Campbell Biology

Bio 151

Custom Edition for Joliet Junior College

| PEARSON COLLECTIONS |

PEARSON

Cover Art: Courtesy of Getty Images

Attention bookstores: For permission to return any unsold stock, contact us at pe-uscustomreturns@pearson.com

Pearson Learning Solutions, 501 Boylston Street, Suite 900, Boston, MA 02116

A Pearson Education Company
www.pearsoned.com

PEARSON

ISBN 10: 1323672214

ISBN 13: 9781323672211

Printed in the USA

2 17

Table of Contents

Brief Contents

About the Authors

Lisa A. Urry is Professor of Biology and Chair of the Biology Department at Mills College. After earning a B.A. at Tufts University, she completed her Ph.D. at the Massachusetts Institute of Technology (MIT). Lisa has conducted research on gene expression during embryonic and larval development in sea urchins. Deeply committed to promoting opportunities in science for women and underrepresented minorities, she has taught courses ranging from introductory and developmental biology to a nonmajors course called Evolution for Future Presidents. Lisa is a coauthor of *Campbell Biology in Focus*.

Michael L. Cain is an ecologist and evolutionary biologist who is now writing full-time. Michael earned an A.B. from Bowdoin College, an M.Sc. from Brown University, and a Ph.D. from Cornell University. As a faculty member at New Mexico State University, he taught introductory biology, ecology, evolution, botany, and conservation biology. Michael is the author of dozens of scientific papers on topics that include foraging behavior in insects and plants, long-distance seed dispersal, and speciation in crickets. He is a coauthor of *Campbell Biology in Focus* and of an ecology textbook.

Steven A. Wasserman is Professor of Biology at the University of California, San Diego (UCSD). He earned an A.B. from Harvard University and a Ph.D. from MIT. Working on the fruit fly *Drosophila*, Steve has done research on developmental biology, reproduction, and immunity. Having taught genetics, development, and physiology to undergraduate, graduate, and medical students, he now focuses on introductory biology, for which he has been honored with UCSD's Distinguished Teaching Award. He is a coauthor of *Campbell Biology in Focus*.

Peter V. Minorsky is Professor of Biology at Mercy College in New York, where he teaches introductory biology, ecology, and botany. He received his A.B. from Vassar College and his Ph.D. from Cornell University. Peter taught at Kenyon College, Union College, Western Connecticut State University, and Vassar College; he is also the science writer for the journal *Plant Physiology*. His research interests concern how plants sense environmental change. Peter received the 2008 Award for Teaching Excellence at Mercy College and is a coauthor of *Campbell Biology in Focus*.

Jane B. Reece, the head of the author team for Editions 8–10 of *CAMPBELL BIOLOGY*, was Neil Campbell's longtime collaborator. Jane taught biology at Middlesex County College and Queensborough Community College. She holds an A.B. from Harvard University, an M.S. from Rutgers University, and a Ph.D. from the University of California, Berkeley. Jane's research as a doctoral student at UC Berkeley and postdoctoral fellow at Stanford University focused on genetic recombination in bacteria. Besides her work on *CAMPBELL BIOLOGY*, Jane has been a coauthor on all the *Campbell* texts.

Neil A. Campbell (1946–2004) earned his M.A. from the University of California, Los Angeles, and his Ph.D. from the University of California, Riverside. His research focused on desert and coastal plants. Neil's 30 years of teaching included introductory biology courses at Cornell University, Pomona College, and San Bernardino Valley College, where he received the college's first Outstanding Professor Award in 1986. For many years he was also a visiting scholar at UC Riverside. Neil was the founding author of *CAMPBELL BIOLOGY*.

Preface

We are honored to present the Eleventh Edition of *Campbell Biology*. For the last three decades, *Campbell Biology* has been the leading college text in the biological sciences. It has been translated into 19 languages and has provided millions of students with a solid foundation in college-level biology. This success is a testament not only to Neil Campbell's original vision but also to the dedication of hundreds of reviewers (listed on pages xxviii–xxxi), who, together with editors, artists, and contributors, have shaped and inspired this work.

Our goals for the Eleventh Edition include:

- **increasing visual literacy** through new figures, questions, and exercises that build students' skills in understanding and creating visual representations of biological structures and processes
- asking students to **practice scientific skills** by applying scientific skills to real-world problems
- **supporting instructors** by providing teaching modules with tools and materials for introducing, teaching, and assessing important and often challenging topics
- **integrating text and media** to engage, guide, and inform students in an active process of inquiry and learning

Our starting point, as always, is our commitment to crafting text and visuals that are accurate, are current, and reflect our passion for teaching biology.

New to This Edition

Here we provide an overview of the new features that we have developed for the Eleventh Edition; we invite you to explore pages ix–xxvi for more information and examples.

- **Visualizing Figures** and **Visual Skills Questions** give students practice in interpreting and creating visual representations in biology. The Visualizing Figures have embedded questions that guide students in exploring how diagrams, photographs, and models represent and reflect biological systems and processes. Assignable questions are also available in **MasteringBiology** to give students practice with the visual skills addressed in the figures.
- **Problem-Solving Exercises** challenge students to apply scientific skills and interpret data in solving real-world problems. These exercises are designed to engage students through compelling case studies and provide practice with data analysis skills. Problem-Solving Exercises have assignable versions in **MasteringBiology**. Some also have more extensive "Solve It" investigations to further explore a given topic.
- **Ready-to-Go Teaching Modules** on key topics provide instructors with assignments to use before and after class, as well as in-class activities that use clickers or Learning Catalytics™ for assessment.
- **Integrated text and media**: Media references in the printed book direct students to the wealth of online self-study resources available to them in the **Study Area** section of **MasteringBiology**. In eText 2.0 (which will be available for Fall 2017 classes), these resources will also be integrated directly into the eText. The new online learning tools include:
  - **Get Ready for This Chapter** questions provide a quick check of student understanding of the background information needed to learn a new chapter's content, with feedback to bolster their preparation.
  - **Figure Walkthroughs** guide students through key figures with narrated explanations, figure markups, and questions that reinforce important points. Additional questions can be assigned in **MasteringBiology**.
  - More than **450 animations and videos** bring biology to life. These include resources from **HHMI BioInteractive** that engage students in topics from the discovery of the double helix to evolution.
  - QR codes and URLs within the Chapter Review provide easy access to **Vocabulary Self-Quizzes** and **Practice Tests** for each chapter that can be used on smartphones, tablets, and computers.
  - **Interviews** from the First Edition through the Eleventh Edition of *Campbell Biology* are referenced in the chapter where they are most relevant. The interviews show students the human side of science by featuring diverse scientists talking about how they became interested in what they study, how they began, and what inspires them.
- The impact of **climate change** at all levels of the biological hierarchy is explored throughout the text, starting with a new figure (Figure 1.12) and discussion in Chapter 1 and concluding with a new Make Connections Figure (Figure 56.30) and expanded coverage on causes and effects of climate change in Chapter 56.
- As in each new edition of *Campbell Biology*, the Eleventh Edition incorporates **new content** and **pedagogical improvements**. These are summarized on pp. vi–viii, following this Preface. Content updates reflect rapid, ongoing changes in technology and knowledge in the fields of genomics, gene editing technology (CRISPR), evolutionary

biology, microbiology, and more. In addition, significant revisions to Unit 8, Ecology, improve the conceptual framework for core ecological topics (such as population growth, species interactions, and community dynamics) and more deeply integrate evolutionary principles.

Our Hallmark Features

Teachers of general biology face a daunting challenge: to help students acquire a conceptual framework for organizing an ever-expanding amount of information. The hallmark features of *CAMPBELL BIOLOGY* provide such a framework, while promoting a deeper understanding of biology and the process of science. As such, they are well aligned with the core competencies outlined by the 2009 **Vision and Change** national conference. Furthermore, the core concepts defined by Vision and Change have close parallels in the unifying themes that are introduced in Chapter 1 and integrated throughout the book. Chief among the themes of both Vision and Change and *CAMPBELL BIOLOGY* is **evolution.** Each chapter of this text includes at least one Evolution section that explicitly focuses on evolutionary aspects of the chapter material, and each chapter ends with an Evolution Connection Question and a Write About a Theme Question.

To help students distinguish the "forest from the trees," each chapter is organized around a framework of three to seven carefully chosen **Key Concepts**. The text, Concept Check Questions, Summary of Key Concepts, and MasteringBiology resources all reinforce these main ideas and essential facts.

Because text and illustrations are equally important for learning biology, **integration of text and figures** has been a hallmark of this text since the First Edition. In addition to the new Visualizing Figures, our popular Exploring Figures and Make Connections Figures epitomize this approach. Each Exploring Figure is a learning unit of core content that brings together related illustrations and text. Make Connections Figures reinforce fundamental conceptual connections throughout biology, helping students overcome tendencies to compartmentalize information. The Eleventh Edition features two new Make Connections Figures. There are also Guided Tour Figures that walk students through complex figures as an instructor would.

To encourage **active reading** of the text, *CAMPBELL BIOLOGY* includes numerous opportunities for students to stop and think about what they are reading, often by putting pencil to paper to draw a sketch, annotate a figure, or graph data. Active reading questions include Visual Skills Questions, Draw It Questions, Make Connections Questions, What If? Questions, Figure Legend Questions, Summary Questions, Synthesize Your Knowledge Questions, and Interpret the Data Questions. Answering these questions requires students to write or draw as well as think and thus helps develop the core competency of communicating science.

Finally, *CAMPBELL BIOLOGY* has always featured **scientific inquiry**, an essential component of any biology course. Complementing stories of scientific discovery in the text narrative and the unit-opening interviews, our standard-setting Inquiry Figures deepen the ability of students to understand how we know what we know. Scientific Inquiry Questions give students opportunities to practice scientific thinking, along with the Problem-Solving Exercises, Scientific Skills Exercises, and Interpret the Data Questions. Together, these activities provide students practice in both applying the process of science and using quantitative reasoning, addressing additional core competencies outlined in Vision and Change.

MasteringBiology®

MasteringBiology, the most widely used online assessment and tutorial program for biology, provides an extensive library of homework assignments that are graded automatically. In addition to the **new Get Ready for This Chapter Questions, Figure Walkthroughs, Problem-Solving Exercises, and Visualizing Figures**, MasteringBiology offers Dynamic Study Modules, Adaptive Follow-Up Assignments, Scientific Skills Exercises, Interpret the Data Questions, Solve It Tutorials, HHMI BioInteractive Short Films, BioFlix® Tutorials with 3-D Animations, Experimental Inquiry Tutorials, Interpreting Data Tutorials, BLAST Tutorials, Make Connections Tutorials, Video Field Trips, Video Tutor Sessions, Get Ready for Biology, Activities, Reading Quiz Questions, Student Misconception Questions, 4,500 Test Bank Questions, and MasteringBiology Virtual Labs. MasteringBiology also includes the *CAMPBELL BIOLOGY* eText, Study Area, Instructor Resources, and Ready-to-Go Teaching Modules. See pages ix–xxiii and www.masteringbiology.com for more details.

Our Partnership with Instructors and Students

A core value underlying our work is our belief in the importance of a partnership with instructors and students. One primary way of serving instructors and students, of course, is providing a text that teaches biology well. In addition, Pearson offers a rich variety of instructor and student resources, in both print and electronic form (see pp. ix–xxiii). In our continuing efforts to improve the book and its supplements, we benefit tremendously from instructor and student feedback, not only in formal reviews from hundreds of scientists, but also via e-mail and other avenues of informal communication.

The real test of any textbook is how well it helps instructors teach and students learn. We welcome comments from both students and instructors. Please address your suggestions to:

Lisa Urry (Chapter 1 and Units 1–3)
lurry@mills.edu
Michael Cain (Units 4, 5, and 8)
mcain@bowdoin.edu
Peter Minorsky (Unit 6)
pminorsky@mercy.edu
Steven Wasserman (Unit 7)
stevenw@ucsd.edu

Highlights of New Content

This section highlights selected new content and pedagogical changes in *Campbell Biology*, Eleventh Edition.

CHAPTER 1 Evolution, the Themes of Biology, and Scientific Inquiry

Chapter 1 opens with a new introduction to a case study on the evolution of coloration in mice. New text and a new photo (Figure 1.12) relate climate change to species survival.

UNIT 1 THE CHEMISTRY OF LIFE

In Unit 1, new content engages students in learning this foundational material. The opening of Chapter 3 and new **Figure 3.7** show organisms affected by **loss of Arctic sea ice**. Chapter 5 has updates on lactose intolerance, *trans* fats, the effects of diet on blood cholesterol, protein sequences and structures, and intrinsically disordered proteins. New Visualizing Figure 5.16 helps students understand various ways proteins are depicted. A new Problem-Solving Exercise engages students by having them compare DNA sequences in a case of possible fish fraud.

Figure 3.7 Effects of climate change on the Arctic.

UNIT 2 THE CELL

Our main goal for this unit was to make the material more accessible and inviting to students. New Visualizing Figure 6.32 shows the profusion of molecules and structures in a cell, all drawn to scale. In Chapter 7, a new figure illustrates levels of LDL receptors in people with and without familial hypercholesterolemia. Chapter 8 includes a beautiful new photo of a geyser with thermophilic bacteria in Figure 8.17, bringing to life the graphs of optimal temperatures for enzyme function. Chapter 10 discusses current research trying to genetically modify rice (a C_3 crop) so that it is capable of carrying out C_4 photosynthesis to increase yields. Chapter 11 includes a new Problem-Solving Exercise that guides students through assessing possible new treatments for bacterial infections by blocking quorum sensing. In Chapter 12, the mechanism of chromosome movement in bacteria has been updated and more cell cycle control checkpoints have been added, including one proposed by researchers in 2014.

UNIT 3 GENETICS

In Chapters 13–17, we have incorporated changes that help students grasp the more abstract concepts of genetics and their chromosomal and molecular underpinnings. For example, a new Visual Skills Question with Figure 13.6 asks students to identify where in the three life cycles haploid cells undergo mitosis, and what type of cells are formed. Chapter 14 includes new information from a 2014 genomic study on the number of genes and genetic variants contributing to height. Figure 14.15b now uses "inability to taste PTC" rather than "attached earlobe." Chapters 14 and 15 are more inclusive, clarifying the meaning of the term "normal" in genetics and explaining that sex is no longer thought to be simply binary. Other updates in Chapter 15 include new research in sex determination and a technique being developed to avoid passing on mitochondrial diseases. New Visualizing Figure 16.7 shows students various ways that DNA is illustrated. Chapter 17 has a new opening photo and story about albino donkeys to pique student interest in gene expression. To help students understand the Beadle and Tatum experiment, new Figure 17.2 explains how they obtained nutritional mutants. A new Problem-Solving Exercise asks students to identify mutations in the insulin gene and predict their effect on the protein.

Chapters 18–21 are extensively updated, driven by exciting new discoveries based on DNA sequencing and gene-editing technology. Chapter 18 has updates on histone modifications, nuclear location and the persistence of transcription factories, chromatin remodeling by ncRNAs, long noncoding RNAs (lncRNAs), the role of master regulatory genes in modifying chromatin structure, and the possible role of *p53* in the low incidence of cancer in elephants. Make Connections Figure 18.27, "Genomics, Cell Signaling, and Cancer," has been expanded to include more information on cell signaling. Chapter 19 features a new section that covers bacterial defenses against bacteriophages and describes the CRISPR-Cas9 system (Figure 19.7); updates include the Ebola, Chikungunya, and Zika viruses (Figure 19.10) and discovery of the largest virus known to date. A discussion has been added of mosquito transmission of diseases and concerns about the effects of global climate change on disease transmission. Chapter 20 has a new photo of next-generation DNA sequencing machines (Figure 20.2) and a new illustration of the widely used technique of RNA sequencing (Figure 20.13). A new section titled Editing Genes and Genomes has been added describing the **CRISPR-Cas9 system (Figure 20.14)** that has been developed to edit genes in living cells. Information has also been

Figure 20.14 Gene editing using the CRISPR-Cas9 system.

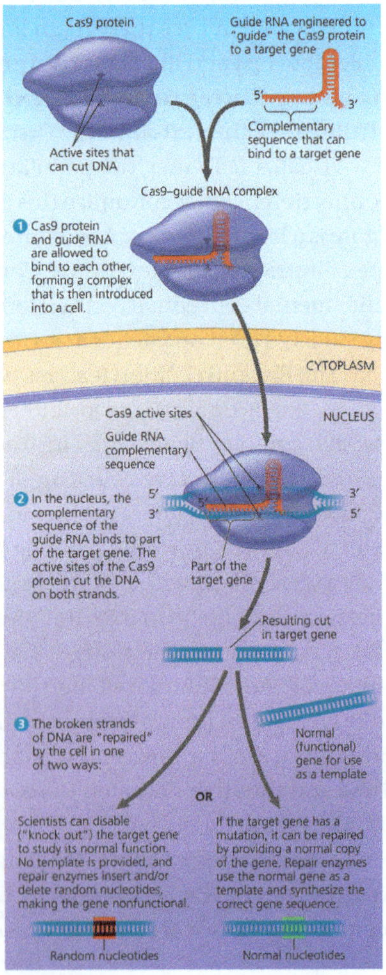

added later in the chapter on use of the CRISPR-Cas9 system, including a study in which a genetic mutation for the disease tyrosinemia was corrected in mice. Finally, the discussion of ethical considerations has been updated to include a recent report of scientists using the CRISPR-Cas9 system to edit a gene in human embryos, along with a discussion of the ethical questions raised by such experiments, such as its usage in the gene drive approach to combat carrying of diseases by mosquitoes. In Chapter 21, in addition to the usual updates of sequence-related data (speed of sequencing, number of species' genomes sequenced, etc.), there are several research updates, including some early results from the new Roadmap Epigenomics Project and results from a 2015 study focusing on 414 important yeast genes.

UNIT 4 MECHANISMS OF EVOLUTION

A major goal for this revision was to strengthen how we help students understand and interpret visual representations of evolutionary data and concepts. Toward this end, we have added a new figure (Figure 25.8), "Visualizing the Scale of Geologic Time," and a new figure **(Figure 23.12)** on **gene flow**. Several figures have been revised to improve the presentation of data, including Figure 24.6 (on reproductive isolation in mosquitofish), Figure 24.10 (on allopolyploid speciation), and Figure 25.25 (on the origin of the insect body plan). The unit also features new material that connects evolutionary concepts and societal problems. Examples include text in Chapter 22 on the 2015 discovery of teixobactin, an antibiotic that is effective against some hard-to-treat pathogens, a new discussion in Chapter 24 on the impact of climate change on hybrid zones, and a new Problem-Solving Exercise in Chapter 24 on how hybridization may have led to the spread of insecticide resistance genes in mosquitoes that transmit malaria. The unit also includes new chapter-opening stories in Chapter 22 (on a moth whose features illustrate the concepts of unity, diversity, and adaptation) and Chapter 25 (on the discovery of whale bones in the Sahara Desert). Additional changes include new text in Concept 22.3 emphasizing how populations can evolve over short periods of time, a new table (Table 23.1) highlighting the five conditions required for a population to be in Hardy-Weinberg equilibrium, and new material in Concept 25.1 describing how researchers recently succeeded for the first time in constructing a "protocell" in which replication of a template strand of RNA could occur.

▼ Figure 23.12 Gene flow and local adaptation in the Lake Erie water snake (*Nerodia sipedon*).

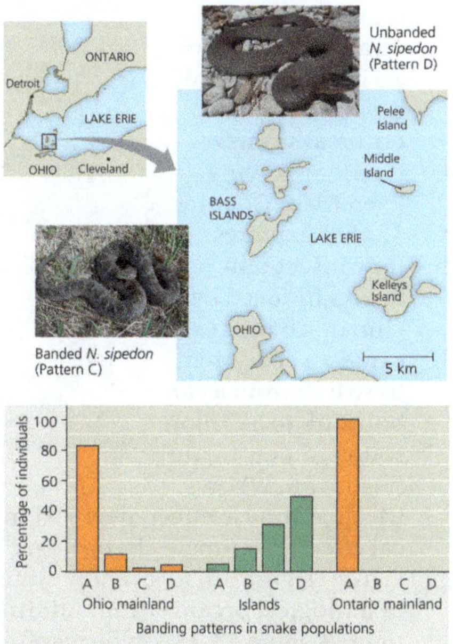

UNIT 5 THE EVOLUTIONARY HISTORY OF BIOLOGICAL DIVERSITY

In keeping with our goal of improving how students interpret and create visual representations in biology, we have added a new figure (Figure 26.5, "Visualizing Phylogenetic Relationships") that introduces the visual conventions used in phylogenetic trees and helps students understand what such trees do and don't convey. Students are also provided many opportunities to practice their visual skills, with more than ten new Visual Skills Questions on topics ranging from interpreting phylogenetic trees to predicting which regions of a bacterial flagellum are hydrophobic. The unit also contains new content on tree thinking, emphasizing such key points as how sister groups provide a clear way to describe evolutionary relationships and how trees do not show a "direction" in evolution. Other major content changes include new text in Concepts 26.6, 27.4, and 28.1 on the 2015 discovery of the Lokiarchaeota, a group of archaea that may represent the sister group of the eukaryotes, new text and a new figure (Figure 26.22) on horizontal gene transfer from prokaryotes to eukaryotes, new text in Concept 27.6 describing the CRISPR-Cas9 system and a new figure (Figure 27.21) that illustrates one example of how CRISPR-Cas 9 technology has opened new avenues of research on HIV, and new material in Concept 29.3 describing how early forests contributed to global climate change (in this case, global cooling). A new Problem-Solving Exercise in Chapter 34 engages students in interpreting data from a study investigating whether frogs can acquire resistance to a fungal pathogen through controlled exposure to it. Other updates include the revision of many phylogenies to reflect recent phylogenomic data, new chapter-opening stories in Chapter 31 (on how mycorrhizae link trees of different species) and Chapter 33 (on the "blue dragon," a mollusc that preys on the highly toxic Portuguese man-of-war), new text and a new figure (Figure 34.37) on the adaptations of the kangaroo rat to its arid environment, and new material in Concept 34.7, including a new figure (Figure 34.52) describing fossil and DNA evidence indicating that humans and Neanderthals interbred, producing viable offspring. The discussion of **human evolution** also includes new text and a new figure **(Figure 34.53)** on *Homo naledi*, the most recently discovered member of the human evolutionary lineage.

▼ Figure 34.53 Fossils of hand and foot bones of *Homo naledi*.

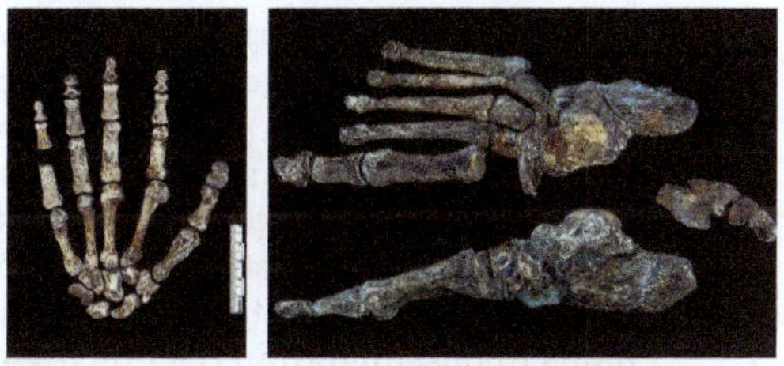

UNIT 6 PLANT FORM AND FUNCTION

A major aim in revising Chapter 35 was to help students better understand how primary and secondary growth are related. New Visualizing Figure 35.11 enables students to picture growth at the cellular level. Also, the terms *protoderm, procambium*, and *ground meristem* have been introduced to underscore the transition of meristematic to mature tissues. A new flowchart (Figure 35.24) summarizes growth in a woody shoot. New text and a figure (Figure 35.26) focus on genome analysis of *Arabidopsis* ecotypes, relating plant morphology to ecology and evolution. In Chapter 36, new Figure 36.8 illustrates the fine branching of leaf veins, and information on phloem-xylem water transfer has been updated. New Make Connections Figure 37.10 highlights mutualism across kingdoms and domains. Figure 37.13 and the related text include new findings on how some soil nitrogen derives from weathering of rocks. New Figure 38.3 clarifies how the terms *carpel* and *pistil* are related. The text on flower structure and the angiosperm life cycle figure identify carpels as megasporophylls and stamens as microsporophylls, correlating with the plant evolution discussion in Unit 5. In Concept 38.3, the current problem of glyphosate-resistant crops is discussed in detail. A revised Figure 39.7 helps students visualize how cells elongate. Figure 39.8 now addresses apical dominance in a Guided Tour format. Information about the role of sugars in controlling apical dominance has been added. In Concept 39.4, a new Problem-Solving Exercise highlights how global climate change affects crop productivity. Figure 39.26 on defense responses against pathogens has been simplified and improved.

UNIT 7 ANIMAL FORM AND FUNCTION

A major goal of the Unit 7 revision was to transform how students interact with and learn from representations of anatomy and physiology. For example, gastrulation is now introduced with a Visualizing Figure (Figure 47.8) that provides a clear and carefully paced introduction to three-dimensional processes that may be difficult for students to grasp. In addition, a number of the new and revised figures help students explore spatial relationships in anatomical contexts, such as the interplay of lymphatic and cardiovascular circulation (Figure 42.15) and the relationship of the limbic system to overall brain structure (Figure 49.14). A new Problem-Solving Exercise in Chapter 45 taps into student interest in medical mysteries through a case study that explores the science behind laboratory testing and diagnosis. Content updates help students appreciate the continued evolution of our understanding of even familiar phenomena, such as the sensation of thirst (Concept 44.4) and the locomotion of kangaroos and jellies (Concept 50.6). Furthermore, new text and figures introduce students to cutting-edge technology relating to such topics as RNA-based antiviral defense in invertebrates (Figure 43.4) and rapid, comprehensive characterization of viral exposure (Figure 43.24), as well as recent discoveries regarding brown fat in adult humans (Figure 40.16), the **microbiome (Figure 41.17)**, parthenogenesis (Concept 46.1), and magnetoreception (Concept 50.1). As always, there is fine-tuning of pedagogy, as in discussions of the complementary roles of inactivation and voltage gating of ion channels during action potential formation (Concept 48.3) and of the experimental characterization of genetic determinants in bird migration (Figure 51.24).

Figure 41.17 Variation in human gut microbiome at different life stages.

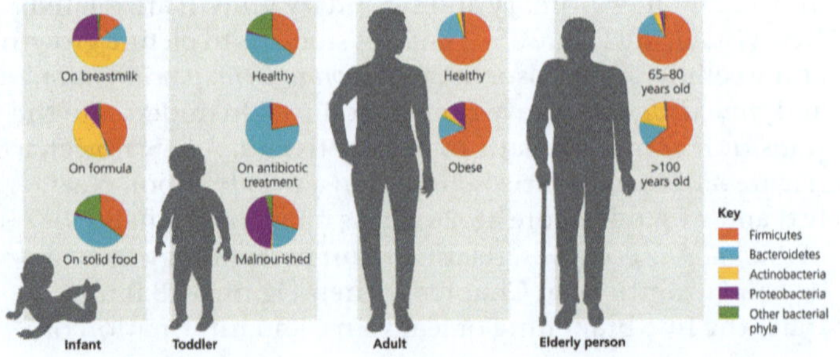

UNIT 8 ECOLOGY

The Ecology Unit has been extensively revised for the Eleventh Edition. We have reorganized and improved the conceptual framework with which students are introduced to the following core ecological topics: life tables, per capita population growth, intrinsic rate of increase ("*r*"), exponential population growth, logistic population growth, density dependence, species interactions (in particular, parasitism, commensalism, and mutualism), and MacArthur and Wilson's island biogeography model. The revision also includes a deeper integration of evolutionary principles, including a new Key Concept (52.5) and two new figures (Figures 52.22 and 52.23) on the reciprocal effects of ecology and evolution, new material in Concept 52.4 on how the geographic distributions of species are shaped by a combination of evolutionary history and ecological factors, and five new Make Connections Questions that ask students to examine how ecological and evolutionary mechanisms interact. In keeping with our goal of expanding and strengthening our coverage of climate change, we have added a new discussion and a new figure (Figure 52.20) on how climate change has affected the distribution of a keystone species, a new section of text in Concept 55.2 on how climate change affects NPP, a new figure **(Figure 55.8)** on how **climate change** has caused an increase in wildfires and insect outbreaks, a new Problem-Solving Exercise in Chapter 55 that explores how insect outbreaks induced by climate change can cause an ecosystem to switch from a carbon sink to a carbon source, a new figure (Figure 56.29) on the greenhouse effect, new text in Concept 56.4 on biological effects of climate change, and a new Make Connections Figure (Figure 56.30) on how climate change affects all levels of biological organization. Additional updates include a new figure (Figure 53.25) on per capita ecological footprints, a new chapter-opening story in Chapter 54 on a seemingly unlikely mutualism between a shrimp and a much larger predatory fish, new text in Concept 54.1 emphasizing that each partner in a mutualism experiences both benefits and costs, new text in Concept 54.1 describing how the outcome of an ecological interaction can change over time, two new figures (Figures 54.29 and 54.30) on the island equilibrium model, a new figure (Figure 54.31) documenting two shrew species as unexpected hosts of the Lyme disease, new text in Concept 56.1 comparing extinction rates today with those typically seen in the fossil record, and a new discussion and figure (Figure 56.22) on the restoration of a degraded urban stream.

Figure 55.8 Climate change, wildfires, and insect outbreaks.

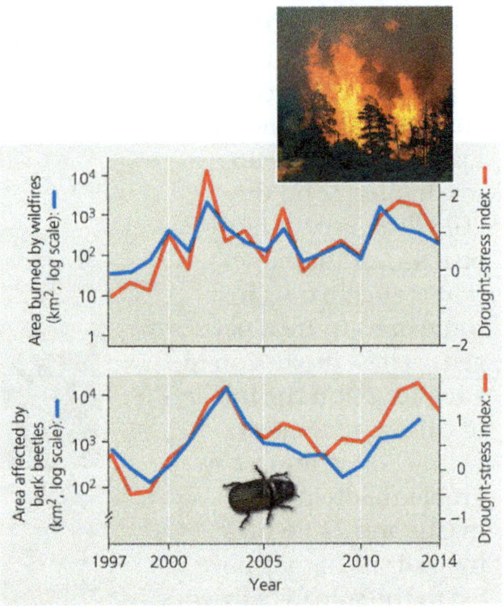

Ready-to-Go Teaching Modules for Instructors

NEW! Ready-to-Go Teaching Modules help instructors efficiently make use of the best teaching tools before, during, and after class.

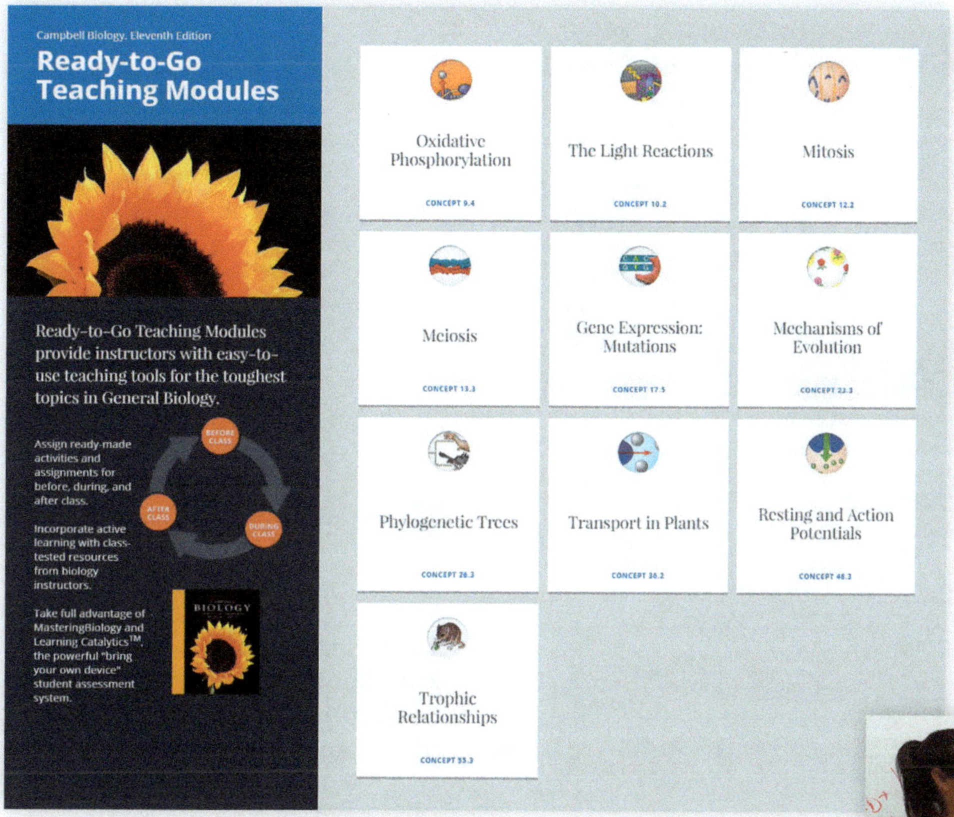

The **Ready-to-Go Teaching Modules** incorporate the best that the text, MasteringBiology, and Learning Catalytics have to offer, along with new ideas for in-class activities. The modules can be accessed through the Instructor Resources area of MasteringBiology.

Instructors can easily incorporate **active learning** into their courses using suggested activity ideas and questions. Videos demonstrate how the activities can be used in class.

Learning Catalytics™ allows students to use their smartphone, tablet, or laptop to respond to questions in class. Visit learningcatalytics.com.

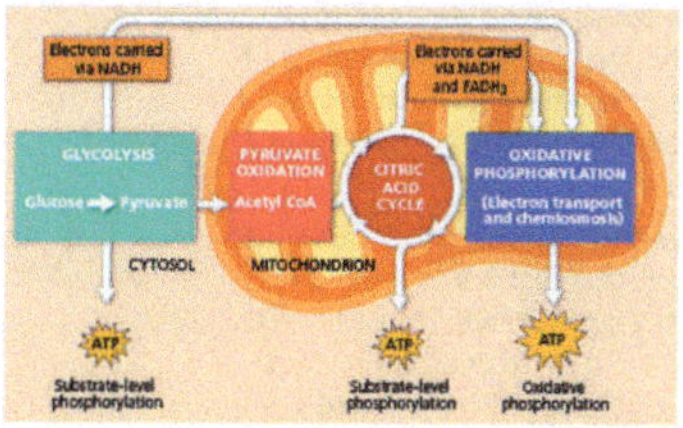
The following conditions were detected in a mutant cell:

The cell is running out of ATP, while ADP is building up to very high levels.
NADH is building up to very high levels, while the level of NAD⁺ is becoming very low.

The amount of protons in the intermembrane space and in the matrix is becoming more equal (the strength of the proton gradient is decreasing/weakening).

Use this information to predict which stage of cellular respiration is not functioning normally in this mutant cell.

A. glycolysis
B. citric acid cycle
C. pyruvate oxidation
D. electron transport chain
E. ATP synthase

See the Big Picture

Each chapter is organized around a framework of 3 to 7 **Key Concepts** that focus on the big picture and provide a context for supporting details.

Every chapter opens with a visually dynamic **photo** accompanied by an **intriguing question** that invites students into the chapter.

The **List of Key Concepts** introduces the big ideas covered in the chapter.

Descent with Modification: A Darwinian View of Life 22

▲ **Figure 22.1** **How is its resemblance to a fallen leaf helpful to this moth?**

KEY CONCEPTS

22.1 **The Darwinian revolution challenged traditional views of a young Earth inhabited by unchanging species**

22.2 **Descent with modification by natural selection explains the adaptations of organisms and the unity and diversity of life**

22.3 **Evolution is supported by an overwhelming amount of scientific evidence**

▼ Juvenile stage (caterpillar) of the dead-leaf moth

Endless Forms Most Beautiful

A hungry bird in the Peruvian rain forest would have to look very closely to spot a "dead-leaf moth" (*Oxytenis modestia*), which blends in well with its forest floor habitat (**Figure 22.1**). This distinctive moth is a member of a diverse group, the more than 120,000 species of lepidopteran insects (moths and butterflies). All lepidopterans have a juvenile stage characterized by a well-developed head and many chewing mouthparts: the ravenous, efficient feeding machines we call caterpillars. (The caterpillar stage of the dead-leaf moth is also protected by its appearance: When threatened, it weaves its head back and forth, resembling a snake about to strike.) As adults, all lepidopterans share other features, such as three pairs of legs and two pairs of wings covered with small scales. But the many lepidopterans also differ from one another. How did there come to be so many different moths and butterflies, and what causes their similarities and differences?

The moth in Figure 22.1 and its many close relatives illustrate three key observations about life:

- the striking ways in which organisms are suited for life in their environments (Here and throughout this text, the term *environment* refers to other organisms as well as to the physical aspects of an organism's surroundings.)
- the many shared characteristics (unity) of life
- the rich diversity of life

When you see this blue icon, log in to **MasteringBiology** and go to the Study Area for digital resources.

Get Ready for This Chapter

466

Get Ready for This Chapter

NEW! Get Ready for This Chapter questions provide a quick check of students' knowledge of basic information needed to learn the new content of a chapter, with feedback.

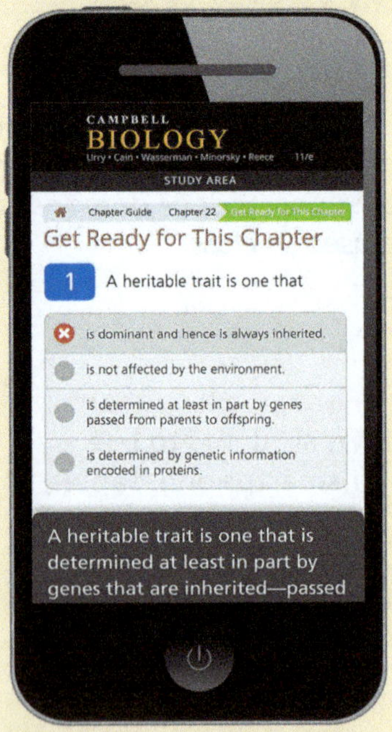

These questions are available as MasteringBiology assignments and as self-study quizzes in the Study Area.

After reading a Key Concept section, students can check their understanding using the **Concept Check Questions.**

Questions throughout the chapter encourage students to **read the text actively.**

What If? Questions ask students to apply what they've learned.

Make Connections Questions ask students to relate content in the chapter to material presented earlier in the course.

CONCEPT CHECK 22.2

1. How does the concept of descent with modification explain both the unity and diversity of life?
2. **WHAT IF?** ➤ If you discovered a fossil of an extinct mammal that lived high in the Andes, would you predict that it would more closely resemble present-day mammals from South American jungles or present-day mammals that live high in Asian mountains? Explain.
3. **MAKE CONNECTIONS** ➤ Review the relationship between genotype and phenotype (see Figures 14.5 and 14.6). Suppose that in a particular pea population, flowers with the white phenotype are favored by natural selection. Predict what would happen over time to the frequency of the *p* allele in the population, and explain your reasoning.

For suggested answers, see Appendix A.

The **Summary of Key Concepts** refocuses students on the main points of the chapter.

VOCAB SELF-QUIZ goo.gl/6u55ks

PRACTICE TEST goo.gl/CUYGKD

NEW! QR codes and URLs at the end of every chapter give students quick access to **Vocabulary Self-Quizzes** and **Practice Tests** on their smartphones, tablets, and computers.

22 Chapter Review

Go to **MasteringBiology**™ for Videos, Animations, Vocab Self-Quiz, Practice Tests, and more in the Study Area.

SUMMARY OF KEY CONCEPTS

CONCEPT 22.1

The Darwinian revolution challenged traditional views of a young Earth inhabited by unchanging species *(pp. 467–469)*

VOCAB SELF-QUIZ

- Darwin proposed that life's diversity arose from ancestral species through natural selection, a departure from prevailing views.
- Cuvier studied **fossils** but denied that evolution occurs; he proposed that sudden catastrophic events in the past caused species to disappear from an area.
- Hutton and Lyell thought that geologic change could result from gradual mechanisms that operated in the past in the same manner as they do today.
- Lamarck hypothesized that species evolve, but the underlying mechanisms he proposed are not supported by evidence.

? *Why was the age of Earth important for Darwin's ideas about evolution?*

CONCEPT 22.2

Descent with modification by natural selection explains the adaptations of organisms and the unity and diversity of life *(pp. 469–474)*

- Darwin's experiences during the voyage of the *Beagle* gave rise to his idea that new species originate from ancestral forms through the accumulation of **adaptations**. He refined his theory for many years and finally published it in 1859 after learning that Wallace had come to the same idea.
- In *The Origin of Species*, Darwin proposed that over long periods of time, descent with modification produced the rich diversity of life through the mechanism of **natural selection**.

Observations

Individuals in a population vary in their heritable characteristics.

Organisms produce more offspring than the environment can support.

Inferences

Individuals that are well suited to their environment tend to leave more offspring than other individuals.

and

Over time, favorable traits accumulate in the population.

? *Describe how overreproduction and heritable variation relate to evolution by natural selection.*

CONCEPT 22.3

Evolution is supported by an overwhelming amount of scientific evidence *(pp. 475–482)*

- Researchers have directly observed natural selection leading to adaptive evolution in many studies, including research on soapberry bug populations and on MRSA.
- Organisms share characteristics because of common descent (**homology**) or because natural selection affects independently evolving species in similar environments in similar ways (**convergent evolution**).
- Fossils show that past organisms differed from living organisms, that many species have become extinct, and that species have evolved over long periods of time; fossils also document the evolutionary origin of new groups of organisms.
- Evolutionary theory can explain some biogeographic patterns.

? *Summarize the different lines of evidence supporting the hypothesis that cetaceans descended from land mammals and are closely related to even-toed ungulates.*

482 UNIT FOUR Mechanisms of Evolution

Summary of Key Concepts Questions check students' understanding of a key idea from each concept.

Summary Figures recap key information visually.

TEST YOUR UNDERSTANDING

Level 1: Knowledge/Comprehension

PRACTICE TEST goo.gl/CUYGKD

1. Which of the following is *not* an observation or inference on which natural selection is based?
   (A) There is heritable variation among individuals.
   (B) Poorly adapted individuals never produce offspring.
   (C) Species produce more offspring than the environment can support.
   (D) Only a fraction of an individual's offspring may survive.
2. Which of the following observations helped Darwin shape his concept of descent with modification?
   (A) Species diversity declines farther from the equator.
   (B) Fewer species live on islands than on the nearest continents.
   (C) Birds live on islands located farther from the mainland than the birds' maximum nonstop flight distance.
   (D) South American temperate plants are more similar to the tropical plants of South America than to the temperate plants of Europe.

Level 2: Application/Analysis

3. Within six months of effectively using methicillin to treat *S. aureus* infections in a community, all new *S. aureus* infections were caused by MRSA. How can this best be explained?
   (A) A patient must have become infected with MRSA from another community.
   (B) In response to the drug, *S. aureus* began making drug-resistant versions of the protein targeted by the drug.
   (C) Some drug-resistant bacteria were present at the start of treatment, and natural selection increased their frequency.
   (D) *S. aureus* evolved to resist vaccines.
4. The upper forelimbs of humans and bats have fairly similar skeletal structures, whereas the corresponding bones in whales have very different shapes and proportions. However, genetic data suggest that all three kinds of organisms diverged from a common ancestor at about the same time. Which of the following is the most likely explanation for these data?
   (A) Forelimb evolution was adaptive in people and bats, but not in whales.
   (B) Natural selection in an aquatic environment resulted in significant changes to whale forelimb anatomy.
   (C) Genes mutate faster in whales than in humans or bats.
   (D) Whales are not properly classified as mammals.
5. DNA sequences in many human genes are very similar to the sequences of corresponding genes in chimpanzees. The most likely explanation for this result is that
   (A) humans and chimpanzees share a relatively recent common ancestor.
   (B) humans evolved from chimpanzees.
   (C) chimpanzees evolved from humans.
   (D) convergent evolution led to the DNA similarities.

Level 3: Synthesis/Evaluation

6. **EVOLUTION CONNECTION** Explain why anatomical and molecular features often fit a similar nested pattern. In addition, describe a process that can cause this not to be the case.
7. **SCIENTIFIC INQUIRY • DRAW IT** Mosquitoes resistant to the pesticide DDT first appeared in India in 1959, but now are found throughout the world. (a) Graph the data in the table below. (b) Examine the graph, then hypothesize why the percentage of mosquitoes resistant to DDT rose rapidly. (c) Suggest an explanation for the global spread of DDT resistance.

Month	0	8	12
Mosquitoes Resistant* to DDT	4%	45%	77%

*Mosquitoes were considered resistant if they were not killed within 1 hour of receiving a dose of 4% DDT.

Data from C. F. Curtis et al., Selection for and against insecticide resistance and possible methods of inhibiting the evolution of resistance in mosquitoes, *Ecological Entomology* 3:273–287 (1978).

8. **WRITE ABOUT A THEME: INTERACTIONS** Write a short essay (about 100–150 words) evaluating whether changes to an organism's physical environment are likely to result in evolutionary change. Use an example to support your reasoning.
9. **SYNTHESIZE YOUR KNOWLEDGE**

This honeypot ant (genus *Myrmecocystus*) can store liquid food inside its expandable abdomen. Consider other ants you are familiar with, and explain how a honeypot ant exemplifies three key features of life: adaptation, unity, and diversity.

For selected answers, see Appendix A.

For additional practice questions, check out the **Dynamic Study Modules** in MasteringBiology. You can use them to study on your smartphone, tablet, or computer anytime, anywhere!

Evolution Connection Questions are included in every Chapter Review.

Synthesize Your Knowledge Questions ask students to apply their understanding of the chapter content to explain an intriguing photo.

Evolution, the fundamental theme of biology, is emphasized throughout. Every chapter has a section explicitly relating the chapter content to evolution:

Evolution of the Genetic Code

EVOLUTION The genetic code is nearly universal, shared by organisms from the simplest bacteria to the most complex plants and animals. The mRNA codon CCG, for instance, translated as the amino acid proline in all organisms whos genetic code has been examined. In laboratory experimen genes can be transcribed and translated after being trans-planted from one species to another, sometimes with quit striking results, as shown in **Figure 17.7**. Bacteria can be programmed by the insertion of human genes to synthesi certain human proteins for medical use, such as insulin. S applications have produced many exciting developments the area of biotechnology (see Concept 20.4).

Figure 17.7 Evidence for evolution: expression of genes from different species. Because diverse forms of life share a common genetic code due to their shared ancestry, one species can be programmed to produce proteins characteristic of a second species by introducing DNA from the second species into the first.

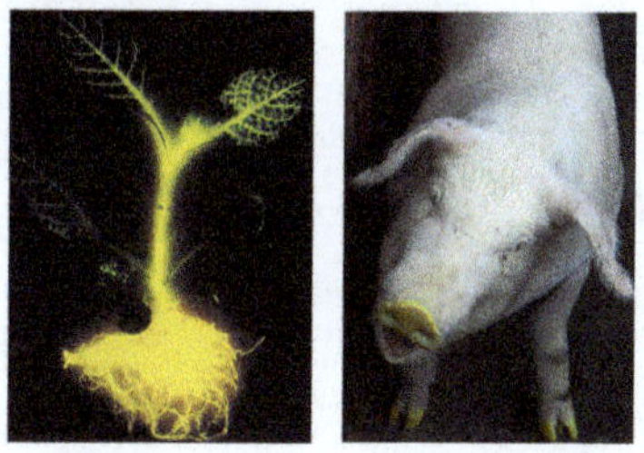

(a) Tobacco plant expressing a firefly gene. The yellow glow is produced by a chemical reaction catalyzed by the protein product of the firefly gene.

(b) Pig expressing a jellyfish gene. Researchers injected a jellyfish gene for a fluorescent protein into fertilized pig eggs. One developed into this fluorescent pig.

Build Visual Skills

NEW! Visualizing Figures teach students how to interpret diagrams and models in biology. Embedded questions give students practice applying visual skills as they read the figure.

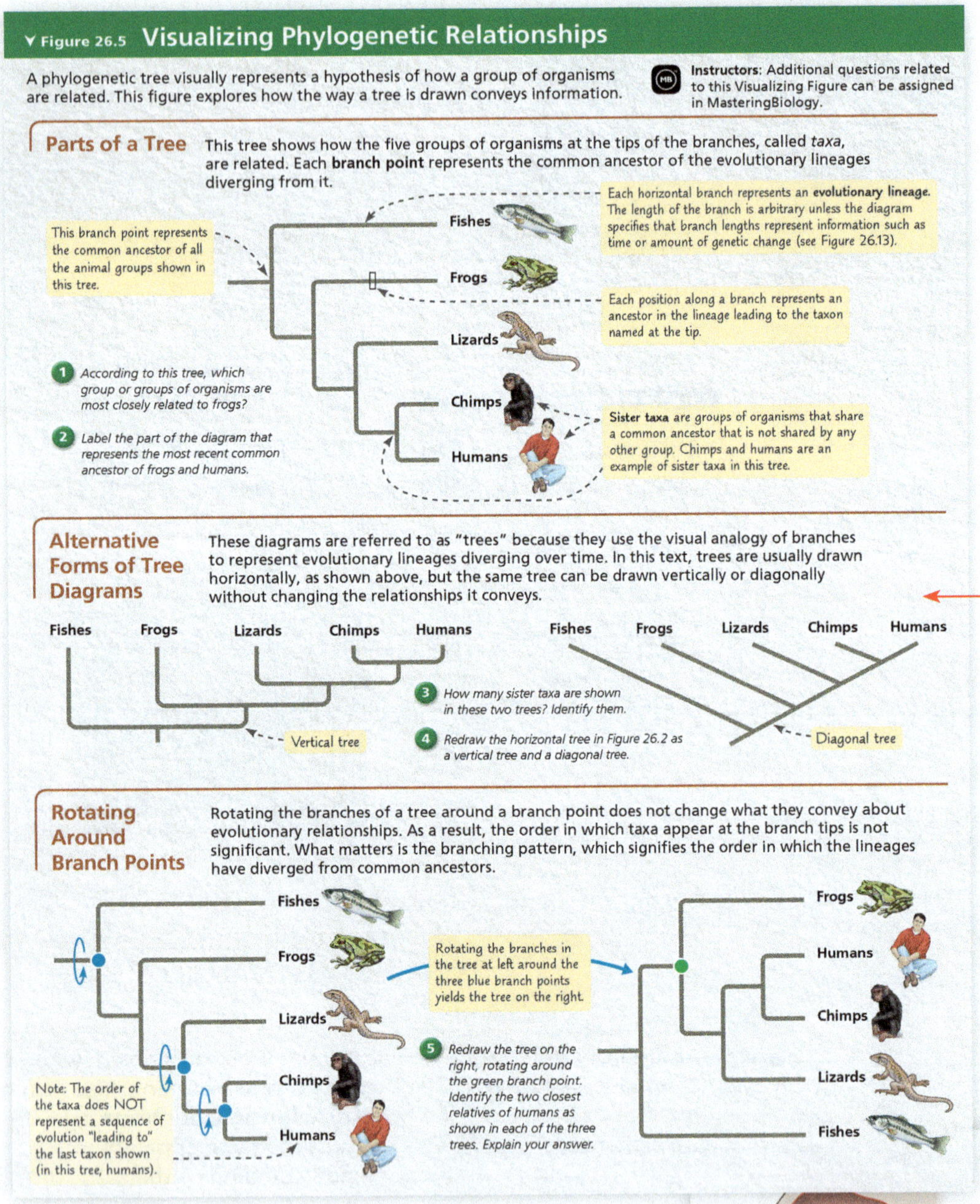
Figure 26.5 Visualizing Phylogenetic Relationships

A phylogenetic tree visually represents a hypothesis of how a group of organisms are related. This figure explores how the way a tree is drawn conveys information.

Instructors: Additional questions related to this Visualizing Figure can be assigned in MasteringBiology.

Parts of a Tree This tree shows how the five groups of organisms at the tips of the branches, called *taxa*, are related. Each **branch point** represents the common ancestor of the evolutionary lineages diverging from it.

1. *According to this tree, which group or groups of organisms are most closely related to frogs?*
2. *Label the part of the diagram that represents the most recent common ancestor of frogs and humans.*

Alternative Forms of Tree Diagrams These diagrams are referred to as "trees" because they use the visual analogy of branches to represent evolutionary lineages diverging over time. In this text, trees are usually drawn horizontally, as shown above, but the same tree can be drawn vertically or diagonally without changing the relationships it conveys.

3. *How many sister taxa are shown in these two trees? Identify them.*
4. *Redraw the horizontal tree in Figure 26.2 as a vertical tree and a diagonal tree.*

Rotating Around Branch Points Rotating the branches of a tree around a branch point does not change what they convey about evolutionary relationships. As a result, the order in which taxa appear at the branch tips is not significant. What matters is the branching pattern, which signifies the order in which the lineages have diverged from common ancestors.

5. *Redraw the tree on the right, rotating around the green branch point. Identify the two closest relatives of humans as shown in each of the three trees. Explain your answer.*

For more practice, each Visualizing Figure is accompanied by an automatically graded assignment in MasteringBiology with feedback for students.

Visualizing Figures include:

NEW! Visual Skills Questions give students practice interpreting illustrations and photos in the text.

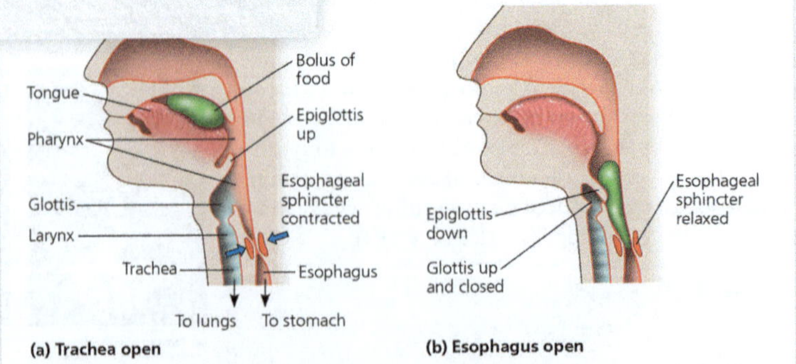

VISUAL SKILLS > *If you laugh while drinking water, the liquid may be ejected from your nostrils. Use this diagram to explain why this happens, taking into account that laughing involves exhaling.*

NEW! Figure Walkthroughs guide students through key figures with narrated explanations, figure markups, and questions that reinforce important points.

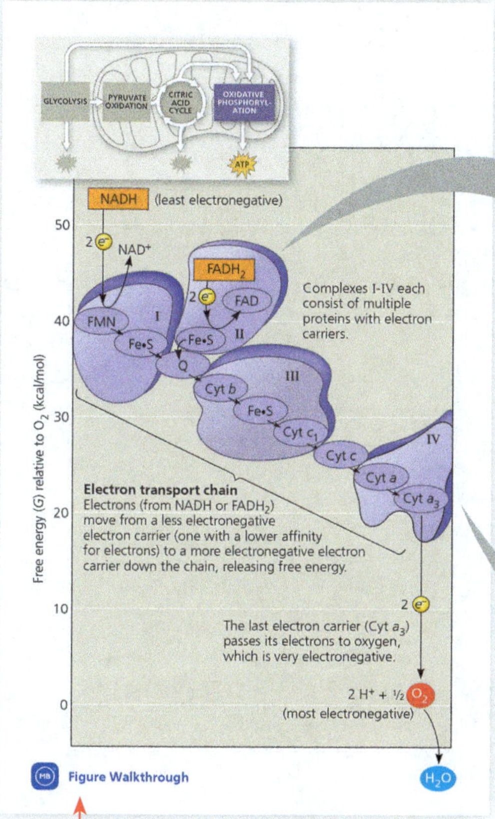

A note in the print book lets students and instructors know when a Figure Walkthrough is available in the Study Area.

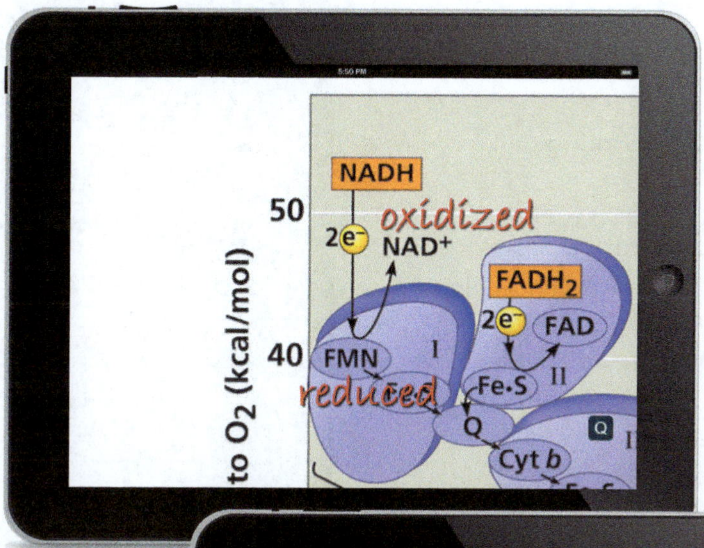

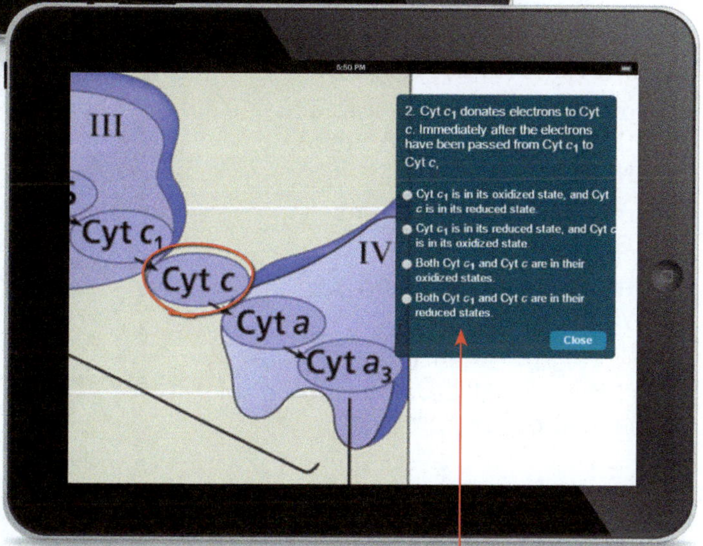

Questions embedded in each Figure Walkthrough encourage students to be active participants in their learning. The Figure Walkthroughs can also be assigned in MasteringBiology with higher-level questions.

➤ Figure 2.17
Photosynthesis: a solar-powered rearrangement of matter. *Elodea*, a freshwater plant, produces sugar by rearranging the atoms of carbon dioxide and water in the chemical process known as photosynthesis, which is powered by sunlight. Much of the sugar is then converted to other food molecules. Oxygen gas (O_2) is a by-product of photosynthesis; notice the bubbles of O_2 gas escaping from the leaves submerged in water.

DRAW IT ➤ *Add labels and arrows on the photo showing the reactants and products of photosynthesis as it takes place in a leaf.*

EXPANDED! Draw It exercises give students practice creating visuals. Students are asked to put pencil to paper and draw a structure, annotate a figure, or graph experimental data.

Make Connections Visually

Eleven **Make Connections Figures** pull together content from different chapters, providing a visual representation of "big picture" relationships.

Make Connections Figures include:

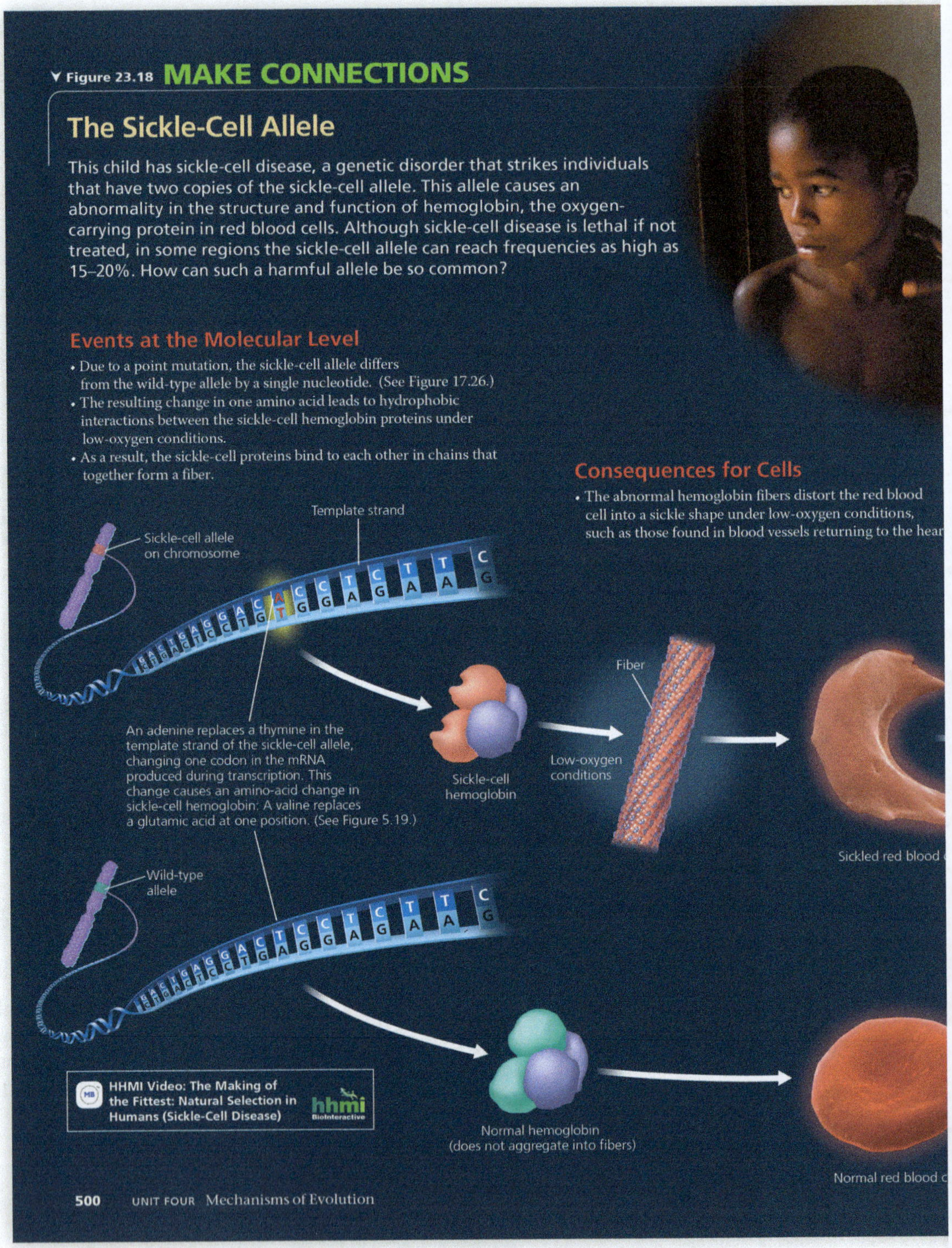

NEW! Media references integrated into the text direct students to digital content in the MasteringBiology Study Area that will help them prepare for class and succeed on exams. Video resources include HHMI BioInteractive Short Films (documentary-quality movies from the Howard Hughes Medical Institute) and much more.

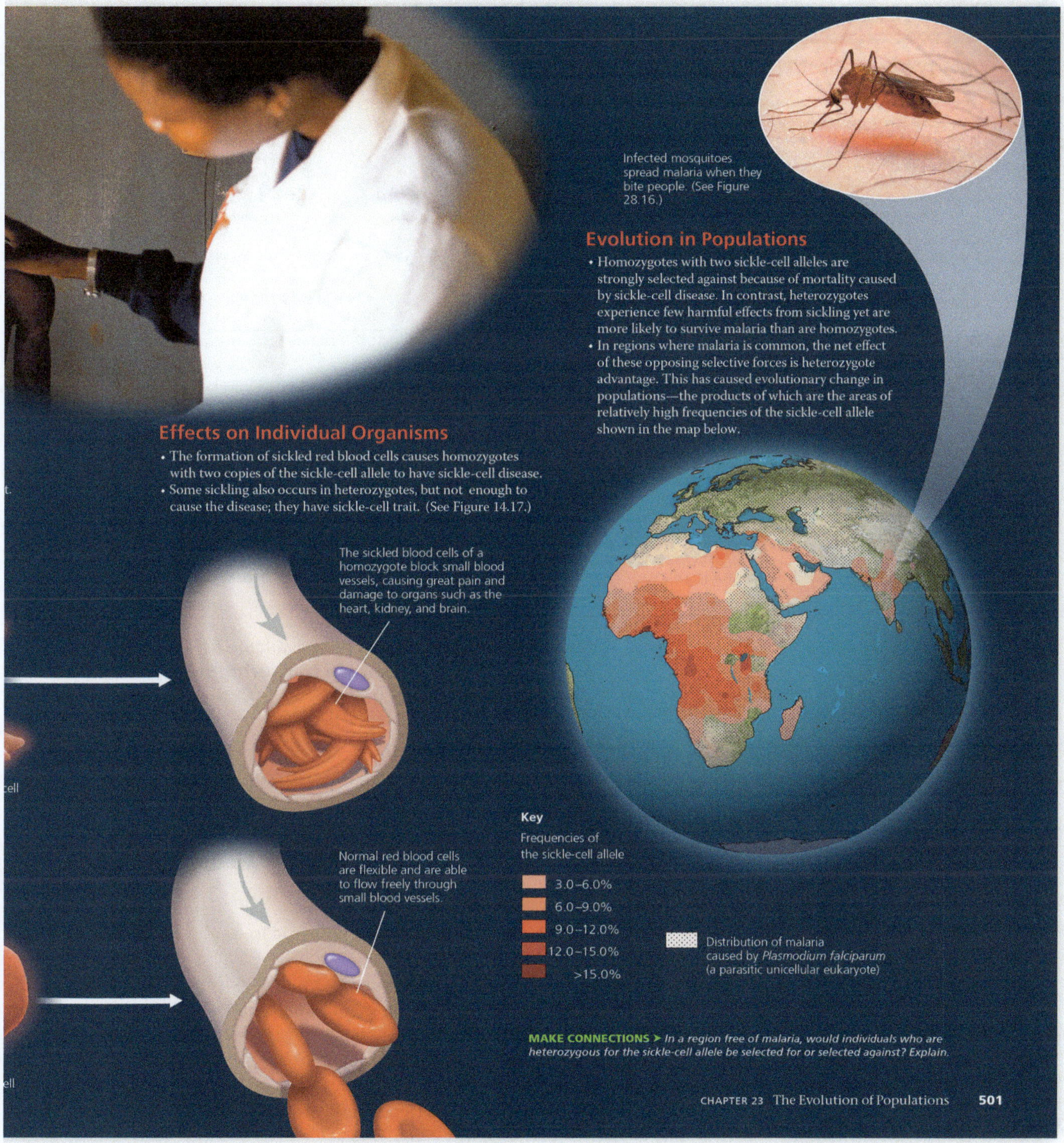

Infected mosquitoes spread malaria when they bite people. (See Figure 28.16.)

Evolution in Populations

- Homozygotes with two sickle-cell alleles are strongly selected against because of mortality caused by sickle-cell disease. In contrast, heterozygotes experience few harmful effects from sickling yet are more likely to survive malaria than are homozygotes.
- In regions where malaria is common, the net effect of these opposing selective forces is heterozygote advantage. This has caused evolutionary change in populations—the products of which are the areas of relatively high frequencies of the sickle-cell allele shown in the map below.

Effects on Individual Organisms

- The formation of sickled red blood cells causes homozygotes with two copies of the sickle-cell allele to have sickle-cell disease.
- Some sickling also occurs in heterozygotes, but not enough to cause the disease; they have sickle-cell trait. (See Figure 14.17.)

MAKE CONNECTIONS ➤ *In a region free of malaria, would individuals who are heterozygous for the sickle-cell allele be selected for or selected against? Explain.*

CHAPTER 23 The Evolution of Populations 501

Make Connections Questions in every chapter ask students to relate content in the chapter to material presented earlier in the course.

Practice Scientific Skills

Scientific Skills Exercises use real data to build key skills needed for biology, including **data analysis**, **graphing**, **experimental design**, and **math skills**.

Each Scientific Skills Exercise is based on an **experiment related to the chapter content.**

Most Scientific Skills Exercises use **data from published research,** which is cited in the exercise.

Questions build in difficulty, walking students through new skills step by step and providing opportunities for higher-level critical thinking.

SCIENTIFIC SKILLS EXERCISE

Interpreting a Scatter Plot with Two Sets of Data

Is Glucose Uptake into Cells Affected by Age? Glucose, an important energy source for animals, is transported into cells by facilitated diffusion using protein carriers. In this exercise, you will interpret a graph with two sets of data from an experiment that examined glucose uptake over time in red blood cells from guinea pigs of different ages. You will determine if the cells' rate of glucose uptake depended on the age of the guinea pigs.

How the Experiment Was Done Researchers incubated guinea pig red blood cells in a 300 m*M* (millimolar) radioactive glucose solution at pH 7.4 at 25°C. Every 10 or 15 minutes, they removed a sample of cells and measured the concentration of radioactive glucose inside those cells. The cells came from either a 15-day-old or a 1-month-old guinea pig.

Data from the Experiment When you have multiple sets of data, it can be useful to plot them on the same graph for comparison. In the graph here, each set of dots (of the same color) forms a *scatter plot*, in which every data point represents two numerical values, one for each variable. For each data set, a curve that best fits the points has been drawn to make it easier to see the trends. (For additional information about graphs, see the Scientific Skills Review in Appendix F.)

Glucose Uptake over Time in Guinea Pig Red Blood Cells

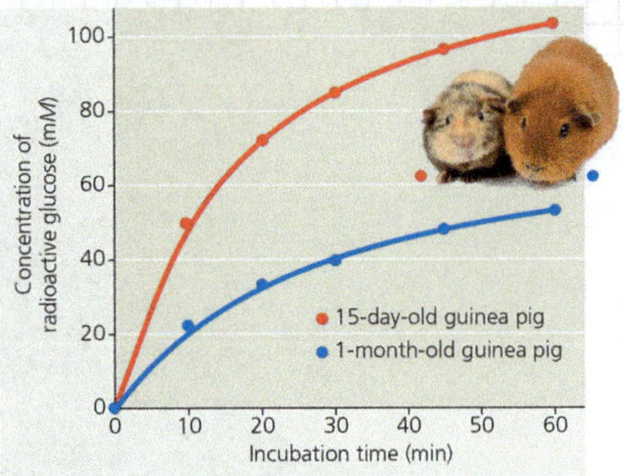

Data from T. Kondo and E. Beutler, Developmental changes in glucose transport of guinea pig erythrocytes, *Journal of Clinical Investigation* 65:1–4 (1980).

INTERPRET THE DATA

1. First make sure you understand the parts of the graph. (a) Which variable is the independent variable—the variable controlled by the researchers? (b) Which variable is the dependent variable—the variable that depended on the treatment and was measured by the researchers? (c) What do the red dots represent? (d) The blue dots?
2. From the data points on the graph, construct a table of the data. Put "Incubation Time (min)" in the left column of the table.
3. What does the graph show? Compare and contrast glucose uptake in red blood cells from 15-day-old and 1-month-old guinea pigs.
4. Develop a hypothesis to explain the difference between glucose uptake in red blood cells from 15-day-old and 1-month-old guinea pigs. (Think about how glucose gets into cells.)
5. Design an experiment to test your hypothesis.

Instructors: A version of this Scientific Skills Exercise can be assigned in MasteringBiology.

All Scientific Skills Exercises are available as **interactive assignments in MasteringBiology** that are automatically graded.

SCIENTIFIC SKILLS EXERCISES are available for every chapter:

Apply Scientific Skills to Solving Problems

NEW! Problem-Solving Exercises guide students in applying scientific skills and interpreting real data in the context of solving a real-world problem.

PROBLEM-SOLVING EXERCISE

Are you a victim of fish fraud?

When buying salmon, perhaps you prefer the more expensive wild-caught Pacific salmon (*Oncorhynchus* species) over farmed Atlantic salmon (*Salmo salar*). But studies reveal that about 40% of the time, you aren't getting the fish you paid for! Watch the video in the MasteringBiology Study Area for more information.

ABC News Video: Fake Fish in Stores and Restaurants

Instructors: A version of this Problem-Solving Exercise can be assigned in Chapter 5 of MasteringBiology. A more extensive investigation is in Chapter 26 of MasteringBiology.

In this exercise, you will investigate whether a piece of salmon has been fraudulently labeled.

Your Approach The principle guiding your investigation is that DNA sequences from within a species or from closely related species are more similar to each other than are sequences from more distantly related species.

Your Data You've been sold a piece of salmon labeled as coho salmon (*Oncorhynchus kisutch*). To see whether your fish was labeled correctly, you will compare a short DNA sequence from your sample to standard sequences from the same gene for three salmon species. The sequences are:

	Sample labeled as *O. kisutch* (coho salmon)	5′-CGGCACCGCCCTAAGTCTCT-3′
Standard sequences	Sequence for *O. kisutch* (coho salmon)	5′-AGGCACCGCCCTAAGTCTAC-3′
	Sequence for *O. keta* (chum salmon)	5′-AGGCACCGCCCTGAGCCTAC-3′
	Sequence for *Salmo salar* (Atlantic salmon)	5′-CGGCACCGCCCTAAGTCTCT-3′

Your Analysis

1. Scan along the standard sequences (*O. kisutch*, *O. keta*, and *S. salar*), base by base, circling any bases that do not match the sequence from your fish sample.
2. How many bases differ between (a) *O. kisutch* and your fish sample? (b) *O. keta* and the sample? (c) *S. salar* and the sample?
3. For each standard, what percentage of its bases are identical to your sample?
4. Based on these data alone, state a hypothesis for the species identity of your sample. What is your reasoning?

A version of each Problem-Solving Exercise can also be assigned in MasteringBiology.

Problem-Solving Exercises include:

* Available only in MasteringBiology. All other Scientific Skills Exercises are in the print book, eText, and MasteringBiology.

Bring Biology to Life

NEW! Over 450 carefully chosen and edited **videos and animations** have been integrated into the print book, MasteringBiology Study Area, and eText 2.0 at point of use to help students learn biology visually. (eText 2.0 will be live for Fall 2017 classes.)

Media references in the print book direct students to digital resources in the Study Area and eText 2.0:

- **NEW!** Get Ready for This Chapter questions
- **NEW!** Figure Walkthroughs
- **NEW!** HHMI BioInteractive Videos and Animations
- **NEW!** QR codes and URLs for easy access to Vocabulary Self-Quizzes and Practice Tests
- BioFlix Animations
- ABC News Videos
- Campbell Interviews

and much more

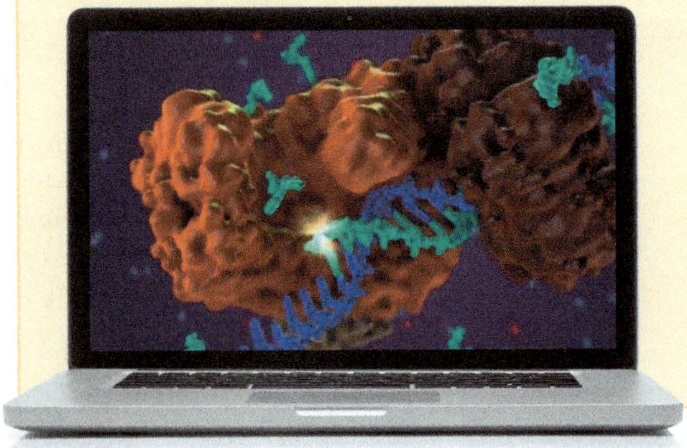

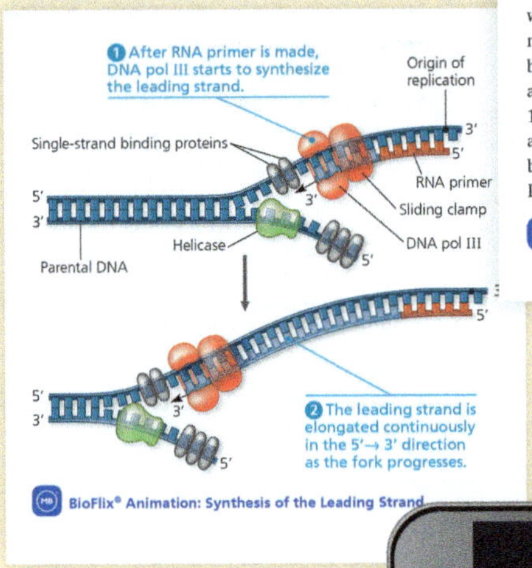

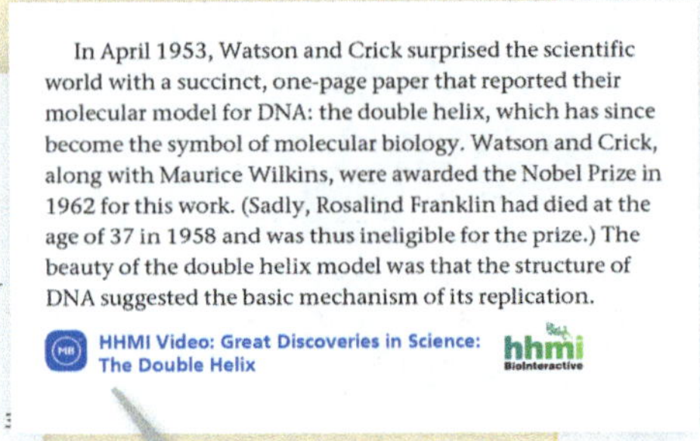
In April 1953, Watson and Crick surprised the scientific world with a succinct, one-page paper that reported their molecular model for DNA: the double helix, which has since become the symbol of molecular biology. Watson and Crick, along with Maurice Wilkins, were awarded the Nobel Prize in 1962 for this work. (Sadly, Rosalind Franklin had died at the age of 37 in 1958 and was thus ineligible for the prize.) The beauty of the double helix model was that the structure of DNA suggested the basic mechanism of its replication.

HHMI Video: Great Discoveries in Science: The Double Helix

Access the complete textbook online!
The Campbell eText includes powerful interactive and customization functions, such as instructor and student note-taking, highlighting, bookmarking, search, and links to glossary terms.

eText 1.0 will be available for Spring 2017 classes.

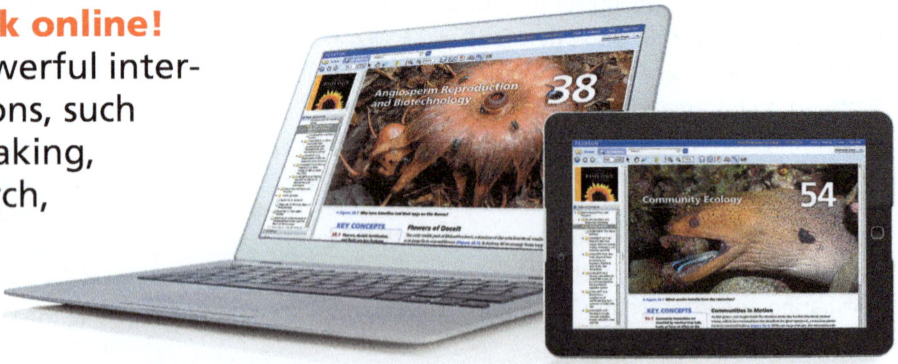

NEW! eText 2.0 will be available for Fall 2017 classes with all of the above plus links to animations, videos, and quizzes, all accessible from a smartphone, tablet, and computer.

Succeed with MasteringBiology

MasteringBiology improves results by engaging students before, during, and after class.

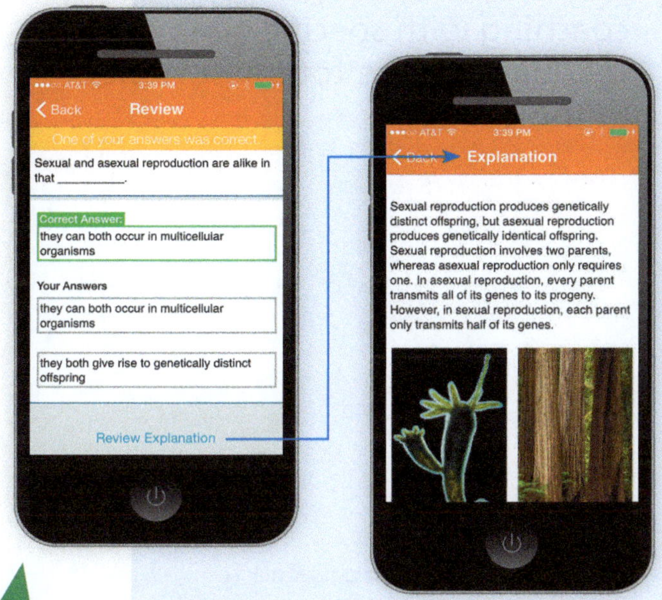

Before Class

Dynamic Study Modules provide students with multiple sets of questions with extensive feedback so that they can **test, learn, and retest** until they achieve mastery of the textbook material.

NEW! Get Ready for This Chapter quizzes help students review content they need to understand from previous chapters (see p. x).

Pre-Class Reading Quizzes help students pinpoint concepts that they understand and concepts that they need to review.

During Class

NEW! For ideas for in-class activities, see the **Ready-to-Go Teaching Modules** (see p. ix).

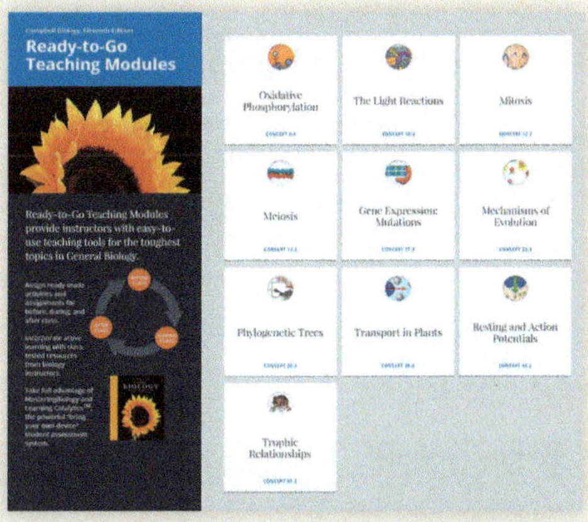

After Class

Hundreds of self-paced tutorials and coaching activities provide students with individualized coaching with specific hints and feedback on the toughest topics in the course.

Optional Adaptive Follow-up Assignments are based on each student's performance on the original MasteringBiology assignment and provide additional questions and activities tailored to each student's needs.

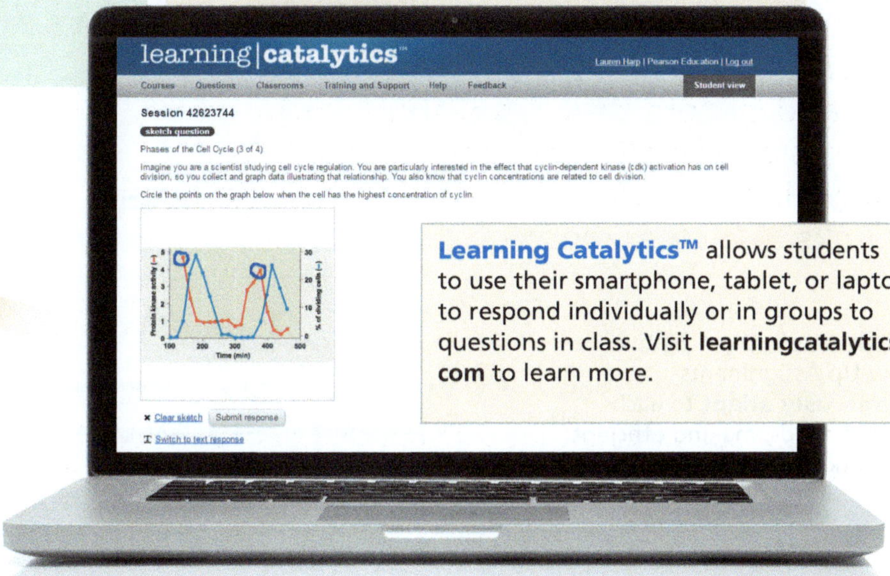

Learning Catalytics™ allows students to use their smartphone, tablet, or laptop to respond individually or in groups to questions in class. Visit **learningcatalytics.com** to learn more.

Personalized Coaching in MasteringBiology

Instructors can assign self-paced **MasteringBiology tutorials** that provide students with individualized coaching with specific hints and feedback on the toughest topics in the course.

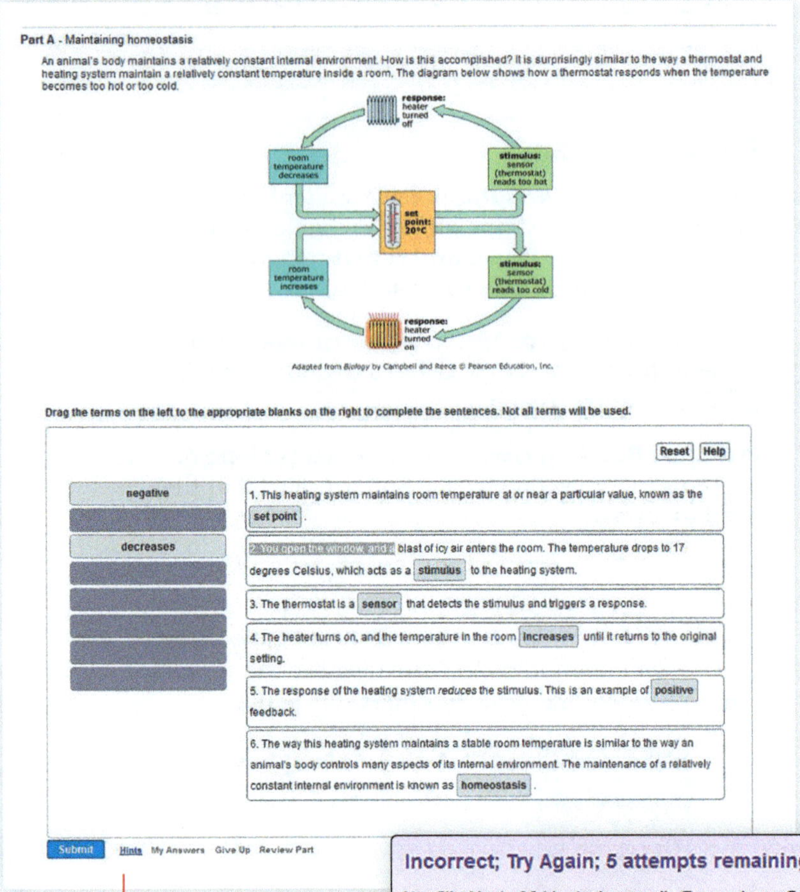

1. If a student gets stuck...
2. Specific wrong-answer **feedback** appears in the purple feedback box.

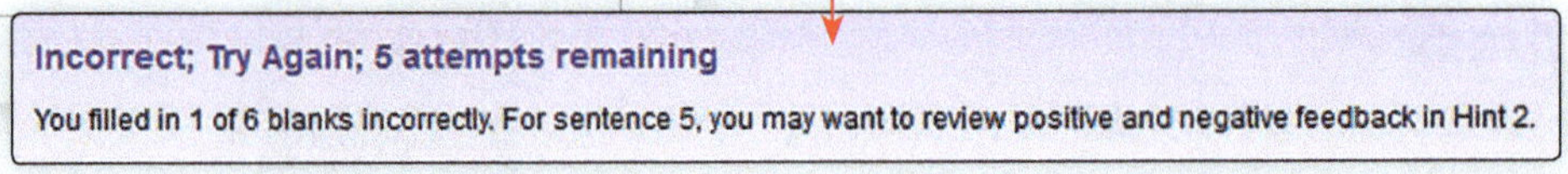

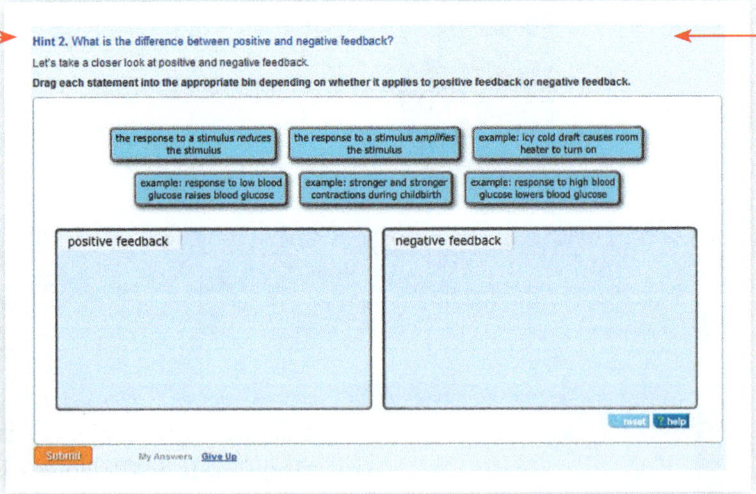

3. **Hints** coach students to the correct response.

4. Optional **Adaptive Follow-Up Assignments** are based on the original homework assignment and provide additional coaching and practice as needed.

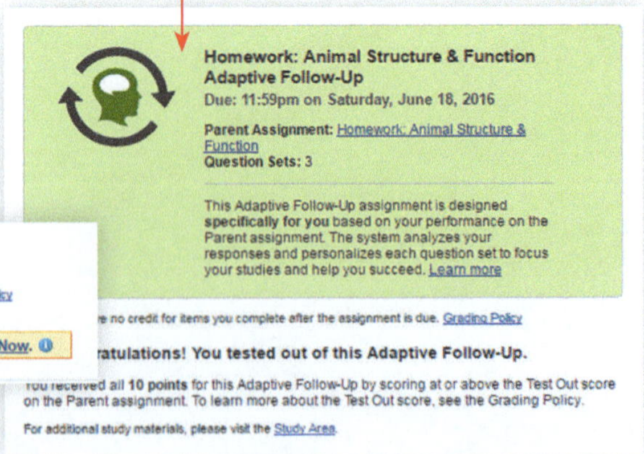

Question sets in the Adaptive Follow-Up Assignments **continuously adapt** to each student's needs, making efficient use of study time.

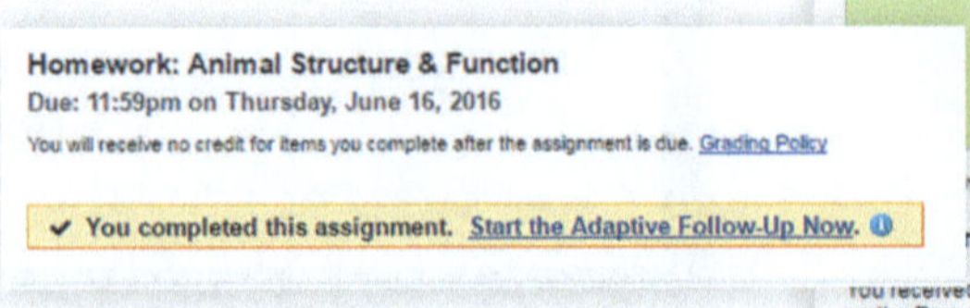

MasteringBiology offers thousands of tutorials, activities, and questions that can be assigned as homework. A few examples are shown below.

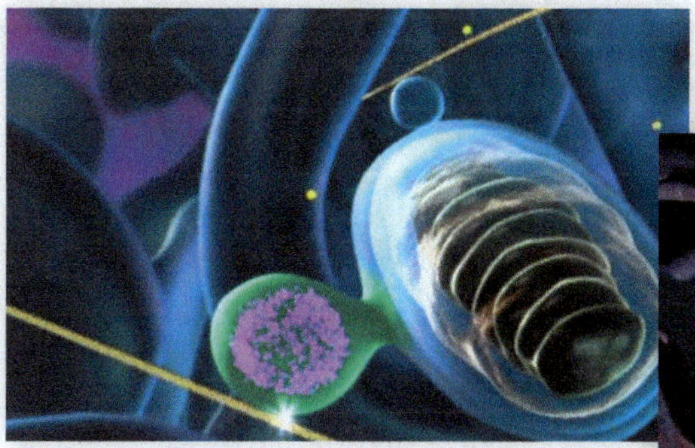

BioFlix Tutorials use 3-D, movie-quality animations and coaching exercises to help students master tough topics outside of class. Animations are also available in the Study Area and eText 2.0, and can be shown in class.

EXPANDED! HHMI BioInteractive Short Films, documentary-quality movies from the Howard Hughes Medical Institute, engage students in topics from the discovery of the double helix to evolution, with assignable questions.

NEW! Galápagos Evolution Video Activities, filmed on the Galápagos Islands by Peter and Rosemary Grant, bring to life the dynamic evolutionary processes that impact Darwin's finches on Daphne Major Island. Videos explore important concepts and data from the Grants' field research, with assignable activities.

The **MasteringBiology Gradebook** provides instructors with quick results and easy-to-interpret insights into student performance. Every assignment is automatically graded. Shades of red highlight vulnerable students and challenging assignments.

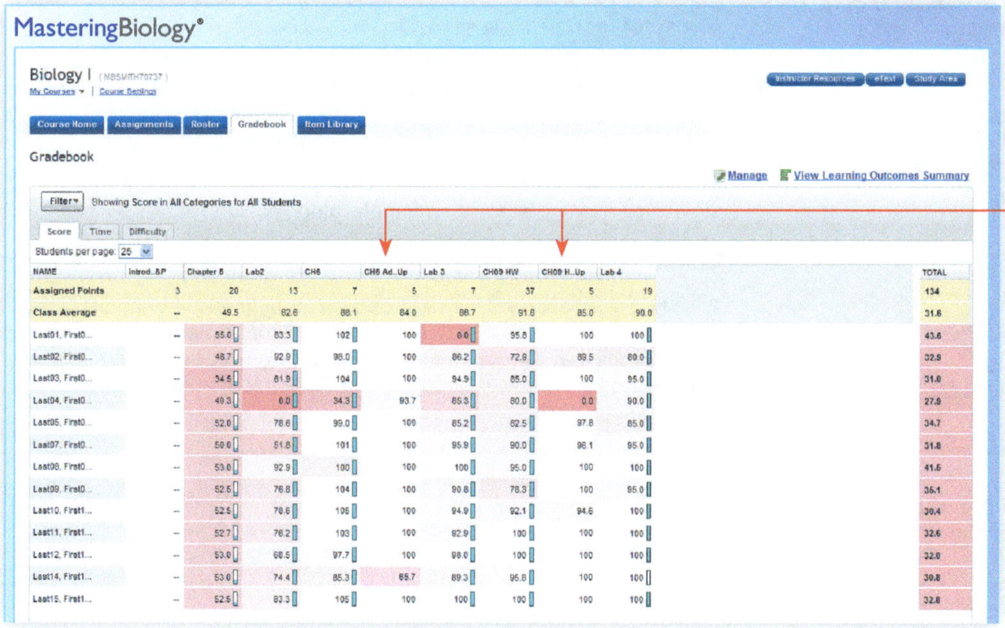

NAME	Introd..&P	Chapter 5	Lab2	CH6	CH6 Ad..Up	Lab 3	CH09 HW	CH09 H..Up	Lab 4	TOTAL
Assigned Points	3	20	13	7	5	7	37	5	19	134
Class Average	--	49.5	82.6	88.1	84.0	88.7	91.6	85.0	90.0	31.6
Last01, First0...	--	55.0	83.3	102	100	0.0	95.6	100	100	43.6
Last02, First0...	--	48.7	92.9	98.0	100	86.2	72.9	89.5	80.0	32.9
Last03, First0...	--	34.5	81.9	104	100	94.9	85.0	100	95.0	31.0
Last04, First0...	--	40.3	0.0	34.3	93.7	65.3	80.0	0.0	90.0	27.9
Last05, First0...	--	52.0	78.6	99.0	100	85.2	82.5	97.8	85.0	34.7
Last07, First0...	--	50.0	51.8	101	100	95.9	90.0	96.1	95.0	31.8
Last08, First0...	--	53.0	92.9	100	100	100	95.0	100	100	41.5
Last09, First0...	--	52.5	76.8	104	100	90.8	78.3	100	95.0	35.1
Last10, First1...	--	52.5	78.6	105	100	94.9	92.1	94.6	100	30.4
Last11, First1...	--	52.7	76.2	103	100	92.9	100	100	100	32.6
Last12, First1...	--	53.0	68.5	97.7	100	98.0	100	100	100	32.0
Last14, First1...	--	53.0	74.4	85.3	65.7	89.3	95.6	100	100	30.8
Last15, First1...	--	52.5	83.3	105	100	100	100	100	100	32.8

Student scores on the optional **Adaptive Follow-Up Assignments** are recorded in the gradebook and offer additional diagnostic information for instructors to monitor learning outcomes and more.

Student and Lab Supplements

For Students

Study Guide, Eleventh Edition

by Martha R. Taylor, Ithaca, New York, and Michael Pollock, Mount Royal University

978-0-134-44377-5 / 0-134-44377-2

This popular study aid provides concept maps, chapter summaries, word roots, and a variety of interactive activities including multiple-choice, short-answer essay, art labeling, and graph interpretation questions.

Inquiry in Action: Interpreting Scientific Papers, Fourth Edition*

by Ruth V. Buskirk, University of Texas at Austin and Christopher M. Gillen, Kenyon College

978-0-134-47861-6 / 0-134-47861-4

This guide helps students learn how to read and understand primary research articles. Part A presents complete articles accompanied by questions that help students analyze the article. Related Inquiry Figures are included in the supplement. Part B covers every part of a research paper, explaining the aim of the sections and how the paper works as a whole.

Practicing Biology: A Student Workbook, Sixth Edition*

by Jean Heitz and Cynthia Giffen, University of Wisconsin, Madison

978-0-134-48603-1 / 0-134-48603-X

This workbook offers a variety of activities to suit different learning styles. Activities such as modeling and concept mapping allow students to visualize and understand biological processes. Other activities focus on basic skills, such as reading and drawing graphs.

Biological Inquiry: A Workbook of Investigative Cases, Fifth Edition*

by Margaret Waterman, Southeast Missouri State University and Ethel Stanley, BioQUEST Curriculum Consortium and Beloit College

978-0-134-48646-8 / 0-134-48646-3

This workbook offers ten investigative cases. Each case study requires students to synthesize information from multiple chapters of the text and apply that knowledge to a real-world scenario as they pose hypotheses, gather new information, analyze evidence, graph data, and draw conclusions. A link to a student website is in the Study Area in MasteringBiology.

Study Card, Eleventh Edition

978-0-134-48648-2 / 0-134-48648-X

This quick-reference card provides students with an overview of the entire field of biology, helping them see the connections between topics.

Spanish Glossary for Biology

By Laura P. Zanello, University of California, Riverside

978-0-32183498-0 / 0-321-83498-4

This resource provides definitions in Spanish for the glossary terms.

Into The Jungle: Great Adventures in the Search for Evolution

by Sean B. Carroll, University of Wisconsin, Madison

978-0-32155671-4 / 0-321-55671-2

These nine short tales vividly depict key discoveries in evolutionary biology and the excitement of the scientific process. Online resources available at www.aw-bc.com/carroll.

Get Ready for Biology

by Lori K. Garrett, Parkland College

978-0-32150057-1 / 0-321-50057-1

This engaging workbook helps students brush up on important math and study skills and get up to speed on biological terminology and the basics of chemistry and cell biology. Also available in MasteringBiology.

A Short Guide to Writing About Biology, Ninth Edition

by Jan A. Pechenik, Tufts University

978-0-13414373-6 / 0-134-14373-6

This best-selling writing guide teaches students to think as biologists and to express ideas clearly and concisely through their writing.

For Lab

Investigating Biology Laboratory Manual, Ninth Edition

by Judith Giles Morgan, Emory University, and M. Eloise Brown Carter, Oxford College of Emory University

978-0-13447346-8 / 0-134-47346-9

With its distinctive investigative approach to learning, this best-selling laboratory manual is now more engaging than ever, with full-color art and photos throughout. The lab manual encourages students to participate in the process of science and develop creative and critical-reasoning skills.

Annotated Instructor's Edition for Investigating Biology Laboratory Manual, Ninth Edition

978-0-13451922-7 / 0-134-51922-1

This Annotated Instructor Edition features teaching information including margin notes with suggestions for lab procedures and answers to questions from the Student Edition. Also included is a detailed Teaching Plan at the end of each lab with specific suggestions for organizing labs, including estimated time allotments for each part of the lab and suggestions for encouraging independent thinking and collaborative discussions.

Preparation Guide for Investigating Biology, Ninth Edition

978-013451801-5 / 0-134-51801-2

Contains materials lists, suggested vendors, instructions for preparing solutions and constructing materials, schedules for planning advance preparation, and more. Available for downloading through the "Instructor Resources" area of MasteringBiology.

Pearson Collections Custom Library gives instructors the power to create custom lab manuals using Pearson content as well as original materials. Learn more at www.pearsonhighered.com/collections.

MasteringBiology® Virtual Labs is an online environment that promotes critical thinking skills using virtual experiments and explorations that can supplement or substitute for existing wet labs in microscopy, molecular biology, genetics, ecology, and systematics.

* An Instructor Guide is available for downloading through the "Instructor Resources" area of MasteringBiology.

Instructor Resources

Instructor's Resource CD/DVD-ROM Set for Campbell Biology, Eleventh Edition

978-0-13445458-0 / 0-134-45458-8

The Instructor's Resource DVD Package consists of a multi-disk set of assets for each chapter. Specific features include:

- Editable figures (art and photos) and tables from the text in PowerPoint®
- Prepared PowerPoint Lecture Presentations for each chapter with lecture notes, editable figures (art and photos), tables, and links to animations and videos
- JPEG images, including labeled and unlabeled art, photos from the text, and extra photos
- Clicker Questions in PowerPoint
- 400 instructor animations and videos, including BioFlix® 3-D animations and ABC News Videos
- Test Bank questions in TestGen® software and Microsoft® Word
- Digital Transparencies
- Quick Reference Guide: An at-a-glance reference to all instructor and student media resources available for each chapter

The Instructor Resources area of MasteringBiology includes:

- **NEW!** Ready-to-Go Teaching Modules help instructors efficiently make use of the available teaching tools for the toughest topics. Before-class assignments, in-class activities, and after-class assignments are provided for ease of use. Instructors can incorporate active learning into their course with the suggested activity ideas and clicker questions or Learning Catalytics questions.

- Editable figures (art and photos) and tables from the text in PowerPoint
- Prepared PowerPoint Lecture Presentations for each chapter with lecture notes, editable figures (art and photos), tables, and links to animations and videos
- JPEG images, including labeled and unlabeled art, photos from the text, and extra photos
- Clicker Questions in PowerPoint
- 400 instructor animations and videos, including BioFlix 3-D animations and ABC News Videos
- Test Bank (in Word and TestGen)
- Digital Transparencies
- Answers to Scientific Skills Exercises, Problem-Solving Exercises, Interpret the Data Questions, and Essay Questions; includes Rubric and Tips for Grading Short-Answer Essays
- Instructor Guides for Supplements: Instructor Guide for Practicing Biology: A Student Workbook, Instructor Guide for Biological Inquiry: A Workbook of Investigative Cases, Answer Key for Inquiry in Action: Interpreting Scientific Papers, Investigating Biology Lab Prep Guide, and Investigating Biology Lab Data Tables
- Quick Reference Guide

Printed Test Bank for Campbell Biology, Eleventh Edition

978-0-13447864-7 / 0-134-47864-9

This invaluable resource contains over 4,500 questions, including scenario-based questions and art, graph, and data interpretation questions. In addition to a print version, the Test Bank is available electronically through the "Instructor Resources" area of MasteringBiology, and in the Instructor's Resource DVD Package.

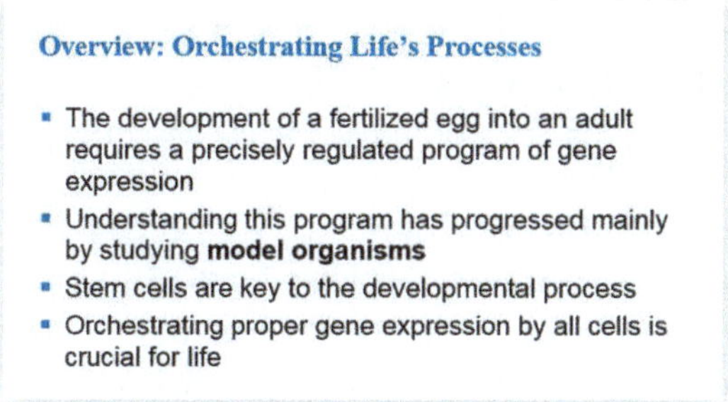

▲ **Customizable PowerPoint Lectures** provide a jumpstart for instruction.

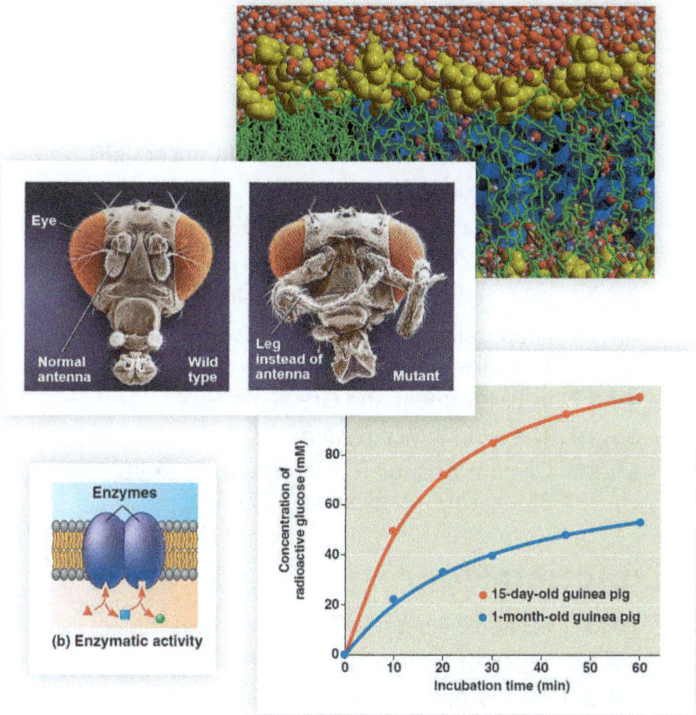

▲ **All of the art, graphs, and photos from the text** are provided with customizable labels. More than 1,600 photos from the text and other sources are included.

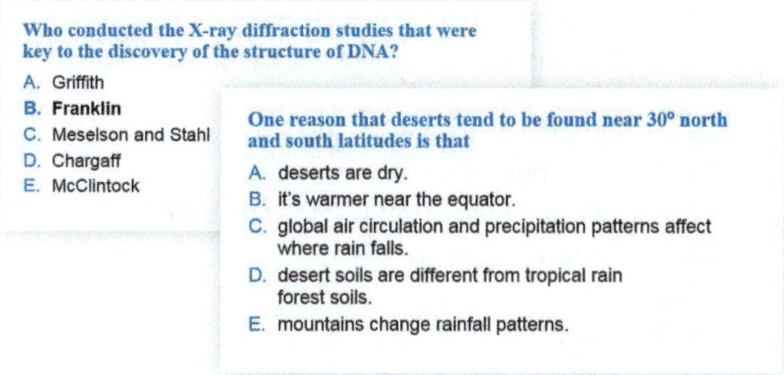

▲ **Clicker Questions** can be used to stimulate effective classroom discussions (for use with or without clickers).

Featured Figures

Visualizing Figures

Make Connections Figures

Exploring Figures

Inquiry Figures

Research Method Figures

*The Inquiry Figure, original research paper, and a worksheet to guide you through the paper are provided in *Inquiry in Action: Interpreting Scientific Papers*, Fourth Edition.
†A related Experimental Inquiry Tutorial can be assigned in MasteringBiology.®

Interviews

UNIT 1 THE CHEMISTRY OF LIFE 27

Lovell Jones
Prairie View A&M University and University of Texas MD Anderson Cancer Center

UNIT 2 THE CELL 92

Elba Serrano
New Mexico State University

UNIT 3 GENETICS 253

Shirley Tilghman
Princeton University

UNIT 4 MECHANISMS OF EVOLUTION 465

Jack Szostak
Harvard University

UNIT 5 THE EVOLUTIONARY HISTORY OF BIOLOGICAL DIVERSITY 550

Nancy Moran
University of Texas at Austin

UNIT 6 PLANT FORM AND FUNCTION 755

Philip Benfey
Duke University

UNIT 7 ANIMAL FORM AND FUNCTION 870

Harald zur Hausen
German Cancer Research Center

UNIT 8 ECOLOGY 1161

Tracy Langkilde
Penn State University

Acknowledgments

The authors wish to express their gratitude to the global community of instructors, researchers, students, and publishing professionals who have contributed to the Eleventh Edition of *CAMPBELL BIOLOGY*.

As authors of this text, we are mindful of the daunting challenge of keeping up to date in all areas of our rapidly expanding subject. We are grateful to the many scientists who helped shape this text by discussing their research fields with us, answering specific questions in their areas of expertise, and sharing their ideas about biology education. We are especially grateful to the following, listed alphabetically: Graham Alexander, John Archibald, Kristian Axelsen, Barbara Bowman, Joanne Chory, Roger Craig, Michael Hothorn, Patrick Keeling, Barrett Klein, Rachel Kramer Green, James Nieh, Kevin Peterson, T.K. Reddy, Andrew Roger, Alastair Simpson, Marty Taylor, and Elisabeth Wade. In addition, the biologists listed on pages xxviii–xxxi provided detailed reviews, helping us ensure the text's scientific accuracy and improve its pedagogical effectiveness. We thank Michael Pollock, author of the Study Guide, for his many contributions to the accuracy, clarity, and consistency of the text; and we thank Carolyn Wetzel, Ruth Buskirk, Joan Sharp, Jennifer Yeh, and Charlene D'Avanzo for their contributions to the Scientific Skills Exercises and Problem-Solving Exercises.

Thanks also to the other professors and students, from all over the world, who contacted the authors directly with useful suggestions. We alone bear the responsibility for any errors that remain, but the dedication of our consultants, reviewers, and other correspondents makes us confident in the accuracy and effectiveness of this text.

Interviews with prominent scientists have been a hallmark of *CAMPBELL BIOLOGY* since its inception, and conducting these interviews was again one of the great pleasures of revising the book. To open the eight units of this edition, we are proud to include interviews with Lovell Jones, Elba Serrano, Shirley Tilghman, Jack Szostak, Nancy Moran, Philip Benfey, Harald zur Hausen, and Tracy Langkilde.

We are especially grateful to Rebecca Orr for her hard work on the digital resources for the eText, Study Area, and Ready-to-Go Teaching Modules. And thanks to the rest of the Ready-to-Go Teaching Modules team: Molly Jacobs, Karen Resendes, Eileen Gregory, Angela Hodgson, Maureen Leupold, Jennifer Metzler, Allison Silveus, Jered Studinski, Sara Tallarovic, Judy Schoonmaker, Michael Pollock, and Chad Brassil. We would also like to extend our sincere appreciation to Carolyn Wetzel, Jennifer Yeh, Matt Lee, and Sherry Seston for their hard work on the Figure Walkthroughs. And our gratitude goes to Katie Cook for keeping these projects so well organized. Thanks also to Kaddee Lawrence for writing the questions that accompany the Visualizing Figures in MasteringBiology and to Mikaela Schmitt-Harsh for converting the Problem-Solving Exercises to MasteringBiology tutorials.

The value of *CAMPBELL BIOLOGY* as a learning tool is greatly enhanced by the supplementary materials that have been created for instructors and students. We recognize that the dedicated authors of these materials are essentially writing mini (and not so mini) books. We appreciate the hard work and creativity of all the authors listed, with their creations, on page xxii. We are also grateful to Kathleen Fitzpatrick and Nicole Tunbridge (PowerPoint® Lecture Presentations); Roberta Batorsky, Douglas Darnowski, James Langeland, and David Knochel (Clicker Questions); Sonish Azam, Kevin Friesen, Murty Kambhampati, Janet Lanza, Ford Lux, Chris Romero, Ruth Sporers, and David Knochel (Test Bank); Natalie Bronstein, Linda Logdberg, Matt McArdle, Ria Murphy, Chris Romero, and Andy Stull (Dynamic Study Modules); and Eileen Gregory, Rebecca Orr, and Elena Pravosudova (Adaptive Follow-up Assignments).

MasteringBiology™ and the other electronic accompaniments for this text are invaluable teaching and learning aids. We thank the hardworking, industrious instructors who worked on the revised and new media: Roberta Batorsky, Beverly Brown, Erica Cline, Willy Cushwa, Tom Kennedy, Tom Owens, Michael Pollock, Frieda Reichsman, Rick Spinney, Dennis Venema, Carolyn Wetzel, Heather Wilson-Ashworth, and Jennifer Yeh. We are also grateful to the many other people—biology instructors, editors, and production experts—who are listed in the credits for these and other elements of the electronic media that accompany the text.

CAMPBELL BIOLOGY results from an unusually strong synergy between a team of scientists and a team of publishing professionals.

Our editorial team at Pearson Education again demonstrated unmatched talents, commitment, and pedagogical insights. Our Courseware Portfolio Management Specialist, Josh Frost, brought publishing savvy, intelligence, and a much-appreciated level head to leading the whole team. The clarity and effectiveness of every page owe much to our extraordinary Supervising Editors Beth Winickoff and Pat Burner, who worked with a top-notch team of Courseware Senior Analysts in John Burner, Mary Ann Murray, Mary Hill, Laura Southworth, and Hilair Chism. Our unsurpassed Courseware Director of Content Development Ginnie Simione Jutson and Courseware Portfolio Management Director Beth Wilbur were indispensable in moving the project in the right direction. We also want to thank Robin Heyden for organizing the annual Biology Leadership Conferences and keeping us in touch with the world of AP Biology.

You would not have this beautiful text if not for the work of the production team: Director of Product Management Services Erin Gregg; Managing Producer Michael Early; Content Producer Lori Newman; Photo Researcher Maureen Spuhler; Copy Editor Joanna Dinsmore; Proofreader Pete Shanks; Rights & Permissions Manager Ben Ferrini; Managing Editor Angel Chavez and the rest of the staff at Integra Software Services, Inc.; Art Production Manager Rebecca Marshall, Artist Kitty Auble, and the rest of the staff at Lachina; Design Manager Marilyn Perry; Text and Cover Designer Elise Lansdon; and Manufacturing Buyer Stacey Weinberger. We also thank those who worked on the text's supplements: Josey Gist, Margaret Young, Kris Langan, Pete Shanks, Crystal Clifton and Jennifer Hastings at Progressive Publishing Alternatives, and Margaret McConnell at Integra. And for creating the wonderful package of electronic media that accompanies the text, we are grateful to Tania Mlawer, Sarah Jensen, Charles Hall, Katie Foley, Laura Tommasi, Lee Ann Doctor, Tod Regan, Libby Reiser, Jackie Jakob, Sarah Young-Dualan, Cady Owens, Caroline Ayres, Katie Cook, and Ziki Dekel as well as VP of Production and Digital Studio Lauren Fogel and Director of Digital Content Development Portfolio Management Stacy Treco.

For their important roles in marketing the text and media, we thank Christy Lesko, Lauren Harp, Kelly Galli, and Jane Campbell. For her market development support, we thank Jessica Moro. We are grateful to Adam Jaworski, VP Portfolio Management, Science, and Paul Corey, Managing Director, Higher Education Courseware, for their enthusiasm, encouragement, and support.

The Pearson sales team, which represents *CAMPBELL BIOLOGY* on campus, is an essential link to the users of the text. They tell us what you like and don't like about the text, communicate the features of the text, and provide prompt service. We thank them for their hard work and professionalism. For representing our text to our international audience, we thank our sales and marketing partners throughout the world. They are all strong allies in biology education.

Finally, we wish to thank our families and friends for their encouragement and patience throughout this long project. Our special thanks to Ross, Lily, and Alex (L.A.U.); Debra and Hannah (M.L.C.); Aaron, Sophie, Noah, and Gabriele (S.A.W.); Natalie (P.V.M.); and Paul, Dan, Maria, Armelle, and Sean (J.B.R.). And, as always, thanks to Rochelle, Allison, Jason, McKay, and Gus.

Lisa A. Urry, Michael L. Cain, Steven A. Wasserman, Peter V. Minorsky, and Jane B. Reece

Reviewers

Eleventh Edition Reviewers

Steve Abedon, *Ohio State University*
John Alcock, *Arizona State University*
Philip Allman, *Florida Gulf Coast College*
Rodney Allrich, *Purdue University*
Jim Barron, *Montana State University Billings*
Stephen Bauer, *Belmont Abbey College*
Aimee Bernard, *University of Colorado Denver*
Teresa Bilinski, *St. Edward's University*
Sarah Bissonnette, *University of California, Berkeley*
Jeffery Bowen, *Bridgewater State University*
Scott Bowling, *Auburn University*
David Broussard, *Lycoming College*
Tessa Burch, *University of Tennessee*
Warren Burggren, *University of North Texas*
Patrick Cafferty, *Emory University*
Michael Campbell, *Penn State University*
Jeffrey Carmichael, *University of North Dakota*
P. Bryant Chase, *Florida State University*
Steve Christenson, *Brigham Young University*
Curt Coffman, *Vincennes University*
Bill Cohen, *University of Kentucky*
Sean Coleman, *University of the Ozarks*
Erin Connolly, *University of South Carolina*
Ron Cooper, *University of California, Los Angeles*
Curtis Daehler, *University of Hawaii at Manoa*
Deborah Dardis, *Southeastern Louisiana University*
Douglas Darnowski, *Indiana University Southeast*
Jeremiah Davie, *D'Youville College*
Melissa Deadmond, *Truckee Meadows Community College*
Jennifer Derkits, *J. Sergeant Reynolds Community College*
Jean DeSaix, *University of Northern Carolina*
Kevin Dixon, *Florida State University*
David Dunbar, *Cabrini College*
Anna Edlund, *Lafayette College*
Rob Erdman, *Florida Gulf Coast College*
Dale Erskine, *Lebanon Valley College*
Susan Erster, *Stony Brook University*
Linda Fergusson-Kolmes, *Portland Community College, Sylvania Campus*
Danilo Fernando, *SUNY College of Environmental Science and Forestry, Syracuse*
Christina Fieber, *Horry-Georgetown Technical College*
Melissa Fierke, *SUNY College of Environmental Science and Forestry*
Mark Flood, *Fairmont State University*
Robert Fowler, *San Jose State University*
Stewart Frankel, *University of Hartford*
Eileen Gregory, *Rollins College*
Gokhan Hacisalihoglu, *Florida A&M University*
Monica Hall-Woods, *St. Charles Community College*
Jean Hardwick, *Ithaca College*
Deborah Harris, *Case Western Reserve University*
Chris Haynes, *Shelton State Community College*
Albert Herrera, *University of Southern California*
Karen Hicks, *Kenyon College*
Elizabeth Hobson, *New Mexico State University*
Mark Holbrook, *University of Iowa*
Erin Irish, *University of Iowa*
Sally Irwin, *University of Hawaii, Maui College*
Jamie Jensen, *Brigham Young University*
Jerry Johnson, *Corban University*
Ann Jorgensen, *University of Hawaii*
Ari Jumpponen, *Kansas State University*
Doug Kane, *Defiance College*
Kasey Karen, *Georgia College & State University*
Paul Kenrick, *Natural History Museum, London*
Stephen T. Kilpatrick, *University of Pittsburgh at Johnstown*
Shannon King, *North Dakota State University*
Karen M. Klein, *Northampton Community College*
Jacob Krans, *Western New England University*
Dubear Kroening, *University of Wisconsin*
Barbara Kuemerle, *Case Western Reserve University*
Jim Langeland, *Kalamazoo College*
Grace Lasker, *Lake Washington Institute of Technology*
Jani Lewis, *State University of New York at Geneseo*
Eric W. Linton, *Central Michigan University*
Tatyana Lobova, *Old Dominion University*
David Longstreth, *Louisiana State University*
Donald Lovett, *College of New Jersey*
Lisa Lyons, *Florida State University*
Mary Martin, *Northern Michigan University*
Scott Meissner, *Cornell University*
Jenny Metzler, *Ball State University*
Grace Miller, *Indiana Wesleyan University*
Jonathan Miller, *Edmonds Community College*
Mill Miller, *Wright State University*
Barbara Nash, *Mercy College*
Karen Neal, *J. Sargeant Reynolds Community College, Richmond*
Shawn Nordell, *Saint Louis University*
Olabisi Ojo, *Southern University at New Orleans*
Fatimata Pale, *Thiel College*
Susan Parrish, *McDaniel College*
Eric Peters, *Chicago State University*
Jarmila Pittermann, *University of California, Santa Cruz*
Jason Porter, *University of the Sciences in Philadelphia*
Elena Pravosudova, *University of Nevada, Reno*
Steven Price, *Virginia Commonwealth University*
Samiksha Raut, *University of Alabama at Birmingham*
Robert Reavis, *Glendale Community College*
Wayne Rickoll, *University of Puget Sound*
Luis Rodriguez, *San Antonio College*
Kara Rosch, *Blinn College*
Scott Russell, *University of Oklahoma*
Jodi Rymer, *College of the Holy Cross*
Per Salvesen, *University of Bergen*
Davison Sangweme, *University of North Georgia*
Karin Scarpinato, *Georgia Southern University*
Cara Schillington, *Eastern Michigan University*
David Schwartz, *Houston Community College*
Carrie Schwarz, *Western Washington University*
Joan Sharp, *Simon Fraser University*
Alison Sherwood, *University of Hawaii at Manoa*
Eric Shows, *Jones County Junior College*
Brian Shmaefsky, *Lone Star College*
John Skillman, *California State University, San Bernardino*
Rebecca Sperry, *Salt Lake Community College*
Clint Springer, *Saint Joseph's University*
Mark Sturtevant, *Oakland University*
Diane Sweeney, *Punahou School*
Kristen Taylor, *Salt Lake Community College*
Rebecca Thomas, *College of St. Joseph*
Martin Vaughan, *Indiana University-Purdue University Indianapolis*
Meena Vijayaraghavan, *Tulane University*
James T. Warren Jr.., *Pennsylvania State University*
Jim Wee, *Loyola University, New Orleans*
Charles Wellman, *Sheffield University*
Christopher Whipps, *State University of New York College of Environmental Science and Forestry*
Philip White, *James Hutton Institute*
Jessica White-Phillip, *Our Lady of the Lake University*
Robert Yost, *Indiana University-Purdue University Indianapolis*
Tia Young, *Pennsylvania State University*

Reviewers of Previous Editions

Kenneth Able, *State University of New York, Albany*; Thomas Adams, *Michigan State University*; Martin Adamson, *University of British Columbia*; Dominique Adriaens, *Ghent University*; Ann Aguanno, *Marymount Manhattan College*; Shylaja Akkaraju, *Bronx Community College of CUNY*; Marc Albrecht, *University of Nebraska*; John Alcock, *Arizona State University*; Eric Alcorn, *Acadia University*; George R. Aliaga, *Tarrant County College*; Rodney Allrich, *Purdue University*; Richard Almon, *State University of New York, Buffalo*; Bonnie Amos, *Angelo State University*; Katherine Anderson, *University of California, Berkeley*; Richard J. Andren, *Montgomery County Community College*; Estry Ang, *University of Pittsburgh, Greensburg*; Jeff Appling, *Clemson University*; J. David Archibald, *San Diego State University*; David Armstrong, *University of Colorado, Boulder*; Howard J. Arnott, *University of Texas, Arlington*; Mary Ashley, *University of Illinois, Chicago*; Angela S. Aspbury, *Texas State University*; Robert Atherton, *University of Wyoming*; Karl Aufderheide, *Texas A&M University*; Leigh Auleb, *San Francisco State University*; Terry Austin, *Temple College*; P. Stephen Baenziger, *University of Nebraska*; Brian Bagatto, *University of Akron*; Ellen Baker, *Santa Monica College*; Katherine Baker, *Millersville University*; Virginia Baker, *Chipola College*; Teri Balser, *University of Wisconsin, Madison*; William Barklow, *Framingham State College*; Susan Barman, *Michigan State University*; Steven Barnhart, *Santa Rosa Junior College*; Andrew Barton, *University of Maine Farmington*; Rebecca A. Bartow, *Western Kentucky University*; Ron Basmajian, *Merced College*; David Bass, *University of Central Oklahoma*; Bonnie Baxter, *Westminster College*; Tim Beagley, *Salt Lake Community College*; Margaret E. Beard, *College of the Holy Cross*; Tom Beatty, *University of British Columbia*; Chris Beck, *Emory University*; Wayne Becker, *University of Wisconsin, Madison*; Patricia Bedinger, *Colorado State University*; Jane Beiswenger, *University of Wyoming*; Anne Bekoff, *University of Colorado, Boulder*; Marc Bekoff, *University of Colorado, Boulder*; Tania Beliz, *College of*

San Mateo; Adrianne Bendich, *Hoffman-La Roche, Inc.*; Marilee Benore, *University of Michigan, Dearborn*; Barbara Bentley, *State University of New York, Stony Brook*; Darwin Berg, *University of California, San Diego*; Werner Bergen, *Michigan State University*; Gerald Bergstrom, *University of Wisconsin, Milwaukee*; Anna W. Berkovitz, *Purdue University*; Dorothy Berner, *Temple University*; Annalisa Berta, *San Diego State University*; Paulette Bierzychudek, *Pomona College*; Charles Biggers, *Memphis State University*; Kenneth Birnbaum, *New York University*; Catherine Black, *Idaho State University*; Michael W. Black, *California Polytechnic State University, San Luis Obispo*; William Blaker, *Furman University*; Robert Blanchard, *University of New Hampshire*; Andrew R. Blaustein, *Oregon State University*; Judy Bluemer, *Morton College*; Edward Blumenthal, *Marquette University*; Robert Blystone, *Trinity University*; Robert Boley, *University of Texas, Arlington*; Jason E. Bond, *East Carolina University*; Eric Bonde, *University of Colorado, Boulder*; Cornelius Bondzi, *Hampton University*; Richard Boohar, *University of Nebraska, Omaha*; Carey L. Booth, *Reed College*; Allan Bornstein, *Southeast Missouri State University*; David Bos, *Purdue University*; Oliver Bossdorf, *State University of New York, Stony Book*; James L. Botsford, *New Mexico State University*; Lisa Boucher, *University of Nebraska, Omaha*; J. Michael Bowes, *Humboldt State University*; Richard Bowker, *Alma College*; Robert Bowker, *Glendale Community College, Arizona*; Scott Bowling, *Auburn University*; Barbara Bowman, *Mills College*; Barry Bowman, *University of California, Santa Cruz*; Deric Bownds, *University of Wisconsin, Madison*; Robert Boyd, *Auburn University*; Sunny Boyd, *University of Notre Dame*; Jerry Brand, *University of Texas, Austin*; Edward Braun, *Iowa State University*; Theodore A. Bremner, *Howard University*; James Brenneman, *University of Evansville*; Charles H. Brenner, *Berkeley, California*; Lawrence Brewer, *University of Kentucky*; Donald P. Briskin, *University of Illinois, Urbana*; Paul Broady, *University of Canterbury*; Chad Brommer, *Emory University*; Judith L. Bronstein, *University of Arizona*; Danny Brower, *University of Arizona*; Carole Browne, *Wake Forest University*; Beverly Brown, *Nazareth College*; Mark Browning, *Purdue University*; David Bruck, *San Jose State University*; Robb T. Brumfield, *Louisiana State University*; Herbert Bruneau, *Oklahoma State University*; Gary Brusca, *Humboldt State University*; Richard C. Brusca, *University of Arizona, Arizona-Sonora Desert Museum*; Alan H. Brush, *University of Connecticut, Storrs*; Howard Buhse, *University of Illinois, Chicago*; Arthur Buikema, *Virginia Tech*; Beth Burch, *Huntington University*; Al Burchsted, *College of Staten Island*; Warren Burggren, *University of North Texas*; Meg Burke, *University of North Dakota*; Edwin Burling, *De Anza College*; Dale Burnside, *Lenoir-Rhyne University*; William Busa, *Johns Hopkins University*; Jorge Busciglio, *University of California, Irvine*; John Bushnell, *University of Colorado*; Linda Butler, *University of Texas, Austin*; David Byres, *Florida Community College, Jacksonville*; Guy A. Caldwell, *University of Alabama*; Jane Caldwell, *West Virginia University*; Kim A. Caldwell, *University of Alabama*; Ragan Callaway, *The University of Montana*; Kenneth M. Cameron, *University of Wisconsin, Madison*; R. Andrew Cameron, *California Institute of Technology*; Alison Campbell, *University of Waikato*; Iain Campbell, *University of Pittsburgh*; Patrick Canary, *Northland Pioneer College*; W. Zacheus Cande, *University of California, Berkeley*; Deborah Canington, *University of California, Davis*; Robert E. Cannon, *University of North Carolina, Greensboro*; Frank Cantelmo, *St. John's University*; John Capeheart, *University of Houston, Downtown*; Gregory Capelli, *College of William and Mary*; Cheryl Keller Capone, *Pennsylvania State University*; Richard Cardullo, *University of California, Riverside*; Nina Caris, *Texas A&M University*; Mickael Cariveau, *Mount Olive College*; Jeffrey Carmichael, *University of North Dakota*; Robert Carroll, *East Carolina University*; Laura L. Carruth, *Georgia State University*; J. Aaron Cassill, *University of Texas, San Antonio*; Karen I. Champ, *Central Florida Community College*; David Champlin, *University of Southern Maine*; Brad Chandler, *Palo Alto College*; Wei-Jen Chang, *Hamilton College*; Bruce Chase, *University of Nebraska, Omaha*; P. Bryant Chase, *Florida State University*; Doug Cheeseman, *De Anza College*; Shepley Chen, *University of Illinois, Chicago*; Giovina Chinchar, *Tougaloo College*; Joseph P. Chinnici, *Virginia Commonwealth University*; Jung H. Choi, *Georgia Institute of Technology*; Steve Christensen, *Brigham Young University, Idaho*; Geoffrey Church, *Fairfield University*; Henry Claman, *University of Colorado Health Science Center*; Anne Clark, *Binghamton University*; Greg Clark, *University of Texas*; Patricia J. Clark, *Indiana University-Purdue University, Indianapolis*; Ross C. Clark, *Eastern Kentucky University*; Lynwood Clemens, *Michigan State University*; Janice J. Clymer, *San Diego Mesa College*; Reggie Cobb, *Nashville Community College*; William P. Coffman, *University of Pittsburgh*; Austin Randy Cohen, *California State University, Northridge*; J. John Cohen, *University of Colorado Health Science Center*; James T. Colbert, *Iowa State University*; Sean Coleman, *University of the Ozarks*; Jan Colpaert, *Hasselt University*; Robert Colvin, *Ohio University*; Jay Comeaux, *McNeese State University*; David Cone, *Saint Mary's University*; Elizabeth Connor, *University of Massachusetts*; Joanne Conover, *University of Connecticut*; Gregory Copenhaver, *University of North Carolina, Chapel Hill*; John Corliss, *University of Maryland*; James T. Costa, *Western Carolina University*; Stuart J. Coward, *University of Georgia*; Charles Creutz, *University of Toledo*; Bruce Criley, *Illinois Wesleyan University*; Norma Criley, *Illinois Wesleyan University*; Joe W. Crim, *University of Georgia*; Greg Crowther, *University of Washington*; Karen Curto, *University of Pittsburgh*; William Cushwa, *Clark College*; Anne Cusic, *University of Alabama, Birmingham*; Richard Cyr, *Pennsylvania State University*; Marymegan Daly, *The Ohio State University*; Deborah Dardis, *Southeastern Louisiana University*; W. Marshall Darley, *University of Georgia*; Cynthia Dassler, *The Ohio State University*; Shannon Datwyler, *California State University, Sacramento*; Marianne Dauwalder, *University of Texas, Austin*; Larry Davenport, *Samford University*; Bonnie J. Davis, *San Francisco State University*; Jerry Davis, *University of Wisconsin, La Crosse*; Michael A. Davis, *Central Connecticut State University*; Thomas Davis, *University of New Hampshire*; Melissa Deadmond, *Truckee Meadows Community College*; John Dearn, *University of Canberra*; Maria E. de Bellard, *California State University, Northridge*; Teresa DeGolier, *Bethel College*; James Dekloe, *University of California, Santa Cruz*; Eugene Delay, *University of Vermont*; Patricia A. DeLeon, *University of Delaware*; Veronique Delesalle, *Gettysburg College*; T. Delevoryas, *University of Texas, Austin*; Roger Del Moral, *University of Washington*; Charles F. Delwiche, *University of Maryland*; Diane C. DeNagel, *Northwestern University*; William L. Dentler, *University of Kansas*; Daniel DerVartanian, *University of Georgia*; Jean DeSaix, *University of North Carolina, Chapel Hill*; Janet De Souza-Hart, *Massachusetts College of Pharmacy & Health Sciences*; Biao Ding, *Ohio State University*; Michael Dini, *Texas Tech University*; Andrew Dobson, *Princeton University*; Stanley Dodson, *University of Wisconsin, Madison*; Jason Douglas, *Angelina College*; Mark Drapeau, *University of California, Irvine*; John Drees, *Temple University School of Medicine*; Charles Drewes, *Iowa State University*; Marvin Druger, *Syracuse University*; Gary Dudley, *University of Georgia*; Susan Dunford, *University of Cincinnati*; Kathryn A. Durham, *Lorain Community College*; Betsey Dyer, *Wheaton College*; Robert Eaton, *University of Colorado*; Robert S. Edgar, *University of California, Santa Cruz*; Anna Edlund, *Lafayette College*; Douglas J. Eernisse, *California State University, Fullerton*; Betty J. Eidemiller, *Lamar University*; Brad Elder, *Doane College*; Curt Elderkin, *College of New Jersey*; William D. Eldred, *Boston University*; Michelle Elekonich, *University of Nevada, Las Vegas*; George Ellmore, *Tufts University*; Mary Ellard-Ivey, *Pacific Lutheran University*; Kurt Elliott, *North West Vista College*; Norman Ellstrand, *University of California, Riverside*; Johnny El-Rady, *University of South Florida*; Dennis Emery, *Iowa State University*; John Endler, *University of California, Santa Barbara*; Rob Erdman, *Florida Gulf Coast College*; Dale Erskine, *Lebanon Valley College*; Margaret T. Erskine, *Lansing Community College*; Gerald Esch, *Wake Forest University*; Frederick B. Essig, *University of South Florida*; Mary Eubanks, *Duke University*; David Evans, *University of Florida*; Robert C. Evans, *Rutgers University, Camden*; Sharon Eversman, *Montana State University*; Olukemi Fadayomi, *Ferris State University*; Lincoln Fairchild, *Ohio State University*; Peter Fajer, *Florida State University*; Bruce Fall, *University of Minnesota*; Sam Fan, *Bradley University*; Lynn Fancher, *College of DuPage*; Ellen H. Fanning, *Vanderbilt University*; Paul Farnsworth, *University of New Mexico*; Larry Farrell, *Idaho State University*; Jerry F. Feldman, *University of California, Santa Cruz*; Lewis Feldman, *University of California, Berkeley*; Myriam Alhadeff Feldman, *Cascadia Community College*; Eugene Fenster, *Longview Community College*; Russell Fernald, *University of Oregon*; Rebecca Ferrell, *Metropolitan State College of Denver*; Kim Finer, *Kent State University*; Milton Fingerman, *Tulane University*; Barbara Finney, *Regis College*; Teresa Fischer, *Indian River Community College*; Frank Fish, *West Chester University*; David Fisher, *University of Hawaii, Manoa*; Jonathan S. Fisher, *St. Louis University*; Steven Fisher, *University of California, Santa Barbara*; David Fitch, *New York University*; Kirk Fitzhugh, *Natural History Museum of Los Angeles County*; Lloyd Fitzpatrick, *University of North Texas*; William Fixsen, *Harvard University*; T. Fleming, *Bradley University*; Abraham Flexer, *Manuscript Consultant, Boulder, Colorado*; Margaret Folsom, *Methodist College*; Kerry Foresman, *University of Montana*; Norma Fowler, *University of Texas, Austin*; Robert G. Fowler, *San Jose State University*; David Fox, *University of Tennessee, Knoxville*; Carl Frankel, *Pennsylvania State University, Hazleton*; Robert Franklin, *College of Charleston*; James Franzen, *University of Pittsburgh*; Art Fredeen, *University of Northern British Columbia*; Kim Fredericks, *Viterbo University*; Bill Freedman, *Dalhousie University*; Matt Friedman, *University of Chicago*; Otto Friesen, *University of Virginia*; Frank Frisch, *Chapman University*; Virginia Fry, *Monterey Peninsula College*; Bernard Frye, *University of Texas, Arlington*; Jed Fuhrman, *University of Southern California*; Alice Fulton, *University of Iowa*; Chandler Fulton, *Brandeis University*; Sara Fultz, *Stanford University*; Berdell Funke, *North Dakota State University*; Anne Funkhouser, *University of the Pacific*; Zofia E. Gagnon, *Marist College*; Michael Gaines, *University of Miami*; Cynthia M. Galloway, *Texas A&M University, Kingsville*; Arthur W. Galston, *Yale University*; Stephen Gammie, *University of Wisconsin, Madison*; Carl Gans, *University of Michigan*; John Gapter, *University of Northern Colorado*; Andrea Gargas, *University of Wisconsin, Madison*; Lauren Garner, *California Polytechnic State University, San Luis Obispo*; Reginald Garrett, *University of Virginia*; Craig Gatto, *Illinois State University*; Kristen Genet, *Anoka Ramsey Community College*; Patricia Gensel, *University of North Carolina*; Chris George, *California Polytechnic State University, San Luis Obispo*; Robert George, *University of Wyoming*; J. Whitfield Gibbons, *University of Georgia*; J. Phil Gibson, *University of Oklahoma*; Frank Gilliam, *Marshall University*; Eric Gillock, *Fort Hayes State University*; Simon Gilroy, *University of Wisconsin, Madison*; Edwin Ginés-Candelaria, *Miami Dade College*; Alan D. Gishlick, *Gustavus Adolphus College*; Todd Gleeson, *University of Colorado*; Jessica Gleffe, *University of California, Irvine*; John Glendinning, *Barnard College*; David Glenn-Lewin, *Wichita State University*; William Glider, *University of Nebraska*; Tricia Glidewell, *Marist School*; Elizabeth A. Godrick, *Boston University*; Jim Goetze, *Laredo Community College*; Lynda Goff, *University of California, Santa Cruz*; Elliott Goldstein, *Arizona State University*; Paul Goldstein, *University of Texas, El Paso*; Sandra Gollnick, *State University of New York, Buffalo*; Roy Golsteyn, *University of Lethbridge*; Anne Good, *University of California, Berkeley*; Judith Goodenough, *University of Massachusetts, Amherst*; Wayne Goodey, *University of British Columbia*; Barbara E. Goodman, *University of South Dakota*; Robert Goodman, *University of Wisconsin, Madison*; Ester Goudsmit, *Oakland University*; Linda Graham, *University of Wisconsin, Madison*; Robert Grammer, *Belmont University*; Joseph Graves, *Arizona State University*; Eileen Gregory, *Rollins College*; Phyllis Griffard, *University of Houston, Downtown*; A. J. F. Griffiths, *University of British Columbia*; Bradley Griggs, *Piedmont Technical College*; William Grimes, *University of Arizona*; David Grise, *Texas A&M University, Corpus Christi*; Mark Gromko, *Bowling Green State University*; Serine Gropper, *Auburn University*; Katherine L. Gross, *Ohio State University*; Gary Gussin, *University of Iowa*; Edward Gruberg, *Temple University*; Carla Guthridge, *Cameron University*; Mark Guyer, *National Human Genome Research Institute*; Ruth Levy Guyer, *Bethesda, Maryland*; Carla Haas, *Pennsylvania State University*; R. Wayne Habermehl, *Montgomery County Community College*; Pryce Pete Haddix, *Auburn University*; Mac Hadley, *University of Arizona*; Joel Hagen, *Radford University*; Jack P. Hailman, *University of Wisconsin*; Leah Haimo, *University of California, Riverside*; Ken Halanych, *Auburn University*; Jody Hall, *Brown University*; Monica Hall-Woods, *St. Charles Community College*; Heather Hallen-Adams, *University of Nebraska, Lincoln*; Douglas Hallett, *Northern Arizona University*; Rebecca Halyard, *Clayton State College*; Devney Hamilton, *Stanford University* (student); E. William Hamilton, *Washington and Lee University*;

Matthew B. Hamilton, *Georgetown University*; Sam Hammer, *Boston University*; Penny Hanchey-Bauer, *Colorado State University*; William F. Hanna, *Massasoit Community College*; Dennis Haney, *Furman University*; Laszlo Hanzely, *Northern Illinois University*; Jeff Hardin, *University of Wisconsin, Madison*; Jean Hardwick, *Ithaca College*; Luke Harmon, *University of Idaho*; Lisa Harper, *University of California, Berkeley*; Jeanne M. Harris, *University of Vermont*; Richard Harrison, *Cornell University*; Stephanie Harvey, *Georgia Southwestern State University*; Carla Hass, *Pennsylvania State University*; Chris Haufler, *University of Kansas*; Bernard A. Hauser, *University of Florida*; Chris Haynes, *Shelton State Community College*; Evan B. Hazard, *Bemidji State University* (emeritus); H. D. Heath, *California State University, East Bay*; George Hechtel, *State University of New York, Stony Brook*; S. Blair Hedges, *Pennsylvania State University*; Brian Hedlund, *University of Nevada, Las Vegas*; David Heins, *Tulane University*; Jean Heitz, *University of Wisconsin, Madison*; Andreas Hejnol, *Sars International Centre for Marine Molecular Biology*; John D. Helmann, *Cornell University*; Colin Henderson, *University of Montana*; Susan Hengeveld, *Indiana University*; Michelle Henricks, *University of California, Los Angeles*; Caroll Henry, *Chicago State University*; Frank Heppner, *University of Rhode Island*; Albert Herrera, *University of Southern California*; Scott Herrick, *Missouri Western State College*; Ira Herskowitz, *University of California, San Francisco*; Paul E. Hertz, *Barnard College*; Chris Hess, *Butler University*; David Hibbett, *Clark University*; R. James Hickey, *Miami University*; Kendra Hill, *San Diego State University*; William Hillenius, *College of Charleston*; Kenneth Hillers, *California Polytechnic State University, San Luis Obispo*; Ralph Hinegardner, *University of California, Santa Cruz*; William Hines, *Foothill College*; Robert Hinrichsen, *Indiana University of Pennsylvania*; Helmut Hirsch, *State University of New York, Albany*; Tuan-hua David Ho, *Washington University*; Carl Hoagstrom, *Ohio Northern University*; Jason Hodin, *Stanford University*; James Hoffman, *University of Vermont*; A. Scott Holaday, *Texas Tech University*; N. Michele Holbrook, *Harvard University*; James Holland, *Indiana State University, Bloomington*; Charles Holliday, *Lafayette College*; Lubbock Karl Holte, *Idaho State University*; Alan R. Holyoak, *Brigham Young University, Idaho*; Laura Hoopes, *Occidental College*; Nancy Hopkins, *Massachusetts Institute of Technology*; Sandra Horikami, *Daytona Beach Community College*; Kathy Hornberger, *Widener University*; Pius F. Horner, *San Bernardino Valley College*; Becky Houck, *University of Portland*; Margaret Houk, *Ripon College*; Laura Houston, *Northeast Lakeview College*; Daniel J. Howard, *New Mexico State University*; Ronald R. Hoy, *Cornell University*; Sandra Hsu, *Skyline College*; Sara Huang, *Los Angeles Valley College*; Cristin Hulslander, *University of Oregon*; Donald Humphrey, *Emory University School of Medicine*; Catherine Hurlbut, *Florida State College, Jacksonville*; Diane Husic, *Moravian College*; Robert J. Huskey, *University of Virginia*; Steven Hutcheson, *University of Maryland, College Park*; Linda L. Hyde, *Gordon College*; Bradley Hyman, *University of California, Riverside*; Jeffrey Ihara, *Mira Costa College*; Mark Iked, *San Bernardino Valley College*; Cheryl Ingram-Smith, *Clemson University*; Harry Itagaki, *Kenyon College*; Alice Jacklet, *State University of New York, Albany*; John Jackson, *North Hennepin Community College*; Thomas Jacobs, *University of Illinois*; Kathy Jacobson, *Grinnell College*; Mark Jaffe, *Nova Southeastern University*; John C. Jahoda, *Bridgewater State College*; Douglas Jensen, *Converse College*; Dan Johnson, *East Tennessee State University*; Lance Johnson, *Midland Lutheran College*; Lee Johnson, *The Ohio State University*; Randall Johnson, *University of California, San Diego*; Roishene Johnson, *Bossier Parish Community College*; Stephen Johnson, *William Penn University*; Wayne Johnson, *Ohio State University*; Kenneth C. Jones, *California State University, Northridge*; Russell Jones, *University of California, Berkeley*; Cheryl Jorcyk, *Boise State University*; Chad Jordan, *North Carolina State University*; Alan Journet, *Southeast Missouri State University*; Walter Judd, *University of Florida*; Thomas W. Jurik, *Iowa State University*; Caroline M. Kane, *University of California, Berkeley*; Thomas C. Kane, *University of Cincinnati*; The-Hui Kao, *Pennsylvania State University*; Tamos Kapros, *University of Missouri*; E. L. Karlstrom, *University of Puget Sound*; Jennifer Katcher, *Pima Community College*; Laura A. Katz, *Smith College*; Judy Kaufman, *Monroe Community College*; Maureen Kearney, *Field Museum of Natural History*; Eric G. Keeling, *Cary Institute of Ecosystem Studies*; Patrick Keeling, *University of British Columbia*; Thomas Keller, *Florida State University*; Elizabeth A. Kellogg, *University of Missouri, St. Louis*; Norm Kenkel, *University of Manitoba*; Chris Kennedy, *Simon Fraser University*; George Khoury, *National Cancer Institute*; Rebecca T. Kimball, *University of Florida*; Mark Kirk, *University of Missouri, Columbia*; Robert Kitchin, *University of Wyoming*; Hillar Klandorf, *West Virginia University*; Attila O. Klein, *Brandeis University*; Daniel Klionsky, *University of Michigan*; Mark Knauss, *Georgia Highlands College*; Janice Knepper, *Villanova University*; Charles Knight, *California Polytechnic State University*; Jennifer Knight, *University of Colorado*; Ned Knight, *Linfield College*; Roger Koeppe, *University of Arkansas*; David Kohl, *University of California, Santa Barbara*; Greg Kopf, *University of Pennsylvania School of Medicine*; Thomas Koppenheffer, *Trinity University*; Peter Kourtev, *Central Michigan University*; Margareta Krabbe, *Uppsala University*; Jacob Krans, *Western New England University*; Anselm Kratochwil, *Universität Osnabrück*; Eliot Krause, *Seton Hall University*; Deborah M. Kristan, *California State University, San Marcos*; Steven Kristoff, *Ivy Tech Community College*; William Kroll, *Loyola University, Chicago*; Janis Kuby, *San Francisco State University*; Barb Kuemerle, *Case Western Reserve University*; Justin P. Kumar, *Indiana University*; Rukmani Kuppuswami, *Laredo Community College*; David Kurijaka, *Ohio University*; Lee Kurtz, *Georgia Gwinnett College*; Michael P. Labare, *United States Military Academy, West Point*; Marc-André Lachance, *University of Western Ontario*; J. A. Lackey, *State University of New York, Oswego*; Elaine Lai, *Brandeis University*; Mohamed Lakrim, *Kingsborough Community College*; Ellen Lamb, *University of North Carolina, Greensboro*; William Lamberts, *College of St Benedict and St John's University*; William L'Amoreaux, *College of Staten Island*; Lynn Lamoreux, *Texas A&M University*; Carmine A. Lanciani, *University of Florida*; Kenneth Lang, *Humboldt State University*; Dominic Lannutti, *El Paso Community College*; Allan Larson, *Washington University*; John Latto, *University of California, Santa Barbara*; Diane K. Lavett, *State University of New York, Cortland, and Emory University*; Charles Leavell, *Fullerton College*; C. S. Lee, *University of Texas*; Daewoo Lee, *Ohio University*; Tali D. Lee, *University of Wisconsin, Eau Claire*; Hugh Lefcort, *Gonzaga University*; Robert Leonard, *University of California, Riverside*; Michael R. Leonardo, *Coe College*; John Lepri, *University of North Carolina, Greensboro*; Donald Levin, *University of Texas, Austin*; Joseph Levine, *Boston College*; Mike Levine, *University of California, Berkeley*; Alcinda Lewis, *University of Colorado, Boulder*; Bill Lewis, *Shoreline Community College*; Jani Lewis, *State University of New York*; John Lewis, *Loma Linda University*; Lorraine Lica, *California State University, East Bay*; Harvey Liftin, *Broward Community College*; Harvey Lillywhite, *University of Florida, Gainesville*; Graeme Lindbeck, *Valencia Community College*; Clark Lindgren, *Grinnell College*; Diana Lipscomb, *George Washington University*; Christopher Little, *The University of Texas, Pan American*; Kevin D. Livingstone, *Trinity University*; Andrea Lloyd, *Middlebury College*; Sam Loker, *University of New Mexico*; Christopher A. Loretz, *State University of New York, Buffalo*; Jane Lubchenco, *Oregon State University*; Douglas B. Luckie, *Michigan State University*; Hannah Lui, *University of California, Irvine*; Margaret A. Lynch, *Tufts University*; Steven Lynch, *Louisiana State University, Shreveport*; Richard Machemer Jr., *St. John Fisher College*; Elizabeth Machunis-Masuoka, *University of Virginia*; James MacMahon, *Utah State University*; Nancy Magill, *Indiana University*; Christine R. Maher, *University of Southern Maine*; Linda Maier, *University of Alabama, Huntsville*; Jose Maldonado, *El Paso Community College*; Richard Malkin, *University of California, Berkeley*; Charles Mallery, *University of Miami*; Keith Malmos, *Valencia Community College, East Campus*; Cindy Malone, *California State University, Northridge*; Mark Maloney, *University of South Mississippi*; Carol Mapes, *Kutztown University of Pennsylvania*; William Margolin, *University of Texas Medical School*; Lynn Margulis, *Boston University*; Julia Marrs, *Barnard College* (student); Kathleen A. Marrs, *Indiana University-Purdue University, Indianapolis*; Edith Marsh, *Angelo State University*; Diane L. Marshall, *University of New Mexico*; Karl Mattox, *Miami University of Ohio*; Joyce Maxwell, *California State University, Northridge*; Jeffrey D. May, *Marshall University*; Mike Mayfield, *Ball State University*; Kamau Mbuthia, *Bowling Green State University*; Lee McClenaghan, *San Diego State University*; Richard McCracken, *Purdue University*; Andrew McCubbin, *Washington State University*; Kerry McDonald, *University of Missouri, Columbia*; Tanya McGhee, *Craven Community College*; Jacqueline McLaughlin, *Pennsylvania State University, Lehigh Valley*; Neal McReynolds, *Texas A&M International*; Darcy Medica, *Pennsylvania State University*; Lisa Marie Meffert, *Rice University*; Susan Meiers, *Western Illinois University*; Michael Meighan, *University of California, Berkeley*; Scott Meissner, *Cornell University*; Paul Melchior, *North Hennepin Community College*; Phillip Meneely, *Haverford College*; John Merrill, *Michigan State University*; Brian Metscher, *University of California, Irvine*; Ralph Meyer, *University of Cincinnati*; James Mickle, *North Carolina State University*; Jan Mikesell, *Gettysburg College*; Roger Milkman, *University of Iowa*; Helen Miller, *Oklahoma State University*; John Miller, *University of California, Berkeley*; Kenneth R. Miller, *Brown University*; Alex Mills, *University of Windsor*; Sarah Milton, *Florida Atlantic University*; Eli Minkoff, *Bates College*; John E. Minnich, *University of Wisconsin, Milwaukee*; Subhash Minocha, *University of New Hampshire*; Michael J. Misamore, *Texas Christian University*; Kenneth Mitchell, *Tulane University School of Medicine*; Ivona Mladenovic, *Simon Fraser University*; Alan Molumby, *University of Illinois, Chicago*; Nicholas Money, *Miami University*; Russell Monson, *University of Colorado, Boulder*; Joseph P. Montoya, *Georgia Institute of Technology*; Frank Moore, *Oregon State University*; Janice Moore, *Colorado State University*; Linda Moore, *Georgia Military College*; Randy Moore, *Wright State University*; William Moore, *Wayne State University*; Carl Moos, *Veterans Administration Hospital, Albany, New York*; Linda Martin Morris, *University of Washington*; Michael Mote, *Temple University*; Alex Motten, *Duke University*; Jeanette Mowery, *Madison Area Technical College*; Deborah Mowshowitz, *Columbia University*; Rita Moyes, *Texas A&M, College Station*; Darrel L. Murray, *University of Illinois, Chicago*; Courtney Murren, *College of Charleston*; John Mutchmor, *Iowa State University*; Elliot Myerowitz, *California Institute of Technology*; Gavin Naylor, *Iowa State University*; Karen Neal, *Reynolds University*; John Neess, *University of Wisconsin, Madison*; Ross Nehm, *Ohio State University*; Tom Neils, *Grand Rapids Community College*; Kimberlyn Nelson, *Pennsylvania State University*; Raymond Neubauer, *University of Texas, Austin*; Todd Newbury, *University of California, Santa Cruz*; James Newcomb, *New England College*; Jacalyn Newman, *University of Pittsburgh*; Harvey Nichols, *University of Colorado, Boulder*; Deborah Nickerson, *University of South Florida*; Bette Nicotri, *University of Washington*; Caroline Niederman, *Tomball College*; Eric Nielsen, *University of Michigan*; Maria Nieto, *California State University, East Bay*; Anders Nilsson, *University of Umeå*; Greg Nishiyama, *College of the Canyons*; Charles R. Noback, *College of Physicians and Surgeons, Columbia University*; Jane Noble-Harvey, *Delaware University*; Mary C. Nolan, *Irvine Valley College*; Kathleen Nolta, *University of Michigan*; Peter Nonacs, *University of California, Los Angeles*; Mohamed A. F. Noor, *Duke University*; Shawn Nordell, *St. Louis University*; Richard S. Norman, *University of Michigan, Dearborn* (emeritus); David O. Norris, *University of Colorado, Boulder*; Steven Norris, *California State University, Channel Islands*; Gretchen North, *Occidental College*; Cynthia Norton, *University of Maine, Augusta*; Steve Norton, *East Carolina University*; Steve Nowicki, *Duke University*; Bette H. Nybakken, *Hartnell College*; Brian O'Conner, *University of Massachusetts, Amherst*; Gerard O'Donovan, *University of North Texas*; Eugene Odum, *University of Georgia*; Mark P. Oemke, *Alma College*; Linda Ogren, *University of California, Santa Cruz*; Patricia O'Hern, *Emory University*; Nathan O. Okia, *Auburn University, Montgomery*; Jeanette Oliver, *St. Louis Community College, Florissant Valley*; Gary P. Olivetti, *University of Vermont*; Margaret Olney, *St. Martin's College*; John Olsen, *Rhodes College*; Laura J. Olsen, *University of Michigan*; Sharman O'Neill, *University of California, Davis*; Wan Ooi, *Houston Community College*; Aharon Oren, *The Hebrew University*; John Oross, *University of California, Riverside*; Rebecca Orr, *Collin College*; Catherine Ortega, *Fort Lewis College*; Charissa Osborne, *Butler University*; Gay Ostarello, *Diablo Valley College*; Henry R. Owen, *Eastern Illinois University*; Thomas G. Owens, *Cornell University*; Penny Padgett, *University of North Carolina, Chapel Hill*; Kevin Padian, *University of California, Berkeley*; Dianna Padilla, *State University of New York, Stony Brook*; Anthony T. Paganini, *Michigan State University*; Barry Palevitz, *University of Georgia*; Michael A. Palladino,

Monmouth University; Matt Palmtag, *Florida Gulf Coast University*; Stephanie Pandolfi, *Michigan State University*; Daniel Papaj, *University of Arizona*; Peter Pappas, *County College of Morris*; Nathalie Pardigon, *Institut Pasteur*; Bulah Parker, *North Carolina State University*; Stanton Parmeter, *Chemeketa Community College*; Cindy Paszkowski, *University of Alberta*; Robert Patterson, *San Francisco State University*; Ronald Patterson, *Michigan State University*; Crellin Pauling, *San Francisco State University*; Kay Pauling, *Foothill Community College*; Daniel Pavuk, *Bowling Green State University*; Debra Pearce, *Northern Kentucky University*; Patricia Pearson, *Western Kentucky University*; Andrew Pease, *Stevenson University*; Nancy Pelaez, *Purdue University*; Shelley Penrod, *North Harris College*; Imara Y. Perera, *North Carolina State University*; Beverly Perry, *Houston Community College*; Irene Perry, *University of Texas of the Permian Basin*; Roger Persell, *Hunter College*; Eric Peters, *Chicago State University*; Larry Peterson, *University of Guelph*; David Pfennig, *University of North Carolina, Chapel Hill*; Mark Pilgrim, *College of Coastal Georgia*; David S. Pilliod, *California Polytechnic State University, San Luis Obispo*; Vera M. Piper, *Shenandoah University*; Deb Pires, *University of California, Los Angeles*; J. Chris Pires, *University of Missouri, Columbia*; Bob Pittman, *Michigan State University*; James Platt, *University of Denver*; Martin Poenie, *University of Texas, Austin*; Scott Poethig, *University of Pennsylvania*; Crima Pogge, *San Francisco Community College*; Michael Pollock, *Mount Royal University*; Roberta Pollock, *Occidental College*; Jeffrey Pommerville, *Texas A&M University*; Therese M. Poole, *Georgia State University*; Angela R. Porta, *Kean University*; Jason Porter, *University of the Sciences, Philadelphia*; Warren Porter, *University of Wisconsin*; Daniel Potter, *University of California, Davis*; Donald Potts, *University of California, Santa Cruz*; Robert Powell, *Avila University*; Andy Pratt, *University of Canterbury*; David Pratt, *University of California, Davis*; Elena Pravosudova, *University of Nevada, Reno*; Halina Presley, *University of Illinois, Chicago*; Eileen Preston, *Tarrant Community College Northwest*; Mary V. Price, *University of California, Riverside*; Mitch Price, *Pennsylvania State University*; Terrell Pritts, *University of Arkansas, Little Rock*; Rong Sun Pu, *Kean University*; Rebecca Pyles, *East Tennessee State University*; Scott Quackenbush, *Florida International University*; Ralph Quatrano, *Oregon State University*; Peter Quinby, *University of Pittsburgh*; Val Raghavan, *Ohio State University*; Deanna Raineri, *University of Illinois, Champaign-Urbana*; David Randall, *City University Hong Kong*; Talitha Rajah, *Indiana University Southeast*; Charles Ralph, *Colorado State University*; Pushpa Ramakrishna, *Chandler-Gilbert Community College*; Thomas Rand, *Saint Mary's University*; Monica Ranes-Goldberg, *University of California, Berkeley*; Robert S. Rawding, *Gannon University*; Robert H. Reavis, *Glendale Community College*; Kurt Redborg, *Coe College*; Ahnya Redman, *Pennsylvania State University*; Brian Reeder, *Morehead State University*; Bruce Reid, *Kean University*; David Reid, *Blackburn College*; C. Gary Reiness, *Lewis & Clark College*; Charles Remington, *Yale University*; Erin Rempala, *San Diego Mesa College*; David Reznick, *University of California, Riverside*; Fred Rhoades, *Western Washington State University*; Douglas Rhoads, *University of Arkansas*; Eric Ribbens, *Western Illinois University*; Christina Richards, *New York University*; Sarah Richart, *Azusa Pacific University*; Christopher Riegle, *Irvine Valley College*; Loren Rieseberg, *University of British Columbia*; Bruce B. Riley, *Texas A&M University*; Todd Rimkus, *Marymount University*; John Rinehart, *Eastern Oregon University*; Donna Ritch, *Pennsylvania State University*; Carol Rivin, *Oregon State University East*; Laurel Roberts, *University of Pittsburgh*; Diane Robins, *University of Michigan*; Kenneth Robinson, *Purdue University*; Thomas Rodella, *Merced College*; Deb Roess, *Colorado State University*; Heather Roffey, *Marianopolis College*; Rodney Rogers, *Drake University*; Suzanne Rogers, *Seton Hill University*; William Roosenburg, *Ohio University*; Mike Rosenzweig, *Virginia Polytechnic Institute and State University*; Wayne Rosing, *Middle Tennessee State University*; Thomas Rost, *University of California, Davis*; Stephen I. Rothstein, *University of California, Santa Barbara*; John Ruben, *Oregon State University*; Albert Ruesink, *Indiana University*; Patricia Rugaber, *College of Coastal Georgia*; Scott Russell, *University of Oklahoma*; Neil Sabine, *Indiana University*; Tyson Sacco, *Cornell University*; Glenn-Peter Saetre, *University of Oslo*; Rowan F. Sage, *University of Toronto*; Tammy Lynn Sage, *University of Toronto*; Sanga Saha, *Harold Washington College*; Don Sakaguchi, *Iowa State University*; Walter Sakai, *Santa Monica College*; Mark F. Sanders, *University of California, Davis*; Kathleen Sandman, *Ohio State University*; Louis Santiago, *University of California, Riverside*; Ted Sargent, *University of Massachusetts, Amherst*; K. Sathasivan, *University of Texas, Austin*; Gary Saunders, *University of New Brunswick*; Thomas R. Sawicki, *Spartanburg Community College*; Inder Saxena, *University of Texas, Austin*; Carl Schaefer, *University of Connecticut*; Andrew Schaffner, *Cal Poly San Luis Obispo*; Maynard H. Schaus, *Virginia Wesleyan College*; Renate Scheibe, *University of Osnabrück*; David Schimpf, *University of Minnesota, Duluth*; William H. Schlesinger, *Duke University*; Mark Schlissel, *University of California, Berkeley*; Christopher J. Schneider, *Boston University*; Thomas W. Schoener, *University of California, Davis*; Robert Schorr, *Colorado State University*; Patricia M. Schulte, *University of British Columbia*; Karen S. Schumaker, *University of Arizona*; Brenda Schumpert, *Valencia Community College*; David J. Schwartz, *Houston Community College*; Christa Schwintzer, *University of Maine*; Erik P. Scully, *Towson State University*; Robert W. Seagull, *Hofstra University*; Edna Seaman, *Northeastern University*; Duane Sears, *University of California, Santa Barbara*; Brent Selinger, *University of Lethbridge*; Orono Shukdeb Sen, *Bethune-Cookman College*; Wendy Sera, *Seton Hill University*; Alison M. Shakarian, *Salve Regina University*; Timothy E. Shannon, *Francis Marion University*; Joan Sharp, *Simon Fraser University*; Victoria C. Sharpe, *Blinn College*; Elaine Shea, *Loyola College, Maryland*; Stephen Sheckler, *Virginia Polytechnic Institute and State University*; Robin L. Sherman, *Nova Southeastern University*; Richard Sherwin, *University of Pittsburgh*; Lisa Shimeld, *Crafton Hills College*; James Shinkle, *Trinity University*; Barbara Shipes, *Hampton University*; Richard M. Showman, *University of South Carolina*; Eric Shows, *Jones County Junior College*; Peter Shugarman, *University of Southern California*; Alice Shuttey, *DeKalb Community College*; James Sidie, *Ursinus College*; Daniel Simberloff, *Florida State University*; Rebecca Simmons, *University of North Dakota*; Anne Simon, *University of Maryland, College Park*; Robert Simons, *University of California, Los Angeles*; Alastair Simpson, *Dalhousie University*; Susan Singer, *Carleton College*; Sedonia Sipes, *Southern Illinois University, Carbondale*; John Skillman, *California State University, San Bernardino*; Roger Sloboda, *Dartmouth University*; John Smarrelli, *Le Moyne College*; Andrew T. Smith, *Arizona State University*; Kelly Smith, *University of North Florida*; Nancy Smith-Huerta, *Miami Ohio University*; John Smol, *Queen's University*; Andrew J. Snope, *Essex Community College*; Mitchell Sogin, *Woods Hole Marine Biological Laboratory*; Doug Soltis, *University of Florida, Gainesville*; Julio G. Soto, *San Jose State University*; Susan Sovonick-Dunford, *University of Cincinnati*; Frederick W. Spiegel, *University of Arkansas*; John Stachowicz, *University of California, Davis*; Joel Stafstrom, *Northern Illinois University*; Alam Stam, *Capital University*; Amanda Starnes, *Emory University*; Karen Steudel, *University of Wisconsin*; Barbara Stewart, *Swarthmore College*; Gail A. Stewart, *Camden County College*; Cecil Still, *Rutgers University, New Brunswick*; Margery Stinson, *Southwestern College*; James Stockand, *University of Texas Health Science Center, San Antonio*; John Stolz, *California Institute of Technology*; Judy Stone, *Colby College*; Richard D. Storey, *Colorado College*; Stephen Strand, *University of California, Los Angeles*; Eric Strauss, *University of Massachusetts, Boston*; Antony Stretton, *University of Wisconsin, Madison*; Russell Stullken, *Augusta College*; Mark Sturtevant, *University of Michigan, Flint*; John Sullivan, *Southern Oregon State University*; Gerald Summers, *University of Missouri*; Judith Sumner, *Assumption College*; Marshall D. Sundberg, *Emporia State University*; Cynthia Surmacz, *Bloomsburg University*; Lucinda Swatzell, *Southeast Missouri State University*; Daryl Sweeney, *University of Illinois, Champaign-Urbana*; Samuel S. Sweet, *University of California, Santa Barbara*; Janice Swenson, *University of North Florida*; Michael A. Sypes, *Pennsylvania State University*; Lincoln Taiz, *University of California, Santa Cruz*; David Tam, *University of North Texas*; Yves Tan, *Cabrillo College*; Samuel Tarsitano, *Southwest Texas State University*; David Tauck, *Santa Clara University*; Emily Taylor, *California Polytechnic State University, San Luis Obispo*; James Taylor, *University of New Hampshire*; John W. Taylor, *University of California, Berkeley*; Martha R. Taylor, *Cornell University*; Franklyn Tan Te, *Miami Dade College*; Thomas Terry, *University of Connecticut*; Roger Thibault, *Bowling Green State University*; Kent Thomas, *Wichita State University*; William Thomas, *Colby-Sawyer College*; Cyril Thong, *Simon Fraser University*; John Thornton, *Oklahoma State University*; Robert Thornton, *University of California, Davis*; William Thwaites, *Tillamook Bay Community College*; Stephen Timme, *Pittsburg State University*; Mike Toliver, *Eureka College*; Eric Toolson, *University of New Mexico*; Leslie Towill, *Arizona State University*; James Traniello, *Boston University*; Paul Q. Trombley, *Florida State University*; Nancy J. Trun, *Duquesne University*; Constantine Tsoukas, *San Diego State University*; Marsha Turell, *Houston Community College*; Victoria Turgeon, *Furman University*; Robert Tuveson, *University of Illinois, Urbana*; Maura G. Tyrrell, *Stonehill College*; Catherine Uekert, *Northern Arizona University*; Claudia Uhde-Stone, *California State University, East Bay*; Gordon Uno, *University of Oklahoma*; Lisa A. Urry, *Mills College*; Saba Valadkhan, *Center for RNA Molecular Biology*; James W. Valentine, *University of California, Santa Barbara*; Joseph Vanable, *Purdue University*; Theodore Van Bruggen, *University of South Dakota*; Kathryn VandenBosch, *Texas A&M University*; Gerald Van Dyke, *North Carolina State University*; Brandi Van Roo, *Framingham State College*; Moira Van Staaden, *Bowling Green State University*; Sarah VanVickle-Chavez, *Washington University, St. Louis*; William Velhagen, *New York University*; Steven D. Verhey, *Central Washington University*; Kathleen Verville, *Washington College*; Sara Via, *University of Maryland*; Frank Visco, *Orange Coast College*; Laurie Vitt, *University of California, Los Angeles*; Neal Voelz, *St. Cloud State University*; Thomas J. Volk, *University of Wisconsin, La Crosse*; Leif Asbjørn Vøllestad, *University of Oslo*; Amy Volmer, *Swarthmore College*; Janice Voltzow, *University of Scranton*; Margaret Voss, *Penn State Erie*; Susan D. Waaland, *University of Washington*; Charles Wade, *C.S. Mott Community College*; William Wade, *Dartmouth Medical College*; John Waggoner, *Loyola Marymount University*; Jyoti Wagle, *Houston Community College*; Edward Wagner, *University of California, Irvine*; D. Alexander Wait, *Southwest Missouri State University*; Claire Walczak, *Indiana University*; Jerry Waldvogel, *Clemson University*; Dan Walker, *San Jose State University*; Robert Lee Wallace, *Ripon College*; Jeffrey Walters, *North Carolina State University*; Linda Walters, *University of Central Florida*; James Wandersee, *Louisiana State University*; Nickolas M. Waser, *University of California, Riverside*; Fred Wasserman, *Boston University*; Margaret Waterman, *University of Pittsburgh*; Charles Webber, *Loyola University of Chicago*; Peter Webster, *University of Massachusetts, Amherst*; Terry Webster, *University of Connecticut, Storrs*; Beth Wee, *Tulane University*; James Wee, *Loyola University*; Andrea Weeks, *George Mason University*; John Weishampel, *University of Central Florida*; Peter Wejksnora, *University of Wisconsin, Milwaukee*; Kentwood Wells, *University of Connecticut*; David J. Westenberg, *University of Missouri, Rolla*; Richard Wetts, *University of California, Irvine*; Matt White, *Ohio University*; Susan Whittemore, *Keene State College*; Murray Wiegand, *University of Winnipeg*; Ernest H. Williams, *Hamilton College*; Kathy Williams, *San Diego State University*; Kimberly Williams, *Kansas State University*; Stephen Williams, *Glendale Community College*; Elizabeth Willott, *University of Arizona*; Christopher Wills, *University of California, San Diego*; Paul Wilson, *California State University, Northridge*; Fred Wilt, *University of California, Berkeley*; Peter Wimberger, *University of Puget Sound*; Robert Winning, *Eastern Michigan University*; E. William Wischusen, *Louisiana State University*; Clarence Wolfe, *Northern Virginia Community College*; Vickie L. Wolfe, *Marshall University*; Janet Wolkenstein, *Hudson Valley Community College*; Robert T. Woodland, *University of Massachusetts Medical School*; Joseph Woodring, *Louisiana State University*; Denise Woodward, *Pennsylvania State University*; Patrick Woolley, *East Central College*; Sarah E. Wyatt, *Ohio University*; Grace Wyngaard, *James Madison University*; Shuhai Xiao, *Virginia Polytechnic Institute*, Ramin Yadegari, *University of Arizona*; Paul Yancey, *Whitman College*; Philip Yant, *University of Michigan*; Linda Yasui, *Northern Illinois University*; Anne D. Yoder, *Duke University*; Hideo Yonenaka, *San Francisco State University*; Gina M. Zainelli, *Loyola University, Chicago*; Edward Zalisko, *Blackburn College*; Nina Zanetti, *Siena College*; Sam Zeveloff, *Weber State University*; Zai Ming Zhao, *University of Texas, Austin*; John Zimmerman, *Kansas State University*; Miriam Zolan, *Indiana University*; Theresa Zucchero, *Methodist University*; Uko Zylstra, *Calvin College*.

Detailed Contents

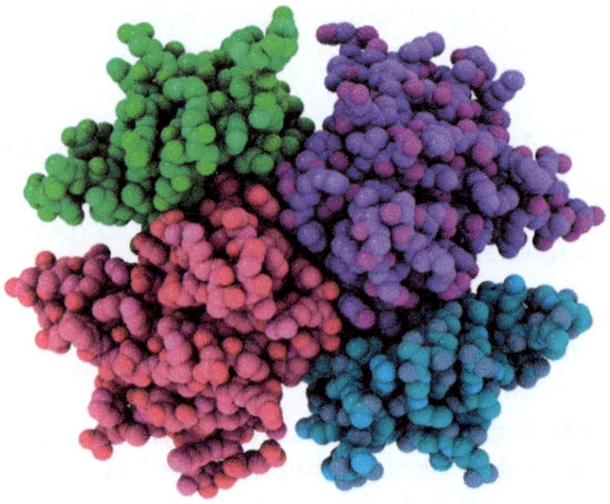

5 The Structure and Function of Large Biological Molecules 66

UNIT 2 THE CELL 92

6 A Tour of the Cell 93

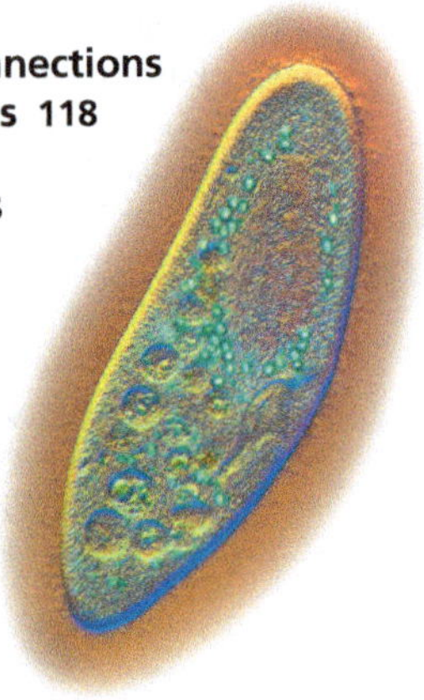

7 Membrane Structure and Function 126

8 An Introduction to Metabolism 143

9 Cellular Respiration and Fermentation 164

10 Photosynthesis 187

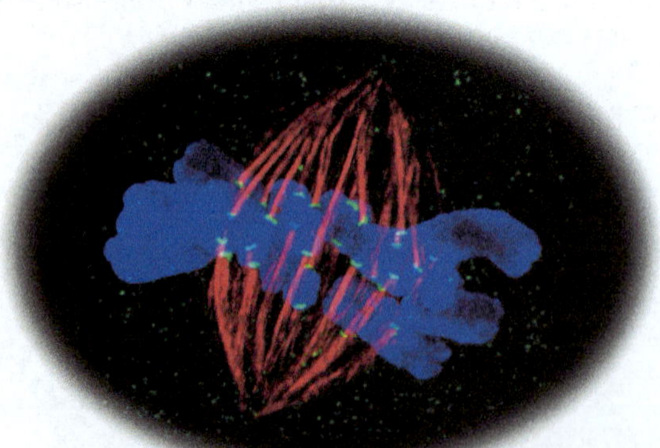

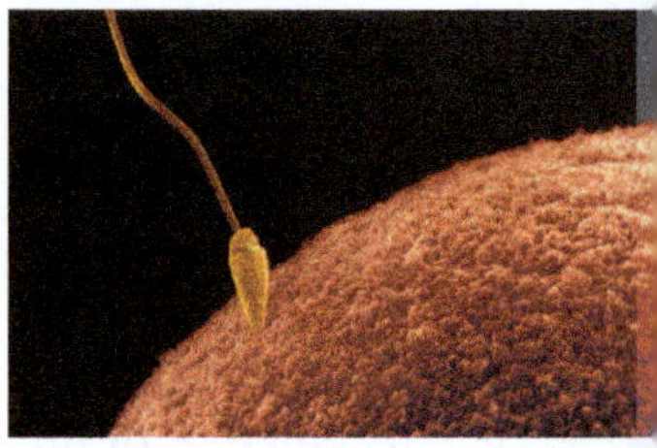

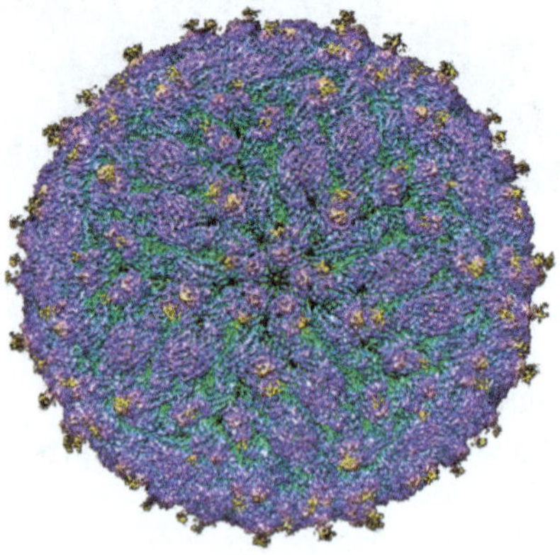

19 Viruses 396

20 DNA Tools and Biotechnology 413

21 Genomes and Their Evolution 440

33 An Introduction to Invertebrates 684

34 The Origin and Evolution of Vertebrates 716

UNIT 6 PLANT FORM AND FUNCTION 755

35 Vascular Plant Structure, Growth, and Development 756

UNIT 7 ANIMAL FORM AND FUNCTION 870

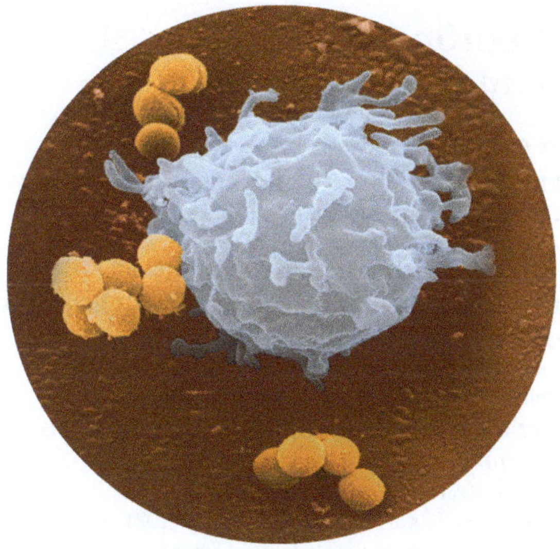

43 The Immune System 950

44 Osmoregulation and Excretion 975

45 Hormones and the Endocrine System 997

46 Animal Reproduction 1017

47 Animal Development 1041

48 Neurons, Synapses, and Signaling 1065

49 Nervous Systems 1083

UNIT 1 THE CHEMISTRY OF LIFE

Born in Baton Rouge, Louisiana, Lovell Jones was one of the first students to integrate the schools in that city. He went on to major in biology at California State University at Hayward (now CSU East Bay). He then earned a Ph.D. in Zoology with an emphasis in endocrinology and tumor biology from the University of California, Berkeley, in 1977. As a postdoctoral fellow at the University of California, San Francisco, he carried out research on the effects of estrogen on breast cancer. In 1980, after his post-doc, he moved to the MD Anderson Cancer Center as an assistant professor in the Department of Gynecology and Biochemistry, where he remained until his retirement in 2013. In addition to making major contributions to our understanding of the role of steroid hormones in reproductive cancers, Dr. Jones is considered a pioneer in addressing health inequities and social justice. He has continued the latter work during his retirement. During his long, distinguished career, Dr. Jones has received numerous awards and has been commended in the U.S. House of Representatives for his lifelong work.

"Studying biochemistry in a holistic way is going to be the key to the answers we need."

▼ Dr. Lovell Jones in his lab, where he has studied the relationship between hormones and cancers.

An Interview with Lovell Jones

What did you study in graduate school?

I got my Ph.D. at UC Berkeley, working with Howard Bern on the effects of hormones on cervical and vaginal development. I developed an interest in the effects of an artificial estrogen called diethylstilbestrol (DES), a medication used to stop spontaneous abortion in pregnant women. Women that used DES during their first trimester gave birth to girls who had an increased incidence of a rare form of cervical cancer at a very early age, as young as 7. At that time, it was thought that DES was acting as a chemical to initiate cancer, but I thought it was a hormonal action because it was mimicking estrogen, and I wanted to test that using mice. By exposing mice to natural estrogen when their cervical and vaginal tracts were developing, I showed that this exposure produced related types of tumors.

How did you continue to work on the relationship between hormones and cancer?

After I got my Ph.D., I worked with Finn Siiteri at UCSF, who was an expert on endocrine-related hormones but had no cancer knowledge. He brought me on because I had shown that hormones could initiate cancer. I started to work on finding a way to measure estrogen levels as a risk factor for developing breast cancer. I found that the level of free estrogen—estrogen that isn't bound to other molecules—was the risk factor because only free estrogen was able to function. At this point, I was recruited to MD Anderson Cancer Center to develop hormonal therapies using agents that blocked estrogens. We developed treatments for lung cancer, head and neck cancers, and salivary cancers. At the same time, I was looking at long-term effects of estrogens on neonatal development and developing a method to use hormone levels to predict breast cancer risk.

Why study biochemistry, and what are the most exciting avenues of biochemical research?

Studying biochemistry in a holistic way is going to be the key to the answers we need. We have to get out of our silos and do what some people call trans-disciplinary research. As far as future research goes, the most exciting area is precision medicine, an approach to treatment that considers a person's genes, environment, and lifestyle. We can now get away from looking at individuals in terms of their race, which is not a biological construct, and instead look at them as members of populations: subgroups that have a particular history together. For example, one of the highest incidences of pre-menopausal breast cancer is among blacks and whites in the Chesapeake Bay area. Why? Well, most white females who came into Maryland during colonial times were Scotch–Irish Catholics who arrived as indentured servants, a higher form of slaves. Roughly a third of them had interactions with African slaves. So a lot of people in that area have genes that came from Africa, and some people who look white have black ancestry. A disease called triple-negative breast cancer, where tumor cells are missing three types of receptors, is spoken of as a "black cancer." But it's really just that a group of genes that arose in one place in Africa leads to early onset of breast cancer. Also, triple-negative breast cancer is a disease of the young, not of black females in particular. So that's why I say precision medicine will need to analyze all of an individual's genes and consider their evolutionary history rather than just assigning them to one race or another.

What is your advice to an undergraduate considering a career in biology?

Find a good mentor. Some of the greatest mentors put their students ahead of themselves. Howard Bern, my Ph.D. advisor, said that the greatest contribution to science is not your papers or grants, but the people you leave behind.

The Chemical Context of Life

2

▲ **Figure 2.1** **What weapon are these wood ants shooting into the air?**

KEY CONCEPTS

2.1 Matter consists of chemical elements in pure form and in combinations called compounds

2.2 An element's properties depend on the structure of its atoms

2.3 The formation and function of molecules depend on chemical bonding between atoms

2.4 Chemical reactions make and break chemical bonds

A Chemical Connection to Biology

Like other animals, ants have structures and mechanisms that defend them from attack. Wood ants live in colonies of hundreds or thousands, and the colony as a whole has a particularly effective way of dealing with enemies. When threatened from above, the ants shoot volleys of formic acid into the air from their abdomens, and the acid bombards the potential predator, such as a hungry bird **(Figure 2.1)**. Formic acid is produced by many species of ants and in fact got its name from the Latin word for ant, *formica*. For quite a few ant species, the formic acid isn't shot out, but probably serves as a disinfectant that protects the ants against microbial parasites. Scientists have long known that chemicals play a major role in insect communication, the attraction of mates, and defense against predators.

Research on ants and other insects is a good example of how relevant chemistry is to the study of life. Unlike college courses, nature is not neatly packaged into individual sciences—biology, chemistry, physics, and so forth. Biologists specialize in the study of life, but organisms and their environments are natural systems to which the concepts of chemistry and physics apply. Biology is multidisciplinary.

This unit of chapters introduces some basic concepts of chemistry that apply to the study of life. Somewhere in the transition from molecules to cells, we will cross the blurry boundary between nonlife and life. This chapter focuses on the chemical components that make up all matter.

When you see this blue icon, log in to **MasteringBiology** and go to the Study Area for digital resources.

MB **Get Ready for This Chapter**

MB **Interview with Deborah Gordon: Investigating ant behavior**

CONCEPT 2.1

Matter consists of chemical elements in pure form and in combinations called compounds

Organisms are composed of **matter**, which is anything that takes up space and has mass.* Matter exists in many forms. Rocks, metals, oils, gases, and living organisms are a few examples of what seems to be an endless assortment of matter.

Elements and Compounds

Matter is made up of elements. An **element** is a substance that cannot be broken down to other substances by chemical reactions. Today, chemists recognize 92 elements occurring in nature; gold, copper, carbon, and oxygen are examples. Each element has a symbol, usually the first letter or two of its name. Some symbols are derived from Latin or German; for instance, the symbol for sodium is Na, from the Latin word *natrium*.

A **compound** is a substance consisting of two or more different elements combined in a fixed ratio. Table salt, for example, is sodium chloride (NaCl), a compound composed of the elements sodium (Na) and chlorine (Cl) in a 1:1 ratio. Pure sodium is a metal, and pure chlorine is a poisonous gas. When chemically combined, however, sodium and chlorine form an edible compound. Water (H_2O), another compound, consists of the elements hydrogen (H) and oxygen (O) in a 2:1 ratio. These are simple examples of organized matter having emergent properties: A compound has characteristics different from those of its elements **(Figure 2.2)**.

▼ Figure 2.2 The emergent properties of a compound. The metal sodium combines with the poisonous gas chlorine, forming the edible compound sodium chloride, or table salt.

Na Sodium + **Cl** Chlorine → **NaCl** Sodium chloride

*In everyday language we tend to substitute the term weight for mass, although the two are not identical. Mass is the amount of matter in an object, whereas the weight of an object is how strongly that mass is pulled by gravity. The weight of an astronaut walking on the moon is approximately $\frac{1}{6}$ the astronaut's weight on Earth, but his or her mass is the same. However, as long as we are earthbound, the weight of an object is a measure of its mass; in everyday language, therefore, we tend to use the terms interchangeably.

The Elements of Life

Of the 92 natural elements, about 20–25% are **essential elements** that an organism needs to live a healthy life and reproduce. The essential elements are similar among organisms, but there is some variation—for example, humans need 25 elements, but plants need only 17.

Just four elements—oxygen (O), carbon (C), hydrogen (H), and nitrogen (N)—make up approximately 96% of living matter. Calcium (Ca), phosphorus (P), potassium (K), sulfur (S), and a few other elements account for most of the remaining 4% or so of an organism's mass. **Trace elements** are required by an organism in only minute quantities. Some trace elements, such as iron (Fe), are needed by all forms of life; others are required only by certain species. For example, in vertebrates (animals with backbones), the element iodine (I) is an essential ingredient of a hormone produced by the thyroid gland. A daily intake of only 0.15 milligram (mg) of iodine is adequate for normal activity of the human thyroid. An iodine deficiency in the diet causes the thyroid gland to grow to abnormal size, a condition called goiter. Consuming seafood or iodized salt reduces the incidence of goiter. Relative amounts of all the elements in the human body are listed in **Table 2.1**.

Some naturally occurring elements are toxic to organisms. In humans, for instance, the element arsenic has been linked to numerous diseases and can be lethal. In some areas of the world, arsenic occurs naturally and can make its way into the groundwater. As a result of using water from drilled wells in southern Asia, millions of people have been inadvertently exposed to arsenic-laden water. Efforts are under way to reduce arsenic levels in their water supply.

Table 2.1 Elements in the Human Body

Element	Symbol	Percentage of Body Mass (including water)	
Oxygen	O	65.0%	96.3%
Carbon	C	18.5%	
Hydrogen	H	9.5%	
Nitrogen	N	3.3%	
Calcium	Ca	1.5%	3.7%
Phosphorus	P	1.0%	
Potassium	K	0.4%	
Sulfur	S	0.3%	
Sodium	Na	0.2%	
Chlorine	Cl	0.2%	
Magnesium	Mg	0.1%	

Trace elements (less than 0.01% of mass): Boron (B), chromium (Cr), cobalt (Co), copper (Cu), fluorine (F), iodine (I), iron (Fe), manganese (Mn), molybdenum (Mo), selenium (Se), silicon (Si), tin (Sn), vanadium (V), zinc (Zn)

INTERPRET THE DATA ➤ *Given the makeup of the human body, what compound do you think accounts for the high percentage of oxygen?*

Figure 2.3 Serpentine plant community. These plants are growing on serpentine soil, which contains elements that are usually toxic to plants. The insets show a close-up of serpentine rock and one of the plants, a Tiburon Mariposa lily (*Calochortus tiburonensis*). This specially-adapted species is found only on this one hill in Tiburon, a peninsula that juts into San Francisco Bay.

Case Study: Evolution of Tolerance to Toxic Elements

EVOLUTION Some species have become adapted to environments containing elements that are usually toxic; an example is serpentine plant communities. Serpentine is a jade-like mineral that contains elevated concentrations of elements such as chromium, nickel, and cobalt. Although most plants cannot survive in soil that forms from serpentine rock, a small number of plant species have adaptations that allow them to do so **(Figure 2.3)**. Presumably, variants of ancestral, nonserpentine species arose that could survive in serpentine soils, and subsequent natural selection resulted in the distinctive array of species we see in these areas today. Researchers are studying whether serpentine-adapted plants could be used to take up toxic heavy metals in contaminated areas, concentrating them for safer disposal.

CONCEPT CHECK 2.1

1. **MAKE CONNECTIONS** > Explain how table salt has emergent properties. (See Concept 1.1.)
2. Is a trace element an essential element? Explain.
3. **WHAT IF?** > In humans, iron is a trace element required for the proper functioning of hemoglobin, the molecule that carries oxygen in red blood cells. What might be the effects of an iron deficiency?
4. **MAKE CONNECTIONS** > Explain how natural selection might have played a role in the evolution of species that are tolerant of serpentine soils. (Review Concept 1.2.)

For suggested answers, see Appendix A.

CONCEPT 2.2

An element's properties depend on the structure of its atoms

Each element consists of a certain type of atom that is different from the atoms of any other element. An **atom** is the smallest unit of matter that still retains the properties of an element. Atoms are so small that it would take about a million of them to stretch across the period printed at the end of this sentence. We symbolize atoms with the same abbreviation used for the element that is made up of those atoms. For example, the symbol C stands for both the element carbon and a single carbon atom.

Subatomic Particles

Although the atom is the smallest unit having the properties of an element, these tiny bits of matter are composed of even smaller parts, called *subatomic particles*. Using high-energy collisions, physicists have produced more than 100 types of particles from the atom, but only three kinds of particles are relevant here: **neutrons**, **protons**, and **electrons**. Protons and electrons are electrically charged. Each proton has one unit of positive charge, and each electron has one unit of negative charge. A neutron, as its name implies, is electrically neutral.

Protons and neutrons are packed together tightly in a dense core, or **atomic nucleus**, at the center of an atom; protons give the nucleus a positive charge. The rapidly moving electrons form a "cloud" of negative charge around the nucleus, and it is the attraction between opposite charges that keeps the electrons in the vicinity of the nucleus. **Figure 2.4** shows

Figure 2.4 Simplified models of a helium (He) atom. The helium nucleus consists of 2 neutrons (brown) and 2 protons (pink). Two electrons (yellow) exist outside the nucleus. These models are not to scale; they greatly overestimate the size of the nucleus in relation to the electron cloud.

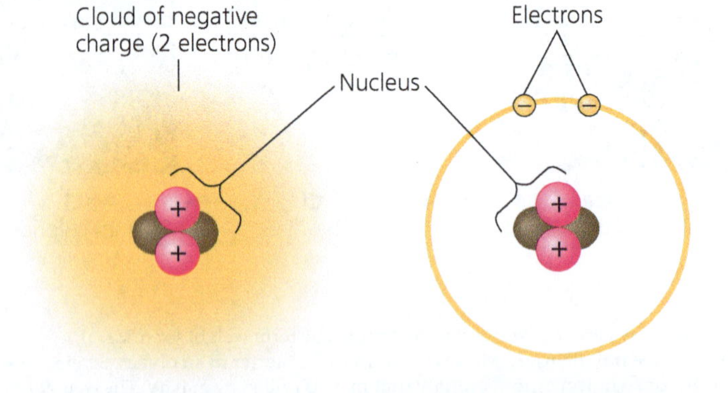

(a) This model represents the two electrons as a cloud of negative charge, a result of their motion around the nucleus.

(b) In this more simplified model, the electrons are shown as two small yellow spheres on a circle around the nucleus.

two commonly used models of the structure of the helium atom as an example.

The neutron and proton are almost identical in mass, each about 1.7×10^{-24} gram (g). Grams and other conventional units are not very useful for describing the mass of objects that are so minuscule. Thus, for atoms and subatomic particles (and for molecules, too), we use a unit of measurement called the **dalton**, in honor of John Dalton, the British scientist who helped develop atomic theory around 1800. (The dalton is the same as the *atomic mass unit*, or *amu*, a unit you may have encountered elsewhere.) Neutrons and protons have masses close to 1 dalton. Because the mass of an electron is only about 1/2,000 that of a neutron or proton, we can ignore electrons when computing the total mass of an atom.

Atomic Number and Atomic Mass

Atoms of the various elements differ in their number of subatomic particles. All atoms of a particular element have the same number of protons in their nuclei. This number of protons, which is unique to that element, is called the **atomic number** and is written as a subscript to the left of the symbol for the element. The abbreviation $_2He$, for example, tells us that an atom of the element helium has 2 protons in its nucleus. Unless otherwise indicated, an atom is neutral in electrical charge, which means that its protons must be balanced by an equal number of electrons. Therefore, the atomic number tells us the number of protons and also the number of electrons in an electrically neutral atom.

We can deduce the number of neutrons from a second quantity, the **mass number**, which is the total number of protons and neutrons in the nucleus of an atom. The mass number is written as a superscript to the left of an element's symbol. For example, we can use this shorthand to write an atom of helium as ^4_2He. Because the atomic number indicates how many protons there are, we can determine the number of neutrons by subtracting the atomic number from the mass number. In our example, the helium atom ^4_2He has 2 neutrons. For sodium (Na):

Mass number = number of protons + neutrons
= 23 for sodium

$^{23}_{11}Na$

Atomic number = number of protons
= number of electrons in a neutral atom
= 11 for sodium

Number of neutrons = mass number − atomic number
= 23 − 11 = 12 for sodium

The simplest atom is hydrogen ^1_1H, which has no neutrons; it consists of a single proton with a single electron.

Because the contribution of electrons to mass is negligible, almost all of an atom's mass is concentrated in its nucleus. Neutrons and protons each have a mass very close to 1 dalton, so the mass number is close to, but slightly different from, the total mass of an atom, called its **atomic mass**. For example, the mass number of sodium ($^{23}_{11}Na$) is 23, but its atomic mass is 22.9898 daltons.

MB Animation: Atomic Number and Atomic Mass

Isotopes

All atoms of a given element have the same number of protons, but some atoms have more neutrons than other atoms of the same element and therefore have greater mass. These different atomic forms of the same element are called **isotopes** of the element. In nature, an element may occur as a mixture of its isotopes. As an example, the element carbon, which has the atomic number 6, has three naturally occurring isotopes. The most common isotope is carbon-12, $^{12}_6C$, which accounts for about 99% of the carbon in nature. The isotope $^{12}_6C$ has 6 neutrons. Most of the remaining 1% of carbon consists of atoms of the isotope $^{13}_6C$, with 7 neutrons. A third, even rarer isotope, $^{14}_6C$, has 8 neutrons. Notice that all three isotopes of carbon have 6 protons; otherwise, they would not be carbon. Although the isotopes of an element have slightly different masses, they behave identically in chemical reactions. (For an element with more than one naturally occurring isotope, the atomic mass is an average of those isotopes, weighted by their abundance. Thus carbon has an atomic mass of 12.01 daltons.)

Both ^{12}C and ^{13}C are stable isotopes, meaning that their nuclei do not have a tendency to lose subatomic particles, a process called decay. The isotope ^{14}C, however, is unstable, or radioactive. A **radioactive isotope** is one in which the nucleus decays spontaneously, giving off particles and energy. When the radioactive decay leads to a change in the number of protons, it transforms the atom to an atom of a different element. For example, when an atom of carbon-14 (^{14}C) decays, it loses a proton, becoming an atom of nitrogen (^{14}N). Radioactive isotopes have many useful applications in biology.

Radioactive Tracers

Radioactive isotopes are often used as diagnostic tools in medicine. Cells can use radioactive atoms just as they would use nonradioactive isotopes of the same element. The radioactive isotopes are incorporated into biologically active molecules, which are then used as tracers to track atoms during metabolism, the chemical processes of an organism. For example, certain kidney disorders are diagnosed by injecting small doses of radioactively labeled substances into the blood and then analyzing the tracer molecules excreted in the urine. Radioactive tracers are also used in combination with sophisticated imaging instruments, such as PET scanners

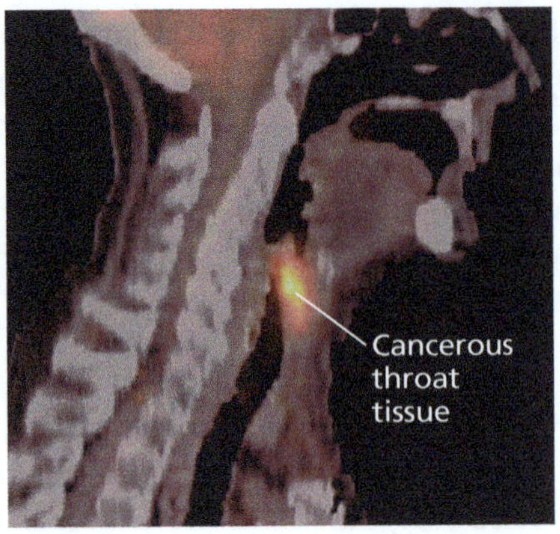

Figure 2.5 A PET scan, a medical use for radioactive isotopes. PET, an acronym for positron-emission tomography, detects locations of intense chemical activity in the body. The bright yellow spot marks an area with an elevated level of radioactively labeled glucose, which in turn indicates high metabolic activity, a hallmark of cancerous tissue.

that can monitor growth and metabolism of cancers in the body **(Figure 2.5)**.

Although radioactive isotopes are very useful in biological research and medicine, radiation from decaying isotopes also poses a hazard to life by damaging cellular molecules. The severity of this damage depends on the type and amount of radiation an organism absorbs. One of the most serious environmental threats is radioactive fallout from nuclear accidents. The doses of most isotopes used in medical diagnosis, however, are relatively safe.

Radiometric Dating

EVOLUTION Researchers measure radioactive decay in fossils to date these relics of past life. Fossils provide a large body of evidence for evolution, documenting differences between organisms from the past and those living at present and giving us insight into species that have disappeared over time. While the layering of fossil beds establishes that deeper fossils are older than more shallow ones, the actual age (in years) of the fossils in each layer cannot be determined by position alone. This is where radioactive isotopes come in.

A "parent" isotope decays into its "daughter" isotope at a fixed rate, expressed as the **half-life** of the isotope—the time it takes for 50% of the parent isotope to decay. Each radioactive isotope has a characteristic half-life that is not affected by temperature, pressure, or any other environmental variable. Using a process called **radiometric dating**, scientists measure the ratio of different isotopes and calculate how many half-lives (in years) have passed since an organism was fossilized or a rock was formed. Half-life values range from very short for some isotopes, measured in seconds or days, to extremely long—uranium-238 has a half-life of 4.5 billion years! Each isotope can best "measure" a particular range of years: Uranium-238 was used to determine that moon rocks are approximately 4.5 billion years old, similar to the estimated age of Earth. In the **Scientific Skills Exercise**, you can work with data from an experiment that used carbon-14 to determine the age of an important fossil. (Figure 25.6 explains more about radiometric dating of fossils.)

The Energy Levels of Electrons

The simplified models of the atom in Figure 2.4 greatly exaggerate the size of the nucleus relative to that of the whole atom. If an atom of helium were the size of a typical football stadium, the nucleus would be the size of a pencil eraser in the center of the field. Moreover, the electrons would be like two tiny gnats buzzing around the stadium. Atoms are mostly empty space. When two atoms approach each other during a chemical reaction, their nuclei do not come close enough to interact. Of the three subatomic particles we have discussed, only electrons are directly involved in chemical reactions.

An atom's electrons vary in the amount of energy they possess. **Energy** is defined as the capacity to cause change—for instance, by doing work. **Potential energy** is the energy that matter possesses because of its location or structure. For example, water in a reservoir on a hill has potential energy because of its altitude. When the gates of the reservoir's dam are opened and the water runs downhill, the energy can be used to do work, such as moving the blades of turbines to generate electricity. Because energy has been expended, the water has less energy at the bottom of the hill than it did in the reservoir. Matter has a natural tendency to move toward the lowest possible state of potential energy; in our example, the water runs downhill. To restore the potential energy of a reservoir, work must be done to elevate the water against gravity.

The electrons of an atom have potential energy due to their distance from the nucleus **(Figure 2.6)**. The negatively charged

Figure 2.6 Energy levels of an atom's electrons. Electrons exist only at fixed levels of potential energy called electron shells.

(a) A ball bouncing down a flight of stairs can come to rest only on each step, not between steps. Similarly, an electron can exist only at certain energy levels, not between levels.

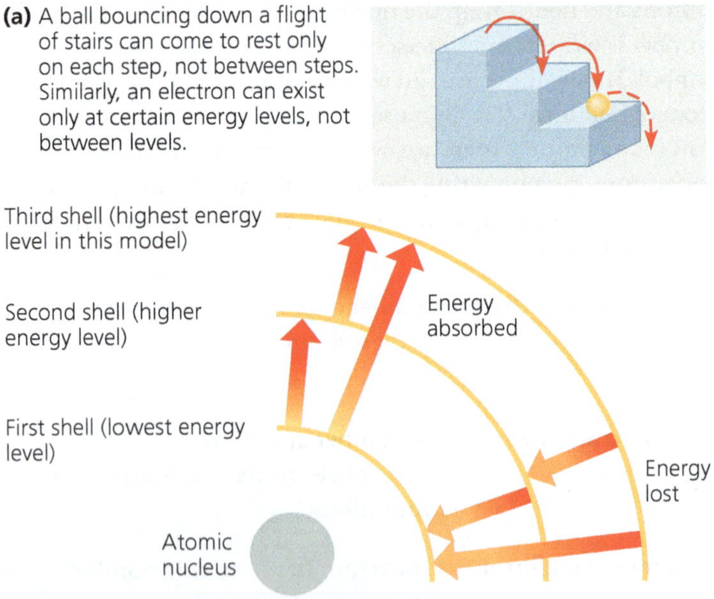

(b) An electron can move from one shell to another only if the energy it gains or loses is exactly equal to the difference in energy between the energy levels of the two shells. Arrows in this model indicate some of the stepwise changes in potential energy that are possible.

Figure Walkthrough

SCIENTIFIC SKILLS EXERCISE

Calibrating a Standard Radioactive Isotope Decay Curve and Interpreting Data

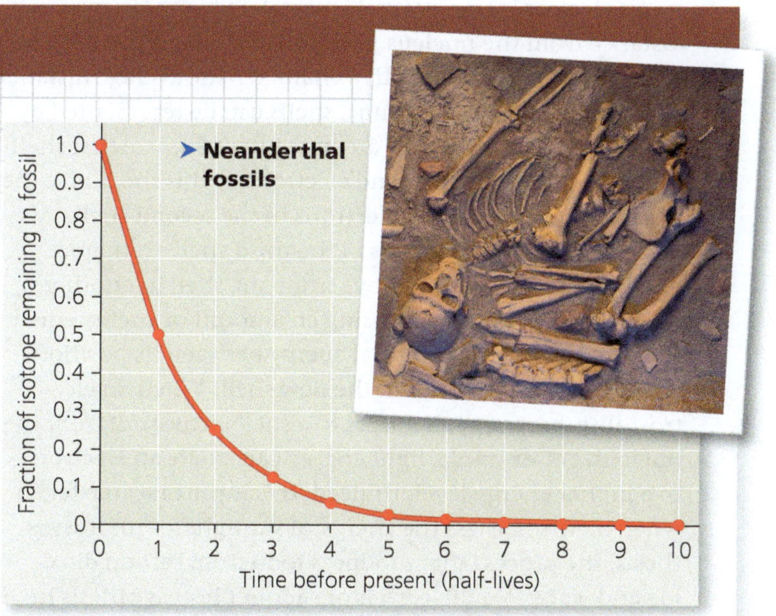

Data from R. Pinhasi et al., Revised age of late Neanderthal occupation and the end of the Middle Paleolithic in the northern Caucasus, *Proceedings of the National Academy of Sciences USA* 147:8611–8616 (2011). doi 10.1073/pnas.1018938108

How Long Might Neanderthals Have Co-Existed with Modern Humans (*Homo sapiens*)? Neanderthals (*Homo neanderthalensis*) were living in Europe by 350,000 years ago and may have coexisted with early *Homo sapiens* in parts of Eurasia for hundreds or thousands of years before Neanderthals became extinct. Researchers sought to more accurately determine the extent of their overlap by pinning down the latest date Neanderthals still lived in the area. They used carbon-14 dating to determine the age of a Neanderthal fossil from the most recent (uppermost) archeological layer containing Neanderthal bones. In this exercise you will calibrate a standard carbon-14 decay curve and use it to determine the age of this Neanderthal fossil. The age will help you approximate the last time the two species may have coexisted at the site where this fossil was collected.

How the Experiment Was Done Carbon-14 (^{14}C) is a radioactive isotope of carbon that decays to ^{14}N at a constant rate. ^{14}C is present in the atmosphere in small amounts at a constant ratio with both ^{13}C and ^{12}C, two other isotopes of carbon. When carbon is taken up from the atmosphere by a plant during photosynthesis, ^{12}C, ^{13}C, and ^{14}C isotopes are incorporated into the plant in the same proportions in which they were present in the atmosphere. These proportions remain the same in the tissues of an animal that eats the plant. While an organism is alive, the ^{14}C in its body constantly decays to ^{14}N but is constantly replaced by new carbon from the environment. Once an organism dies, it stops taking in new ^{14}C but the ^{14}C in its tissues continues to decay, while the ^{12}C in its tissues remains the same because it is not radioactive and does not decay. Thus, scientists can calculate how long the pool of original ^{14}C has been decaying in a fossil by measuring the ratio of ^{14}C to ^{12}C and comparing it to the ratio of ^{14}C to ^{12}C present originally in the atmosphere. The fraction of ^{14}C in a fossil compared to the original fraction of ^{14}C can be converted to years because we know that the half-life of ^{14}C is 5,730 years—in other words, half of the ^{14}C in a fossil decays every 5,730 years.

Data from the Experiment The researchers found that the Neanderthal fossil had approximately 0.0078 (or, in scientific notation, 7.8×10^{-3}) as much ^{14}C as the atmosphere. The following questions will guide you through translating this fraction into the age of the fossil.

INTERPRET THE DATA

1. A standard curve of radioactive isotope decay is shown at the top of the right column. The graph line shows the fraction of the radioactive isotope over time (before present) in units of half-lives. Recall that a half-life is the amount of time it takes for half of the radioactive isotope to decay. Labeling each data point with the corresponding fractions will help orient you to this graph. Draw an arrow to the data point for half-life = 1 and write the fraction of ^{14}C that will remain after one half-life. Calculate the fraction of ^{14}C remaining at each half-life and write the fractions on the graph near arrows pointing to the data points. Convert each fraction to a decimal number and round off to a maximum of three significant digits (zeros at the beginning of the number do not count as significant digits). Also write each decimal number in scientific notation.
2. Recall that ^{14}C has a half-life of 5,730 years. To calibrate the *x*-axis for ^{14}C decay, write the time before present in years below each half-life.
3. The researchers found that the Neanderthal fossil had approximately 0.0078 as much ^{14}C as found originally in the atmosphere. (a) Using the numbers on your graph, determine how many half-lives have passed since the Neanderthal died. (b) Using your ^{14}C calibration on the *x*-axis, what is the approximate age of the Neanderthal fossil in years (round off to the nearest thousand)? (c) Approximately when did Neanderthals become extinct according to this study? (d) The researchers cite evidence that modern humans (*H. sapiens*) became established in the same region as the last Neanderthals approximately 39,000–42,000 years ago. What does this suggest about the overlap of Neanderthals and modern humans?
4. Carbon-14 dating works for fossils up to about 75,000 years old; fossils older than that contain too little ^{14}C to be detected. Most dinosaurs went extinct 65.5 million years ago. (a) Can ^{14}C be used to date dinosaur bones? Explain. (b) Radioactive uranium-235 has a half-life of 704 million years. If it was incorporated into dinosaur bones, could it be used to date the dinosaur fossils? Explain.

MB **Instructors**: A version of this Scientific Skills Exercise can be assigned in MasteringBiology.

electrons are attracted to the positively charged nucleus. It takes work to move a given electron farther away from the nucleus, so the more distant an electron is from the nucleus, the greater its potential energy. Unlike the continuous flow of water downhill, changes in the potential energy of electrons can occur only in steps of fixed amounts. An electron having a certain amount of energy is something like a ball on a staircase (Figure 2.6a). The ball can have different amounts of potential energy, depending on which step it is on, but it cannot spend much time between the steps. Similarly, an electron's potential energy is determined by its energy level. An electron can exist only at certain energy levels, not between them.

An electron's energy level is correlated with its average distance from the nucleus. Electrons are found in different **electron shells**, each with a characteristic average distance and energy level. In diagrams, shells can be represented by concentric circles, as they are in Figure 2.6b. The first shell is closest to the nucleus, and electrons in this shell have the lowest potential energy. Electrons in the second shell have more energy, and electrons in the third shell even more energy. An electron can move from one shell to another, but only by absorbing or losing an amount of energy equal to the difference in potential energy between its position in the old shell and that in the new shell. When an electron absorbs energy, it moves to a shell farther out from the nucleus. For example, light energy can excite an electron to a higher energy level. (Indeed, this is the first step taken when plants harness the energy of sunlight for photosynthesis, the process that produces food from carbon dioxide and water. You'll learn more about photosynthesis in Chapter 10.) When an electron loses energy, it "falls back" to a shell closer to the nucleus, and the lost energy is usually released to the environment as heat. For example, sunlight excites electrons in the surface of a car to higher energy levels. When the electrons fall back to their original levels, the car's surface heats up. This thermal energy can be transferred to the air or to your hand if you touch the car.

Electron Distribution and Chemical Properties

The chemical behavior of an atom is determined by the distribution of electrons in the atom's electron shells. Beginning with hydrogen, the simplest atom, we can imagine building the atoms of the other elements by adding 1 proton and 1 electron at a time (along with an appropriate number of neutrons). **Figure 2.7**, a modified version of what is called the *periodic table of the elements*, shows this distribution of electrons for the first 18 elements, from hydrogen ($_1H$) to argon ($_{18}Ar$). The elements are arranged in three rows, or *periods*, corresponding to the number of electron shells in their atoms. The left-to-right sequence of elements in each row corresponds to the sequential addition of electrons and protons. (See Appendix B for the complete periodic table.)

▼ Figure 2.7 Electron distribution diagrams for the first 18 elements in the periodic table. In a standard periodic table (see Appendix B), information for each element is presented as shown for helium in the inset. In the diagrams in this table, electrons are represented as yellow dots and electron shells as concentric circles. These diagrams are a convenient way to picture the distribution of an atom's electrons among its electron shells, but these simplified models do not accurately represent the shape of the atom or the location of its electrons. The elements are arranged in rows, each representing the filling of an electron shell. As electrons are added, they occupy the lowest available shell.

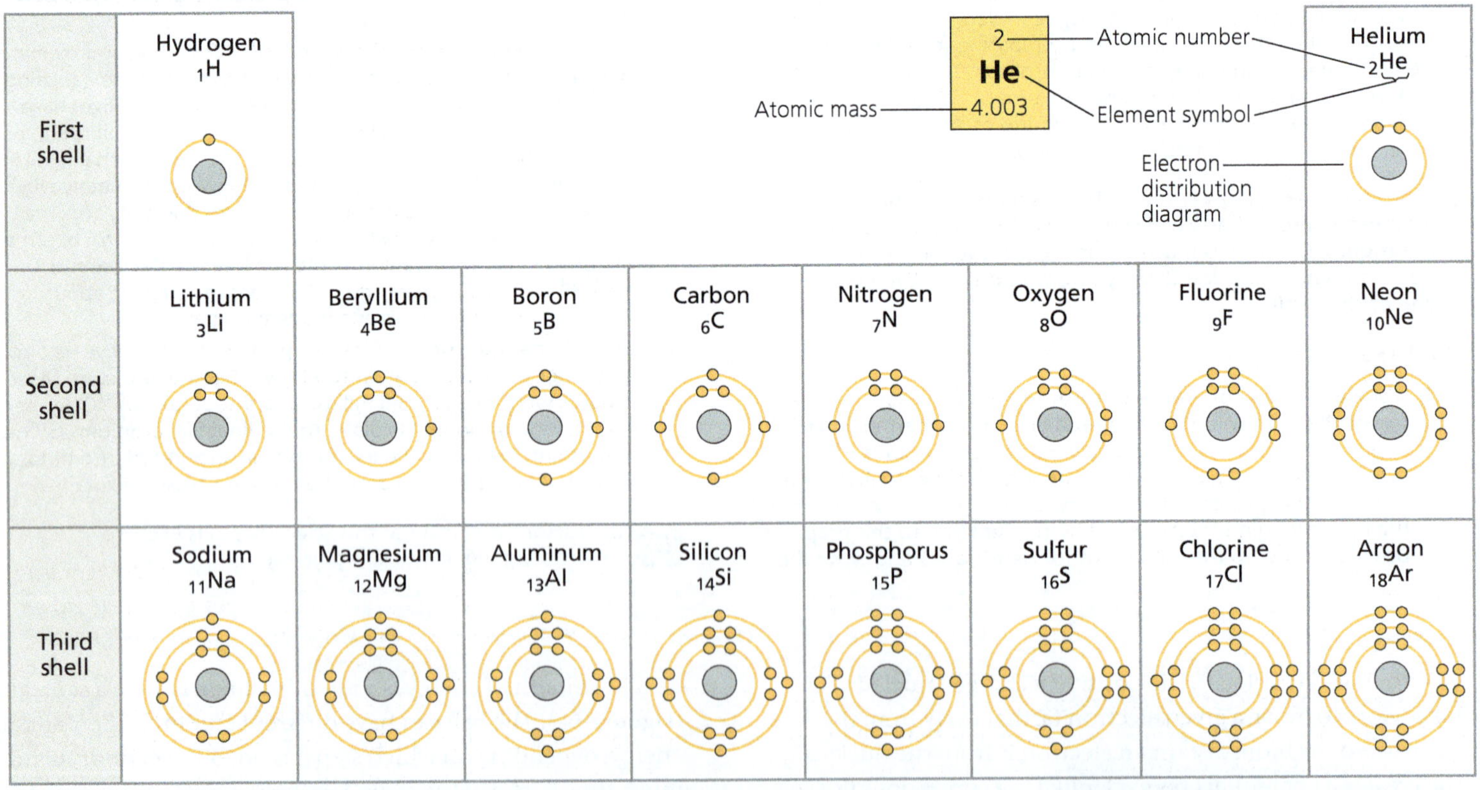

VISUAL SKILLS ➤ *Looking at the depictions of atoms in this chart, what is the atomic number of magnesium? How many protons and electrons does it have? How many electron shells? How many valence electrons?*

Animation: Electron Distribution Diagrams

Hydrogen's 1 electron and helium's 2 electrons are located in the first shell. Electrons, like all matter, tend to exist in the lowest available state of potential energy. In an atom, this state is in the first shell. However, the first shell can hold no more than 2 electrons; thus, hydrogen and helium are the only elements in the first row of the table. In an atom with more than 2 electrons, the additional electrons must occupy higher shells because the first shell is full. The next element, lithium, has 3 electrons. Two of these electrons fill the first shell, while the third electron occupies the second shell. The second shell holds a maximum of 8 electrons. Neon, at the end of the second row, has 8 electrons in the second shell, giving it a total of 10 electrons.

The chemical behavior of an atom depends mostly on the number of electrons in its *outermost* shell. We call those outer electrons **valence electrons** and the outermost electron shell the **valence shell**. In the case of lithium, there is only 1 valence electron, and the second shell is the valence shell. Atoms with the same number of electrons in their valence shells exhibit similar chemical behavior. For example, fluorine (F) and chlorine (Cl) both have 7 valence electrons, and both form compounds when combined with the element sodium (Na): Sodium fluoride (NaF) is commonly added to toothpaste to prevent tooth decay, and, as described earlier, NaCl is table salt (see Figure 2.2). An atom with a completed valence shell is unreactive; that is, it will not interact readily with other atoms. At the far right of the periodic table are helium, neon, and argon, the only three elements shown in Figure 2.7 that have full valence shells. These elements are said to be *inert*, meaning chemically unreactive. All the other atoms in Figure 2.7 are chemically reactive because they have incomplete valence shells.

Electron Orbitals

In the early 1900s, the electron shells of an atom were visualized as concentric paths of electrons orbiting the nucleus, somewhat like planets orbiting the sun. It is still convenient to use two-dimensional concentric-circle diagrams, as in Figure 2.7, to symbolize three-dimensional electron shells. However, you need to remember that each concentric circle represents only the *average* distance between an electron in that shell and the nucleus. Accordingly, the concentric-circle diagrams do not give a real picture of an atom. In reality, we can never know the exact location of an electron. What we can do instead is describe the space in which an electron spends most of its time. The three-dimensional space where an electron is found 90% of the time is called an **orbital**.

Each electron shell contains electrons at a particular energy level, distributed among a specific number of orbitals of distinctive shapes and orientations. **Figure 2.8** shows the orbitals of neon as an example, with its electron distribution diagram for reference. You can think of an orbital as a component of an electron shell. The first electron shell has only one spherical *s* orbital (called 1*s*), but the second shell has

▼ Figure 2.8 Electron orbitals.

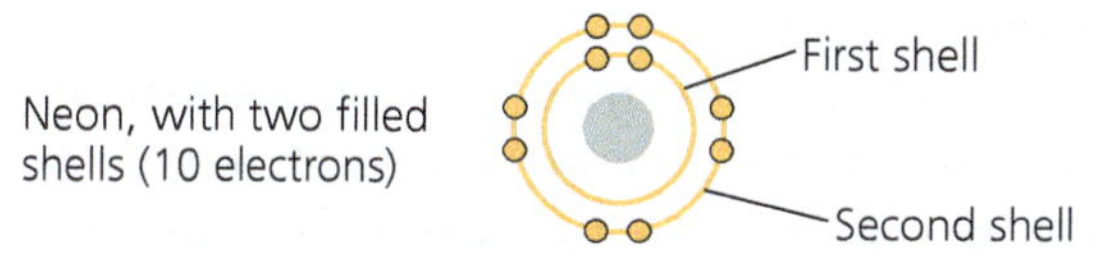

(a) Electron distribution diagram. An electron distribution diagram is shown here for a neon atom, which has a total of 10 electrons. Each concentric circle represents an electron shell, which can be subdivided into electron orbitals.

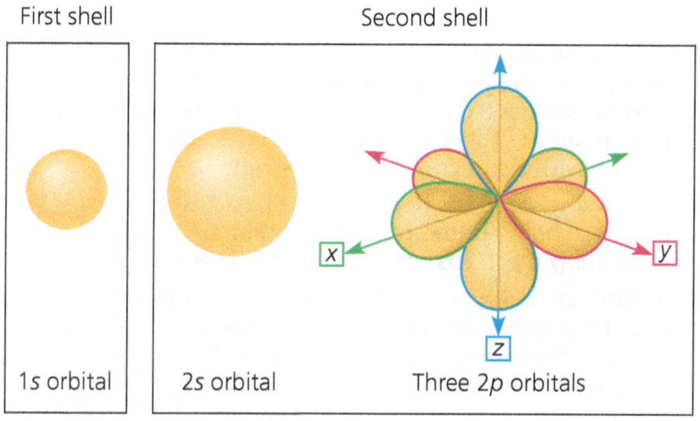

(b) Separate electron orbitals. The three-dimensional shapes represent electron orbitals—the volumes of space where the electrons of an atom are most likely to be found. Each orbital holds a maximum of 2 electrons. The first electron shell, on the left, has one spherical (*s*) orbital, designated 1*s*. The second shell, on the right, has one larger *s* orbital (designated 2*s* for the second shell) plus three dumbbell-shaped orbitals called *p* orbitals (2*p* for the second shell). The three 2*p* orbitals lie at right angles to one another along imaginary *x*-, *y*-, and *z*-axes of the atom. Each 2*p* orbital is outlined here in a different color.

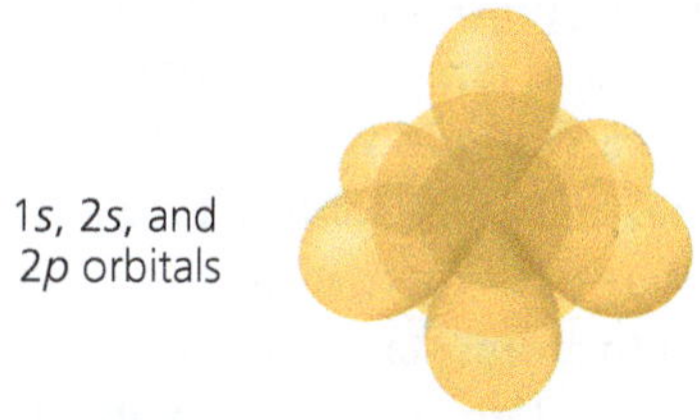

(c) Superimposed electron orbitals. To reveal the complete picture of the electron orbitals of neon, we superimpose the 1*s* orbital of the first shell and the 2*s* and three 2*p* orbitals of the second shell.

four orbitals: one large spherical *s* orbital (called 2*s*) and three dumbbell-shaped *p* orbitals (called 2*p* orbitals). (The third shell and other higher electron shells also have *s* and *p* orbitals, as well as orbitals of more complex shapes.)

No more than 2 electrons can occupy a single orbital. The first electron shell can therefore accommodate up to 2 electrons in its *s* orbital. The lone electron of a hydrogen atom occupies the 1*s* orbital, as do the 2 electrons of a helium atom. The four orbitals of the second electron shell can hold up to 8 electrons,

2 in each orbital. Electrons in each of the four orbitals in the second shell have nearly the same energy, but they move in different volumes of space.

The reactivity of an atom arises from the presence of unpaired electrons in one or more orbitals of the atom's valence shell. As you will see in the next section, atoms interact in a way that completes their valence shells. When they do so, it is the *unpaired* electrons that are involved.

CONCEPT CHECK 2.2

1. A lithium atom has 3 protons and 4 neutrons. What is its mass number?
2. A nitrogen atom has 7 protons, and the most common isotope of nitrogen has 7 neutrons. A radioactive isotope of nitrogen has 8 neutrons. Write the atomic number and mass number of this radioactive nitrogen as a chemical symbol with a subscript and superscript.
3. How many electrons does fluorine have? How many electron shells? Name the orbitals that are occupied. How many electrons are needed to fill the valence shell?
4. **VISUAL SKILLS >** In Figure 2.7, if two or more elements are in the same row, what do they have in common? If two or more elements are in the same column, what do they have in common?

For suggested answers, see Appendix A.

CONCEPT 2.3

The formation and function of molecules depend on chemical bonding between atoms

Now that we have looked at the structure of atoms, we can move up the hierarchy of organization and see how atoms combine to form molecules and ionic compounds. Atoms with incomplete valence shells can interact with certain other atoms in such a way that each partner atom completes its valence shell: The atoms either share or transfer valence electrons. These interactions usually result in atoms staying close together, held by attractions called **chemical bonds**. The strongest kinds of chemical bonds are covalent bonds and ionic bonds in dry ionic compounds. (Ionic bonds in aqueous, or water-based, solutions are weak interactions, as we will see later.)

MB **Animation: Introduction to Chemical Bonds**

Covalent Bonds

A **covalent bond** is the sharing of a pair of valence electrons by two atoms. For example, let's consider what happens when two hydrogen atoms approach each other. Recall that hydrogen has 1 valence electron in the first shell, but the shell's capacity is 2 electrons. When the two hydrogen atoms come close enough for their 1*s* orbitals to overlap, they can share their electrons **(Figure 2.9)**. Each hydrogen atom is now associated with 2 electrons in what amounts to a completed valence shell. Two or more atoms held together by covalent bonds constitute a **molecule**, in this case a hydrogen molecule.

Figure 2.9 Formation of a covalent bond.

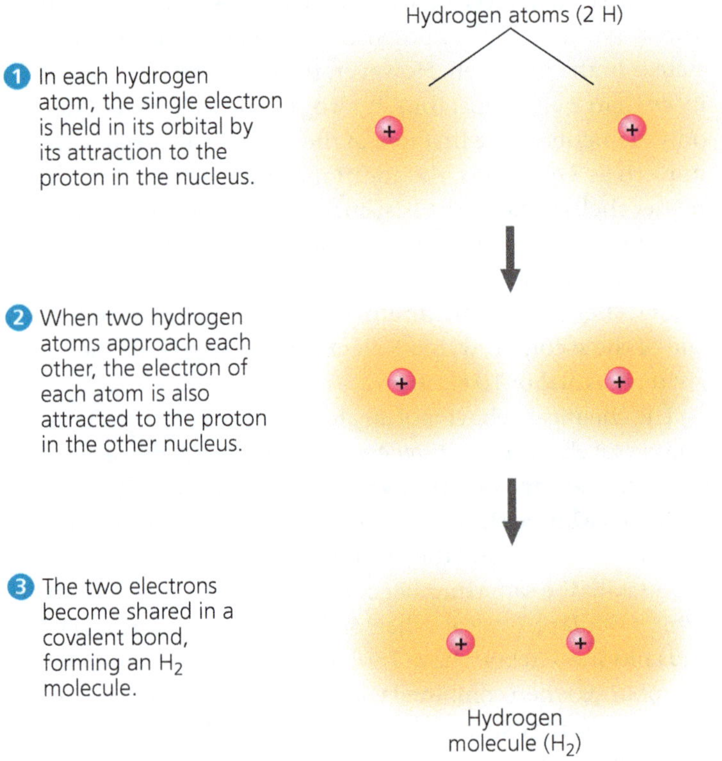

Figure 2.10a shows several ways of representing a hydrogen molecule. Its *molecular formula*, H_2, simply indicates that the molecule consists of two atoms of hydrogen. Electron sharing can be depicted by an electron distribution diagram or by a *Lewis dot structure*, in which element symbols are surrounded by dots that represent the valence electrons (H:H). We can also use a *structural formula*, H—H, where the line represents a **single bond**, a pair of shared electrons. A *space-filling model* comes closest to representing the actual shape of the molecule. (You may also be familiar with ball-and-stick models, which are shown in Figure 2.15.)

Oxygen has 6 electrons in its second electron shell and therefore needs 2 more electrons to complete its valence shell. Two oxygen atoms form a molecule by sharing *two* pairs of valence electrons **(Figure 2.10b)**. The atoms are thus joined by what is called a **double bond** (O=O).

Each atom that can share valence electrons has a bonding capacity corresponding to the number of covalent bonds the atom can form. When the bonds form, they give the atom a full complement of electrons in the valence shell. The bonding capacity of oxygen, for example, is 2. This bonding capacity is called the atom's **valence** and usually equals the number of unpaired electrons required to complete the atom's outermost (valence) shell. See if you can determine the valences of hydrogen, oxygen, nitrogen, and carbon by studying the electron distribution diagrams in Figure 2.7. You can see that the valence of hydrogen is 1; oxygen, 2; nitrogen, 3; and carbon, 4. The situation is more complicated

Figure 2.10 Covalent bonding in four molecules. The number of electrons required to complete an atom's valence shell generally determines how many covalent bonds that atom will form. This figure shows several ways of indicating covalent bonds.

Name and Molecular Formula	Electron Distribution Diagram	Lewis Dot Structure and Structural Formula	Space-Filling Model
(a) Hydrogen (H_2). Two hydrogen atoms share one pair of electrons, forming a single bond.	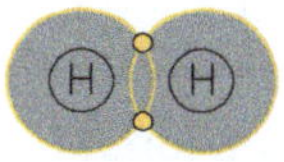	H:H H—H	
(b) Oxygen (O_2). Two oxygen atoms share two pairs of electrons, forming a double bond.	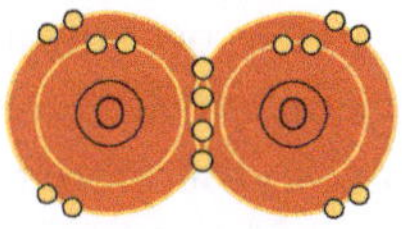	Ö::Ö O=O	
(c) Water (H_2O). Two hydrogen atoms and one oxygen atom are joined by single bonds, forming a molecule of water.	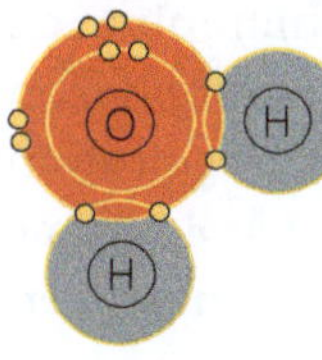	:Ö:H H O—H | H	
(d) Methane (CH_4). Four hydrogen atoms can satisfy the valence of one carbon atom, forming methane.	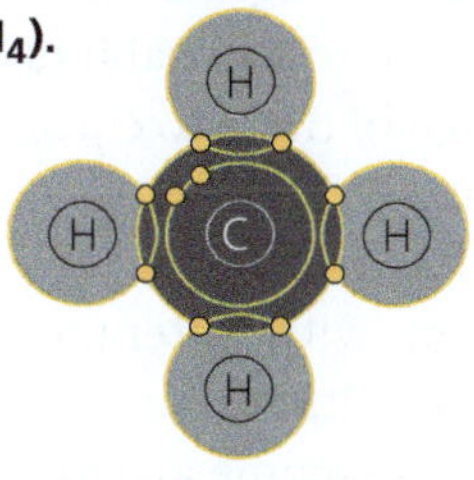	H H:C:H H H | H—C—H | H	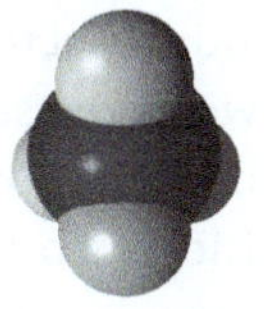

Animation: Covalent Bonds

for phosphorus, in the third row of the periodic table, which can have a valence of 3 or 5 depending on the combination of single and double bonds it makes.

The molecules H_2 and O_2 are pure elements rather than compounds because a compound is a combination of two or more *different* elements. Water, with the molecular formula H_2O, is a compound. Two atoms of hydrogen are needed to satisfy the valence of one oxygen atom. **Figure 2.10c** shows the structure of a water molecule. (Water is so important to life that Chapter 3 is devoted entirely to its structure and behavior.)

Methane, the main component of natural gas, is a compound with the molecular formula CH_4. It takes four hydrogen atoms, each with a valence of 1, to complement one atom of carbon, with its valence of 4 **(Figure 2.10d)**. (We will look at many other compounds of carbon in Chapter 4.)

Atoms in a molecule attract shared bonding electrons to varying degrees, depending on the element. The attraction of a particular atom for the electrons of a covalent bond is called its

Figure 2.11 Polar covalent bonds in a water molecule.

Because oxygen (O) is more electronegative than hydrogen (H), shared electrons are pulled more toward oxygen.

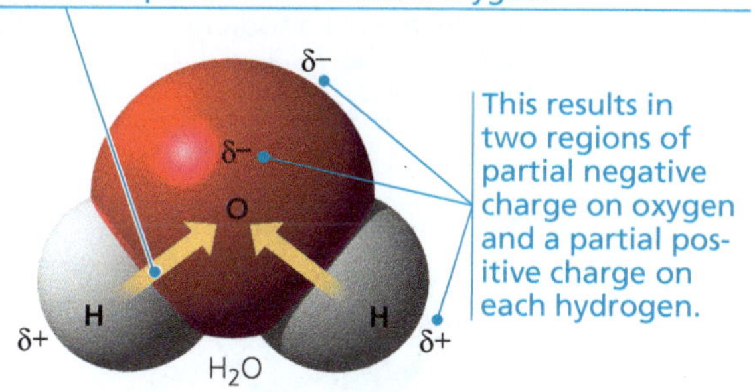

Animation: Nonpolar and Polar Molecules

electronegativity. The more electronegative an atom is, the more strongly it pulls shared electrons toward itself. In a covalent bond between two atoms of the same element, the electrons are shared equally because the two atoms have the same electronegativity—the tug-of-war is at a standoff. Such a bond is called a **nonpolar covalent bond**. For example, the single bond of H_2 is nonpolar, as is the double bond of O_2. However, when an atom is bonded to a more electronegative atom, the electrons of the bond are not shared equally. This type of bond is called a **polar covalent bond**. Such bonds vary in their polarity, depending on the relative electronegativity of the two atoms. For example, the bonds between the oxygen and hydrogen atoms of a water molecule are quite polar **(Figure 2.11)**.

Oxygen is one of the most electronegative elements, attracting shared electrons much more strongly than hydrogen does. In a covalent bond between oxygen and hydrogen, the electrons spend more time near the oxygen nucleus than near the hydrogen nucleus. Because electrons have a negative charge and are pulled toward oxygen in a water molecule, the oxygen atom has two regions of partial negative charge (each indicated by the Greek letter δ with a minus sign, δ−, "delta minus"), and each hydrogen atom has a partial positive charge (δ+, "delta plus"). In contrast, the individual bonds of methane (CH_4) are much less polar because the electronegativities of carbon and hydrogen are similar.

Ionic Bonds

In some cases, two atoms are so unequal in their attraction for valence electrons that the more electronegative atom strips an electron completely away from its partner. The two resulting oppositely charged atoms (or molecules) are called **ions**. A positively charged ion is called a **cation**, while a negatively charged ion is called an **anion**. Because of their opposite charges, cations and anions attract each other; this attraction is called an **ionic bond**. Note that the transfer of an electron is not, by itself, the formation of a bond; rather, it allows a bond to form because it results in two ions of opposite charge. Any two ions of opposite charge can form an ionic bond. The ions do not need to have acquired their charge by an electron transfer with each other.

Figure 2.12 Electron transfer and ionic bonding. The attraction between oppositely charged atoms, or ions, is an ionic bond. An ionic bond can form between any two oppositely charged ions, even if they have not been formed by transfer of an electron from one to the other.

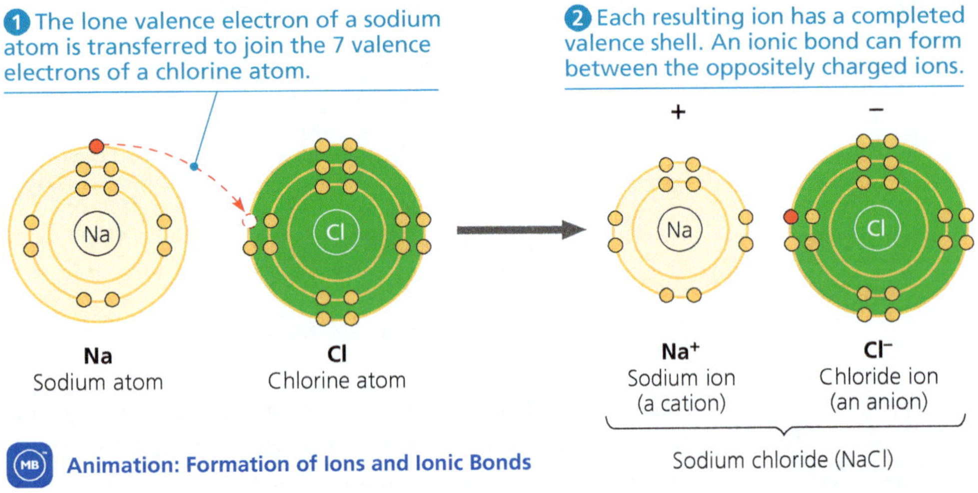

MB Animation: Formation of Ions and Ionic Bonds

This is what happens when an atom of sodium ($_{11}Na$) encounters an atom of chlorine ($_{17}Cl$) **(Figure 2.12)**. A sodium atom has a total of 11 electrons, with its single valence electron in the third electron shell. A chlorine atom has a total of 17 electrons, with 7 electrons in its valence shell. When these two atoms meet, the lone valence electron of sodium is transferred to the chlorine atom, and both atoms end up with their valence shells complete. (Because sodium no longer has an electron in the third shell, the second shell is now the valence shell.) The electron transfer between the two atoms moves one unit of negative charge from sodium to chlorine. Sodium, now with 11 protons but only 10 electrons, has a net electrical charge of 1+; the sodium atom has become a cation. Conversely, the chlorine atom, having gained an extra electron, now has 17 protons and 18 electrons, giving it a net electrical charge of 1−; it has become a chloride ion—an anion.

Compounds formed by ionic bonds are called **ionic compounds**, or **salts**. We know the ionic compound sodium chloride (NaCl) as table salt **(Figure 2.13)**. Salts are often found in nature as crystals of various sizes and shapes. Each salt crystal is an aggregate of vast numbers of cations and anions bonded by their electrical attraction and arranged in a three-dimensional lattice. Unlike a covalent compound, which consists of molecules having a definite size and number of atoms, an ionic compound does not consist of molecules. The formula for an ionic compound, such as NaCl, indicates only the ratio of elements in a crystal of the salt. "NaCl" by itself is not a molecule.

Figure 2.13 A sodium chloride (NaCl) crystal. The sodium ions (Na^+) and chloride ions (Cl^-) are held together by ionic bonds. The formula NaCl tells us that the ratio of Na^+ to Cl^- is 1:1.

Not all salts have equal numbers of cations and anions. For example, the ionic compound magnesium chloride ($MgCl_2$) has two chloride ions for each magnesium ion. Magnesium ($_{12}Mg$) must lose 2 outer electrons if the atom is to have a complete valence shell, so it has a tendency to become a cation with a net charge of 2+ (Mg^{2+}). One magnesium cation can therefore form ionic bonds with two chloride anions (Cl^-).

The term *ion* also applies to entire molecules that are electrically charged. In the salt ammonium chloride (NH_4Cl), for instance, the anion is a single chloride ion (Cl^-), but the cation is ammonium (NH_4^+), a nitrogen atom covalently bonded to four hydrogen atoms. The whole ammonium ion has an electrical charge of 1+ because it has given up 1 electron and thus is 1 electron short.

Environment affects the strength of ionic bonds. In a dry salt crystal, the bonds are so strong that it takes a hammer and chisel to break enough of them to crack the crystal in two. If the same salt crystal is dissolved in water, however, the ionic bonds are much weaker because each ion is partially shielded by its interactions with water molecules. Most drugs are manufactured as salts because they are quite stable when dry but can dissociate (come apart) easily in water. (In Concept 3.2, you will learn how water dissolves salts.)

Weak Chemical Interactions

In organisms, most of the strongest chemical bonds are covalent bonds, which link atoms to form a cell's molecules. But weaker interactions within and between molecules are also indispensable, contributing greatly to the emergent properties of life. Many large biological molecules are held in their functional form by weak interactions. In addition, when two molecules in the cell make contact, they may adhere temporarily by weak interactions. The reversibility of weak interactions can be an advantage: Two molecules can come together, affect one another in some way, and then separate.

Several types of weak chemical interactions are important in organisms. One is the ionic bond as it exists between ions dissociated in water, which we just discussed. Hydrogen bonds and van der Waals interactions are also crucial to life.

Hydrogen Bonds

Among weak chemical interactions, hydrogen bonds are so central to the chemistry of life that they deserve special

Figure 2.14 A hydrogen bond.

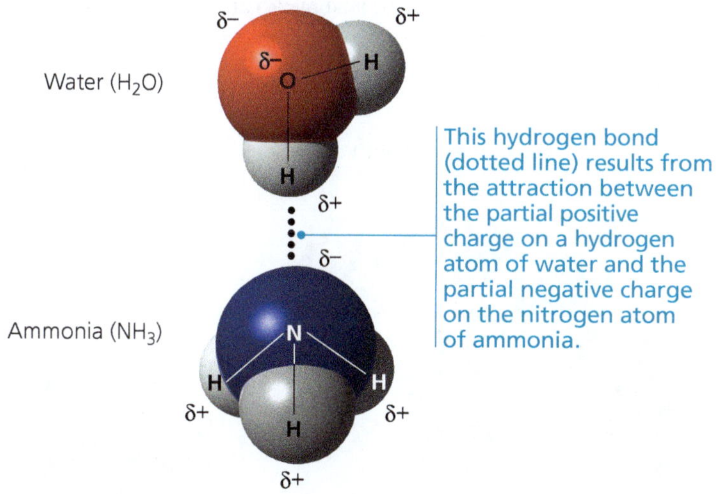

DRAW IT ➤ *Draw one water molecule surrounded by four other water molecules, arranged so that they can make hydrogen bonds with each other. Use simple outlines of space-filling models. Draw the partial charges on the water molecules and use dots for the hydrogen bonds.*

MB Animation: Hydrogen Bonds

attention. When a hydrogen atom is covalently bonded to an electronegative atom, the hydrogen atom has a partial positive charge that allows it to be attracted to a different electronegative atom nearby. This attraction between a hydrogen and an electronegative atom is called a **hydrogen bond**. In living cells, the electronegative partners are usually oxygen or nitrogen atoms. Refer to **Figure 2.14** to examine the simple case of hydrogen bonding between water (H_2O) and ammonia (NH_3).

Van der Waals Interactions

Even a molecule with nonpolar covalent bonds may have positively and negatively charged regions. Electrons are not always evenly distributed; at any instant, they may accumulate by chance in one part of a molecule or another. The results are ever-changing regions of positive and negative charge that enable all atoms and molecules to stick to one another. These **van der Waals interactions** are individually weak and occur only when atoms and molecules are very close together. When many such interactions occur simultaneously, however, they can be powerful: Van der Waals interactions allow the gecko lizard shown here to walk straight up a wall! The anatomy of the gecko's foot—including many minuscule hairlike projections from the toes and strong tendons underlying the skin—strikes a balance between maximum surface contact with the wall and necessary stiffness of the foot. The van der Waals interactions between the foot molecules and the molecules of the wall's surface are so numerous that despite their individual weakness, together they can support the gecko's body weight. This discovery has inspired development of an artificial adhesive called Geckskin: A patch the size of an index card can hold a 700-pound weight to a wall!

Van der Waals interactions, hydrogen bonds, ionic bonds in water, and other weak interactions may form not only between molecules but also between parts of a large molecule, such as a protein. The cumulative effect of weak interactions is to reinforce the three-dimensional shape of the molecule. (You will learn more about the very important biological roles of weak interactions in Chapter 5.)

Molecular Shape and Function

A molecule has a characteristic size and shape, which are key to its function in the living cell. A molecule consisting of two atoms, such as H_2 or O_2, is always linear, but most molecules with more than two atoms have more complicated shapes. These shapes are determined by the positions of the atoms' orbitals **(Figure 2.15)**. When an atom forms covalent bonds,

Figure 2.15 Molecular shapes due to hybrid orbitals.

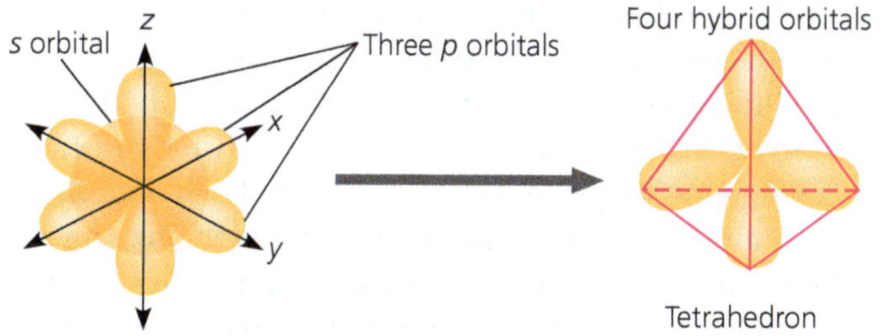

(a) Hybridization of orbitals. The single *s* and three *p* orbitals of a valence shell involved in covalent bonding combine to form four teardrop-shaped hybrid orbitals. These orbitals extend to the four corners of an imaginary tetrahedron (outlined in pink).

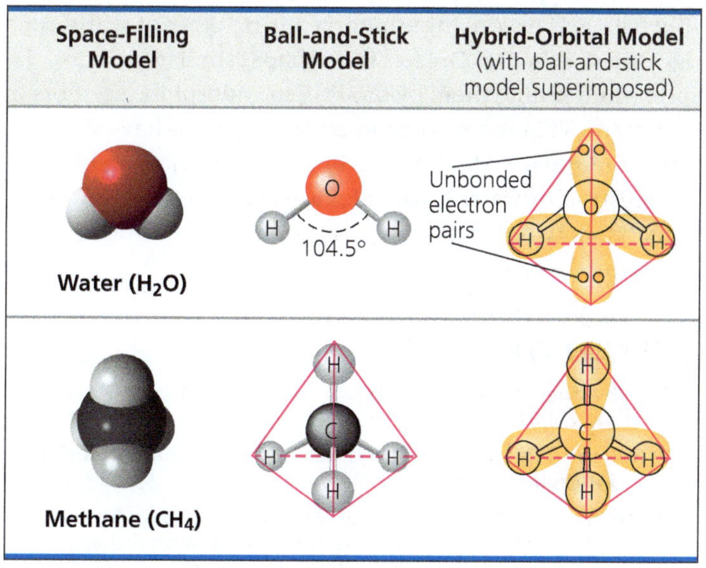

(b) Molecular-shape models. Three models representing molecular shape are shown for water and methane. The positions of the hybrid orbitals determine the shapes of the molecules.

the orbitals in its valence shell undergo rearrangement. For atoms with valence electrons in both *s* and *p* orbitals (review Figure 2.8), the single *s* and three *p* orbitals form four new hybrid orbitals shaped like identical teardrops extending from the region of the atomic nucleus, as shown in Figure 2.15a. If we connect the larger ends of the teardrops with lines, we have the outline of a geometric shape called a tetrahedron, a pyramid with a triangular base.

For water molecules (H_2O), two of the hybrid orbitals in the oxygen's valence shell are shared with hydrogens (see Figure 2.15b). The result is a molecule shaped roughly like a V, with its two covalent bonds at an angle of 104.5°.

The methane molecule (CH_4) has the shape of a completed tetrahedron because all four hybrid orbitals of the carbon atom are shared with hydrogen atoms (see Figure 2.15b). The carbon nucleus is at the center, with its four covalent bonds radiating to hydrogen nuclei at the corners of the tetrahedron. Larger molecules containing multiple carbon atoms, including many of the molecules that make up living matter, have more complex overall shapes. However, the tetrahedral shape of a carbon atom bonded to four other atoms is often a repeating motif within such molecules.

Molecular shape is crucial: It determines how biological molecules recognize and respond to one another with specificity. Biological molecules often bind temporarily to each other by forming weak interactions, but only if their shapes are complementary. Consider the effects of opiates, drugs such as morphine and heroin derived from opium. Opiates relieve pain and alter mood by weakly binding to specific receptor molecules on the surfaces of brain cells. Why would brain cells carry receptors for opiates, compounds that are not made by the body? In 1975, the discovery of endorphins answered this question. Endorphins are signaling molecules made by the pituitary gland that bind to the receptors, relieving pain and producing euphoria during times of stress, such as intense exercise. Opiates have shapes similar to endorphins and mimic them by binding to endorphin receptors in the brain. That is why opiates and endorphins have similar effects **(Figure 2.16)**. The role of molecular shape in brain chemistry illustrates how biological organization leads to a match between structure and function, one of biology's unifying themes.

CONCEPT CHECK 2.3

1. Why does the structure H—C=C—H fail to make sense chemically?
2. What holds the atoms together in a crystal of magnesium chloride ($MgCl_2$)?
3. **WHAT IF?** ➤ If you were a pharmaceutical researcher, why would you want to learn the three-dimensional shapes of naturally occurring signaling molecules?

For suggested answers, see Appendix A.

▼ Figure 2.16 A molecular mimic. Morphine affects pain perception and emotional state by mimicking the brain's natural endorphins.

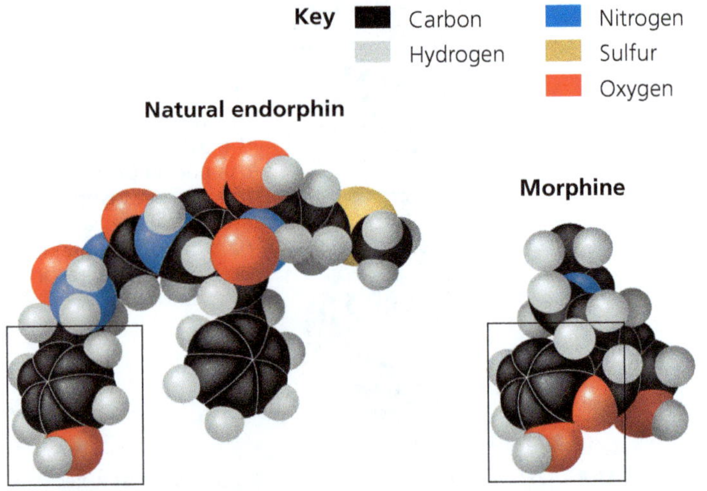

(a) Structures of endorphin and morphine. The boxed portion of the endorphin molecule (left) binds to receptor molecules on target cells in the brain. The boxed portion of the morphine molecule (right) is a close match.

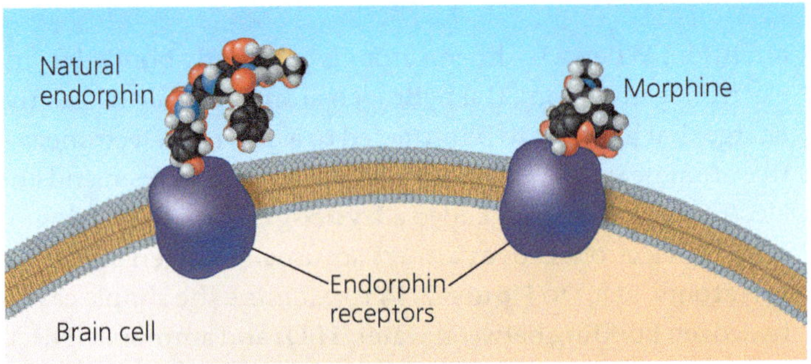

(b) Binding to endorphin receptors. Both endorphin and morphine can bind to endorphin receptors on the surface of a brain cell.

Interview with Candace Pert: Discovering opiate receptors in the brain

CONCEPT 2.4

Chemical reactions make and break chemical bonds

The making and breaking of chemical bonds, leading to changes in the composition of matter, are called **chemical reactions**. An example is the reaction between hydrogen and oxygen molecules that forms water:

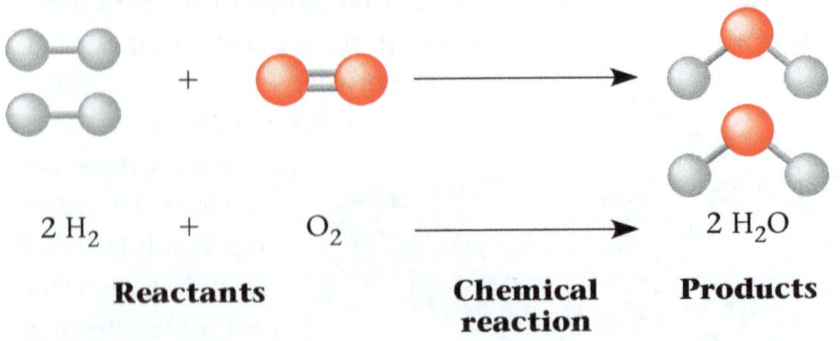

This reaction breaks the covalent bonds of H_2 and O_2 and forms the new bonds of H_2O. When we write the equation for a chemical reaction, we use an arrow to indicate the conversion of the starting materials, called the **reactants**, to the resulting materials, or **products**. The coefficients indicate the number of molecules involved; for example, the coefficient 2 before the H_2 means that the reaction starts with two molecules of hydrogen. Notice that all atoms of the reactants must be accounted for in the products. Matter is conserved in a chemical reaction: Reactions cannot create or destroy atoms but can only rearrange (redistribute) the electrons among them.

Photosynthesis, which takes place within the cells of green plant tissues, is an important biological example of how chemical reactions rearrange matter. Humans and other animals ultimately depend on photosynthesis for food and oxygen, and this process is at the foundation of almost all ecosystems. The following summarizes the process of photosynthesis:

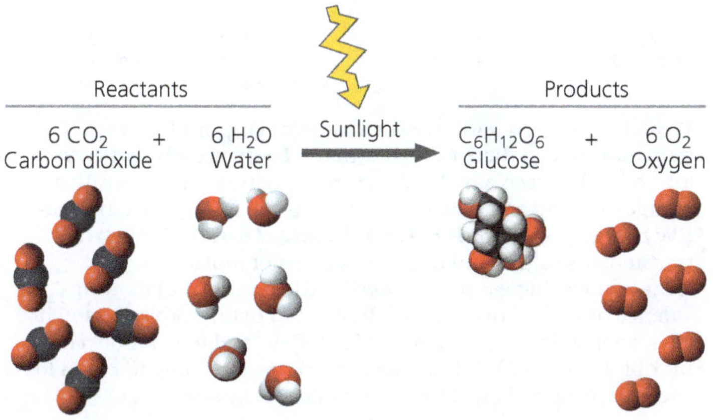

The raw materials of photosynthesis are carbon dioxide (CO_2) and water (H_2O), which land plants absorb from the air and soil, respectively. Within the plant cells, sunlight powers the conversion of these ingredients to a sugar called glucose ($C_6H_{12}O_6$) and oxygen molecules (O_2), a by-product that can be seen when released by a water plant **(Figure 2.17)**. Although photosynthesis is actually a sequence of many chemical reactions, we still end up with the same number and types of atoms that we had when we started. Matter has simply been rearranged, with an input of energy provided by sunlight.

All chemical reactions are theoretically reversible, with the products of the forward reaction becoming the reactants for the reverse reaction. For example, hydrogen and nitrogen molecules can combine to form ammonia, but ammonia can also decompose to regenerate hydrogen and nitrogen:

$$3\,H_2 + N_2 \rightleftharpoons 2\,NH_3$$

The two opposite-headed arrows indicate that the reaction is reversible.

One of the factors affecting the rate of a reaction is the concentration of reactants. The greater the concentration of reactant molecules, the more frequently they collide with one another and have an opportunity to react and form products.

➤ Figure 2.17 Photosynthesis: a solar-powered rearrangement of matter. *Elodea*, a freshwater plant, produces sugar by rearranging the atoms of carbon dioxide and water in the chemical process known as photosynthesis, which is powered by sunlight. Much of the sugar is then converted to other food molecules. Oxygen gas (O_2) is a by-product of photosynthesis; notice the bubbles of O_2 gas escaping from the leaves submerged in water.

DRAW IT ➤ *Add labels and arrows on the photo showing the reactants and products of photosynthesis as it takes place in a leaf.*

The same holds true for products. As products accumulate, collisions resulting in the reverse reaction become more frequent. Eventually, the forward and reverse reactions occur at the same rate, and the relative concentrations of products and reactants stop changing. The point at which the reactions offset one another exactly is called **chemical equilibrium**. This is a dynamic equilibrium; reactions are still going on in both directions, but with no net effect on the concentrations of reactants and products. Equilibrium does *not* mean that the reactants and products are equal in concentration, but only that their concentrations have stabilized at a particular ratio. The reaction involving ammonia reaches equilibrium when ammonia decomposes as rapidly as it forms. In some chemical reactions, the equilibrium point may lie so far to the right that these reactions go essentially to completion; that is, virtually all the reactants are converted to products.

We will return to the subject of chemical reactions after more detailed study of the various types of molecules that are important to life. In the next chapter, we focus on water, the substance in which all the chemical processes of organisms occur.

CONCEPT CHECK 2.4

1. **MAKE CONNECTIONS ➤** Consider the reaction between hydrogen and oxygen that forms water, shown with ball-and-stick models at the beginning of Concept 2.4. After studying Figure 2.10, draw and label the Lewis dot structures representing this reaction.
2. Which type of chemical reaction, if any, occurs faster at equilibrium: the formation of products from reactants or that of reactants from products?
3. **WHAT IF? ➤** Write an equation that uses the products of photosynthesis as reactants and the reactants of photosynthesis as products. Add energy as another product. This new equation describes a process that occurs in your cells. Describe this equation in words. How does this equation relate to breathing?

For suggested answers, see Appendix A.

2 Chapter Review

Go to **MasteringBiology**™ for Videos, Animations, Vocab Self-Quiz, Practice Tests, and more in the Study Area.

SUMMARY OF KEY CONCEPTS

CONCEPT 2.1

Matter consists of chemical elements in pure form and in combinations called compounds *(pp. 29–30)*

VOCAB SELF-QUIZ goo.gl/6u55ks

- **Elements** cannot be broken down chemically to other substances. A **compound** contains two or more different elements in a fixed ratio. Oxygen, carbon, hydrogen, and nitrogen make up approximately 96% of living matter.

? *Compare an element and a compound.*

CONCEPT 2.2

An element's properties depend on the structure of its atoms *(pp. 30–36)*

- An **atom**, the smallest unit of an element, has the following components:

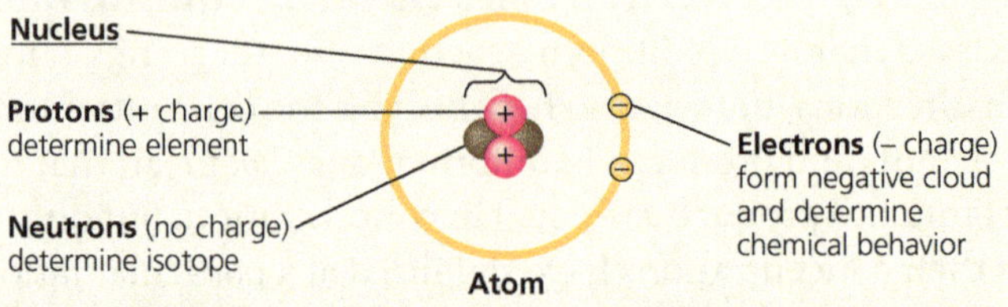

- An electrically neutral atom has equal numbers of electrons and protons; the number of protons determines the **atomic number**. The **atomic mass** is measured in **daltons** and is roughly equal to the **mass number**, the sum of protons plus neutrons. **Isotopes** of an element differ from each other in neutron number and therefore mass. Unstable isotopes give off particles and energy as radioactivity.
- In an atom, electrons occupy specific **electron shells**; the electrons in a shell have a characteristic energy level. Electron distribution in shells determines the chemical behavior of an atom. An atom that has an incomplete outer shell, the **valence shell**, is reactive.
- Electrons exist in **orbitals**, three-dimensional spaces with specific shapes that are components of electron shells.

DRAW IT ➤ *Draw the electron distribution diagrams for neon ($_{10}$Ne) and argon ($_{18}$Ar). Use these diagrams to explain why these elements are chemically unreactive.*

CONCEPT 2.3

The formation and function of molecules depend on chemical bonding between atoms *(pp. 36–40)*

- **Chemical bonds** form when atoms interact and complete their valence shells. **Covalent bonds** form when pairs of electrons are shared:

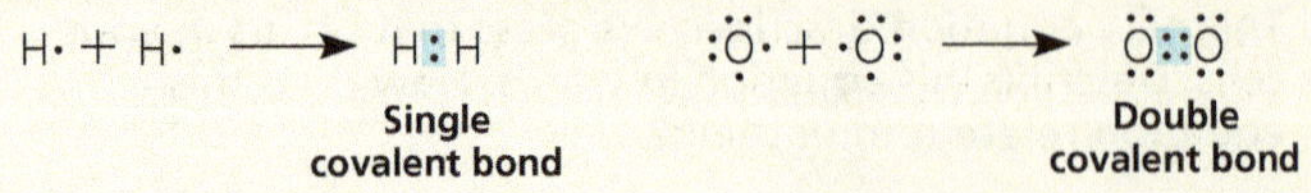

- **Molecules** consist of two or more covalently bonded atoms. The attraction of an atom for the electrons of a covalent bond is its **electronegativity**. If both atoms are the same, they have the same electronegativity and share a **nonpolar covalent bond**. Electrons of a **polar covalent bond** are pulled closer to the more electronegative atom, such as the oxygen in H_2O.
- An **ion** forms when an atom or molecule gains or loses an electron and becomes charged. An **ionic bond** is the attraction between two oppositely charged ions:

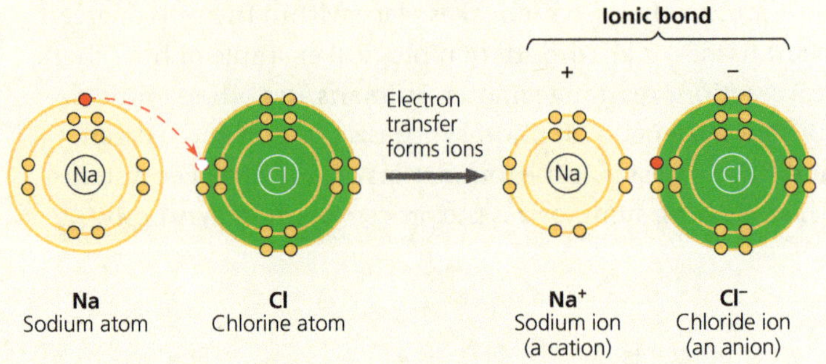

- Weak interactions reinforce the shapes of large molecules and help molecules adhere to each other. A **hydrogen bond** is an attraction between a hydrogen atom carrying a partial positive charge (δ+) and an electronegative atom carrying a partial negative charge (δ–). **Van der Waals interactions** occur between transiently positive and negative regions of molecules.
- A molecule's shape is determined by the positions of its atoms' valence orbitals. Covalent bonds result in hybrid orbitals, which are responsible for the shapes of H_2O, CH_4, and many more complex biological molecules. Molecular shape is usually the basis for the recognition of one biological molecule by another.

? *In terms of electron sharing between atoms, compare nonpolar covalent bonds, polar covalent bonds, and the formation of ions.*

CONCEPT 2.4

Chemical reactions make and break chemical bonds *(pp. 40–41)*

- **Chemical reactions** change **reactants** into **products** while conserving matter. All chemical reactions are theoretically reversible. **Chemical equilibrium** is reached when the forward and reverse reaction rates are equal.

? *What would happen to the concentration of products if more reactants were added to a reaction that was in chemical equilibrium? How would this addition affect the equilibrium?*

TEST YOUR UNDERSTANDING

Level 1: Knowledge/Comprehension

PRACTICE TEST goo.gl/CUYGKD

1. In the term *trace element*, the adjective *trace* means that
   (A) the element is required in very small amounts.
   (B) the element can be used as a label to trace atoms through an organism's metabolism.
   (C) the element is very rare on Earth.
   (D) the element enhances health but is not essential for the organism's long-term survival.

2. Compared with ^{31}P, the radioactive isotope ^{32}P has
   (A) a different atomic number.
   (B) one more proton.
   (C) one more electron.
   (D) one more neutron.

3. The reactivity of an atom arises from
   (A) the average distance of the outermost electron shell from the nucleus.
   (B) the existence of unpaired electrons in the valence shell.
   (C) the sum of the potential energies of all the electron shells.
   (D) the potential energy of the valence shell.

4. Which statement is true of all atoms that are anions?
   (A) The atom has more electrons than protons.
   (B) The atom has more protons than electrons.
   (C) The atom has fewer protons than does a neutral atom of the same element.
   (D) The atom has more neutrons than protons.

5. Which of the following statements correctly describes any chemical reaction that has reached equilibrium?
   (A) The concentrations of products and reactants are equal.
   (B) The reaction is now irreversible.
   (C) Both forward and reverse reactions have halted.
   (D) The rates of the forward and reverse reactions are equal.

Level 2: Application/Analysis

6. We can represent atoms by listing the number of protons, neutrons, and electrons—for example, $2p^+$, $2n^0$, $2e^-$ for helium. Which of the following represents the ^{18}O isotope of oxygen?
   (A) $7p^+$, $2n^0$, $9e^-$
   (B) $8p^+$, $10n^0$, $8e^-$
   (C) $9p^+$, $9n^0$, $9e^-$
   (D) $10p^+$, $8n^0$, $9e^-$

7. The atomic number of sulfur is 16. Sulfur combines with hydrogen by covalent bonding to form a compound, hydrogen sulfide. Based on the number of valence electrons in a sulfur atom, predict the molecular formula of the compound.
   (A) HS (C) H_2S
   (B) HS_2 (D) H_4S

8. What coefficients must be placed in the following blanks so that all atoms are accounted for in the products?

$$C_6H_{12}O_6 \rightarrow \_\_\_\_ C_2H_6O + \_\_\_\_ CO_2$$

   (A) 2; 1 (C) 1; 3
   (B) 3; 1 (D) 2; 2

9. **DRAW IT** Draw Lewis dot structures for each hypothetical molecule shown below, using the correct number of valence electrons for each atom. Determine which molecule makes sense because each atom has a complete valence shell and each bond has the correct number of electrons. Explain what makes the other molecule nonsensical, considering the number of bonds each type of atom can make.

(a)
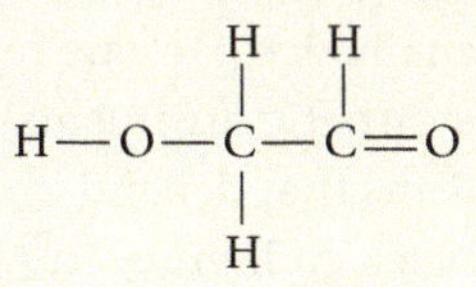

(b)
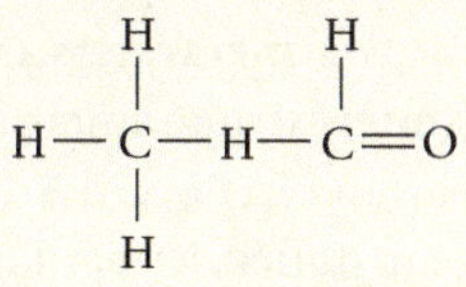

Level 3: Synthesis/Evaluation

10. **EVOLUTION CONNECTION** The percentages of naturally occurring elements making up the human body (see Table 2.1) are similar to the percentages of these elements found in other organisms. How could you account for this similarity among organisms?

11. **SCIENTIFIC INQUIRY** Female luna moths (*Actias luna*) attract males by emitting chemical signals that spread through the air. A male hundreds of meters away can detect these molecules and fly toward their source. The sensory organs responsible for this behavior are the comblike antennae visible in the photograph shown here. Each filament of an antenna is equipped with thousands of receptor cells that detect the sex attractant. Based on what you learned in this chapter, propose a hypothesis to account for the ability of the male moth to detect a specific molecule in the presence of many other molecules in the air. What predictions does your hypothesis make? Design an experiment to test one of these predictions.

12. **WRITE ABOUT A THEME: ORGANIZATION** While waiting at an airport, Neil Campbell once overheard this claim: "It's paranoid and ignorant to worry about industry or agriculture contaminating the environment with their chemical wastes. After all, this stuff is just made of the same atoms that were already present in our environment." Drawing on your knowledge of electron distribution, bonding, and emergent properties (see Concept 1.1), write a short essay (100–150 words) countering this argument.

13. **SYNTHESIZE YOUR KNOWLEDGE**

This bombardier beetle is spraying a boiling hot liquid that contains irritating chemicals, used as a defense mechanism against its enemies. The beetle stores two sets of chemicals separately in its glands. Using what you learned about chemistry in this chapter, propose a possible explanation for why the beetle is not harmed by the chemicals it stores and what causes the explosive discharge.

For selected answers, see Appendix A.

For additional practice questions, check out the **Dynamic Study Modules** in MasteringBiology. You can use them to study on your smartphone, tablet, or computer anytime, anywhere!

▲ **Figure 3.1** **How does life on Earth depend on the chemistry of water?**

KEY CONCEPTS

3.1 **Polar covalent bonds in water molecules result in hydrogen bonding**

3.2 **Four emergent properties of water contribute to Earth's suitability for life**

3.3 **Acidic and basic conditions affect living organisms**

▲ Black guillemots, threatened by climate change

The Molecule That Supports All of Life

Life on Earth began in water and evolved there for 3 billion years before spreading onto land. Water is the substance that makes life possible as we know it here on Earth, and possibly on other planets as well. All organisms familiar to us are made mostly of water and live in an environment dominated by water.

Three-quarters of Earth's surface is covered by water. Although most of this water is in liquid form, water is also present on Earth as a solid (ice) and a gas (water vapor). Water is the only common substance on Earth to exist in the natural environment in all three physical states of matter. Furthermore, the solid form of water floats on the liquid form, a rare property emerging from the chemistry of the water molecule.

As the Earth is warming from climate change (see Concept 1.1), the ratio of ice to liquid water is changing. Arctic sea ice and glaciers are melting, affecting life on, under, and around them. In the Arctic, warmer waters and the smaller ice pack are resulting in blooms of phytoplankton (microscopic aquatic photosynthetic organisms), seen from space as the "cloudy" seawater in **Figure 3.1**. Organisms that depend on Arctic ice, however, are suffering. For instance, a population of black guillemots in Alaska is declining due to the warming climate and reduction of Arctic sea ice.

In this chapter, you will learn how the structure of a water molecule allows it to interact with other molecules, including other water molecules. This ability leads to water's unique emergent properties that help make Earth suitable for life.

When you see this blue icon, log in to **MasteringBiology** and go to the Study Area for digital resources.

MB **Get Ready for This Chapter**

CONCEPT 3.1

Polar covalent bonds in water molecules result in hydrogen bonding

Water is so familiar to us that it is easy to overlook its many extraordinary qualities. Following the theme of emergent properties, we can trace water's unique behavior to the structure and interactions of its molecules.

Studied on its own, the water molecule is deceptively simple. It is shaped like a wide V, with its two hydrogen atoms joined to the oxygen atom by single covalent bonds. Oxygen is more electronegative than hydrogen, so the electrons of the covalent bonds spend more time closer to oxygen than to hydrogen; these are **polar covalent bonds** (see Figure 2.11). This unequal sharing of electrons and water's V-like shape make it a **polar molecule**, meaning that its overall charge is unevenly distributed. In water, the oxygen of the molecule has two regions of partial negative charge (δ–), and each hydrogen has a partial positive charge (δ+).

The properties of water arise from attractions between oppositely charged atoms of different water molecules: The partially positive hydrogen of one molecule is attracted to the partially negative oxygen of a nearby molecule. The two molecules are thus held together by a hydrogen bond **(Figure 3.2)**. When water is in its liquid form, its hydrogen bonds are very fragile, each only about 1/20 as strong as a covalent bond. The hydrogen bonds form, break, and re-form with great frequency. Each lasts only a few trillionths of a second, but the molecules are constantly forming new hydrogen bonds with a succession of partners. Therefore, at any instant, most of the water molecules are hydrogen-bonded to their neighbors. The extraordinary properties of water emerge from this hydrogen bonding, which organizes water molecules into a higher level of structural order.

▼ Figure 3.2 Hydrogen bonds between water molecules.

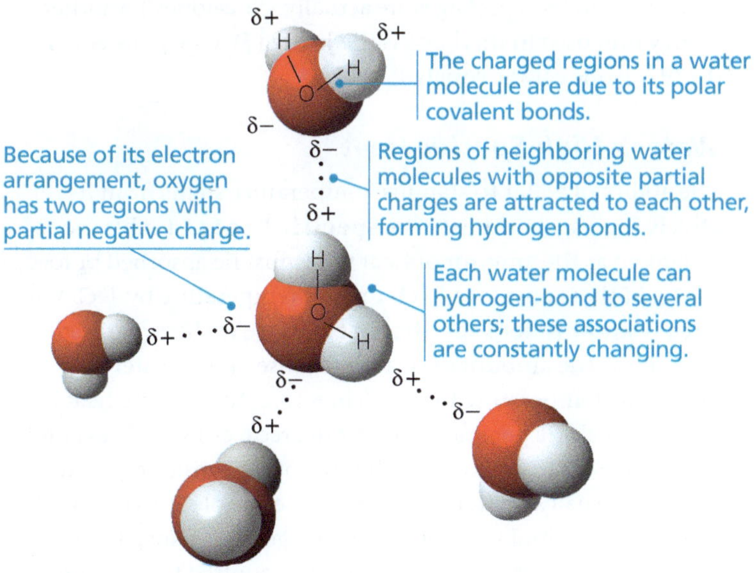

DRAW IT ➤ *Draw partial charges on the water molecule at the far left, and draw three more water molecules hydrogen-bonded to it.*

Animation: Polarity of Water

CONCEPT CHECK 3.1

1. **MAKE CONNECTIONS ➤** What is electronegativity, and how does it affect interactions between water molecules? (Review Figure 2.11.)
2. **VISUAL SKILLS ➤** Look at Figure 3.2 and explain why the central water molecule can hydrogen bond to four (rather than three or five) other water molecules.
3. Why is it unlikely that two neighboring water molecules would be arranged like this?

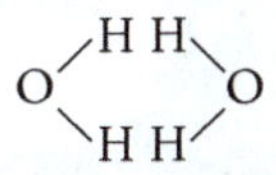

4. **WHAT IF? ➤** What would be the effect on the properties of the water molecule if oxygen and hydrogen had equal electronegativity?

For suggested answers, see Appendix A.

CONCEPT 3.2

Four emergent properties of water contribute to Earth's suitability for life

We will examine four emergent properties of water that contribute to Earth's suitability as an environment for life: cohesive behavior, ability to moderate temperature, expansion upon freezing, and versatility as a solvent.

Cohesion of Water Molecules

Water molecules stay close to each other as a result of hydrogen bonding. Although the arrangement of molecules in a sample of liquid water is constantly changing, at any given moment many of the molecules are linked by multiple hydrogen bonds. These linkages make water more structured than most other liquids. Collectively, the hydrogen bonds hold the substance together, a phenomenon called **cohesion**.

Cohesion due to hydrogen bonding contributes to the transport of water and dissolved nutrients against gravity in plants. Water from the roots reaches the leaves through a network of water-conducting cells **(Figure 3.3)**. As water evaporates from a leaf, hydrogen bonds cause water molecules leaving the veins to tug on molecules farther down, and the upward pull is transmitted through the water-conducting cells all the way to the roots. **Adhesion**, the clinging of one substance to another, also plays a role. Adhesion of water by hydrogen bonds to the molecules of cell walls helps counter the downward pull of gravity (see Figure 3.3).

Related to cohesion is **surface tension**, a measure of how difficult it is to stretch or break the surface of a liquid. At the interface between water and air is an ordered arrangement of water molecules, hydrogen-bonded to one another and to the water below, but not to the air above. This asymmetry gives

▼ Figure 3.3 Water transport in plants. Because of the properties of cohesion and adhesion, the tallest trees can transport water more than 100 m upward—approximately one-quarter the height of the Empire State Building in New York City.

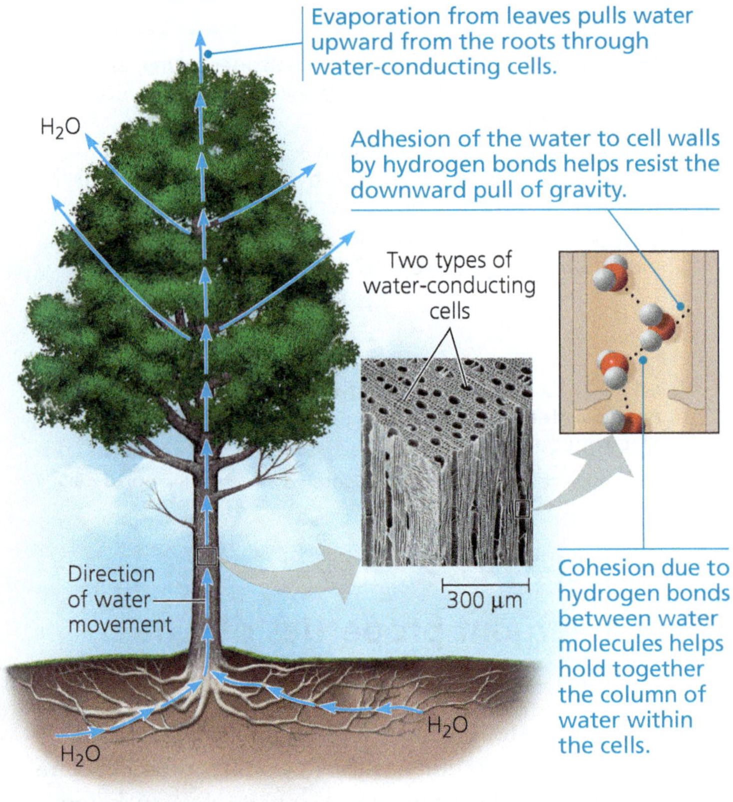

BioFlix® Animation: Adhesion and Cohesion in Plants
Animation: Cohesion of Water

water an unusually high surface tension, making it behave as though it were coated with an invisible film. You can observe the surface tension of water by slightly overfilling a drinking glass; the water will stand above the rim. The spider in **Figure 3.4** takes advantage of the surface tension of water to walk across a pond without breaking the surface.

Moderation of Temperature by Water

Water moderates air temperature by absorbing heat from air that is warmer and releasing the stored heat to air that is cooler. Water is effective as a heat bank because it can absorb or release a relatively large amount of heat with only a slight change in its own temperature. To understand this capability of water, let's first look at temperature and heat.

◀ Figure 3.4 Walking on water. The high surface tension of water, resulting from the collective strength of its hydrogen bonds, allows this raft spider to walk on the surface of a pond.

Temperature and Heat

Anything that moves has **kinetic energy**, the energy of motion. Atoms and molecules have kinetic energy because they are always moving, although not necessarily in any particular direction. The faster a molecule moves, the greater its kinetic energy. The kinetic energy associated with the random movement of atoms or molecules is called **thermal energy**. Thermal energy is related to temperature, but they are not the same thing. **Temperature** represents the *average* kinetic energy of the molecules in a body of matter, regardless of volume, whereas the thermal energy of a body of matter reflects the *total* kinetic energy, and thus depends on the matter's volume. When water is heated in a coffeemaker, the average speed of the molecules increases, and the thermometer records this as a rise in temperature of the liquid. The total amount of thermal energy also increases in this case. Note, however, that although the pot of coffee has a much higher temperature than, say, the water in a swimming pool, the swimming pool contains more thermal energy because of its much greater volume.

Whenever two objects of different temperature are brought together, thermal energy passes from the warmer to the cooler object until the two are the same temperature. Molecules in the cooler object speed up at the expense of the thermal energy of the warmer object. An ice cube cools a drink not by adding coldness to the liquid, but by absorbing thermal energy from the liquid as the ice itself melts. Thermal energy in transfer from one body of matter to another is defined as **heat**.

One convenient unit of heat used in this book is the **calorie (cal)**. A calorie is the amount of heat it takes to raise the temperature of 1 g of water by 1°C. Conversely, a calorie is also the amount of heat that 1 g of water releases when it cools by 1°C. A **kilocalorie (kcal)**, 1,000 cal, is the quantity of heat required to raise the temperature of 1 kilogram (kg) of water by 1°C. (The "Calories" on food packages are actually kilocalories.) Another energy unit used in this book is the **joule (J)**. One joule equals 0.239 cal; one calorie equals 4.184 J.

Water's High Specific Heat

The ability of water to stabilize temperature stems from its relatively high specific heat. The **specific heat** of a substance is defined as the amount of heat that must be absorbed or lost for 1 g of that substance to change its temperature by 1°C. We already know water's specific heat because we have defined a calorie as the amount of heat that causes 1 g of water to change its temperature by 1°C. Therefore, the specific heat of water is 1 calorie per gram and per degree Celsius, abbreviated as 1 cal/(g · °C). Compared with most other substances, water has an unusually high specific heat. For example, ethyl alcohol, the type of alcohol in alcoholic beverages, has a specific heat of 0.6 cal/(g · °C); that is, only 0.6 cal is required to raise the temperature of 1 g of ethyl alcohol by 1°C.

Because of the high specific heat of water relative to other materials, water will change its temperature less than other liquids when it absorbs or loses a given amount of heat. The reason you can burn your fingers by touching the side of an iron pot on the stove when the water in the pot is still lukewarm is that the specific heat of water is ten times greater than that of iron. In other words, the same amount of heat will raise the temperature of 1 g of the iron much faster than it will raise the temperature of 1 g of the water. Specific heat can be thought of as a measure of how well a substance resists changing its temperature when it absorbs or releases heat. Water resists changing its temperature; when it does change its temperature, it absorbs or loses a relatively large quantity of heat for each degree of change.

We can trace water's high specific heat, like many of its other properties, to hydrogen bonding. Heat must be absorbed in order to break hydrogen bonds; by the same token, heat is released when hydrogen bonds form. A calorie of heat causes a relatively small change in the temperature of water because much of the heat is used to disrupt hydrogen bonds before the water molecules can begin moving faster. And when the temperature of water drops slightly, many additional hydrogen bonds form, releasing a considerable amount of energy in the form of heat.

What is the relevance of water's high specific heat to life on Earth? A large body of water can absorb and store a huge amount of heat from the sun in the daytime and during summer while warming up only a few degrees. At night and during winter, the gradually cooling water can warm the air. This capability of water serves to moderate air temperatures in coastal areas **(Figure 3.5)**. The high specific heat of water also tends to stabilize ocean temperatures, creating a favorable environment for marine life. Thus, because of its high specific heat, the water that covers most of Earth keeps temperature fluctuations on land and in water within limits that permit life. Also, because organisms are made primarily of water, they are better able to resist changes in their own temperature than if they were made of a liquid with a lower specific heat.

Evaporative Cooling

Molecules of any liquid stay close together because they are attracted to one another. Molecules moving fast enough to overcome these attractions can depart the liquid and enter the air as a gas (vapor). This transformation from a liquid to a gas is called vaporization, or evaporation. Recall that the speed of molecular movement varies and that temperature is the *average* kinetic energy of molecules. Even at low temperatures, the speediest molecules can escape into the air. Some evaporation occurs at any temperature; a glass of water at room temperature, for example, will eventually evaporate completely. If a liquid is heated, the average kinetic energy of molecules increases and the liquid evaporates more rapidly.

Heat of vaporization is the quantity of heat a liquid must absorb for 1 g of it to be converted from the liquid to the gaseous state. For the same reason that water has a high specific heat, it also has a high heat of vaporization relative to most other liquids. To evaporate 1 g of water at 25°C, about 580 cal of heat is needed—nearly double the amount needed to vaporize a gram of alcohol or ammonia. Water's high heat of vaporization is another emergent property resulting from the strength of its hydrogen bonds, which must be broken before the molecules can exit from the liquid in the form of water vapor.

The high amount of energy required to vaporize water has a wide range of effects. On a global scale, for example, it helps moderate Earth's climate. A considerable amount of solar heat absorbed by tropical seas is consumed during the evaporation of surface water. Then, as moist tropical air circulates poleward, it releases heat as it condenses and forms rain. On an organismal level, water's high heat of vaporization accounts for the severity of steam burns. These burns are caused by the heat energy released when steam condenses into liquid on the skin.

As a liquid evaporates, the surface of the liquid that remains behind cools down (its temperature decreases). This **evaporative cooling** occurs because the "hottest" molecules, those with the greatest kinetic energy, are the most likely to leave as gas. It is as if the 100 fastest runners at a college transferred to another school; the average speed of the remaining students would decline.

Evaporative cooling of water contributes to the stability of temperature in lakes and ponds and also provides a mechanism that prevents terrestrial organisms from overheating. For example, evaporation of water from the leaves of a plant helps keep the tissues in the leaves from becoming too warm in the sunlight. Evaporation of sweat from human skin dissipates body heat and helps prevent overheating on a hot day or when excess heat is generated by strenuous activity. High humidity on a hot day increases discomfort because the high concentration of water vapor in the air inhibits the evaporation of sweat from the body.

▼ Figure 3.5 Temperatures for the Pacific Ocean and Southern California on an August day.

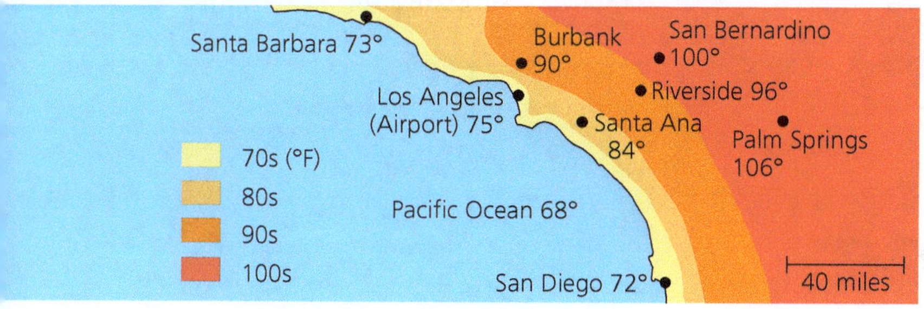

INTERPRET THE DATA ➤ *Explain the pattern of temperatures shown in this diagram.*

Floating of Ice on Liquid Water

Water is one of the few substances that are less dense as a solid than as a liquid. In other words, ice floats on liquid water. While other materials contract and become denser when they solidify, water expands. The cause of this exotic behavior is, once again,

➤ Figure 3.6 Ice: crystalline structure and floating barrier. In ice, each molecule is hydrogen-bonded to four neighbors in a three-dimensional crystal. Because the crystal is spacious, ice has fewer molecules than an equal volume of liquid water. In other words, ice is less dense than liquid water. Floating ice becomes a barrier that insulates the liquid water below from the colder air. The marine organism shown here is a type of shrimp called krill; it was photographed beneath floating ice in the Southern Ocean near Antarctica.

WHAT IF? ➤ *If water did not form hydrogen bonds, what would happen to the shrimp's habitat, shown here?*

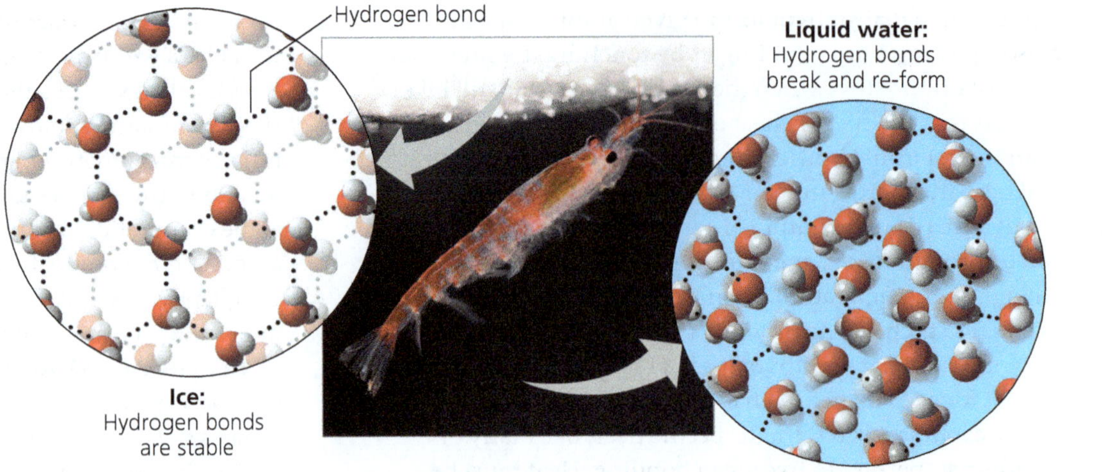

hydrogen bonding. At temperatures above 4°C, water behaves like other liquids, expanding as it warms and contracting as it cools. As the temperature falls from 4°C to 0°C, water begins to freeze because more and more of its molecules are moving too slowly to break hydrogen bonds. At 0°C, the molecules become locked into a crystalline lattice, each water molecule hydrogen-bonded to four partners **(Figure 3.6)**. The hydrogen bonds keep the molecules at "arm's length," far enough apart to make ice about 10% less dense (10% fewer molecules in the same volume) than liquid water at 4°C. When ice absorbs enough heat for its temperature to rise above 0°C, hydrogen bonds between molecules are disrupted. As the crystal collapses, the ice melts and molecules have fewer hydrogen bonds, allowing them to slip closer together. Water reaches its greatest density at 4°C and then begins to expand as the molecules move faster. Even in liquid water, many of the molecules are connected by hydrogen bonds, though only transiently: The hydrogen bonds are constantly breaking and re-forming.

The ability of ice to float due to its lower density is an important factor in the suitability of the environment for life. If ice sank, then eventually all ponds, lakes, and even oceans would freeze solid, making life as we know it impossible on Earth. During summer, only the upper few inches of the ocean would thaw. Instead, when a deep body of water cools, the floating ice insulates the liquid water below, preventing it from freezing and allowing life to exist under the frozen surface, as shown in the photo in Figure 3.6. Besides insulating the water below, ice also provides a solid habitat for some animals, such as polar bears and seals.

Many scientists are worried that these bodies of ice are at risk of disappearing. Global warming, which is caused by carbon dioxide and other "greenhouse" gases in the atmosphere (see Figure 56.28), is having a profound effect on icy environments around the globe. In the Arctic, the average air temperature has risen 2.2°C just since 1961. This temperature increase has affected the seasonal balance between Arctic sea ice and liquid water, causing ice to form later in the year, to melt earlier, and to cover a smaller area. The rate at which glaciers and Arctic sea ice are disappearing is posing an extreme challenge to animals that depend on ice for their survival **(Figure 3.7)**.

▼ Figure 3.7 Effects of climate change on the Arctic. Warmer temperatures in the Arctic cause more sea ice to melt in the summer, benefiting some organisms and harming others.

Interview with Susan Solomon: Understanding climate change

Water: The Solvent of Life

A sugar cube placed in a glass of water will dissolve with a little stirring. The glass will then contain a uniform mixture of sugar and water; the concentration of dissolved sugar will be the same everywhere in the mixture. A liquid that is a completely homogeneous mixture of two or more substances is called a **solution**. The dissolving agent of a solution is the **solvent**, and the substance that is dissolved is the **solute**. In this case, water is the solvent and sugar is the solute. An **aqueous solution** is one in which the solute is dissolved in water; water is the solvent.

Water is a very versatile solvent, a quality we can trace to the polarity of the water molecule. Suppose, for example, that a spoonful of table salt, the ionic compound sodium chloride (NaCl), is placed in water **(Figure 3.8)**. At the surface of each crystal of salt, the sodium and chloride ions are exposed to the solvent. These ions and regions of the water molecules are attracted to each other due to their opposite charges. The oxygens of the water molecules have regions of partial negative charge that are attracted to sodium cations. The hydrogen regions are partially positively charged and are attracted to chloride anions. As a result, water molecules surround the individual sodium and chloride ions, separating and shielding them from one another. The sphere of water molecules around each dissolved ion is called a **hydration shell**. Working inward from the surface of each salt crystal, water eventually dissolves all the ions. The result is a solution of two solutes, sodium cations and chloride anions, homogeneously mixed with water, the solvent. Other ionic compounds also dissolve in water. Seawater, for instance, contains a great variety of dissolved ions, as do living cells.

A compound does not need to be ionic to dissolve in water; many compounds made up of nonionic polar molecules, such as the sugar in the sugar cube mentioned earlier, are also water-soluble. Such compounds dissolve when water molecules surround each of the solute molecules, forming hydrogen bonds with them. Even molecules as large as proteins can dissolve in water if they have ionic and polar regions on their surface **(Figure 3.9)**. Many different kinds of polar compounds are dissolved (along with ions) in the water of such biological fluids as blood, the sap of plants, and the liquid within all cells. Water is the solvent of life.

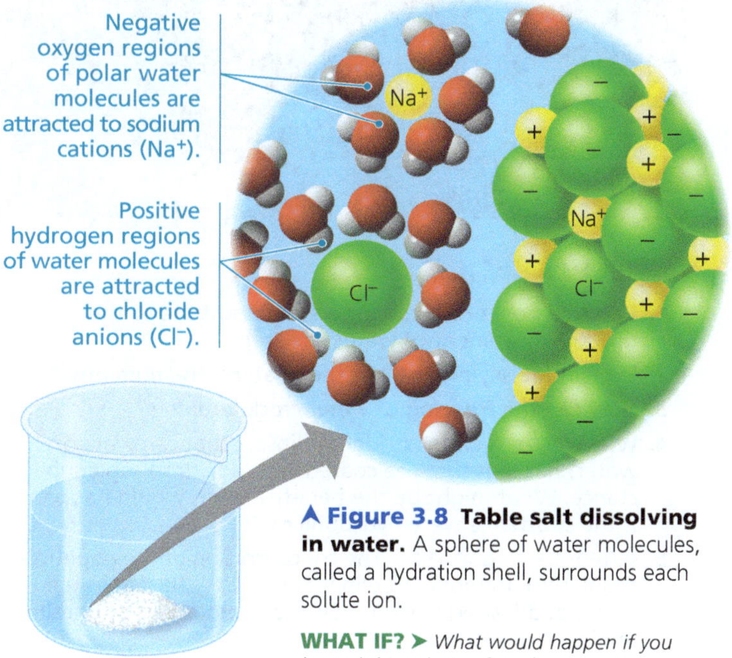

▲ Figure 3.8 Table salt dissolving in water. A sphere of water molecules, called a hydration shell, surrounds each solute ion.

WHAT IF? ➤ *What would happen if you heated this solution for a long time?*

▼ Figure 3.9 A water-soluble protein. Human lysozyme is a protein found in tears and saliva that has antibacterial action (see Figure 5.16). This model shows the lysozyme molecule (purple) in an aqueous environment. Ionic and polar regions on the protein's surface attract the partially charged regions on water molecules.

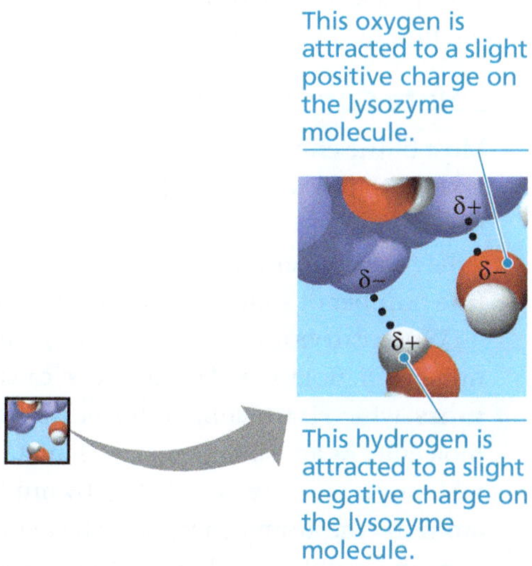

Hydrophilic and Hydrophobic Substances

Any substance that has an affinity for water is said to be **hydrophilic** (from the Greek *hydro*, water, and *philos*, loving). In some cases, substances can be hydrophilic without actually dissolving. For example, some molecules in cells are so large that they do not dissolve. Another example of a hydrophilic substance that does not dissolve is cotton, a plant product. Cotton consists of giant molecules of cellulose, a compound with numerous regions of partial positive and partial negative charges that can form hydrogen bonds with water. Water adheres to the cellulose fibers. Thus, a cotton towel does a great job of drying the body, yet it does not dissolve in the washing machine. Cellulose is also present in the walls of water-conducting cells in a plant; you read earlier how the adhesion of water to these hydrophilic walls helps water move up the plant against gravity.

There are, of course, substances that do not have an affinity for water. Substances that are nonionic and nonpolar (or otherwise cannot form hydrogen bonds) actually seem to repel water; these substances are said to be **hydrophobic** (from the Greek *phobos*, fearing). An example from the kitchen is vegetable oil, which, as you know, does not mix stably with water-based substances such as vinegar. The hydrophobic behavior of the oil molecules results from a prevalence of relatively nonpolar covalent bonds, in this case bonds between carbon and hydrogen, which share electrons almost equally. Hydrophobic molecules related to oils are major ingredients of cell membranes. (Imagine what would happen to a cell if its membrane dissolved!)

Solute Concentration in Aqueous Solutions

Most of the chemical reactions in organisms involve solutes dissolved in water. To understand such reactions, we must know how many atoms and molecules are involved and calculate the concentration of solutes in an aqueous solution (the number of solute molecules in a volume of solution).

When carrying out experiments, we use mass to calculate the number of molecules. We must first calculate the **molecular mass**, which is the sum of the masses of all the atoms in a molecule. As an example, let's calculate the molecular mass of table sugar (sucrose), $C_{12}H_{22}O_{11}$, by multiplying the number of atoms by the atomic mass of each element (see Appendix B). In round numbers of daltons, the mass of a carbon atom is 12, the mass of a hydrogen atom is 1, and the mass of an oxygen atom is 16. Thus, sucrose has a molecular mass of $(12 \times 12) + (22 \times 1) + (11 \times 16) = 342$ daltons. Because we can't weigh out small numbers of molecules, we usually measure substances in units called moles. Just as a dozen always means 12 objects, a **mole (mol)** represents an exact number of objects: 6.02×10^{23}, which is called Avogadro's number. Because of the way in which Avogadro's number and the unit *dalton* were originally defined, there are 6.02×10^{23} daltons in 1 g. Once we determine the molecular mass of a molecule such as sucrose, we can use the same number (342), but with the unit *gram*, to represent the mass of 6.02×10^{23} molecules of sucrose, or 1 mol of sucrose (this is sometimes called the *molar mass*). To obtain 1 mol of sucrose in the lab, therefore, we weigh out 342 g.

The practical advantage of measuring a quantity of chemicals in moles is that a mole of one substance has exactly the same number of molecules as a mole of any other substance. If the molecular mass of substance A is 342 daltons and that of substance B is 10 daltons, then 342 g of A will have the same number of molecules as 10 g of B. A mole of ethyl alcohol (C_2H_6O) also contains 6.02×10^{23} molecules, but its mass is only 46 g because the mass of a molecule of ethyl alcohol is less than that of a molecule of sucrose. Measuring in moles makes it convenient for scientists working in the laboratory to combine substances in fixed ratios of molecules.

How would we make a liter (L) of solution consisting of 1 mol of sucrose dissolved in water? We would measure out 342 g of sucrose and then gradually add water, while stirring, until the sugar was completely dissolved. We would then add enough water to bring the total volume of the solution up to 1 L. At that point, we would have a 1-molar (1 *M*) solution of sucrose. **Molarity**—the number of moles of solute per liter of solution—is the unit of concentration most often used by biologists for aqueous solutions.

Water's capacity as a versatile solvent complements the other properties discussed in this chapter. Since these remarkable properties allow water to support life on Earth so well, scientists who seek life elsewhere in the universe look for water as a sign that a planet might sustain life.

MB MP3 Tutor: The Properties of Water

Possible Evolution of Life on Other Planets

EVOLUTION Biologists who look for life elsewhere in the universe (known as *astrobiologists*) have concentrated their search on planets that might have water. More than 800 planets have been found outside our solar system, and there is evidence for the presence of water vapor on a few of them. In our own solar system, Mars has been a focus of study. Like Earth, Mars has an ice cap at both poles. Images from spacecraft sent to Mars showed that ice is present just under the surface of Mars and enough water vapor exists in its atmosphere for frost to form. In 2015, scientists found evidence of water flowing on Mars **(Figure 3.10)**, and other studies suggested conditions existed that could have supported microorganismal life. Drilling below the surface may be the next step in the search for signs of life on Mars. If any life-forms or fossils are found, their study will shed light on the process of evolution from an entirely new perspective.

▼ **Figure 3.10 Evidence for liquid water on Mars.** Water appears to have helped form these dark streaks that run downhill on Mars during the summer. NASA scientists also found evidence of hydrated salts, indicating water is present. (This digitally treated photograph was taken by the Mars Reconnaissance Orbiter.)

CONCEPT CHECK 3.2

1. Describe how properties of water contribute to the upward movement of water in a tree.
2. Explain the saying "It's not the heat; it's the humidity."
3. How can the freezing of water crack boulders?
4. **WHAT IF?** ➤ A water strider (an insect that can walk on water) has legs that are coated with a hydrophobic substance. What might be the benefit? What would happen if the substance were hydrophilic?
5. **INTERPRET THE DATA** ➤ The concentration of the appetite-regulating hormone ghrelin is about 1.3×10^{-10} *M* in the blood of a fasting person. How many molecules of ghrelin are in 1 L of blood?

For selected answers, see Appendix A.

CONCEPT 3.3

Acidic and basic conditions affect living organisms

Occasionally, a hydrogen atom participating in a hydrogen bond between two water molecules shifts from one molecule to the other. When this happens, the hydrogen atom leaves its electron behind, and what is actually transferred is a **hydrogen ion** (H^+), a single proton with a charge of 1+. The water molecule that lost a proton is now a **hydroxide ion** (OH^-), which has a charge of 1−. The proton binds to the other water molecule, making that molecule a **hydronium ion** (H_3O^+). We can picture the chemical reaction as follows:

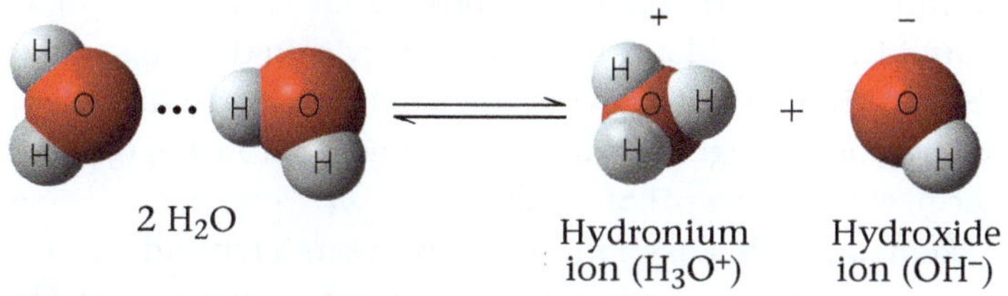

MB Animation: Dissociation of Water Molecules

By convention, H^+ (the hydrogen ion) is used to represent H_3O^+ (the hydronium ion), and we follow that practice in this book. Keep in mind, though, that H^+ does not exist on its own in an aqueous solution. It is always associated with a water molecule in the form of H_3O^+.

As indicated by the double arrows, this is a reversible reaction that reaches a state of dynamic equilibrium when water molecules dissociate at the same rate that they are being reformed from H^+ and OH^-. At this equilibrium point, the concentration of water molecules greatly exceeds the concentrations of H^+ and OH^-. In pure water, only one water molecule in every 554 million is dissociated; the concentration of H^+ and of OH^- in pure water is therefore 10^{-7} *M* (at 25°C). This means there is only one ten-millionth of a mole of hydrogen ions per liter of pure water and an equal number of hydroxide ions. (Even so, this is a huge number—over 60,000 *trillion*—of each ion in a liter of pure water.)

Although the dissociation of water is reversible and statistically rare, it is exceedingly important in the chemistry of life. H^+ and OH^- are very reactive. Changes in their concentrations can drastically affect a cell's proteins and other complex molecules. As we have seen, the concentrations of H^+ and OH^- are equal in pure water, but adding certain kinds of solutes, called acids and bases, disrupts this balance. Biologists use something called the pH scale to describe how acidic or basic (the opposite of acidic) a solution is. In the remainder of this chapter, you will learn about acids, bases, and pH and why changes in pH can adversely affect organisms.

Acids and Bases

What would cause an aqueous solution to have an imbalance in H^+ and OH^- concentrations? When acids dissolve in water, they donate additional H^+ to the solution. An **acid** is a substance that increases the hydrogen ion concentration of a solution. For example, when hydrochloric acid (HCl) is added to water, hydrogen ions dissociate from chloride ions:

$$HCl \rightarrow H^+ + Cl^-$$

This source of H^+ (dissociation of water is the other source) results in an acidic solution—one having more H^+ than OH^-.

A substance that reduces the hydrogen ion concentration of a solution is called a **base**. Some bases reduce the H^+ concentration directly by accepting hydrogen ions. Ammonia (NH_3), for instance, acts as a base when the unshared electron pair in nitrogen's valence shell attracts a hydrogen ion from the solution, resulting in an ammonium ion (NH_4^+):

$$NH_3 + H^+ \rightleftharpoons NH_4^+$$

Other bases reduce the H^+ concentration indirectly by dissociating to form hydroxide ions, which combine with hydrogen ions and form water. One such base is sodium hydroxide (NaOH), which in water dissociates into its ions:

$$NaOH \rightarrow Na^+ + OH^-$$

In either case, the base reduces the H^+ concentration. Solutions with a higher concentration of OH^- than H^+ are known as basic solutions. A solution in which the H^+ and OH^- concentrations are equal is said to be neutral.

Notice that single arrows were used in the reactions for HCl and NaOH. These compounds dissociate completely when mixed with water, so hydrochloric acid is called a strong acid and sodium hydroxide a strong base. In contrast, ammonia is a weak base. The double arrows in the reaction for ammonia indicate that the binding and release of hydrogen ions are reversible reactions, although at equilibrium there will be a fixed ratio of NH_4^+ to NH_3.

Weak acids are acids that reversibly release and accept back hydrogen ions. An example is carbonic acid:

H_2CO_3	$\rightleftharpoons$	HCO_3^-	+	H^+
Carbonic acid		Bicarbonate ion		Hydrogen ion

Here the equilibrium so favors the reaction in the left direction that when carbonic acid is added to pure water, only 1% of the molecules are dissociated at any particular time. Still, that is enough to shift the balance of H^+ and OH^- from neutrality.

The pH Scale

In any aqueous solution at 25°C, the product of the H^+ and OH^- concentrations is constant at 10^{-14}. This can be written

$$[H^+][OH^-] = 10^{-14}$$

(The brackets indicate molar concentration.) As previously mentioned, in a neutral solution at 25°C, $[H^+] = 10^{-7}$ and $[OH^-] = 10^{-7}$. Therefore, the product of $[H^+]$ and $[OH^-]$ in a neutral solution at 25°C is 10^{-14}. If enough acid is added to a solution to increase [H^+] to 10^{-5} *M*, then [OH^-] will decline by an equivalent factor to 10^{-9} *M* (note that $10^{-5} \times 10^{-9} = 10^{-14}$). This constant relationship expresses the behavior of acids and bases in an aqueous solution. An acid not only adds hydrogen ions to a solution, but also removes hydroxide ions because of the tendency for H^+ to combine with OH^-, forming water. A base has the opposite effect, increasing OH^- concentration but also reducing H^+ concentration by the formation of water. If enough of a base is added to raise the OH^- concentration to 10^{-4} *M*, it will cause the H^+ concentration to drop to 10^{-10} *M*. Whenever we know the concentration of either H^+ or OH^- in an aqueous solution, we can deduce the concentration of the other ion.

Because the H^+ and OH^- concentrations of solutions can vary by a factor of 100 trillion or more, scientists have developed a way to express this variation more conveniently than in moles per liter. The pH scale **(Figure 3.11)** compresses the range of H^+ and OH^- concentrations by employing logarithms.

▼ Figure 3.11 The pH scale and pH values of some aqueous solutions.

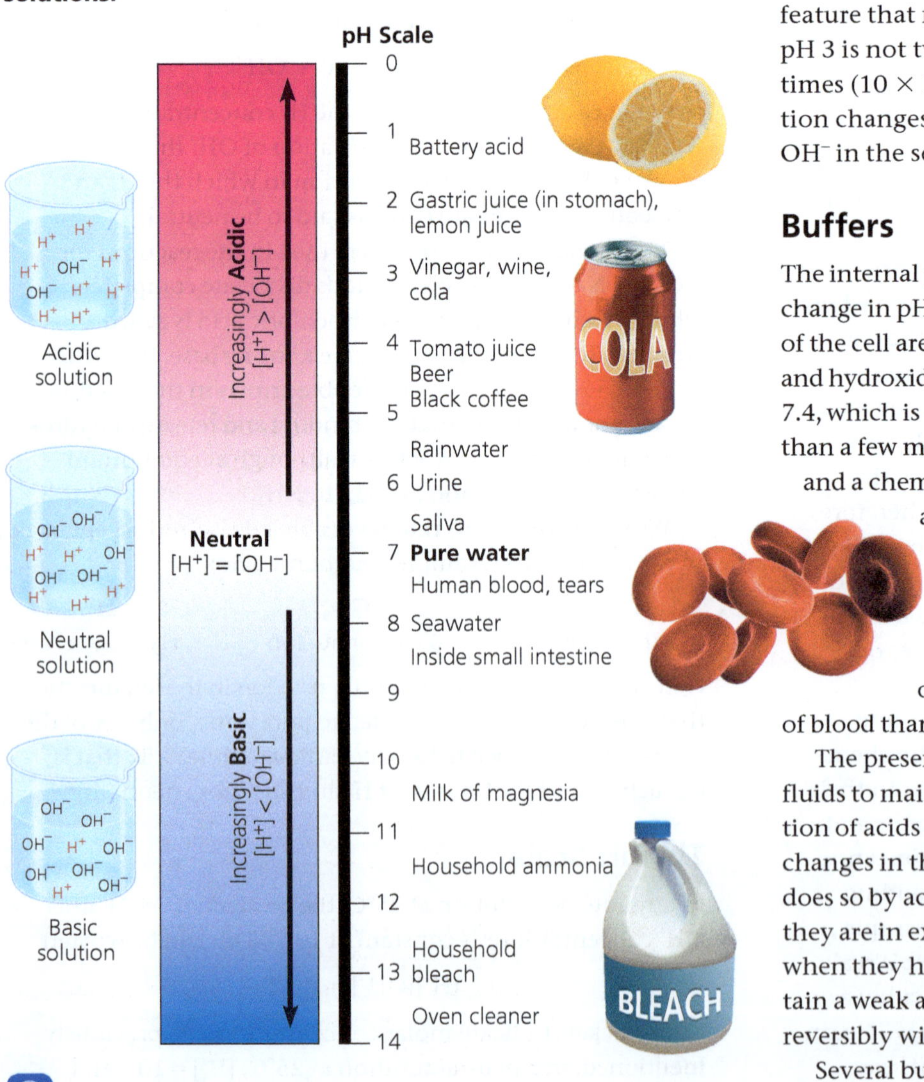

Animation: Acids, Bases, and pH

The **pH** of a solution is defined as the negative logarithm (base 10) of the hydrogen ion concentration:

$$pH = -\log [H^+]$$

For a neutral aqueous solution, [H^+] is 10^{-7} *M*, giving us

$$-\log 10^{-7} = -(-7) = 7$$

Notice that pH *decreases* as H^+ concentration *increases* (see Figure 3.11). Notice, too, that although the pH scale is based on H^+ concentration, it also implies OH^- concentration. A solution of pH 10 has a hydrogen ion concentration of 10^{-10} *M* and a hydroxide ion concentration of 10^{-4} *M*.

The pH of a neutral aqueous solution at 25°C is 7, the midpoint of the pH scale. A pH value less than 7 denotes an acidic solution; the lower the number, the more acidic the solution. The pH for basic solutions is above 7. Most biological fluids, such as blood and saliva, are within the range of pH 6–8. There are a few exceptions, however, including the strongly acidic digestive juice of the human stomach (gastric juice), which has a pH of about 2.

Remember that each pH unit represents a tenfold difference in H^+ and OH^- concentrations. It is this mathematical feature that makes the pH scale so compact. A solution of pH 3 is not twice as acidic as a solution of pH 6, but 1,000 times ($10 \times 10 \times 10$) more acidic. When the pH of a solution changes slightly, the actual concentrations of H^+ and OH^- in the solution change substantially.

Buffers

The internal pH of most living cells is close to 7. Even a slight change in pH can be harmful because the chemical processes of the cell are very sensitive to the concentrations of hydrogen and hydroxide ions. The pH of human blood is very close to 7.4, which is slightly basic. A person cannot survive for more than a few minutes if the blood pH drops to 7 or rises to 7.8, and a chemical system exists in the blood that maintains a stable pH. If 0.01 mol of a strong acid is added to a liter of pure water, the pH drops from 7.0 to 2.0. If the same amount of acid is added to a liter of blood, however, the pH decrease is only from 7.4 to 7.3. Why does the addition of acid have so much less of an effect on the pH of blood than it does on the pH of water?

The presence of substances called buffers allows biological fluids to maintain a relatively constant pH despite the addition of acids or bases. A **buffer** is a substance that minimizes changes in the concentrations of H^+ and OH^- in a solution. It does so by accepting hydrogen ions from the solution when they are in excess and donating hydrogen ions to the solution when they have been depleted. Most buffer solutions contain a weak acid and its corresponding base, which combine reversibly with hydrogen ions.

Several buffers contribute to pH stability in human blood and many other biological solutions. One of these is carbonic acid (H_2CO_3), which is formed when CO_2 reacts with water in blood plasma. As mentioned earlier, carbonic acid dissociates to yield a bicarbonate ion (HCO_3^-) and a hydrogen ion (H^+):

$$\underset{\substack{H^+\text{ donor}\\ \text{(acid)}}}{H_2CO_3} \underset{\text{Response to a drop in pH}}{\overset{\text{Response to a rise in pH}}{\rightleftharpoons}} \underset{\substack{H^+\text{ acceptor}\\ \text{(base)}}}{HCO_3^-} + \underset{\substack{\text{Hydrogen}\\ \text{ion}}}{H^+}$$

The chemical equilibrium between carbonic acid and bicarbonate acts as a pH regulator, the reaction shifting left or right as other processes in the solution add or remove hydrogen ions. If the H^+ concentration in blood begins to fall (that is, if pH rises), the reaction proceeds to the right and more carbonic acid dissociates, replenishing hydrogen ions. But when the H^+ concentration in blood begins to rise (when pH drops), the reaction proceeds to the left, with HCO_3^- (the base) removing the hydrogen ions from the solution and forming H_2CO_3. Thus, the carbonic acid–bicarbonate buffering system consists of an acid and a base in equilibrium with each other. Most other buffers are also acid-base pairs.

Acidification: A Threat to Our Oceans

Among the many threats to water quality posed by human activities is the burning of fossil fuels, which releases CO_2 into the atmosphere. The resulting increase in atmospheric CO_2 levels has caused global warming and other aspects of climate change (see Concept 56.4). In addition, about 25% of human-generated CO_2 is absorbed by the oceans. In spite of the huge volume of water in the oceans, scientists worry that the absorption of so much CO_2 will harm marine ecosystems.

Recent data have shown that such fears are well founded. When CO_2 dissolves in seawater, it reacts with water to form carbonic acid, which lowers ocean pH. This process, known as **ocean acidification**, alters the delicate balance of conditions for life in the oceans **(Figure 3.12)**. Based on measurements of CO_2 levels in air bubbles trapped in ice over thousands of years, scientists calculate that the pH of the oceans is 0.1 pH unit lower now than at any time in the past 420,000 years. Recent studies predict that it will drop another 0.3–0.5 pH unit by the end of this century.

As seawater acidifies, the extra hydrogen ions combine with carbonate ions (CO_3^{2-}) to form bicarbonate ions (HCO_3^-), thereby reducing the carbonate ion concentration (see Figure 3.12). Scientists predict that ocean acidification will cause the carbonate ion concentration to decrease by 40% by the year 2100. This is of great concern because carbonate ions are required for calcification, the production of calcium carbonate ($CaCO_3$) by many marine organisms, including reef-building corals and animals that build shells. The **Scientific Skills Exercise** allows you to work with data from an experiment examining the effect of carbonate ion concentration on coral reefs. Coral reefs are sensitive ecosystems that act as havens for a great diversity of marine life.

▼ Figure 3.12 Atmospheric CO_2 from human activities and its fate in the ocean.

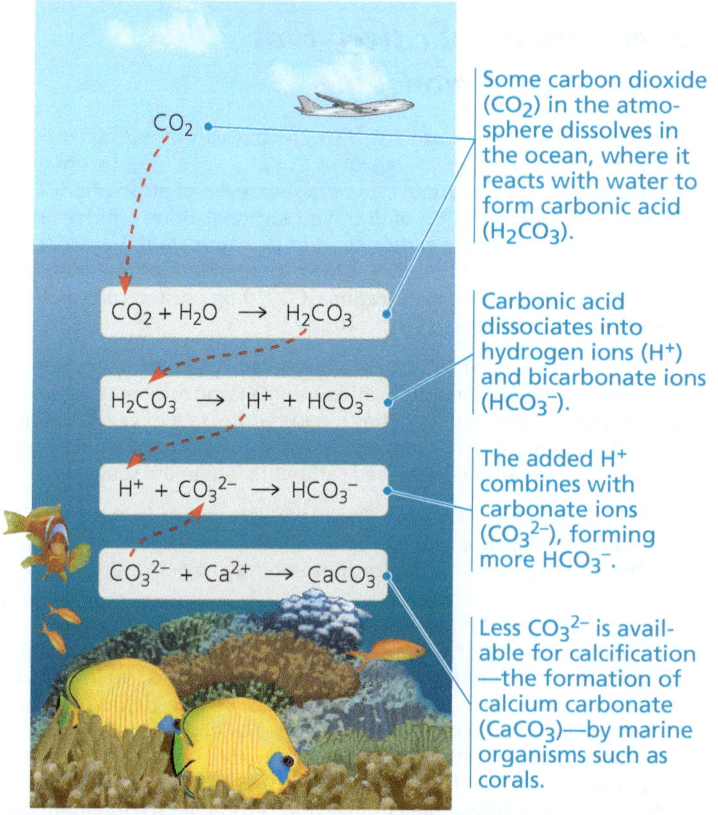

VISUAL SKILLS ➤ *Looking at all the chemical equations above, summarize the effect of adding excess CO_2 to the oceans on the calcification process in the final equation.*

ABC News Video: Ocean Acidification

The disappearance of coral reef ecosystems would be a tragic loss of biological diversity.

If there is any reason for optimism about the future quality of water resources on our planet, it is that we have made progress in learning about the delicate chemical balances in oceans, lakes, and rivers. Continued progress can come only from the actions of informed individuals, like yourselves, who are concerned about environmental quality. This requires understanding the crucial role that water plays in the suitability of the environment for continued life on Earth.

CONCEPT CHECK 3.3

1. Compared with a basic solution at pH 9, the same volume of an acidic solution at pH 4 has ________ times as many hydrogen ions (H^+).
2. HCl is a strong acid that dissociates in water: $HCl \rightarrow H^+ + Cl^-$. What is the pH of 0.01 *M* HCl?
3. Acetic acid (CH_3COOH) can be a buffer, similar to carbonic acid. Write the dissociation reaction, identifying the acid, base, H^+ acceptor, and H^+ donor.
4. **WHAT IF? ➤** Given a liter of pure water and a liter solution of acetic acid, what would happen to the pH, in general, if you added 0.01 mol of a strong acid to each? Use the reaction from question 3 to explain the result.

For suggested answers, see Appendix A.

SCIENTIFIC SKILLS EXERCISE

Interpreting a Scatter Plot with a Regression Line

How Does the Carbonate Ion Concentration of Seawater Affect the Calcification Rate of a Coral Reef? Scientists predict that acidification of the ocean due to higher levels of atmospheric CO_2 will lower the concentration of dissolved carbonate ions, which living corals use to build calcium carbonate reef structures. In this exercise, you will analyze data from a controlled experiment that examined the effect of carbonate ion concentration ($[CO_3^{2-}]$) on calcium carbonate deposition, a process called calcification.

How the Experiment Was Done For several years, scientists conducted research on ocean acidification using a large coral reef aquarium at Biosphere 2 in Arizona. They measured the rate of calcification by the reef organisms and examined how the calcification rate changed with differing amounts of dissolved carbonate ions in the seawater.

Data from the Experiment The black data points in the graph form a scatter plot. The red line, known as a linear regression line, is the best-fitting straight line for these points.

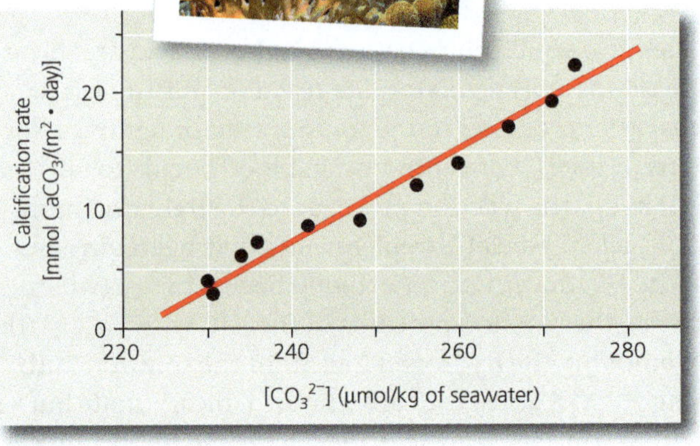

Data from C. Langdon et al., Effect of calcium carbonate saturation state on the calcification rate of an experimental coral reef, *Global Biogeochemical Cycles* 14:639–654 (2000).

INTERPRET THE DATA

1. When presented with a graph of experimental data, the first step in analysis is to determine what each axis represents. (a) In words, explain what is being shown on the *x*-axis. Be sure to include the units. (b) What is being shown on the *y*-axis (including units)? (c) Which variable is the independent variable—the variable that was *manipulated* by the researchers? (d) Which variable is the dependent variable—the variable that responded to or depended on the treatment, which was *measured* by the researchers? (For additional information about graphs, see the Scientific Skills Review in Appendix F.)
2. Based on the data shown in the graph, describe in words the relationship between carbonate ion concentration and calcification rate.
3. (a) If the seawater carbonate ion concentration is 270 μmol/kg, what is the approximate rate of calcification, and approximately how many days would it take 1 square meter of reef to accumulate 30 mmol of calcium carbonate ($CaCO_3$)? (b) If the seawater carbonate ion concentration is 250 μmol/kg, what is the approximate rate of calcification, and approximately how many days would it take 1 square meter of reef to accumulate 30 mmol of calcium carbonate? (c) If carbonate ion concentration decreases, how does the calcification rate change, and how does that affect the time it takes coral to grow?
4. (a) Referring to the equations in Figure 3.12, determine which step of the process is measured in this experiment. (b) Are the results of this experiment consistent with the hypothesis that increased atmospheric $[CO_2]$ will slow the growth of coral reefs? Why or why not?

MB **Instructors**: A version of this Scientific Skills Exercise can be assigned in MasteringBiology.

3 Chapter Review

MB Go to **MasteringBiology**™ for Videos, Animations, Vocab Self-Quiz, Practice Tests, and more in the Study Area.

SUMMARY OF KEY CONCEPTS

CONCEPT 3.1

Polar covalent bonds in water molecules result in hydrogen bonding *(p. 45)*

VOCAB SELF-QUIZ goo.gl/6u55ks

- Water is a **polar molecule**. A hydrogen bond forms when a partially negatively charged region on the oxygen of one water molecule is attracted to the partially positively charged hydrogen of a nearby water molecule. Hydrogen bonding between water molecules is the basis for water's properties.

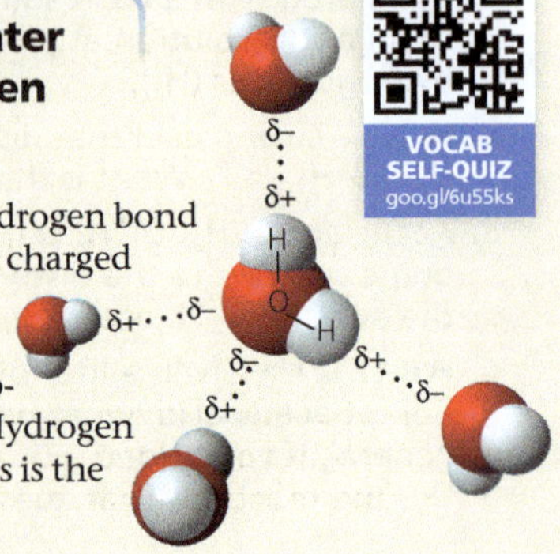

DRAW IT ➤ *Label a hydrogen bond and a polar covalent bond in the diagram of five water molecules. Is a hydrogen bond a covalent bond? Explain.*

CONCEPT 3.2

Four emergent properties of water contribute to Earth's suitability for life *(pp. 45–50)*

- Hydrogen bonding keeps water molecules close to each other, giving water **cohesion**. Hydrogen bonding is also responsible for water's **surface tension**.
- Water has a high **specific heat**: Heat is absorbed when hydrogen bonds break and is released when hydrogen bonds form. This helps keep **temperatures** relatively steady, within limits that permit life. **Evaporative cooling** is based on water's high **heat of vaporization**. The evaporative loss of the most energetic water molecules cools a surface.

- Ice floats because it is less dense than liquid water. This property allows life to exist under the frozen surfaces of lakes and polar seas.
- Water is an unusually versatile **solvent** because its polar molecules are attracted to ions and polar substances that can form hydrogen bonds. **Hydrophilic** substances have an affinity for water; **hydrophobic** substances do not. **Molarity**, the number of moles of **solute** per liter of **solution**, is used as a measure of solute concentration in solutions. A **mole** is a certain number of molecules of a substance. The mass of a mole of a substance in grams is the same as the **molecular mass** in daltons.
- The emergent properties of water support life on Earth and may contribute to the potential for life to have evolved on other planets.

? *Describe how different types of solutes dissolve in water. Explain what a solution is.*

CONCEPT 3.3

Acidic and basic conditions affect living organisms *(pp. 51–54)*

- A water molecule can transfer an H^+ to another water molecule to form H_3O^+ (represented simply by H^+) and OH^-.
- The concentration of H^+ is expressed as **pH**; pH = $-\log\,[H^+]$. A **buffer** consists of an acid-base pair that combines reversibly with hydrogen ions, allowing it to resist pH changes.
- The burning of fossil fuels increases the amount of CO_2 in the atmosphere. Some CO_2 dissolves in the oceans, causing **ocean acidification**, which has potentially grave consequences for marine organisms that rely on calcification.

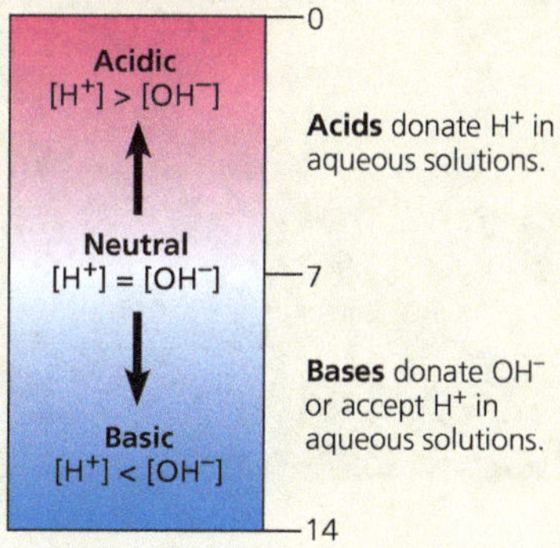

? *Explain what happens to the concentration of hydrogen ions in an aqueous solution when you add a base and cause the concentration of OH^- to rise to 10^{-3}. What is the pH of this solution?*

TEST YOUR UNDERSTANDING

Level 1: Knowledge/Comprehension

PRACTICE TEST goo.gl/CUYGKD

1. Which of the following is a hydrophobic material?
   (A) paper (B) table salt
   (C) wax (D) sugar
2. We can be sure that a mole of table sugar and a mole of vitamin C are equal in their
   (A) mass. (B) volume.
   (C) number of atoms. (D) number of molecules.
3. Measurements show that the pH of a particular lake is 4.0. What is the hydrogen ion concentration of the lake?
   (A) 4.0 M (B) 10^{-10} M
   (C) 10^{-4} M (D) 10^{4} M
4. What is the *hydroxide* ion concentration of the lake described in question 3?
   (A) 10^{-10} M (B) 10^{-4} M
   (C) 10^{-7} M (D) 10.0 M

Level 2: Application/Analysis

5. A slice of pizza has 500 kcal. If we could burn the pizza and use all the heat to warm a 50-L container of cold water, what would be the approximate increase in the temperature of the water? (*Note*: A liter of cold water weighs about 1 kg.)
   (A) 50°C (B) 5°C
   (C) 100°C (D) 10°C
6. **DRAW IT** Draw the hydration shells that form around a potassium ion and a chloride ion when potassium chloride (KCl) dissolves in water. Label the positive, negative, and partial charges on the atoms.

Level 3: Synthesis/Evaluation

7. In agricultural areas, farmers pay close attention to the weather forecast. Right before a predicted overnight freeze, farmers spray water on crops to protect the plants. Use the properties of water to explain how this method works. Be sure to mention why hydrogen bonds are responsible for this phenomenon.
8. **MAKE CONNECTIONS** What do climate change (see Concept 1.1 and Concept 3.2) and ocean acidification have in common?
9. **EVOLUTION CONNECTION** This chapter explains how the emergent properties of water contribute to the suitability of the environment for life. Until fairly recently, scientists assumed that other physical requirements for life included a moderate range of temperature, pH, atmospheric pressure, and salinity, as well as low levels of toxic chemicals. That view has changed with the discovery of organisms known as extremophiles, which have been found flourishing in hot, acidic sulfur springs, around hydrothermal vents deep in the ocean, and in soils with high levels of toxic metals. Why would astrobiologists be interested in studying extremophiles? What does the existence of life in such extreme environments say about the possibility of life on other planets?
10. **SCIENTIFIC INQUIRY** Design a controlled experiment to test the hypothesis that water acidification caused by acidic rain would inhibit the growth of *Elodea*, a freshwater plant (see Figure 2.17).
11. **WRITE ABOUT A THEME: ORGANIZATION** Several emergent properties of water contribute to the suitability of the environment for life. In a short essay (100–150 words), describe how the ability of water to function as a versatile solvent arises from the structure of water molecules.
12. **SYNTHESIZE YOUR KNOWLEDGE**

How do cats drink? Scientists using high-speed video have shown that cats use an interesting technique to drink aqueous substances like water and milk. Four times a second, the cat touches the tip of its tongue to the water and draws a column of water up into its mouth (as you can see in the photo), which then shuts before gravity can pull the water back down. Describe how the properties of water allow cats to drink in this fashion, including how water's molecular structure contributes to the process.

For selected answers, see Appendix A.

For additional practice questions, check out the **Dynamic Study Modules** in MasteringBiology. You can use them to study on your smartphone, tablet, or computer anytime, anywhere!

Carbon and the Molecular Diversity of Life

4

Figure 4.1 What properties make carbon the basis of all life?

KEY CONCEPTS

4.1 Organic chemistry is the study of carbon compounds

4.2 Carbon atoms can form diverse molecules by bonding to four other atoms

4.3 A few chemical groups are key to molecular function

Carbon can bond to four other atoms or groups of atoms, making a large variety of molecules possible.

Carbon: The Backbone of Life

Living organisms, such as the plants and the Qinling golden snub-nosed monkeys shown in **Figure 4.1**, are made up of chemicals based mostly on the element carbon. Carbon enters the biosphere through the action of producers—plants and other photosynthetic organisms that use solar energy to transform atmospheric CO_2 into the carbon-based molecules of life. These molecules are then taken up by consumers, which feed on other organisms.

Of all the chemical elements, carbon is unparalleled in its ability to form molecules that are large, complex, and varied, making possible the diversity of organisms that have evolved on Earth. Proteins, DNA, carbohydrates, and other molecules that distinguish living matter from inanimate material are all composed of carbon atoms bonded to one another and to atoms of other elements. Hydrogen (H), oxygen (O), nitrogen (N), sulfur (S), and phosphorus (P) are other common ingredients of these compounds, but it is the element carbon (C) that accounts for the enormous variety of biological molecules.

Large biological molecules, such as proteins, are the main focus of Chapter 5. In this chapter, we investigate the properties of smaller molecules. We will use these small molecules to illustrate concepts of molecular architecture that help explain why carbon is so important to life, while highlighting the theme that emergent properties arise from the organization of matter in living organisms.

When you see this blue icon, log in to **MasteringBiology** and go to the Study Area for digital resources.

MB **Get Ready for This Chapter**

CONCEPT 4.1

Organic chemistry is the study of carbon compounds

For historical reasons, compounds containing carbon are said to be organic, and their study is called **organic chemistry**. By the early 1800s, chemists had learned to make simple compounds in the laboratory by combining elements under the right conditions. Artificial synthesis of the complex molecules extracted from living matter seemed impossible, however. Organic compounds were thought to arise only in living organisms, which were believed to contain a life force beyond the jurisdiction of physical and chemical laws.

Chemists began to chip away at this notion when they learned to synthesize organic compounds in the laboratory. In 1828, Friedrich Wöhler, a German chemist, tried to make an "inorganic" salt, ammonium cyanate, by mixing solutions of ammonium ions (NH_4^+) and cyanate ions (CNO^-). Wöhler was astonished to find that instead he had made urea, an organic compound present in the urine of animals.

The next few decades saw laboratory synthesis of increasingly complex organic compounds, supporting the view that physical and chemical laws govern the processes of life. Organic chemistry was redefined as the study of carbon compounds, regardless of origin. Organic compounds range from simple molecules, such as methane (CH_4), to colossal ones, such as proteins, with thousands of atoms.

Organic Molecules and the Origin of Life on Earth

EVOLUTION In 1953, Stanley Miller, a graduate student of Harold Urey at the University of Chicago, helped bring the abiotic (nonliving) synthesis of organic compounds into the context of evolution. Study **Figure 4.2** to learn about his classic experiment. From his results, Miller concluded that complex organic molecules could arise spontaneously under conditions thought at that time to have existed on the early Earth. You can work with the data from a related experiment in the **Scientific Skills Exercise**. These experiments support the idea that abiotic synthesis of organic compounds, perhaps near volcanoes, could have been an early stage in the origin of life (see Figure 25.2).

The overall percentages of the major elements of life—C, H, O, N, S, and P—are quite uniform from one organism to another, reflecting the common evolutionary origin of all life. Because of carbon's ability to form four bonds, however, this limited assortment of atomic building blocks can be used to build an inexhaustible variety of organic molecules. Different species of organisms, and different individuals within a species, are distinguished by variations in the types of organic molecules they make. In a sense, the great diversity

Figure 4.2

Inquiry Can organic molecules form under conditions estimated to simulate those on the early Earth?

Experiment In 1953, Stanley Miller set up a closed system to mimic conditions thought at that time to have existed on the early Earth. A flask of water simulated the primeval sea. The water was heated so that some vaporized and moved into a second, higher flask containing the "atmosphere"—a mixture of gases. Sparks were discharged in the synthetic atmosphere to mimic lightning.

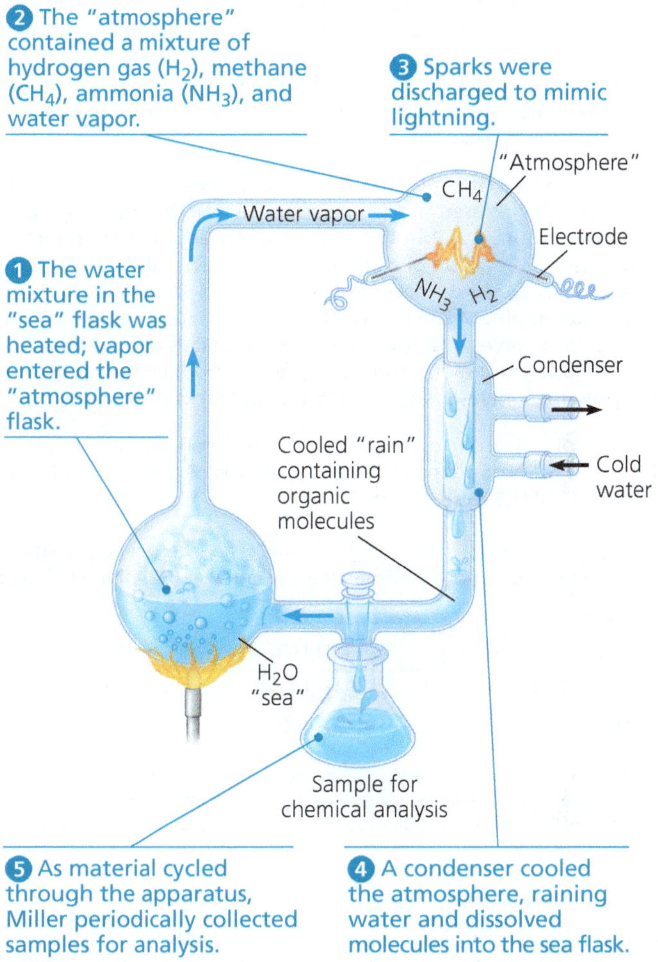

Results Miller identified a variety of organic molecules that are common in organisms. These included simple compounds, such as formaldehyde (CH_2O) and hydrogen cyanide (HCN), and more complex molecules, such as amino acids and long chains of carbon and hydrogen known as hydrocarbons.

Conclusion Organic molecules, a first step in the origin of life, may have been synthesized abiotically on the early Earth. Although new evidence indicates that the early Earth's atmosphere was different from the "atmosphere" used by Miller in this experiment, recent experiments using the revised list of chemicals also produced organic molecules. (We will explore this hypothesis in more detail in Concept 25.1.)

Data from S. L. Miller, A production of amino acids under possible primitive Earth conditions, *Science* 117:528–529 (1953).

WHAT IF ➤ *If Miller had increased the concentration of NH_3 in his experiment, how might the relative amounts of the products HCN and CH_2O have differed?*

MB **Interview with Stanley Miller: Investigating the origin of life**

SCIENTIFIC SKILLS EXERCISE

Working with Moles and Molar Ratios

Could the First Biological Molecules Have Formed Near Volcanoes on Early Earth? In 2007, Jeffrey Bada, a former graduate student of Stanley Miller, discovered some vials of samples that had never been analyzed from an experiment performed by Miller in 1958. In that experiment, Miller used hydrogen sulfide gas (H_2S) as one of the gases in the reactant mixture. Since H_2S is released by volcanoes, the H_2S experiment was designed to mimic conditions near volcanoes on early Earth. In 2011, Bada and colleagues published the results of their analysis of these "lost" samples. In this exercise, you will make calculations using the molar ratios of reactants and products from the H_2S experiment.

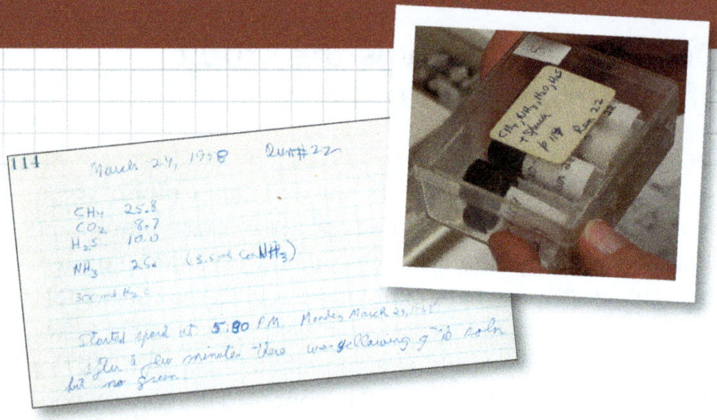

▲ Some of Stanley Miller's notes from his 1958 hydrogen sulfide (H_2S) experiment along with his original vials.

How the Experiment Was Done According to his laboratory notebook, Miller used the same apparatus as in his original experiment (see Figure 4.2), but the mixture of gaseous reactants included methane (CH_4), carbon dioxide (CO_2), hydrogen sulfide (H_2S), and ammonia (NH_3). After three days of simulated volcanic activity, he collected samples of the liquid, partially purified the chemicals, and sealed the samples in sterile vials. In 2011, Bada's research team used modern analytical methods to analyze the products in the vials for the presence of amino acids, the building blocks of proteins.

Data from the Experiment The table below shows 4 of the 23 amino acids detected in the 2011 analysis of the samples from Miller's 1958 H_2S experiment.

Product Compound	Molecular Formula	Molar Ratio (Relative to Glycine)
Glycine	$C_2H_5NO_2$	1.0
Serine	$C_3H_7NO_3$	3.0×10^{-2}
Methionine	$C_5H_{11}NO_2S$	1.8×10^{-3}
Alanine	$C_3H_7NO_2$	1.1

Data from E. T. Parker et al., Primordial synthesis of amines and amino acids in a 1958 Miller H_2S-rich spark discharge experiment, *Proceedings of the National Academy of Sciences USA* 108:5526-5531 (2011). www.pnas.org/cgi/doi/10.1073/pnas.1019191108.

INTERPRET THE DATA

1. A *mole* is the number of particles of a substance with a mass equivalent to its molecular (or atomic) mass in daltons. There are 6.02×10^{23} molecules (or atoms) in 1.0 mole (Avogadro's number; see Concept 3.2). The data table shows the "molar ratios" of some of the products from the Miller H_2S experiment. In a molar ratio, each unitless value is expressed relative to a standard for that experiment. Here, the standard is the number of moles of the amino acid glycine, which is set to a value of 1.0. For instance, serine has a molar ratio of 3.0×10^{-2}, meaning that for every mole of glycine, there is 3.0×10^{-2} mole of serine. (a) Give the molar ratio of methionine to glycine and explain what it means. (b) How many molecules of glycine are present in 1.0 mole? (c) For every 1.0 mole of glycine in the sample, how many molecules of methionine are present? (Recall that to multiply two numbers with exponents, you add their exponents; to divide them, you subtract the exponent in the denominator from that in the numerator.)
2. (a) Which amino acid is present in higher amounts than glycine? (b) How many more molecules of that amino acid are present than the number of molecules in 1.0 mole of glycine?
3. The synthesis of products is limited by the amount of reactants. (a) If one mole each of CH_4, NH_3, H_2S, and CO_2 is added to 1 liter of water (= 55.5 moles of H_2O) in a flask, how many moles of hydrogen, carbon, oxygen, nitrogen, and sulfur are in the flask? (b) Looking at the molecular formula in the table, how many moles of each element would be needed to make 1.0 mole of glycine? (c) What is the maximum number of moles of glycine that could be made in that flask, with the specified ingredients, if no other molecules were made? Explain. (d) If serine or methionine were made individually, which element(s) would be used up first for each? How much of each product could be made?
4. The earlier published experiment carried out by Miller did not include H_2S in the reactants (see Figure 4.2). Which of the compounds shown in the data table can be made in the H_2S experiment but could not be made in the earlier experiment?

(MB) **Instructors**: A version of this Scientific Skills Exercise can be assigned in MasteringBiology.

of living organisms we see on the planet (and in fossil remains) is made possible by the unique chemical versatility of the element carbon.

CONCEPT CHECK 4.1

1. Why was Wöhler astonished to find he had made urea?
2. **VISUAL SKILLS ➤** See Figure 4.2. Miller carried out a control experiment without discharging sparks and found no organic compounds. What might explain this result?

For suggested answers, see Appendix A.

CONCEPT 4.2

Carbon atoms can form diverse molecules by bonding to four other atoms

The key to an atom's chemical characteristics is its electron configuration. This configuration determines the kinds and number of bonds an atom will form with other atoms. Recall that it is the valence electrons, those in

Figure 4.3 The shapes of three simple organic molecules.

Molecule and Molecular Shape	Molecular Formula	Structural Formula	Ball-and-Stick Model (molecular shape in pink)	Space-Filling Model
(a) Methane. When a carbon atom has four single bonds to other atoms, the molecule is tetrahedral.	CH_4	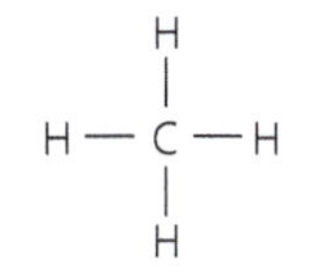	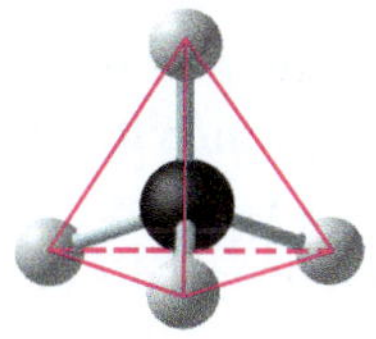	
(b) Ethane. A molecule may have more than one tetrahedral group of single-bonded atoms. (Ethane consists of two such groups.)	C_2H_6	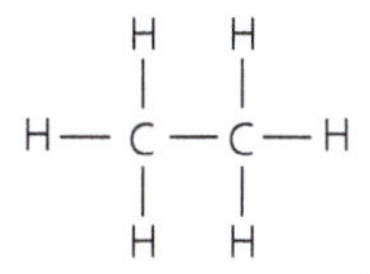	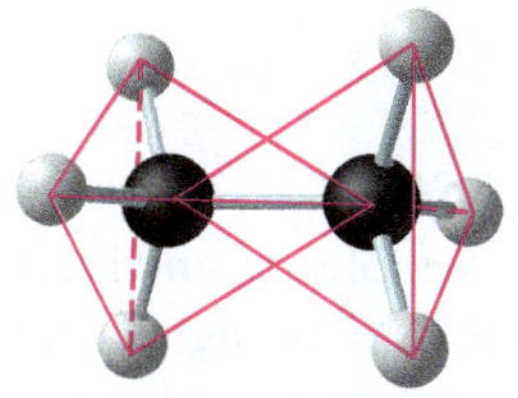	
(c) Ethene (ethylene). When two carbon atoms are joined by a double bond, all atoms attached to those carbons are in the same plane, and the molecule is flat.	C_2H_4	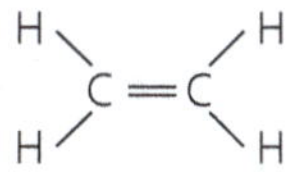	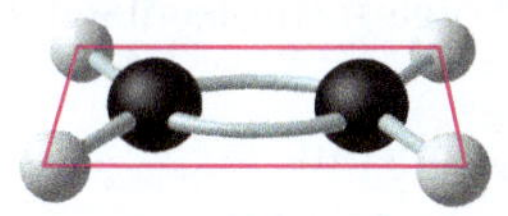	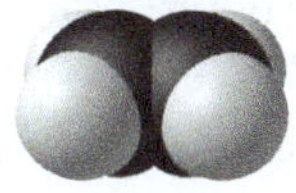

the outermost shell, that are available to form bonds with other atoms.

The Formation of Bonds with Carbon

Carbon has 6 electrons, with 2 in the first electron shell and 4 in the second shell; thus, it has 4 valence electrons in a shell that can hold up to 8 electrons. A carbon atom usually completes its valence shell by sharing its 4 electrons with other atoms so that 8 electrons are present. Each pair of shared electrons constitutes a covalent bond (see Figure 2.10d). In organic molecules, carbon usually forms single or double covalent bonds. Each carbon atom acts as an intersection point from which a molecule can branch off in as many as four directions. This enables carbon to form large, complex molecules.

When a carbon atom forms four single covalent bonds, the arrangement of its four hybrid orbitals causes the bonds to angle toward the corners of an imaginary tetrahedron. The bond angles in methane (CH_4) are 109.5° **(Figure 4.3a)**, and they are roughly the same in any group of atoms where carbon has four single bonds. For example, ethane (C_2H_6) is shaped like two overlapping tetrahedrons **(Figure 4.3b)**. In molecules with more carbons, every grouping of a carbon bonded to four other atoms has a tetrahedral shape. But when two carbon atoms are joined by a double bond, as in ethene (C_2H_4), the bonds from both carbons are all in the same plane, so the atoms joined to those carbons are in the same plane as well **(Figure 4.3c)**. We find it convenient to write molecules as structural formulas, as if the molecules being represented are two-dimensional,

Figure 4.4 Valences of the major elements of organic molecules. Valence, the number of covalent bonds an atom can form, is generally equal to the number of electrons required to complete the valence shell. All electrons are shown in the electron distribution diagrams (top), but only valence shell electrons are shown in the Lewis dot structures (bottom). Note that carbon can form four bonds.

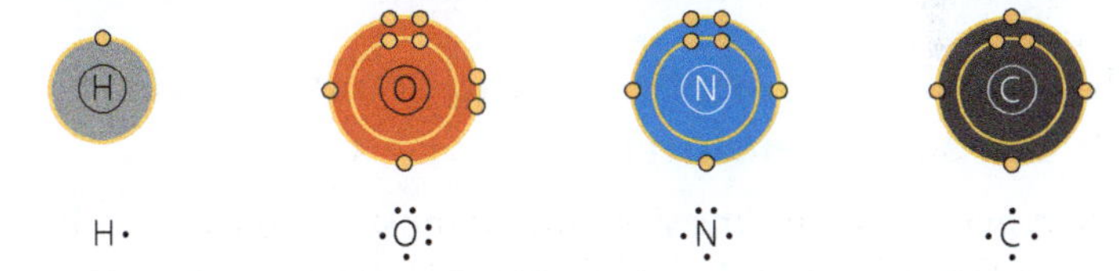

MAKE CONNECTIONS *Draw the Lewis dot structures for sodium, phosphorus, sulfur, and chlorine. (Refer to Figure 2.7.)*

but keep in mind that molecules are three-dimensional and that the shape of a molecule is central to its function.

The number of unpaired electrons in the valence shell of an atom is generally equal to the atom's **valence**, the number of covalent bonds it can form. **Figure 4.4** shows the valences of carbon and its most frequent bonding partners—hydrogen, oxygen, and nitrogen. These are the four main atoms in organic molecules.

The electron configuration of carbon gives it covalent compatibility with many different elements. Let's consider how valence and the rules of covalent bonding apply to carbon atoms with partners other than hydrogen. We'll look at two examples, the simple molecules carbon dioxide and urea.

In the carbon dioxide molecule (CO_2), a single carbon atom is joined to two atoms of oxygen by double covalent bonds. The structural formula for CO_2 is shown here:

$$O{=}C{=}O$$

Each line in a structural formula represents a pair of shared electrons. Thus, the two double bonds in CO_2 have the same number of shared electrons as four single bonds. The arrangement completes the valence shells of all atoms in the molecule:

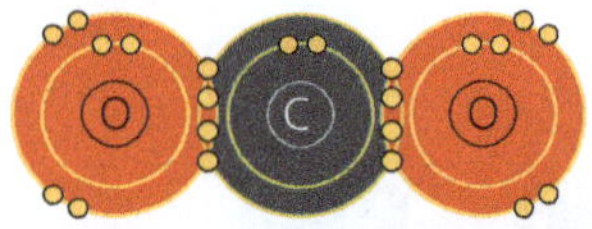

Because CO_2 is a very simple molecule and lacks hydrogen, it is often considered inorganic, even though it contains carbon. Whether we call CO_2 organic or inorganic, however, it is clearly important to the living world as the source of carbon, via photosynthetic organisms, for all organic molecules in organisms (see Concept 2.4).

Urea, $CO(NH_2)_2$, is the organic compound found in urine that Wöhler synthesized in the early 1800s. Again, each atom has the required number of covalent bonds. In this case, one carbon atom participates in both single and double bonds.

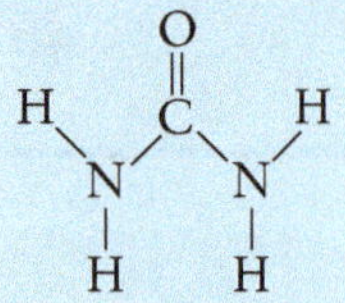

Urea

Urea and carbon dioxide are molecules with only one carbon atom. But as Figure 4.3 shows, a carbon atom can also use one or more valence electrons to form covalent bonds to other carbon atoms, each of which can also form four bonds. Thus, the atoms can be linked into chains of seemingly infinite variety.

Molecular Diversity Arising from Variation in Carbon Skeletons

Carbon chains form the skeletons of most organic molecules. The skeletons vary in length and may be straight, branched, or arranged in closed rings **(Figure 4.5)**. Some carbon skeletons have double bonds, which vary in number and location. Such variation in carbon skeletons is one important source of the molecular complexity and diversity that characterize living matter. In addition, atoms of other elements can be bonded to the skeletons at available sites.

Hydrocarbons

All of the molecules that are shown in Figures 4.3 and 4.5 are **hydrocarbons**, organic molecules consisting of only carbon and hydrogen. Atoms of hydrogen are attached to the carbon skeleton wherever electrons are available for covalent bonding. Hydrocarbons are the major components of petroleum, which is called a fossil fuel because it consists of the partially decomposed remains of organisms that lived millions of years ago.

Figure 4.5 Four ways that carbon skeletons can vary.

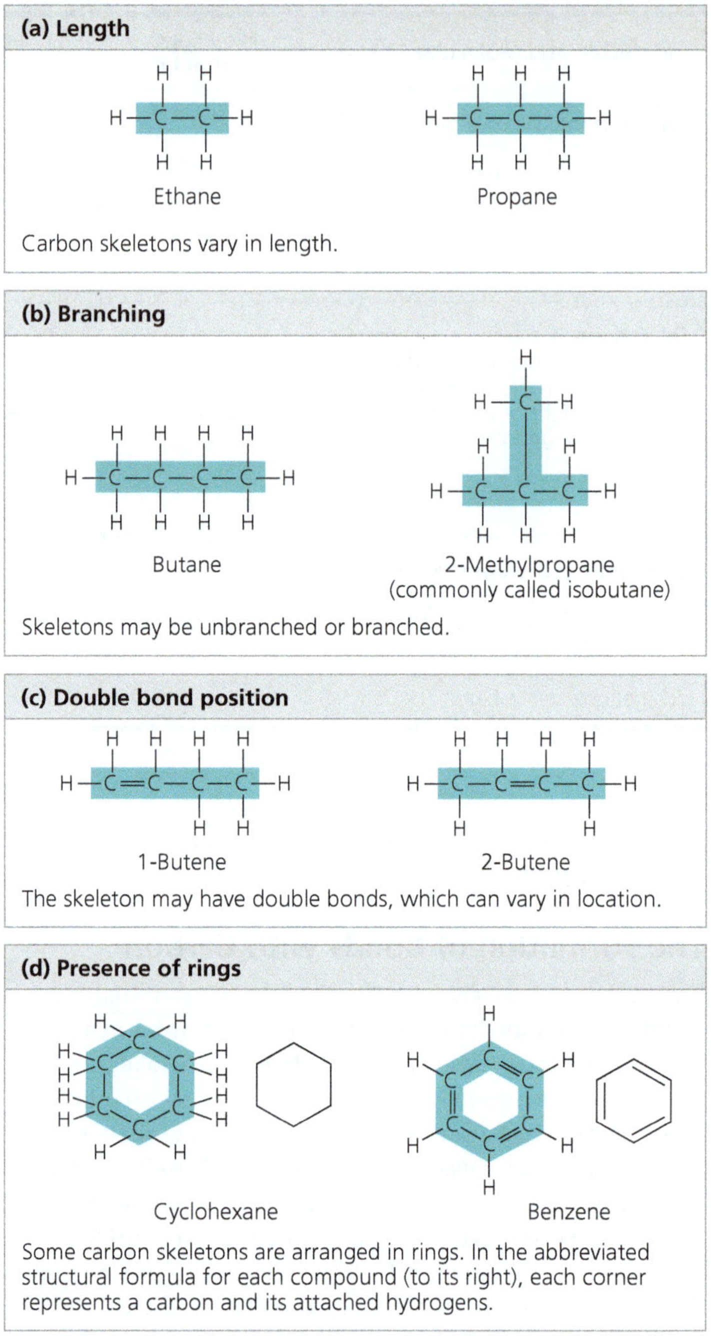

MB Animation: Diversity of Carbon-Based Molecules

Although hydrocarbons are not prevalent in most living organisms, many of a cell's organic molecules have regions consisting of only carbon and hydrogen. For example, the molecules known as fats have long hydrocarbon tails attached to a nonhydrocarbon component **(Figure 4.6)**. Neither petroleum nor fat dissolves in water; both are hydrophobic compounds because the great majority of their bonds are relatively nonpolar carbon-to-hydrogen linkages. Another characteristic of hydrocarbons is that they can undergo reactions that release a relatively large amount of energy. The gasoline that fuels a car consists of hydrocarbons, and the hydrocarbon tails of fats serve as stored fuel for plant embryos (seeds) and animals.

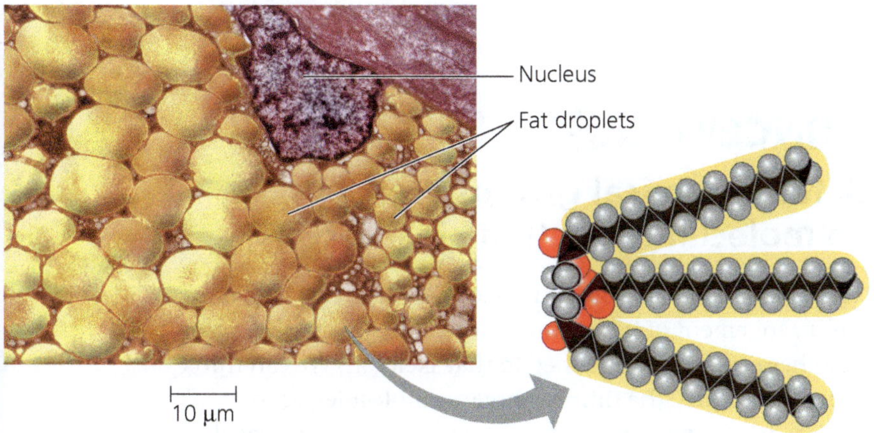

(a) Part of a human adipose cell **(b) A fat molecule**

▲ Figure 4.6 The role of hydrocarbons in fats.
(a) Mammalian adipose cells stockpile fat molecules as a fuel reserve. This colorized micrograph shows part of a human adipose cell with many fat droplets, each containing a large number of fat molecules. **(b)** A fat molecule consists of a small, nonhydrocarbon component joined to three hydrocarbon tails that account for the hydrophobic behavior of fats. The tails can be broken down to provide energy. (Black = carbon; gray = hydrogen; red = oxygen.)

MAKE CONNECTIONS ➤ *How do the tails account for the hydrophobic nature of fats? (See Concept 3.2.)*

Isomers

Variation in the architecture of organic molecules can be seen in **isomers**, compounds that have the same numbers of atoms of the same elements but different structures and hence different properties. We will examine three types of isomers: structural isomers, *cis-trans* isomers, and enantiomers.

Structural isomers differ in the covalent arrangements of their atoms. Compare, for example, the two five-carbon compounds in **Figure 4.7a**. Both have the molecular formula C_5H_{12}, but they differ in the covalent arrangement of their carbon skeletons. The skeleton is straight in one compound but branched in the other. The number of possible isomers increases tremendously as carbon skeletons increase in size. There are only three forms of C_5H_{12} (two of which are shown in Figure 4.7a), but there are 18 variants of C_8H_{18} and 366,319 possible structural isomers of $C_{20}H_{42}$. Structural isomers may also differ in the location of double bonds.

In ***cis-trans* isomers**, carbons have covalent bonds to the same atoms, but these atoms differ in their spatial arrangements due to the inflexibility of double bonds. Single bonds allow the atoms they join to rotate freely about the bond axis without changing the compound. In contrast, double bonds do not permit such rotation. If a double bond joins two carbon atoms, and each C also has two different atoms (or groups of atoms) attached to it, then two distinct *cis-trans* isomers are possible. Consider a simple molecule with two double-bonded carbons, each of which has an H and an X attached to it **(Figure 4.7b)**. The arrangement with both Xs on the same side of the double bond is called a *cis isomer*, and that with the Xs on opposite sides is called a *trans isomer*. The subtle difference in shape between such isomers can have a dramatic effect on the biological activities of organic molecules. For example, the biochemistry of vision involves a light-induced change of retinal, a chemical compound in the eye, from the *cis* isomer to the *trans* isomer (see Figure 50.17). Another example involves *trans* fats, harmful fats formed during food processing that are discussed in Concept 5.3.

Enantiomers are isomers that are mirror images of each other and that differ in shape due to the presence of an *asymmetric carbon*, one that is attached to four different atoms or groups of atoms. (See the middle carbon in the ball-and-stick models shown in **Figure 4.7c**.) The four groups can

▼ Figure 4.7 Three types of isomers. Isomers are compounds that have the same molecular formula but different structures.

(a) Structural isomers

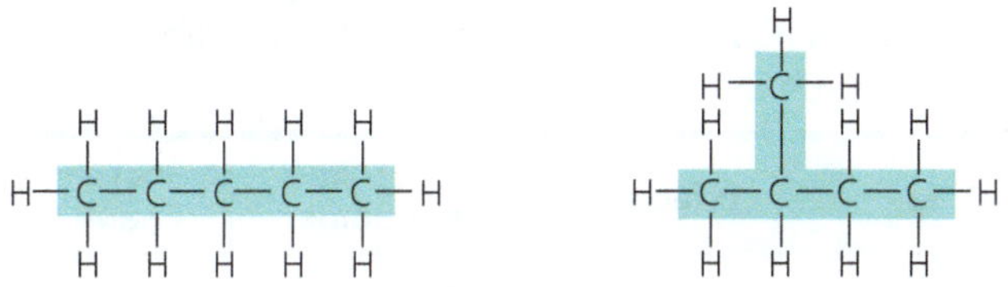

Pentane 2-Methylbutane

Structural isomers differ in covalent partners, as shown in this example of two isomers of C_5H_{12}.

(b) *Cis-trans* isomers

cis isomer: The two Xs are on the same side.

trans isomer: The two Xs are on opposite sides.

Cis-trans isomers differ in arrangement about a double bond. In these diagrams, X represents an atom or group of atoms attached to a double-bonded carbon.

(c) Enantiomers

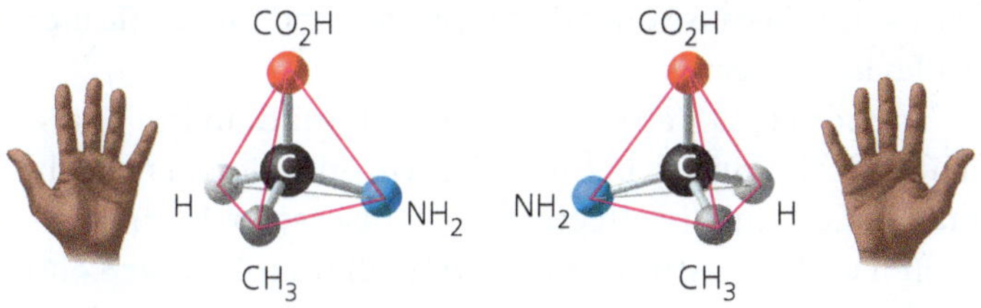

L isomer D isomer

Enantiomers differ in spatial arrangement around an asymmetric carbon, resulting in molecules that are mirror images, like left and right hands. The two isomers here are designated the L and D isomers from the Latin for "left" and "right" (*levo* and *dextro*). Enantiomers cannot be superimposed on each other.

DRAW IT ➤ *There are three structural isomers of C_5H_{12}; draw the one not shown in (a).*

Animation: Isomers

Figure 4.8 The pharmacological importance of enantiomers. Ibuprofen and albuterol are drugs whose enantiomers have different effects. (*S* and *R* are used here to distinguish between enantiomers.) Ibuprofen is commonly sold as a mixture of the two enantiomers; the *S* enantiomer is 100 times more effective than the *R* form. Albuterol is synthesized and sold only as the *R* form of that particular drug; the *S* form counteracts the active *R* form.

Drug	Effects	Effective Enantiomer	Ineffective Enantiomer
Ibuprofen	Reduces inflammation and pain	 *S*-Ibuprofen	 *R*-Ibuprofen
Albuterol	Relaxes bronchial (airway) muscles, improving airflow in asthma patients	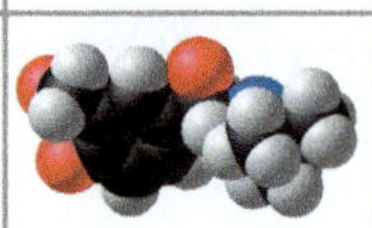 *R*-Albuterol	 *S*-Albuterol

be arranged in space around the asymmetric carbon in two different ways that are mirror images. Enantiomers are, in a way, left-handed and right-handed versions of the molecule. Just as your right hand won't fit into a left-handed glove, a "right-handed" molecule won't fit into the same space as the "left-handed" version. Usually, only one isomer is biologically active because only that form can bind to specific molecules in an organism.

The concept of enantiomers is important in the pharmaceutical industry because the two enantiomers of a drug may not be equally effective, as is the case for both ibuprofen and the asthma medication albuterol (**Figure 4.8**). Methamphetamine also occurs in two enantiomers that have very different effects. One enantiomer is the highly addictive stimulant drug known as "crank," sold illegally in the street drug trade. The other has a much weaker effect and is the active ingredient in an over-the-counter vapor inhaler for treatment of nasal congestion. The differing effects of enantiomers in the body demonstrate that organisms are sensitive to even the subtlest variations in molecular architecture. Once again, we see that molecules have emergent properties that depend on the specific arrangement of their atoms.

CONCEPT CHECK 4.2

1. **DRAW IT** ➤ (a) Draw a structural formula for C_2H_4. (b) Draw the *trans* isomer of $C_2H_2Cl_2$.
2. **VISUAL SKILLS** ➤ Which two pairs of molecules in Figure 4.5 are isomers? For each pair, identify the type of isomer.
3. How are gasoline and fat chemically similar?
4. **VISUAL SKILLS** ➤ See Figures 4.5a and 4.7. Can propane (C_3H_8) form isomers? Explain.

For suggested answers, see Appendix A.

CONCEPT 4.3

A few chemical groups are key to molecular function

The properties of an organic molecule depend not only on the arrangement of its carbon skeleton but also on the various chemical groups attached to that skeleton. We can think of hydrocarbons, the simplest organic molecules, as the underlying framework for more complex organic molecules. A number of chemical groups can replace one or more hydrogens of the hydrocarbon. These groups may participate in chemical reactions or may contribute to function indirectly by their effects on molecular shape; they help give each molecule its unique properties.

The Chemical Groups Most Important in the Processes of Life

Consider the differences between estradiol (a type of estrogen) and testosterone. These compounds are female and male sex hormones, respectively, in humans and other vertebrates. Both are steroids, organic molecules with a common carbon skeleton in the form of four fused rings. They differ only in the chemical groups attached to the rings (shown here in abbreviated form); the distinctions in molecular architecture are shaded in blue:

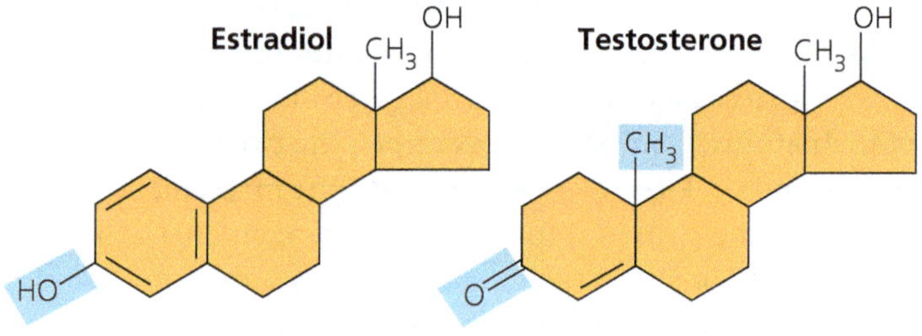

The different actions of these two molecules on many targets throughout the body are the basis of gender, producing the contrasting features of male and female vertebrates. In this case, the chemical groups are important because they affect molecular shape, contributing to function.

In other cases, chemical groups are directly involved in chemical reactions; such groups are known as **functional groups**. Each has certain properties, such as shape and charge, that cause it to participate in chemical reactions in a characteristic way.

The seven chemical groups most important in biological processes are the hydroxyl, carbonyl, carboxyl, amino, sulfhydryl, phosphate, and methyl groups. The first six groups can be chemically reactive; of these six, all except the sulfhydryl group are also hydrophilic and thus increase the solubility of organic compounds in water. The methyl group is not reactive, but instead often serves as a recognizable tag on biological molecules. Study **Figure 4.9** to become familiar with these biologically important chemical groups.

Figure 4.9 Some biologically important chemical groups.

Chemical Group	Group Properties and Compound Name	Examples
Hydroxyl group (—OH)  (may be written HO—)	Is polar due to electronegative oxygen. Forms hydrogen bonds with water, helping dissolve compounds such as sugars. Compound name: **Alcohol** (specific name usually ends in *-ol*)	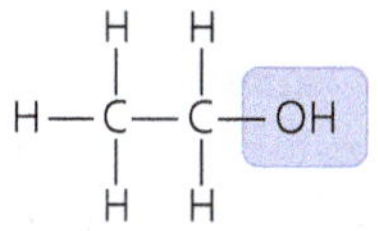 **Ethanol**, the alcohol present in alcoholic beverages
Carbonyl group (>C=O) 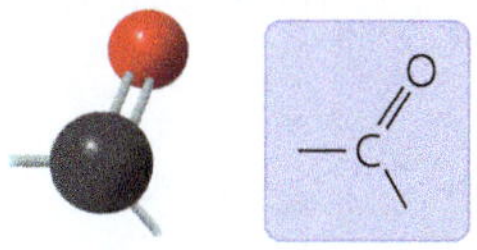	Sugars with ketone groups are called ketoses; those with aldehydes are called aldoses. Compound name: **Ketone** (carbonyl group is within a carbon skeleton) or **aldehyde** (carbonyl group is at the end of a carbon skeleton)	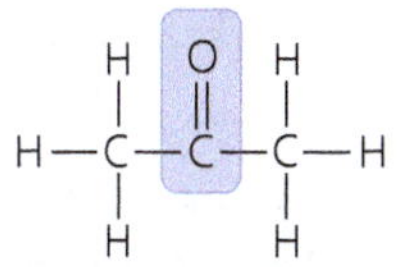 **Acetone**, the simplest ketone 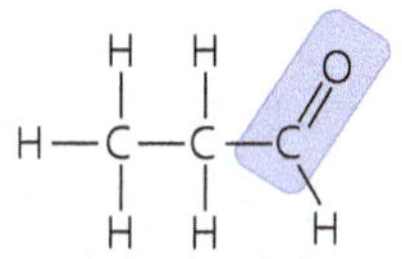 **Propanal**, an aldehyde
Carboxyl group (—COOH) 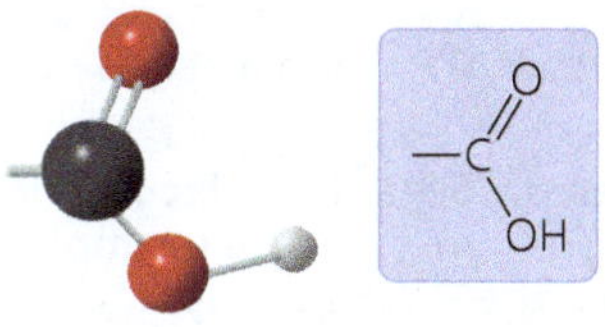	Acts as an acid (can donate H^+) because the covalent bond between oxygen and hydrogen is so polar. Compound name: **Carboxylic acid**, or **organic acid**	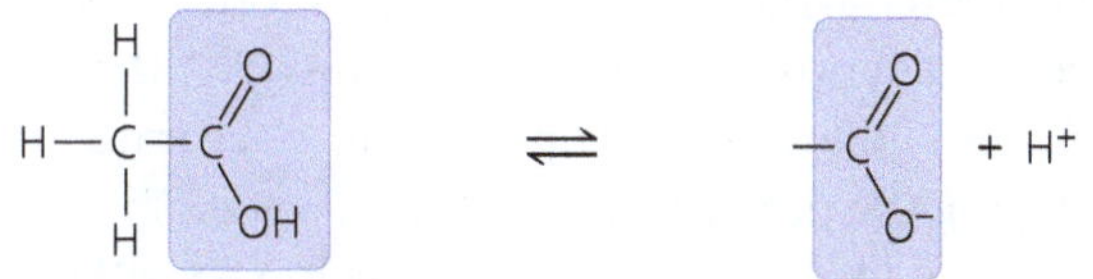 **Acetic acid**, which gives vinegar its sour taste Ionized form of —COOH (carboxylate ion), found in cells
Amino group (—NH_2) 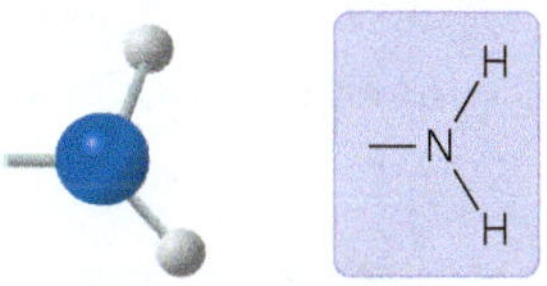	Acts as a base; can pick up an H^+ from the surrounding solution (water, in living organisms). Compound name: **Amine**	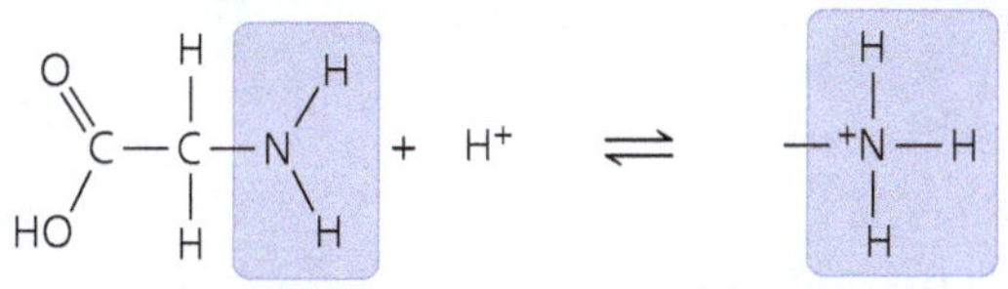 **Glycine**, an amino acid (note its carboxyl group) Ionized form of —NH_2, found in cells
Sulfhydryl group (—SH) 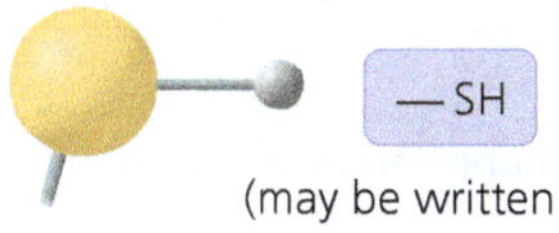 (may be written HS—)	Two —SH groups can react, forming a "cross-link" that helps stabilize protein structure. Hair protein cross-links maintain the straightness or curliness of hair; in hair salons, permanent treatments break cross-links, then re-form them while the hair is in the desired shape. Compound name: **Thiol**	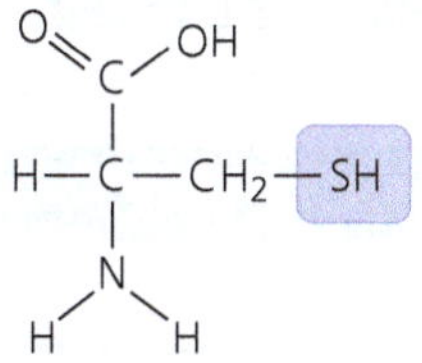 **Cysteine**, a sulfur-containing amino acid
Phosphate group (—OPO_3^{2-}) 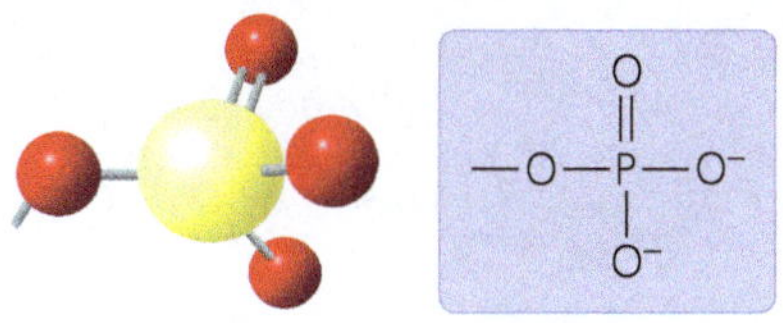	Contributes negative charge (1– when positioned inside a chain of phosphates; 2– when at the end). When attached, confers on a molecule the ability to react with water, releasing energy. Compound name: **Organic phosphate**	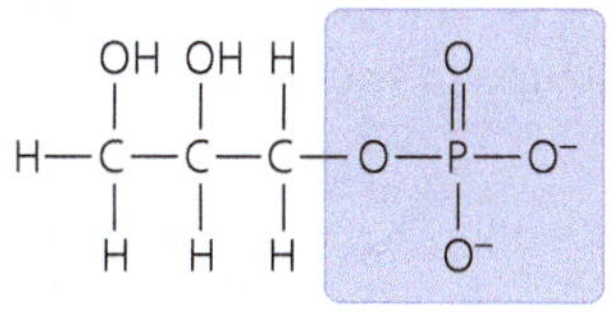 **Glycerol phosphate**, which takes part in many important chemical reactions in cells
Methyl group (—CH_3) 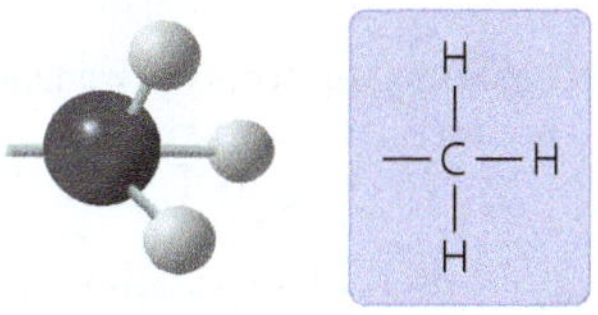	Affects the expression of genes when on DNA or on proteins bound to DNA. Affects the shape and function of male and female sex hormones. Compound name: **Methylated compound**	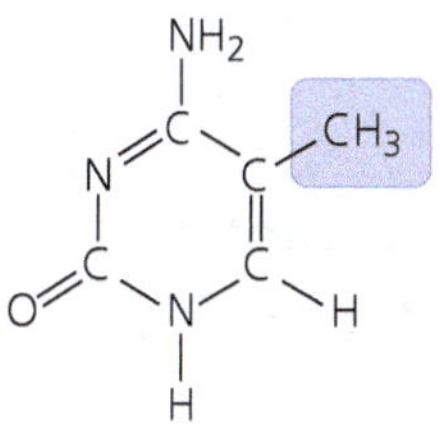 **5-Methylcytosine**, a component of DNA that has been modified by addition of a methyl group

Animation: Functional Groups

ATP: An Important Source of Energy for Cellular Processes

The "Phosphate group" row in Figure 4.9 shows a simple example of an organic phosphate molecule. A more complicated organic phosphate, **adenosine triphosphate,** or **ATP**, is worth mentioning here because its function in the cell is so important. ATP consists of an organic molecule called adenosine attached to a string of three phosphate groups:

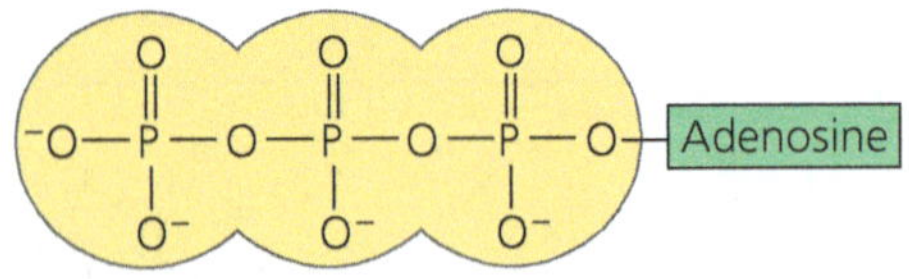

Where three phosphates are present in series, as in ATP, one phosphate may be split off as a result of a reaction with water. This inorganic phosphate ion, $HOPO_3^{2-}$, is often abbreviated Ⓟ$_i$ in this book, and a phosphate group in an organic molecule is often written as Ⓟ. Having lost one phosphate, ATP becomes adenosine *di*phosphate, or ADP. Although ATP is sometimes said to store energy, it is more accurate to think of it as storing the potential to react with water. This reaction releases energy that can be used by the cell. You will learn about this in more detail in Concept 8.3.

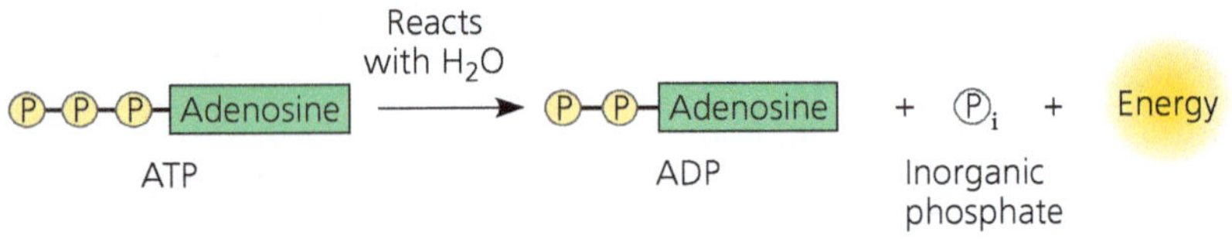

The Chemical Elements of Life: *A Review*

Living matter, as you have learned, consists mainly of carbon, oxygen, hydrogen, and nitrogen, with smaller amounts of sulfur and phosphorus. These elements all form strong covalent bonds, an essential characteristic in the architecture of complex organic molecules. Of all these elements, carbon is the virtuoso of the covalent bond. The versatility of carbon makes possible the great diversity of organic molecules, each with particular properties that emerge from the unique arrangement of its carbon skeleton and the chemical groups appended to that skeleton. This variation at the molecular level provides the foundation for the rich biological diversity found on our planet.

CONCEPT CHECK 4.3

1. **VISUAL SKILLS >** What does the term *amino acid* signify about the structure of such a molecule? See Figure 4.9.
2. What chemical change occurs to ATP when it reacts with water and releases energy?
3. **DRAW IT >** Suppose you had an organic molecule such as cysteine (see Figure 4.9, sulfhydryl group example), and you chemically removed the —NH_2 group and replaced it with —COOH. Draw this structure. How would this change the chemical properties of the molecule? Is the central carbon asymmetric before the change? After?

For suggested answers, see Appendix A.

4 Chapter Review

Go to **MasteringBiology**™ for Videos, Animations, Vocab Self-Quiz, Practice Tests, and more in the Study Area.

SUMMARY OF KEY CONCEPTS

CONCEPT 4.1

Organic chemistry is the study of carbon compounds *(pp. 57–58)*

VOCAB SELF-QUIZ goo.gl/6u55ks

- **Organic** compounds, once thought to arise only within living organisms, were finally synthesized in the laboratory.
- Living matter is made mostly of carbon, oxygen, hydrogen, and nitrogen. Biological diversity results from carbon's ability to form a huge number of molecules with particular shapes and properties.

? *How did Stanley Miller's experiments support the idea that, even at life's origins, physical and chemical laws govern the processes of life?*

CONCEPT 4.2

Carbon atoms can form diverse molecules by bonding to four other atoms *(pp. 58–62)*

- Carbon, with a valence of 4, can bond to various other atoms, including O, H, and N. Carbon can also bond to other carbon atoms, forming the carbon skeletons of organic compounds. These skeletons vary in length and shape and have bonding sites for atoms of other elements.
- **Hydrocarbons** consist of carbon and hydrogen.
- **Isomers** are compounds that have the same molecular formula but different structures and therefore different properties. Three types of isomers are **structural isomers, *cis-trans* isomers**, and **enantiomers**.

VISUAL SKILLS > *Refer back to Figure 4.9. What type of isomers are acetone and propanal? How many asymmetric carbons are present in acetic acid, glycine, and glycerol phosphate? Can these three molecules exist as forms that are enantiomers?*

CONCEPT 4.3

A few chemical groups are key to molecular function *(pp. 62–64)*

- Chemical groups attached to the carbon skeletons of organic molecules participate in chemical reactions (**functional groups**) or contribute to function by affecting molecular shape (see Figure 4.9).
- **ATP (adenosine triphosphate)** consists of adenosine attached to three phosphate groups. ATP can react with water, forming

ADP (adenosine diphosphate) and inorganic phosphate. This reaction releases energy that can be used by the cell.

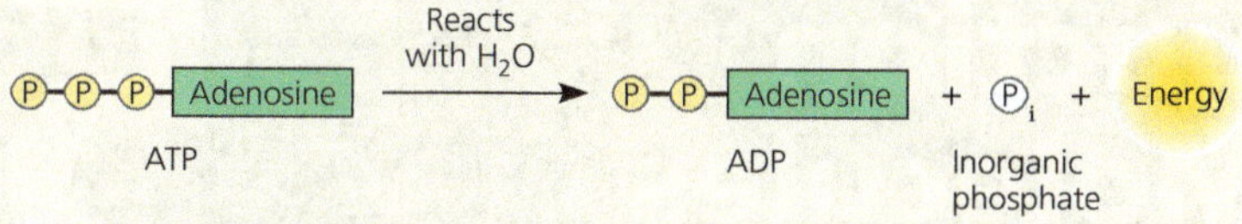

In what ways does a methyl group differ chemically from the other six important chemical groups shown in Figure 4.9?

TEST YOUR UNDERSTANDING

Level 1: Knowledge/Comprehension

PRACTICE TEST goo.gl/CUYGKD

1. Organic chemistry is currently defined as
   (A) the study of compounds made only by living cells.
   (B) the study of carbon compounds.
   (C) the study of natural (as opposed to synthetic) compounds.
   (D) the study of hydrocarbons.

2. **VISUAL SKILLS** Which functional group is *not* present in this molecule?
   (A) carboxyl
   (B) sulfhydryl
   (C) hydroxyl
   (D) amino

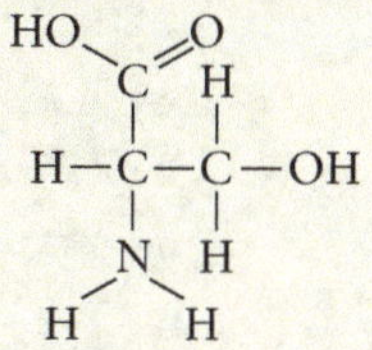

3. **MAKE CONNECTIONS** Which chemical group is most likely to be responsible for an organic molecule behaving as a base (see Concept 3.3)?
   (A) hydroxyl
   (B) carbonyl
   (C) amino
   (D) phosphate

Level 2: Application/Analysis

4. **VISUAL SKILLS** Visualize the structural formula of each of the following hydrocarbons. Which hydrocarbon has a double bond in its carbon skeleton?
   (A) C_3H_8
   (B) C_2H_6
   (C) C_2H_4
   (D) C_2H_2

5. **VISUAL SKILLS** Choose the term that correctly describes the relationship between these two sugar molecules:
   (A) structural isomers
   (B) *cis-trans* isomers
   (C) enantiomers
   (D) isotopes

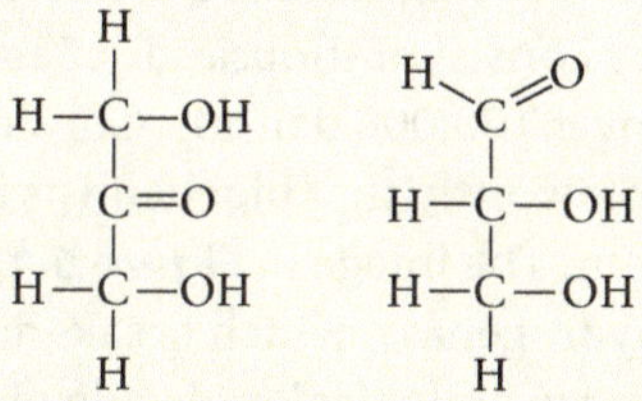

6. **VISUAL SKILLS** Identify the asymmetric carbon in this molecule:

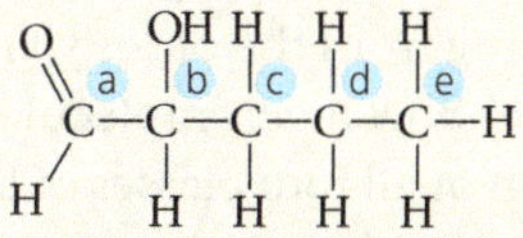

7. Which action could produce a carbonyl group?
   (A) the replacement of the —OH of a carboxyl group with hydrogen
   (B) the addition of a thiol to a hydroxyl
   (C) the addition of a hydroxyl to a phosphate
   (D) the replacement of the nitrogen of an amine with oxygen

8. **VISUAL SKILLS** Which of the molecules shown in question 5 has an asymmetric carbon? Which carbon is asymmetric?

Level 3: Synthesis/Evaluation

9. **EVOLUTION CONNECTION • DRAW IT** Some scientists think that life elsewhere in the universe might be based on the element silicon, rather than on carbon, as on Earth. Look at the electron distribution diagram for silicon in Figure 2.7 and draw the Lewis dot structure for silicon. What properties does silicon share with carbon that would make silicon-based life more likely than, say, neon-based life or aluminum-based life?

10. **SCIENTIFIC INQUIRY** 50 years ago, pregnant women who were prescribed thalidomide for morning sickness gave birth to children with birth defects. Thalidomide is a mixture of two enantiomers; one reduces morning sickness, but the other causes severe birth defects. Today, the FDA has approved this drug for non-pregnant individuals with Hansen's disease (leprosy) or newly diagnosed multiple myeloma, a blood and bone marrow cancer. The beneficial enantiomer can be synthesized and given to patients, but over time, both the beneficial *and* the harmful enantiomer can be detected in the body. Propose a possible explanation for the presence of the harmful enantiomer.

11. **WRITE ABOUT A THEME: ORGANIZATION** In 1918, an epidemic of sleeping sickness caused an unusual rigid paralysis in some survivors, similar to symptoms of advanced Parkinson's disease. Years later, L-dopa (below, left), a chemical used to treat Parkinson's disease, was given to some of these patients. L-dopa was remarkably effective at eliminating the paralysis, at least temporarily. However, its enantiomer, D-dopa (right), was subsequently shown to have no effect at all, as is the case for Parkinson's disease. In a short essay (100–150 words), discuss how the effectiveness of one enantiomer and not the other illustrates the theme of structure and function.

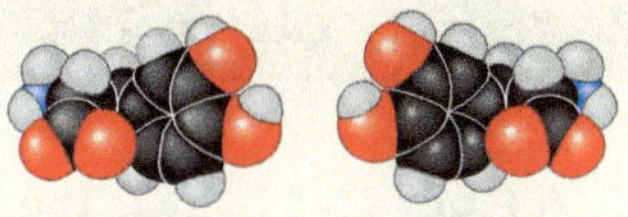

L-dopa **D-dopa**

12. **SYNTHESIZE YOUR KNOWLEDGE**

Explain how the chemical structure of the carbon atom accounts for the differences between the male and female lions seen in the photo.

For selected answers, see Appendix A.

For additional practice questions, check out the **Dynamic Study Modules** in MasteringBiology. You can use them to study on your smartphone, tablet, or computer anytime, anywhere!

The Structure and Function of Large Biological Molecules

5

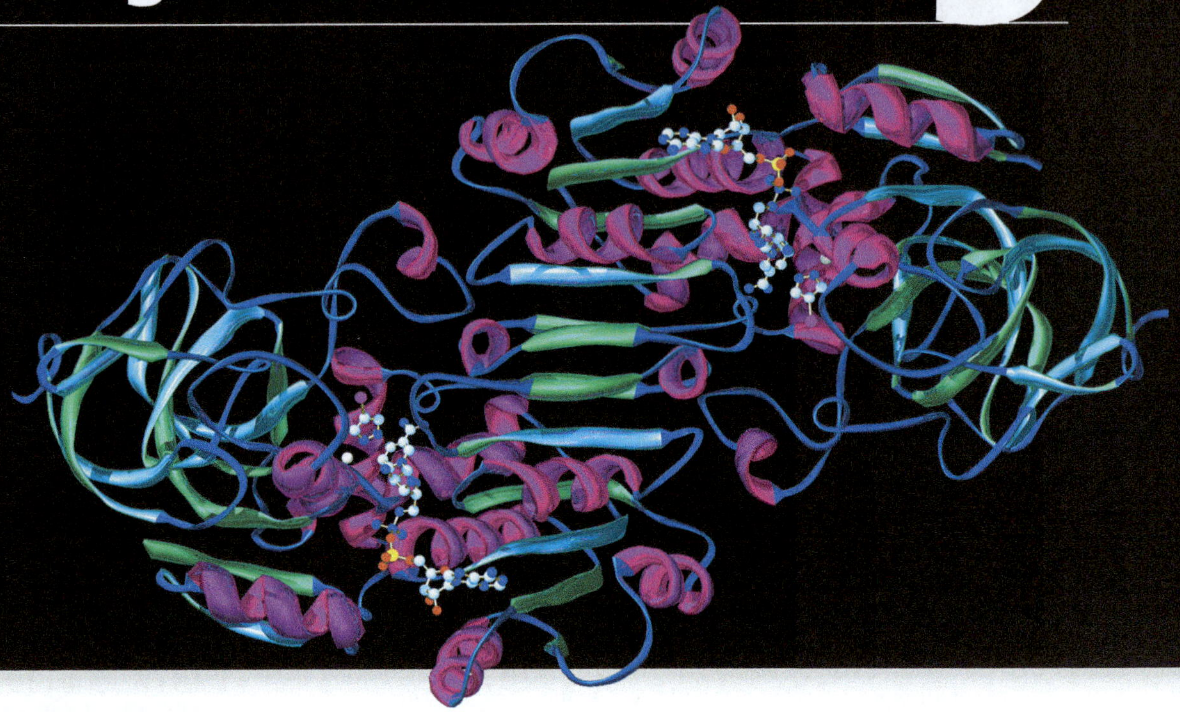

▲ **Figure 5.1 Why is the structure of a protein important for its function?**

KEY CONCEPTS

5.1 Macromolecules are polymers, built from monomers

5.2 Carbohydrates serve as fuel and building material

5.3 Lipids are a diverse group of hydrophobic molecules

5.4 Proteins include a diversity of structures, resulting in a wide range of functions

5.5 Nucleic acids store, transmit, and help express hereditary information

5.6 Genomics and proteomics have transformed biological inquiry and applications

◀ The scientist in the foreground is using 3-D glasses to help her visualize the structure of the protein displayed on her screen.

The Molecules of Life

Given the rich complexity of life on Earth, it might surprise you that the most important large molecules found in all living things—from bacteria to elephants—can be sorted into just four main classes: carbohydrates, lipids, proteins, and nucleic acids. On the molecular scale, members of three of these classes—carbohydrates, proteins, and nucleic acids—are huge and are therefore called **macromolecules**. For example, a protein may consist of thousands of atoms that form a molecular colossus with a mass well over 100,000 daltons. Considering the size and complexity of macromolecules, it is noteworthy that biochemists have determined the detailed structure of so many of them. The image in **Figure 5.1** is a molecular model of a protein called alcohol dehydrogenase, which breaks down alcohol in the body.

The architecture of a large biological molecule plays an essential role in its function. Like water and simple organic molecules, large biological molecules exhibit unique emergent properties arising from the orderly arrangement of their atoms. In this chapter, we'll first consider how macromolecules are built. Then we'll examine the structure and function of all four classes of large biological molecules: carbohydrates, lipids, proteins, and nucleic acids.

When you see this blue icon, log in to **MasteringBiology** and go to the Study Area for digital resources.

MB **Get Ready for This Chapter**

CONCEPT 5.1

Macromolecules are polymers, built from monomers

Large carbohydrates, proteins, and nucleic acids are chain-like molecules called polymers (from the Greek *polys*, many, and *meros*, part). A **polymer** is a long molecule consisting of many similar or identical building blocks linked by covalent bonds, much as a train consists of a chain of cars. The repeating units that serve as the building blocks of a polymer are smaller molecules called **monomers** (from the Greek *monos*, single). In addition to forming polymers, some monomers have functions of their own.

The Synthesis and Breakdown of Polymers

Although each class of polymer is made up of a different type of monomer, the chemical mechanisms by which cells make and break down polymers are basically the same in all cases. In cells, these processes are facilitated by **enzymes**, specialized macromolecules that speed up chemical reactions. The reaction connecting monomers is a good example of a **dehydration reaction**, a reaction in which two molecules are covalently bonded to each other with the loss of a water molecule **(Figure 5.2a)**. When a bond forms between two monomers, each monomer contributes part of the water molecule that is released during the reaction: One monomer provides a hydroxyl group (—OH), while the other provides a hydrogen (—H). This reaction is repeated as monomers are added to the chain one by one, making a polymer (also called polymerization).

Polymers are disassembled to monomers by **hydrolysis**, a process that is essentially the reverse of the dehydration reaction **(Figure 5.2b)**. Hydrolysis means water breakage (from the Greek *hydro*, water, and *lysis*, break). The bond between monomers is broken by the addition of a water molecule, with a hydrogen from water attaching to one monomer and the hydroxyl group attaching to the other. An example of hydrolysis within our bodies is the process of digestion. The bulk of the organic material in our food is in the form of polymers that are much too large to enter our cells. Within the digestive tract, various enzymes attack the polymers, speeding up hydrolysis. Released monomers are then absorbed into the bloodstream for distribution to all body cells. Those cells can then use dehydration reactions to assemble the monomers into new, different polymers that can perform specific functions required by the cell. (Dehydration reactions and hydrolysis can also be involved in the formation and breakdown of molecules that are not polymers, such as some lipids.)

The Diversity of Polymers

A cell has thousands of different macromolecules; the collection varies from one type of cell to another. The inherited differences between close relatives, such as human siblings, reflect small variations in polymers, particularly DNA and proteins. Molecular differences between unrelated individuals are more extensive, and those between species greater still. The diversity of macromolecules in the living world is vast, and the possible variety is effectively limitless.

What is the basis for such diversity in life's polymers? These molecules are constructed from only 40 to 50 common monomers and some others that occur rarely. Building a huge variety of polymers from such a limited number of monomers is analogous to constructing hundreds of thousands of words from only 26 letters of the alphabet. The key is arrangement—the particular linear sequence that the units follow. However, this analogy falls far short of describing the great diversity of macromolecules because most biological polymers have many more monomers than the number of letters in even the longest word. Proteins, for example, are built from 20 kinds of amino acids arranged in chains that are typically hundreds of amino acids long. The molecular logic of life is simple but elegant: Small molecules common to all organisms act as building blocks that are ordered into unique macromolecules.

Despite this immense diversity, molecular structure and function can still be grouped roughly by class. Let's examine each of the four major classes of large biological molecules. For each class, the large molecules have emergent properties not found in their individual components.

Figure 5.2 The synthesis and breakdown of polymers.

(a) Dehydration reaction: synthesizing a polymer

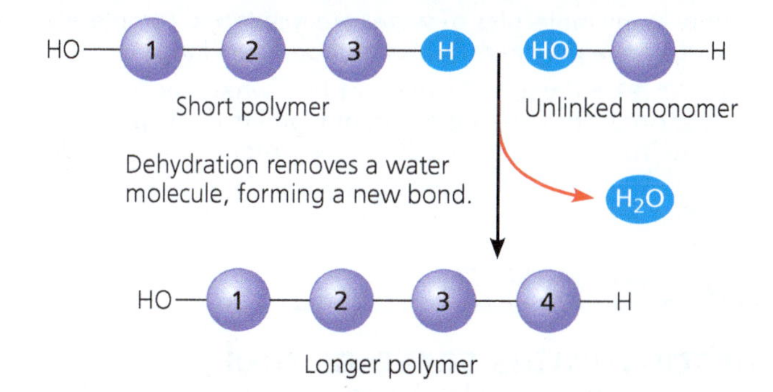

(b) Hydrolysis: breaking down a polymer

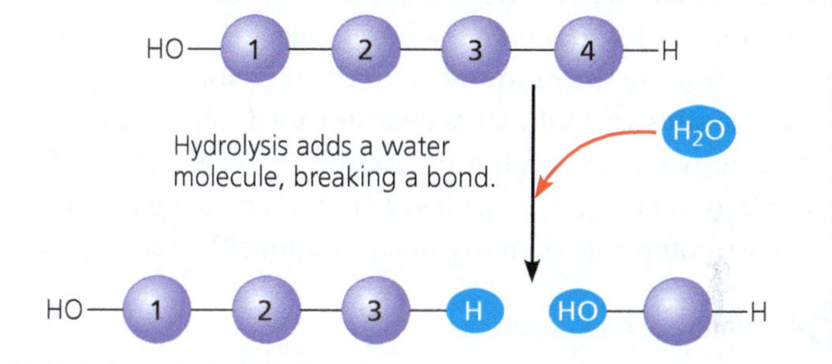

MB **Animation: Making and Breaking Polymers**

CONCEPT CHECK 5.1

1. What are the four main classes of large biological molecules? Which class does not consist of polymers?
2. How many molecules of water are needed to completely hydrolyze a polymer that is ten monomers long?
3. **WHAT IF?** ➤ If you eat a piece of fish, what reactions must occur for the amino acid monomers in the protein of the fish to be converted to new proteins in your body?

For suggested answers, see Appendix A.

CONCEPT 5.2

Carbohydrates serve as fuel and building material

Carbohydrates include sugars and polymers of sugars. The simplest carbohydrates are the monosaccharides, or simple sugars; these are the monomers from which more complex carbohydrates are built. Disaccharides are double sugars, consisting of two monosaccharides joined by a covalent bond. Carbohydrate macromolecules are polymers called polysaccharides, composed of many sugar building blocks.

Animation: Carbohydrates

Sugars

Monosaccharides (from the Greek *monos*, single, and *sacchar*, sugar) generally have molecular formulas that are some multiple of the unit CH_2O. Glucose ($C_6H_{12}O_6$), the most common monosaccharide, is of central importance in the chemistry of life. In the structure of glucose, we can see the trademarks of a sugar: The molecule has a carbonyl group, $\rangle C{=}O$, and multiple hydroxyl groups, —OH **(Figure 5.3)**. Depending on the location of the carbonyl group, a sugar is either an aldose (aldehyde sugar) or a ketose (ketone sugar). Glucose, for example, is an aldose; fructose, an isomer of glucose, is a ketose. (Most names for sugars end in *-ose*.) Another criterion for classifying sugars is the size of the carbon skeleton, which ranges from three to seven carbons long. Glucose, fructose, and other sugars that have six carbons are called hexoses. Trioses (three-carbon sugars) and pentoses (five-carbon sugars) are also common.

Still another source of diversity for simple sugars is in the way their parts are arranged spatially around asymmetric carbons. (Recall that an asymmetric carbon is a carbon attached to four different atoms or groups of atoms.) Glucose and galactose, for example, differ only in the placement of parts around one asymmetric carbon (see the purple boxes in Figure 5.3). What seems like a small difference is significant enough to give the two sugars distinctive shapes and binding activities, thus different behaviors.

Although it is convenient to draw glucose with a linear carbon skeleton, this representation is not completely accurate. In aqueous solutions, glucose molecules, as well as most other

▼ Figure 5.3 The structure and classification of some monosaccharides. Sugars vary in the location of their carbonyl groups (orange), the length of their carbon skeletons, and the way their parts are arranged spatially around asymmetric carbons (compare, for example, the purple portions of glucose and galactose).

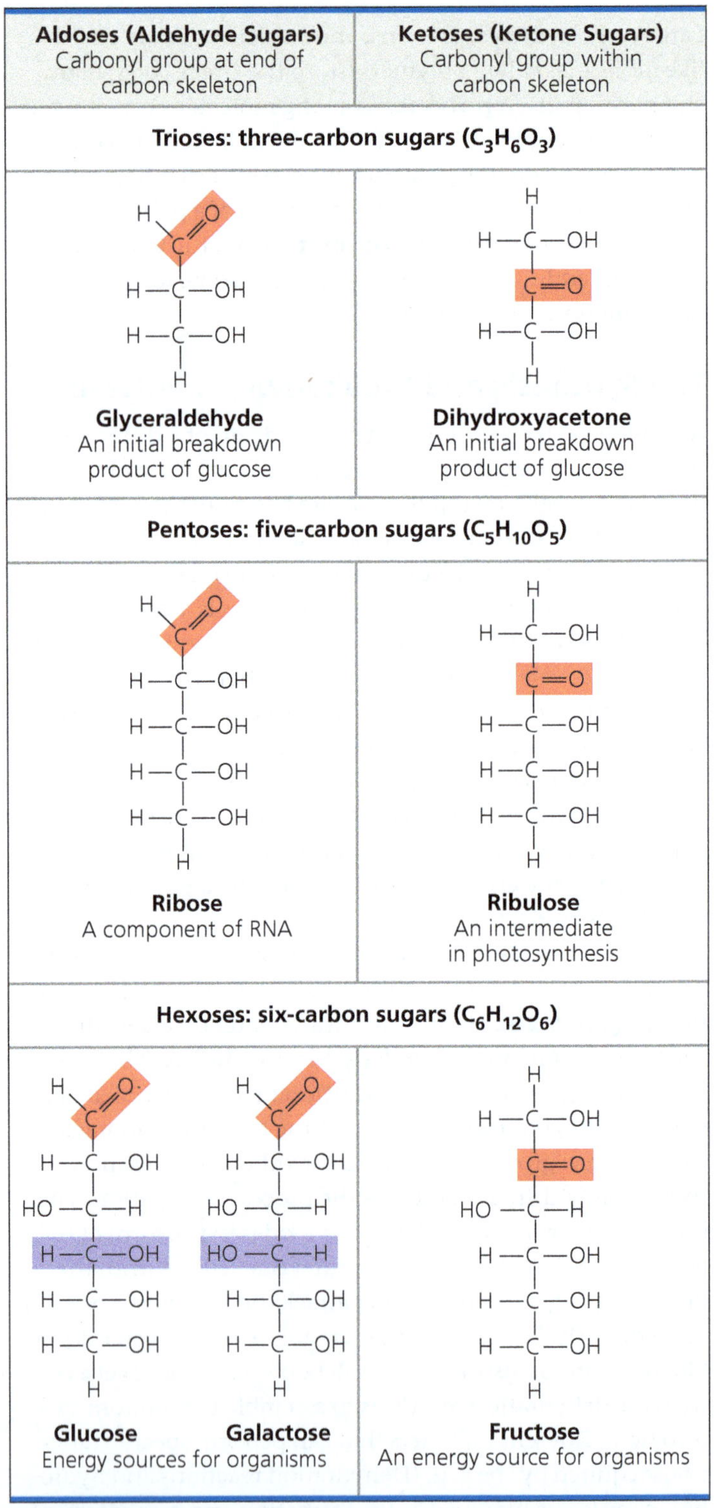

MAKE CONNECTIONS ➤ *In the 1970s, a process was developed that converts the glucose in corn syrup to its sweeter-tasting isomer, fructose. High-fructose corn syrup, a common ingredient in soft drinks and processed food, is a mixture of glucose and fructose. What type of isomers are glucose and fructose? (See Figure 4.7.)*

Animation: Monosaccharides

Figure 5.4 Linear and ring forms of glucose.

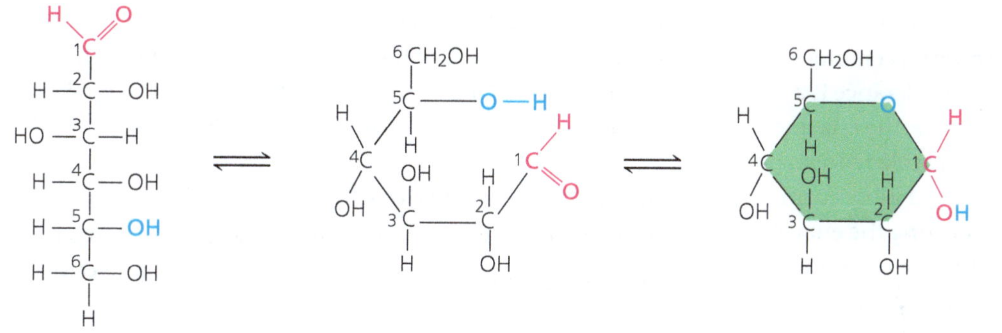

(a) Linear and ring forms. Chemical equilibrium between the linear and ring structures greatly favors the formation of rings. The carbons of the sugar are numbered 1 to 6, as shown. To form the glucose ring, carbon 1 (magenta) bonds to the oxygen (blue) attached to carbon 5.

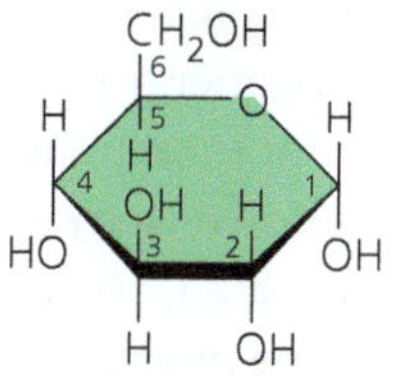

(b) Abbreviated ring structure. Each unlabeled corner represents a carbon. The ring's thicker edge indicates that you are looking at the ring edge-on; the components attached to the ring lie above or below the plane of the ring.

DRAW IT ➤ *Start with the linear form of fructose (see Figure 5.3) and draw the formation of the fructose ring in two steps, as shown in (a). First, number the carbons starting at the top of the linear structure. Then draw the molecule in a ringlike orientation, attaching carbon 5 via its oxygen to carbon 2. Compare the number of carbons in the fructose and glucose rings.*

five- and six-carbon sugars, form rings, because they are the most stable form of these sugars under physiological conditions **(Figure 5.4)**.

Monosaccharides, particularly glucose, are major nutrients for cells. In the process known as cellular respiration, cells extract energy from glucose molecules by breaking them down in a series of reactions. Not only are simple-sugar molecules a major fuel for cellular work, but their carbon skeletons also serve as raw material for the synthesis of other types of small organic molecules, such as amino acids and fatty acids. Sugar molecules that are not immediately used in these ways are generally incorporated as monomers into disaccharides or polysaccharides, discussed next.

A **disaccharide** consists of two monosaccharides joined by a **glycosidic linkage**, a covalent bond formed between two monosaccharides by a dehydration reaction (*glyco* refers to carbohydrate). For example, maltose is a disaccharide formed by the linking of two molecules of glucose **(Figure 5.5a)**. Also known as malt sugar, maltose is an ingredient used in brewing beer. The most prevalent disaccharide is sucrose, or table sugar. Its two monomers are glucose and fructose **(Figure 5.5b)**. Plants generally transport carbohydrates from leaves to roots

Figure 5.5 Examples of disaccharide synthesis.

(a) Dehydration reaction in the synthesis of maltose. The bonding of two glucose units forms maltose. The 1–4 glycosidic linkage joins the number 1 carbon of one glucose to the number 4 carbon of the second glucose. Joining the glucose monomers in a different way would result in a different disaccharide.

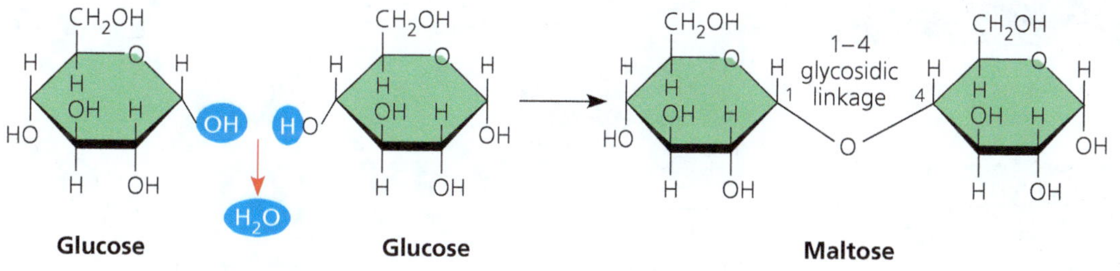

(b) Dehydration reaction in the synthesis of sucrose. Sucrose is a disaccharide formed from glucose and fructose. Notice that fructose forms a five-sided ring, though it is a hexose like glucose.

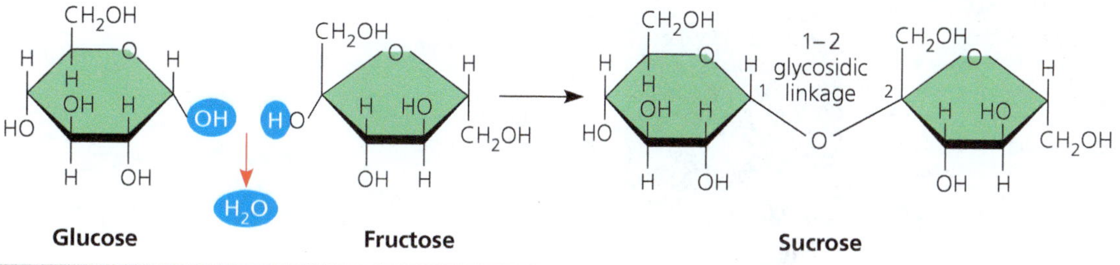

DRAW IT ➤ *Referring to Figures 5.3 and 5.4, number the carbons in each sugar in this figure. How does the name of each linkage relate to the numbers?*

MB Animation: Synthesis of Sucrose

and other nonphotosynthetic organs in the form of sucrose. Lactose, the sugar present in milk, is another disaccharide, in this case a glucose molecule joined to a galactose molecule. Disaccharides must be broken down into monosaccharides to be used for energy by organisms. Lactose intolerance is a common condition in humans who lack lactase, the enzyme that breaks down lactose. The sugar is instead broken down by intestinal bacteria, causing formation of gas and subsequent cramping. The problem may be avoided by taking the enzyme lactase when eating or drinking dairy products or consuming dairy products that have already been treated with lactase to break down the lactose.

Polysaccharides

Polysaccharides are macromolecules, polymers with a few hundred to a few thousand monosaccharides joined by glycosidic linkages. Some polysaccharides serve as storage material, hydrolyzed as needed to provide sugar for cells. Other polysaccharides serve as building material for structures that protect the cell or the whole organism. The architecture and function of a polysaccharide are determined by its sugar monomers and by the positions of its glycosidic linkages.

Storage Polysaccharides

Both plants and animals store sugars for later use in the form of storage polysaccharides **(Figure 5.6)**. Plants store **starch**, a polymer of glucose monomers, as granules within cellular structures known as plastids. (Plastids include chloroplasts.) Synthesizing starch enables the plant to stockpile surplus glucose. Because glucose is a major cellular fuel, starch represents stored energy. The sugar can later be withdrawn by the plant from this carbohydrate "bank" by hydrolysis, which breaks the

Figure 5.6 Polysaccharides of plants and animals. (a) Starch stored in plant cells, **(b)** glycogen stored in muscle cells, and **(c)** structural cellulose fibers in plant cell walls are all polysaccharides composed entirely of glucose monomers (green hexagons). In starch and glycogen, the polymer chains tend to form helices in unbranched regions because of the angle of the linkages between glucose molecules. There are two kinds of starch: amylose and amylopectin. Cellulose, with a different kind of glucose linkage, is always unbranched.

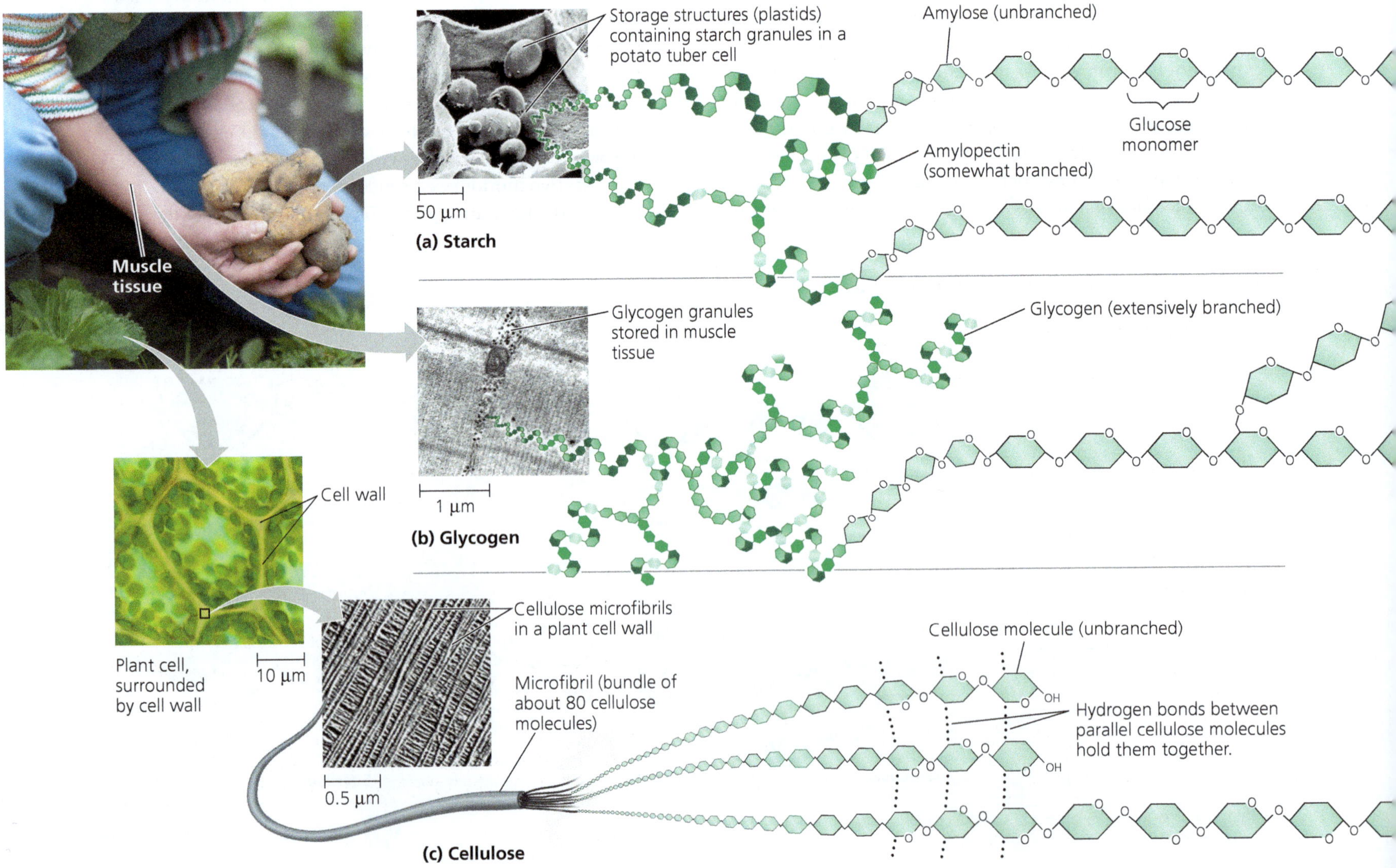

bonds between the glucose monomers. Most animals, including humans, also have enzymes that can hydrolyze plant starch, making glucose available as a nutrient for cells. Potato tubers and grains—the fruits of wheat, maize (corn), rice, and other grasses—are the major sources of starch in the human diet.

Most of the glucose monomers in starch are joined by 1–4 linkages (number 1 carbon to number 4 carbon), like the glucose units in maltose (see Figure 5.5a). The simplest form of starch, amylose, is unbranched. Amylopectin, a more complex starch, is a branched polymer with 1–6 linkages at the branch points. Both of these starches are shown in **Figure 5.6a**.

Animals store a polysaccharide called **glycogen**, a polymer of glucose that is like amylopectin but more extensively branched **(Figure 5.6b)**. Vertebrates store glycogen mainly in liver and muscle cells. Hydrolysis of glycogen in these cells releases glucose when the demand for sugar increases. (The extensively branched structure of glycogen fits its function: More free ends are available for hydrolysis.) This stored fuel cannot sustain an animal for long, however. In humans, for example, glycogen stores are depleted in about a day unless they are replenished by eating. This is an issue of concern in low-carbohydrate diets, which can result in weakness and fatigue.

Structural Polysaccharides

Organisms build strong materials from structural polysaccharides. For example, the polysaccharide called **cellulose** is a major component of the tough walls that enclose plant cells **(Figure 5.6c)**. Globally, plants produce almost 10^{14} kg (100 billion tons) of cellulose per year; it is the most abundant organic compound on Earth.

Like starch, cellulose is a polymer of glucose with 1–4 glycosidic linkages, but the linkages in these two polymers differ. The difference is based on the fact that there are actually two slightly different ring structures for glucose **(Figure 5.7a)**. When glucose forms a ring, the hydroxyl group attached to the number 1 carbon is positioned either below or above the plane of the ring. These two ring forms for glucose are called alpha (α) and beta (β), respectively. (Greek letters are often used as a "numbering" system for different versions of biological structures, much as we use the letters a, b, c, and so on for the parts of a question or a figure.) In starch, all the glucose monomers are in the α configuration **(Figure 5.7b)**, the arrangement we saw in Figures 5.4 and 5.5. In contrast, the glucose monomers of cellulose are all in the β configuration, making every glucose monomer "upside down" with respect to its neighbors (**Figure 5.7c**; see also Figure 5.6c).

The differing glycosidic linkages in starch and cellulose give the two molecules distinct three-dimensional shapes. Certain starch molecules are largely helical, fitting their function of efficiently storing glucose units. Conversely, a cellulose molecule is straight. Cellulose is never branched, and some hydroxyl groups on its glucose monomers are free to hydrogen-bond with the hydroxyls of other cellulose molecules lying parallel to it. In plant cell walls, parallel cellulose molecules held together in this way are grouped into units called microfibrils (see Figure 5.6c). These cable-like microfibrils are a strong building material for plants and an important substance for humans because cellulose is the major constituent of paper and the only component of cotton. The unbranched structure of cellulose thus fits its function: imparting strength to parts of the plant.

Enzymes that digest starch by hydrolyzing its α linkages are unable to hydrolyze the β linkages of cellulose due to the different shapes of these two molecules. In fact,

▼ Figure 5.7 Starch and cellulose structures.

(a) α and β glucose ring structures. These two interconvertible forms of glucose differ in the placement of the hydroxyl group (highlighted in blue) attached to the number 1 carbon.

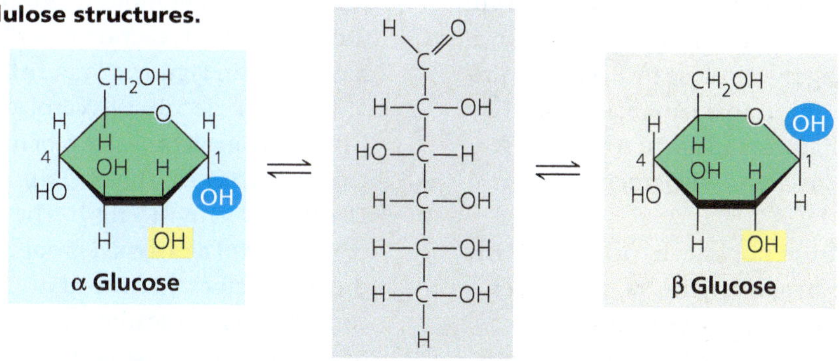

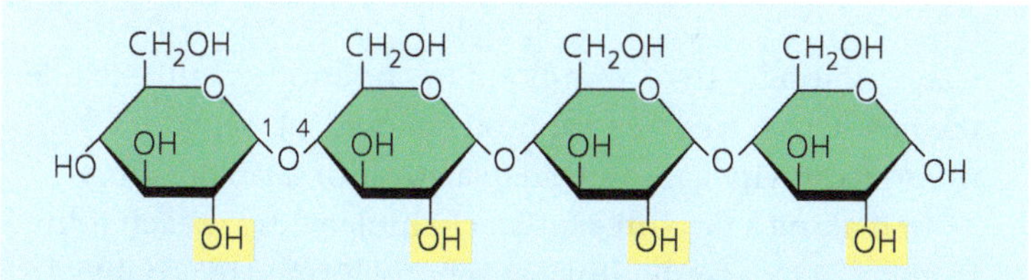

(b) Starch: 1–4 linkage of α glucose monomers. All monomers are in the same orientation. Compare the positions of the —OH groups highlighted in yellow with those in cellulose (c).

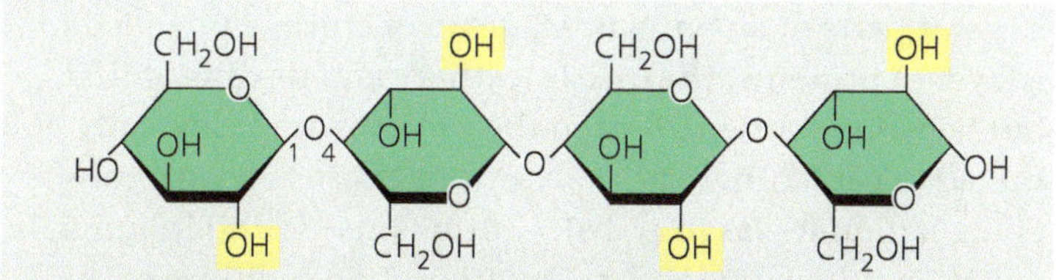

(c) Cellulose: 1–4 linkage of β glucose monomers. In cellulose, every β glucose monomer is upside down with respect to its neighbors. (See the highlighted —OH groups.)

Animation: Starch, Cellulose, and Glycogen Structures

Figure 5.8 Chitin, a structural polysaccharide.

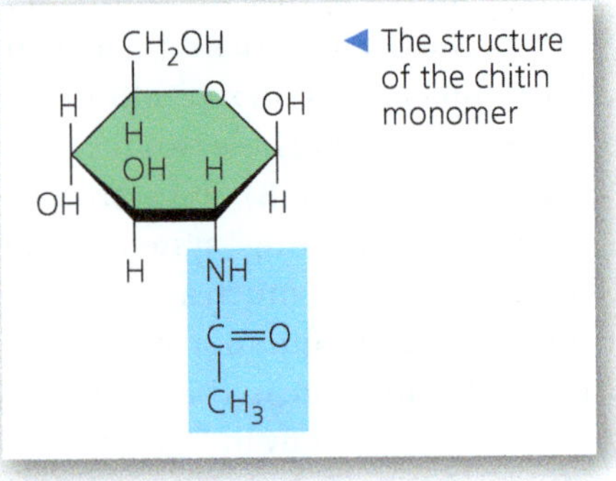

The structure of the chitin monomer

Chitin, embedded in proteins, forms the exoskeleton of arthropods. This emperor dragonfly (*Anax imperator*) is molting—shedding its old exoskeleton (brown) and emerging upside down in adult form.

few organisms possess enzymes that can digest cellulose. Almost all animals, including humans, do not; the cellulose in our food passes through the digestive tract and is eliminated with the feces. Along the way, the cellulose abrades the wall of the digestive tract and stimulates the lining to secrete mucus, which aids in the smooth passage of food through the tract. Thus, although cellulose is not a nutrient for humans, it is an important part of a healthful diet. Most fruits, vegetables, and whole grains are rich in cellulose. On food packages, "insoluble fiber" refers mainly to cellulose.

Some microorganisms can digest cellulose, breaking it down into glucose monomers. A cow harbors cellulose-digesting prokaryotes and protists in its gut. These microbes hydrolyze the cellulose of hay and grass and convert the glucose to other compounds that nourish the cow. Similarly, a termite, which is unable to digest cellulose by itself, has prokaryotes or protists living in its gut that can make a meal of wood. Some fungi can also digest cellulose in soil and elsewhere, thereby helping recycle chemical elements within Earth's ecosystems.

Another important structural polysaccharide is **chitin**, the carbohydrate used by arthropods (insects, spiders, crustaceans, and related animals) to build their exoskeletons **(Figure 5.8)**. An exoskeleton is a hard case that surrounds the soft parts of an animal. Made up of chitin embedded in a layer of proteins, the case is leathery and flexible at first, but becomes hardened when the proteins are chemically linked to each other (as in insects) or encrusted with calcium carbonate (as in crabs). Chitin is also found in fungi, which use this polysaccharide rather than cellulose as the building material for their cell walls. Chitin is similar to cellulose, with β linkages, except that the glucose monomer of chitin has a nitrogen-containing attachment (see Figure 5.8).

CONCEPT CHECK 5.2

1. Write the formula for a monosaccharide that has three carbons.
2. A dehydration reaction joins two glucose molecules to form maltose. The formula for glucose is $C_6H_{12}O_6$. What is the formula for maltose?
3. **WHAT IF?** ➤ After a cow is given antibiotics to treat an infection, a vet gives the animal a drink of "gut culture" containing various prokaryotes. Why is this necessary?

For suggested answers, see Appendix A.

CONCEPT 5.3

Lipids are a diverse group of hydrophobic molecules

Lipids are the one class of large biological molecules that does not include true polymers, and they are generally not big enough to be considered macromolecules. The compounds called **lipids** are grouped with each other because they share one important trait: They mix poorly, if at all, with water. The hydrophobic behavior of lipids is based on their molecular structure. Although they may have some polar bonds associated with oxygen, lipids consist mostly of hydrocarbon regions. Lipids are varied in form and function. They include waxes and certain pigments, but we will focus on the types of lipids that are most important biologically: fats, phospholipids, and steroids.

MB Animation: Lipids

Fats

Although fats are not polymers, they are large molecules assembled from smaller molecules by dehydration reactions, like the dehydration reaction described for the polymerization of monomers in Figure 5.2a. A **fat** is constructed from two kinds of smaller molecules: glycerol and fatty acids **(Figure 5.9a)**. Glycerol is an alcohol; each of its three carbons bears a hydroxyl group. A **fatty acid** has a long carbon skeleton, usually 16 or 18 carbon atoms in length. The carbon at one end of the skeleton is part of a carboxyl group, the functional group that gives these molecules the name fatty *acid*. The rest of the skeleton consists of a hydrocarbon chain. The relatively nonpolar C—H bonds in the hydrocarbon chains of fatty acids are the reason fats are hydrophobic. Fats separate from water because the water molecules hydrogen-bond to one another and exclude the fats. This is why vegetable oil (a liquid fat) separates from the aqueous vinegar solution in a bottle of salad dressing.

In making a fat, three fatty acid molecules are each joined to glycerol by an ester linkage, a bond formed by a dehydration reaction between a hydroxyl group and a carboxyl group. The resulting fat, also called a **triacylglycerol**, thus consists of three fatty acids linked to one glycerol molecule. (Still another name for a fat is *triglyceride*, a word often found

Figure 5.9 The synthesis and structure of a fat, or triacylglycerol. The molecular building blocks of a fat are one molecule of glycerol and three molecules of fatty acids. **(a)** One water molecule is removed for each fatty acid joined to the glycerol. **(b)** A fat molecule with three fatty acid units, two of them identical. The carbons of the fatty acids are arranged zigzag to suggest the actual orientations of the four single bonds extending from each carbon (see Figures 4.3a and 4.6b).

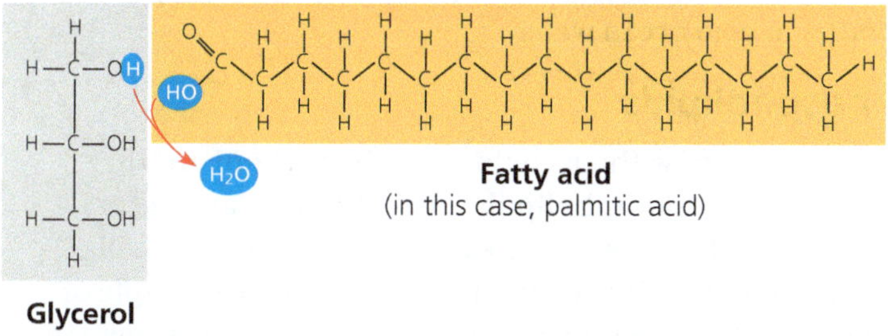

(a) One of three dehydration reactions in the synthesis of a fat

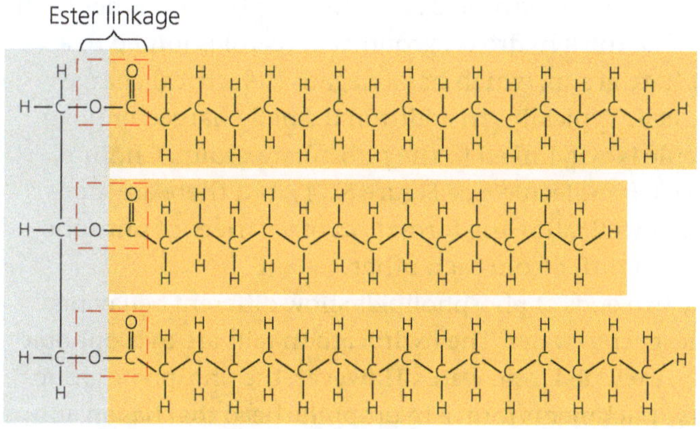

(b) Fat molecule (triacylglycerol)

in the list of ingredients on packaged foods.) The fatty acids in a fat can all be the same, or they can be of two or three different kinds, as in **Figure 5.9b**.

The terms *saturated* fats and *unsaturated* fats are commonly used in the context of nutrition **(Figure 5.10)**. These terms refer to the structure of the hydrocarbon chains of the fatty acids. If there are no double bonds between carbon atoms composing a chain, then as many hydrogen atoms as possible are bonded to the carbon skeleton. Such a structure is said to be *saturated* with hydrogen, and the resulting fatty acid is therefore called a **saturated fatty acid (Figure 5.10a)**. An **unsaturated fatty acid** has one or more double bonds, with one fewer hydrogen atom on each double-bonded carbon. Nearly every double bond in naturally occurring fatty acids is a *cis* double bond, which creates a kink in the hydrocarbon chain wherever it occurs **(Figure 5.10b)**. (See Figure 4.7b to remind yourself about *cis* and *trans* double bonds.)

A fat made from saturated fatty acids is called a saturated fat. Most animal fats are saturated: The hydrocarbon chains of their fatty acids—the "tails" of the fat molecules—lack double bonds, and their flexibility allows the fat molecules to pack together tightly. Saturated animal fats—such as lard

Figure 5.10 Saturated and unsaturated fats and fatty acids.

(a) Saturated fat

At room temperature, the molecules of a saturated fat, such as the fat in butter, are packed closely together, forming a solid.

Structural formula of a saturated fat molecule (Each hydrocarbon chain is represented as a zigzag line, where each bend represents a carbon atom; hydrogens are not shown.)

Space-filling model of stearic acid, a saturated fatty acid (red = oxygen, black = carbon, gray = hydrogen)

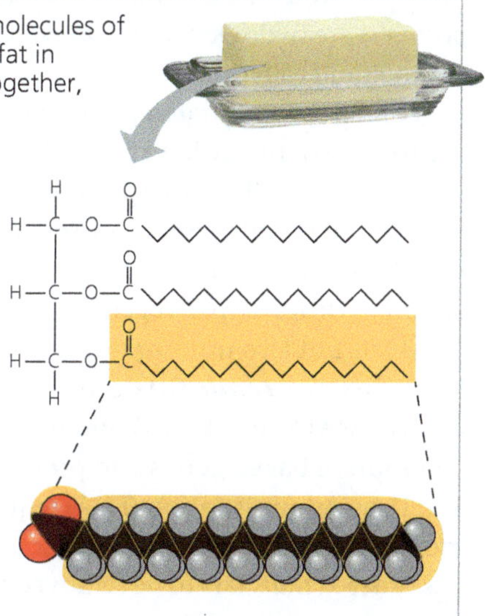

(b) Unsaturated fat

At room temperature, the molecules of an unsaturated fat such as olive oil cannot pack together closely enough to solidify because of the kinks in some of their fatty acid hydrocarbon chains.

Structural formula of an unsaturated fat molecule

Space-filling model of oleic acid, an unsaturated fatty acid

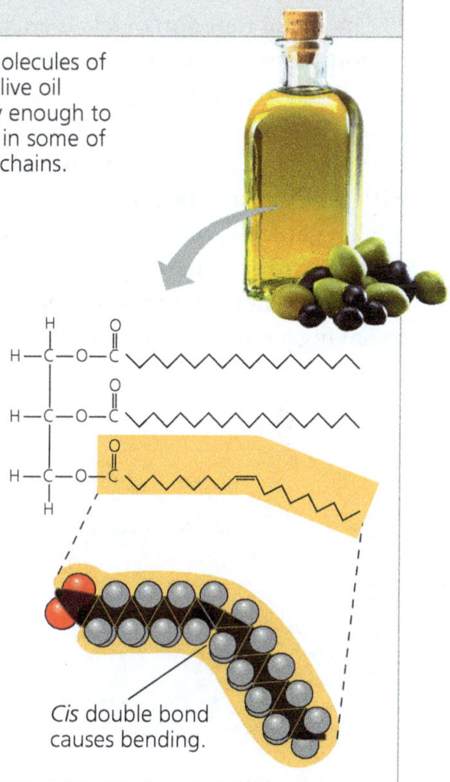

and butter—are solid at room temperature. In contrast, the fats of plants and fishes are generally unsaturated, meaning that they are built of one or more types of unsaturated fatty acids. Usually liquid at room temperature, plant and fish fats are referred to as oils—olive oil and cod liver oil are examples. The kinks where the *cis* double bonds are located prevent the molecules from packing together closely enough to solidify at room temperature. The phrase "hydrogenated vegetable oils" on food labels means that unsaturated fats have been synthetically converted to saturated fats by adding hydrogen, allowing them to solidify. Peanut butter, margarine, and many other products are hydrogenated to prevent lipids from separating out in liquid (oil) form.

A diet rich in saturated fats is one of several factors that may contribute to the cardiovascular disease known as atherosclerosis. In this condition, deposits called plaques develop within the walls of blood vessels, causing inward bulges that impede blood flow and reduce the resilience of the vessels. The process of hydrogenating vegetable oils produces not only saturated fats but also unsaturated fats with *trans* double bonds. It appears that ***trans* fats** can contribute to coronary heart disease (see Concept 42.4). Because *trans* fats are especially common in baked goods and processed foods, the U.S. Food and Drug Administration (FDA) requires nutritional labels to include information on *trans* fat content. In addition, the FDA has ordered *trans* fats to be removed from the U.S. food supply by 2018. Some countries, such as Denmark and Switzerland, have already banned *trans* fats in foods.

The major function of fats is energy storage. The hydrocarbon chains of fats are similar to gasoline molecules and just as rich in energy. A gram of fat stores more than twice as much energy as a gram of a polysaccharide, such as starch. Because plants are relatively immobile, they can function with bulky energy storage in the form of starch. (Vegetable oils are generally obtained from seeds, where more compact storage is an asset to the plant.) Animals, however, must carry their energy stores with them, so there is an advantage to having a more compact reservoir of fuel—fat. Humans and other mammals stock their long-term food reserves in adipose cells (see Figure 4.6a), which swell and shrink as fat is deposited and withdrawn from storage. In addition to storing energy, adipose tissue also cushions such vital organs as the kidneys, and a layer of fat beneath the skin insulates the body. This subcutaneous layer is especially thick in whales, seals, and most other marine mammals, insulating their bodies in cold ocean water.

Phospholipids

Cells as we know them could not exist without another type of lipid—phospholipids. Phospholipids are essential for cells because they are major constituents of cell membranes. Their structure provides a classic example of how form fits function at the molecular level. As shown in **Figure 5.11**, a **phospholipid** is similar to a fat molecule but has only two fatty acids attached to glycerol rather than three. The third hydroxyl group of glycerol is joined to a phosphate group, which has a negative electrical charge in the cell. Typically, an additional small charged or polar molecule is also linked to the phosphate group. Choline is one such molecule (see Figure 5.11), but there are many others as well, allowing formation of a variety of phospholipids that differ from each other.

The two ends of phospholipids show different behaviors with respect to water. The hydrocarbon tails are hydrophobic and are excluded from water. However, the phosphate group and its attachments form a hydrophilic head that has an affinity for water. When phospholipids are added to water, they self-assemble into a double-layered sheet called a "bilayer"

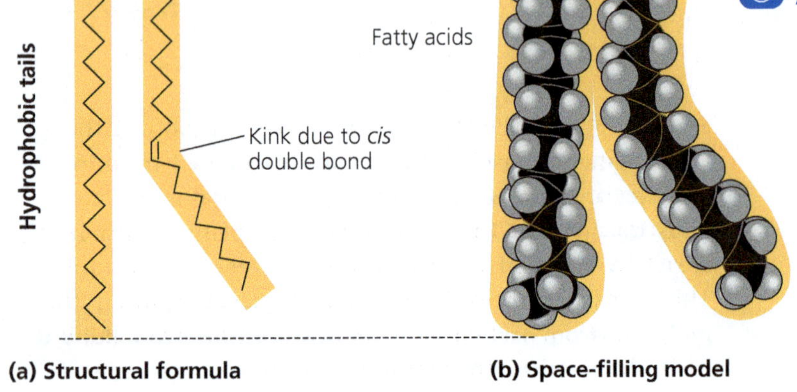

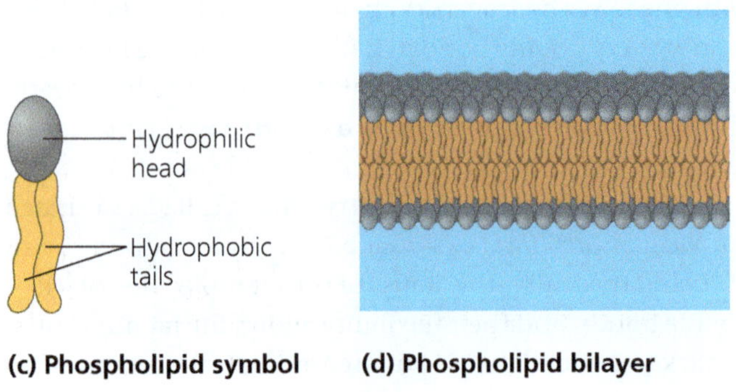

▼ Figure 5.11 The structure of a phospholipid. A phospholipid has a hydrophilic (polar) head and two hydrophobic (nonpolar) tails. This particular phospholipid, called a phosphatidylcholine, has a choline attached to a phosphate group. Shown here are **(a)** the structural formula, **(b)** the space-filling model (yellow = phosphorus, blue = nitrogen), **(c)** the symbol for a phospholipid that will appear throughout this book, and **(d)** the bilayer structure formed by self-assembly of phospholipids in an aqueous environment.

DRAW IT ➤ *Draw an oval around the hydrophilic head of the space-filling model.*

Figure Walkthrough
Animation: Space-Filling Model of a Phospholipid

that shields their hydrophobic fatty acid tails from water (Figure 5.11d).

At the surface of a cell, phospholipids are arranged in a similar bilayer. The hydrophilic heads of the molecules are on the outside of the bilayer, in contact with the aqueous solutions inside and outside of the cell. The hydrophobic tails point toward the interior of the bilayer, away from the water. The phospholipid bilayer forms a boundary between the cell and its external environment and establishes separate compartments within eukaryotic cells; in fact, the existence of cells depends on the properties of phospholipids.

Steroids

Steroids are lipids characterized by a carbon skeleton consisting of four fused rings. Different steroids are distinguished by the particular chemical groups attached to this ensemble of rings. **Cholesterol**, a type of steroid, is a crucial molecule in animals **(Figure 5.12)**. It is a common component of animal cell membranes and is also the precursor from which other steroids, such as the vertebrate sex hormones, are synthesized. In vertebrates, cholesterol is synthesized in the liver and is also obtained from the diet. A high level of cholesterol in the blood may contribute to atherosclerosis, although some researchers are questioning the roles of cholesterol and saturated fats in the development of this condition.

▼ Figure 5.12 Cholesterol, a steroid. Cholesterol is the molecule from which other steroids, including the sex hormones, are synthesized. Steroids vary in the chemical groups attached to their four interconnected rings (shown in gold).

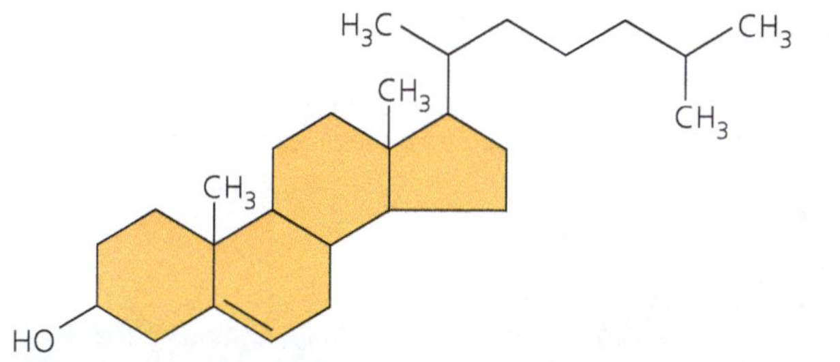

MAKE CONNECTIONS ➤ *Compare cholesterol with the sex hormones shown in the figure at the beginning of Concept 4.3. Circle the chemical groups that cholesterol has in common with estradiol; put a square around the chemical groups that cholesterol has in common with testosterone.*

Interview with Lovell Jones: Investigating the effects of sex hormones on cancer (see the interview before Chapter 2)

CONCEPT CHECK 5.3

1. Compare the structure of a fat (triglyceride) with that of a phospholipid.
2. Why are human sex hormones considered lipids?
3. **WHAT IF? ➤** Suppose a membrane surrounded an oil droplet, as it does in the cells of plant seeds and in some animal cells. Describe and explain the form it might take.

For suggested answers, see Appendix A.

CONCEPT 5.4

Proteins include a diversity of structures, resulting in a wide range of functions

Nearly every dynamic function of a living being depends on proteins. In fact, the importance of proteins is underscored by their name, which comes from the Greek word *proteios*, meaning "first," or "primary." Proteins account for more than 50% of the dry mass of most cells, and they are instrumental in almost everything organisms do. Some proteins speed up chemical reactions, while others play a role in defense, storage, transport, cellular communication, movement, or structural support. **Figure 5.13** shows examples of proteins with these functions, which you'll learn more about in later chapters.

Life would not be possible without enzymes, most of which are proteins. Enzymatic proteins regulate metabolism by acting as **catalysts**, chemical agents that selectively speed up chemical reactions without being consumed in the reaction. Because an enzyme can perform its function over and over again, these molecules can be thought of as workhorses that keep cells running by carrying out the processes of life.

A human has tens of thousands of different proteins, each with a specific structure and function; proteins, in fact, are the most structurally sophisticated molecules known. Consistent with their diverse functions, they vary extensively in structure, each type of protein having a unique three-dimensional shape.

Proteins are all constructed from the same set of 20 amino acids, linked in unbranched polymers. The bond between amino acids is called a *peptide bond*, so a polymer of amino acids is called a **polypeptide**. A **protein** is a biologically functional molecule made up of one or more polypeptides, each folded and coiled into a specific three-dimensional structure.

Amino Acid Monomers

All amino acids share a common structure. An **amino acid** is an organic molecule with both an amino group and a carboxyl group (see Figure 4.9); the small figure shows the general formula for an amino acid. At the center of the amino acid is an asymmetric carbon atom called the *alpha* (α) *carbon*. Its four different partners are an amino group, a carboxyl group, a hydrogen atom, and a variable group symbolized by R. The R group, also called the side chain, differs with each amino acid.

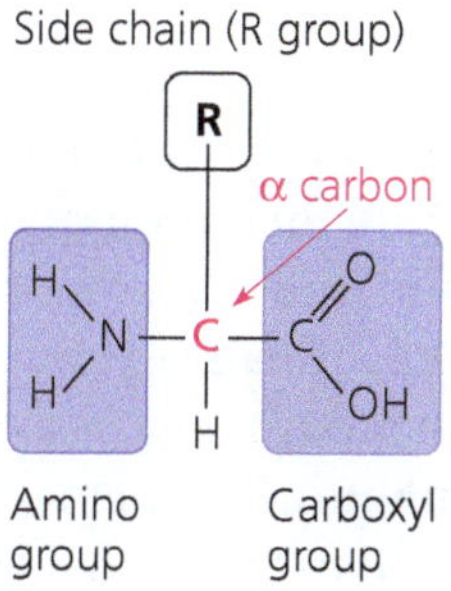

Figure 5.13 An overview of protein functions.

Animation: Protein Functions

Enzymatic proteins

Function: Selective acceleration of chemical reactions

Example: Digestive enzymes catalyze the hydrolysis of bonds in food molecules.

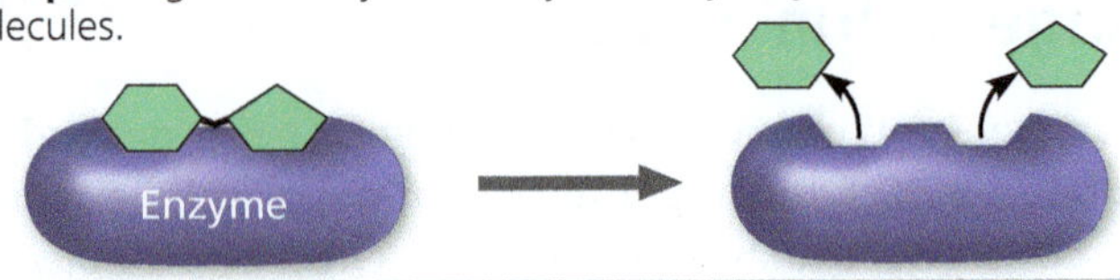

Defensive proteins

Function: Protection against disease

Example: Antibodies inactivate and help destroy viruses and bacteria.

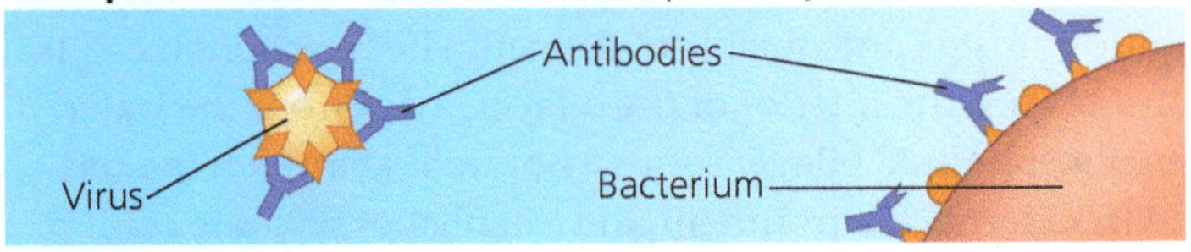

Storage proteins

Function: Storage of amino acids

Examples: Casein, the protein of milk, is the major source of amino acids for baby mammals. Plants have storage proteins in their seeds. Ovalbumin is the protein of egg white, used as an amino acid source for the developing embryo.

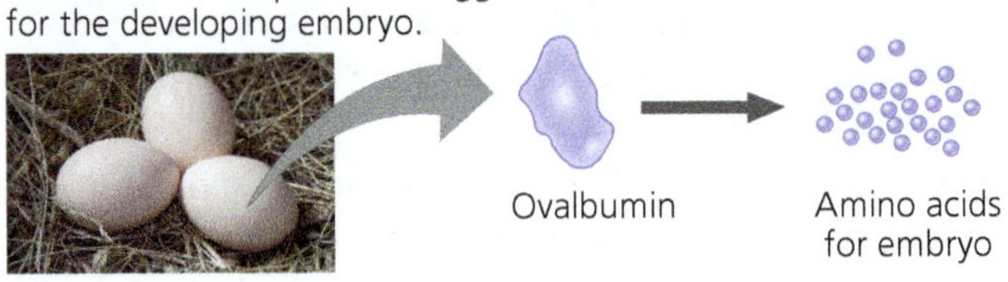

Transport proteins

Function: Transport of substances

Examples: Hemoglobin, the iron-containing protein of vertebrate blood, transports oxygen from the lungs to other parts of the body. Other proteins transport molecules across membranes, as shown here.

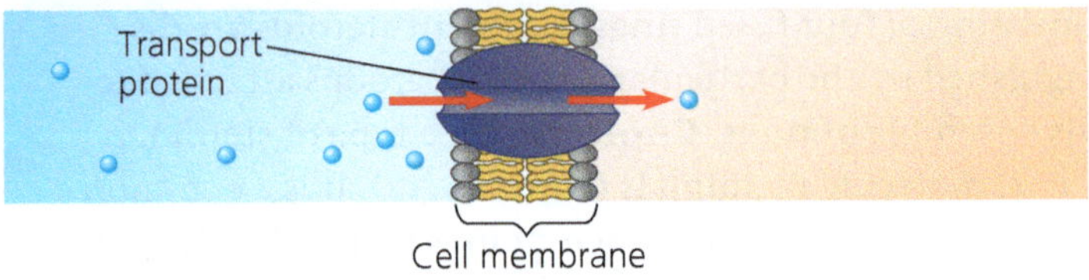

Hormonal proteins

Function: Coordination of an organism's activities

Example: Insulin, a hormone secreted by the pancreas, causes other tissues to take up glucose, thus regulating blood sugar concentration.

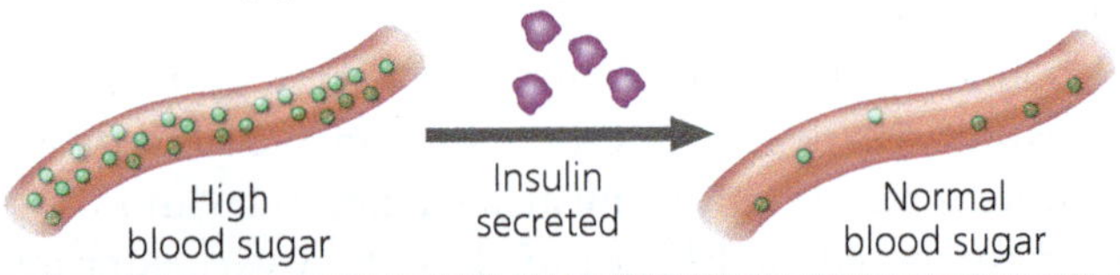

Receptor proteins

Function: Response of cell to chemical stimuli

Example: Receptors built into the membrane of a nerve cell detect signaling molecules released by other nerve cells.

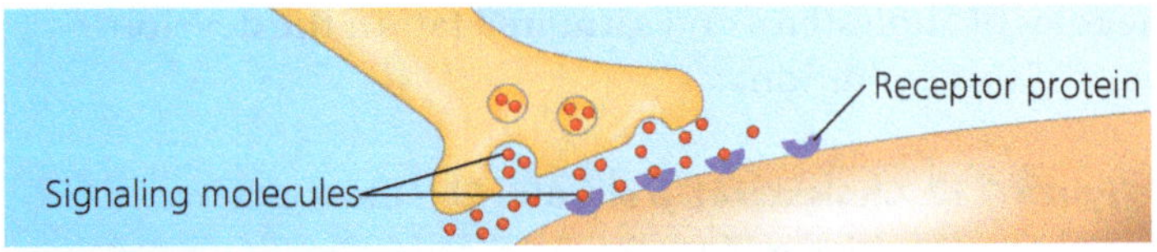

Contractile and motor proteins

Function: Movement

Examples: Motor proteins are responsible for the undulations of cilia and flagella. Actin and myosin proteins are responsible for the contraction of muscles.

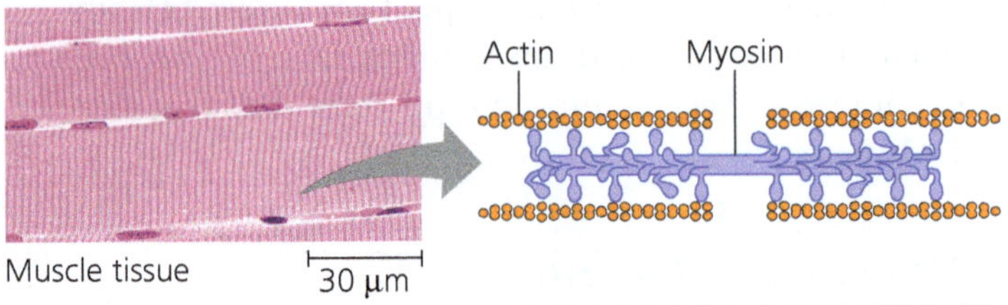

Structural proteins

Function: Support

Examples: Keratin is the protein of hair, horns, feathers, and other skin appendages. Insects and spiders use silk fibers to make their cocoons and webs, respectively. Collagen and elastin proteins provide a fibrous framework in animal connective tissues.

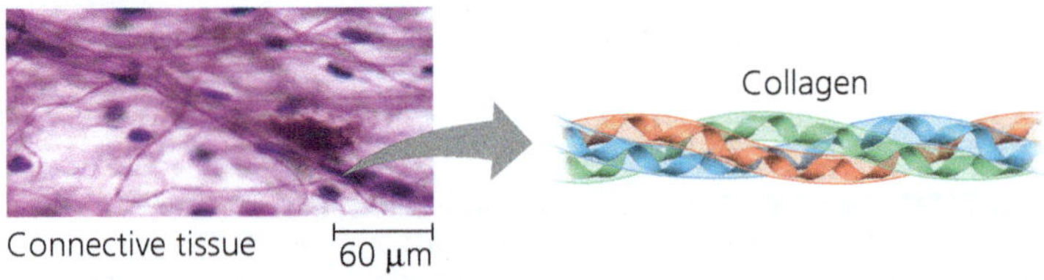

Figure 5.14 shows the 20 amino acids that cells use to build their thousands of proteins. Here the amino groups and carboxyl groups are all depicted in ionized form, the way they usually exist at the pH found in a cell. The side chain (R group) may be as simple as a hydrogen atom, as in the amino acid glycine, or it may be a carbon skeleton with various functional groups attached, as in glutamine.

The physical and chemical properties of the side chain determine the unique characteristics of a particular amino acid, thus affecting its functional role in a polypeptide. In Figure 5.14, the amino acids are grouped according to the properties of their side chains. One group consists of amino acids with nonpolar side chains, which are hydrophobic. Another group consists of amino acids with polar side chains, which are hydrophilic. Acidic amino acids have side chains that are generally negative in charge due to the presence of a carboxyl group, which is usually dissociated (ionized) at cellular pH. Basic amino acids have amino groups in their side chains that are generally positive in charge. (Notice that *all* amino acids have carboxyl groups and amino groups; the terms *acidic* and *basic* in this context refer only to groups in the side chains.) Because they are charged, acidic and basic side chains are also hydrophilic.

21 - Selenocysteine (new - when ?)

Figure 5.14 The 20 amino acids of proteins. The amino acids are grouped here according to the properties of their side chains (R groups) and shown in their prevailing ionic forms at pH 7.2, the pH within a cell. The three-letter and one-letter abbreviations for the amino acids are in parentheses. All of the amino acids used in proteins are L enantiomers (see Figure 4.7c).

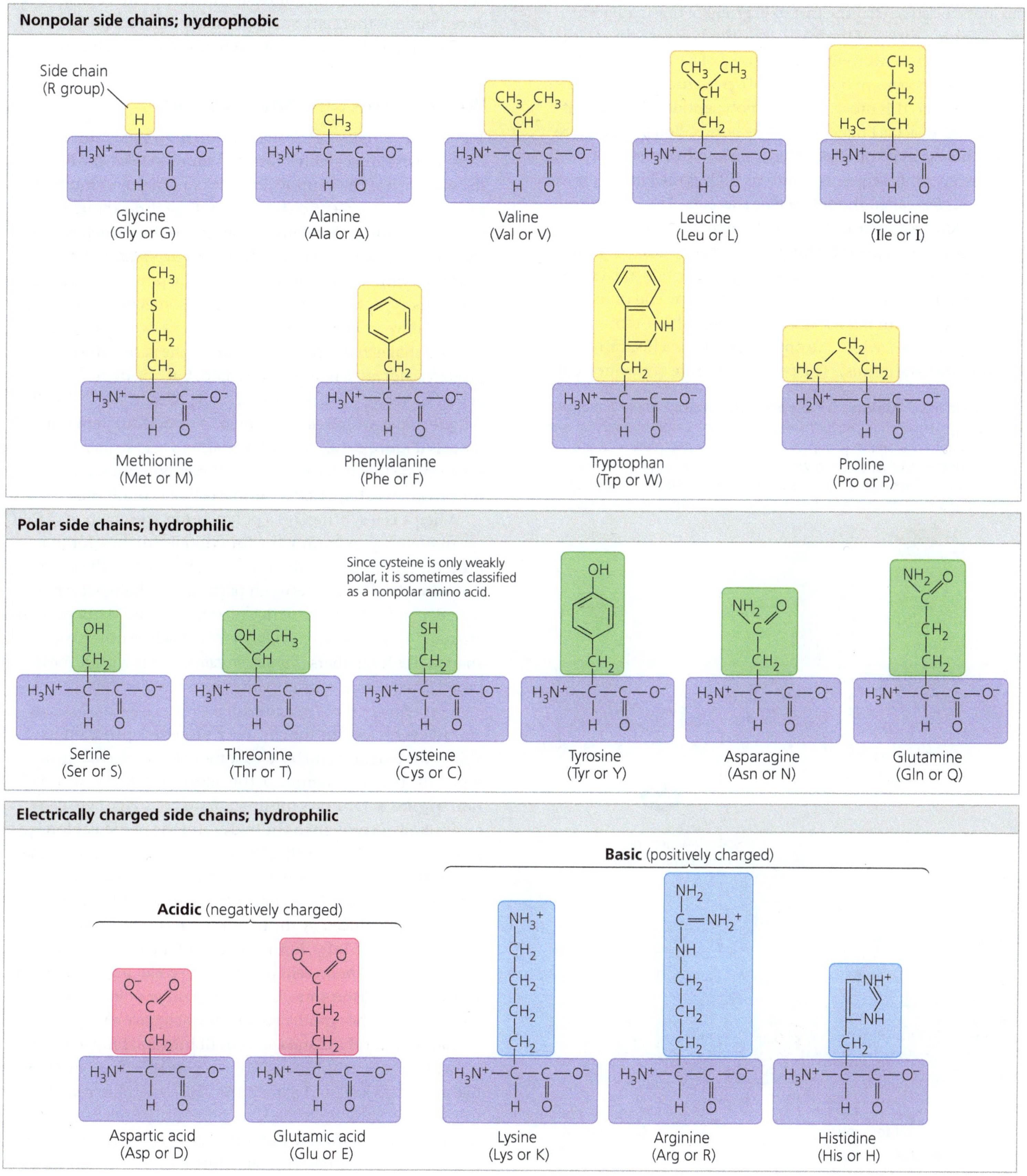

Polypeptides (Amino Acid Polymers)

Now that we have examined amino acids, let's see how they are linked to form polymers **(Figure 5.15)**. When two amino acids are positioned so that the carboxyl group of one is adjacent to the amino group of the other, they can become joined by a dehydration reaction, with the removal of a water molecule. The resulting covalent bond is called a **peptide bond**. Repeated over and over, this process yields a polypeptide, a polymer of many amino acids linked by peptide bonds. You'll learn more about how cells synthesize polypeptides in Concept 17.4.

The repeating sequence of atoms highlighted in purple in Figure 5.15 is called the *polypeptide backbone*. Extending from this backbone are the different side chains (R groups) of the amino acids. Polypeptides range in length from a few amino acids to 1,000 or more. Each specific polypeptide has a unique linear sequence of amino acids. Note that one end of the polypeptide chain has a free amino group (the N-terminus of the polypeptide), while the opposite end has a free carboxyl group (the C-terminus). The chemical nature of the molecule as a whole is determined by the kind and sequence of the side chains, which determine how a polypeptide folds and thus its final shape and chemical characteristics. The immense variety of polypeptides in nature illustrates an important concept introduced earlier—that cells can make many different polymers by linking a limited set of monomers into diverse sequences.

▼ Figure 5.15 Making a polypeptide chain. Peptide bonds are formed by dehydration reactions, which link the carboxyl group of one amino acid to the amino group of the next. The peptide bonds are formed one at a time, starting with the amino acid at the amino end (N-terminus). The polypeptide has a repetitive backbone (purple) to which the amino acid side chains (yellow and green) are attached.

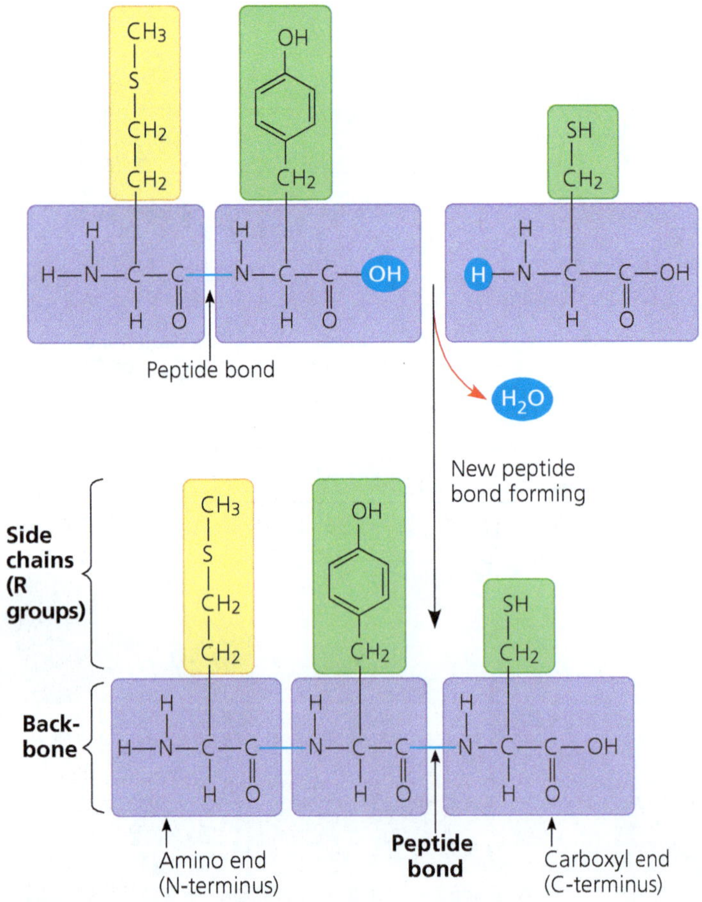

DRAW IT ➤ *Label the three amino acids in the upper part of the figure using three-letter and one-letter codes. Circle and label the carboxyl and amino groups that will form the new peptide bond.*

Protein Structure and Function

The specific activities of proteins result from their intricate three-dimensional architecture, the simplest level of which is the sequence of their amino acids. What can the amino acid sequence of a polypeptide tell us about the three-dimensional structure (commonly referred to simply as the "structure") of the protein and its function? The term *polypeptide* is not synonymous with the term *protein*. Even for a protein consisting of a single polypeptide, the relationship is somewhat analogous to that between a long strand of yarn and a sweater of particular size and shape that can be knitted from the yarn. A functional protein is not *just* a polypeptide chain, but one or more polypeptides precisely twisted, folded, and coiled into a molecule of unique shape, which can be shown in several different types of models **(Figure 5.16)**. And it is the amino acid sequence of each polypeptide that determines what three-dimensional structure the protein will have under normal cellular conditions.

When a cell synthesizes a polypeptide, the chain may fold spontaneously, assuming the functional structure for that protein. This folding is driven and reinforced by the formation of various bonds between parts of the chain, which in turn depends on the sequence of amino acids. Many proteins are roughly spherical (*globular proteins*), while others are shaped like long fibers (*fibrous proteins*). Even within these broad categories, countless variations exist.

A protein's specific structure determines how it works. In almost every case, the function of a protein depends on its ability to recognize and bind to some other molecule. In an especially striking example of the marriage of form and function, **Figure 5.17** shows the exact match of shape between an antibody (a protein in the body) and the particular foreign substance on a flu virus that the antibody binds to and marks for destruction. Also, you may recall another example of molecules with matching shapes from Concept 2.3: endorphin molecules (produced by the body) and morphine molecules (a manufactured drug), both of which fit into receptor proteins on the surface of brain cells in humans, producing euphoria and relieving pain. Morphine, heroin, and other opiate drugs are able to mimic endorphins because they all have a shape similar to that of endorphins and can thus fit into and bind to endorphin receptors in the brain. This fit is very specific, something like a lock and key (see Figure 2.16). The endorphin receptor, like other receptor molecules, is a protein. The function of a protein—for instance, the ability of a receptor protein to bind to a particular pain-relieving signaling molecule—is an emergent property resulting from exquisite molecular order.

▼ Figure 5.16 Visualizing Proteins

Proteins can be represented in different ways, depending on the goal of the illustration.

Structural Models

Using data from structural studies of proteins, computers can generate various types of models. Each model emphasizes a different aspect of the protein's structure, but no model can show what a protein actually looks like. These three models depict lysozyme, a protein in tears and saliva that helps prevent infection by binding to target molecules on bacteria.

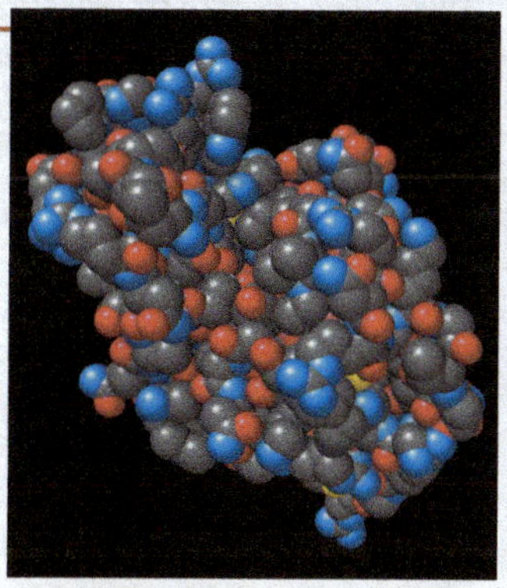

Space-filling model: Shows all the atoms of the protein (except hydrogen), emphasizing the overall globular shape. The atoms are color-coded: gray = carbon, red = oxygen, blue = nitrogen, and yellow = sulfur.

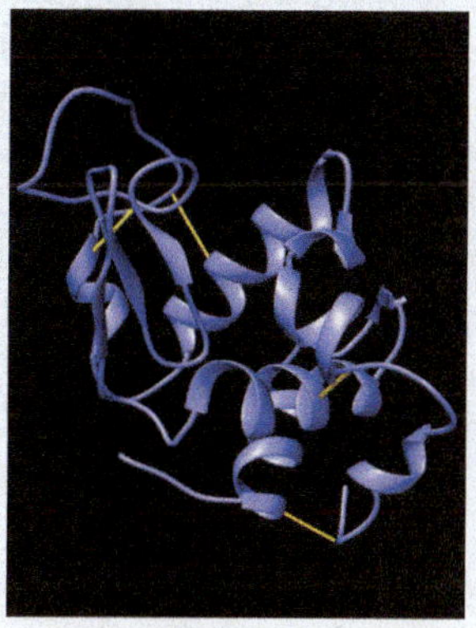

Ribbon model: Shows only the backbone of the polypeptide, emphasizing how it folds and coils to form a 3-D shape, in this case stabilized by disulfide bridges (yellow lines).

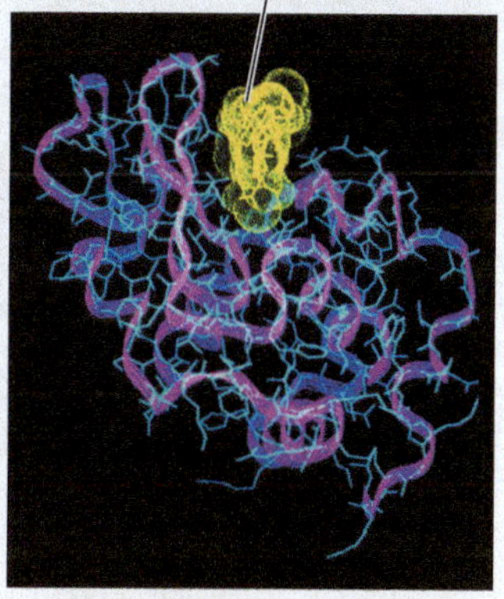

Wireframe model (blue): Shows the backbone of the polypeptide chain with side chains (R groups) extending from it (see Figure 5.15). A ribbon model (purple) is superimposed on the wireframe model.

1 *In which model is it easiest to follow the polypeptide backbone?*

MB **Instructors:** The tutorial "Molecular Model: Lysozyme," in which students rotate 3-D models of lysozyme, can be assigned in MasteringBiology.

Simplified Diagrams

It isn't always necessary to use a detailed computer model; simplified diagrams are useful when the focus of the figure is on the function of the protein, not the structure.

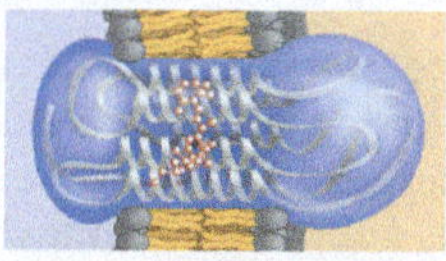

In this diagram of the protein rhodopsin, a simple transparent shape is drawn around the contours of a ribbon model, showing the overall shape of the molecule as well as some internal details.

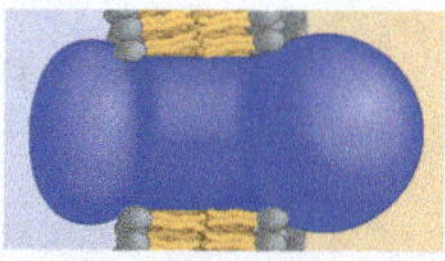

When structural details are not needed, a solid shape can be used to represent a protein.

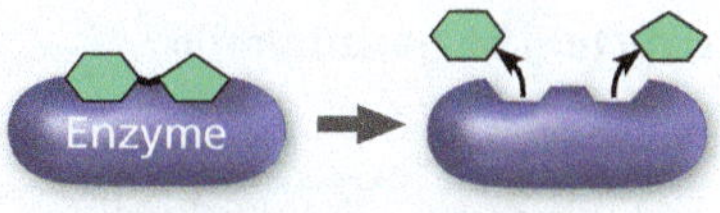

A simple shape is used here to represent a generic enzyme because the diagram focuses on enzyme action in general.

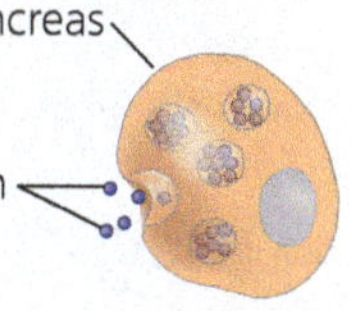

Sometimes a protein is represented simply as a dot, as shown here for insulin.

2 *Draw a simple version of lysozyme that shows its overall shape, based on the molecular models in the top section of the figure.*

3 *Why is it unnecessary to show the actual shape of insulin here?*

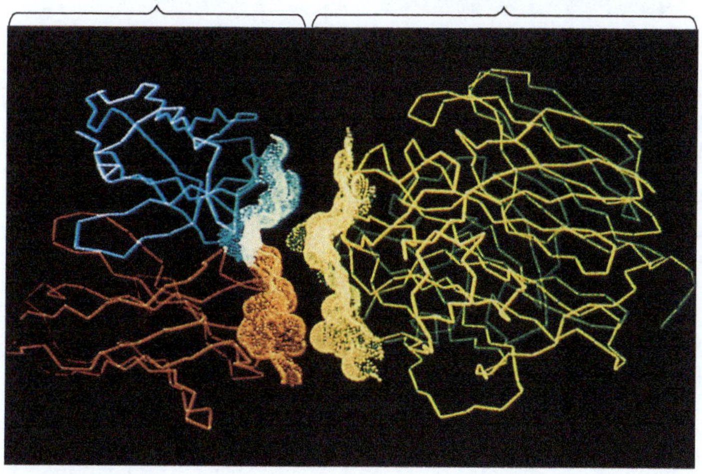

➤ Figure 5.17 Complementarity of shape between two protein surfaces. A technique called X-ray crystallography was used to generate a computer model of an antibody protein (blue and orange, left) bound to a flu virus protein (yellow and green, right). This is a wireframe model modified by adding an "electron density map" in the region where the two proteins meet. Computer software was then used to back the images away from each other slightly.

VISUAL SKILLS ➤ *What do these computer models allow you to see about the two proteins?*

Four Levels of Protein Structure

In spite of their great diversity, proteins share three superimposed levels of structure, known as primary, secondary, and tertiary structure. A fourth level, quaternary structure, arises when a protein consists of two or more polypeptide chains. **Figure 5.18** describes these four levels of protein structure. Be sure to study this figure thoroughly before going on to the next section.

MB **Animation: Protein Structure**

▼ Figure 5.18 Exploring Levels of Protein Structure

Primary Structure

Linear chain of amino acids

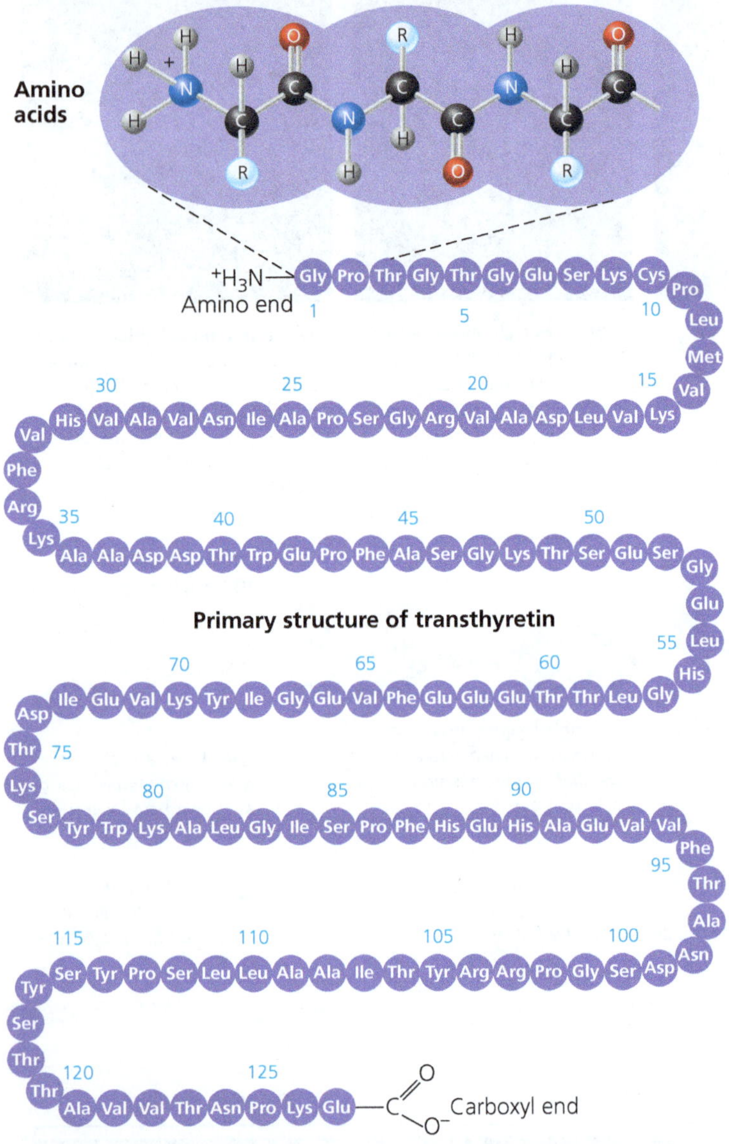

The **primary structure** of a protein is its sequence of amino acids. As an example, let's consider transthyretin, a globular blood protein that transports vitamin A and one of the thyroid hormones throughout the body. Transthyretin is made up of four identical polypeptide chains, each composed of 127 amino acids. Shown here is one of these chains unraveled for a closer look at its primary structure. Each of the 127 positions along the chain is occupied by one of the 20 amino acids, indicated here by its three-letter abbreviation.

The primary structure is like the order of letters in a very long word. If left to chance, there would be 20^{127} different ways of making a polypeptide chain 127 amino acids long. However, the precise primary structure of a protein is determined not by the random linking of amino acids, but by inherited genetic information. The primary structure in turn dictates secondary and tertiary structure, due to the chemical nature of the backbone and the side chains (R groups) of the amino acids along the polypeptide.

Secondary Structure

Regions stabilized by hydrogen bonds between atoms of the polypeptide backbone

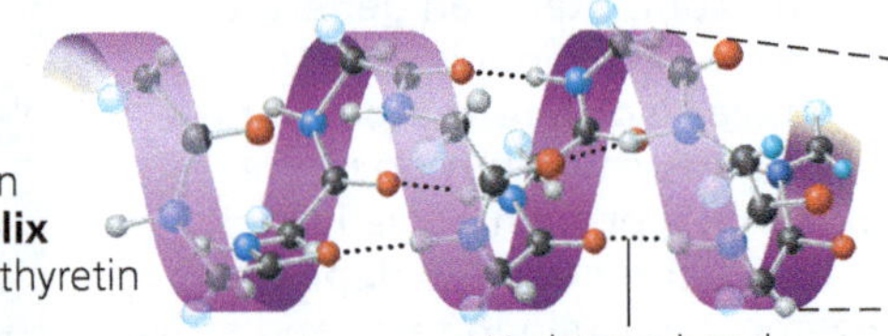

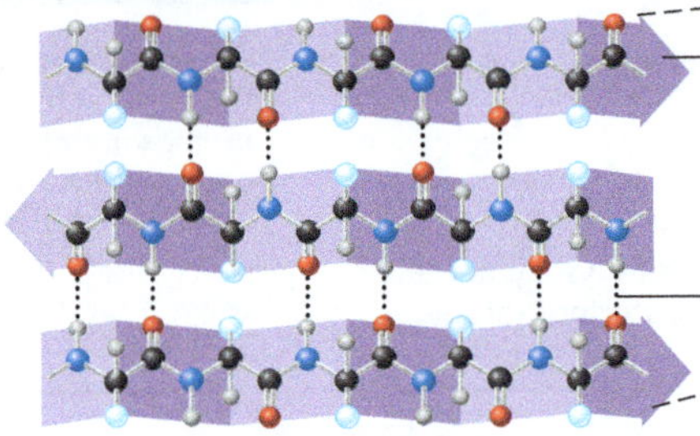

Most proteins have segments of their polypeptide chains repeatedly coiled or folded in patterns that contribute to the protein's overall shape. These coils and folds, collectively referred to as **secondary structure**, are the result of hydrogen bonds between the repeating constituents of the polypeptide backbone (not the amino acid side chains). Within the backbone, the oxygen atoms have a partial negative charge, and the hydrogen atoms attached to the nitrogens have a partial positive charge (see Figure 2.14); therefore, hydrogen bonds can form between these atoms. Individually, these hydrogen bonds are weak, but because they are repeated many times over a relatively long region of the polypeptide chain, they can support a particular shape for that part of the protein.

One such secondary structure is the **α helix**, a delicate coil held together by hydrogen bonding between every fourth amino acid, as shown above. Although each transthyretin polypeptide has only one α helix region (see the Tertiary Structure section), other globular proteins have multiple stretches of α helix separated by nonhelical regions (see hemoglobin in the Quaternary Structure section). Some fibrous proteins, such as α-keratin, the structural protein of hair, have the α helix formation over most of their length.

The other main type of secondary structure is the **β pleated sheet**. As shown above, in this structure two or more segments of the polypeptide chain lying side by side (called β strands) are connected by hydrogen bonds between parts of the two parallel segments of polypeptide backbone. β pleated sheets make up the core of many globular proteins, as is the case for transthyretin (see Tertiary Structure), and dominate some fibrous proteins, including the silk protein of a spider's web. The teamwork of so many hydrogen bonds makes each spider silk fiber stronger than a steel strand of the same weight.

▶ Spiders secrete silk fibers made of a structural protein containing β pleated sheets, which allow the spider web to stretch and recoil.

Tertiary Structure

Three-dimensional shape stabilized by interactions between side chains

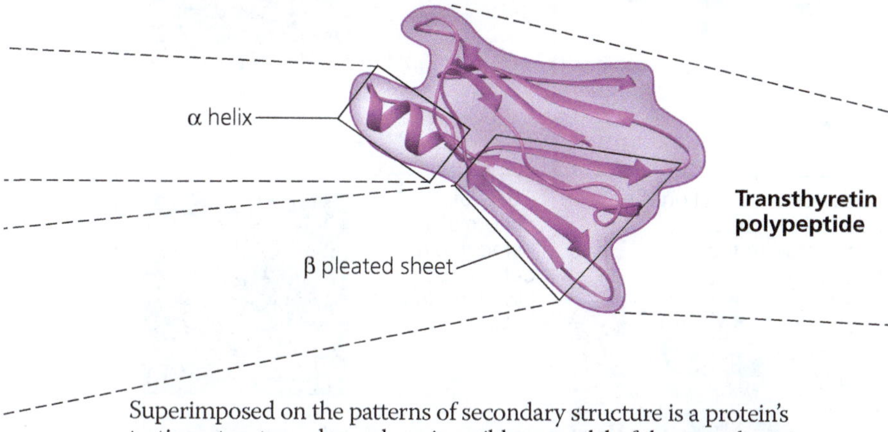

Superimposed on the patterns of secondary structure is a protein's tertiary structure, shown here in a ribbon model of the transthyretin polypeptide. While secondary structure involves interactions between backbone constituents, **tertiary structure** is the overall shape of a polypeptide resulting from interactions between the side chains (R groups) of the various amino acids. One type of interaction that contributes to tertiary structure is called—somewhat misleadingly—a **hydrophobic interaction**. As a polypeptide folds into its functional shape, amino acids with hydrophobic (nonpolar) side chains usually end up in clusters at the core of the protein, out of contact with water. Thus, a "hydrophobic interaction" is actually caused by the exclusion of nonpolar substances by water molecules. Once nonpolar amino acid side chains are close together, van der Waals interactions help hold them together. Meanwhile, hydrogen bonds between polar side chains and ionic bonds between positively and negatively charged side chains also help stabilize tertiary structure. These are all weak interactions in the aqueous cellular environment, but their cumulative effect helps give the protein a unique shape.

Covalent bonds called **disulfide bridges** may further reinforce the shape of a protein. Disulfide bridges form where two cysteine monomers, which have sulfhydryl groups (—SH) on their side chains (see Figure 4.9), are brought close together by the folding of the protein. The sulfur of one cysteine bonds to the sulfur of the second, and the disulfide bridge (—S—S—) rivets parts of the protein together (see yellow lines in Figure 5.16 ribbon model). All of these different kinds of interactions can contribute to the tertiary structure of a protein, as shown here in a small part of a hypothetical protein:

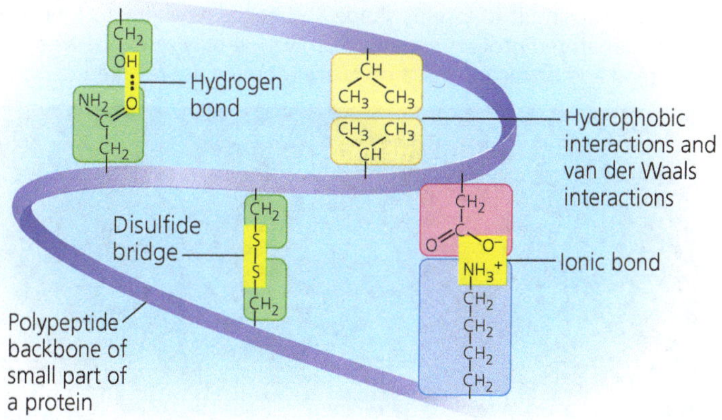

Quaternary Structure

Association of two or more polypeptides (some proteins only)

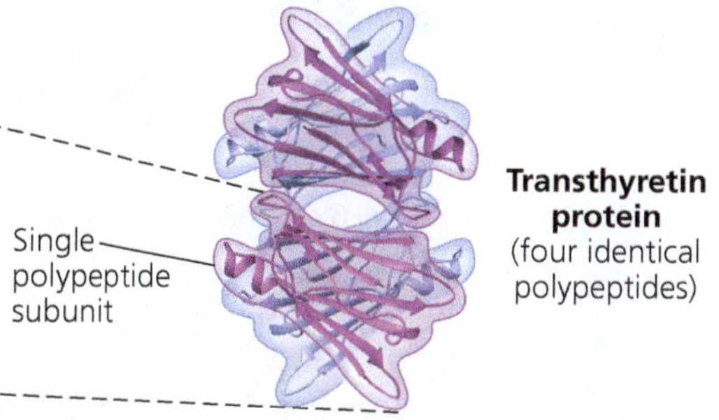

Some proteins consist of two or more polypeptide chains aggregated into one functional macromolecule. **Quaternary structure** is the overall protein structure that results from the aggregation of these polypeptide subunits. For example, shown above is the complete globular transthyretin protein, made up of its four polypeptides.

Another example is collagen, which is a fibrous protein that has three identical helical polypeptides intertwined into a larger triple helix, giving the long fibers great strength. This suits collagen fibers to their function as the girders of connective tissue in skin, bone, tendons, ligaments, and other body parts. (Collagen accounts for 40% of the protein in a human body.)

Hemoglobin, the oxygen-binding protein of red blood cells, is another example of a globular protein with quaternary structure. It consists of four polypeptide subunits, two of one kind (α) and two of another kind (β). Both α and β subunits consist primarily of α-helical secondary structure. Each subunit has a nonpolypeptide component, called heme, with an iron atom that binds oxygen.

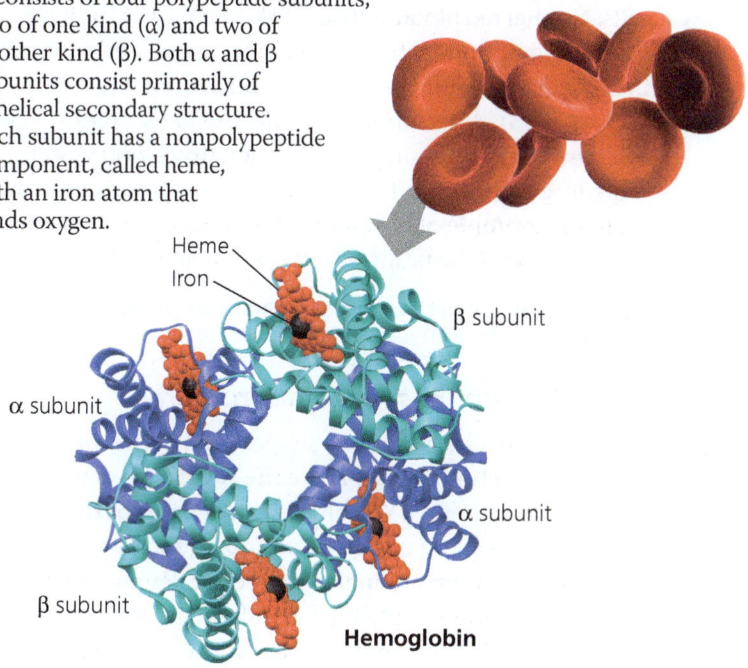

Animation: Proteins
MP3 Tutor: Protein Structure and Function

Figure 5.19 A single amino acid substitution in a protein causes sickle-cell disease.

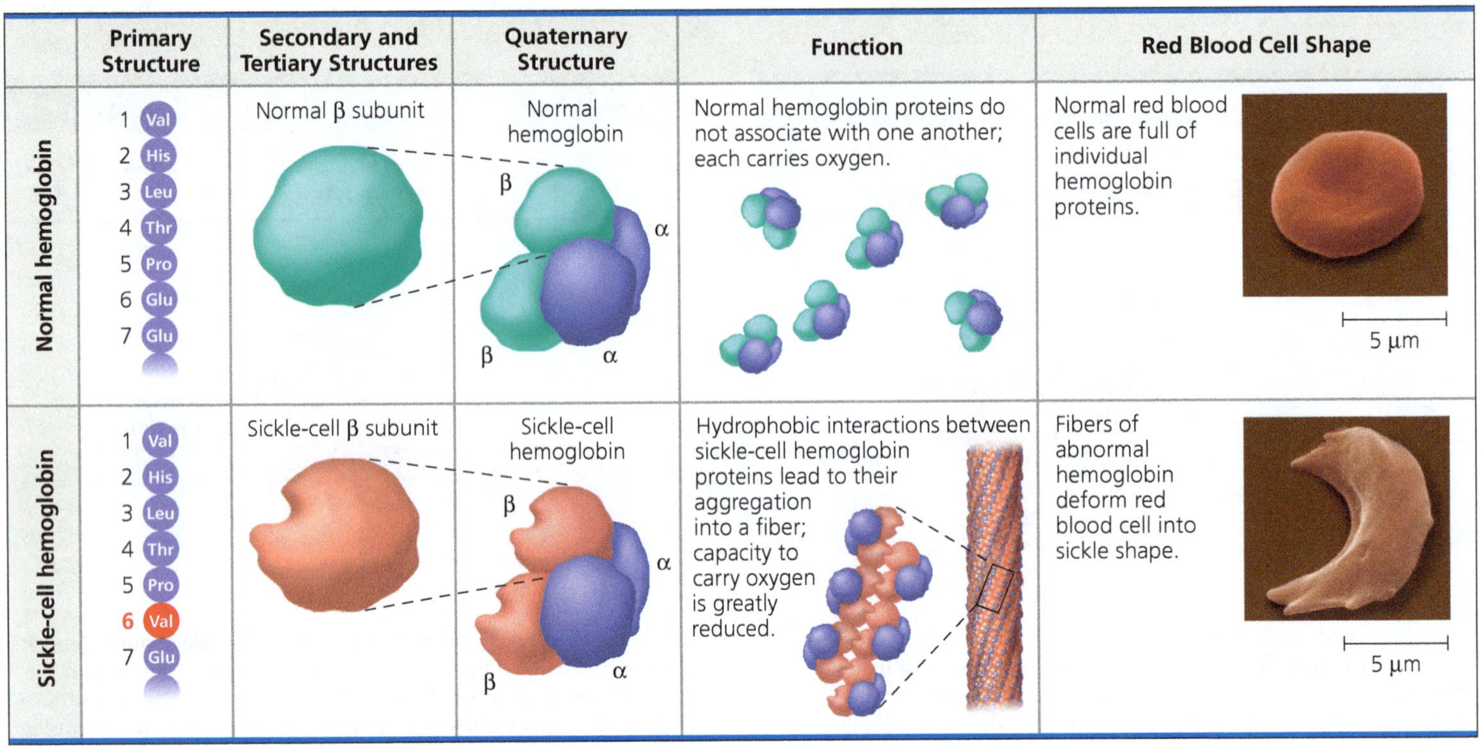

MAKE CONNECTIONS ➤ *Considering the chemical characteristics of the amino acids valine and glutamic acid (see Figure 5.14), propose a possible explanation for the dramatic effect on protein function that occurs when valine is substituted for glutamic acid.*

MB **HHMI Animation: Sickle-Cell Disease** hhmi BioInteractive

Sickle-Cell Disease: A Change in Primary Structure

Even a slight change in primary structure can affect a protein's shape and ability to function. For instance, **sickle-cell disease**, an inherited blood disorder, is caused by the substitution of one amino acid (valine) for the normal one (glutamic acid) at the position of the sixth amino acid in the primary structure of hemoglobin, the protein that carries oxygen in red blood cells. Normal red blood cells are disk-shaped, but in sickle-cell disease, the abnormal hemoglobin molecules tend to aggregate into chains, deforming some of the cells into a sickle shape **(Figure 5.19)**. A person with the disease has periodic "sickle-cell crises" when the angular cells clog tiny blood vessels, impeding blood flow. The toll taken on such patients is a dramatic example of how a simple change in protein structure can have devastating effects on protein function.

MB **Interview with Linus Pauling: Winner of the Nobel Prize in Chemistry and the Nobel Peace Prize**

What Determines Protein Structure?

You've learned that a unique shape endows each protein with a specific function. But what are the key factors determining protein structure? You already know most of the answer: A polypeptide chain of a given amino acid sequence can be arranged into a three-dimensional shape determined by the interactions responsible for secondary and tertiary structure. This folding normally occurs as the protein is being synthesized in the crowded environment within a cell, aided by other proteins. However, protein structure also depends on the physical and chemical conditions of the protein's environment. If the pH, salt concentration, temperature, or other aspects of its environment are altered, the weak chemical bonds and interactions within a protein may be destroyed, causing the protein to unravel and lose its native shape, a change called **denaturation (Figure 5.20)**. Because it is misshapen, the denatured protein is biologically inactive.

Most proteins become denatured if they are transferred from an aqueous environment to a nonpolar solvent, such as ether or chloroform; the polypeptide chain refolds so that

Figure 5.20 Denaturation and renaturation of a protein. High temperatures or various chemical treatments will denature a protein, causing it to lose its shape and hence its ability to function. If the denatured protein remains dissolved, it may renature when the chemical and physical aspects of its environment are restored to normal.

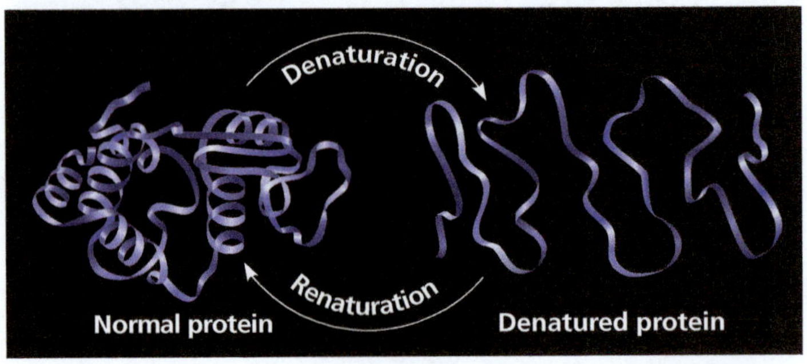

its hydrophobic regions face outward toward the solvent. Other denaturation agents include chemicals that disrupt the hydrogen bonds, ionic bonds, and disulfide bridges that maintain a protein's shape. Denaturation can also result from excessive heat, which agitates the polypeptide chain enough to overpower the weak interactions that stabilize the structure. The white of an egg becomes opaque during cooking because the denatured proteins are insoluble and solidify. This also explains why excessively high fevers can be fatal: Proteins in the blood tend to denature at very high body temperatures.

When a protein in a test-tube solution has been denatured by heat or chemicals, it can sometimes return to its functional shape when the denaturing agent is removed. (Sometimes this is not possible: For example, a fried egg will not become liquefied when placed back into the refrigerator!) We can conclude that the information for building specific shape is intrinsic to the protein's primary structure; this is often the case for small proteins. The sequence of amino acids determines the protein's shape—where an α helix can form, where β pleated sheets can exist, where disulfide bridges are located, where ionic bonds can form, and so on. But how does protein folding occur in the cell?

Protein Folding in the Cell

Biochemists now know the amino acid sequence for about 65 million proteins, with roughly 1.5 million added each month, and the three-dimensional shape for almost 35,000. Researchers have tried to correlate the primary structure of many proteins with their three-dimensional structure to discover the rules of protein folding. Unfortunately, however, the protein-folding process is not that simple. Most proteins probably go through several intermediate structures on their way to a stable shape, and looking at the mature structure does not reveal the stages of folding required to achieve that form. However, biochemists have developed methods for tracking a protein through such stages and learning more about this important process.

Misfolding of polypeptides in cells is a serious problem that has come under increasing scrutiny by medical researchers. Many diseases—such as cystic fibrosis, Alzheimer's, Parkinson's, and mad cow disease—are associated with an accumulation of misfolded proteins. In fact, misfolded versions of the transthyretin protein featured in Figure 5.18 have been implicated in several diseases, including one form of senile dementia.

Even when scientists have a correctly folded protein in hand, determining its exact three-dimensional structure is not simple, for a single protein has thousands of atoms. The method most commonly used to determine the 3-D structure of a protein is **X-ray crystallography**, which depends on the diffraction of an X-ray beam by the atoms of a crystallized molecule. Using this technique, scientists can build a 3-D model that shows the exact position of every atom in a protein molecule **(Figure 5.21)**. Nuclear magnetic resonance (NMR) spectroscopy and bioinformatics (see Concept 5.6) are complementary approaches to understanding protein structure and function.

Figure 5.21

Research Method X-Ray Crystallography

Application Scientists use X-ray crystallography to determine the three-dimensional (3-D) structure of macromolecules such as nucleic acids and proteins.

Technique Researchers aim an X-ray beam through a crystallized protein or nucleic acid. The atoms of the crystal diffract (bend) the X-rays into an orderly array that a digital detector records as a pattern of spots called an X-ray diffraction pattern, an example of which is shown here.

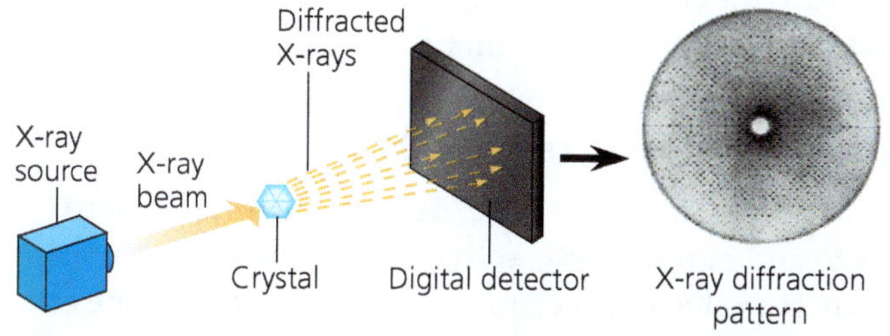

Results Using data from X-ray diffraction patterns and the sequence of monomers determined by chemical methods, researchers can build a 3-D computer model of the macromolecule being studied, such as the four-subunit protein transthyretin (see Figure 5.18) shown here.

The structure of some proteins is difficult to determine for a simple reason: A growing body of biochemical research has revealed that a significant number of proteins, or regions of proteins, do not have a distinct 3-D structure until they interact with a target protein or other molecule. Their flexibility and indefinite structure are important for their function, which may require binding with different targets at different times. These proteins, which may account for 20–30% of mammalian proteins, are called *intrinsically disordered proteins* and are the focus of current research.

CONCEPT CHECK 5.4

1. What parts of a polypeptide participate in the bonds that hold together secondary structure? Tertiary structure?
2. Thus far in the chapter, the Greek letters α and β have been used to specify at least three different pairs of structures. Name and briefly describe them.
3. **WHAT IF?** ➤ Where would you expect a polypeptide region rich in the amino acids valine, leucine, and isoleucine to be located in a folded polypeptide? Explain.

For suggested answers, see Appendix A.

CONCEPT 5.5

Nucleic acids store, transmit, and help express hereditary information

If the primary structure of polypeptides determines a protein's shape, what determines primary structure? The amino acid sequence of a polypeptide is programmed by a discrete unit of inheritance known as a **gene**. Genes consist of DNA, which belongs to the class of compounds called nucleic acids. **Nucleic acids** are polymers made of monomers called nucleotides.

The Roles of Nucleic Acids

The two types of nucleic acids, **deoxyribonucleic acid (DNA)** and **ribonucleic acid (RNA)**, enable living organisms to reproduce their complex components from one generation to the next. Unique among molecules, DNA provides directions for its own replication. DNA also directs RNA synthesis and, through RNA, controls protein synthesis; this entire process is called **gene expression** (Figure 5.22).

DNA is the genetic material that organisms inherit from their parents. Each chromosome contains one long DNA molecule, usually carrying several hundred or more genes. When a cell reproduces itself by dividing, its DNA molecules are copied and passed along from one generation of cells to the next. The information that programs all the cell's activities is encoded in the structure of the DNA. The DNA, however, is not directly involved in running the operations of the cell, any more than computer software by itself can read the bar code on a box of cereal. Just as a scanner is needed to read a bar code, proteins are required to implement genetic programs. The molecular hardware of the cell—the tools that carry out biological functions—consists mostly of proteins. For example, the oxygen carrier in red blood cells is the protein hemoglobin that you saw earlier (see Figure 5.18), not the DNA that specifies its structure.

How does RNA, the other type of nucleic acid, fit into gene expression, the flow of genetic information from DNA to proteins? A given gene along a DNA molecule can direct synthesis of a type of RNA called *messenger RNA* (*mRNA*). The mRNA molecule interacts with the cell's protein-synthesizing machinery to direct production of a polypeptide, which folds into all or part of a protein. We can summarize the flow of genetic information as DNA → RNA → protein (see Figure 5.22). The sites of protein synthesis are cellular structures called ribosomes. In a eukaryotic cell, ribosomes are in the cytoplasm—the region between the nucleus and the plasma membrane, the cell's outer boundary—but DNA resides in the nucleus. Messenger RNA conveys genetic instructions for building proteins from the nucleus to the cytoplasm. Prokaryotic cells lack nuclei but still use mRNA to convey a message from the DNA to ribosomes and other cellular equipment that translate the coded information into amino acid sequences. Later in the book, you'll read about other functions of some recently discovered RNA molecules; the stretches of DNA that direct synthesis of these RNAs are also considered genes (see Concept 18.3).

▼ Figure 5.22 Gene expression: DNA → RNA → protein. In a eukaryotic cell, DNA in the nucleus programs protein production in the cytoplasm by dictating synthesis of messenger RNA (mRNA).

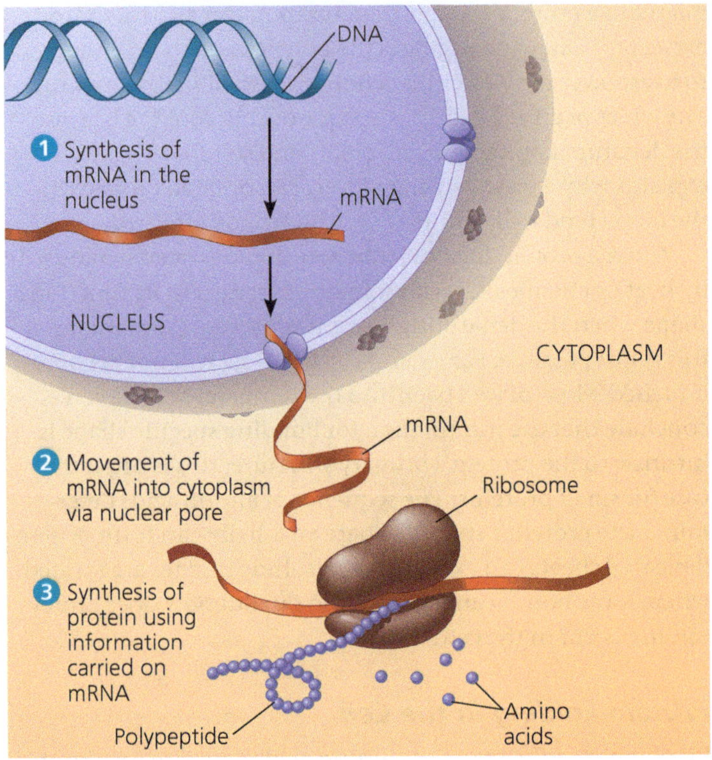

MB BioFlix® Animation: Gene Expression

The Components of Nucleic Acids

Nucleic acids are macromolecules that exist as polymers called **polynucleotides** (Figure 5.23a). As indicated by the name, each polynucleotide consists of monomers called **nucleotides**. A nucleotide, in general, is composed of three parts: a five-carbon sugar (a pentose), a nitrogen-containing (nitrogenous) base, and one to three phosphate groups (Figure 5.23b). The beginning monomer used to build a polynucleotide has three phosphate groups, but two are lost during the polymerization process. The portion of a nucleotide without any phosphate groups is called a *nucleoside*.

To understand the structure of a single nucleotide, let's first consider the nitrogenous bases (Figure 5.23c). Each nitrogenous base has one or two rings that include nitrogen atoms. (They are called nitrogenous *bases* because the nitrogen atoms tend to take up H^+ from solution, thus acting as bases.) There are two families of nitrogenous bases: pyrimidines and purines. A **pyrimidine** has one six-membered ring of carbon and nitrogen atoms. The members of the pyrimidine family are cytosine (C), thymine (T), and uracil (U). **Purines** are larger, with a six-membered ring fused to a five-membered ring. The purines are adenine (A) and guanine (G). The specific pyrimidines and purines differ in the chemical groups attached to the rings. Adenine, guanine, and cytosine are found in both DNA and RNA; thymine is found only in DNA and uracil only in RNA.

Now let's add the sugar to which the nitrogenous base is attached. In DNA the sugar is **deoxyribose**; in RNA it is **ribose** (see Figure 5.23c). The only difference between these two sugars is that deoxyribose lacks an oxygen atom on the second carbon in the ring, hence the name *deoxy*ribose.

So far, we have built a nucleoside (base plus sugar). To complete the construction of a nucleotide, we attach one to three phosphate groups to the 5′ carbon of the sugar (the carbon numbers in the sugar include ′, the prime symbol; see Figure 5.23b). With one phosphate, this is a nucleoside monophosphate, more often called a nucleotide.

Nucleotide Polymers

The linkage of nucleotides into a polynucleotide involves a dehydration reaction. (You will learn the details in Concept 16.2.) In the polynucleotide, adjacent nucleotides are joined by a phosphodiester linkage, which consists of a phosphate group that links the sugars of two nucleotides. This bonding results in a repeating pattern of sugar-phosphate units called the *sugar-phosphate backbone* (see Figure 5.23a). (Note that the nitrogenous bases are not part of the backbone.) The two free ends of the polymer are distinctly different from each other. One end has a phosphate attached to a 5′ carbon, and the other end has a hydroxyl group on a 3′ carbon; we refer to these as the *5′ end* and the *3′ end*, respectively. We can say that a polynucleotide has a built-in directionality along its sugar-phosphate backbone, from 5′ to 3′, somewhat like a one-way street. The bases are attached all along the sugar-phosphate backbone.

The sequence of bases along a DNA (or mRNA) polymer is unique for each gene and provides very specific information to the cell. Because genes are hundreds to thousands of nucleotides long, the number of possible base sequences is effectively limitless. The information carried by the gene is encoded in its specific sequence of the four DNA bases. For example, the sequence 5′-AGGTAACTT-3′ means one thing, whereas the sequence 5′-CGCTTTAAC-3′ has a different meaning. (Entire genes, of course, are much longer.) The linear order of bases in a gene specifies the amino acid sequence—the primary structure—of a protein, which in turn specifies that protein's 3-D structure, thus enabling its function in the cell.

▼ **Figure 5.23 Components of nucleic acids. (a)** A polynucleotide has a sugar-phosphate backbone with variable appendages, the nitrogenous bases. **(b)** In a polynucleotide, each nucleotide monomer includes a nitrogenous base, a sugar, and a phosphate group. Note that carbon numbers in the sugar include primes (′). **(c)** A nucleoside includes a nitrogenous base (purine or pyrimidine) and a five-carbon sugar (deoxyribose or ribose).

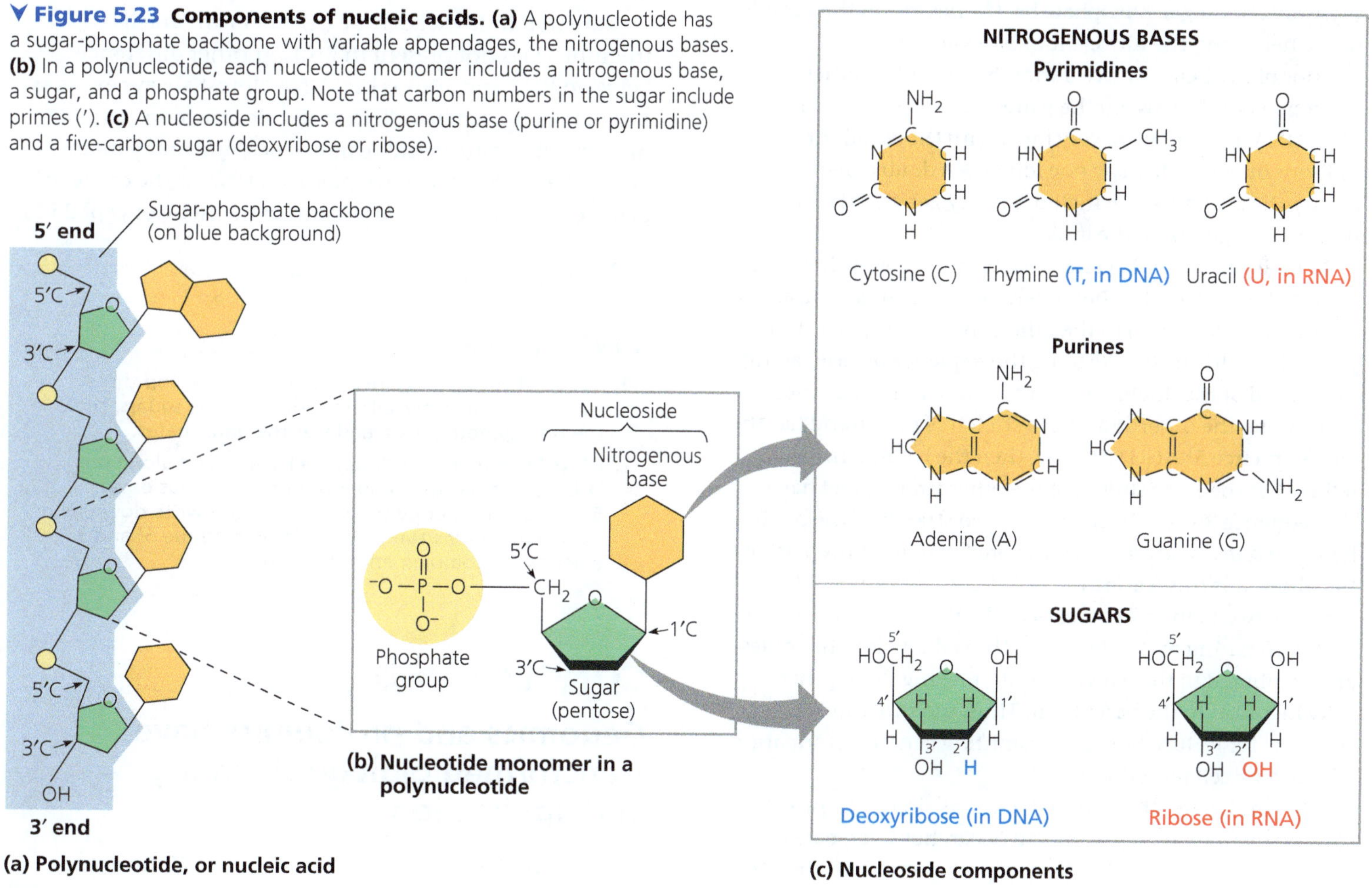

HHMI Animation: The Chemical Structure of DNA

Figure 5.24 The structures of DNA and tRNA molecules. **(a)** The DNA molecule is usually a double helix, with the sugar-phosphate backbones of the antiparallel polynucleotide strands (symbolized here by blue ribbons) on the outside of the helix. Hydrogen bonds between pairs of nitrogenous bases hold the two strands together. As illustrated here with symbolic shapes for the bases, adenine (A) can pair only with thymine (T), and guanine (G) can pair only with cytosine (C). Each DNA strand in this figure is the structural equivalent of the polynucleotide diagrammed in Figure 5.23a. **(b)** A tRNA molecule has a roughly L-shaped structure due to complementary base pairing of antiparallel stretches of RNA. In RNA, A pairs with U.

HHMI Animation: Paired DNA Strands

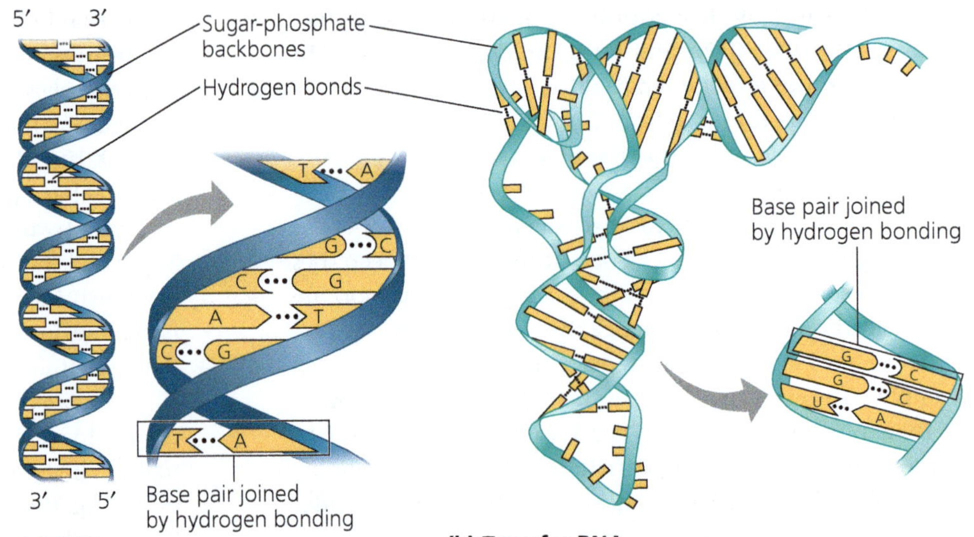

(a) DNA **(b) Transfer RNA**

The Structures of DNA and RNA Molecules

DNA molecules have two polynucleotides, or "strands," that wind around an imaginary axis, forming a **double helix** **(Figure 5.24a)**. The two sugar-phosphate backbones run in opposite $5' \rightarrow 3'$ directions from each other; this arrangement is referred to as **antiparallel**, somewhat like a divided highway. The sugar-phosphate backbones are on the outside of the helix, and the nitrogenous bases are paired in the interior of the helix. The two strands are held together by hydrogen bonds between the paired bases (see Figure 5.24a). Most DNA molecules are very long, with thousands or even millions of base pairs. The one long DNA double helix in a eukaryotic chromosome includes many genes, each one a particular segment of the molecule.

In base pairing, only certain bases in the double helix are compatible with each other. Adenine (A) in one strand always pairs with thymine (T) in the other, and guanine (G) always pairs with cytosine (C). Reading the sequence of bases along one strand of the double helix would tell us the sequence of bases along the other strand. If a stretch of one strand has the base sequence 5′-AGGTCCG-3′, then the base-pairing rules tell us that the same stretch of the other strand must have the sequence 3′-TCCAGGC-5′ The two strands of the double helix are *complementary*, each the predictable counterpart of the other. It is this feature of DNA that makes it possible to generate two identical copies of each DNA molecule in a cell that is preparing to divide. When the cell divides, the copies are distributed to the daughter cells, making them genetically identical to the parent cell. Thus, the structure of DNA accounts for its function of transmitting genetic information whenever a cell reproduces.

RNA molecules, by contrast, exist as single strands. Complementary base pairing can occur, however, between regions of two RNA molecules or even between two stretches of nucleotides in the *same* RNA molecule. In fact, base pairing within an RNA molecule allows it to take on the particular three-dimensional shape necessary for its function. Consider, for example, the type of RNA called *transfer RNA* (*tRNA*), which brings amino acids to the ribosome during the synthesis of a polypeptide. A tRNA molecule is about 80 nucleotides in length. Its functional shape results from base pairing between nucleotides where complementary stretches of the molecule can run antiparallel to each other **(Figure 5.24b)**.

Note that in RNA, adenine (A) pairs with uracil (U); thymine (T) is not present in RNA. Another difference between RNA and DNA is that DNA almost always exists as a double helix, whereas RNA molecules are more variable in shape. RNAs are versatile molecules, and many biologists believe RNA may have preceded DNA as the carrier of genetic information in early forms of life (see Concept 25.1).

Animation: Nucleic Acid Structure

CONCEPT CHECK 5.5

1. **DRAW IT** ➤ Go to Figure 5.23a and, for the top three nucleotides, number all the carbons in the sugars, circle the nitrogenous bases, and star the phosphates.
2. **DRAW IT** ➤ In a DNA double helix, a region along one DNA strand has this sequence of nitrogenous bases: 5′-TAGGCCT-3′. Copy this sequence, and write down its complementary strand, clearly indicating the 5′ and 3′ ends of the complementary strand.

For suggested answers, see Appendix A.

CONCEPT 5.6

Genomics and proteomics have transformed biological inquiry and applications

Experimental work in the first half of the 20th century established the role of DNA as the bearer of genetic information, passed from generation to generation, that specified the functioning of living cells and organisms. Once the structure of the DNA molecule was described in 1953, and the linear sequence of nucleotide bases was understood to specify the amino acid sequence of proteins, biologists sought to "decode" genes by learning their nucleotide sequences (often called "base sequences").

The first chemical techniques for *DNA sequencing*, or determining the sequence of nucleotides along a DNA strand, one by one, were developed in the 1970s. Researchers began to study gene sequences, gene by gene, and the more they learned, the more questions they had: How was expression of genes regulated? Genes and their protein products clearly interacted with each other, but how? What was the function, if any, of the DNA that is not part of genes? To fully understand the genetic functioning of a living organism, the entire sequence of the full complement of DNA, the organism's *genome*, would be most enlightening. In spite of the apparent impracticality of this idea, in the late 1980s several prominent biologists put forth an audacious proposal to launch a project that would sequence the entire human genome—all 3 billion bases of it! This endeavor began in 1990 and was effectively completed in the early 2000s.

An unplanned but profound side benefit of this project—the Human Genome Project—was the rapid development of faster and less expensive methods of sequencing. This trend has continued: The cost for sequencing 1 million bases in 2001, well over $5,000, has decreased to less than $0.02 in 2016. And a human genome, the first of which took over 10 years to sequence, could be completed at today's pace in just a few days **(Figure 5.25)**. The number of genomes that have been fully sequenced has burgeoned, generating reams of data and prompting development of **bioinformatics**, the use of computer software and other computational tools that can handle and analyze these large data sets.

The reverberations of these developments have transformed the study of biology and related fields. Biologists often look at problems by analyzing large sets of genes or even comparing whole genomes of different species, an approach called **genomics**. A similar analysis of large sets of proteins, including their sequences, is called **proteomics**. (Protein sequences can be determined either by using biochemical techniques or by translating the DNA sequences that code for them.) These approaches permeate all fields of biology, some examples of which are shown in **Figure 5.26**.

◀ Figure 5.25 Automatic DNA sequencing machines and abundant computing power enable rapid sequencing of genes and genomes.

Perhaps the most significant impact of genomics and proteomics on the field of biology as a whole has been their contributions to our understanding of evolution. In addition to confirming evidence for evolution from the study of fossils and characteristics of currently existing species, genomics has helped us tease out relationships among different groups of organisms that had not been resolved by previous types of evidence, and thus infer evolutionary history.

DNA and Proteins as Tape Measures of Evolution

EVOLUTION We are accustomed to thinking of shared traits, such as hair and milk production in mammals, as evidence of shared ancestry. Because DNA carries heritable information in the form of genes, sequences of genes and their protein products document the hereditary background of an organism. The linear sequences of nucleotides in DNA molecules are passed from parents to offspring; these sequences determine the amino acid sequences of proteins. As a result, siblings have greater similarity in their DNA and proteins than do unrelated individuals of the same species.

Given our evolutionary view of life, we can extend this concept of "molecular genealogy" to relationships between species: We would expect two species that appear to be closely related based on anatomical evidence (and possibly fossil evidence) to also share a greater proportion of their DNA and protein sequences than do less closely related species. In fact, that is the case. An example is the comparison of the β polypeptide chain of human hemoglobin with the corresponding hemoglobin polypeptide in other vertebrates. In this chain of 146 amino acids, humans and gorillas differ in just 1 amino acid, while humans and frogs, more distantly related, differ in 67 amino acids. In the **Scientific Skills Exercise**, you can apply this sort of reasoning to additional species. And this conclusion holds true as well when comparing whole genomes: The human genome is 95–98% identical to that of the chimpanzee, but only roughly 85% identical to that of the mouse, a more distant evolutionary relative. Molecular biology has added a new tape measure to the toolkit biologists use to assess evolutionary kinship.

Comparing genomic sequences has practical applications as well. In the **Problem-Solving Exercise**, you can see how this type of genomic analysis can help you detect consumer fraud.

CONCEPT CHECK 5.6

1. How would sequencing the entire genome of an organism help scientists to understand how that organism functioned?
2. Given the function of DNA, why would you expect two species with very similar traits to also have very similar genomes?

For suggested answers, see Appendix A.

▼ **Figure 5.26** **MAKE CONNECTIONS**

Contributions of Genomics and Proteomics to Biology

Nucleotide sequencing and the analysis of large sets of genes and proteins can be done rapidly and inexpensively due to advances in technology and information processing. Taken together, genomics and proteomics have advanced our understanding of biology across many different fields.

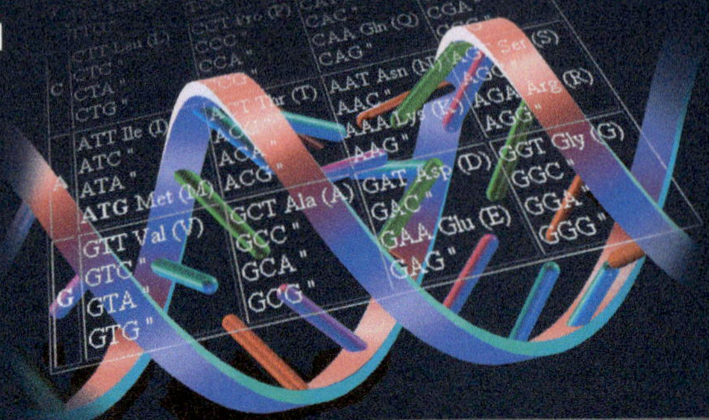

Paleontology

New DNA sequencing techniques have allowed decoding of minute quantities of DNA found in ancient tissues from our extinct relatives, the Neanderthals (*Homo neanderthalensis*). Sequencing the Neanderthal genome has informed our understanding of their physical appearance (as in this reconstruction), as well as their relationship with modern humans. (See Figures 34.51 and 34.52.)

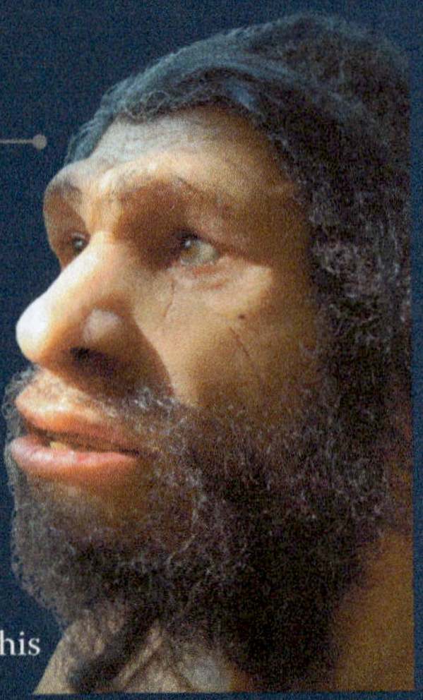

Medical Science

Identifying the genetic basis for human diseases like cancer helps researchers focus their search for potential future treatments. Currently, sequencing the sets of genes expressed in an individual's tumor can allow a more targeted approach to treating the cancer, a type of "personalized medicine." (See Figures 12.20 and 18.27.)

Evolution

A major aim of evolutionary biology is to understand the relationships among species, both living and extinct. For example, genome sequence comparisons have identified the hippopotamus as the land mammal sharing the most recent common ancestor with whales. (See Figure 22.20.)

Hippopotamus

Short-finned pilot whale

Conservation Biology

The tools of molecular genetics and genomics are increasingly used by forensic ecologists to identify which species of animals and plants are killed illegally. In one case, genomic sequences of DNA from illegal shipments of elephant tusks were used to track down poachers and pinpoint the territory where they were operating. (See Figure 56.9.)

Species Interactions

Most plant species exist in a mutually beneficial partnership with fungi (right) and bacteria associated with the plants' roots; these interactions improve plant growth. Genome sequencing and analysis of gene expression have allowed characterization of plant-associated communities. Such studies will help advance our understanding of such interactions and may improve agricultural practices. (See the Chapter 31 Scientific Skills Exercise and Figure 37.11.)

MAKE CONNECTIONS ➤ *Considering the examples provided here, describe how the approaches of genomics and proteomics help us to address a variety of biological questions.*

MB **HHMI Video: The Making of the Fittest: The Birth and Death of Genes (Icefish)** hhmi BioInteractive

SCIENTIFIC SKILLS EXERCISE

Analyzing Polypeptide Sequence Data

Human

Rhesus monkey

Gibbon

Are Rhesus Monkeys or Gibbons More Closely Related to Humans? In this exercise, you will look at amino acid sequence data for the β polypeptide chain of hemoglobin, often called β-globin. You will then interpret the data to hypothesize whether the monkey or the gibbon is more closely related to humans.

How Such Experiments Are Done Researchers can isolate the polypeptide of interest from an organism and then determine the amino acid sequence. More frequently, the DNA of the relevant gene is sequenced, and the amino acid sequence of the polypeptide is deduced from the DNA sequence of its gene.

Data from the Experiments In the data below, the letters give the sequence of the 146 amino acids in β-globin from humans, rhesus monkeys, and gibbons. Because a complete sequence would not fit on one line here, the sequences are broken into three segments. The sequences for the three different species are aligned so that you can compare them easily. For example, you can see that for all three species, the first amino acid is V (valine) and the 146th amino acid is H (histidine).

Species	Alignment of Amino Acid Sequences of β-globin
Human	1 VHLTPEEKSA VTALWGKVNV DEVGGEALGR LLVVYPWTQR FFESFGDLST
Monkey	1 VHLTPEEKNA VTTLWGKVNV DEVGGEALGR LLLVYPWTQR FFESFGDLSS
Gibbon	1 VHLTPEEKSA VTALWGKVNV DEVGGEALGR LLVVYPWTQR FFESFGDLST
Human	51 PDAVMGNPKV KAHGKKVLGA FSDGLAHLDN LKGTFATLSE LHCDKLHVDP
Monkey	51 PDAVMGNPKV KAHGKKVLGA FSDGLNHLDN LKGTFAQLSE LHCDKLHVDP
Gibbon	51 PDAVMGNPKV KAHGKKVLGA FSDGLAHLDN LKGTFAQLSE LHCDKLHVDP
Human	101 ENFRLLGNVL VCVLAHHFGK EFTPPVQAAY QKVVAGVANA LAHKYH
Monkey	101 ENFKLLGNVL VCVLAHHFGK EFTPQVQAAY QKVVAGVANA LAHKYH
Gibbon	101 ENFRLLGNVL VCVLAHHFGK EFTPQVQAAY QKVVAGVANA LAHKYH

INTERPRET THE DATA

1. Scan the monkey and gibbon sequences, letter by letter, circling any amino acids that do not match the human sequence. (a) How many amino acids differ between the monkey and the human sequences? (b) Between the gibbon and human?
2. For each nonhuman species, what percent of its amino acids are identical to the human sequence of β-globin?
3. Based on these data alone, state a hypothesis for which of these two species is more closely related to humans. What is your reasoning?
4. What other evidence could you use to support your hypothesis?

Data from Human: http://www.ncbi.nlm.nih.gov/protein/AAA21113.1; rhesus monkey: http://www.ncbi.nlm.nih.gov/protein/122634; gibbon: http://www.ncbi.nlm.nih.gov/protein/122616

MB **Instructors:** A version of this Scientific Skills Exercise can be assigned in MasteringBiology.

PROBLEM-SOLVING EXERCISE

Are you a victim of fish fraud?

When buying salmon, perhaps you prefer the more expensive wild-caught Pacific salmon (*Oncorhynchus* species) over farmed Atlantic salmon (*Salmo salar*). But studies reveal that about 40% of the time, you aren't getting the fish you paid for! Watch the video in the MasteringBiology Study Area for more information.

MB **ABC News Video: Fake Fish in Stores and Restaurants**

MB **Instructors:** A version of this Problem-Solving Exercise can be assigned in Chapter 5 of MasteringBiology. A more extensive investigation is in Chapter 26 of MasteringBiology.

In this exercise, you will investigate whether a piece of salmon has been fraudulently labeled.

Your Approach The principle guiding your investigation is that DNA sequences from within a species or from closely related species are more similar to each other than are sequences from more distantly related species.

Your Data You've been sold a piece of salmon labeled as coho salmon (*Oncorhynchus kisutch*). To see whether your fish was labeled correctly, you will compare a short DNA sequence from your sample to standard sequences from the same gene for three salmon species. The sequences are:

	Sample labeled as *O. kisutch* (coho salmon)	5′-CGGCACCGCCCTAAGTCTCT-3′
Standard sequences	Sequence for *O. kisutch* (coho salmon)	5′-AGGCACCGCCCTAAGTCTAC-3′
	Sequence for *O. keta* (chum salmon)	5′-AGGCACCGCCCTGAGCCTAC-3′
	Sequence for *Salmo salar* (Atlantic salmon)	5′-CGGCACCGCCCTAAGTCTCT-3′

Your Analysis

1. Scan along the standard sequences (*O. kisutch, O. keta*, and *S. salar*), base by base, circling any bases that do not match the sequence from your fish sample.
2. How many bases differ between (a) *O. kisutch* and your fish sample? (b) *O. keta* and the sample? (c) *S. salar* and the sample?
3. For each standard, what percentage of its bases are identical to your sample?
4. Based on these data alone, state a hypothesis for the species identity of your sample. What is your reasoning?

5 Chapter Review

Go to **MasteringBiology**™ for Videos, Animations, Vocab Self-Quiz, Practice Tests, and more in the Study Area.

SUMMARY OF KEY CONCEPTS

VOCAB SELF-QUIZ goo.gl/6u55ks

CONCEPT 5.1

Macromolecules are polymers, built from monomers *(pp. 67–68)*

- Large carbohydrates (polysaccharides), proteins, and nucleic acids are **polymers**, which are chains of **monomers**. The components of lipids vary. Monomers form larger molecules by **dehydration reactions**, in which water molecules are released. Polymers can disassemble by the reverse process, **hydrolysis**. An immense variety of polymers can be built from a small set of monomers.

? *What is the fundamental basis for the differences between large carbohydrates, proteins, and nucleic acids?*

Large Biological Molecules	Components	Examples	Functions
CONCEPT 5.2 **Carbohydrates serve as fuel and building material** *(pp. 68–72)* ? *Compare the composition, structure, and function of starch and cellulose. What role do starch and cellulose play in the human body?*	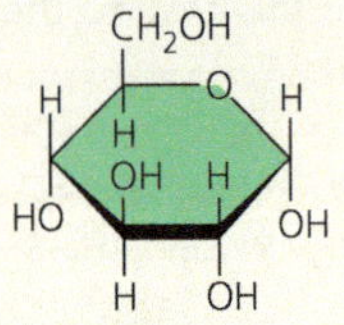 Monosaccharide monomer	**Monosaccharides:** glucose, fructose **Disaccharides:** lactose, sucrose **Polysaccharides:** ▪ Cellulose (plants) ▪ Starch (plants) ▪ Glycogen (animals) ▪ Chitin (animals and fungi)	Fuel; carbon sources that can be converted to other molecules or combined into polymers ▪ Strengthens plant cell walls ▪ Stores glucose for energy ▪ Stores glucose for energy ▪ Strengthens exoskeletons and fungal cell walls
CONCEPT 5.3 **Lipids are a diverse group of hydrophobic molecules** *(pp. 72–75)* ? *Why are lipids not considered to be polymers or macromolecules?*	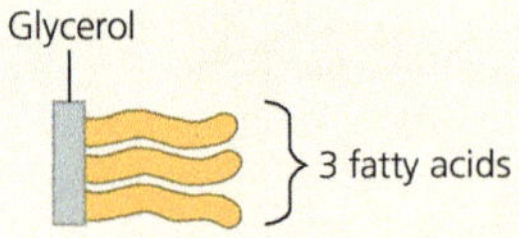	**Triacylglycerols** (fats or oils): glycerol + three fatty acids	Important energy source 
	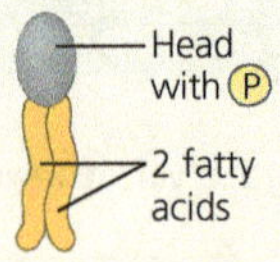	**Phospholipids:** glycerol + phosphate group + two fatty acids	Lipid bilayers of membranes 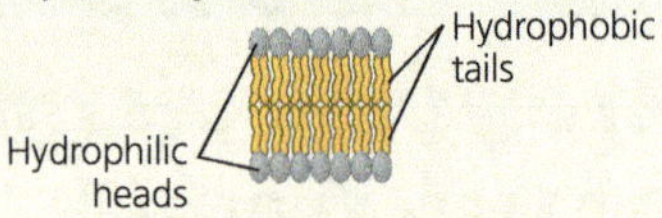
	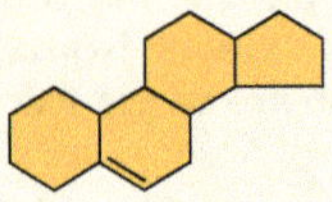 Steroid backbone	**Steroids:** four fused rings with attached chemical groups	▪ Component of cell membranes (cholesterol) ▪ Signaling molecules that travel through the body (hormones)
CONCEPT 5.4 **Proteins include a diversity of structures, resulting in a wide range of functions** *(pp. 75–83)* ? *Explain the basis for the great diversity of proteins.*	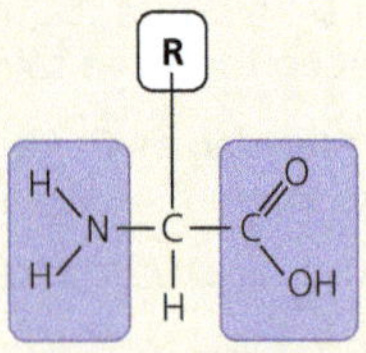 Amino acid monomer (20 types)	▪ Enzymes ▪ Defensive proteins ▪ Storage proteins ▪ Transport proteins ▪ Hormones ▪ Receptor proteins ▪ Motor proteins ▪ Structural proteins	▪ Catalyze chemical reactions ▪ Protect against disease ▪ Store amino acids ▪ Transport substances ▪ Coordinate organismal responses ▪ Receive signals from outside cell ▪ Function in cell movement ▪ Provide structural support
CONCEPT 5.5 **Nucleic acids store, transmit, and help express hereditary information** *(pp. 84–86)* ? *What role does complementary base pairing play in the functions of nucleic acids?*	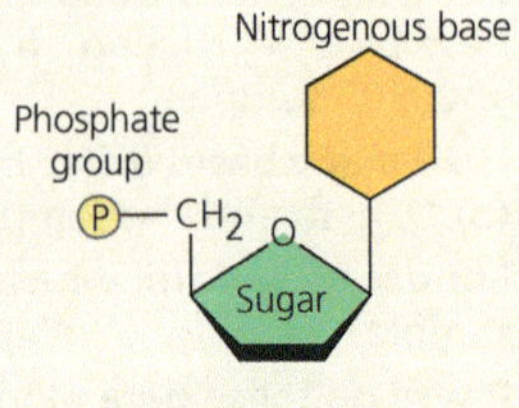 Nucleotide (monomer of a polynucleotide)	**DNA:**  ▪ Sugar = deoxyribose ▪ Nitrogenous bases = C, G, A, T ▪ Usually double-stranded	Stores hereditary information
		RNA:  ▪ Sugar = ribose ▪ Nitrogenous bases = C, G, A, U ▪ Usually single-stranded	Various functions in gene expression, including carrying instructions from DNA to ribosomes

CONCEPT 5.6

Genomics and proteomics have transformed biological inquiry and applications *(pp. 86–89)*

- Recent technological advances in DNA sequencing have given rise to **genomics**, an approach that analyzes large sets of genes or whole genomes, and **proteomics**, a similar approach for large sets of proteins. **Bioinformatics** is the use of computational tools and computer software to analyze these large data sets.
- The more closely two species are related evolutionarily, the more similar their DNA sequences are. DNA sequence data confirm models of evolution based on fossils and anatomical evidence.

? *Given the sequences of a particular gene in fruit flies, fish, mice, and humans, predict the relative similarity of the human sequence to that of each of the other species.*

TEST YOUR UNDERSTANDING

Level 1: Knowledge/Comprehension

PRACTICE TEST
goo.gl/CUYGKD

1. Which of the following categories includes all others in the list?
(A) disaccharide
(B) polysaccharide
(C) starch
(D) carbohydrate

2. The enzyme amylase can break glycosidic linkages between glucose monomers only if the monomers are in the α form. Which of the following could amylase break down?
(A) glycogen, starch, and amylopectin
(B) glycogen and cellulose
(C) cellulose and chitin
(D) starch, chitin, and cellulose

3. Which of the following is true of *unsaturated* fats?
(A) They are more common in animals than in plants.
(B) They have double bonds in their fatty acid chains.
(C) They generally solidify at room temperature.
(D) They contain more hydrogen than do saturated fats having the same number of carbon atoms.

4. The structural level of a protein *least* affected by a disruption in hydrogen bonding is the
(A) primary level.
(B) secondary level.
(C) tertiary level.
(D) quaternary level.

5. Enzymes that break down DNA catalyze the hydrolysis of the covalent bonds that join nucleotides together. What would happen to DNA molecules treated with these enzymes?
(A) The two strands of the double helix would separate.
(B) The phosphodiester linkages of the polynucleotide backbone would be broken.
(C) The pyrimidines would be separated from the deoxyribose sugars.
(D) All bases would be separated from the deoxyribose sugars.

Level 2: Application/Analysis

6. The molecular formula for glucose is $C_6H_{12}O_6$. What would be the molecular formula for a polymer made by linking ten glucose molecules together by dehydration reactions?
(A) $C_{60}H_{120}O_{60}$
(B) $C_{60}H_{102}O_{51}$
(C) $C_{60}H_{100}O_{50}$
(D) $C_{60}H_{111}O_{51}$

7. Which of the following pairs of base sequences could form a short stretch of a normal double helix of DNA?
(A) 5′-AGCT-3′ with 5′-TCGA-3′
(B) 5′-GCGC-3′ with 5′-TATA-3′
(C) 5′-ATGC-3′ with 5′-GCAT-3′
(D) All of these pairs are correct.

8. Construct a table that organizes the following terms, and label the columns and rows.

Monosaccharides	Polypeptides	Phosphodiester linkages
Fatty acids	Triacylglycerols	Peptide bonds
Amino acids	Polynucleotides	Glycosidic linkages
Nucleotides	Polysaccharides	Ester linkages

9. **DRAW IT** Copy the polynucleotide strand in Figure 5.23a and label the bases G, T, C, and T, starting from the 5′ end. Assuming this is a DNA polynucleotide, now draw the complementary strand, using the same symbols for phosphates (circles), sugars (pentagons), and bases. Label the bases. Draw arrows showing the 5′ → 3′ direction of each strand. Use the arrows to make sure the second strand is antiparallel to the first. *Hint*: After you draw the first strand vertically, turn the paper upside down; it is easier to draw the second strand from the 5′ toward the 3′ direction as you go from top to bottom.

Level 3: Synthesis/Evaluation

10. **EVOLUTION CONNECTION** Comparisons of amino acid sequences can shed light on the evolutionary divergence of related species. If you were comparing two living species, would you expect all proteins to show the same degree of divergence? Why or why not? Justify your answer.

11. **SCIENTIFIC INQUIRY** Suppose you are a research assistant in a lab studying DNA-binding proteins. You have been given the amino acid sequences of all the proteins encoded by the genome of a certain species and have been asked to find candidate proteins that could bind DNA. What type of amino acids would you expect to see in the DNA-binding regions of such proteins? Explain your thinking.

12. **WRITE ABOUT A THEME: ORGANIZATION** Proteins, which have diverse functions in a cell, are all polymers of the same kinds of monomers—amino acids. Write a short essay (100–150 words) that discusses how the structure of amino acids allows this one type of polymer to perform so many functions.

13. **SYNTHESIZE YOUR KNOWLEDGE**

Given that the function of egg yolk is to nourish and support the developing chick, explain why egg yolks are so high in fat, protein, and cholesterol.

For selected answers, see Appendix A.

MB For additional practice questions, check out the **Dynamic Study Modules** in MasteringBiology. You can use them to study on your smartphone, tablet, or computer anytime, anywhere!

UNIT 2 THE CELL

Elba Serrano was born in Old San Juan, Puerto Rico. Because her father was a career sergeant in the U.S. military, she had a nomadic childhood, tempered by her family's strong Hispanic culture. She majored in physics at the University of Rochester, one of two women among 80 majors. She then earned a Ph.D. from Stanford University, studying the permeability of nerve cells to ions, and was a postdoctoral fellow at UCLA. There, she started working on the auditory system, as a way of understanding how cells allow an organism to respond to changes in the environment. Since 1992, Dr. Serrano has been a faculty member at New Mexico State University, where she has made many discoveries about the development of the auditory system. Dr. Serrano is also an outspoken advocate and mentor for women and underrepresented groups in science.

An Interview with Elba Serrano

How did you get interested in science, and in biology in particular?

When you're a military kid, you're asked to pick out what you want for Christmas from catalogs, and I remember always gravitating to the telescope and the chemistry set. To this day, I love math—I solve math problems for fun. In college I studied in a medical school library, and I'd take breaks from studying physics by reading about biology. I became very interested in biochemistry, and I said, "It looks like there is a field called biophysics—I think I'll go to grad school for that." I ended up at Stanford, and I really struggled—I was pretty much ready to drop out at the end of the first year. Then I was assigned a book called *Nerve, Muscle, and Synapse,* by Bernard Katz. I started reading this book at 8 AM and didn't stop—I didn't eat lunch. I finished it by the end of the day and knew "this is what I want to do." It was basically membrane channel biophysics, and there were equations! I thought, "This is the science I recognize."

What was your dissertation research?

I worked on neurons, and I focused on the flow of ions like potassium through the cell membrane. Membranes are interesting because they both define a cell—they're barriers—and at the same time they are a conduit. Membranes are like borders—very dynamic places where a lot of things happen. Within the membrane there are special proteins called ion channels, a kind of gatekeeper that permits the passage of ions. Neurons are the basic cells of the brain. The way the brain does its job is through the coordinated work of many neurons, and these cells are similar in some ways but also extremely diverse. So my dissertation question explored the diversity of neuron cell membranes: How are the proteins on different neurons different?

"Membranes are like borders—very dynamic places where a lot of things happen."

How did you become interested in the auditory system, and what questions are you asking?

I always want to use the concepts of physics to understand biological systems. Biophysicists love the inner ear: it has sensory surfaces with beautiful cells called mechanosensory hair cells (see photo). These are similar but diverse in their function; the ear's capacity to pick up all sound frequencies is because all the little hair cells are slightly different from each other. The mechanosensory hair cells we work on in my lab are marvelous, beautiful, very special cells. We study the structure and positioning of these cells, and at the same time we ask genetics questions. We are interested in identifying the genes for the ion channel proteins that are important in the electrical response of these cells—how are they organized in the genome, how are they turned on and off? The ear is a powerful system that has both redundancy—many similar cells—and diversity—each cell is unique.

▼ Inner ear "hair" cells with bundles of rod-like protrusions that detect sound waves. Damage to these bundles by noise or aging can lead to deafness or loss of balance. (SEM)

What is your advice to an undergraduate who is considering a career in biology?

First, don't limit yourself in your possibilities for your future. Next, get a variety of experiences—part of life's journey is about finding out what gives us satisfaction. If you're choosing a career in biology, there are many ways you can be a biologist. You can work in labs or in field settings, among others. Finally, understanding biology today requires understanding math, physics, and chemistry, so it is important to invest in those areas. Even if you feel you will not be able to excel in them, do not let that deter you, because everything we do in science can be learned—it just requires dedication.

6 A Tour of the Cell

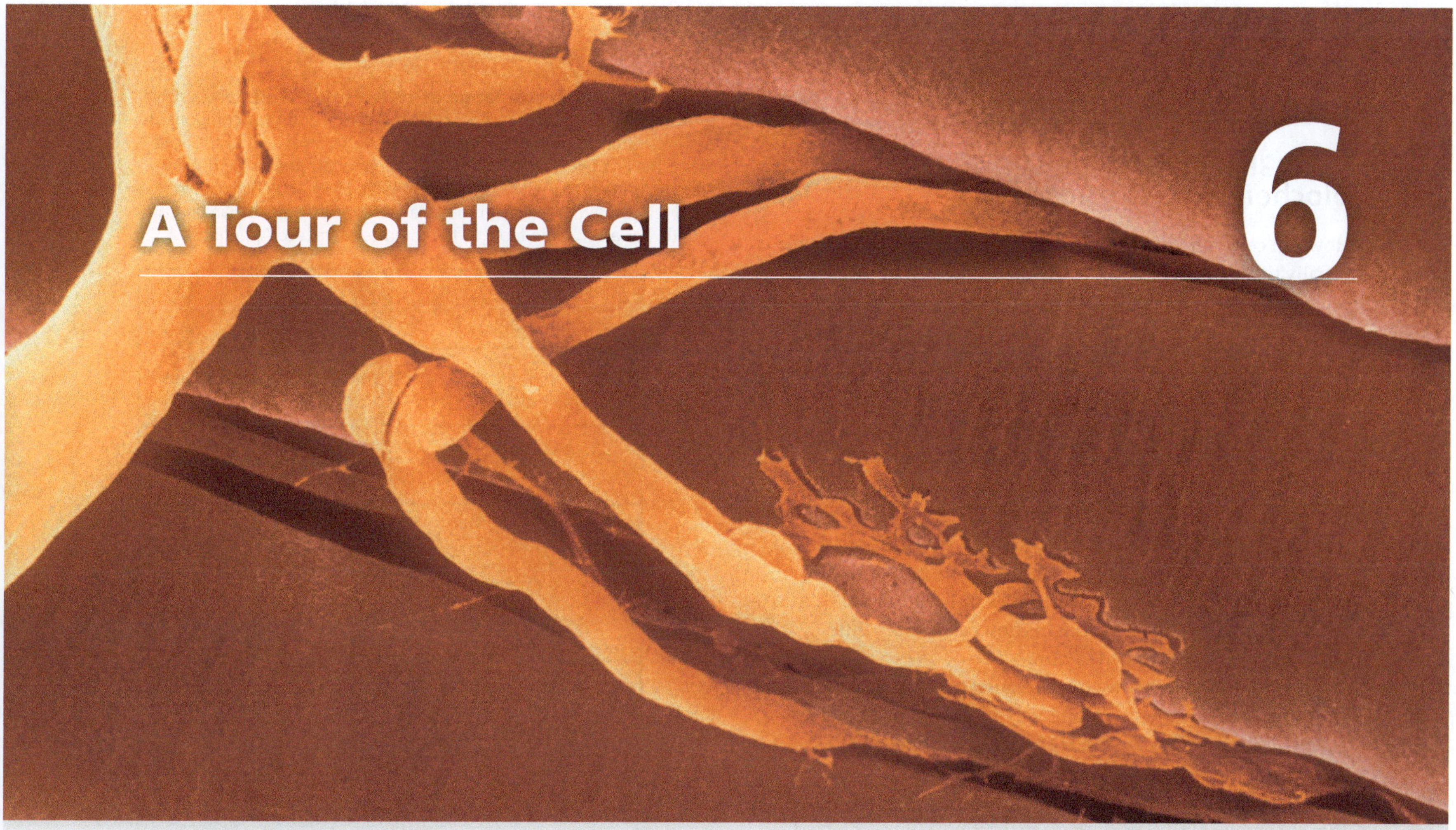

Figure 6.1 How do your cells help you learn about biology?

KEY CONCEPTS

6.1 Biologists use microscopes and biochemistry to study cells

6.2 Eukaryotic cells have internal membranes that compartmentalize their functions

6.3 The eukaryotic cell's genetic instructions are housed in the nucleus and carried out by the ribosomes

6.4 The endomembrane system regulates protein traffic and performs metabolic functions

6.5 Mitochondria and chloroplasts change energy from one form to another

6.6 The cytoskeleton is a network of fibers that organizes structures and activities in the cell

6.7 Extracellular components and connections between cells help coordinate cellular activities

6.8 A cell is greater than the sum of its parts

The Fundamental Units of Life

Cells are as fundamental to the living systems of biology as the atom is to chemistry. Many different types of cells are working for you right now. The contraction of muscle cells moves your eyes as you read this sentence. **Figure 6.1** shows extensions from a nerve cell (orange) making contact with muscle cells (red). The words on the page are translated into signals that nerve cells carry to your brain, where they are passed on to other nerve cells. As you study, your cells make connections between nerve cells that solidify memories and permit learning to occur.

All organisms are made of cells. In the hierarchy of biological organization, the cell is the simplest collection of matter that can be considered a living entity. Indeed, many forms of life exist as single-celled organisms, like the *Paramecium* shown here, a eukaryote that lives in pond water. Larger, more complex organisms, including plants and animals, are multicellular; their bodies are associations of many kinds of specialized cells that could not survive for long on their own. Even when cells are arranged into higher levels of organization, such as tissues and organs, the cell remains the organism's basic unit of structure and function.

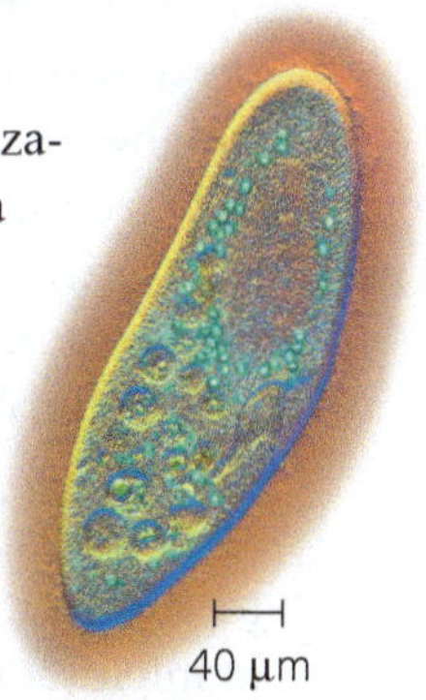

All cells are related by their descent from earlier cells. During the long evolutionary history of life on Earth, cells have been modified in many different ways. But although cells can differ substantially from one another, they share common features. In this chapter, we'll first examine the tools and techniques that allow us to understand cells, then tour the cell and become acquainted with its components.

When you see this blue icon, log in to **MasteringBiology** and go to the Study Area for digital resources.

MB Get Ready for This Chapter

CONCEPT 6.1

Biologists use microscopes and biochemistry to study cells

How can cell biologists investigate the inner workings of a cell, usually too small to be seen by the unaided eye? Before we tour the cell, it will be helpful to learn how cells are studied.

Microscopy

The development of instruments that extend the human senses allowed the discovery and early study of cells. Microscopes were invented in 1590 and further refined during the 1600s. Cell walls were first seen by Robert Hooke in 1665 as he looked through a microscope at dead cells from the bark of an oak tree. But it took the wonderfully crafted lenses of Antoni van Leeuwenhoek to visualize living cells. Imagine Hooke's excitement when he visited van Leeuwenhoek in 1674 and the world of microorganisms—what his host called "very little animalcules"—was revealed to him.

The microscopes first used by Renaissance scientists, as well as the microscopes you are likely to use in the laboratory, are all light microscopes. In a **light microscope (LM)**, visible light is passed through the specimen and then through glass lenses. The lenses refract (bend) the light in such a way that the image of the specimen is magnified as it is projected into the eye or into a camera (see Appendix D).

Three important parameters in microscopy are magnification, resolution, and contrast. *Magnification* is the ratio of an object's image size to its real size. Light microscopes can magnify effectively to about 1,000 times the actual size of the specimen; at greater magnifications, additional details cannot be seen clearly. *Resolution* is a measure of the clarity of the image; it is the minimum distance two points can be separated and still be distinguished as separate points. For example, what appears to the unaided eye as one star in the sky may be resolved as twin stars with a telescope, which has a higher resolving ability than the eye. Similarly, using standard techniques, the light microscope cannot resolve detail finer than about 0.2 micrometer (μm), or 200 nanometers (nm), regardless of the magnification **(Figure 6.2)**. The third parameter, *contrast*, is the difference in brightness between the light and dark areas of an image. Methods for enhancing contrast include staining or labeling cell components to stand out visually. **Figure 6.3** shows some different types of microscopy; study this figure as you read this section.

Until recently, the resolution barrier prevented cell biologists from using standard light microscopy when studying **organelles**, the membrane-enclosed structures within eukaryotic cells. To see these structures in any detail required the development of a new instrument. In the 1950s, the electron microscope was introduced to biology. Rather than focusing light, the **electron microscope (EM)** focuses a beam of electrons through the specimen or onto its surface (see Appendix D). Resolution is inversely related to the wavelength of the light (or electrons) a microscope uses for imaging, and electron beams have much shorter wavelengths than visible light. Modern electron microscopes can theoretically achieve a resolution of about 0.002 nm, though in practice they usually cannot resolve structures smaller than about 2 nm across. Still, this is a 100-fold improvement over the standard light microscope.

The **scanning electron microscope (SEM)** is especially useful for detailed study of the topography

▼ Figure 6.2 The size range of cells. Most cells are between 1 and 100 μm in diameter (yellow region of chart) and their components are even smaller (see Figure 6.32), as are viruses. Notice that the scale along the left side is logarithmic, to accommodate the range of sizes shown. Starting at the top of the scale with 10 m and going down, each reference measurement marks a tenfold decrease in diameter or length. For a complete table of the metric system, see Appendix C.

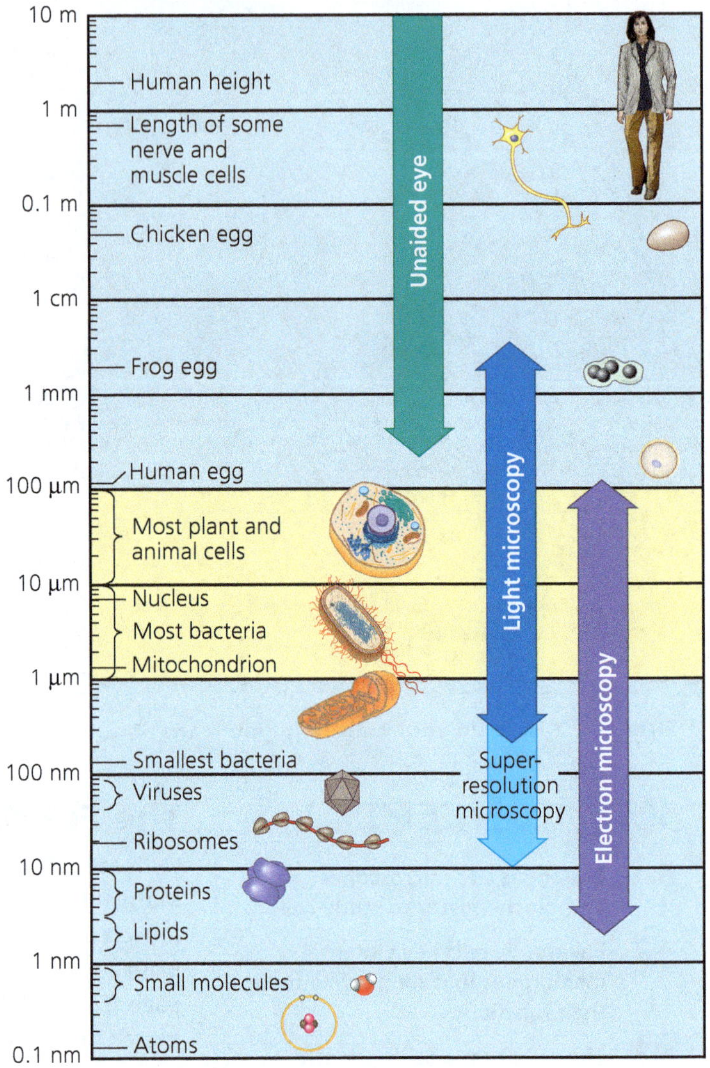

1 centimeter (cm) = 10^{-2} meter (m) = 0.4 inch
1 millimeter (mm) = 10^{-3} m
1 micrometer (μm) = 10^{-3} mm = 10^{-6} m
1 nanometer (nm) = 10^{-3} μm = 10^{-9} m

MB **Animation: Metric System Review**

Figure 6.3 Exploring Microscopy

Light Microscopy (LM)

Brightfield (unstained specimen). Light passes directly through the specimen. Unless the cell is naturally pigmented or artificially stained, the image has little contrast. (The first four light micrographs show human cheek epithelial cells; the scale bar pertains to all four micrographs.)

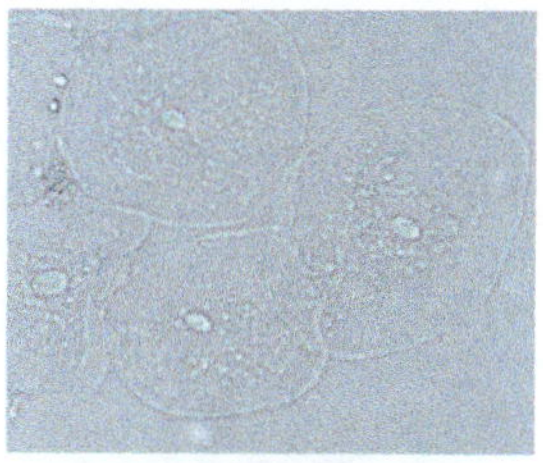

Brightfield (stained specimen). Staining with various dyes enhances contrast. Most staining procedures require that cells be fixed (preserved), thereby killing them.

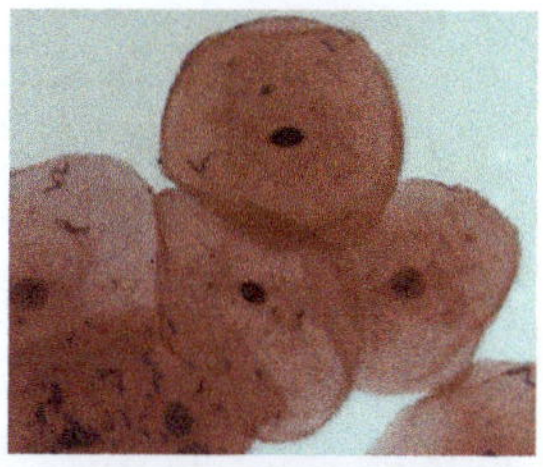

Phase-contrast. Variations in density within the specimen are amplified to enhance contrast in unstained cells; this is especially useful for examining living, unpigmented cells.

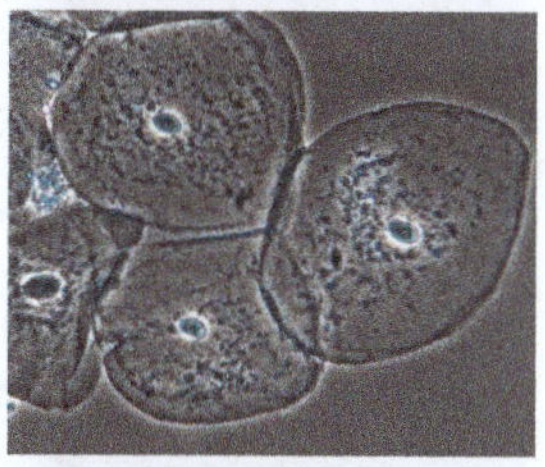

Differential interference contrast (Nomarski). As in phase-contrast microscopy, optical modifications are used to exaggerate differences in density; the image appears almost 3-D.

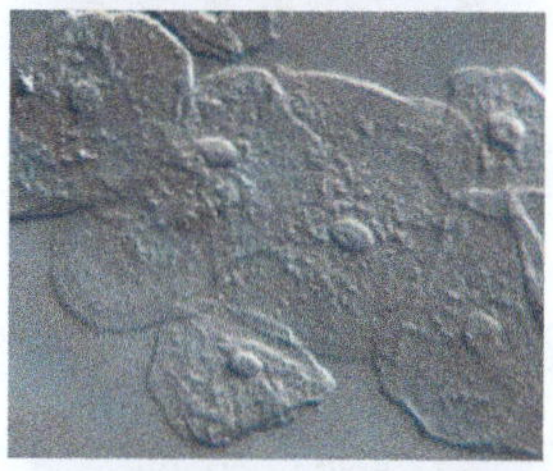

Fluorescence. The locations of specific molecules in the cell can be revealed by labeling the molecules with fluorescent dyes or antibodies; some cells have molecules that fluoresce on their own. Fluorescent substances absorb ultraviolet radiation and emit visible light. In this fluorescently labeled uterine cell, nuclear material is blue, organelles called mitochondria are orange, and the cell's "skeleton" is green.

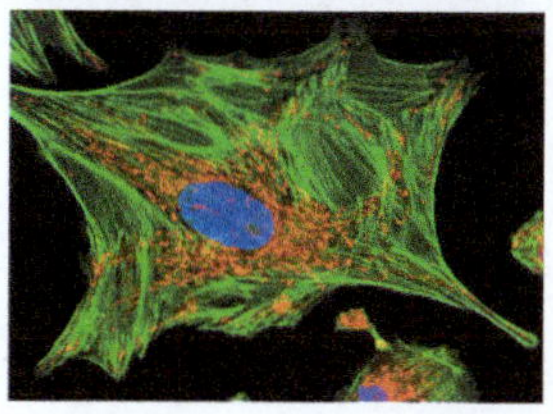

Confocal. The top image is a standard fluorescence micrograph of fluorescently labeled nervous tissue (nerve cells are green, support cells are orange, and regions of overlap are yellow); below it is a confocal image of the same tissue. Using a laser, this "optical sectioning" technique eliminates out-of-focus light from a thick sample, creating a single plane of fluorescence in the image. By capturing sharp images at many different planes, a 3-D reconstruction can be created. The standard image is blurry because out-of-focus light is not excluded.

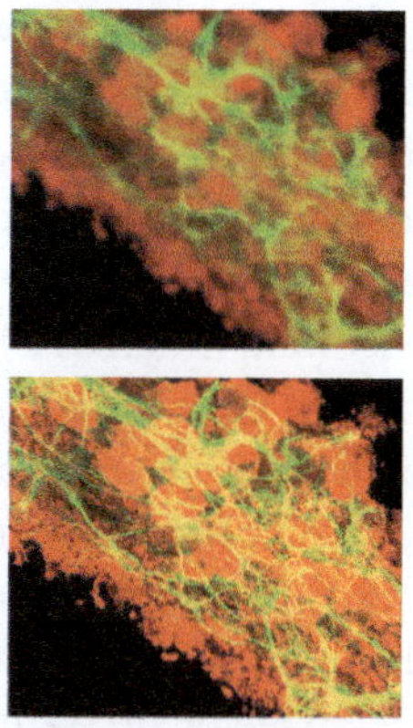

Deconvolution. The top of this split image is a compilation of standard fluorescence micrographs through the depth of a white blood cell. Below is an image of the same cell reconstructed from many blurry images at different planes, each of which was processed using deconvolution software. This process digitally removes out-of-focus light and reassigns it to its source, creating a much sharper 3-D image.

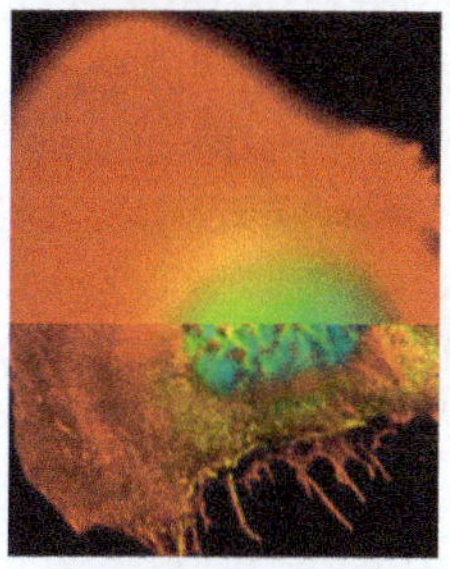

Super-resolution. On the top is a confocal image of part of a nerve cell, using a fluorescent label that binds to a molecule clustered in small sacs in the cell (vesicles) that are 40 nm in diameter. The greenish-yellow spots are blurry because 40 nm is below the 200-nm limit of resolution for standard light microscopy. Below is an image of the same part of the cell, seen using a new super-resolution technique. Sophisticated equipment is used to light up individual fluorescent molecules and record their position. Combining information from many molecules in different places "breaks" the limit of resolution, resulting in the sharp greenish-yellow dots seen here. (Each dot is a 40-nm vesicle.)

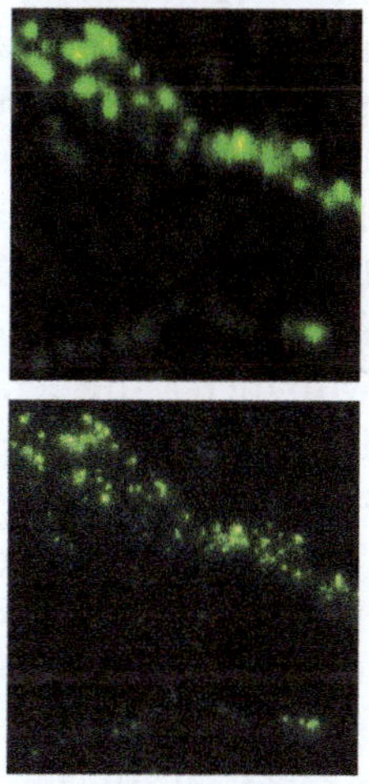

Electron Microscopy (EM)

Scanning electron microscopy (SEM). Micrographs taken with a scanning electron microscope show a 3-D image of the surface of a specimen. This SEM shows the surface of a cell from a trachea (windpipe) covered with cell projections called cilia. Electron micrographs are black and white but are often artificially colorized to highlight particular structures, as has been done with both electron micrographs shown here.

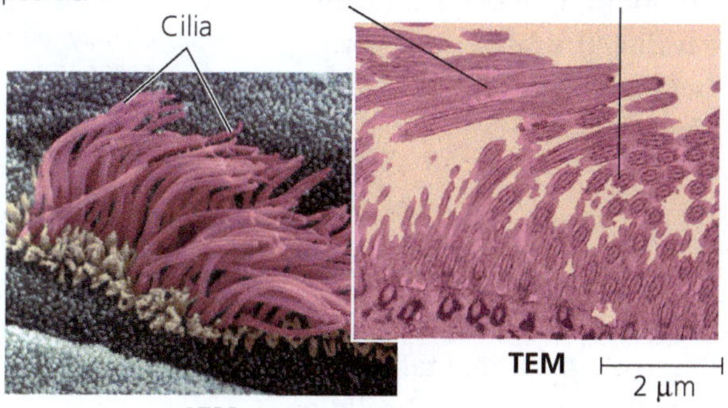

Transmission electron microscopy (TEM). A transmission electron microscope profiles a thin section of a specimen. This TEM shows a section through a tracheal cell, revealing its internal structure. In preparing the specimen, some cilia were cut along their lengths, creating longitudinal sections, while other cilia were cut straight across, creating cross sections.

Abbreviations used in figure legends in this text:
LM = Light Micrograph
SEM = Scanning Electron Micrograph
TEM = Transmission Electron Micrograph

VISUAL SKILLS ➤ *When the tissue was sliced for the TEM, what was the orientation of the cilia in the upper left? On the right? Explain how the orientation determined the type of section we see.*

of a specimen (see Figure 6.3). The electron beam scans the surface of the sample, usually coated with a thin film of gold. The beam excites electrons on the surface, and these secondary electrons are detected by a device that translates the pattern of electrons into an electronic signal sent to a video screen. The result is an image of the specimen's surface that appears three-dimensional.

The **transmission electron microscope (TEM)** is used to study the internal structure of cells (see Figure 6.3). The TEM aims an electron beam through a very thin section of the specimen, much as a light microscope aims light through a sample on a slide. For the TEM, the specimen has been stained with atoms of heavy metals, which attach to certain cellular structures, thus enhancing the electron density of some parts of the cell more than others. The electrons passing through the specimen are scattered more in the denser regions, so fewer are transmitted. The image displays the pattern of transmitted electrons. Instead of using glass lenses, both the SEM and TEM use electromagnets as lenses to bend the paths of the electrons, ultimately focusing the image onto a monitor for viewing.

Electron microscopes have revealed many subcellular structures that were impossible to resolve with the light microscope. But the light microscope offers advantages, especially in studying living cells. A disadvantage of electron microscopy is that the methods used to prepare the specimen kill the cells. Specimen preparation for any type of microscopy can introduce artifacts, structural features seen in micrographs that do not exist in the living cell.

In the past several decades, light microscopy has been revitalized by major technical advances (see Figure 6.3). Labeling individual cellular molecules or structures with fluorescent markers has made it possible to see such structures with increasing detail. In addition, both confocal and deconvolution microscopy have produced sharper images of three-dimensional tissues and cells. Finally, a group of new techniques and labeling molecules developed in recent years has allowed researchers to "break" the resolution barrier and distinguish subcellular structures even as small as 10–20 nm across. As this *super-resolution microscopy* becomes more widespread, the images we see of living cells are proving as exciting for us as van Leeuwenhoek's were for Robert Hooke 350 years ago.

Microscopes are the most important tools of *cytology*, the study of cell structure. Understanding the function of each structure, however, required the integration of cytology and *biochemistry*, the study of the chemical processes (metabolism) of cells.

Cell Fractionation

A useful technique for studying cell structure and function is **cell fractionation (Figure 6.4)**, which takes cells apart and separates major organelles and other subcellular

▼ Figure 6.4

Research Method Cell Fractionation

Application Cell fractionation is used to separate (fractionate) cell components based on size and density.

Technique Cells are homogenized in a blender to break them up. The resulting mixture (homogenate) is centrifuged. The supernatant (the liquid above the pellet) is poured into another tube and centrifuged at a higher speed for a longer period. This process is repeated several times. This "differential centrifugation" results in a series of pellets, each containing different cell components.

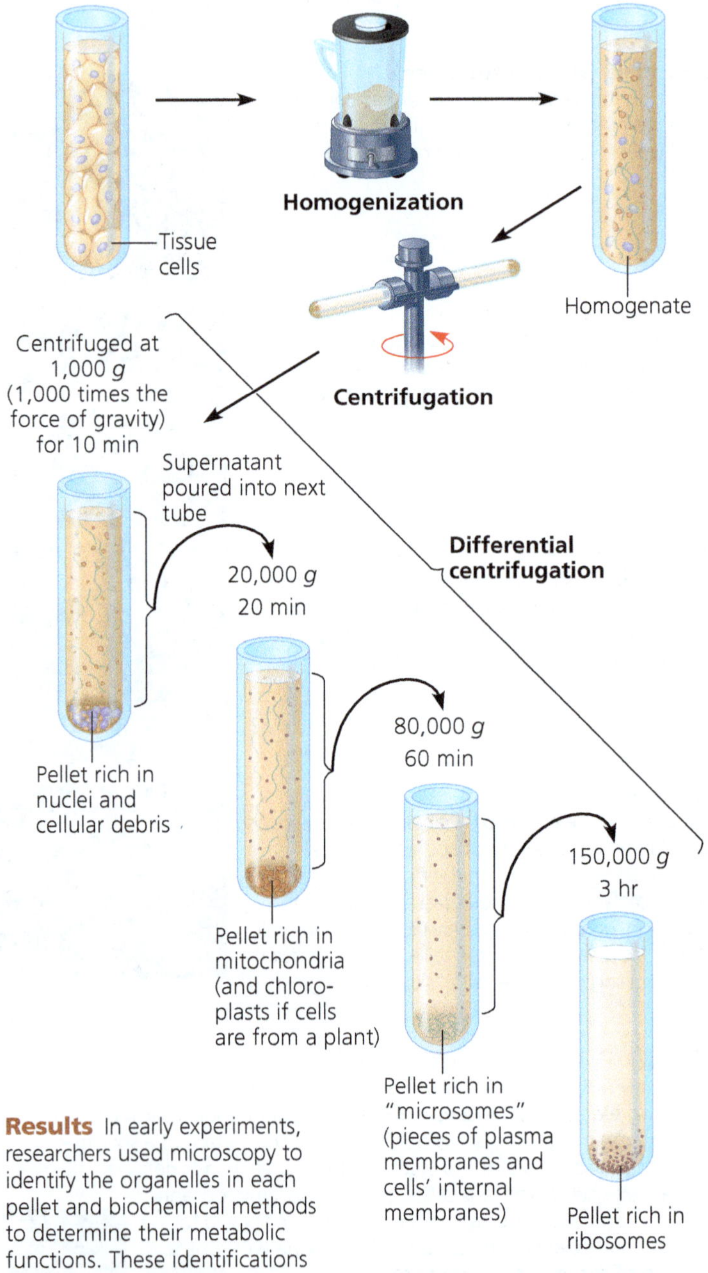

Results In early experiments, researchers used microscopy to identify the organelles in each pellet and biochemical methods to determine their metabolic functions. These identifications established a baseline for this method, enabling today's researchers to know which cell fraction they should collect in order to isolate and study particular organelles.

MAKE CONNECTIONS ➤ *If you wanted to study the process of translation of proteins from mRNA, which part of which fraction would you use? (See Figure 5.22.)*

structures from one another. The piece of equipment that is used for this task is the centrifuge, which spins test tubes holding mixtures of disrupted cells at a series of increasing speeds. At each speed, the resulting force causes a subset of the cell components to settle to the bottom of the tube, forming a pellet. At lower speeds, the pellet consists of larger components, and higher speeds result in a pellet with smaller components.

Cell fractionation enables researchers to prepare specific cell components in bulk and identify their functions, a task not usually possible with intact cells. For example, on one of the cell fractions, biochemical tests showed the presence of enzymes involved in cellular respiration, while electron microscopy revealed large numbers of the organelles called mitochondria. Together, these data helped biologists determine that mitochondria are the sites of cellular respiration. Biochemistry and cytology thus complement each other in correlating cell function with structure.

CONCEPT CHECK 6.1

1. How do stains used for light microscopy compare with those used for electron microscopy?
2. **WHAT IF?** ➤ Which type of microscope would you use to study (a) the changes in shape of a living white blood cell and (b) the details of surface texture of a hair?

For suggested answers, see Appendix A.

CONCEPT 6.2

Eukaryotic cells have internal membranes that compartmentalize their functions

Cells—the basic structural and functional units of every organism—are of two distinct types: prokaryotic and eukaryotic. Organisms of the domains Bacteria and Archaea consist of prokaryotic cells. Protists, fungi, animals, and plants all consist of eukaryotic cells. ("Protist" is an informal term referring to a diverse group of mostly unicellular eukaryotes.)

Comparing Prokaryotic and Eukaryotic Cells

All cells share certain basic features: They are all bounded by a selective barrier, called the *plasma membrane* (also referred to as the cell membrane). Inside all cells is a semifluid, jellylike substance called **cytosol**, in which subcellular components are suspended. All cells contain *chromosomes*, which carry genes in the form of DNA. And all cells have *ribosomes*, tiny complexes that make proteins according to instructions from the genes.

A major difference between prokaryotic and eukaryotic cells is the location of their DNA. In a **eukaryotic cell**, most of the DNA is in an organelle called the *nucleus*, which is bounded by a double membrane (see Figure 6.8). In a **prokaryotic cell**, the DNA is concentrated in a region that is not membrane-enclosed, called the **nucleoid** (**Figure 6.5**).

Figure 6.5 A prokaryotic cell. Lacking a true nucleus and the other membrane-enclosed organelles of the eukaryotic cell, the prokaryotic cell appears much simpler in internal structure. Prokaryotes include bacteria and archaea; the general cell structure of these two domains is quite similar.

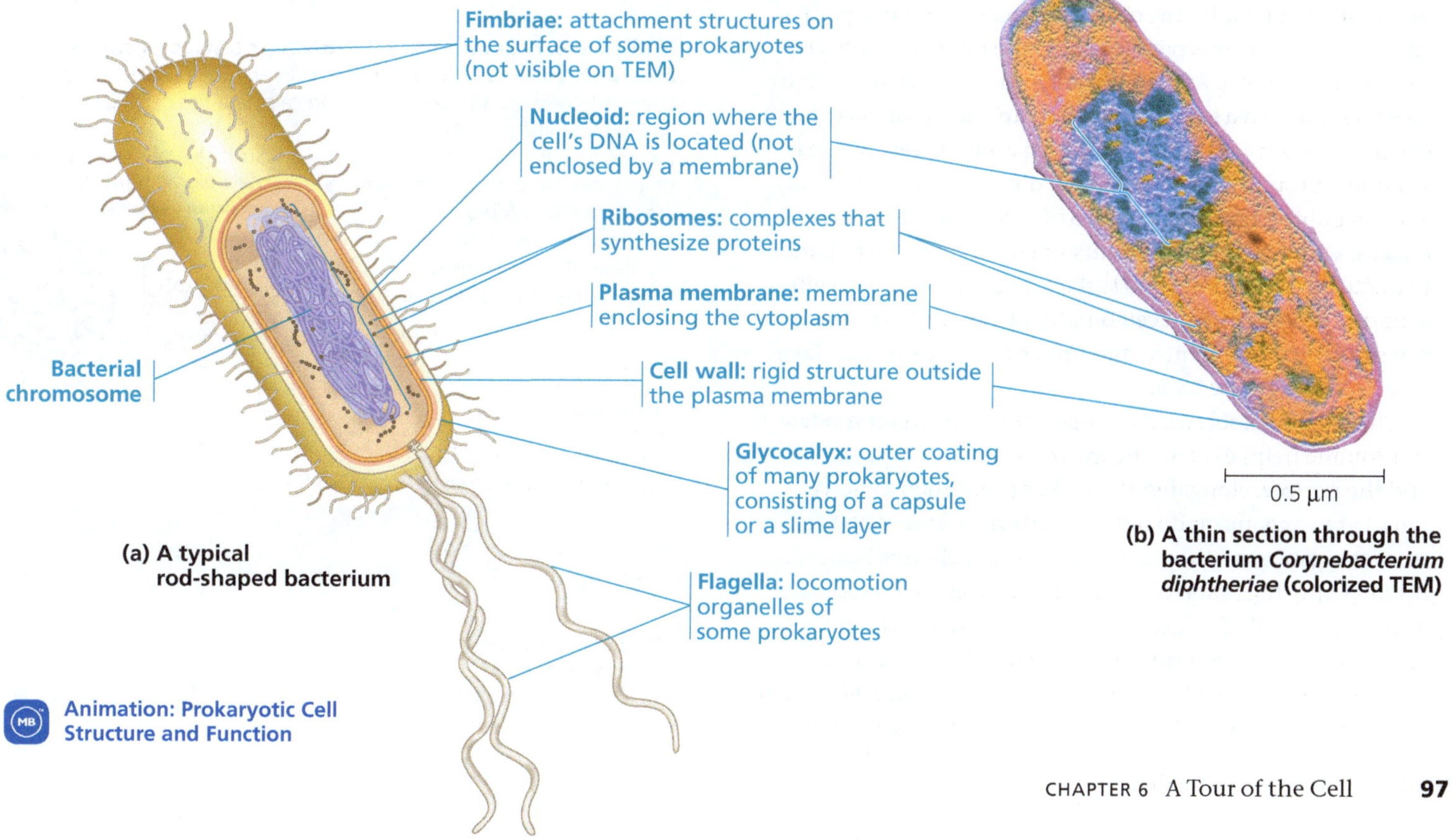

(a) A typical rod-shaped bacterium

(b) A thin section through the bacterium *Corynebacterium diphtheriae* (colorized TEM)

Animation: Prokaryotic Cell Structure and Function

Eukaryotic means "true nucleus" (from the Greek *eu*, true, and *karyon*, kernel, referring to the nucleus), and *prokaryotic* means "before nucleus" (from the Greek *pro*, before), reflecting the earlier evolution of prokaryotic cells.

The interior of either type of cell is called the **cytoplasm**; in eukaryotic cells, this term refers only to the region between the nucleus and the plasma membrane. Within the cytoplasm of a eukaryotic cell, suspended in cytosol, are a variety of organelles of specialized form and function. These membrane-bounded structures are absent in almost all prokaryotic cells, another distinction between prokaryotic and eukaryotic cells. In spite of the absence of organelles, though, the prokaryotic cytoplasm is not a formless soup. For example, some prokaryotes contain regions surrounded by proteins (not membranes), within which specific reactions take place.

Eukaryotic cells are generally much larger than prokaryotic cells (see Figure 6.2). Size is a general feature of cell structure that relates to function. The logistics of carrying out cellular metabolism sets limits on cell size. At the lower limit, the smallest cells known are bacteria called mycoplasmas, which have diameters between 0.1 and 1.0 µm. These are perhaps the smallest packages with enough DNA to program metabolism and enough enzymes and other cellular equipment to carry out the activities necessary for a cell to sustain itself and reproduce. Typical bacteria are 1–5 µm in diameter, about ten times the size of mycoplasmas. Eukaryotic cells are typically 10–100 µm in diameter.

Metabolic requirements also impose theoretical upper limits on the size that is practical for a single cell. At the boundary of every cell, the **plasma membrane** functions as a selective barrier that allows passage of enough oxygen, nutrients, and wastes to service the entire cell **(Figure 6.6)**. For each square micrometer of membrane, only a limited amount of a particular substance can cross per second, so the ratio of surface area to volume is critical. As a cell (or any other object) increases in size, its surface area grows proportionately less than its volume. (Area is proportional to a linear dimension squared, whereas volume is proportional to the linear dimension cubed.) Thus, a smaller object has a greater ratio of surface area to volume **(Figure 6.7)**. The **Scientific Skills Exercise** gives you a chance to calculate the volumes and surface areas of two actual cells—a mature yeast cell and a cell budding from it. To see different ways organisms maximize the surface area of cells, see Make Connections Figure 33.9.

The need for a surface area large enough to accommodate the volume helps explain the microscopic size of most cells and the narrow, elongated shapes of others, such as nerve cells. Larger organisms do not generally have *larger* cells than smaller organisms—they simply have *more* cells (see Figure 6.7). A sufficiently high ratio of surface area to volume is especially important in cells that exchange a lot of material with their surroundings, such as intestinal cells. Such cells may have many long, thin projections from their surface called *microvilli*, which increase surface area without an appreciable increase in volume.

▼ Figure 6.6 The plasma membrane. The plasma membrane and the membranes of organelles consist of a double layer (bilayer) of phospholipids with various proteins attached to or embedded in it. The hydrophobic parts of phospholipids and membrane proteins are found in the interior of the membrane, while the hydrophilic parts are in contact with aqueous solutions on either side. Carbohydrate side chains may be attached to proteins or lipids on the outer surface of the plasma membrane.

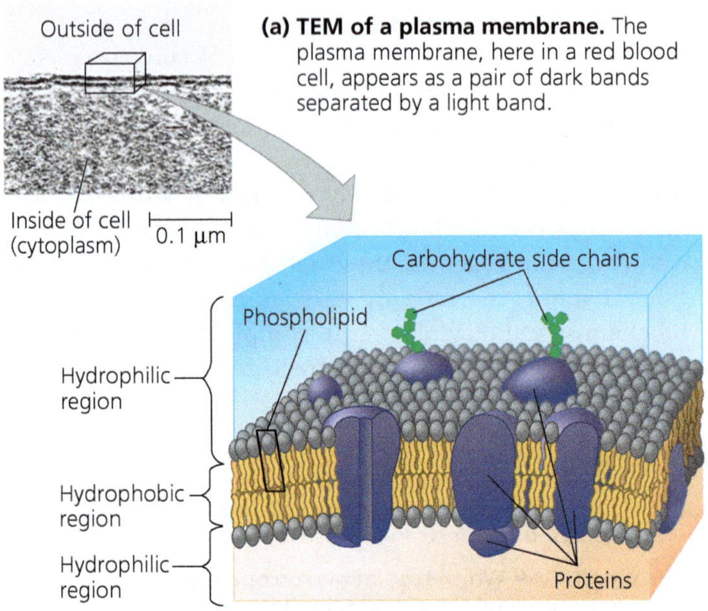

(a) TEM of a plasma membrane. The plasma membrane, here in a red blood cell, appears as a pair of dark bands separated by a light band.

(b) Structure of the plasma membrane

VISUAL SKILLS ➤ *What parts of the membrane diagram in (b) correspond to the dark bands, and which to the light band, in the TEM in (a)? (Review Figure 5.11.)*

BioFlix® Animation: Membranes

▼ Figure 6.7 Geometric relationships between surface area and volume. In this diagram, cells are represented as boxes. Using arbitrary units of length, we can calculate the cell's surface area (in square units, or units^2), volume (in cubic units, or units^3), and ratio of surface area to volume. A high surface-to-volume ratio facilitates the exchange of materials between a cell and its environment.

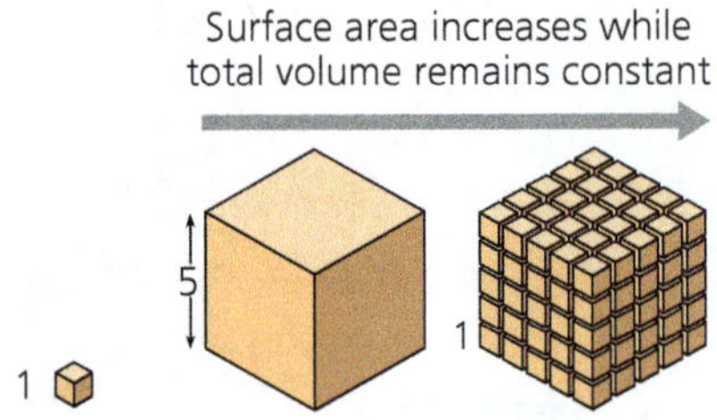

Total surface area [sum of the surface areas (height × width) of all box sides × number of boxes]	6	150	750
Total volume [height × width × length × number of boxes]	1	125	125
Surface-to-volume (S-to-V) ratio [surface area ÷ volume]	6	1.2	6

SCIENTIFIC SKILLS EXERCISE

Using a Scale Bar to Calculate Volume and Surface Area of a Cell

How Much New Cytoplasm and Plasma Membrane Are Made by a Growing Yeast Cell? The unicellular yeast *Saccharomyces cerevisiae* divides by budding off a small new cell that then grows to full size (see the yeast cells at the bottom of Figure 6.8). During its growth, the new cell synthesizes new cytoplasm, which increases its volume, and new plasma membrane, which increases its surface area. In this exercise, you will use a scale bar to determine the sizes of a mature parent yeast cell and a cell budding from it. You will then calculate the volume and surface area of each cell. You will use your calculations to determine how much cytoplasm and plasma membrane the new cell needs to synthesize to grow to full size.

How the Experiment Was Done Yeast cells were grown under conditions that promoted division by budding. The cells were then viewed with a differential interference contrast light microscope and photographed.

Data from the Experiment This light micrograph shows a budding yeast cell about to be released from the mature parent cell:

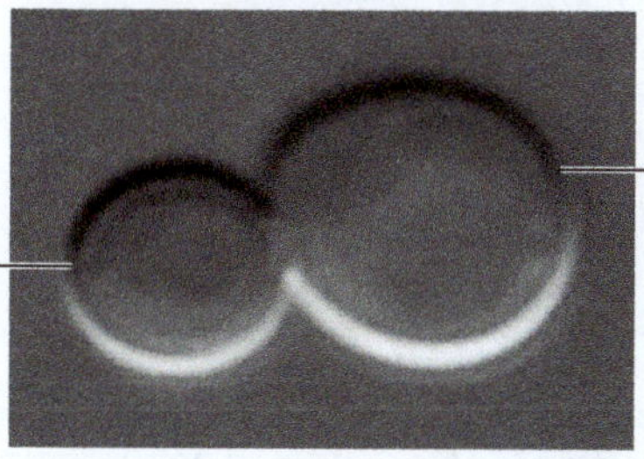

Micrograph from Kelly Tatchell, using yeast cells grown for experiments described in L. Kozubowski et al., Role of the septin ring in the asymmetric localization of proteins at the mother-bud neck in *Saccharomyces cerevisiae, Molecular Biology of the Cell* 16:3455–3466 (2005).

INTERPRET THE DATA

1. Examine the micrograph of the yeast cells. The scale bar under the photo is labeled 1 µm. The scale bar works in the same way as a scale on a map, where, for example, 1 inch equals 1 mile. In this case the bar represents one thousandth of a millimeter. Using the scale bar as a basic unit, determine the diameter of the mature parent cell and the new cell. Start by measuring the scale bar and the diameter of each cell. The units you use are irrelevant, but working in millimeters is convenient. Divide each diameter by the length of the scale bar and then multiply by the scale bar's length value to give you the diameter in micrometers.
2. The shape of a yeast cell can be approximated by a sphere. (a) Calculate the volume of each cell using the formula for the volume of a sphere:

$$V = \frac{4}{3}\pi r^3$$

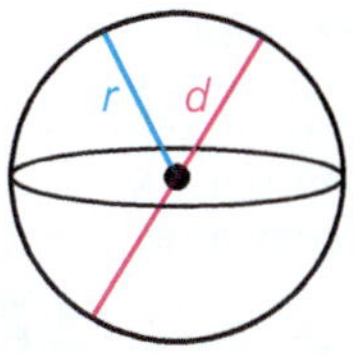

Note that π (the Greek letter pi) is a constant with an approximate value of 3.14, *d* stands for diameter, and *r* stands for radius, which is half the diameter. (b) What volume of new cytoplasm will the new cell have to synthesize as it matures? To determine this, calculate the difference between the volume of the full-sized cell and the volume of the new cell.

3. As the new cell grows, its plasma membrane needs to expand to contain the increased volume of the cell. (a) Calculate the surface area of each cell using the formula for the surface area of a sphere: $A = 4\pi r^2$. (b) How much area of new plasma membrane will the new cell have to synthesize as it matures?
4. When the new cell matures, it will be approximately how many times greater in volume and how many times greater in surface area than its current size?

Instructors: A version of this Scientific Skills Exercise can be assigned in MasteringBiology.

The evolutionary relationships between prokaryotic and eukaryotic cells will be discussed later in this chapter, and prokaryotic cells will be described in detail elsewhere (see Chapter 27). Most of the discussion of cell structure that follows in this chapter applies to eukaryotic cells.

A Panoramic View of the Eukaryotic Cell

In addition to the plasma membrane at its outer surface, a eukaryotic cell has extensive, elaborately arranged internal membranes that divide the cell into compartments—the organelles mentioned earlier. The cell's compartments provide different local environments that support specific metabolic functions, so incompatible processes can occur simultaneously in a single cell. The plasma membrane and organelle membranes also participate directly in the cell's metabolism because many enzymes are built right into the membranes.

The basic fabric of most biological membranes is a double layer of phospholipids and other lipids. Embedded in this lipid bilayer or attached to its surfaces are diverse proteins (see Figure 6.6). However, each type of membrane has a unique composition of lipids and proteins suited to that membrane's specific functions. For example, enzymes embedded in the membranes of the organelles called mitochondria function in cellular respiration. Because membranes are so fundamental to the organization of the cell, Chapter 7 will discuss them in detail.

Before continuing with this chapter, examine the eukaryotic cells in **Figure 6.8**. The generalized diagrams of an animal cell and a plant cell introduce the various organelles and show the key differences between animal and plant cells. The micrographs at the bottom of the figure give you a glimpse of cells from different types of eukaryotic organisms.

Figure 6.8 Exploring Eukaryotic Cells

Animal Cell (cutaway view of generalized cell)

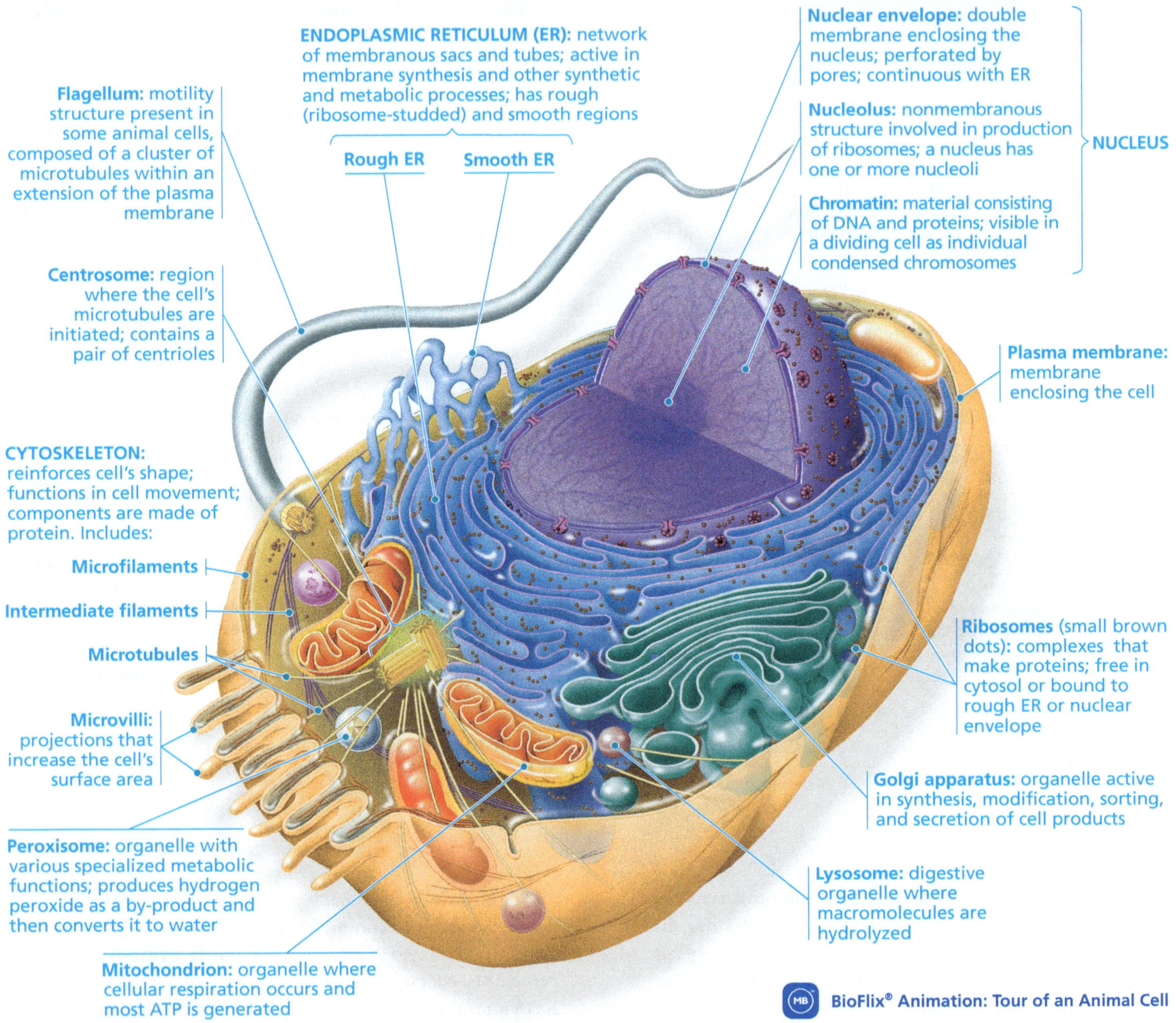

BioFlix® Animation: Tour of an Animal Cell

Animal Cells

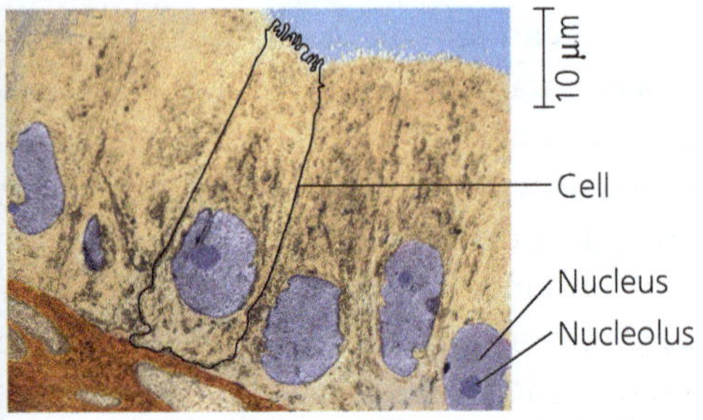

Human cells from lining of uterus (colorized TEM)

Fungal Cells

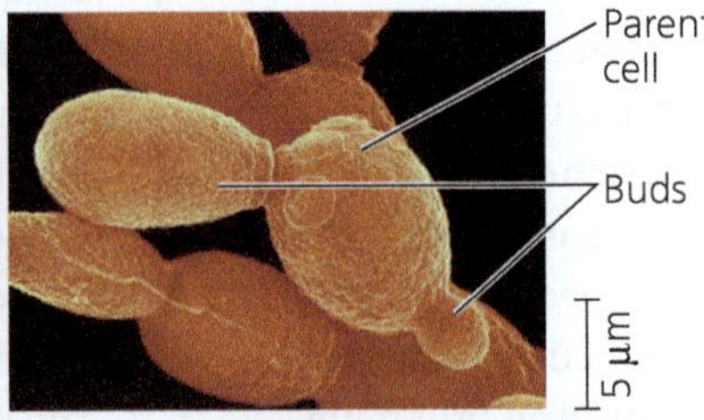

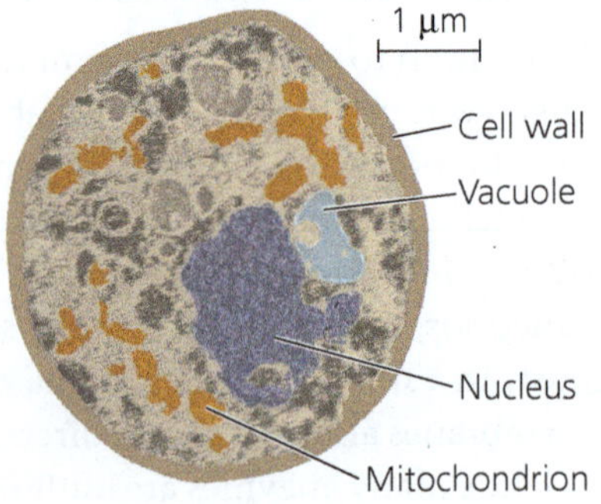

Yeast cells: reproducing by budding (above, colorized SEM) and a single cell (right, colorized TEM)

Plant Cell (cutaway view of generalized cell)

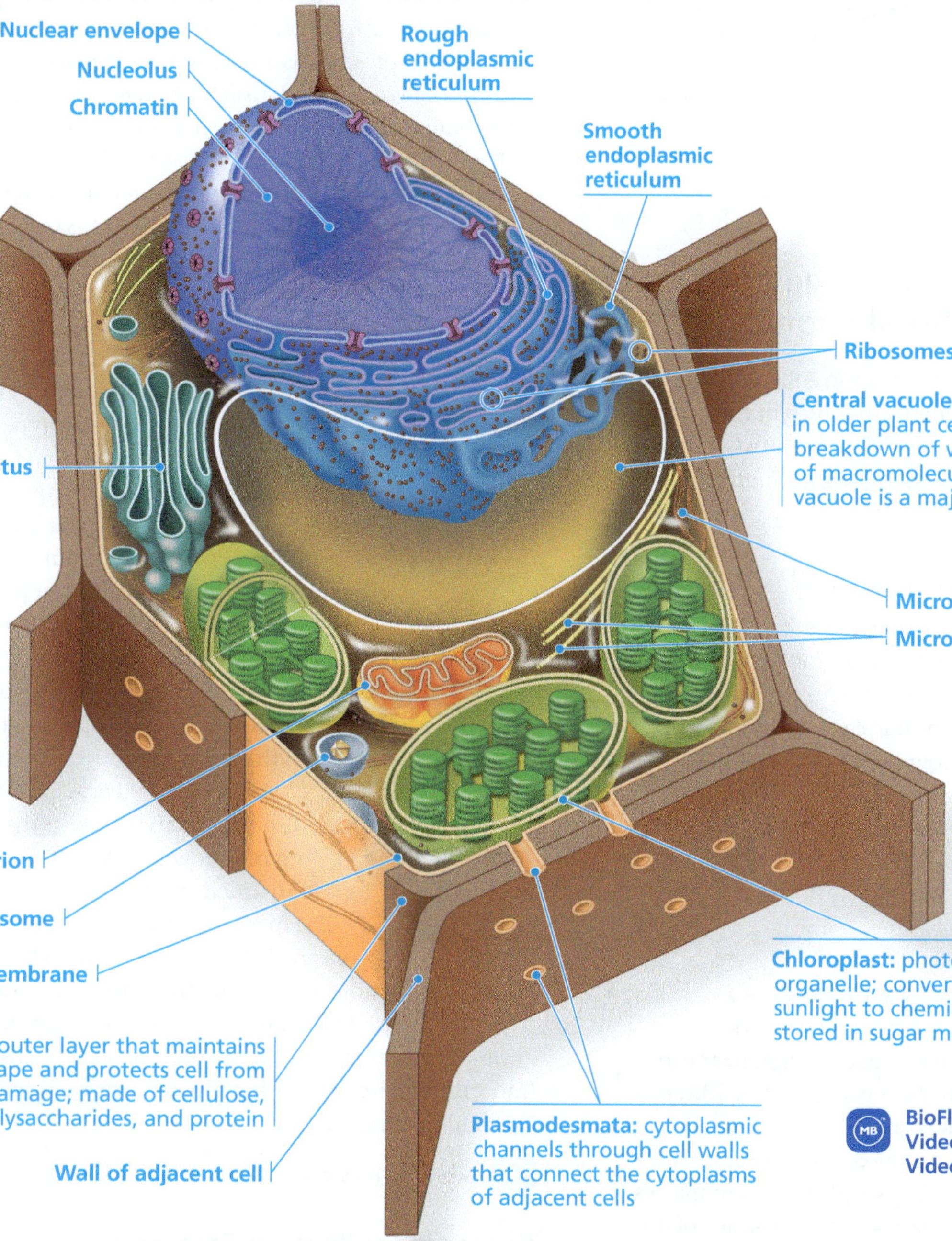

BioFlix® Animation: Tour of a Plant Cell
Video: Turgid *Elodea*
Video: *Chlamydomonas*

Plant Cells

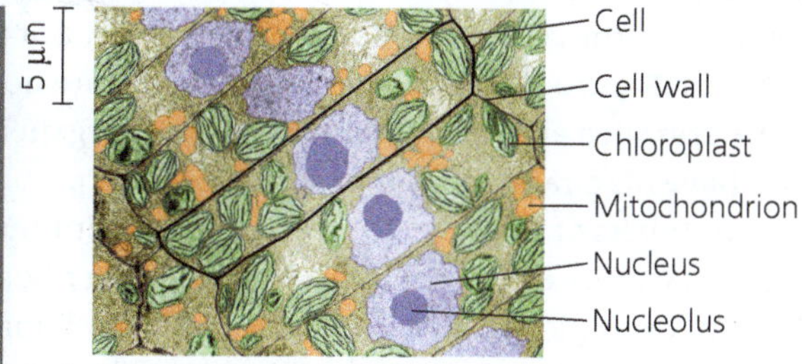

Cells from duckweed (*Spirodela oligorrhiza*), a floating plant (colorized TEM)

Unicellular Eukaryotes

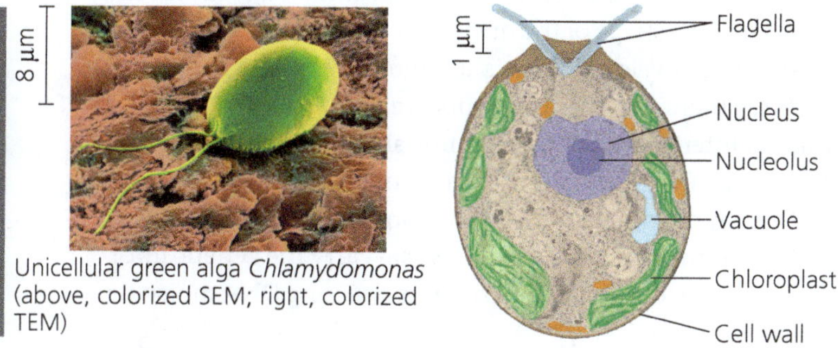

Unicellular green alga *Chlamydomonas* (above, colorized SEM; right, colorized TEM)

CONCEPT CHECK 6.2

1. Briefly describe the structure and function of the nucleus, the mitochondrion, the chloroplast, and the endoplasmic reticulum.
2. **DRAW IT ➤** Draw a simplified elongated cell that measures 125 × 1 × 1 arbitrary units. A nerve cell would be roughly this shape. Predict how its surface-to-volume ratio would compare with those in Figure 6.7. Then calculate the ratio and check your prediction.

For suggested answers, see Appendix A.

CONCEPT 6.3

The eukaryotic cell's genetic instructions are housed in the nucleus and carried out by the ribosomes

On the first stop of our detailed tour of the eukaryotic cell, let's look at two cellular components involved in the genetic control of the cell: the nucleus, which houses most of the cell's DNA, and the ribosomes, which use information from the DNA to make proteins.

The Nucleus: Information Central

The **nucleus** contains most of the genes in the eukaryotic cell. (Some genes are located in mitochondria and chloroplasts.) It is usually the most conspicuous organelle (see the purple structure in the fluorescence micrograph), averaging about 5 μm in diameter. The **nuclear envelope** encloses the nucleus **(Figure 6.9)**, separating its contents from the cytoplasm.

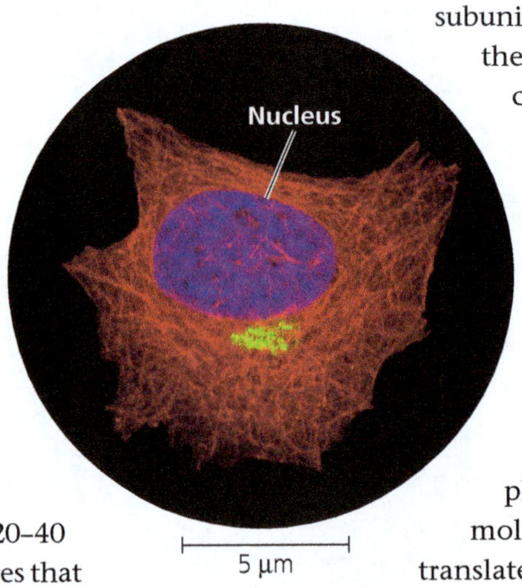

The nuclear envelope is a *double* membrane. The two membranes, each a lipid bilayer with associated proteins, are separated by a space of 20–40 nm. The envelope is perforated by pore structures that are about 100 nm in diameter. At the lip of each pore, the inner and outer membranes of the nuclear envelope are continuous. An intricate protein structure called a *pore complex* lines each pore and plays an important role in the cell by regulating the entry and exit of proteins and RNAs, as well as large complexes of macromolecules. Except at the pores, the nuclear side of the envelope is lined by the **nuclear lamina**, a netlike array of protein filaments (in animal cells, called *intermediate filaments*) that maintains the shape of the nucleus by mechanically supporting the nuclear envelope. There is also much evidence for a *nuclear matrix*, a framework of protein fibers extending throughout the nuclear interior. The nuclear lamina and matrix may help organize the genetic material so it functions efficiently.

Within the nucleus, the DNA is organized into discrete units called **chromosomes**, structures that carry the genetic information. Each chromosome contains one long DNA molecule associated with many proteins. Some of the proteins help coil the DNA molecule of each chromosome, reducing its length and allowing it to fit into the nucleus. The complex of DNA and proteins making up chromosomes is called **chromatin**. When a cell is not dividing, stained chromatin appears as a diffuse mass in micrographs, and the chromosomes cannot be distinguished from one another, even though discrete chromosomes are present. As a cell prepares to divide, however, the chromosomes coil (condense) further, becoming thick enough to be distinguished under a microscope as separate structures. Each eukaryotic species has a characteristic number of chromosomes. For example, a typical human cell has 46 chromosomes in its nucleus; the exceptions are the sex cells (eggs and sperm), which have only 23 chromosomes in humans. A fruit fly cell has 8 chromosomes in most cells and 4 in the sex cells.

A prominent structure within the nondividing nucleus is the **nucleolus** (plural, *nucleoli*), which appears through the electron microscope as a mass of densely stained granules and fibers adjoining part of the chromatin. Here a type of RNA called *ribosomal RNA* (rRNA) is synthesized from instructions in the DNA. Also in the nucleolus, proteins imported from the cytoplasm are assembled with rRNA into large and small subunits of ribosomes. These subunits then exit the nucleus through the nuclear pores to the cytoplasm, where a large and a small subunit can assemble into a ribosome. Sometimes there are two or more nucleoli; the number depends on the species and the stage in the cell's reproductive cycle.

As we saw in Figure 5.22, the nucleus directs protein synthesis by synthesizing messenger RNA (mRNA) according to instructions provided by the DNA. The mRNA is then transported to the cytoplasm via the nuclear pores. Once an mRNA molecule reaches the cytoplasm, ribosomes translate the mRNA's genetic message into the primary structure of a specific polypeptide. (This process of transcribing and translating genetic information is described in detail in Chapter 17.)

MB BioFlix® Animation: Nucleus and Ribosomes

Ribosomes: Protein Factories

Ribosomes, which are complexes made of ribosomal RNAs and proteins, are the cellular components that carry out protein synthesis **(Figure 6.10)**. (Note that ribosomes are not membrane bounded and thus are not considered organelles.) Cells that have high rates of protein synthesis have particularly large numbers of ribosomes as well as prominent nucleoli, which makes sense, given the role of nucleoli in ribosome assembly. For example, a human pancreas cell, which makes many digestive enzymes, has a few million ribosomes.

Figure 6.9 The nucleus and its envelope. Within the nucleus are the chromosomes, which appear as a mass of chromatin (DNA and associated proteins) and one or more nucleoli (singular, *nucleolus*), which function in ribosome synthesis. The nuclear envelope, which consists of two membranes separated by a narrow space, is perforated with pores and lined by the nuclear lamina.

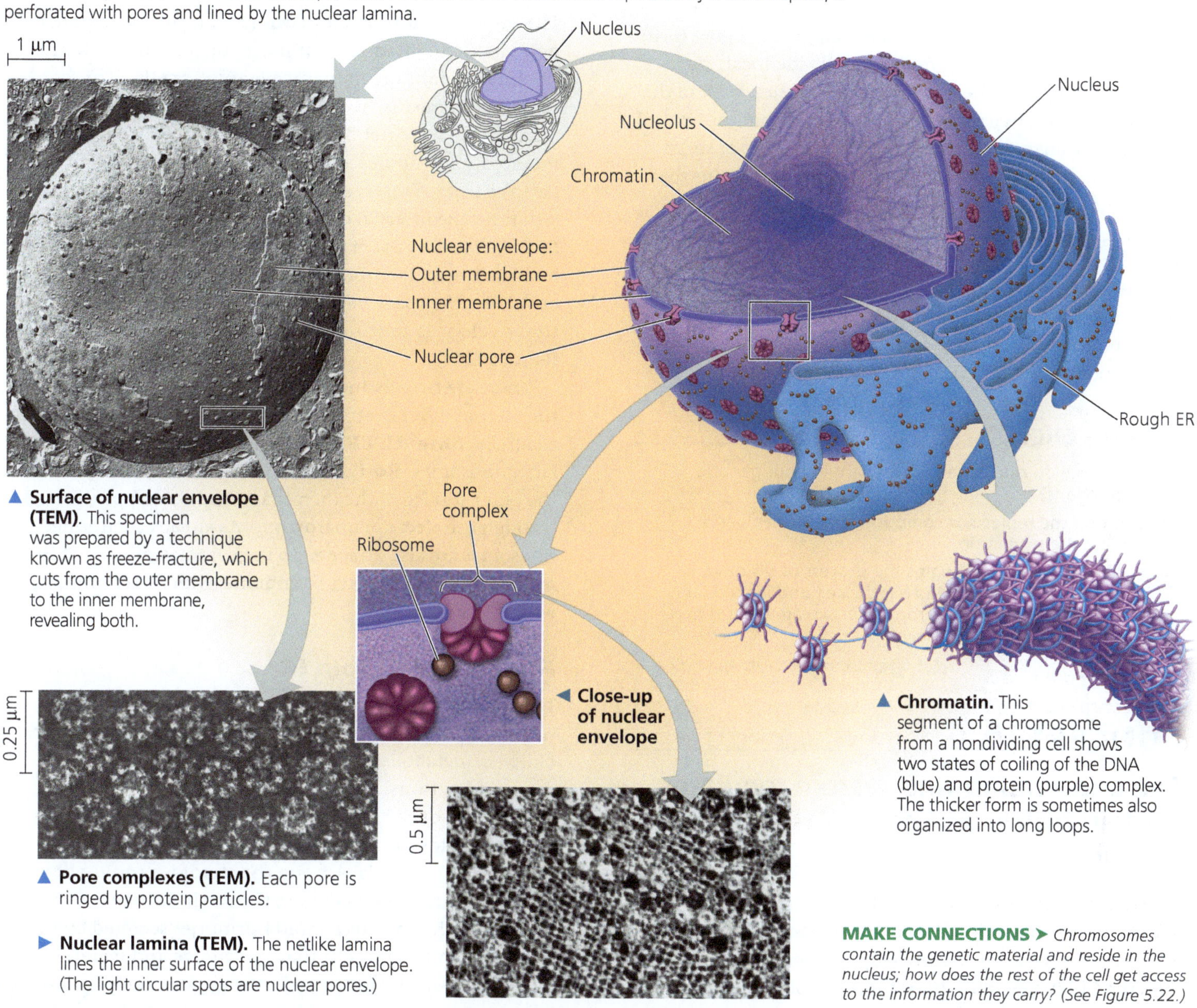

▲ **Surface of nuclear envelope (TEM).** This specimen was prepared by a technique known as freeze-fracture, which cuts from the outer membrane to the inner membrane, revealing both.

◀ **Close-up of nuclear envelope**

▲ **Chromatin.** This segment of a chromosome from a nondividing cell shows two states of coiling of the DNA (blue) and protein (purple) complex. The thicker form is sometimes also organized into long loops.

▲ **Pore complexes (TEM).** Each pore is ringed by protein particles.

▶ **Nuclear lamina (TEM).** The netlike lamina lines the inner surface of the nuclear envelope. (The light circular spots are nuclear pores.)

MAKE CONNECTIONS ➤ *Chromosomes contain the genetic material and reside in the nucleus; how does the rest of the cell get access to the information they carry? (See Figure 5.22.)*

Figure 6.10 Ribosomes. This electron micrograph of a pancreas cell shows both free and bound ribosomes. The simplified diagram and computer model show the two subunits of a ribosome.

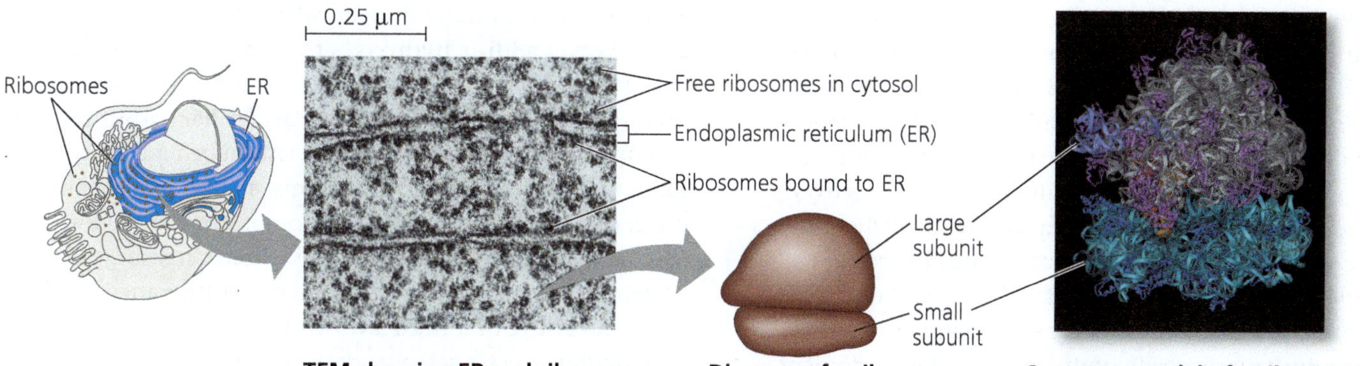

TEM showing ER and ribosomes **Diagram of a ribosome** **Computer model of a ribosome**

DRAW IT ➤ *After you have read the section on ribosomes, circle a ribosome in the micrograph that might be making a protein that will be secreted.*

Interview with Venki Ramakrishnan: Studying ribosome structure

Ribosomes build proteins in two cytoplasmic locales. At any given time, *free ribosomes* are suspended in the cytosol, while *bound ribosomes* are attached to the outside of the endoplasmic reticulum or nuclear envelope (see Figure 6.10). Bound and free ribosomes are structurally identical, and ribosomes can play either role at different times. Most of the proteins made on free ribosomes function within the cytosol; examples are enzymes that catalyze the first steps of sugar breakdown. Bound ribosomes generally make proteins that are destined for insertion into membranes, for packaging within certain organelles such as lysosomes (see Figure 6.8), or for export from the cell (secretion). Cells that specialize in protein secretion—for instance, the cells of the pancreas that secrete digestive enzymes—frequently have a high proportion of bound ribosomes. (You will learn more about ribosome structure and function in Concept 17.4.)

CONCEPT CHECK 6.3

1. What role do ribosomes play in carrying out genetic instructions?
2. Describe the molecular composition of nucleoli and explain their function.
3. **WHAT IF? >** As a cell begins the process of dividing, its chromosomes become shorter, thicker, and individually visible in an LM (light micrograph). Explain what is happening at the molecular level.

For suggested answers, see Appendix A.

CONCEPT 6.4

The endomembrane system regulates protein traffic and performs metabolic functions

Many of the different membrane-bounded organelles of the eukaryotic cell are part of the **endomembrane system**, which includes the nuclear envelope, the endoplasmic reticulum, the Golgi apparatus, lysosomes, various kinds of vesicles and vacuoles, and the plasma membrane. This system carries out a variety of tasks in the cell, including synthesis of proteins, transport of proteins into membranes and organelles or out of the cell, metabolism and movement of lipids, and detoxification of poisons. The membranes of this system are related either through direct physical continuity or by the transfer of membrane segments as tiny **vesicles** (sacs made of membrane). Despite these relationships, the various membranes are not identical in structure and function. Moreover, the thickness, molecular composition, and types of chemical reactions carried out in a given membrane are not fixed, but may be modified several times during the membrane's life. Having already discussed the nuclear envelope, we will now focus on the endoplasmic reticulum and the other endomembranes to which the endoplasmic reticulum gives rise.

The Endoplasmic Reticulum: Biosynthetic Factory

The **endoplasmic reticulum (ER)** is such an extensive network of membranes that it accounts for more than half the total membrane in many eukaryotic cells. (The word *endoplasmic* means "within the cytoplasm," and *reticulum* is Latin for "little net.") The ER consists of a network of membranous tubules and sacs called cisternae (from the Latin *cisterna*, a reservoir for a liquid). The ER membrane separates the internal compartment of the ER, called the *ER lumen* (cavity) or cisternal space, from the cytosol. And because the ER membrane is continuous with the nuclear envelope, the space between the two membranes of the envelope is continuous with the lumen of the ER **(Figure 6.11)**.

There are two distinct, though connected, regions of the ER that differ in structure and function: smooth ER and rough ER. **Smooth ER** is so named because its outer surface lacks ribosomes. **Rough ER** is studded with ribosomes on the outer surface of the membrane and thus appears rough through the electron microscope. As already mentioned, ribosomes are also attached to the cytoplasmic side of the nuclear envelope's outer membrane, which is continuous with rough ER.

Functions of Smooth ER

The smooth ER functions in diverse metabolic processes, which vary with cell type. These processes include synthesis of lipids, metabolism of carbohydrates, detoxification of drugs and poisons, and storage of calcium ions.

Enzymes of the smooth ER are important in the synthesis of lipids, including oils, steroids, and new membrane phospholipids. Among the steroids produced by the smooth ER in animal cells are the sex hormones of vertebrates and the various steroid hormones secreted by the adrenal glands. The cells that synthesize and secrete these hormones—in the testes and ovaries, for example—are rich in smooth ER, a structural feature that fits the function of these cells.

Other enzymes of the smooth ER help detoxify drugs and poisons, especially in liver cells. Detoxification usually involves adding hydroxyl groups to drug molecules, making them more soluble and easier to flush from the body. The sedative phenobarbital and other barbiturates are examples of drugs metabolized in this manner by smooth ER in liver cells. In fact, barbiturates, alcohol, and many other drugs induce the proliferation of smooth ER and its associated detoxification enzymes, thus increasing the rate of detoxification. This, in turn, increases tolerance to the drugs, meaning that higher doses are required to achieve a particular effect, such as sedation. Also, because some of the detoxification enzymes have relatively broad action, the proliferation of smooth ER in response to one drug can increase the need for higher dosages of other drugs as well. Barbiturate abuse, for example, can decrease the effectiveness of certain antibiotics and other useful drugs.

▼ Figure 6.11 Endoplasmic reticulum (ER). A membranous system of interconnected tubules and flattened sacs called cisternae, the ER is also continuous with the nuclear envelope, as shown in the cutaway diagram at the top. The membrane of the ER encloses a continuous compartment called the ER lumen (or cisternal space). Rough ER, which is studded on its outer surface with ribosomes, can be distinguished from smooth ER in the electron micrograph (TEM). Transport vesicles bud off from a region of the rough ER called transitional ER and travel to the Golgi apparatus and other destinations.

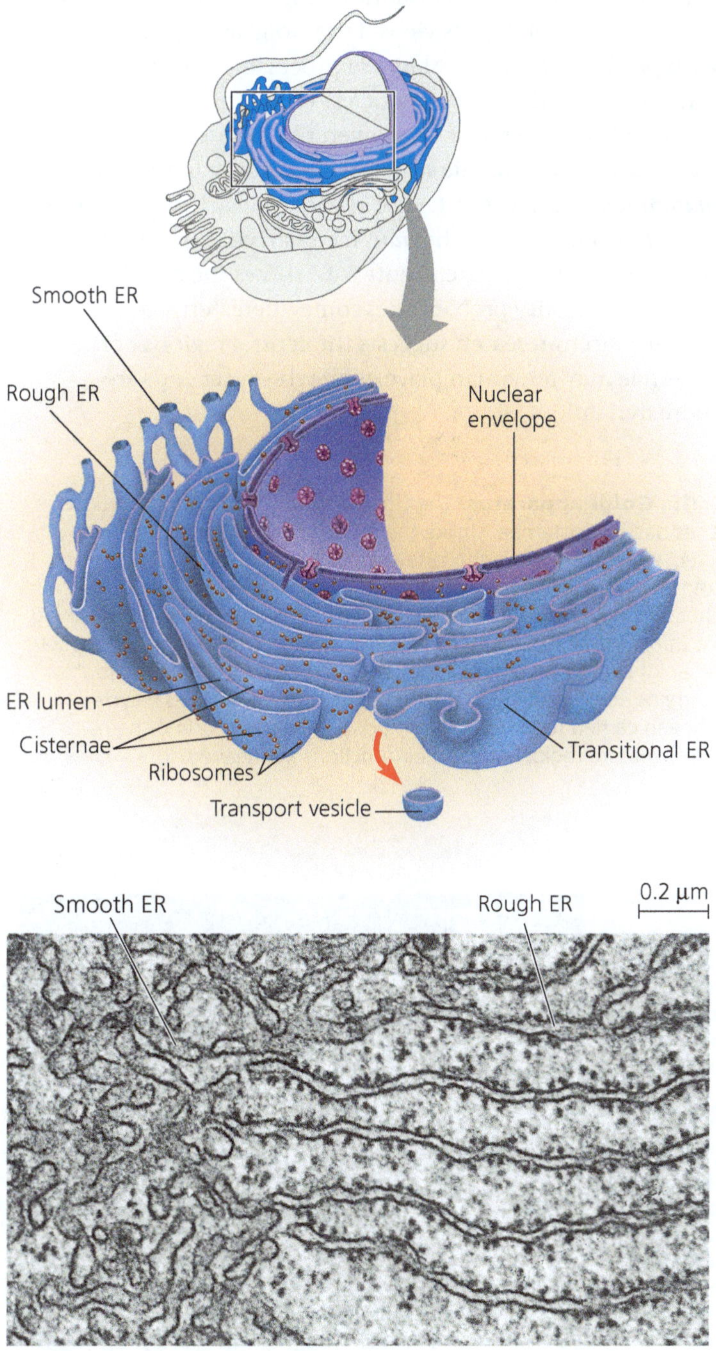

The smooth ER also stores calcium ions. In muscle cells, for example, the smooth ER membrane pumps calcium ions from the cytosol into the ER lumen. When a muscle cell is stimulated by a nerve impulse, calcium ions rush back across the ER membrane into the cytosol and trigger contraction of the muscle cell. In other cell types, release of calcium ions from the smooth ER triggers different responses, such as secretion of vesicles carrying newly synthesized proteins.

Functions of Rough ER

Many cells secrete proteins that are produced by ribosomes attached to rough ER. For example, certain pancreatic cells synthesize the protein insulin in the ER and secrete this hormone into the bloodstream. As a polypeptide chain grows from a bound ribosome, the chain is threaded into the ER lumen through a pore formed by a protein complex in the ER membrane. The new polypeptide folds into its functional shape as it enters the ER lumen. Most secretory proteins are **glycoproteins**, proteins with carbohydrates covalently bonded to them. The carbohydrates are attached to the proteins in the ER lumen by enzymes built into the ER membrane.

After secretory proteins are formed, the ER membrane keeps them separate from proteins in the cytosol, which are produced by free ribosomes. Secretory proteins depart from the ER wrapped in the membranes of vesicles that bud like bubbles from a specialized region called transitional ER (see Figure 6.11). Vesicles in transit from one part of the cell to another are called **transport vesicles**; we will discuss their fate shortly.

In addition to making secretory proteins, rough ER is a membrane factory for the cell; it grows in place by adding membrane proteins and phospholipids to its own membrane. As polypeptides destined to be membrane proteins grow from the ribosomes, they are inserted into the ER membrane itself and anchored there by their hydrophobic portions. Like the smooth ER, the rough ER also makes membrane phospholipids; enzymes built into the ER membrane assemble phospholipids from precursors in the cytosol. The ER membrane expands, and portions of it are transferred in the form of transport vesicles to other components of the endomembrane system.

The Golgi Apparatus: Shipping and Receiving Center

After leaving the ER, many transport vesicles travel to the **Golgi apparatus**. We can think of the Golgi as a warehouse for receiving, sorting, shipping, and even some manufacturing. Here, products of the ER, such as proteins, are modified and stored and then sent to other destinations. Not surprisingly, the Golgi apparatus is especially extensive in cells specialized for secretion.

The Golgi apparatus consists of a group of associated, flattened membranous sacs—cisternae—looking like a stack of pita bread **(Figure 6.12)**. A cell may have many, even hundreds, of these stacks. The membrane of each cisterna in a stack separates its internal space from the cytosol. Vesicles concentrated in the vicinity of the Golgi apparatus are engaged in the transfer of material between parts of the Golgi and other structures.

A Golgi stack has a distinct structural directionality, with the membranes of cisternae on opposite sides of the stack differing in thickness and molecular composition. The two sides of a Golgi stack are referred to as the *cis* face and the *trans* face; these act, respectively, as the receiving and shipping departments of the Golgi apparatus. The term *cis* means "on the same side," and the *cis* face is usually located near the ER. Transport vesicles move material from the ER to the Golgi apparatus. A vesicle that buds from the ER can add its membrane and the contents of its lumen to the *cis* face by fusing with a Golgi membrane on that side. The *trans* face ("on the opposite side") gives rise to vesicles that pinch off and travel to other sites.

Products of the endoplasmic reticulum are usually modified during their transit from the *cis* region to the *trans* region of the Golgi apparatus. For example, glycoproteins formed in the ER have their carbohydrates modified, first in the ER itself, and then as they pass through the Golgi. The Golgi removes some sugar monomers and substitutes others, producing a large variety of carbohydrates. Membrane phospholipids may also be altered in the Golgi.

In addition to its finishing work, the Golgi apparatus also manufactures some macromolecules. Many polysaccharides secreted by cells are Golgi products. For example, pectins and certain other noncellulose polysaccharides are made in the Golgi of plant cells and then incorporated along with cellulose into their cell walls. Like secretory proteins, nonprotein Golgi products that will be secreted depart from the *trans* face of the Golgi inside transport vesicles that eventually fuse with the plasma membrane.

The Golgi manufactures and refines its products in stages, with different cisternae containing unique teams of enzymes. Until recently, biologists viewed the Golgi as a static structure, with products in various stages of processing transferred from one cisterna to the next by vesicles. While this may occur, research from several labs has given rise to a new model of the Golgi as a more dynamic structure. According to the *cisternal maturation model*, the cisternae of the Golgi actually progress forward from the *cis* to the *trans* face, carrying and modifying their cargo as they move. Figure 6.12 shows the details of this model. The reality probably lies somewhere between the two models; recent research suggests the central regions of the cisternae may remain in place, while the outer ends are more dynamic.

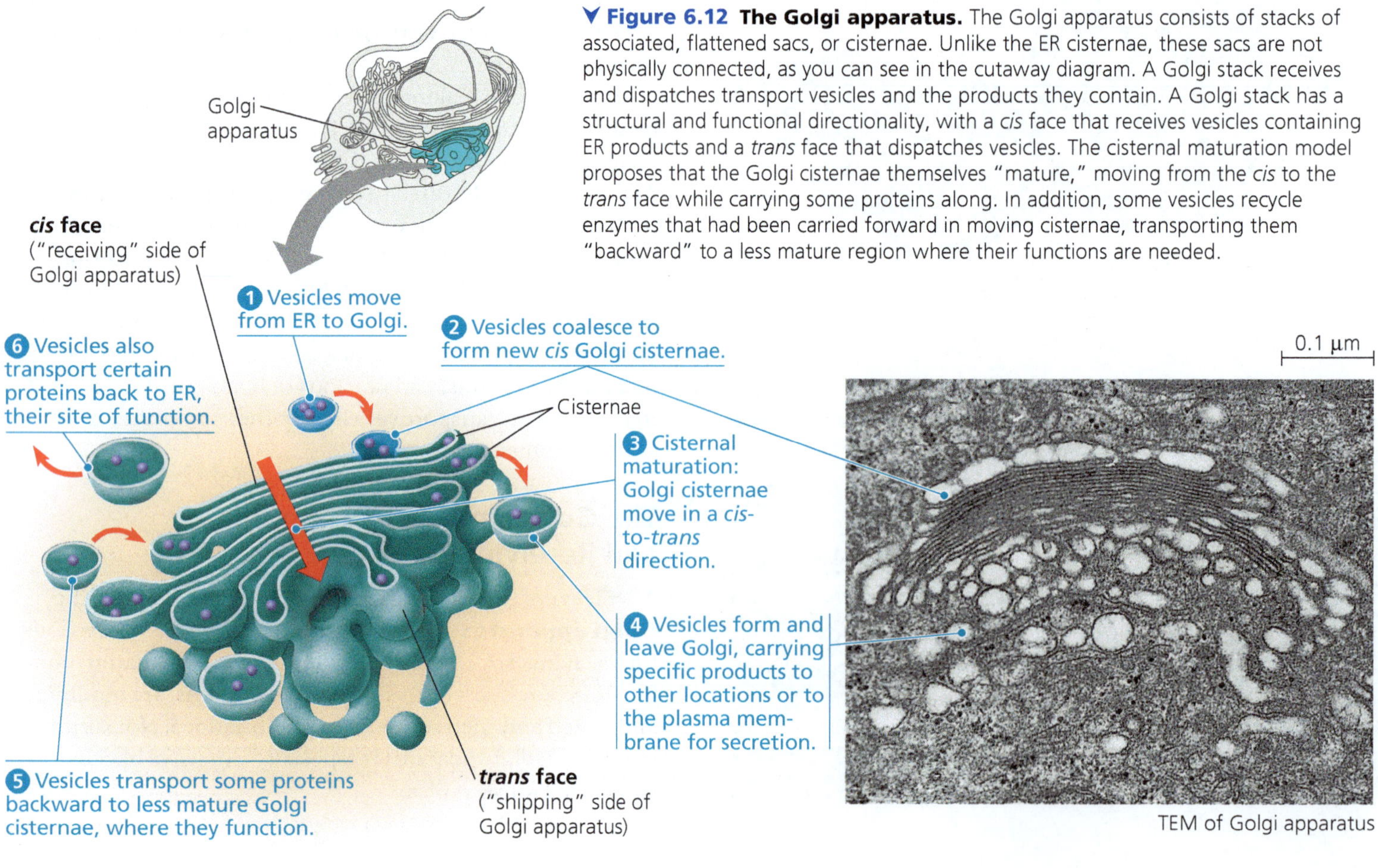

▼ **Figure 6.12 The Golgi apparatus.** The Golgi apparatus consists of stacks of associated, flattened sacs, or cisternae. Unlike the ER cisternae, these sacs are not physically connected, as you can see in the cutaway diagram. A Golgi stack receives and dispatches transport vesicles and the products they contain. A Golgi stack has a structural and functional directionality, with a *cis* face that receives vesicles containing ER products and a *trans* face that dispatches vesicles. The cisternal maturation model proposes that the Golgi cisternae themselves "mature," moving from the *cis* to the *trans* face while carrying some proteins along. In addition, some vesicles recycle enzymes that had been carried forward in moving cisternae, transporting them "backward" to a less mature region where their functions are needed.

Before a Golgi stack dispatches its products by budding vesicles from the *trans* face, it sorts these products and targets them for various parts of the cell. Molecular identification tags, such as phosphate groups added to the Golgi products, aid in sorting by acting like zip codes on mailing labels. Finally, transport vesicles budded from the Golgi may have external molecules on their membranes that recognize "docking sites" on the surface of specific organelles or on the plasma membrane, thus targeting the vesicles appropriately.

Lysosomes: Digestive Compartments

A **lysosome** is a membranous sac of hydrolytic enzymes that many eukaryotic cells use to digest (hydrolyze) macromolecules. Lysosomal enzymes work best in the acidic environment found in lysosomes. If a lysosome breaks open or leaks its contents, the released enzymes are not very active because the cytosol has a near-neutral pH. However, excessive leakage from a large number of lysosomes can destroy a cell by self-digestion.

Hydrolytic enzymes and lysosomal membrane are made by rough ER and then transferred to the Golgi apparatus for further processing. At least some lysosomes probably arise by budding from the *trans* face of the Golgi apparatus (see Figure 6.12). How are the proteins of the inner surface of the lysosomal membrane and the digestive enzymes themselves spared from destruction? Apparently, the three-dimensional shapes of these proteins protect vulnerable bonds from enzymatic attack.

Lysosomes carry out intracellular digestion in a variety of circumstances. Amoebas and many other unicellular eukaryotes eat by engulfing smaller organisms or food particles, a process called **phagocytosis** (from the Greek *phagein*, to eat, and *kytos*, vessel, referring here to the cell). The *food vacuole* formed in this way then fuses with a lysosome, whose enzymes digest the food (**Figure 6.13a**, bottom). Digestion products, including simple sugars, amino acids, and other monomers, pass into the cytosol and become nutrients for the cell. Some human cells also carry out phagocytosis. Among them are macrophages, a type of white blood cell that helps defend the body by engulfing and destroying bacteria and other invaders (see Figure 6.13a, top, and Figure 6.31).

Figure 6.13 Lysosomes.

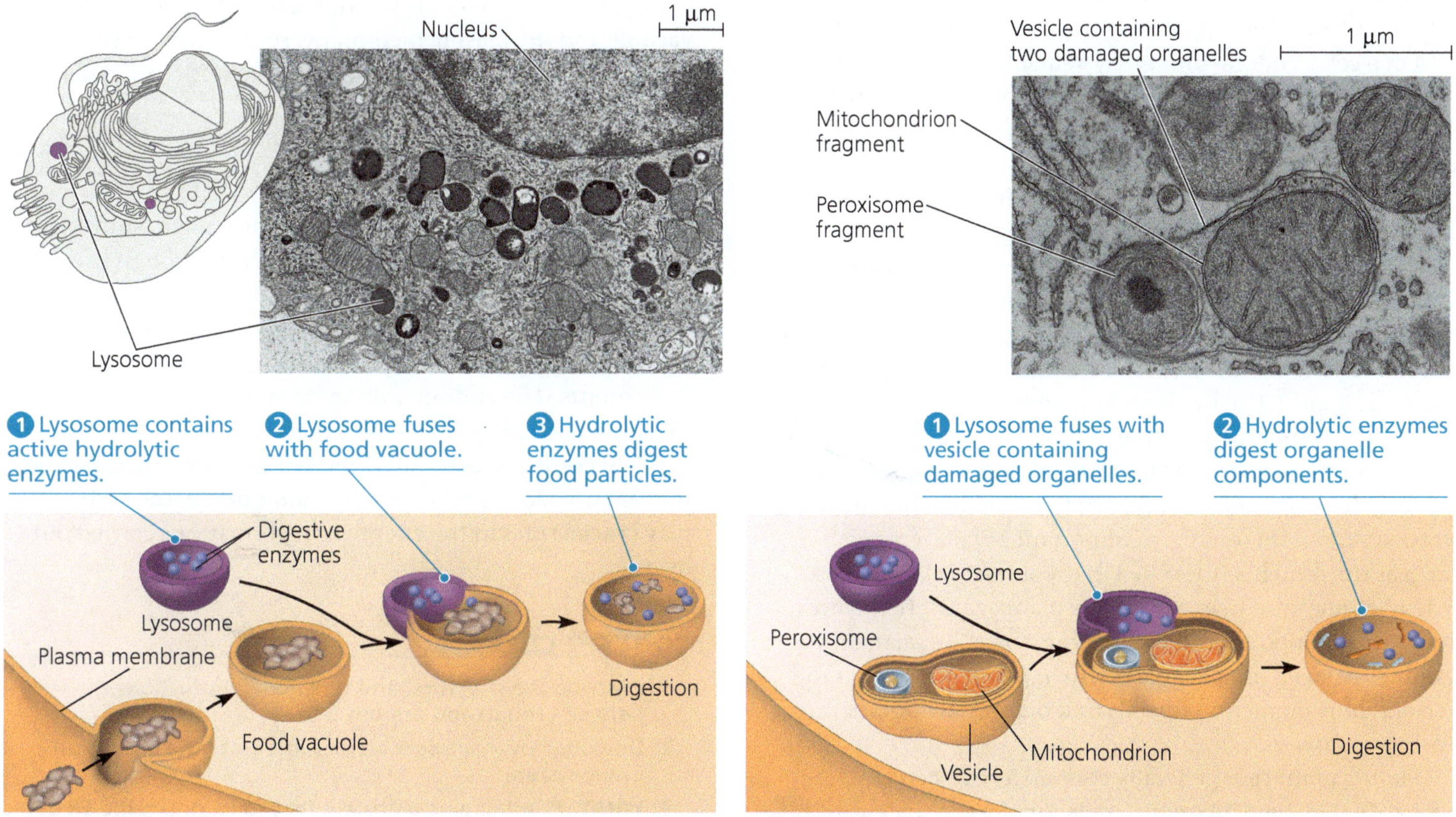

(a) Phagocytosis. In phagocytosis, lysosomes digest (hydrolyze) materials taken into the cell. *Top:* In this macrophage (a type of white blood cell) from a rat, the lysosomes are very dark because of a stain that reacts with one of the products of digestion inside the lysosome (TEM). Macrophages ingest bacteria and viruses and destroy them using lysosomes. *Bottom:* This diagram shows a lysosome fusing with a food vacuole during the process of phagocytosis by a unicellular eukaryote.

(b) Autophagy. In autophagy, lysosomes recycle intracellular materials. *Top:* In the cytoplasm of this rat liver cell is a vesicle containing two disabled organelles (TEM). The vesicle will fuse with a lysosome in the process of autophagy, which recycles intracellular materials. *Bottom:* This diagram shows fusion of such a vesicle with a lysosome. This type of vesicle has a double membrane of unknown origin. The outer membrane fuses with the lysosome, and the inner membrane is degraded along with the damaged organelles.

Lysosomes also use their hydrolytic enzymes to recycle the cell's own organic material, a process called *autophagy*. During autophagy, a damaged organelle or small amount of cytosol becomes surrounded by a double membrane (of unknown origin), and a lysosome fuses with the outer membrane of this vesicle **(Figure 6.13b)**. The lysosomal enzymes dismantle the inner membrane with the enclosed material, and the resulting small organic compounds are released to the cytosol for reuse. With the help of lysosomes, the cell continually renews itself. A human liver cell, for example, recycles half of its macromolecules each week.

The cells of people with inherited lysosomal storage diseases lack a functioning hydrolytic enzyme normally present in lysosomes. The lysosomes become engorged with indigestible material, which begins to interfere with other cellular activities. In Tay-Sachs disease, for example, a lipid-digesting enzyme is missing or inactive, and the brain becomes impaired by an accumulation of lipids in the cells. Fortunately, lysosomal storage diseases are rare in the general population.

Vacuoles: Diverse Maintenance Compartments

Vacuoles are large vesicles derived from the endoplasmic reticulum and Golgi apparatus. Thus, vacuoles are an integral part of a cell's endomembrane system. Like all cellular membranes, the vacuolar membrane is selective in transporting solutes; as a result, the solution inside a vacuole differs in composition from the cytosol.

Vacuoles perform a variety of functions in different kinds of cells. **Food vacuoles**, formed by phagocytosis, have already been mentioned (see Figure 6.13a). Many unicellular eukaryotes living in fresh water have **contractile vacuoles** that pump excess water out of the cell, thereby maintaining a suitable concentration of ions and molecules inside the cell (see Figure 7.13). In plants and fungi, certain vacuoles carry out enzymatic hydrolysis, a function shared by lysosomes in animal cells. (In fact, some biologists consider these hydrolytic vacuoles to be a type of lysosome.) In plants, small vacuoles can hold reserves of important organic compounds, such as the proteins stockpiled in the storage cells in seeds. Vacuoles may also help protect the plant against herbivores by storing compounds that are poisonous or unpalatable to animals. Some plant vacuoles contain pigments, such as the red and blue pigments of petals that help attract pollinating insects to flowers.

Mature plant cells generally contain a large **central vacuole (Figure 6.14)**, which develops by the coalescence of smaller vacuoles. The solution inside the central vacuole, called cell sap, is the plant cell's main repository of inorganic ions, including potassium and chloride. The central vacuole plays a major role in the growth of plant cells, which enlarge as the vacuole absorbs water, enabling the cell to become larger with a minimal investment in new cytoplasm. The cytosol often occupies only a thin layer between the central vacuole and the plasma membrane, so the ratio of plasma membrane surface to cytosolic volume is sufficient, even for a large plant cell.

▼ Figure 6.14 The plant cell vacuole. The central vacuole is usually the largest compartment in a plant cell; the rest of the cytoplasm is often confined to a narrow zone between the vacuolar membrane and the plasma membrane (TEM).

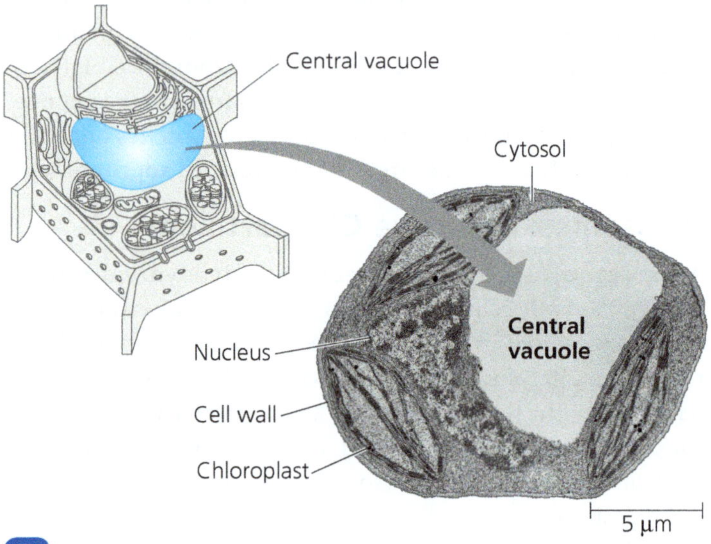

BioFlix® Animation: Central Vacuole

The Endomembrane System: *A Review*

Figure 6.15 reviews the endomembrane system, showing the flow of membrane lipids and proteins through the various organelles. As the membrane moves from the ER to the Golgi and then elsewhere, its molecular composition and metabolic functions are modified, along with those of its contents. The endomembrane system is a complex and dynamic player in the cell's compartmental organization.

We'll continue our tour of the cell with some organelles that are not closely related to the endomembrane system but play crucial roles in the energy transformations carried out by cells.

CONCEPT CHECK 6.4

1. Describe the structural and functional distinctions between rough and smooth ER.
2. Describe how transport vesicles integrate the endomembrane system.
3. **WHAT IF?** ➤ Imagine a protein that functions in the ER but requires modification in the Golgi apparatus before it can achieve that function. Describe the protein's path through the cell, starting with the mRNA molecule that specifies the protein.

For suggested answers, see Appendix A.

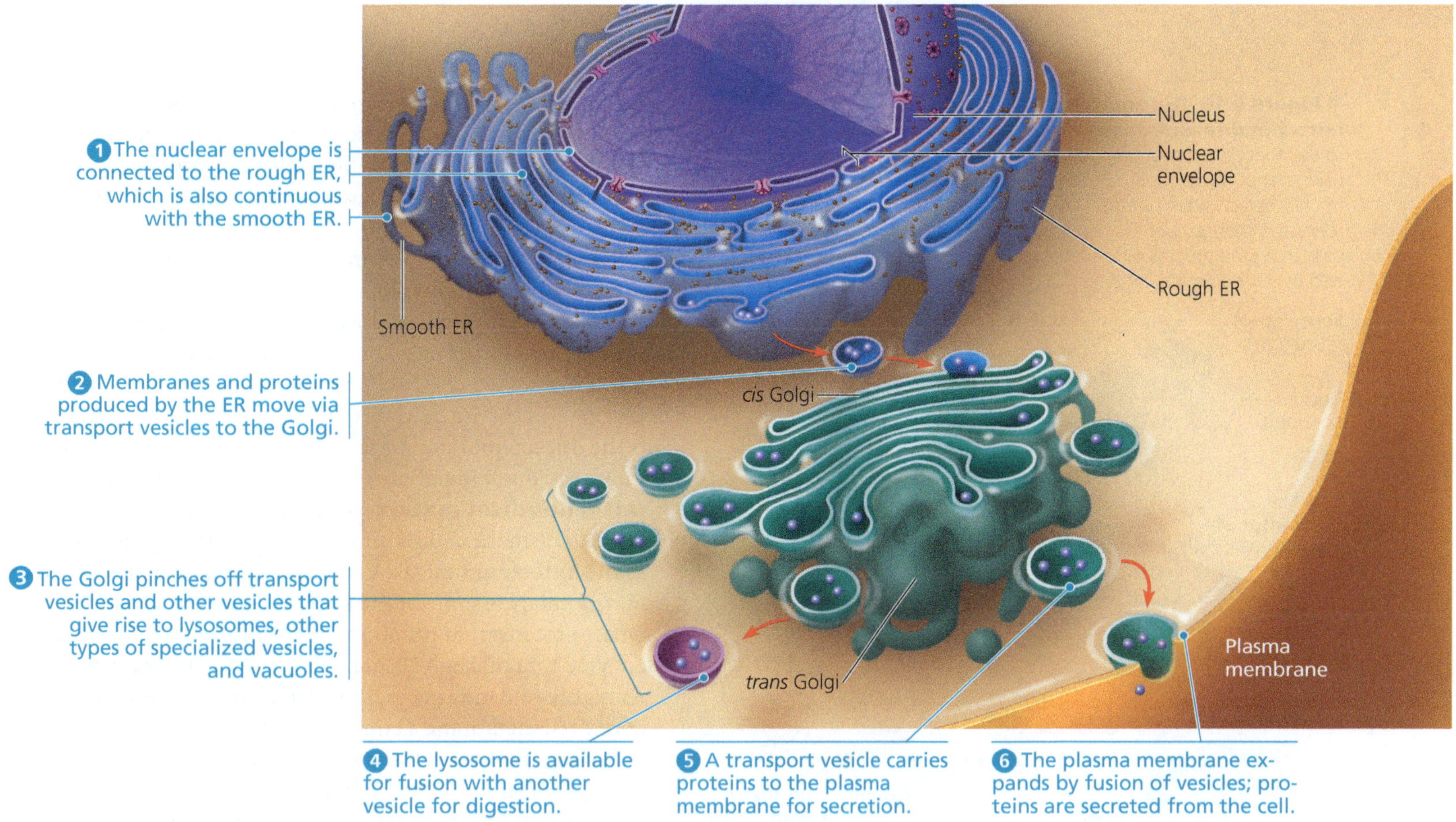

▲ **Figure 6.15 Review: relationships among organelles of the endomembrane system.** The red arrows show some of the migration pathways for membranes and the materials they enclose.

MB **BioFlix® Animation: Endomembrane System**

CONCEPT 6.5

Mitochondria and chloroplasts change energy from one form to another

Organisms transform the energy they acquire from their surroundings. In eukaryotic cells, mitochondria and chloroplasts are the organelles that convert energy to forms that cells can use for work. **Mitochondria** (singular, *mitochondrion*) are the sites of cellular respiration, the metabolic process that uses oxygen to drive the generation of ATP by extracting energy from sugars, fats, and other fuels. **Chloroplasts**, found in plants and algae, are the sites of photosynthesis. This process in chloroplasts converts solar energy to chemical energy by absorbing sunlight and using it to drive the synthesis of organic compounds such as sugars from carbon dioxide and water.

In addition to having related functions, mitochondria and chloroplasts share similar evolutionary origins, which we'll discuss briefly before describing their structures. In this section, we will also consider the peroxisome, an oxidative organelle. The evolutionary origin of the peroxisome, as well as its relation to other organelles, is still a matter of some debate.

The Evolutionary Origins of Mitochondria and Chloroplasts

EVOLUTION Mitochondria and chloroplasts display similarities with bacteria that led to the **endosymbiont theory**, illustrated in **Figure 6.16**. This theory states that an early ancestor of eukaryotic cells engulfed an oxygen-using nonphotosynthetic prokaryotic cell. Eventually, the engulfed cell formed a relationship with the host cell in which it was enclosed, becoming an *endosymbiont* (a cell living within another cell). Indeed, over the course of evolution, the host cell and its endosymbiont merged into a single organism, a eukaryotic cell with a mitochondrion. At least one of these cells may have then taken up a photosynthetic prokaryote, becoming the ancestor of eukaryotic cells that contain chloroplasts.

This is a widely accepted theory, which we will discuss in more detail in Concept 25.3. This theory is consistent with many structural features of mitochondria and chloroplasts. First, rather than being bounded by a single membrane like organelles of the endomembrane system, mitochondria and typical chloroplasts have two membranes surrounding them. (Chloroplasts also have an internal system of membranous sacs.) There is evidence that the ancestral engulfed

▼ **Figure 6.16 The endosymbiont theory of the origins of mitochondria and chloroplasts in eukaryotic cells.** According to this theory, the proposed ancestors of mitochondria were oxygen-using nonphotosynthetic prokaryotes, while the proposed ancestors of chloroplasts were photosynthetic prokaryotes. The large arrows represent change over evolutionary time; the small arrows inside the cells show the process of the endosymbiont becoming an organelle, also over long periods of time.

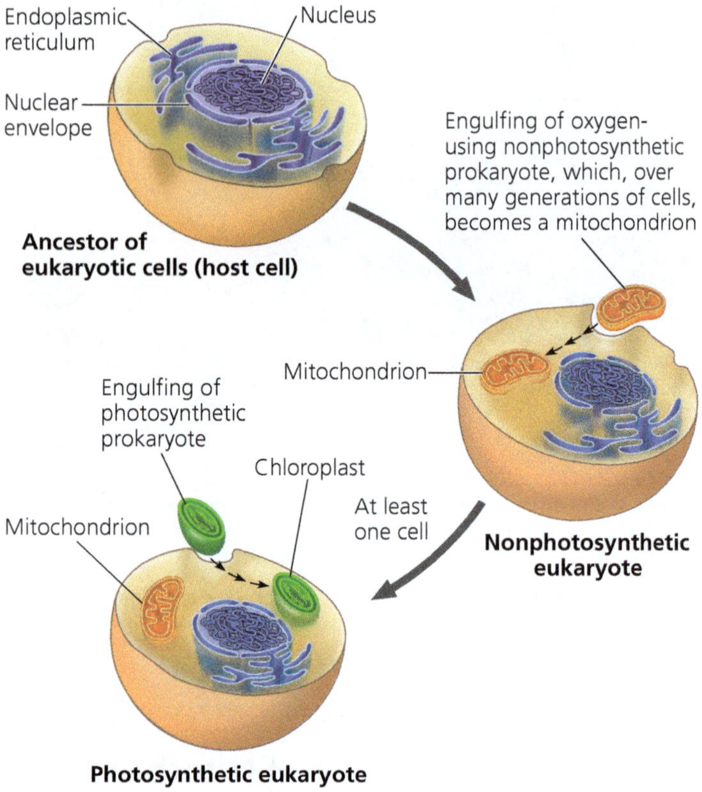

prokaryotes had two outer membranes, which became the double membranes of mitochondria and chloroplasts. Second, like prokaryotes, mitochondria and chloroplasts contain ribosomes, as well as circular DNA molecules—like bacterial chromosomes—associated with their inner membranes. The DNA in these organelles programs the synthesis of some organelle proteins on ribosomes that have been synthesized and assembled there as well. Third, also consistent with their probable evolutionary origins as cells, mitochondria and chloroplasts are autonomous (somewhat independent) organelles that grow and reproduce within the cell.

Next, we focus on the structures of mitochondria and chloroplasts, while providing an overview of their structures and functions. (In Chapters 9 and 10, we will examine their roles as energy transformers.)

Mitochondria: Chemical Energy Conversion

Mitochondria are found in nearly all eukaryotic cells, including those of plants, animals, fungi, and most unicellular eukaryotes. Some cells have a single large mitochondrion, but more often a cell has hundreds or even thousands of mitochondria; the number correlates with the cell's level of metabolic activity. For example, cells that move or contract have proportionally more mitochondria per volume than less active cells.

Each of the two membranes enclosing the mitochondrion is a phospholipid bilayer with a unique collection of embedded proteins **(Figure 6.17)**. The outer membrane is smooth, but the inner membrane is convoluted, with infoldings called **cristae**. The inner membrane divides the mitochondrion into two internal compartments. The first is the intermembrane space, the narrow region between the inner and outer membranes. The second compartment, the **mitochondrial matrix**, is enclosed by the inner membrane. The matrix contains many different enzymes as well as the mitochondrial DNA and ribosomes. Enzymes in the matrix catalyze some of the steps of cellular respiration. Other proteins that function in respiration, including the enzyme that makes ATP, are built into the inner membrane. As highly folded surfaces, the cristae give the inner mitochondrial membrane a large surface area, thus enhancing the productivity of cellular respiration. This is another example of structure fitting function.

Mitochondria are generally in the range of 1–10 μm long. Time-lapse films of living cells reveal mitochondria moving around, changing their shapes, and fusing or dividing in two, unlike the static structures seen in electron micrographs of dead cells. These studies helped cell biologists understand that mitochondria in a living cell form a branched tubular network, seen in a whole cell in Figure 6.17b, that is in a dynamic state of flux.

Chloroplasts: Capture of Light Energy

Chloroplasts contain the green pigment chlorophyll, along with enzymes and other molecules that function in the photosynthetic production of sugar. These lens-shaped organelles, about 3–6 μm in length, are found in leaves and other green organs of plants and in algae **(Figure 6.18**; see also Figure 6.26c).

The contents of a chloroplast are partitioned from the cytosol by an envelope consisting of two membranes separated by a very narrow intermembrane space. Inside the chloroplast is another membranous system in the form of flattened, interconnected sacs called **thylakoids**. In some regions, thylakoids are stacked like poker chips; each stack is called a **granum** (plural, *grana*). The fluid outside the thylakoids is the **stroma**, which contains the chloroplast DNA and ribosomes as well as many enzymes. The membranes of the chloroplast divide the chloroplast space into three compartments: the intermembrane space, the stroma, and the thylakoid space. This compartmental organization enables the chloroplast to convert light energy to chemical energy

Figure 6.17 The mitochondrion, site of cellular respiration. (a) The inner and outer membranes of the mitochondrion are evident in the drawing and electron micrograph (TEM). The cristae are infoldings of the inner membrane, which increase its surface area. The cutaway drawing shows the two compartments bounded by the membranes: the intermembrane space and the mitochondrial matrix. Many respiratory enzymes are found in the inner membrane and the matrix. Free ribosomes are also present in the matrix. The circular DNA molecules are associated with the inner mitochondrial membrane. **(b)** The light micrograph shows an entire unicellular eukaryote (*Euglena gracilis*) at a much lower magnification than the TEM. The mitochondrial matrix has been stained green. The mitochondria form a branched tubular network. The nuclear DNA is stained red; molecules of mitochondrial DNA appear as bright yellow spots.

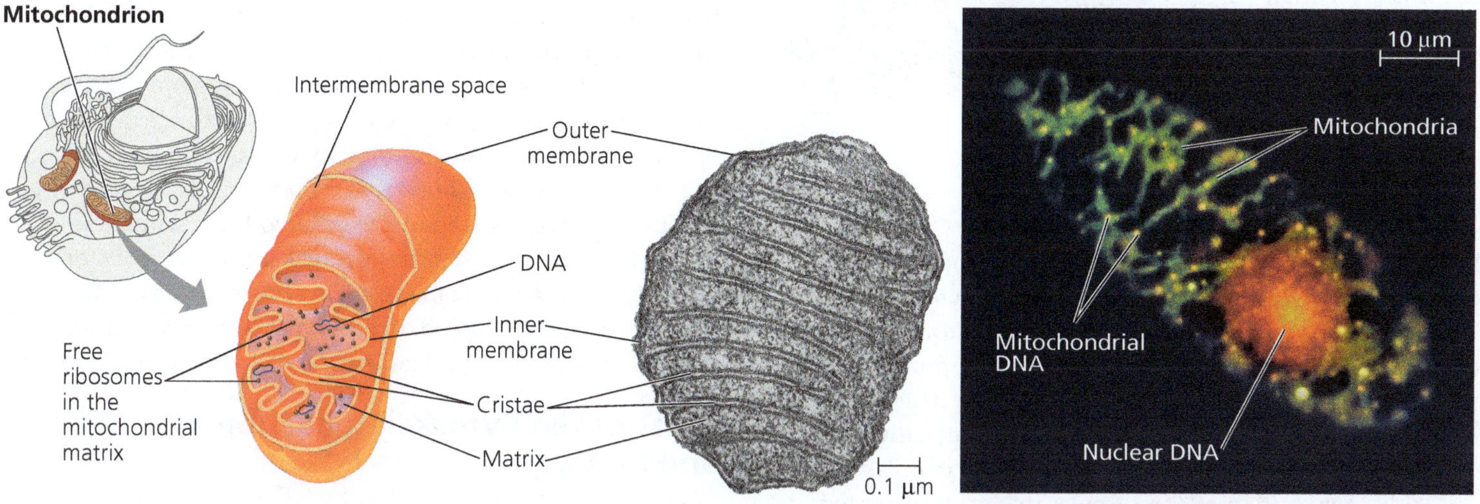

(a) Diagram and TEM of mitochondrion

(b) Network of mitochondria in *Euglena* (LM)

BioFlix® Animation: Mitochondria

during photosynthesis. (You will learn more about photosynthesis in Chapter 10.)

As with mitochondria, the static and rigid appearance of chloroplasts in micrographs or schematic diagrams cannot accurately depict their dynamic behavior in the living cell. Their shape is changeable, and they grow and occasionally pinch in two, reproducing themselves. They are mobile and, with mitochondria and other organelles, move around the cell along tracks of the cytoskeleton, a structural network we will consider in Concept 6.6.

The chloroplast is a specialized member of a family of closely related plant organelles called **plastids**. One type of plastid, the *amyloplast*, is a colorless organelle that stores starch (amylose), particularly in roots and tubers. Another is the *chromoplast*, which has pigments that give fruits and flowers their orange and yellow hues.

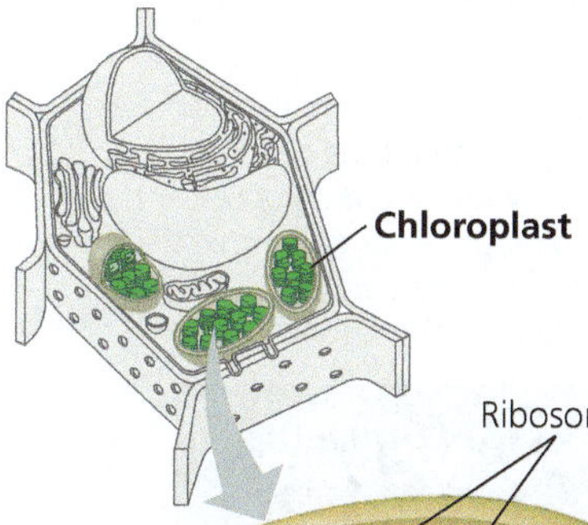

Figure 6.18 The chloroplast, site of photosynthesis. (a) Many plants have lens-shaped chloroplasts, as shown here in a diagram and a TEM. A chloroplast has three compartments: the intermembrane space, the stroma, and the thylakoid space. Free ribosomes are present in the stroma, as are copies of chloroplast DNA molecules. **(b)** This fluorescence micrograph, at a much lower magnification than the TEM, shows a whole cell of the green alga *Spirogyra crassa*, which is named for its spiral chloroplasts. Under natural light the chloroplasts appear green, but under ultraviolet light they naturally fluoresce red, as shown here.

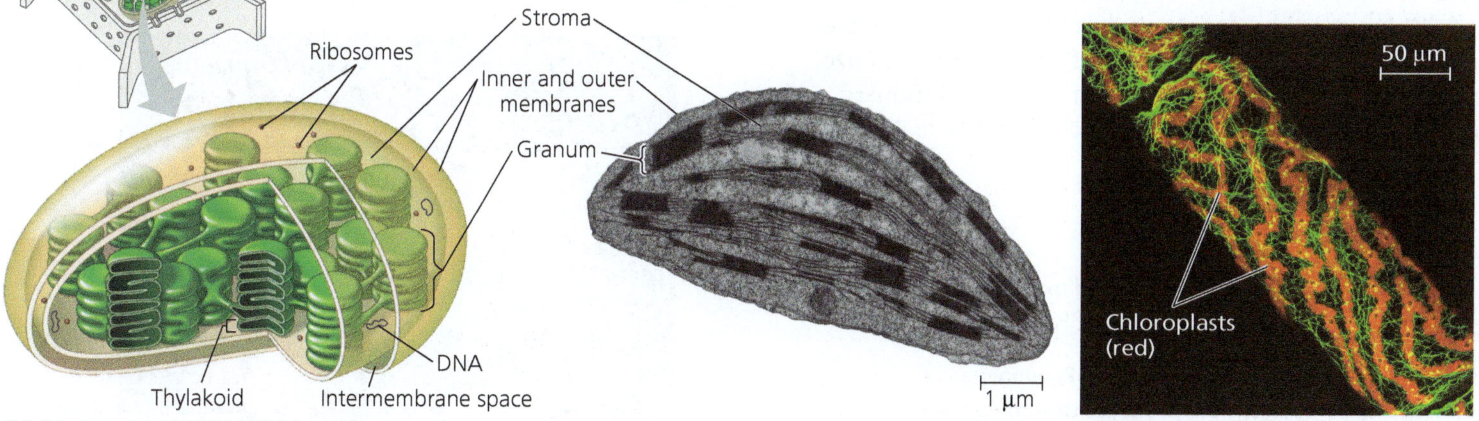

(a) Diagram and TEM of chloroplast

(b) Chloroplasts in an algal cell

BioFlix® Animation: Chloroplasts and Mitochondria

Figure 6.19 **A peroxisome.** Peroxisomes are roughly spherical and often have a granular or crystalline core that is thought to be a dense collection of enzyme molecules. Chloroplasts and mitochondria cooperate with peroxisomes in certain metabolic functions (TEM).

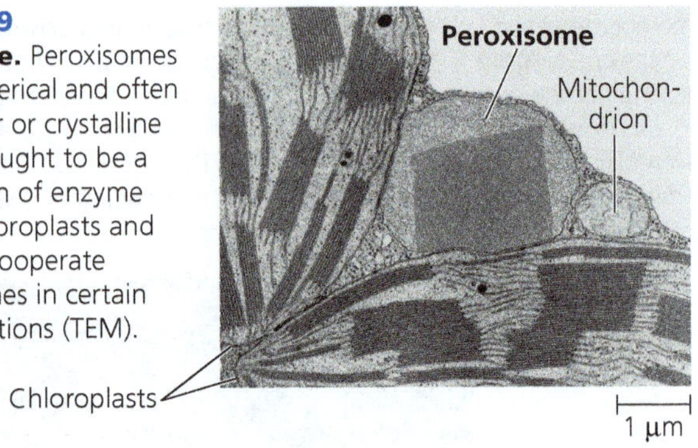

Peroxisomes: Oxidation

The **peroxisome** is a specialized metabolic compartment bounded by a single membrane **(Figure 6.19)**. Peroxisomes contain enzymes that remove hydrogen atoms from various substrates and transfer them to oxygen (O_2), producing hydrogen peroxide (H_2O_2) as a by-product (from which the organelle derives its name). These reactions have many different functions. Some peroxisomes use oxygen to break fatty acids down into smaller molecules that are transported to mitochondria and used as fuel for cellular respiration. Peroxisomes in the liver detoxify alcohol and other harmful compounds by transferring hydrogen from the poisonous compounds to oxygen. The H_2O_2 formed by peroxisomes is itself toxic, but the organelle also contains an enzyme that converts H_2O_2 to water. This is an excellent example of how the cell's compartmental structure is crucial to its functions: The enzymes that produce H_2O_2 and those that dispose of this toxic compound are sequestered away from other cellular components that could be damaged.

Specialized peroxisomes called *glyoxysomes* are found in the fat-storing tissues of plant seeds. These organelles contain enzymes that initiate the conversion of fatty acids to sugar, which the emerging seedling uses as a source of energy and carbon until it can produce its own sugar by photosynthesis.

How peroxisomes are related to other organelles is still an open question. They grow larger by incorporating proteins made in the cytosol and ER, as well as lipids made in the ER and within the peroxisome itself. Peroxisomes may increase in number by splitting in two when they reach a certain size, sparking the suggestion of an endosymbiotic evolutionary origin, but others argue against this scenario. Discussion of this issue is ongoing.

CONCEPT CHECK 6.5

1. Describe two characteristics shared by chloroplasts and mitochondria. Consider both function and membrane structure.
2. Do plant cells have mitochondria? Explain.
3. **WHAT IF?** ➤ A classmate proposes that mitochondria and chloroplasts should be classified in the endomembrane system. Argue against the proposal.

For suggested answers, see Appendix A.

CONCEPT 6.6

The cytoskeleton is a network of fibers that organizes structures and activities in the cell

In the early days of electron microscopy, biologists thought that the organelles of a eukaryotic cell floated freely in the cytosol. But improvements in both light microscopy and electron microscopy have revealed the **cytoskeleton**, a network of fibers extending throughout the cytoplasm **(Figure 6.20)**. Bacterial cells also have fibers that form a type of cytoskeleton, constructed of proteins similar to eukaryotic ones, but here we will concentrate on eukaryotes. The eukaryotic cytoskeleton, which plays a major role in organizing the structures and activities of the cell, is composed of three types of molecular structures: microtubules, microfilaments, and intermediate filaments.

Roles of the Cytoskeleton: Support and Motility

The most obvious function of the cytoskeleton is to give mechanical support to the cell and maintain its shape. This is especially important for animal cells, which lack walls. The remarkable strength and resilience of the cytoskeleton as a whole are based on its architecture. Like a dome tent, the cytoskeleton is stabilized by a balance between opposing forces exerted by its elements. And just as the skeleton of an animal helps fix the positions of other body parts, the cytoskeleton provides anchorage for many organelles and even cytosolic enzyme molecules. The cytoskeleton is more dynamic than an animal skeleton, however. It can be quickly dismantled in one part of the cell and reassembled in a new location, changing the shape of the cell.

Figure 6.20 **The cytoskeleton.** As shown in this fluorescence micrograph, the cytoskeleton extends throughout the cell. The cytoskeletal elements have been tagged with different fluorescent molecules: green for microtubules and reddish-orange for microfilaments. A third component of the cytoskeleton, intermediate filaments, is not evident here. (The blue color tags the DNA in the nucleus.)

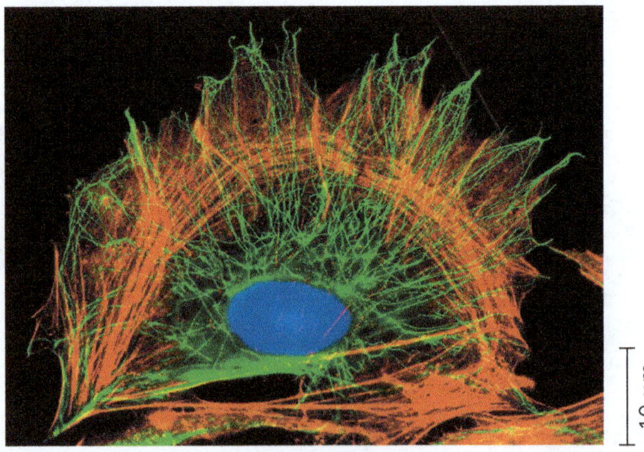

BioFlix® Animation: Cytoskeleton

Some types of cell motility (movement) also involve the cytoskeleton. The term *cell motility* includes both changes in cell location and movements of cell parts. Cell motility generally requires interaction of the cytoskeleton with **motor proteins**. There are many such examples: Cytoskeletal elements and motor proteins work together with plasma membrane molecules to allow whole cells to move along fibers outside the cell. Inside the cell, vesicles and other organelles often use motor protein "feet" to "walk" to their destinations along a track provided by the cytoskeleton. For example, this is how vesicles containing neurotransmitter molecules migrate to the tips of axons, the long extensions of nerve cells that release these molecules as chemical signals to adjacent nerve cells **(Figure 6.21)**. The cytoskeleton also manipulates the plasma membrane, bending it inward to form food vacuoles or other phagocytic vesicles.

MB **Interview with George Langford: Investigating how motor proteins move cargo**

Components of the Cytoskeleton

Now let's look more closely at the three main types of fibers that make up the cytoskeleton: *Microtubules* are the thickest of the three types; *microfilaments* (also called actin filaments) are the thinnest; and *intermediate filaments* are fibers with diameters in a middle range **(Table 6.1)**.

▼ Figure 6.21 Motor proteins and the cytoskeleton.

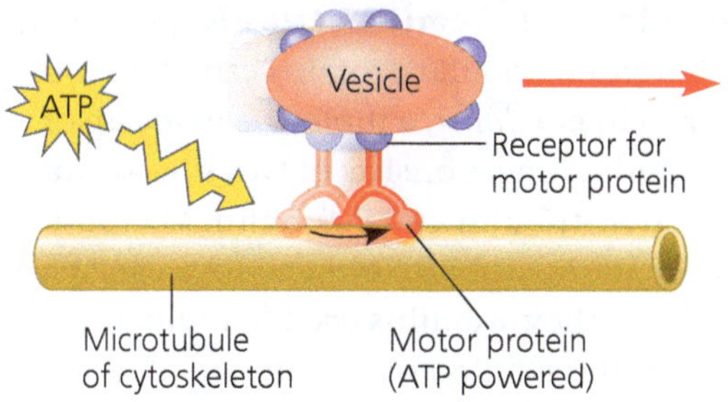

(a) Motor proteins that attach to receptors on vesicles can "walk" the vesicles along microtubules or, in some cases, along microfilaments.

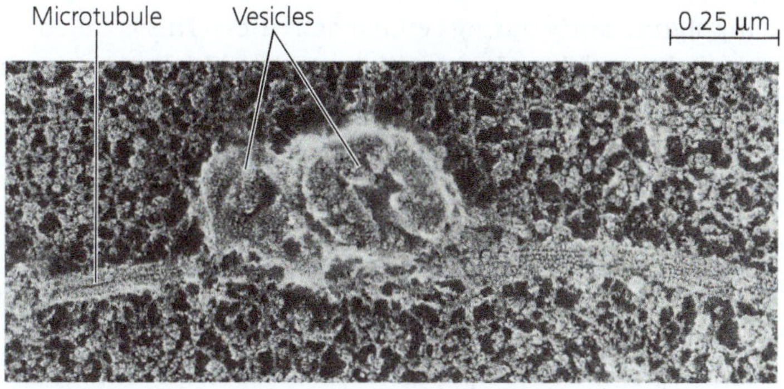

(b) Two vesicles containing neurotransmitters move along a microtubule toward the tip of a nerve cell extension called an axon (SEM).

Table 6.1 The Structure and Function of the Cytoskeleton

Property	Microtubules (Tubulin Polymers)	Microfilaments (Actin Filaments)	Intermediate Filaments
Structure	Hollow tubes	Two intertwined strands of actin	Fibrous proteins coiled into cables
Diameter	25 nm with 15-nm lumen	7 nm	8–12 nm
Protein subunits	Tubulin, a dimer consisting of α-tubulin and β-tubulin	Actin	One of several different proteins (such as keratins)
Main functions	Maintenance of cell shape (compression-resisting "girders"); cell motility (as in cilia or flagella); chromosome movements in cell division; organelle movements	Maintenance of cell shape (tension-bearing elements); changes in cell shape; muscle contraction; cytoplasmic streaming in plant cells; cell motility (as in amoeboid movement); division of animal cells	Maintenance of cell shape (tension-bearing elements); anchorage of nucleus and certain other organelles; formation of nuclear lamina
Fluorescence micrographs of fibroblasts. Fibroblasts are a favorite cell type for cell biology studies because they spread out flat and their internal structures are easy to see. In each, the structure of interest has been tagged with fluorescent molecules. The DNA in the nucleus has also been tagged in the first micrograph (blue) and third micrograph (orange).	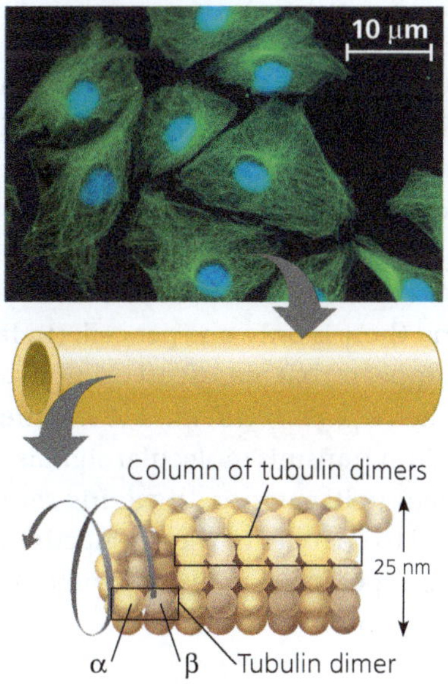	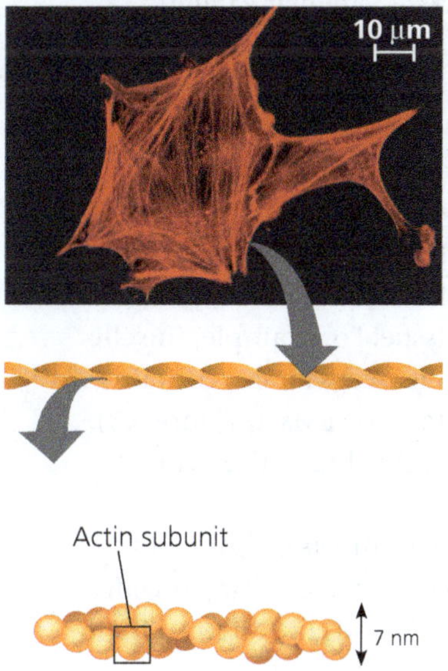	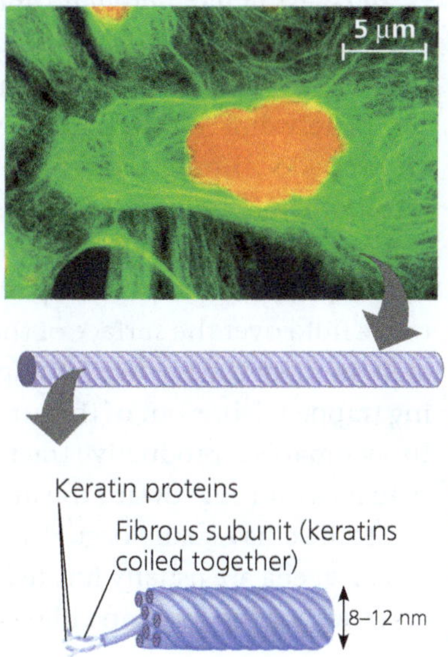

Microtubules

All eukaryotic cells have **microtubules**, hollow rods constructed from globular proteins called tubulins. Each tubulin protein is a *dimer*, a molecule made up of two subunits. A tubulin dimer consists of two slightly different polypeptides, α-tubulin and β-tubulin. Microtubules grow in length by adding tubulin dimers; they can also be disassembled and their tubulins used to build microtubules elsewhere in the cell. Because of the orientation of tubulin dimers, the two ends of a microtubule are slightly different. One end can accumulate or release tubulin dimers at a much higher rate than the other, thus growing and shrinking significantly during cellular activities. (This is called the "plus end," not because it can only add tubulin proteins but because it's the end where both "on" and "off" rates are much higher.)

Microtubules shape and support the cell and also serve as tracks along which organelles equipped with motor proteins can move. In addition to the example in Figure 6.21, microtubules guide vesicles from the ER to the Golgi apparatus and from the Golgi to the plasma membrane. Microtubules are also involved in the separation of chromosomes during cell division, as shown in Figure 12.7.

Centrosomes and Centrioles In animal cells, microtubules grow out from a **centrosome**, a region that is often located near the nucleus. These microtubules function as compression-resisting girders of the cytoskeleton. Within the centrosome is a pair of **centrioles**, each composed of nine sets of triplet microtubules arranged in a ring **(Figure 6.22)**. Although centrosomes with centrioles may help organize microtubule assembly in animal cells, many other eukaryotic cells lack centrosomes with centrioles and instead organize microtubules by other means.

Cilia and Flagella In eukaryotes, a specialized arrangement of microtubules is responsible for the beating of **flagella** (singular, *flagellum*) and **cilia** (singular, *cilium*), microtubule-containing extensions that project from some cells. (The bacterial flagellum, shown in Figure 6.5, has a completely different structure.) Many unicellular eukaryotes are propelled through water by cilia or flagella that act as locomotor appendages, and the sperm of animals, algae, and some plants have flagella. When cilia or flagella extend from cells that are held in place as part of a tissue layer, they can move fluid over the surface of the tissue. For example, the ciliated lining of the trachea (windpipe) sweeps mucus containing trapped debris out of the lungs (see the EMs in Figure 6.3). In a woman's reproductive tract, the cilia lining the oviducts help move an egg toward the uterus.

Motile cilia usually occur in large numbers on the cell surface. Flagella are usually limited to just one or a few per cell, and they are longer than cilia. Flagella and cilia differ in their beating patterns. A flagellum has an undulating motion like the tail of a fish. In contrast, cilia have alternating power and recovery strokes, much like the oars of a racing crew boat **(Figure 6.23)**.

A cilium may also act as a signal-receiving "antenna" for the cell. Cilia that have this function are generally nonmotile, and there is only one per cell. (In fact, in vertebrate animals, it appears that almost all cells have such a cilium, which is called a *primary cilium*.) Membrane proteins on this kind of cilium transmit molecular signals from the cell's environment to its interior, triggering signaling pathways that may lead to changes in the cell's activities. Cilium-based signaling appears to be crucial to brain function and to embryonic development.

▼ **Figure 6.22 Centrosome containing a pair of centrioles.** Most animal cells have a centrosome, a region near the nucleus where the cell's microtubules are initiated. Within the centrosome is a pair of centrioles, each about 250 nm (0.25 μm) in diameter. The two centrioles are at right angles to each other, and each is made up of nine sets of three microtubules. The blue portions of the drawing represent nontubulin proteins that connect the microtubule triplets.

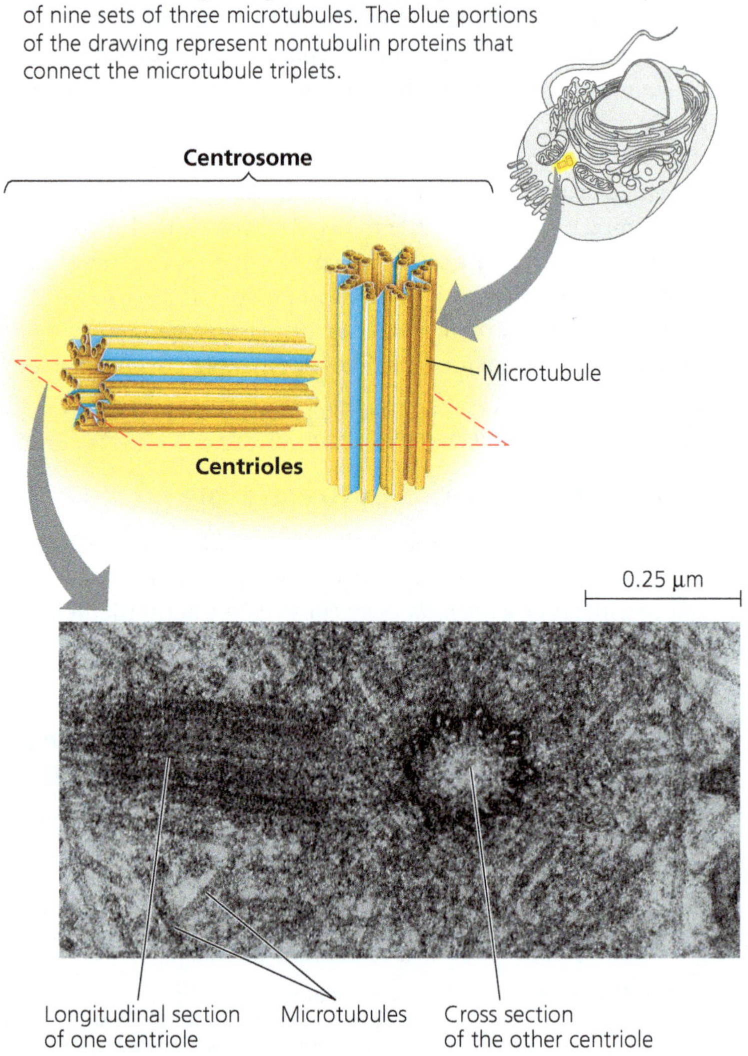

VISUAL SKILLS ➤ *How many microtubules are in a centrosome? In the drawing, circle and label one microtubule and describe its structure. Circle and label a triplet.*

Figure 6.23 A comparison of the beating of flagella and motile cilia.

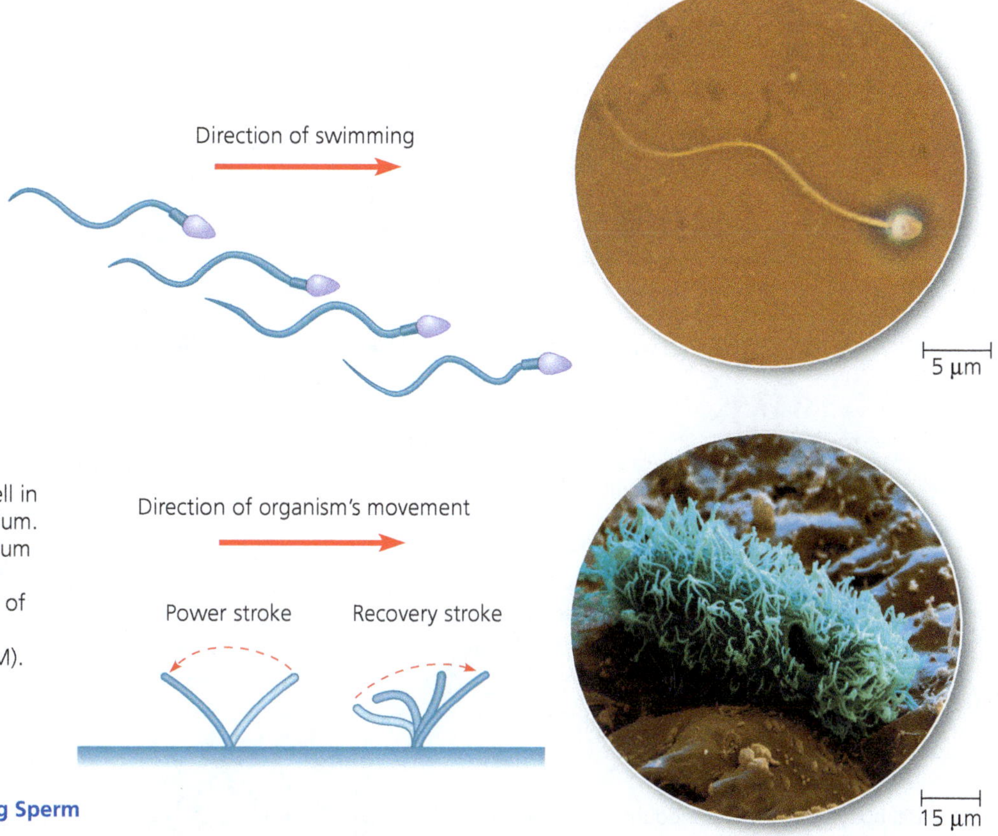

(a) Motion of flagella. A flagellum usually undulates, its snakelike motion driving a cell in the same direction as the axis of the flagellum. Propulsion of a human sperm cell is an example of flagellate locomotion (LM).

(b) Motion of cilia. Cilia have a back-and-forth motion. The rapid power stroke moves the cell in a direction perpendicular to the axis of the cilium. Then, during the slower recovery stroke,the cilium bends and sweeps sideways, closer to the cell surface. A dense nap of cilia, beating at a rate of about 40 to 60 strokes a second, covers this *Colpidium*, a freshwater protist (colorized SEM).

Video: Flagellum Movement in Swimming Sperm
Video: Flagellum Beating with ATP
Video: *Paramecium* Cilia

Though different in length, number per cell, and beating pattern, motile cilia and flagella share a common structure. Each motile cilium or flagellum has a group of microtubules sheathed in an extension of the plasma membrane **(Figure 6.24a)**. Nine doublets of microtubules are arranged in a ring with two single microtubules in its center **(Figure 6.24b)**. This arrangement, referred to as the "9 + 2" pattern, is found in nearly all eukaryotic flagella and motile cilia. (Nonmotile primary cilia have a "9 + 0" pattern, lacking the central pair of microtubules.) The microtubule assembly of a cilium or flagellum is anchored in the cell by a **basal body**, which is structurally very similar to a centriole, with microtubule triplets in a "9 + 0" pattern **(Figure 6.24c)**. In fact, in many animals (including humans), the basal body of the fertilizing sperm's flagellum enters the egg and becomes a centriole.

How does the microtubule assembly produce the bending movements of flagella and motile cilia? Bending involves large motor proteins called **dyneins** (red in the diagram in Figure 6.24) that are attached along each outer microtubule doublet. A typical dynein protein has two "feet" that "walk" along the microtubule of the adjacent doublet, using ATP for energy. One foot maintains contact, while the other releases and reattaches one step farther along the microtubule (see Figure 6.21). The outer doublets and two central microtubules are held together by flexible cross-linking proteins (blue in the diagram in Figure 6.24), and the walking movement is coordinated so that it happens on one side of the circle at a time. If the doublets were not held in place, the walking action would make them slide past each other. Instead, the movements of the dynein feet cause the microtubules—and the organelle as a whole—to bend.

Microfilaments (Actin Filaments)

Microfilaments are thin solid rods. They are also called actin filaments because they are built from molecules of **actin**, a globular protein. A microfilament is a twisted double chain of actin subunits (see Table 6.1). Besides occurring as linear filaments, microfilaments can form structural networks when certain proteins bind along the side of such a filament and allow a new filament to extend as a branch. Like microtubules, microfilaments seem to be present in all eukaryotic cells.

In contrast to the compression-resisting role of microtubules, the structural role of microfilaments in the cytoskeleton is to bear tension (pulling forces). A three-dimensional

Figure 6.24 Structure of a flagellum or motile cilium.

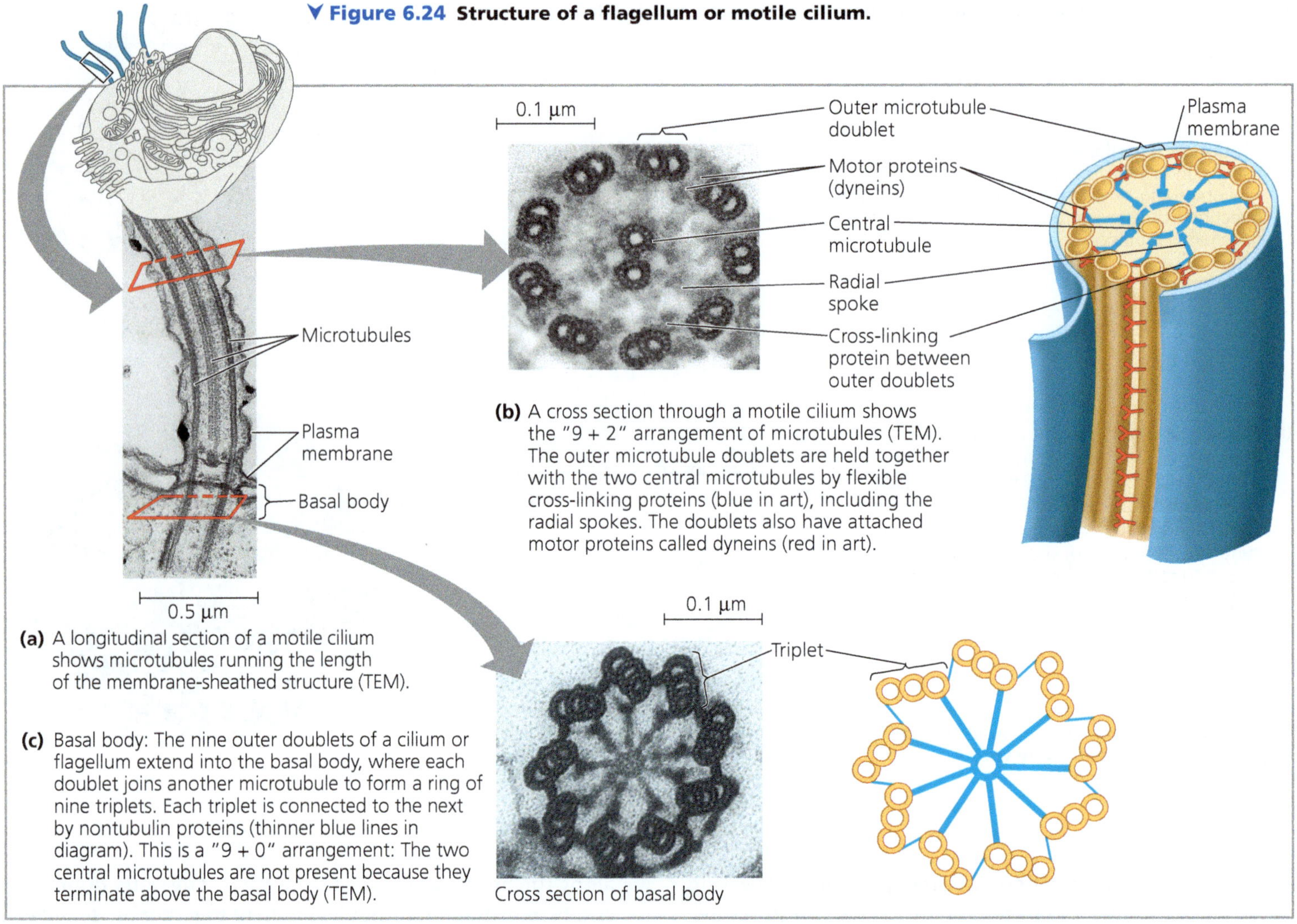

(a) A longitudinal section of a motile cilium shows microtubules running the length of the membrane-sheathed structure (TEM).

(b) A cross section through a motile cilium shows the "9 + 2" arrangement of microtubules (TEM). The outer microtubule doublets are held together with the two central microtubules by flexible cross-linking proteins (blue in art), including the radial spokes. The doublets also have attached motor proteins called dyneins (red in art).

(c) Basal body: The nine outer doublets of a cilium or flagellum extend into the basal body, where each doublet joins another microtubule to form a ring of nine triplets. Each triplet is connected to the next by nontubulin proteins (thinner blue lines in diagram). This is a "9 + 0" arrangement: The two central microtubules are not present because they terminate above the basal body (TEM).

Cross section of basal body

DRAW IT ➤ *In (a) and (b), circle and label the central pair of microtubules. In (a), show where they terminate, and explain why they aren't seen in the cross section of the basal body in (c).*

Animation: Cilia and Flagella

network formed by microfilaments just inside the plasma membrane (*cortical microfilaments*) helps support the cell's shape (see Figure 6.8). This network gives the outer cytoplasmic layer of a cell, called the **cortex**, the semisolid consistency of a gel, in contrast with the more fluid state of the interior cytoplasm. In some kinds of animal cells, such as nutrient-absorbing intestinal cells, bundles of microfilaments make up the core of microvilli, delicate projections that increase the cell's surface area **(Figure 6.25)**.

Microfilaments are well known for their role in cell motility. Thousands of actin filaments and thicker filaments

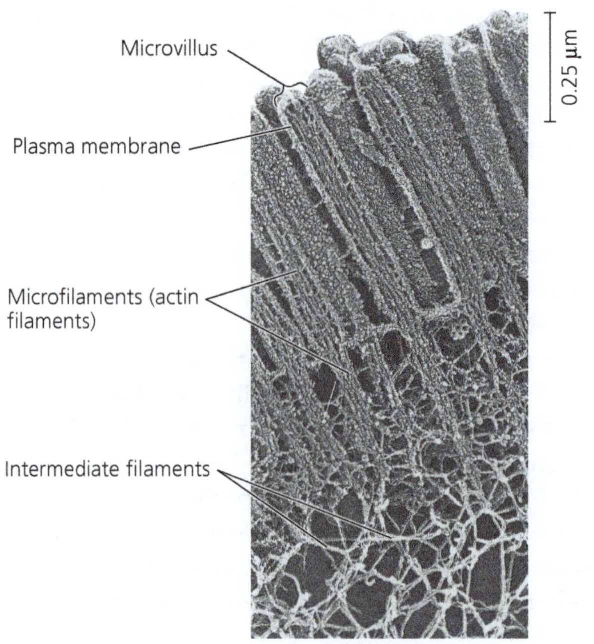

➤ Figure 6.25 A structural role of microfilaments. The surface area of this nutrient-absorbing intestinal cell is increased by its many microvilli (singular, *microvillus*), cellular extensions reinforced by bundles of microfilaments. These actin filaments are anchored to a network of intermediate filaments (TEM).

made of a protein called **myosin** interact to cause contraction of muscle cells **(Figure 6.26a)**; muscle contraction is described in detail in Concept 50.5. In the unicellular eukaryote *Amoeba* and some of our white blood cells, localized contractions brought about by actin and myosin are involved in the amoeboid (crawling) movement of the cells. The cell crawls along a surface by extending cellular extensions called **pseudopodia** (from the Greek *pseudes*, false, and *pod*, foot) and moving toward them **(Figure 6.26b)**. In plant cells, actin-protein interactions contribute to **cytoplasmic streaming**, a circular flow of cytoplasm within cells **(Figure 6.26c)**. This movement, which is especially common in large plant cells, speeds the movement of organelles and the distribution of materials within the cell.

Intermediate Filaments

Intermediate filaments are named for their diameter, which is larger than the diameter of microfilaments but smaller than that of microtubules (see Table 6.1). While microtubules and microfilaments are found in all eukaryotic cells, intermediate filaments are only found in the cells of some animals, including vertebrates. Specialized for bearing tension (like microfilaments), intermediate filaments are a diverse class of cytoskeletal elements. Each type is constructed from a particular molecular subunit belonging to a family of proteins whose members include the keratins. Microtubules and microfilaments, in contrast, are consistent in diameter and composition in all eukaryotic cells.

Intermediate filaments are more permanent fixtures of cells than are microfilaments and microtubules, which are often disassembled and reassembled in various parts of a cell. Even after cells die, intermediate filament networks often persist; for example, the outer layer of our skin consists of dead skin cells full of keratin filaments. Chemical treatments that remove microfilaments and microtubules from the cytoplasm of living cells leave a web of intermediate filaments that retains its original shape. Such experiments suggest that intermediate filaments are especially sturdy and that they play an important role in reinforcing the shape of a cell and fixing the position of certain organelles. For instance, the nucleus typically sits within a cage made of intermediate filaments, fixed in location by branches of the filaments that extend into the cytoplasm. Other intermediate filaments make up the nuclear lamina, which lines the interior of the nuclear envelope (see Figure 6.9). By supporting a cell's shape, intermediate filaments help the cell carry out its specific function. For example, the network of intermediate filaments shown in Figure 6.25 anchors the microfilaments supporting the intestinal microvilli. Thus, the various kinds of intermediate filaments may function together as the permanent framework of the entire cell.

▼ **Figure 6.26 Microfilaments and motility.** In these three examples, interactions between actin filaments and motor proteins bring about cell movement.

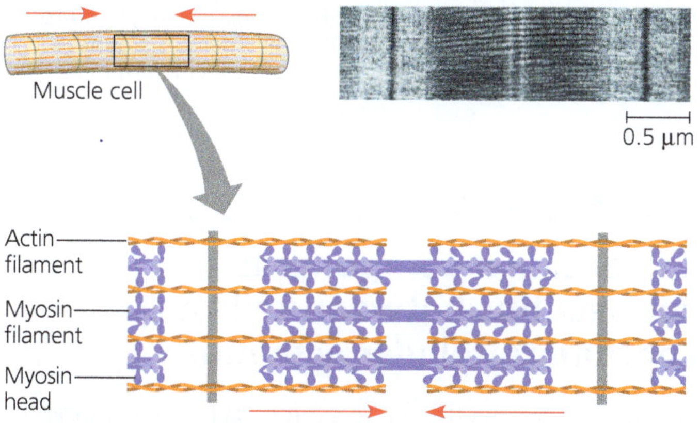

(a) Myosin motors in muscle cell contraction. The "walking" of myosin projections (the so-called heads) drives the parallel myosin and actin filaments past each other so that the actin filaments approach each other in the middle (red arrows). This shortens the muscle cell. Muscle contraction involves the shortening of many muscle cells at the same time (TEM).

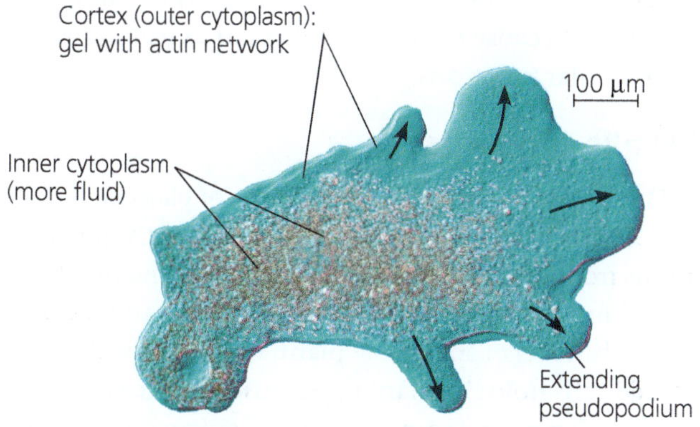

(b) Amoeboid movement. Interaction of actin filaments with myosin causes contraction of the cell, pulling the cell's trailing end (at left) forward (to the right) (LM).

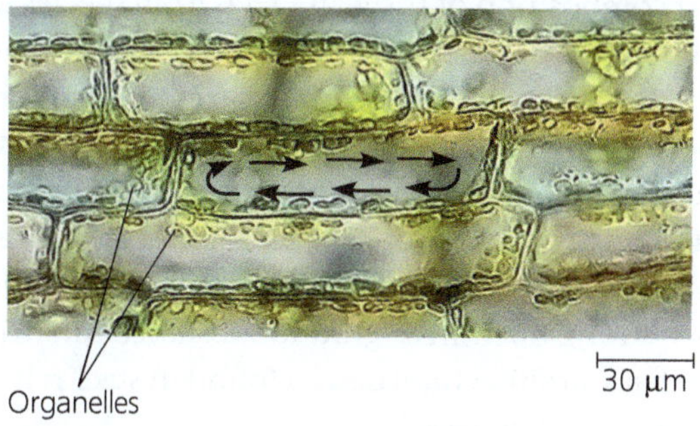

(c) Cytoplasmic streaming in plant cells. A layer of cytoplasm cycles around the cell, moving over tracks of actin filaments. Myosin motors attached to some organelles drive the streaming by interacting with the actin (LM).

BioFlix® Animation: Actin and Myosin in Muscle Contraction
Video: *Amoeba* Pseudopodia
Video: Cytoplasmic Streaming

CONCEPT CHECK 6.6

1. Describe how cilia and flagella bend.
2. **WHAT IF?** ➤ Males afflicted with Kartagener's syndrome are sterile because of immotile sperm, and they tend to suffer from lung infections. This disorder has a genetic basis. Suggest what the underlying defect might be.

For suggested answers, see Appendix A.

CONCEPT 6.7

Extracellular components and connections between cells help coordinate cellular activities

Having crisscrossed the cell to explore its interior components, we complete our tour of the cell by returning to the surface of this microscopic world, where there are additional structures with important functions. The plasma membrane is usually regarded as the boundary of the living cell, but most cells synthesize and secrete materials extracellularly (to the outside of the cell). Although these materials and the structures they form are outside the cell, their study is important to cell biology because they are involved in a great many essential cellular functions.

Cell Walls of Plants

The **cell wall** is an extracellular structure of plant cells (**Figure 6.27**). This is one of the features that distinguishes plant cells from animal cells. The wall protects the plant cell, maintains its shape, and prevents excessive uptake of water. On the level of the whole plant, the strong walls of specialized cells hold the plant up against the force of gravity. Prokaryotes, fungi, and some unicellular eukaryotes also have cell walls, as you saw in Figures 6.5 and 6.8, but we will postpone discussion of them until Unit Five.

Plant cell walls are much thicker than the plasma membrane, ranging from 0.1 μm to several micrometers. The exact chemical composition of the wall varies from species to species and even from one cell type to another in the same plant, but the basic design of the wall is consistent. Microfibrils made of the polysaccharide cellulose (see Figure 5.6) are synthesized by an enzyme called cellulose synthase and secreted to the extracellular space, where they become embedded in a matrix of other polysaccharides and proteins. This combination of materials, strong fibers in a "ground substance" (matrix), is the same basic architectural design found in steel-reinforced concrete and in fiberglass.

A young plant cell first secretes a relatively thin and flexible wall called the **primary cell wall** (see the micrograph in Figure 6.27). Between primary walls of adjacent cells is the **middle lamella**, a thin layer rich in sticky polysaccharides called pectins. The middle lamella glues adjacent cells together. (Pectin is used in cooking as a thickening agent in jams and jellies.) When the cell matures and stops growing, it strengthens its wall. Some plant cells do this simply by secreting hardening substances into the primary wall. Other cells add a **secondary cell wall** between the plasma membrane and the primary wall. The secondary wall, often deposited in several laminated layers, has a strong and durable matrix that affords the cell protection and support. Wood, for example, consists mainly of secondary walls. Plant cell walls are usually perforated by channels between adjacent cells called plasmodesmata, which will be discussed shortly.

▼ **Figure 6.27 Plant cell walls.** The drawing shows several cells, each with a cell wall, large vacuole, a nucleus, and several chloroplasts and mitochondria. The TEM shows the cell walls where two cells come together. The multilayered partition between plant cells consists of adjoining walls individually secreted by the cells.

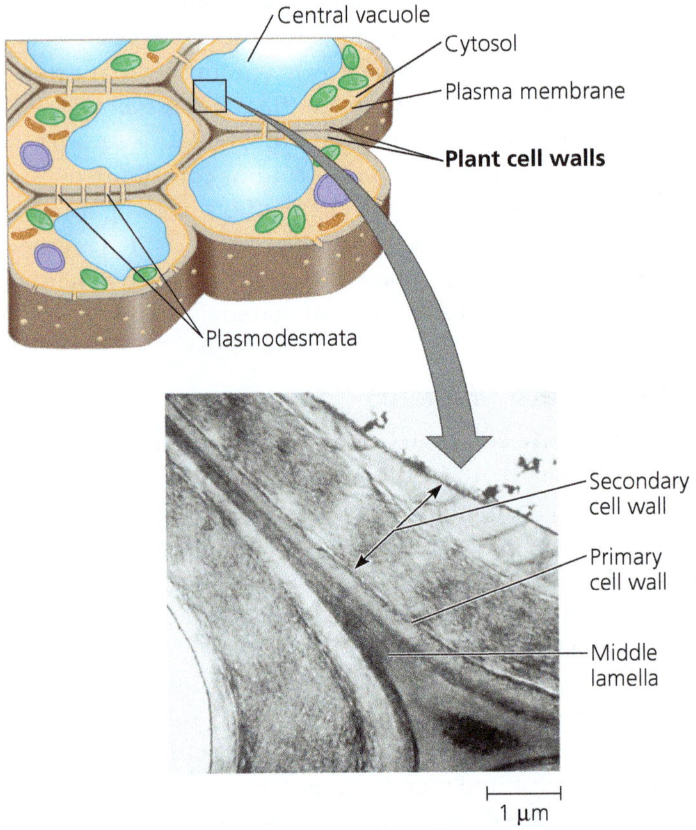

The Extracellular Matrix (ECM) of Animal Cells

Although animal cells lack walls akin to those of plant cells, they do have an elaborate **extracellular matrix (ECM)**. The main ingredients of the ECM are glycoproteins and other carbohydrate-containing molecules secreted by the cells. (Recall that glycoproteins are proteins with covalently bonded carbohydrates, usually short chains of sugars.) The most abundant glycoprotein in the ECM of most animal cells is **collagen**, which forms strong fibers outside the cells (see Figure 5.18). In fact, collagen accounts for about 40% of the total protein in the human body. The collagen fibers are embedded in a network woven out of **proteoglycans**

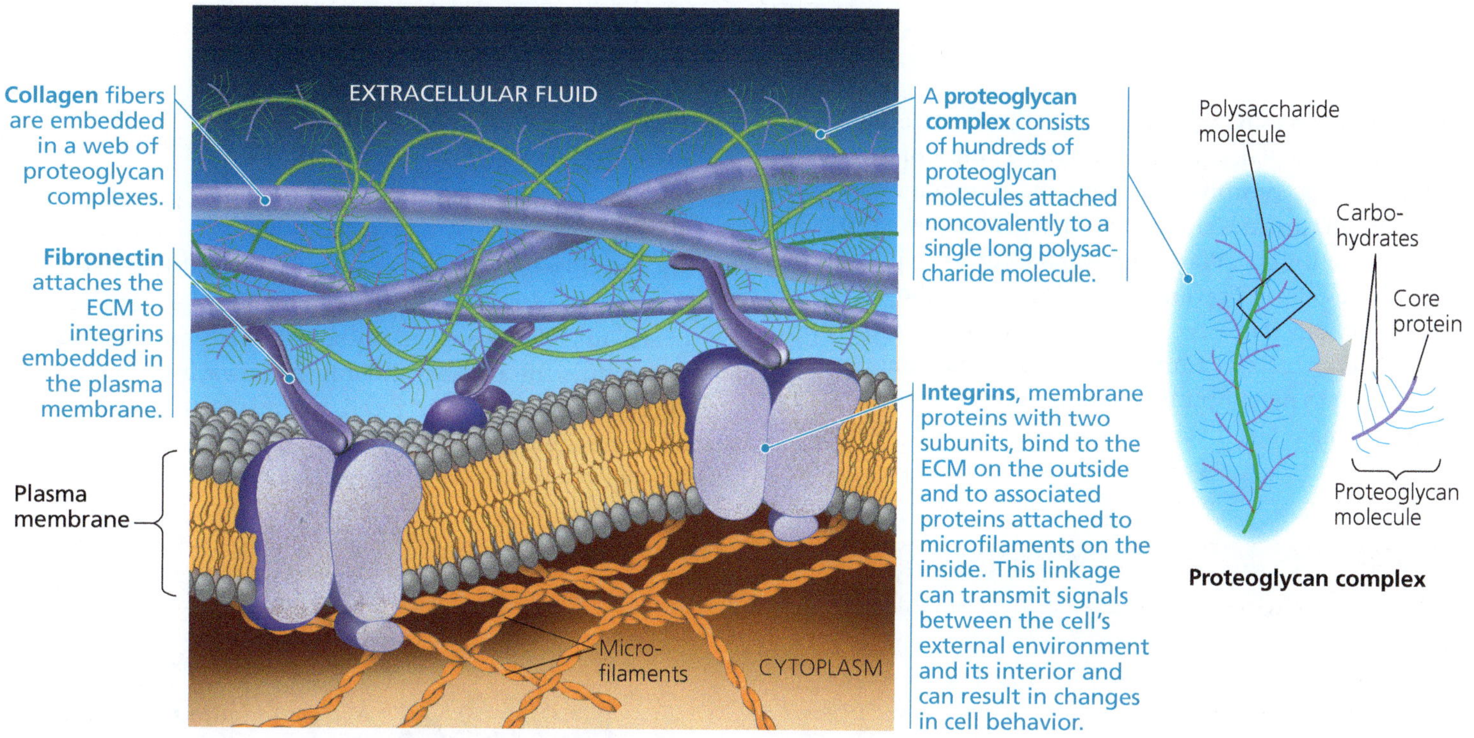

▲ **Figure 6.28 Extracellular matrix (ECM) of an animal cell.** The molecular composition and structure of the ECM vary from one cell type to another. In this example, three different types of ECM molecules are present: collagen, fibronectin, and proteoglycans.

secreted by cells **(Figure 6.28)**. A proteoglycan molecule consists of a small core protein with many carbohydrate chains covalently attached, so that it may be up to 95% carbohydrate. Large proteoglycan complexes can form when hundreds of proteoglycan molecules become noncovalently attached to a single long polysaccharide molecule, as shown in Figure 6.28. Some cells are attached to the ECM by ECM glycoproteins such as **fibronectin**. Fibronectin and other ECM proteins bind to cell-surface receptor proteins called **integrins** that are built into the plasma membrane. Integrins span the membrane and bind on their cytoplasmic side to associated proteins attached to microfilaments of the cytoskeleton. The name *integrin* is based on the word *integrate*: Integrins are in a position to transmit signals between the ECM and the cytoskeleton and thus to integrate changes occurring outside and inside the cell.

Current research on fibronectin, other ECM molecules, and integrins reveals the influential role of the ECM in the lives of cells. By communicating with a cell through integrins, the ECM can regulate a cell's behavior. For example, some cells in a developing embryo migrate along specific pathways by matching the orientation of their microfilaments to the "grain" of fibers in the extracellular matrix. Researchers have also learned that the extracellular matrix around a cell can influence the activity of genes in the nucleus. Information about the ECM probably reaches the nucleus by a combination of mechanical and chemical signaling pathways. Mechanical signaling involves fibronectin, integrins, and microfilaments of the cytoskeleton. Changes in the cytoskeleton may in turn trigger signaling pathways inside the cell, leading to changes in the set of proteins being made by the cell and therefore changes in the cell's function. In this way, the extracellular matrix of a particular tissue may help coordinate the behavior of all the cells of that tissue. Direct connections between cells also function in this coordination, as we discuss next.

Cell Junctions

Cells in an animal or plant are organized into tissues, organs, and organ systems. Neighboring cells often adhere, interact, and communicate via sites of direct physical contact.

Plasmodesmata in Plant Cells

It might seem that the nonliving cell walls of plants would isolate plant cells from one another. But in fact, as shown in **Figure 6.29**, cell walls are perforated with **plasmodesmata** (singular, *plasmodesma*; from the Greek *desma*, bond), channels that connect cells. Cytosol passing through the

▼ **Figure 6.29 Plasmodesmata between plant cells.** The cytoplasm of one plant cell is continuous with the cytoplasm of its neighbors via plasmodesmata, cytoplasmic channels through the cell walls (TEM).

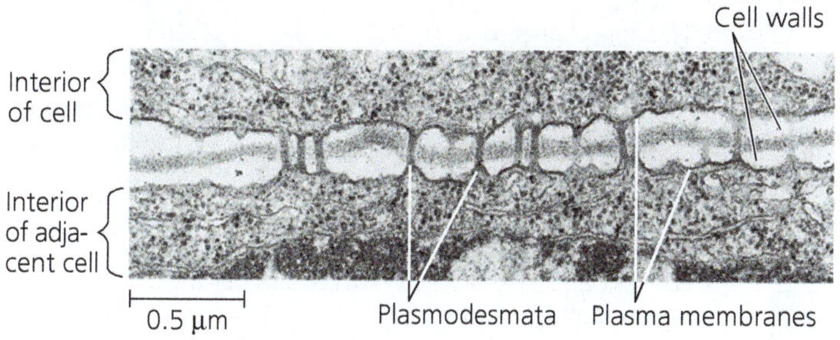

plasmodesmata joins the internal chemical environments of adjacent cells. These connections unify most of the plant into one living continuum. The plasma membranes of adjacent cells line the channel of each plasmodesma and thus are continuous. Water and small solutes can pass freely from cell to cell, and several experiments have shown that in some circumstances, certain proteins and RNA molecules can do this as well (see Concept 36.6). The macromolecules transported to neighboring cells appear to reach the plasmodesmata by moving along fibers of the cytoskeleton.

Tight Junctions, Desmosomes, and Gap Junctions in Animal Cells

In animals, there are three main types of cell junctions: *tight junctions, desmosomes*, and *gap junctions*. (Gap junctions are most like the plasmodesmata of plants, although gap junction pores are not lined with membrane.) All three types of cell junctions are especially common in epithelial tissue, which lines the external and internal surfaces of the body. **Figure 6.30** uses epithelial cells of the intestinal lining to illustrate these junctions.

▼ Figure 6.30 Exploring Cell Junctions in Animal Tissues

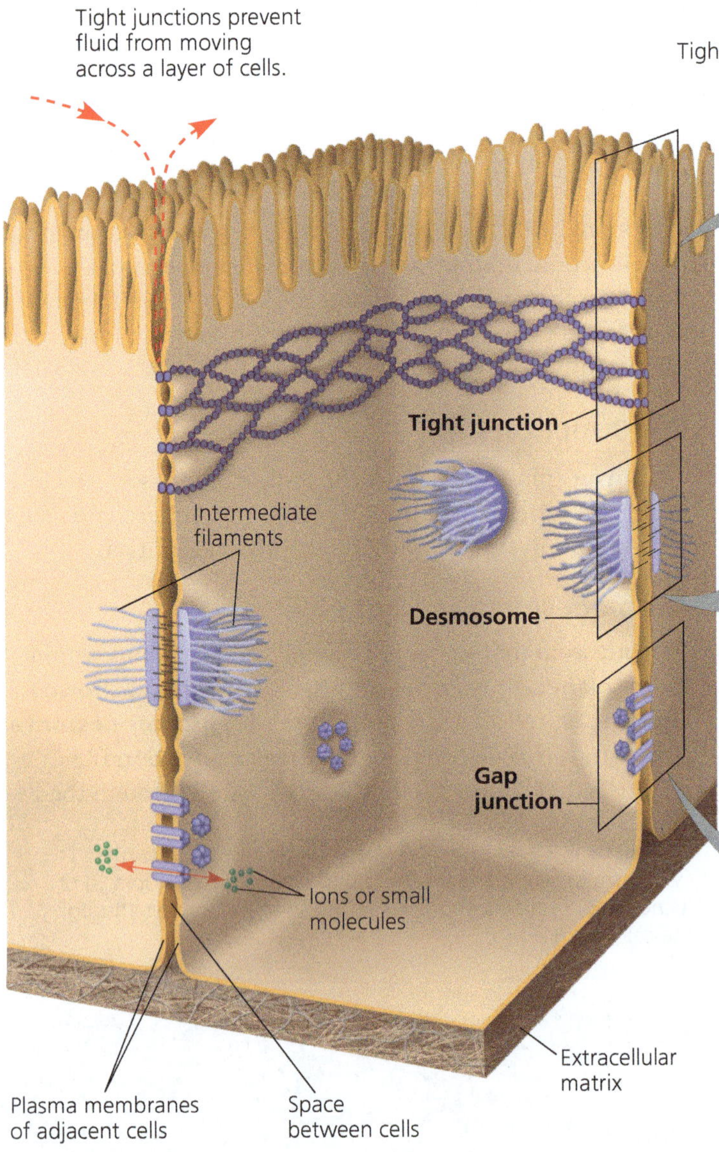

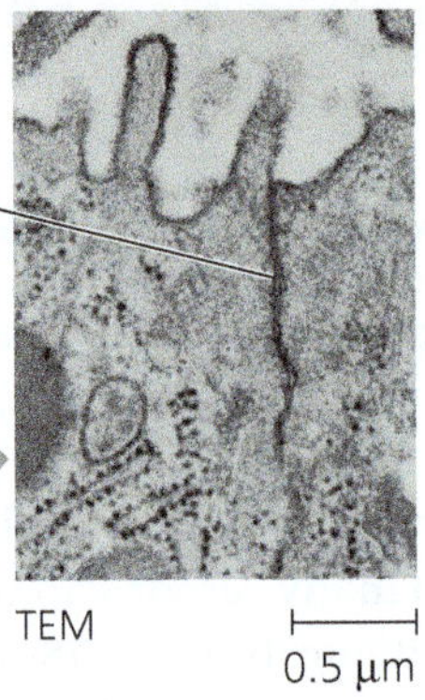

Tight Junctions

At **tight junctions**, the plasma membranes of neighboring cells are very tightly pressed against each other, bound together by specific proteins. Forming continuous seals around the cells, tight junctions establish a barrier that prevents leakage of extracellular fluid across a layer of epithelial cells (see red dashed arrow). For example, tight junctions between skin cells make us watertight.

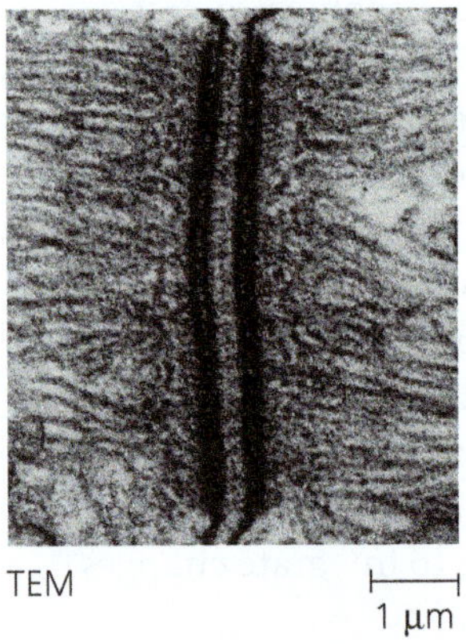

Desmosomes

Desmosomes (one type of *anchoring junction*) function like rivets, fastening cells together into strong sheets. Intermediate filaments made of sturdy keratin proteins anchor desmosomes in the cytoplasm. Desmosomes attach muscle cells to each other in a muscle. Some "muscle tears" involve the rupture of desmosomes.

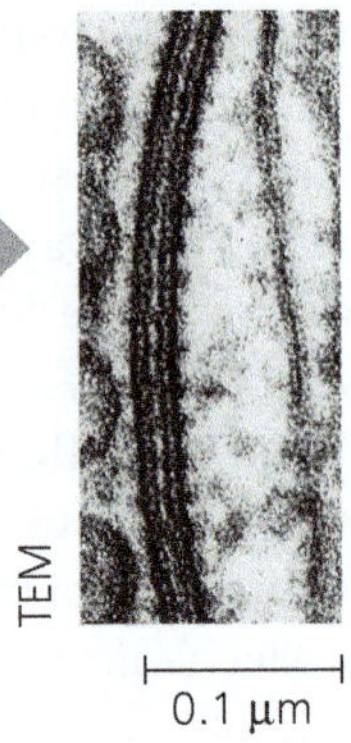

Gap Junctions

Gap junctions (also called *communicating junctions*) provide cytoplasmic channels from one cell to an adjacent cell and in this way are similar in their function to the plasmodesmata in plants. Gap junctions consist of membrane proteins that surround a pore through which ions, sugars, amino acids, and other small molecules may pass. Gap junctions are necessary for communication between cells in many types of tissues, such as heart muscle, and in animal embryos.

MB **Animation: Cell Junctions**

CONCEPT CHECK 6.7

1. In what way are the cells of plants and animals structurally different from single-celled eukaryotes?
2. **WHAT IF? >** If the plant cell wall or the animal extracellular matrix were impermeable, what effect would this have on cell function?
3. **MAKE CONNECTIONS >** The polypeptide chain that makes up a tight junction weaves back and forth through the membrane four times, with two extracellular loops, and one loop plus short C-terminal and N-terminal tails in the cytoplasm. Looking at Figure 5.14, what would you predict about the amino acid sequence of the tight junction protein?

For suggested answers, see Appendix A.

CONCEPT 6.8

A cell is greater than the sum of its parts

From our panoramic view of the cell's compartmental organization to our close-up inspection of each organelle's architecture, this tour of the cell has provided many opportunities to correlate structure with function. (This would be a good time to review cell structure by returning to Figure 6.8.)

BioFlix® Animation: Tour of an Animal Cell
BioFlix® Animation: Tour of a Plant Cell

Even as we dissect the cell, remember that none of its components works alone. As an example of cellular integration, consider the microscopic scene in **Figure 6.31**. The large cell is a macrophage (see Figure 6.13a). It helps defend the mammalian body against infections by ingesting bacteria (the smaller cells) into phagocytic vesicles. The macrophage crawls along a surface and reaches out to the bacteria with thin pseudopodia (specifically, filopodia). Actin filaments interact with other elements of the cytoskeleton in these movements. After the macrophage engulfs the bacteria, they are destroyed by lysosomes produced by the elaborate endomembrane system. The digestive enzymes of the lysosomes and the proteins of the cytoskeleton are all made by ribosomes. And the synthesis of these proteins is programmed by genetic messages dispatched from the DNA in the nucleus. All these processes require energy, which mitochondria supply in the form of ATP.

Cellular functions arise from cellular order: The cell is a living unit greater than the sum of its parts. The cell in Figure 6.31 is a good example of integration of cellular processes, seen from the outside. But what about the internal organization of a cell? As you proceed in your study of biology to consider different cellular processes, it will be helpful to try to visualize the architecture and furnishings inside a cell. **Figure 6.32** is designed to help you get a sense of the relative sizes and organization of important biological molecules and macromolecules, along with cellular structures and organelles. As you study this figure, see if you can shrink yourself down to the size of a protein and contemplate your surroundings.

CONCEPT CHECK 6.8

1. *Colpidium colpoda* is a unicellular eukaryote that lives in freshwater, eats bacteria, and moves by cilia (see Figure 6.23b). Describe how the parts of this cell work together in the functioning of *C. colpoda*, including as many organelles and other cell structures as you can.

For suggested answers, see Appendix A.

Figure 6.31 The emergence of cellular functions. The ability of this macrophage (brown) to recognize, apprehend, and destroy *Staphylococcus* bacteria (orange) is a coordinated activity of the whole cell. Its cytoskeleton, lysosomes, and plasma membrane are among the components that function in phagocytosis (colorized SEM).

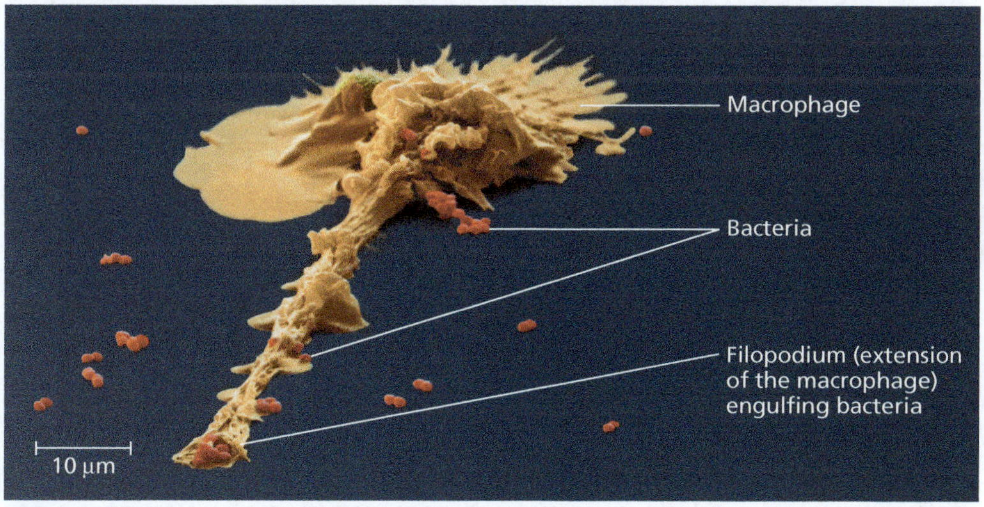

Animation: Review of Animal Cell Structure and Function

▼ Figure 6.32 Visualizing the Scale of the Molecular Machinery in a Cell

A slice of a plant cell's interior is illustrated in the center panel, with all structures and molecules drawn to scale. Selected molecules and structures are shown above and below, all enlarged by the same factor so you can compare their sizes. All protein and nucleic acid structures are based on data from the Protein Data Bank; regions whose structure has not yet been determined are shown in gray.

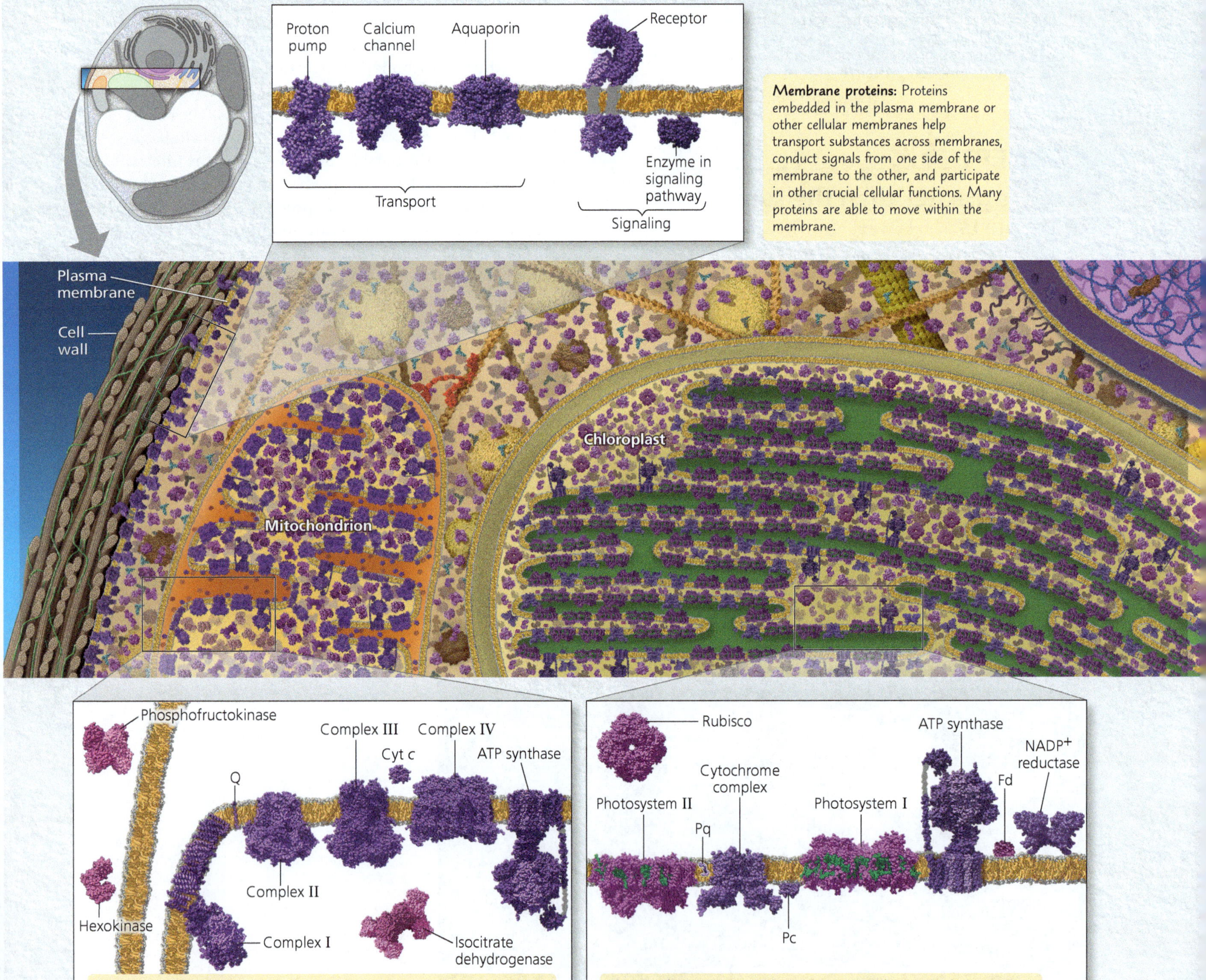

Membrane proteins: Proteins embedded in the plasma membrane or other cellular membranes help transport substances across membranes, conduct signals from one side of the membrane to the other, and participate in other crucial cellular functions. Many proteins are able to move within the membrane.

Cellular respiration: Cellular respiration includes many steps, some carried out by individual proteins or protein complexes in the cytoplasm and the mitochondrial matrix. Other proteins and protein complexes, involved in generating ATP from food molecules, form a "chain" in the inner mitochondrial membrane.

Photosynthesis: Large complexes of proteins with associated nonprotein molecules are embedded in chloroplast membranes. Together, they can trap light energy in molecules that are later used by other proteins inside the chloroplast to make sugars. This is the basis for all life on the planet.

Transcription: In the nucleus, the information contained in a DNA sequence is transferred to messenger RNA (mRNA) by an enzyme called RNA polymerase. After their synthesis, mRNA molecules leave the nucleus via nuclear pores.

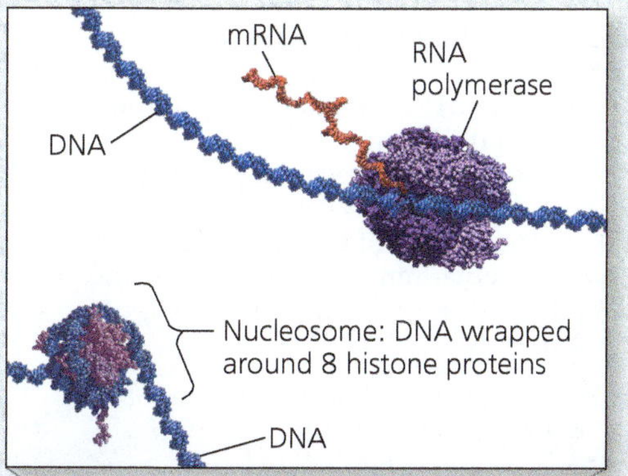

Nuclear pore: The nuclear pore complex regulates molecular traffic in and out of the nucleus, which is bounded by a double membrane. Among the largest structures that pass through the pore are the ribosomal subunits, which are built in the nucleus.

Translation: In the cytoplasm, the information in mRNA is used to assemble a polypeptide with a specific sequence of amino acids. Both transfer RNA (tRNA) molecules and a ribosome play a role. The eukaryotic ribosome, which includes a large subunit and a small subunit, is a colossal complex composed of four large ribosomal RNA (rRNA) molecules and more than 80 proteins. Through transcription and translation, the nucleotide sequence of DNA determines the amino acid sequence of a polypeptide, via the intermediary mRNA.

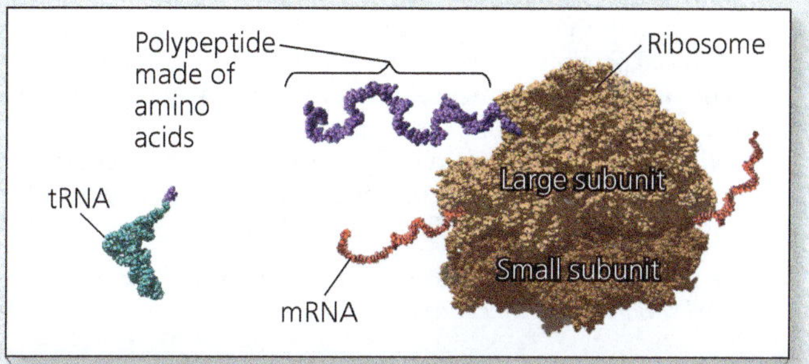

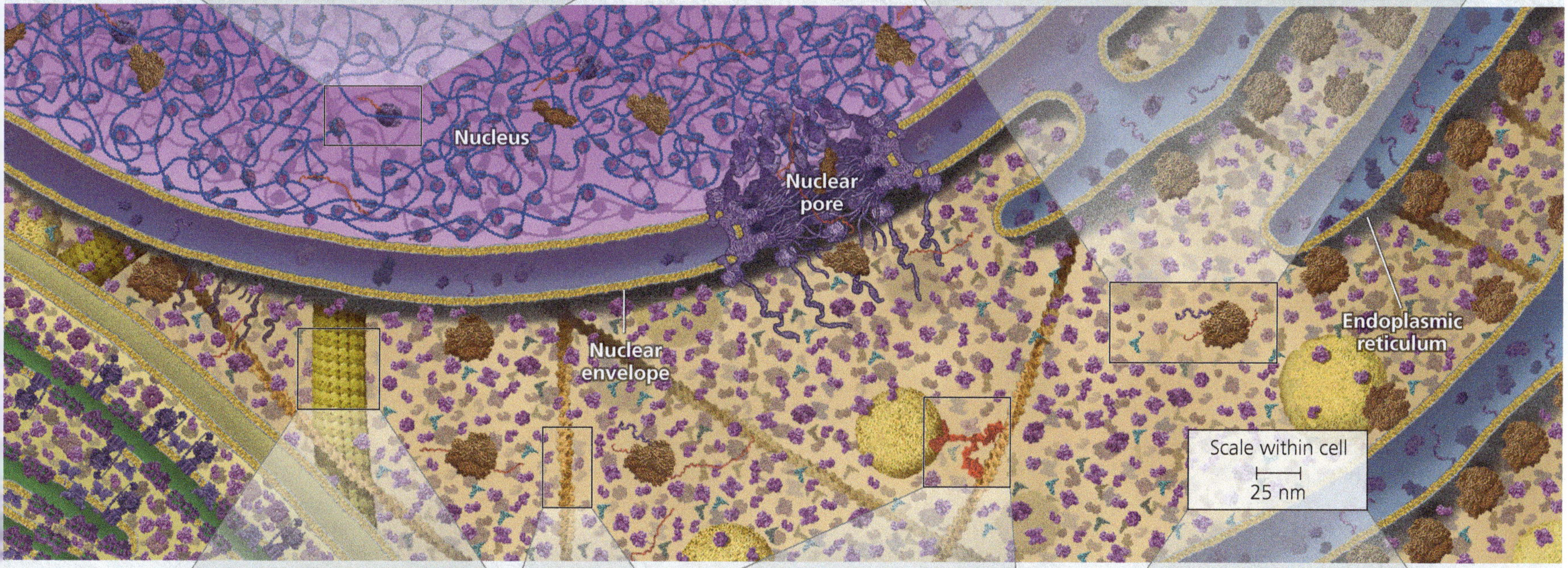

Cytoskeleton: Cytoskeletal structures are polymers of protein subunits. Microtubules are hollow structural rods made of tubulin protein subunits, while microfilaments are cables that have two chains of actin proteins wound around each other.

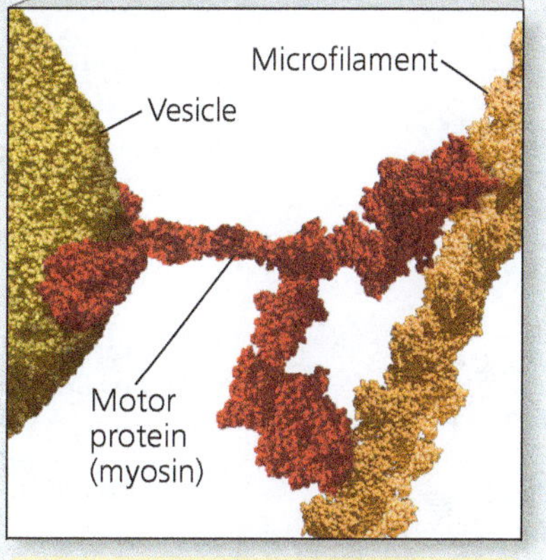

Motor proteins: Responsible for transport of vesicles and movement of organelles within the cell. This requires energy, often provided by ATP hydrolysis.

25 nm
Scale of enlarged structures

1. *List the following structures from largest to smallest: proton pump, nuclear pore, Cyt c, ribosome.*
2. *Considering the structures of a nucleosome and of RNA polymerase, speculate about what must happen before RNA polymerase can transcribe the DNA wrapped around the histone proteins of a nucleosome.*
3. *Find another myosin motor protein walking on a microfilament in this figure. What organelle is being moved by that myosin protein?*

Instructors: Additional questions related to this Visualizing Figure can be assigned in MasteringBiology.

6 Chapter Review

Go to **MasteringBiology**™ for Videos, Animations, Vocab Self-Quiz, Practice Tests, and more in the Study Area.

SUMMARY OF KEY CONCEPTS

VOCAB SELF-QUIZ goo.gl/6u55ks

CONCEPT 6.1

Biologists use microscopes and biochemistry to study cells *(pp. 94–97)*

- Improvements in microscopy that affect the parameters of magnification, resolution, and contrast have catalyzed progress in the study of cell structure. **Light microscopy** (LM) and **electron microscopy** (EM), as well as other types, remain important tools.
- Cell biologists can obtain pellets enriched in particular cellular components by centrifuging disrupted cells at sequential speeds, a process known as **cell fractionation**.

? *How do microscopy and biochemistry complement each other to reveal cell structure and function?*

CONCEPT 6.2

Eukaryotic cells have internal membranes that compartmentalize their functions
(pp. 97–102)

- All cells are bounded by a **plasma membrane**.
- **Prokaryotic cells** lack nuclei and other membrane-enclosed **organelles**, while **eukaryotic cells** have internal membranes that compartmentalize cellular functions.
- The surface-to-volume ratio is an important parameter affecting cell size and shape.
- Plant and animal cells have most of the same organelles: a nucleus, endoplasmic reticulum, Golgi apparatus, and mitochondria. Chloroplasts are present only in cells of photosynthetic eukaryotes.

? *Explain how the compartmental organization of a eukaryotic cell contributes to its biochemical functioning.*

	Cell Component	Structure	Function
CONCEPT 6.3 **The eukaryotic cell's genetic instructions are housed in the nucleus and carried out by the ribosomes** *(pp. 102–104)* ? *Describe the relationship between the nucleus and ribosomes.*	Nucleus 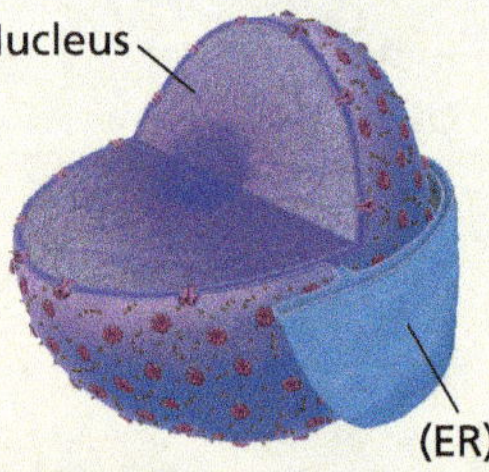 	Surrounded by nuclear envelope (double membrane) perforated by nuclear pores; nuclear envelope continuous with endoplasmic reticulum (ER)	Houses chromosomes, which are made of chromatin (DNA and proteins); contains nucleoli, where ribosomal subunits are made; pores regulate entry and exit of materials
	Ribosome 	Two subunits made of ribosomal RNAs and proteins; can be free in cytosol or bound to ER	Protein synthesis
CONCEPT 6.4 **The endomembrane system regulates protein traffic and performs metabolic functions** *(pp. 104–109)* ? *Describe the key role played by transport vesicles in the endomembrane system.*	Endoplasmic reticulum (ER) 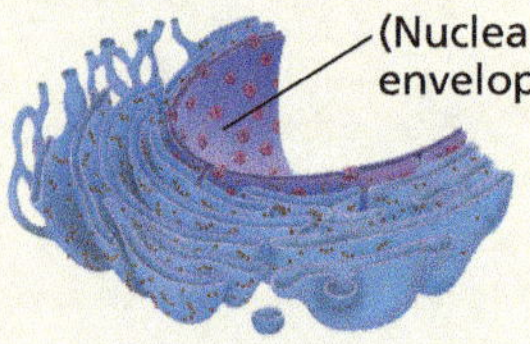 	Extensive network of membrane-bounded tubules and sacs; membrane separates lumen from cytosol; continuous with nuclear envelope	Smooth ER: synthesis of lipids, metabolism of carbohydrates, Ca^{2+} storage, detoxification of drugs and poisons Rough ER: aids in synthesis of secretory and other proteins on bound ribosomes; adds carbohydrates to proteins to make glycoproteins; produces new membrane
	Golgi apparatus 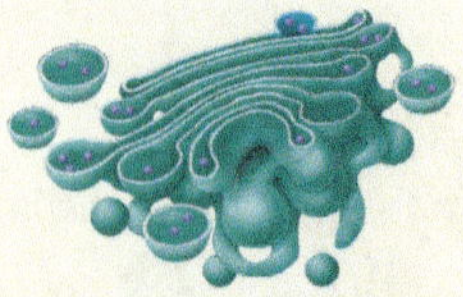	Stacks of flattened membranous sacs; has polarity (*cis* and *trans* faces)	Modification of proteins, carbohydrates on proteins, and phospholipids; synthesis of many polysaccharides; sorting of Golgi products, which are then released in vesicles
	Lysosome 	Membranous sac of hydrolytic enzymes (in animal cells)	Breakdown of ingested substances, cell macromolecules, and damaged organelles for recycling
	Vacuole 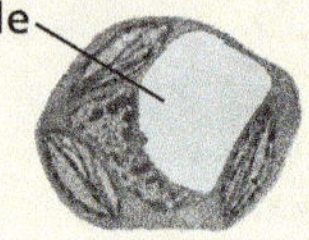	Large membrane-bounded vesicle	Digestion, storage, waste disposal, water balance, cell growth, and protection

	Cell Component	Structure	Function
CONCEPT 6.5 **Mitochondria and chloroplasts change energy from one form to another** *(pp. 109–112)* ? *What does the endosymbiont theory propose as the origin for mitochondria and chloroplasts? Explain.*	Mitochondrion 	Bounded by double membrane; inner membrane has infoldings	Cellular respiration
	Chloroplast 	Typically two membranes around fluid stroma, which contains thylakoids stacked into grana	Photosynthesis (chloroplasts are in cells of photosynthetic eukaryotes, including plants)
	Peroxisome 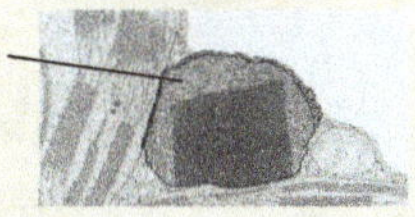	Specialized metabolic compartment bounded by a single membrane	Contains enzymes that transfer H atoms from substrates to oxygen, producing H_2O_2 (hydrogen peroxide), which is converted to H_2O.

CONCEPT 6.6

The cytoskeleton is a network of fibers that organizes structures and activities in the cell *(pp. 112–118)*

- The **cytoskeleton** functions in structural support for the cell and in motility and signal transmission.
- **Microtubules** shape the cell, guide organelle movement, and separate chromosomes in dividing cells. **Cilia** and **flagella** are motile appendages containing microtubules. Primary cilia also play sensory and signaling roles. **Microfilaments** are thin rods that function in muscle contraction, amoeboid movement, **cytoplasmic streaming**, and support of microvilli. **Intermediate filaments** support cell shape and fix organelles in place.

? *Describe the role of motor proteins inside the eukaryotic cell and in whole-cell movement.*

CONCEPT 6.7

Extracellular components and connections between cells help coordinate cellular activities *(pp. 118–121)*

- Plant **cell walls** are made of cellulose fibers embedded in other polysaccharides and proteins.
- Animal cells secrete glycoproteins and proteoglycans that form the **extracellular matrix (ECM)**, which functions in support, adhesion, movement, and regulation.
- Cell junctions connect neighboring cells. Plants have **plasmodesmata** that pass through adjoining cell walls. Animal cells have **tight junctions, desmosomes,** and **gap junctions.**

? *Compare the structure and functions of a plant cell wall and the extracellular matrix of an animal cell.*

CONCEPT 6.8

A cell is greater than the sum of its parts *(pp. 121–123)*

- Many components work together in a functioning cell.

? *When a cell ingests a bacterium, what role does the nucleus play?*

TEST YOUR UNDERSTANDING

Level 1: Knowledge/Comprehension

PRACTICE TEST goo.gl/CUYGKD

1. Which structure is *not* part of the endomembrane system?
   (A) nuclear envelope (C) Golgi apparatus
   (B) chloroplast (D) plasma membrane

2. Which structure is common to plant *and* animal cells?
   (A) chloroplast (C) mitochondrion
   (B) central vacuole (D) centriole

3. Which of the following is present in a prokaryotic cell?
   (A) mitochondrion (C) nuclear envelope
   (B) ribosome (D) chloroplast

Level 2: Application/Analysis

4. Cyanide binds to at least one molecule involved in producing ATP. If a cell is exposed to cyanide, most of the cyanide will be found within the
   (A) mitochondria. (C) peroxisomes.
   (B) ribosomes. (D) lysosomes.

5. Which cell would be best for studying lysosomes?
   (A) muscle cell (C) bacterial cell
   (B) nerve cell (D) phagocytic white blood cell

6. **DRAW IT** From memory, draw two eukaryotic cells. Label the structures listed here and show any physical connections between the internal structures of each cell: nucleus, rough ER, smooth ER, mitochondrion, centrosome, chloroplast, vacuole, lysosome, microtubule, cell wall, ECM, microfilament, Golgi apparatus, intermediate filament, plasma membrane, peroxisome, ribosome, nucleolus, nuclear pore, vesicle, flagellum, microvilli, plasmodesma.

Level 3: Synthesis/Evaluation

7. **EVOLUTION CONNECTION** (a) What cell structures best reveal evolutionary unity? (b) Provide an example of diversity related to specialized cellular modifications.

8. **SCIENTIFIC INQUIRY** Imagine protein X, destined to span the plasma membrane. Assume that the mRNA carrying the genetic message for protein X has already been translated by ribosomes in a cell culture. If you fractionate the cells (see Figure 6.4), in which fraction would you find protein X? Explain by describing its transit through the cell.

9. **WRITE ABOUT A THEME: ORGANIZATION** Considering some of the characteristics that define life and drawing on your knowledge of cellular structures and functions, write a short essay (100–150 words) that discusses this statement: Life is an emergent property that appears at the level of the cell. (See Concept 1.1.)

10. **SYNTHESIZE YOUR KNOWLEDGE**

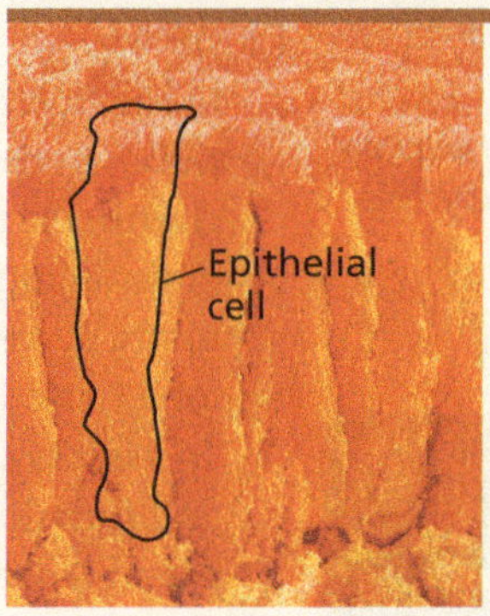

The cells in this SEM are epithelial cells from the small intestine. Discuss how aspects of their structure contribute to their specialized functions of nutrient absorption and as a barrier between the intestinal contents and the blood supply on the other side of the sheet of epithelial cells.

For selected answers, see Appendix A.

MB For additional practice questions, check out the **Dynamic Study Modules** in MasteringBiology. You can use them to study on your smartphone, tablet, or computer anytime, anywhere!

7 Membrane Structure and Function

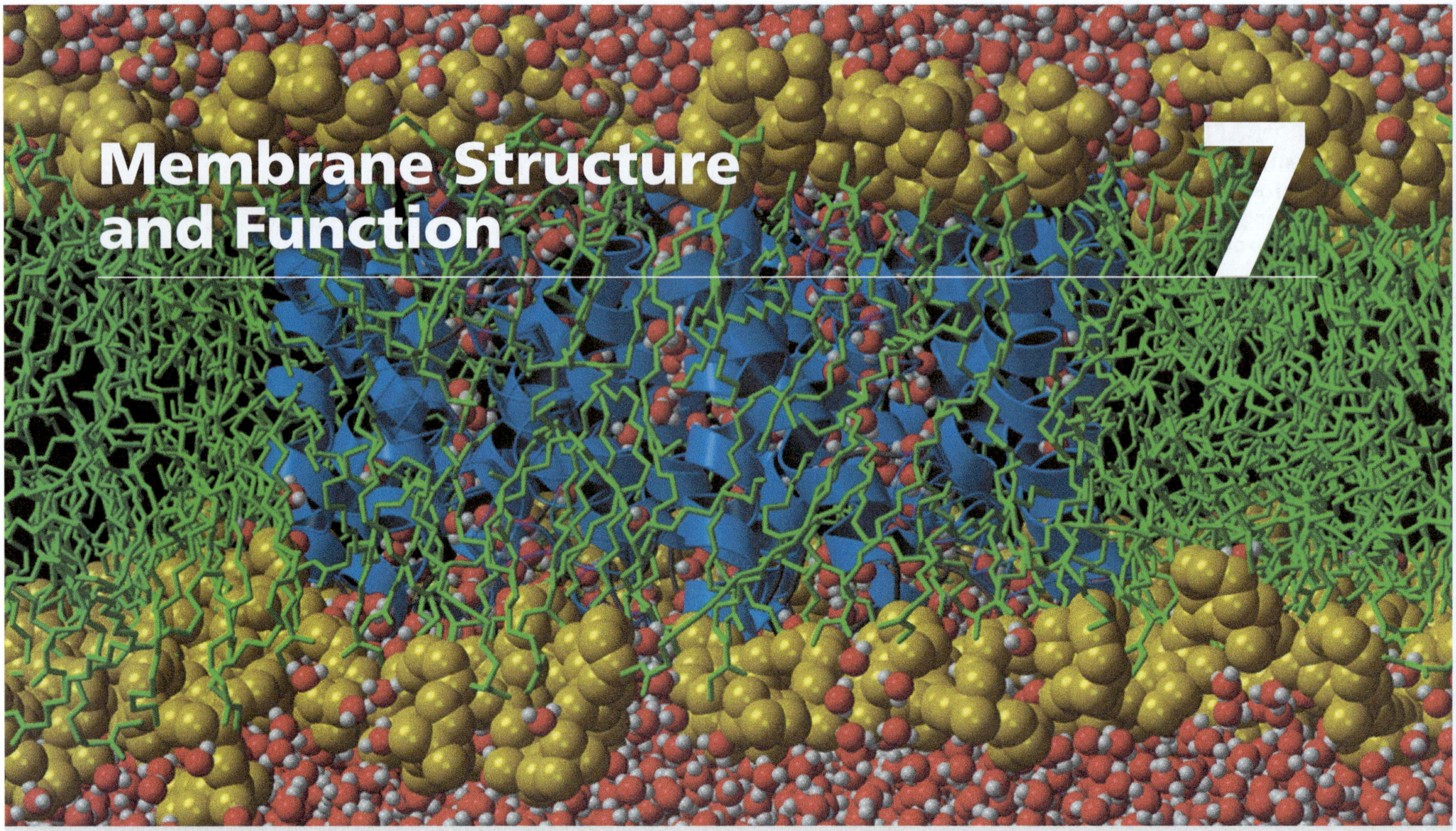

▲ **Figure 7.1 How do cell membrane proteins like this aquaporin (blue ribbons) help regulate chemical traffic?**

KEY CONCEPTS

7.1 Cellular membranes are fluid mosaics of lipids and proteins

7.2 Membrane structure results in selective permeability

7.3 Passive transport is diffusion of a substance across a membrane with no energy investment

7.4 Active transport uses energy to move solutes against their gradients

7.5 Bulk transport across the plasma membrane occurs by exocytosis and endocytosis

▼ **Potassium ion channel protein**

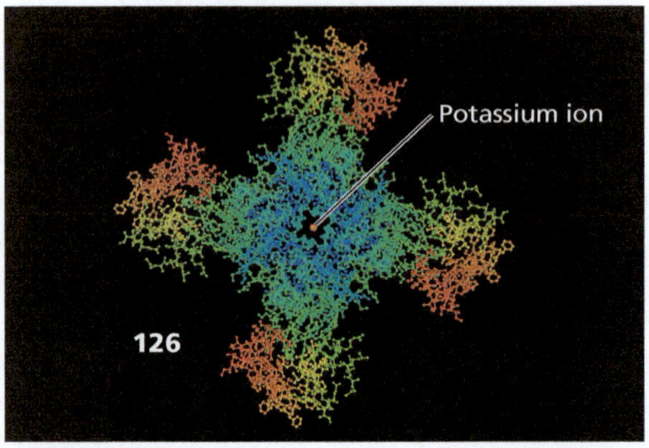

Life at the Edge

The plasma membrane that surrounds the cell can be considered the edge of life, the boundary that separates a living cell from its surroundings and controls all inbound and outbound traffic. Like all biological membranes, the plasma membrane exhibits **selective permeability**; that is, it allows some substances to cross it more easily than others. The ability of the cell to discriminate in its chemical exchanges is fundamental to life, and it is the plasma membrane and its component molecules that make this selectivity possible.

In this chapter, you will learn how cellular membranes control the passage of substances, often with transport proteins. For example, the image in **Figure 7.1** shows a computer model of a short section of the phospholipid bilayer of a membrane (hydrophilic heads are yellow, and hydrophobic tails are green). The blue ribbons within the lipid bilayer represent helical regions of a membrane transport channel protein called an aquaporin. One molecule of this protein enables billions of water molecules (red and gray) to pass through the membrane every second, many more than could cross on their own. Another type of transport protein is the ion channel shown here; it allows potassium ions to pass through the membrane. To understand how the plasma membrane and its proteins enable cells to survive and function, we begin by examining membrane structure, then explore how plasma membranes control transport into and out of cells.

When you see this blue icon, log in to **MasteringBiology** and go to the Study Area for digital resources.

MB **Get Ready for This Chapter**

MB **Animation: Membrane in Motion**
Interview with Peter Agre: Discovering aquaporins

CONCEPT 7.1

Cellular membranes are fluid mosaics of lipids and proteins

Lipids and proteins are the staple ingredients of membranes, although carbohydrates are also important. The most abundant lipids in most membranes are phospholipids. The ability of phospholipids to form membranes is inherent in their molecular structure. A phospholipid is an **amphipathic** molecule, meaning it has both a hydrophilic ("water-loving") region and a hydrophobic ("water-fearing") region (see Figure 5.11). Other types of membrane lipids are also amphipathic. A phospholipid bilayer can exist as a stable boundary between two aqueous compartments because the molecular arrangement shelters the hydrophobic tails of the phospholipids from water while exposing the hydrophilic heads to water **(Figure 7.2)**.

Like membrane lipids, most membrane proteins are amphipathic. Such proteins can reside in the phospholipid bilayer with their hydrophilic regions protruding. This molecular orientation maximizes contact of hydrophilic regions of proteins with water in the cytosol and extracellular fluid, while providing their hydrophobic parts with a nonaqueous environment. **Figure 7.3** shows the currently accepted model of the arrangement of molecules in the plasma membrane. In this **fluid mosaic model**, the membrane is a mosaic of protein molecules bobbing in a fluid bilayer of phospholipids.

▼ Figure 7.2 Phospholipid bilayer (cross section).

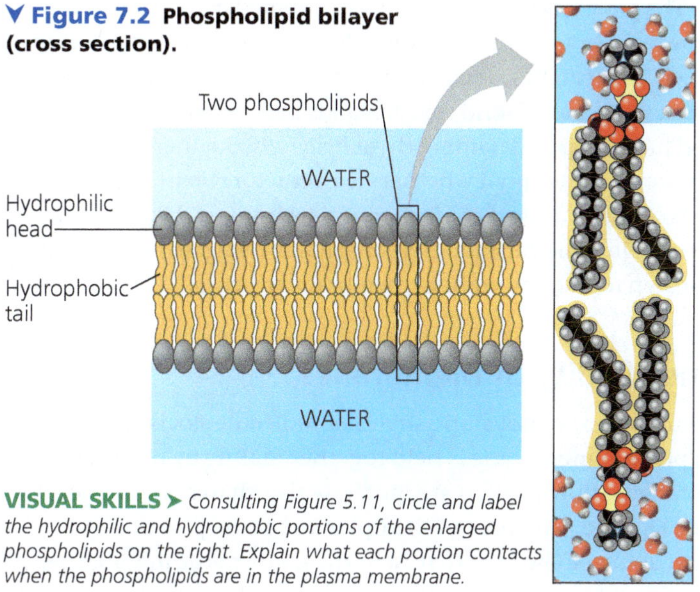

VISUAL SKILLS ➤ *Consulting Figure 5.11, circle and label the hydrophilic and hydrophobic portions of the enlarged phospholipids on the right. Explain what each portion contacts when the phospholipids are in the plasma membrane.*

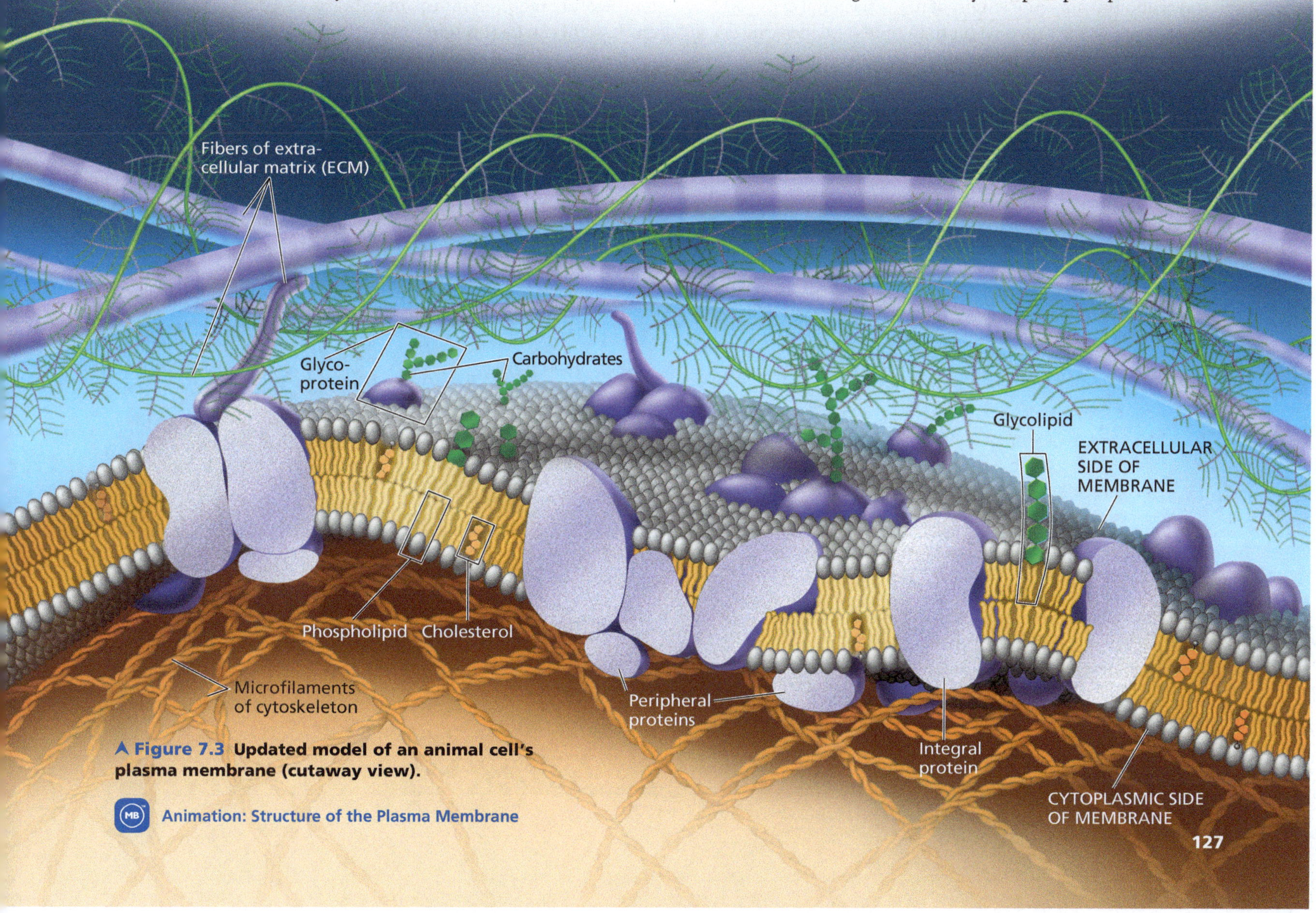

▲ Figure 7.3 Updated model of an animal cell's plasma membrane (cutaway view).

MB Animation: Structure of the Plasma Membrane

The proteins are not randomly distributed in the membrane, however. Groups of proteins are often associated in long-lasting, specialized patches, where they carry out common functions. Researchers have found specific lipids in these patches as well and have proposed naming them *lipid rafts*, but there is ongoing controversy about whether such structures exist in living cells or are an artifact of biochemical techniques. Like all models, the fluid mosaic model is continually being refined as new research reveals more about membrane structure.

The Fluidity of Membranes

Membranes are not static sheets of molecules locked rigidly in place. A membrane is held together mainly by hydrophobic interactions, which are much weaker than covalent bonds (see Figure 5.18). Most of the lipids and some proteins can shift about sideways—that is, in the plane of the membrane, like partygoers elbowing their way through a crowded room. Very rarely, also, a lipid may flip-flop across the membrane, switching from one phospholipid layer to the other.

The sideways movement of phospholipids within the membrane is rapid. Adjacent phospholipids switch positions about 10^7 times per second, which means that a phospholipid can travel about 2 μm—the length of many bacterial cells—in 1 second. Proteins are much larger than lipids and move more slowly, but some membrane proteins do drift, as shown in a classic experiment described in **Figure 7.4**. Some membrane proteins seem to move in a highly directed manner, perhaps driven along cytoskeletal fibers in the cell by motor proteins connected to the membrane proteins' cytoplasmic regions. However, many other membrane proteins seem to be held immobile by their attachment to the cytoskeleton or to the extracellular matrix (see Figure 7.3).

A membrane remains fluid as temperature decreases until the phospholipids settle into a closely packed arrangement and the membrane solidifies, much as bacon grease forms lard when it cools. The temperature at which a membrane solidifies depends on the types of lipids it is made of. As the temperature decreases, the membrane remains fluid to a lower temperature if it is rich in phospholipids with unsaturated hydrocarbon tails (see Figures 5.10 and 5.11). Because of kinks in the tails where double bonds are located, unsaturated hydrocarbon tails cannot pack together as closely as saturated hydrocarbon tails, making the membrane more fluid **(Figure 7.5a)**.

The steroid cholesterol, which is wedged between phospholipid molecules in the plasma membranes of animal cells, has different effects on membrane fluidity at different temperatures **(Figure 7.5b)**. At relatively high temperatures—at 37°C, the body temperature of humans, for example—cholesterol makes the membrane less fluid by restraining phospholipid movement. However, because cholesterol also hinders the close packing of phospholipids, it lowers the temperature required for the membrane to solidify. Thus, cholesterol can be thought of as a "fluidity buffer" for the membrane, resisting changes in membrane fluidity that can be caused by changes in temperature. Compared to animals, plants have very low levels of cholesterol; rather, related steroid lipids buffer membrane fluidity in plant cells.

Figure 7.4

Inquiry Do membrane proteins move?

Experiment Larry Frye and Michael Edidin, at Johns Hopkins University, labeled the plasma membrane proteins of a mouse cell and a human cell with two different markers and fused the cells. Using a microscope, they observed the markers on the hybrid cell.

Results

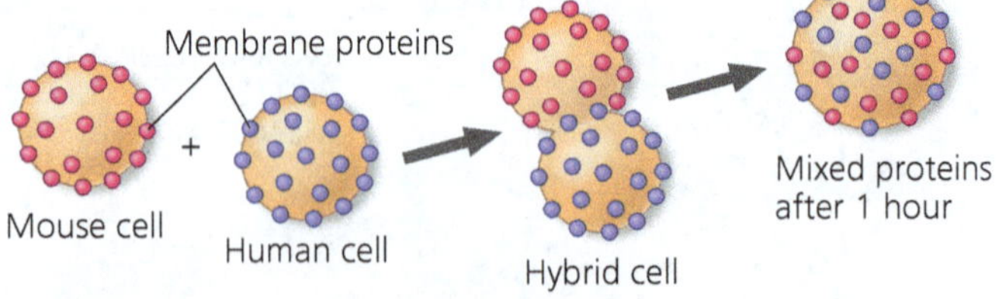

Conclusion The mixing of the mouse and human membrane proteins indicates that at least some membrane proteins move sideways within the plane of the plasma membrane.

Data from L. D. Frye and M. Edidin, The rapid intermixing of cell surface antigens after formation of mouse-human heterokaryons, *Journal of Cell Science* 7:319 (1970).

WHAT IF? ➤ *Suppose the proteins did not mix in the hybrid cell, even many hours after fusion. Would you be able to conclude that proteins don't move within the membrane? What other explanation could there be?*

Figure 7.5 Factors that affect membrane fluidity.

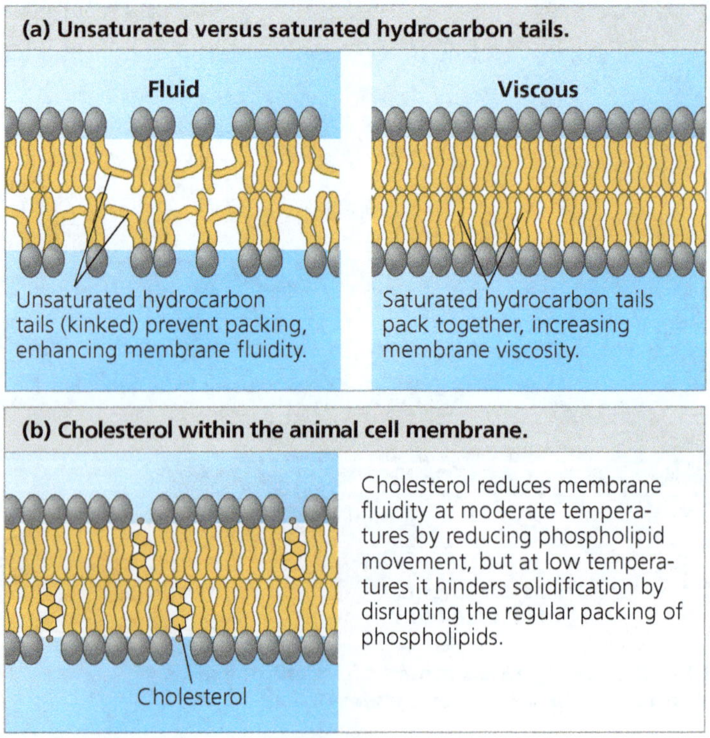

Membranes must be fluid to work properly; the fluidity of a membrane affects both its permeability and the ability of membrane proteins to move to where their function is needed. Usually, membranes are about as fluid as salad oil. When a membrane solidifies, its permeability changes, and enzymatic proteins in the membrane may become inactive if their activity requires movement within the membrane. However, membranes that are too fluid cannot support protein function either. Therefore, extreme environments pose a challenge for life, resulting in evolutionary adaptations that include differences in membrane lipid composition.

Evolution of Differences in Membrane Lipid Composition

EVOLUTION Variations in the cell membrane lipid compositions of many species appear to be evolutionary adaptations that maintain the appropriate membrane fluidity under specific environmental conditions. For instance, fishes that live in extreme cold have membranes with a high proportion of unsaturated hydrocarbon tails, enabling their membranes to remain fluid (see Figure 7.5a). At the other extreme, some bacteria and archaea thrive at temperatures greater than 90°C (194°F) in thermal hot springs and geysers. Their membranes include unusual lipids that may prevent excessive fluidity at such high temperatures.

The ability to change the lipid composition of cell membranes in response to changing temperatures has evolved in organisms that live where temperatures vary. In many plants that tolerate extreme cold, such as winter wheat, the percentage of unsaturated phospholipids increases in autumn, an adjustment that keeps the membranes from solidifying during winter. Certain bacteria and archaea can also change the proportion of unsaturated phospholipids in their cell membranes, depending on the temperature at which they are growing. Overall, natural selection has apparently favored organisms whose mix of membrane lipids ensures an appropriate level of membrane fluidity for their environment.

Membrane Proteins and Their Functions

Now we come to the *mosaic* aspect of the fluid mosaic model. Somewhat like a tile mosaic (shown here), a membrane is a collage of different proteins, often clustered together in groups, embedded in the fluid matrix of the lipid bilayer (see Figure 7.3). In the plasma membrane of red blood cells alone, for example, more than 50 kinds of proteins have been found so far. Phospholipids form the main fabric of the membrane, but proteins determine most of the membrane's functions. Different types of cells contain different sets of membrane proteins, and the various membranes within a cell each have a unique collection of proteins.

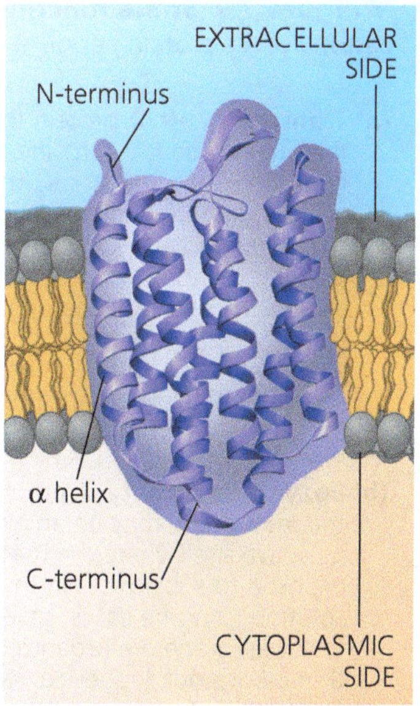

Figure 7.6 The structure of a transmembrane protein. Bacteriorhodopsin (a bacterial transport protein) has a distinct orientation in the membrane, with its N-terminus outside the cell and its C-terminus inside. This ribbon model highlights the secondary structure of the hydrophobic parts, including seven transmembrane α helices, which lie mostly within the hydrophobic interior of the membrane. The nonhelical hydrophilic segments are in contact with the aqueous solutions on the extracellular and cytoplasmic sides of the membrane.

Notice in Figure 7.3 that there are two major populations of membrane proteins: integral proteins and peripheral proteins. **Integral proteins** penetrate the hydrophobic interior of the lipid bilayer. The majority are *transmembrane proteins*, which span the membrane; other integral proteins extend only partway into the hydrophobic interior. The hydrophobic regions of an integral protein consist of one or more stretches of nonpolar amino acids (see Figure 5.14), typically 20–30 amino acids in length, usually coiled into α helices **(Figure 7.6)**. The hydrophilic parts of the molecule are exposed to the aqueous solutions on either side of the membrane. Some proteins also have one or more hydrophilic channels that allow passage through the membrane of hydrophilic substances (even of water itself; see Figure 7.1). **Peripheral proteins** are not embedded in the lipid bilayer at all; they are loosely bound to the surface of the membrane, often to exposed parts of integral proteins (see Figure 7.3).

On the cytoplasmic side of the plasma membrane, some membrane proteins are held in place by attachment to the cytoskeleton. And on the extracellular side, certain membrane proteins may attach to materials outside the cell. For example, in animal cells, membrane proteins may be attached to fibers of the extracellular matrix (see Figure 6.28; *integrins* are one type of integral, transmembrane protein). These attachments combine to give animal cells a stronger framework than the plasma membrane alone could provide.

A single cell may have cell-surface membrane proteins that carry out several different functions, such as transport through the cell membrane, enzymatic activity, or attaching a cell to either a neighboring cell or the extracellular matrix. Furthermore, a single membrane protein may itself carry out multiple functions. Thus, the membrane is not only a structural mosaic, with many proteins embedded in the membrane, but also a functional mosaic, carrying out a range

Figure 7.7 Some functions of membrane proteins. In many cases, a single protein performs multiple tasks.

(a) Transport. *Left:* A protein that spans the membrane may provide a hydrophilic channel across the membrane that is selective for a particular solute. *Right:* Other transport proteins shuttle a substance from one side to the other by changing shape (see Figure 7.14b). Some of these proteins hydrolyze ATP as an energy source to actively pump substances across the membrane.

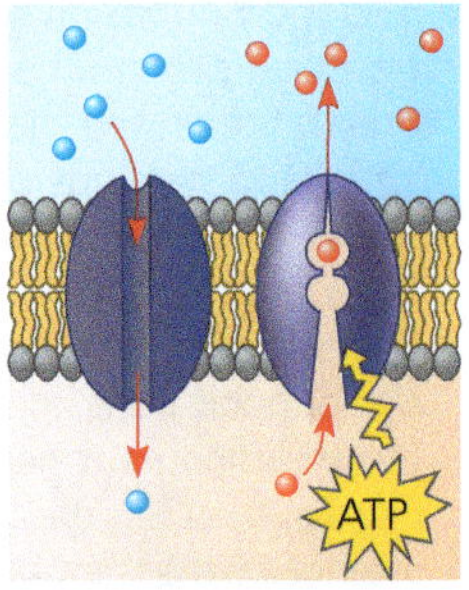

(b) Enzymatic activity. A protein built into the membrane may be an enzyme with its active site (where the reactant binds) exposed to substances in the adjacent solution. In some cases, several enzymes in a membrane are organized as a team that carries out sequential steps of a metabolic pathway.

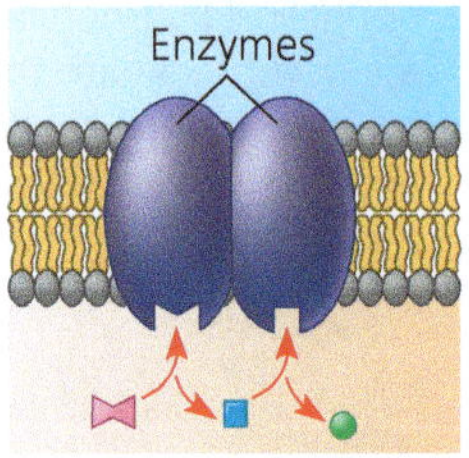

(c) Signal transduction. A membrane protein (receptor) may have a binding site with a specific shape that fits the shape of a chemical messenger, such as a hormone. The external messenger (signaling molecule) may cause the protein to change shape, allowing it to relay the message to the inside of the cell, usually by binding to a cytoplasmic protein (see Figure 11.6).

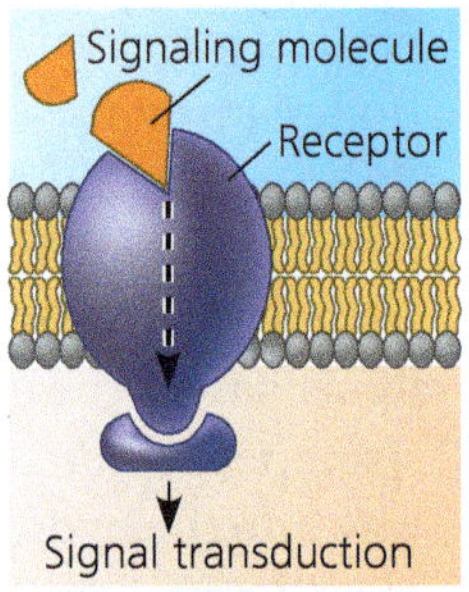

(d) Cell-cell recognition. Some glycoproteins serve as identification tags that are specifically recognized by membrane proteins of other cells. This type of cell-cell binding is usually short-lived compared to that shown in (e).

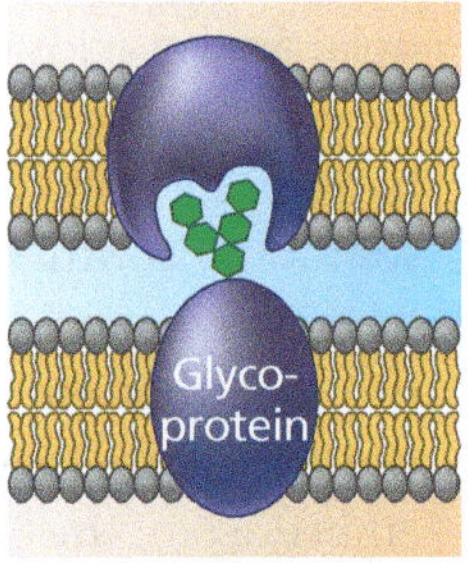

(e) Intercellular joining. Membrane proteins of adjacent cells may hook together in various kinds of junctions, such as gap junctions or tight junctions (see Figure 6.30). This type of binding is more long-lasting than that shown in (d).

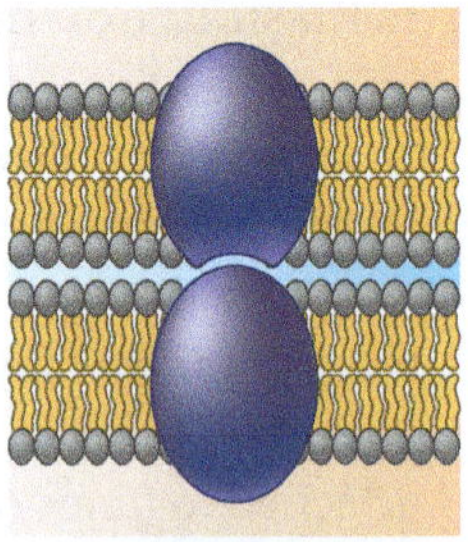

(f) Attachment to the cytoskeleton and extracellular matrix (ECM). Microfilaments or other elements of the cytoskeleton may be noncovalently bound to membrane proteins, a function that helps maintain cell shape and stabilizes the location of certain membrane proteins. Proteins that can bind to ECM molecules can coordinate extracellular and intracellular changes (see Figure 6.28).

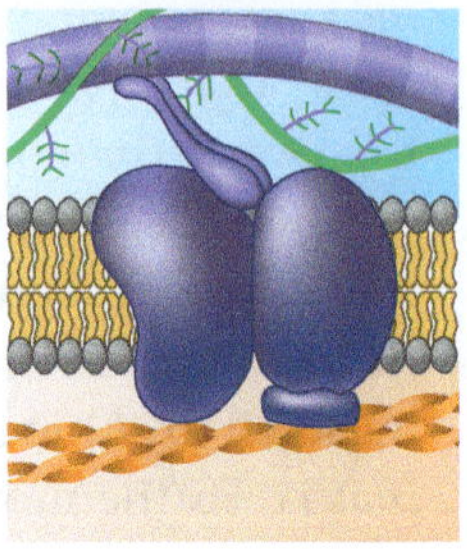

VISUAL SKILLS ➤ *Some transmembrane proteins can bind to a particular ECM molecule and, when bound, transmit a signal into the cell. Use the proteins shown in (c) and (f) to explain how this might occur.*

of functions. **Figure 7.7** illustrates six major functions performed by proteins of the plasma membrane.

Animation: Functions of the Plasma Membrane

Proteins on a cell's surface are important in the medical field. For example, a protein called CD4 on the surface of immune cells helps the human immunodeficiency virus (HIV) infect these cells, leading to acquired immune deficiency syndrome (AIDS). Despite multiple exposures to HIV, however, a small number of people do not develop AIDS and show no evidence of HIV-infected cells. Comparing their genes with the genes of infected individuals, researchers learned that resistant people have an unusual form of a gene that codes for an immune cell-surface protein called CCR5. Further work showed that although CD4 is the main HIV receptor, HIV must also bind to CCR5 as a "co-receptor" to infect most cells **(Figure 7.8a)**. An absence of CCR5 on the cells of resistant individuals, due to the gene alteration, prevents the virus from entering the cells **(Figure 7.8b)**.

This information has been key to developing a treatment for HIV infection. Interfering with CD4 causes dangerous side effects because of its many important functions in cells. Discovery of the CCR5 co-receptor provided a safer target for development of drugs that mask this protein and block HIV entry. One such drug, maraviroc (brand name Selzentry), was approved for treatment of HIV in 2007 and is now being tested to determine whether this drug might also work to prevent HIV infection in uninfected, at-risk patients.

The Role of Membrane Carbohydrates in Cell-Cell Recognition

Cell-cell recognition, a cell's ability to distinguish one type of neighboring cell from another, is crucial to the functioning of an organism. It is important, for example, in the sorting of cells into tissues and organs in an animal embryo. It is also

Figure 7.8 The genetic basis for HIV resistance.

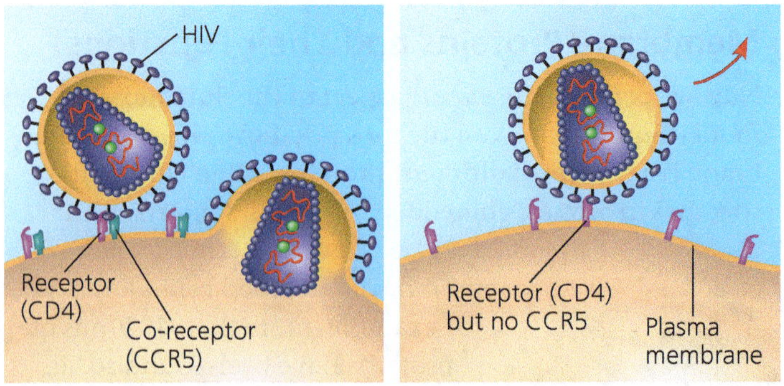

(a) HIV can infect a cell with CCR5 on its surface, as in most people.

(b) HIV cannot infect a cell lacking CCR5 on its surface, as in resistant individuals.

MAKE CONNECTIONS ➤ *Study Figures 2.16 and 5.17; each shows pairs of molecules binding to each other. What would you predict about CCR5 that would allow HIV to bind to it? How could a drug molecule interfere with this binding?*

the basis for the rejection of foreign cells by the immune system, an important line of defense in vertebrate animals (see Concept 43.1). Cells recognize other cells by binding to molecules, often containing carbohydrates, on the extracellular surface of the plasma membrane (see Figure 7.7d).

Membrane carbohydrates are usually short, branched chains of fewer than 15 sugar units. Some are covalently bonded to lipids, forming molecules called **glycolipids**. (Recall that *glyco* refers to carbohydrate.) However, most are covalently bonded to proteins, which are thereby **glycoproteins** (see Figure 7.3).

The carbohydrates on the extracellular side of the plasma membrane vary from species to species, among individuals of the same species, and even from one cell type to another in a single individual. The diversity of the molecules and their location on the cell's surface enable membrane carbohydrates to function as markers that distinguish one cell from another. For example, the four human blood types designated A, B, AB, and O reflect variation in the carbohydrate part of glycoproteins on the surface of red blood cells.

Synthesis and Sidedness of Membranes

Membranes have distinct inside and outside faces. The two lipid layers may differ in lipid composition, and each protein has directional orientation in the membrane (see Figure 7.6). **Figure 7.9** shows how membrane sidedness arises: The asymmetrical arrangement of proteins, lipids, and their associated carbohydrates in the plasma membrane is determined as the membrane is being built by the endoplasmic reticulum (ER) and Golgi apparatus, components of the endomembrane system (see Figure 6.15).

CONCEPT CHECK 7.1

1. **VISUAL SKILLS ➤** Carbohydrates are attached to plasma membrane proteins in the ER (see Figure 7.9). On which side of the vesicle membrane are the carbohydrates during transport to the cell surface?
2. **WHAT IF? ➤** How would the membrane lipid composition of a native grass found in very warm soil around hot springs compare with that of a native grass found in cooler soil? Explain.

For suggested answers, see Appendix A.

CONCEPT 7.2

Membrane structure results in selective permeability

The biological membrane is an exquisite example of a supramolecular structure—many molecules ordered into a higher level of organization—with emergent properties beyond those of the individual molecules. The remainder of this chapter focuses on one of those properties: the ability to regulate transport across cellular boundaries, a function essential to the cell's existence. We will see once again that form fits

▼ Figure 7.9 Synthesis of membrane components and their orientation in the membrane. The cytoplasmic (orange) face of the plasma membrane differs from the extracellular (aqua) face. The latter arises from the inside face of ER, Golgi, and vesicle membranes.

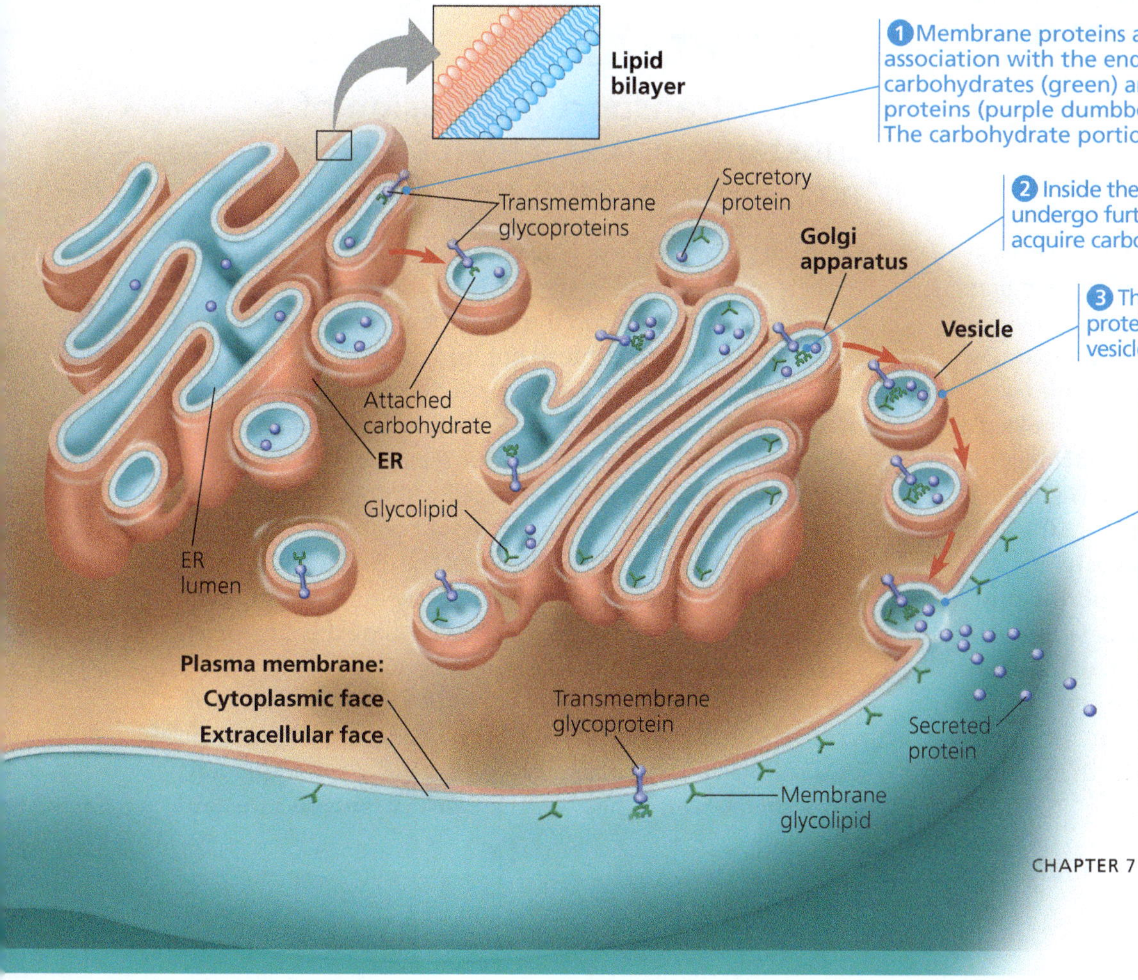

1 Membrane proteins and lipids are synthesized in association with the endoplasmic reticulum (ER). In the ER, carbohydrates (green) are added to the transmembrane proteins (purple dumbbells), making them glycoproteins. The carbohydrate portions may then be modified.

2 Inside the Golgi apparatus, the glycoproteins undergo further carbohydrate modification, and lipids acquire carbohydrates, becoming glycolipids.

3 The glycoproteins, glycolipids, and secretory proteins (purple spheres) are transported in vesicles to the plasma membrane.

4 As vesicles fuse with the plasma membrane, the outside face of the vesicle becomes continuous with the inside (cytoplasmic) face of the plasma membrane. This releases the secretory proteins from the cell, a process called *exocytosis*, and positions the carbohydrates of membrane glycoproteins and glycolipids on the outside (extracellular) face of the plasma membrane.

DRAW IT ➤ *Draw an integral membrane protein extending from partway through the ER membrane into the ER lumen. Next, draw the protein where it would be located in a series of numbered steps ending at the plasma membrane. Would the protein contact the cytoplasm or the extracellular fluid? Explain.*

function: The fluid mosaic model helps explain how membranes regulate the cell's molecular traffic.

A steady traffic of small molecules and ions moves across the plasma membrane in both directions. Consider the chemical exchanges between a muscle cell and the extracellular fluid that bathes it. Sugars, amino acids, and other nutrients enter the cell, and metabolic waste products leave it. The cell takes in O_2 for use in cellular respiration and expels CO_2. Also, the cell regulates its concentrations of inorganic ions, such as Na^+, K^+, Ca^{2+}, and Cl^-, by shuttling them one way or the other across the plasma membrane. Although the heavy traffic through them may seem to suggest otherwise, cell membranes are selectively permeable, and substances do not cross the barrier indiscriminately. The cell is able to take up some small molecules and ions and exclude others.

The Permeability of the Lipid Bilayer

Nonpolar molecules, such as hydrocarbons, CO_2, and O_2, are hydrophobic, as are lipids. They can all therefore dissolve in the lipid bilayer of the membrane and cross it easily, without the aid of membrane proteins. However, the hydrophobic interior of the membrane impedes direct passage through the membrane of ions and polar molecules, which are hydrophilic. Polar molecules such as glucose and other sugars pass only slowly through a lipid bilayer, and even water, a very small polar molecule, does not cross rapidly relative to nonpolar molecules. A charged atom or molecule and its surrounding shell of water (see Figure 3.8) are even less likely to penetrate the hydrophobic interior of the membrane. Furthermore, the lipid bilayer is only one aspect of the gatekeeper system responsible for a cell's selective permeability. Proteins built into the membrane play key roles in regulating transport.

Transport Proteins

Specific ions and a variety of polar molecules can't move through cell membranes on their own. However, these hydrophilic substances can avoid contact with the lipid bilayer by passing through **transport proteins** that span the membrane.

Some transport proteins, called *channel proteins*, function by having a hydrophilic channel that certain molecules or atomic ions use as a tunnel through the membrane (see Figure 7.7a, left). For example, the passage of water molecules through the membrane in certain cells is greatly facilitated by channel proteins known as **aquaporins** (see Figure 7.1). Each aquaporin allows entry of up to *3 billion* (3×10^9) water molecules per second, passing single file through its central channel, which fits ten at a time. Without aquaporins, only a tiny fraction of these water molecules would pass through the same area of the cell membrane in a second, so the channel protein brings about a tremendous increase in rate. Other transport proteins, called *carrier proteins*, hold onto their passengers and change shape in a way that shuttles them across the membrane (see Figure 7.7a, right).

A transport protein is specific for the substance it translocates (moves), allowing only a certain substance (or a small group of related substances) to cross the membrane. For example, a specific carrier protein in the plasma membrane of red blood cells transports glucose across the membrane 50,000 times faster than glucose can pass through on its own. This "glucose transporter" is so selective that it even rejects fructose, a structural isomer of glucose. Thus, the selective permeability of a membrane depends on both the discriminating barrier of the lipid bilayer and the specific transport proteins built into the membrane.

MB **Animation: Selective Permeability of Membranes**

What establishes the *direction* of traffic across a membrane? And what mechanisms drive molecules across membranes? We will address these questions next as we explore two modes of membrane traffic: passive transport and active transport.

CONCEPT CHECK 7.2

1. What property allows O_2 and CO_2 to cross a lipid bilayer without the aid of membrane proteins?
2. **VISUAL SKILLS ➤** Examine Figure 7.2. Why is a transport protein needed to move many water molecules rapidly across a membrane?
3. **MAKE CONNECTIONS ➤** Aquaporins exclude passage of hydronium ions (H_3O^+), but some aquaporins allow passage of glycerol, a three-carbon alcohol (see Figure 5.9), as well as H_2O. Since H_3O^+ is closer in size to water than glycerol is, yet cannot pass through, what might be the basis of this selectivity?

For suggested answers, see Appendix A.

CONCEPT 7.3

Passive transport is diffusion of a substance across a membrane with no energy investment

Molecules have a type of energy called thermal energy, due to their constant motion (see Concept 3.2). One result of this motion is **diffusion**, the movement of particles of any substance so that they spread out into the available space. Each molecule moves randomly, yet diffusion of a *population* of molecules may be directional. To understand this process, let's imagine a synthetic membrane separating pure water from a solution of a dye in water. Study **Figure 7.10a** carefully to appreciate how diffusion would result in both solutions having equal concentrations of the dye molecules. Once that point is reached, there will be a dynamic equilibrium, with roughly as many dye molecules crossing the membrane each second in one direction as in the other.

We can now state a simple rule of diffusion: In the absence of other forces, a substance will diffuse from where it is more concentrated to where it is less concentrated. Put another way, any substance will diffuse down its **concentration gradient**, the region along which the density of a chemical substance increases or decreases (in this case, decreases). No work must be done to make this happen; diffusion is a spontaneous process, needing no input of energy. Note that each substance diffuses down its *own* concentration gradient, unaffected by the concentration gradients of other substances **(Figure 7.10b)**.

Much of the traffic across cell membranes occurs by diffusion. When a substance is more concentrated on one side of a membrane than on the other, there is a tendency for the substance to diffuse across the membrane down its concentration gradient (assuming that the membrane is permeable to that substance). One important example is the uptake of oxygen by a cell performing cellular respiration. Dissolved oxygen diffuses into the cell across the plasma membrane. As long as cellular respiration consumes the O_2 as it enters, diffusion into the cell will continue because the concentration gradient favors movement in that direction.

The diffusion of a substance across a biological membrane is called **passive transport** because the cell does not have to expend energy to make it happen. The concentration gradient itself represents potential energy (see Concept 2.2 and Figure 8.5b) and drives diffusion. Remember, however, that membranes are selectively permeable and therefore have different effects on the rates of diffusion of various molecules. In the case of water, the presence of aquaporin proteins allows water to diffuse very rapidly across the membranes of certain cells compared with diffusion in the absence of aquaporins. As we'll see next, the movement of water across the plasma membrane has important consequences for cells.

Figure 7.10 The diffusion of solutes across a synthetic membrane. Each of the large arrows under the diagrams shows the net diffusion of the dye molecules of that color.

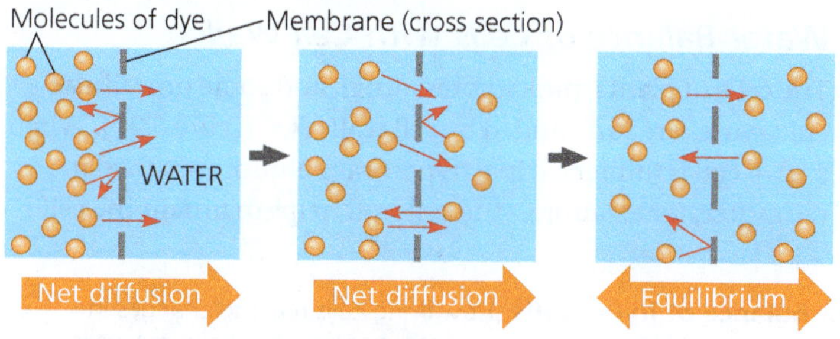

(a) Diffusion of one solute. The membrane has pores large enough for molecules of dye to pass through. Random movement of dye molecules will cause some to pass through the pores; this will happen more often on the side with more dye molecules. The dye diffuses from where it is more concentrated to where it is less concentrated (called diffusing down a concentration gradient). This leads to a dynamic equilibrium: The solute molecules continue to cross the membrane, but at roughly equal rates in both directions.

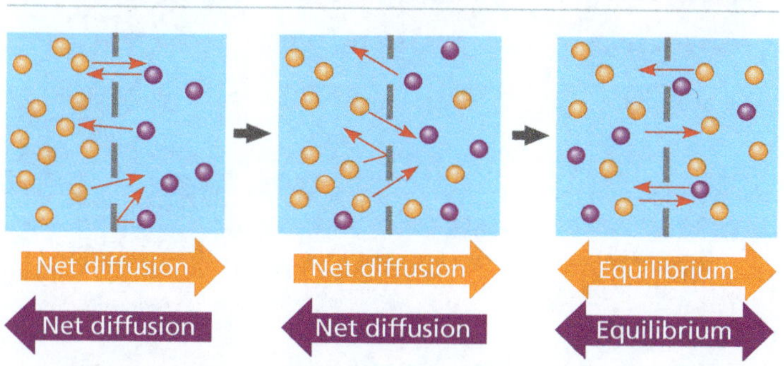

(b) Diffusion of two solutes. Solutions of two different dyes are separated by a membrane that is permeable to both. Each dye diffuses down its own concentration gradient. There will be a net diffusion of the purple dye toward the left, even though the *total* solute concentration was initially greater on the left side.

Animation: Diffusion

Effects of Osmosis on Water Balance

To see how two solutions with different solute concentrations interact, picture a U-shaped glass tube with a selectively permeable artificial membrane separating two sugar solutions **(Figure 7.11)**. Pores in this synthetic membrane are too small

Figure 7.11 Osmosis. Two sugar solutions of different concentrations are separated by a membrane that the solvent (water) can pass through but the solute (sugar) cannot. Water molecules move randomly and may cross in either direction, but overall, water diffuses from the solution with less concentrated solute to that with more concentrated solute. This passive transport of water, or osmosis, makes the sugar concentrations on both sides more nearly equal. (The concentrations are prevented from being exactly equal due to the effect of water pressure on the higher side, which is not discussed here, for simplicity.)

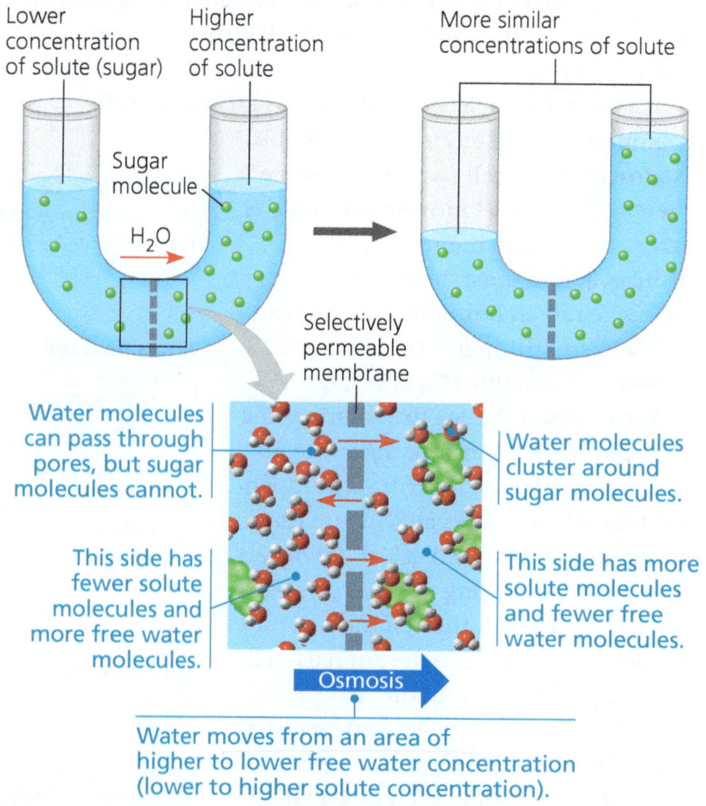

VISUAL SKILLS ➤ *If an orange dye capable of passing through the membrane was added to the left side of the tube above, how would it be distributed at the end of the experiment? (See Figure 7.10.) Would the final solution levels in the tube be affected?*

Figure Walkthrough

for sugar molecules to pass through but large enough for water molecules. However, tight clustering of water molecules around the hydrophilic solute molecules makes some of the water unavailable to cross the membrane. As a result, the solution with a higher solute concentration has a lower *free* water concentration. Water diffuses across the membrane from the region of higher free water concentration (lower solute concentration) to that of lower free water concentration (higher solute concentration) until the solute concentrations on both sides of the membrane are more nearly equal. The diffusion of free water across a selectively permeable membrane, whether artificial or cellular, is called **osmosis**. The movement of water across cell membranes and the balance of water between the cell and its environment are crucial to organisms. Let's now apply what we've learned about osmosis in this system to living cells.

Water Balance of Cells Without Cell Walls

To explain the behavior of a cell in a solution, we must consider both solute concentration and membrane permeability. Both factors are taken into account in the concept of **tonicity**, the ability of a surrounding solution to cause a cell to gain or lose water. The tonicity of a solution depends in part on its concentration of solutes that cannot cross the membrane (nonpenetrating solutes) relative to that inside the cell. If there is a higher concentration of nonpenetrating solutes in the surrounding solution, water will tend to leave the cell, and vice versa.

If a cell without a cell wall, such as an animal cell, is immersed in an environment that is **isotonic** to the cell (*iso* means "same"), there will be no *net* movement of water across the plasma membrane. Water diffuses across the membrane, but at the same rate in both directions. In an isotonic environment, the volume of an animal cell is stable **(Figure 7.12a)**.

Let's transfer the cell to a solution that is **hypertonic** to the cell (*hyper* means "more," in this case referring to nonpenetrating solutes). The cell will lose water, shrivel, and probably die. This is why an increase in the salinity (saltiness) of a lake can kill the animals there; if the lake water becomes hypertonic to the animals' cells, they might shrivel and die. However, taking up too much water can be just as hazardous as losing water. If we place the cell in a solution that is **hypotonic** to the cell (*hypo* means "less"), water will enter the cell faster than it leaves, and the cell will swell and lyse (burst) like an overfilled water balloon.

A cell without rigid cell walls can tolerate neither excessive uptake nor excessive loss of water. This problem of water balance is automatically solved if such a cell lives in isotonic surroundings. Seawater is isotonic to many marine invertebrates. The cells of most terrestrial (land-dwelling) animals are bathed in an extracellular fluid that is isotonic to the cells. In hypertonic or hypotonic environments, however, organisms that lack rigid cell walls must have other adaptations for **osmoregulation**, the control of solute concentrations and water balance. For example, the unicellular eukaryote *Paramecium* lives in pond water, which is hypotonic to the cell. *Paramecium* has a plasma membrane that is much less permeable to water than the membranes of most other cells, but this only slows the uptake of water, which continually enters the cell. The *Paramecium* cell doesn't burst because it is also equipped with a contractile vacuole, an organelle that functions as a bilge pump to force water out of the cell as fast as it enters by osmosis **(Figure 7.13)**. In contrast, the bacteria and archaea that live in hypersaline (excessively salty) environments (see Figure 27.1) have cellular mechanisms that balance the internal and external solute concentrations to ensure that water does not move out of the cell. We will examine other evolutionary adaptations for osmoregulation by animals in Concept 44.1.

Water Balance of Cells with Cell Walls

The cells of plants, prokaryotes, fungi, and some unicellular eukaryotes are surrounded by cell walls (see Figure 6.27). When such a cell is immersed in a hypotonic solution—bathed in rainwater, for example—the cell wall helps maintain the cell's

▼ Figure 7.12 The water balance of living cells. How living cells react to changes in the solute concentration of their environment depends on whether or not they have cell walls. **(a)** Animal cells, such as this red blood cell, do not have cell walls. **(b)** Plant cells do have cell walls. (Arrows indicate net water movement after the cells were first placed in these solutions.)

(a) Animal cell. An animal cell fares best in an isotonic environment unless it has special adaptations that offset the osmotic uptake or loss of water.

(b) Plant cell. Plant cells are turgid (firm) and generally healthiest in a hypotonic environment, where the uptake of water is eventually balanced by the wall pushing back on the cell.

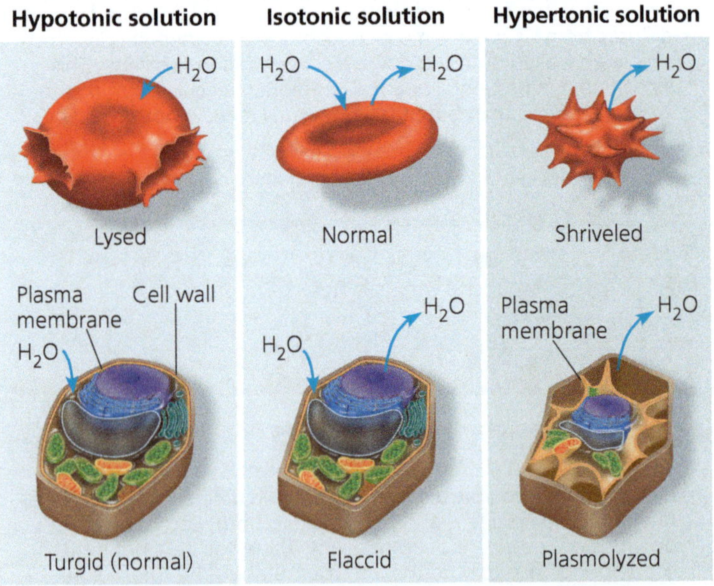

MB **Animation: Osmosis and Water Balance in Cells**
Video: Turgid *Elodea*
Video: Plasmolysis in *Elodea*

Figure 7.13 The contractile vacuole of *Paramecium*. The vacuole collects fluid from canals in the cytoplasm. When full, the vacuole and canals contract, expelling fluid from the cell (LM).

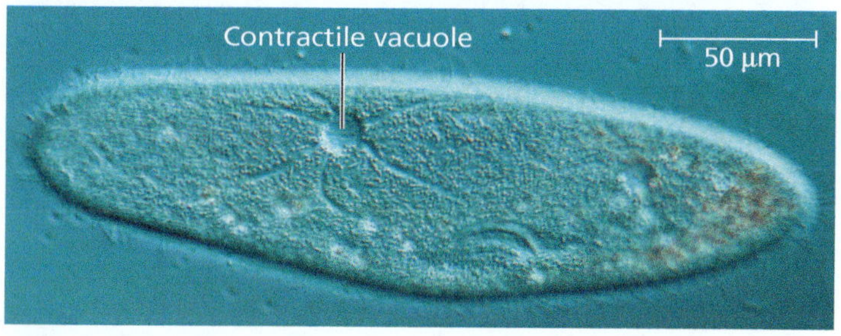

Video: *Paramecium* Vacuole

water balance. Consider a plant cell. Like an animal cell, the plant cell swells as water enters by osmosis **(Figure 7.12b)**. However, the relatively inelastic cell wall will expand only so much before it exerts a back pressure on the cell, called *turgor pressure*, that opposes further water uptake. At this point, the cell is **turgid** (very firm), the healthy state for most plant cells. Plants that are not woody, such as most houseplants, depend for mechanical support on cells kept turgid by a surrounding hypotonic solution. If a plant's cells and surroundings are isotonic, there is no net tendency for water to enter and the cells become **flaccid** (limp); the plant wilts.

However, a cell wall is of no advantage if the cell is immersed in a hypertonic environment. In this case, a plant cell, like an animal cell, will lose water to its surroundings and shrink. As the plant cell shrivels, its plasma membrane pulls away from the cell wall at multiple places. This phenomenon, called **plasmolysis**, causes the plant to wilt and can lead to plant death. The walled cells of bacteria and fungi also plasmolyze in hypertonic environments.

Facilitated Diffusion: Passive Transport Aided by Proteins

Let's look more closely at how water and certain hydrophilic solutes cross a membrane. As mentioned earlier, many polar molecules and ions impeded by the lipid bilayer of the membrane diffuse passively with the help of transport proteins that span the membrane. This phenomenon is called **facilitated diffusion**. Cell biologists are still trying to learn exactly how various transport proteins facilitate diffusion. Most transport proteins are very specific: They transport some substances but not others.

As mentioned earlier, the two types of transport proteins are channel proteins and carrier proteins. Channel proteins simply provide corridors that allow specific molecules or ions to cross the membrane **(Figure 7.14a)**. The hydrophilic passageways provided by these proteins can allow water molecules or small ions to diffuse very quickly from one side of the membrane to the other. Aquaporins, the water channel proteins, facilitate the massive levels of diffusion of water (osmosis) that occur in plant cells and in animal cells such as

Figure 7.14 Two types of transport proteins that carry out facilitated diffusion. In both cases, the protein can transport the solute in either direction, but the net movement is down the concentration gradient of the solute.

(a) A channel protein has a channel through which water molecules or a specific solute can pass.

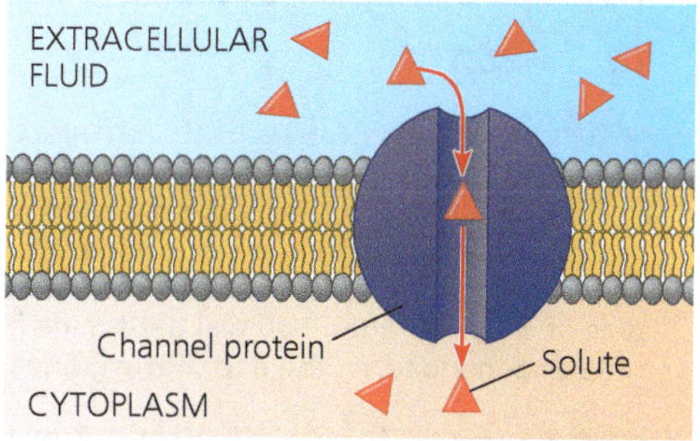

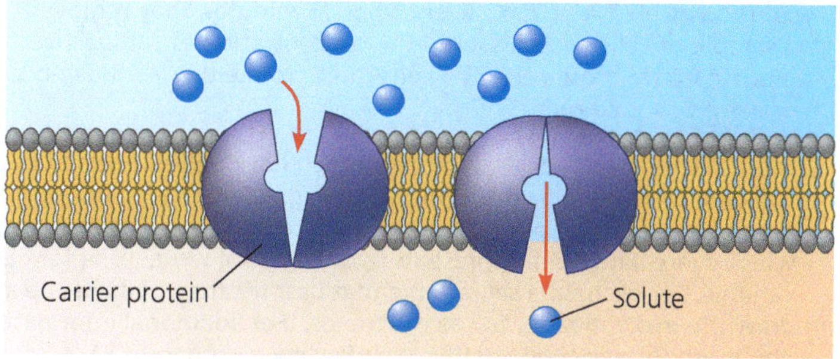

(b) A carrier protein alternates between two shapes, moving a solute across the membrane during the shape change.

Animation: Facilitated Diffusion

red blood cells (see Figure 7.12). Certain kidney cells also have a high number of aquaporins, allowing them to reclaim water from urine before it is excreted. If the kidneys did not perform this function, you would excrete about 180 L of urine per day—and have to drink an equal volume of water!

Channel proteins that transport ions are called **ion channels**. Many ion channels function as **gated channels**, which open or close in response to a stimulus. For some gated channels, the stimulus is electrical. In a nerve cell, for example, an ion channel opens in response to an electrical stimulus, allowing a stream of potassium ions to leave the cell. (See the potassium ion channel at the beginning of this chapter.) This restores the cell's ability to fire again. Other gated channels open or close when a specific substance other than the one to be transported binds to the channel. These gated channels are also important in the functioning of the nervous system, as you'll learn in Concepts 48.2 and 48.3.

Interview with Elba Serrano: Investigating how ion channels enable you to hear (see the interview before Chapter 6)

Carrier proteins, such as the glucose transporter mentioned earlier, seem to undergo a subtle change in shape that somehow translocates the solute-binding site across the membrane **(Figure 7.14b)**. Such a change in shape may be triggered by the binding and release of the transported molecule. Like ion channels, carrier proteins involved in facilitated diffusion result in the net movement of a substance down its concentration gradient.

SCIENTIFIC SKILLS EXERCISE

Interpreting a Scatter Plot with Two Sets of Data

Is Glucose Uptake into Cells Affected by Age? Glucose, an important energy source for animals, is transported into cells by facilitated diffusion using protein carriers. In this exercise, you will interpret a graph with two sets of data from an experiment that examined glucose uptake over time in red blood cells from guinea pigs of different ages. You will determine if the cells' rate of glucose uptake depended on the age of the guinea pigs.

How the Experiment Was Done Researchers incubated guinea pig red blood cells in a 300 m*M* (millimolar) radioactive glucose solution at pH 7.4 at 25°C. Every 10 or 15 minutes, they removed a sample of cells and measured the concentration of radioactive glucose inside those cells. The cells came from either a 15-day-old or a 1-month-old guinea pig.

Data from the Experiment When you have multiple sets of data, it can be useful to plot them on the same graph for comparison. In the graph here, each set of dots (of the same color) forms a *scatter plot*, in which every data point represents two numerical values, one for each variable. For each data set, a curve that best fits the points has been drawn to make it easier to see the trends. (For additional information about graphs, see the Scientific Skills Review in Appendix F.)

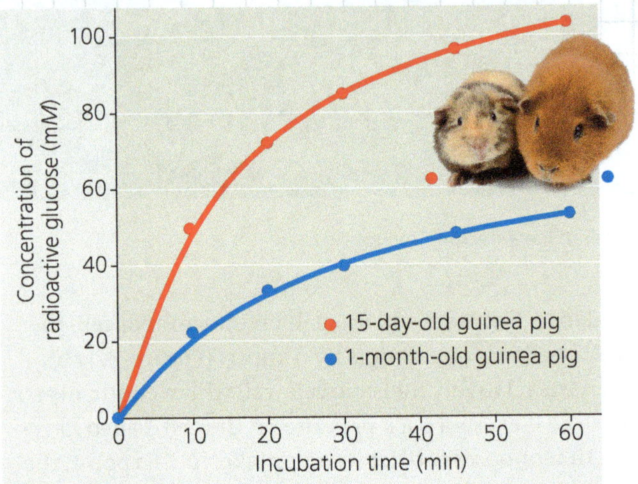

Data from T. Kondo and E. Beutler, Developmental changes in glucose transport of guinea pig erythrocytes, *Journal of Clinical Investigation* 65:1–4 (1980).

INTERPRET THE DATA

1. First make sure you understand the parts of the graph. (a) Which variable is the independent variable—the variable controlled by the researchers? (b) Which variable is the dependent variable—the variable that depended on the treatment and was measured by the researchers? (c) What do the red dots represent? (d) The blue dots?
2. From the data points on the graph, construct a table of the data. Put "Incubation Time (min)" in the left column of the table.
3. What does the graph show? Compare and contrast glucose uptake in red blood cells from 15-day-old and 1-month-old guinea pigs.
4. Develop a hypothesis to explain the difference between glucose uptake in red blood cells from 15-day-old and 1-month-old guinea pigs. (Think about how glucose gets into cells.)
5. Design an experiment to test your hypothesis.

Instructors: A version of this Scientific Skills Exercise can be assigned in MasteringBiology.

No energy input is thus required: This is passive transport. The **Scientific Skills Exercise** gives you an opportunity to work with data from an experiment related to glucose transport.

BioFlix® Animation: Passive Transport

CONCEPT CHECK 7.3

1. How do you think a cell performing cellular respiration rids itself of the resulting CO_2?
2. **WHAT IF?** ➤ If a *Paramecium* swims from a hypotonic to an isotonic environment, will its contractile vacuole become more active or less? Why?

For suggested answers, see Appendix A.

CONCEPT 7.4

Active transport uses energy to move solutes against their gradients

Despite the help of transport proteins, facilitated diffusion is considered passive transport because the solute is moving down its concentration gradient, a process that requires no energy. Facilitated diffusion speeds transport of a solute by providing efficient passage through the membrane, but it does not alter the direction of transport. Some other transport proteins, however, can move solutes against their concentration gradients, across the plasma membrane from the side where they are less concentrated (whether inside or outside) to the side where they are more concentrated.

The Need for Energy in Active Transport

To pump a solute across a membrane against its gradient requires work; the cell must expend energy. Therefore, this type of membrane traffic is called **active transport**. The transport proteins that move solutes against their concentration gradients are all carrier proteins rather than channel proteins. This makes sense because when channel proteins are open, they merely allow solutes to diffuse down their concentration gradients rather than picking them up and transporting them against their gradients.

Active transport enables a cell to maintain internal concentrations of small solutes that differ from concentrations in its environment. For example, compared with its

Figure 7.15 The sodium-potassium pump: a specific case of active transport. This transport system pumps ions against steep concentration gradients: Sodium ion concentration ([Na^+]) is high outside the cell and low inside, while potassium ion concentration ([K^+]) is low outside the cell and high inside. The pump oscillates between two shapes in a cycle that moves three Na^+ out of the cell (steps 1–3) for every two K^+ pumped into the cell (steps 4–6). The two shapes have different binding affinities for Na^+ and K^+. ATP hydrolysis powers the shape change by transferring a phosphate group to the transport protein (phosphorylating the protein).

Animation: Active Transport

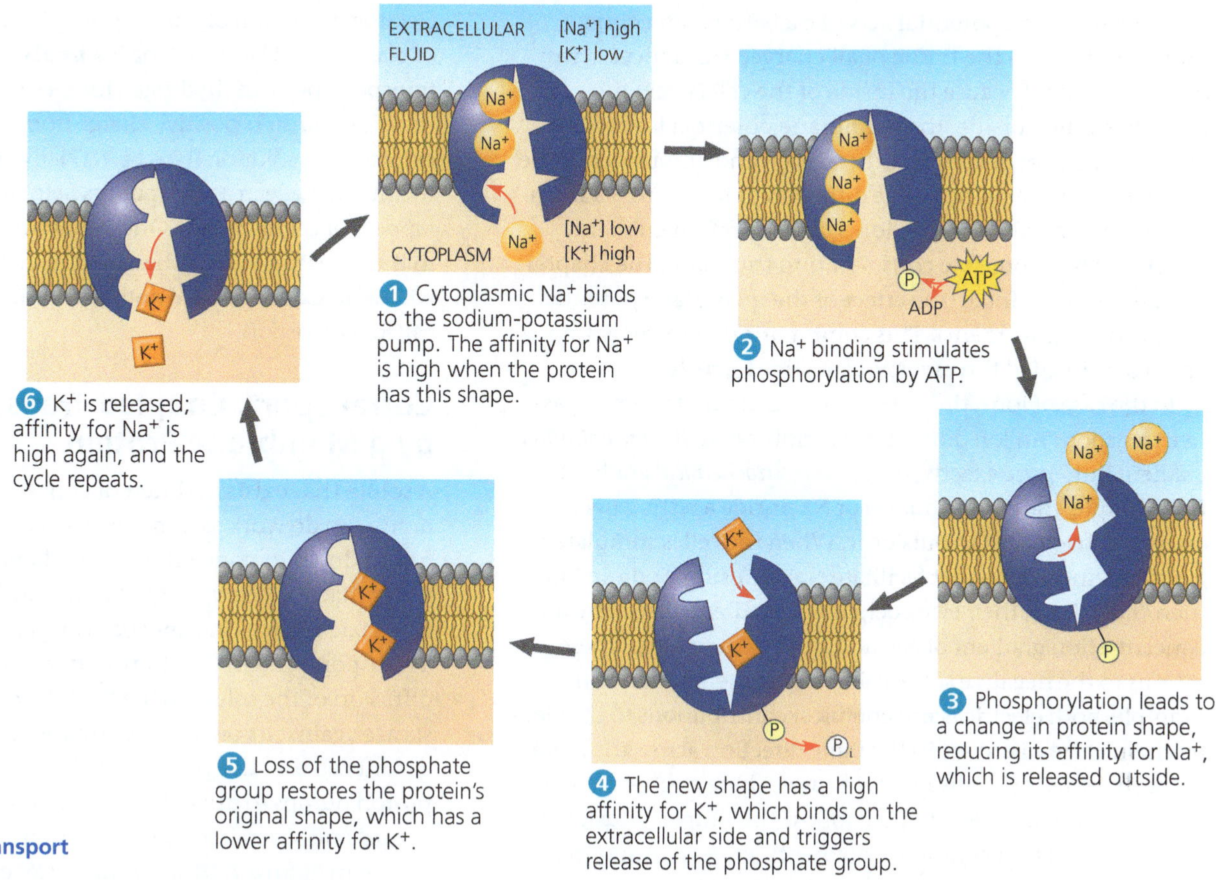

surroundings, an animal cell has a much higher concentration of potassium ions (K^+) and a much lower concentration of sodium ions (Na^+). The plasma membrane helps maintain these steep gradients by pumping Na^+ out of the cell and K^+ into the cell.

As in other types of cellular work, ATP hydrolysis supplies the energy for most active transport. One way ATP can power active transport is when its terminal phosphate group is transferred directly to the transport protein. This can induce the protein to change its shape in a manner that translocates a solute bound to the protein across the membrane. One transport system that works this way is the **sodium-potassium pump**, which exchanges Na^+ for K^+ across the plasma membrane of animal cells **(Figure 7.15)**. The distinction between passive transport and active transport is reviewed in **Figure 7.16**.

How Ion Pumps Maintain Membrane Potential

All cells have voltages across their plasma membranes. Voltage is electrical potential energy (see Concept 2.2)—a separation of opposite charges. The cytoplasmic side of the membrane is negative in charge relative to the extracellular side because of an unequal distribution of anions and cations on the two sides. The voltage across a membrane, called a **membrane potential**, ranges from about −50 to −200 millivolts (mV). (The minus sign indicates that the inside of the cell is negative relative to the outside.)

Figure 7.16 Review: passive and active transport.

Passive transport. Substances diffuse spontaneously down their concentration gradients, crossing a membrane with no expenditure of energy by the cell. The rate of diffusion can be greatly increased by transport proteins in the membrane.

Active transport. Some transport proteins act as pumps, moving substances across a membrane against their concentration (or electrochemical) gradients. Energy is usually supplied by ATP hydrolysis.

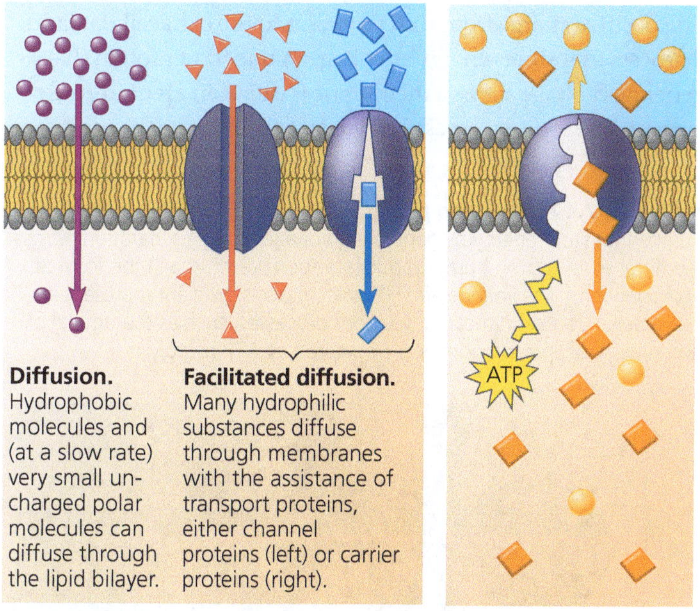

VISUAL SKILLS *For each solute in the right panel, describe its direction of movement, and state whether it is moving with or against its concentration gradient.*

The membrane potential acts like a battery, an energy source that affects the traffic of all charged substances across the membrane. Because the inside of the cell is negative compared with the outside, the membrane potential favors the passive transport of cations into the cell and anions out of the cell. Thus, *two* forces drive the diffusion of ions across a membrane: a chemical force (the ion's concentration gradient, which has been our sole consideration thus far in the chapter) and an electrical force (the effect of the membrane potential on the ion's movement). This combination of forces acting on an ion is called the **electrochemical gradient**.

In the case of ions, then, we must refine our concept of passive transport: An ion diffuses not simply down its *concentration* gradient but, more exactly, down its *electrochemical* gradient. For example, the concentration of Na^+ inside a resting nerve cell is much lower than outside it. When the cell is stimulated, gated channels open that facilitate Na^+ diffusion. Sodium ions then "fall" down their electrochemical gradient, driven by the concentration gradient of Na^+ and by the attraction of these cations to the negative side (inside) of the membrane. In this example, both electrical and chemical contributions to the electrochemical gradient act in the same direction across the membrane, but this is not always so. In cases where electrical forces due to the membrane potential oppose the simple diffusion of an ion down its concentration gradient, active transport may be necessary. In Concepts 48.2 and 48.3, you'll learn about the importance of electrochemical gradients and membrane potentials in the transmission of nerve impulses.

Some membrane proteins that actively transport ions contribute to the membrane potential. An example is the sodium-potassium pump. Notice in Figure 7.15 that the pump does not translocate Na^+ and K^+ one for one, but pumps three sodium ions out of the cell for every two potassium ions it pumps into the cell. With each "crank" of the pump, there is a net transfer of one positive charge from the cytoplasm to the extracellular fluid, a process that stores energy as voltage. A transport protein that generates voltage across a membrane is called an **electrogenic pump**. The sodium-potassium pump appears to be the major electrogenic pump of animal cells. The main electrogenic pump of plants, fungi, and bacteria is a **proton pump**, which actively transports protons (hydrogen ions, H^+) out of the cell. The pumping of H^+ transfers positive charge from the cytoplasm to the extracellular solution **(Figure 7.17)**. By generating voltage across membranes, electrogenic pumps help store energy that can be tapped for cellular work. One important use of proton gradients in the cell is for ATP synthesis during cellular respiration, as you will see in Concept 9.4. Another is a type of membrane traffic called cotransport.

Figure 7.17 A proton pump. Proton pumps are electrogenic pumps that store energy by generating voltage (charge separation) across membranes. A proton pump translocates positive charge in the form of hydrogen ions. The voltage and H^+ concentration gradient represent a dual energy source that can drive other processes, such as the uptake of nutrients. Most proton pumps are powered by ATP hydrolysis.

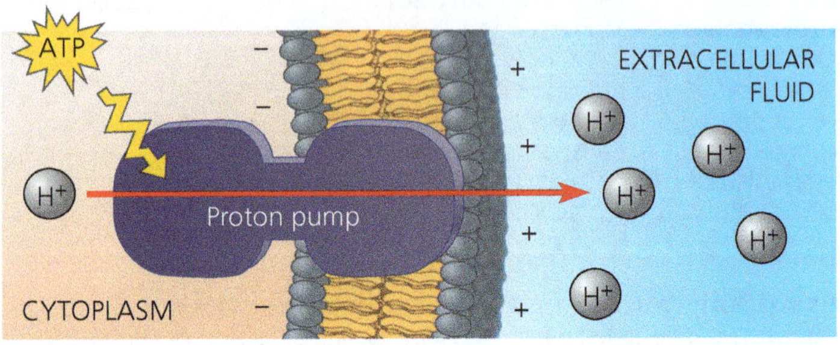

Cotransport: Coupled Transport by a Membrane Protein

A solute that exists in different concentrations across a membrane can do work as it moves across that membrane by diffusion down its concentration gradient. This is analogous to water that has been pumped uphill and performs work as it flows back down. In a mechanism called **cotransport**, a transport protein (a cotransporter) can couple the "downhill" diffusion of the solute to the "uphill" transport of a second substance against its own concentration gradient. For instance, a plant cell uses the gradient of H^+ generated by its ATP-powered proton pumps to drive the active transport of amino acids, sugars, and several other nutrients into the cell. In the example shown in **Figure 7.18**, a cotransporter couples the return of H^+ to the transport of sucrose into the cell. This protein can translocate sucrose into the cell against its concentration gradient, but only if the sucrose molecule travels in the company of an H^+.

Figure 7.18 Cotransport: active transport driven by a concentration gradient. A carrier protein, such as this H^+/sucrose cotransporter in a plant cell (top), is able to use the diffusion of H^+ down its electrochemical gradient into the cell to drive the uptake of sucrose. (The cell wall is not shown.) Although not technically part of the cotransport process, an ATP-driven proton pump is shown here (bottom), which concentrates H^+ outside the cell. The resulting H^+ gradient represents potential energy that can be used for active transport—of sucrose, in this case. Thus, ATP hydrolysis indirectly provides the energy necessary for cotransport.

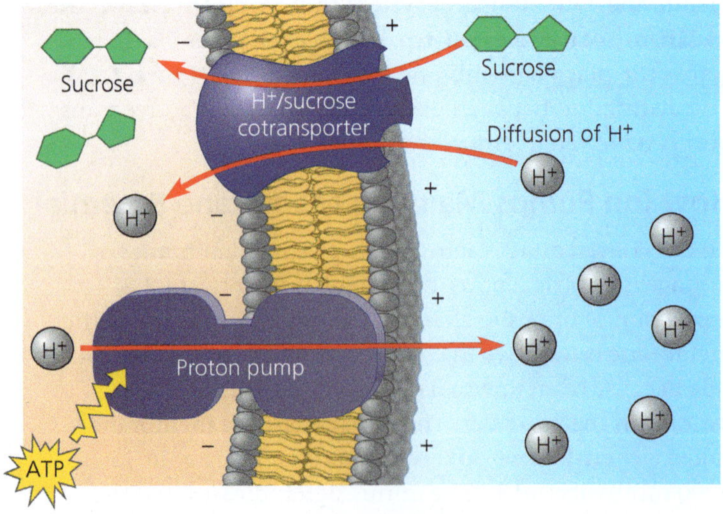

The H^+ uses the transport protein as an avenue to diffuse down its own electrochemical gradient, which is maintained by the proton pump. Plants use H^+/sucrose cotransport to load sucrose produced by photosynthesis into cells in the veins of leaves. The vascular tissue of the plant can then distribute the sugar to roots and other nonphotosynthetic organs that do not make their own food.

What we know about cotransport proteins in animal cells has helped us find more effective treatments for diarrhea, a serious problem in developing countries. Normally, sodium in waste is reabsorbed in the colon, maintaining constant levels in the body, but diarrhea expels waste so rapidly that reabsorption is not possible, and sodium levels fall precipitously. To treat this life-threatening condition, patients are given a solution to drink containing high concentrations of salt (NaCl) and glucose. The solutes are taken up by sodium-glucose cotransporters on the surface of intestinal cells and passed through the cells into the blood. This simple treatment has lowered infant mortality worldwide.

MB BioFlix® Animation: Active Transport

CONCEPT CHECK 7.4

1. Sodium-potassium pumps help nerve cells establish a voltage across their plasma membranes. Do these pumps use ATP or produce ATP? Explain.
2. **VISUAL SKILLS ➤** Compare the sodium-potassium pump in Figure 7.15 with the cotransporter in Figure 7.18. Explain why the sodium-potassium pump would not be considered a cotransporter.
3. **MAKE CONNECTIONS ➤** Review the characteristics of the lysosome in Concept 6.4. Given the internal environment of a lysosome, what transport protein might you expect to see in its membrane?

For suggested answers, see Appendix A.

CONCEPT 7.5

Bulk transport across the plasma membrane occurs by exocytosis and endocytosis

Water and small solutes enter and leave the cell by diffusing through the lipid bilayer of the plasma membrane or by being pumped or moved across the membrane by transport proteins. However, large molecules—such as proteins and polysaccharides, as well as larger particles—generally cross the membrane in bulk, packaged in vesicles. Like active transport, these processes require energy.

MB BioFlix® Animation: Exocytosis and Endocytosis

Exocytosis

As you saw in Figure 6.15, the cell secretes certain molecules by the fusion of vesicles with the plasma membrane; this process is called **exocytosis**. A transport vesicle that has budded from the Golgi apparatus moves along microtubules of the cytoskeleton to the plasma membrane. When the vesicle membrane and plasma membrane come into contact, specific proteins rearrange the lipid molecules of the two bilayers so that the two membranes fuse. The contents of the vesicle spill out of the cell, and the vesicle membrane becomes part of the plasma membrane (see Figure 7.9, step 4).

Many secretory cells use exocytosis to export products. For example, cells in the pancreas that make insulin secrete it into the extracellular fluid by exocytosis. In another example, nerve cells use exocytosis to release neurotransmitters that signal other neurons or muscle cells. When plant cells are making cell walls, exocytosis delivers some of the necessary proteins and carbohydrates from Golgi vesicles to the outside of the cell.

Endocytosis

In **endocytosis**, the cell takes in molecules and particulate matter by forming new vesicles from the plasma membrane. Although the proteins involved in the processes are different, the events of endocytosis look like the reverse of exocytosis. First, a small area of the plasma membrane sinks inward to form a pocket. Then, as the pocket deepens, it pinches in, forming a vesicle containing material that had been outside the cell. Study **Figure 7.19** carefully to understand the three types of endocytosis: phagocytosis ("cellular eating"), pinocytosis ("cellular drinking"), and receptor-mediated endocytosis.

Human cells use receptor-mediated endocytosis to take in cholesterol for membrane synthesis and the synthesis of other steroids. Cholesterol travels in the blood in particles called low-density lipoproteins (LDLs), each a complex of lipids and a protein. LDLs bind to LDL receptors on plasma membranes and then enter the cells by endocytosis. In the inherited disease familial hypercholesterolemia, characterized by a very high level of cholesterol in the blood, LDLs cannot enter cells because the LDL receptor proteins are defective or missing:

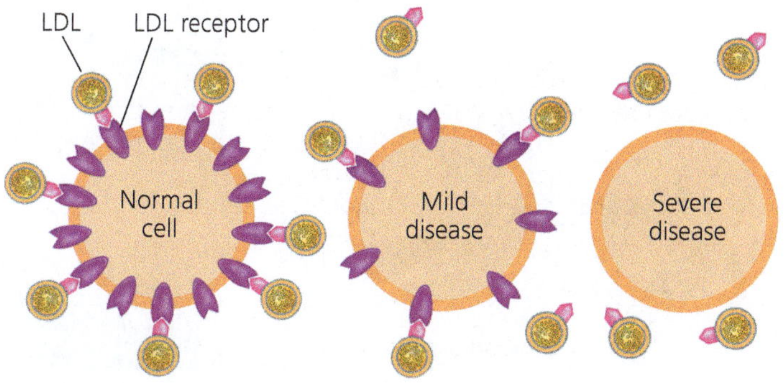

Consequently, cholesterol accumulates in the blood, where it contributes to early atherosclerosis, the buildup of lipid deposits within the walls of blood vessels. This buildup narrows the space in the vessels and impedes blood flow, potentially resulting in heart damage and stroke.

▼ Figure 7.19 Exploring Endocytosis in Animal Cells

Phagocytosis

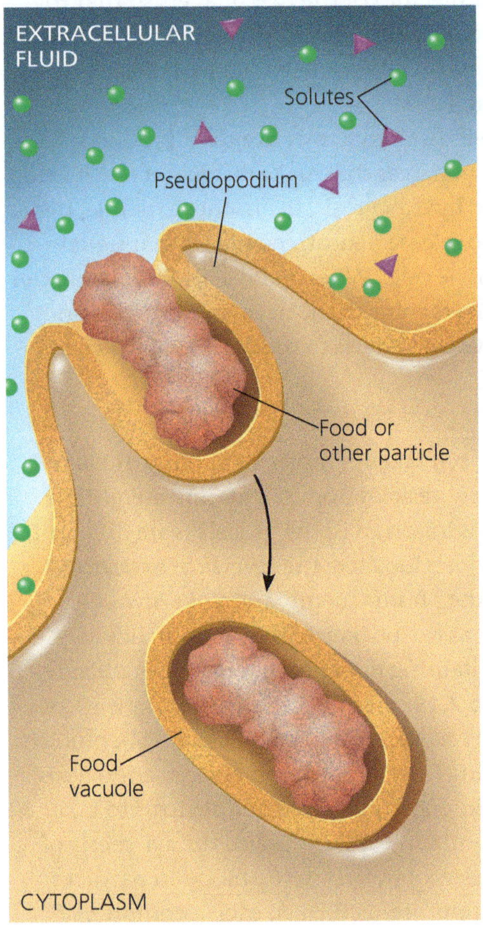

In **phagocytosis**, a cell engulfs a particle by extending pseudopodia (singular, *pseudopodium*) around it and packaging it within a membranous sac called a food vacuole. The particle will be digested after the food vacuole fuses with a lysosome containing hydrolytic enzymes (see Figure 6.13).

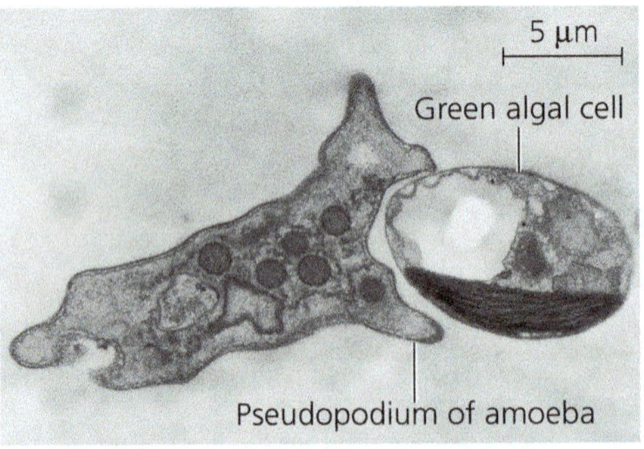

An amoeba engulfing a green algal cell via phagocytosis (TEM).

Pinocytosis

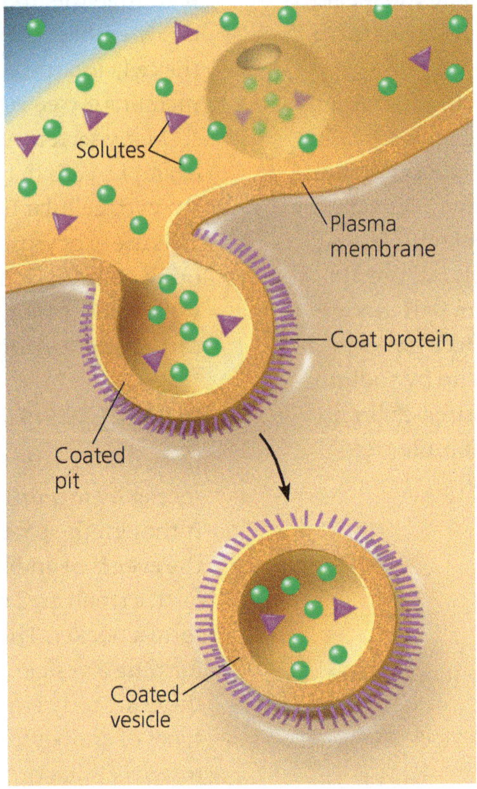

In **pinocytosis**, a cell continually "gulps" droplets of extracellular fluid into tiny vesicles, formed by infoldings of the plasma membrane. In this way, the cell obtains molecules dissolved in the droplets. Because any and all solutes are taken into the cell, pinocytosis as shown here is nonspecific for the substances it transports. In many cases, as above, the parts of the plasma membrane that form vesicles are lined on their cytoplasmic side by a fuzzy layer of coat protein; the "pits" and resulting vesicles are said to be "coated."

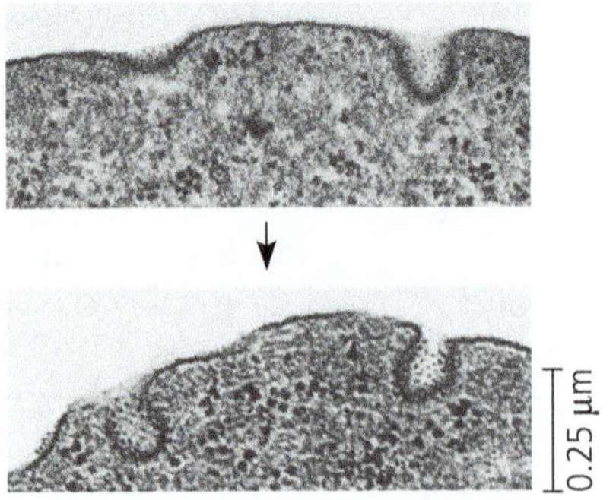

Pinocytotic vesicles forming (TEMs).

Receptor-Mediated Endocytosis

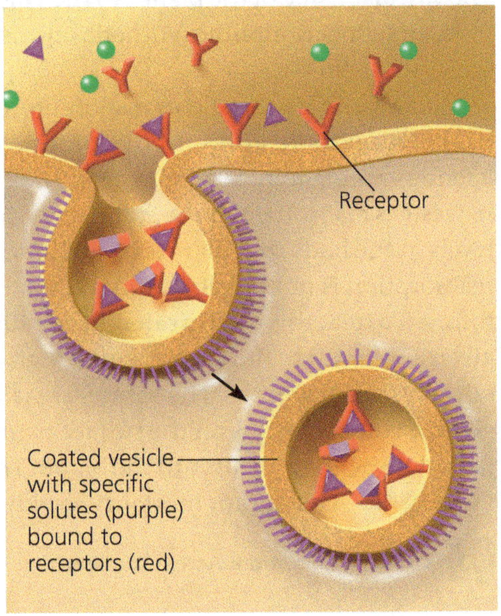

Receptor-mediated endocytosis is a specialized type of pinocytosis that enables the cell to acquire bulk quantities of specific substances, even though those substances may not be very concentrated in the extracellular fluid. Embedded in the plasma membrane are proteins with receptor sites exposed to the extracellular fluid. Specific solutes bind to the receptors. The receptor proteins then cluster in coated pits, and each coated pit forms a vesicle containing the bound molecules. The diagram shows only bound molecules (purple triangles) inside the vesicle, but other molecules from the extracellular fluid are also present.
After the ingested material is liberated from the vesicle, the emptied receptors are recycled to the plasma membrane by the same vesicle (not shown).

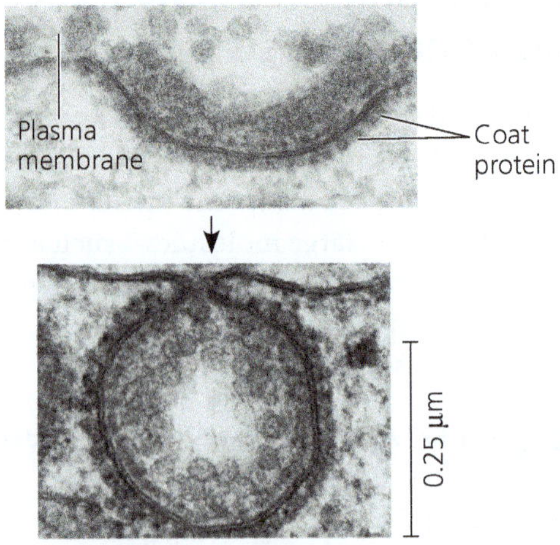

Top: A coated pit. *Bottom*: A coated vesicle forming during receptor-mediated endocytosis (TEMs).

VISUAL SKILLS ➤ *Use the scale bars to estimate the diameters of (a) the food vacuole that will form around the algal cell (left micrograph) and (b) the coated vesicle (lower right micrograph). (c) Which is larger, and by what factor?*

MB **Animation: Exocytosis and Endocytosis**

Endocytosis and exocytosis also provide mechanisms for rejuvenating or remodeling the plasma membrane. These processes occur continually in most eukaryotic cells, yet the amount of plasma membrane in a nongrowing cell remains fairly constant. The addition of membrane by one process appears to offset the loss of membrane by the other.

Energy and cellular work have figured prominently in our study of membranes. In the next three chapters, you will learn more about how cells acquire chemical energy to do the work of life.

MB BioFlix® Animation: Membrane Transport

CONCEPT CHECK 7.5

1. As a cell grows, its plasma membrane expands. Does this involve endocytosis or exocytosis? Explain.
2. **DRAW IT >** Return to Figure 7.9, and circle a patch of plasma membrane that is coming from a vesicle involved in exocytosis.
3. **MAKE CONNECTIONS >** In Concept 6.7, you learned that animal cells make an extracellular matrix (ECM). Describe the cellular pathway of synthesis and deposition of an ECM glycoprotein.

For suggested answers, see Appendix A.

7 Chapter Review

MB Go to **MasteringBiology**™ for Videos, Animations, Vocab Self-Quiz, Practice Tests, and more in the Study Area.

SUMMARY OF KEY CONCEPTS

CONCEPT 7.1

Cellular membranes are fluid mosaics of lipids and proteins *(pp. 127–131)*

VOCAB SELF-QUIZ goo.gl/6u55ks

- In the **fluid mosaic model, amphipathic** proteins are embedded in the phospholipid bilayer.
- Phospholipids and some proteins move sideways within the membrane. The unsaturated hydrocarbon tails of some phospholipids keep membranes fluid at lower temperatures, while cholesterol helps membranes resist changes in fluidity caused by temperature changes.
- Membrane proteins function in transport, enzymatic activity, signal transduction, cell-cell recognition, intercellular joining, and attachment to the cytoskeleton and extracellular matrix. Short chains of sugars linked to proteins (in **glycoproteins**) and lipids (in **glycolipids**) on the exterior side of the plasma membrane interact with surface molecules of other cells.
- Membrane proteins and lipids are synthesized in the ER and modified in the ER and Golgi apparatus. The inside and outside faces of membranes differ in molecular composition.

? *In what ways are membranes crucial to life?*

CONCEPT 7.2

Membrane structure results in selective permeability *(pp. 131–132)*

- A cell must exchange molecules and ions with its surroundings, a process controlled by the **selective permeability** of the plasma membrane. Hydrophobic substances are soluble in lipids and pass through membranes rapidly, whereas polar molecules and ions generally require specific **transport proteins**.

? *How do aquaporins affect the permeability of a membrane?*

CONCEPT 7.3

Passive transport is diffusion of a substance across a membrane with no energy investment *(pp. 132–136)*

- **Diffusion** is the spontaneous movement of a substance down its **concentration gradient**. Water diffuses out through the permeable membrane of a cell (**osmosis**) if the solution outside has a higher solute concentration (**hypertonic**) than the cytosol; water enters the cell if the solution has a lower solute concentration (**hypotonic**). If the concentrations are equal (**isotonic**), no net osmosis occurs. Cell survival depends on balancing water uptake and loss.
- In **facilitated diffusion**, a transport protein speeds the movement of water or a solute across a membrane down its concentration gradient. **Ion channels** facilitate the diffusion of ions across a membrane. Carrier proteins can undergo changes in shape that translocate bound solutes across the membrane.

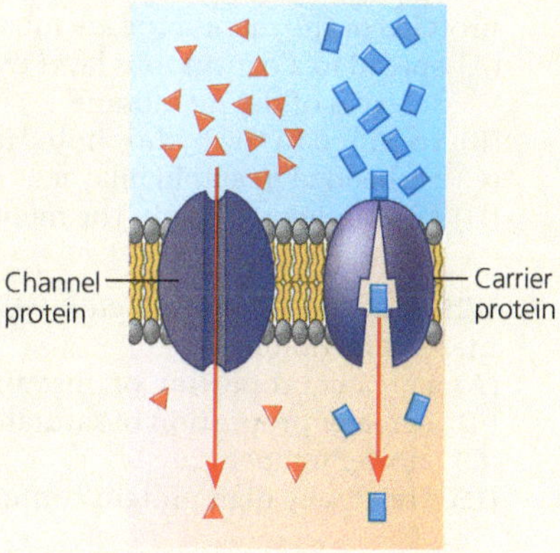

? *What happens to a cell placed in a hypertonic solution? Describe the free water concentration inside and out.*

CONCEPT 7.4

Active transport uses energy to move solutes against their gradients *(pp. 136–139)*

- Specific membrane proteins use energy, usually in the form of ATP, to do the work of **active transport**.
- Ions can have both a concentration (chemical) gradient and an electrical gradient (voltage). These gradients combine in the **electrochemical gradient**, which determines the net direction of ionic diffusion.
- **Cotransport** of two solutes occurs when a membrane protein enables the "downhill" diffusion of one solute to drive the "uphill" transport of the other.

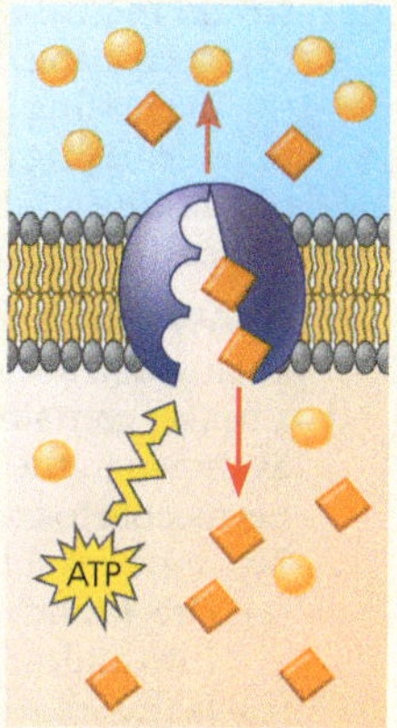

? *ATP is not directly involved in the functioning of a cotransporter. Why, then, is cotransport considered active transport?*

CONCEPT 7.5

Bulk transport across the plasma membrane occurs by exocytosis and endocytosis *(pp. 139–141)*

- In **exocytosis**, transport vesicles migrate to the plasma membrane, fuse with it, and release their contents. In **endocytosis**, molecules enter cells within vesicles that pinch inward from the plasma membrane. The three types of endocytosis are **phagocytosis, pinocytosis**, and **receptor-mediated endocytosis**.

? *Which type of endocytosis involves the binding of specific substances in the extracellular fluid to membrane proteins? What does this type of transport enable a cell to do?*

TEST YOUR UNDERSTANDING

Level 1: Knowledge/Comprehension

PRACTICE TEST
goo.gl/CUYGKD

1. In what way do the membranes of a eukaryotic cell vary?
   (A) Phospholipids are found only in certain membranes.
   (B) Certain proteins are unique to each membrane.
   (C) Only certain membranes of the cell are selectively permeable.
   (D) Only certain membranes are constructed from amphipathic molecules.
2. According to the fluid mosaic model of membrane structure, proteins of the membrane are mostly
   (A) spread in a continuous layer over the inner and outer surfaces of the membrane.
   (B) confined to the hydrophobic interior of the membrane.
   (C) embedded in a lipid bilayer.
   (D) randomly oriented in the membrane, with no fixed inside-outside polarity.
3. Which of the following factors would tend to increase membrane fluidity?
   (A) a greater proportion of unsaturated phospholipids
   (B) a greater proportion of saturated phospholipids
   (C) a lower temperature
   (D) a relatively high protein content in the membrane

Level 2: Application/Analysis

4. Which of the following processes includes all the others?
   (A) osmosis
   (B) diffusion of a solute across a membrane
   (C) passive transport
   (D) transport of an ion down its electrochemical gradient
5. Based on Figure 7.18, which of these experimental treatments would increase the rate of sucrose transport into a plant cell?
   (A) decreasing extracellular sucrose concentration
   (B) decreasing extracellular pH
   (C) decreasing cytoplasmic pH
   (D) adding a substance that makes the membrane more permeable to hydrogen ions
6. **DRAW IT** An artificial "cell" consisting of an aqueous solution enclosed in a selectively permeable membrane is immersed in a beaker containing a different solution, the "environment," as shown in the accompanying diagram. The membrane is permeable to water and to the simple sugars glucose and fructose but impermeable to the disaccharide sucrose.
   (a) Draw solid arrows to indicate the net movement of solutes into and/or out of the cell.
   (b) Is the solution outside the cell isotonic, hypotonic, or hypertonic?
   (c) Draw a dashed arrow to show the net osmosis, if any.
   (d) Will the artificial cell become more flaccid, more turgid, or stay the same?
   (e) Eventually, will the two solutions have the same or different solute concentrations?

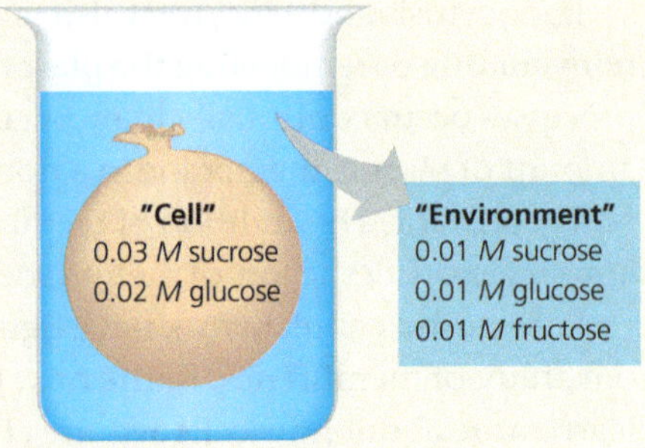

Level 3: Synthesis/Evaluation

7. **EVOLUTION CONNECTION** *Paramecium* and other unicellular eukaryotes that live in hypotonic environments have cell membranes that limit water uptake, while those living in isotonic environments have membranes that are more permeable to water. Describe what water regulation adaptations might have evolved in unicellular eukaryotes in hypertonic habitats such as the Great Salt Lake and in habitats with changing salt concentration.
8. **SCIENTIFIC INQUIRY** An experiment is designed to study the mechanism of sucrose uptake by plant cells. Cells are immersed in a sucrose solution, and the pH of the solution is monitored. Samples of the cells are taken at intervals, and their sucrose concentration is measured. The pH is observed to decrease until it reaches a steady, slightly acidic level, and then sucrose uptake begins. (a) Evaluate these results and propose a hypothesis to explain them. (b) Predict what would happen if an inhibitor of ATP regeneration by the cell were added to the beaker once the pH was at a steady level. Explain.
9. **SCIENCE, TECHNOLOGY, AND SOCIETY** Extensive irrigation in arid regions causes salts to accumulate in the soil. (When water evaporates, salts that were dissolved in the water are left behind in the soil.) Based on what you learned about water balance in plant cells, explain why increased soil salinity (saltiness) might be harmful to crops.
10. **WRITE ABOUT A THEME: INTERACTIONS** A human pancreatic cell obtains O_2—and necessary molecules such as glucose, amino acids, and cholesterol—from its environment, and it releases CO_2 as a waste product. In response to hormonal signals, the cell secretes digestive enzymes. It also regulates its ion concentrations by exchange with its environment. Based on what you have just learned about the structure and function of cellular membranes, write a short essay (100–150 words) to describe how such a cell accomplishes these interactions with its environment.
11. **SYNTHESIZE YOUR KNOWLEDGE**

In the supermarket, lettuce and other produce are often sprayed with water. Explain why this makes vegetables crisp.

For selected answers, see Appendix A.

For additional practice questions, check out the **Dynamic Study Modules** in MasteringBiology. You can use them to study on your smartphone, tablet, or computer anytime, anywhere!

An Introduction to Metabolism

8

▲ **Figure 8.1 What causes these breaking waves to glow?**

KEY CONCEPTS

8.1 An organism's metabolism transforms matter and energy, subject to the laws of thermodynamics

8.2 The free-energy change of a reaction tells us whether or not the reaction occurs spontaneously

8.3 ATP powers cellular work by coupling exergonic reactions to endergonic reactions

8.4 Enzymes speed up metabolic reactions by lowering energy barriers

8.5 Regulation of enzyme activity helps control metabolism

▼ **Firefly**

The Energy of Life

The living cell is a chemical factory in miniature, where thousands of reactions occur within a microscopic space. Sugars can be converted to amino acids that are linked together into proteins when needed. Conversely, when food is digested, proteins are dismantled into amino acids that can be converted to sugars. In multicellular organisms, many cells export chemical products that are used in other parts of the organism. The process called cellular respiration drives this cellular economy by extracting the energy stored in sugars and other fuels. Cells apply this energy to perform various types of work, such as the transport of solutes across the plasma membrane, which we discussed in Concept 7.4.

In a more exotic example, the ocean waves shown in **Figure 8.1** are brightly illuminated from within by free-floating, single-celled marine organisms called dinoflagellates. These dinoflagellates convert the energy stored in certain organic molecules to light, a process called bioluminescence. Most bioluminescent organisms are found in the oceans, but some exist on land, such as the bioluminescent firefly shown in the small photo. Bioluminescence and other metabolic activities carried out by a cell are precisely coordinated and controlled. In its complexity, its efficiency, and its responsiveness to subtle changes, the cell is peerless as a chemical factory. The concepts of metabolism that you learn in this chapter will help you understand how matter and energy flow during life's processes and how that flow is regulated.

When you see this blue icon, log in to **MasteringBiology** and go to the Study Area for digital resources.

MB **Get Ready for This Chapter**

CONCEPT 8.1

An organism's metabolism transforms matter and energy, subject to the laws of thermodynamics

The totality of an organism's chemical reactions is called **metabolism** (from the Greek *metabole*, change). Metabolism is an emergent property of life that arises from orderly interactions between molecules.

Organization of the Chemistry of Life into Metabolic Pathways

We can picture a cell's metabolism as an elaborate road map of the thousands of chemical reactions that occur in a cell, arranged as intersecting metabolic pathways. A **metabolic pathway** begins with a specific molecule, which is then altered in a series of defined steps, resulting in a certain product. Each step of the pathway is catalyzed by a specific enzyme:

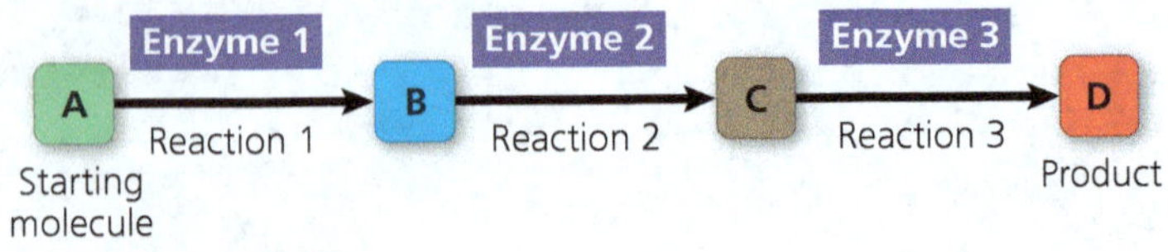

There are mechanisms that regulate these enzymes, thus balancing metabolic supply and demand, analogous to the red, yellow, and green stoplights that control the flow of automobile traffic.

Metabolism as a whole manages the material and energy resources of the cell. Some metabolic pathways release energy by breaking down complex molecules to simpler compounds. These degradative processes are called **catabolic pathways**, or breakdown pathways. A major pathway of catabolism is cellular respiration, in which the sugar glucose and other organic fuels are broken down in the presence of oxygen to carbon dioxide and water. (Pathways can have more than one starting molecule and/or product.) Energy that was stored in the organic molecules becomes available to do the work of the cell, such as ciliary beating or membrane transport. **Anabolic pathways**, in contrast, consume energy to build complicated molecules from simpler ones; they are sometimes called biosynthetic pathways. Examples of anabolism are the synthesis of an amino acid from simpler molecules and the synthesis of a protein from amino acids. Catabolic and anabolic pathways are the "downhill" and "uphill" avenues of the metabolic landscape. Energy released from the downhill reactions of catabolic pathways can be stored and then used to drive the uphill reactions of anabolic pathways.

In this chapter, we will focus on mechanisms common to metabolic pathways. Because energy is fundamental to all metabolic processes, a basic knowledge of energy is necessary to understand how the living cell works. Although we will use some nonliving examples to study energy, the concepts demonstrated by these examples also apply to **bioenergetics**, the study of how energy flows through living organisms.

Forms of Energy

Energy is the capacity to cause change. In everyday life, energy is important because some forms of energy can be used to do work—that is, to move matter against opposing forces, such as gravity and friction. Put another way, energy is the ability to rearrange a collection of matter. For example, you expend energy to turn the pages of this book, and your cells expend energy in transporting certain substances across membranes. Energy exists in various forms, and the work of life depends on the ability of cells to transform energy from one form to another.

Energy can be associated with the relative motion of objects; this energy is called **kinetic energy**. Moving objects can perform work by imparting motion to other matter: A pool player uses the motion of the cue stick to push the cue ball, which in turn moves the other balls; water gushing through a dam turns turbines; and the contraction of leg muscles pushes bicycle pedals. **Thermal energy** is kinetic energy associated with the random movement of atoms or molecules; thermal energy in transfer from one object to another is called **heat**. Light is also a type of energy that can be harnessed to perform work, such as powering photosynthesis in green plants.

An object not presently moving may still possess energy. Energy that is not kinetic is called **potential energy**; it is energy that matter possesses because of its location or structure. Water behind a dam, for instance, possesses energy because of its altitude above sea level. Molecules possess energy because of the arrangement of electrons in the bonds between their atoms. **Chemical energy** is a term used by biologists to refer to the potential energy available for release in a chemical reaction. Recall that catabolic pathways release energy by breaking down complex molecules. Biologists say that these complex molecules, such as glucose, are high in chemical energy. During a catabolic reaction, some bonds are broken and others are formed, releasing energy and resulting in lower-energy breakdown products. This transformation also occurs in the engine of a car when the hydrocarbons of gasoline react explosively with oxygen, releasing the energy that pushes the pistons and producing exhaust. Although less explosive, a similar reaction of food molecules with oxygen provides chemical energy in biological systems, producing carbon dioxide and water as waste products. Biochemical pathways, carried out in the context of cellular structures, enable cells to release chemical energy from food molecules and use the energy to power life processes.

How is energy converted from one form to another? Consider **Figure 8.2**. The woman climbing the ladder to the diving platform is releasing chemical energy from the food she ate for lunch and using some of that energy to perform

Figure 8.2 Transformations between potential and kinetic energy.

Animation: Energy Transformations

the work of climbing. The kinetic energy of muscle movement is thus being transformed into potential energy due to her increasing height above the water. The man diving is converting his potential energy to kinetic energy, which is then transferred to the water as he enters it, resulting in splashing, noise, and increased movement of water molecules. A small amount of energy is lost as heat due to friction.

Now let's consider the original source of the organic food molecules that provided the necessary chemical energy for these divers to climb the steps. This chemical energy was itself derived from light energy absorbed by plants during photosynthesis. Organisms are energy transformers.

The Laws of Energy Transformation

The study of the energy transformations that occur in a collection of matter is called **thermodynamics**. Scientists use the word *system* to denote the matter under study; they refer to the rest of the universe—everything outside the system—as the *surroundings*. An *isolated system*, such as that approximated by liquid in a thermos bottle, is unable to exchange either energy or matter with its surroundings outside the thermos. In an *open system*, energy and matter can be transferred between the system and its surroundings. Organisms are open systems. They absorb energy—for instance, light energy or chemical energy in the form of organic molecules—and release heat and metabolic waste products, such as carbon dioxide, to the surroundings. Two laws of thermodynamics govern energy transformations in organisms and all other collections of matter.

The First Law of Thermodynamics

According to the **first law of thermodynamics**, the energy of the universe is constant: *Energy can be transferred and transformed, but it cannot be created or destroyed.* The first law is also known as the *principle of conservation of energy*. The electric company does not make energy, but merely converts it to a form that is convenient for us to use. By converting sunlight to chemical energy, a plant acts as an energy transformer, not an energy producer.

The brown bear in Figure 8.3a will convert the chemical energy of the organic molecules in its food to kinetic and other forms of energy as it carries out biological processes. What happens to this energy after it has performed work? The second law of thermodynamics helps to answer this question.

Figure 8.3 The two laws of thermodynamics.

(a) First law of thermodynamics: Energy can be transferred or transformed but neither created nor destroyed. For example, chemical reactions in this brown bear will convert the chemical (potential) energy in the fish into the kinetic energy of running.

(b) Second law of thermodynamics: Every energy transfer or transformation increases the disorder (entropy) of the universe. For example, as the bear runs, disorder is increased around its body by the release of heat and small molecules that are the by-products of metabolism. A brown bear can run at speeds up to 35 miles per hour (56 km/hr)—as fast as a racehorse.

MP3 Tutor: Basic Energy Concepts

The Second Law of Thermodynamics

If energy cannot be destroyed, why can't organisms simply recycle their energy over and over again? It turns out that during every energy transfer or transformation, some energy becomes unavailable to do work. In most energy transformations, the more usable forms of energy are at least partly converted to thermal energy and released as heat. Only a small fraction of the chemical energy from the food in Figure 8.3a is transformed into the motion of the brown bear shown in **Figure 8.3b**; most is lost as heat, which dissipates rapidly through the surroundings.

In the process of carrying out chemical reactions that perform various kinds of work, living cells unavoidably convert other forms of energy to heat. A system can put this energy to work only when there is a temperature difference that results in thermal energy flowing as heat from a warmer location to a cooler one. If temperature is uniform, as it is in a living cell, then the heat generated during a chemical reaction will simply warm a body of matter, such as the organism. (This can make a room crowded with people uncomfortably warm, as each person is carrying out a multitude of chemical reactions!)

A consequence of the loss of usable energy as heat to the surroundings is that each energy transfer or transformation makes the universe more disordered. We are all familiar with the word "disorder" in the sense of a messy room or a rundown building. The word "disorder" as used by scientists, however, has a specific molecular definition related to how dispersed the energy is in a system, and how many different energy levels are present. For simplicity, we use "disorder" in the following discussion because our common understanding (the messy room) is a good analogy for molecular disorder.

Scientists use a quantity called **entropy** as a measure of molecular disorder, or randomness. The more randomly arranged a collection of matter is, the greater its entropy. We can now state the **second law of thermodynamics**: *Every energy transfer or transformation increases the entropy of the universe.* Although order can increase locally, there is an unstoppable trend toward randomization of the universe as a whole.

The physical disintegration of a system's organized structure is a good analogy for an increase in entropy. For example, you can observe the gradual decay of an unmaintained building over time. Much of the increasing entropy of the universe is more abstract, however, because it takes the form of increasing amounts of heat and less ordered forms of matter. As the bear in Figure 8.3b converts chemical energy to kinetic energy, it is also increasing the disorder of its surroundings by producing heat and small molecules, such as the CO_2 it exhales, that are the breakdown products of food.

The concept of entropy helps us understand why certain processes are energetically favorable and occur on their own. It turns out that if a given process, by itself, leads to an increase in entropy, that process can proceed without requiring an input of energy. Such a process is called a **spontaneous process**. Note that as we're using it here, the word *spontaneous* does not imply that the process would occur quickly; rather, the word signifies that it is energetically favorable. (In fact, it may be helpful for you to think of the phrase "energetically favorable" when you read the formal term "spontaneous," the word favored by chemists.) Some spontaneous processes, such as an explosion, may be virtually instantaneous, while others, such as the rusting of an old car over time, are much slower.

A process that, on its own, leads to a decrease in entropy is said to be nonspontaneous: It will happen only if energy is supplied. We know from experience that certain events occur spontaneously and others do not. For instance, we know that water flows downhill spontaneously but moves uphill only with an input of energy, such as when a machine pumps the water against gravity. Some energy is inevitably lost as heat, increasing entropy in the surroundings, so usage of energy means that a nonspontaneous process also leads to an increase in the entropy of the universe as a whole.

Biological Order and Disorder

Living systems increase the entropy of their surroundings, as predicted by thermodynamic law. It is true that cells create ordered structures from less organized starting materials. For example, simpler molecules are ordered into the more complex structure of an amino acid, and amino acids are ordered into polypeptide chains. At the organismal level as well, complex and beautifully ordered structures result from biological processes that use simpler starting materials **(Figure 8.4)**.

▼ **Figure 8.4 Order as a characteristic of life.** Order is evident in the detailed structures of the biscuit star and the agave plant shown here. As open systems, organisms can increase their order as long as the order of their surroundings decreases, with an overall increase in entropy in the universe.

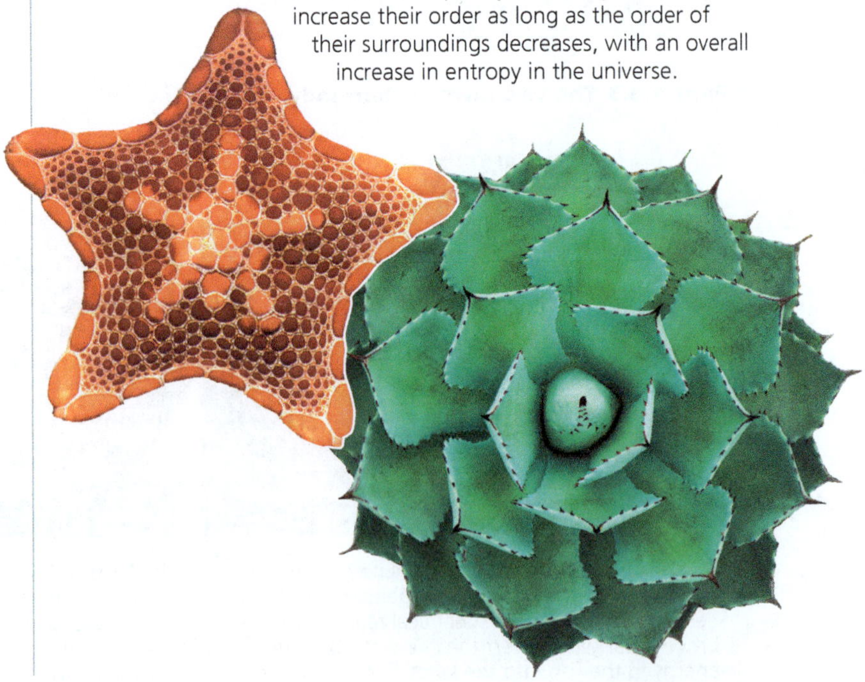

However, an organism also takes in organized forms of matter and energy from the surroundings and replaces them with less ordered forms. For example, an animal obtains starch, proteins, and other complex molecules from the food it eats. As catabolic pathways break these molecules down, the animal releases carbon dioxide and water—small molecules that possess less chemical energy than the food did (see Figure 8.3b). The depletion of chemical energy is accounted for by heat generated during metabolism. On a larger scale, energy flows into most ecosystems in the form of light and exits in the form of heat (see Figure 1.11).

During the early history of life, complex organisms evolved from simpler ancestors. For instance, we can trace the ancestry of the plant kingdom from much simpler organisms called green algae to more complex flowering plants. However, this increase in organization over time in no way violates the second law. The entropy of a particular system, such as an organism, may actually decrease as long as the total entropy of the *universe*—the system plus its surroundings—increases. Thus, organisms are islands of low entropy in an increasingly random universe. The evolution of biological order is perfectly consistent with the laws of thermodynamics.

CONCEPT CHECK 8.1

1. **MAKE CONNECTIONS ➤** How does the second law of thermodynamics help explain the diffusion of a substance across a membrane? (See Figure 7.10.)
2. Describe the forms of energy found in an apple as it grows on a tree, then falls, then is digested by someone who eats it.
3. **WHAT IF? ➤** If you place a teaspoon of sugar in the bottom of a glass of water, it will dissolve completely over time. Left longer, eventually the water will disappear and the sugar crystals will reappear. Explain these observations in terms of entropy.

For suggested answers, see Appendix A.

CONCEPT 8.2

The free-energy change of a reaction tells us whether or not the reaction occurs spontaneously

The laws of thermodynamics that we've just discussed apply to the universe as a whole. As biologists, we want to understand the chemical reactions of life—for example, which reactions occur spontaneously and which ones require some input of energy from outside. But how can we know this without assessing the energy and entropy changes in the entire universe for each separate reaction?

Free-Energy Change, ΔG

Recall that the universe is really equivalent to "the system" plus "the surroundings." In 1878, J. Willard Gibbs, a professor at Yale, defined a very useful function called the Gibbs free energy of a system (without considering its surroundings), symbolized by the letter G. We'll refer to the Gibbs free energy simply as free energy. **Free energy** is the portion of a system's energy that can perform work when temperature and pressure are uniform throughout the system, as in a living cell. Let's consider how we determine the free-energy change that occurs when a system changes—for example, during a chemical reaction.

The change in free energy, ΔG, can be calculated for a chemical reaction by applying the following equation:

$$\Delta G = \Delta H - T\Delta S$$

This equation uses only properties of the system (the reaction) itself: ΔH symbolizes the change in the system's *enthalpy* (in biological systems, equivalent to total energy); ΔS is the change in the system's entropy; and T is the absolute temperature in Kelvin (K) units (K = °C + 273; see Appendix C).

Using chemical methods, we can measure ΔG for any reaction. (The value will depend on conditions such as pH, temperature, and concentrations of reactants and products.) Once we know the value of ΔG for a process, we can use it to predict whether the process will be spontaneous (that is, whether it is energetically favorable and will occur without an input of energy). More than a century of experiments has shown that only processes with a negative ΔG are spontaneous. For ΔG to be negative, ΔH must be negative (the system gives up enthalpy and H decreases) or $T\Delta S$ must be positive (the system gives up order and S increases), or both: When ΔH and $T\Delta S$ are tallied, ΔG has a negative value ($\Delta G < 0$) for all spontaneous processes. In other words, every spontaneous process decreases the system's free energy, and processes that have a positive or zero ΔG are never spontaneous.

This information is immensely interesting to biologists, for it allows us to predict which kinds of change can happen without an input of energy. Such spontaneous changes can be harnessed to perform work. This principle is very important in the study of metabolism, where a major goal is to determine which reactions can supply energy for cellular work.

Free Energy, Stability, and Equilibrium

As we saw in the previous section, when a process occurs spontaneously in a system, we can be sure that ΔG is negative. Another way to think of ΔG is to realize that it represents the difference between the free energy of the final state and the free energy of the initial state:

$$\Delta G = G_{\text{final state}} - G_{\text{initial state}}$$

Thus, ΔG can be negative only when the process involves a loss of free energy during the change from initial state to final state. Because it has less free energy, the system in its final state is less likely to change and is therefore more stable than it was previously.

We can think of free energy as a measure of a system's instability—its tendency to change to a more stable state. Unstable systems (higher G) tend to change in such a way that they become more stable (lower G). For example, a diver on top of a platform is less stable (more likely to fall) than when floating in the water; a drop of concentrated dye is less stable (more likely to disperse) than when the dye is spread randomly through the liquid; and a glucose molecule is less stable (more likely to break down) than the simpler molecules into which it can be split **(Figure 8.5)**. Unless something prevents it, each of these systems will move toward greater stability: The diver falls, the solution becomes uniformly colored, and the glucose molecule is broken down into smaller molecules.

Another term that describes a state of maximum stability is *equilibrium*, which you learned about in Concept 2.4 in connection with chemical reactions. There is an important relationship between free energy and equilibrium, including chemical equilibrium. Recall that most chemical reactions are reversible and proceed to a point at which the forward and backward reactions occur at the same rate. The reaction is then said to be at chemical equilibrium, and there is no further net change in the relative concentration of products and reactants.

As a reaction proceeds toward equilibrium, the free energy of the mixture of reactants and products decreases. Free energy increases when a reaction is somehow pushed away from equilibrium, perhaps by removing some of the products (and thus changing their concentration relative to that of the reactants). For a system at equilibrium, G is at its lowest possible value in that system. We can think of the equilibrium state as a free-energy valley. Any change from the equilibrium position will have a positive ΔG and will not be spontaneous. For this reason, systems never spontaneously move away from equilibrium. Because a system at equilibrium cannot spontaneously change, it can do no work. *A process is spontaneous and can perform work only when it is moving toward equilibrium.*

Free Energy and Metabolism

We can now apply the free-energy concept more specifically to the chemistry of life's processes.

Exergonic and Endergonic Reactions in Metabolism

Based on their free-energy changes, chemical reactions can be classified as either exergonic ("energy outward") or endergonic ("energy inward"). An **exergonic reaction** proceeds with a net release of free energy **(Figure 8.6a)**. Because the chemical mixture loses free energy (G decreases), ΔG is negative for an exergonic reaction. Using ΔG as a standard for spontaneity, exergonic reactions are those that occur spontaneously. (Remember, the word *spontaneous* implies that

▼ Figure 8.5 The relationship of free energy to stability, work capacity, and spontaneous change. Unstable systems (top) are rich in free energy, G. They have a tendency to change spontaneously to a more stable state (bottom), and it is possible to harness this "downhill" change to perform work.

(a) Gravitational motion. Objects move spontaneously from a higher altitude to a lower one.

(b) Diffusion. Molecules in a drop of dye diffuse until they are randomly dispersed.

(c) Chemical reaction. In a cell, a glucose molecule is broken down into simpler molecules.

MAKE CONNECTIONS ▶ *Compare the redistribution of molecules shown in (b) to the transport of hydrogen ions (H^+) across a membrane by a proton pump, creating a concentration gradient, as shown in Figure 7.17. Which process(es) result(s) in higher free energy? Which system(s) can do work?*

Figure 8.6 Free energy changes (ΔG) in exergonic and endergonic reactions.

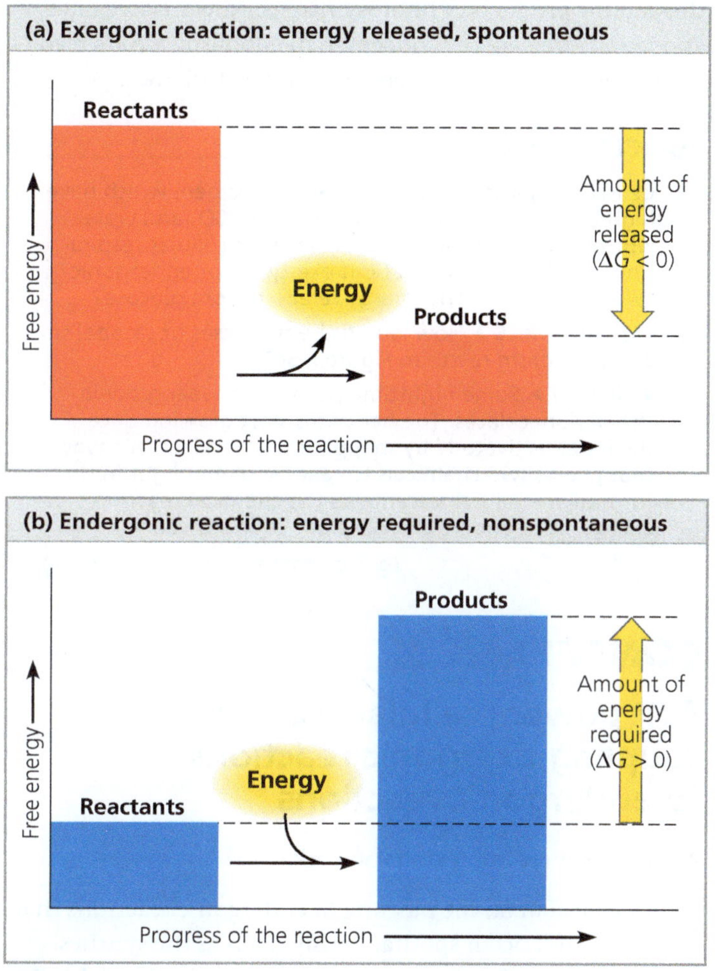

Animation: Exergonic and Endergonic Reactions

it is energetically favorable, not that it will occur rapidly.) The magnitude of ΔG for an exergonic reaction represents the maximum amount of work the reaction can perform.* The greater the decrease in free energy, the greater the amount of work that can be done.

We can use the overall reaction for cellular respiration as an example:

$$C_6H_{12}O_6 + 6\,O_2 \rightarrow 6\,CO_2 + 6\,H_2O$$
$$\Delta G = -686 \text{ kcal/mol } (-2{,}870 \text{ kJ/mol})$$

For each mole (180 g) of glucose broken down by respiration under what are called "standard conditions" (1 *M* of each reactant and product, 25°C, pH 7), 686 kcal (2,870 kJ) of energy is made available for work. Because energy must be conserved, the chemical *products* of respiration store 686 kcal less free energy per mole than the *reactants*. The products are, in a sense, the spent exhaust of a process that tapped the free energy stored in the bonds of the sugar molecules.

It is important to realize that the breaking of bonds does not release energy; on the contrary, as you will soon see, it requires energy. The phrase "energy stored in bonds" is shorthand for the potential energy that can be released when new bonds are formed after the original bonds break, as long as the products are of lower free energy than the reactants.

An **endergonic reaction** is one that absorbs free energy from its surroundings **(Figure 8.6b)**. Because this kind of reaction essentially *stores* free energy in molecules (*G* increases), ΔG is positive. Such reactions are nonspontaneous, and the magnitude of ΔG is the quantity of energy required to drive the reaction. If a chemical process is exergonic (downhill), releasing energy in one direction, then the reverse process must be endergonic (uphill), using energy. A reversible process cannot be downhill in both directions. If $\Delta G = -686$ kcal/mol for respiration, which converts glucose and oxygen to carbon dioxide and water, then the reverse process—the conversion of carbon dioxide and water to glucose and oxygen—must be strongly endergonic, with $\Delta G = +686$ kcal/mol. Such a reaction would never happen by itself.

How, then, do plants make the sugar that organisms use for energy? Plants get the required energy—686 kcal to make a mole of glucose—from the environment by capturing light and converting its energy to chemical energy. Next, in a long series of exergonic steps, they gradually spend that chemical energy to assemble glucose molecules.

Equilibrium and Metabolism

Reactions in an isolated system eventually reach equilibrium and can then do no work, as illustrated by the isolated hydroelectric system in **Figure 8.7**. The chemical reactions of metabolism are reversible, and they, too, would reach equilibrium if they occurred in the isolation of a test tube. Because systems at equilibrium are at a minimum of *G* and can do no work, a cell that has reached metabolic equilibrium is dead! *The fact that metabolism as a whole is never at equilibrium is one of the defining features of life.*

Figure 8.7 Equilibrium and work in an isolated hydroelectric system. Water flowing downhill turns a turbine that drives a generator providing electricity to a lightbulb, but only until the system reaches equilibrium.

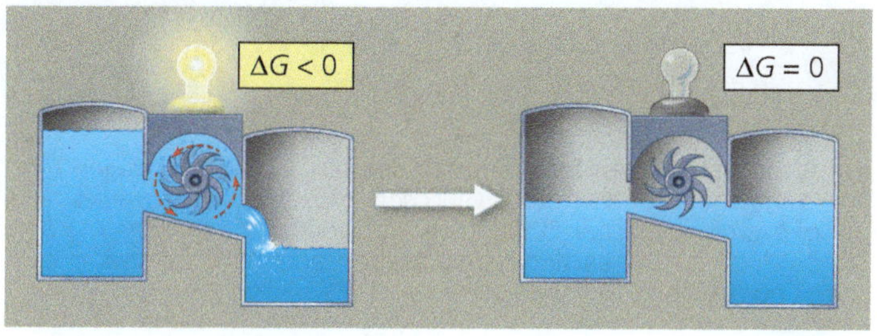

*The word *maximum* qualifies this statement, because some of the free energy is released as heat and cannot do work. Therefore, ΔG represents a theoretical upper limit of available energy.

Like most systems, a living cell is not in equilibrium. The constant flow of materials in and out of the cell keeps the metabolic pathways from ever reaching equilibrium, and the cell continues to do work throughout its life. This principle is illustrated by the open (and more realistic) hydroelectric system in **Figure 8.8a**. However, unlike this simple system in which water flowing downhill turns a single turbine, a catabolic pathway in a cell releases free energy in a series of reactions. An example is cellular respiration, illustrated by analogy in **Figure 8.8b**. Some of the reversible reactions of respiration are constantly "pulled" in one direction—that is, they are kept out of equilibrium. The key to maintaining this lack of equilibrium is that the product of a reaction does not accumulate but instead becomes a reactant in the next step; finally, waste products are expelled from the cell. The overall sequence of reactions is kept going by the huge free-energy difference between glucose and oxygen at the top of the energy "hill" and carbon dioxide and water at the "downhill" end. As long as our cells have a steady supply of glucose or other fuels and oxygen and are able to expel waste products to the surroundings, their metabolic pathways never reach equilibrium and can continue to do the work of life.

Stepping back to look at the big picture, we can see once again how important it is to think of organisms as open systems. Sunlight provides a daily source of free energy for an ecosystem's plants and other photosynthetic organisms.

Figure 8.8 Equilibrium and work in open systems.

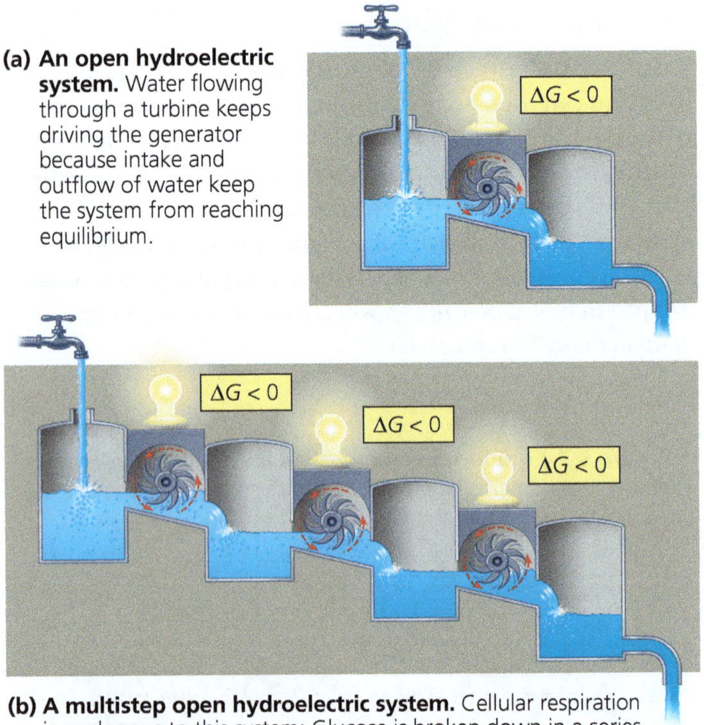

(a) An open hydroelectric system. Water flowing through a turbine keeps driving the generator because intake and outflow of water keep the system from reaching equilibrium.

(b) A multistep open hydroelectric system. Cellular respiration is analogous to this system: Glucose is broken down in a series of exergonic reactions that power the work of the cell. The product of each reaction is used as the reactant for the next, so no reaction reaches equilibrium.

Animals and other nonphotosynthetic organisms in an ecosystem must have a source of free energy in the form of the organic products of photosynthesis. Now that we have applied the free-energy concept to metabolism, we are ready to see how a cell actually performs the work of life.

CONCEPT CHECK 8.2

1. Cellular respiration uses glucose and oxygen, which have high levels of free energy, and releases CO_2 and water, which have low levels of free energy. Is cellular respiration spontaneous or not? Is it exergonic or endergonic? What happens to the energy released from glucose?
2. **VISUAL SKILLS ➤** How would the processes of catabolism and anabolism relate to Figure 8.5c?
3. **WHAT IF? ➤** Some nighttime partygoers wear glow-in-the-dark necklaces. The necklaces start glowing once they are "activated" by snapping the necklace in a way that allows two chemicals to react and emit light in the form of chemiluminescence. Is the chemical reaction exergonic or endergonic? Explain your answer.

For suggested answers, see Appendix A.

CONCEPT 8.3

ATP powers cellular work by coupling exergonic reactions to endergonic reactions

A cell does three main kinds of work:

- *Chemical work*, the pushing of endergonic reactions that would not occur spontaneously, such as the synthesis of polymers from monomers (chemical work will be discussed further here; examples are shown in Chapters 9 and 10)
- *Transport work*, the pumping of substances across membranes against the direction of spontaneous movement (see Concept 7.4)
- *Mechanical work*, such as the beating of cilia (see Concept 6.6), the contraction of muscle cells, and the movement of chromosomes during cellular reproduction

A key feature in the way cells manage their energy resources to do this work is **energy coupling**, the use of an exergonic process to drive an endergonic one. ATP is responsible for mediating most energy coupling in cells, and in most cases it acts as the immediate source of energy that powers cellular work.

MB Animation: Energy Coupling

The Structure and Hydrolysis of ATP

ATP (adenosine triphosphate) was introduced when we discussed the phosphate group as a functional group (see Concept 4.3). ATP contains the sugar ribose, with the nitrogenous base adenine and a chain of three phosphate groups (the triphosphate group) bonded to it **(Figure 8.9a)**. In addition to its role in energy coupling, ATP is also one of the nucleoside triphosphates used to make RNA (see Figure 5.23).

The bonds between the phosphate groups of ATP can be broken by hydrolysis. When the terminal phosphate bond is broken by addition of a water molecule, a molecule of inorganic phosphate ($HOPO_3^{2-}$, abbreviated Ⓟi throughout this book) leaves the ATP, which becomes adenosine diphosphate, or ADP **(Figure 8.9b)**. The reaction is exergonic and releases 7.3 kcal of energy per mole of ATP hydrolyzed:

$$ATP + H_2O \rightarrow ADP + Ⓟ_i$$
$$\Delta G = -7.3 \text{ kcal/mol } (-30.5 \text{ kJ/mol})$$

This is the free-energy change measured under standard conditions. In the cell, conditions do not conform to standard

Figure 8.9 The structure and hydrolysis of adenosine triphosphate (ATP). Throughout this book, the chemical structure of the triphosphate group seen in **(a)** will be represented by the three joined yellow circles shown in **(b)**.

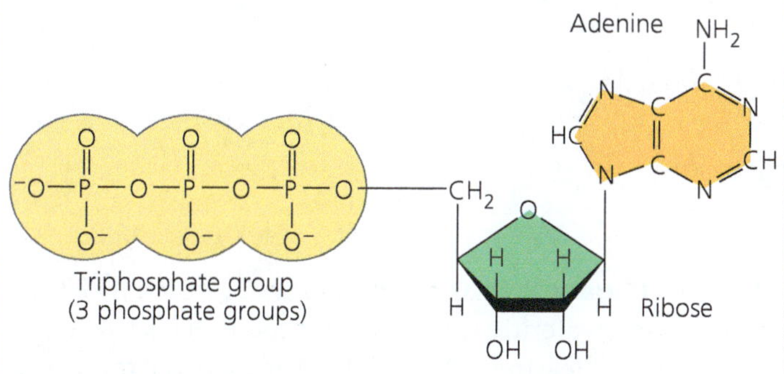

(a) The structure of ATP. In the cell, most hydroxyl groups of phosphates are ionized (—O⁻).

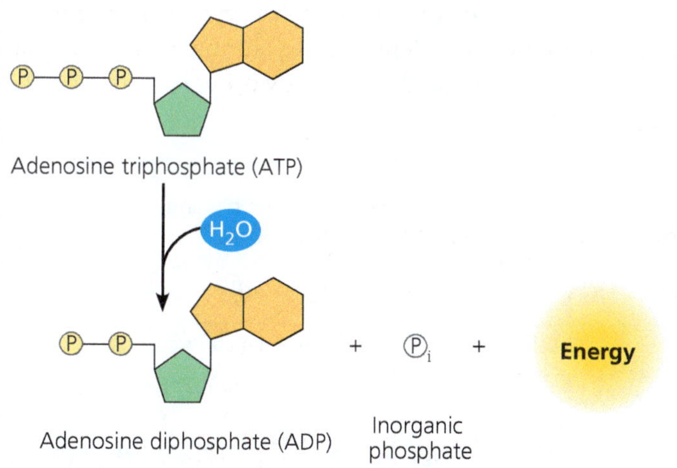

(b) The hydrolysis of ATP. The reaction of ATP and water yields ADP and inorganic phosphate (Ⓟi) and releases energy.

Animation: The Structure of ATP
Animation: Space-Filling Model of ATP
Animation: Stick Model of ATP

conditions, primarily because reactant and product concentrations differ from 1 *M*. For example, when ATP hydrolysis occurs under cellular conditions, the actual ΔG is about −13 kcal/mol, 78% greater than the energy released by ATP hydrolysis under standard conditions.

Because their hydrolysis releases energy, the phosphate bonds of ATP are sometimes referred to as high-energy phosphate bonds, but the term is misleading. The phosphate bonds of ATP are not unusually strong bonds, as "high-energy" may imply; rather, the reactants (ATP and water) themselves have high energy relative to the energy of the products (ADP and Ⓟi). The release of energy during the hydrolysis of ATP comes from the chemical change of the system to a state of lower free energy, not from the phosphate bonds themselves.

ATP is useful to the cell because the energy it releases on losing a phosphate group is somewhat greater than the energy most other molecules could deliver. But why does this hydrolysis release so much energy? If we reexamine the ATP molecule in Figure 8.9a, we can see that all three phosphate groups are negatively charged. These like charges are crowded together, and their mutual repulsion contributes to the instability of this region of the ATP molecule. The triphosphate tail of ATP is the chemical equivalent of a compressed spring.

How the Hydrolysis of ATP Performs Work

When ATP is hydrolyzed in a test tube, the release of free energy merely heats the surrounding water. In an organism, this same generation of heat can sometimes be beneficial. For instance, the process of shivering uses ATP hydrolysis during muscle contraction to warm the body. In most cases in the cell, however, the generation of heat alone would be an inefficient (and potentially dangerous) use of a valuable energy resource. Instead, the cell's proteins harness the energy released during ATP hydrolysis in several ways to perform the three types of cellular work—chemical, transport, and mechanical.

For example, with the help of specific enzymes, the cell is able to use the energy released by ATP hydrolysis directly to drive chemical reactions that, by themselves, are endergonic. If the ΔG of an endergonic reaction is less than the amount of energy released by ATP hydrolysis, then the two reactions can be coupled so that, overall, the coupled reactions are exergonic. This usually involves phosphorylation, the transfer of a phosphate group from ATP to some other molecule, such as the reactant. The recipient molecule with the phosphate group covalently bonded to it is then called a **phosphorylated intermediate**. The key to coupling exergonic and endergonic reactions is the formation of this phosphorylated intermediate, which is more

Figure 8.10 How ATP drives chemical work: Energy coupling using ATP hydrolysis. In this example, the exergonic process of ATP hydrolysis is used to drive an endergonic process—the cellular synthesis of the amino acid glutamine from glutamic acid and ammonia.

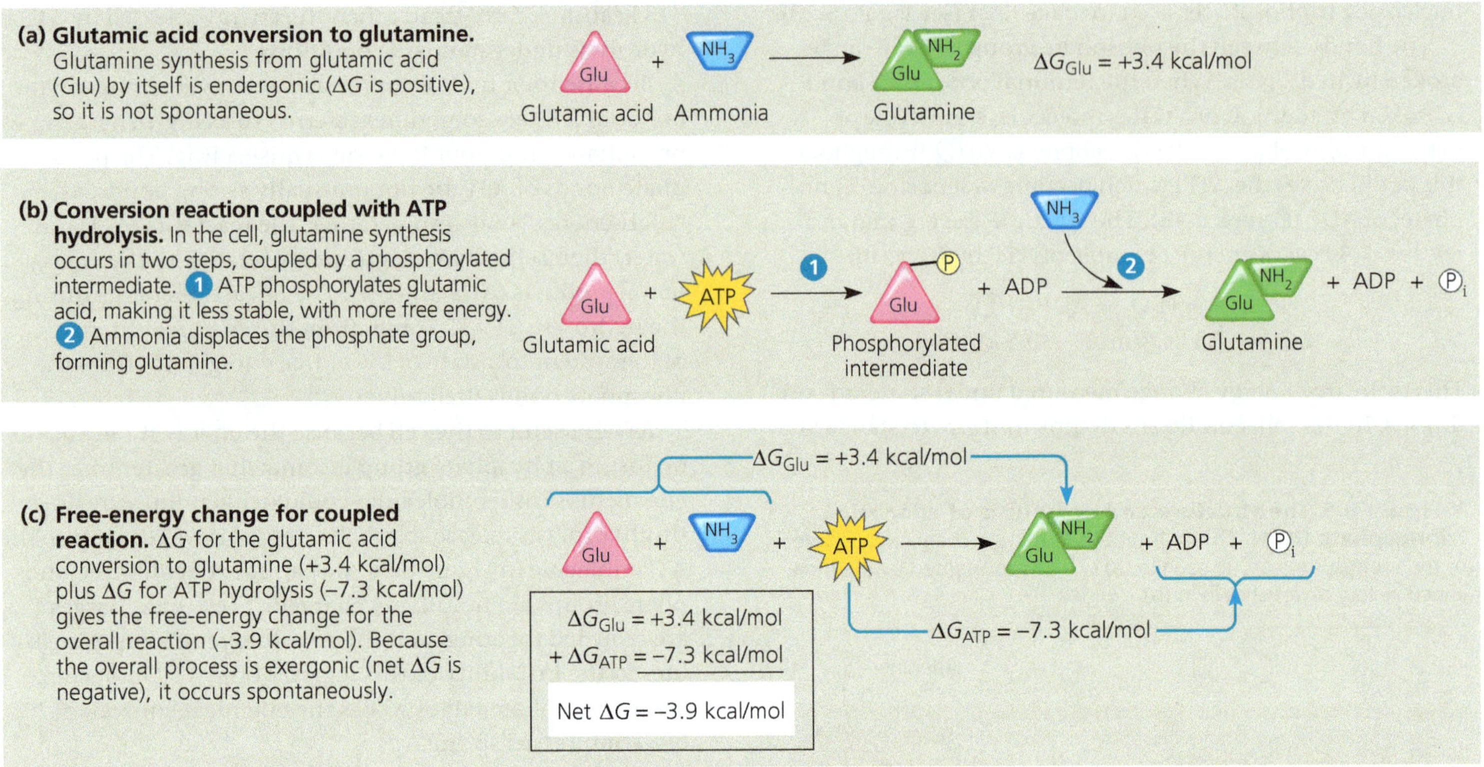

MAKE CONNECTIONS ➤ *Referring to Figure 5.14, explain why glutamine (Gln) is diagrammed as a glutamic acid (Glu) with an amino group attached.*

reactive (less stable, with more free energy) than the original unphosphorylated molecule **(Figure 8.10)**.

Transport and mechanical work in the cell are also nearly always powered by the hydrolysis of ATP. In these cases, ATP hydrolysis leads to a change in a protein's shape and often its ability to bind another molecule. Sometimes this occurs via a phosphorylated intermediate, as seen for the transport protein in **Figure 8.11a**. In most instances of mechanical work involving motor proteins "walking" along cytoskeletal elements **(Figure 8.11b)**, a cycle occurs in which ATP is first bound noncovalently to the motor protein. Next, ATP is hydrolyzed, releasing ADP and Ⓟ$_i$. Another ATP molecule can then bind. At each stage, the motor protein changes its shape and ability to bind the cytoskeleton, resulting in movement of the protein along the cytoskeletal track. Phosphorylation and dephosphorylation promote crucial protein shape changes during many other important cellular processes as well.

Figure 8.11 How ATP drives transport and mechanical work. ATP hydrolysis causes changes in the shapes and binding affinities of proteins. This can occur either **(a)** directly, by phosphorylation, as shown for a membrane protein carrying out active transport of a solute (see also Figure 7.15), or **(b)** indirectly, via noncovalent binding of ATP and its hydrolytic products, as is the case for motor proteins that move vesicles (and other organelles) along cytoskeletal "tracks" in the cell (see also Figure 6.21).

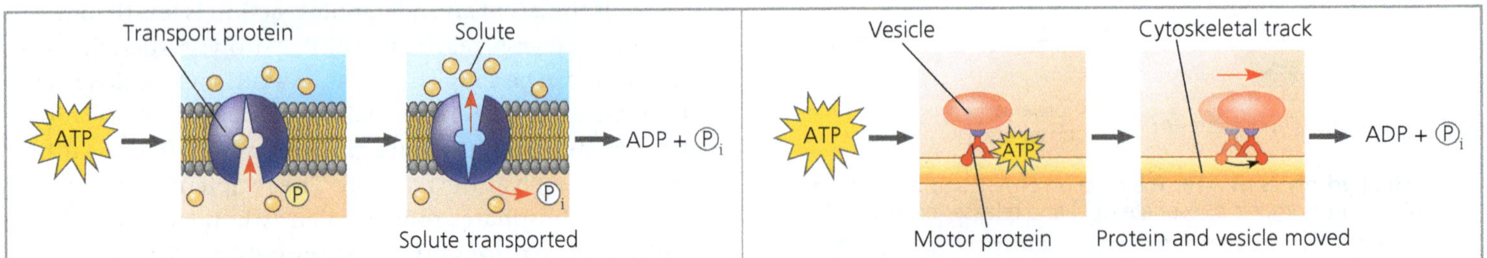

(a) Transport work: ATP phosphorylates transport proteins, causing a shape change that allows transport of solutes.

(b) Mechanical work: ATP binds noncovalently to motor proteins and then is hydrolyzed, causing a shape change that walks the motor protein forward.

Figure 8.12 The ATP cycle. Energy released by breakdown reactions (catabolism) in the cell is used to phosphorylate ADP, regenerating ATP. Chemical potential energy stored in ATP drives most cellular work.

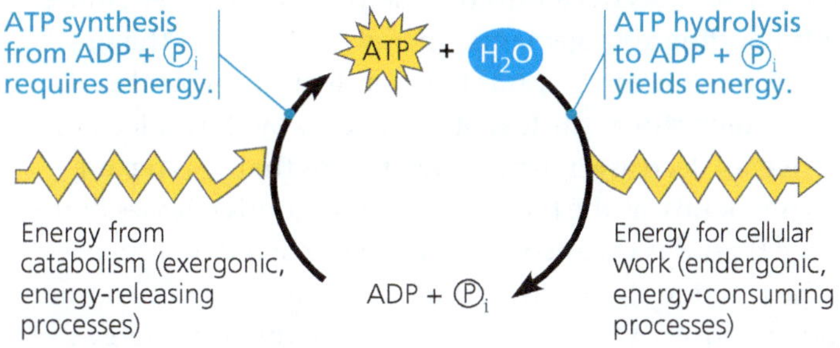

Animation: Metabolism Overview

The Regeneration of ATP

An organism at work uses ATP continuously, but ATP is a renewable resource that can be regenerated by the addition of phosphate to ADP **(Figure 8.12)**. The free energy required to phosphorylate ADP comes from exergonic breakdown reactions (catabolism) in the cell. This shuttling of inorganic phosphate and energy is called the ATP cycle, and it couples the cell's energy-yielding (exergonic) processes to the energy-consuming (endergonic) ones. The ATP cycle proceeds at an astonishing pace. For example, a working muscle cell recycles its entire pool of ATP in less than a minute. That turnover represents 10 million molecules of ATP consumed and regenerated per second per cell. If ATP could not be regenerated by the phosphorylation of ADP, humans would use up nearly their body weight in ATP each day.

Because both directions of a reversible process cannot be downhill, the regeneration of ATP is necessarily endergonic:

$$\text{ADP} + \text{Ⓟ}_i \rightarrow \text{ATP} + H_2O$$
$$\Delta G = +7.3 \text{ kcal/mol } (+30.5 \text{ kJ/mol}) \text{ (standard conditions)}$$

Since ATP formation from ADP and Ⓟ_i is not spontaneous, free energy must be spent to make it occur. Catabolic (exergonic) pathways, especially cellular respiration, provide the energy for the endergonic process of making ATP. Plants also use light energy to produce ATP. Thus, the ATP cycle is a revolving door through which energy passes during its transfer from catabolic to anabolic pathways.

CONCEPT CHECK 8.3

1. How does ATP typically transfer energy from an exergonic to an endergonic reaction in the cell?
2. Which of the following has more free energy: glutamic acid + ammonia + ATP or glutamine + ADP + Ⓟ_i? Explain your answer.
3. **MAKE CONNECTIONS** ➤ Does Figure 8.11a show passive or active transport? Explain. (See Concepts 7.3 and 7.4.)

For suggested answers, see Appendix A.

CONCEPT 8.4

Enzymes speed up metabolic reactions by lowering energy barriers

The laws of thermodynamics tell us what will and will not happen under given conditions but say nothing about the rate of these processes. A spontaneous chemical reaction occurs without any requirement for outside energy, but it may occur so slowly that it is imperceptible. For example, even though the hydrolysis of sucrose (table sugar) to glucose and fructose is exergonic, occurring spontaneously with a release of free energy ($\Delta G = -7$ kcal/mol), a solution of sucrose dissolved in sterile water will sit for years at room temperature with no appreciable hydrolysis. However, if we add a small amount of the enzyme sucrase to the solution, then all the sucrose may be hydrolyzed within seconds, as shown here:

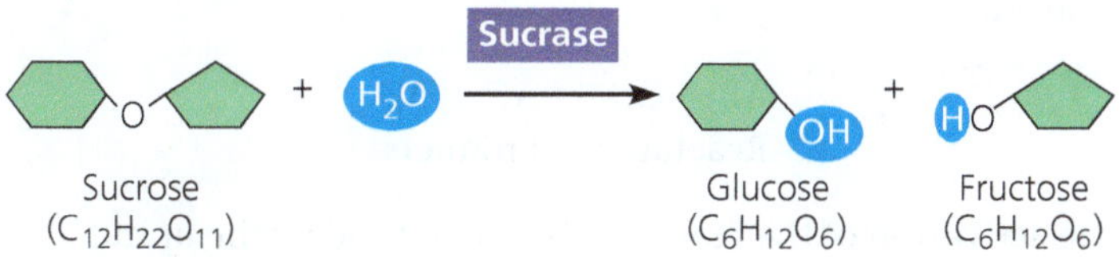

How does the enzyme do this?

An **enzyme** is a macromolecule that acts as a **catalyst**, a chemical agent that speeds up a reaction without being consumed by the reaction. In this chapter, we are focusing on enzymes that are proteins. (Some RNA molecules, called ribozymes, can function as enzymes; these will be discussed in Concepts 17.3 and 25.1.) Without regulation by enzymes, chemical traffic through the pathways of metabolism would become terribly congested because many chemical reactions would take such a long time. In the next two sections, we will see why spontaneous reactions can be slow and how an enzyme changes the situation.

The Activation Energy Barrier

Every chemical reaction between molecules involves both bond breaking and bond forming. For example, the hydrolysis of sucrose involves breaking the bond between glucose and fructose and one of the bonds of a water molecule and then forming two new bonds, as shown above. Changing one molecule into another generally involves contorting the starting molecule into a highly unstable state before the reaction can proceed. This contortion can be compared to the bending of a metal key ring when you pry it open to add a new key. The key ring is highly unstable in its opened form but returns to a stable state once the key is threaded all the way onto the ring. To reach the contorted state where bonds can change, reactant molecules must absorb energy from their surroundings. When the new bonds of the product molecules form, energy is released as heat, and the molecules return to stable shapes with lower energy than the contorted state.

The initial investment of energy for starting a reaction—the energy required to contort the reactant molecules so the bonds can break—is known as the *free energy of activation*, or **activation energy**, abbreviated E_A in this book. We can think of activation energy as the amount of energy needed to push the reactants to the top of an energy barrier, or uphill, so that the "downhill" part of the reaction can begin. Activation energy is often supplied by heat in the form of thermal energy that the reactant molecules absorb from the surroundings. The absorption of thermal energy accelerates the reactant molecules, so they collide more often and more forcefully. It also agitates the atoms within the molecules, making the breakage of bonds more likely. When the molecules have absorbed enough energy for the bonds to break, the reactants are in an unstable condition known as the *transition state*.

Figure 8.13 graphs the energy changes for a hypothetical exergonic reaction that swaps portions of two reactant molecules:

$$\underset{\text{Reactants}}{AB + CD} \rightarrow \underset{\text{Products}}{AC + BD}$$

The activation of the reactants is represented by the uphill portion of the graph, in which the free-energy content of the reactant molecules is increasing. At the summit, when energy equivalent to E_A has been absorbed, the reactants are in the transition state: They are activated, and their bonds can be broken. As the atoms then settle into their new, more stable bonding arrangements, energy is released to the surroundings. This corresponds to the downhill part of the curve, which shows the loss of free energy by the molecules. The overall decrease in free energy means that E_A is repaid with dividends, as the formation of new bonds releases more energy than was invested in the breaking of old bonds.

▼ Figure 8.13 Energy profile of an exergonic reaction. The "molecules" are hypothetical, with A, B, C, and D representing portions of the molecules. Thermodynamically, this is an exergonic reaction, with a negative ΔG, and the reaction occurs spontaneously. However, the activation energy (E_A) provides a barrier that determines the rate of the reaction.

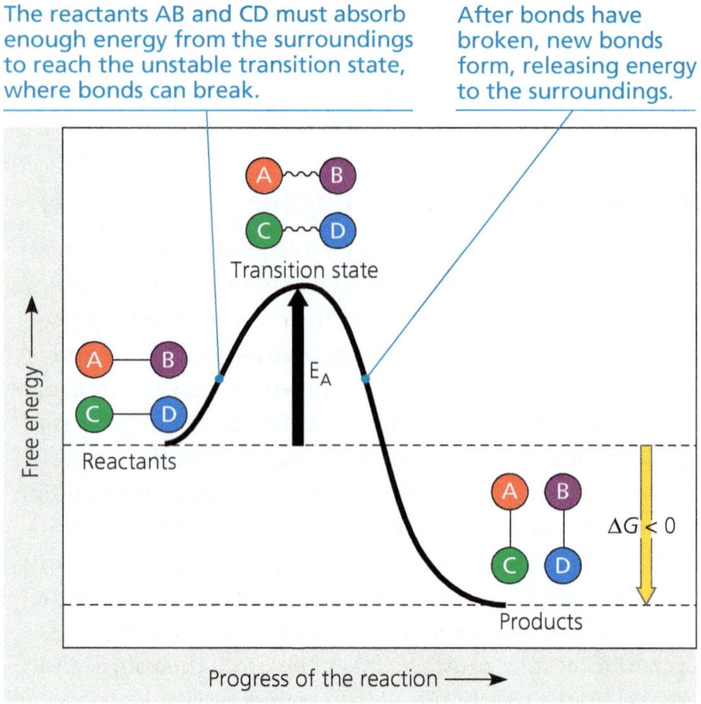

DRAW IT ➤ *Graph the progress of an endergonic reaction in which EF and GH form products EG and FH, assuming that the reactants must pass through a transition state.*

The reaction shown in Figure 8.13 is exergonic and occurs spontaneously ($\Delta G < 0$). However, the activation energy provides a barrier that determines the rate of the reaction. The reactants must absorb enough energy to reach the top of the activation energy barrier before the reaction can occur. For some reactions, E_A is modest enough that even at room temperature there is sufficient thermal energy for many of the reactant molecules to reach the transition state in a short time. In most cases, however, E_A is so high and the transition state is reached so rarely that the reaction will hardly proceed at all. In these cases, the reaction will occur at a noticeable rate only if energy is provided, usually by heat. For example, the reaction of gasoline and oxygen is exergonic and will occur spontaneously, but energy is required for the molecules to reach the transition state and react. Only when the spark plugs fire in an automobile engine can there be the explosive release of energy that pushes the pistons. Without a spark, a mixture of gasoline hydrocarbons and oxygen will not react because the E_A barrier is too high.

How Enzymes Speed Up Reactions

Proteins, DNA, and other complex cellular molecules are rich in free energy and have the potential to decompose spontaneously; that is, the laws of thermodynamics favor their breakdown. These molecules only persist because at temperatures typical for cells, few molecules can make it over the hump of activation energy. The barriers for selected reactions must occasionally be surmounted, however, for cells to carry out the processes needed for life. Heat can increase the rate of a reaction by allowing reactants to attain the transition state more often, but this would not work well in biological systems. First, high temperature denatures proteins and kills cells. Second, heat would speed up *all* reactions, not just those that are needed. Instead of heat, organisms carry out **catalysis**, a process by which a catalyst (for example, an enzyme) selectively speeds up a reaction without itself being consumed. (You learned about catalysts in Concept 5.4.)

An enzyme catalyzes a reaction by lowering the E_A barrier **(Figure 8.14)**, enabling the reactant molecules to absorb enough energy to reach the transition state even at moderate temperatures, as we'll discuss shortly. It is crucial to note that *an enzyme cannot change the ΔG for a reaction; it cannot make an endergonic reaction exergonic.* Enzymes can only hasten reactions

Figure 8.14 The effect of an enzyme on activation energy. Without affecting the free-energy change (ΔG) for a reaction, an enzyme speeds the reaction by reducing its activation energy (E_A).

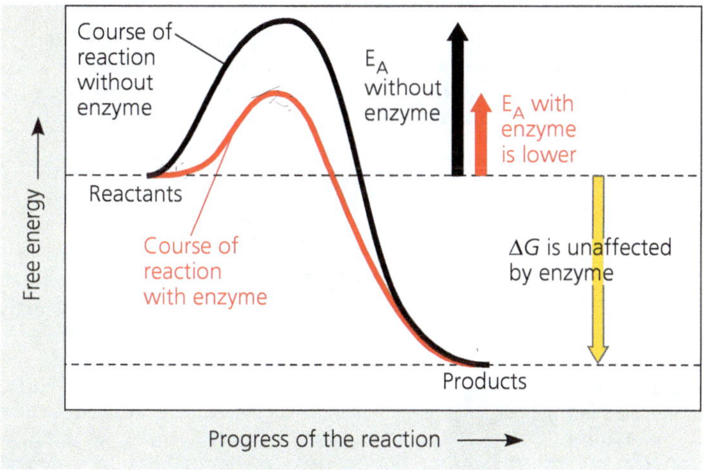

Animation: How Enzymes Work

that would eventually occur anyway, but this enables the cell to have a dynamic metabolism, routing chemicals smoothly through metabolic pathways. Also, enzymes are very specific for the reactions they catalyze, so they determine which chemical processes will be going on in the cell at any given time.

Substrate Specificity of Enzymes

The reactant an enzyme acts on is referred to as the enzyme's **substrate**. The enzyme binds to its substrate (or substrates, when there are two or more reactants), forming an **enzyme-substrate complex**. While enzyme and substrate are joined, the catalytic action of the enzyme converts the substrate to the product (or products) of the reaction. The overall process can be summarized as follows:

$$\text{Enzyme} + \text{Substrate(s)} \rightleftharpoons \text{Enzyme-substrate complex} \rightleftharpoons \text{Enzyme} + \text{Product(s)}$$

Most enzyme names end in *-ase*. For example, the enzyme sucrase catalyzes the hydrolysis of the disaccharide sucrose into its two monosaccharides, glucose and fructose (see the diagram at the beginning of Concept 8.4):

$$\text{Sucrase} + \text{Sucrose} + H_2O \rightleftharpoons \text{Sucrase-sucrose-}H_2O\text{ complex} \rightleftharpoons \text{Sucrase} + \text{Glucose} + \text{Fructose}$$

The reaction catalyzed by each enzyme is very specific; an enzyme can recognize its specific substrate even among closely related compounds. For instance, sucrase will act only on sucrose and will not bind to other disaccharides, such as maltose. What accounts for this molecular recognition? Recall that most enzymes are proteins, and proteins are macromolecules with unique three-dimensional configurations. The specificity of an enzyme results from its shape, which is a consequence of its amino acid sequence.

Only a restricted region of the enzyme molecule actually binds to the substrate. This region, called the **active site**, is typically a pocket or groove on the surface of the enzyme where catalysis occurs (**Figure 8.15a**; see also Figure 5.16). Usually, the active site is formed by only a few of the enzyme's amino acids, with the rest of the protein molecule providing a framework that determines the shape of the active site. The specificity of an enzyme is attributed to a complementary fit between the shape of its active site and the shape of the substrate.

An enzyme is not a stiff structure locked into a given shape. In fact, recent work by biochemists has shown that enzymes (and other proteins) seem to "dance" between subtly different shapes in a dynamic equilibrium, with slight differences in free energy for each "pose." The shape that best fits the substrate isn't necessarily the one with the lowest energy, but during the very short time the enzyme takes on this shape, its active site can bind to the substrate. The active site itself is also not a rigid receptacle for the substrate. As the substrate enters the active site, the enzyme changes shape slightly due to interactions between the substrate's chemical groups and chemical groups on the side chains of the amino acids that form the active site. This shape change makes the active site fit even more snugly around the substrate **(Figure 8.15b)**.

Figure 8.15 Induced fit between an enzyme and its substrate.

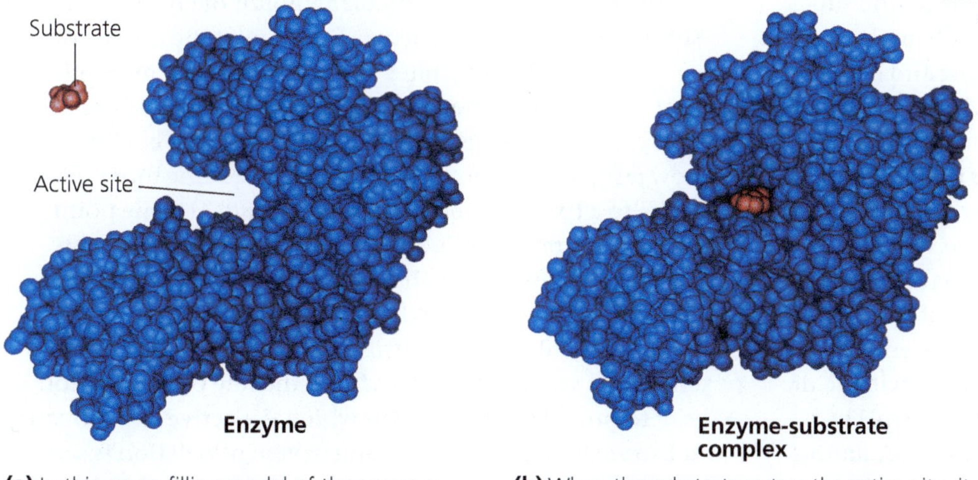

(a) In this space-filling model of the enzyme hexokinase (blue), the active site forms a groove on the surface. The enzyme's substrate is glucose (red).

(b) When the substrate enters the active site, it forms weak bonds with the enzyme, inducing a change in the shape of the protein. This change allows additional weak bonds to form, causing the active site to enfold the substrate and hold it in place.

The tightening of the binding after initial contact—called **induced fit**—is like a clasping handshake. Induced fit brings chemical groups of the active site into positions that enhance their ability to catalyze the chemical reaction.

Catalysis in the Enzyme's Active Site

In most enzymatic reactions, the substrate is held in the active site by so-called weak interactions, such as hydrogen bonds and ionic bonds. The R groups of a few of the amino acids that make up the active site catalyze the conversion of substrate to product, and the product departs from the active site. The enzyme is then free to take another substrate molecule into its active site. The entire cycle happens so fast that a single enzyme molecule typically acts on about 1,000 substrate molecules per second, and some enzymes are even faster. Enzymes, like other catalysts, emerge from the reaction in their original form. Therefore, very small amounts of enzyme can have a huge metabolic impact by functioning over and over again in catalytic cycles. **Figure 8.16** shows a catalytic cycle involving two substrates and two products.

Most metabolic reactions are reversible, and an enzyme can catalyze either the forward or the reverse reaction, depending on which direction has a negative ΔG. This in turn depends mainly on the relative concentrations of reactants and products. The net effect is always in the direction of equilibrium.

Enzymes use a variety of mechanisms that lower activation energy and speed up a reaction (see Figure 8.16, step 3):

- When there are two or more reactants, the active site provides a template on which the substrates can come together in the proper orientation for a reaction to occur between them.
- As the active site of an enzyme clutches the bound substrates, the enzyme may stretch the substrate molecules toward their transition state form, stressing and bending critical chemical bonds to be broken during the reaction. Because E_A is proportional to the difficulty of breaking the bonds, distorting the substrate helps it approach the transition state and reduces the amount of free energy that must be absorbed to achieve that state.
- The active site may also provide a microenvironment that is more conducive to a particular type of reaction than the solution itself would be without the enzyme. For example, if the active site has amino acids with acidic R groups, the active site may be a pocket of low pH in an otherwise neutral cell. In such cases, an acidic amino acid may facilitate H^+ transfer to the substrate as a key step in catalyzing the reaction.
- Amino acids in the active site directly participate in the chemical reaction. Sometimes this process even involves brief covalent bonding between the substrate and the side chain of an amino acid of the enzyme. Subsequent steps of the reaction restore the side chains to their original states so that the active site is the same after the reaction as it was before.

▼ Figure 8.16 The active site and catalytic cycle of an enzyme. An enzyme can convert one or more reactant molecules to one or more product molecules. The enzyme shown here converts two substrate molecules to two product molecules.

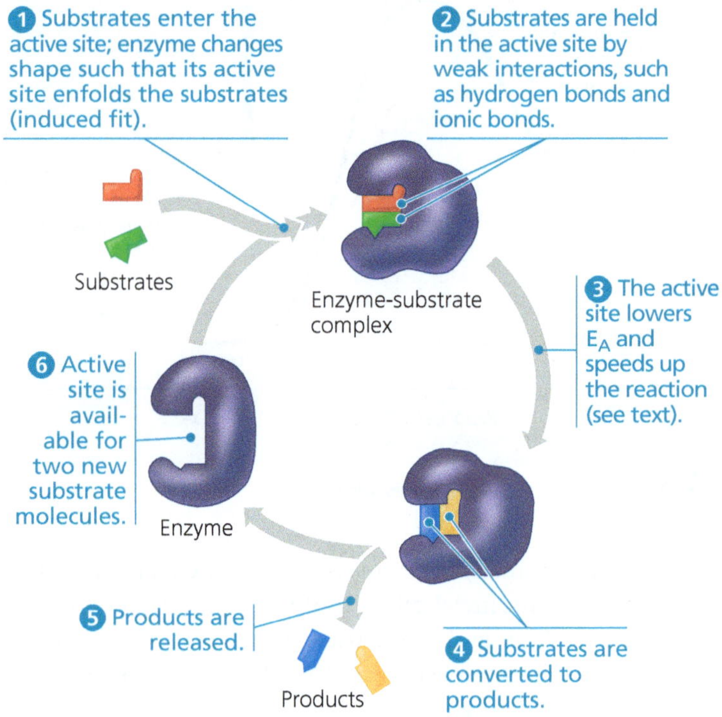

DRAW IT ➤ *The enzyme-substrate complex passes through a transition state (see Figure 8.13). Label the part of the cycle where the transition state occurs.*

MB Animation: Enzymes: Steps in a Reaction

The rate at which a particular amount of enzyme converts substrate to product is partly a function of the initial concentration of the substrate: The more substrate molecules that are available, the more frequently they access the active sites of the enzyme molecules. However, there is a limit to how fast the reaction can be pushed by adding more substrate to a fixed concentration of enzyme. At some point, the concentration of substrate will be high enough that all enzyme molecules will have their active sites engaged. As soon as the product exits an active site, another substrate molecule enters. At this substrate concentration, the enzyme is said to be *saturated*, and the rate of the reaction is determined by the speed at which the active site converts substrate to product. When an enzyme population is saturated, the only way to increase the rate of product formation is to add more enzyme. Cells often increase the rate of a reaction by producing more enzyme molecules. You can graph the overall progress of an enzymatic reaction in the **Scientific Skills Exercise**.

SCIENTIFIC SKILLS EXERCISE

Making a Line Graph and Calculating a Slope

Does the Rate of Glucose 6-Phosphatase Activity Change over Time in Isolated Liver Cells? Glucose 6-phosphatase, which is found in mammalian liver cells, is a key enzyme in control of blood glucose levels. The enzyme catalyzes the breakdown of glucose 6-phosphate into glucose and inorganic phosphate (Ⓟ$_i$). These products are transported out of liver cells into the blood, increasing blood glucose levels. In this exercise, you will graph data from a time-course experiment that measured Ⓟ$_i$ concentration in the buffer outside isolated liver cells, thus indirectly measuring glucose 6-phosphatase activity inside the cells.

How the Experiment Was Done Isolated rat liver cells were placed in a dish with buffer at physiological conditions (pH 7.4, 37°C). Glucose 6-phosphate (the substrate) was added so it could be taken up by cells and broken down by glucose 6-phosphatase. A sample of buffer was removed every 5 minutes and the concentration of Ⓟ$_i$ that had been transported out of the cells was determined

Data from the Experiment

Time (min)	Concentration of Ⓟ$_i$ (μmol/mL)
0	0
5	10
10	90
15	180
20	270
25	330
30	355
35	355
40	355

Data from S. R. Commerford et al., Diets enriched in sucrose or fat increase gluconeogenesis and G-6-Pase but not basal glucose production in rats, *American Journal of Physiology - Endocrinology and Metabolism* 283:E545–E555 (2002).

INTERPRET THE DATA

1. To see patterns in the data from a time-course experiment like this, it is helpful to graph the data. First, determine which set of data goes on each axis. (a) What did the researchers intentionally vary in the experiment? This is the independent variable, which goes on the *x*-axis. (b) What are the units (abbreviated) for the independent variable? Explain in words what the abbreviation stands for. (c) What was measured by the researchers? This is the dependent variable, which goes on the *y*-axis. (d) What does the units abbreviation stand for? Label each axis, including the units.
2. Next, you'll want to mark off the axes with just enough evenly spaced tick marks to accommodate the full set of data. Determine the range of data values for each axis. (a) What is the largest value to go on the *x*-axis? What is a reasonable spacing for the tick marks, and what should be the highest one? (b) What is the largest value to go on the *y*-axis? What is a reasonable spacing for the tick marks, and what should be the highest one?
3. Plot the data points on your graph. Match each *x*-value with its partner *y*-value and place a point on the graph at that coordinate. Draw a line that connects the points. (For additional information about graphs, see the Scientific Skills Review in Appendix F.)
4. Examine your graph and look for patterns in the data. (a) Does the concentration of Ⓟ$_i$ increase evenly through the course of the experiment? To answer this question, describe the pattern you see in the graph. (b) What part of the graph shows the highest rate of enzyme activity? Consider that the rate of enzyme activity is related to the slope of the line, $\Delta y/\Delta x$ (the "rise" over the "run"), in μmol/(mL · min), with the steepest slope indicating the highest rate of enzyme activity. Calculate the rate of enzyme activity (slope) where the graph is steepest. (c) Can you think of a biological explanation for the pattern you see?
5. If your blood sugar level is low from skipping lunch, what reaction (discussed in this exercise) will occur in your liver cells? Write out the reaction and put the name of the enzyme over the reaction arrow. How will this reaction affect your blood sugar level?

Instructors: A version of this Scientific Skills Exercise can be assigned in MasteringBiology.

Effects of Local Conditions on Enzyme Activity

The activity of an enzyme—how efficiently the enzyme functions—is affected by general environmental factors, such as temperature and pH. It can also be affected by chemicals that specifically influence that enzyme. In fact, researchers have learned much about enzyme function by employing such chemicals.

Effects of Temperature and pH

Recall from Figure 5.20 that the three-dimensional structures of proteins are sensitive to their environment. As a consequence, each enzyme works better under some conditions than under other conditions, because these *optimal conditions* favor the most active shape for the enzyme.

Temperature and pH are environmental factors important in the activity of an enzyme. Up to a point, the rate of an enzymatic reaction increases with increasing temperature, partly because substrates collide with active sites more frequently when the molecules move rapidly. Above that temperature, however, the speed of the enzymatic reaction drops sharply. The thermal agitation of the enzyme molecule disrupts the hydrogen bonds, ionic bonds, and other weak interactions that stabilize the active shape of the enzyme, and the protein molecule eventually denatures. Each enzyme has an optimal temperature at which its reaction rate is greatest. Without denaturing the enzyme, this temperature allows the greatest number of molecular collisions and the fastest conversion of the reactants to product molecules. Most human enzymes have optimal temperatures of about 35–40°C (close to human body temperature). The thermophilic bacteria that live in hot springs contain enzymes with optimal temperatures of 70°C or higher **(Figure 8.17a)**.

Just as each enzyme has an optimal temperature, it also has a pH at which it is most active. The optimal pH values for most enzymes fall in the range of pH 6–8, but there are exceptions. For example, pepsin, a digestive enzyme in the human stomach, works best at a very low pH. Such an acidic environment denatures most enzymes, but pepsin is adapted to maintain its functional three-dimensional structure in the acidic environment of the stomach. In contrast, trypsin, a digestive enzyme residing in the more alkaline environment of the human intestine would be denatured in the stomach **(Figure 8.17b)**.

▼ Figure 8.17 Environmental factors affecting enzyme activity. Each enzyme has an optimal **(a)** temperature and **(b)** pH that favor the most active shape of the protein molecule.

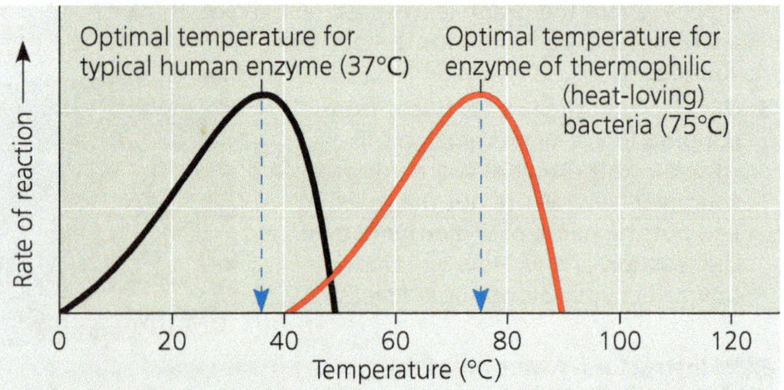

(a) The photo shows thermophilic cyanobacteria (green) thriving in the hot water of a Nevada geyser. The graph compares the optimal temperatures for an enzyme from the thermophilic bacterium *Thermus oshimai* (75°C) and human enzymes (body temperature, 37°C).

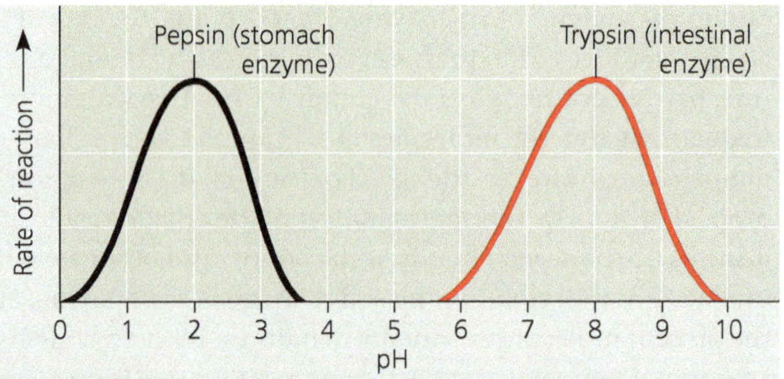

(b) This graph shows the rate of reaction for two digestive enzymes over a range of pH values.

INTERPRET THE DATA ➤ *Looking at the graph in (b), what is the optimal pH for pepsin activity? Explain why natural selection might have resulted in the optimal pH for pepsin, a stomach enzyme (see Figure 3.11). What is the optimal pH for trypsin?*

Cofactors

Many enzymes require nonprotein helpers for catalytic activity, often for chemical processes like electron transfers that cannot easily be carried out by the amino acids in proteins. These adjuncts, called **cofactors**, may be bound tightly to the enzyme as permanent residents, or they may bind loosely and reversibly along with the substrate. The cofactors of some enzymes are inorganic, such as the metal atoms zinc, iron, and copper in ionic form. If the cofactor is an organic molecule, it is referred to, more specifically, as a **coenzyme**. Most vitamins are important in nutrition because they act as coenzymes or raw materials from which coenzymes are made.

Enzyme Inhibitors

Certain chemicals selectively inhibit the action of specific enzymes. Sometimes the inhibitor attaches to the enzyme by covalent bonds, in which case the inhibition is usually irreversible. Many enzyme inhibitors, however, bind to the enzyme by weak interactions, and when this occurs the inhibition is reversible. Some reversible inhibitors resemble the normal substrate molecule and compete for admission into the active site **(Figure 8.18a** and **b)**. These mimics, called **competitive inhibitors**, reduce the productivity of enzymes by blocking substrates from entering active sites. This kind of inhibition can be overcome by increasing the concentration of substrate so that as active sites become available, more substrate molecules than inhibitor molecules are around to gain entry to the sites.

In contrast, **noncompetitive inhibitors** do not directly compete with the substrate to bind to the enzyme at the active site **(Figure 8.18c)**. Instead, they impede enzymatic reactions by binding to another part of the enzyme. This interaction causes the enzyme molecule to change its shape in such a way that the active site becomes much less effective at catalyzing the conversion of substrate to product.

Toxins and poisons are often irreversible enzyme inhibitors. An example is sarin, a nerve gas. In the mid-1990s, terrorists released sarin in the Tokyo subway, killing several people and injuring many others. This small molecule binds covalently to the R group on the amino acid serine, which is found in the active site of acetylcholinesterase, an enzyme important in the nervous system. Other examples include the pesticides DDT and parathion, inhibitors of key enzymes in the nervous system. Finally, many antibiotics are inhibitors of specific

Figure 8.18 Inhibition of enzyme activity.

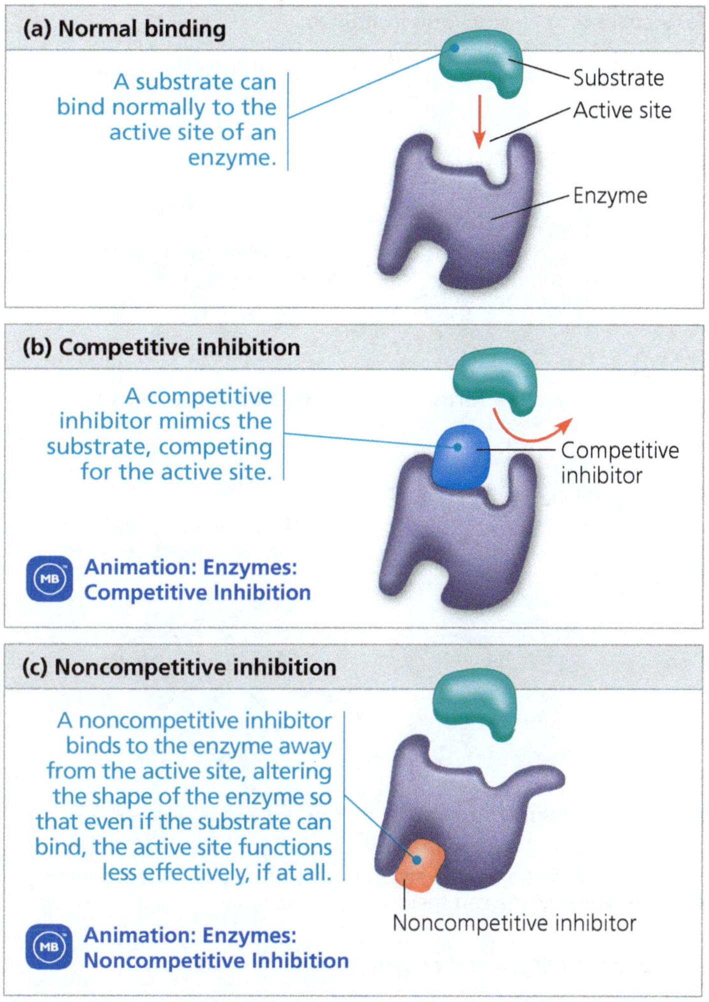

enzymes in bacteria. For instance, penicillin blocks the active site of an enzyme that many bacteria use to make cell walls.

Citing enzyme inhibitors that are metabolic poisons may give the impression that enzyme inhibition is generally abnormal and harmful. In fact, molecules naturally present in the cell often regulate enzyme activity by acting as inhibitors. Such regulation—selective inhibition—is essential to the control of cellular metabolism, as we will discuss in Concept 8.5.

The Evolution of Enzymes

EVOLUTION Thus far, biochemists have identified more than 4,000 different enzymes in various species, most likely a very small fraction of all enzymes. How did this grand profusion of enzymes arise? Recall that most enzymes are proteins, and proteins are encoded by genes. A permanent change in a gene, known as a *mutation*, can result in a protein with one or more changed amino acids. In the case of an enzyme, if the changed amino acids are in the active site or some other crucial region, the altered enzyme might have a novel activity or might bind to a different substrate. Under environmental conditions where the new function benefits the organism, natural selection would tend to favor the mutated form of the gene, causing it to persist in the population.

Figure 8.19 Mimicking evolution of an enzyme with a new function. Researchers tested whether the function of an enzyme called β-galactosidase, which breaks down the sugar lactose, could change over time in populations of the bacterium *Escherichia coli*. After seven rounds of mutation and selection in the lab, β-galactosidase evolved into an enzyme specialized for breaking down a sugar other than lactose. This ribbon model shows one subunit of the altered enzyme; six amino acids were different.

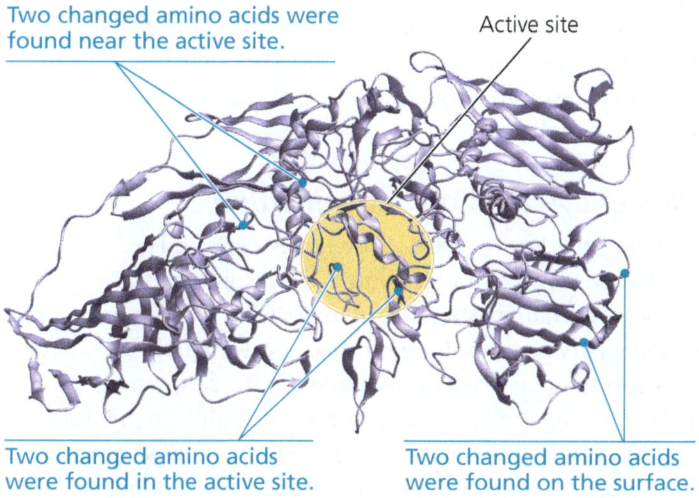

This simplified model is generally accepted as the main way in which the multitude of different enzymes arose over the past few billion years of life's history. Data supporting this model have been collected by researchers using a lab procedure that mimics evolution in natural populations **(Figure 8.19)**.

CONCEPT CHECK 8.4

1. Many spontaneous reactions occur very slowly. Why don't all spontaneous reactions occur instantly?
2. Why do enzymes act only on very specific substrates?
3. **WHAT IF?** ➤ Malonate is an inhibitor of the enzyme succinate dehydrogenase. How would you determine whether malonate is a competitive or noncompetitive inhibitor?
4. **DRAW IT** ➤ A mature lysosome has an internal pH of around 4.5. Using Figure 8.17b as a guide, draw a graph showing what you would predict for the rate of reaction for a lysosomal enzyme. Label its optimal pH, assuming its optimal pH matches its environment.

For suggested answers, see Appendix A.

CONCEPT 8.5

Regulation of enzyme activity helps control metabolism

Chemical chaos would result if all of a cell's metabolic pathways were operating simultaneously. Intrinsic to life's processes is a cell's ability to tightly regulate its metabolic pathways by controlling when and where its various enzymes are active. It does this either by switching on and off the genes that encode specific enzymes (as we will discuss in Unit Three) or, as we discuss here, by regulating the activity of enzymes once they are made.

Allosteric Regulation of Enzymes

In many cases, the molecules that naturally regulate enzyme activity in a cell behave something like reversible noncompetitive inhibitors (see Figure 8.18c): These regulatory molecules change an enzyme's shape and the functioning of its active site by binding to a site elsewhere on the molecule, via noncovalent interactions. **Allosteric regulation** is the term used to describe any case in which a protein's function at one site is affected by the binding of a regulatory molecule to a separate site. It may result in either inhibition or stimulation of an enzyme's activity.

Allosteric Activation and Inhibition

Most enzymes known to be allosterically regulated are constructed from two or more subunits, each composed of a polypeptide chain with its own active site. The entire complex oscillates between two different shapes, one catalytically active and the other inactive **(Figure 8.20a)**. In the simplest kind of allosteric regulation, an activating or inhibiting regulatory molecule binds to a regulatory site (sometimes called an allosteric site), often located where subunits join. The binding of an *activator* to a regulatory site stabilizes the shape that has functional active sites, whereas the binding of an *inhibitor* stabilizes the inactive form of the enzyme. The subunits of an allosteric enzyme fit together in such a way that a shape change in one subunit is transmitted to all others. Through this interaction of subunits, a single activator or inhibitor molecule that binds to one regulatory site will affect the active sites of all subunits.

Fluctuating concentrations of regulators can cause a sophisticated pattern of response in the activity of cellular enzymes. The products of ATP hydrolysis (ADP and Ⓟi), for example, play a complex role in balancing the flow of traffic between anabolic and catabolic pathways by their effects on key enzymes. ATP binds to several catabolic enzymes allosterically, lowering their affinity for substrate and thus inhibiting their activity. ADP, however, functions as an activator of the same enzymes. This is logical because catabolism functions in regenerating ATP. If ATP production lags behind its use, ADP accumulates and activates the enzymes that speed up catabolism, producing more ATP. If the supply of ATP exceeds demand, then catabolism slows down as ATP molecules accumulate and bind to the same enzymes, inhibiting them. (You'll see specific examples of this type of regulation when you learn about cellular respiration in the next chapter; see, for example, Figure 9.20.) ATP, ADP, and other related molecules also affect key enzymes in anabolic pathways. In this way, allosteric enzymes control the rates of important reactions in both sorts of metabolic pathways.

In another kind of allosteric activation, a *substrate* molecule binding to one active site in a multisubunit enzyme triggers a shape change in all the subunits, thereby increasing catalytic activity at the other active sites **(Figure 8.20b)**. Called **cooperativity**, this mechanism amplifies the response of enzymes to substrates: One substrate molecule primes an enzyme to act on additional substrate molecules more readily. Cooperativity is considered allosteric regulation because, even though substrate is binding to an active site, its binding affects catalysis in another active site.

▼ Figure 8.20 Allosteric regulation of enzyme activity.

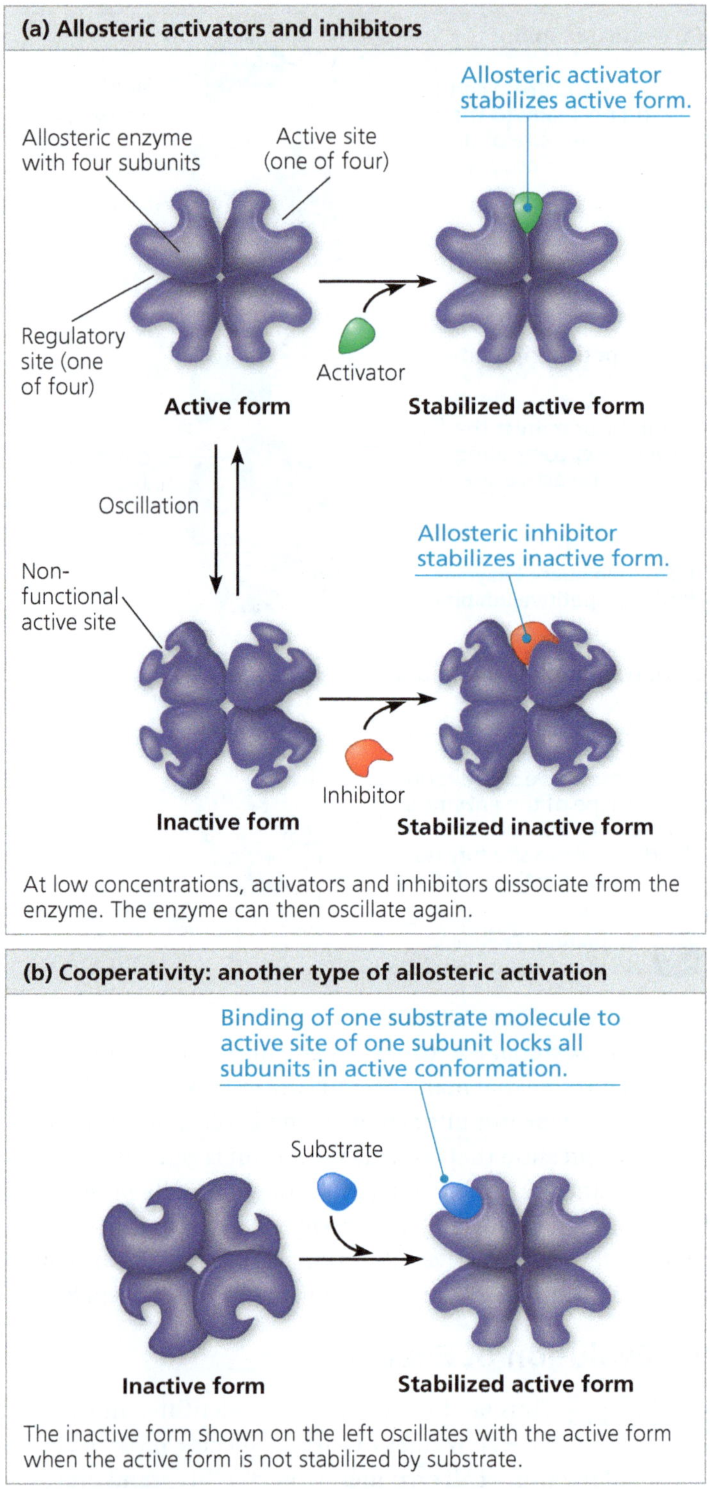

Although hemoglobin is not an enzyme (it carries O_2 rather than catalyzing a reaction), classic studies of hemoglobin have elucidated the principle of cooperativity. Hemoglobin is made up of four subunits, each with an oxygen-binding site (see Figure 5.18). The binding of an oxygen molecule to one binding site increases the affinity for oxygen of the remaining binding sites. Thus, where oxygen is at high levels, such as in the lungs or gills, hemoglobin's affinity for oxygen increases as more binding sites are filled. In oxygen-deprived tissues, however, the release of each oxygen molecule decreases the oxygen affinity of the other binding sites, resulting in the release of oxygen where it is most needed. Cooperativity works similarly in multisubunit enzymes that have been studied.

Feedback Inhibition

Earlier, we discussed the allosteric inhibition of an enzyme in an ATP-generating pathway by ATP itself. This is a common mode of metabolic control, called **feedback inhibition**, in which a metabolic pathway is halted by the inhibitory binding of its end product to an enzyme that acts early in the pathway. **Figure 8.21** shows an example of feedback inhibition operating on an anabolic pathway. Some cells use this five-step pathway to synthesize the amino acid isoleucine from threonine, another amino acid. As isoleucine accumulates, it slows down its own synthesis by allosterically inhibiting the enzyme for the first step of the pathway. Feedback inhibition thereby prevents the cell from making more isoleucine than is necessary and thus wasting chemical resources.

▼ Figure 8.21 Feedback inhibition in isoleucine synthesis.

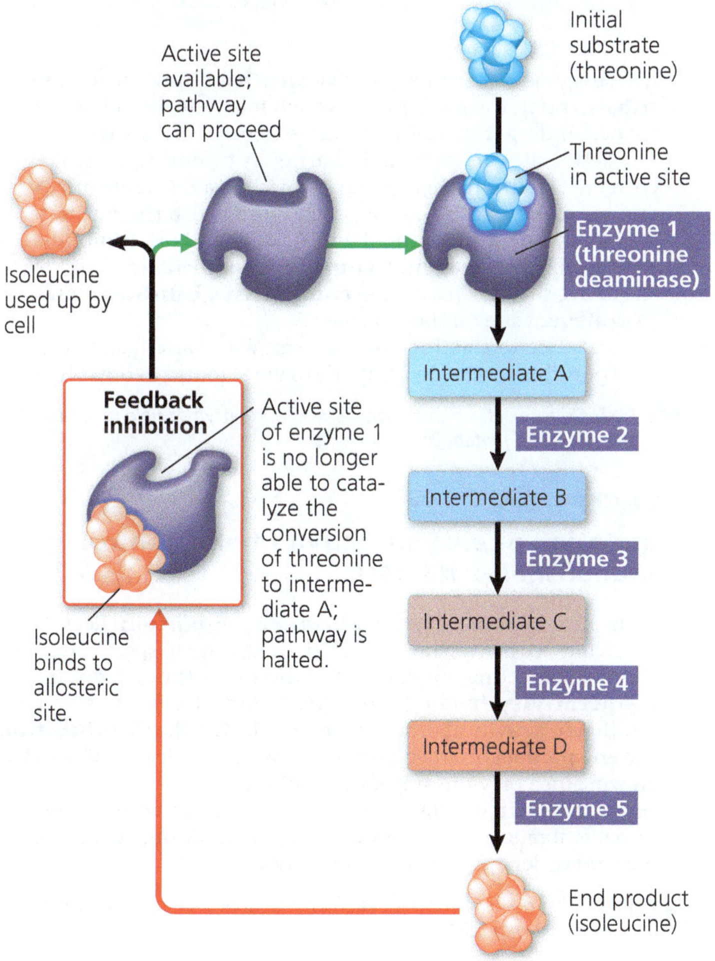

▼ Figure 8.22 Organelles and structural order in metabolism. Organelles such as the mitochondrion (TEM) contain enzymes that carry out specific functions, in this case the second and third stages of cellular respiration.

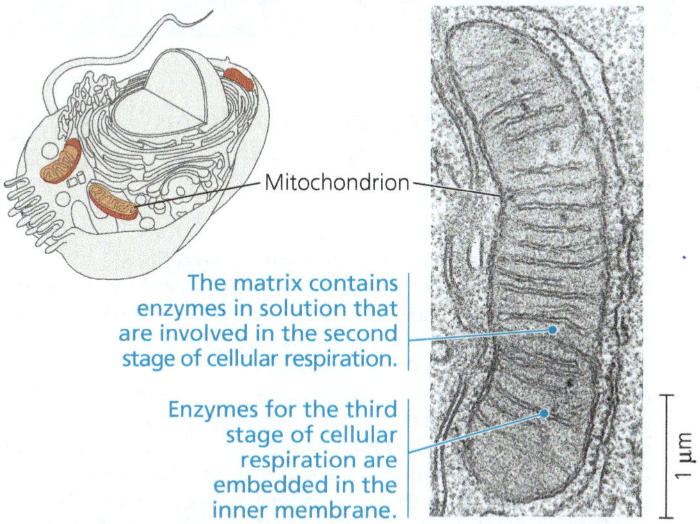

Localization of Enzymes Within the Cell

The cell is not just a bag of chemicals with thousands of different kinds of enzymes and substrates in a random mix. The cell is compartmentalized, and cellular structures help bring order to metabolic pathways. In some cases, a team of enzymes for several steps of a metabolic pathway are assembled into a multienzyme complex. The arrangement facilitates the sequence of reactions, with the product from the first enzyme becoming the substrate for an adjacent enzyme in the complex, and so on, until the end product is released. Some enzymes and enzyme complexes have fixed locations within the cell and act as structural components of particular membranes. Others are in solution within particular membrane-enclosed eukaryotic organelles, each with its own internal chemical environment. For example, in eukaryotic cells, the enzymes for the second and third stages of cellular respiration reside in specific locations within mitochondria **(Figure 8.22)**.

In this chapter, you have learned about the laws of thermodynamics that govern metabolism, the intersecting set of chemical pathways characteristic of life. We have explored the bioenergetics of breaking down and building up biological molecules. To continue the theme of bioenergetics, we will next examine cellular respiration, the major catabolic pathway that breaks down organic molecules and releases energy that can be used for the crucial processes of life.

CONCEPT CHECK 8.5

1. How do an activator and an inhibitor have different effects on an allosterically regulated enzyme?
2. **WHAT IF? ➤** Regulation of isoleucine synthesis is an example of feedback inhibition of an anabolic pathway. With that in mind, explain how ATP might be involved in feedback inhibition of a catabolic pathway.

For suggested answers, see Appendix A.

8 Chapter Review

Go to **MasteringBiology**™ for Videos, Animations, Vocab Self-Quiz, Practice Tests, and more in the Study Area.

SUMMARY OF KEY CONCEPTS

VOCAB SELF-QUIZ
goo.gl/6u5Sks

CONCEPT 8.1

An organism's metabolism transforms matter and energy, subject to the laws of thermodynamics *(pp. 144–147)*

- **Metabolism** is the collection of chemical reactions that occur in an organism. Enzymes catalyze reactions in intersecting **metabolic pathways**, which may be **catabolic** (breaking down molecules, releasing energy) or **anabolic** (building molecules, consuming energy). **Bioenergetics** is the study of the flow of energy through living organisms.
- **Energy** is the capacity to cause change; some forms of energy do work by moving matter. **Kinetic energy** is associated with motion and includes **thermal energy** associated with random motion of atoms or molecules. **Heat** is thermal energy in transfer from one object to another. **Potential energy** is related to the location or structure of matter and includes **chemical energy** possessed by a molecule due to its structure.
- **The first law of thermodynamics**, conservation of energy, states that energy cannot be created or destroyed, only transferred or transformed. The **second law of thermodynamics** states that **spontaneous processes**, those requiring no outside input of energy, increase the **entropy** (molecular disorder) of the universe.

? *Explain how the highly ordered structure of a cell does not conflict with the second law of thermodynamics.*

CONCEPT 8.2

The free-energy change of a reaction tells us whether or not the reaction occurs spontaneously *(pp. 147–150)*

- A living system's **free energy** is energy that can do work under cellular conditions. The change in free energy (ΔG) during a biological process is related directly to enthalpy change (ΔH) and to the change in entropy (ΔS): $\Delta G = \Delta H - T\Delta S$. Organisms live at the expense of free energy. A spontaneous process occurs with no energy input; during such a process, free energy decreases and the stability of a system increases. At maximum stability, the system is at equilibrium and can do no work.
- In an **exergonic** (spontaneous) chemical reaction, the products have less free energy than the reactants ($-\Delta G$). **Endergonic** (nonspontaneous) reactions require an input of energy ($+\Delta G$). The addition of starting materials and the removal of end products prevent metabolism from reaching equilibrium.

? *Explain the meaning of each component in the equation for the change in free energy of a spontaneous chemical reaction. Why are spontaneous reactions important in the metabolism of a cell?*

CONCEPT 8.3

ATP powers cellular work by coupling exergonic reactions to endergonic reactions *(pp. 150–153)*

- **ATP** is the cell's energy shuttle. Hydrolysis of its terminal phosphate yields ADP and Ⓟ$_i$ and releases free energy.
- Through **energy coupling**, the exergonic process of ATP hydrolysis drives endergonic reactions by transfer of a phosphate group to specific reactants, forming a **phosphorylated intermediate** that is more reactive. ATP hydrolysis (sometimes with protein phosphorylation) also causes changes in the shape and binding affinities of transport and motor proteins.
- Catabolic pathways drive regeneration of ATP from ADP + Ⓟ$_i$.

? *Describe the ATP cycle: How is ATP used and regenerated in a cell?*

CONCEPT 8.4

Enzymes speed up metabolic reactions by lowering energy barriers *(pp. 153–159)*

- In a chemical reaction, the energy necessary to break the bonds of the reactants is the **activation energy**, E_A.
- **Enzymes** lower the E_A barrier:

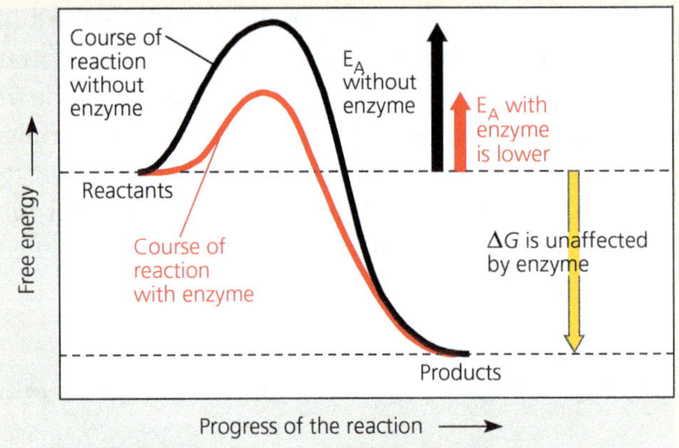

- Each enzyme has a unique **active site** that binds one or more **substrate(s)**, the reactants on which it acts. It then changes shape, binding the substrate(s) more tightly (**induced fit**).
- The active site can lower an E_A barrier by orienting substrates correctly, straining their bonds, providing a favorable microenvironment, or even covalently bonding with the substrate.
- Each enzyme has an optimal temperature and pH. Inhibitors reduce enzyme function. A **competitive inhibitor** binds to the active site, whereas a **noncompetitive inhibitor** binds to a different site on the enzyme.
- Natural selection, acting on organisms with variant enzymes, is responsible for the diversity of enzymes found in organisms.

? *How do both activation energy barriers and enzymes help maintain the structural and metabolic order of life?*

CONCEPT 8.5

Regulation of enzyme activity helps control metabolism *(pp. 159–161)*

- Many enzymes are subject to **allosteric regulation**: Regulatory molecules, either activators or inhibitors, bind to specific regulatory sites, affecting the shape and function of the enzyme. In **cooperativity**, binding of one substrate molecule can stimulate binding or activity at other active sites. In **feedback inhibition**, the end product of a metabolic pathway allosterically inhibits the enzyme for a previous step in the pathway.
- Some enzymes are grouped into complexes, some are incorporated into membranes, and some are contained inside organelles, increasing the efficiency of metabolic processes.

? *What roles do allosteric regulation and feedback inhibition play in the metabolism of a cell?*

TEST YOUR UNDERSTANDING

Level 1: Knowledge/Comprehension

PRACTICE TEST
goo.gl/CUYGKD

1. Choose the pair of terms that correctly completes this sentence: Catabolism is to anabolism as ________ is to ________.
   (A) exergonic; spontaneous
   (B) exergonic; endergonic
   (C) free energy; entropy
   (D) work; energy
2. Most cells cannot harness heat to perform work because
   (A) heat does not involve a transfer of energy.
   (B) cells do not have much thermal energy; they are relatively cool.
   (C) temperature is usually uniform throughout a cell.
   (D) heat can never be used to do work.
3. Which of the following metabolic processes can occur without a net influx of energy from some other process?
   (A) $ADP + Ⓟ_i \rightarrow ATP + H_2O$
   (B) $C_6H_{12}O_6 + 6\ O_2 \rightarrow 6\ CO_2 + 6\ H_2O$
   (C) $6\ CO_2 + 6\ H_2O \rightarrow C_6H_{12}O_6 + 6\ O_2$
   (D) Amino acids → Protein
4. If an enzyme in solution is saturated with substrate, the most effective way to obtain a faster yield of products is to
   (A) add more of the enzyme.
   (B) heat the solution to 90°C.
   (C) add more substrate.
   (D) add a noncompetitive inhibitor.
5. Some bacteria are metabolically active in hot springs because
   (A) they are able to maintain a lower internal temperature.
   (B) high temperatures make catalysis unnecessary.
   (C) their enzymes have high optimal temperatures.
   (D) their enzymes are completely insensitive to temperature.

Level 2: Application/Analysis

6. If an enzyme is added to a solution where its substrate and product are in equilibrium, what will occur?
   (A) Additional substrate will be formed.
   (B) The reaction will change from endergonic to exergonic.
   (C) The free energy of the system will change.
   (D) Nothing; the reaction will stay at equilibrium.

Level 3: Synthesis/Evaluation

7. **DRAW IT** Using a series of arrows, draw the branched metabolic reaction pathway described by the following statements, and then answer the question at the end. Use red arrows and minus signs to indicate inhibition.

   L can form either M or N.

   M can form O.

   O can form either P or R.

   P can form Q.

   R can form S.

   O inhibits the reaction of L to form M.

   Q inhibits the reaction of O to form P.

   S inhibits the reaction of O to form R.

   Which reaction would prevail if both Q and S were present in the cell in high concentrations?
   (A) L → M
   (B) M → O
   (C) L → N
   (D) O → P
8. **EVOLUTION CONNECTION** Some people argue that biochemical pathways are too complex to have evolved because all intermediate steps in a given pathway must be present to produce the final product. Critique this argument. How could you use the diversity of metabolic pathways that produce the same or similar products to support your case?
9. **SCIENTIFIC INQUIRY • DRAW IT** A researcher has developed an assay to measure the activity of an important enzyme present in pancreatic cells growing in culture. She adds the enzyme's substrate to a dish of cells and then measures the appearance of reaction products. The results are graphed as the amount of product on the y-axis versus time on the x-axis. The researcher notes four sections of the graph. For a short period of time, no products appear (section A). Then (section B) the reaction rate is quite high (the slope of the line is steep). Next, the reaction gradually slows down (section C). Finally, the graph line becomes flat (section D). Draw and label the graph, and propose a model to explain the molecular events occurring at each stage of this reaction profile.
10. **WRITE ABOUT A THEME: ENERGY AND MATTER** Life requires energy. In a short essay (100–150 words), describe the basic principles of bioenergetics in an animal cell. How is the flow and transformation of energy different in a photosynthesizing cell? Include the role of ATP and enzymes in your discussion.
11. **SYNTHESIZE YOUR KNOWLEDGE**

Explain what is happening in this photo in terms of kinetic energy and potential energy. Include the energy conversions that occur when the penguins eat fish and climb back up on the glacier. Describe the role of ATP and enzymes in the underlying molecular processes, including what happens to the free energy of some of the molecules involved.

For selected answers, see Appendix A.

For additional practice questions, check out the **Dynamic Study Modules** in MasteringBiology. You can use them to study on your smartphone, tablet, or computer anytime, anywhere!

Cellular Respiration and Fermentation

9

▲ **Figure 9.1** **How does food, like the sand eels captured by this puffin, power the work of life?**

KEY CONCEPTS

9.1 Catabolic pathways yield energy by oxidizing organic fuels

9.2 Glycolysis harvests chemical energy by oxidizing glucose to pyruvate

9.3 After pyruvate is oxidized, the citric acid cycle completes the energy-yielding oxidation of organic molecules

9.4 During oxidative phosphorylation, chemiosmosis couples electron transport to ATP synthesis

9.5 Fermentation and anaerobic respiration enable cells to produce ATP without the use of oxygen

9.6 Glycolysis and the citric acid cycle connect to many other metabolic pathways

BioFlix® Animation: Introduction to Cellular Respiration

Life Is Work

Living cells require transfusions of energy from outside sources to perform their many tasks—for example, assembling polymers, pumping substances across membranes, moving, and reproducing. The puffin in **Figure 9.1** obtains energy for its cells by feeding upon sand eels and other aquatic organisms; many other animals obtain energy by feeding on photosynthetic organisms such as plants and algae.

The energy stored in the organic molecules of food ultimately comes from the sun. Energy flows into an ecosystem as sunlight and exits as heat; in contrast, the chemical elements essential to life are recycled **(Figure 9.2)**. Photosynthesis generates oxygen, as well as organic molecules used by the mitochondria of eukaryotes as fuel for cellular respiration. Respiration breaks this fuel down, using oxygen and generating ATP. The waste products of this type of respiration, carbon dioxide and water, are the raw materials for photosynthesis.

In this chapter, we consider how cells harvest the chemical energy stored in organic molecules and use it to generate ATP, the molecule that drives most cellular work. After presenting some basic information about respiration, we'll focus on three key pathways of respiration: glycolysis, pyruvate oxidation and the citric acid cycle, and oxidative phosphorylation. We'll also consider fermentation, a somewhat simpler pathway coupled to glycolysis that has deep evolutionary roots.

When you see this blue icon, log in to **MasteringBiology** and go to the Study Area for digital resources.

Get Ready for This Chapter

Figure 9.2 Energy flow and chemical recycling in ecosystems. Energy flows into an ecosystem as sunlight and ultimately leaves as heat, while the chemical elements essential to life are recycled.

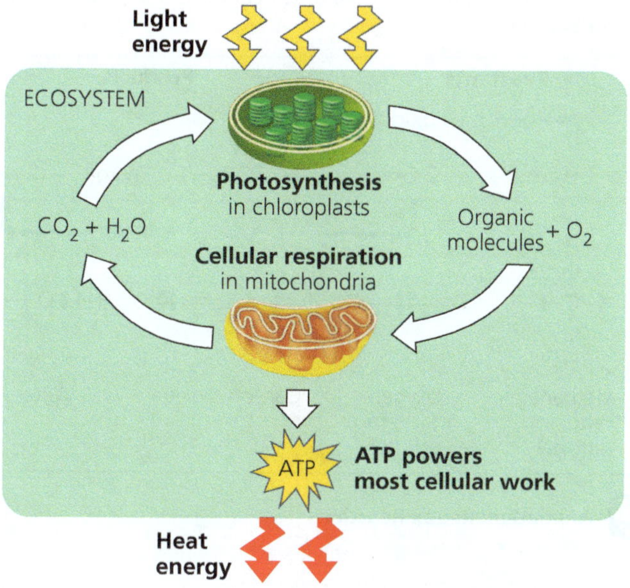

Animation: Energy Flow and Chemical Recycling

CONCEPT 9.1

Catabolic pathways yield energy by oxidizing organic fuels

Metabolic pathways that release stored energy by breaking down complex molecules are called catabolic pathways (see Concept 8.1). Transfer of electrons from fuel molecules (like glucose) to other molecules plays a major role in these pathways. In this section, we consider these processes, which are central to cellular respiration.

Catabolic Pathways and Production of ATP

Organic compounds possess potential energy as a result of the arrangement of electrons in the bonds between their atoms. Compounds that can participate in exergonic reactions can act as fuels. Through the activity of enzymes, a cell systematically degrades complex organic molecules that are rich in potential energy to simpler waste products that have less energy. Some of the energy taken out of chemical storage can be used to do work; the rest is dissipated as heat.

One catabolic process, **fermentation**, is a partial degradation of sugars or other organic fuel that occurs without the use of oxygen. However, the most efficient catabolic pathway is **aerobic respiration**, in which oxygen is consumed as a reactant along with the organic fuel (*aerobic* is from the Greek *aer*, air, and *bios*, life). The cells of most eukaryotic and many prokaryotic organisms can carry out aerobic respiration. Some prokaryotes use substances other than oxygen as reactants in a similar process that harvests chemical energy without oxygen; this process is called *anaerobic respiration* (the prefix *an-* means "without"). Technically, the term **cellular respiration** includes both aerobic and anaerobic processes. However, it originated as a synonym for aerobic respiration because of the relationship of that process to organismal respiration, in which an animal breathes in oxygen. Thus, *cellular respiration* is often used to refer to the aerobic process, a practice we follow in most of this chapter.

Although very different in mechanism, aerobic respiration is in principle similar to the combustion of gasoline in an automobile engine after oxygen is mixed with the fuel (hydrocarbons). Food provides the fuel for respiration, and the exhaust is carbon dioxide and water. The overall process can be summarized as follows:

$$\text{Organic compounds} + \text{Oxygen} \rightarrow \text{Carbon dioxide} + \text{Water} + \text{Energy}$$

Carbohydrates, fats, and protein molecules from food can all be processed and consumed as fuel, as we will discuss later in the chapter. In animal diets, a major source of carbohydrates is starch, a storage polysaccharide that can be broken down into glucose ($C_6H_{12}O_6$) subunits. Here, we will learn the steps of cellular respiration by tracking the degradation of the sugar glucose:

$$C_6H_{12}O_6 + 6\,O_2 \rightarrow 6\,CO_2 + 6\,H_2O + \text{Energy (ATP + heat)}$$

This breakdown of glucose is exergonic, having a free-energy change of −686 kcal (2,870 kJ) per mole of glucose decomposed ($\Delta G = -686$ kcal/mol). Recall that a negative ΔG ($\Delta G < 0$) indicates that the products of the chemical process store less energy than the reactants and that the reaction can happen spontaneously—in other words, without an input of energy.

Catabolic pathways do not directly move flagella, pump solutes across membranes, polymerize monomers, or perform other cellular work. Catabolism is linked to work by a chemical drive shaft—ATP (see Concept 8.3). To keep working, the cell must regenerate its supply of ATP from ADP and $Ⓟ_i$ (see Figure 8.12). To understand how cellular respiration accomplishes this, let's examine the fundamental chemical processes known as oxidation and reduction.

Redox Reactions: Oxidation and Reduction

How do the catabolic pathways that decompose glucose and other organic fuels yield energy? The answer is based on the transfer of electrons during the chemical reactions. The relocation of electrons releases energy stored in organic molecules, and this energy ultimately is used to synthesize ATP.

The Principle of Redox

In many chemical reactions, there is a transfer of one or more electrons (e^-) from one reactant to another. These electron transfers are called oxidation-reduction reactions, or **redox reactions** for short. In a redox reaction, the loss of electrons from one substance is called **oxidation**, and the addition of electrons to another substance is known as **reduction**. (Note that *adding* electrons is called *reduction*; adding negatively charged electrons to an atom *reduces* the amount of positive charge of that atom.)

To take a simple, nonbiological example, consider the reaction between the elements sodium (Na) and chlorine (Cl) that forms table salt:

becomes oxidized (loses electron): Na → Na^+

$$Na + Cl \longrightarrow Na^+ + Cl^-$$

becomes reduced (gains electron): Cl → Cl^-

We could generalize a redox reaction this way:

becomes oxidized: Xe^- → X

$$Xe^- + Y \longrightarrow X + Ye^-$$

becomes reduced: Y → Ye^-

In the generalized reaction, substance Xe^-, the electron donor, is called the **reducing agent**; it reduces Y, which accepts the donated electron. Substance Y, the electron acceptor, is the **oxidizing agent**; it oxidizes Xe^- by removing its electron. Because an electron transfer requires both an electron donor and an acceptor, oxidation and reduction always go hand in hand.

Not all redox reactions involve the complete transfer of electrons from one substance to another; some change the degree of electron sharing in covalent bonds. Methane combustion, shown in **Figure 9.3**, is an example. The covalent electrons in methane are shared nearly equally between the bonded atoms because carbon and hydrogen have about the same affinity for valence electrons; they are about equally electronegative (see Concept 2.3). But when methane reacts with oxygen, forming carbon dioxide, electrons end up shared less equally between the carbon atom and its new covalent partners, the oxygen atoms, which are very electronegative. In effect, the carbon atom has partially "lost" its shared electrons; thus, methane has been oxidized.

Now let's examine the fate of the reactant O_2. The two atoms of the oxygen molecule (O_2) share their electrons equally. But when oxygen reacts with the hydrogen from methane, forming water, the electrons of the covalent bonds spend more time near the oxygen (see Figure 9.3). In effect, each oxygen atom has partially "gained" electrons, so the oxygen molecule has been reduced. Because oxygen is so electronegative, it is one of the most powerful of all oxidizing agents.

Energy must be added to pull an electron away from an atom, just as energy is required to push a ball uphill. The more electronegative the atom (the stronger its pull on electrons), the more energy is required to take an electron away from it. An electron loses potential energy when it shifts from a less electronegative atom toward a more electronegative one, just as a ball loses potential energy when it rolls downhill. A redox reaction that moves electrons closer to oxygen, such as the burning (oxidation) of methane, therefore releases chemical energy that can be put to work.

▼ Figure 9.3 Methane combustion as an energy-yielding redox reaction. The reaction releases energy to the surroundings because the electrons lose potential energy when they end up being shared unequally, spending more time near electronegative atoms such as oxygen.

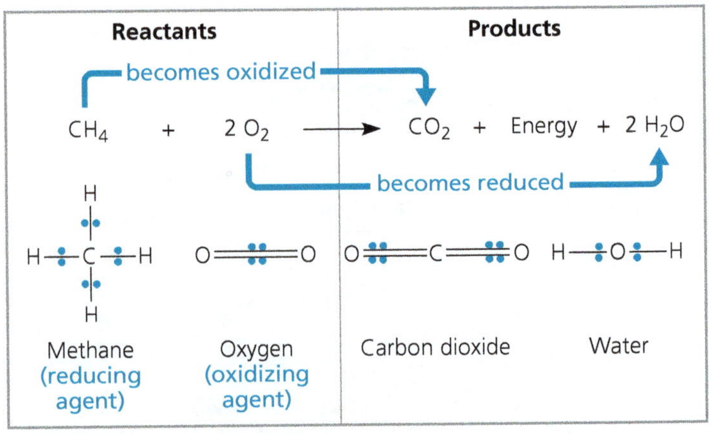

Animation: Redox Reactions

Oxidation of Organic Fuel Molecules During Cellular Respiration

The oxidation of methane by oxygen is the main combustion reaction that occurs at the burner of a gas stove. The combustion of gasoline in an automobile engine is also a redox reaction; the energy released pushes the pistons. But the energy-yielding redox process of greatest interest to biologists is respiration: the oxidation of glucose and other molecules in food. Examine again the summary equation for cellular respiration, but this time think of it as a redox process:

becomes oxidized: $C_6H_{12}O_6$ → $6\ CO_2$

$$C_6H_{12}O_6 + 6\ O_2 \longrightarrow 6\ CO_2 + 6\ H_2O + \text{Energy}$$

becomes reduced: $6\ O_2$ → $6\ H_2O$

As in the combustion of methane or gasoline, the fuel (glucose) is oxidized and oxygen is reduced. The electrons lose potential energy along the way, and energy is released.

In general, organic molecules that have an abundance of hydrogen are excellent fuels because their bonds are a source of "hilltop" electrons, whose energy may be released as these electrons "fall" down an energy gradient during their transfer to oxygen. The summary equation for respiration indicates that hydrogen is transferred from glucose to oxygen. But the important point, not visible in the summary equation, is that the energy state of the electron changes as hydrogen (with its electron) is transferred to oxygen. In respiration, the oxidation of glucose transfers electrons to a lower energy state, liberating energy that becomes available for ATP synthesis. So, in general, we see fuels with multiple C—H bonds oxidized into products with multiple C—O bonds.

The main energy-yielding foods—carbohydrates and fats—are reservoirs of electrons associated with hydrogen, often in the form of C—H bonds. Only the barrier of activation energy holds back the flood of electrons to a lower energy state (see Figure 8.13). Without this barrier, a food substance like glucose would combine almost instantaneously with O_2. If we supply the activation energy by igniting glucose, it burns in air, releasing 686 kcal (2,870 kJ) of heat per mole of glucose (about 180 g). Body temperature is not high enough to initiate burning, of course. Instead, if you swallow some glucose, enzymes in your cells will lower the barrier of activation energy, allowing the sugar to be oxidized in a series of steps.

Stepwise Energy Harvest via NAD+ and the Electron Transport Chain

If energy is released from a fuel all at once, it cannot be harnessed efficiently for constructive work. For example, if a gasoline tank explodes, it cannot drive a car very far. Cellular respiration does not oxidize glucose (or any other organic fuel) in a single explosive step either. Rather, glucose is broken down in a series of steps, each one catalyzed by an enzyme. At key steps, electrons are stripped from the glucose. As is often the case in oxidation reactions, each electron travels with a proton—thus, as a hydrogen atom. The hydrogen atoms are not transferred directly to oxygen, but instead are usually passed first to an electron carrier, a coenzyme called nicotinamide adenine dinucleotide, a derivative of the vitamin niacin. This coenzyme is well suited as an electron carrier because it can cycle easily between its oxidized form, **NAD^+**, and its reduced form, **NADH**. As an electron acceptor, NAD^+ functions as an oxidizing agent during respiration.

How does NAD^+ trap electrons from glucose and the other organic molecules in food? Enzymes called dehydrogenases remove a pair of hydrogen atoms (2 electrons and 2 protons) from the substrate (glucose, in the preceding example), thereby oxidizing it. The enzyme delivers the 2 electrons along with 1 proton to its coenzyme, NAD^+, forming NADH (**Figure 9.4**). The other proton is released as a hydrogen ion (H^+) into the surrounding solution:

$$\mathrm{H{-}\overset{|}{\underset{|}{C}}{-}OH} + \mathrm{NAD^+} \xrightarrow{\text{Dehydrogenase}} \mathrm{\overset{|}{\underset{|}{C}}{=}O} + \mathrm{NADH} + \mathrm{H^+}$$

By receiving 2 negatively charged electrons but only 1 positively charged proton, the nicotinamide portion of NAD^+ has its charge neutralized when NAD^+ is reduced to NADH. The name NADH shows the hydrogen that has been received in the reaction. NAD^+ is the most versatile electron acceptor in cellular respiration and functions in several of the redox steps during the breakdown of glucose.

Electrons lose very little of their potential energy when they are transferred from glucose to NAD^+. Each NADH molecule formed during respiration represents stored energy. This energy can be tapped to make ATP when the electrons complete their "fall" in a series of steps down an energy gradient from NADH to oxygen.

How do electrons that are extracted from glucose and stored as potential energy in NADH finally reach oxygen? It will help to compare the redox chemistry of cellular respiration to a much simpler reaction: the reaction between hydrogen and oxygen to form water (**Figure 9.5a**). Mix H_2 and O_2, provide a spark for activation energy, and the gases combine explosively. In fact, combustion of liquid H_2 and O_2 is harnessed to help power the rocket engines that boost satellites into orbit and launch spacecraft. The explosion represents a release of energy as the electrons of hydrogen "fall" closer to the electronegative oxygen atoms. Cellular respiration also brings hydrogen and oxygen together to form water, but there are two important differences. First, in cellular respiration, the hydrogen that reacts with oxygen

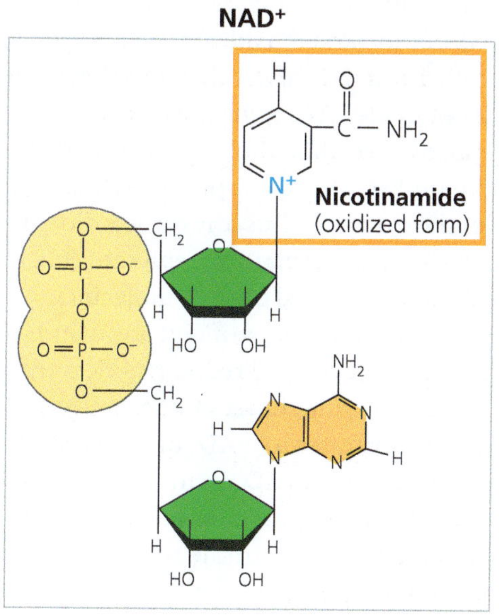

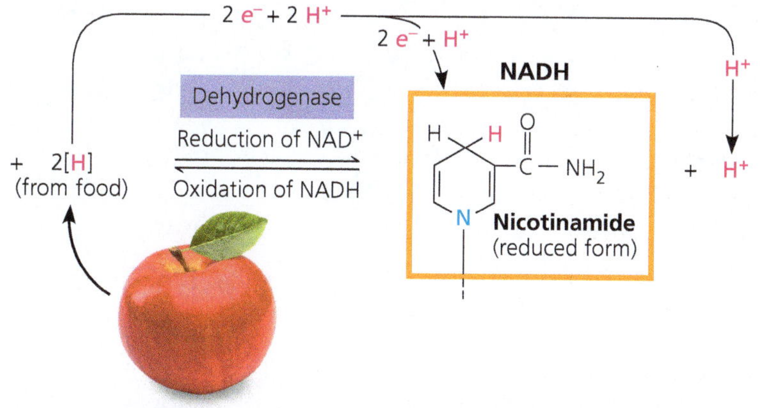

VISUAL SKILLS ➤ *Describe the structural differences between the oxidized form and the reduced form of nicotinamide.*

▲ **Figure 9.4 NAD^+ as an electron shuttle.** The full name for NAD^+, nicotinamide adenine dinucleotide, describes its structure—the molecule consists of two nucleotides joined together at their phosphate groups (shown in yellow). (Nicotinamide is a nitrogenous base, although not one that is present in DNA or RNA; see Figure 5.23.) The enzymatic transfer of 2 electrons and 1 proton (H^+) from an organic molecule in food to NAD^+ reduces the NAD^+ to NADH: Most of the electrons removed from food are transferred initially to NAD^+, forming NADH.

is derived from organic molecules rather than H_2. Second, instead of occurring in one explosive reaction, respiration uses an electron transport chain to break the fall of electrons to oxygen into several energy-releasing steps **(Figure 9.5b)**. An **electron transport chain** consists of a number of molecules, mostly proteins, built into the inner membrane of the mitochondria of eukaryotic cells (and the plasma membrane of respiring prokaryotes). Electrons removed from glucose are shuttled by NADH to the "top," higher-energy end of the chain. At the "bottom," lower-energy end, O_2 captures these electrons along with hydrogen nuclei (H^+), forming water. (Anaerobically respiring prokaryotes have an electron acceptor at the end of the chain that is different from O_2.)

Electron transfer from NADH to oxygen is an exergonic reaction with a free-energy change of −53 kcal/mol (−222 kJ/mol). Instead of this energy being released and wasted in a single explosive step, electrons cascade down the chain from one carrier molecule to the next in a series of redox reactions, losing a small amount of energy with each step until they finally reach oxygen, the terminal electron acceptor, which has a very great affinity for electrons. Each "downhill" carrier is more electronegative than, and thus capable of oxidizing, its "uphill" neighbor, with oxygen at the bottom of the chain. Therefore, the electrons transferred from glucose to NAD^+, which is thus reduced to NADH, fall down an energy gradient in the electron transport chain to a far more stable location in the electronegative oxygen atom. Put another way, oxygen pulls electrons down the chain in an energy-yielding tumble analogous to gravity pulling objects downhill.

In summary, during cellular respiration, most electrons travel the following "downhill" route: glucose → NADH → electron transport chain → oxygen. Later in this chapter, you will learn more about how the cell uses the energy released from this exergonic electron fall to regenerate its supply of ATP. For now, having covered the basic redox mechanisms of cellular respiration, let's look at the entire process by which energy is harvested from organic fuels.

Figure 9.5 An introduction to electron transport chains.

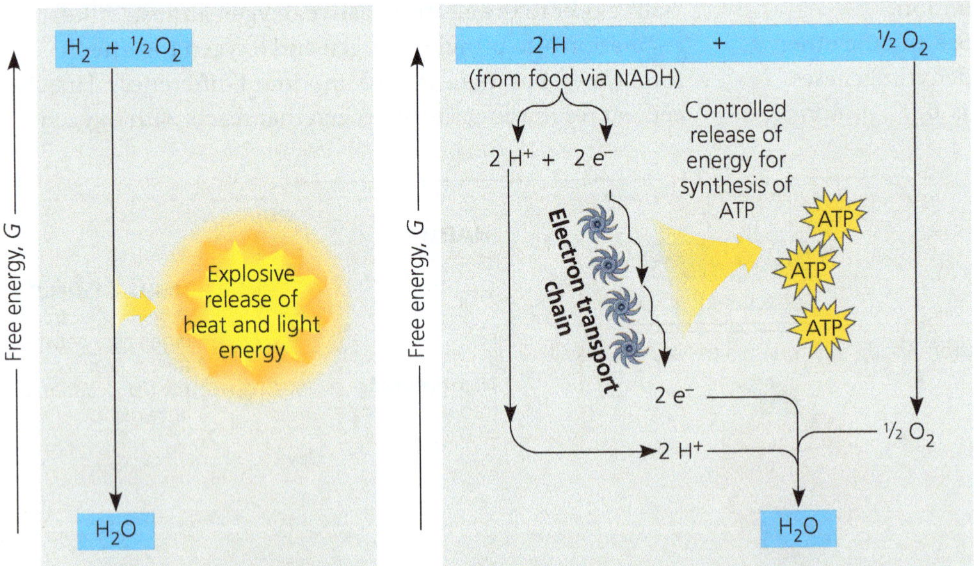

(a) Uncontrolled reaction. The one-step exergonic reaction of hydrogen with oxygen to form water releases a large amount of energy in the form of heat and light: an explosion.

(b) Cellular respiration. In cellular respiration, the same reaction occurs in stages: An electron transport chain breaks the "fall" of electrons in this reaction into a series of smaller steps and stores some of the released energy in a form that can be used to make ATP. (The rest of the energy is released as heat.)

The Stages of Cellular Respiration: *A Preview*

The harvesting of energy from glucose by cellular respiration is a cumulative function of three metabolic stages. We list them here along with a color-coding scheme we will use throughout the chapter to help you keep track of the big picture:

1. GLYCOLYSIS (color-coded blue throughout the chapter)
2. PYRUVATE OXIDATION and the CITRIC ACID CYCLE (color-coded light orange and dark orange)
3. OXIDATIVE PHOSPHORYLATION: Electron transport and chemiosmosis (color-coded purple)

Biochemists usually reserve the term *cellular respiration* for stages 2 and 3 together. In this text, however, we include glycolysis as a part of cellular respiration because most respiring cells deriving energy from glucose use glycolysis to produce the starting material for the citric acid cycle.

As diagrammed in **Figure 9.6**, glycolysis and then pyruvate oxidation and the citric acid cycle are the catabolic pathways that break down glucose and other organic fuels. **Glycolysis**, which occurs in the cytosol, begins the degradation process by breaking glucose into two molecules of a compound called pyruvate. In eukaryotes, pyruvate enters the mitochondrion and is oxidized to a compound called acetyl CoA, which enters the **citric acid cycle**. There, the breakdown of glucose to carbon dioxide is completed. (In prokaryotes, these processes take place in the cytosol.) Thus, the carbon dioxide produced by respiration represents fragments of oxidized organic molecules.

Some of the steps of glycolysis and the citric acid cycle are redox reactions in which dehydrogenases transfer electrons from substrates to NAD^+ or the related electron carrier FAD, forming NADH or $FADH_2$. (You'll learn more about FAD and $FADH_2$ later.) In the third stage of respiration, the electron transport chain

▶ **Figure 9.6 An overview of cellular respiration.** During glycolysis, each glucose molecule is broken down into two molecules of pyruvate. In eukaryotic cells, as shown here, the pyruvate enters the mitochondrion. There it is oxidized to acetyl CoA, which will be further oxidized to CO_2 in the citric acid cycle. The electron carriers NADH and $FADH_2$ transfer electrons derived from glucose to electron transport chains. During oxidative phosphorylation, electron transport chains convert the chemical energy to a form used for ATP synthesis in the process called chemiosmosis. (During earlier steps of cellular respiration, a few molecules of ATP are synthesized in a process called substrate-level phosphorylation.) To visualize these processes in their cellular context, see Figure 6.32.

Animation: Overview of Cellular Respiration

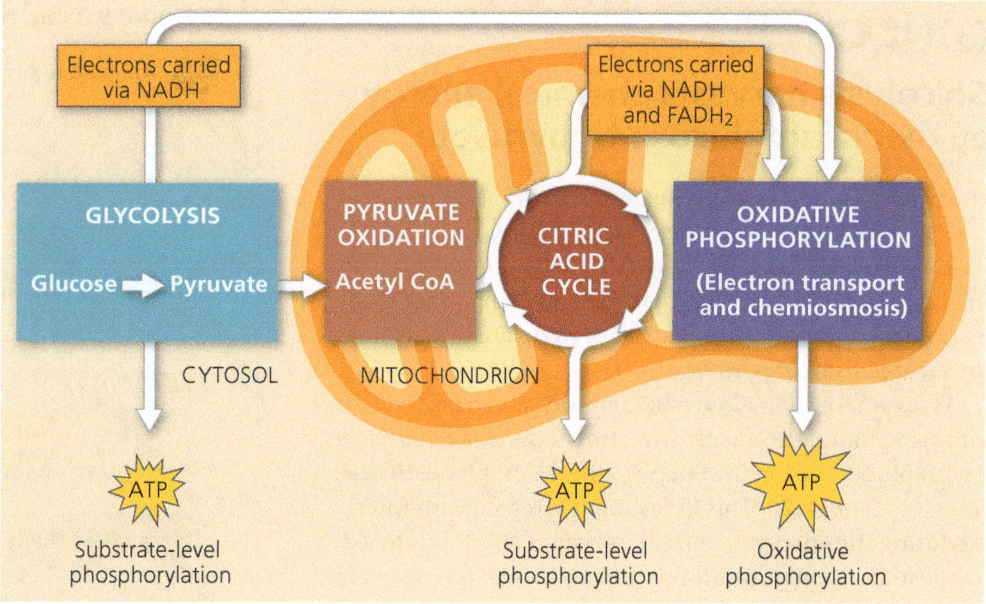

accepts electrons from NADH or $FADH_2$ generated during the first two stages and passes these electrons down the chain. At the end of the chain, the electrons are combined with molecular oxygen and hydrogen ions (H^+), forming water (see Figure 9.5b). The energy released at each step of the chain is stored in a form the mitochondrion (or prokaryotic cell) can use to make ATP from ADP. This mode of ATP synthesis is called **oxidative phosphorylation** because it is powered by the redox reactions of the electron transport chain.

In eukaryotic cells, the inner membrane of the mitochondrion is the site of electron transport and another process called *chemiosmosis*, together making up oxidative phosphorylation. (In prokaryotes, these processes take place in the plasma membrane.) Oxidative phosphorylation accounts for almost 90% of the ATP generated by respiration. A smaller amount of ATP is formed directly in a few reactions of glycolysis and the citric acid cycle by a mechanism called **substrate-level phosphorylation** (**Figure 9.7**). This mode of ATP synthesis

▼ **Figure 9.7 Substrate-level phosphorylation.** Some ATP is made by direct transfer of a phosphate group from an organic substrate to ADP by an enzyme. (For examples in glycolysis, see Figure 9.9, steps 7 and 10.)

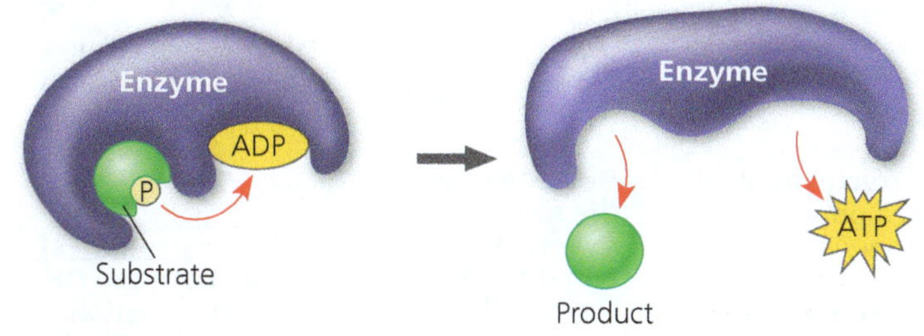

MAKE CONNECTIONS ▶ *Review Figure 8.9. Do you think the potential energy is higher for the reactants or the products in the reaction shown above? Explain.*

occurs when an enzyme transfers a phosphate group from a substrate molecule to ADP, rather than adding an inorganic phosphate to ADP as in oxidative phosphorylation. "Substrate molecule" here refers to an organic molecule generated as an intermediate during the catabolism of glucose. You'll see examples of substrate-level phosphorylation later in the chapter, in both glycolysis and the citric acid cycle.

You can think of the whole process this way: When you withdraw a relatively large sum of money from an ATM, it is not delivered to you in a single bill of larger denomination. Instead, a number of smaller-denomination bills are dispensed that you can spend more easily. This is analogous to ATP production during cellular respiration. For each molecule of glucose degraded to carbon dioxide and water by respiration, the cell makes up to about 32 molecules of ATP, each with 7.3 kcal/mol of free energy. Respiration cashes in the large denomination of energy banked in a single molecule of glucose (686 kcal/mol under standard conditions) for the small change of many molecules of ATP, which is more practical for the cell to spend on its work.

This preview has introduced you to how glycolysis, the citric acid cycle, and oxidative phosphorylation fit into the process of cellular respiration. We are now ready to take a closer look at each of these three stages of respiration.

CONCEPT CHECK 9.1

1. Compare and contrast aerobic and anaerobic respiration, including the processes involved.
2. **WHAT IF? ▶** If the following redox reaction occurred, which compound would be oxidized? Reduced?

$$C_4H_6O_5 + NAD^+ \rightarrow C_4H_4O_5 + NADH + H^+$$

For suggested answers, see Appendix A.

CONCEPT 9.2

Glycolysis harvests chemical energy by oxidizing glucose to pyruvate

The word *glycolysis* means "sugar splitting," and that is exactly what happens during this pathway. Glucose, a six-carbon sugar, is split into two three-carbon sugars. These smaller sugars are then oxidized and their remaining atoms rearranged to form two molecules of pyruvate. (Pyruvate is the ionized form of pyruvic acid.)

As summarized in **Figure 9.8**, glycolysis can be divided into two phases: the energy investment phase and the energy payoff phase. During the energy investment phase, the cell actually spends ATP. This investment is repaid with interest during the energy payoff phase, when ATP is produced by substrate-level phosphorylation and NAD^+ is reduced to NADH by electrons released from the oxidation of glucose. The net energy yield from glycolysis, per glucose molecule, is 2 ATP plus 2 NADH. The ten steps of the glycolytic pathway are shown in **Figure 9.9**.

All of the carbon originally present in glucose is accounted for in the two molecules of pyruvate; no carbon is released as CO_2 during glycolysis. Glycolysis occurs whether or not O_2 is present. However, if O_2 *is* present, the chemical energy stored in pyruvate and NADH can be extracted by pyruvate oxidation, the citric acid cycle, and oxidative phosphorylation.

Figure 9.8 The energy input and output of glycolysis.

Animation: Glycolysis

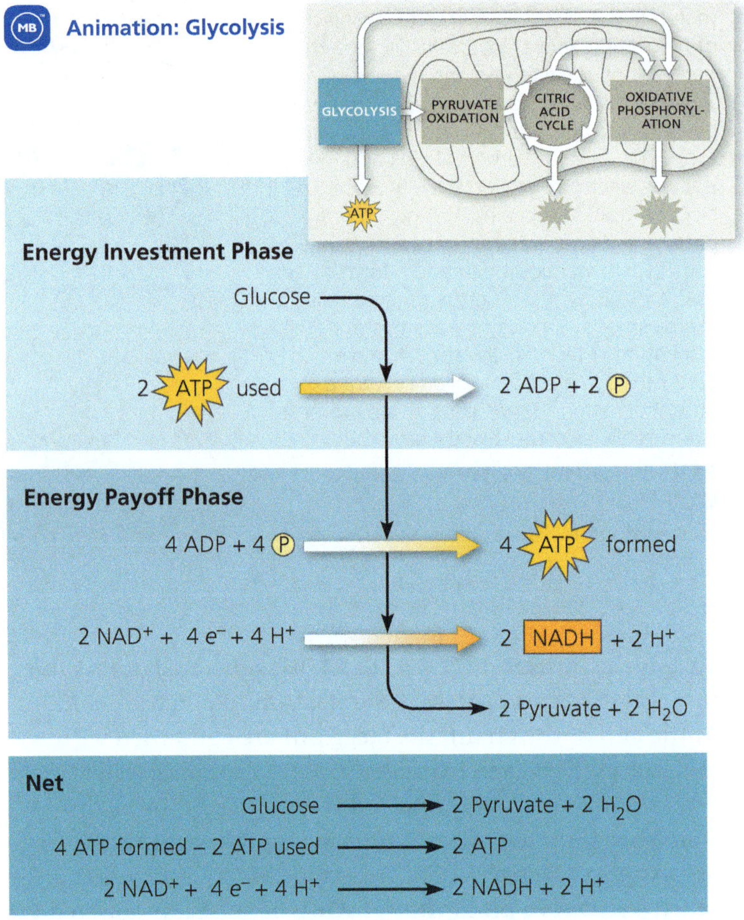

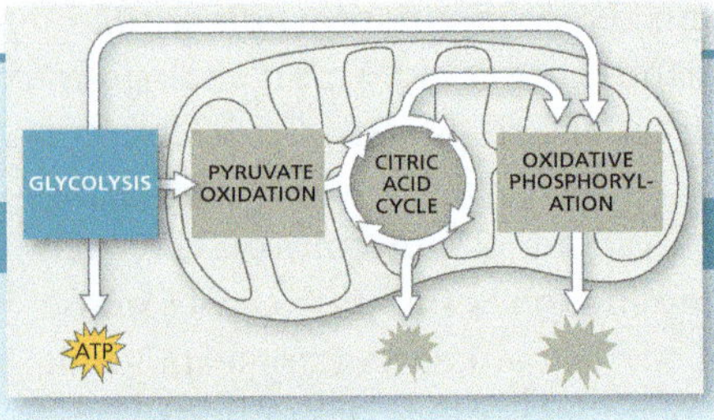

Figure 9.9 A closer look at glycolysis. Note that glycolysis is a source of ATP and NADH.

GLYCOLYSIS: Energy Investment Phase

WHAT IF? *What would happen if you removed the dihydroxyacetone phosphate generated in step 4 as fast as it was produced?*

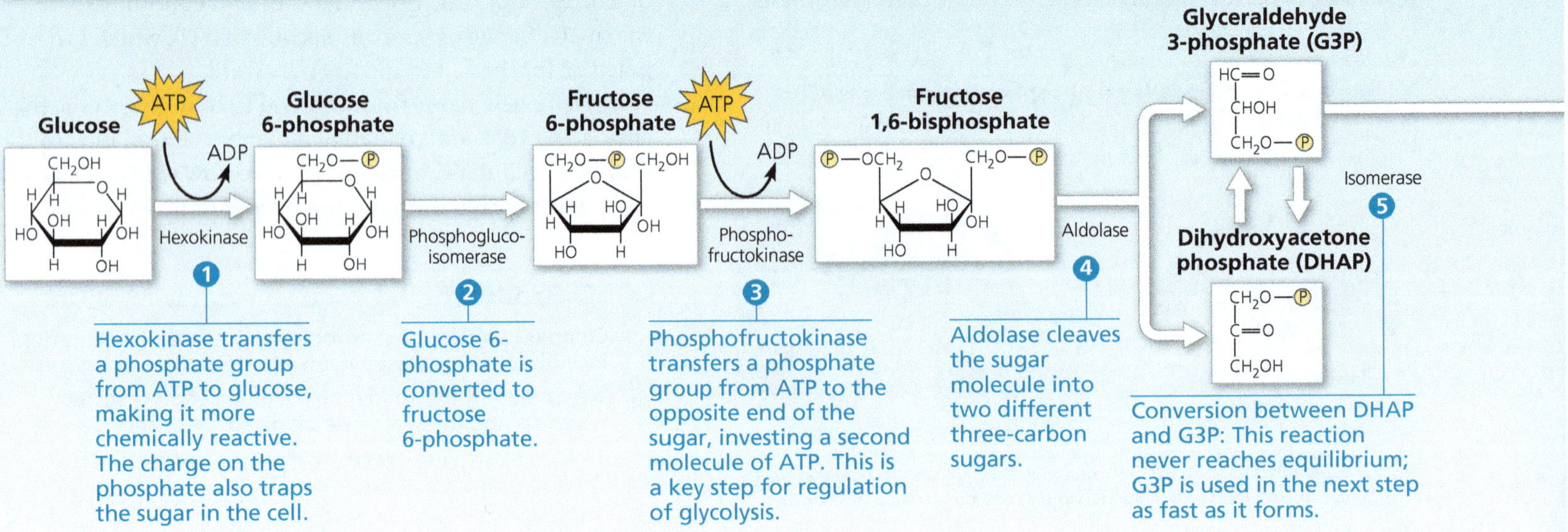

CONCEPT CHECK 9.2

1. **VISUAL SKILLS ➤** During the redox reaction in glycolysis (see step 6 in Figure 9.9), which molecule acts as the oxidizing agent? The reducing agent?

For suggested answers, see Appendix A.

CONCEPT 9.3

After pyruvate is oxidized, the citric acid cycle completes the energy-yielding oxidation of organic molecules

Glycolysis releases less than a quarter of the chemical energy in glucose that can be harvested by cells; most of the energy remains stockpiled in the two molecules of pyruvate. When O_2 is present, the pyruvate in eukaryotic cells enters a mitochondrion, where the oxidation of glucose is completed. In aerobically respiring prokaryotic cells, this process occurs in the cytosol. (Later in the chapter, we'll discuss what happens to pyruvate when O_2 is unavailable or in a prokaryote that is unable to use O_2.)

Oxidation of Pyruvate to Acetyl CoA

Upon entering the mitochondrion via active transport, pyruvate is first converted to a compound called acetyl coenzyme A, or **acetyl CoA** (**Figure 9.10**). This step, linking glycolysis and the citric acid cycle, is carried out by a multienzyme complex that catalyzes three reactions: ❶ Pyruvate's carboxyl

▼ Figure 9.10 Oxidation of pyruvate to acetyl CoA, the step before the citric acid cycle. Pyruvate is a charged molecule, so in eukaryotic cells it must enter the mitochondrion via active transport, with the help of a transport protein. Next, a complex of several enzymes (the pyruvate dehydrogenase complex) catalyzes the three numbered steps, which are described in the text. The acetyl group of acetyl CoA will enter the citric acid cycle. The CO_2 molecule will diffuse out of the cell. By convention, coenzyme A is abbreviated S-CoA when it is attached to a molecule, emphasizing the sulfur atom (S). (The gene encoding the transport protein was finally identified several years ago, after 40 years of research.)

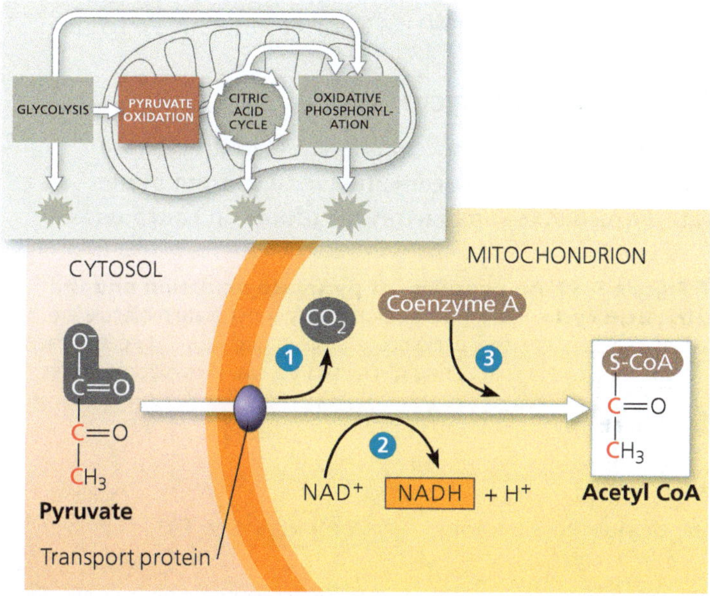

BioFlix® Animation: Acetyl CoA

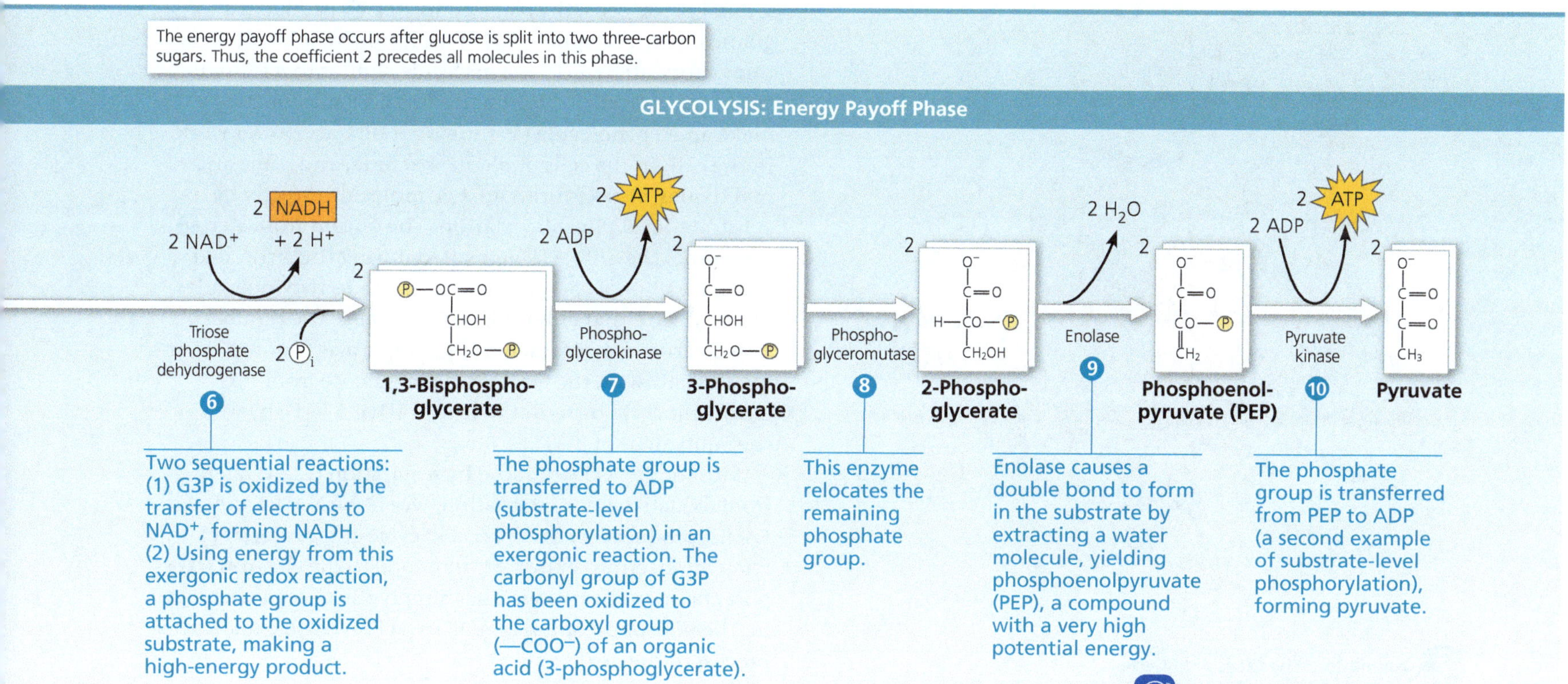

BioFlix® Animation: Glycolysis

group (—COO^-), already somewhat oxidized and thus carrying little chemical energy, is now fully oxidized and given off as a molecule of CO_2. This is the first step in which CO_2 is released during respiration. ❷ Next, the remaining two-carbon fragment is oxidized and the electrons transferred to NAD^+, storing energy in the form of NADH. ❸ Finally, coenzyme A (CoA), a sulfur-containing compound derived from a B vitamin, is attached via its sulfur atom to the two-carbon intermediate, forming acetyl CoA. Acetyl CoA has a high potential energy, which is used to transfer the acetyl group to a molecule in the citric acid cycle, a reaction that is therefore highly exergonic.

The Citric Acid Cycle

The citric acid cycle functions as a metabolic furnace that further oxidizes organic fuel derived from pyruvate. **Figure 9.11** summarizes the inputs and outputs as

▼ Figure 9.11 An overview of pyruvate oxidation and the citric acid cycle. The inputs and outputs per pyruvate molecule are shown. To calculate on a per-glucose basis, multiply by 2, because each glucose molecule is split during glycolysis into two pyruvate molecules.

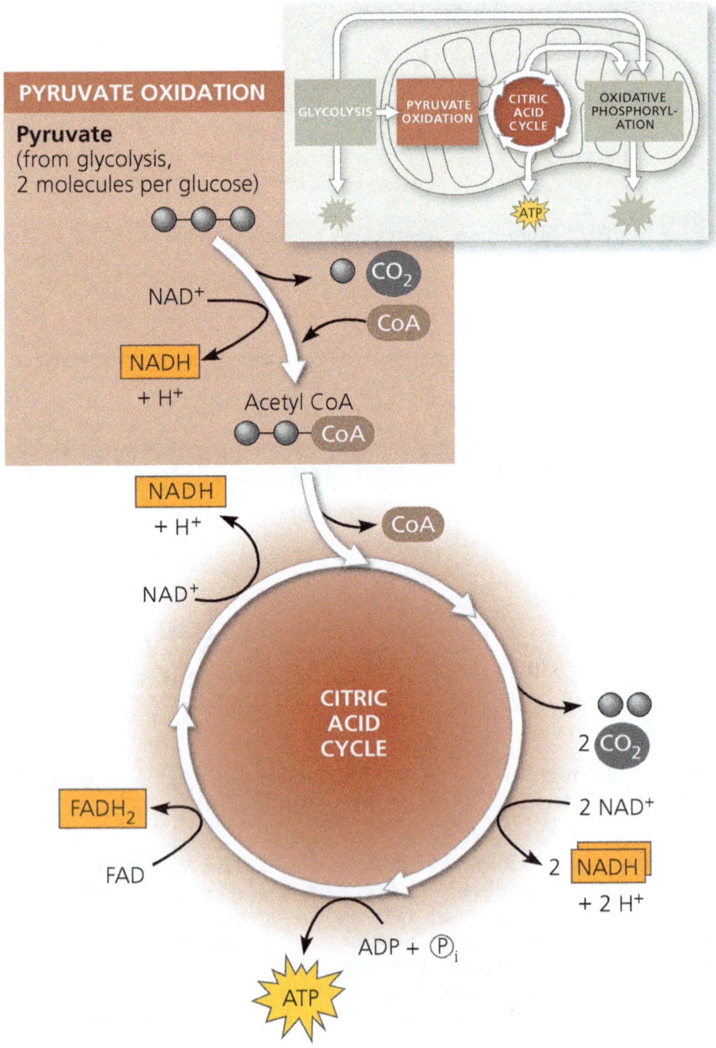

Animation: The Citric Acid Cycle

pyruvate is broken down to three CO_2 molecules, including the molecule of CO_2 released during the conversion of pyruvate to acetyl CoA. The cycle generates 1 ATP per turn by substrate-level phosphorylation, but most of the chemical energy is transferred to NAD^+ and FAD during the redox reactions. The reduced coenzymes, NADH and $FADH_2$, shuttle their cargo of high-energy electrons into the electron transport chain. The citric acid cycle is also called the tricarboxylic acid cycle or the Krebs cycle, the latter honoring Hans Krebs, the German-British scientist who was largely responsible for working out the pathway in the 1930s.

Now let's look at the citric acid cycle in more detail. The cycle has eight steps, each catalyzed by a specific enzyme. You can see in **Figure 9.12** that for each turn of the citric acid cycle, two carbons (red) enter in the relatively reduced form of an acetyl group (step ❶), and two different carbons (blue) leave in the completely oxidized form of CO_2 molecules (steps ❸ and ❹). The acetyl group of acetyl CoA joins the cycle by combining with the compound oxaloacetate, forming citrate (step ❶). Citrate is the ionized form of citric acid, for which the cycle is named. The next seven steps decompose the citrate back to oxaloacetate. It is this regeneration of oxaloacetate that makes the process a *cycle*.

Referring to Figure 9.12, we can tally the energy-rich molecules produced by the citric acid cycle. For each acetyl group entering the cycle, 3 NAD^+ are reduced to NADH (steps ❸, ❹, and ❽). In step ❻, electrons are transferred not to NAD^+, but to FAD, which accepts 2 electrons and 2 protons to become $FADH_2$. In many animal tissue cells, the reaction in step ❺ produces a guanosine triphosphate (GTP) molecule by substrate-level phosphorylation. GTP is a molecule similar to ATP in its structure and cellular function. This GTP may be used to make an ATP molecule (as shown) or directly power work in the cell. In the cells of plants, bacteria, and some animal tissues, step ❺ forms an ATP molecule directly by substrate-level phosphorylation. The output from step ❺ represents the only ATP generated during the citric acid cycle. Recall that each glucose gives rise to two molecules of acetyl CoA that enter the cycle. Because the numbers noted earlier are obtained from a single acetyl group entering the pathway, the total yield per glucose from the citric acid cycle turns out to be 6 NADH, 2 $FADH_2$, and the equivalent of 2 ATP.

Most of the ATP produced by respiration results later, from oxidative phosphorylation, when the NADH and $FADH_2$ produced by the citric acid cycle and earlier steps relay the electrons extracted from food to the electron transport chain. In the process, they supply the necessary energy for the phosphorylation of ADP to ATP. We will explore this process in the next section.

Figure 9.12 A closer look at the citric acid cycle. In the chemical structures, red type traces the fate of the two carbon atoms that enter the cycle via acetyl CoA (step 1), and blue type indicates the two carbons that exit the cycle as CO_2 in steps 3 and 4. (The red type goes only through step 5 because the succinate molecule is symmetrical; the two ends cannot be distinguished from each other.) Notice that the carbon atoms that enter the cycle from acetyl CoA do not leave the cycle in the same turn. They remain in the cycle, occupying a different location in the molecules on their next turn, after another acetyl group is added. Therefore, the oxaloacetate regenerated at step 8 is made up of different carbon atoms each time around. In eukaryotic cells, all the citric acid cycle enzymes are located in the mitochondrial matrix except for the enzyme that catalyzes step 6, which resides in the inner mitochondrial membrane. Carboxylic acids are represented in their ionized forms, as —COO^-, because the ionized forms prevail at the pH within the mitochondrion.

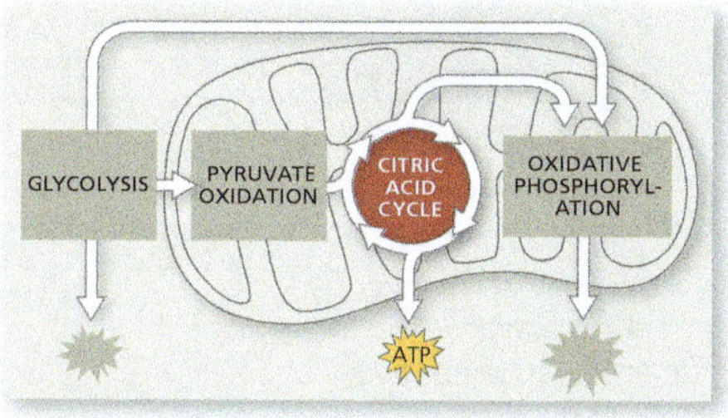

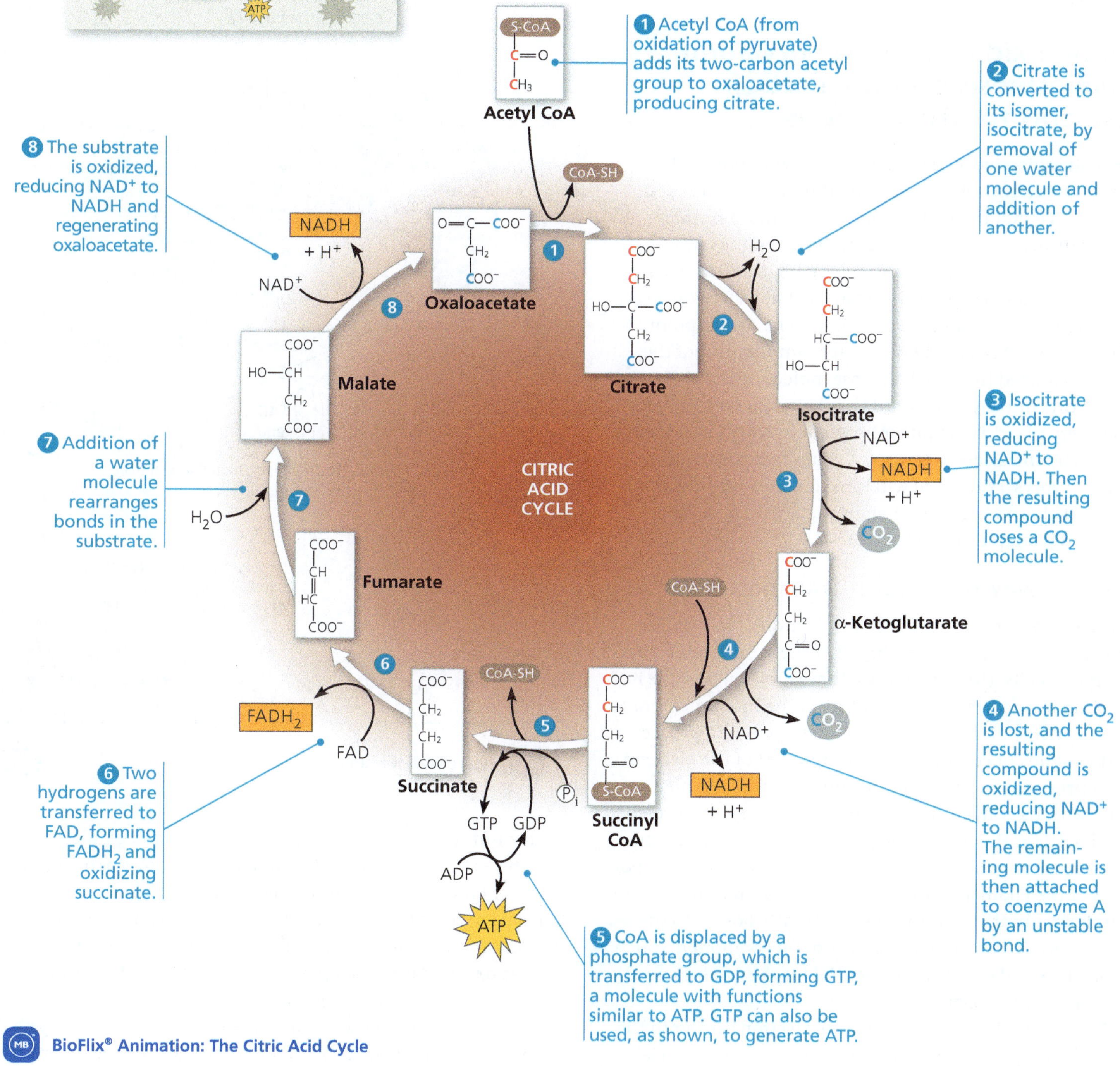

MB BioFlix® Animation: The Citric Acid Cycle

CONCEPT CHECK 9.3

1. **VISUAL SKILLS >** Name the molecules that conserve most of the energy from the redox reactions of the citric acid cycle (see Figure 9.12). How is this energy converted to a form that can be used to make ATP?
2. What processes in your cells produce the CO_2 that you exhale?
3. **VISUAL SKILLS >** The conversions shown in Figure 9.10 and step 4 of Figure 9.12 are each catalyzed by a large multienzyme complex. What similarities are there in the reactions that occur in these two cases?

For suggested answers, see Appendix A.

CONCEPT 9.4

During oxidative phosphorylation, chemiosmosis couples electron transport to ATP synthesis

Our main objective in this chapter is to learn how cells harvest the energy of glucose and other nutrients in food to make ATP. But the metabolic components of respiration we have examined so far, glycolysis and the citric acid cycle, produce only 4 ATP molecules per glucose molecule, all by substrate-level phosphorylation: 2 net ATP from glycolysis and 2 ATP from the citric acid cycle. At this point, molecules of NADH (and $FADH_2$) account for most of the energy extracted from each glucose molecule. These electron escorts link glycolysis and the citric acid cycle to the machinery of oxidative phosphorylation, which uses energy released by the electron transport chain to power ATP synthesis. In this section, you will learn first how the electron transport chain works and then how electron flow down the chain is coupled to ATP synthesis.

The Pathway of Electron Transport

The electron transport chain is a collection of molecules embedded in the inner membrane of the mitochondrion in eukaryotic cells. (In prokaryotes, these molecules reside in the plasma membrane.) The folding of the inner membrane to form cristae increases its surface area, providing space for thousands of copies of each component of the electron transport chain in a mitochondrion. The infolded membrane with its concentration of electron carrier molecules is well-suited for the series of sequential redox reactions that take place along the electron transport chain. Most components of the chain are proteins, which exist in multiprotein complexes numbered I through IV. Tightly bound to these proteins are *prosthetic groups*, nonprotein components such as cofactors and coenzymes essential for the catalytic functions of certain enzymes.

Figure 9.13 shows the sequence of electron carriers in the electron transport chain and the drop in free energy as electrons travel down the chain. During this electron transport,

Figure 9.13 Free-energy change during electron transport. The overall energy drop (ΔG) for electrons traveling from NADH to oxygen is 53 kcal/mol, but this "fall" is broken up into a series of smaller steps by the electron transport chain. (An oxygen atom is represented here as $\frac{1}{2}$ O_2 to emphasize that the electron transport chain reduces molecular oxygen, O_2, not individual oxygen atoms.)

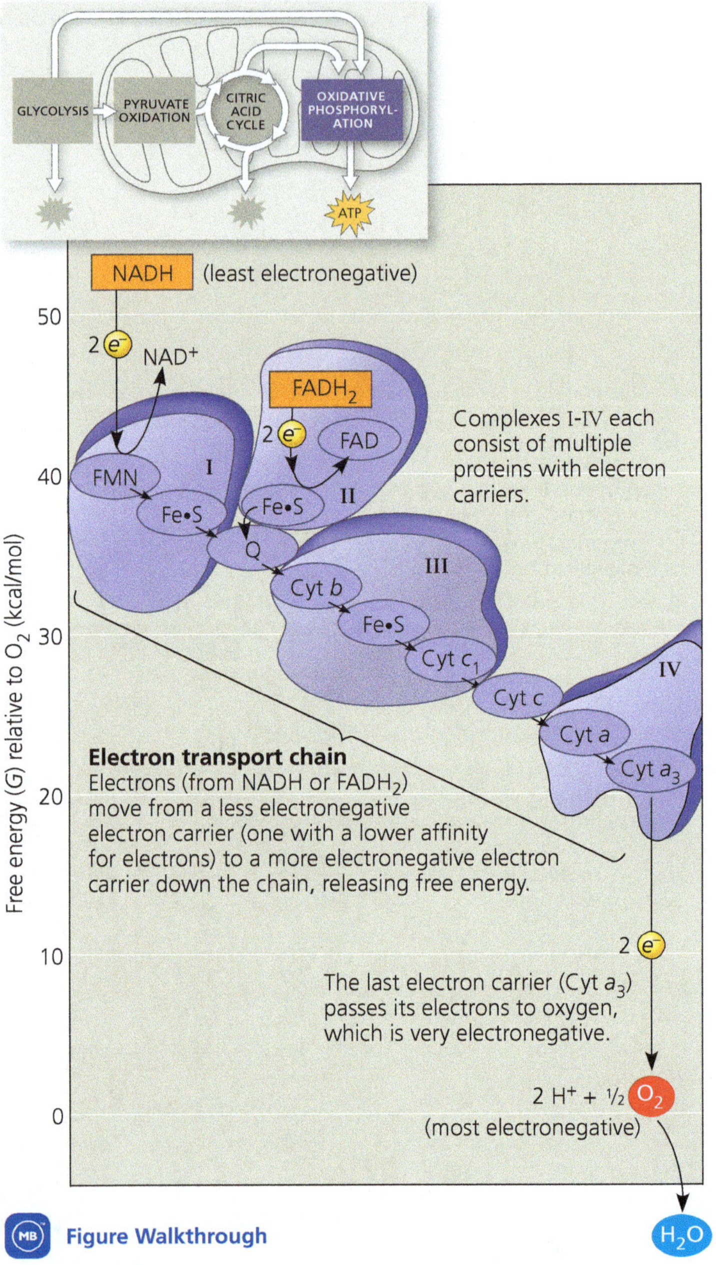

MB Figure Walkthrough

electron carriers alternate between reduced and oxidized states as they accept and then donate electrons. Each component of the chain becomes reduced when it accepts electrons from its "uphill" neighbor, which has a lower affinity for electrons (in other words, is less electronegative). It then returns to its oxidized form as it passes electrons to its "downhill," more electronegative neighbor.

Now let's take a closer look at the electrons as they drop in energy level, passing through the components of the electron transport chain in Figure 9.13. We'll first describe the passage of electrons through complex I in some detail as an illustration of the general principles involved in electron transport. Electrons acquired from glucose by NAD^+ during glycolysis and the citric acid cycle are transferred from NADH to the first molecule of the electron transport chain in complex I. This molecule is a flavoprotein, so named because it has a prosthetic group called flavin mononucleotide (FMN). In the next redox reaction, the flavoprotein returns to its oxidized form as it passes electrons to an iron-sulfur protein (Fe•S in complex I), one of a family of proteins with both iron and sulfur tightly bound. The iron-sulfur protein then passes the electrons to a compound called ubiquinone (Q in Figure 9.13). This electron carrier is a small hydrophobic molecule, the only member of the electron transport chain that is not a protein. Ubiquinone is individually mobile within the membrane rather than residing in a particular complex. (Another name for ubiquinone is coenzyme Q, or CoQ; you may have seen it sold as a nutritional supplement.)

Most of the remaining electron carriers between ubiquinone and oxygen are proteins called **cytochromes**. Their prosthetic group, called a heme group, has an iron atom that accepts and donates electrons. (The heme group in a cytochrome is similar to the heme group in hemoglobin, the protein of red blood cells, except that the iron in hemoglobin carries oxygen, not electrons.) The electron transport chain has several types of cytochromes, each named "cyt" with a letter and number to distinguish it as a different protein with a slightly different electron-carrying heme group. The last cytochrome of the chain, Cyt a_3, passes its electrons to oxygen, which is *very* electronegative. Each oxygen atom also picks up a pair of hydrogen ions (protons) from the aqueous solution, neutralizing the −2 charge of the added electrons and forming water.

Another source of electrons for the electron transport chain is $FADH_2$, the other reduced product of the citric acid cycle. Notice in Figure 9.13 that $FADH_2$ adds its electrons from within complex II, at a lower energy level than NADH does. Consequently, although NADH and $FADH_2$ each donate an equivalent number of electrons (2) for oxygen reduction, the electron transport chain provides about one-third less energy for ATP synthesis when the electron donor is $FADH_2$ rather than NADH. We'll see why in the next section.

The electron transport chain makes no ATP directly. Instead, it eases the fall of electrons from food to oxygen, breaking a large free-energy drop into a series of smaller steps that release energy in manageable amounts, step by step. How does the mitochondrion (or the plasma membrane in prokaryotes) couple this electron transport and energy release to ATP synthesis? The answer is a mechanism called chemiosmosis.

Chemiosmosis: The Energy-Coupling Mechanism

Populating the inner membrane of the mitochondrion or the prokaryotic plasma membrane are many copies of a protein complex called **ATP synthase**, the enzyme that makes ATP from ADP and inorganic phosphate **(Figure 9.14)**. ATP synthase works like an ion pump running in reverse. Ion pumps usually use ATP as an energy source to transport ions against their gradients. Enzymes can catalyze a reaction in either direction, depending on the ΔG for the reaction, which is affected by the local concentrations of reactants and products (see Concepts 8.2 and 8.3). Rather than hydrolyzing ATP to pump protons against their concentration gradient, under the conditions of cellular respiration ATP synthase uses the energy of an existing ion gradient to power ATP synthesis. The power source for ATP synthase is a difference in the concentration of H^+ on opposite sides of the inner mitochondrial membrane. This process, in which energy stored in the form of a hydrogen ion gradient across a membrane is used to drive cellular work such as the synthesis of ATP, is called **chemiosmosis** (from

Figure 9.14 ATP synthase, a molecular mill. Multiple ATP synthases reside in eukaryotic mitochondrial and chloroplast membranes and in prokaryotic plasma membranes.

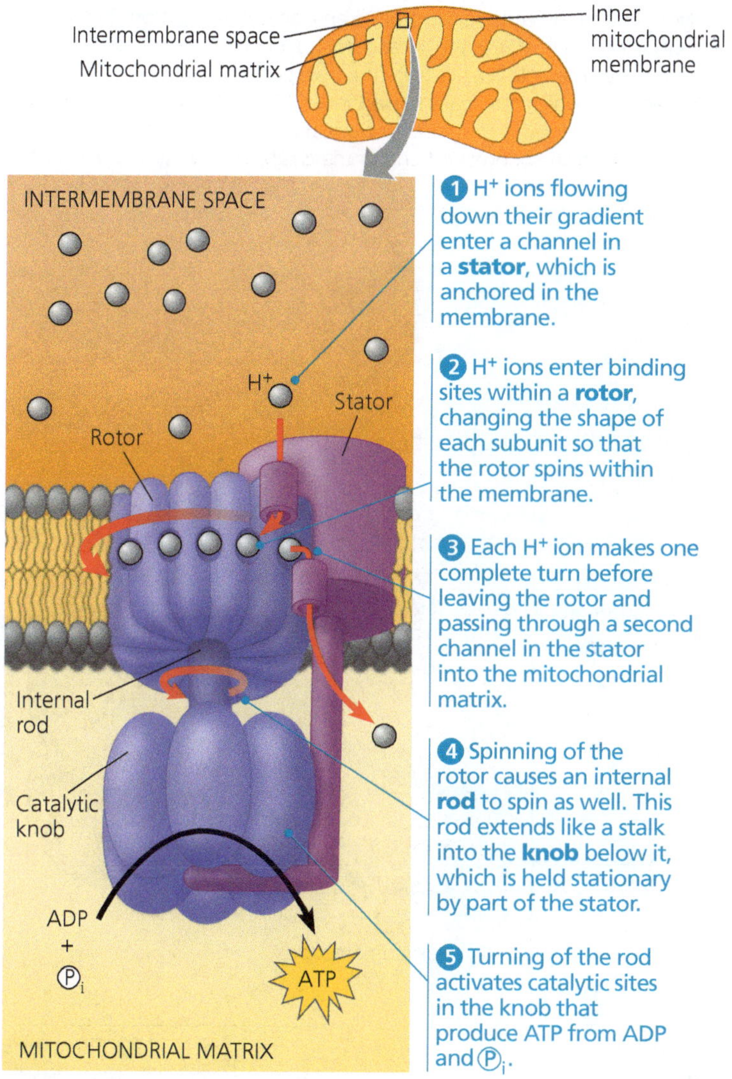

BioFlix® Animation: ATP Synthase
Animation: Rotating ATP Synthase

Figure 9.15 Chemiosmosis couples the electron transport chain to ATP synthesis. ❶ NADH and $FADH_2$ shuttle high-energy electrons extracted from food during glycolysis and the citric acid cycle into an electron transport chain built into the inner mitochondrial membrane. The gold arrows trace the transport of electrons, which are finally passed to a terminal acceptor (O_2, in the case of aerobic respiration) at the "downhill" end of the chain, forming water. Most of the electron carriers of the chain are grouped into four complexes (I–IV). Two mobile carriers, ubiquinone (Q) and cytochrome *c* (Cyt *c*), move rapidly, ferrying electrons between the large complexes. As the complexes shuttle electrons, they pump protons from the mitochondrial matrix into the intermembrane space. $FADH_2$ deposits its electrons via complex II and so results in fewer protons being pumped into the intermembrane space than occurs with NADH. Chemical energy originally harvested from food is transformed into a proton-motive force, a gradient of H^+ across the membrane. ❷ During chemiosmosis, the protons flow back down their gradient via ATP synthase, which is built into the membrane nearby. The ATP synthase harnesses the proton-motive force to phosphorylate ADP, forming ATP. Together, electron transport and chemiosmosis make up oxidative phosphorylation.

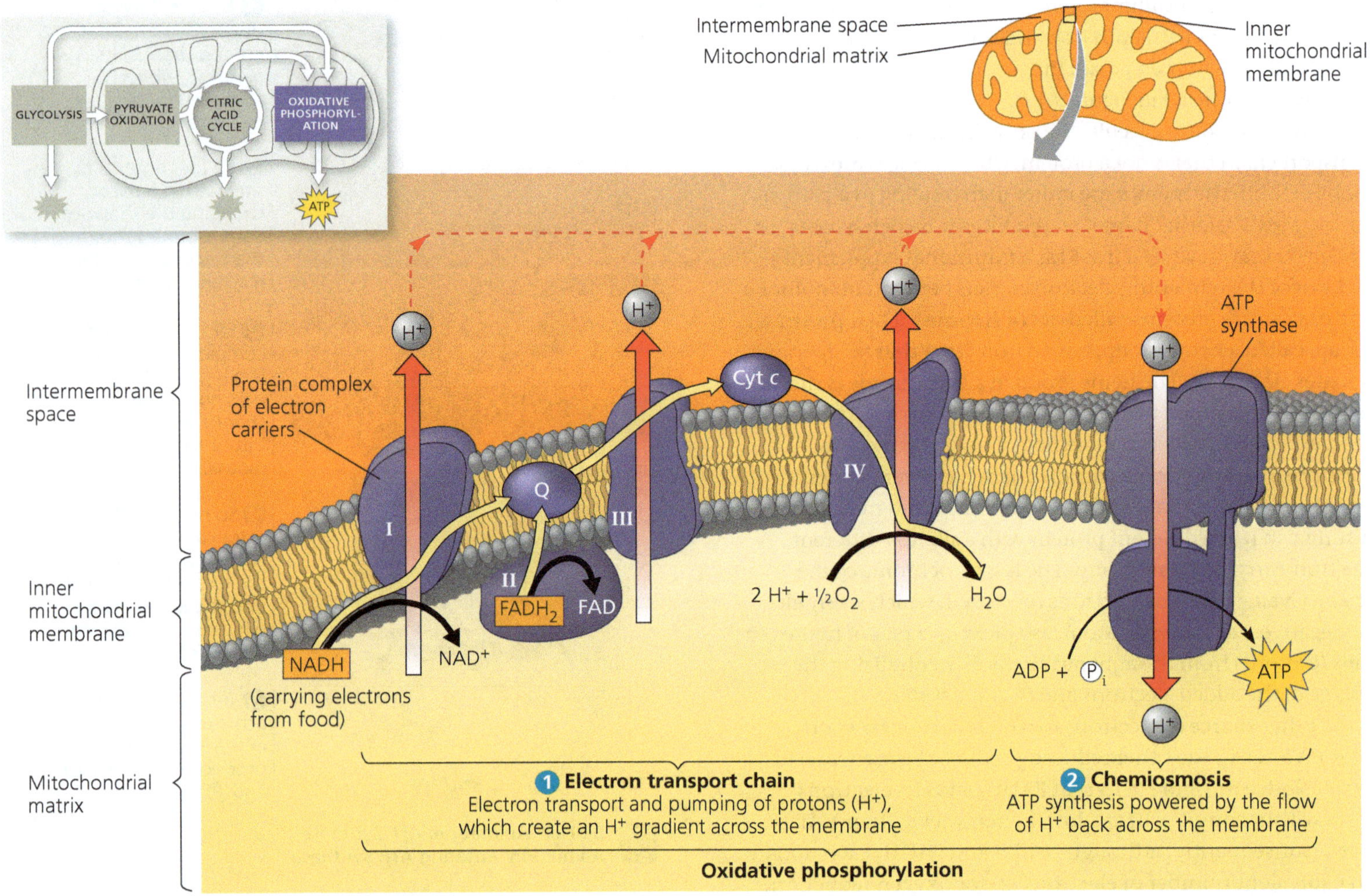

WHAT IF ➤ *If complex IV were nonfunctional, could chemiosmosis produce any ATP, and if so, how would the rate of synthesis differ?*

Animation: Electron Transport
BioFlix® Animation: Electron Transport

the Greek *osmos*, push). We have previously used the word *osmosis* in discussing water transport, but here it refers to the flow of H^+ across a membrane.

From studying the structure of ATP synthase, scientists have learned how the flow of H^+ through this large enzyme powers ATP generation. ATP synthase is a multisubunit complex with four main parts, each made up of multiple polypeptides (see Figure 9.14). Protons move one by one into binding sites on one of the parts (the rotor), causing it to spin in a way that catalyzes ATP production from ADP and inorganic phosphate. The flow of protons thus behaves somewhat like a rushing stream that turns a waterwheel.

How does the inner mitochondrial membrane or the prokaryotic plasma membrane generate and maintain the H^+ gradient that drives ATP synthesis by the ATP synthase protein complex? Establishing the H^+ gradient is a major function of the electron transport chain, which is shown in its mitochondrial location in **Figure 9.15**. The chain is an energy converter that uses the exergonic flow of electrons from NADH and $FADH_2$ to pump H^+ across the membrane, from the mitochondrial matrix into the intermembrane space. The H^+ has a tendency to move back across the membrane, diffusing down its gradient. And the ATP synthases are the only sites that provide a route through the membrane for H^+. As we described previously, the passage of H^+ through ATP synthase uses the exergonic flow of H^+ to drive the phosphorylation of ADP. Thus, the energy stored in an H^+ gradient across a membrane couples the redox reactions of the electron transport chain to ATP synthesis.

At this point, you may be wondering how the electron transport chain pumps hydrogen ions. Researchers have found that certain members of the electron transport chain accept and release protons (H^+) along with electrons. (The aqueous solutions inside and surrounding the cell are a ready source of H^+.) At certain steps along the chain, electron transfers cause H^+ to be taken up and released into the surrounding solution. In eukaryotic cells, the electron carriers are spatially arranged in the inner mitochondrial membrane in such a way that H^+ is accepted from the mitochondrial matrix and deposited in the intermembrane space (see Figure 9.15). The H^+ gradient that results is referred to as a **proton-motive force**, emphasizing the capacity of the gradient to perform work. The force drives H^+ back across the membrane through the H^+ channels provided by ATP synthases.

In general terms, *chemiosmosis is an energy-coupling mechanism that uses energy stored in the form of an H^+ gradient across a membrane to drive cellular work*. In mitochondria, the energy for gradient formation comes from exergonic redox reactions along the electron transport chain, and ATP synthesis is the work performed. But chemiosmosis also occurs elsewhere and in other variations. Chloroplasts use chemiosmosis to generate ATP during photosynthesis; in these organelles, light (rather than chemical energy) drives both electron flow down an electron transport chain and the resulting H^+ gradient formation. Prokaryotes, as already mentioned, generate H^+ gradients across their plasma membranes. They then tap the proton-motive force not only to make ATP inside the cell but also to rotate their flagella and to pump nutrients and waste products across the membrane. Because of its central importance to energy conversions in prokaryotes and eukaryotes, chemiosmosis has helped unify the study of bioenergetics. Peter Mitchell was awarded the Nobel Prize in 1978 for originally proposing the chemiosmotic model.

An Accounting of ATP Production by Cellular Respiration

In the last few sections, we have looked rather closely at the key processes of cellular respiration. Now let's take a step back and remind ourselves of its overall function: harvesting the energy of glucose for ATP synthesis.

During respiration, most energy flows in this sequence: glucose → NADH → electron transport chain → proton-motive force → ATP. We can do some bookkeeping to calculate the ATP profit when cellular respiration oxidizes a molecule of glucose to six molecules of carbon dioxide. The three main departments of this metabolic enterprise are glycolysis, pyruvate oxidation and the citric acid cycle, and the electron transport chain, which drives oxidative phosphorylation. **Figure 9.16** gives a detailed accounting of the ATP yield for each glucose molecule that is oxidized. The tally adds the 4 ATP produced directly by substrate-level phosphorylation during glycolysis and the citric acid cycle to the many more molecules of ATP generated by oxidative

Figure 9.16 ATP yield per molecule of glucose at each stage of cellular respiration.

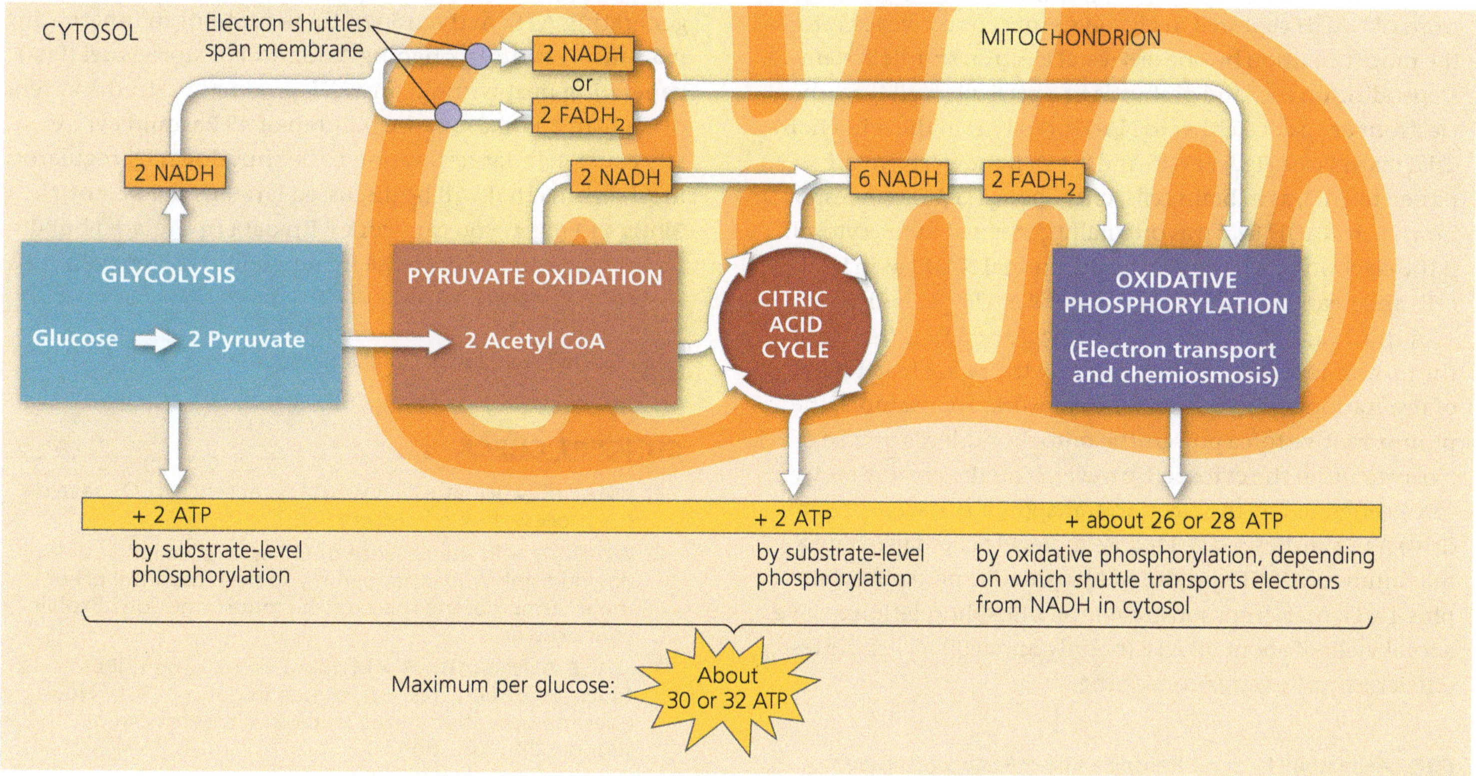

VISUAL SKILLS ➤ *Explain exactly how the total of 26 or 28 ATP from oxidative phosphorylation was calculated (see the yellow bar in the figure).*

Animation: ATP Yield from Cellular Respiration

phosphorylation. Each NADH that transfers a pair of electrons from glucose to the electron transport chain contributes enough to the proton-motive force to generate a maximum of about 3 ATP.

Why are the numbers in Figure 9.16 inexact? There are three reasons we cannot state an exact number of ATP molecules generated by the breakdown of one molecule of glucose. First, phosphorylation and the redox reactions are not directly coupled to each other, so the ratio of the number of NADH molecules to the number of ATP molecules is not a whole number. We know that 1 NADH results in 10 H^+ being transported out across the inner mitochondrial membrane, but the exact number of H^+ that must reenter the mitochondrial matrix via ATP synthase to generate 1 ATP has long been debated. Based on experimental data, however, most biochemists now agree that the most accurate number is 4 H^+. Therefore, a single molecule of NADH generates enough proton-motive force for the synthesis of 2.5 ATP. The citric acid cycle also supplies electrons to the electron transport chain via $FADH_2$, but since its electrons enter later in the chain, each molecule of this electron carrier is responsible for transport of only enough H^+ for the synthesis of 1.5 ATP. These numbers also take into account the slight energetic cost of moving the ATP formed in the mitochondrion out into the cytosol, where it will be used.

Second, the ATP yield varies slightly depending on the type of shuttle used to transport electrons from the cytosol into the mitochondrion. The mitochondrial inner membrane is impermeable to NADH, so NADH in the cytosol is segregated from the machinery of oxidative phosphorylation. The 2 electrons of NADH captured in glycolysis must be conveyed into the mitochondrion by one of several electron shuttle systems. Depending on the kind of shuttle in a particular cell type, the electrons are passed either to NAD^+ or to FAD in the mitochondrial matrix (see Figures 9.15 and 9.16). If the electrons are passed to FAD, as in brain cells, only about 1.5 ATP can result from each NADH that was originally generated in the cytosol. If the electrons are passed to mitochondrial NAD^+, as in liver cells and heart cells, the yield is about 2.5 ATP per NADH.

A third variable that reduces the yield of ATP is the use of the proton-motive force generated by the redox reactions of respiration to drive other kinds of work. For example, the proton-motive force powers the mitochondrion's uptake of pyruvate from the cytosol. However, if *all* the proton-motive force generated by the electron transport chain were used to drive ATP synthesis, one glucose molecule could generate a maximum of 28 ATP produced by oxidative phosphorylation plus 4 ATP (net) from substrate-level phosphorylation to give a total yield of about 32 ATP (or only about 30 ATP if the less efficient shuttle were functioning).

MB Animation: Electron Transport Chain: Factors Affecting ATP Yield

We can now roughly estimate the efficiency of respiration—that is, the percentage of chemical energy in glucose that has been transferred to ATP. Recall that the complete oxidation of a mole of glucose releases 686 kcal of energy under standard conditions ($\Delta G = -686$ kcal/mol). Phosphorylation of ADP to form ATP stores at least 7.3 kcal per mole of ATP. Therefore, the efficiency of respiration is 7.3 kcal per mole of ATP times 32 moles of ATP per mole of glucose divided by 686 kcal per mole of glucose, which equals 0.34. Thus, about 34% of the potential chemical energy in glucose has been transferred to ATP; the actual percentage is bound to vary as ΔG varies under different cellular conditions. Cellular respiration is remarkably efficient in its energy conversion. By comparison, even the most efficient automobile converts only about 25% of the energy stored in gasoline to energy that moves the car.

The rest of the energy stored in glucose is lost as heat. We humans use some of this heat to maintain our relatively high body temperature (37°C), and we dissipate the rest through sweating and other cooling mechanisms.

Surprisingly, perhaps, it may be beneficial under certain conditions to reduce the efficiency of cellular respiration. A remarkable adaptation is shown by hibernating mammals, which overwinter in a state of inactivity and lowered metabolism. Although their internal body temperature is lower than normal, it still must be kept significantly higher than the external air temperature. One type of tissue, called brown fat, is made up of cells packed full of mitochondria. The inner mitochondrial membrane contains a channel protein called the uncoupling protein that allows protons to flow back down their concentration gradient without generating ATP. Activation of these proteins in hibernating mammals results in ongoing oxidation of stored fuel (fats), generating heat without any ATP production. In the absence of such an adaptation, the buildup of ATP would eventually cause cellular respiration to be shut down by regulatory mechanisms that will be discussed later. In the **Scientific Skills Exercise**, you can work with data in a related but different case where a decrease in metabolic efficiency in cells is used to generate heat.

MB BioFlix® Animation: Cellular Respiration

CONCEPT CHECK 9.4

1. **WHAT IF? >** What effect would an absence of O_2 have on the process shown in Figure 9.15?
2. **WHAT IF? >** In the absence of O_2, as in question 1, what do you think would happen if you decreased the pH of the intermembrane space of the mitochondrion? Explain your answer.
3. **MAKE CONNECTIONS >** Membranes must be fluid to function properly (as you learned in Concept 7.1). How does the operation of the electron transport chain support that assertion?

For suggested answers, see Appendix A.

SCIENTIFIC SKILLS EXERCISE

Making a Bar Graph and Evaluating a Hypothesis

Does Thyroid Hormone Level Affect Oxygen Consumption in Cells? Some animals, such as mammals and birds, maintain a relatively constant body temperature, above that of their environment, by using heat produced as a by-product of metabolism. When the core temperature of these animals drops below an internal set point, their cells are triggered to reduce the efficiency of ATP production by the electron transport chains in mitochondria. At lower efficiency, extra fuel must be consumed to produce the same number of ATPs, generating additional heat. Because this response is moderated by the endocrine system, researchers hypothesized that thyroid hormone might trigger this cellular response. In this exercise, you will use a bar graph to visualize data from an experiment that compared the metabolic rates (by measuring oxygen consumption) in mitochondria of cells from animals with different levels of thyroid hormone.

How the Experiment Was Done Liver cells were isolated from sibling rats that had low, normal, or elevated thyroid hormone levels. The oxygen consumption rate due to activity of the mitochondrial electron transport chains of each type of cell was measured under controlled conditions.

Data from the Experiment

Thyroid Hormone Level	Oxygen Consumption Rate [nmol O_2/(min · mg cells)]
Low	4.3
Normal	4.8
Elevated	8.7

Data from M. E. Harper and M. D. Brand, The quantitative contributions of mitochondrial proton leak and ATP turnover reactions to the changed respiration rates of hepatocytes from rats of different thyroid status, *Journal of Biological Chemistry* 268:14850–14860 (1993).

INTERPRET THE DATA

1. To visualize any differences in oxygen consumption between cell types, it will be useful to graph the data in a bar graph. First, set up the axes. (a) What is the independent variable (intentionally varied by the researchers), which goes on the *x*-axis? List the categories along the *x*-axis; because they are discrete rather than continuous, you can list them in any order. (b) What is the dependent variable (measured by the researchers), which goes on the *y*-axis? (c) What units (abbreviated) should go on the *y*-axis? Label the *y*-axis, including the units specified in the data table. Determine the range of values of the data that will need to go on the *y*-axis. What is the largest value? Draw evenly spaced tick marks and label them, starting with 0 at the bottom.

2. Graph the data for each sample. Match each *x*-value with its *y*-value and place a mark on the graph at that coordinate, then draw a bar from the *x*-axis up to the correct height for each sample. Why is a bar graph more appropriate than a scatter plot or line graph? (For additional information about graphs, see the Scientific Skills Review in Appendix F.)

3. Examine your graph and look for a pattern in the data. (a) Which cell type had the highest rate of oxygen consumption, and which had the lowest? (b) Does this support the researchers' hypothesis? Explain. (c) Based on what you know about mitochondrial electron transport and heat production, predict which rats had the highest, and which had the lowest, body temperature.

(MB) **Instructors**: A version of this Scientific Skills Exercise can be assigned in MasteringBiology.

CONCEPT 9.5

Fermentation and anaerobic respiration enable cells to produce ATP without the use of oxygen

Because most of the ATP generated by cellular respiration is due to the work of oxidative phosphorylation, our estimate of ATP yield from aerobic respiration depends on an adequate supply of oxygen to the cell. Without the electronegative oxygen to pull electrons down the transport chain, oxidative phosphorylation eventually ceases. However, there are two general mechanisms by which certain cells can oxidize organic fuel and generate ATP *without* the use of oxygen: anaerobic respiration and fermentation. The distinction between these two is that an electron transport chain is used in anaerobic respiration but not in fermentation. (The electron transport chain is also called the respiratory chain because of its role in both types of cellular respiration.)

We have already mentioned anaerobic respiration, which takes place in certain prokaryotic organisms that live in environments without oxygen. These organisms have an electron transport chain but do not use oxygen as a final electron acceptor at the end of the chain. Oxygen performs this function very well because it is extremely electronegative, but other, less electronegative substances can also serve as final electron acceptors. Some "sulfate-reducing" marine bacteria, for instance, use the sulfate ion (SO_4^{2-}) at the end of their respiratory chain. Operation of the chain builds up a proton-motive force used to produce ATP, but H_2S (hydrogen sulfide) is made as a by-product rather than water. The rotten-egg odor you may have smelled while walking through a salt marsh or a mudflat signals the presence of sulfate-reducing bacteria.

Fermentation is a way of harvesting chemical energy without using either oxygen or any electron transport chain—in other words, without cellular respiration. How can food be oxidized without cellular respiration? Remember, oxidation simply refers to the loss of electrons to an electron acceptor, so it does not need to involve oxygen. Glycolysis oxidizes glucose to two molecules of pyruvate. The oxidizing agent of glycolysis is NAD^+, and neither oxygen nor any electron transfer chain is involved. Overall, glycolysis is exergonic, and some of the energy made available is used to produce 2 ATP (net) by substrate-level phosphorylation. If oxygen *is* present, then additional ATP is made by oxidative phosphorylation when NADH passes electrons removed from glucose to the electron transport chain. But glycolysis generates 2 ATP whether oxygen is present or not—that is, whether conditions are aerobic or anaerobic.

As an alternative to respiratory oxidation of organic nutrients, fermentation is an extension of glycolysis that allows continuous generation of ATP by the substrate-level phosphorylation of glycolysis. For this to occur, there must be a sufficient supply of NAD^+ to accept electrons during the oxidation step of glycolysis. Without some mechanism to recycle NAD^+ from NADH, glycolysis would soon deplete the cell's pool of NAD^+ by reducing it all to NADH and would shut itself down for lack of an oxidizing agent. Under aerobic conditions, NAD^+ is recycled from NADH by the transfer of electrons to the electron transport chain. An anaerobic alternative is to transfer electrons from NADH to pyruvate, the end product of glycolysis.

Types of Fermentation

Fermentation consists of glycolysis plus reactions that regenerate NAD^+ by transferring electrons from NADH to pyruvate or derivatives of pyruvate. The NAD^+ can then be reused to oxidize sugar by glycolysis, which nets two molecules of ATP by substrate-level phosphorylation. There are many types of fermentation, differing in the end products formed from pyruvate. Two types are alcohol fermentation and lactic acid fermentation, and both are harnessed by humans for food and industrial production.

In **alcohol fermentation** **(Figure 9.17a)**, pyruvate is converted to ethanol (ethyl alcohol) in two steps. The first step releases carbon dioxide from the pyruvate, which is converted to the two-carbon compound acetaldehyde. In the second step, acetaldehyde is reduced by NADH to ethanol. This regenerates the supply of NAD^+ needed for the continuation of glycolysis. Many bacteria carry out alcohol fermentation under anaerobic conditions. Yeast (a fungus), in addition to aerobic respiration, also carries out alcohol fermentation. For thousands of years, humans have used yeast in brewing, winemaking, and baking. The CO_2 bubbles generated by baker's yeast during alcohol fermentation allow bread to rise.

▼ **Figure 9.17 Fermentation.** In the absence of oxygen, many cells use fermentation to produce ATP by substrate-level phosphorylation. NAD^+ is regenerated for use in glycolysis when pyruvate, the end product of glycolysis, serves as an electron acceptor for oxidizing NADH. Two of the common end products formed from fermentation are **(a)** ethanol and **(b)** lactate, the ionized form of lactic acid.

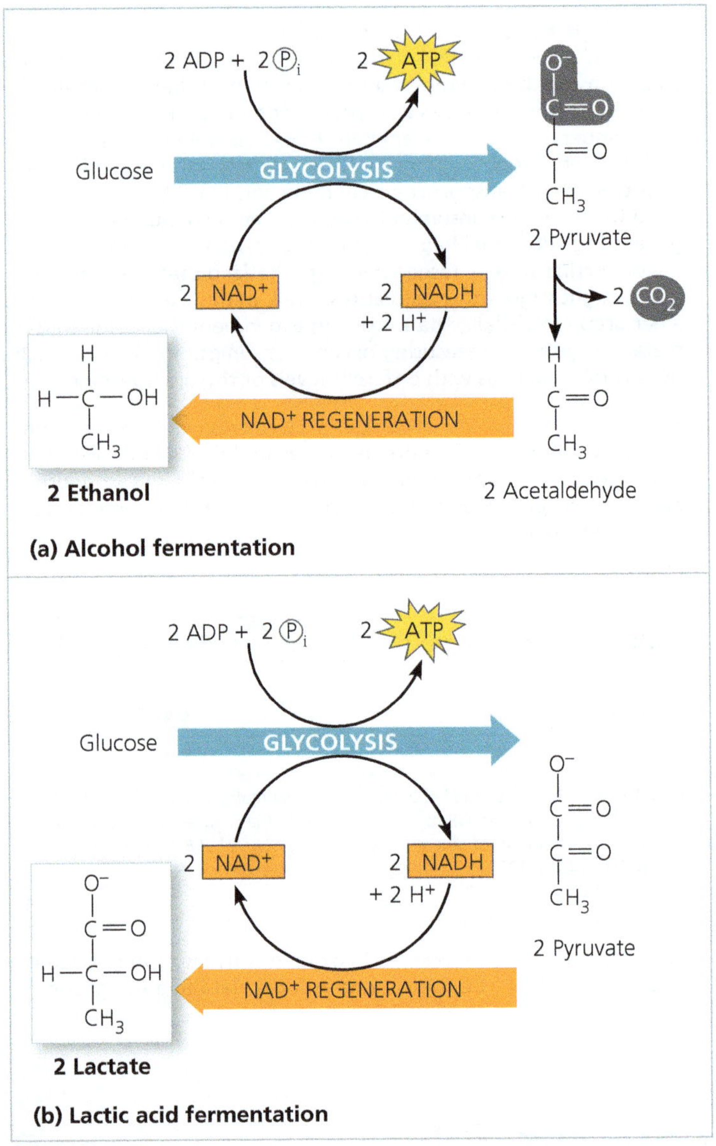

MB **Animation: Fermentation**

During **lactic acid fermentation** **(Figure 9.17b)**, pyruvate is reduced directly by NADH to form lactate as an end product, regenerating NAD^+ with no release of CO_2. (Lactate is the ionized form of lactic acid.) Lactic acid fermentation by certain fungi and bacteria is used in the dairy industry to make cheese and yogurt.

Human muscle cells make ATP by lactic acid fermentation when oxygen is scarce. This occurs during strenuous exercise, when sugar catabolism for ATP production outpaces the muscle's supply of oxygen from the blood. Under these conditions, the cells switch from aerobic respiration to fermentation. The lactate that accumulates was previously thought to cause the muscle fatigue and pain that occurs a day or so after intense exercise. However, evidence shows that within an hour, blood carries the excess lactate from the muscles to the liver, where it is converted back to pyruvate by liver cells. Because oxygen is available, this pyruvate can then enter the mitochondria in liver cells and complete cellular respiration. (Next-day muscle soreness is more likely caused by trauma to small muscle fibers, which leads to inflammation and pain.)

Comparing Fermentation with Anaerobic and Aerobic Respiration

Fermentation, anaerobic respiration, and aerobic respiration are three alternative cellular pathways for producing ATP by harvesting the chemical energy of food. All three use glycolysis to oxidize glucose and other organic fuels to pyruvate, with a net production of 2 ATP by substrate-level phosphorylation. And in all three pathways, NAD^+ is the oxidizing agent that accepts electrons from food during glycolysis.

A key difference is the contrasting mechanisms for oxidizing NADH back to NAD^+, which is required to sustain glycolysis. In fermentation, the final electron acceptor is an organic molecule such as pyruvate (lactic acid fermentation) or acetaldehyde (alcohol fermentation). In cellular respiration, by contrast, electrons carried by NADH are transferred to an electron transport chain, which regenerates the NAD^+ required for glycolysis.

Another major difference is the amount of ATP produced. Fermentation yields two molecules of ATP, produced by substrate-level phosphorylation. In the absence of an electron transport chain, the energy stored in pyruvate is unavailable. In cellular respiration, however, pyruvate is completely oxidized in the mitochondrion. Most of the chemical energy from this process is shuttled by NADH and $FADH_2$ in the form of electrons to the electron transport chain. There, the electrons move stepwise down a series of redox reactions to a final electron acceptor. (In aerobic respiration, the final electron acceptor is oxygen; in anaerobic respiration, the final acceptor is another molecule that is electronegative, although less so than oxygen.) Stepwise electron transport drives oxidative phosphorylation, yielding ATP. Thus, cellular respiration harvests much more energy from each sugar molecule than fermentation can. In fact, aerobic respiration yields up to 32 molecules of ATP per glucose molecule—up to 16 times as much as does fermentation.

Some organisms, called **obligate anaerobes**, carry out only fermentation or anaerobic respiration. In fact, these organisms cannot survive in the presence of oxygen, some forms of which can actually be toxic if protective systems are

Figure 9.18 Pyruvate as a key juncture in catabolism. Glycolysis is common to fermentation and cellular respiration. The end product of glycolysis, pyruvate, represents a fork in the catabolic pathways of glucose oxidation. In a facultative anaerobe or a muscle cell, which are capable of both aerobic cellular respiration and fermentation, pyruvate is committed to one of those two pathways, usually depending on whether or not oxygen is present.

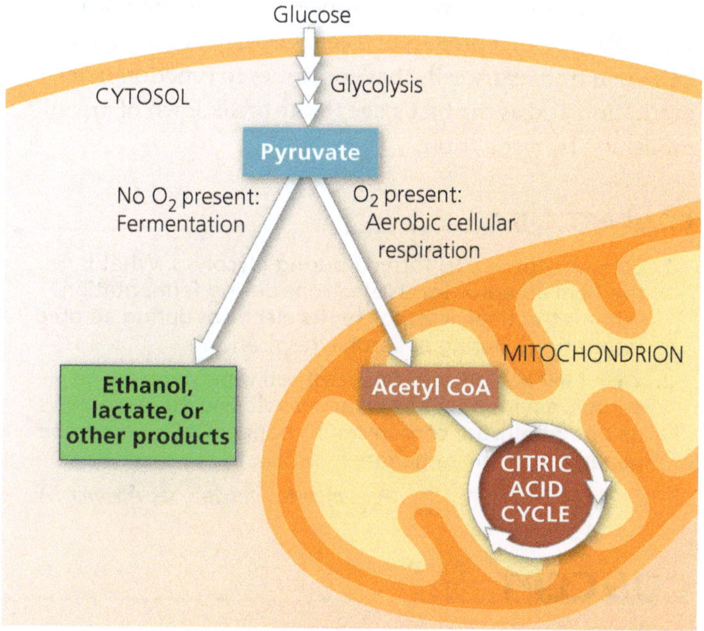

not present in the cell. A few cell types, such as cells of the vertebrate brain, can carry out only aerobic oxidation of pyruvate, not fermentation. Other organisms, including yeasts and many bacteria, can make enough ATP to survive using either fermentation or respiration. Such species are called **facultative anaerobes**. On the cellular level, our muscle cells behave as facultative anaerobes. In such cells, pyruvate is a fork in the metabolic road that leads to two alternative catabolic routes **(Figure 9.18)**. Under aerobic conditions, pyruvate can be converted to acetyl CoA, and oxidation continues in the citric acid cycle via aerobic respiration. Under anaerobic conditions, lactic acid fermentation occurs: Pyruvate is diverted from the citric acid cycle, serving instead as an electron acceptor to recycle NAD^+. To make the same amount of ATP, a facultative anaerobe has to consume sugar at a much faster rate when fermenting than when respiring.

The Evolutionary Significance of Glycolysis

EVOLUTION The role of glycolysis in both fermentation and respiration has an evolutionary basis. Ancient prokaryotes are thought to have used glycolysis to make ATP long before oxygen was present in Earth's atmosphere. The oldest known fossils of bacteria date back 3.5 billion years, but appreciable quantities of oxygen probably did not begin to accumulate in the atmosphere until about 2.7 billion years ago. Cyanobacteria produced this O_2 as a by-product of photosynthesis. Therefore, early prokaryotes may have generated ATP exclusively from glycolysis. The fact that glycolysis is today the most widespread metabolic pathway among Earth's organisms suggests that it evolved very early in the history of life. The cytosolic location of glycolysis also implies great antiquity; the pathway does not require any of the membrane-enclosed organelles of the eukaryotic cell, which evolved approximately 1 billion years after the first prokaryotic cell. Glycolysis is a metabolic heirloom from early cells that continues to function in fermentation and as the first stage in the breakdown of organic molecules by respiration.

CONCEPT CHECK 9.5

1. Consider the NADH formed during glycolysis. What is the final acceptor for its electrons during fermentation? What is the final acceptor for its electrons during aerobic respiration? During anaerobic respiration?
2. **WHAT IF?** ➤ A glucose-fed yeast cell is moved from an aerobic environment to an anaerobic one. How would its rate of glucose consumption change if ATP were to be generated at the same rate?

For suggested answers, see Appendix A.

CONCEPT 9.6

Glycolysis and the citric acid cycle connect to many other metabolic pathways

So far, we have treated the oxidative breakdown of glucose in isolation from the cell's overall metabolic economy. In this section, you will learn that glycolysis and the citric acid cycle are major intersections of the cell's catabolic (breakdown) and anabolic (biosynthetic) pathways.

The Versatility of Catabolism

Throughout this chapter, we have used glucose as an example of a fuel for cellular respiration. But free glucose molecules are not common in the diets of humans and other animals. We obtain most of our calories in the form of fats, proteins, and carbohydrates such as sucrose and other disaccharides, and starch, a polysaccharide. All these organic molecules in food can be used by cellular respiration to make ATP **(Figure 9.19)**.

Glycolysis can accept a wide range of carbohydrates for catabolism. In the digestive tract, starch is hydrolyzed to glucose, which can then be broken down in the cells by glycolysis and the citric acid cycle. Similarly, glycogen, the polysaccharide that humans and many other animals store in their liver and muscle cells, can be hydrolyzed to glucose between meals as fuel for respiration. The digestion of disaccharides, including sucrose, provides glucose and other monosaccharides as fuel for respiration.

▼ Figure 9.19 The catabolism of various molecules from food. Carbohydrates, fats, and proteins can all be used as fuel for cellular respiration. Monomers of these molecules enter glycolysis or the citric acid cycle at various points. Glycolysis and the citric acid cycle are catabolic funnels through which electrons from all kinds of organic molecules flow on their exergonic fall to oxygen.

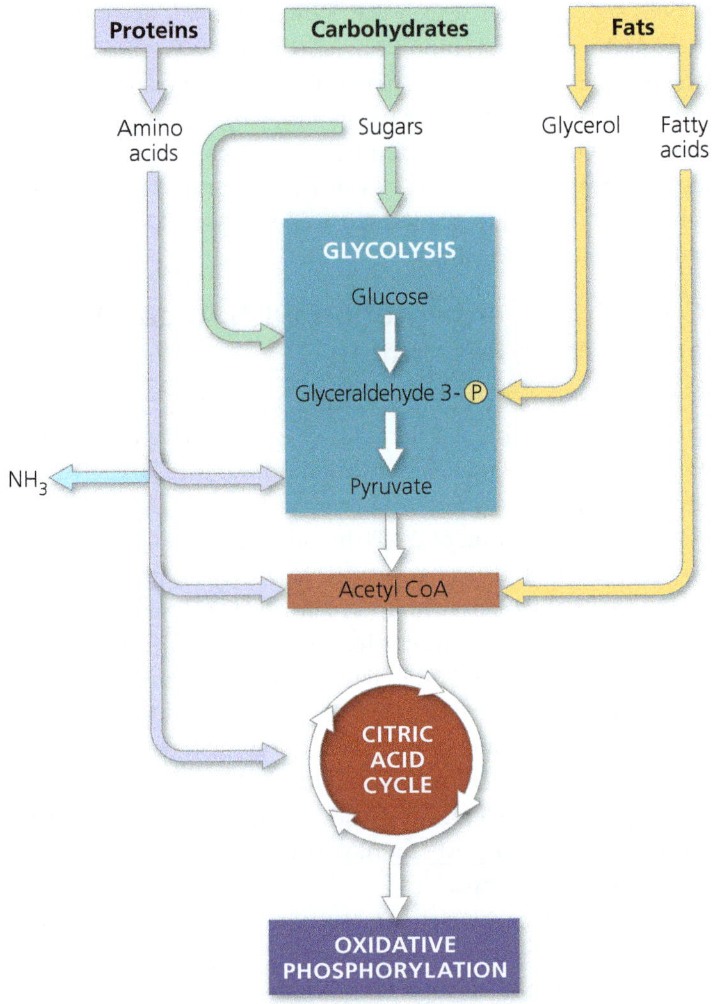

Proteins can also be used for fuel, but first they must be digested to their constituent amino acids. Many of the amino acids are used by the organism to build new proteins. Amino acids present in excess are converted by enzymes to intermediates of glycolysis and the citric acid cycle. Before amino acids can feed into glycolysis or the citric acid cycle, their amino groups must be removed, a process called *deamination*. The nitrogenous waste is excreted from the animal in the form of ammonia (NH_3), urea, or other waste products.

Catabolism can also harvest energy stored in fats obtained either from food or from fat cells in the body. After fats are digested to glycerol and fatty acids, the glycerol is converted to glyceraldehyde 3-phosphate, an intermediate of glycolysis. Most of the energy of a fat is stored in the fatty acids. A metabolic sequence called **beta oxidation** breaks the fatty acids down to two-carbon fragments, which enter the citric acid cycle as acetyl CoA. NADH and $FADH_2$ are also generated during beta oxidation; they can enter the electron transport chain, leading to further ATP production. Fats make excellent fuels, in large part due to their chemical structure and the high energy level of their electrons (present in many C—H bonds, equally shared between carbon and hydrogen) compared to those of carbohydrates. A gram of fat oxidized by respiration produces more than twice as much ATP as a gram of carbohydrate. Unfortunately, this also means that a person trying to lose weight must work hard to use up fat stored in the body because so many calories are stockpiled in each gram of fat.

Biosynthesis (Anabolic Pathways)

Cells need substance as well as energy. Not all the organic molecules of food are destined to be oxidized as fuel to make ATP. In addition to calories, food must also provide the carbon skeletons that cells require to make their own molecules. Some organic monomers obtained from digestion can be used directly. For example, as previously mentioned, amino acids from the hydrolysis of proteins in food can be incorporated into the organism's own proteins. Often, however, the body needs specific molecules that are not present as such in food. Compounds formed as intermediates of glycolysis and the citric acid cycle can be diverted into anabolic pathways as precursors from which the cell can synthesize the molecules it requires. For example, humans can make about half of the 20 amino acids in proteins by modifying compounds siphoned away from the citric acid cycle; the rest are "essential amino acids" that must be obtained in the diet. Also, glucose can be made from pyruvate, and fatty acids can be synthesized from acetyl CoA. Of course, these anabolic, or biosynthetic, pathways do not generate ATP, but instead consume it.

In addition, glycolysis and the citric acid cycle function as metabolic interchanges that enable our cells to convert some kinds of molecules to others as we need them. For example, an intermediate compound generated during glycolysis, dihydroxyacetone phosphate (see Figure 9.9, step 5), can be converted to one of the major precursors of fats. If we eat more food than we need, we store fat even if our diet is fat-free. Metabolism is remarkably versatile and adaptable.

Regulation of Cellular Respiration via Feedback Mechanisms

Basic principles of supply and demand regulate the metabolic economy. The cell does not waste energy making more of a particular substance than it needs. If there is a surplus of a certain amino acid, for example, the anabolic pathway that synthesizes that amino acid from an intermediate of the citric acid cycle is switched off. The most common mechanism for this control is feedback inhibition: The end product of the anabolic pathway inhibits the enzyme that catalyzes an early step of the pathway (see Figure 8.21). This prevents the needless diversion of key metabolic intermediates from uses that are more urgent.

The cell also controls its catabolism. If the cell is working hard and its ATP concentration begins to drop, respiration speeds up. When there is plenty of ATP to meet demand, respiration slows down, sparing valuable organic molecules for other functions. Again, control is based mainly on regulating the activity of enzymes at strategic points in the catabolic pathway. As shown in **Figure 9.20**, one important switch is phosphofructokinase, the enzyme that catalyzes step 3 of glycolysis (see Figure 9.9). That is the first step that commits the substrate irreversibly to the glycolytic pathway. By controlling the rate of this step, the cell can speed up or slow down the entire catabolic process. Phosphofructokinase can thus be considered the pacemaker of respiration.

Figure 9.20 The control of cellular respiration. Allosteric enzymes at certain points in the respiratory pathway respond to inhibitors and activators that help set the pace of glycolysis and the citric acid cycle. Phosphofructokinase, which catalyzes an early step in glycolysis (see Figure 9.9, step 3), is one such enzyme. It is stimulated by AMP (derived from ADP) but is inhibited by ATP and by citrate. This feedback regulation adjusts the rate of respiration as the cell's catabolic and anabolic demands change.

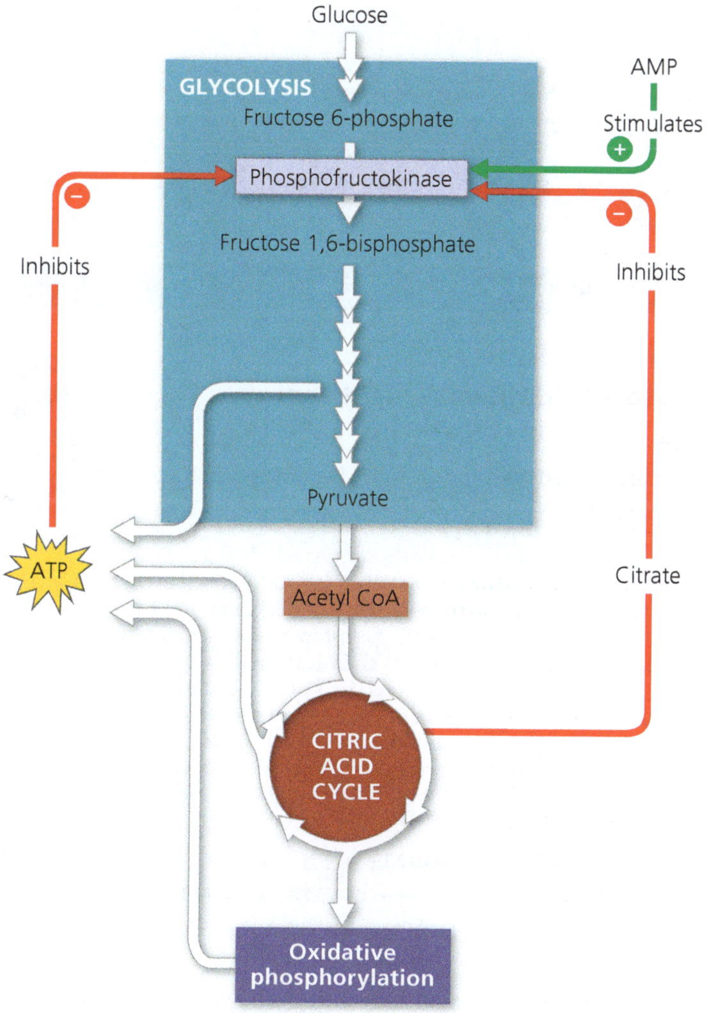

Phosphofructokinase is an allosteric enzyme with receptor sites for specific inhibitors and activators. It is inhibited by ATP and stimulated by AMP (adenosine monophosphate), which the cell derives from ADP. As ATP accumulates, inhibition of the enzyme slows down glycolysis. The enzyme becomes active again as cellular work converts ATP to ADP (and AMP) faster than ATP is being regenerated. Phosphofructokinase is also sensitive to citrate, the first product of the citric acid cycle. If citrate accumulates in mitochondria, some of it passes into the cytosol and inhibits phosphofructokinase. This mechanism helps synchronize the rates of glycolysis and the citric acid cycle. As citrate accumulates, glycolysis slows down, and the supply of pyruvate and thus acetyl groups to the citric acid cycle decreases. If citrate consumption increases, either because of a demand for more ATP or because anabolic pathways are draining off intermediates of the citric acid cycle, glycolysis accelerates and meets the demand. Metabolic balance is augmented by the control of enzymes that catalyze other key steps of glycolysis and the citric acid cycle. Cells are thrifty, expedient, and responsive in their metabolism.

Cellular respiration and metabolic pathways play a role of central importance in organisms. Examine Figure 9.2 again to put cellular respiration into the broader context of energy flow and chemical cycling in ecosystems. The energy that keeps us alive is *released*, not *produced*, by cellular respiration. We are tapping energy that was stored in food by photosynthesis. In the next chapter, you will learn how photosynthesis captures light and converts it to chemical energy.

CONCEPT CHECK 9.6

1. **MAKE CONNECTIONS >** Compare the structure of a fat (see Figure 5.9) with that of a carbohydrate (see Figure 5.3). What features of their structures make fat a much better fuel?
2. Under what circumstances might your body synthesize fat molecules?
3. **VISUAL SKILLS >** What will happen in a muscle cell that has used up its supply of oxygen and ATP? (Review Figures 9.18 and 9.20.)
4. **VISUAL SKILLS >** During intense exercise, can a muscle cell use fat as a concentrated source of chemical energy? Explain. (Review Figures 9.18 and 9.19.)

For suggested answers, see Appendix A.

9 Chapter Review

Go to **MasteringBiology**™ for Videos, Animations, Vocab Self-Quiz, Practice Tests, and more in the Study Area.

SUMMARY OF KEY CONCEPTS

CONCEPT 9.1

Catabolic pathways yield energy by oxidizing organic fuels *(pp. 165–169)*

VOCAB SELF-QUIZ goo.gl/6u55ks

- Cells break down glucose and other organic fuels to yield chemical energy in the form of ATP. **Fermentation** is a process that results in the partial degradation of glucose without the use of oxygen. The process of **cellular respiration** is a more complete breakdown of glucose. In **aerobic respiration**, oxygen is used as a reactant; in *anaerobic respiration*, other substances are used as reactants in a similar process that harvests chemical energy without oxygen.
- The cell taps the energy stored in food molecules through **redox reactions**, in which one substance partially or totally shifts electrons to another. **Oxidation** is the total or partial loss of electrons, while **reduction** is the total or partial addition of electrons. During aerobic respiration, glucose ($C_6H_{12}O_6$) is oxidized to CO_2, and O_2 is reduced to H_2O:

$$\underbrace{C_6H_{12}O_6 \rightarrow 6\,CO_2}_{\text{becomes oxidized}} \qquad C_6H_{12}O_6 + 6\,O_2 \longrightarrow 6\,CO_2 + 6\,H_2O + \text{Energy} \qquad \underbrace{6\,O_2 \rightarrow 6\,H_2O}_{\text{becomes reduced}}$$

- Electrons lose potential energy during their transfer from glucose or other organic compounds to oxygen. Electrons are usually passed first to **NAD**$^+$, reducing it to NADH, and are then passed from NADH to an **electron transport chain**, which conducts the electrons to O_2 in energy-releasing steps. The energy that is released is used to make ATP.
- Aerobic respiration occurs in three stages: (1) **glycolysis**, (2) pyruvate oxidation and the **citric acid cycle**, and (3) **oxidative phosphorylation** (electron transport and chemiosmosis).

? *Describe the difference between the two processes in cellular respiration that produce ATP: oxidative phosphorylation and substrate-level phosphorylation.*

CONCEPT 9.2

Glycolysis harvests chemical energy by oxidizing glucose to pyruvate *(pp. 170–171)*

- Glycolysis ("splitting of sugar") is a series of reactions that breaks down glucose into two pyruvate molecules, which may go on to enter the citric acid cycle, and nets 2 ATP and 2 NADH per glucose molecule.

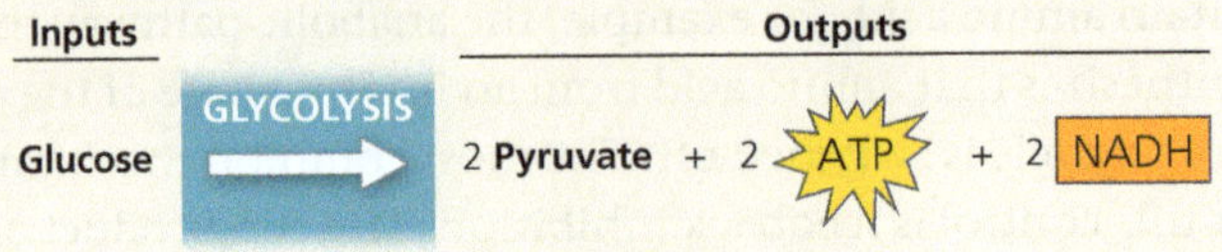

? *Which reactions in glycolysis are the source of energy for the formation of ATP and NADH?*

CONCEPT 9.3

After pyruvate is oxidized, the citric acid cycle completes the energy-yielding oxidation of organic molecules *(pp. 171–174)*

- In eukaryotic cells, pyruvate enters the mitochondrion and is oxidized to **acetyl CoA**, which is further oxidized in the citric acid cycle.

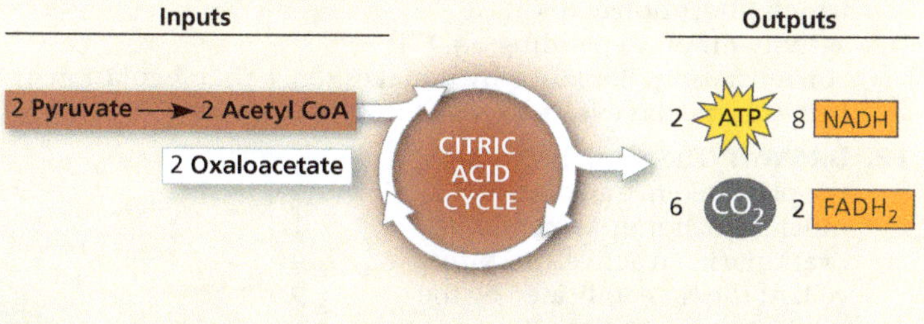

? *What molecular products indicate the complete oxidation of glucose during cellular respiration?*

CONCEPT 9.4

During oxidative phosphorylation, chemiosmosis couples electron transport to ATP synthesis *(pp. 174–179)*

- NADH and $FADH_2$ transfer electrons to the electron transport chain. Electrons move down the chain, losing energy in several energy-releasing steps. Finally, electrons are passed to O_2, reducing it to H_2O.

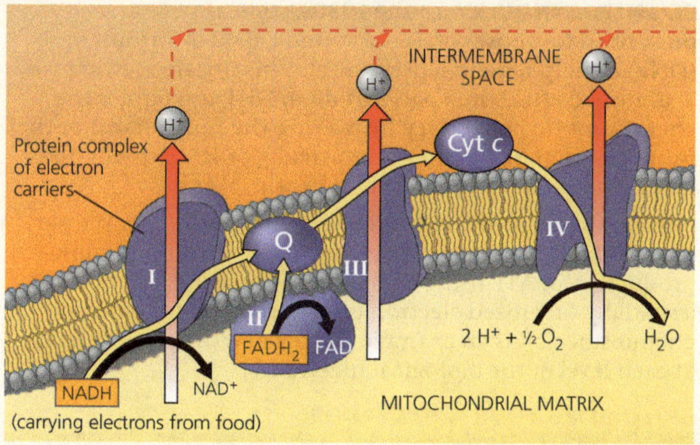

- Along the electron transport chain, electron transfer causes protein complexes to move H^+ from the mitochondrial matrix (in eukaryotes) to the intermembrane space, storing energy as a **proton-motive force** (H^+ gradient). As H^+ diffuses back into the matrix through **ATP synthase**, its passage drives the phosphorylation of ADP to form ATP, called **chemiosmosis**.
- About 34% of the energy stored in a glucose molecule is transferred to ATP during cellular respiration, producing a maximum of about 32 ATP.

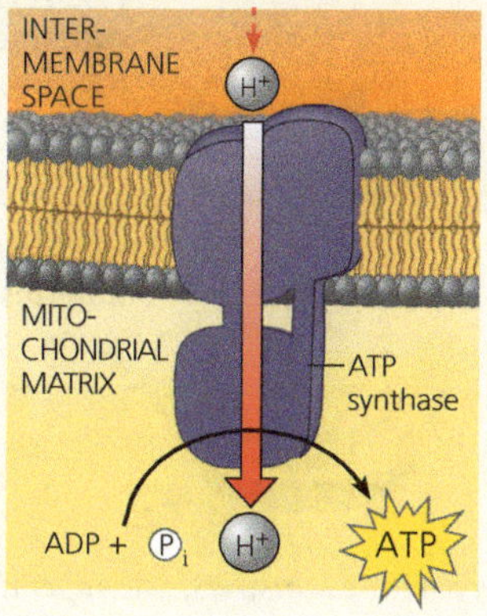

? *Briefly explain the mechanism by which ATP synthase produces ATP. List three locations in which ATP synthases are found.*

CONCEPT 9.5

Fermentation and anaerobic respiration enable cells to produce ATP without the use of oxygen *(pp. 179–182)*

- Glycolysis nets 2 ATP by substrate-level phosphorylation, whether oxygen is present or not. Under anaerobic conditions, anaerobic respiration or fermentation can take place. In anaerobic respiration, an electron transport chain is present with a final electron acceptor other than oxygen. In fermentation, the electrons from NADH are passed to pyruvate or a derivative of pyruvate, regenerating the NAD^+ required to oxidize more glucose. Two common types of fermentation are **alcohol fermentation** and **lactic acid fermentation**.
- Fermentation and anaerobic or aerobic respiration all use glycolysis to oxidize glucose, but they differ in their final electron acceptor and whether an electron transport chain is used (respiration) or not (fermentation). Respiration yields more ATP; aerobic respiration, with O_2 as the final electron acceptor, yields about 16 times as much ATP as does fermentation.
- Glycolysis occurs in nearly all organisms and is thought to have evolved in ancient prokaryotes before there was O_2 in the atmosphere.

? *Which process yields more ATP, fermentation or anaerobic respiration? Explain.*

CONCEPT 9.6

Glycolysis and the citric acid cycle connect to many other metabolic pathways *(pp. 182–184)*

- Catabolic pathways funnel electrons from many kinds of organic molecules into cellular respiration. Many carbohydrates can enter glycolysis, most often after conversion to glucose. Amino acids of proteins must be deaminated before being oxidized. The fatty acids of fats undergo **beta oxidation** to two-carbon fragments and then enter the citric acid cycle as acetyl CoA. Anabolic pathways can use small molecules from food directly or build other substances using intermediates of glycolysis or the citric acid cycle.
- Cellular respiration is controlled by allosteric enzymes at key points in glycolysis and the citric acid cycle.

? *Describe how the catabolic pathways of glycolysis and the citric acid cycle intersect with anabolic pathways in the metabolism of a cell.*

TEST YOUR UNDERSTANDING

Level 1: Knowledge/Comprehension

PRACTICE TEST
goo.gl/CUYGKD

1. The *immediate* energy source that drives ATP synthesis by ATP synthase during oxidative phosphorylation is the
   (A) oxidation of glucose and other organic compounds.
   (B) flow of electrons down the electron transport chain.
   (C) H^+ concentration gradient across the membrane holding ATP synthase.
   (D) transfer of phosphate to ADP.
2. Which metabolic pathway is common to both fermentation and cellular respiration of a glucose molecule?
   (A) the citric acid cycle
   (B) the electron transport chain
   (C) glycolysis
   (D) reduction of pyruvate to lactate

3. The final electron acceptor of the electron transport chain that functions in aerobic oxidative phosphorylation is
   (A) oxygen.
   (B) water.
   (C) NAD^+.
   (D) pyruvate.

4. In mitochondria, exergonic redox reactions
   (A) are the source of energy driving prokaryotic ATP synthesis.
   (B) provide the energy that establishes the proton gradient.
   (C) reduce carbon atoms to carbon dioxide.
   (D) are coupled via phosphorylated intermediates to endergonic processes.

Level 2: Application/Analysis

5. What is the oxidizing agent in the following reaction?

$$\text{Pyruvate} + \text{NADH} + H^+ \rightarrow \text{Lactate} + NAD^+$$

   (A) oxygen
   (B) NADH
   (C) lactate
   (D) pyruvate

6. When electrons flow along the electron transport chains of mitochondria, which of the following changes occurs?
   (A) The pH of the matrix increases.
   (B) ATP synthase pumps protons by active transport.
   (C) The electrons gain free energy.
   (D) NAD^+ is oxidized.

7. Most CO_2 from catabolism is released during
   (A) glycolysis.
   (B) the citric acid cycle.
   (C) lactate fermentation.
   (D) electron transport.

8. **MAKE CONNECTIONS** Step 3 in Figure 9.9 is a major point of regulation of glycolysis. The enzyme phosphofructokinase is allosterically regulated by ATP and related molecules (see Concept 8.5). Considering the overall result of glycolysis, would you expect ATP to inhibit or stimulate activity of this enzyme? Explain. (Hint: Make sure you consider the role of ATP as an allosteric regulator, not as a substrate of the enzyme.)

9. **MAKE CONNECTIONS** The proton pump shown in Figures 7.17 and 7.18 is a type of ATP synthase (see Figure 9.14). Compare the processes shown in the two figures, and say whether they are involved in active or passive transport (see Concepts 7.3 and 7.4).

10. **VISUAL SKILLS** This computer model shows the four parts of ATP synthase, each part consisting of a number of polypeptide subunits (the structure in gray is still an area of active research). Using Figure 9.14 as a guide, label the rotor, stator, internal rod, and catalytic knob of this molecular motor.

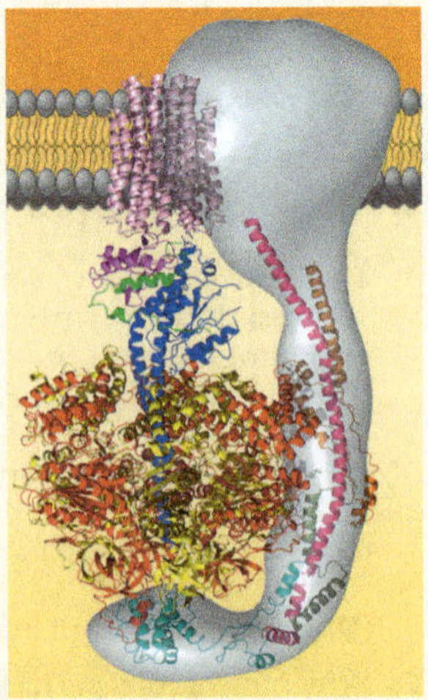

Level 3: Synthesis/Evaluation

11. **INTERPRET THE DATA** Phosphofructokinase is an enzyme that acts on fructose 6-phosphate at an early step in glucose breakdown. Regulation of this enzyme controls whether the sugar will continue on in the glycolytic pathway. Considering this graph, under which condition is phosphofructokinase more active? Given what you know about glycolysis and regulation of metabolism by this enzyme, explain the mechanism by which phosphofructokinase activity differs depending on ATP concentration. Explain why it makes sense that regulation of this enzyme has evolved so that it works this way.

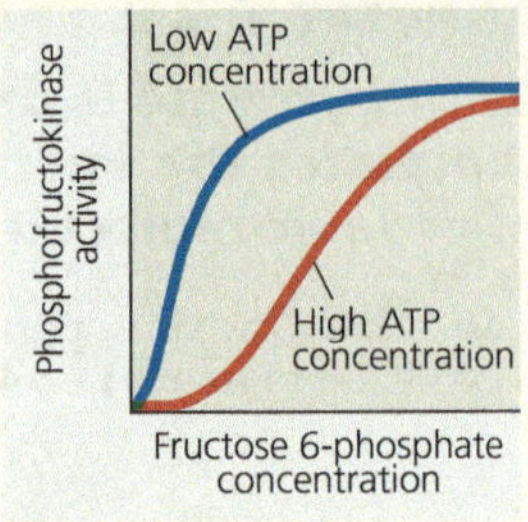

12. **DRAW IT** The graph here shows the pH difference across the inner mitochondrial membrane over time in an actively respiring cell. At the time indicated by the vertical arrow, a metabolic poison is added that specifically and completely inhibits all function of mitochondrial ATP synthase. Draw what you would expect to see for the rest of the graphed line, and explain your graph.

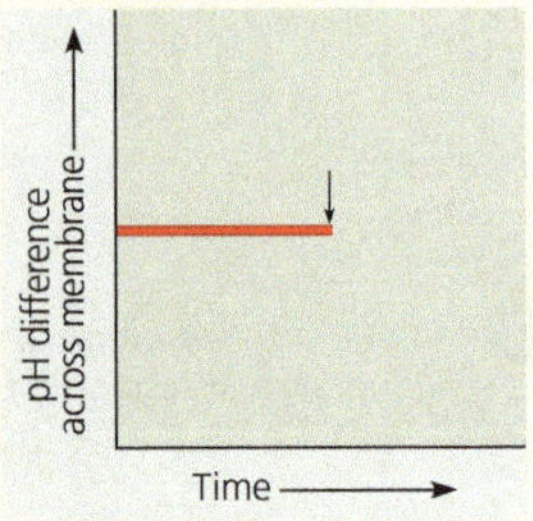

13. **EVOLUTION CONNECTION** ATP synthases are found in the prokaryotic plasma membrane and in mitochondria and chloroplasts. (a) Propose a hypothesis to account for an evolutionary relationship of these eukaryotic organelles and prokaryotes. (b) Explain how the amino acid sequences of the ATP synthases from the different sources could be used to support or refute your hypothesis.

14. **SCIENTIFIC INQUIRY** In the 1930s, some physicians prescribed low doses of a compound called dinitrophenol (DNP) to help patients lose weight. This unsafe method was abandoned after some patients died. DNP uncouples the chemiosmotic machinery by making the lipid bilayer of the inner mitochondrial membrane leaky to H^+. Explain how this could cause weight loss and death.

15. **WRITE ABOUT A THEME: ORGANIZATION** In a short essay (100–150 words), explain how oxidative phosphorylation—production of ATP using energy from the redox reactions of a spatially organized electron transport chain followed by chemiosmosis—is an example of how new properties emerge at each level of the biological hierarchy.

16. **SYNTHESIZE YOUR KNOWLEDGE**

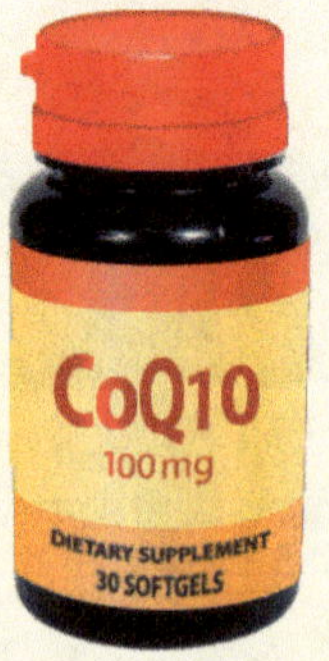

Coenzyme Q (CoQ) is sold as a nutritional supplement. One company uses this marketing slogan for CoQ: "Give your heart the fuel it craves most." Considering the role of coenzyme Q, critique this claim. How do you think this product might function to benefit the heart? Is CoQ used as a "fuel" during cellular respiration?

For selected answers, see Appendix A.

For additional practice questions, check out the **Dynamic Study Modules** in MasteringBiology. You can use them to study on your smartphone, tablet, or computer anytime, anywhere!

▲ **Figure 10.1 How does sunlight help build the trunk, branches, and leaves of this broadleaf tree?**

KEY CONCEPTS

10.1 Photosynthesis converts light energy to the chemical energy of food

10.2 The light reactions convert solar energy to the chemical energy of ATP and NADPH

10.3 The Calvin cycle uses the chemical energy of ATP and NADPH to reduce CO_2 to sugar

10.4 Alternative mechanisms of carbon fixation have evolved in hot, arid climates

10.5 Life depends on photosynthesis

▼ **Other organisms also benefit from photosynthesis.**

The Process That Feeds the Biosphere

Life on Earth is solar powered. Plants and other photosynthetic organisms contain cellular organelles called **chloroplasts**. Specialized molecular complexes in chloroplasts capture light energy that has traveled 150 million km from the sun and convert it to chemical energy that is stored in sugar and other organic molecules. This conversion process is called **photosynthesis**. Let's begin by placing photosynthesis in its ecological context.

Photosynthesis nourishes almost the entire living world directly or indirectly. An organism acquires the organic compounds it uses for energy and carbon skeletons by one of two major modes: autotrophic nutrition or heterotrophic nutrition. **Autotrophs** are "self-feeders" (*auto-* means "self," and *trophos* means "feeder"); they sustain themselves without eating anything derived from other living beings. Autotrophs produce their organic molecules from CO_2 and other inorganic raw materials obtained from the environment. They are the ultimate sources of organic compounds for all nonautotrophic organisms, and for this reason, biologists refer to autotrophs as the *producers* of the biosphere.

Almost all plants are autotrophs; the only nutrients they require are water and minerals from the soil and carbon dioxide from the air. Specifically, plants are *photo*autotrophs, organisms that use light as a source of energy to synthesize organic substances (**Figure 10.1**). Photosynthesis also occurs in algae, certain

When you see this blue icon, log in to **MasteringBiology** and go to the Study Area for digital resources.

MB **Get Ready for This Chapter**

unicellular eukaryotes, and some prokaryotes **(Figure 10.2)**. In this chapter, we will touch on these other groups in passing, but our emphasis will be on plants. Variations in autotrophic nutrition that occur in prokaryotes and algae will be described in Concept 27.3.

Figure 10.2 Photoautotrophs. These organisms use light energy to drive the synthesis of organic molecules from carbon dioxide and (in most cases) water. They feed themselves and the entire living world. On land, **(a)** plants are the predominant producers of food. In aquatic environments, photoautotrophs include unicellular and **(b)** multicellular algae, such as this kelp; **(c)** some non-algal unicellular eukaryotes, such as *Euglena*; **(d)** the prokaryotes called cyanobacteria; and **(e)** other photosynthetic prokaryotes, such as these purple sulfur bacteria, which produce sulfur (the yellow globules within the cells) (c–e, LMs).

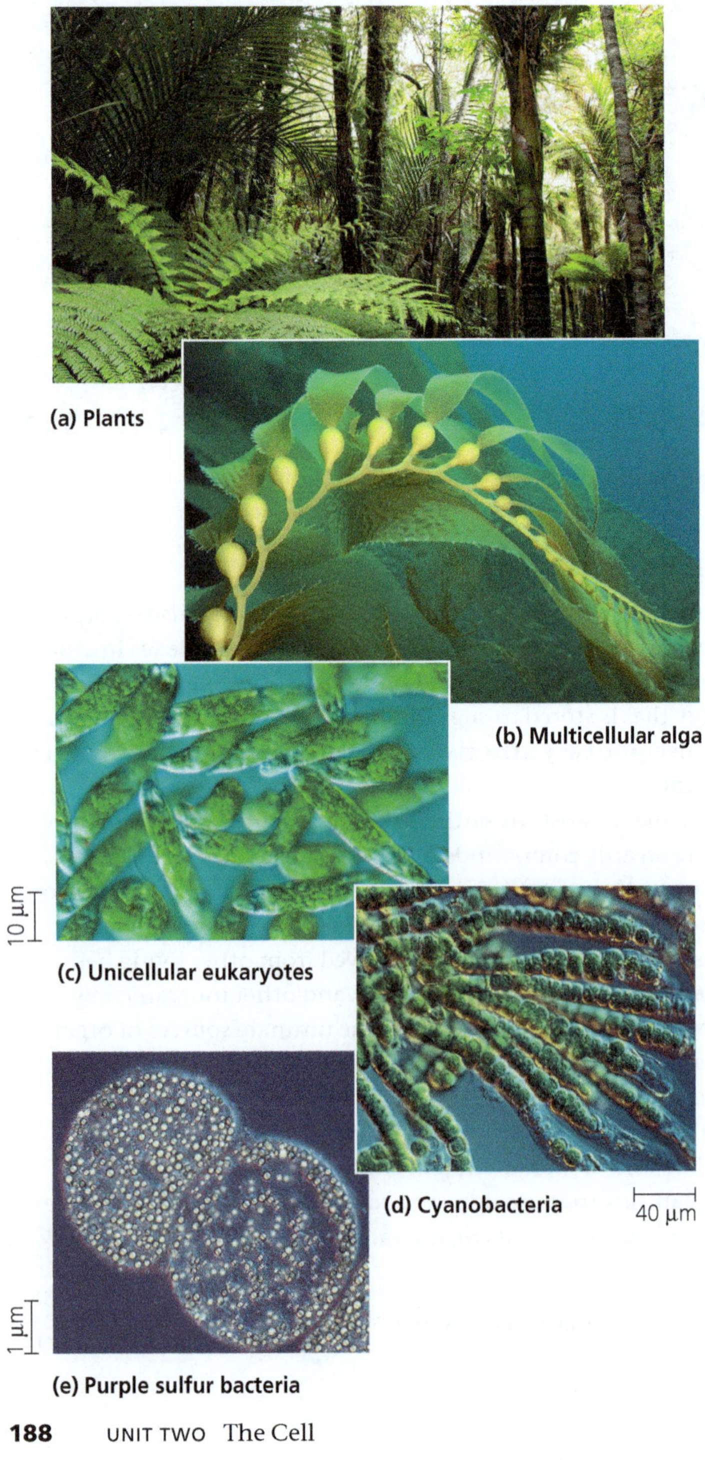

Heterotrophs obtain organic material by the second major mode of nutrition. Unable to make their own food, they live on compounds produced by other organisms (*hetero-* means "other"). Heterotrophs are the biosphere's *consumers*. The most obvious "other-feeding" occurs when an animal eats plants or other organisms. But heterotrophic nutrition may be more subtle. Some heterotrophs consume the remains of other organisms by decomposing and feeding on organic litter such as dead organisms, feces, and fallen leaves; these types of heterotrophs are known as *decomposers*. Most fungi and many types of prokaryotes get their nourishment this way. Almost all heterotrophs, including humans, are completely dependent, either directly or indirectly, on photoautotrophs for food—and also for oxygen, a by-product of photosynthesis.

MB **BioFlix® Animation: The Flow of Carbon Atoms in Producers, Consumers, and Decomposers**

The Earth's supply of fossil fuels was formed from remains of organisms that died hundreds of millions of years ago. In a sense, then, fossil fuels represent stores of the sun's energy from the distant past. Because these resources are being used at a much higher rate than they are replenished, researchers are exploring methods of capitalizing on the photosynthetic process to provide alternative fuels **(Figure 10.3)**.

In this chapter, you'll learn how photosynthesis works. After discussing general principles of photosynthesis, we'll consider the two stages of photosynthesis: the light reactions, which capture solar energy and transform it into chemical energy; and the Calvin cycle, which uses that chemical

Figure 10.3 Alternative fuels from algae. The power of sunlight can be tapped to generate a sustainable alternative to fossil fuels. Species of unicellular algae that are prolific producers of plant oils can be cultured in long, transparent tanks called photobioreactors, such as the one shown here at Arizona State University. A simple chemical process can yield "biodiesel," which can be mixed with gasoline or used alone to power vehicles.

WHAT IF? ➤ *The main product of fossil fuel combustion is CO_2, and this is the source of the increase in atmospheric CO_2 concentration. Scientists have proposed strategically situating containers of these algae near industrial plants or near highly congested city streets. Considering the process of photosynthesis, how does this arrangement make sense?*

MB **ABC News Video: Turning Algae into Biofuel**

energy to make the organic molecules of food. Finally, we will consider some aspects of photosynthesis from an evolutionary perspective.

CONCEPT 10.1

Photosynthesis converts light energy to the chemical energy of food

The remarkable ability of an organism to harness light energy and use it to drive the synthesis of organic compounds emerges from structural organization in the cell: Photosynthetic enzymes and other molecules are grouped together in a biological membrane, enabling the necessary series of chemical reactions to be carried out efficiently. The process of photosynthesis most likely originated in a group of bacteria that had infolded regions of the plasma membrane containing clusters of such molecules. In existing photosynthetic bacteria, infolded photosynthetic membranes function similarly to the internal membranes of the chloroplast, a eukaryotic organelle. According to what has come to be known as the endosymbiont theory, the original chloroplast was a photosynthetic prokaryote that lived inside an ancestor of eukaryotic cells. (You learned about this theory in Concept 6.5, and it will be described more fully in Concept 25.3.) Chloroplasts are present in a variety of photosynthesizing organisms (see some examples in Figure 10.2), but here we focus on chloroplasts in plants.

Chloroplasts: The Sites of Photosynthesis in Plants

All green parts of a plant, including green stems and unripened fruit, have chloroplasts, but the leaves are the major sites of photosynthesis in most plants **(Figure 10.4)**. There are about half a million chloroplasts in a chunk of leaf with a top surface area of 1 mm^2. Chloroplasts are found mainly in the cells of the **mesophyll**, the tissue in the interior of the leaf. Carbon dioxide enters the leaf, and oxygen exits, by way of microscopic pores called **stomata** (singular, *stoma*; from the Greek, meaning "mouth"). Water absorbed by the roots is delivered to the leaves in veins. Leaves also use veins to export sugar to roots and other nonphotosynthetic parts of the plant.

A typical mesophyll cell has about 30–40 chloroplasts, each measuring about 2–4 μm by 4–7 μm. A chloroplast has two membranes surrounding a dense fluid called the **stroma**. Suspended within the stroma is a third membrane system, made up of sacs called **thylakoids**, which segregates the stroma from the *thylakoid space* inside these sacs. In some places, thylakoid sacs are stacked in columns called *grana* (singular, *granum*). **Chlorophyll**, the green pigment that gives leaves their color, resides in the thylakoid membranes of the chloroplast. (The internal photosynthetic membranes

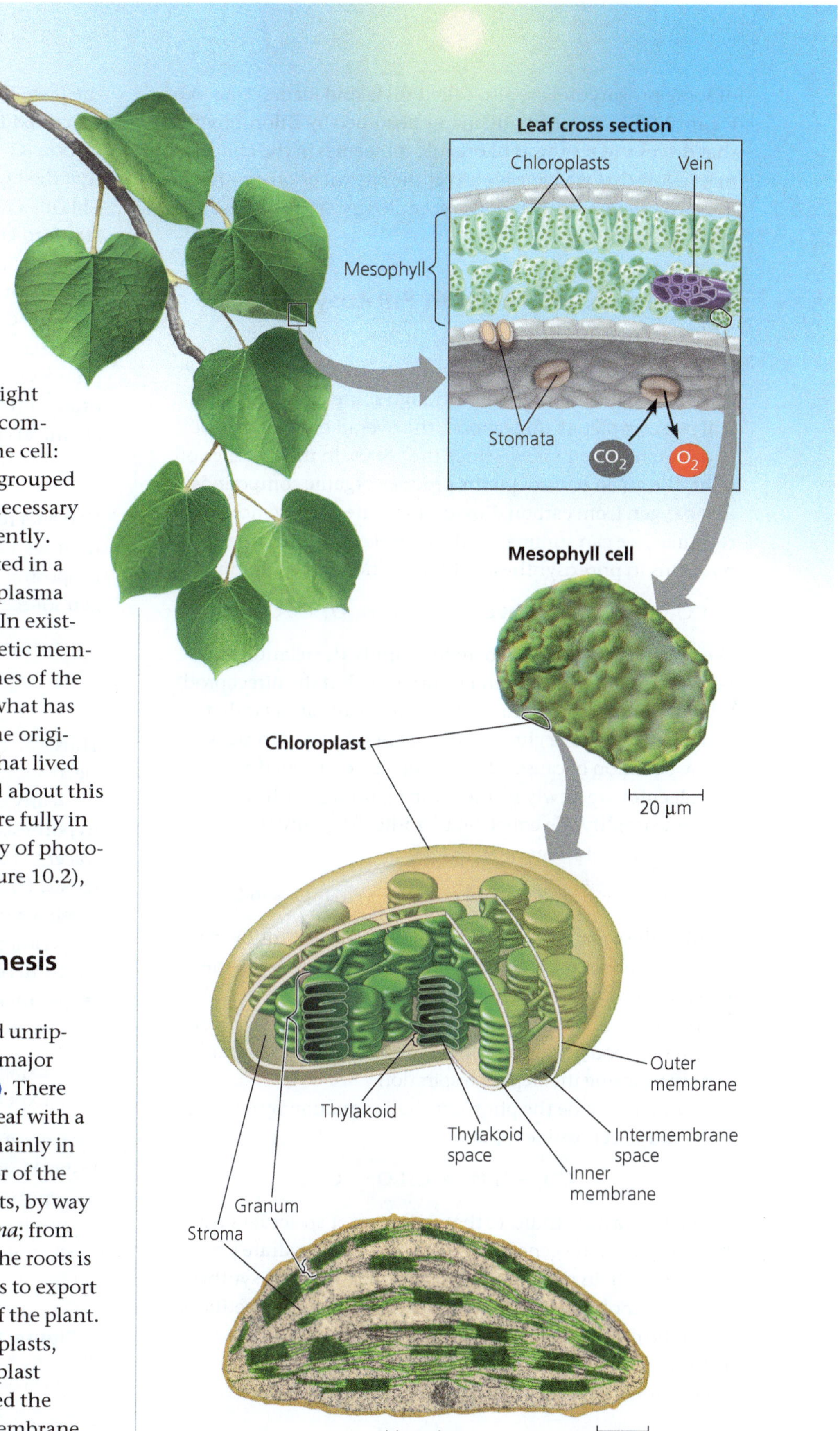

▲ **Figure 10.4 Zooming in on the location of photosynthesis in a plant.** Leaves are the major organs of photosynthesis in plants. These images take you into a leaf, then into a cell, and finally into a chloroplast, the organelle where photosynthesis occurs (middle, LM; bottom, TEM).

Video: Chloroplasts in Motion

of some prokaryotes are also called thylakoid membranes; see Figure 27.8b.) It is the light energy absorbed by chlorophyll that drives the synthesis of organic molecules in the chloroplast. Now that we have looked at the sites of photosynthesis in plants, we are ready to look more closely at the process of photosynthesis.

Tracking Atoms Through Photosynthesis: *Scientific Inquiry*

Scientists have tried for centuries to piece together the process by which plants make food. Although some of the steps are still not completely understood, the overall photosynthetic equation has been known since the 1800s: In the presence of light, the green parts of plants produce organic compounds and oxygen from carbon dioxide and water. Using molecular formulas, we can summarize the complex series of chemical reactions in photosynthesis with this chemical equation:

$$6\ CO_2 + 12\ H_2O + \text{Light energy} \rightarrow C_6H_{12}O_6 + 6\ O_2 + 6\ H_2O$$

We use glucose ($C_6H_{12}O_6$) here to simplify the relationship between photosynthesis and respiration, but the direct product of photosynthesis is actually a three-carbon sugar that can be used to make glucose. Water appears on both sides of the equation because 12 molecules are consumed and 6 molecules are newly formed during photosynthesis. We can simplify the equation by indicating only the net consumption of water:

$$6\ CO_2 + 6\ H_2O + \text{Light energy} \rightarrow C_6H_{12}O_6 + 6\ O_2$$

Writing the equation in this form, we can see that the overall chemical change during photosynthesis is the reverse of the one that occurs during cellular respiration (see Concept 9.1). Both of these metabolic processes occur in plant cells. However, as you will soon learn, chloroplasts do not synthesize sugars by simply reversing the steps of respiration.

Now let's divide the photosynthetic equation by 6 to put it in its simplest possible form:

$$CO_2 + H_2O \rightarrow [CH_2O] + O_2$$

Here, the brackets indicate that CH_2O is not an actual sugar but represents the general formula for a carbohydrate (see Concept 5.2). In other words, we are imagining the synthesis of a sugar molecule one carbon at a time (with six repetitions theoretically adding up to a glucose molecule—$C_6H_{12}O_6$). Let's now see how researchers tracked the elements C, H, and O from the reactants of photosynthesis to the products.

The Splitting of Water

One of the first clues to the mechanism of photosynthesis came from the discovery that the O_2 given off by plants is derived from H_2O and not from CO_2. The chloroplast splits water into hydrogen and oxygen. Before this discovery, the prevailing hypothesis was that photosynthesis split carbon dioxide ($CO_2 \rightarrow C + O_2$) and then added water to the carbon ($C + H_2O \rightarrow [CH_2O]$). This hypothesis predicted that the O_2 released during photosynthesis came from CO_2. This idea was challenged in the 1930s by C. B. van Niel, of Stanford University. Van Niel was investigating photosynthesis in bacteria that make their carbohydrate from CO_2 but do not release O_2. He concluded that, at least in these bacteria, CO_2 is not split into carbon and oxygen. One group of bacteria used hydrogen sulfide (H_2S) rather than water for photosynthesis, forming yellow globules of sulfur as a waste product (these globules are visible in Figure 10.2e). Here is the chemical equation for photosynthesis in these sulfur bacteria:

$$CO_2 + 2\ H_2S \rightarrow [CH_2O] + H_2O + 2\ S$$

Van Niel reasoned that the bacteria split H_2S and used the hydrogen atoms to make sugar. He then generalized that idea, proposing that all photosynthetic organisms require a hydrogen source but that the source varies:

$$\text{Sulfur bacteria: } CO_2 + 2\ H_2S \rightarrow [CH_2O] + H_2O + 2\ S$$
$$\text{Plants: } CO_2 + 2\ H_2O \rightarrow [CH_2O] + H_2O + O_2$$
$$\text{General: } CO_2 + 2\ H_2X \rightarrow [CH_2O] + H_2O + 2\ X$$

Thus, van Niel hypothesized that plants split H_2O as a source of electrons from hydrogen atoms, releasing O_2 as a by-product.

Nearly 20 years later, scientists confirmed van Niel's hypothesis by using oxygen-18 (^{18}O), a heavy isotope, as a tracer to follow the path of oxygen atoms during photosynthesis. The experiments showed that the O_2 from plants was labeled with ^{18}O only if water was the source of the tracer (experiment 1). If the ^{18}O was introduced to the plant in the form of CO_2, the label did not turn up in the released O_2 (experiment 2). In the following summary, magenta denotes labeled atoms of oxygen (^{18}O):

$$\text{Experiment 1: } CO_2 + 2\ H_2O \rightarrow [CH_2O] + H_2O + O_2$$
$$\text{Experiment 2: } CO_2 + 2\ H_2O \rightarrow [CH_2O] + H_2O + O_2$$

A significant result of the shuffling of atoms during photosynthesis is the extraction of hydrogen from water and its incorporation into sugar. The waste product of photosynthesis, O_2, is released to the atmosphere. **Figure 10.5** shows the fates of all atoms in photosynthesis.

Figure 10.5 Tracking atoms through photosynthesis. The atoms from CO_2 are shown in magenta, and the atoms from H_2O are shown in blue.

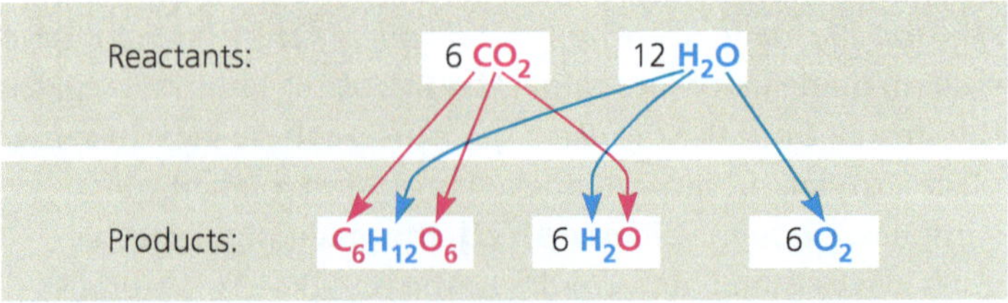

Photosynthesis as a Redox Process

Let's briefly compare photosynthesis with cellular respiration. Both processes involve redox reactions. During cellular respiration, energy is released from sugar when electrons associated with hydrogen are transported by carriers to oxygen, forming water as a by-product (see Concept 9.1). The electrons lose potential energy as they "fall" down the electron transport chain toward electronegative oxygen, and the mitochondrion harnesses that energy to synthesize ATP (see Figure 9.15). Photosynthesis reverses the direction of electron flow. Water is split, and its electrons are transferred along with hydrogen ions (H^+) from the water to carbon dioxide, reducing it to sugar.

$$\text{Energy} + 6\,CO_2 + 6\,H_2O \longrightarrow C_6H_{12}O_6 + 6\,O_2$$

becomes reduced (CO_2 → $C_6H_{12}O_6$); becomes oxidized (H_2O → O_2)

Because the electrons increase in potential energy as they move from water to sugar, this process requires energy—in other words, it is endergonic. This energy boost that occurs during photosynthesis is provided by light.

The Two Stages of Photosynthesis: *A Preview*

The equation for photosynthesis is a deceptively simple summary of a very complex process. Actually, photosynthesis is not a single process, but two processes, each with multiple steps. These two stages of photosynthesis are known as the **light reactions** (the *photo* part of photosynthesis) and the **Calvin cycle** (the *synthesis* part) **(Figure 10.6)**.

The light reactions are the steps of photosynthesis that convert solar energy to chemical energy. Water is split, providing a source of electrons and protons (hydrogen ions, H^+) and giving off O_2 as a by-product. Light absorbed by chlorophyll drives a transfer of the electrons and hydrogen ions from water to an acceptor called **NADP$^+$** (nicotinamide adenine dinucleotide phosphate), where they are temporarily stored. (The electron acceptor NADP$^+$ is first cousin to NAD$^+$, which functions as an electron carrier in cellular respiration; the two molecules differ only by the presence of an extra phosphate group in the NADP$^+$ molecule.) The light reactions use solar energy to reduce NADP$^+$ to **NADPH** by adding a pair of electrons along with an H^+. The light reactions also generate ATP, using chemiosmosis to power the addition of a phosphate group to ADP, a process called **photophosphorylation**. Thus, light energy is initially converted to chemical energy in the form of two compounds: NADPH and ATP. NADPH, a source of electrons, acts as "reducing power" that can be passed along to an electron acceptor, reducing it, while ATP is the versatile energy currency of cells. Notice that the light reactions produce no sugar; that happens in the second stage of photosynthesis, the Calvin cycle.

The Calvin cycle is named for Melvin Calvin, who, along with his colleagues James Bassham and Andrew Benson, began to elucidate its steps in the late 1940s. The cycle begins by incorporating CO_2 from the air into organic molecules already present in the chloroplast. This initial incorporation of carbon into organic compounds is known as **carbon fixation**. The Calvin cycle then reduces the fixed carbon to carbohydrate by the addition of electrons. The reducing power is provided by NADPH, which acquired its cargo of electrons in the light reactions. To convert CO_2 to carbohydrate, the Calvin cycle also requires chemical energy

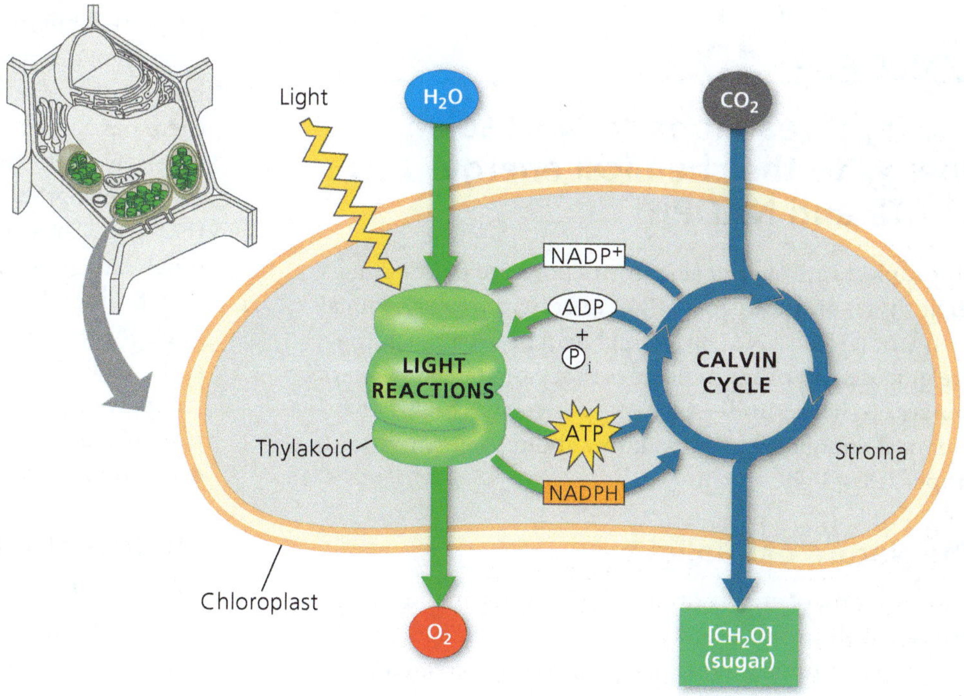

➤ **Figure 10.6 An overview of photosynthesis: cooperation of the light reactions and the Calvin cycle.** In the chloroplast, the thylakoid membranes (green) are the sites of the light reactions, whereas the Calvin cycle occurs in the stroma (gray). The light reactions use solar energy to make ATP and NADPH, which supply chemical energy and reducing power, respectively, to the Calvin cycle. The Calvin cycle incorporates CO_2 into organic molecules, which are converted to sugar. (Recall that most simple sugars have formulas that are some multiple of CH_2O.) To visualize these processes in their cellular context, see Figure 6.32.

MB **BioFlix® Animation: Photosynthesis**

in the form of ATP, which is also generated by the light reactions. Thus, it is the Calvin cycle that makes sugar, but it can do so only with the help of the NADPH and ATP produced by the light reactions. The metabolic steps of the Calvin cycle are sometimes referred to as the dark reactions, or light-independent reactions, because none of the steps requires light *directly*. Nevertheless, the Calvin cycle in most plants occurs during daylight, for only then can the light reactions provide the NADPH and ATP that the Calvin cycle requires. In essence, the chloroplast uses light energy to make sugar by coordinating the two stages of photosynthesis.

As Figure 10.6 indicates, the thylakoids of the chloroplast are the sites of the light reactions, while the Calvin cycle occurs in the stroma. On the outside of the thylakoids, molecules of $NADP^+$ and ADP pick up electrons and phosphate, respectively, and NADPH and ATP are then released to the stroma, where they play crucial roles in the Calvin cycle. The two stages of photosynthesis are treated in this figure as metabolic modules that take in ingredients and crank out products. In the next two sections, we'll look more closely at how the two stages work, beginning with the light reactions.

CONCEPT CHECK 10.1

1. **MAKE CONNECTIONS ➤** How do the CO_2 molecules used in photosynthesis reach and enter the chloroplasts inside leaf cells? (See Concept 7.2.)
2. Explain how the use of an oxygen isotope helped elucidate the chemistry of photosynthesis.
3. **WHAT IF? ➤** The Calvin cycle requires ATP and NADPH, products of the light reactions. If a classmate asserted that the light reactions don't depend on the Calvin cycle and, with continual light, could just keep on producing ATP and NADPH, how would you respond?

For suggested answers, see Appendix A.

CONCEPT 10.2

The light reactions convert solar energy to the chemical energy of ATP and NADPH

Chloroplasts are chemical factories powered by the sun. Their thylakoids transform light energy into the chemical energy of ATP and NADPH, which will be used to synthesize glucose and other molecules that can be used as energy sources. To better understand the conversion of light to chemical energy, we need to know about some important properties of light.

The Nature of Sunlight

Light is a form of energy known as electromagnetic energy, also called electromagnetic radiation. Electromagnetic energy travels in rhythmic waves analogous to those created by

▼ Figure 10.7 The electromagnetic spectrum. White light is a mixture of all wavelengths of visible light. A prism can sort white light into its component colors by bending light of different wavelengths at different angles. (Droplets of water in the atmosphere can act as prisms, causing a rainbow to form.) Visible light drives photosynthesis.

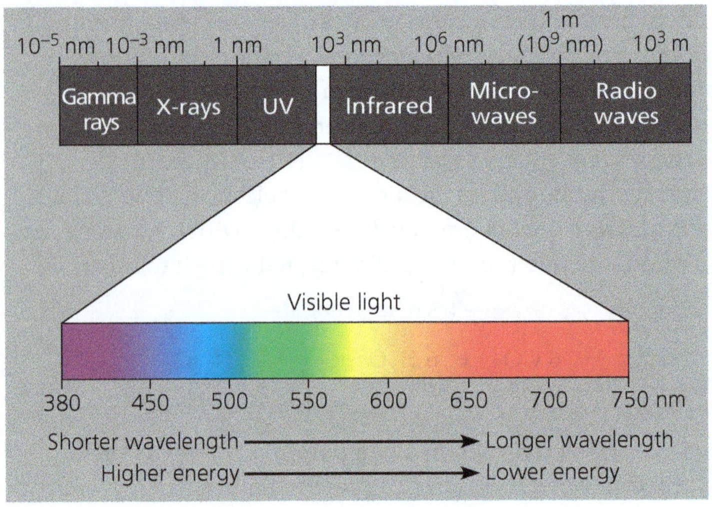

dropping a pebble into a pond. Electromagnetic waves, however, are disturbances of electric and magnetic fields rather than disturbances of a material medium such as water.

The distance between the crests of electromagnetic waves is called the **wavelength**. Wavelengths range from less than a nanometer (for gamma rays) to more than a kilometer (for radio waves). This entire range of radiation is known as the **electromagnetic spectrum** (**Figure 10.7**). The segment most important to life is the narrow band from about 380 nm to 750 nm in wavelength. This radiation is known as **visible light** because it can be detected as various colors by the human eye.

The model of light as waves explains many of light's properties, but in certain respects light behaves as though it consists of discrete particles, called **photons**. Photons are not tangible objects, but they act like objects in that each of them has a fixed quantity of energy. The amount of energy is inversely related to the wavelength of the light: The shorter the wavelength, the greater the energy of each photon of that light. Thus, a photon of violet light packs nearly twice as much energy as a photon of red light (see Figure 10.7).

Although the sun radiates the full spectrum of electromagnetic energy, the atmosphere acts like a selective window, allowing visible light to pass through while screening out a substantial fraction of other radiation. The part of the spectrum we can see—visible light—is also the radiation that drives photosynthesis.

Photosynthetic Pigments: The Light Receptors

When light meets matter, it may be reflected, transmitted, or absorbed. Substances that absorb visible light are known as

▼ Figure 10.8 Why leaves are green: interaction of light with chloroplasts. The chlorophyll molecules of chloroplasts absorb violet-blue and red light (the colors most effective in driving photosynthesis) and reflect or transmit green light. This is why leaves appear green.

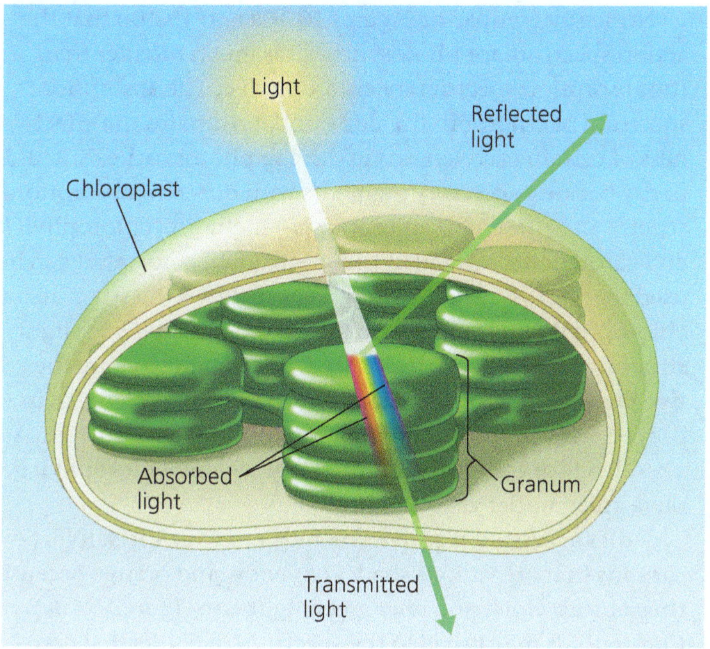

Animation: Light Energy and Pigments

pigments. Different pigments absorb light of different wavelengths, and the wavelengths that are absorbed disappear. If a pigment is illuminated with white light, the color we see is the color most reflected or transmitted by the pigment. (If a pigment absorbs all wavelengths, it appears black.) We see green when we look at a leaf because chlorophyll absorbs violet-blue and red light while transmitting and reflecting green light **(Figure 10.8)**. The ability of a pigment to absorb various wavelengths of light can be measured with an instrument called a **spectrophotometer**. This machine directs beams of light of different wavelengths through a solution of the pigment and measures the fraction of the light transmitted at each wavelength. A graph plotting a pigment's light absorption versus wavelength is called an **absorption spectrum** **(Figure 10.9)**.

The absorption spectra of chloroplast pigments provide clues to the relative effectiveness of different wavelengths for driving photosynthesis, since light can perform work in chloroplasts only if it is absorbed. **Figure 10.10a** shows the absorption spectra of three types of pigments in chloroplasts: **chlorophyll *a***, the key light-capturing pigment that participates directly in the light reactions; the accessory pigment **chlorophyll *b***; and a separate group of accessory pigments called carotenoids. The spectrum of chlorophyll *a* suggests that violet-blue and red light work best for photosynthesis, since they are absorbed, while green is the least effective color. This is confirmed by an **action spectrum** for photosynthesis **(Figure 10.10b)**, which profiles the relative effectiveness of different wavelengths of radiation in driving the process. An action spectrum is prepared by illuminating chloroplasts with light of different colors and then plotting wavelength against some measure of photosynthetic rate, such as CO_2 consumption or O_2 release. The action spectrum for photosynthesis was first demonstrated by Theodor W. Engelmann, a German botanist, in 1883. Before equipment for measuring O_2 levels had even been invented, Engelmann

▼ Figure 10.9

Research Method Determining an Absorption Spectrum

Application An absorption spectrum is a visual representation of how well a particular pigment absorbs different wavelengths of visible light. Absorption spectra of various chloroplast pigments help scientists decipher the role of each pigment in a plant.

Technique A spectrophotometer measures the relative amounts of light of different wavelengths absorbed and transmitted by a pigment solution.

1. White light is separated into colors (wavelengths) by a prism.
2. One by one, the different colors of light are passed through the sample (chlorophyll in this example). Green light and blue light are shown here.
3. The transmitted light strikes a photoelectric tube, which converts the light energy to electricity.
4. The electric current is measured by a galvanometer. The meter indicates the fraction of light transmitted through the sample, from which we can determine the amount of light absorbed.

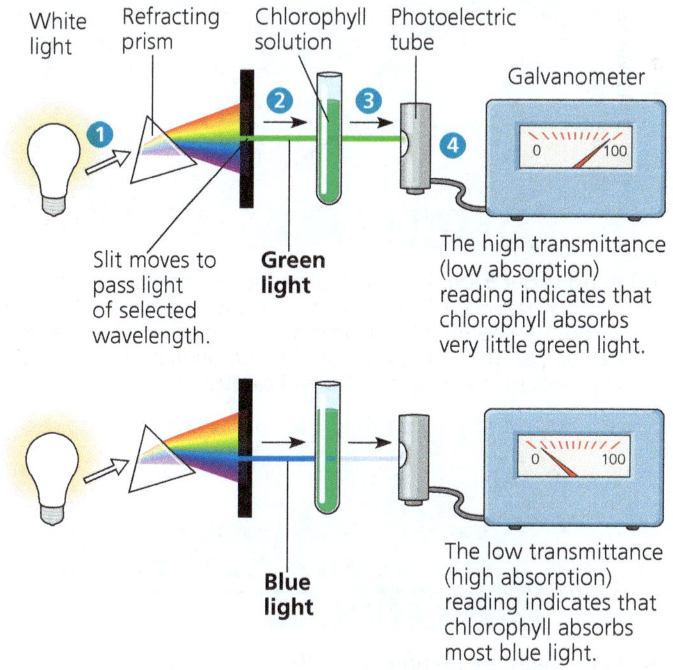

Results See Figure 10.10a for absorption spectra of three types of chloroplast pigments.

⮟ Figure 10.10

Inquiry Which wavelengths of light are most effective in driving photosynthesis?

Experiment Absorption and action spectra, along with a classic experiment by Theodor W. Engelmann, reveal which wavelengths of light are photosynthetically important.

Results

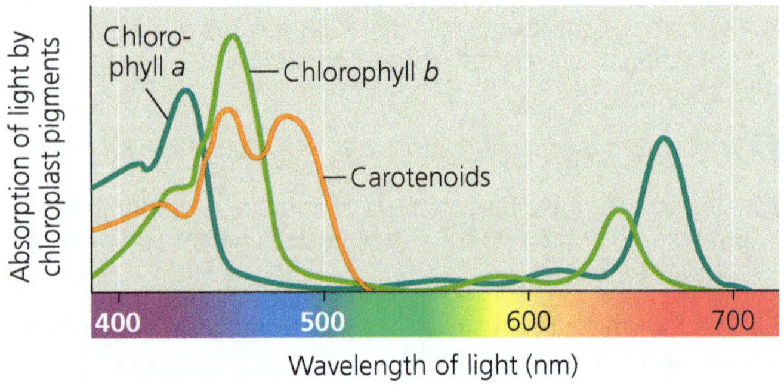

(a) Absorption spectra. The three curves show the wavelengths of light best absorbed by three types of chloroplast pigments.

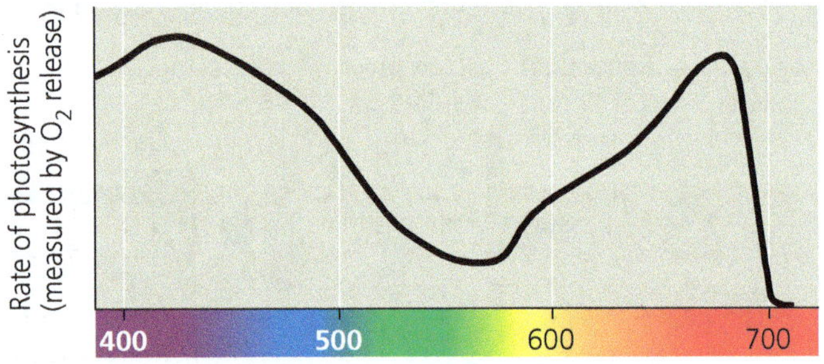

(b) Action spectrum. This graph plots the rate of photosynthesis versus wavelength. The resulting action spectrum resembles the absorption spectrum for chlorophyll *a* but does not match exactly (see part a). This is partly due to the absorption of light by accessory pigments such as chlorophyll *b* and carotenoids.

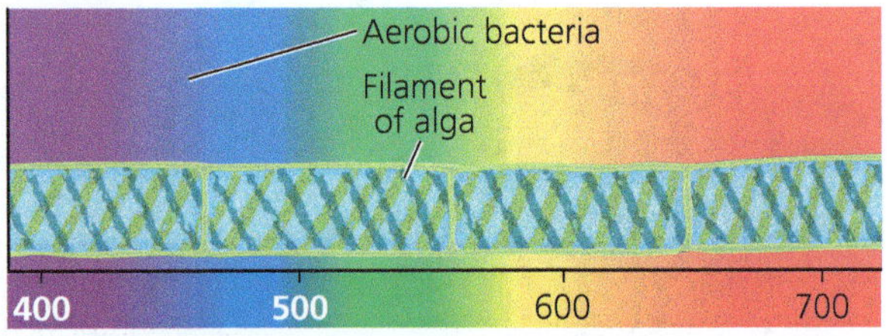

(c) Engelmann's experiment. In 1883, Theodor W. Engelmann illuminated a filamentous alga with light that had been passed through a prism, exposing different segments of the alga to different wavelengths. He used aerobic bacteria, which congregate near an oxygen source, to determine which segments of the alga were releasing the most O_2 and thus photosynthesizing most. Bacteria congregated in greatest numbers around the parts of the alga illuminated with violet-blue or red light.

Conclusion The action spectra, confirmed by Engelmann's experiment, show which portions of the spectrum are most effective in driving photosynthesis.

Data from T. W. Engelmann, *Bacterium photometricum*. Ein Beitrag zur vergleichenden Physiologie des Licht-und Farbensinnes, *Archiv. für Physiologie* 30:95–124 (1883).

MB **Instructors:** A related Experimental Inquiry Tutorial can be assigned in MasteringBiology.

INTERPRET THE DATA ➤ *According to the graph, which wavelengths of light drive the highest rates of photosynthesis?*

performed a clever experiment in which he used bacteria to measure rates of photosynthesis in filamentous algae **(Figure 10.10c)**. His results are a striking match to the modern action spectrum shown in Figure 10.10b.

Notice by comparing Figures 10.10a and 10.10b that the action spectrum for photosynthesis is much broader than the absorption spectrum of chlorophyll *a*. The absorption spectrum of chlorophyll *a* alone underestimates the effectiveness of certain wavelengths in driving photosynthesis. This is partly because accessory pigments with different absorption spectra also present in chloroplasts—including chlorophyll *b* and carotenoids—broaden the spectrum of colors that can be used for photosynthesis. **Figure 10.11** shows the structure of chlorophyll *a* compared with that of chlorophyll *b*. A slight structural difference between them is enough to cause the two pigments to absorb at slightly different wavelengths in the red and blue parts of the spectrum (see Figure 10.10a). As a result, chlorophyll *a* appears blue green and chlorophyll *b* olive green under visible light.

Other accessory pigments include **carotenoids**, hydrocarbons that are various shades of yellow and orange because they absorb violet and blue-green light (see Figure 10.10a). Carotenoids may broaden the spectrum of colors that can

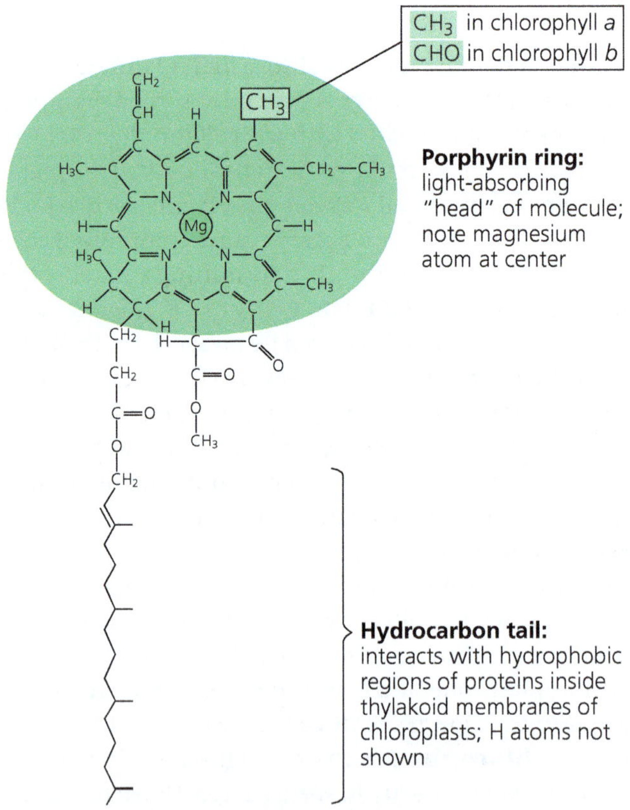

▲ **Figure 10.11 Structure of chlorophyll molecules in chloroplasts of plants.** Chlorophyll *a* and chlorophyll *b* differ in only one of the functional groups bonded to the porphyrin ring. (Also see the space-filling model of chlorophyll in Figure 1.3.)

MB **Animation: Space-Filling Model of Chlorophyll**

drive photosynthesis. However, a more important function of at least some carotenoids seems to be *photoprotection*: These compounds absorb and dissipate excessive light energy that would otherwise damage chlorophyll or interact with oxygen, forming reactive oxidative molecules that are dangerous to the cell. Interestingly, carotenoids similar to the photoprotective ones in chloroplasts have a photoprotective role in the human eye. (Carrots, known for aiding night vision, are rich in carotenoids.) These and related molecules are, of course, found naturally in many vegetables and fruits. They are also often advertised in health food products as "phytochemicals" (from the Greek *phyton*, plant), some of which have antioxidant properties. Plants can synthesize all the antioxidants they require, but humans and other animals must obtain some of them from their diets.

Excitation of Chlorophyll by Light

What exactly happens when chlorophyll and other pigments absorb light? The colors corresponding to the absorbed wavelengths disappear from the spectrum of the transmitted and reflected light, but energy cannot disappear. When a molecule absorbs a photon of light, one of the molecule's electrons is elevated to an orbital where it has more potential energy (see Figure 2.6b). When the electron is in its normal orbital, the pigment molecule is said to be in its ground state. Absorption of a photon boosts an electron to an orbital of higher energy, and the pigment molecule is then said to be in an excited state. The only photons absorbed are those whose energy is exactly equal to the energy difference between the ground state and an excited state, and this energy difference varies from one kind of molecule to another. Thus, a particular compound absorbs only photons corresponding to specific wavelengths, which is why each pigment has a unique absorption spectrum.

Once absorption of a photon raises an electron to an excited state, the electron cannot stay there long. The excited state, like all high-energy states, is unstable. Generally, when isolated pigment molecules absorb light, their excited electrons drop back down to the ground-state orbital in a billionth of a second, releasing their excess energy as heat. This conversion of light energy to heat is what makes the top of an automobile so hot on a sunny day. (White cars are coolest because their paint reflects all wavelengths of visible light.) In isolation, some pigments, including chlorophyll, emit light as well as heat after absorbing photons. As excited electrons fall back to the ground state, photons are given off, an afterglow called fluorescence. An illuminated solution of chlorophyll isolated from chloroplasts will fluoresce in the red part of the spectrum and also give off heat **(Figure 10.12)**. This is best seen by illuminating with ultraviolet light, which chlorophyll can also absorb (see Figures 10.7 and 10.10a). Viewed under visible light, the fluorescence would be difficult to see against the green of the solution.

A Photosystem: A Reaction-Center Complex Associated with Light-Harvesting Complexes

Chlorophyll molecules excited by the absorption of light energy produce very different results in an intact chloroplast than they do in isolation (see Figure 10.12). In their native environment of the thylakoid membrane, chlorophyll molecules are organized along with other small organic molecules and proteins into complexes called photosystems.

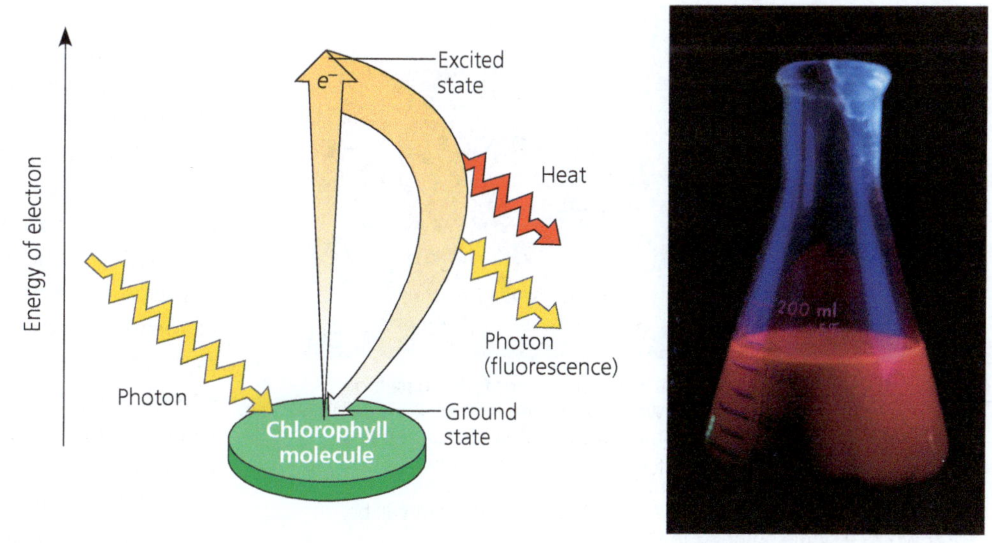

(a) Excitation of isolated chlorophyll molecule

(b) Fluorescence

➤ **Figure 10.12 Excitation of isolated chlorophyll by light. (a)** Absorption of a photon causes a transition of the chlorophyll molecule from its ground state to its excited state. The photon boosts an electron to an orbital where it has more potential energy. If the illuminated molecule exists in isolation, its excited electron immediately drops back down to the ground-state orbital, and its excess energy is given off as heat and fluorescence (light). **(b)** A chlorophyll solution excited with ultraviolet light fluoresces with a red-orange glow.

WHAT IF? ➤ *If a leaf containing the same concentration of chlorophyll as in the solution was exposed to the same ultraviolet light, no fluorescence would be seen. Propose an explanation for the difference in fluorescence emission between the solution and the leaf.*

A **photosystem** is composed of a reaction-center complex surrounded by several light-harvesting complexes **(Figure 10.13)**. The **reaction-center complex** is an organized association of proteins holding a special pair of chlorophyll *a* molecules and a primary electron acceptor. Each **light-harvesting complex** consists of various pigment molecules (which may include chlorophyll *a*, chlorophyll *b*, and multiple carotenoids) bound to proteins. The number and variety of pigment molecules enable a photosystem to harvest light over a larger surface area and a larger portion of the spectrum than could any single pigment molecule alone. Together, these light-harvesting complexes act as an antenna for the reaction-center complex. When a pigment molecule absorbs a photon, the energy is transferred from pigment molecule to pigment molecule within a light-harvesting complex, like a human "wave" at a sports arena, until it is passed to the pair of chlorophyll *a* molecules in the reaction-center complex. The pair of chlorophyll *a* molecules in the reaction-center complex are special because their molecular environment—their location and the other molecules with which they are associated—enables them to use the energy from light not only to boost one of their electrons to a higher energy level, but also to transfer it to a different molecule—the **primary electron acceptor**, which is a molecule capable of accepting electrons and becoming reduced.

▼ Figure 10.13 The structure and function of a photosystem.

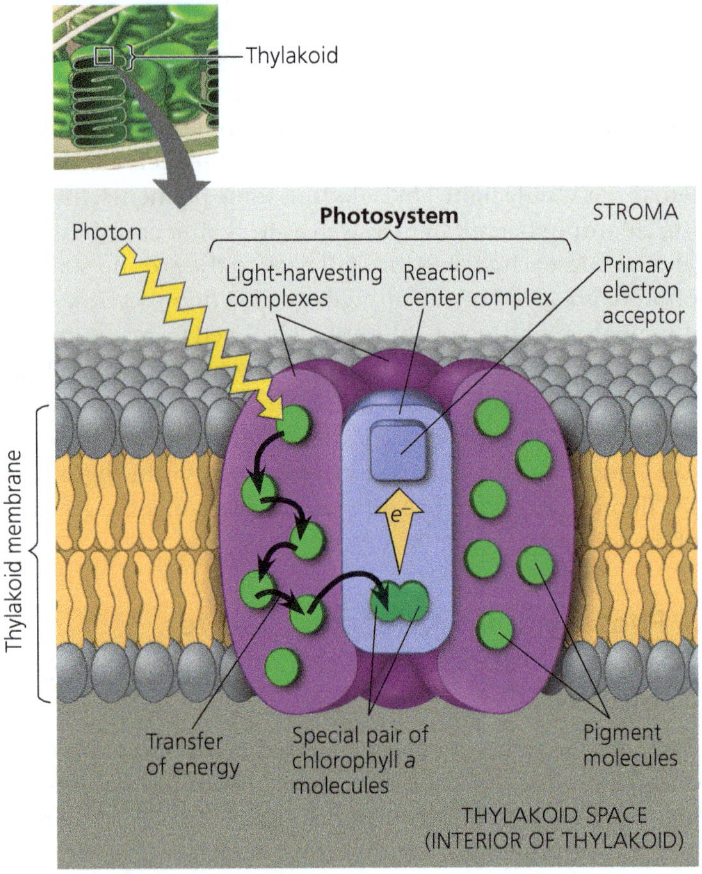

(a) How a photosystem harvests light. When a photon strikes a pigment molecule in a light-harvesting complex, the energy is passed from molecule to molecule until it reaches the reaction-center complex. Here, an excited electron from the special pair of chlorophyll *a* molecules is transferred to the primary electron acceptor.

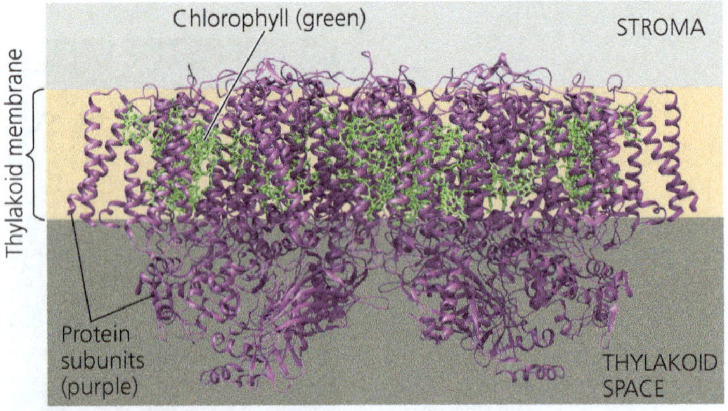

(b) Structure of a photosystem. This computer model, based on X-ray crystallography, shows two photosystem complexes side by side. Chlorophyll molecules (bright green ball-and-stick models within the membrane; the tails are not shown) are interspersed with protein subunits (purple ribbons; notice the many α helices spanning the membrane). For simplicity, a photosystem will be shown as a single complex in the rest of the chapter.

The solar-powered transfer of an electron from the reaction-center chlorophyll *a* pair to the primary electron acceptor is the first step of the light reactions. As soon as the chlorophyll electron is excited to a higher energy level, the primary electron acceptor captures it; this is a redox reaction. In the flask shown in Figure 10.12b, isolated chlorophyll fluoresces because there is no electron acceptor, so electrons of photoexcited chlorophyll drop right back to the ground state. In the structured environment of a chloroplast, however, an electron acceptor is readily available, and the potential energy represented by the excited electron is not dissipated as light and heat. Thus, each photosystem—a reaction-center complex surrounded by light-harvesting complexes—functions in the chloroplast as a unit. It converts light energy to chemical energy, which will ultimately be used for the synthesis of sugar.

The thylakoid membrane is populated by two types of photosystems that cooperate in the light reactions of photosynthesis: **photosystem II (PS II)** and **photosystem I (PS I)**. (They were named in order of their discovery, but photosystem II functions first in the light reactions.) Each has a characteristic reaction-center complex—a particular kind of primary electron acceptor next to a special pair of chlorophyll *a* molecules associated with specific proteins. The reaction-center chlorophyll *a* of photosystem II is known as P680 because this pigment is best at absorbing light having a wavelength of 680 nm (in the red part of the spectrum). The chlorophyll *a* at the reaction-center complex of photosystem I is called P700 because it most effectively absorbs light of wavelength 700 nm (in the far-red part of the spectrum). These two pigments, P680 and P700, are nearly identical chlorophyll *a* molecules. However, their association with different proteins in the thylakoid membrane affects the electron distribution in the two pigments and accounts for the slight differences in their light-absorbing properties. Now let's see how the two types of photosystems work together in using light energy to generate ATP and NADPH, the two main products of the light reactions.

Linear Electron Flow

Light drives the synthesis of ATP and NADPH by energizing the two types of photosystems embedded in the thylakoid membranes of chloroplasts. The key to this energy transformation is a flow of electrons through the photosystems and other molecular components built into the thylakoid membrane. This is called **linear electron flow**, and it occurs during the light reactions of photosynthesis, as shown in **Figure 10.14**. The numbered steps in the text correspond to the numbered steps in the figure.

1. A photon of light strikes one of the pigment molecules in a light-harvesting complex of PS II, boosting one of its electrons to a higher energy level. As this electron falls back to its ground state, an electron in a nearby pigment molecule is simultaneously raised to an excited state. The process continues, with the energy being relayed to other pigment molecules until it reaches the P680 pair of chlorophyll *a* molecules in the PS II reaction-center complex. It excites an electron in this pair of chlorophylls to a higher energy state.
2. This electron is transferred from the excited P680 to the primary electron acceptor. We can refer to the resulting form of P680, missing the negative charge of an electron, as $P680^+$.
3. An enzyme catalyzes the splitting of a water molecule into two electrons, two hydrogen ions (H^+), and an oxygen atom. The electrons are supplied one by one to the $P680^+$ pair, each electron replacing one transferred to the primary electron acceptor. ($P680^+$ is the strongest biological oxidizing agent known; its electron "hole" must be filled. This greatly facilitates the transfer of electrons from the split water molecule.) The H^+ are released into the thylakoid space (interior of the thylakoid). The oxygen atom

▼ Figure 10.14 How linear electron flow during the light reactions generates ATP and NADPH. The gold arrows trace the flow of light-driven electrons from water to NADPH. The black arrows trace the transfer of energy from pigment molecule to pigment molecule.

MB Figure Walkthrough

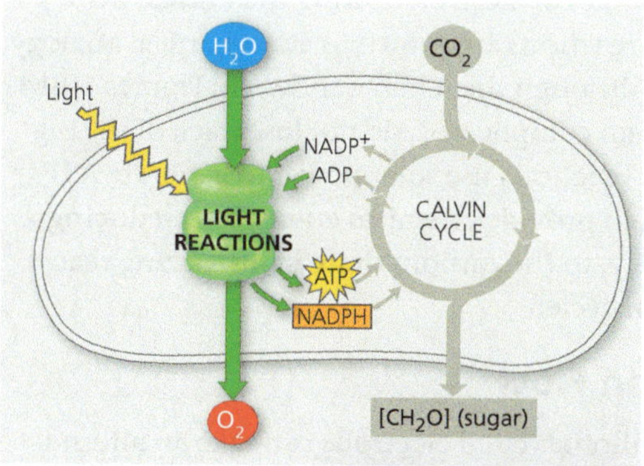

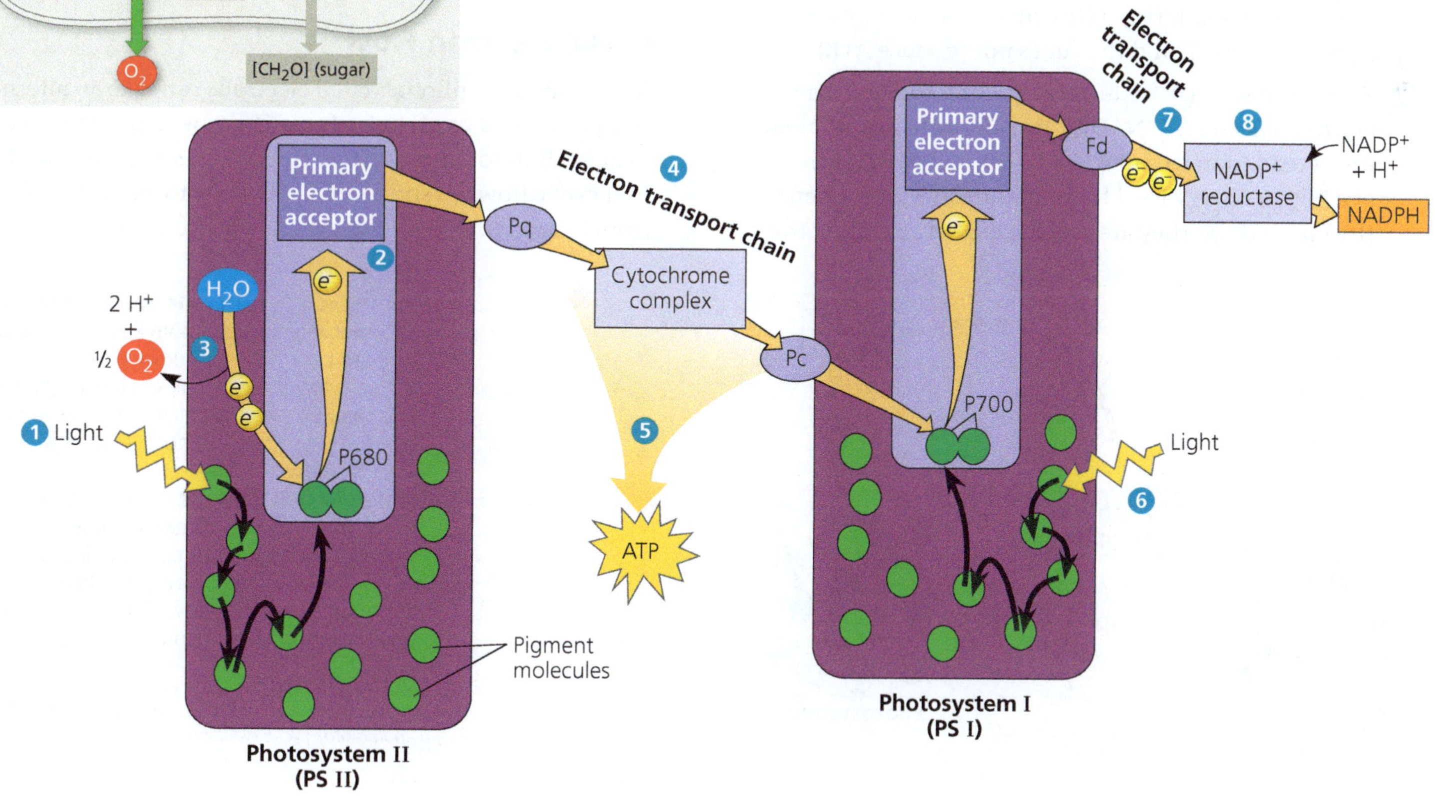

immediately combines with an oxygen atom generated by the splitting of another water molecule, forming O_2.

4 Each photoexcited electron passes from the primary electron acceptor of PS II to PS I via an electron transport chain, the components of which are similar to those of the electron transport chain that functions in cellular respiration. The electron transport chain between PS II and PS I is made up of the electron carrier plastoquinone (Pq), a cytochrome complex, and a protein called plastocyanin (Pc). Each component carries out redox reactions as electrons flow down the electron transport chain, releasing free energy that is used to pump protons (H^+) into the thylakoid space, contributing to a proton gradient across the thylakoid membrane.

5 The potential energy stored in the proton gradient is used to make ATP in a process called chemiosmosis, to be discussed shortly.

6 Meanwhile, light energy has been transferred via light-harvesting complex pigments to the PS I reaction-center complex, exciting an electron of the P700 pair of chlorophyll *a* molecules located there. The photoexcited electron is then transferred to PS I's primary electron acceptor, creating an electron "hole" in the P700—which we now can call $P700^+$. In other words, $P700^+$ can now act as an electron acceptor, accepting an electron that reaches the bottom of the electron transport chain from PS II.

7 Photoexcited electrons are passed in a series of redox reactions from the primary electron acceptor of PS I down a second electron transport chain through the protein ferredoxin (Fd). (This chain does not create a proton gradient and thus does not produce ATP.)

8 The enzyme $NADP^+$ reductase catalyzes the transfer of electrons from Fd to $NADP^+$. Two electrons are required for its reduction to NADPH. Electrons in NADPH are at a higher energy level than they are in water (where they started), so they are more readily available for the reactions of the Calvin cycle. This process also removes an H^+ from the stroma.

Figure 10.15 A mechanical analogy for linear electron flow during the light reactions.

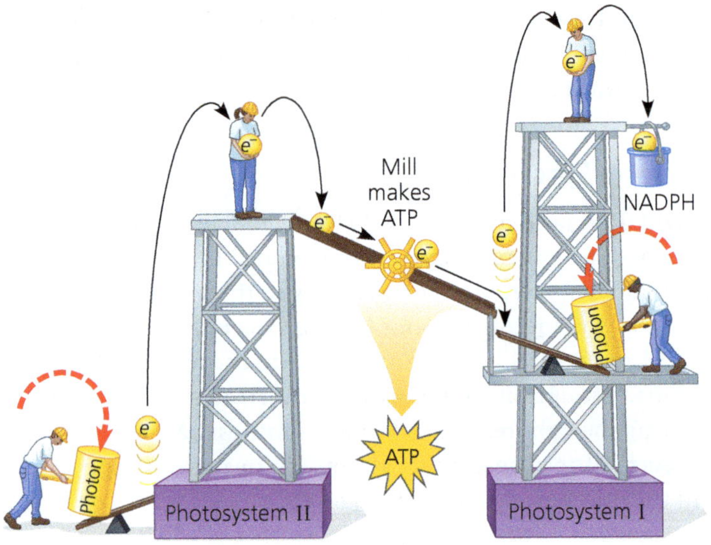

The energy changes of electrons during their linear flow through the light reactions are shown in a mechanical analogy in **Figure 10.15**. Although the scheme shown in Figures 10.14 and 10.15 may seem complicated, do not lose track of the big picture: The light reactions use solar power to generate ATP and NADPH, which provide chemical energy and reducing power, respectively, to the carbohydrate-synthesizing reactions of the Calvin cycle.

Cyclic Electron Flow

In certain cases, photoexcited electrons can take an alternative path called **cyclic electron flow**, which uses photosystem I but not photosystem II. You can see in **Figure 10.16** that cyclic flow is a short circuit: The electrons cycle back from ferredoxin (Fd) to the cytochrome complex, then

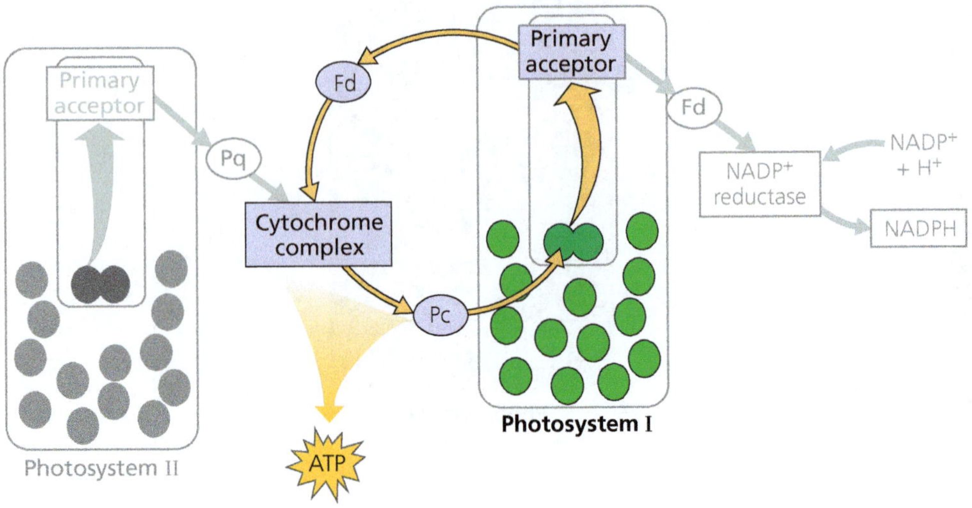

Figure 10.16 Cyclic electron flow. Photoexcited electrons from PS I are occasionally shunted back from ferredoxin (Fd) to chlorophyll via the cytochrome complex and plastocyanin (Pc). This electron shunt supplements the supply of ATP (via chemiosmosis) but produces no NADPH. The "shadow" of linear electron flow is included in the diagram for comparison with the cyclic route. The two Fd molecules in this diagram are actually one and the same—the final electron carrier in the electron transport chain of PS I—although it is depicted twice to clearly show its role in two parts of the process.

VISUAL SKILLS ➤ *Look at Figure 10.15 and explain how you would alter it to show a mechanical analogy for cyclic electron flow.*

via a plastocyanin molecule (Pc) to a P700 chlorophyll in the PS I reaction-center complex. There is no production of NADPH and no release of oxygen that results from this process. On the other hand, cyclic flow does generate ATP.

Rather than having both PS II and PS I, several of the currently existing groups of photosynthetic bacteria are known to have a single photosystem related to either PS II or PS I. For these species, which include the purple sulfur bacteria (see Figure 10.2e) and the green sulfur bacteria, cyclic electron flow is the one and only means of generating ATP during the process of photosynthesis. Evolutionary biologists hypothesize that these bacterial groups are descendants of ancestral bacteria in which photosynthesis first evolved, in a form similar to cyclic electron flow.

Cyclic electron flow can also occur in photosynthetic species that possess both photosystems; this includes some prokaryotes, such as the cyanobacteria shown in Figure 10.2d, as well as the eukaryotic photosynthetic species that have been tested thus far. Although the process is probably in part an "evolutionary leftover," research suggests it plays at least one beneficial role for these organisms. Plants with mutations that render them unable to carry out cyclic electron flow are capable of growing well in low light, but do not grow well where light is intense. This is evidence for the idea that cyclic electron flow may be photoprotective. Later you'll learn more about cyclic electron flow as it relates to a particular adaptation of photosynthesis (C_4 plants; see Concept 10.4).

Whether ATP synthesis is driven by linear or cyclic electron flow, the actual mechanism is the same. Before we move on to consider the Calvin cycle, let's review chemiosmosis, the process that uses membranes to couple redox reactions to ATP production.

A Comparison of Chemiosmosis in Chloroplasts and Mitochondria

Chloroplasts and mitochondria generate ATP by the same basic mechanism: chemiosmosis (see Figure 9.15). An electron transport chain pumps protons (H^+) across a membrane as electrons are passed through a series of carriers that are progressively more electronegative. Thus, electron transport chains transform redox energy to a proton-motive force, potential energy stored in the form of an H^+ gradient across a membrane. An ATP synthase complex in the same membrane couples the diffusion of hydrogen ions down their gradient to the phosphorylation of ADP, forming ATP.

Some of the electron carriers, including the iron-containing proteins called cytochromes, are very similar in chloroplasts and mitochondria. The ATP synthase complexes of the two organelles are also quite similar. But there are noteworthy differences between photophosphorylation in chloroplasts and oxidative phosphorylation in mitochondria. Both work by way of chemiosmosis, but in chloroplasts, the high-energy electrons dropped down the transport chain come from water, while in mitochondria, they are extracted from organic molecules (which are thus oxidized). Chloroplasts do not need molecules from food to make ATP; their photosystems capture light energy and use it to drive the electrons from water to the top of the transport chain. In other words, mitochondria use chemiosmosis to transfer chemical energy from food molecules to ATP, whereas chloroplasts use it to transform light energy into chemical energy in ATP.

Although the spatial organization of chemiosmosis differs slightly between chloroplasts and mitochondria, it is easy to see similarities in the two **(Figure 10.17)**. Electron transport chain

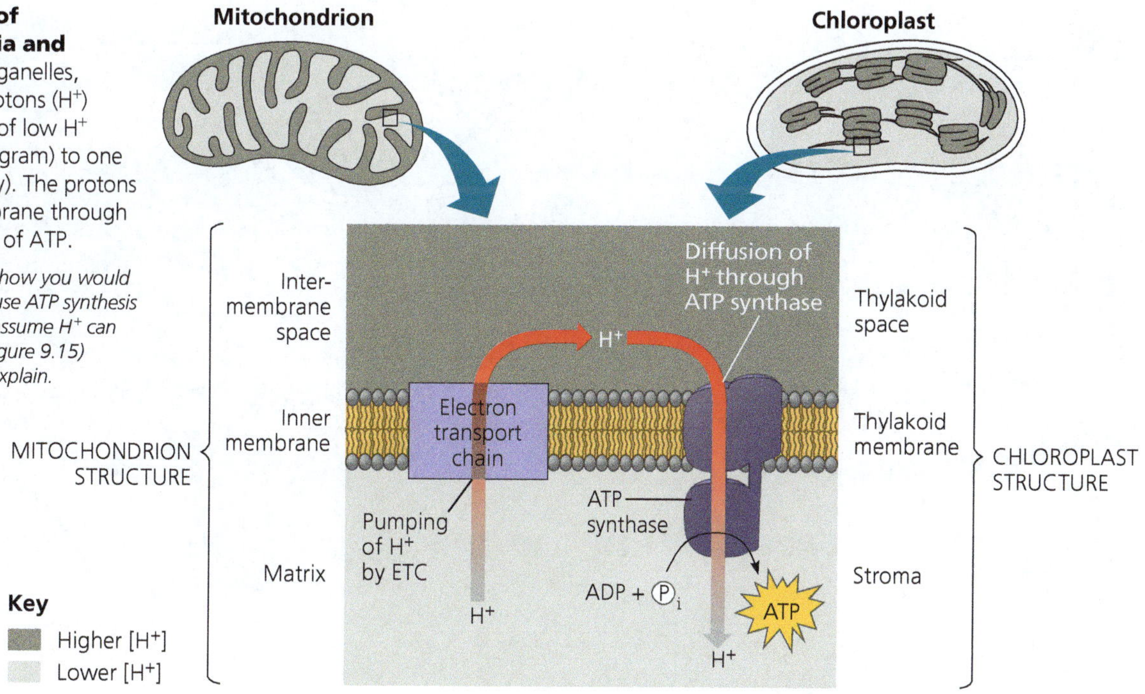

➤ **Figure 10.17 Comparison of chemiosmosis in mitochondria and chloroplasts.** In both kinds of organelles, electron transport chains pump protons (H^+) across a membrane from a region of low H^+ concentration (light gray in this diagram) to one of high H^+ concentration (dark gray). The protons then diffuse back across the membrane through ATP synthase, driving the synthesis of ATP.

MAKE CONNECTIONS ➤ *Describe how you would change the pH in order to artificially cause ATP synthesis (a) outside an isolated mitochondrion (assume H^+ can freely cross the outer membrane; see Figure 9.15) and (b) in the stroma of a chloroplast. Explain.*

proteins in the inner membrane of the mitochondrion pump protons from the mitochondrial matrix out to the intermembrane space, which then serves as a reservoir of hydrogen ions. Similarly, electron transport chain proteins in the thylakoid membrane of the chloroplast pump protons from the stroma into the thylakoid space (interior of the thylakoid), which functions as the H^+ reservoir. If you imagine the cristae of mitochondria pinching off from the inner membrane, this may help you see how the thylakoid space and the intermembrane space are comparable spaces in the two organelles, while the mitochondrial matrix is analogous to the stroma of the chloroplast.

In the mitochondrion, protons diffuse down their concentration gradient from the intermembrane space through ATP synthase to the matrix, driving ATP synthesis. In the chloroplast, ATP is synthesized as the hydrogen ions diffuse from the thylakoid space back to the stroma through ATP synthase complexes **(Figure 10.18)**, whose catalytic knobs are on the stroma side of the membrane. Thus, ATP forms in the

▼ Figure 10.18 The light reactions and chemiosmosis: Current model of the organization of the thylakoid membrane. The gold arrows track the linear electron flow outlined in Figure 10.14. At least three steps in the light reactions contribute to the H^+ gradient across the thylakoid membrane: 1 Water is split by photosystem II on the side of the membrane facing the thylakoid space; 2 as plastoquinone (Pq) transfers electrons to the cytochrome complex, four protons are translocated across the membrane into the thylakoid space; and 3 a hydrogen ion is removed from the stroma when it is taken up by $NADP^+$. Notice that in step 2, hydrogen ions are being pumped from the stroma into the thylakoid space, as in Figure 10.17. The diffusion of H^+ from the thylakoid space back to the stroma (along the H^+ concentration gradient) powers the ATP synthase. These light-driven reactions store chemical energy in NADPH and ATP, which shuttle the energy to the carbohydrate-producing Calvin cycle.

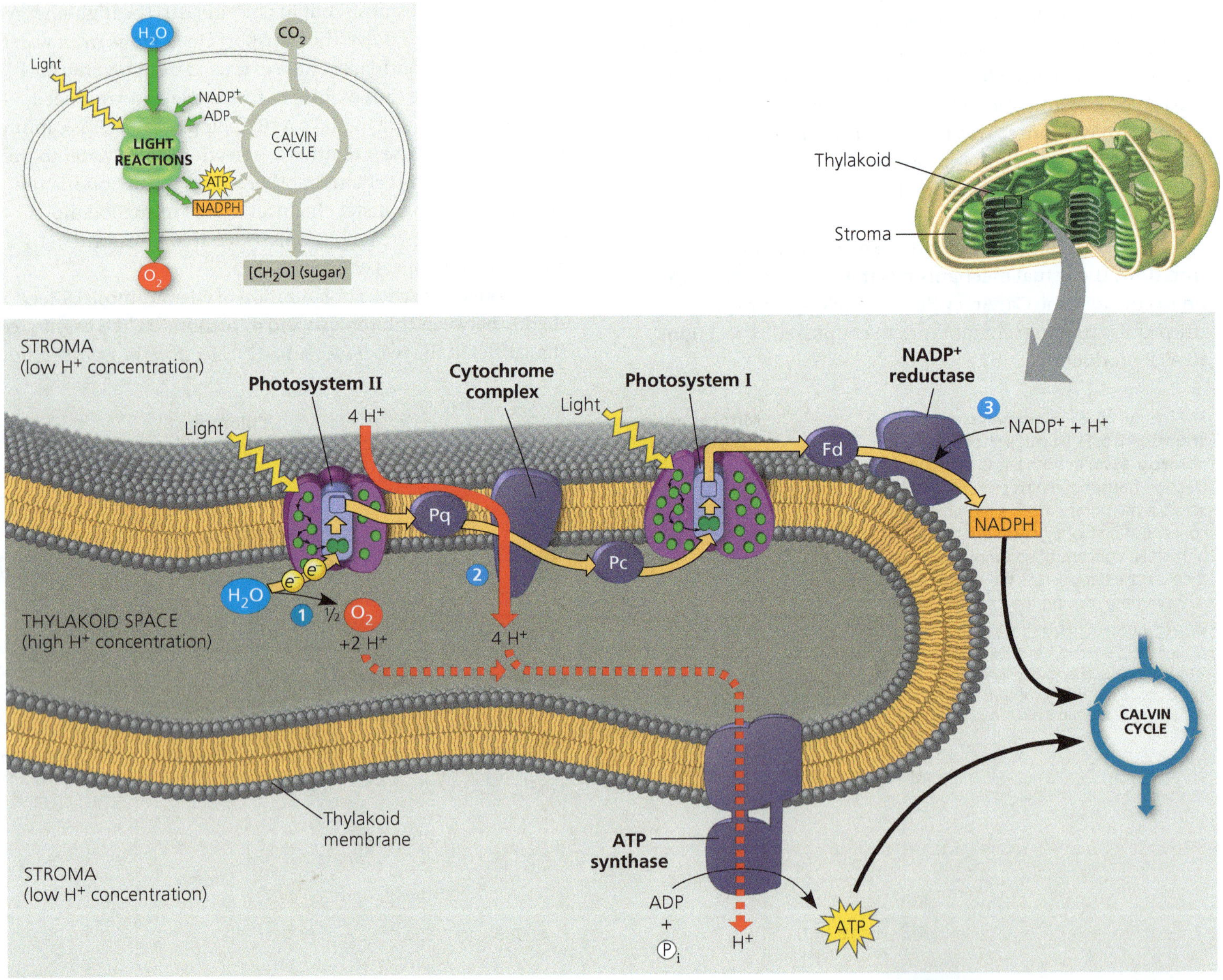

Animation: The Light Reactions

stroma, where it is used to help drive sugar synthesis during the Calvin cycle.

The proton (H^+) gradient, or pH gradient, across the thylakoid membrane is substantial. When chloroplasts in an experimental setting are illuminated, the pH in the thylakoid space drops to about 5 (the H^+ concentration increases), and the pH in the stroma increases to about 8 (the H^+ concentration decreases). This gradient of three pH units corresponds to a thousandfold difference in H^+ concentration. If the lights are then turned off, the pH gradient is abolished, but it can quickly be restored by turning the lights back on. Experiments such as this provided strong evidence in support of the chemiosmotic model.

The currently accepted model for the organization of the light-reaction "machinery" within the thylakoid membrane is based on several research studies. Each of the molecules and molecular complexes in the figure is present in numerous copies in each thylakoid. Notice that NADPH, like ATP, is produced on the side of the membrane facing the stroma, where the Calvin cycle reactions take place.

Let's summarize the light reactions. Electron flow pushes electrons from water, where they are at a state of low potential energy, ultimately to NADPH, where they are stored at a state of high potential energy. The light-driven electron flow also generates ATP. Thus, the equipment of the thylakoid membrane converts light energy to chemical energy stored in ATP and NADPH. Oxygen is produced as a by-product. Let's now see how the Calvin cycle uses the products of the light reactions to synthesize sugar from CO_2.

MB BioFlix® Animation: The Light Reactions

CONCEPT CHECK 10.2

1. What color of light is *least* effective in driving photosynthesis? Explain.
2. In the light reactions, what is the initial electron donor? Where do the electrons finally end up?
3. **WHAT IF?** ➤ In an experiment, isolated chloroplasts placed in an illuminated solution with the appropriate chemicals can carry out ATP synthesis. Predict what would happen to the rate of synthesis if a compound is added to the solution that makes membranes freely permeable to hydrogen ions.

For suggested answers, see Appendix A.

CONCEPT 10.3

The Calvin cycle uses the chemical energy of ATP and NADPH to reduce CO_2 to sugar

The Calvin cycle is similar to the citric acid cycle in that a starting material is regenerated after some molecules enter and others exit the cycle. However, the citric acid cycle is catabolic, oxidizing acetyl CoA and using the energy to synthesize ATP, while the Calvin cycle is anabolic, building carbohydrates from smaller molecules and consuming energy. Carbon enters the Calvin cycle in the form of CO_2 and leaves in the form of sugar. The cycle spends ATP as an energy source and consumes NADPH as reducing power for adding high-energy electrons to make the sugar.

As we mentioned in Concept 10.1, the carbohydrate produced directly from the Calvin cycle is not glucose. It is actually a three-carbon sugar called **glyceraldehyde 3-phosphate (G3P)**. For the net synthesis of *one* molecule of G3P, the cycle must take place three times, fixing *three* molecules of CO_2—one per turn of the cycle. (Recall that the term *carbon fixation* refers to the initial incorporation of CO_2 into organic material.) As we trace the steps of the Calvin cycle, keep in mind that we are following three molecules of CO_2 through the reactions. **Figure 10.19** divides the Calvin cycle into three phases: carbon fixation, reduction, and regeneration of the CO_2 acceptor.

Phase 1: Carbon fixation. The Calvin cycle incorporates each CO_2 molecule, one at a time, by attaching it to a five-carbon sugar named ribulose bisphosphate (abbreviated RuBP). The enzyme that catalyzes this first step is RuBP carboxylase-oxygenase, or **rubisco**. (This is the most abundant protein in chloroplasts and is also thought to be the most abundant protein on Earth.) The product of the reaction is a six-carbon intermediate that is short-lived because it is so energetically unstable that it immediately splits in half, forming two molecules of 3-phosphoglycerate (for each CO_2 fixed).

Phase 2: Reduction. Each molecule of 3-phosphoglycerate receives an additional phosphate group from ATP, becoming 1,3-bisphosphoglycerate. Next, a pair of electrons donated from NADPH reduces 1,3-bisphosphoglycerate, which also loses a phosphate group in the process, becoming glyceraldehyde 3-phosphate (G3P). Specifically, the electrons from NADPH reduce a carboyxl group on 1,3-bisphosphoglycerate to the aldehyde group of G3P, which stores more potential energy. G3P is a sugar—the same three-carbon sugar formed in glycolysis by the splitting of glucose (see Figure 9.9). Notice in Figure 10.19 that for every *three* molecules of CO_2 that enter the cycle, there are *six* molecules of G3P formed. But only one molecule of this three-carbon sugar can be counted as a net gain of carbohydrate because the rest are required to complete the cycle. The cycle began with 15 carbons' worth of carbohydrate in the form of three molecules of the five-carbon sugar RuBP. Now there are 18 carbons' worth of carbohydrate in the form of six molecules of G3P. One molecule exits the cycle to be used by the plant cell, but the other five molecules must be recycled to regenerate the three molecules of RuBP.

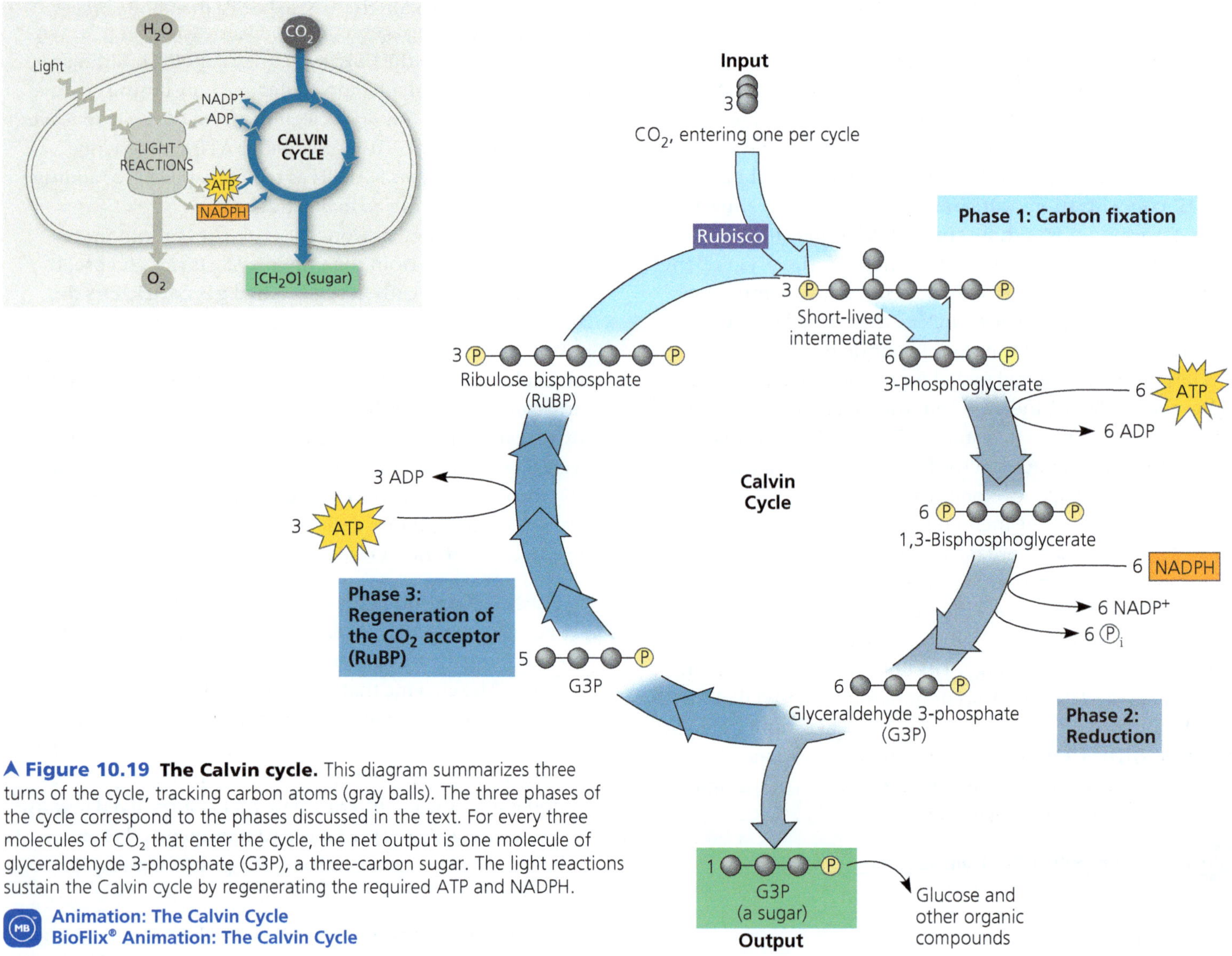

▲ **Figure 10.19 The Calvin cycle.** This diagram summarizes three turns of the cycle, tracking carbon atoms (gray balls). The three phases of the cycle correspond to the phases discussed in the text. For every three molecules of CO_2 that enter the cycle, the net output is one molecule of glyceraldehyde 3-phosphate (G3P), a three-carbon sugar. The light reactions sustain the Calvin cycle by regenerating the required ATP and NADPH.

Animation: The Calvin Cycle
BioFlix® Animation: The Calvin Cycle

Phase 3: Regeneration of the CO_2 acceptor (RuBP). In a complex series of reactions, the carbon skeletons of five molecules of G3P are rearranged by the last steps of the Calvin cycle into three molecules of RuBP. To accomplish this, the cycle spends three more molecules of ATP. The RuBP is now prepared to receive CO_2 again, and the cycle continues.

For the net synthesis of one G3P molecule, the Calvin cycle consumes a total of nine molecules of ATP and six molecules of NADPH. The light reactions regenerate the ATP and NADPH. The G3P spun off from the Calvin cycle becomes the starting material for metabolic pathways that synthesize other organic compounds, including glucose (from two molecules of G3P), the disaccharide sucrose, and other carbohydrates. Neither the light reactions nor the Calvin cycle alone can make sugar from CO_2. Photosynthesis is an emergent property of the intact chloroplast, which integrates the two stages of photosynthesis.

CONCEPT CHECK 10.3

1. To synthesize one glucose molecule, the Calvin cycle uses ________ molecules of CO_2, ________ molecules of ATP, and ________ molecules of NADPH.
2. How are the large numbers of ATP and NADPH molecules used during the Calvin cycle consistent with the high value of glucose as an energy source?
3. **WHAT IF? ➤** Explain why a poison that inhibits an enzyme of the Calvin cycle will also inhibit the light reactions.
4. **DRAW IT ➤** Redraw the cycle in Figure 10.19 using numerals to indicate the numbers of carbons instead of gray balls, multiplying at each step to ensure that you have accounted for all the carbons. In what forms do the carbon atoms enter and leave the cycle?
5. **MAKE CONNECTIONS ➤** Review Figures 9.9 and 10.19. Discuss the roles of intermediate and product played by glyceraldehyde 3-phosphate (G3P) in the two processes shown in these figures.

For suggested answers, see Appendix A.

CONCEPT 10.4

Alternative mechanisms of carbon fixation have evolved in hot, arid climates

EVOLUTION Ever since plants first moved onto land about 475 million years ago, they have been adapting to the problems of terrestrial life, particularly the problem of dehydration. In Concept 36.4, we will consider anatomical adaptations that help plants conserve water, while in this chapter we are concerned with metabolic adaptations. The solutions often involve trade-offs. An important example is the balance between photosynthesis and the prevention of excessive water loss from the plant. The CO_2 required for photosynthesis enters a leaf (and the resulting O_2 exits) via stomata, the pores on the leaf surface (see Figure 10.4). However, stomata are also the main avenues of transpiration, the evaporative loss of water from leaves. On a hot, dry day, most plants close their stomata, a response that conserves water but also reduces CO_2 levels. With stomata even partially closed, CO_2 concentrations begin to decrease in the air spaces within the leaf, and the concentration of O_2 released from the light reactions begins to increase. These conditions within the leaf favor an apparently wasteful process called photorespiration.

Photorespiration: An Evolutionary Relic?

In most plants, initial fixation of carbon occurs via rubisco, the Calvin cycle enzyme that adds CO_2 to ribulose bisphosphate. Such plants are called **C_3 plants** because the first organic product of carbon fixation is a three-carbon compound, 3-phosphoglycerate (see Figure 10.19). Rice, wheat, and soybeans are C_3 plants that are important in agriculture. When their stomata partially close on hot, dry days, C_3 plants produce less sugar because the declining level of CO_2 in the leaf starves the Calvin cycle. In addition, rubisco is capable of binding O_2 in place of CO_2. As CO_2 becomes scarce within the air spaces of the leaf and O_2 builds up, rubisco adds O_2 to the Calvin cycle instead of CO_2. The product splits, and a two-carbon compound leaves the chloroplast. Peroxisomes and mitochondria within the plant cell rearrange and split this compound, releasing CO_2. The process is called **photorespiration** because it occurs in the light (*photo*) and consumes O_2 while producing CO_2 (*respiration*). However, unlike normal cellular respiration, photorespiration uses ATP rather than generating it. And unlike photosynthesis, photorespiration produces no sugar. In fact, photorespiration *decreases* photosynthetic output by siphoning organic material from the Calvin cycle and releasing CO_2 that would otherwise be fixed. This CO_2 can eventually be fixed if it is still in the leaf once the CO_2 concentration builds up to a high enough level. In the meantime, though, the process is energetically costly, much like a hamster running on its wheel.

How can we explain the existence of a metabolic process that seems to be counterproductive for the plant? According to one hypothesis, photorespiration is evolutionary baggage—a metabolic relic from a much earlier time when the atmosphere had less O_2 and more CO_2 than it does today. In the ancient atmosphere that prevailed when rubisco first evolved, the ability of the enzyme's active site to bind O_2 would have made little difference. The hypothesis suggests that modern rubisco retains some of its chance affinity for O_2, which is now so concentrated in the atmosphere that a certain amount of photorespiration is inevitable. There is also some evidence that photorespiration may provide protection against the damaging products of the light reactions, which build up when the Calvin cycle slows due to low CO_2.

In many types of plants—including a significant number of crop plants—photorespiration drains away as much as 50% of the carbon fixed by the Calvin cycle. Indeed, if photorespiration could be reduced in certain plant species without otherwise affecting photosynthetic productivity, crop yields and food supplies might increase.

In some plant species, alternate modes of carbon fixation have evolved that minimize photorespiration and optimize the Calvin cycle—even in hot, arid climates. The two most important of these photosynthetic adaptations are C_4 photosynthesis and crassulacean acid metabolism (CAM).

C_4 Plants

The **C_4 plants** are so named because they preface the Calvin cycle with an alternate mode of carbon fixation that forms a four-carbon compound as its first product. The C_4 pathway is believed to have evolved independently at least 45 separate times and is used by several thousand species in at least 19 plant families. Among the C_4 plants important to agriculture are sugarcane and corn, members of the grass family.

The anatomy of a C_4 leaf is correlated with the mechanism of C_4 photosynthesis. In C_4 plants, there are two distinct types of photosynthetic cells: bundle-sheath cells and mesophyll cells. **Bundle-sheath cells** are arranged into tightly packed sheaths around the veins of the leaf **(Figure 10.20)**. Between the bundle sheath and the leaf surface are the more loosely arranged mesophyll cells, which, in C_4 leaves, are closely associated and never more than two to three cells away from the bundle-sheath cells. The Calvin cycle is confined to the chloroplasts of the bundle-sheath cells. However, the Calvin cycle is preceded by incorporation of CO_2 into organic compounds in the mesophyll cells. See the numbered steps in Figure 10.20, which are also described here:

1. The first step is carried out by an enzyme present only in mesophyll cells called **PEP carboxylase**. This enzyme adds CO_2 to phosphoenolpyruvate (PEP), forming the four-carbon product oxaloacetate. PEP carboxylase has a much higher affinity for CO_2 than does rubisco and

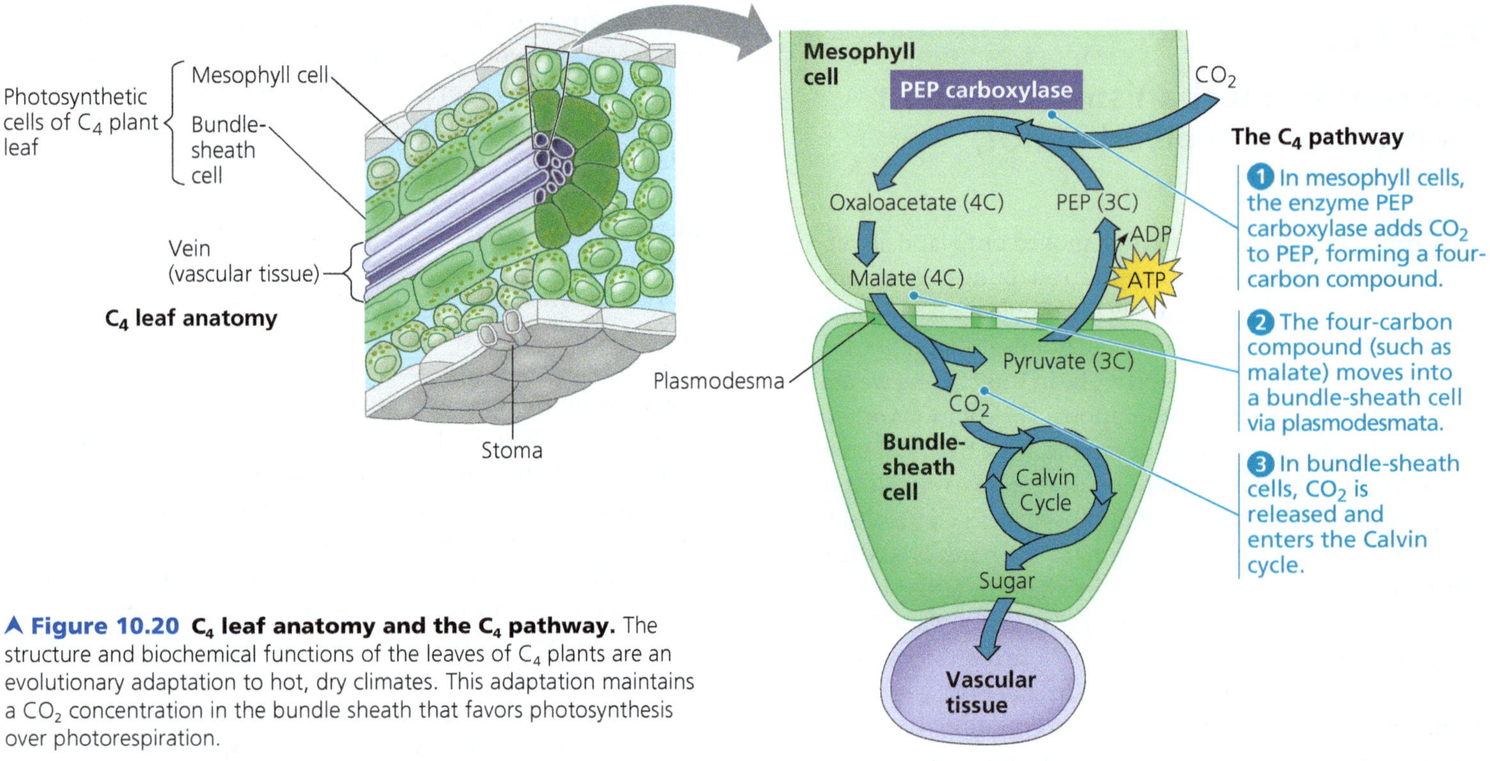

▲ **Figure 10.20 C_4 leaf anatomy and the C_4 pathway.** The structure and biochemical functions of the leaves of C_4 plants are an evolutionary adaptation to hot, dry climates. This adaptation maintains a CO_2 concentration in the bundle sheath that favors photosynthesis over photorespiration.

no affinity for O_2. Therefore, PEP carboxylase can fix carbon efficiently when rubisco cannot—that is, when it is hot and dry and stomata are partially closed, causing CO_2 concentration in the leaf to be lower and O_2 concentration to be relatively higher.

2. After the CO_2 is fixed in the mesophyll cells, the four-carbon products (malate in the example shown in Figure 10.20) are exported to bundle-sheath cells through plasmodesmata (see Figure 6.29).

3. Within the bundle-sheath cells, the four-carbon compounds release CO_2, which is re-fixed into organic material by rubisco and the Calvin cycle. The same reaction regenerates pyruvate, which is transported to mesophyll cells. There, ATP is used to convert pyruvate to PEP, which can accept addition of another CO_2, allowing the reaction cycle to continue. This ATP can be thought of, in a sense, as the "price" of concentrating CO_2 in the bundle-sheath cells. To generate this extra ATP, bundle-sheath cells carry out cyclic electron flow, the process described earlier in this chapter (see Figure 10.16). In fact, these cells contain PS I but no PS II, so cyclic electron flow is their only photosynthetic mode of generating ATP.

In effect, the mesophyll cells of a C_4 plant pump CO_2 into the bundle-sheath cells, keeping the CO_2 concentration in those cells high enough for rubisco to bind CO_2 rather than O_2. The cyclic series of reactions involving PEP carboxylase and the regeneration of PEP can be thought of as an ATP-powered pump that concentrates CO_2. In this way, C_4 photosynthesis spends ATP energy to minimize photorespiration and enhance sugar production. This adaptation is especially advantageous in hot regions with intense sunlight, where stomata partially close during the day, and it is in such environments that C_4 plants evolved and thrive today.

The concentration of CO_2 in the atmosphere has drastically increased since the Industrial Revolution began in the 1800s, and it continues to rise today due to human activities such as the burning of fossil fuels. The resulting global climate change, including an increase in average temperatures around the planet, may have far-reaching effects on plant species. Scientists are concerned that increasing CO_2 concentration and temperature may affect C_3 and C_4 plants differently, thus changing the relative abundance of these species in a given plant community.

Which type of plant would stand to gain more from increasing CO_2 levels? Recall that in C_3 plants, the binding of O_2 rather than CO_2 by rubisco leads to photorespiration, lowering the efficiency of photosynthesis. C_4 plants overcome this problem by concentrating CO_2 in the bundle-sheath cells at the cost of ATP. Rising CO_2 levels should benefit C_3 plants by lowering the amount of photorespiration that occurs. At the same time, rising temperatures have the opposite effect, increasing photorespiration. (Other factors such as water availability may also come into play.) In contrast, many C_4 plants could be largely unaffected by increasing CO_2 levels or temperature. Researchers have investigated aspects of this question in several studies; you can work with data from one such experiment in the **Scientific Skills Exercise**. In different regions, the particular combination of CO_2 concentration and temperature is likely to alter the balance of C_3 and C_4 plants in varying ways. The effects of such a widespread and variable change in community structure are unpredictable and thus a cause of legitimate concern.

SCIENTIFIC SKILLS EXERCISE

Making Scatter Plots with Regression Lines

Does Atmospheric CO_2 Concentration Affect the Productivity of Agricultural Crops? The atmospheric concentration of CO_2 has been rising globally, and scientists wondered whether this would affect C_3 and C_4 plants differently. In this exercise, you will make a scatter plot to examine the relationship between CO_2 concentration and growth of both corn (maize), a C_4 crop plant, and velvetleaf, a C_3 weed found in cornfields.

➤ Corn plant surrounded by invasive velvetleaf plants

How the Experiment Was Done Researchers grew corn and velvetleaf plants under controlled conditions for 45 days, where all plants received the same amounts of water and light. The plants were divided into three groups, and each was exposed to a different concentration of CO_2 in the air: 350, 600, or 1,000 ppm (parts per million).

Data from the Experiment The table shows the dry mass (in grams) of corn and velvetleaf plants grown at the three concentrations of CO_2. The dry mass values are averages of the leaves, stems, and roots of eight plants.

	350 ppm CO_2	600 ppm CO_2	1,000 ppm CO_2
Average dry mass of one corn plant (g)	91	89	80
Average dry mass of one velvetleaf plant (g)	35	48	54

Data from D. T. Patterson and E. P. Flint, Potential effects of global atmospheric CO_2 enrichment on the growth and competitiveness of C_3 and C_4 weed and crop plants, *Weed Science* 28(1):71–75 (1980).

INTERPRET THE DATA

1. To explore the relationship between the two variables, it is useful to graph the data in a scatter plot, and then draw a regression line. (a) First, place labels for the dependent and independent variables on the appropriate axes. Explain your choices. (b) Now plot the data points for corn and velvetleaf using different symbols for each set of data, and add a key for the two symbols. (For additional information about graphs, see the Scientific Skills Review in Appendix F.)
2. Draw a "best-fit" line for each set of points. A best-fit line does not necessarily pass through all or even most points. Instead, it is a straight line that passes as close as possible to all data points from that set. Draw a best-fit line for each set of data. Because placement of the line is a matter of judgment, two individuals may draw two slightly different lines for a given set of points. The line that actually fits best, a regression line, can be identified by squaring the distances of all points to any candidate line, then selecting the line that minimizes the sum of the squares. (See the graph in the Scientific Skills Exercise in Chapter 3 for an example of a linear regression line.) Excel or other software programs, including those on a graphing calculator, can plot a regression line once data points are entered. Using either Excel or a graphing calculator, enter the data points for each data set and have the program draw the two regression lines. Compare them to the lines you drew.
3. Describe the trends shown by the regression lines in your scatter plot. (a) Compare the relationship between increasing concentration of CO_2 and the dry mass of corn to that for velvetleaf. (b) Considering that velvetleaf is a weed invasive to cornfields, predict how increased CO_2 concentration may affect interactions between the two species.
4. Based on the data in the scatter plot, estimate the percentage change in dry mass of corn and velvetleaf plants if atmospheric CO_2 concentration increased from 390 ppm (current levels) to 800 ppm. (a) What is the estimated dry mass of corn and velvetleaf plants at 390 ppm? 800 ppm? (b) To calculate the percentage change in mass for each plant, subtract the mass at 390 ppm from the mass at 800 ppm (change in mass), divide by the mass at 390 ppm (initial mass), and multiply by 100. What is the estimated percentage change in dry mass for corn? For velvetleaf? (c) Do these results support the conclusion from other experiments that C_3 plants grow better than C_4 plants under increased CO_2 concentration? Why or why not?

MB **Instructors:** A version of this Scientific Skills Exercise can be assigned in MasteringBiology.

C_4 photosynthesis is considered more efficient than C_3 photosynthesis because it uses less water and resources. On our planet today, the world population and demand for food are rapidly increasing. At the same time, the amount of land suitable for growing crops is decreasing due to the effects of global climate change, which include an increase in sea level as well as a hotter, drier climate in many regions. To address issues of food supply, scientists in the Philippines have been working on genetically modifying rice—an important food staple that is a C_3 crop—so that it can instead carry out C_4 photosynthesis. Results so far seem promising, and these researchers estimate that the yield of C_4 rice might be 30–50% higher than C_3 rice with the same input of water and resources.

CAM Plants

A second photosynthetic adaptation to arid conditions has evolved in many succulent (water-storing) plants, numerous cacti, pineapples, and representatives of several other plant families. These plants open their stomata during the night and close them during the day, just the reverse of how other plants behave. Closing stomata during the day helps desert plants conserve water, but it also prevents CO_2 from entering the leaves. During the night, when their stomata are open, these plants take up CO_2 and incorporate it into a variety of organic acids. This mode of carbon fixation is called **crassulacean acid metabolism**, or **CAM**, after the plant family Crassulaceae, the succulents in which the process was first discovered. The mesophyll cells of **CAM plants** store the organic acids they make during the night in their vacuoles until morning, when the stomata close. During the day, when the light reactions can supply ATP and NADPH for the Calvin cycle, CO_2 is released from the organic acids made the night before to become incorporated into sugar in the chloroplasts.

Notice in **Figure 10.21** that the CAM pathway is similar to the C_4 pathway in that CO_2 is first incorporated into organic intermediates before it enters the Calvin cycle. The difference is that in C_4 plants, the initial steps of carbon fixation are separated structurally from the Calvin cycle, whereas in CAM plants, the two steps occur within the same cell but at separate times. (Keep in mind that CAM, C_4, and C_3 plants all eventually use the Calvin cycle to make sugar from carbon dioxide.)

Figure 10.21 C_4 and CAM photosynthesis compared. The C_4 and CAM pathways are two evolutionary solutions to the problem of maintaining photosynthesis with stomata partially or completely closed on hot, dry days. Both adaptations are characterized by (1) preliminary incorporation of CO_2 into organic acids, followed by (2) transfer of CO_2 to the Calvin cycle.

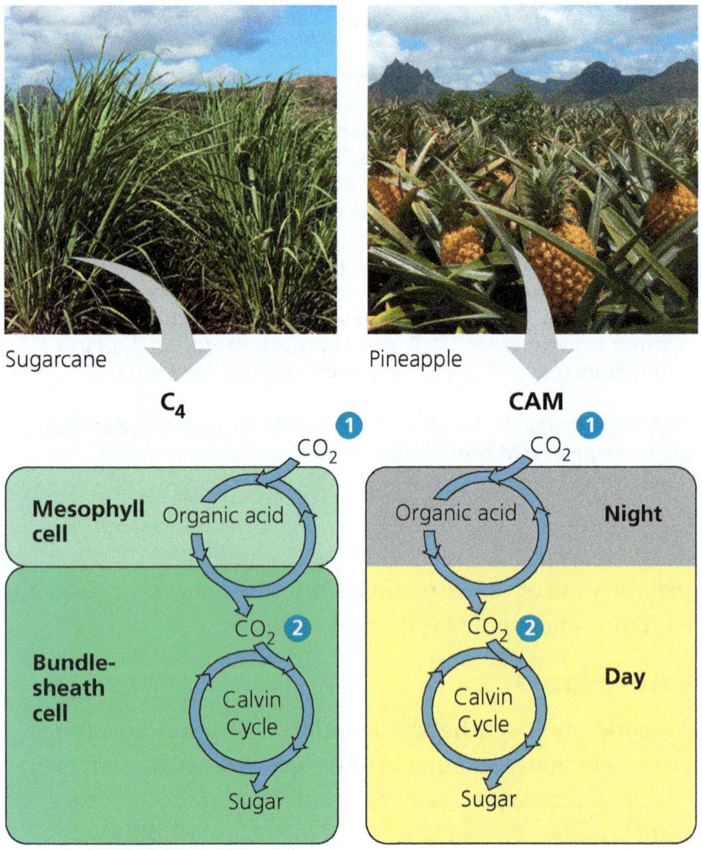

(a) Spatial separation of steps. In C_4 plants, carbon fixation and the Calvin cycle occur in different types of cells.

(b) Temporal separation of steps. In CAM plants, carbon fixation and the Calvin cycle occur in the same cell at different times.

MB Animation: Photosynthesis in Dry Climates

CONCEPT CHECK 10.4

1. Describe how photorespiration lowers photosynthetic output for plants.
2. The presence of only PS I, not PS II, in the bundle-sheath cells of C_4 plants has an effect on O_2 concentration. What is that effect, and how might that benefit the plant?
3. **MAKE CONNECTIONS ➤** Refer to the discussion of ocean acidification in Concept 3.3. Ocean acidification and changes in the distribution of C_3 and C_4 plants may seem to be two very different problems, but what do they have in common? Explain.
4. **WHAT IF? ➤** How would you expect the relative abundance of C_3 versus C_4 and CAM species to change in a geographic region whose climate becomes much hotter and drier, with no change in CO_2 concentration?

For suggested answers, see Appendix A.

CONCEPT 10.5

Life depends on photosynthesis

The Importance of Photosynthesis: *A Review*

In this chapter, we have followed photosynthesis from photons to food. The light reactions capture solar energy and use it to make ATP and transfer electrons from water to $NADP^+$, forming NADPH. The Calvin cycle uses the ATP and NADPH to produce sugar from carbon dioxide. The energy that enters the chloroplasts as sunlight becomes stored as chemical energy in organic compounds. The entire process is reviewed visually in **Figure 10.22**, where photosynthesis is also put in its natural context.

As for the fates of photosynthetic products, enzymes in the chloroplast and cytosol convert the G3P made in the Calvin cycle to many other organic compounds. In fact, the sugar made in the chloroplasts supplies the entire plant with chemical energy and carbon skeletons for the synthesis of all the major organic molecules of plant cells. About 50% of the organic material made by photosynthesis is consumed as fuel for cellular respiration in plant cell mitochondria.

Technically, green cells are the only autotrophic parts of the plant. The rest of the plant depends on organic molecules exported from leaves via veins (see Figure 10.22, top). In most plants, carbohydrate is transported out of the leaves to the rest of the plant in the form of sucrose, a disaccharide. After arriving at nonphotosynthetic cells, the sucrose provides raw material for cellular respiration and a multitude of anabolic pathways that synthesize proteins, lipids, and other products. A considerable amount of sugar in the form of glucose is linked together to make the polysaccharide cellulose (see Figure 5.6c), especially in plant cells that are still growing and maturing. Cellulose, the main ingredient of cell walls, is the most abundant organic molecule in the plant—and probably on the surface of the planet.

Most plants and other photosynthesizers make more organic material each day than they need to use as respiratory fuel and precursors for biosynthesis. They stockpile the extra sugar by synthesizing starch, storing some in the chloroplasts themselves and some in storage cells of roots, tubers, seeds, and fruits. In accounting for the consumption of the food molecules produced by photosynthesis, let's not forget that most plants lose leaves, roots, stems, fruits, and sometimes their entire bodies to heterotrophs, including humans.

On a global scale, photosynthesis is the process responsible for the presence of oxygen in our atmosphere. Furthermore, although each chloroplast is minuscule, their collective productivity in terms of food production is prodigious: Photosynthesis makes an estimated 150 billion metric tons of carbohydrate per year (a metric ton is 1,000 kg, about 1.1 tons). That's organic matter equivalent in mass to a stack of about 60 trillion biology textbooks—and the stack would reach 17 times the distance from Earth to the sun! No chemical process is more important than photosynthesis to the welfare of life on Earth.

In Chapters 5 through 10, you have learned about many activities of cells. **Figure 10.23** integrates these cellular processes into the context of a working plant cell. As you study the figure, reflect on how each process fits into the big picture: As the most basic unit of living organisms, a cell performs all functions characteristic of life.

CONCEPT CHECK 10.5

1. **MAKE CONNECTIONS ➤** Can plants use the sugar they produce during photosynthesis to directly power the work of the cell? Explain. (See Figures 8.10, 8.11, and 9.6.).

For suggested answers, see Appendix A.

▼ Figure 10.22 A review of photosynthesis. This diagram shows the main reactants and products of photosynthesis as they move through the tissues of a tree (left) and a chloroplast (right).

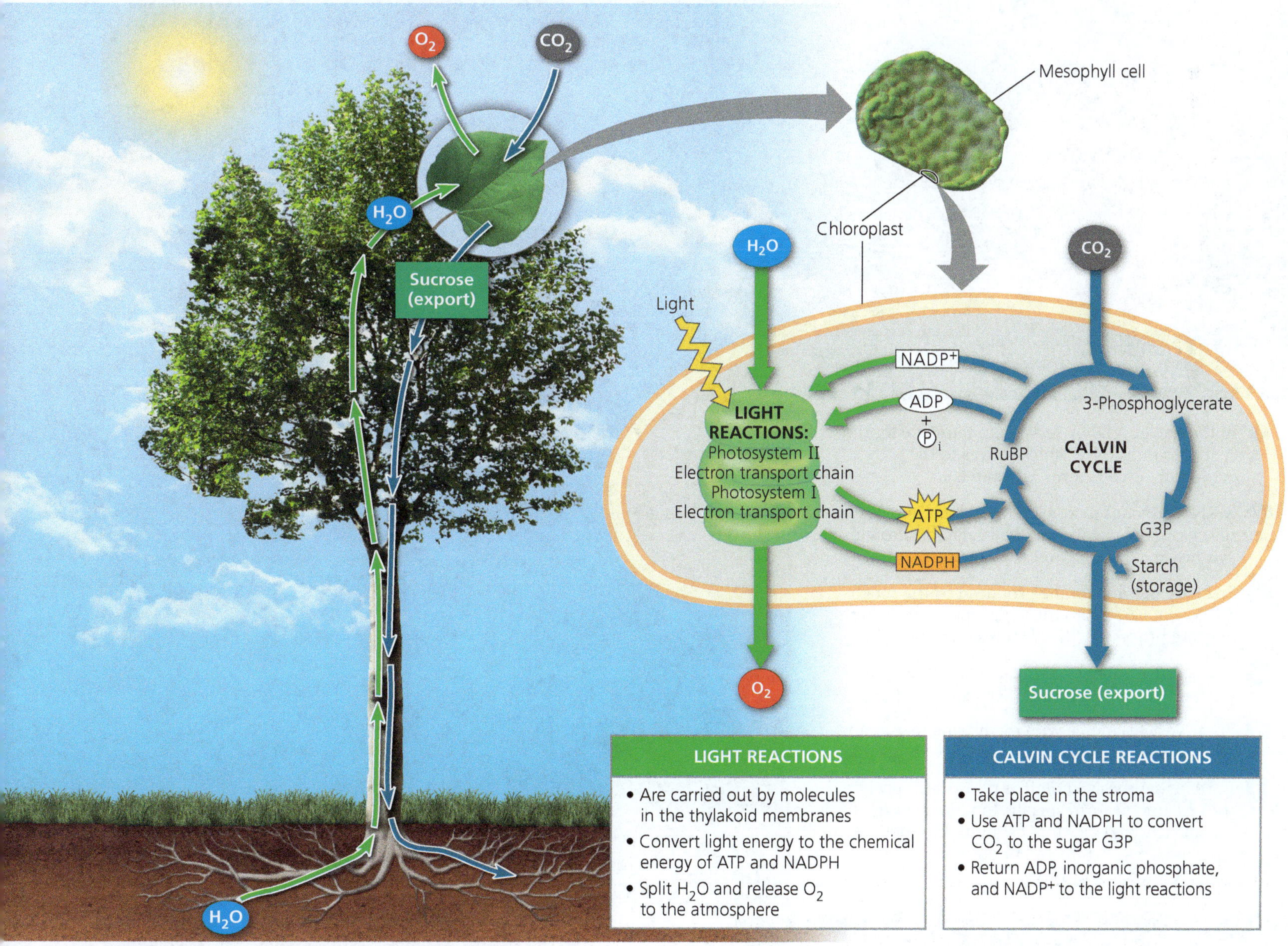

LIGHT REACTIONS	CALVIN CYCLE REACTIONS
• Are carried out by molecules in the thylakoid membranes • Convert light energy to the chemical energy of ATP and NADPH • Split H_2O and release O_2 to the atmosphere	• Take place in the stroma • Use ATP and NADPH to convert CO_2 to the sugar G3P • Return ADP, inorganic phosphate, and NADP+ to the light reactions

BioFlix® Animation: Photosynthesis
MP3 Tutor: Photosynthesis

Figure 10.23 MAKE CONNECTIONS

The Working Cell

This figure illustrates how a generalized plant cell functions, integrating the cellular activities you learned about in Chapters 5–10.

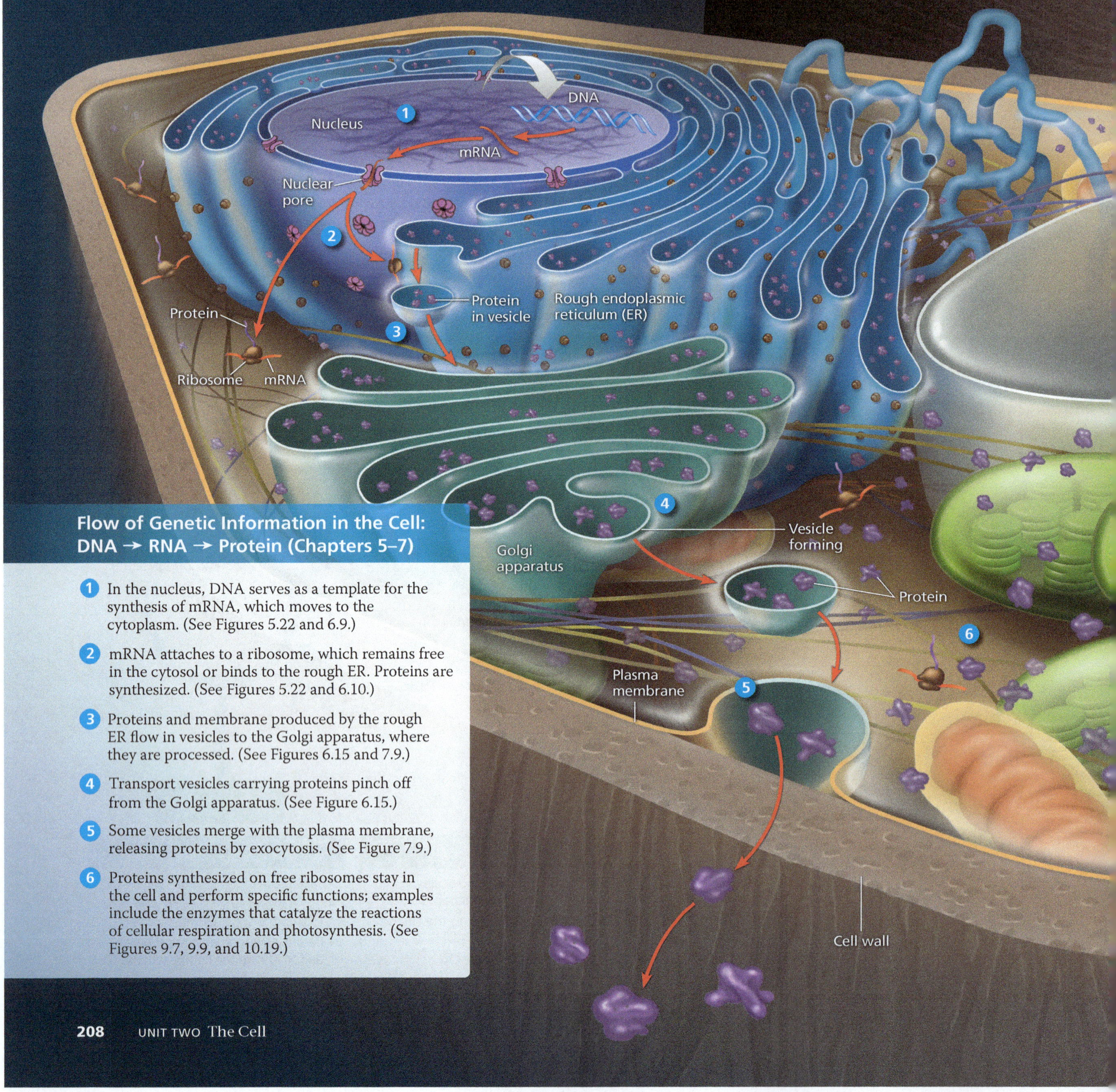

Flow of Genetic Information in the Cell: DNA → RNA → Protein (Chapters 5–7)

1. In the nucleus, DNA serves as a template for the synthesis of mRNA, which moves to the cytoplasm. (See Figures 5.22 and 6.9.)
2. mRNA attaches to a ribosome, which remains free in the cytosol or binds to the rough ER. Proteins are synthesized. (See Figures 5.22 and 6.10.)
3. Proteins and membrane produced by the rough ER flow in vesicles to the Golgi apparatus, where they are processed. (See Figures 6.15 and 7.9.)
4. Transport vesicles carrying proteins pinch off from the Golgi apparatus. (See Figure 6.15.)
5. Some vesicles merge with the plasma membrane, releasing proteins by exocytosis. (See Figure 7.9.)
6. Proteins synthesized on free ribosomes stay in the cell and perform specific functions; examples include the enzymes that catalyze the reactions of cellular respiration and photosynthesis. (See Figures 9.7, 9.9, and 10.19.)

Energy Transformations in the Cell: Photosynthesis and Cellular Respiration (Chapters 8–10)

7 In chloroplasts, the process of photosynthesis uses the energy of light to convert CO_2 and H_2O to organic molecules, with O_2 as a by-product. (See Figure 10.22.)

8 In mitochondria, organic molecules are broken down by cellular respiration, capturing energy in molecules of ATP, which are used to power the work of the cell, such as protein synthesis and active transport. CO_2 and H_2O are by-products. (See Figures 8.9–8.11, 9.2, and 9.16.)

Movement Across Cell Membranes (Chapter 7)

9 Water diffuses into and out of the cell directly through the plasma membrane and by facilitated diffusion through aquaporins. (See Figure 7.1.)

10 By passive transport, the CO_2 used in photosynthesis diffuses into the cell and the O_2 formed as a by-product of photosynthesis diffuses out of the cell. Both solutes move down their concentration gradients. (See Figures 7.10 and 10.22.)

11 In active transport, energy (usually supplied by ATP) is used to transport a solute against its concentration gradient. (See Figure 7.16.)

Exocytosis (shown in step 5) and endocytosis move larger materials out of and into the cell. (See Figures 7.9 and 7.19.)

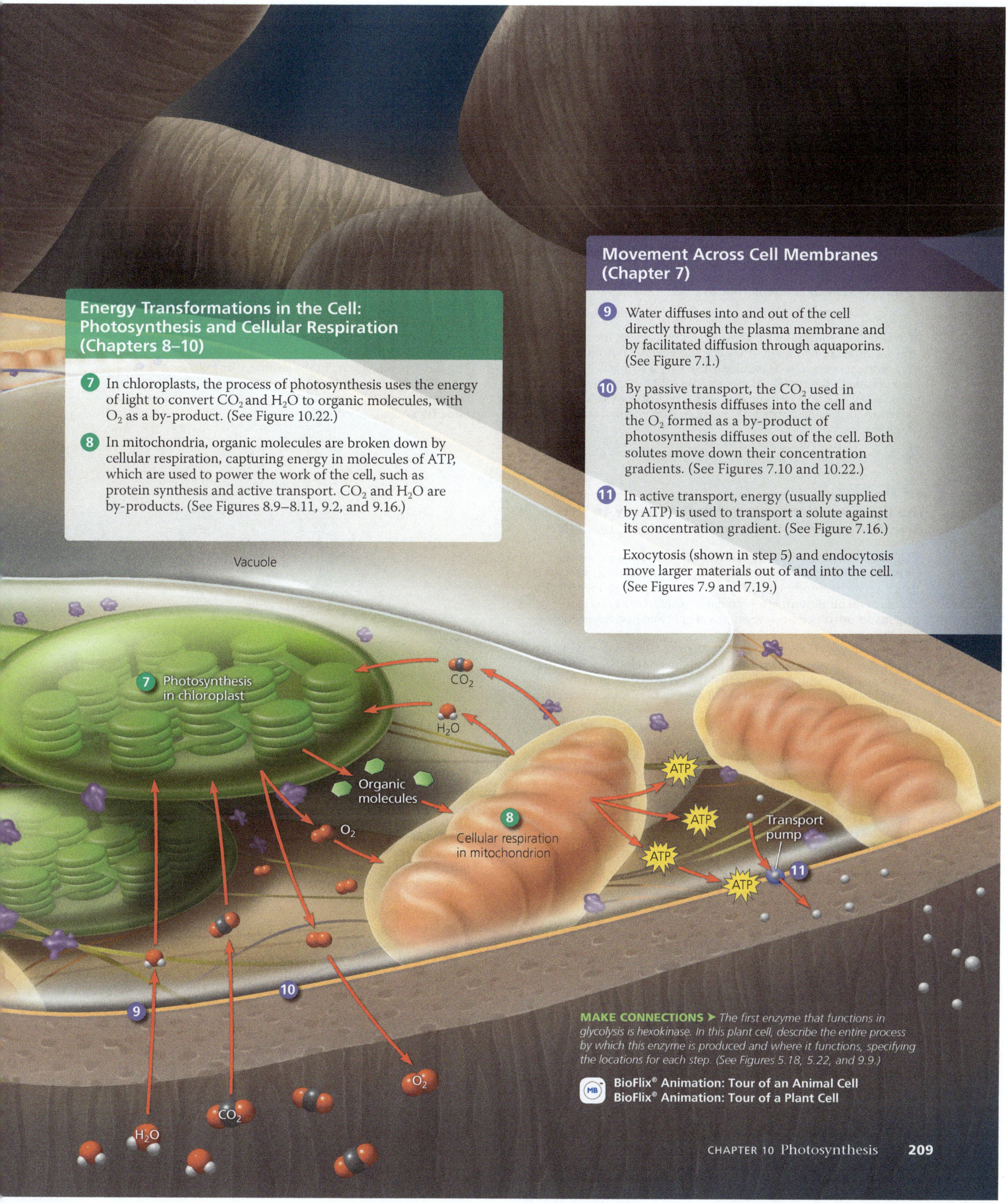

MAKE CONNECTIONS ➤ *The first enzyme that functions in glycolysis is hexokinase. In this plant cell, describe the entire process by which this enzyme is produced and where it functions, specifying the locations for each step. (See Figures 5.18, 5.22, and 9.9.)*

BioFlix® Animation: Tour of an Animal Cell
BioFlix® Animation: Tour of a Plant Cell

10 Chapter Review

Go to **MasteringBiology**™ for Videos, Animations, Vocab Self-Quiz, Practice Tests, and more in the Study Area.

SUMMARY OF KEY CONCEPTS

CONCEPT 10.1

Photosynthesis converts light energy to the chemical energy of food *(pp. 189–192)*

VOCAB SELF-QUIZ goo.gl/6u55ks

- In eukaryotes that are **autotrophs**, photosynthesis occurs in **chloroplasts**, organelles containing **thylakoids**. Stacks of thylakoids form grana. **Photosynthesis** is summarized as

$$6\,CO_2 + 12\,H_2O + \text{Light energy} \rightarrow C_6H_{12}O_6 + 6\,O_2 + 6\,H_2O.$$

Chloroplasts split water into hydrogen and oxygen, incorporating the electrons of hydrogen into sugar molecules. Photosynthesis is a redox process: H_2O is oxidized, and CO_2 is reduced. The **light reactions** in the thylakoid membranes split water, releasing O_2, producing ATP, and forming **NADPH**. The **Calvin cycle** in the **stroma** forms sugar from CO_2, using ATP for energy and NADPH for reducing power.

? *Compare the roles of CO_2 and H_2O in cellular respiration and photosynthesis.*

CONCEPT 10.2

The light reactions convert solar energy to the chemical energy of ATP and NADPH *(pp. 192–201)*

- Light is a form of electromagnetic energy. The colors we see as **visible light** include those **wavelengths** that drive photosynthesis. A pigment absorbs light of specific wavelengths; **chlorophyll *a*** is the main photosynthetic pigment in plants. Other accessory pigments absorb different wavelengths of light and pass the energy on to chlorophyll *a*.
- A pigment goes from a ground state to an excited state when a **photon** of light boosts one of the pigment's electrons to a higher-energy orbital. This excited state is unstable. Electrons from isolated pigments tend to fall back to the ground state, giving off heat and/or light.
- A **photosystem** is composed of a **reaction-center complex** surrounded by **light-harvesting complexes** that funnel the energy of photons to the reaction-center complex. When a special pair of reaction-center chlorophyll *a* molecules absorbs energy, one of its electrons is boosted to a higher energy level and transferred to the **primary electron acceptor. Photosystem II** contains P680 chlorophyll *a* molecules in the reaction-center complex; **photosystem I** contains P700 molecules.
- **Linear electron flow** during the light reactions uses both photosystems and produces NADPH, ATP, and oxygen:

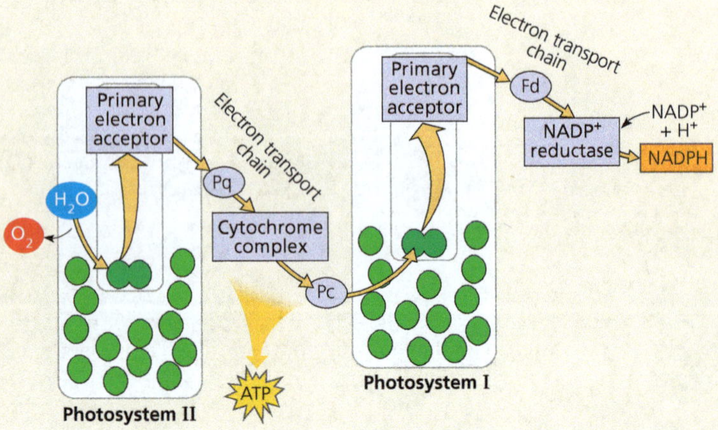

- **Cyclic electron flow** employs only one photosystem, producing ATP but no NADPH or O_2.
- During chemiosmosis in both mitochondria and chloroplasts, electron transport chains generate an H^+ gradient across a membrane. ATP synthase uses this proton-motive force to make ATP.

? *The absorption spectrum of chlorophyll a differs from the action spectrum of photosynthesis. Explain this observation.*

CONCEPT 10.3

The Calvin cycle uses the chemical energy of ATP and NADPH to reduce CO_2 to sugar *(pp. 201–202)*

- The Calvin cycle occurs in the stroma, using electrons from NADPH and energy from ATP. One molecule of **G3P** exits the cycle per three CO_2 molecules fixed and is converted to glucose and other organic molecules.

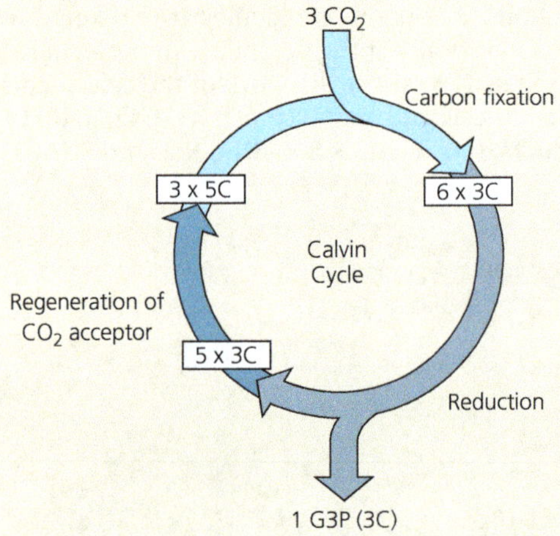

DRAW IT ➤ *On the diagram above, draw where ATP and NADPH are used and where rubisco functions. Describe these steps.*

CONCEPT 10.4

Alternative mechanisms of carbon fixation have evolved in hot, arid climates *(pp. 203–206)*

- On dry, hot days, **C_3 plants** close their stomata, conserving water but keeping CO_2 out and O_2 in. Under these conditions, **photorespiration** can occur: **Rubisco** binds O_2 instead of CO_2, consuming ATP and releasing CO_2 without producing ATP or carbohydrate. Photorespiration may be an evolutionary relic, and it may play a photoprotective role.
- **C_4 plants** minimize the cost of photorespiration by incorporating CO_2 into four-carbon compounds in mesophyll cells. These compounds are exported to **bundle-sheath cells**, where they release carbon dioxide for use in the Calvin cycle.
- **CAM plants** open their stomata at night, incorporating CO_2 into organic acids, which are stored in mesophyll cells. During the day, the stomata close, and the CO_2 is released from the organic acids for use in the Calvin cycle.
- Organic compounds produced by photosynthesis provide the energy and building material for Earth's ecosystems.

? *Why are C_4 and CAM photosynthesis more energetically expensive than C_3 photosynthesis? What climate conditions would favor C_4 and CAM plants?*

CONCEPT 10.5

Life depends on photosynthesis *(pp. 206–209)*

- Organic compounds produced by photosynthesis provide the energy and building material for Earth's ecosystems.

? *Explain how all life depends on photosynthesis.*

TEST YOUR UNDERSTANDING

Level 1: Knowledge/Comprehension

PRACTICE TEST goo.gl/CUYGKD

1. The light reactions of photosynthesis supply the Calvin cycle with
(A) light energy.
(B) CO_2 and ATP.
(C) H_2O and NADPH.
(D) ATP and NADPH.

2. Which of the following sequences correctly represents the flow of electrons during photosynthesis?
(A) NADPH $\rightarrow O_2 \rightarrow CO_2$
(B) $H_2O \rightarrow$ NADPH $\rightarrow$ Calvin cycle
(C) $H_2O \rightarrow$ photosystem I $\rightarrow$ photosystem II
(D) NADPH $\rightarrow$ electron transport chain $\rightarrow O_2$

3. How is photosynthesis similar in C_4 plants and CAM plants?
(A) In both cases, only photosystem I is used.
(B) Both types of plants make sugar without the Calvin cycle.
(C) In both cases, rubisco is not used to fix carbon initially.
(D) Both types of plants make most of their sugar in the dark.

4. Which of the following statements is a correct distinction between autotrophs and heterotrophs?
(A) Autotrophs, but not heterotrophs, can nourish themselves beginning with CO_2 and other nutrients that are inorganic.
(B) Only heterotrophs require chemical compounds from the environment.
(C) Cellular respiration is unique to heterotrophs.
(D) Only heterotrophs have mitochondria.

5. Which of the following does *not* occur during the Calvin cycle?
(A) carbon fixation
(B) oxidation of NADPH
(C) release of oxygen
(D) regeneration of the CO_2 acceptor

Level 2: Application/Analysis

6. In mechanism, photophosphorylation is most similar to
(A) substrate-level phosphorylation in glycolysis.
(B) oxidative phosphorylation in cellular respiration.
(C) carbon fixation.
(D) reduction of $NADP^+$.

7. Which process is most directly driven by light energy?
(A) creation of a pH gradient by pumping protons across the thylakoid membrane
(B) reduction of $NADP^+$ molecules
(C) transfer of energy from pigment molecule to pigment molecule
(D) ATP synthesis

Level 3: Synthesis/Evaluation

8. SCIENCE, TECHNOLOGY, AND SOCIETY Scientific evidence indicates that the CO_2 added to the air by the burning of wood and fossil fuels is contributing to global warming, a rise in global temperature. Tropical rain forests are estimated to be responsible for approximately 20% of global photosynthesis, yet the consumption of large amounts of CO_2 by living trees is thought to make little or no *net* contribution to reduction of global warming. Explain why this might be the case. (*Hint*: What processes in both living and dead trees produce CO_2?)

9. EVOLUTION CONNECTION Photorespiration can decrease soybeans' photosynthetic output by about 50%. Would you expect this figure to be higher or lower in wild relatives of soybeans? Why?

10. SCIENTIFIC INQUIRY • MAKE CONNECTIONS The following diagram represents an experiment with isolated thylakoids. The thylakoids were first made acidic by soaking them in a solution at pH 4. After the thylakoid space reached pH 4, the thylakoids were transferred to a basic solution at pH 8. The thylakoids then made ATP in the dark. (See Concept 3.3 to review pH.)

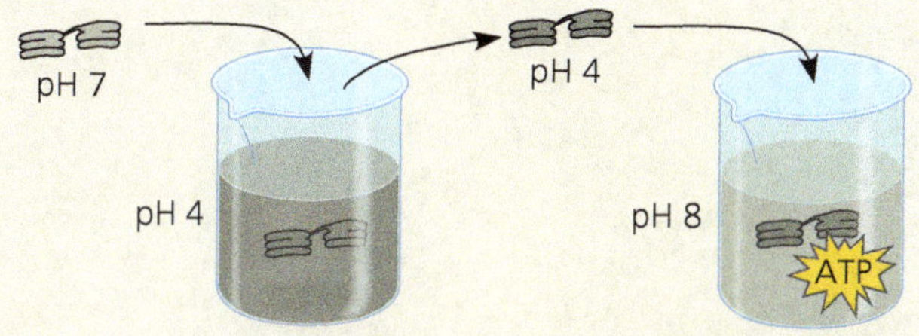

Draw an enlargement of part of the thylakoid membrane in the beaker with the solution at pH 8. Draw ATP synthase. Label the areas of high H^+ concentration and low H^+ concentration. Show the direction protons flow through the enzyme, and show the reaction where ATP is synthesized. Would ATP end up in the thylakoid or outside of it? Explain why the thylakoids in the experiment were able to make ATP in the dark.

11. WRITE ABOUT A THEME: ENERGY AND MATTER Life is solar powered. Almost all the producers of the biosphere depend on energy from the sun to produce the organic molecules that supply the energy and carbon skeletons needed for life. In a short essay (100–150 words), describe how the process of photosynthesis in the chloroplasts of plants transforms the energy of sunlight into the chemical energy of sugar molecules.

12. SYNTHESIZE YOUR KNOWLEDGE

"Watermelon snow" in Antarctica is caused by a species of photosynthetic green algae that thrives in subzero temperatures (*Chlamydomonas nivalis*). These algae are also found in high-altitude year-round snowfields. In both locations, UV light levels tend to be high. Based on what you learned in this chapter, propose an explanation for why this photosynthetic alga appears reddish-pink.

For selected answers, see Appendix A.

For additional practice questions, check out the **Dynamic Study Modules** in MasteringBiology. You can use them to study on your smartphone, tablet, or computer anytime, anywhere!

11 Cell Communication

▲ **Figure 11.1** **How does cell signaling trigger the desperate flight of this impala?**

KEY CONCEPTS

11.1 External signals are converted to responses within the cell

11.2 Reception: A signaling molecule binds to a receptor protein, causing it to change shape

11.3 Transduction: Cascades of molecular interactions relay signals from receptors to target molecules in the cell

11.4 Response: Cell signaling leads to regulation of transcription or cytoplasmic activities

11.5 Apoptosis integrates multiple cell-signaling pathways

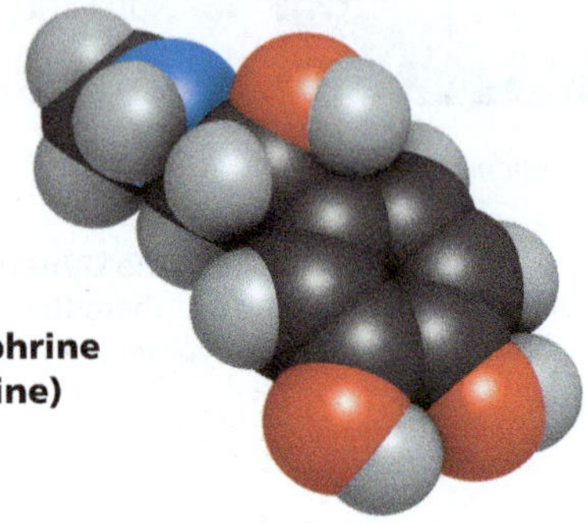

➤ **Epinephrine (adrenaline)**

Cellular Messaging

The impala in **Figure 11.1** flees for its life, racing to escape the predatory cheetah nipping at its heels. The impala is breathing rapidly, its heart pounding and its legs pumping furiously. These physiological functions are all part of the impala's "fight-or-flight" response, driven by hormones released from its adrenal glands at times of stress—in this case, upon sensing the cheetah. What systems allow the trillions of cells in the impala to "talk" to each other, coordinating their activities?

Cells can signal to each other and interpret the signals they receive from other cells and the environment. The signals may include light and touch, but are most often chemicals. The flight response shown here is triggered by a signaling molecule called epinephrine (also called adrenaline; see the space-filling model included here). In studying cell communication, biologists have discovered ample evidence for the evolutionary relatedness of all life. The same set of cell-signaling mechanisms shows up again and again in diverse species, in processes ranging from bacterial signaling to embryonic development to cancer. In this chapter, we focus on the main mechanisms by which cells receive, process, and respond to chemical signals sent from other cells. We will also consider *apoptosis*, a mechanism of programmed cell death that integrates input from multiple signaling pathways.

When you see this blue icon, log in to **MasteringBiology** and go to the Study Area for digital resources.

MB **Get Ready for This Chapter**

CONCEPT 11.1

External signals are converted to responses within the cell

What does a cell that is "talking" say to a "listening" cell, and how does the latter cell respond to the message—that is, how do cells communicate? Let's approach these questions by first looking at communication among microorganisms.

Evolution of Cell Signaling

EVOLUTION One topic cells communicate about is sex. Cells of the unicellular yeast *Saccharomyces cerevisiae*—which are used to make bread, wine, and beer—identify their sexual mates by chemical signaling. There are two sexes, or mating types, called **a** and **α** **(Figure 11.2)**. Each type secretes a specific factor that binds only to receptors on the other type of cell. When exposed to each other's mating factors, a pair of cells of opposite type change shape, grow toward each other, and fuse (mate). The new **a**/**α** cell contains all the genes of both original cells, a combination of genetic resources that provides advantages to the cell's descendants, which arise by subsequent cell divisions.

The unique match between mating factor and receptor is key to ensuring mating only among cells of the same species of yeast. Recently, researchers were able to genetically engineer yeast cells with both receptors and mating factors altered so that the altered proteins would bind to each other but not to the original proteins of the parent cells. The genetically engineered cells were thus able to mate with one another but not with cells of the parent population. This evidence supports a model in which changes in the genes encoding receptor and mating factor proteins can lead to the establishment of new species.

How does the binding of a mating factor by the yeast cell surface receptor initiate a signal that brings about the cellular response of mating? This occurs in a series of steps called a *signal transduction pathway*. Many such pathways exist in both yeast and animal cells. In fact, the molecular details of signal transduction in yeasts and mammals are strikingly similar, even though it's been over a billion years since they shared a common ancestor. This suggests that early versions of cell-signaling mechanisms evolved hundreds of millions of years before the first multicellular creatures appeared on Earth.

Scientists think that signaling mechanisms first evolved in ancient prokaryotes and single-celled eukaryotes like yeasts and then were adopted for new uses by their multicellular descendants. Cell signaling is also critical among prokaryotes. For example, bacterial cells secrete molecules that can be detected by other bacterial cells **(Figure 11.3)**. Sensing the concentration of such signaling molecules allows bacteria to monitor their own local cell density, a phenomenon called *quorum sensing*.

Quorum sensing allows bacterial populations to coordinate their behaviors in activities that require a given number of cells

Figure 11.2 Communication between mating yeast cells. *Saccharomyces cerevisiae* cells use chemical signaling to identify cells of opposite mating type and initiate the mating process. The two mating types and their corresponding chemical signaling molecules, or mating factors, are called **a** and **α**.

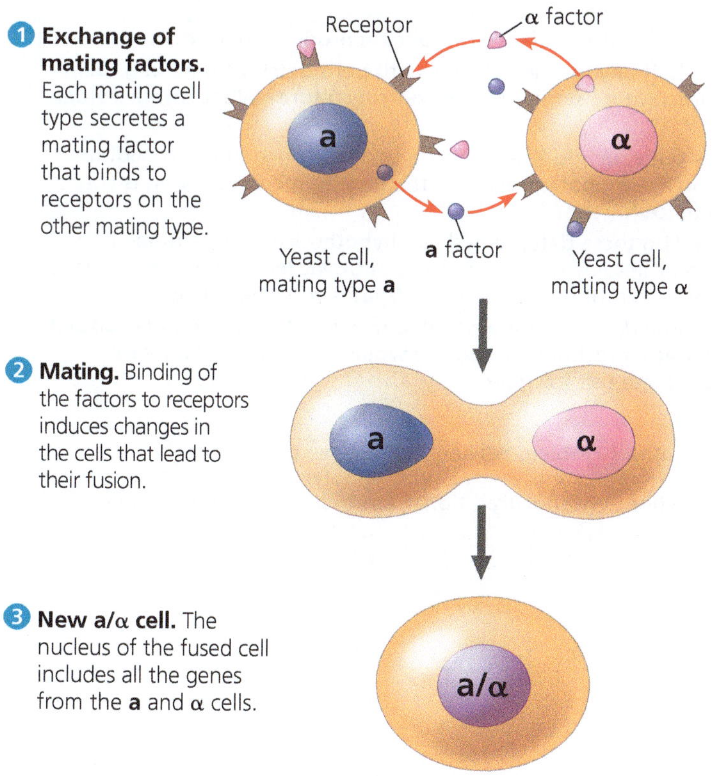

Figure 11.3 Communication among bacteria. Soil-dwelling bacteria called myxobacteria ("slime bacteria") use chemical signals to share information about nutrient availability. When food is scarce, starving cells secrete a signaling molecule that stimulates neighboring cells to aggregate. The cells form a structure called a fruiting body that produces spores, thick-walled cells capable of surviving until the environment improves. The myxobacteria shown here are the species *Myxococcus xanthus* (steps 1–3, SEMs; lower photo, LM).

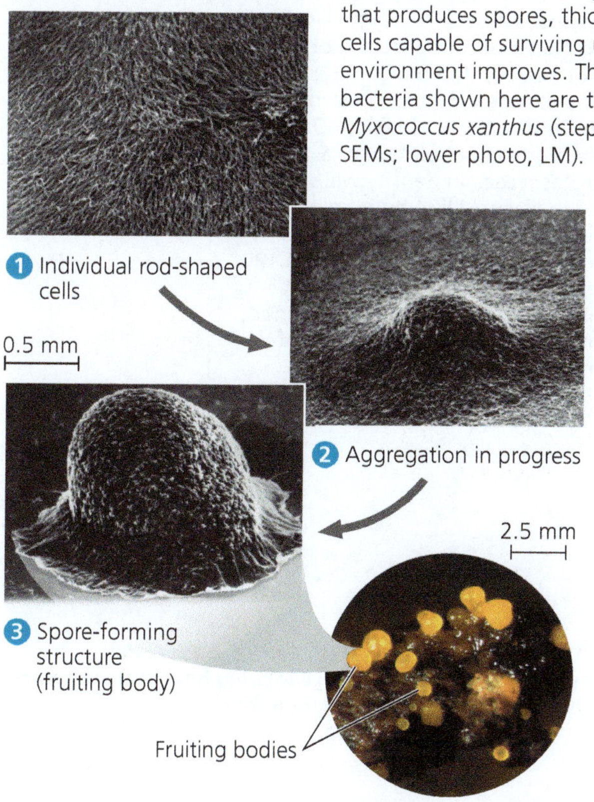

PROBLEM-SOLVING EXERCISE

Can a skin wound turn deadly?

"That scrape I got at the game last week looks infected. I wonder if I should go to the doctor?" Contact sports can be hard on your body even if you are in top physical condition. "Contact" in many cases leads to skin wounds that can become infected—and even deadly, if infected with antibiotic-resistant bacteria.

Watch the video in the Mastering-Biology Study Area to see what happened when a strain of antibiotic-resistant bacteria called MRSA infected at least one high school student. MRSA stands for methicillin-resistant *Staphylococcus aureus*, a strain of bacteria that is resistant to several types of antibiotics, not just methicillin. Most "staph" infections are not antibiotic-resistant and can be treated with antibiotics.

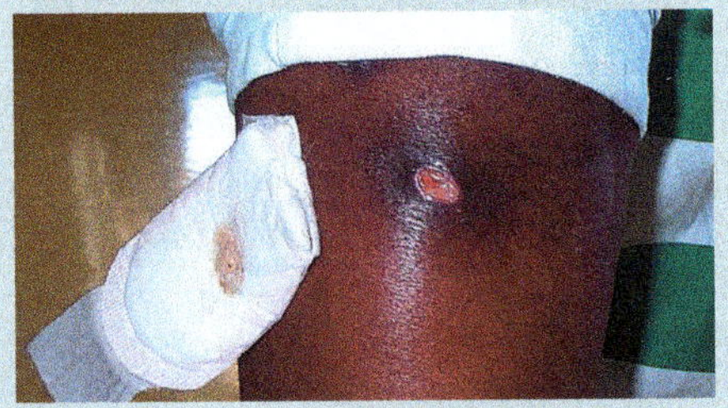

ABC News Video: MRSA Outbreak

Staphylococcus aureus (S. aureus) is a common bacterial species found on the surface of healthy skin that can turn into a serious pathogen if introduced into tissue through a cut or abrasion. Once inside the body, a population of *S. aureus* that reaches a certain density will start to secrete a toxin, killing body cells and contributing significantly to inflammation and damage. Because about 1 in 100 people carry a strain of *S. aureus* that is resistant to common antibiotics, a minor infection can turn permanently harmful or even deadly.

In this exercise, you will investigate the mechanism by which cells sense their own population density (so-called *quorum sensing*) to analyze whether blocking it can stop *S. aureus* from producing toxin.

Your Approach

The facts you have in hand for your investigation are that quorum sensing in *S. aureus* involves two separate signal transduction pathways that can lead to toxin production. Two candidate synthetic peptides (short proteins), called peptides 1 and 2, have been proposed to interfere with the *S. aureus* quorum-sensing pathways. Your job is to test these two potential inhibitors of quorum sensing to see if they block either or both of the pathways that lead to toxin production.

For your experiment, you grow four cultures of *S. aureus* to a standardized high density and measure the concentration of toxin in the culture. The control culture contains no peptide. The other cultures have one or both candidate inhibitory peptides mixed into the growth medium before starting the cultures.

Your Data

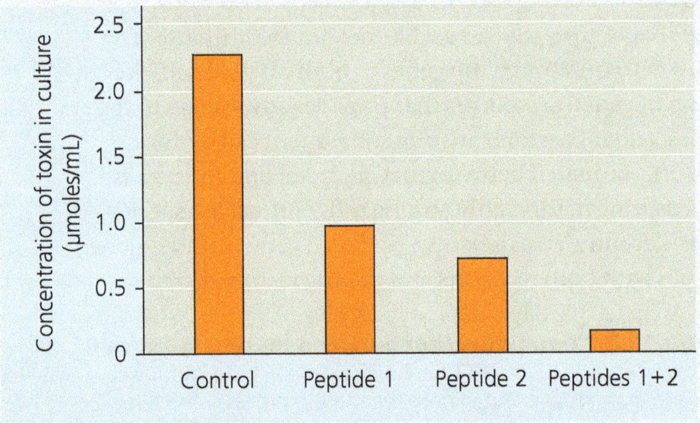

Data from N. Balaban et al., Treatment of *Staphylococcus aureus* biofilm infection by the quorum-sensing inhibitor RIP, *Antimicrobial Agents and Chemotherapy* 51(6):2226–2229 (2007).

Your Analysis

1. Rank the cultures according to toxin production, from most to least.
2. Which, if any, of the cultures with peptide(s) resulted in a toxin concentration similar to the control culture? What is your evidence for this?
3. Was there an additive effect on toxin production when peptides 1 and 2 were both present in the growth medium? What is your evidence for this?
4. Based on these data, would you hypothesize that peptides 1 and 2 act on the same quorum-sensing pathway leading to toxin production or on two different pathways? What is your reasoning?
5. Do these data suggest a possible treatment for antibiotic-resistant *S. aureus* infections? What else would you want to know to investigate this further?

Instructors: A version of this Problem-Solving Exercise can be assigned in MasteringBiology. Or a more extensive investigation called "Solve It: Is It Possible to Treat Bacterial Infections Without Traditional Antibiotics?" can be assigned.

BBC Video: Brushing Your Teeth Can Save Your Life
Interview with Bonnie Bassler: Exploring how bacteria communicate with each other
HHMI Video: Interview with Bonnie Bassler

acting synchronously. One example is formation of a *biofilm*, an aggregation of bacterial cells adhered to a surface. The cells in the biofilm often derive nutrition from the surface they are on. You have probably encountered biofilms many times, perhaps without realizing it. The slimy coating on a fallen log or on leaves lying on a forest path, and even the film on your teeth each morning, are examples of bacterial biofilms. In fact, tooth-brushing and flossing disrupt biofilms that would otherwise cause cavities and gum disease.

Another example of bacterial behavior coordinated by quorum sensing is one that has serious medical implications: the secretion of toxins by infectious bacteria. Sometimes treatment by antibiotics doesn't work with such infections due to antibiotic resistance that has evolved in a particular strain of bacteria. Interfering with the signaling pathways used in quorum sensing represents a promising approach as an alternative treatment. In the **Problem-Solving Exercise**, you can participate in the process of scientific thinking involved in this novel approach.

Local and Long-Distance Signaling

Like bacteria or yeast cells, cells in a multicellular organism usually communicate via signaling molecules targeted for cells that may or may not be immediately adjacent. As we saw in Concepts 6.7 and 7.1, eukaryotic cells may communicate by direct contact, which is one type of local signaling **(Figure 11.4)**. Both animals and plants have cell junctions that, where present, directly connect the cytoplasms of adjacent cells **(Figure 11.4a)**. In these cases, signaling substances dissolved in the cytosol can pass freely between neighboring cells. Moreover, animal cells may communicate via direct contact between membrane-bound cell-surface molecules,

▼ Figure 11.4 Communication by direct contact between cells.

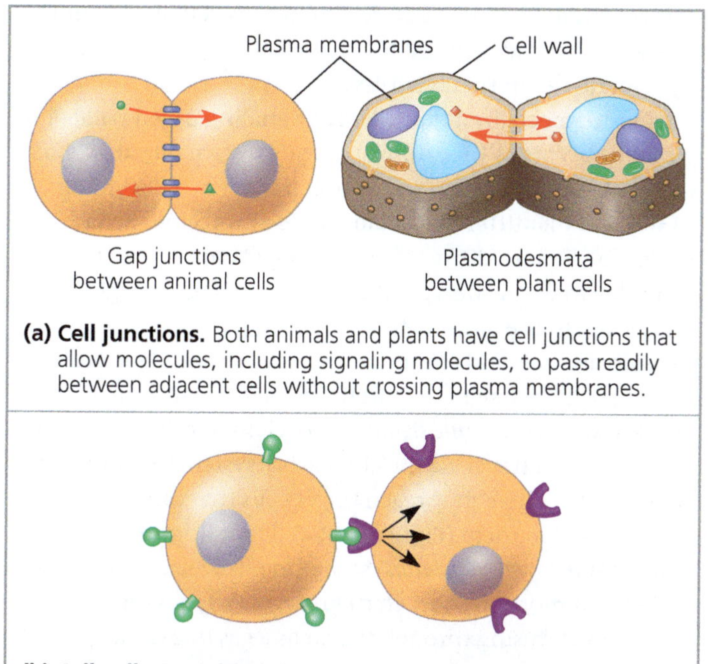

(a) Cell junctions. Both animals and plants have cell junctions that allow molecules, including signaling molecules, to pass readily between adjacent cells without crossing plasma membranes.

(b) Cell-cell recognition. Two cells in an animal may communicate by interaction between molecules protruding from their surfaces.

in a process called cell-cell recognition **(Figure 11.4b)**. This sort of local signaling is especially important in embryonic development and the immune response.

In many other cases of local signaling, signaling molecules are secreted by the signaling cell. Some molecules travel only short distances; such local regulators influence cells in the vicinity. This type of local signaling in animals is called *paracrine signaling* **(Figure 11.5a)**. One class of local regulators in

▼ Figure 11.5 Local and long-distance cell signaling by secreted molecules in animals. In both local and long-distance signaling, only specific target cells that can recognize a given signaling molecule will respond to it.

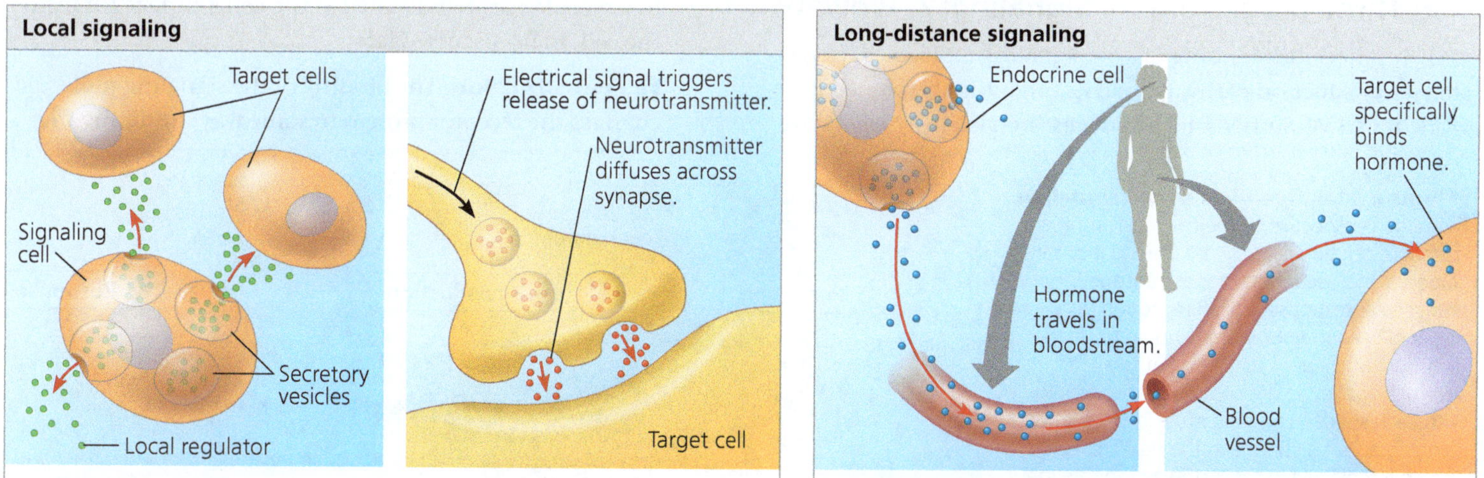

(a) Paracrine signaling. A signaling cell acts on nearby target cells by secreting molecules of a local regulator (a growth factor, for example).

(b) Synaptic signaling. A nerve cell releases neurotransmitter molecules into a synapse, stimulating the target cell, such as a muscle or another nerve cell.

(c) Endocrine (hormonal) signaling. Specialized endocrine cells secrete hormones into body fluids, often blood. Hormones reach most body cells, but are bound by and affect only some cells.

animals, *growth factors*, are compounds that stimulate nearby target cells to grow and divide. Numerous cells can simultaneously receive and respond to the growth factors produced by a single cell in their vicinity.

A more specialized type of local signaling called *synaptic signaling* occurs in the animal nervous system **(Figure 11.5b)**. An electrical signal along a nerve cell triggers the secretion of neurotransmitter molecules. These molecules act as chemical signals, diffusing across the synapse—the narrow space between the nerve cell and its target cell—triggering a response in the target cell.

Both animals and plants use molecules called **hormones** for long-distance signaling. In hormonal signaling in animals, also known as *endocrine signaling*, specialized cells release hormones, which travel via the circulatory system to other parts of the body, where they reach target cells that can recognize and respond to them **(Figure 11.5c)**. Plant hormones (often called *plant growth regulators*) sometimes travel in plant vessels (tubes) but more often reach their targets by moving through cells or by diffusing through the air as a gas (see Concept 39.2). Like local regulators, hormones vary widely in size and type. For instance, the plant hormone ethylene, a gas that promotes fruit ripening and helps regulate growth, is a hydrocarbon of only six atoms (C_2H_4), small enough to pass through cell walls. In contrast, the mammalian hormone insulin, which regulates sugar levels in the blood, is a protein with thousands of atoms.

What happens when a potential target cell is exposed to a secreted signaling molecule? The ability of a cell to respond is determined by whether it has a specific receptor molecule that can bind to the signaling molecule. The information conveyed by this binding, the signal, must then be changed into another form—transduced—inside the cell before the cell can respond. The remainder of the chapter discusses this process, primarily as it occurs in animal cells.

The Three Stages of Cell Signaling: *A Preview*

Our current understanding of how signaling molecules act via signal transduction pathways had its origins in the pioneering work of Earl W. Sutherland, whose research led to a Nobel Prize in 1971. Sutherland and his colleagues at Vanderbilt University were investigating how the animal hormone epinephrine (also called adrenaline) triggers the "fight-or-flight" response in animals by stimulating the breakdown of the storage polysaccharide glycogen within liver cells and skeletal muscle cells. Glycogen breakdown releases the sugar glucose 1-phosphate, which the cell converts to glucose 6-phosphate. The liver or muscle cell can then use this compound, an early intermediate in glycolysis, for energy production (see Figure 9.9). Alternatively, the compound can be stripped of phosphate and released from the cell into the blood as glucose, which can fuel cells throughout the body. Thus, one effect of epinephrine is the mobilization of fuel reserves, which can be used by the animal to either defend itself (fight) or escape whatever elicited a scare (flight), as the impala in Figure 11.1 is clearly doing.

Sutherland's research team discovered that epinephrine stimulates glycogen breakdown by somehow activating a cytosolic enzyme, glycogen phosphorylase. However, when epinephrine was added to a cell-free mixture containing the enzyme and its substrate, glycogen, no breakdown occurred. Glycogen phosphorylase could be activated by epinephrine only when the hormone was added to *intact* cells. This result told Sutherland two things. First, epinephrine does not interact directly with the enzyme responsible for glycogen breakdown; an intermediate step or series of steps must be occurring in the cell. Second, an intact, membrane-bound cell must be present for transmission of the signal to take place.

Sutherland's work suggested that the process going on at the receiving end of a cellular communication can be dissected into three stages: reception, transduction, and response **(Figure 11.6)**:

1. **Reception.** Reception is the target cell's detection of a signaling molecule coming from outside the cell. A chemical signal is "detected" when the signaling molecule binds to a receptor protein located at the cell's surface (or inside the cell, to be discussed later).

2. **Transduction.** The binding of the signaling molecule changes the receptor protein in some way, initiating the

➤ Figure 11.6 Overview of cell signaling. From the perspective of the cell receiving the message, cell signaling can be divided into three stages: signal reception, signal transduction, and cellular response. When reception occurs at the plasma membrane, as shown here, the transduction stage is usually a pathway of several steps (three are shown as an example), with each specific relay molecule in the pathway bringing about a change in the next molecule. The final molecule in the pathway triggers the cell's response.

VISUAL SKILLS ➤ *Where would the epinephrine in Sutherland's experiment fit into this diagram of cell signaling?*

Animation: Overview of Cell Signaling

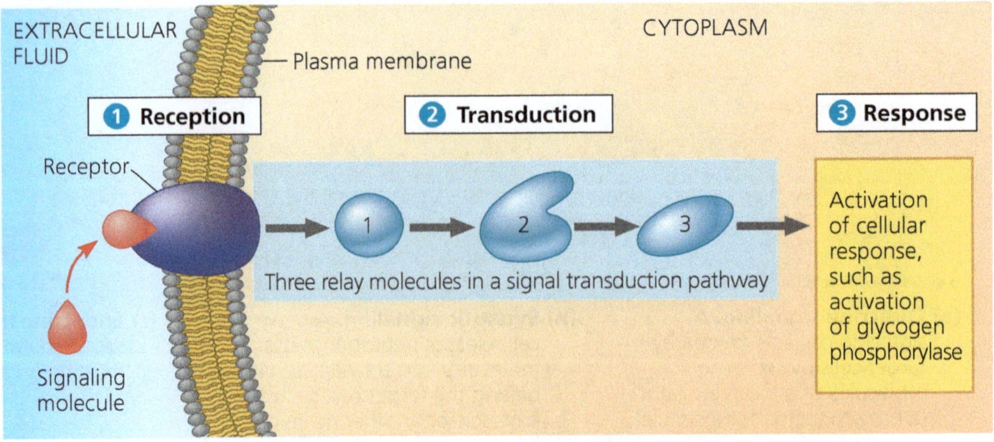

process of transduction. The transduction stage converts the signal to a form that can bring about a specific cellular response. In Sutherland's system, the binding of epinephrine to a receptor protein in a liver cell's plasma membrane leads to activation of glycogen phosphorylase in the cytosol. Transduction sometimes occurs in a single step but more often requires a sequence of changes in a series of different molecules—a **signal transduction pathway**. The molecules in the pathway are often called relay molecules; three are shown as an example.

3 **Response**. In the third stage of cell signaling, the transduced signal finally triggers a specific cellular response. The response may be almost any imaginable cellular activity—such as catalysis by an enzyme (for example, glycogen phosphorylase), rearrangement of the cytoskeleton, or activation of specific genes in the nucleus. The cell-signaling process helps ensure that crucial activities like these occur in the right cells, at the right time, and in proper coordination with the activities of other cells of the organism. We'll now explore the mechanisms of cell signaling in more detail, including a discussion of regulation and termination of the process.

CONCEPT CHECK 11.1

1. Explain how signaling is involved in ensuring that yeast cells fuse only with cells of the opposite mating type.
2. In liver cells, glycogen phosphorylase acts in which of the three stages of the signaling pathway associated with an epinephrine-initiated signal?
3. **WHAT IF?** ➤ If epinephrine were mixed with glycogen phosphorylase and glycogen in a cell-free mixture in a test tube, would glucose 1-phosphate be generated? Why or why not?

For suggested answers, see Appendix A.

CONCEPT 11.2

Reception: A signaling molecule binds to a receptor protein, causing it to change shape

A wireless router may broadcast its network signal indiscriminately, but often it can be joined only by computers with the correct password: Reception of the signal depends on the receiver. Similarly, the signals emitted by an **a** mating type yeast cell are "heard" only by its prospective mates, α cells. In the case of the epinephrine circulating throughout the bloodstream of the impala in Figure 11.1, the hormone encounters many types of cells, but only certain target cells detect and react to the epinephrine molecule. A receptor protein on or in the target cell allows the cell to "hear" the signal and respond to it. The signaling molecule is complementary in shape to a specific site on the receptor and attaches there, like a hand in a glove. The signaling molecule acts as a **ligand**, the term for a molecule that specifically binds to another (often larger) molecule. Ligand binding generally causes a receptor protein to undergo a change in shape. For many receptors, this shape change directly activates the receptor, enabling it to interact with other cellular molecules. For other kinds of receptors, the immediate effect of ligand binding is to cause the aggregation of two or more receptor proteins, which leads to further molecular events inside the cell. Most signal receptors are plasma membrane proteins, but others are located inside the cell. We discuss both of these types next.

MB Animation: Reception

Receptors in the Plasma Membrane

Cell-surface transmembrane receptors play crucial roles in the biological systems of animals. The largest family of human cell-surface receptors is the G protein-coupled receptors (GPCRs). There are more than 800 GPCRs; an example is shown in **Figure 11.7**. Another example is the co-receptor hijacked by HIV to enter immune cells (see Figure 7.8); this GPCR is the target of the drug maraviroc, which has shown some success at treating AIDS.

Most water-soluble signaling molecules bind to specific sites on transmembrane receptor proteins that transmit information from the extracellular environment to the inside of the cell. We can see how cell-surface transmembrane receptors work by looking at three major types: G protein-coupled receptors (GPCRs), receptor tyrosine kinases, and ion channel receptors. These receptors are discussed and illustrated in **Figure 11.8**; study this figure before going on.

▼ Figure 11.7 The structure of a G protein-coupled receptor (GPCR). Shown here is a model of the human β_2-adrenergic receptor, which binds adrenaline (epinephrine) and was able to be crystallized in the presence of both a molecule that mimics adrenaline (green in the model) and cholesterol in the membrane (orange). Two receptor molecules (blue) are shown as ribbon models in a side view. Caffeine can also bind to this receptor; see question 10 at the end of the chapter.

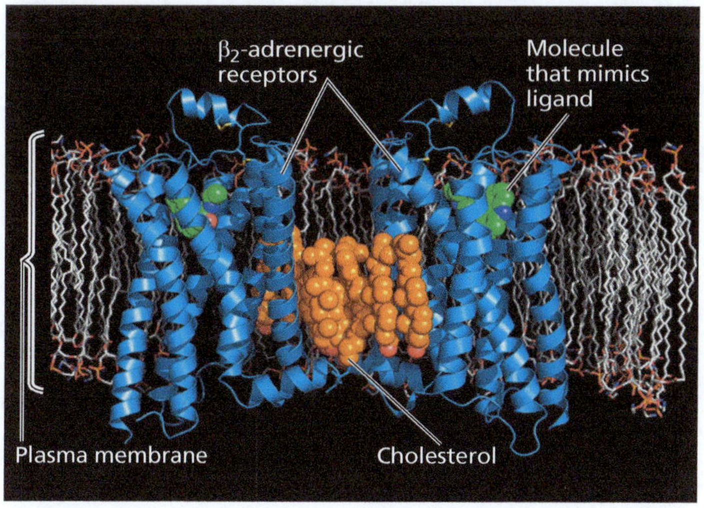

▼ Figure 11.8 Exploring Cell-Surface Transmembrane Receptors

G Protein-Coupled Receptors

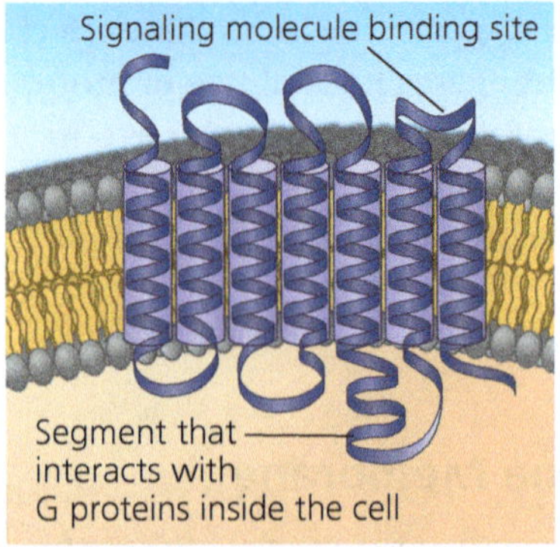

G protein-coupled receptor

A **G protein-coupled receptor** (GPCR) is a cell-surface transmembrane receptor that works with the help of a **G protein**, a protein that binds the energy-rich molecule GTP. Many different signaling molecules—including yeast mating factors, neurotransmitters, and epinephrine (adrenaline) and many other hormones—use GPCRs.

G protein-coupled receptors vary in the binding sites for their ligands and also for different types of G proteins inside the cell. Nevertheless, GPCR proteins are all remarkably similar in structure. In fact, they make up a large family of eukaryotic receptor proteins with a secondary structure in which the single polypeptide, represented here in a ribbon model, has seven transmembrane α helices (outlined with cylinders and depicted in a row for clarity). Specific loops between the helices (here, the loops on the right) form binding sites for signaling molecules (outside the cell) and G proteins (on the cytoplasmic side).

GPCR-based signaling systems are extremely widespread and diverse in their functions, including roles in embryonic development and sensory reception. In humans, for example, vision, smell, and taste depend on GPCRs (see Concept 50.4). Similarities in structure in G proteins and GPCRs in diverse organisms suggest that G proteins and their associated receptors evolved very early among eukaryotes.

Malfunctions of the associated G proteins themselves are involved in many human diseases, including bacterial infections. The bacteria that cause cholera, pertussis (whooping cough), and botulism, among others, make their victims ill by producing toxins that interfere with G protein function. Up to 60% of all medicines used today exert their effects by influencing G protein pathways.

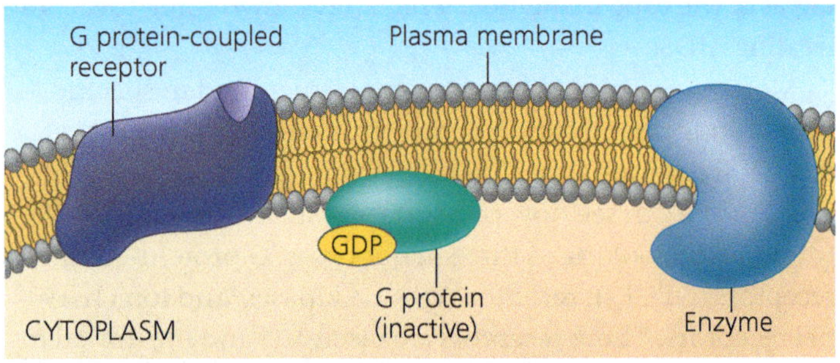

❶ Attached but able to move along the cytoplasmic side of the membrane, a G protein functions as a molecular switch that is either on or off, depending on whether GDP or GTP is attached—hence the term *G protein*. (GTP, or guanosine triphosphate, is similar to ATP.) When GDP is bound to the G protein, as shown above, the G protein is inactive. The receptor and G protein work together with another protein, usually an enzyme.

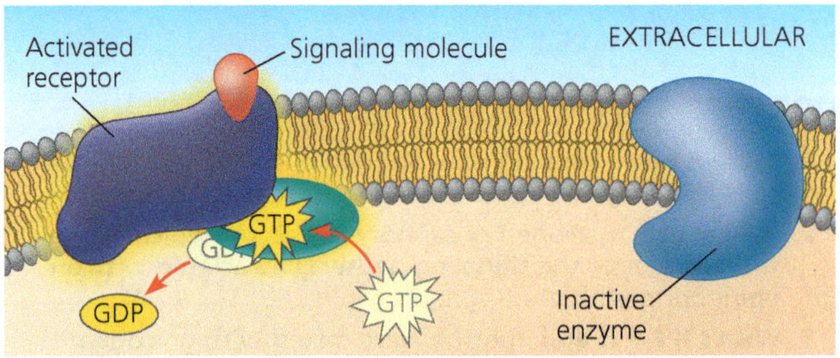

❷ When the appropriate signaling molecule binds to the extracellular side of the receptor, the receptor is activated and changes shape. Its cytoplasmic side then binds an inactive G protein, causing a GTP to displace the GDP. This activates the G protein.

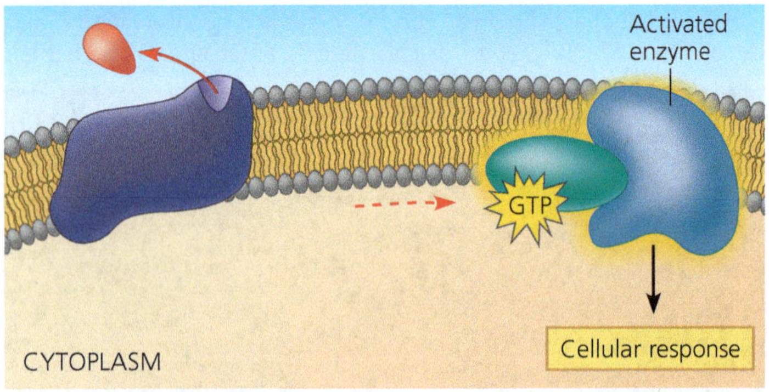

❸ The activated G protein dissociates from the receptor, diffuses along the membrane, and then binds to an enzyme, altering the enzyme's shape and activity. Once activated, the enzyme can trigger the next step leading to a cellular response. Binding of signaling molecules is reversible: Like other ligands, they bind and dissociate many times. The ligand concentration outside the cell determines how often a ligand is bound and initiates signaling.

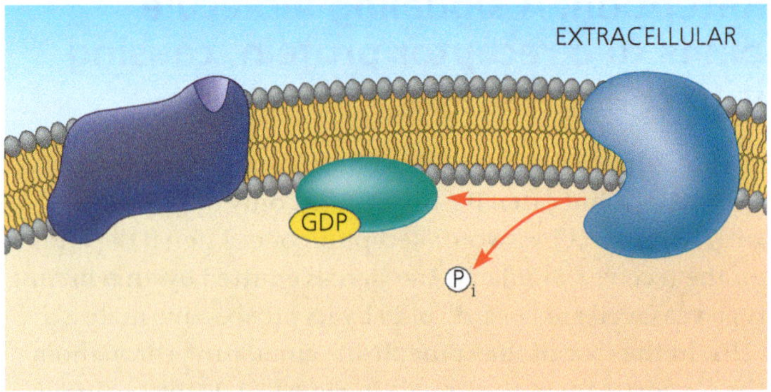

❹ The changes in the enzyme and G protein are only temporary because the G protein also functions as a GTPase enzyme—in other words, it then hydrolyzes its bound GTP to GDP and Ⓟ$_i$. Now inactive again, the G protein leaves the enzyme, which returns to its original state. The G protein is now available for reuse. The GTPase function of the G protein allows the pathway to shut down rapidly when the signaling molecule is no longer present.

Receptor Tyrosine Kinases

Receptor tyrosine kinases (RTKs) belong to a major class of plasma membrane receptors characterized by having enzymatic activity. An RTK is a *protein kinase*—an enzyme that catalyzes the transfer of phosphate groups from ATP to another protein. The part of the receptor protein extending into the cytoplasm functions more specifically as a tyrosine kinase, an enzyme that catalyzes the transfer of a phosphate group from ATP to the amino acid tyrosine of a substrate protein. Thus, RTKs are membrane receptors that attach phosphates to tyrosines.

Upon binding a ligand such as a growth factor, one RTK may activate ten or more different transduction pathways and cellular responses. Often, more than one signal transduction pathway can be triggered at once, helping the cell regulate and coordinate many aspects of cell growth and cell reproduction. The ability of a single ligand-binding event to trigger so many pathways is a key difference between RTKs and GPCRs, which generally activate a single transduction pathway. Abnormal RTKs that function even in the absence of signaling molecules are associated with many kinds of cancer.

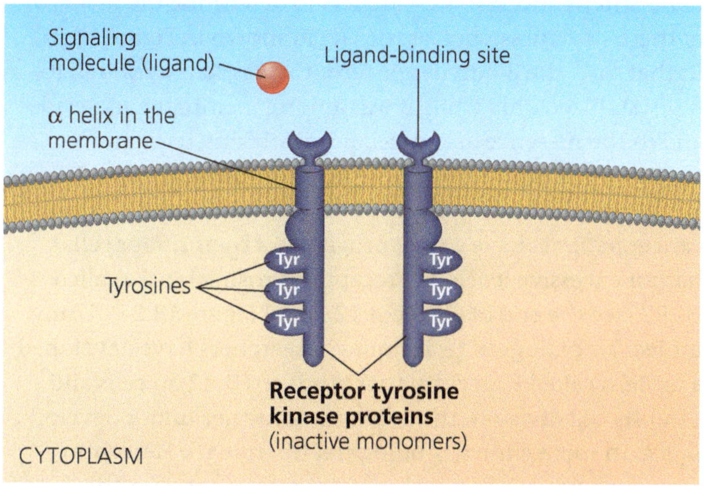

1 Many receptor tyrosine kinases have the structure depicted schematically here. Before the signaling molecule binds, the receptors exist as individual units referred to as monomers. Notice that each monomer has an extracellular ligand-binding site, an α helix spanning the membrane, and an intracellular tail containing multiple tyrosines.

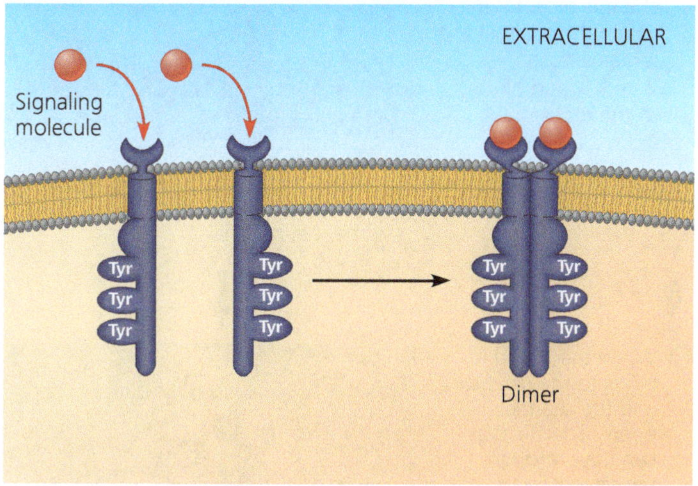

2 The binding of a signaling molecule (such as a growth factor) causes two receptor monomers to associate closely with each other, forming a complex known as a dimer, in a process called dimerization. (In some cases, larger clusters form. The details of monomer association are a focus of current research.)

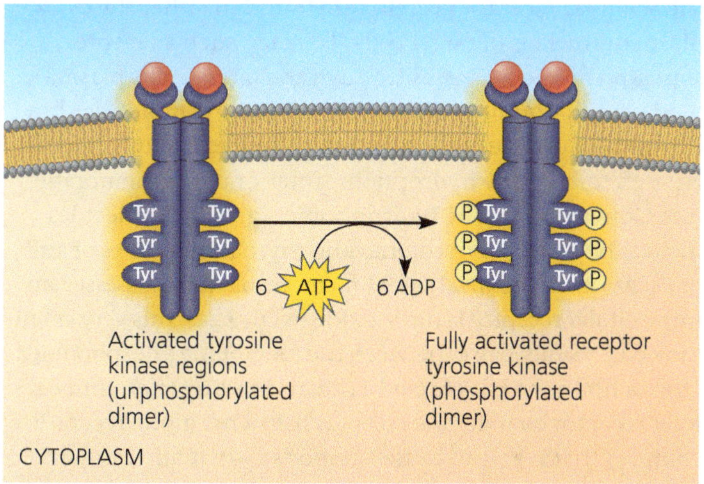

3 Dimerization activates the tyrosine kinase region of each monomer; each tyrosine kinase adds a phosphate from an ATP molecule to a tyrosine that is part of the tail of the other monomer.

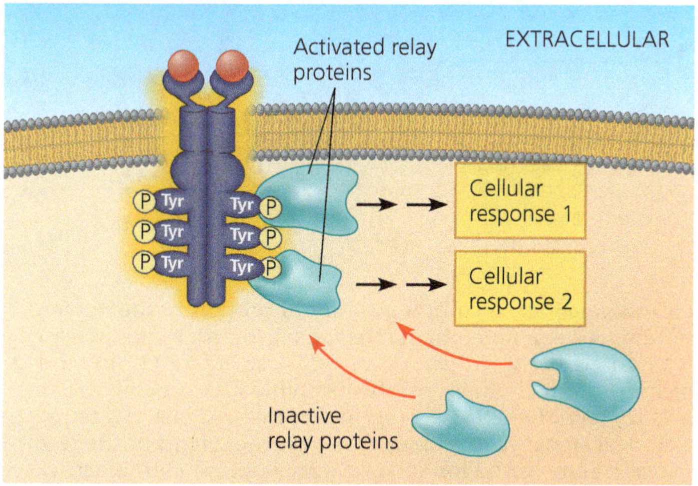

4 Now that the receptor is fully activated, it is recognized by specific relay proteins inside the cell. Each such protein binds to a specific phosphorylated tyrosine, undergoing a resulting structural change that activates the bound relay protein. Each activated protein triggers a transduction pathway, leading to a cellular response.

Continued on next page

▼ **Figure 11.8 (continued)**

Ion Channel Receptors

A **ligand-gated ion channel** is a type of membrane channel receptor containing a region that can act as a "gate," opening or closing the channel when the receptor changes shape. When a signaling molecule binds as a ligand to the channel receptor, the channel opens or closes, allowing or blocking the flow of specific ions, such as Na^+ or Ca^{2+}. Like the other receptors we have discussed, these proteins bind the ligand at a specific site on their extracellular sides.

1 Here we show a ligand-gated ion channel receptor in which the channel remains closed until a ligand binds to the receptor.

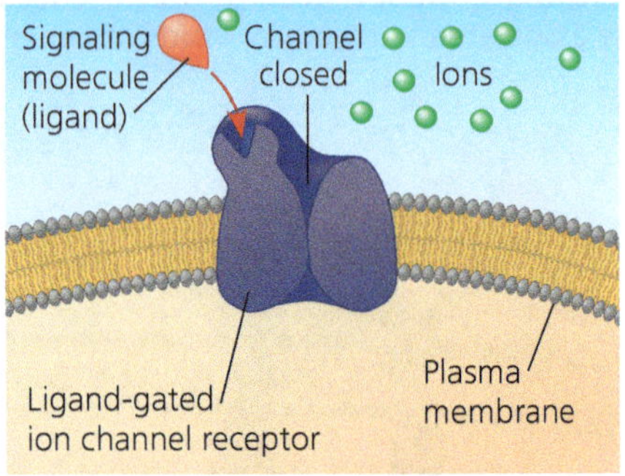

2 When the ligand binds to the receptor and the channel opens, specific ions can flow through the channel and rapidly change the concentration of that particular ion inside the cell. This change may directly affect the activity of the cell in some way.

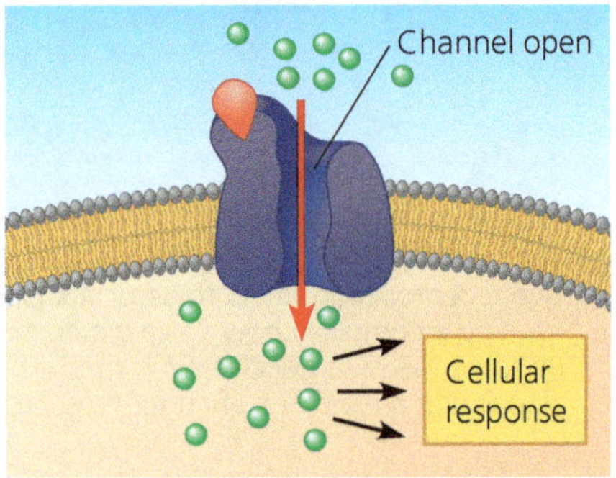

3 When the ligand dissociates from this receptor, the channel closes and ions no longer enter the cell.

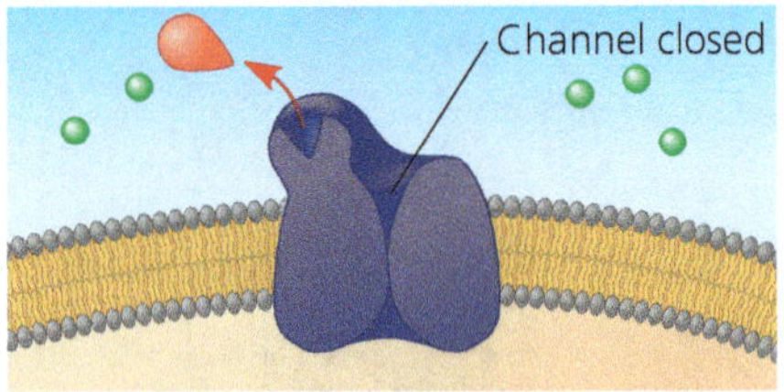

Ligand-gated ion channels are very important in the nervous system. For example, the neurotransmitter molecules released at a synapse between two nerve cells (see Figure 11.5b) bind as ligands to ion channels on the receiving cell, causing the channels to open. Ions flow in (or, in some cases, out), triggering an electrical signal that propagates down the length of the receiving cell. Some gated ion channels are controlled by electrical signals instead of ligands; these voltage-gated ion channels are also crucial to the functioning of the nervous system, as we will discuss in Chapter 48. Some ion channels are present on membranes of organelles, such as the ER.

MAKE CONNECTIONS ➤ *Is the flow of ions through a ligand-gated channel an example of active or passive transport? (Review Concepts 7.3 and 7.4.)*

Animation: Acetylcholine Receptor

Given the many important functions of cell-surface receptors, it is not surprising that their malfunctions are associated with many human diseases, including cancer, heart disease, and asthma. To better understand and treat these conditions, a major focus of both university research teams and the pharmaceutical industry has been to analyze the structure of these receptors.

Although cell-surface receptors represent 30% of all human proteins, determining their structures has proved challenging: They make up only 1% of the proteins whose structures have been determined by X-ray crystallography (see Figure 5.21). For one thing, cell-surface receptors tend to be flexible and inherently unstable, thus difficult to crystallize. It took years of persistent efforts for researchers to determine the first few of these structures, such as the GPCR shown in Figure 11.7. In that case, the β-adrenergic receptor was stable enough to be crystallized only while it was among membrane molecules and in the presence of a molecule mimicking its ligand.

Abnormal functioning of receptor tyrosine kinases (RTKs) is associated with many types of cancers. For example, breast cancer patients have a poor prognosis if their tumor cells harbor excessive levels of a receptor tyrosine kinase called HER2 (see the end of Concept 12.3 and Figure 18.27). Using molecular biological techniques, researchers have developed a protein called Herceptin that binds to HER2 on cells and inhibits cell division, thus thwarting further tumor development. In some clinical studies, treatment with Herceptin improved patient survival rates by more than one-third. One goal of ongoing research into these cell-surface receptors and other cell-signaling proteins is development of additional successful treatments.

Intracellular Receptors

Intracellular receptor proteins are found in either the cytoplasm or nucleus of target cells. To reach such a receptor, a signaling molecule passes through the target cell's plasma membrane. A number of important signaling molecules can do this because they are either hydrophobic enough or small enough to cross the hydrophobic interior of the membrane (see Concept 7.1). The hydrophobic signaling molecules include both steroid hormones and thyroid hormones of animals. Another chemical signaling molecule that possesses an intracellular receptor is nitric oxide (NO), a gas; this very small molecule readily passes between the membrane phospholipids. Once a hormone has entered a cell, its binding to an intracellular receptor changes the receptor into a hormone-receptor complex that is able to cause a response—in many cases, the turning on or off of particular genes.

The behavior of aldosterone is a representative example of how steroid hormones work. This hormone is secreted by cells of the adrenal gland, a gland that lies above the kidney. Aldosterone then travels through the blood and enters cells all over the body. However, a response occurs only in kidney cells, which contain receptor molecules for aldosterone. In these cells, the hormone binds to and activates the receptor protein. With aldosterone attached, the active form of the receptor protein then enters the nucleus and turns on specific genes that control water and sodium flow in kidney cells, ultimately affecting blood volume **(Figure 11.9)**.

How does the activated hormone-receptor complex turn on genes? Recall that the genes in a cell's DNA function by being transcribed and processed into messenger RNA (mRNA), which leaves the nucleus and is translated into a specific protein by ribosomes in the cytoplasm (see Figure 5.22). Special proteins called *transcription factors* control which genes are turned on—that is, which genes are transcribed into mRNA—in a particular cell at a particular time. When the aldosterone receptor is activated, it acts as a transcription factor that turns on specific genes. (You'll learn more about transcription factors in Chapters 17 and 18.)

By acting as a transcription factor, the aldosterone receptor itself carries out the transduction part of the signaling pathway. Most other intracellular receptors function in the same way, although many of them, such as the thyroid hormone receptor, are already in the nucleus before the signaling molecule reaches them. Interestingly, many of these intracellular receptor proteins are structurally similar, suggesting an evolutionary kinship.

▼ Figure 11.9 Steroid hormone interacting with an intracellular receptor.

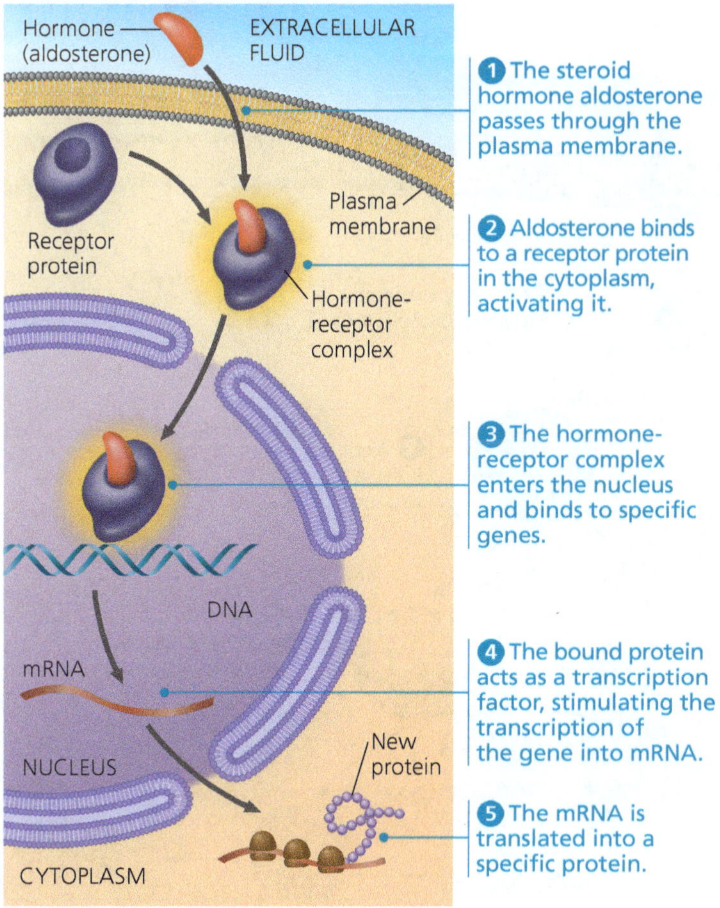

MAKE CONNECTIONS ➤ *Why is a cell-surface receptor protein not required for this steroid hormone to enter the cell? (See Concept 7.2.)*

CONCEPT CHECK 11.2

1. Nerve growth factor (NGF) is a water-soluble signaling molecule. Would you expect the receptor for NGF to be intracellular or in the plasma membrane? Why?
2. **WHAT IF? ➤** What would the effect be if a cell made defective receptor tyrosine kinase proteins that were unable to dimerize?
3. **MAKE CONNECTIONS ➤** How is ligand binding similar to the process of allosteric regulation of enzymes? (See Figure 8.20.)

For suggested answers, see Appendix A.

CONCEPT 11.3

Transduction: Cascades of molecular interactions relay signals from receptors to target molecules in the cell

When receptors for signaling molecules are plasma membrane proteins, like most of those we have discussed, the transduction stage of cell signaling is usually a multistep pathway involving many molecules. Steps often include activation of proteins by addition or removal of phosphate groups or release of other small molecules or ions that act as signaling molecules. One benefit of multiple steps is the possibility of greatly amplifying a signal. If each molecule transmits the signal to numerous molecules at the next step in the series, the result is a geometric increase in the number of activated molecules by the end (see Figure 11.16). Moreover, multistep pathways provide more opportunities for coordination and control than do simpler systems. This allows regulation of the response, as we'll discuss later in the chapter.

Signal Transduction Pathways

The binding of a specific signaling molecule to a receptor in the plasma membrane triggers the first step in the signal transduction pathway—the chain of molecular interactions that leads to a particular response within the cell. Like falling dominoes, the signal-activated receptor activates another molecule, which activates yet another molecule, and so on, until the protein that produces the final cellular response is activated. The molecules that relay a signal from receptor to response, which we call relay molecules in this book, are often proteins. Protein-protein interactions are a major theme of cell signaling—indeed, a unifying theme of all cellular activities.

MB **Animation: Signal Transduction Pathways**

Keep in mind that the original signaling molecule is not physically passed along a signaling pathway; in most cases, it never even enters the cell. When we say that the signal is relayed along a pathway, we mean that certain information is passed on. At each step, the signal is transduced into a different form, commonly a shape change in the next protein. Very often, the shape change is brought about by phosphorylation.

Protein Phosphorylation and Dephosphorylation

Previous chapters introduced the concept of activating a protein by adding one or more phosphate groups to it (see Figure 8.11a). In Figure 11.8, you have already seen how phosphorylation is involved in the activation of receptor tyrosine kinases. In fact, the phosphorylation and dephosphorylation of proteins is a widespread cellular mechanism for regulating protein activity. An enzyme that transfers phosphate groups from ATP to a protein is generally known as a **protein kinase**. Recall that a receptor tyrosine kinase is a specific kind of protein kinase that phosphorylates tyrosines on the other receptor tyrosine kinase in a dimer. Most cytoplasmic protein kinases, however, act on proteins different from themselves. Another distinction is that most cytoplasmic protein kinases phosphorylate either of two other amino acids, serine or threonine, rather than tyrosine. Serine/threonine kinases are widely involved in signaling pathways in animals, plants, and fungi.

Many of the relay molecules in signal transduction pathways are protein kinases, and they often act on other protein kinases in the pathway. **Figure 11.10** depicts a hypothetical pathway containing two different protein kinases that create a **phosphorylation cascade**. The sequence of steps shown in the figure is similar to many known pathways, including those triggered in yeast by mating factors and in animal cells by many growth factors. The signal is transmitted by a cascade of protein phosphorylations, each causing a shape change in the phosphorylated protein. The shape change results from the interaction of the newly added phosphate groups with charged or polar amino acids on the protein being phosphorylated (see Figure 5.14). The shape change in turn alters the function of the protein, most often activating it. In some cases, though, phosphorylation instead *decreases* the activity of the protein.

About 2% of our own genes are thought to code for protein kinases, a significant percentage. A single cell may have hundreds of different kinds, each specific for a different substrate protein. Together, protein kinases probably regulate the activity of a large proportion of the thousands of proteins in a cell.

➤ Figure 11.10 A phosphorylation cascade. In a phosphorylation cascade, a series of different proteins in a pathway are phosphorylated in turn, each protein adding a phosphate group to the next one in line. Here, phosphorylation activates each protein, and dephosphorylation returns it to its inactive form. The active and inactive forms of each protein are represented by different shapes to remind you that activation is usually associated with a change in molecular shape.

WHAT IF? ➤ *What would happen if a mutation in protein kinase 2 made it incapable of being phosphorylated?*

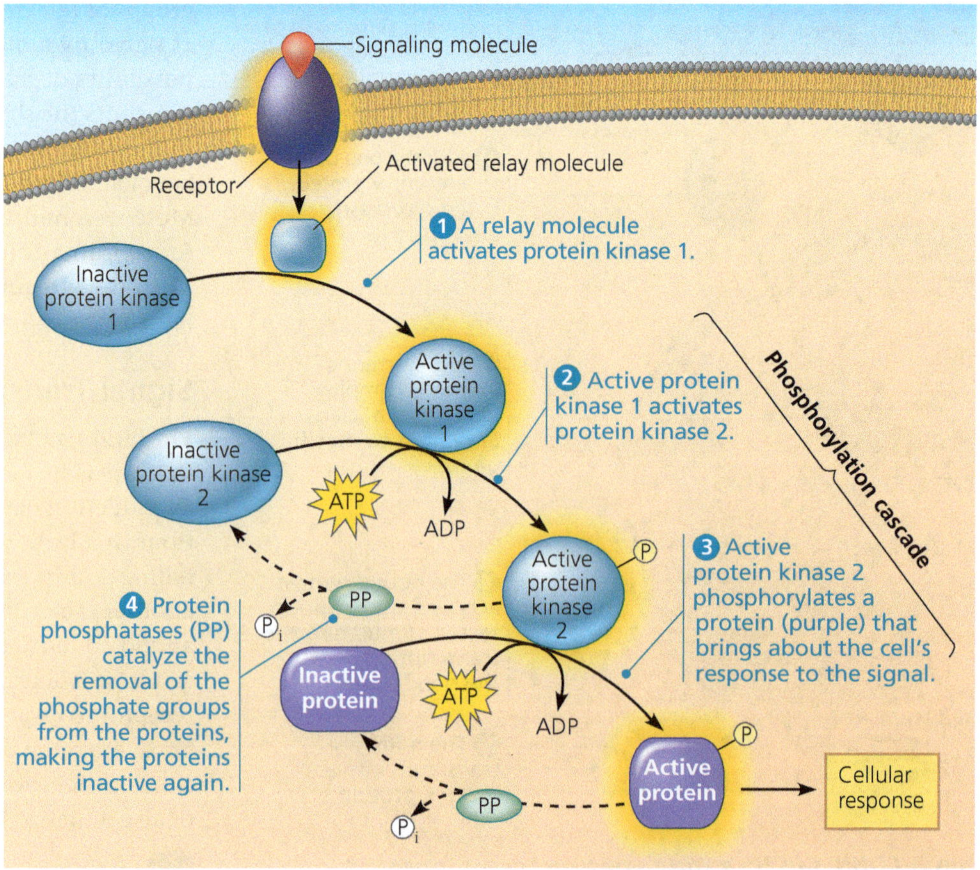

Among these are most of the proteins that, in turn, regulate cell division. Abnormal activity of such a kinase can cause abnormal cell division and contribute to the development of cancer.

Equally important in the phosphorylation cascade are the **protein phosphatases**, enzymes that can rapidly remove phosphate groups from proteins, a process called dephosphorylation. By dephosphorylating and thus inactivating protein kinases, phosphatases provide the mechanism for turning off the signal transduction pathway when the initial signal is no longer present. Phosphatases also make the protein kinases available for reuse, enabling the cell to respond again to an extracellular signal. The phosphorylation-dephosphorylation system acts as a molecular switch in the cell, turning activities on or off, or up or down, as required. At any given moment, the activity of a protein regulated by phosphorylation depends on the balance in the cell between active kinase molecules and active phosphatase molecules.

Small Molecules and Ions as Second Messengers

Not all components of signal transduction pathways are proteins. Many signaling pathways also involve small, nonprotein, water-soluble molecules or ions called **second messengers**. (The pathway's "first messenger" is considered to be the extracellular signaling molecule—the ligand—that binds to the membrane receptor.) Because second messengers are small and also water-soluble, they can readily spread throughout the cell by diffusion. For example, as we'll see shortly, a second messenger called cyclic AMP carries the signal initiated by epinephrine from the plasma membrane of a liver or muscle cell into the cell's interior, where the signal eventually brings about glycogen breakdown. Second messengers participate in pathways that are initiated by both G protein-coupled receptors and receptor tyrosine kinases. The two most widely used second messengers are cyclic AMP and calcium ions, Ca^{2+}. A large variety of relay proteins are sensitive to changes in the cytosolic concentration of one or the other of these second messengers.

Cyclic AMP

As discussed previously, Earl Sutherland established that, without passing through the plasma membrane, epinephrine somehow causes glycogen breakdown within cells. This discovery prompted him to search for a second messenger that transmits the signal from the plasma membrane to the metabolic machinery in the cytoplasm.

Sutherland found that the binding of epinephrine to the plasma membrane of a liver cell elevates the cytosolic concentration of **cyclic AMP** (**cAMP**; cyclic adenosine monophosphate). As shown in **Figure 11.11**, an enzyme embedded in the plasma membrane, **adenylyl cyclase** (also known as adenylate cyclase), converts ATP to cAMP in response to an extracellular signal—in this case, provided by epinephrine. But epinephrine doesn't stimulate adenylyl cyclase directly. When epinephrine outside the cell binds to a G protein-coupled receptor, the protein activates adenylyl cyclase, which in turn can catalyze the synthesis of many molecules of cAMP. In this way, the normal cellular concentration of cAMP can be boosted 20-fold in a matter of seconds. The cAMP broadcasts the signal to the cytoplasm. It does not persist for long in the absence of the hormone because a different enzyme, called phosphodiesterase, converts cAMP to AMP. Another surge of epinephrine is needed to boost the cytosolic concentration of cAMP again.

Subsequent research has revealed that epinephrine and many other signaling molecules lead to activation of adenylyl cyclase by G proteins and formation of cAMP **(Figure 11.12)**. The immediate effect of an elevation in cAMP levels is usually the activation of a serine/threonine kinase called *protein kinase A*. The activated protein kinase A then phosphorylates various other proteins, depending on the cell type. (The complete pathway for epinephrine's stimulation of glycogen breakdown is shown later, in Figure 11.16.)

Figure 11.11 Cyclic AMP. The second messenger cyclic AMP (cAMP) is made from ATP by adenylyl cyclase, an enzyme embedded in the plasma membrane. Note that the phosphate group in cAMP is attached to both the 5′ and the 3′ carbons; this cyclic arrangement is the basis for the molecule's name. Cyclic AMP is inactivated by phosphodiesterase, an enzyme that converts it to AMP.

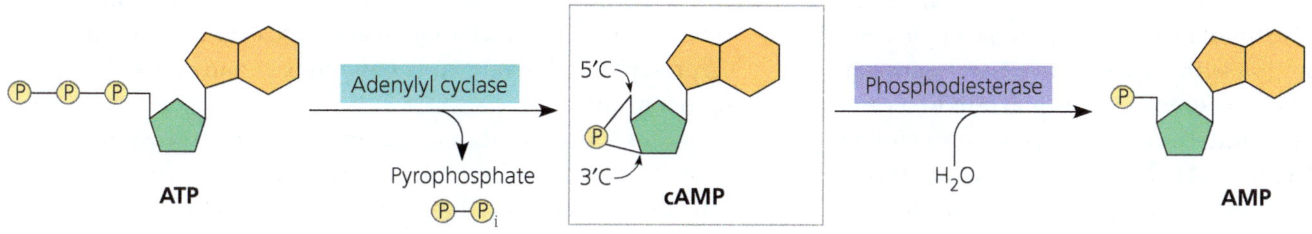

WHAT IF? ➤ *What would happen if a molecule that inactivated phosphodiesterase were introduced into the cell?*

Figure 11.12 cAMP as a second messenger in a G protein signaling pathway.

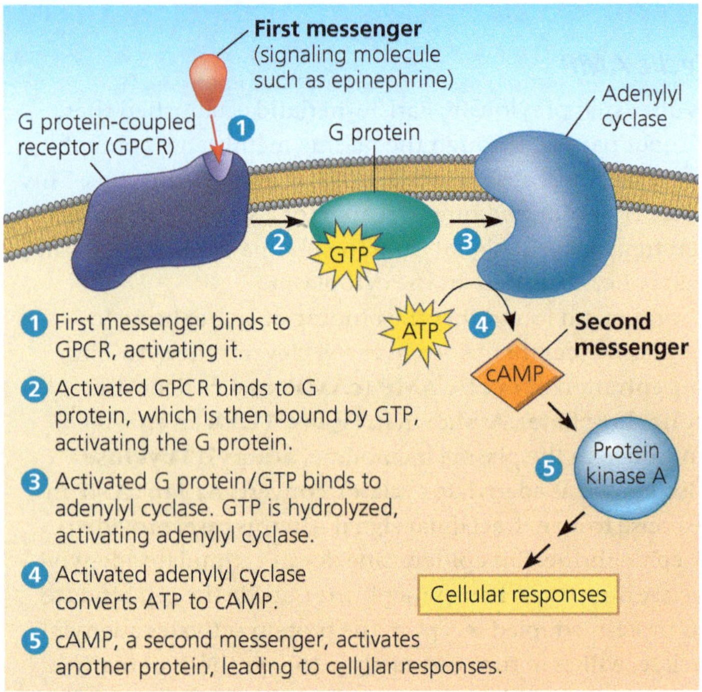

DRAW IT *The bacterium that causes the disease cholera produces a toxin that locks the G protein in its activated state. Review Figure 11.8 then draw this figure as it would be if cholera toxin were present. (You do not need to draw the cholera toxin molecule.)*

Further regulation of cell metabolism is provided by other G protein systems that *inhibit* adenylyl cyclase. In these systems, a different signaling molecule activates a different receptor, which in turn activates an *inhibitory* G protein that blocks activation of adenylyl cyclase.

Now that we know about the role of cAMP in G protein signaling pathways, we can explain in molecular detail how certain microbes cause disease. Consider cholera, a disease that is frequently epidemic in places where the water supply is contaminated with human feces. People acquire the cholera bacterium, *Vibrio cholerae*, by drinking contaminated water. The bacteria form a biofilm on the lining of the small intestine and produce a toxin. The cholera toxin is an enzyme that chemically modifies a G protein involved in regulating salt and water secretion. Because the modified G protein is unable to hydrolyze GTP to GDP, it remains stuck in its active form, continuously stimulating adenylyl cyclase to make cAMP (see the question with Figure 11.12). The resulting high concentration of cAMP causes the intestinal cells to secrete large amounts of salts into the intestines, with water following by osmosis. An infected person quickly develops profuse diarrhea and if left untreated can soon die from the loss of water and salts.

Our understanding of signaling pathways involving cyclic AMP or related messengers has allowed us to develop treatments for certain conditions in humans. In one pathway, a molecule similar to cAMP called *cyclic GMP (cGMP)* is produced by a muscle cell in response to the gas nitric oxide (NO) after it is released by a neighboring cell. cGMP then acts as a second messenger that causes relaxation of muscles, such as those in the walls of arteries. A compound that inhibits the hydrolysis of cGMP to GMP, thus prolonging the signal, was originally prescribed for chest pains because it relaxed blood vessels and increased blood flow to the heart muscle. Under the trade name Viagra, this compound is now widely used as a treatment for erectile dysfunction in human males. Because Viagra leads to dilation of blood vessels, it also allows increased blood flow to the penis, optimizing physiological conditions for penile erections.

Calcium Ions and Inositol Trisphosphate (IP_3)

Many of the signaling molecules that function in animals—including neurotransmitters, growth factors, and some hormones—induce responses in their target cells via signal transduction pathways that increase the cytosolic concentration of calcium ions (Ca^{2+}). Calcium is even more widely used than cAMP as a second messenger. Increasing the cytosolic concentration of Ca^{2+} causes many responses in animal cells, including muscle cell contraction, exocytosis of molecules (secretion), and cell division. In plant cells, a wide range of hormonal and environmental stimuli can cause brief increases in cytosolic Ca^{2+} concentration, triggering various signaling pathways, such as the pathway for greening in response to light (see Figure 39.4). Cells use Ca^{2+} as a second messenger in pathways triggered by both G protein-coupled receptors and receptor tyrosine kinases.

Although cells always contain some Ca^{2+}, this ion can function as a second messenger because its concentration in the cytosol is normally much lower than the concentration outside the cell **(Figure 11.13)**. In fact, the level of Ca^{2+} in the blood and extracellular fluid of an animal is often more than 10,000 times higher than that in the cytosol. Calcium ions are actively transported out of the cell and are actively imported from the cytosol into the endoplasmic reticulum (and, under some conditions, into mitochondria and chloroplasts) by various protein pumps. As a result, the calcium concentration in the ER is usually much higher than that in the cytosol. Because the cytosolic calcium level is low, a small change in absolute numbers of ions represents a relatively large percentage change in calcium concentration.

In response to a signal relayed by a signal transduction pathway, the cytosolic calcium level may rise, usually by a mechanism that releases Ca^{2+} from the cell's ER. The pathways leading to calcium release involve two other

Figure 11.13 The maintenance of calcium ion concentrations in an animal cell. The Ca^{2+} concentration in the cytosol is usually much lower (beige) than in the extracellular fluid and ER (green). Protein pumps in the plasma membrane and the ER membrane, driven by ATP, move Ca^{2+} from the cytosol into the extracellular fluid and into the lumen of the ER. Mitochondrial pumps, driven by chemiosmosis (see Concept 9.4), move Ca^{2+} into mitochondria when the calcium level in the cytosol rises significantly.

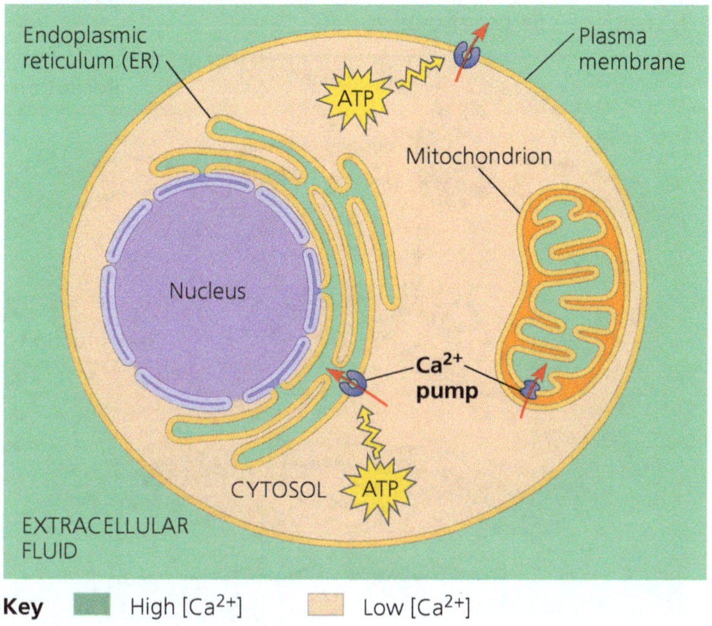

second messengers, **inositol trisphosphate (IP_3)** and **diacylglycerol (DAG)**. These two messengers are produced by cleavage of a certain kind of phospholipid in the plasma membrane. **Figure 11.14** shows the complete picture of how a signal causes IP_3 to stimulate the release of calcium from the ER. Because IP_3 acts before calcium in these pathways, calcium could be considered a "*third* messenger." However, scientists use the term *second messenger* for all small, nonprotein components of signal transduction pathways.

CONCEPT CHECK 11.3

1. What is a protein kinase, and what is its role in a signal transduction pathway?
2. When a signal transduction pathway involves a phosphorylation cascade, how does the cell's response get turned off?
3. What is the actual "signal" that is being transduced in any signal transduction pathway, such as those shown in Figures 11.6 and 11.10? In what way is this information being passed from the exterior to the interior of the cell?
4. **WHAT IF?** ➤ If you exposed a cell to a ligand that binds to a receptor and activates phospholipase C, predict the effect the IP_3-gated calcium channel would have on Ca^{2+} concentration in the cytosol.

For suggested answers, see Appendix A.

Figure 11.14 Calcium and IP_3 in signaling pathways. Calcium ions (Ca^{2+}) and inositol trisphosphate (IP_3) function as second messengers in many signal transduction pathways. In this figure, the process is initiated by the binding of a signaling molecule to a G protein-coupled receptor. A receptor tyrosine kinase could also initiate this pathway by activating phospholipase C.

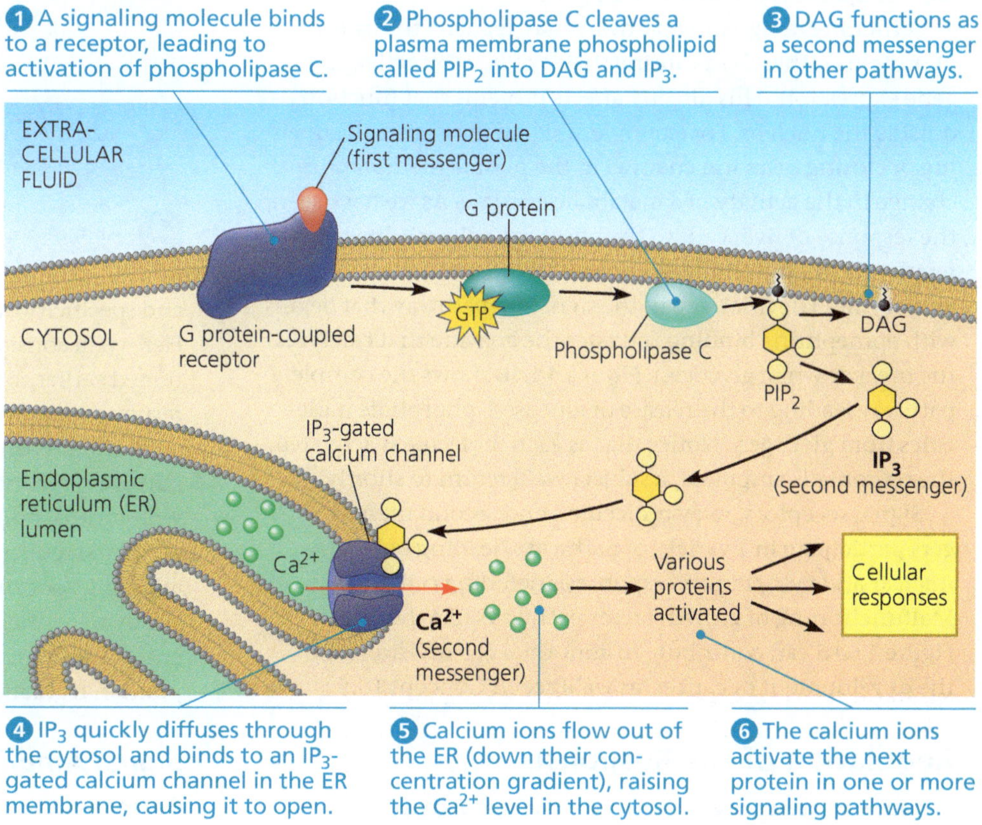

CONCEPT 11.4

Response: Cell signaling leads to regulation of transcription or cytoplasmic activities

We now take a closer look at the cell's subsequent response to an extracellular signal—what some researchers call the "output response." What is the nature of the final step in a signaling pathway?

Nuclear and Cytoplasmic Responses

Ultimately, a signal transduction pathway leads to the regulation of one or more cellular activities. The response at the end of the pathway may occur in the nucleus of the cell or in the cytoplasm.

Many signaling pathways ultimately regulate protein synthesis, usually by turning specific genes on or off in the nucleus. Like an activated steroid receptor (see Figure 11.9), the final activated molecule in a signaling pathway may function as a transcription factor. **Figure 11.15** shows an example in which a signaling pathway activates a transcription factor that turns a gene on: The response to this growth factor signal is transcription, the synthesis of one or more specific mRNAs, which will be translated in the cytoplasm into specific proteins. In other cases, the transcription factor might regulate a gene by turning it off. Often a transcription factor regulates several different genes.

Sometimes a signaling pathway may regulate the *activity* of proteins rather than causing their *synthesis* by activating gene expression. This directly affects proteins that function outside the nucleus. For example, a signal may cause the opening or closing of an ion channel in the plasma membrane or a change in the activity of a metabolic enzyme. As we have seen, the response of liver cells to the hormone epinephrine helps regulate cellular energy metabolism by affecting the activity of an enzyme. The final step in the signaling pathway that begins with epinephrine binding activates the enzyme that catalyzes the breakdown of glycogen. **Figure 11.16** shows the complete pathway leading to the release of glucose 1-phosphate molecules from glycogen. Notice that as each molecule is activated, the response is amplified, a subject we'll return to shortly.

Signal receptors, relay molecules, and second messengers participate in a variety of pathways, leading to both nuclear and cytoplasmic responses, including cell division. Malfunctioning of growth factor pathways like the one in Figure 11.15 can contribute to abnormal cell division and the development of cancer, as we'll see in Concept 18.5.

Regulation of the Response

Whether the response occurs in the nucleus or in the cytoplasm, it is not simply turned "on" or "off." Rather, the extent

Figure 11.15 Nuclear responses to a signal: the activation of a specific gene by a growth factor. This diagram shows a typical signaling pathway that leads to regulation of gene activity in the cell nucleus. The initial signaling molecule, in this case a growth factor, triggers a phosphorylation cascade, as in Figure 11.10. (The ATP molecules and phosphate groups are not shown.) Once phosphorylated, the last kinase in the sequence enters the nucleus and activates a transcription factor, which stimulates transcription of a specific gene (or genes). The resulting mRNAs then direct the synthesis of a particular protein.

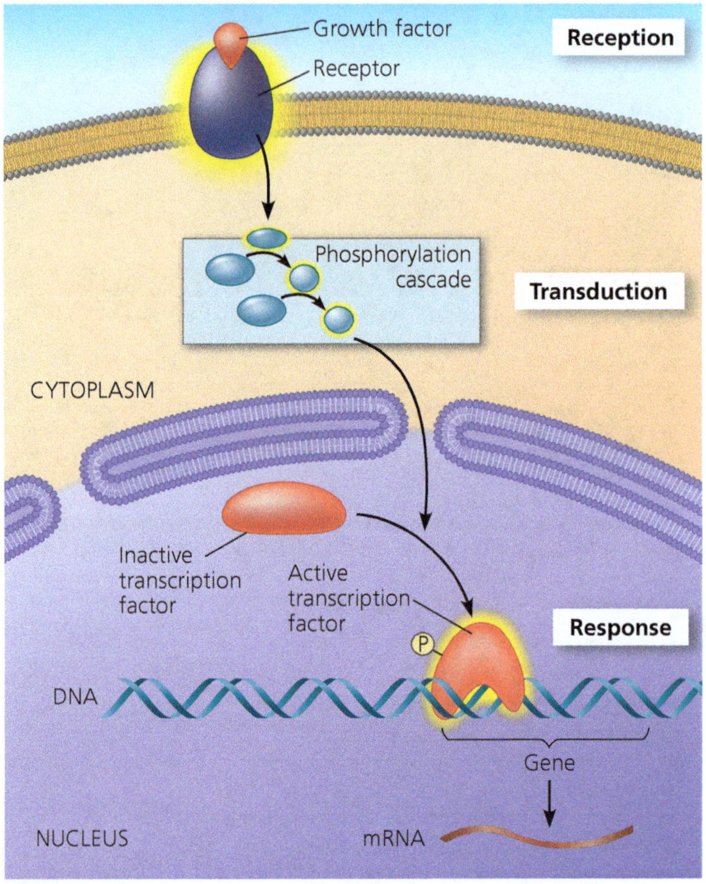

Animation: Nuclear Response: Activating a Gene

and specificity of the response are regulated in multiple ways. Here we'll consider four aspects of this regulation. First, as mentioned earlier, signaling pathways generally amplify the cell's response to a single signaling event. The degree of amplification depends on the function of the specific molecules in the pathway. Second, the many steps in a multistep pathway provide control points at which the cell's response can be further regulated, contributing to the specificity of the response and allowing coordination with other signaling pathways. Third, the overall efficiency of the response is enhanced by the presence of proteins known as scaffolding proteins. Finally, a crucial point in regulating the response is the termination of the signal.

Signal Amplification

Elaborate enzyme cascades amplify the cell's response to a signal. At each catalytic step in the cascade, the number

Figure 11.16 Cytoplasmic response to a signal: the stimulation of glycogen breakdown by epinephrine (adrenaline). In this signaling system, the hormone epinephrine acts through a G protein-coupled receptor to activate a succession of relay molecules, including cAMP and two protein kinases (see also Figure 11.12). The final protein activated is the enzyme glycogen phosphorylase, which uses inorganic phosphate to release glucose monomers from glycogen in the form of glucose 1-phosphate molecules. This pathway amplifies the hormonal signal: One receptor protein can activate approximately 100 molecules of G protein, and each enzyme in the pathway, once activated, can act on many molecules of its substrate, the next molecule in the cascade. The number of activated molecules given for each step is approximate.

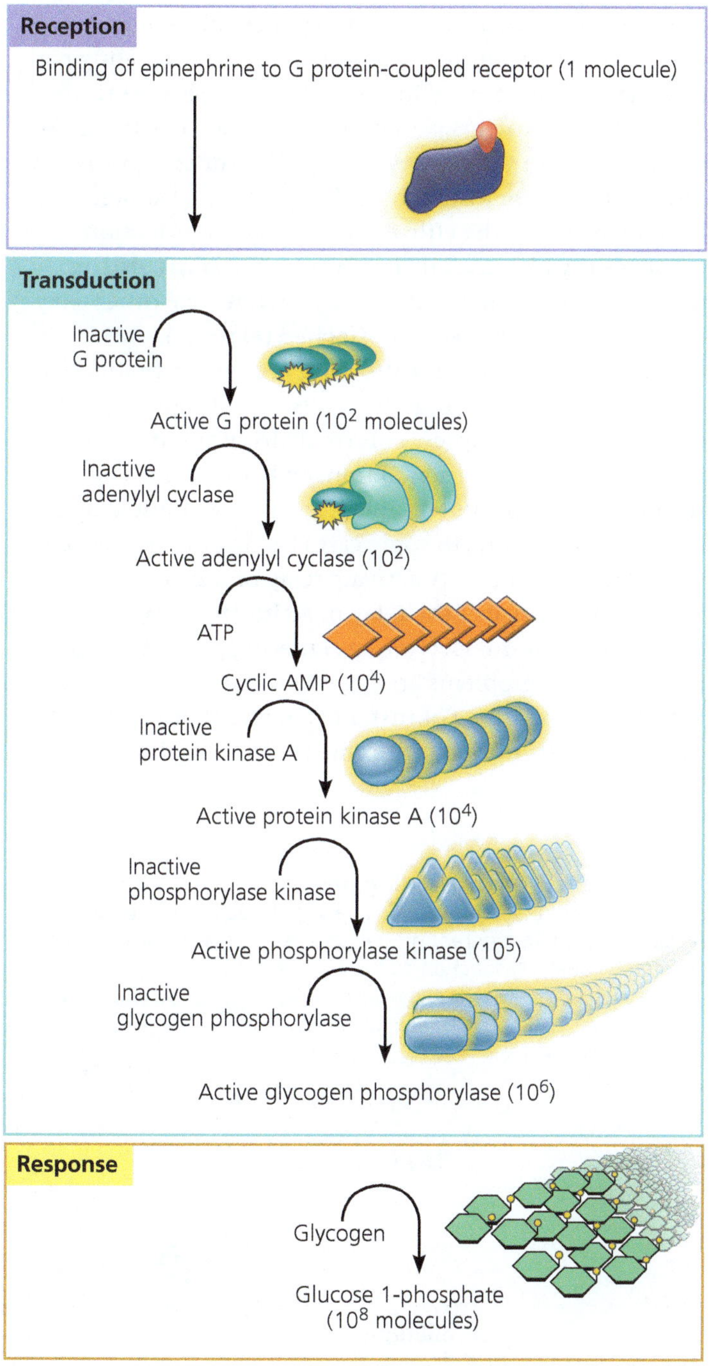

VISUAL SKILLS ➤ *In the figure, how many glucose 1-phosphate molecules are released in response to one signaling molecule? Calculate the factor by which the response is amplified in going from each step to the next.*

Animation: Cytoplasmic Response: Glycogen Breakdown

of activated products can be much greater than in the preceding step. For example, in the epinephrine-triggered pathway in Figure 11.16, each adenylyl cyclase molecule catalyzes the formation of 100 or so cAMP molecules, each molecule of protein kinase A phosphorylates about 10 molecules of the next kinase in the pathway, and so on. The amplification effect stems from the fact that these proteins persist in their active form long enough to process multiple molecules of substrate before they become inactive again. As a result of the signal's amplification, a small number of epinephrine molecules binding to receptors on the surface of a liver cell or muscle cell can lead to the release of hundreds of millions of glucose molecules from glycogen.

The Specificity of Cell Signaling and Coordination of the Response

Consider two different cells in your body—a liver cell and a heart muscle cell, for example. Both are in contact with your bloodstream and are therefore constantly exposed to many different hormone molecules, as well as to local regulators secreted by nearby cells. Yet the liver cell responds to some signals but ignores others, and the same is true for the heart cell. And some kinds of signals trigger responses in both cells—but different responses. For instance, epinephrine stimulates the liver cell to break down glycogen, but the main response of the heart cell to epinephrine is contraction, leading to a more rapid heartbeat. How do we account for this difference?

The explanation for the specificity exhibited in cellular responses to signals is the same as the basic explanation for virtually all differences between cells: Because different kinds of cells turn on different sets of genes, *different kinds of cells have different collections of proteins*. The response of a cell to a signal depends on its particular collection of signal receptor proteins, relay proteins, and proteins needed to carry out the response. A liver cell, for example, is poised to respond appropriately to epinephrine by having the proteins listed in Figure 11.16 as well as those needed to manufacture glycogen.

Thus, two cells that respond differently to the same signal differ in one or more proteins that respond to the signal. Notice in **Figure 11.17** that different pathways may have some molecules in common. For example, cells A, B, and C all use the same receptor protein for the red signaling molecule; differences in other proteins account for their differing responses. In cell D, a different receptor protein is used for the same signaling molecule, leading to yet another response. In cell B, a pathway triggered by one signal diverges to produce two responses; such branched pathways often involve receptor tyrosine kinases (which can activate multiple relay proteins) or second messengers (which can regulate numerous proteins). In cell C, two pathways triggered by separate signals converge to modulate a single response. Branching of pathways and "cross-talk" (interaction) between pathways are important in regulating and coordinating a cell's responses to information coming in from different sources in the body. (You'll learn more about this coordination in Concept 11.5.) Moreover, the use of some of the same proteins in more than one pathway allows the cell to economize on the number of different proteins it must make.

Figure 11.17 The specificity of cell signaling. The particular proteins a cell possesses determine what signaling molecules it responds to and the nature of the response. The four cells in these diagrams respond to the same signaling molecule (red) in different ways because each has a different set of proteins (purple and teal). Note, however, that the same kinds of molecules can participate in more than one pathway.

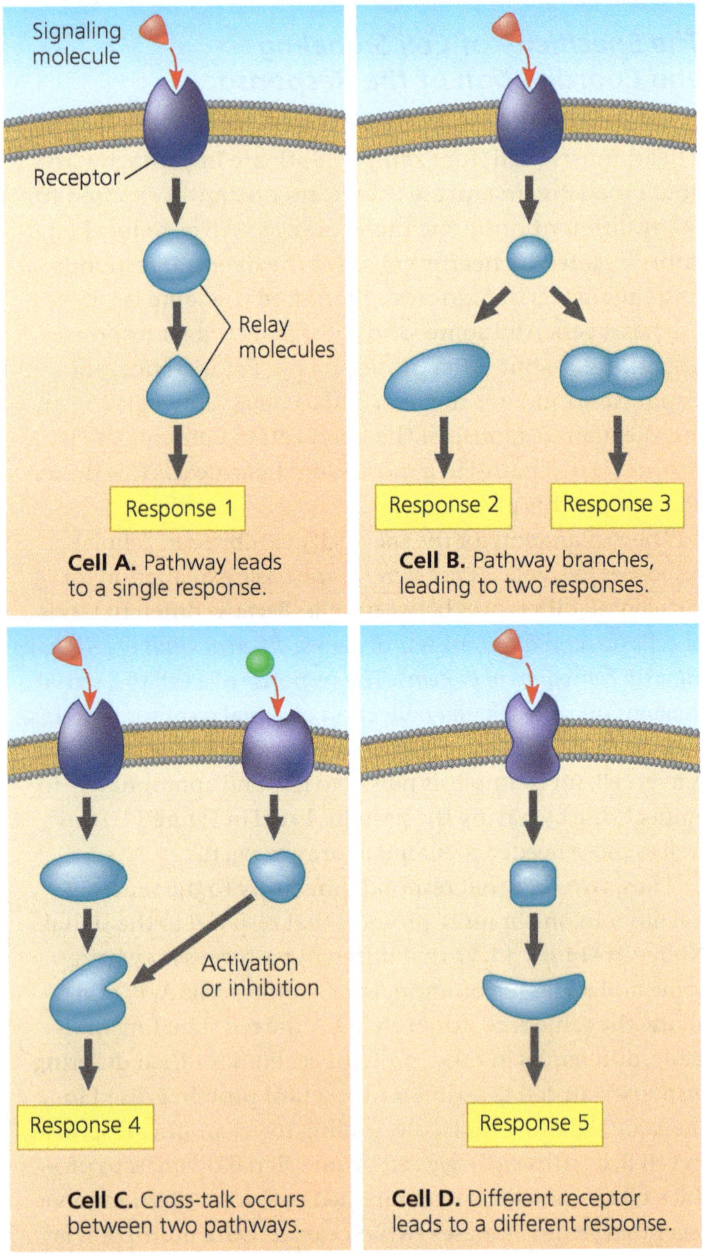

VISUAL SKILLS ➤ *Study the signaling pathway shown in Figure 11.14, and explain how the situation pictured for cell B in Figure 11.17 could apply to that pathway.*

Instructors: The Scientific Skills Exercise "Using Experiments to Test a Model" can be assigned in MasteringBiology. It explores the cellular response of a yeast cell to the signal initiated by a mating factor from a cell of the opposite mating type.

Signaling Efficiency: Scaffolding Proteins and Signaling Complexes

The illustrations of signaling pathways in Figure 11.17 (as well as diagrams of other pathways in this chapter) are greatly simplified. The diagrams show only a few relay molecules and, for clarity's sake, display these molecules spread out in the cytosol. If this were true in the cell, signaling pathways would operate very inefficiently because most relay molecules are proteins, and proteins are too large to diffuse quickly through the viscous cytosol. How does a given protein kinase, for instance, find its protein substrate?

In many cases, the efficiency of signal transduction is apparently increased by the presence of **scaffolding proteins**, large relay proteins to which several other relay proteins are simultaneously attached **(Figure 11.18)**. Researchers have found scaffolding proteins in brain cells that *permanently* hold together networks of signaling pathway proteins at synapses. This hardwiring enhances the speed and accuracy of signal transfer between cells because the rate of protein-protein interaction is not limited by diffusion. Furthermore, in some cases the scaffolding proteins themselves may directly activate relay proteins.

The importance of the relay proteins that serve as points of branching or intersection in signaling pathways is highlighted by the problems arising when these proteins are defective or missing. For instance, in an inherited disorder called Wiskott-Aldrich syndrome (WAS), the absence of

Figure 11.18 A scaffolding protein. The scaffolding protein shown here simultaneously binds to a specific activated membrane receptor and three different protein kinases. This physical arrangement facilitates signal transduction by these molecules.

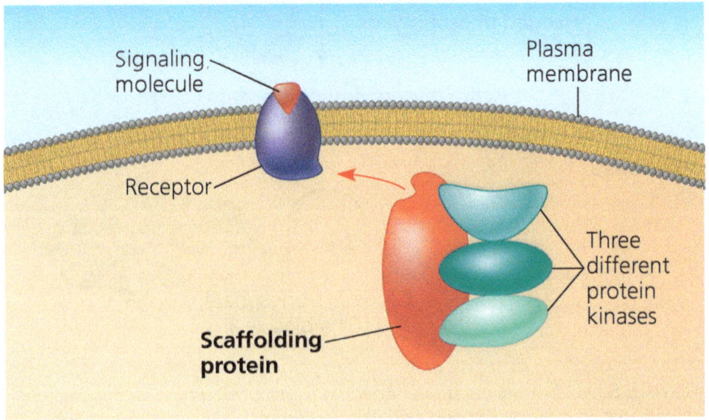

a single relay protein leads to such diverse effects as abnormal bleeding, eczema, and a predisposition to infections and leukemia. These symptoms are thought to arise primarily from the absence of the protein in cells of the immune system. By studying normal cells, scientists found that the WAS protein is located just beneath the immune cell surface. The protein interacts both with microfilaments of the cytoskeleton and with several different components of signaling pathways that relay information from the cell surface, including pathways regulating immune cell proliferation. This multifunctional relay protein is thus both a branch point and an important intersection point in a complex signal transduction network that controls immune cell behavior. When the WAS protein is absent, the cytoskeleton is not properly organized and signaling pathways are disrupted, leading to the WAS symptoms.

Termination of the Signal

In the interest of keeping Figure 11.17 simple, we did not show the *inactivation* mechanisms that are an essential aspect of any cell-signaling pathway. For a cell of a multicellular organism to remain capable of responding to incoming signals, each molecular change in its signaling pathways must last only a short time. As we saw in the cholera example, if a signaling pathway component becomes locked into one state, whether active or inactive, consequences for the organism can be serious.

The ability of a cell to receive new signals depends on reversibility of the changes produced by prior signals. The binding of signaling molecules to receptors is reversible. As the external concentration of signaling molecules falls, fewer receptors are bound at any given moment, and the unbound receptors revert to their inactive form. The cellular response occurs only when the concentration of receptors with bound signaling molecules is above a certain threshold. When the number of active receptors falls below that threshold, the cellular response ceases. Then, by a variety of means, the relay molecules return to their inactive forms: The GTPase activity intrinsic to a G protein hydrolyzes its bound GTP; the enzyme phosphodiesterase converts cAMP to AMP; protein phosphatases inactivate phosphorylated kinases and other proteins; and so forth. As a result, the cell is soon ready to respond to a fresh signal.

MB BioFlix Animation: Mechanism of Hormone Action: Second Messenger cAMP

In this section, we explored the complexity of signaling initiation and termination in a single pathway, and we saw the potential for pathways to intersect with each other. In the next section, we'll consider one especially important network of interacting pathways in the cell.

CONCEPT CHECK 11.4

1. How can a target cell's response to a single hormone molecule result in a response that affects a million other molecules?
2. **WHAT IF?** ➤ If two cells have different scaffolding proteins, explain how they might behave differently in response to the same signaling molecule.
3. **WHAT IF?** ➤ Some human diseases are associated with malfunctioning protein phosphatases. How would such proteins affect signaling pathways? (Review the discussion of protein phosphatases in Concept 11.3, and see Figure 11.10.)

For suggested answers, see Appendix A.

CONCEPT 11.5

Apoptosis integrates multiple cell-signaling pathways

When signaling pathways were first discovered, they were thought to be linear, independent pathways. Our understanding of cellular communication has benefited from the realization that signaling pathway components interact with each other in various ways. For a cell to carry out the appropriate response, cellular proteins often must integrate multiple signals. Let's consider an important cellular process—cellular suicide—as an example.

Cells that are infected, are damaged, or have reached the end of their functional life span often undergo "programmed cell death" **(Figure 11.19)**. The best-understood type of this controlled cell suicide is **apoptosis** (from the Greek, meaning "falling off," and used in a classic Greek poem to refer to leaves falling from a tree). During this process, cellular agents chop up

▼ Figure 11.19 Apoptosis of a human white blood cell. On the left is a normal white blood cell, while on the right is a white blood cell undergoing apoptosis. The apoptotic cell is shrinking and forming lobes ("blebs"), which eventually are shed as membrane-bounded cell fragments (colorized SEMs).

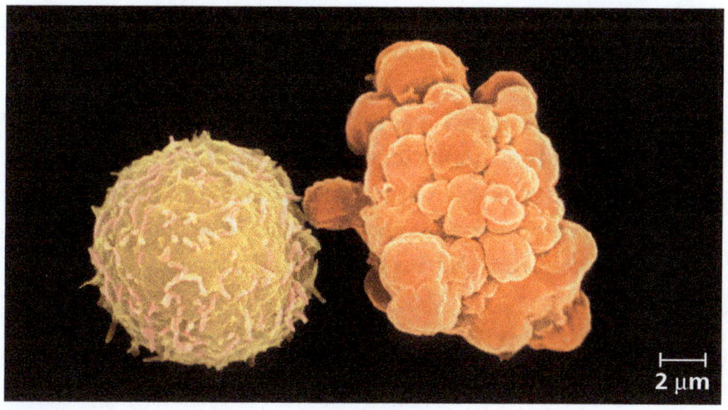

MB Video: Phosphate-Induced Apoptosis

the DNA and fragment the organelles and other cytoplasmic components. The cell shrinks and becomes lobed (a change called "blebbing"), and the cell's parts are packaged up in vesicles that are engulfed and digested by specialized scavenger cells, leaving no trace. Apoptosis protects neighboring cells from damage that they would otherwise suffer if a dying cell merely leaked out all its contents, including its many digestive enzymes.

The signal that triggers apoptosis can come from either outside or inside the cell. Outside the cell, signaling molecules released from other cells can initiate a signal transduction pathway that activates the genes and proteins responsible for carrying out cell death. Within a cell whose DNA has been irretrievably damaged, a series of protein-protein interactions can pass along a signal that similarly triggers cell death. Considering some examples of apoptosis can help us to see how signaling pathways are integrated in cells.

Apoptosis in the Soil Worm *Caenorhabditis elegans*

The molecular mechanisms of apoptosis were worked out by researchers studying embryonic development of a small soil worm, a nematode called *Caenorhabditis elegans*. Because the adult worm has only about 1,000 cells, the researchers were able to work out the entire ancestry of each cell. The timely suicide of cells occurs exactly 131 times during normal development of *C. elegans*, at precisely the same points in the cell lineage of each worm. In worms and other species, apoptosis is triggered by signals that activate a cascade of "suicide" proteins in the cells destined to die.

Genetic research on *C. elegans* initially revealed two key apoptosis genes, called *ced-3* and *ced-4* (*ced* stands for "cell death"), which encode proteins essential for apoptosis. The proteins are called Ced-3 and Ced-4, respectively. These and most other proteins involved in apoptosis are continually present in cells, but in inactive form; thus, regulation in this case occurs at the level of protein activity rather than through gene activity and protein synthesis. In *C. elegans*, a protein in the outer mitochondrial membrane, called Ced-9 (the product of the *ced-9* gene), serves as a master regulator of apoptosis, acting as a brake in the absence of a signal promoting apoptosis **(Figure 11.20)**. When a death signal is received by the cell, signal transduction involves a change in Ced-9 that disables the brake, and the apoptotic pathway activates proteases and nucleases, enzymes that cut up the proteins and DNA of the cell. The main proteases of apoptosis are called *caspases*; in the nematode, the chief caspase is the Ced-3 protein.

▼ Figure 11.20 Molecular basis of apoptosis in *C. elegans*. Three proteins, Ced-3, Ced-4, and Ced-9, are critical to apoptosis and its regulation in the nematode. Apoptosis is more complicated in mammals but involves proteins similar to those in *C. elegans*.

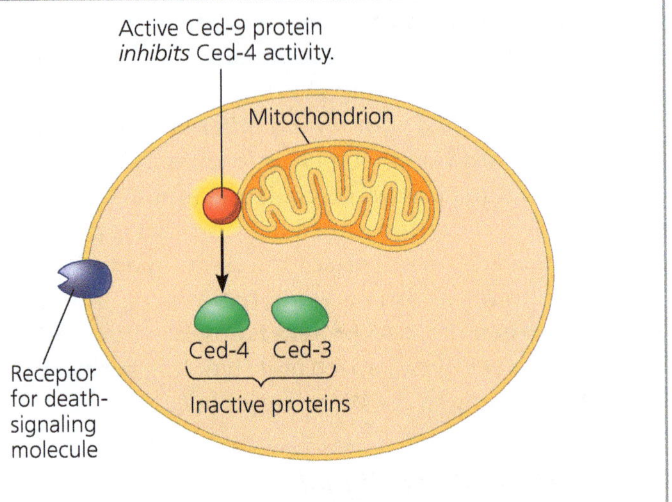

(a) No death signal. As long as Ced-9, located in the outer mitochondrial membrane, is active, apoptosis is inhibited, and the cell remains alive.

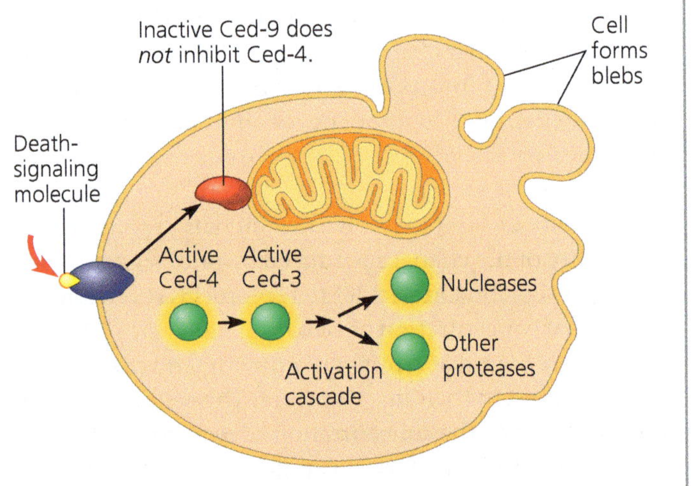

(b) Death signal. When a cell receives a death signal, Ced-9 is inactivated, relieving its inhibition of Ced-4. Active Ced-4 activates Ced-3, a protease, which triggers a cascade of reactions leading to the activation of nucleases and other proteases. The action of these enzymes causes the changes seen in apoptotic cells and eventually leads to cell death.

Apoptotic Pathways and the Signals That Trigger Them

In humans and other mammals, several different pathways, involving about 15 different caspases, can carry out apoptosis. The pathway that is used depends on the type of cell and on the particular signal that initiates apoptosis. One major pathway involves certain mitochondrial proteins that are triggered to form molecular pores in the mitochondrial outer membrane, causing it to leak and release other proteins that promote apoptosis. Perhaps surprisingly, these latter include cytochrome *c*, which functions in mitochondrial electron transport in healthy cells (see Figure 9.15) but acts as a cell death factor when released from mitochondria. The process of mitochondrial apoptosis in mammals uses proteins similar to the nematode proteins Ced-3, Ced-4, and Ced-9. These can be thought of as relay proteins capable of transducing the apoptotic signal.

Animation: Apoptosis

At key gateways into the apoptotic program, relay proteins integrate signals from several different sources and can send a cell down an apoptotic pathway. Often, the signal originates outside the cell, like the death-signaling molecule depicted in Figure 11.20b, which presumably was released by a neighboring cell. When a death-signaling ligand occupies a cell-surface receptor, this binding leads to activation of caspases and other enzymes that carry out apoptosis, without involving the mitochondrial pathway. This process of signal reception, transduction, and response is similar to what we have discussed throughout this chapter. In a twist on the classic scenario, two other types of alarm signals that can lead to apoptosis originate from *inside* the cell rather than from a cell-surface receptor. One signal comes from the nucleus, generated when the DNA has suffered irreparable damage, and a second comes from the endoplasmic reticulum when excessive protein misfolding occurs. Mammalian cells make life-or-death "decisions" by somehow integrating the death signals and life signals they receive from these external and internal sources.

A built-in cell suicide mechanism is essential to development and maintenance in all animals. The similarities between apoptosis genes in nematodes and those in mammals, as well as the observation that apoptosis occurs in multicellular fungi and even in single-celled yeasts, indicate that the basic mechanism evolved early in the evolution of eukaryotes. In vertebrates, apoptosis is essential for normal development of the nervous system, for normal operation of the immune system, and for normal morphogenesis of hands and feet in humans and paws in other mammals **(Figure 11.21)**. The level of apoptosis between the developing digits is lower in the webbed feet of ducks and other water birds than in the nonwebbed feet of land birds, such as chickens. In the case of humans, the failure of appropriate apoptosis can result in webbed fingers and toes.

Significant evidence points to the involvement of apoptosis in certain degenerative diseases of the nervous system, such as Parkinson's disease and Alzheimer's disease. In Alzheimer's disease, an accumulation of aggregated proteins in neuronal cells activates an enzyme that triggers apoptosis, resulting in the loss of brain function seen in these patients. Furthermore, cancer can result from a failure of cell suicide; some cases of human melanoma, for example, have been linked to faulty forms of the human version of the *C. elegans* Ced-4 protein. It is not surprising, therefore, that the signaling pathways feeding into apoptosis are quite elaborate. After all, the life-or-death question is the most fundamental one imaginable for a cell.

This chapter has introduced you to many of the general mechanisms of cell communication, such as ligand binding, protein-protein interactions and shape changes, cascades of interactions, and protein phosphorylation. Throughout your study of biology, you will encounter numerous examples of cell signaling.

CONCEPT CHECK 11.5

1. Give an example of apoptosis during embryonic development, and explain its function in the developing embryo.
2. **WHAT IF?** ➤ If apoptosis occurred when it should not, what types of protein defects might be the cause? What types could result in apoptosis not occurring when it should?

For suggested answers, see Appendix A.

▼ Figure 11.21 Effect of apoptosis during paw development in the mouse. In mice, humans, other mammals, and land birds, the embryonic region that develops into feet or hands initially has a solid, platelike structure. Apoptosis eliminates the cells in the interdigital regions, thus forming the digits. The embryonic mouse paws shown in these fluorescence light micrographs are stained so that cells undergoing apoptosis appear a bright yellowish green. Apoptosis of cells begins at the margin of each interdigital region (left), peaks as the tissue in these regions is reduced (middle), and is no longer visible when the interdigital tissue has been eliminated (right).

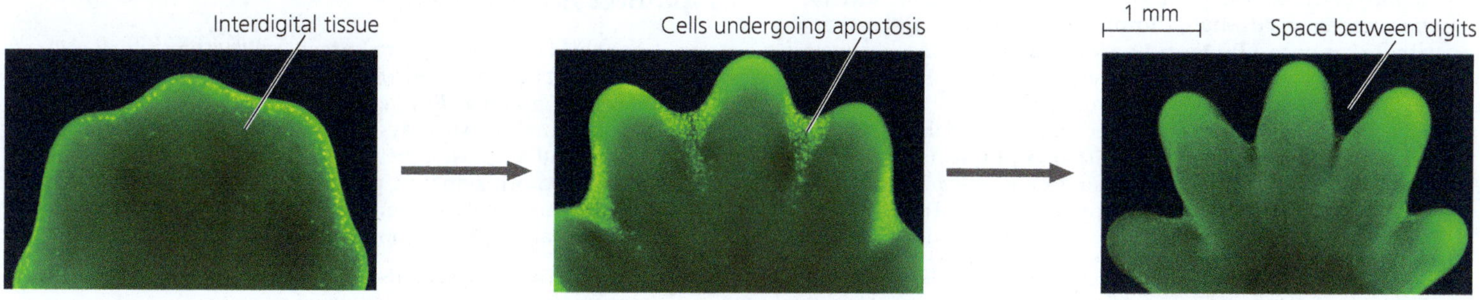

11 Chapter Review

Go to **MasteringBiology**™ for Videos, Animations, Vocab Self-Quiz, Practice Tests, and more in the Study Area.

SUMMARY OF KEY CONCEPTS

VOCAB SELF-QUIZ goo.gl/6u55ks

CONCEPT 11.1

External signals are converted to responses within the cell *(pp. 213–217)*

- **Signal transduction pathways** are crucial for many processes. Signaling during yeast cell mating has much in common with processes in multicellular organisms, suggesting an early evolutionary origin of signaling mechanisms. Bacterial cells can sense the local density of bacterial cells (quorum sensing).
- Local signaling by animal cells involves direct contact or the secretion of local regulators. For long-distance signaling, animal and plant cells use **hormones**; animals also pass signals electrically.
- Like epinephrine, other hormones that bind to membrane receptors trigger a three-stage cell-signaling pathway:

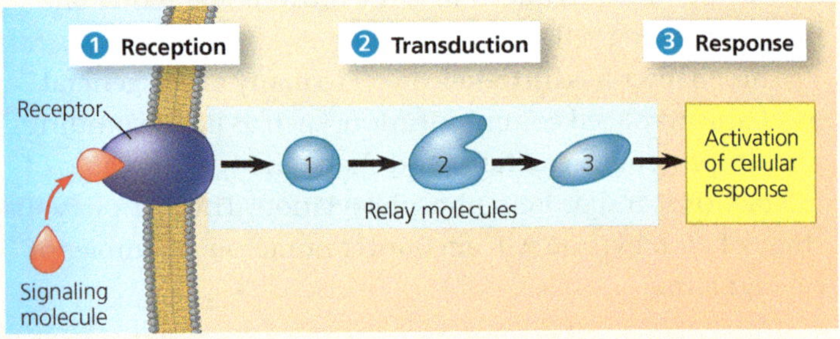

? *What determines whether a cell responds to a hormone such as epinephrine? What determines how a cell responds to such a hormone?*

CONCEPT 11.2

Reception: A signaling molecule binds to a receptor protein, causing it to change shape *(pp. 217–221)*

- The binding between signaling molecule (**ligand**) and receptor is highly specific. A specific shape change in a receptor is often the initial transduction of the signal.
- There are three major types of cell-surface transmembrane receptors: (1) **G protein-coupled receptors (GPCRs)** work with cytoplasmic **G proteins**. Ligand binding activates the receptor, which then activates a specific G protein, which activates yet another protein, thus propagating the signal. (2) **Receptor tyrosine kinases (RTKs)** react to the binding of signaling molecules by forming dimers and then adding phosphate groups to tyrosines on the cytoplasmic part of the other monomer making up the dimer. Relay proteins in the cell can then be activated by binding to different phosphorylated tyrosines, allowing this receptor to trigger several pathways at once. (3) **Ligand-gated ion channels** open or close in response to binding by specific signaling molecules, regulating the flow of specific ions across the membrane.
- The activity of all three types of receptors is crucial; abnormal GPCRs and RTKs are associated with many human diseases.
- Intracellular receptors are cytoplasmic or nuclear proteins. Signaling molecules that are hydrophobic or small enough to cross the plasma membrane bind to these receptors inside the cell.

? *How are the structures of a GPCR and an RTK similar? How does initiation of signal transduction differ for these two types of receptors?*

CONCEPT 11.3

Transduction: Cascades of molecular interactions relay signals from receptors to target molecules in the cell *(pp. 221–225)*

- At each step in a signal transduction pathway, the signal is transduced into a different form, which commonly involves a shape change in a protein. Many signal transduction pathways include **phosphorylation cascades**, in which a series of **protein kinases** each add a phosphate group to the next one in line, activating it. Enzymes called **protein phosphatases** remove the phosphate groups. The balance between phosphorylation and dephosphorylation regulates the activity of proteins involved in the sequential steps of a signal transduction pathway.
- **Second messengers**, such as the small molecule **cyclic AMP (cAMP)** and the ion Ca^{2+}, diffuse readily through the cytosol and thus help broadcast signals quickly. Many G proteins activate **adenylyl cyclase**, which makes cAMP from ATP. Cells use Ca^{2+} as a second messenger in both GPCR and RTK pathways. The tyrosine kinase pathways can also involve two other second messengers, **diacylglycerol (DAG)** and **inositol trisphosphate (IP_3)**. IP_3 can trigger a subsequent increase in Ca^{2+} levels.

? *What is the difference between a protein kinase and a second messenger? Can both operate in the same signal transduction pathway?*

CONCEPT 11.4

Response: Cell signaling leads to regulation of transcription or cytoplasmic activities *(pp. 226–229)*

- Some pathways lead to a nuclear response: Specific genes are turned on or off by activated transcription factors. In others, the response involves cytoplasmic regulation.
- Cellular responses are not simply on or off; they are regulated at many steps. Each protein in a signaling pathway amplifies the signal by activating multiple copies of the next component; for long pathways, the total amplification may be over a millionfold. The combination of proteins in a cell confers specificity in the signals it detects and the responses it carries out. **Scaffolding proteins** increase signaling efficiency. Pathway branching further helps the cell coordinate signals and responses. Signal response can be terminated quickly because ligand binding is reversible.

? *What mechanisms in the cell terminate its response to a signal and maintain its ability to respond to new signals?*

CONCEPT 11.5

Apoptosis integrates multiple cell-signaling pathways *(pp. 229–231)*

- **Apoptosis** is a type of programmed cell death in which cell components are disposed of in an orderly fashion. Studies of the soil worm *Caenorhabditis elegans* clarified molecular details of the relevant signaling pathways. A death signal leads to activation of caspases and nucleases, the main enzymes involved in apoptosis.
- Several apoptotic signaling pathways exist in the cells of humans and other mammals, triggered in different ways. Signals eliciting apoptosis can originate from outside or inside the cell.

? *What is an explanation for the similarities between genes in yeasts, nematodes, and mammals that control apoptosis?*

TEST YOUR UNDERSTANDING

Level 1: Knowledge/Comprehension

PRACTICE TEST goo.gl/CUYGKD

1. Binding of a signaling molecule to which type of receptor leads directly to a change in the distribution of substances on opposite sides of the membrane?
   (A) intracellular receptor
   (B) G protein-coupled receptor
   (C) phosphorylated receptor tyrosine kinase dimer
   (D) ligand-gated ion channel

2. The activation of receptor tyrosine kinases is characterized by
   (A) dimerization and phosphorylation.
   (B) dimerization and IP_3 binding.
   (C) a phosphorylation cascade.
   (D) GTP hydrolysis.

3. Lipid-soluble signaling molecules, such as aldosterone, cross the membranes of all cells but affect only target cells because
   (A) only target cells retain the appropriate DNA segments.
   (B) intracellular receptors are present only in target cells.
   (C) only target cells have enzymes that break down aldosterone.
   (D) only in target cells is aldosterone able to initiate the phosphorylation cascade that turns genes on.

4. Consider this pathway: epinephrine → G protein-coupled receptor → G protein → adenylyl cyclase → cAMP. Identify the second messenger.
   (A) cAMP
   (B) G protein
   (C) GTP
   (D) adenylyl cyclase

5. Apoptosis involves all but which of the following?
   (A) fragmentation of the DNA
   (B) cell-signaling pathways
   (C) lysis of the cell
   (D) digestion of cellular contents by scavenger cells

Level 2: Application/Analysis

6. Which observation suggested to Sutherland the involvement of a second messenger in epinephrine's effect on liver cells?
   (A) Enzymatic activity was proportional to the amount of calcium added to a cell-free extract.
   (B) Receptor studies indicated that epinephrine was a ligand.
   (C) Glycogen breakdown was observed only when epinephrine was administered to intact cells.
   (D) Glycogen breakdown was observed only when epinephrine and glycogen phosphorylase were mixed.

7. Protein phosphorylation is commonly involved with all of the following *except*
   (A) activation of receptor tyrosine kinases.
   (B) activation of protein kinase molecules.
   (C) activation of G protein-coupled receptors.
   (D) regulation of transcription by signaling molecules.

Level 3: Synthesis/Evaluation

8. **DRAW IT** Draw the following apoptotic pathway, which operates in human immune cells. A death signal is received when a molecule called Fas binds its cell-surface receptor. The binding of many Fas molecules to receptors causes receptor clustering. The intracellular regions of the receptors, when together, bind proteins called adaptor proteins. These in turn bind to inactive molecules of caspase-8, which become activated and then activate caspase-3. Once activated, caspase-3 initiates apoptosis.

9. **EVOLUTION CONNECTION** Identify the evolutionary mechanisms that might account for the origin and persistence of cell-to-cell signaling systems in prokaryotes.

10. **SCIENTIFIC INQUIRY** Epinephrine initiates a signal transduction pathway that produces cyclic AMP (cAMP) and leads to the breakdown of glycogen to glucose, a major energy source for cells. But glycogen breakdown is only part of the fight-or-flight response that epinephrine brings about; the overall effect on the body includes an increase in heart rate and alertness, as well as a burst of energy. Given that caffeine blocks the activity of cAMP phosphodiesterase, propose a mechanism by which caffeine ingestion leads to heightened alertness and sleeplessness.

11. **SCIENCE, TECHNOLOGY, AND SOCIETY** The aging process is thought to be initiated at the cellular level. Among the changes that can occur after a certain number of cell divisions is the loss of a cell's ability to respond to growth factors and other signals. Much research into aging is aimed at understanding such losses, with the ultimate goal of extending the human life span. Not everyone, however, agrees that this is a desirable goal. If life expectancy were greatly increased, discuss what might be the social and ecological consequences.

12. **WRITE ABOUT A THEME: ORGANIZATION** The properties of life emerge at the biological level of the cell. The highly regulated process of apoptosis is not simply the destruction of a cell; it is also an emergent property. Write a short essay (about 100–150 words) that briefly explains the role of apoptosis in the development and proper functioning of an animal, and describe how this form of programmed cell death is a process that emerges from the orderly integration of signaling pathways.

13. **SYNTHESIZE YOUR KNOWLEDGE**

There are five basic tastes—sour, salty, sweet, bitter, and "umami." Salt is detected when the concentration of salt outside of a taste bud cell is higher than that inside of it, and ion channels allow the passive leakage of Na^+ into the cell. The resulting change in membrane potential (see Concept 7.4) sends the "salty" signal to the brain. Umami is a savory taste generated by glutamate (glutamic acid, found in monosodium glutamate, or MSG), which is used as a flavor enhancer in foods such as taco-flavored tortilla chips. The glutamate receptor is a GPCR, which, when bound, initiates a signaling pathway that ends with a cellular response, perceived by you as "taste." If you eat a regular potato chip and then rinse your mouth, you will no longer taste salt. But if you eat a flavored tortilla chip and then rinse, the taste persists. (Try it!) Propose a possible explanation for this difference.

For selected answers, see Appendix A.

MB For additional practice questions, check out the **Dynamic Study Modules** in MasteringBiology. You can use them to study on your smartphone, tablet, or computer anytime, anywhere!

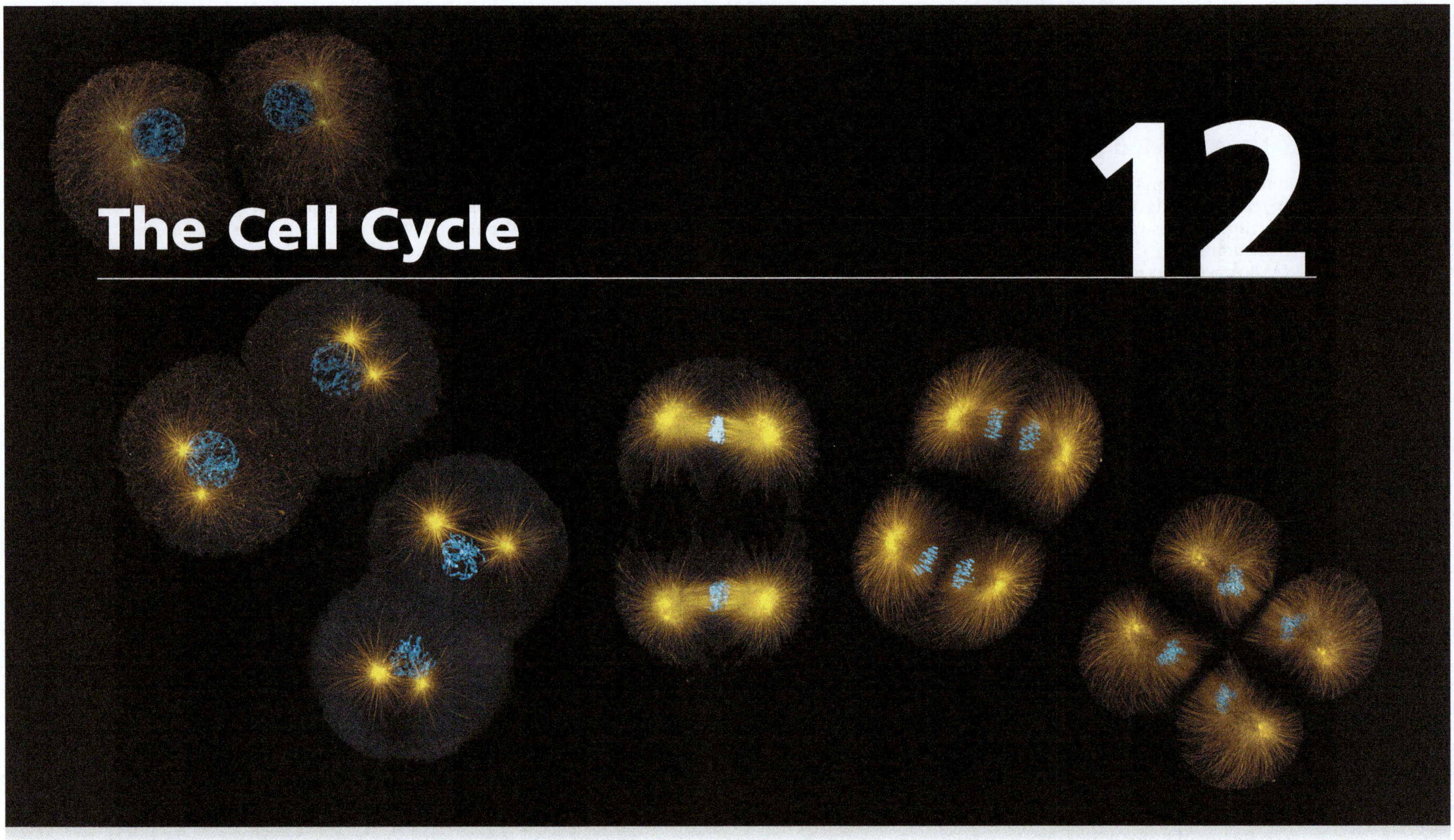

The Cell Cycle

12

▲ **Figure 12.1 How do dividing cells distribute chromosomes to daughter cells?**

KEY CONCEPTS

12.1 Most cell division results in genetically identical daughter cells

12.2 The mitotic phase alternates with interphase in the cell cycle

12.3 The eukaryotic cell cycle is regulated by a molecular control system

▼ Chromosomes (blue) are attached by specific proteins (green) to cell machinery (red) and are moved during division of a rat kangaroo cell.

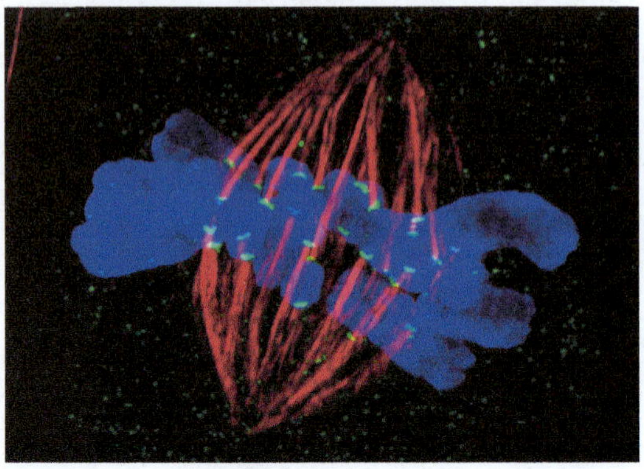

The Key Roles of Cell Division

The ability of organisms to produce more of their own kind is the one characteristic that best distinguishes living things from nonliving matter. This unique capacity to procreate, like all biological functions, has a cellular basis. In 1855, Rudolf Virchow, a German physician, put it this way: "Where a cell exists, there must have been a preexisting cell, just as the animal arises only from an animal and the plant only from a plant." He summarized this concept with the Latin saying "*Omnis cellula e cellula*," meaning "Every cell from a cell." The continuity of life is based on the reproduction of cells, or **cell division**. The series of confocal fluorescence micrographs in **Figure 12.1**, starting at the upper left and reading from left to right, follows the events of cell division as the cells of a two-celled marine worm embryo become four.

Cell division plays several important roles in life. When a prokaryotic cell divides, it is actually reproducing, since the process gives rise to a new organism (another cell). The same is true of any unicellular eukaryote, such as the amoeba shown in **Figure 12.2a**. As for multicellular eukaryotes, cell division enables each of these organisms to develop from a single cell—the fertilized egg. A two-celled embryo, the first stage in this process, is shown in **Figure 12.2b**. And cell division continues to function in renewal and repair in fully grown multicellular eukaryotes, replacing cells that die from accidents or normal wear and tear. For example, dividing cells in your bone marrow continuously make new blood cells **(Figure 12.2c)**.

When you see this blue icon, log in to **MasteringBiology** and go to the Study Area for digital resources.

MB **Get Ready for This Chapter**

Figure 12.2 The functions of cell division.

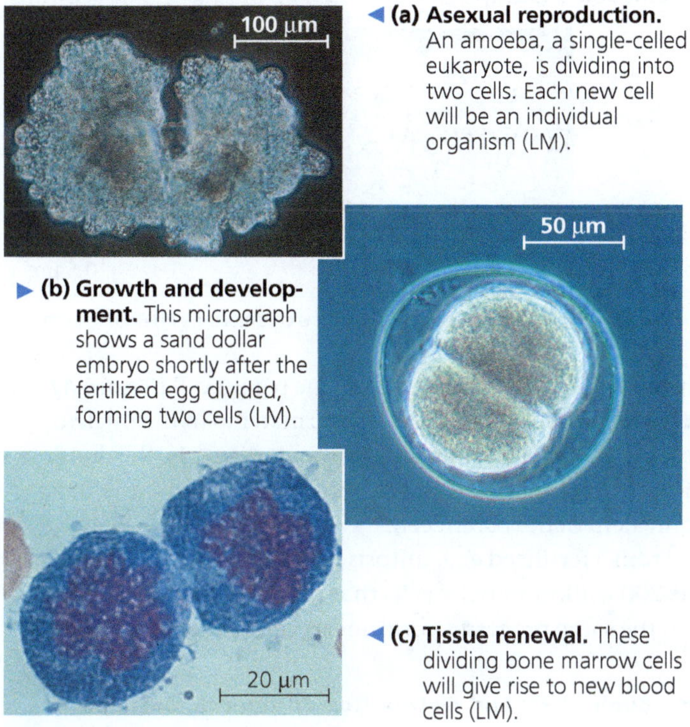

(a) Asexual reproduction. An amoeba, a single-celled eukaryote, is dividing into two cells. Each new cell will be an individual organism (LM).

(b) Growth and development. This micrograph shows a sand dollar embryo shortly after the fertilized egg divided, forming two cells (LM).

(c) Tissue renewal. These dividing bone marrow cells will give rise to new blood cells (LM).

MB Cell Division in a Sea Urchin Embryo

The cell division process is an integral part of the **cell cycle**, the life of a cell from the time it is first formed during division of a parent cell until its own division into two daughter cells. (Biologists use the words *daughter* or *sister* in relation to cells, but this is not meant to imply gender.) Passing identical genetic material to cellular offspring is a crucial function of cell division. In this chapter, you will learn how this process occurs in the context of the cell cycle. After studying the cellular mechanics of cell division in eukaryotes and bacteria, you will learn about the molecular control system that regulates progress through the eukaryotic cell cycle and what happens when the control system malfunctions. Because a breakdown in cell cycle control plays a major role in the development of cancer, this aspect of cell biology is an active area of research.

CONCEPT 12.1

Most cell division results in genetically identical daughter cells

The reproduction of a cell, with all of its complexity, cannot occur by a mere pinching in half; a cell is not like a soap bubble that simply enlarges and splits in two. In both prokaryotes and eukaryotes, most cell division involves the distribution of identical genetic material—DNA—to two daughter cells. (The exception is meiosis, the special type of eukaryotic cell division that can produce sperm and eggs.) What is most remarkable about cell division is the accuracy with which the DNA is passed from one generation of cells to the next. A dividing cell replicates its DNA, distributes the two copies to opposite ends of the cell, and then splits into daughter cells.

Cellular Organization of the Genetic Material

A cell's DNA, its genetic information, is called its **genome**. Although a prokaryotic genome is often a single DNA molecule, eukaryotic genomes usually consist of a number of DNA molecules. The overall length of DNA in a eukaryotic cell is enormous. A typical human cell, for example, has about 2 m of DNA—a length about 250,000 times greater than the cell's diameter. Before the cell can divide to form genetically identical daughter cells, all of this DNA must be copied, or replicated, and then the two copies must be separated so that each daughter cell ends up with a complete genome.

The replication and distribution of so much DNA are manageable because the DNA molecules are packaged into structures called **chromosomes** (from the Greek *chroma*, color, and *soma*, body), so named because they take up certain dyes used in microscopy **(Figure 12.3)**. Each eukaryotic chromosome consists of one very long, linear DNA molecule associated with many proteins (see Figure 6.9). The DNA molecule carries several hundred to a few thousand genes, the units of information that specify an organism's inherited traits. The associated proteins maintain the structure of the chromosome and help control the activity of the genes. Together, the entire complex of DNA and proteins that is the building material of chromosomes is referred to as **chromatin**. As you will soon see, the chromatin of a chromosome varies in its degree of condensation during the process of cell division.

Every eukaryotic species has a characteristic number of chromosomes in each cell's nucleus. For example, the nuclei of human **somatic cells** (all body cells except the reproductive cells) each contain 46 chromosomes, made up of two sets

Figure 12.3 Eukaryotic chromosomes. Chromosomes (stained purple) are visible within the nucleus of this cell from an African blood lily. The thinner red threads in the surrounding cytoplasm are the cytoskeleton. The cell is preparing to divide (LM).

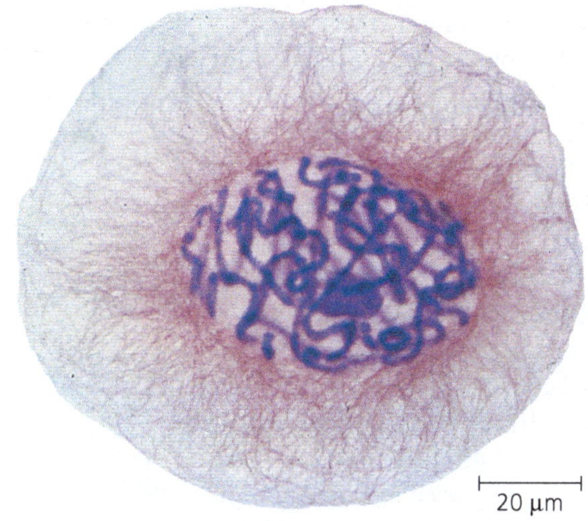

of 23, one set inherited from each parent. Reproductive cells, or **gametes**—such as sperm and eggs—have half as many chromosomes as somatic cells; in our example, human gametes have one set of 23 chromosomes. The number of chromosomes in somatic cells varies widely among species: 18 in cabbage plants, 48 in chimpanzees, 56 in elephants, 90 in hedgehogs, and 148 in one species of alga. We'll now consider how these chromosomes behave during cell division.

Figure 12.4 A highly condensed, duplicated human chromosome (SEM).

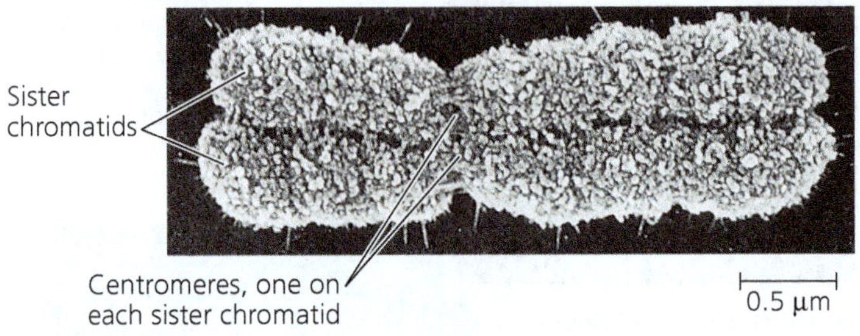

DRAW IT ➤ *Circle one sister chromatid of the chromosome in this micrograph.*

Distribution of Chromosomes During Eukaryotic Cell Division

When a cell is not dividing, and even as it replicates its DNA in preparation for cell division, each chromosome is in the form of a long, thin chromatin fiber. After DNA replication, however, the chromosomes condense as a part of cell division: Each chromatin fiber becomes densely coiled and folded, making the chromosomes much shorter and so thick that we can see them with a light microscope.

Each duplicated chromosome consists of two **sister chromatids**, which are joined copies of the original chromosome **(Figure 12.4)**. The two chromatids, each containing an identical DNA molecule, are typically attached all along their lengths by protein complexes called *cohesins*; this attachment is known as *sister chromatid cohesion*. Each sister chromatid has a **centromere**, a region made up of repetitive sequences in the chromosomal DNA where the chromatid is attached most closely to its sister chromatid. This attachment is mediated by proteins that recognize and bind to the centromeric DNA; other bound proteins condense the DNA, giving the duplicated chromosome a narrow "waist." The portion of a chromatid to either side of the centromere is referred to as an *arm* of the chromatid. (An unduplicated chromosome has a single centromere, distinguished by the proteins that bind there, and two arms.)

Later in the cell division process, the two sister chromatids of each duplicated chromosome separate and move into two new nuclei, one forming at each end of the cell. Once the sister chromatids separate, they are no longer called sister chromatids but are considered individual chromosomes; this is the step that essentially doubles the number of chromosomes during cell division. Thus, each new nucleus receives a collection of chromosomes identical to that of the parent cell **(Figure 12.5)**. **Mitosis**, the division of the genetic material in the nucleus, is usually followed immediately by **cytokinesis**, the division of the cytoplasm. One cell has become two, each the genetic equivalent of the parent cell.

From a fertilized egg, mitosis and cytokinesis produced the 200 trillion somatic cells that now make up your body, and the same processes continue to generate new cells to

Figure 12.5 Chromosome duplication and distribution during cell division.

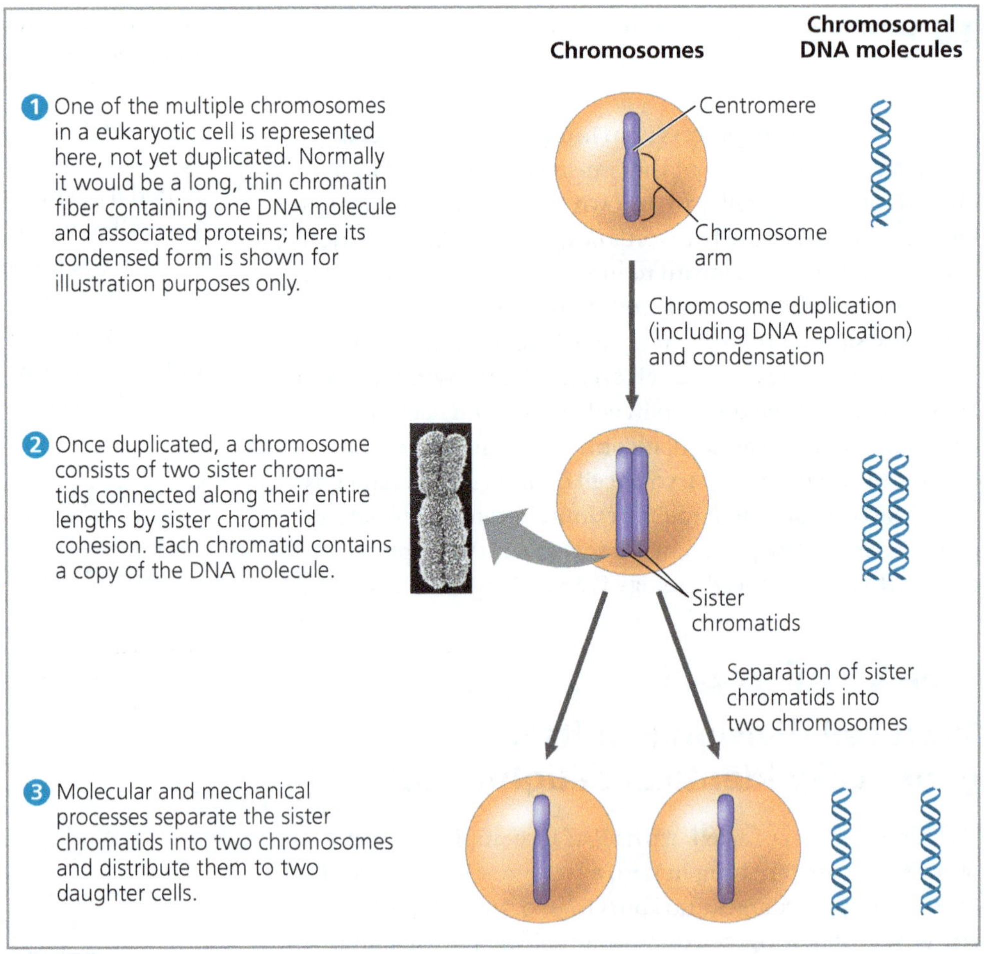

? *How many chromatid arms does the chromosome in* 2 *have? Identify the point in the figure where one chromosome becomes two.*

BioFlix® Animation: Chromosome Duplication

replace dead and damaged ones. In contrast, you produce gametes—eggs or sperm—by a variation of cell division called *meiosis*, which yields daughter cells with only one set of chromosomes, half as many chromosomes as the parent cell. Meiosis in humans occurs only in special cells in the ovaries or testes (the gonads). Generating gametes, meiosis reduces the chromosome number from 46 (two sets) to 23 (one set). Fertilization fuses two gametes together and returns the chromosome number to 46 (two sets). Mitosis then conserves that number in every somatic cell nucleus of the new human individual. In Chapter 13, we will examine the role of meiosis in reproduction and inheritance in more detail. In the remainder of this chapter, we focus on mitosis and the rest of the cell cycle in eukaryotes.

CONCEPT CHECK 12.1

1. How many chromosomes are drawn in each part of Figure 12.5? (Ignore the micrograph in step 2.)
2. **WHAT IF?** ➤ A chicken has 78 chromosomes in its somatic cells. How many chromosomes did the chicken inherit from each parent? How many chromosomes are in each of the chicken's gametes? How many chromosomes will be in each somatic cell of the chicken's offspring?

For suggested answers, see Appendix A.

CONCEPT 12.2

The mitotic phase alternates with interphase in the cell cycle

In 1882, a German anatomist named Walther Flemming developed dyes that allowed him to observe, for the first time, the behavior of chromosomes during mitosis and cytokinesis. (In fact, Flemming coined the terms *mitosis* and *chromatin*.) During the period between one cell division and the next, it appeared to Flemming that the cell was simply growing larger. But we now know that many critical events occur during this stage in the life of a cell.

Phases of the Cell Cycle

Mitosis is just one part of the cell cycle **(Figure 12.6)**. In fact, the **mitotic (M) phase**, which includes both mitosis and cytokinesis, is usually the shortest part of the cell cycle. The mitotic phase alternates with a much longer stage called **interphase**, which often accounts for about 90% of the cycle. Interphase can be divided into three phases: the **G_1 phase** ("first gap"), the **S phase** ("synthesis"), and the **G_2 phase** ("second gap"). The G phases were misnamed as "gaps" when they were first observed because the cells appeared inactive, but we now know that intense metabolic activity and growth occur throughout interphase. During all three phases of interphase, in fact, a cell grows by producing proteins and cytoplasmic organelles such as mitochondria and endoplasmic reticulum. Duplication of

▼ Figure 12.6 The cell cycle. In a dividing cell, the mitotic (M) phase alternates with interphase, a growth period. The first part of interphase (G_1) is followed by the S phase, when the chromosomes duplicate; G_2 is the last part of interphase. In the M phase, mitosis distributes the daughter chromosomes to daughter nuclei, and cytokinesis divides the cytoplasm, producing two daughter cells.

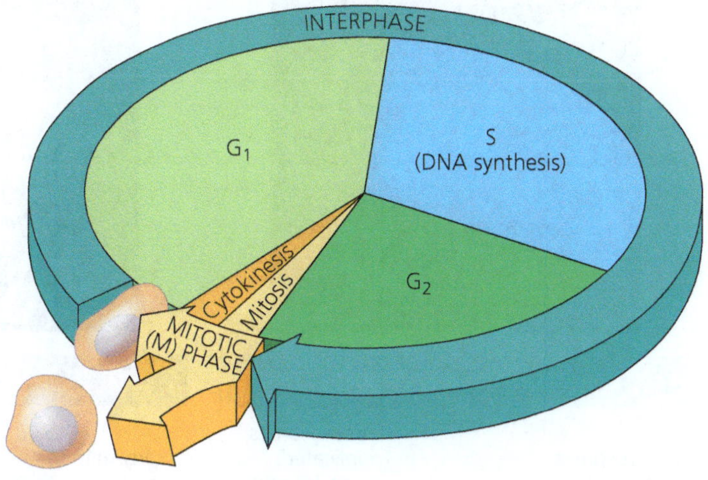

MB Animation: The Cell Cycle

the chromosomes, crucial for eventual division of the cell, occurs entirely during the S phase. (We will discuss synthesis, or replication, of DNA in Concept 16.2.) Thus, a cell grows (G_1), continues to grow as it copies its chromosomes (S), grows more as it completes preparations for cell division (G_2), and divides (M). The daughter cells may then repeat the cycle.

A particular human cell might undergo one division in 24 hours. Of this time, the M phase would occupy less than 1 hour, while the S phase might occupy 10–12 hours, or about half the cycle. The rest of the time would be apportioned between the G_1 and G_2 phases. The G_2 phase usually takes 4–6 hours; in our example, G_1 would occupy about 5–6 hours. G_1 is the most variable in length in different types of cells. Some cells in a multicellular organism divide very infrequently or not at all. These cells spend their time in G_1 (or a related phase called G_0, to be discussed later in the chapter) doing their job in the organism—a cell of the pancreas secretes digestive enzymes, for example.

Mitosis is conventionally broken down into five stages: **prophase**, **prometaphase**, **metaphase**, **anaphase**, and **telophase**. Overlapping with the latter stages of mitosis, cytokinesis completes the mitotic phase. **Figure 12.7** describes these stages in an animal cell. Study this figure thoroughly before progressing to the next two sections, which examine mitosis and cytokinesis more closely.

The Mitotic Spindle: *A Closer Look*

Many of the events of mitosis depend on the **mitotic spindle**, which begins to form in the cytoplasm during prophase. This structure consists of fibers made of microtubules and associated

∨ Figure 12.7 Exploring Mitosis in an Animal Cell

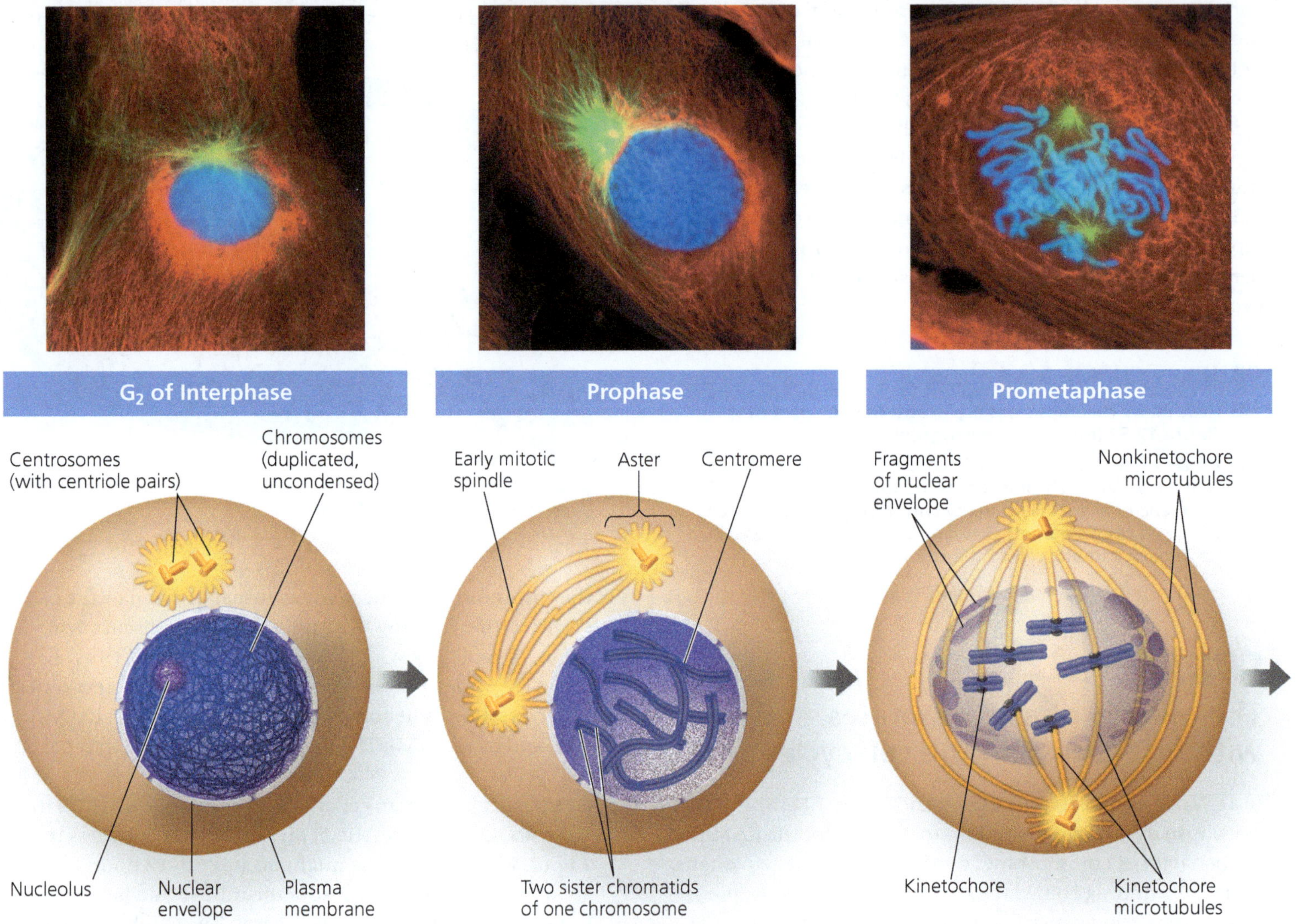

G$_2$ of Interphase

- A nuclear envelope encloses the nucleus.
- The nucleus contains one or more nucleoli (singular, *nucleolus*).
- Two centrosomes have formed by duplication of a single centrosome. Centrosomes are regions in animal cells that organize the microtubules of the spindle. Each centrosome contains two centrioles.
- Chromosomes, duplicated during S phase, cannot be seen individually because they have not yet condensed.

The fluorescence micrographs show dividing lung cells from a newt; this species has 22 chromosomes. Chromosomes appear blue, microtubules green, and intermediate filaments red. For simplicity, the drawings show only 6 chromosomes.

Prophase

- The chromatin fibers become more tightly coiled, condensing into discrete chromosomes observable with a light microscope.
- The nucleoli disappear.
- Each duplicated chromosome appears as two identical sister chromatids joined at their centromeres and, in some species, all along their arms by cohesins (sister chromatid cohesion).
- The mitotic spindle (named for its shape) begins to form. It is composed of the centrosomes and the microtubules that extend from them. The radial arrays of shorter microtubules that extend from the centrosomes are called asters ("stars").
- The centrosomes move away from each other, propelled partly by the lengthening microtubules between them.

Prometaphase

- The nuclear envelope fragments.
- The microtubules extending from each centrosome can now invade the nuclear area.
- The chromosomes have become even more condensed.
- A kinetochore, a specialized protein structure, has now formed at the centromere of each chromatid (thus, two per chromosome).
- Some of the microtubules attach to the kinetochores, becoming "kinetochore microtubules," which jerk the chromosomes back and forth.
- Nonkinetochore microtubules interact with those from the opposite pole of the spindle, lengthening the cell.

? *How many molecules of DNA are in the prometaphase drawing? How many molecules per chromosome? How many double helices are there per chromosome? Per chromatid?*

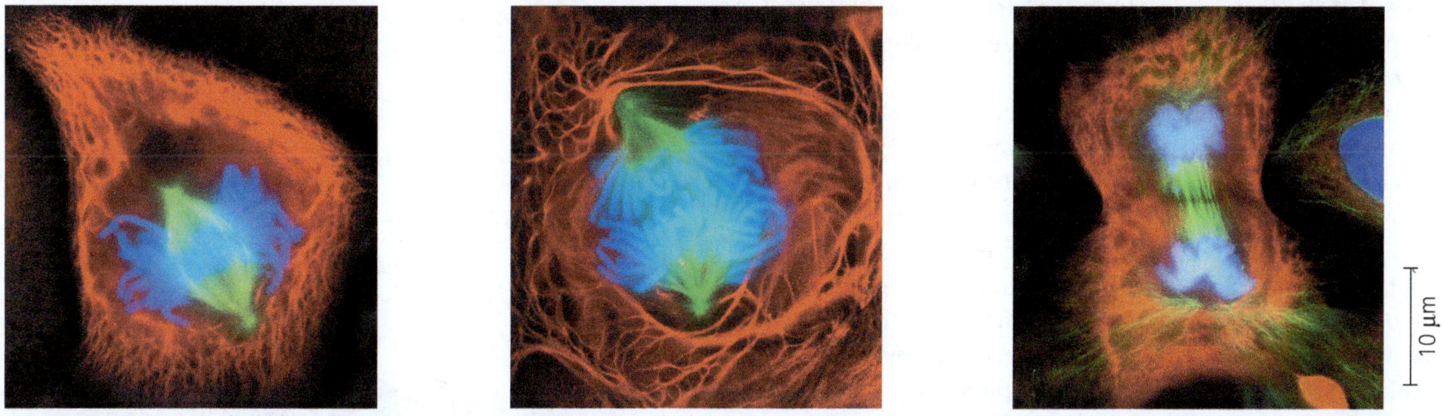

Metaphase | **Anaphase** | **Telophase and Cytokinesis**

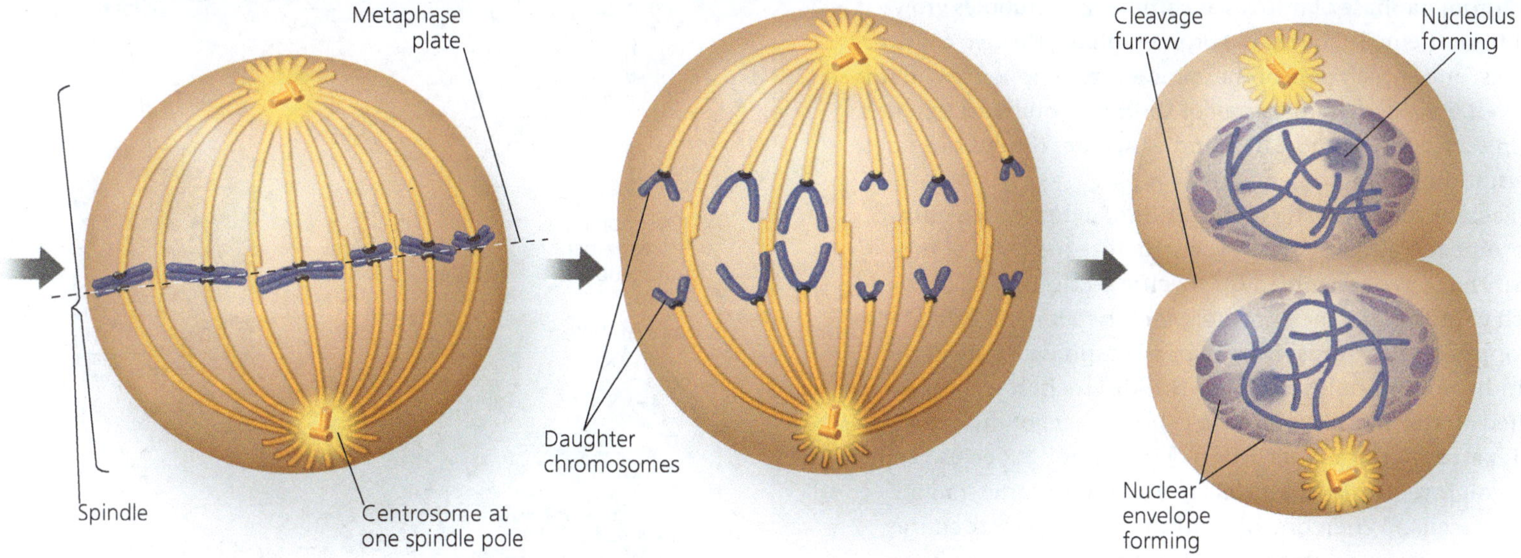

Metaphase

- The centrosomes are now at opposite poles of the cell.
- The chromosomes have all arrived at the *metaphase plate*, a plane that is equidistant between the spindle's two poles. The chromosomes' centromeres lie at the metaphase plate.
- For each chromosome, the kinetochores of the sister chromatids are attached to kinetochore microtubules coming from opposite poles.

BioFlix® Animation: Mitosis
Video: Animal Mitosis (time-lapse)

Anaphase

- Anaphase is the shortest stage of mitosis, often lasting only a few minutes.
- Anaphase begins when the cohesin proteins are cleaved. This allows the two sister chromatids of each pair to part suddenly. Each chromatid thus becomes an independent chromosome.
- The two new daughter chromosomes begin moving toward opposite ends of the cell as their kinetochore microtubules shorten. Because these microtubules are attached at the centromere region, the centromeres are pulled ahead of the arms, moving at a rate of about 1 µm/min.
- The cell elongates as the nonkinetochore microtubules lengthen.
- By the end of anaphase, the two ends of the cell have equivalent—and complete—collections of chromosomes.

Telophase

- Two daughter nuclei form in the cell. Nuclear envelopes arise from the fragments of the parent cell's nuclear envelope and other portions of the endomembrane system.
- Nucleoli reappear.
- The chromosomes become less condensed.
- Any remaining spindle microtubules are depolymerized.
- Mitosis, the division of one nucleus into two genetically identical nuclei, is now complete.

Cytokinesis

- The division of the cytoplasm is usually well under way by late telophase, so the two daughter cells appear shortly after the end of mitosis.
- In animal cells, cytokinesis involves the formation of a cleavage furrow, which pinches the cell in two.

proteins. While the mitotic spindle assembles, the other microtubules of the cytoskeleton partially disassemble, providing the material used to construct the spindle. The spindle microtubules elongate (polymerize) by incorporating more subunits of the protein tubulin (see Table 6.1) and shorten (depolymerize) by losing subunits.

In animal cells, the assembly of spindle microtubules starts at the **centrosome**, a subcellular region containing material that functions throughout the cell cycle to organize the cell's microtubules. (It is also a type of *microtubule-organizing center.*) A pair of centrioles is located at the center of the centrosome, but they are not essential for cell division: If the centrioles are destroyed with a laser microbeam, a spindle nevertheless forms during mitosis. In fact, centrioles are not even present in plant cells, which do form mitotic spindles.

During interphase in animal cells, the single centrosome duplicates, forming two centrosomes, which remain near the nucleus. The two centrosomes move apart during prophase and prometaphase of mitosis as spindle microtubules grow out from them. By the end of prometaphase, the two centrosomes, one at each pole of the spindle, are at opposite ends of the cell. An *aster*, a radial array of short microtubules, extends from each centrosome. The spindle includes the centrosomes, the spindle microtubules, and the asters.

Each of the two sister chromatids of a duplicated chromosome has a **kinetochore**, a structure made up of proteins that have assembled on specific sections of DNA at each centromere. The chromosome's two kinetochores face in opposite directions. During prometaphase, some of the spindle microtubules attach to the kinetochores; these are called *kinetochore microtubules*. (The number of microtubules attached to a kinetochore varies among species, from one microtubule in yeast cells to 40 or so in some mammalian cells.) When one of a chromosome's kinetochores is "captured" by microtubules, the chromosome begins to move toward the pole from which those microtubules extend. However, this movement comes to a halt as soon as microtubules from the opposite pole attach to the kinetochore on the other chromatid. What happens next is like a tug-of-war that ends in a draw. The chromosome moves first in one direction and then in the other, back and forth, finally settling midway between the two ends of the cell. At metaphase, the centromeres of all the duplicated chromosomes are on a plane midway between the spindle's two poles. This plane is called the **metaphase plate**, which is an imaginary plate rather than an actual cellular structure **(Figure 12.8)**. Meanwhile, microtubules that do not attach to kinetochores have been elongating, and by metaphase they overlap and interact with other *nonkinetochore microtubules* from the opposite pole of the spindle. By metaphase, the microtubules of the asters have also grown and are in contact with the plasma membrane. The spindle is now complete.

▼ **Figure 12.8 The mitotic spindle at metaphase.** The kinetochores of each chromosome's two sister chromatids face in opposite directions. Here, each kinetochore is attached to a cluster of kinetochore microtubules extending from the nearest centrosome. Nonkinetochore microtubules overlap at the metaphase plate (TEMs).

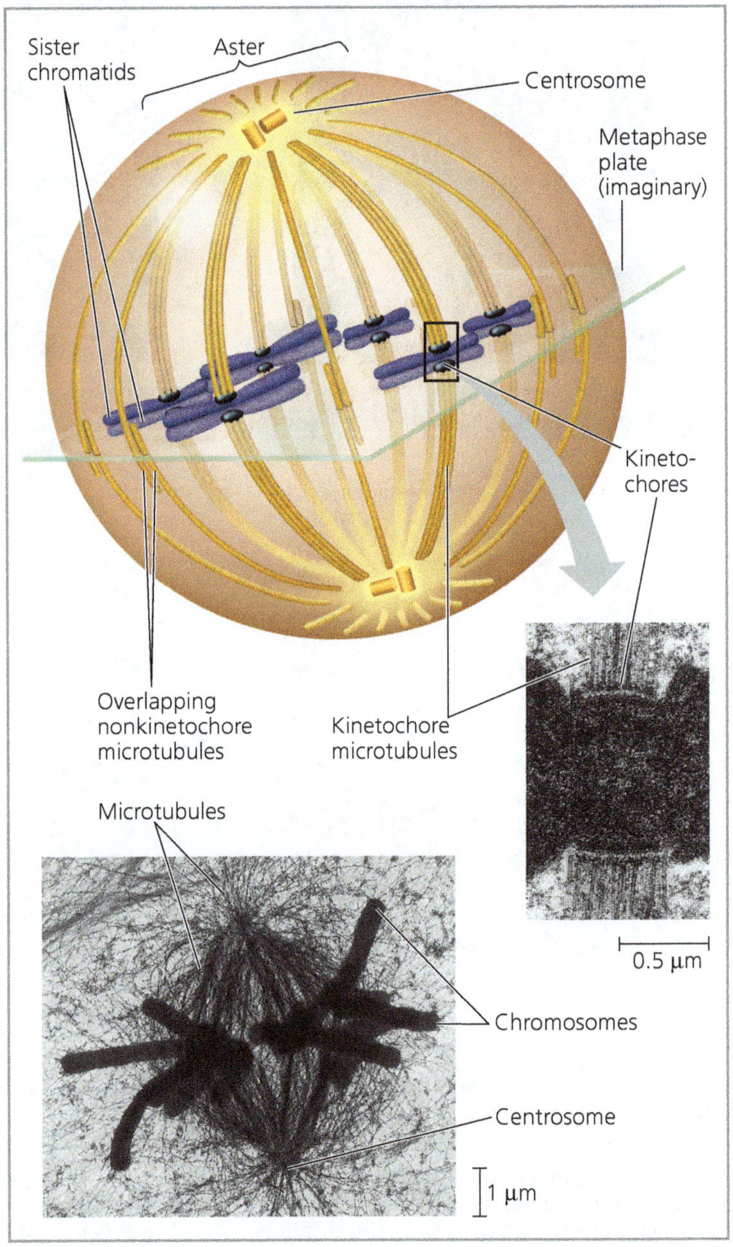

DRAW IT ➤ *On the lower micrograph, draw a line indicating the position of the metaphase plate. Circle an aster. Draw arrows indicating the directions of chromosome movement once anaphase begins.*

Video: Spindle Formation During Mitosis
Animation: Mitosis

The structure of the spindle correlates well with its function during anaphase. Anaphase begins suddenly when the cohesins holding together the sister chromatids of each chromosome are cleaved by an enzyme called *separase*. Once separated, the chromatids become individual chromosomes that move toward opposite ends of the cell.

How do the kinetochore microtubules function in this poleward movement of chromosomes? Apparently, two mechanisms are in play, both involving motor proteins. (To review how motor proteins move an object along a microtubule, see Figure 6.21.) Results of a cleverly designed experiment suggested that motor proteins on the kinetochores "walk" the chromosomes along the microtubules, which depolymerize at their kinetochore ends after the motor proteins have passed **(Figure 12.9)**. (This is referred to as the "Pac-man" mechanism because of its resemblance to the arcade game character that moves by eating all the dots in its path.) However, other researchers, working with different cell types or cells from other species, have shown that chromosomes are "reeled in" by motor proteins at the spindle poles and that the microtubules depolymerize after they pass by these motor proteins at the poles. The general consensus now is that both mechanisms are used and that their relative contributions vary among cell types.

In a dividing animal cell, the nonkinetochore microtubules are responsible for elongating the whole cell during anaphase. Nonkinetochore microtubules from opposite poles overlap each other extensively during metaphase (see Figure 12.8). During anaphase, the region of overlap is reduced as motor proteins attached to the microtubules walk them away from one another, using energy from ATP. As the microtubules push apart from each other, their spindle poles are pushed apart, elongating the cell. At the same time, the microtubules lengthen somewhat by the addition of tubulin subunits to their overlapping ends. As a result, the microtubules continue to overlap.

At the end of anaphase, duplicate groups of chromosomes have arrived at opposite ends of the elongated parent cell. Nuclei re-form during telophase. Cytokinesis generally begins during anaphase or telophase, and the spindle eventually disassembles by depolymerization of microtubules.

Cytokinesis: *A Closer Look*

In animal cells, cytokinesis occurs by a process known as **cleavage**. The first sign of cleavage is the appearance of a **cleavage furrow**, a shallow groove in the cell surface near the old metaphase plate **(Figure 12.10a)**. On the cytoplasmic side of the furrow is a contractile ring of actin microfilaments associated with molecules of the protein myosin. The actin microfilaments interact with the myosin molecules, causing the ring to contract. The contraction of the dividing cell's ring of microfilaments is like the pulling of a drawstring. The cleavage furrow deepens until the parent cell is pinched in two, producing two completely separated cells, each with its own nucleus and its own share of cytosol, organelles, and other subcellular structures.

Cytokinesis in plant cells, which have cell walls, is markedly different. There is no cleavage furrow. Instead, during telophase, vesicles derived from the Golgi apparatus move along microtubules to the middle of the cell, where they coalesce, producing a **cell plate**. Cell wall materials

Figure 12.9

Inquiry At which end do kinetochore microtubules shorten during anaphase?

Experiment Gary Borisy and colleagues at the University of Wisconsin wanted to determine whether kinetochore microtubules depolymerize at the kinetochore end or the pole end as chromosomes move toward the poles during mitosis. First they labeled the microtubules of a pig kidney cell in early anaphase with a yellow fluorescent dye. (Nonkinetochore microtubules are not shown.)

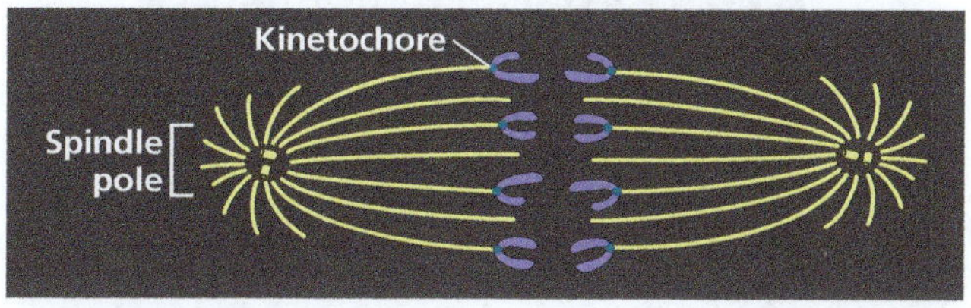

Then they marked a region of the kinetochore microtubules between one spindle pole and the chromosomes by using a laser to eliminate the fluorescence from that region, while leaving the microtubules intact (see below). As anaphase proceeded, they monitored the changes in microtubule length on either side of the mark.

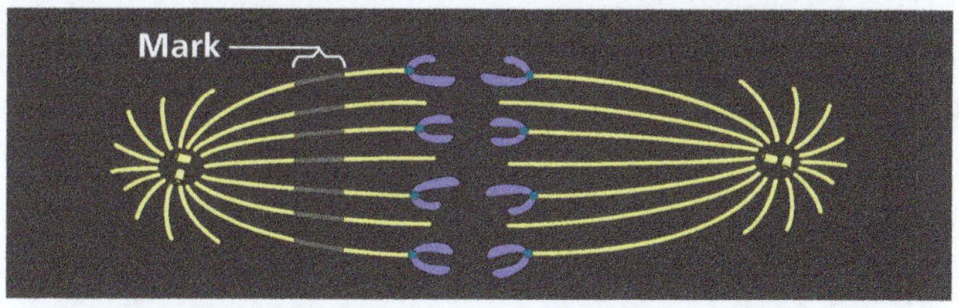

Results As the chromosomes moved poleward, the microtubule segments on the kinetochore side of the mark shortened, while those on the spindle pole side stayed the same length.

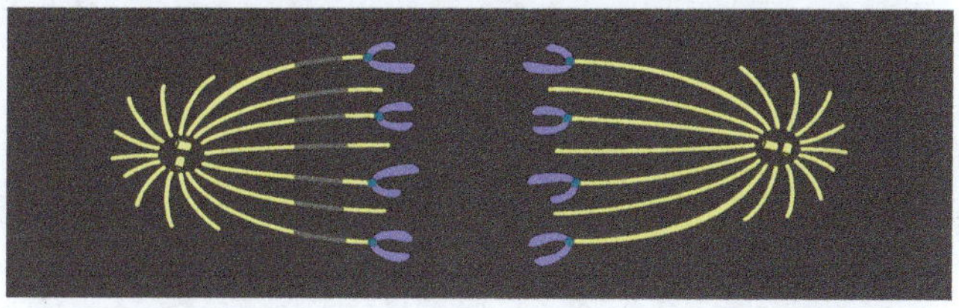

Conclusion During anaphase in this cell type, chromosome movement is correlated with kinetochore microtubules shortening at their kinetochore ends and not at their spindle pole ends. This experiment supports the hypothesis that during anaphase, a chromosome is walked along a microtubule as the microtubule depolymerizes at its kinetochore end, releasing tubulin subunits.

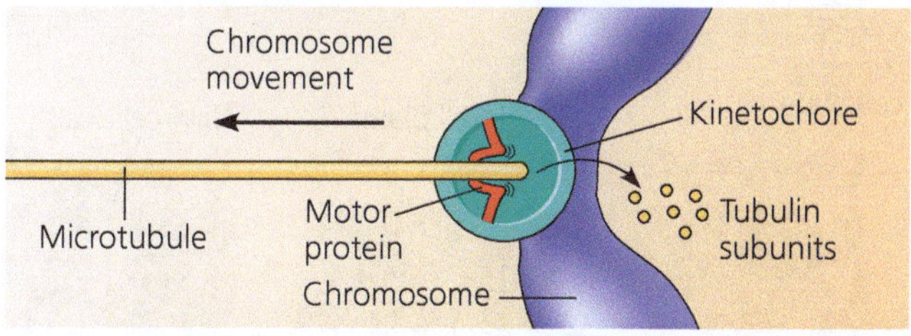

Data from G. J. Gorbsky, P. J. Sammak, and G. G. Borisy, Chromosomes move poleward in anaphase along stationary microtubules that coordinately disassemble from their kinetochore ends, *Journal of Cell Biology* 104:9–18 (1987).

WHAT IF? ➤ *If this experiment had been done on a cell type in which "reeling in" at the poles was the main cause of chromosome movement, how would the mark have moved relative to the poles? How would the microtubule portions on either side of the mark have changed?*

Animation: Microtubule Depolymerization

Figure 12.10 Cytokinesis in animal and plant cells.

(a) Cleavage of an animal cell (SEM)

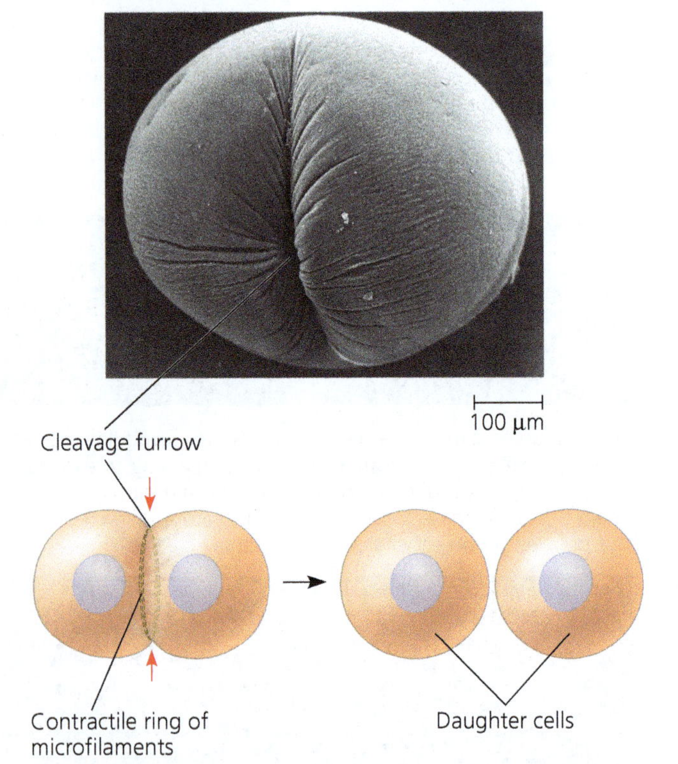

(b) Cell plate formation in a plant cell (TEM)

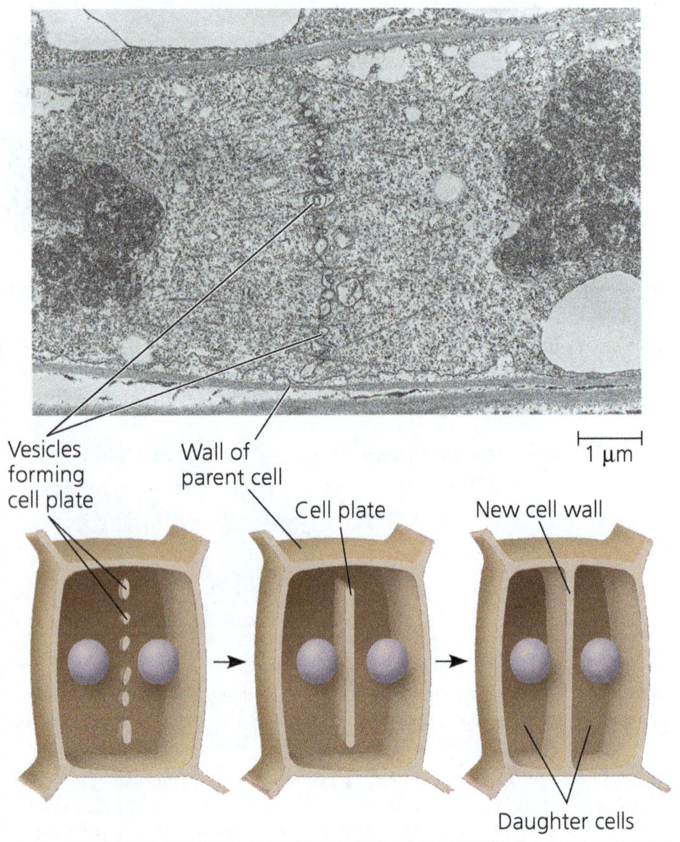

Animation: Cytokinesis
Video: Cytokinesis in an Animal Cell

carried in the vesicles collect inside the cell plate as it grows **(Figure 12.10b)**. The cell plate enlarges until its surrounding membrane fuses with the plasma membrane along the perimeter of the cell. Two daughter cells result, each with its own plasma membrane. Meanwhile, a new cell wall arising from the contents of the cell plate forms between the daughter cells.

Figure 12.11 is a series of micrographs of a dividing plant cell. Examining this figure will help you review mitosis and cytokinesis.

Binary Fission in Bacteria

Prokaryotes (bacteria and archaea) can undergo a type of reproduction in which the cell grows to roughly double its size and then divides to form two cells. The term **binary fission**, meaning "division in half," refers to this process and to the asexual reproduction of single-celled eukaryotes, such as the amoeba in Figure 12.2a. However, the process in eukaryotes involves mitosis, while that in prokaryotes does not.

In bacteria, most genes are carried on a single bacterial chromosome that consists of a circular DNA molecule and associated proteins. Although bacteria are smaller and simpler than eukaryotic cells, the challenge of replicating their genomes in an orderly fashion and distributing the copies equally to two daughter cells is still formidable. For example, when it is fully stretched out, the chromosome of the bacterium *Escherichia coli* is about 500 times as long as the cell. For such a long chromosome to fit within the cell, it must be highly coiled and folded.

In some bacteria, the process of cell division is initiated when the DNA of the bacterial chromosome begins to replicate at a specific place on the chromosome called the **origin of replication**, producing two origins. As the chromosome continues to replicate, one origin moves rapidly toward the opposite end of the cell **(Figure 12.12)**. While the chromosome is replicating, the cell elongates. When replication is complete and the bacterium has reached about twice its initial size, proteins cause its plasma membrane to pinch inward, dividing the parent bacterial cell into two daughter cells. In this way, each cell inherits a complete genome.

Using the techniques of modern DNA technology to tag the origins of replication with molecules that glow green in fluorescence microscopy (see Figure 6.3), researchers have directly observed the movement of bacterial chromosomes. This movement is reminiscent of the poleward movements of the centromere regions of eukaryotic chromosomes during anaphase of mitosis, but bacteria don't have visible mitotic spindles or even microtubules. In most bacterial species studied, the two origins of replication end up at opposite ends of the cell or in some other very specific location, possibly anchored there by one or more proteins. How bacterial chromosomes move and how their specific location is established and maintained are active areas of research. Several proteins that play important roles have been identified. Polymerization of one protein resembling eukaryotic actin

Figure 12.11 Mitosis in a plant cell. These light micrographs show mitosis in cells of an onion root.

1 **Prophase.** The chromosomes are condensing and the nucleolus is beginning to disappear. Although not yet visible in the micrograph, the mitotic spindle is starting to form.

2 **Prometaphase.** Discrete chromosomes are now visible; each consists of two aligned, identical sister chromatids. Later in prometaphase, the nuclear envelope will fragment.

3 **Metaphase.** The spindle is complete, and the chromosomes, attached to microtubules at their kinetochores, are all at the metaphase plate.

4 **Anaphase.** The chromatids of each chromosome have separated, and the daughter chromosomes are moving to the ends of the cell as their kinetochore microtubules shorten.

5 **Telophase.** Daughter nuclei are forming. Meanwhile, cytokinesis has started: The cell plate, which will divide the cytoplasm in two, is growing toward the perimeter of the parent cell.

Figure 12.12 Bacterial cell division by binary fission. The bacterium shown here has a single, circular chromosome.

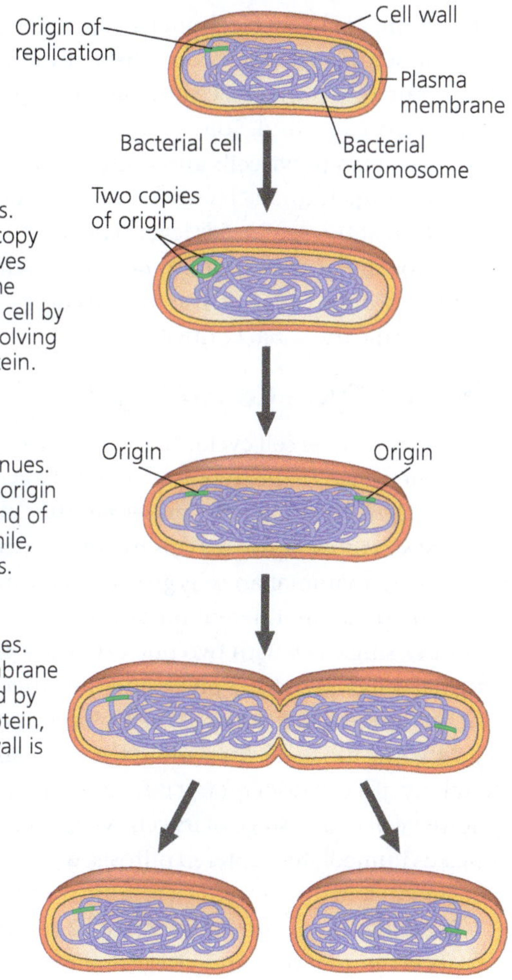

1 Chromosome replication begins. Soon after, one copy of the origin moves rapidly toward the other end of the cell by a mechanism involving an actin-like protein.

2 Replication continues. One copy of the origin is now at each end of the cell. Meanwhile, the cell elongates.

3 Replication finishes. The plasma membrane is pinched inward by a tubulin-like protein, and a new cell wall is deposited.

4 Two daughter cells result.

Animation: Cell Division in Bacteria

apparently functions in bacterial chromosome movement during cell division, and another protein that is related to tubulin helps pinch the plasma membrane inward, separating the two bacterial daughter cells.

The Evolution of Mitosis

EVOLUTION Given that prokaryotes preceded eukaryotes on Earth by more than a billion years, we might hypothesize that mitosis evolved from simpler prokaryotic mechanisms of cell reproduction. The fact that some of the proteins involved in bacterial binary fission are related to eukaryotic proteins that function in mitosis supports that hypothesis.

As eukaryotes with nuclear envelopes and larger genomes evolved, the ancestral process of binary fission, seen today in bacteria, somehow gave rise to mitosis. Variations on cell division exist in different groups of organisms. These variant processes may be similar to mechanisms used by ancestral species and thus may resemble steps in the evolution of mitosis from a binary fission-like process presumably carried out by very early bacteria. Possible intermediate stages are suggested by two unusual types of nuclear division found today in certain unicellular eukaryotes—dinoflagellates, diatoms, and some yeasts **(Figure 12.13)**. These two modes of nuclear division are thought to be cases where ancestral mechanisms have remained relatively unchanged over evolutionary time. In both types, the nuclear envelope remains intact, in contrast to what happens in most eukaryotic cells. Keep in mind, however, that we can't observe cell division in cells of extinct species. This hypothesis uses only currently existing species as examples and must ignore any potential intermediate mechanisms used by species that disappeared long ago.

Figure 12.13 Mechanisms of cell division in several groups of organisms. Some unicellular eukaryotes existing today have mechanisms of cell division that may resemble intermediate steps in the evolution of mitosis. Except for (a), cell walls are not shown.

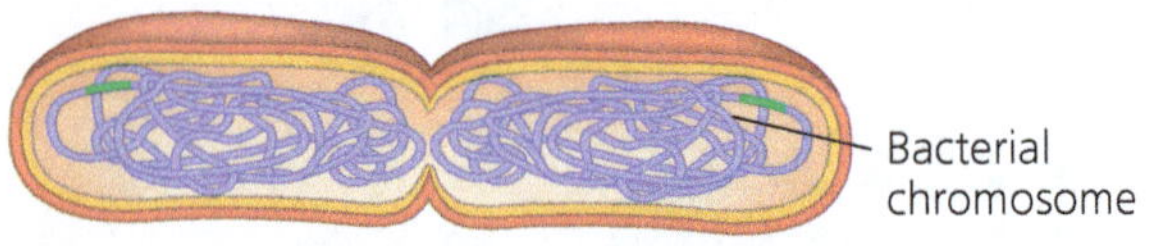

(a) Bacteria. During binary fission in bacteria, the origins of the daughter chromosomes move to opposite ends of the cell. The mechanism involves polymerization of actin-like molecules, and possibly proteins that may anchor the daughter chromosomes to specific sites on the plasma membrane.

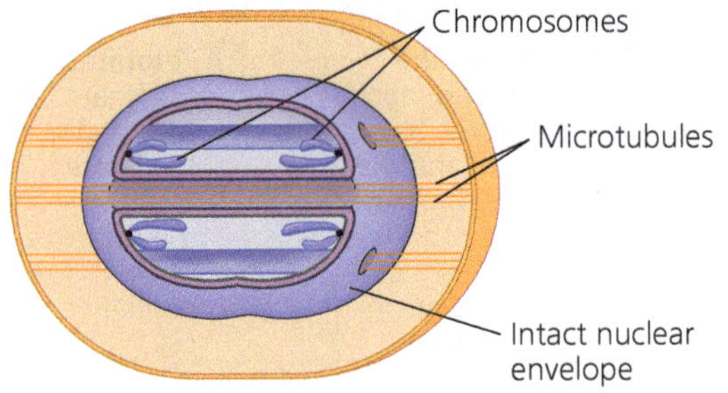

(b) Dinoflagellates. In unicellular protists called dinoflagellates, the chromosomes attach to the nuclear envelope, which remains intact during cell division. Microtubules pass through the nucleus inside cytoplasmic tunnels, reinforcing the spatial orientation of the nucleus, which then divides in a process reminiscent of bacterial binary fission.

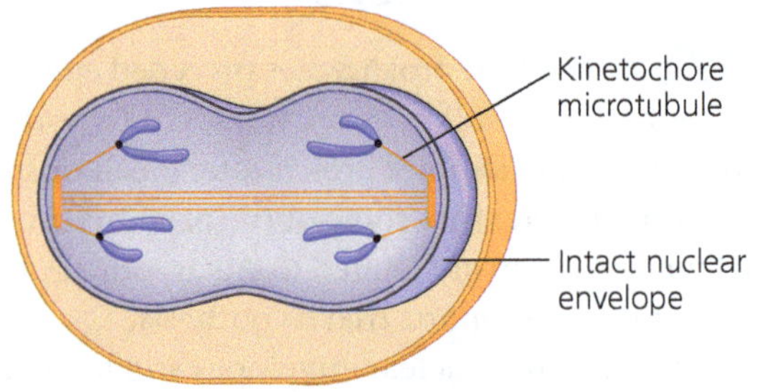

(c) Diatoms and some yeasts. In these two other groups of unicellular eukaryotes, the nuclear envelope also remains intact during cell division. In these organisms, the microtubules form a spindle *within* the nucleus. Microtubules separate the chromosomes, and the nucleus splits into two daughter nuclei.

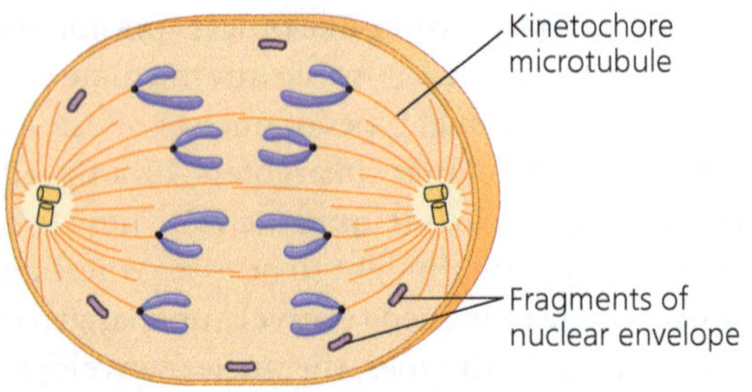

(d) Most eukaryotes. In most other eukaryotes, including plants and animals, the spindle forms outside the nucleus, and the nuclear envelope breaks down during mitosis. Microtubules separate the chromosomes, and two nuclear envelopes then form.

Video: Nuclear Envelope Breakdown and Formation During Mitosis in *C. elegans*, a Eukaryote

CONCEPT CHECK 12.2

1. How many chromosomes are shown in the illustration in Figure 12.8? Are they duplicated? How many chromatids are shown?
2. Compare cytokinesis in animal cells and plant cells.
3. During which stages of the cell cycle does a chromosome consist of two identical chromatids?
4. Compare the roles of tubulin and actin during eukaryotic cell division with the roles of tubulin-like and actin-like proteins during bacterial binary fission.
5. A kinetochore has been compared to a coupling device that connects a motor to the cargo that it moves. Explain.
6. **MAKE CONNECTIONS ➤** What other functions do actin and tubulin carry out? Name the proteins they interact with to do so. (Review Figures 6.21a and 6.26a.)

For suggested answers, see Appendix A.

CONCEPT 12.3

The eukaryotic cell cycle is regulated by a molecular control system

The timing and rate of cell division in different parts of a plant or animal are crucial to normal growth, development, and maintenance. The frequency of cell division varies with the type of cell. For example, human skin cells divide frequently throughout life, whereas liver cells maintain the ability to divide but keep it in reserve until an appropriate need arises—say, to repair a wound. Some of the most specialized cells, such as fully formed nerve cells and muscle cells, do not divide at all in a mature human. These cell cycle differences result from regulation at the molecular level. The mechanisms of this regulation are of great interest, not only to understand the life cycles of normal cells but also to learn how cancer cells manage to escape the usual controls.

The Cell Cycle Control System

What controls the cell cycle? In the early 1970s, a variety of experiments led to the hypothesis that the cell cycle is driven by specific signaling molecules present in the cytoplasm. Some of the first strong evidence for this hypothesis came from experiments with mammalian cells grown in culture. In these experiments, two cells in different phases of the cell cycle were fused to form a single cell with two nuclei **(Figure 12.14)**. If one of the original cells was in the S phase and the other was in G_1, the G_1 nucleus immediately entered the S phase, as though stimulated by signaling molecules present in the cytoplasm of the first cell. Similarly, if a cell undergoing mitosis (M phase) was fused with another cell in any stage of its cell cycle, even G_1, the second nucleus immediately entered mitosis, with condensation of the chromatin and formation of a mitotic spindle.

The experiment shown in Figure 12.14 and other experiments on animal cells and yeasts demonstrated that the sequential events of the cell cycle are directed by a distinct **cell cycle control system**, a cyclically operating set of

▼ Figure 12.14

Inquiry Do molecular signals in the cytoplasm regulate the cell cycle?

Experiment Researchers at the University of Colorado wondered whether a cell's progression through the cell cycle is controlled by cytoplasmic molecules. They induced cultured mammalian cells at different phases of the cell cycle to fuse. Two experiments are shown.

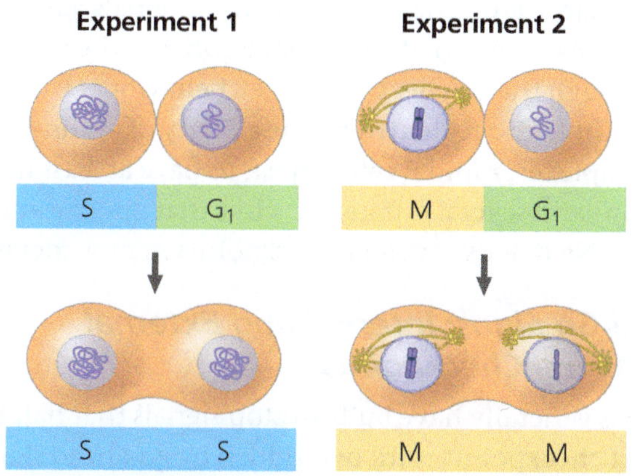

When a cell in the S phase was fused with a cell in G_1, the G_1 nucleus immediately entered the S phase—DNA was synthesized.

When a cell in the M phase was fused with a cell in G_1, the G_1 nucleus immediately began mitosis—a spindle formed and the chromosomes condensed, even though the chromosomes had not been duplicated.

Conclusion The results of fusing a G_1 cell with a cell in the S or M phase of the cell cycle suggest that molecules present in the cytoplasm during the S or M phase control the progression to those phases.

Data from R. T. Johnson and P. N. Rao, Mammalian cell fusion: Induction of premature chromosome condensation in interphase nuclei, *Nature* 226:717–722 (1970).

WHAT IF? ➤ *If the progression of phases did not depend on cytoplasmic molecules and, instead, each phase automatically began when the previous one was complete, how would the results have differed?*

molecules in the cell that both triggers and coordinates key events in the cell cycle **(Figure 12.15)**. The cell cycle control system has been compared to the control device of a washing machine. Like the washer's timing device, the cell cycle control system proceeds on its own, according to a built-in clock. However, just as a washer's cycle is subject to both internal control (such as the sensor that detects when the tub is filled with water) and external adjustment (such as starting or stopping the machine), the cell cycle is regulated at certain checkpoints by both internal and external signals. A **checkpoint** in the cell cycle is a control point where stop and go-ahead signals can regulate the cycle. Three important checkpoints are found in the G_1, G_2, and M phases (red gates in Figure 12.15), which will be discussed shortly.

To understand how cell cycle checkpoints work, we'll first identify some of the molecules that make up the cell cycle control system (the molecular basis for the cell cycle clock) and describe how a cell progresses through

▼ Figure 12.15 Mechanical analogy for the cell cycle control system. In this diagram, the flat "stepping stones" around the perimeter represent sequential events. Like the control device of a washing machine, the cell cycle control system proceeds on its own, driven by a built-in clock. However, the system is subject to internal and external regulation at various checkpoints; three important checkpoints are shown (red).

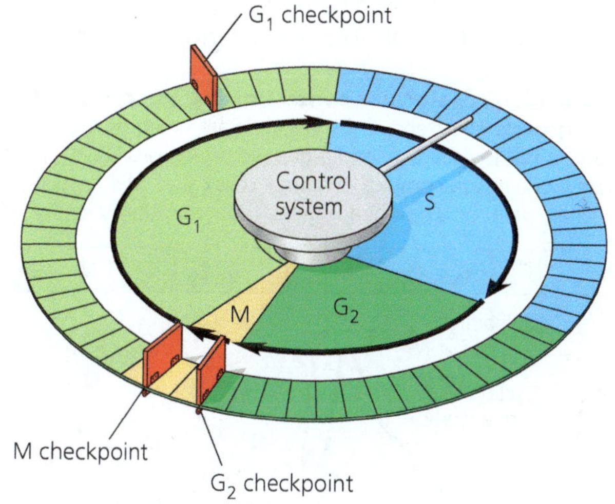

Animation: Control of the Cell Cycle

the cycle. We'll then consider the internal and external checkpoint signals that can make the clock either pause or continue.

The Cell Cycle Clock: Cyclins and Cyclin-Dependent Kinases

Rhythmic fluctuations in the abundance and activity of cell cycle control molecules pace the sequential events of the cell cycle. These regulatory molecules are mainly proteins of two types: protein kinases and cyclins. Protein kinases are enzymes that activate or inactivate other proteins by phosphorylating them (see Concept 11.3).

Many of the kinases that drive the cell cycle are actually present at a constant concentration in the growing cell, but much of the time they are in an inactive form. To be active, such a kinase must be attached to a **cyclin**, a protein that gets its name from its cyclically fluctuating concentration in the cell. Because of this requirement, these kinases are called **cyclin-dependent kinases,** or **Cdks**. The activity of a Cdk rises and falls with changes in the concentration of its cyclin partner. **Figure 12.16a** shows the fluctuating activity of **MPF**, the cyclin-Cdk complex that was discovered first (in frog eggs). Note that the peaks of MPF activity correspond to the peaks of cyclin concentration. The cyclin level rises during the S and G_2 phases and then falls abruptly during M phase.

The initials MPF stand for "maturation-promoting factor," but we can think of MPF as "M-phase-promoting factor" because it triggers the cell's passage into the M phase, past the G_2 checkpoint. When cyclins that accumulate during G_2 associate with Cdk molecules, the resulting MPF complex is active—it phosphorylates a variety of proteins, initiating mitosis **(Figure 12.16b)**. MPF acts both directly as a kinase and indirectly by activating other kinases. For example, MPF causes phosphorylation of various proteins of the nuclear lamina (see Figure 6.9), which promotes fragmentation of the nuclear envelope during prometaphase of mitosis. There is also evidence that MPF contributes to molecular events required for chromosome condensation and spindle formation during prophase.

Figure 12.16 Molecular control of the cell cycle at the G_2 checkpoint. The steps of the cell cycle are timed by rhythmic fluctuations in the activity of cyclin-dependent kinases (Cdks). Here we focus on a cyclin-Cdk complex in animal cells called MPF, which acts at the G_2 checkpoint as a go-ahead signal, triggering the events of mitosis.

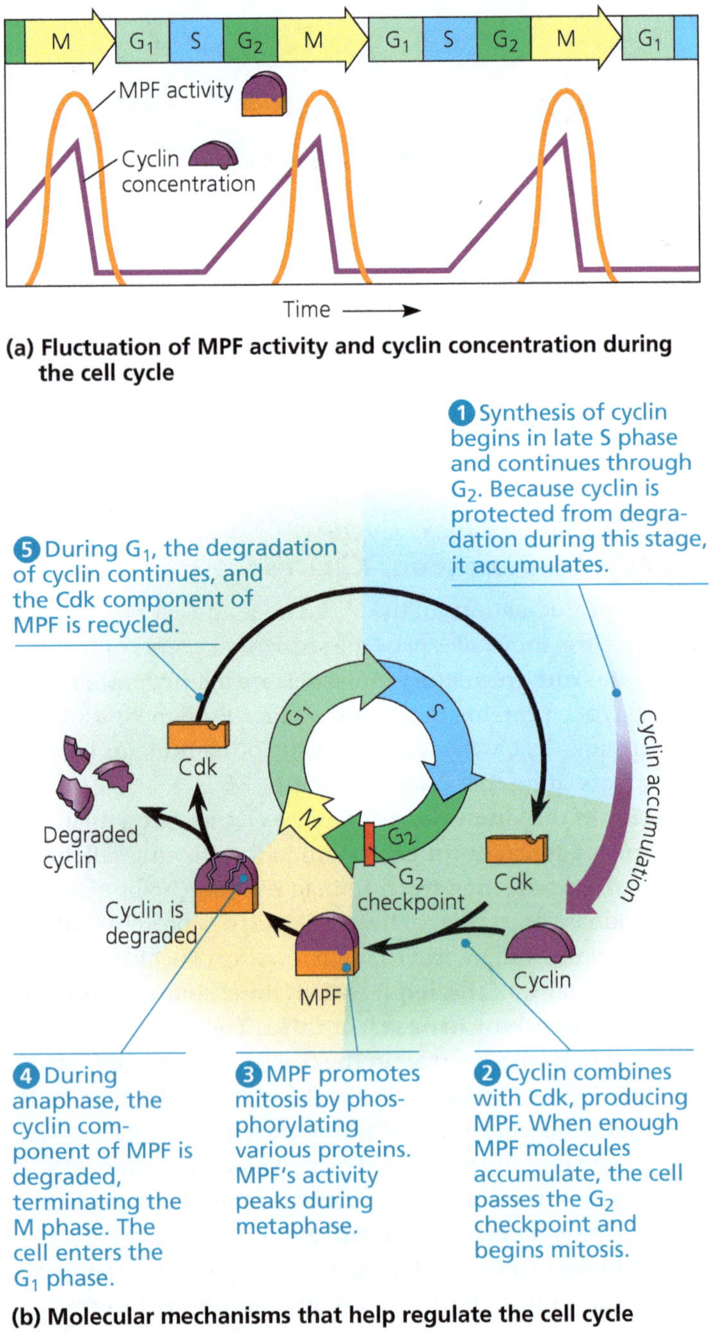

(a) Fluctuation of MPF activity and cyclin concentration during the cell cycle

1 Synthesis of cyclin begins in late S phase and continues through G_2. Because cyclin is protected from degradation during this stage, it accumulates.

2 Cyclin combines with Cdk, producing MPF. When enough MPF molecules accumulate, the cell passes the G_2 checkpoint and begins mitosis.

3 MPF promotes mitosis by phosphorylating various proteins. MPF's activity peaks during metaphase.

4 During anaphase, the cyclin component of MPF is degraded, terminating the M phase. The cell enters the G_1 phase.

5 During G_1, the degradation of cyclin continues, and the Cdk component of MPF is recycled.

(b) Molecular mechanisms that help regulate the cell cycle

VISUAL SKILLS ➤ *Explain how the events in the diagram in (b) are related to the "Time" axis of the graph in (a), beginning at the left.*

MB **Interview with Paul Nurse: Discovering how protein kinases control the cell cycle**

During anaphase, MPF helps switch itself off by initiating a process that leads to the destruction of its own cyclin. The noncyclin part of MPF, the Cdk, persists in the cell, inactive until it becomes part of MPF again by associating with new cyclin molecules synthesized during the S and G_2 phases of the next round of the cycle.

The fluctuating activities of different cyclin-Cdk complexes are of major importance in controlling all the stages of the cell cycle; they also give the go-ahead signals at some checkpoints. As mentioned above, MPF controls the cell's passage through the G_2 checkpoint. Cell behavior at the G_1 checkpoint is also regulated by the activity of cyclin-Cdk protein complexes. Animal cells appear to have at least three Cdk proteins and several different cyclins that operate at this checkpoint. Next, let's consider checkpoints in more detail.

Stop and Go Signs: Internal and External Signals at the Checkpoints

Animal cells generally have built-in stop signals that halt the cell cycle at checkpoints until overridden by go-ahead signals. (The signals are transmitted within the cell by the kinds of signal transduction pathways discussed in Chapter 11.) Many signals registered at checkpoints come from cellular surveillance mechanisms inside the cell. These signals report whether crucial cellular processes that should have occurred by that point have in fact been completed correctly and thus whether or not the cell cycle should proceed. Checkpoints also register signals from outside the cell.

Three important checkpoints are those in the G_1, G_2, and M phases (see Figure 12.15). For many cells, the G_1 checkpoint seems to be the most important. If a cell receives a go-ahead signal at the G_1 checkpoint, it will usually complete the G_1, S, G_2, and M phases and divide. If it does not receive a go-ahead signal at that point, it may exit the cycle, switching into a nondividing state called the **G_0 phase (Figure 12.17a)**. Most cells of the human body are actually in the G_0 phase. As mentioned earlier, mature nerve cells and muscle cells never divide. Other cells, such as liver cells, can be "called back" from the G_0 phase to the cell cycle by external cues, such as growth factors released during injury.

Biologists are currently working out the pathways that link signals originating inside and outside the cell with the responses by cyclin-dependent kinases and other proteins. An example of an internal signal occurs at the third important checkpoint, the M checkpoint **(Figure 12.17b)**. Anaphase, the separation of sister chromatids, does not begin until all the chromosomes are properly attached to the spindle at the metaphase plate. Researchers have learned that as long as some kinetochores are unattached to spindle microtubules, the sister chromatids remain together, delaying anaphase. Only when the kinetochores of all the chromosomes are properly attached to the spindle does the appropriate regulatory protein complex become activated. (In this case, the regulatory molecule

➤ Figure 12.17 Two important checkpoints. At certain checkpoints in the cell cycle (red gates), cells do different things depending on the signals they receive. Events of the **(a)** G_1 and **(b)** M checkpoints are shown. In **(b)**, the G_2 checkpoint has already been passed by the cell.

WHAT IF? ➤ *What might be the result if the cell ignored either checkpoint and progressed through the cell cycle?*

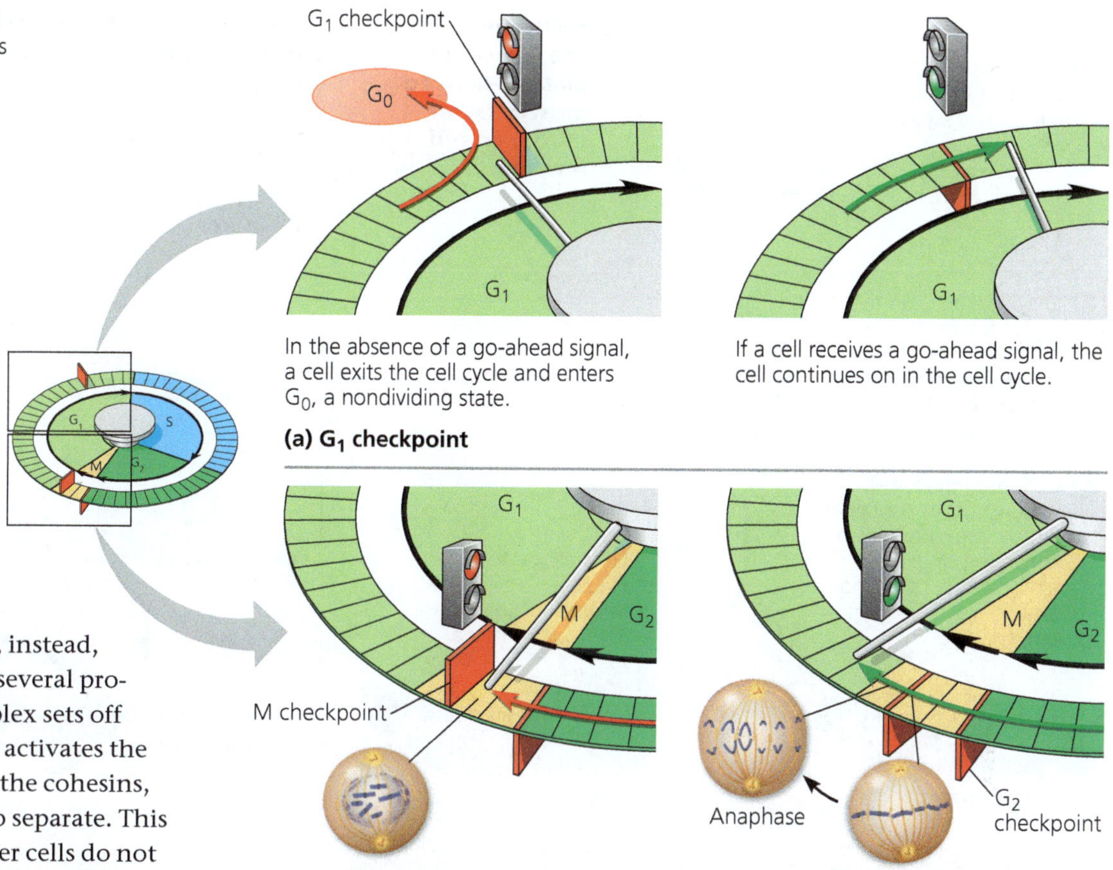

In the absence of a go-ahead signal, a cell exits the cell cycle and enters G_0, a nondividing state.

If a cell receives a go-ahead signal, the cell continues on in the cell cycle.

(a) G_1 checkpoint

A cell in mitosis receives a stop signal when any of its chromosomes are not attached to spindle fibers.

When all chromosomes are attached to spindle fibers from both poles, a go-ahead signal allows the cell to proceed into anaphase.

(b) M checkpoint

is not a cyclin-Cdk complex but, instead, a different complex made up of several proteins.) Once activated, the complex sets off a chain of molecular events that activates the enzyme separase, which cleaves the cohesins, allowing the sister chromatids to separate. This mechanism ensures that daughter cells do not end up with missing or extra chromosomes.

There are checkpoints in addition to those in G_1, G_2, and M. For instance, a checkpoint in S phase stops cells with DNA damage from proceeding in the cell cycle. And, in 2014, researchers presented evidence for another checkpoint between anaphase and telophase that ensures anaphase is completed and the chromosomes are well separated before cytokinesis can begin, thus avoiding chromosomal damage.

What about the stop and go-ahead signals themselves—what are the signaling molecules? Studies using animal cells in culture have led to the identification of many external factors, both chemical and physical, that can influence cell division. For example, cells fail to divide if an essential nutrient is lacking in the culture medium. (This is analogous to trying to run a washing machine without the water supply hooked up; an internal sensor won't allow the machine to continue past the point where water is needed.) And even if all other conditions are favorable, most types of mammalian cells divide in culture only if the growth medium includes specific growth factors. As mentioned in Concept 11.1, a **growth factor** is a protein released by certain cells that stimulates other cells to divide. Different cell types respond specifically to different growth factors or combinations of growth factors.

Consider, for example, *platelet-derived growth factor* (*PDGF*), which is made by blood cell fragments called platelets. The experiment illustrated in **Figure 12.18** demonstrates that PDGF is required for the division of cultured fibroblasts, a type of connective tissue cell. Fibroblasts have PDGF receptors on their plasma membranes. The binding of PDGF molecules to these receptors (which are receptor tyrosine kinases; see Figure 11.8) triggers a signal transduction pathway that allows the cells to pass the G_1 checkpoint and divide. PDGF stimulates fibroblast division not only in the artificial conditions of cell culture but also in an animal's body. When an injury occurs, platelets release PDGF in the vicinity. The resulting proliferation of fibroblasts helps heal the wound.

The effect of an external physical factor on cell division is clearly seen in **density-dependent inhibition**, a phenomenon in which crowded cells stop dividing **(Figure 12.19a)**. As first observed many years ago, cultured cells normally divide until they form a single layer of cells on the inner surface of a culture flask, at which point the cells stop dividing. If some cells are removed, those bordering the open space begin dividing again and continue until the vacancy is filled. Follow-up studies revealed that the binding of a cell-surface protein to its counterpart on an adjoining cell sends a signal to both cells that inhibits cell division, preventing them from moving forward in the cell cycle, even in the presence of growth factors.

Most animal cells also exhibit **anchorage dependence** (see Figure 12.19a). To divide, they must be attached to a substratum, such as the inside of a culture flask or the

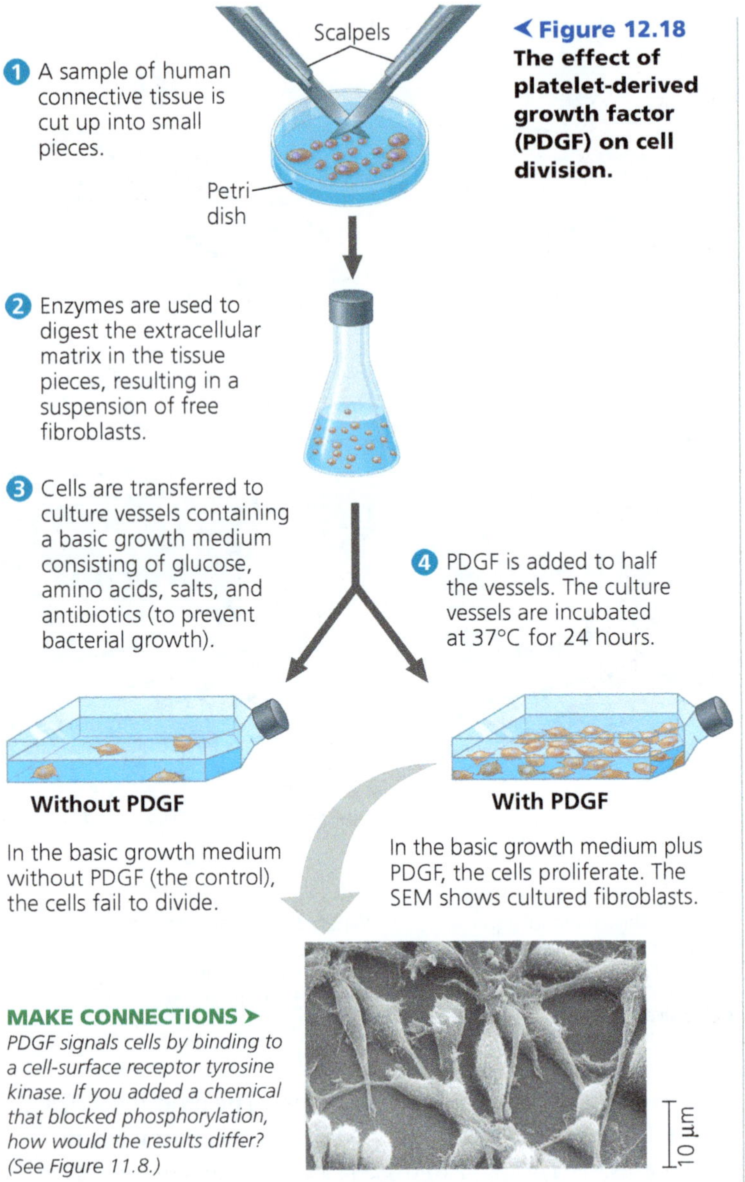

◀ Figure 12.18 The effect of platelet-derived growth factor (PDGF) on cell division.

MAKE CONNECTIONS ▶ *PDGF signals cells by binding to a cell-surface receptor tyrosine kinase. If you added a chemical that blocked phosphorylation, how would the results differ? (See Figure 11.8.)*

▼ Figure 12.19 Density-dependent inhibition and anchorage dependence of cell division. Individual cells are shown disproportionately large in the drawings.

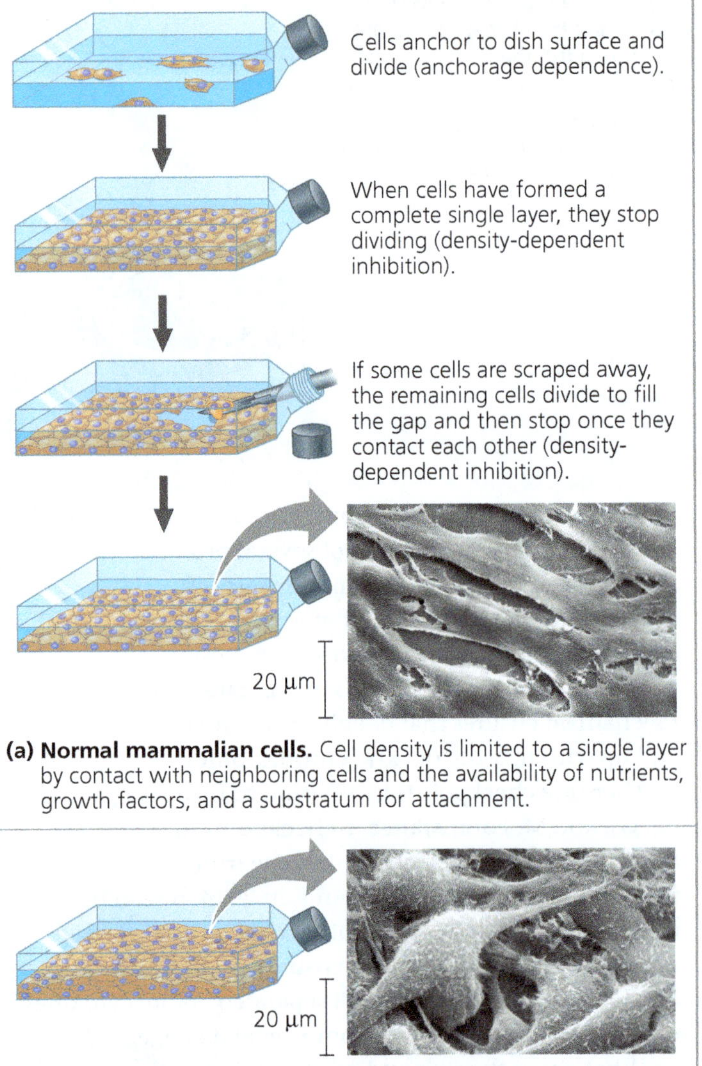

(a) Normal mammalian cells. Cell density is limited to a single layer by contact with neighboring cells and the availability of nutrients, growth factors, and a substratum for attachment.

(b) Cancer cells. Cancer cells usually continue to divide well beyond a single layer, forming a clump of overlapping cells. They do not exhibit anchorage dependence or density-dependent inhibition.

extracellular matrix of a tissue. Experiments suggest that like cell density, anchorage is signaled to the cell cycle control system via pathways involving plasma membrane proteins and elements of the cytoskeleton linked to them.

Density-dependent inhibition and anchorage dependence appear to function not only in cell culture but also in the body's tissues, checking the growth of cells at some optimal density and location during embryonic development and throughout an organism's life. Cancer cells, which we discuss next, exhibit neither density-dependent inhibition nor anchorage dependence **(Figure 12.19b)**.

Loss of Cell Cycle Controls in Cancer Cells

Cancer cells do not heed the normal signals that regulate the cell cycle. In culture, they do not stop dividing when growth factors are depleted. A logical hypothesis is that cancer cells do not need growth factors in their culture medium to grow and divide. They may make a required growth factor themselves, or they may have an abnormality in the signaling pathway that conveys the growth factor's signal to the cell cycle control system even in the absence of that factor. Another possibility is an abnormal cell cycle control system. In these scenarios, the underlying basis of the abnormality is almost always a change in one or more genes (for example, a mutation) that alters the function of their protein products, resulting in faulty cell cycle control.

There are other important differences between normal cells and cancer cells that reflect derangements of the cell cycle. If and when they stop dividing, cancer cells do so at random points in the cycle, rather than at the normal checkpoints. Moreover, cancer cells can go on dividing indefinitely in culture if they are given a continual supply of nutrients; in essence, they are "immortal." A striking example is a cell line that has been reproducing in culture since 1951. Cells of this line are called HeLa cells because their original source was a tumor removed from a woman named *He*nrietta *La*cks. Cells in culture that acquire the ability to divide indefinitely are said to have undergone **transformation**, the process that causes them to behave like cancer cells. By contrast, nearly all normal, nontransformed mammalian cells growing in culture divide only about 20 to 50 times before they stop dividing, age, and die. Finally, cancer cells evade the normal controls that trigger a cell to undergo apoptosis when something is wrong—for example, when an irreparable mistake has occurred during DNA replication preceding mitosis.

MB **ABC News Video: Henrietta Lacks' Cells**

Abnormal cell behavior in the body can be catastrophic. The problem begins when a single cell in a tissue undergoes the first of many steps that converts a normal cell to a cancer cell. Such a cell often has altered proteins on its surface, and the body's immune system normally recognizes the cell as "nonself"—an insurgent—and destroys it. However, if the cell evades destruction, it may proliferate and form a tumor, a mass of abnormal cells within otherwise normal tissue. The abnormal cells may remain at the original site if they have too few genetic and cellular changes to survive at another site. In that case, the tumor is called a **benign tumor**. Most benign tumors do not cause serious problems (depending on their location) and can be removed by surgery. In contrast, a **malignant tumor** includes cells whose genetic and cellular changes enable them to spread to new tissues and impair the functions of one or more organs; these cells are also sometimes called *transformed* cells (although usage of this term is generally restricted to cells in culture). An individual with a malignant tumor is said to have cancer **(Figure 12.20)**.

The changes that have occurred in cells of malignant tumors show up in many ways besides excessive proliferation. These cells may have unusual numbers of chromosomes, though whether this is a cause or an effect of tumor-related changes is a topic of debate. Their metabolism may be altered, and they may cease to function in any constructive way. Abnormal changes on the cell surface cause cancer cells to lose attachments to neighboring cells and the extracellular matrix, allowing them to spread into nearby tissues. Cancer cells may also secrete signaling molecules that cause blood vessels to grow toward the tumor. A few tumor cells may separate from the original tumor, enter blood vessels and lymph vessels, and travel to other parts of the body. There, they may proliferate and form a new tumor. This spread of cancer cells to locations distant from their original site is called **metastasis** (see Figure 12.20).

A tumor that appears to be localized may be treated with high-energy radiation, which damages DNA in cancer cells much more than DNA in normal cells, apparently because the majority of cancer cells have lost the ability to repair such damage. To treat known or suspected metastatic tumors, chemotherapy is used, in which drugs that are toxic to actively dividing cells are administered through the circulatory system. As you might expect, chemotherapeutic drugs interfere with specific steps in the cell cycle. For example, the drug Taxol freezes the mitotic spindle by preventing microtubule depolymerization, which stops actively dividing cells from proceeding past metaphase and leads to their destruction. The side effects of chemotherapy are due to the effects of the drugs on normal cells that divide frequently, due to the function of that cell type in the organism. For example, nausea results from chemotherapy's effects on intestinal cells, hair loss from effects on hair follicle cells, and susceptibility to infection from effects on immune system cells. You'll work with data from an experiment involving a potential chemotherapeutic agent in the **Scientific Skills Exercise**.

▼ **Figure 12.20 The growth and metastasis of a malignant breast tumor.** A series of genetic and cellular changes contribute to a tumor becoming malignant (cancerous). The cells of malignant tumors grow in an uncontrolled way and can spread to neighboring tissues and, via lymph and blood vessels, to other parts of the body. The spread of cancer cells beyond their original site is called metastasis.

MB **Video: Tumor Growth in a Living Mouse**

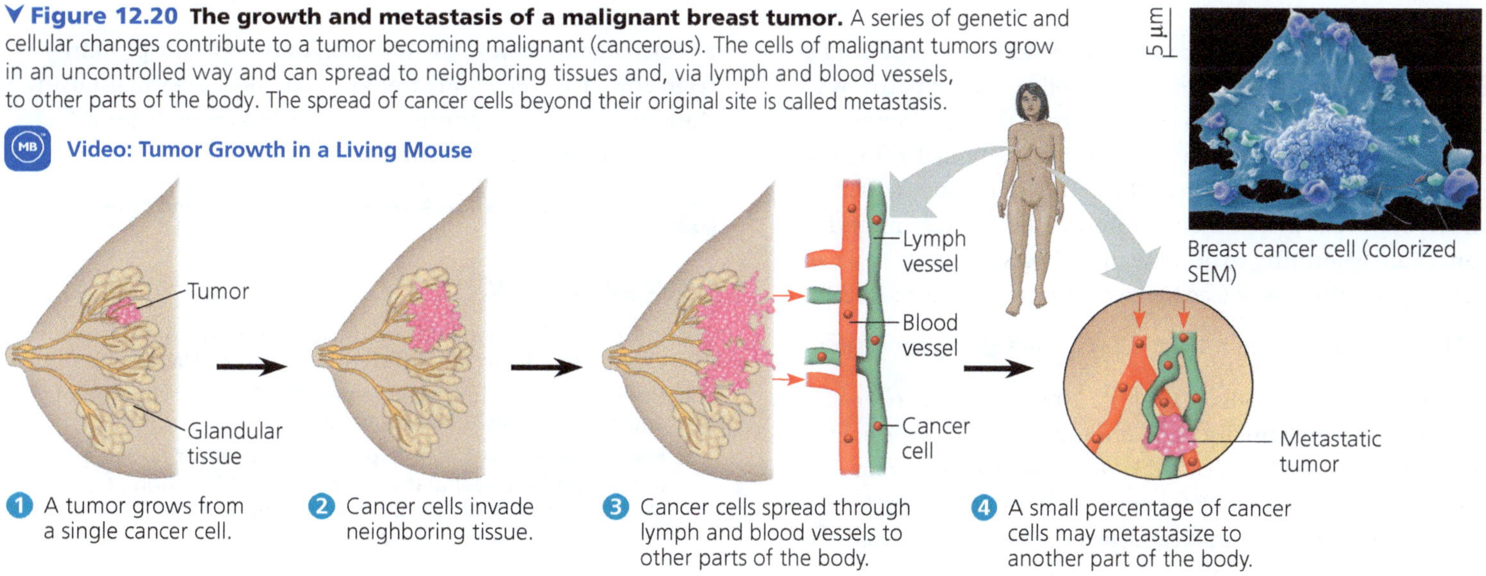

SCIENTIFIC SKILLS EXERCISE

Interpreting Histograms

At What Phase Is the Cell Cycle Arrested by an Inhibitor? Many medical treatments are aimed at stopping cancer cell proliferation by blocking the cell cycle of cancerous tumor cells. One potential treatment is a cell cycle inhibitor derived from human umbilical cord stem cells. In this exercise, you will compare two histograms to determine where in the cell cycle the inhibitor blocks the division of cancer cells.

How the Experiment Was Done In the treated sample, human glioblastoma (brain cancer) cells were grown in tissue culture in the presence of the inhibitor, while a control sample of glioblastoma cells was grown in its absence. After 72 hours of growth, the two cell samples were harvested. To get a "snapshot" of the phase of the cell cycle each cell was in at that time, the samples were treated with a fluorescent chemical that binds to DNA and then run through a flow cytometer, an instrument that records the fluorescence level of each cell. Computer software then graphed the number of cells in each sample with a particular fluorescence level, as shown below.

Data from the Experiment

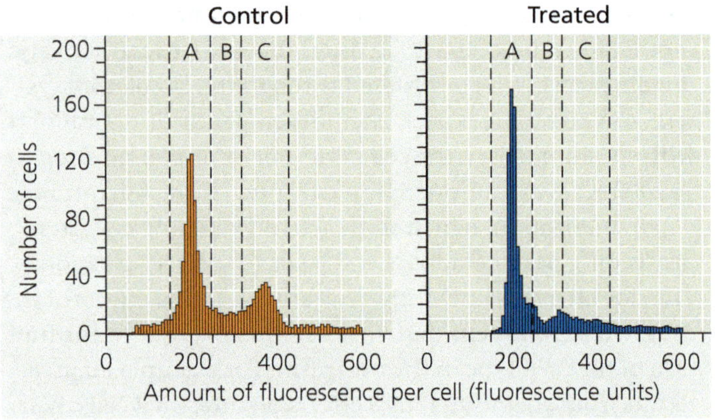

The data are plotted in a type of graph called a histogram (above), which groups the values for a numeric variable on the *x*-axis into intervals. A histogram allows you to see how all the experimental subjects (cells, in this case) are distributed along a continuous variable (amount of fluorescence). In these histograms, the bars

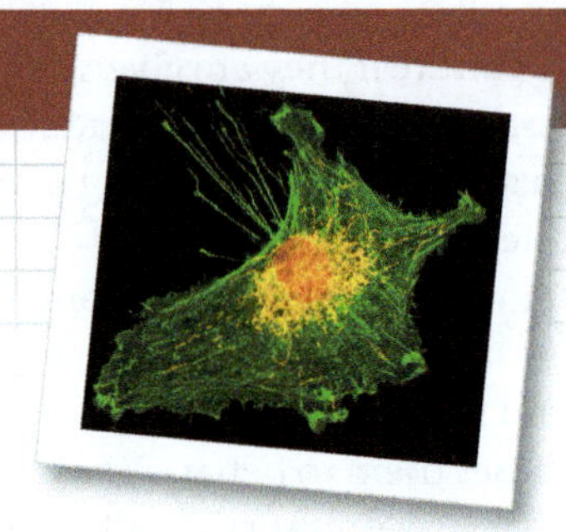

are so narrow that the data appear to follow a curve for which you can detect peaks and dips. Each narrow bar represents the number of cells observed to have a level of fluorescence in the range of that interval. This in turn indicates the relative amount of DNA in those cells. Overall, comparing the two histograms allows you to see how the DNA content of this cell population is altered by the treatment.

INTERPRET THE DATA

1. Study the data in the histograms. (a) Which axis indirectly shows the relative amount of DNA per cell? Explain your answer. (b) In the control sample, compare the first peak in the histogram (in region A) to the second peak (in region C). Which peak shows the population of cells with the higher amount of DNA per cell? Explain. (For additional information about graphs, see the Scientific Skills Review in Appendix F.)
2. (a) In the control sample histogram, identify the phase of the cell cycle (G_1, S, or G_2) of the population of cells in each region delineated by vertical lines. Label the histogram with these phases and explain your answer. (b) Does the S phase population of cells show a distinct peak in the histogram? Why or why not?
3. The histogram representing the treated sample shows the effect of growing the cancer cells alongside human umbilical cord stem cells that produce the potential inhibitor. (a) Label the histogram with the cell cycle phases. Which phase of the cell cycle has the greatest number of cells in the treated sample? Explain. (b) Compare the distribution of cells among G_1, S, and G_2 phases in the control and treated samples. What does this tell you about the cells in the treated sample? (c) Based on what you learned in Concept 12.3, propose a mechanism by which the stem cell–derived inhibitor might arrest the cancer cell cycle at this stage. (More than one answer is possible.)

MB **Instructors**: A version of this Scientific Skills Exercise can be assigned in MasteringBiology.

Data from K. K. Velpula et al., Regulation of glioblastoma progression by cord blood stem cells is mediated by downregulation of cyclin D1, *PLoS ONE* 6(3):e18017 (2011).

Over the past several decades, researchers have produced a flood of valuable information about cell-signaling pathways and how their malfunction contributes to the development of cancer through effects on the cell cycle. Coupled with new molecular techniques, such as the ability to rapidly sequence the DNA of cells in a particular tumor, medical treatments for cancer are beginning to become more "personalized" to a particular patient's tumor (see Make Connections Figure 18.27).

MB **BBC Video: Are Fruit Flies the Key in the Fight Against Cancer?**

For example, the cells of roughly 20% of breast cancer tumors show abnormally high amounts of a cell-surface receptor tyrosine kinase called HER2, and many show an increase in the number of estrogen receptor (ER) molecules, intracellular receptors that can trigger cell division. Based on lab findings, a physician can prescribe chemotherapy with a molecule that blocks the function of the specific protein (Herceptin for HER2 and tamoxifen for ERs). Treatment using these agents, when appropriate, has led to increased survival rates and fewer cancer recurrences.

MB **Interview with Bruce Alberts: Cancer control and careers in science**

CONCEPT CHECK 12.3

1. In Figure 12.14, why do the nuclei resulting from experiment 2 contain different amounts of DNA?
2. How does MPF allow a cell to pass the G_2 phase checkpoint and enter mitosis? (See Figure 12.16.)
3. **MAKE CONNECTIONS ➤** Explain how receptor tyrosine kinases and intracellular receptors might function in triggering cell division. (Review Figures 11.8 and 11.9 and Concept 11.2.)

For suggested answers, see Appendix A.

12 Chapter Review

Go to **MasteringBiology™** for Videos, Animations, Vocab Self-Quiz, Practice Tests, and more in the Study Area.

SUMMARY OF KEY CONCEPTS

- Unicellular organisms reproduce by **cell division**; multicellular organisms depend on cell division for their development from a fertilized egg and for growth and repair. Cell division is part of the **cell cycle**, an ordered sequence of events in the life of a cell.

VOCAB SELF-QUIZ goo.gl/6u55ks

CONCEPT 12.1

Most cell division results in genetically identical daughter cells *(pp. 235–237)*

- The genetic material (DNA) of a cell—its **genome**—is partitioned among **chromosomes**. Each eukaryotic chromosome consists of one DNA molecule associated with many proteins. Together, the complex of DNA and associated proteins is called **chromatin**. The chromatin of a chromosome exists in different states of condensation at different times. In animals, **gametes** have one set of chromosomes and **somatic cells** have two sets.
- Cells replicate their genetic material before they divide, each daughter cell receiving a copy of the DNA. Prior to cell division, chromosomes are duplicated. Each one then consists of two identical **sister chromatids** joined along their lengths by sister chromatid cohesion and held most tightly together at a constricted region at the **centromeres**. When this cohesion is broken, the chromatids separate during cell division, becoming the chromosomes of the daughter cells. Eukaryotic cell division consists of **mitosis** (division of the nucleus) and **cytokinesis** (division of the cytoplasm).

? *Differentiate between these terms: chromosome, chromatin, and chromatid.*

CONCEPT 12.2

The mitotic phase alternates with interphase in the cell cycle *(pp. 237–244)*

- Between divisions, a cell is in **interphase**: the **G_1, S**, and **G_2** phases. The cell grows throughout interphase, with DNA being replicated only during the synthesis (S) phase. Mitosis and cytokinesis make up the **mitotic (M) phase** of the cell cycle.

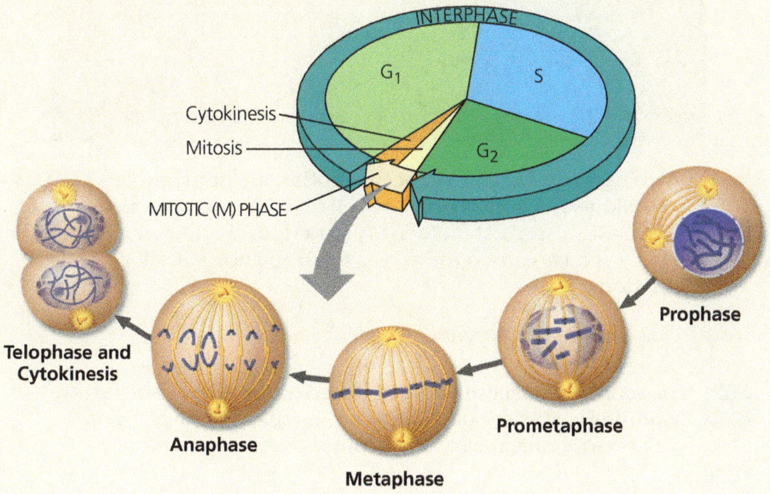

- The **mitotic spindle**, made up of microtubules, controls chromosome movement during mitosis. In animal cells, it arises from the **centrosomes** and includes spindle microtubules and asters. Some spindle microtubules attach to the **kinetochores** of chromosomes and move the chromosomes to the **metaphase plate**. After sister chromatids separate, motor proteins move them along kinetochore microtubules toward opposite ends of the cell. The cell elongates when motor proteins push nonkinetochore microtubules from opposite poles away from each other.
- Mitosis is usually followed by cytokinesis. Animal cells carry out cytokinesis by **cleavage**, and plant cells form a **cell plate**.
- During **binary fission** in bacteria, the chromosome replicates and the daughter chromosomes actively move apart. Some of the proteins involved in bacterial binary fission are related to eukaryotic actin and tubulin.
- Since prokaryotes preceded eukaryotes by more than a billion years, it is likely that mitosis evolved from prokaryotic cell division. Certain unicellular eukaryotes exhibit mechanisms of cell division that may be similar to those of ancestors of existing eukaryotes. Such mechanisms might represent intermediate steps in the evolution of mitosis.

? *In which of the three phases of interphase and the stages of mitosis do chromosomes exist as single DNA molecules?*

CONCEPT 12.3

The eukaryotic cell cycle is regulated by a molecular control system *(pp. 244–250)*

- Signaling molecules present in the cytoplasm regulate progress through the cell cycle.
- The **cell cycle control system** is molecularly based. Cyclic changes in regulatory proteins work as a cell cycle clock. The key molecules are **cyclins** and **cyclin-dependent kinases (Cdks)**. The clock has specific **checkpoints** where the cell cycle stops until a go-ahead signal is received; important checkpoints occur in G_1, G_2, and M phases. Cell culture has enabled researchers to study the molecular details of cell division. Both internal signals and external signals control the cell cycle checkpoints via signal transduction pathways. Most cells exhibit **density-dependent inhibition** of cell division as well as **anchorage dependence**.
- Cancer cells elude normal cell cycle regulation and divide unchecked, forming tumors. **Malignant tumors** invade nearby tissues and can undergo **metastasis**, exporting cancer cells to other sites, where they may form secondary tumors. Recent cell cycle and cell signaling research, and new techniques for sequencing DNA, have led to improved cancer treatments.

? *Explain the significance of the G_1, G_2, and M checkpoints and the go-ahead signals involved in the cell cycle control system.*

TEST YOUR UNDERSTANDING

Level 1: Knowledge/Comprehension

PRACTICE TEST goo.gl/CUYGKD

1. Through a microscope, you can see a cell plate beginning to develop across the middle of a cell and nuclei forming on either side of the cell plate. This cell is most likely
   (A) an animal cell in the process of cytokinesis.
   (B) a plant cell in the process of cytokinesis.
   (C) a bacterial cell dividing.
   (D) a plant cell in metaphase.

2. Vinblastine is a standard chemotherapeutic drug used to treat cancer. Because it interferes with the assembly of microtubules, its effectiveness must be related to
   (A) disruption of mitotic spindle formation.
   (B) suppression of cyclin production.
   (C) myosin denaturation and inhibition of cleavage furrow formation.
   (D) inhibition of DNA synthesis.
3. One difference between cancer cells and normal cells is that cancer cells
   (A) are unable to synthesize DNA.
   (B) are arrested at the S phase of the cell cycle.
   (C) continue to divide even when they are tightly packed together.
   (D) cannot function properly because they are affected by density-dependent inhibition.
4. The decline of MPF activity at the end of mitosis is due to
   (A) the destruction of the protein kinase Cdk.
   (B) decreased synthesis of Cdk.
   (C) the degradation of cyclin.
   (D) the accumulation of cyclin.
5. In the cells of some organisms, mitosis occurs without cytokinesis. This will result in
   (A) cells with more than one nucleus.
   (B) cells that are unusually small.
   (C) cells lacking nuclei.
   (D) cell cycles lacking an S phase.
6. Which of the following does *not* occur during mitosis?
   (A) condensation of the chromosomes
   (B) replication of the DNA
   (C) separation of sister chromatids
   (D) spindle formation

Level 2: Application/Analysis

7. Cell A has half as much DNA as cells B, C, and D in a mitotically active tissue. Cell A is most likely in
   (A) G_1.
   (B) G_2.
   (C) prophase.
   (D) metaphase.
8. The drug cytochalasin B blocks the function of actin. Which of the following aspects of the animal cell cycle would be most disrupted by cytochalasin B?
   (A) spindle formation
   (B) spindle attachment to kinetochores
   (C) cell elongation during anaphase
   (D) cleavage furrow formation and cytokinesis
9. **VISUAL SKILLS** The light micrograph shows dividing cells near the tip of an onion root. Identify a cell in each of the following stages: prophase, prometaphase, metaphase, anaphase, and telophase. Describe the major events occurring at each stage.

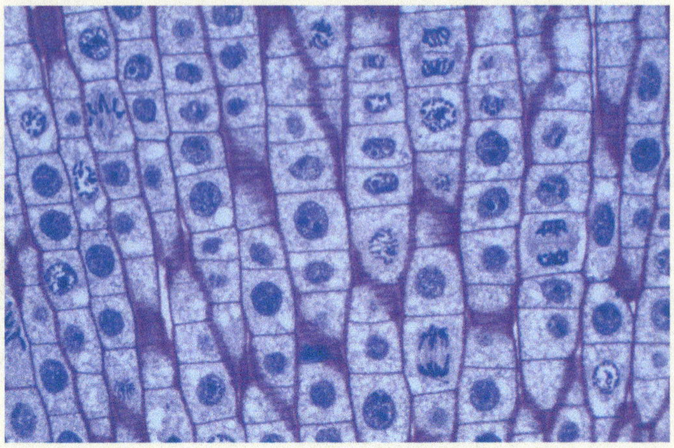

10. **DRAW IT** Draw one eukaryotic chromosome as it would appear during interphase, during each of the stages of mitosis, and during cytokinesis. Also draw and label the nuclear envelope and any microtubules attached to the chromosome(s).

Level 3: Synthesis/Evaluation

11. **EVOLUTION CONNECTION** The result of mitosis is that the daughter cells end up with the same number of chromosomes that the parent cell had. Another potential way to maintain the number of chromosomes would be to carry out cell division first and then duplicate the chromosomes in each daughter cell. Assess whether this would be an equally good way of organizing the cell cycle. Explain why evolution has not led to this alternative.
12. **SCIENTIFIC INQUIRY** Although both ends of a microtubule can gain or lose subunits, one end (called the plus end) polymerizes and depolymerizes at a higher rate than the other end (the minus end). For spindle microtubules, the plus ends are in the center of the spindle, and the minus ends are at the poles. Motor proteins that move along microtubules specialize in walking either toward the plus end or toward the minus end; the two types are called plus end–directed and minus end–directed motor proteins, respectively. Given what you know about chromosome movement and spindle changes during anaphase, predict which type of motor proteins would be present on (a) kinetochore microtubules and (b) nonkinetochore microtubules.
13. **WRITE ABOUT A THEME: INFORMATION** The continuity of life is based on heritable information in the form of DNA. In a short essay (100–150 words), explain how the process of mitosis faithfully parcels out exact copies of this heritable information in the production of genetically identical daughter cells.
14. **SYNTHESIZE YOUR KNOWLEDGE**

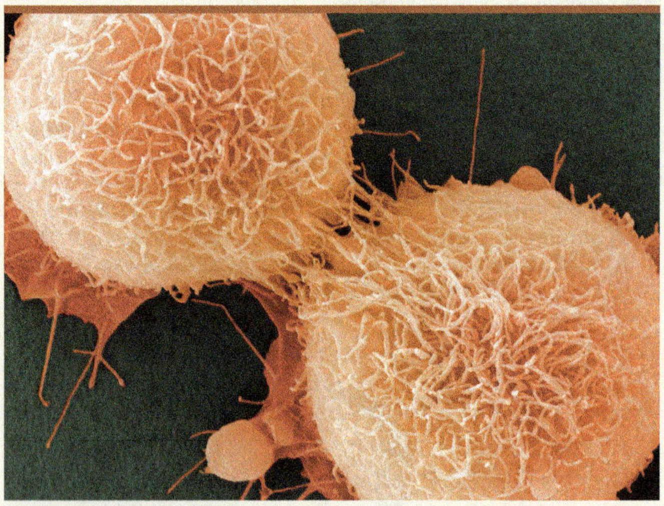

Shown here are two HeLa cancer cells that are just completing cytokinesis. Explain how the cell division of cancer cells like these is misregulated. Identify genetic and other changes that might have caused these cells to escape normal cell cycle regulation.

For selected answers, see Appendix A.

For additional practice questions, check out the **Dynamic Study Modules** in MasteringBiology. You can use them to study on your smartphone, tablet, or computer anytime, anywhere!

UNIT 3 GENETICS

Born in Toronto, Canada, Shirley Tilghman majored in chemistry and biochemistry at Queen's University in Kingston, Ontario. She earned a Ph.D. in Biochemistry from Temple University in 1975. As a postdoctoral fellow at the National Institutes of Health, she helped clone the first mammalian gene (β-globin) and began to ask questions about the molecular basis for embryonic development. After her postdoc, she was a researcher at what is now the Fox Chase Cancer Center in Philadelphia. In 1986, she moved to Princeton University, where she made seminal discoveries about "imprinted" genes, genes that are expressed differently depending on the parent from which they are inherited. She was the founding director of the Lewis-Sigler Institute for Integrative Genomics. Dr. Tilghman was the president of Princeton from 2001 to 2013. She is a member of the National Academy of Sciences and serves as a trustee for multiple institutions.

"If you've seen the children's book *Little Bunny Follows His Nose*, that's how I think of the *H19* gene."

▼ Dr. Tilghman and graduate student Ekaterina Semenova examine a mouse tissue preparation.

An Interview with Shirley Tilghman

How did you get interested in science, and in biology in particular?

It began with loving math. My father claims that when I was five, I wanted to do mental math puzzles instead of having stories read to me! In high school this evolved into a strong interest in chemistry because I loved the logic of chemistry—the rules. But in college I realized I was unlikely to be a good chemist because it wasn't coming naturally to me. I went looking for something more compatible with the way my brain worked and happily stumbled upon the Meselson-Stahl experiment [see Figure 16.11], which I thought was the coolest thing I had ever read. I barely knew what DNA was, but I knew I wanted to learn about it! I wanted to be a molecular biologist.

When you set up your own molecular biology lab, how did you choose what to study?

I wanted to study how genes get turned on and off during early mammalian development, so I went looking through the literature for a messenger RNA that was expressed very early at very high levels and was regulated. I stumbled on the α-fetoprotein gene, the *Afp* gene. It's turned on early in fetal development and is one of the most abundant mRNAs, but is completely shut off after birth. We learned that it is the enhancers of this gene—the DNA sequences that transcription factors bind to—that make such high levels of gene expression possible during specific stages of development.

How did your work on regulation of *Afp* lead you to study the *H19* gene?

This was a classic case of serendipity. If you've seen the children's book *Little Bunny Follows His Nose*, that's how I think of the *H19* gene. When we were working on *Afp*, we wanted to find other genes regulated in the same way, and *H19* was the most dramatic example. The first thing you did then when you got a gene was to sequence it, and that's what my graduate student Vassilis Pachnis did. He brought the sequence to me, and I could see that there was no region in it that coded for a protein, so I asked him to sequence it again. That poor guy sequenced *H19* probably ten times! What finally convinced us that the gene's product was a "noncoding RNA"—an RNA that didn't code for a protein—was sequencing the same gene from multiple mammals. We saw that the sequence was very similar in all of them, but none of them had a sequence that coded for a protein. *H19* was the first gene for a noncoding RNA ever found!

What else did you find out about the *H19* gene?

When we were working on *H19*, I had a classic science experience. I went to a scientific meeting and met a researcher who had reported "imprinting" of the *Igf2* gene, which is near *H19*. Imprinting means that only the copy of the gene inherited from one particular parent is expressed. There were many parallels between *Igf2* and *H19*, so I ran back to the lab and convinced a postdoc named Marisa Bartolomei to do an experiment to determine whether *H19* is imprinted. We discovered that only the copy of the *H19* gene inherited from the mother is expressed—in other words, *H19* was imprinted also. We spent the next ten years working out the details. This was the first imprinted gene where the mechanism was unraveled. It turns out that imprinting involves adding methyl groups to the DNA that affect expression of the gene—the methyl groups are only added in the gamete from one parent, either the mother or the father. At least 100 genes are now known to be imprinted, and there are even more candidate genes to analyze.

Meiosis and Sexual Life Cycles

13

▲ **Figure 13.1 What accounts for family resemblance?**

KEY CONCEPTS

13.1 Offspring acquire genes from parents by inheriting chromosomes

13.2 Fertilization and meiosis alternate in sexual life cycles

13.3 Meiosis reduces the number of chromosome sets from diploid to haploid

13.4 Genetic variation produced in sexual life cycles contributes to evolution

▼ **A sperm fertilizing an egg.**

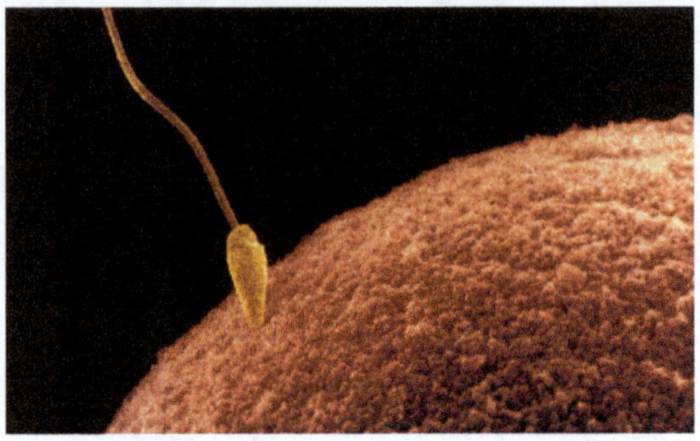

Variations on a Theme

We all know that offspring resemble their parents more than they do unrelated individuals. If you examine the family members shown in **Figure 13.1**, you can pick out some similar features among them. The transmission of traits from one generation to the next is called inheritance, or **heredity** (from the Latin *heres*, heir). However, sons and daughters are not identical copies of either parent or of their siblings. Along with inherited similarity, there is also **variation**. What are the biological mechanisms leading to the "family resemblance" evident among the family members in the photo? A detailed answer to this question eluded biologists until the advance of genetics in the 20th century.

Genetics is the scientific study of heredity and inherited variation. In this unit, you'll learn about genetics at multiple levels, from organisms to cells to molecules. We begin by examining how chromosomes pass from parents to offspring in sexually reproducing organisms. The processes of meiosis (a special type of cell division) and fertilization (the fusion of sperm and egg, as seen in the small photo) maintain a species' chromosome count during the sexual life cycle. We will describe the cellular mechanics of meiosis and explain how this process differs from mitosis. Finally, we will consider how both meiosis and fertilization contribute to genetic variation, such as that seen in Figure 13.1.

When you see this blue icon, log in to **MasteringBiology** and go to the Study Area for digital resources.

MB **Get Ready for This Chapter**

CONCEPT 13.1

Offspring acquire genes from parents by inheriting chromosomes

Family friends may tell you that you have your mother's nose or your father's eyes. Of course, parents do not, in any literal sense, give their children a nose, eyes, hair, or any other traits. What, then, *is* actually inherited?

Inheritance of Genes

Parents endow their offspring with coded information in the form of hereditary units called **genes**. The genes we inherit from our mothers and fathers are our genetic link to our parents, and they account for family resemblances such as shared eye color or freckles. Our genes program specific traits that emerge as we develop from fertilized eggs into adults.

The genetic program is written in the language of DNA, the polymer of four different nucleotides you learned about in Concepts 1.1 and 5.5. Inherited information is passed on in the form of each gene's specific sequence of DNA nucleotides, much as printed information is communicated in the form of meaningful sequences of letters. In both cases, the language is symbolic. Just as your brain translates the word *apple* into a mental image of the fruit, cells translate genes into freckles and other features. Most genes program cells to synthesize specific enzymes and other proteins, whose cumulative action produces an organism's inherited traits. The programming of these traits in the form of DNA is one of the unifying themes of biology.

The transmission of hereditary traits has its molecular basis in the replication of DNA, which produces copies of genes that can be passed from parents to offspring. In animals and plants, reproductive cells called **gametes** are the vehicles that transmit genes from one generation to the next. During fertilization, male and female gametes (sperm and eggs) unite, passing on genes of both parents to their offspring.

Except for small amounts of DNA in mitochondria and chloroplasts, the DNA of a eukaryotic cell is packaged into chromosomes within the nucleus. Every species has a characteristic number of chromosomes. For example, humans have 46 chromosomes in their **somatic cells**—all cells of the body except the gametes and their precursors. Each chromosome consists of a single long DNA molecule, elaborately coiled in association with various proteins. One chromosome includes several hundred to a few thousand genes, each of which is a precise sequence of nucleotides along the DNA molecule. A gene's specific location along the length of a chromosome is called the gene's **locus** (plural, *loci*; from the Latin, meaning "place"). Our genetic endowment (our genome) consists of the genes and other DNA that make up the chromosomes we inherited from our parents.

Comparison of Asexual and Sexual Reproduction

Only organisms that reproduce asexually have offspring that are exact genetic copies of themselves. In **asexual reproduction**, a single individual (like a yeast cell or an amoeba; see Figure 12.2a) is the sole parent and passes copies of all its genes to its offspring without the fusion of gametes. For example, single-celled eukaryotic organisms can reproduce asexually by mitotic cell division, in which DNA is copied and allocated equally to two daughter cells. The genomes of the offspring are virtually exact copies of the parent's genome. Some multicellular organisms are also capable of reproducing asexually **(Figure 13.2)**. Because the cells of the offspring arise via mitosis in the parent, the offspring is usually genetically identical to its parent. An individual that reproduces asexually gives rise to a **clone**, a group of genetically identical individuals. Genetic differences occasionally arise in asexually reproducing organisms as a result of changes in the DNA called mutations, which we will discuss in Concept 17.5.

In **sexual reproduction**, two parents give rise to offspring that have unique combinations of genes inherited from the two parents. In contrast to a clone, offspring of sexual reproduction vary genetically from their siblings and both parents: They are variations on a common theme of family resemblance, not exact replicas. Genetic variation like that shown in Figure 13.1 is an important consequence of sexual reproduction. What mechanisms generate this genetic variation? The key is the behavior of chromosomes during the sexual life cycle.

▼ Figure 13.2 Asexual reproduction in two multicellular organisms. (a) This relatively simple animal, a hydra, reproduces by budding. The bud, a localized mass of mitotically dividing cells, develops into a small hydra, which detaches from the parent (LM). **(b)** All the trees in this circle of redwoods arose asexually from a single parent tree, whose stump is in the center of the circle.

(a) Hydra **(b) Redwoods**

Video: Hydra Budding
Animation: Asexual Reproduction

CONCEPT CHECK 13.1

1. **MAKE CONNECTIONS** ➤ Using what you know of gene expression in a cell, explain what causes the traits of parents (such as hair color) to show up in their offspring. (See Concept 5.5.)
2. How does an asexually reproducing eukaryotic organism produce offspring that are genetically identical to each other and to their parents?
3. **WHAT IF?** ➤ A horticulturalist breeds orchids, trying to obtain a plant with a unique combination of desirable traits. After many years, she finally succeeds. To produce more plants like this one, should she crossbreed it with another plant or clone it? Why?

For suggested answers, see Appendix A.

CONCEPT 13.2

Fertilization and meiosis alternate in sexual life cycles

A **life cycle** is the generation-to-generation sequence of stages in the reproductive history of an organism, from conception to production of its own offspring. In this section, we use humans as an example to track the behavior of chromosomes through the sexual life cycle. We begin by considering the chromosome count in human somatic cells and gametes. We will then explore how the behavior of chromosomes relates to the human life cycle and other types of sexual life cycles.

Sets of Chromosomes in Human Cells

In humans, each somatic cell has 46 chromosomes. During mitosis, the chromosomes become condensed enough to be visible under a light microscope. At this point, they can be distinguished from one another by their size, the position of their centromeres, and the pattern of colored bands produced by certain chromatin-binding stains.

Careful examination of a micrograph of the 46 human chromosomes from a single cell in mitosis reveals that there are two chromosomes of each of 23 types. This becomes clear when images of the chromosomes are arranged in pairs, starting with the longest chromosomes. The resulting ordered display is called a **karyotype** (Figure 13.3). The two chromosomes of a pair have the same length, centromere position, and staining pattern: These are called **homologous chromosomes** (or **homologs**). Both chromosomes of each pair carry genes controlling the same inherited characters. For example, if a gene for eye color is situated at a particular locus on a certain chromosome, then its homologous chromosome (its homolog) will also have a version of the eye-color gene at the equivalent locus.

The two chromosomes referred to as X and Y are an important exception to the general pattern of homologous chromosomes in human somatic cells. Typically, human females have a homologous pair of X chromosomes (XX), while males have one X and one Y chromosome (XY; see Figure 13.3). Only small parts of the X and Y are homologous. Most of the

▼ Figure 13.3

Research Method **Preparing a Karyotype**

Application A karyotype is a display of condensed chromosomes arranged in pairs. Karyotyping can be used to screen for defective chromosomes or abnormal numbers of chromosomes associated with certain congenital disorders, such as Down syndrome.

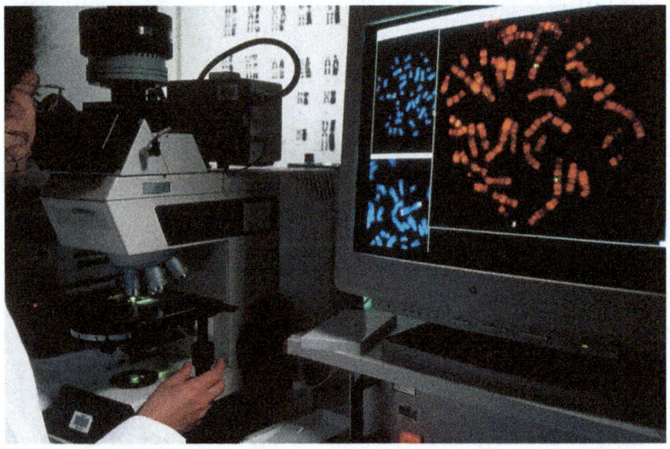

Technique Karyotypes are prepared from isolated somatic cells, which are treated with a drug to stimulate mitosis and then grown in culture for several days. Cells arrested when the chromosomes are most highly condensed—at metaphase—are stained and then viewed with a microscope equipped with a digital camera. An image of the chromosomes is displayed on a computer monitor, and digital software is used to arrange them in pairs according to their appearance.

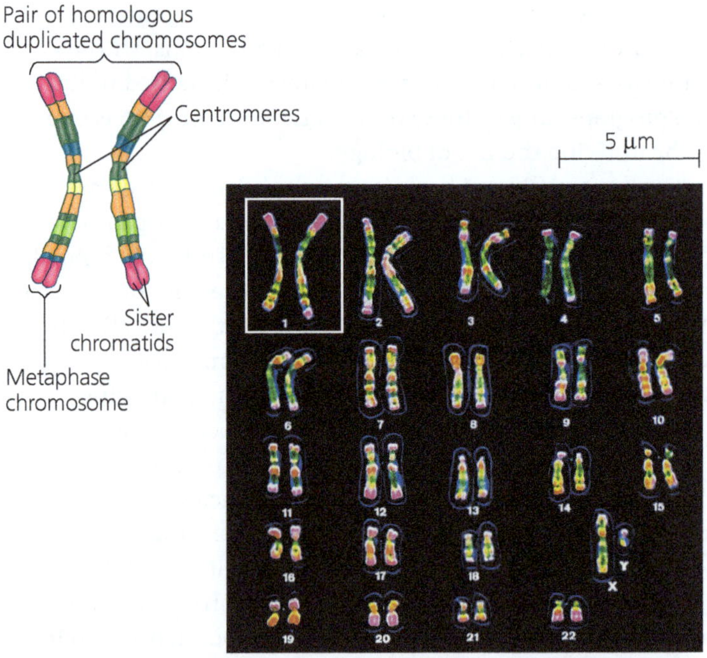

Results This karyotype shows the chromosomes from a human male (as seen by the presence of the XY chromosome pair), colored to emphasize their chromosome banding patterns. The size of the chromosome, position of the centromere, and pattern of stained bands help identify specific chromosomes. Although difficult to discern in the karyotype, each metaphase chromosome consists of two closely attached sister chromatids (see the diagram of the first pair of homologous duplicated chromosomes).

genes carried on the X chromosome do not have counterparts on the tiny Y, and the Y chromosome has genes lacking on the X. Due to their role in sex determination, the X and Y chromosomes are called **sex chromosomes**. The other chromosomes are called **autosomes**.

The occurrence of pairs of homologous chromosomes in each human somatic cell is a consequence of our sexual origins. We inherit one chromosome of a pair from each parent. Thus, the 46 chromosomes in our somatic cells are actually two sets of 23 chromosomes—a maternal set (from our mother) and a paternal set (from our father). The number of chromosomes in a single set is represented by n. Any cell with two chromosome sets is called a **diploid cell** and has a diploid number of chromosomes, abbreviated $2n$. For humans, the diploid number is 46 ($2n = 46$), the number of chromosomes in our somatic cells. In a cell in which DNA synthesis has occurred, all the chromosomes are duplicated, and therefore each consists of two identical sister chromatids, associated closely at the centromere and along the arms. (Even though the chromosomes are duplicated, we still say the cell is diploid, or $2n$. This because it has only two sets of information regardless of the number of chromatids, which are merely copies of the information in one set.) **Figure 13.4** helps clarify the various terms that we use to describe duplicated chromosomes in a diploid cell.

Unlike somatic cells, gametes contain a single set of chromosomes. Such cells are called **haploid cells**, and each has a haploid number of chromosomes (n). For humans, the haploid number is 23 ($n = 23$). The set of 23 consists of the 22 autosomes plus a single sex chromosome. An unfertilized egg contains an X chromosome; a sperm contains either an X or a Y chromosome.

Each sexually reproducing species has a characteristic diploid and haploid number. For example, the fruit fly *Drosophila melanogaster* has a diploid number ($2n$) of 8 and a haploid number (n) of 4, while for dogs, $2n$ is 78 and n is 39. The chromosome number generally does not correlate with the size or complexity of a species' genome; it simply reflects how many linear pieces of DNA make up the genome, which is a function of the evolutionary history of that species (see Concept 21.5). Now let's consider chromosome behavior during sexual life cycles. We'll use the human life cycle as an example.

Figure 13.4 Describing chromosomes. A cell from an organism with a diploid number of 6 ($2n = 6$) is depicted here following chromosome duplication and condensation. Each of the six duplicated chromosomes consists of two sister chromatids associated closely along their lengths. Each homologous pair is composed of one chromosome from the maternal set (red) and one from the paternal set (blue). Each set is made up of three chromosomes in this example (long, medium, and short). Together, one maternal and one paternal chromatid in a pair of homologous chromosomes are called nonsister chromatids.

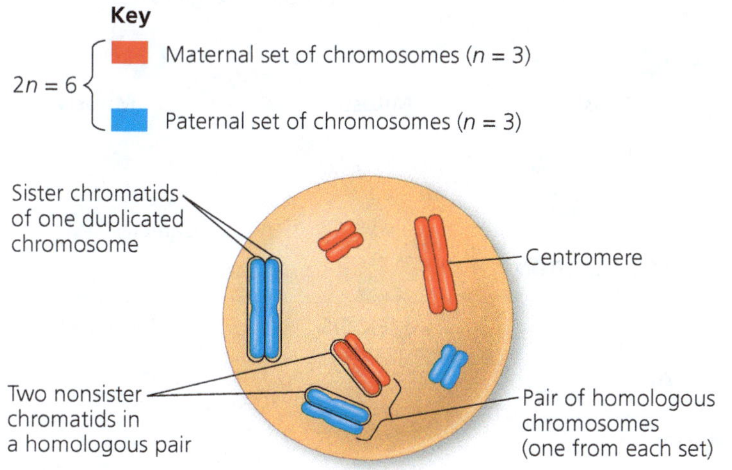

VISUAL SKILLS ➤ *How many sets of chromosomes are present in this diagram? How many pairs of homologous chromosomes are present?*

BioFlix® Animation: Chromosomes

Behavior of Chromosome Sets in the Human Life Cycle

The human life cycle begins when a haploid sperm from the father fuses with a haploid egg from the mother **(Figure 13.5)**.

Figure 13.5 The human life cycle. In each generation, the number of chromosome sets is halved during meiosis but doubles at fertilization. For humans, the number of chromosomes in a haploid cell is 23, consisting of one set ($n = 23$); the number of chromosomes in the diploid zygote and all somatic cells arising from it is 46, consisting of two sets ($2n = 46$).

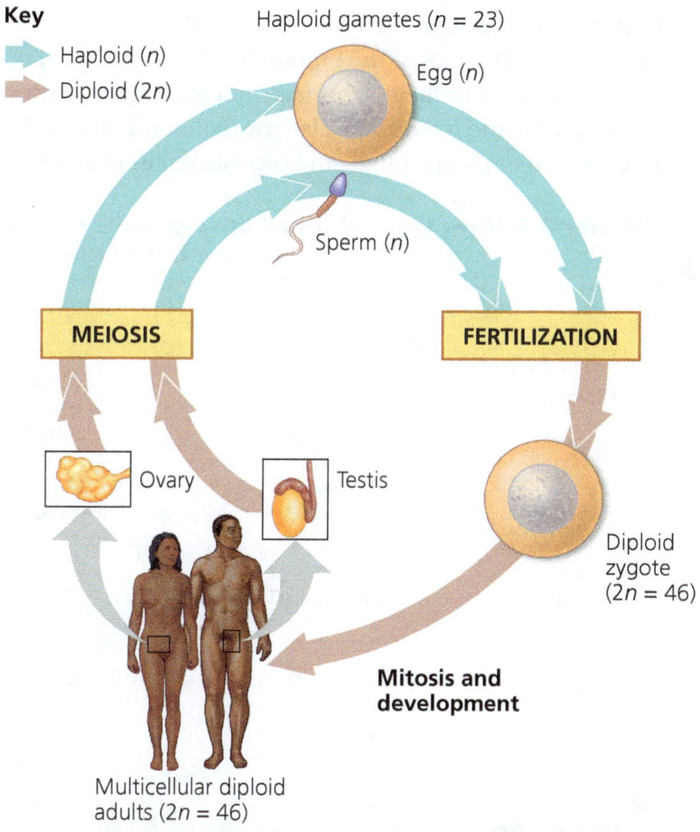

This figure introduces a color code that will be used for other life cycles later in this book. The aqua arrows identify haploid stages of a life cycle, and the tan arrows identify diploid stages.

Animation: The Human Life Cycle

This union of gametes, culminating in fusion of their nuclei, is called **fertilization**. The resulting fertilized egg, or **zygote**, is diploid because it contains two haploid sets of chromosomes bearing genes representing the maternal and paternal family lines. As a human develops into a sexually mature adult, mitosis of the zygote and its descendant cells generates all the somatic cells of the body. Both chromosome sets in the zygote and all the genes they carry are passed with precision to the somatic cells.

The only cells of the human body not produced by mitosis are the gametes, which develop from specialized cells called *germ cells* in the gonads—ovaries in females and testes in males (see Figure 13.5). Imagine what would happen if human gametes were made by mitosis: They would be diploid like the somatic cells. At the next round of fertilization, when two gametes fused, the normal chromosome number of 46 would double to 92, and each subsequent generation would double the number of chromosomes yet again. This does not happen, however, because in sexually reproducing organisms, gamete formation involves a type of cell division called **meiosis**. This type of cell division reduces the number of sets of chromosomes from two to one in the gametes, counterbalancing the doubling that occurs at fertilization. As a result of meiosis, each human sperm and egg is haploid ($n = 23$). Fertilization restores the diploid condition by combining two sets of chromosomes, and the human life cycle is repeated, generation after generation (see Figure 13.5).

In general, the steps of the human life cycle are typical of many sexually reproducing animals. Indeed, the processes of fertilization and meiosis are also the hallmarks of sexual reproduction in plants, fungi, and protists just as in animals. Fertilization and meiosis alternate in sexual life cycles, maintaining a constant number of chromosomes in each species from one generation to the next.

The Variety of Sexual Life Cycles

Although the alternation of meiosis and fertilization is common to all organisms that reproduce sexually, the timing of these two events in the life cycle varies, depending on the species. These variations can be grouped into three main types of life cycles. In the type that occurs in humans and most other animals, gametes are the only haploid cells **(Figure 13.6a)**. Meiosis occurs in germ cells during the production of gametes, which undergo no further cell division prior to fertilization. After fertilization, the diploid zygote divides by mitosis, producing a multicellular organism that is diploid.

Plants and some species of algae exhibit a second type of life cycle called **alternation of generations (Figure 13.6b)**. This type includes both diploid and haploid stages that are multicellular. The multicellular diploid stage is called the *sporophyte*. Meiosis in the sporophyte produces haploid cells called *spores*. Unlike a gamete, a haploid spore doesn't fuse with another cell but divides mitotically, generating a multicellular haploid stage called the *gametophyte*. Cells of the gametophyte give rise to gametes by mitosis. Fusion of two haploid gametes at fertilization results in a diploid zygote, which develops into the next sporophyte generation. Therefore, in this type of life cycle, the sporophyte generation produces a gametophyte as its offspring, and the gametophyte generation produces the next sporophyte generation (see Figure 13.6b). The term *alternation of generations* fits well as a name for this type of life cycle.

Figure 13.6 Three types of sexual life cycles. The common feature of all three cycles is the alternation of meiosis and fertilization, key events that contribute to genetic variation among offspring. The cycles differ in the timing of these two key events. (Small circles are cells; large circles are organisms.)

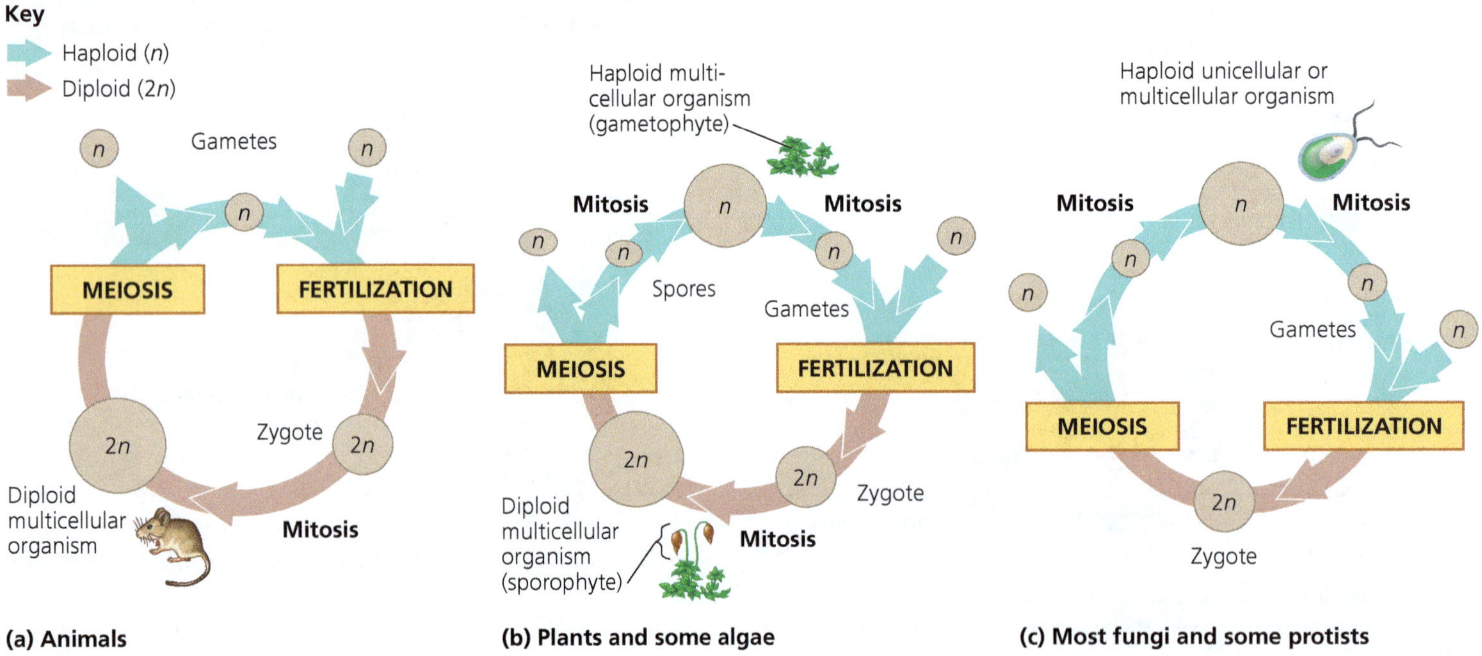

VISUAL SKILLS ➤ *For each type of life cycle, indicate whether haploid cells undergo mitosis, and if they do, describe the cells that are formed.*

A third type of life cycle occurs in most fungi and some protists, including some algae **(Figure 13.6c)**. After gametes fuse and form a diploid zygote, meiosis occurs without a multicellular diploid offspring developing. Meiosis produces not gametes but haploid cells that then divide by mitosis and give rise to either unicellular descendants or a haploid multicellular adult organism. Subsequently, the haploid organism carries out further mitoses, producing the cells that develop into gametes. The only diploid stage found in these species is the single-celled zygote.

Note that *either* haploid or diploid cells can divide by mitosis, depending on the type of life cycle. Only diploid cells, however, can undergo meiosis because haploid cells have only a single set of chromosomes that cannot be further reduced. Though the three types of sexual life cycles differ in the timing of meiosis and fertilization, they share a fundamental result: genetic variation among offspring.

CONCEPT CHECK 13.2

1. **MAKE CONNECTIONS ➤** In Figure 13.4, how many DNA molecules (double helices) are present (see Figure 12.5)? What is the haploid number of this cell? Is a set of chromosomes haploid or diploid?
2. **VISUAL SKILLS ➤** In the karyotype shown in Figure 13.3, how many pairs of chromosomes are present? How many sets?
3. **WHAT IF? ➤** A certain eukaryote lives as a unicellular organism, but during environmental stress, it produces gametes. The gametes fuse, and the resulting zygote undergoes meiosis, generating new single cells. What type of organism could this be?

For suggested answers, see Appendix A.

CONCEPT 13.3

Meiosis reduces the number of chromosome sets from diploid to haploid

Several steps of meiosis closely resemble corresponding steps in mitosis. Meiosis, like mitosis, is preceded by the duplication of chromosomes. However, this single duplication is followed by not one but two consecutive cell divisions, called **meiosis I** and **meiosis II**. These two divisions result in four daughter cells (rather than the two daughter cells of mitosis), each with only half as many chromosomes as the parent cell—one set, rather than two.

The Stages of Meiosis

The overview of meiosis in **Figure 13.7** shows, for a single pair of homologous chromosomes in a diploid cell, that both members of the pair are duplicated and the copies sorted into four haploid daughter cells. Recall that sister chromatids are two copies of *one* chromosome, closely associated all along their lengths; this association is called *sister chromatid cohesion*. Together, the sister chromatids make up one duplicated chromosome (see Figure 13.4). In contrast, the two chromosomes of a homologous pair are individual chromosomes that were inherited from each parent. Homologs appear alike in the microscope, but they may have different versions of genes at corresponding loci; each version is called an *allele* of that gene (see Figure 14.4). Homologs are not associated with each other in any obvious way except during meiosis.

Figure 13.8 describes in detail the stages of the two divisions of meiosis for an animal cell whose diploid number is 6. Study this figure thoroughly before going on.

▼ Figure 13.7 Overview of meiosis: how meiosis reduces chromosome number. After the chromosomes duplicate in interphase, the diploid cell divides *twice*, yielding four haploid daughter cells. This overview tracks just one pair of homologous chromosomes, which for the sake of simplicity are drawn in the condensed state throughout.

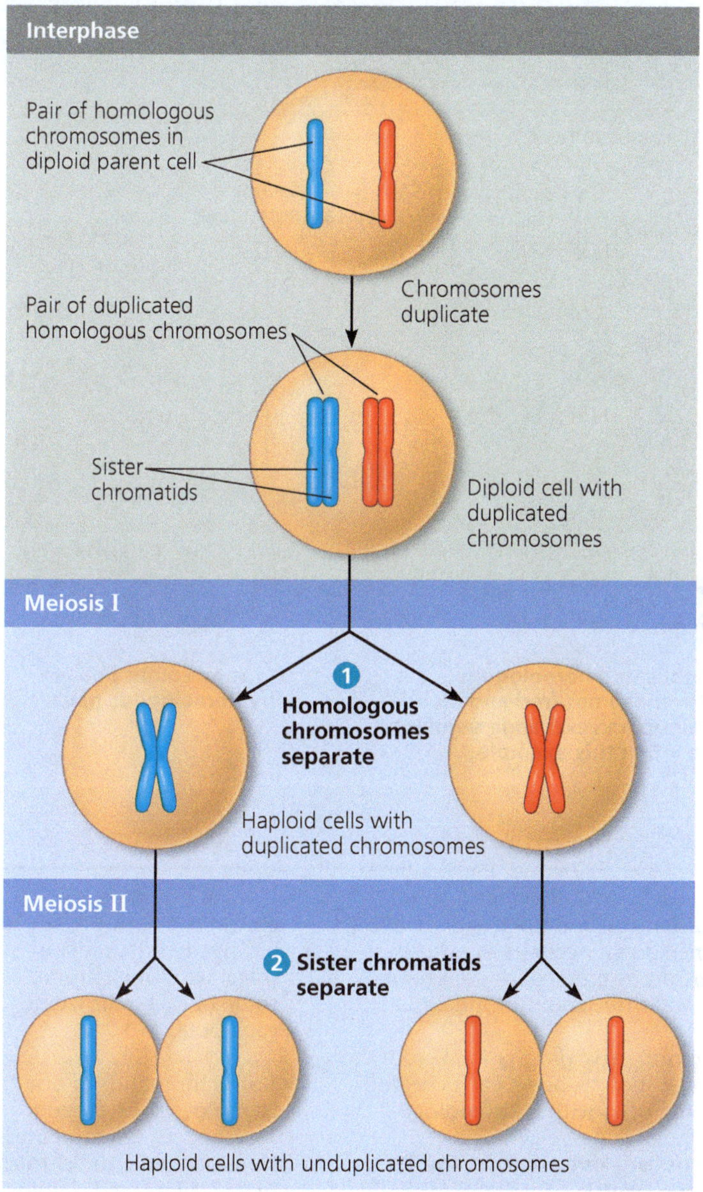

DRAW IT ➤ *Redraw the cells in this figure using a simple double helix to represent each DNA molecule.*

MB **Animation: Overview of Meiosis**

▼ Figure 13.8 Exploring Meiosis in an Animal Cell

MEIOSIS I: Separates homologous chromosomes

Prophase I	Metaphase I	Anaphase I	Telophase I and Cytokinesis

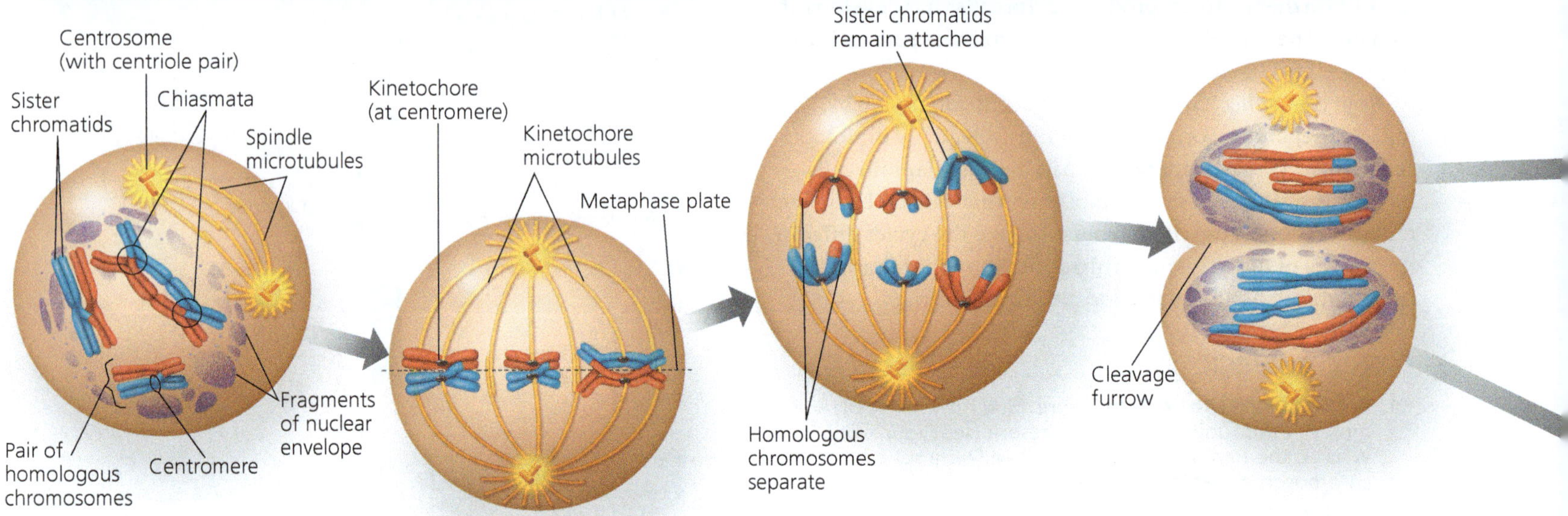

Duplicated homologous chromosomes (red and blue) pair up and exchange segments; $2n = 6$ in this example.

Chromosomes line up by homologous pairs.

The two homologous chromosomes of each pair separate.

Two haploid cells form; each chromosome still consists of two sister chromatids.

Prophase I

- Centrosome movement, spindle formation, and nuclear envelope breakdown occur as in mitosis. Chromosomes condense progressively throughout prophase I.
- During early prophase I, before the stage shown above, each chromosome pairs with its homolog, aligned gene by gene, and **crossing over** occurs: The DNA molecules of nonsister chromatids are broken (by proteins) and are rejoined to each other.
- At the stage shown above, each homologous pair has one or more X-shaped regions called **chiasmata** (singular, *chiasma*), where crossovers have occurred.
- Later in prophase I, microtubules from one pole or the other attach to the kinetochores, one at the centromere of each homolog. (The two kinetochores on the sister chromatids of a homolog are linked together by proteins and act as a single kinetochore.) Microtubules move the homologous pairs toward the metaphase plate (see the metaphase I diagram).

Metaphase I

- Pairs of homologous chromosomes are now arranged at the metaphase plate, with one chromosome of each pair facing each pole.
- Both chromatids of one homolog are attached to kinetochore microtubules from one pole; the chromatids of the other homolog are attached to microtubules from the opposite pole.

Anaphase I

- Breakdown of proteins that are responsible for sister chromatid cohesion along chromatid arms allows homologs to separate.
- The homologs move toward opposite poles, guided by the spindle apparatus.
- Sister chromatid cohesion persists at the centromere, causing chromatids to move as a unit toward the same pole.

Telophase I and Cytokinesis

- When telophase I begins, each half of the cell has a complete haploid set of duplicated chromosomes. Each chromosome is composed of two sister chromatids; one or both chromatids include regions of nonsister chromatid DNA.
- Cytokinesis (division of the cytoplasm) usually occurs simultaneously with telophase I, forming two haploid daughter cells.
- In animal cells like these, a cleavage furrow forms. (In plant cells, a cell plate forms.)
- In some species, chromosomes decondense and nuclear envelopes form.
- No chromosome duplication occurs between meiosis I and meiosis II.

MB Video: Meiosis I in Sperm Formation

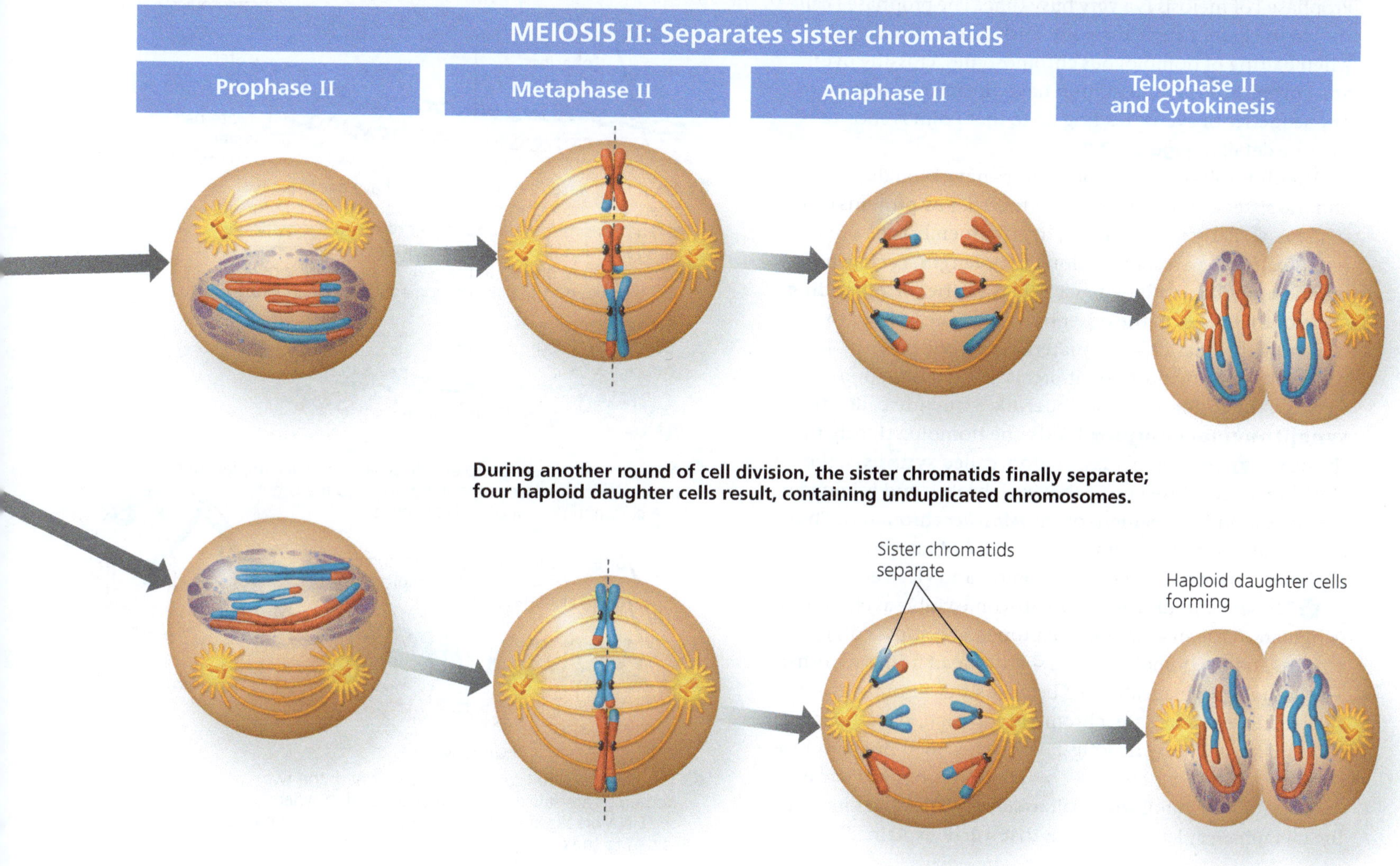

Prophase II

- A spindle apparatus forms.
- In late prophase II (not shown here), chromosomes, each still composed of two chromatids associated at the centromere, are moved by microtubules toward the metaphase II plate.

Metaphase II

- The chromosomes are positioned at the metaphase plate as in mitosis.
- Because of crossing over in meiosis I, the two sister chromatids of each chromosome are not genetically identical.
- The kinetochores of sister chromatids are attached to microtubules extending from opposite poles.

Anaphase II

- Breakdown of proteins holding the sister chromatids together at the centromere allows the chromatids to separate. The chromatids move toward opposite poles as individual chromosomes.

Telophase II and Cytokinesis

- Nuclei form, the chromosomes begin decondensing, and cytokinesis occurs.
- The meiotic division of one parent cell produces four daughter cells, each with a haploid set of (unduplicated) chromosomes.
- The four daughter cells are genetically distinct from one another and from the parent cell.

MAKE CONNECTIONS ➤ *Look at Figure 12.7 and imagine the two daughter cells undergoing another round of mitosis, yielding four cells. Compare the number of chromosomes in each of those four cells, after mitosis, with the number in each cell in Figure 13.8, after meiosis. What is it about the process of meiosis that accounts for this difference, even though meiosis also includes two cell divisions?*

BioFlix® Animation: Meiosis
Animation: Meiosis

Crossing Over and Synapsis During Prophase I

Prophase I of meiosis is a very busy time. The prophase I cell shown in Figure 13.8 is at a point fairly late in prophase I, when pairing of homologous chromosomes, crossing over, and chromosome condensation have already taken place. The sequence of events leading up to that point is shown in more detail in **Figure 13.9**.

After interphase, the chromosomes have been duplicated and the sister chromatids are held together by proteins called *cohesins*. ❶ Early in prophase I, the two members of a homologous pair associate loosely along their length. Each gene on one homolog is aligned precisely with the corresponding allele of that gene on the other homolog. The DNA of two nonsister chromatids—one maternal and one paternal—is broken by specific proteins at precisely matching points. ❷ Next, the formation of a zipper-like structure called the **synaptonemal complex** holds one homolog tightly to the other. ❸ During this association, called **synapsis**, the DNA breaks are closed up so that each broken end is joined to the corresponding segment of the *nonsister* chromatid. Thus, a paternal chromatid is joined to a piece of maternal chromatid beyond the crossover point, and vice versa.

❹ These points of crossing over become visible as chiasmata (singular, *chiasma*) after the synaptonemal complex disassembles and the homologs move slightly apart from each other. The homologs remain attached because sister chromatids are still held together by sister chromatid cohesion, even though some of the DNA may no longer be attached to its original chromosome. At least one crossover per chromosome must occur in order for the homologous pair to stay together as it moves to the metaphase I plate, for reasons that will be explained shortly.

A Comparison of Mitosis and Meiosis

Figure 13.10 summarizes the key differences between meiosis and mitosis in diploid cells. Basically, meiosis reduces the number of chromosome sets from two (diploid) to one (haploid), whereas mitosis conserves the number of chromosome sets. Therefore, meiosis produces cells that differ genetically from their parent cell and from each other, whereas mitosis produces daughter cells that are genetically identical to their parent cell and to each other.

Three events unique to meiosis occur during meiosis I:

1. **Synapsis and crossing over.** During prophase I, duplicated homologs pair up and crossing over occurs, as described previously and in Figure 13.9. Synapsis and crossing over do not occur during prophase of mitosis.
2. **Alignment of homologous pairs at the metaphase plate.** At metaphase I of meiosis, pairs of homologs are positioned at the metaphase plate, rather than individual chromosomes, as in metaphase of mitosis.

▼ Figure 13.9 Crossing over and synapsis in prophase I: a closer look.

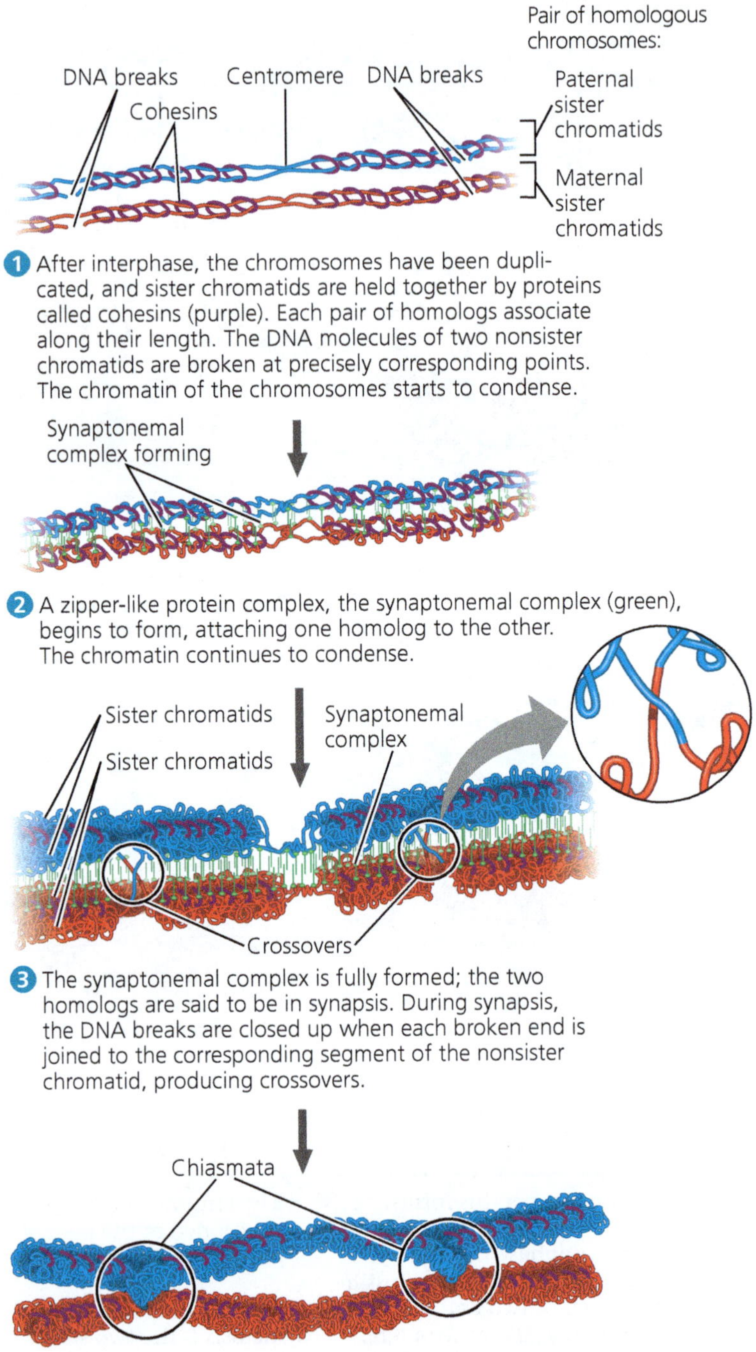

❶ After interphase, the chromosomes have been duplicated, and sister chromatids are held together by proteins called cohesins (purple). Each pair of homologs associate along their length. The DNA molecules of two nonsister chromatids are broken at precisely corresponding points. The chromatin of the chromosomes starts to condense.

❷ A zipper-like protein complex, the synaptonemal complex (green), begins to form, attaching one homolog to the other. The chromatin continues to condense.

❸ The synaptonemal complex is fully formed; the two homologs are said to be in synapsis. During synapsis, the DNA breaks are closed up when each broken end is joined to the corresponding segment of the nonsister chromatid, producing crossovers.

❹ After the synaptonemal complex disassembles, the homologs move slightly apart from each other but remain attached because of sister chromatid cohesion, even though some of the DNA may no longer be attached to its original chromosome. The points of attachment where crossovers have occurred show up as chiasmata. The chromosomes continue to condense as they move toward the metaphase plate.

3. **Separation of homologs.** At anaphase I of meiosis, the duplicated chromosomes of each homologous pair move toward opposite poles, but the sister chromatids of each duplicated chromosome remain attached. In anaphase of mitosis, by contrast, sister chromatids separate.

Figure 13.10 A comparison of mitosis and meiosis.

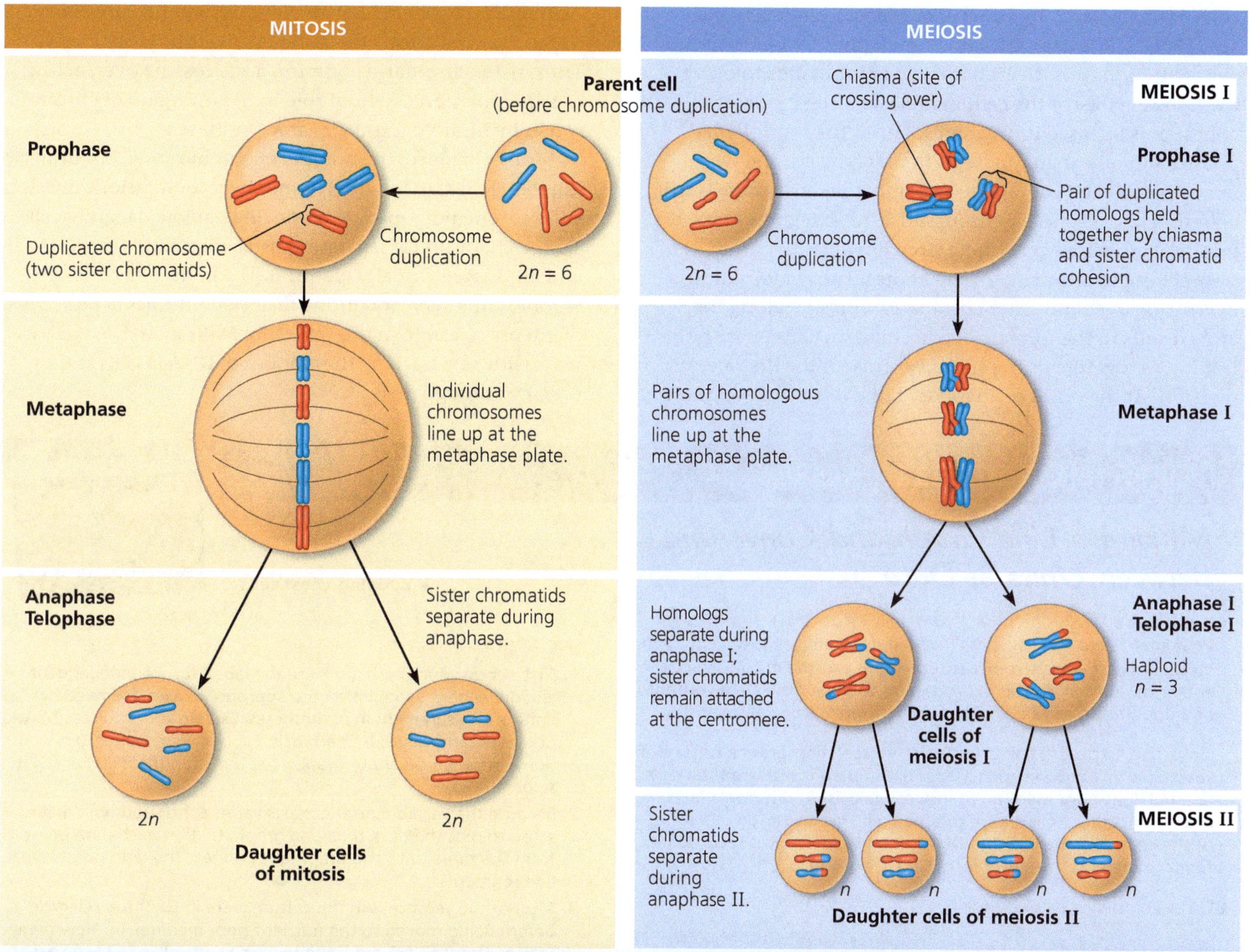

SUMMARY

Property	**Mitosis** (occurs in both diploid and haploid cells)	**Meiosis** (can only occur in diploid cells)
DNA replication	Occurs during interphase, before mitosis begins	Occurs during interphase before meiosis I but not meiosis II
Number of divisions	One, including prophase, prometaphase, metaphase, anaphase, and telophase	Two, each including prophase, metaphase, anaphase, and telophase
Synapsis of homologous chromosomes	Does not occur	Occurs during prophase I along with crossing over between nonsister chromatids; resulting chiasmata hold pairs together due to sister chromatid cohesion
Number of daughter cells and genetic composition	Two, each genetically identical to the parent cell, with the same number of chromosomes	Four, each haploid (n); genetically different from the parent cell and from each other
Role in animals, fungi, and plants	Enables multicellular animal, fungus, or plant (gametophyte or sporophyte) to arise from a single cell; produces cells for growth, repair, and, in some species, asexual reproduction; produces gametes in the plant gametophyte	Produces gametes (in animals) or spores (in fungi and in plant sporophytes); reduces number of chromosome sets by half and introduces genetic variability among the gametes or spores

DRAW IT ➤ *Could any other combinations of chromosomes be generated during meiosis II from the specific cells shown in telophase I? Explain. (Hint: Draw the cells as they would appear in metaphase II.)*

Sister chromatids stay together due to sister chromatid cohesion, mediated by cohesin proteins. In mitosis, this attachment lasts until the end of metaphase, when enzymes cleave the cohesins, freeing the sister chromatids to move to opposite poles of the cell. In meiosis, sister chromatid cohesion is released in two steps, one at the start of anaphase I and one at anaphase II. In metaphase I, the two homologs of each pair are held together because there is still cohesion between sister chromatid arms in regions beyond points of crossing over, where stretches of sister chromatids now belong to different chromosomes. The combination of crossing over and sister chromatid cohesion along the arms results in the formation of a chiasma. Chiasmata hold homologs together as the spindle forms for the first meiotic division. At the onset of anaphase I, the release of cohesion along sister chromatid *arms* allows homologs to separate. At anaphase II, the release of sister chromatid cohesion at the *centromeres* allows the sister chromatids to separate. Thus, sister chromatid cohesion and crossing over, acting together, play an essential role in the lining up of chromosomes by homologous pairs at metaphase I.

Meiosis I reduces the number of chromosome sets: from two (diploid) to one (haploid). During the second meiotic division, sister chromatids separate, producing haploid daughter cells. The mechanisms for separating sister chromatids in meiosis II and mitosis are virtually identical. The molecular basis of chromosome behavior during meiosis continues to be a focus of intense research. In the **Scientific Skills Exercise**, you can work with data tracking the amount of DNA in cells as they progress through meiosis.

SCIENTIFIC SKILLS EXERCISE

Making a Line Graph and Converting Between Units of Data

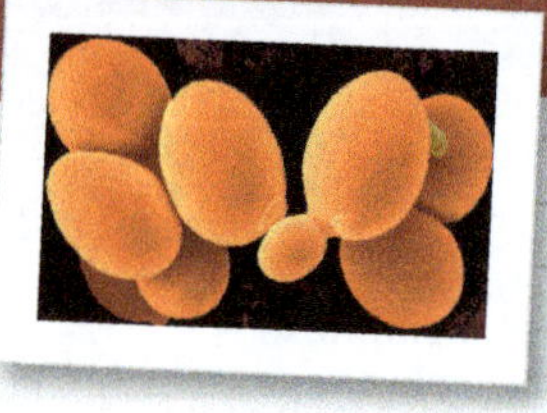

➤ **Budding yeast cells**

How Does DNA Content Change as Budding Yeast Cells Proceed Through Meiosis? When nutrients are low, cells of the budding yeast (*Saccharomyces cerevisiae*) exit the mitotic cell cycle and enter meiosis. In this exercise, you will track the DNA content of a population of yeast cells as they progress through meiosis.

How the Experiment Was Done Researchers grew a culture of yeast cells in a nutrient-rich medium and then transferred the cells to a nutrient-poor medium to induce meiosis. At different times after induction, the DNA content per cell was measured in a sample of the cells, and the average DNA content per cell was recorded in femtograms (fg; 1 femtogram = 1×10^{-15} gram).

Data from the Experiment

Time After Induction (hours)	Average Amount of DNA per Cell (fg)
0.0	24.0
1.0	24.0
2.0	40.0
3.0	47.0
4.0	47.5
5.0	48.0
6.0	48.0
7.0	47.5
7.5	25.0
8.0	24.0
9.0	23.5
9.5	14.0
10.0	13.0
11.0	12.5
12.0	12.0
13.0	12.5
14.0	12.0

INTERPRET THE DATA

1. First, set up your graph. (a) Place the labels for the independent and dependent variables on the appropriate axes, followed by units of measurement in parentheses. Explain your choices. (b) Add tick marks and values for each axis. Explain your choices. (For additional information about graphs, see the Scientific Skills Review in Appendix F.)
2. Because the variable on the *x*-axis varies continuously, it makes sense to plot the data on a line graph. (a) Plot each data point from the table onto the graph. (b) Connect the data points with line segments.
3. Most of the yeast cells in the culture were in G_1 of the cell cycle before being moved to the nutrient-poor medium. (a) How many femtograms of DNA are there in each yeast cell in G_1? Estimate this value from the data in your graph. (b) How many femtograms of DNA should be present in each cell in G_2? (See Concept 12.2 and Figure 12.6.) At the end of meiosis I (MI)? At the end of meiosis II (MII)? (See Figure 13.7.) (c) Using these values as a guideline, distinguish the different phases by inserting vertical dashed lines in the graph between phases and label each phase (G_1, S, G_2, MI, MII). You can figure out where to put the dividing lines based on what you know about the DNA content of each phase (see Figure 13.7). (d) Think carefully about the point where the line at the highest value begins to slope downward. What specific point of meiosis does this "corner" represent? What stage(s) correspond to the downward sloping line?
4. Given the fact that 1 fg of DNA = 9.78×10^5 base pairs (on average), you can convert the amount of DNA per cell to the length of DNA in numbers of base pairs. (a) Calculate the number of base pairs of DNA in the haploid yeast genome. Express your answer in millions of base pairs (Mb), a standard unit for expressing genome size. Show your work. (b) How many base pairs per minute were synthesized during the S phase of these yeast cells?

(MB) **Instructors:** A version of this Scientific Skills Exercise can be assigned in MasteringBiology.

Further Reading G. Simchen, Commitment to meiosis: what determines the mode of division in budding yeast? *BioEssays* 31:169–177 (2009).

CONCEPT CHECK 13.3

1. **MAKE CONNECTIONS >** Compare the chromosomes in a cell at metaphase of mitosis with those in a cell at metaphase II. (See Figures 12.7 and 13.8.)
2. **WHAT IF? >** After the synaptonemal complex disappears, how would any pair of homologous chromosomes be associated if crossing over did not occur? What effect might this have on gamete formation?

For suggested answers, see Appendix A.

CONCEPT 13.4

Genetic variation produced in sexual life cycles contributes to evolution

How do we account for the genetic variation of the family members in Figure 13.1? As you will learn more about in later chapters, mutations are the original source of genetic diversity. These changes in an organism's DNA create the different versions of genes, known as alleles. Once these differences arise, reshuffling of the alleles during sexual reproduction produces the variation that results in each member of a sexually reproducing population having a unique combination of traits.

Origins of Genetic Variation Among Offspring

In species that reproduce sexually, the behavior of chromosomes during meiosis and fertilization is responsible for most of the variation that arises in each generation. Three mechanisms contribute to the genetic variation arising from sexual reproduction: independent assortment of chromosomes, crossing over, and random fertilization.

Independent Assortment of Chromosomes

One aspect of sexual reproduction that generates genetic variation is the random orientation of pairs of homologous chromosomes at metaphase of meiosis I. At metaphase I, the homologous pairs, each consisting of one maternal and one paternal chromosome, are situated at the metaphase plate. (Note that the terms *maternal* and *paternal* refer, respectively, to whether the chromosome in question was contributed by the mother or the father of the individual whose cells are undergoing meiosis.) Each pair may orient with either its maternal or paternal homolog closer to a given pole—its orientation is as random as the flip of a coin. Thus, there is a 50% chance that a particular daughter cell of meiosis I will get the maternal chromosome of a certain homologous pair and a 50% chance that it will get the paternal chromosome.

Because each pair of homologous chromosomes is positioned independently of the other pairs at metaphase I, the first meiotic division results in each pair sorting its maternal and paternal homologs into daughter cells independently of every other pair. This is called *independent assortment*. Each daughter cell represents one outcome of all possible combinations of maternal and paternal chromosomes. As shown in **Figure 13.11**, the number of combinations possible for daughter cells formed by meiosis of a diploid cell with two pairs of homologous chromosomes ($n = 2$) is four: two possible arrangements for the first pair times two possible arrangements for the second pair. Note that only two of the four combinations of daughter cells shown in the figure would result from meiosis of a *single* diploid cell, because a single parent cell would have one or the other possible chromosomal arrangement at metaphase I, but not both. However, the population of daughter cells resulting from meiosis of a large number of diploid cells contains all four types in approximately equal numbers. In the case of $n = 3$, eight combinations ($2 \times 2 \times 2 = 2^3$) of chromosomes are possible for daughter cells. More generally, the number of possible combinations when chromosomes sort independently during meiosis is 2^n, where n is the haploid number of the organism.

In the case of humans ($n = 23$), the number of possible combinations of maternal and paternal chromosomes in the resulting gametes is 2^{23}, or about 8.4 million. Each gamete that you produce in your lifetime contains one of roughly 8.4 million possible combinations of chromosomes. This is an underestimate, because it doesn't take into account crossing over, which we'll consider next.

Crossing Over

As a consequence of the independent assortment of chromosomes during meiosis, each of us produces a collection

▼ Figure 13.11 The independent assortment of homologous chromosomes in meiosis.

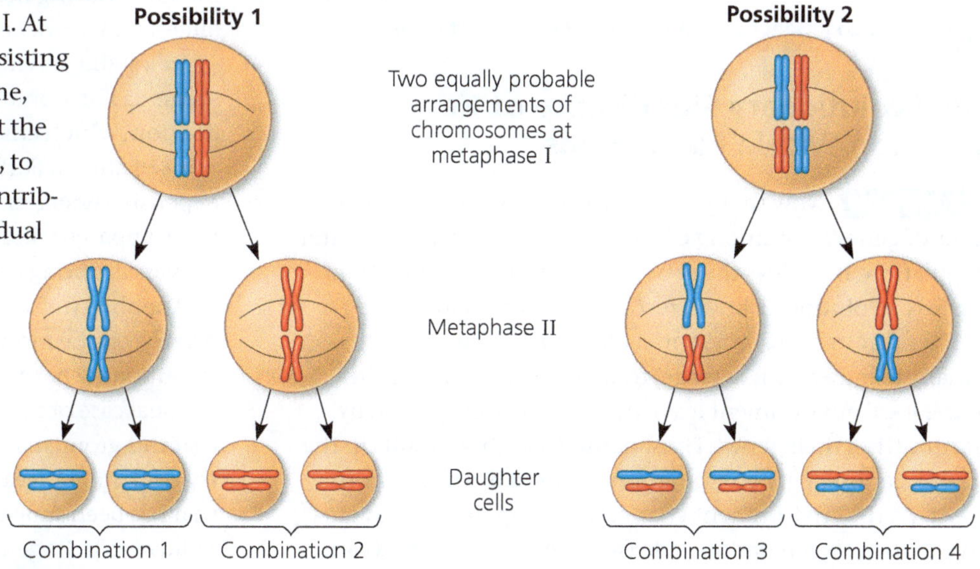

Animation: Genetic Variation from Independent Assortment of Chromosomes

of gametes differing greatly in their combinations of the chromosomes we inherited from our two parents. Figure 13.11 suggests that each chromosome in a gamete is exclusively maternal or paternal in origin. In fact, this is *not* the case, because crossing over produces **recombinant chromosomes**, individual chromosomes that carry genes (DNA) from two different parents **(Figure 13.12)**. In meiosis in humans, an average of one to three crossover events occurs per chromosome pair, depending on the size of the chromosomes and the position of their centromeres.

As you learned in Figure 13.9, crossing over produces chromosomes with new combinations of maternal and paternal alleles. At metaphase II, chromosomes that contain one or more recombinant chromatids can be oriented in two alternative, nonequivalent ways with respect to other chromosomes because their sister chromatids are no longer identical (see Figure 13.12). The different possible arrangements of nonidentical sister chromatids during meiosis II further increase the number of genetic types of daughter cells that can result from meiosis.

You'll learn more about crossing over in Chapter 15. The important point for now is that crossing over, by combining DNA inherited from two parents into a single chromosome, is an important source of genetic variation in sexual life cycles.

Random Fertilization

The random nature of fertilization adds to the genetic variation arising from meiosis. In humans, each male and female gamete represents one of about 8.4 million (2^{23}) possible chromosome combinations due to independent assortment. The fusion of a male gamete with a female gamete during fertilization will produce a zygote with any of about 70 trillion ($2^{23} \times 2^{23}$) diploid combinations. If we factor in the variation brought about by crossing over, the number of possibilities is truly astronomical. It may sound trite, but you really *are* unique.

MB **Animation: Genetic Variation from Random Fertilization**

The Evolutionary Significance of Genetic Variation Within Populations

EVOLUTION Now that you've learned how new combinations of genes arise among offspring in a sexually reproducing population, how does the genetic variation in a population relate to evolution? Darwin recognized that a population evolves through the differential reproductive success of its variant members. On average, those individuals best suited to the local environment leave the most offspring, thereby transmitting their genes. Thus, natural selection results in the accumulation of genetic variations favored by the environment. As the environment changes, the population may survive if, in each generation, at least some of its members can cope effectively with the new conditions. Mutations are the original source of different alleles, which are then mixed and matched during meiosis. New and different combinations of alleles may work better than those that previously prevailed.

Figure 13.12 The results of crossing over during meiosis.

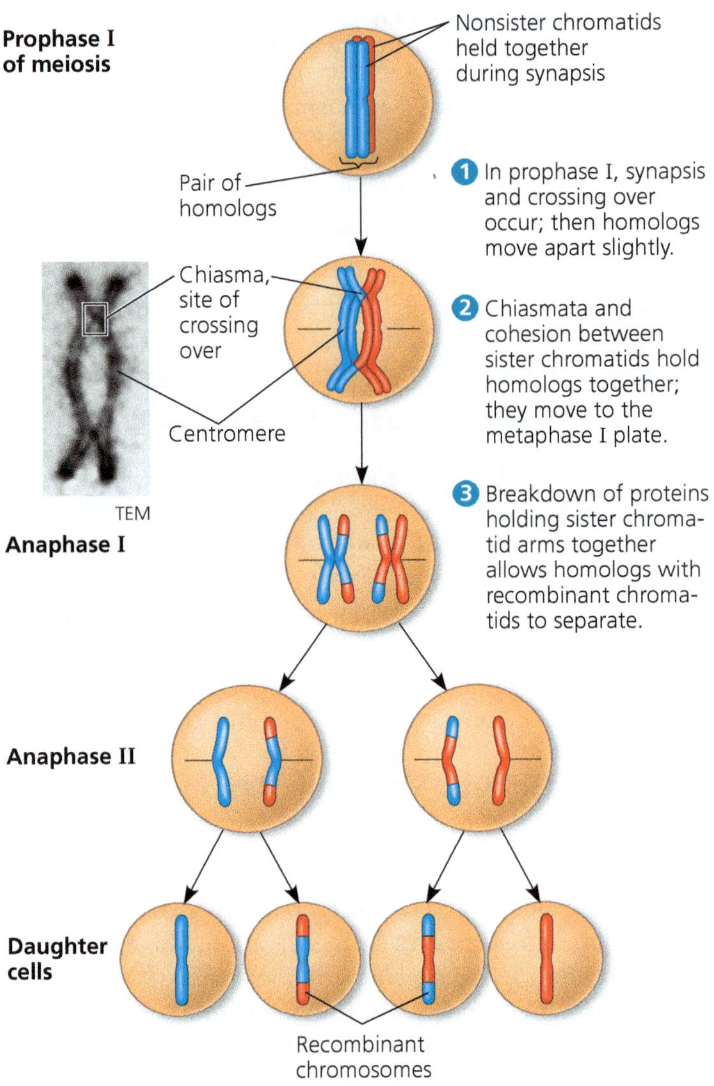

MB **Animation: Genetic Variation from Crossing Over**

In a stable environment, though, sexual reproduction seems as if it would be less advantageous than asexual reproduction, which ensures perpetuation of successful combinations of alleles. Furthermore, sexual reproduction is more expensive energetically than asexual reproduction. In spite of these apparent disadvantages, sexual reproduction is almost universal among animals. Why is this?

The ability of sexual reproduction to generate genetic diversity is the most commonly proposed explanation for the evolutionary persistence of this process. However, consider the unusual case of the bdelloid rotifer **(Figure 13.13)**. It appears that this group may not have reproduced sexually for more than 50 million years of their evolutionary history, a model that has been supported by recent analysis of the genetic sequences in its genome. Does this mean that genetic diversity is not advantageous in this species? It turns out that bdelloid rotifers are an exception to the "rule" that sex alone generates genetic diversity: Bdelloids have mechanisms other than sexual reproduction for generating genetic diversity. For example, they live in environments that can dry up for long periods of time, during which they can enter a state of suspended animation. In this state, their cell membranes may crack in places, allowing entry of DNA from other rotifer species and even from more distantly related species. Evidence suggests that this foreign DNA can become incorporated into the genome of the bdelloid, leading to increased genetic diversity. In fact, the genomic analysis shows that bdelloid rotifers pick up non-bdelloid DNA at a much higher rate than most other species pick up foreign DNA. The conclusion that bdelloid rotifers have developed other ways of achieving genetic diversity supports the idea that genetic diversity is advantageous, but that sexual reproduction is not the only way of generating such diversity.

▼ Figure 13.13 A bdelloid rotifer, an animal that reproduces only asexually.

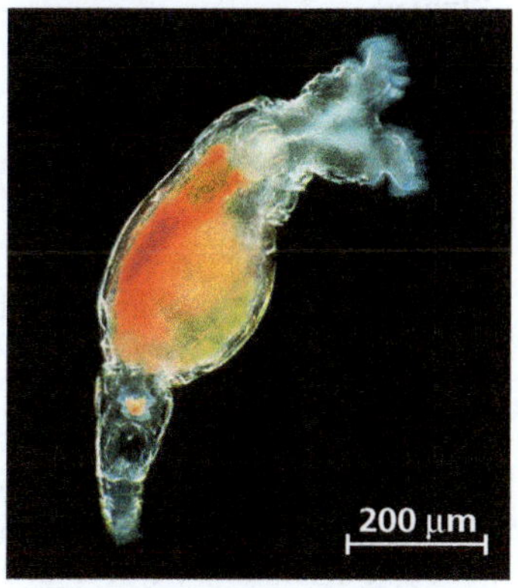

In this chapter, we have seen how sexual reproduction greatly increases the genetic variation present in a population. Although Darwin realized that heritable variation is what makes evolution possible, he could not explain why offspring resemble—but are not identical to—their parents. Ironically, Gregor Mendel, a contemporary of Darwin, published a theory of inheritance that helps explain genetic variation, but his discoveries had no impact on biologists until 1900, more than 15 years after Darwin (1809–1882) and Mendel (1822–1884) had died. In the next chapter, you'll learn how Mendel discovered the basic rules governing the inheritance of specific traits.

CONCEPT CHECK 13.4

1. What is the original source of variation among the different alleles of a gene?
2. The diploid number for fruit flies is 8, and the diploid number for grasshoppers is 46. If no crossing over took place, would the genetic variation among offspring from a given pair of parents be greater in fruit flies or grasshoppers? Explain.
3. **WHAT IF? ➤** If maternal and paternal chromatids have the same two alleles for every gene, will crossing over lead to genetic variation?

For suggested answers, see Appendix A.

13 Chapter Review

Go to **MasteringBiology**™ for Videos, Animations, Vocab Self-Quiz, Practice Tests, and more in the Study Area.

SUMMARY OF KEY CONCEPTS

CONCEPT 13.1

Offspring acquire genes from parents by inheriting chromosomes *(pp. 255–256)*

VOCAB SELF-QUIZ goo.gl/6u55ks

- Each **gene** in an organism's DNA exists at a specific **locus** on a certain chromosome.
- In **asexual reproduction**, a single parent produces genetically identical offspring by mitosis. **Sexual reproduction** combines genes from two parents, leading to genetically diverse offspring.

? *Explain why human offspring resemble their parents but are not identical to them.*

CONCEPT 13.2

Fertilization and meiosis alternate in sexual life cycles *(pp. 256–259)*

- Normal human **somatic cells** are **diploid**. They have 46 chromosomes made up of two sets of 23, one set from each parent. Human diploid cells have 22 pairs of **homologs** that are **autosomes**, and one pair of **sex chromosomes**; the latter typically determines whether the person is female (XX) or male (XY).
- In humans, ovaries and testes produce **haploid gametes** by **meiosis**, each gamete containing a single set of 23 chromosomes ($n = 23$). During **fertilization**, an egg and sperm unite, forming a diploid ($2n = 46$) single-celled **zygote**, which develops into a multicellular organism by mitosis.
- Sexual life cycles differ in the timing of meiosis relative to fertilization and in the point(s) of the cycle at which a multicellular organism is produced by mitosis.

? *Compare the life cycles of animals and plants, mentioning their similarities and differences.*

CONCEPT 13.3

Meiosis reduces the number of chromosome sets from diploid to haploid *(pp. 259–265)*

- The two cell divisions of meiosis, **meiosis I** and **meiosis II**, produce four haploid daughter cells. The number of chromosome sets is reduced from two (diploid) to one (haploid) during meiosis I.
- Meiosis is distinguished from mitosis by three events of meiosis I:

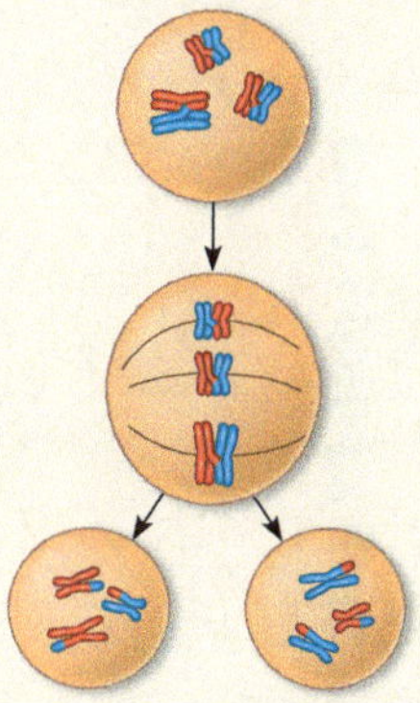

Prophase I: Each pair of homologous chromosomes undergoes **synapsis** and **crossing over** between nonsister chromatids with the subsequent appearance of **chiasmata**.

Metaphase I: Chromosomes line up as homologous pairs on the metaphase plate.

Anaphase I: Homologs separate from each other; sister chromatids remain joined at the centromere.

Meiosis II separates the sister chromatids.

- Sister chromatid cohesion and crossing over allow chiasmata to hold homologs together until anaphase I. Cohesins are cleaved along the arms at anaphase I, allowing homologs to separate, and at the centromeres in anaphase II, releasing sister chromatids.

? *In prophase I, homologous chromosomes pair up and undergo synapsis and crossing over. Can this also occur during prophase II? Explain.*

CONCEPT 13.4

Genetic variation produced in sexual life cycles contributes to evolution *(pp. 265–267)*

- Three events in sexual reproduction contribute to genetic variation in a population: independent assortment of chromosomes during meiosis I, crossing over during meiosis I, and random fertilization of egg cells by sperm. During crossing over, DNA of nonsister chromatids in a homologous pair is broken and rejoined.
- Genetic variation is the raw material for evolution by natural selection. Mutations are the original source of this variation; recombination of variant genes generates additional genetic diversity.

? *Explain how three processes unique to sexual reproduction generate a great deal of genetic variation.*

TEST YOUR UNDERSTANDING

Level 1: Knowledge/Comprehension

PRACTICE TEST
goo.gl/CUYGKD

1. A human cell containing 22 autosomes and a Y chromosome is
(A) a sperm.
(B) an egg.
(C) a zygote.
(D) a somatic cell of a male.

2. The two homologs of a pair move toward opposite poles of dividing cell during
(A) mitosis.
(B) meiosis I.
(C) meiosis II.
(D) fertilization.

Level 2: Application/Analysis

3. Meiosis II is similar to mitosis in that
(A) sister chromatids separate during anaphase.
(B) DNA replicates before the division.
(C) the daughter cells are diploid.
(D) homologous chromosomes synapse.

4. If the DNA content of a diploid cell in the G_1 phase of the cell cycle is x, then the DNA content of the same cell at metaphase of meiosis I will be
(A) $0.25x$. (B) $0.5x$. (C) x. (D) $2x$.

5. If we continue to follow the cell lineage from question 4, then the DNA content of a single cell at metaphase of meiosis II will be
(A) $0.25x$. (B) $0.5x$. (C) x. (D) $2x$.

6. **DRAW IT** The diagram shows a cell in meiosis.

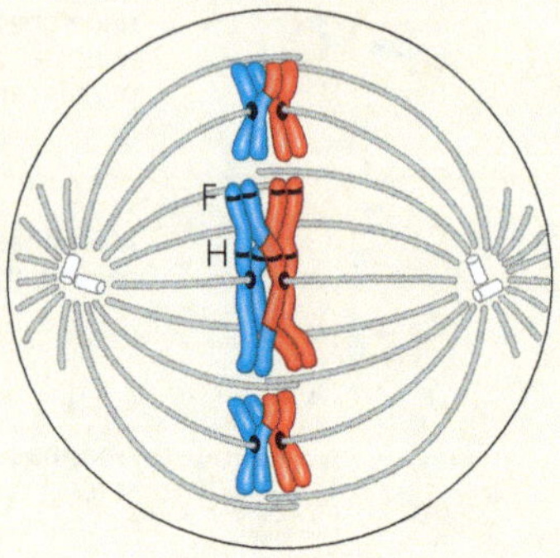

(a) Label the appropriate structures with these terms: chromosome (label as duplicated or unduplicated), centromere, kinetochore, sister chromatids, nonsister chromatids, homologous pair (use a bracket when labeling), homolog (label each one), chiasma, sister chromatid cohesion, and gene loci, labeling the alleles of the F and H genes.
(b) Describe the makeup of a haploid set and a diploid set.
(c) Identify the stage of meiosis shown.

Level 3: Synthesis/Evaluation

7. Explain how you can tell that the cell in question 6 is undergoing meiosis, not mitosis.

8. **EVOLUTION CONNECTION** Many species can reproduce either asexually or sexually. Explain what you think might be the evolutionary significance of the switch from asexual to sexual reproduction that occurs in some organisms when the environment becomes unfavorable.

9. **SCIENTIFIC INQUIRY** The diagram in question 6 represents just a few of the chromosomes of a meiotic cell in a certain person. Assume the freckles gene is located at the locus marked F, and the hair-color gene is located at the locus marked H, both on the long chromosome. The individual from whom this cell was taken has inherited different alleles for each gene ("freckles" and "black hair" from one parent, and "no freckles" and "blond hair" from the other). Predict allele combinations in the gametes resulting from this meiotic event. (It will help if you draw out the rest of meiosis and label the alleles by name.) List other possible combinations of these alleles in this individual's gametes.

10. **WRITE ABOUT A THEME: INFORMATION** The continuity of life is based on heritable information in the form of DNA. In a short essay (100–150 words), explain how chromosome behavior during sexual reproduction in animals ensures perpetuation of parental traits in offspring and, at the same time, genetic variation among offspring.

11. **SYNTHESIZE YOUR KNOWLEDGE**

The Cavendish banana, the world's most popular fruit, is currently threatened by extinction due to a fungus. This banana variety is "triploid" ($3n$, with three sets of chromosomes) and can only reproduce through cloning by cultivators. Given what you know about meiosis, explain how the banana's triploid number accounts for its inability to form normal gametes. Considering genetic diversity, discuss how the absence of sexual reproduction might make this domesticated species vulnerable to infectious agents.

For selected answers, see Appendix A.

For additional practice questions, check out the **Dynamic Study Modules** in MasteringBiology. You can use them to study on your smartphone, tablet, or computer anytime, anywhere!

Mendel and the Gene Idea

14

▲ **Figure 14.1 What principles of inheritance did Gregor Mendel discover by breeding pea plants?**

KEY CONCEPTS

14.1 Mendel used the scientific approach to identify two laws of inheritance

14.2 Probability laws govern Mendelian inheritance

14.3 Inheritance patterns are often more complex than predicted by simple Mendelian genetics

14.4 Many human traits follow Mendelian patterns of inheritance

Drawing from the Deck of Genes

The crowd at a soccer match attests to the marvelous variety and diversity of humankind. Brown, blue, or gray eyes; black, brown, or blond hair—these are just a few examples of heritable variations that we may observe. What principles account for the transmission of such traits from parents to offspring?

The explanation of heredity most widely in favor during the 1800s was the "blending" hypothesis, the idea that genetic material contributed by the two parents mixes just as blue and yellow paints blend to make green. This hypothesis predicts that over many generations, a freely mating population will give rise to a uniform population of individuals, something we don't see. The blending hypothesis also fails to explain how traits can reappear after they've skipped a generation.

An alternative to the blending model is a "particulate" hypothesis of inheritance: the gene idea. In this model, parents pass on discrete heritable units—genes—that retain their separate identities in offspring. An organism's collection of genes is more like a deck of cards than a pail of paint. Like cards, genes can be shuffled and passed along, generation after generation, in undiluted form.

Modern genetics had its genesis in an abbey garden, where a monk named Gregor Mendel documented a particulate mechanism for inheritance using pea plants (**Figure 14.1**). Mendel developed his theory of inheritance several decades

When you see this blue icon, log in to **MasteringBiology** and go to the Study Area for digital resources.

MB Get Ready for This Chapter

◀ **Mendel (third from right, holding a sprig of fuchsia) with his fellow monks.**

"self-pollinate"? see p. 271 usage

before chromosomes were observed under the microscope and the significance of their behavior during mitosis or meiosis was understood. In this chapter, we'll step into Mendel's garden to re-create his experiments and explain how he arrived at his theory of inheritance. We'll also explore inheritance patterns more complex than those observed by Mendel in garden peas. Finally, we will see how the Mendelian model applies to the inheritance of human variations, including hereditary disorders such as sickle-cell disease.

CONCEPT 14.1

Mendel used the scientific approach to identify two laws of inheritance

Mendel discovered the basic principles of heredity by breeding garden peas in carefully planned experiments. As we retrace his work, you will recognize the key elements of the scientific process that were introduced in Chapter 1.

Mendel's Experimental, Quantitative Approach

Mendel grew up on his parents' small farm in a region of Austria that is now part of the Czech Republic. In this agricultural area, Mendel and the other children received agricultural training in school along with their basic education. As an adolescent, Mendel overcame financial hardship and illness to excel in high school and, later, at the Olmutz Philosophical Institute.

In 1843, at the age of 21, Mendel entered an Augustinian monastery, a reasonable choice at that time for someone who valued the life of the mind. He considered becoming a teacher but failed the necessary examination. In 1851, he left the monastery to pursue two years of study in physics and chemistry at the University of Vienna. These were very important years for Mendel's development as a scientist, in large part due to the strong influence of two professors. One was the physicist Christian Doppler, who encouraged his students to learn science through experimentation and trained Mendel to use mathematics to help explain natural phenomena. The other was a botanist named Franz Unger, who aroused Mendel's interest in the causes of variation in plants.

After attending the university, Mendel returned to the monastery and was assigned to teach at a local school, where several other instructors were enthusiastic about scientific research. In addition, his fellow monks shared a long-standing fascination with the breeding of plants. Around 1857, Mendel began breeding garden peas in the abbey garden to study inheritance. Although the question of heredity had long been a focus of curiosity at the monastery, Mendel's fresh approach allowed him to deduce principles that had remained elusive to others.

One reason Mendel probably chose to work with peas is that there are many varieties. For example, one variety has purple flowers, while another variety has white flowers. A heritable feature that varies among individuals, such as flower color, is called a **character**. Each variant for a character, such as purple or white color for flowers, is called a **trait**.

Other advantages of using peas are their short generation time and the large number of offspring from each mating. Furthermore, Mendel could strictly control mating between plants (**Figure 14.2**). Each pea flower has both pollen-producing organs (stamens) and an egg-bearing organ (carpel). In nature, pea plants usually self-fertilize: Pollen grains from the stamens land on the carpel of the same flower, and sperm released

Figure 14.2

Research Method **Crossing Pea Plants**

Application By crossing (mating) two true-breeding varieties of an organism, scientists can study patterns of inheritance. In this example, Mendel crossed pea plants that varied in flower color.

Technique

Results When pollen from a white flower was transferred to a purple flower, the first-generation hybrids all had purple flowers. The result was the same for the reciprocal cross, which involved the transfer of pollen from purple flowers to white flowers.

from the pollen grains fertilize eggs present in the carpel.*
To achieve cross-pollination of two plants, Mendel removed the immature stamens of a plant before they produced pollen and then dusted pollen from another plant onto the altered flowers (see Figure 14.2). Each resulting zygote then developed into a plant embryo encased in a seed (pea). His method allowed Mendel to always be sure of the parentage of new seeds.

Mendel chose to track only those characters that occurred in two distinct, alternative forms, such as purple or white flower color. He also made sure that he started his experiments with varieties that were **true-breeding**—that is, over many generations of self-pollination, these plants had produced only the same variety as the parent plant. For example, a plant with purple flowers is true-breeding if the seeds produced by self-pollination in successive generations all give rise to plants that also have purple flowers.

In a typical breeding experiment, Mendel cross-pollinated two contrasting, true-breeding pea varieties—for example, purple-flowered plants and white-flowered plants (see Figure 14.2). This mating, or *crossing*, of two true-breeding varieties is called **hybridization**. The true-breeding parents are referred to as the **P generation** (parental generation), and their hybrid offspring are the **F_1 generation** (first filial generation, the word *filial* from the Latin word for "son"). Allowing these F_1 hybrids to self-pollinate (or to cross-pollinate with other F_1 hybrids) produces an **F_2 generation** (second filial generation). Mendel usually followed traits for at least the P, F_1, and F_2 generations. Had Mendel stopped his experiments with the F_1 generation, the basic patterns of inheritance would have eluded him. Mendel's quantitative analysis of the F_2 plants from thousands of genetic crosses like these allowed him to deduce two fundamental principles of heredity, now called the law of segregation and the law of independent assortment.

The Law of Segregation

If the blending model of inheritance were correct, the F_1 hybrids from a cross between purple-flowered and white-flowered pea plants would have pale purple flowers, a trait intermediate between those of the P generation. Notice in Figure 14.2 that the experiment produced a very different result: All the F_1 offspring had flowers of the same color as the purple-flowered parents. What happened to the white-flowered plants' genetic contribution to the hybrids? If it were lost, then the F_1 plants could produce only purple-flowered offspring in the F_2 generation. But when Mendel allowed the F_1 plants to self- or cross-pollinate and planted their seeds, the white-flower trait reappeared in the F_2 generation.

Mendel used very large sample sizes and kept accurate records of his results: 705 of the F_2 plants had purple flowers, and 224 had white flowers. These data fit a ratio of approximately three purple to one white **(Figure 14.3)**. Mendel reasoned that the heritable factor for white flowers did not disappear in the F_1 plants but was somehow hidden, or masked, when the purple-flower factor was present. In Mendel's terminology, purple flower color is a *dominant* trait, and white flower color is a *recessive* trait. The reappearance of white-flowered plants in the F_2 generation was evidence

▼ Figure 14.3

Inquiry When F_1 hybrid pea plants self- or cross-pollinate, which traits appear in the F_2 generation?

Experiment Mendel crossed true-breeding purple-flowered plants and white-flowered plants (crosses are symbolized by ×). The resulting F_1 hybrids were allowed to self-pollinate or were cross-pollinated with other F_1 hybrids. The F_2 generation plants were then observed for flower color.

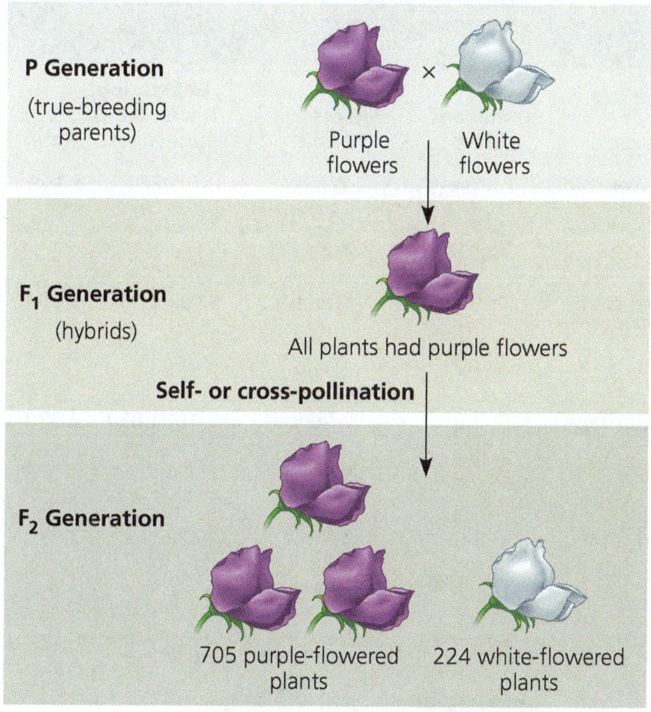

Results Both purple-flowered and white-flowered plants appeared in the F_2 generation, in a ratio of approximately 3:1.

Conclusion The "heritable factor" for the recessive trait (white flowers) had not been destroyed, deleted, or "blended" in the F_1 generation but was merely masked by the presence of the factor for purple flowers, which is the dominant trait.

Data from G. Mendel, Experiments in plant hybridization, *Proceedings of the Natural History Society of Brünn* 4:3–47 (1866).

WHAT IF? ➤ *If you mated two purple-flowered plants from the P generation, what ratio of traits would you expect to observe in the offspring? Explain. What might Mendel have concluded if he stopped his experiment after the F_1 generation?*

*As you learned in Figure 13.6b, meiosis in plants produces spores, not gametes. In flowering plants like the pea, each spore develops into a microscopic haploid gametophyte that contains only a few cells and is located on the parent plant. The gametophyte produces sperm, in pollen grains, and eggs, in the carpel. For simplicity, we will not include the gametophyte stage in our discussion of fertilization in plants.

that the heritable factor causing white flowers had not been diluted or destroyed by coexisting with the purple-flower factor in the F_1 hybrids. Instead, it had been hidden when in the presence of the purple-flower factor.

Mendel observed the same pattern of inheritance in six other characters, each represented by two distinctly different traits **(Table 14.1)**. For example, when Mendel crossed a true-breeding variety that produced smooth, round pea seeds with one that produced wrinkled seeds, all the F_1 hybrids produced round seeds; this is the dominant trait for seed shape. In the F_2 generation, approximately 75% of the seeds were round and 25% were wrinkled—a 3:1 ratio, as in Figure 14.3. Now let's see how Mendel deduced the law of segregation from his experimental results. In the discussion that follows, we will use modern terms instead of some of the terms used by Mendel. (For example, we'll use "gene" instead of Mendel's "heritable factor.")

Table 14.1 The Results of Mendel's F_1 Crosses for Seven Characters in Pea Plants

Character	Dominant Trait	×	Recessive Trait	F_2 Generation Dominant: Recessive	Ratio
Flower color	Purple 	×	White 	705:224	3.15:1
Seed color	Yellow 	×	Green 	6,022:2,001	3.01:1
Seed shape	Round 	×	Wrinkled 	5,474:1,850	2.96:1
Pod color	Green 	×	Yellow 	428:152	2.82:1
Pod shape	Inflated 	×	Constricted 	882:299	2.95:1
Flower position	Axial 	×	Terminal 	651:207	3.14:1
Stem length	Tall 	×	Dwarf 	787:277	2.84:1

Mendel's Model

Mendel developed a model to explain the 3:1 inheritance pattern that he consistently observed among the F_2 offspring in his pea experiments. We describe four related concepts making up this model, the fourth of which is the law of segregation.

First, *alternative versions of genes account for variations in inherited characters.* The gene for flower color in pea plants, for example, exists in two versions, one for purple flowers and the other for white flowers. These alternative versions of a gene are called **alleles**. Today, we can relate this concept to chromosomes and DNA. As shown in **Figure 14.4**, each gene is a sequence of nucleotides at a specific place, or locus, along a particular chromosome. The DNA at that locus, however, can vary slightly in its nucleotide sequence. This variation in information content can affect the function of the encoded protein and thus an inherited character of the organism. The purple-flower allele and the white-flower allele are two DNA sequence variations possible at the flower-color locus on a pea plant's chromosomes. The purple-flower allele sequence allows synthesis of purple pigment, and the white-flower allele sequence does not.

Second, *for each character, an organism inherits two copies (that is, two alleles) of a gene, one from each parent.* Remarkably, Mendel made this deduction without knowing about the role, or even the existence, of chromosomes. Each somatic cell in a diploid organism has two sets of chromosomes, one set inherited from each parent (see Figure 13.4). Thus, a genetic locus is actually represented twice in a diploid cell, once on each homolog of a specific pair of chromosomes. The two alleles at a particular locus may be identical, as in the true-breeding plants of Mendel's P generation. Or the alleles may differ, as in the F_1 hybrids (see Figure 14.4).

Third, *if the two alleles at a locus differ, then one, the* ***dominant allele****, determines the organism's appearance; the other, the* ***recessive allele****, has no noticeable effect on the organism's appearance*. Accordingly, Mendel's F_1 plants had purple flowers because the allele for that trait is dominant and the allele for white flowers is recessive.

The fourth and final part of Mendel's model, the **law of segregation**, states that *the two alleles for a heritable character segregate (separate from each other) during gamete formation and end up in different gametes*. Thus, an egg or a sperm gets only one of the two alleles that are present in the somatic cells of the organism making the gamete. In terms of chromosomes, this segregation corresponds to the distribution of copies of the two members of a pair of homologous chromosomes to different gametes in meiosis (see Figure 13.7). Note that if an organism has identical alleles for a particular character then that allele is present in all gametes. Because it is the only allele that can be passed on to offspring, the offspring always look like their parents; this explains why these plants are true-breeding. But if different alleles are present, as in the F_1 hybrids, then 50% of the gametes receive the dominant allele and 50% receive the recessive allele.

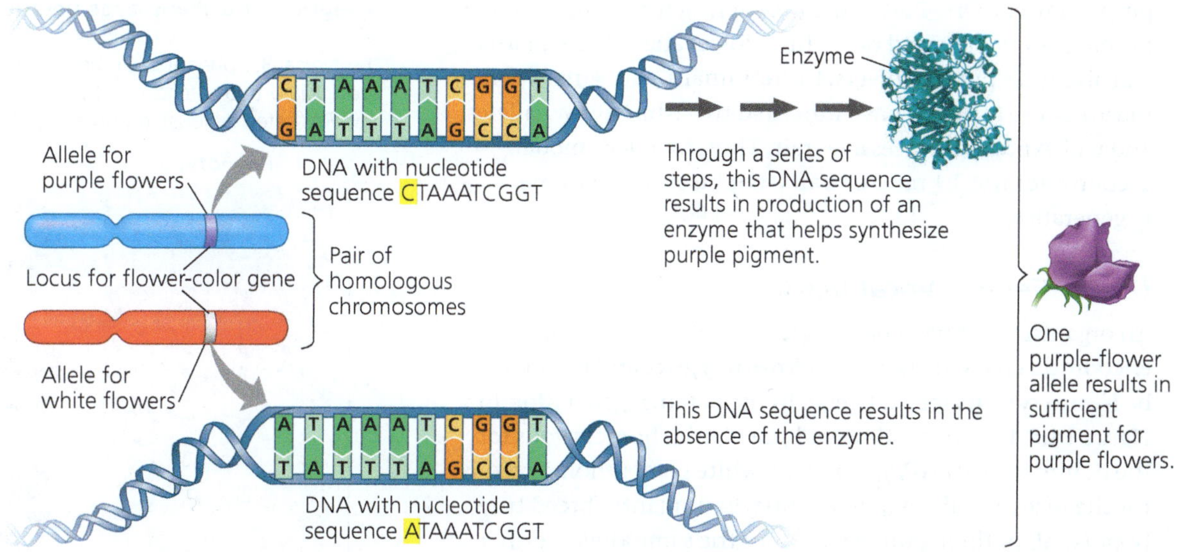

➤ Figure 14.4 Alleles, alternative versions of a gene. This diagram shows a pair of homologous chromosomes in an F_1 hybrid pea plant, with the actual DNA sequence from the flower-color allele of each chromosome. The paternally inherited chromosome (blue) has an allele for purple flowers, which codes for a protein that indirectly controls synthesis of purple pigment. The maternally inherited chromosome (red) has an allele for white flowers, which results in no functional protein being made.

Does Mendel's segregation model account for the 3:1 ratio he observed in the F_2 generation of his numerous crosses? For the flower-color character, the model predicts that the two different alleles present in an F_1 individual will segregate into gametes such that half the gametes will have the purple-flower allele and half will have the white-flower allele. During self-pollination, gametes of each class unite randomly. An egg with a purple-flower allele has an equal chance of being fertilized by a sperm with a purple-flower allele or by one with a white-flower allele. Since the same is true for an egg with a white-flower allele, there are four equally likely combinations of sperm and egg. **Figure 14.5** illustrates these combinations using a **Punnett square**, a handy diagrammatic device for predicting the allele composition of offspring from a cross between individuals of known genetic makeup. Notice that we use a capital letter to symbolize a dominant allele and a lowercase letter for a recessive allele. In our example, *P* is the purple-flower allele, and *p* is the white-flower allele; it is often useful as well to be able to refer to the gene itself as the *P*/*p* gene.

In the F_2 offspring, what color will the flowers be? One-fourth of the plants have inherited two purple-flower alleles; clearly, these plants will have

▼ Figure 14.5 Mendel's law of segregation. This diagram shows the genetic makeup of the generations in Figure 14.3. It illustrates Mendel's model for inheritance of the alleles of a single gene. Each plant has two alleles for the gene controlling flower color, one allele inherited from each of the plant's parents. To construct a Punnett square that predicts the F_2 generation offspring, we list all the possible gametes from one parent (here, the F_1 female) along the left side of the square and all the possible gametes from the other parent (here, the F_1 male) along the top. The boxes represent the offspring resulting from all the possible unions of male and female gametes.

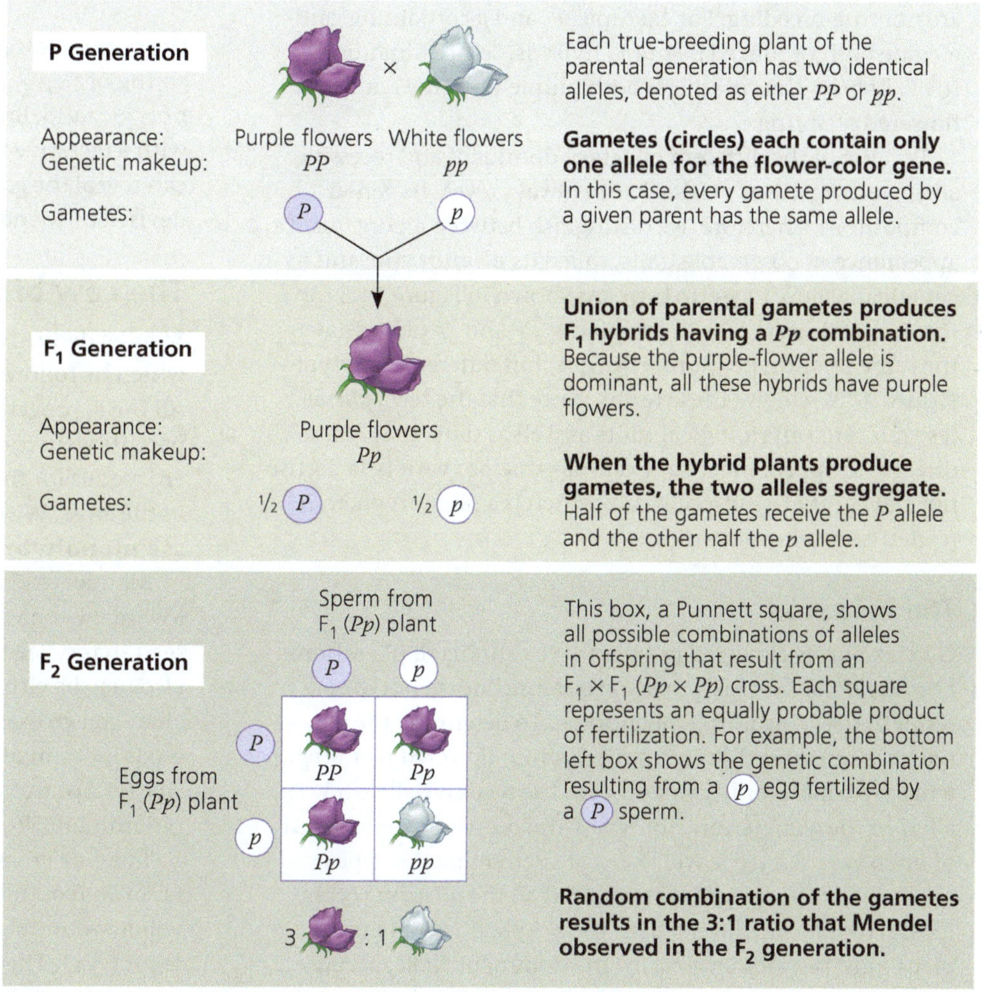

MB Animation: Mendel's Cross of One Character: Flower Color
Animation: Simplified Crosses of One Character in Humans
Animation: Cross of One Character in "MendAliens"

purple flowers. One-half of the F_2 offspring have inherited one purple-flower allele and one white-flower allele; these plants will also have purple flowers, the dominant trait. Finally, one-fourth of the F_2 plants have inherited two white-flower alleles and will express the recessive trait. Thus, Mendel's model accounts for the 3:1 ratio of traits that he observed in the F_2 generation.

Useful Genetic Vocabulary

An organism that has a pair of identical alleles for a gene encoding a character is called a **homozygote** and is said to be **homozygous** for that gene. In the parental generation in Figure 14.5, the purple-flowered pea plant is homozygous for the dominant allele (*PP*), while the white plant is homozygous for the recessive allele (*pp*). Homozygous plants "breed true" because all of their gametes contain the same allele—either *P* or *p* in this example. If we cross dominant homozygotes with recessive homozygotes, every offspring will have two different alleles—*Pp* in the case of the F_1 hybrids of our flower-color experiment (see Figure 14.5). An organism that has two different alleles for a gene is called a **heterozygote** and is said to be **heterozygous** for that gene. Unlike homozygotes, heterozygotes produce gametes with different alleles, so they are not true-breeding. For example, *P*- and *p*-containing gametes are both produced by our F_1 hybrids. Self-pollination of the F_1 hybrids thus produces both purple-flowered and white-flowered offspring.

Because of the different effects of dominant and recessive alleles, an organism's traits do not always reveal its genetic composition. Therefore, we distinguish between an organism's appearance or observable traits, called its **phenotype**, and its genetic makeup, its **genotype**. As shown in Figure 14.5 for the case of flower color in pea plants, *PP* and *Pp* plants have the same phenotype (purple flowers) but different genotypes. **Figure 14.6** reviews these terms. Note that the term *phenotype* refers to physiological traits as well as traits that relate directly to appearance. For example, one pea variety lacks the normal ability to self-pollinate, which is a phenotypic trait (called non-self-pollination).

The Testcross

Given a purple-flowered pea plant, we cannot tell if it is homozygous (*PP*) or heterozygous (*Pp*) because both genotypes result in the same purple phenotype. To determine the genotype, we can cross this plant with a white-flowered plant (*pp*), which will make only gametes with the recessive allele (*p*). The allele in the gamete contributed by the purple-flowered plant of unknown genotype will therefore determine the appearance of the offspring **(Figure 14.7)**. If all the offspring of the cross have purple flowers, then the purple-flowered mystery plant must be homozygous for the dominant allele, because a *PP* × *pp* cross produces all *Pp* offspring. But if both the purple and the white phenotypes appear among the offspring, then the purple-flowered parent must be heterozygous. The offspring of a *Pp* × *pp* cross will be expected to have a 1:1 phenotypic ratio. Breeding an organism of unknown genotype with a recessive homozygote is called a **testcross** because it can reveal the genotype of that organism. The testcross was devised by Mendel and continues to be used by geneticists.

▼ Figure 14.6 Phenotype versus genotype. Grouping F_2 offspring from a cross for flower color according to phenotype results in the typical 3:1 phenotypic ratio. In terms of genotype, however, there are actually two categories of purple-flowered plants, *PP* (homozygous) and *Pp* (heterozygous), giving a 1:2:1 genotypic ratio.

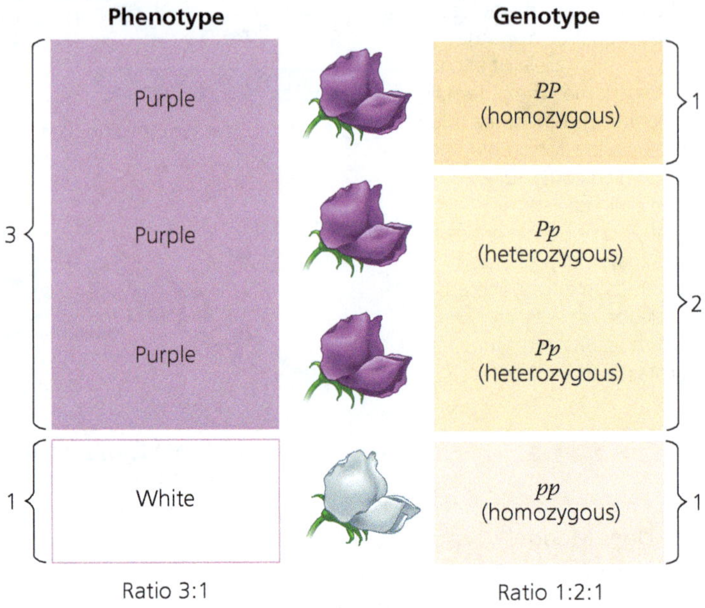

The Law of Independent Assortment

Mendel derived the law of segregation from experiments in which he followed only a *single* character, such as flower color. All the F_1 progeny produced in his crosses of true-breeding parents were **monohybrids**, meaning that they were heterozygous for the one particular character being followed in the cross. We refer to a cross between such heterozygotes as a **monohybrid cross**.

Mendel worked out the second law of inheritance by following *two* characters at the same time, such as seed color and seed shape. Seeds (peas) may be either yellow or green. They also may be either round (smooth) or wrinkled. From single-character crosses, Mendel knew that the allele for yellow seeds is dominant (*Y*), and the allele for green seeds is recessive (*y*). For the seed-shape character, the allele for round is dominant (*R*), and the allele for wrinkled is recessive (*r*).

Imagine crossing two true-breeding pea varieties that differ in *both* of these characters—a cross between a plant with yellow-round seeds (*YYRR*) and a plant with green-wrinkled seeds (*yyrr*). The F_1 plants will be **dihybrids**, individuals heterozygous for the two characters being followed in the cross (*YyRr*). But are these two characters transmitted from parents

Figure 14.7

Research Method **The Testcross**

Application An organism that shows a dominant trait in its phenotype, such as purple flowers in pea plants, can be either homozygous for the dominant allele or heterozygous. To determine the organism's genotype, geneticists can perform a testcross.

Technique In a testcross, the individual with the unknown genotype is crossed with a homozygous individual expressing the recessive trait (white flowers in this example), and Punnett squares are used to predict the possible outcomes.

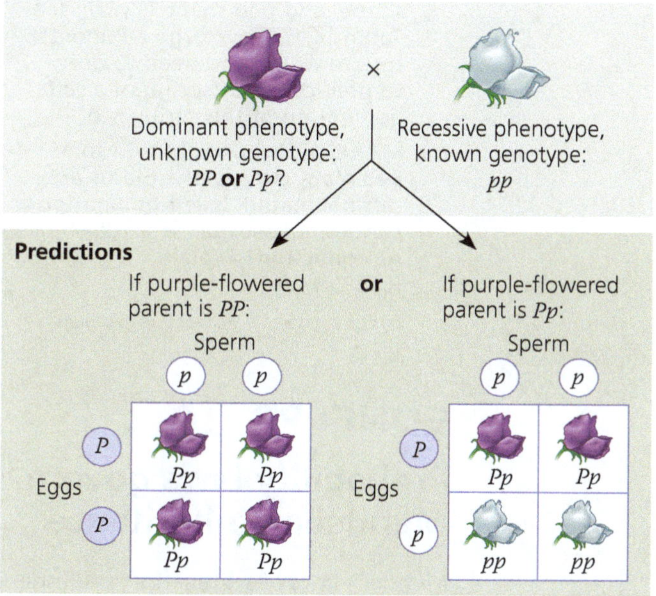

Results Matching the results to either prediction identifies the unknown parental genotype (either *PP* or *Pp* in this example). In this testcross, we transferred pollen from a white-flowered plant to the carpels of a purple-flowered plant; the opposite (reciprocal) cross would have led to the same results.

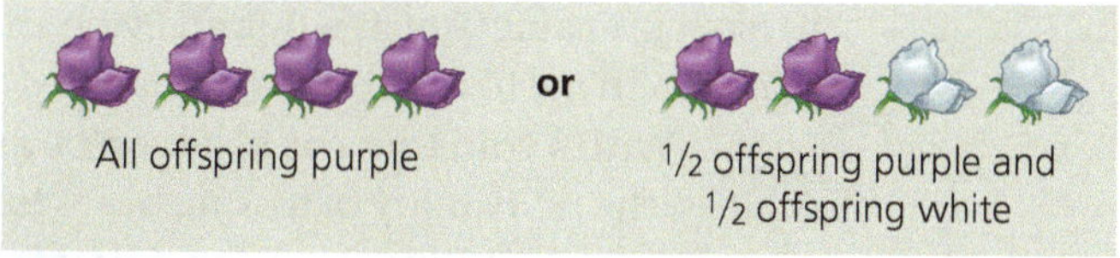

Animation: Testcross in "MendAliens"

to offspring as a package? That is, will the *Y* and *R* alleles always stay together, generation after generation? Or are seed color and seed shape inherited independently? **Figure 14.8** shows how a **dihybrid cross**, a cross between F_1 dihybrids, can determine which of these two hypotheses is correct.

The F_1 plants, of genotype *YyRr*, exhibit both dominant phenotypes, yellow seeds with round shapes, no matter which hypothesis is correct. The key step in the experiment is to see what happens when F_1 plants self-pollinate and produce F_2 offspring. If the hybrids must transmit their alleles in the same combinations in which the alleles were inherited from the P generation, then the F_1 hybrids will produce only two classes of gametes: *YR* and *yr*. As shown on the left side of Figure 14.8, this "dependent assortment" hypothesis predicts that the phenotypic ratio of the F_2 generation will be 3:1, just as in a monohybrid cross:

The alternative hypothesis is that the two pairs of alleles segregate independently of each other. In other words, genes are packaged into gametes in all possible allelic combinations, as long as each gamete has one allele for each gene (see Figure 13.11). In our example, an F_1 plant will produce four classes of gametes in equal quantities: *YR, Yr, yR,* and *yr*. If sperm of the four classes fertilize eggs of the four classes, there will be 16 (4 × 4) equally probable ways in which the alleles can combine in the F_2 generation, as shown on the right side of Figure 14.8. These combinations result in four phenotypic categories with a ratio of 9:3:3:1 (nine yellow round to three green round to three yellow wrinkled to one green wrinkled):

When Mendel did the experiment and classified the F_2 offspring, his results were close to the predicted 9:3:3:1 phenotypic ratio, supporting the hypothesis that the alleles for one gene—controlling seed color, for example—segregate into gametes independently of the alleles of any other gene, such as seed shape.

Mendel tested his seven pea characters in various dihybrid combinations and always observed a 9:3:3:1 phenotypic ratio in the F_2 generation. Is this consistent with the 3:1 phenotypic ratio seen for the monohybrid cross shown in Figure 14.5? To explore this question, count the number of yellow and green peas, ignoring shape, and calculate the ratio. The results of Mendel's dihybrid experiments are the basis for what we now call the **law of independent assortment**, which states that *two or more genes assort independently—that is, each pair of alleles segregates independently of any other pair of alleles—during gamete formation.*

This law applies only to genes (allele pairs) located on different chromosomes (that is, on chromosomes that are not homologous) or, alternatively, to genes that are very far apart on the same chromosome. (This will be explained in Concept 15.3, along with the more complex inheritance patterns of genes located near each other, alleles of which tend to be inherited together.) All the pea characters Mendel chose for analysis were controlled by genes on different chromosomes or were far apart on the same chromosome; this situation greatly simplified interpretation of his multicharacter pea crosses. All the examples we consider in the rest of this chapter involve genes located on different chromosomes.

Animation: Independent Assortment

▼ Figure 14.8

Inquiry Do the alleles for one character segregate into gametes dependently or independently of the alleles for a different character?

Experiment To follow the characters of seed color and seed shape through the F_2 generation, Mendel crossed a true-breeding plant with yellow round seeds with a true-breeding plant with green wrinkled seeds, producing dihybrid F_1 plants. Self-pollination of the F_1 dihybrids produced the F_2 generation. The two hypotheses (dependent and independent "assortment" of the two genes) predict different phenotypic ratios.

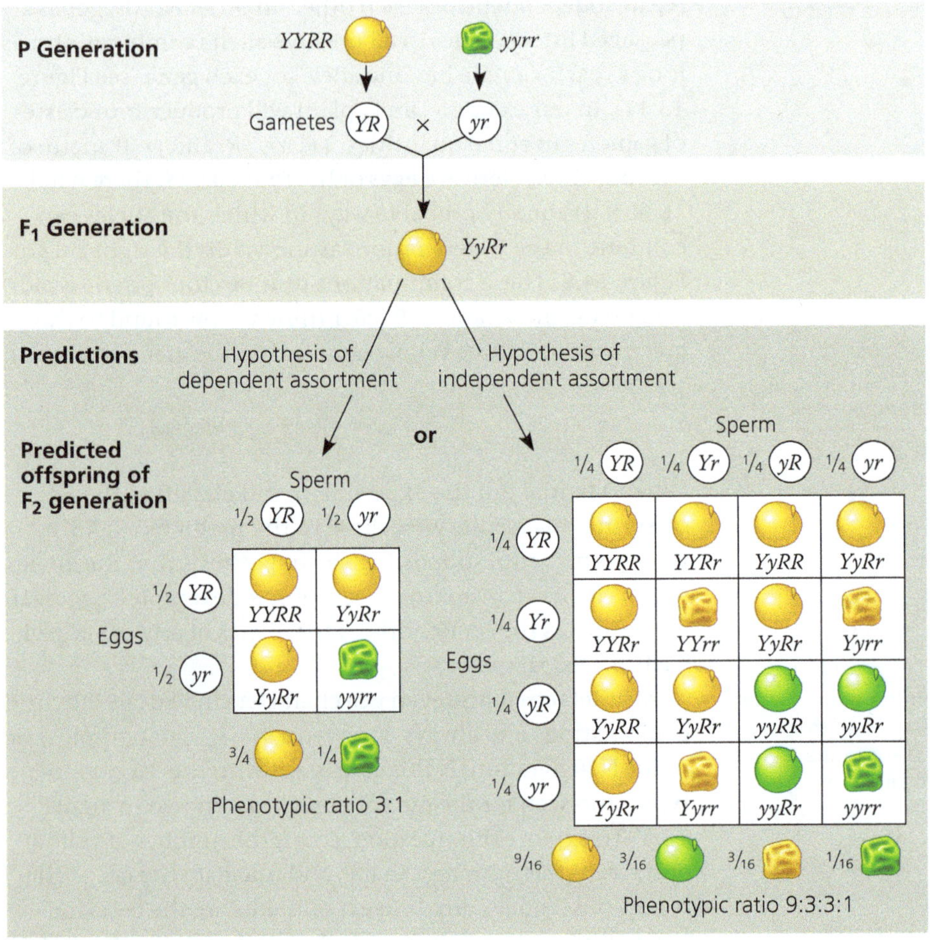

Results

Conclusion The results support the hypothesis of independent assortment, the only one that predicts two newly observed phenotypes: green round seeds and yellow wrinkled seeds (see the right-hand Punnett square). The alleles for each gene segregate independently of those of the other, and the two genes are said to assort independently.

Data from G. Mendel, Experiments in plant hybridization, *Proceedings of the Natural History Society of Brünn* 4:3–47 (1866).

WHAT IF? ➤ *Suppose Mendel had transferred pollen from an F_1 plant to the carpel of a plant that was homozygous recessive for both genes. Set up the cross and draw Punnett squares that predict the offspring for both hypotheses. Would this cross have supported the hypothesis of independent assortment equally well?*

Animation: Mendel's Cross of Two Characters: Seed Shape and Seed Color
Animation: A Simplified Cross of Two Characters in Humans
Animation: Crosses of Two Characters in "MendAliens"

CONCEPT CHECK 14.1

1. **DRAW IT ➤** Pea plants heterozygous for flower position and stem length (*AaTt*) are allowed to self-pollinate, and 400 of the resulting seeds are planted. Draw a Punnett square for this cross. How many offspring would be predicted to have terminal flowers and be dwarf? (See Table 14.1.)
2. **WHAT IF? ➤** List all gametes that could be made by a pea plant heterozygous for seed color, seed shape, and pod shape (*YyRrIi*; see Table 14.1). How large a Punnett square would you need to draw to predict the offspring of a self-pollination of this "trihybrid"?
3. **MAKE CONNECTIONS ➤** In some pea plant crosses, the plants are self-pollinated. Is self-pollination considered asexual or sexual reproduction? Explain. (See Concept 13.1.)

For suggested answers, see Appendix A.

CONCEPT 14.2

Probability laws govern Mendelian inheritance

Mendel's laws of segregation and independent assortment reflect the same rules of probability that apply to tossing coins, rolling dice, and drawing cards from a deck. The probability scale ranges from 0 to 1. An event that is certain to occur has a probability of 1, while an event that is certain *not* to occur has a probability of 0. With a coin that has heads on both sides, the probability of tossing heads is 1, and the probability of tossing tails is 0. With a normal coin, the chance of tossing heads is $\frac{1}{2}$, and the chance of tossing tails is $\frac{1}{2}$. The probability of drawing the ace of spades from a 52-card deck is $\frac{1}{52}$. The probabilities of all possible outcomes for an event must add up to 1. With a deck of cards, the chance of picking a card other than the ace of spades is $\frac{51}{52}$.

Tossing a coin illustrates an important lesson about probability. For every toss, the probability of heads is $\frac{1}{2}$. The outcome of any particular toss is unaffected by what has happened on previous trials. We refer to phenomena such as coin tosses as independent events. Each toss of a coin, whether done sequentially with one coin or simultaneously with many, is independent of every other toss. And like two separate coin tosses, the alleles of one gene segregate into gametes independently of another gene's alleles (the law of independent assortment). We'll now look at two basic rules of probability that help us predict the outcome of the fusion of such gametes in simple monohybrid crosses and more complicated crosses as well.

The Multiplication and Addition Rules Applied to Monohybrid Crosses

How do we determine the probability that two or more independent events will occur together in some specific combination? For example, what is the chance that two coins tossed simultaneously will both land heads up? The **multiplication rule** states that to determine this probability, we multiply the probability of one event (one coin coming up heads) by the probability of the other event (the other coin coming up heads). By the multiplication rule, then, the probability that both coins will land heads up is $\frac{1}{2} \times \frac{1}{2} = \frac{1}{4}$.

We can apply the same reasoning to an F_1 monohybrid cross. With seed shape in pea plants as the heritable character, the genotype of F_1 plants is *Rr*. Segregation in a heterozygous plant is like flipping a coin in terms of calculating the probability of each outcome: Each egg produced has a $\frac{1}{2}$ chance of carrying the dominant allele (*R*) and a $\frac{1}{2}$ chance of carrying the recessive allele (*r*). The same odds apply to each sperm cell produced. For a particular F_2 plant to have wrinkled seeds, the recessive trait, both the egg and the sperm that come together must carry the *r* allele. The probability that an *r* allele will be present in both gametes at fertilization is found by multiplying $\frac{1}{2}$ (the probability that the egg will have an *r*) $\times$ $\frac{1}{2}$ (the probability that the sperm will have an *r*). Thus, the multiplication rule tells us that the probability of an F_2 plant having wrinkled seeds (*rr*) is $\frac{1}{4}$ **(Figure 14.9)**. Likewise, the probability of an F_2 plant carrying both dominant alleles for seed shape (*RR*) is $\frac{1}{4}$.

To figure out the probability that an F_2 plant from a monohybrid cross will be heterozygous rather than homozygous, we need to invoke a second rule. Notice in Figure 14.9 that the dominant allele can come from the egg and the recessive allele from the sperm, or vice versa. That is, F_1 gametes can combine to produce *Rr* offspring in two *mutually exclusive* ways: For any particular heterozygous F_2 plant, the dominant allele can come from the egg *or* the sperm, but not from both. According to the **addition rule**, the probability that any one of two or more mutually exclusive events will occur is calculated by adding their individual probabilities. As we have just seen, the multiplication rule gives us the individual probabilities that we will now add together. The probability for one possible way of obtaining an F_2 heterozygote—the dominant allele from the egg and the recessive allele from the sperm—is $\frac{1}{4}$. The probability for the other possible way—the recessive allele from the egg and the dominant allele from the sperm—is also $\frac{1}{4}$ (see Figure 14.9). Using the rule of addition, then, we can calculate the probability of an F_2 heterozygote as $\frac{1}{4} + \frac{1}{4} = \frac{1}{2}$.

▼ Figure 14.9 Segregation of alleles and fertilization as chance events. When a heterozygote (*Rr*) forms gametes, whether a particular gamete ends up with an *R* or an *r* is like the toss of a coin. We can determine the probability for any genotype among the offspring of two heterozygotes by multiplying together the individual probabilities of an egg and sperm having a particular allele (*R* or *r* in this example).

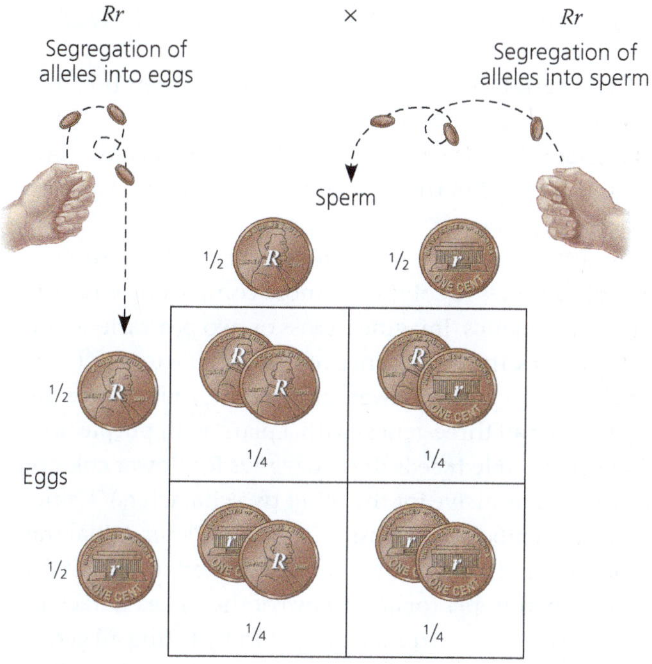

Solving Complex Genetics Problems with the Rules of Probability

We can also apply the rules of probability to predict the outcome of crosses involving multiple characters. Recall that each allelic pair segregates independently during gamete formation (the law of independent assortment). Thus, a dihybrid or other multicharacter cross is equivalent to two or more independent monohybrid crosses occurring simultaneously. By applying what we have learned about monohybrid crosses, we can determine the probability of specific genotypes occurring in the F_2 generation without having to construct unwieldy Punnett squares.

Consider the dihybrid cross between *YyRr* heterozygotes shown in Figure 14.8. We will focus first on the seed-color character. For a monohybrid cross of *Yy* plants, we can use a simple Punnett square to determine that the probabilities of the offspring genotypes are $\frac{1}{4}$ for *YY*, $\frac{1}{2}$ for *Yy*, and $\frac{1}{4}$ for *yy*. We can draw a second Punnett square to determine that the same probabilities apply to the offspring genotypes for seed shape: $\frac{1}{4}$ *RR*, $\frac{1}{2}$ *Rr*, and $\frac{1}{4}$ *rr*. Knowing these probabilities, we can simply use the multiplication rule to determine the probability of each of the genotypes in the F_2 generation.

To give two examples, the calculations for finding the probabilities of two of the possible F_2 genotypes (*YYRR* and *YyRR*) are shown below:

Probability of *YYRR* = $\frac{1}{4}$(probability of *YY*) × $\frac{1}{4}$(*RR*) = $\frac{1}{16}$

Probability of *YyRR* = $\frac{1}{2}$(*Yy*) × $\frac{1}{4}$(*RR*) = $\frac{1}{8}$

The *YYRR* genotype corresponds to the upper left box in the larger Punnett square in Figure 14.8 (one box = $\frac{1}{16}$). Looking closely at the larger Punnett square in Figure 14.8, you will see that 2 of the 16 boxes ($\frac{1}{8}$) correspond to the *YyRR* genotype.

Now let's see how we can combine the multiplication and addition rules to solve even more complex problems in Mendelian genetics. Imagine a cross of two pea varieties in which we track the inheritance of three characters. Let's cross a trihybrid with purple flowers and yellow, round seeds (heterozygous for all three genes) with a plant with purple flowers and green, wrinkled seeds (heterozygous for flower color but homozygous recessive for the other two characters). Using Mendelian symbols, our cross is *PpYyRr* × *Ppyyrr*. What fraction of offspring from this cross are predicted to exhibit the recessive phenotypes for *at least two* of the three characters?

To answer this question, we can start by listing all genotypes we could get that fulfill this condition: *ppyyRr, ppYyrr, Ppyyrr, PPyyrr,* and *ppyyrr*. (Because the condition is *at least two* recessive traits, it includes the last genotype, which shows all three recessive traits.) Next, we calculate the probability for each of these genotypes resulting from our *PpYyRr* × *Ppyyrr* cross by multiplying together the individual probabilities for the allele pairs, just as we did in our dihybrid example. Note that in a cross involving heterozygous and homozygous allele pairs (for example, *Yy* × *yy*), the probability of heterozygous offspring is $\frac{1}{2}$ and the probability of homozygous offspring is $\frac{1}{2}$. Finally, we use the addition rule to add the probabilities for all the different genotypes that fulfill the condition of at least two recessive traits resulting from our *PpYyRr* × *Ppyyrr* cross, as shown below:

ppyyRr	$\frac{1}{4}$ (probability of *pp*) × $\frac{1}{2}$ (*yy*) × $\frac{1}{2}$ (*Rr*)	= $\frac{1}{16}$
ppYyrr	$\frac{1}{4} \times \frac{1}{2} \times \frac{1}{2}$	= $\frac{1}{16}$
Ppyyrr	$\frac{1}{2} \times \frac{1}{2} \times \frac{1}{2}$	= $\frac{2}{16}$
PPyyrr	$\frac{1}{4} \times \frac{1}{2} \times \frac{1}{2}$	= $\frac{1}{16}$
ppyyrr	$\frac{1}{4} \times \frac{1}{2} \times \frac{1}{2}$	= $\frac{1}{16}$
Chance of *at least two* recessive traits		= $\frac{6}{16}$ or $\frac{3}{8}$

In time, you'll be able to solve genetics problems faster by using the rules of probability than by filling in Punnett squares.

We cannot predict with certainty the exact numbers of progeny of different genotypes resulting from a genetic cross. But the rules of probability give us the *likelihood* of various outcomes. Usually, the larger the sample size, the closer the results will conform to our predictions. Mendel understood this statistical feature of inheritance and had a keen sense of the rules of chance. It was for this reason that he set up his experiments so as to generate, and then count, large numbers of offspring from his crosses.

CONCEPT CHECK 14.2

1. For any gene with a dominant allele *A* and recessive allele *a*, what proportions of the offspring from an *AA* × *Aa* cross are expected to be homozygous dominant, homozygous recessive, and heterozygous?
2. Two organisms, with genotypes *BbDD* and *BBDd*, are mated. Assuming independent assortment of the *B*/*b* and *D*/*d* genes, write the genotypes of all possible offspring from this cross and use the rules of probability to calculate the chance of each genotype occurring.
3. **WHAT IF?** ➤ Three characters (flower color, seed color, and pod shape) are considered in a cross between two pea plants: *PpYyIi* × *ppYyii*. What fraction of offspring is predicted to be homozygous recessive for at least two of the three characters?

For suggested answers, see Appendix A.

CONCEPT 14.3

Inheritance patterns are often more complex than predicted by simple Mendelian genetics

In the 20th century, geneticists extended Mendelian principles not only to diverse organisms, but also to patterns of inheritance more complex than those described by Mendel. For the work that led to his two laws of inheritance, Mendel chose pea plant characters that turn out to have a relatively simple genetic basis: Each character is determined by one gene, for which there are only two alleles, one completely dominant and the other completely recessive. (There is one exception: Mendel's pod shape character is actually determined by two genes.) Not all heritable characters are determined so simply, and the relationship between genotype and phenotype is rarely so straightforward. Mendel himself realized that he could not explain the more complicated patterns he observed in crosses involving other pea characters or other plant species. This does not diminish the utility of Mendelian genetics, however, because the basic principles of segregation and independent assortment apply even to more complex patterns of inheritance. In this section, we will extend Mendelian genetics to hereditary patterns that were not reported by Mendel.

Extending Mendelian Genetics for a Single Gene

The inheritance of characters determined by a single gene deviates from simple Mendelian patterns when alleles are not completely dominant or recessive, when a particular gene has more than two alleles, or when a single gene produces multiple phenotypes. We will describe examples of each of these situations in this section.

Degrees of Dominance

Alleles can show different degrees of dominance and recessiveness in relation to each other. In Mendel's classic pea crosses, the F_1 offspring always looked like one of the two parental varieties because one allele in a pair showed **complete dominance** over the other. In such situations, the phenotypes of the heterozygote and the dominant homozygote are indistinguishable (see Figure 14.6).

For some genes, however, neither allele is completely dominant, and the F_1 hybrids have a phenotype somewhere between those of the two parental varieties. This phenomenon, called **incomplete dominance**, is seen when red snapdragons are crossed with white snapdragons: All the F_1 hybrids have pink flowers **(Figure 14.10)**. This third, intermediate phenotype results from flowers of the heterozygotes having less red pigment than the red homozygotes. (This is unlike the case of Mendel's pea plants, where the *Pp* heterozygotes make enough pigment for the flowers to be purple, indistinguishable from those of *PP* plants.)

At first glance, incomplete dominance of either allele seems to provide evidence for the blending hypothesis of inheritance, which would predict that the red or white trait could never reappear among offspring of the pink hybrids. In fact, interbreeding F_1 hybrids produces F_2 offspring with a phenotypic ratio of one red to two pink to one white. (Because heterozygotes have a separate phenotype, the genotypic and phenotypic ratios for the F_2 generation are the same, 1:2:1.) The segregation of the red-flower and white-flower alleles in the gametes produced by the pink-flowered plants confirms that the alleles for flower color are heritable factors that maintain their identity in the hybrids; that is, inheritance is particulate.

Another variation on dominance relationships between alleles is called **codominance**; in this variation, the two alleles each affect the phenotype in separate, distinguishable ways. For example, the human MN blood group is determined by codominant alleles for two specific molecules located on the surface of red blood cells, the M and N molecules. A single gene locus, at which two allelic variations are possible, determines the phenotype of this blood group. Individuals homozygous for the *M* allele (*MM*) have red blood cells with only M molecules; individuals homozygous for the *N* allele (*NN*) have red blood cells with only N molecules. But *both* M and N molecules are present on the red blood cells of individuals heterozygous for the *M* and *N* alleles (*MN*). Note that the MN phenotype is *not* intermediate between the M and N phenotypes, which distinguishes codominance from incomplete dominance. Rather, *both* M and N phenotypes are exhibited by heterozygotes, since both molecules are present.

▼ Figure 14.10 Incomplete dominance in snapdragon color. When red snapdragons are crossed with white ones, the F_1 hybrids have pink flowers. Segregation of alleles into gametes of the F_1 plants results in an F_2 generation with a 1:2:1 ratio for both genotype and phenotype. Neither allele is dominant, so rather than using upper- and lowercase letters, we use the letter *C* with a superscript to indicate an allele for flower color: C^R for red and C^W for white.

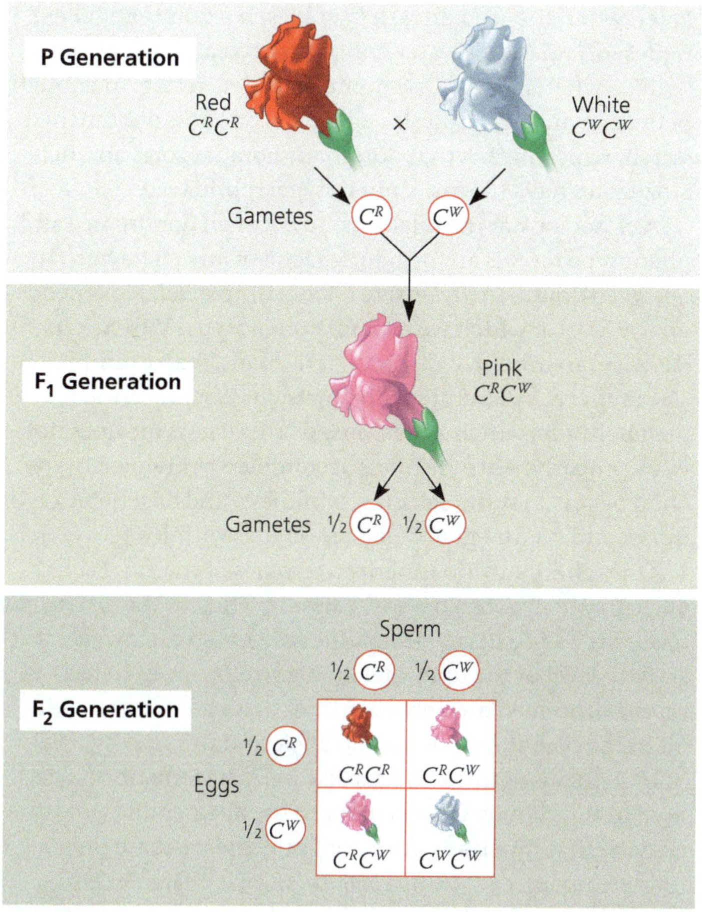

? *Suppose a classmate argues that this figure supports the blending hypothesis for inheritance. What might your classmate say, and how would you respond?*

MB **Animation: Incomplete Dominance in "MendAliens"**

The Relationship Between Dominance and Phenotype

We've now seen that the relative effects of two alleles range from complete dominance of one allele, to incomplete dominance of either allele, to codominance of both alleles. It is important to understand that an allele is called *dominant* because it is seen in the phenotype, not because it somehow subdues a recessive allele. Alleles are simply variations in a gene's nucleotide sequence (see Figure 14.4). When a dominant allele coexists with a recessive allele in a heterozygote, they do not actually interact at all. It is in the pathway from genotype to phenotype that dominance and recessiveness come into play.

To illustrate the relationship between dominance and phenotype, we can use one of the characters Mendel studied—round versus wrinkled pea seed shape. The dominant allele (round) codes for an enzyme that helps convert an unbranched form of starch to a branched form in the seed. The recessive allele (wrinkled) codes for a defective form of this enzyme, leading to an accumulation of unbranched starch, which causes excess water to enter the seed by osmosis. Later, when the seed dries, it wrinkles. If a dominant allele is present, no excess water enters the seed and it does not wrinkle when it dries. One dominant allele results in enough of the enzyme to synthesize adequate amounts of branched starch, which means that dominant homozygotes and heterozygotes have the same phenotype: round seeds.

A closer look at the relationship between dominance and phenotype reveals an intriguing fact: For any character, the observed dominant/recessive relationship of alleles depends on the level at which we examine phenotype. **Tay-Sachs disease**, an inherited disorder in humans, is an example. The brain cells of a child with Tay-Sachs disease cannot metabolize certain lipids because a crucial enzyme does not work properly. As these lipids accumulate in brain cells, the child begins to suffer seizures, blindness, and degeneration of motor and mental performance and dies within a few years.

Only children who inherit two copies of the Tay-Sachs allele (homozygotes) have the disease. Thus, at the *organismal* level, the Tay-Sachs allele qualifies as recessive. However, the activity level of the lipid-metabolizing enzyme in heterozygotes is intermediate between the activity level in individuals homozygous for the normal allele and the activity level in individuals with Tay-Sachs disease. (The term normal is used in the genetic sense to refer to the allele coding for the enzyme that functions properly.) The intermediate phenotype observed at the *biochemical* level is characteristic of incomplete dominance of either allele. Fortunately, the heterozygote condition does not lead to disease symptoms, apparently because half the normal enzyme activity is sufficient to prevent lipid accumulation in the brain. Extending our analysis to yet another level, we find that heterozygous individuals produce equal numbers of normal and dysfunctional enzyme molecules. Thus, at the *molecular* level, the normal allele and the Tay-Sachs allele are codominant. As you can see, whether alleles appear to be completely dominant, incompletely dominant, or codominant depends on the level at which the phenotype is analyzed.

Frequency of Dominant Alleles Although you might assume that the dominant allele for a particular character would be more common than the recessive allele, this is not always the case. For an example of a rare dominant allele, about one baby out of 400 in the United States is born with extra fingers or toes, a condition known as polydactyly. Some cases are caused by the presence of a dominant allele. The low frequency of polydactyly indicates that the recessive allele, which results in five digits per appendage, is far more prevalent than the dominant allele in the population. In Concept 23.3, you will learn how relative frequencies of alleles in a population are affected by natural selection.

Multiple Alleles

Only two alleles exist for the pea characters that Mendel studied, but most genes exist in more than two allelic forms. The ABO blood groups in humans, for instance, are determined by that person's two alleles of the blood group gene; there are three possible alleles: I^A, I^B, and i. A person's blood group may be one of four types: A, B, AB, or O. These letters refer to two carbohydrates—A and B—that may be found attached to specific cell-surface molecules on red blood cells. An individual's blood cells may have carbohydrate A (type A blood), carbohydrate B (type B), both (type AB), or neither (type O), as shown in **Figure 14.11**. Matching compatible blood groups is critical for safe blood transfusions (see Concept 43.3).

Figure 14.11 Multiple alleles for the ABO blood groups. The four blood groups result from different combinations of three alleles.

(a) The three alleles for the ABO blood groups and their carbohydrates. Each allele codes for an enzyme that may add a specific carbohydrate (designated by the superscript on the allele and shown as a triangle or circle) to red blood cells.

Allele	I^A	I^B	i
Carbohydrate	A △	B ○	none

(b) Blood group genotypes and phenotypes. There are six possible genotypes, resulting in four different phenotypes.

Genotype	I^AI^A or I^Ai	I^BI^B or I^Bi	I^AI^B	ii
Red blood cell with surface carbohydrates		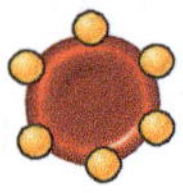		
Phenotype (blood group)	A	B	AB	O

VISUAL SKILLS ➤ *Based on the surface carbohydrate phenotype in (b), what are the dominance relationships among the alleles?*

Pleiotropy

So far, we have treated Mendelian inheritance as though each gene affects only one phenotypic character. Most genes, however, have multiple phenotypic effects, a property called **pleiotropy** (from the Greek *pleion*, more). In humans, for example, pleiotropic alleles are responsible for the multiple symptoms associated with certain hereditary diseases, such as cystic fibrosis and sickle-cell disease, discussed later in this chapter. In the garden pea, the gene that determines flower color also affects the color of the coating on the outer surface of the seed, which can be gray or white. Given the intricate molecular and cellular interactions responsible for an organism's development and physiology, it isn't surprising that a single gene can affect a number of characters.

Extending Mendelian Genetics for Two or More Genes

Dominance relationships, multiple alleles, and pleiotropy all have to do with the effects of the alleles of a single gene. We now consider two situations in which two or more genes are involved in determining a particular phenotype. In the first case, called epistasis, one gene affects the phenotype of another because the two gene products interact; in the second case, called polygenic inheritance, multiple genes independently affect a single trait.

Epistasis

In **epistasis** (from the Greek for "standing upon"), the phenotypic expression of a gene at one locus alters that of a gene at a second locus. An example will help clarify this concept. In Labrador retrievers (commonly called "Labs"), black coat color is dominant to brown. Let's designate *B* and *b* as the two alleles for this character. For a Lab to have brown fur, its genotype must be *bb*; these dogs are called chocolate Labs. But there is more to the story. A second gene determines whether or not pigment will be deposited in the hair. The dominant allele, symbolized by *E*, results in the deposition of either black or brown pigment, depending on the genotype at the first locus. But if the Lab is homozygous recessive for the second locus (*ee*), then the coat is yellow, regardless of the genotype at the black/brown locus (yellow Labs). In this case, the gene for pigment deposition (*E*/*e*) is said to be epistatic to the gene that codes for black or brown pigment (*B*/*b*).

What happens if we mate black Labs that are heterozygous for both genes (*BbEe*)? Although the two genes affect the same phenotypic character (coat color), they follow the law of independent assortment. Thus, our breeding experiment represents an F_1 dihybrid cross, like those that produced a 9:3:3:1 ratio in Mendel's experiments. We can use a Punnett square to represent the genotypes of the F_2 offspring **(Figure 14.12)**. As a result of epistasis, the phenotypic ratio among the F_2 offspring is 9 black to 3 chocolate to 4 yellow Labs. Other types of epistatic interactions produce different ratios, but all are modified versions of 9:3:3:1.

Figure 14.12 An example of epistasis. This Punnett square illustrates the genotypes and phenotypes predicted for offspring of matings between two black Labrador retrievers of genotype *BbEe*. The *E*/*e* gene, which is epistatic to the *B*/*b* gene coding for hair pigment, controls whether or not pigment of any color will be deposited in the hair.

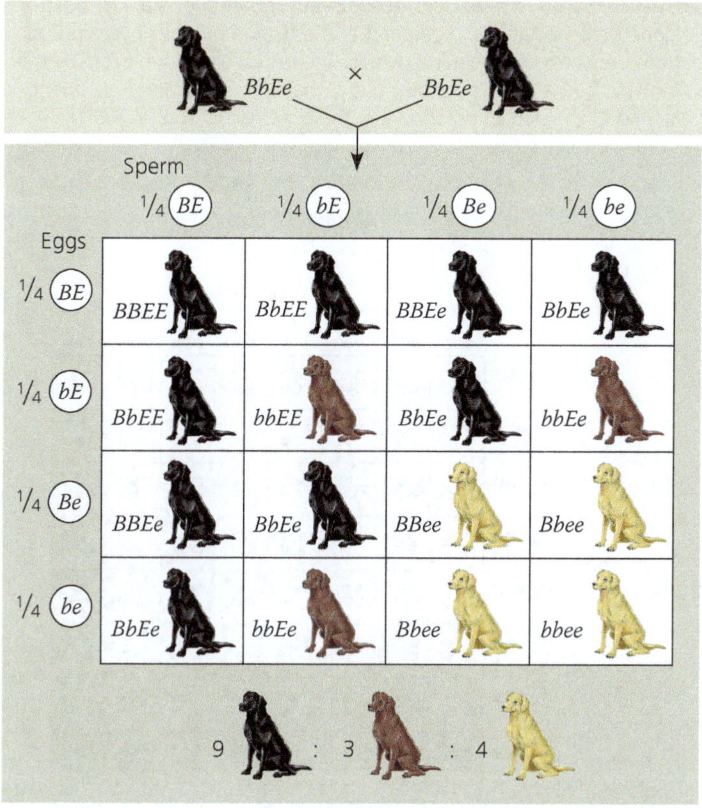

VISUAL SKILLS > *Compare the four squares in the lower right of this Punnett square with those in Figure 14.8. Explain the genetic basis for the difference between the ratio (9:3:4) of phenotypes seen in this cross and the 9:3:3:1 ratio seen in Figure 14.8.*

Polygenic Inheritance

Mendel studied characters that could be classified on an either-or basis, such as purple versus white flower color. But many characters, such as human skin color and height, are not one of two discrete characters, but instead vary in the population in gradations along a continuum. These are called **quantitative characters**. Quantitative variation usually indicates **polygenic inheritance**, an additive effect of two or more genes on a single phenotypic character. (In a way, this is the converse of pleiotropy, where a single gene affects several phenotypic characters.) Height is a good example of polygenic inheritance: In 2014, a genomic study of over 250,000 people found almost 700 genetic variations associated with over 180 genes that affect height. Many variations were in or near genes involved in biochemical pathways affecting growth of the skeleton, but others were associated with genes not obviously related to growth.

Skin pigmentation in humans is also controlled by many separately inherited genes. Here, we'll simplify the story in order to understand the concept of polygenic inheritance. Let's consider three genes, with a dark-skin allele for each gene (*A*, *B*, or *C*) contributing one "unit" of darkness (also a simplification) to the phenotype and being incompletely dominant to the other allele (*a*, *b*, or *c*). In our model, an *AABBCC* person would be very dark, whereas an *aabbcc* individual would be very light. An *AaBbCc* person would have skin of an intermediate shade. Because the alleles have a cumulative effect, the genotypes *AaBbCc* and *AABbcc* would make the same genetic contribution (three units) to skin

Figure 14.13 A simplified model for polygenic inheritance of skin color. In this model, three separately inherited genes affect skin color. The heterozygous individuals (*AaBbCc*) represented by the two rectangles at the top of this figure each carry three dark-skin alleles (black circles, which represent *A*, *B*, or *C*) and three light-skin alleles (white circles, which represent *a*, *b*, or *c*). The Punnett square shows all the possible genetic combinations in gametes and offspring of many hypothetical matings between these heterozygotes. The results are summarized by the phenotypic frequencies (fractions) under the Punnett square. (The phenotypic ratio of the skin colors shown in the boxes is 1:6:15:20:15:6:1.)

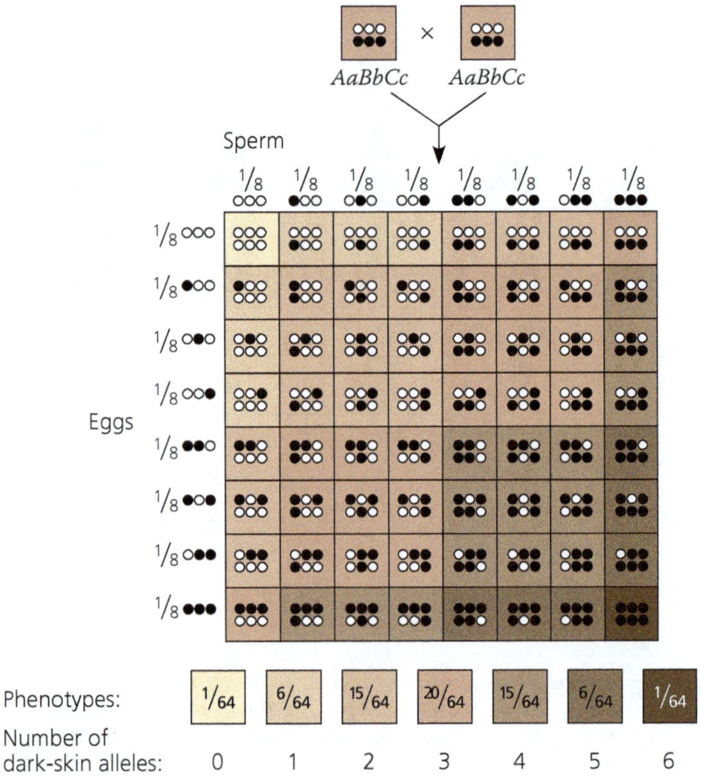

HHMI Video: The Biology of Skin Color

darkness. There are seven skin color phenotypes that could result from a mating between *AaBbCc* heterozygotes, as shown in **Figure 14.13**. In a large number of such matings, the majority of offspring would be expected to have intermediate phenotypes (skin color in the middle range). You can graph the predictions from the Punnett square in the **Scientific Skills Exercise**. Environmental factors, such as exposure to the sun, also affect the skin color phenotype.

Nature and Nurture: The Environmental Impact on Phenotype

Another departure from simple Mendelian genetics arises when the phenotype for a character depends on environment as well as genotype. A single tree, locked into its inherited genotype, has leaves that vary in size, shape, and greenness, depending on their exposure to wind and sun. For humans, nutrition influences height, exercise alters build, sun-tanning darkens the skin, and experience improves performance on intelligence tests. Even identical twins, who are genetic equals, accumulate phenotypic differences as a result of their unique experiences.

Whether human characters are more influenced by genes or the environment—in everyday terms, nature versus nurture—is a debate that we will not attempt to settle here. We can say, however, that a genotype generally is not associated with a rigidly defined phenotype, but rather with a range of phenotypic possibilities due to environmental influences **(Figure 14.14)**. For some characters, such as the ABO blood group system, the phenotypic range has no breadth whatsoever; that is, a given genotype mandates a very specific phenotype. Other characters, such as a person's blood count of red and white cells, vary quite a bit, depending on such factors as the altitude, the customary level of physical activity, and the presence of infectious agents.

Figure 14.14 The effect of environment on phenotype. The outcome of a genotype lies within a phenotypic range that depends on the environment in which the genotype is expressed. For example, the acidity and free aluminum content of the soil affect the color of hydrangea flowers, which range from pink (basic soil) to blue-violet (acidic soil). Free aluminum is necessary for bluer colors.

(a) Hydrangeas grown in basic soil

(b) Hydrangeas of the same genetic variety grown in acidic soil with free aluminum

Generally, the phenotypic range is broadest for polygenic characters. Environment contributes to the quantitative nature of these characters, as we have seen in the continuous variation of skin color. Geneticists refer to such characters as **multifactorial**, meaning that many factors, both genetic and environmental, collectively influence phenotype.

BBC Video: Genetics vs. Environment

A Mendelian View of Heredity and Variation

We have now broadened our view of Mendelian inheritance by exploring degrees of dominance as well as multiple alleles, pleiotropy, epistasis, polygenic inheritance, and the phenotypic impact of the environment. How can we integrate these refinements into a comprehensive theory of Mendelian genetics? The key is to make the transition from the reductionist emphasis on single genes and phenotypic characters to the emergent properties of the organism as a whole, one of the themes of this book.

The term *phenotype* can refer not only to specific characters, such as flower color and blood group, but also to an organism in its entirety—*all* aspects of its physical

SCIENTIFIC SKILLS EXERCISE

Making a Histogram and Analyzing a Distribution Pattern

What Is the Distribution of Phenotypes Among Offspring of Two Parents Who Are Both Heterozygous for Three Additive Genes? Human skin color is a polygenic trait that is determined by the additive effects of many different genes. In this exercise, you will work with a simplified model of skin-color genetics where only three genes are assumed to affect the darkness of skin color and where each gene has two alleles—dark or light (see Figure 14.13). In this model, each dark allele contributes equally to the darkness of skin color, and each pair of alleles segregates independently of any other pair. Using a type of graph called a histogram, you will determine the distribution of phenotypes of offspring with different numbers of dark-skin alleles. (For additional information about graphs, see the Scientific Skills Review in Appendix F.)

How This Model Is Analyzed To predict the phenotypes of the offspring of parents heterozygous for the three genes in our simplified model, we can use the Punnett square in Figure 14.13. The heterozygous individuals (*AaBbCc*) represented by the two rectangles at the top of that figure each carry three dark-skin alleles (black circles, which represent *A, B,* or *C*) and three light-skin alleles (white circles, which represent *a, b,* or *c*). The Punnett square shows all the possible genetic combinations in gametes and in offspring of a large number of hypothetical matings between these heterozygotes.

Predictions from the Punnett Square If we assume that each square in the Punnett square represents one offspring of the heterozygous *AaBbCc* parents, then the squares below show the possible skin-color phenotypes and their predicted frequencies. Below the squares is the number of dark-skin alleles for each phenotype.

Phenotypes:	$1/64$	$6/64$	$15/64$	$20/64$	$15/64$	$6/64$	$1/64$
Number of dark-skin alleles:	0	1	2	3	4	5	6

INTERPRET THE DATA

1. A histogram is a bar graph that shows the distribution of numeric data (here, the number of dark-skin alleles). To make a histogram of the allele distribution, put skin color (as the number of dark-skin alleles) along the *x*-axis and predicted number of offspring (out of 64) with each phenotype on the *y*-axis. There are no gaps in these allele data, so draw the bars next to each other with no space in between.
2. You can see that the skin-color phenotypes are not distributed uniformly. (a) Which phenotype has the highest frequency? Draw a vertical dashed line through that bar. (b) Distributions of values like this one tend to show one of several common patterns. Sketch a rough curve that approximates the values and look at its shape. Is it symmetrically distributed around a central peak value (a "normal distribution," sometimes called a bell curve); is it skewed to one end of the *x*-axis or the other (a "skewed distribution"); or does it show two apparent groups of frequencies (a "bimodal distribution")? Explain the reason for the curve's shape. (It will help to read the text description that supports Figure 14.13.)

(MB) **Instructors:** A version of this Scientific Skills Exercise can be assigned in MasteringBiology.

Further Reading R. A. Sturm, A golden age of human pigmentation genetics, *Trends in Genetics* 22:464–468 (2006).

appearance, internal anatomy, physiology, and behavior. Similarly, the term *genotype* can refer to an organism's entire genetic makeup, not just its alleles for a single genetic locus. In most cases, a gene's impact on phenotype is affected by other genes and by the environment. In this integrated view of heredity and variation, an organism's phenotype reflects its overall genotype and unique environmental history.

Considering all that can occur in the pathway from genotype to phenotype, it is indeed impressive that Mendel could uncover the fundamental principles governing the transmission of individual genes from parents to offspring. Mendel's laws of segregation and of independent assortment explain heritable variations in terms of alternative forms of genes (hereditary "particles," now known as the alleles of genes) that are passed along, generation after generation, according to simple rules of probability. This theory of inheritance is equally valid for peas, flies, fishes, birds, and human beings—indeed, for any organism with a sexual life cycle. Furthermore, by extending the principles of segregation and independent assortment to help explain such hereditary patterns as epistasis and quantitative characters, we begin to see how broadly Mendelian genetics applies. From Mendel's abbey garden came a theory of particulate inheritance that anchors modern genetics. In the last section of this chapter, we will apply Mendelian genetics to human inheritance, with emphasis on the transmission of hereditary diseases.

CONCEPT CHECK 14.3

1. *Incomplete dominance* and *epistasis* are both terms that define genetic relationships. What is the most basic distinction between these terms?
2. If a man with type AB blood marries a woman with type O, what blood types would you expect in their children? What fraction would you expect of each type?
3. **WHAT IF?** ➤ A rooster with gray feathers and a hen of the same phenotype produce 15 gray, 6 black, and 8 white chicks. What is the simplest explanation for the inheritance of these colors in chickens? What phenotypes would you expect in the offspring of a cross between a gray rooster and a black hen?

For suggested answers, see Appendix A.

CONCEPT 14.4

Many human traits follow Mendelian patterns of inheritance

Peas are convenient subjects for genetic research, but humans are not. The human generation span is long—about 20 years—and human parents produce many fewer offspring than peas and most other species. Even more important, it wouldn't be ethical to ask pairs of humans to breed so that the phenotypes of their offspring could be analyzed! In spite of these constraints, the study of human genetics continues, spurred on by our desire to understand our own inheritance and to develop treatments and cures for human genetically based diseases. New molecular biological techniques have led to many breakthrough discoveries, as we will see in Concept 20.4, but basic Mendelian genetics endures as the foundation of human genetics.

Pedigree Analysis

Unable to manipulate the mating patterns of people, geneticists instead analyze the results of matings that have already occurred. They do so by collecting information about a family's history for a particular trait and assembling this information into a family tree describing the traits of parents and children across the generations—a family **pedigree**.

Figure 14.15a shows a three-generation pedigree that traces the occurrence of a pointed contour of the hairline on the forehead. This trait, called a widow's peak, is due to a dominant allele, *W*. Because the widow's-peak allele is dominant, all individuals who lack a widow's peak must be homozygous recessive (*ww*). The two grandparents with widow's peaks must have the *Ww* genotype, since some of their offspring are homozygous recessive. The offspring in the second generation who *do* have widow's peaks must also be heterozygous, because they are the products of *Ww* × *ww* matings. The third generation in this pedigree consists of two sisters. The one who has a widow's peak could be either homozygous (*WW*) or heterozygous (*Ww*), given what we know about the genotypes of her parents (both *Ww*).

Figure 14.15b is a pedigree of the same family, but this time we focus on a recessive trait, the inability of individuals to taste a chemical called PTC (phenylthiocarbamide). Compounds similar to PTC are found in broccoli, brussels sprouts, and related vegetables and account for the bitter taste some people report when eating these foods. We'll use *t* for the recessive allele and *T* for the dominant allele, which results in the ability to taste PTC. As you work your way through the pedigree, notice once again that you can apply what you have learned about Mendelian inheritance to understand the genotypes shown for the family members.

An important application of a pedigree is to help us calculate the probability that a future child will have a particular

▼ Figure 14.15 Pedigree analysis. Each of these pedigrees traces a trait through three generations of the same family. The two traits have different inheritance patterns, as shown by the pedigrees. (Note: While most traits are not determined by a single gene, that is generally agreed to be the case for these two traits.)

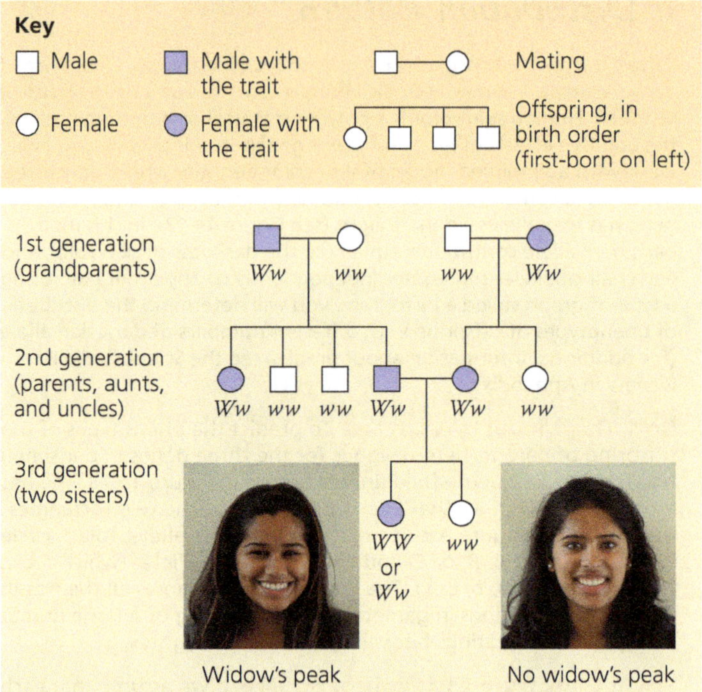

(a) Is a widow's peak a dominant or recessive trait?
Tips for pedigree analysis: Notice in the third generation that the second-born daughter lacks a widow's peak, although both of her parents had the trait. Such a pattern indicates that the trait is due to a dominant allele. If it were due to a *recessive* allele, and both parents had the recessive phenotype (straight hairline), *all* of their offspring would also have the recessive phenotype.

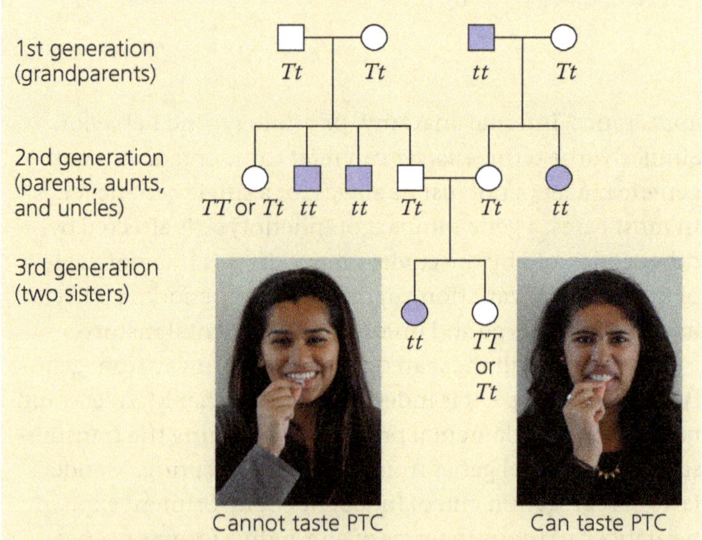

(b) Is the inability to taste a chemical called PTC a dominant or recessive trait?
Tips for pedigree analysis: Notice that the first-born daughter in the third generation has the trait (is unable to taste PTC), although both parents lack that trait (they *can* taste PTC). Such a pattern is explained if the non-taster phenotype is due to a recessive allele. (If it were due to a *dominant* allele, then at least one parent would also have had the trait.)

genotype and phenotype. Suppose that the couple represented in the second generation of Figure 14.15 decides to have one more child. What is the probability that the child will have a widow's peak? This is equivalent to a Mendelian F_1 monohybrid cross ($Ww \times Ww$), and therefore the probability that a child will inherit a dominant allele and have a widow's peak is $\frac{3}{4}$ ($\frac{1}{4}$ *WW* + $\frac{1}{2}$ *Ww*). What is the probability that the child will be unable to taste PTC? We can also treat this as a monohybrid cross ($Tt \times Tt$), but this time we want to know the chance that the offspring will be homozygous recessive (*tt*). That probability is $\frac{1}{4}$. Finally, what is the chance that the child will have a widow's peak *and* be unable to taste PTC? Assuming that the genes for these two characters are on different chromosomes, the two pairs of alleles will assort independently in this dihybrid cross ($WwTt \times WwTt$). Therefore, we can use the multiplication rule: $\frac{3}{4}$ (chance of widow's peak) $\times$ $\frac{1}{4}$ (chance of inability to taste PTC) $=$ $\frac{3}{16}$ (chance of widow's peak and inability to taste PTC).

Pedigrees are a more serious matter when the alleles in question cause disabling or deadly diseases instead of innocuous human variations such as hairline or inability to taste an innocuous chemical. However, for disorders inherited as simple Mendelian traits, the same techniques of pedigree analysis apply.

Recessively Inherited Disorders

Thousands of genetic disorders are known to be inherited as simple recessive traits. These disorders range in severity from relatively mild, such as albinism (lack of pigmentation, which results in susceptibility to skin cancers and vision problems), to life-threatening, such as cystic fibrosis.

The Behavior of Recessive Alleles

How can we account for the behavior of alleles that cause recessively inherited disorders? Recall that genes code for proteins of specific function. An allele that causes a genetic disorder (let's call it allele *a*) codes for either a malfunctioning protein or no protein at all. In the case of disorders classified as recessive, heterozygotes (*Aa*) typically have the normal phenotype because one copy of the normal allele (*A*) produces a sufficient amount of the specific protein. Thus, a recessively inherited disorder shows up only in the homozygous individuals (*aa*) who inherit a recessive allele from each parent. Although phenotypically normal with regard to the disorder, heterozygotes may transmit the recessive allele to their offspring and thus are called **carriers**. **Figure 14.16** illustrates these ideas using albinism as an example.

Figure 14.16 Albinism: a recessive trait. One of the two sisters shown here has normal coloration; the other has albinism. Most recessive homozygotes are born to parents who are carriers of the disorder but themselves have a normal phenotype, the case shown in the Punnett square.

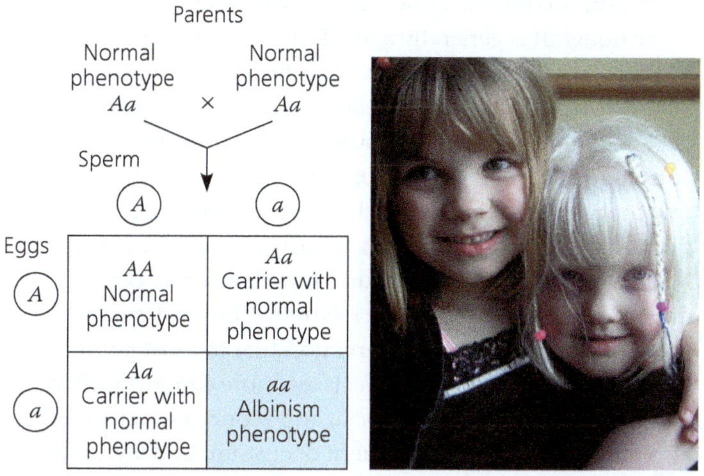

What is the probability that the sister with normal coloration is a carrier of the albinism allele?

Most people who have recessive disorders are born to parents who are carriers of the disorder but have a normal phenotype, as is the case shown in the Punnett square in Figure 14.16. A mating between two carriers corresponds to a Mendelian F_1 monohybrid cross, so the predicted genotypic ratio for the offspring is 1 *AA* : 2 *Aa* : 1 *aa*. Thus, each child has a $\frac{1}{4}$ chance of inheriting a double dose of the recessive allele; in the case of albinism, such a child will have albinism. From the genotypic ratio, we also can see that out of three offspring with the *normal* phenotype (one *AA* plus two *Aa*), two are predicted to be heterozygous carriers, a $\frac{2}{3}$ chance. Recessive homozygotes could also result from $Aa \times aa$ and $aa \times aa$ matings, but if the disorder is lethal before reproductive age or results in sterility (neither of which is true for albinism), no *aa* individuals will reproduce. Even if recessive homozygotes are able to reproduce, such matings will occur relatively rarely because these individuals account for a much smaller percentage of the population than heterozygous carriers (for reasons we'll examine in Concept 23.2).

In general, genetic disorders are not evenly distributed among all groups of people. For example, the incidence of Tay-Sachs disease, which we described earlier in this chapter, is disproportionately high among Ashkenazic Jews, Jewish people whose ancestors lived in central Europe. In that population, Tay-Sachs disease occurs in one out of 3,600 births, an incidence about 100 times greater than that among non-Jews or Mediterranean (Sephardic) Jews. This uneven distribution results from the different genetic histories of the world's peoples during less technological times, when populations were more geographically (and hence genetically) isolated.

When a disease-causing recessive allele is rare, it is relatively unlikely that two carriers of the same harmful allele will meet and mate. The probability of passing on recessive traits increases greatly, however, if the man and woman are close relatives (for example, siblings or first cousins). This is because people with recent common ancestors are more likely to carry the same recessive alleles than are unrelated people.

Thus, these consanguineous ("same blood") matings, indicated in pedigrees by double lines, are more likely to produce offspring homozygous for recessive traits—including harmful ones. Such effects can be observed in many types of domesticated and zoo animals that have become inbred.

Although it is generally agreed that inbreeding causes an increase in autosomal recessive conditions compared to those resulting from matings between unrelated parents, there is debate among geneticists about exactly how much human consanguinity increases the risk of inherited diseases. For one thing, many harmful alleles have such severe effects that a homozygous embryo spontaneously aborts long before birth. Most societies and cultures have laws or taboos forbidding marriages between close relatives. These rules may have evolved out of empirical observation that in most populations, stillbirths and birth defects are more common when parents are closely related. Social and economic factors have also influenced the development of customs and laws against consanguineous marriages.

Cystic Fibrosis

The most common lethal genetic disease in the United States is **cystic fibrosis**, which strikes one out of every 2,500 people of European descent but is much rarer in other groups. Among people of European descent, one out of 25 (4%) are carriers of the cystic fibrosis allele. The normal allele for this gene codes for a membrane protein that functions in the transport of chloride ions between certain cells and the extracellular fluid. These chloride transport channels are defective or absent in the plasma membranes of children who inherit two recessive alleles for cystic fibrosis. The result is an abnormally high concentration of intracellular chloride, which causes an uptake of water due to osmosis. This in turn causes the mucus that coats certain cells to become thicker and stickier than normal. The mucus builds up in the pancreas, lungs, digestive tract, and other organs, leading to multiple (pleiotropic) effects, including poor absorption of nutrients from the intestines, chronic bronchitis, and recurrent bacterial infections.

Untreated, cystic fibrosis can cause death by the age of 5. Daily doses of antibiotics to stop infection, gentle pounding on the chest to clear mucus from clogged airways, and other therapies can prolong life. In the United States, more than half of those with cystic fibrosis now survive into their 30s and beyond.

Sickle-Cell Disease: A Genetic Disorder with Evolutionary Implications

EVOLUTION The most common inherited disorder among people of African descent is **sickle-cell disease**, which affects one out of 400 African-Americans. Sickle-cell disease is caused by the substitution of a single amino acid in the hemoglobin protein of red blood cells; in homozygous individuals, all hemoglobin is of the sickle-cell (abnormal) variety. When the oxygen content of an affected individual's blood is low (at high altitudes or under physical stress, for instance), the sickle-cell hemoglobin proteins aggregate into long fibers that deform the red cells into a sickle shape (see Figure 5.19). Sickled cells may clump and clog small blood vessels, often leading to other symptoms throughout the body, including physical weakness, pain, organ damage, and even stroke and paralysis. Regular blood transfusions can ward off brain damage in children with sickle-cell disease, and new drugs can help prevent or treat other problems. There is currently no widely available cure, but the disease is the target of ongoing gene therapy research.

Although two sickle-cell alleles are necessary for an individual to manifest full-blown sickle-cell disease and thus the condition is considered a recessive one, the presence of one sickle-cell allele can affect the phenotype. Thus, at the organismal level, the normal allele is incompletely dominant to the sickle-cell allele **(Figure 14.17)**. At the molecular level, the two alleles are codominant; both normal and abnormal (sickle-cell) hemoglobins are made in heterozygotes (carriers), who are said to have *sickle-cell trait*. (Here the word "trait" is used to distinguish this condition from full-blown sickle-cell disease, thus it is used differently from its definition earlier in the chapter—any variant of a phenotypic character.) Heterozygotes are usually healthy but may suffer some symptoms during long periods of reduced blood oxygen.

Figure 14.17 Sickle-cell disease and sickle-cell trait.

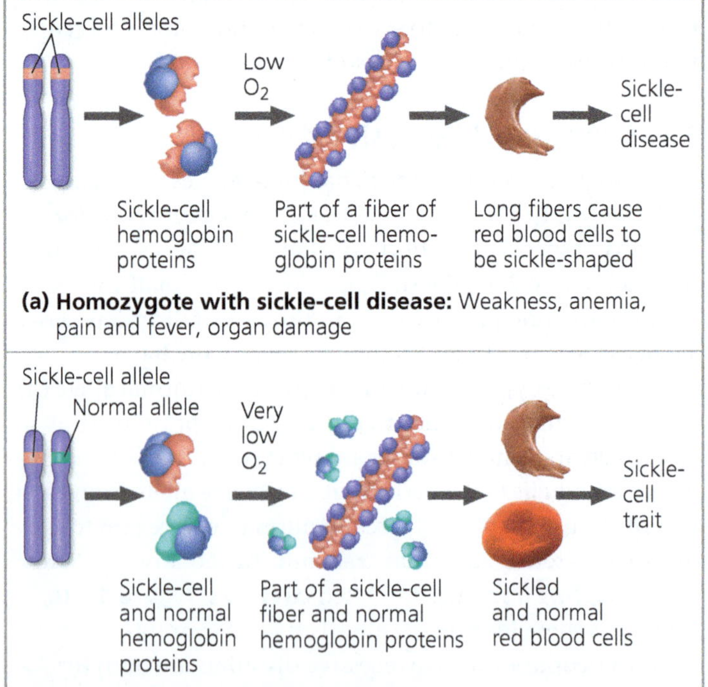

(a) Homozygote with sickle-cell disease: Weakness, anemia, pain and fever, organ damage

(b) Heterozygote with sickle-cell trait: Some symptoms when blood oxygen is very low; reduction of malaria symptoms

HHMI Animation: Sickle-Cell Disease

About one out of ten African-Americans have sickle-cell trait, an unusually high frequency of heterozygotes for an allele with severe detrimental effects in homozygotes. Why haven't evolutionary processes resulted in the disappearance of the allele among this population? One explanation is that having a single copy of the sickle-cell allele reduces the frequency and severity of malaria attacks, especially among young children. The malaria parasite spends part of its life cycle in red blood cells (see Figure 28.16), and the presence of even heterozygous amounts of sickle-cell hemoglobin results in lower parasite densities and hence reduced malaria symptoms. Thus, in tropical Africa, where infection with the malaria parasite is common, the sickle-cell allele confers an advantage to heterozygotes even though it is harmful in the homozygous state. (The balance between these two effects will be discussed in Concept 23.4; see Make Connections Figure 23.18.) The relatively high frequency of African-Americans with sickle-cell trait is a vestige of their African ancestry.

HHMI Video: The Making of the Fittest: Natural Selection in Humans (Sickle-Cell Disease)

Dominantly Inherited Disorders

Although many harmful alleles are recessive, a number of human disorders are due to dominant alleles. One example is *achondroplasia*, a form of dwarfism that occurs in one of every 25,000 people. Heterozygous individuals have the dwarf phenotype **(Figure 14.18)**. Therefore, all people who do not have achondroplasia—99.99% of the population—are homozygous for the recessive allele. Like the presence of extra fingers or toes mentioned earlier, achondroplasia is a trait for which the recessive allele is much more prevalent than the corresponding dominant allele.

Figure 14.18 Achondroplasia: a dominant trait. Dr. Michael C. Ain has achondroplasia, a form of dwarfism caused by a dominant allele. This has inspired his work: He is a specialist in the repair of bone defects caused by achondroplasia and other disorders. The dominant allele (*D*) might have arisen as a mutation in the egg or sperm of a parent or could have been inherited from an affected parent, as shown for an affected father in the Punnett square.

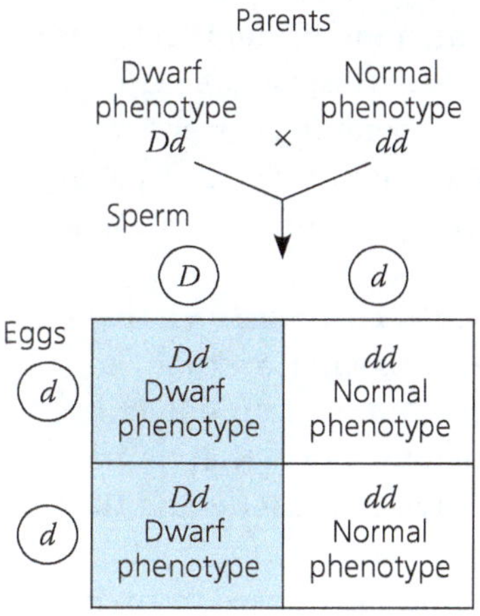

Unlike achondroplasia, which is relatively harmless, some dominant alleles cause lethal diseases. Those that do are much less common than recessive alleles that have lethal effects. A lethal recessive allele is only lethal when homozygous; it can be passed from one generation to the next by heterozygous carriers because the carriers themselves have normal phenotypes. A lethal dominant allele, however, often causes the death of afflicted individuals before they can mature and reproduce, and in this case the allele is not passed on to future generations.

A lethal dominant allele may be passed on, though, if the lethal disease symptoms first appear after reproductive age. In these cases, the individual may already have transmitted the allele to his or her children. For example, a degenerative disease of the nervous system, called **Huntington's disease**, is caused by a lethal dominant allele that has no obvious phenotypic effect until the individual is about 35 to 45 years old. Once the deterioration of the nervous system begins, it is irreversible and inevitably fatal. As with other dominant traits, a child born to a parent with the Huntington's disease allele has a 50% chance of inheriting the allele and the disorder (see the Punnett square in Figure 14.18). In the United States, this disease afflicts about one in 10,000 people.

At one time, the onset of symptoms was the only way to know if a person had inherited the Huntington's allele, but this is no longer the case. By analyzing DNA samples from a large family with a high incidence of the disorder, geneticists tracked the Huntington's allele to a locus near the tip of chromosome 4, and the gene was sequenced in 1993. This information led to the development of a test that could detect the presence of the Huntington's allele in an individual's genome. (The methods that make such tests possible are discussed in Concepts 20.1 and 20.4.) The availability of this test poses an agonizing dilemma for those with a family history of Huntington's disease. Some individuals may want to be tested for this disease, whereas others may decide it would be too stressful to find out.

Interview with Nancy Wexler: Mapping the gene that causes Huntington's disease

Multifactorial Disorders

The hereditary diseases we have discussed so far are sometimes described as simple Mendelian disorders because they result from abnormality of one or both alleles at a single genetic locus. Many more people are susceptible to diseases that have a multifactorial basis—a genetic component plus a significant environmental influence. Heart disease, diabetes, cancer, alcoholism, certain mental illnesses such as schizophrenia and bipolar disorder, and many other diseases are multifactorial. In these cases, the hereditary component is polygenic. For example, many genes affect cardiovascular health, making some of us more prone than others to heart attacks and strokes. No matter what our genotype, however, our lifestyle has a tremendous effect on phenotype for cardiovascular health and other multifactorial characters. Exercise, a healthful diet, abstinence from smoking, and an ability to handle stressful situations all reduce our risk of heart disease and some types of cancer.

Genetic Testing and Counseling

Avoiding simple Mendelian disorders is possible when the risk of a particular genetic disorder can be assessed before a child is conceived or during the early stages of the pregnancy. Many hospitals have genetic counselors who can provide information to prospective parents concerned about a family history for a specific disease. Fetal and newborn testing can also reveal genetic disorders.

Counseling Based on Mendelian Genetics and Probability Rules

Consider the case of a hypothetical couple, John and Carol. Each had a brother who died from the same recessively inherited lethal disease. Before conceiving their first child, John and Carol seek genetic counseling to determine the risk of having a child with the disease. From the information about their brothers, we know that both parents of John and both parents of Carol must have been carriers of the recessive allele. Thus, John and Carol are both products of $Aa \times Aa$ crosses, where a symbolizes the allele that causes this particular disease. We also know that John and Carol are not homozygous recessive (aa), because they do not have the disease. Therefore, their genotypes are either AA or Aa.

Given a genotypic ratio of 1 AA : 2 Aa : 1 aa for offspring of an $Aa \times Aa$ cross, John and Carol each have a $2/3$ chance of being carriers (Aa). According to the rule of multiplication, the overall probability of their firstborn having the disorder is $2/3$ (the chance that John is a carrier) times $2/3$ (the chance that Carol is a carrier) times $1/4$ (the chance of two carriers having a child with the disease), which equals $1/9$. Suppose that Carol and John decide to have a child—after all, there is an $8/9$ chance that their baby will not have the disorder. If, despite these odds, their child is born with the disease, then we would know that *both* John and Carol are, in fact, carriers (Aa genotype). If both John and Carol are carriers, there is a $1/4$ chance that any subsequent child this couple has will have the disease. The probability is higher for subsequent children because the diagnosis of the disease in the first child established that both parents are carriers, not because the genotype of the first child affects in any way that of future children.

When we use Mendel's laws to predict possible outcomes of matings, it is important to remember that each child represents an independent event in the sense that its genotype is unaffected by the genotypes of older siblings. Suppose that John and Carol have three more children, and *all three* have the hypothetical hereditary disease. There is only one chance in 64 ($1/4 \times 1/4 \times 1/4$) that such an outcome will occur. Despite this run of misfortune, the chance that a fourth child of this couple will have the disease remains $1/4$.

Tests for Identifying Carriers

Most children with recessive disorders are born to parents with normal phenotypes. The key to accurately assessing the genetic risk for a particular disease is therefore to find out whether the prospective parents are heterozygous carriers of the recessive allele. For an increasing number of heritable disorders, tests are available that can distinguish individuals of normal phenotype who are dominant homozygotes from those who are heterozygous carriers. There are now tests that can identify carriers of the alleles for Tay-Sachs disease, sickle-cell disease, and the most common form of cystic fibrosis. A program testing for carriers of Tay-Sachs disease that began in the 1980s has successfully reduced the rate of babies born with this disease.

These tests for identifying carriers enable people with family histories of genetic disorders to make informed decisions about having children, including whether to do genetic testing of the fetus, should they decide to become pregnant. The tests also raise other issues: Could carriers be denied health or life insurance or lose the jobs providing those benefits, even though they themselves are healthy? The Genetic Information Nondiscrimination Act, signed into law in the United States in 2008, allays these concerns by prohibiting discrimination in employment or insurance coverage based on genetic test results. A question that remains is whether sufficient genetic counseling is available to help large numbers of individuals understand their genetic test results. Even when test results are clearly understood, affected individuals may still face difficult decisions. Advances in biotechnology offer the potential to reduce human suffering, but along with them come ethical issues that require conscientious deliberation.

Fetal Testing

Suppose a couple expecting a child learns that they are both carriers of the Tay-Sachs allele. One of the tests that can be done to determine whether the developing fetus has Tay-Sachs disease is **amniocentesis**, which can be performed starting at the 15th week of pregnancy **(Figure 14.19a)**. In this procedure, a physician inserts a needle into the uterus and extracts about 10 mL of amniotic fluid, the liquid that bathes the fetus. Some genetic disorders can be detected from the presence of certain molecules in the amniotic fluid itself. Tests for other disorders, including Tay-Sachs disease, are performed on the DNA of cells cultured in the laboratory, descendants of fetal cells sloughed off into the amniotic fluid. A karyotype of these cultured cells can also identify certain chromosomal defects (see Figure 13.3).

In an alternative technique called **chorionic villus sampling (CVS)**, a physician inserts a narrow tube through the cervix into the uterus and suctions out a tiny sample of tissue from the placenta, the organ that transmits nutrients and fetal wastes between the fetus and the mother **(Figure 14.19b)**. The cells of the chorionic villi of the placenta—the portion sampled—are derived from the

Figure 14.19 Testing a fetus for genetic disorders. Biochemical tests may detect substances associated with particular disorders, and genetic testing can detect many genetic abnormalities. Karyotyping shows whether the chromosomes of the fetus are normal in number and appearance.

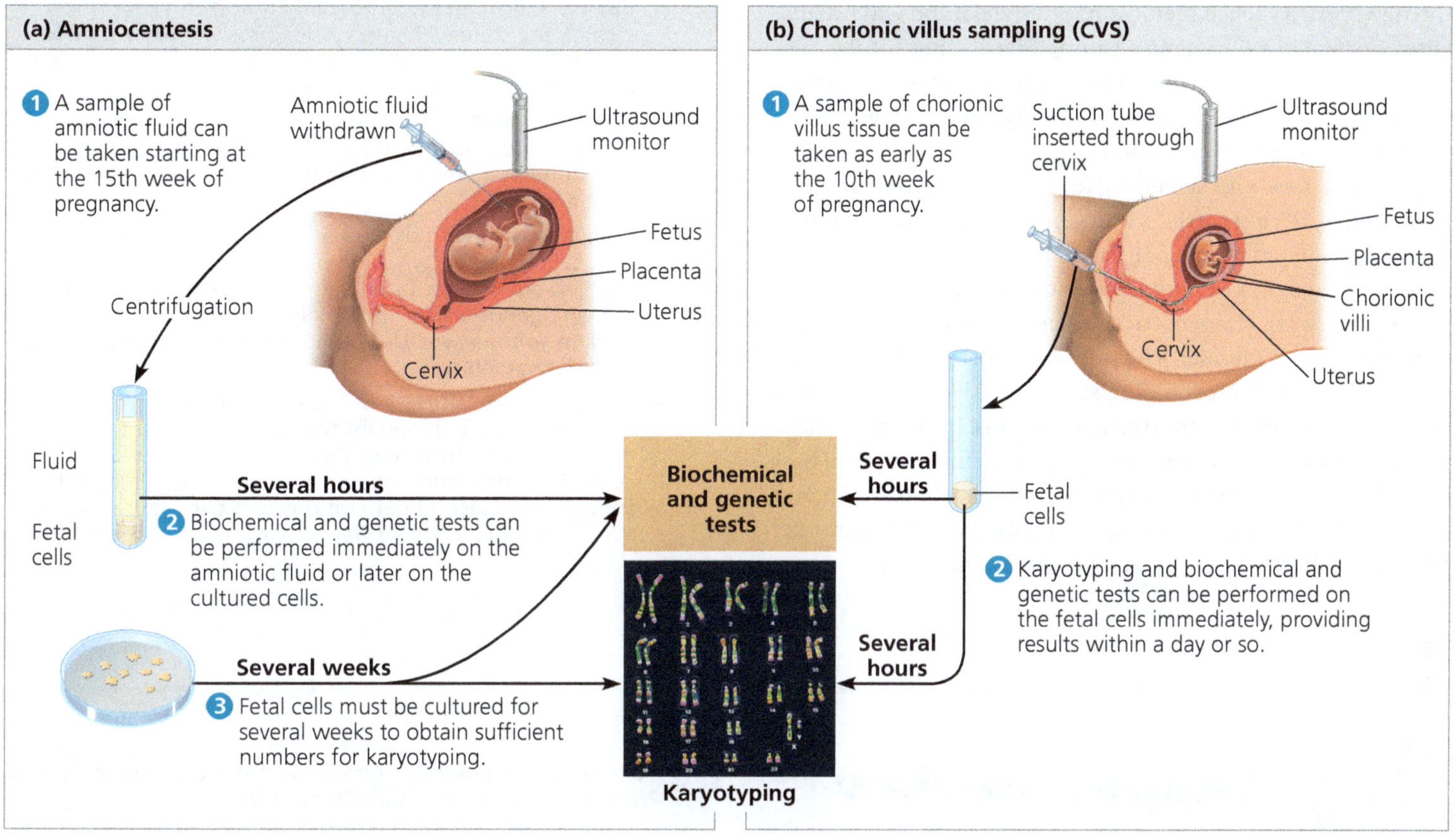

fetus and have the same genotype and DNA sequence as the new individual. These cells are proliferating rapidly enough to allow karyotyping to be carried out immediately. This rapid analysis represents an advantage over amniocentesis, in which the cells must be cultured for several weeks before karyotyping. Another advantage of CVS is that it can be performed as early as the 10th week of pregnancy.

Medical scientists have also developed methods for isolating fetal cells, or even fetal DNA, that have escaped into the mother's blood. Although very few are present, the cells can be cultured and tested, and the fetal DNA can be analyzed. In 2012, researchers were able to analyze the entire genome of a fetus, comparing sequences of samples obtained from both parents and fetal DNA found in the mother's blood. Cell-free fetal DNA tests and other blood tests are increasingly being used as noninvasive prenatal screening tests for certain disorders; a positive result indicates to the parents that further diagnostic testing, such as amniocentesis or CVS, should be considered.

Imaging techniques allow a physician to examine a fetus directly for major anatomical abnormalities that might not show up in genetic tests. In the *ultrasound* technique, for example, reflected sound waves are used to produce an image of the fetus by a simple noninvasive procedure.

Ultrasound and isolation of fetal cells or DNA from maternal blood pose no known risk to either mother or fetus, while the other procedures can cause complications in a small percentage of cases. Amniocentesis or CVS for diagnostic testing is generally offered to women over age 35, due to their increased risk of bearing a child with Down syndrome, and may also be offered to younger women if there are known concerns. If the fetal tests reveal a serious disorder like Tay-Sachs, the parents face the difficult choice of either terminating the pregnancy or preparing to care for a child with a genetic disorder, one that might even be fatal. Parental and fetal screening for Tay-Sachs alleles done since 1980 has reduced the number of children born with this incurable disease by 90%. In 2008, the Chinese government initiated a program of fetal testing to detect a harmful genetic blood disorder called β-thalassemia. This effort resulted in a reduction in the rate of this disorder from over 21 births per 1000 in 2008 to just under 13 in 2011.

Newborn Screening

Some genetic disorders can be detected at birth by simple biochemical tests that are now routinely performed in most hospitals in the United States. One common screening program is for phenylketonuria (PKU), a recessively inherited disorder that occurs in about one out of every 10,000–15,000 births in the United States. Children with this disease cannot properly metabolize the amino acid phenylalanine. This compound and its by-product, phenylpyruvate, can accumulate to toxic levels in the blood, causing severe intellectual disability (mental retardation). However, if PKU is detected in the newborn, a special diet low in phenylalanine will usually allow normal development. (Among many other substances, this diet excludes the artificial sweetener aspartame, which contains phenylalanine.) Unfortunately, few other genetic disorders are treatable at present.

Fetal and newborn screening for serious inherited diseases, tests for identifying carriers, and genetic counseling all rely on the Mendelian model of inheritance. We owe the "gene idea"—the concept of heritable factors transmitted according to simple rules of chance—to the elegant quantitative experiments of Gregor Mendel. The importance of his discoveries was overlooked by most biologists until early in the 20th century, decades after he reported his findings. In the next chapter, you will learn how Mendel's laws have their physical basis in the behavior of chromosomes during sexual life cycles and how the synthesis of Mendelian genetics and a chromosome theory of inheritance catalyzed progress in genetics.

CONCEPT CHECK 14.4

1. Beth and Tom each have a sibling with cystic fibrosis, but neither Beth nor Tom nor any of their parents have the disease. Calculate the probability that if this couple has a child, the child will have cystic fibrosis. What would be the probability if a test revealed that Tom is a carrier but Beth is not? Explain your answers.
2. **MAKE CONNECTIONS ➤** Explain how the change of a single amino acid in hemoglobin leads to the aggregation of hemoglobin into long fibers. (Review Figures 5.14, 5.18, and 5.19.)
3. Joan was born with six toes on each foot, a dominant trait called polydactyly. Two of her five siblings and her mother, but not her father, also have extra digits. What is Joan's genotype for the number-of-digits character? Explain your answer. Use *D* and *d* to symbolize the alleles for this character.
4. **MAKE CONNECTIONS ➤** In Table 14.1, note the phenotypic ratio of the dominant to recessive trait in the F_2 generation for the monohybrid cross involving flower color. Then determine the phenotypic ratio for the offspring of the second-generation couple in Figure 14.15b. What accounts for the difference in the two ratios?

For suggested answers, see Appendix A.

14 Chapter Review

Go to **MasteringBiology**™ for Videos, Animations, Vocab Self-Quiz, Practice Tests, and more in the Study Area.

SUMMARY OF KEY CONCEPTS

CONCEPT 14.1

Mendel used the scientific approach to identify two laws of inheritance
(pp. 270–276)

VOCAB SELF-QUIZ goo.gl/6u55ks

- Gregor Mendel formulated a theory of inheritance based on experiments with garden peas, proposing that parents pass on to their offspring discrete genes that retain their identity through generations. This theory includes two "laws."
- The **law of segregation** states that genes have alternative forms, or **alleles**. In a diploid organism, the two alleles of a gene segregate (separate) during meiosis and gamete formation; each sperm or egg carries only one allele of each pair. This law explains the 3:1 ratio of F_2 phenotypes observed when **monohybrids** self-pollinate. Each organism inherits one allele for each gene from each parent. In **heterozygotes**, the two alleles are different; expression of the **dominant allele** masks the phenotypic effect of the **recessive allele. Homozygotes** have identical alleles of a given gene and are therefore **true-breeding**.
- The **law of independent assortment** states that the pair of alleles for a given gene segregates into gametes independently of the pair of alleles for any other gene. In a cross between **dihybrids** (individuals heterozygous for two genes), the offspring have four phenotypes in a 9:3:3:1 ratio.

? *When Mendel did crosses of true-breeding purple- and white-flowered pea plants, the white-flowered trait disappeared from the F_1 generation but reappeared in the F_2 generation. Use genetic terms to explain why that happened.*

CONCEPT 14.2

Probability laws govern Mendelian inheritance
(pp. 276–278)

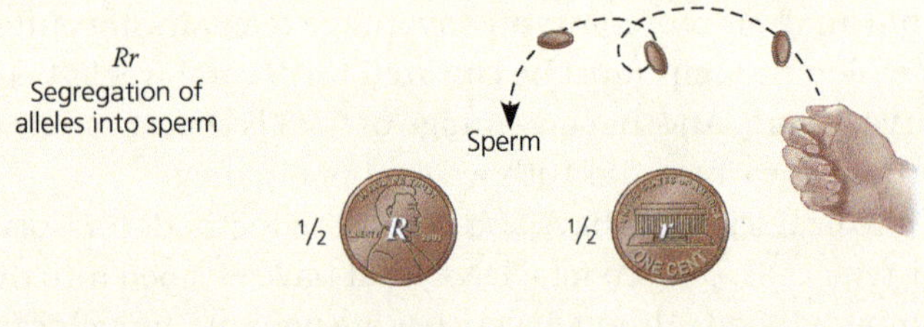

- The **multiplication rule** states that the probability of two or more events occurring together is equal to the product of the individual probabilities of the independent single events. The **addition rule** states that the probability of an event that can occur in two or more independent, mutually exclusive ways is the sum of the individual probabilities.
- The rules of probability can be used to solve complex genetics problems. A dihybrid or other multicharacter cross is equivalent to two or more independent monohybrid crosses occurring simultaneously. In calculating the chances of the various offspring genotypes from such crosses, each character is first considered separately and then the individual probabilities are multiplied.

DRAW IT ➤ *Redraw the Punnett square on the right side of Figure 14.8 as two smaller monohybrid Punnett squares, one for each gene. Below each square, list the fractions of each phenotype produced. Use the rule of multiplication to compute the overall fraction of each possible dihybrid phenotype. What is the phenotypic ratio?*

CONCEPT 14.3

Inheritance patterns are often more complex than predicted by simple Mendelian genetics *(pp. 278–283)*

- Extensions of Mendelian genetics for a single gene:

Relationship among alleles of a single gene	Description	Example
Complete dominance of one allele	Heterozygous phenotype same as that of homozygous dominant	
Incomplete dominance of either allele	Heterozygous phenotype intermediate between the two homozygous phenotypes	
Codominance	Both phenotypes expressed in heterozygotes	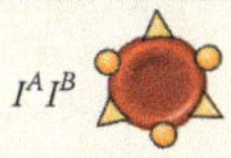
Multiple alleles	In the population, some genes have more than two alleles	ABO blood group alleles I^A, I^B, i
Pleiotropy	One gene affects multiple phenotypic characters	Sickle-cell disease

- Extensions of Mendelian genetics for two or more genes:

Relationship among two or more genes	Description	Example
Epistasis	The phenotypic expression of one gene affects the expression of another gene	
Polygenic inheritance	A single phenotypic character is affected by two or more genes	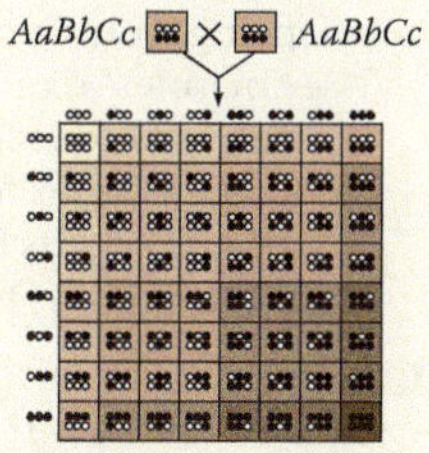

- The expression of a genotype can be affected by environmental influences, resulting in a range of phenotypes. Polygenic characters that are also influenced by the environment are called **multifactorial** characters.
- An organism's overall phenotype, including its physical appearance, internal anatomy, physiology, and behavior, reflects its overall genotype and unique environmental history. Even in more complex inheritance patterns, Mendel's fundamental laws still apply.

? *Which genetic relationships listed in the first column of the two tables above are demonstrated by the inheritance pattern of the ABO blood group alleles? For each genetic relationship, explain why this inheritance pattern is or is not an example.*

CONCEPT 14.4

Many human traits follow Mendelian patterns of inheritance *(pp. 284–290)*

- Analysis of family **pedigrees** can be used to deduce the possible genotypes of individuals and make predictions about future offspring. Such predictions are statistical probabilities rather than certainties.

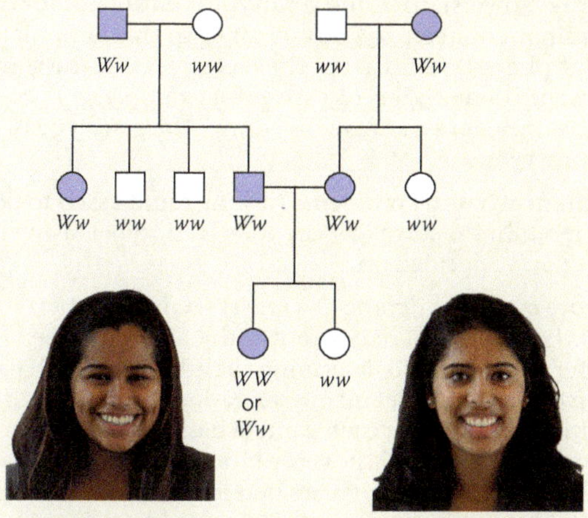

- Many genetic disorders are inherited as simple recessive traits. Most affected (homozygous recessive) individuals are children of phenotypically normal, heterozygous **carriers**.
- The sickle-cell allele has probably persisted for evolutionary reasons: Homozygotes have sickle-cell disease, but heterozygotes have an advantage because one copy of the sickle-cell allele reduces both the frequency and severity of malaria attacks.

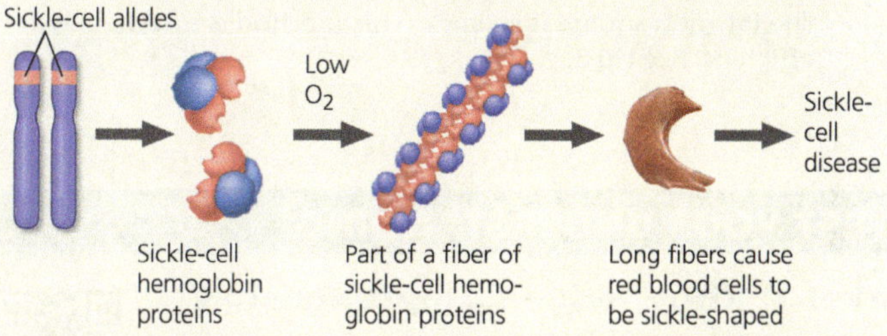

- Lethal dominant alleles are eliminated from the population if affected people die before reproducing. Nonlethal dominant alleles and lethal ones that strike relatively late in life can be inherited in a Mendelian way.
- Many human diseases are multifactorial—that is, they have both genetic and environmental components and do not follow simple Mendelian inheritance patterns.
- Using family histories, genetic counselors help couples determine the probability that their children will have genetic disorders. Genetic testing of prospective parents to reveal whether they are carriers of recessive alleles associated with specific disorders has become widely available. Blood tests can screen for certain disorders in a fetus. **Amniocentesis** and **chorionic villus sampling** can indicate whether a suspected genetic disorder is present in a fetus. Other genetic tests can be performed after birth.

? *Both members of a couple know that they are carriers of the cystic fibrosis allele. None of their three children has cystic fibrosis, but any one of them might be a carrier. The couple would like to have a fourth child but are worried that he or she would very likely have the disease, since the first three do not. What would you tell the couple? Would it remove some uncertainty from their prediction if they could find out whether the three children are carriers?*

TIPS FOR GENETICS PROBLEMS

1. Write down symbols for the alleles. (These may be given in the problem.) When represented by single letters, the dominant allele is uppercase and the recessive is lowercase.
2. Write down the possible genotypes, as determined by the phenotype.
   a. If the phenotype is that of the dominant trait (for example, purple flowers), then the genotype is either homozygous dominant or heterozygous (*PP* or *Pp* in this example).
   b. If the phenotype is that of the recessive trait, the genotype must be homozygous recessive (for example, *pp*).
   c. If the problem says "true-breeding," the genotype is homozygous.
3. Determine what the problem is asking for. If asked to do a cross, write it out in the form [genotype] × [genotype], using the alleles you've decided on.
4. To figure out the outcome of a cross, set up a Punnett square.
   a. Put the gametes of one parent at the top and those of the other on the left. To determine the allele(s) in each gamete for a given genotype, set up a systematic way to list all the possibilities. (Remember, each gamete has one allele of each gene.) Note that there are 2^n possible types of gametes, where n is the number of gene loci that are heterozygous. For example, an individual with genotype *AaBbCc* would produce $2^3 = 8$ types of gametes. Write the genotypes of the gametes in circles above the columns and to the left of the rows.
   b. Fill in the Punnett square as if each possible sperm were fertilizing each possible egg, making all of the possible offspring. In a cross of *AaBbCc* × *AaBbCc*, for example, the Punnett square would have 8 columns and 8 rows, so there are 64 different offspring; you would know the genotype of each and thus the phenotype. Count genotypes and phenotypes to obtain the genotypic and phenotypic ratios. Because the Punnett square is so large, this method is not the most efficient. See tip 5.
5. You can use the rules of probability if a Punnett square would be too big. (For example, see the question at the end of the summary for Concept 14.2 and question 7 below.) You can consider each gene separately (see the section Solving Complex Genetics Problems with the Rules of Probability in Concept 14.2).
6. If the problem gives you the phenotypic ratios of offspring but not the genotypes of the parents in a given cross, the phenotypes can help you deduce the parents' unknown genotypes.
   a. For example, if $1/2$ of the offspring have the recessive phenotype and $1/2$ the dominant, you know that the cross was between a heterozygote and a homozygous recessive.
   b. If the ratio is 3:1, the cross was between two heterozygotes.
   c. If two genes are involved and you see a 9:3:3:1 ratio in the offspring, you know that each parent is heterozygous for both genes. Caution: Don't assume that the reported numbers will exactly equal the predicted ratios. For example, if there are 13 offspring with the dominant trait and 11 with the recessive, assume that the ratio is one dominant to one recessive.
7. For pedigree problems, use the tips in Figure 14.15 and below to determine what kind of trait is involved.
   a. If parents without the trait have offspring with the trait, the trait must be recessive and the parents both carriers.
   b. If the trait is seen in every generation, it is most likely dominant (see the next possibility, though).
   c. If both parents have the trait, then in order for it to be recessive, all offspring must show the trait.
   d. To determine the likely genotype of a certain individual in a pedigree, first label the genotypes of all the family members you can. Even if some of the genotypes are incomplete, label what you do know. For example, if an individual has the dominant phenotype, the genotype must be *AA* or *Aa*; you can write this as *A*–. Try different possibilities to see which fits the results. Use the rules of probability to calculate the probability of each possible genotype being the correct one.

TEST YOUR UNDERSTANDING

Level 1: Knowledge/Comprehension

PRACTICE TEST
goo.gl/CUYGKD

1. **DRAW IT** Two pea plants heterozygous for the characters of pod color and pod shape are crossed. Draw a Punnett square to determine the phenotypic ratios of the offspring.
2. A man with type A blood marries a woman with type B blood. Their child has type O blood. What are the genotypes of these three individuals? What genotypes, and in what frequencies, would you expect in future offspring from this marriage?
3. A man has six fingers on each hand and six toes on each foot. His wife and their daughter have the normal number of digits. Remember that extra digits is a dominant trait. What fraction of this couple's children would be expected to have extra digits?
4. **DRAW IT** A pea plant heterozygous for inflated pods (*Ii*) is crossed with a plant homozygous for constricted pods (*ii*). Draw a Punnett square for this cross to predict genotypic and phenotypic ratios. Assume that pollen comes from the *ii* plant.

Level 2: Application/Analysis

5. Flower position, stem length, and seed shape are three characters that Mendel studied. Each is controlled by an independently assorting gene and has dominant and recessive expression as indicated in Table 14.1. If a plant that is heterozygous for all three characters is allowed to self-fertilize, what proportion of the offspring would you expect to be each of the following? (*Note:* Use the rules of probability instead of a huge Punnett square.)
   (a) homozygous for the three dominant traits
   (b) homozygous for the three recessive traits
   (c) heterozygous for all three characters
   (d) homozygous for axial and tall, heterozygous for seed shape
6. Hemochromatosis is an inherited disease caused by a recessive allele. If a woman and her husband, who are both carriers, have three children, what is the probability of each of the following?
   (a) All three children are of normal phenotype.
   (b) One or more of the three children have the disease.
   (c) All three children have the disease.
   (d) At least one child is phenotypically normal.

   (*Note:* It will help to remember that the probabilities of all possible outcomes always add up to 1.)
7. The genotype of F_1 individuals in a tetrahybrid cross is *AaBbCcDd*. Assuming independent assortment of these four genes, what are the probabilities that F_2 offspring will have the following genotypes?
   (a) *aabbccdd*
   (b) *AaBbCcDd*
   (c) *AABBCCDD*
   (d) *AaBBccDd*
   (e) *AaBBCCdd*

8. What is the probability that each of the following pairs of parents will produce the indicated offspring? (Assume independent assortment of all gene pairs.)
(a) *AABBCC* × *aabbcc* → *AaBbCc*
(b) *AABbCc* × *AaBbCc* → *AAbbCC*
(c) *AaBbCc* × *AaBbCc* → *AaBbCc*
(d) *aaBbCC* × *AABbcc* → *AaBbCc*

9. Karen and Steve each have a sibling with sickle-cell disease. Neither Karen nor Steve nor any of their parents have the disease, and none of them have been tested to see if they carry the sickle-cell allele. Based on this incomplete information, calculate the probability that if this couple has a child, the child will have sickle-cell disease.

10. In 1981, a stray black cat with unusual rounded, curled-back ears was adopted by a family in California. Hundreds of descendants of the cat have since been born, and cat fanciers hope to develop the curl cat into a show breed. Suppose you owned the first curl cat and wanted to develop a true-breeding variety. How would you determine whether the curl allele is dominant or recessive? How would you obtain true-breeding curl cats? How could you be sure they are true-breeding?

11. In tigers, a recessive allele of a particular gene causes both an absence of fur pigmentation (a white tiger) and a cross-eyed condition. If two phenotypically normal tigers that are heterozygous at this locus are mated, what percentage of their offspring will be cross-eyed? What percentage of cross-eyed tigers will be white?

12. In maize (corn) plants, a dominant allele *I* inhibits kernel color, while the recessive allele *i* permits color when homozygous. At a different locus, the dominant allele *P* causes purple kernel color, while the homozygous recessive genotype *pp* causes red kernels. If plants heterozygous at both loci are crossed, what will be the phenotypic ratio of the offspring?

13. The pedigree below traces the inheritance of alkaptonuria, a biochemical disorder. Affected individuals, indicated here by the colored circles and squares, are unable to metabolize a substance called alkapton, which colors the urine and stains body tissues. Does alkaptonuria appear to be caused by a dominant allele or by a recessive allele? Fill in the genotypes of the individuals whose genotypes can be deduced. What genotypes are possible for each of the other individuals?

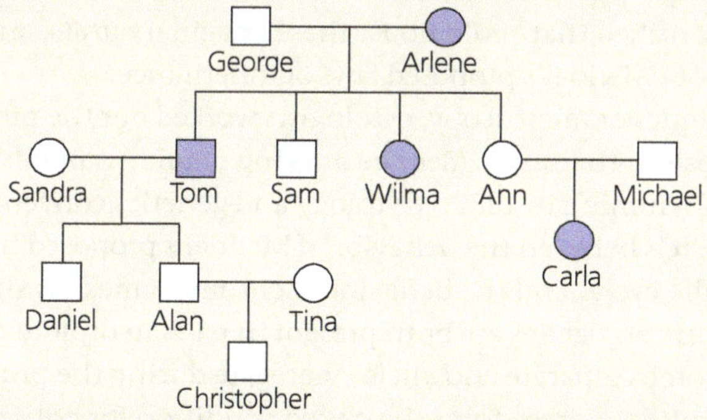

14. Imagine that you are a genetic counselor, and a couple planning to start a family comes to you for information. Charles was married once before, and he and his first wife had a child with cystic fibrosis. The brother of his current wife, Elaine, died of cystic fibrosis. What is the probability that Charles and Elaine will have a baby with cystic fibrosis? (Neither Charles, Elaine, nor their parents have cystic fibrosis.)

Level 3: Synthesis/Evaluation

15. EVOLUTION CONNECTION Over the past half century, there has been a trend in the United States and other developed countries for people to marry and start families later in life than did their parents and grandparents. What effects might this trend have on the incidence (frequency) of late-acting dominant lethal alleles in the population?

16. SCIENTIFIC INQUIRY You are handed a mystery pea plant with tall stems and axial flowers and asked to determine its genotype as quickly as possible. You know that the allele for tall stems (*T*) is dominant to that for dwarf stems (*t*) and that the allele for axial flowers (*A*) is dominant to that for terminal flowers (*a*).
(a) Identify all the possible genotypes for your mystery plant.
(b) Describe the one cross you would do, out in your garden, to determine the exact genotype of your mystery plant.
(c) While waiting for the results of your cross, you predict the results for each possible genotype listed in part a. Explain how you do this and why this is not called "performing a cross."
(d) Explain how the results of your cross and your predictions will help you learn the genotype of your mystery plant.

17. WRITE ABOUT A THEME: INFORMATION The continuity of life is based on heritable information in the form of DNA. In a short essay (100–150 words), explain how the passage of genes from parents to offspring, in the form of particular alleles, ensures perpetuation of parental traits in offspring and, at the same time, genetic variation among offspring. Use genetic terms in your explanation.

18. SYNTHESIZE YOUR KNOWLEDGE

Just for fun, imagine that "shirt-striping" is a phenotypic character caused by a single gene. Construct a genetic explanation for the appearance of the family in the above photograph, consistent with their "shirt phenotypes." Include in your answer the presumed allele combinations for "shirt-striping" in each family member. Identify the inheritance pattern shown by the child.

For selected answers, see Appendix A.

For additional practice questions, check out the **Dynamic Study Modules** in MasteringBiology. You can use them to study on your smartphone, tablet, or computer anytime, anywhere!

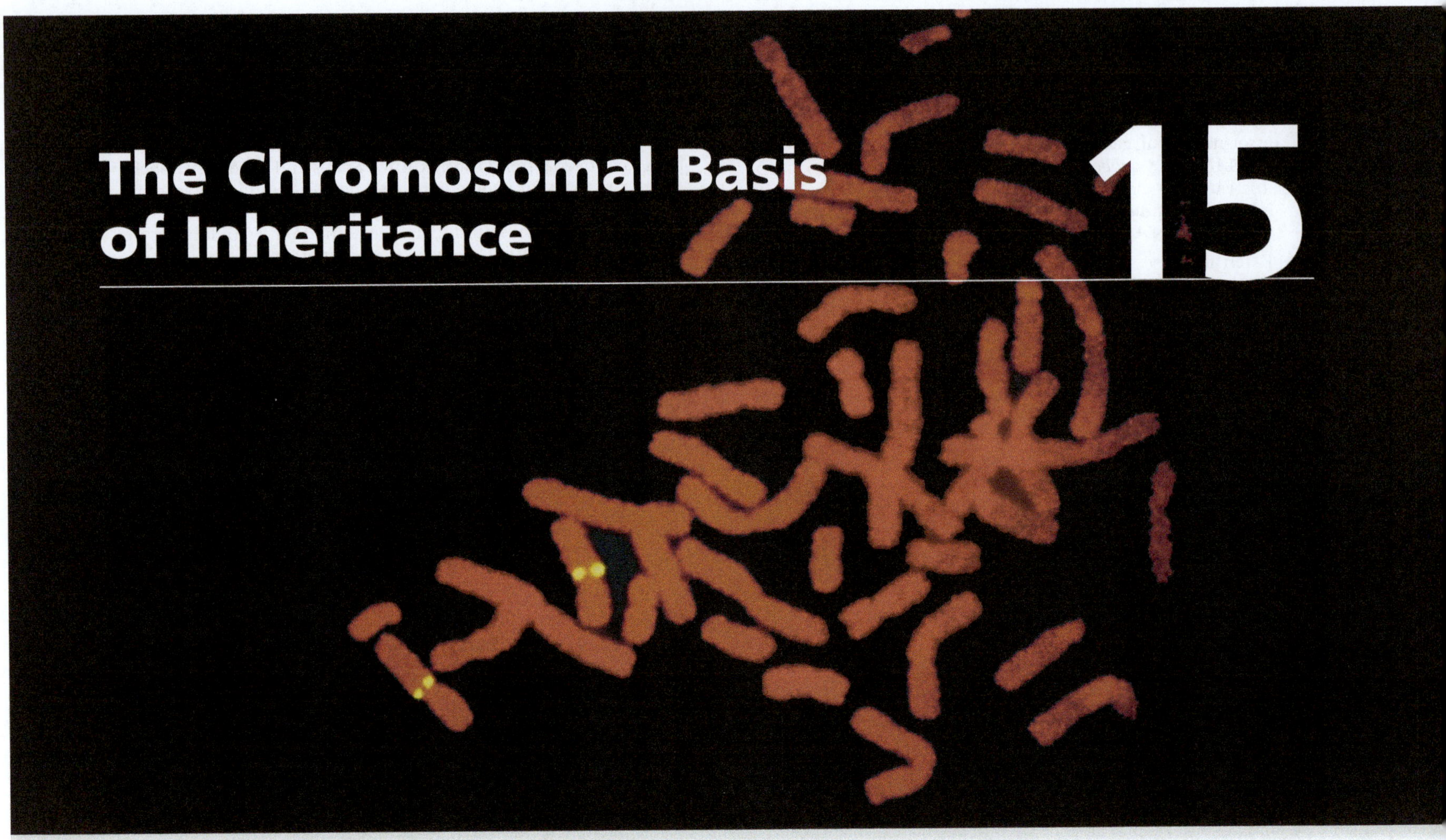

The Chromosomal Basis of Inheritance

15

▲ **Figure 15.1** **Where in the cell are Mendel's hereditary factors located?**

KEY CONCEPTS

15.1 Morgan showed that Mendelian inheritance has its physical basis in the behavior of chromosomes: *scientific inquiry*

15.2 Sex-linked genes exhibit unique patterns of inheritance

15.3 Linked genes tend to be inherited together because they are located near each other on the same chromosome

15.4 Alterations of chromosome number or structure cause some genetic disorders

15.5 Some inheritance patterns are exceptions to standard Mendelian inheritance

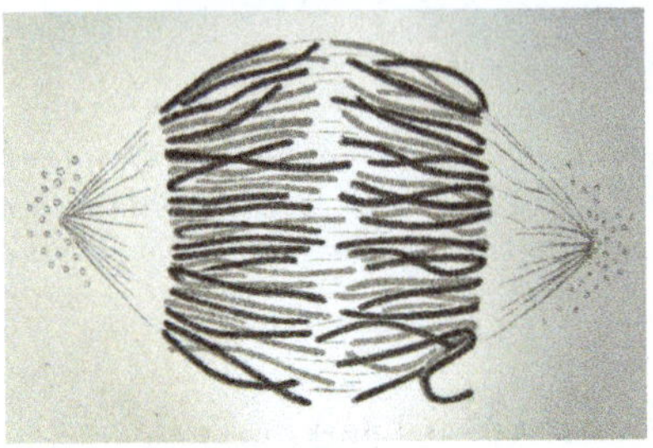

◀ **Drawing of mitosis (Flemming, 1882)**

Locating Genes Along Chromosomes

Today, we know that genes—Mendel's "factors"—are segments of DNA located along chromosomes. We can see the location of a particular gene by tagging chromosomes with a fluorescent dye that highlights that gene. For example, the four yellow dots in **Figure 15.1** mark a specific gene on a pair of homologous human chromosomes. (The chromosomes have duplicated, so each chromosome has two copies of the allele, one on each sister chromatid.) However, Gregor Mendel's "hereditary factors" were purely an abstract concept when he proposed their existence in 1860. At that time, no cellular structures had been identified that could house these imaginary units, and most biologists were skeptical about Mendel's proposed laws of inheritance.

Using improved techniques of microscopy, cytologists worked out the process of mitosis in 1875 and meiosis in the 1890s. (See the drawing of mitosis published by German biologist Walther Flemming in 1882.) Cytology and genetics converged as biologists began to see parallels between the behavior of Mendel's proposed hereditary factors during sexual life cycles and the behavior of chromosomes: As shown in **Figure 15.2**, chromosomes and genes are both present in pairs in diploid cells, and homologous chromosomes separate and alleles segregate during the process of meiosis. After meiosis, fertilization restores the paired condition for both chromosomes and genes. Around 1902, Walter S. Sutton, Theodor Boveri, and others

When you see this blue icon, log in to **MasteringBiology** and go to the Study Area for digital resources.

MB **Get Ready for This Chapter**

Figure 15.2 The chromosomal basis of Mendel's laws. Here we correlate the results of one of Mendel's dihybrid crosses (see Figure 14.8) with the behavior of chromosomes during meiosis (see Figure 13.8). The arrangement of chromosomes at metaphase I of meiosis and their movement during anaphase I account for, respectively, the independent assortment and segregation of the alleles for seed color and shape. Each cell that undergoes meiosis in an F_1 plant produces two kinds of gametes. If we count the results for all cells, however, each F_1 plant produces equal numbers of all four kinds of gametes because the alternative chromosome arrangements at metaphase I are equally likely.

MP3 Tutor: The Chromosomal Basis of Inheritance
Animation: The Chromosomal Basis of Independent Assortment

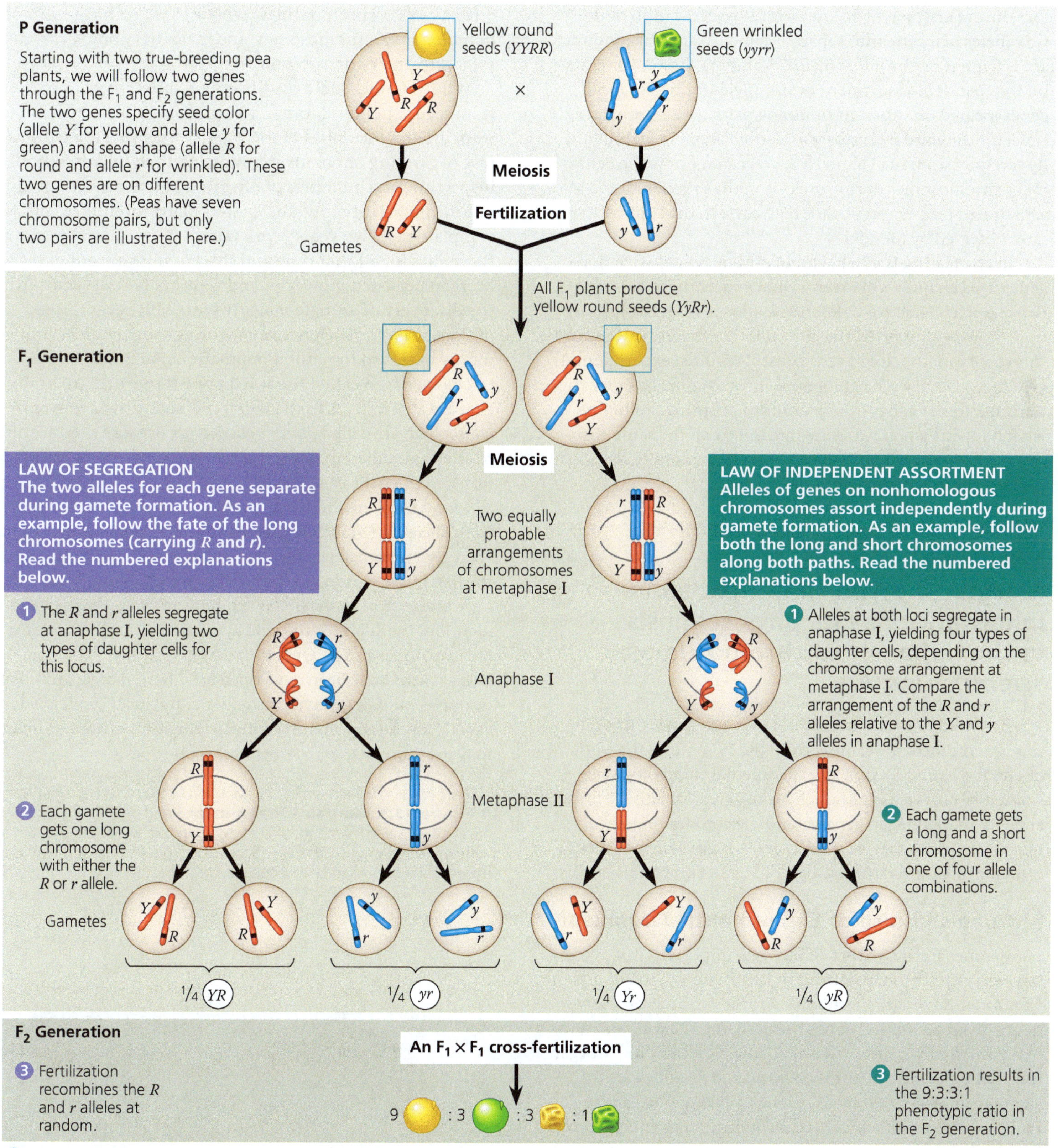

If you crossed an F_1 plant with a plant that was homozygous recessive for both genes (yyrr), how would the phenotypic ratio of the offspring compare with the 9:3:3:1 ratio seen here?

independently noted these parallels and began to develop the **chromosome theory of inheritance**. According to this theory, Mendelian genes have specific loci (positions) along chromosomes, and it is the chromosomes that undergo segregation and independent assortment.

As you can see in Figure 15.2, the separation of homologs during anaphase I accounts for the segregation of the two alleles of a gene into separate gametes, and the random arrangement of chromosome pairs at metaphase I accounts for independent assortment of the alleles for two or more genes located on different homolog pairs. This figure traces the same dihybrid pea cross you learned about in Figure 14.8. By carefully studying Figure 15.2, you can see how the behavior of chromosomes during meiosis in the F_1 generation and subsequent random fertilization give rise to the F_2 phenotypic ratio observed by Mendel.

In correlating the behavior of chromosomes with that of genes, this chapter will extend what you learned in the past two chapters. First, we'll describe evidence from the fruit fly that strongly supported the chromosome theory. (Although this theory made a lot of sense, it still required experimental evidence.) Next, we'll explore the chromosomal basis for the transmission of genes from parents to offspring, including what happens when two genes are linked on the same chromosome. Finally, we will discuss some important exceptions to the standard mode of inheritance.

CONCEPT 15.1

Morgan showed that Mendelian inheritance has its physical basis in the behavior of chromosomes: *scientific inquiry*

The first solid evidence associating a specific gene with a specific chromosome came early in the 1900s from the work of Thomas Hunt Morgan, an experimental embryologist at Columbia University. Although Morgan was initially skeptical about both Mendelian genetics and the chromosome theory, his early experiments provided convincing evidence that chromosomes are indeed the location of Mendel's heritable factors.

Morgan's Choice of Experimental Organism

Many times in the history of biology, important discoveries have come to those insightful or lucky enough to choose an experimental organism suitable for the research problem being tackled. Mendel chose the garden pea because a number of distinct varieties were available. For his work, Morgan selected a species of fruit fly, *Drosophila melanogaster*, a common insect that feeds on the fungi growing on fruit. Fruit flies are prolific breeders; a single mating will produce hundreds of offspring, and a new generation can be bred every two weeks. Morgan's laboratory began using this convenient organism for genetic studies in 1907 and soon became known as "the fly room."

Another advantage of the fruit fly is that it has only four pairs of chromosomes, which are easily distinguishable with a light microscope. There are three pairs of autosomes and one pair of sex chromosomes. Female fruit flies have a pair of homologous X chromosomes, and males have one X chromosome and one Y chromosome.

While Mendel could readily obtain different pea varieties from seed suppliers, Morgan was probably the first person to want different varieties of the fruit fly. He faced the tedious task of carrying out many matings and then microscopically inspecting large numbers of offspring in search of naturally occurring variant individuals. After many months of this, he complained, "Two years' work wasted. I have been breeding those flies for all that time and I've got nothing out of it." Morgan persisted, however, and was finally rewarded with the discovery of a single male fly with white eyes instead of the usual red. The phenotype for a character most commonly observed in natural populations, such as red eyes in *Drosophila*, is called the **wild type** (**Figure 15.3**). Traits that are alternatives to the wild type, such as white eyes in *Drosophila*, are called *mutant phenotypes* because they are due to alleles assumed to have originated as changes, or mutations, in the wild-type allele.

Morgan and his students invented a notation for symbolizing alleles in *Drosophila* that is still widely used for fruit flies. For a given character in flies, the gene takes its symbol from the first mutant (non–wild type) discovered. Thus, the allele for white eyes in *Drosophila* is symbolized by *w*. A superscript + identifies the allele for the wild-type trait: w^+ for the allele for red eyes, for example. Over the years, a variety of gene notation systems have been developed for different organisms. For example, human genes are usually written in all capitals, such as *HTT* for the gene involved in Huntington's disease. (Alleles may use more than one letter, as this one does.)

▼ Figure 15.3 Morgan's first mutant. Wild-type *Drosophila* flies have red eyes (left). Among his flies, Morgan discovered a mutant male with white eyes (right). This variation made it possible for Morgan to trace a gene for eye color to a specific chromosome.

Correlating Behavior of a Gene's Alleles with Behavior of a Chromosome Pair

Morgan mated his white-eyed male fly with a red-eyed female. All the F_1 offspring had red eyes, suggesting that the wild-type allele is dominant. When Morgan bred the F_1 flies to each other, he observed the classical 3:1 phenotypic ratio among the F_2 offspring. However, there was a surprising additional result: The white-eye trait showed up only in males. All the F_2 females had red eyes, while half the males had red eyes and half had white eyes. Therefore, Morgan concluded that somehow a fly's eye color was linked to its sex. (If the eye color gene were unrelated to sex, half of the white-eyed flies would have been female.)

Recall that a female fly has two X chromosomes (XX), while a male fly has an X and a Y (XY). The correlation between the trait of white eye color and the male sex of the affected F_2 flies suggested to Morgan that the gene involved in his white-eyed mutant was located exclusively on the X chromosome, with no corresponding allele present on the Y chromosome. His reasoning can be followed in **Figure 15.4**. For a male, a single copy of the mutant allele would confer white eyes; since a male has only one X chromosome, there can be no wild-type allele (w^+) present to mask the recessive allele. However, a female could have white eyes only if both her X chromosomes carried the recessive mutant allele (w). This was impossible for the F_2 females in Morgan's experiment because all the F_1 fathers had red eyes, so each F_2 female received a w^+ allele on the X chromosome inherited from her father.

Morgan's finding of the correlation between a particular trait and an individual's sex provided support for the chromosome theory of inheritance: namely, that a specific gene is carried on a specific chromosome (in this case, an eye color gene on the X chromosome). In addition, Morgan's work indicated that genes located on a sex chromosome exhibit unique inheritance patterns, which we will discuss in the next section. Recognizing the importance of Morgan's early work, many bright students were attracted to his fly room.

CONCEPT CHECK 15.1

1. Which one of Mendel's laws describes the inheritance of alleles for a single character? Which law relates to the inheritance of alleles for two characters in a dihybrid cross?
2. **MAKE CONNECTIONS ➤** Review the description of meiosis (see Figure 13.8) and Mendel's laws of segregation and independent assortment (see Concept 14.1). What is the physical basis for each of Mendel's laws?
3. **WHAT IF? ➤** Propose a possible reason that the first naturally occurring mutant fruit fly Morgan saw involved a gene on a sex chromosome and was found in a male.

For suggested answers, see Appendix A.

▼ Figure 15.4

Inquiry In a cross between a wild-type female fruit fly and a mutant white-eyed male, what color eyes will the F_1 and F_2 offspring have?

Experiment Thomas Hunt Morgan wanted to analyze the behavior of two alleles of a fruit fly eye color gene. In crosses similar to those done by Mendel with pea plants, Morgan and his colleagues mated a wild-type (red-eyed) female with a mutant white-eyed male.

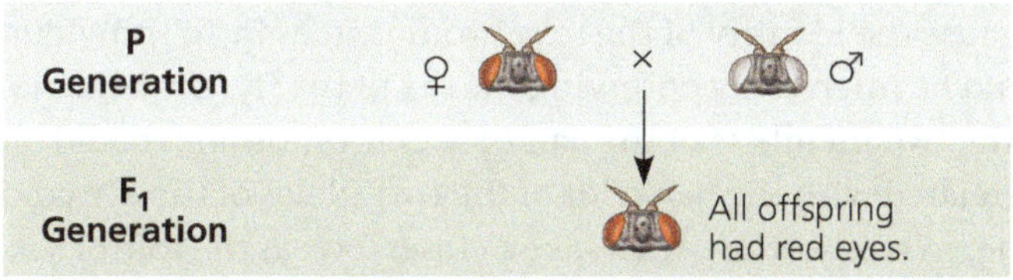

Morgan then bred an F_1 red-eyed female to an F_1 red-eyed male to produce the F_2 generation.

Results The F_2 generation showed a typical Mendelian ratio of 3 red-eyed flies: 1 white-eyed fly. However, all white-eyed flies were males; no females displayed the white-eye trait.

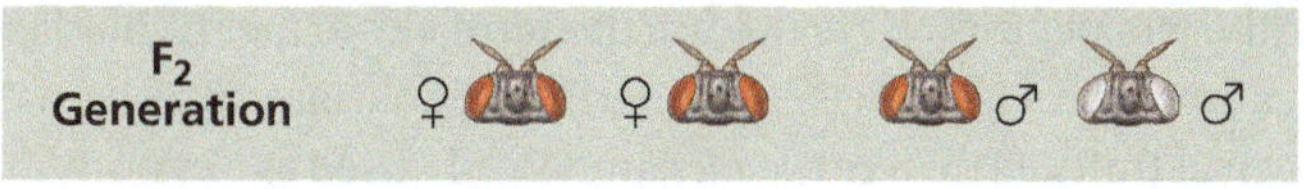

Conclusion All F_1 offspring had red eyes, so the mutant white-eye trait (w) must be recessive to the wild-type red-eye trait (w^+). Since the recessive trait—white eyes—was expressed only in males in the F_2 generation, Morgan deduced that this eye color gene is located on the X chromosome and that there is no corresponding locus on the Y chromosome.

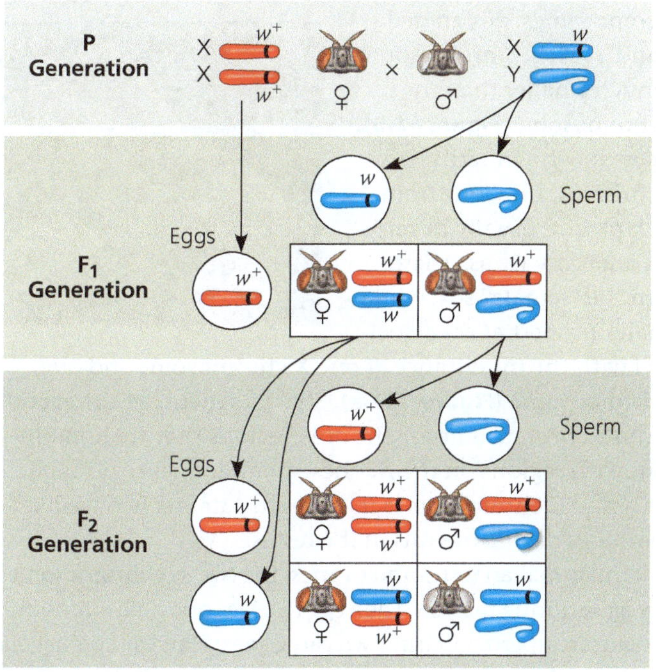

Data from T. H. Morgan, Sex-limited inheritance in *Drosophila, Science* 32:120–122 (1910).

Instructors: A related Experimental Inquiry Tutorial can be assigned in MasteringBiology.

WHAT IF? ➤ *Suppose this eye color gene were located on an autosome. Predict the phenotypes (including gender) of the F_2 flies in this hypothetical cross. (Hint: Draw a Punnett square.)*

CONCEPT 15.2

Sex-linked genes exhibit unique patterns of inheritance

As you just learned, Morgan's discovery of a trait (white eyes) that correlated with the sex of flies was a key episode in the development of the chromosome theory of inheritance. Because the identity of the sex chromosomes in an individual could be inferred by observing the sex of the fly, the behavior of the two members of the pair of sex chromosomes could be correlated with the behavior of the two alleles of the eye color gene. In this section, we'll take a closer look at the role of sex chromosomes in inheritance.

The Chromosomal Basis of Sex

Although sex has traditionally been described as a pair of binary categories, we are coming to understand that sex classifications may be less distinct. Here, we use the term sex to mean the classification into a group with a shared set of anatomical and physiological traits. (The term *gender*, previously used as a synonym of sex, is now more often used to refer to an individual's own cultural experience of identifying as male, female, or otherwise.) In this sense, sex is determined largely by chromosomes.

Humans and other mammals have two types of sex chromosomes, designated X and Y. The Y chromosome is much smaller than the X chromosome **(Figure 15.5)**. A person who inherits two X chromosomes, one from each parent, usually develops anatomy we associate with the "female" sex, while "male" properties are associated with the inheritance of one X chromosome and one Y chromosome **(Figure 15.6a)**. Short segments at either end of the Y chromosome are the only regions that are homologous with regions on the X. These homologous regions allow the X and Y chromosomes in males to pair and behave like homologs during meiosis in the testes.

▼ Figure 15.5 Human sex chromosomes.

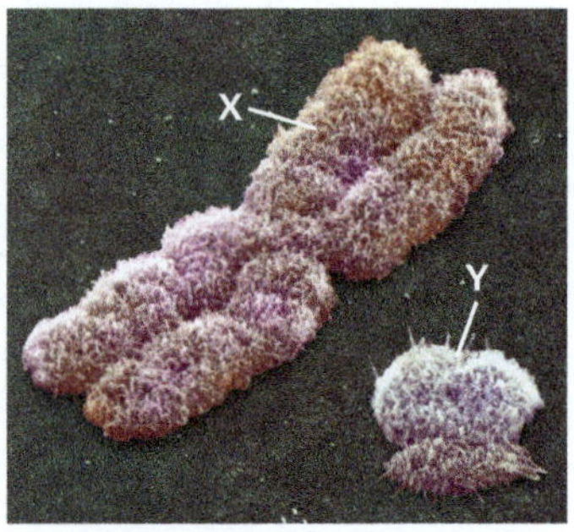

In mammalian testes and ovaries, the two sex chromosomes segregate during meiosis. Each egg receives one X chromosome. In contrast, sperm fall into two categories: Half the sperm cells a male produces receive an X chromosome, and half receive a Y chromosome. We can trace the sex of each offspring to the events of conception: If a sperm cell bearing an X chromosome fertilizes an egg, the zygote is XX, a female; if a sperm cell containing a Y chromosome fertilizes an egg, the zygote is XY, a male (see Figure 15.6a). Thus, in general, sex determination is a matter of chance—a fifty-fifty chance. Note that the mammalian X-Y system isn't the only chromosomal system for determining sex. **Figure 15.6b–d** illustrates three other systems.

▼ Figure 15.6 Some chromosomal systems of sex determination. Numerals indicate the number of autosomes in the species pictured. In *Drosophila*, males are XY, but sex depends on the ratio of the number of X chromosomes to the number of autosome sets, not simply on the presence of a Y chromosome.

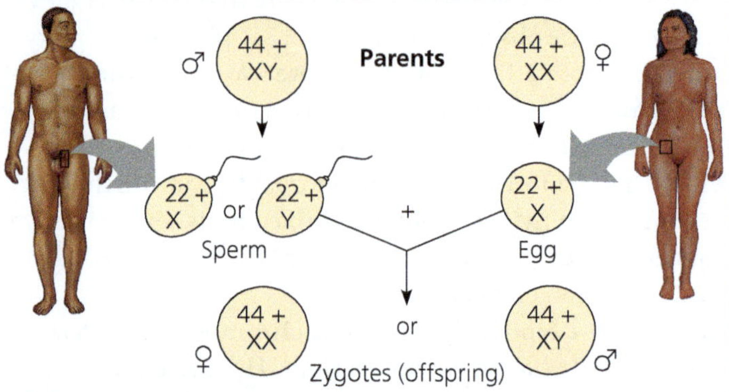

(a) The X-Y system. In mammals, the sex of an offspring depends on whether the sperm cell contains an X chromosome or a Y.

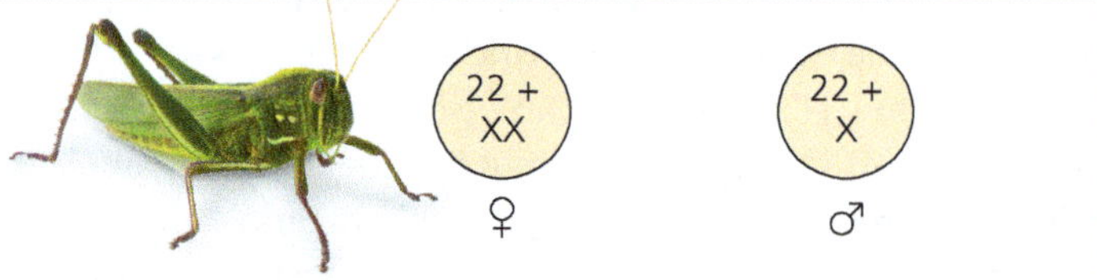

(b) The X-0 system. In grasshoppers, cockroaches, and some other insects, there is only one type of sex chromosome, the X. Females are XX; males have only one sex chromosome (X0). Sex of the offspring is determined by whether the sperm cell contains an X chromosome or no sex chromosome.

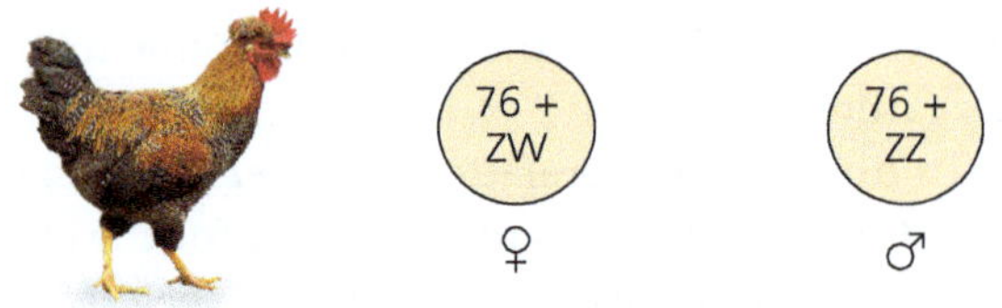

(c) The Z-W system. In birds, some fishes, and some insects, the sex chromosomes present in the egg (not the sperm) determine the sex of offspring. The sex chromosomes are designated Z and W. Females are ZW and males are ZZ.

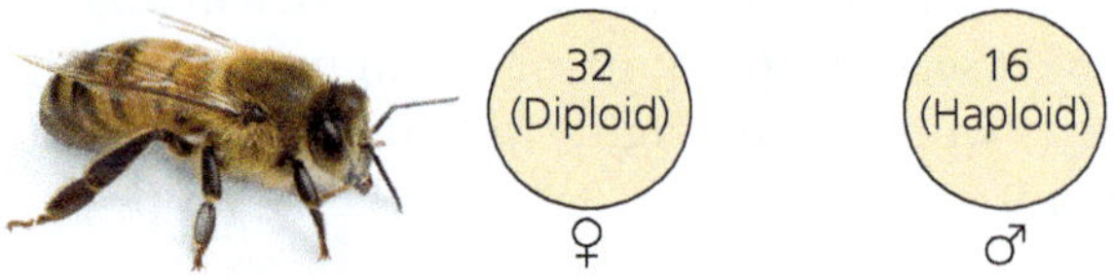

(d) The haplo-diploid system. There are no sex chromosomes in most species of bees and ants. Females develop from fertilized eggs and are thus diploid. Males develop from unfertilized eggs and are haploid; they have no fathers.

In humans, the anatomical signs of sex begin to emerge when the embryo is about 2 months old. Before then, the rudiments of the gonads are generic—they can develop into either testes or ovaries, depending on whether or not a Y chromosome is present, and depending on what genes are active. A gene on the Y chromosome—called *SRY*, for sex-determining region of Y—is required for the development of testes. In the absence of *SRY*, the gonads develop into ovaries, even in an XY embryo.

A gene located on either sex chromosome is called a **sex-linked gene**. The human X chromosome contains approximately 1,100 genes, which are called **X-linked genes**, while genes located on the Y chromosome are called *Y-linked genes.* On the human Y chromosome, researchers have identified 78 genes that code for about 25 proteins (some genes are duplicates). About half of these genes are expressed only in the testis, and some are required for normal testicular functioning and the production of normal sperm. The Y chromosome is passed along virtually intact from a father to all his sons. Because there are so few Y-linked genes, very few disorders are transferred from father to son on the Y chromosome.

The development of female gonads requires a gene called *WNT4* (on chromosome 1, an autosome), which encodes a protein that promotes ovary development. An embryo that is XY but has extra copies of the *WNT4* gene can develop rudimentary female gonads. Overall, sex is determined by the interactions of a network of gene products like these.

The biochemical, physiological, and anatomical features associated with "males" and "females" are turning out to be more complicated than previously thought, with many genes involved in their development. Because of the complexity of this process, many variations exist. Some individuals are born with intermediate sexual ("intersex") characteristics, or even with anatomical features that do not match an individual's sense of their own gender ("transgender" individuals). Sex determination is an active area of research that should yield a more sophisticated understanding in years to come.

Inheritance of X-Linked Genes

The fact that males and females inherit a different number of X chromosomes leads to a pattern of inheritance different from that produced by genes located on autosomes. While there are very few Y-linked genes and many help determine sex, the X chromosomes have genes for many characters unrelated to sex. X-linked genes in humans follow the same pattern of inheritance that Morgan observed for the eye color locus he studied in *Drosophila* (see Figure 15.4). Fathers pass X-linked alleles to all of their daughters but to none of their sons. In contrast, mothers can pass X-linked alleles to both sons and daughters, as shown in **Figure 15.7** for the inheritance of a mild X-linked disorder, red-green color blindness.

If an X-linked trait is due to a recessive allele, a female will express the phenotype only if she is homozygous for that allele. Because males have only one locus, the terms *homozygous* and *heterozygous* lack meaning for describing their X-linked genes; the term *hemizygous* is used in such cases. Any male receiving the recessive allele from his mother will express the trait. For this reason, far more males than females have X-linked recessive disorders. However, even though the chance of a female inheriting a double dose of the mutant allele is much less than the probability of a male inheriting a single dose, there *are* females with X-linked disorders. For instance, color blindness is almost always inherited as an X-linked trait. A color-blind daughter may be born to a color-blind father whose mate is a carrier (see Figure 15.7c). Because the X-linked allele for color blindness is relatively rare, though, the probability that such a man and woman will mate is low.

A number of human X-linked disorders are much more serious than color blindness, such as **Duchenne muscular dystrophy**, which affects about one out of 3,500 males born in the United States. The disease is characterized by a progressive weakening of the muscles and loss of coordination. Affected individuals rarely live past their early 20s. Researchers have traced the disorder to the absence of a key muscle protein called dystrophin and have mapped the gene for this protein to a specific locus on the X chromosome.

▼ Figure 15.7 The transmission of X-linked recessive traits. In this diagram, red-green color blindness is used as an example. The superscript N represents the dominant allele for normal color vision carried on the X chromosome, while n represents the recessive allele, which has a mutation causing color blindness. White boxes indicate unaffected individuals, light orange boxes indicate carriers, and dark orange boxes indicate color-blind individuals.

If a color-blind woman married a man who had normal color vision, what would be the probable phenotypes of their children?

Animation: X-Linked Genes

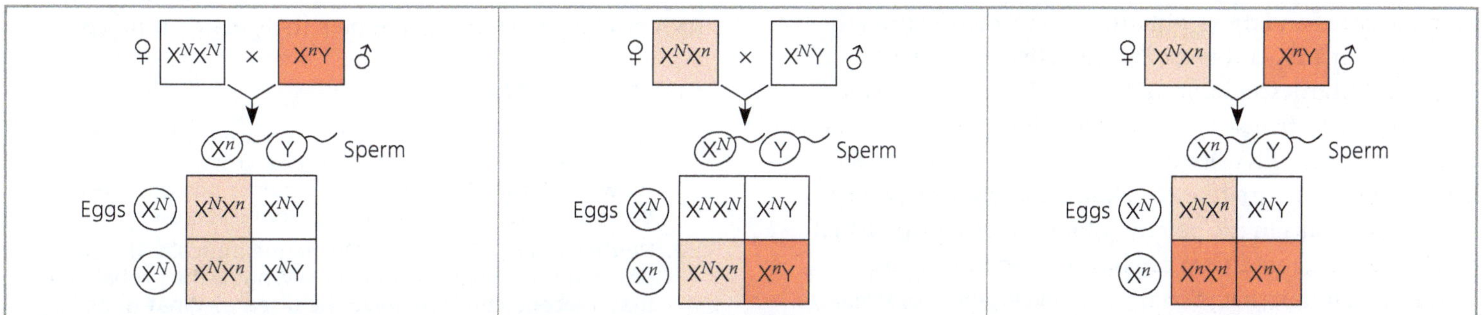

(a) A color-blind father will transmit the mutant allele to all daughters but to no sons. When the mother is a dominant homozygote, the daughters will have the normal phenotype but will be carriers of the mutation.

(b) If a carrier mates with a male who has normal color vision, there is a 50% chance that each daughter will be a carrier like her mother and a 50% chance that each son will have the disorder.

(c) If a carrier mates with a color-blind male, there is a 50% chance that each child born to them will have the disorder, regardless of sex. Daughters who have normal color vision will be carriers, whereas males who have normal color vision will be free of the recessive allele.

Hemophilia is an X-linked recessive disorder defined by the absence of one or more of the proteins required for blood clotting. When a person with hemophilia is injured, bleeding is prolonged because a firm clot is slow to form. Small cuts in the skin are usually not a problem, but bleeding in the muscles or joints can be painful and can lead to serious damage. In the 1800s, hemophilia was widespread among the royal families of Europe. Queen Victoria of England is known to have passed the allele to several of her descendants. Subsequent intermarriage with royal family members of other nations, such as Spain and Russia, further spread this X-linked trait, and its incidence is well documented in royal pedigrees. A few years ago, new genomic techniques allowed sequencing of DNA from tiny amounts isolated from the buried remains of royal family members. The genetic basis of the mutation, and how it resulted in a nonfunctional blood-clotting factor, is now understood. Today, people with hemophilia are treated as needed with intravenous injections of the protein that is missing.

X Inactivation in Female Mammals

Female mammals, including human females, inherit two X chromosomes—twice the number inherited by males—so you may wonder whether females make twice as many of the proteins encoded by X-linked genes as males. In fact, almost all of one X chromosome in each cell in female mammals becomes inactivated during early embryonic development. As a result, the cells of females and males have the same effective dose (one active copy) of most X-linked genes. The inactive X in each cell of a female condenses into a compact object called a **Barr body** (discovered by Canadian anatomist Murray Barr), which lies along the inside of the nuclear envelope. Most of the genes of the X chromosome that forms the Barr body are not expressed. In the ovaries, however, Barr body chromosomes are reactivated in the cells that give rise to eggs, resulting in every female gamete (egg) having an active X after meiosis.

British geneticist Mary Lyon demonstrated that the selection of which X chromosome will form the Barr body occurs randomly and independently in each embryonic cell present at the time of X inactivation. As a consequence, females consist of a *mosaic* of two types of cells: those with the active X derived from the father and those with the active X derived from the mother. After an X chromosome is inactivated in a particular cell, all mitotic descendants of that cell have the same inactive X. Thus, if a female is heterozygous for a sex-linked trait, about half of her cells will express one allele, while the others will express the alternate allele. **Figure 15.8** shows how this mosaicism results in the patchy coloration of a tortoiseshell cat. In humans, mosaicism can be observed in a recessive X-linked mutation that prevents the development of sweat glands. A woman who is heterozygous for this trait has patches of normal skin and patches of skin lacking sweat glands.

Inactivation of an X chromosome involves modification of the DNA and proteins bound to it called histones, including attachment of methyl groups (—CH_3) to DNA nucleotides.

▼ Figure 15.8 X inactivation and the tortoiseshell cat. The tortoiseshell gene is on the X chromosome, and the tortoiseshell phenotype requires the presence of two different alleles, one for orange fur and one for black fur. Normally, only females can have both alleles, because only they have two X chromosomes. If a female cat is heterozygous for the tortoiseshell gene, she is tortoiseshell. Orange patches are formed by populations of cells in which the X chromosome with the orange allele is active; black patches have cells in which the X chromosome with the black allele is active. ("Calico" cats also have white areas, which are determined by another gene.)

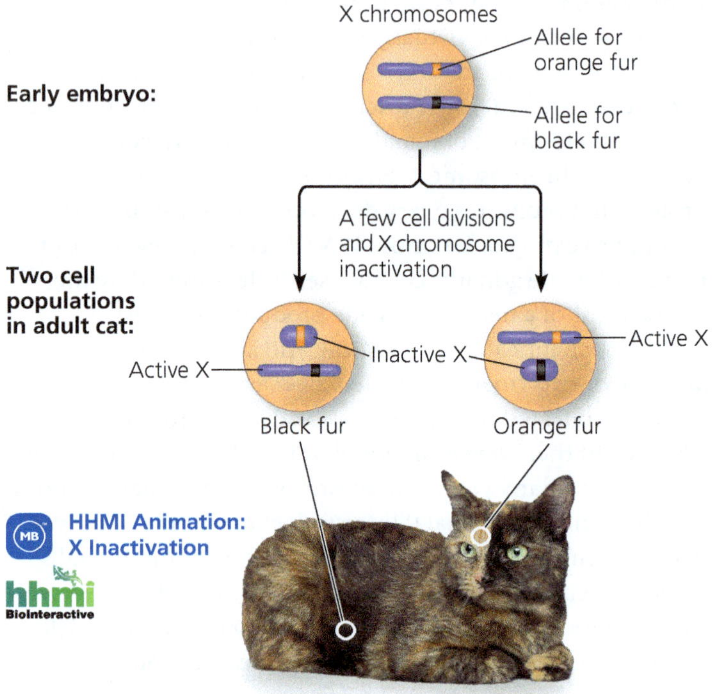

(The regulatory role of DNA methylation is discussed in Concept 18.2.) A particular region of each X chromosome contains several genes involved in the inactivation process. The two regions, one on each X chromosome, associate briefly with each other in each cell at an early stage of embryonic development. Then one of the genes, called *XIST* (for X-inactive specific transcript), becomes active *only* on the chromosome that will become the Barr body. Multiple copies of the RNA product of this gene apparently attach to the X chromosome on which they are made, eventually almost covering it. Interaction of this RNA with the chromosome initiates X inactivation, and the RNA products of nearby genes help to regulate the process.

CONCEPT CHECK 15.2

1. A white-eyed female *Drosophila* is mated with a red-eyed (wild-type) male, the reciprocal cross of the one shown in Figure 15.4. What phenotypes and genotypes do you predict for the offspring from this cross?
2. Neither Tim nor Rhoda has Duchenne muscular dystrophy, but their firstborn son does. What is the probability that a second child will have the disease? What is the probability if the second child is a boy? A girl?
3. **MAKE CONNECTIONS ➤** Consider what you learned about dominant and recessive alleles in Concept 14.1. If a disorder were caused by a dominant X-linked allele, how would the inheritance pattern differ from what we see for recessive X-linked disorders?

For suggested answers, see Appendix A.

CONCEPT 15.3

Linked genes tend to be inherited together because they are located near each other on the same chromosome

The number of genes in a cell is far greater than the number of chromosomes; in fact, each chromosome (except the Y) has hundreds or thousands of genes. Genes located near each other on the same chromosome tend to be inherited together in genetic crosses; such genes are said to be genetically linked and are called **linked genes**. When geneticists follow linked genes in breeding experiments, the results deviate from those expected from Mendel's law of independent assortment.

How Linkage Affects Inheritance

To see how linkage between genes affects the inheritance of two different characters, let's examine another of Morgan's *Drosophila* experiments. In this case, the characters are body color and wing size, each with two different phenotypes. Wild-type flies have gray bodies and normal-sized wings. In addition to these flies, Morgan had managed to obtain, through breeding, doubly mutant flies with black bodies and wings much smaller than normal, called vestigial wings. The mutant alleles are recessive to the wild-type alleles, and neither gene is on a sex chromosome. In his investigation of these two genes, Morgan carried out the crosses shown in **Figure 15.9**. The first was a P generation cross to generate F_1 dihybrid flies, and the second was a testcross.

▼ Figure 15.9

Inquiry How does linkage between two genes affect inheritance of characters?

Experiment Morgan wanted to know whether the genes for body color and wing size are genetically linked and, if so, how this affects their inheritance. The alleles for body color are b^+ (gray) and b (black), and those for wing size are vg^+ (normal) and vg (vestigial).

Morgan mated true-breeding P (parental) generation flies—wild-type flies with black, vestigial-winged flies—to produce heterozygous F_1 dihybrids ($b^+ b\ vg^+ vg$), all of which are wild-type in appearance.

He then mated wild-type F_1 dihybrid females with homozygous recessive males. This testcross will reveal the genotype of the eggs made by the dihybrid female.

The male's sperm contributes only recessive alleles, so the phenotype of the offspring reflects the genotype of the female's eggs.

Note: Although only females (with pointed abdomens) are shown, half the offspring in each class would be males (with rounded abdomens).

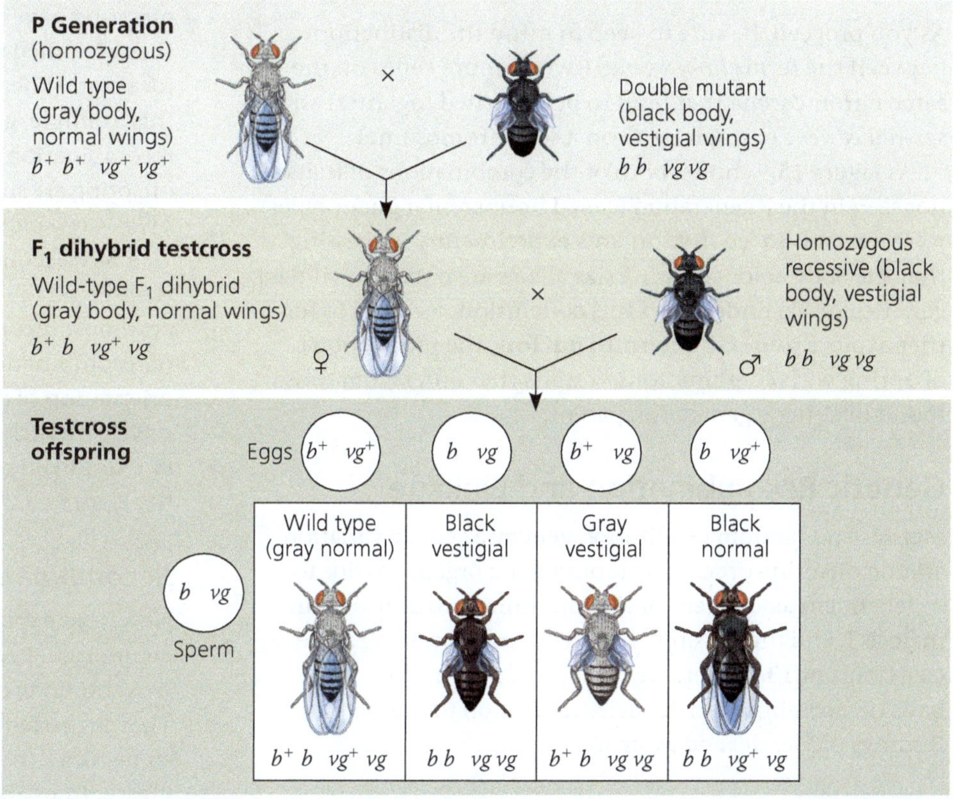

Predicted ratios of testcross offspring	if genes are located on different chromosomes:	1	1	1	1
	if genes are located on the same chromosome *and* parental alleles are always inherited together:	1	1	0	0
Results	Data from Morgan's experiment:	965	944	206	185

Conclusion Since most offspring had a parental (P generation) phenotype, Morgan concluded that the genes for body color and wing size are genetically linked on the same chromosome. However, the production of a relatively small number of offspring with nonparental phenotypes indicated that some mechanism occasionally breaks the linkage between specific alleles of genes on the same chromosome.

Data from T. H. Morgan and C. J. Lynch, The linkage of two factors in *Drosophila* that are not sex-linked, *Biological Bulletin* 23:174–182 (1912).

WHAT IF? ➤ *If the parental (P generation) flies had been true-breeding for gray body with vestigial wings and black body with normal wings, which phenotypic class(es) would be largest among the testcross offspring?*

The resulting flies had a much higher proportion of the combinations of traits seen in the P generation flies (called parental phenotypes) than would be expected if the two genes assorted independently. Morgan thus concluded that body color and wing size are usually inherited together in specific (parental) combinations because the genes are linked; they are near each other on the same chromosome:

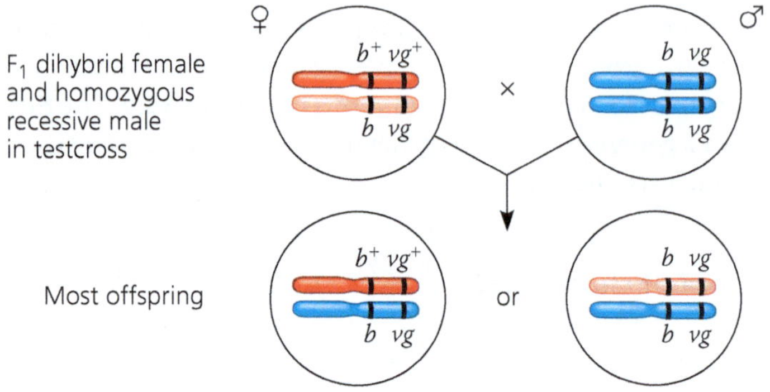

As you proceed, be sure to keep in mind the distinction between the terms *linked genes* (two or more genes on the same chromosome that tend to be inherited together) and *sex-linked gene* (a single gene on a sex chromosome).

As Figure 15.9 shows, both of the combinations of traits not seen in the P generation (called nonparental phenotypes) were also produced in Morgan's experiments, suggesting that the body color and wing size alleles are not always linked genetically. To understand this conclusion, we need to further explore **genetic recombination**, the production of offspring with combinations of traits that differ from those found in either P generation parent.

Genetic Recombination and Linkage

Meiosis and random fertilization generate genetic variation among offspring of sexually reproducing organisms due to independent assortment of chromosomes, crossing over in meiosis I, and the possibility of any sperm fertilizing any egg (see Concept 13.4). Here we'll examine the chromosomal basis of recombination of alleles in relation to the genetic findings of Mendel and Morgan.

Recombination of Unlinked Genes: Independent Assortment of Chromosomes

Mendel learned from crosses in which he followed two characters that some offspring have combinations of traits that do not match those of either parent. For example, consider a cross of a dihybrid pea plant with yellow round seeds, heterozygous for both seed color and seed shape (*YyRr*), with a plant homozygous for both recessive alleles (with green wrinkled seeds, *yyrr*). (This cross acts as a testcross because the results will reveal the genotype of the gametes made in the dihybrid *YyRr* plant.) Let's represent the cross by the following Punnett square:

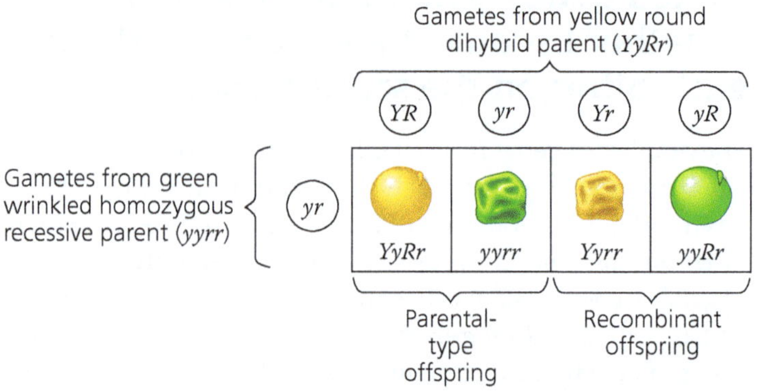

Notice in this Punnett square that one-half of the offspring are expected to inherit a phenotype that matches either of the phenotypes of the P (parental) generation originally crossed to produce the F_1 dihybrid (see Figure 15.2). These matching offspring are called **parental types**. But two nonparental phenotypes are also found among the offspring. Because these offspring have new combinations of seed shape and color, they are called **recombinant types**, or **recombinants** for short. When 50% of all offspring are recombinants, as in this example, geneticists say that there is a 50% frequency of recombination. The predicted phenotypic ratios among the offspring are similar to what Mendel actually found in his *YyRr* × *yyrr* crosses.

A 50% frequency of recombination in such testcrosses is observed for any two genes that are located on different chromosomes and thus cannot be linked. The physical basis of recombination between unlinked genes is the random orientation of homologous chromosomes at metaphase I of meiosis, which leads to the independent assortment of the two unlinked genes (see Figure 13.11 and the question in the Figure 15.2 legend).

Recombination of Linked Genes: Crossing Over

Now, let's explain the results of the *Drosophila* testcross in Figure 15.9. Recall that most of the offspring from the testcross for body color and wing size had parental phenotypes. That suggested that the two genes were on the same chromosome, since the occurrence of parental types with a frequency greater than 50% indicates that the genes are linked. About 17% of offspring, however, were recombinants.

Seeing these results, Morgan proposed that some process must occasionally break the physical connection between specific alleles of genes on the same chromosome. Later experiments showed that this process, now called **crossing over**, accounts for the recombination of linked genes. In crossing over, which occurs while replicated homologous chromosomes are paired during prophase of meiosis I, a set of proteins orchestrates an exchange of corresponding segments of one maternal and one paternal chromatid (see Figure 13.9).

Figure 15.10 Chromosomal basis for recombination of linked genes. In these diagrams re-creating the testcross in Figure 15.9, we track chromosomes as well as genes. The maternal chromosomes (present in the wild-type F_1 dihybrid) are color-coded red and pink to distinguish one homolog from the other before any meiotic crossing over has occurred. Because crossing over between the b^+/b and vg^+/vg loci occurs in some, but not all, egg-producing cells, more eggs with parental-type chromosomes than with recombinant ones are produced in the mating females. Fertilization of the eggs by sperm of genotype *b vg* gives rise to some recombinant offspring. The recombination frequency is the percentage of recombinant flies in the total pool of offspring.

DRAW IT *Suppose, as in the question at the bottom of Figure 15.9, the parental (P generation) flies were true-breeding for gray body with vestigial wings and black body with normal wings. Draw the chromosomes in each of the four possible kinds of eggs from an F_1 female, and label each chromosome as "parental" or "recombinant."*

Animation: Linked Genes and Crossing Over

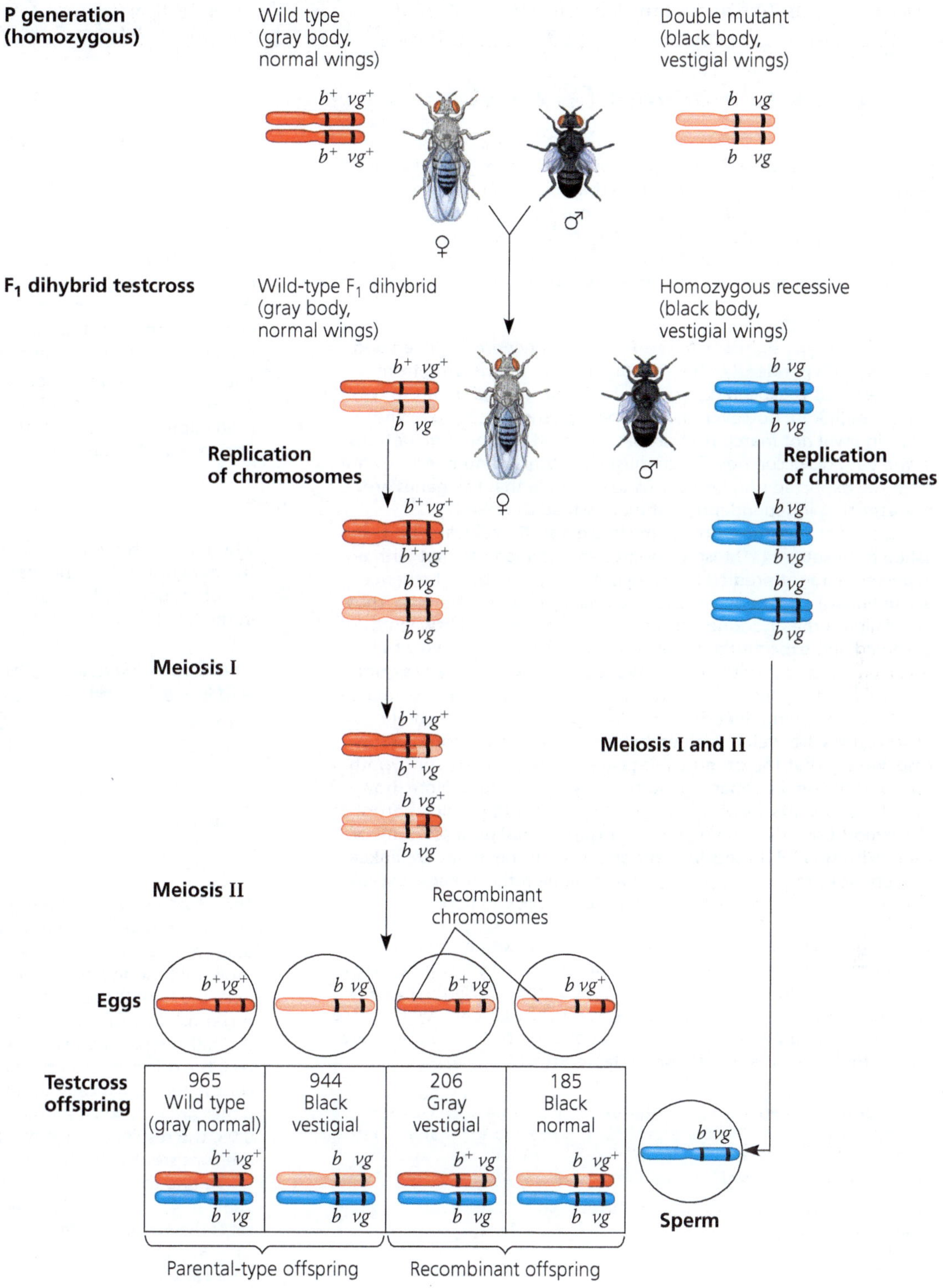

$$\text{Recombination frequency} = \frac{391 \text{ recombinants}}{2{,}300 \text{ total offspring}} \times 100 = 17\%$$

In effect, when a single crossover occurs, end portions of two nonsister chromatids trade places.

Figure 15.10 shows how crossing over in a dihybrid female fly resulted in recombinant eggs and ultimately recombinant offspring in Morgan's testcross. Most eggs had a chromosome with either the $b^+ vg^+$ or *b vg* parental genotype, but some had a recombinant chromosome ($b^+ vg$ or $b\ vg^+$). Fertilization of all classes of eggs by homozygous recessive sperm (*b vg*) produced an offspring population in which 17% exhibited a nonparental, recombinant phenotype, reflecting combinations of alleles not seen before in either P generation parent. In the **Scientific Skills Exercise**, you can use a statistical test to analyze the results from an F_1 dihybrid testcross and see whether the two genes assort independently or are linked.

SCIENTIFIC SKILLS EXERCISE

Using the Chi-Square (χ^2) Test

➤ **Cosmos plants**

Are Two Genes Linked or Unlinked? Genes that are in close proximity on the same chromosome will result in the linked alleles being inherited together more often than not. But how can you tell if certain alleles are inherited together due to linkage or whether they just happen to assort together? In this exercise, you will use a simple statistical test, the chi-square (χ^2) test, to analyze phenotypes of F_1 testcross progeny in order to see whether two genes are linked or unlinked.

How These Experiments Are Done If genes are unlinked and assorting independently, the phenotypic ratio of offspring from an F_1 testcross is expected to be 1:1:1:1 (see Figure 15.9). If the two genes are linked, however, the observed phenotypic ratio of the offspring will not match that ratio. Given that random fluctuations in the data do occur, how much must the observed numbers deviate from the expected numbers for us to conclude that the genes are not assorting independently but may instead be linked?

To answer this question, scientists use a statistical test. This test, called a chi-square (χ^2) test, compares an observed data set with an expected data set predicted by a hypothesis (here, that the genes are unlinked) and measures the discrepancy between the two, thus determining the "goodness of fit." If the discrepancy between the observed and expected data sets is so large that it is unlikely to have occurred by random fluctuation, we say there is statistically significant evidence against the hypothesis (or, more specifically, evidence for the genes being linked). If the discrepancy is small, then our observations are well explained by random variation alone. In this case, we say that the observed data are consistent with our hypothesis, or that the discrepancy is statistically insignificant. Note, however, that consistency with our hypothesis is not the same as proof of our hypothesis. Also, the size of the experimental data set is important: With small data sets like this one, even if the genes are linked, discrepancies might be small by chance alone if the linkage is weak. For simplicity, we overlook the effect of sample size here.

Data from the Simulated Experiment In cosmos plants, purple stem (A) is dominant to green stem (a), and short petals (B) is dominant to long petals (b). In a simulated cross, $AABB$ plants were crossed with $aabb$ plants to generate F_1 dihybrids ($AaBb$), which were then testcrossed ($AaBb \times aabb$). A total of 900 offspring plants were scored for stem color and flower petal length.

Offspring from test-cross of $AaBb$ (F_1) × $aabb$	Purple stem/short petals (A–B–)	Green stem/short petals (aaB–)	Purple stem/long petals (A–bb)	Green stem/long petals ($aabb$)
Expected ratio if the genes are unlinked	1	1	1	1
Expected number of offspring (of 900)				
Observed number of offspring (of 900)	220	210	231	239

INTERPRET THE DATA

1. The results in the data table are from a simulated F_1 dihybrid testcross. The hypothesis that the two genes are unlinked predicts that the offspring phenotypic ratio will be 1:1:1:1. Using this ratio, calculate the expected number of each phenotype out of the 900 total offspring, and enter the values in that data table.
2. The goodness of fit is measured by χ^2. This statistic measures the amounts by which the observed values differ from their respective predictions to indicate how closely the two sets of values match. The formula for calculating this value is

$$\chi^2 = \sum \frac{(o - e)^2}{e}$$

where o = observed and e = expected. Calculate the χ^2 value for the data using the table below. Fill out that table, carrying out the operations indicated in the top row. Then add up the entries in the last column to find the χ^2 value.

Testcross Offspring	Expected (e)	Observed (o)	Deviation ($o - e$)	$(o - e)^2$	$(o - e)^2/e$
(A–B–)		220			
(aaB–)		210			
(A–bb)		231			
($aabb$)		239			
				χ^2 = Sum	

3. The χ^2 value means nothing on its own—it is used to find the probability that, assuming the hypothesis is true, the observed data set could have resulted from random fluctuations. A low probability suggests that the observed data are not consistent with the hypothesis and thus the hypothesis should be rejected. A standard cutoff point used by biologists is a probability of 0.05 (5%). If the probability corresponding to the χ^2 value is 0.05 or less, the differences between observed and expected values are considered statistically significant and the hypothesis (that the genes are unlinked) should be rejected. If the probability is above 0.05, the results are not statistically significant; the observed data are consistent with the hypothesis.

To find the probability, locate your χ^2 value in the χ^2 Distribution Table in Appendix F. The "degrees of freedom" (df) of your data set is the number of categories (here, 4 phenotypes) minus 1, so df = 3. (a) Determine which values on the df = 3 line of the table your calculated χ^2 value lies between. (b) The column headings for these values show the probability range for your χ^2 number. Based on whether there are nonsignificant ($p \leq 0.05$) or significant ($p > 0.05$) differences between the observed and expected values, are the data consistent with the hypothesis that the two genes are unlinked and assorting independently, or is there enough evidence to reject this hypothesis?

Instructors: A version of this Scientific Skills Exercise can be assigned in MasteringBiology.

New Combinations of Alleles: Variation for Natural Selection

EVOLUTION The physical behavior of chromosomes during meiosis contributes to the generation of variation in offspring (see Concept 13.4). Each pair of homologous chromosomes lines up independently of other pairs during metaphase I, and crossing over prior to that, during prophase I, can mix and match parts of maternal and paternal homologs. Mendel's elegant experiments show that the behavior of the abstract entities known as genes—or, more concretely, alleles of genes—also leads to variation in offspring (see Concept 14.1). Now, putting these different ideas together, you can see that the recombinant chromosomes resulting from crossing over may bring alleles together in new combinations, and the subsequent events of meiosis distribute to gametes the recombinant chromosomes in a multitude of combinations, such as the new variants seen in Figures 15.9 and 15.10. Random fertilization then increases even further the number of variant allele combinations that can be created.

This abundance of genetic variation provides the raw material on which natural selection works. If the traits conferred by particular combinations of alleles are better suited for a given environment, organisms possessing those genotypes will be expected to thrive and leave more offspring, ensuring the continuation of their genetic complement. In the next generation, of course, the alleles will be shuffled anew. Ultimately, the interplay between environment and phenotype (and thus genotype) will determine which genetic combinations persist over time.

Mapping the Distance Between Genes Using Recombination Data: *Scientific Inquiry*

The discovery of linked genes and recombination due to crossing over motivated one of Morgan's students, Alfred H. Sturtevant, to work out a method for constructing a **genetic map**, an ordered list of the genetic loci along a particular chromosome.

Sturtevant hypothesized that the percentage of recombinant offspring, the *recombination frequency*, calculated from experiments like the one in Figures 15.9 and 15.10, depends on the distance between genes on a chromosome. He assumed that crossing over is a random event, with the chance of crossing over approximately equal at all points along a chromosome. Based on these assumptions, Sturtevant predicted that *the farther apart two genes are, the higher the probability that a crossover will occur between them and therefore the higher the recombination frequency*. His reasoning was simple: The greater the distance between two genes, the more points there are between them where crossing over can occur. Using recombination data from various fruit fly crosses, Sturtevant proceeded to assign relative positions to genes on the same chromosomes—that is, to *map* genes.

A genetic map based on recombination frequencies is called a **linkage map**. **Figure 15.11** shows Sturtevant's linkage map of three genes: the body color (*b*) and wing size (*vg*) genes depicted in Figure 15.10 and a third gene, called cinnabar (*cn*). Cinnabar is one of many *Drosophila* genes affecting eye color. Cinnabar eyes, a mutant phenotype, are a brighter red than the wild-type color. The recombination frequency between *cn* and *b* is 9%; that between *cn* and *vg*, 9.5%; and that between *b* and *vg*, 17%. In other words, crossovers between *cn* and *b* and between *cn* and *vg* are about half as frequent as crossovers between *b* and *vg*. Only a map that locates *cn* about midway between *b* and *vg* is consistent with these data, as you can prove to yourself by drawing alternative maps. Sturtevant expressed the distances between genes in **map units**, defining one map unit as equivalent to a 1% recombination frequency.

In practice, the interpretation of recombination data is more complicated than this example suggests. Some genes on a chromosome are so far from each other that a crossover between them is virtually certain. The observed frequency of recombination in crosses involving two such genes can have

▼ Figure 15.11

Research Method **Constructing a Linkage Map**

Application A linkage map shows the relative locations of genes along a chromosome.

Technique A linkage map is based on the assumption that the probability of a crossover between two genetic loci is proportional to the distance separating the loci. The recombination frequencies used to construct a linkage map for a particular chromosome are obtained from experimental crosses, such as the cross depicted in Figures 15.9 and 15.10. The distances between genes are expressed as map units, with one map unit equivalent to a 1% recombination frequency. Genes are arranged on the chromosome in the order that best fits the data.

Results In this example, the observed recombination frequencies between three *Drosophila* gene pairs (*b* and *cn*, 9%; *cn* and *vg*, 9.5%; and *b* and *vg*, 17%) best fit a linear order in which *cn* is positioned about halfway between the other two genes:

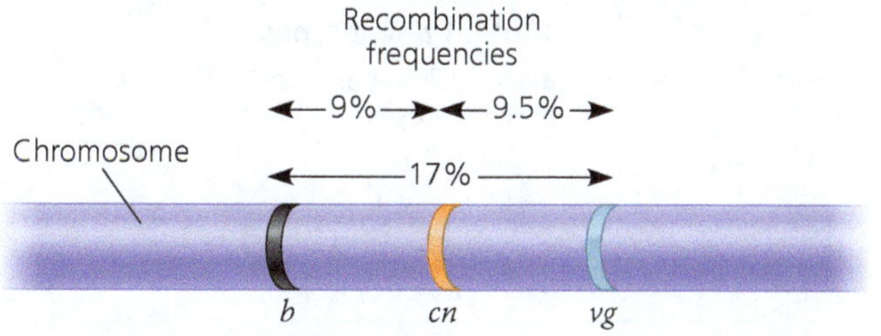

The *b*-*vg* recombination frequency (17%) is slightly less than the sum of the *b*-*cn* and *cn*-*vg* frequencies (9 + 9.5 = 18.5%) because of the few times that one crossover occurs between *b* and *cn* and another crossover occurs between *cn* and *vg*. The second crossover would "cancel out" the first, reducing the observed *b*-*vg* recombination frequency while contributing to the frequency between each of the closer pairs of genes. The value of 18.5% (18.5 map units) is closer to the actual distance between the genes. In practice, a geneticist would add the smaller distances in constructing a map.

a maximum value of 50%, a result indistinguishable from that for genes on different chromosomes. In this case, the physical connection between genes on the same chromosome is not reflected in the results of genetic crosses. Despite being on the same chromosome and thus being *physically connected*, the genes are *genetically unlinked*; alleles of such genes assort independently, as if they were on different chromosomes. In fact, at least two of the genes for pea characters that Mendel studied are now known to be on the same chromosome, but the distance between them is so great that linkage is not observed in genetic crosses. Consequently, the two genes behaved as if they were on different chromosomes in Mendel's experiments. Genes located far apart on a chromosome are mapped by adding the recombination frequencies from crosses involving closer pairs of genes lying between the two distant genes.

Using recombination data, Sturtevant and his colleagues were able to map numerous *Drosophila* genes in linear arrays. They found that the genes clustered into four groups of linked genes (*linkage groups*). Light microscopy had revealed four pairs of chromosomes in *Drosophila*, so the linkage map provided additional evidence that genes are located on chromosomes. Each chromosome has a linear array of specific genes, each gene with its own locus **(Figure 15.12)**.

Because a linkage map is based strictly on recombination frequencies, it gives only an approximate picture of a chromosome. The frequency of crossing over is not actually uniform over the length of a chromosome, as Sturtevant assumed, and therefore map units do not correspond to actual physical distances (in nanometers, for instance). A linkage map does portray the order of genes along a chromosome, but it does not accurately portray the precise locations of those genes. Other methods enable geneticists to construct *cytogenetic maps* of chromosomes, which locate genes with respect to chromosomal features, such as stained bands, that can be seen in the microscope. Technical advances over the last two decades have enormously increased the rate and affordability of DNA sequencing. Therefore, today, most researchers sequence whole genomes to map the locations of genes of a given species. The entire nucleotide sequence is the ultimate physical map of a chromosome, revealing the physical distances in DNA nucleotides between gene loci (see Concept 21.1). Comparing a linkage map with such a physical map or with a cytogenetic map of the same chromosome, we find that the linear order of genes is identical in all the maps, but the spacing between genes is not.

▼ Figure 15.12 A partial genetic (linkage) map of a *Drosophila* chromosome. This simplified map shows just seven of the genes that have been mapped on *Drosophila* chromosome II. (DNA sequencing has revealed over 9,000 genes on that chromosome.) The number at each gene locus indicates the number of map units between that locus and the locus for arista length (left). Notice that more than one gene can affect a given phenotypic characteristic, such as eye color.

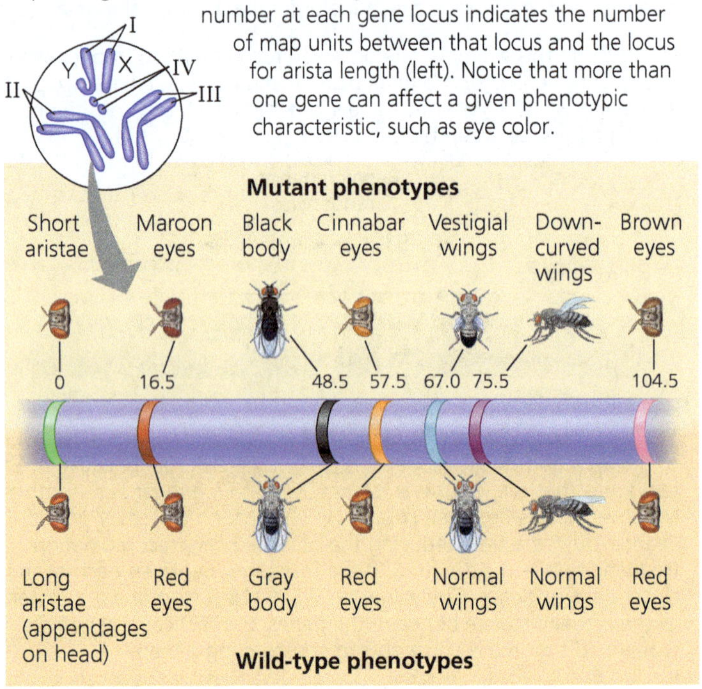

CONCEPT CHECK 15.3

1. When two genes are located on the same chromosome, what is the physical basis for the production of recombinant offspring in a testcross between a dihybrid parent and a double-mutant (recessive) parent?
2. **VISUAL SKILLS ➤** For each type of offspring of the testcross in Figure 15.9, explain the relationship between its phenotype and the alleles contributed by the female parent. (It will be useful to draw out the chromosomes of each fly and follow the alleles throughout the cross.)
3. **WHAT IF? ➤** Genes *A*, *B*, and *C* are located on the same chromosome. Testcrosses show that the recombination frequency between *A* and *B* is 28% and that between *A* and *C* is 12%. Can you determine the linear order of these genes? Explain.

For suggested answers, see Appendix A.

CONCEPT 15.4

Alterations of chromosome number or structure cause some genetic disorders

As you have learned so far in this chapter, the phenotype of an organism can be affected by small-scale changes involving individual genes. Random mutations are the source of all new alleles, which can lead to new phenotypic traits.

Large-scale chromosomal changes can also affect an organism's phenotype. Physical and chemical disturbances, as well as errors during meiosis, can damage chromosomes in major ways or alter their number in a cell. Large-scale chromosomal alterations in humans and other mammals often lead to spontaneous abortion (miscarriage) of a fetus, and individuals born with these types of genetic defects commonly exhibit various developmental disorders. Plants appear to tolerate such genetic defects better than animals do.

Interview with Terry Orr-Weaver: Why are humans so bad at meiosis?!

Abnormal Chromosome Number

Ideally, the meiotic spindle distributes chromosomes to daughter cells without error. But there is an occasional mishap, called a **nondisjunction**, in which the members of a pair of homologous chromosomes do not move apart properly during meiosis I or sister chromatids fail to separate during meiosis II **(Figure 15.13)**. In nondisjunction, one gamete receives two of the same type of chromosome and another gamete receives no copy. The other chromosomes are usually distributed normally.

If either of the aberrant gametes unites with a normal one at fertilization, the zygote will also have an abnormal number of a particular chromosome, a condition known as **aneuploidy**. Fertilization involving a gamete that has no copy of a particular chromosome will lead to a missing chromosome in the zygote (so that the cell has $2n - 1$ chromosomes); the aneuploid zygote is said to be **monosomic** for that chromosome. If a chromosome is present in triplicate in the zygote (so that the cell has $2n + 1$ chromosomes), the aneuploid cell is **trisomic** for that chromosome. Mitosis will subsequently transmit the anomaly to all embryonic cells. Monosomy and trisomy are estimated to occur in 10–25% of human conceptions and are the main reason for pregnancy loss. If the organism survives, it usually has a set of traits caused by the abnormal dose of the genes associated with the extra or missing chromosome. Down syndrome is an example of trisomy in humans that will be discussed later. Nondisjunction can also occur during mitosis. If such an error takes place early in embryonic development, then the aneuploid condition is passed along by mitosis to a large number of cells and is likely to have a substantial effect on the organism.

Figure 15.13 Meiotic nondisjunction. Gametes with an abnormal chromosome number can arise by nondisjunction in either meiosis I or meiosis II. For simplicity, the figure does not show the spores formed by meiosis in plants. Ultimately, spores form gametes that have the defects shown. (See Figure 13.6.)

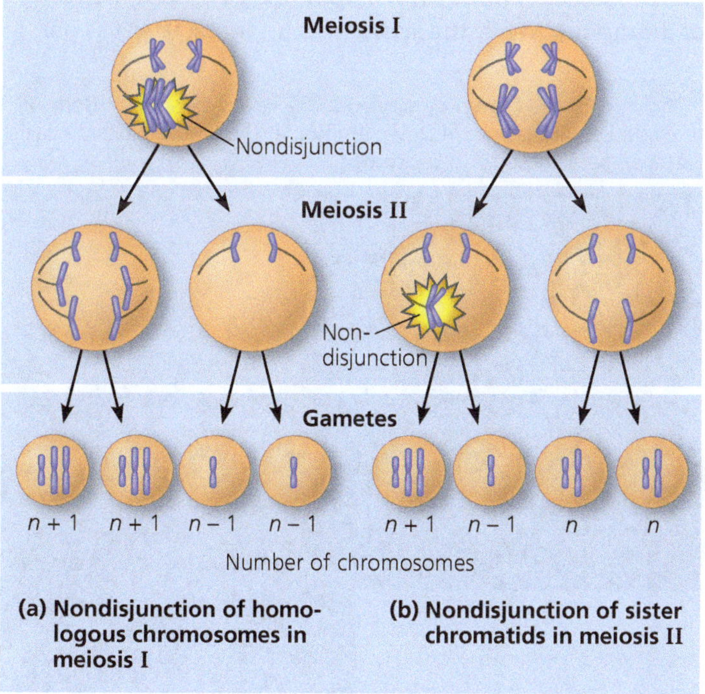

Some organisms have more than two complete chromosome sets in all somatic cells. The general term for this chromosomal alteration is **polyploidy**; the specific terms *triploidy* ($3n$) and *tetraploidy* ($4n$) indicate three and four chromosomal sets, respectively. One way a triploid cell may arise is by the fertilization of an abnormal diploid egg produced by nondisjunction of all its chromosomes. Tetraploidy could result from the failure of a $2n$ zygote to divide after replicating its chromosomes. Subsequent normal mitotic divisions would then produce a $4n$ embryo.

Polyploidy is fairly common in the plant kingdom. The spontaneous origin of polyploid individuals plays an important role in plant evolution (see Concept 24.2). Many species we eat are polyploid: Bananas are triploid, wheat hexaploid ($6n$), and strawberries octoploid ($8n$). Polyploid animal species are much less common, but there are a few fishes and amphibians known to be polyploid. In general, polyploids are more nearly normal in appearance than aneuploids. One extra (or missing) chromosome apparently disrupts genetic balance more than does an entire extra set of chromosomes.

MB Animation: Polyploid Plants

Alterations of Chromosome Structure

Errors in meiosis or damaging agents such as radiation can cause breakage of a chromosome, which can lead to four types of changes in chromosome structure **(Figure 15.14)**. A **deletion** occurs when a chromosomal fragment is lost. The affected chromosome is then missing certain genes. The "deleted" fragment may become attached as an extra segment to a sister or nonsister chromatid, producing a **duplication** of a portion of that chromosome. A chromosomal fragment may also reattach to the original chromosome but in the reverse orientation, producing an **inversion**. A fourth possible result of chromosomal breakage is for the fragment to join a nonhomologous chromosome, a rearrangement called a **translocation**.

Deletions and duplications are especially likely to occur during meiosis. In crossing over, nonsister chromatids sometimes exchange unequal-sized segments of DNA, so that one partner gives up more genes than it receives (see Figure 21.13). The products of such an unequal crossover are one chromosome with a deletion and one chromosome with a duplication.

A diploid embryo that is homozygous for a large deletion (or has a single X chromosome with a large deletion, in a male) is usually missing a number of essential genes, a condition that is typically lethal. Duplications and translocations also tend to be harmful. In reciprocal translocations, in which segments are exchanged between nonhomologous chromosomes, and in

Figure 15.14 Alterations of chromosome structure. Red arrows indicate breakage points. Dark purple highlights the chromosomal parts affected by the rearrangements.

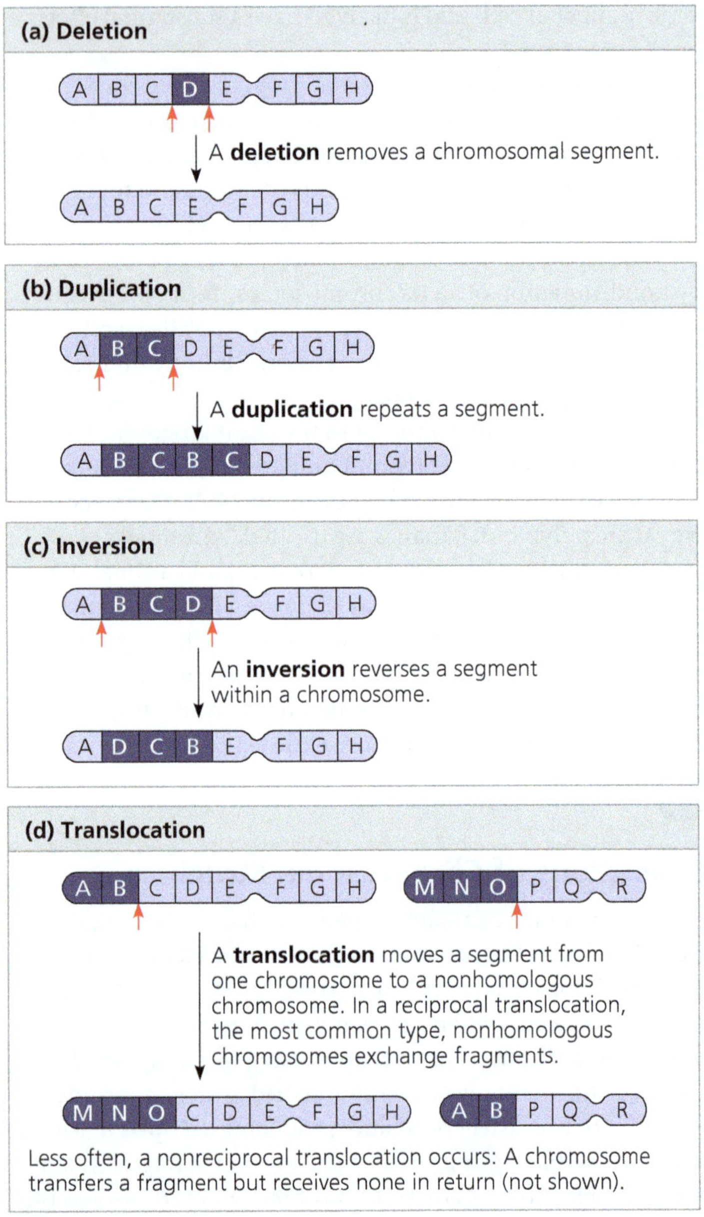

Animation: Alterations of Chromosome Structure

inversions, the balance of genes is not abnormal—all genes are present in their normal doses. Nevertheless, translocations and inversions can alter phenotype because a gene's expression can be influenced by its location among neighboring genes, which can have devastating effects.

Human Disorders Due to Chromosomal Alterations

Alterations of chromosome number and structure are associated with a number of serious human disorders. As described earlier, nondisjunction in meiosis results in aneuploidy in gametes and any resulting zygotes. Although the frequency of aneuploid zygotes may be quite high in humans, most of these chromosomal alterations are so disastrous to development that the affected embryos are spontaneously aborted long before birth. However, some types of aneuploidy appear to upset the genetic balance less than others, so individuals with certain aneuploid conditions can survive to birth and beyond. These individuals have a set of traits—a *syndrome*—characteristic of the type of aneuploidy. Genetic disorders caused by aneuploidy can be diagnosed before birth by fetal testing (see Figure 14.19).

Down Syndrome (Trisomy 21)

One aneuploid condition, **Down syndrome**, affects approximately one out of every 830 children born in the United States **(Figure 15.15)**. Down syndrome is usually the result of an extra chromosome 21, so that each body cell has a total of 47 chromosomes. Because the cells are trisomic for chromosome 21, Down syndrome is often called *trisomy 21*. Down syndrome includes characteristic facial features, short stature, correctable heart defects, and developmental delays. Individuals with Down syndrome have an increased chance of developing leukemia and Alzheimer's disease but have a lower rate of high blood pressure, atherosclerosis (hardening of the arteries), stroke, and many types of solid tumors. Although people with Down syndrome, on average, have a life span shorter than normal, most, with proper medical treatment, live to middle age and beyond. Many live independently or at home with their families, are employed, and are valuable contributors to their communities. Almost all males and about half of females with Down syndrome are sexually underdeveloped and sterile.

The frequency of Down syndrome increases with the age of the mother. While the disorder occurs in just 0.04% (4 out

Figure 15.15 Down syndrome. The karyotype shows trisomy 21, the most common cause of Down syndrome. The child exhibits the facial features characteristic of this disorder.

of 10,000 births) of children born to women under age 30, the risk climbs to 0.92% (92 out of 10,000) for mothers at age 40 and is even higher for older mothers. The correlation of Down syndrome with maternal age has not yet been explained. Most cases result from nondisjunction during meiosis I, and some research points to an age-dependent abnormality in meiosis. Trisomies of some other chromosomes also increase in incidence with maternal age, although infants with other autosomal trisomies rarely survive for long. Medical experts recommend that prenatal screening for trisomies in the embryo be offered to all pregnant women, due to its low risk and useful results. In 2008, a law was passed stipulating that medical practitioners give accurate, up-to-date information about any prenatal or postnatal diagnosis received by parents and that they connect parents with appropriate support services.

Aneuploidy of Sex Chromosomes

Aneuploid conditions involving sex chromosomes appear to upset the genetic balance less than those involving autosomes. This may be because the Y chromosome carries relatively few genes. Also, extra copies of the X chromosome simply become inactivated as Barr bodies.

An extra X chromosome in a male, producing XXY, occurs approximately once in every 500 to 1,000 live male births. People with this disorder, called *Klinefelter syndrome*, have male sex organs, but the testes are abnormally small and the man is sterile. Even though the extra X is inactivated, some breast enlargement and other female body characteristics are common. Affected individuals may have subnormal intelligence. About 1 of every 1,000 males is born with an extra Y chromosome (XYY). These males undergo normal sexual development and do not exhibit any well-defined syndrome, but tend to be taller than average.

Females with trisomy X (XXX), which occurs once in approximately 1,000 live female births, are healthy and have no unusual physical features other than being slightly taller than average. Triple-X females are at risk for learning disabilities but are fertile. Monosomy X, which is called *Turner syndrome*, occurs about once in every 2,500 female births and is the only known viable monosomy in humans. Although these X0 individuals are phenotypically female, they are sterile because their sex organs do not mature. When provided with estrogen replacement therapy, girls with Turner syndrome do develop secondary sex characteristics. Most have normal intelligence.

Disorders Caused by Structurally Altered Chromosomes

Many deletions in human chromosomes, even in a heterozygous state, cause severe problems. One such syndrome, known as *cri du chat* ("cry of the cat"), results from a specific deletion in chromosome 5. A child born with this deletion is severely intellectually disabled, has a small head with unusual facial features, and has a cry that sounds like the mewing of a distressed cat. Such individuals usually die in infancy or early childhood.

Chromosomal translocations can also occur during mitosis; some have been implicated in certain cancers, including *chronic myelogenous leukemia* (*CML*). This disease occurs when a reciprocal translocation happens during mitosis of cells that are precursors of white blood cells. The exchange of a large portion of chromosome 22 with a small fragment from a tip of chromosome 9 produces a much shortened, easily recognized chromosome 22, called the *Philadelphia chromosome* (**Figure 15.16**). Such an exchange causes cancer by creating a new "fused" gene that leads to uncontrolled cell cycle progression. (The mechanism of gene activation will be discussed in Chapter 18.)

Figure 15.16 Translocation associated with chronic myelogenous leukemia (CML). The cancerous cells in nearly all CML patients contain an abnormally short chromosome 22, the so-called Philadelphia chromosome, and an abnormally long chromosome 9. These altered chromosomes result from the reciprocal translocation shown here, which presumably occurred in a single white blood cell precursor undergoing mitosis and was then passed along to all descendant cells.

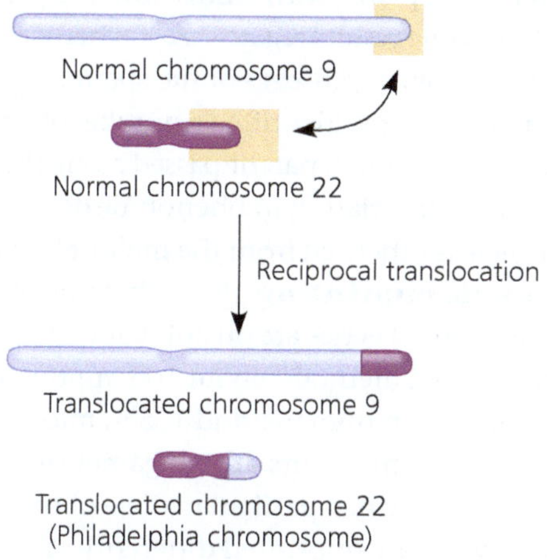

CONCEPT CHECK 15.4

1. About 5% of individuals with Down syndrome have a chromosomal translocation in which a third copy of chromosome 21 is attached to chromosome 14. If this translocation occurred in a parent's gonad, how could it lead to Down syndrome in a child?
2. **MAKE CONNECTIONS** ➤ The ABO blood type locus has been mapped on chromosome 9. A father who has type AB blood and a mother who has type O blood have a child with trisomy 9 and type A blood. Using this information, can you tell in which parent the nondisjunction occurred? Explain your answer. (See Figures 14.11 and 15.13.)
3. **MAKE CONNECTIONS** ➤ The gene that is activated on the Philadelphia chromosome codes for an intracellular tyrosine kinase. Review the discussion of cell cycle control in Concept 12.3, and explain how the activation of this gene could contribute to the development of cancer.

For suggested answers, see Appendix A.

CONCEPT 15.5

Some inheritance patterns are exceptions to standard Mendelian inheritance

In the previous section, you learned about deviations from the usual patterns of chromosomal inheritance due to abnormal events in meiosis and mitosis. We conclude this chapter by describing two normally occurring exceptions to Mendelian genetics, one involving genes located in the nucleus and the other involving genes located outside the nucleus. In both cases, the sex of the parent contributing an allele is a factor in the pattern of inheritance.

Genomic Imprinting

Throughout our discussions of Mendelian genetics and the chromosomal basis of inheritance, we have assumed that a given allele will have the same effect whether it was inherited from the mother or the father. This is probably a safe assumption most of the time. For example, when Mendel crossed purple-flowered pea plants with white-flowered pea plants, he observed the same results regardless of whether the purple-flowered parent supplied the eggs or the sperm. In recent years, however, geneticists have identified a number of traits in mammals that depend on which parent passed along the alleles for those traits. Such variation in phenotype depending on whether an allele is inherited from the male or female parent is called **genomic imprinting**. (Note that unlike sex-linked genes, most imprinted genes are on autosomes.) Using newer DNA sequence-based methods, about 100 imprinted genes have been identified in humans, and 125 in mice.

Genomic imprinting occurs during gamete formation and results in the silencing of a particular allele of certain genes. Because these genes are imprinted differently in sperm and eggs, the offspring expresses only one allele of an imprinted gene, the one that has been inherited from a specific parent—either the female parent or the male parent, depending on the particular gene. The imprints are then transmitted to all body cells during development. In each generation, the old imprints are "erased" in gamete-producing cells, and the chromosomes of the developing gametes are newly imprinted according to the sex of the individual forming the gametes. In a given species, the imprinted genes are always imprinted in the same way. For instance, a gene imprinted for maternal allele expression is always imprinted this way, generation after generation.

Consider, for example, the mouse gene for insulin-like growth factor 2 (*Igf2*), one of the first imprinted genes to be identified. Although this growth factor is required for normal prenatal growth, only the paternal allele is expressed **(Figure 15.17a)**. Evidence that the *Igf2* gene is imprinted came initially from crosses between normal-sized (wild-type) mice and dwarf (mutant) mice homozygous for a recessive mutation in the *Igf2* gene. The phenotypes of heterozygous offspring (with one normal allele and one mutant) differed depending on whether the mutant allele came from the mother or the father **(Figure 15.17b)**.

▼ Figure 15.17 Genomic imprinting of the mouse *Igf2* gene.

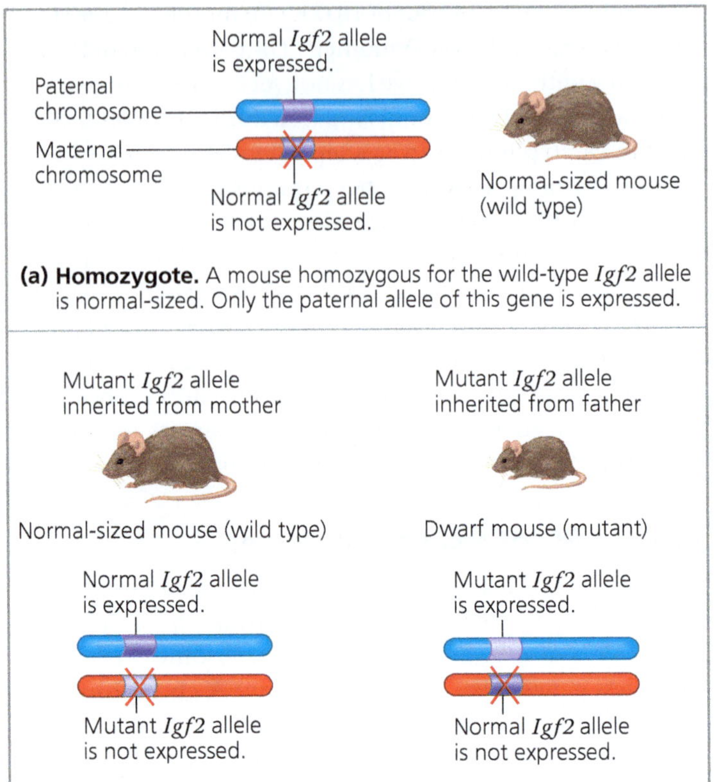

(a) Homozygote. A mouse homozygous for the wild-type *Igf2* allele is normal-sized. Only the paternal allele of this gene is expressed.

(b) Heterozygotes. Heterozygotes have different phenotypes depending on which allele is contributed by each parent. Matings between wild-type mice and those homozygous for the recessive mutant *Igf2* allele produce heterozygous offspring. Since the maternal *Igf2* allele is not expressed, the dwarf (mutant) phenotype is seen only when the father contributes the mutant allele.

What exactly is a genomic imprint? It turns out that imprinting can involve either silencing an allele in one type of gamete (egg or sperm) or activating it in the other. In many cases, the imprint seems to consist of methyl (—CH_3) groups that are added to cytosine nucleotides of one of the alleles. Such methylation may silence the allele, an effect consistent with evidence that heavily methylated genes are usually inactive (see Concept 18.2). However, for a few genes, methylation has been shown to *activate* expression of the allele. This is the case for the *Igf2* gene: Methylation of certain cytosines on the paternal chromosome leads to expression of the paternal *Igf2* allele, by an indirect mechanism involving chromatin structure and protein-DNA interactions.

Genomic imprinting may affect only a small fraction of the genes in mammalian genomes, but most of the known imprinted genes are critical for embryonic development. In experiments with mice, embryos engineered to inherit both copies of certain chromosomes from the same parent usually die before birth, whether that parent is male or female. A few years ago, however, scientists in Japan combined the genetic material from two eggs in a zygote while allowing expression of the *Igf2* gene, along with several other imprinted genes, from only one of the egg nuclei. The zygote developed into an apparently healthy mouse. Normal development seems to require that embryonic cells have exactly one active copy—not zero, not two—of certain genes. The association of improper imprinting with abnormal development and certain cancers has stimulated ongoing studies of how different genes are imprinted.

Interview with Shirley Tilghman: Discovering imprinted genes (see the interview before Chapter 13)

Inheritance of Organelle Genes

Although our focus in this chapter has been on the chromosomal basis of inheritance, we end with an important amendment: Not all of a eukaryotic cell's genes are located on nuclear chromosomes, or even in the nucleus; some genes are located in organelles in the cytoplasm. Because they are outside the nucleus, these genes are sometimes called *extranuclear genes* or *cytoplasmic genes*. Mitochondria, as well as chloroplasts and other plastids in plants, contain small circular DNA molecules that carry a number of genes. These organelles reproduce themselves and transmit their genes to daughter organelles. Genes on organelle DNA are not distributed to offspring according to the same rules that direct the distribution of nuclear chromosomes during meiosis, so they do not display Mendelian inheritance.

The first hint that extranuclear genes exist came from studies by the German scientist Carl Correns on the inheritance of yellow or white patches on the leaves of an otherwise green plant. In 1909, he observed that the coloration of the offspring was determined only by the maternal parent (the source of eggs) and not by the paternal parent (the source of sperm). Subsequent research showed that such coloration patterns, or variegation, are due to mutations in plastid genes that control pigmentation **(Figure 15.18)**. In most plants, a zygote receives all its plastids from the cytoplasm of the egg and none from the sperm, which contributes little more than a haploid set of chromosomes. An egg may contain plastids with different alleles for a pigmentation gene. As the zygote develops, plastids containing wild-type or mutant pigmentation genes are distributed randomly to daughter cells. The pattern of leaf coloration exhibited by a plant depends on the ratio of wild-type to mutant plastids in its various tissues.

Similar maternal inheritance is also the rule for mitochondrial genes in most animals and plants, because the mitochondria passed on to a zygote come from the cytoplasm of the egg. (The few mitochondria contributed by the sperm appear to be destroyed in the egg by autophagy; see Figure 6.13.) The products of most mitochondrial genes help make up some of the protein complexes of the electron transport chain and ATP synthase (see Figure 9.15). Defects in one or more of these proteins, therefore, reduce the amount of ATP the cell can make and have been shown to cause a number of human disorders in as many as one out of every 5,000 births. Because the parts of the body most susceptible to energy deprivation are the nervous system and the muscles, most mitochondrial diseases primarily affect these systems. For example, *mitochondrial myopathy* causes weakness, intolerance of exercise, and muscle deterioration. Another mitochondrial disorder is *Leber's hereditary optic neuropathy*, which can produce sudden blindness in people as young as their 20s or 30s. The four mutations found thus far to cause this disorder affect oxidative phosphorylation during cellular respiration, a crucial function for the cell (see Concept 9.4).

◀ Figure 15.18 A painted nettle coleus plant. The variegated (patterned) leaves on this coleus plant (*Plectranthus scutellarioides*) result from mutations that affect expression of pigment genes located in plastids, which generally are inherited from the maternal parent.

The fact that mitochondrial disorders are inherited only from the mother suggests a way to avoid passing along these disorders. The chromosomes from the egg of an affected mother could be transferred to an egg of a healthy donor that has had its own chromosomes removed. This "two-mother" egg could then be fertilized by a sperm from the prospective father and transplanted into the womb of the prospective mother, becoming an embryo with three parents. After optimizing conditions for this approach in monkeys, researchers reported in 2013 that they have successfully carried out this procedure on human eggs. More research will be necessary to optimize experimental conditions for the health of the embryo, and eventual use of this procedure would require approval by the relevant federal agencies.

In addition to the rarer diseases clearly caused by defects in mitochondrial DNA, mitochondrial mutations inherited from a person's mother may contribute to at least some types of diabetes and heart disease, as well as to other disorders that commonly debilitate the elderly, such as Alzheimer's disease. In the course of a lifetime, new mutations gradually accumulate in our mitochondrial DNA, and some researchers think that these mutations play a role in the normal aging process.

CONCEPT CHECK 15.5

1. Gene dosage—the number of copies of a gene that are actively being expressed—is important to proper development. Identify and describe two processes that establish the proper dosage of certain genes.
2. Reciprocal crosses between two primrose varieties, A and B, produced the following results: A female × B male → offspring with all green (nonvariegated) leaves; B female × A male → offspring with patterned (variegated) leaves. Explain these results.
3. **WHAT IF?** ➤ Mitochondrial genes are critical to the energy metabolism of cells, but mitochondrial disorders caused by mutations in these genes are generally not lethal. Why not?

For suggested answers, see Appendix A.

15 Chapter Review

Go to **MasteringBiology**™ for Videos, Animations, Vocab Self-Quiz, Practice Tests, and more in the Study Area.

SUMMARY OF KEY CONCEPTS

VOCAB SELF-QUIZ goo.gl/6u55ks

CONCEPT 15.1

Morgan showed that Mendelian inheritance has its physical basis in the behavior of chromosomes: *scientific inquiry* *(pp. 296–297)*

- Morgan's work with an eye color gene in *Drosophila* led to the **chromosome theory of inheritance**, which states that genes are located on chromosomes and that the behavior of chromosomes during meiosis accounts for Mendel's laws.

? *What characteristic of the sex chromosomes allowed Morgan to correlate their behavior with that of the alleles of the eye color gene?*

CONCEPT 15.2

Sex-linked genes exhibit unique patterns of inheritance *(pp. 298–300)*

- Sex is often chromosomally based. Humans and other mammals have an X-Y system in which sex is largely determined by whether a Y chromosome is present. Other systems are found in birds, fishes, and insects.
- The sex chromosomes carry **sex-linked genes**, virtually all of which are on the X chromosome (**X-linked**). Any male who inherits a recessive X-linked allele (from his mother) will express the trait, such as color blindness.
- In mammalian females, one of the two X chromosomes in each cell is randomly inactivated during early embryonic development, becoming highly condensed into a **Barr body**.

? *Why are males affected by X-linked disorders much more often than females?*

CONCEPT 15.3

Linked genes tend to be inherited together because they are located near each other on the same chromosome *(pp. 301–306)*

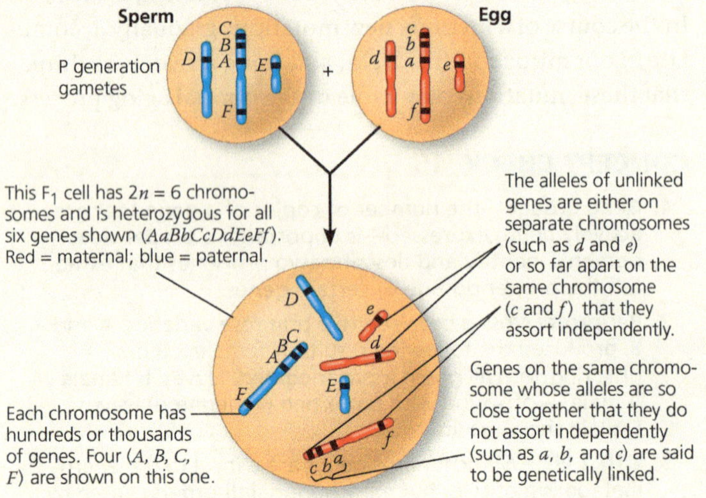

- An F_1 dihybrid testcross yields **parental types** with the same combination of traits as those in the P generation parents and **recombinant types** (**recombinants**) with new combinations of traits not seen in either P generation parent. Because of the independent assortment of chromosomes, unlinked genes exhibit a 50% frequency of recombination in the gametes. For genetically **linked genes, crossing over** between nonsister chromatids during meiosis I accounts for the observed recombinants, always less than 50%.
- The order of genes on a chromosome and the relative distances between them can be deduced from recombination frequencies observed in genetic crosses. These data allow construction of a **linkage map** (a type of **genetic map**). The farther apart genes are, the more likely their allele combinations will be recombined during crossing over.

? *Why are specific alleles of two distant genes more likely to show recombination than those of two closer genes?*

CONCEPT 15.4

Alterations of chromosome number or structure cause some genetic disorders *(pp. 306–309)*

- **Aneuploidy**, an abnormal chromosome number, can result from **nondisjunction** during meiosis. When a normal gamete unites with one containing two copies or no copies of a particular chromosome, the resulting zygote and its descendant cells either have one extra copy of that chromosome (**trisomy**, $2n + 1$) or are missing a copy (**monosomy**, $2n - 1$). **Polyploidy** (extra sets of chromosomes) can result from complete nondisjunction.
- Chromosome breakage can result in alterations of chromosome structure: **deletions, duplications, inversions,** and **translocations**. Translocations can be reciprocal or nonreciprocal.
- Changes in the number of chromosomes per cell or in the structure of individual chromosomes can affect the phenotype and, in some cases, lead to disorders. Such alterations cause **Down syndrome** (usually due to trisomy of chromosome 21), certain cancers associated with chromosomal translocations that occur during mitosis, and various other human disorders.

? *Why are inversions and reciprocal translocations less likely to be lethal than are aneuploidy, duplications, deletions, and nonreciprocal translocations?*

CONCEPT 15.5

Some inheritance patterns are exceptions to standard Mendelian inheritance *(pp. 310–311)*

- In mammals, the phenotypic effects of a small number of particular genes depend on which allele is inherited from each parent, a phenomenon called **genomic imprinting**. Imprints are formed during gamete production, with the result that one allele (either maternal or paternal) is not expressed in offspring.
- The inheritance of traits controlled by the genes present in mitochondria and plastids depends solely on the maternal parent because the zygote's cytoplasm containing these organelles comes from the egg. Some diseases affecting the nervous and muscular systems are caused by defects in mitochondrial genes that prevent cells from making enough ATP.

? *Explain how genomic imprinting and inheritance of mitochondrial and chloroplast DNA are exceptions to standard Mendelian inheritance.*

TEST YOUR UNDERSTANDING

Level 1: Knowledge/Comprehension

PRACTICE TEST
goo.gl/CUYGKD

1. A man with hemophilia (a recessive, sex-linked condition) has a daughter without the condition. She marries a man who does not have hemophilia. What is the probability that their daughter will have hemophilia? Their son? If they have four sons, what is the probability that all will be affected?

2. Pseudohypertrophic muscular dystrophy is an inherited disorder that causes gradual deterioration of the muscles. It is seen almost exclusively in boys born to apparently unaffected parents and usually results in death in the early teens. Is this disorder caused by a dominant or a recessive allele? Is its inheritance sex-linked or autosomal? How do you know? Explain why this disorder is almost never seen in girls.

3. A wild-type fruit fly (heterozygous for gray body color and normal wings) is mated with a black fly with vestigial wings. The offspring have the following phenotypic distribution: wild-type, 778; black vestigial, 785; black normal, 158; gray vestigial, 162. What is the recombination frequency between these genes for body color and wing size? Is this consistent with the results of the experiment in Figure 15.9?

4. A planet is inhabited by creatures that reproduce with the same hereditary patterns seen in humans. Three phenotypic characters are height (T = tall, t = dwarf), head appendages (A = antennae, a = no antennae), and nose morphology (S = upturned snout, s = downturned snout). Since the creatures are not "intelligent," Earth scientists are able to do some controlled breeding experiments using various heterozygotes in testcrosses. For tall heterozygotes with antennae, the offspring are tall antennae, 46; dwarf antennae, 7; dwarf no antennae, 42; tall no antennae, 5. For heterozygotes with antennae and an upturned snout, the offspring are antennae upturned snout, 47; antennae downturned snout, 2; no antennae downturned snout, 48; no antennae upturned snout, 3. Calculate the recombination frequencies for both experiments.

Level 2: Application/Analysis

5. Using the information from problem 4, scientists do a further testcross using a heterozygote for height and nose morphology. The offspring are tall upturned snout, 40; dwarf upturned snout, 9; dwarf downturned snout, 42; tall downturned snout, 9. Calculate the recombination frequency from these data, and then use your answer from problem 4 to determine the correct order of the three linked genes.

6. A wild-type fruit fly (heterozygous for gray body color and red eyes) is mated with a black fruit fly with purple eyes. The offspring are wild-type, 721; black purple, 751; gray purple, 49; black red, 45. What is the recombination frequency between these genes for body color and eye color? Using information from problem 3, what fruit flies (genotypes and phenotypes) would you mate to determine the order of the body color, wing size, and eye color genes on the chromosome?

7. Assume that genes A and B are on the same chromosome and are 50 map units apart. An animal heterozygous at both loci is crossed with one that is homozygous recessive at both loci. What percentage of the offspring will show recombinant phenotypes resulting from crossovers? Without knowing these genes are on the same chromosome, how would you interpret the results of this cross?

8. Two genes of a flower, one controlling blue (B) versus white (b) petals and the other controlling round (R) versus oval (r) stamens, are linked and are 10 map units apart. You cross a homozygous blue oval plant with a homozygous white round plant. The resulting F_1 progeny are crossed with homozygous white oval plants, and 1,000 offspring plants are obtained. How many plants of each of the four phenotypes do you expect?

9. You design *Drosophila* crosses to provide recombination data for gene a, which is located on the chromosome shown in Figure 15.12. Gene a has recombination frequencies of 14% with the vestigial wing locus and 26% with the brown eye locus. Approximately where is a located along the chromosome?

Level 3: Synthesis/Evaluation

10. Banana plants, which are triploid, are seedless and therefore sterile. Propose a possible explanation.

11. **EVOLUTION CONNECTION** Crossing over is thought to be evolutionarily advantageous because it continually shuffles genetic alleles into novel combinations. Until recently, it was thought that the genes on the Y chromosome might degenerate because they lack homologous genes on the X chromosome with which to pair up prior to crossing over. However, when the Y chromosome was sequenced, eight large regions were found to be internally homologous to each other, and quite a few of the 78 genes represent duplicates. (Y chromosome researcher David Page has called it a "hall of mirrors.") Explain what might be a benefit of these regions.

12. **SCIENTIFIC INQUIRY • DRAW IT** Assume you are mapping genes *A, B, C,* and *D* in *Drosophila*. You know that these genes are linked on the same chromosome, and you determine the recombination frequencies between each pair of genes to be as follows: *A–B*, 8%; *A–C*, 28%; *A–D*, 25%; *B–C*, 20%; *B–D*, 33%.
    (a) Describe how you determined the recombination frequency for each pair of genes.
    (b) Draw a chromosome map based on your data.

13. **WRITE ABOUT A THEME: INFORMATION** The continuity of life is based on heritable information in the form of DNA. In a short essay (100–150 words), relate the structure and behavior of chromosomes to inheritance in both asexually and sexually reproducing species.

14. **SYNTHESIZE YOUR KNOWLEDGE**

Butterflies have an X-Y sex determination system that is different from that of flies or humans. Female butterflies may be either XY or X0, while butterflies with two or more X chromosomes are males. This photograph shows a tiger swallowtail *gynandromorph*, which is half male (left side) and half female (right side). Given that the first division of the zygote divides the embryo into the future right and left halves of the butterfly, propose a hypothesis that explains how nondisjunction during the first mitosis might have produced this unusual-looking butterfly.

For selected answers, see Appendix A.

MB For additional practice questions, check out the **Dynamic Study Modules** in MasteringBiology. You can use them to study on your smartphone, tablet, or computer anytime, anywhere!

The Molecular Basis of Inheritance 16

▲ **Figure 16.1 What is the structure of DNA?**

KEY CONCEPTS

16.1 DNA is the genetic material

16.2 Many proteins work together in DNA replication and repair

16.3 A chromosome consists of a DNA molecule packed together with proteins

▲ James Watson (left) and Francis Crick with their DNA model.

Life's Operating Instructions

The elegant double-helical structure of deoxyribonucleic acid, or DNA, has become an icon of modern biology **(Figure 16.1)**. James Watson and Francis Crick shook the scientific world in April 1953 with their DNA model, which they constructed from sheet metal and wire, shown in the small photo. Gregor Mendel's heritable factors and Thomas Hunt Morgan's genes on chromosomes are, in fact, composed of DNA. Chemically speaking, your genetic endowment is the DNA you inherited from your parents. DNA, the substance of inheritance, is the most celebrated molecule of our time.

Of all nature's molecules, nucleic acids are unique in their ability to direct their own replication from monomers. Indeed, the resemblance of offspring to their parents has its basis in the accurate replication of DNA and its transmission from one generation to the next. Hereditary information in DNA directs the development of your biochemical, anatomical, physiological, and, to some extent, behavioral traits. In this chapter, you will discover how biologists deduced that DNA is the genetic material and how Watson and Crick worked out its structure. You will also learn how a molecule of DNA is copied during **DNA replication** and how cells repair their DNA. Finally, you will explore how a molecule of DNA is packaged together with proteins in a chromosome.

When you see this blue icon, log in to **MasteringBiology** and go to the Study Area for digital resources.

MB **Get Ready for This Chapter**

MB **Interview with James Watson: Co-discoverer of the structure of DNA**

CONCEPT 16.1

DNA is the genetic material

Today, even schoolchildren have heard of DNA, and scientists routinely manipulate DNA in the laboratory. Early in the 20th century, however, identifying the molecules of inheritance loomed as a major challenge to biologists.

The Search for the Genetic Material: *Scientific Inquiry*

Once T. H. Morgan's group showed that genes exist as parts of chromosomes (described in Concept 15.1), the two chemical components of chromosomes—DNA and protein—emerged as the leading candidates for the genetic material. Until the 1940s, the case for proteins seemed stronger: Biochemists had identified proteins as a class of macromolecules with great heterogeneity and specificity of function, essential requirements for the hereditary material. Moreover, little was known about nucleic acids, whose physical and chemical properties seemed far too uniform to account for the multitude of specific inherited traits exhibited by every organism. This view gradually changed as the role of DNA in heredity was worked out in studies of bacteria and the viruses that infect them, systems far simpler than fruit flies or humans. Let's trace the search for the genetic material as a case study in scientific inquiry.

Evidence That DNA Can Transform Bacteria

In 1928, a British medical officer named Frederick Griffith was trying to develop a vaccine against pneumonia. He was studying *Streptococcus pneumoniae*, a bacterium that causes pneumonia in mammals. Griffith had two strains (varieties) of the bacterium, one pathogenic (disease-causing) and one nonpathogenic (harmless). He was surprised to find that when he killed the pathogenic bacteria with heat and then mixed the cell remains with living bacteria of the nonpathogenic strain, some of the living cells became pathogenic **(Figure 16.2)**. Furthermore, this newly acquired trait of pathogenicity was inherited by all the descendants of the transformed bacteria. Apparently, some chemical component of the dead pathogenic cells caused this heritable change, although the identity of the substance was not known. Griffith called the phenomenon **transformation**, now defined as a change in genotype and phenotype due to the assimilation of external DNA by a cell. Later work by Oswald Avery, Maclyn McCarty, and Colin MacLeod identified the transforming substance as DNA.

Scientists remained skeptical, however, since many still viewed proteins as better candidates for the genetic material. Also, many biologists were not convinced that bacterial genes would be similar in composition and function to those of more complex organisms. But the major reason for the continued doubt was that so little was known about DNA.

▼ Figure 16.2

Inquiry Can a genetic trait be transferred between different bacterial strains?

Experiment Frederick Griffith studied two strains of the bacterium *Streptococcus pneumoniae*. The S (smooth) strain can cause pneumonia in mice; it is pathogenic because the cells have an outer capsule that protects them from an animal's immune system. Cells of the R (rough) strain lack a capsule and are nonpathogenic. To test for the trait of pathogenicity, Griffith injected mice with the two strains:

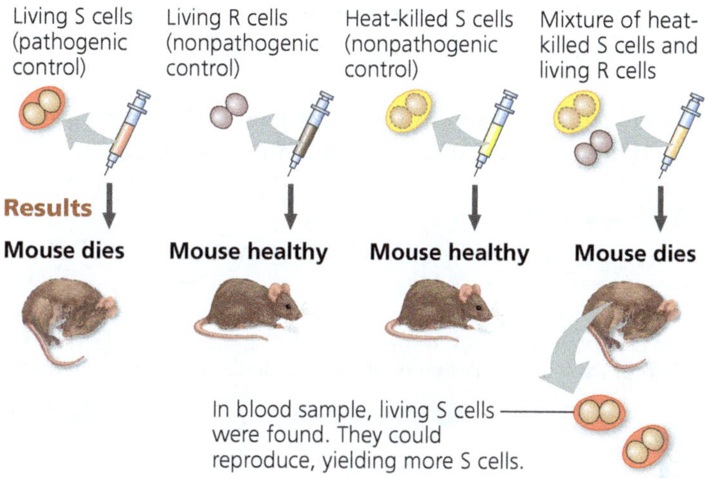

Conclusion The living R bacteria had been transformed into pathogenic S bacteria by an unknown, heritable substance from the dead S cells that enabled the R cells to make capsules.

Data from F. Griffith, The significance of pneumococcal types, *Journal of Hygiene* 27:113–159 (1928).

WHAT IF? ➤ *How did this experiment rule out the possibility that the R cells simply used the dead S cells' capsules to become pathogenic?*

Evidence That Viral DNA Can Program Cells

Additional evidence that DNA was the genetic material came from studies of viruses that infect bacteria **(Figure 16.3)**. These viruses are called **bacteriophages** (meaning "bacteria-eaters"), or **phages** for short. Viruses are much simpler than cells. A **virus** is little more than DNA (or sometimes RNA) enclosed by a protective coat, which is often simply protein. To produce more viruses, a virus must infect a cell and take over the cell's metabolic machinery.

Phages have been widely used as tools by researchers in molecular genetics. In 1952, Alfred Hershey and Martha Chase performed

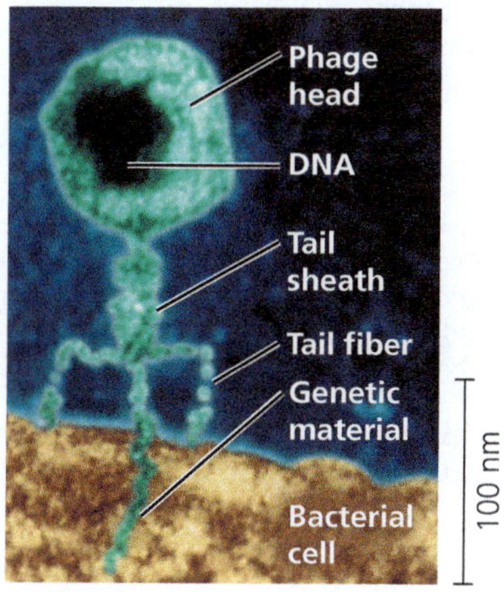

➤ Figure 16.3 A virus infecting a bacterial cell. A phage called T2 attaches to a host cell and injects its genetic material through the plasma membrane, while the head and tail parts remain on the outer bacterial surface (colorized TEM).

experiments showing that DNA is the genetic material of a phage known as T2. This is one of many phages that infect *Escherichia coli* (*E. coli*), a bacterium that normally lives in the intestines of mammals and is a model organism for molecular biologists. At that time, biologists already knew that T2, like many other phages, was composed almost entirely of DNA and protein. They also knew that the T2 phage could quickly turn an *E. coli* cell into a T2-producing factory that released many copies of new phages when the cell ruptured. Somehow, T2 could reprogram its host cell to produce viruses. But which viral component—protein or DNA—was responsible?

Hershey and Chase answered this question by devising an experiment showing that only one of the two components of T2 actually enters the *E. coli* cell during infection **(Figure 16.4)**. In their experiment, they used a radioactive isotope of sulfur to tag protein in one batch of T2 and a radioactive isotope of phosphorus to tag DNA in a second batch. Because protein, but not DNA, contains sulfur, radioactive sulfur atoms were incorporated only into the protein of the phage. In a similar way, the atoms of radioactive phosphorus labeled only the DNA, not the protein, because nearly all the phage's phosphorus is in its DNA. In the experiment, separate samples of nonradioactive *E. coli* cells were infected with the protein-labeled and DNA-labeled batches of T2. The researchers then tested the two samples shortly after the onset of infection to see which type of molecule—protein or DNA—had entered the bacterial cells and would therefore be capable of reprogramming them.

▼ Figure 16.4

Inquiry Is protein or DNA the genetic material of phage T2?

Experiment Alfred Hershey and Martha Chase used radioactive sulfur and phosphorus to trace the fates of protein and DNA, respectively, of T2 phages that infected bacterial cells. They wanted to see which of these molecules entered the cells and could reprogram them to make more phages.

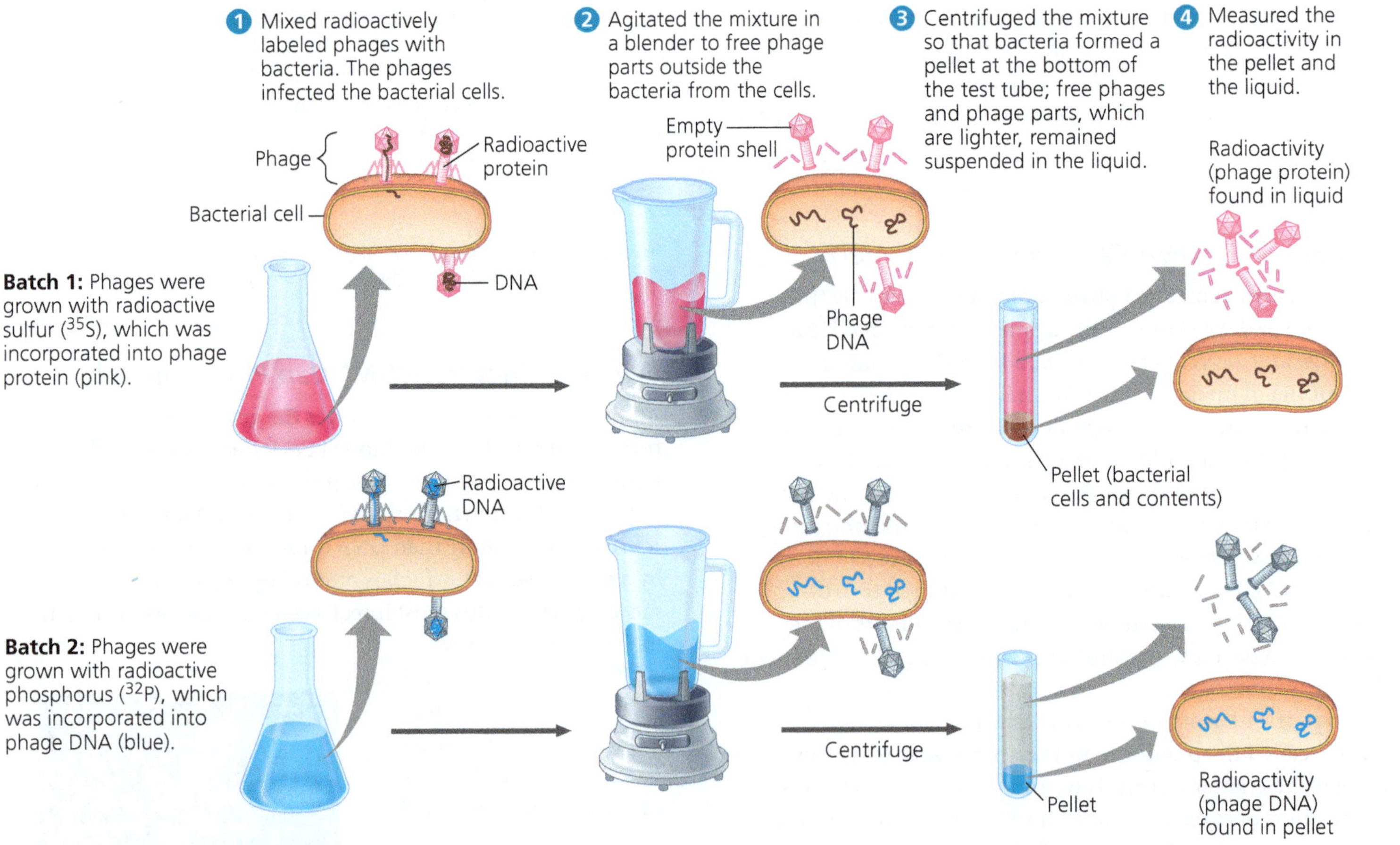

Results When proteins were labeled (batch 1), radioactivity remained outside the cells, but when DNA was labeled (batch 2), radioactivity was found inside the cells. Cells containing radioactive phage DNA released new phages with some radioactive phosphorus.

Conclusion Phage DNA entered bacterial cells, but phage proteins did not. Hershey and Chase concluded that DNA, not protein, functions as the genetic material of phage T2.

Data from A. D. Hershey and M. Chase, Independent functions of viral protein and nucleic acid in growth of bacteriophage, *Journal of General Physiology* 36:39–56 (1952).

WHAT IF? ➤ *How would the results have differed if proteins carried the genetic information?*

Animation: The Hershey-Chase Experiment

Hershey and Chase found that the phage DNA entered the host cells but the phage protein did not. Moreover, when these bacteria were returned to a culture medium and the infection ran its course, the *E. coli* released phages that contained some radioactive phosphorus. This result further showed that the DNA inside the cell played an ongoing role during the infection process. They concluded that the DNA injected by the phage must be the molecule carrying the genetic information that makes the cells produce new viral DNA and proteins. The Hershey-Chase experiment was a landmark study because it provided powerful evidence that nucleic acids, rather than proteins, are the hereditary material, at least for certain viruses.

Additional Evidence That DNA Is the Genetic Material

Further evidence that DNA is the genetic material came from the laboratory of biochemist Erwin Chargaff. DNA was known to be a polymer of nucleotides, each having three components: a nitrogenous (nitrogen-containing) base, a pentose sugar called deoxyribose, and a phosphate group **(Figure 16.5)**. The base can be adenine (A), thymine (T), guanine (G), or cytosine (C). Chargaff analyzed the base composition of DNA from a number of different organisms. In 1950, he reported that the base composition of DNA varies from one species to another. For example, he found that 32.8% of sea urchin DNA nucleotides have the base A, whereas 30.4% of human DNA nucleotides have the base A and only 24.7% of the DNA nucleotides from the bacterium *E. coli* have the base A. Chargaff's evidence of molecular diversity among species, which most scientists had presumed to be absent from DNA, made DNA a more credible candidate for the genetic material.

Chargaff also noticed a peculiar regularity in the ratios of nucleotide bases. In the DNA of each species he studied, the number of adenines approximately equaled the number of thymines, and the number of guanines approximately equaled the number of cytosines. In sea urchin DNA, for example, Chargaff's analysis found the four bases in these percentages: A = 32.8% and T = 32.1%; G = 17.7% and C = 17.3%. (The percentages are not exactly the same because of limitations in Chargaff's techniques.)

These two findings became known as *Chargaff's rules*: (1) DNA base composition varies between species, and (2) for each species, the percentages of A and T bases are roughly equal, as are those of G and C bases. In the **Scientific Skills Exercise**, you can use Chargaff's rules to predict percentages of nucleotide bases. The basis for these rules remained unexplained until the discovery of the double helix.

Building a Structural Model of DNA: *Scientific Inquiry*

Once most biologists were convinced that DNA was the genetic material, the challenge was to determine how the structure of DNA could account for its role in inheritance.

▼ Figure 16.5 The structure of a DNA strand. Each DNA nucleotide monomer consists of a nitrogenous base (T, A, C, or G), the sugar deoxyribose (blue), and a phosphate group (yellow). The phosphate group of one nucleotide is attached to the sugar of the next by a covalent bond, forming a "backbone" of alternating phosphates and sugars from which the bases project. A polynucleotide strand has directionality, from the 5′ end (with the phosphate group) to the 3′ end (with the —OH group of the sugar). 5′ and 3′ refer to the numbers assigned to the carbons in the sugar ring.

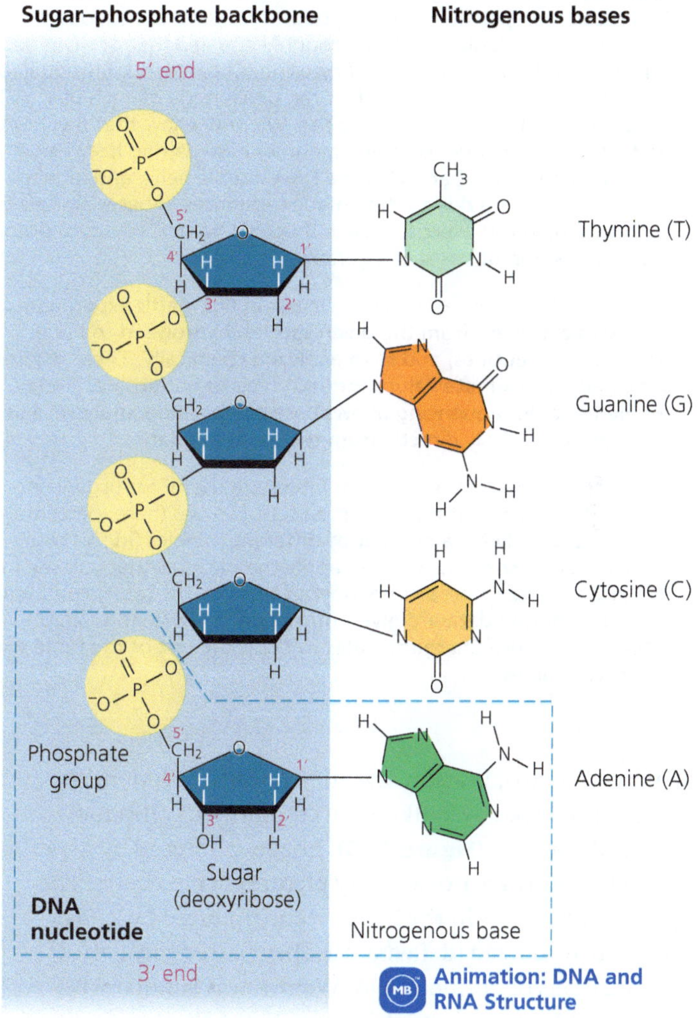

Animation: DNA and RNA Structure

By the early 1950s, the arrangement of covalent bonds in a nucleic acid polymer was well established (see Figure 16.5), and researchers focused on discovering the three-dimensional structure of DNA. Among the scientists working on the problem were Linus Pauling, at the California Institute of Technology, and Maurice Wilkins and Rosalind Franklin, at King's College in London. First to come up with the complete answer, however, were two scientists who were relatively unknown at the time—the American James Watson and the Englishman Francis Crick.

The brief but celebrated partnership that solved the puzzle of DNA structure began soon after Watson journeyed to Cambridge University, where Crick was studying protein structure with a technique called X-ray crystallography (see Figure 5.21). While visiting the laboratory of

SCIENTIFIC SKILLS EXERCISE

Working with Data in a Table

Given the Percentage Composition of One Nucleotide in a Genome, Can We Predict the Percentages of the Other Three Nucleotides? Even before the structure of DNA was elucidated, Erwin Chargaff and his co-workers noticed a pattern in the base composition of nucleotides from different species: The percentage of adenine (A) bases roughly equaled that of thymine (T) bases, and the percentage of cytosine (C) bases roughly equaled that of guanine (G) bases. Further, the percentage of each pair (A/T or C/G) varied from species to species. We now know that the 1:1 A/T and C/G ratios are due to complementary base pairing between A and T and between C and G in the DNA double helix, and interspecies differences are due to the unique sequences of bases along a DNA strand. In this exercise, you will apply Chargaff's rules to predict the composition of bases in a genome.

How the Experiments Were Done In Chargaff's experiments, DNA was extracted from the given species, hydrolyzed to break apart the nucleotides, and then analyzed chemically. These studies provided approximate values for each type of nucleotide. (Today, whole-genome sequencing allows base composition analysis to be done more precisely directly from the sequence data.)

Data from the Experiments Tables are useful for organizing sets of data representing a common set of values (here, percentages of A, G, C, and T) for a number of different samples (in this case, from different species). You can apply the patterns that you see in the known data to predict unknown values. In the table, complete base distribution data are given for sea urchin DNA and salmon DNA; you will use Chargaff's rules to fill in the rest of the table with predicted values.

	Base Percentage			
Source of DNA	**Adenine**	**Guanine**	**Cytosine**	**Thymine**
Sea urchin	32.8	17.7	17.3	32.1
Salmon	29.7	20.8	20.4	29.1
Wheat	28.1	21.8	22.7	
E. coli	24.7	26.0		
Human	30.4			30.1
Ox	29.0			
Average %				

Data from several papers by Chargaff: for example, E. Chargaff et al., Composition of the desoxypentose nucleic acids of four genera of sea-urchin, *Journal of Biological Chemistry* 195:155–160 (1952).

➤ **Sea urchin**

INTERPRET THE DATA

1. Explain how the sea urchin and salmon data demonstrate both of Chargaff's rules.
2. Using Chargaff's rules, fill in the table with your predictions of the missing percentages of bases, starting with the wheat genome and proceeding through *E. coli*, human, and ox. Show how you arrived at your answers.
3. If Chargaff's rule—that the amount of A equals the amount of T and the amount of C equals the amount of G—is valid, then hypothetically we could extrapolate this to the combined DNA of all species on Earth (like one huge Earth genome). To see whether the data in the table support this hypothesis, calculate the average percentage for each base in your completed table by averaging the values in each column. Does Chargaff's equivalence rule still hold true?

Instructors: A version of this Scientific Skills Exercise can be assigned in MasteringBiology.

Maurice Wilkins, Watson saw an X-ray diffraction image of DNA produced by Wilkins's accomplished colleague Rosalind Franklin **(Figure 16.6)**. Images produced by X-ray crystallography are not actually pictures of molecules. The spots and smudges in the image were produced by X-rays that were diffracted (deflected) as they passed through aligned fibers of purified DNA. Watson was familiar with the type of X-ray diffraction pattern that helical molecules produce, and an examination of the photo that Wilkins showed him confirmed that DNA was helical in shape. The photo also augmented earlier data obtained by Franklin and others suggesting the width of the helix and the spacing of

◀ Figure 16.6 Rosalind Franklin and her X-ray diffraction photo of DNA.

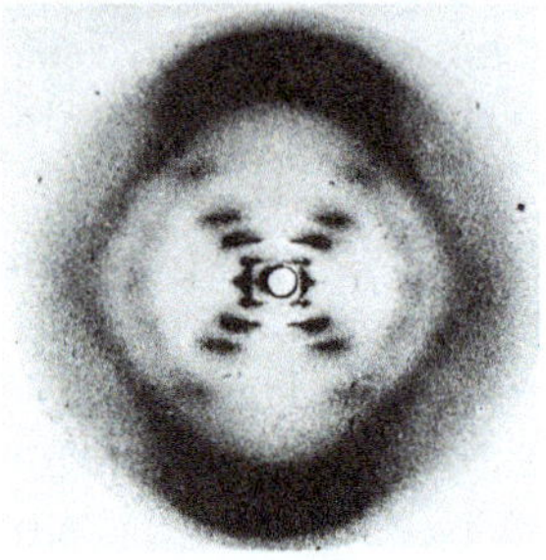

the nitrogenous bases along it. The pattern in this photo implied that the helix was made up of two strands, contrary to a three-stranded model that Linus Pauling had proposed a short time earlier. The presence of two strands accounts for the now-familiar term **double helix**. DNA is shown in some of its many different representations in **Figure 16.7**.

Watson and Crick began building models of a double helix that would conform to the X-ray measurements and what was then known about the chemistry of DNA, including Chargaff's rule of base equivalences. Having also read an unpublished annual report summarizing Franklin's work, they knew she had concluded that the sugar-phosphate backbones were on the outside of the DNA molecule, contrary to their working model. Franklin's arrangement was appealing because it put the negatively charged phosphate groups facing the aqueous surroundings, while the relatively hydrophobic nitrogenous bases were hidden in the interior. Watson constructed such a model, shown in the small photo on the first page of this chapter. In this model, the two sugar-phosphate backbones are **antiparallel**—that is, their subunits run in opposite directions (see Figure 16.7). You can imagine the overall arrangement as a rope ladder with rigid rungs. The side ropes represent the sugar-phosphate backbones, and the rungs represent pairs of nitrogenous bases. Now imagine twisting the

Figure 16.7 Visualizing DNA

DNA can be illustrated in many ways, but all diagrams represent the same basic structure. The level of detail shown depends on the process or the type of information being conveyed.

Structural Images

These structural images show the three-dimensional shape of the DNA double helix (left) and chemical details of DNA's structure (right). Both images use the same colors for phosphate groups (yellow), deoxyribose sugars (blue), and nitrogenous bases (shades of green and orange).

The DNA double helix is right-handed, as shown in this computer-generated space-filling model. Use your right hand as shown to follow the sugar-phosphate backbone up the helix (red arrow) and around to the back. (It won't work with your left hand.)

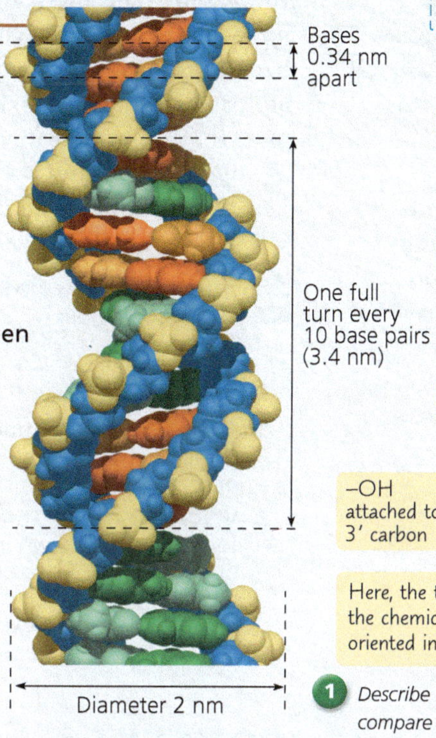

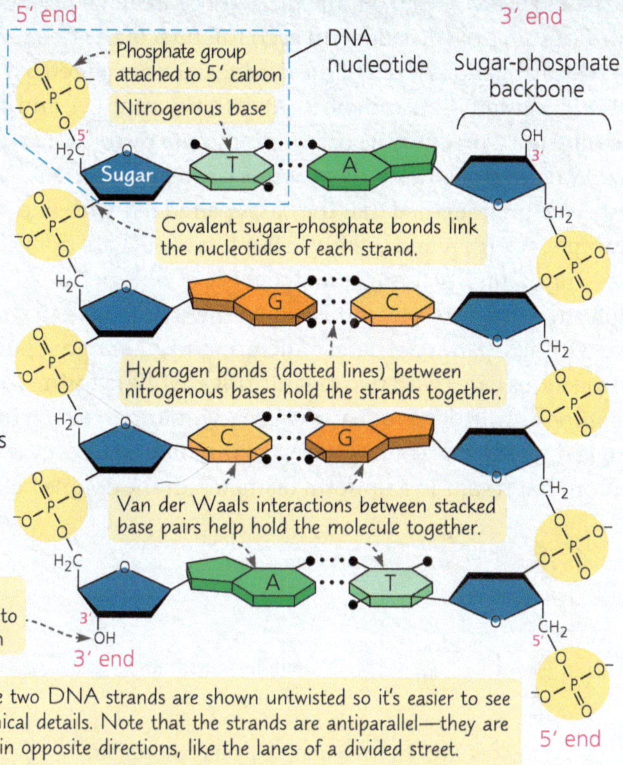

Here, the two DNA strands are shown untwisted so it's easier to see the chemical details. Note that the strands are antiparallel—they are oriented in opposite directions, like the lanes of a divided street.

1 *Describe the bonds that hold together the nucleotides in one DNA strand. Then compare them with the bonds that hold the two DNA strands together.*

Simplified Images

When molecular detail is not necessary, DNA is portrayed in a range of simplified diagrams, depending on the focus of the figure.

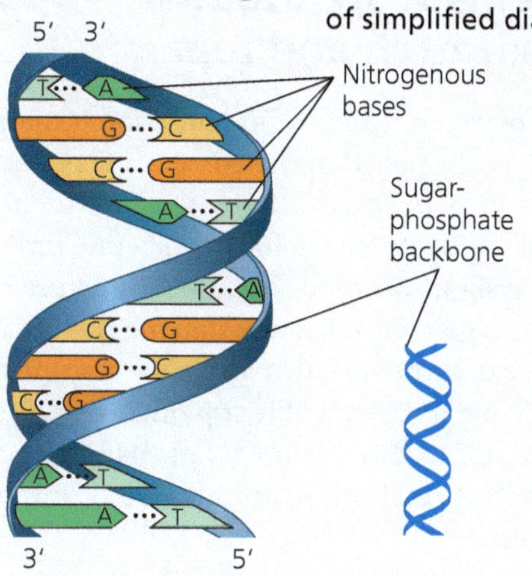

The "ribbons" in these simplified double helix diagrams represent the sugar-phosphate backbones.

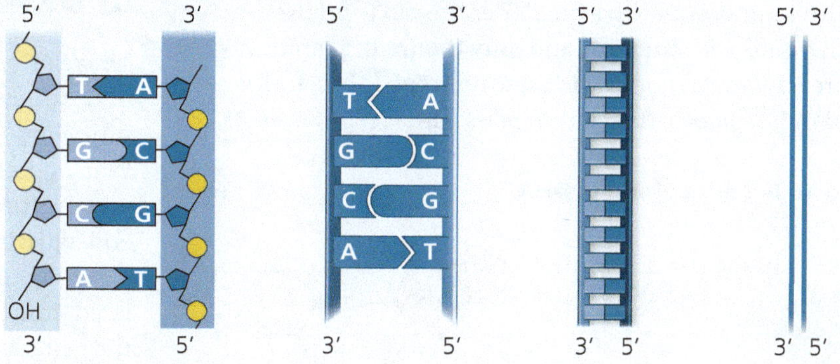

These flattened "ladder style" diagrams of DNA depict the sugar-phosphate backbones like the side rails of a ladder, with the base pairs as rungs. Light blue is used to indicate the more recently synthesized strand.

Sometimes the double-stranded DNA molecule is shown simply as two straight lines.

2 *Compare the information conveyed in the three ladder diagrams.*

DNA Sequences

Genetic information is carried in DNA as a linear sequence of nucleotides that may be transcribed into mRNA and translated into a polypeptide. When focusing on the DNA sequence, each nucleotide can be represented simply as the letter of its base: A, T, C, or G.

3′- A C G T A A G C G G T T A A T -5′
5′- T G C A T T C G C C A A T T A -3′

Instructors: Additional questions related to this Visualizing Figure can be assigned in MasteringBiology.

Animation: DNA Double Helix

ladder to form a helix. Franklin's X-ray data indicated that the helix makes one full turn every 3.4 nm along its length. With the bases stacked just 0.34 nm apart, there are ten layers of base pairs, or rungs of the ladder, in each full turn of the helix.

The nitrogenous bases of the double helix are paired in specific combinations: adenine (A) with thymine (T), and guanine (G) with cytosine (C). It was mainly by trial and error that Watson and Crick arrived at this key feature of DNA. At first, Watson imagined that the bases paired like with like—for example, A with A and C with C. But this model did not fit the X-ray data, which suggested that the double helix had a uniform diameter. Why is this requirement inconsistent with like-with-like pairing of bases? Adenine and guanine are purines, nitrogenous bases with two organic rings, while cytosine and thymine are nitrogenous bases called pyrimidines, which have a single ring. Pairing a purine with a pyrimidine is the only combination that results in a uniform diameter for the double helix:

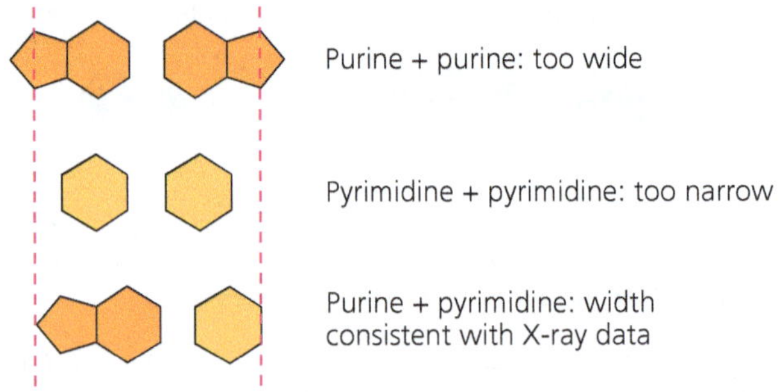

Watson and Crick reasoned that there must be additional specificity of pairing dictated by the structure of the bases. Each base has chemical side groups that can form hydrogen bonds with its appropriate partner: Adenine can form two hydrogen bonds with thymine and only thymine; guanine forms three hydrogen bonds with cytosine and only cytosine. In shorthand, A pairs with T, and G pairs with C **(Figure 16.8)**.

▼ Figure 16.8 Base pairing in DNA.

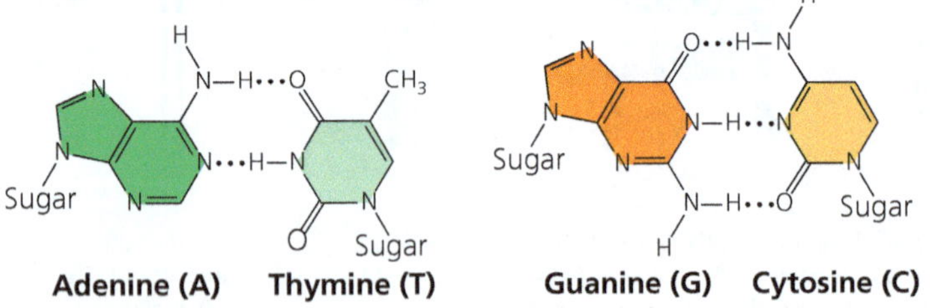

The Watson-Crick model took into account Chargaff's ratios and ultimately explained them. Wherever one strand of a DNA molecule has an A, the partner strand has a T. Similarly, a G in one strand is always paired with a C in the complementary strand. Therefore, in the DNA of any organism, the amount of adenine equals the amount of thymine, and the amount of guanine equals the amount of cytosine. (Modern DNA sequencing techniques have confirmed that the amounts are exactly equal.) Although the base-pairing rules dictate the combinations of nitrogenous bases that form the "rungs" of the double helix, they do not restrict the sequence of nucleotides *along* each DNA strand. The linear sequence of the four bases can be varied in countless ways, and each gene has a unique base sequence.

In April 1953, Watson and Crick surprised the scientific world with a succinct, one-page paper that reported their molecular model for DNA: the double helix, which has since become the symbol of molecular biology. Watson and Crick, along with Maurice Wilkins, were awarded the Nobel Prize in 1962 for this work. (Sadly, Rosalind Franklin had died at the age of 37 in 1958 and was thus ineligible for the prize.) The beauty of the double helix model was that the structure of DNA suggested the basic mechanism of its replication.

HHMI Video: Great Discoveries in Science: The Double Helix

CONCEPT CHECK 16.1

1. Given a polynucleotide sequence such as GAATTC, explain what further information you would need in order to identify which is the 5′ end. (See Figure 16.5.)
2. **VISUAL SKILLS >** Griffith was trying to develop a vaccine for *S. pneumonia* when he was surprised to discover the phenomenon of bacterial transformation. Look at the second and third panels of Figure 16.2. Based on these results, what result was he expecting in the fourth panel? Explain.

For suggested answers, see Appendix A.

CONCEPT 16.2

Many proteins work together in DNA replication and repair

The relationship between structure and function is manifest in the double helix. The idea that there is specific pairing of nitrogenous bases in DNA was the flash of inspiration that led Watson and Crick to the double helix. At the same time, they saw the functional significance of the base-pairing rules. They ended their classic paper with this wry statement: "It has not escaped our notice that the specific pairing we have postulated immediately suggests a possible copying mechanism for the genetic material."* In this section, you will learn about the basic principle of DNA replication, as well as some important details of the process.

The Basic Principle: Base Pairing to a Template Strand

In a second paper, Watson and Crick stated their hypothesis for how DNA replicates: "Now our model for deoxyribonucleic acid is, in effect, a pair of templates, each of which is complementary to the other. We imagine that prior to duplication the hydrogen bonds are broken, and the two chains unwind and separate. Each chain then acts as a template for

*J. D. Watson and F. H. C. Crick, Molecular structure of nucleic acids: a structure for deoxyribose nucleic acids, *Nature* 171:737–738 (1953).

Figure 16.9 A model for DNA replication: the basic concept. In this simplified illustration, a short segment of DNA has been untwisted. Simple shapes symbolize the four kinds of bases. Dark blue represents DNA strands present in the parental molecule; light blue represents newly synthesized DNA.

Animation: DNA Replication: An Overview

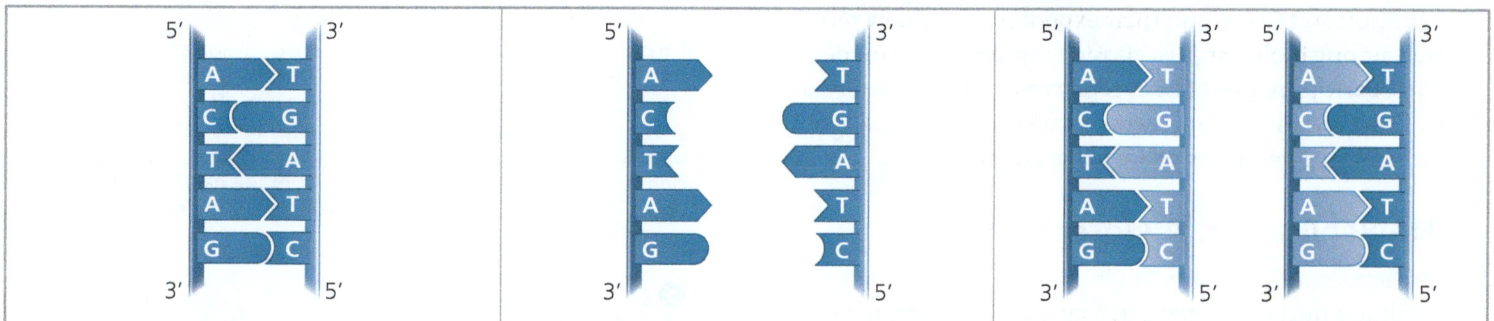

(a) The parental molecule has two complementary strands of DNA. Each base is paired by hydrogen bonding with its specific partner, A with T and G with C.

(b) First, the two DNA strands are separated. Each parental strand can now serve as a template for a new, complementary strand.

(c) Nucleotides complementary to the parental (dark blue) strand are connected to form the sugar-phosphate backbones of the new "daughter" (light blue) strands.

the formation on to itself of a new companion chain, so that eventually we shall have two pairs of chains, where we only had one before. Moreover, the sequence of the pairs of bases will have been duplicated exactly."[†]

Figure 16.9 illustrates Watson and Crick's basic idea. To make it easier to follow, we show only a short section of double helix in untwisted form. Notice that if you cover one of the two DNA strands of Figure 16.9a, you can still determine its linear sequence of nucleotides by referring to the uncovered strand and applying the base-pairing rules. The two strands are complementary; each stores the information necessary to reconstruct the other. When a cell copies a DNA molecule, each strand serves as a template for ordering nucleotides into a new, complementary strand. Nucleotides line up along the template strand according to the base-pairing rules and are linked to form the new strands. Where there was one double-stranded DNA molecule at the beginning of the process, there are soon two, each an exact replica of the "parental" molecule. The copying mechanism is analogous to using a photographic negative to make a positive image, which can in turn be used to make another negative, and so on.

This model of DNA replication remained untested for several years following publication of the DNA structure. The requisite experiments were simple in concept but difficult to perform. Watson and Crick's model predicts that when a double helix replicates, each of the two daughter molecules will have one old strand, from the parental molecule, and one newly made strand. This **semiconservative model** can be distinguished from a conservative model of replication, in which the two parental strands somehow come back together after the process (that is, the parental molecule is conserved). In yet a third model, called the dispersive model, all four strands of DNA following replication have a mixture of old and new DNA **(Figure 16.10)**.

[†]J. D. Watson and F. H. C. Crick, Genetical implications of the structure of deoxyribonucleic acid, *Nature* 171:964–967 (1953).

Figure 16.10 DNA replication: three alternative models. Each short segment of double helix symbolizes the DNA within a cell. Beginning with a parent cell, we follow the DNA for two more generations of cells—two rounds of DNA replication. Parental DNA is dark blue; newly made DNA is light blue.

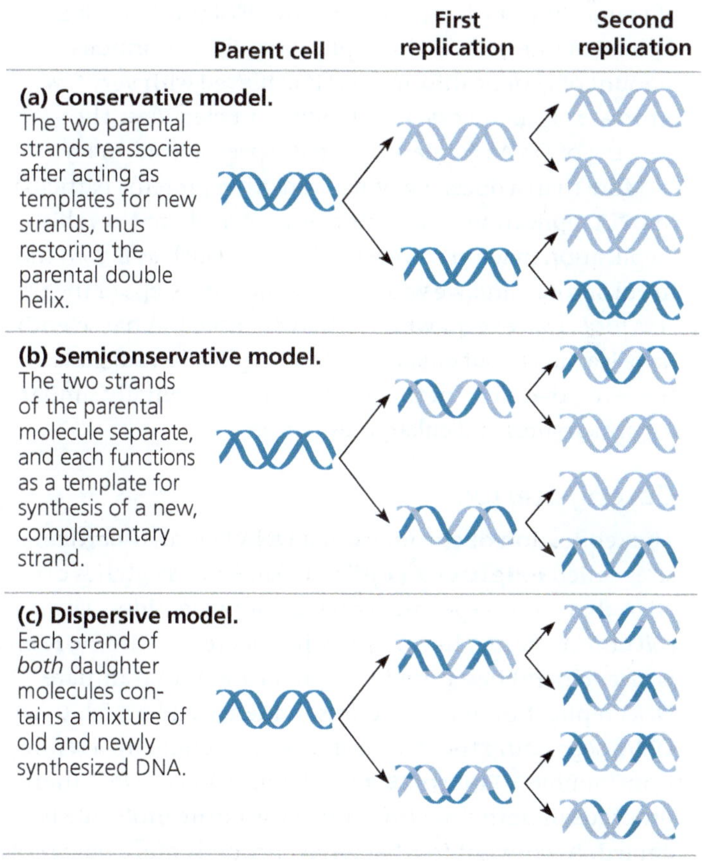

Although mechanisms for conservative or dispersive DNA replication are not easy to devise, these models remained possibilities until they could be ruled out. After two years of preliminary work at the California Institute of Technology in the late 1950s, Matthew Meselson and Franklin Stahl devised a clever experiment that distinguished between the three models, described in **Figure 16.11**. Their results supported the semiconservative model of DNA replication, as predicted by Watson and Crick, and their experiment is widely recognized among biologists as a classic example of elegant design.

The basic principle of DNA replication is conceptually simple. However, the actual process involves some complicated biochemical gymnastics, as we will now see.

DNA Replication: *A Closer Look*

The bacterium *E. coli* has a single chromosome of about 4.6 million nucleotide pairs. In a favorable environment, an *E. coli* cell can copy all of this DNA and divide to form two genetically identical daughter cells in considerably less than an hour. Each of *your* somatic cells has 46 DNA molecules in its nucleus, one long double-helical molecule per chromosome. In all, that represents about 6 billion nucleotide pairs, or over 1,000 times more DNA than is found in most bacterial cells. If we were to print the one-letter symbols for these bases (A, G, C, and T) the size of the type you are now reading, the 6 billion nucleotide pairs of information in a diploid human cell would fill about 1,400 biology textbooks. Yet it takes one of your cells just a few hours to copy all of this DNA during S phase of interphase. This replication of an enormous amount of genetic information is achieved with very few errors—only about one per 10 billion nucleotides. The copying of DNA is remarkable in its speed and accuracy.

More than a dozen enzymes and other proteins participate in DNA replication. Much more is known about how this "replication machine" works in bacteria (such as *E. coli*) than in eukaryotes, and we will describe the basic steps of the process for *E. coli*, except where otherwise noted. What scientists have learned about eukaryotic DNA replication suggests, however, that most of the process is fundamentally similar for prokaryotes and eukaryotes.

Getting Started

The replication of chromosomal DNA begins at particular sites called **origins of replication**, short stretches of DNA that have a specific sequence of nucleotides. The *E. coli* chromosome, like many other bacterial chromosomes, is circular and has a single origin. Proteins that initiate DNA replication recognize this sequence and attach to the DNA, separating the two strands and opening up a replication "bubble" **(Figure 16.12a)**. Replication of DNA then proceeds in both directions until the entire molecule is copied. In contrast to a bacterial chromosome, a eukaryotic chromosome may have hundreds or even a few thousand replication origins. Multiple replication bubbles form and eventually fuse, thus speeding up the copying of the very long DNA molecules **(Figure 16.12b)**. As in bacteria, eukaryotic DNA replication proceeds in both directions from each origin.

▼ Figure 16.11

Inquiry Does DNA replication follow the conservative, semiconservative, or dispersive model?

Experiment Matthew Meselson and Franklin Stahl cultured *E. coli* for several generations in a medium containing nucleotide precursors labeled with a heavy isotope of nitrogen, ^{15}N. They then transferred the bacteria to a medium with only ^{14}N, a lighter isotope. They took one sample after the first DNA replication and another after the second replication. They extracted DNA from the bacteria in the samples and then centrifuged each DNA sample to separate DNA of different densities.

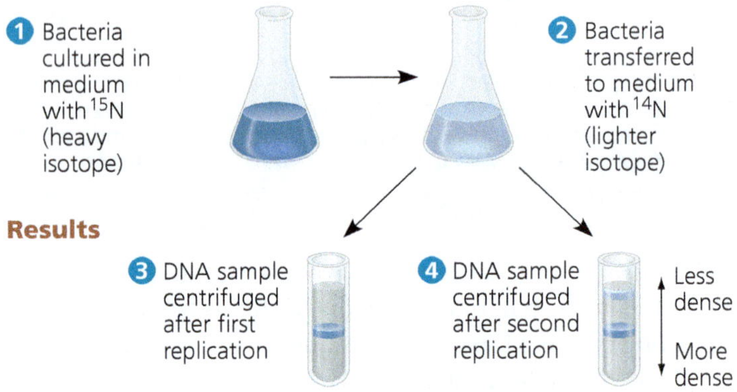

Conclusion Meselson and Stahl compared their results to those predicted by each of the three models in Figure 16.10, as shown below. The first replication in the ^{14}N medium produced a band of hybrid (^{15}N-^{14}N) DNA. This result eliminated the conservative model. The second replication produced both light and hybrid DNA, a result that refuted the dispersive model and supported the semiconservative model. They therefore concluded that DNA replication is semiconservative.

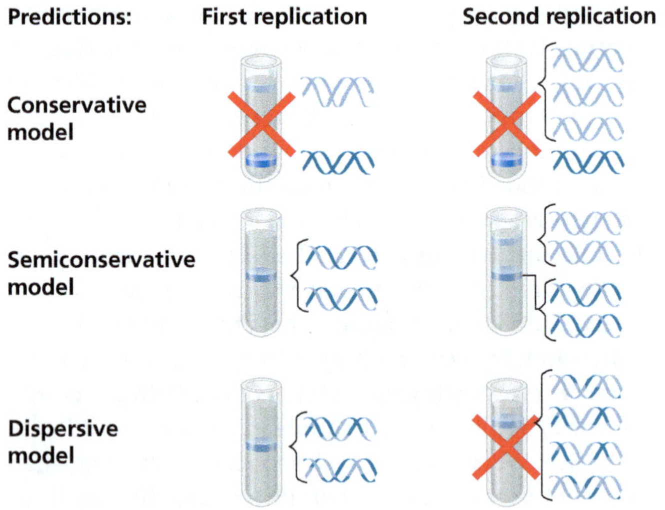

Data from M. Meselson and F. W. Stahl, The replication of DNA in *Escherichia coli*, *Proceedings of the National Academy of Sciences USA* 44:671–682 (1958).

Inquiry in Action Read and analyze the original paper in *Inquiry in Action: Interpreting Scientific Papers.*

MB **Instructors:** A related Experimental Inquiry Tutorial can be assigned in MasteringBiology.

WHAT IF? ➤ *If Meselson and Stahl had first grown the cells in ^{14}N-containing medium and then moved them into ^{15}N-containing medium before taking samples, what would have been the result after each of the two replications?*

Figure 16.12 Origins of replication in *E. coli* and eukaryotes. The red arrows indicate the movement of the replication forks and thus the overall directions of DNA replication within each bubble.

(a) Origin of replication in an *E. coli* cell

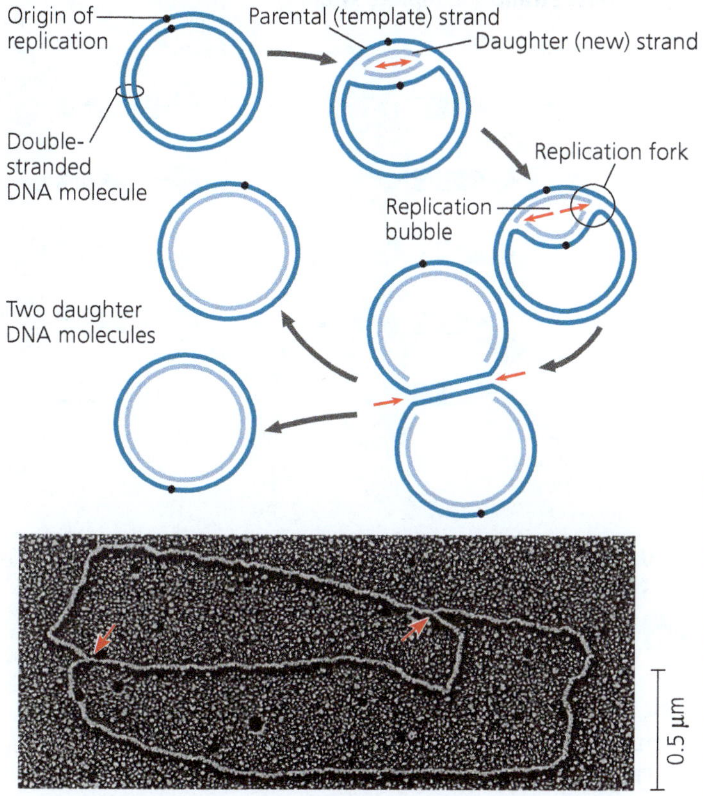

The circular chromosome of *E. coli* and other bacteria has only one origin of replication. The parental strands separate there, forming a replication bubble with two forks (red arrows). Replication proceeds in both directions until the forks meet on the other side, resulting in two daughter DNA molecules. The TEM shows a bacterial chromosome with a replication bubble.

(b) Origins of replication in a eukaryotic cell

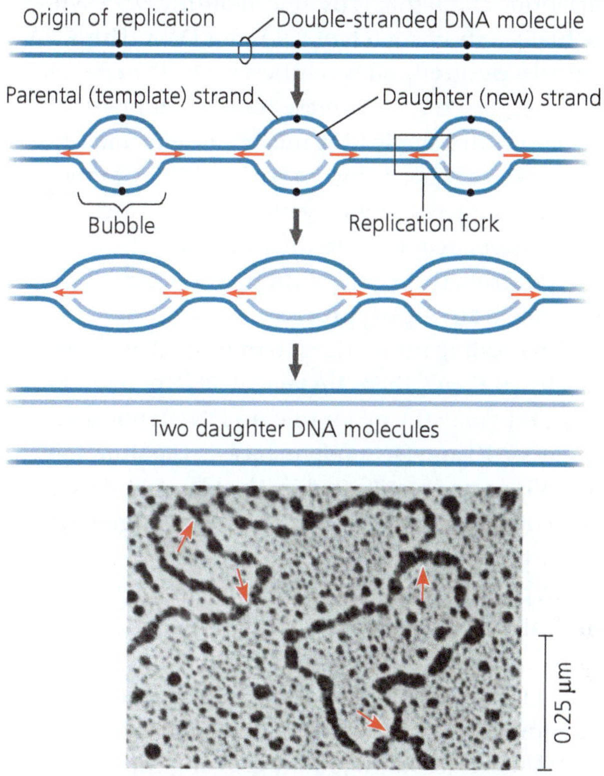

In a linear chromosome of a eukaryote, replication bubbles form at many sites along the giant DNA molecule during S phase of interphase. The bubbles expand as replication proceeds in both directions (red arrows). Eventually, the bubbles fuse and synthesis of the daughter strands is complete. The TEM shows three replication bubbles along the DNA of a cultured Chinese hamster cell.

DRAW IT ➤ *In the TEM, add arrows in the forks of the third bubble.*

BioFlix® Animation: The Replication Fork in *E. coli*

At each end of a replication bubble is a **replication fork**, a Y-shaped region where the parental strands of DNA are being unwound. Several kinds of proteins participate in the unwinding **(Figure 16.13)**. **Helicases** are enzymes that untwist the double helix at the replication forks, separating the two parental strands and making them available as template strands. After the parental strands separate, **single-strand binding proteins** bind to the unpaired DNA strands, keeping them from re-pairing. The untwisting of the double helix causes tighter twisting and strain ahead of the replication fork. **Topoisomerase** is an enzyme that helps relieve this strain by breaking, swiveling, and rejoining DNA strands.

Synthesizing a New DNA Strand

The unwound sections of parental DNA strands are now available to serve as templates for the synthesis of new complementary DNA strands. However, the enzymes that synthesize

Figure 16.13 Some of the proteins involved in the initiation of DNA replication. The same proteins function at both replication forks in a replication bubble. For simplicity, only the left-hand fork is shown, and the DNA bases are drawn much larger in relation to the proteins than they are in reality.

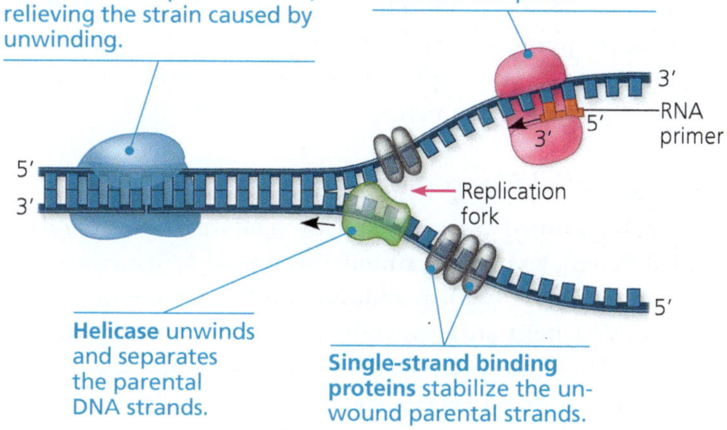

DNA cannot *initiate* the synthesis of a polynucleotide; they can only add DNA nucleotides to the end of an already existing chain that is base-paired with the template strand. The initial nucleotide chain that is produced during DNA synthesis is actually a short stretch of RNA, not DNA. This RNA chain is called a **primer** and is synthesized by the enzyme **primase** (see Figure 16.13). Primase starts a complementary RNA chain with a single RNA nucleotide and adds RNA nucleotides one at a time, using the parental DNA strand as a template. The completed primer, generally 5–10 nucleotides long, is thus base-paired to the template strand. The new DNA strand will start from the 3′ end of the RNA primer.

Enzymes called **DNA polymerases** catalyze the synthesis of new DNA by adding nucleotides to the 3′ end of a preexisting chain. In *E. coli*, there are several DNA polymerases, but two appear to play the major roles in DNA replication: DNA polymerase III and DNA polymerase I. The situation in eukaryotes is more complicated, with at least 11 different DNA polymerases discovered so far, although the general principles are the same.

Most DNA polymerases require a primer and a DNA template strand, along which complementary DNA nucleotides are lined up. In *E. coli*, DNA polymerase III (abbreviated DNA pol III) adds a DNA nucleotide to the RNA primer and then continues adding DNA nucleotides, complementary to the parental DNA template strand, to the growing end of the new DNA strand. The rate of elongation is about 500 nucleotides per second in bacteria and 50 per second in human cells.

Each nucleotide to be added to a growing DNA strand consists of a sugar attached to a base and to three phosphate groups. You have already encountered such a molecule—ATP (adenosine triphosphate; see Figure 8.9). The only difference between the ATP of energy metabolism and dATP, the adenine nucleotide used to make DNA, is the sugar component, which is deoxyribose in the building block of DNA but ribose in ATP. Like ATP, the nucleotides used for DNA synthesis are chemically reactive, partly because their triphosphate tails have an unstable cluster of negative charge. DNA polymerase catalyzes the addition of each monomer via a dehydration reaction (see Figure 5.2a). As each monomer is joined to the growing end of a DNA strand, two phosphate groups are lost as a molecule of pyrophosphate (Ⓟ—Ⓟ$_i$). Subsequent hydrolysis of the pyrophosphate to two molecules of inorganic phosphate (Ⓟ$_i$) is a coupled exergonic reaction that helps drive the polymerization reaction **(Figure 16.14)**.

Antiparallel Elongation

As we have noted previously, the two ends of a DNA strand are different, giving each strand directionality, like a one-way street (see Figure 16.5). In addition, the two strands of DNA in a double helix are antiparallel, meaning that they are oriented in opposite directions to each other, like the two sides of a divided street (see Figure 16.14). Therefore, the two new strands formed during DNA replication must also be antiparallel to their template strands.

▼ Figure 16.14 Addition of a nucleotide to a DNA strand. DNA polymerase catalyzes the addition of a nucleotide to the 3′ end of a growing DNA strand, with the release of two phosphates.

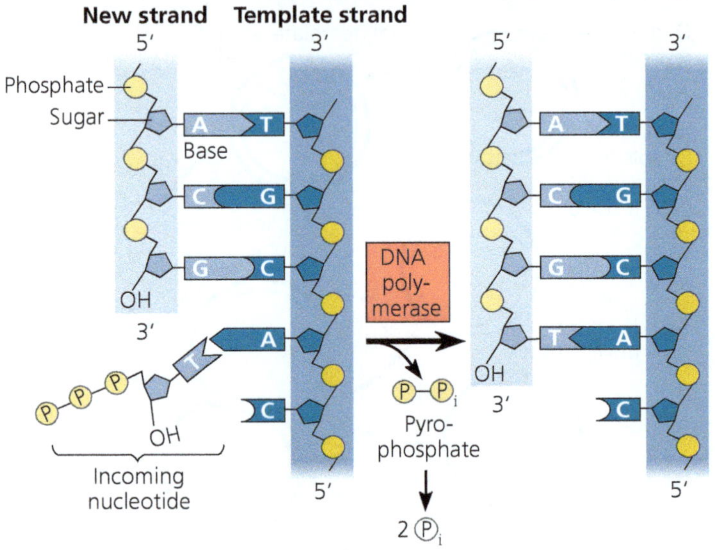

VISUAL SKILLS ➤ *Use this diagram to explain what we mean when we say that each DNA strand has directionality.*

Figure Walkthrough

The antiparallel arrangement of the double helix, together with a property of DNA polymerases, has an important effect on how replication occurs. Because of their structure, DNA polymerases can add nucleotides only to the free 3′ end of a primer or growing DNA strand, never to the 5′ end (see Figure 16.14). Thus, a new DNA strand can elongate only in the 5′ → 3′ direction. With this in mind, let's examine one of the two replication forks in a bubble **(Figure 16.15)**. Along one template strand, DNA polymerase III can synthesize a complementary strand continuously by elongating the new DNA in the mandatory 5′ → 3′ direction. DNA pol III remains in the replication fork on that template strand and continuously adds nucleotides to the new complementary strand as the fork progresses. The DNA strand made by this mechanism is called the **leading strand**. Only one primer is required for DNA pol III to synthesize the entire leading strand (see Figure 16.15).

To elongate the other new strand of DNA in the mandatory 5′ → 3′ direction, DNA pol III must work along the other template strand in the direction *away from* the replication fork. The DNA strand elongating in this direction is called the **lagging strand**. In contrast to the leading strand, which elongates continuously, the lagging strand is synthesized discontinuously, as a series of segments. These segments of the lagging strand are called **Okazaki fragments**, after Reiji Okazaki, the Japanese scientist who discovered them. The fragments are about 1,000–2,000 nucleotides long in *E. coli* and 100–200 nucleotides long in eukaryotes.

Figure 16.15 Synthesis of the leading strand during DNA replication. This diagram focuses on the left replication fork shown in the overview box. DNA polymerase III (DNA pol III), shaped like a cupped hand, is shown closely associated with a protein called the "sliding clamp" that encircles the newly synthesized double helix like a doughnut. The sliding clamp moves DNA pol III along the DNA template strand.

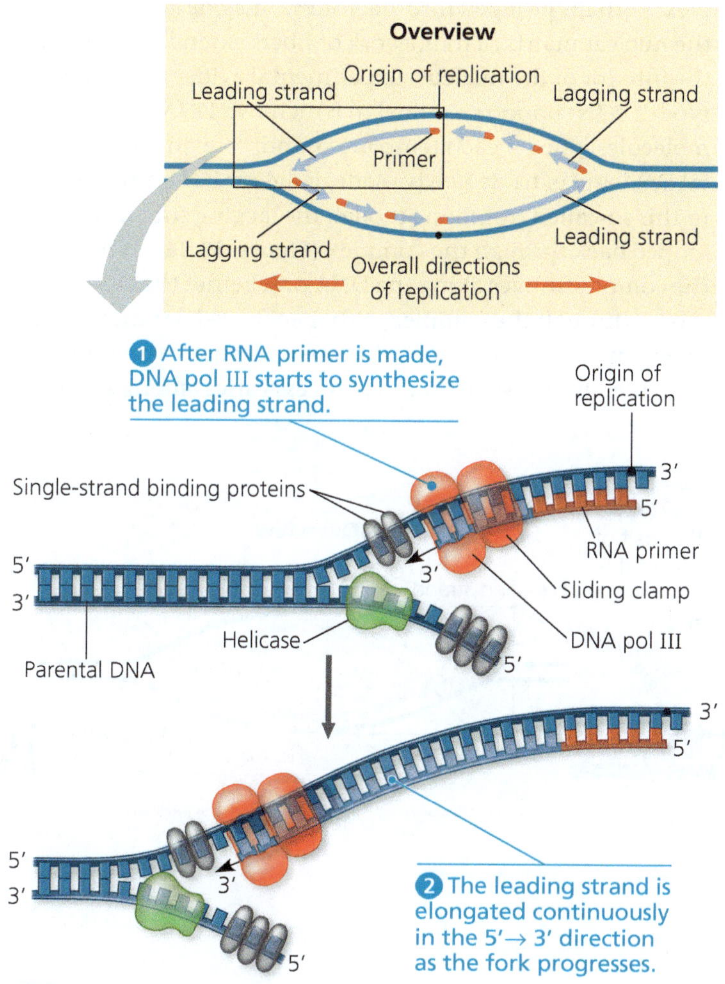

BioFlix® Animation: Synthesis of the Leading Strand

Figure 16.16 illustrates the steps in the synthesis of the lagging strand at one fork. Whereas only one primer is required on the leading strand, each Okazaki fragment on the lagging strand must be primed separately (steps 1 and 4). After DNA pol III forms an Okazaki fragment (steps 2 to 4), another DNA polymerase, DNA pol I, replaces the RNA nucleotides of the adjacent primer with DNA nucleotides one at a time (step 5). But DNA pol I cannot join the final nucleotide of this replacement DNA segment to the first DNA nucleotide of the adjacent Okazaki fragment. Another enzyme, **DNA ligase**, accomplishes this task, joining the sugar-phosphate backbones of all the Okazaki fragments into a continuous DNA strand (step 6).

BioFlix® Animation: Synthesis of the Lagging Strand

Figure 16.16 Synthesis of the lagging strand.

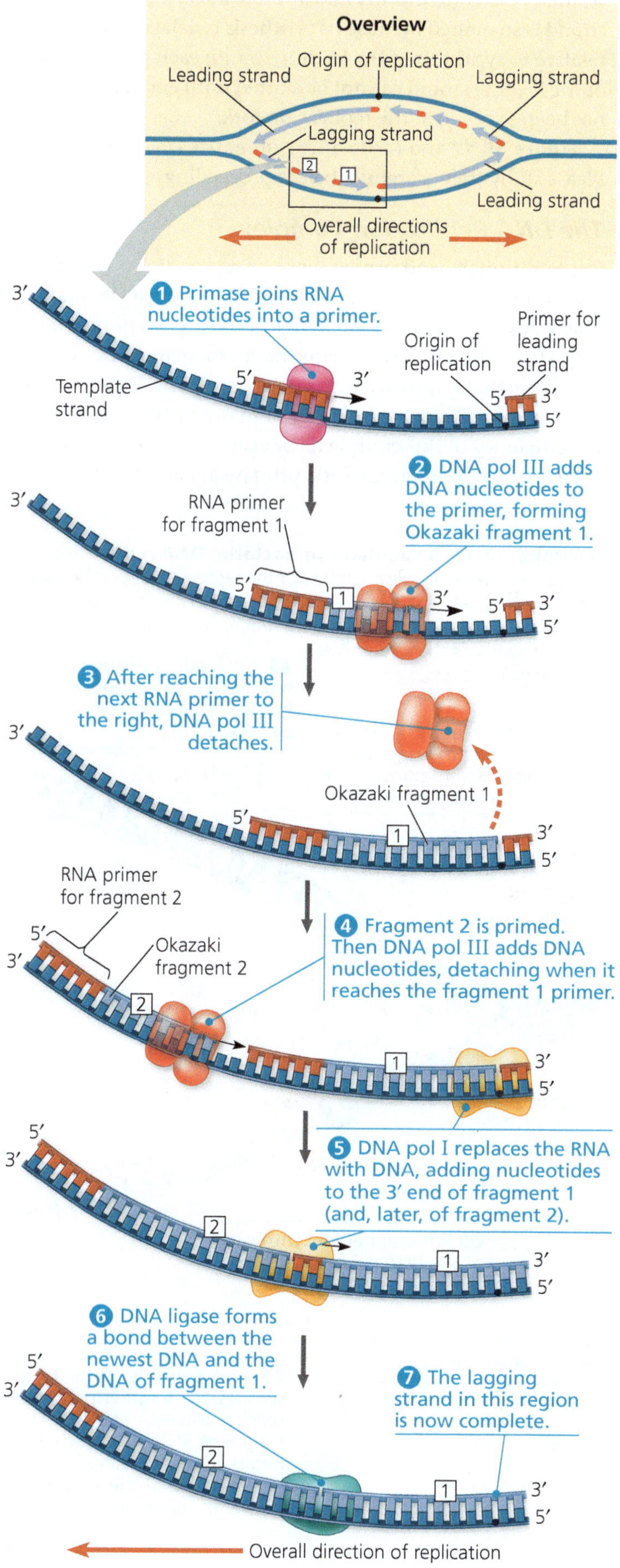

Synthesis of the leading strand and synthesis of the lagging strand occur concurrently and at the same rate. The lagging strand is so named because its synthesis is delayed slightly relative to synthesis of the leading strand; each new fragment of the lagging strand cannot be started until enough template has been exposed at the replication fork.

Figure 16.17 and **Table 16.1** summarize DNA replication. Please study them carefully before proceeding.

The DNA Replication Complex

It is traditional—and convenient—to represent DNA polymerase molecules as locomotives moving along a DNA railroad track, but such a model is inaccurate in two important ways. First, the various proteins that participate in DNA replication actually form a single large complex, a "DNA replication machine." Many protein-protein interactions facilitate the efficiency of this complex. For example, by interacting with other proteins at the fork, primase apparently acts as a molecular brake, slowing progress of the replication fork and coordinating the placement of primers and the rates of replication on the leading and lagging strands. Second, the DNA replication complex may not move along the DNA; rather, the DNA may move through the complex during the replication process. In eukaryotic cells, multiple copies of the complex, perhaps grouped into "factories," may be anchored to the nuclear matrix, a framework of fibers extending through the interior of the nucleus. Experimental evidence in some type of cells supports a model in which two DNA polymerase molecules, one on each template strand, "reel in" the parental DNA and extrude newly made daughter DNA molecules. In this so-called trombone model, the lagging strand is also looped back through the complex **(Figure 16.18)**. Whether the complex moves along the DNA or whether the DNA moves through the complex, either anchored or not, are still open, unresolved questions that are under active investigation. It is also possible that the process varies among species.

▼ **Figure 16.17 A summary of bacterial DNA replication.** The detailed diagram shows the left-hand replication fork of the replication bubble shown in the overview (upper right). Viewing each daughter strand in its entirety in the overview, you can see that half of it is made continuously as the leading strand, while the other half (on the other side of the origin) is synthesized in fragments as the lagging strand.

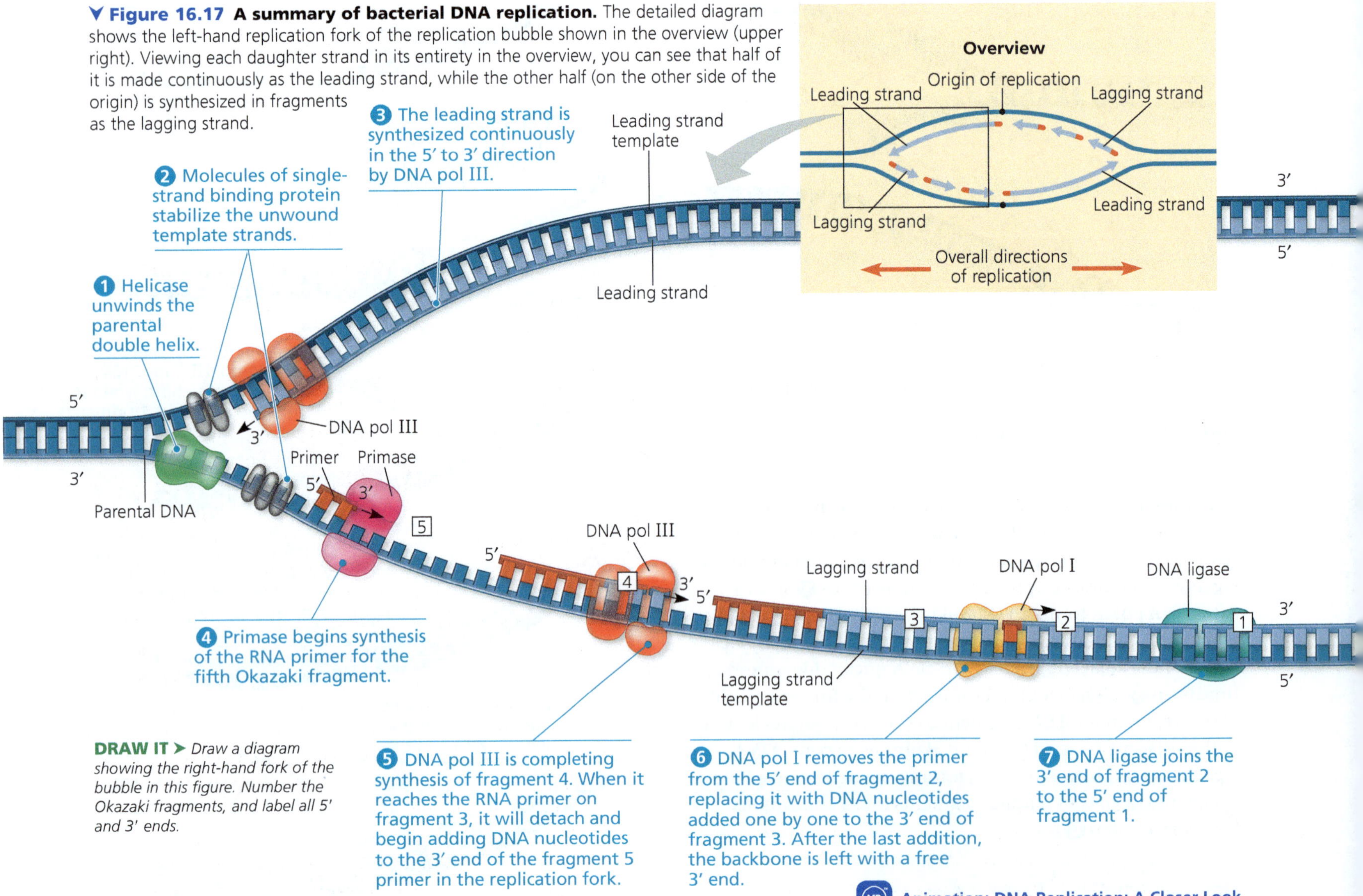

DRAW IT ➤ *Draw a diagram showing the right-hand fork of the bubble in this figure. Number the Okazaki fragments, and label all 5′ and 3′ ends.*

Animation: DNA Replication: A Closer Look

Table 16.1 Bacterial DNA Replication Proteins and Their Functions

Protein	Function
Helicase 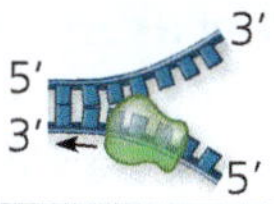	Unwinds parental double helix at replication forks
Single-strand binding protein 	Binds to and stabilizes single-stranded DNA until it is used as a template
Topoisomerase 	Relieves overwinding strain ahead of replication forks by breaking, swiveling, and rejoining DNA strands
Primase 	Synthesizes an RNA primer at 5′ end of leading strand and at 5′ end of each Okazaki fragment of lagging strand
DNA pol III 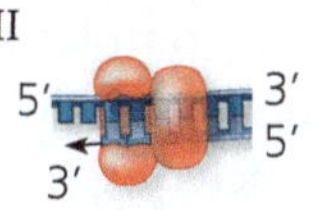	Using parental DNA as a template, synthesizes new DNA strand by adding nucleotides to an RNA primer or a pre-existing DNA strand
DNA pol I 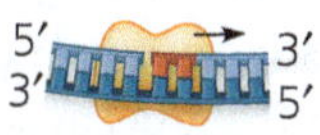	Removes RNA nucleotides of primer from 5′ end and replaces them with DNA nucleotides added to 3′ end of adjacent fragment
DNA ligase 	Joins Okazaki fragments of lagging strand; on leading strand, joins 3′ end of DNA that replaces primer to rest of leading strand DNA

Proofreading and Repairing DNA

We cannot attribute the accuracy of DNA replication solely to the specificity of base pairing. Initial pairing errors between incoming nucleotides and those in the template strand occur at a rate of one in 10^5 nucleotides. However, errors in the completed DNA molecule amount to only one in 10^{10} (10 billion) nucleotides, an error rate that is 100,000 times lower. This is because during DNA replication, many DNA polymerases proofread each nucleotide against its template as soon as it is covalently bonded to the growing strand. Upon finding an incorrectly paired nucleotide, the polymerase removes the nucleotide and then resumes synthesis. (This action is similar to fixing a texting error by deleting the wrong letter and then entering the correct one.)

Mismatched nucleotides sometimes evade proofreading by a DNA polymerase. In **mismatch repair**, other enzymes remove and replace incorrectly paired nucleotides that have resulted from replication errors. Researchers highlighted the importance of such repair enzymes when they found that a hereditary defect in one of them is associated with a form of colon cancer. Apparently, this defect allows cancer-causing errors to accumulate in the DNA faster than normal.

▼ **Figure 16.18 The "trombone" model of the DNA replication complex.** In this proposed model, two molecules of DNA polymerase III work together in a complex, one on each strand, with helicase and other proteins. The lagging strand template DNA loops through the complex, resembling the slide of a trombone.

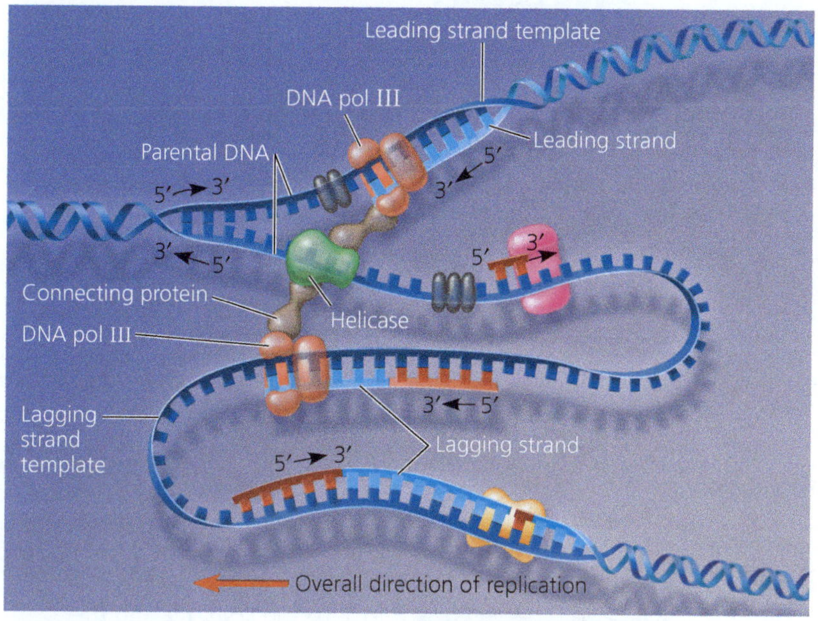

DRAW IT ➤ *Draw a line tracing the lagging strand template along the entire stretch of DNA shown here.*

BioFlix® Animation: The DNA Replication Complex

Incorrectly paired or altered nucleotides can also arise after replication. In fact, maintenance of the genetic information encoded in DNA requires frequent repair of various kinds of damage to existing DNA. DNA molecules are constantly subjected to potentially harmful chemical and physical agents, such as X-rays, as we'll discuss in Concept 17.5. In addition, DNA bases may undergo spontaneous chemical changes under normal cellular conditions. However, these changes in DNA are usually corrected before they become permanent changes—*mutations*—perpetuated through successive replications. Each cell continuously monitors and repairs its genetic material. Because repair of damaged DNA is so important to the survival of an organism, it is no surprise that many different DNA repair enzymes have evolved. Almost 100 are known in *E. coli*, and about 170 have been identified so far in humans.

Most cellular systems for repairing incorrectly paired nucleotides, whether they are due to DNA damage or to replication errors, use a mechanism that takes advantage of the base-paired structure of DNA. In many cases, a segment of the strand containing the damage is cut out (excised) by a DNA-cutting enzyme—a **nuclease**—and the resulting gap is then filled in with nucleotides, using the undamaged strand as a template. The enzymes involved in filling the gap are a DNA polymerase and DNA ligase. One such DNA repair system is called **nucleotide excision repair** (Figure 16.19).

Figure 16.19 Nucleotide excision repair of DNA damage.

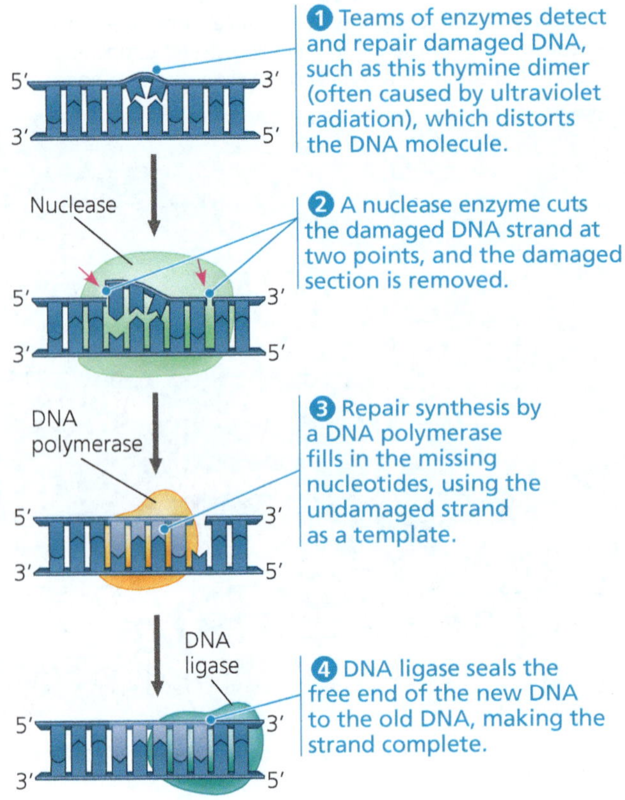

An important function of the DNA repair enzymes in our skin cells is to repair genetic damage caused by the ultraviolet rays of sunlight. One type of damage, shown in Figure 16.19, is the covalent linking of thymine bases that are adjacent on a DNA strand. Such *thymine dimers* cause the DNA to buckle and interfere with DNA replication. The importance of repairing this kind of damage is underscored by a disorder called xeroderma pigmentosum (XP), which in most cases is caused by an inherited defect in a nucleotide excision repair enzyme. Individuals with XP are hypersensitive to sunlight; mutations in their skin cells caused by ultraviolet light are left uncorrected, often resulting in skin cancer. The effects are extreme: Without sun protection, children who have XP can develop skin cancer by age 10.

Evolutionary Significance of Altered DNA Nucleotides

EVOLUTION Faithful replication of the genome and repair of DNA damage are important for the functioning of the organism and for passing on a complete, accurate genome to the next generation. The error rate after proofreading and repair is extremely low, but rare mistakes do slip through. Once a mismatched nucleotide pair is replicated, the sequence change is permanent in the daughter molecule that has the incorrect nucleotide as well as in any subsequent copies. As we mentioned earlier, a permanent change in the DNA sequence is called a mutation.

Mutations can change the phenotype of an organism (as you'll learn in Concept 17.5). And if they occur in germ cells, which give rise to gametes, mutations can be passed on from generation to generation. The vast majority of such changes either have no effect or are harmful, but a very small percentage can be beneficial. In either case, mutations are the original source of the variation on which natural selection operates during evolution and are ultimately responsible for the appearance of new species. (You'll learn more about this process in Unit Four.) The balance between complete fidelity of DNA replication or repair and a low mutation rate has resulted in new proteins that contribute to different phenotypes. Ultimately, over long periods of time, this process leads to new species and thus to the rich diversity of life we see on Earth today.

Replicating the Ends of DNA Molecules

For linear DNA, such as the DNA of eukaryotic chromosomes, the usual replication machinery cannot complete the 5′ ends of daughter DNA strands. (This is another consequence of the fact that a DNA polymerase can add nucleotides only to the 3′ end of a preexisting polynucleotide.) Even if an Okazaki fragment can be started with an RNA primer hydrogen-bonded to the very end of the template strand, once that primer is removed, it cannot be replaced with DNA because there is no 3′ end available for nucleotide addition **(Figure 16.20)**. As a result, repeated rounds of replication produce shorter and shorter DNA molecules with uneven ("staggered") ends.

Most prokaryotes have a circular chromosome, with no ends, so the shortening of DNA does not occur. But what protects the genes of linear eukaryotic chromosomes from being eroded away during successive rounds of DNA replication? Eukaryotic chromosomal DNA molecules have special nucleotide sequences called **telomeres** at their ends **(Figure 16.21)**. Telomeres do not contain genes; instead, the DNA typically consists of multiple repetitions of one short nucleotide sequence. In each human telomere, for example, the six-nucleotide sequence TTAGGG is repeated between 100 and 1,000 times.

Telomeres have two protective functions. First, specific proteins associated with telomeric DNA prevent the staggered ends of the daughter molecule from activating the cell's systems for monitoring DNA damage. (Staggered ends of a DNA molecule, which often result from double-strand breaks, can trigger signal transduction pathways leading to cell cycle arrest or cell death.) Second, telomeric DNA acts as a kind of buffer zone that provides some protection against the organism's genes shortening, somewhat like how the plastic-wrapped ends of a shoelace slow down its unraveling. Telomeres do not prevent the erosion of genes near the ends of chromosomes; they merely postpone it.

As shown in Figure 16.20, telomeres become shorter during every round of replication. Thus, as expected, telomeric DNA tends to be shorter in dividing somatic cells of older individuals and in cultured cells that have divided many

Figure 16.20 Shortening of the ends of linear DNA molecules. Here we follow the left end of one DNA molecule through two rounds of replication. After the first round, the new lagging strand is shorter than its template. After a second round, both the leading and lagging strands have become shorter than the original parental DNA. Although not shown here, the other ends of these chromosomal DNA molecules (not shown) also become shorter.

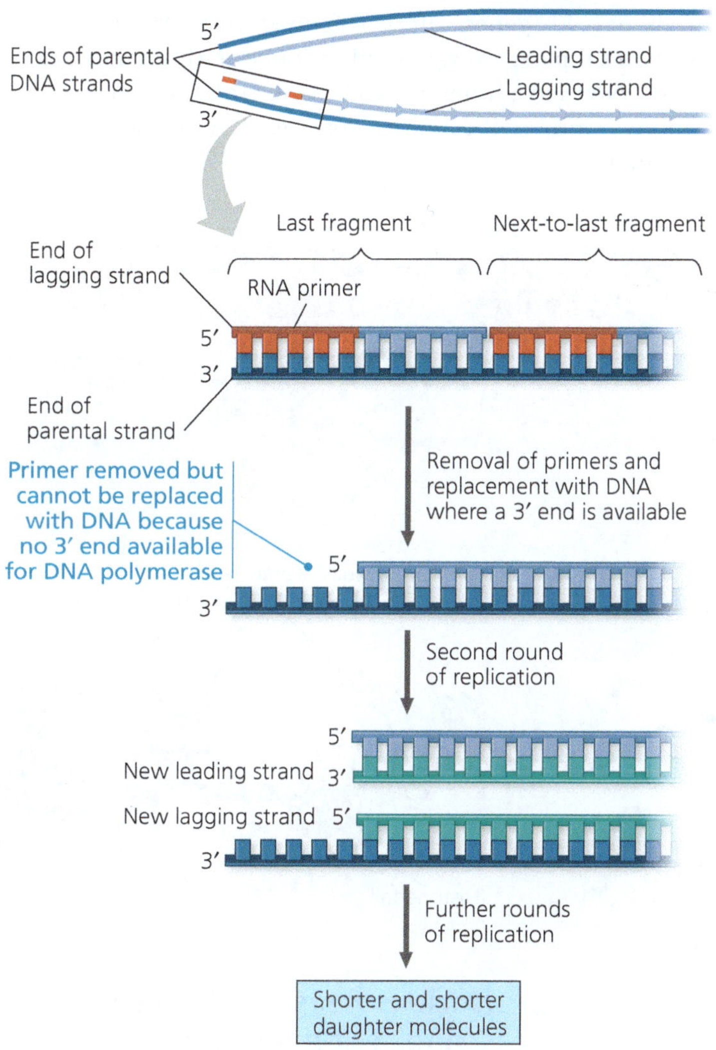

times. It has been proposed that shortening of telomeres is somehow connected to the aging process of certain tissues and even to aging of the organism as a whole.

ABC News Video: The Effect of Exercise on Cells

But what about cells whose genome must persist virtually unchanged from an organism to its offspring over many generations? If the chromosomes of germ cells became shorter in every cell cycle, essential genes would eventually be missing from the gametes they produce. However, this does not occur: An enzyme called *telomerase* catalyzes the lengthening of telomeres in eukaryotic germ cells, thus restoring their original length and compensating for the shortening that occurs during DNA replication. This enzyme contains its own RNA molecule that it uses as a template to artificially

Figure 16.21 Telomeres. Eukaryotes have repetitive, noncoding sequences called telomeres at the ends of their DNA. Telomeres are stained orange in these mouse chromosomes (LM).

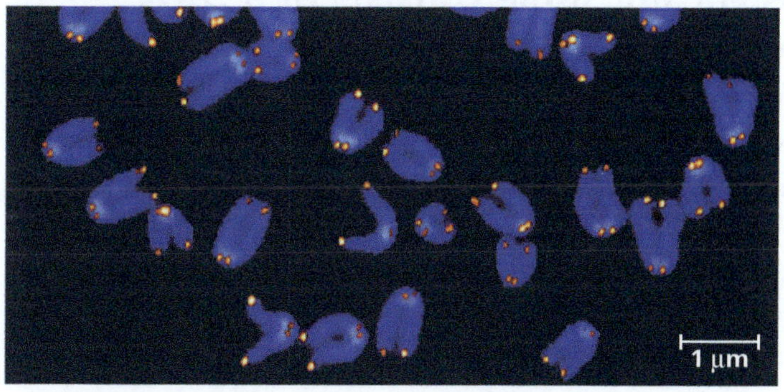

"extend" the leading strand, allowing the lagging strand to maintain a given length. Telomerase is not active in most human somatic cells, but its activity varies from tissue to tissue. The activity of telomerase in germ cells results in telomeres of maximum length in the zygote.

BBC Video: Can Humans Live Forever?

Normal shortening of telomeres may protect organisms from cancer by limiting the number of divisions that somatic cells can undergo. Cells from large tumors often have unusually short telomeres, as we would expect for cells that have undergone many cell divisions. Further shortening would presumably lead to self-destruction of the tumor cells. Telomerase activity is abnormally high in cancerous somatic cells, suggesting that its ability to stabilize telomere length may allow these cancer cells to persist. Many cancer cells do seem capable of unlimited cell division, as do immortal strains of cultured cells (see Concept 12.3). For several years, researchers have studied inhibition of telomerase as a possible cancer therapy. While studies that inhibited telomerase in mice with tumors have led to the death of cancer cells, eventually the cells have restored the length of their telomeres by an alternative pathway. This is an area of ongoing research that may eventually yield useful cancer treatments.

CONCEPT CHECK 16.2

1. What role does complementary base pairing play in the replication of DNA?
2. Identify two major functions of DNA pol III in DNA replication.
3. **MAKE CONNECTIONS** ➤ What is the relationship between DNA replication and the S phase of the cell cycle? See Figure 12.6.
4. **VISUAL SKILLS** ➤ If the DNA pol I in a given cell were nonfunctional, how would that affect the synthesis of a *leading* strand? In the overview box in Figure 16.17, point out where DNA pol I would normally function on the top leading strand.

For suggested answers, see Appendix A.

CONCEPT 16.3

A chromosome consists of a DNA molecule packed together with proteins

Now that you have learned about the structure and replication of DNA, let's take a step back and examine how DNA is packaged into chromosomes, the structures that carry genetic information. The main component of the genome in most bacteria is one double-stranded, circular DNA molecule that is associated with a small amount of protein. Although we refer to this structure as the bacterial chromosome, it is very different from a eukaryotic chromosome, which consists of one linear DNA molecule associated with a large amount of protein. In *E. coli*, the chromosomal DNA consists of about 4.6 million nucleotide pairs, representing about 4,400 genes. This is 100 times more DNA than is found in a typical virus, but only about one-thousandth as much DNA as in a human somatic cell. Still, that is a tremendous amount of DNA to be packaged in such a small container.

Figure 16.22 Exploring Chromatin Packing in a Eukaryotic Chromosome

This illustration, accompanied by transmission electron micrographs, depicts a current model for the progressive levels of DNA coiling and folding. The illustration zooms out from a single molecule of DNA to a metaphase chromosome, which is large enough to be seen with a light microscope.

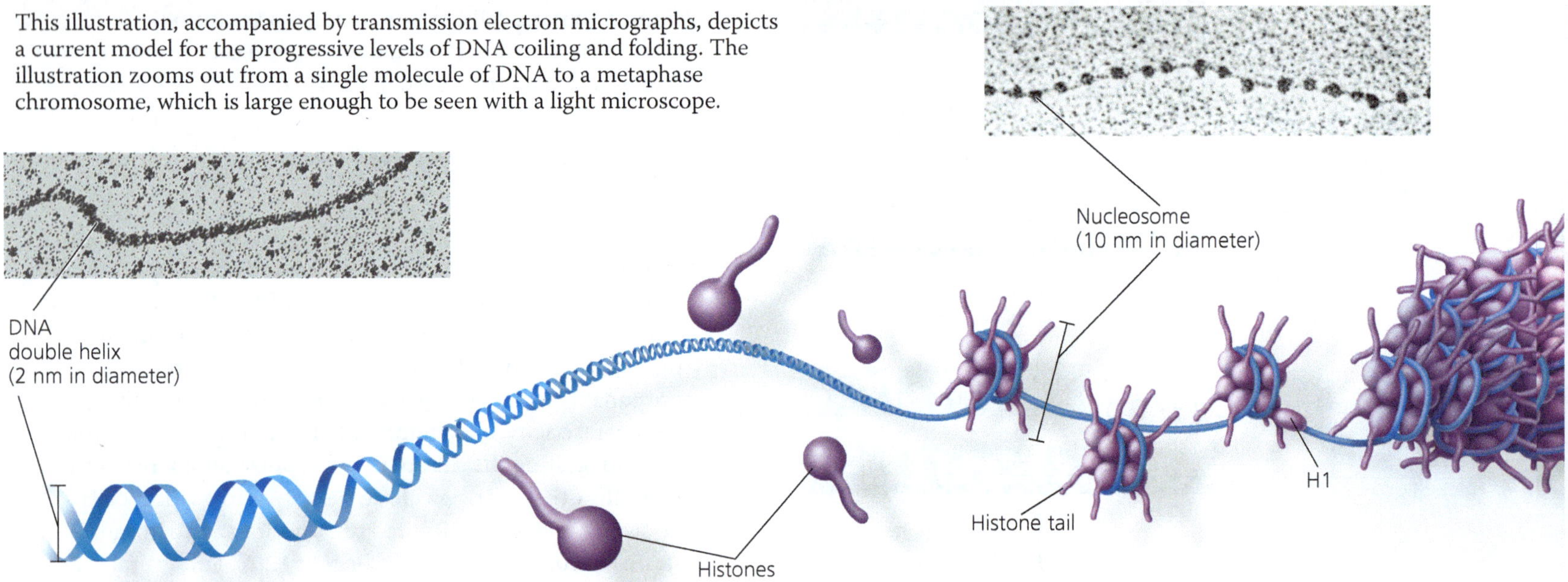

DNA, the double helix

Shown above is a ribbon model of DNA, with each ribbon representing one of the polynucleotide strands. Recall that each phosphate group along the backbone contributes a negative charge along the outside of each strand. The TEM shows a molecule of naked (protein-free) DNA; the double helix alone is 2 nm across.

Histones

Proteins called **histones** are responsible for the first level of DNA packing in chromatin. Although each histone is small—containing only about 100 amino acids—the total mass of histone in chromatin roughly equals the mass of DNA. More than a fifth of a histone's amino acids are positively charged (lysine or arginine) and therefore bind tightly to the negatively charged DNA.

Four types of histones are most common in chromatin. The histones are very similar among eukaryotes; for example, histones of the same type in cows and pea plants differ by only two amino acids. The apparent conservation of histone genes during evolution probably reflects the important role of histones in organizing DNA within cells.

These four types of histones are critical to the next level of DNA packing. (A fifth type of histone is involved in a further stage of packing.)

Nucleosomes, or "beads on a string" (10-nm fiber)

In electron micrographs, unfolded chromatin is 10 nm in diameter (the *10-nm fiber*). Such chromatin resembles beads on a string (see the TEM). Each "bead" is a **nucleosome**, the basic unit of DNA packing; the "string" between beads is called *linker* DNA.

A nucleosome consists of DNA wound twice around a protein core of eight histones, two each of the main histone types. The amino end (N-terminus) of each histone (the *histone tail*) extends outward from the nucleosome.

In the cell cycle, the histones leave the DNA only briefly during DNA replication. Generally, they do the same during the process of transcription, which also requires access to the DNA by the cell's molecular machinery. Nucleosomes, and in particular their histone tails, are involved in the regulation of gene expression.

MB Animation: DNA Packing

Stretched out, the DNA of an *E. coli* cell would measure about a millimeter in length, which is 500 times longer than the cell. Within a bacterium, however, certain proteins cause the chromosome to coil and "supercoil," densely packing it so that it fills only part of the cell. Unlike the nucleus of a eukaryotic cell, this dense region of DNA in a bacterium, called the nucleoid, is not bounded by membrane (see Figure 6.5).

Each eukaryotic chromosome contains a single linear DNA double helix that, in humans, averages about 1.5×10^8 nucleotide pairs. This is an enormous amount of DNA relative to a chromosome's condensed length. If completely stretched out, such a DNA molecule would be about 4 cm long, thousands of times the diameter of a cell nucleus—and that's not even considering the DNA of the other 45 human chromosomes!

In the cell, eukaryotic DNA is precisely combined with a large amount of protein. Together, this complex of DNA and protein, called **chromatin**, fits into the nucleus through an elaborate, multilevel system of packing. Our current view of the successive levels of DNA packing in a chromosome is outlined in **Figure 16.22**. Study this figure carefully before reading further.

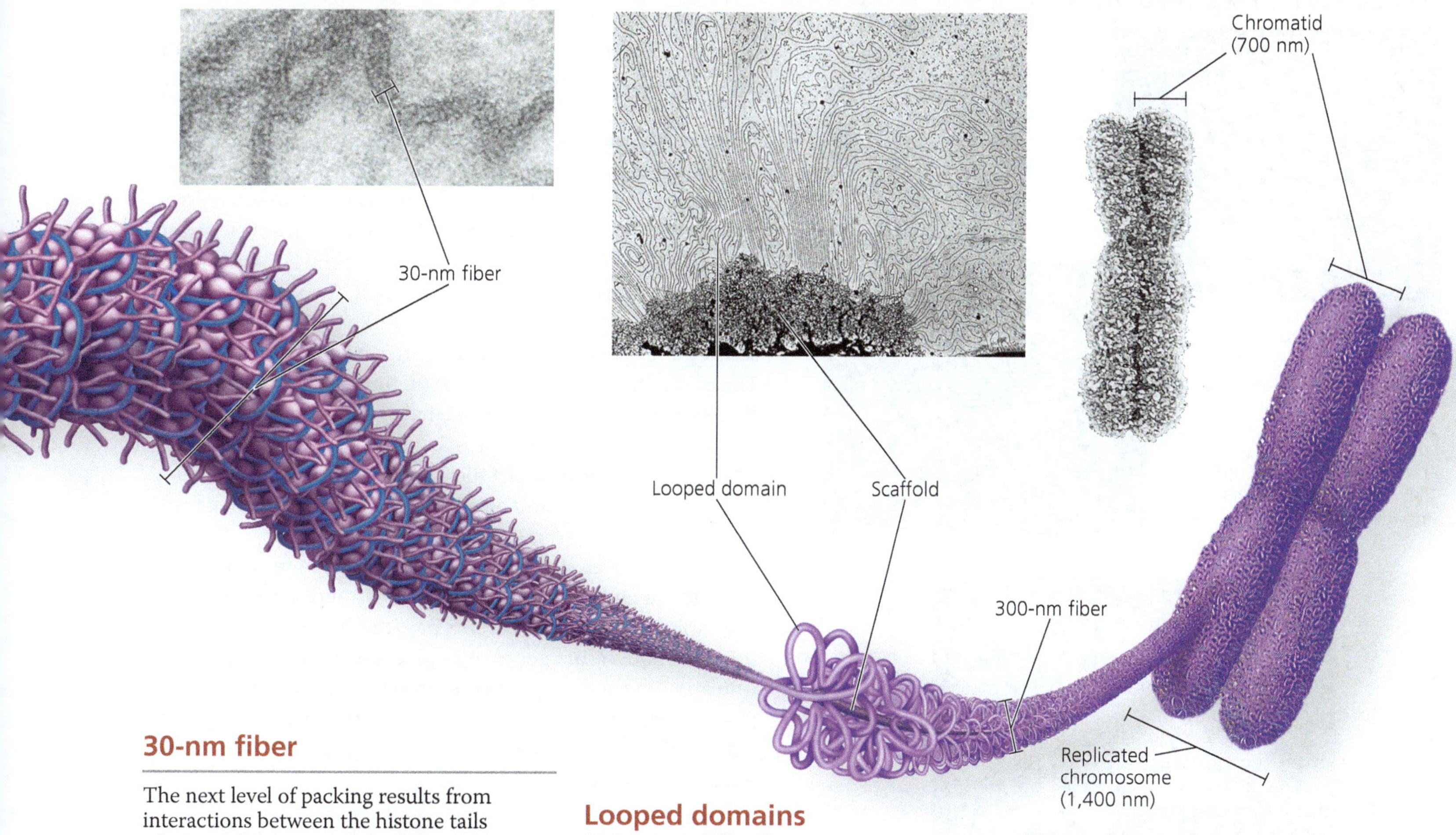

30-nm fiber

The next level of packing results from interactions between the histone tails of one nucleosome and the linker DNA and nucleosomes on either side. The fifth type of histone is involved at this level. These interactions cause the extended 10-nm fiber to coil or fold, forming a chromatin fiber roughly 30 nm in thickness, the *30-nm fiber*. Although the 30-nm fiber is quite prevalent in the interphase nucleus, the packing arrangement of nucleosomes in this form of chromatin is still a matter of some debate.

Looped domains (300-nm fiber)

The 30-nm fiber, in turn, forms loops called *looped* domains attached to a chromosome scaffold composed of proteins, thus making up a *300-nm fiber*. The scaffold is rich in one type of topoisomerase.

Metaphase chromosome

In a mitotic chromosome, the looped domains themselves coil and fold in a manner not yet fully understood, further compacting all the chromatin to produce the characteristic metaphase chromosome (also shown in the micrograph above). The width of one chromatid is 700 nm. Particular genes always end up located at the same places in metaphase chromosomes, indicating that the packing steps are highly specific and precise.

Chromatin undergoes striking changes in its degree of packing during the course of the cell cycle (see Figure 12.7). In interphase cells stained for light microscopy, the chromatin usually appears as a diffuse mass within the nucleus, suggesting that the chromatin is highly extended. As a cell prepares for mitosis, its chromatin coils and folds up (condenses), eventually forming a characteristic number of short, thick metaphase chromosomes that are distinguishable from each other with the light microscope **(Figure 16.23a)**.

Though interphase chromatin is generally much less condensed than the chromatin of mitotic chromosomes, it shows several of the same levels of higher-order packing.

▼ Figure 16.23 "Painting" chromosomes. Researchers can treat ("paint") human chromosomes with molecular tags that cause each chromosome pair to appear a different color.

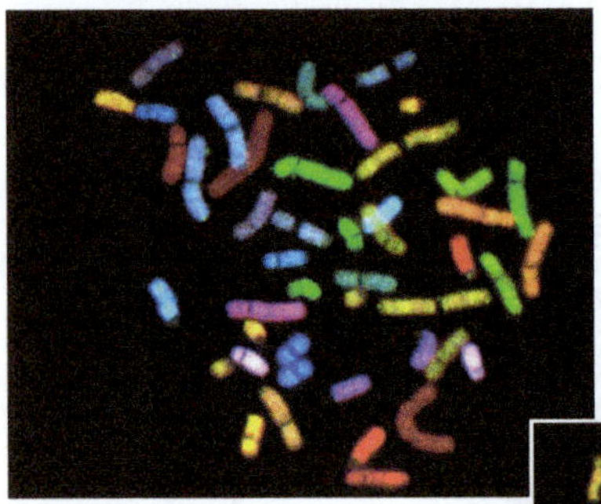

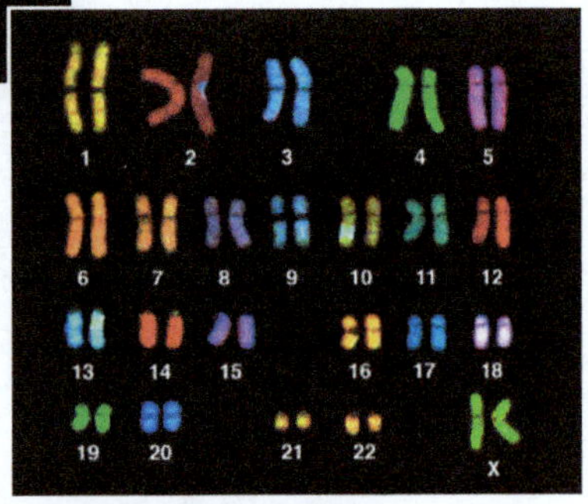

(a) These metaphase chromosomes have been "painted" so that the two homologs of a pair are the same color. Above is a spread of treated chromosomes; on the right, they have been organized into a karyotype.

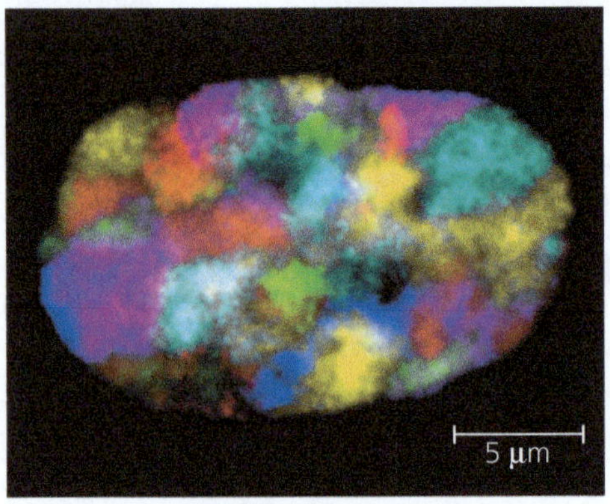

(b) The ability to visually distinguish among chromosomes makes it possible to see how the chromosomes are arranged in the interphase nucleus. Each chromosome appears to occupy a specific territory during interphase. In general, the two homologs of a pair are not located together.

MAKE CONNECTIONS ➤ *If you arrested a human cell in metaphase I of meiosis and applied this technique, what would you observe? How would this differ from what you would see in metaphase of mitosis? Review Figure 13.8 and Figure 12.7.*

Some of the chromatin comprising a chromosome seems to be present as a 10-nm fiber, but much is compacted into a 30-nm fiber, which in some regions is further folded into looped domains. Early on, biologists assumed that interphase chromatin was a tangled mass in the nucleus, like a bowl of spaghetti, but this is far from the case. Although an interphase chromosome lacks an obvious scaffold, its looped domains appear to be attached to the nuclear lamina, on the inside of the nuclear envelope, and perhaps also to fibers of the nuclear matrix. These attachments may help organize regions of chromatin where genes are active. The chromatin of each chromosome occupies a specific restricted area within the interphase nucleus, and the chromatin fibers of different chromosomes do not appear to be entangled **(Figure 16.23b)**.

Even during interphase, the centromeres and telomeres of chromosomes, as well as other chromosomal regions in some cells, exist in a highly condensed state similar to that seen in a metaphase chromosome. This type of interphase chromatin, visible as irregular clumps with a light microscope, is called **heterochromatin**, to distinguish it from the less compacted, more dispersed **euchromatin** ("true chromatin"). Because of its compaction, heterochromatic DNA is largely inaccessible to the machinery in the cell responsible for transcribing the genetic information coded in the DNA, a crucial early step in gene expression. In contrast, the looser packing of euchromatin makes its DNA accessible to this machinery, so the genes present in euchromatin can be transcribed. The chromosome is a dynamic structure that is condensed, loosened, modified, and remodeled as necessary for various cell processes, including mitosis, meiosis, and gene activity. Chemical modifications of histones affect the state of chromatin condensation and also have multiple effects on gene activity, as you'll see in Concept 18.2.

In this chapter, you have learned how DNA molecules are arranged in chromosomes and how DNA replication provides the copies of genes that parents pass to offspring. However, it is not enough that genes be copied and transmitted; the information they carry must be used by the cell. In other words, genes must also be expressed. In the next chapter, we will examine how the cell expresses the genetic information encoded in DNA.

CONCEPT CHECK 16.3

1. Describe the structure of a nucleosome, the basic unit of DNA packing in eukaryotic cells.
2. What two properties, one structural and one functional, distinguish heterochromatin from euchromatin?
3. **MAKE CONNECTIONS ➤** Interphase chromosomes appear to be attached to the nuclear lamina and perhaps also the nuclear matrix. Describe these two structures. See Figure 6.9 and the associated text.

For suggested answers, see Appendix A.

16 Chapter Review

Go to **MasteringBiology**™ for Videos, Animations, Vocab Self-Quiz, Practice Tests, and more in the Study Area.

SUMMARY OF KEY CONCEPTS

VOCAB SELF-QUIZ goo.gl/6u55ks

CONCEPT 16.1

DNA is the genetic material *(pp. 315–320)*

- Experiments with bacteria and with **phages** provided the first strong evidence that the genetic material is DNA.
- Watson and Crick deduced that DNA is a **double helix** and built a structural model. Two **antiparallel** sugar-phosphate chains wind around the outside of the molecule; the nitrogenous bases project into the interior, where they hydrogen-bond in specific pairs, A with T, G with C.

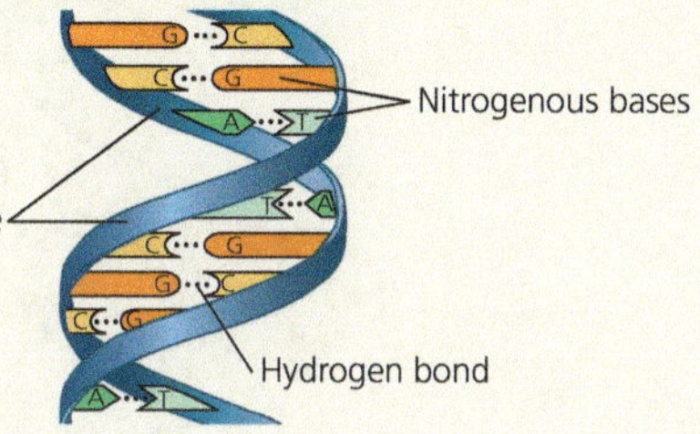

? *What does it mean when we say that the two DNA strands in the double helix are antiparallel? What would an end of the double helix look like if the strands were parallel?*

CONCEPT 16.2

Many proteins work together in DNA replication and repair *(pp. 320–329)*

- The Meselson-Stahl experiment showed that **DNA replication** is **semiconservative**: The parental molecule unwinds, and each strand then serves as a template for the synthesis of a new strand according to base-pairing rules.
- DNA replication at one **replication fork** is summarized here:

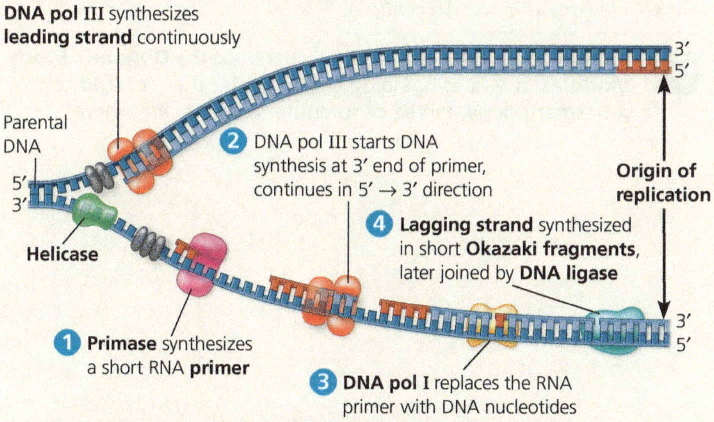

- **DNA polymerases** proofread new DNA, replacing incorrect nucleotides. In **mismatch repair**, enzymes correct errors that persist. **Nucleotide excision repair** is a process by which **nucleases** cut out and other enzymes replace damaged stretches of DNA.
- The ends of eukaryotic chromosomal DNA get shorter with each round of replication. The presence of **telomeres**, repetitive sequences at the ends of linear DNA molecules, postpones the erosion of genes. Telomerase catalyzes the lengthening of telomeres in germ cells.

? *Compare DNA replication on the leading and lagging strands, including both similarities and differences.*

CONCEPT 16.3

A chromosome consists of a DNA molecule packed together with proteins *(pp. 330–332)*

- The chromosome of most bacterial species is a circular DNA molecule with some associated proteins, making up the nucleoid. The **chromatin** making up a eukaryotic chromosome is composed of DNA, **histones**, and other proteins. The histones bind to each other and to the DNA to form **nucleosomes**, the most basic units of DNA packing. Histone tails extend outward from each bead-like nucleosome core. Additional coiling and folding lead ultimately to the highly condensed chromatin of the metaphase chromosome.
- Chromosomes occupy restricted areas in the interphase nucleus. In interphase cells, most chromatin is less compacted (**euchromatin**), but some remains highly condensed (**heterochromatin**). Euchromatin, but not heterochromatin, is generally accessible for transcription of genes.

? *Describe the levels of chromatin packing you'd expect to see in an interphase nucleus.*

TEST YOUR UNDERSTANDING

Level 1: Knowledge/Comprehension

PRACTICE TEST goo.gl/CUYGKD

1. In his work with pneumonia-causing bacteria and mice, Griffith found that
   (A) the protein coat from pathogenic cells was able to transform nonpathogenic cells.
   (B) heat-killed pathogenic cells caused pneumonia.
   (C) some substance from pathogenic cells was transferred to nonpathogenic cells, making them pathogenic.
   (D) the polysaccharide coat of bacteria caused pneumonia.
2. What is the basis for the difference in how the leading and lagging strands of DNA molecules are synthesized?
   (A) The origins of replication occur only at the 5′ end.
   (B) Helicases and single-strand binding proteins work at the 5′ end.
   (C) DNA polymerase can join new nucleotides only to the 3′ end of a pre-existing strand, and the strands are antiparallel.
   (D) DNA ligase works only in the 3′ → 5′ direction.
3. In analyzing the number of different bases in a DNA sample, which result would be consistent with the base-pairing rules?
   (A) A = G
   (B) A + G = C + T
   (C) A + T = G + C
   (D) A = C
4. The elongation of the leading strand during DNA synthesis
   (A) progresses away from the replication fork.
   (B) occurs in the 3′ → 5′ direction.
   (C) produces Okazaki fragments.
   (D) depends on the action of DNA polymerase.
5. In a nucleosome, the DNA is wrapped around
   (A) histones.
   (B) ribosomes.
   (C) polymerase molecules.
   (D) a thymine dimer.

Level 2: Application/Analysis

6. *E. coli* cells grown on ^{15}N medium are transferred to ^{14}N medium and allowed to grow for two more generations (two rounds of DNA replication). DNA extracted from these cells is centrifuged. What density distribution of DNA would you expect in this experiment?
   (A) one high-density and one low-density band
   (B) one intermediate-density band
   (C) one high-density and one intermediate-density band
   (D) one low-density and one intermediate-density band

7. A biochemist isolates, purifies, and combines in a test tube a variety of molecules needed for DNA replication. When she adds some DNA to the mixture, replication occurs, but each DNA molecule consists of a normal strand paired with numerous segments of DNA a few hundred nucleotides long. What has she probably left out of the mixture?
   (A) DNA polymerase
   (B) DNA ligase
   (C) Okazaki fragments
   (D) primase

8. The spontaneous loss of amino groups from adenine in DNA results in hypoxanthine, an uncommon base, opposite thymine. What combination of proteins could repair such damage?
   (A) nuclease, DNA polymerase, DNA ligase
   (B) telomerase, primase, DNA polymerase
   (C) telomerase, helicase, single-strand binding protein
   (D) DNA ligase, replication fork proteins, adenylyl cyclase

9. **MAKE CONNECTIONS** Although the proteins that cause the *E. coli* chromosome to coil are not histones, what property would you expect them to share with histones, given their ability to bind to DNA (see Figure 5.14)?

Level 3: Synthesis/Evaluation

10. **EVOLUTION CONNECTION** Some bacteria may be able to respond to environmental stress by increasing the rate at which mutations occur during cell division. How might this be accomplished? Might there be an evolutionary advantage to this ability? Explain.

11. **SCIENTIFIC INQUIRY**

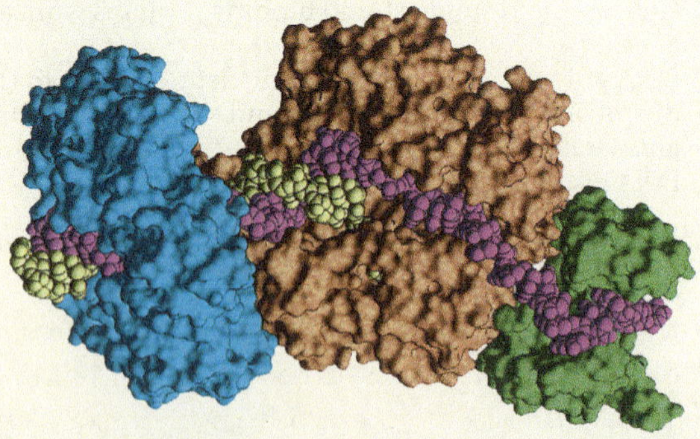

DRAW IT Model building can be an important part of the scientific process. The illustration shown above is a computer-generated model of a DNA replication complex. The parental and newly synthesized DNA strands are color-coded differently, as are each of the following three proteins: DNA pol III, the sliding clamp, and single-strand binding protein.
   (a) Using what you've learned in this chapter to clarify this model, label each DNA strand and each protein.
   (b) Draw an arrow to indicate the overall direction of DNA replication.

12. **WRITE ABOUT A THEME: INFORMATION** The continuity of life is based on heritable information in the form of DNA, and structure and function are correlated at all levels of biological organization. In a short essay (100–150 words), describe how the structure of DNA is correlated with its role as the molecular basis of inheritance.

13. **SYNTHESIZE YOUR KNOWLEDGE**

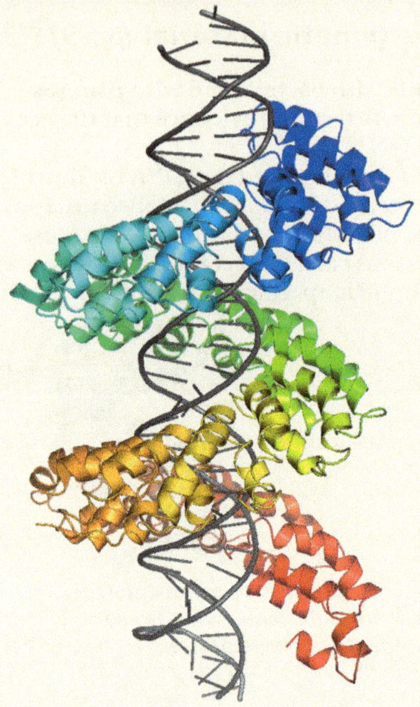

This image shows DNA (gray) interacting with a computer-generated model of a TAL protein (multicolored), one of a family of proteins found only in a species of the bacterium *Xanthomonas*. The bacterium uses proteins like this one to find specific gene sequences in cells of the organisms it infects, such as tomatoes, rice, and citrus fruits. Given what you know about DNA structure and considering the image above, discuss how the TAL protein's structure suggests that it functions.

For selected answers, see Appendix A.

For additional practice questions, check out the **Dynamic Study Modules** in MasteringBiology. You can use them to study on your smartphone, tablet, or computer anytime, anywhere!

Gene Expression: From Gene to Protein

17

▲ **Figure 17.1** **How does a single faulty gene result in the dramatic appearance of these albino donkeys?**

KEY CONCEPTS

17.1 Genes specify proteins via transcription and translation

17.2 Transcription is the DNA-directed synthesis of RNA: *a closer look*

17.3 Eukaryotic cells modify RNA after transcription

17.4 Translation is the RNA-directed synthesis of a polypeptide: *a closer look*

17.5 Mutations of one or a few nucleotides can affect protein structure and function

▼ **An albino raccoon.**

The Flow of Genetic Information

The island of Asinara lies off the coast of the Italian island of Sardinia. The name Asinara probably originated from the Latin work *sinuaria*, which means "sinus-shaped." A second meaning of Asinara is "donkey-inhabited," which is perhaps even more appropriate because Asinara is home to a wild population of albino donkeys **(Figure 17.1)**. What factors are responsible for the albino phenotype?

Inherited traits are determined by genes, and the trait of albinism is caused by a recessive allele of a pigmentation gene (see Concept 14.4). The information content of genes is in the form of specific sequences of nucleotides along strands of DNA, the genetic material. The albino donkey has a faulty version of a key protein, an enzyme required for pigment synthesis, and this protein is faulty because the gene that codes for it contains incorrect information.

This example illustrates the main point of this chapter: The DNA inherited by an organism leads to specific traits by dictating the synthesis of proteins and of RNA molecules involved in protein synthesis. In other words, proteins are the link between genotype and phenotype. **Gene expression** is the process by which DNA directs the synthesis of proteins (or, in some cases, just RNAs). The expression of genes that code for proteins includes two stages: transcription and translation. This chapter describes the flow of information from gene to protein and explains how genetic mutations affect organisms through their proteins. Understanding the processes of gene expression will allow us to revisit the concept of the gene in more detail at the end of the chapter.

When you see this blue icon, log in to **MasteringBiology** and go to the Study Area for digital resources.

MB Get Ready for This Chapter

CONCEPT 17.1

Genes specify proteins via transcription and translation

Before going into the details of how genes direct protein synthesis, let's step back and examine how the fundamental relationship between genes and proteins was discovered.

Evidence from Studying Metabolic Defects

In 1902, British physician Archibald Garrod was the first to suggest that genes dictate phenotypes through enzymes, proteins that catalyze specific chemical reactions in the cell. He postulated that the symptoms of an inherited disease reflect an inability to make a particular enzyme. He later referred to such diseases as "inborn errors of metabolism." For example, people with a disease called alkaptonuria have black urine because it contains a chemical called alkapton, which darkens upon exposure to air. Garrod reasoned that most people have an enzyme that breaks down alkapton, whereas people with alkaptonuria have inherited an inability to make that enzyme, so alkapton is expelled in their urine.

Several decades later, research supported Garrod's hypothesis that a gene dictates the production of a specific enzyme, later named the *one gene–one enzyme hypothesis*. Biochemists learned that cells synthesize and degrade most organic molecules via metabolic pathways, in which each chemical reaction in a sequence is catalyzed by a specific enzyme (see Concept 8.1). Such metabolic pathways lead, for instance, to the synthesis of the pigments that give the brown donkey in Figure 17.1 its fur color or fruit flies (*Drosophila*) their eye color (see Figure 15.3). In the 1930s, the American biochemist and geneticist George Beadle and his French colleague Boris Ephrussi speculated that in *Drosophila*, each mutation affecting eye color blocks pigment synthesis at a specific step by preventing production of the enzyme that catalyzes that step. But neither the chemical reactions nor the enzymes that catalyze them were known at the time.

Nutritional Mutants in Neurospora: Scientific Inquiry

A breakthrough came a few years later at Stanford University, where Beadle and Edward Tatum began working with a bread mold, *Neurospora crassa*, a haploid species. To observe a change in a mutant's phenotype, Beadle and Tatum needed to disable just one allele (rather than two, as in a diploid species) of a protein-coding gene required for a specific metabolic activity. They bombarded *Neurospora* with X-rays, known to cause mutations, and looked among the survivors for mutants that differed in their nutritional needs from the wild-type bread mold.

Wild-type *Neurospora* has modest food requirements. It can grow in the laboratory on a simple solution containing minimal nutrients for growth—inorganic salts, glucose, and the vitamin biotin—incorporated into agar, a support medium. From this so-called *minimal medium*, wild-type mold cells use their metabolic pathways to produce all the other molecules they need for growth, dividing repeatedly and forming visible colonies of genetically identical cells. As shown in **Figure 17.2**, Beadle and Tatum generated different "nutritional mutants" of *Neurospora* cells, each of which was unable to synthesize a particular essential nutrient. Such cells could not grow on minimal medium but could grow on *complete medium*, which contains all nutrients needed for growth. For *Neurospora*, the complete medium consists of the minimal medium supplemented with all 20 amino acids and a few other nutrients. Beadle and Tatum hypothesized that in each nutritional mutant, the gene for the enzyme that synthesizes a particular nutrient had been disabled.

This approach resulted in a valuable collection of mutant strains of *Neurospora*, catalogued by their defect in a particular pathway. Two colleagues of theirs, Adrian Srb and Norman Horowitz, used a collection of arginine-requiring mutants to investigate the biochemical pathway for arginine synthesis in *Neurospora* **(Figure 17.3)**. Srb and Horowitz pinned down each

▼ Figure 17.2 Beadle and Tatum's experimental approach. To obtain nutritional mutants, Beadle and Tatum exposed *Neurospora* cells to X-rays, inducing mutations, then screened mutants that had new nutritional requirements, such as arginine, as shown here.

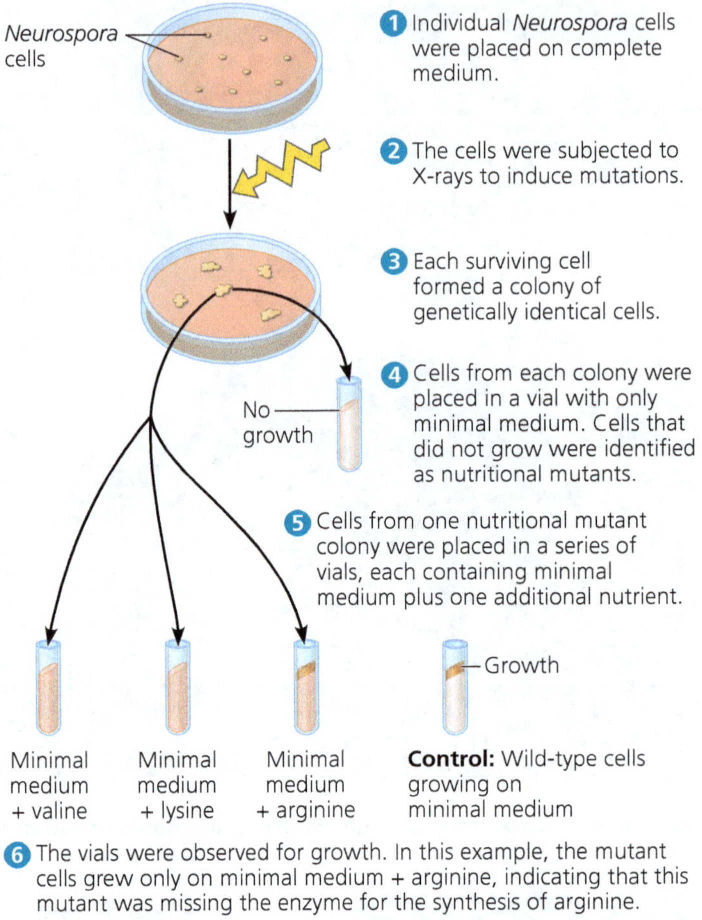

Figure 17.3

Inquiry Do individual genes specify the enzymes that function in a biochemical pathway?

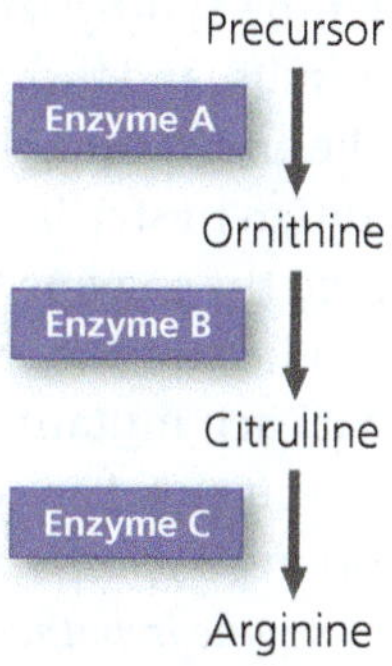

Experiment Working with the mold *Neurospora crassa*, Adrian Srb and Norman Horowitz, then at Stanford University, used Beadle and Tatum's experimental approach (see Figure 17.2) to isolate mutants that required arginine in their growth medium. The researchers showed that these mutants fell into three classes, each defective in a different gene. From studies by others on mammalian liver cells, they suspected that the metabolic pathway of arginine biosynthesis involved a precursor nutrient and the intermediate molecules ornithine and citrulline, as shown in the diagram on the right.

Their most famous experiment, shown here, tested both the *one gene–one enzyme hypothesis* and their postulated arginine-synthesizing pathway. In this experiment, they grew their three classes of mutants under the four different conditions shown in the Results Table below. They included minimal medium (MM) as a control, knowing that wild-type cells could grow on MM but mutant cells could not. (See test tubes below.)

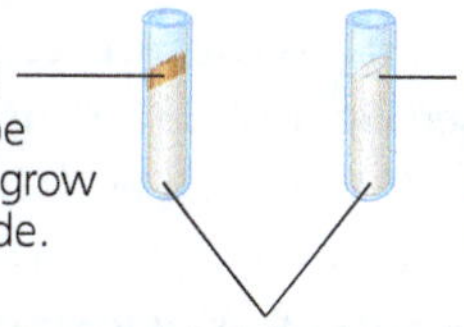

Control: Minimal medium

Results As shown in the table on the right, the wild-type strain was capable of growth under all experimental conditions, requiring only the minimal medium. The three classes of mutants each had a specific set of growth requirements. For example, class II mutants could not grow when ornithine alone was added but could grow when either citrulline or arginine was added.

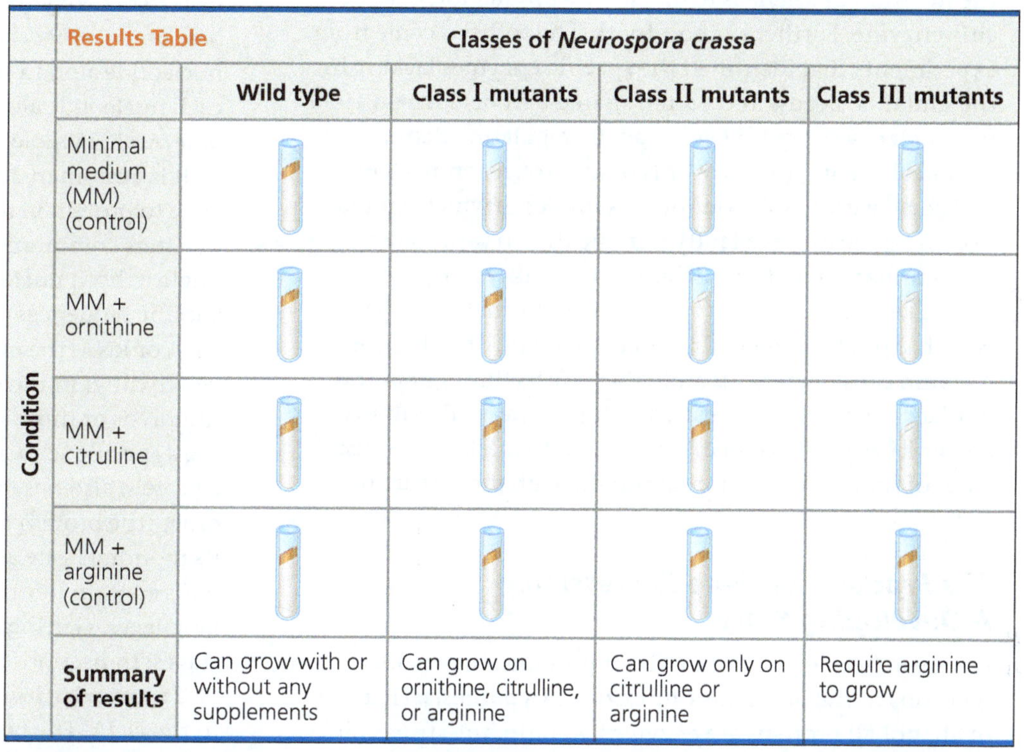

Results Table

Condition	Classes of *Neurospora crassa*: Wild type	Class I mutants	Class II mutants	Class III mutants
Minimal medium (MM) (control)				
MM + ornithine				
MM + citrulline				
MM + arginine (control)				
Summary of results	Can grow with or without any supplements	Can grow on ornithine, citrulline, or arginine	Can grow only on citrulline or arginine	Require arginine to grow

Conclusion From the growth requirements of the mutants, Srb and Horowitz deduced that each class of mutant was unable to carry out one step in the pathway for synthesizing arginine, presumably because it lacked the necessary enzyme, as shown in the table on the right. Because each of their mutants was mutated in a single gene, they concluded that each mutated gene must normally dictate the production of one enzyme. Their results supported the one gene–one enzyme hypothesis, proposed by Beadle and Tatum, and also confirmed that the arginine pathway described in the mammalian liver also operates in *Neurospora*. (Notice in the Results Table that a mutant can grow only if supplied with a compound made *after* the defective step because this bypasses the defect.)

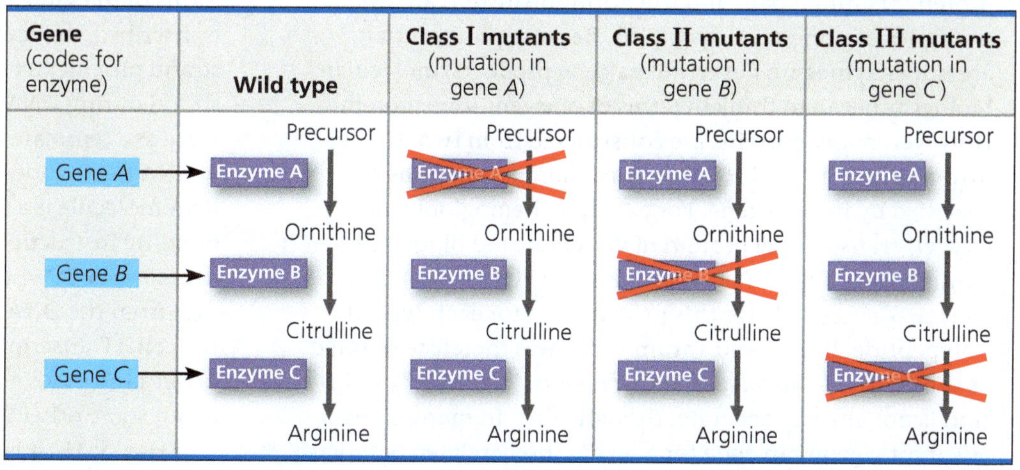

Gene (codes for enzyme)	**Wild type**	**Class I mutants** (mutation in gene *A*)	**Class II mutants** (mutation in gene *B*)	**Class III mutants** (mutation in gene *C*)
Gene *A* Gene *B* Gene *C*	Precursor Enzyme A Ornithine Enzyme B Citrulline Enzyme C Arginine	Precursor Enzyme A Ornithine Enzyme B Citrulline Enzyme C Arginine	Precursor Enzyme A Ornithine Enzyme B Citrulline Enzyme C Arginine	Precursor Enzyme A Ornithine Enzyme B Citrulline Enzyme C Arginine

Data from A. M. Srb and N. H. Horowitz, The ornithine cycle in *Neurospora* and its genetic control, *Journal of Biological Chemistry* 154:129–139 (1944).

WHAT IF? *Suppose the experiment had shown that class I mutants could grow only in MM supplemented by ornithine or arginine and that class II mutants could grow in MM supplemented by citrulline, ornithine, or arginine. What conclusions would the researchers have drawn from those results regarding the biochemical pathway and the defect in class I and class II mutants?*

mutant's defect more specifically, using additional tests to distinguish among three classes of arginine-requiring mutants. Mutants in each class required a different set of compounds along the arginine-synthesizing pathway, which has three steps. These results, and those of many similar experiments done by Beadle and Tatum, suggested that each class was blocked at a different step in this pathway because mutants in that class lacked the enzyme that catalyzes the blocked step.

Because Beadle and Tatum set up their experimental conditions so that each mutant was defective in a single gene, the collected results, taken together, provided strong support for a working hypothesis they had proposed earlier. The *one gene–one enzyme hypothesis*, as they dubbed it, states that the function of a gene is to dictate the production of a specific enzyme. Further support for this hypothesis came from experiments that identified the specific enzymes lacking in the mutants. Beadle and Tatum shared a Nobel Prize in 1958 for "their discovery that genes act by regulating definite chemical events" (in the words of the Nobel committee).

Today, we know of countless examples in which a mutation in a gene causes a faulty enzyme that in turn leads to an identifiable condition. The albino donkey in Figure 17.1 lacks a key enzyme called tyrosinase in the metabolic pathway that produces melanin, a dark pigment. The absence of melanin causes white fur and other effects throughout the donkey's body. Its nose, ears, and hooves, as well as its eyes, are pink because no melanin is present to mask the reddish color of the blood vessels that run through those structures.

The Products of Gene Expression: A Developing Story

As researchers learned more about proteins, they made revisions to the one gene–one enzyme hypothesis. First of all, not all proteins are enzymes. Keratin, the structural protein of animal hair, and the hormone insulin are two examples of nonenzyme proteins. Because proteins that are not enzymes are nevertheless gene products, molecular biologists began to think in terms of one gene–one protein. However, many proteins are constructed from two or more different polypeptide chains, and each polypeptide is specified by its own gene. For example, hemoglobin—the oxygen-transporting protein of vertebrate red blood cells—contains two kinds of polypeptides (see Figure 5.18), and thus two genes code for this protein, one for each type of polypeptide. Beadle and Tatum's idea was therefore restated as the *one gene–one polypeptide hypothesis*. Even this description is not entirely accurate, though. First, in many cases, a eukaryotic gene can code for a set of closely related polypeptides via a process called alternative splicing, which you will learn about later in this chapter. Second, quite a few genes code for RNA molecules that have important functions in cells even though they are never translated into protein. For now, we will focus on genes that do code for polypeptides. (Note that it is common to refer to these gene products as proteins—a practice you will encounter in this book—rather than more precisely as polypeptides.)

Basic Principles of Transcription and Translation

Genes provide the instructions for making specific proteins. But a gene does not build a protein directly. The bridge between DNA and protein synthesis is the nucleic acid RNA. RNA is chemically similar to DNA except that it contains ribose instead of deoxyribose as its sugar and has the nitrogenous base uracil rather than thymine (see Figure 5.23). Thus, each nucleotide along a DNA strand has A, G, C, or T as its base, and each nucleotide along an RNA strand has A, G, C, or U as its base. An RNA molecule usually consists of a single strand.

It is customary to describe the flow of information from gene to protein in linguistic terms. Just as specific sequences of letters communicate information in a language such as English, both nucleic acids and proteins are polymers with specific sequences of monomers that convey information. In DNA or RNA, the monomers are the four types of nucleotides, which differ in their nitrogenous bases. Genes are typically hundreds or thousands of nucleotides long, each gene having a specific sequence of nucleotides. Each polypeptide of a protein also has monomers arranged in a particular linear order (the protein's primary structure; see Figure 5.18), but its monomers are amino acids. Thus, nucleic acids and proteins contain information written in two different chemical languages. Getting from DNA to protein requires two major stages: transcription and translation.

Transcription is the synthesis of RNA using information in the DNA. The two nucleic acids are written in different forms of the same language, and the information is simply transcribed, or "rewritten," from DNA to RNA. Just as a DNA strand provides a template for making a new complementary strand during DNA replication (see Concept 16.2), it also can serve as a template for assembling a complementary sequence of RNA nucleotides. For a protein-coding gene, the resulting RNA molecule is a faithful transcript of the gene's protein-building instructions. This type of RNA molecule is called **messenger RNA (mRNA)** because it carries a genetic message from the DNA to the protein-synthesizing machinery of the cell. (Transcription is the general term for the synthesis of *any* kind of RNA on a DNA template. Later, you will learn about some other types of RNA produced by transcription.)

Translation is the synthesis of a polypeptide using the information in the mRNA. During this stage, there is a change in language: The cell must translate the nucleotide sequence of an mRNA molecule into the amino acid sequence of a polypeptide. The sites of translation are **ribosomes**, molecular complexes that facilitate the orderly linking of amino acids into polypeptide chains.

Transcription and translation occur in all organisms. Because most studies have involved bacteria and eukaryotic cells, they are our main focus in this chapter. Our understanding of transcription and translation in archaea lags behind, but we do know that archaeal cells share some features of gene expression with bacteria and others with eukaryotes.

The basic mechanics of transcription and translation are similar for bacteria and eukaryotes, but there is an important difference in the flow of genetic information within the cells. Bacteria do not have nuclei. Therefore, nuclear membranes do not separate bacterial DNA and mRNA from ribosomes and the other protein-synthesizing equipment **(Figure 17.4a)**. As you'll see later, this lack of compartmentalization allows translation of an mRNA to begin while its transcription is still in progress. By contrast, eukaryotic cells have nuclei. The presence of a nuclear envelope separates transcription from translation in space and time **(Figure 17.4b)**. Transcription occurs in the nucleus, but the mRNA must be transported to the cytoplasm for translation. Before eukaryotic RNA transcripts from protein-coding genes can leave the nucleus, they are modified in various ways to produce the final, functional mRNA. The transcription of a protein-coding eukaryotic gene results in *pre-mRNA*, and further processing yields the finished mRNA. The initial RNA transcript from any gene, including those specifying RNA that is not translated into protein, is more generally called a **primary transcript**.

▼ Figure 17.4 Overview: the roles of transcription and translation in the flow of genetic information. In a cell, inherited information flows from DNA to RNA to protein. The two main stages of information flow are transcription and translation. A miniature version of part **(a)** or **(b)** accompanies several figures later in the chapter as an orientation diagram to help you see where a particular figure fits into the overall scheme of gene expression.

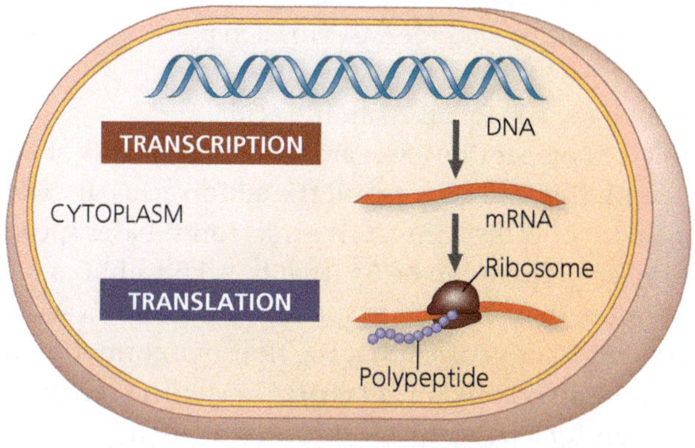

(a) Bacterial cell. In a bacterial cell, which lacks a nucleus, mRNA produced by transcription is immediately translated without additional processing.

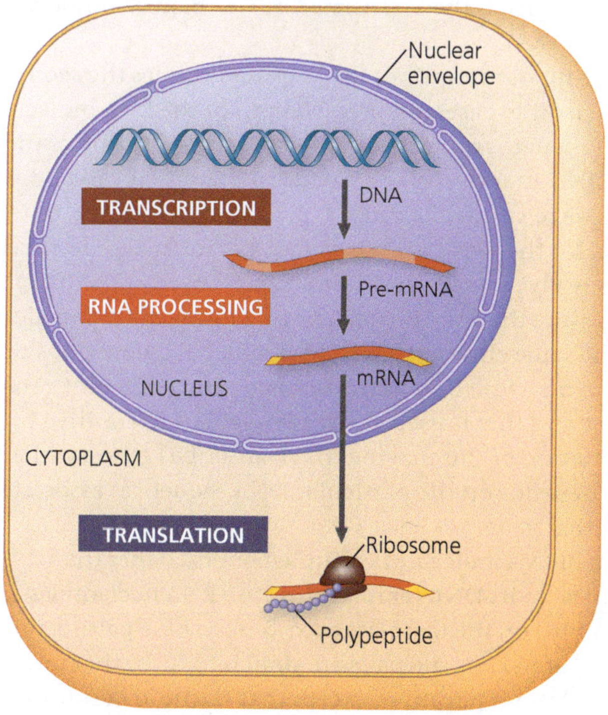

(b) Eukaryotic cell. The nucleus provides a separate compartment for transcription. The original RNA transcript, called pre-mRNA, is processed in various ways before leaving the nucleus as mRNA.

MB **Animation: Overview of Protein Synthesis in Bacteria**
Animation: Overview of Protein Synthesis in Eukaryotes

To summarize: Genes program protein synthesis via genetic messages in the form of messenger RNA. Put another way, cells are governed by a molecular chain of command with a directional flow of genetic information:

This concept was dubbed the *central dogma* by Francis Crick in 1956. But in the 1970s, scientists were surprised to discover some enzymes that use RNA molecules as templates for DNA synthesis (which we'll cover in Concept 19.2). However, these exceptions do not invalidate the idea that, in general, genetic information flows from DNA to RNA to protein. In the next section, we discuss how the instructions for assembling amino acids into a specific order are encoded in nucleic acids.

The Genetic Code

When biologists began to suspect that the instructions for protein synthesis were encoded in DNA, they recognized a problem: There are only four nucleotide bases to specify 20 amino acids. Thus, the genetic code cannot be a language like Chinese, where each written symbol corresponds to a word. How many nucleotides, then, would turn out to correspond to an amino acid?

Codons: Triplets of Nucleotides

If each kind of nucleotide base were translated into an amino acid, only four amino acids could be specified, one per nucleotide base. Would a language of two-letter code words suffice? The two-nucleotide sequence AG, for example, could specify one amino acid, and GT could specify another. Since there are four possible nucleotide bases in each position, this would give us 16 (that is, 4×4, or 4^2) possible arrangements—still not enough to code for all 20 amino acids.

Triplets of nucleotide bases are the smallest units of uniform length that can code for all the amino acids. If each arrangement of three consecutive nucleotide bases specifies an amino acid, there can be 64 (that is, 4^3) possible code words—more than enough to specify all the amino acids. Experiments have verified that the flow of information from gene to protein is based on a **triplet code**: The genetic instructions for a polypeptide chain are written in the DNA as a series of nonoverlapping, three-nucleotide words. The series of words in a gene is transcribed into a complementary series of nonoverlapping, three-nucleotide words in mRNA, which is then translated into a chain of amino acids **(Figure 17.5)**.

During transcription, the gene determines the sequence of nucleotide bases along the length of the RNA molecule that is being synthesized. For each gene, only one of the two DNA strands is transcribed. This strand is called the **template strand** because it provides the pattern, or template, for the sequence of nucleotides in an RNA transcript. For any given gene, the same strand is used as the template every time the gene is transcribed. However, farther along on the same chromosomal DNA molecule, the opposite strand may function as the template for a different gene. The strand that is used as the template is determined by the orientation of the enzyme that transcribes genes, which in turn depends on the particular DNA sequences associated with that gene.

An mRNA molecule is complementary rather than identical to its DNA template because RNA nucleotides are assembled on the template according to base-pairing rules (see Figure 17.5). The pairs are similar to those that form during DNA replication, except that U (the RNA substitute for T) pairs with A and the mRNA nucleotides contain ribose instead of deoxyribose. Like a new strand of DNA, the RNA molecule is synthesized in an antiparallel direction to the template strand of DNA. (To review what is meant by "antiparallel" and the 5′ and 3′ ends of a nucleic acid chain, see Figure 16.7.) In the example in Figure 17.5, the nucleotide triplet ACC along the DNA template strand (written as 3′-ACC-5′) provides a template for 5′-UGG-3′ in the mRNA molecule. The mRNA nucleotide triplets are called **codons**, and they are customarily written in the $5' \rightarrow 3'$ direction. In our example, UGG is the codon for the amino acid tryptophan (abbreviated Trp, or W). The term *codon* is also used for the DNA nucleotide triplets along the *nontemplate* strand. These codons are complementary to the template strand and thus identical in sequence to the mRNA, except that they have a T wherever there is a U in the mRNA. For this reason, the nontemplate DNA strand is often called the **coding strand**; by convention, the sequence of the coding strand is used when a gene's sequence is reported.

During translation, the sequence of codons along an mRNA molecule is decoded, or translated, into a sequence of amino acids making up a polypeptide chain. The codons are read by the translation machinery in the $5' \rightarrow 3'$ direction along the mRNA. Each codon specifies which one of the 20 amino acids will be incorporated at the corresponding position along a polypeptide. Because codons are nucleotide triplets, the number of nucleotides making up a genetic message must be three times the number of amino acids in the protein product. For example, it takes 300 nucleotides along an mRNA strand to code for the amino acids in a polypeptide that is 100 amino acids long.

▼ Figure 17.5 The triplet code. For each gene, one DNA strand functions as a template for transcription of RNAs, such as mRNA. The base-pairing rules for DNA synthesis also guide transcription, except that RNA is made with uracil (U) instead of thymine (T). During translation, the mRNA is read as a sequence of nucleotide triplets, called codons. Each codon specifies an amino acid to be added to the growing polypeptide chain. The mRNA is read in the $5' \rightarrow 3'$ direction.

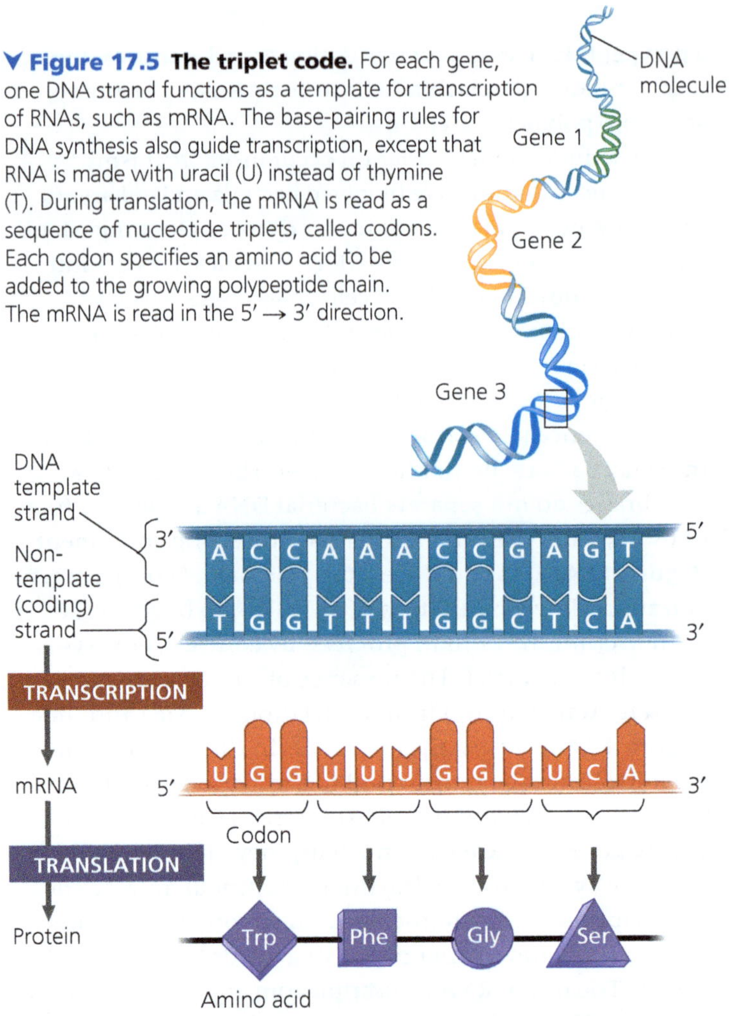

VISUAL SKILLS ➤ *By convention, the nontemplate strand, also called the coding strand, is used to represent a DNA sequence. Write the sequence of the mRNA strand and the nontemplate strand—in both cases reading from 5′ to 3′—and compare them. Why do you think this convention was adopted? (Hint: Why is this called the coding strand?)*

MP3 Tutor: DNA to RNA to Protein

Cracking the Code

Molecular biologists cracked the genetic code of life in the early 1960s when a series of elegant experiments disclosed the amino acid translations of each of the RNA codons. The first codon was deciphered in 1961 by Marshall Nirenberg, of the National Institutes of Health, along with his colleagues. Nirenberg synthesized an artificial mRNA by linking together many identical RNA nucleotides containing uracil as their base. No matter where the genetic message started or stopped, it could contain only one codon (UUU) over and over. Nirenberg added this "poly-U" polynucleotide to a test-tube mixture containing amino acids, ribosomes, and the other components required for protein synthesis. His artificial system translated the poly-U mRNA into a polypeptide containing many units of the amino acid phenylalanine (Phe, or F), strung together as a long polyphenylalanine chain. Thus, Nirenberg determined that the mRNA codon UUU specifies the amino acid phenylalanine. Soon, the amino acids specified by the codons AAA, GGG, and CCC were also identified.

Although more elaborate techniques were required to decode mixed triplets such as AUA and CGA, all 64 codons were deciphered by the mid-1960s. As **Figure 17.6** shows, 61 of the 64 triplets code for amino acids. The three codons that do not designate amino acids are "stop" signals, or termination codons, marking the end of translation. Notice that the codon AUG has a dual function: It codes for the amino acid methionine (Met, or M) and also functions as a "start" signal, or initiation codon. Genetic messages usually begin with the mRNA codon AUG, which signals the protein-synthesizing machinery to begin translating the mRNA at that location. (Because AUG also stands for methionine, polypeptide chains begin with methionine when they are synthesized. However, an enzyme may subsequently remove this starter amino acid from the chain.)

Notice in Figure 17.6 that there is redundancy in the genetic code, but no ambiguity. For example, although codons GAA and GAG both specify glutamic acid (redundancy), neither of them ever specifies any other amino acid (no ambiguity). The redundancy in the code is not altogether random. In many cases, codons that are synonyms for a particular amino acid differ only in the third nucleotide base of the triplet. We will consider the significance of this redundancy later in the chapter.

Our ability to extract the intended message from a written language depends on reading the symbols in the correct groupings—that is, in the correct **reading frame**. Consider this statement: "The red dog ate the bug." Group the letters incorrectly by starting at the wrong point, and the result will probably be gibberish: for example, "her edd oga tet heb ug." The reading frame is also important in the molecular language of cells. The short stretch of polypeptide shown in Figure 17.5, for instance, will be made

Figure 17.6 The codon table for mRNA. The three nucleotide bases of an mRNA codon are designated here as the first, second, and third bases, reading in the 5′ → 3′ direction along the mRNA. The codon AUG not only stands for the amino acid methionine (Met, or M) but also functions as a "start" signal for ribosomes to begin translating the mRNA at that point. Three of the 64 codons function as "stop" signals, marking where ribosomes end translation. Both one- and three-letter codes are shown for the amino acids; see Figure 5.14 for their full names.

First mRNA base (5′ end of codon)	Second mRNA base: U	Second mRNA base: C	Second mRNA base: A	Second mRNA base: G	Third mRNA base (3′ end of codon)
U	UUU Phe (F)	UCU Ser (S)	UAU Tyr (Y)	UGU Cys (C)	U
U	UUC	UCC	UAC	UGC	C
U	UUA Leu (L)	UCA	UAA Stop	UGA Stop	A
U	UUG	UCG	UAG Stop	UGG Trp (W)	G
C	CUU Leu (L)	CCU Pro (P)	CAU His (H)	CGU Arg (R)	U
C	CUC	CCC	CAC	CGC	C
C	CUA	CCA	CAA Gln (Q)	CGA	A
C	CUG	CCG	CAG	CGG	G
A	AUU Ile (I)	ACU Thr (T)	AAU Asn (N)	AGU Ser (S)	U
A	AUC	ACC	AAC	AGC	C
A	AUA	ACA	AAA Lys (K)	AGA Arg (R)	A
A	AUG Met (M) or start	ACG	AAG	AGG	G
G	GUU Val (V)	GCU Ala (A)	GAU Asp (D)	GGU Gly (G)	U
G	GUC	GCC	GAC	GGC	C
G	GUA	GCA	GAA Glu (E)	GGA	A
G	GUG	GCG	GAG	GGG	G

VISUAL SKILLS ➤ *A segment in the middle of an mRNA has the sequence 5′-AGAGAACCGCGA-3′. Using the codon table, translate this sequence, assuming the first three nucleotides are a codon.*

(MB) **Animation: Translation: The Genetic Code**

correctly only if the mRNA nucleotides are read from left to right (5′ → 3′) in the groups of three shown in the figure: UGG UUU GGC UCA. Although a genetic message is written with no spaces between the codons, the cell's protein-synthesizing machinery reads the message as a series of nonoverlapping three-letter words. The message is *not* read as a series of overlapping words—UGGUUU, and so on—which would convey a very different message.

Evolution of the Genetic Code

EVOLUTION The genetic code is nearly universal, shared by organisms from the simplest bacteria to the most complex plants and animals. The mRNA codon CCG, for instance, is translated as the amino acid proline in all organisms whose genetic code has been examined. In laboratory experiments, genes can be transcribed and translated after being transplanted from one species to another, sometimes with quite

Figure 17.7 Evidence for evolution: expression of genes from different species. Because diverse forms of life share a common genetic code due to their shared ancestry, one species can be programmed to produce proteins characteristic of a second species by introducing DNA from the second species into the first.

(a) Tobacco plant expressing a firefly gene. The yellow glow is produced by a chemical reaction catalyzed by the protein product of the firefly gene.

(b) Pig expressing a jellyfish gene. Researchers injected a jellyfish gene for a fluorescent protein into fertilized pig eggs. One developed into this fluorescent pig.

MB **Video: GFP Transgenic Mice**

striking results, as shown in **Figure 17.7**. Bacteria can be programmed by the insertion of human genes to synthesize certain human proteins for medical use, such as insulin. Such applications have produced many exciting developments in the area of biotechnology (see Concept 20.4).

Despite a small number of exceptions, the evolutionary significance of the code's near universality is clear. A language shared by all living things must have been operating very early in the history of life—early enough to be present in the common ancestor of all present-day organisms. A shared genetic vocabulary is a reminder of the kinship of all life.

CONCEPT CHECK 17.1

1. **MAKE CONNECTIONS ➤** In a research article about alkaptonuria published in 1902, Garrod suggested that humans inherit two "characters" (alleles) for a particular enzyme and that both parents must contribute a faulty version for the offspring to have alkaptonuria. Today, would this disorder be called dominant or recessive? (See Concept 14.4.)
2. What polypeptide product would you expect from a poly-G mRNA that is 30 nucleotides long?
3. **DRAW IT ➤** T The template strand of a gene contains the sequence 3′-TTCAGTCGT-5′. Imagine that the nontemplate sequence was transcribed instead of the template sequence. Draw the mRNA sequence and translate it using Figure 17.6. (Be sure to pay attention to the 5′ and 3′ ends.) Predict how well the protein synthesized from the nontemplate strand would function, if at all.

For suggested answers, see Appendix A.

CONCEPT 17.2

Transcription is the DNA-directed synthesis of RNA: *a closer look*

Now that we have considered the linguistic logic and evolutionary significance of the genetic code, we are ready to reexamine transcription, the first stage of gene expression, in greater detail.

Molecular Components of Transcription

Messenger RNA, the carrier of information from DNA to the cell's protein-synthesizing machinery, is transcribed from the template strand of a gene. An enzyme called an **RNA polymerase** pries the two strands of DNA apart and joins together RNA nucleotides complementary to the DNA template strand, thus elongating the RNA polynucleotide **(Figure 17.8)**. Like the DNA polymerases that function in DNA replication, RNA polymerases can assemble a polynucleotide only in its $5' \rightarrow 3'$ direction, adding onto its $3'$ end. Unlike DNA polymerases, however, RNA polymerases are able to start a chain from scratch; they don't need to add the first nucleotide onto a pre-existing primer.

Specific sequences of nucleotides along the DNA mark where transcription of a gene begins and ends. The DNA sequence where RNA polymerase attaches and initiates transcription is known as the **promoter**; in bacteria, the sequence that signals the end of transcription is called the **terminator**. (The termination mechanism is different in eukaryotes; we'll describe it later.) Molecular biologists refer to the direction of transcription as "downstream" and the other direction as "upstream." These terms are also used to describe the positions of nucleotide sequences within the DNA or RNA. Thus, the promoter sequence in DNA is said to be upstream from the terminator. The stretch of DNA downstream from the promoter that is transcribed into an RNA molecule is called a **transcription unit**.

Bacteria have a single type of RNA polymerase that synthesizes not only mRNA but also other types of RNA that function in protein synthesis, such as ribosomal RNA. In contrast, eukaryotes have at least three types of RNA polymerase in their nuclei; the one used for pre-mRNA synthesis is called RNA polymerase II. The other RNA polymerases transcribe RNA molecules that are not translated into protein. In the discussion that follows, we start with the features of mRNA synthesis common to both bacteria and eukaryotes and then describe some key differences.

Synthesis of an RNA Transcript

The three stages of transcription, as shown in Figure 17.8 and described next, are initiation, elongation, and termination of the RNA chain. Study Figure 17.8 to familiarize yourself with the stages and the terms used to describe them.

RNA Polymerase Binding and Initiation of Transcription

The promoter of a gene includes within it the transcription **start point**—the nucleotide where RNA polymerase actually begins synthesis of the mRNA—and typically extends several dozen or so nucleotide pairs upstream from the start point **(Figure 17.9)**. Based on interactions with proteins that will be covered shortly, RNA polymerase binds in a precise location and orientation on the promoter. This in turn determines where transcription starts and which of the two strands of the DNA helix is used as the template.

Figure 17.8 The stages of transcription: initiation, elongation, and termination. This general depiction of transcription applies to both bacteria and eukaryotes, but the details of termination differ, as described in the text. Also, in a bacterium, the RNA transcript is immediately usable as mRNA; in a eukaryote, the RNA transcript must first undergo processing.

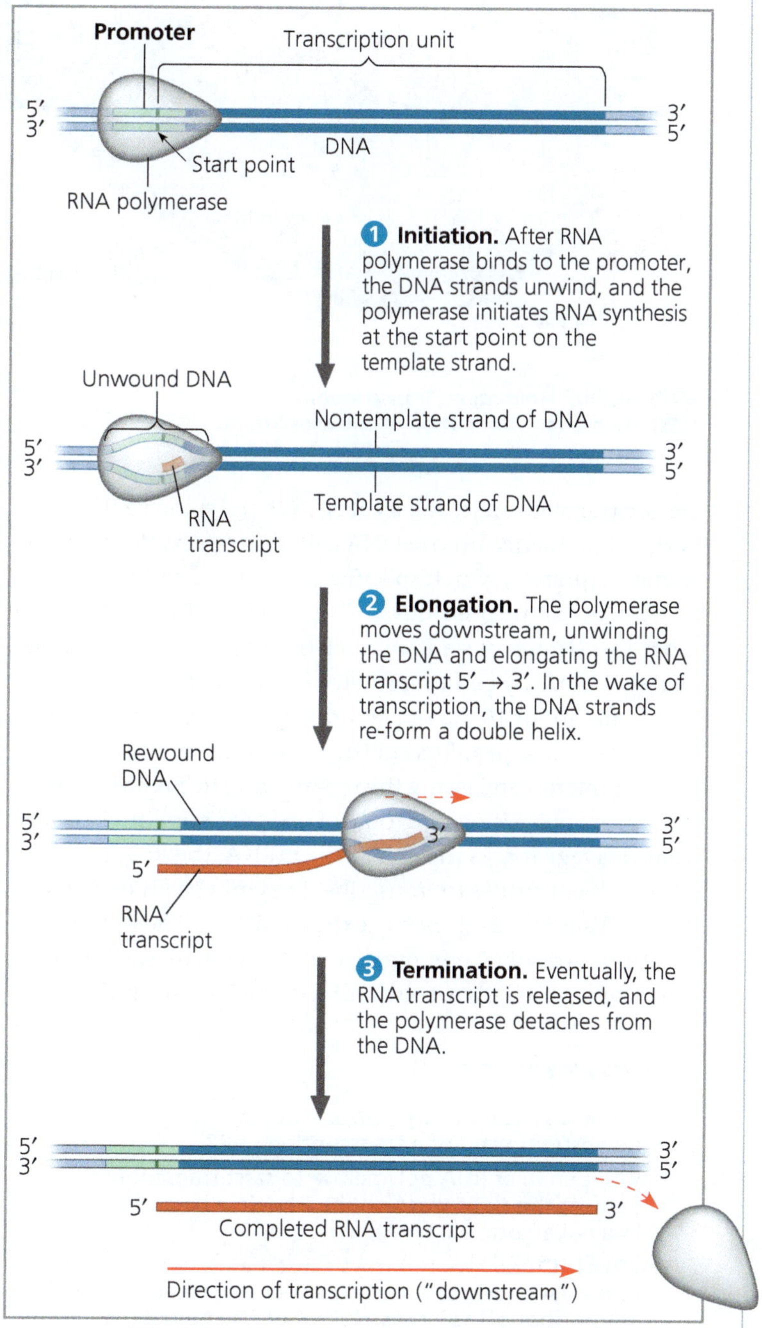

MAKE CONNECTIONS ➤ *Compare the use of a template strand during transcription and replication. See Figure 16.17.*

Animation: Overview of Transcription
Animation: Overview of Transcription in Bacteria

Figure 17.9 The initiation of transcription at a eukaryotic promoter. In eukaryotic cells, proteins called transcription factors mediate the initiation of transcription by RNA polymerase II.

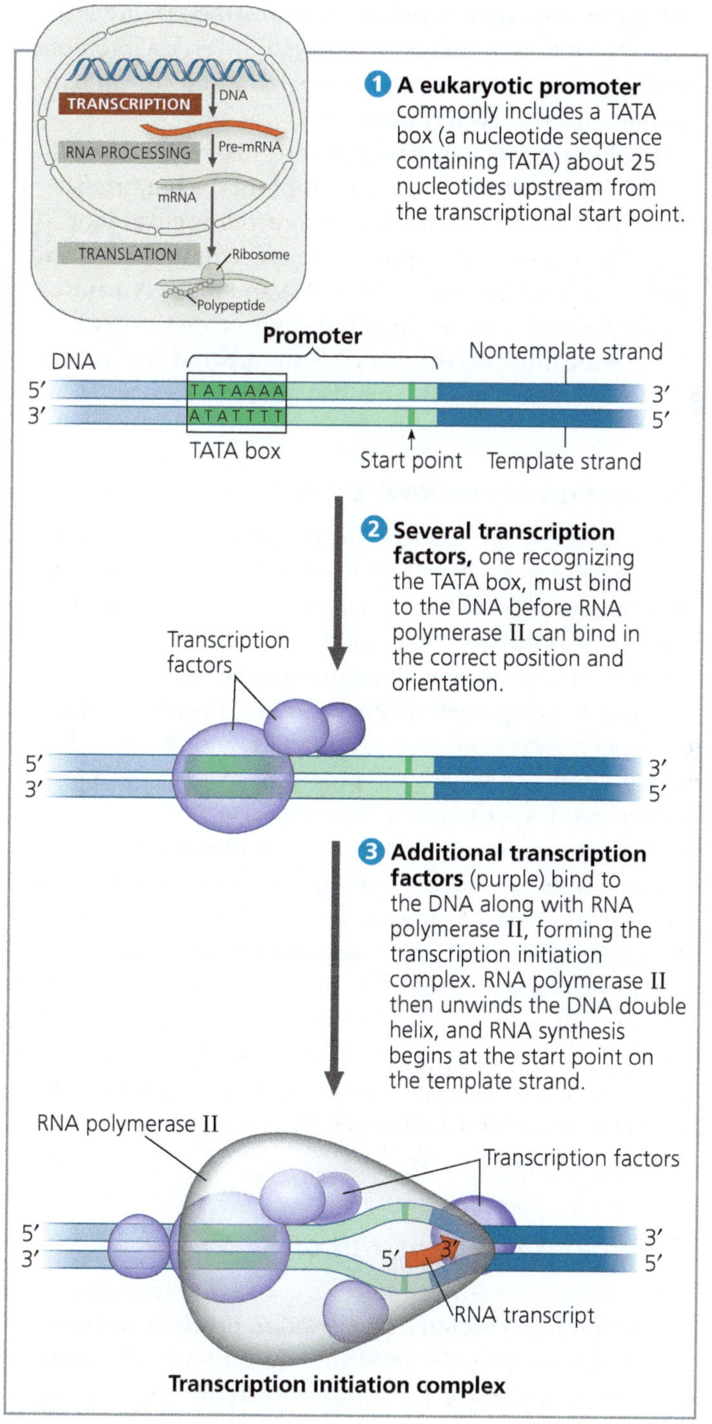

? *Explain how the interaction of RNA polymerase with the promoter would differ if the figure showed transcription initiation for bacteria.*

Certain sections of a promoter are especially important for binding RNA polymerase in a way that ensures that transcription will begin at the right place. In bacteria, part of the RNA polymerase itself specifically recognizes and binds to the promoter. In eukaryotes, a collection of proteins called **transcription factors** mediate the binding of RNA polymerase and the initiation of transcription. Only after transcription factors are attached to the promoter does RNA polymerase II bind to it. The whole complex of transcription factors and RNA polymerase II bound to the promoter is called a **transcription initiation complex**. Figure 17.9 shows the role of transcription factors and a crucial promoter DNA sequence called the **TATA box** in forming the initiation complex at a eukaryotic promoter.

The interaction between eukaryotic RNA polymerase II and transcription factors is an example of the importance of protein-protein interactions in controlling eukaryotic transcription. Once the appropriate transcription factors are firmly attached to the promoter DNA and the polymerase is bound to them in the correct orientation on the DNA, the enzyme unwinds the two DNA strands and begins transcribing the template strand at the start point.

Elongation of the RNA Strand

As RNA polymerase moves along the DNA, it untwists the double helix, exposing about 10–20 DNA nucleotides at a time for pairing with RNA nucleotides **(Figure 17.10)**. The enzyme adds nucleotides to the 3′ end of the growing RNA molecule as it continues along the double helix. In the wake of this advancing wave of RNA synthesis, the new RNA molecule peels away from its DNA template, and the DNA double helix re-forms. Transcription progresses at a rate of about 40 nucleotides per second in eukaryotes.

A single gene can be transcribed simultaneously by several molecules of RNA polymerase following each other like trucks in a convoy. A growing strand of RNA trails off from each polymerase, with the length of each new strand reflecting how far along the template the enzyme has traveled from the start point (see the mRNA molecules in Figure 17.23). The congregation of many polymerase molecules simultaneously transcribing a single gene increases the amount of mRNA transcribed from it, which helps the cell make the encoded protein in large amounts.

▼ Figure 17.10 Transcription elongation. RNA polymerase moves along the DNA template strand, joining complementary RNA nucleotides to the 3′ end of the growing RNA transcript. Behind the polymerase, the new RNA peels away from the template strand, which re-forms a double helix with the nontemplate strand.

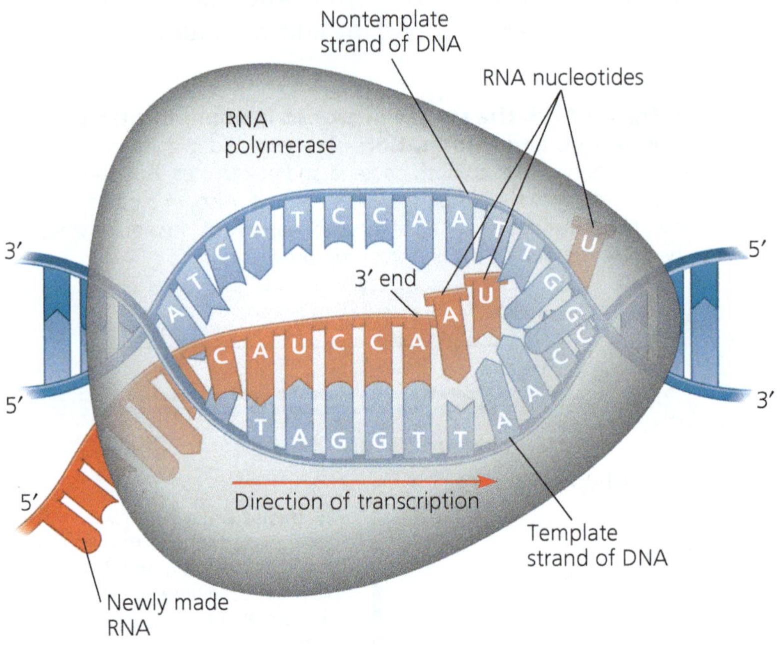

BioFlix® Animation: Transcription
Animation: Elongation of the RNA Strand

Termination of Transcription

Bacteria and eukaryotes differ in the way they terminate transcription. In bacteria, transcription proceeds through a terminator sequence in the DNA. The transcribed terminator (an RNA sequence) functions as the termination signal, causing the polymerase to detach from the DNA and release the transcript, which requires no further modification before translation. In eukaryotes, RNA polymerase II transcribes a sequence on the DNA called the polyadenylation signal sequence, which specifies a polyadenylation signal (AAUAAA) in the pre-mRNA. This is called a "signal" because once this stretch of six RNA nucleotides appears, it is immediately bound by certain proteins in the nucleus. Then, at a point about 10–35 nucleotides downstream from the AAUAAA, these proteins cut the RNA transcript free from the polymerase, releasing the pre-mRNA. The pre-mRNA then undergoes processing, the topic of the next section. Although that cleavage marks the end of the mRNA, the RNA polymerase II continues to transcribe. Enzymes begin to degrade the RNA starting at its newly exposed 5′ end. The polymerase continues transcribing, pursued by the enzymes, until they catch up to the polymerase and it dissociates from the DNA.

CONCEPT CHECK 17.2

1. What is a promoter? Is it located at the upstream or downstream end of a transcription unit?
2. What enables RNA polymerase to start transcribing a gene at the right place on the DNA in a bacterial cell? In a eukaryotic cell?
3. **WHAT IF?** ➤ Suppose X-rays caused a sequence change in the TATA box of a particular gene's promoter. How would that affect transcription of the gene? (See Figure 17.9.)

For suggested answers, see Appendix A.

Figure 17.11 RNA processing: addition of the 5′ cap and poly-A tail. Enzymes modify the two ends of a eukaryotic pre-mRNA molecule. The modified ends may promote the export of mRNA from the nucleus, and they help protect the mRNA from degradation. When the mRNA reaches the cytoplasm, the modified ends, in conjunction with certain cytoplasmic proteins, facilitate ribosome attachment. The 5′ cap and poly-A tail are not translated into protein, nor are the regions called the 5′ untranslated region (5′ UTR) and 3′ untranslated region (3′ UTR). The pink segments are introns, which will be described shortly (see Figure 17.12).

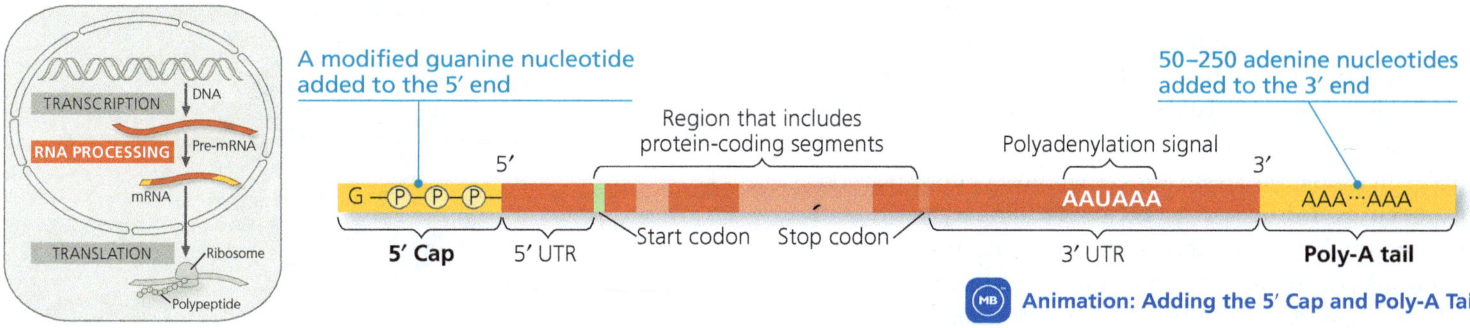

Animation: Adding the 5′ Cap and Poly-A Tail

CONCEPT 17.3

Eukaryotic cells modify RNA after transcription

Enzymes in the eukaryotic nucleus modify pre-mRNA in specific ways before the genetic message is dispatched to the cytoplasm. During this **RNA processing**, both ends of the primary transcript are altered. Also, in most cases, certain interior sections of the RNA molecule are cut out and the remaining parts spliced together. These modifications produce an mRNA molecule ready for translation.

Alteration of mRNA Ends

Each end of a pre-mRNA molecule is modified in a particular way **(Figure 17.11)**. The 5′ end, which is synthesized first, receives a **5′ cap**, a modified form of a guanine (G) nucleotide added onto the 5′ end after transcription of the first 20–40 nucleotides. The 3′ end of the pre-mRNA molecule is also modified before the mRNA exits the nucleus. Recall that the pre-mRNA is cut and released soon after the polyadenylation signal, AAUAAA, is transcribed. At the 3′ end, an enzyme then adds 50–250 more adenine (A) nucleotides, forming a **poly-A tail**. The 5′ cap and poly-A tail share several important functions. First, they seem to facilitate the export of the mature mRNA from the nucleus. Second, they help protect the mRNA from degradation by hydrolytic enzymes. And third, they help ribosomes attach to the 5′ end of the mRNA once the mRNA reaches the cytoplasm. Figure 17.11 also shows the untranslated regions (UTRs) at the 5′ and 3′ ends of the mRNA (referred to as the 5′ UTR and 3′ UTR). The UTRs are parts of the mRNA that will not be translated into protein, but they have other functions, such as ribosome binding.

Split Genes and RNA Splicing

A remarkable stage of RNA processing in the eukaryotic nucleus is **RNA splicing (Figure 17.12)**, where large portions of the RNA molecules are removed and the remaining portions

Figure 17.12 RNA processing: RNA splicing. The RNA molecule shown here codes for β-globin, one of the polypeptides of hemoglobin. The numbers under the RNA refer to codons; β-globin is 146 amino acids long. The β-globin gene and its pre-mRNA transcript have three exons, corresponding to sequences that will leave the nucleus as mRNA. (The 5′ UTR and 3′ UTR are parts of exons because they are included in the mRNA; however, they do not code for protein.) During RNA processing, the introns are cut out and the exons spliced together. In many genes, the introns are much longer than the exons.

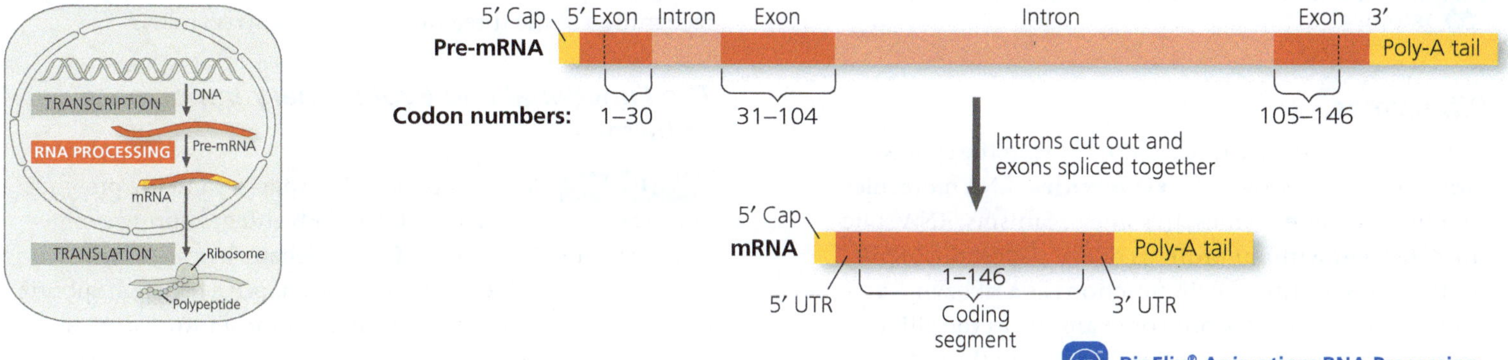

BioFlix® Animation: RNA Processing

DRAW IT ➤ *On the mRNA, indicate the sites of the start and stop codons.*

are reconnected. This cut-and-paste job is similar to editing a movie. The average length of a transcription unit along a human DNA molecule is about 27,000 nucleotide pairs, so the primary RNA transcript is also that long. However, the average-sized protein of 400 amino acids requires only 1,200 nucleotides in RNA to code for it. (Remember, each amino acid is encoded by a *triplet* of nucleotides.) This is because most eukaryotic genes and their RNA transcripts have long noncoding stretches of nucleotides, regions that are not translated. Even more surprising is that most of these noncoding sequences are interspersed between coding segments of the gene and thus between coding segments of the pre-mRNA. In other words, the sequence of DNA nucleotides that codes for a eukaryotic polypeptide is usually not continuous; it is split into segments. The noncoding segments of nucleic acid that lie between coding regions are called *int*ervening sequences, or **introns**. The other regions are called **exons**, because they are eventually *ex*pressed, usually by being translated into amino acid sequences. (Exceptions include the UTRs of the exons at the ends of the RNA, which make up part of the mRNA but are not translated into protein. Because of these exceptions, you may prefer to think of exons as sequences of RNA that *ex*it the nucleus.) The terms *intron* and *exon* are used for both RNA sequences and the DNA sequences that specify them.

In making a primary transcript from a gene, RNA polymerase II transcribes both introns and exons from the DNA, but the mRNA molecule that enters the cytoplasm is an abridged version. In RNA splicing, the introns are cut out from the molecule and the exons joined together, forming an mRNA molecule with a continuous coding sequence.

How is pre-mRNA splicing carried out? The removal of introns is accomplished by a large complex made of proteins and small RNAs called a **spliceosome**. This complex binds to several short nucleotide sequences along an intron, including key sequences at each end **(Figure 17.13)**. The intron is then released (and rapidly degraded), and the spliceosome joins together the two exons that flanked the intron. It turns out that the small RNAs in the spliceosome not only participate in spliceosome assembly and splice site recognition, but also catalyze the splicing reaction.

Interview with Joan Steitz: Studying RNA, her "favorite molecule"

Ribozymes

The idea of a catalytic role for the RNAs in the spliceosome arose from the discovery of **ribozymes**, RNA molecules that function as enzymes. In some organisms, RNA splicing can occur without proteins or even additional RNA molecules: The intron RNA functions as a ribozyme and catalyzes its own excision! For example, in the ciliate protist *Tetrahymena*, self-splicing occurs in the production of ribosomal RNA (rRNA), a component of the organism's

Figure 17.13 A spliceosome splicing a pre-mRNA. The diagram shows a portion of a pre-mRNA transcript, with an intron (pink) flanked by two exons (red). Small RNAs within the spliceosome base-pair with nucleotides at specific sites along the intron. Next, small spliceosome RNAs catalyze cutting of the pre-mRNA and the splicing together of the exons, releasing the intron for rapid degradation.

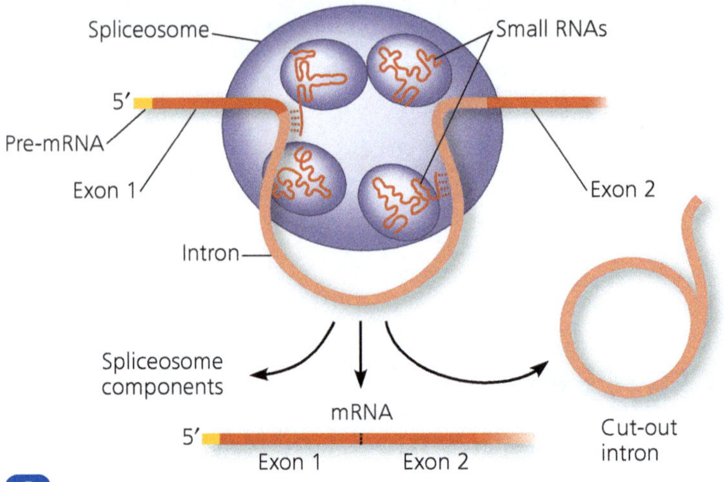

Animation: A Spliceosome

ribosomes. The pre-rRNA actually removes its own introns. The discovery of ribozymes rendered obsolete the idea that all biological catalysts are proteins.

Three properties of RNA enable some RNA molecules to function as enzymes. First, because RNA is single-stranded, a region of an RNA molecule may base-pair, in an antiparallel arrangement, with a complementary region elsewhere in the same molecule; this gives the molecule a particular three-dimensional structure. A specific structure is essential to the catalytic function of ribozymes, just as it is for enzymatic proteins. Second, like certain amino acids in an enzymatic protein, some of the bases in RNA contain functional groups that can participate in catalysis. Third, the ability of RNA to hydrogen-bond with other nucleic acid molecules (either RNA or DNA) adds specificity to its catalytic activity. For example, complementary base pairing between the RNA of the spliceosome and the RNA of a primary RNA transcript precisely locates the region where the ribozyme catalyzes splicing. Later in this chapter, you will see how these properties of RNA also allow it to perform important noncatalytic roles in the cell, such as recognition of the three-nucleotide codons on mRNA.

The Functional and Evolutionary Importance of Introns

EVOLUTION Whether or not RNA splicing and the presence of introns have provided selective advantages during evolutionary history is a matter of some debate. In any case, it is informative to consider their possible adaptive benefits. Specific functions have not been identified for most introns, but at least some contain sequences that regulate gene expression, and many affect gene products.

One important consequence of the presence of introns in genes is that a single gene can encode more than one kind of polypeptide. Many genes are known to give rise to two or more different polypeptides, depending on which segments are treated as exons during RNA processing; this is called **alternative RNA splicing** (see Figure 18.13). Results from the Human Genome Project (discussed in Concept 21.1) suggest that alternative RNA splicing is one reason humans can get along with about the same number of genes as a nematode (roundworm). Because of alternative splicing, the number of different protein products an organism produces can be much greater than its number of genes.

Proteins often have a modular architecture consisting of discrete structural and functional regions called **domains**. One domain of an enzyme, for example, might include the active site, while another might allow the enzyme to bind to a cellular membrane. In quite a few cases, different exons code for the different domains of a protein **(Figure 17.14)**.

The presence of introns in a gene may facilitate the evolution of new and potentially beneficial proteins as a result of a process known as *exon shuffling* (see Figure 21.16). Introns increase the probability of crossing over between the exons of alleles of a gene—simply by providing more terrain for crossovers without interrupting coding sequences. This might result in new combinations of exons and proteins with altered structure and function. We can also imagine the occasional mixing and matching of exons between completely different (nonallelic) genes. Exon shuffling of either sort could lead to new proteins with novel combinations of functions. While most of the shuffling would result in nonbeneficial changes, occasionally a beneficial variant might arise.

Figure 17.14 Correspondence between exons and protein domains.

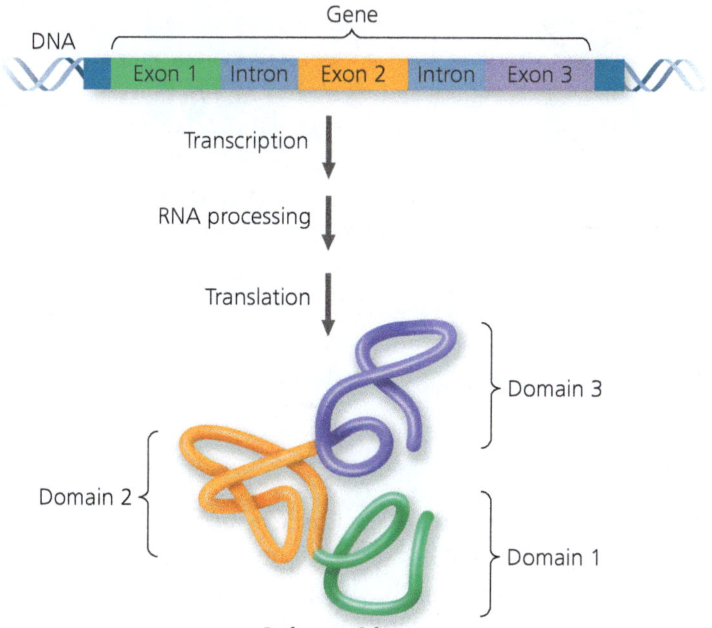

CONCEPT CHECK 17.3

1. There are about 20,000 human protein-coding genes. How can human cells make 75,000–100,000 different proteins?
2. How is RNA splicing similar to how you would watch a recorded television show? What would introns be?
3. **WHAT IF?** ➤ What would be the effect of treating cells with an agent that removed the cap from mRNAs?

For suggested answers, see Appendix A.

CONCEPT 17.4

Translation is the RNA-directed synthesis of a polypeptide: *a closer look*

We will now examine how genetic information flows from mRNA to protein—the process of translation **(Figure 17.15)**. We'll focus on the basic steps of translation that occur in both bacteria and eukaryotes, while pointing out key differences.

Figure 17.15 Translation: the basic concept. As a molecule of mRNA is moved through a ribosome, codons are translated into amino acids, one by one. The translators, or interpreters, are tRNA molecules, each type with a specific anticodon at one end and a corresponding amino acid at the other end. A tRNA adds its amino acid cargo to a growing polypeptide chain when the anticodon hydrogen-bonds to the complementary codon on the mRNA.

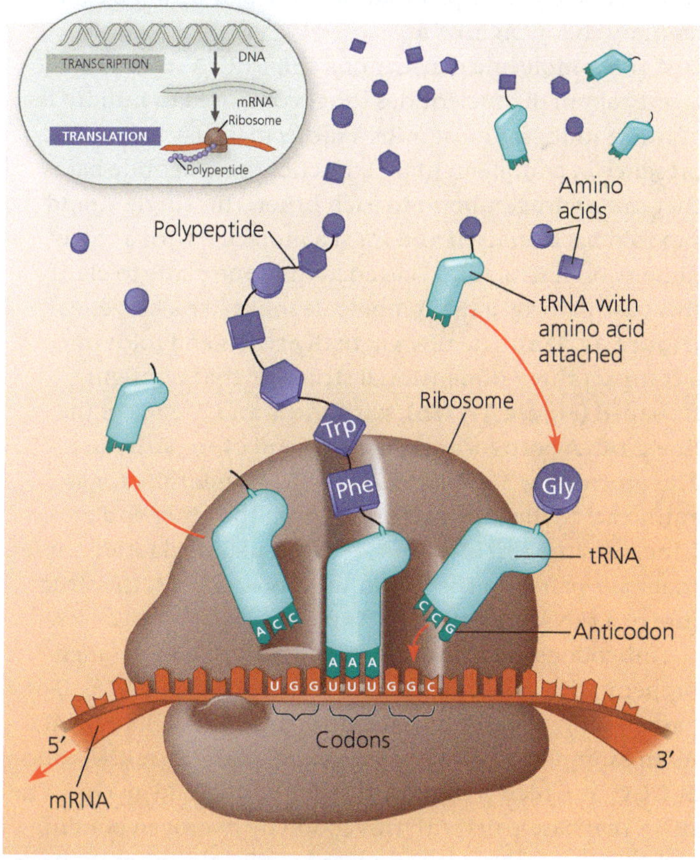

BioFlix® Animation: Translation
Animation: Overview of Translation

Molecular Components of Translation

In the process of translation, a cell "reads" a genetic message and builds a polypeptide accordingly. The message is a series of codons along an mRNA molecule, and the translator is called a **transfer RNA (tRNA)**. The function of a tRNA is to transfer an amino acid from the cytoplasmic pool of amino acids to a growing polypeptide in a ribosome. A cell keeps its cytoplasm stocked with all 20 amino acids, either by synthesizing them from other compounds or by taking them up from the surrounding solution. The ribosome, a structure made of proteins and RNAs, adds each amino acid brought to it by a tRNA to the growing end of a polypeptide chain (see Figure 17.15).

Translation is simple in principle but complex in its biochemistry and mechanics, especially in the eukaryotic cell. In dissecting translation, we'll focus on the slightly less complicated version of the process that occurs in bacteria. We'll first look at the major players in this process.

The Structure and Function of Transfer RNA

The key to translating a genetic message into a specific amino acid sequence is the fact that each tRNA molecule enables translation of a given mRNA codon into a certain amino acid. This is possible because a tRNA bears a specific amino acid at one end of its three-dimensional structure, while at the other end is a nucleotide triplet that can base-pair with the complementary codon on mRNA.

A tRNA molecule consists of a single RNA strand that is only about 80 nucleotides long (compared to hundreds of nucleotides for most mRNA molecules). Because of the presence of complementary stretches of nucleotide bases that can hydrogen-bond to each other, this single strand can fold back on itself and form a molecule with a three-dimensional structure. Flattened into one plane to clarify this base pairing, a tRNA molecule looks like a cloverleaf **(Figure 17.16a)**. The tRNA actually twists and folds into a compact three-dimensional structure that is roughly L-shaped **(Figure 17.16b)**, with the 5′ and 3′ ends of the linear tRNA both located near one end of the structure. The protruding 3′ end acts as the attachment site for an amino acid. The loop extending from the other end of the L includes the **anticodon**, the particular nucleotide triplet that base-pairs to a specific mRNA codon. Thus, the structure of a tRNA molecule fits its function.

Anticodons are conventionally written 3′ → 5′ to align properly with codons written 5′ → 3′ (see Figure 17.15). (For base pairing, RNA strands must be antiparallel, like DNA.) As an example of how tRNAs work, consider the mRNA codon 5′-GGC-3′, which is translated as the amino acid glycine. The tRNA that base-pairs with this codon by hydrogen bonding has 3′-CCG-5′ as its anticodon and carries glycine at its other end (see the incoming tRNA approaching the ribosome in Figure 17.15). As an mRNA molecule is moved through a ribosome, glycine will be added to the polypeptide chain whenever the codon 5′-GGC-3′ is presented for translation. Codon by codon, the genetic message is translated as tRNAs

▼ **Figure 17.16 The structure of transfer RNA (tRNA).**

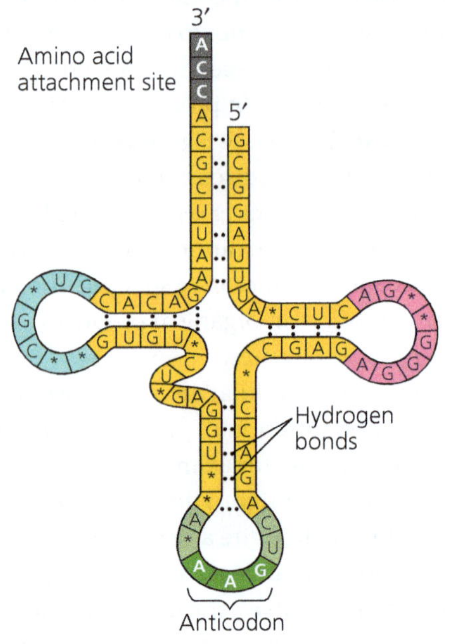

(a) Two-dimensional structure. The four base-paired regions and three loops are characteristic of all tRNAs, as is the base sequence of the amino acid attachment site at the 3′ end. The anticodon triplet is unique to each tRNA type, as are some sequences in the other two loops. (The asterisks mark bases that have been chemically modified, a characteristic of tRNA. The modified bases contribute to tRNA function in a way that is not yet understood.)

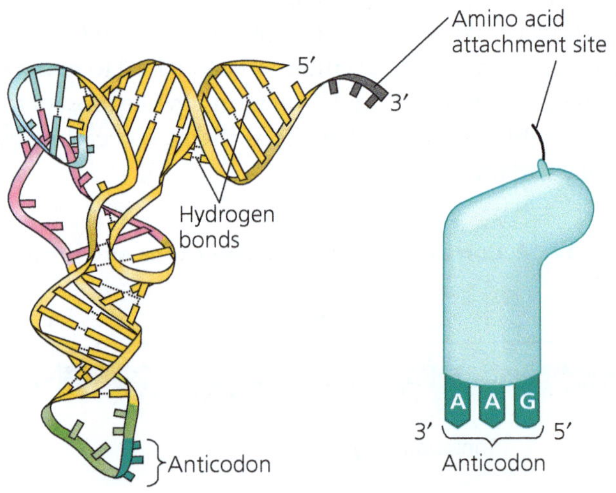

(b) Three-dimensional structure

(c) Symbol used in this book

VISUAL SKILLS ➤ *Look at the tRNA shown in this figure. Based on its anticodon, identify the codon it would bind to, as well as the amino acid that it would carry.*

HHMI Video: RNA Folding hhmi BioInteractive

position each amino acid in the order prescribed, and the ribosome adds that amino acid onto the growing polypeptide chain. The tRNA molecule is a translator in the sense that, in the context of the ribosome, it can read a nucleic acid word (the mRNA codon) and interpret it as a protein word (the amino acid).

Like mRNA and other types of cellular RNA, transfer RNA molecules are transcribed from DNA templates. In a eukaryotic cell, tRNA, like mRNA, is made in the nucleus and then travels to the cytoplasm, where it will participate in the process of translation. In both bacterial and eukaryotic cells, each tRNA molecule is used repeatedly, picking up its designated amino acid in the cytosol, depositing this cargo onto a polypeptide chain at the ribosome, and then leaving the ribosome, ready to pick up another of the same amino acid.

The accurate translation of a genetic message requires two instances of molecular recognition. First, a tRNA that binds to an mRNA codon specifying a particular amino acid must carry that amino acid, and no other, to the ribosome. The correct matching up of tRNA and amino acid is carried out by a family of related enzymes that are aptly named **aminoacyl-tRNA synthetases** **(Figure 17.17)**. The active site of each type of aminoacyl-tRNA synthetase fits only a specific combination of amino acid and tRNA. There are 20 different synthetases, one for each amino acid. A synthetase joins a given amino acid to an appropriate tRNA; one synthetase is able to bind to all the different tRNAs for its particular amino acid. The synthetase catalyzes the covalent attachment of the amino acid to its tRNA in a process driven by the hydrolysis of ATP. The resulting aminoacyl tRNA, also called a charged tRNA, is released from the enzyme and is then available to deliver its amino acid to a growing polypeptide chain on a ribosome.

The second instance of molecular recognition is the pairing of the tRNA anticodon with the appropriate mRNA codon. If one tRNA variety existed for each mRNA codon specifying an amino acid, there would be 61 tRNAs (see Figure 17.6). In bacteria, however, there are only about 45 tRNAs, signifying that some tRNAs must be able to bind to more than one codon. Such versatility is possible because the rules for base pairing between the third nucleotide base of a codon and the corresponding base of a tRNA anticodon are relaxed compared to those at other codon positions. For example, the nucleotide base U at the 5′ end of a tRNA anticodon can pair with either A or G in the third position (at the 3′ end) of an mRNA codon. The flexible base pairing at this codon position is called **wobble**. Wobble explains why the synonymous codons for a given amino acid most often differ in their third nucleotide base. Accordingly, a tRNA with the anticodon 3′-UCU-5′ can base-pair with either the mRNA codon 5′-AGA-3′ or 5′-AGG-3′, both of which code for arginine (see Figure 17.6).

▼ Figure 17.17 Aminoacyl-tRNA synthetases provide specificity in joining amino acids to their tRNAs. Linkage of a tRNA to its amino acid is an endergonic process that occurs at the expense of ATP, which loses two phosphate groups, becoming AMP.

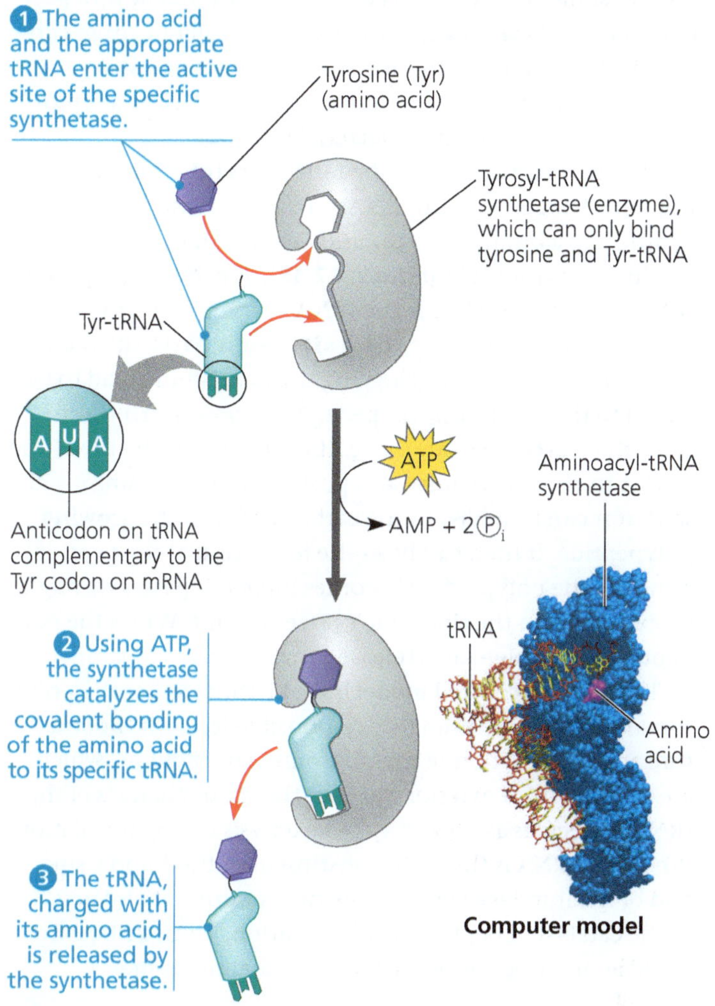

MB Animation: Transfer RNA

The Structure and Function of Ribosomes

Ribosomes facilitate the specific coupling of tRNA anticodons with mRNA codons during protein synthesis. A ribosome consists of a large subunit and a small subunit, each made up of proteins and one or more **ribosomal RNAs (rRNAs)**. In eukaryotes, the subunits are made in the nucleolus. Ribosomal RNA genes are transcribed, and the RNA is processed and assembled with proteins imported from the cytoplasm. Completed ribosomal subunits are then exported via nuclear pores to the cytoplasm. In both bacteria and eukaryotes, a large and a small subunit join to form a functional ribosome only when attached to an mRNA molecule. About one-third of the mass of a ribosome is made up of proteins; the rest consists of three rRNA molecules (in bacteria) or four (in eukaryotes). Because most cells contain thousands of ribosomes, rRNA is the most abundant type of cellular RNA.

Although the ribosomes of bacteria and eukaryotes are very similar in structure and function, eukaryotic ribosomes are slightly larger, as well as differing somewhat from bacterial ribosomes in their molecular composition. The differences are medically significant. Certain antibiotic drugs can inactivate bacterial ribosomes without affecting eukaryotic ribosomes. These drugs, including tetracycline and streptomycin, are used to combat bacterial infections.

The structure of a ribosome reflects its function of bringing mRNA together with tRNAs carrying amino acids. In addition to a binding site for mRNA, each ribosome has three binding sites for tRNA **(Figure 17.18)**. The **P site** (peptidyl-tRNA binding site) holds the tRNA carrying the growing polypeptide chain, while the **A site** (aminoacyl-tRNA binding site) holds the tRNA carrying the next amino acid to be added to the chain. Discharged tRNAs leave the ribosome from the **E site** (exit site). The ribosome holds the tRNA and mRNA in close proximity and positions the new amino acid so that it can be added to the carboxyl end of the growing polypeptide. It then catalyzes the formation of the peptide bond. As the polypeptide becomes longer, it passes through an *exit tunnel* in the ribosome's large subunit. When the polypeptide is complete, it is released through the exit tunnel.

The widely accepted model is that rRNAs, rather than ribosomal proteins, are primarily responsible for both the structure and the function of the ribosome. The proteins, which are largely on the exterior, support the shape changes of the rRNA molecules as they carry out catalysis during translation. Ribosomal RNA is the main constituent of the A and P sites and of the interface between the two subunits; it also acts as the catalyst of peptide bond formation. Thus, a ribosome could actually be considered one colossal ribozyme!

Building a Polypeptide

We can divide translation, the synthesis of a polypeptide, into three stages: initiation, elongation, and termination. All three require protein "factors" that aid in the translation process. Some steps of initiation and elongation also require energy, provided by the hydrolysis of guanosine triphosphate (GTP).

Ribosome Association and Initiation of Translation

In either bacteria or eukaryotes, the start codon (AUG) signals the start of translation; this is important because it establishes the codon reading frame for the mRNA. In the first step of translation, a small ribosomal subunit binds to both the mRNA and a specific initiator tRNA, which carries the amino acid methionine. In bacteria, the small subunit can bind the two in either order; it binds the mRNA at a specific RNA sequence, just upstream of the AUG start codon. In the **Scientific Skills Exercise**, you can work with DNA sequences encoding the ribosomal binding sites on the mRNAs of a group of *Eschericia coli*

Figure 17.18 The anatomy of a functioning ribosome.

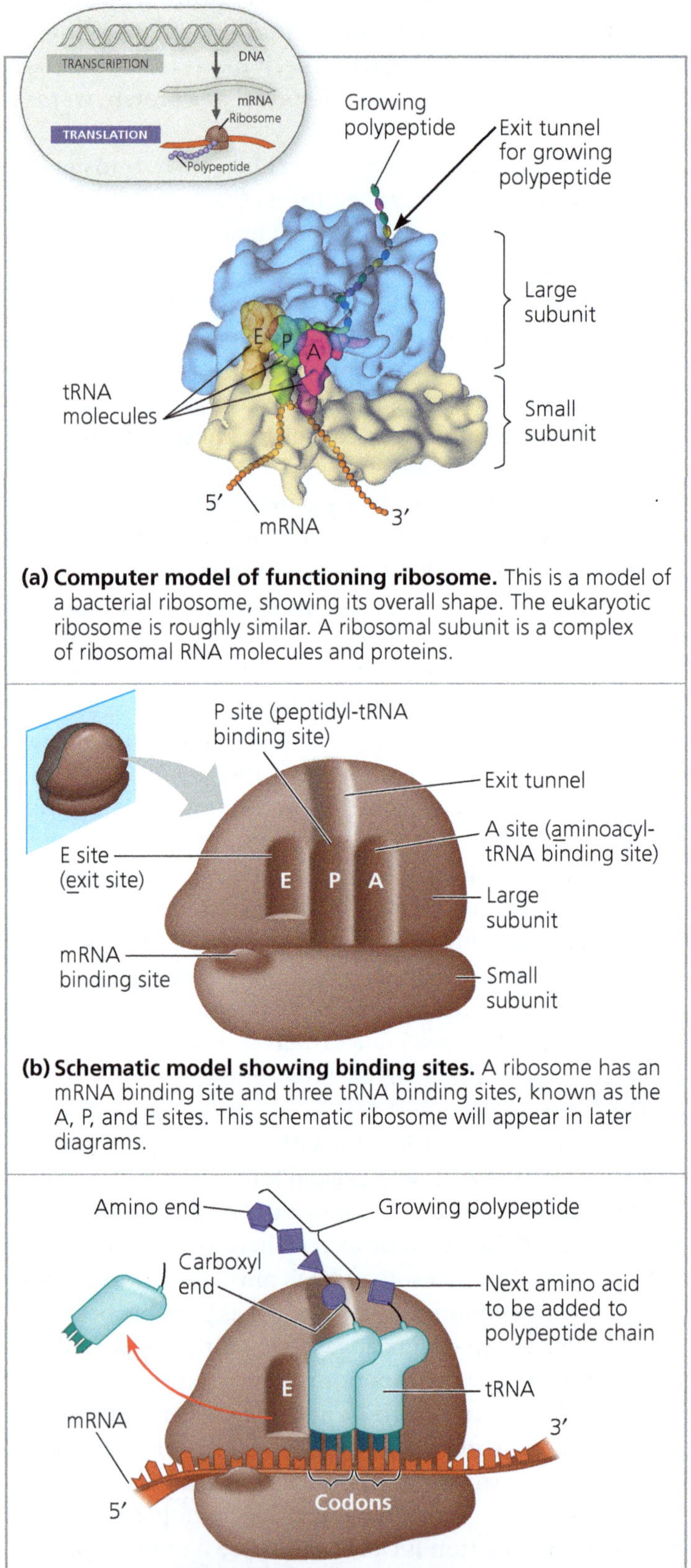

(a) Computer model of functioning ribosome. This is a model of a bacterial ribosome, showing its overall shape. The eukaryotic ribosome is roughly similar. A ribosomal subunit is a complex of ribosomal RNA molecules and proteins.

(b) Schematic model showing binding sites. A ribosome has an mRNA binding site and three tRNA binding sites, known as the A, P, and E sites. This schematic ribosome will appear in later diagrams.

(c) Schematic model with mRNA and tRNA. A tRNA fits into a binding site when its anticodon base-pairs with an mRNA codon. The P site holds the tRNA attached to the growing polypeptide. The A site holds the tRNA carrying the next amino acid to be added to the polypeptide chain. Discharged tRNAs leave from the E site. The polypeptide grows at its carboxyl end.

Animation: Ribosomes

SCIENTIFIC SKILLS EXERCISE

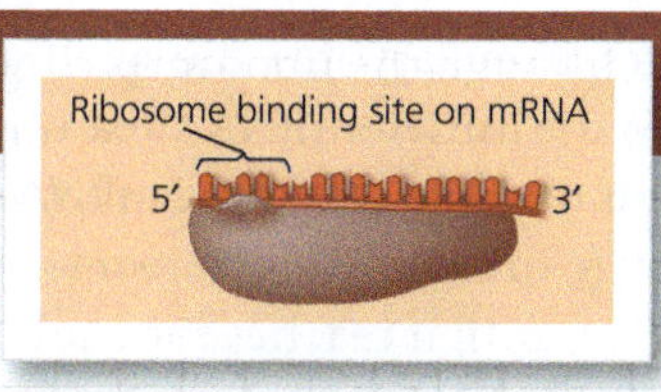

Interpreting a Sequence Logo

How Can a Sequence Logo Be Used to Identify Ribosome Binding Sites on Bacterial mRNAs? When initiating translation, ribosomes bind to an mRNA at a ribosome binding site upstream of the AUG start codon. Because mRNAs from different genes all bind to a ribosome, the genes encoding these mRNAs are likely to have a similar base sequence where the ribosomes bind. Therefore, candidate ribosome binding sites on mRNA can be identified by comparing DNA sequences (and thus the mRNA sequences) of multiple genes in a species, searching the region upstream of the start codon for shared (conserved) stretches of bases. In this exercise, you will analyze DNA sequences from multiple such genes, represented by a visual graphic called a sequence logo.

How the Experiment Was Done The DNA sequences of 149 genes from the *E. coli* genome were aligned using computer software. The aim was to identify similar base sequences—at the appropriate location in each gene—as potential ribosome binding sites. Rather than presenting the data as a series of 149 sequences aligned in a column (a sequence alignment), the researchers used a sequence logo.

Data from the Experiment To show how sequence logos are made, the potential ribosome binding regions from 10 *E. coli* genes are shown below in a sequence alignment, followed by the sequence logo derived from the aligned sequences. Note that the DNA shown is the nontemplate (coding) strand, which is how DNA sequences are typically presented.

thrA	G G T A A C G A G G T A A C A A C C A T G C G A G T G
lacA	C A T A A C G G A G T G A T C G C A T T G A A C A T G
lacY	C G C G T A A G G A A A T C C A T T A T G T A C T A T
lacZ	T T C A C A C A G G A A A C A G C T A T G A C C A T G
lacI	C A A T T C A G G G T G G T G A A T G T G A A A C C A
recA	G G C A T G A C A G G A G T A A A A A T G G C T A T C
galR	A C C C A C T A A G G T A T T T T C A T G G C G A C C
metJ	A A G A G G A T T A A G T A T C T C A T G G C T G A A
lexA	A T A C A C C C A G G G G G C G G A A T G A A A G C G
trpR	T A A C A A T G G C G A C A T A T T A T G G C C C A A
5′ … 3′	−18 −17 −16 −15 −14 −13 −12 −11 −10 −9 −8 −7 −6 −5 −4 −3 −2 −1 0 1 2 3 4 5 6 7 8

▲ **Sequence alignment**

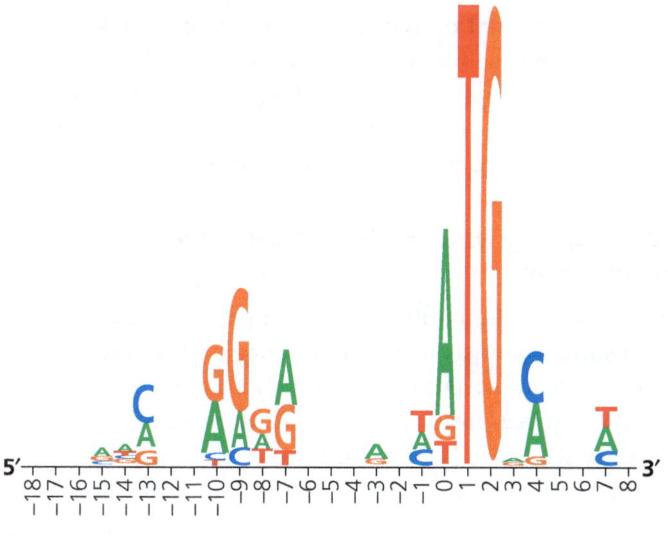

▲ **Sequence logo**

INTERPRET THE DATA

1. In the sequence logo (bottom, left), the horizontal axis shows the primary sequence of the DNA by nucleotide position. Letters for each base are stacked on top of each other according to their relative frequency at that position among the aligned sequences, with the most common base as the largest letter at the top of the stack. The height of each letter represents the relative frequency of that base *at that position*. (a) In the sequence alignment, count the number of each base at position −9 and order them from most to least frequent. Compare this to the size and placement of each base at −9 in the logo. (b) Do the same for positions 0 and 1.
2. The height of a stack of letters in a logo indicates the predictive power of that stack (determined statistically). If the stack is tall, we can be more confident in predicting what base will be in that position if a new sequence is added to the logo. For example, at position 2 in the sequence alignment, all 10 sequences have a G; the probability of finding a G there in a new sequence is very high, as is the stack in the sequence logo. For short stacks, the bases all have about the same frequency, so it's hard to predict a base at those positions. (a) Looking at the sequence logo, which two positions have the most predictable bases? What bases do you predict would be at those positions in a newly sequenced gene? (b) Which 12 positions have the least predictable bases? How do you know? How does this reflect the relative frequencies of the bases shown at these positions in the sequence alignment? Use the two leftmost positions of the 12 as examples in your answer.
3. In the actual experiment, the researchers used 149 sequences to build their sequence logo, which is shown below. There is a stack at each position, even if short, because the sequence logo includes more data. (a) Which three positions in this sequence logo have the most predictable bases? Name the most frequent base at each. (b) Which four positions have the least predictable bases? How can you tell?

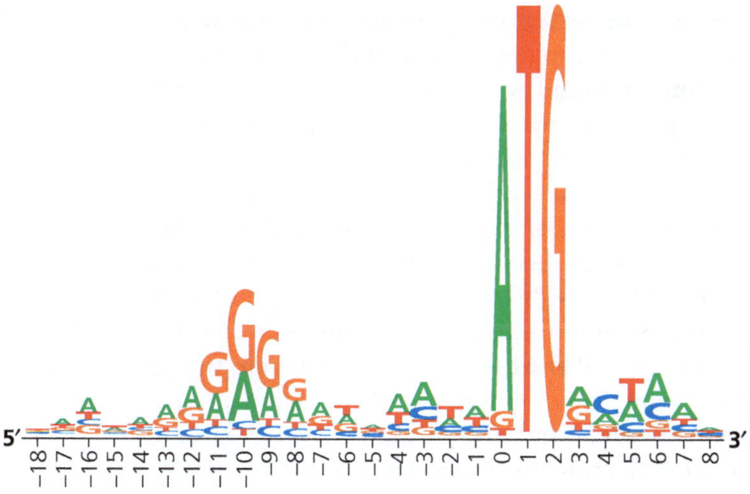

4. A consensus sequence identifies the base occurring most often at each position in the set of sequences. (a) Write out the consensus sequence of this (the nontemplate) strand. In any position where the base can't be determined, put a dash. (b) Which provides more information—the consensus sequence or the sequence logo? What is lost in the less informative method?
5. (a) Based on the logo, what five adjacent base positions in the 5′ UTR region are most likely to be involved in ribosome binding? Explain. (b) What is represented by the bases in positions 0–2?

MB **Instructors**: A version of this Scientific Skills Exercise can be assigned in MasteringBiology.

Further Reading T. D. Schneider and R. M. Stephens, Sequence logos: A new way to display consensus sequences, *Nucleic Acids Research* 18:6097–6100 (1990).

genes. In eukaryotes, the small subunit, with the initiator tRNA already bound, binds to the 5′ cap of the mRNA and then moves, or *scans*, downstream along the mRNA until it reaches the start codon; the initiator tRNA then hydrogen-bonds to the AUG start codon.

Thus, the first components to associate with each other during the initiation stage of translation are mRNA, a tRNA bearing the first amino acid of the polypeptide, and the small ribosomal subunit **(Figure 17.19)**. This is followed by the attachment of a large ribosomal subunit, completing the *translation initiation complex*. Proteins called *initiation factors* are required to bring all these components together. The cell also expends energy obtained by hydrolysis of a GTP molecule to form the initiation complex. At the completion of the initiation process, the initiator tRNA sits in the P site of the ribosome, and the vacant A site is ready for the next aminoacyl tRNA. Note that a polypeptide is always synthesized in one direction, from the initial methionine at the amino end, also called the N-terminus, toward the final amino acid at the carboxyl end, also called the C-terminus (see Figure 5.15).

▼ Figure 17.19 The initiation of translation. Animation: Initiation of Translation

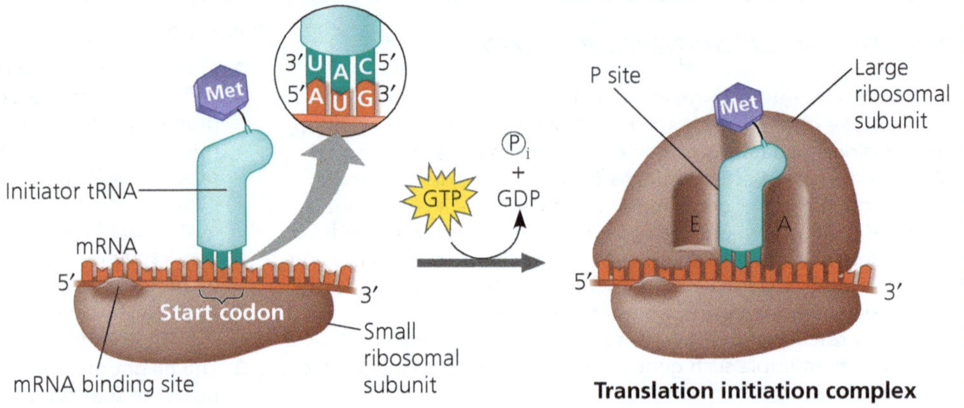

❶ A small ribosomal subunit binds to a molecule of mRNA. In a bacterial cell, the mRNA binding site on this subunit recognizes a specific nucleotide sequence on the mRNA just upstream of the start codon. An initiator tRNA, with the anticodon UAC, base-pairs with the start codon, AUG. This tRNA carries the amino acid methionine (Met).

❷ The arrival of a large ribosomal subunit completes the initiation complex. Proteins called initiation factors (not shown) are required to bring all the translation components together. Hydrolysis of GTP provides the energy for the assembly. The initiator tRNA is in the P site; the A site is available to the tRNA bearing the next amino acid.

Elongation of the Polypeptide Chain

In the elongation stage of translation, amino acids are added one by one to the previous amino acid at the C-terminus of the growing chain. Each addition involves several proteins called *elongation factors* and occurs in a three-step cycle described in **Figure 17.20**. Energy expenditure occurs in the first and third steps. Codon recognition requires hydrolysis of one molecule of GTP, which increases the accuracy and efficiency of this step. One more GTP is hydrolyzed to provide energy for the translocation step.

The mRNA is moved through the ribosome in one direction only, 5′ end first; this is equivalent to the ribosome moving 5′ → 3′ on the mRNA. The main point is that the ribosome and the mRNA move relative to each other, unidirectionally, codon by codon. The elongation cycle takes less than a tenth of a second in bacteria and is repeated as each amino acid is added until the polypeptide is complete. The empty tRNAs that are released from the E site return to the cytoplasm, where they will be reloaded with the appropriate amino acid (see Figure 17.17).

Termination of Translation

The final stage of translation is termination **(Figure 17.21)**. Elongation continues until a stop codon in the mRNA reaches the A site. The nucleotide base triplets UAG, UAA, and UGA (all written 5′ → 3′) do not code for amino acids but instead act as signals to stop translation. A *release factor*, a protein shaped like an aminoacyl tRNA, binds directly to the stop codon in the A site. The release factor causes the addition of a water molecule instead of an amino acid to the polypeptide chain. (Water molecules are abundant in the cytosol.) This reaction breaks (hydrolyzes) the bond between the completed polypeptide and the tRNA in the P site, releasing the polypeptide through the exit tunnel of the ribosome's large subunit. The remainder of the translation assembly then comes apart in a multistep process, aided by other protein factors. Breakdown of the translation assembly requires the hydrolysis of two more GTP molecules.

Completing and Targeting the Functional Protein

The process of translation is often not sufficient to make a functional protein. In this section, you will learn about modifications that polypeptide chains undergo after the translation process as well as some of the mechanisms used to target completed proteins to specific sites in the cell.

Protein Folding and Post-Translational Modifications

During its synthesis, a polypeptide chain begins to coil and fold spontaneously as a consequence of its amino acid sequence (primary structure), forming a protein with a specific shape: a three-dimensional molecule with secondary and tertiary structure (see Figure 5.18). Thus, a gene determines primary structure, which in turn determines shape.

Figure 17.20 The elongation cycle of translation. The hydrolysis of GTP plays an important role in the elongation process; elongation factors are not shown.

Animation: Elongation Cycle of Translation

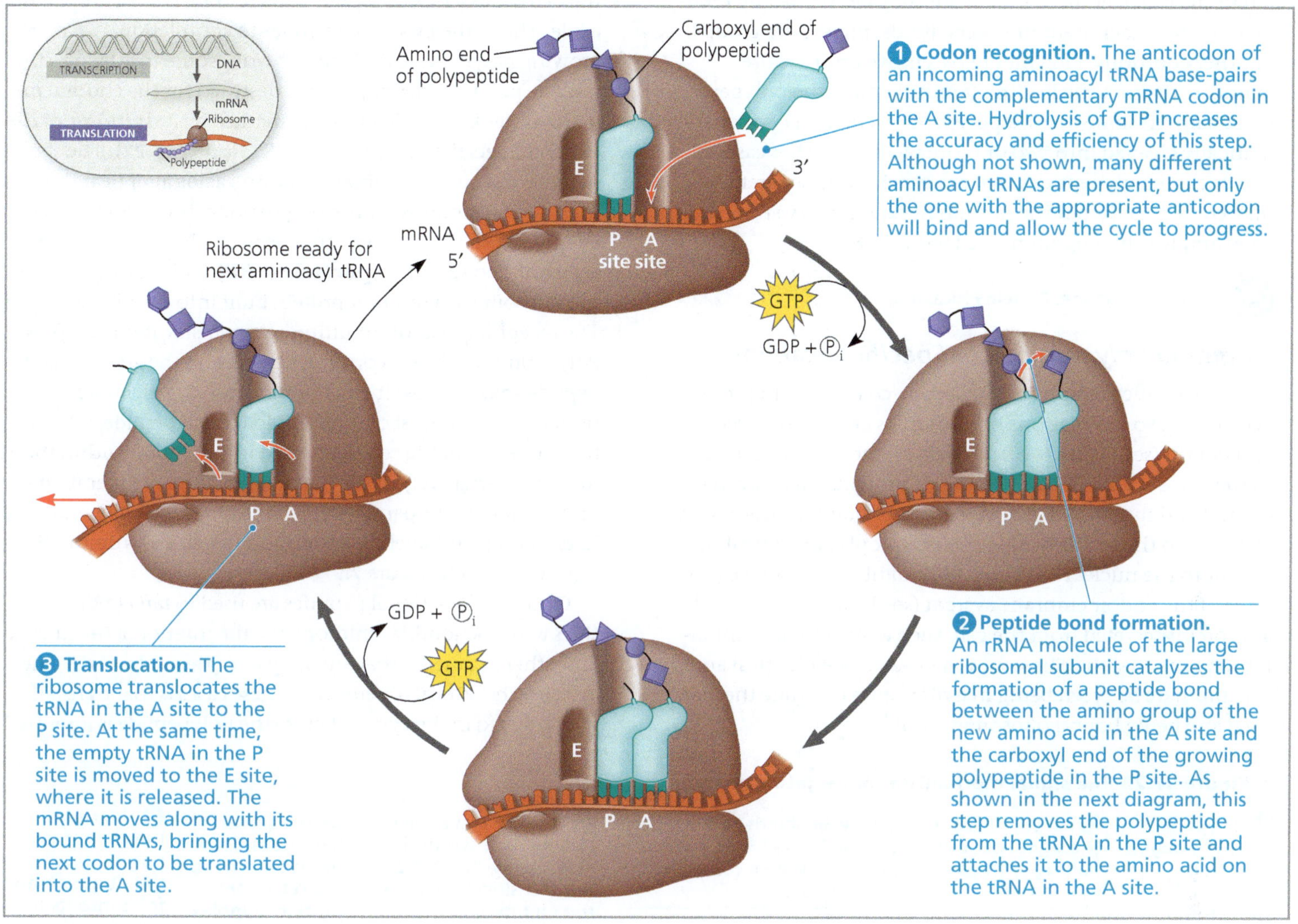

1 Codon recognition. The anticodon of an incoming aminoacyl tRNA base-pairs with the complementary mRNA codon in the A site. Hydrolysis of GTP increases the accuracy and efficiency of this step. Although not shown, many different aminoacyl tRNAs are present, but only the one with the appropriate anticodon will bind and allow the cycle to progress.

2 Peptide bond formation. An rRNA molecule of the large ribosomal subunit catalyzes the formation of a peptide bond between the amino group of the new amino acid in the A site and the carboxyl end of the growing polypeptide in the P site. As shown in the next diagram, this step removes the polypeptide from the tRNA in the P site and attaches it to the amino acid on the tRNA in the A site.

3 Translocation. The ribosome translocates the tRNA in the A site to the P site. At the same time, the empty tRNA in the P site is moved to the E site, where it is released. The mRNA moves along with its bound tRNAs, bringing the next codon to be translated into the A site.

Figure 17.21 The termination of translation. Like elongation, termination requires GTP hydrolysis as well as additional protein factors, which are not shown here.

Animation: Termination of Translation

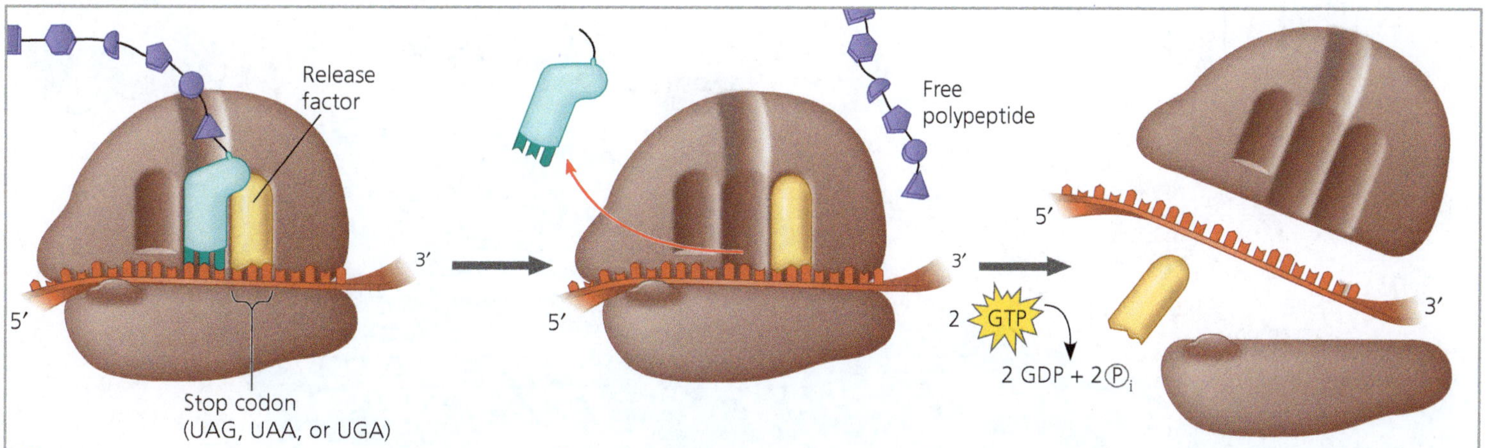

1 When a ribosome reaches a stop codon on mRNA, the A site of the ribosome accepts a "release factor," a protein shaped like a tRNA, instead of an aminoacyl tRNA.

2 The release factor promotes hydrolysis of the bond between the tRNA in the P site and the last amino acid of the polypeptide, thus freeing the polypeptide from the ribosome.

3 The two ribosomal subunits and the other components of the assembly dissociate with the hydrolysis of two GTP molecules.

Additional steps—*post-translational modifications*—may be required before the protein can begin doing its particular job in the cell. Certain amino acids may be chemically modified by the attachment of sugars, lipids, phosphate groups, or other additions. Enzymes may remove one or more amino acids from the leading (amino) end of the polypeptide chain. In some cases, a polypeptide chain may be enzymatically cleaved into two or more pieces. In other cases, two or more polypeptides that are synthesized separately may come together, if the protein has quaternary structure; an example is hemoglobin (see Figure 5.18).

BioFlix® Animation: Protein Processing

Targeting Polypeptides to Specific Locations

In electron micrographs of eukaryotic cells active in protein synthesis, two populations of ribosomes are evident: free and bound (see Figure 6.10). Free ribosomes are suspended in the cytosol and mostly synthesize proteins that stay in the cytosol and function there. In contrast, bound ribosomes are attached to the cytosolic side of the endoplasmic reticulum (ER) or to the nuclear envelope. Bound ribosomes make proteins of the endomembrane system (see Figure 6.15) as well as proteins secreted from the cell, such as insulin. It is important to note that the ribosomes themselves are identical and can alternate between being free ribosomes one time they are used and being bound ribosomes the next.

What determines whether a ribosome is free in the cytosol or bound to rough ER? Polypeptide synthesis always begins in the cytosol as a free ribosome starts to translate an mRNA molecule. There, the process continues to completion—*unless* the growing polypeptide itself cues the ribosome to attach to the ER. The polypeptides of proteins destined for the endomembrane system or for secretion are marked by a **signal peptide**, which targets the protein to the ER **(Figure 17.22)**. The signal peptide, a sequence of about 20 amino acids at or near the leading end (N-terminus) of the polypeptide, is recognized as it emerges from the ribosome by a protein-RNA complex called a **signal-recognition particle (SRP)**. This particle escorts the ribosome to a receptor protein built into the ER membrane. The receptor is part of a multiprotein translocation complex. Polypeptide synthesis continues there, and the growing polypeptide snakes across the membrane into the ER lumen via a protein pore. The rest of the completed polypeptide, if it is to be secreted from the cell, is released into solution within the ER lumen. Alternatively, if the polypeptide is to be a membrane protein, it remains partially embedded in the ER membrane. In either case, it travels in a transport vesicle to its destination (see, for example, Figure 7.9).

Other kinds of signal peptides are used to target polypeptides to mitochondria, chloroplasts, the interior of the nucleus, and other organelles that are not part of the endomembrane system. The critical difference in these cases is that translation is completed in the cytosol before the polypeptide is imported

▼ Figure 17.22 The signal mechanism for targeting proteins to the ER.

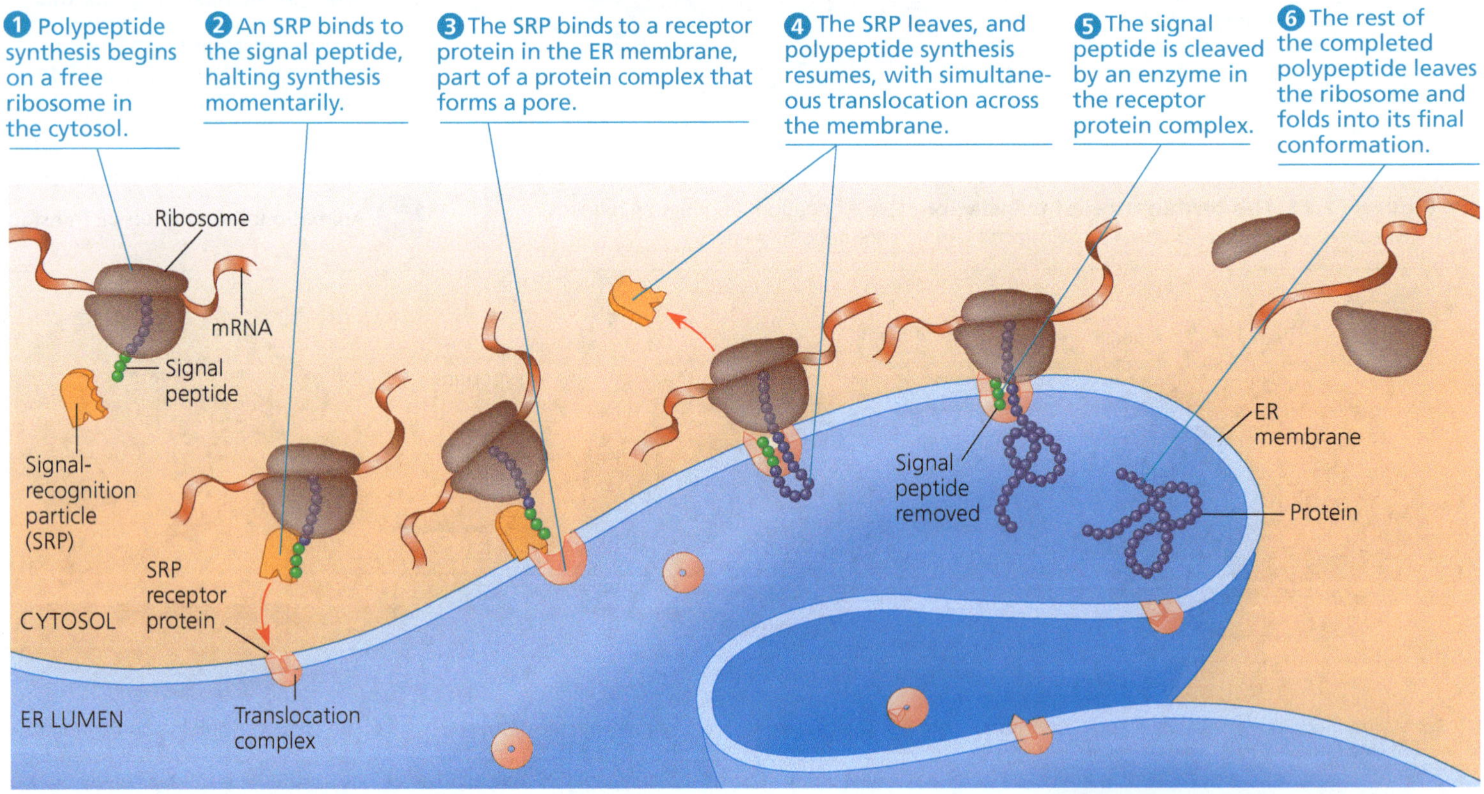

MAKE CONNECTIONS ➤ *If this protein were destined for secretion, what would happen to it after its synthesis was completed? See Figure 7.9.*

Video: Cotranslational Translocation

into the organelle. Translocation mechanisms also vary, but in all cases studied to date, the "postal zip codes" that address proteins for secretion or to cellular locations are signal peptides of some sort. Bacteria also employ signal peptides to target proteins to the plasma membrane for secretion.

Making Multiple Polypeptides in Bacteria and Eukaryotes

In previous sections, you learned how a single polypeptide is synthesized using the information encoded in an mRNA molecule. When a polypeptide is required in a cell, though, the need is for many copies, not just one.

A single ribosome can make an average-sized polypeptide in less than a minute. In both bacteria and eukaryotes, however, multiple ribosomes translate an mRNA at the same time **(Figure 17.23)**; that is, a single mRNA is used to make many copies of a polypeptide simultaneously. Once a ribosome is far enough past the start codon, a second ribosome can attach to the mRNA, eventually resulting in a number of ribosomes trailing along the mRNA. Such strings of ribosomes, called **polyribosomes** (or **polysomes**), can be seen with an electron microscope; they can be either free or bound. They enable a cell to rapidly make many copies of a polypeptide.

Another way both bacteria and eukaryotes augment the number of copies of a polypeptide is by transcribing multiple mRNAs from the same gene. However, the coordination of the two processes—transcription and translation—differs in the two groups. The most important differences between bacteria and eukaryotes arise from the bacterial cell's lack of compartmental organization. Like a one-room workshop, a bacterial cell ensures a streamlined operation by coupling the two processes. In the absence of a nucleus, it can simultaneously transcribe and translate the same gene **(Figure 17.24)**, and the newly made protein can quickly diffuse to its site of function.

In contrast, the eukaryotic cell's nuclear envelope segregates transcription from translation and provides a compartment for extensive RNA processing. This processing stage includes additional steps, discussed earlier, the regulation of which can help coordinate the eukaryotic cell's elaborate activities. **Figure 17.25** summarizes the path from gene to polypeptide in a eukaryotic cell.

▼ Figure 17.23 Polyribosomes.

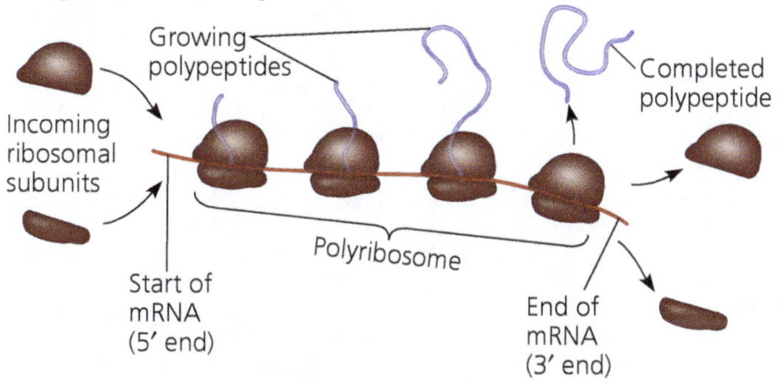

(a) An mRNA molecule is generally translated simultaneously by several ribosomes in clusters called polyribosomes.

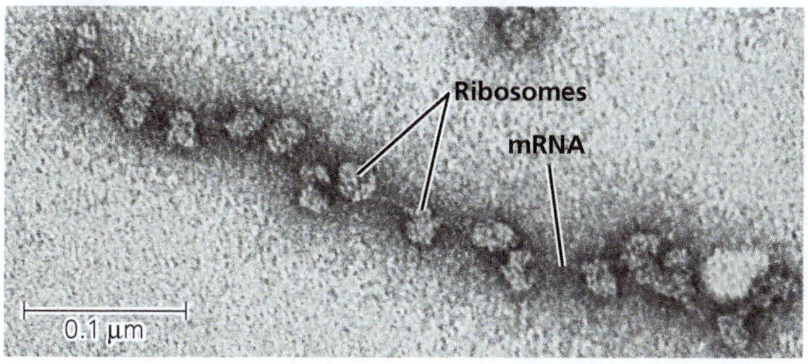

(b) This micrograph shows a large polyribosome in a bacterial cell. Growing polypeptides are not visible here (TEM).

▼ Figure 17.24 Coupled transcription and translation in bacteria. In bacterial cells, the translation of mRNA can begin as soon as the leading (5′) end of the mRNA molecule peels away from the DNA template. The micrograph (TEM) shows a strand of *E. coli* DNA being transcribed by RNA polymerase molecules. Attached to each RNA polymerase molecule is a growing strand of mRNA, which is already being translated by ribosomes. The newly synthesized polypeptides are not visible in the micrograph but are shown in the diagram.

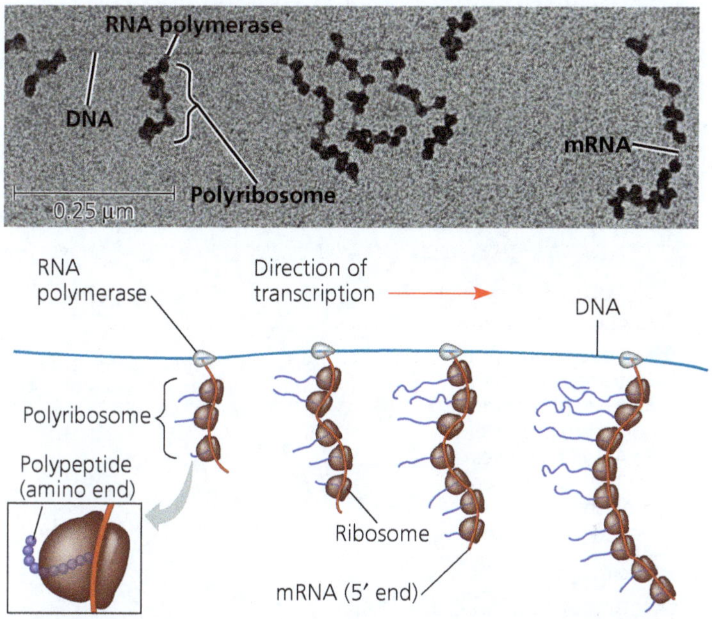

VISUAL SKILLS ➤ *Which one of the mRNA molecules started being transcribed first? On that mRNA, which ribosome started translating the mRNA first?*

CONCEPT CHECK 17.4

1. What two processes ensure that the correct amino acid is added to a growing polypeptide chain?
2. Describe how a polypeptide to be secreted reaches the endomembrane system.
3. **WHAT IF? DRAW IT ➤** Draw a tRNA with the anticodon 3′-CGU-5′. What two different codons could it bind to? Draw each codon on an mRNA, labeling all 5′ and 3′ ends, the tRNA, and the amino acid it carries.
4. **WHAT IF? ➤** In eukaryotic cells, mRNAs have been found to have a circular arrangement in which proteins hold the poly-A tail near the 5′ cap. How might this increase translation efficiency?

For suggested answers, see Appendix A.

Figure 17.25 A summary of transcription and translation in a eukaryotic cell. This diagram shows the path from one gene to one polypeptide. Each gene in the DNA can be transcribed repeatedly into many identical RNA molecules and each mRNA can be translated repeatedly to yield many identical polypeptide molecules. (Also, remember that the final products of some genes are not polypeptides but RNA molecules that don't get translated, including tRNA and rRNA.) In general, the steps of transcription and translation are similar in bacterial, archaeal, and eukaryotic cells. The major difference is the occurrence of RNA processing in the eukaryotic nucleus. Other significant differences are found in the initiation stages of both transcription and translation and in the termination of transcription. To visualize these processes in their cellular context, see Figure 6.32.

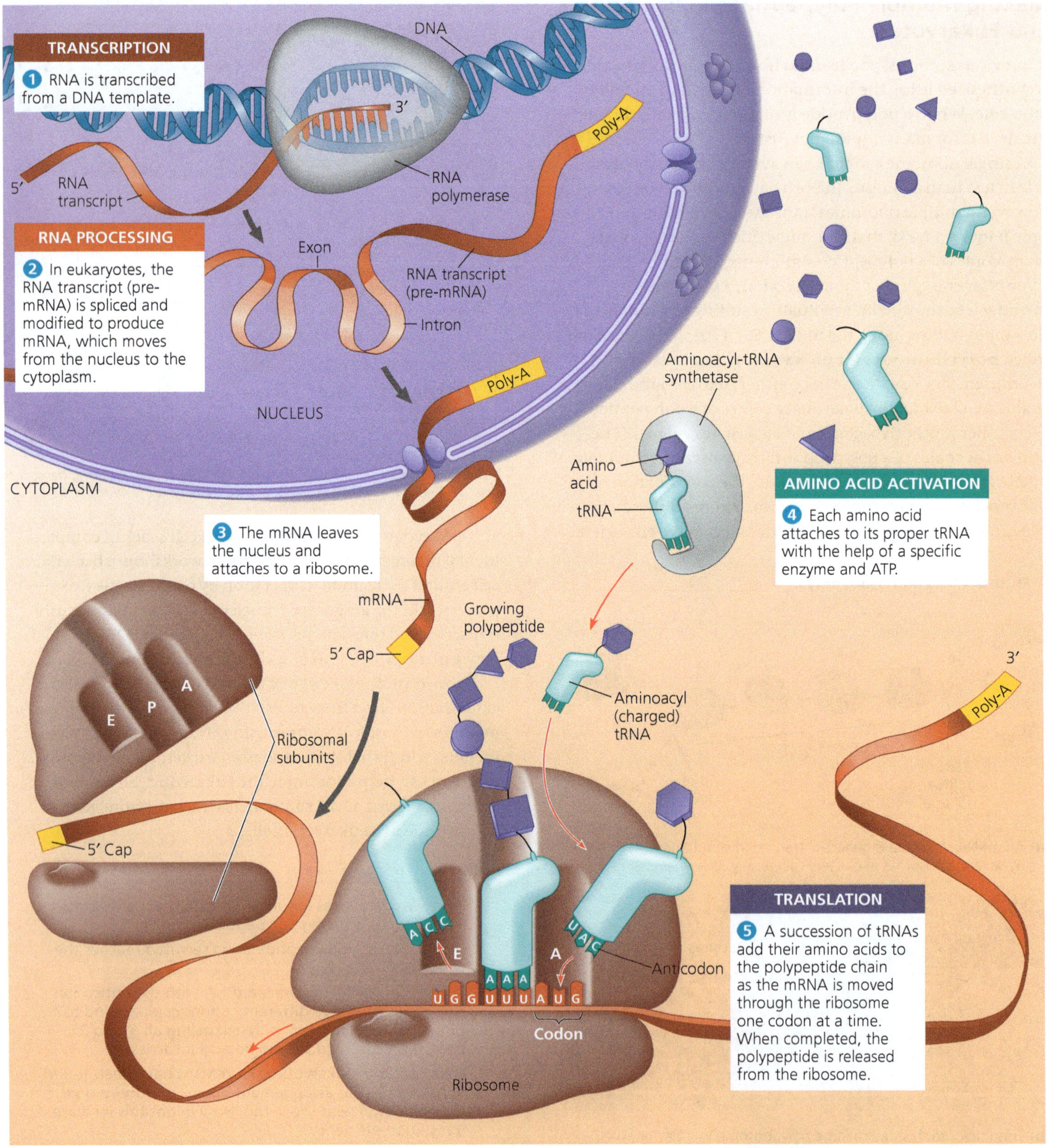

BioFlix® Animation: Protein Synthesis

CONCEPT 17.5

Mutations of one or a few nucleotides can affect protein structure and function

Now that you have explored the process of gene expression, you are ready to understand the effects of changes to the genetic information of a cell. These changes, called **mutations**, are responsible for the huge diversity of genes found among organisms because mutations are the ultimate source of new genes. Earlier, we considered chromosomal rearrangements that affect long segments of DNA (see Figure 15.14); these are considered large-scale mutations. Here we examine small-scale mutations of one or a few nucleotide pairs, including **point mutations**, changes in a single nucleotide pair of a gene.

If a point mutation occurs in a gamete or in a cell that gives rise to gametes, it may be transmitted to offspring and to future generations. If the mutation has an adverse effect on the phenotype of a person, the mutant condition is referred to as a genetic disorder or hereditary disease. For example, we can trace the genetic basis of sickle-cell disease to the mutation of a single nucleotide pair in the gene that encodes the β-globin polypeptide of hemoglobin. The change of a single nucleotide in the DNA's template strand leads to an altered mRNA and the production of an abnormal protein (**Figure 17.26**; also see Figure 5.19). In individuals who are homozygous for the mutant allele, the sickling of red blood cells caused by the altered hemoglobin produces the multiple symptoms associated with sickle-cell disease (see Concept 14.4 and Figure 23.18). Another disorder caused by a point mutation is a heart condition called familial cardiomyopathy, which is responsible for some of the tragic incidents of sudden death in young athletes. Point mutations in several genes encoding muscle proteins have been identified, any of which can lead to this disorder.

▼ Figure 17.26 The molecular basis of sickle-cell disease: a point mutation. The allele that causes sickle-cell disease differs from the wild-type (normal) allele by a single DNA nucleotide pair. The micrographs are SEMs of a normal red blood cell (on the left) and a sickled red blood cell (right) from individuals homozygous for wild-type and mutant alleles, respectively.

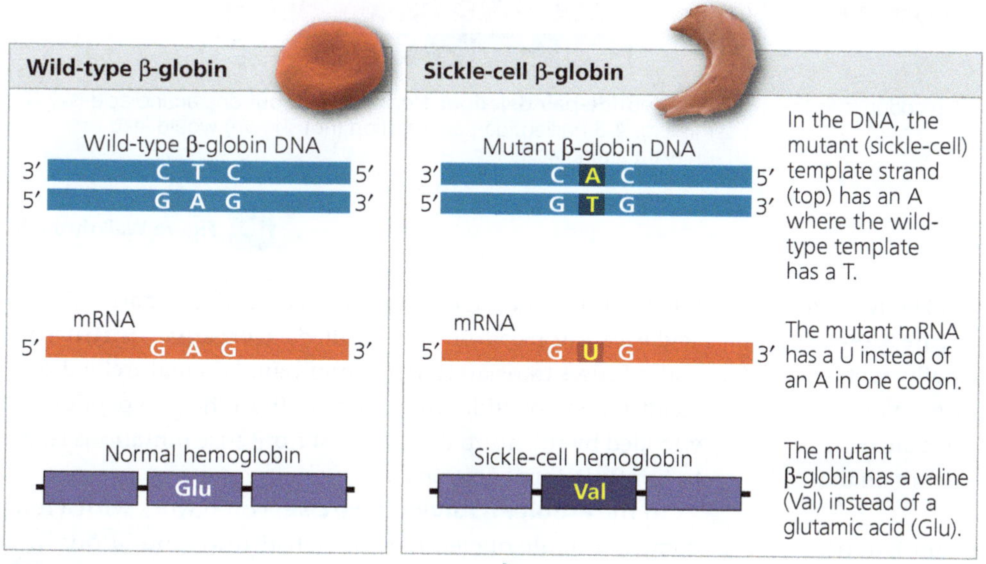

MB HHMI Animation: Sickle-Cell Disease
Animation: Mutation Types

Types of Small-Scale Mutations

Let's now consider how small-scale mutations affect proteins. We should first note that many mutations occur in regions outside of protein-coding genes, and any potential effect they have on the phenotype of the organism may be subtle and hard to detect. For this reason, here we'll concentrate on mutations within protein-coding genes. Small-scale mutations within a gene can be divided into two general categories: (1) single nucleotide-pair substitutions and (2) nucleotide-pair insertions or deletions. Insertions and deletions can involve one or more nucleotide pairs.

Substitutions

A **nucleotide-pair substitution** is the replacement of one nucleotide and its partner with another pair of nucleotides (**Figure 17.27a**). Some substitutions have no effect on the encoded protein, owing to the redundancy of the genetic code. For example, if 3′-CCG-5′ on the template strand mutated to 3′-CCA-5′, the mRNA codon that used to be GGC would become GGU, but a glycine would still be inserted at the proper location in the protein (see Figure 17.6). In other words, a change in a nucleotide pair may transform one codon into another that is translated into the same amino acid. Such a change is an example of a **silent mutation**, which has no observable effect on the phenotype. (Silent mutations can occur outside genes as well.) Interestingly, there is evidence that some silent mutations may indirectly affect where or at what level the gene gets expressed, even though the actual protein is the same.

Substitutions that change one amino acid to another one are called **missense mutations**. Such a mutation may have little effect on the protein: The new amino acid may have properties similar to those of the amino acid it replaces, or it may be in a region of the protein where the exact sequence of amino acids is not essential to the protein's function.

However, the nucleotide-pair substitutions of greatest interest are those that cause a major change in a protein. The alteration of a single amino acid in a crucial area of a protein—such as in the part

Figure 17.27 Types of small-scale mutations that affect mRNA sequence. All but one of the types shown here also affect the amino acid sequence of the encoded polypeptide.

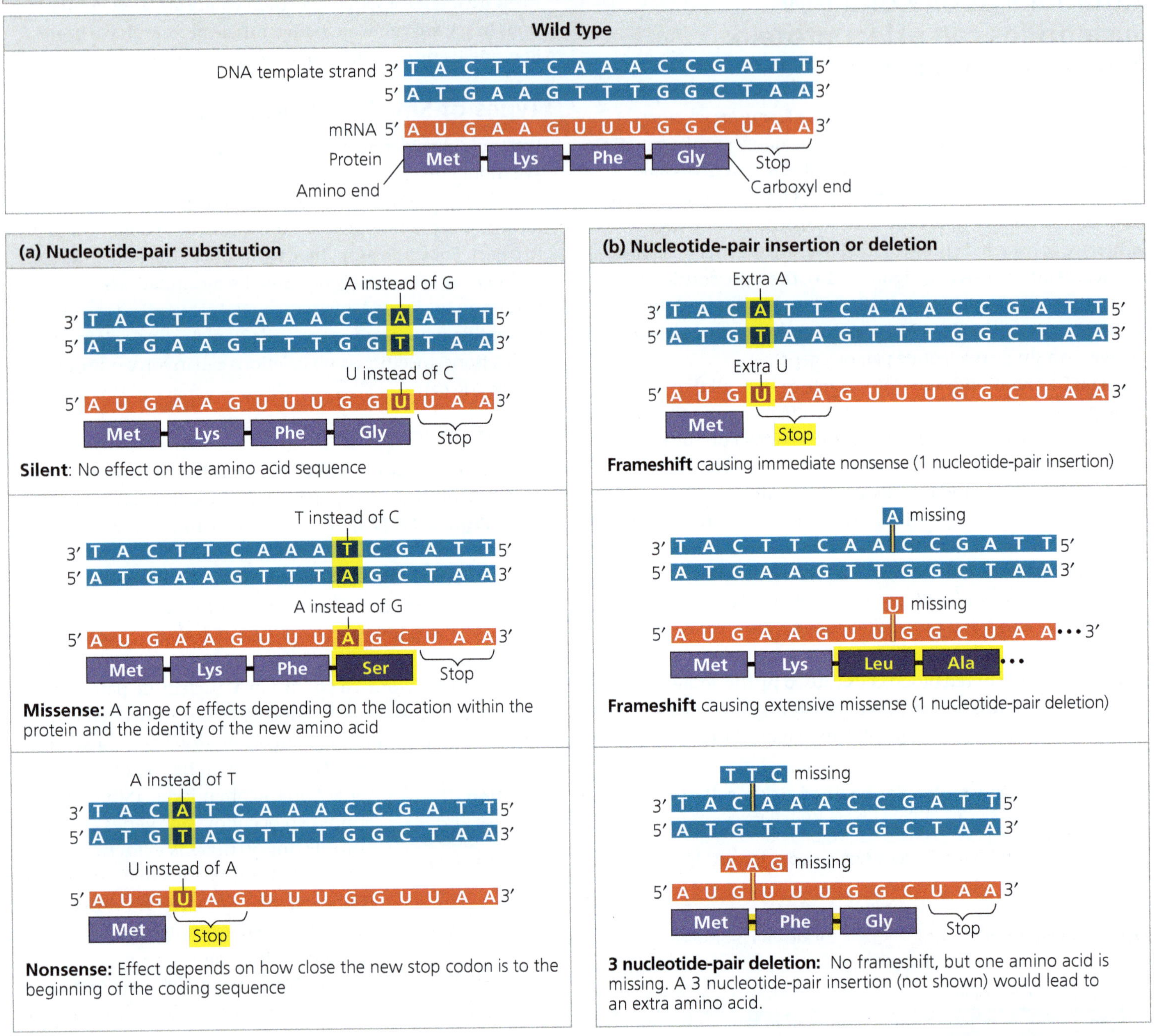

Figure Walkthrough

of the β-globin subunit of hemoglobin shown in Figure 17.26 or in the active site of an enzyme as shown in Figure 8.19—can significantly alter protein activity. Occasionally, such a mutation leads to an improved protein or one with novel capabilities, but much more often such mutations are neutral or detrimental, leading to a useless or less active protein that impairs cellular function.

Substitution mutations are usually missense mutations; that is, the altered codon still codes for an amino acid and thus makes sense, although not necessarily the *right* sense. But a point mutation can also change a codon for an amino acid into a stop codon. This is called a **nonsense mutation**, and it causes translation to be terminated prematurely; the resulting polypeptide will be shorter than the polypeptide encoded by the normal gene. Most nonsense mutations lead to nonfunctional proteins.

In the **Problem-Solving Exercise**, you'll work with a few common single nucleotide-pair substitution mutations in the gene encoding insulin, some or all of which may lead to diabetes. You will classify these mutations into one of

PROBLEM-SOLVING EXERCISE

Are insulin mutations the cause of three infants' neonatal diabetes?

Insulin is a hormone that acts as a key regulator of blood glucose level. In some cases of neonatal diabetes, the gene coding for the insulin protein has a nucleotide-pair substitution mutation that alters the protein structure enough to cause it to malfunction. How can you identify a nucleotide-pair substitution and determine its effect on the amino acid sequence?

Now that it's possible to sequence an individual's whole genome, doctors can use that DNA sequence information to diagnose diseases and identify new treatments. For example, the insulin gene sequence of a patient with neonatal diabetes can be analyzed to determine if it has a mutation and, if so, its effect.

Watch the video in the Mastering-Biology Study Area to see how genome sequencing is changing medicine.

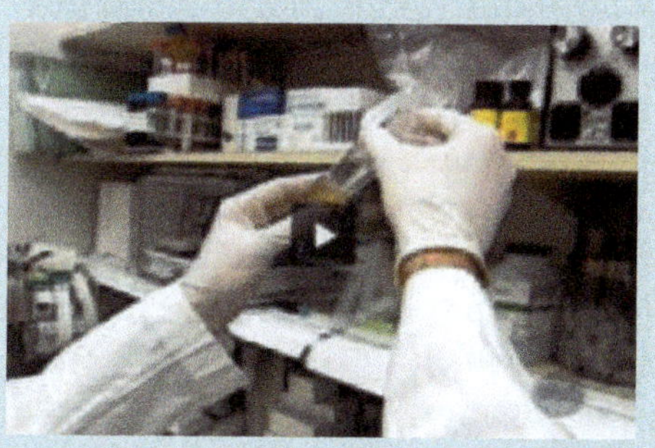

MB ABC News Video: Using Genome Sequencing to Diagnose Gene-Based Diseases

In this exercise, you will determine the effect of mutations present in a portion of diabetes patients' insulin gene sequences.

Your Approach Suppose you are a medical geneticist presented with three infant patients, all of whom have a nucleotide-pair substitution in their insulin gene. It is your job to analyze each mutation to figure out the effect of the mutation on the amino acid sequence of the insulin protein. To identify the mutation in each patient, you will compare his or her individual insulin complementary DNA (cDNA) sequence to that of the wild-type cDNA. (cDNA is a double-stranded DNA molecule that is based on the mRNA sequence and thus contains only the portion of a gene that is translated—introns are not included. cDNA sequences are commonly used to compare the coding regions of genes.) Identifying the codons that have been changed will tell you which, if any, amino acids are altered in the patient's insulin protein.

Your Data You will analyze the cDNA codons for amino acids 35–54 (of the 110 amino acids) of each patient's insulin protein, so the start codon (AUG) is not present. The sequences of the wild-type cDNA and the patients' cDNA are shown below, arranged in codons.

Wild-type cDNA	5′-CTG GTG GAA GCT CTC TAC CTA GTG TGC GGG GAA CGA GGC TTC TTC TAC ACA CCC AAG ACC-3′
Patient 1 cDNA	5′-CTG GTG GAA GCT CTC TAC CTA GTG TGC GGG GAA CGA GGC TGC TTC TAC ACA CCC AAG ACC-3′
Patient 2 cDNA	5′-CTG GTG GAA GCT CTC TAC CTA GTG TGC GGG GAA CGA GGC TCC TTC TAC ACA CCC AAG ACC-3′
Patient 3 cDNA	5′-CTG GTG GAA GCT CTC TAC CTA GTG TGC GGG GAA CGA GGC TTC TCG TAC ACA CCC AAG ACC-3′

Data from N. Nishi and K. Nanjo, Insulin gene mutations and diabetes, *Journal of Diabetes Investigation* 2:92–100 (2011).

Your Analysis

1. Comparing each patient's cDNA sequence to the wild-type cDNA sequence, circle the codons where a nucleotide-pair substitution mutation has occurred.
2. Use a codon table (see Figure 17.6) to identify the amino acid that will be made by the codon with the mutation in each patient's insulin sequence, and compare it to the amino acid made by the codon in the corresponding wild-type sequence. As is standard practice with DNA sequences, the cDNA *coding* (nontemplate) strand has been provided, so to convert it to mRNA for use with the codon table, you just need to change T to U. Classify each patient's nucleotide-pair substitution mutation: Is it a silent, missense, or nonsense mutation? Explain, for each answer.
3. Compare the structure of the amino acid you identified in each patient's insulin sequence to that of the corresponding amino acid in the wild-type insulin sequence (see Figure 5.14). Given that each patient has neonatal diabetes, discuss how the change of amino acid in each might have affected the insulin protein and thus resulted in the disease.

MB **Instructors**: A version of this Problem-Solving Exercise can be assigned in MasteringBiology. Or a more extensive investigation called "Solve It: Which Insulin Mutations May Result in Disease?" can be assigned.

the types we just described and characterize the change in amino acid sequence.

Insertions and Deletions

Insertions and **deletions** are additions or losses of nucleotide pairs in a gene **(Figure 17.27b)**. These mutations have a disastrous effect on the resulting protein more often than substitutions do. Insertion or deletion of nucleotides may alter the reading frame of the genetic message, the triplet grouping of nucleotides on the mRNA that is read during translation. Such a mutation, called a **frameshift mutation**, occurs whenever the number of nucleotides inserted or deleted is not a multiple of three. All nucleotides downstream of the deletion or insertion will be improperly grouped into codons; the result will be extensive missense mutations, usually ending sooner or later in a nonsense mutation that leads to premature termination. Unless the frameshift is very near the end of the gene, the protein is almost certain to be nonfunctional. Insertions and deletions also occur outside of coding regions; these are not called frameshift mutations, but can have effects on the phenotype—for instance, they can affect how a gene is expressed.

New Mutations and Mutagens

Mutations can arise in a number of ways. Errors during DNA replication or recombination can lead to nucleotide-pair substitutions, insertions, or deletions, as well as to mutations affecting longer stretches of DNA. If an incorrect nucleotide is added to a growing chain during replication, for example, the base on that nucleotide will then be mismatched with the nucleotide base on the other strand. In many cases, the error will be corrected by DNA proofreading and repair systems (see Concept 16.2). Otherwise, the incorrect base will be used as a template in the next round of replication, resulting in a mutation. Such mutations are called *spontaneous mutations*. It is difficult to calculate the rate at which such mutations occur. Rough estimates have been made of the rate of mutation during DNA replication for both *E. coli* and eukaryotes, and the numbers are similar: About one nucleotide in every 10^{10} is altered, and the change is passed on to the next generation of cells.

A number of physical and chemical agents, called **mutagens**, interact with DNA in ways that cause mutations. In the 1920s, Hermann Muller discovered that X-rays caused genetic changes in fruit flies, and he used X-rays to make *Drosophila* mutants for his genetic studies. But he also recognized an alarming implication of his discovery: X-rays and other forms of high-energy radiation pose hazards to the genetic material of people as well as laboratory organisms. Mutagenic radiation, a physical mutagen, includes ultraviolet (UV) light, which can cause disruptive thymine dimers in DNA (see Figure 16.19).

Chemical mutagens fall into several categories. Nucleotide analogs are chemicals similar to normal DNA nucleotides but that pair incorrectly during DNA replication. Other chemical mutagens interfere with correct DNA replication by inserting themselves into the DNA and distorting the double helix. Still other mutagens cause chemical changes in bases that change their pairing properties.

Researchers have developed a variety of methods to test the mutagenic activity of chemicals. A major application of these tests is the preliminary screening of chemicals to identify those that may cause cancer. This approach makes sense because most carcinogens (cancer-causing chemicals) are mutagenic, and conversely, most mutagens are carcinogenic.

What Is a Gene? *Revisiting the Question*

Our definition of a gene has evolved over the past few chapters, as it has through the history of genetics. We began with the Mendelian concept of a gene as a discrete unit of inheritance that affects a phenotypic character (Chapter 14). We saw that Morgan and his colleagues assigned such genes to specific loci on chromosomes (Chapter 15). We went on to view a gene as a region of specific nucleotide sequence along the length of the DNA molecule of a chromosome (Chapter 16). Finally, in this chapter, we have considered a functional definition of a gene as a DNA sequence that codes for a specific polypeptide chain or a functional RNA molecule, such as a tRNA. All these definitions are useful, depending on the context in which genes are being studied.

We have noted that merely saying a gene codes for a polypeptide is an oversimplification. Most eukaryotic genes contain noncoding segments (such as introns), so large portions of these genes have no corresponding segments in polypeptides. Molecular biologists also often include promoters and certain other regulatory regions of DNA within the boundaries of a gene. These DNA sequences are not transcribed, but they can be considered part of the functional gene because they must be present for transcription to occur. Our definition of a gene must also be broad enough to include the DNA that is transcribed into rRNA, tRNA, and other RNAs that are not translated. These genes have no polypeptide products but play crucial roles in the cell. Thus, we arrive at the following definition: *A gene is a region of DNA that can be expressed to produce a final functional product that is either a polypeptide or an RNA molecule.*

When considering phenotypes, however, it is often useful to start by focusing on genes that code for polypeptides. In this chapter, you have learned in molecular terms how a typical gene is expressed—by transcription into RNA and then translation into a polypeptide that forms a protein of specific structure and function. Proteins, in turn, bring about an organism's observable phenotype.

A given type of cell expresses only a subset of its genes. This is an essential feature in multicellular organisms: You'd be in trouble if the lens cells in your eyes started expressing the genes for hair proteins, which are normally expressed only in hair follicle cells! Gene expression is precisely regulated, which we'll explore in the next chapter, beginning with the simpler case of bacteria and continuing with eukaryotes.

CONCEPT CHECK 17.5

1. What happens when one nucleotide pair is lost from the middle of the coding sequence of a gene?
2. **MAKE CONNECTIONS** ➤ Individuals heterozygous for the sickle-cell allele are generally healthy but show phenotypic effects of the allele under some circumstances (see Figure 14.17). Explain in terms of gene expression.
3. **WHAT IF? DRAW IT** ➤ The template strand of a gene includes this sequence:
   3′-TACTTGTCCGATATC-5′. It is mutated to
   3′-TACTTGTCCAATATC-5′. For both wild-type and mutant sequences, draw the double-stranded DNA, the resulting mRNA, and the amino acid sequence each encodes. What is the effect of the mutation on the amino acid sequence?

For suggested answers, see Appendix A.

17 Chapter Review

Go to **MasteringBiology**™ for Videos, Animations, Vocab Self-Quiz, Practice Tests, and more in the Study Area.

SUMMARY OF KEY CONCEPTS

CONCEPT 17.1

Genes specify proteins via transcription and translation *(pp. 336–342)*

VOCAB SELF-QUIZ goo.gl/6u55ks

- Beadle and Tatum's studies of mutant strains of *Neurospora* led to the one gene–one polypeptide hypothesis. During **gene expression**, the information encoded in genes is used to make specific polypeptide chains (enzymes and other proteins) or RNA molecules.
- **Transcription** is the synthesis of RNA complementary to a **template strand** of DNA. **Translation** is the synthesis of a polypeptide whose amino acid sequence is specified by the nucleotide sequence in **messenger RNA (mRNA)**.
- Genetic information is encoded as a sequence of nonoverlapping nucleotide triplets, or **codons**. A codon in mRNA either is translated into an amino acid (61 of the 64 codons) or serves as a stop signal (3 codons). Codons must be read in the correct **reading frame**.

? *Describe the process of gene expression, by which a gene affects the phenotype of an organism.*

CONCEPT 17.2

Transcription is the DNA-directed synthesis of RNA: *a closer look* *(pp. 342–344)*

- RNA synthesis is catalyzed by **RNA polymerase**, which links together RNA nucleotides complementary to a DNA template strand. This process follows the same base-pairing rules as DNA replication, except that in RNA, uracil substitutes for thymine.

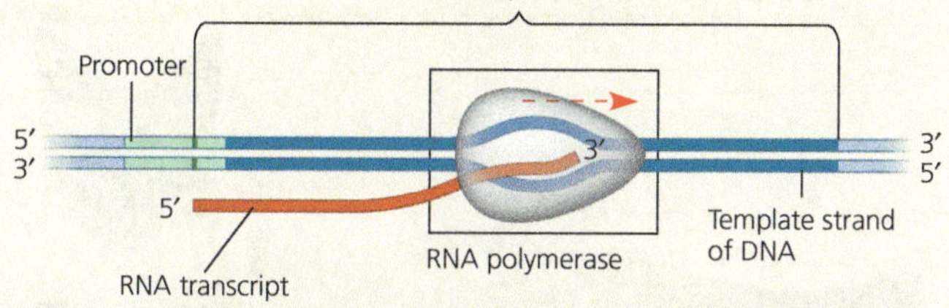

- The three stages of transcription are initiation, elongation, and termination. A **promoter**, often including a **TATA box** in eukaryotes, establishes where RNA synthesis is initiated. **Transcription factors** help eukaryotic RNA polymerase recognize promoter sequences, forming a **transcription initiation complex**. Termination differs in bacteria and eukaryotes.

? *What are the similarities and differences in the initiation of gene transcription in bacteria and eukaryotes?*

CONCEPT 17.3

Eukaryotic cells modify RNA after transcription *(pp. 345–347)*

- Eukaryotic mRNAs undergo **RNA processing**, which includes RNA splicing, the addition of a modified nucleotide **5′ cap** to the 5′ end, and the addition of a **poly-A tail** to the 3′ end. The processed mRNA includes an untranslated region (5′ UTR or 3′ UTR) at each end of the coding segment.
- Most eukaryotic genes are split into segments: They have **introns** interspersed among the **exons** (the regions included in the mRNA). In **RNA splicing**, introns are removed and exons joined. RNA splicing is typically carried out by **spliceosomes**, but in some cases, RNA alone catalyzes its own splicing. The catalytic ability of some RNA molecules, called **ribozymes**, derives from the inherent properties of RNA. The presence of introns allows for **alternative RNA splicing**.

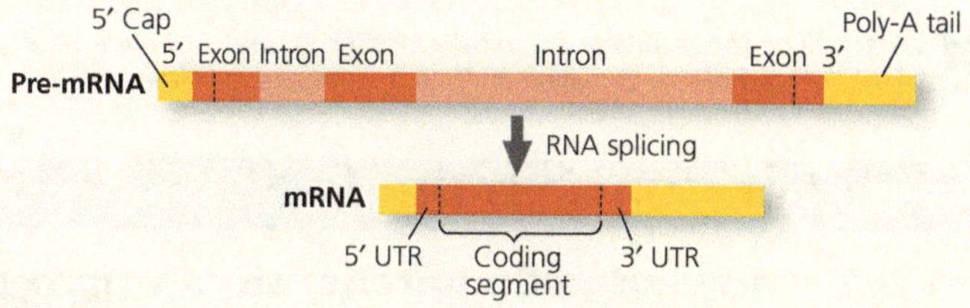

? *What function do the 5′ cap and the poly-A tail serve on a eukaryotic mRNA?*

CONCEPT 17.4

Translation is the RNA-directed synthesis of a polypeptide: *a closer look* *(pp. 347–356)*

- A cell translates an mRNA message into protein using **transfer RNAs (tRNAs)**. After being bound to a specific amino acid by an **aminoacyl-tRNA synthetase**, a tRNA lines up via its **anticodon** at the complementary codon on mRNA. A **ribosome**, made up of **ribosomal RNAs (rRNAs)** and proteins, facilitates this coupling with binding sites for mRNA and tRNA.
- Ribosomes coordinate the three stages of translation: initiation, elongation, and termination. The formation of peptide bonds between amino acids is catalyzed by rRNAs as tRNAs move through the **A** and **P sites** and exit through the **E site**.

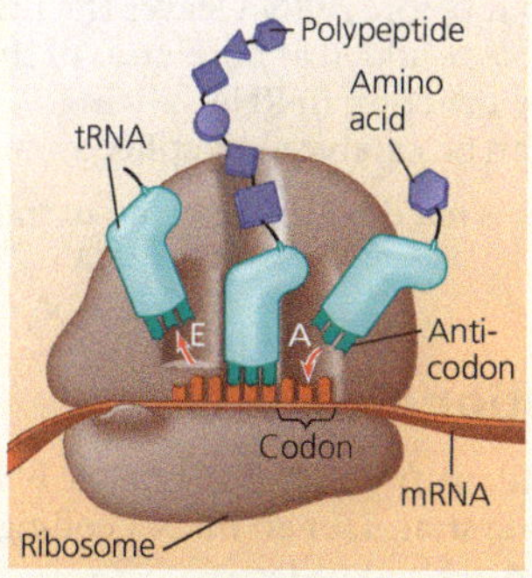

- After translation, during protein processing, proteins may be modified by cleavage or by attachment of sugars, lipids, phosphates, or other chemical groups.
- Free ribosomes in the cytosol initiate synthesis of all proteins, but proteins with a **signal peptide** are synthesized on the ER.
- A gene can be transcribed by multiple RNA polymerases simultaneously. Also, a single mRNA molecule can be translated simultaneously by a number of ribosomes, forming a **polyribosome**. In bacteria, these processes are coupled, but in eukaryotes they are separated in space and time by the nuclear membrane.

? *Describe how tRNAs function in the context of the ribosome in building a polypeptide.*

CONCEPT 17.5

Mutations of one or a few nucleotides can affect protein structure and function *(pp. 357–360)*

- Small-scale **mutations** include **point mutations**, changes in one DNA nucleotide pair, which may lead to production of non-functional proteins. **Nucleotide-pair substitutions** can cause **missense** or **nonsense mutations**. Nucleotide-pair **insertions** or **deletions** may produce **frameshift mutations**.
- Spontaneous mutations can occur during DNA replication and recombination. Chemical and physical **mutagens** cause DNA damage that can alter genes.

? *What will be the results of chemically modifying one nucleotide base of a gene? What role is played by DNA repair systems in the cell?*

TEST YOUR UNDERSTANDING

Level 1: Knowledge/Comprehension

PRACTICE TEST
goo.gl/CUYGKD

1. In eukaryotic cells, transcription cannot begin until
   (A) the two DNA strands have completely separated and exposed the promoter.
   (B) several transcription factors have bound to the promoter.
   (C) the 5′ caps are removed from the mRNA.
   (D) the DNA introns are removed from the template.
2. Which of the following is *not* true of a codon?
   (A) It may code for the same amino acid as another codon.
   (B) It never codes for more than one amino acid.
   (C) It extends from one end of a tRNA molecule.
   (D) It is the basic unit of the genetic code.
3. The anticodon of a particular tRNA molecule is
   (A) complementary to the corresponding mRNA codon.
   (B) complementary to the corresponding triplet in rRNA.
   (C) the part of tRNA that bonds to a specific amino acid.
   (D) catalytic, making the tRNA a ribozyme.
4. Which of the following is *not* true of RNA processing?
   (A) Exons are cut out before mRNA leaves the nucleus.
   (B) Nucleotides may be added at both ends of the RNA.
   (C) Ribozymes may function in RNA splicing.
   (D) RNA splicing can be catalyzed by spliceosomes.
5. Which component is *not* directly involved in translation?
   (A) GTP
   (B) DNA
   (C) tRNA
   (D) ribosomes

Level 2: Application/Analysis

6. Using Figure 17.6, identify a 5′ → 3′ sequence of nucleotides in the DNA template strand for an mRNA coding for the polypeptide sequence Phe-Pro-Lys.
   (A) 5′-UUUCCCAAA-3′
   (B) 5′-GAACCCCTT-3′
   (C) 5′-CTTCGGGAA-3′
   (D) 5′-AAACCCUUU-3′
7. Which of the following mutations would be *most* likely to have a harmful effect on an organism?
   (A) a deletion of three nucleotides near the middle of a gene
   (B) a single nucleotide deletion in the middle of an intron
   (C) a single nucleotide deletion near the end of the coding sequence
   (D) a single nucleotide insertion downstream of, and close to, the start of the coding sequence
8. Would the coupling of the processes shown in Figure 17.24 be found in a eukaryotic cell? Explain why or why not.
9. Complete the following table:

Type of RNA	Functions
Messenger RNA (mRNA)	
Transfer RNA (tRNA)	
	In a ribosome, plays a structural role; as a ribozyme, plays a catalytic role (catalyzes peptide bond formation)
Primary transcript	
Small RNAs in the spliceosome	

Level 3: Synthesis/Evaluation

10. **EVOLUTION CONNECTION** Most amino acids are coded for by a set of similar codons (see Figure 17.6). Propose at least one evolutionary explanation to account for this pattern.
11. **SCIENTIFIC INQUIRY** Knowing that the genetic code is almost universal, a scientist uses molecular biological methods to insert the human β-globin gene (shown in Figure 17.12) into bacterial cells, hoping the cells will express it and synthesize functional β-globin protein. Instead, the protein produced is nonfunctional and is found to contain many fewer amino acids than does β-globin made by a eukaryotic cell. Explain why.
12. **WRITE ABOUT A THEME: INFORMATION** Evolution accounts for the unity and diversity of life, and the continuity of life is based on heritable information in the form of DNA. In a short essay (100–150 words), discuss how the fidelity with which DNA is inherited is related to the processes of evolution. (Review the discussion of proofreading and DNA repair in Concept 16.2.)
13. **SYNTHESIZE YOUR KNOWLEDGE**

Some mutations result in proteins that function well at one temperature but are nonfunctional at a different (usually higher) temperature. Siamese cats have such a "temperature-sensitive" mutation in a gene encoding an enzyme that makes dark pigment in the fur. The mutation results in the breed's distinctive point markings and lighter body color (see the photo). Using this information and what you learned in the chapter, explain the pattern of the cat's fur pigmentation.

For selected answers, see Appendix A.

For additional practice questions, check out the **Dynamic Study Modules** in MasteringBiology. You can use them to study on your smartphone, tablet, or computer anytime, anywhere!

Regulation of Gene Expression 18

▲ **Figure 18.1** **How can this fish's eyes see equally well in both air and water?**

KEY CONCEPTS

18.1 Bacteria often respond to environmental change by regulating transcription

18.2 Eukaryotic gene expression is regulated at many stages

18.3 Noncoding RNAs play multiple roles in controlling gene expression

18.4 A program of differential gene expression leads to the different cell types in a multicellular organism

18.5 Cancer results from genetic changes that affect cell cycle control

Beauty in the Eye of the Beholder

The fish in **Figure 18.1** is keeping an eye out for predators—or, more precisely, both halves of each eye! *Anableps anableps*, commonly known as "cuatro ojos" ("four eyes"), glides through freshwater lakes and ponds in Central and South America with the upper half of each eye protruding from the water. The eye's upper half is particularly well-suited for aerial vision and the lower half for aquatic vision. The molecular basis of this specialization has recently been revealed: The cells of the two parts of the eye express a slightly different set of genes involved in vision, even though these two groups of cells are quite similar and contain identical genomes. What is the biological mechanism underlying the difference in gene expression that makes this remarkable feat possible?

A hallmark of prokaryotic and eukaryotic cells alike—from a bacterium to the cells of a fish—is their intricate and precise regulation of gene expression. In this chapter, we first explore how bacteria regulate expression of their genes in response to different environmental conditions. We then examine the general mechanisms by which eukaryotes regulate gene expression, including the many roles played by RNA molecules. In the final two sections, we explore the role of gene regulation in both embryonic development, as the ultimate example of proper gene regulation, and cancer, as an illustration of what happens when regulation goes awry. Orchestrating proper gene expression by all cells is crucial to the functions of life.

When you see this blue icon, log in to **MasteringBiology** and go to the Study Area for digital resources.

MB **Get Ready for This Chapter**

CONCEPT 18.1

Bacteria often respond to environmental change by regulating transcription

Bacterial cells that can conserve resources and energy have a selective advantage over cells that are unable to do so. Thus, natural selection has favored bacteria that express only the genes whose products are needed by the cell.

Consider, for instance, an individual *Escherichia coli* (*E. coli*) cell living in the erratic environment of a human colon, dependent for its nutrients on the whimsical eating habits of its host. If the environment is lacking in the amino acid tryptophan, which the bacterium needs to survive, the cell responds by activating a metabolic pathway that makes tryptophan from another compound. If the human host later eats a tryptophan-rich meal, the bacterial cell stops producing tryptophan, thus avoiding wasting resources to produce a substance that is readily available from the surrounding solution.

A metabolic pathway can be controlled on two levels, as shown for the synthesis of tryptophan in **Figure 18.2**. First, cells can adjust the activity of enzymes already present. This is a fairly rapid response, which relies on the sensitivity of many enzymes to chemical cues that increase or decrease their catalytic activity (see Concept 8.5). The activity of the first enzyme in the pathway is inhibited by the pathway's end product—tryptophan, in this case **(Figure 18.2a)**. Thus, if tryptophan accumulates in a cell, it shuts down the synthesis of more tryptophan by inhibiting enzyme activity. Such *feedback inhibition*, typical of anabolic (biosynthetic) pathways, allows a cell to adapt to short-term fluctuations in the supply of a substance it needs (see Figure 8.21).

Second, cells can adjust the production level of certain enzymes; that is, they can regulate the expression of the genes encoding the enzymes. If, in our example, the environment provides all the tryptophan the cell needs, the cell stops making the enzymes that catalyze the synthesis of tryptophan **(Figure 18.2b)**. In this case, the control of enzyme production occurs at the level of transcription, the synthesis of messenger RNA from the genes that code for these enzymes.

Regulation of the tryptophan synthesis pathway is just one example of how bacteria tune their metabolism to changing environments. Many genes of the bacterial genome are switched on or off by changes in the metabolic status of the cell. One basic mechanism for this control of gene expression in bacteria, described as the *operon model*, was discovered in 1961 by François Jacob and Jacques Monod at the Pasteur Institute in Paris. Let's see what an operon is and how it works.

Operons: The Basic Concept

E. coli synthesizes the amino acid tryptophan from a precursor molecule in the three-step pathway shown in Figure 18.2.

Figure 18.2 Regulation of a metabolic pathway. In the pathway for tryptophan synthesis, an abundance of tryptophan can both **(a)** inhibit the activity of the first enzyme in the pathway (feedback inhibition), a rapid response, and **(b)** repress expression of the genes encoding all subunits of the enzymes in the pathway, a longer-term response. Genes *trpE* and *trpD* encode the two subunits of enzyme 1, and genes *trpB* and *trpA* encode the two subunits of enzyme 3. (The genes were named before the order in which they functioned in the pathway was determined.) The symbol ⊖ stands for inhibition.

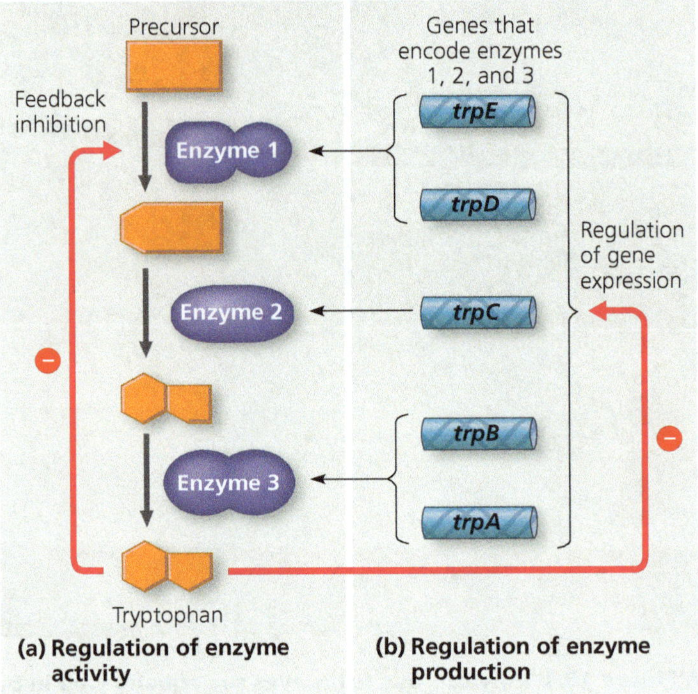

Each reaction in the pathway is catalyzed by a specific enzyme, and the five genes that code for the subunits of these enzymes are clustered together on the bacterial chromosome. A single promoter serves all five genes, which together constitute a transcription unit. (Recall that a promoter is a site where RNA polymerase can bind to DNA and begin transcription; see Figure 17.8.) Thus, transcription gives rise to one long mRNA molecule that codes for the five polypeptides making up the enzymes in the tryptophan pathway **(Figure 18.3a)**. The cell can translate this one mRNA into five separate polypeptides because the mRNA is punctuated with start and stop codons that signal where the coding sequence for each polypeptide begins and ends.

A key advantage of grouping genes of related function into one transcription unit is that a single "on-off switch" can control the whole cluster of functionally related genes; in other words, these genes are *coordinately controlled*. When an *E. coli* cell must make tryptophan for itself because its surrounding environment lacks this amino acid, all the enzymes for the metabolic pathway are synthesized at the same time. The on-off switch is a segment of DNA called an **operator**. Both its location and name suit its function: Positioned within the promoter or, in some cases, between the promoter and the enzyme-coding genes, the operator controls the access

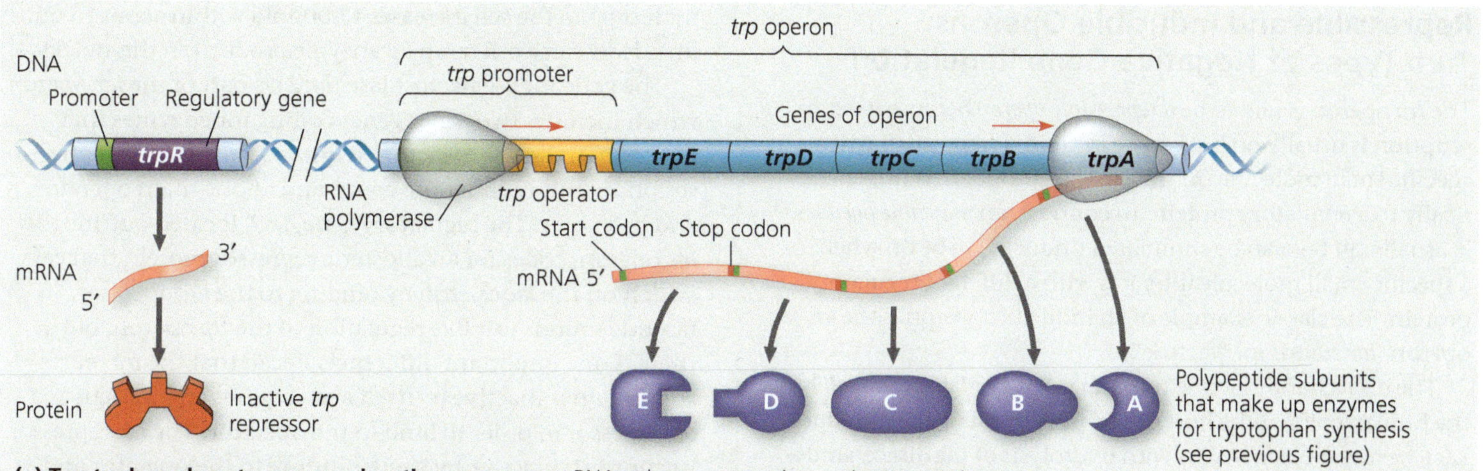

(a) Tryptophan absent, repressor inactive, operon on. RNA polymerase attaches to the DNA at the operon's promoter and transcribes the operon's genes. Enzymes for tryptophan synthesis are made.

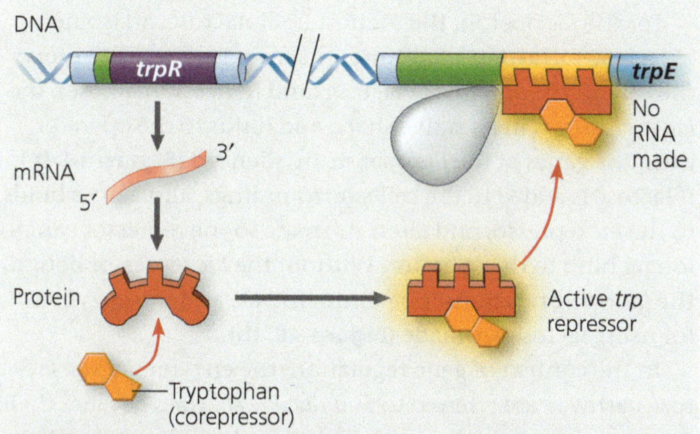

(b) Tryptophan present, repressor active, operon off. As tryptophan accumulates, it inhibits its own production by activating the repressor protein, which binds to the operator, blocking transcription. Enzymes for tryptophan synthesis are not made.

▲ Figure 18.3 The *trp* operon in *E. coli*: regulated synthesis of repressible enzymes. Tryptophan is an amino acid produced by an anabolic pathway catalyzed by three enzymes (see Figure 18.2). **(a)** The five genes encoding the polypeptide subunits of the enzymes in this pathway are grouped, along with a promoter, into the *trp* operon. The *trp* operator (the repressor binding site) is located within the *trp* promoter (the RNA polymerase binding site). **(b)** Accumulation of tryptophan, the end product of the pathway, represses transcription of the *trp* operon, thus blocking synthesis of all the enzymes in the pathway and shutting down tryptophan production.

VISUAL SKILLS ▶ *Describe what happens to the* trp *operon as the cell uses up its store of tryptophan.*

of RNA polymerase to the genes. Together, the operator, the promoter, and the genes they control—the entire stretch of DNA required for enzyme production for the tryptophan pathway—constitute an **operon**. The *trp* operon (*trp* for tryptophan) is one of many operons in the *E. coli* genome (see Figure 18.3a).

If the operator is the operon's switch for controlling transcription, how does this switch work? By itself, the *trp* operon is turned on; that is, RNA polymerase can bind to the promoter and transcribe the genes of the operon. The *trp* operon can be switched off by a protein that is called the *trp* repressor. A **repressor** binds to the operator and blocks attachment of RNA polymerase to the promoter, preventing transcription of the genes **(Figure 18.3b)**. A repressor protein is specific for the operator of a particular operon. For example, the *trp* repressor, which switches off the *trp* operon by binding to the *trp* operator, has no effect on other operons in the *E. coli* genome.

A repressor protein is encoded by a **regulatory gene**—in this case, a gene called *trpR*; *trpR* is located some distance from the *trp* operon and has its own promoter. Regulatory genes are expressed continuously, although at a low rate, and a few *trp* repressor molecules are always present in *E. coli* cells. Why, then, is the *trp* operon not switched off permanently? First, the binding of repressors to operators is reversible. An operator alternates between two states: one with the repressor bound and one without the repressor bound. The relative duration of the repressor-bound state is higher when more active repressor molecules are present. Second, the *trp* repressor, like most regulatory proteins, is an allosteric protein, with two alternative shapes: active and inactive (see Figure 8.20). The *trp* repressor is synthesized in the inactive form, which has little affinity for the *trp* operator. Only when a tryptophan molecule binds to the *trp* repressor at an allosteric site does the repressor protein change to the active form that can attach to the operator, turning the operon off.

Tryptophan functions in this system as a **corepressor**, a small molecule that cooperates with a repressor protein to switch an operon off. As tryptophan accumulates, more tryptophan molecules associate with *trp* repressor molecules, which can then bind to the *trp* operator and shut down production of the tryptophan pathway enzymes. If the cell's tryptophan level drops, many fewer *trp* repressor proteins would have tryptophan bound, rendering them inactive; they would dissociate from the operator, allowing transcription of the operon's genes to resume. The *trp* operon is one example of how gene expression can respond to changes in the cell's internal and external environment.

Repressible and Inducible Operons: Two Types of Negative Gene Regulation

The *trp* operon is said to be a *repressible operon* because its transcription is usually on but can be inhibited (repressed) when a specific small molecule (in this case, tryptophan) binds allosterically to a regulatory protein. In contrast, an *inducible operon* is usually off but can be stimulated (induced) to be on when a specific small molecule interacts with a different regulatory protein. The classic example of an inducible operon is the *lac* operon (*lac* stands for "lactose").

The disaccharide lactose (milk sugar) is available to *E. coli* in the human colon if the host drinks milk or eats a dairy product. Lactose metabolism begins with hydrolysis of the disaccharide into its component monosaccharides (glucose and galactose), a reaction catalyzed by the enzyme β-galactosidase. Only a few molecules of this enzyme are present in an *E. coli* cell growing in the absence of lactose. If lactose is added to the bacterium's environment, however, the number of β-galactosidase molecules in the cell increases 1,000-fold within about 15 minutes. How can a cell ramp up enzyme production this quickly?

The gene for β-galactosidase (*lacZ*) is part of the *lac* operon, which includes two other genes coding for enzymes that function in the use of lactose **(Figure 18.4)**. The entire transcription unit is under the command of one main operator and promoter. The regulatory gene, *lacI*, located outside the *lac* operon, codes for an allosteric repressor protein that can switch off the *lac* operon by binding to the *lac* operator. So far, this sounds just like regulation of the *trp* operon, but there is one important difference. Recall that the *trp* repressor protein is inactive by itself and requires tryptophan as a corepressor in order to bind to the operator. The *lac* repressor, in contrast, is active by itself, binding to the operator and switching the *lac* operon off. In this case, a specific small molecule, called an **inducer**, *inactivates* the repressor.

For the *lac* operon, the inducer is allolactose, an isomer of lactose formed in small amounts from lactose that enters the cell. In the absence of lactose (and hence allolactose), the *lac* repressor is in its active shape and binds to the operator; thus, the genes of the *lac* operon are silenced **(Figure 18.4a)**. If lactose is added to the cell's surroundings, allolactose binds to the *lac* repressor and alters its shape so the repressor can no longer bind to the operator. Without the *lac* repressor bound, the *lac* operon is transcribed into mRNA, and the enzymes for using lactose are made **(Figure 18.4b)**.

In the context of gene regulation, the enzymes of the lactose pathway are referred to as *inducible enzymes* because their synthesis is induced by a chemical signal (allolactose, in this case). Analogously, the enzymes for tryptophan synthesis are said to be repressible. *Repressible enzymes* generally function in anabolic pathways, which synthesize essential end products from raw materials (precursors). By suspending production

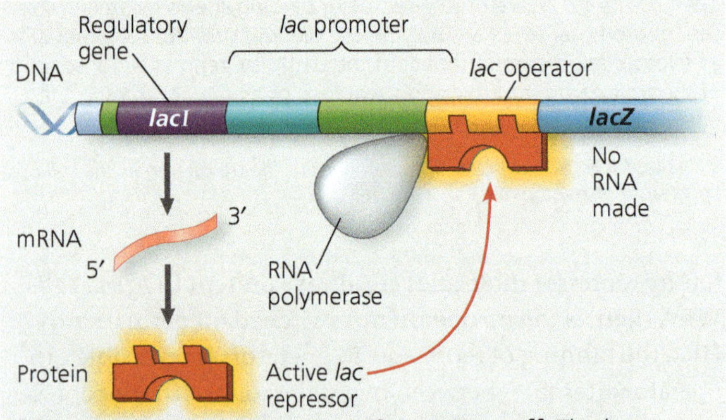

(a) Lactose absent, repressor active, operon off. The *lac* repressor is innately active, and in the absence of lactose it switches off the operon by binding to the operator. The enzymes for using lactose are not made.

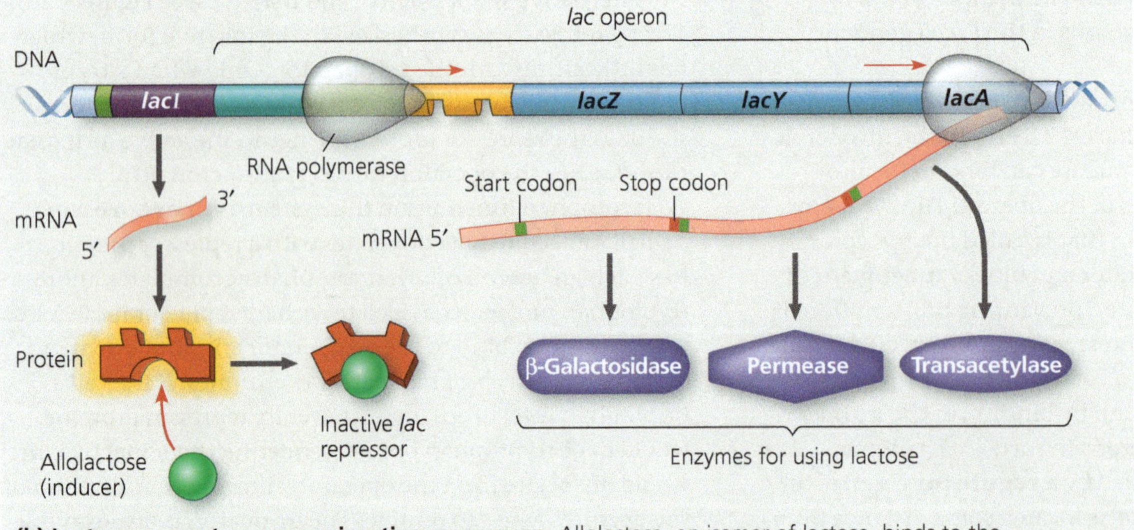

(b) Lactose present, repressor inactive, operon on. Allolactose, an isomer of lactose, binds to the repressor, inactivating it and "derepressing" the operon. The inactive repressor cannot bind to the operator, and so the genes of the *lac* operon are transcribed, and the enzymes for using lactose are made.

MB Animation: The *lac* Operon in *E. coli*

◀ Figure 18.4 The *lac* operon in *E. coli*: regulated synthesis of inducible enzymes. *E. coli* uses three enzymes to take up and metabolize lactose, the genes for which are clustered in the *lac* operon. The first gene, *lacZ*, codes for β-galactosidase, which hydrolyzes lactose to glucose and galactose. The second, *lacY*, codes for a permease, the membrane protein that transports lactose into the cell. The third, *lacA*, codes for transacetylase, an enzyme that detoxifies other molecules entering the cell via the permease. Unusually, the gene for the *lac* repressor, *lacI*, is adjacent to the *lac* operon. The function of the teal region within the promoter will be revealed in Figure 18.5.

of an end product when it is already present in sufficient quantity, the cell can allocate its organic precursors and energy for other uses. In contrast, inducible enzymes usually function in catabolic pathways, which break down a nutrient to simpler molecules. By producing the appropriate enzymes only when the nutrient is available, the cell avoids wasting energy and precursors making proteins that are not needed.

Regulation of both the *trp* and *lac* operons involves the *negative* control of genes because the operons are switched off by the active form of their respective repressor protein. It may be easier to see this for the *trp* operon, but it is also true for the *lac* operon. In the case of the *lac* operon, allolactose induces enzyme synthesis not by directly activating the *lac* operon, but by freeing it from the negative effect of the repressor (see Figure 18.4b). Gene regulation is said to be *positive* only when a regulatory protein interacts directly with the genome to switch transcription on.

Positive Gene Regulation

When glucose and lactose are both present in its environment, *E. coli* preferentially uses glucose. The enzymes for glucose breakdown in glycolysis (see Figure 9.9) are continually present. Only when lactose is present *and* glucose is in short supply does *E. coli* use lactose as an energy source, and only then does it synthesize appreciable quantities of the enzymes for lactose breakdown.

How does the *E. coli* cell sense the glucose concentration and relay this information to the *lac* operon? Again, the mechanism depends on the interaction of an allosteric regulatory protein with a small organic molecule, **cyclic AMP (cAMP)** in this case, which accumulates when glucose is scarce (see Figure 11.11 for the structure of cAMP). The regulatory protein, called *cAMP receptor protein* (*CRP*), is an **activator**, a protein that binds to DNA and stimulates transcription of a gene. When cAMP binds to this regulatory protein, CRP assumes its active shape and can attach to a specific site at the upstream end of the *lac* promoter **(Figure 18.5a)**. This attachment increases the affinity of RNA polymerase for the *lac* promoter, which is actually rather low even when no *lac* repressor is bound to the operator. By facilitating the binding of RNA polymerase to the promoter and thereby increasing the rate of transcription of the *lac* operon, the attachment of CRP to the promoter directly stimulates gene expression. Therefore, this mechanism qualifies as positive regulation.

If the amount of glucose in the cell increases, the cAMP concentration falls, and without cAMP, CRP detaches from the *lac* operon. Because CRP is inactive, RNA polymerase binds less efficiently to the promoter, and transcription of the *lac* operon proceeds only at a low level, even when lactose is present **(Figure 18.5b)**. Thus, the *lac* operon is under dual control: negative control by the *lac* repressor and positive control by CRP. The state of the *lac* repressor (with allolactose bound or without it) determines whether or not transcription of the *lac* operon's genes occurs at all; the state of CRP (with bound cAMP or without it) controls the *rate* of transcription if the operon is repressor-free. It is as though the operon has both an on-off switch and a volume control.

In addition to regulating the *lac* operon, CRP helps regulate other operons that encode enzymes used in catabolic pathways. All told, it may affect the expression of more than 100 genes in *E. coli*. When glucose is plentiful and CRP is inactive, the synthesis of enzymes that catabolize compounds other than glucose generally slows down. The ability to catabolize other compounds, such as lactose, enables a cell deprived of glucose to survive. The compounds present in any given cell at the moment determine which operons are switched on—the result of simple interactions of activator and repressor proteins with the promoters of the genes in question.

Figure 18.5 Positive control of the lac operon by cAMP receptor protein (CRP). RNA polymerase has high affinity for the *lac* promoter only when CRP is bound to a DNA site at the upstream end of the promoter. CRP, in turn, attaches to its DNA site only when associated with cyclic AMP (cAMP), whose concentration in the cell rises when the glucose concentration falls. Thus, when glucose is present, even if lactose is also available, the cell preferentially catabolizes glucose and makes very little of the enzymes for using lactose.

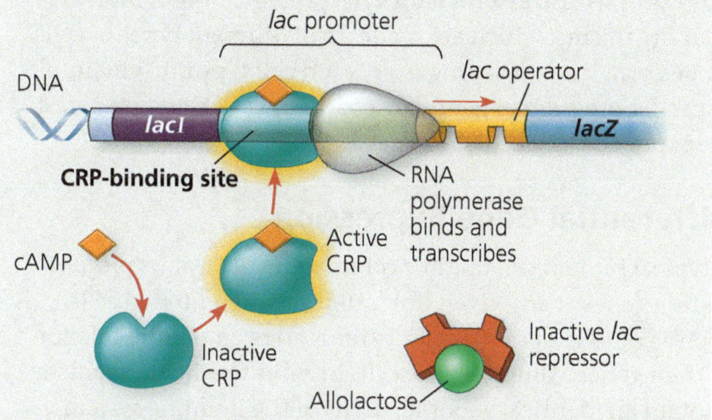

(a) Lactose present, glucose scarce (cAMP level high): abundant *lac* mRNA synthesized. If glucose is scarce, the high level of cAMP activates CRP, which binds to the promoter and increases RNA polymerase binding there. The *lac* operon produces large amounts of mRNA coding for the enzymes that the cell needs for use of lactose.

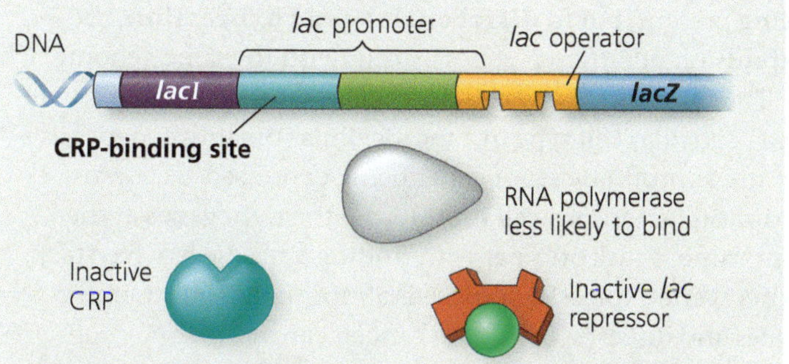

(b) Lactose present, glucose present (cAMP level low): little *lac* mRNA synthesized. When glucose is present, cAMP is scarce, and CRP is unable to stimulate transcription at a significant rate, even though no repressor is bound.

CONCEPT CHECK 18.1

1. How does binding of the *trp* corepressor to the *trp* repressor alter repressor function and transcription? What about the binding of the *lac* inducer to the *lac* repressor?
2. Describe the binding of RNA polymerase, repressors, and activators to the *lac* operon when both lactose and glucose are scarce. What is the effect of these scarcities on transcription of the *lac* operon?
3. **WHAT IF? >** A certain mutation in *E. coli* changes the *lac* operator so that the active repressor cannot bind. How would this affect the cell's production of β-galactosidase?

For suggested answers, see Appendix A.

CONCEPT 18.2

Eukaryotic gene expression is regulated at many stages

All organisms, whether prokaryotes or eukaryotes, must regulate which genes are expressed at any given time. Both unicellular organisms and the cells of multicellular organisms continually turn genes on and off in response to signals from their external and internal environments. Regulation of gene expression is also essential for cell specialization in multicellular organisms, which are made up of different types of cells. To perform its own distinct role, each cell type must maintain a specific program of gene expression in which certain genes are expressed and others are not.

Differential Gene Expression

A typical human cell might express about 20% of its protein-coding genes at any given time. Highly differentiated cells, such as muscle or nerve cells, express an even smaller fraction of their genes. Almost all the cells in a multicellular organism contain an identical genome. (Cells of the immune system are one exception, as you will see in Chapter 43.) A subset of genes is expressed in each cell type; some of these are "housekeeping" genes, expressed by many cell types, while others are unique to that cell type. The uniquely expressed genes allow these cells to carry out their specific function. The differences between cell types, therefore, are due not to different genes being present, but to **differential gene expression**, the expression of different genes by cells with the same genome.

The function of any cell, whether a single-celled eukaryote or a particular cell type in a multicellular organism, depends on the appropriate set of genes being expressed. The transcription factors of a cell must locate the right genes at the right time, a task on a par with finding a needle in a haystack. When gene expression proceeds abnormally, serious imbalances and diseases, including cancer, can arise.

Figure 18.6 summarizes the process of gene expression in a eukaryotic cell, highlighting key stages in the expression of a protein-coding gene. Each stage depicted in Figure 18.6

Figure 18.6 Stages in gene expression that can be regulated in eukaryotic cells. In this diagram, the colored boxes indicate the processes most often regulated; each color indicates the type of molecule that is affected (blue = DNA, red/orange = RNA, purple = protein). The nuclear envelope separating transcription from translation in eukaryotic cells offers an opportunity for post-transcriptional control in the form of RNA processing that is absent in prokaryotes. In addition, eukaryotes have a greater variety of control mechanisms operating before transcription and after translation. A miniature version of this figure accompanies several figures later in the chapter as an orientation diagram.

Animation: Control of Gene Expression

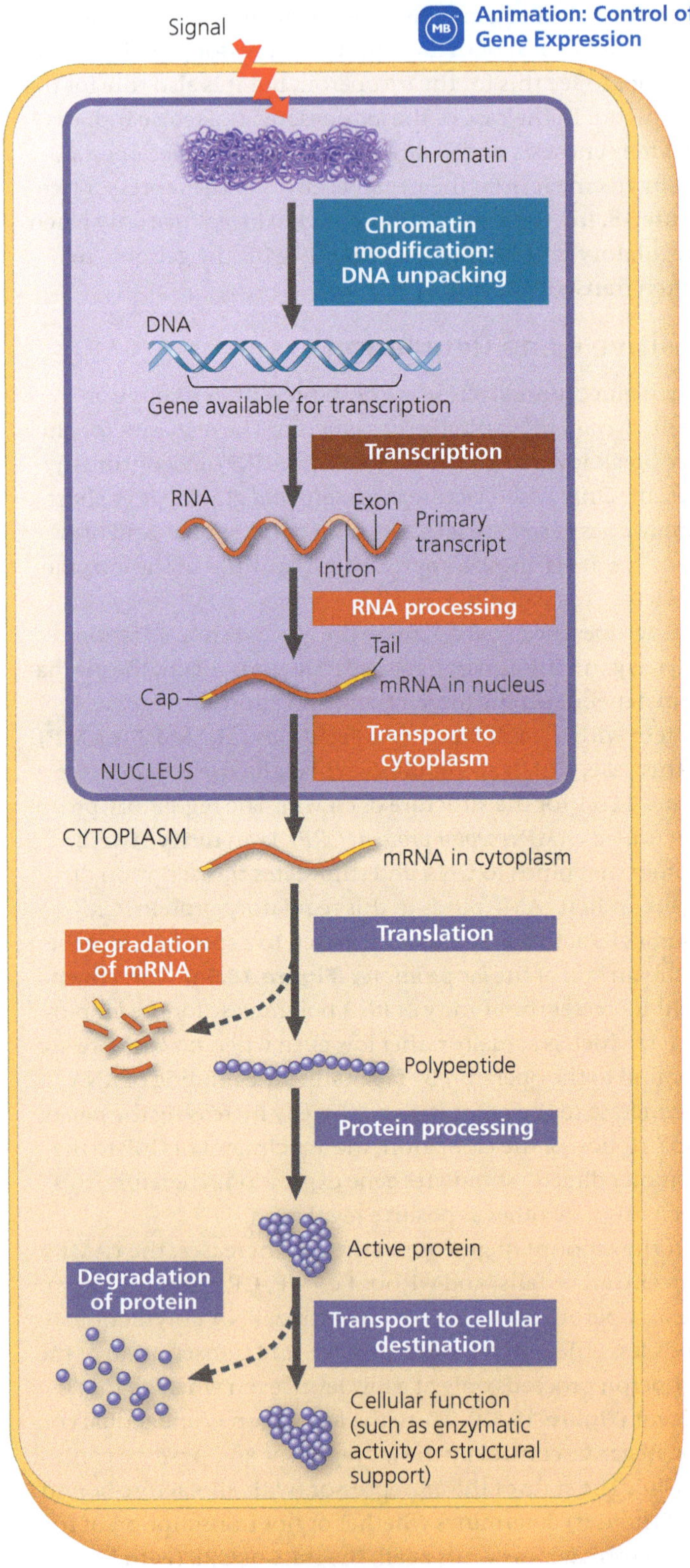

is a potential control point at which gene expression can be turned on or off, accelerated, or slowed down.

Fifty or so years ago, an understanding of the mechanisms that control gene expression in eukaryotes seemed almost hopelessly out of reach. Since then, new research methods, notably advances in DNA technology (see Chapter 20), have enabled molecular biologists to uncover many details of eukaryotic gene regulation. In all organisms, gene expression is commonly controlled at transcription; regulation at this stage often occurs in response to signals coming from outside the cell, such as hormones or other signaling molecules. For this reason, the term *gene expression* is often equated with transcription for both bacteria and eukaryotes. While this may often be the case for bacteria, the greater complexity of eukaryotic cell structure and function provides opportunities for regulating gene expression at many additional stages (see Figure 18.6). In the remainder of this section, we'll examine some of the important control points of eukaryotic gene expression more closely.

Regulation of Chromatin Structure

Recall that the DNA of eukaryotic cells is packaged with proteins in an elaborate complex known as chromatin, the basic unit of which is the nucleosome (see Figure 16.22). The structural organization of chromatin not only packs a cell's DNA into a compact form that fits inside the nucleus, but also helps regulate gene expression in several ways. The location of a gene's promoter, relative to both placement of nucleosomes and the sites where the DNA attaches to the chromosome scaffold, can affect whether the gene is transcribed. In addition, genes within heterochromatin, which is highly condensed, are usually not expressed. Lastly, certain chemical modifications to chromatin—both to the histone proteins of the nucleosomes around which DNA is wrapped and to the nucleotides that make up that DNA—can influence chromatin structure and gene expression. Here we examine the effects of these modifications, which are catalyzed by specific enzymes.

Histone Modifications and DNA Methylation

There is abundant evidence that chemical modifications to histones, found in all eukaryotic organisms, play a direct role in the regulation of gene transcription. The N-terminus of each histone protein in a nucleosome protrudes outward from the nucleosome **(Figure 18.7a)**. These so-called *histone tails* are accessible to various modifying enzymes that catalyze the addition or removal of specific chemical groups, such as acetyl (—$COCH_3$), methyl, and phosphate groups (see Figure 4.9). Generally, **histone acetylation**—the addition of an acetyl group to an amino acid in a histone tail—appears to promote transcription by opening up the chromatin structure **(Figure 18.7b)**, while the addition of methyl groups to histones can lead to the condensation of chromatin and reduced transcription. Often, the addition of a particular chemical group may create a new binding site for enzymes that further modify chromatin structure in various ways.

Rather than modifying histone proteins, a different set of enzymes can methylate the DNA itself on certain bases, usually cytosine. Such **DNA methylation** occurs in most plants, animals, and fungi. Long stretches of inactive DNA, such as that of inactivated mammalian X chromosomes (see Figure 15.8), are generally more methylated than regions of actively transcribed DNA (although there are exceptions). On a smaller scale, the DNA of individual genes is usually more heavily methylated in cells in which those genes are not expressed. Removal of the extra methyl groups can turn on some of these genes.

Figure 18.7 A simple model of histone tails and the effect of histone acetylation. In addition to acetylation, histones can undergo several other types of modifications, such as methylation or phosphorylation. These also help determine the chromatin configuration in a region, sometimes by establishing binding sites for chromatin-modifying enzymes.

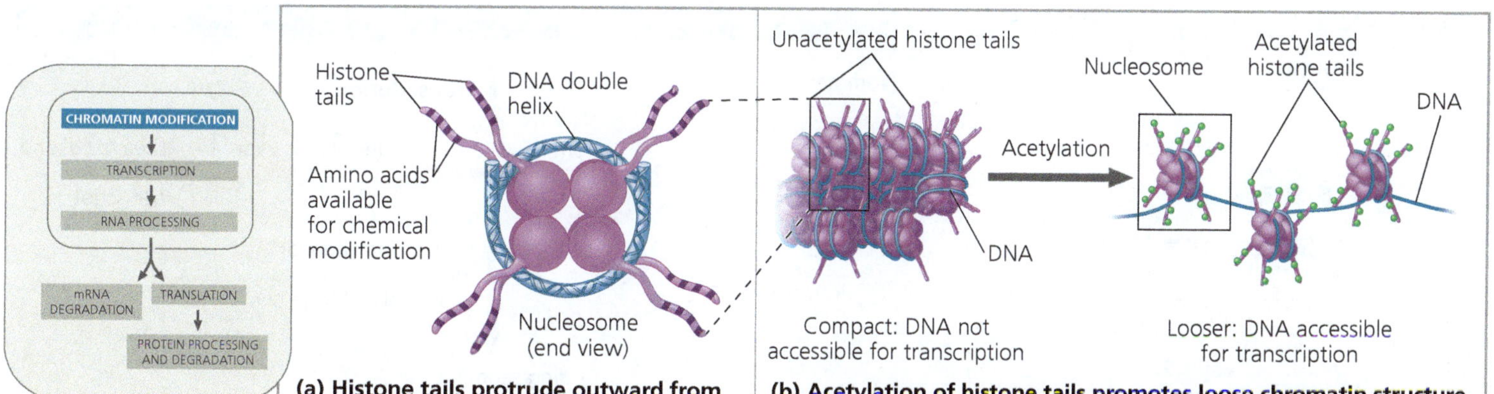

(a) Histone tails protrude outward from a nucleosome. The amino acids in the histone tails are accessible for chemical modification.

(b) Acetylation of histone tails promotes loose chromatin structure that permits transcription. A region of chromatin in which nucleosomes are unacetylated forms a compact structure (left) in which the DNA is not transcribed. When nucleosomes become highly acetylated (right), the chromatin becomes less condensed, and the DNA is accessible for transcription.

Once methylated, genes usually stay that way through successive cell divisions in a given individual. At DNA sites where one strand is already methylated, enzymes methylate the correct daughter strand after each round of DNA replication. Methylation patterns are thus passed on to daughter cells, and cells forming specialized tissues keep a chemical record of what occurred during embryonic development. A methylation pattern maintained in this way also accounts for *genomic imprinting* in mammals, where methylation permanently regulates expression of either the maternal or paternal allele of particular genes at the start of development (see Figure 15.17).

Epigenetic Inheritance

The chromatin modifications that we just discussed do not change the DNA sequence, yet they still may be passed along to future generations of cells. Inheritance of traits transmitted by mechanisms not involving the nucleotide sequence itself is called **epigenetic inheritance**. Whereas mutations in the DNA are permanent changes, modifications to the chromatin can be reversed. For example, DNA methylation patterns are largely erased and reestablished during gamete formation.

Researchers are amassing more and more evidence for the importance of epigenetic information in the regulation of gene expression. Epigenetic variations might help explain why one identical twin acquires a genetically based disease, such as schizophrenia, but the other does not, despite their identical genomes. Alterations in normal patterns of DNA methylation are seen in some cancers, where they are associated with inappropriate gene expression. Evidently, enzymes that modify chromatin structure are integral parts of the eukaryotic cell's machinery for regulating transcription.

Regulation of Transcription Initiation

Chromatin-modifying enzymes provide initial control of gene expression by making a region of DNA either more or less able to bind the transcription machinery. Once the chromatin of a gene is optimally modified for expression, the initiation of transcription is the next major step at which gene expression is regulated. As in bacteria, the regulation of transcription initiation in eukaryotes involves proteins that bind to DNA and either facilitate or inhibit binding of RNA polymerase. The process is more complicated in eukaryotes, however. Before looking at how eukaryotic cells control their transcription, let's review the structure of a eukaryotic gene.

Organization of a Typical Eukaryotic Gene and Its Transcript

A eukaryotic gene and the DNA elements (segments) that control it are typically organized as shown in **Figure 18.8**, which extends what you learned about eukaryotic genes in Chapter 17. Recall that a cluster of proteins called a *transcription initiation complex* assembles on the promoter sequence

▼ Figure 18.8 A eukaryotic gene and its transcript. Each eukaryotic gene has a promoter—a DNA sequence where RNA polymerase binds and starts transcription, proceeding "downstream." A number of control elements (gold) are involved in regulating the initiation of transcription; these are DNA sequences located near (proximal to) or far from (distal to) the promoter. Distal control elements can be grouped together as enhancers, one of which is shown for this gene. At the other end of the gene, a polyadenylation (poly-A) signal sequence in the last exon of the gene is transcribed into an RNA sequence that signals where the transcript is cleaved and the poly-A tail added. Transcription may continue for hundreds of nucleotides beyond the poly-A signal before terminating. RNA processing of the primary transcript into a functional mRNA involves three steps: addition of the 5′ cap, addition of the poly-A tail, and splicing. In the cell, the 5′ cap is added soon after transcription is initiated, and splicing occurs while transcription is still under way (see Figure 17.11).

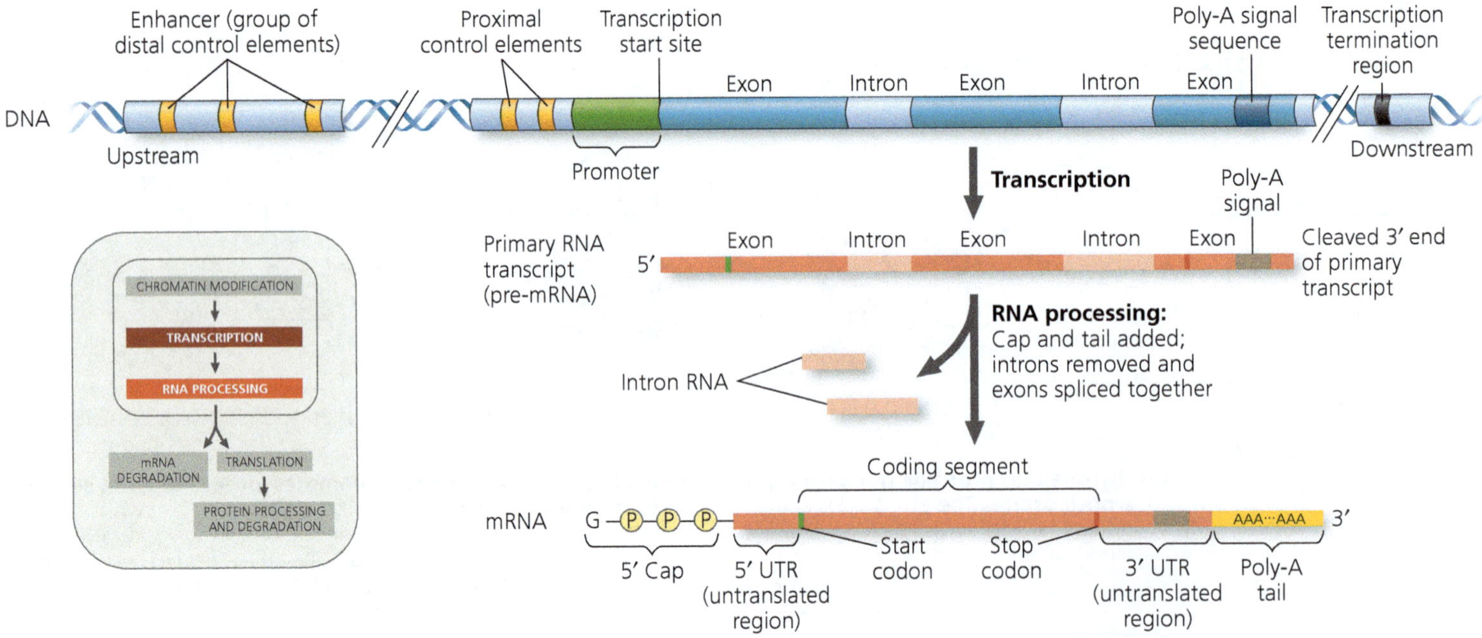

at the "upstream" end of the gene (see Figure 17.9). One of these proteins, RNA polymerase II, then proceeds to transcribe the gene, synthesizing a primary RNA transcript (pre-mRNA). RNA processing includes enzymatic addition of a 5′ cap and a poly-A tail, as well as splicing out of introns, to yield a mature mRNA. Associated with most eukaryotic genes are multiple **control elements**, segments of noncoding DNA that serve as binding sites for the proteins called transcription factors, which bind to the control elements and regulate transcription. Control elements on the DNA and the transcription factors that bind to them are critical to the precise regulation of gene expression seen in different cell types.

The Roles of General and Specific Transcription Factors

There are two types of transcription factors: General transcription factors act at the promoter of all genes, while some genes require specific transcription factors that bind to control elements that may be close to or farther away from the promoter.

General Transcription Factors at the Promoter To initiate transcription, eukaryotic RNA polymerase requires the assistance of transcription factors. Some transcription factors (such as those illustrated in Figure 17.9) are essential for the transcription of *all* protein-coding genes; therefore, they are often called *general transcription factors*. A few general transcription factors bind to a DNA sequence such as the TATA box within the promoter, but many bind to proteins, including other transcription factors and RNA polymerase II. Protein-protein interactions are crucial to the initiation of eukaryotic transcription. Only when the complete initiation complex has assembled can the polymerase begin to move along the DNA template strand, producing a complementary strand of RNA.

The interaction of general transcription factors and RNA polymerase II with a promoter usually leads to a low rate of initiation and production of few RNA transcripts from genes that are not expressed all the time, but instead are regulated. In eukaryotes, high levels of transcription of these particular genes at the appropriate time and place depend on the interaction of control elements with another set of proteins, which can be thought of as *specific transcription factors*.

Enhancers and Specific Transcription Factors As you can see in Figure 18.8, some control elements, named *proximal control elements*, are located close to the promoter. (Although some biologists consider proximal control elements part of the promoter, in this text we do not.) The more distant *distal control elements*, groupings of which are called **enhancers**, may be thousands of nucleotides upstream or downstream of a gene or even within an intron. A given gene may have multiple enhancers, each active at a different time or in a different cell type or location in the organism. Each enhancer, however, is generally associated with only that gene and no other.

▼ Figure 18.9 The structure of MyoD, a transcriptional activator. The MyoD protein is made up of two polypeptide subunits (purple and salmon) with extensive regions of α helix. Each subunit has one DNA-binding domain (lower half) and one activation domain (upper half). The latter includes binding sites for the other subunit and for other proteins. MyoD is involved in muscle development in vertebrate embryos (see Concept 18.4).

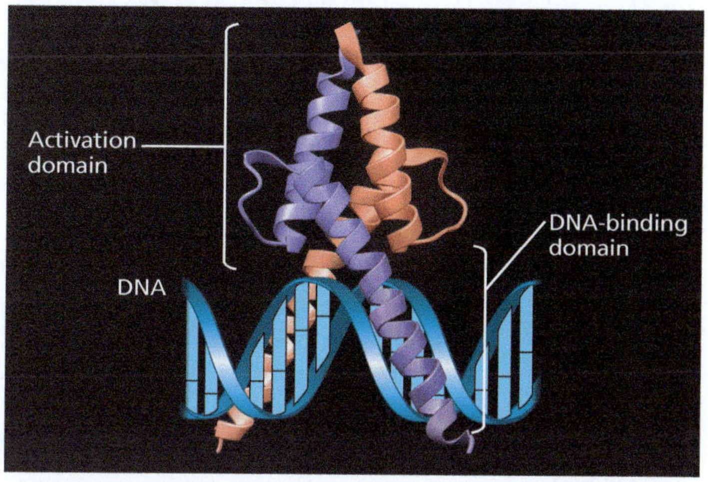

VISUAL SKILLS ➤ *Describe how the two functional domains of the MyoD protein relate to the two polypeptide subunits.*

In eukaryotes, the rate of gene expression can be strongly increased or decreased by the binding of specific transcription factors, either activators or repressors, to the control elements of enhancers. Hundreds of transcription activators have been discovered in eukaryotes; the structure of one example is shown in **Figure 18.9**. Researchers have identified two types of structural domains that are commonly found in a large number of activator proteins: a *DNA-binding domain*—a part of the protein's three-dimensional structure that binds to DNA—and one or more *activation domains*. Activation domains bind other regulatory proteins or components of the transcription machinery, facilitating a series of protein-protein interactions that result in enhanced transcription of a given gene.

How can binding of activators to an enhancer located far from the promoter influence transcription? One study shows that the proteins regulating a mouse globin gene contact both the gene's promoter and an enhancer located about 50,000 nucleotides upstream. This and many other studies support the currently accepted model, in which protein-mediated bending of the DNA is thought to bring the bound activators into contact with a group of *mediator proteins*, which in turn interact with general transcription factors at the promoter **(Figure 18.10)**. These protein-protein interactions help assemble and position the initiation complex on the promoter, and allow the promoter and enhancer to come together in a very specific fashion, in spite of what is often a large number of nucleotide pairs between them. In the **Scientific Skills Exercise**, you can work with data from an experiment that identified the control elements in an enhancer of a particular human gene.

Figure 18.10 A model for the action of enhancers and transcription activators. Bending of the DNA by a protein enables enhancers to influence a promoter hundreds or even thousands of nucleotides away. Specific transcription factors called activators bind to the enhancer DNA sequences and then to a group of mediator proteins. These in turn bind to general transcription factors and then RNA polymerase II, thus assembling the transcription initiation complex. These protein-protein interactions lead to correct positioning of the complex on the promoter and the initiation of RNA synthesis. Only one enhancer (with three gold control elements) is shown here, but a gene may have several enhancers that act at different times or in different cell types.

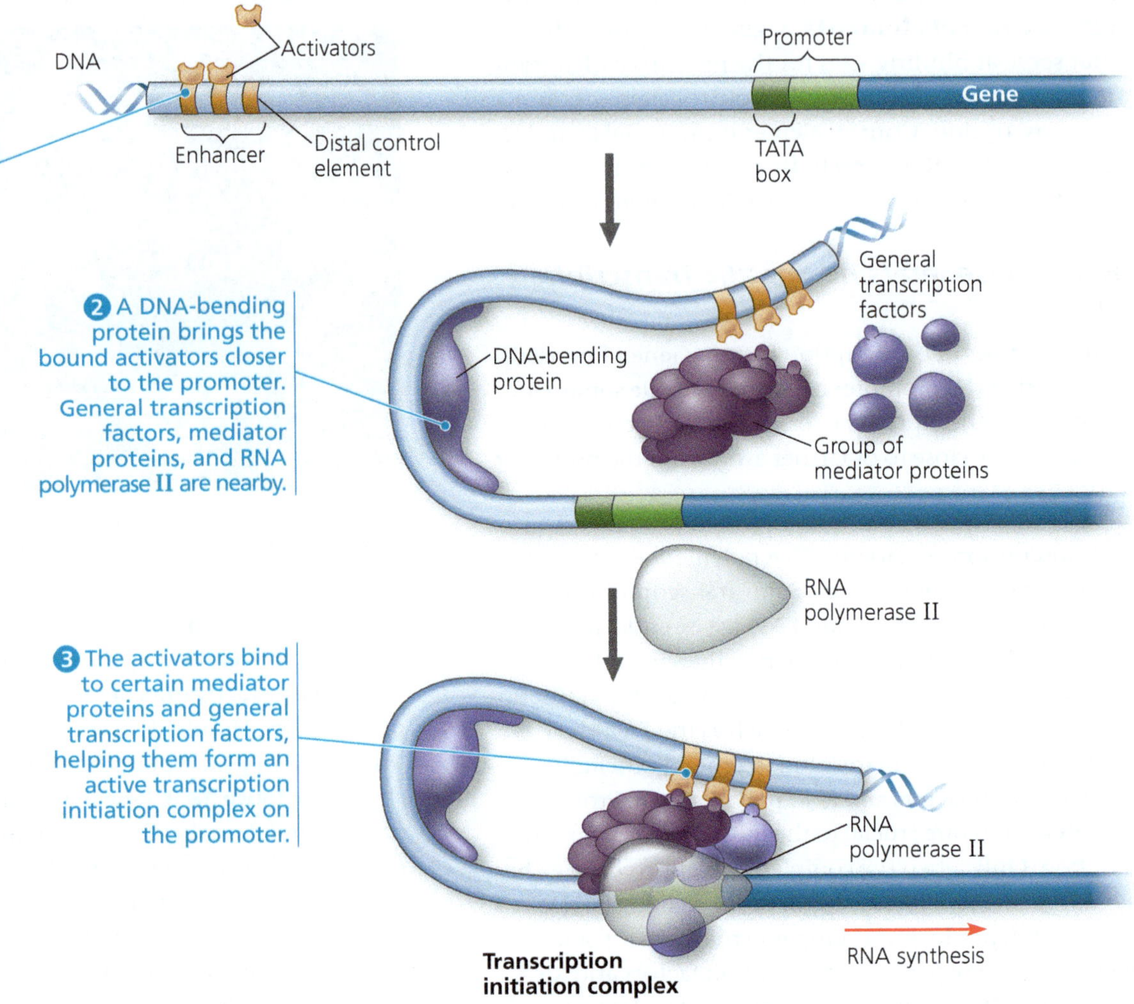

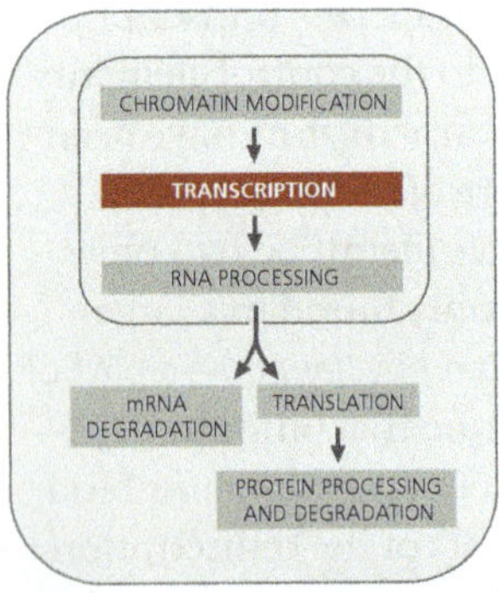

Animation: Regulation of Transcription by Transcriptional Activators and Enhancers

Specific transcription factors that function as repressors can inhibit gene expression in several different ways. Some repressors bind directly to control element DNA (in enhancers or elsewhere), blocking activator binding. Other repressors interfere with the activator itself so it can't bind the DNA.

In addition to influencing transcription directly, some activators and repressors act indirectly by affecting chromatin structure. Studies using yeast and mammalian cells show that some activators recruit proteins that acetylate histones near the promoters of specific genes, thus promoting transcription (see Figure 18.7). Similarly, some repressors recruit proteins that remove acetyl groups from histones, leading to reduced transcription, a phenomenon referred to as *silencing*. Indeed, recruitment of chromatin-modifying proteins seems to be the most common mechanism of repression in eukaryotic cells.

Combinatorial Control of Gene Activation In eukaryotes, the precise control of transcription depends largely on the binding of activators to DNA control elements. Considering the great number of genes that must be regulated in a typical animal or plant cell, the number of completely different nucleotide sequences found in control elements is surprisingly small. A dozen or so short nucleotide sequences appear again and again in the control elements for different genes. On average, each enhancer is composed of about ten control elements, each of which can

SCIENTIFIC SKILLS EXERCISE

Analyzing DNA Deletion Experiments

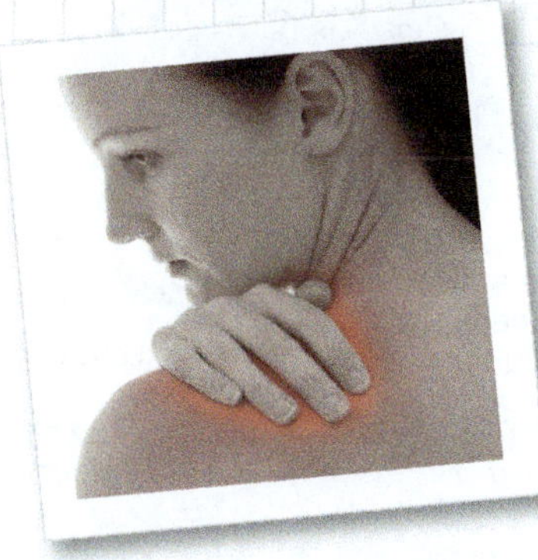

What Control Elements Regulate Expression of the *mPGES-1* Gene? The promoter of a gene includes the DNA immediately upstream of the transcription start site, but the control elements that regulate the level of transcription of the gene (grouped in an enhancer) may be thousands of base pairs upstream of the promoter. Because the distance and spacing of control elements make them difficult to identify, scientists begin by deleting possible control elements and measuring the effect on gene expression. In this exercise, you will analyze data obtained from DNA deletion experiments that tested possible control elements for the human gene *mPGES-1*. This gene codes for an enzyme that synthesizes a type of prostaglandin, a chemical made during tissue inflammation.

How the Experiment Was Done The researchers hypothesized that there were three possible control elements in an enhancer region located 8–9 kilobases upstream of the *mPGES-1* gene. Control elements regulate whatever gene is in the appropriate downstream location. Thus, to test the activity of the possible elements, researchers first synthesized molecules of DNA ("constructs") that had the intact enhancer region upstream of a "reporter gene," a gene whose mRNA product could be easily measured experimentally. Next, they made three more DNA constructs, with one of the three proposed control elements deleted in each (see the left side of the figure). The researchers then introduced each DNA construct into a separate human cell culture, where the cells took up the DNA constructs. After 48 hours, the amount of reporter gene mRNA made by the cells was measured. Comparing these amounts allowed researchers to determine if any of the deletions had an effect on expression of the reporter gene, mimicking the effect of deletions on *mPGES-1* gene expression. (The *mPGES-1* gene itself couldn't be used to measure expression levels because the cells express their own *mPGES-1* gene, and the mRNA from that gene would confuse the results.)

Data from the Experiment The diagrams on the left side of the figure show the intact DNA sequence (top) and the three experimental DNA constructs. A red X is located on the possible control element (1, 2, or 3) that was deleted in each experimental DNA construct. The area between the slashes represents the approximately 8 kilobases of DNA located between the promoter and the enhancer region. The horizontal bar graph on the right shows the amount of reporter gene mRNA that was present in each cell culture after 48 hours relative to the amount that was in the culture containing the intact enhancer region (top bar = 100%).

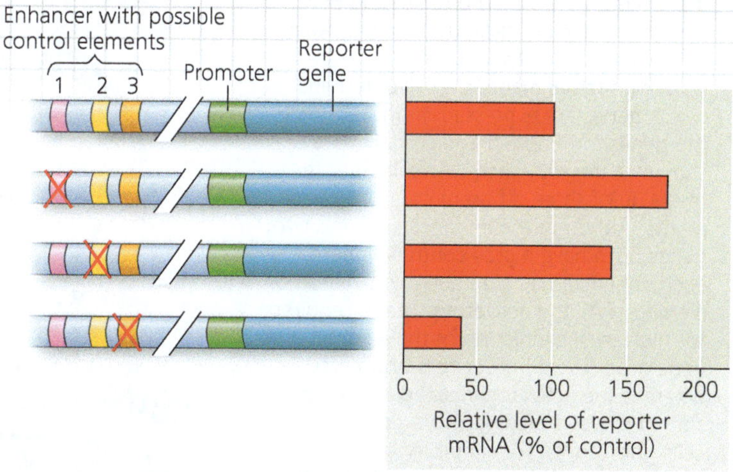

Data from J. N. Walters et al., Regulation of human microsomal prostaglandin E synthase-1 by IL-1b requires a distal enhancer element with a unique role for C/EBPb, *Biochemical Journal* 443:561–571 (2012).

INTERPRET THE DATA

1. (a) What is the independent variable in the graph? (b) What is the dependent variable? (c) What was the control treatment in this experiment? Label it on the diagram.
2. Do the data suggest that any of these possible control elements are actual control elements? Explain.
3. (a) Did deletion of any of the possible control elements cause a *reduction* in reporter gene expression? If so, which one(s), and how can you tell? (b) If loss of a control element causes a reduction in gene expression, what must be the normal role of that control element? Provide a biological explanation for how the loss of such a control element could lead to a reduction in gene expression.
4. (a) Did deletion of any of the possible control elements cause an *increase* in reporter gene expression relative to the control? If so, which one(s), and how can you tell? (b) If loss of a control element causes an increase in gene expression, what must be the normal role of that control element? Propose a biological explanation for how the loss of such a control element could lead to an increase in gene expression.

Instructors: A version of this Scientific Skills Exercise can be assigned in MasteringBiology.

bind only one or two specific transcription factors. It is the particular *combination* of control elements in an enhancer associated with a gene, rather than the presence of a single unique control element, that is important in regulating transcription of the gene.

Even with only a dozen control element sequences available, a very large number of combinations is possible. Each combination of control elements is able to activate transcription only when the appropriate activator proteins are present, which may occur at a precise time during development or in a particular cell type. **Figure 18.11** illustrates how the use of different combinations of just a few control elements can allow differential regulation of transcription in two representative cell types—liver cells and lens cells. This can occur because each cell type contains a different group of activator proteins. How cell types come to differ during this process, even though they all arise from one cell (the fertilized egg), will be explored in Concept 18.4.

➤ **Figure 18.11 Cell type–specific transcription.** Both liver cells and lens cells have the genes for making the proteins albumin and crystallin, but only liver cells make albumin (a blood protein) and only lens cells make crystallin (the main protein of the lens of the eye). The specific transcription factors made in a cell determine which genes are expressed. In this example, the genes for albumin and crystallin are shown at the top, each with an enhancer made up of three different control elements. Although the enhancers for the two genes both have a gray control element, each enhancer has a unique *combination* of elements. All the activator proteins required for high-level expression of the albumin gene are present in liver cells only (left), whereas the activators needed for expression of the crystallin gene are present in lens cells only (right). For simplicity, we consider only the role of specific transcription factors that are activators here, although repressors may also influence transcription in certain cell types.

VISUAL SKILLS ➤ *Describe the enhancer for the albumin gene in each type of cell. How would the nucleotide sequence of this enhancer in the liver cell compare with that in the lens cell?*

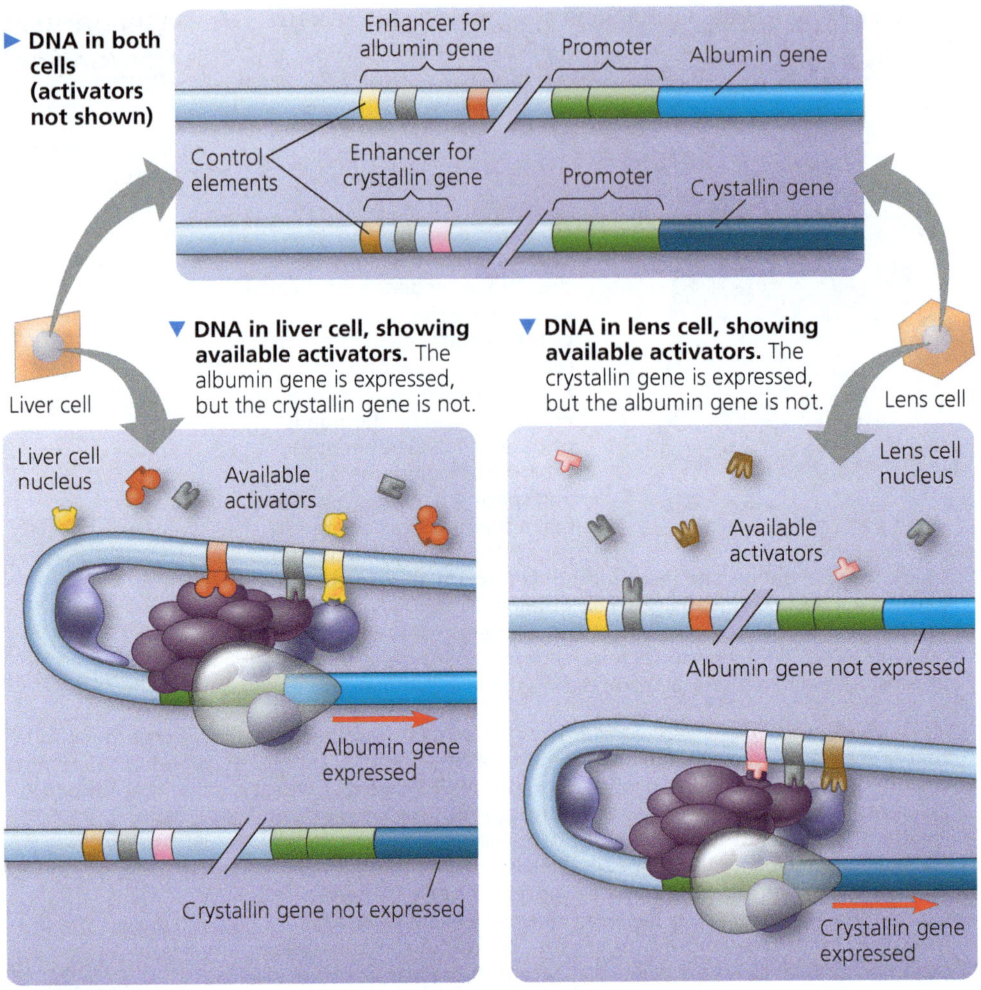

Coordinately Controlled Genes in Eukaryotes

How does the eukaryotic cell deal with a group of genes of related function that need to be turned on or off at the same time? Earlier in this chapter, you learned that in bacteria, such *coordinately controlled* genes are often clustered into an operon, which is regulated by a single promoter and transcribed into a single mRNA molecule. Thus, the genes are expressed together, and the encoded proteins are produced concurrently. With a few exceptions, operons that work in this way have *not* been found in eukaryotic cells.

Eukaryotic genes that are co-expressed, such as genes coding for the enzymes of a metabolic pathway, are typically scattered over different chromosomes. Here, coordinate gene expression depends on every gene of a dispersed group having a specific combination of control elements. Activator proteins in the nucleus that recognize the control elements bind to them, promoting simultaneous transcription of the genes, no matter where they are in the genome.

Coordinate control of dispersed genes in a eukaryotic cell often occurs in response to chemical signals from outside the cell. A steroid hormone, for example, enters a cell and binds to a specific intracellular receptor protein, forming a hormone-receptor complex that serves as a transcription activator (see Figure 11.9). Every gene whose transcription is stimulated by a given steroid hormone, regardless of its chromosomal location, has a control element recognized by that hormone-receptor complex. This is how estrogen activates a group of genes that stimulate cell division in uterine cells, preparing the uterus for pregnancy.

Many signaling molecules, such as nonsteroid hormones and growth factors, bind to receptors on a cell's surface and never actually enter the cell. Such molecules can control gene expression indirectly by triggering signal transduction pathways that activate particular transcription factors (see Figure 11.15). Coordinate regulation in such pathways is the same as for steroid hormones: Genes with the same sets of control elements are activated by the same chemical signals. Because this system for coordinating gene regulation is so widespread, biologists think that it probably arose early in evolutionary history.

Nuclear Architecture and Gene Expression

You saw in Figure 16.23b that each chromosome in the interphase nucleus occupies a distinct territory. The chromosomes are not completely isolated, however. Recently, *chromosome conformation capture (3C)* techniques have been developed that allow researchers to cross-link and identify regions of

Figure 18.12 Chromosomal interactions in the interphase nucleus. Although each chromosome has its own territory (see Figure 16.23b), loops of chromatin may extend into other sites in the nucleus. Some of these sites are transcription factories that are occupied by multiple chromatin loops from the same chromosome (blue loops) or other chromosomes (red and green loops).

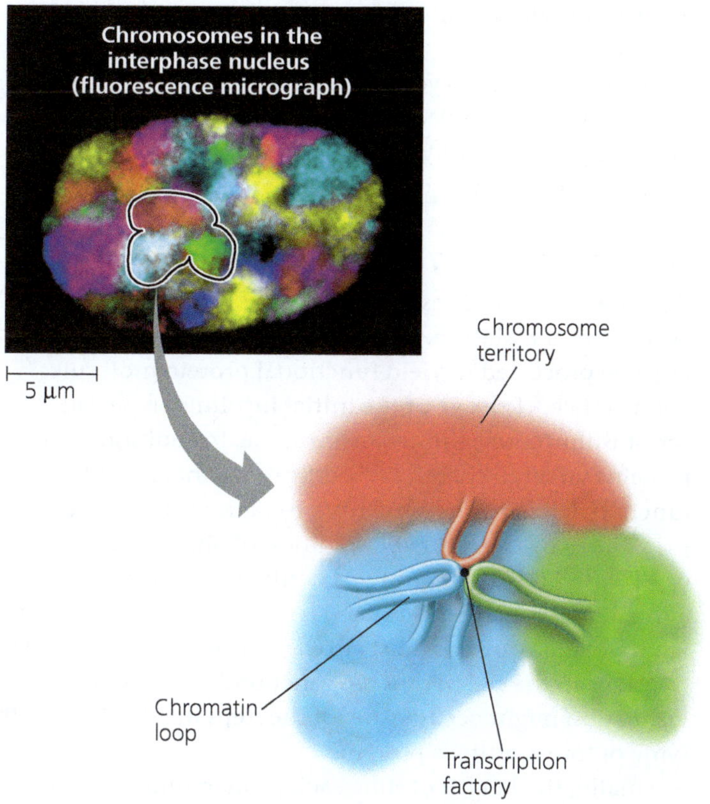

chromosomes that associate with each other during interphase. These studies reveal that loops of chromatin extend from individual chromosomal territories into specific sites in the nucleus **(Figure 18.12)**. Different loops from the same chromosome and loops from other chromosomes may congregate in such sites, some of which are rich in RNA polymerases and other transcription-associated proteins. Like a recreation center that draws members from many different neighborhoods, these so-called *transcription factories* are thought to be areas specialized for a common function.

The old view that the nuclear contents are like a bowl of amorphous chromosomal spaghetti has given way to a new model of a nucleus with a defined architecture and regulated movements of chromatin. Several lines of evidence suggest that genes that are not being expressed are located in the outer edges of the nucleus, while those that are being expressed are found in its interior region. Relocation of particular genes from their chromosomal territories to transcription factories in the interior may be part of the process of readying genes for transcription. How long an individual transcription factory may last has not yet been established. In 2014, the National Institutes of Health announced funding for a new "4D Nucleome" program, which aims to investigate the many fascinating questions addressed by this exciting area of current research.

Mechanisms of Post-Transcriptional Regulation

Transcription alone does not constitute gene expression. The expression of a protein-coding gene is ultimately measured by the amount of functional protein a cell makes, and much happens between the synthesis of the RNA transcript and the activity of the protein in the cell. Many regulatory mechanisms operate at the various stages after transcription (see Figure 18.6). These mechanisms allow a cell to rapidly fine-tune gene expression in response to environmental changes without altering its transcription patterns. Here we discuss how cells can regulate gene expression once a gene has been transcribed.

RNA Processing

RNA processing in the nucleus and the export of mature RNA to the cytoplasm provide opportunities for regulating gene expression not available in prokaryotes. One example of regulation at the RNA-processing level is **alternative RNA splicing**, in which different mRNA molecules are produced from the same primary transcript, depending on which RNA segments are treated as exons and which as introns. Regulatory proteins specific to a cell type control intron-exon choices by binding to regulatory sequences within the primary transcript.

A simple example of alternative RNA splicing is shown in **Figure 18.13** for the troponin T gene, which encodes two different (though related) proteins. Other genes code for many more possible products. For instance, researchers have found a *Drosophila* gene with enough alternatively spliced exons to

Figure 18.13 Alternative RNA splicing of the troponin T gene. The primary transcript of this gene can be spliced in more than one way, generating different mRNA molecules. Notice that one mRNA molecule has ended up with exon 3 (green) and the other with exon 4 (purple). These two mRNAs are translated into different but related muscle proteins.

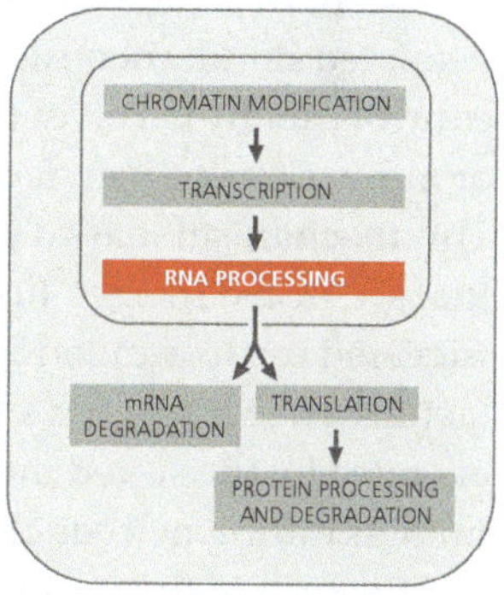

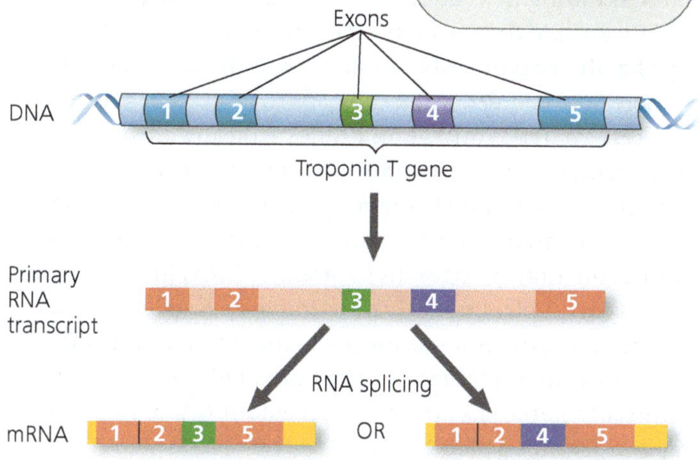

generate about 19,000 membrane proteins that have different extracellular domains. At least 17,500 (94%) of the alternative mRNAs are actually synthesized. Each developing nerve cell in the fly appears to synthesize a different form of the protein, which acts as unique identifier on the cell surface and helps prevent excessive overlap of nerve cells during development of the nervous system.

It is clear that alternative RNA splicing can significantly expand the repertoire of a eukaryotic genome. In fact, alternative splicing was proposed as one explanation for the surprisingly low number of human genes counted when the human genome was sequenced. The number of human genes was found to be similar to that of a soil worm (nematode), a mustard plant, or a sea anemone. This discovery prompted questions about what, if not the number of genes, accounts for the more complex morphology (external form) of humans. It turns out that more than 90% of human protein-coding genes probably undergo alternative splicing. Thus, the extent of alternative splicing greatly multiplies the number of possible human proteins, which may be better correlated with complexity of form.

Initiation of Translation and mRNA Degradation

Translation is another opportunity for regulating gene expression; it occurs most commonly at the initiation stage (see Figure 17.19). For some mRNAs, the initiation of translation can be blocked by regulatory proteins that bind to specific sequences or structures within the untranslated region (UTR) at the 5′ or 3′ end, preventing the attachment of ribosomes. (Recall from Concept 17.3 that both the 5′ cap and the poly-A tail of an mRNA molecule are important for ribosome binding.)

Alternatively, translation of *all* the mRNAs in a cell may be regulated simultaneously. In a eukaryotic cell, such "global" control usually involves the activation or inactivation of one or more of the protein factors required to initiate translation. This mechanism plays a role in starting translation of mRNAs that are stored in eggs. Just after fertilization, translation is triggered by the sudden activation of translation initiation factors. The response is a burst of synthesis of the proteins encoded by the stored mRNAs. Some plants and algae store mRNAs during periods of darkness; light then triggers the reactivation of the translational apparatus.

The life span of mRNA molecules in the cytoplasm is important in determining the pattern of protein synthesis in a cell. Bacterial mRNA molecules typically are degraded by enzymes within a few minutes of their synthesis. This short life span of mRNAs is one reason bacteria can change their patterns of protein synthesis so quickly in response to environmental changes. In contrast, mRNAs in multicellular eukaryotes typically survive for hours, days, or even weeks. For instance, the mRNAs for the hemoglobin polypeptides (α-globin and β-globin) in developing red blood cells are unusually stable, and these long-lived mRNAs are translated repeatedly in red blood cells.

Nucleotide sequences that affect how long an mRNA remains intact are often found in the untranslated region at the 3′ end of the molecule (see Figure 18.8). In one experiment, researchers transferred such a sequence from the short-lived mRNA for a growth factor to the 3′ end of a normally stable globin mRNA. The globin mRNA was quickly degraded.

During the past few years, other mechanisms that degrade or block expression of mRNA molecules have come to light. They involve a group of newly discovered RNA molecules that regulate gene expression at several levels, as we'll discuss shortly.

Protein Processing and Degradation

The final opportunities for controlling gene expression occur after translation. Often, eukaryotic polypeptides must be processed to yield functional protein molecules. For instance, cleavage of the initial insulin polypeptide (pro-insulin) forms the active hormone. In addition, many proteins undergo chemical modifications that make them functional. Regulatory proteins are commonly activated or inactivated by the reversible addition of phosphate groups (see Figure 11.10), and proteins destined for the surface of animal cells acquire sugars (see Figure 6.12). Cell-surface proteins and many others must also be transported to target destinations in the cell in order to function (see Figure 17.22). Regulation might occur at any of the steps involved in modifying or transporting a protein.

Finally, the length of time each protein functions in the cell is strictly regulated by selective degradation. Many proteins, such as the cyclins involved in regulating the cell cycle, must be relatively short-lived if the cell is to function appropriately (see Figure 12.16). To mark a protein for destruction, the cell commonly attaches molecules of a small protein called ubiquitin to the protein. Giant protein complexes called proteasomes then recognize the ubiquitin-tagged proteins and degrade them.

MB **Animation: Post-Transcriptional Control Mechanisms**

CONCEPT CHECK 18.2

1. In general, what are the effects of histone acetylation and DNA methylation on gene expression?
2. **MAKE CONNECTIONS ➤** Speculate about whether the same enzyme could methylate both a histone and a DNA base. (See Concept 5.4.)
3. Compare the roles of general and specific transcription factors in regulating gene expression.
4. Once mRNA encoding a particular protein reaches the cytoplasm, what are four mechanisms that can regulate the amount of the protein that is active in the cell?
5. **WHAT IF? ➤** Suppose you compared the nucleotide sequences of the distal control elements in the enhancers of three genes that are expressed only in muscle cells. What would you expect to find? Why?

For suggested answers, see Appendix A.

CONCEPT 18.3

Noncoding RNAs play multiple roles in controlling gene expression

Genome sequencing has revealed that protein-coding DNA accounts for only 1.5% of the human genome and a similarly small percentage of the genomes of many other multicellular eukaryotes. A very small fraction of the non-protein-coding DNA consists of genes for RNAs such as ribosomal RNA and transfer RNA. Until recently, scientists assumed that most of the remaining DNA was not transcribed, thinking that since it didn't specify proteins or the few known types of RNA, such DNA didn't contain meaningful genetic information—in fact, it was called "junk DNA." However, a flood of recent data has contradicted this idea. For example, a massive study of the entire human genome showed that roughly 75% of the genome is transcribed at some point in any given cell. Introns account for only a fraction of this transcribed, nontranslated RNA. These and other results suggest that a significant amount of the genome may be transcribed into non-protein-coding RNAs—also called *noncoding RNAs*, or *ncRNAs*—including a variety of small RNAs. Researchers are uncovering more evidence of the biological roles of these ncRNAs every day.

Biologists are excited about these discoveries, which have revealed a large and diverse population of RNA molecules in the cell that play crucial roles in regulating gene expression—but have gone largely unnoticed until fairly recently. Our long-standing view that because mRNAs code for proteins, they are the most important RNAs functioning in the cell demands revision. This represents a major shift in the thinking of biologists, one that you are witnessing as students entering this field of study.

Effects on mRNAs by MicroRNAs and Small Interfering RNAs

Regulation by both small and large ncRNAs occurs at several points in the pathway of gene expression, including mRNA translation and chromatin modification. We'll examine two types of small ncRNAs, the importance of which was acknowledged when their discovery was the focus of the 2006 Nobel Prize in Physiology or Medicine, which was awarded for work completed only eight years earlier.

Since 1993, a number of research studies have uncovered **microRNAs (miRNAs)**—small, single-stranded RNA molecules capable of binding to complementary sequences in mRNA molecules. A longer RNA precursor is processed by cellular enzymes into an miRNA, a single-stranded RNA of about 22 nucleotides that forms a complex with one or more proteins **(Figure 18.14)**. The miRNA allows the complex to bind to any mRNA molecule with at least 7 or 8 nucleotides of complementary sequence. The miRNA-protein complex then degrades the target mRNA or, less often, simply blocks its translation. There are approximately 1,500 genes for miRNAs in the human genome, and biologists estimate that expression of at least one-half of all human genes may be regulated by miRNAs, a remarkable figure given that the existence of miRNAs was unknown 25 years ago.

Another class of small RNAs, similar in size and function to miRNAs, is called **small interfering RNAs (siRNAs)**. Both miRNAs and siRNAs can associate with the same proteins, producing similar results. In fact, if siRNA precursor RNA molecules are injected into a cell, the cell's machinery can process them into siRNAs that turn off expression of genes with related sequences, similarly to how miRNAs function. The distinction between miRNAs and siRNAs is based on subtle differences in the structure of their precursors, which in both cases are RNA molecules that are mostly double-stranded. The blocking of gene expression by siRNAs is referred to as **RNA interference (RNAi)**, and it is used in the laboratory as a means of disabling specific genes to investigate their function.

How did the RNAi pathway evolve? As you will learn in Concept 19.2, some viruses have double-stranded RNA genomes. Given that the cellular RNAi pathway can process double-stranded RNAs into homing devices that lead to destruction of related RNAs, some scientists think that this pathway may have evolved as a natural defense against

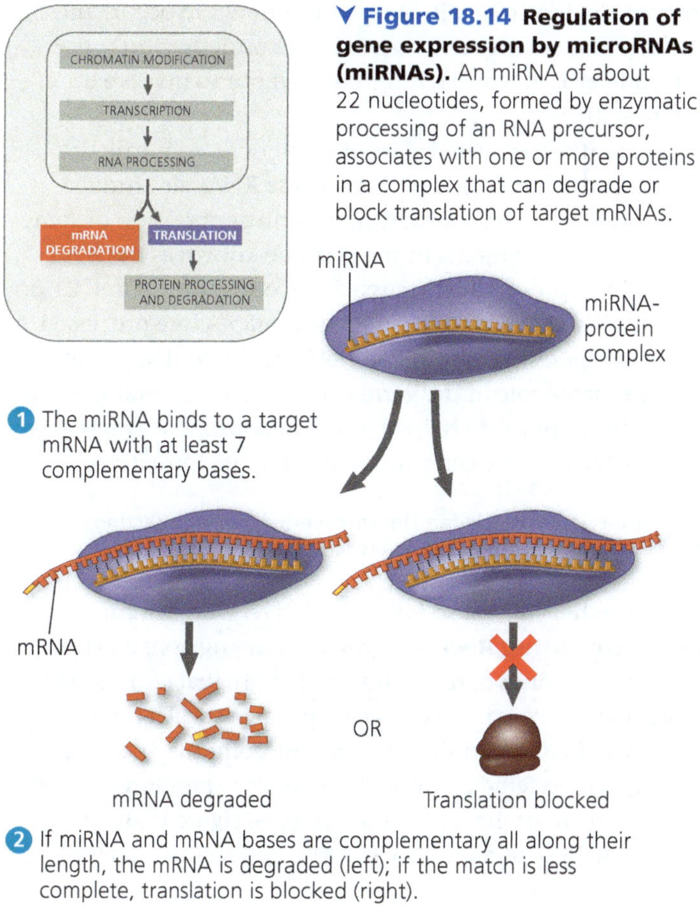

Figure 18.14 Regulation of gene expression by microRNAs (miRNAs). An miRNA of about 22 nucleotides, formed by enzymatic processing of an RNA precursor, associates with one or more proteins in a complex that can degrade or block translation of target mRNAs.

infection by such viruses. However, the fact that RNAi can also affect the expression of nonviral cellular genes may reflect a different evolutionary origin for the RNAi pathway. Moreover, many species, including mammals, apparently produce their own long, double-stranded RNA precursors to small RNAs such as siRNAs. Once produced, these RNAs can interfere with gene expression at stages other than translation, as we'll discuss next.

Chromatin Remodeling and Effects on Transcription by ncRNAs

In addition to regulating mRNAs, some ncRNAs act to bring about remodeling of chromatin structure. One example occurs during formation of heterochromatin at the centromere, as studied in a species of yeast.

In the S phase of the cell cycle, the centromeric regions of DNA must be loosened for chromosomal replication and then re-condensed into heterochromatin in preparation for mitosis. In some yeasts, siRNAs produced by the yeast cells themselves are required to re-form the heterochromatin at the centromeres. A model for how this happens is shown in **Figure 18.15**. Exactly how the process starts is still under debate, but biologists agree on the general idea: The siRNA system in yeast interacts with other ncRNAs and with chromatin-modifying enzymes to condense the centromere chromatin into heterochromatin. Under most conditions in mammalian cells, siRNAs are not known to occur, and the mechanism for centromere DNA condensation is not yet understood. However, it may turn out to involve other small ncRNAs.

A recently discovered class of small ncRNAs is called *piwi-interacting RNAs*, or *piRNAs*. These RNAs also induce formation of heterochromatin, blocking expression of some parasitic DNA elements in the genome known as transposons. (Transposons are discussed in Concepts 21.4 and 21.5.) Usually 24–31 nucleotides in length, piRNAs are processed from a longer, single-stranded RNA precursor. They play an indispensable role in the germ cells of many animal species, where they appear to help reestablish appropriate methylation patterns in the genome during gamete formation.

MB **Interview with Haifan Lin: Discovering piwi-interacting RNAs and connections to testicular cancer**

Researchers have also found a relatively large number of **long noncoding RNAs (lncRNAs)**, ranging from 200 to hundreds of thousands of nucleotides in length, that are expressed at significant levels in specific cell types at particular times. One such lncRNA is responsible for X chromosome inactivation, which, in most female mammals, prevents expression of genes located on one of the X chromosomes (see Figure 15.8). In this case, lncRNAs—transcripts of the *XIST* gene located on the

Figure 18.15 Condensation of chromatin at the centromere. In one type of yeast, siRNAs and longer noncoding RNAs cooperate in the pathway that leads to re-formation of highly condensed heterochromatin at the centromere of each chromatid after DNA replication.

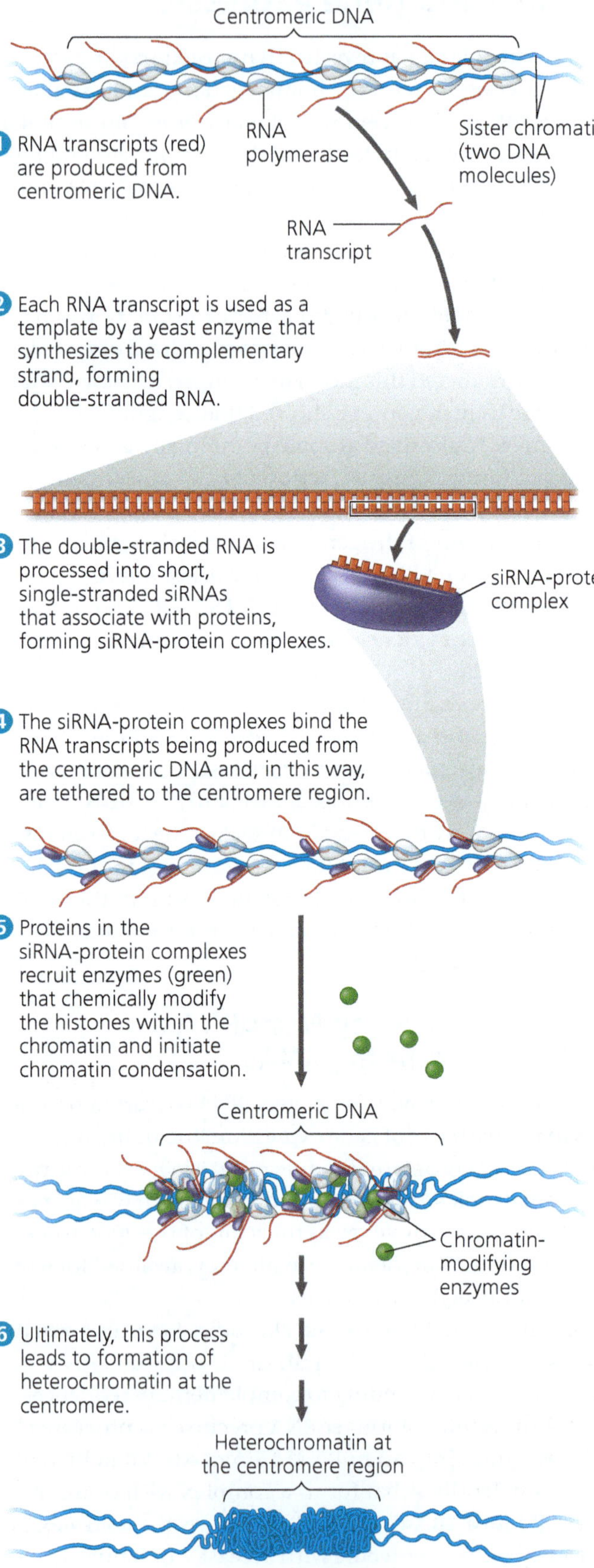

chromosome to be inactivated—bind back to and coat that chromosome, and this binding leads to condensation of the entire chromosome into heterochromatin.

The cases described above involve chromatin remodeling in large regions of the chromosome. Because chromatin structure affects transcription and thus gene expression, RNA-based regulation of chromatin structure is sure to play an important role in gene regulation. Additionally, some experimental evidence supports the idea of an alternate role for lncRNAs in which they can act as a scaffold, bringing together DNA, proteins, and other RNAs into complexes. These associations may act either to condense chromatin or, in some cases, to help bring the enhancer of a gene together with mediator proteins and the gene's promoter, activating gene expression in a more direct fashion.

The Evolutionary Significance of Small ncRNAs

EVOLUTION Small ncRNAs can regulate gene expression at multiple steps and in many ways. While this section has focused on ncRNAs in eukaryotes, small ncRNAs are also used by bacteria as a defense system, called the CRISPR-Cas9 system, against viruses that infect them. (You'll learn more about this in Concept 19.2.) The use of ncRNAs thus evolved long ago, but we don't yet know how bacterial ncRNAs are related to those of eukaryotes.

What about the evolutionary significance of small eukaryotic ncRNAs? In general, extra levels of gene regulation might allow evolution of a higher degree of complexity of form. The versatility of miRNA regulation has therefore led some biologists to hypothesize that an increase in the number of different miRNAs specified by the genome of a given species has allowed morphological complexity to increase over evolutionary time. While this hypothesis is still being evaluated, it is logical to expand the discussion to include all small ncRNAs. Exciting new techniques for rapidly sequencing genomes have allowed biologists to begin asking how many genes for ncRNAs are present in the genome of any given species. A survey of different species supports the notion that siRNAs evolved first, followed by miRNAs and later piRNAs, which are found only in animals. And while there are hundreds of types of miRNAs, there appear to be 60,000 or so types of piRNAs, allowing the potential for very sophisticated gene regulation by piRNAs.

Given the extensive functions of ncRNAs, it is not surprising that many of the ncRNAs characterized thus far play important roles in embryonic development—the topic we turn to in the next section. Embryonic development is perhaps the ultimate example of precisely regulated gene expression.

CONCEPT CHECK 18.3

1. Compare miRNAs and siRNAs, including their functions.
2. **WHAT IF?** ➤ Suppose the mRNA being degraded in Figure 18.14 coded for a protein that promotes cell division in a multicellular organism. What would happen if a mutation disabled the gene for the miRNA that triggers this degradation?
3. **MAKE CONNECTIONS** ➤ Inactivation of one of the X chromosomes in female mammals involves lncRNA called *XIST* RNA, mentioned in this section and in Concept 15.2. Describe transcription and binding of *XIST* RNA, then suggest a model for how it initiates Barr body formation.

For suggested answers, see Appendix A.

CONCEPT 18.4

A program of differential gene expression leads to the different cell types in a multicellular organism

In the embryonic development of multicellular organisms, a fertilized egg (a zygote) gives rise to cells of many different types, each with a different structure and corresponding function. Typically, cells are organized into tissues, tissues into organs, organs into organ systems, and organ systems into the whole organism. Thus, any developmental program must produce cells of different types that form higher-level structures arranged in a particular way in three dimensions. The processes that occur during development in plants and animals are detailed in Chapters 35 and 47, respectively. In this chapter, we focus on the program of regulation of gene expression that orchestrates development, using a few animal species as examples.

A Genetic Program for Embryonic Development

The photos in **Figure 18.16** illustrate the dramatic difference between a frog zygote (fertilized egg) and the tadpole it

Figure 18.16 From fertilized egg to animal: What a difference four days makes. It takes just four days for cell division, differentiation, and morphogenesis to transform each of the fertilized frog eggs shown in **(a)** into a tadpole like the one in **(b)**.

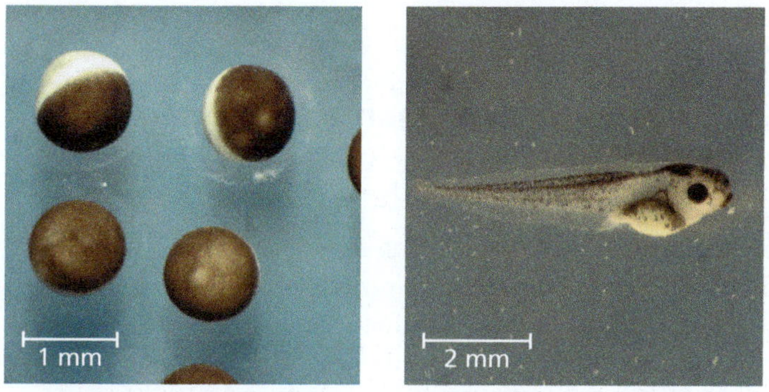

(a) Fertilized eggs of a frog **(b) Newly hatched tadpole**

becomes. This remarkable transformation results from three interrelated processes: cell division, cell differentiation, and morphogenesis. Through a succession of mitotic cell divisions, the zygote gives rise to a large number of cells. Cell division alone, however, would merely produce a great ball of identical cells, nothing like a tadpole. During embryonic development, cells not only increase in number, but also undergo cell **differentiation**, the process by which cells become specialized in structure and function. Moreover, the different kinds of cells are not randomly distributed but are organized into tissues and organs in a particular three-dimensional arrangement. The physical processes that give an organism its shape constitute **morphogenesis**, the development of the form of an organism and its structures.

All three processes are rooted in cellular behavior. Even morphogenesis, the shaping of the organism, can be traced back to changes in the shape, motility, and other characteristics of the cells that make up various regions of the embryo. As you have seen, the activities of a cell depend on the genes it expresses and the proteins it produces. Almost all cells in an organism have the same genome; therefore, differential gene expression results from the genes being regulated differently in each cell type.

In Figure 18.11, you saw a simplified view of how differential gene expression occurs in two cell types, a liver cell and a lens cell. Each of these fully differentiated cells has a particular mix of specific activators that turn on the collection of genes whose products are required in the cell. The fact that both cells arose through a series of mitoses from a common fertilized egg inevitably leads to a question: How do different sets of activators come to be present in the two cells?

It turns out that materials placed into the egg by maternal cells set up a sequential program of gene regulation that is carried out as embryonic cells divide, and this program coordinates cell differentiation during embryonic development. To understand how this works, we will consider two basic developmental processes. First, we'll explore how cells that arise from early embryonic mitoses develop the differences that start each cell along its own differentiation pathway. Second, we'll see how cellular differentiation leads to one particular cell type, using muscle development as an example.

Cytoplasmic Determinants and Inductive Signals

What generates the first differences among cells in an early embryo? And what controls the differentiation of all the various cell types as development proceeds? By this point in the chapter, you can probably deduce the answer: The specific genes expressed in any particular cell of a developing organism determine its path. Two sources of information, used to varying extents in different species, "tell" a cell which genes to express at any given time during embryonic development.

One important source of information early in development is the egg's cytoplasm, which contains both RNA and proteins encoded by the mother's DNA. The cytoplasm of an unfertilized egg is not homogeneous. Messenger RNA, proteins, other substances, and organelles are distributed unevenly in the unfertilized egg, and this unevenness has a profound impact on the development of the future embryo in many species. Maternal substances in the egg that influence the course of early development are called **cytoplasmic determinants** (Figure 18.17a). After fertilization, early mitotic divisions distribute the zygote's

Figure 18.17 Sources of developmental information for the early embryo.

(a) Cytoplasmic determinants in the egg

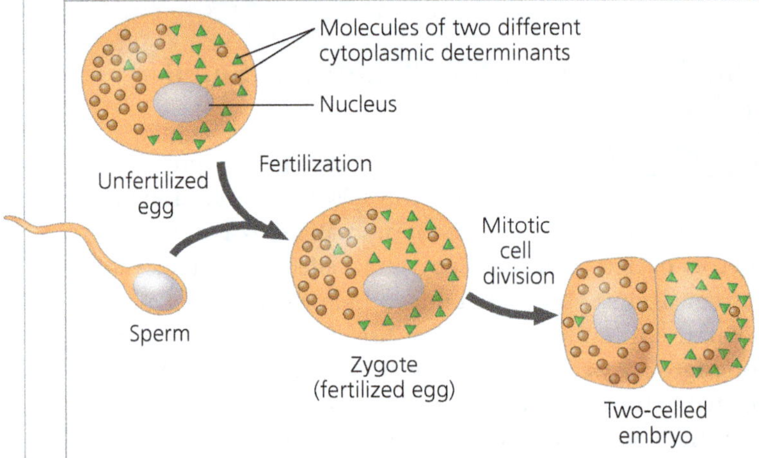

The unfertilized egg has molecules in its cytoplasm, encoded by the mother's genes, that influence development. Many of these cytoplasmic determinants, like the two shown here, are unevenly distributed in the egg. After fertilization and mitotic division, the cell nuclei of the embryo are exposed to different sets of cytoplasmic determinants and, as a result, express different genes.

(b) Induction by nearby cells

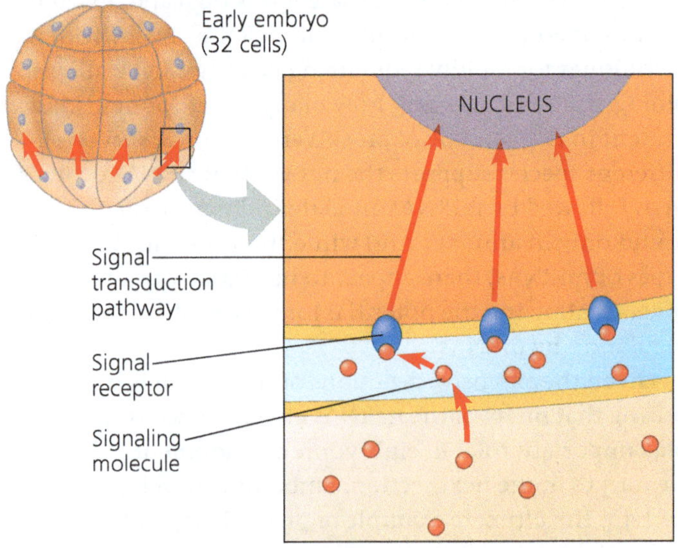

Cells at the bottom of the early embryo are releasing molecules that signal (induce) nearby cells to change their gene expression.

cytoplasm into separate cells. The nuclei of these cells may thus be exposed to different cytoplasmic determinants, depending on which portions of the zygotic cytoplasm a cell received. The combination of cytoplasmic determinants in a cell helps determine its developmental fate by regulating expression of the cell's genes during the course of cell differentiation.

The other major source of developmental information, which becomes increasingly important as the number of embryonic cells increases, is the environment around a particular cell. Most influential are the signals conveyed to an embryonic cell from other embryonic cells in the vicinity, including contact with cell-surface molecules on neighboring cells and the binding of growth factors secreted by neighboring cells (see Concept 11.1). Such signals cause changes in the target cells, a process called **induction** **(Figure 18.17b)**. The molecules passing along these signals within the target cell are cell-surface receptors and other signaling pathway proteins. In general, the signaling molecules send a cell down a specific developmental path by causing changes in its gene expression that eventually result in observable cellular changes. Thus, interactions between embryonic cells help induce differentiation into the many specialized cell types making up a new organism.

Sequential Regulation of Gene Expression During Cellular Differentiation

The earliest changes that set a cell on its path to specialization are subtle ones, showing up only at the molecular level. Before biologists knew much about the molecular changes occurring in embryos, they coined the term **determination** to refer to the point at which an embryonic cell is irreversibly committed to becoming a particular cell type. Once it has undergone determination, an embryonic cell can be experimentally placed in another location in the embryo and it will still differentiate into the cell type that is its normal fate. Differentiation, then, is the process by which a cell attains its determined fate. As the tissues and organs of an embryo develop and their cells differentiate, the cells become more noticeably different in structure and function.

Today we understand determination in terms of molecular changes. The outcome of determination, observable cell differentiation, is marked by the expression of genes for *tissue-specific proteins*. These proteins are found only in a specific cell type and give the cell its characteristic structure and function. The first evidence of differentiation is the appearance of mRNAs for these proteins. Eventually, differentiation is observable with a microscope as changes in cellular structure. On the molecular level, different sets of genes are sequentially expressed in a regulated manner as new cells arise from division of their precursors. A number of the steps in gene expression may be regulated during differentiation, transcription being the most common. In the fully differentiated cell, transcription remains the principal regulatory point for maintaining appropriate gene expression.

Differentiated cells are specialists at making tissue-specific proteins. For example, as a result of transcriptional regulation, liver cells specialize in making albumin, and lens cells specialize in making crystallin (see Figure 18.11). Skeletal muscle cells in vertebrates are another instructive example. Each of these cells is a long fiber containing many nuclei within a single plasma membrane. Skeletal muscle cells have high concentrations of muscle-specific versions of the contractile proteins myosin and actin, as well as membrane receptor proteins that detect signals from nerve cells.

Muscle cells develop from embryonic precursor cells that have the potential to develop into a number of cell types, including cartilage cells and fat cells, but particular conditions commit them to becoming muscle cells. Although the committed cells appear unchanged under the microscope, determination has occurred, and they are now *myoblasts*. Eventually, myoblasts start to churn out large amounts of muscle-specific proteins and fuse to form mature, elongated, multinucleate skeletal muscle cells.

Researchers have worked out what happens at the molecular level during muscle cell determination. To do so, they grew embryonic precursor cells in culture and analyzed them using molecular techniques you will learn about in Concepts 20.1 and 20.2. In a series of experiments, they isolated different genes, caused each to be expressed in a separate embryonic precursor cell, and then looked for differentiation into myoblasts and muscle cells. In this way, they identified several so-called "master regulatory genes" whose protein products commit the cells to becoming skeletal muscle. Thus, in the case of muscle cells, the molecular basis of determination is the expression of one or more of these master regulatory genes.

To understand more about how determination occurs in muscle cell differentiation, let's focus on the master regulatory gene called *myoD*. The *myoD* gene deserves its designation as a master regulatory gene. Researchers have shown that the MyoD protein it encodes is capable of changing some kinds of fully differentiated nonmuscle cells, such as fat cells and liver cells, into muscle cells. Why doesn't MyoD work on *all* kinds of cells? One likely explanation is that activation of muscle-specific genes is not solely dependent on MyoD but requires a particular *combination* of regulatory proteins, some of which are lacking in cells that do not respond to MyoD. The determination and differentiation of other kinds of tissues may play out in a similar fashion. A growing body of experimental evidence supports the idea that master regulatory proteins like MyoD might actually function by opening the chromatin in particular regions. This allows access to transcription machinery for activation of the next set of cell-type-specific genes.

What is the molecular basis for muscle cell differentiation? The MyoD protein is a transcription factor (see Figure 18.9) that binds to specific control elements in the enhancers of various target genes and stimulates their expression **(Figure 18.18)**. Some target genes for MyoD encode still other muscle-specific transcription factors. MyoD also stimulates expression of the

Figure 18.18 Determination and differentiation of muscle cells. Skeletal muscle cells arise from embryonic cells as a result of changes in gene expression. (In this depiction, the process of gene activation is greatly simplified.)

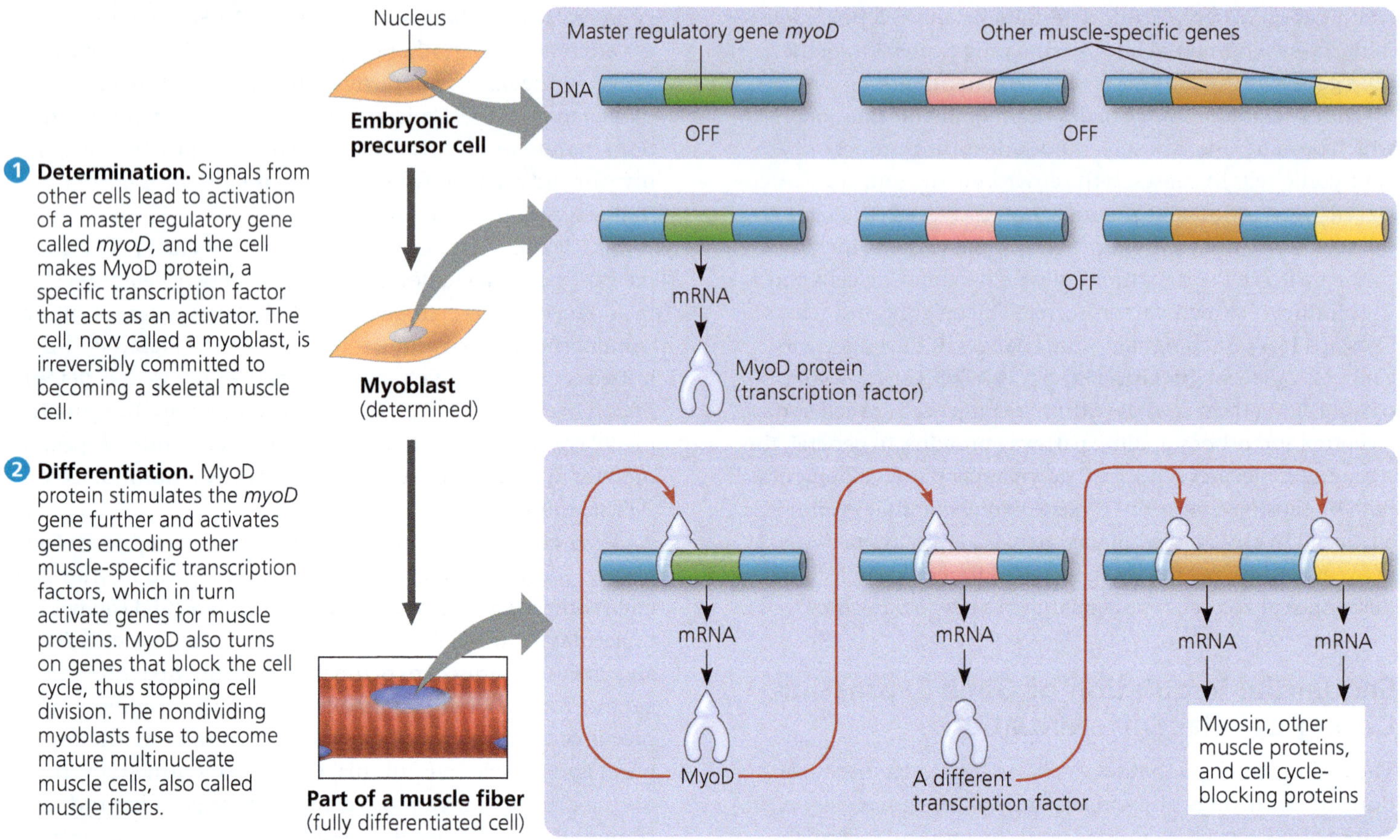

WHAT IF? ➤ *What would happen if a mutation in the* myoD *gene resulted in the production of an altered MyoD protein that could not activate the* myoD *gene?*

myoD gene itself, an example of positive feedback that perpetuates MyoD's effect in maintaining the cell's differentiated state. Presumably, all the genes activated by MyoD have enhancer control elements recognized by MyoD and are thus coordinately controlled. Finally, the secondary transcription factors activate the genes for proteins such as myosin and actin that confer the unique properties of skeletal muscle cells.

We have now seen how different programs of gene expression that are activated in the fertilized egg can result in differentiated cells and tissues. But for the tissues to function effectively in the organism as a whole, the organism's *body plan*—its overall three-dimensional arrangement—must be established and superimposed on the differentiation process. Next we'll investigate the molecular basis for the establishment of the body plan, using the well-studied fruit fly *Drosophila melanogaster* as an example.

Pattern Formation: Setting Up the Body Plan

Cytoplasmic determinants and inductive signals both contribute to the development of a spatial organization in which the tissues and organs of an organism are all in their characteristic places. This process is referred to as **pattern formation**.

Just as the locations of the front, back, and sides of a new building are determined before construction begins, pattern formation in animals begins in the early embryo, when the major axes of an animal are established. In a bilaterally symmetrical animal, the relative positions of head and tail, right and left sides, and back and front—the three major body axes—are set up before the organs appear. The molecular cues that control pattern formation, collectively called **positional information**, are provided by cytoplasmic determinants and inductive signals (see Figure 18.17). These cues tell a cell its location relative to the body axes and to neighboring cells and determine how the cell and its descendants will respond to future molecular signals.

During the first half of the 20th century, classical embryologists made detailed anatomical observations of embryonic development in a number of species and performed experiments in which they manipulated embryonic tissues. Although this research laid the groundwork for understanding the mechanisms of development, it did not reveal the specific molecules that guide development or determine how patterns are established.

In the 1940s, scientists began using the genetic approach—the study of mutants—to investigate *Drosophila* development. That approach has had spectacular success. These studies have established that genes control development and have led to an understanding of the key roles that specific molecules play in defining position and directing differentiation. By combining anatomical, genetic, and biochemical approaches to the study of *Drosophila* development, researchers have discovered developmental principles common to many other species, including humans.

The Life Cycle of Drosophila

Fruit flies and other arthropods have a modular construction, an ordered series of segments. These segments make up the body's three major parts: the head, the thorax (the mid-body, from which the wings and legs extend), and the abdomen **(Figure 18.19a)**. Like other bilaterally symmetrical animals, *Drosophila* has an anterior-posterior (head-to-tail) axis, a dorsal-ventral (back-to-belly) axis, and a right-left axis. In *Drosophila*, cytoplasmic determinants that are localized in the unfertilized egg provide positional information for the placement of anterior-posterior and dorsal-ventral axes even before fertilization. We'll focus here on the molecules involved in establishing the anterior-posterior axis.

The *Drosophila* egg develops in one of the female's ovaries, adjacent to ovarian cells called nurse cells and surrounded by so-called follicle cells (**Figure 18.19b**, top). These support cells supply the egg with nutrients, mRNAs, and other substances needed for development and make the egg shell. After fertilization and laying of the egg, embryonic development results in the formation of a segmented larva, which goes through three larval stages. Then, in a process much like that by which a caterpillar becomes a butterfly, the fly larva forms a pupa in which it metamorphoses into the adult fly pictured in Figure 18.19a.

Genetic Analysis of Early Development: Scientific Inquiry

Edward B. Lewis was a visionary American biologist who, in the 1940s, first showed the value of the genetic approach to studying embryonic development in *Drosophila*. Lewis studied bizarre mutant flies with developmental defects that led to extra wings or legs in the wrong place **(Figure 18.20)**. He located the mutations on the fly's genetic map, thus connecting the developmental abnormalities to specific genes. This research supplied the first concrete evidence that genes somehow direct the developmental processes studied by embryologists. The genes Lewis discovered, called **homeotic genes**, are regulatory genes that control pattern formation in the late embryo, larva, and adult.

▼ Figure 18.19 Key events in the *Drosophila* life cycle.

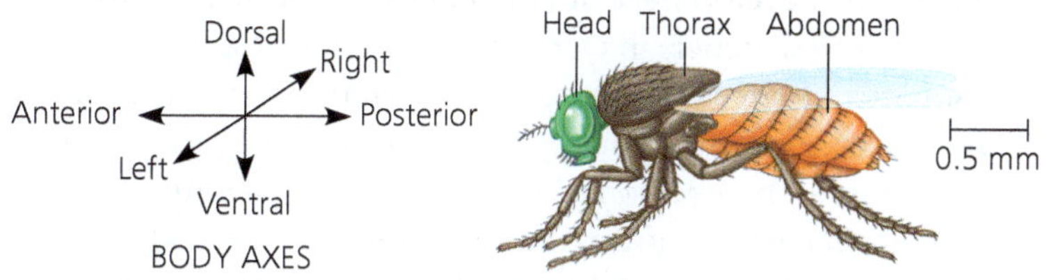

(a) Adult. The adult fly is segmented, and multiple segments make up each of the three main body parts—head, thorax, and abdomen. The body axes are shown by arrows.

1 Developing egg within one ovarian follicle (among many in an ovary). The egg (yellow) is surrounded by support cells (follicle cells).

2 Mature, unfertilized egg. The developing egg enlarges as nutrients and mRNAs are supplied to it by other support cells (nurse cells), which shrink. Eventually, the mature egg fills the egg shell that is secreted by the follicle cells.

3 Fertilized egg. The egg is fertilized within the mother and then laid.

4 Segmented embryo. The egg develops into a segmented embryo.

5 Larva. The embryo develops into a larva, which has three stages. The third stage forms a pupa (not shown), within which the larva metamorphoses into the adult shown in (a).

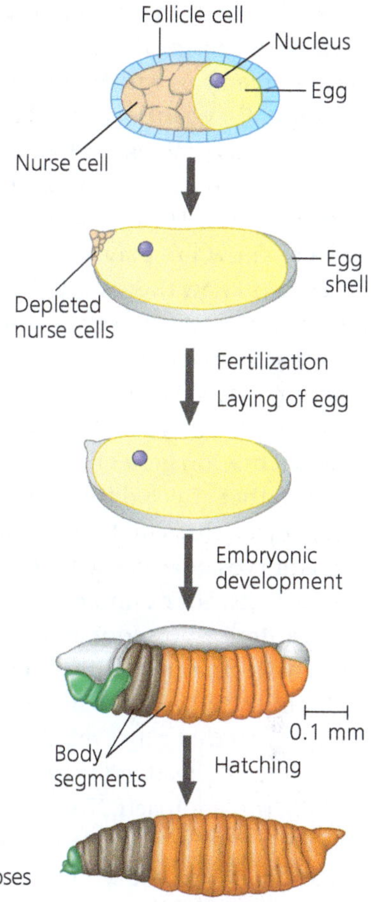

(b) Development from egg to larva.

▼ Figure 18.20 Abnormal pattern formation in *Drosophila*. Mutations in homeotic genes cause abnormal placement of structures in an animal, such as the legs extending from the mutant fly's head in place of antenna. (Colorized SEMs)

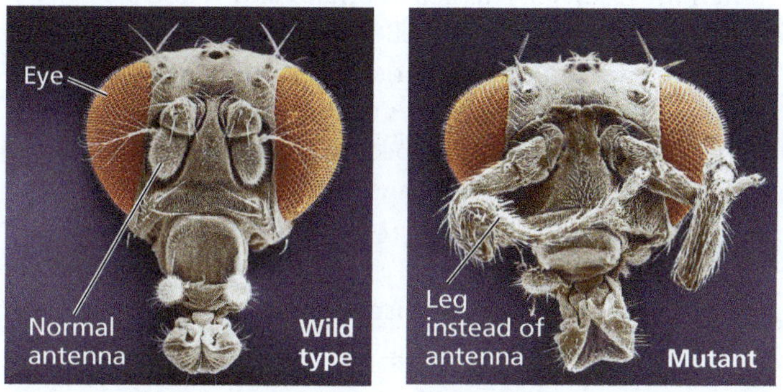

Further insight into pattern formation during early embryonic development did not come for another 30 years, when two researchers in Germany, Christiane Nüsslein-Volhard and Eric Wieschaus, set out to identify *all* the genes that affect segment formation in *Drosophila*. The project was daunting for three reasons. The first was the sheer number of *Drosophila* genes, now known to total about 14,000. The genes affecting segmentation might be just a few needles in a haystack or might be so numerous and varied that the scientists would be unable to make sense of them. Second, mutations affecting a process as fundamental as segmentation would surely be **embryonic lethals**, mutations with phenotypes causing death at the embryonic or larval stage. Because organisms with embryonic lethal mutations never reproduce, they cannot be bred for study. The researchers dealt with this problem by looking for recessive mutations, which can be propagated in heterozygous flies that act as genetic carriers. Third, cytoplasmic determinants in the egg were known to play a role in axis formation, so the researchers knew they would have to study the mother's genes as well as those of the embryo. It is the mother's genes that we will discuss further as we focus on how the anterior-posterior body axis is set up in the developing egg.

Nüsslein-Volhard and Wieschaus began their search for segmentation genes by exposing flies to a mutagenic chemical that affected the flies' gametes. They mated the mutagenized flies and then scanned their descendants for dead embryos or larvae with abnormal segmentation or other defects. For example, to find genes that might set up the anterior-posterior axis, they looked for embryos or larvae with abnormal ends, such as two heads or two tails, predicting that such abnormalities would arise from mutations in maternal genes required for correctly setting up the offspring's head or tail end.

Using this approach, Nüsslein-Volhard and Wieschaus eventually identified about 1,200 genes essential for pattern formation during embryonic development. Of these, about 120 were essential for normal segmentation. Over several years, the researchers were able to group these segmentation genes by general function, to find them on the fly's chromosomes, and to isolate many of them for further study in the lab. The result was a detailed molecular understanding of the early steps in pattern formation in *Drosophila*.

When the results of Nüsslein-Volhard and Wieschaus were combined with Lewis's earlier work, a coherent picture of *Drosophila* development emerged. In recognition of their discoveries, the three researchers were awarded a Nobel Prize in 1995. Next, let's consider a specific example of the genes that Nüsslein-Volhard, Wieschaus, and co-workers found.

Axis Establishment

As we mentioned earlier, cytoplasmic determinants in the egg are the substances that initially establish the axes of the *Drosophila* body. These substances are encoded by genes of the mother, fittingly called maternal effect genes. A gene classified as a **maternal effect gene** is one that, when mutant in the mother, results in a mutant phenotype in the offspring, regardless of the offspring's own genotype. In fruit fly development, the mRNA or protein products of maternal effect genes are placed in the egg while it is still in the mother's ovary. When the mother has a mutation in such a gene, she makes a defective gene product (or none at all), and her eggs are defective; when these eggs are fertilized, they fail to develop properly.

Because they control the orientation (polarity) of the egg and consequently that of the fly, these maternal effect genes are also called *egg-polarity genes*. One group of these genes sets up the anterior-posterior axis of the embryo, while a second group establishes the dorsal-ventral axis. Like mutations in segmentation genes, mutations in maternal effect genes are generally embryonic lethals.

Bicoid: A Morphogen That Determines Head Structures To see how maternal effect genes determine the body axes of the offspring, we will focus on one such gene, called ***bicoid***, a term meaning "two-tailed." An embryo or larva whose mother has two mutant *bicoid* alleles lacks the front half of its body and has posterior structures at both ends **(Figure 18.21)**. This phenotype suggested to Nüsslein-Volhard and her colleagues that the product of the mother's *bicoid* gene is essential for setting up the anterior end of the fly and might be concentrated at the future anterior end of the embryo. This hypothesis is an example of the *morphogen gradient hypothesis* first proposed

▼ Figure 18.21 Effect of the *bicoid* gene on *Drosophila* development. A wild-type fruit fly larva has a head, three thoracic (T) segments, eight abdominal (A) segments, and a tail. A larva whose mother has two mutant alleles of the *bicoid* gene has two tails and lacks all anterior structures (LMs).

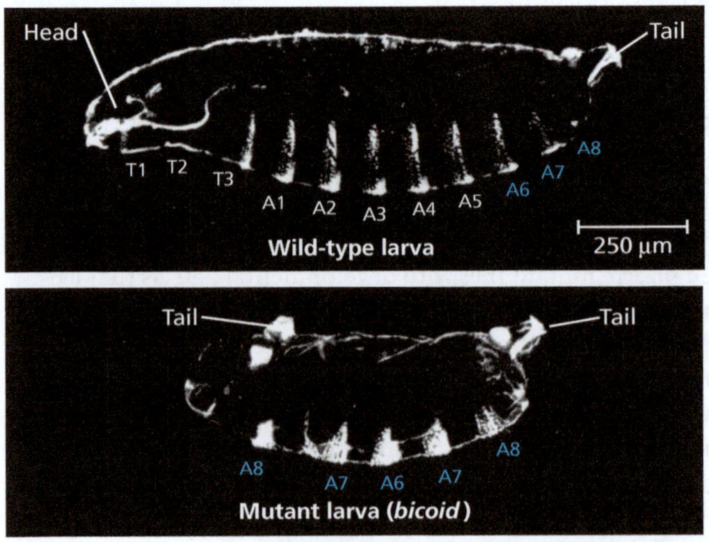

by embryologists a century ago, in which gradients of substances called **morphogens** establish an embryo's axes and other features of its form.

DNA technology and other modern biochemical methods enabled the researchers to test whether the *bicoid* product, a protein called Bicoid, is in fact a morphogen that determines the anterior end of the fly. The first question they asked was whether the mRNA and protein products of this gene are located in the egg in a position consistent with the hypothesis. They found that *bicoid* mRNA is highly concentrated at the extreme anterior end of the mature egg **(Figure 18.22)**. After the egg is fertilized, the mRNA is translated into protein. The Bicoid protein then diffuses from the anterior end toward the posterior, resulting in a gradient of protein within the early embryo, with the highest concentration at the anterior end. These results are consistent with the hypothesis that Bicoid protein specifies the fly's anterior end. To test the hypothesis more specifically, scientists injected pure *bicoid* mRNA into various regions of early embryos. The protein that resulted from its translation caused anterior structures to form at the injection sites.

The *bicoid* research was groundbreaking for several reasons. First, it led to the identification of a specific protein required for some of the earliest steps in pattern formation. It thus helped us understand how different regions of the egg can give rise to cells that go down different developmental pathways. Second, it increased our understanding of the mother's critical role in the initial phases of embryonic development. Finally, the principle that a gradient of morphogens can determine polarity and position has proved to be a key developmental concept for a number of species, just as early embryologists had hypothesized.

Maternal mRNAs are crucial during development of many species. In *Drosophila*, gradients of specific proteins encoded by maternal mRNAs not only determine the posterior and anterior ends but also establish the dorsal-ventral axis. As the fly embryo grows, it reaches a point when the embryonic program of gene expression takes over, and the maternal mRNAs must be destroyed. (This process involves miRNAs in *Drosophila* and other species.) Later, positional information encoded by the embryo's genes, operating on an ever finer scale, establishes a specific number of correctly oriented segments and triggers the formation of each segment's characteristic structures. When the genes operating in this final step are abnormal, the pattern of the adult is abnormal, as you saw in Figure 18.20.

MB **Interview with Nancy Hopkins: Studying the genetic basis of development in zebrafish**

Evolutionary Developmental Biology ("Evo-Devo")

EVOLUTION The fly with legs emerging from its head in Figure 18.20 is the result of a single mutation in one gene, a homeotic gene. The gene does not encode any antenna protein, however. Instead, it encodes a transcription factor that regulates other genes, and its malfunction leads to misplaced structures, such as legs instead of antennae. The observation that a change in gene regulation during development could lead to such a fantastic change in body form prompted some scientists to consider whether these types of mutations could contribute to evolution by generating novel body shapes. Ultimately, this line of inquiry gave rise to the field of evolutionary developmental biology, so-called "evo-devo," which will be further discussed in Concept 21.6.

In this section, we have seen how a carefully orchestrated program of sequential gene regulation controls the

▼ Figure 18.22

Inquiry Could Bicoid be a morphogen that determines the anterior end of a fruit fly?

Experiment Using a genetic approach to study *Drosophila* development, Christiane Nüsslein-Volhard and colleagues at two research institutions in Germany analyzed expression of the *bicoid* gene. The researchers hypothesized that *bicoid* normally codes for a morphogen that specifies the head (anterior) end of the embryo. To begin to test this hypothesis, they used molecular techniques to determine whether the mRNA and protein encoded by this gene were found in the anterior end of the fertilized egg and early embryo of wild-type flies.

Results *Bicoid* mRNA (dark blue in the light micrographs and drawings) was confined to the anterior end of the unfertilized egg. Later in development, Bicoid protein (dark orange) was seen to be concentrated in cells at the anterior end of the embryo.

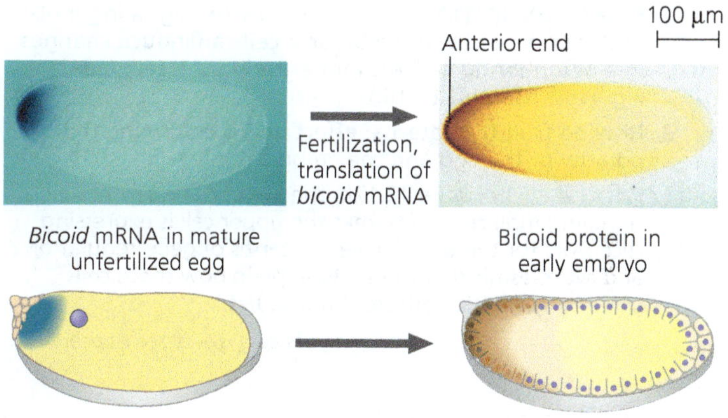

Conclusion The location of *bicoid* mRNA and the diffuse gradient of Bicoid protein seen later are consistent with the hypothesis that Bicoid protein is a morphogen specifying formation of head-specific structures.

Further Reading C. Nüsslein-Volhard et al., Determination of anteroposterior polarity in *Drosophila, Science* 238:1675–1681 (1987); W. Driever and C. Nüsslein-Volhard, A gradient of Bicoid protein in *Drosophila* embryos, *Cell* 54:83–93 (1988); T. Berleth et al., The role of localization of *bicoid* RNA in organizing the anterior pattern of the *Drosophila* embryo, *EMBO Journal* 7:1749–1756 (1988).

WHAT IF? ➤ *The researchers needed further evidence, so they injected bicoid mRNA into the anterior end of an egg from a female with a mutation disabling the bicoid gene. Given that the hypothesis was supported, what must their results have been?*

MB **Animation: Role of *bicoid* Gene in *Drosophila* Development** transformation of a fertilized egg into a multicellular organism. The program is carefully balanced between turning on the genes for differentiation in the right place and turning off other genes. Even when an organism is fully developed, gene expression is regulated in a similarly fine-tuned manner. In the final section of the chapter, we'll consider how fine this tuning is by looking at how specific changes in expression of just a few genes can lead to the development of cancer.

CONCEPT CHECK 18.4

1. **MAKE CONNECTIONS ➤** As you learned in Chapter 12, mitosis gives rise to two daughter cells that are genetically identical to the parent cell. Yet you, the product of many mitotic divisions, are not composed of identical, zygote-like cells. Why?
2. **MAKE CONNECTIONS ➤** Explain how the signaling molecules released by an embryonic cell can induce changes in a neighboring cell without entering the cell. (See Figures 11.15 and 11.16.)
3. How do fruit fly maternal effect genes determine the polarity of the egg and the embryo?
4. **WHAT IF? ➤** In Figure 18.17b, the lower cell is synthesizing signaling molecules, whereas the upper cell is expressing receptors for these molecules. In terms of gene regulation and cytoplasmic determinants, explain how these cells came to synthesize different molecules.

For suggested answers, see Appendix A.

CONCEPT 18.5

Cancer results from genetic changes that affect cell cycle control

In Concept 12.3, we considered cancer as a type of disease in which cells escape from the control mechanisms that normally limit their growth. Now that we have discussed the molecular basis of gene expression and its regulation, we are ready to look at cancer more closely. The gene regulation systems that go wrong during cancer turn out to be the very same systems that play important roles in embryonic development, the immune response, and many other biological processes. Thus, research into the molecular basis of cancer has both benefited from and informed many other fields of biology.

Types of Genes Associated with Cancer

The genes that normally regulate cell growth and division during the cell cycle include genes for growth factors, their receptors, and the intracellular molecules of signaling pathways. (To review cell signaling, see Concept 11.2; for regulation of the cell cycle, see Concept 12.3.) Mutations that alter any of these genes in somatic cells can lead to cancer. The agent of such change can be random spontaneous mutation. However, it is also likely that many cancer-causing mutations result from environmental influences, such as chemical carcinogens, X-rays and other high-energy radiation, and some viruses.

Cancer research led to the discovery of cancer-causing genes called **oncogenes** (from the Greek *onco*, tumor) in certain types of viruses. Subsequently, close counterparts of viral oncogenes were found in the genomes of humans and other animals. The normal versions of the cellular genes, called **proto-oncogenes**, code for proteins that stimulate normal cell growth and division.

How might a proto-oncogene—a gene that has an essential function in normal cells—become an oncogene, a cancer-causing gene? In general, an oncogene arises from a genetic change that leads to an increase either in the amount of the proto-oncogene's protein product or in the intrinsic activity of each protein molecule. The genetic changes that convert proto-oncogenes to oncogenes fall into three main categories: movement of DNA within the genome, amplification of a proto-oncogene, and point mutations in a control element or in the proto-oncogene itself **(Figure 18.23)**.

▼ Figure 18.23 Genetic changes that can turn proto-oncogenes into oncogenes.

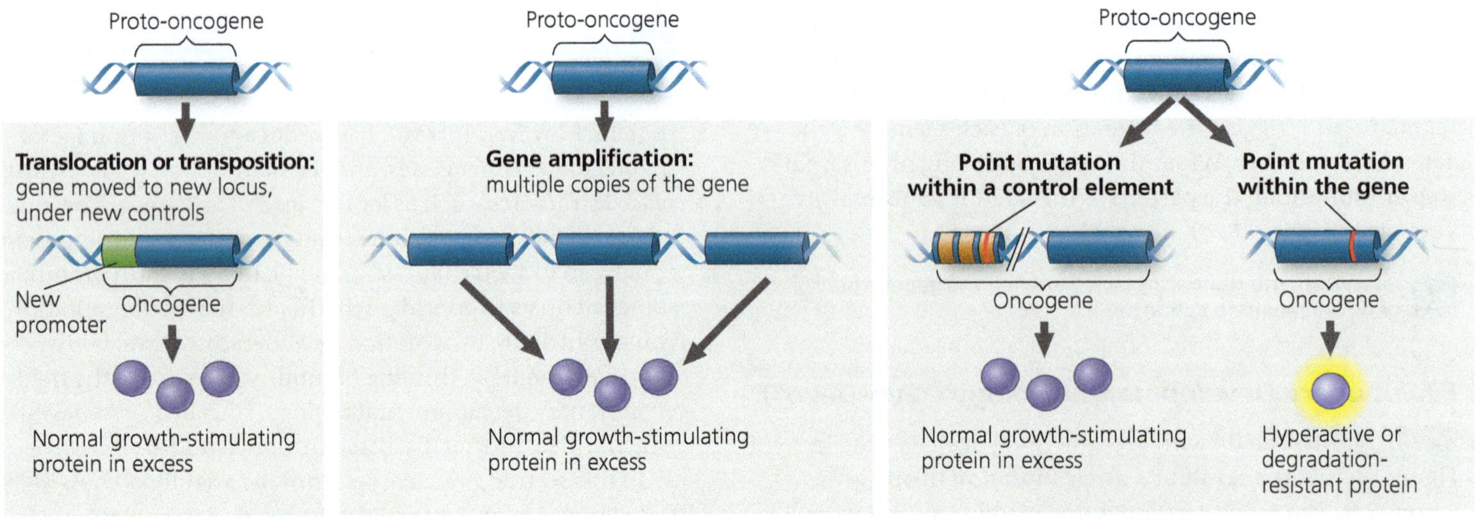

Cancer cells are frequently found to contain chromosomes that have broken and rejoined incorrectly, translocating fragments from one chromosome to another (see Figure 15.14). Having learned how gene expression is regulated, you can now see the possible consequences of such translocations. If a translocated proto-oncogene ends up near an especially active promoter (or other control element), its transcription may increase, making it an oncogene. The second main type of genetic change, amplification, increases the number of copies of the proto-oncogene in the cell through repeated gene duplication (discussed in Concept 21.5). The third possibility is a point mutation either in the promoter or an enhancer that controls a proto-oncogene, causing an increase in its expression, or in the coding sequence of the proto-oncogene, changing the gene's product to a protein that is more active or more resistant to degradation than the normal protein. These mechanisms can lead to abnormal stimulation of the cell cycle and put the cell on the path to becoming a cancer cell.

In addition to genes whose products normally promote cell division, cells contain genes whose normal products *inhibit* cell division. Such genes are called **tumor-suppressor genes** since the proteins they encode help prevent uncontrolled cell growth. Any mutation that decreases the normal activity of a tumor-suppressor protein may contribute to the onset of cancer, in effect stimulating growth through the absence of suppression.

The protein products of tumor-suppressor genes have various functions. Some repair damaged DNA, a function that prevents the cell from accumulating cancer-causing mutations. Other tumor-suppressor proteins control the adhesion of cells to each other or to the extracellular matrix; proper cell anchorage is crucial in normal tissues—and is often absent in cancers. Still other tumor-suppressor proteins are components of cell-signaling pathways that inhibit the cell cycle.

Interference with Normal Cell-Signaling Pathways

The proteins encoded by many proto-oncogenes and tumor-suppressor genes are components of cell-signaling pathways. Let's take a closer look at how such proteins function in normal cells and what goes wrong with their function in cancer cells. We will focus on the products of two key genes, the *ras* proto-oncogene and the *p53* tumor-suppressor gene. Mutations in *ras* occur in about 30% of human cancers, and mutations in *p53* in more than 50%.

The Ras protein, encoded by the ***ras* gene** (named for rat sarcoma, a connective tissue cancer), is a G protein that relays a signal from a growth factor receptor on the plasma membrane to a cascade of protein kinases (see Figures 11.8 and 11.10). The cellular response at the end of the pathway is the synthesis of a protein that stimulates the cell cycle **(Figure 18.24a)**. Normally, such a pathway will not operate unless triggered by the appropriate growth factor. But certain mutations in the *ras* gene can lead to production of a hyperactive Ras protein that triggers the kinase cascade even in

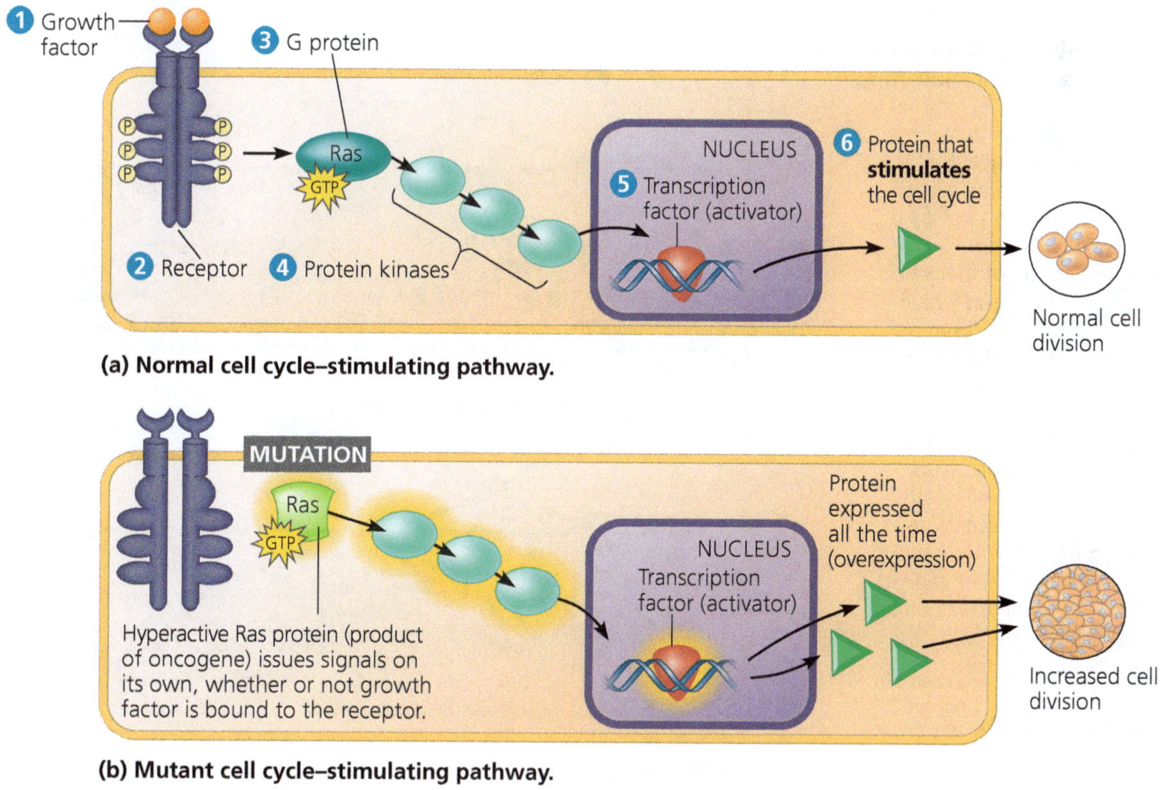

➤ **Figure 18.24 Normal and mutant cell cycle–stimulating pathway.** **(a)** The normal pathway is triggered by 1 a growth factor that binds to 2 its receptor in the plasma membrane. The signal is relayed to 3 a G protein called Ras. Like all G proteins, Ras is active when GTP is bound to it. Ras passes the signal to 4 a series of protein kinases. The last kinase activates 5 a transcription factor (activator) that turns on one or more genes for 6 a protein that stimulates the cell cycle. **(b)** If a mutation makes Ras or any other pathway component abnormally active, excessive cell division and cancer may result.

the absence of growth factor, resulting in increased cell division **(Figure 18.24b)**. In fact, hyperactive versions or excess amounts of any of the pathway's components can have the same outcome: excessive cell division.

Figure 18.25a shows a pathway in which an intracellular signal leads to the synthesis of a protein that suppresses the cell cycle. In this case, the signal is damage to the cell's DNA, perhaps as the result of exposure to ultraviolet light. Operation of this signaling pathway blocks the cell cycle until the damage has been repaired. Otherwise, the damage might contribute to tumor formation by causing mutations or chromosomal abnormalities. Thus, the genes for the components of the pathway act as tumor-suppressor genes. The ***p53* gene**, named for the 53,000-dalton molecular weight of its protein product, is a tumor-suppressor gene. The protein it encodes is a specific transcription factor that promotes the synthesis of cell cycle–inhibiting proteins. That is why a mutation that knocks out the *p53* gene, like a mutation that leads to a hyperactive Ras protein, can lead to excessive cell growth and cancer **(Figure 18.25b)**.

The *p53* gene has been called the "guardian angel of the genome." Once the gene is activated—for example, by DNA damage—the p53 protein functions as an activator for several other genes. Often it activates a gene called *p21*, whose product halts the cell cycle by binding to cyclin-dependent kinases, allowing time for the cell to repair the DNA. Researchers recently showed that p53 also activates expression of a group of miRNAs, which in turn inhibit the cell cycle. In addition, the p53 protein can turn on genes directly involved in DNA repair. Finally, when DNA damage is irreparable, p53 activates "suicide" genes, whose protein products bring about programmed cell death (apoptosis; see Figure 11.20). Thus, p53 acts in several ways to prevent a cell from passing on mutations due to DNA damage. If mutations do accumulate and the cell survives through many divisions—as is more likely if the *p53* tumor-suppressor gene is defective or missing—cancer may ensue. The many functions of p53 suggest a complex picture of regulation in normal cells, one that we do not yet fully understand.

A recent study may underscore the protective role of *p53* while illuminating a long-standing research question: Why is cancer so rare among elephants? The incidence of cancer among elephants in zoo-based studies has been estimated at about 3%, compared to closer to 30% for humans. Genome sequencing revealed that elephants have 20 copies of the *p53* gene, compared to one copy in humans, other mammals, and even manatees, elephants' closest living relatives. There are undoubtedly other underlying reasons, but the correlation between low cancer rate and extra copies of the *p53* gene bears further investigation.

For the present, the diagrams in Figure 18.24 and Figure 18.25 are an accurate view of how mutations can contribute to cancer, but we still don't know exactly how a particular cell becomes a cancer cell. As we discover previously unknown

➤ Figure 18.25 Normal and mutant cell cycle–inhibiting pathway. (a) In the normal pathway, ① DNA damage is an intracellular signal that is passed via ② protein kinases, leading to activation of ③ p53. Activated p53 promotes ④ transcription of the gene for ⑤ a protein that inhibits the cell cycle. The resulting suppression of cell division ensures that the damaged DNA is not replicated. If the DNA damage is irreparable, then the p53 signal leads to programmed cell death (apoptosis). **(b)** Mutations causing deficiencies in any pathway component can contribute to the development of cancer.

? *Explain whether a cancer-causing mutation in a tumor-suppressor gene, such as p53, is more likely to be a recessive or a dominant mutation.*

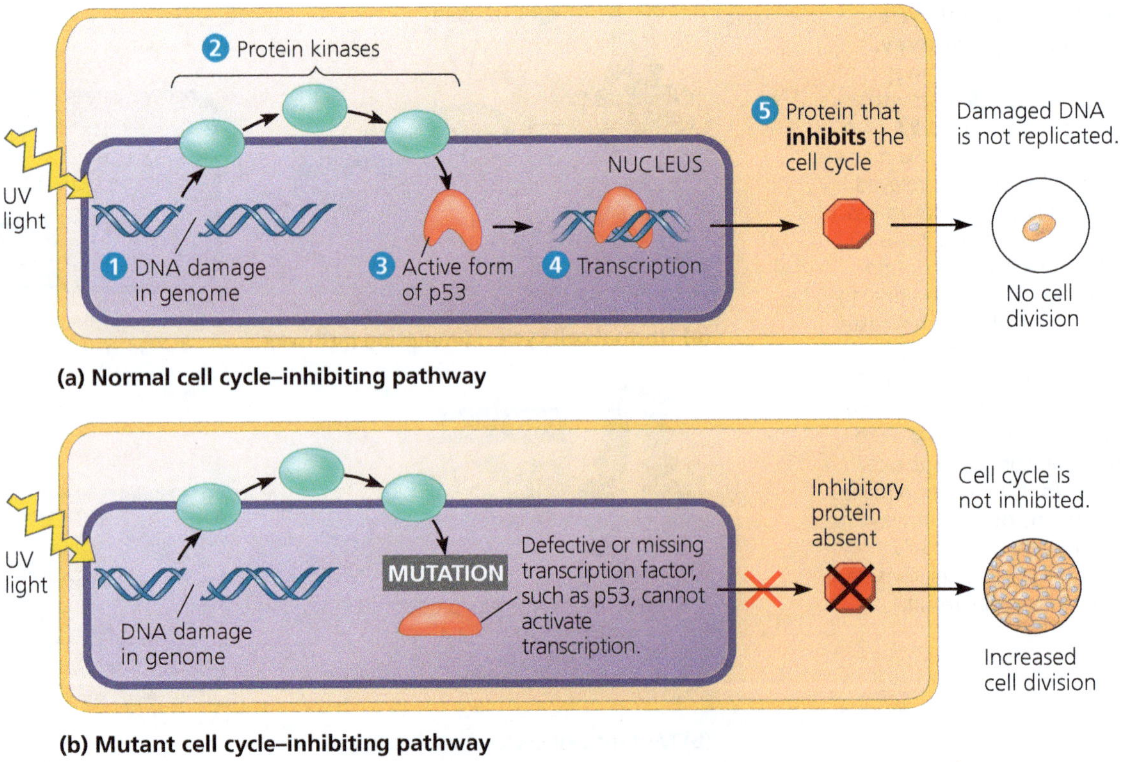

aspects of gene regulation, it is informative to study their role in the onset of cancer. Such studies have shown, for instance, that DNA methylation and histone modification patterns differ in normal and cancer cells and that miRNAs probably participate in cancer development. While we've learned a lot about cancer by studying cell-signaling pathways, there is still much more to learn.

The Multistep Model of Cancer Development

More than one somatic mutation is generally needed to produce all the changes characteristic of a full-fledged cancer cell. This may help explain why the incidence of cancer increases greatly with age. If cancer results from an accumulation of mutations and if mutations occur throughout life, then the longer we live, the more likely we are to develop cancer.

The model of a multistep path to cancer is well supported by studies of one of the best-understood types of human cancer: colorectal cancer, which affects the colon and/or rectum. About 140,000 new cases of colorectal cancer are diagnosed each year in the United States, and the disease causes 50,000 deaths per year. Like most cancers, colorectal cancer develops gradually **(Figure 18.26)**. The first sign is often a polyp, a small, benign growth in the colon lining. The cells of the polyp look normal, although they divide unusually frequently. The tumor grows and may eventually become malignant, invading other tissues. The development of a malignant tumor is paralleled by a gradual accumulation of mutations that convert proto-oncogenes to oncogenes and knock out tumor-suppressor genes. A *ras* oncogene and a mutated *p53* tumor-suppressor gene are often involved.

About half a dozen changes must occur at the DNA level for a cell to become fully cancerous. These changes usually include the appearance of at least one active oncogene and the mutation or loss of several tumor-suppressor genes. Furthermore, since mutant tumor-suppressor alleles are usually recessive, in most cases mutations must knock out *both* alleles in a cell's genome to block tumor suppression. (Most oncogenes, on the other hand, behave as dominant alleles.)

Since we understand the progression of this type of cancer, routine screenings (colonoscopies, for example) are recommended to identify and remove any suspicious polyps. The colorectal cancer mortality rate has been declining for the past 20 years due to increased screening and improved treatments. Treatments for other cancers have improved as well. Advances in the sequencing of DNA and mRNA allow medical researchers to compare the genes expressed by different types of tumors and by the same type in different people. These comparisons have led to personalized treatments based on the molecular characteristics of a person's tumor.

Breast cancer is the second most common form of cancer in the United States, and the first among women. Each year, this cancer strikes over 230,000 women (and some men) in the United States and kills 40,000 (450,000 worldwide). A major problem with understanding breast cancer is its heterogeneity: Tumors differ in significant ways. Identifying differences between types of breast cancer is expected to improve treatment and decrease the mortality rate. In 2012, The Cancer Genome Atlas Network, sponsored by the National Institutes of Health, published the results of a multi-team effort that used a genomics approach to profile subtypes of breast cancer based on their molecular signatures. Four major types of breast cancer were identified **(Figure 18.27)**. It is now routine to screen for the presence of particular signaling receptors in any breast cancer tumors, and individuals with breast cancer, along with their physicians, can now make more informed decisions about their treatments.

▼ Figure 18.26 A multistep model for the development of colorectal cancer. This type of cancer is one of the best understood. Changes in a tumor parallel a series of genetic changes, including mutations affecting several tumor-suppressor genes (such as *p53*) and the *ras* proto-oncogene. Mutations of tumor-suppressor genes often entail loss (deletion) of the gene. *APC* stands for adenomatous polyposis coli, and *SMAD4* is a gene involved in signaling that results in apoptosis.

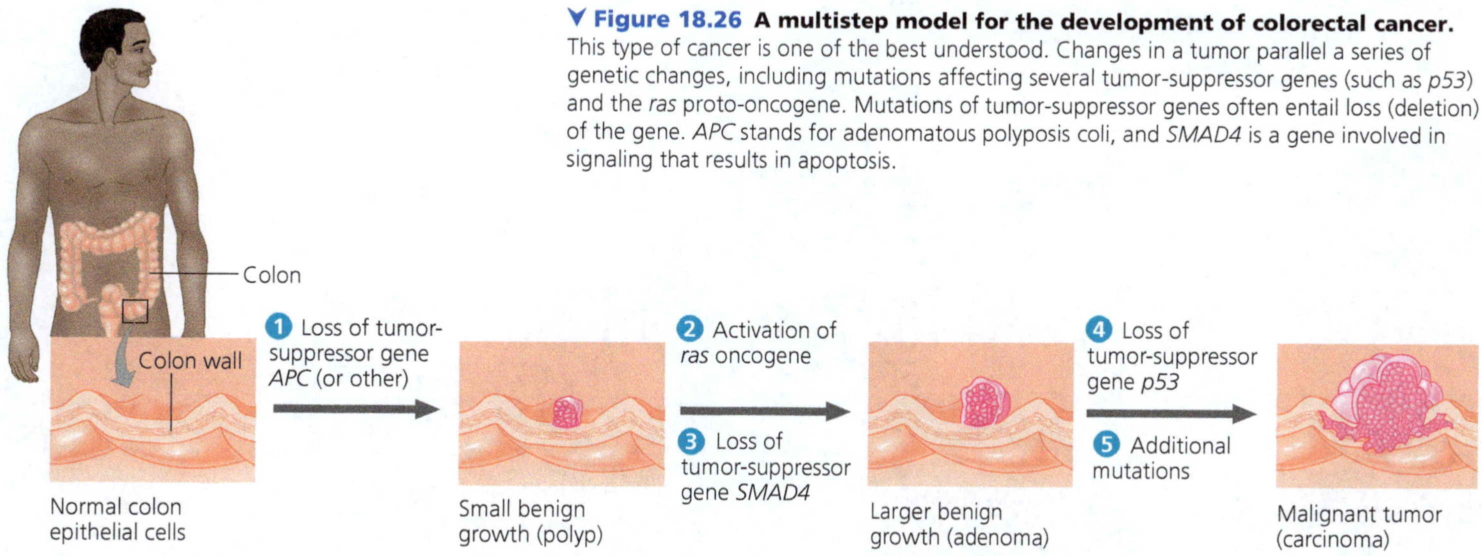

Figure 18.27 MAKE CONNECTIONS

Genomics, Cell Signaling, and Cancer

Modern medicine that melds genome-wide molecular studies with cell-signaling research is transforming the treatment of many diseases, such as breast cancer. Using microarray analysis (*see Figure 20.13*) and other techniques, researchers measured the relative levels of mRNA transcripts for every gene in hundreds of breast cancer tumor samples. They identified four major subtypes of breast cancer that differ in their expression of three signal receptors involved in regulating cell growth and division:

- Estrogen receptor alpha (ERα)
- Progesterone receptor (PR)
- HER2, a type of receptor called a receptor tyrosine kinase (*see Figure 11.8*)

(ERα and PR are steroid receptors; *see Figure 11.9*.) The absence or excess expression of these receptors can cause aberrant cell signaling, leading in some cases to inappropriate cell division, which may contribute to cancer (*see Figure 18.24*).

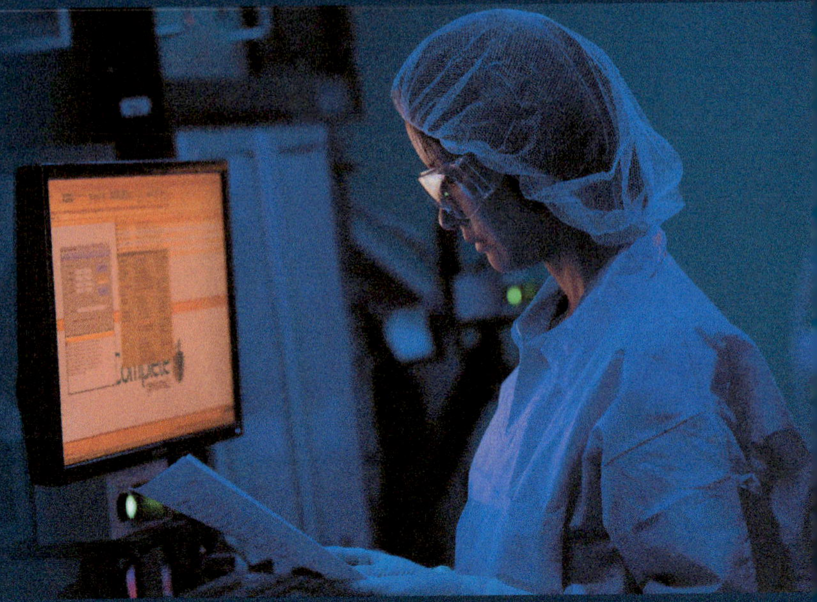

▲ **A research scientist examines DNA sequencing data from breast cancer samples.**

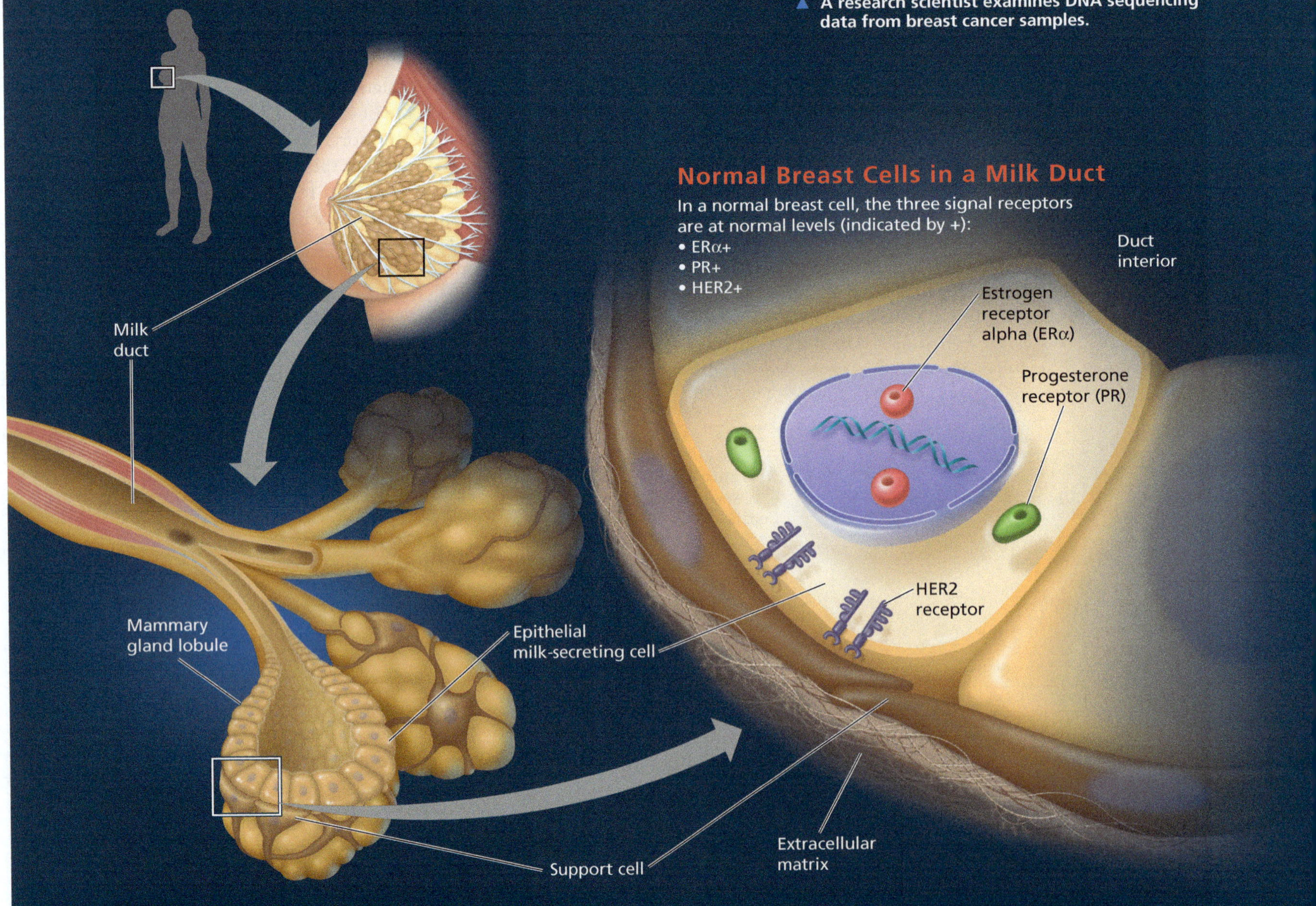

Breast Cancer Subtypes

Each breast cancer subtype is characterized by the overexpression (indicated by ++ or +++) or absence (–) of three signal receptors: ERα, PR, and HER2. Breast cancer treatments are becoming more effective because they can be tailored to the specific cancer subtype.

Luminal A

Luminal B

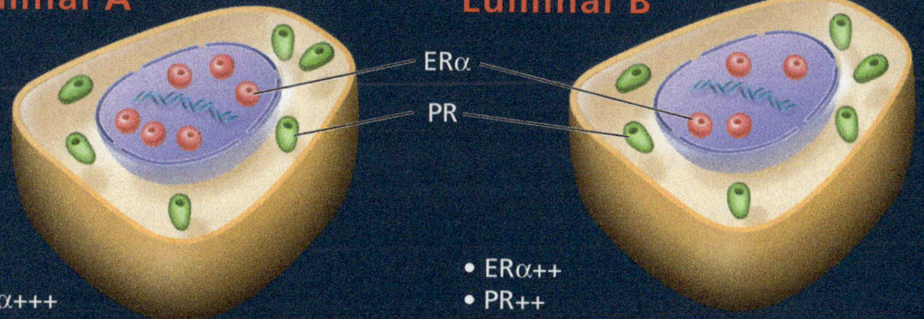

Luminal A

- ERα+++
- PR++
- HER2–
- 40% of breast cancers
- Best prognosis

Luminal B

- ERα++
- PR++
- HER2– (shown here); some HER2++
- 15–20% of breast cancers
- Poorer prognosis than luminal A subtype

Both luminal subtypes overexpress ERα (luminal A more than luminal B) and PR, and usually lack expression of HER2. Both can be treated with drugs that target ERα and inactivate it, the most well-known drug being tamoxifen. These subtypes can also be treated with drugs that inhibit estrogen synthesis.

Basal-like

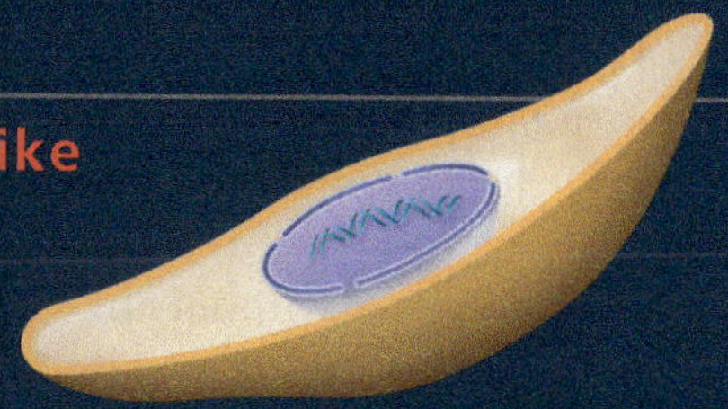

- ERα–
- PR–
- HER2–
- 15–20% of breast cancers
- More aggressive; poorer prognosis than other subtypes

The basal-like subtype is "triple negative"—it does not express ERα, PR, or HER2. It often has a mutation in the tumor-suppressor gene *BRCA1* (*see Concept 18.5*). Treatments that target ER, PR, or HER2 are not effective, but new treatments are being developed. Currently, patients are treated with cytotoxic chemotherapy, which selectively kills fast-growing cells.

HER2

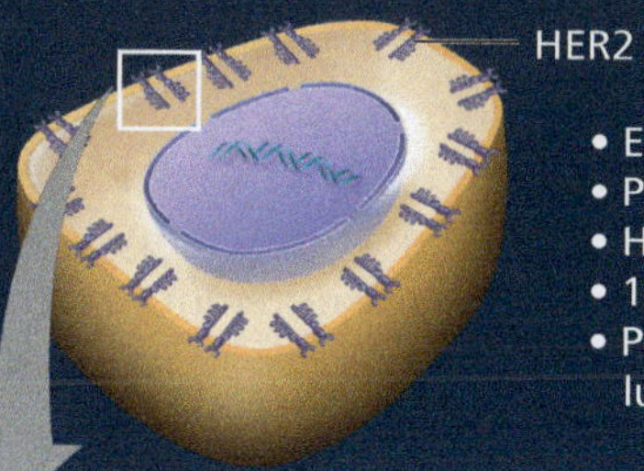

- ERα–
- PR–
- HER2++
- 10–15% of breast cancers
- Poorer prognosis than luminal A subtype

The HER2 subtype overexpresses HER2. Because it does not express either ERα or PR at normal levels, the cells are unresponsive to therapies that target those two receptors. However, patients with the HER2 subtype can be treated with Herceptin, an antibody protein that inactivates HER2 (*see Concept 12.3*).

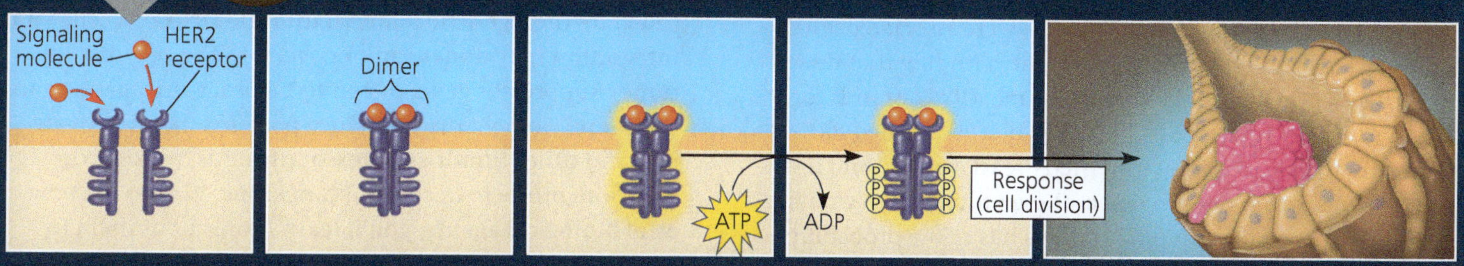

1 Signaling molecules (such as a growth factor) bind to HER2 receptor monomers (single receptor proteins).

2 Binding of signaling molecules causes two receptor monomers to associate closely with each other, forming a dimer.

3 Formation of a dimer activates each monomer.

4 Each monomer adds phosphate from ATP to the other monomer, triggering a signal transduction pathway.

5 The signal is transduced through the cell, which leads to a cellular response—in this case, turning on genes that trigger cell division. HER2 cells have up to 100 times as many HER2 receptors as normal cells, so they undergo uncontrolled cell division.

Treatment with Herceptin for the HER2 subtype

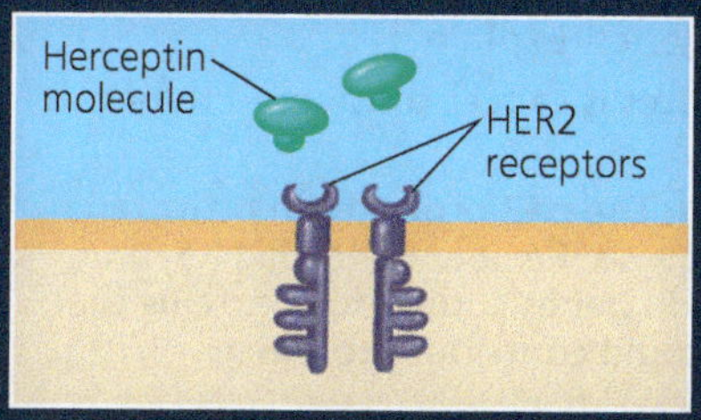

1 The drug Herceptin binds to HER2 receptors in place of the usual signaling molecules.

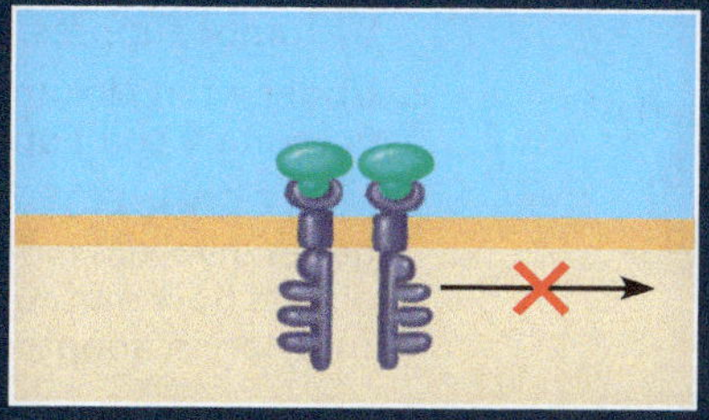

2 In certain patients with the HER2 subtype, signaling is blocked and excessive cell division does not occur.

MAKE CONNECTIONS ➤ *When researchers compared gene expression in normal breast cells and cells from breast cancers, they found that the genes showing the most significant differences in expression encoded signal receptors, as shown here. Given what you learned in Chapters 11, 12, and this chapter, explain why this result is not surprising.*

Inherited Predisposition and Environmental Factors Contributing to Cancer

The fact that multiple genetic changes are required to produce a cancer cell helps explain the observation that cancers can run in families. An individual inheriting an oncogene or a mutant allele of a tumor-suppressor gene is one step closer to accumulating the necessary mutations for cancer to develop than is an individual without any such mutations.

Geneticists are working to identify inherited cancer alleles so that predisposition to certain cancers can be detected early in life. About 15% of colorectal cancers, for example, involve inherited mutations. Many of these affect the tumor-suppressor gene called *adenomatous polyposis coli*, or *APC* (see Figure 18.26). This gene has multiple functions in the cell, including regulation of cell migration and adhesion. Even in patients with no family history of the disease, the *APC* gene is mutated in 60% of colorectal cancers. In these individuals, new mutations must occur in both *APC* alleles before the gene's function is lost. Since only 15% of colorectal cancers are associated with known inherited mutations, researchers continue to try to identify "markers" that could predict the risk of developing this type of cancer.

Given the prevalence and significance of breast cancer, it is not surprising that it was one of the first cancers for which the role of inheritance was investigated. It turns out that for 5–10% of patients with breast cancer, there is evidence of a strong inherited predisposition. Geneticist Mary-Claire King began working on this problem in the mid-1970s. After 16 years of research, she convincingly demonstrated that mutations in one gene—*BRCA1*—were associated with increased susceptibility to breast cancer, a finding that flew in the face of medical opinion at the time. (*BRCA* stands for <u>br</u>east <u>ca</u>ncer.) Mutations in that gene or a gene called *BRCA2* are found in at least half of inherited breast cancers, and tests using DNA sequencing can detect these mutations. A woman who inherits one mutant *BRCA1* allele has a 60% probability of developing breast cancer before the age of 50, compared with only a 2% probability for an individual homozygous for the normal allele.

BRCA1 and *BRCA2* are considered tumor-suppressor genes because their wild-type alleles protect against breast cancer and their mutant alleles are recessive. (Note that mutations in *BRCA1* are commonly found in the genomes of cells from basal-like breast cancers; see Figure 18.27.) The BRCA1 and BRCA2 proteins both appear to function in the cell's DNA damage repair pathway. More is known about BRCA2, which, along with another protein, helps repair breaks that occur in both strands of DNA; this repair function is crucial for maintaining undamaged DNA.

MB Interview with Mary-Claire King: Discovering breast cancer genes

Because DNA breakage can contribute to cancer, it makes sense that the risk of cancer can be lowered by minimizing exposure to DNA-damaging agents, such as the ultraviolet radiation in sunlight and chemicals found in cigarette smoke. Novel genomics-based analyses of specific cancers, such as the approach described in Figure 18.27, are contributing to both early diagnosis and development of treatments that interfere with expression of key genes in tumors. Ultimately, such approaches are expected to lower the death rate from cancer.

The Role of Viruses in Cancer

The study of genes associated with cancer, inherited or not, increases our basic understanding of how disruption of normal gene regulation can result in this disease. In addition to the mutations and other genetic alterations described in this section, a number of *tumor viruses* can cause cancer in various animals, including humans. In fact, one of the earliest breakthroughs in understanding cancer came in 1911, when Peyton Rous, an American pathologist, discovered a virus that causes cancer in chickens. The Epstein-Barr virus, which causes infectious mononucleosis, has been linked to several types of cancer in humans, notably Burkitt's lymphoma. Papillomaviruses are associated with cancer of the cervix, and a virus called HTLV-1 causes a type of adult leukemia. Viruses play a role in about 15% of the cases of human cancer.

Viruses may at first seem very different from mutations as a cause of cancer. However, we now know that viruses can interfere with gene regulation in several ways if they integrate their genetic material into the DNA of a cell. Viral integration may donate an oncogene to the cell, disrupt a tumor-suppressor gene, or convert a proto-oncogene to an oncogene. Some viruses produce proteins that inactivate p53 and other tumor-suppressor proteins, making the cell more prone to becoming cancerous. Viruses are powerful biological agents; you'll learn more about them in Chapter 19.

MB Animation: Causes of Cancer

CONCEPT CHECK 18.5

1. Cancer-promoting mutations are likely to have different effects on the activity of proteins encoded by proto-oncogenes than they do on proteins encoded by tumor-suppressor genes. Explain.
2. Under what circumstances is cancer considered to have a hereditary component?
3. **MAKE CONNECTIONS ➤** The p53 protein can activate genes involved in apoptosis. Review Concept 11.5, and discuss how mutations in genes coding for proteins that function in apoptosis could contribute to cancer.

For suggested answers, see Appendix A.

18 Chapter Review

Go to **MasteringBiology**™ for Videos, Animations, Vocab Self-Quiz, Practice Tests, and more in the Study Area.

SUMMARY OF KEY CONCEPTS

CONCEPT 18.1

Bacteria often respond to environmental change by regulating transcription *(pp. 364–368)*

VOCAB SELF-QUIZ goo.gl/6u55ks

- Cells control metabolism by regulating enzyme activity or the expression of genes coding for enzymes. In bacteria, genes are often clustered into **operons**, with one promoter serving several adjacent genes. An **operator** site on the DNA switches the operon on or off, resulting in coordinate regulation of the genes.

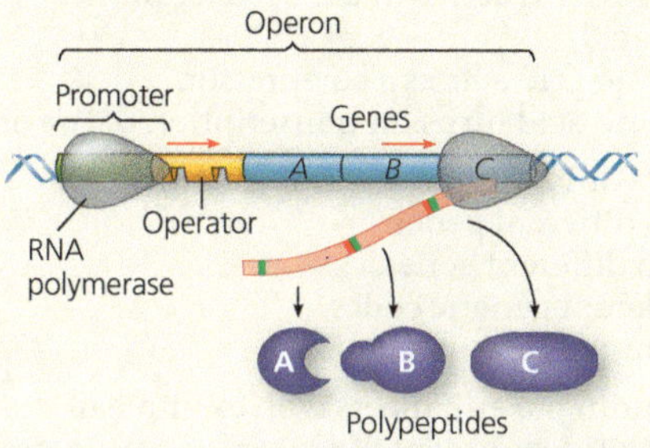

- Both repressible and inducible operons are examples of negative gene regulation. In either type of operon, binding of a specific **repressor** protein to the operator shuts off transcription. (The repressor is encoded by a separate **regulatory gene**.) In a repressible operon, the repressor is active when bound to a **corepressor**, usually the end product of an anabolic pathway.

Repressible operon:

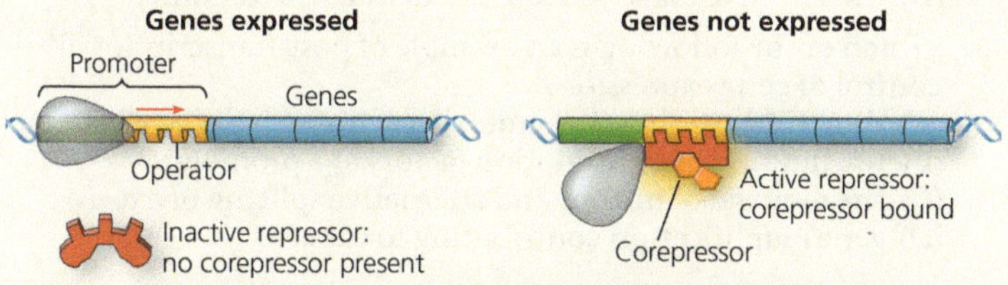

- In an inducible operon, binding of an **inducer** to an innately active repressor inactivates the repressor and turns on transcription. Inducible enzymes usually function in catabolic pathways.

Inducible operon:

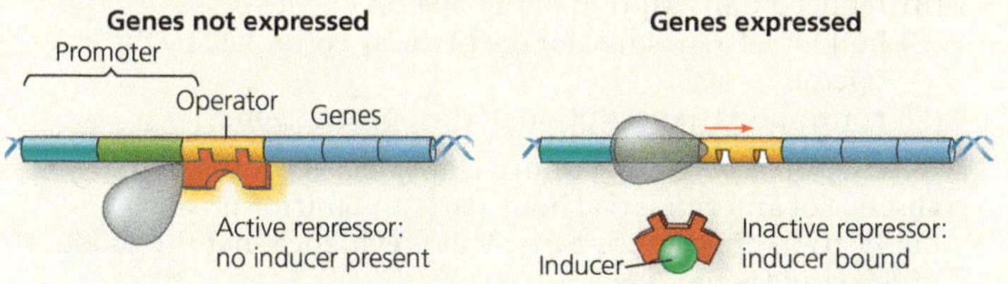

- Some operons are also subject to positive gene regulation via a stimulatory **activator** protein, such as cAMP receptor protein (CRP), which, when activated by **cyclic AMP**, binds to a site within the promoter and stimulates transcription.

? *Compare and contrast the roles of a corepressor and an inducer in negative regulation of an operon.*

CONCEPT 18.2

Eukaryotic gene expression is regulated at many stages *(pp. 368–376)*

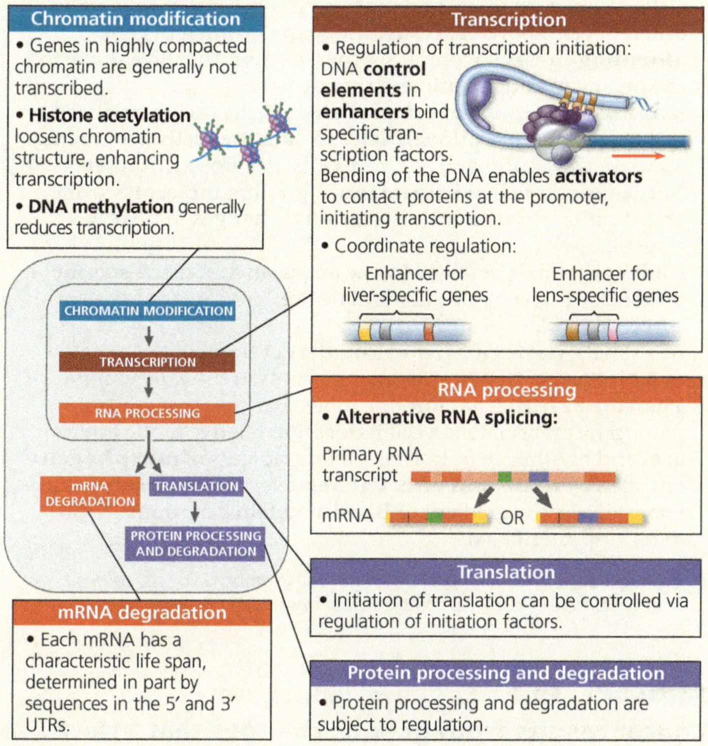

? *Describe what must happen in a cell for a gene specific to that cell type to be transcribed.*

CONCEPT 18.3

Noncoding RNAs play multiple roles in controlling gene expression *(pp. 377–379)*

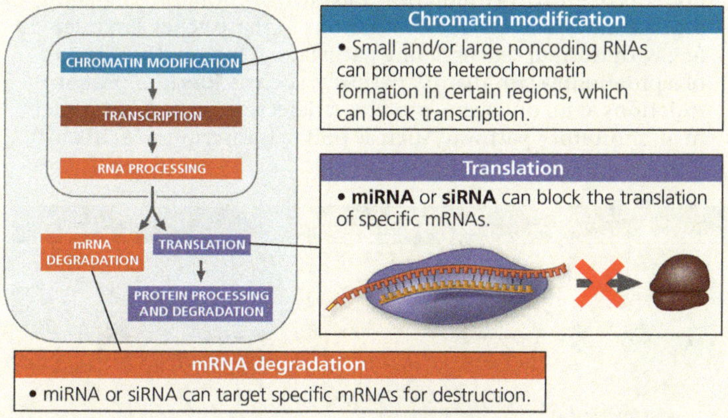

? *Why are miRNAs called noncoding RNAs? Explain how they participate in gene regulation.*

CONCEPT 18.4

A program of differential gene expression leads to the different cell types in a multicellular organism *(pp. 379–386)*

- Embryonic cells become committed to a certain fate (**determination**), and undergo **differentiation**, becoming specialized in structure and function for their determined fate. Cells differ in structure and function not because they contain different genomes but because they express different genes. **Morphogenesis** encompasses the processes that give shape to the organism and its various structures.
- Localized **cytoplasmic determinants** in the unfertilized egg are distributed differentially to daughter cells, where they regulate the expression of those cells' developmental fate. In the process called **induction**, signaling molecules from embryonic cells cause transcriptional changes in nearby target cells.
- Differentiation is heralded by the appearance of tissue-specific proteins, which enable differentiated cells to carry out their specialized roles.
- In animals, **pattern formation**, the development of a spatial organization of tissues and organs, begins in the early embryo. **Positional information**, the molecular cues that control pattern formation, tells a cell its location relative to the body's axes and to other cells. In *Drosophila*, gradients of **morphogens** encoded by **maternal effect genes** determine the body axes. For example, the gradient of **Bicoid** protein determines the anterior-posterior axis.

? *Describe the two main processes that cause embryonic cells to head down different pathways to their final fates.*

CONCEPT 18.5

Cancer results from genetic changes that affect cell cycle control *(pp. 386–392)*

- The products of **proto-oncogenes** and **tumor-suppressor genes** control cell division. A DNA change that makes a proto-oncogene excessively active converts it to an **oncogene**, which may promote excessive cell division and cancer. A tumor-suppressor gene encodes a protein that inhibits abnormal cell division. A mutation that reduces the activity of its protein product may lead to excessive cell division and cancer.
- Many proto-oncogenes and tumor-suppressor genes encode components of growth-stimulating and growth-inhibiting signaling pathways, respectively, and mutations in these genes can interfere with normal cell-signaling pathways. A hyperactive version of a protein in a stimulatory pathway, such as **Ras** (a G protein), functions as an oncogene protein. A defective version of a protein in an inhibitory pathway, such as **p53** (a transcription activator), fails to function as a tumor suppressor.

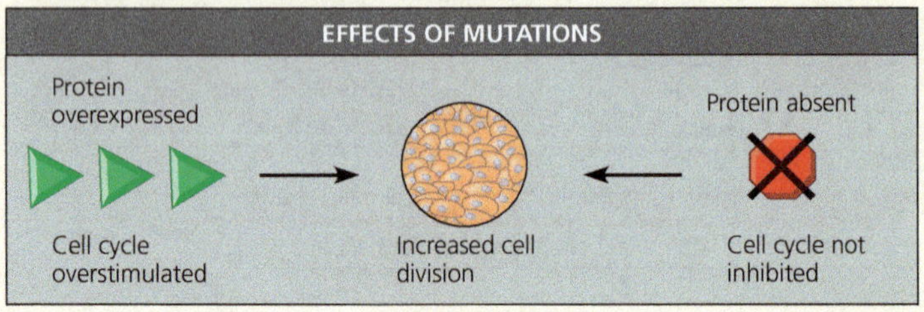

- In the multistep model of cancer development, normal cells are converted to cancer cells by the accumulation of mutations affecting proto-oncogenes and tumor-suppressor genes. Technical advances in DNA and mRNA sequencing are enabling cancer treatments that are more individually based.
- Genomics-based studies have resulted in researchers proposing four subtypes of breast cancer, based on expression of genes by tumor cells.
- Individuals who inherit a mutant allele of a proto-oncogene or tumor-suppressor gene have a predisposition to develop a particular cancer. Certain viruses promote cancer by integration of viral DNA into a cell's genome.

? *Compare the usual functions of proteins encoded by proto-oncogenes with those of proteins encoded by tumor-suppressor genes.*

TEST YOUR UNDERSTANDING

Level 1: Knowledge/Comprehension

PRACTICE TEST
goo.gl/CUYGKD

1. If a particular operon encodes enzymes for making an essential amino acid and is regulated like the *trp* operon, then
(A) the amino acid inactivates the repressor.
(B) the repressor is active in the absence of the amino acid.
(C) the amino acid acts as a corepressor.
(D) the amino acid turns on transcription of the operon.

2. Muscle cells differ from nerve cells mainly because they
(A) express different genes.
(B) contain different genes.
(C) use different genetic codes.
(D) have unique ribosomes.

3. The functioning of enhancers is an example of
(A) a eukaryotic equivalent of prokaryotic promoter functioning.
(B) transcriptional control of gene expression.
(C) the stimulation of translation by initiation factors.
(D) post-translational control that activates certain proteins.

4. Cell differentiation always involves
(A) transcription of the *myoD* gene.
(B) the movement of cells.
(C) the production of tissue-specific proteins.
(D) the selective loss of certain genes from the genome.

5. Which of the following is an example of post-transcriptional control of gene expression?
(A) the addition of methyl groups to cytosine bases of DNA
(B) the binding of transcription factors to a promoter
(C) the removal of introns and alternative splicing of exons
(D) gene amplification contributing to cancer

Level 2: Application/Analysis

6. What would occur if the repressor of an inducible operon were mutated so it could not bind the operator?
(A) irreversible binding of the repressor to the promoter
(B) reduced transcription of the operon's genes
(C) buildup of a substrate for the pathway controlled by the operon
(D) continuous transcription of the operon's genes

7. Absence of *bicoid* mRNA from a *Drosophila* egg leads to the absence of anterior larval body parts and mirror-image duplication of posterior parts. This is evidence that the product of the *bicoid* gene
(A) normally leads to formation of head structures.
(B) normally leads to formation of tail structures.
(C) is transcribed in the early embryo.
(D) is a protein present in all head structures.

8. Which of the following statements about the DNA in one of your brain cells is true?
(A) Most of the DNA codes for protein.
(B) The majority of genes are likely to be transcribed.
(C) It is the same as the DNA in one of your liver cells.
(D) Each gene lies immediately adjacent to an enhancer.

9. Within a cell, the amount of protein made using a given mRNA molecule depends partly on
(A) the degree of DNA methylation.
(B) the rate at which the mRNA is degraded.
(C) the number of introns present in the mRNA.
(D) the types of ribosomes present in the cytoplasm.

10. Proto-oncogenes can change into oncogenes that cause cancer. Which of the following best explains the presence of these potential time bombs in eukaryotic cells?
(A) Proto-oncogenes first arose from viral infections.
(B) Proto-oncogenes are mutant versions of normal genes.
(C) Proto-oncogenes are genetic "junk."
(D) Proto-oncogenes normally help regulate cell division.

Level 3: Synthesis/Evaluation

11. **DRAW IT** The diagram below shows five genes, including their enhancers, from the genome of a certain species. Imagine that yellow, blue, green, black, red, and purple activator proteins exist that can bind to the appropriately color-coded control elements in the enhancers of these genes.

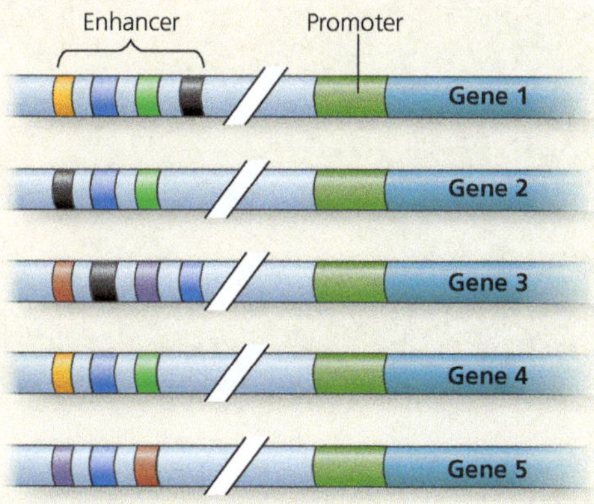

(a) Draw an X above enhancer elements (of all the genes) that would have activators bound in a cell in which only gene 5 is transcribed. Identify which colored activators would be present.
(b) Draw a dot above all enhancer elements that would have activators bound in a cell in which the green, blue, and yellow activators are present. Identify which gene(s) would be transcribed.
(c) Imagine that genes 1, 2, and 4 code for nerve-specific proteins, and genes 3 and 5 are skin specific. Identify which activators would have to be present in each cell type to ensure transcription of the appropriate genes.

12. **EVOLUTION CONNECTION** DNA sequences can act as "tape measures of evolution" (see Concept 5.6). Scientists analyzing the human genome sequence were surprised to find that some of the regions of the human genome that are most highly conserved (similar to comparable regions in other species) don't code for proteins. Propose a possible explanation for this observation.

13. **SCIENTIFIC INQUIRY** Prostate cells usually require testosterone and other androgens to survive. But some prostate cancer cells thrive despite treatments that eliminate androgens. One hypothesis is that estrogen, often considered a female hormone, may be activating genes normally controlled by an androgen in these cancer cells. Describe one or more experiments to test this hypothesis. (See Figure 11.9 to review the action of these steroid hormones.)

14. **SCIENCE, TECHNOLOGY, AND SOCIETY** Trace amounts of dioxin were present in Agent Orange, a defoliant sprayed on vegetation during the Vietnam War. Animal tests suggest that dioxin can cause birth defects, cancer, liver and thymus damage, and immune system suppression, sometimes leading to death. But the animal tests are equivocal; a hamster is not affected by a dose that can kill a guinea pig. Dioxin acts like a steroid hormone, entering a cell and binding to a cytoplasmic receptor that then binds the cell's DNA.
(a) Discuss how this mechanism might help explain the variety of dioxin's effects on different body systems and in different animals.
(b) Discuss how you might determine whether a type of illness is related to dioxin exposure. Next, discuss how you might determine whether a particular individual became ill as a result of exposure to dioxin. Which would be more difficult to demonstrate? Why?

15. **WRITE ABOUT A THEME: INTERACTIONS** In a short essay (100–150 words), discuss how the processes shown in Figure 18.2 are examples of feedback mechanisms regulating biological systems in bacterial cells.

16. **SYNTHESIZE YOUR KNOWLEDGE**

The flashlight fish has an organ under its eye that emits light, which serves to startle predators and attract prey, and allows the fish to communicate with other fish. Some species can rotate the organ inside and then out, so the light appears to flash on and off. The light is actually emitted by bacteria (of the genus *Vibrio*) that live in the organ in a mutualistic relationship with the fish. (The bacteria receive nutrients from the fish.) The bacteria must multiply until they reach a certain density in the organ (a "quorum"; see Concept 11.1), at which point they all begin emitting light at the same time. There is a group of six or so genes, called *lux* genes, whose gene products are necessary for light formation. Given that these bacterial genes are regulated together, propose a hypothesis for how the genes are organized and regulated.

For selected answers, see Appendix A.

For additional practice questions, check out the **Dynamic Study Modules** in MasteringBiology. You can use them to study on your smartphone, tablet, or computer anytime, anywhere!

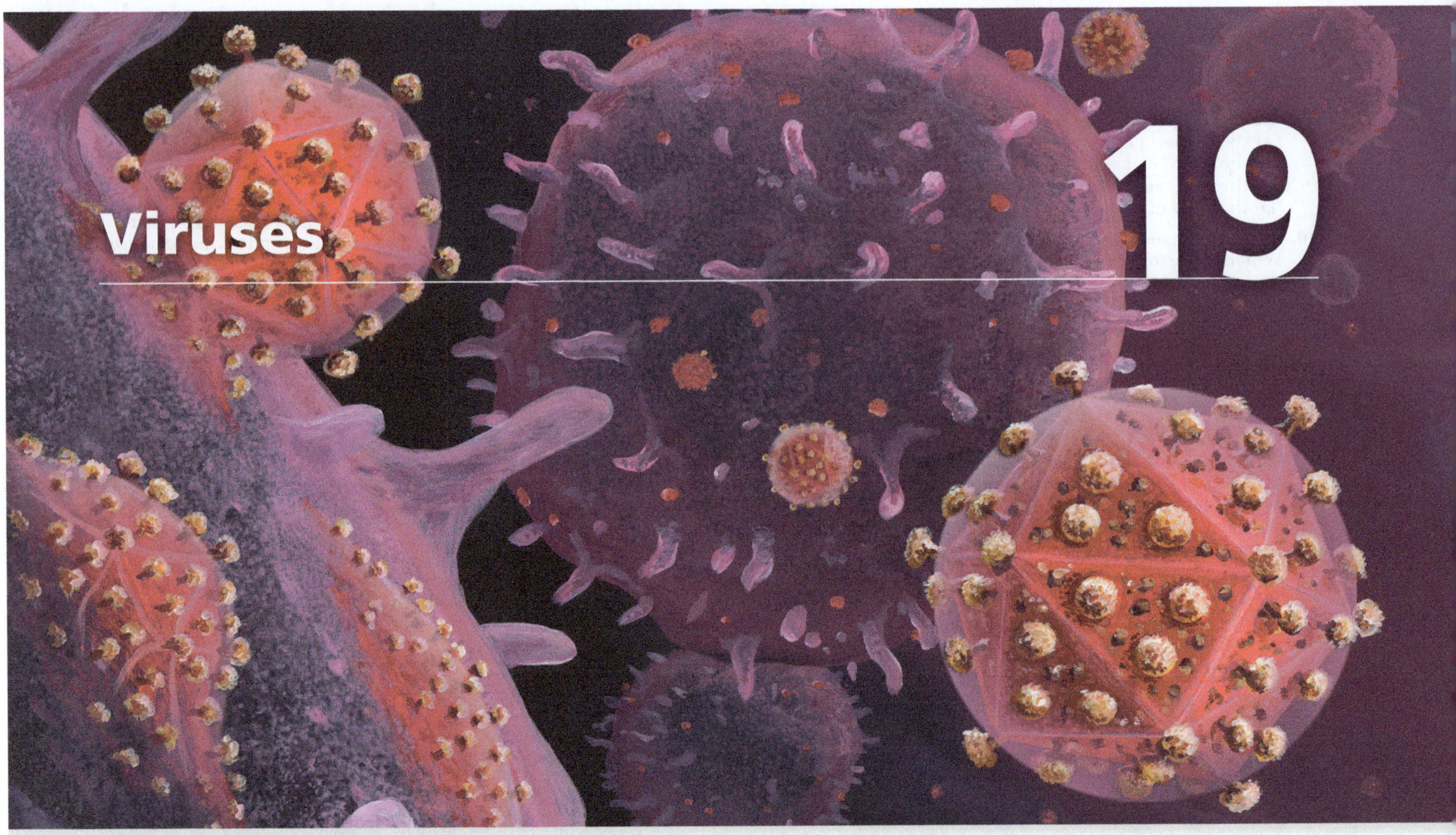

▲ **Figure 19.1** **Are the viruses (reddish-purple) budding from these cells alive?**

KEY CONCEPTS

19.1 A virus consists of a nucleic acid surrounded by a protein coat

19.2 Viruses replicate only in host cells

19.3 Viruses and prions are formidable pathogens in animals and plants

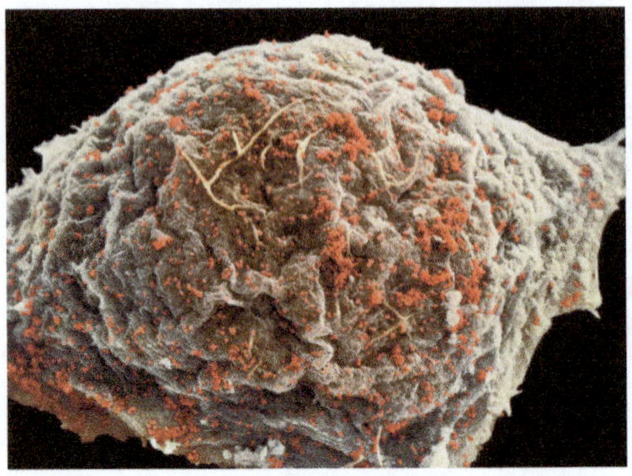

▲ A human immune cell infected with HIV. New viruses (red) are budding from the plasma membrane (colorized SEM).

A Borrowed Life

The illustration in **Figure 19.1** shows a remarkable event: Human immune cells (purple) infected by human immunodeficiency viruses (HIV) are releasing more HIV viruses. These viruses (red, surrounded by a protein-studded purple membrane from the immune cell) will infect other cells. (The SEM below shows one infected cell.) By injecting its genetic information into a cell, a virus hijacks the cell, recruiting cellular machinery to manufacture many new viruses and promote further infection. Left untreated, HIV causes acquired immunodeficiency syndrome (AIDS) by destroying vital immune system cells.

Compared with eukaryotic and even prokaryotic cells, viruses are much smaller and simpler in structure. Lacking the structures and metabolic machinery found in a cell, a **virus** is an infectious particle consisting of little more than genes packaged in a protein coat.

Are viruses living or nonliving? Early on, they were considered biological chemicals; the Latin root for virus means "poison." Viruses are capable of causing a wide variety of diseases, so researchers in the late 1800s saw a parallel with bacteria and proposed that viruses were the simplest of living forms. However, viruses cannot reproduce or carry out metabolic activities outside of a host cell. Most biologists studying viruses today would probably agree that they are not alive but exist in a shady area between life-forms and chemicals. The simple phrase used recently by two researchers describes them aptly enough: Viruses lead "a kind of borrowed life."

When you see this blue icon, log in to **MasteringBiology** and go to the Study Area for digital resources.

MB **Get Ready for This Chapter**

To a large extent, molecular biology was born in the laboratories of biologists studying viruses that infect bacteria. Experiments with these viruses provided evidence that genes are made of nucleic acids, and they were critical in working out the molecular mechanisms of the fundamental processes of DNA replication, transcription, and translation.

In this chapter, we will explore the biology of viruses, beginning with their structure and then describing how they replicate. Next, we will discuss the role of viruses as disease-causing agents, or pathogens, and conclude by considering some even simpler infectious agents called prions.

CONCEPT 19.1

A virus consists of a nucleic acid surrounded by a protein coat

Scientists were able to detect viruses indirectly long before they were actually able to see them. The story of how viruses were discovered begins near the end of the 19th century.

The Discovery of Viruses: *Scientific Inquiry*

Tobacco mosaic disease stunts the growth of tobacco plants and gives their leaves a mottled, or mosaic, coloration. In 1883, Adolf Mayer, a German scientist, discovered that he could transmit the disease from plant to plant by rubbing sap extracted from diseased leaves onto healthy plants. After an unsuccessful search for an infectious microbe in the sap, Mayer suggested that the disease was caused by unusually small bacteria that were invisible under a microscope. This hypothesis was tested a decade later by Dmitri Ivanowsky, a Russian biologist who passed sap from infected tobacco leaves through a filter designed to remove bacteria. After filtration, the sap still produced mosaic disease.

But Ivanowsky clung to the hypothesis that bacteria caused tobacco mosaic disease. Perhaps, he reasoned, the bacteria were small enough to pass through the filter or made a toxin that could do so. The second possibility was ruled out when the Dutch botanist Martinus Beijerinck carried out a classic series of experiments that showed that the infectious agent in the filtered sap could replicate **(Figure 19.2)**.

In fact, the pathogen replicated only within the host it infected. In further experiments, Beijerinck showed that unlike bacteria used in the lab at that time, the mysterious agent of mosaic disease could not be cultivated on nutrient media in test tubes or petri dishes. Beijerinck imagined a replicating particle much smaller and simpler than a bacterium, and he is generally credited with being the first scientist to voice the concept of a virus. His suspicions were confirmed in 1935 when the American scientist Wendell Stanley crystallized the infectious particle, now known as tobacco mosaic virus (TMV). Subsequently, TMV and many other viruses were actually seen with the help of the electron microscope.

Figure 19.2

Inquiry **What causes tobacco mosaic disease?**

Experiment In the late 1800s, Martinus Beijerinck, of the Technical School in Delft, the Netherlands, investigated the properties of the agent that causes tobacco mosaic disease (then called spot disease).

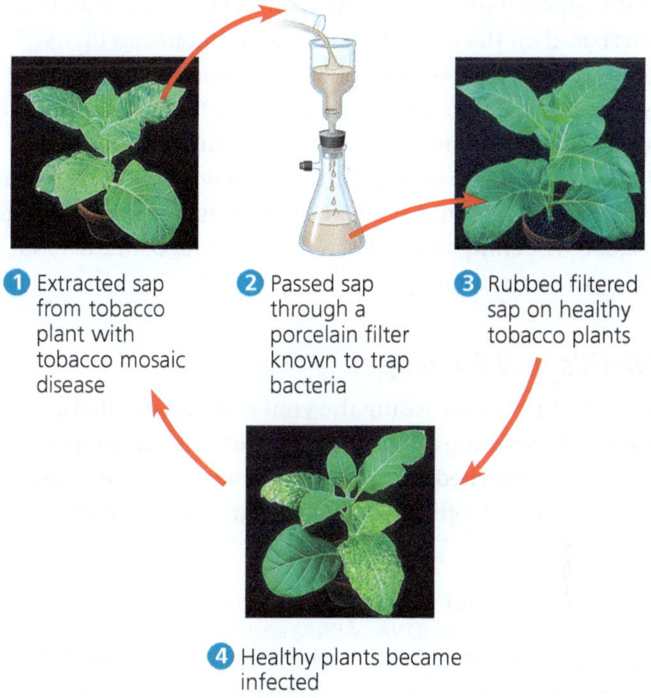

Results When the filtered sap was rubbed on healthy plants, they became infected. Their sap, extracted and filtered, could then act as a source of infection for another group of plants. Each successive group of plants developed the disease to the same extent as earlier groups.

Conclusion The infectious agent was apparently not a bacterium because it could pass through a bacterium-trapping filter. The pathogen must have been replicating in the plants because its ability to cause disease was undiluted after several transfers from plant to plant.

Data from M. J. Beijerinck, Concerning a *contagium vivum fluidum* as cause of the spot disease of tobacco leaves, *Verhandelingen der Koninkyke akademie Wettenschappen te Amsterdam* 65:3–21 (1898). Translation published in English as Phytopathological Classics Number 7 (1942), American Phytopathological Society Press, St. Paul, MN.

WHAT IF? *If Beijerinck had observed that the infection of each group was weaker than that of the previous group and that ultimately the sap could no longer cause disease, what might he have concluded?*

Structure of Viruses

The tiniest viruses are only 20 nm in diameter—smaller than a ribosome. Millions could easily fit on a pinhead. Even the largest known virus, which has a diameter of 1,500 nanometers (1.5 μm), is barely visible under the light microscope. Stanley's discovery that some viruses could be crystallized was exciting and puzzling news. Not even the simplest of cells can aggregate into regular crystals. But if viruses are not cells, then what are they? Examining the structure of a virus more closely reveals that it is an infectious particle consisting of nucleic acid enclosed in a protein coat and, for some viruses, surrounded by a membranous envelope.

Viral Genomes

We usually think of genes as being made of double-stranded DNA, but many viruses defy this convention. Their genomes may consist of double-stranded DNA, single-stranded DNA, double-stranded RNA, or single-stranded RNA, depending on the type of virus. A virus is called a DNA virus or an RNA virus based on the kind of nucleic acid that makes up its genome. In either case, the genome is usually organized as a single linear or circular molecule of nucleic acid, although the genomes of some viruses consist of multiple molecules of nucleic acid. The smallest viruses known have only three genes in their genome, while the largest have several hundred to 2,000. For comparison, bacterial genomes contain about 200 to a few thousand genes.

Capsids and Envelopes

The protein shell enclosing the viral genome is called a **capsid**. Depending on the type of virus, the capsid may be rod-shaped, polyhedral, or more complex in shape. Capsids are built from a large number of protein subunits called *capsomeres*, but the number of different *kinds* of proteins in a capsid is usually small. Tobacco mosaic virus has a rigid, rod-shaped capsid made from over 1,000 molecules of a single type of protein arranged in a helix; rod-shaped viruses are commonly called *helical viruses* for this reason **(Figure 19.3a)**. Adenoviruses, which infect the respiratory tracts of animals, have 252 identical protein molecules arranged in a polyhedral capsid with 20 triangular facets—an icosahedron; thus, these and other similarly shaped viruses are referred to as *icosahedral viruses* **(Figure 19.3b)**.

Some viruses have accessory structures that help them infect their hosts. For instance, a membranous envelope surrounds the capsids of influenza viruses and many other viruses found in animals **(Figure 19.3c)**. These **viral envelopes**, which are derived from the membranes of the host cell, contain host cell phospholipids and membrane proteins. They also contain proteins and glycoproteins of viral origin. (Glycoproteins are proteins with carbohydrates covalently attached.) Some viruses carry a few viral enzyme molecules within their capsids.

Many of the most complex capsids are found among the viruses that infect bacteria, called **bacteriophages**,

▼ **Figure 19.3 Viral structure.** Viruses are made up of nucleic acid (DNA or RNA) enclosed in a protein coat (the capsid) and sometimes further wrapped in a membranous envelope. The individual protein subunits making up the capsid are called capsomeres. Although diverse in size and shape, viruses have many common structural features. (All micrographs are colorized TEMs.)

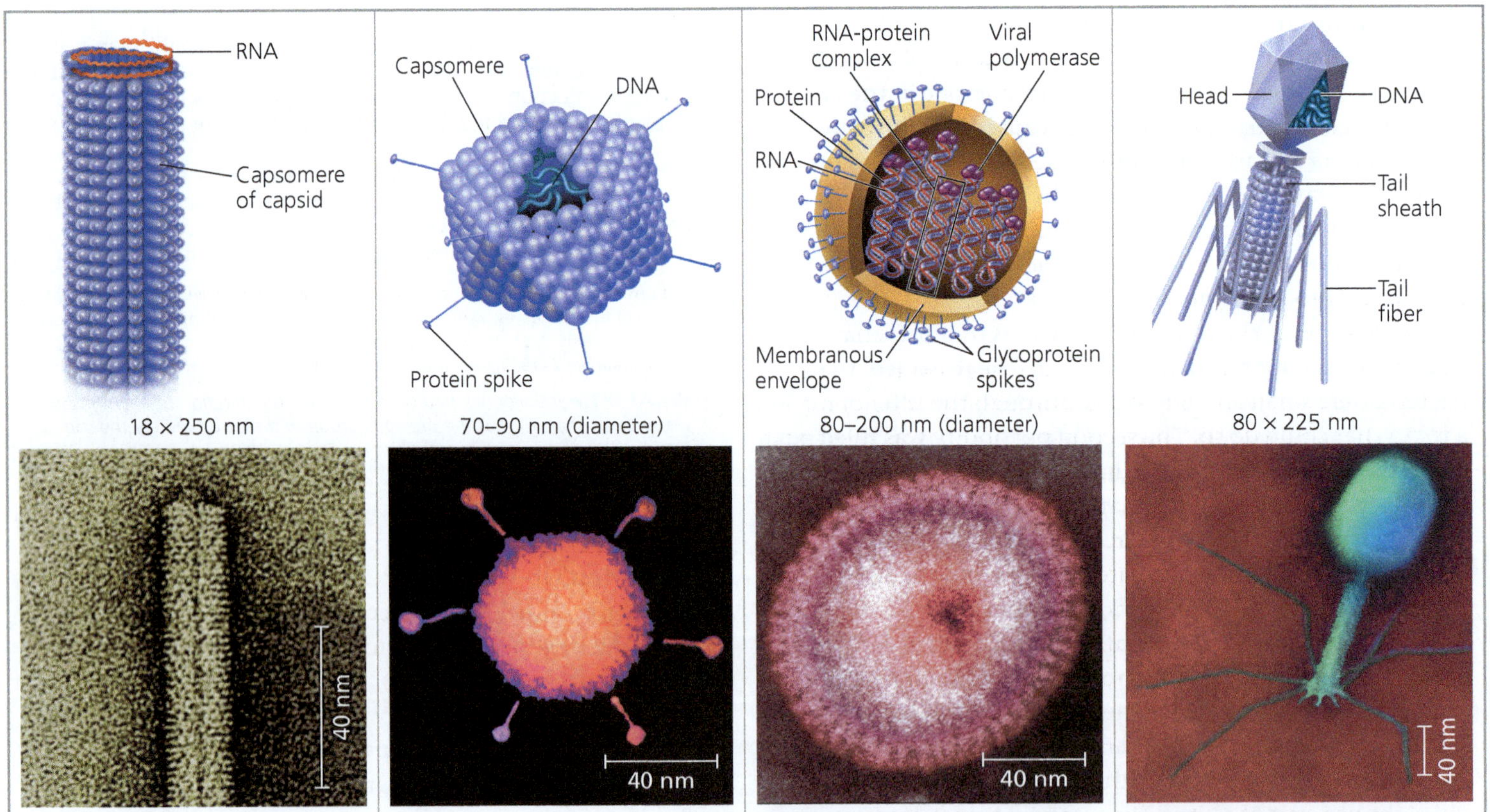

(a) Tobacco mosaic virus has a helical capsid surrounding a single helical RNA molecule, with the overall shape of a rigid rod.

(b) Adenoviruses have an icosahedral capsid with a protein spike at each vertex (corner). The capsid, made up of capsomeres, contains double-stranded DNA.

(c) Influenza viruses have an outer envelope, studded with glycoprotein spikes, that contains eight double-helical RNA-protein complexes, each associated with a viral polymerase.

(d) Bacteriophage T4, like other "T-even" phages, has a complex capsid consisting of an icosahedral head and a tail apparatus. T4's double-stranded DNA is enclosed in the head.

or simply **phages**. The first phages studied included seven that infect *Escherichia coli*. These seven phages were named type 1 (T1), type 2 (T2), and so forth, in the order of their discovery. The three "T-even" phages (T2, T4, and T6) turned out to be very similar in structure. Their capsids have elongated icosahedral heads enclosing their DNA. Attached to the head is a protein tail piece with fibers by which the phages attach to a bacterial cell **(Figure 19.3d)**. In the next section, we'll examine how these few viral parts function together with cellular components to produce large numbers of viral progeny.

CONCEPT CHECK 19.1

1. **VISUAL SKILLS ➤** Compare the structures of tobacco mosaic virus (TMV) and influenza virus (see Figure 19.3).
2. **MAKE CONNECTIONS ➤** Bacteriophages were used to provide evidence that DNA carries genetic information (see Figure 16.4). Briefly describe the experiment carried out by Hershey and Chase, including in your description why the researchers chose to use phages.

For suggested answers, see Appendix A.

CONCEPT 19.2

Viruses replicate only in host cells

Viruses lack metabolic enzymes and equipment for making proteins, such as ribosomes. They are obligate intracellular parasites; in other words, they can replicate only within a host cell. It is fair to say that viruses in isolation are merely packaged sets of genes in transit from one host cell to another.

Each particular virus can infect cells of only a limited number of host species, called the **host range** of the virus. This host specificity results from the evolution of recognition systems by the virus. Viruses usually identify host cells by a "lock-and-key" fit between viral surface proteins and specific receptor molecules on the outside of cells. According to one model, such receptor molecules originally carried out functions that benefited the host cell but were co-opted later by viruses as portals of entry. Some viruses have broad host ranges. For example, West Nile virus and equine encephalitis virus are distinctly different viruses that can each infect mosquitoes, birds, horses, and humans. Other viruses have host ranges so narrow that they infect only a single species. Measles virus, for instance, can infect only humans. Furthermore, viral infection of multicellular eukaryotes is usually limited to particular tissues. Human cold viruses infect only the cells lining the upper respiratory tract, and the HIV seen in Figure 19.1 binds to receptors present only on certain types of immune cells.

General Features of Viral Replicative Cycles

A viral infection begins when a virus binds to a host cell and the viral genome makes its way inside **(Figure 19.4)**. The mechanism of genome entry depends on the type of virus and the type of host cell. For example, T-even phages use their elaborate tail apparatus to inject DNA into a bacterium (see Figure 19.3d). Other viruses are taken up by endocytosis or, in the case of enveloped viruses, by fusion of the viral envelope with the host's plasma membrane. Once the viral genome is inside, the proteins it encodes can commandeer the host, reprogramming the cell to copy the viral genome and manufacture viral proteins. The host provides the nucleotides for making viral nucleic acids, as well as enzymes, ribosomes, tRNAs, amino acids, ATP, and other components needed for making the viral proteins. Many DNA viruses use the DNA polymerases of the host cell to synthesize new genomes along the templates provided by the viral DNA. In contrast, to replicate their genomes, RNA viruses use virally encoded RNA polymerases that can use RNA as a template. (Uninfected cells generally make no enzymes for carrying out this process.)

▼ Figure 19.4 A simplified viral replicative cycle. A virus is an intracellular parasite that uses the equipment and small molecules of its host cell to replicate. In this simplest of viral cycles, the parasite is a DNA virus with a capsid consisting of a single type of protein.

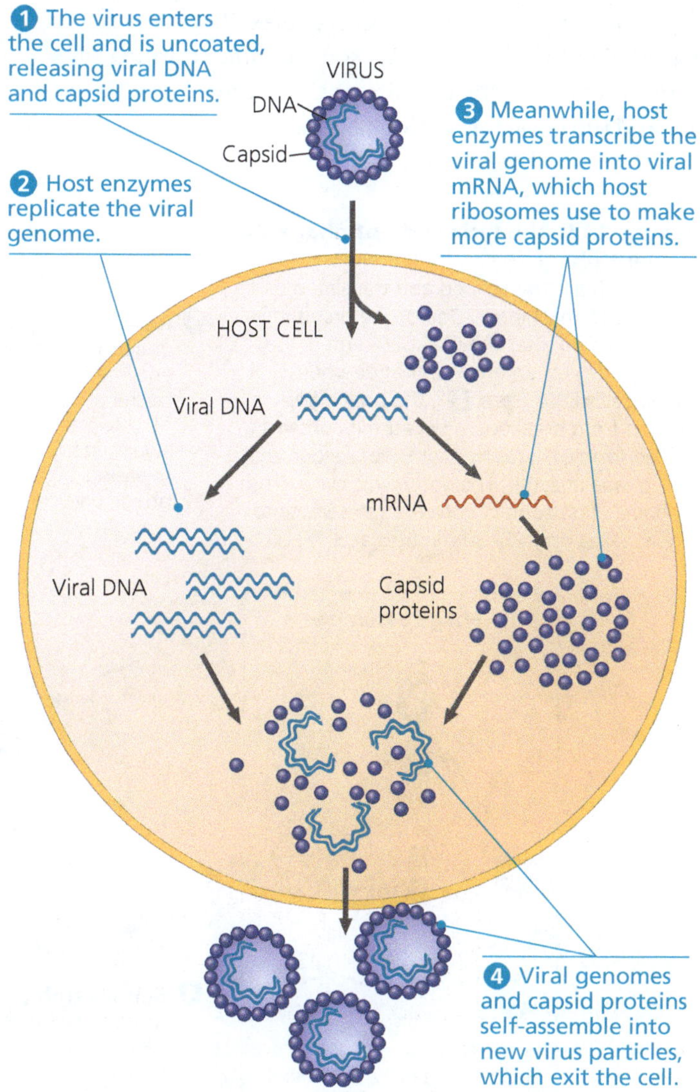

MAKE CONNECTIONS ➤ *Label each of the straight gray arrows with one word representing the name of the process that is occurring. Review Figure 17.25.*

Animation: Simplified Viral Replicative Cycle

After the viral nucleic acid molecules and capsomeres are produced, they spontaneously self-assemble into new viruses. In fact, researchers can separate the RNA and capsomeres of TMV and then reassemble complete viruses simply by mixing the components together under the right conditions. The simplest type of viral replicative cycle ends with the exit of hundreds or thousands of viruses from the infected host cell, a process that often damages or destroys the cell. Such cellular damage and death, as well as the body's responses to this destruction, cause many of the symptoms associated with viral infections. The viral progeny that exit a cell have the potential to infect additional cells, spreading the viral infection.

There are many variations on the simplified viral replicative cycle we have just described. We will now take a look at some of these variations in bacterial viruses (phages) and animal viruses; later in the chapter, we will consider plant viruses.

Replicative Cycles of Phages

Phages are the best understood of all viruses, although some of them are also among the most complex. Research on phages led to the discovery that some double-stranded DNA viruses can replicate by two alternative mechanisms: the lytic cycle and the lysogenic cycle.

The Lytic Cycle

A phage replicative cycle that culminates in death of the host cell is known as a **lytic cycle**. The term refers to the last stage of infection, during which the bacterium lyses (breaks open) and releases the phages that were produced within the cell. Each of these phages can then infect a healthy cell, and a few successive lytic cycles can destroy an entire bacterial population in just a few hours. A phage that replicates only by a lytic cycle is a **virulent phage**. **Figure 19.5** illustrates the major steps in the lytic cycle of T4, a typical virulent phage.

The Lysogenic Cycle

Instead of lysing their host cells, many phages coexist with them in a state called lysogeny. In contrast to the lytic cycle, which kills the host cell, the **lysogenic cycle** allows replication of the phage genome without destroying the host. Phages capable of using both modes of replicating within a bacterium are called **temperate phages**. A temperate phage called lambda, written with the Greek letter λ, has been widely used in biological research. Phage λ resembles T4, but its tail has only one short tail fiber.

Infection of an *E. coli* cell by phage λ begins when the phage binds to the surface of the cell and injects its linear DNA

➤ Figure 19.5 The lytic cycle of phage T4, a virulent phage. Phage T4 has almost 300 genes, which are transcribed and translated using the host cell's machinery. One of the first phage genes translated after the viral DNA enters the host cell codes for an enzyme that degrades the host cell's DNA (step 2); the phage DNA is protected from breakdown because it contains a modified form of cytosine that is not recognized by the phage enzyme. The entire lytic cycle, from the phage's first contact with the cell surface to cell lysis, takes only 20–30 minutes at 37°C.

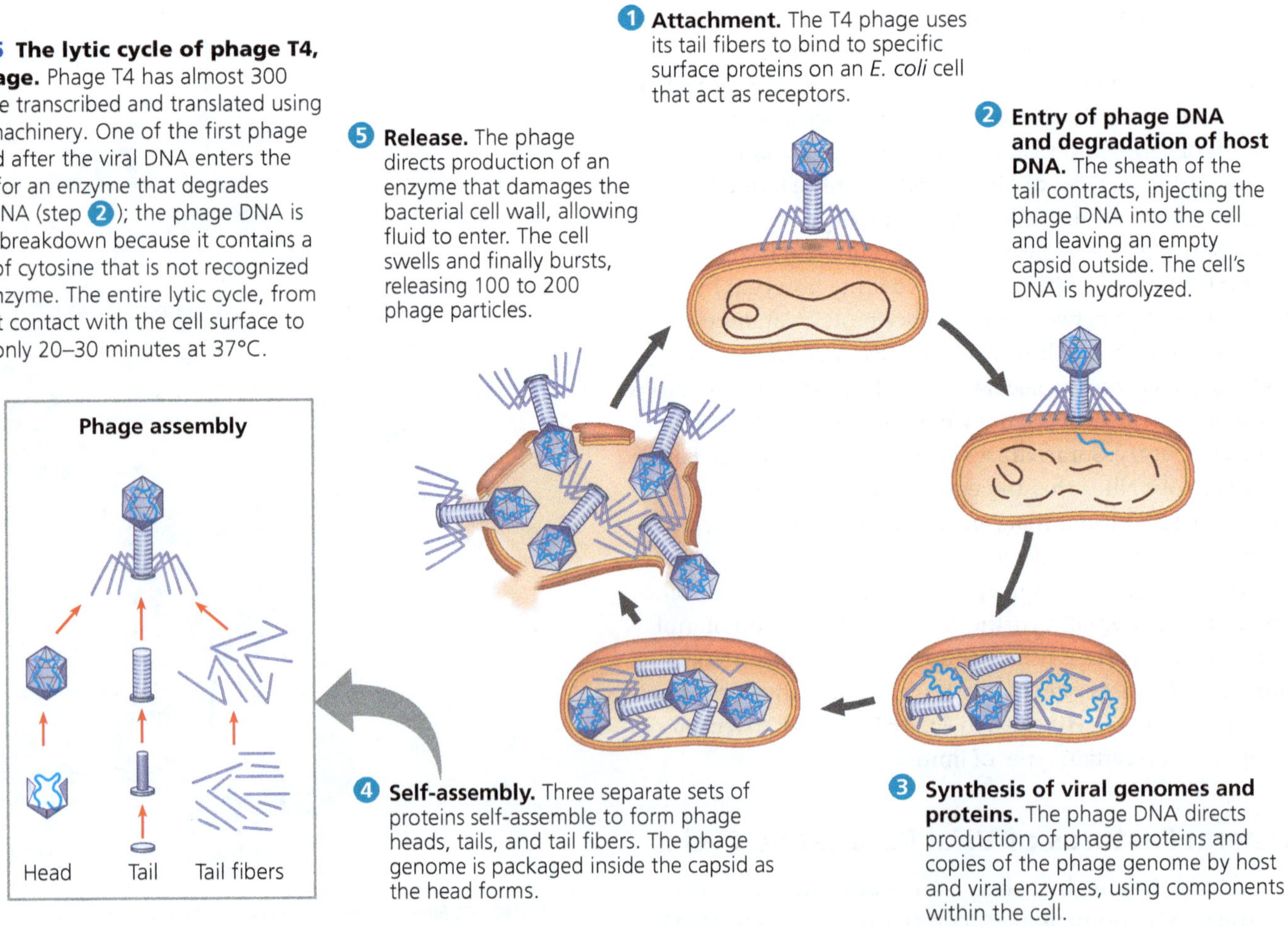

MB Animation: Phage Lytic Cycle

genome **(Figure 19.6)**. Within the host, the λ DNA molecule forms a circle. What happens next depends on the replicative mode: lytic cycle or lysogenic cycle. During a lytic cycle, the viral genes immediately turn the host cell into a λ-producing factory, and the cell soon lyses and releases its virus progeny. During a lysogenic cycle, however, the λ DNA molecule is incorporated into a specific site on the *E. coli* chromosome by viral proteins that break both circular DNA molecules and join them to each other. When integrated into the bacterial chromosome in this way, the viral DNA is known as a **prophage**. One prophage gene codes for a protein that prevents transcription of most of the other prophage genes. Thus, the phage genome is mostly silent within the bacterium. Every time the *E. coli* cell prepares to divide, it replicates the phage DNA along with its own chromosome such that each daughter cell inherits a prophage. A single infected cell can quickly give rise to a large population of bacteria carrying the virus in prophage form. This mechanism enables viruses to propagate without killing the host cells on which they depend.

The term *lysogenic* signifies that prophages are capable of generating active phages that lyse their host cells. This occurs when the λ genome (or that of another temperate phage) is induced to exit the bacterial chromosome and initiate a lytic cycle. An environmental signal, such as a certain chemical or high-energy radiation, usually triggers the switchover from the lysogenic to the lytic mode.

In addition to the gene for the viral protein that prevents transcription, a few other prophage genes may be expressed during lysogeny. Expression of these genes may alter the host's phenotype, a phenomenon that can have important medical significance. For example, the three species of bacteria that cause the human diseases diphtheria, botulism, and scarlet fever would not be so harmful to humans without certain prophage genes that cause the host bacteria to make toxins. And the difference between the *E. coli* strain in our intestines and the O157:H7 strain that has caused several deaths by food poisoning appears to be the presence of toxin genes of prophages in the O157:H7 strain.

Bacterial Defenses Against Phages

After reading about the lytic cycle, you may have wondered why phages haven't exterminated all bacteria. Lysogeny is one major reason why bacteria have been spared from extinction caused by phages. Bacteria also have their own defenses against phages. First, natural selection favors bacterial mutants with surface proteins that are no longer recognized as receptors by a particular type of phage. Second, when phage DNA does enter a bacterium, the DNA

▼ **Figure 19.6 The lytic and lysogenic cycles of phage λ, a temperate phage.** After entering the bacterial cell and circularizing, the λ DNA can immediately initiate the production of a large number of progeny phages (lytic cycle) or integrate into the bacterial chromosome (lysogenic cycle). In most cases, phage λ follows the lytic pathway, which is similar to that detailed in Figure 19.5. However, once a lysogenic cycle begins, the prophage may be carried in the host cell's chromosome for many generations. Phage λ has one main tail fiber, which is short.

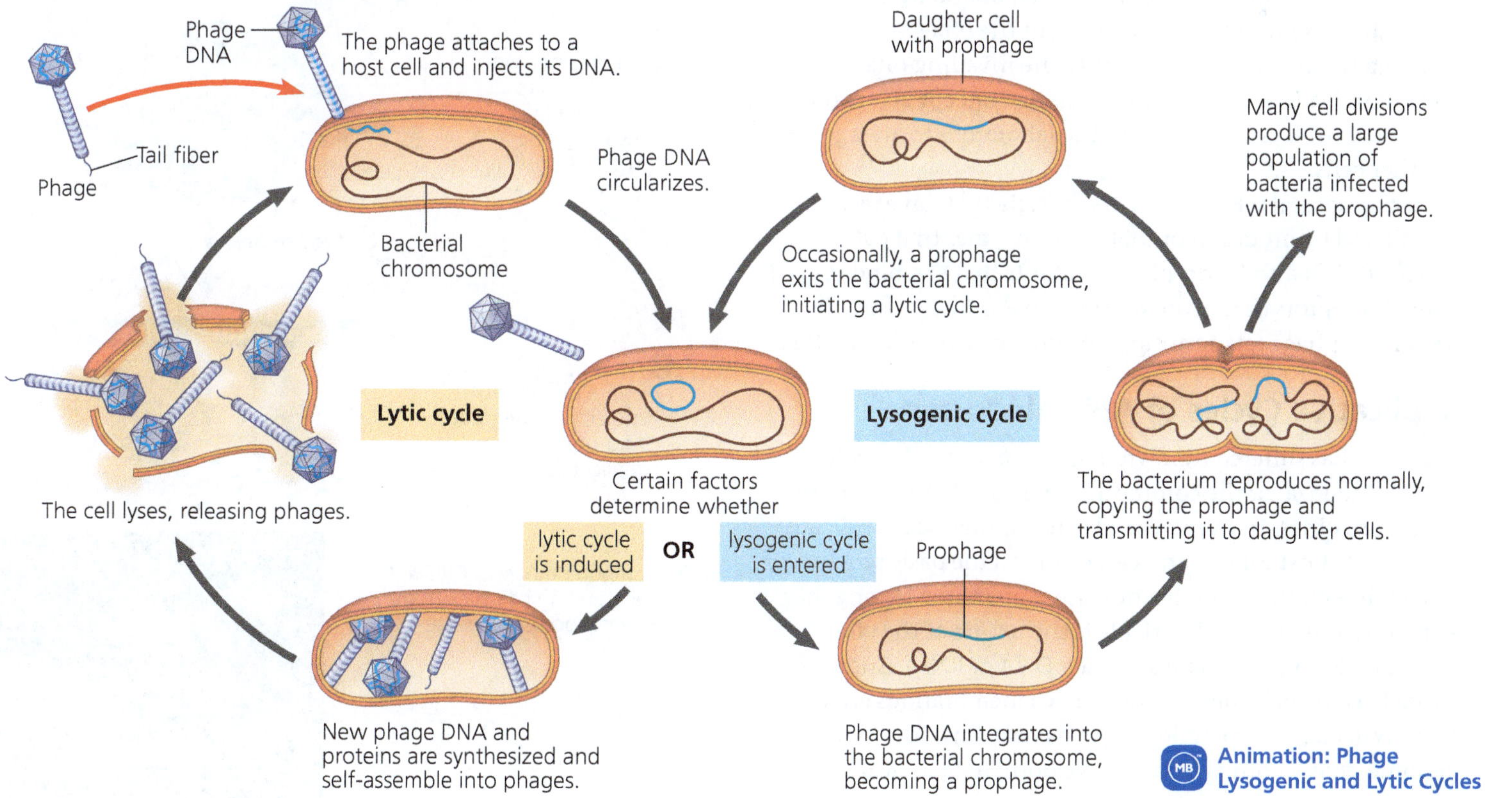

MB **Animation: Phage Lysogenic and Lytic Cycles**

often is identified as foreign and cut up by cellular enzymes called **restriction enzymes**, which are so named because they *restrict* a phage's ability to replicate within the bacterium. (Restriction enzymes are used in molecular biology and DNA cloning techniques; see Concept 20.1.) The bacterium's own DNA is methylated in a way that prevents attack by its own restriction enzymes. A third defense is a system present in both bacteria and archaea called the *CRISPR-Cas system*.

The CRISPR-Cas system was discovered during a study of repetitive DNA sequences present in the genomes of many prokaryotes. These sequences, which puzzled scientists, were named clustered regularly interspaced short palindromic repeats (CRISPRs) because each sequence read the same forward and backward (a palindrome), with different stretches of "spacer DNA" in between the repeats. At first, scientists assumed the spacer DNA sequences were random and meaningless, but analysis by several research groups showed that each spacer sequence corresponded to DNA from a particular phage that had infected the cell. Further studies revealed that particular nuclease proteins interact with the CRISPR region. These nucleases, called Cas (CRISPR-associated) proteins, can identify and cut phage DNA, thereby defending the bacterium against phage infection.

When a phage infects a bacterial cell that has the CRISPR-Cas system, the DNA of the invading phage is integrated into the genome between two repeat sequences. If the cell survives the infection, any further attempt by the same type of phage to infect this cell (or its offspring) triggers transcription of the CRISPR region into RNA molecules **(Figure 19.7)**. These RNAs are cut into pieces and then bound by Cas proteins. The Cas protein uses a portion of the phage-related RNA as a homing device to identify the invading phage DNA and cut it, leading to its destruction. In Concept 20.1, you'll learn how this system is used in the laboratory to alter genes in other cells.

Just as natural selection favors bacteria that have receptors altered by mutation or that have enzymes that cut phage DNA, it also favors phage mutants that can bind to altered receptors or that are resistant to enzymes. Thus, the bacterium-phage relationship is in constant evolutionary flux.

Replicative Cycles of Animal Viruses

Everyone has suffered from viral infections, whether cold sores, influenza, or the common cold. Like all viruses, those that cause illness in humans and other animals can replicate only inside host cells. Many variations on the basic scheme of viral infection and replication are represented among the animal viruses. One key variable is the nature of the viral genome (double- or single-stranded DNA or RNA). Another variable is the presence or absence of a membranous envelope. Whereas few bacteriophages have an envelope or RNA genome, many animal viruses have both. In fact, nearly all

▼ Figure 19.7 The CRISPR-Cas system: a type of bacterial immune system.

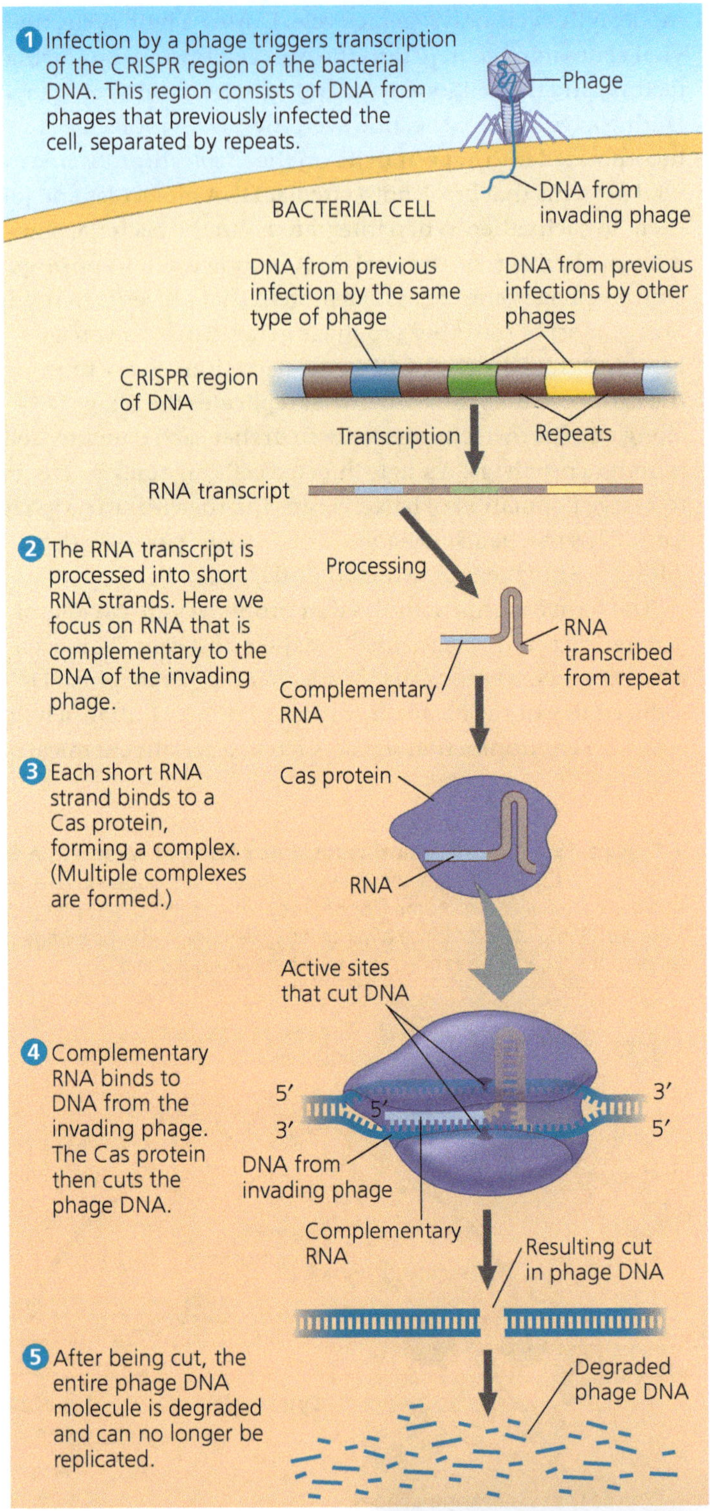

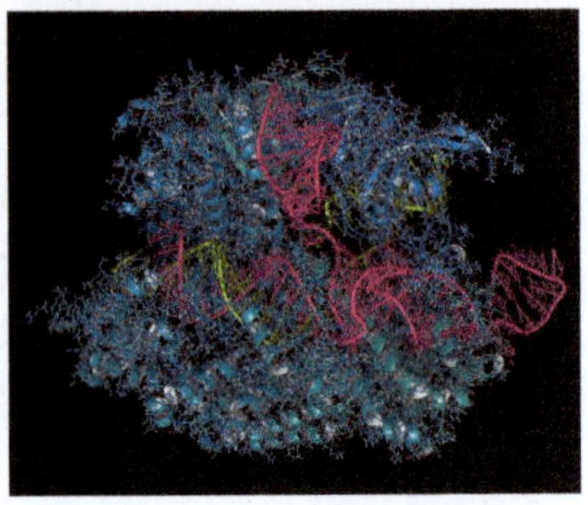

▶ Computer model of CRISPR-Cas9 gene editing complex from *Streptococcus pyogenes*

animal viruses with RNA genomes have an envelope, as do some with DNA genomes. Rather than consider all the mechanisms of viral infection and replication, we will focus first on the roles of viral envelopes and then on the functioning of RNA as the genetic material of many animal viruses.

Viral Envelopes

An animal virus equipped with an envelope—that is, a membranous outer layer—uses it to enter the host cell. Protruding from the outer surface of this envelope are viral glycoproteins that bind to specific receptor molecules on the surface of a host cell. **Figure 19.8** outlines the events in the replicative cycle of an enveloped virus with an RNA genome. Ribosomes bound to the endoplasmic reticulum (ER) of the host cell make the protein parts of the envelope glycoproteins; cellular enzymes in the ER and Golgi apparatus then add the sugars. The resulting viral glycoproteins, embedded in membrane derived from the host cell, are transported to the cell surface. In a process much like exocytosis, new viral capsids are wrapped in membrane as they bud from the cell. In other words, the viral envelope is usually derived from the host cell's plasma membrane, although all or most of the molecules of this membrane are specified by viral genes. The enveloped viruses are now free to infect other cells. This replicative cycle does not necessarily kill the host cell, in contrast to the lytic cycles of phages.

Some viruses have envelopes that are not derived from plasma membrane. Herpesviruses, for example, are temporarily cloaked in membrane derived from the nuclear envelope of the host; they then shed this membrane in the cytoplasm and acquire a new envelope made from membrane of the Golgi apparatus. These viruses have a double-stranded DNA genome and replicate within the host cell nucleus, using a combination of viral and cellular enzymes to replicate and transcribe their DNA. In the case of herpesviruses, copies of the viral DNA can remain behind as mini-chromosomes in the nuclei of certain nerve cells. There they remain latent until some sort of physical or emotional stress triggers a new round of active virus production. The infection of other cells by these new viruses causes the blisters characteristic of herpes, such as cold sores or genital sores. Once someone acquires a herpesvirus infection, flare-ups may recur throughout the person's life.

▼ Figure 19.8 The replicative cycle of an enveloped RNA virus. Shown here is a virus with a single-stranded RNA genome that functions as a template for synthesis of mRNA. Some enveloped viruses enter the host cell by fusion of the envelope with the cell's plasma membrane; others enter by endocytosis. For all enveloped RNA viruses, formation of new envelopes for progeny viruses occurs by the mechanism depicted in this figure.

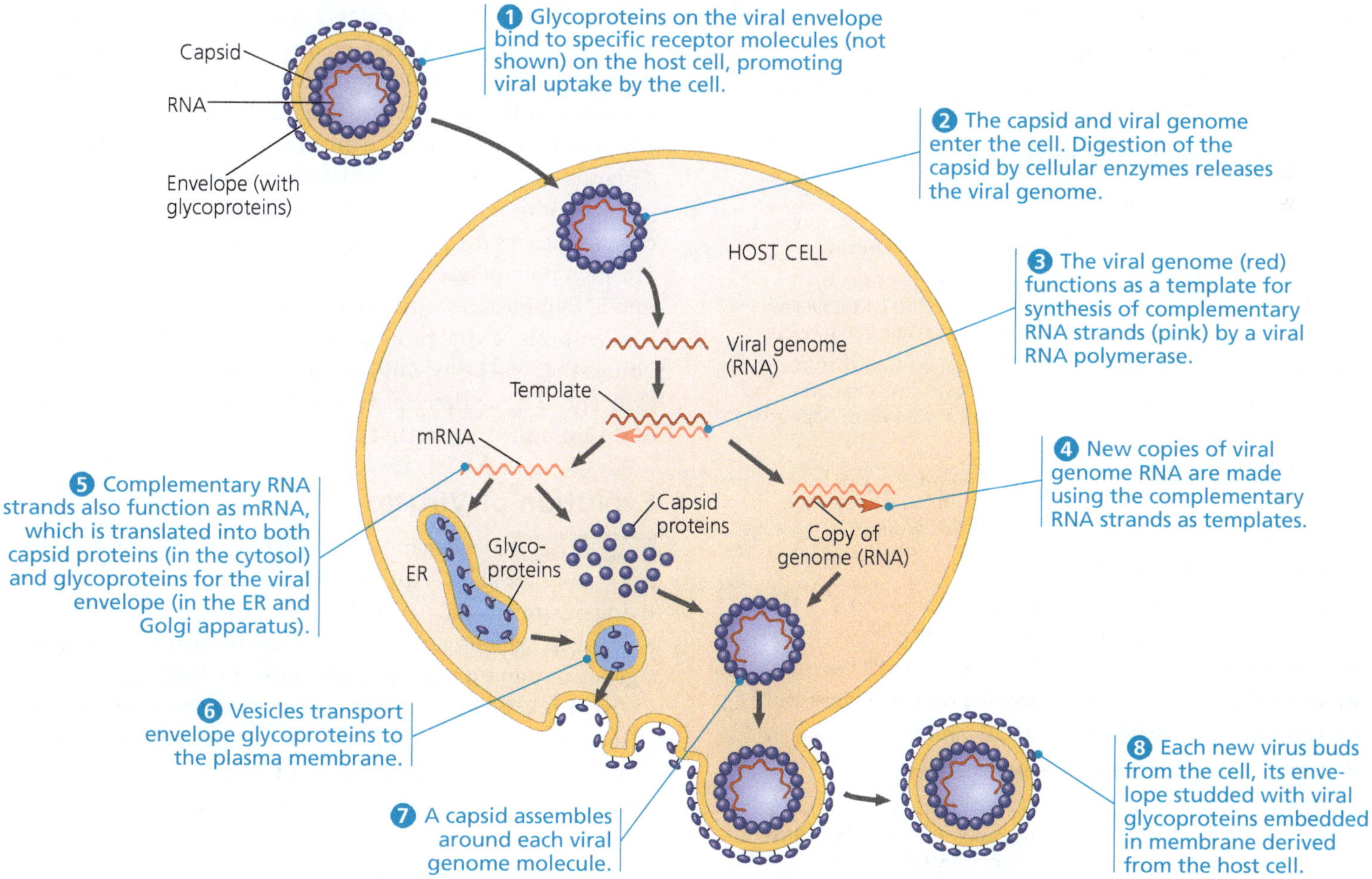

Viral Genetic Material

Table 19.1 shows the common classification system for animal viruses, which is based on their genetic material: double- or single-stranded DNA, or double- or single-stranded RNA. Although some phages and most plant viruses are RNA viruses, the broadest variety of RNA genomes is found among the viruses that infect animals. There are three types of single-stranded RNA genomes found in animal viruses (classes IV–VI in Table 19.1). The genome of class IV viruses can directly serve as mRNA and thus can be translated into viral protein immediately after infection. Figure 19.8 shows a virus of class V, in which the RNA genome serves instead as a *template* for mRNA synthesis. The RNA genome is transcribed into complementary RNA strands, which function both as mRNA and as templates for the synthesis of additional copies of genomic RNA. All viruses that use an RNA genome as a template for mRNA transcription require RNA → RNA synthesis. These viruses use a viral enzyme capable of carrying out this process; there are no such enzymes in most cells. The enzyme used in this process is encoded by the viral genome, and after its synthesis the protein is packaged during viral self-assembly with the genome inside the viral capsid.

Table 19.1 Classes of Animal Viruses

Class/Family	Envelope?	Examples That Cause Human Diseases
I. Double-Stranded DNA (dsDNA)		
Adenovirus (see Figure 19.3b)	No	Respiratory viruses
Papillomavirus	No	Warts, cervical cancer
Polyomavirus	No	Tumors
Herpesvirus	Yes	Herpes simplex I and II (cold sores, genital sores); varicella zoster (shingles, chicken pox); Epstein-Barr virus (mononucleosis, Burkitt's lymphoma)
Poxvirus	Yes	Smallpox virus; cowpox virus
II. Single-Stranded DNA (ssDNA)		
Parvovirus	No	B19 parvovirus (mild rash)
III. Double-Stranded RNA (dsRNA)		
Reovirus	No	Rotavirus (diarrhea); Colorado tick fever virus
IV. Single-Stranded RNA (ssRNA); Serves as mRNA		
Picornavirus	No	Rhinovirus (common cold); poliovirus; hepatitis A virus; other intestinal viruses
Coronavirus	Yes	Severe acute respiratory syndrome (SARS); Middle East respiratory syndrome (MERS)
Flavivirus	Yes	Zika virus (see Figure 19.10c); yellow fever virus; dengue virus; West Nile virus; hepatitis C virus
Togavirus	Yes	Chikungunya virus (see Figure 19.10b); rubella virus; equine encephalitis viruses
V. ssRNA; Serves as Template for mRNA Synthesis		
Filovirus	Yes	Ebola virus (hemorrhagic fever; see Figure 19.10a)
Orthomyxovirus	Yes	Influenza virus (see Figure 19.3c)
Paramyxovirus	Yes	Measles virus; mumps virus
Rhabdovirus	Yes	Rabies virus
VI. ssRNA; Serves as Template for DNA Synthesis		
Retrovirus	Yes	Human immunodeficiency virus (HIV/AIDS; see Figure 19.9); RNA tumor viruses (leukemia)

The RNA animal viruses with the most complicated replicative cycles are the **retroviruses** (class VI). These viruses have an enzyme called **reverse transcriptase** that transcribes an RNA template into DNA, an RNA → DNA information flow that is the opposite of the usual direction. This unusual phenomenon is the source of the name retroviruses (*retro* means "backward"). Of particular medical importance is **HIV (human immunodeficiency virus)**, the retrovirus shown in Figure 19.1 that causes **AIDS (acquired immunodeficiency syndrome)**. HIV and other retroviruses are enveloped viruses that contain two identical molecules of single-stranded RNA and two molecules of reverse transcriptase.

The HIV replicative cycle (traced in **Figure 19.9**) is typical of a retrovirus. After HIV enters a host cell, its reverse transcriptase molecules are released into the cytoplasm, where they catalyze synthesis of viral DNA. The newly made viral DNA then enters the cell's nucleus and integrates into the DNA of a chromosome. The integrated viral DNA, called a **provirus**, never leaves the host's genome, remaining a permanent resident of the cell. (Recall that a prophage, in contrast, leaves the host's genome at the start of a lytic cycle.) The RNA polymerase of the host transcribes the proviral DNA into RNA molecules, which can function both as mRNA for the synthesis of viral proteins and as genomes for the new viruses that will be assembled and released from the cell. In Concept 43.4, we describe how HIV causes the deterioration of the immune system that occurs in AIDS.

Evolution of Viruses

EVOLUTION We began this chapter by asking whether or not viruses are alive. Viruses do not really fit our definition of living organisms. An isolated virus is biologically inert, unable to replicate its genes or regenerate its own ATP. Yet it has a genetic program written in the universal language of life. Do we think of viruses as nature's most complex associations of molecules or as the simplest forms of life? Either way, we must bend our usual definitions. Although viruses cannot replicate or carry out metabolic activities independently, their use of the genetic code makes it hard to deny their evolutionary connection to the living world.

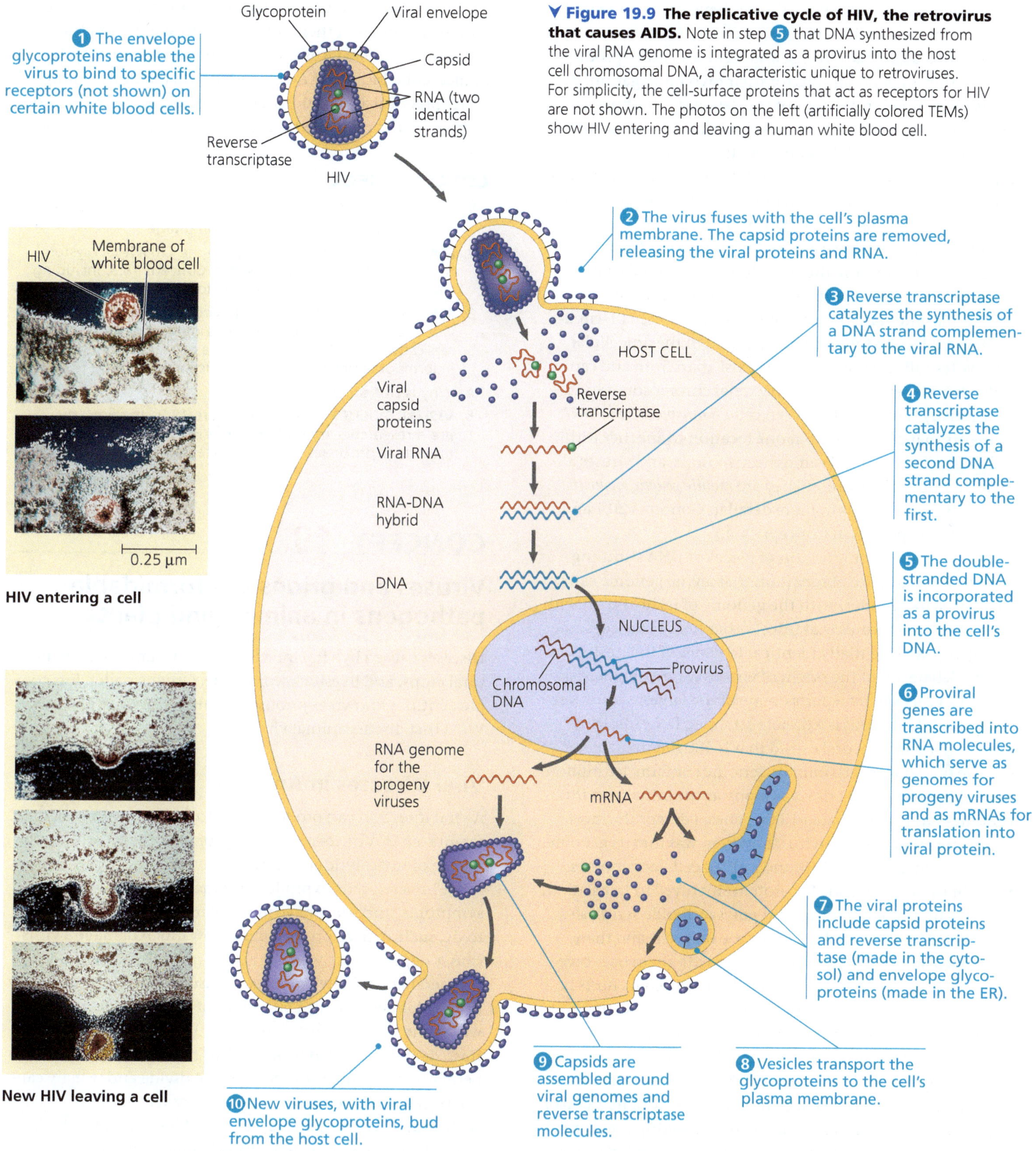

Figure 19.9 The replicative cycle of HIV, the retrovirus that causes AIDS. Note in step 5 that DNA synthesized from the viral RNA genome is integrated as a provirus into the host cell chromosomal DNA, a characteristic unique to retroviruses. For simplicity, the cell-surface proteins that act as receptors for HIV are not shown. The photos on the left (artificially colored TEMs) show HIV entering and leaving a human white blood cell.

MAKE CONNECTIONS ➤ *Describe what is known about binding of HIV to immune system cells. (See Figure 7.8.) How was this discovered?*

Animation: Retrovirus (HIV) Replicative Cycle

How did viruses originate? Viruses have been found that infect every form of life—not just bacteria, animals, and plants, but also archaea, fungi, and algae and other protists. Because they depend on cells for their own propagation, it seems likely that viruses are not the descendants of precellular forms of life but evolved—possibly multiple times—*after* the first cells appeared. Most molecular biologists favor the hypothesis that viruses originated from naked bits of cellular nucleic acids that moved from one cell to another, perhaps via injured cell surfaces. The evolution of genes coding for capsid proteins may have allowed viruses to bind cell membranes, thus facilitating the infection of uninjured cells.

Candidates for the original sources of viral genomes include plasmids and transposons. *Plasmids* are small, circular DNA molecules found in bacteria and in the unicellular eukaryotes called yeasts. Plasmids exist apart from and can replicate independently of the bacterial chromosome and are occasionally transferred between cells. *Transposons* are DNA segments that can move from one location to another within a cell's genome. Thus, plasmids, transposons, and viruses all share an important feature: They are *mobile genetic elements*. (We'll discuss plasmids in more detail in Concepts 20.1 and 27.2 and transposons in Concept 21.4.)

Consistent with this notion of pieces of DNA shuttling from cell to cell is the observation that a viral genome can have more in common with the genome of its host than with the genomes of viruses that infect other hosts. Indeed, some viral genes are essentially identical to genes of the host.

The debate about the origin of viruses was reinvigorated about 15 years ago by reports of one of the largest viruses yet discovered: Mimivirus is a double-stranded DNA (dsDNA) virus with an icosahedral capsid that is 400 nm in diameter, the size of a small bacterium. Its genome contains 1.2 million bases (Mb)—about 100 times as many as the influenza virus genome—and an estimated 1,000 genes. Perhaps the most surprising aspect of mimivirus, however, was that its genome included genes previously found only in cellular genomes. Some of these genes code for proteins involved in translation, DNA repair, protein folding, and polysaccharide synthesis. Whether mimivirus evolved *before* the first cells and then developed an exploitative relationship with them or evolved more recently and simply scavenged genes from its hosts is not yet settled. Since 2013 several even larger viruses have been discovered that cannot be classified with any existing known virus. One such virus is 1 μm (1,000 nm) in diameter, with a dsDNA genome of around 2–2.5 Mb, larger than that of some small eukaryotes. What's more, over 90% of its 2,000 or so genes are unrelated to cellular genes, inspiring the name it was given, pandoravirus. A second virus, called *Pithovirus sibericum*, with a diameter of 1.5 μm and 500 genes, was discovered in permanently frozen soil in Siberia. This virus, once thawed, was able to infect an amoeba after being frozen for 30,000 years! How these and all other viruses fit in the tree of life is an intriguing, unresolved question.

The ongoing evolutionary relationship between viruses and the genomes of their host cells is an association that continues to make viruses very useful experimental systems in molecular biology. Knowledge about viruses also allows many practical applications, since viruses have a tremendous impact on all organisms through their ability to cause disease.

CONCEPT CHECK 19.2

1. Compare the effect on the host cell of a lytic (virulent) phage and a lysogenic (temperate) phage.
2. **MAKE CONNECTIONS ➤** Compare the CRISPR-Cas system to the miRNA system discussed in Concept 18.3, including their mechanisms and their functions.
3. **MAKE CONNECTIONS ➤** The RNA virus in Figure 19.8 has a viral RNA polymerase that functions in step 3 of the virus's replicative cycle. Compare this with a cellular RNA polymerase in terms of template and overall function (see Figure 17.10).
4. Why is HIV called a retrovirus?
5. **VISUAL SKILLS ➤** Looking at Figure 19.9, imagine you are a researcher trying to combat HIV infection. What molecular processes could you attempt to block?

For suggested answers, see Appendix A.

CONCEPT 19.3

Viruses and prions are formidable pathogens in animals and plants

Diseases caused by viral infections afflict humans, agricultural crops, and livestock worldwide. Other smaller, less complex entities known as prions also cause disease in animals. We'll first discuss animal viruses.

Viral Diseases in Animals

A viral infection can produce symptoms by a number of different routes. Viruses may damage or kill cells by causing the release of hydrolytic enzymes from lysosomes. Some viruses cause infected cells to produce toxins that lead to disease symptoms, and some have molecular components that are toxic, such as envelope proteins. How much damage a virus causes depends partly on the ability of the infected tissue to regenerate by cell division. People usually recover completely from colds because the epithelium of the respiratory tract, which the viruses infect, can efficiently repair itself. In contrast, damage inflicted by poliovirus to mature nerve cells is permanent because these cells do not divide and usually cannot be replaced. Many of the temporary symptoms associated with viral infections, such as fever and body aches, actually result from the body's own efforts to defend itself against infection rather than from cell death caused by the virus.

The immune system is a critical part of the body's natural defenses (see Chapter 43). It is also the basis for the major medical tool used to prevent viral infections—vaccines. A **vaccine** is a harmless derivative of a pathogen that stimulates the immune system to mount defenses against the harmful pathogen. Smallpox, a viral disease that was once a devastating scourge in many parts of the world, was eradicated by a vaccination program carried out by the World Health Organization (WHO). The very narrow host range of the smallpox virus—it infects only humans—was a critical factor in the success of this program. Similar worldwide vaccination campaigns are currently under way to eradicate polio and measles. Effective vaccines are also available to protect against rubella, mumps, hepatitis B, and a number of other viral diseases.

Although vaccines can prevent some viral illnesses, medical care can do little, at present, to cure most viral infections once they occur. The antibiotics that help us recover from bacterial infections are powerless against viruses. Antibiotics kill bacteria by inhibiting enzymes specific to bacteria but have no effect on eukaryotic or virally encoded enzymes. However, the few enzymes that are encoded only by viruses have provided targets for other drugs. Most antiviral drugs resemble nucleosides and thus interfere with viral nucleic acid synthesis. One such drug is acyclovir, which impedes herpesvirus replication by inhibiting the viral polymerase that synthesizes viral DNA but not the eukaryotic one. Similarly, azidothymidine (AZT) curbs HIV replication by interfering with the synthesis of DNA by reverse transcriptase. In the past 20 years, much effort has gone into developing drugs to treat HIV. Currently, multidrug treatments, sometimes called "cocktails," are considered to be most effective. Such treatments commonly include a combination of two nucleoside mimics and a protease inhibitor, which interferes with an enzyme required for assembly of the viruses. Another effective treatment involves a drug called maraviroc, which blocks a protein on the surface of human immune cells that helps bind the HIV virus (see Figure 7.8). This drug has also been used successfully to prevent infection in individuals who either have been exposed to, or are at risk of exposure to, HIV.

MB BBC Video: Know Your Enemy: Bacteria vs. Viruses

Emerging Viruses

Viruses that suddenly become apparent are often referred to as *emerging viruses*. HIV, the AIDS virus, is a classic example: This virus appeared in San Francisco in the early 1980s, seemingly out of nowhere, although later studies uncovered a case in the Belgian Congo in 1959. A number of other dangerous emerging viruses cause encephalitis, inflammation of the brain. One example is the West Nile virus, which appeared in North America in 1999 and has spread to all 48 contiguous states in the United States, by now resulting in over 40,000 cases and almost 2,000 deaths.

MB HHMI Video: Interview with Katie Walter

The deadly Ebola virus **(Figure 19.10a)**, recognized initially in 1976 in central Africa, is one of several emerging viruses that cause *hemorrhagic fever*, an often fatal illness characterized by fever, vomiting, massive bleeding, and circulatory system collapse. In 2014, a widespread outbreak of Ebola virus in western Africa caused the World Health Organization to declare an international health emergency. By mid-2015 the outbreak, centered in Guinea, Sierra Leone, and Liberia, had caused over 27,000 illnesses and 11,000 deaths.

Another example is the mosquito-borne virus called chikungunya **(Figure 19.10b)**, which causes an acute illness with fever, rashes, and persistent joint pain. Chikungunya has long been considered a tropical virus, but it has now appeared in northern Italy and southeastern France. A more recently emerging virus is the Zika virus **(Figure 19.10c)**, which caused an outbreak of disease in spring 2015 in Brazil. Although symptoms of Zika are often mild, the outbreak was noticed because infection of pregnant women was correlated with a striking increase in the number of babies born with abnormally small brains, a condition called microcephaly. Zika is a mosquito-borne flavivirus (like West Nile virus) that infects neural cells, posing a particular danger to fetal brain development. Because of the neurological defects associated with Zika and its spread to 28 other countries by early 2016, the World Health Organization declared Zika an international health emergency.

Types of influenza often emerge as outbreaks of illness. In 2009, a widespread outbreak, or **epidemic**, of a flu-like illness appeared in Mexico and the United States. The infectious agent was quickly identified as an influenza virus related to viruses that cause the seasonal flu. This particular virus was named H1N1 for reasons that will be explained shortly. The illness spread rapidly, prompting WHO to declare a global epidemic, or **pandemic**, shortly thereafter. Half a year later, the disease had reached 207 countries, infecting over 600,000 people and killing almost 8,000.

▼ Figure 19.10 Emerging viruses.

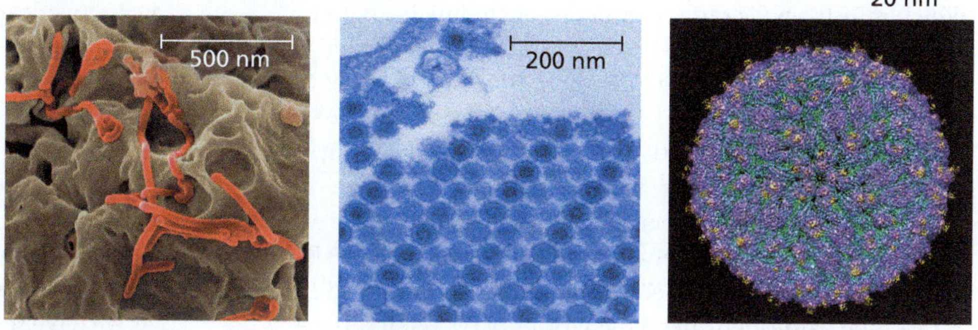

(a) Ebola viruses budding from a monkey cell (colorized SEM).

(b) Chikungunya viruses emerging from a cell in the upper left and packing together (colorized TEM).

(c) Computer-generated image of a **Zika virus**, based on a technique called cryo-electron microscopy.

How do such viruses burst on the human scene, giving rise to harmful diseases that were previously rare or even unknown? Three processes contribute to the emergence of viral diseases. The first, and perhaps most important, is the mutation of existing viruses. RNA viruses tend to have an unusually high rate of mutation because viral RNA polymerases do not proofread and correct errors in replicating their RNA genomes. Some mutations change existing viruses into new genetic varieties (strains) that can cause disease, even in individuals who are immune to the ancestral virus. For instance, seasonal flu epidemics are caused by new strains of influenza virus genetically different enough from earlier strains that people have little immunity to them. You'll see an example of this process in the **Scientific Skills Exercise**, where you'll analyze genetic changes in variants of the H1N1 flu virus and correlate them with spread of the disease.

A second process that can lead to the emergence of viral diseases is the dissemination of a viral disease from a small, isolated human population. For instance, AIDS went unnamed and virtually unnoticed for decades before it began to spread around the world. In this case, technological and social factors, including affordable international travel, blood transfusions, sexual promiscuity, and the abuse of intravenous drugs, allowed a previously rare human disease to become a global scourge.

Interview with David Satcher: The role of the CDC in recognizing AIDS and in public health

A third source of new viral diseases in humans is the spread of existing viruses from other animals. Scientists estimate that about three-quarters of new human diseases originate in this way. Animals that harbor and can transmit a particular virus but are generally unaffected by it are said to act as a natural reservoir for that virus. For example, the H1N1 virus that caused the 2009 flu pandemic mentioned earlier was likely passed to humans from pigs; for this reason, the disease it caused was originally called "swine flu."

In general, flu epidemics provide an instructive example of the effects of viruses moving between species. There are three types of influenza virus: types B and C, which infect only humans and have never caused an epidemic, and type A, which infects a wide range of animals, including birds, pigs, horses, and humans. Influenza A strains have caused four major flu epidemics among humans in the last 100 years. The worst was the first one, the "Spanish flu" pandemic of 1918–1919, which killed 40–50 million people, including many World War I soldiers.

Different strains of influenza A are given standardized names; for example, both the strain that caused the 1918 flu and the one that caused the 2009 pandemic flu are called H1N1. The name identifies which forms of two viral surface proteins are present: hemagglutinin (HA) and neuraminidase (NA). There are 16 different types of hemagglutinin, a protein that helps the flu virus attach to host cells, and 9 types of neuraminidase, an enzyme that helps release new virus particles from infected cells. Waterbirds have been found that carry viruses with all possible combinations of HA and NA. Variations of the hemagglutinin protein are used each year to generate vaccines against the strains predicted most likely to occur the next year.

A likely scenario for the 1918 pandemic and others is that the virus mutated as it passed from one host species to another. When an animal like a pig or a bird is infected with more than one strain of flu virus, the different strains can undergo genetic recombination if the RNA molecules making up their genomes mix and match during viral assembly. Pigs were probably the main hosts for recombination that led to the 2009 flu virus, which turns out to contain sequences from bird, pig, and human flu viruses. Coupled with mutation, these reassortments can lead to the emergence of a viral strain capable of infecting human cells. People who have never been exposed to that particular strain before will lack immunity, and the recombinant virus has the potential to be highly pathogenic. If such a flu virus recombines with viruses that circulate widely among humans, it may acquire the ability to spread easily from person to person, dramatically increasing the potential for a major human outbreak.

The many avian flu viruses carried by wild and domestic birds pose a potential long-term threat. A case in point is an H5N1 virus; the first transmission of H5N1 from birds to humans was documented in Hong Kong in 1997. Since then, the overall mortality rate due to H5N1 has been greater than 50% of those infected—an alarming number. Also, the host range of H5N1 is expanding, which provides increasing chances for reassortment between different strains. If the H5N1 avian flu virus evolves so that it can spread easily from person to person, it could represent a major global health threat akin to that of the 1918 pandemic.

How easily could this happen? In 2011, scientists working with ferrets, small mammals that are animal models for human flu, found out that only a few mutations of the avian flu virus would allow infection of cells in the human nasal cavity and windpipe. Furthermore, when the scientists transferred nasal swabs serially from ferret to ferret, the virus became transmissible through the air. Reports of this startling discovery at a scientific conference ignited a firestorm of debate about whether to publish the results and led to an ongoing reevaluation of the federal policies governing this type of experiment in the United States. The risks of doing this type of research (what if the new virus escapes or the procedure falls into the hands of bioterrorists?) must be considered in relation to the risks of not doing it—the possibility that we will be unable to combat new, more transmissible viruses because we lack an understanding of how they develop.

As we have seen, emerging viruses are generally not new; rather, they are existing viruses that mutate, disseminate more widely in the current host species, or spread to new host species. Changes in host behavior or environmental changes

SCIENTIFIC SKILLS EXERCISE

Analyzing a Sequence-Based Phylogenetic Tree to Understand Viral Evolution

How Can Sequence Data Be Used to Track Flu Virus Evolution? In 2009, an influenza A H1N1 virus caused a pandemic, and the virus has continued to resurface in outbreaks across the world. Researchers in Taiwan were curious about why the virus kept appearing despite widespread flu vaccine initiatives. They hypothesized that newly evolved variant strains of the H1N1 virus were able to evade human immune system defenses. To test this hypothesis, they needed to determine if each wave of flu infection was caused by a different H1N1 variant strain.

How the Experiment Was Done Scientists obtained the genome sequences for 4,703 virus isolates collected from patients with H1N1 flu in Taiwan. They compared the sequences in different strains for the viral hemagglutinin (HA) gene, and based on mutations that had occurred, arranged the isolates into a phylogenetic tree (see Figure 26.5 for information on how to read phylogenetic trees).

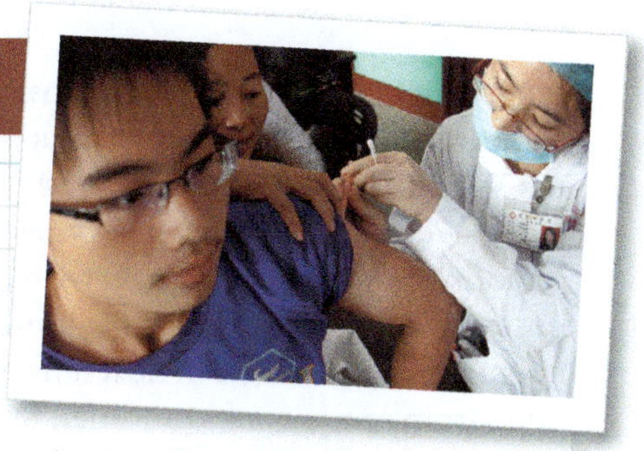

▲ **H1N1 flu vaccination.**

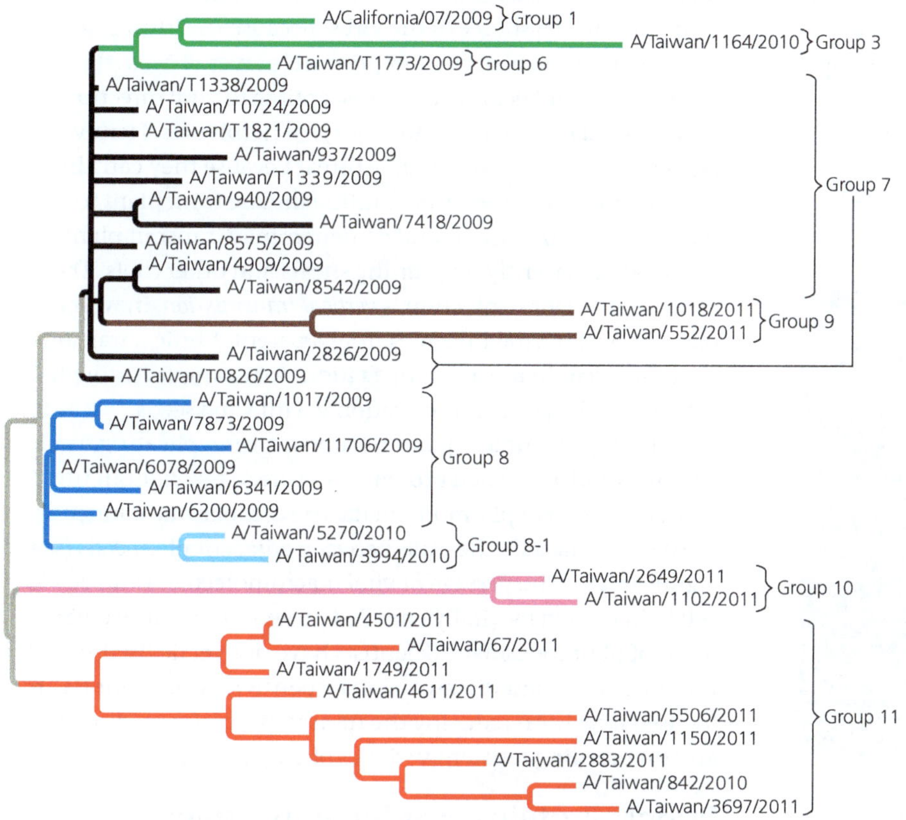

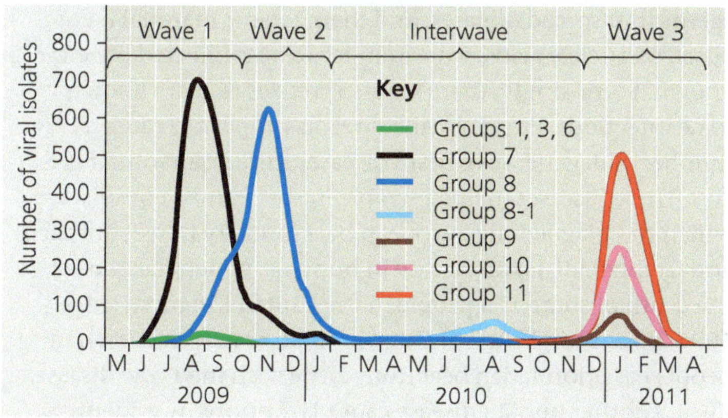

▲ Scientists graphed the number of isolates by the month and year of isolate collection to show the period in which each viral variant was actively causing illness in people.

Data from the Experiment In the phylogenetic tree, each branch tip is one variant strain of the H1N1 virus with a unique HA gene sequence. The tree is a way to visualize a working hypothesis about the evolutionary relationships between H1N1 variants.

INTERPRET THE DATA

1. The phylogenetic tree shows the hypothesized evolutionary relationship between the variant strains of H1N1 virus. The more closely connected two variants are, the more alike they are in terms of HA gene sequence. Each fork in a branch, called a node, shows where two lineages separate due to different accumulated mutations. The length of the branches is a measure of how many sequence differences there are between the variants, indicating how distantly related they are. Referring to the phylogenetic tree, which variants are more closely related to each other: A/Taiwan/1018/2011 and A/Taiwan/552/2011 or A/Taiwan/1018/2011 and A/Taiwan/8542/2009? Explain your answer.
2. The scientists arranged the branches into groups made up of one ancestral variant and all of its descendant, mutated variants. They are color-coded in the figure. Using group 11 as an example, trace the lineage of its variants. (a) Do all of the nodes have the same number of branches or branch tips? (b) Are all of the branches in the group the same length? (c) What do these results indicate?
3. The graph at the lower left shows the number of isolates collected (each from an ill patient) on the *y*-axis and the month and year that the isolates were collected on the *x*-axis. Each group of variants is plotted separately with a line color that matches the tree diagram. (a) Which group of variants was the earliest to cause the first wave of H1N1 flu in over 100 patients in Taiwan? (b) After a group of variants had a peak number of infections, did members of that same group cause another (later) wave of infection? (c) One variant in group 1 (green, uppermost branch) was used to make a vaccine that was distributed very early in the pandemic. Based on the graphed data, does it look like the vaccine was effective?
4. Groups 9, 10, and 11 all had H1N1 variants that caused a large number of infections at the same time in Taiwan. Does this mean that the scientists' hypothesis, that new variants cause new waves of infection, was incorrect? Explain your answer.

(MB) **Instructors**: A version of this Scientific Skills Exercise can be assigned in MasteringBiology.

Data from J.-R. Yang et al., New variants and age shift to high fatality groups contribute to severe successive waves in the 2009 influenza pandemic in Taiwan, *PLoS ONE* 6(11): e28288 (2011).

can increase the viral traffic responsible for emerging diseases. For instance, new roads built through remote areas can allow viruses to spread between previously isolated human populations. Also, the destruction of forests to expand cropland can bring humans into contact with other animals that may host viruses capable of infecting humans. Finally, genetic mutations and changes in host ranges can allow viruses to jump from one species to another. Many viruses, including chikungunya, mentioned earlier, can be transmitted by mosquitoes. A dramatic expansion of the disease caused by chikungunya occurred in the mid-2000s when a mutation in the virus allowed it to infect not only the mosquito species *Aedes aegypti*, but also the related *Aedes albopictus*. Promotion of the use of insecticides and mosquito netting over beds are crucial tools in public health attempts to prevent diseases carried by mosquitoes **(Figure 19.11)**.

▼ Figure 19.11 Mosquitoes as vectors for disease. Mosquitoes transmit viruses when they feed on infected blood from one person and then bite other people. Mosquito netting is an important means of preventing infection in affected areas.

Recently, scientists have become concerned about the possible effects of climate change on worldwide viral transmission. Dengue fever, also mosquito-borne, has appeared in Florida and Portugal, regions where it had not been seen before. The possibility that global climate change has allowed mosquito species carrying these viruses to expand their ranges and interact more is troubling because of the increased chance of a mutation allowing a virus species to jump to a new host. This is an area of active research by scientists applying climate change models to what is known about the habitat requirements of mosquito species.

Viral Diseases in Plants

More than 2,000 types of viral diseases of plants are known, and together they account for an estimated annual loss of $15 billion worldwide due to their destruction of agricultural and horticultural crops. Common signs of viral infection include bleached or brown spots on leaves and fruits **(Figure 19.12)**, stunted growth, and damaged flowers or roots, all of which can diminish the yield and quality of crops.

➤ Figure 19.12 Immature tomato infected by a virus.

Plant viruses have the same basic structure and mode of replication as animal viruses. Most plant viruses discovered thus far, including tobacco mosaic virus (TMV), have an RNA genome. Many have a helical capsid, like TMV, while others have an icosahedral capsid (see Figure 19.3b).

Viral diseases of plants spread by two major routes. In the first route, called *horizontal transmission*, a plant is infected from an external source of the virus. Because the invading virus must get past the plant's outer protective layer of cells (the epidermis), a plant becomes more susceptible to viral infections if it has been damaged by wind, injury, or herbivores. Herbivores, especially insects, pose a double threat because they can also act as carriers of viruses, transmitting disease from plant to plant. Moreover, farmers and gardeners may transmit plant viruses inadvertently on pruning shears and other tools. The other route of viral infection is *vertical transmission*, in which a plant inherits a viral infection from a parent. Vertical transmission can occur in asexual propagation (for example, through cuttings) or in sexual reproduction via infected seeds.

Once a virus enters a plant cell and begins replicating, viral genomes and associated proteins can spread throughout the plant by means of plasmodesmata, the cytoplasmic connections that penetrate the walls between adjacent plant cells (see Figure 36.19). The passage of viral macromolecules from cell to cell is facilitated by virally encoded proteins that cause enlargement of plasmodesmata. Scientists have not yet devised cures for most viral plant diseases. Consequently, research efforts are focused largely on reducing the transmission of such diseases and on breeding resistant varieties of crop plants.

Prions: Proteins as Infectious Agents

The viruses we have discussed in this chapter are infectious agents that spread diseases, and their genetic material is composed of nucleic acids, whose ability to be replicated is well known. Surprisingly, there are also *proteins* that are known to be infectious. Proteins called **prions** appear to cause a number of degenerative brain diseases in various animal species. These diseases include scrapie in sheep; mad cow disease, which has plagued the European beef industry in recent years; and Creutzfeldt-Jakob disease in humans, which has caused the death of some 175 people in the United Kingdom since 1996. Prions can be transmitted in food, as may occur when people eat prion-laden beef from cattle with mad cow disease. Kuru, another human disease caused by prions, was identified in the early 1900s among the South Fore natives of New Guinea. A kuru epidemic peaked there in the 1960s, puzzling scientists, who at first thought the disease had a genetic basis.

➤ **Figure 19.13 Model for how prions propagate.** Prions are misfolded versions of normal brain proteins. When a prion contacts a normally folded version of the same protein, it may induce the normal protein to assume the abnormal shape. The resulting chain reaction may continue until high levels of prion aggregation cause cellular malfunction and eventual degeneration of the brain.

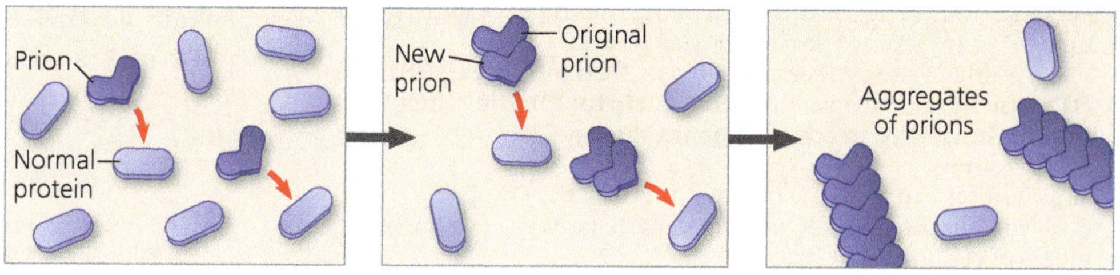

MB **Animation: Prions: Characteristics**
Animation: Prions: Diseases

Eventually, however, anthropological investigations ferreted out how the disease was spread: ritual cannibalism, a widespread practice among South Fore natives at that time.

Two characteristics of prions are especially alarming. First, prions act very slowly, with an incubation period of at least ten years before symptoms develop. The lengthy incubation period prevents sources of infection from being identified until long after the first cases appear, allowing many more infections to occur. Second, prions are virtually indestructible; they are not destroyed or deactivated by heating to normal cooking temperatures. To date, there is no known cure for prion diseases, and the only hope for developing effective treatments lies in understanding the process of infection.

How can a protein, which cannot replicate itself, be a transmissible pathogen? According to the leading model, a prion is a misfolded form of a protein normally present in brain cells. When the prion gets into a cell containing the normal form of the protein, the prion somehow converts normal protein molecules to the misfolded prion versions. Several prions then aggregate into a complex that can convert other normal proteins to prions, which join the chain **(Figure 19.13)**. Prion aggregation interferes with normal cellular functions and causes disease symptoms. This model was greeted with much skepticism when it was first proposed by Stanley Prusiner in the early 1980s, but it is now widely accepted. Prusiner was awarded the Nobel Prize in 1997 for his work on prions. He has recently proposed that prions are also involved in neurodegenerative diseases such as Alzheimer's and Parkinson's disease. There are many outstanding questions about these small infectious agents.

CONCEPT CHECK 19.3

1. Describe two ways in which a preexisting virus can become an emerging virus.
2. Contrast horizontal and vertical transmission of viruses in plants.
3. **WHAT IF?** ➤ TMV has been isolated from virtually all commercial tobacco products. Why, then, is TMV infection not an additional hazard for smokers?

For suggested answers, see Appendix A.

19 Chapter Review

MB Go to **MasteringBiology**™ for Videos, Animations, Vocab Self-Quiz, Practice Tests, and more in the Study Area.

SUMMARY OF KEY CONCEPTS

CONCEPT 19.1

A virus consists of a nucleic acid surrounded by a protein coat *(pp. 397–399)*

VOCAB SELF-QUIZ goo.gl/6u55ks

- Researchers discovered viruses in the late 1800s by studying a plant disease, tobacco mosaic disease.
- A **virus** is a small nucleic acid genome enclosed in a protein **capsid** and sometimes a membranous **viral envelope**. The genome may be single- or double-stranded DNA or RNA.

? *Are viruses generally considered living or nonliving? Explain.*

CONCEPT 19.2

Viruses replicate only in host cells *(pp. 399–406)*

- Viruses use enzymes, ribosomes, and small molecules of host cells to synthesize progeny viruses during replication.
- Each type of virus has a characteristic **host range**, affected by whether cell-surface proteins are present that viral surface proteins can bind to.
- **Phages** (viruses that infect bacteria) can replicate by two alternative mechanisms: the **lytic cycle** and the **lysogenic cycle**.

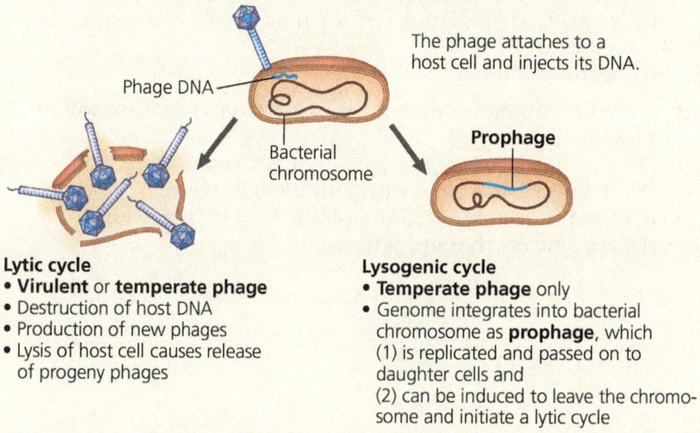

- Bacteria have various ways of defending themselves against phage infections, including the CRISPR-Cas system.
- Many animal viruses have an envelope. **Retroviruses** (such as **HIV**) use the enzyme **reverse transcriptase** to copy their RNA genome into DNA, which can be integrated into the host genome as a **provirus**.
- Since viruses can replicate only within cells, they probably evolved after the first cells appeared, perhaps as packaged fragments of cellular nucleic acid.

? *Describe enzymes that are not found in most cells but are necessary for the replication of certain types of viruses.*

CONCEPT 19.3

Viruses and prions are formidable pathogens in animals and plants *(pp. 406–411)*

- Symptoms of viral diseases may be caused by direct viral harm to cells or by the body's immune response. **Vaccines** stimulate the immune system to defend the host against specific viruses.
- An **epidemic**, a widespread outbreak of a disease, can become a **pandemic**, a global epidemic.
- Outbreaks of emerging viral diseases in humans are usually not new, but rather are caused by existing viruses that expand their host territory. The H1N1 2009 flu virus was a new combination of pig, human, and avian viral genes that caused a pandemic. The H5N1 avian flu virus has the potential to cause a high-mortality flu pandemic.
- Viruses enter plant cells through damaged cell walls (horizontal transmission) or are inherited from a parent (vertical transmission).
- **Prions** are slow-acting, virtually indestructible infectious proteins that cause brain diseases in mammals.

? *What aspect of an RNA virus makes it more likely than a DNA virus to become an emerging virus?*

TEST YOUR UNDERSTANDING

Level 1: Knowledge/Comprehension

PRACTICE TEST
goo.gl/CUYGKD

1. Which of the following characteristics, structures, or processes is common to both bacteria and viruses?
(A) metabolism
(B) ribosomes
(C) genetic material composed of nucleic acid
(D) cell division

2. Emerging viruses arise by
(A) mutation of existing viruses.
(B) the spread of existing viruses to new host species.
(C) the spread of existing viruses more widely within their host species.
(D) all of the above.

3. To cause a human pandemic, the H5N1 avian flu virus would have to
(A) spread to primates such as chimpanzees.
(B) develop into a virus with a different host range.
(C) become capable of human-to-human transmission.
(D) become much more pathogenic.

Level 2: Application/Analysis

4. A bacterium is infected with an experimentally constructed bacteriophage composed of the T2 phage protein coat and T4 phage DNA. The new phages produced would have
(A) T2 protein and T4 DNA.
(B) T4 protein and T2 DNA.
(C) T2 protein and T2 DNA.
(D) T4 protein and T4 DNA.

5. RNA viruses require their own supply of certain enzymes because
(A) host cells rapidly destroy the viruses.
(B) host cells lack enzymes that can replicate the viral genome.
(C) these enzymes translate viral mRNA into proteins.
(D) these enzymes penetrate host cell membranes.

6. **DRAW IT** Redraw Figure 19.8 to show the replicative cycle of a virus with a single-stranded genome that can function as mRNA (a class IV virus).

Level 3: Synthesis/Evaluation

7. **EVOLUTION CONNECTION** The success of some viruses lies in their ability to evolve rapidly within the host. Such viruses evade the host's defenses by mutating and producing many altered progeny viruses before the body can mount an attack. Thus, the viruses present late in infection differ from those that initially infected the body. Discuss this as an example of evolution in microcosm. Which viral lineages tend to predominate?

8. **SCIENTIFIC INQUIRY** When bacteria infect an animal, the number of bacteria in the body increases in an exponential fashion (graph A). After infection by a virulent animal virus with a lytic replicative cycle, there is no evidence of infection for a while. Then the number of viruses rises suddenly and subsequently increases in a series of steps (graph B). Explain the difference in the curves.

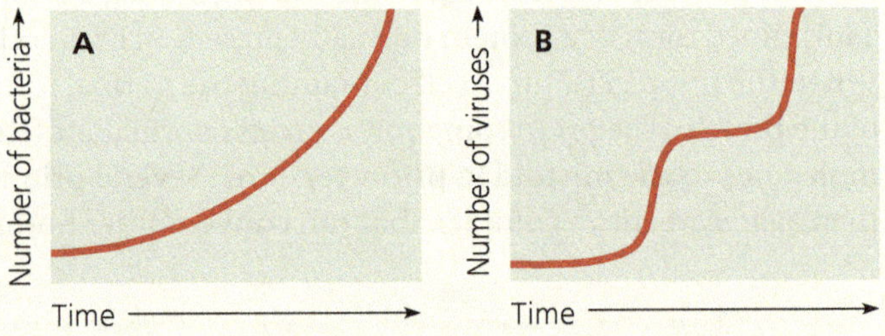

9. **WRITE ABOUT A THEME: ORGANIZATION** While viruses are considered by most scientists to be nonliving, they do show some characteristics of life, including the correlation of structure and function. In a short essay (100–150 words), discuss how the structure of a virus correlates with its function.

10. **SYNTHESIZE YOUR KNOWLEDGE**

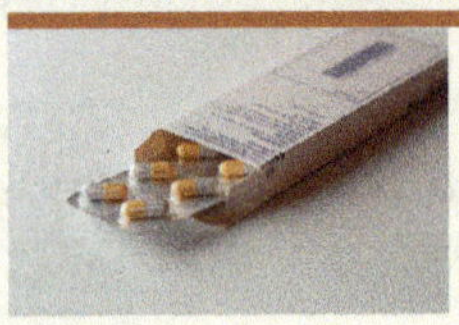

Oseltamivir (Tamiflu), an antiviral drug prescribed for influenza, inhibits the enzyme neuraminidase. Explain how this drug could prevent infection in someone exposed to the flu or could shorten the course of flu in an infected patient (the reasons for which it is prescribed).

For selected answers, see Appendix A.

For additional practice questions, check out the **Dynamic Study Modules** in MasteringBiology. You can use them to study on your smartphone, tablet, or computer anytime, anywhere!

20 DNA Tools and Biotechnology

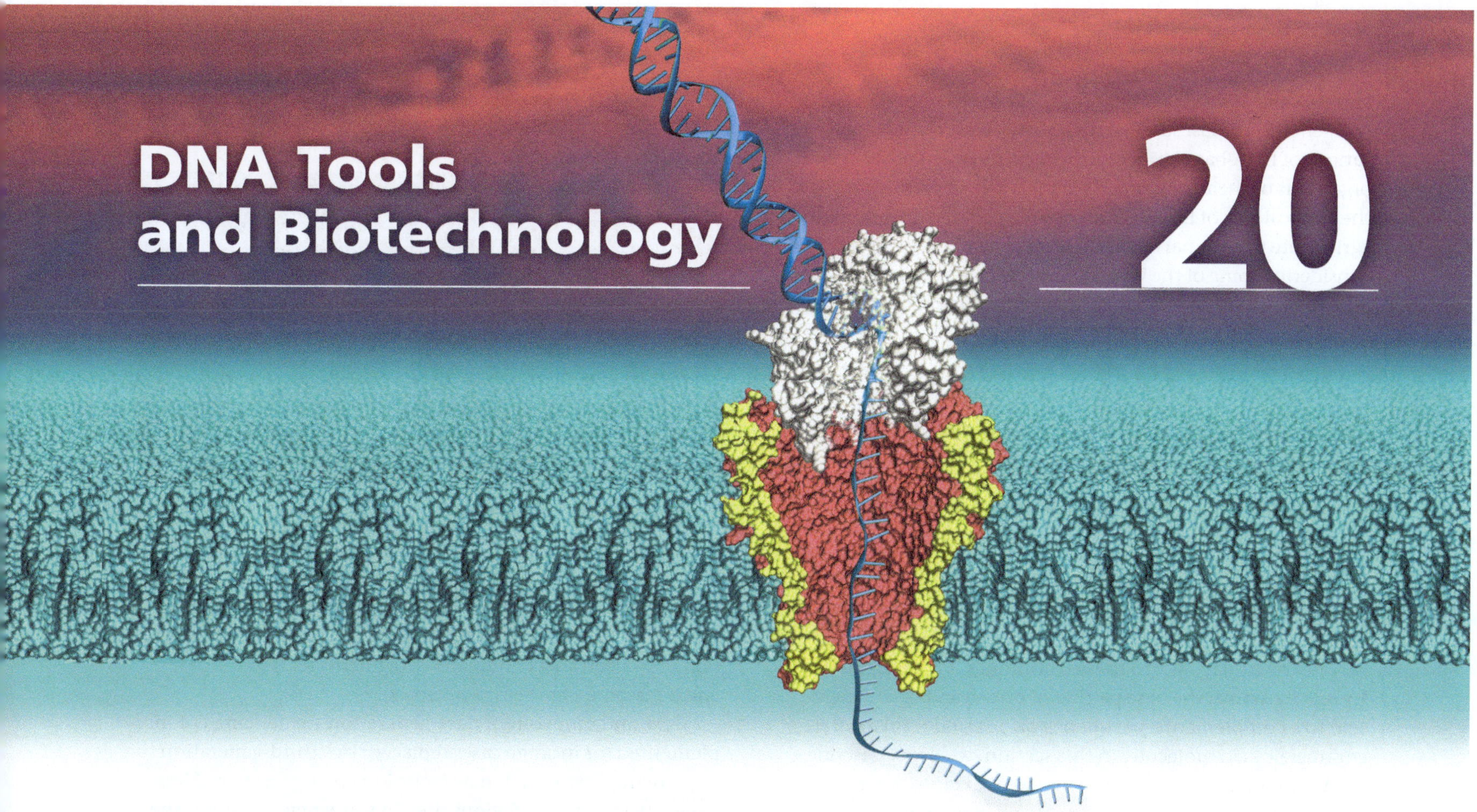

▲ **Figure 20.1 How can the technique shown in this model advance our sequencing of genomes?**

KEY CONCEPTS

20.1 DNA sequencing and DNA cloning are valuable tools for genetic engineering and biological inquiry

20.2 Biologists use DNA technology to study gene expression and function

20.3 Cloned organisms and stem cells are useful for basic research and other applications

20.4 The practical applications of DNA-based biotechnology affect our lives in many ways

◄ Woolly mammoth, an extinct organism whose genome was sequenced using mummified remains

The DNA Toolbox

The last decade or so has seen some extraordinary feats in biology, among them determination of the complete DNA sequences of several extinct species, including woolly mammoths (see below), Neanderthals, and a 700,000-year-old horse. Pivotal to those discoveries was the sequencing of the human genome, essentially completed in 2003. This endeavor marked a turning point in biology because it sparked remarkable technological advances in DNA sequencing.

The first human genome sequence took several years at a cost of 1 billion dollars; the time and cost of sequencing a genome have been in free fall since then. **Figure 20.1** shows a model of a sequencing technique in which the nucleotides of a single strand of DNA are passed one by one through a very small pore in a membrane, and the resulting tiny changes in an electrical current are used to determine the nucleotide sequence. Developers of this technique, which you will learn more about later in the chapter, claim that ultimately we will be able to sequence a human genome in about 6 hours on a $900 device the size of a pack of gum.

In this chapter, we'll first describe the main techniques for sequencing and manipulating DNA—**DNA technology**—and for using these DNA tools to analyze gene expression. Next, we'll explore advances in cloning organisms and producing stem cells, techniques that have both expanded our basic understanding of biology

When you see this blue icon, log in to **MasteringBiology** and go to the Study Area for digital resources.

MB **Get Ready for This Chapter**

and enhanced our ability to apply that understanding to global problems. In the last section, we'll survey the practical applications of DNA-based **biotechnology**, the manipulation of organisms or their components to make useful products. Today, the applications of DNA technology affect everything from agriculture to criminal law to medical research. We will end by considering some of the important social and ethical issues that arise as biotechnology becomes more pervasive in our lives.

CONCEPT 20.1

DNA sequencing and DNA cloning are valuable tools for genetic engineering and biological inquiry

The discovery of the structure of the DNA molecule, and specifically the recognition that its two strands are complementary to each other, opened the door for the development of DNA sequencing and other techniques used in biological research today. Key to these techniques is **nucleic acid hybridization**, the base pairing of one strand of a nucleic acid to a complementary sequence on a strand from a different nucleic acid molecule. In this section, we'll first describe DNA-sequencing techniques. Then we'll explore other important methods used in **genetic engineering**, the direct manipulation of genes for practical purposes.

DNA Sequencing

Researchers can exploit the principle of complementary base pairing to determine the complete nucleotide sequence of a DNA molecule, a process called **DNA sequencing**. The DNA is first cut into fragments, and then each fragment is sequenced. The first automated procedure used a technique called *dideoxyribonucleotide* (or *dideoxy*) *chain termination sequencing*. In this technique, one strand of a DNA fragment is used as a template for synthesis of a nested set of complementary fragments; these are further analyzed to yield the sequence. Biochemist Frederick Sanger received the Nobel Prize in 1980 for developing this method. Dideoxy sequencing is still used for routine small-scale sequencing jobs.

MB HHMI Video: Sanger Method of DNA Sequencing hhmi BioInteractive

In the last 15 years, "next-generation sequencing" techniques have been developed that are much faster **(Figure 20.2)**. DNA fragments are amplified (copied) to yield an enormous number of identical fragments **(Figure 20.3)**. A specific strand of each fragment is immobilized, and the complementary strand is synthesized, one nucleotide at a time. A chemical technique enables electronic monitors to identify in real time which of the four nucleotides is added; this method is thus called *sequencing by synthesis*. Thousands or hundreds of thousands of fragments, each about 300 nucleotides long, are sequenced in parallel in machines like those shown in Figure 20.2, accounting for the high rate of nucleotides sequenced per hour. This is an example of "high-throughput" DNA technology and is currently the method of choice for studies where massive numbers of DNA samples—even a set of numerous fragments representing an entire genome—are being sequenced.

▼ Figure 20.2 Next-generation DNA sequencing machines. These machines use sequencing by synthesis and can sequence 70–90 million nucleotides per hour.

More and more often, next-generation sequencing is complemented (or in some cases replaced) by "third-generation sequencing," with each new technique being faster and less expensive than the previous one. In some new methods, the DNA is neither cut into fragments nor amplified. Instead, a single, very long DNA molecule is sequenced on its own. Several groups have developed techniques that move a single strand of a DNA molecule through a very small pore (a *nanopore*) in a membrane, identifying the bases one by one by the distinct way each interrupts an electrical current. One model of this concept is shown in Figure 20.1, in which the pore is a protein channel embedded in a lipid membrane. (Other researchers are using artificial membranes and nanopores.) The idea is that each type of base interrupts the electrical current for a slightly different length of time. In 2015, after a year of use and review by scientists, the first nanopore sequencer went on the market; this device is the size of a small candy bar and connects to a computer via a USB port. Associated software allows immediate identification and analysis of the sequence. This is only one of many approaches to further increase the rate and cut the cost of sequencing, while also allowing the methodology to move out of the laboratory and into the field.

Improved DNA-sequencing techniques have transformed the way in which we can explore fundamental biological questions about evolution and how life works (see Make Connections Figure 5.26). Little more than 15 years after the human genome sequence was announced, researchers had completed the sequencing of thousands of genomes, with tens of thousands in progress. Complete genome sequences have been determined for cells from several cancers, for ancient humans, and for the many bacteria that live in the

Figure 20.3

Research Method Sequencing by Synthesis: Next-Generation Sequencing

Application In current next-generation sequencing techniques, each fragment is about 300 nucleotides long; by sequencing the fragments in parallel, about 2 billion nucleotides can be sequenced in 24 hours.

Technique See numbered steps and diagrams.

Results Each of the 2,000,000 wells in the multiwell plate, which holds a different fragment, yields a different sequence. The results for one fragment are shown below as a "flow-gram." The sequences of the entire set of fragments are analyzed using computer software, which "stitches" them together into a whole sequence—here, an entire genome.

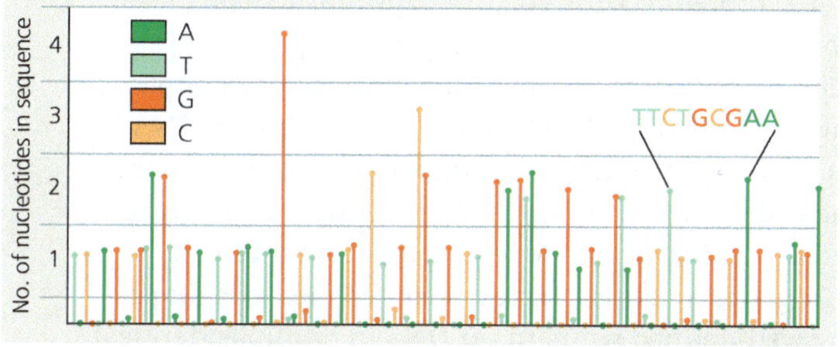

INTERPRET THE DATA ➤ *If the template strand has two or more identical nucleotides in a row, their complementary nucleotides will be added one after the other in the same flow step. How are two or more of the same nucleotide (in a row) detected in the flow-gram? (See sample on the right.) Write out the sequence of the first 25 nucleotides in the flow-gram above, starting from the left. (Ignore the very short lines.)*

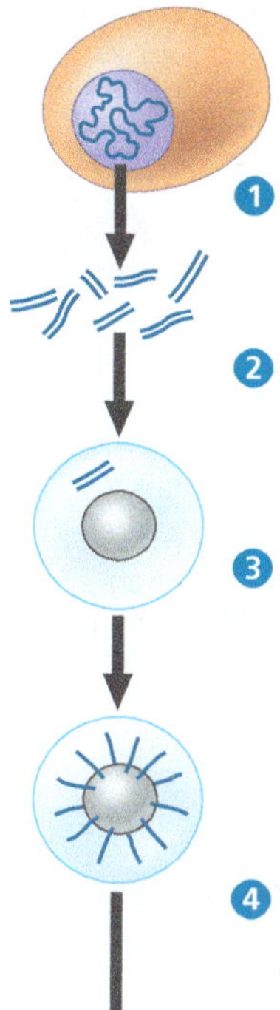

1. Genomic DNA is fragmented, and fragments of 300 base pairs are selected.

2. Each fragment is isolated with a bead in a droplet of aqueous solution.

3. The fragment is copied over and over by a technique called PCR (to be described later). All the 5′ ends of one strand are specifically "captured" by the bead. Eventually, 10^6 identical copies of the same single strand, which will be used as a template strand, are attached to the bead.

4. The bead is placed into a small well along with DNA polymerases and primers that can hybridize to the 3′ end of the single (template) strand.

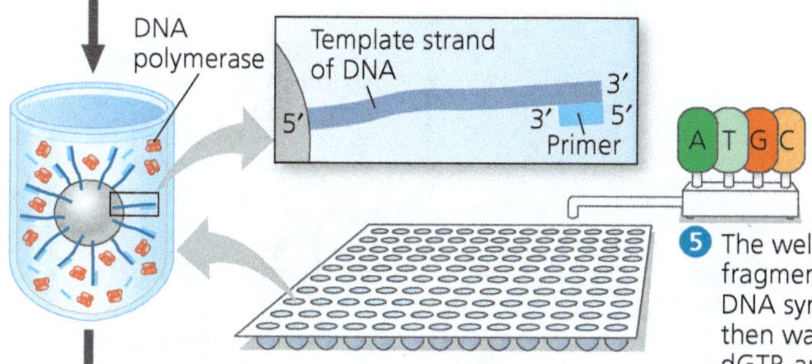

5. The well is one of 2 million on a multiwell plate, each containing a different DNA fragment to be sequenced. A solution of one of the four nucleotides required for DNA synthesis (deoxynucleoside triphosphates, or dNTPs) is added to all wells and then washed off. This is done sequentially for all four nucleotides: dATP, dTTP, dGTP, and then dCTP. The entire process is then repeated.

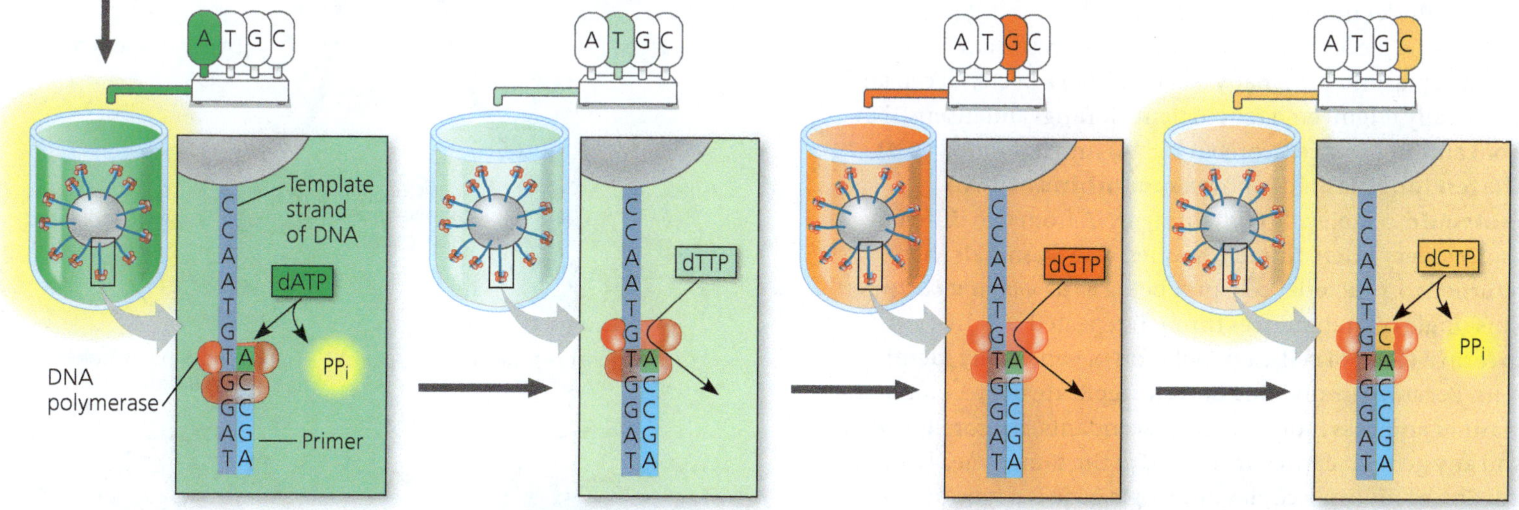

6. In each well, if the next base on the template strand (T in this example) is complementary to the added nucleotide (A, here), the nucleotide is joined to the growing strand, releasing PP_i, which causes a flash of light that is recorded.

7. The nucleotide is washed off and a different nucleotide (dTTP, here) is added. If the nucleotide is not complementary to the next template base (G, here), it is not joined to the strand and there is no flash.

8. The process of adding and washing off the four nucleotides is repeated until every fragment has a complete complementary strand. The pattern of flashes reveals the sequence of the original fragment in each well.

human intestine. In Chapter 21, you'll learn more about how this rapid acceleration of sequencing technology has revolutionized our study of the evolution of species and the genome itself. Now, let's consider how individual genes are studied.

Making Multiple Copies of a Gene or Other DNA Segment

A molecular biologist studying a particular gene or group of genes faces a challenge. Naturally occurring DNA molecules are very long, and a single molecule usually carries hundreds or even thousands of genes. Moreover, in many eukaryotic genomes, protein-coding genes occupy only a small proportion of the chromosomal DNA, the rest being noncoding nucleotide sequences. A single human gene, for example, might constitute only 1/100,000 of a chromosomal DNA molecule. As a further complication, the distinctions between a gene and the surrounding DNA are subtle, consisting only of differences in nucleotide sequence. To work directly with specific genes, scientists have developed methods for preparing well-defined segments of DNA in multiple identical copies, a process called **DNA cloning**.

Most methods for cloning pieces of DNA in the laboratory share certain general features. One common approach uses bacteria, most often *Escherichia coli*. Recall from Figure 16.12 that the *E. coli* chromosome is a large, circular molecule of DNA. In addition, *E. coli* and many other bacteria also have **plasmids**, small, circular DNA molecules that are replicated separately. A plasmid has only a small number of genes; these genes may be useful when the bacterium is in a particular environment but may not be required for survival or reproduction under most conditions.

To clone pieces of DNA using bacteria, researchers first obtain a plasmid (originally isolated from a bacterial cell and genetically engineered for efficient cloning) and insert DNA from another source ("foreign" DNA) into it **(Figure 20.4)**. The resulting plasmid is now a **recombinant DNA molecule**, a molecule containing DNA from two different sources, very often different species. The plasmid is then returned to a bacterial cell, producing a *recombinant bacterium*. This single cell reproduces through repeated cell divisions to form a clone of cells, a population of genetically identical cells. Because the dividing bacteria replicate the recombinant plasmid and pass it on to their descendants, the foreign DNA and any genes it carries are cloned at the same time. The production of multiple copies of a single gene is a type of DNA cloning called **gene cloning**.

In Figure 20.4, the plasmid acts as a **cloning vector**, a DNA molecule that can carry foreign DNA into a host cell and be replicated there. Bacterial plasmids are widely used as cloning vectors for several reasons: They can be readily obtained from commercial suppliers, manipulated to form recombinant plasmids by insertion of foreign DNA in a test

▼ Figure 20.4 Gene cloning and some uses of cloned genes. In this simplified diagram of gene cloning, we start with a plasmid (originally isolated from a bacterial cell) and a gene of interest from another organism. Only one plasmid and one copy of the gene of interest are shown at the top of the figure, but the starting materials would include many of each.

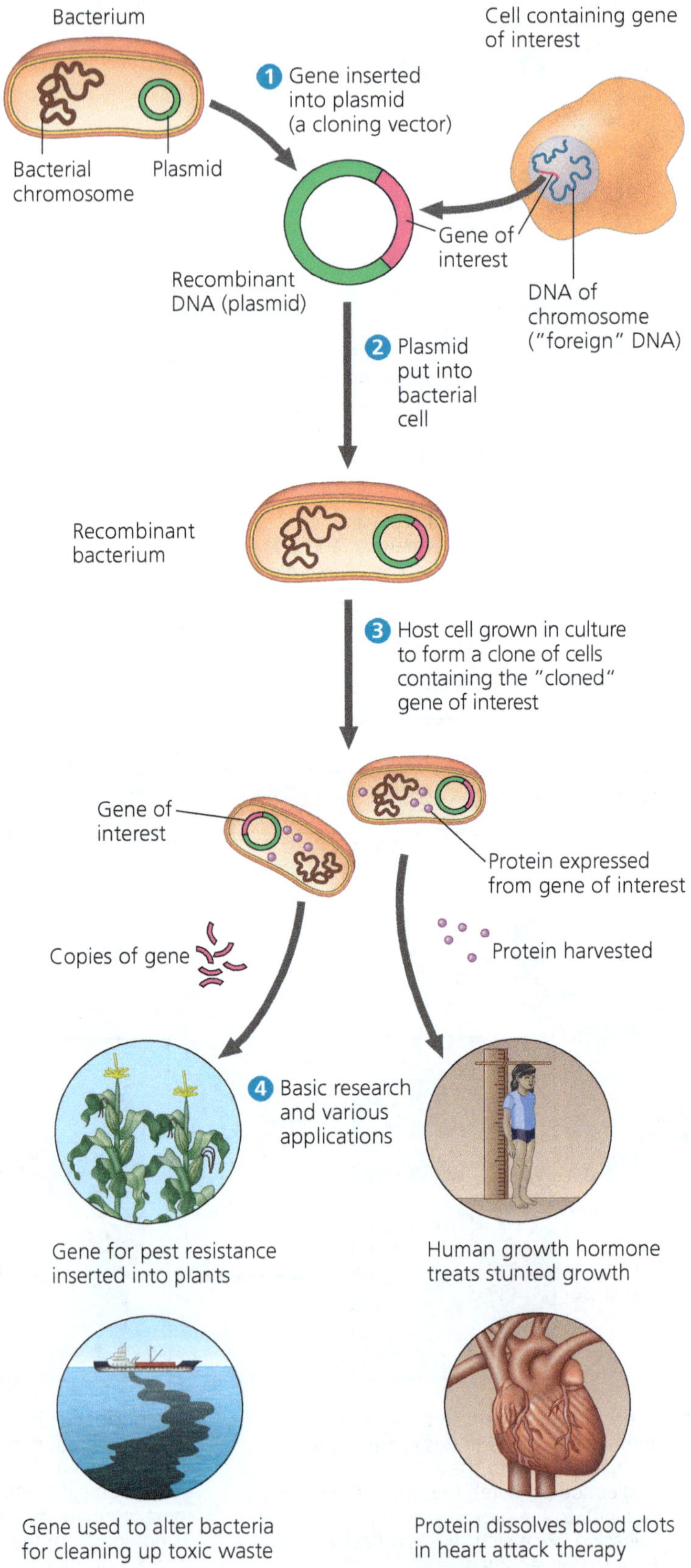

MB HHMI Video: Genetic Engineering

tube (referred to as *in vitro*, from the Latin meaning "in glass"), and then easily introduced into bacterial cells. The foreign DNA in Figure 20.4 is a gene from a eukaryotic cell; we will describe in more detail how the foreign DNA segment was obtained later in this section.

Gene cloning is useful for two basic purposes: to make many copies of, or *amplify*, a particular gene and to produce a protein product from it (see Figure 20.4). Researchers can isolate copies of a cloned gene from bacteria for use in basic research or to endow another organism with a new metabolic capability, such as pest resistance. For example, a resistance gene present in one crop species might be cloned and transferred into plants of another species. (Such organisms are called *genetically modified*, or *GM* for short; they will be discussed later in the chapter.) Alternatively, a protein with medical uses, such as human growth hormone, can be harvested in large quantities from cultures of bacteria carrying a cloned gene for the protein. (We'll describe the techniques for expressing cloned genes later.) Since one gene is only a very small part of the total DNA in a cell, the ability to amplify such rare DNA fragments is crucial for any application involving a single gene.

Using Restriction Enzymes to Make a Recombinant DNA Plasmid

Gene cloning and genetic engineering generally rely on the use of enzymes that cut DNA molecules at a limited number of specific locations. These enzymes, called restriction endonucleases, or **restriction enzymes**, were discovered in the late 1960s by biologists doing basic research on bacteria. Restriction enzymes protect the bacterial cell by cutting up foreign DNA from other organisms or phages (see Concept 19.2).

Hundreds of different restriction enzymes have been identified and isolated. Each restriction enzyme is very specific, recognizing a particular short DNA sequence, or **restriction site**, and cutting both DNA strands at precise points within this restriction site. The DNA of a bacterial cell is protected from the cell's own restriction enzymes by the addition of methyl groups ($—CH_3$) to adenines or cytosines within the sequences recognized by the enzymes.

MB **Animation: Restriction Enzymes**

Figure 20.5 shows how restriction enzymes are used to clone a foreign DNA fragment into a bacterial plasmid. At the top is a bacterial plasmid (like the one in Figure 20.4) that has a single restriction site recognized by a particular restriction enzyme from *E. coli*. As shown in this example, most restriction sites are symmetrical. That is, the sequence of nucleotides is the same on both strands when read in the $5' \rightarrow 3'$ direction. The most commonly used restriction enzymes recognize sequences containing four to eight nucleotide pairs. Because any sequence this short usually occurs (by chance)

Figure 20.5 Using a restriction enzyme and DNA ligase to make a recombinant DNA plasmid. The restriction enzyme in this example (called *Eco*RI) recognizes a single six-base-pair restriction site present in this plasmid. It makes staggered cuts in the sugar-phosphate backbones, producing fragments with "sticky ends." Foreign DNA fragments with complementary sticky ends can base-pair with the plasmid ends; the ligated product is a recombinant plasmid. (If the two plasmid sticky ends base-pair, the original nonrecombinant plasmid is reformed.)

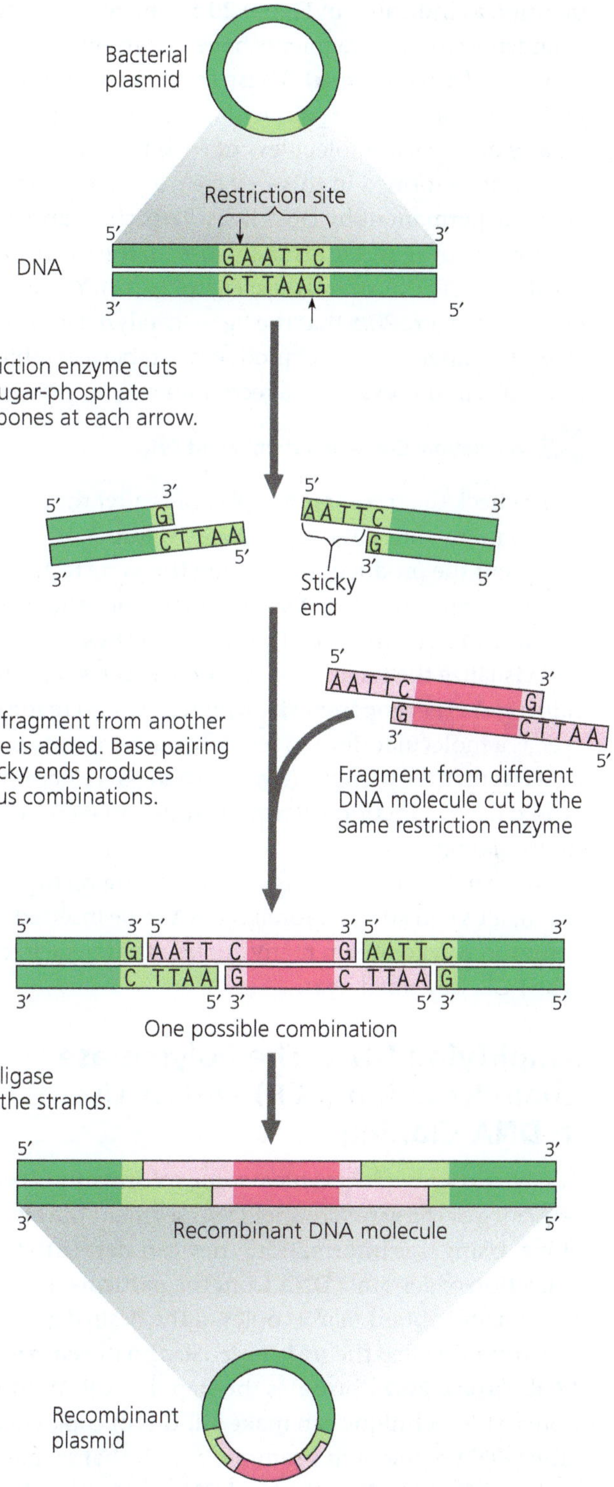

DRAW IT ➤ *The restriction enzyme Hind*III *recognizes the sequence 5′-AAGCTT-3′, cutting between the two A's. Draw the double-stranded sequence before and after the enzyme cuts it.*

MB **Animation: Recombinant DNA**

many times in a long DNA molecule, a restriction enzyme will make many cuts in such a DNA molecule, yielding a set of **restriction fragments**. All copies of a given DNA molecule always yield the same set of restriction fragments when exposed to the same restriction enzyme.

The most useful restriction enzymes cleave the sugar-phosphate backbones in the two DNA strands in a staggered manner, as indicated in Figure 20.5. The resulting double-stranded restriction fragments have at least one single-stranded end, called a **sticky end**. These short extensions can form hydrogen-bonded base pairs with complementary sticky ends on any other DNA molecules cut with the same enzyme. The associations formed in this way are only temporary but can be made permanent by DNA ligase, which catalyzes the formation of covalent bonds that close up the sugar-phosphate backbones of DNA strands (see Figure 16.16). You can see at the bottom of Figure 20.5 that the ligase-catalyzed joining of DNA from two different sources produces a stable recombinant DNA molecule, in this example, a recombinant plasmid.

MB Animation: Creating Recombinant DNA

To check the recombinant plasmids after they have been copied many times in host cells (see Figure 20.4), a researcher might cut the products again using the same restriction enzyme, expecting two DNA fragments, one the size of the plasmid and one the size of the inserted DNA. To separate and visualize the fragments, researchers carry out a technique called **gel electrophoresis**, which uses a gel made of a polymer as a molecular sieve to separate out a mixture of nucleic acid fragments by length **(Figure 20.6)**. Gel electrophoresis is used in conjunction with many different techniques in molecular biology.

Now that we have discussed the cloning vector in some detail, let's consider the foreign DNA to be inserted. The most common way to obtain many copies of the gene to be cloned is by PCR, described next.

Amplifying DNA: The Polymerase Chain Reaction (PCR) and Its Use in DNA Cloning

Today, most researchers have some information about the sequence of the gene or other DNA segment they want to clone. Using this information, they can start with the entire collection of genomic DNA from the particular species of interest and obtain many copies of the desired gene by using a technique called the **polymerase chain reaction**, or **PCR**. **Figure 20.7** illustrates the steps in PCR. Within a few hours, this technique can make billions of copies of a specific target DNA segment in a sample, even if that segment makes up less than 0.001% of the total DNA in the sample.

MB HHMI Video: Polymerase Chain Reaction (PCR) hhmi BioInteractive

▼ **Figure 20.6 Gel electrophoresis.** A gel made of a polymer acts as a molecular sieve to separate nucleic acids or proteins differing in size, electrical charge, or other physical properties as they move in an electric field. In the example shown here, DNA molecules are separated by length in a gel made of a polysaccharide called agarose.

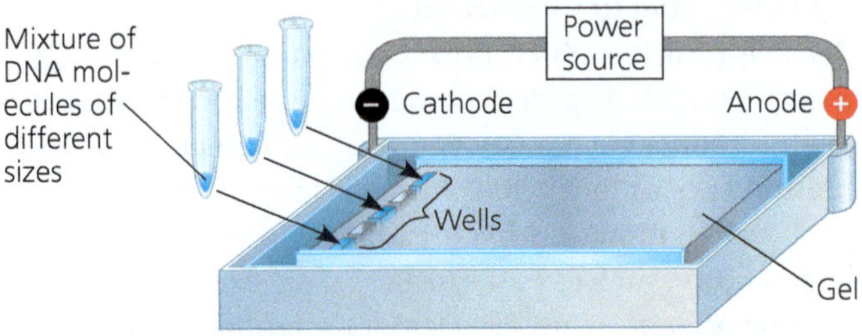

(a) Each sample, a mixture of different DNA molecules, is placed in a separate well near one end of a thin slab of agarose gel. The gel is set into a small plastic support and immersed in an aqueous, buffered solution in a tray with electrodes at each end. The current is then turned on, causing the negatively charged DNA molecules to move toward the positive electrode.

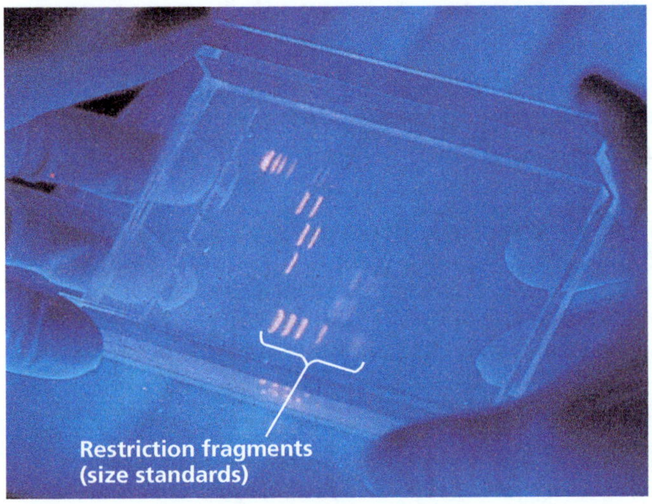

(b) Shorter molecules are slowed down less than longer ones, so they move faster through the gel. After the current is turned off, a DNA-binding dye is added that fluoresces pink in UV light. Each pink band corresponds to many thousands of DNA molecules of the same length. The horizontal ladder of bands at the bottom of the gel is a set of restriction fragments of known sizes for comparison with samples of unknown length.

MB Animation: Gel Electrophoresis of DNA

In the PCR procedure, a three-step cycle brings about a chain reaction that produces an exponentially growing population of identical DNA molecules. During each cycle, the reaction mixture is heated to denature (separate) the strands of the double-stranded DNA and then cooled to allow annealing (hydrogen bonding) of short, single-stranded DNA primers complementary to sequences on opposite strands at each end of the target sequence; finally, a heat-stable DNA polymerase extends the primers in the $5' \rightarrow 3'$ direction. If a standard DNA polymerase were used, the protein would be denatured along with the DNA during the first heating step and would have to be replaced after each cycle. The key to

Figure 20.7

Research Method The Polymerase Chain Reaction (PCR)

Application With PCR, any specific segment—the so-called target sequence—in a DNA sample can be copied many times (amplified) within a test tube.

Technique PCR requires double-stranded DNA containing the target sequence, a heat-resistant DNA polymerase, all four nucleotides, and two 15- to 20-nucleotide single DNA strands that serve as primers. One primer is complementary to one end of the target sequence on one strand; the second primer is complementary to the other end of the sequence on the other strand.

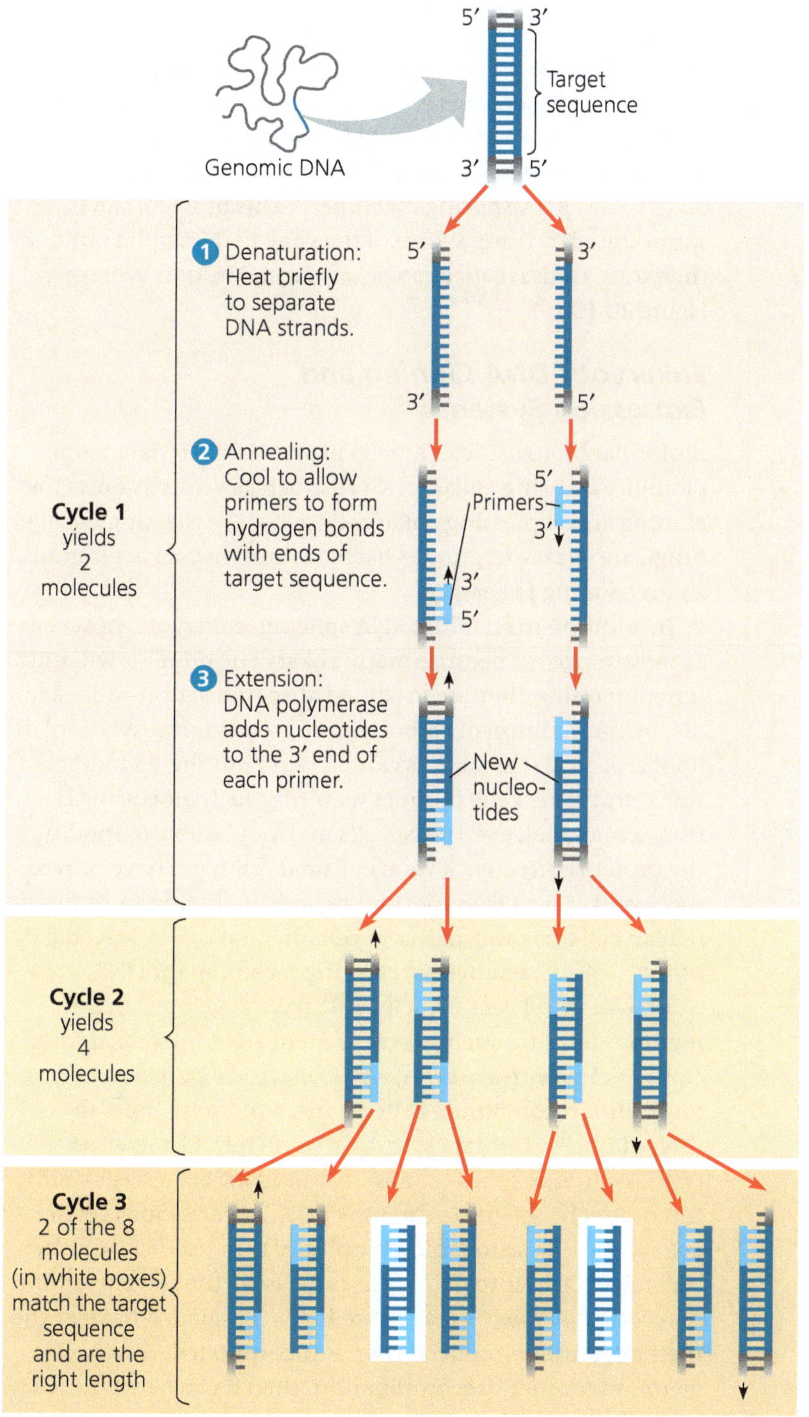

Results After three cycles, two molecules match the target sequence exactly. After 30 more cycles, over 1 billion (10^9) molecules match the target sequence.

Animation: Copying DNA through PCR

automating PCR was the discovery of an unusual heat-stable DNA polymerase called *Taq* polymerase, named after the bacterial species from which it was first isolated. This bacterial species, *Thermus aquaticus*, lives in hot springs, and the stability of its DNA polymerase at high temperatures is an evolutionary adaptation that enables the enzyme to function at temperatures up to 95°C. Today, researchers also use a DNA polymerase from the archaean species *Pyrococcus furiosus*. This enzyme, called *Pfu* polymerase, is more accurate and stable but more expensive than *Taq* polymerase.

PCR is speedy and very specific. Only a minuscule amount of DNA need be present in the starting material, and this DNA can be partially degraded, as long as there are a few copies of the complete target sequence. The key to the high specificity is the pair of primers used for each PCR amplification. The primer sequences are chosen so they hybridize *only* to sequences at opposite ends of the target segment, one on the 3′ end of each strand. (For high specificity, the primers must be at least 15 or so nucleotides long.) With each successive cycle, the number of target segment molecules of the correct length doubles, so the number of molecules equals 2^n, where n is the number of cycles. After 30 or so cycles, about a billion copies of the target sequence are present!

Despite its speed and specificity, PCR amplification cannot substitute for gene cloning in cells to make large amounts of a gene. This is because occasional errors during PCR replication limit the number of good copies and the length of DNA fragments that can be copied. Instead, PCR is used to provide the specific DNA fragment for cloning. PCR primers are synthesized to include a restriction site at each end of the DNA fragment that matches the site in the cloning vector, and the fragment and vector are cut and ligated together **(Figure 20.8)**. The resulting plasmids are sequenced so that those with error-free inserts can be selected.

Devised in 1985, PCR has had a major impact on biological research and genetic engineering. PCR has been used to amplify DNA from a wide variety of sources: a 40,000-year-old frozen woolly mammoth (see the photo on the first page of this chapter); fingerprints or tiny amounts of blood, tissue, or semen found at crime scenes; single embryonic cells for rapid prenatal diagnosis of genetic disorders (see Figure 14.19); and cells infected with viruses that are difficult to detect, such as HIV. (To test for HIV, viral genes are amplified.) We'll return to applications of PCR later.

Figure 20.8 Use of a restriction enzyme and PCR in gene cloning. In a closer look at the process shown at the top of Figure 20.4, PCR is used to produce the DNA fragment or gene of interest that will be ligated into a cloning vector, in this case a bacterial plasmid. The ends of the fragments have the same restriction site as the cloning vector. The plasmid and the DNA fragments are cut with the same restriction enzyme and combined so the sticky ends can hybridize and be ligated together. The resulting plasmids are then introduced into bacterial host cells. The plasmid also contains an antibiotic resistance gene that allows only cells with a plasmid to survive when the antibiotic is present. Other genetic engineering techniques are used to ensure that cells with nonrecombinant plasmids can be eliminated.

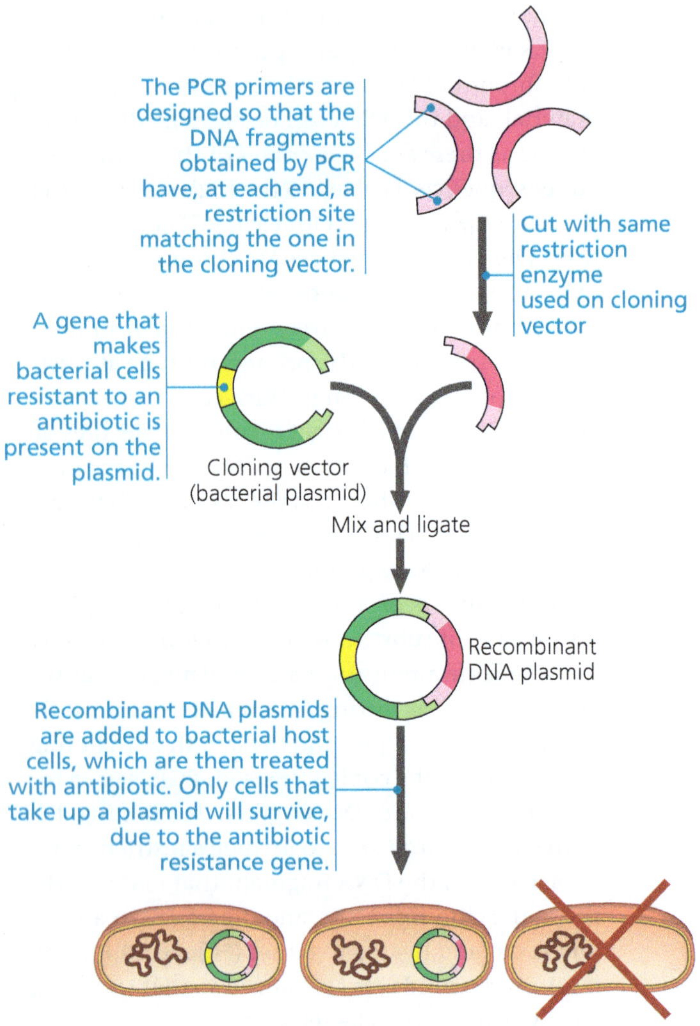

Expressing Cloned Eukaryotic Genes

Once a gene has been cloned in host cells, its protein product can be expressed in large amounts for research or for practical applications, which we'll explore in Concept 20.4. Cloned genes can be expressed in either bacterial or eukaryotic cells; each option has advantages and disadvantages.

Bacterial Expression Systems

Getting a cloned eukaryotic gene to function in bacterial host cells can be difficult because certain aspects of gene expression are different in eukaryotes and bacteria. To overcome differences in promoters and other DNA control sequences, scientists usually employ an **expression vector**, a cloning vector that contains a highly active bacterial promoter just upstream of a restriction site where the eukaryotic gene can be inserted in the correct reading frame. The bacterial host cell will recognize the promoter and proceed to express the foreign gene now linked to that promoter. Such expression vectors allow the synthesis of many eukaryotic proteins in bacterial cells.

Another problem with expressing cloned eukaryotic genes in bacteria is the presence of noncoding regions (introns) in most eukaryotic genes (see Concept 17.3). Introns can make a eukaryotic gene very long and unwieldy, and they prevent correct expression of the gene by bacterial cells, which do not have RNA-splicing machinery. This problem can be surmounted by using a form of the gene that includes only the exons. (This is called *complementary DNA,* or *cDNA*; see Figure 20.10.)

Eukaryotic DNA Cloning and Expression Systems

Molecular biologists can avoid eukaryotic-bacterial incompatibility by using eukaryotic cells such as yeasts as hosts for cloning and expressing eukaryotic genes. Yeasts, single-celled fungi, are as easy to grow as bacteria, and they have plasmids, a rarity among eukaryotes.

In addition to enabling RNA splicing, eukaryotic host cells are advantageous because many eukaryotic proteins will not function unless they are modified after translation—for example, by the addition of carbohydrate groups (glycosylation) or lipid groups. Bacterial cells cannot carry out these modifications, and if the gene product requiring such processing is from a mammal, even yeast cells may not be able to modify the protein correctly. Several cultured cell types have proved successful as host cells for this purpose, including some mammalian cell lines and an insect cell line that can be infected by a virus (called baculovirus) carrying recombinant DNA.

Besides using vectors, scientists have developed other methods for introducing recombinant DNA into eukaryotic cells. In **electroporation**, a brief electrical pulse applied to a solution containing cells creates temporary holes in their plasma membranes, through which DNA can enter. (This technique is now commonly used for bacteria as well.) Alternatively, scientists can inject DNA directly into single eukaryotic cells using microscopically thin needles. Another way to get DNA into plant cells is by using the soil bacterium *Agrobacterium tumefaciens*, as we'll discuss later. Whatever the method, if the introduced DNA is incorporated into a cell's genome by genetic recombination, then it can be expressed by the cell. Expressing different versions of genes in cells allows researchers to study protein function, a topic we'll return to in Concept 20.2.

Cross-Species Gene Expression and Evolutionary Ancestry

EVOLUTION The ability to express eukaryotic proteins in bacteria (even if the proteins aren't modified properly) is quite remarkable when we consider how different eukaryotic and bacterial cells are. In fact, examples abound of genes that are taken from one species and function perfectly well when transferred into another very different species, such as a firefly gene in a tobacco plant and a jellyfish gene in a pig (see Figure 17.7). These observations underscore the shared evolutionary ancestry of species living today.

One example involves a gene called *Pax-6*, which has been found in animals as diverse as vertebrates and fruit flies. The vertebrate *Pax-6* gene product (the PAX-6 protein) triggers a complex program of gene expression resulting in formation of the vertebrate eye, which has a single lens. Expression of the fly *Pax-6* gene leads to formation of the compound fly eye, which is quite different from the vertebrate eye. When the mouse *Pax-6* gene was cloned and introduced into a fly embryo so that it replaced the fly's own *Pax-6* gene, researchers were surprised to see that the mouse version of the gene led to formation of a compound fly eye (see Figure 50.16). Conversely, when the fly *Pax-6* gene was transferred into a vertebrate embryo—a frog, in this case—a frog eye formed. Although the genetic programs triggered in vertebrates and flies generate very different eyes, the two versions of the *Pax-6* gene can substitute for each other to trigger lens development, evidence of their evolution from a gene in a very ancient common ancestor. Because of their ancient evolutionary roots, all living organisms share the same basic mechanisms of gene expression. This commonality is the basis of many recombinant DNA techniques described in this chapter.

CONCEPT CHECK 20.1

1. **MAKE CONNECTIONS >** The restriction site for an enzyme called *Pvu*I is the following sequence:

   5′-CGATCG-3′
   3′-GCTAGC-5′

   Staggered cuts are made between the T and C on each strand. What type of bonds are being cleaved? (See Concept 5.5.)
2. **DRAW IT >** One strand of a DNA molecule has the following sequence:

   5′-CTTGACGATCGTTACCG-3′

   Draw the other strand. Will *Pvu*I (see question 1) cut this molecule? If so, draw the products.
3. What are some potential difficulties in using plasmid vectors and bacterial host cells to produce large quantities of proteins from cloned eukaryotic genes?
4. **VISUAL SKILLS >** Compare Figure 20.7 with Figure 16.20. How does replication of DNA ends during PCR proceed without shortening the fragments each time?

For suggested answers, see Appendix A.

CONCEPT 20.2

Biologists use DNA technology to study gene expression and function

To see how a biological system works, scientists seek to understand the functioning of the system's component parts. Analysis of when and where a gene or group of genes is expressed can provide important clues about their function.

Analyzing Gene Expression

Biologists driven to understand the assorted cell types of a multicellular organism, cancer cells, or the developing tissues of an embryo first try to discover which genes are expressed by the cells of interest. The most straightforward way to do this is usually to identify the mRNAs being made. We'll first examine techniques that look for patterns of expression of specific individual genes. Next, we'll explore ways to characterize groups of genes being expressed by cells or tissues of interest. As you will see, all of these procedures depend in some way on base pairing between complementary nucleotide sequences.

Studying the Expression of Single Genes

Suppose we have cloned a gene that we suspect plays an important role in the embryonic development of *Drosophila melanogaster* (the fruit fly). The first thing we might want to know is which embryonic cells express the gene—in other words, where in the embryo is the corresponding mRNA found? We can detect the mRNA by nucleic acid hybridization with molecules of complementary sequence that we can follow in some way. The complementary molecule, a short, single-stranded nucleic acid that can be either RNA or DNA, is called a **nucleic acid probe**. Using our cloned gene as a template, we can synthesize a probe complementary to the mRNA. For example, if part of the sequence on the mRNA were

5′ ···C U C A U C A C C G G C··· 3′

then we would synthesize this single-stranded DNA probe:

3′ G A G T A G T G G C C G 5′

Each probe molecule is labeled during synthesis with a fluorescent tag so we can follow it. A solution containing probe molecules is applied to *Drosophila* embryos, allowing the probe to hybridize specifically with any complementary sequences on the many mRNAs in embryonic cells in which the gene is being transcribed. Because this technique allows us to see the mRNA in place (or *in situ*, in Latin) in the intact organism, this technique is called ***in situ* hybridization**. Different probes can be labeled with different fluorescent dyes, sometimes with strikingly beautiful results **(Figure 20.9)**.

Figure 20.9 Determining where single genes are expressed by *in situ* hybridization analysis. A *Drosophila* embryo was incubated in a solution containing DNA probes for five different mRNAs, each probe labeled with a different fluorescently colored tag. The embryo was then viewed using fluorescence microscopy; the resulting fluorescent micrograph is shown. Each color marks where a specific gene is expressed as mRNA. The arrows from the groups of yellow and blue cells above the micrograph show a magnified view of nucleic acid hybridization of the appropriately colored probe to the mRNA. Yellow cells (expressing the *wg* gene) interact with blue cells (expressing the *en* gene); their interaction helps establish the pattern in a body segment. The diagram at the bottom clarifies the eight segments visible in this view.

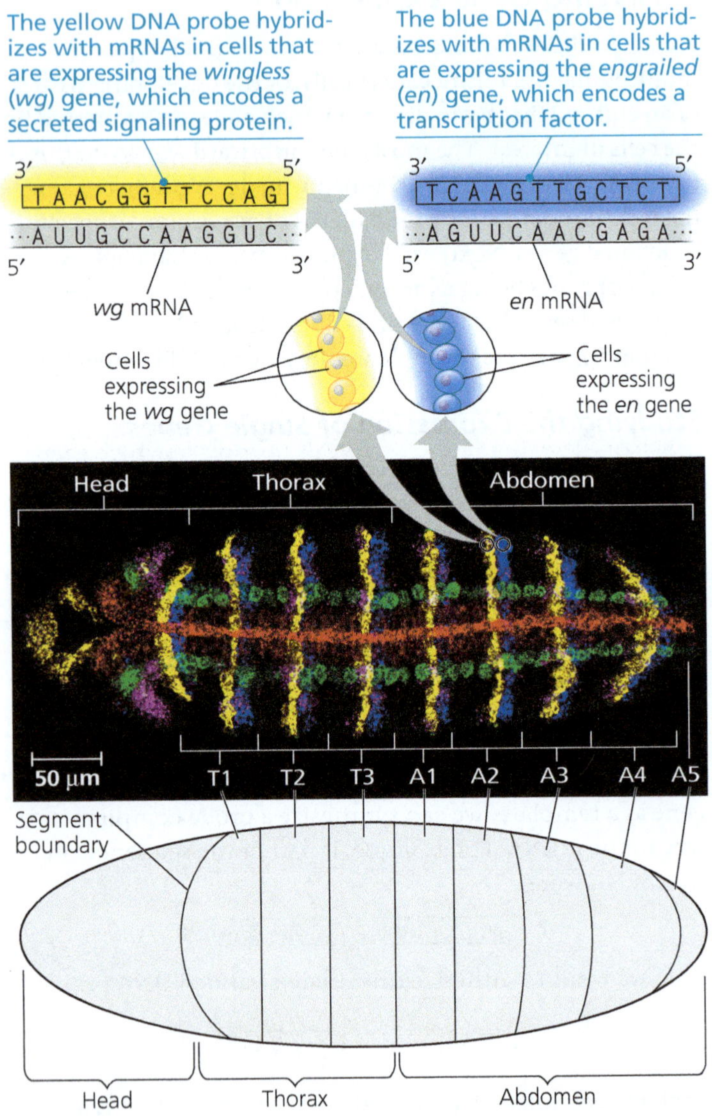

Figure 20.10 Making complementary DNA (cDNA) from eukaryotic genes. Complementary DNA is made in a test tube using mRNA as a template for the first strand. Only one mRNA is shown here, but the final collection of cDNAs would reflect all the mRNAs present in the cell.

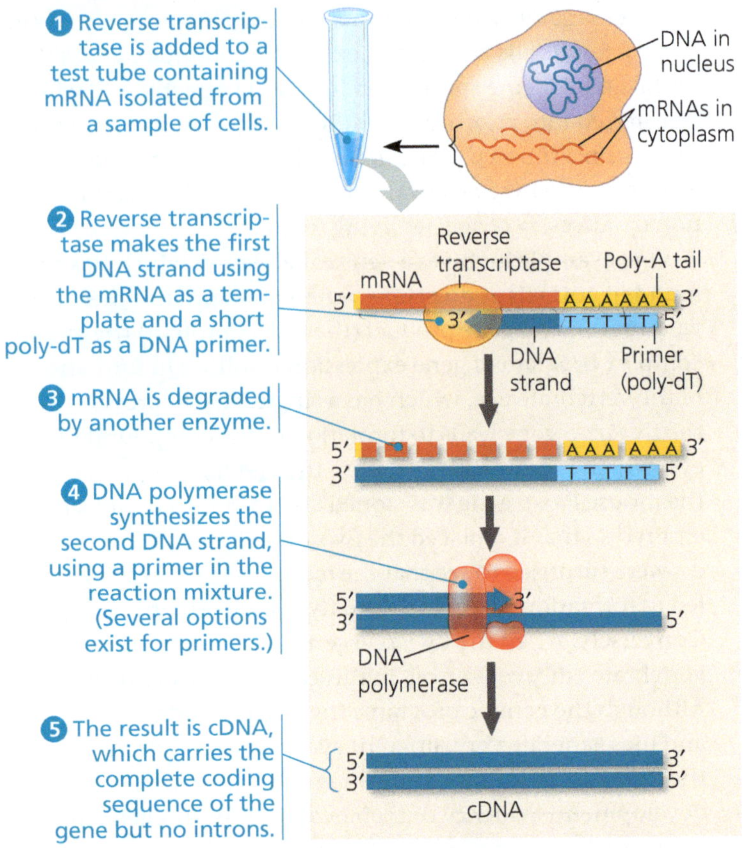

Other mRNA detection techniques may be preferable for comparing the amounts of a specific mRNA in several samples at the same time—for example, in different cell types or in embryos at different stages of development. One method that is widely used is called the **reverse transcriptase polymerase chain reaction**, or **RT-PCR**.

RT-PCR begins by turning sample sets of mRNAs into double-stranded DNAs with the corresponding sequences. First, the enzyme reverse transcriptase (from a retrovirus; see Figure 19.9) is used to synthesize a complementary DNA copy of each mRNA in the sample, called a *reverse transcript* **(Figure 20.10)**. Recall that the 3′ end of an mRNA has a stretch of adenine (A) nucleotides called a poly-A tail. This allows a short complementary strand of thymine deoxyribonucleotides (poly-dT) to be added and used as a primer for synthesis of this DNA strand. Following enzymatic degradation of the mRNA, a second DNA strand, complementary to the first, is synthesized by DNA polymerase. The resulting double-stranded DNA is called **complementary DNA (cDNA)**. (Made from mRNA, cDNA lacks introns and can be used for protein expression in bacteria, as mentioned earlier.) To analyze the timing of expression of the *Drosophila* gene of interest, for example, we would first isolate all the mRNAs from different stages of *Drosophila* embryos and make cDNA from each stage **(Figure 20.11)**.

Next in RT-PCR is the PCR step (see Figure 20.7). As you will recall, PCR is a way of rapidly making many copies of one specific stretch of double-stranded DNA, using primers that hybridize to the opposite ends of the segment of interest. In our case, we would add primers corresponding to a segment of our *Drosophila* gene, using the cDNA from each embryonic stage as a template for PCR amplification in separate samples.

Figure 20.11

Research Method RT-PCR Analysis of the Expression of Single Genes

Application RT-PCR uses the enzyme reverse transcriptase (RT) in combination with PCR and gel electrophoresis. RT-PCR can be used to compare gene expression between samples—for instance, in different embryonic stages, in different tissues, or in the same type of cell under different conditions.

Technique In this example, samples containing mRNAs from six embryonic stages of *Drosophila* were analyzed for a specific mRNA as shown below. (In steps 1 and 2, the mRNA from only one stage is shown.)

1. **cDNA synthesis** is carried out by incubating the mRNAs with reverse transcriptase and other necessary components.
2. **PCR amplification** of the sample is performed using primers specific to the *Drosophila* gene of interest.
3. **Gel electrophoresis** will reveal amplified DNA products only in samples that contained mRNA transcribed from the specific *Drosophila* gene.

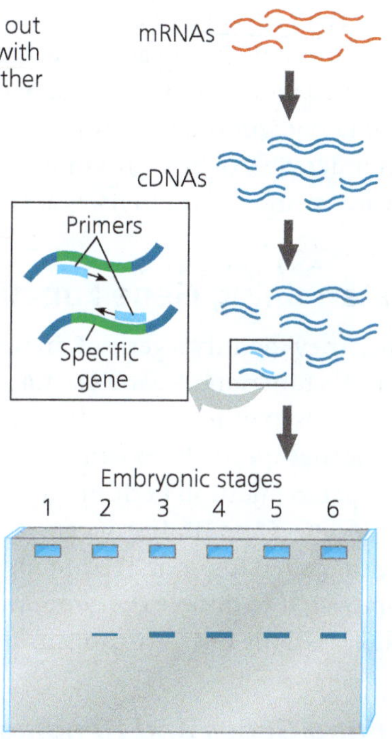

Results The mRNA for this gene first is expressed at stage 2 and continues to be expressed through stage 6. The size of the amplified fragment (shown by its position on the gel) depends on the distance between the primers that were used (not on the size of the mRNA).

When the products are analyzed on a gel, copies of the amplified region will be observed as bands only in samples that originally contained mRNA from the gene of interest. An enhancement called *quantitative RT-PCR (qRT-PCR)* uses a fluorescent dye that fluoresces only when bound to a double-stranded PCR product. The newer quantitative PCR machines can detect the light and measure the PCR product, thus avoiding the need for electrophoresis while also providing quantitative data, a distinct advantage. RT-PCR or qRT-PCR can also be carried out with mRNAs collected from different tissues at one time to discover which tissue is producing a specific mRNA.

Instructors: The **Scientific Skills Exercise** "Analyzing Quantitative and Spatial Gene Expression Data" can be assigned in MasteringBiology. Students work with data from an experiment that investigated mRNA expression using both *in situ* hybridization and quantitative RT-PCR.

Studying the Expression of Interacting Groups of Genes

A major goal of biologists is to learn how genes act together to produce and maintain a functioning organism. Now that the genomes of a number of species have been sequenced, it is possible to study the expression of large groups of genes—the so-called *systems approach.* Researchers use what is known about the whole genome to investigate which genes are transcribed in different tissues or at different stages of development. One aim is to identify networks of gene expression across an entire genome.

Genome-wide expression studies can be carried out using **DNA microarray assays.** A DNA microarray consists of tiny amounts of a large number of single-stranded DNA fragments representing different genes fixed to a glass slide in a tightly spaced array, or grid, of dots. (The microarray is also called a *DNA chip* by analogy to a computer chip.) Ideally, these fragments represent all the genes of an organism. The mRNAs from cells under study are reverse-transcribed into cDNAs (see Figure 20.10), and a fluorescent label is added so the cDNAs can be used as probes on the microarray. Different fluorescent labels are used for different cell samples so that multiple samples can be tested in the same experiment. The resulting pattern of colored dots, shown in an actual-size microarray in **Figure 20.12**, reveals the dots to which each probe was bound and thus the genes that are expressed in the cell samples being tested. Microarray technology started taking off after several papers about it were published in 1995;

Figure 20.12 Use of microarrays to analyze expression of many genes. In this DNA microarray assay, researchers extracted mRNAs from two different human tissues and synthesized two sets of cDNAs, fluorescently labeled red (tissue #1) or green (tissue #2). Labeled cDNAs were hybridized with a microarray containing 5,760 human genes (about 25% of human genes), part of which is shown in the enlargement. Red indicates that the gene in that well was expressed in tissue #1, green in tissue #2, yellow in both, and black in neither. The fluorescence intensity at each spot indicates the relative expression of the gene.

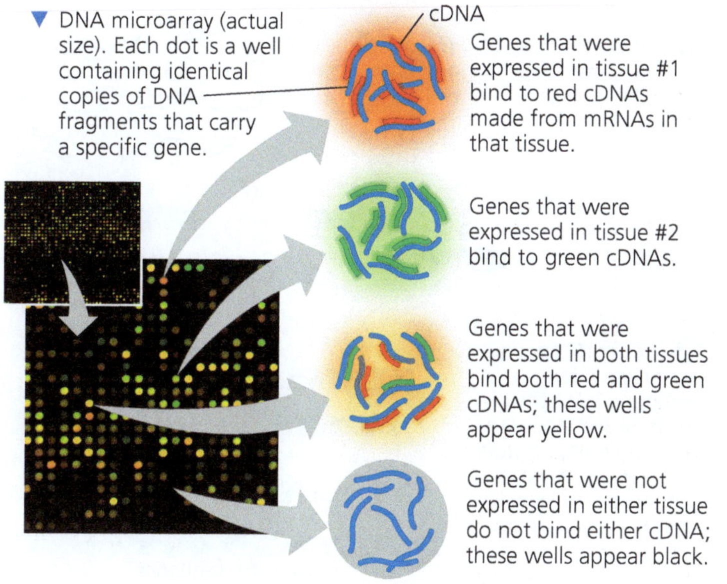

since then more sophisticated applications have been developed and are in use.

Increasingly, with the advent of rapid, inexpensive DNA sequencing methods, microarray usage is decreasing: Researchers can now afford to simply sequence the cDNA samples from different tissues or different embryonic stages in order to discover which genes are expressed. This straightforward method is called **RNA sequencing**, or **RNA-seq** (pronounced "RNA-seek"), even though it is the cDNA that is actually sequenced. In RNA-seq, the mRNA (or other RNA) samples are isolated, cut into shorter, similar-sized fragments, and converted into cDNAs **(Figure 20.13)**. These short cDNA stretches are sequenced, and a computer program reassembles them, either mapping them onto the genome of the species in question (when available) or simply ordering them from scratch based on overlapping sequences of multiple RNAs. RNA-seq has several advantages over microarrays. First, the procedure is not based on hybridization with a labeled probe, so it doesn't depend on having genomic sequences in hand (although they are usually available). Second, it can measure levels of expression over a very wide range, unlike microarrays, which cannot accurately measure either very low or very high levels. Third, a careful analysis provides a wealth of information about expression of a particular gene, such as relative levels of alternatively spliced mRNAs. As the price of DNA sequencing plummets, RNA-seq is becoming more widely used for many applications. In most cases, however, expression of individual genes still needs to be confirmed by RT-PCR.

Figure 20.13 Use of RNA sequencing (RNA-seq) to analyze expression of many genes. RNA-seq yields a wide range of information about expression of genes, including their level of expression.

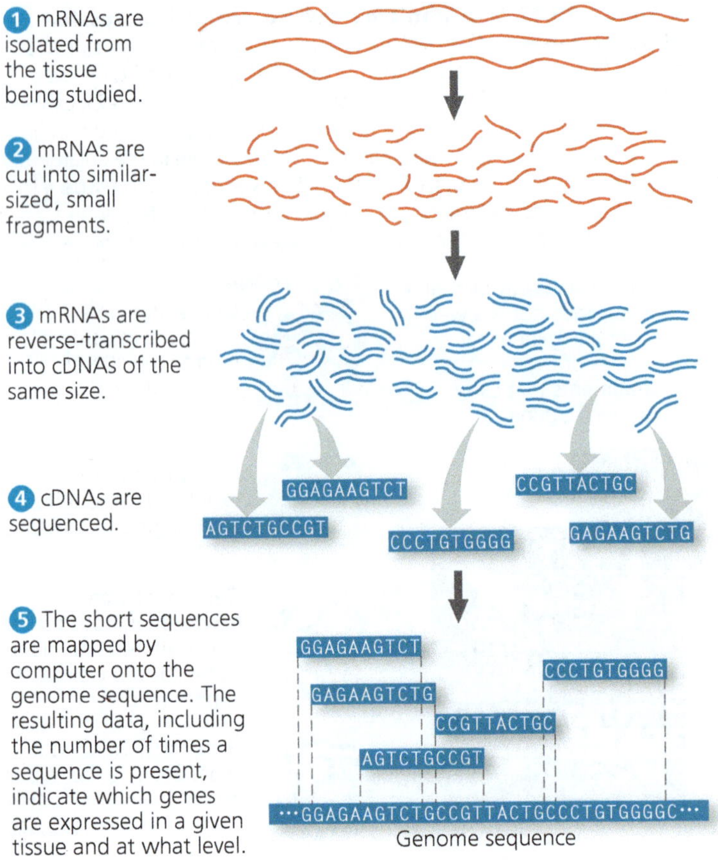

Scientists can now measure the expression of thousands of genes at one time. DNA technology makes such studies possible; with automation, they are easily performed on a large scale. By uncovering gene interactions and providing clues to gene function, DNA microarray assays and RNA-seq may contribute to a better understanding of diseases and suggest new diagnostic techniques or therapies. For instance, comparing patterns of gene expression in breast cancer tumors and noncancerous breast tissue has already resulted in more informed and effective treatment protocols (see Figure 18.27). Ultimately, information from these methods should provide a grander view of how ensembles of genes interact to form an organism and maintain its vital systems.

Determining Gene Function

Once they identify a gene of interest, how do scientists determine its function? A gene's sequence can be compared with sequences in other species. If the function of a similar gene in another species is known, one might suspect that the gene product in question performs a comparable task. Data about the location and timing of gene expression may reinforce the suggested function. To obtain stronger evidence, one approach is to disable the gene and then observe the consequences in the cell or organism.

Editing Genes and Genomes

Molecular biologists have long sought techniques for altering, or editing, the genetic material of cells or organisms in a predictable way. In one such technique, called ***in vitro* mutagenesis**, specific mutations are introduced into a cloned gene, and the mutated gene is returned to a cell in such a way that it disables ("knocks out") the normal cellular copies of the same gene. If the introduced mutations alter or destroy the function of the gene product, the phenotype of the mutant cell may help reveal the function of the missing normal protein. Using molecular and genetic techniques worked out in the 1980s, researchers can generate mice with any given gene disabled in order to study the role of that gene in development and in the adult. Mario Capecchi, Martin Evans, and Oliver Smithies received the Nobel Prize in 2007 for developing this technique.

Over the past 10 years, biologists have developed a powerful new technique for gene editing in living cells and organisms, called the **CRISPR-Cas9 system**, that is taking the field of genetic engineering by storm. Cas9 is a bacterial protein that helps defend bacteria against bacteriophage infections in a system worked out by Jennifer Doudna and Emmanuelle Charpentier. In bacterial cells, Cas9 acts together with a "guide RNA" made from the CRISPR region of the bacterial system (see Figure 19.7).

Similar to the restriction enzymes described earlier, Cas9 is a nuclease that cuts double-stranded DNA molecules. However, while a given restriction enzyme recognizes only one particular DNA sequence, the Cas9 protein will cut any sequence to which it is directed. Cas9 takes its marching orders from a guide RNA molecule that it binds and uses as a homing device, cutting both strands of any DNA sequence that is exactly complementary to the guide RNA. Scientists have been able to exploit the function of Cas9 by introducing a Cas9–guide RNA complex into a cell they wish to alter **(Figure 20.14)**. The guide RNA in the complex is engineered to be complementary to the "target" gene. Cas9 cuts both strands of the target DNA, and the resulting broken ends of DNA trigger a DNA repair system (similar to that shown in Figure 16.19). When there is no undamaged DNA for the enzymes of the repair system to use as a template, as shown at the bottom left of Figure 20.14, the repair enzymes rejoin the ends, sometimes introducing or removing nucleotides. If the cut is directed to a coding portion of the gene, the rejoining process often alters the DNA sequence so that the gene no longer works properly.

This technique is a highly effective way for researchers to knock out a given gene in order to study what that gene does, and it has already been used in many organisms, including bacteria, fish, mice, insects, human cells, and various crop plants. Researchers have also modified the technique so that CRISPR-Cas9 can be used to repair a gene that has a mutation (see bottom right of Figure 20.14). They introduce a segment from the normal (functional) gene along with the CRISPR-Cas9 system. After Cas9 cuts the target DNA, repair enzymes can use the normal DNA segment as a template to repair the target DNA at the break point. This approach is used for gene therapy, which will be discussed later in the chapter.

In another application of CRISPR-Cas9, scientists are attempting to address the global problem of insect-borne diseases by altering genes in the insect so that, for example, it cannot transmit disease. An extra twist to this approach is engineering the new allele so that it is much more highly favored for inheritance than is the wild-type allele. This is called a **gene drive** because the biased inheritance of the engineered gene during reproduction rapidly "drives" the new allele through the population.

Other Methods for Studying Gene Function

Another method for silencing expression of selected genes doesn't alter the genome; instead, it exploits the phenomenon of **RNA interference (RNAi)**, described in Concept 18.3. This experimental approach uses synthetic double-stranded RNA molecules matching the sequence of a particular gene to trigger breakdown of the gene's messenger RNA or to block its translation. In organisms such as the nematode

▼ Figure 20.14 Gene editing using the CRISPR-Cas9 system.

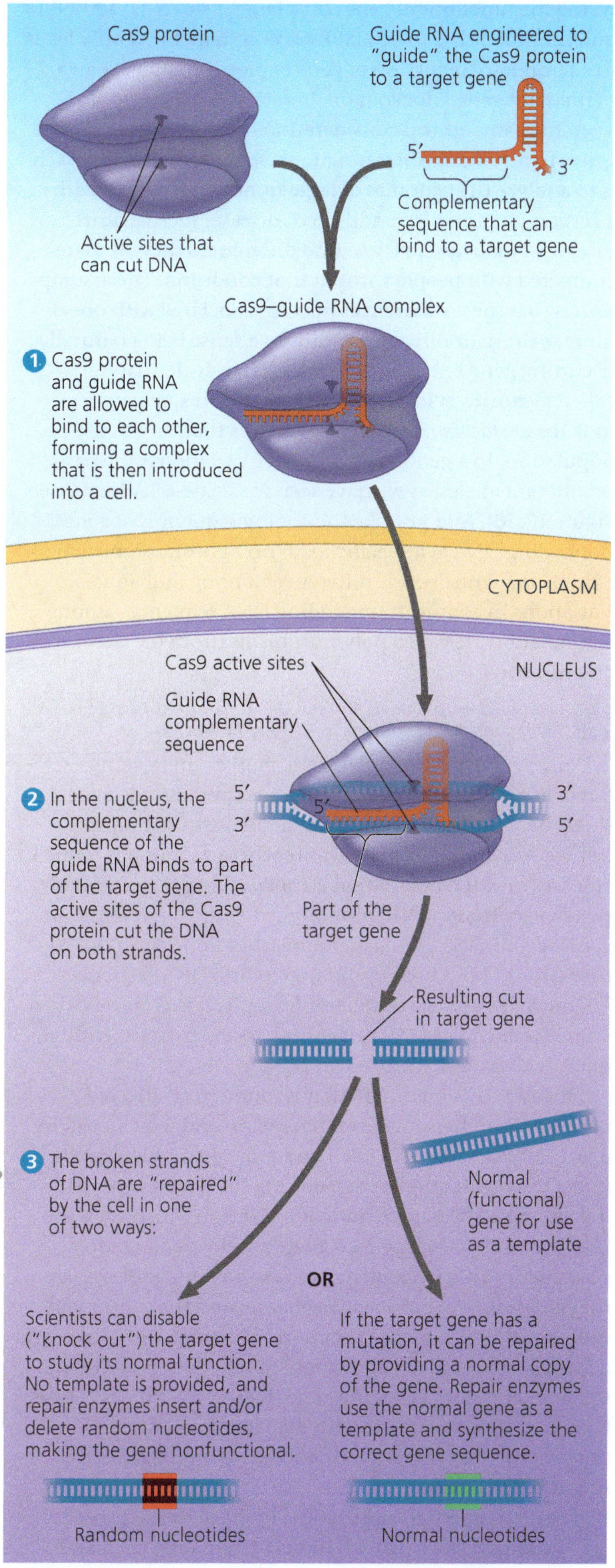

and the fruit fly, RNAi has already proved valuable for analyzing the functions of genes on a large scale. This method is quicker than using the CRISPR-Cas9 system, but it only leads to a temporary reduction of gene expression rather than a permanent gene knockout or alteration.

In humans, ethical considerations prohibit knocking out genes to determine their functions. An alternative approach is to analyze the genomes of large numbers of people with a certain phenotypic condition or disease, such as heart disease or diabetes, to try to find differences they all share compared with people without that condition. The assumption is that these differences may be associated with one or more malfunctioning genes, thus in a sense being naturally occurring gene knockouts. In these large-scale analyses, called **genome-wide association studies**, researchers look for *genetic markers*, DNA sequences that vary in the population. In a gene, such sequence variation is the basis of different alleles, as we have seen for sickle-cell disease (see Figure 17.26). And just like the coding sequences of genes, noncoding DNA at a specific locus on a chromosome may exhibit small nucleotide differences among individuals. Variations in coding or noncoding DNA sequences among a population are called polymorphisms (from the Greek for "many forms").

MB Interview with Charles Rotimi: Using genomics to study health-related conditions in African-Americans

Among the most useful genetic markers in tracking down genes that contribute to diseases and disorders are single base-pair variations in the genomes of the human population. A single base-pair site where variation is found in at least 1% of the population is called a **single nucleotide polymorphism** (**SNP**, pronounced "snip"). A few million SNPs occur in the human genome, about once in 100–300 base pairs of both coding and noncoding DNA sequences. To find SNPs in large numbers of people, it isn't necessary to sequence their DNA; SNPs can be detected by very sensitive microarray assays, RNA-seq, or PCR.

Once a SNP is identified that is found in all affected people, researchers focus on that region and sequence it. In nearly all cases, the SNP itself does not contribute directly to the disease in question by altering the encoded protein; in fact, most SNPs are in noncoding regions. Instead, if the SNP and a disease-associated allele are close enough to be genetically linked, scientists can take advantage of the fact that crossing over between the marker and the gene is very unlikely during gamete formation. Therefore, the marker and gene will almost always be inherited together, even though the marker is not part of the gene (**Figure 20.15**). SNPs have been found that correlate with diabetes, heart disease, and several types of cancer, and the search is on for genes that might be involved.

The experimental approaches you have learned about thus far focused on working with molecules, mainly DNA and proteins. In a parallel line of inquiry, biologists have been developing powerful techniques for cloning whole multicellular organisms. One aim of this work is to obtain special types of cells, called stem cells, that can give rise to all types of tissues. Being able to manipulate stem cells would allow scientists to use the DNA-based methods previously discussed to alter stem cells for the treatment of diseases. Methods involving the cloning of organisms and production of stem cells are the subject of the next section.

Figure 20.15 Single nucleotide polymorphisms (SNPs) as genetic markers for disease-associated alleles. This diagram depicts the same region of the genome from two groups of individuals, one group having a particular disease or condition with a genetic basis. Unaffected people have an A/T pair at a given SNP locus, while affected people have a C/G pair there. Once the allele is confirmed as being associated with the disease in question, the SNP that varies in this way can be used as a marker for the disease-associated allele.

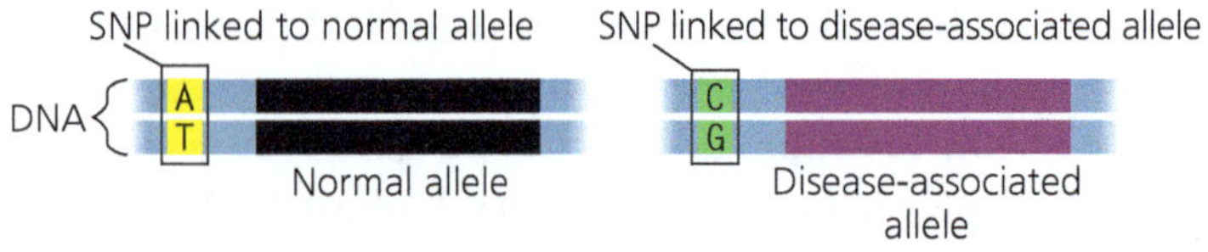

CONCEPT CHECK 20.2

1. Describe the role of complementary base pairing during RT-PCR, DNA microarray analysis, RNA sequencing, and CRISPR-Cas9 editing.
2. **VISUAL SKILLS** ➤ Consider the microarray in Figure 20.12. If a sample from normal tissue is labeled with a green fluorescent dye and a sample from cancerous tissue is labeled red, what color spots would represent genes you would be interested in if you were studying cancer? Explain.

For suggested answers, see Appendix A.

CONCEPT 20.3

Cloned organisms and stem cells are useful for basic research and other applications

Along with advances in DNA technology, scientists have been developing and refining methods for cloning whole multicellular organisms from single cells. In this context, cloning produces one or more organisms that are genetically identical to the "parent" that donated the single cell. This is often called *organismal cloning* to differentiate it from gene cloning and, more significantly, from cell cloning—the division of an asexually reproducing cell such as a bacterium into a group of genetically identical cells. (The common theme is that the product is genetically identical to the parent. In fact, the word *clone* comes from the Greek *klon*, meaning "twig.") The current interest in organismal cloning arises primarily from its ability to generate stem cells. A **stem cell** is a relatively unspecialized cell that can both reproduce itself indefinitely and, under appropriate conditions, differentiate into specialized cells of one or more types. Stem cells have great potential for regenerating damaged tissues.

The cloning of plants and animals was first attempted over 50 years ago in experiments designed to answer basic biological questions. For example, researchers wondered if all the cells of an organism have the same genes or whether cells lose genes during the process of differentiation (see Concept 18.4). One way to answer this question is to see whether a differentiated cell can generate a whole organism—in other words, whether cloning an organism is possible. Let's discuss these early experiments before we consider more recent progress in organismal cloning and procedures for producing stem cells.

Cloning Plants: Single-Cell Cultures

The successful cloning of whole plants from single differentiated cells was accomplished during the 1950s by F. C. Steward and his students at Cornell University, who worked with carrot plants. They found that differentiated cells taken from the root (the carrot) and incubated in culture medium could grow into normal adult plants, each genetically identical to the parent plant. These results showed that differentiation does not necessarily involve irreversible changes in the DNA. In plants, mature cells can "dedifferentiate" and then give rise to all the specialized cell types of the organism. Any cell with this potential is said to be **totipotent**.

Plant cloning is used extensively in agriculture. For plants such as orchids, cloning is the only commercially practical means of reproducing plants. In other cases, cloning has been used to reproduce a plant with valuable characteristics, such as resistance to plant pathogens. In fact, you yourself may be a plant cloner: If you have ever grown a new plant from a cutting, you have practiced cloning!

Cloning Animals: Nuclear Transplantation

Differentiated cells from animals generally do not divide in culture, much less develop into the multiple cell types of a new organism. Therefore, early researchers had to use a different approach to answer the question of whether differentiated animal cells are totipotent. Their approach was to remove the nucleus of an egg (creating an *enucleated* egg) and replace it with the nucleus of a differentiated cell, a procedure called *nuclear transplantation*, now more commonly called *somatic cell nuclear transfer*. If the nucleus from the differentiated donor cell retains its full genetic capability, then it should be able to direct development of the recipient cell into all the tissues and organs of an organism. Such experiments were conducted on one species of frog (*Rana pipiens*) by Robert Briggs and Thomas King in the 1950s and on another frog species (*Xenopus laevis*) by John Gurdon in the 1970s **(Figure 20.16)**. These researchers transplanted a nucleus from an embryonic or tadpole cell into an enucleated egg of the same species. In

▼ Figure 20.16

Inquiry Can the nucleus from a differentiated animal cell direct development of an organism?

Experiment John Gurdon and colleagues at Oxford University, in England, destroyed the nuclei of frog (*Xenopus laevis*) eggs by exposing the eggs to ultraviolet light. They then transplanted nuclei from cells of frog embryos and tadpoles into the enucleated eggs.

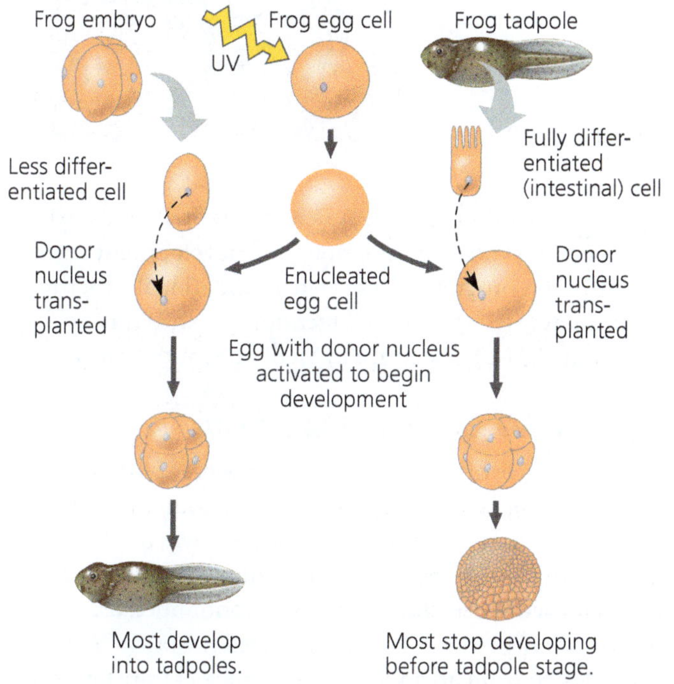

Results When the transplanted nuclei came from an early embryo, the cells of which are relatively undifferentiated, most of the recipient eggs developed into tadpoles. But when the nuclei came from the fully differentiated intestinal cells of a tadpole, fewer than 2% of the eggs developed into normal tadpoles, and most of the embryos stopped developing at a much earlier stage.

Conclusion The nucleus from a differentiated frog cell can direct development of a tadpole. However, its ability to do so decreases as the donor cell becomes more differentiated, presumably because of changes in the nucleus.

Data from J. B. Gurdon et al., The developmental capacity of nuclei transplanted from keratinized cells of adult frogs, *Journal of Embryology and Experimental Morphology* 34:93–112 (1975).

WHAT IF? ➤ *If each cell in a four-cell embryo were already so specialized that it was not totipotent, what results would you predict for the experiment on the left side of the figure?*

MB **HHMI Video: Somatic Cell Nuclear Transfer**

Gurdon's experiments, the transplanted nucleus was often able to support normal development of the egg into a tadpole. However, he found that the potential of a transplanted nucleus to direct normal development was inversely related to the age of the donor: The older the donor nucleus, the lower the percentage of normal tadpoles (see Figure 20.16).

From these results, Gurdon concluded that something in the nucleus *does* change as animal cells differentiate. In

➤ Figure 20.17 Reproductive cloning of a mammal by nuclear transfer. Dolly, shown here as a lamb, has a very different appearance from her surrogate mother, standing beside her.

frogs and most other animals, nuclear potential tends to be restricted more and more as embryonic development and cell differentiation progress. These were foundational experiments that ultimately led to stem cell technology, and Gurdon received the 2012 Nobel Prize in Medicine for this work.

Reproductive Cloning of Mammals

In addition to cloning frogs, researchers were able to clone mammals using early embryonic cells as a source of donor nuclei. Until about 20 years ago, though, it was not known whether a nucleus from a fully differentiated cell could be reprogrammed successfully to act as a donor nucleus. In 1997, researchers in Scotland announced the birth of Dolly, a lamb cloned from an adult sheep by nuclear transfer from a differentiated mammary gland cell **(Figure 20.17)**. Using a technique related to that in 20.16, the researchers implanted early embryos into surrogate mothers. Out of several hundred embryos, one successfully completed normal development, and Dolly was born, a genetic clone of the nucleus donor. At the age of 6, Dolly suffered complications from a lung infection often seen in sheep kept indoors and was euthanized. Another cloned sheep from the same experiment developed an unusual lung disease. This led to speculation that this sheep's cells were in some way not quite as healthy as those of a normal sheep, possibly reflecting incomplete reprogramming of the original transplanted nucleus. Reprogramming involves epigenetic changes that lead to changes in chromatin structure (see Concept 18.2), to be discussed shortly.

Since that time, researchers have cloned numerous other mammals, including mice, cats, cows, horses, pigs, dogs, and monkeys. In most cases, their goal has been the production of new individuals; this is known as *reproductive cloning*. We have already learned a lot from such experiments. For example, cloned animals of the same species do *not* always look or behave identically. In a herd of cows cloned from the same line of cultured cells, certain cows are dominant in behavior and others are more submissive. Another example of nonidentity in clones is the first cloned cat, named CC for Carbon Copy **(Figure 20.18)**. She has a calico coat, like her single

⮟ Figure 20.18 CC ("Carbon Copy"), the first cloned cat (right), and her single parent. Rainbow (left) donated the nucleus in a cloning procedure that resulted in CC. However, the two cats are not identical: Rainbow is a classic calico cat with orange patches on her fur and has a "reserved personality," while CC has a gray and white coat and is more playful.

female parent, but the color and pattern are different because of random X chromosome inactivation, which is a normal occurrence during embryonic development (see Figure 15.8). And identical human twins, which are naturally occurring "clones," are always slightly different. Clearly, environmental influences and random phenomena play a significant role during development.

Faulty Gene Regulation in Cloned Animals Due to Epigenetic Differences

In most nuclear transplantation studies thus far, only a small percentage of cloned embryos develop normally to birth. And like Dolly, many cloned animals exhibit defects. Cloned mice, for instance, are prone to obesity, pneumonia, liver failure, and premature death. Scientists assert that even cloned animals that appear normal are likely to have subtle defects.

Researchers have uncovered some reasons for the low efficiency of cloning and the high incidence of abnormalities. In the nuclei of fully differentiated cells, a small subset of genes is turned on and expression of the rest of the genes is repressed. This regulation often is the result of epigenetic changes in chromatin, such as acetylation of histones or methylation of DNA (see Figure 18.7). During the nuclear transfer procedure, many of these changes must be reversed in the later-stage nucleus from a donor animal for genes to be expressed or repressed appropriately in early stages of development. Researchers have found that the DNA in cells from cloned embryos, like that of differentiated cells, often has more methyl groups than does the DNA in equivalent cells from normal embryos of the same species. This finding suggests that the reprogramming of donor nuclei requires more accurate and complete chromatin restructuring than occurs during cloning procedures. Because DNA methylation helps regulate gene expression, misplaced or extra methyl groups in the DNA of donor nuclei may interfere with the pattern of gene expression necessary for normal embryonic development. In fact, the success of a cloning attempt may depend in large part on whether or not the chromatin in the donor nucleus can be artificially modified to resemble that of a newly fertilized egg.

Stem Cells of Animals

Progress in cloning mammalian embryos, including primates, has heightened speculation about the cloning of humans, which has not yet been achieved past very early embryonic stages. The main reason researchers have been trying to clone human embryos is not for reproduction, but for the production of stem cells to treat human diseases. Recall that a stem cell is a relatively unspecialized cell that can both reproduce itself indefinitely and, under appropriate conditions, differentiate into specialized cells of one or more types **(Figure 20.19)**. Thus, stem cells are able to both replenish their own population and generate cells that travel down specific differentiation pathways.

HHMI Animation: Somatic Cell Nuclear Transfer

Embryonic and Adult Stem Cells

Many early animal embryos contain stem cells capable of giving rise to differentiated cells of any type. Stem cells can be isolated from early embryos at a stage called the blastula stage or its human equivalent, the blastocyst stage. In culture, these *embryonic stem (ES) cells* reproduce indefinitely, and depending on culture conditions, they can be made to differentiate

Figure 20.19 How stem cells maintain their own population and generate differentiated cells.

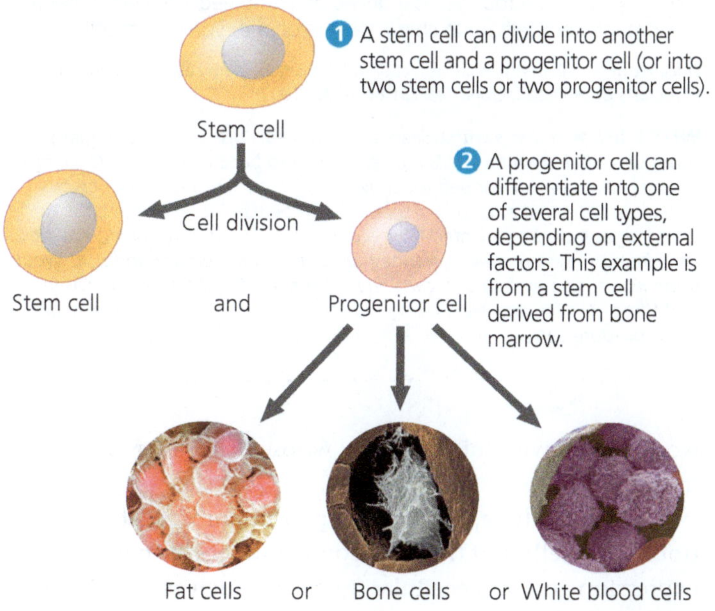

HHMI Video: Cultured Human Embryonic Stem Cells

Figure 20.20 Working with stem cells. Animal stem cells, which can be isolated from early embryos or adult tissues and grown in culture, are self-perpetuating, relatively undifferentiated cells. Embryonic stem cells are easier to grow than adult stem cells and can theoretically give rise to *all* types of cells in an organism. The range of cell types that can arise from adult stem cells is not yet fully understood.

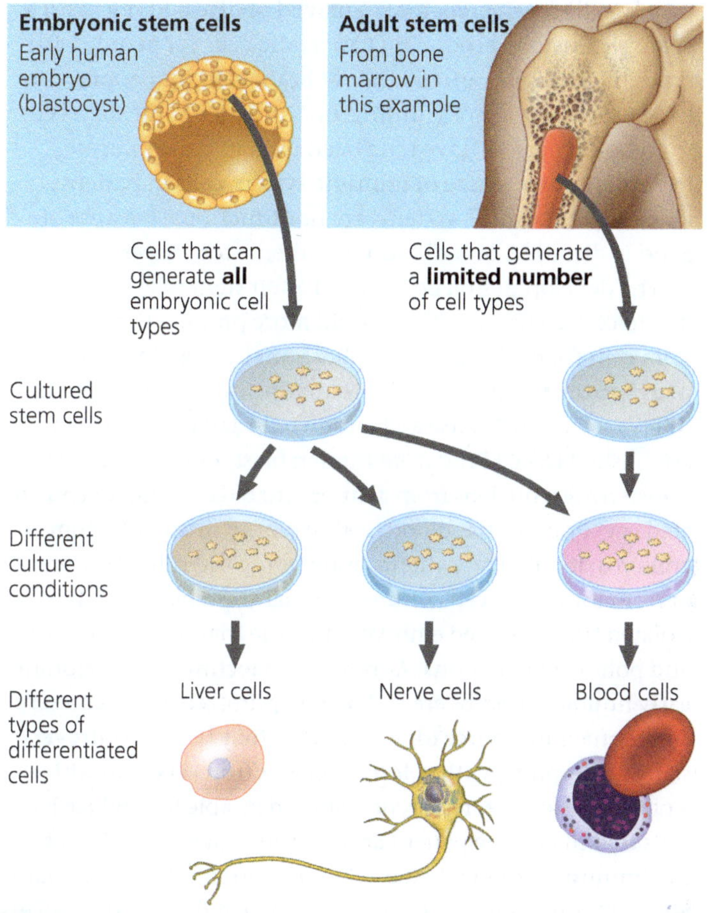

HHMI Animation: Creating Embryonic Stem Cell Lines

into a wide variety of specialized cells **(Figure 20.20)**, including even eggs and sperm.

The adult body also has stem cells, which serve to replace nonreproducing specialized cells as needed. In contrast to ES cells, *adult stem cells* are not able to give rise to all cell types in the organism, though they can generate several defined types. For example, one of the several types of stem cells in bone marrow can generate all the different kinds of blood cells (see Figure 20.20), and another type of bone marrow stem cell can differentiate into bone, cartilage, fat, muscle, and the linings of blood vessels. To the surprise of many, the adult brain has been found to contain stem cells that continue to produce certain kinds of nerve cells there. Researchers have also reported finding stem cells in skin, hair, eyes, and dental pulp. Although adult animals have only tiny numbers of stem cells, scientists are learning to identify and isolate these cells from various tissues and, in some cases, to grow them in culture. With the right culture conditions (for instance, the addition of specific growth factors), cultured stem cells from adult animals have been made to differentiate into various defined types of specialized cells, although none are as versatile as ES cells.

Research with embryonic or adult stem cells is a source of valuable data about differentiation and has enormous potential for medical applications. The ultimate aim is to supply cells for the repair of damaged or diseased organs: for example, insulin-producing pancreatic cells for people with type 1 diabetes or certain kinds of brain cells for people with Parkinson's disease or Huntington's disease. Adult stem cells from bone marrow have long been used in bone marrow transplants as a source of immune system cells in patients whose own immune systems are nonfunctional because of genetic disorders or radiation treatments for cancer.

The developmental potential of adult stem cells is limited to certain tissues. ES cells hold more promise than adult stem cells for most medical applications because ES cells are **pluripotent**, capable of differentiating into many different cell types. In 2013, a research group reported that they had established ES cell lines from human blastocysts produced by transferring a nucleus from a differentiated cell into an enucleated egg. Prior to that report, cells were obtained only from embryos donated by patients undergoing infertility treatments or from long-term cell cultures originally established with cells isolated from donated embryos, an issue that prompts ethical and political discussions. Although the techniques for cloning early human embryos are still being optimized, they represent a potential new source for ES cells that may be less controversial. Furthermore, with a donor nucleus from a person with a particular disease, researchers should be able to produce ES cells that match the patient and are thus not rejected by his or her immune system when used for treatment. When the main aim of cloning is to produce ES cells to treat disease, the process is called *therapeutic cloning*. Although most people believe that reproductive cloning of humans is unethical, opinions vary about the morality of therapeutic cloning.

Induced Pluripotent Stem (iPS) Cells

Resolving the debate now seems less urgent because researchers have learned to turn back the clock in fully differentiated cells, reprogramming them to act like ES cells. The accomplishment of this feat, which posed formidable obstacles, was announced in 2007, first by labs using mouse skin cells and then by additional groups using cells from human skin and other organs or tissues. In all these cases, researchers transformed the differentiated cells into a type of ES cell by using a retrovirus to introduce extra, cloned copies of four "stem cell" master regulatory genes. The "deprogrammed" cells are known as *induced pluripotent stem* (*iPS*) cells because, in using this fairly simple laboratory technique to return them to their undifferentiated state, pluripotency has been restored. The experiments that first transformed human differentiated cells into iPS cells are described in **Figure 20.21**. Shinya Yamanaka received the 2012 Nobel Prize in Medicine for this work, shared with John Gurdon, whose work you read about in Figure 20.16.

By many criteria, iPS cells can perform most of the functions of ES cells, but there are some differences in gene expression and other cellular functions, such as cell division. At least until these differences are fully understood, the study

▼ Figure 20.21

Inquiry Can a fully differentiated human cell be "deprogrammed" to become a stem cell?

Experiment Shinya Yamanaka and colleagues at Kyoto University, in Japan, used a retroviral vector to introduce four genes into fully differentiated human skin fibroblast cells. The cells were then cultured in a medium that would support growth of stem cells.

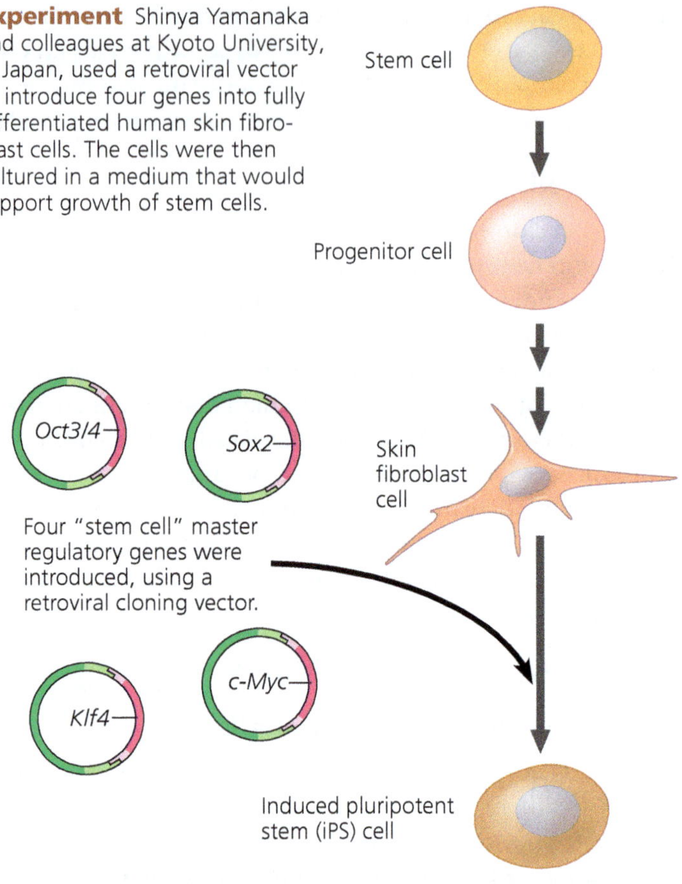

Results Two weeks later, the cells resembled embryonic stem cells in appearance and were actively dividing. Their gene expression patterns, gene methylation patterns, and other characteristics were also consistent with those of embryonic stem cells. The iPS cells were able to differentiate into heart muscle cells, as well as other cell types.

Conclusion The four genes induced differentiated skin cells to become pluripotent stem cells, with characteristics of embryonic stem cells.

Data from K. Takahashi et al., Induction of pluripotent stem cells from adult human fibroblasts by defined factors, *Cell* 131:861–872 (2007).

WHAT IF? ➤ *Patients with diseases such as heart disease or Alzheimer's could have their own skin cells reprogrammed to become iPS cells. Once procedures have been developed for converting iPS cells into heart or nervous system cells, the patients' own iPS cells might be used to treat their disease. When organs are transplanted from a donor to a diseased recipient, the recipient's immune system may reject the transplant, a dangerous condition. Would using iPS cells be expected to carry the same risk? Why or why not? Given that these cells are actively dividing, undifferentiated cells, what risks might this procedure carry?*

of ES cells will continue to make important contributions to the development of stem cell therapies. (In fact, it is likely that ES cells will always be a focus of basic research as well.) In the meantime, work is proceeding using the iPS cells that have been experimentally produced.

There are two major potential uses for human iPS cells. First, cells from patients suffering from diseases have been reprogrammed to become iPS cells, which act as model cells for studying the disease and potential treatments. Human iPS cell lines have already been developed from individuals with type 1 diabetes, Parkinson's disease, Huntington's disease, Down syndrome, and many other diseases. Second, in the field of regenerative medicine, a patient's own cells could be reprogrammed into iPS cells and then used to replace nonfunctional tissues, such as insulin-producing cells of the pancreas. In fact, in 2014 two research groups described successful methods for growing insulin-producing cells from both iPS cells and ES cells. Before these cells can be used in patients, however, researchers will have to develop a way to ensure the cells are not destroyed by the patients' immune system (the original cause of type 1 diabetes, in which the immune system malfunctions).

In another surprising development, researchers have been able to identify genes that can directly reprogram a differentiated cell into another type of differentiated cell without it passing through a pluripotent state. In the first reported example, one type of cell in the pancreas was transformed into another type. However, the two types of cells do not need to be very closely related: Another research group has been able to directly reprogram a skin fibroblast into a nerve cell. Development techniques that direct iPS cells or even fully differentiated cells to become specific cell types for regenerative medicine is an area of intense research, one that has already seen some success. The iPS cells created in this way could eventually provide tailor-made "replacement" cells for patients without using any human eggs or embryos, thus circumventing most ethical objections.

CONCEPT CHECK 20.3

1. Based on current knowledge, how would you explain the difference in the percentage of tadpoles that developed from the two kinds of donor nuclei in Figure 20.16?
2. A few companies in China and South Korea provide the service of cloning dogs, using cells from their clients' pets to provide nuclei in procedures like that in Figure 20.17. Should their clients expect the clone to look identical to their original pet? Why or why not? What ethical questions does this bring up?
3. **MAKE CONNECTIONS** ➤ Based on what you know about muscle differentiation (see Figure 18.18) and genetic engineering, propose the first experiment you might try if you wanted to direct an embryonic stem cell or iPS cell to develop into a muscle cell.

For suggested answers, see Appendix A.

CONCEPT 20.4

The practical applications of DNA-based biotechnology affect our lives in many ways

DNA technology is in the news almost every day. Most often, the topic is a new and promising application in medicine, but this is just one of numerous fields benefiting from DNA technology and genetic engineering.

Medical Applications

One important use of DNA technology is the identification of human genes whose mutation plays a role in genetic diseases. These discoveries may lead to ways of diagnosing, treating, and even preventing such conditions. DNA technology is also contributing to our understanding of "nongenetic" diseases, from arthritis to AIDS, since a person's genes influence susceptibility to these diseases. Furthermore, diseases of all sorts involve changes in gene expression within the affected cells and often within the patient's immune system. By using RNA-seq and DNA microarray assays (see Figures 20.12 and 20.13) or other techniques to compare gene expression in healthy and diseased tissues, researchers are finding genes that are turned on or off in particular diseases. These genes and their products are potential targets for prevention or therapy.

Diagnosis and Treatment of Diseases

A new chapter in the diagnosis of infectious diseases has been opened by DNA technology, in particular the use of PCR and labeled nucleic acid probes to track down pathogens. For example, because the sequence of the RNA genome of HIV is known, RT-PCR can be used to amplify, and thus detect, HIV RNA in blood or tissue samples (see Figure 20.11). RT-PCR is often the best way to detect an otherwise elusive infective agent.

Medical scientists can now diagnose hundreds of human genetic disorders by using PCR with primers that target the genes associated with these disorders. The amplified DNA product is then sequenced to reveal the presence or absence of the disease-causing mutation. Among the genes for human diseases that have been identified are those for sickle-cell disease, hemophilia, cystic fibrosis, Huntington's disease, and Duchenne muscular dystrophy. Individuals with such diseases can often be identified before the onset of symptoms, even before birth (see Figure 14.19). PCR can also be used to identify symptomless carriers of potentially harmful recessive alleles.

As you learned earlier, genome-wide association studies have pinpointed SNPs (single nucleotide polymorphisms) that are linked to disease-associated alleles (see Figure 20.15).

Individuals can be tested by PCR and sequencing for a SNP that is correlated with the abnormal allele. The presence of particular SNPs is correlated with increased risk for conditions such as heart disease, Alzheimer's disease, and some types of cancer. Companies that offer individual genetic testing for risk factors like these are looking for previously identified, linked SNPs. It may be helpful for individuals to learn about their health risks, with the understanding that such genetic tests merely reflect correlations and do not make predictions.

The techniques described in this chapter have also prompted improvements in disease treatments. By analyzing the expression of many genes in breast cancer patients, researchers have been able to refine their understanding of the different subtypes of breast cancer (see Figure 18.27). Knowing the expression levels of particular genes can help physicians determine the likelihood that the cancer will recur, thus helping them design an appropriate treatment. Given that some low-risk patients have a 96% survival rate over a ten-year period with no treatment, gene expression analysis allows doctors and patients access to valuable information when they are considering treatment options.

Many envision a future of "personalized medicine" where each person's genetic profile can inform them about diseases or conditions for which they are especially at risk and help them make treatment choices. As we will discuss later in the chapter, a *genetic profile* is currently taken to mean a set of genetic markers such as SNPs. Ultimately, however, it will likely mean the complete DNA sequence of an individual—once sequencing becomes inexpensive enough. Our ability to sequence a person's genome rapidly and inexpensively is advancing faster than our development of appropriate treatments for the conditions we are characterizing. Still, the identification of genes involved in these conditions provides us with good targets for therapeutic interventions.

Human Gene Therapy and Gene Editing

Gene therapy—the introduction of genes into an afflicted individual for therapeutic purposes—holds great potential for treating the relatively small number of disorders traceable to a single defective gene. The aim of this approach is to insert a normal allele of the defective gene into the somatic cells of the tissue affected by the disorder.

BBC Video: Cystic Fibrosis and the Promise of Gene Therapy

For gene therapy of somatic cells to be permanent, the cells that have the normal allele must be cells that multiply throughout the patient's life. Bone marrow cells, which include the stem cells that give rise to all the cells of the blood and immune system, are prime candidates. **Figure 20.22** outlines one procedure for gene therapy in an individual whose bone marrow cells do not produce a vital

▼ Figure 20.22 Gene therapy using a retroviral vector. A retrovirus that has been rendered harmless is used as a vector in this procedure, which exploits the ability of a retrovirus to insert a DNA transcript of its RNA genome into the chromosomal DNA of its host cell (see Figure 19.9). If the foreign gene carried by the retroviral vector is expressed, the cell and its descendants will possess the gene product. Cells that reproduce throughout life, such as bone marrow cells, are ideal candidates for gene therapy.

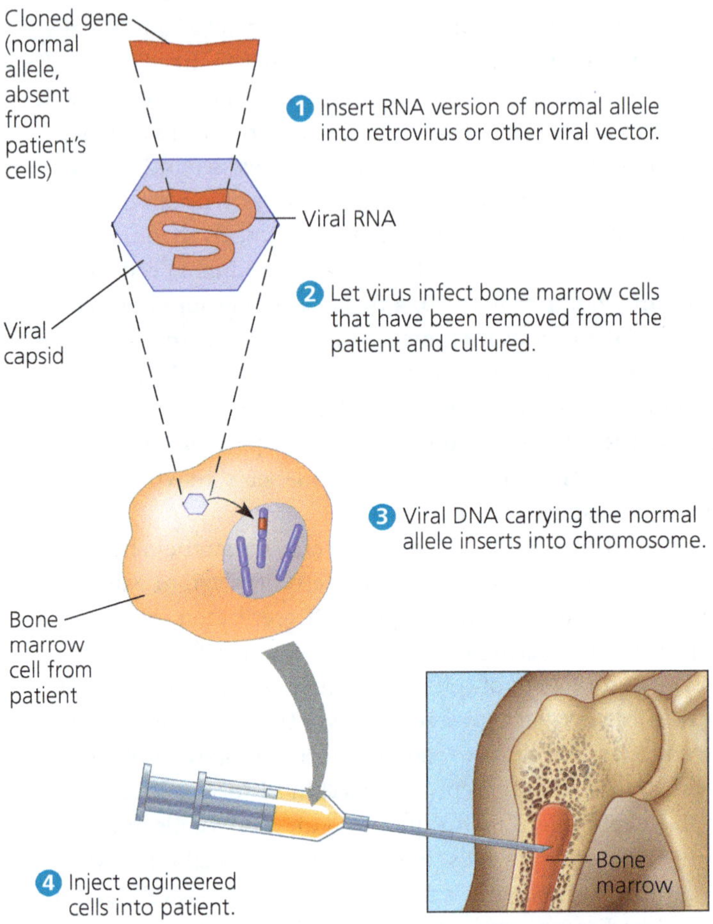

enzyme because of a single defective gene. One type of severe combined immunodeficiency (SCID) is caused by this kind of defect. If the treatment is successful, the patient's bone marrow cells will begin producing the missing protein, and the patient may be cured.

The procedure shown in Figure 20.22 was used in gene therapy trials for SCID in France in 2000. In that trial, ten young children with SCID were treated by the same procedure. Nine of these patients showed significant, definitive improvement after two years, the first indisputable success of gene therapy. However, three of the patients subsequently developed leukemia, a type of blood cell cancer, and one of them died. Researchers have concluded it is likely that the insertion of the retroviral vector occurred near a gene that triggers the proliferation of blood cells. Using a viral vector that does not come from a retrovirus, clinical researchers have treated at least three other genetic diseases somewhat successfully with gene therapy: a type of progressive blindness (see Concept 50.3), a degenerative disease of the nervous system, and a blood disorder involving the β-globin gene.

Gene therapy raises many technical issues. For example, how can the activity of the transferred gene be controlled so that cells make appropriate amounts of the gene product at the right time and in the right place? How can we be sure that the insertion of the therapeutic gene does not harm some other necessary cell function? As more is learned about DNA control elements and gene interactions, researchers may be able to answer such questions.

A more direct approach that avoids the complications of using a viral vector in gene therapy is made possible by gene editing, especially given the development of the CRISPR-Cas9 system described earlier. In this approach the existing defective gene is edited to correct the mutation. As shown in Figure 20.14, the CRISPR-Cas9 system is capable of doing this.

In 2014, a group of researchers reported correcting a genetic defect in mice using CRISPR-Cas9 technology. The lab mice had been genetically engineered to have a mutation in a gene encoding a liver enzyme that metabolizes the amino acid tyrosine, mimicking a fatal genetic disorder in humans called tyrosinemia. A guide RNA molecule complementary to the mutated region of the gene was introduced into the mouse along with the Cas9 protein and a segment of DNA from the same region of the normal gene for use as a template. Subsequent analysis indicated that the faulty gene had been corrected in enough of the liver cells that the amount of functional enzyme made was sufficient to alleviate the disease symptoms. There are still hurdles to overcome before this approach can be used in clinical trials in humans, but the CRISPR technology is sparking widespread excitement among researchers and physicians alike.

In addition to technical challenges, gene therapy and gene editing provoke ethical questions. Some critics believe that tampering with human genes in any way is immoral or unethical. Other observers see no fundamental difference between the transplantation of genes into somatic cells and the transplantation of organs. You might wonder whether scientists are considering engineering human germ-line cells in the hope of correcting a defect in future generations. Such genetic engineering is now routinely done in laboratory mice, and, in fact, conditions that would allow genetic engineering of human embryos have been worked out.

The development of the CRISPR-Cas9 system has engendered much debate about the ethics of gene editing, related to applications both potential and real. In March of 2015, an editorial was published by leading scientists working with CRISPR-Cas9 calling for the research community to "strongly discourage" any experimental work on human eggs or embryos. A month later, however, scientists in China reported using CRISPR-Cas9 technology to edit a gene in human embryos. (They used "nonviable" fertilized eggs—zygotes—that would not develop all the way but could form blastocysts.) The researchers were attempting to edit the β-thalassemia gene, mutations in which cause a blood disease of the same name. They injected 86 zygotes, only four of which showed the gene to be edited properly. In many of the other embryos, there were effects on genes other than the β-thalassemia gene—at a much higher level than had been seen in mouse embryos or human cell lines. This study highlighted problems with the technique, at least in human embryos, and at the same time accelerated concern about ethical considerations. Under what circumstances, if any, should we alter the genomes of human germ lines? Would this inevitably lead to the practice of eugenics, a deliberate effort to control the genetic makeup of human populations? While we may not have to resolve these questions immediately, considering them is imperative because they will likely come to the fore at some point in the near future.

Pharmaceutical Products

The pharmaceutical industry derives significant benefit from advances in DNA technology and genetic research, applying them to the development of useful drugs to treat diseases. Pharmaceutical products are synthesized using methods of either organic chemistry or biotechnology, depending on the nature of the product.

Synthesis of Small Molecules for Use as Drugs

Determining the sequence and structure of proteins crucial for tumor cell survival has led to the identification of small molecules that combat certain cancers by blocking the function of these proteins. One drug, imatinib (trade name Gleevec), is a small molecule that inhibits one tyrosine kinase (see Figure 11.8). The overexpression of this kinase, resulting from a chromosomal translocation, is instrumental in causing chronic myelogenous leukemia (CML; see Figure 15.16). Patients in the early stages of CML who are treated with imatinib have exhibited nearly complete, sustained remission from the cancer. Drugs that work in a similar way have also been used with success to treat a few types of lung and breast cancers. This approach is feasible only for cancers for which the molecular basis is fairly well understood.

MB HHMI Video: Gleevec hhmi BioInteractive

In many cases of such drug-treated tumors, though, cells later arise that are resistant to the new drug. In one study, the whole genome of the tumor cells was sequenced both before and after the appearance of drug resistance. Comparison of the sequences showed genetic changes that allowed the tumor cells to "get around" the drug-inhibited protein. Here, we can see that cancer cells demonstrate the principles of evolution: Certain tumor cells have a random mutation that allows them to survive in the presence of a particular drug, and as a consequence of natural selection in the presence of the drug, these are the cells that survive and reproduce.

Protein Production in Cell Cultures Pharmaceutical products that are proteins are commonly synthesized on a large scale using cell cultures. You learned earlier in the chapter about DNA cloning and gene expression systems for producing large quantities of a chosen protein that is present naturally in only minute amounts. The host cells used in such expression systems can even be engineered to secrete a protein as it is made, thereby simplifying the task of purifying it by traditional biochemical methods.

Among the first pharmaceutical products manufactured in this way were human insulin and human growth hormone (HGH). Some 2 million people with diabetes in the United States depend on insulin treatment to control their disease. Human growth hormone has been a boon to children born with a form of dwarfism caused by inadequate amounts of HGH, as well as helping AIDS patients gain weight. Another important pharmaceutical product produced by genetic engineering is tissue plasminogen activator (TPA). If administered shortly after a heart attack, TPA helps dissolve blood clots and reduces the risk of subsequent heart attacks.

Protein Production by "Pharm" Animals In some cases, instead of using cell systems to produce large quantities of protein products, pharmaceutical scientists can use whole animals. They can introduce a gene (or other DNA) from an animal into the genome of another individual, often of a different species. This individual is then called a **transgenic** animal. To do this, they first remove eggs from a female of the recipient species and fertilize them *in vitro*. Meanwhile, they have cloned the desired gene from the donor organism. They then inject the cloned DNA directly into the nuclei of the fertilized eggs. Some of the cells integrate the foreign DNA, the *transgene*, into their genome and are able to express the foreign gene. The engineered embryos that arise from these zygotes are then surgically implanted in a surrogate mother. If an embryo develops successfully, the result is a transgenic animal that expresses its new, "foreign" gene.

Assuming that the introduced gene encodes a protein desired in large quantities, these transgenic animals can act as pharmaceutical "factories." For example, a transgene for a human blood protein such as antithrombin, which prevents blood clots, can be inserted into the genome of a goat in such a way that the transgene's product is secreted in the animal's milk **(Figure 20.23)**. The protein is then purified from the milk (which is easier than purification from a cell culture). Such proteins must be tested to ensure that they (or contaminants from the farm animals) will not cause allergic reactions or other adverse effects in patients who receive them.

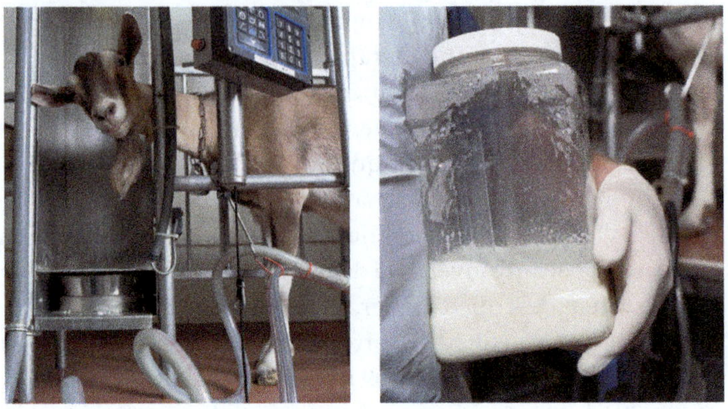

▼ **Figure 20.23 Goats as "pharm" animals.** This transgenic goat carries a gene for a human blood protein, antithrombin, which she secretes in her milk. Patients with a rare hereditary disorder in which this protein is lacking suffer from formation of blood clots in their blood vessels. Easily purified from the goat's milk, the protein is used to prevent blood clots in these patients during surgery or childbirth.

Forensic Evidence and Genetic Profiles

In violent crimes, body fluids or small pieces of tissue may be left at the scene or on the clothes or other possessions of the victim or assailant. If enough blood, semen, or tissue is available, forensic laboratories can determine the blood type or tissue type by using antibodies to detect specific cell-surface proteins. However, such tests require fairly fresh samples in relatively large amounts. Also, because many people have the same blood or tissue type, this approach can only exclude a suspect; it cannot provide strong evidence of guilt.

DNA testing, on the other hand, can identify the guilty individual with a high degree of certainty because the DNA sequence of every person is unique (except for identical twins). Genetic markers that vary in the population can be analyzed for a given person to determine that individual's unique set of genetic markers, or **genetic profile**. (This term is preferred over "DNA fingerprint" by forensic scientists, who want to emphasize the heritable aspect of these markers rather than the fact that they produce a pattern on a gel that, like a fingerprint, is visually recognizable.) The FBI started applying DNA technology to forensics in 1988, using a method involving gel electrophoresis and nucleic acid hybridization to detect similarities and differences in DNA samples. This method required much smaller samples of blood or tissue than earlier methods—only about 1,000 cells.

MB **Animation: Genetic Profiles**

Today, forensic scientists use an even more sensitive method that takes advantage of variations in length of genetic markers called **short tandem repeats (STRs)**. These are tandemly repeated units of two- to five-nucleotide sequences in specific regions of the genome. The number of repeats present in these regions is highly variable from person to person (polymorphic), and even for a single individual, the two alleles of an STR may differ from each other. For example, one individual may have the sequence ACAT repeated

30 times at one genome locus and 15 times at the same locus on the other homolog, whereas another individual may have 18 repeats at this locus on each homolog. (These two genotypes can be expressed by the two repeat numbers: 30,15 and 18,18.) PCR is used to amplify particular STRs, using sets of primers that are labeled with different-colored fluorescent tags; the length of the region, and thus the number of repeats, can then be determined by electrophoresis. The PCR step allows use of this method even when the DNA is in poor condition or available only in minute quantities: A tissue sample containing as few as 20 cells can be sufficient.

In a murder case, for example, this method can be used to compare DNA samples from the suspect, the victim, and a small amount of blood found at the crime scene. The forensic scientist tests only a few selected portions of the DNA—usually 13 STR markers. However, even this small set of markers can provide a forensically useful genetic profile because the probability that two people (who are not identical twins) would have exactly the same set of STR markers is vanishingly small. The Innocence Project, a nonprofit organization dedicated to overturning wrongful convictions, uses STR analysis of archived samples from crime scenes to revisit old cases. As of 2016, more than 340 innocent people have been released from prison as a result of forensic and legal work by this group **(Figure 20.24)**.

Genetic profiles can also be useful for other purposes. A comparison of the DNA of a mother, her child, and the purported father can conclusively settle a question of paternity. Sometimes paternity is of historical interest: Genetic profiles provided strong evidence that Thomas Jefferson or one of his close male relatives fathered at least one of the children of his slave Sally Hemings. Genetic profiles can also identify victims of mass casualties. The largest such effort occurred after the attack on the World Trade Center in 2001; more than 10,000 samples of victims' remains were compared with DNA samples from personal items, such as toothbrushes, provided by families. Ultimately, forensic scientists succeeded in identifying almost 3,000 victims using these methods.

Just how reliable is a genetic profile? The greater the number of markers examined in a DNA sample, the more likely it is that the profile is unique to one individual. In forensic cases using STR analysis with 13 markers, the probability of two people having identical DNA profiles is somewhere between one chance in 10 billion and one in several trillion. (For comparison, the world's population is between 7 and 8 billion.) The exact probability depends on the frequency of those markers in the general population. Information on how common various markers are in different ethnic groups is critical because these marker frequencies may vary considerably among ethnic groups and between a particular ethnic group and the population as a whole. With the increasing availability of frequency data, forensic scientists can make extremely accurate statistical calculations. Thus, despite problems that can still arise from insufficient data, human error, or flawed evidence, genetic profiles are now accepted as compelling evidence by legal experts and scientists alike.

▼ Figure 20.24 STR analysis used to release an innocent man from prison.

(a) In 1984, Earl Washington was convicted and sentenced to death for the 1982 rape and murder of Rebecca Williams. His sentence was commuted to life in prison in 1993 due to new doubts about the evidence. In 2000, STR analysis by forensic scientists associated with the Innocence Project showed conclusively that he was innocent. This photo shows Washington just before his release in 2001, after 17 years in prison.

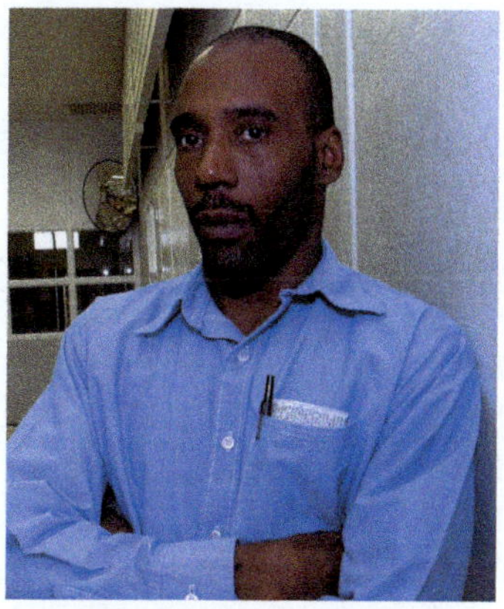

Source of sample	STR marker 1	STR marker 2	STR marker 3
Semen on victim	17,19	13,16	12,12
Earl Washington	16,18	14,15	11,12
Kenneth Tinsley	17,19	13,16	12,12

(b) In STR analysis, selected STR markers in a DNA sample are amplified by PCR, and the PCR products are separated by electrophoresis. The procedure reveals how many repeats are present for each STR locus in the sample. An individual has two alleles per STR locus, each with a certain number of repeats. This table shows the number of repeats for three STR markers in three samples: from semen found on the victim, from Washington, and from another man (Kenneth Tinsley), who was in prison because of an unrelated conviction. These and other STR data (not shown) exonerated Washington and led Tinsley to plead guilty to the murder.

Environmental Cleanup

Increasingly, the diverse abilities of certain microorganisms to transform chemicals are being exploited for environmental cleanup. If the growth needs of such microorganisms make them unsuitable for direct use, scientists can now transfer the genes for their valuable metabolic capabilities into other microorganisms, which can then be used to treat environmental problems. For example, many bacteria can extract heavy metals, such as copper, lead, and nickel, from their environments and incorporate the metals into compounds such as copper sulfate or lead sulfate, which are readily recoverable. Genetically engineered microorganisms may become important in both mining (especially as ore reserves are depleted) and cleaning up highly toxic mining wastes. Biotechnologists are also trying to engineer microorganisms that can degrade chlorinated hydrocarbons and other harmful compounds.

These microorganisms could be used in wastewater treatment plants or by manufacturers before the compounds are ever released into the environment.

Agricultural Applications

Scientists are working to learn more about the genomes of agriculturally important plants and animals. For a number of years, they have been using DNA technology in an effort to improve agricultural productivity. The selective breeding of both livestock (animal husbandry) and crops has exploited naturally occurring mutations and genetic recombination for thousands of years.

As we described earlier, DNA technology enables scientists to produce transgenic animals, which speeds up the selective breeding process. The goals of creating a transgenic animal are often the same as the goals of traditional breeding—for instance, to make a sheep with better quality wool, a pig with leaner meat, or a cow that will mature in a shorter time. Scientists might, for example, identify and clone a gene that causes the development of larger muscles (muscles make up most of the meat we eat) in one breed of cattle and transfer it to other cattle or even to sheep. However, health problems are not uncommon among farm animals carrying genes from other species, and modification of the animal's own genes using the CRISPR-Cas9 system will likely emerge as a more useful technique. Animal health and welfare are important issues to consider when genetically altering animals.

Agricultural scientists have already endowed a number of crop plants with genes for desirable traits, such as delayed ripening and resistance to spoilage, disease, and drought. Modifications can also add value to food crops, giving them a longer shelf life or improved flavor or nutritional value. For many plant species, a single tissue cell grown in culture can give rise to an adult plant. Thus, genetic manipulations can be performed on an ordinary somatic cell and the cell then used to generate a plant with new traits.

Genetic engineering is rapidly replacing traditional plant-breeding programs, especially for useful traits, such as herbicide or pest resistance, determined by one or a few genes. Crops engineered with a bacterial gene making the plants resistant to an herbicide can grow while weeds are destroyed, and genetically engineered crops that can resist destructive insects reduce the need for chemical insecticides. In India, the insertion of a salinity resistance gene from a coastal mangrove plant into the genomes of several rice varieties has resulted in rice plants that can grow in water three times as salty as seawater. The research foundation that carried out this feat of genetic engineering estimates that one-third of all irrigated land has high salinity owing to overirrigation and intensive use of chemical fertilizers, representing a serious threat to the food supply. Thus, salinity-resistant crop plants would be enormously valuable worldwide.

Safety and Ethical Questions Raised by DNA Technology

Early concerns about potential dangers associated with recombinant DNA technology focused on the possibility that hazardous new pathogens might be created. What might happen, for instance, if in a research study cancer cell genes were transferred into bacteria or viruses? To guard against such rogue microorganisms, scientists developed a set of guidelines that were adopted as formal government regulations in the United States and some other countries. One safety measure is a set of strict laboratory procedures designed to prevent engineered microorganisms from infecting researchers or accidentally leaving the laboratory. In addition, strains of microorganisms to be used in recombinant DNA experiments are genetically crippled to ensure that they cannot survive outside the laboratory. Finally, certain obviously dangerous experiments have been banned.

Today, most public concern about possible hazards centers not on recombinant microorganisms but on **genetically modified organisms (GMOs)** used as food. A GMO is a transgenic organism, one that has acquired by artificial means one or more genes from another species or even from another variety of the same species. Some salmon, for example, have been genetically modified by addition of a more active salmon growth hormone gene. However, the majority of the GMOs that contribute to our food supply are not animals, but crop plants.

GM crops are widespread in the United States, Argentina, and Brazil; together, these countries account for over 80% of the world's acreage devoted to such crops. In the United States, most corn, soybean, and canola crops are genetically modified, and a recent law requires labeling of GM products. The same foods are an ongoing subject of controversy in Europe, where the GM revolution has been met with strong opposition. Many Europeans are concerned about the safety of GM foods and the possible environmental consequences of growing GM plants. Although a small number of GM crops have been grown on European soil, the European Union established a comprehensive legal framework regarding GMOs in 2015. Among other regulations, individual member states may ban either the growing or importing of GM crops, which must be clearly labeled. The high degree of consumer distrust in Europe makes the future of GM crops there uncertain.

Advocates of a cautious approach toward GM crops fear that transgenic plants might pass their new genes to close relatives in nearby wild areas. We know that lawn and crop grasses, for example, commonly exchange genes with wild relatives via pollen transfer. If crop plants carrying genes for resistance to herbicides, diseases, or insect pests pollinated wild ones, the offspring might become "super weeds" that are very difficult to control. Another worry involves possible risks to human health from GM foods. Some people fear that the protein products of transgenes might lead to allergic reactions. Although there is some evidence that this could happen, advocates claim that these proteins could be tested in advance to avoid producing ones that cause allergic reactions. (For further discussion of plant biotechnology and GM crops, see Concept 38.3.)

Today, governments and regulatory agencies throughout the world are grappling with how to facilitate the use of biotechnology in agriculture, industry, and medicine while ensuring that new products and procedures are safe. In the United States, such applications of biotechnology must be evaluated for potential risks by various regulatory agencies, including the Food and Drug Administration, the Environmental Protection Agency, the National Institutes of Health, and the Department of Agriculture. Meanwhile, these same agencies and the public must consider the ethical implications of biotechnology.

MB **ABC News Video: Genetically Altered Salmon**

Advances in biotechnology have allowed us to obtain complete genome sequences for humans and many other species, providing a vast treasure trove of information about genes. We can ask how certain genes differ from species to species, as well as how genes and, ultimately, entire genomes have evolved. (These are the subjects of Chapter 21.) At the same time, the increasing speed and falling cost of sequencing the genomes of individuals are raising significant ethical questions. Who should have the right to examine someone else's genetic information? How should that information be used? Should a person's genome be a factor in determining eligibility for a job or insurance? Ethical considerations, as well as concerns about potential environmental and health hazards, will likely slow some applications of biotechnology. There is always a danger that too much regulation will stifle basic research and its potential benefits. On the other hand, genetic engineering—especially gene editing with the CRISPR-Cas system—enables us to profoundly and rapidly alter species that have been evolving for millennia. A good example is the potential use of a gene drive that would eliminate the ability of mosquito species to carry diseases or even eradicate certain mosquito species. There would probably be health benefits to this approach, at least initially, but unforeseen problems could easily arise. Given the tremendous power of DNA technology, we must proceed with humility and caution.

MB **Interview with David Suzuki: Exploring ethical issues related to DNA technology**

CONCEPT CHECK 20.4

1. What is the advantage of using stem cells for gene therapy or gene editing?
2. List at least three different properties that have been acquired by crop plants via genetic engineering.
3. **WHAT IF?** ➤ As a physician, you have a patient with symptoms that suggest a hepatitis A infection, but you have not been able to detect viral proteins in the blood. Knowing that hepatitis A is an RNA virus, what lab tests could you perform to support your diagnosis? Explain the results that would support your hypothesis.

For suggested answers, see Appendix A.

20 Chapter Review

MB Go to **MasteringBiology**™ for Videos, Animations, Vocab Self-Quiz, Practice Tests, and more in the Study Area.

SUMMARY OF KEY CONCEPTS

CONCEPT 20.1

DNA sequencing and DNA cloning are valuable tools for genetic engineering and biological inquiry *(pp. 414–421)*

VOCAB SELF-QUIZ goo.gl/6u55ks

- **Nucleic acid hybridization**, the base pairing of one strand of a nucleic acid to the complementary sequence on a strand from another nucleic acid molecule, is widely used in **DNA technology**.
- **DNA sequencing** can be carried out using the dideoxy chain termination method in automated sequencing machines.
- Next-generation (high-throughput) techniques for sequencing DNA are based on sequencing by synthesis: DNA polymerase is used to synthesize a stretch of DNA from a single-stranded template, and the order in which nucleotides are added reveals the sequence. Third-generation sequencing methods, including nanopore technology, sequence long DNA molecules one at a time by distinguishing the nucleotide bases as they pass through a pore in a membrane.
- **Gene cloning** (or **DNA cloning**) produces multiple copies of a gene (or DNA segment) that can be used to manipulate and analyze DNA and to produce useful new products or organisms with beneficial traits.
- In **genetic engineering**, bacterial **restriction enzymes** are used to cut DNA molecules within short, specific nucleotide sequences (**restriction sites**), yielding a set of double-stranded **restriction fragments** with single-stranded **sticky ends**:

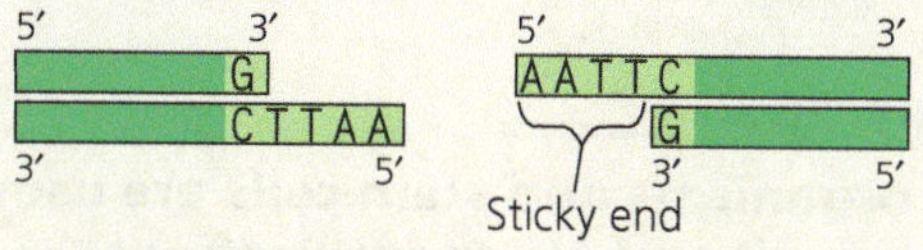

- The sticky ends on restriction fragments from one DNA source can base-pair with complementary sticky ends on fragments from other DNA molecules. Sealing the base-paired fragments with DNA ligase produces **recombinant DNA molecules**.
- DNA restriction fragments of different lengths can be separated by **gel electrophoresis**.

- The **polymerase chain reaction (PCR)** can produce many copies of (amplify) a specific target segment of DNA *in vitro*, using primers that bracket the desired sequence and a heat-resistant DNA polymerase.
- To clone a eukaryotic gene:

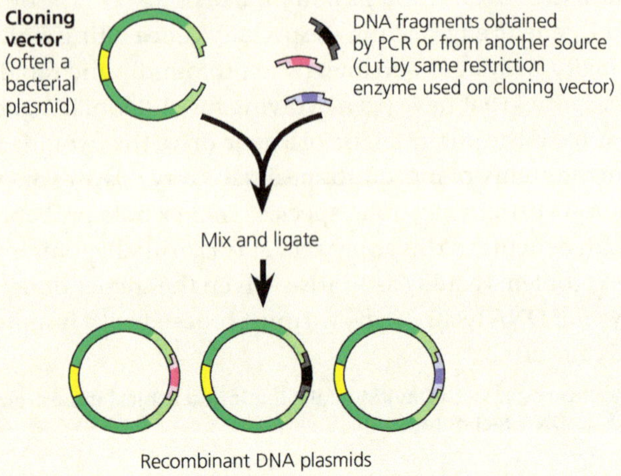

Recombinant plasmids are returned to host cells, each of which divides to form a clone of cells.
- Several technical difficulties hinder the expression of cloned eukaryotic genes in bacterial host cells. The use of cultured eukaryotic cells as host cells, coupled with appropriate **expression vectors**, helps avoid these problems.

? *Describe how the process of gene cloning results in a cell clone containing a recombinant plasmid.*

CONCEPT 20.2

Biologists use DNA technology to study gene expression and function *(pp. 421–426)*

- Several techniques use hybridization of a **nucleic acid probe** to detect the presence of specific mRNAs.
- ***In situ* hybridization** and **RT-PCR** can detect the presence of a given mRNA in a tissue or an RNA sample, respectively.
- **DNA microarrays** are used to identify sets of genes co-expressed by a group of cells. Increasingly, instead, **RNA sequencing (RNA-seq)** is used to sequence the **cDNAs** corresponding to RNAs from the cells.
- For a gene of unknown function, experimental inactivation of the gene (a gene knockout) and observation of the resulting phenotypic effects can provide clues to its function. The **CRISPR-Cas9 system** allows researchers to edit genes in living cells in a specific, desired way. The new alleles can be altered so that they are inherited in a biased way through a population (**gene drive**). In humans, **genome-wide association studies** identify and use **single nucleotide polymorphisms (SNPs)** as genetic markers for alleles that are associated with particular conditions.

? *What useful information is obtained by detecting expression of specific genes?*

CONCEPT 20.3

Cloned organisms and stem cells are useful for basic research and other applications *(pp. 426–431)*

- The question of whether all the cells in an organism have the same genome prompted the first attempts at organismal cloning.
- Single differentiated cells from plants are often **totipotent**: capable of generating all the tissues of a complete new plant.
- Transplantation of the nucleus from a differentiated animal cell into an enucleated egg can sometimes give rise to a new animal.
- Certain embryonic **stem cells** (ES cells) from animal embryos and particular adult stem cells from adult tissues can reproduce and differentiate both in the lab and in the organism, offering the potential for medical use. ES cells are **pluripotent** but difficult to acquire. Induced pluripotent stem (iPS) cells resemble ES cells in their capacity to differentiate; they can be generated by reprogramming differentiated cells. iPS cells hold promise for medical research and regenerative medicine.

? *Describe how, using mice, a researcher could carry out (1) organismal cloning, (2) production of ES cells, and (3) generation of iPS cells, focusing on how the cells are reprogrammed. (The procedures are basically the same in humans and mice.)*

CONCEPT 20.4

The practical applications of DNA-based biotechnology affect our lives in many ways *(pp. 431–437)*

- DNA technology, including the analysis of genetic markers such as SNPs, is increasingly being used in the diagnosis of genetic and other diseases and offers potential for better treatment of genetic disorders or even permanent cures through **gene therapy**, or gene editing with the CRISPR-Cas9 system. It also enables more informed cancer therapies. DNA technology is used with cell cultures in the large-scale production of protein hormones and other proteins with therapeutic uses. Some therapeutic proteins are being produced in **transgenic** "pharm" animals.
- Analysis of genetic markers such as **short tandem repeats (STRs)** in DNA isolated from tissue or body fluids found at crime scenes leads to a **genetic profile**. Use of genetic profiles can provide definitive evidence that a suspect is innocent or strong evidence of guilt. Such analysis is also useful in parenthood disputes and in identifying the remains of crime victims.
- Genetically engineered microorganisms can be used to extract minerals from the environment or degrade various types of toxic waste materials.
- The aims of developing transgenic plants and animals are to improve agricultural productivity and food quality.
- The potential benefits of genetic engineering must be carefully weighed against the potential for harm to humans or the environment.

? *What factors affect whether a given genetic disease would be a good candidate for successful gene therapy?*

TEST YOUR UNDERSTANDING

Level 1: Knowledge/Comprehension

PRACTICE TEST goo.gl/CUYGKD

1. In DNA technology, the term *vector* can refer to
   (A) the enzyme that cuts DNA into restriction fragments.
   (B) the sticky end of a DNA fragment.
   (C) a SNP marker.
   (D) a plasmid used to transfer DNA into a living cell.
2. Which of the following tools of DNA technology is incorrectly paired with its use?
   (A) electrophoresis—separation of DNA fragments
   (B) DNA ligase—cutting DNA, creating sticky ends of restriction fragments
   (C) DNA polymerase—polymerase chain reaction to amplify sections of DNA
   (D) reverse transcriptase—production of cDNA from mRNA

3. Plants are more readily manipulated by genetic engineering than are animals because
   (A) plant genes do not contain introns.
   (B) more vectors are available for transferring recombinant DNA into plant cells.
   (C) a somatic plant cell can often give rise to a complete plant.
   (D) plant cells have larger nuclei.

4. A paleontologist has recovered a bit of tissue from the 400-year-old preserved skin of an extinct dodo (a bird). To compare a specific region of the DNA from a sample with DNA from living birds, which of the following would be most useful for increasing the amount of dodo DNA available for testing?
   (A) SNP analysis
   (B) polymerase chain reaction (PCR)
   (C) electroporation
   (D) gel electrophoresis

5. DNA technology has many medical applications. Which of the following is *not* done routinely at present?
   (A) production of hormones for treating diabetes and dwarfism
   (B) analysis of gene expression for more informed cancer treatments
   (C) gene editing by the CRISPR-Cas9 system in viable human embryos to correct genetic diseases
   (D) prenatal identification of genetic disease alleles

Level 2: Application/Analysis

6. Which of the following is *not* true of cDNA produced using human brain tissue as the starting material?
   (A) It can be amplified by the polymerase chain reaction.
   (B) It was produced from pre-mRNA using reverse transcriptase.
   (C) It can be labeled and used as a probe to detect genes expressed in the brain.
   (D) It lacks the introns of the pre-mRNA.

7. Expression of a cloned eukaryotic gene in a bacterial cell involves many challenges. The use of mRNA and reverse transcriptase is part of a strategy to solve the problem of
   (A) post-transcriptional processing.
   (B) post-translational processing.
   (C) nucleic acid hybridization.
   (D) restriction fragment ligation.

8. Which of the following sequences in double-stranded DNA is most likely to be recognized as a cutting site for a restriction enzyme?
   (A) AAGG
   TTCC
   (B) GGCC
   CCGG
   (C) ACCA
   TGGT
   (D) AAAA
   TTTT

Level 3: Synthesis/Evaluation

9. **MAKE CONNECTIONS** Imagine you want to study one of the human crystallins, proteins present in the lens of the eye (see Figure 1.8). To obtain a sufficient amount of the protein of interest, you decide to clone the gene that codes for it. Assume you know the sequence of this gene. Explain how you would go about this.

10. **MAKE CONNECTIONS** Looking at Figure 20.15, what does it mean for a SNP to be "linked" to a disease-associated allele? How does this allow the SNP to be used as a genetic marker? (See Concept 15.3.)

11. **DRAW IT** You are cloning an aardvark gene, using a bacterial plasmid as a vector. The green diagram shows the plasmid, which contains the restriction site for the enzyme used in Figure 20.5. Above the plasmid is a segment of linear aardvark DNA that was synthesized using PCR. Diagram your cloning procedure, and show what would happen to these two molecules during each step. Use one color for the aardvark DNA and its bases and another color for those of the plasmid. Label each step and all 5′ and 3′ ends.

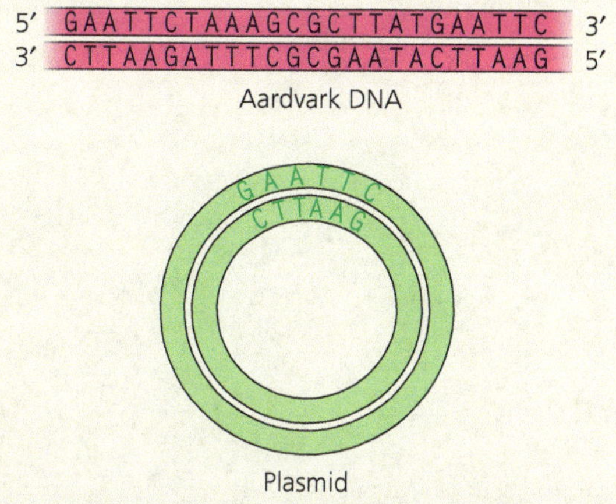

12. **EVOLUTION CONNECTION** Ethical considerations aside, if DNA-based technologies became widely used, discuss how they might change the way evolution proceeds, as compared with the natural evolutionary mechanisms that have operated for the past 4 billion years.

13. **SCIENTIFIC INQUIRY** You hope to study a gene that codes for a neurotransmitter protein produced in human brain cells. You know the amino acid sequence of the protein. Explain how you might (a) identify what genes are expressed in a specific type of brain cell, (b) identify (and isolate) the neurotransmitter gene, (c) produce multiple copies of the gene for study, and (d) produce large quantities of the neurotransmitter for evaluation as a potential medication.

14. **WRITE ABOUT A THEME: INFORMATION** In a short essay (100–150 words), discuss how the genetic basis of life plays a central role in biotechnology.

15. **SYNTHESIZE YOUR KNOWLEDGE**

The water in the Yellowstone National Park hot springs shown here is around 160°F (70°C). Biologists assumed that no species of organisms could live in water above about 130°F (55°C), so they were surprised to find several species of bacteria there, now called *thermophiles* ("heat-lovers"). You've learned in this chapter how an enzyme from one species, *Thermus aquaticus*, made feasible one of the most important DNA-based techniques used in labs today. Identify the enzyme, and indicate the value of its being isolated from a thermophile. Suggest other reasons why enzymes from this bacterium (or other thermophiles) might also be valuable.

For selected answers, see Appendix A.

Genomes and Their Evolution

21

Figure 21.1 What genomic information distinguishes a human from a chimpanzee?

KEY CONCEPTS

- **21.1** The Human Genome Project fostered development of faster, less expensive sequencing techniques
- **21.2** Scientists use bioinformatics to analyze genomes and their functions
- **21.3** Genomes vary in size, number of genes, and gene density
- **21.4** Multicellular eukaryotes have a lot of noncoding DNA and many multigene families
- **21.5** Duplication, rearrangement, and mutation of DNA contribute to genome evolution
- **21.6** Comparing genome sequences provides clues to evolution and development

Elephant shark (*Callorhinchus milii*)

Reading the Leaves from the Tree of Life

The chimpanzee (*Pan troglodytes*) is our closest living relative on the evolutionary tree of life. The boy in **Figure 21.1** and his chimpanzee companion are intently studying the same leaf, but only one of them is able to talk about it. What accounts for this difference between two primates that share so much of their evolutionary history? With advances in sequencing technology, we are now addressing the genetic basis of such intriguing questions. Later in the chapter, you'll learn about the *FOXP2* gene involved in vocalization, which differs between the two species.

The chimpanzee genome was sequenced two years after sequencing of the human genome. Now that we can compare our genome, base by base, with that of the chimpanzee, we can tackle the more general issue of what differences in genetic information account for the distinct characteristics of these two species of primates.

In addition to determining the sequences of the human and chimpanzee genomes, researchers have obtained complete genome sequences for *Escherichia coli* (*E. coli*) and numerous other prokaryotes, as well as many eukaryotes, including *Zea mays* (corn), *Drosophila melanogaster* (fruit fly), *Octopus bimaculoides* (California two-spot octopus), and *Callorhinchus milii* (elephant shark; see the small photo). In 2014, a high-quality sequence was announced for the genome of *Homo neanderthalensis*, (Neanderthals), an extinct species closely related to present-day humans. These genomes are of great interest in their own right, but they also provide important insights into evolution as well as other biological processes. Broadening the

When you see this blue icon, log in to **MasteringBiology** and go to the Study Area for digital resources.

MB **Get Ready for This Chapter**

human-chimpanzee comparison to the genomes of other primates and more distantly related animals should reveal the sets of genes that control group-defining characteristics. Beyond that, comparisons with the genomes of bacteria, archaea, fungi, protists, and plants should enlighten us about the long evolutionary history of the ancient genes we all share.

With the genomes of many species fully sequenced, scientists can study whole sets of genes and their interactions, an approach called **genomics**. The sequencing efforts that feed this approach have generated, and continue to generate, enormous volumes of data. The need to deal with this ever-increasing flood of information has spawned the field of **bioinformatics**, the application of computational methods to store and analyze biological data.

We will begin this chapter by discussing two approaches to genome sequencing and some of the advances in bioinformatics and its applications. We will then summarize what has been learned from the genomes that have been sequenced thus far. Next, we will describe the composition of the human genome as a representative genome of a complex multicellular eukaryote. Finally, we will explore current ideas about how genomes evolve and about how the evolution of developmental mechanisms could have generated the great diversity of life on Earth today.

CONCEPT 21.1

The Human Genome Project fostered development of faster, less expensive sequencing techniques

Sequencing of the human genome, an ambitious undertaking, officially began as the **Human Genome Project** in 1990. Organized by an international, publicly funded consortium of scientists at universities and research institutes, the project involved 20 large sequencing centers in six countries plus a host of other labs working on smaller parts of the project.

After the human genome sequence was largely completed in 2003, the sequence of each chromosome was analyzed and described in a series of papers, the last of which covered chromosome 1 and was published in 2006. At that point, the sequencing was considered "virtually complete."

The ultimate goal in mapping any genome is to determine the complete nucleotide sequence of each chromosome. For the human genome, this was accomplished by sequencing machines using the dideoxy chain termination method mentioned in Concept 20.1. Even with this initial use of automation, though, the sequencing of all 3 billion base pairs in a haploid set of human chromosomes presented a formidable challenge. In fact, a major thrust of the Human Genome Project was the development of technology for faster sequencing (see Concept 20.1). Improvements over the years chipped away at each time-consuming step, enabling the rate of sequencing to accelerate impressively: Whereas a productive lab could typically sequence 1,000 base pairs a day in the 1980s, by the year 2000 each research center working on the Human Genome Project was sequencing 1,000 base pairs *per second*. As of 2016, the most widely used automated machines can sequence nearly 25 million base pairs per second, while developers of some newer techniques claim they can achieve a rate of 66 billion base pairs per second. Methods that can analyze biological materials very rapidly and produce enormous volumes of data are said to be "high-throughput." Sequencing machines are an example of high-throughput devices.

Two approaches complemented each other in obtaining the complete sequence. The initial approach was a methodical one that built on an earlier storehouse of human genetic information. In 1998, however, molecular biologist J. Craig Venter set up a company (Celera Genomics) and declared his intention to sequence the entire human genome using an alternative strategy. The **whole-genome shotgun approach** starts with the cloning and sequencing of DNA fragments from randomly cut DNA. Powerful computer programs then assemble the resulting very large number of overlapping short sequences into a single continuous sequence **(Figure 21.2)**.

▼ **Figure 21.2 Whole-genome shotgun approach to sequencing.** In this approach, developed by J. Craig Venter and colleagues at Celera Genomics, random DNA fragments are cloned (see Figure 20.4), sequenced, and then ordered relative to each other.

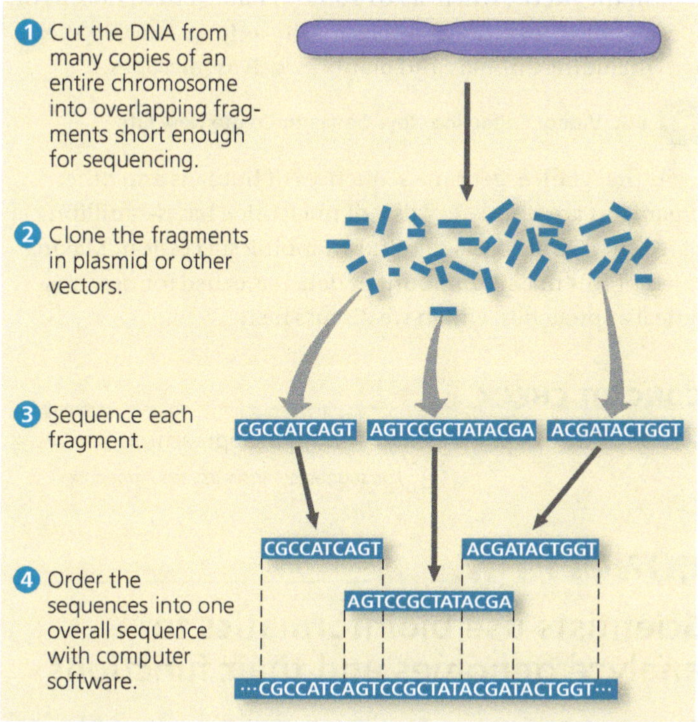

VISUAL SKILLS ➤ *The fragments in step 2 of this figure are depicted as scattered, rather than arranged in an ordered array. How does this depiction reflect the approach?*

HHMI Animation: Shotgun Sequencing

Today, the whole-genome shotgun approach is still used, although newer, "next-generation" sequencing techniques (see Figure 20.3) have resulted in massive increases in speed and decreases in the cost of sequencing entire genomes. In these new techniques, many very small DNA fragments (each about 300 base pairs long) are sequenced at the same time, and computer software rapidly assembles the complete sequence. Because of the sensitivity of these techniques, the fragments can be sequenced directly; the cloning step (2 in Figure 21.2) is unnecessary. Whereas sequencing the first human genome took 13 years and cost $100 million, the genome of James Watson (co-discoverer of DNA structure) was sequenced using newer techniques in four months in 2007 for about $1 million, and as of 2016, an individual's genome can be sequenced in a day or so for about $1,000.

These technological advances have also facilitated an approach called **metagenomics** (from the Greek *meta*, beyond), in which DNA from an entire community of species (a *metagenome*) is collected from an environmental sample and sequenced. Again, computer software sorts out the partial sequences and assembles them into the individual specific genomes. An advantage of this technique is the ability to sequence the DNA of mixed microbial populations, which eliminates the need to culture each species separately in the lab, a difficulty that has limited the study of microbes. So far, this approach has been applied to communities found in environments as diverse as the human intestine and ancient soils in the Arctic, where a 2014 study characterized dozens of species living together as a community as long as 50,000 years ago, including animals and plants as well as microbes.

MB BBC Video: Collecting New Genes to Create New Life

At first glance, genome sequences of humans and other organisms are simply dry lists of nucleotide bases—millions of A's, T's, C's, and G's in mind-numbing succession. Making sense of this massive amount of data has called for new analytical approaches, which we discuss next.

CONCEPT CHECK 21.1

1. Describe the whole-genome shotgun approach.

For suggested answers, see Appendix A.

CONCEPT 21.2

Scientists use bioinformatics to analyze genomes and their functions

Each of the 20 or so sequencing centers around the world working on the Human Genome Project churned out voluminous amounts of DNA sequence day after day. As the data began to accumulate, the need to coordinate efforts to keep track of all the sequences became clear. Thanks to the foresight of research scientists and government officials involved in the Human Genome Project, its goals included establishing centralized databases and refining analytical software, all to be made readily accessible on the Internet.

MB HHMI Video: Leading Edge Bioinformatics hhmi BioInteractive

Centralized Resources for Analyzing Genome Sequences

Making bioinformatics resources available to researchers worldwide and speeding up the dissemination of information served to accelerate progress in DNA sequence analysis. For example, in 1988, in preparation for the Human Genome Project in the United States, the National Library of Medicine (NLM) and the National Institutes of Health (NIH) joined forces to create an organization called the National Center for Biotechnology Information (NCBI), which today maintains a website (www.ncbi.nlm.nih.gov) with extensive resources useful for bioinformatics. On this site are links to databases, software, and a wealth of information about genomics and related topics. Similar websites have also been established by three genome centers with which the NCBI collaborates: the European Molecular Biology Laboratory, the DNA Data Bank of Japan, and BGI (formerly known as the Beijing Genomics Institute) in Shenzhen, China. These large, comprehensive websites are complemented by others maintained by individual or small groups of laboratories. Smaller websites often provide databases and software designed for a narrower purpose, such as studying genetic and genomic changes in one particular type of cancer.

The NCBI database of sequences is called GenBank. As of June 2016, it included the sequences of 194 million fragments of genomic DNA, totaling 213 billion base pairs! GenBank is constantly updated, and the amount of data it contains increases rapidly. Any sequence in the database can be retrieved and analyzed using software from the NCBI website or elsewhere.

One very widely used software program available on the NCBI website, called BLAST, allows the user to compare a DNA sequence with every sequence in GenBank, base by base. A researcher might search for similar regions in other genes of the same species or among the genes of other species. Another program allows comparison of protein sequences. Yet a third can search any protein sequence for *conserved* (common) stretches of amino acids (domains) for which a function is known or suspected, and it can show a three-dimensional model of the domain alongside other relevant information **(Figure 21.3)**. There is even a software program that can align and compare a collection of sequences, either nucleic acids or polypeptides, and diagram them in the form of an evolutionary tree based on the sequence relationships. (One such diagram is shown in Figure 21.17.)

Two research institutions, Rutgers University and the University of California, San Diego, also maintain a worldwide database of all three-dimensional protein structures that

Figure 21.3 Bioinformatics tools that are available on the Internet. A website maintained by the National Center for Biotechnology Information (NCBI) allows scientists and the public to access DNA and protein sequences and other stored data. The site includes a link to a protein structure database (Conserved Domain Database, CDD) that can find and describe similar domains in related proteins, as well as software (Cn3D, "See in 3-D") that displays models of domains. Some results are shown from a search for regions of proteins similar to an amino acid sequence in a muskmelon protein. The WD40 domain is very common in proteins encoded by eukaryotic genomes. It often plays a key role in molecular interactions during signal transduction.

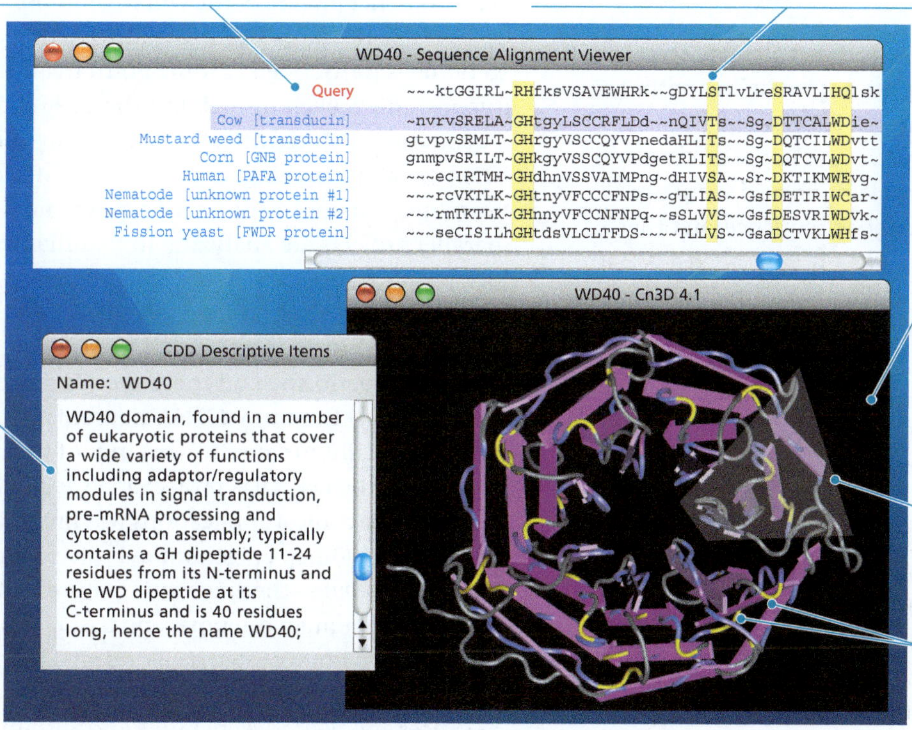

have been experimentally determined, called the Protein Data Bank (www.wwpdb.org). These structures can be rotated by the viewer to show all sides of the protein. Throughout this book, you'll find images of protein structures that have been obtained from the Protein Data Bank.

There is a vast array of resources available for researchers anywhere in the world to use free of charge. Let us now consider the types of questions scientists can address using these resources.

Instructors: BLAST Data Analysis Tutorials, which teach students how to work with real data from the BLAST database, can be assigned in MasteringBiology.

Identifying Protein-Coding Genes and Understanding Their Functions

Using available DNA sequences, geneticists can study genes directly, rather than taking the classical genetic approach, which requires determining the function of an unknown gene from the phenotype. But this more recent approach poses a new challenge: What does the gene actually do? Given a long DNA sequence from a database like GenBank, scientists aim to identify all protein-coding genes in the sequence and ultimately their functions. This process, called **gene annotation**, uses three lines of evidence to identify a gene.

First, computers are utilized in a search for patterns that indicate the presence of genes. The usual approach is to use software to scan the stored sequences for those that represent transcriptional and translational start and stop signals, RNA-splicing sites, and other telltale signs of protein-coding genes. The software also looks for certain short sequences that specify known mRNAs. Thousands of such sequences, called *expressed sequence tags*, or *ESTs*, have been collected from cDNA sequences and are cataloged in computer databases. This type of analysis identifies sequences that may turn out to be previously unknown protein-coding genes.

Although the identities of about half of the human genes were known before the Human Genome Project began, the other genes, previously unknown, were revealed by DNA sequence analysis. Once such suspected genes are identified, the second step is to obtain clues about their identities and functions by using software to compare their sequences with those of known genes from other organisms. Due to redundancy in the genetic code, the DNA sequence itself may vary more among species than the protein sequence does. Thus, scientists interested in proteins often compare the predicted amino acid sequence of a protein to that of other proteins. Third, the identities of these genes must then be confirmed by using RNA-seq (see Figure 20.13) or some other method to show that the relevant RNA is actually expressed from the proposed gene.

Animation: The Human Genome Project: Genes on Human Chromosome 17

Sometimes a newly identified sequence will match, at least partially, the sequence of a gene or protein in another species whose function is well known. For example, a plant researcher working on signaling pathways in the muskmelon would be excited to see that a partial amino acid sequence from a gene she had identified matched sequences in other species encoding a functional part of a protein called a WD40 domain (see Figure 21.3). WD40 domains are present in many eukaryotic proteins and are known to function in signal transduction pathways. Alternatively, a new gene sequence might be similar to a previously encountered sequence whose function is still unknown. Another possibility is that the sequence is entirely unlike anything ever seen before. This was true for about a third of the genes of *E. coli* when its genome was sequenced. In such cases, protein function is usually deduced through a combination of biochemical and functional studies. The biochemical approach aims to determine the three-dimensional structure of the protein as well as other attributes, such as potential binding sites for other molecules. Functional studies usually involve *knocking out* (blocking or disabling) the gene in an organism to see how the phenotype is affected. The CRISPR-Cas 9 system, described in Figure 20.14, is an example of an experimental technique used to block gene function.

Understanding Genes and Gene Expression at the Systems Level

The impressive computational power provided by the tools of bioinformatics allows the study of whole sets of genes and their interactions, as well as the comparison of genomes from different species. Genomics is a rich source of new insights into fundamental questions about genome organization, regulation of gene expression, embryonic development, and evolution.

One informative approach was taken by a long-term research project called ENCODE (Encyclopedia of DNA Elements), which ran from 2003 to 2012. The aim of the project was to learn everything possible about the functionally important elements in the human genome using multiple experimental techniques on different types of cultured cells. Investigators sought to identify protein-coding genes and genes for non-coding RNAs, along with sequences that regulate gene expression, such as enhancers and promoters. In addition, they extensively characterized DNA and histone modifications and chromatin structure—features termed "epigenetic" since they affect gene expression without changing the sequence of nucleotide bases (see Concept 18.3). The second phase of the project, involving more than 440 scientists in 32 research groups, culminated in 2012 with the simultaneous publication of 30 papers describing over 1,600 large data sets. The considerable power of this project is that it provides the opportunity to compare results from specific projects with each other, yielding a much richer picture of the whole genome.

Perhaps the most striking finding is that about 75% of the genome is transcribed at some point in at least one of the cell types studied, even though less than 2% codes for proteins. Furthermore, biochemical functions have been assigned to DNA elements making up at least 80% of the genome. To learn more about the different types of functional elements, parallel projects are analyzing in a similar way the genomes of two model organisms, the soil nematode *Caenorhabditis elegans* and the fruit fly *Drosophila melanogaster*. Because genetic and biochemical experiments using DNA technology can be performed on these species, testing the activities of potentially functional DNA elements in their genomes is expected to illuminate the workings of the human genome.

Because the ENCODE project analyzed cells in culture, its potential for clinical applications was limited. A related project called the Roadmap Epigenomics Project set out to characterize the *epigenome*—the epigenetic features of the genome—of hundreds of human cell types and tissues. The aim was to focus on the epigenomes of stem cells, normal tissues from mature adults, and relevant tissues from individuals with diseases such as cancer and neurodegenerative and autoimmune disorders. In 2015, a series of papers reported on the results from 111 tissues. One of the most useful findings was that the original tissue in which a cancer arose can be identified in cells of a secondary tumor based on characterization of their epigenomes.

Systems Biology

The scientific progress resulting from sequencing genomes and studying large sets of genes has encouraged scientists to attempt similar systematic studies of sets of proteins and their properties (such as their abundance, chemical modifications, and interactions), an approach called **proteomics**. (A **proteome** is the entire set of proteins expressed by a cell or group of cells.) Proteins, not the genes that encode them, carry out most of the activities of the cell. Therefore, we must study when and where proteins are produced in an organism, as well as how they interact in networks, if we are to understand the functioning of cells and organisms.

Genomics and proteomics enable molecular biologists to approach the study of life from an increasingly global perspective. Using the tools we have described, biologists have begun to compile catalogs of genes and proteins— listings of all the "parts" that contribute to the operation of cells, tissues, and organisms. With such catalogs in hand, researchers have shifted their attention from the individual parts—the genes and proteins—to their functional integration in biological systems. As you may recall, Concept 1.1 discussed this approach, called **systems biology**, which aims to model the dynamic behavior of whole biological systems based on the study of the interactions among the system's parts. Because of the vast amounts of data generated in these types of studies, advances in computer technology and bioinformatics are crucial to studying systems biology.

One important use of the systems biology approach is to define gene and protein interaction networks. To map the protein interaction network in the yeast *Saccharomyces cerevisiae,* for instance, researchers used sophisticated techniques to knock out pairs of genes, one pair at a time, creating doubly mutant cells. They then compared the fitness of each double mutant (based in part on the size of the cell colony it formed) to that predicted from the fitness of each of the two single mutants. The researchers reasoned that if the observed fitness matched the prediction, then the products of the two genes didn't interact with each other, but if the observed fitness was greater or less than predicted, then the gene products interacted in the cell. They then used computer software to build a graphic model by "mapping" the gene products to certain locations in the model, based on the similarity of their interactions. This resulted in the network-like "functional map" of protein interactions shown in **Figure 21.4**. Processing the vast number of protein-protein interactions generated by this experiment and integrating them into the completed map required powerful computers, mathematical tools, and newly developed software.

Application of Systems Biology to Medicine

The Cancer Genome Atlas is another example of systems biology in which many interacting genes and gene products are analyzed together, as a group. This project, under the joint leadership of the National Cancer Institute and the NIH, aims to determine how changes in biological systems lead to cancer. A three-year pilot project that ended in 2010 set out to find all the common mutations in three types of cancer—lung cancer, ovarian cancer, and glioblastoma of the brain—by comparing gene sequences and patterns of gene expression in cancer cells with those in normal cells. Work on glioblastoma confirmed the role of several suspected genes and identified a few previously unknown ones, suggesting possible new targets for therapies. The approach proved so fruitful for these three types of cancer that it has been extended to ten other types, chosen because they are common and often lethal in humans.

As high-throughput techniques become more rapid and less expensive, they are being increasingly applied to the problem of cancer, as seen with the Roadmap Epigenomics Project described earlier. Rather than sequencing only protein-coding genes, sequencing the whole genomes of many tumors of a particular type allows scientists to uncover common chromosomal abnormalities, as well as any other consistent changes in these aberrant genomes.

Figure 21.4 The systems biology approach to protein interactions. This global protein interaction map shows the likely interactions (lines) among about 4,500 gene products (dots) in *Saccharomyces cerevisiae,* the budding yeast. Dots of the same color represent gene products involved in one of the 13 similarly colored cellular functions listed around the map. The white dots represent proteins that haven't been assigned to any color-coded function. The expanded area shows additional details of one map region where the gene products (blue dots) carry out amino acid biosynthesis, uptake, and related functions.

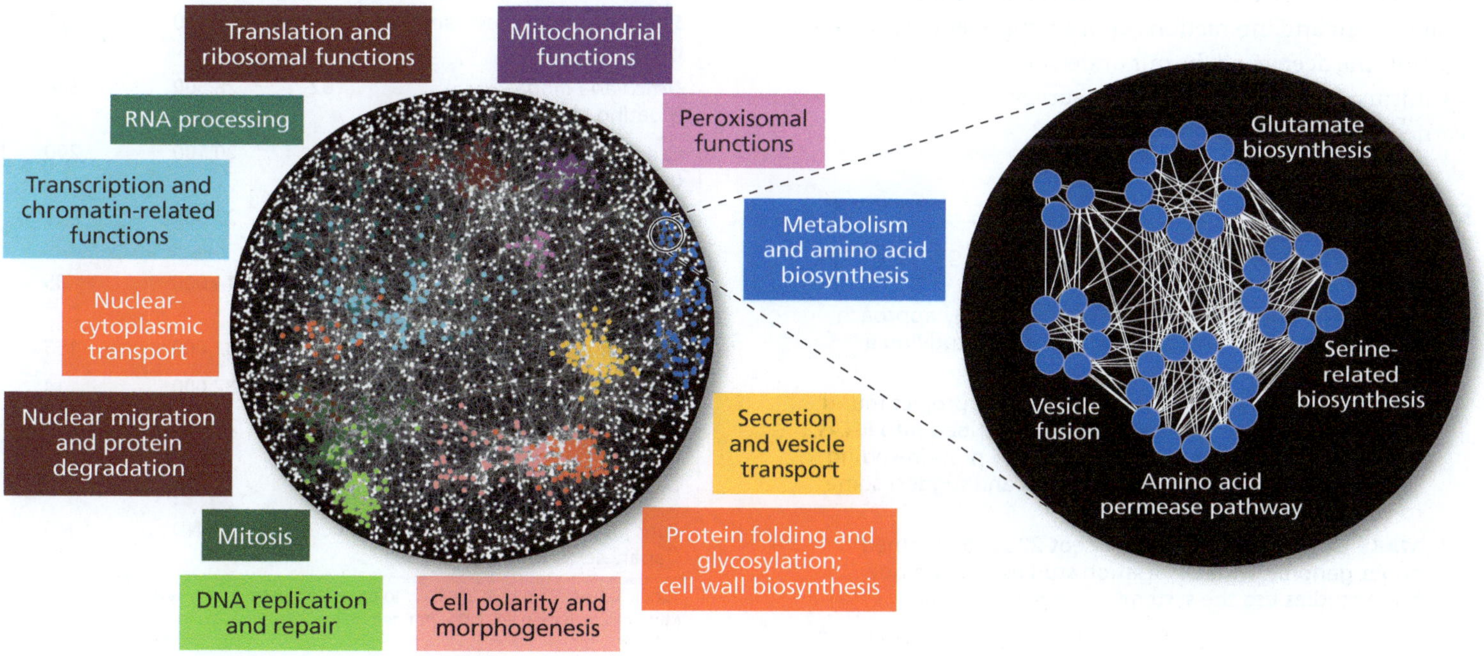

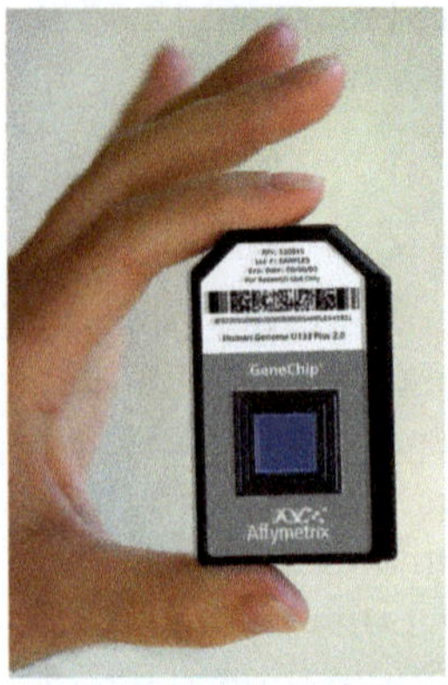

Figure 21.5 A human gene microarray chip. Tiny spots of DNA arranged in a grid on this silicon wafer represent almost all of the genes in the human genome. Using this chip, researchers can analyze expression patterns for all these genes at the same time (see Figure 20.12).

HHMI Animation: Gene Chip Manufacturing

In addition to whole-genome sequencing, silicon and glass "chips" that hold a microarray of most of the known human genes are now used to analyze gene expression patterns in patients suffering from various cancers and other diseases **(Figure 21.5)**. Increasingly, RNA-seq (see Figure 20.13) is replacing microarray analysis. Analyzing which genes are over- or underexpressed in a particular cancer may allow physicians to tailor patients' treatment to their unique genetic makeup and the specifics of their cancers. This approach has been used to characterize subsets of particular cancers, enabling more refined treatments. Breast cancer is one example (see Figure 18.27).

Ultimately, medical records may include an individual's DNA sequence, a sort of genetic bar code, with regions highlighted that predispose the person to specific diseases. The use of such sequences for personalized medicine—disease prevention and treatment—has great potential.

Systems biology is a very efficient way to study emergent properties at the molecular level. Novel properties arise at each successive level of biological complexity as a result of the arrangement of building blocks at the underlying level (see Concept 1.1). The more we can learn about the arrangement and interactions of the components of genetic systems, the deeper will be our understanding of whole organisms. The rest of this chapter surveys what we've learned from genomic studies.

CONCEPT CHECK 21.2

1. What role does the Internet play in current genomics and proteomics research?
2. Explain the advantage of the systems biology approach to studying cancer versus the approach of studying a single gene at a time.
3. **MAKE CONNECTIONS** ➤ The ENCODE pilot project found that at least 75% of the genome is transcribed into RNAs, far more than could be accounted for by protein-coding genes. Review Concepts 17.3 and 18.3 and suggest some roles that these RNAs might play.
4. **MAKE CONNECTIONS** ➤ In Concept 20.2, you learned about genome-wide association studies. Explain how these studies use the systems biology approach.

For suggested answers, see Appendix A.

CONCEPT 21.3

Genomes vary in size, number of genes, and gene density

The sequences of thousands of genomes have been completed, with tens of thousands of genomes either in progress or considered permanent drafts (because they require more work than it would be worth to complete them). Among the sequences in progress are roughly 3,400 metagenomes. In the completely sequenced group, about 5,000 are genomes of bacteria, and 242 are archaeal genomes. There are 283 completed eukaryotic species, along with 2,635 permanent drafts. Among these are vertebrates, invertebrates, protists, fungi, and plants. Next, we'll discuss what we've learned about genome size, number of genes, and gene density, focusing on general trends.

Genome Size

Comparing the three domains (Bacteria, Archaea, and Eukarya), we find a general difference in genome size between prokaryotes and eukaryotes **(Table 21.1)**. While there are

Table 21.1 Genome Sizes and Estimated Numbers of Genes*

Organism	Haploid Genome Size (Mb)	Number of Genes	Genes per Mb
Bacteria			
Haemophilus influenzae	1.8	1,700	940
Escherichia coli	4.6	4,400	950
Archaea			
Archaeoglobus fulgidus	2.2	2,500	1,130
Methanosarcina barkeri	4.8	3,600	750
Eukaryotes			
Saccharomyces cerevisiae (yeast, a fungus)	12	6,300	525
Urticularia gibba (floating bladderwort)	82	28,500	348
Caenorhabditis elegans (nematode)	100	20,100	200
Arabidopsis thaliana (mustard family plant)	120	27,000	225
Drosophila melanogaster (fruit fly)	165	14,000	85
Daphnia pulex (water flea)	200	31,000	155
Zea mays (corn)	2,300	32,000	14
Ailuropoda melanoleuca (giant panda)	2,400	21,000	9
Homo sapiens (human)	3,000	$<21,000$	7
Paris japonica (Japanese canopy plant)	149,000	ND	ND

*Some values given here are likely to be revised as genome analysis continues. Mb = million base pairs. ND = not determined.

some exceptions, most bacterial genomes have between 1 and 6 million base pairs (Mb); the genome of *E. coli*, for instance, has 4.6 Mb. Genomes of archaea are, for the most part, within the size range of bacterial genomes. (Keep in mind, however, that many fewer archaeal genomes have been completely sequenced, so this picture may change.) Eukaryotic genomes tend to be larger: The genome of the single-celled yeast *Saccharomyces cerevisiae* (a fungus) has about 12 Mb, while most animals and plants, which are multicellular, have genomes of at least 100 Mb. There are 165 Mb in the fruit fly genome, while humans have 3,000 Mb, about 500 to 3,000 times as many as a typical bacterium.

Aside from this general difference between prokaryotes and eukaryotes, a comparison of genome sizes among eukaryotes fails to reveal any systematic relationship between genome size and the organism's phenotype. For instance, the genome of *Paris japonica*, the Japanese canopy plant, contains 149 billion base pairs (149,000 Mb), while that of another plant, *Urticularia gibba*, a bladderwort, contains only 82 Mb. Even more striking, there is a single-celled amoeba, *Polychaos dubium*, whose genome size has been estimated at 670 billion base pairs (670,000 Mb). (This genome has not yet been sequenced.) On a finer scale, comparing two insect species, the cricket (*Anabrus simplex*) genome turns out to have 11 times as many base pairs as the fruit fly (*Drosophila melanogaster*) genome. There is a wide range of genome sizes within the groups of unicellular eukaryotes, insects, amphibians, and plants, and less of a range within mammals and reptiles.

Number of Genes

The number of genes also varies between prokaryotes and eukaryotes: Bacteria and archaea, in general, have fewer genes than eukaryotes. Free-living bacteria and archaea have from 1,500 to 7,500 genes, while the number of genes in eukaryotes ranges from about 5,000 for unicellular fungi (yeasts) to at least 40,000 for some multicellular eukaryotes.

Within the eukaryotes, the number of genes in a species is often lower than expected from considering simply the size of its genome. Looking at Table 21.1, you can see that the genome of the nematode *C. elegans* is 100 Mb in size and contains roughly 20,100 genes. In comparison, the genome of *Drosophila melanogaster* is much bigger (165 Mb) but has only about two-thirds the number of genes—14,000 genes.

Considering an example closer to home, we noted that the human genome contains 3,000 Mb, well over ten times the size of either the *D. melanogaster* or *C. elegans* genome. At the outset of the Human Genome Project, biologists expected somewhere between 50,000 and 100,000 genes to be identified in the completed sequence, based on the number of known human proteins. As the project progressed, the estimate was revised downward several times, and the ENCODE project discussed above has established the number to be fewer than 21,000. This relatively low number, similar to the number of genes in the nematode *C. elegans*, surprised biologists, who had been expecting many more human genes.

Interview with Eric Lander: Exploring the Human Genome

What genetic attributes allow humans (and other vertebrates) to get by with no more genes than nematodes? An important factor is that vertebrate genomes "get more bang for the buck" from their coding sequences because of extensive alternative splicing of RNA transcripts. Recall that this process generates more than one polypeptide from a single gene (see Figure 18.13). A typical human gene contains about ten exons, and an estimated 90% or more of these multi-exon genes are spliced in at least two different ways. Some genes are expressed in hundreds of alternatively spliced forms, others in just two. Scientists have not yet catalogued all of the different forms, but it is clear that the number of different proteins encoded in the human genome far exceeds the proposed number of genes.

Additional polypeptide diversity could result from post-translational modifications such as cleavage or the addition of carbohydrate groups in different cell types or at different developmental stages. Finally, the discovery of miRNAs and other small RNAs that play regulatory roles has added a new variable to the mix (see Concept 18.3). Some scientists think that this added level of regulation, when present, may contribute to greater organismal complexity for some genes.

Gene Density and Noncoding DNA

We can take both genome size and number of genes into account by comparing gene density in different species. In other words, we can ask: How many genes are in a given length of DNA? When we compare the genomes of bacteria, archaea, and eukaryotes, we see that eukaryotes generally have larger genomes but fewer genes in a given number of base pairs. Humans have hundreds or thousands of times as many base pairs in their genome as most bacteria, as we already noted, but only 5 to 15 times as many genes; thus, gene density is lower in humans (see Table 21.1). Even unicellular eukaryotes, such as yeasts, have fewer genes per million base pairs than bacteria and archaea. Among the genomes that have been sequenced completely, humans and other mammals have the lowest gene density.

In all bacterial genomes studied so far, most of the DNA consists of genes for protein, tRNA, or rRNA; the small amount remaining consists mainly of nontranscribed regulatory sequences, such as promoters. The sequence of nucleotides along a bacterial protein-coding gene is not interrupted by noncoding sequences (introns). In eukaryotic genomes, by contrast, most of the DNA neither encodes protein nor is transcribed into RNA molecules of known function, and the DNA includes more complex regulatory sequences. In fact, humans have 10,000 times as much noncoding DNA as bacteria. Some of this DNA in multicellular eukaryotes is present as introns within genes. Indeed, introns account for most of the difference in average length between human genes (27,000 base pairs) and bacterial genes (1,000 base pairs).

In addition to introns, multicellular eukaryotes have a vast amount of non-protein-coding DNA between genes. In the next section, we will describe the composition and arrangement of these great stretches of DNA in the human genome.

CONCEPT CHECK 21.3

1. The best estimate is that the human genome contains fewer than 21,000 genes. However, there is evidence that human cells produce many more than 21,000 different polypeptides. What processes might account for this discrepancy?
2. The Genomes Online Database (GOLD) website of the Joint Genome Institute has information about genome sequencing projects. Scroll through the page at https://gold.jgi.doe.gov/statistics and describe the information you find. What percent of bacterial genome projects have medical relevance?
3. **WHAT IF?** ➤ What evolutionary processes might account for prokaryotes having smaller genomes than eukaryotes?

For suggested answers, see Appendix A.

CONCEPT 21.4

Multicellular eukaryotes have a lot of noncoding DNA and many multigene families

We have spent most of this chapter, and indeed this unit, focusing on genes that code for proteins. Yet the coding regions of these genes and the genes for noncoding RNA products such as rRNA, tRNA, and miRNA make up only a small portion of the genomes of most multicellular eukaryotes. For example, once the sequencing of the human genome was completed, it became clear that only a tiny part—about 1.5%—codes for proteins or is transcribed into rRNAs or tRNAs. **Figure 21.6** shows what is known about the makeup of the remaining 98.5% of the genome.

Gene-related regulatory sequences and introns account, respectively, for 5% and about 20% of the human genome. The rest, located between functional genes, includes some unique (single-copy) noncoding DNA, such as gene fragments and **pseudogenes**, former genes that have accumulated mutations over a long time and no longer produce functional proteins. (The genes that produce small noncoding RNAs are a tiny percentage of the genome, distributed between the 20% introns and the 15% unique noncoding DNA.) Most of the DNA between functional genes, however, is **repetitive DNA**, which consists of sequences that are present in multiple copies in the genome.

▼ Figure 21.6 Types of DNA sequences in the human genome. The gene sequences that code for proteins or are transcribed into rRNA or tRNA molecules make up only about 1.5% of the human genome (dark purple in the pie chart), while introns and regulatory sequences associated with genes (light purple) make up about a quarter. The vast majority of the human genome does not code for proteins (although much of it gives rise to RNAs), and a large amount is repetitive DNA (dark and light green and teal).

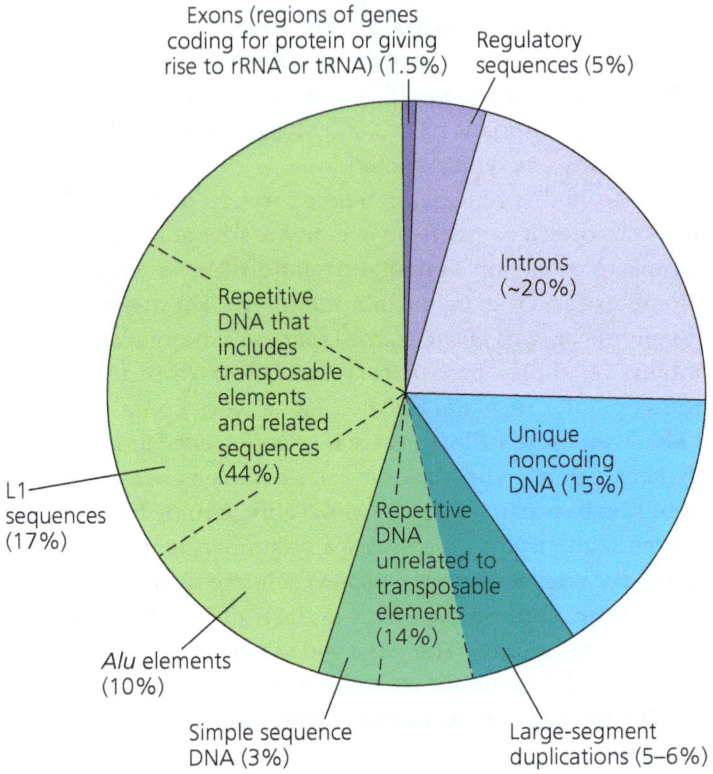

The bulk of many eukaryotic genomes consists of DNA sequences that neither code for proteins nor are transcribed to produce RNAs with known functions; this noncoding DNA was often described in the past as "junk DNA." However, genome comparisons over the past 10 years have revealed the persistence of this DNA in diverse genomes over many hundreds of generations. For example, the genomes of humans, rats, and mice contain almost 500 regions of noncoding DNA that are *identical* in sequence in all three species. This is a higher level of sequence conservation than is seen for protein-coding regions in these species, strongly suggesting that the noncoding regions have important functions. The results of the ENCODE project discussed earlier have underscored the key roles played by much of this noncoding DNA. In the next few pages, we examine how genes and noncoding DNA sequences are organized within genomes of multicellular eukaryotes, using the human genome as our main example. Genome organization tells us a lot about how genomes have evolved and continue to evolve, as we'll discuss in Concept 21.5.

Transposable Elements and Related Sequences

Both prokaryotes and eukaryotes have stretches of DNA that can move from one location to another within the genome. These stretches are known as *transposable genetic elements*, or simply **transposable elements**. During the process called *transposition*, a transposable element moves from one site in a cell's DNA to a different target site by a type of recombination process. Transposable elements are sometimes called "jumping genes," but actually they never completely detach from the cell's DNA. Instead, the original and new DNA sites are brought very close together by enzymes and other proteins that bend the DNA. Surprisingly, about 75% of human repetitive DNA (44% of the entire human genome) is made up of transposable elements and sequences related to them.

The first evidence for wandering DNA segments came from American geneticist Barbara McClintock's breeding experiments with Indian corn (maize) in the 1940s and 1950s **(Figure 21.7)**. As she tracked corn plants through multiple generations, McClintock identified changes in the color of corn kernels that made sense only if she postulated the existence of genetic elements capable of moving from other locations in the genome into the genes for kernel color, disrupting the genes so that the kernel color was changed. McClintock's discovery was met with great skepticism and virtually discounted at the time. Her careful work and insightful ideas were finally validated many years later when transposable elements were found in bacteria. In 1983, at the age of 81, McClintock received the Nobel Prize for her pioneering research.

Movement of Transposons and Retrotransposons

Eukaryotic transposable elements are of two types. The first type, **transposons**, move within a genome by means of a

◀ Figure 21.7 The effect of transposable elements on corn kernel color. Barbara McClintock first proposed the idea of mobile genetic elements after observing variegations in the color of the kernels on a corn cob (top right).

▼ Figure 21.8 Transposon movement. Movement of transposons by either the copy-and-paste mechanism (shown here) or the cut-and-paste mechanism involves a double-stranded DNA intermediate that is inserted into the genome.

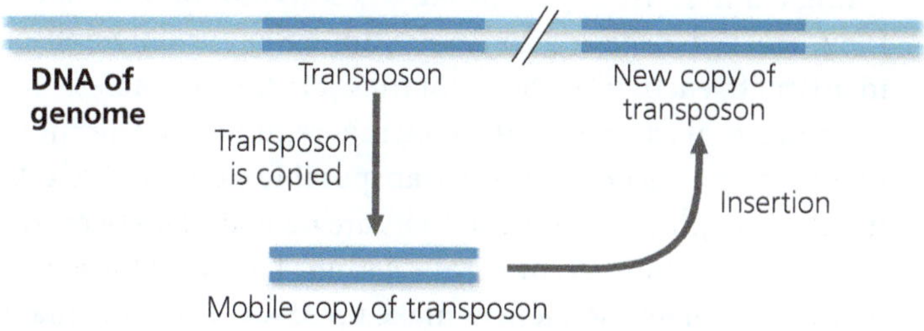

VISUAL SKILLS ▶ *How would this figure differ if it showed the cut-and-paste mechanism?*

MB Interview with Virginia Walbot: Plant genetics and development

DNA intermediate. Transposons can move by a "cut-and-paste" mechanism, which removes the element from the original site, or by a "copy-and-paste" mechanism, which leaves a copy behind **(Figure 21.8)**. Both mechanisms require an enzyme called *transposase*, which is generally encoded by the transposon.

Most transposable elements in eukaryotic genomes are of the second type, **retrotransposons**, which move by means of an RNA intermediate that is a transcript of the retrotransposon DNA. Thus, retrotransposons always leave a copy at the original site during transposition **(Figure 21.9)**. To insert at another site, the RNA intermediate is first converted back to DNA by reverse transcriptase, an enzyme encoded by the retrotransposon. (Reverse transcriptase is also encoded by retroviruses, as you learned in Concept 19.2. In fact, retroviruses may have evolved from retrotransposons.) Another cellular enzyme catalyzes insertion of the reverse-transcribed DNA at a new site.

▼ Figure 21.9 Retrotransposon movement. Movement begins with synthesis of a single-stranded RNA intermediate. The remaining steps are essentially identical to part of the retrovirus replicative cycle (see Figure 19.8).

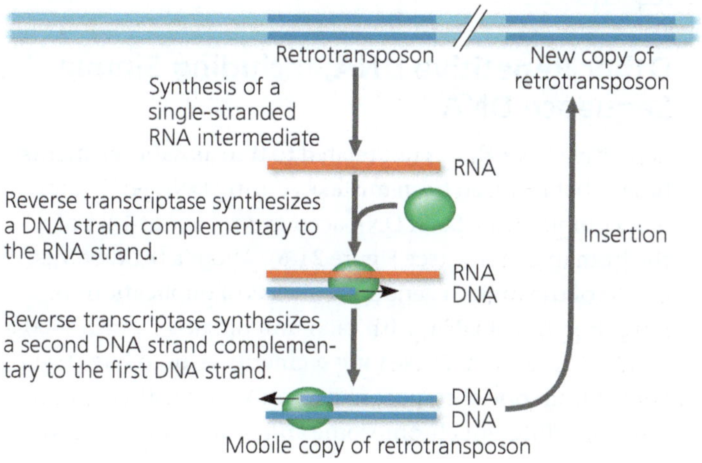

Sequences Related to Transposable Elements

Multiple copies of transposable elements and sequences related to them are scattered throughout eukaryotic genomes. A single unit is usually hundreds to thousands of base pairs long, and the dispersed copies are similar but usually not identical to each other. Some of these are transposable elements that can move; the enzymes required for this movement may be encoded by any transposable element, including the one that is moving. Others are related sequences that have lost the ability to move altogether. Transposable elements and related sequences make up 25–50% of most mammalian genomes (see Figure 21.6) and even higher percentages in amphibians and many plants. In fact, the very large size of some plant genomes is accounted for by extra transposable elements rather than by extra genes. For example, transposable elements make up 85% of the corn genome!

In humans and other primates, a large portion of transposable element-related DNA consists of a family of similar sequences called *Alu elements*. These sequences alone account for approximately 10% of the human genome. *Alu* elements are about 300 nucleotides long, much shorter than most functional transposable elements, and they do not code for any protein. However, many *Alu* elements are transcribed into RNA, and at least some of these RNAs are thought to help regulate gene expression.

An even larger percentage (17%) of the human genome is made up of a type of retrotransposon called *LINE-1*, or *L1*. These sequences are much longer than *Alu* elements—about 6,500 base pairs—and typically have a very low rate of transposition. However, researchers working with rats have found L1 retrotransposons to be more active in cells of the developing brain. They have proposed that different effects on gene expression of L1 retrotransposition in developing neurons may contribute to the great diversity of neuronal cell types (see Concept 48.1).

Although many transposable elements encode proteins, these proteins do not carry out normal cellular functions. Therefore, transposable elements are usually included in the "noncoding" DNA category, along with other repetitive sequences.

Other Repetitive DNA, Including Simple Sequence DNA

Repetitive DNA that is not related to transposable elements has probably arisen from mistakes during DNA replication or recombination. Such DNA accounts for about 14% of the human genome (see Figure 21.6). About a third of this (5–6% of the human genome) consists of duplications of long stretches of DNA, with each unit ranging from 10,000 to 300,000 base pairs. These long segments seem to have been copied from one chromosomal location to another site on the same or a different chromosome and probably include some functional genes.

In contrast to scattered copies of long sequences, stretches of DNA known as **simple sequence DNA** contain many copies of tandemly repeated short sequences, as in the following example (showing one DNA strand only):

. . . GTTACGTTACGTTACGTTACGTTACGTTAC . . .

In this case, the repeated unit (GTTAC) consists of 5 nucleotides. Repeated units may contain as many as 500 nucleotides, but often contain fewer than 15 nucleotides, as in this example. When the unit contains 2–5 nucleotides, the series of repeats is called a **short tandem repeat**, or **STR**; we discussed the use of STR analysis in preparing genetic profiles in Concept 20.4 (see Figure 20.24). The number of copies of the repeated unit can vary from site to site within a given genome. There could be as many as several hundred thousand repetitions of the GTTAC unit at one site, but only half that number at another. STR analysis is performed on sites selected because they have relatively few repeats. The repeat number varies from person to person, and since humans are diploid, each person has two alleles per site, which can differ in repeat number. This diversity produces the variation represented in the genetic profiles that result from STR analysis. Altogether, simple sequence DNA makes up 3% of the human genome.

Much of a genome's simple sequence DNA is located at chromosomal telomeres and centromeres, suggesting that this DNA plays a structural role for chromosomes. The DNA at centromeres is essential for the separation of chromatids in cell division (see Concept 12.2). Centromeric DNA, along with simple sequence DNA located elsewhere, may also help organize the chromatin within the interphase nucleus. The simple sequence DNA located at telomeres—the tips of chromosomes—prevents genes from being lost as the DNA shortens with each round of replication (see Concept 16.2). Telomeric DNA also binds proteins that protect the ends of a chromosome from degradation and from joining to other chromosomes.

Short repetitive sequences like those described here provide a challenge for whole-genome shotgun sequencing because the presence of many short repeats hinders accurate reassembly of fragment sequences by computers. Regions of simple sequence DNA account for much of the uncertainty present in estimates of whole-genome sizes and are the reason some sequences are considered "permanent drafts."

Genes and Multigene Families

We finish our discussion of the various types of DNA sequences in eukaryotic genomes with a closer look at genes. Recall that DNA sequences that code for proteins or give rise to tRNA or rRNA compose a mere 1.5% of the human genome (see Figure 21.6). If we include introns and regulatory sequences associated with genes, the total amount of DNA that is gene-related—coding and noncoding—constitutes about 25% of the human genome. Put another way, only about 6% (1.5% out of 25%) of the length of the average gene is represented in the final gene product.

Like the genes of bacteria, many eukaryotic genes are present as unique sequences, with only one copy per haploid set of chromosomes. But in the human genome and the genomes of many other animals and plants, these unique genes make up less than half of the total gene-related DNA. The rest occur in **multigene families**, collections of two or more identical or very similar genes.

In multigene families that consist of *identical* DNA sequences, those sequences are usually clustered tandemly and, with the notable exception of the genes for histone proteins, have RNAs as their final products. An example is the family of identical DNA sequences that each include the genes for the three largest rRNA molecules **(Figure 21.10a)**. These rRNA molecules are transcribed from a single transcription unit that is repeated tandemly hundreds to thousands of times in one or several clusters in the genome of a multicellular eukaryote. The many copies of this rRNA transcription unit help cells to quickly make the millions of ribosomes needed for active protein synthesis. The primary transcript is cleaved to yield three rRNA molecules, which combine with proteins and one other kind of rRNA (5*S* rRNA) to form ribosomal subunits.

The classic examples of multigene families of *nonidentical* genes are two related families of genes that encode globins, a group of proteins that include the α and β polypeptide subunits of hemoglobin. One family, located on chromosome 16 in humans, encodes various forms of α-globin; the other, on chromosome 11, encodes forms of β-globin **(Figure 21.10b)**. The different forms of each globin subunit are expressed at different times in development, allowing hemoglobin to function effectively in the changing environment of the developing animal. In humans, for example, the embryonic and fetal forms of hemoglobin have a higher affinity for oxygen than the adult forms, ensuring the efficient transfer of oxygen from mother to fetus. Also found in the globin gene family clusters are several pseudogenes.

In Concept 21.5, we'll consider the evolution of these two globin gene families as we explore how arrangements of genes provide insight into the evolution of genomes. We'll also examine some processes that have shaped the genomes of different species over evolutionary time.

CONCEPT CHECK 21.4

1. Discuss the characteristics of mammalian genomes that make them larger than prokaryotic genomes.
2. **VISUAL SKILLS ➤** Which of the three mechanisms described in Figures 21.8 and 21.9 result(s) in a copy remaining at the original site as well as a copy appearing in a new location?
3. Contrast the organizations of the rRNA gene family and the globin gene families. For each, explain how the existence of a family of genes benefits the organism.
4. **MAKE CONNECTIONS ➤** Assign each DNA segment at the top of Figure 18.8 to a sector in the pie chart in Figure 21.6.

For suggested answers, see Appendix A.

▼ Figure 21.10 Gene families.

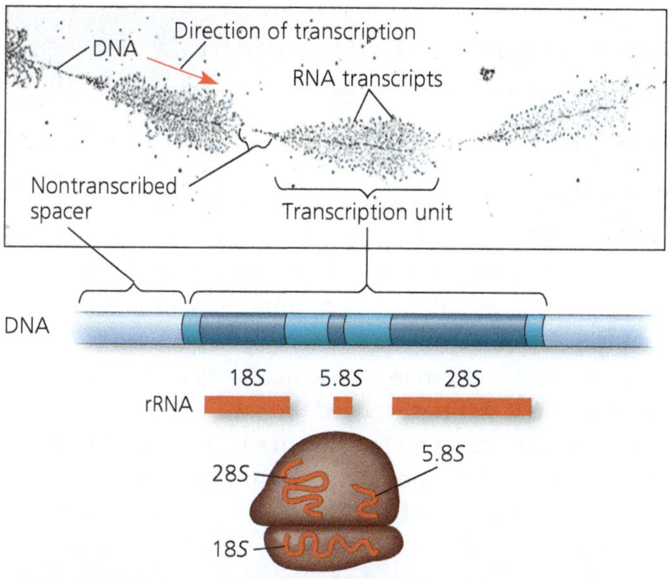

(a) Part of the ribosomal RNA gene family. The TEM at the top shows three of the hundreds of copies of rRNA transcription units in the rRNA gene family of a salamander genome. Each "feather" corresponds to a single unit being transcribed by about 100 molecules of RNA polymerase (dark dots along the DNA), moving left to right (red arrow). The growing RNA transcripts extend from the DNA, accounting for the feather-like appearance. In the diagram of a transcription unit below the TEM, the genes for three types of rRNA (darker blue) are adjacent to regions that are transcribed but later removed (medium blue). A single transcript is processed to yield one of each of the three rRNAs (red), key components of the ribosome.

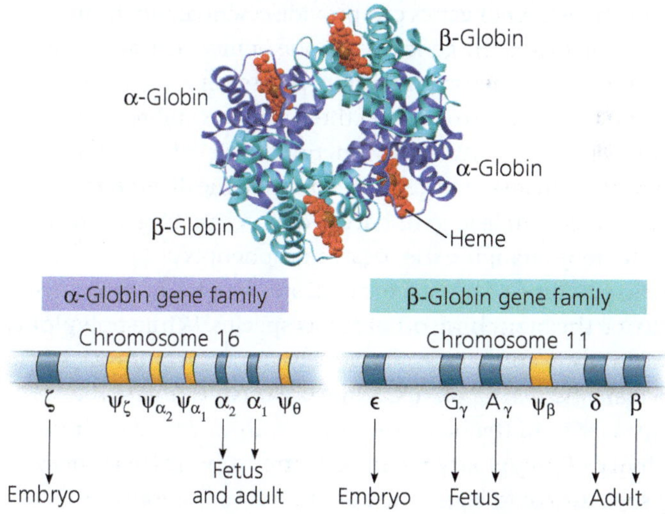

(b) The human α-globin and β-globin gene families. Adult hemoglobin is composed of two α-globin and two β-globin polypeptide subunits, as shown in the molecular model. The genes (darker blue) encoding α- and β-globins are found in two families, organized as shown here. The noncoding DNA (light blue) separating the functional genes within each family includes pseudogenes (ψ; gold), versions of the functional genes that no longer encode functional polypeptides. Genes and pseudogenes are named with Greek letters, as you have seen previously for the α- and β-globins. Some genes are expressed only in the embryo or fetus.

VISUAL SKILLS ➤ *In the TEM at the top of part (a), how could you determine the direction of transcription if it weren't indicated by the red arrow?*

CONCEPT 21.5

Duplication, rearrangement, and mutation of DNA contribute to genome evolution

EVOLUTION Now that we have explored the makeup of the human genome, let's see what its composition reveals about how the genome evolved. The basis of change at the genomic level is mutation, which underlies much of genome evolution. It seems likely that the earliest forms of life had a minimal number of genes—those necessary for survival and reproduction. If this were indeed the case, one aspect of evolution must have been an increase in the size of the genome, with the extra genetic material providing the raw material for gene diversification. In this section, we'll first describe how extra copies of all or part of a genome can arise and then consider subsequent processes that can lead to the evolution of proteins (or RNA products) with slightly different or entirely new functions.

Duplication of Entire Chromosome Sets

An accident in meiosis, such as failure to separate homologs during meiosis I, can result in one or more extra sets of chromosomes, a condition known as polyploidy (see Concept 15.4). Although such accidents would most often be lethal, in rare cases they could facilitate the evolution of genes. In a polyploid organism, one set of genes can provide essential functions for the organism. The genes in the one or more extra sets can diverge by accumulating mutations; these variations may persist if the organism carrying them survives and reproduces. In this way, genes with novel functions can evolve. As long as one copy of an essential gene is expressed, the divergence of another copy can lead to its encoded protein acting in a novel way, thereby changing the organism's phenotype.

The outcome of this accumulation of mutations may eventually be the branching off of a new species. While polyploidy is rare among animals, it is relatively common among plants, especially flowering plants. Some botanists estimate that as many as 80% of the plant species that are alive today show evidence of polyploidy having occurred among their ancestral species. You'll learn more about how polyploidy leads to plant speciation in Concept 24.2.

Alterations of Chromosome Structure

With the recent explosion in genomic sequence information, we can now compare the chromosomal organizations of many different species in detail. This information allows us to make inferences about the evolutionary processes that shape chromosomes and may drive speciation. For example, scientists have long known that sometime in the last 6 million years, when the ancestors of humans and chimpanzees diverged as species, the fusion of two ancestral chromosomes in the human line led to different haploid numbers for humans ($n = 23$) and chimpanzees ($n = 24$). The banding patterns in stained chromosomes suggested that the ancestral versions of current chimpanzee chromosomes 12 and 13 fused end to end, forming chromosome 2 in an ancestor of the human lineage. Sequencing and analysis of human chromosome 2 during the Human Genome Project provided very strong supporting evidence for the model we have just described **(Figure 21.11)**.

Figure 21.11 Human and chimpanzee chromosomes. The positions of telomere-like and centromere-like sequences on human chromosome 2 (left) match those of telomeres on chimpanzee chromosomes 12 and 13 and the centromere on chimpanzee chromosome 13 (right). This suggests that chromosomes 12 and 13 in a human ancestor fused end to end to form human chromosome 2. The centromere from ancestral chromosome 12 remained functional on human chromosome 2, while the one from ancestral chromosome 13 did not.

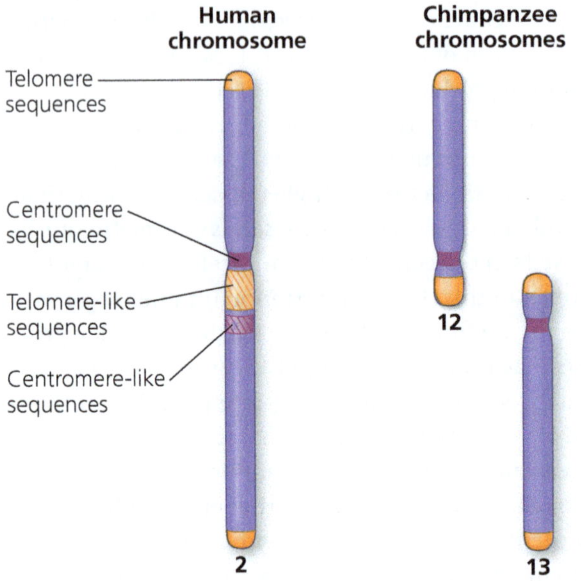

In another study of broader scope, researchers compared the DNA sequence of each human chromosome with the whole-genome sequence of the mouse **(Figure 21.12)**. One part of their study showed that large blocks of genes on human chromosome 16 are found on four mouse chromosomes, indicating that the genes in each block stayed together in both the mouse and the human lineages during their divergent evolution from a common ancestor.

Performing the same comparison of chromosomes of humans and six other mammalian species allowed the researchers to reconstruct the evolutionary history of chromosomal rearrangements in these eight species. They found many duplications and inversions of large portions of chromosomes, the result of errors during meiotic recombination in which the DNA was broken and rejoined incorrectly. The rate of these events seems to have begun accelerating about 100 million years ago, around 35 million years before large

Figure 21.12 Human and mouse chromosomes. Here, we can see that DNA sequences very similar to large blocks of human chromosome 16 (colored areas in this diagram) are found on mouse chromosomes 7, 8, 16, and 17. This finding suggests that the DNA sequence in each block has stayed together in the mouse and human lineages since the time they diverged from a common ancestor.

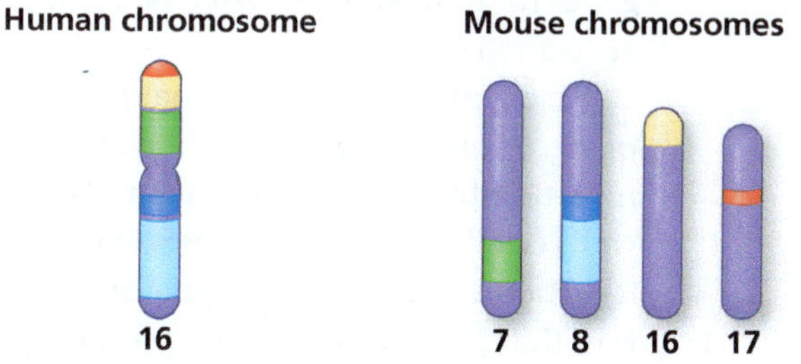

dinosaurs became extinct and the number of mammalian species began rapidly increasing. The apparent coincidence is interesting because chromosomal rearrangements are thought to contribute to the generation of new species. Although two individuals with different arrangements could still mate and produce offspring, the offspring would have two nonequivalent sets of chromosomes, making meiosis inefficient or even impossible. Thus, chromosomal rearrangements would lead to two populations that could not successfully mate with each other, a step on the way to their becoming two separate species. (You'll learn more about this in Concept 24.2.)

The same study also unearthed a pattern with medical relevance. Analysis of the chromosomal breakage points associated with the rearrangements showed that specific sites were used over and over again. A number of these recombination "hot spots" correspond to locations of chromosomal rearrangements within the human genome that are associated with congenital diseases (see Concept 15.4).

Duplication and Divergence of Gene-Sized Regions of DNA

Errors during meiosis can also lead to the duplication of chromosomal regions that are smaller than the ones we've just discussed, including segments the length of individual genes. Unequal crossing over during prophase I of meiosis, for instance, can result in one chromosome with a deletion and another with a duplication of a particular gene. Transposable elements can provide homologous sites where nonsister chromatids can cross over, even when other chromatid regions are not correctly aligned **(Figure 21.13)**.

Also, slippage can occur during DNA replication, such that the template shifts with respect to the new complementary strand, and a part of the template strand is either skipped by the replication machinery or used twice as a template. As a result, a segment of DNA is deleted or duplicated. It is easy to imagine how such errors could occur in regions of repeats. The variable number of repeated units of simple sequence

Figure 21.13 Gene duplication due to unequal crossing over. One mechanism by which a gene (or other DNA segment) can be duplicated is recombination during meiosis between copies of a transposable element (yellow) flanking the gene (blue). Such recombination between misaligned nonsister chromatids of homologous chromosomes produces one chromatid with two copies of the gene and one chromatid with no copy. (Genes and transposable elements are shown only in the region of interest.)

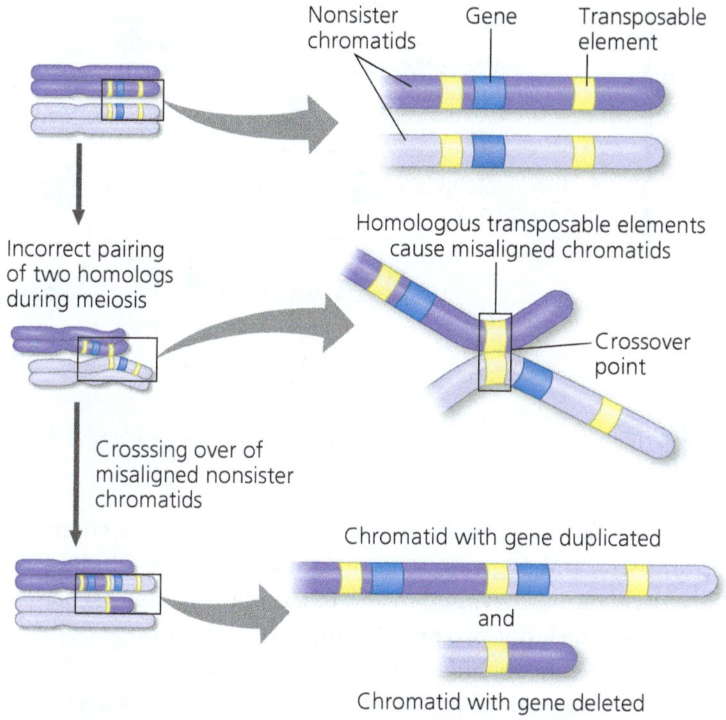

MAKE CONNECTIONS *Examine how crossing over occurs in Figure 13.9. In the middle panel above, draw a line through the portions that result in the upper chromatid in the bottom panel. Use a different color to do the same for the other chromatid.*

DNA at a given site, used for STR analysis, is probably due to errors like these. Evidence that unequal crossing over and template slippage during DNA replication lead to duplication of genes is found in the existence of multigene families, such as the globin family.

Evolution of Genes with Related Functions: The Human Globin Genes

In Figure 21.10b, you saw the organization of the α-globin and β-globin gene families as they exist in the human genome today. Now, let's consider how events such as duplications can lead to the evolution of genes with related functions like the globin genes. A comparison of gene sequences within a multigene family can suggest the order in which the genes arose. Re-creating the evolutionary history of the globin genes using this approach indicates that they all evolved from one common ancestral globin gene that underwent duplication and divergence into the α-globin and β-globin ancestral genes about 450–500 million years ago. Each of these genes was later duplicated several times, and the copies then diverged from each other in sequence, yielding the

Figure 21.14 A proposed model for the sequence of events in the evolution of the human α-globin and β-globin gene families from a single ancestral globin gene.

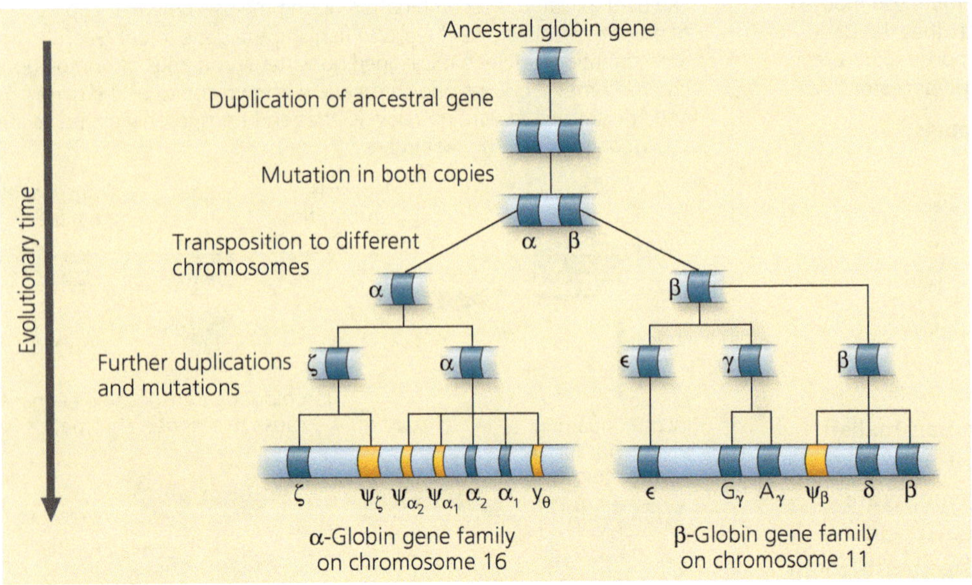

? *The gold elements are pseudogenes. Explain how they could have arisen after gene duplication.*

current family members **(Figure 21.14)**. In fact, the common ancestral globin gene also gave rise to the oxygen-binding muscle protein myoglobin and to the plant protein leghemoglobin. The latter two proteins function as monomers, and their genes are included in a "globin superfamily."

After the duplication events, the differences between the genes in the globin families undoubtedly arose from mutations that accumulated in the gene copies over many generations. The current model is that the necessary function provided by an α-globin protein, for example, was fulfilled by one gene, while other copies of the α-globin gene accumulated random mutations. Many mutations may have had an adverse effect on the organism, and others may have had no effect. However, a few mutations must have altered the function of the protein product in a way that benefitted the organism at a particular life stage without substantially changing the protein's oxygen-carrying function. Presumably, natural selection acted on these altered genes, maintaining them in the population.

In the **Scientific Skills Exercise**, you can compare amino acid sequences of the globin family proteins and see how such comparisons were used to generate the model for globin gene evolution shown in Figure 21.14. The existence of several pseudogenes among the functional globin genes provides additional evidence for this model: Random mutations in these "genes" over evolutionary time have destroyed their function.

Evolution of Genes with Novel Functions

In the evolution of the globin gene families, gene duplication and subsequent divergence produced family members whose protein products performed functions similar to each other (oxygen transport). However, an alternative scenario is that one copy of a duplicated gene can undergo alterations that lead to a completely new function for the protein product. The genes for lysozyme and α-lactalbumin are a good example.

Lysozyme is an enzyme that helps protect animals against bacterial infection by hydrolyzing bacterial cell walls (see Visualizing Figure 5.16); α-lactalbumin is a nonenzymatic protein that plays a role in milk production in mammals. The two proteins are quite similar in their amino acid sequences and three-dimensional structures **(Figure 21.15)**. Both genes are found in mammals, but only the lysozyme gene is present in birds. These findings suggest that at some time after the lineages leading to mammals and birds had separated, the lysozyme gene was duplicated in the mammalian lineage but not in the avian lineage. Subsequently, one copy of the duplicated lysozyme gene evolved into a gene encoding α-lactalbumin, a protein with a completely new function associated with a key characteristic of mammals—milk production. In a recent study, evolutionary biologists searched vertebrate genomes for genes with similar sequences. There appear to be at least eight members of the lysozyme family, distributed widely among mammalian species. The functions of all the encoded gene products are not yet known, but it will be exciting to discover whether they are as different as the functions of lysozyme and α-lactalbumin.

Besides the duplication and divergence of whole genes, rearrangement of existing DNA sequences within genes has also contributed to genome evolution. The presence of introns may have promoted the evolution of new proteins by facilitating the duplication or shuffling of exons, as we'll discuss next.

Rearrangements of Parts of Genes: Exon Duplication and Exon Shuffling

Recall from Concept 17.3 that an exon often codes for a protein domain, a distinct structural and functional region of a protein molecule. We've already seen that unequal crossing over during meiosis can lead to duplication of a gene on one chromosome and its loss from the homologous chromosome (see Figure 21.13). By a similar process, a particular exon within a gene could be duplicated on one chromosome and deleted from the other. The gene with the duplicated exon would code for a protein containing a second copy

Figure 21.15 Comparison of lysozyme and α-lactalbumin proteins. Computer-generated ribbon models of the similar structures of **(a)** lysozyme and **(b)** α-lactalbumin are shown, along with **(c)** a comparison of the amino acid sequences of the two proteins. The amino acids are arranged in groups of ten for ease of reading, and single-letter amino acid codes are used (see Figure 5.14). Identical amino acids are highlighted in yellow, and dashes indicate gaps in one sequence that have been introduced by the software to optimize the alignment.

MAKE CONNECTIONS ➤ *Even though two amino acids are not identical, they may be structurally and chemically similar and therefore behave similarly. Using Figure 5.14 as a reference, examine the nonidentical amino acids in positions 1–30 and note cases where the amino acids in the two sequences are similarly acidic or basic.*

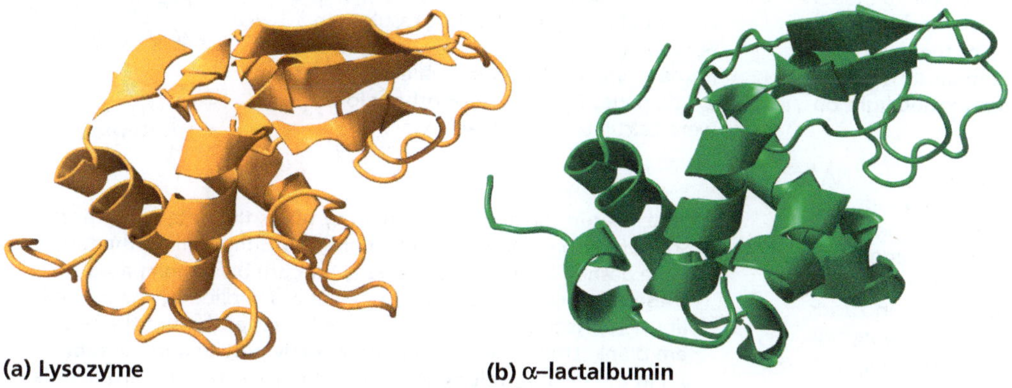

(a) Lysozyme **(b) α–lactalbumin**

Lysozyme	**1**	KVFERCELAR	TLKRLGMDGY	RGISLANWMC	LAKWESGYNT	RATNYNAGDR
α–lactalbumin	**1**	KQFTKCELSQ	LLK--DIDGY	GGIALPELIC	TMFHTSGYDT	QAIVENN--E
Lysozyme	**51**	STDYGIFQIN	SRYWCNDGKT	PGAVNACHLS	CSALLQDNIA	DAVACAKRVV
α–lactalbumin	**51**	STEYGLFQIS	NKLWCKSSQV	PQSRNICDIS	CDKFLDDDIT	DDIMCAKKIL
Lysozyme	**101**	RDPQGIRAWV	AWRNRCQ-NR	DVRQYVQGCG	V	
α–lactalbumin	**101**	D-IKGIDYWL	AHKALCT--E	KLEQWLCEKL	-	

(c) Amino acid sequence alignments of lysozyme and α–lactalbumin

of the encoded domain. This change in the protein's structure might augment its function by increasing its stability, enhancing its ability to bind a particular ligand, or altering some other property. Quite a few protein-coding genes have multiple copies of related exons, which presumably arose by duplication and then diverged. The gene encoding the extracellular matrix protein collagen is a good example. Collagen is a structural protein (see Figure 5.18) with a highly repetitive amino acid sequence, which reflects the repetitive pattern of exons in the collagen gene.

As an alternative possibility, we can imagine the occasional mixing and matching of different exons either within a gene or between two different (nonallelic) genes owing to errors in meiotic recombination. This process, termed *exon shuffling*, could lead to new proteins with novel combinations of functions. As an example, let's consider the gene for tissue plasminogen activator (TPA). The TPA protein is an extracellular protein that helps control blood clotting. It has four domains of three types, each encoded by an exon, and one of those exons is present in two copies. Because each type of exon is also found in other proteins, the current version of the gene for TPA is thought to have arisen by several instances of exon shuffling during errors in meiotic recombination and subsequent duplication **(Figure 21.16)**.

Figure 21.16 Evolution of a new gene by exon shuffling. Meiotic errors could have moved exons, each encoding a particular domain, from ancestral forms of the genes for epidermal growth factor, fibronectin, and plasminogen (left) into the evolving gene for tissue plasminogen activator, TPA (right). Subsequent duplication of the "kringle" exon (K) from the plasminogen gene after its movement into the TPA gene could account for the two copies of this exon in the TPA gene existing today.

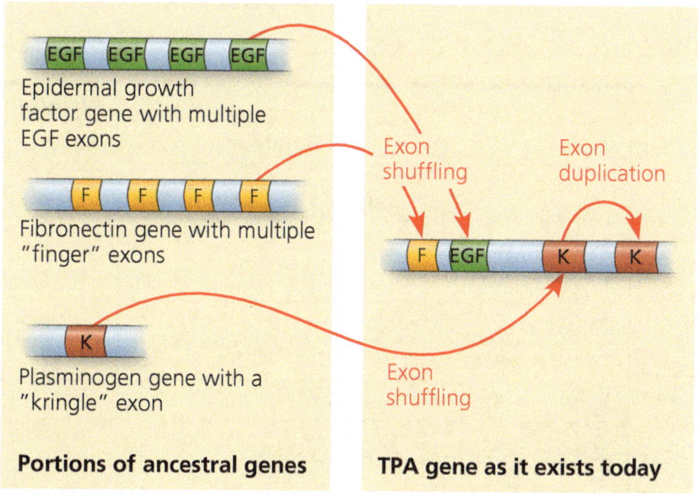

VISUAL SKILLS ➤ *Looking at Figure 21.13, describe the steps by which transposable elements within introns might have facilitated the exon shuffling shown here.*

SCIENTIFIC SKILLS EXERCISE

Reading an Amino Acid Sequence Identity Table

How Have Amino Acid Sequences of Human Globin Genes Diverged During Their Evolution? To build a model of the evolutionary history of the globin genes (see Figure 21.14), researchers compared the amino acid sequences of the polypeptides they encode. In this exercise, you will analyze comparisons of the amino acid sequences of the globin polypeptides to shed light on their evolutionary relationships.

How the Experiment Was Done Scientists obtained the DNA sequences for each of the eight globin genes and "translated" them into amino acid sequences. They then used a computer program to align the sequences (with dashes indicating gaps in one sequence) and calculate a percent identity value for each pair of globins. The percent identity reflects the number of positions with identical amino acids relative to the total number of amino acids in a globin polypeptide. The data were displayed in a table to show the pairwise comparisons.

Data from the Experiment The following table shows an example of a pairwise alignment—that of the α_1-globin (alpha-1 globin) and ζ-globin (zeta globin) amino acid sequences—using the standard single-letter symbols for amino acids. To the left of each line of amino acid sequence is the number of the first amino acid in that line. The percent identity value for the α_1- and ζ-globin amino acid sequences was calculated by counting the number of matching amino acids (86, highlighted in yellow), dividing by the total number of amino acid positions (143), and then multiplying by 100. This resulted in a 60% identity value for the α_1-ζ pair, as shown in the amino acid identity table at the bottom of the page. The values for other globin pairs were calculated in the same way.

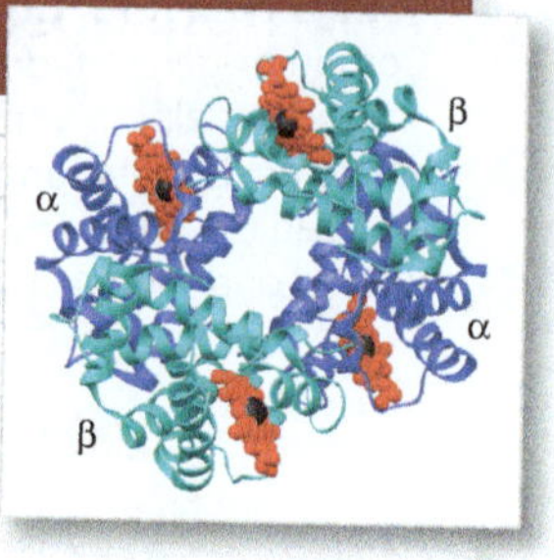

Hemoglobin

Globin	Alignment of Globin Amino Acid Sequences
α_1	1 MVLSPADKTNVKAAWGKVGAHAGEYGAEAL
ζ	1 MSLTKTERTIIVSMWAKISTQADTIGTETL
α_1	31 ERMFLSFPTTKTYFPHFDLSH-GSAQVKGH
ζ	31 ERLFLSHPQTKTYFPHFDL-HPGSAQLRAH
α_1	61 GKKVADALTNAVAHVDDMPNALSALSDLHA
ζ	61 GSKVVAAVGDAVKSIDDIGGALSKLSELHA
α_1	91 HKLRVDPVNFKLLSHCLLVTLAAHLPAEFT
ζ	91 YILRVDPVNFKLLSHCLLVTLAARFPADFT
α_1	121 PAVHASLDKFLASVSTVLTSKYR
ζ	121 AEAHAAWDKFLSVVSSVLTEKYR

INTERPRET THE DATA

1. Note that in the amino acid identity table, the data are arranged so each globin pair can be compared. (a) Some cells in the table have dashed lines. Given the pairs that are being compared for these cells, what percent identity value is implied by the dashed lines? (b) Notice that the cells in the lower left half of the table are blank. Using the information already provided in the table, fill in the missing values. Why does it make sense that these cells were left blank in the table?
2. The earlier that two genes arose from a duplicated gene, the more their nucleotide sequences can have diverged, which may result in amino acid differences in the protein products. (a) Based on that premise, identify which two genes are most divergent from each other. What is the percent amino acid identity between their polypeptides? (b) Using the same approach, identify which two globin genes are the most recently duplicated. What is the percent identity between them?
3. The model of globin gene evolution shown in Figure 21.14 suggests that an ancestral gene duplicated and mutated to become α- and β-globin genes, and then each one was further duplicated and mutated. What features of the data set support the model?
4. Make an ordered list of all the percent identity values from the table, starting with 100% at the top. Next to each number write the globin pair(s) with that percent identity value. Use one color for the globins from the α family and a different color for the globins from the β family. (a) Compare the order of pairs on your list with their positions in the model shown in Figure 21.14. Does the order of pairs describe the same relative "closeness" of globin family members seen in the model? (b) Compare the percent identity values for pairs within the α or β group to the values for between-group pairs.

MB **Instructors:** A version of this Scientific Skills Exercise can be assigned in MasteringBiology.

Further Reading R. C. Hardison, Globin genes on the move, *Journal of Biology* 7:35.1–35.5 (2008).

Amino Acid Identity Table

		α Family			β Family				
		α_1 (alpha 1)	α_2 (alpha 2)	ζ (zeta)	β (beta)	δ (delta)	ϵ (epsilon)	A_γ (gamma A)	G_γ (gamma G)
α Family	α_1	-----	100	60	45	44	39	42	42
	α_2		-----	60	45	44	39	42	42
	ζ			-----	38	40	41	41	41
β Family	β				-----	93	76	73	73
	δ					-----	73	71	72
	ϵ						-----	80	80
	A_γ							-----	99
	G_γ								-----

Compiled using data from the National Center for Biotechnology Information (NCBI).

How Transposable Elements Contribute to Genome Evolution

The persistence of transposable elements as a large fraction of some eukaryotic genomes is consistent with the idea that they play an important role in shaping a genome over evolutionary time. These elements can contribute to the evolution of the genome in several ways. They can promote recombination, disrupt cellular genes or control elements, and carry entire genes or individual exons to new locations.

Transposable elements of similar sequence scattered throughout the genome facilitate recombination between different (nonhomologous) chromosomes by providing homologous regions for crossing over (see Figure 21.13). Most such recombination events are probably detrimental, causing chromosomal translocations and other changes in the genome that may be lethal to the organism. But over the course of evolutionary time, an occasional recombination event of this sort may be advantageous to the organism. (For the change to be heritable, of course, it must happen in a cell that will give rise to a gamete.)

The movement of a transposable element can have a variety of consequences. For instance, a transposable element that "jumps" into a protein-coding sequence will prevent the production of a normal transcript of the gene. (Introns provide a sort of "safety zone" that does not affect the transcript because the transposable element will be spliced out.) If a transposable element inserts within a regulatory sequence, the transposition may lead to increased or decreased production of one or more proteins. Transposition caused both types of effects on the genes coding for pigment-synthesizing enzymes in McClintock's corn kernels. Again, while such changes are usually harmful, in the long run some may provide a survival advantage. A possible example was mentioned earlier: At least some of the *Alu* transposable elements in the human genome are known to produce RNAs that regulate expression of human genes.

During transposition, a transposable element may carry along a gene or even a group of genes to a new position in the genome. This mechanism probably accounts for the location of the α-globin and β-globin gene families on different human chromosomes, as well as the dispersion of the genes of certain other gene families. By a similar tag-along process, an exon from one gene may be inserted into another gene in a mechanism similar to that of exon shuffling during recombination. For example, an exon may be inserted by transposition into the intron of a protein-coding gene. If the inserted exon is retained in the RNA transcript during RNA splicing, the protein that is synthesized will have an additional domain, which may confer a new function on the protein.

Most often, the processes discussed in this section produce harmful effects, which may be lethal, or have no effect at all. In a few cases, however, small heritable changes that are beneficial may occur. Over many generations, the resulting genetic diversity provides valuable raw material for natural selection. Diversification of genes and their products is an important factor in the evolution of new species. Thus, the accumulation of changes in the genome of each species provides a record of its evolutionary history. To read this record, we must be able to identify genomic changes. Comparing the genomes of different species allows us to do that, increasing our understanding of how genomes evolve. You will learn more about these topics next.

CONCEPT CHECK 21.5

1. Describe three examples of errors in cellular processes that lead to DNA duplications.
2. Explain how multiple exons might have arisen in the ancestral EGF and fibronectin genes shown on the left side of Figure 21.16.
3. What are three ways that transposable elements are thought to contribute to genome evolution?
4. **WHAT IF?** ➤ In 2005, Icelandic scientists reported finding a large chromosomal inversion present in 20% of northern Europeans, and they noted that Icelandic women with this inversion had significantly more children than women without it. What would you expect to happen to the frequency of this inversion in the Icelandic population in future generations?

For suggested answers, see Appendix A.

CONCEPT 21.6

Comparing genome sequences provides clues to evolution and development

EVOLUTION One researcher has likened the current state of biology to the Age of Exploration in the 1400s, which occurred soon after major improvements in navigation and ship design. In the last 30 years, we have seen rapid advances in genome sequencing and data collection, new techniques for assessing gene activity across the whole genome, and refined approaches for understanding how genes and their products work together in complex systems. In the field of biology, we are truly poised on the brink of a new world.

Comparisons of genome sequences from different species reveal a lot about the evolutionary history of life, from very ancient to more recent. Similarly, comparative studies of the genetic programs that direct embryonic development in different species are beginning to clarify the mechanisms that generated the great diversity of life-forms present today. In this final section of the chapter, we will discuss what has been learned from these two approaches.

Comparing Genomes

The more similar in sequence the genes and genomes of two species are, the more closely related those species are in their evolutionary history—because not enough time has passed for mutations and other changes to accumulate. Comparing genomes of closely related species sheds light on more recent evolutionary events, whereas comparing genomes of very distantly related species helps us understand ancient evolutionary history. In either case, learning about characteristics that are shared or divergent between groups enhances our picture of the evolution of organisms and biological processes. As you learned in Concept 1.2, the evolutionary relationships between species can be represented by a diagram in the form of a tree, where each branch point marks the divergence of two lineages. **Figure 21.17** shows the evolutionary relationships of some groups and species we will discuss.

Comparing Distantly Related Species

Determining which genes have remained similar—that is, are *highly conserved*—in distantly related species can help clarify evolutionary relationships among species that diverged from each other long ago. Indeed, comparisons of the specific gene sequences of bacteria, archaea, and eukaryotes indicate that these three groups diverged between 2 and 4 billion years ago and strongly support the theory that they are the fundamental domains of life (see Figure 21.17).

In addition to their value in evolutionary biology, comparative genomic studies confirm the relevance of research on model organisms to our understanding of biology in general and human biology in particular. Very ancient genes can still be surprisingly similar in disparate species. A 2015 study tested the ability of the human version of each of 414 important yeast genes to function equivalently in yeast cells. Remarkably, the researchers concluded that 47% of these yeast genes could be replaced by the human gene. This striking result underscores the common origin of yeasts and humans—two distantly related species.

▼ Figure 21.17 Evolutionary relationships of the three domains of life. The tree diagram at the top shows the ancient divergence of bacteria, archaea, and eukaryotes. A portion of the eukaryote lineage is expanded in the inset to show the more recent divergence of three mammalian species discussed in this chapter.

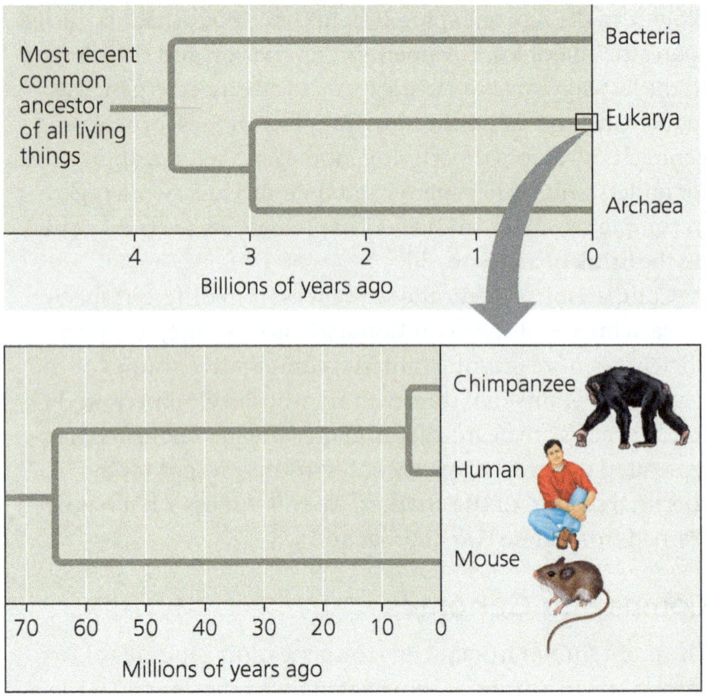

Comparing Closely Related Species

The genomes of two closely related species are likely to be organized similarly because of their relatively recent divergence. Their long shared history also means that only a small number of gene differences are found when their genomes are compared. These genetic differences can thus be more easily correlated with phenotypic differences between the two species. An exciting application of this type of analysis is seen as researchers compare the human genome with the genomes of the chimpanzee, mouse, rat, and other mammals. Identifying the genes shared by all of these species but not by nonmammals gives us clues about what it takes to make a mammal, while finding the genes shared by chimpanzees and humans but not by rodents tells us something about primates. And, of course, comparing the human genome with that of the chimpanzee helps us answer the tantalizing question we asked at the beginning of the chapter: What genomic information defines a human or a chimpanzee?

An analysis of the overall composition of the human and chimpanzee genomes, which are thought to have diverged only about 6 million years ago (see Figure 21.17), reveals some general differences. Considering single nucleotide substitutions, the two genomes differ by only 1.2%. When researchers looked at longer stretches of DNA, however, they were surprised to find a further 2.7% difference due to insertions or deletions of larger regions in the genome of one or the other species; many of the insertions were duplications or other repetitive DNA. In fact, a third of the human duplications are not present in the chimpanzee genome, and some of these duplications contain regions associated with human diseases. There are more *Alu* elements in the human genome than in the chimpanzee genome, and the latter contains many copies of a retroviral provirus not present in humans. All of these observations provide clues to the forces that might have swept the two genomes along different paths, but we don't have a complete picture yet.

Along with chimpanzees, bonobos are the other African ape species that are the closest living relatives to humans. The sequencing of the bonobo genome, completed in 2012, revealed that in some regions, human sequences were more closely related to either chimpanzee or bonobo sequences than chimpanzee or bonobo sequences were to each other. Such a fine-grained comparison of three closely related species allows even more detail to be worked out in reconstructing their related evolutionary history.

We also don't know how the genetic differences revealed by genome sequencing might account for the distinct characteristics of each species. To discover the basis for the phenotypic differences between chimpanzees and humans, biologists are studying specific genes and types of genes that differ between the two species and comparing them with their counterparts in other mammals. This approach has revealed a number of genes that are apparently changing (evolving) faster in the human than in either the chimpanzee or the mouse. Among them are genes involved in defense against malaria and tuberculosis as well as at least one gene that regulates brain size. When genes are classified by function, the genes that seem to be evolving the fastest are those that code for transcription factors. This discovery makes sense because transcription factors regulate gene expression and thus play a key role in orchestrating the overall genetic program.

One transcription factor whose gene shows evidence of rapid change in the human lineage is called FOXP2 (**Figure 21.18**). Several lines of evidence suggest that the *FOXP2* gene product regulates genes that function in

▼ Figure 21.18

Inquiry What is the function of a gene (*FOXP2*) that is rapidly evolving in the human lineage?

Experiment Several lines of evidence support a role for the *FOXP2* gene in the development of speech and language in humans and of vocalization in other vertebrates. In 2005, Joseph Buxbaum and collaborators at the Mount Sinai School of Medicine and several other institutions tested the function of *FOXP2*. They used the mouse, a model organism in which genes can be easily knocked out, as a representative vertebrate that vocalizes: Mice produce ultrasonic squeaks (whistles) to communicate stress. The researchers used genetic engineering to produce mice in which one or both copies of *FOXP2* were disrupted.

Wild type: two normal copies of *FOXP2*

Heterozygote: one copy of *FOXP2* disrupted

Homozygote: both copies of *FOXP2* disrupted

They then compared the phenotypes of these mice. Two of the characters they examined are included here: brain anatomy and vocalization.

Experiment 1: Researchers cut thin sections of brain and stained them with reagents that allow visualization of brain anatomy in a UV fluorescence microscope.

Experiment 2: To induce stress, researchers separated each newborn mouse pup from its mother and recorded the number of ultrasonic whistles produced by the pup.

Results

Experiment 1 Results: Disruption of both copies of *FOXP2* led to brain abnormalities in which the cells were disorganized. Phenotypic effects on the brain of heterozygotes, with one disrupted copy, were less severe. (Each color in the micrographs below reveals a different cell or tissue type.)

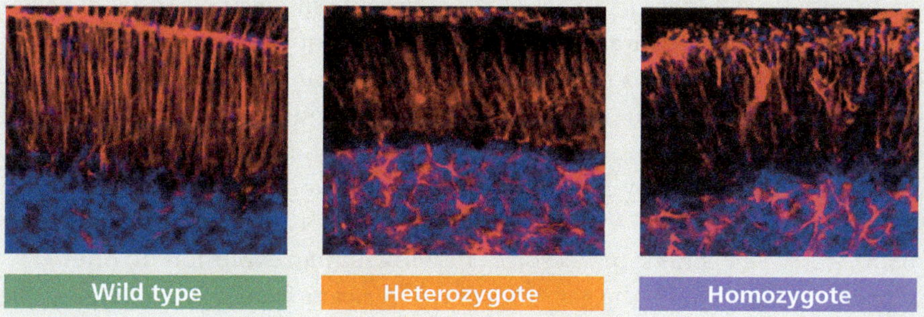

Experiment 2 Results: Disruption of both copies of *FOXP2* led to an absence of ultrasonic vocalization in response to stress. The effect on vocalization in the heterozygote was also extreme.

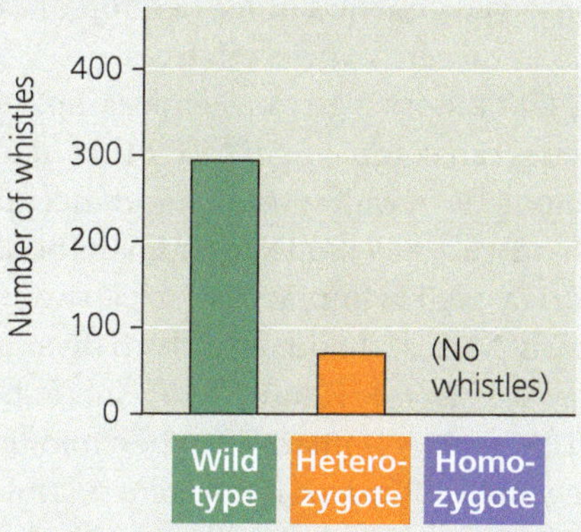

Conclusion *FOXP2* plays a significant role in the development of functional communication systems in mice. The results augment evidence from studies of birds and humans, supporting the hypothesis that *FOXP2* may act similarly in diverse organisms.

Data from W. Shu et al., Altered ultrasonic vocalization in mice with a disruption in the *Foxp2* gene, *Proceedings of the National Academy of Sciences USA* 102:9643–9648 (2005).

WHAT IF? ➤ *Since the results support a role for mouse FOXP2 in vocalization, you might wonder whether the human FOXP2 protein is a key regulator of speech. If you were given the amino acid sequences of wild-type and mutant human FOXP2 proteins and the wild-type chimpanzee FOXP2 protein, how would you investigate this question? What further clues could you obtain by comparing these sequences to that of the mouse FOXP2 protein?*

vocalization in vertebrates. First, mutations in this gene can produce severe speech and language impairment in humans. Moreover, the *FOXP2* gene is expressed in the brains of zebra finches and canaries at the time when these songbirds are learning their songs. But perhaps the strongest evidence comes from a "knockout" experiment in which researchers disrupted the *FOXP2* gene in mice and analyzed the resulting phenotype (see Figure 21.18). The homozygous mutant mice had malformed brains and failed to emit normal ultrasonic vocalizations, and mice with one faulty copy of the gene also showed significant problems with vocalization. These results support the idea that the *FOXP2* gene product turns on genes involved in vocalization.

Expanding on this analysis, another research group more recently replaced the *FOXP2* gene in mice with a "humanized" copy coding for the human versions of two amino acids that differ between human and chimpanzee; these are the changes potentially responsible for a human's ability to speak. Although the mice were generally healthy, they had subtly different vocalizations and showed changes in brain cells in circuits associated with speech in human brains.

In 2010, the Neanderthal genome was sequenced from a very small amount of preserved genomic DNA, and a high-quality sequence was completed in 2014. Neanderthals (*Homo neanderthalensis*) are members of the same genus to which humans (*Homo sapiens*) belong (see Concept 34.7). A reconstruction of their evolutionary history based on genomic comparisons between the two species suggests that some groups of humans and Neanderthals co-existed and interbred for a period of time before Neanderthals went extinct about 30,000 years ago. While Neanderthals have sometimes been portrayed as primitive beings that could only grunt, their *FOXP2* gene sequence encodes a protein identical to that of humans. This suggests that Neanderthals may have been capable of speech of some type and, along with other observed genetic similarities, forces us to reevaluate our image of our recent extinct relatives.

The *FOXP2* story is an excellent example of how different approaches can complement each other in uncovering biological phenomena of widespread importance. The *FOXP2* experiments used mice as a model for humans because it would be unethical (as well as impractical) to carry out such experiments in humans. Mice and humans, which diverged about 65.5 million years ago (see Figure 21.17), share about 85% of their genes. This genetic similarity can be exploited in studying human genetic disorders. If researchers know the organ or tissue that is affected by a particular genetic disorder, they can look for genes that are expressed in these locations in mice.

Even though more distantly related to humans, fruit flies have also been a useful model species for study of such human disorders as Parkinson's disease and alcoholism, while nematodes (soil worms) have yielded a wealth of information about aging. Further research efforts are under way to extend genomic studies to many more species, including neglected species from diverse branches of the tree of life. These studies will advance our understanding of evolution, of course, as well as all aspects of biology, from human health to ecology.

Comparing Genomes Within a Species

Another exciting consequence of our ability to analyze genomes is our growing understanding of the spectrum of genetic variation in humans. Because the history of the human species is so short—probably about 200,000 years—the amount of DNA variation among humans is small compared to that of many other species. Much of our diversity seems to be in the form of single nucleotide polymorphisms (SNPs). SNPs are single base-pair sites where genetic variation is found in at least 1% of the population (see Concept 20.2); they are usually detected by DNA sequencing. In the human genome, SNPs occur on average about once in 100–300 base pairs. Scientists have already identified the locations of several million SNP sites in the human genome and continue to find additional SNPs. These are stored in databases around the world, one of which is run by the National Center for Biotechnology Information (NCBI) and can be accessed at http://www.ncbi.nlm.nih.gov/SNP/.

In the course of this search, they have also found other variations—including chromosomal regions with inversions, deletions, and duplications. The most surprising discovery has been the widespread occurrence of *copy-number variants* (*CNVs*), loci where some individuals have one or multiple copies of a particular gene or genetic region rather than the standard two copies (one on each homolog). CNVs result from regions of the genome being duplicated or deleted inconsistently within the population. One study of 40 people found more than 8,000 CNVs involving 13% of the genes in the genome, and these CNVs probably represent just a small subset of the total. Since these variants encompass much longer stretches of DNA than the single nucleotides of SNPs, CNVs are more likely to have phenotypic consequences and to play a role in complex diseases and disorders. At the very least, the high incidence of copy-number variation blurs the meaning of the phrase "a normal human genome."

Copy-number variants, SNPs, and variations in repetitive DNA such as short tandem repeats (STRs) are useful genetic markers for studying human evolution. In one study, the genomes of two Africans from different communities were sequenced: Archbishop Desmond Tutu, the South African civil rights advocate and a member of the Bantu tribe, the majority population in southern Africa, and !Gubi, a hunter-gatherer from the Khoisan community in Namibia, a minority African population that is probably the human group with the oldest known lineage. The comparison revealed many differences, as you might expect. The analysis was then broadened to compare the protein-coding regions of !Gubi's genome with those of three other

Khoisan community members (self-identified Bushmen) living nearby. Remarkably, the four African genomes differed more from each other than a European would from an Asian. These data highlight the extensive diversity among African genomes. Extending this approach will help us answer important questions about the differences between human populations and the migratory routes of human populations throughout history.

Widespread Conservation of Developmental Genes Among Animals

Biologists in the field of evolutionary developmental biology, or **evo-devo** as it is often called, compare developmental processes of different multicellular organisms. Their aim is to understand how these processes have evolved and how changes in them can modify existing organismal features or lead to new ones. With the advent of molecular techniques and the recent flood of genomic information, we are beginning to realize that the genomes of related species with strikingly different forms may have only minor differences in gene sequence or, perhaps more importantly, in gene regulation. Discovering the molecular basis of these differences in turn helps us understand the origins of the myriad diverse forms that cohabit this planet, thus informing our study of the evolution of life.

In Concept 18.4, you learned about the homeotic genes in *Drosophila melanogaster* (see Figure 18.20), which encode transcription factors that regulate gene expression and specify the identity of body segments in the fruit fly. Molecular analysis of the homeotic genes in *Drosophila* has shown that they all include a 180-nucleotide sequence called a **homeobox**, which codes for a 60-amino-acid *homeodomain* in the encoded proteins. An identical or very similar nucleotide sequence has been discovered in the homeotic genes of many invertebrates and vertebrates. In fact, the nucleotide sequences in humans and fruit flies are so similar that one researcher has whimsically referred to flies as "little people with wings." The resemblance even extends to the organization of these genes: The vertebrate genes homologous to the homeotic genes of fruit flies have kept the same chromosomal arrangement **(Figure 21.19)**. Homeobox-containing sequences have also been found in regulatory genes of much more distantly related eukaryotes, including plants and yeasts. From these similarities, we can deduce that the homeobox DNA sequence evolved very early in the history of life and was sufficiently valuable to organisms to have been conserved in animals and plants virtually unchanged for hundreds of millions of years.

Homeotic genes in animals were named *Hox* genes, short for homeobox-containing genes, because homeotic genes were the first genes found to have this sequence. Other homeobox-containing genes were later found that do not act as homeotic genes; that is, they do not directly control

Figure 21.19 Conservation of homeotic genes in a fruit fly and a mouse. Homeotic genes that control the form of anterior and posterior structures of the body occur in the same linear sequence on chromosomes in *Drosophila* and mice. Each colored band on the chromosomes shown here represents a homeotic gene. In fruit flies, all homeotic genes are found on one chromosome. The mouse and other mammals have the same or similar sets of genes on four chromosomes. The color code indicates the parts of the embryos in which these genes are expressed and the adult body regions that result. All of these genes are essentially identical in flies and mice, except for those represented by black bands, which are less similar in the two animals.

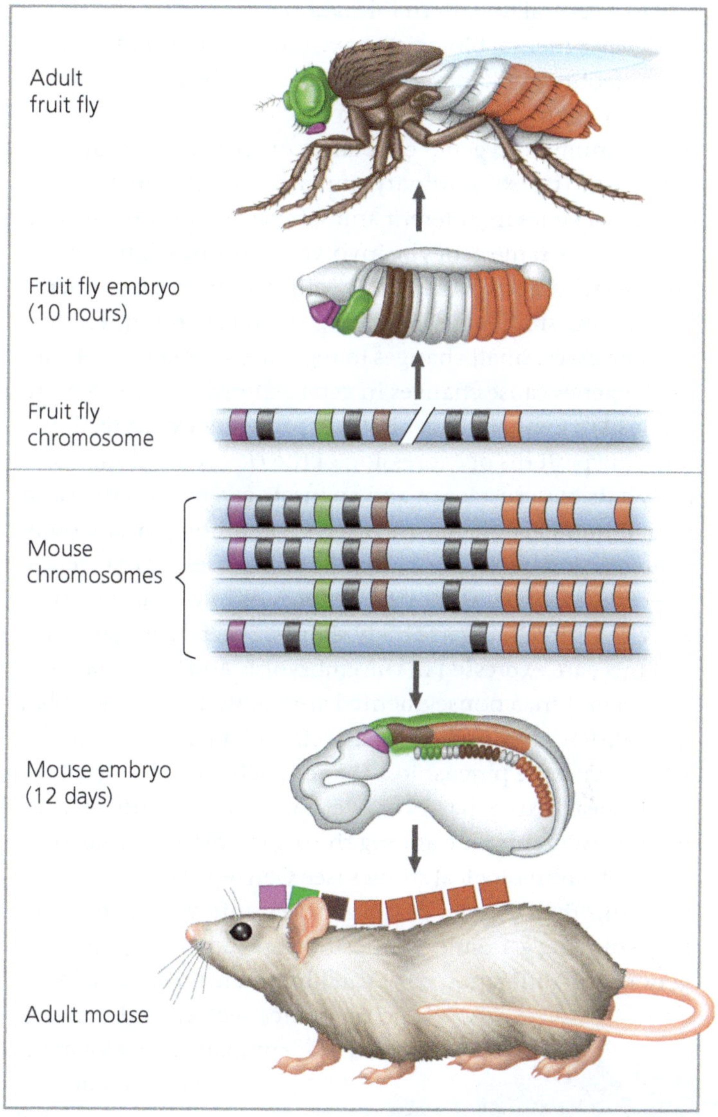

the identity of body parts. However, most of these genes, in animals at least, are associated with development, suggesting their ancient and fundamental importance in that process. In *Drosophila*, for example, homeoboxes are present not only in the homeotic genes but also in the egg-polarity gene *bicoid* (see Figures 18.21 and 18.22), in several of the segmentation genes, and in a master regulatory gene for eye development.

Researchers have discovered that the homeobox-encoded homeodomain binds to DNA when the protein functions as a transcription factor. Elsewhere in the protein, domains that are more variable interact with other transcription factors, allowing the homeodomain-containing protein to recognize specific enhancers and regulate the associated genes. Proteins with homeodomains probably regulate development by coordinating the transcription of batteries of developmental genes, switching them on or off. In embryos of *Drosophila* and other animal species, different combinations of homeobox genes are active in different parts of the embryo. This selective expression of regulatory genes, varying over time and space, is central to pattern formation.

Developmental biologists have found that in addition to homeotic genes, many other genes involved in development are highly conserved from species to species. These include numerous genes encoding components of signaling pathways. The extraordinary similarity among some developmental genes in different animal species raises a question: How can the same genes be involved in the development of animals whose forms are so very different from each other?

Ongoing studies are suggesting answers to this question. In some cases, small changes in regulatory sequences of particular genes cause changes in gene expression patterns that can lead to major changes in body form. For example, the differing patterns of expression of the *Hox* genes along the body axis in a crustacean and an insect can explain the variation in number of leg-bearing segments among these closely related animals **(Figure 21.20)**. In other cases, similar genes direct different developmental processes in various organisms, resulting in diverse body shapes. Several *Hox* genes, for instance, are expressed in the embryonic and larval stages of the sea urchin, a nonsegmented animal that has a body plan quite different from those of insects and mice. Sea urchin adults make the pincushion-shaped shells you may have seen on the beach; two species of live sea urchins are shown in the photo. Sea urchins are among the organisms long used in classical embryological studies (see Concept 47.2).

In this final chapter of the genetics unit, you have learned how studying genomic composition and comparing the genomes of different species can illuminate the process by which genomes evolve. Furthermore, comparing developmental programs, we can see that

Figure 21.20 Effect of differences in *Hox* gene expression in a crustacean and an insect. Changes in the expression patterns of *Hox* genes have occurred over evolutionary time since insects diverged from a crustacean ancestor. These changes account in part for the different body plans of **(a)** the brine shrimp *Artemia*, a crustacean, and **(b)** the grasshopper, an insect. Shown here are regions of the adult body color-coded for expression of four *Hox* genes that determine the formation of particular body parts during embryonic development. Each color represents a specific *Hox* gene.

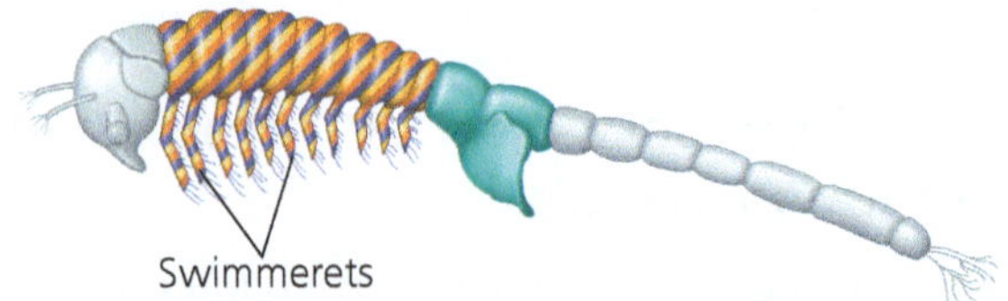

(a) Expression of four *Hox* genes in the brine shrimp *Artemia*. Three of the *Hox* genes are expressed together in one region (indicated by stripes), specifying the identity of the segments that have swimmerets. The fourth (teal) specifies the identity of the genital segments.

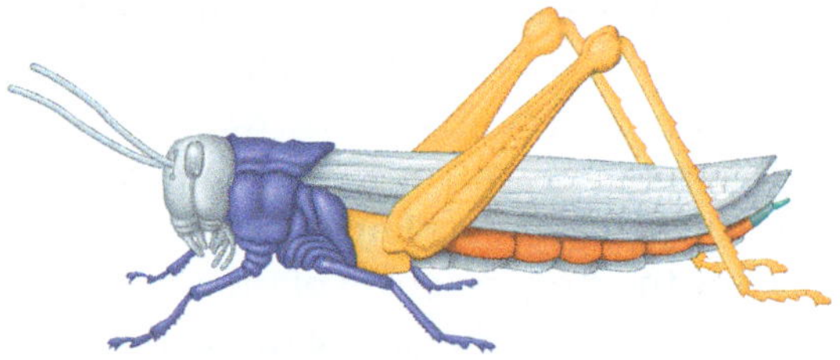

(b) Expression of the grasshopper versions of the same four *Hox* genes. In the grasshopper, each *Hox* gene is expressed in a discrete region and specifies the identity of that region.

the unity of life is reflected in the similarity of molecular and cellular mechanisms used to establish body pattern, although the genes directing development may differ among organisms. The similarities between genomes reflect the common ancestry of life on Earth. But the differences are also crucial, for they have created the huge diversity of organisms that have evolved. In the remaining chapters, we expand our perspective beyond the level of molecules, cells, and genes to explore this diversity on the organismal level.

CONCEPT CHECK 21.6

1. Would you expect the genome of the macaque (a monkey) to be more like that of a mouse or that of a human? Explain.
2. DNA sequences called homeoboxes help homeotic genes in animals direct development. Given that they are common to flies and mice, explain why these animals are so different.
3. **WHAT IF?** ➤ There are three times as many *Alu* elements in the human genome as in the chimpanzee genome. How do you think these extra *Alu* elements arose in the human genome? Propose a role they might have played in the divergence of these two species.

For suggested answers, see Appendix A.

21 Chapter Review

Go to **MasteringBiology**™ for Videos, Animations, Vocab Self-Quiz, Practice Tests, and more in the Study Area.

SUMMARY OF KEY CONCEPTS

VOCAB SELF-QUIZ goo.gl/6u55ks

CONCEPT 21.1

The Human Genome Project fostered development of faster, less expensive sequencing techniques *(pp. 441–442)*

- The **Human Genome Project** was largely completed in 2003, aided by major advances in sequencing technology.
- In the **whole-genome shotgun** approach, the whole genome is cut into many small, overlapping fragments that are sequenced; computer software then assembles the genome sequence.

? *How did the Human Genome Project result in more rapid, less expensive DNA-sequencing technology?*

CONCEPT 21.2

Scientists use bioinformatics to analyze genomes and their functions *(pp. 442–446)*

- Computer analysis of genome sequences aids **gene annotation**, the identification of protein-coding sequences. Methods to determine gene function include comparing sequences of newly discovered genes with those of known genes in other species and observing the effects of experimentally inactivating the genes.
- In **systems biology**, scientists use the computer-based tools of **bioinformatics** to compare genomes and study sets of genes and proteins as whole systems (**genomics** and **proteomics**). Studies include large-scale analyses of protein interactions, functional DNA elements, and genes contributing to medical conditions.

? *What has been the most significant finding of the ENCODE project? Why was the project expanded to include non-human species?*

CONCEPT 21.3

Genomes vary in size, number of genes, and gene density *(pp. 446–448)*

	Bacteria	Archaea	Eukarya
Genome size	Most are 1–6 Mb		Most are 10–4,000 Mb, but a few are much larger
Number of genes	1,500–7,500		Most are 5,000–45,000
Gene density	Higher than in eukaryotes		Lower than in prokaryotes (Within eukaryotes, lower density is correlated with larger genomes.)
Introns	None in protein-coding genes	Present in some genes	Present in most genes of multicellular eukaryotes, but only in some genes of unicellular eukaryotes
Other noncoding DNA	Very little		Can exist in large amounts; generally more repetitive noncoding DNA in multicellular eukaryotes

? *Compare genome size, gene number, and gene density (a) in the three domains and (b) among eukaryotes.*

CONCEPT 21.4

Multicellular eukaryotes have a lot of noncoding DNA and many multigene families *(pp. 448–451)*

- Only 1.5% of the human genome codes for proteins or gives rise to rRNAs or tRNAs; the rest is noncoding DNA, including **pseudogenes** and **repetitive DNA** of unknown function.
- The most abundant type of repetitive DNA in multicellular eukaryotes consists of **transposable elements** and related sequences. In eukaryotes, there are two types of transposable elements: **transposons**, which move via a DNA intermediate, and **retrotransposons**, which are more prevalent and move via an RNA intermediate.
- Other repetitive DNA includes short, noncoding sequences that are tandemly repeated thousands of times (**simple sequence DNA**, which includes **STRs**); these sequences are especially prominent in centromeres and telomeres, where they probably play structural roles in the chromosome.
- Though many eukaryotic genes are present in one copy per haploid chromosome set, others (most, in some species) are members of a gene family, such as the human globin gene families:

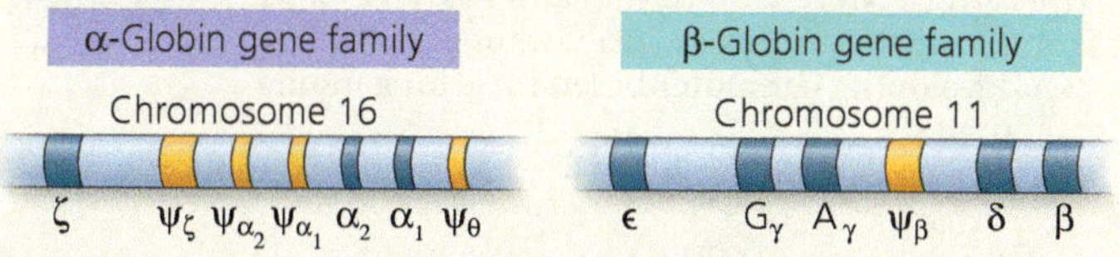

? *Explain how the function of transposable elements might account for their prevalence in human noncoding DNA.*

CONCEPT 21.5

Duplication, rearrangement, and mutation of DNA contribute to genome evolution *(pp. 452–457)*

- Errors in cell division can lead to extra copies of all or part of entire chromosome sets, which may then diverge if one set accumulates sequence changes. Polyploidy occurs more often among plants than animals and contributes to speciation.
- The chromosomal organization of genomes can be compared among species, providing information about evolutionary relationships. Within a given species, rearrangements of chromosomes are thought to contribute to the emergence of new species.
- The genes encoding the various related but different globin proteins evolved from one common ancestral globin gene, which duplicated and diverged into α-globin and β-globin ancestral genes. Subsequent duplication and random mutation gave rise to the present globin genes, all of which code for oxygen-binding proteins. The copies of some duplicated genes have diverged so much that the functions of their encoded proteins (such as lysozyme and α-lactalbumin) are now substantially different.
- Rearrangement of exons within and between genes during evolution has led to genes containing multiple copies of similar exons and/or several different exons derived from other genes.
- Movement of transposable elements or recombination between copies of the same element can generate new sequence combinations that are beneficial to the organism. These may alter the functions of genes or their patterns of expression and regulation.

? *How could chromosomal rearrangements lead to the emergence of new species?*

CONCEPT 21.6

Comparing genome sequences provides clues to evolution and development *(pp. 457–462)*

- Comparisons of genomes from widely divergent and closely related species provide valuable information about ancient and more recent evolutionary history, respectively. Analysis of single nucleotide polymorphisms (SNPs) and copy-number variants (CNVs) among individuals in a species can also shed light on the evolution of that species.
- Evolutionary developmental (**evo-devo**) biologists have shown that homeotic genes and some other genes associated with animal development contain a **homeobox** region whose sequence is highly conserved among diverse species. Related sequences are present in the genes of plants and yeasts.

? *What type of information can be obtained by comparing the genomes of closely related species? Of very distantly related species?*

TEST YOUR UNDERSTANDING

Level 1: Knowledge/Comprehension

PRACTICE TEST goo.gl/CUYGKD

1. Bioinformatics includes all of the following except
   (A) using computer programs to align DNA sequences.
   (B) using DNA technology to combine DNA from two different sources in a test tube.
   (C) developing computer-based tools for genome analysis.
   (D) using mathematical tools to make sense of biological systems.

2. Homeotic genes
   (A) encode transcription factors that control the expression of genes responsible for specific anatomical structures.
   (B) are found only in *Drosophila* and other arthropods.
   (C) are the only genes that contain the homeobox domain.
   (D) encode proteins that form anatomical structures in the fly.

Level 2: Application/Analysis

3. Two eukaryotic proteins have one domain in common but are otherwise very different. Which of the following processes is most likely to have contributed to this similarity?
   (A) gene duplication
   (B) alternative splicing
   (C) exon shuffling
   (D) random point mutations

4. **DRAW IT** Below are the amino acid sequences (using the single-letter code; see Figure 5.14) of four short segments of the FOXP2 protein from six species: chimpanzee (C), orangutan (O), gorilla (G), rhesus macaque (R), mouse (M), and human (H). These segments contain all of the amino acid differences between the FOXP2 proteins of these species.

   1. ATETI...PKSSD...TSSTT...NARRD
   2. ATETI...PKSSE...TSSTT...NARRD
   3. ATETI...PKSSD...TSSTT...NARRD
   4. ATETI...PKSSD...TSSNT... SARRD
   5. ATETI...PKSSD...TSSTT...NARRD
   6. VTETI...PKSSD...TSSTT...NARRD

   Use a highlighter to color any amino acid that varies among the species. (Color that amino acid in all sequences.)
   (a) The C, G, R sequences are identical. Identify which lines correspond to those sequences.
   (b) The H sequence differs from that of the C, G, R species at two amino acids. Underline the two differences in the H sequence.
   (c) The O sequence differs from the C, G, R sequences at one amino acid (having V instead of A) and from the H sequence at three amino acids. Identify the O sequence.
   (d) In the M sequence, circle the amino acid(s) that differ from the C, G, R sequences, and draw a square around those that differ from the H sequence.
   (e) Primates and rodents diverged between 60 and 100 million years ago, and chimpanzees and humans about 6 million years ago. Compare the amino acid differences between the mouse and the C, G, R species with those between the human and the C, G, R species. What can you conclude?

Level 3: Synthesis/Evaluation

5. **EVOLUTION CONNECTION** Genes important in the embryonic development of animals, such as homeobox-containing genes, have been relatively well conserved during evolution; that is, they are more similar among different species than are many other genes. Explain why this is.

6. **SCIENTIFIC INQUIRY** The scientists mapping the SNPs in the human genome noticed that groups of SNPs tended to be inherited together, in blocks known as haplotypes, ranging in length from 5,000 to 200,000 base pairs. There are as few as four or five commonly occurring combinations of SNPs per haplotype. Integrating what you've learned throughout this chapter and this unit, propose an explanation for this observation.

7. **WRITE ABOUT A THEME: INFORMATION** The continuity of life is based on heritable information in the form of DNA. In a short essay (100–150 words), explain how mutations in protein-coding genes and regulatory DNA contribute to evolution.

8. **SYNTHESIZE YOUR KNOWLEDGE**

Insects have three thoracic (trunk) segments. While researchers have found insect fossils with pairs of wings on all three segments, modern insects have wings or related structures on only the second and third segment. It turns out that in modern insects, *Hox* gene products act to inhibit wing formation on the first segment. The treehopper insect (above) is somewhat of an exception. In addition to having wings on its second segment, the treehopper's first segment has an ornate helmet that resembles a set of thorns, which a recent study has found to be a modified, fused pair of "wings." The thorn-like structure helps to camouflage the treehopper in tree branches, thus reducing its risk of predation. Explain how changes in gene regulation could have led to the evolution of such a structure.

For selected answers, see Appendix A.

For additional practice questions, check out the **Dynamic Study Modules** in MasteringBiology. You can use them to study on your smartphone, tablet, or computer anytime, anywhere!

UNIT 4 MECHANISMS OF EVOLUTION

A Nobel Prize laureate, Jack W. Szostak is Professor of Genetics at Harvard Medical School and Professor of Chemistry and Chemical Biology at Harvard University and is the Alexander Rich Distinguished Investigator at Massachusetts General Hospital. Dr. Szostak grew up in Canada and received a B.Sc. from McGill University and a Ph.D. in biochemistry from Cornell University. He has made pioneering contributions to genetics, including identifying the role of programmed DNA double-strand breaks in meiotic recombination and the discovery of how chromosomes are protected by telomeres. A member of the National Academy of Sciences, Dr. Szostak is now conducting groundbreaking research on the origin of life on Earth.

An Interview with Jack Szostak

You won a Nobel Prize for your work on telomeres. How did you transition from that topic to studying the origin of life?

After working on telomeres and DNA damage repair in yeast, I started to look around for something new to study. At that time, Sid Altman and Tom Cech had just discovered ribozymes, which are RNA molecules that can function as enzyme-like catalysts. I found this really exciting, and so I switched the focus of my lab to RNA biochemistry. At first, we looked at molecular evolution. But it's one thing to study evolution in the lab, where you, the experimenter, control the conditions under which an existing ribozyme can evolve. I started to wonder: How in the first place did RNA form and replicate itself on early Earth? I became more and more interested in that question, and now my whole lab is working on it.

What approaches do you use to study the origin of life?

We're using chemical experiments to try to figure out the conditions under which a primitive cell could form, grow, and divide. Much work from other laboratories has contributed to our current understanding of how the chemical building blocks of biology could have been made on early Earth. This has set the stage for exploring the next question: How did these inanimate chemicals assemble into larger structures and start acting like a living cell?

What hurdles remain to answer this question?

The first living cells must have been extremely simple. But they also must have had some of the universal features of cells today, such as a cell membrane and a genetic material like RNA. We've learned how to assemble cell-like structures with a membrane that surrounds RNA. The membranes of these structures can grow and produce "offspring" similar to themselves. The big remaining hurdle is to discover conditions under which RNA can replicate itself in a way that is compatible with a primitive cell membrane.

"All present-day life shares many features, including the same underlying biochemistry. We're trying to find explanations for such shared features."

What can you learn about present-day cells by studying the origin of life?

All present-day life shares many features, including the same underlying biochemistry. We're trying to find explanations for such shared features. We've learned that the essential role of RNA seems to emerge naturally from chemical conditions like those on early Earth. But questions remain. For example, all cells maintain certain concentration gradients across their membranes, such as higher potassium levels inside the cell than out, and lower sodium levels inside than out. Cells spend a lot of energy maintaining these gradients—why is that? We don't have a good answer for that yet, but we hope to learn more about such puzzling features of life today by studying its origins.

▼ Model of a protocell containing nucleotides and short bits of RNA

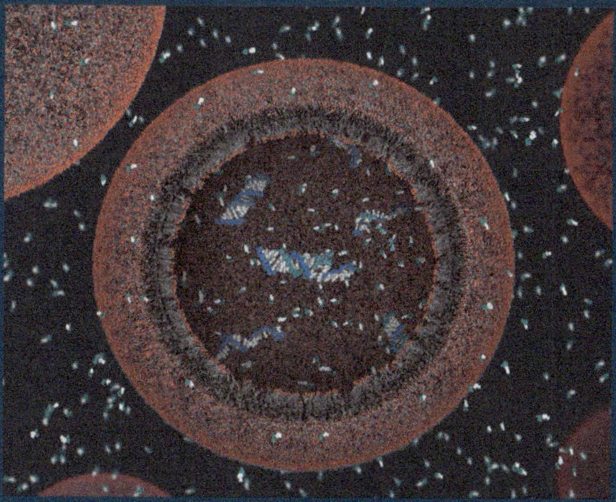

What do you find most rewarding about your job?

I especially enjoy talking with people about new experiments. The lab is filled with talented students and post-docs who have different backgrounds and different scientific interests. Everyone has to learn how to talk with each other. This makes the lab an exciting and fun place where we all are working together to solve a big scientific problem.

22 Descent with Modification: A Darwinian View of Life

▲ **Figure 22.1 How is its resemblance to a fallen leaf helpful to this moth?**

KEY CONCEPTS

22.1 The Darwinian revolution challenged traditional views of a young Earth inhabited by unchanging species

22.2 Descent with modification by natural selection explains the adaptations of organisms and the unity and diversity of life

22.3 Evolution is supported by an overwhelming amount of scientific evidence

▼ Juvenile stage (caterpillar) of the dead-leaf moth

Endless Forms Most Beautiful

A hungry bird in the Peruvian rain forest would have to look very closely to spot a "dead-leaf moth" (*Oxytenis modestia*), which blends in well with its forest floor habitat **(Figure 22.1)**. This distinctive moth is a member of a diverse group, the more than 120,000 species of lepidopteran insects (moths and butterflies). All lepidopterans have a juvenile stage characterized by a well-developed head and many chewing mouthparts: the ravenous, efficient feeding machines we call caterpillars. (The caterpillar stage of the dead-leaf moth is also protected by its appearance: When threatened, it weaves its head back and forth, resembling a snake about to strike.) As adults, all lepidopterans share other features, such as three pairs of legs and two pairs of wings covered with small scales. But the many lepidopterans also differ from one another. How did there come to be so many different moths and butterflies, and what causes their similarities and differences?

The moth in Figure 22.1 and its many close relatives illustrate three key observations about life:

- the striking ways in which organisms are suited for life in their environments (Here and throughout this text, the term *environment* refers to other organisms as well as to the physical aspects of an organism's surroundings.)
- the many shared characteristics (unity) of life
- the rich diversity of life

When you see this blue icon, log in to **MasteringBiology** and go to the Study Area for digital resources.

MB **Get Ready for This Chapter**

More than a century and a half ago, Charles Darwin was inspired to develop a scientific explanation for these three broad observations. When he published his hypothesis in his book *The Origin of Species*, Darwin ushered in a scientific revolution—the era of evolutionary biology.

For now, we will define **evolution** as *descent with modification*, a phrase Darwin used in proposing that Earth's many species are descendants of ancestral species that were different from the present-day species. Evolution can also be defined as a change in the genetic composition of a population from generation to generation (see Concept 23.3).

We can also view evolution in two related but different ways: as a pattern and as a process. The *pattern* of evolutionary change is revealed by data from many scientific disciplines, including biology, geology, physics, and chemistry. These data are facts—they are observations about the natural world—and these observations show that life has evolved over time. The *process* of evolution consists of the mechanisms that cause the observed pattern of change. These mechanisms represent natural causes of the natural phenomena we observe. Indeed, the power of evolution as a unifying theory is its ability to explain and connect a vast array of observations about the living world.

As with all general theories in science, we continue to test our understanding of evolution by examining whether it can account for new observations and experimental results. In this and the following chapters, we'll examine how ongoing discoveries shape what we know about the pattern and process of evolution. To set the stage, we'll first retrace Darwin's quest to explain the adaptations, unity, and diversity of what he called life's "endless forms most beautiful."

CONCEPT 22.1

The Darwinian revolution challenged traditional views of a young Earth inhabited by unchanging species

What impelled Darwin to challenge the prevailing views about Earth and its life? Darwin developed his revolutionary proposal over time, influenced by the work of others and by his travels **(Figure 22.2)**. As we'll see, his ideas also had deep historical roots.

Figure 22.2 The intellectual context of Darwin's ideas.

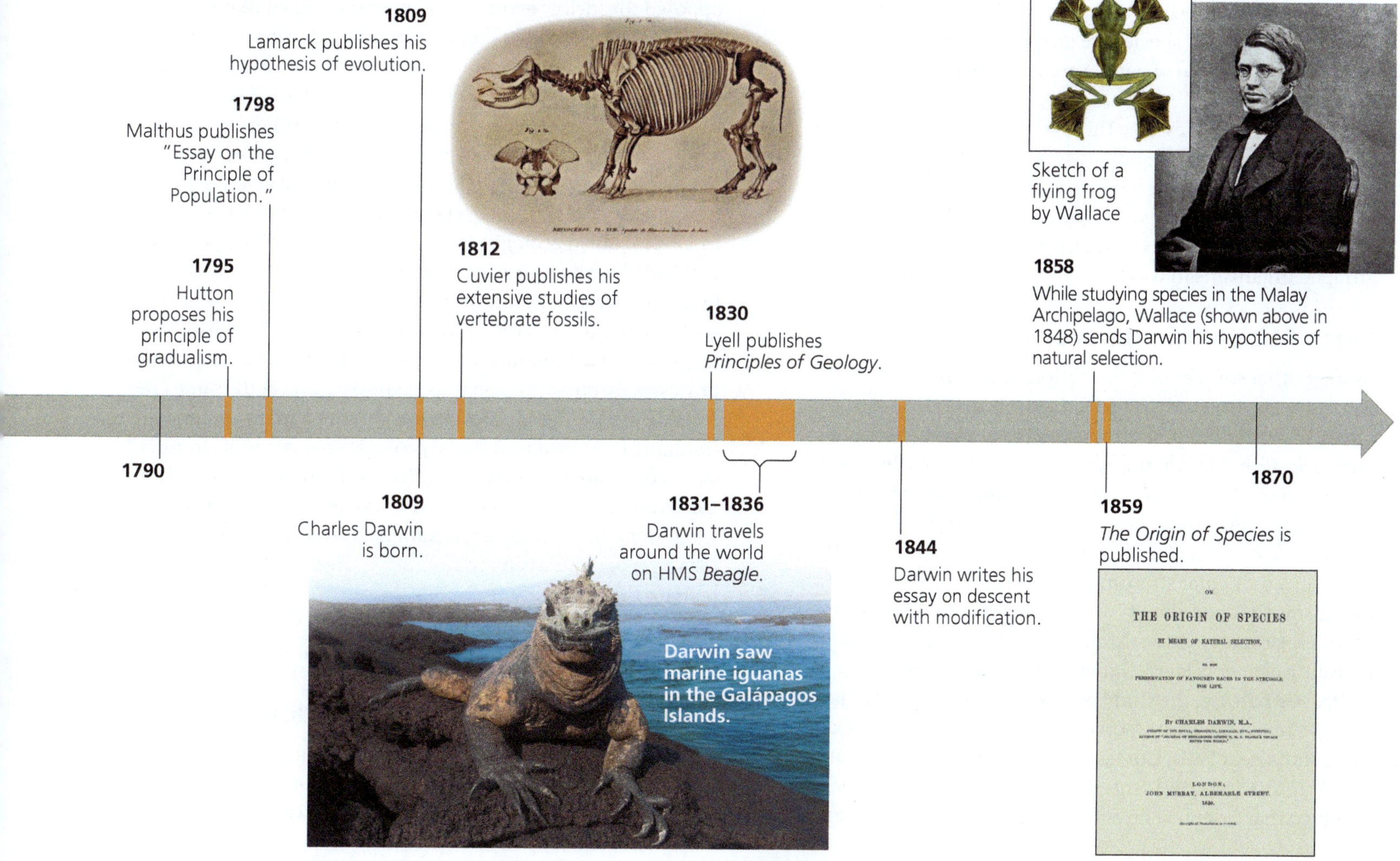

Scala Naturae and Classification of Species

Long before Darwin was born, several Greek philosophers suggested that life might have changed gradually over time. But one philosopher who greatly influenced early Western science, Aristotle (384–322 BCE), viewed species as fixed (unchanging). Through his observations of nature, Aristotle recognized certain "affinities" among organisms. He concluded that life-forms could be arranged on a ladder, or scale, of increasing complexity, later called the *scala naturae* ("scale of nature"). Each form of life, perfect and permanent, had its allotted rung on this ladder.

These ideas were generally consistent with the Old Testament account of creation, which holds that species were individually designed by God and therefore perfect. In the 1700s, many scientists interpreted the often remarkable match of organisms to their environment as evidence that the Creator had designed each species for a particular purpose.

One such scientist was Carolus Linnaeus (1707–1778), a Swedish physician and botanist who sought to classify life's diversity, in his words, "for the greater glory of God." In the 1750s, Linnaeus developed the two-part, or *binomial*, format for naming species (such as *Homo sapiens* for humans) that is still used today. In contrast to the linear hierarchy of the *scala naturae*, Linnaeus adopted a nested classification system, grouping similar species into increasingly general categories. For example, similar species are grouped in the same genus, similar genera (plural of genus) are grouped in the same family, and so on.

Linnaeus did not ascribe the resemblances among species to evolutionary kinship, but rather to the pattern of their creation. A century later, however, Darwin argued that classification should be based on evolutionary relationships. He also noted that scientists using the Linnaean system often grouped organisms in ways that reflected those relationships.

Ideas About Change over Time

Among other sources of information, Darwin drew from the work of scientists studying **fossils**, the remains or traces of organisms from the past. Many fossils are found in sedimentary rocks formed from the sand and mud that settle to the bottom of seas, lakes, and swamps **(Figure 22.3)**. New layers of sediment cover older ones and compress them into superimposed layers of rock called **strata** (singular, *stratum*). The fossils in a particular stratum provide a glimpse of some of the organisms that populated Earth at the time that layer formed. Later, erosion may carve through upper (younger) strata, revealing deeper (older) strata that had been buried.

Paleontology, the study of fossils, was developed in large part by French scientist Georges Cuvier (1769–1832). In examining strata near Paris, Cuvier noted that the older the stratum, the more dissimilar its fossils were to current life-forms. He also observed that from one layer to the next, some new species

▼ Figure 22.3 Formation of sedimentary strata with fossils.

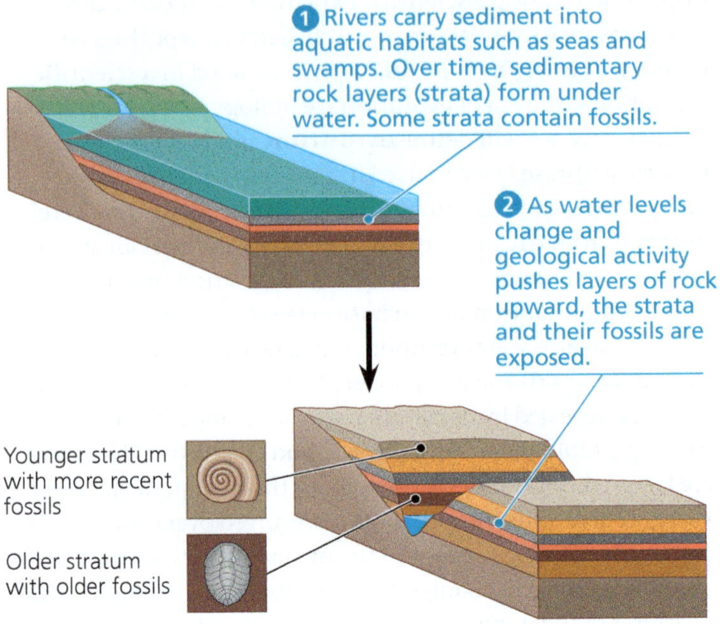

MB Video: Grand Canyon

appeared while others disappeared. He inferred that extinctions must have been a common occurrence, but he staunchly opposed the idea of evolution. Cuvier speculated that each boundary between strata represented a sudden catastrophic event, such as a flood, that had destroyed many of the species living in that area. Such regions, he reasoned, were later repopulated by different species immigrating from other areas.

In contrast to Cuvier's emphasis on sudden events, other scientists suggested that profound change could take place through the cumulative effect of slow but continuous processes. In 1795, Scottish geologist James Hutton (1726–1797) proposed that Earth's geologic features could be explained by gradual mechanisms, such as valleys being formed by rivers. The leading geologist of Darwin's time, Charles Lyell (1797–1875), incorporated Hutton's thinking into his proposal that the same geologic processes are operating today as in the past, and at the same rate.

Hutton's and Lyell's ideas strongly influenced Darwin's thinking. Darwin agreed that if geologic change results from slow, continuous actions rather than from sudden events, then Earth must be much older than the widely accepted age of a few thousand years. It would, for example, take a very long time for a river to carve a canyon by erosion. He later reasoned that perhaps similarly slow and subtle processes could produce substantial biological change. However, Darwin was not the first to apply the idea of gradual change to biological evolution.

Lamarck's Hypothesis of Evolution

Although some 18th-century naturalists suggested that life evolves as environments change, only one proposed a mechanism for *how* life changes over time: French biologist

Jean-Baptiste de Lamarck (1744–1829). Alas, Lamarck is primarily remembered today *not* for his visionary recognition that evolutionary change explains patterns in fossils and the match of organisms to their environments, but for the incorrect mechanism he proposed.

Lamarck published his hypothesis in 1809, the year Darwin was born. By comparing living species with fossil forms, Lamarck had found what appeared to be several lines of descent, each a chronological series of older to younger fossils leading to a living species. He explained his findings using two principles that were widely accepted at the time. The first was *use and disuse*, the idea that parts of the body that are used extensively become larger and stronger, while those that are not used deteriorate. Among many examples, he cited a giraffe stretching its neck to reach leaves on high branches. The second principle, *inheritance of acquired characteristics*, stated that an organism could pass these modifications to its offspring. Lamarck reasoned that the long, muscular neck of the living giraffe had evolved over many generations as giraffes stretched their necks ever higher.

Lamarck also thought that evolution happens because organisms have an innate drive to become more complex. Darwin rejected this idea, but he, too, thought that variation was introduced into the evolutionary process in part through inheritance of acquired characteristics. Today, however, our understanding of genetics refutes this mechanism: Experiments show that traits acquired by use during an individual's life are not inherited in the way proposed by Lamarck (**Figure 22.4**).

➤ **Figure 22.4 Acquired traits cannot be inherited.** This bonsai tree was "trained" to grow as a dwarf by pruning and shaping. However, seeds from this tree would produce offspring of normal size.

Lamarck was vilified in his own time, especially by Cuvier, who denied that species ever evolve. In retrospect, however, Lamarck did recognize that the fact that organisms are well-suited for life in their environments can be explained by gradual evolutionary change, and he did propose a testable explanation for how this change occurs.

CONCEPT CHECK 22.1

1. How did Hutton's and Lyell's ideas influence Darwin's thinking about evolution?
2. **MAKE CONNECTIONS ➤** Scientific hypotheses must be testable (see Concept 1.3). Applying this criterion, are Cuvier's explanation of the fossil record and Lamarck's hypothesis of evolution scientific? Explain your answer in each case.

For suggested answers, see Appendix A.

CONCEPT 22.2

Descent with modification by natural selection explains the adaptations of organisms and the unity and diversity of life

As the 19th century dawned, it was generally thought that species had remained unchanged since their creation. A few clouds of doubt about the permanence of species were beginning to gather, but no one could have forecast the thundering storm just beyond the horizon. How did Charles Darwin become the lightning rod for a revolutionary view of life?

Darwin's Research

Charles Darwin (1809–1882) was born in Shrewsbury, in western England. Even as a boy, he had a consuming interest in nature. When he was not reading nature books, he was fishing, hunting, riding, and collecting insects. However, Darwin's father, a physician, could see no future for his son as a naturalist and sent him to medical school in Edinburgh. But Charles found medicine boring and surgery before the days of anesthesia horrifying. He quit medical school and enrolled at Cambridge University, intending to become a clergyman. (At that time, many scholars of science belonged to the clergy.)

At Cambridge, Darwin became the protégé of John Henslow, a botany professor. Soon after Darwin graduated, Henslow recommended him to Captain Robert FitzRoy, who was preparing the survey ship HMS *Beagle* for a long voyage around the world. Darwin would pay his own way and serve as a conversation partner to the young captain. FitzRoy, who was himself an accomplished scientist, accepted Darwin because he was a skilled naturalist and because they were of similar age and social class.

The Voyage of the Beagle

Darwin embarked from England on the *Beagle* in December 1831. The primary mission of the voyage was to chart poorly known stretches of the South American coastline. Darwin, however, spent most of his time on shore, observing and collecting thousands of plants and animals. He described features of organisms that made them well suited to such diverse environments as the humid jungles of Brazil, the expansive grasslands of Argentina, and the towering peaks of the Andes. He also noted that the plants and animals in temperate regions of South America more closely resembled species living in the South American tropics than species living in temperate regions of Europe. Furthermore, the fossils he found, though clearly different from living species, distinctly resembled the living organisms of South America.

Darwin also spent much time thinking about geology. Despite repeated bouts of seasickness, he read Lyell's *Principles of Geology* during the voyage. He experienced geologic change firsthand when a violent earthquake shook the coast of Chile, and he observed afterward that rocks along the coast had been thrust upward by several meters. Finding fossils of ocean organisms high in the Andes, Darwin inferred that the rocks containing the fossils must have been raised there by many similar earthquakes. These observations reinforced what he had learned from Lyell: Physical evidence did not support the traditional view that Earth was only a few thousand years old.

Darwin's interest in the species (or fossils) found in an area was further stimulated by the *Beagle*'s stop at the Galápagos, a group of volcanic islands located near the equator about 900 km west of South America **(Figure 22.5)**. Darwin was fascinated by the unusual organisms there. The birds he collected included several kinds of mockingbirds. These mockingbirds, though similar to each other, seemed to be different species. Some were unique to individual islands, while others lived on two or more adjacent islands. Furthermore, although the animals on the Galápagos resembled species living on the South American mainland, most of the Galápagos species were not known from anywhere else in the world. Darwin hypothesized that the Galápagos had been colonized by organisms that had strayed from South America and then diversified, giving rise to new species on the various islands.

Darwin's Focus on Adaptation

During the voyage of the *Beagle*, Darwin observed many examples of **adaptations**, inherited characteristics of organisms that enhance their survival and reproduction in specific environments. Later, as he reassessed his observations, he began to perceive adaptation to the environment and the origin of new species as closely related processes. Could a new species arise from an ancestral form by the gradual accumulation of adaptations to a different environment? From studies made years after Darwin's voyage, biologists have concluded that this is indeed what happened to a diverse group of finches found on the Galápagos Islands (see Figure 1.20). The finches' various beaks and behaviors are adapted to the specific foods available on their home

▼ Figure 22.5 The voyage of HMS *Beagle* (December 1831–October 1836).

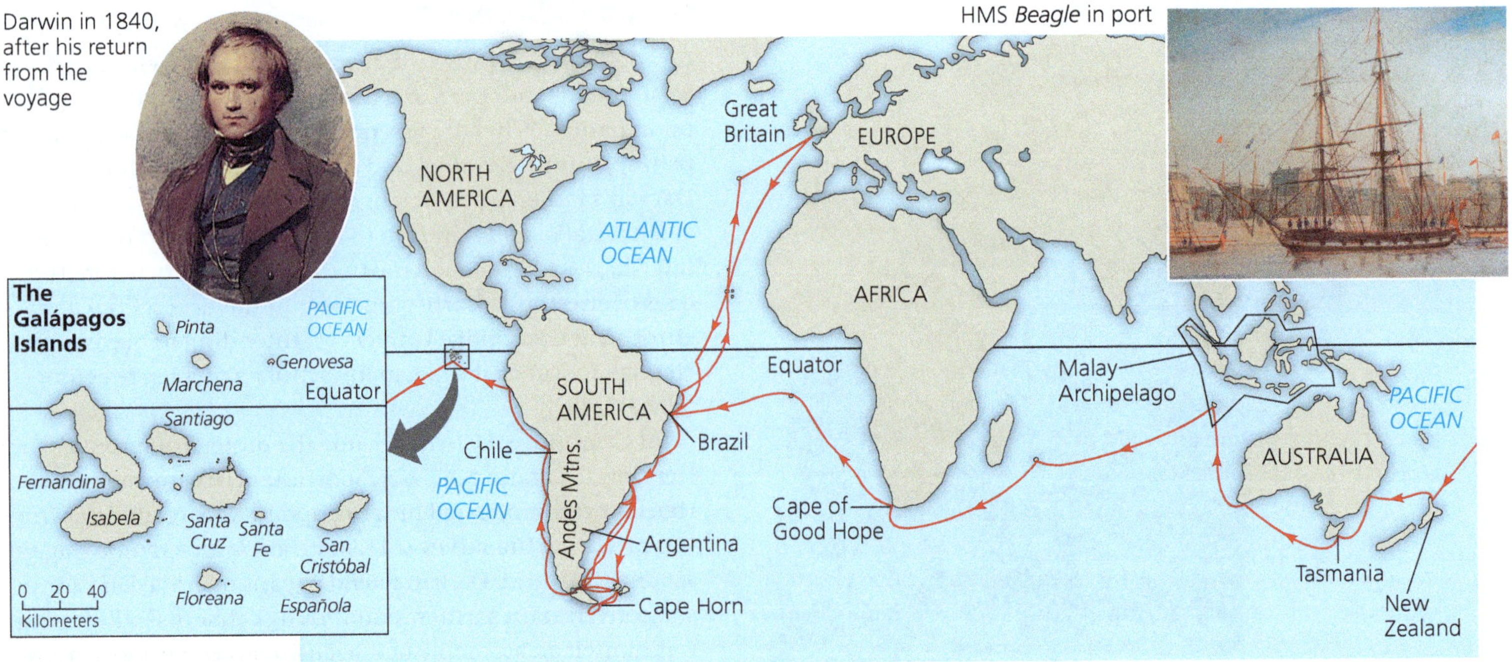

MB ABC News Video: Protecting the Galápagos Islands

Figure 22.6 Three examples of beak variation in Galápagos finches. The Galápagos Islands are home to more than a dozen species of closely related finches, some found only on a single island. A striking difference among them is their beaks, which are adapted for specific diets.

(a) Cactus-eater. The long, sharp beak of the common cactus finch (*Geospiza scandens*) helps it tear and eat cactus flowers and pulp.

(b) Insect-eater. The green warbler finch (*Certhidea olivacea*) uses its narrow, pointed beak to grasp insects.

(c) Seed-eater. The large ground finch (*Geospiza magnirostris*) has a large beak adapted for cracking seeds found on the ground.

MAKE CONNECTIONS ➤ *Review Figure 1.20. Circle the most recent common ancestor shared by the three species that eat insects. Are all of the descendants of that ancestor insect-eaters?*

islands **(Figure 22.6)**. Darwin realized that explaining such adaptations was essential to understanding evolution. His explanation of how adaptations arise centered on **natural selection**, a process in which individuals that have certain inherited traits tend to survive and reproduce at higher rates than do other individuals *because of* those traits.

By the early 1840s, Darwin had worked out the major features of his hypothesis. He set these ideas on paper in 1844, when he wrote a long essay on descent with modification and its underlying mechanism, natural selection. Yet he was still reluctant to publish his ideas, in part because he anticipated the uproar they would cause. During this time, Darwin continued to compile evidence in support of his hypothesis. By the mid-1850s, he had described his ideas to Lyell and a few others. Lyell, who was not yet convinced of evolution, nevertheless urged Darwin to publish on the subject before someone else came to the same conclusions and published first.

In June 1858, Lyell's prediction came true. Darwin received a manuscript from Alfred Russel Wallace (1823–1913), a British naturalist working in the South Pacific islands of the Malay Archipelago (see Figure 22.2). Wallace had developed a hypothesis of natural selection nearly identical to Darwin's. He asked Darwin to evaluate his paper and forward it to Lyell if it merited publication. Darwin complied, writing to Lyell: "Your words have come true with a vengeance. . . . I never saw a more striking coincidence . . . so all my originality, whatever it may amount to, will be smashed." On July 1, 1858, Lyell and a colleague presented Wallace's paper, along with extracts from Darwin's unpublished 1844 essay, to the Linnean Society of London. Darwin quickly finished his book, titled *On the Origin of Species by Means of Natural Selection* (commonly referred to as *The Origin of Species*), and published it the next year. Although Wallace had submitted his ideas for publication first, he admired Darwin and thought that Darwin had developed and tested the idea of natural selection so extensively that he should be known as its main architect.

Within a decade, Darwin's book and its proponents had convinced most scientists of the time that life's diversity is the product of evolution. Darwin succeeded where previous evolutionists had failed, mainly by presenting a plausible scientific mechanism with immaculate logic and an avalanche of supporting evidence.

MB **HHMI Video: The Origin of Species: The Making of a Theory**

Ideas from *The Origin of Species*

In his book, Darwin amassed evidence that descent with modification by natural selection explains three broad observations about nature—the unity of life, the diversity of life, and the striking ways in which organisms are suited for life in their environments.

Descent with Modification

In the first edition of *The Origin of Species*, Darwin never used the word *evolution* (although the final word of the book is "evolved"). Rather, he discussed *descent with modification*, a phrase that summarized his view of life. Organisms share many characteristics, leading Darwin to perceive unity in life. He attributed the unity of life to the descent of all organisms from an ancestor that lived in the remote past. He also thought that as the descendants of that ancestral organism lived in various habitats, they gradually accumulated diverse modifications, or adaptations, that fit them to specific ways of life. Darwin reasoned that over a long period of time, descent with modification eventually led to the rich diversity of life we see today.

Figure 22.7 "I think . . . ." In this 1837 sketch, Darwin envisioned the branching pattern of evolution. Branches that end in twigs labeled A–D represent particular groups of living organisms; all other branches represent extinct groups.

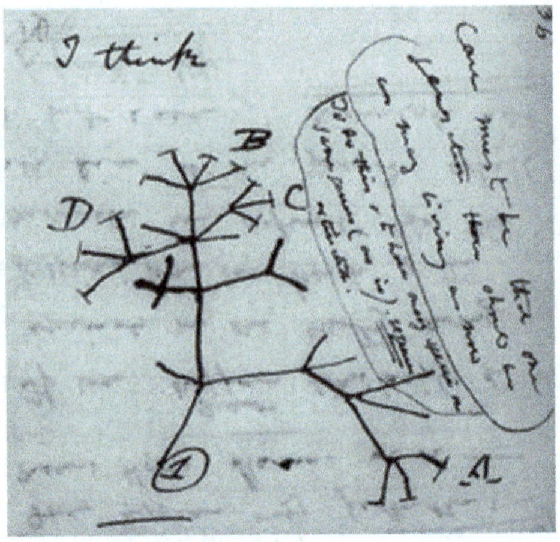

Darwin viewed the history of life as a tree, with multiple branchings from a common trunk out to the tips of the youngest twigs **(Figure 22.7)**. In his diagram, the tips of the twigs that are labeled A–D represent several groups of organisms living in the present day, while the unlabeled branches represent groups that are extinct. Each fork of the tree represents the most recent common ancestor of all the lines of evolution that subsequently branch from that point.

Darwin thought that such a branching process, along with past extinction events, could explain the large morphological gaps that sometimes exist between related groups of organisms. As an example, let's consider the three living species of elephants: the Asian elephant (*Elephas maximus*) and two species of African elephants (*Loxodonta africana* and *L. cyclotis*). These closely related species are very similar because they shared the same line of descent until a relatively recent split from their common ancestor, as shown in the tree diagram in **Figure 22.8**. Note that seven lineages related to elephants have become extinct over the past 32 million years. As a result, there are no living species that fill the morphological gap between the elephants and their nearest relatives today, the hyraxes and manatees.

Extinctions like those in Figure 22.8 are not uncommon. In fact, many evolutionary branches, even some major ones, are dead ends: Scientists estimate that over 99% of all species that have ever lived are now extinct. As in Figure 22.8, fossils of extinct species can document the divergence of present-day groups by "filling in" gaps between them.

Artificial Selection, Natural Selection, and Adaptation

Darwin proposed the mechanism of natural selection to explain the observable patterns of evolution. He crafted his argument carefully, hoping to persuade even the most skeptical readers. First he discussed familiar examples of selective breeding of domesticated plants and animals. Humans have modified other species over many generations by selecting and breeding individuals that possess desired traits, a process

Figure 22.8 Descent with modification. This evolutionary tree of elephants and their relatives is based mainly on fossils—their anatomy, order of appearance in strata, and geographic distribution. Note that most branches of descent ended in extinction (denoted by the dagger symbol, †). (Time line not to scale.)

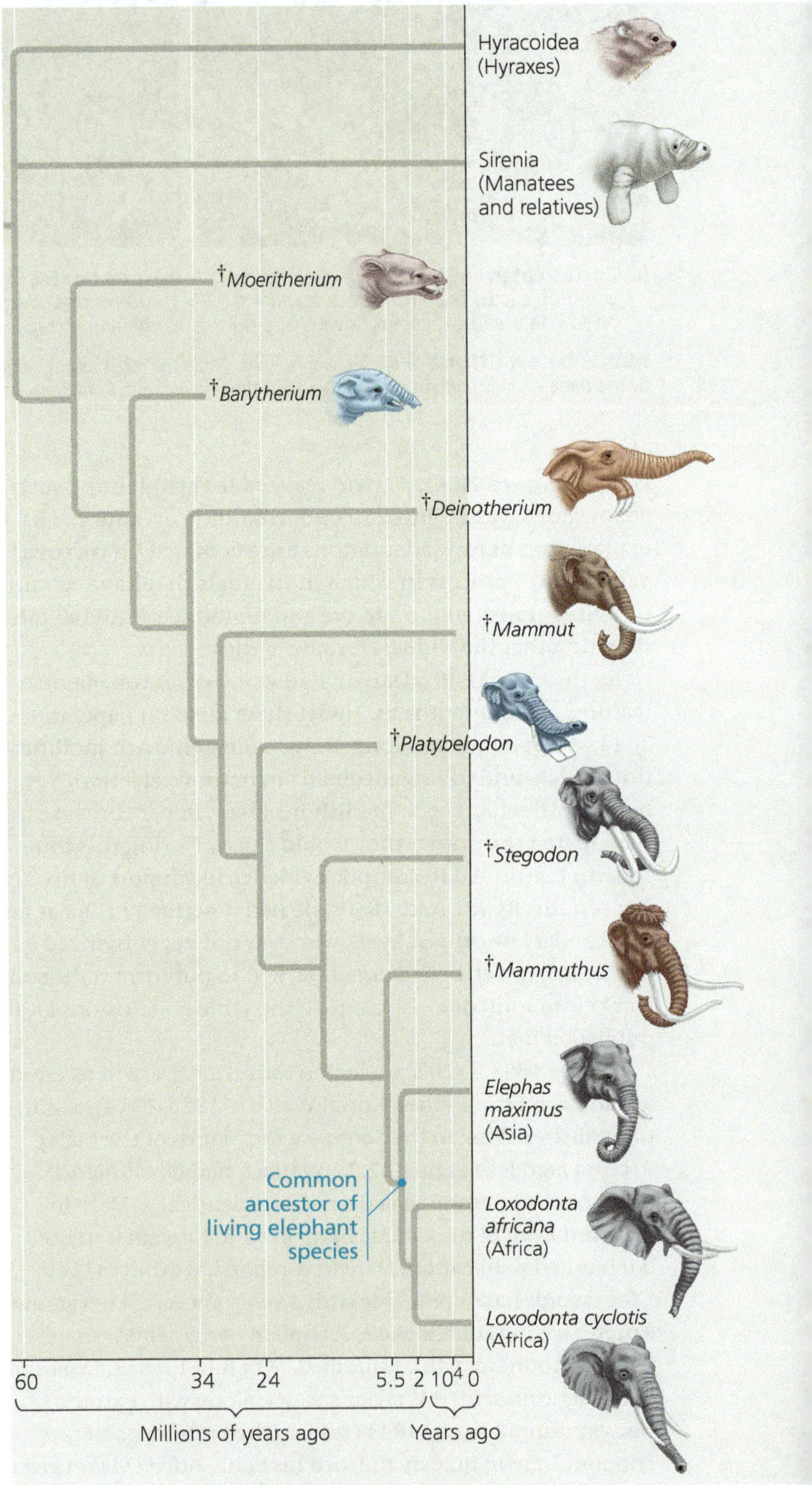

VISUAL SKILLS ➤ *Based on this tree, approximately when did the most recent ancestor shared by* Mammuthus *(woolly mammoths), Asian elephants, and African elephants live?*

Figure 22.9 Artificial selection. These different vegetables have all been selected from one species of wild mustard. By selecting variations in different parts of the plant, breeders have obtained these divergent results.

called **artificial selection** (Figure 22.9). As a result of artificial selection, crops, livestock animals, and pets often bear little resemblance to their wild ancestors.

Darwin then argued that a similar process occurs in nature. He based his argument on two observations, from which he drew two inferences:

Observation #1: Members of a population often vary in their inherited traits (Figure 22.10).

Observation #2: All species can produce more offspring than their environment can support (Figure 22.11), and many of these offspring fail to survive and reproduce.

Inference #1: Individuals whose inherited traits give them a higher probability of surviving and reproducing in a given environment tend to leave more offspring than do other individuals.

Inference #2: This unequal ability of individuals to survive and reproduce will lead to the accumulation of favorable traits in the population over generations.

As these two inferences suggest, Darwin saw an important connection between natural selection and the capacity of organisms to "overreproduce." He began to make this connection after reading an essay by economist Thomas Malthus, who contended that much of human suffering—disease, famine, and war—resulted from the human population's potential to increase faster than food supplies and other resources. Similarly, Darwin realized that the capacity to overreproduce was characteristic of all species. Of the many eggs laid, young born, and seeds spread, only a tiny fraction complete their development and leave offspring of their own. The rest are eaten, starved, diseased, unmated, or unable to tolerate physical conditions of the environment such as salinity or temperature.

Figure 22.10 Variation in a population. Individuals in this population of Asian ladybird beetles vary in color and spot pattern. Natural selection may act on these variations only if (1) they are heritable and (2) they affect the beetles' ability to survive and reproduce.

Figure 22.11 Overproduction of offspring. A single puffball fungus can produce billions of spores that give rise to offspring. If all of these offspring and their descendants survived to maturity, they would carpet the surrounding land surface.

An organism's heritable traits can influence not only its own performance, but also how well its offspring cope with environmental challenges. For example, an organism might have a trait that gives its offspring an advantage in escaping predators, obtaining food, or tolerating physical conditions. When such advantages increase the number of offspring that survive and reproduce, the traits that are favored will likely appear at a greater frequency in the next generation. Thus, over time, natural selection resulting from factors such as predators, lack of food, or adverse physical conditions can lead to an increase in the proportion of favorable traits in a population.

How rapidly do such changes occur? Darwin reasoned that if artificial selection can bring about dramatic change in a relatively short period of time, then natural selection should be capable of substantial modification of species over many hundreds of generations. Even if the advantages of some heritable traits over others are slight, the advantageous variations will gradually accumulate in the population, and less favorable variations will diminish. Over time, this process will increase the frequency of individuals with favorable adaptations, hence increasing the degree to which organisms are well suited for life in their environment.

Key Features of Natural Selection

Let's now recap the main ideas of natural selection:

- Natural selection is a process in which individuals that have certain heritable traits survive and reproduce at a higher rate than do other individuals because of those traits.
- Over time, natural selection can increase the frequency of adaptations that are favorable in a given environment **(Figure 22.12)**.
- If an environment changes, or if individuals move to a new environment, natural selection may result in adaptation to these new conditions, sometimes giving rise to new species.

Text 474

One subtle but important point is that although natural selection occurs through interactions between individual organisms and their environment, *individuals do not evolve*. Rather, it is the population that evolves over time.

A second key point is that natural selection can amplify or diminish only those heritable traits that differ among the individuals in a population. Thus, even if a trait is heritable, if all the individuals in a population are genetically identical for that trait, evolution by natural selection cannot occur.

Third, remember that environmental factors vary from place to place and over time. A trait that is favorable in one place or time may be useless—or even detrimental—in other places or times. Natural selection is always operating, but which traits are favored depends on the context in which a species lives and mates.

▼ Figure 22.12 Camouflage as an example of evolutionary adaptation. Related species of the insects called mantises have diverse shapes and colors that evolved in different environments, as seen in this South African flower-eyed mantis (*Pseudocreobotra wahlbergi*; top) and Malaysian orchid mantis (*Hymenopus coronatus*; bottom).

VISUAL SKILLS ➤ *Use evidence from these two images to explain how these mantises demonstrate the three key observations about life introduced at the beginning of this chapter: the unity and diversity of life and the match between organisms and their environments.*

Next, we'll survey the wide range of observations that support a Darwinian view of evolution by natural selection.

CONCEPT CHECK 22.2

1. How does the concept of descent with modification explain both the unity and diversity of life?
2. **WHAT IF? ➤** If you discovered a fossil of an extinct mammal that lived high in the Andes, would you predict that it would more closely resemble present-day mammals from South American jungles or present-day mammals that live high in Asian mountains? Explain.
3. **MAKE CONNECTIONS ➤** Review the relationship between genotype and phenotype (see Figures 14.5 and 14.6). Suppose that in a particular pea population, flowers with the white phenotype are favored by natural selection. Predict what would happen over time to the frequency of the *p* allele in the population, and explain your reasoning.

For suggested answers, see Appendix A.

CONCEPT 22.3

Evolution is supported by an overwhelming amount of scientific evidence

In *The Origin of Species*, Darwin marshaled a broad range of evidence to support the concept of descent with modification. Still—as he readily acknowledged—there were instances in which key evidence was lacking. For example, Darwin referred to the origin of flowering plants as an "abominable mystery," and he lamented the lack of fossils showing how earlier groups of organisms gave rise to new groups.

In the last 150 years, new discoveries have filled many of the gaps that Darwin identified. The origin of flowering plants, for example, is much better understood (see Concept 30.3), and many fossils have been discovered that signify the origin of new groups of organisms (see Concept 25.2). In this section, we'll consider four types of data that document the pattern of evolution and illuminate how it occurs: direct observations, homology, the fossil record, and biogeography.

Direct Observations of Evolutionary Change

Biologists have documented evolutionary change in thousands of scientific studies. We'll examine many such studies throughout this unit, but let's look at two examples here.

Natural Selection in Response to Introduced Species

Animals that eat plants, called herbivores, often have adaptations that help them feed efficiently on their primary food sources. What happens when herbivores switch to a new food source with different characteristics?

An opportunity to study this question in nature is provided by soapberry bugs, which use their "beak"—a hollow, needlelike mouthpart—to feed on seeds located within the fruits of various plants. In southern Florida, the soapberry bug (*Jadera haematoloma*) feeds on the seeds of a native plant, the balloon vine (*Cardiospermum corindum*). In central Florida, however, balloon vines have become rare. Instead, soapberry bugs in that region now feed on the seeds of the goldenrain tree (*Koelreuteria elegans*), a species recently introduced from Asia.

Soapberry bugs feed most effectively when the length of their beak is similar to the depth at which seeds are found within the fruit. Goldenrain tree fruit consists of three flat lobes, and its seeds are much closer to the fruit surface than are the seeds of the plump, round fruit of the native balloon vine. These differences led researchers to predict that in populations that feed on goldenrain tree, natural selection would result in beaks that are *shorter* than those in populations that feed on balloon vine **(Figure 22.13)**. Indeed, beak lengths are shorter in the populations that feed on goldenrain tree.

▼ Figure 22.13

Inquiry **Can a change in a population's food source result in evolution by natural selection?**

Field Study Soapberry bugs feed most effectively when the length of their "beak" is similar to the depth of the seeds within the fruit. Scott Carroll and his colleagues measured beak lengths in soapberry bug populations feeding on the native balloon vine. They also measured beak lengths in populations feeding on the introduced goldenrain tree. The researchers then compared the measurements with those of museum specimens collected in the two areas before the goldenrain tree was introduced.

Soapberry bug with beak inserted in balloon vine fruit

Results Beak lengths were shorter in populations feeding on the introduced species than in populations feeding on the native species, in which the seeds are buried more deeply. The average beak length in museum specimens from each population (indicated by red arrows) was similar to beak lengths in populations feeding on native species.

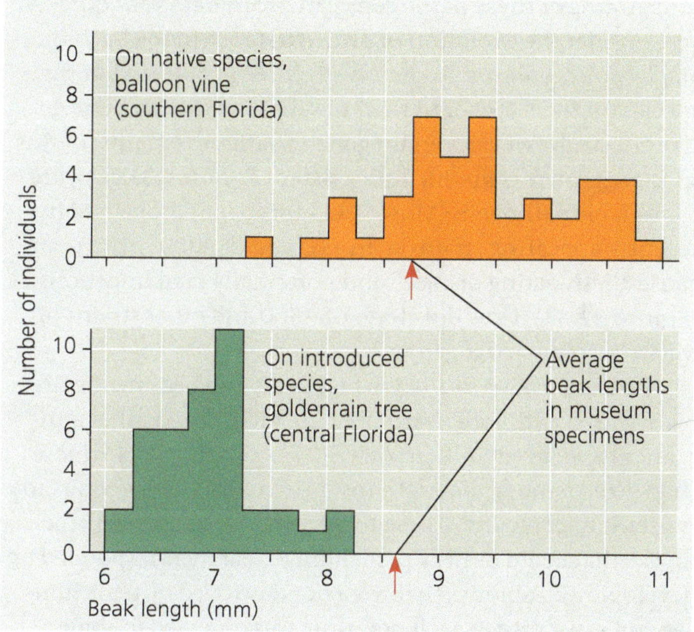

Conclusion Museum specimens and contemporary data suggest that a change in the size of the soapberry bug's food source can result in evolution by natural selection for a corresponding change in beak size.

Data from S. P. Carroll and C. Boyd, Host race radiation in the soapberry bug: natural history with the history, *Evolution* 46:1052–1069 (1992).

WHAT IF? ➤ *Data from additional studies showed that when soapberry bug eggs from a population that fed on balloon vine fruits were then reared on goldenrain tree fruits (or vice versa), the beak lengths of the adult insects were most similar to those in the population from which the eggs were initially obtained. Interpret these results.*

Researchers have also studied beak length evolution in soapberry bug populations that feed on plants introduced to Louisiana, Oklahoma, and Australia. In each of these locations, the fruit of the introduced plants is larger than the fruit of the native plant. Thus, in populations feeding on introduced species in these regions, researchers predicted that natural selection would result in the evolution of *longer* beaks. Again, data collected in field studies upheld this prediction.

The observed changes in beak lengths had important consequences: In Australia, for example, the increase in beak length nearly doubled the success with which soapberry bugs could eat the seeds of the introduced species. Furthermore, since historical data show that the goldenrain tree reached central Florida just 35 years before the scientific studies were initiated, the results demonstrate that natural selection can cause rapid evolution in a wild population.

The Evolution of Drug-Resistant Bacteria

An example of ongoing natural selection that dramatically affects humans is the evolution of drug-resistant pathogens (disease-causing organisms and viruses). This is a particular problem with bacteria and viruses because they can produce new generations in a short period of time; as a result, resistant strains of these pathogens can proliferate very quickly.

Consider the evolution of drug resistance in the bacterium *Staphylococcus aureus*. About one in three people harbor this species on their skin or in their nasal passages with no negative effects. However, certain genetic varieties (strains) of this species, known as methicillin-resistant *S. aureus* (MRSA), are formidable pathogens. Most MRSA infections are caused by recently appearing strains such as clone USA300, which can cause "flesh-eating disease" and potentially fatal infections **(Figure 22.14)**. How did clone USA300 and other strains of MRSA become so dangerous?

The story begins in 1943, when penicillin became the first widely used antibiotic. Since then, penicillin and other antibiotics have saved millions of lives. However, by 1945, more than 20% of the *S. aureus* strains seen in hospitals were already resistant to penicillin. These bacteria had an enzyme, penicillinase, that could destroy penicillin. Researchers responded by developing antibiotics that were not destroyed by penicillinase, but resistance to each new drug was observed in some *S. aureus* populations within a few years.

Then, in 1959, doctors began using a promising new antibiotic, methicillin. But within two years, methicillin-resistant strains of *S. aureus* were observed. How did these resistant strains emerge? Methicillin works by deactivating an enzyme that bacteria use to synthesize their cell walls. However, some *S. aureus* populations included individuals that were able to synthesize their cell walls using a different enzyme that was not affected by methicillin. These individuals survived the methicillin treatments and reproduced at higher rates than did other individuals. Over time, these resistant individuals became increasingly common, leading to the spread of MRSA.

▼ Figure 22.14 Clone USA300: a virulent strain of methicillin-resistant *Staphylococcus aureus* (MRSA). Resistant to multiple antibiotics and highly contagious, this strain and its close relatives can cause lethal infections of the skin, lungs, and blood. As shown here, researchers have identified key areas of the USA300 genome that code for adaptations that cause its virulent properties.

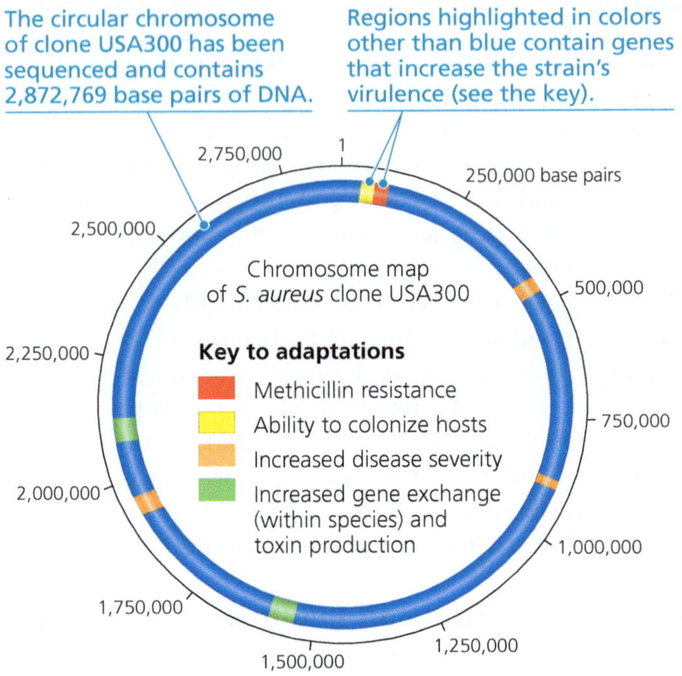

WHAT IF? ➤ *Some drugs being developed specifically target and kill only* S. aureus; *others slow the growth of MRSA but do not kill it. Based on how natural selection works and on the fact that bacterial species can exchange genes, explain why each of these strategies might be effective.*

Initially, MRSA could be controlled by antibiotics that work differently from the way methicillin works. But this has become less effective because some MRSA strains are resistant to multiple antibiotics—probably because bacteria can exchange genes with members of their own and other species. Thus, the multidrug-resistant strains of today may have emerged over time as MRSA strains that were resistant to different antibiotics exchanged genes.

Finally, it is important to note that *S. aureus* is not the only pathogenic bacterium that has evolved resistance to multiple antibiotics. Furthermore, in recent decades, antibiotic resistance has spread much faster than new antibiotics have been discovered—a problem of great public health concern. Hope may loom on the horizon, however. For example, in 2015, scientists reported the discovery of "teixobactin," a new antibiotic that shows promise for treating MRSA and other pathogens. In addition, as we'll describe in the Scientific Skills Exercise in Chapter 27, the methods used in the discovery of teixobactin may lead to the discovery of other new antibiotics as well.

The *S. aureus* and soapberry bug examples highlight three key points about natural selection. First, natural selection is a process of editing, not a creative mechanism. A drug does not *create* resistant pathogens; it *selects for* resistant individuals that are already present in the population. Second, in species that produce new generations in short periods of time, evolution by natural selection can occur rapidly—in just a few years (*S. aureus*) or decades (soapberry bugs). Third, natural selection depends on time and place. It favors those characteristics in a genetically variable population that provide an advantage in the current, local environment. What is beneficial in one situation may be useless or even harmful in another. Beak lengths suitable for the size of the typical fruit eaten by members of a particular soapberry bug population are favored by natural selection. However, a beak length suitable for fruit of one size can be disadvantageous when the bug is feeding on fruit of another size.

MB **BBC Video: A Future Without Antibiotics**

Figure 22.15 Mammalian forelimbs: homologous structures. Even though they have become adapted for different functions, the forelimbs of all mammals are constructed from the same basic skeletal elements: one large bone (purple), attached to two smaller bones (orange and tan), attached to several small bones (gold), attached to several metacarpals (green), attached to approximately five digits, each of which is composed of multiple phalanges (blue).

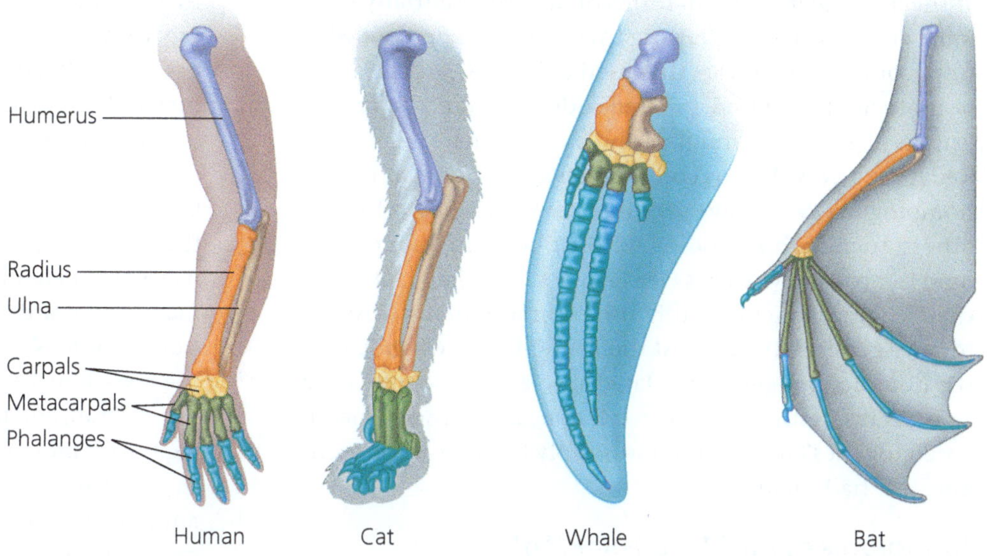

Homology

A second type of evidence for evolution comes from analyzing similarities among different organisms. As we've discussed, evolution is a process of descent with modification: Characteristics present in an ancestral organism are altered (by natural selection) in its descendants over time as they face different environmental conditions. As a result, related species can have characteristics that have an underlying similarity yet function differently. Similarity resulting from common ancestry is known as **homology**. As we'll describe in this section, an understanding of homology can be used to make testable predictions and explain observations that are otherwise puzzling.

Anatomical and Molecular Homologies

The view of evolution as a remodeling process leads to the prediction that closely related species should share similar features—and they do. Of course, closely related species share the features used to determine their relationship, but they also share many other features. Some of these shared features make little sense except in the context of evolution. For example, the forelimbs of all mammals, including humans, cats, whales, and bats, show the same arrangement of bones from the shoulder to the tips of the digits, even though the appendages have very different functions: lifting, walking, swimming, and flying **(Figure 22.15)**. Such striking anatomical resemblances would be highly unlikely if these structures had arisen anew in each species. Rather, the underlying skeletons of the arms, forelegs, flippers, and wings of different mammals are **homologous structures** that represent variations on a structural theme that was present in their common ancestor.

Comparing early stages of development in different animal species reveals additional anatomical homologies not visible in adult organisms. For example, at some point in their development, all vertebrate embryos have a tail located posterior to (behind) the anus, as well as structures called pharyngeal (throat) arches **(Figure 22.16)**. These homologous throat arches ultimately develop into structures with very different functions, such as gills in fishes and parts of the ears and throat in humans and other mammals.

Some of the most intriguing homologies concern "leftover" structures of marginal, if any, importance to the organism. These **vestigial structures** are remnants of features that served a function in the organism's ancestors. For instance, the skeletons of some snakes retain vestiges of the pelvis and leg bones of walking ancestors. Another example is provided by eye remnants that are buried under scales in blind species

Figure 22.16 Anatomical similarities in vertebrate embryos. At some stage in their embryonic development, all vertebrates have a tail located posterior to the anus (referred to as a post-anal tail), as well as pharyngeal (throat) arches. Descent from a common ancestor can explain such similarities.

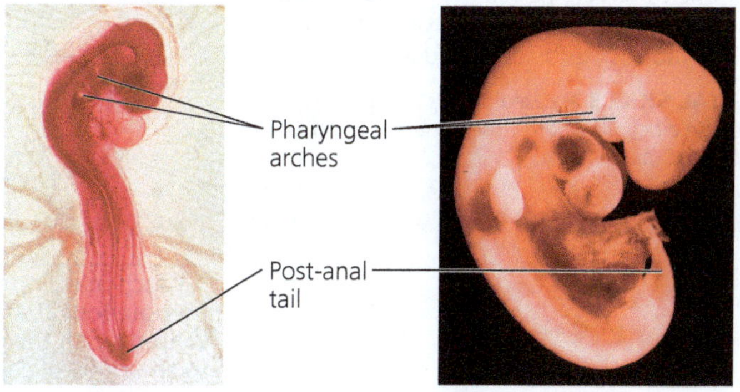

Chick embryo (LM)   Human embryo

of cave fishes. We would not expect to see these vestigial structures if snakes and blind cave fishes had origins separate from those of other vertebrate animals.

Biologists also observe similarities among organisms at the molecular level. All forms of life use essentially the same genetic code, suggesting that all species descended from common ancestors that used this code. But molecular homologies go beyond a shared code. For example, organisms as dissimilar as humans and bacteria share genes inherited from a very distant common ancestor. Some of these homologous genes have acquired new functions, while others, such as those coding for the ribosomal subunits used in protein synthesis (see Figure 17.18), have retained their original functions. It is also common for organisms to have genes that have lost their function, even though the homologous genes in related species may be fully functional. Like vestigial structures, it appears that such inactive "pseudogenes" may be present simply because a common ancestor had them.

Homologies and "Tree Thinking"

Some homologous characteristics, such as the genetic code, are shared by all species because they date to the deep ancestral past. In contrast, homologous characteristics that evolved more recently are shared only within smaller groups of organisms. Consider the *tetrapods* (from the Greek *tetra*, four, and *pod*, foot), the vertebrate group that consists of amphibians, mammals, and reptiles. Like all vertebrates, tetrapods have a backbone. But unlike other vertebrates, tetrapods also have limbs with digits (see Figure 22.15). As suggested by this example, homologous characteristics form a nested pattern: All life shares the deepest layer (in this case, all vertebrates have a backbone), and each successive smaller group adds its own homologies to those it shares with larger groups (in this case, all tetrapods have a backbone *and* limbs with digits). This nested pattern is exactly what we would expect to result from descent with modification from a common ancestor.

Biologists often represent the pattern of descent from common ancestors with an **evolutionary tree**, a diagram that reflects evolutionary relationships among groups of organisms. We will explore evolutionary trees in more detail in Chapter 26, but for now, let's consider how we can interpret and use such trees.

Figure 22.17 is an evolutionary tree of tetrapods and their closest living relatives, the lungfishes. In this diagram, each branch point represents the most recent common ancestor of the two lineages diverging from that point. For example, lungfishes and all tetrapods descended from ancestor (1), whereas mammals, lizards and snakes, crocodiles, and birds all descended from ancestor (3). As expected, the three homologies shown on the tree—limbs with digits, the amnion (a protective embryonic membrane), and feathers—form a nested pattern. Limbs with digits were present in common ancestor (2) and hence are found in all of the descendants of that ancestor (the tetrapods). The amnion was present only in ancestor (3) and hence is shared only by some tetrapods (mammals and reptiles). Feathers were present only in ancestor (6) and hence are found only in birds.

To explore "tree thinking" further, note that in Figure 22.17, mammals are positioned closer to amphibians than to birds. As a result, you might conclude that mammals are more closely related to amphibians than they are to birds. However, mammals are actually more closely related to birds than to amphibians because mammals and birds share a more recent common ancestor (ancestor (3)) than do mammals and amphibians (ancestor (2)). Ancestor (2) is also the most recent common ancestor of birds and amphibians, making mammals and birds equally related to amphibians. Finally, note that the tree in Figure 22.17 shows the relative timing of events but not their actual dates. Thus, we can conclude that ancestor (2) lived before ancestor (3), but we do not know when that was.

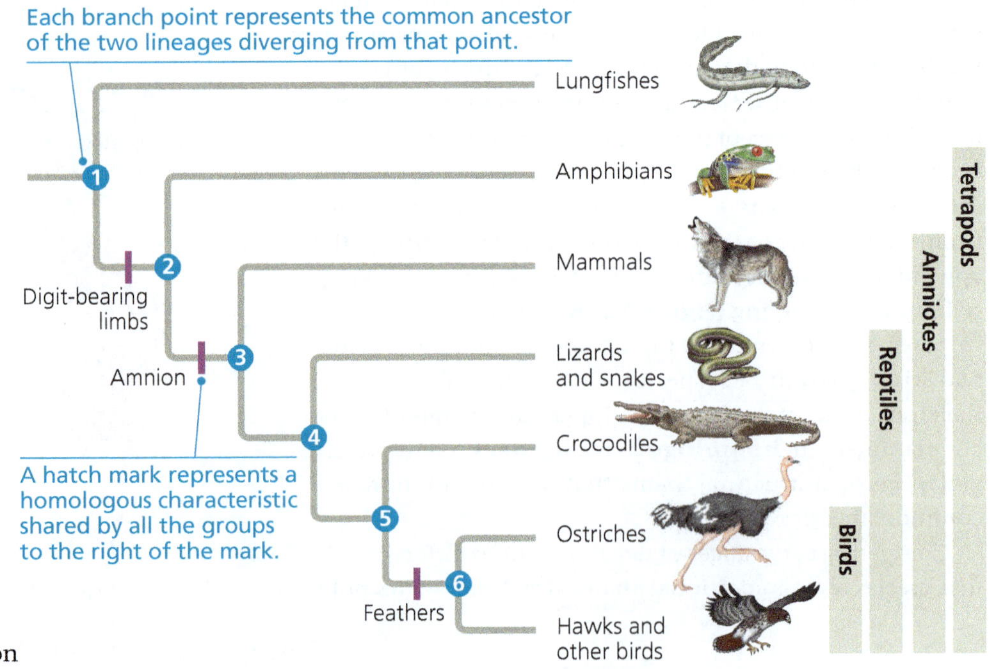

➤ Figure 22.17 Tree thinking: information provided in an evolutionary tree. This evolutionary tree for tetrapods and their closest living relatives, the lungfishes, is based on anatomical and DNA sequence data. The purple bars indicate the origin of three important homologies, each of which evolved only once. Birds are nested within and evolved from reptiles; hence, the group of organisms called "reptiles" technically includes birds.

VISUAL SKILLS ➤ *Based on this evolutionary tree, are crocodiles more closely related to lizards or birds? Explain.*

MB **Figure Walkthrough**

Evolutionary trees are hypotheses that summarize our current understanding of patterns of descent. Our confidence in these relationships, as with any hypothesis, depends on the strength of the supporting data. In the case of Figure 22.17, the tree is supported by many different data sets, including both anatomical and DNA sequence data. As a result, biologists are confident that it accurately reflects evolutionary history. Scientists can use such well-supported evolutionary trees to make specific and sometimes surprising predictions about organisms (see Figure 26.17).

▲ **Figure 22.18 Convergent evolution.** The ability to glide through the air evolved independently in these two distantly related mammals.

A Different Cause of Resemblance: Convergent Evolution

Although organisms that are closely related share characteristics because of common descent, distantly related organisms can resemble one another for a different reason: **convergent evolution**, the independent evolution of similar features in different lineages. Consider marsupial mammals, many of which live in Australia. Marsupials are distinct from another group of mammals—the placental mammals, or eutherians—few of which live in Australia. (Eutherians complete their embryonic development in the uterus, whereas marsupials are born as embryos and complete their development in an external pouch.) Some Australian marsupials have eutherian look-alikes with superficially similar adaptations. For instance, a forest-dwelling Australian marsupial called the sugar glider is superficially very similar to flying squirrels, gliding eutherians that live in North American forests **(Figure 22.18)**. But the sugar glider has many other characteristics that make it a marsupial, much more closely related to kangaroos and other Australian marsupials than to flying squirrels or other eutherians. Once again, our understanding of evolution can explain these observations. Although they evolved independently from different ancestors, these two mammals have adapted to similar environments in similar ways. In such examples in which species share features because of convergent evolution, the resemblance is said to be **analogous**, not homologous. Analogous features share similar function, but not common ancestry, while homologous features share common ancestry, but not necessarily similar function.

The Fossil Record

A third type of evidence for evolution comes from fossils. The fossil record documents the pattern of evolution, showing that past organisms differed from present-day organisms and that many species have become extinct. Fossils also show the evolutionary changes that have occurred in various groups of organisms. To give one of hundreds of possible examples, researchers found that over several thousand years, the pelvic bone in fossil stickleback fish became greatly reduced in size. The consistent nature of this change over time suggests that the reduction in the size of the pelvic bone may have been driven by natural selection.

Fossils can also shed light on the origins of new groups of organisms. An example is the fossil record of cetaceans, the mammalian order that includes whales, dolphins, and porpoises. Some of these fossils **(Figure 22.19)** provided strong support for a hypothesis based on DNA sequence data: that cetaceans are closely related to even-toed ungulates, a group that includes hippopotamuses, pigs, deer, and cows.

What else can fossils tell us about cetacean origins? The earliest cetaceans lived 50–60 million years ago. The fossil record indicates that prior to that time, most mammals were terrestrial. Although scientists had long realized that whales and other cetaceans originated from land mammals, few fossils

▼ **Figure 22.19 Ankle bones: one piece of the puzzle.** Comparing fossils and present-day examples of the astragalus (a type of ankle bone) indicates that cetaceans are closely related to even-toed ungulates. **(a)** In most mammals, the astragalus is shaped like that of a dog, with a double hump on one end (red arrows) but not at the opposite end (blue arrow). **(b)** Fossils show that the early cetacean *Pakicetus* had an astragalus with double humps at both ends, a shape otherwise found only in pigs **(c)**, deer **(d)**, and all other even-toed ungulates.

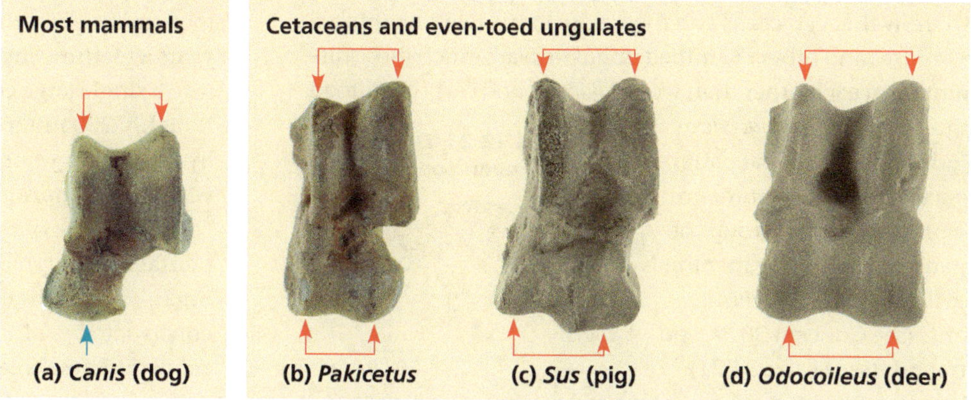

Figure 22.20 The transition to life in the sea. Multiple lines of evidence support the hypothesis that cetaceans (highlighted in yellow) evolved from terrestrial mammals. Fossils document the reduction over time in the pelvis and hind limb bones of extinct (†) cetacean ancestors, including *Pakicetus, Rodhocetus,* and *Dorudon*. DNA sequence data support the hypothesis that cetaceans are most closely related to hippopotamuses.

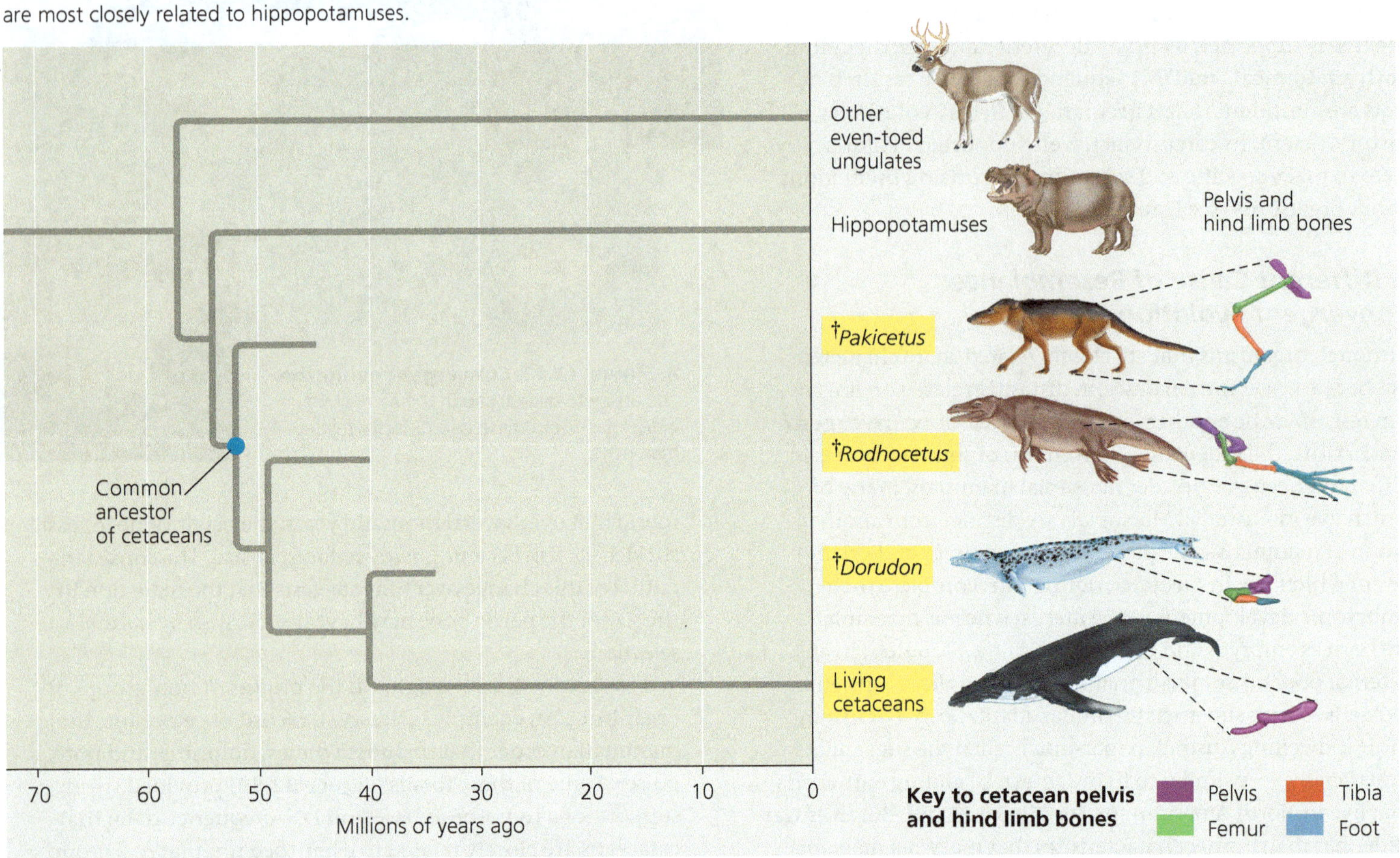

VISUAL SKILLS ➤ *Use the diagram to determine which happened first during the evolution of cetaceans: changes in hind limb structure or the origin of tail flukes. Explain.*

had been found that revealed how cetacean limb structure had changed over time, leading eventually to the loss of hind limbs and the development of flukes (the lobes on a whale's tail) and flippers. In the past few decades, however, a series of remarkable fossils have been discovered in Pakistan, Egypt, and North America. These fossils document steps in the transition from life on land to life in the sea, filling in some of the gaps between ancestral and living cetaceans **(Figure 22.20)**.

Collectively, the recent fossil discoveries document the origin of a group of mammals, the cetaceans. These discoveries also show that cetaceans and their close living relatives (hippopotamuses and other even-toed ungulates) are much more different from each other than were *Pakicetus* and early even-toed ungulates, such as *Diacodexis* **(Figure 22.21)**. Similar patterns are seen in fossils documenting the origins of other groups of organisms, including mammals (see Figure 25.7), flowering plants (see Concept 30.3), and tetrapods (see Figure 34.21).

Figure 22.21 *Diacodexis*, an early even-toed ungulate.

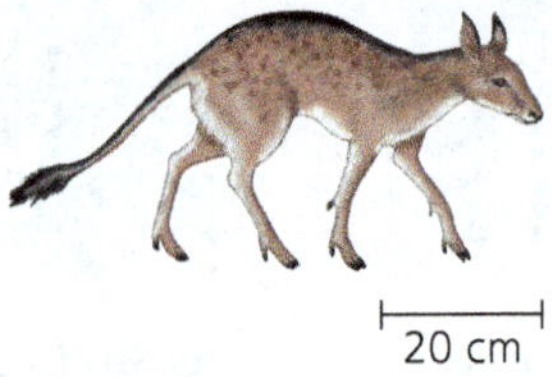

In each of these cases, the fossil record shows that over time, descent with modification produced increasingly large differences among related groups of organisms, ultimately resulting in the diversity of life we see today.

Biogeography

A fourth type of evidence for evolution comes from the field of **biogeography**, the scientific study of the geographic distributions of species. The geographic distributions of organisms are influenced by many factors, including *continental drift*, the slow movement of Earth's continents over time. About 250 million years ago, these movements united all of Earth's landmasses into a single large continent called **Pangaea** (see Figure 25.16). Roughly 200 million years ago, Pangaea began to break apart; by 20 million years ago, the continents we know today were within a few hundred kilometers of their present locations.

We can use our understanding of evolution and continental drift to predict where fossils of different groups of organisms might be found. For example, scientists have constructed evolutionary trees for horses based on anatomical data. These trees and the ages of fossils of horse ancestors suggest that the genus that includes present-day horses (*Equus*) originated 5 million years ago in North America. Geologic evidence indicates that at that time, North and South America were not yet connected, making it difficult for horses to travel between them. Thus, we would predict that the oldest *Equus* fossils should be found only on the continent on which the group originated—North America. This prediction and others like it for different groups of organisms have been upheld, providing more evidence for evolution.

We can also use our understanding of evolution to explain biogeographic data. For example, islands generally have many plant and animal species that are **endemic** (found nowhere else in the world). Yet, as Darwin described in *The Origin of Species*, most island species are closely related to species from the nearest mainland or a neighboring island. He explained this observation by suggesting that islands are colonized by species from the nearest mainland. These colonists eventually give rise to new species as they adapt to their new environments. Such a process also explains why two islands with similar environments in distant parts of the world tend to be populated not by species that are closely related to each other, but rather by species related to those of the nearest mainland, where the environment is often quite different.

What Is Theoretical About Darwin's View of Life?

Some people dismiss Darwin's ideas as "just a theory." However, as we have seen, the *pattern* of evolution—the observation that life has evolved over time—has been documented directly and is supported by a great deal of evidence. In addition, Darwin's explanation of the *process* of evolution—that natural selection is the primary cause of the observed pattern of evolutionary change—makes sense of massive amounts of data. The effects of natural selection also can be observed and tested in nature. One such experiment is described in the **Scientific Skills Exercise**.

SCIENTIFIC SKILLS EXERCISE

Making and Testing Predictions

Can Predation Result in Natural Selection for Color Patterns in Guppies? Our understanding of evolution changes constantly as new observations lead to new hypotheses—and hence to new ways to test our understanding of evolutionary theory. Consider the wild guppies (*Poecilia reticulata*) that live in pools connected by streams on the Caribbean island of Trinidad. Male guppies have highly varied color patterns that are controlled by genes that are only expressed in adult males. Female guppies choose males with bright color patterns as mates more often than they choose males with drab coloring. But the bright colors that attract females also can make the males more conspicuous to predators. Researchers observed that in pools with few predator species, the benefits of bright colors appear to "win out," and males are more brightly colored than in pools where predation is more intense.

One guppy predator, the killifish, preys on juvenile guppies that have not yet displayed their adult coloration. Researchers predicted that if adult guppies with drab colors were transferred to a pool with only killifish, eventually the descendants of these guppies would be more brightly colored (because of the female preference for brightly colored males).

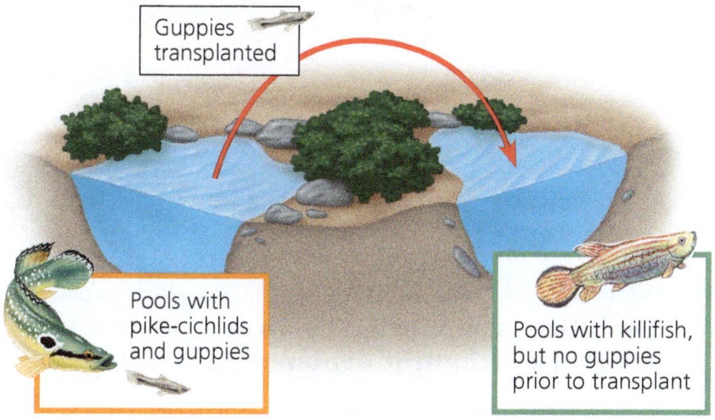

How the Experiment Was Done Researchers transplanted 200 guppies from pools containing pike-cichlid fish, intense predators of adult guppies, to pools containing killifish, less active predators that prey mainly on juvenile guppies. They tracked the number of bright-colored spots and the total area of those spots on male guppies in each generation.

Data from the Experiment After 22 months (15 generations), researchers compared the color pattern data for guppies from the source and transplanted populations.

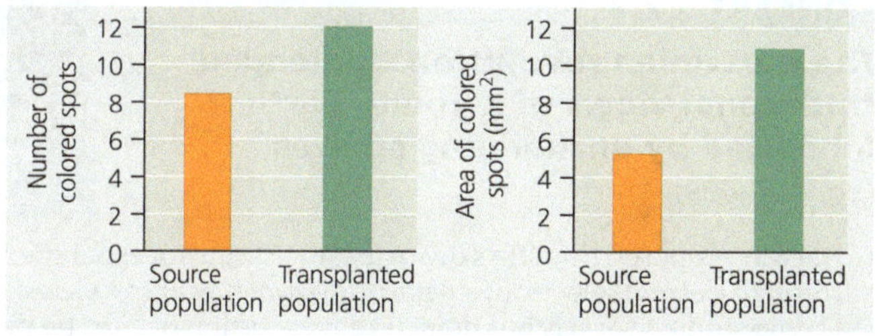

Data from J. A. Endler, Natural selection on color patterns in *Poecilia reticulata*, *Evolution* 34:76–91 (1980).

INTERPRET THE DATA

1. Identify the following elements of hypothesis-based science in this example: (a) question, (b) hypothesis, (c) prediction, (d) control group, and (e) experimental group. (For additional information about hypothesis-based science, see Chapter 1 and the Scientific Skills Review in Appendix F and the Study Area of MasteringBiology.)
2. Explain how the types of data the researchers chose to collect enabled them to test their prediction.
3. What conclusion do you draw from the data presented above?
4. Predict what would happen if, after 22 months, guppies from the transplanted population were returned to the source pool. Describe an experiment to test your prediction.

MB **Instructors**: A version of this Scientific Skills Exercise can be assigned in MasteringBiology.

What, then, is theoretical about evolution? Keep in mind that the scientific meaning of the term *theory* is very different from its meaning in everyday use. The colloquial use of the word *theory* comes close to what scientists mean by a hypothesis. In science, a theory is more comprehensive than a hypothesis. A theory, such as the theory of evolution by natural selection, accounts for many observations and explains and integrates a great variety of phenomena. Such a unifying theory does not become widely accepted unless its predictions stand up to thorough and continual testing by experiment and additional observation (see Concept 1.3). As the rest of this unit demonstrates, this has certainly been the case with the theory of evolution by natural selection.

The skepticism of scientists as they continue to test theories prevents these ideas from becoming dogma. For example, although Darwin thought that evolution was a very slow process, we now know that this isn't always true. Populations can evolve rapidly, and new species can form in relatively short periods of time: a few thousand years or less. Furthermore, evolutionary biologists now recognize that natural selection is not the only mechanism responsible for evolution. Indeed, the study of evolution today is livelier than ever as scientists use a wide range of experimental approaches and genetic analyses to test predictions based on natural selection and other evolutionary mechanisms.

Although Darwin's theory attributes life's diversity to natural processes, the diverse products of evolution are nevertheless elegant and inspiring. As Darwin wrote in the final sentence of *The Origin of Species*, "There is grandeur in this view of life . . . [in which] endless forms most beautiful and most wonderful have been, and are being, evolved."

CONCEPT CHECK 22.3

1. Explain how the following statement is inaccurate: "Antibiotics have created drug resistance in MRSA."
2. How does evolution account for (a) the similar mammalian forelimbs with different functions shown in Figure 22.15 and (b) the similar forms of the two distantly related mammals shown in Figure 22.18?
3. **WHAT IF?** ➤ Fossils show that dinosaurs originated 200–250 million years ago. Would you expect the geographic distribution of early dinosaur fossils to be broad (on many continents) or narrow (on one or a few continents only)? Explain.

For suggested answers, see Appendix A.

22 Chapter Review

Go to **MasteringBiology**™ for Videos, Animations, Vocab Self-Quiz, Practice Tests, and more in the Study Area.

SUMMARY OF KEY CONCEPTS

VOCAB SELF-QUIZ
goo.gl/6u5Sks

CONCEPT 22.1

The Darwinian revolution challenged traditional views of a young Earth inhabited by unchanging species *(pp. 467–469)*

- Darwin proposed that life's diversity arose from ancestral species through natural selection, a departure from prevailing views.
- Cuvier studied **fossils** but denied that evolution occurs; he proposed that sudden catastrophic events in the past caused species to disappear from an area.
- Hutton and Lyell thought that geologic change could result from gradual mechanisms that operated in the past in the same manner as they do today.
- Lamarck hypothesized that species evolve, but the underlying mechanisms he proposed are not supported by evidence.

? *Why was the age of Earth important for Darwin's ideas about evolution?*

CONCEPT 22.2

Descent with modification by natural selection explains the adaptations of organisms and the unity and diversity of life *(pp. 469–474)*

- Darwin's experiences during the voyage of the *Beagle* gave rise to his idea that new species originate from ancestral forms through the accumulation of **adaptations**. He refined his theory for many years and finally published it in 1859 after learning that Wallace had come to the same idea.
- In *The Origin of Species*, Darwin proposed that over long periods of time, descent with modification produced the rich diversity of life through the mechanism of **natural selection**.

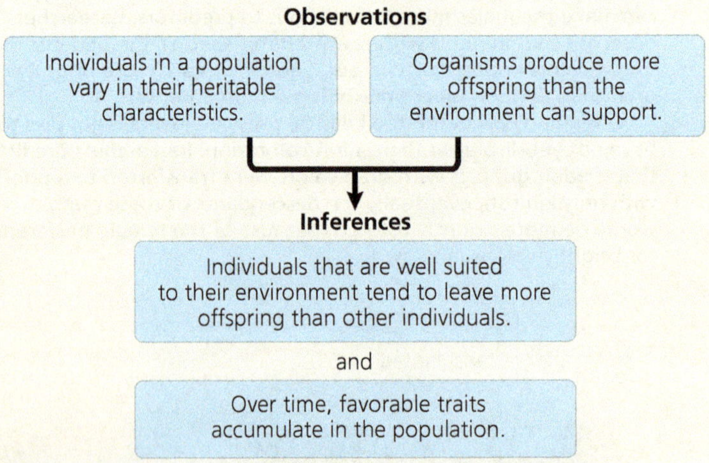

? *Describe how overreproduction and heritable variation relate to evolution by natural selection.*

CONCEPT 22.3

Evolution is supported by an overwhelming amount of scientific evidence *(pp. 475–482)*

- Researchers have directly observed natural selection leading to adaptive evolution in many studies, including research on soapberry bug populations and on MRSA.

- Organisms share characteristics because of common descent (**homology**) or because natural selection affects independently evolving species in similar environments in similar ways (**convergent evolution**).
- Fossils show that past organisms differed from living organisms, that many species have become extinct, and that species have evolved over long periods of time; fossils also document the evolutionary origin of new groups of organisms.
- Evolutionary theory can explain some biogeographic patterns.

Summarize the different lines of evidence supporting the hypothesis that cetaceans descended from land mammals and are closely related to even-toed ungulates.

TEST YOUR UNDERSTANDING

Level 1: Knowledge/Comprehension

PRACTICE TEST
goo.gl/CUYGKD

1. Which of the following is *not* an observation or inference on which natural selection is based?
   (A) There is heritable variation among individuals.
   (B) Poorly adapted individuals never produce offspring.
   (C) Species produce more offspring than the environment can support.
   (D) Only a fraction of an individual's offspring may survive.
2. Which of the following observations helped Darwin shape his concept of descent with modification?
   (A) Species diversity declines farther from the equator.
   (B) Fewer species live on islands than on the nearest continents.
   (C) Birds live on islands located farther from the mainland than the birds' maximum nonstop flight distance.
   (D) South American temperate plants are more similar to the tropical plants of South America than to the temperate plants of Europe.

Level 2: Application/Analysis

3. Within six months of effectively using methicillin to treat *S. aureus* infections in a community, all new *S. aureus* infections were caused by MRSA. How can this best be explained?
   (A) A patient must have become infected with MRSA from another community.
   (B) In response to the drug, *S. aureus* began making drug-resistant versions of the protein targeted by the drug.
   (C) Some drug-resistant bacteria were present at the start of treatment, and natural selection increased their frequency.
   (D) *S. aureus* evolved to resist vaccines.
4. The upper forelimbs of humans and bats have fairly similar skeletal structures, whereas the corresponding bones in whales have very different shapes and proportions. However, genetic data suggest that all three kinds of organisms diverged from a common ancestor at about the same time. Which of the following is the most likely explanation for these data?
   (A) Forelimb evolution was adaptive in people and bats, but not in whales.
   (B) Natural selection in an aquatic environment resulted in significant changes to whale forelimb anatomy.
   (C) Genes mutate faster in whales than in humans or bats.
   (D) Whales are not properly classified as mammals.
5. DNA sequences in many human genes are very similar to the sequences of corresponding genes in chimpanzees. The most likely explanation for this result is that
   (A) humans and chimpanzees share a relatively recent common ancestor.
   (B) humans evolved from chimpanzees.
   (C) chimpanzees evolved from humans.
   (D) convergent evolution led to the DNA similarities.

Level 3: Synthesis/Evaluation

6. **EVOLUTION CONNECTION** Explain why anatomical and molecular features often fit a similar nested pattern. In addition, describe a process that can cause this not to be the case.
7. **SCIENTIFIC INQUIRY • DRAW IT** Mosquitoes resistant to the pesticide DDT first appeared in India in 1959, but now are found throughout the world. (a) Graph the data in the table below. (b) Examine the graph, then hypothesize why the percentage of mosquitoes resistant to DDT rose rapidly. (c) Suggest an explanation for the global spread of DDT resistance.

Month	0	8	12
Mosquitoes Resistant* to DDT	4%	45%	77%

*Mosquitoes were considered resistant if they were not killed within 1 hour of receiving a dose of 4% DDT.

Data from C. F. Curtis et al., Selection for and against insecticide resistance and possible methods of inhibiting the evolution of resistance in mosquitoes, *Ecological Entomology* 3:273–287 (1978).

8. **WRITE ABOUT A THEME: INTERACTIONS** Write a short essay (about 100–150 words) evaluating whether changes to an organism's physical environment are likely to result in evolutionary change. Use an example to support your reasoning.
9. **SYNTHESIZE YOUR KNOWLEDGE**

This honeypot ant (genus *Myrmecocystus*) can store liquid food inside its expandable abdomen. Consider other ants you are familiar with, and explain how a honeypot ant exemplifies three key features of life: adaptation, unity, and diversity.

For selected answers, see Appendix A.

For additional practice questions, check out the **Dynamic Study Modules** in MasteringBiology. You can use them to study on your smartphone, tablet, or computer anytime, anywhere!

The Evolution of Populations 23

Figure 23.1 Is this finch evolving?

KEY CONCEPTS

23.1 Genetic variation makes evolution possible

23.2 The Hardy-Weinberg equation can be used to test whether a population is evolving

23.3 Natural selection, genetic drift, and gene flow can alter allele frequencies in a population

23.4 Natural selection is the only mechanism that consistently causes adaptive evolution

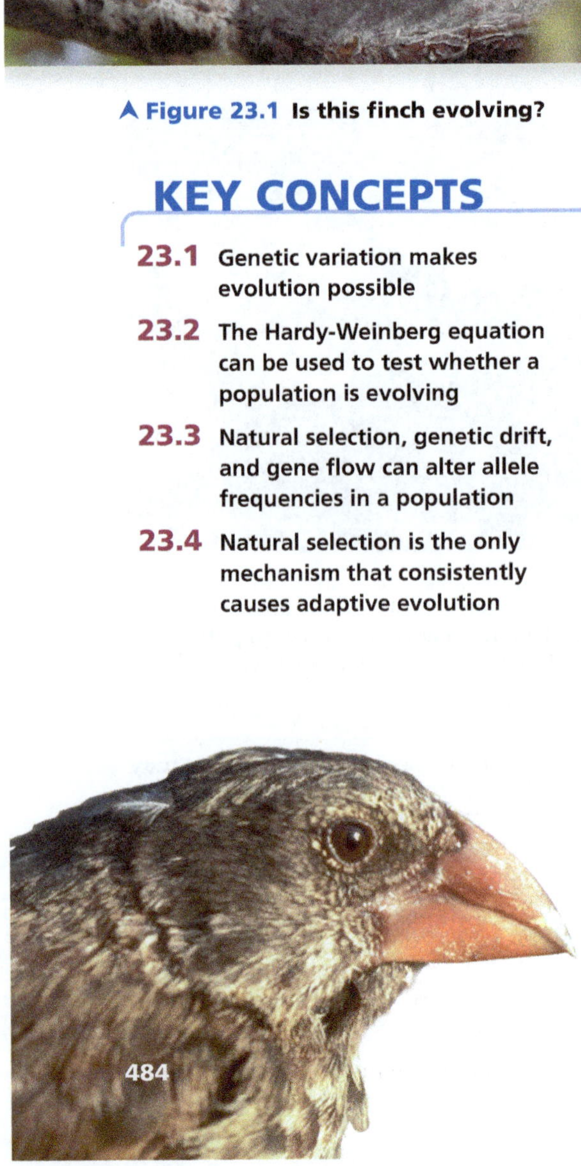

The Smallest Unit of Evolution

One common misconception about evolution is that individual organisms evolve. It is true that natural selection acts on individuals: Each organism's traits affect its survival and reproductive success compared with those of other individuals. But the evolutionary impact of natural selection is only apparent in how a *population* of organisms changes over time.

Consider the medium ground finch (*Geospiza fortis*), a seed-eating bird that inhabits the Galápagos Islands **(Figure 23.1)**. In 1977, the *G. fortis* population on the island of Daphne Major was decimated by a long period of drought: Of some 1,200 birds, only 180 survived. Researchers Peter and Rosemary Grant observed that during the drought, small, soft seeds were in short supply. The finches mostly fed on large, hard seeds that were more plentiful. Birds with larger, deeper beaks were better able to crack and eat these larger seeds, and they survived at a higher rate than did finches with smaller beaks. Since beak depth is an inherited trait in these birds, the offspring of surviving birds also tended to have deep beaks. As a result, the average beak depth in the next generation of *G. fortis* was greater than it had been in the pre-drought population **(Figure 23.2)**. The finch population had evolved by natural selection. However, the *individual* finches did not evolve. Each bird had a beak of a particular size, which did not grow larger during the drought. Rather, the proportion of large

When you see this blue icon, log in to **MasteringBiology** and go to the Study Area for digital resources.

MB **Get Ready for This Chapter**

➤ Figure 23.2 Evidence of selection by food source. The data represent adult beak depth measurements of medium ground finches hatched in the generations before and after the 1977 drought. In one generation, natural selection resulted in a larger average beak size in the population.

Instructors: A related Experimental Inquiry Tutorial can be assigned in MasteringBiology.

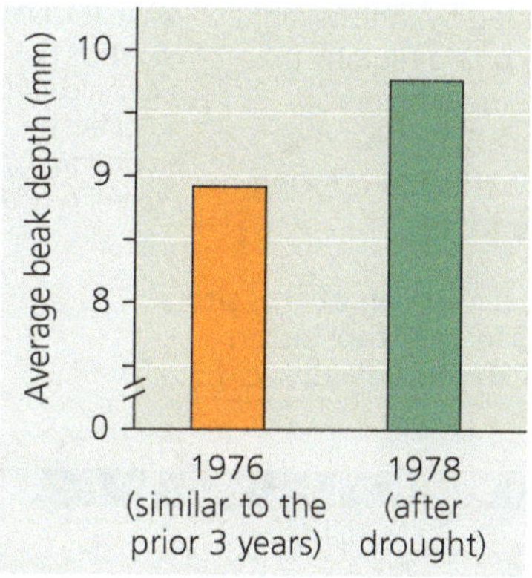

▲ Figure 23.3 Phenotypic variation in horses. In horses, coat color varies along a continuum and is influenced by multiple genes.

beaks in the population increased from generation to generation: The population evolved, not its individual members.

Focusing on evolutionary change in populations, we can define evolution on its smallest scale, called **microevolution**, as a change in allele frequencies in a population over generations. As you will see in this chapter, natural selection is not the only cause of microevolution. In fact, there are three main mechanisms that can cause allele frequency change: natural selection, genetic drift (chance events that alter allele frequencies), and gene flow (the transfer of alleles between populations). Each of these mechanisms has distinctive effects on the genetic composition of populations. However, only natural selection consistently improves the degree to which organisms are well suited for life in their environment (adaptation). Before we examine natural selection and adaptation more closely, let's revisit a prerequisite for these processes in a population: genetic variation.

Interview with Peter and Rosemary Grant: Studying evolution in the Galápagos finches

CONCEPT 23.1

Genetic variation makes evolution possible

In *The Origin of Species*, Darwin provided abundant evidence that life on Earth has evolved over time, and he proposed natural selection as the primary mechanism for that change. He observed that individuals differ in their inherited traits and that selection acts on such differences, leading to evolutionary change. Although Darwin realized that variation in heritable traits is a prerequisite for evolution, he did not know precisely how organisms pass heritable traits to their offspring.

Just a few years after Darwin published *The Origin of Species*, Gregor Mendel wrote a groundbreaking paper on inheritance in pea plants (see Concept 14.1). In that paper, Mendel proposed a model of inheritance in which organisms transmit discrete heritable units (now called genes) to their offspring. Although Darwin did not know about genes, Mendel's paper set the stage for understanding the genetic differences on which evolution is based. Here we'll examine such genetic differences and how they are produced.

Genetic Variation

Individuals within all species vary in their phenotypic traits. Among humans, for example, you can easily observe phenotypic variation in facial features, height, and voice. Indeed, individual variation occurs in all species. And though you cannot identify a person's blood group (A, B, AB, or O) from his or her appearance, this and many other molecular traits also vary extensively among individuals.

Such phenotypic variations often reflect **genetic variation**, differences among individuals in the composition of their genes or other DNA sequences. Some heritable phenotypic differences occur on an "either-or" basis, such as the flower colors of Mendel's pea plants: Each plant had flowers that were either purple or white (see Figure 14.3). Characters that vary in this way are typically determined by a single gene locus, with different alleles producing distinct phenotypes. In contrast, other phenotypic differences vary in gradations along a continuum. Such variation usually results from the influence of two or more genes on a single phenotypic character. In fact, many phenotypic characters are influenced by multiple genes, including coat color in horses **(Figure 23.3)**, seed number in maize (corn), and height in humans.

How much do genes and other DNA sequences vary from one individual to another? Genetic variation at the whole-gene level (*gene variability*) can be quantified as the average percentage of loci that are heterozygous. (Recall that a heterozygous individual has two different alleles for a given locus, whereas a homozygous individual has two identical alleles for that locus.) As an example, on average the fruit fly *Drosophila melanogaster* is heterozygous for about 1,920 of its 13,700 loci (14%) and homozygous for all the rest.

Considerable genetic variation can also be measured at the molecular level of DNA (*nucleotide variability*). But little of this variation results in phenotypic variation. Why? Many nucleotide variations occur within *introns*, noncoding

Figure 23.4 Extensive genetic variation at the molecular level. This diagram summarizes data from a study comparing the DNA sequence of the alcohol dehydrogenase (*Adh*) gene in several fruit flies (*Drosophila melanogaster*). The *Adh* gene has four exons (dark blue) separated by introns (light blue); the exons include the coding regions that are ultimately translated into the amino acids of the *Adh* enzyme (see Figure 5.1). Only one substitution has a phenotypic effect, producing a different form of the *Adh* enzyme.

MAKE CONNECTIONS ➤ *Review Figures 17.6 and 17.11. Explain how a base-pair substitution that alters a coding region of the* Adh *locus could have no effect on amino acid sequence. Then explain how an insertion in an exon could have no effect on the protein produced.*

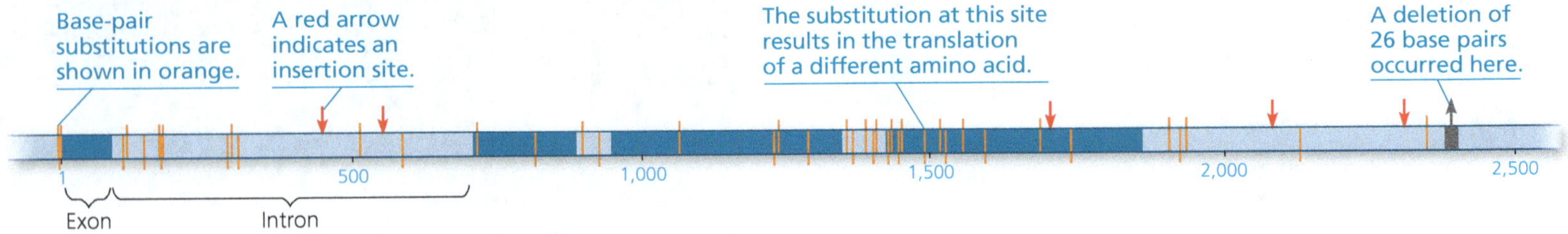

segments of DNA lying between *exons*, the regions retained in mRNA after RNA processing (see Figure 17.12). And of the variations that occur within exons, most do not cause a change in the amino acid sequence of the protein encoded by the gene. For example, in the sequence comparison shown in **Figure 23.4**, there are 43 nucleotide sites with variable base pairs (where substitutions have occurred), as well as several sites where insertions or deletions have occurred. Although 18 variable sites occur within the four exons of the *Adh* gene, only one of these variations (at site 1,490) results in an amino acid change. Note, however, that this single variable site is enough to cause genetic variation at the level of the gene—and hence two different forms of the *Adh* enzyme are produced.

It is important to bear in mind that some phenotypic variation does not result from genetic differences among individuals **(Figure 23.5** shows a striking example in a caterpillar of the southwestern United States). Phenotype is the product of an inherited genotype and many environmental influences (see Concept 14.3). In a human example, bodybuilders alter their phenotypes dramatically but do not pass their huge muscles on to the next generation. In general, only the genetically determined part of phenotypic variation can have evolutionary consequences. As such, genetic variation provides the raw material for evolutionary change: Without genetic variation, evolution cannot occur.

Figure 23.5 Nonheritable variation. These caterpillars of the moth *Nemoria arizonaria* owe their different appearances to chemicals in their diets, not to differences in their genotypes. **(a)** Caterpillars raised on a diet of oak flowers resemble the flowers, whereas **(b)** their siblings raised on oak leaves resemble oak twigs.

Sources of Genetic Variation

The genetic variation on which evolution depends originates when mutation, gene duplication, or other processes produce new alleles and new genes. Genetic variants can be produced rapidly in organisms with short generation times. Sexual reproduction can also result in genetic variation as existing genes are arranged in new ways.

Formation of New Alleles

New alleles can arise by *mutation*, a change in the nucleotide sequence of an organism's DNA. A change of as little as one base in a gene—a "point mutation"—can have a significant impact on phenotype, as in sickle-cell disease (see Figure 17.26). We might expect that this would be the case: Organisms reflect many generations of past selection, and hence their phenotypes tend to be suited for life in their environments. As a result, most new mutations that alter a phenotype are at least slightly harmful.

In some cases, natural selection quickly removes such harmful alleles. In diploid organisms, however, harmful alleles that are recessive can be hidden from selection. Indeed, a harmful recessive allele can persist for generations by propagation in heterozygous individuals (where its harmful effects can be masked by the more favorable dominant allele). Such "heterozygote protection" maintains a huge pool of alleles that might not be favored under present conditions, but that could be beneficial if the environment changes.

While many mutations are harmful, many others are not. Recall that much of the DNA in eukaryotic genomes does not encode proteins (see Figure 21.6). Point mutations in these noncoding regions generally result in **neutral variation**, differences in DNA sequence that do not confer a selective advantage or disadvantage. The redundancy in the genetic code is another source of neutral variation: Even a point mutation in a gene that encodes a protein will have no effect on the protein's function if the amino acid composition is not changed. And even where there is a change in the amino acid, it may not affect the protein's shape and function. Moreover, as you will see later in this chapter, a mutant allele may on rare occasions actually make its bearer better suited to the environment, enhancing reproductive success.

Finally, note that in multicellular organisms, only mutations in cell lines that produce gametes can be passed to offspring. In plants and fungi, this is not as limiting as it may sound, since many different cell lines can produce gametes. But in most animals, the majority of mutations occur in somatic cells and are not passed to offspring.

Altering Gene Number or Position

Chromosomal changes that delete, disrupt, or rearrange many loci are usually harmful. However, when such large-scale changes leave genes intact, they may not affect the organisms' phenotype. In rare cases, chromosomal rearrangements may even be beneficial. For example, the translocation of part of one chromosome to a different chromosome could link genes in a way that produces a positive effect.

A key potential source of variation is the duplication of genes due to errors in meiosis (such as unequal crossing over), slippage during DNA replication, or the activities of transposable elements (see Concept 21.5). Duplications of large chromosome segments, like other chromosomal aberrations, are often harmful, but the duplication of smaller pieces of DNA may not be. Gene duplications that do not have severe effects can persist over generations, allowing mutations to accumulate. The result is an expanded genome with new genes that may take on new functions.

Such increases in gene number appear to have played a major role in evolution. For example, the remote ancestors of mammals had a single gene for detecting odors that has since been duplicated many times. As a result, humans today have about 380 functional olfactory receptor genes, and mice have about 1,200. This dramatic proliferation of olfactory genes probably helped early mammals, enabling them to detect faint odors and to distinguish among many different smells.

Rapid Reproduction

Mutation rates tend to be low in plants and animals, averaging about one mutation in every 100,000 genes per generation, and they are often even lower in prokaryotes. But prokaryotes have many more generations per unit of time, so mutations can quickly generate genetic variation in their populations. The same is true of viruses. For instance, HIV has a generation time of about two days (that is, it takes two days for a newly formed virus to produce the next generation of viruses). HIV also has an RNA genome, which has a much higher mutation rate than a typical DNA genome because of the lack of RNA repair mechanisms in host cells (see Concept 19.2). For this reason, single-drug treatments are unlikely to be effective against HIV: Mutant forms of the virus that are resistant to a particular drug would tend to proliferate in relatively short order. The most effective AIDS treatments to date have been drug "cocktails" that combine several medications. This approach has worked well because it is less likely that a set of mutations that together confer resistance to *all* the drugs will occur in a short time period.

Sexual Reproduction

In organisms that reproduce sexually, most of the genetic variation in a population results from the unique combination of alleles that each individual receives from its parents. Of course, at the nucleotide level, all the differences among these alleles have originated from past mutations. Sexual reproduction then shuffles existing alleles and deals them at random to produce individual genotypes.

Three mechanisms contribute to this shuffling: crossing over, independent assortment of chromosomes, and fertilization (see Concept 13.4). During meiosis, homologous chromosomes, one inherited from each parent, trade some of their alleles by crossing over. These homologous chromosomes and the alleles they carry are then distributed at random into gametes. Then, because myriad possible mating combinations exist in a population, fertilization typically brings together gametes that have different genetic backgrounds. The combined effects of these three mechanisms ensure that sexual reproduction rearranges existing alleles into fresh combinations each generation, providing much of the genetic variation that makes evolution possible.

MB Animation: Origins of Genetic Variation

CONCEPT CHECK 23.1

1. Explain why genetic variation within a population is a prerequisite for evolution.
2. Of all the mutations that occur in a population, why do only a small fraction become widespread?
3. **MAKE CONNECTIONS** ➤ If a population stopped reproducing sexually (but still reproduced asexually), how would its genetic variation be affected over time? Explain. (See Concept 13.4.)

For suggested answers, see Appendix A.

CONCEPT 23.2

The Hardy-Weinberg equation can be used to test whether a population is evolving

Although the individuals in a population must differ genetically for evolution to occur, the presence of genetic variation does not guarantee that a population will evolve. For that to happen, one or more factors that cause evolution must be at work. In this section, we'll explore one way to test whether evolution is occurring in a population. First, let's clarify what we mean by a population.

Figure 23.6 One species, two populations. These two caribou populations in the Yukon are not totally isolated; they sometimes share the same area. Still, members of either population are most likely to breed within their own population.

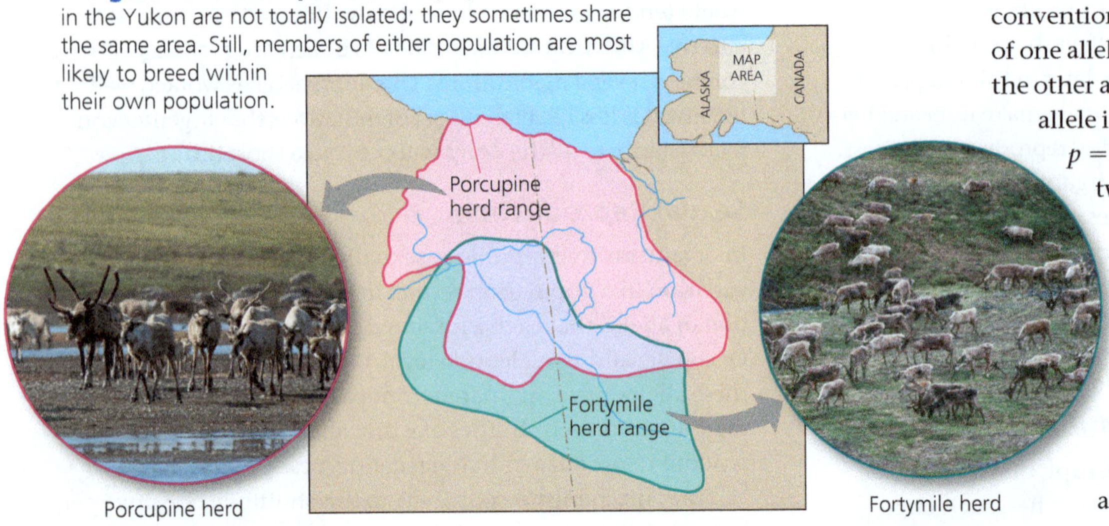

When studying a locus with two alleles, the convention is to use p to represent the frequency of one allele and q to represent the frequency of the other allele. Thus, p, the frequency of the C^R allele in the gene pool of this population, is $p = 0.8$ (80%). And because there are only two alleles for this gene, the frequency of the C^W allele, represented by q, must be $q = 1 - p = 0.2$ (20%). For loci that have more than two alleles, the sum of all allele frequencies must still equal 1 (100%).

Next we'll see how allele and genotype frequencies can be used to test whether evolution is occurring in a population.

Gene Pools and Allele Frequencies

A **population** is a group of individuals of the same species that live in the same area and interbreed, producing fertile offspring. Different populations of a species may be isolated geographically from one another, exchanging genetic material only rarely. Such isolation is common for species that live on widely separated islands or in different lakes. But not all populations are isolated (**Figure 23.6**). Still, members of a population typically breed with one another and thus on average are more closely related to each other than to members of other populations.

We can characterize a population's genetic makeup by describing its **gene pool**, which consists of all copies of every type of allele at every locus in all members of the population. If only one allele exists for a particular locus in a population, that allele is said to be *fixed* in the gene pool, and all individuals are homozygous for that allele. But if there are two or more alleles for a particular locus in a population, individuals may be either homozygous or heterozygous.

For example, imagine a population of 500 wildflower plants with two alleles, C^R and C^W, for a locus that codes for flower pigment. These alleles show incomplete dominance; thus, each genotype has a distinct phenotype. Plants homozygous for the C^R allele (C^RC^R) produce red pigment and have red flowers; plants homozygous for the C^W allele (C^WC^W) produce no red pigment and have white flowers; and heterozygotes (C^RC^W) produce some red pigment and have pink flowers.

Each allele has a frequency (proportion) in the population. For example, suppose our population has 320 plants with red flowers, 160 with pink flowers, and 20 with white flowers. Because these are diploid organisms, these 500 individuals have a total of 1,000 copies of the gene for flower color. The C^R allele accounts for 800 of these copies ($320 \times 2 = 640$ for C^RC^R plants, plus $160 \times 1 = 160$ for C^RC^W plants). Thus, the frequency of the C^R allele is $800/1{,}000 = 0.8$ (80%).

The Hardy-Weinberg Equation

One way to assess whether natural selection or other factors are causing evolution at a particular locus is to determine what the genetic makeup of a population would be if it were *not* evolving at that locus. We can then compare that scenario with the data we actually observed for the population. If there are no differences, we can conclude that the population is not evolving. If there are differences, this suggests that the population may be evolving—and then we can try to figure out why.

Hardy-Weinberg Equilibrium

In a population that is not evolving, allele and genotype frequencies will remain constant from generation to generation, provided that only Mendelian segregation and recombination of alleles are at work. Such a population is said to be in **Hardy-Weinberg equilibrium**, named for the British mathematician and German physician, respectively, who independently developed this idea in 1908.

To determine whether a population is in Hardy-Weinberg equilibrium, it is helpful to think about genetic crosses in a new way. Previously, we used Punnett squares to determine the genotypes of offspring in a genetic cross (see Figure 14.5). Here, instead of considering the possible allele combinations from one cross, we'll consider the combination of alleles in *all* of the crosses in a population.

Imagine that all the alleles for a given locus from all the individuals in a population are placed in a large bin (**Figure 23.7**). We can think of this bin as holding the population's gene pool for that locus. "Reproduction" occurs by selecting alleles at random from the bin; somewhat similar events occur in nature when fish release sperm and eggs into the water or when pollen (containing plant sperm) is blown about by the wind. By viewing reproduction as a process of randomly selecting and combining alleles from the bin (the gene pool), we are in effect assuming that mating occurs at random—that is, that all male-female matings are equally likely.

Figure 23.7 Selecting alleles at random from a gene pool.

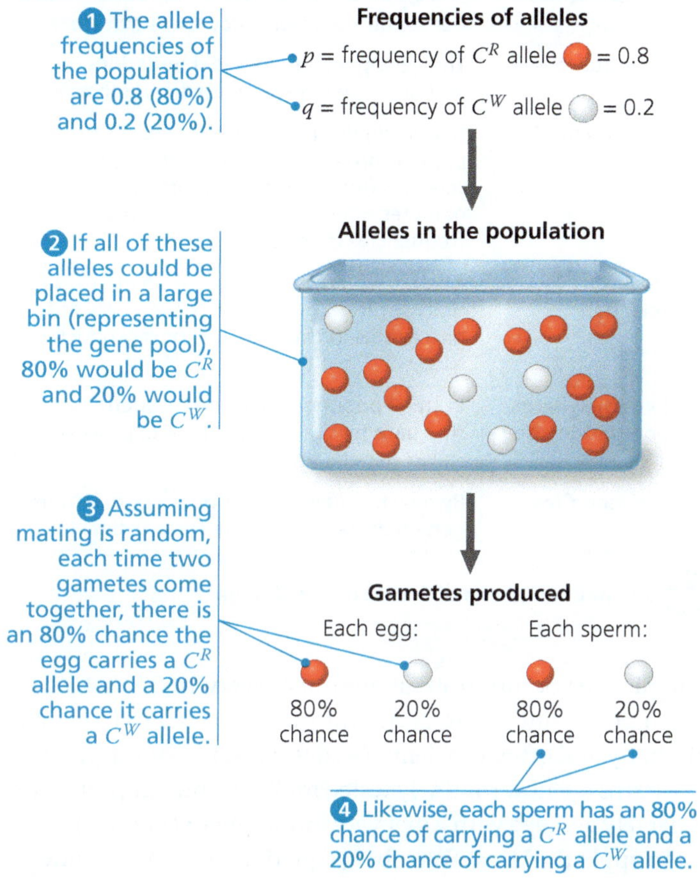

DRAW IT ➤ *Draw a similar bin that contains six white balls instead of four. For the frequency of C^R in the bin to remain equal to 0.8, how many red balls should the bin contain?*

BioFlix® Animation: Allele Frequencies

Let's apply the bin analogy to the hypothetical wildflower population discussed earlier. In that population of 500 flowers, the frequency of the allele for red flowers (C^R) is $p = 0.8$, and the frequency of the allele for white flowers (C^W) is $q = 0.2$. This implies that a bin holding all 1,000 copies of the flower-color gene in the population would contain 800 C^R alleles and 200 C^W alleles. Assuming that gametes are formed by selecting alleles at random from the bin, the probability that an egg or sperm contains a C^R or C^W allele is equal to the frequency of these alleles in the bin. Thus, as shown in Figure 23.7, each egg has an 80% chance of containing a C^R allele and a 20% chance of containing a C^W allele; the same is true for each sperm.

Using the rule of multiplication (see Figure 14.9), we can now calculate the frequencies of the three possible genotypes, assuming random unions of sperm and eggs. The probability that two C^R alleles will come together is $p \times p = p^2 = 0.8 \times 0.8 = 0.64$. Thus, about 64% of the plants in the next generation will have the genotype C^RC^R. The frequency of C^WC^W individuals is expected to be about $q \times q = q^2 = 0.2 \times 0.2 = 0.04$, or 4%. C^RC^W heterozygotes can arise in two different ways. If the sperm provides the C^R allele and the egg provides the C^W allele, the resulting heterozygotes will be $p \times q = 0.8 \times 0.2 = 0.16$, or 16% of the total. If the sperm provides the C^W allele and the egg the C^R allele, the heterozygous offspring will make up $q \times p = 0.2 \times 0.8 = 0.16$, or 16%. The frequency of heterozygotes is thus the sum of these possibilities: $pq + qp = 2pq = 0.16 + 0.16 = 0.32$, or 32%.

As shown in **Figure 23.8**, the genotype frequencies in the next generation must add up to 1 (100%). Thus, the equation for Hardy-Weinberg equilibrium states that at a locus with

Figure 23.8 Hardy-Weinberg equilibrium. In our wildflower population, the gene pool remains constant from one generation to the next. Mendelian processes alone do not alter frequencies of alleles or genotypes.

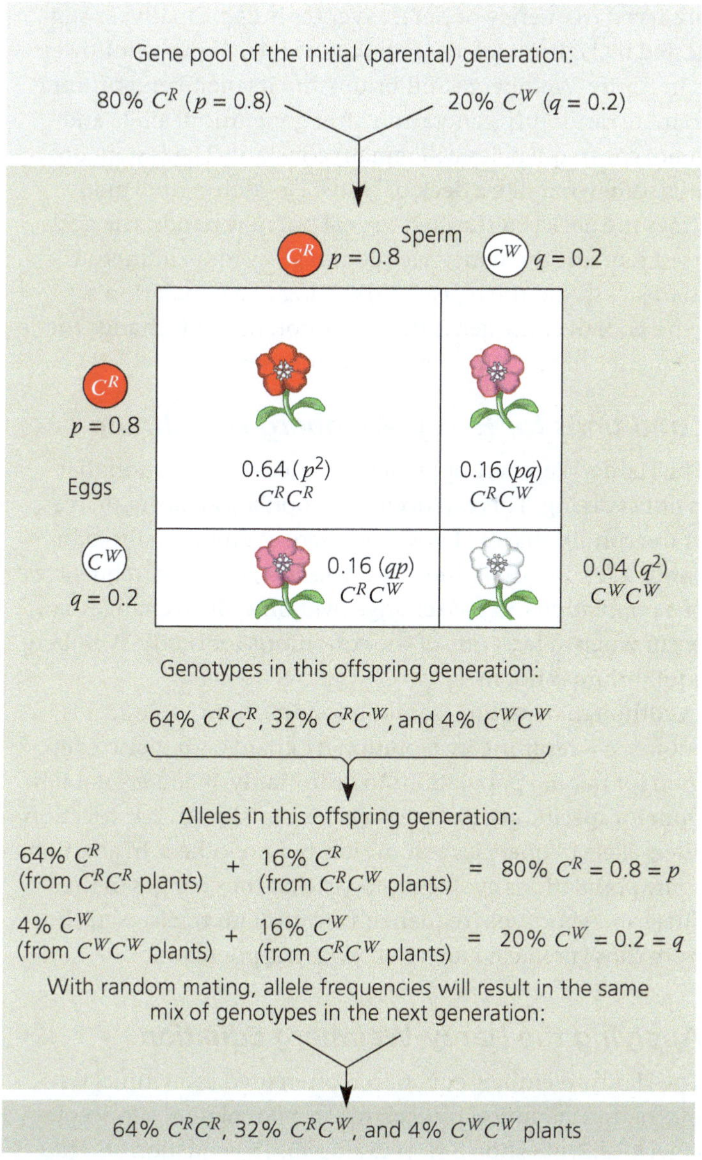

WHAT IF? ➤ *If the frequency of the C^R allele were 0.6, predict the frequencies of the C^RC^R, C^RC^W, and C^WC^W genotypes.*

Figure Walkthrough

two alleles, the three genotypes will appear in the following proportions:

p^2	+	$2pq$	+	q^2	=	1
Expected frequency of genotype C^RC^R		Expected frequency of genotype C^RC^W		Expected frequency of genotype C^WC^W		

Note that for a locus with two alleles, only three genotypes are possible (in this case, C^RC^R, C^RC^W, and C^WC^W). As a result, the sum of the frequencies of the three genotypes must equal 1 (100%) in *any* population—regardless of whether the population is in Hardy-Weinberg equilibrium. The key point is that a population is in Hardy-Weinberg equilibrium only if the observed genotype frequency of one homozygote is p^2, the observed frequency of the other homozygote is q^2, and the observed frequency of heterozygotes is $2pq$. Finally, as suggested by Figure 23.8, if a population such as our wildflowers is in Hardy-Weinberg equilibrium and its members continue to mate randomly generation after generation, allele and genotype frequencies will remain constant. The system operates somewhat like a deck of cards: No matter how many times the deck is reshuffled to deal out new hands, the deck itself remains the same. Aces do not grow more numerous than jacks. And the repeated shuffling of a population's gene pool over the generations cannot, in itself, change the frequency of one allele relative to another.

Conditions for Hardy-Weinberg Equilibrium

The Hardy-Weinberg approach describes a population that is not evolving. This can occur if a population meets all five of the conditions for Hardy-Weinberg equilibrium listed in **Table 23.1**. But in nature, the allele and genotype frequencies of a population often *do* change over time. Such changes can occur when at least one of the conditions for Hardy-Weinberg equilibrium is not met.

Although departure from the conditions in Table 23.1 is common—resulting in evolutionary change—it is also common for natural populations to be in Hardy-Weinberg equilibrium for specific genes. One way this can happen is if selection alters allele frequencies at some loci but not others. In addition, some populations evolve so slowly that the changes in their allele and genotype frequencies are difficult to distinguish from those predicted for a non-evolving population.

Table 23.1 Conditions for Hardy-Weinberg Equilibrium

Condition	Consequence if Condition Does Not Hold
1. No mutations	The gene pool is modified if mutations occur or if entire genes are deleted or duplicated.
2. Random mating	If individuals mate within a subset of the population, such as near neighbors or close relatives (inbreeding), random mixing of gametes does not occur and genotype frequencies change.
3. No natural selection	Allele frequencies change when individuals with different genotypes show consistent differences in their survival or reproductive success.
4. Extremely large population size	In small populations, allele frequencies fluctuate by chance over time (a process called genetic drift).
5. No gene flow	By moving alleles into or out of populations, gene flow can alter allele frequencies.

MB Animation: Causes of Evolutionary Change

Applying the Hardy-Weinberg Equation

The Hardy-Weinberg equation is often used as an initial test of whether evolution is occurring in a population (Concept Check 23.2, question 3 is an example). The equation also has medical applications, such as estimating the percentage of a population carrying the allele for an inherited disease. For example, consider phenylketonuria (PKU), a metabolic disorder that results from homozygosity for a recessive allele. This disorder occurs in about one out of every 10,000 babies born in the United States. Left untreated, PKU results in mental disability and other problems. (As described in Concept 14.4, newborns are now routinely tested for PKU, and symptoms can be largely avoided with a diet very low in phenylalanine.)

To apply the Hardy-Weinberg equation, we must assume that no new PKU mutations are being introduced into the population (condition 1) and that people neither choose their mates on the basis of whether or not they carry this gene nor generally mate with close relatives (condition 2). We must also ignore any effects of differential survival and reproductive success among PKU genotypes (condition 3) and assume that there are no effects of genetic drift (condition 4) or of gene flow from other populations into the United States (condition 5). These assumptions are reasonable: The mutation rate for the PKU gene is low, inbreeding and other forms of nonrandom mating are not common in the United States, selection occurs only against the rare homozygotes (and then only if dietary restrictions are not followed), the U.S. population is very large, and populations outside the country have PKU allele frequencies similar to those seen in the United States.

If all these assumptions hold, then the frequency of individuals in the population born with PKU will correspond to q^2 in the Hardy-Weinberg equation (q^2 = frequency of homozygotes). Because the allele is recessive, we must estimate the number of heterozygotes rather than counting them directly as we did with the pink flowers. Recall that there is one PKU occurrence per 10,000 births, which indicates that $q^2 = 0.0001$. Thus, the frequency (q) of the recessive allele for PKU is

$$q = \sqrt{0.0001} = 0.01$$

and the frequency of the dominant allele is

$$p = 1 - q = 1 - 0.01 = 0.99$$

The frequency of carriers, heterozygous people who do not have PKU but may pass the PKU allele to offspring, is

$$2pq = 2 \times 0.99 \times 0.01 = 0.0198$$
(approximately 2% of the U.S. population)

Remember, the assumption of Hardy-Weinberg equilibrium yields an approximation; the real number of carriers may differ. Still, our calculations suggest that harmful recessive alleles at this and other loci can be concealed in a population because they are carried by healthy heterozygotes. The **Scientific Skills Exercise** provides another opportunity for you to apply the Hardy-Weinberg equation to allele data.

CONCEPT CHECK 23.2

1. A population has 700 individuals, 85 of genotype *AA*, 320 of genotype *Aa*, and 295 of genotype *aa*. What are the frequencies of alleles *A* and *a*?
1. The frequency of allele *a* is 0.45 for a population in Hardy-Weinberg equilibrium. What are the expected frequencies of genotypes *AA*, *Aa*, and *aa*?
1. **WHAT IF?** ➤ A locus that affects susceptibility to a degenerative brain disease has two alleles, *V* and *v*. In a population, 16 people have genotype *VV*, 92 have genotype *Vv*, and 12 have genotype *vv*. Is this population evolving? Explain.

For suggested answers, see Appendix A.

CONCEPT 23.3

Natural selection, genetic drift, and gene flow can alter allele frequencies in a population

Note again the five conditions required for a population to be in Hardy-Weinberg equilibrium (see Table 23.1). A deviation from any of these conditions is a potential cause of evolution. New mutations (violation of condition 1) can alter allele frequencies, but because mutations are rare, the change from one generation to the next is likely to be very small. Nonrandom mating (violation of condition 2) can affect the frequencies of homozygous and heterozygous genotypes but by itself has no effect on allele frequencies in the gene pool. (Allele frequencies can change if individuals with certain inherited traits are more likely than other individuals to obtain mates. However, such a situation not only causes a deviation from random mating, but also violates condition 3, no natural selection.)

For the rest of this section we will focus on the three mechanisms that alter allele frequencies directly and cause most evolutionary change: natural selection, genetic drift, and gene flow (violations of conditions 3–5).

SCIENTIFIC SKILLS EXERCISE

Using the Hardy-Weinberg Equation to Interpret Data and Make Predictions

Is Evolution Occurring in a Soybean Population? One way to test whether evolution is occurring in a population is to compare the observed genotype frequencies at a locus with those expected for a non-evolving population based on the Hardy-Weinberg equation. In this exercise, you'll test whether a soybean population is evolving at a locus with two alleles, C^G and C^Y, that affect chlorophyll production and hence leaf color.

How the Experiment Was Done Students planted soybean seeds and then counted the number of seedlings of each genotype at day 7 and again at day 21. Seedlings of each genotype could be distinguished visually because the C^G and C^Y alleles show incomplete dominance: C^GC^G seedlings have green leaves, C^GC^Y seedlings have green-yellow leaves, and C^YC^Y seedlings have yellow leaves.

Data from the Experiment

Time (days)	Number of Seedlings			
	Green (C^GC^G)	Green-yellow (C^GC^Y)	Yellow (C^YC^Y)	Total
7	49	111	56	216
21	47	106	20	173

INTERPRET THE DATA

1. Use the observed genotype frequencies from the day 7 data to calculate the frequencies of the C^G allele (p) and the C^Y allele (q).
2. Next, use the Hardy-Weinberg equation ($p^2 + 2pq + q^2 = 1$) to calculate the day 7 expected frequencies of genotypes C^GC^G, C^GC^Y, and C^YC^Y for a population in Hardy-Weinberg equilibrium.
3. Calculate the observed frequencies of genotypes C^GC^G, C^GC^Y, and C^YC^Y at day 7. Compare these frequencies to the expected frequencies calculated in question 2. Is the seedling population in Hardy-Weinberg equilibrium at day 7, or is evolution occurring? Explain your reasoning and identify which genotypes, if any, appear to be selected for or against.
4. Calculate the observed frequencies of genotypes C^GC^G, C^GC^Y, and C^YC^Y at day 21. Compare these frequencies to the expected frequencies calculated in question 2 and to the observed frequencies at day 7. Is the seedling population in Hardy-Weinberg equilibrium at day 21, or is evolution occurring? Explain your reasoning and identify which genotypes, if any, appear to be selected for or against.
5. Homozygous C^YC^Y individuals cannot produce chlorophyll. The ability to photosynthesize becomes more critical as seedlings age and begin to exhaust the supply of food that was stored in the seed from which they emerged. Develop a hypothesis that explains the data for days 7 and 21. Based on this hypothesis, predict how the frequencies of the C^G and C^Y alleles will change beyond day 21.

Instructors: A version of this Scientific Skills Exercise can be assigned in MasteringBiology.

Natural Selection

The concept of natural selection is based on differential success in survival and reproduction: Individuals in a population exhibit variations in their heritable traits, and those with traits that are better suited to their environment tend to produce more offspring than those with traits that are not as well suited.

In genetic terms, selection results in alleles being passed to the next generation in proportions that differ from those in the present generation. For example, the fruit fly *D. melanogaster* has an allele that confers resistance to several insecticides, including DDT. This allele has a frequency of 0% in laboratory strains of *D. melanogaster* established from flies collected in the wild in the early 1930s, prior to DDT use. However, in strains established from flies collected after 1960 (following 20 or more years of DDT use), the allele frequency is 37%. We can infer that this allele either arose by mutation between 1930 and 1960 or was present in 1930, but very rare. In any case, the rise in frequency of this allele most likely occurred because DDT is a powerful poison that is a strong selective force in exposed fly populations.

As the *D. melanogaster* example suggests, an allele that confers resistance to an insecticide will increase in frequency in a population exposed to that insecticide. Such changes are not coincidental. By consistently favoring some alleles over others, natural selection can cause **adaptive evolution**, a process in which traits that enhance survival or reproduction tend to increase in frequency over time. We'll explore this process in more detail later in this chapter.

Genetic Drift

If you flip a coin 1,000 times, a result of 700 heads and 300 tails might make you suspicious about that coin. But if you flip a coin only 10 times, an outcome of 7 heads and 3 tails would not be surprising. The smaller the number of coin flips, the more likely it is that chance alone will cause a deviation from the predicted result. (In this case, the prediction is an equal number of heads and tails.) Chance events can also cause allele frequencies to fluctuate unpredictably from one generation to the next, especially in small populations—a process called **genetic drift**.

Figure 23.9 models how genetic drift might affect a small population of our wildflowers. In this example, drift leads to the loss of an allele from the gene pool, but it is a matter of chance that the C^W allele is lost and not the C^R allele. Such unpredictable changes in allele frequencies can be caused by chance events associated with survival and reproduction. Perhaps a large animal such as a moose stepped on the three C^WC^W individuals in generation 2, killing them and increasing the chance that only the C^R allele would be passed to the next generation. Allele frequencies can also be affected by chance events that occur during fertilization. For example, suppose two individuals of genotype C^RC^W had a small number of offspring. By chance alone, every egg and sperm pair that generated offspring could happen to have carried the C^R allele and not the C^W allele.

Certain circumstances can result in genetic drift having a significant impact on a population. Two examples are the founder effect and the bottleneck effect.

▼ **Figure 23.9 Genetic drift.** This small wildflower population has a stable size of ten plants. Suppose that by chance only five plants of generation 1 (those highlighted in yellow) produce fertile offspring. (This could occur, for example, if only those plants happened to grow in a location that provided enough nutrients to support the production of offspring.) Again by chance, only two plants of generation 2 leave fertile offspring.

VISUAL SKILLS ➤ *Based on this diagram, summarize how the frequency of the* C^W *allele changes over time.*

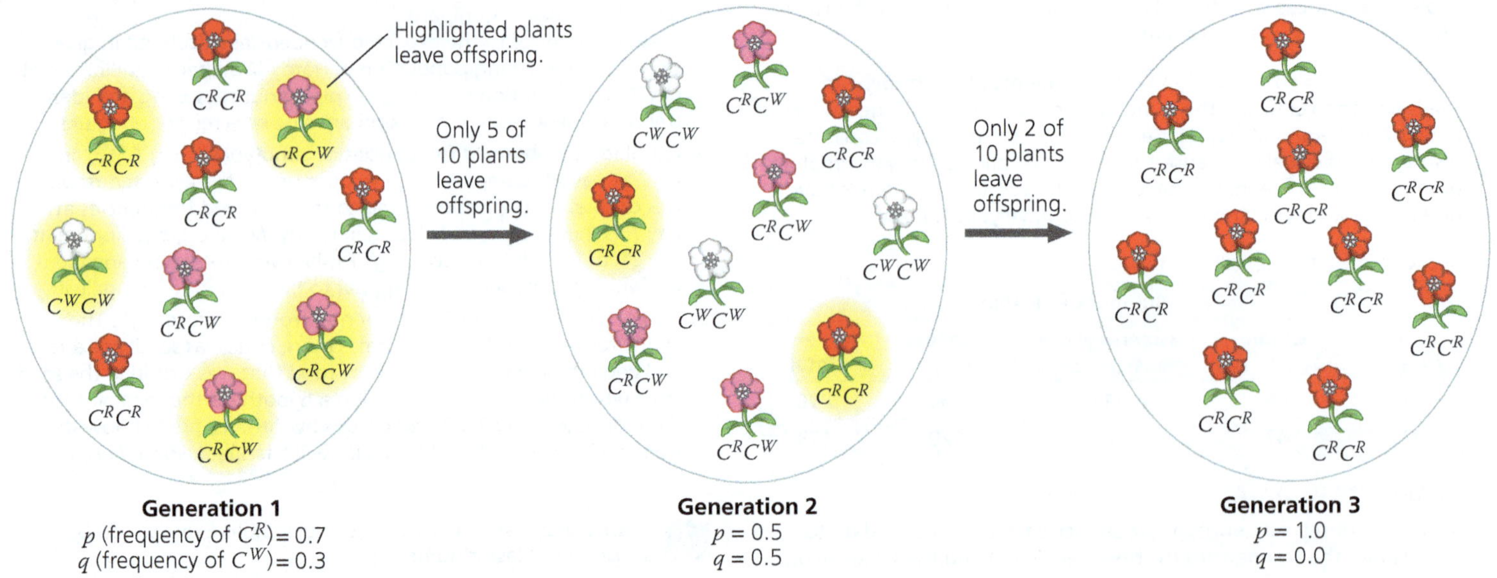

The Founder Effect

When a few individuals become isolated from a larger population, this smaller group may establish a new population whose gene pool differs from the source population; this is called the **founder effect**. The founder effect might occur, for example, when a few members of a population are blown by a storm to a new island. Genetic drift, in which chance events alter allele frequencies, can occur in such a case because the storm indiscriminately transports some individuals (and their alleles), but not others, from the source population.

The founder effect probably accounts for the relatively high frequency of certain inherited disorders among isolated human populations. For example, in 1814, 15 British colonists founded a settlement on Tristan da Cunha, a group of small islands in the Atlantic Ocean midway between Africa and South America. Apparently, one of the colonists carried a recessive allele for retinitis pigmentosa, a progressive form of blindness that afflicts homozygous individuals. Of the founding colonists' 240 descendants on the island in the late 1960s, four had retinitis pigmentosa. The frequency of the allele that causes this disease is ten times higher on Tristan da Cunha than in the populations from which the founders came.

The Bottleneck Effect

A sudden change in the environment, such as a fire or flood, may drastically reduce the size of a population. A severe drop in population size can cause the **bottleneck effect**, so named because the population has passed through a "bottleneck" that reduces its size **(Figure 23.10)**. By chance alone, certain alleles may be overrepresented among the survivors, others may be underrepresented, and some may be absent altogether. Ongoing genetic drift is likely to have substantial effects on the gene pool until the population becomes large enough that chance events have less impact. But even if a population that has passed through a bottleneck ultimately recovers in size, it may have low levels of genetic variation for a long period of time—a legacy of the genetic drift that occurred when the population was small.

Human actions sometimes create severe bottlenecks for other species, as the following example shows.

▼ Figure 23.10 The bottleneck effect. Shaking just a few marbles through the narrow neck of a bottle is analogous to a drastic reduction in the size of a population. By chance, blue marbles are overrepresented in the surviving population and gold marbles are absent.

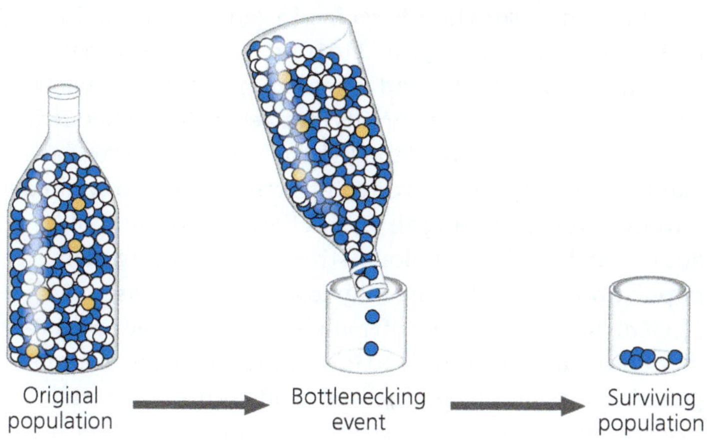

Case Study: Impact of Genetic Drift on the Greater Prairie Chicken

Millions of greater prairie chickens (*Tympanuchus cupido*) once lived on the prairies of Illinois. As these prairies were converted to farmland and other uses during the 19th and 20th centuries, the number of greater prairie chickens plummeted **(Figure 23.11a)**. By 1993 fewer than 50 birds remained. These few surviving birds had low levels of genetic variation, and less than 50% of their eggs hatched, compared with much higher hatching rates of the larger populations in Kansas and Nebraska **(Figure 23.11b)**.

▼ Figure 23.11 Genetic drift and loss of genetic variation.

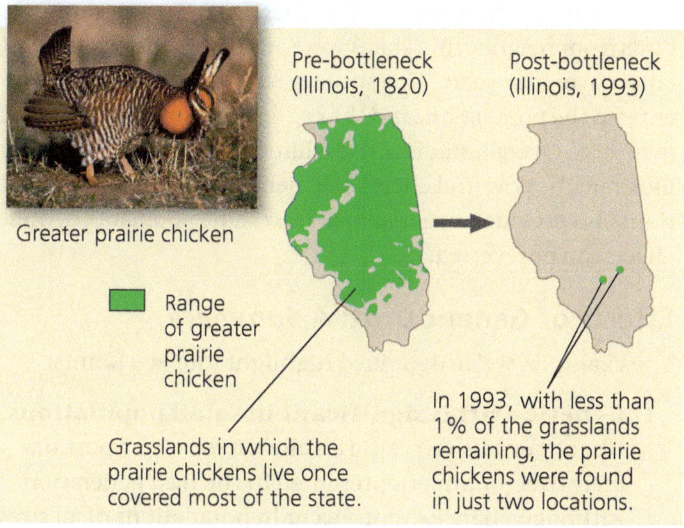

(a) The Illinois population of greater prairie chickens dropped from millions of birds in the 1800s to fewer than 50 birds in 1993.

Location	Population size	Number of alleles per locus	Percentage of eggs hatched
Illinois			
1930–1960s	1,000–25,000	5.2	93
1993	<50	3.7	<50
Kansas, 1998 (no bottleneck)	750,000	5.8	99
Nebraska, 1998 (no bottleneck)	75,000–200,000	5.8	96

(b) In the small Illinois population, genetic drift led to decreases in the number of alleles per locus and the percentage of eggs hatched.

These data suggest that genetic drift during the bottleneck may have led to a loss of genetic variation and an increase in the frequency of harmful alleles. To investigate this hypothesis, researchers extracted DNA from 15 museum specimens of Illinois greater prairie chickens. Of the 15 birds, 10 had been collected in the 1930s, when there were 25,000 greater prairie chickens in Illinois, and 5 had been collected in the 1960s, when there were 1,000 greater prairie chickens in Illinois. By studying the DNA of these specimens, the researchers were able to obtain a minimum, baseline estimate of how much genetic variation was present in the Illinois population *before* the population shrank to extremely low numbers. This baseline estimate is a key piece of information that is not usually available in cases of population bottlenecks.

The researchers surveyed six loci and found that the 1993 population had fewer alleles per locus than the pre-bottleneck Illinois or the current Kansas and Nebraska populations (see Figure 23.11b). Thus, as predicted, drift had reduced the genetic variation of the small 1993 population. Drift may also have increased the frequency of harmful alleles, leading to the low egg-hatching rate. To counteract these negative effects, 271 birds from neighboring states were added to the Illinois population over four years. This strategy succeeded: New alleles entered the population, and the egg-hatching rate improved to over 90%. Overall, studies on the Illinois greater prairie chicken illustrate the powerful effects of genetic drift in small populations and provide hope that in at least some populations, these effects can be reversed.

Effects of Genetic Drift: *A Summary*

The examples we've described highlight four key points:

1. **Genetic drift is significant in small populations.** Chance events can cause an allele to be disproportionately over- or underrepresented in the next generation. Although chance events occur in populations of all sizes, they tend to alter allele frequencies substantially only in small populations.
2. **Genetic drift can cause allele frequencies to change at random.** Because of genetic drift, an allele may increase in frequency one year, then decrease the next; the change from year to year is not predictable. Thus, unlike natural selection, which in a given environment consistently favors some alleles over others, genetic drift causes allele frequencies to change at random over time.
3. **Genetic drift can lead to a loss of genetic variation within populations.** By causing allele frequencies to fluctuate randomly over time, genetic drift can eliminate alleles from a population. Because evolution depends on genetic variation, such losses can influence how effectively a population can adapt to a change in the environment.
4. **Genetic drift can cause harmful alleles to become fixed.** Alleles that are neither harmful nor beneficial can be lost or become fixed (reach a frequency of 100%) by chance through genetic drift. In very small populations, genetic drift can also cause alleles that are slightly harmful to become fixed. When this occurs, the population's survival can be threatened (as in greater prairie chickens).

Gene Flow

Natural selection and genetic drift are not the only phenomena affecting allele frequencies. Allele frequencies can also change by **gene flow**, the transfer of alleles into or out of a population due to the movement of fertile individuals or their gametes. For example, suppose that near our original hypothetical wildflower population there is another population consisting primarily of white-flowered individuals ($C^W C^W$). Insects carrying pollen from these plants may fly to and pollinate plants in our original population. The introduced C^W alleles would modify our original population's allele frequencies in the next generation. Because alleles are transferred between populations, gene flow tends to reduce the genetic differences between populations. In fact, if it is extensive enough, gene flow can result in two populations combining into a single population with a common gene pool.

Alleles transferred by gene flow can also affect how well populations are adapted to local environmental conditions. For instance, mainland and island populations of the Lake Erie water snake (*Nerodia sipedon*) differ in their color patterns: Nearly all snakes from the Ohio or Ontario mainlands are strongly banded, whereas the majority of snakes from islands are unbanded or intermediate **(Figure 23.12)**. Banding coloration is an inherited trait, determined by a few loci (with alleles that encode bands being dominant to alleles that encode the absence of bands). On islands, water snakes live along rocky shorelines, while on the mainland, they live in marshes. Snakes without bands are more well camouflaged in island habitats than are snakes with bands. Hence, on islands, snakes without bands survive at higher rates than do snakes with bands.

These data indicate that snakes without bands are favored by natural selection in island populations. Thus, we might expect that *all* snakes on islands would lack bands. Why is this not the case? The answer lies in gene flow from the mainland. In any given year, 3 to10 snakes from the mainland swim to the islands and join the populations there. As a result, each year such migrants transfer alleles for banded coloration from the mainland (where nearly all snakes have bands) to the islands. This ongoing gene flow has prevented selection from removing all of the alleles for banded coloration from island populations—thereby preventing island populations from adapting fully to local conditions.

Gene flow can also transfer alleles that improve the ability of populations to adapt to local conditions. For example, gene flow has resulted in the worldwide spread of several insecticide resistance alleles in the mosquito *Culex pipiens*,

Figure 23.12 Gene flow and local adaptation in the Lake Erie water snake (*Nerodia sipedon*). Researchers assigned letters to variations in coloration in *N. sipedon* populations. Color pattern A is strong banding, patterns B and C are intermediate banding, and pattern D is no banding. Banding is advantageous for camouflage in mainland environments, whereas having no bands is advantageous in island environments. However, gene flow from the mainland causes banding to persist in island populations.

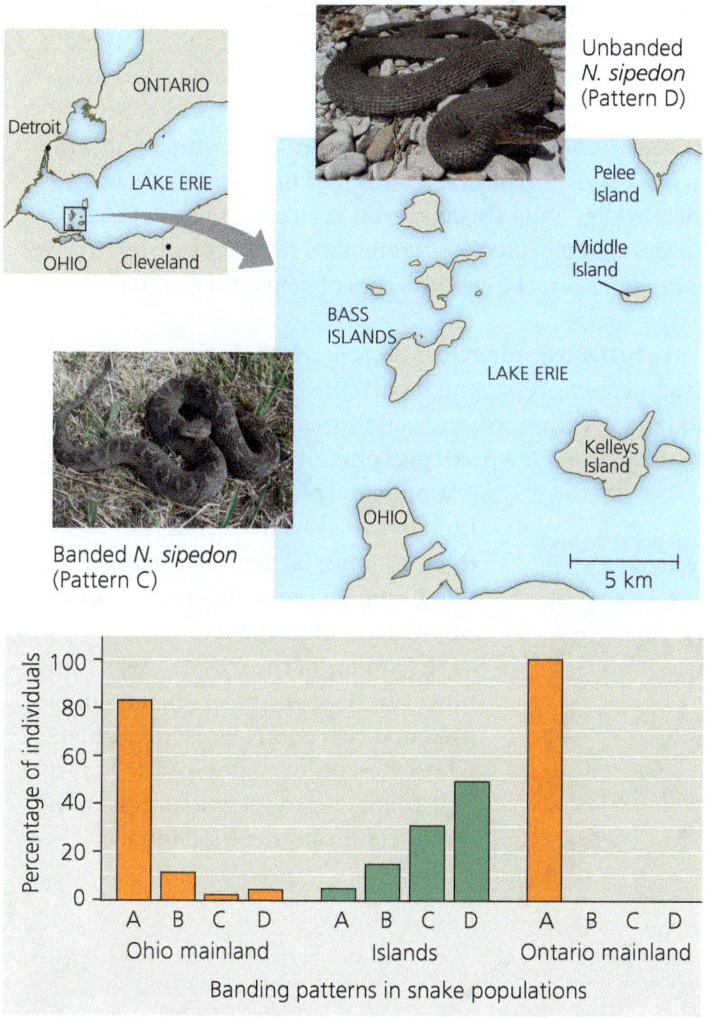

WHAT IF? ➤ *Suppose a severe weather event caused island populations to decrease in size but did not affect the size of mainland populations. Predict how gene flow from the mainland would affect color patterns in island populations. Explain.*

a vector of West Nile virus and other diseases. Each of these alleles has a unique genetic signature that allowed researchers to document that it arose by mutation in only one or a few geographic locations. In their population of origin, these alleles increased because they provided insecticide resistance. These alleles were then transferred to new populations, where again, their frequencies increased as a result of natural selection.

Finally, gene flow has become an increasingly important agent of evolutionary change in human populations. Humans today move much more freely about the world than in the past. As a result, mating is more common between members of populations that previously had very little contact, leading to an exchange of alleles and fewer genetic differences between those populations.

BioFlix® Animation: Natural Selection, Genetic Drift, and Gene Flow

CONCEPT CHECK 23.3

1. In what sense is natural selection more "predictable" than genetic drift?
2. Distinguish genetic drift from gene flow in terms of (a) how they occur and (b) their implications for future genetic variation in a population.
3. **WHAT IF?** ➤ Suppose two plant populations exchange pollen and seeds. In one population, individuals of genotype *AA* are most common (9,000 *AA*, 900 *Aa*, 100 *aa*), while the opposite is true in the other population (100 *AA*, 900 *Aa*, 9,000 *aa*). If neither allele has a selective advantage, what will happen over time to the allele and genotype frequencies of these populations?

For suggested answers, see Appendix A.

CONCEPT 23.4

Natural selection is the only mechanism that consistently causes adaptive evolution

Evolution by natural selection is a blend of chance and "sorting": chance in the creation of new genetic variations (as in mutation) and sorting as natural selection favors some alleles over others. Because of this favoring process, the outcome of natural selection is *not* random. Instead, natural selection consistently increases the frequencies of alleles that provide reproductive advantage, thus leading to adaptive evolution.

Natural Selection: *A Closer Look*

To see how natural selection can cause adaptive evolution, we'll begin with the concept of relative fitness and the different ways that selection acts on an organism's phenotype.

Relative Fitness

The phrases "struggle for existence" and "survival of the fittest" are commonly used to describe natural selection, but these expressions are misleading if always taken to mean direct competitive contests among individuals. There *are* animal species in which individuals, usually the males, lock horns or otherwise do combat to determine mating privilege. But reproductive success is generally more subtle and depends on many factors besides outright battle. For example, a barnacle that is more efficient at collecting food than its neighbors may have greater stores of energy and hence be able to produce a larger number of eggs. A moth may have more offspring than other moths in the same population because its body colors more effectively conceal it from predators, improving its chance of surviving long enough to produce more offspring. These examples illustrate how in a given environment, certain traits can lead to greater **relative fitness**: the contribution an individual makes to the gene pool of the next generation *relative to* the contributions of other individuals.

Although we often refer to the relative fitness of a genotype, remember that the entity that is subjected to natural selection is the whole organism, not the underlying genotype. Thus, selection acts more directly on the phenotype than on the genotype; it acts on the genotype indirectly, via how the genotype affects the phenotype.

Directional, Disruptive, and Stabilizing Selection

Natural selection can alter the frequency distribution of heritable traits in three ways, depending on which phenotypes in a population are favored: through directional selection, disruptive selection, and stabilizing selection.

Directional selection occurs when conditions favor individuals exhibiting one extreme of a phenotypic range, thereby shifting a population's frequency curve for the phenotypic character in one direction or the other **(Figure 23.13a)**. Directional selection is common when a population's environment changes or when members of a population migrate to a new (and different) habitat. For instance, an increase in the relative abundance of large seeds over small seeds led to an increase in beak depth in a population of Galápagos finches (see Figure 23.2).

Disruptive selection (Figure 23.13b) occurs when conditions favor individuals at both extremes of a phenotypic range over individuals with intermediate phenotypes. One example is a population of black-bellied seedcracker finches in Cameroon whose members display two distinctly different beak sizes. Small-billed birds feed mainly on soft seeds, whereas large-billed birds specialize in cracking hard seeds. It appears that birds with intermediate-sized bills are relatively inefficient at cracking both types of seeds and thus have lower relative fitness.

Stabilizing selection (Figure 23.13c) acts against both extreme phenotypes and favors intermediate variants. This mode of selection reduces variation and tends to maintain the status quo for a particular phenotypic character. For example, the birth weights of most human babies lie in the range of 3–4 kg (6.6–8.8 pounds); babies who are either much smaller or much larger suffer higher rates of mortality.

Regardless of the mode of selection, however, the basic mechanism remains the same. Selection favors individuals whose heritable phenotypic traits provide higher reproductive success than do the traits of other individuals.

▼ Figure 23.13 Modes of selection. These cases describe three ways in which a hypothetical deer mouse population with heritable variation in fur coloration might evolve. The graphs show how the frequencies of individuals with different fur colors change over time. The large white arrows symbolize selective pressures against certain phenotypes.

MAKE CONNECTIONS ➤ *Review Figure 22.13. Which mode of selection has occurred in soapberry bug populations that feed on the introduced goldenrain tree? Explain.*

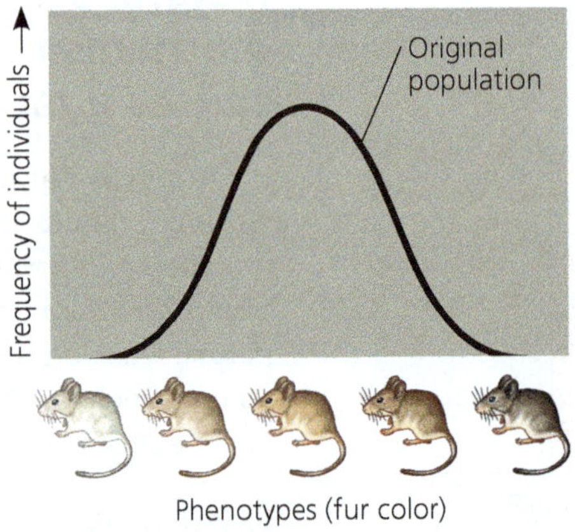

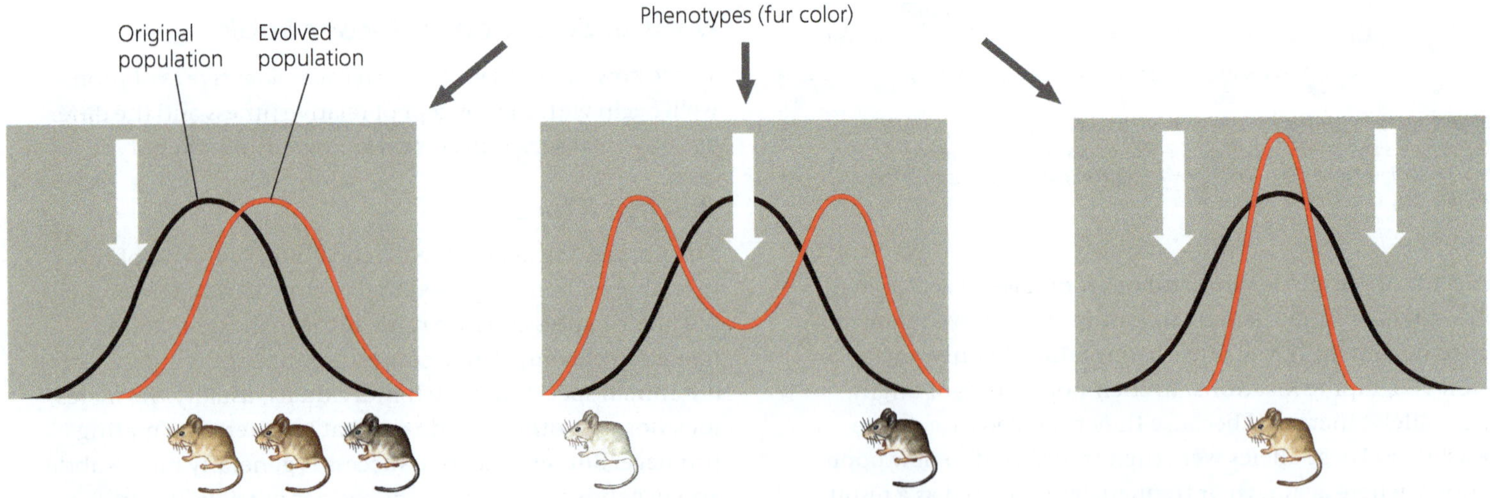

(a) Directional selection shifts the overall makeup of the population by favoring variants that are at one extreme of the distribution. In this case, lighter mice are selected against because they live among dark rocks, making it harder for them to hide from predators.

(b) Disruptive selection favors variants at both ends of the distribution. These mice have colonized a patchy habitat made up of light and dark rocks, with the result that mice of an intermediate color are selected against.

(c) Stabilizing selection removes extreme variants from the population and preserves intermediate types. If the environment consists of rocks of an intermediate color, both light and dark mice will be selected against.

The Key Role of Natural Selection in Adaptive Evolution

The adaptations of organisms include many striking examples. Certain octopuses, for example, have the ability to change color rapidly, enabling them to blend into different backgrounds. Another example is the remarkable jaws of snakes **(Figure 23.14)**, which allow them to swallow prey much larger than their own head (a feat analogous to a person swallowing a whole watermelon). Other adaptations, such as a version of an enzyme that shows improved function in cold environments, may be less visually dramatic but just as important for survival and reproduction.

Such adaptations can arise gradually over time as natural selection increases the frequencies of alleles that enhance survival or reproduction. As the proportion of individuals that have favorable traits increases, the degree to which a species is well suited for life in its environment improves; that is, adaptive evolution occurs. However, the physical and biological components of an organism's environment may change over time. As a result, what constitutes a "good match" between an organism and its environment can be a moving target, making adaptive evolution a continuous, dynamic process. Environmental conditions can also differ from place to place, causing different alleles to be favored in different locations. When this occurs, natural selection can cause the populations of a species to differ genetically from one another.

And what about genetic drift and gene flow? Both can, in fact, increase the frequencies of alleles that enhance survival or reproduction, but neither does so consistently. Genetic drift can cause the frequency of a slightly beneficial allele to increase, but it also can cause the frequency of such an allele to decrease. Similarly, gene flow may introduce alleles that are advantageous or ones that are disadvantageous. Natural selection is the only evolutionary mechanism that consistently leads to adaptive evolution.

▼ Figure 23.14 Movable jaw bones in snakes.

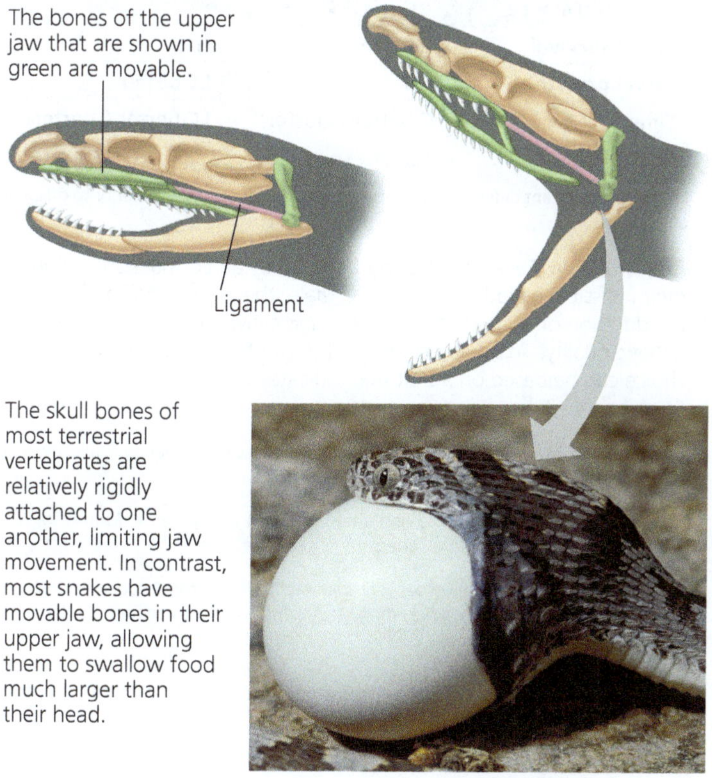

HHMI Video: Got Lactase? The Co-evolution of Genes and Culture

Sexual Selection

Charles Darwin was the first to explore the implications of **sexual selection**, a process in which individuals with certain inherited characteristics are more likely than other individuals of the same sex to obtain mates. Sexual selection can result in **sexual dimorphism**, a difference in secondary sexual characteristics between males and females of the same species **(Figure 23.15)**. These distinctions include differences in size, color, ornamentation, and behavior.

How does sexual selection operate? There are several ways. In **intrasexual selection**, meaning selection within the same sex, individuals of one sex compete directly for mates of the opposite sex. In many species, intrasexual selection occurs among males. For example, a single male may patrol a group of females and prevent other males from mating with them. The patrolling male may defend his status by defeating smaller, weaker, or less fierce males in combat. More often, this male is the psychological victor in ritualized displays that discourage would-be competitors but do not risk injury that would reduce his own fitness (see Figure 51.16). Intrasexual selection also occurs among females in a variety of species, including ring-tailed lemurs and broadnosed pipefish.

In **intersexual selection**, also called *mate choice*, individuals of one sex (usually the females) are choosy in selecting their mates from the other sex. In many cases, the female's choice depends on the showiness of the male's appearance

▼ Figure 23.15 Sexual dimorphism and sexual selection. A peacock (left) and a peahen (right) show extreme sexual dimorphism. There is intrasexual selection between competing males, followed by intersexual selection when the females choose among the showiest males.

or behavior (see Figure 23.15). What intrigued Darwin about mate choice is that male showiness may not seem adaptive in any other way and may in fact pose some risk. For example, bright plumage may make male birds more visible to predators. But if such characteristics help a male gain a mate, and if this benefit outweighs the risk from predation, then both the bright plumage and the female preference for it will be reinforced because they enhance overall reproductive success.

How do female preferences for certain male characteristics evolve in the first place? One hypothesis is that females prefer male traits that are correlated with "good genes." If the trait preferred by females is indicative of a male's overall genetic quality, both the male trait and female preference for it should increase in frequency. **Figure 23.16** describes one experiment testing this hypothesis in gray tree frogs.

Other researchers have shown that in several bird species, the traits preferred by females are related to overall male health. Here, too, female preference appears to be based on traits that reflect "good genes," in this case, alleles indicative of a robust immune system.

Balancing Selection

As we've seen, genetic variation is often found at loci affected by selection. What prevents natural selection from reducing the variation at those loci by culling all unfavorable alleles? As mentioned earlier, in diploid organisms, many unfavorable recessive alleles persist because they are hidden from selection when in heterozygous individuals. In addition, selection itself may preserve variation at some loci, thus maintaining two or more phenotypic forms in a population. Known as **balancing selection**, this type of selection includes frequency-dependent selection and heterozygote advantage.

Frequency-Dependent Selection

In **frequency-dependent selection**, the fitness of a phenotype depends on how common it is in the population. Consider the scale-eating fish (*Perissodus microlepis*) of Lake Tanganyika, in Africa. These fish attack other fish from behind, darting in to remove a few scales from the flank of their prey. Of interest here is a peculiar feature of the scale-eating fish: Some are "left-mouthed" and some are "right-mouthed." This trait is determined by two alleles and simple Mendelian inheritance. Hence, all individuals in a population are either left-mouthed or right-mouthed, and the frequencies of these two phenotypes must add up to 100%.

Because their mouth twists to the left, left-mouthed fish always attack their prey's right flank **(Figure 23.17)**. (To see why, twist your lower jaw and lips to the left and imagine trying to take a bite from the left side of a fish, approaching it from behind.) Similarly, right-mouthed fish always attack from the left. Prey species guard against attack from whatever phenotype of scale-eating fish is most common in the lake.

▼ Figure 23.16

Inquiry Do females select mates based on traits indicative of "good genes"?

Experiment Female gray tree frogs (*Hyla versicolor*) prefer to mate with males that give long mating calls. Allison Welch and colleagues, at the University of Missouri, tested whether the genetic makeup of long-calling (LC) males is superior to that of short-calling (SC) males. The researchers fertilized half the eggs of each female with sperm from an LC male and fertilized the remaining eggs with sperm from an SC male. In two separate experiments (one in 1995, the other in 1996), the resulting half-sibling offspring were raised in a common environment and their survival and growth were monitored.

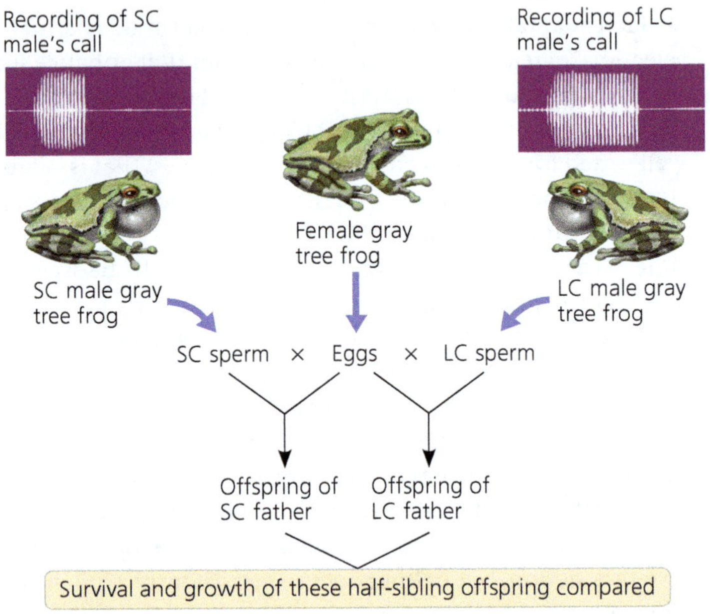

Results

Offspring Performance	1995	1996
Larval survival	LC better	NSD
Larval growth	NSD	LC better
Time to metamorphosis	LC better (shorter)	LC better (shorter)

NSD = no significant difference; LC better = offspring of LC males superior to offspring of SC males.

Conclusion Because offspring fathered by an LC male outperformed their half-siblings fathered by an SC male, the team concluded that the duration of a male's mating call is indicative of the male's overall genetic quality. This result supports the hypothesis that female mate choice can be based on a trait that indicates whether the male has "good genes."

Data from A. M. Welch et al., Call duration as an indicator of genetic quality in male gray tree frogs, *Science* 280:1928–1930 (1998).

INQUIRY IN ACTION Read and analyze the original paper in *Inquiry in Action: Interpreting Scientific Papers.*

WHAT IF? *Why did the researchers split each female frog's eggs into two batches for fertilization by different males? Why didn't they mate each female with a single male frog?*

▼ **Figure 23.17 Frequency-dependent selection.** In a population of the scale-eating fish *Perissodus microlepis*, the frequency of left-mouthed individuals (red data points) rises and falls in a regular manner. The frequency of left-mouthed individuals among adults that reproduced was also recorded in three sample years (green data points).

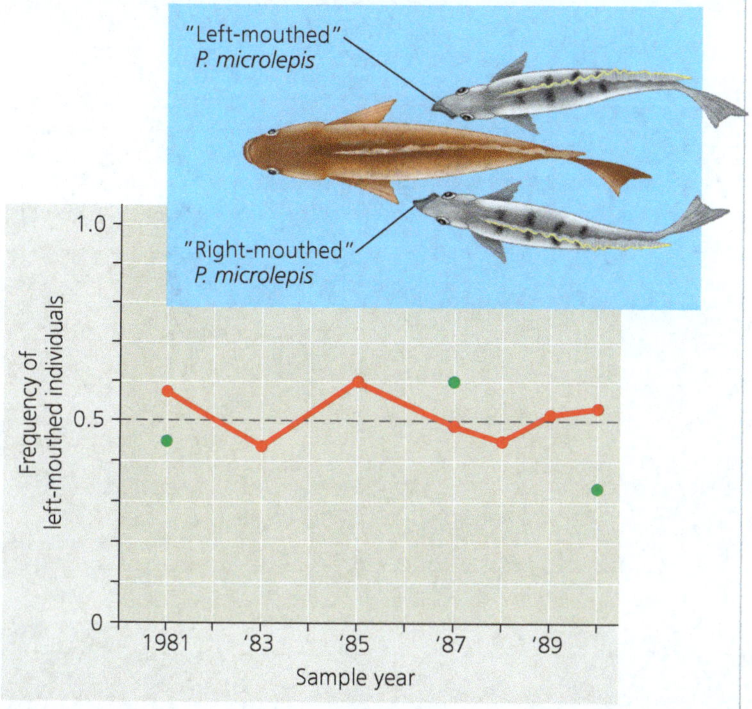

INTERPRET THE DATA ➤ *For 1981, 1987, and 1990, compare the frequency of left-mouthed individuals among breeding adults to the frequency of left-mouthed individuals in the entire population. What do the data indicate about when natural selection favors left-mouthed individuals over right-mouthed individuals (or vice versa)? Explain.*

Thus, from year to year, selection favors whichever mouth phenotype is least common. As a result, the frequency of left- and right-mouthed fish oscillates over time, and balancing selection (due to frequency dependence) keeps the frequency of each phenotype close to 50%.

Heterozygote Advantage

If individuals who are heterozygous at a particular locus have greater fitness than do both kinds of homozygotes, they exhibit **heterozygote advantage**. In such a case, natural selection tends to maintain two or more alleles at that locus. Note that heterozygote advantage is defined in terms of *genotype*, not phenotype. Thus, whether heterozygote advantage represents stabilizing or directional selection depends on the relationship between the genotype and the phenotype. For example, if the phenotype of a heterozygote is intermediate to the phenotypes of both homozygotes, heterozygote advantage is a form of stabilizing selection.

An example of heterozygote advantage occurs at the locus in humans that codes for the β polypeptide subunit of hemoglobin, the oxygen-carrying protein of red blood cells. In homozygous individuals, a recessive allele at that locus causes sickle-cell disease. The red blood cells of people with sickle-cell disease become distorted in shape, or *sickled*, under low-oxygen conditions (see Figure 5.19), as occurs in the capillaries. These sickled cells can clump together and block the flow of blood in the capillaries, damaging organs such as the kidney, heart, and brain. Although some red blood cells become sickled in heterozygotes, not enough become sickled to cause sickle-cell disease.

Heterozygotes for the sickle-cell allele are protected against the most severe effects of malaria, a disease caused by a parasite that infects red blood cells (see Figure 28.16). One reason for this partial protection is that the body destroys sickled red blood cells rapidly, killing the parasites they harbor. Malaria is a major killer in some tropical regions. In such regions, selection favors heterozygotes over homozygous dominant individuals, who are more vulnerable to the effects of malaria, and also over homozygous recessive individuals, who develop sickle-cell disease. As described in **Figure 23.18**, these selective pressures have caused the frequency of the sickle-cell allele to reach relatively high levels in areas where the malaria parasite is common.

Why Natural Selection Cannot Fashion Perfect Organisms

Though natural selection leads to adaptation, nature abounds with examples of organisms that are less than ideally suited for their lifestyles. There are several reasons why.

1. **Selection can act only on existing variations.** Natural selection favors only the fittest phenotypes among those currently in the population, which may not be the ideal traits. New advantageous alleles do not arise on demand.
2. **Evolution is limited by historical constraints.** Each species has a legacy of descent with modification from ancestral forms. Evolution does not scrap the ancestral anatomy and build each new complex structure from scratch; rather, evolution co-opts existing structures and adapts them to new situations. We could imagine that if a terrestrial animal were to adapt to an environment in which flight would be advantageous, it might be best just to grow an extra pair of limbs that would serve as wings. However, evolution does not work this way; instead, it operates on the traits an organism already has. Thus, in birds and bats, an existing pair of limbs took on new functions for flight as these organisms evolved from nonflying ancestors.
3. **Adaptations are often compromises.** Each organism must do many different things. A seal spends part of its time on rocks; it could probably walk better if it had legs instead of flippers, but then it would not swim nearly as well. We humans owe much of our versatility and athleticism to our prehensile hands and flexible limbs, but these also make us prone to sprains, torn ligaments, and dislocations: Structural reinforcement has been compromised for agility.

▼ **Figure 23.18** **MAKE CONNECTIONS**

The Sickle-Cell Allele

This child has sickle-cell disease, a genetic disorder that strikes individuals that have two copies of the sickle-cell allele. This allele causes an abnormality in the structure and function of hemoglobin, the oxygen-carrying protein in red blood cells. Although sickle-cell disease is lethal if not treated, in some regions the sickle-cell allele can reach frequencies as high as 15–20%. How can such a harmful allele be so common?

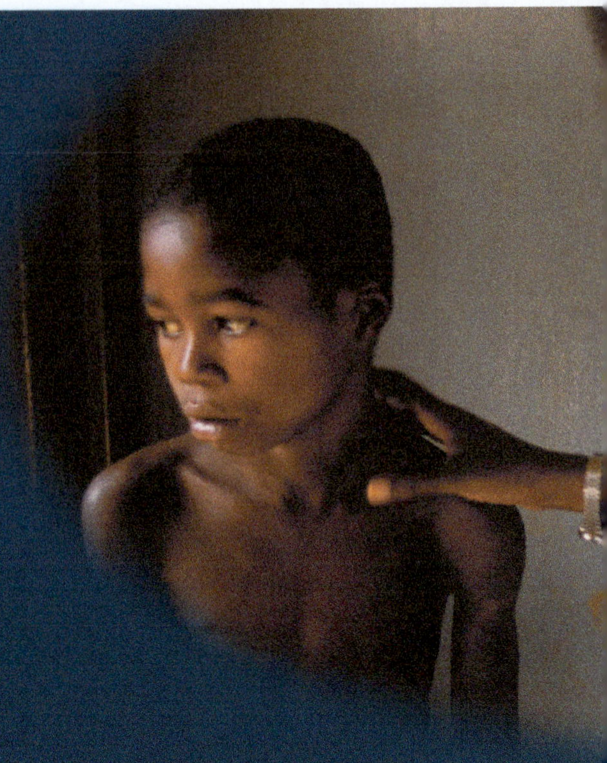

Events at the Molecular Level

- Due to a point mutation, the sickle-cell allele differs from the wild-type allele by a single nucleotide. (See Figure 17.26.)
- The resulting change in one amino acid leads to hydrophobic interactions between the sickle-cell hemoglobin proteins under low-oxygen conditions.
- As a result, the sickle-cell proteins bind to each other in chains that together form a fiber.

Consequences for Cells

- The abnormal hemoglobin fibers distort the red blood cell into a sickle shape under low-oxygen conditions, such as those found in blood vessels returning to the heart.

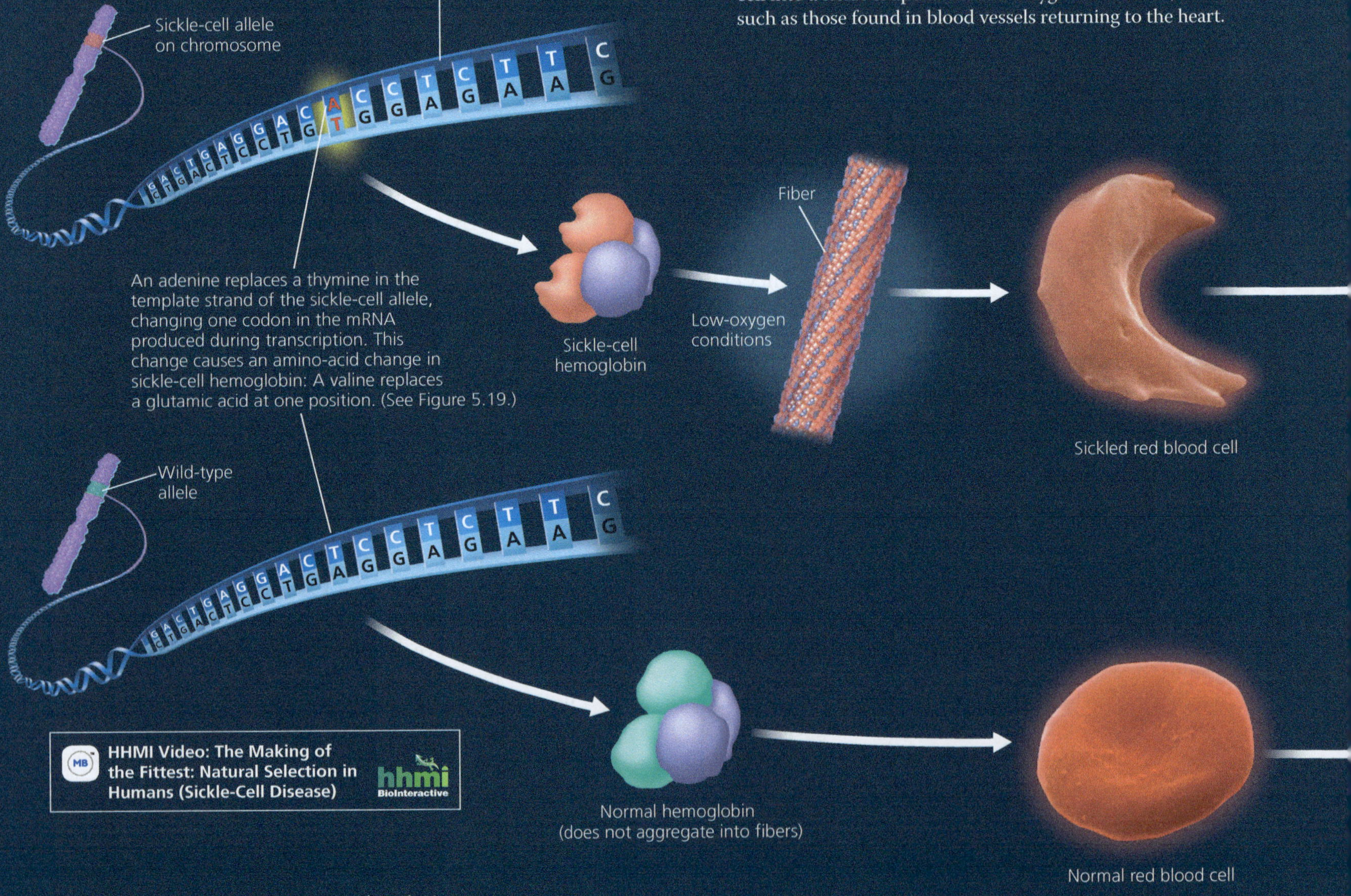

HHMI Video: The Making of the Fittest: Natural Selection in Humans (Sickle-Cell Disease)

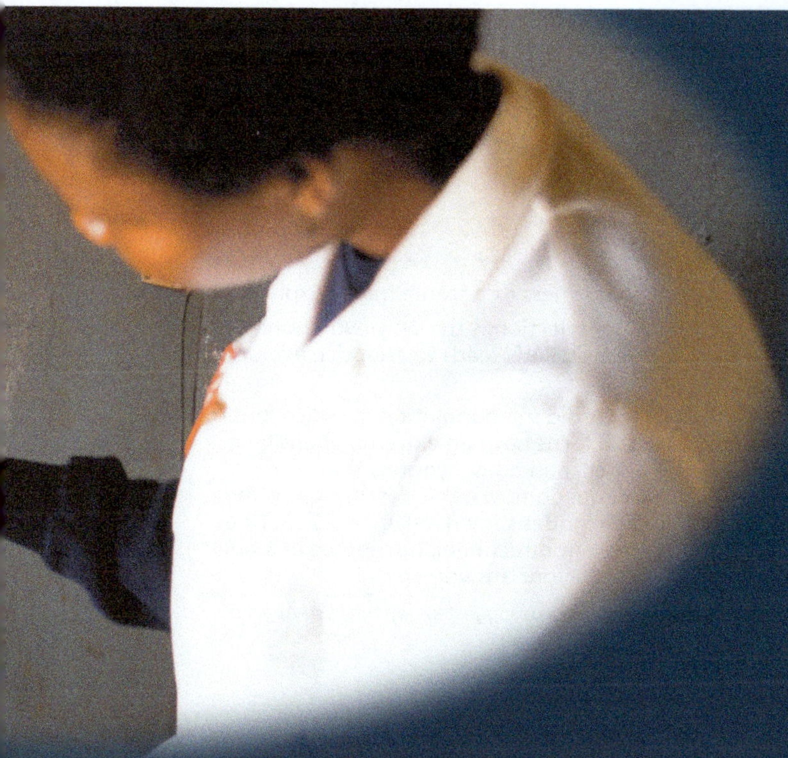

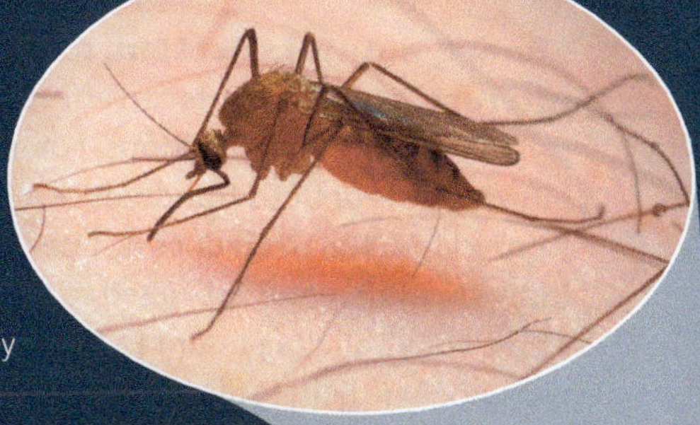

Infected mosquitoes spread malaria when they bite people. (See Figure 28.16.)

Evolution in Populations

- Homozygotes with two sickle-cell alleles are strongly selected against because of mortality caused by sickle-cell disease. In contrast, heterozygotes experience few harmful effects from sickling yet are more likely to survive malaria than are homozygotes.
- In regions where malaria is common, the net effect of these opposing selective forces is heterozygote advantage. This has caused evolutionary change in populations—the products of which are the areas of relatively high frequencies of the sickle-cell allele shown in the map below.

Effects on Individual Organisms

- The formation of sickled red blood cells causes homozygotes with two copies of the sickle-cell allele to have sickle-cell disease.
- Some sickling also occurs in heterozygotes, but not enough to cause the disease; they have sickle-cell trait. (See Figure 14.17.)

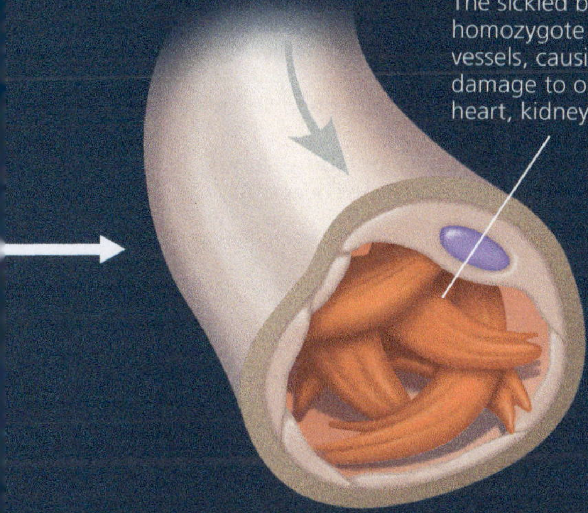

The sickled blood cells of a homozygote block small blood vessels, causing great pain and damage to organs such as the heart, kidney, and brain.

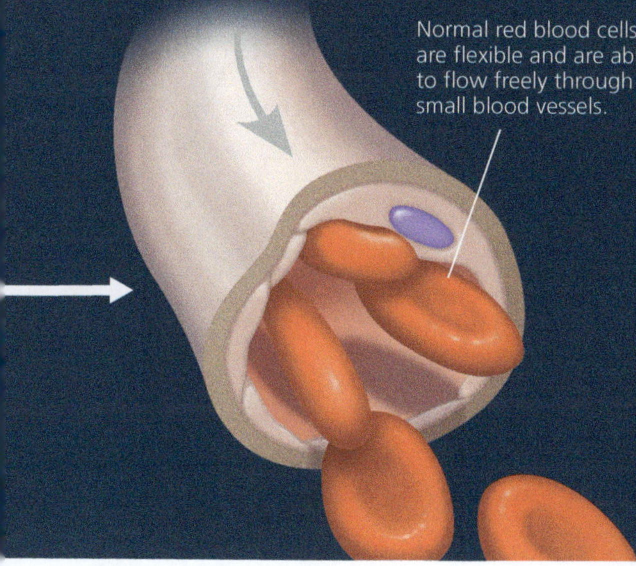

Normal red blood cells are flexible and are able to flow freely through small blood vessels.

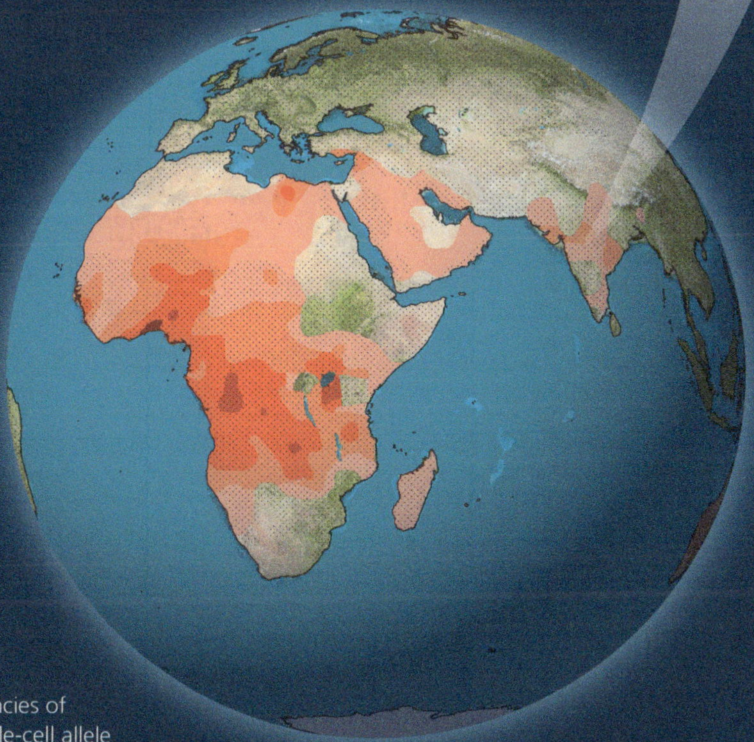

Key

Frequencies of the sickle-cell allele

- 3.0–6.0%
- 6.0–9.0%
- 9.0–12.0%
- 12.0–15.0%
- >15.0%

Distribution of malaria caused by *Plasmodium falciparum* (a parasitic unicellular eukaryote)

MAKE CONNECTIONS ➤ *In a region free of malaria, would individuals who are heterozygous for the sickle-cell allele be selected for or selected against? Explain.*

4. **Chance, natural selection, and the environment interact.** Chance events can affect the subsequent evolutionary history of populations. For instance, when a storm blows insects or birds hundreds of kilometers over an ocean to an island, the wind does not necessarily transport those individuals that are best suited to the new environment. Thus, not all alleles present in the founding population's gene pool are better suited to the new environment than the alleles that are "left behind." In addition, the environment at a particular location may change unpredictably from year to year, again limiting the extent to which adaptive evolution results in organisms being well suited for current environmental conditions.

With these four constraints, evolution does not tend to craft perfect organisms. Natural selection operates on a "better than" basis. We can, in fact, see evidence for evolution in the many imperfections of the organisms it produces.

CONCEPT CHECK 23.4

1. What is the relative fitness of a sterile mule? Explain.
2. Explain why natural selection is the only evolutionary mechanism that consistently leads to adaptive evolution in a population.
3. **VISUAL SKILLS ➤** Consider a population in which heterozygotes at a certain locus have an extreme phenotype (such as being larger than homozygotes) that confers a selective advantage. Compare this description to the models of selection modes shown in Figure 23.13. Does this situation represent directional, disruptive, or stabilizing selection? Explain your answer.

For suggested answers, see Appendix A.

23 Chapter Review

Go to **MasteringBiology**™ for Videos, Animations, Vocab Self-Quiz, Practice Tests, and more in the Study Area.

SUMMARY OF KEY CONCEPTS

CONCEPT 23.1

Genetic variation makes evolution possible *(pp. 485–487)*

VOCAB SELF-QUIZ goo.gl/6u55ks

- **Genetic variation** refers to genetic differences among individuals within a population.
- The nucleotide differences that provide the basis of genetic variation originate when mutation and gene duplication produce new alleles and new genes. New genetic variants are produced rapidly in organisms with short generation times. In sexually reproducing organisms, most of the genetic differences among individuals result from crossing over, the independent assortment of chromosomes, and fertilization.

? *Typically, most of the nucleotide variability that occurs within a genetic locus does not affect the phenotype. Explain why.*

CONCEPT 23.2

The Hardy-Weinberg equation can be used to test whether a population is evolving *(pp. 487–491)*

- A **population**, a localized group of organisms belonging to one species, is united by its **gene pool**, the aggregate of all the alleles in the population.
- For a population in **Hardy-Weinberg equilibrium**, the allele and genotype frequencies will remain constant if the population is large, mating is random, mutation is negligible, there is no gene flow, and there is no natural selection. For such a population, if p and q represent the frequencies of the only two possible alleles at a particular locus, then p^2 is the frequency of one kind of homozygote, q^2 is the frequency of the other kind of homozygote, and $2pq$ is the frequency of the heterozygous genotype.

? *Is it circular reasoning to calculate p and q from observed genotype frequencies and then use those values of p and q to test if the population is in Hardy-Weinberg equilibrium? Explain your answer.*

CONCEPT 23.3

Natural selection, genetic drift, and gene flow can alter allele frequencies in a population *(pp. 491–495)*

- In natural selection, individuals that have certain inherited traits tend to survive and reproduce at higher rates than other individuals *because of* those traits.
- In **genetic drift**, chance fluctuations in allele frequencies over generations tend to reduce genetic variation.
- **Gene flow**, the transfer of alleles between populations, tends to reduce genetic differences between populations over time.

? *Would two small, geographically isolated populations in very different environments be likely to evolve in similar ways? Explain.*

CONCEPT 23.4

Natural selection is the only mechanism that consistently causes adaptive evolution *(pp. 495–502)*

- One organism has greater **relative fitness** than another organism if it leaves more fertile descendants. The modes of natural selection differ in their effect on phenotype:

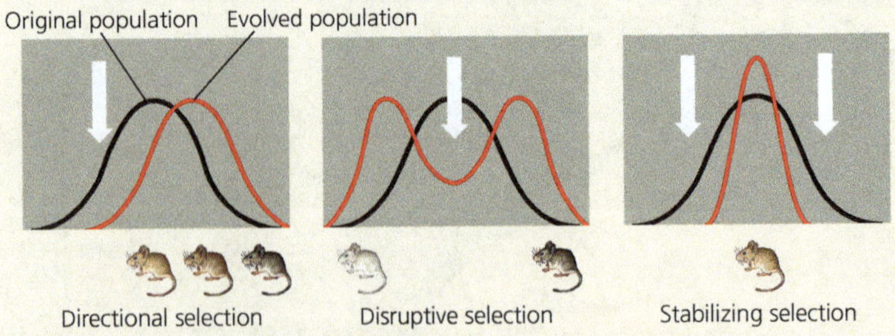

- Unlike genetic drift and gene flow, natural selection consistently increases the frequencies of alleles that enhance survival and reproduction, thus improving the degree to which organisms are well-suited for life in their environment.
- **Sexual selection** can result in secondary sex characteristics that can give individuals advantages in mating.

- **Balancing selection** occurs when natural selection maintains two or more forms in a population.
- There are constraints to evolution: Natural selection can act only on available variation; structures result from modified ancestral anatomy; adaptations are often compromises; and chance, natural selection, and the environment interact.

? *How might secondary sex characteristics in males differ from those in females in a species in which females compete for mates?*

TEST YOUR UNDERSTANDING

Level 1: Knowledge/Comprehension

PRACTICE TEST goo.gl/CUYGKD

1. Natural selection changes allele frequencies because some ________ survive and reproduce better than others.
(A) alleles
(B) loci
(C) species
(D) individuals

2. No two people are genetically identical, except for identical twins. The main source of genetic variation among humans is
(A) new mutations that occurred in the preceding generation.
(B) genetic drift.
(C) the reshuffling of alleles in sexual reproduction.
(D) environmental effects.

Level 2: Application/Analysis

3. If the nucleotide variability of a locus equals 0%, what is the gene variability and number of alleles at that locus?
(A) gene variability = 0%; number of alleles = 0
(B) gene variability = 0%; number of alleles = 1
(C) gene variability = 0%; number of alleles = 2
(D) gene variability > 0%; number of alleles = 2

4. There are 25 individuals in population 1, all with genotype *AA*, and there are 40 individuals in population 2, all with genotype *aa*. Assume that these populations are located far from each other and that their environmental conditions are very similar. Based on the information given here, the observed genetic variation most likely resulted from
(A) genetic drift.
(B) gene flow.
(C) nonrandom mating.
(D) directional selection.

5. A fruit fly population has a gene with two alleles, *A1* and *A2*. Tests show that 70% of the gametes produced in the population contain the *A1* allele. If the population is in Hardy-Weinberg equilibrium, what proportion of the flies carry both *A1* and *A2*?
(A) 0.7
(B) 0.49
(C) 0.42
(D) 0.21

Level 3: Synthesis/Evaluation

6. **EVOLUTION CONNECTION** Using at least two examples, explain how the process of evolution is revealed by the imperfections of living organisms.

7. **SCIENTIFIC INQUIRY • INTERPRET THE DATA** Researchers studied genetic variation in the marine mussel *Mytilus edulis* around Long Island, New York. They measured the frequency of a particular allele (*lap*^{94}) for an enzyme involved in regulating the mussel's internal saltwater balance. The researchers presented their data as a series of pie charts linked to sampling sites within Long Island Sound, where the salinity is highly variable, and along the coast of the open ocean, where salinity is constant. (a) Create a data table for the 11 sampling sites by estimating the frequency of *lap*^{94} from the pie charts. (*Hint*: Think of each pie chart as a clock face to help you estimate the proportion of the shaded area.) (b) Graph the frequencies for sites 1–8 to show how the frequency of this allele changes with increasing salinity in Long Island Sound (from southwest to northeast). Evaluate how the data from sites 9–11 compare with the data from the sites within the Sound. (c) Considering the various mechanisms that can alter allele frequency, construct a hypothesis that explains the patterns you observe in the data and that accounts for the following observations: (1) The *lap*^{94} allele helps mussels maintain osmotic balance in water with a high salt concentration but is costly to use in less salty water; and (2) mussels produce larvae that can disperse long distances before they settle on rocks and grow into adults.

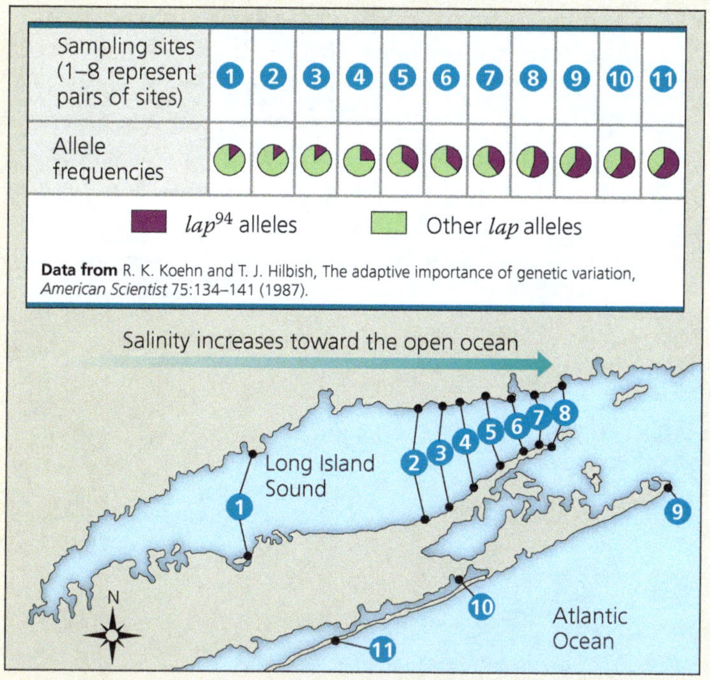

Data from R. K. Koehn and T. J. Hilbish, The adaptive importance of genetic variation, *American Scientist* 75:134–141 (1987).

8. **WRITE ABOUT A THEME: ORGANIZATION** Heterozygotes at the sickle-cell locus produce both normal and abnormal (sickle-cell) hemoglobin (see Concept 14.4). When hemoglobin molecules are packed into a heterozygote's red blood cells, some cells receive relatively large quantities of abnormal hemoglobin, making these cells prone to sickling. In a short essay (approximately 100–150 words), explain how these molecular and cellular events lead to emergent properties at the individual and population levels of biological organization.

9. **SYNTHESIZE YOUR KNOWLEDGE**

This kettle lake formed 14,000 years ago when a glacier that covered the surrounding area melted. Initially devoid of animal life, over time the lake was colonized by invertebrates and other animals. Hypothesize how mutation, natural selection, genetic drift, and gene flow may have affected populations that colonized the lake.

For selected answers, see Appendix A.

MB For additional practice questions, check out the **Dynamic Study Modules** in MasteringBiology. You can use them to study on your smartphone, tablet, or computer anytime, anywhere!

The Origin of Species 24

▲ **Figure 24.1 How did this flightless bird come to live on the isolated Galápagos Islands?**

KEY CONCEPTS

24.1 The biological species concept emphasizes reproductive isolation

24.2 Speciation can take place with or without geographic separation

24.3 Hybrid zones reveal factors that cause reproductive isolation

24.4 Speciation can occur rapidly or slowly and can result from changes in few or many genes

▼ **Galápagos giant tortoise, another species unique to the islands**

That "Mystery of Mysteries"

When Darwin came to the Galápagos Islands, he noted that these volcanic islands were teeming with plants and animals found nowhere else in the world **(Figure 24.1)**. Later he realized that these species had formed relatively recently. He wrote in his diary, "Both in space and time, we seem to be brought somewhat near to that great fact—that mystery of mysteries—the first appearance of new beings on this Earth."

The "mystery of mysteries" that captivated Darwin is **speciation**, the process by which one species splits into two or more species. Speciation fascinated Darwin (and many biologists since) because it has produced the tremendous diversity of life, repeatedly yielding new species that differ from existing ones. Later, Darwin realized that speciation also helps to explain the many features that organisms share (the unity of life): When one species splits into two, the species that result share many characteristics because they are descended from this common ancestor. At the DNA sequence level, for example, such similarities indicate that the flightless cormorant (*Phalacrocorax harrisi*) in Figure 24.1 is closely related to flying cormorants found in the Americas. This suggests that the flightless cormorant originated from an ancestral cormorant species that flew from the mainland to the Galápagos.

Speciation also forms a conceptual bridge between **microevolution**, changes over time in allele frequencies in a population, and **macroevolution**, the broad pattern of evolution above the species level. An example of

When you see this blue icon, log in to **MasteringBiology** and go to the Study Area for digital resources.

MB **Get Ready for This Chapter**

macroevolutionary change is the origin of new groups of organisms, such as mammals or flowering plants, through a series of speciation events. We examined microevolutionary mechanisms in Chapter 23, and we'll turn to macroevolution in Chapter 25. In this chapter, we'll explore the "bridge" between microevolution and macroevolution—the mechanisms by which new species originate from existing ones. First, let's establish what we actually mean by a "species."

MB HHMI Video: The Origin of Species: The Beak of the Finch

CONCEPT 24.1

The biological species concept emphasizes reproductive isolation

The word *species* is Latin for "kind" or "appearance." In daily life, we commonly distinguish between various "kinds" of organisms—dogs and cats, for instance—based on differences in their appearance. But are organisms truly divided into the discrete units we call species? To answer this question, biologists compare not only the morphology (body form) of different groups of organisms but also less obvious differences in physiology, biochemistry, and DNA sequences. The results generally confirm that morphologically distinct species are indeed discrete groups, differing in many ways besides their body forms.

The Biological Species Concept

The primary definition of species used in this textbook is the **biological species concept**. According to this concept, a **species** is a group of populations whose members have the potential to interbreed in nature and produce viable, fertile offspring—but do not produce viable, fertile offspring with members of other such groups (**Figure 24.2**). Thus, the members of a biological species are united by being reproductively compatible, at least potentially. All human beings, for example, belong to the same species. A businesswoman in Manhattan may be unlikely to meet a dairy farmer in Mongolia, but if the two should happen to meet and mate, they could have viable babies who develop into fertile adults. In contrast, humans and chimpanzees remain distinct biological species, even where they live in the same region, because many factors keep them from interbreeding and producing fertile offspring.

What holds the gene pool of a species together, causing its members to resemble each other more than they resemble members of other species? Recall the evolutionary mechanism of *gene flow*, the transfer of alleles between populations (see Concept 23.3). Typically, gene flow occurs between the different populations of a species. This ongoing exchange of alleles tends to hold the populations together genetically. But as we'll explore in this chapter, a reduction or lack of gene flow can play a key role in the formation of new species.

(a) Similarity between different species. The eastern meadowlark (*Sturnella magna*, left) and the western meadowlark (*Sturnella neglecta*, right) have similar body shapes and colorations. Nevertheless, they are distinct biological species because their songs and other behaviors are different enough to prevent interbreeding should they meet in the wild.

(b) Diversity within a species. As diverse as we may be in appearance, all humans belong to a single biological species (*Homo sapiens*), defined by our capacity to interbreed successfully.

▲ **Figure 24.2 The biological species concept is based on the potential to interbreed, not on physical similarity.**

Reproductive Isolation

Because biological species are defined in terms of reproductive compatibility, the formation of a new species hinges on **reproductive isolation**—the existence of biological factors (barriers) that impede members of two species from interbreeding and producing viable, fertile offspring. Such barriers block gene flow between the species and limit the formation of **hybrids**, offspring that result from an interspecific mating. Although a single barrier may not prevent all gene flow, a combination of several barriers can effectively isolate a species' gene pool.

Clearly, a fly cannot mate with a frog or a fern, but the reproductive barriers between more closely related species are not so obvious. As described in **Figure 24.3**,

▼ Figure 24.3 Exploring Reproductive Barriers

Prezygotic barriers impede mating or hinder fertilization if mating does occur

Habitat Isolation

Two species that occupy different habitats within the same area may encounter each other rarely, if at all, even though they are not isolated by obvious physical barriers, such as mountain ranges.

Example: These two fly species in the genus *Rhagoletis* occur in the same geographic areas, but the apple maggot fly (*Rhagoletis pomonella*) feeds and mates on hawthorns and apples **(a)** while its close relative, the blueberry maggot fly (*R. mendax*), mates and lays its eggs only on blueberries **(b)**.

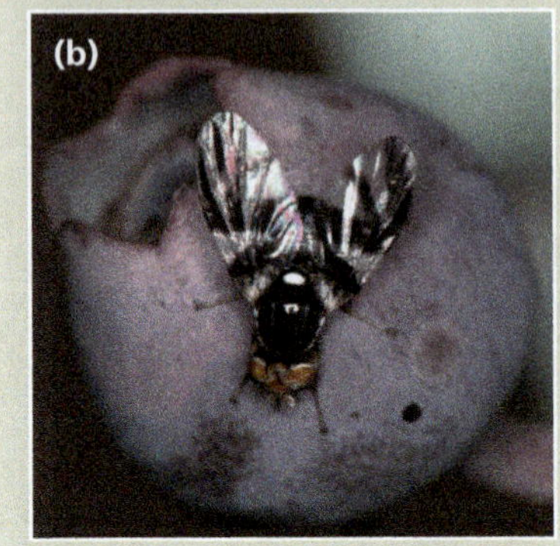

Temporal Isolation

Species that breed during different times of the day, different seasons, or different years cannot mix their gametes.

Example: In North America, the geographic ranges of the western spotted skunk (*Spilogale gracilis*) **(c)** and the eastern spotted skunk (*Spilogale putorius*) **(d)** overlap, but *S. gracilis* mates in late summer and *S. putorius* mates in late winter.

Behavioral Isolation

Courtship rituals that attract mates and other behaviors unique to a species are effective reproductive barriers, even between closely related species. Such behavioral rituals enable *mate recognition*—a way to identify potential mates of the same species.

Example: Blue-footed boobies, inhabitants of the Galápagos, mate only after a courtship display unique to their species. Part of the "script" calls for the male to high-step **(e)**, a behavior that calls the female's attention to his bright blue feet.

Video: Blue-Footed Boobies Courtship Ritual

Mechanical Isolation

Mating is attempted, but morphological differences prevent its successful completion.

Example: The shells of two species of snails in the genus *Bradybaena* spiral in different directions: Moving inward to the center, one spirals in a counterclockwise direction (**f**, left), the other in a clockwise direction (**f**, right). As a result, the snails' genital openings (indicated by arrows) are not aligned, and mating cannot be completed.

Postzygotic barriers prevent a hybrid zygote from developing into a viable, fertile adult

Gametic Isolation

Sperm of one species may not be able to fertilize the eggs of another species. For instance, sperm may not be able to survive in the reproductive tract of females of the other species, or biochemical mechanisms may prevent the sperm from penetrating the membrane surrounding the other species' eggs.

Example: Gametic isolation separates certain closely related species of aquatic animals, such as sea urchins **(g)**. Sea urchins release their sperm and eggs into the surrounding water, where they fuse and form zygotes. It is difficult for gametes of different species, such as the red and purple urchins shown here, to fuse because proteins on the surfaces of the eggs and sperm bind very poorly to each other.

Reduced Hybrid Viability

The genes of different parent species may interact in ways that impair the hybrid's development or survival in its environment.

Example: Some salamander subspecies of the genus *Ensatina* live in the same regions and habitats, where they may occasionally hybridize. But most of the hybrids do not complete development, and those that do are frail **(h)**.

Reduced Hybrid Fertility

Even if hybrids are vigorous, they may be sterile. If the chromosomes of the two parent species differ in number or structure, meiosis in the hybrids may fail to produce normal gametes. Since the infertile hybrids cannot produce offspring when they mate with either parent species, genes cannot flow freely between the species.

Example: The hybrid offspring of a male donkey **(i)** and a female horse **(j)** is a mule **(k)**, which is robust but sterile. A "hinny" (not shown), the offspring of a female donkey and a male horse, is also sterile.

Hybrid Breakdown

Some first-generation hybrids are viable and fertile, but when they mate with one another or with either parent species, offspring of the next generation are feeble or sterile.

Example: Strains of cultivated rice have accumulated different mutant recessive alleles at two loci in the course of their divergence from a common ancestor. Hybrids between them are vigorous and fertile (**l**, left and right), but plants in the next generation that carry too many of these recessive alleles are small and sterile (**l**, center). Although these rice strains are not yet considered different species, they have begun to be separated by postzygotic barriers.

these barriers can be classified according to whether they contribute to reproductive isolation before or after fertilization. **Prezygotic barriers** ("before the zygote") block fertilization from occurring. Such barriers typically act in one of three ways: by impeding members of different species from attempting to mate, by preventing an attempted mating from being completed successfully, or by hindering fertilization if mating is completed successfully. If a sperm cell from one species overcomes prezygotic barriers and fertilizes an ovum from another species, a variety of **postzygotic barriers** ("after the zygote") may contribute to reproductive isolation after the hybrid zygote is formed. Developmental errors may reduce survival among hybrid embryos. Or problems after birth may cause hybrids to be infertile or decrease their chance of surviving long enough to reproduce.

MB **Video: Prezygotic Barriers to Mating in Galápagos Finches by Peter and Rosemary Grant**

Limitations of the Biological Species Concept

One strength of the biological species concept is that it directs our attention to a way by which speciation can occur: by the evolution of reproductive isolation. However, the number of species to which this concept can be usefully applied is limited. There is, for example, no way to evaluate the reproductive isolation of fossils. The biological species concept also does not apply to organisms that reproduce asexually all or most of the time, such as prokaryotes. (Many prokaryotes do transfer genes among themselves, as we will discuss in Concept 27.2, but this is not part of their reproductive process.) Furthermore, in the biological species concept, species are designated by the *absence* of gene flow. But there are many pairs of species that are morphologically and ecologically distinct, and yet gene flow occurs between them. An example is the grizzly bear (*Ursus arctos*) and polar bear (*Ursus maritimus*), whose hybrid offspring have been dubbed "grolar bears" **(Figure 24.4)**. As we'll discuss, natural selection can cause such species to remain distinct even though some gene flow occurs between them. Because of the limitations to the biological species concept, alternative species concepts are useful in certain situations.

Other Definitions of Species

While the biological species concept emphasizes the *separateness* of different species due to reproductive barriers, several other definitions emphasize the *unity within* a species. For example, the **morphological species concept** distinguishes a species by body shape and other structural features. The morphological species concept can be applied to asexual and sexual organisms, and it can be useful even without information on the extent of gene flow. In practice, scientists often distinguish species using morphological criteria. A disadvantage of this approach, however, is that it relies on

▲ **Figure 24.4 Hybridization between two species of bears in the genus *Ursus*.**

subjective criteria; researchers may disagree on which structural features distinguish a species.

The **ecological species concept** defines a species in terms of its ecological niche, the sum of how members of the species interact with the nonliving and living parts of their environment (see Concept 54.1). For example, two species of oak trees might differ in their size or in their ability to tolerate dry conditions, yet still occasionally interbreed. Because they occupy different ecological niches, these oaks would be considered separate species even though they are connected by some gene flow. Unlike the biological species concept, the ecological species concept can accommodate asexual as well as sexual species. It also emphasizes the role of disruptive natural selection as organisms adapt to different environments.

In addition to those discussed here, more than 20 other species definitions have been proposed. The usefulness of each definition depends on the situation and the research questions being asked. For our purposes of studying how species originate, the biological species concept, with its focus on reproductive barriers, is particularly helpful.

CONCEPT CHECK 24.1

1. (a) Which species concept(s) could you apply to both asexual and sexual species? (b) Which would be most useful for identifying species in the field? Explain.
2. **WHAT IF? ➤** Suppose two bird species live in a forest and are not known to interbreed. One species feeds and mates in the treetops and the other on the ground. But in captivity, the birds can interbreed and produce viable, fertile offspring. What type of reproductive barrier most likely keeps these species separate in nature? Explain.

For suggested answers, see Appendix A.

CONCEPT 24.2

Speciation can take place with or without geographic separation

Having discussed what constitutes a unique species, let's return to the process by which such species arise from existing species. We'll describe this process by focusing on the geographic setting in which gene flow is interrupted between populations of the existing species—in allopatric speciation the populations are geographically isolated, while in sympatric speciation they are not **(Figure 24.5)**.

Allopatric ("Other Country") Speciation

In **allopatric speciation** (from the Greek *allos*, other, and *patra*, homeland), gene flow is interrupted when a population is divided into geographically isolated subpopulations. For example, the water level in a lake may subside, resulting in two or more smaller lakes that are now home to separated populations (see Figure 24.5a). Or a river may change course and divide a population of animals that cannot cross it. Allopatric speciation can also occur without geologic remodeling, such as when individuals colonize a remote area and their descendants become geographically isolated from the parent population. The flightless cormorant shown in Figure 24.1 probably originated in this way from an ancestral flying species that reached the Galápagos Islands.

The Process of Allopatric Speciation

How formidable must a geographic barrier be to promote allopatric speciation? The answer depends on the ability of the organisms to move about. Birds, mountain lions, and coyotes can cross rivers and canyons—as can the windblown pollen of pine trees and the seeds of some flowering plants. In contrast, small rodents may find a wide river or deep canyon a formidable barrier.

Once geographic isolation has occurred, the separated gene pools may diverge. Different mutations arise, and natural selection and genetic drift may alter allele frequencies in different ways in the separated populations. Reproductive isolation may then evolve as a by-product of the genetic divergence that results from selection or drift.

Figure 24.6 describes an example. On Andros Island, in the Bahamas, populations of the mosquitofish *Gambusia hubbsi* colonized a series of ponds that later became isolated from one another. Genetic analyses indicate that little or no gene flow currently occurs between the ponds. The environments of these ponds are very similar except that some contain predatory fishes, while others do not. In ponds with predatory fishes, selection has favored the evolution of a mosquitofish body shape that enables rapid bursts of speed (Figure 24.6). In ponds without predatory fishes, selection has favored a different body shape, one that improves the ability to swim for long periods of time. How have these different selective pressures affected the evolution of reproductive barriers? Researchers studied this question by bringing together mosquitofish from the two types of ponds. They found that female mosquitofish prefer to mate with males whose body shape is similar to their own. This preference establishes a behavioral barrier to reproduction between mosquitofish from ponds with predators and those

▼ Figure 24.5 The geography of speciation.

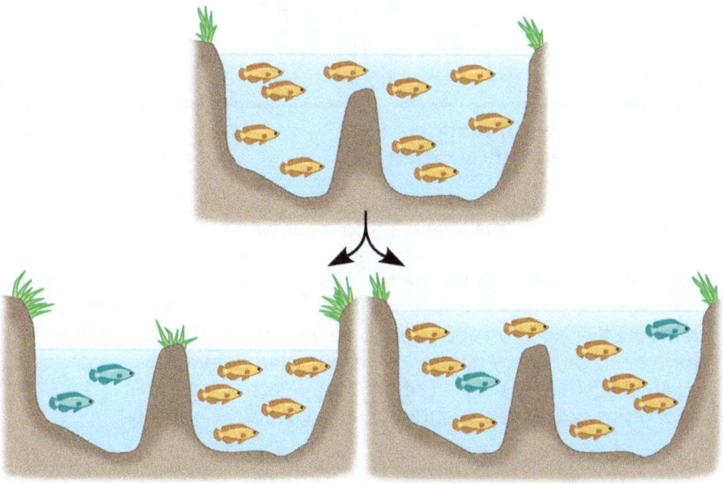

(a) Allopatric speciation. A population forms a new species while geographically isolated from its parent population.

(b) Sympatric speciation. A subset of a population forms a new species without geographic separation.

▼ Figure 24.6 Evolution in mosquitofish populations. Different body shapes have evolved in mosquitofish populations from ponds with and without predators. These differences affect how quickly the fish can accelerate to escape and their survival rate when exposed to predators.

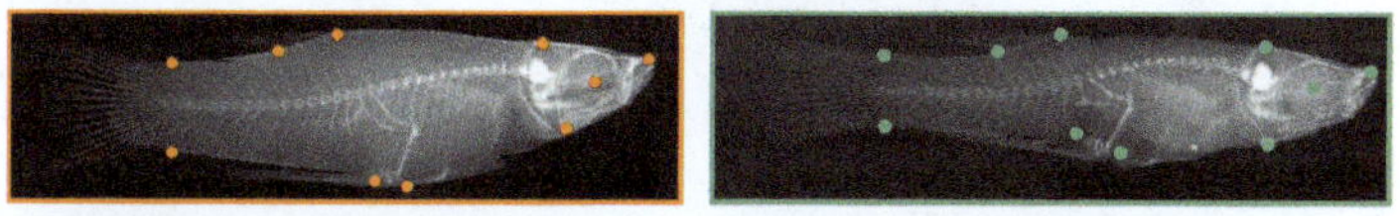

In ponds with predatory fishes, the mosquitofish's head is streamlined and the tail is powerful, enabling rapid bursts of speed.

In ponds without predatory fishes, mosquitofish have a different body shape that favors long, steady swimming.

(a) Differences in body shape

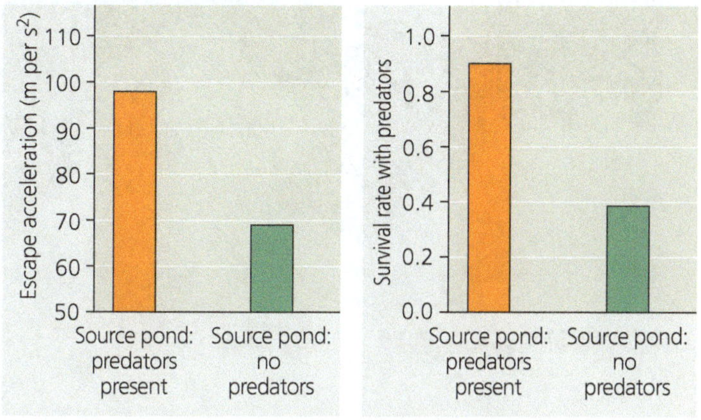

(b) Differences in escape acceleration and survival

from ponds without predators. Thus, as a by-product of selection for avoiding predators, reproductive barriers have formed in these allopatric populations.

Evidence of Allopatric Speciation

Many studies provide evidence that speciation can occur in allopatric populations. For example, laboratory studies show that reproductive barriers can develop when populations are isolated experimentally and subjected to different environmental conditions **(Figure 24.7)**.

Field studies indicate that allopatric speciation also can occur in nature. Consider the 30 species of snapping shrimp in the genus *Alpheus* that live off the Isthmus of Panama, the land bridge that connects South and North America **(Figure 24.8)**. Fifteen of these species live on the Atlantic side of the isthmus, while the other 15 live on the Pacific side. Before the isthmus formed, gene flow could occur between the Atlantic and Pacific populations of snapping shrimp. Did the species on different sides of the isthmus originate by allopatric speciation? Morphological and genetic data group these shrimp into 15 pairs of *sister species*, pairs whose member species are each other's closest relative.

Figure 24.8 Allopatric speciation in snapping shrimp (*Alpheus*). The shrimp pictured are just 2 of the 15 pairs of sister species that arose as populations were divided by the formation of the Isthmus of Panama. The color-coded type indicates the sister species.

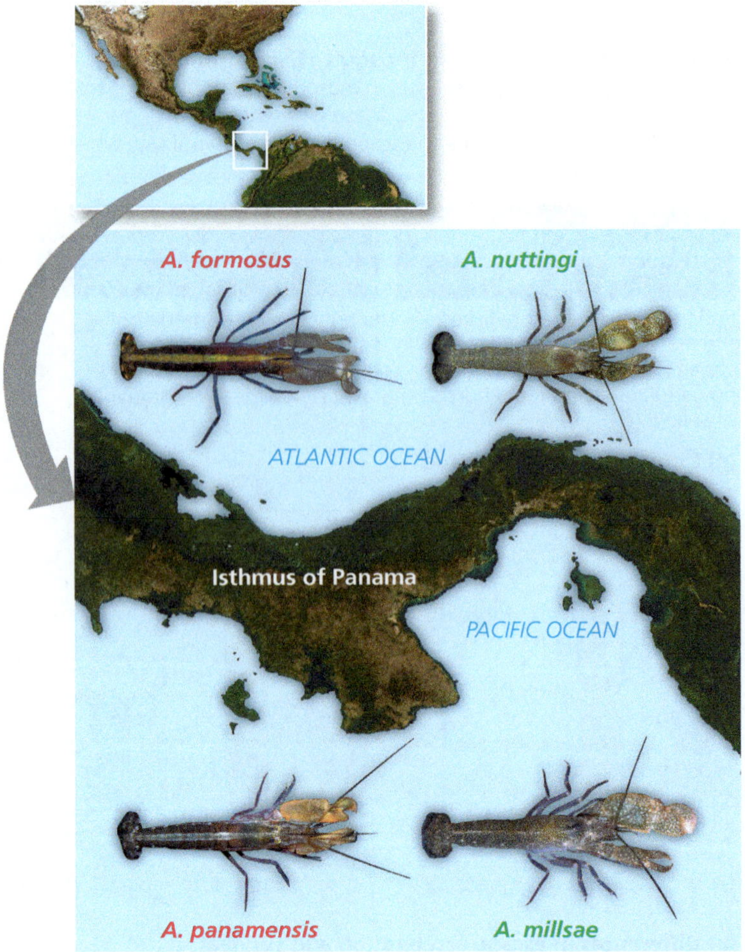

Figure 24.7

Inquiry Can divergence of allopatric populations lead to reproductive isolation?

Experiment A researcher divided a laboratory population of the fruit fly *Drosophila pseudoobscura*, raising some flies on a starch medium and others on a maltose medium. After one year (about 40 generations), natural selection resulted in divergent evolution: Populations raised on starch digested starch more efficiently, while those raised on maltose digested maltose more efficiently. The researcher then put flies from the same or different populations in mating cages and measured mating frequencies. All flies used in the mating preference tests were reared for one generation on a standard cornmeal medium.

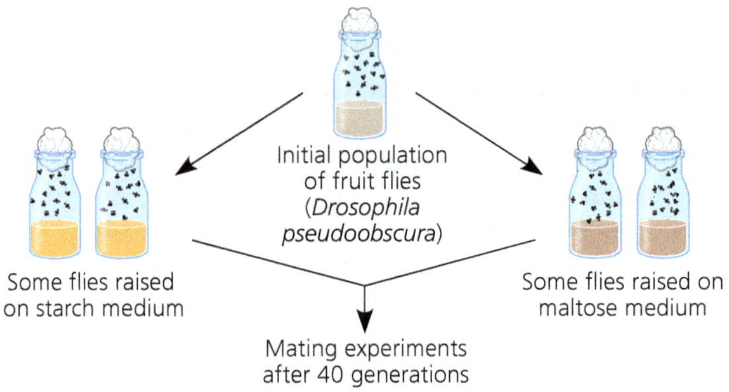

Results Mating patterns among populations of flies raised on different media are shown below. When flies from "starch populations" were mixed with flies from "maltose populations," the flies tended to mate with like partners. But in the control group (shown on the right), flies from different populations adapted to starch were about as likely to mate with each other as with flies from their own population; similar results were obtained for control groups adapted to maltose.

		Female	
		Starch	Maltose
Male	Starch	22	9
	Maltose	8	20

Number of matings in experimental group

		Female	
		Starch population 1	Starch population 2
Male	Starch population 1	18	15
	Starch population 2	12	15

Number of matings in control group

Conclusion In the experimental group, the strong preference of "starch flies" and "maltose flies" to mate with like-adapted flies indicates that a reproductive barrier was forming between these fly populations. Although this barrier was not absolute (some mating between starch flies and maltose flies did occur), after 40 generations reproductive isolation appeared to be increasing. This barrier may have been caused by differences in courtship behavior that arose as an incidental by-product of differing selective pressures as these allopatric populations adapted to different sources of food.

Data from D. M. B. Dodd, Reproductive isolation as a consequence of adaptive divergence in *Drosophila pseudoobscura*, *Evolution* 43:1308–1311 (1989).

WHAT IF? ➤ *Why were all flies used in the mating preference tests reared on a standard medium (rather than on starch or maltose)?*

In each of these 15 pairs, one of the sister species lives on the Atlantic side of the isthmus, while the other lives on the Pacific side. This fact strongly suggests that the two species arose as a consequence of geographic separation. Furthermore, genetic analyses indicate that the *Alpheus* species originated from 9 to 3 million years ago, with the sister species that live in the deepest water diverging first. These divergence times are consistent with geologic evidence that the isthmus formed gradually, starting 10 million years ago, and closing completely about 3 million years ago.

The importance of allopatric speciation is also suggested by the fact that regions that are isolated or highly subdivided by barriers typically have more species than do otherwise similar regions that lack such features. For example, many unique plants and animals are found on the geographically isolated Hawaiian Islands (we'll return to the origin of Hawaiian species in Concept 25.4). Field studies also show that reproductive isolation between two populations generally increases as the geographic distance between them increases, a finding consistent with allopatric speciation. In the **Scientific Skills Exercise**, you will analyze data from one such study that examined reproductive isolation in geographically separated salamander populations.

Note that while geographic isolation prevents interbreeding between members of allopatric populations, physical separation is not a biological barrier to reproduction. Biological reproductive barriers such as those described in Figure 24.3 are intrinsic to the organisms themselves. Hence, it is biological barriers that can prevent interbreeding when members of different populations come into contact with one another.

Sympatric ("Same Country") Speciation

In **sympatric speciation** (from the Greek *syn*, together), speciation occurs in populations that live in the same geographic area (see Figure 24.5b). How can reproductive barriers form between sympatric populations while their members remain in contact with each other? Although such contact (and the ongoing gene flow that results) makes sympatric speciation less common than allopatric speciation, sympatric speciation can occur if gene flow is reduced by such factors as polyploidy, sexual selection, and habitat differentiation. (Note that these factors can also promote allopatric speciation.)

SCIENTIFIC SKILLS EXERCISE

Identifying Independent and Dependent Variables, Making a Scatter Plot, and Interpreting Data

Does Distance Between Salamander Populations Increase Their Reproductive Isolation? Allopatric speciation begins when populations become geographically isolated, preventing mating between individuals in different populations and thus stopping gene flow. It is logical that as distance between populations increases, so will their degree of reproductive isolation. To test this hypothesis, researchers studied populations of the dusky salamander (*Desmognathus ochrophaeus*) living on different mountain ranges in the southern Appalachians.

How the Experiment Was Done The researchers tested the reproductive isolation of pairs of salamander populations by leaving one male and one female together and later checking the females for the presence of sperm. Four mating combinations were tested for each pair of populations (A and B)—two *within* the same population (female A with male A and female B with male B) and two *between* populations (female A with male B and female B with male A).

Data from the Experiment The researchers used an index of reproductive isolation that ranged from a value of 0 (no isolation) to a value of 2 (full isolation). The proportion of successful matings for each mating combination was measured, with 100% success = 1 and no success = 0. The reproductive isolation value for two populations is the sum of the proportion of successful matings of each type within populations (AA + BB) minus the sum of the proportion of successful matings of each type between populations (AB + BA). The table provides distance and reproductive isolation data for 27 pairs of dusky salamander populations.

INTERPRET THE DATA

1. State the researchers' hypothesis, and identify the independent and dependent variables in this study. Explain why the researchers used four mating combinations for each pair of populations.
2. Calculate the value of the reproductive isolation index if (a) *all* of the matings within a population were successful, but *none* of the matings between populations were successful; (b) salamanders are equally successful in mating with members of their own population and members of another population.
3. Make a scatter plot to help you visualize any patterns that might indicate a relationship between the variables. Plot the independent variable on the *x*-axis and the dependent variable on the *y*-axis. (For additional information about graphs, see the Scientific Skills Review in Appendix F and the Study Area of MasteringBiology.)
4. Interpret your graph by (a) explaining in words any pattern indicating a possible relationship between the variables and (b) hypothesizing the possible cause of such a relationship.

MB **Instructors**: A version of this Scientific Skills Exercise can be assigned in MasteringBiology.

Data from S. G. Tilley, A. Verrell, and S. J. Arnold, Correspondence between sexual isolation and allozyme differentiation: a test in the salamander *Desmognathus ochrophaeus*, *Proceedings of the National Academy of Sciences USA* 87:2715–2719 (1990).

Geographic Distance (km)	15	32	40	47	42	62	63	81	86	107	107	115	137	147
Reproductive Isolation Value	0.32	0.54	0.50	0.50	0.82	0.37	0.67	0.53	1.15	0.73	0.82	0.81	0.87	0.87
Distance (continued)	137	150	165	189	219	239	247	53	55	62	105	179	169	
Isolation (continued)	0.50	0.57	0.91	0.93	1.5	1.22	0.82	0.99	0.21	0.56	0.41	0.72	1.15	

Polyploidy

A species may originate from an accident during cell division that results in extra sets of chromosomes, a condition called **polyploidy**. Polyploid speciation occasionally occurs in animals; for example, the gray tree frog *Hyla versicolor* (see Figure 23.16) is thought to have originated in this way. However, polyploidy is far more common in plants. In fact, botanists estimate that more than 80% of the plant species alive today are descended from ancestors that formed by polyploid speciation.

Figure 24.9 Sympatric speciation by autopolyploidy.

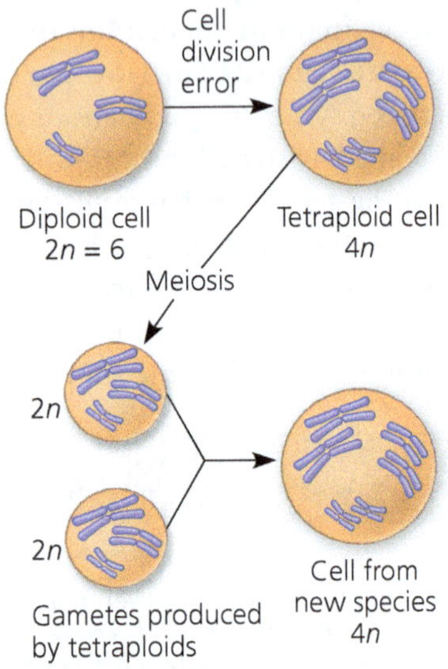

Two distinct forms of polyploidy have been observed in plant (and a few animal) populations. An **autopolyploid** (from the Greek *autos*, self) is an individual that has more than two chromosome sets that are all derived from a single species. In plants, for example, a failure of cell division could double a cell's chromosome number from the original number ($2n$) to a tetraploid number ($4n$) **(Figure 24.9)**.

A tetraploid can produce fertile tetraploid offspring by self-pollinating or by mating with other tetraploids. In addition, the tetraploids are reproductively isolated from $2n$ plants of the original population, because the triploid ($3n$) offspring of such unions have reduced fertility. Thus, in just one generation, autopolyploidy can generate reproductive isolation without any geographic separation.

A second form of polyploidy can occur when two different species interbreed and produce hybrid offspring. Most such hybrids are sterile because the set of chromosomes from one species cannot pair during meiosis with the set of chromosomes from the other species. However, an infertile hybrid may be able to propagate itself asexually (as many plants can do). In subsequent generations, various mechanisms can change a sterile hybrid into a fertile polyploid called an **allopolyploid (Figure 24.10)**. The allopolyploids are fertile when mating with each other but cannot interbreed with either parent species; thus, they represent a new biological species.

Although it can be challenging to study speciation in the field, scientists have documented at least five new plant species that have originated by polyploid speciation since 1850. One of these examples involves the origin of a new species of goatsbeard plant (genus *Tragopogon*) in the Pacific Northwest. *Tragopogon* first arrived in the region when humans introduced three European species in the early 1900s: *T. pratensis, T. dubius,* and *T. porrifolius.* These three species are now common weeds

Figure 24.10 One mechanism for allopolyploid speciation in plants. Most hybrids are sterile because their chromosomes are not homologous and cannot pair during meiosis. However, such a hybrid may be able to reproduce asexually. This diagram traces one mechanism that can produce fertile hybrids (allopolyploids) that are members of a new species. The new species has a diploid chromosome number equal to the sum of the diploid chromosome numbers of the two parent species.

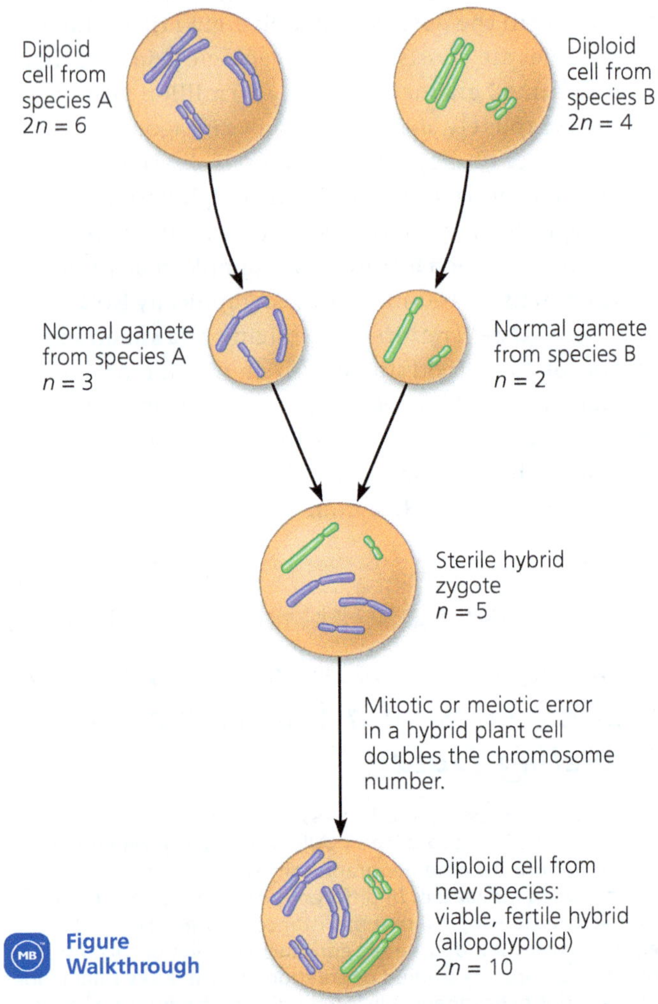

MB Figure Walkthrough

in abandoned parking lots and other urban sites. In 1950, a new *Tragopogon* species was discovered near the Idaho-Washington border, a region where all three European species also were found. Genetic analyses revealed that this new species, *Tragopogon miscellus,* is a hybrid of two of the European species **(Figure 24.11)**. Although the *T. miscellus* population grows mainly by reproduction of its own members, additional episodes of hybridization between the parent species continue to add new members to the *T. miscellus* population. Later, scientists discovered another new *Tragopogon* species, *T. mirus*—this one a hybrid of *T. dubius* and *T. porrifolius* (see Figure 24.11). The *Tragopogon* story is just one of several well-studied examples in which scientists have observed speciation in progress.

Many important agricultural crops—such as oats, cotton, potatoes, tobacco, and wheat—are polyploids. For example, the wheat used for bread, *Triticum aestivum,* is an allohexaploid (six sets of chromosomes, two sets from each of three different species). The first of the polyploidy events that eventually led

Figure 24.11 Allopolyploid speciation in *Tragopogon*. The gray boxes indicate the three parent species. The diploid chromosome number of each species is shown in parentheses.

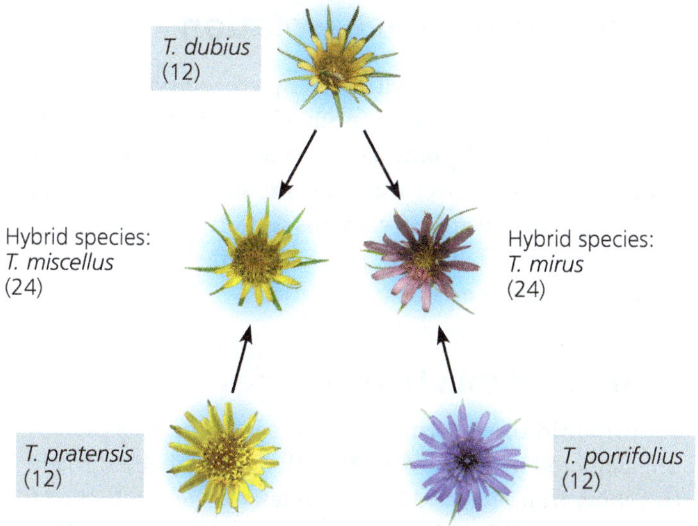

to modern wheat probably occurred about 8,000 years ago in the Middle East as a spontaneous hybrid of an early cultivated wheat species and a wild grass. Today, plant geneticists generate new polyploids in the laboratory by using chemicals that induce meiotic and mitotic errors. By harnessing the evolutionary process, researchers can produce new hybrid species with desired qualities, such as a hybrid that combines the high yield of wheat with the hardiness of rye.

MB **Animation: Speciation by Changes in Ploidy**

Sexual Selection

There is evidence that sympatric speciation can also be driven by sexual selection. Clues to how this can occur have been found in cichlid fish from one of Earth's hot spots of animal speciation, East Africa's Lake Victoria. This lake was once home to as many as 600 species of cichlids. Genetic data indicate that these species originated within the last 100,000 years from a small number of colonizing species that arrived from other lakes and rivers. How did so many species—more than double the number of freshwater fish species known in all of Europe—originate within a single lake?

One hypothesis is that subgroups of the original cichlid populations adapted to different food sources and the resulting genetic divergence contributed to speciation in Lake Victoria. But sexual selection, in which (typically) females select males based on their appearance (see Concept 23.4), may also have been a factor. Researchers have studied two closely related sympatric species of cichlids that differ mainly in the coloration of breeding males: Breeding *Pundamilia pundamilia* males have a blue-tinged back, whereas breeding *Pundamilia nyererei* males have a red-tinged back **(Figure 24.12)**. Their results suggest that mate choice based on male breeding coloration can act as a reproductive barrier that keeps the gene pools of these two species separate.

Figure 24.12

Inquiry Does sexual selection in cichlids result in reproductive isolation?

Experiment Researchers placed males and females of *Pundamilia pundamilia* and *P. nyererei* together in two aquarium tanks, one with natural light and one with a monochromatic orange lamp. Under normal light, the two species are noticeably different in male breeding coloration; under monochromatic orange light, the two species are very similar in color. The researchers then observed the mate choices of the females in each tank.

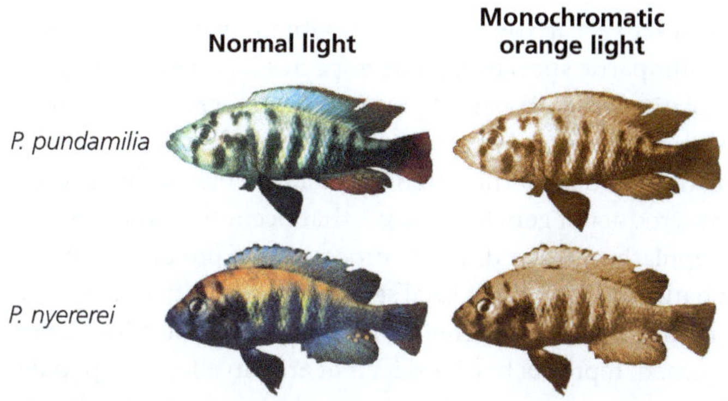

Results Under normal light, females of each species strongly preferred males of their own species. But under orange light, females of each species responded indiscriminately to males of both species. The resulting hybrids were viable and fertile.

Conclusion The researchers concluded that mate choice by females based on male breeding coloration can act as a reproductive barrier that keeps the gene pools of these two species separate. Since the species can still interbreed when this prezygotic behavioral barrier is breached in the laboratory, the genetic divergence between the species is likely to be small. This suggests that speciation in nature has occurred relatively recently.

Data from O. Seehausen and J. J. M. van Alphen, The effect of male coloration on female mate choice in closely related Lake Victoria cichlids (*Haplochromis nyererei* complex), *Behavioral Ecology and Sociobiology* 42:1–8 (1998).

WHAT IF? ➤ *Suppose that female cichlids living in the murky waters of a polluted lake could not distinguish colors well. In such waters, how might the gene pools of these species change over time?*

Habitat Differentiation

Sympatric speciation can also occur when a subpopulation exploits a habitat or resource not used by the parent population. Consider the North American apple maggot fly (*Rhagoletis pomonella*), a pest of apples. The fly's original habitat was the native hawthorn tree (see Figure 24.3a), but about 200 years ago, some populations colonized apple trees that had been introduced by European settlers. Apple maggot flies usually mate on or near their host plant. This results in a prezygotic barrier (habitat isolation) between populations that feed on apples and populations that feed on hawthorns. Furthermore, because apples mature more quickly than hawthorn fruit, natural selection has favored apple-feeding flies with rapid development. These apple-feeding populations now show temporal isolation from the hawthorn-feeding *R. pomonella*, providing a second prezygotic barrier to gene flow between the two populations. Researchers also have identified alleles that benefit the flies that use one host plant but harm the flies that use the other host plant. Natural selection operating on these alleles has provided a postzygotic barrier to reproduction, further limiting gene flow. Altogether, although the two populations are still classified as subspecies rather than separate species, sympatric speciation appears to be well under way.

Allopatric and Sympatric Speciation: *A Review*

Now let's recap the processes by which new species form. In allopatric speciation, a new species forms in geographic isolation from its parent population. Geographic isolation severely restricts gene flow. Intrinsic barriers to reproduction with members of the parent population may then arise as a by-product of genetic changes that occur within the isolated population. Many different processes can produce such genetic changes, including natural selection under different environmental conditions, genetic drift, and sexual selection. Once formed, reproductive barriers that arise in allopatric populations can prevent interbreeding with the parent population even if the populations come back into contact.

Sympatric speciation, in contrast, requires the emergence of a reproductive barrier that isolates a subset of a population from the remainder of the population in the same area. Though rarer than allopatric speciation, sympatric speciation can occur when gene flow to and from the isolated subpopulation is blocked. This can occur as a result of polyploidy, a condition in which an organism has extra sets of chromosomes. Sympatric speciation also can result from sexual selection. Finally, sympatric speciation can occur when a subset of a population becomes reproductively isolated because of natural selection that results from a switch to a habitat or food source not used by the parent population.

Having reviewed the geographic context in which species originate, we'll next explore in more detail what can happen when new or partially formed species come into contact.

MB **HHMI Video: Anole Lizards: An Example of Speciation** hhmi BioInteractive

CONCEPT CHECK 24.2

1. Summarize key differences between allopatric and sympatric speciation. Which type of speciation is more common, and why?
2. Describe two mechanisms that can decrease gene flow in sympatric populations, thereby making sympatric speciation more likely to occur.
3. **WHAT IF?** ➤ Is allopatric speciation more likely to occur on an island close to a mainland or on a more isolated island of the same size? Explain your prediction.
4. **MAKE CONNECTIONS** ➤ Review the process of meiosis in Figure 13.8. Describe how an error during meiosis could lead to polyploidy.

For suggested answers, see Appendix A.

CONCEPT 24.3

Hybrid zones reveal factors that cause reproductive isolation

What happens if species with incomplete reproductive barriers come into contact with one another? One possible outcome is the formation of a **hybrid zone**, a region in which members of different species meet and mate, producing at least some offspring of mixed ancestry. In this section, we'll explore hybrid zones and what they reveal about factors that cause the evolution of reproductive isolation.

Patterns Within Hybrid Zones

Some hybrid zones form as narrow bands, such as the one depicted in **Figure 24.13** for the yellow-bellied toad (*Bombina variegata*) and its close relative, the fire-bellied toad (*B. bombina*). This hybrid zone, represented by the red line on the map, extends for 4,000 km but is less than 10 km wide in most places. The hybrid zone occurs where the higher-altitude habitat of the yellow-bellied toad meets the lowland habitat of the fire-bellied toad. Across a given "slice" of the zone, the frequency of alleles specific to yellow-bellied toads typically decreases from close to 100% at the edge where only yellow-bellied toads are found to around 50% in the central portion of the zone to close to 0% at the edge where only fire-bellied toads are found.

What causes such a pattern of allele frequencies across a hybrid zone? We can infer that there is an obstacle to gene flow—otherwise, alleles from one parent species would also be common in the gene pool of the other parent species. Are geographic barriers reducing gene flow? Not in this case, since the toads can move throughout the hybrid zone. A more important factor is that hybrid toads have increased rates of embryonic mortality and a variety of morphological abnormalities, including ribs that are fused to the spine and malformed tadpole mouthparts. Because the hybrids have poor survival and reproduction, they produce few viable offspring with members of the parent species. As a result, hybrid individuals rarely serve as a stepping-stone from which alleles are passed from one species to the other. Outside the hybrid zone, additional obstacles to gene flow may be provided by natural selection in the different environments in which the parent species live.

Hybrid zones typically are located wherever the habitats of the interbreeding species meet. Those regions often resemble a group of isolated patches scattered across the landscape—more like the complex pattern of spots on a Dalmatian than the continuous band shown in Figure 24.13. But regardless of whether they have complex or simple spatial patterns, hybrid zones form when two species lacking complete barriers to reproduction come into contact. What happens when the habitats of the interbreeding species change over time?

Figure 24.13 A narrow hybrid zone for *Bombina* toads in Europe. The graph shows species-specific allele frequencies across the width of the zone near Krakow, Poland, averaged over six genetic loci. A value of 1.0 indicates that all individuals were yellow-bellied toads, 0 indicates that all individuals were fire-bellied toads, and intermediate frequencies indicate that some individuals were of mixed ancestry.

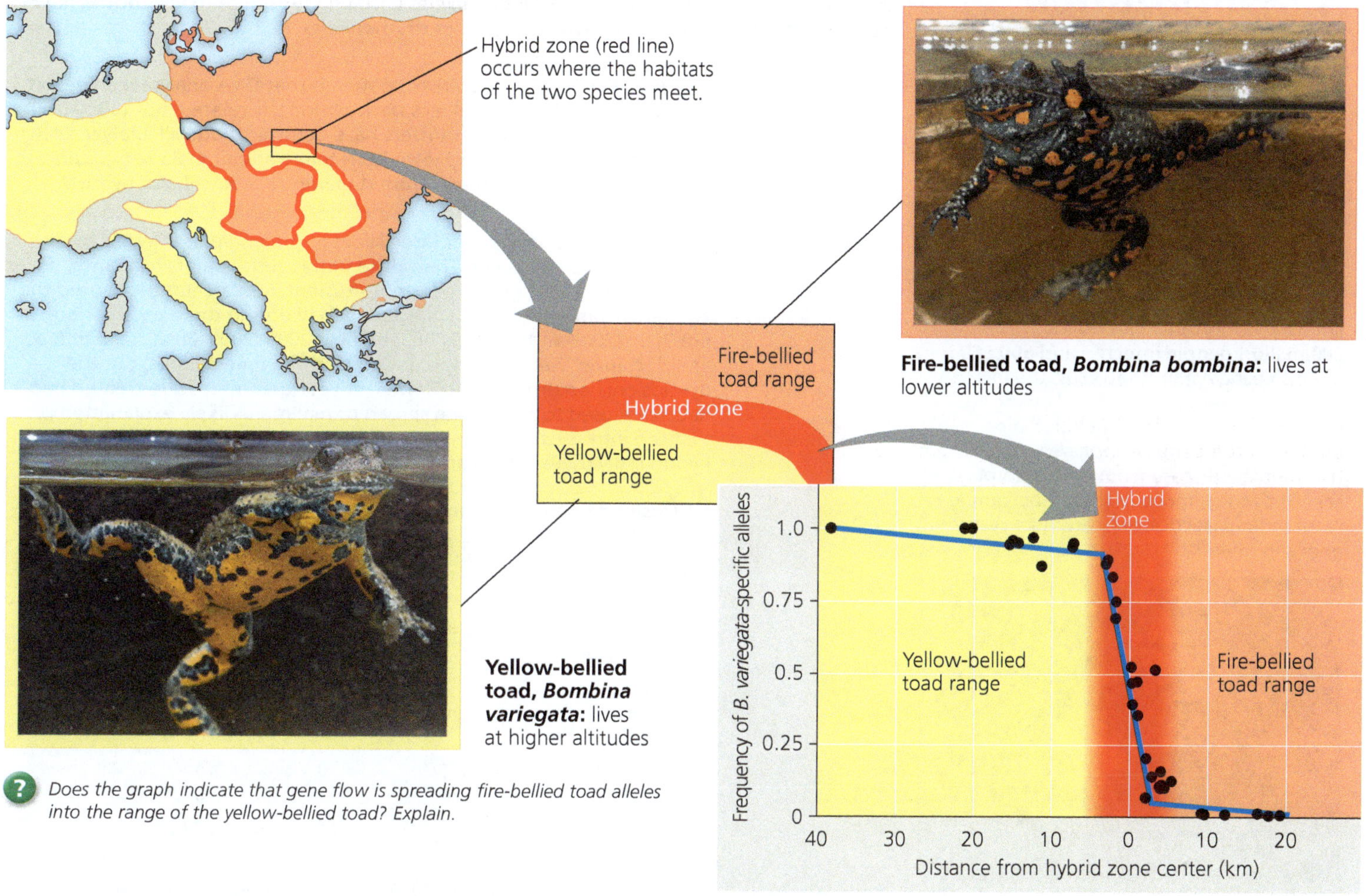

Does the graph indicate that gene flow is spreading fire-bellied toad alleles into the range of the yellow-bellied toad? Explain.

Hybrid Zones and Environmental Change

A change in environmental conditions can alter where the habitats of interbreeding species meet. When this happens, an existing hybrid zone can move to a new location, or a novel hybrid zone may form.

For example, black-capped chickadees (*Poecile atricapillus*) and Carolina chickadees (*P. carolinensis*) interbreed in a narrow hybrid zone that runs from New Jersey to Kansas. Recent studies have shown that the location of this hybrid zone has shifted northward as the climate has warmed. In another example, a series of warm winters prior to 2003 enabled the southern flying squirrel (*Glaucomys volans*) to expand northward into the range of the northern flying squirrel, *G. sabrinus*. Previously, the ranges of these two species had not overlapped. Genetic analyses showed that these flying squirrels began to hybridize where their ranges came into contact, thereby forming a novel hybrid zone induced by climate change.

Finally, note that a hybrid zone can be a source of novel genetic variation that improves the ability of one or both parent species to cope with changing environmental conditions. This can occur when an allele found only in one parent species is transferred first to hybrid individuals, and then to the other parent species when hybrids breed with the second parent species. Recent genetic analyses have shown that hybridization has been a source for such novel genetic variation in various insect, bird, and plant species. In the **Problem-Solving Exercise**, you can examine one such example: a case in which hybridization may have led to the transfer of insecticide-resistance alleles between mosquitoes that transmit malaria.

Hybrid Zones over Time

Studying a hybrid zone is like observing a naturally occurring experiment on speciation. Will the hybrids become reproductively isolated from their parents and form a new species, as occurred by polyploidy in the goatsbeard plant of the Pacific Northwest? If not, there are three other common outcomes for the hybrid zone over time: reinforcement of

PROBLEM-SOLVING EXERCISE

Is hybridization promoting insecticide resistance in mosquitoes that transmit malaria?

Malaria is a leading cause of human illness and mortality worldwide, with 200 million people infected and 600,000 deaths each year. In the 1960s, the incidence of malaria was reduced owing to the use of insecticides that killed mosquitoes in the genus *Anopheles*, which transmit the disease from person to person. But today, mosquitoes are becoming resistant to insecticides—causing a resurgence in malaria.

▲ Insecticide-treated bed nets have helped reduce cases of malaria in many countries, but resistance to insecticides is rising in mosquito populations.

Instructors: A version of this Problem-Solving Exercise can be assigned in MasteringBiology.

In this exercise, you will investigate whether alleles encoding resistance to insecticides have been transferred between closely related species of *Anopheles*.

Your Approach The principle guiding your investigation is that DNA analyses can detect the transfer of resistance alleles between closely related mosquito species. To find out whether such transfers have occurred, you will analyze DNA results from two species of mosquitoes that transmit malaria (*Anopheles gambiae* and *A. coluzzii*) and from *A. gambiae* × *A. coluzzii* hybrids.

Your Data Resistance to DDT and other insecticides in *Anopheles* is affected by a sodium channel gene, *kdr*. The *r* allele of this gene confers resistance, while the wild type (+/+) genotype is not resistant. Researchers sequenced the *kdr* gene from mosquitoes collected in Mali during three time periods: pre-2006 (2002 and 2004), 2006, and post-2006 (2009–2012). *A. gambiae* and *A. coluzzii* were collected during all three time periods, but their hybrids only occurred in 2006, the first year that insecticide-treated bed nets were used to reduce the spread of malaria. A likely explanation is that the introduction of the treated bed nets may have briefly favored hybrid individuals, which are usually at a selective disadvantage.

Observed numbers of mosquitoes by *kdr* genotype	+/+	+/*r*	*r*/*r*
A. gambiae			
Pre-2006	3	5	2
2006	8	8	7
Post-2006	3	3	57
Hybrids			
2006	10	7	0
A. coluzzii			
Pre-2006	226	0	0
2006	70	7	0
Post-2006	79	127	94

Your Analysis

1. How did the frequencies of *kdr* genotypes change over time in *A. gambiae*? Describe a hypothesis that accounts for these observations.
2. How did the frequencies of *kdr* genotypes change over time in *A. coluzzii*? Describe a hypothesis that accounts for these observations.
3. Do these results indicate that hybridization can lead to the transfer of adaptive alleles? Explain.
4. Predict how the transfer of the *r* allele to *A. coluzzii* populations could affect the number of malaria cases.

barriers, fusion of species, or stability **(Figure 24.14)**. Let's examine what studies suggest about these possibilities.

Reinforcement: Strengthening Reproductive Barriers

Hybrids often are less fit than members of their parent species. In such cases, natural selection should strengthen prezygotic barriers to reproduction, reducing the formation of unfit hybrids. Because this process involves *reinforcing* reproductive barriers, it is called **reinforcement**. If reinforcement is occurring, a logical prediction is that barriers to reproduction between species should be stronger for sympatric populations than for allopatric populations.

As an example, let's consider two species of European flycatcher, the pied flycatcher (*Ficedula hypoleuca*) and the collared flycatcher (*Ficedula albicollis*). In allopatric populations of these birds, males of the two species closely resemble one another, while in sympatric populations, the males look very different. Female flycatchers do not select males of the other species when given a choice between males from sympatric populations, but they frequently do make mistakes when selecting between males from allopatric populations. Thus, barriers to reproduction are stronger in birds from sympatric populations than in birds from allopatric populations, as you would predict if reinforcement were occurring. Similar results have been observed in a number of organisms, including fishes, insects, plants, and other birds.

Figure 24.14 Formation of a hybrid zone and possible outcomes for hybrids over time. The thick gray arrows represent the passage of time.

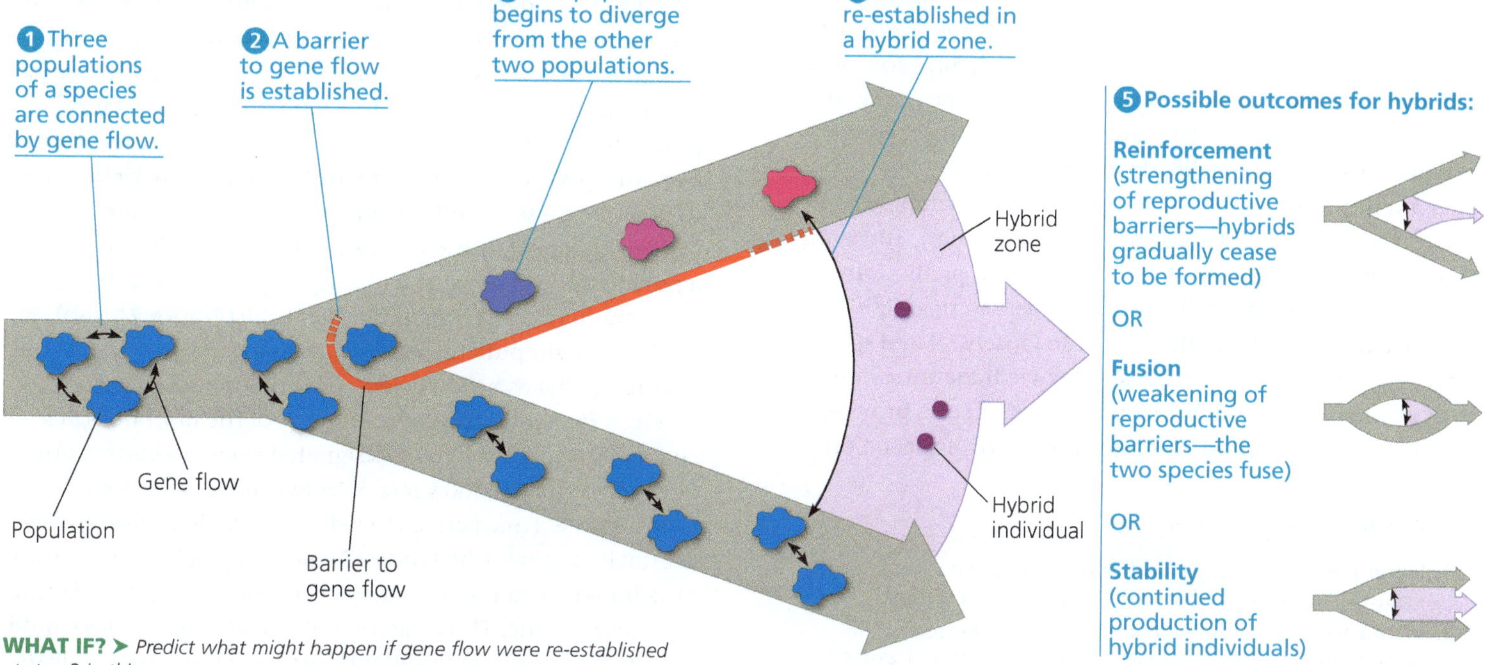

WHAT IF? ➤ *Predict what might happen if gene flow were re-established at step 3 in this process.*

Fusion: Weakening Reproductive Barriers

Barriers to reproduction may be weak when two species meet in a hybrid zone. Indeed, so much gene flow may occur that reproductive barriers weaken further and the gene pools of the two species become increasingly alike. In effect, the speciation process reverses, eventually causing the two hybridizing species to fuse into a single species.

For example, genetic and morphological evidence indicate that the recent loss of the large tree finch from the Galápagos island of Floreana resulted from extensive hybridization with another finch species on that island. Such a situation also may be occurring among Lake Victoria cichlids. Many pairs of ecologically similar cichlid species are reproductively isolated because the females of one species prefer to mate with males of one color, while females of the other species prefer to mate with males of a different color (see Figure 24.12). Results from field and laboratory studies indicate that murky waters caused by pollution have reduced the ability of females to use color to distinguish males of their own species from males of closely related species. In some polluted waters, many hybrids have been produced, leading to fusion of the parent species' gene pools and a loss of species **(Figure 24.15)**.

Figure 24.15 Fusion: the breakdown of reproductive barriers. Increasingly cloudy water in Lake Victoria over the past several decades may have weakened reproductive barriers between *P. nyererei* and *P. pundamilia*. In areas of cloudy water, the two species have hybridized extensively, causing their gene pools to fuse.

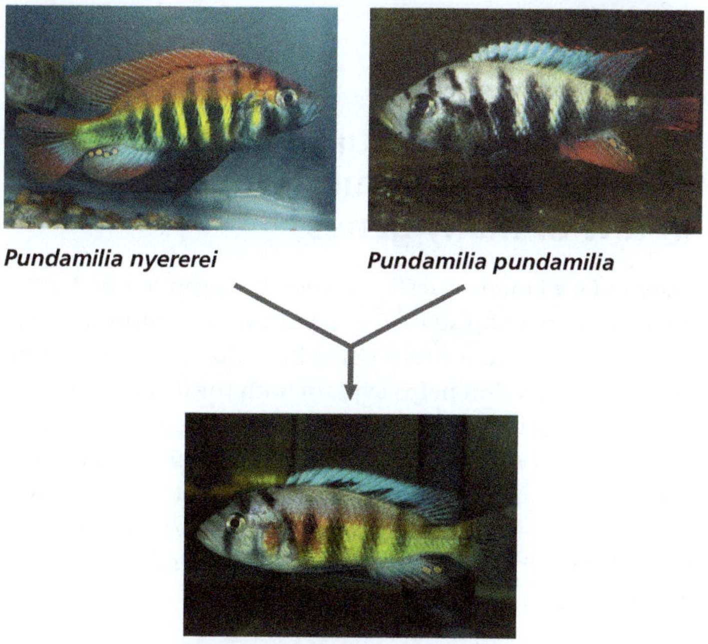

Pundamilia nyererei ***Pundamilia pundamilia***

Pundamilia "turbid water," hybrid offspring from a location with turbid water

Stability: Continued Formation of Hybrid Individuals

Many hybrid zones are stable in the sense that hybrids continue to be produced. In some cases, this occurs because the hybrids survive or reproduce better than members of either parent species, at least in certain habitats or years. But stable hybrid zones have also been observed in cases where the hybrids are selected *against*—an unexpected result.

For example, hybrids continue to form in the *Bombina* hybrid zone even though they are strongly selected against.

One explanation relates to the narrowness of the *Bombina* hybrid zone (see Figure 24.13). Evidence suggests that members of both parent species migrate into the zone from the parent populations located outside the zone, thus leading to the continued production of hybrids. If the hybrid zone were wider, this would be less likely to occur, since the center of the zone would receive little gene flow from distant parent populations located outside the hybrid zone.

Sometimes the outcomes in hybrid zones match our predictions (European flycatchers and cichlid fishes), and sometimes they don't (*Bombina*). But whether our predictions are upheld or not, events in hybrid zones can shed light on how barriers to reproduction between closely related species change over time. In the next section, we'll examine how interactions between hybridizing species can also provide a glimpse into the speed and genetic control of speciation.

CONCEPT CHECK 24.3

1. What are hybrid zones, and why can they be viewed as "natural laboratories" in which to study speciation?
2. **WHAT IF?** ➤ Consider two species that diverged while geographically separated but resumed contact before reproductive isolation was complete. Predict the outcome over time if the two species mated indiscriminately and (a) hybrid offspring survived and reproduced more poorly than offspring from intraspecific matings or (b) hybrid offspring survived and reproduced as well as offspring from intraspecific matings.

For suggested answers, see Appendix A.

CONCEPT 24.4

Speciation can occur rapidly or slowly and can result from changes in few or many genes

Darwin faced many questions when he began to ponder that "mystery of mysteries"—speciation. He found answers to some of those questions when he realized that evolution by natural selection helps explain both the diversity of life and the adaptations of organisms (see Concept 22.2). But biologists since Darwin have continued to ask fundamental questions about speciation. How long does it take for new species to form? And how many genes change when one species splits into two? Answers to these questions are also emerging.

The Time Course of Speciation

We can gather information about how long it takes new species to form from broad patterns in the fossil record and from studies that use morphological data (including fossils) or molecular data to assess the time interval between speciation events in particular groups of organisms.

Patterns in the Fossil Record

The fossil record includes many episodes in which new species appear suddenly in a geologic stratum, persist essentially unchanged through several strata, and then disappear. For example, there are dozens of species of marine invertebrates that make their debut in the fossil record with novel morphologies, but then change little for millions of years before becoming extinct. The term **punctuated equilibria** is used to describe these periods of apparent stasis punctuated by sudden change **(Figure 24.16a)**. Other species do not show a punctuated pattern; instead, they appear to have changed more gradually over long periods of time **(Figure 24.16b)**.

What might punctuated and gradual patterns tell us about how long it takes new species to form? Suppose that a species survived for 5 million years, but most of the morphological changes that caused it to be designated a new species occurred during the first 50,000 years of its existence—just 1% of its total lifetime. Time periods this short (in geologic terms) often cannot be distinguished in fossil strata, in part because the rate of sediment accumulation may be too slow to separate layers this close in time. Thus, based on its fossils, the species would seem to have appeared suddenly and then lingered with little or no change before becoming extinct. Even though such a species may have originated more slowly than its fossils suggest (in this case taking up to 50,000 years), a punctuated pattern indicates that speciation occurred relatively rapidly. For species whose fossils changed much more gradually, we also cannot tell exactly when a new biological species formed, since information about reproductive isolation does not fossilize.

▼ Figure 24.16 Two models for the tempo of speciation.

(a) Punctuated model. New species change most as they branch from a parent species and then change little for the rest of their existence.

(b) Gradual model. Species diverge from one another more slowly and steadily over time.

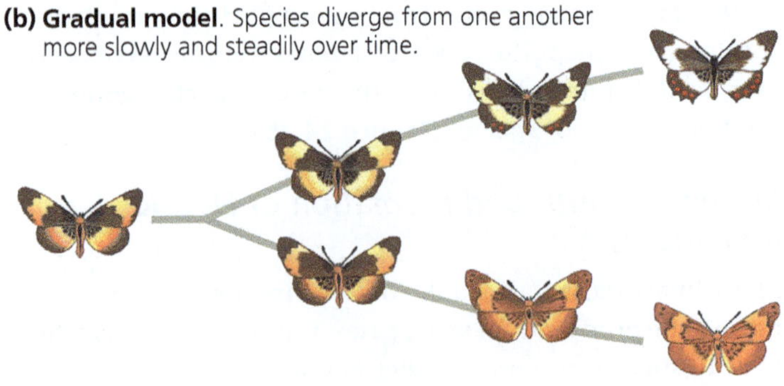

However, it is likely that speciation in such groups occurred relatively slowly, perhaps taking millions of years.

MB **Interview with Stephen Jay Gould: An "architect" of the concept of punctuated equilibria**

Speciation Rates

The existence of fossils that display a punctuated pattern suggests that once the process of speciation begins, it can be completed relatively rapidly—a suggestion supported by a growing number of studies.

For example, rapid speciation appears to have produced the wild sunflower *Helianthus anomalus*. Genetic evidence indicates that this species originated by the hybridization of two other sunflower species, *H. annuus* and *H. petiolaris*. The hybrid species *H. anomalus* is ecologically distinct and reproductively isolated from both parent species **(Figure 24.17)**. Unlike the outcome of allopolyploid speciation, in which there is a change in chromosome number after hybridization, in these sunflowers the two parent species and the hybrid all have the same number of chromosomes ($2n = 34$). How, then, did speciation occur? To study this question, researchers performed an experiment designed to mimic events in nature **(Figure 24.18)**. Their results indicated that natural selection could produce extensive genetic changes in hybrid populations over short periods of time. These changes appear to have caused the hybrids to diverge reproductively from their parents and form a new species, *H. anomalus*.

The sunflower example, along with the apple maggot fly, Lake Victoria cichlid, and fruit fly examples discussed earlier, suggests that new species can arise rapidly *once divergence begins*. But what is the total length of time between speciation events? This interval consists of the time that elapses before populations of a newly formed species start to diverge from one another plus the time it takes for speciation to be complete once divergence begins. It turns out that the total time between speciation events varies considerably. In a survey of data from 84 groups of plants and animals, speciation intervals ranged from 4,000 years (in cichlids of Lake Nabugabo,

▼ Figure 24.17 A hybrid sunflower species and its dry sand dune habitat. The wild sunflower *Helianthus anomalus* shown here originated via the hybridization of two other sunflowers, *H. annuus* and *H. petiolaris*, which live in nearby but moister environments.

▼ Figure 24.18

Inquiry How does hybridization lead to speciation in sunflowers?

Experiment Loren Rieseberg and his colleagues crossed the two parent sunflower species, *H. annuus* and *H. petiolaris*, to produce experimental hybrids in the laboratory (for each gamete, only two of the $n = 17$ chromosomes are shown).

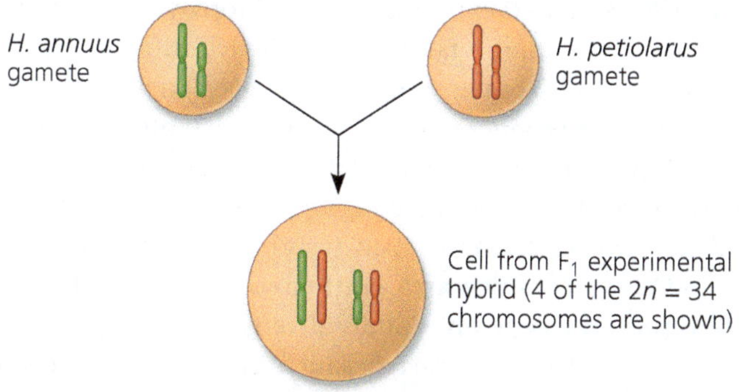

Note that in the first (F_1) generation, each chromosome of the experimental hybrids consisted entirely of DNA from one or the other parent species. The researchers then tested whether the F_1 and subsequent generations of experimental hybrids were fertile. They also used species-specific genetic markers to compare the chromosomes in the experimental hybrids with the chromosomes in the naturally occurring hybrid *H. anomalus*.

Results Although only 5% of the F_1 experimental hybrids were fertile, after just four more generations the hybrid fertility rose to more than 90%. The chromosomes of individuals from this fifth hybrid generation differed from those in the F_1 generation (see above) but were similar to those in *H. anomalus* individuals from natural populations:

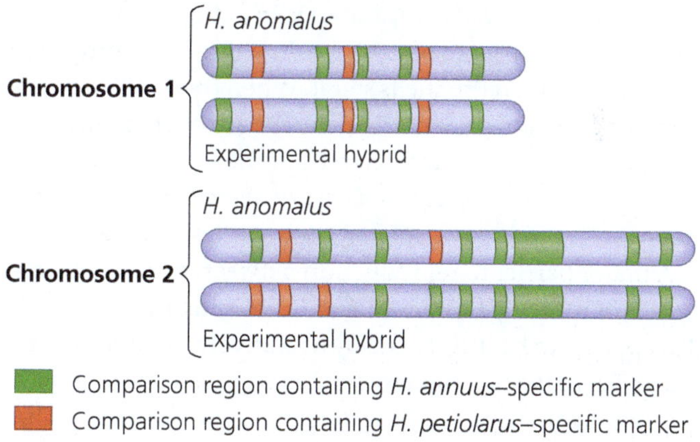

Conclusion Over time, the chromosomes in the population of experimental hybrids became similar to the chromosomes of *H. anomalus* individuals from natural populations. This suggests that the observed rise in the fertility of the experimental hybrids may have occurred as selection eliminated regions of DNA from the parent species that were not compatible with one another. Overall, it appeared that the initial steps of the speciation process occurred rapidly and could be mimicked in a laboratory experiment.

Data from L. H. Rieseberg et al., Role of gene interactions in hybrid speciation: evidence from ancient and experimental hybrids, *Science* 272:741–745 (1996).

WHAT IF? *The increased fertility of the experimental hybrids could have resulted from natural selection for thriving under laboratory conditions. Evaluate this alternative explanation for the result.*

Uganda) to 40 million years (in some beetles). Overall, the time between speciation events averaged 6.5 million years and was rarely less than 500,000 years.

These data suggest that on average, millions of years may pass before a newly formed plant or animal species will itself give rise to another new species. As you'll read in Concept 25.4, this finding has implications for how long it takes life on Earth to recover from mass extinction events. Moreover, the extreme variability in the time it takes new species to form indicates that organisms do not have an internal "speciation clock" that causes them to produce new species at regular intervals. Instead, speciation begins only after gene flow between populations is interrupted, perhaps by changing environmental conditions or by unpredictable events, such as a storm that transports a few individuals to a new area. Furthermore, once gene flow is interrupted, the populations must diverge genetically to such an extent that they become reproductively isolated—all before other events cause gene flow to resume, possibly reversing the speciation process (see Figure 24.15).

Studying the Genetics of Speciation

Studies of ongoing speciation (as in hybrid zones) can reveal traits that cause reproductive isolation. By identifying the genes that control those traits, scientists can explore a fundamental question of evolutionary biology: How many genes influence the formation of new species?

In some cases, the evolution of reproductive isolation results from the effects of a single gene. For example, in Japanese snails of the genus *Euhadra*, a change in a single gene results in a mechanical barrier to reproduction. This gene controls the direction in which the shells spiral. When their shells spiral in different directions, the snails' genitalia are oriented in a manner that prevents mating (Figure 24.3f shows a similar example). Recent genetic analyses have uncovered other single genes that cause reproductive isolation in fruit flies or mice.

A major barrier to reproduction between two closely related species of monkey flower, *Mimulus cardinalis* and *M. lewisii*, also appears to be influenced by a relatively small number of genes. These two species are isolated by several prezygotic and postzygotic barriers. Of these, one prezygotic barrier, pollinator choice, accounts for most of the isolation: In a hybrid zone between *M. cardinalis* and *M. lewisii*, nearly 98% of pollinator visits were restricted to one species or the other.

The two monkey flower species are visited by different pollinators: Hummingbirds prefer the red-flowered *M. cardinalis*, and bumblebees prefer the pink-flowered *M. lewisii*. Pollinator choice is affected by at least two loci in the monkey flowers, one of which, the "yellow upper," or *yup*, locus, influences flower color **(Figure 24.19)**. By crossing the two parent species to produce F_1 hybrids and then performing repeated backcrosses of these F_1 hybrids to each parent species, researchers succeeded in transferring the *M. cardinalis* allele at this locus into *M. lewisii*, and vice versa. In a field experiment, *M. lewisii* plants with the *M. cardinalis yup* allele received 68-fold more visits from hummingbirds than did wild-type *M. lewisii*. Similarly, *M. cardinalis* plants with the *M. lewisii yup* allele received 74-fold more visits from bumblebees than did wild-type *M. cardinalis*. Thus, a mutation at a single locus can influence pollinator preference and hence contribute to reproductive isolation in monkey flowers.

In other organisms, the speciation process is influenced by larger numbers of genes and gene interactions. For example, hybrid sterility between two subspecies of the fruit fly *Drosophila pseudoobscura* results from gene interactions among at least four loci, and postzygotic isolation in the sunflower hybrid zone discussed earlier is influenced by at least 26 chromosome segments (and an unknown number of genes). Overall, studies suggest that few or many genes can influence the evolution of reproductive isolation and hence the emergence of a new species.

▼ **Figure 24.19 A locus that influences pollinator choice.** Pollinator preferences provide a strong barrier to reproduction between *Mimulus lewisii* and *M. cardinalis*. After transferring the *M. lewisii* allele for a flower-color locus into *M. cardinalis* and vice versa, researchers observed a shift in some pollinators' preferences.

(a) Typical *Mimulus lewisii*

(b) *M. lewisii* with an *M. cardinalis* flower-color allele

(c) Typical *Mimulus cardinalis*

(d) *M. cardinalis* with an *M. lewisii* flower-color allele

WHAT IF ➤ *If* M. cardinalis *individuals that had the* M. lewisii yup *allele were planted in an area that housed both monkey flower species, how might the production of hybrid offspring be affected?*

From Speciation to Macroevolution

As you've seen, speciation may begin with differences as small as the color on a cichlid's back. However, as speciation occurs again and again, such differences can accumulate and become more pronounced, eventually leading to the formation of new groups of organisms that differ greatly from their ancestors (as in the origin of whales from terrestrial mammals; see Figure 22.20). Moreover, as one group of organisms increases in size by producing many new species, another group of organisms may shrink, losing species to extinction. The cumulative effects of many such speciation and extinction events have helped shape the sweeping evolutionary changes that are documented in the fossil record. In the next chapter, we turn to such large-scale evolutionary changes as we begin our study of macroevolution.

CONCEPT CHECK 24.4

1. Speciation can occur rapidly between diverging populations, yet the time between speciation events is often more than a million years. Explain this apparent contradiction.
2. Summarize evidence that the *yup* locus acts as a prezygotic barrier to reproduction in two species of monkey flowers. Do these results demonstrate that the *yup* locus alone controls barriers to reproduction between these species? Explain.
3. **MAKE CONNECTIONS >** Compare Figure 13.12 with Figure 24.18. What cellular process could cause the hybrid chromosomes in Figure 24.18 to contain DNA from both parent species? Explain.

For suggested answers, see Appendix A.

24 Chapter Review

Go to **MasteringBiology**™ for Videos, Animations, Vocab Self-Quiz, Practice Tests, and more in the Study Area.

SUMMARY OF KEY CONCEPTS

CONCEPT 24.1

The biological species concept emphasizes reproductive isolation *(pp. 505–508)*

VOCAB SELF-QUIZ goo.gl/6u55ks

- A biological **species** is a group of populations whose individuals may interbreed and produce viable, fertile offspring with each other but not with members of other species.
- The **biological species concept** emphasizes reproductive isolation through **prezygotic** and **postzygotic barriers** that separate gene pools.

? *Explain the role of gene flow in the biological species concept.*

CONCEPT 24.2

Speciation can take place with or without geographic separation *(pp. 509–514)*

- In **allopatric speciation**, gene flow is reduced when two populations of one species become geographically separated from each other. One or both populations may undergo evolutionary change during the period of separation, resulting in the establishment of barriers to reproduction.

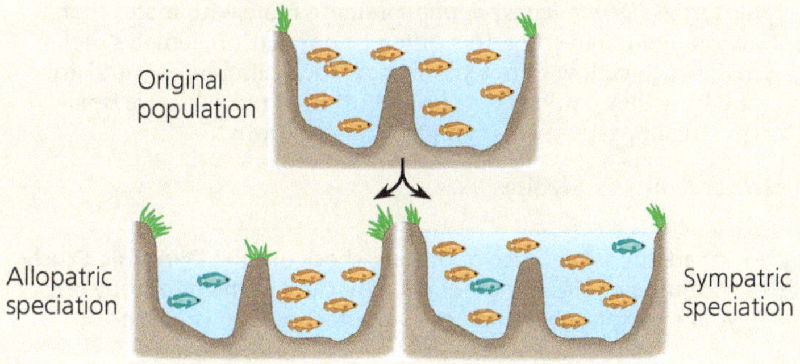

- In **sympatric speciation**, a new species originates while remaining in the same geographic area as the parent species. Plant species (and, more rarely, animal species) have evolved sympatrically through **polyploidy**. Sympatric speciation can also result from sexual selection and habitat shifts.

? *Can factors that cause sympatric speciation also cause allopatric speciation? Explain.*

CONCEPT 24.3

Hybrid zones reveal factors that cause reproductive isolation *(pp. 514–518)*

- Many groups of organisms form **hybrid zones** in which members of different species meet and mate, producing at least some offspring of mixed ancestry.
- Many hybrid zones are *stable,* in that hybrid offspring continue to be produced over time. In others, **reinforcement** strengthens prezygotic barriers to reproduction, thus decreasing the formation of unfit hybrids. In still other hybrid zones, barriers to reproduction may weaken over time, resulting in the *fusion* of the species' gene pools (reversing the speciation process).

? *What factors can support the long-term stability of a hybrid zone if the parent species live in different environments?*

CONCEPT 24.4

Speciation can occur rapidly or slowly and can result from changes in few or many genes *(pp. 518–521)*

- New species can form rapidly once divergence begins—but it can take millions of years for that to happen. The time interval between speciation events varies considerably, from a few thousand years to tens of millions of years.
- Researchers have identified particular genes involved in some cases of speciation. Speciation can be driven by few or many genes.

? *Is speciation something that happened only in the distant past, or are new species continuing to arise today? Explain.*

TEST YOUR UNDERSTANDING

Level 1: Knowledge/Comprehension

PRACTICE TEST
goo.gl/CUYGKD

1. The *largest* unit within which gene flow can readily occur is a
   (A) population.
   (B) species.
   (C) genus.
   (D) hybrid.

2. Males of different species of the fruit fly *Drosophila* that live in the same parts of the Hawaiian Islands have different elaborate courtship rituals. These rituals involve fighting other males and making stylized movements that attract females. What type of reproductive isolation does this represent?
   (A) habitat isolation
   (B) temporal isolation
   (C) behavioral isolation
   (D) gametic isolation

3. According to the punctuated equilibria model,
   (A) given enough time, most existing species will branch gradually into new species.
   (B) most new species accumulate their unique features relatively rapidly as they come into existence, then change little for the rest of their duration as a species.
   (C) most evolution occurs in sympatric populations.
   (D) speciation is usually due to a single mutation.

Level 2: Application/Analysis

4. Bird guides once listed the myrtle warbler and Audubon's warbler as distinct species. Recently, these birds have been classified as eastern and western forms of a single species, the yellow-rumped warbler. Which of the following pieces of evidence, if true, would be cause for this reclassification?
   (A) The two forms interbreed often in nature, and their offspring survive and reproduce well.
   (B) The two forms live in similar habitats and have similar food requirements.
   (C) The two forms have many genes in common.
   (D) The two forms are very similar in appearance.

5. Which of the following factors would *not* contribute to allopatric speciation?
   (A) The separated population is small, and genetic drift occurs.
   (B) The isolated population is exposed to different selection pressures than the ancestral population.
   (C) Different mutations begin to distinguish the gene pools of the separated populations.
   (D) Gene flow between the two populations is extensive.

6. Plant species A has a diploid chromosome number of 12. Plant species B has a diploid number of 16. A new species, C, arises as an allopolyploid from A and B. The diploid number for species C would probably be
   (A) 14. (B) 16. (C) 28. (D) 56.

Level 3: Synthesis/Evaluation

7. **EVOLUTION CONNECTION** Explain the biological basis for assigning all human populations to a single species. Can you think of a scenario by which a second human species could originate in the future?

8. **SCIENTIFIC INQUIRY • DRAW IT** In this chapter, you read that bread wheat (*Triticum aestivum*) is an allohexaploid, containing two sets of chromosomes from each of three different parent species. Genetic analysis suggests that the three species pictured following this question each contributed chromosome sets to *T. aestivum.* (The capital letters here represent sets of chromosomes rather than individual genes, and the diploid chromosome number for each species is shown in parentheses.) Evidence also indicates that the first polyploidy event was a spontaneous hybridization of the early cultivated wheat species *T. monococcum* and a wild *Triticum* grass species. Based on this information, draw a diagram of one possible chain of events that could have produced the allohexaploid *T. aestivum.*

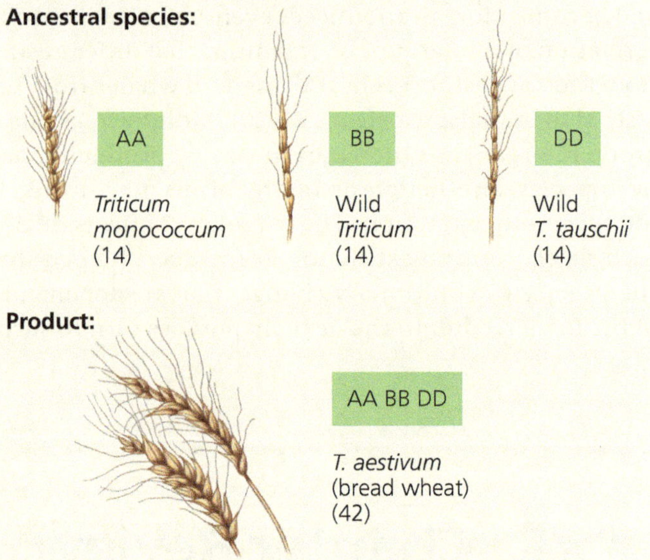

9. **WRITE ABOUT A THEME: INFORMATION** In sexually reproducing species, each individual inherits DNA from both parent organisms. In a short essay (100–150 words), apply this idea to what occurs when organisms of two species that have homologous chromosomes mate and produce (F_1) hybrid offspring. What percentage of the DNA in the F_1 hybrids' chromosomes comes from each parent species? As the hybrids mate and produce F_2 and later-generation hybrid offspring, describe how recombination and natural selection may affect whether the DNA in hybrid chromosomes is derived from one parent species or the other.

10. **SYNTHESIZE YOUR KNOWLEDGE**

Suppose that females of one population of strawberry poison dart frogs (*Dendrobates pumilio*) prefer to mate with males that are orange-red in color. In a different population, females prefer males with yellow skin. Explain how such differences could arise and how they could affect the evolution of reproductive isolation in allopatric versus sympatric populations.

For selected answers, see Appendix A.

For additional practice questions, check out the **Dynamic Study Modules** in MasteringBiology. You can use them to study on your smartphone, tablet, or computer anytime, anywhere!

25 The History of Life on Earth

▲ **Figure 25.1 Would you have expected to find whale bones buried here?**

KEY CONCEPTS

25.1 Conditions on early Earth made the origin of life possible

25.2 The fossil record documents the history of life

25.3 Key events in life's history include the origins of unicellular and multicellular organisms and the colonization of land

25.4 The rise and fall of groups of organisms reflect differences in speciation and extinction rates

25.5 Major changes in body form can result from changes in the sequences and regulation of developmental genes

25.6 Evolution is not goal oriented

▼ Fossil of *Dorudon atrox*, an ancient whale

A Surprise in the Desert

With its dry, wind-sculpted sands and searing heat, the Sahara Desert seems an unlikely place to discover the bones of whales. But starting in the 1870s, researchers uncovered fossils of ancient whales at several locations that once were covered by an ancient sea **(Figure 25.1)**. For example, a nearly complete skeleton of *Dorudon atrox*, an extinct whale that lived 35 million years ago, was discovered in a region that came to be called Wadi Hitan, the "Valley of Whales." Collectively, the whale fossils found in the Sahara were spectacular not only for where they were found, but also for documenting early steps in the transition from life on land to life in the sea.

Fossils discovered in other parts of the world tell a similar story: Past organisms were very different from those presently living. The sweeping changes in life on Earth as revealed by fossils illustrate **macroevolution**, the broad pattern of evolution above the species level. Examples of macroevolutionary change include the emergence of terrestrial vertebrates through a series of speciation events, the impact of mass extinctions on biodiversity, and the origin of key adaptations such as flight.

Taken together, such changes provide a grand view of the evolutionary history of life. We'll examine that history in this chapter, beginning with a discussion of hypotheses regarding the origin of life. This is the most speculative topic of the entire unit, for no fossil evidence of that seminal episode exists. We will then turn to evidence from the fossil record about major events in the history of life and the factors that have shaped the rise and fall of different groups of organisms over time.

When you see this blue icon, log in to **MasteringBiology** and go to the Study Area for digital resources.

MB **Get Ready for This Chapter**

CONCEPT 25.1

Conditions on early Earth made the origin of life possible

Direct evidence of life on early Earth comes from fossils of microorganisms that lived 3.5 billion years ago. But how did the first living cells appear? Observations and experiments in chemistry, geology, and physics have led scientists to propose one scenario that we'll examine here. They hypothesize that chemical and physical processes could have produced simple cells through a sequence of four main stages:

1. The abiotic (nonliving) synthesis of small organic molecules, such as amino acids and nitrogenous bases
2. The joining of these small molecules into macromolecules, such as proteins and nucleic acids
3. The packaging of these molecules into **protocells**, droplets with membranes that maintained an internal chemistry different from that of their surroundings
4. The origin of self-replicating molecules that eventually made inheritance possible

Though speculative, this scenario leads to predictions that can be tested in the laboratory. In this section, we'll examine some of the evidence for each stage.

Synthesis of Organic Compounds on Early Earth

Our planet formed 4.6 billion years ago, condensing from a vast cloud of dust and rocks that surrounded the young sun. For its first few hundred million years, Earth was bombarded by huge chunks of rock and ice left over from the formation of the solar system. The collisions generated so much heat that all of the available water was vaporized, preventing the formation of seas and lakes.

This massive bombardment ended 4 billion years ago, setting the stage for the origin of life. The first atmosphere had little oxygen and was likely thick with water vapor, along with compounds released by volcanic eruptions, such as nitrogen and its oxides, carbon dioxide, methane, ammonia, and hydrogen. As Earth cooled, the water vapor condensed into oceans, and much of the hydrogen escaped into space.

During the 1920s, Russian chemist A. I. Oparin and British scientist J. B. S. Haldane independently hypothesized that Earth's early atmosphere was a reducing (electron-adding) environment, in which organic compounds could have formed from simpler molecules. The energy for this synthesis could have come from lightning and UV radiation. Haldane suggested that the early oceans were a solution of organic molecules, a "primitive soup" from which life arose.

In 1953, Stanley Miller and Harold Urey, working at the University of Chicago, tested the Oparin-Haldane hypothesis by creating laboratory conditions comparable to those that scientists at the time thought existed on early Earth (see Figure 4.2). His apparatus yielded a variety of amino acids found in organisms today, along with other organic compounds. Many laboratories have since repeated Miller's classic experiment using different recipes for the atmosphere, some of which also produced organic compounds.

▼ Figure 25.2 Amino acid synthesis in a simulated volcanic eruption. In addition to his classic 1953 study, Miller also conducted an experiment simulating a volcanic eruption. In a 2008 reanalysis of those results, researchers found that far more amino acids were produced under simulated volcanic conditions than were produced in the conditions of the original 1953 experiment.

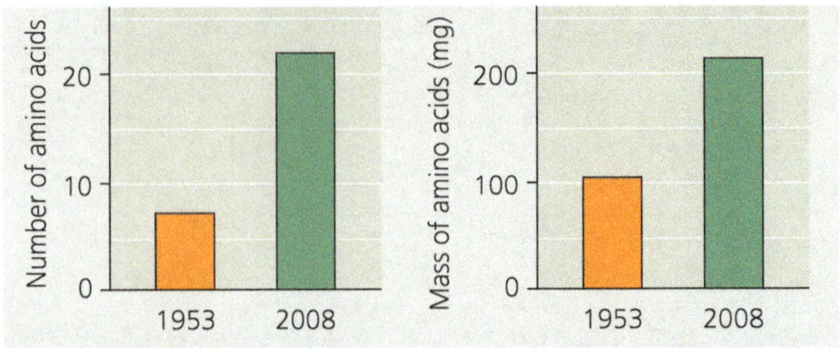

MAKE CONNECTIONS ➤ *Explain how more than 20 amino acids could have been produced in the 2008 experiment. (See Concept 5.4.)*

However, some evidence suggests that the early atmosphere was made up primarily of nitrogen and carbon dioxide and was neither reducing nor oxidizing (electron removing). Recent Miller/Urey-type experiments using such "neutral" atmospheres have also produced organic molecules. In addition, small pockets of the early atmosphere, such as those near the openings of volcanoes, may have been reducing. Perhaps the first organic compounds formed near volcanoes. In 2008, researchers used modern equipment to reanalyze molecules that Miller had saved from one of his experiments. The 2008 paper found that numerous amino acids had formed under conditions that simulated a volcanic eruption **(Figure 25.2)**.

Another hypothesis is that organic compounds were first produced in deep-sea **hydrothermal vents**, areas on the seafloor where heated water and minerals gush from Earth's interior into the ocean. Some of these vents, known as "black smokers," release water so hot (300–400°C) that organic compounds formed there may have been unstable. But other deep-sea vents, called **alkaline vents**, release water that has a high pH (9–11) and is warm (40–90°C) rather than hot, an environment that may have been more suitable for the origin of life **(Figure 25.3)**.

Studies related to the volcanic-atmosphere and alkaline-vent hypotheses show that the abiotic synthesis of organic molecules is possible under various conditions. Another source of organic molecules may have been meteorites. For example, fragments of the Murchison meteorite, a 4.5-billion-year-old rock that landed in Australia in 1969, contain more than 80 amino acids, some in large amounts. These amino

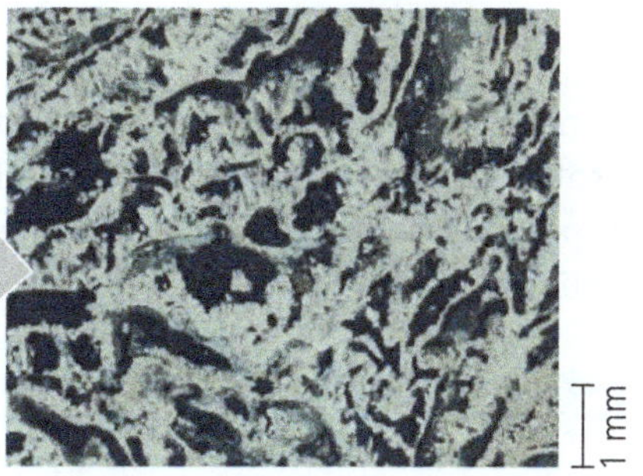

◀ **Figure 25.3 Did life originate in deep-sea alkaline vents?** The first organic compounds may have arisen in warm alkaline vents similar to this one from the 40,000-year-old "Lost City" vent field in the mid-Atlantic Ocean. These vents contain hydrocarbons and are full of tiny pores (inset) lined with iron and other catalytic minerals. Early oceans were acidic, so a pH gradient would have formed between the interior of the vents and the surrounding ocean water. Energy for the synthesis of organic compounds could have been harnessed from this pH gradient.

acids cannot be contaminants from Earth because they consist of an equal mix of D and L isomers (see Figure 4.7). Organisms make and use only L isomers, with a few rare exceptions. Recent studies have shown that the Murchison meteorite also contained other key organic molecules, including lipids, simple sugars, and nitrogenous bases such as uracil.

Abiotic Synthesis of Macromolecules

The presence of small organic molecules, such as amino acids and nitrogenous bases, is not sufficient for the emergence of life as we know it. Every cell has many types of macromolecules, including enzymes and other proteins and the nucleic acids needed for self-replication. Could such macromolecules have formed on early Earth? A 2009 study demonstrated that one key step, the abiotic synthesis of RNA monomers, can occur spontaneously from simple precursor molecules. In addition, by dripping solutions of amino acids or RNA nucleotides onto hot sand, clay, or rock, researchers have produced polymers of these molecules. The polymers formed spontaneously, without the help of enzymes or ribosomes. Unlike proteins, the amino acid polymers are a complex mix of linked and cross-linked amino acids. Still, it is possible that such polymers acted as weak catalysts for a variety of chemical reactions on early Earth.

Protocells

All organisms must be able to carry out both reproduction and energy processing (metabolism). DNA molecules carry genetic information, including the instructions needed to replicate themselves accurately during reproduction. But DNA replication requires elaborate enzymatic machinery, along with an abundant supply of nucleotide building blocks provided by the cell's metabolism. This suggests that self-replicating molecules and a metabolic source of building blocks may have appeared together in early protocells. The necessary conditions may have been met in *vesicles*, fluid-filled compartments enclosed by a membrane-like structure. Recent experiments show that abiotically produced vesicles can exhibit certain properties of life, including simple reproduction and metabolism, as well as the maintenance of an internal chemical environment different from that of their surroundings **(Figure 25.4)**.

For example, vesicles can form spontaneously when lipids or other organic molecules are added to water. When this occurs, molecules that have both a hydrophobic region and a hydrophilic region can organize into a bilayer similar to the lipid bilayer of a plasma membrane. Adding substances such as *montmorillonite*, a soft mineral clay produced by the weathering of volcanic ash, greatly increases the rate of vesicle self-assembly (see Figure 25.4a). This clay, which is thought to have been common on early Earth, provides surfaces on which organic molecules become concentrated, increasing the likelihood that the molecules will react with each other and form vesicles. Abiotically produced vesicles can "reproduce" on their own (see Figure 25.4b), and they can increase

▼ **Figure 25.4 Features of abiotically produced vesicles.**

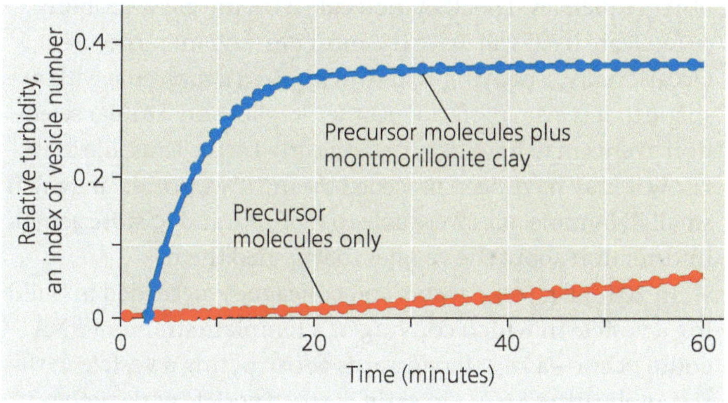

(a) Self-assembly. The presence of montmorillonite clay greatly increases the rate of vesicle self-assembly.

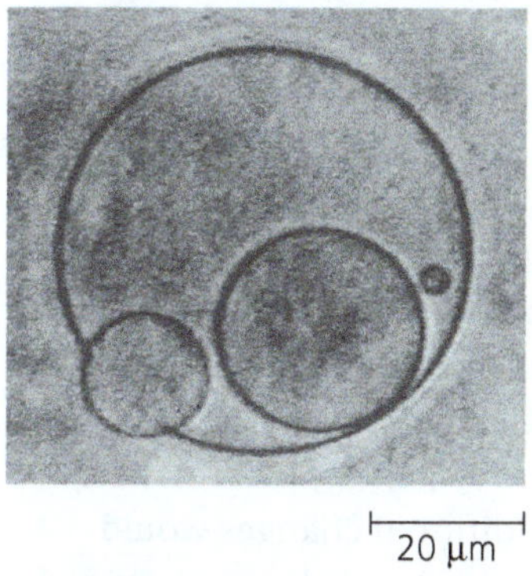

(b) Reproduction. Vesicles can divide on their own, as in this vesicle "giving birth" to smaller vesicles (LM).

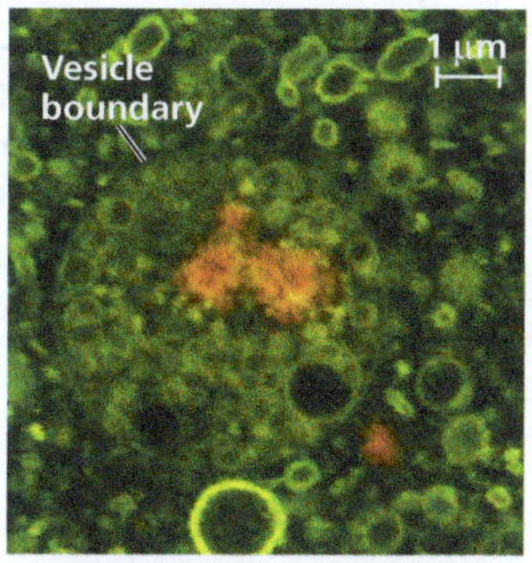

(c) Absorption of RNA. This vesicle has incorporated montmorillonite clay particles coated with RNA (orange).

MAKE CONNECTIONS ▶ *Explain how molecules with both a hydrophobic region and a hydrophilic region can self-assemble into a bilayer when in water. (See Concept 5.3.)*

in size ("grow") without dilution of their contents. Vesicles also can absorb montmorillonite particles, including those on which RNA and other organic molecules have become attached (see Figure 25.4c). Finally, experiments have shown that some vesicles have a selectively permeable bilayer and can perform metabolic reactions using an external source of reagents—another important prerequisite for life.

Self-Replicating RNA

The first genetic material was most likely RNA, not DNA. RNA plays a central role in protein synthesis, but it can also function as an enzyme-like catalyst (see Concept 17.3). Such RNA catalysts are called **ribozymes**. Some ribozymes can make complementary copies of short pieces of RNA, provided that they are supplied with nucleotide building blocks.

Natural selection on the molecular level has produced ribozymes capable of self-replication in the laboratory. How does this occur? Unlike double-stranded DNA, which takes the form of a uniform helix, single-stranded RNA molecules assume a variety of specific three-dimensional shapes mandated by their nucleotide sequences. In a given environment, RNA molecules with certain nucleotide sequences may have shapes that enable them to replicate faster and with fewer errors than other sequences. The RNA molecule with the greatest ability to replicate itself will leave the most descendant molecules. Occasionally, a copying error will result in a molecule with a shape that is even more adept at self-replication. Similar selection events may have occurred on early Earth. Thus, life as we know it may have been preceded by an "RNA world," in which small RNA molecules were able to replicate and to store genetic information about the vesicles that carried them.

In 2013, Dr. Jack Szostak and colleagues succeeded in building a vesicle in which copying of a template strand of RNA could occur—a key step towards constructing a vesicle with self-replicating RNA. On early Earth, a vesicle with such self-replicating, catalytic RNA would differ from its many neighbors that lacked such molecules. If that vesicle could grow, split, and pass its RNA molecules to its "daughters," the daughters would be protocells. Although the first such protocells likely carried only limited amounts of genetic information, specifying only a few properties, their inherited characteristics could have been acted on by natural selection. The most successful of the early protocells would have increased in number because they could exploit their resources effectively and pass their abilities on to subsequent generations.

Once RNA sequences that carried genetic information appeared in protocells, many additional changes would have been possible. For example, RNA could have provided the template on which DNA nucleotides were assembled. Double-stranded DNA is a more chemically stable repository for genetic information than is the more fragile RNA. DNA also can be replicated more accurately. Accurate replication was advantageous as genomes grew larger through gene duplication and other processes and as more properties of the protocells became coded in genetic information. Once DNA appeared, the stage was set for a blossoming of new forms of life—a change we see documented in the fossil record.

MB Interview with Jack Szostak: Studying the origin of life (see the interview before Chapter 22)

CONCEPT CHECK 25.1

1. What hypothesis did Miller test in his classic experiment?
2. How would the appearance of protocells have represented a key step in the origin of life?
3. **MAKE CONNECTIONS ➤** In changing from an "RNA world" to today's "DNA world," genetic information must have flowed from RNA to DNA. After reviewing Figures 17.4 and 19.9, suggest how this could have occurred. Does such a flow occur today?

For suggested answers, see Appendix A.

CONCEPT 25.2

The fossil record documents the history of life

Starting with the earliest traces of life, the fossil record opens a window into the world of long ago and provides glimpses of the evolution of life over billions of years. In this section, we'll examine fossils as a form of scientific evidence: how fossils form, how scientists date and interpret them, and what they can and cannot tell us about changes in the history of life.

The Fossil Record

Sedimentary rocks are the richest source of fossils. As a result, the fossil record is based primarily on the sequence in which fossils have accumulated in sedimentary rock layers, called *strata* (see Figure 22.3). Useful information is also provided by other types of fossils, such as insects preserved in amber (fossilized tree sap) and mammals frozen in ice.

The fossil record shows that there have been great changes in the kinds of organisms on Earth at different points in time **(Figure 25.5)**. Many past organisms were unlike organisms living today, and many organisms that once were common are now extinct. As we'll see later in this section, fossils also document how new groups of organisms arose from previously existing ones.

As substantial and significant as the fossil record is, keep in mind that it is an incomplete chronicle of evolution. Many of Earth's organisms did not die in the right place and time to be preserved as fossils. Of those fossils that were formed, many were destroyed by later geologic processes, and only a fraction of the others have been discovered. As a result, the known fossil record is biased in favor of species that existed for a long time, were abundant and widespread in certain kinds of environments, and had hard shells, skeletons, or other parts that

Figure 25.5 Documenting the history of life. These fossils illustrate representative organisms from different points in time. Although prokaryotes and unicellular eukaryotes are shown only at the base of the diagram, these organisms continue to thrive today. In fact, most organisms on Earth are unicellular.

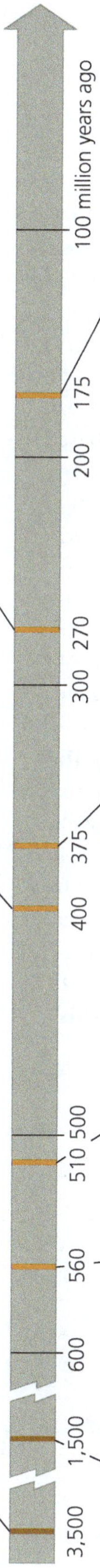

▼ *Rhomaleosaurus victor*, a plesiosaur. These large marine reptiles were important predators from 200 million to 66 million years ago.

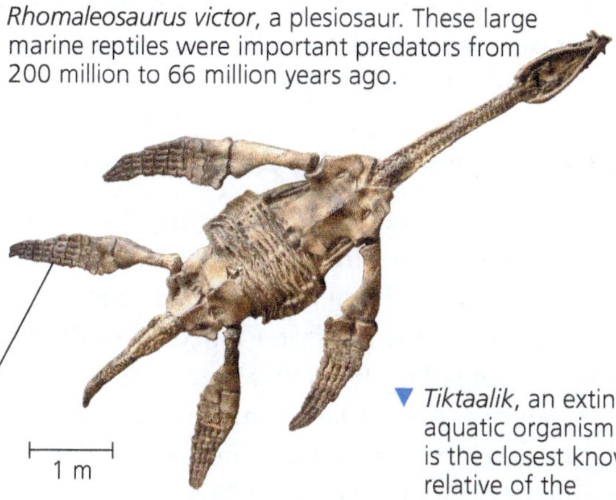

▼ *Dimetrodon*, the largest known carnivore of its day, was more closely related to mammals than to reptiles. The spectacular "sail" on its back may have functioned in temperature regulation or as an ornament that served to attract mates.

▼ *Tiktaalik*, an extinct aquatic organism that is the closest known relative of the four-legged vertebrates that went on to colonize land

▲ *Coccosteus cuspidatus*, a placoderm (fishlike vertebrate) that had a bony shield covering its head and front end

▶ *Hallucigenia*, a member of a morphologically diverse group of animals found in the Burgess Shale fossil bed in the Canadian Rockies

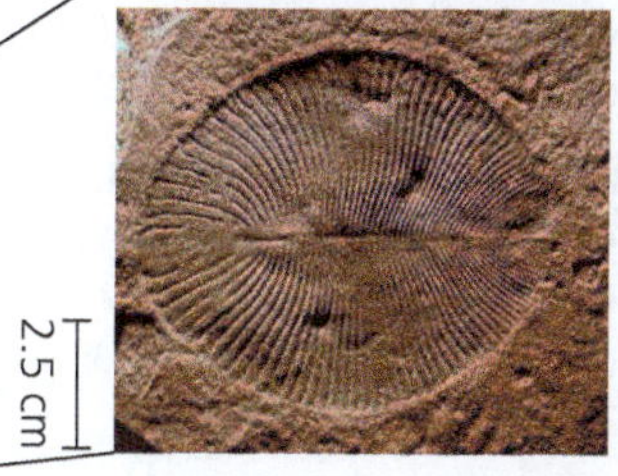

◀ *Dickinsonia costata*, a member of the Ediacaran biota, an extinct group of soft-bodied organisms

▲ Some prokaryotes bind thin films of sediments together, producing layered rocks called stromatolites. Present-day stromatolites are found in a few shallow marine bays, such as Shark Bay, Australia, shown here.

▲ A section through a fossilized stromatolite

▶ *Tappania*, a unicellular eukaryote thought to be either an alga or a fungus

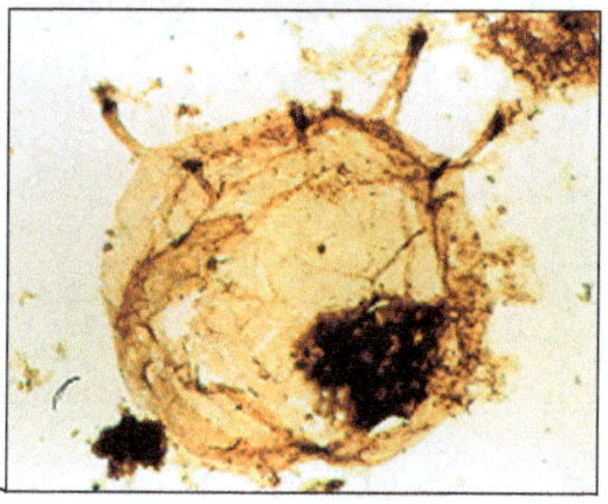

facilitated their fossilization. Even with its limitations, however, the fossil record is a remarkably detailed account of biological change over the vast scale of geologic time. Furthermore, as shown by the recently unearthed fossils of whale ancestors with hind limbs (see Figures 22.19, 22.20, and 25.1), gaps in the fossil record continue to be filled by new discoveries.

How Rocks and Fossils Are Dated

Fossils are valuable data for reconstructing the history of life, but only if we can determine where they fit in that unfolding story. While the order of fossils in rock strata tells us the sequence in which the fossils were laid down—their relative ages—it does not tell us their actual ages. Examining the relative positions of fossils is like peeling off layers of wallpaper in an old house. You can infer the sequence in which the layers were applied, but not the year each layer was added.

How can we determine the age of a fossil? One of the most common techniques is **radiometric dating**, which is based on the decay of radioactive isotopes (see Concept 2.2). In this process, a radioactive "parent" isotope decays to a "daughter" isotope at a characteristic rate. The rate of decay is expressed by the **half-life**, the time required for 50% of the parent isotope to decay **(Figure 25.6)**. Each type of radioactive isotope has a characteristic half-life, which is not affected by temperature, pressure, or other environmental variables. For example, carbon-14 decays relatively quickly; its half-life is 5,730 years. Uranium-238 decays slowly; its half-life is 4.5 billion years.

Fossils contain isotopes of elements that accumulated in the organisms when they were alive. For example, a living organism contains the most common carbon isotope, carbon-12, as well as a radioactive isotope, carbon-14. When the organism dies, it stops accumulating carbon, and the amount of carbon-12 in its tissues does not change over time. However, the carbon-14 that it contains at the time of death slowly decays into another element, nitrogen-14. Thus, by measuring the ratio of carbon-14 to carbon-12 in a fossil, we can determine the fossil's age. This method works for fossils up to about 75,000 years old; fossils

▼ Figure 25.6 Radiometric dating. In this diagram, each unit of time represents one half-life of a radioactive isotope.

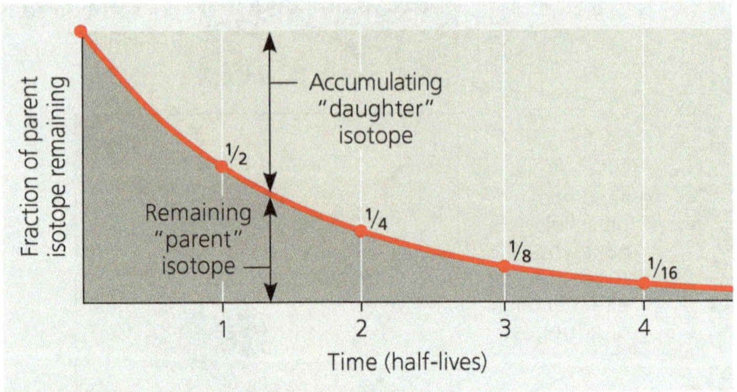

DRAW IT ➤ *Relabel the x-axis of this graph in years to illustrate the radioactive decay of uranium-238 (half-life = 4.5 billion years).*

older than that contain too little carbon-14 to be detected with current techniques. Radioactive isotopes with longer half-lives are used to date older fossils.

Determining the age of these older fossils in sedimentary rocks can be challenging. Organisms do not use radioisotopes with long half-lives, such as uranium-238, to build their bones or shells. In addition, sedimentary rocks are often composed of sediments of differing ages. Although we cannot date these older fossils directly, an indirect method can be used to infer the age of fossils that are sandwiched between two layers of volcanic rock. As lava cools into volcanic rock, radioisotopes from the surrounding environment become trapped in the newly formed rock. Some of the trapped radioisotopes have long half-lives, allowing geologists to estimate the ages of ancient volcanic rocks. If two volcanic layers surrounding fossils are found to be 525 million and 535 million years old, for example, then the fossils are roughly 530 million years old.

The Origin of New Groups of Organisms

Some fossils provide a detailed look at the origin of new groups of organisms. Such fossils are central to our understanding of evolution; they illustrate how new features arise and how long it takes for such changes to occur. We'll examine one such case here: the origin of mammals.

Along with amphibians and reptiles, mammals belong to the group of animals called *tetrapods* (from the Greek *tetra*, four, and *pod*, foot), named for having four limbs. Mammals have a number of unique anatomical features that fossilize readily, allowing scientists to trace their origin. For example, the lower jaw is composed of one bone (the dentary) in mammals but several bones in other tetrapods. In addition, the lower and upper jaws in mammals hinge between a different set of bones than in other tetrapods. Mammals also have a unique set of three bones that transmit sound in the middle ear, the hammer, anvil, and stirrup, whereas other tetrapods have only one such bone, the stirrup (see Concept 34.6). Finally, the teeth of mammals are differentiated into incisors (for tearing), canines (for piercing), and the multi-pointed premolars and molars (for crushing and grinding). In contrast, the teeth of other tetrapods usually consist of a row of undifferentiated, single-pointed teeth.

As detailed in **Figure 25.7**, the fossil record shows that the unique features of mammalian jaws and teeth evolved gradually over time, in a series of steps. As you study Figure 25.7, bear in mind that it includes just a few examples of the fossil skulls that document the origin of mammals. If all the known fossils in the sequence were arranged by shape and placed side by side, their features would blend smoothly from one group to the next. Some of these fossils would reflect how the features of a group that dominates life today, the mammals, gradually arose in a previously existing group, the cynodonts. Others would reveal side branches on the tree of life—groups of organisms that thrived for millions of years but ultimately left no descendants that survive today.

⮟ Figure 25.7 Exploring The Origin of Mammals

Over the course of 120 million years, mammals originated gradually from a group of tetrapods called synapsids. Shown here are a few of the many fossil organisms whose morphological features represent intermediate steps between living mammals and their early synapsid ancestors. The evolutionary context of the origin of mammals is shown in the tree diagram at right (the dagger symbol † indicates extinct lineages).

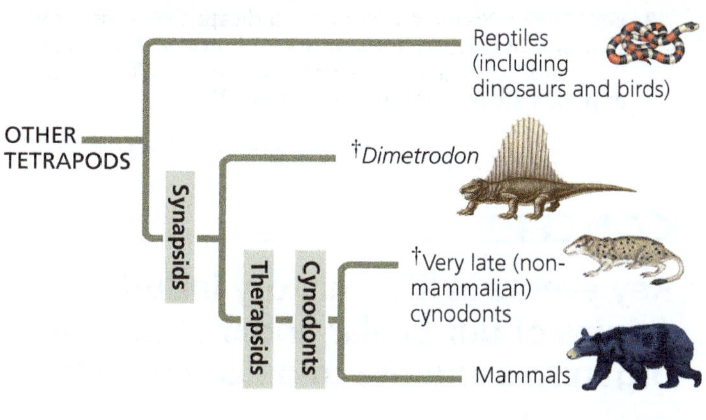

Key to skull bones

- Articular
- Quadrate
- Dentary
- Squamosal

Synapsid (300 mya)

Early synapsids had multiple bones in the lower jaw and single-pointed teeth. The jaw hinge was formed by the articular and quadrate bones. Early synapsids also had an opening called the *temporal fenestra* behind the eye socket. Powerful cheek muscles for closing the jaws probably passed through the temporal fenestra. Over time, this opening enlarged and moved in front of the hinge between the lower and upper jaws, thereby increasing the power and precision with which the jaws could be closed (much as moving a doorknob away from the hinge makes a door easier to close).

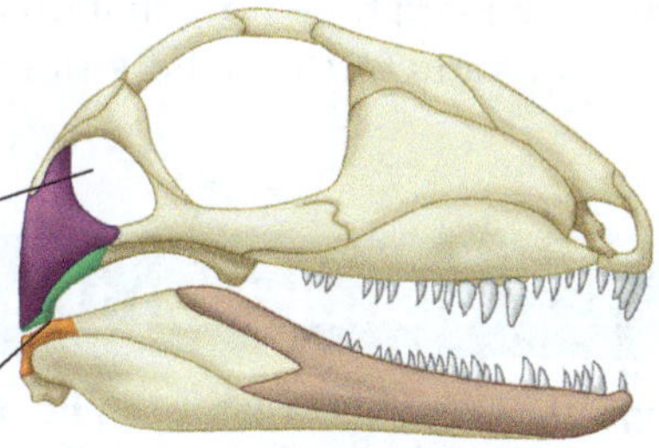

Therapsid (280 mya)

Later, a group of synapsids called therapsids appeared. Therapsids had large dentary bones, long faces, and the first examples of specialized teeth, large canines. These trends continued in a group of therapsids called cynodonts.

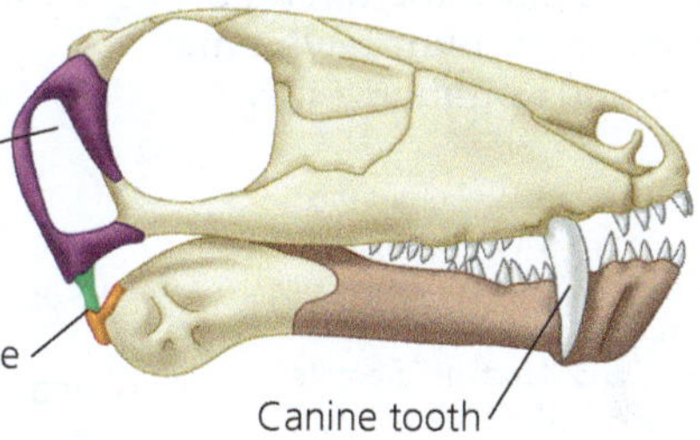

Early cynodont (260 mya)

In early cynodont therapsids, the dentary was the largest bone in the lower jaw, the temporal fenestra was large and positioned forward of the jaw hinge, and teeth with several cusps first appeared (not visible in the diagram). As in earlier synapsids, the jaw had an articular-quadrate hinge.

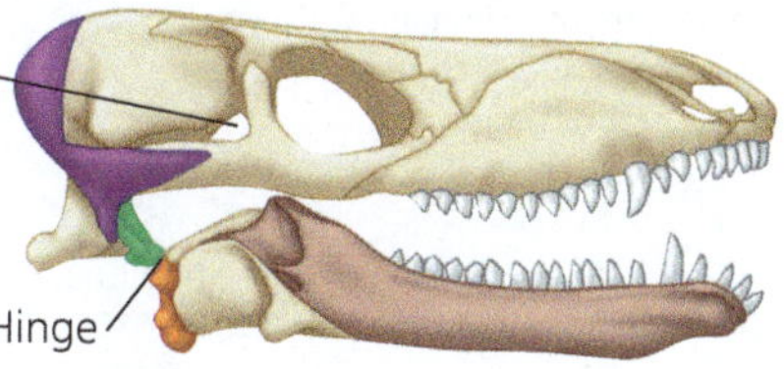

Later cynodont (220 mya)

Later cynodonts had teeth with complex cusp patterns, and their lower and upper jaws hinged in two locations: They retained the original articular-quadrate hinge and formed a new, second hinge between the dentary and squamosal bones. (The temporal fenestra is not visible in this or the below cynodont skull at the angles shown.)

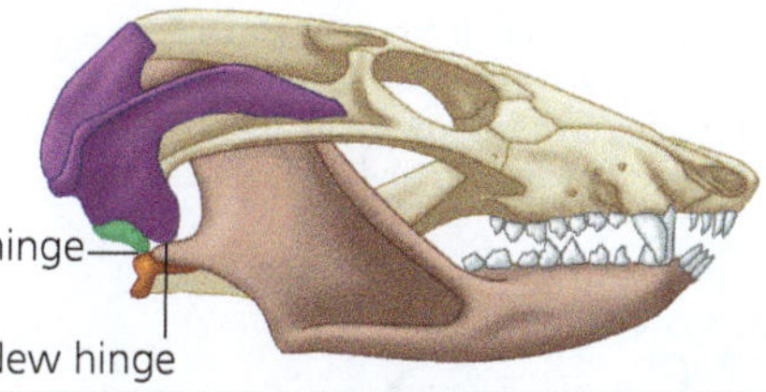

Very late cynodont (195 mya)

In some very late (nonmammalian) cynodonts and early mammals, the original articular-quadrate hinge was lost, leaving the dentary-squamosal hinge as the only hinge between the lower and upper jaws, as in living mammals. The articular and quadrate bones migrated into the ear region (not shown), where they functioned in transmitting sound. In the mammal lineage, these two bones later evolved into the familiar hammer (malleus) and anvil (incus) bones of the ear.

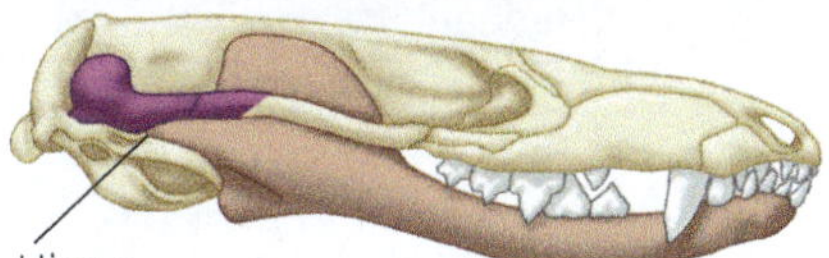

CONCEPT CHECK 25.2

1. Describe an example from the fossil record that shows how life has changed over time.
2. **WHAT IF?** ➤ Your measurements indicate that a fossilized skull you unearthed has a carbon-14/carbon-12 ratio about $\frac{1}{16}$ that of the skulls of present-day animals. What is the approximate age of the fossilized skull?

For suggested answers, see Appendix A.

CONCEPT 25.3

Key events in life's history include the origins of unicellular and multicellular organisms and the colonization of land

The study of fossils has helped geologists establish a **geologic record**: a standard time scale that divides Earth's history into four eons and further subdivisions **(Table 25.1)**. The first three eons—the Hadean, Archaean, and Proterozoic—together lasted about 4 billion years. The Phanerozoic eon, roughly the last half billion years, encompasses most of the time that animals have existed on Earth. It is divided into three eras: the Paleozoic, Mesozoic, and Cenozoic. Each era represents a distinct age in the history of Earth and its life. For example, the Mesozoic era is sometimes called the "age of reptiles" because of its abundance of reptilian fossils, including those of dinosaurs. The boundaries between the eras correspond to major extinction events, when many forms of life disappeared and were replaced by forms that evolved from the survivors.

As we've seen, the fossil record provides a sweeping overview of the history of life over geologic time. Here we will focus on a few major events in that history, returning to study the details in Unit Five. **Figure 25.8** will help you visualize how long ago these key events occurred against the vast backdrop of geologic time.

▼ Figure 25.8 Visualizing the Scale of Geologic Time

Geologic time is so vast that it can be difficult to visualize when key events in the history of life on Earth occurred. This figure introduces two common representations that help place the timing of those events in context: a countdown timer and a horizontal time line.

Instructors: Additional questions related to this Visualizing Figure can be assigned in MasteringBiology.

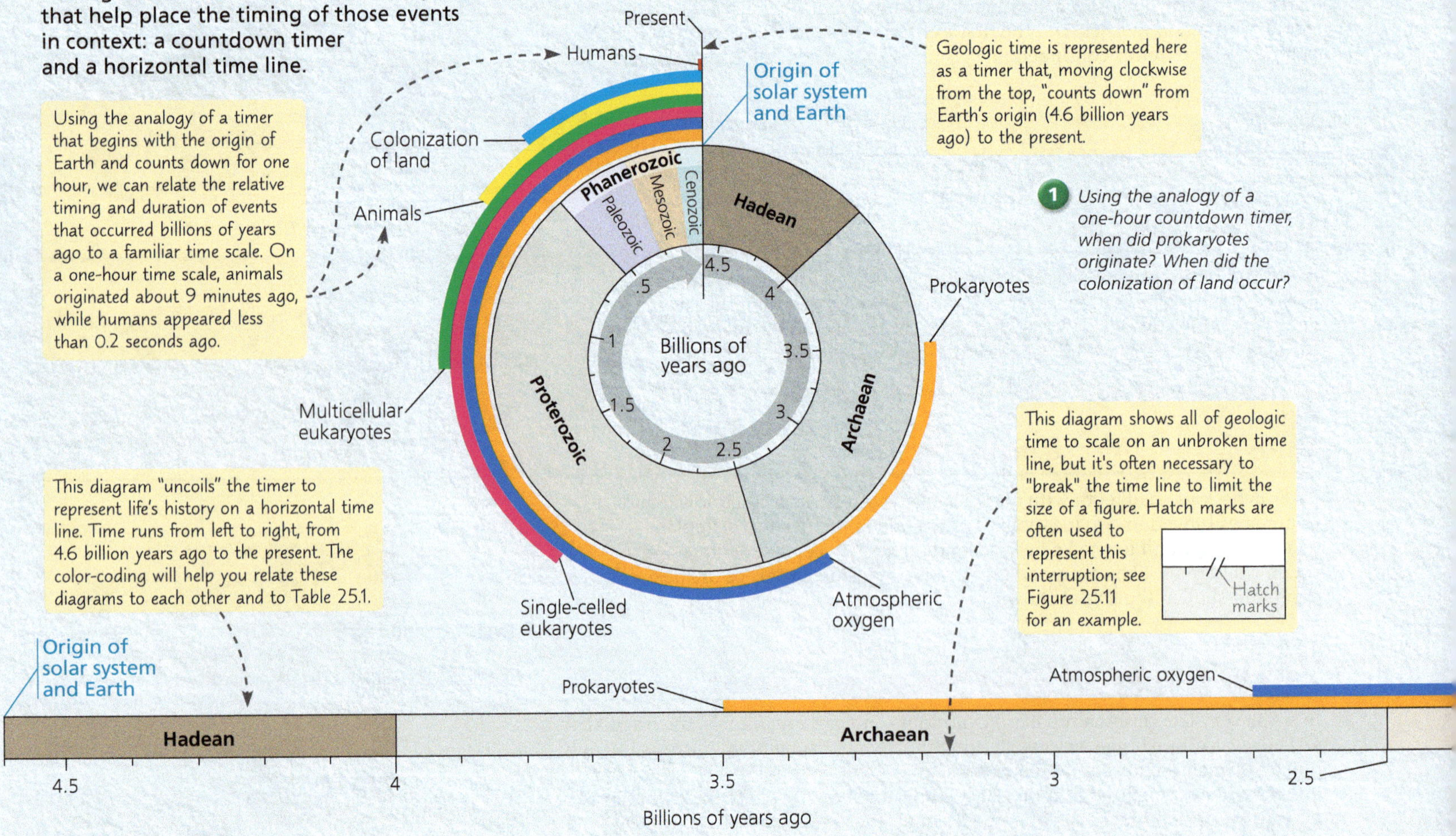

Table 25.1 The Geologic Record

Eons (duration not to scale)	Era	Period	Epoch	Age (Millions of Years Ago)	Some Important Events in the History of Life
Phanerozoic	Cenozoic	Quaternary	Holocene	0.01	Historical time
			Pleistocene	2.6	Ice ages; origin of genus *Homo*
		Neogene	Pliocene	5.3	Appearance of bipedal human ancestors
			Miocene	23	Continued radiation of mammals and angiosperms; earliest direct human ancestors
		Paleogene	Oligocene	34	Origins of many primate groups
			Eocene	56	Angiosperm dominance increases; continued radiation of most present-day mammalian orders
			Paleocene	66	Major radiation of mammals, birds, and pollinating insects
	Mesozoic	Cretaceous		145	Flowering plants (angiosperms) appear and diversify; many groups of organisms, including most dinosaurs, become extinct at end of period
		Jurassic		201	Gymnosperms continue as dominant plants; dinosaurs abundant and diverse
		Triassic		252	Cone-bearing plants (gymnosperms) dominate landscape; dinosaurs evolve and radiate; origin of mammals
	Paleozoic	Permian		299	Radiation of reptiles; origin of most present-day groups of insects; extinction of many marine and terrestrial organisms at end of period
		Carboniferous		359	Extensive forests of vascular plants form; first seed plants appear; origin of reptiles; amphibians dominant
		Devonian		419	Diversification of bony fishes; first tetrapods and insects appear
		Silurian		444	Diversification of early vascular plants
		Ordovician		485	Marine algae abundant; colonization of land by diverse fungi, plants, and animals
		Cambrian		541	Sudden increase in diversity of many animal phyla (Cambrian explosion)
Proterozoic	Neoproterozoic	Ediacaran		635	Diverse algae and soft-bodied invertebrate animals appear
				1,000	
				1,800	Oldest fossils of eukaryotic cells appear
				2,500	
Archaean				2,700	Concentration of atmospheric oxygen begins to increase
				3,500	Oldest fossils of cells (prokaryotes) appear
				4,000	Oldest known rocks on Earth's surface
Hadean				Approx. 4,600	Origin of Earth

2 *Once life began, what types of organisms lived on Earth for the next two billion years? Where did these organisms live?*

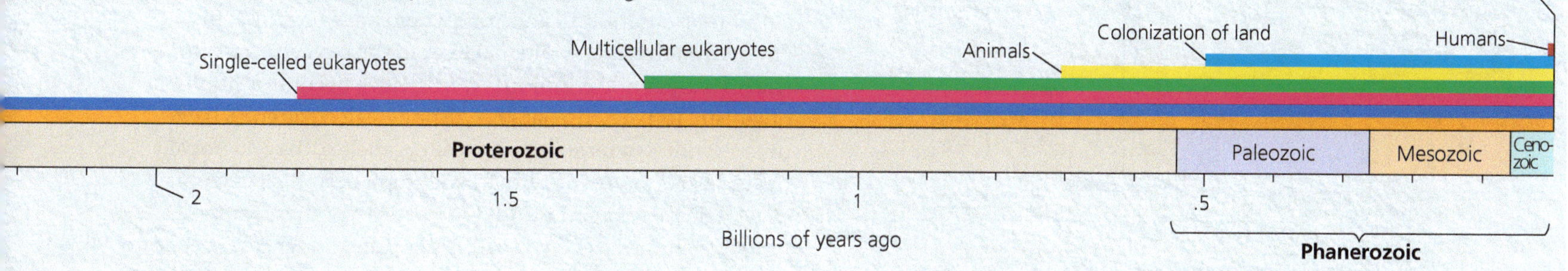

The First Single-Celled Organisms

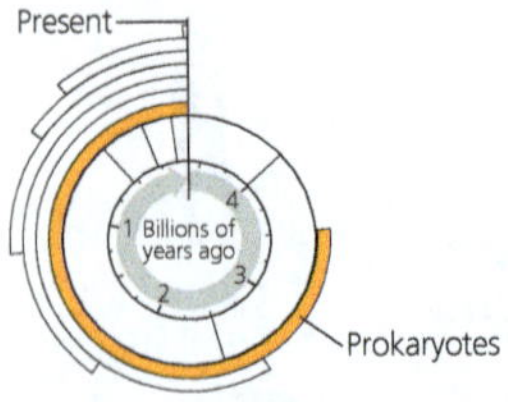

The earliest direct evidence of life, dating from 3.5 billion years ago, comes from fossilized stromatolites (see Figure 25.5). **Stromatolites** are layered rocks that form when certain prokaryotes bind thin films of sediment together. Stromatolites and other early prokaryotes were Earth's sole inhabitants for about 1.5 billion years. As we will see, these prokaryotes transformed life on our planet.

Photosynthesis and the Oxygen Revolution

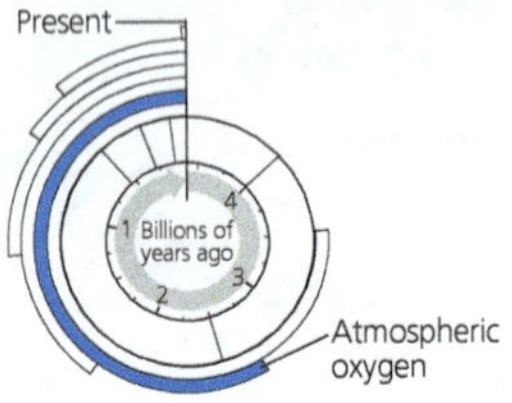

Most atmospheric oxygen gas (O_2) is of biological origin, produced during the water-splitting step of photosynthesis. When oxygenic photosynthesis first evolved—in photosynthetic prokaryotes— the free O_2 it produced probably dissolved in the surrounding water until it reached a high enough concentration to react with elements dissolved in water, including iron. This would have caused the iron to precipitate as iron oxide, which accumulated as sediments. These sediments were compressed into banded iron formations, red layers of rock containing iron oxide that are a source of iron ore today. Once all of the dissolved iron had precipitated, additional O_2 dissolved in the water until the seas and lakes became saturated with O_2. After this occurred, the O_2 finally began to "gas out" of the water and enter the atmosphere. This change left its mark in the rusting of iron-rich terrestrial rocks, a process that began about 2.7 billion years ago. This chronology implies that bacteria similar to today's cyanobacteria (oxygen-releasing, photosynthetic bacteria) originated before 2.7 billion years ago.

As shown in **Figure 25.9**, the amount of atmospheric O_2 increased gradually from about 2.7 to 2.4 billion years ago,

▼ Figure 25.9 The rise of atmospheric oxygen. Chemical analyses of ancient rocks have enabled this reconstruction of atmospheric oxygen levels during Earth's history.

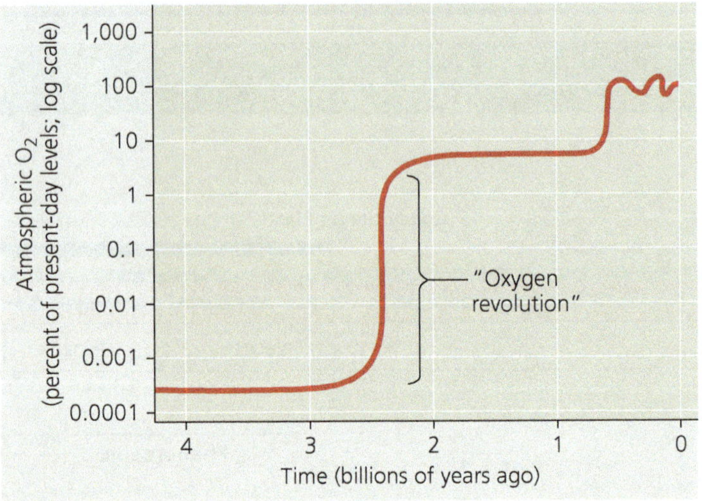

but then shot up relatively rapidly to between 1% and 10% of its present level. This "oxygen revolution" had an enormous impact on life. In some of its chemical forms, oxygen attacks chemical bonds and can inhibit enzymes and damage cells. As a result, the rising concentration of atmospheric O_2 probably doomed many prokaryotic groups. Some species survived in habitats that remained anaerobic, where we find their descendants living today (see Concept 27.4). Among other survivors, diverse adaptations to the changing atmosphere evolved, including cellular respiration, which uses O_2 in the process of harvesting the energy stored in organic molecules.

The rise in atmospheric O_2 levels left a huge imprint on the history of life. A few hundred million years later, another fundamental change occurred: the origin of the eukaryotic cell.

The First Eukaryotes

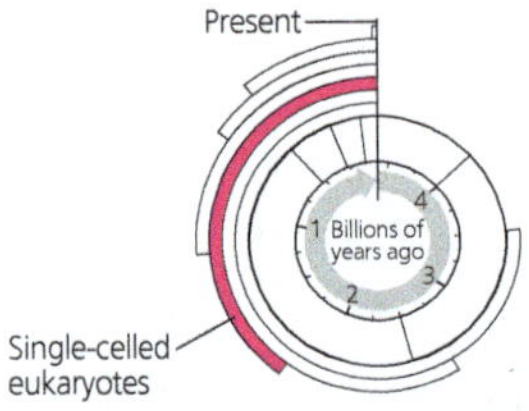

The oldest widely accepted fossils of eukaryotic organisms are 1.8 billion years old. Recall that eukaryotic cells have more complex organization than prokaryotic cells: Eukaryotic cells have a nuclear envelope, mitochondria, endoplasmic reticulum, and other internal structures that prokaryotes lack. Also, unlike prokaryotic cells, eukaryotic cells have a well-developed cytoskeleton, a feature that enables eukaryotic cells to change their shape and thereby surround and engulf other cells.

How did the eukaryotes evolve from their prokaryotic ancestors? Current evidence indicates that the eukaryotes originated by **endosymbiosis** when a prokaryotic cell engulfed a small cell that would evolve into an organelle found in all eukaryotes, the mitochondrion. The small, engulfed cell is an example of an *endosymbiont*, a cell that lives within another cell, called the *host cell*. The prokaryotic ancestor of the mitochondrion probably entered the host cell as undigested prey or an internal parasite. Though such a process may seem unlikely, scientists have directly observed cases in which endosymbionts that began as prey or parasites developed a mutually beneficial relationship with the host in as little as five years.

By whatever means the relationship began, we can hypothesize how the symbiosis could have become beneficial. For example, in a world that was becoming increasingly aerobic, a host that was itself an anaerobe would have benefited from endosymbionts that could make use of the oxygen. Over time, the host and endosymbionts would have become a single organism, its parts inseparable. Although all eukaryotes have mitochondria or remnants of these organelles, they do not all have plastids (a general term for chloroplasts and related organelles). Thus, the **serial endosymbiosis** hypothesis supposes that mitochondria evolved before plastids through a sequence of endosymbiotic events. As shown in **Figure 25.10**, both

Figure 25.10 A hypothesis for the origin of mitochondria and plastids through serial endosymbiosis. The proposed host was an archaean or a close relative of the archaeans. The proposed ancestors of mitochondria were aerobic, heterotrophic bacteria, while the proposed ancestors of plastids were photosynthetic bacteria. In this figure, the arrows represent change over evolutionary time.

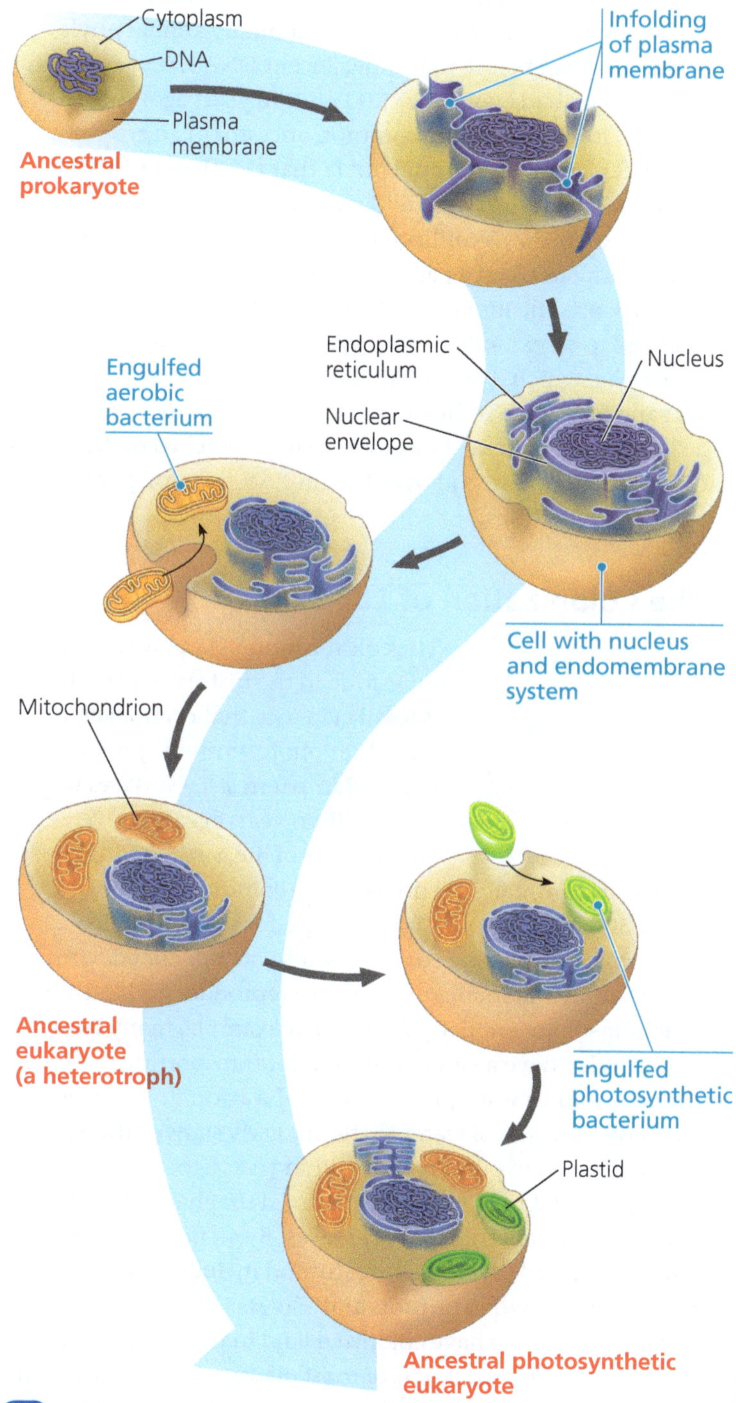

Figure Walkthrough

mitochondria and plastids are thought to have descended from bacterial cells. The original host—the cell that engulfed the bacterium whose descendants gave rise to the mitochondrion—is thought to have been an archaean or a close relative of the archaeans.

A great deal of evidence supports the endosymbiotic origin of mitochondria and plastids:

- The inner membranes of both organelles have enzymes and transport systems that are homologous to those found in the plasma membranes of living bacteria.
- Mitochondria and plastids replicate by a splitting process that is similar to that of certain bacteria. In addition, each of these organelles contains circular DNA molecules that, like the chromosomes of bacteria, are not associated with histones or large amounts of other proteins.
- As might be expected of organelles descended from free-living organisms, mitochondria and plastids also have the cellular machinery (including ribosomes) needed to transcribe and translate their DNA into proteins.
- Finally, in terms of size, RNA sequences, and sensitivity to certain antibiotics, the ribosomes of mitochondria and plastids are more similar to bacterial ribosomes than they are to the cytoplasmic ribosomes of eukaryotic cells.

In Chapter 28, we'll return to the origin of eukaryotes, focusing on what genomic data have revealed about the prokaryotic lineages that gave rise to the host and endosymbiont cells.

The Origin of Multicellularity

An orchestra can play a greater variety of musical compositions than a violin soloist can; the increased complexity of the orchestra makes more variations possible. Likewise, the origin of structurally complex eukaryotic cells sparked the evolution of greater morphological diversity than was possible for the simpler prokaryotic cells. After the first eukaryotes appeared, a great range of unicellular forms evolved, giving rise to the diversity of single-celled eukaryotes that continue to flourish today. Another wave of diversification also occurred: Some single-celled eukaryotes gave rise to multicellular forms, whose descendants include a variety of algae, plants, fungi, and animals.

Early Multicellular Eukaryotes

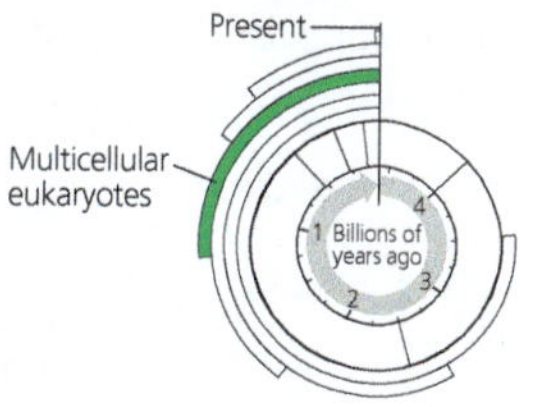

The oldest known fossils of multicellular eukaryotes that can be resolved taxonomically are of relatively small red algae that lived 1.2 billion years ago; even older fossils, dating to 1.8 billion years ago, may also be of small, multicellular eukaryotes. Larger and more diverse multicellular eukaryotes do not appear in the fossil record until about 600 million years ago (see Figure 25.5). These fossils, referred to as the Ediacaran biota, were of soft-bodied organisms—some over 1 m long—that lived from 635 to 541 million years ago. The Ediacaran biota included both algae and animals, along with various organisms of unknown taxonomic affinity.

The rise of large eukaryotes in the Ediacaran period represents an enormous change in the history of life. Before that time, Earth was a microbial world: Its only inhabitants were single-celled prokaryotes and eukaryotes, along with an assortment of microscopic, multicellular eukaryotes. As the diversification of the Ediacaran biota came to a close about 541 million years ago, the stage was set for another, even more spectacular burst of evolutionary change: the "Cambrian explosion."

The Cambrian Explosion

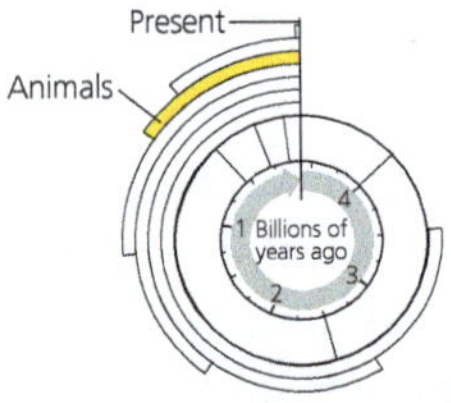

Many present-day animal phyla appear suddenly in fossils formed 535–525 million years ago, early in the Cambrian period. This phenomenon is referred to as the **Cambrian explosion**. Fossils of several animal groups—sponges, cnidarians (sea anemones and their relatives), and molluscs (snails, clams, and their relatives)—appear in even older rocks dating from the late Proterozoic **(Figure 25.11)**.

Prior to the Cambrian explosion, all large animals were soft-bodied. The fossils of large pre-Cambrian animals reveal little evidence of predation. Instead, these animals appear to have been grazers (feeding on algae), filter feeders, or scavengers, not hunters. The Cambrian explosion changed all of that. In a relatively short period of time (10 million years), predators over 1 m in length emerged that had claws and other features for capturing prey; simultaneously, new defensive adaptations, such as sharp spines and heavy body armor, appeared in their prey (see Figure 25.5).

Although the Cambrian explosion had an enormous impact on life on Earth, it appears that many animal phyla originated long before that time. Recent DNA analyses suggest that sponges had evolved by 700 million years ago; such analyses also indicate that the common ancestor of arthropods, chordates, and other animal phyla that radiated during the Cambrian explosion lived 670 million years ago. Researchers have unearthed 710-million-year-old sediments containing steroids indicative of a particular group of sponges—a finding that supports the molecular data. In contrast, the oldest fossil assigned to an extant animal phylum is that of the mollusc *Kimberella*, which lived 560 million years ago. Overall, molecular and fossil data indicate that the Cambrian explosion had a "long fuse"—at least 25 million years long based on the age of *Kimberella* fossils, and over 100 million years long based on some DNA analyses.

▼ Figure 25.11 Appearance of selected animal groups. The white bars indicate earliest appearances of these animal groups in the fossil record.

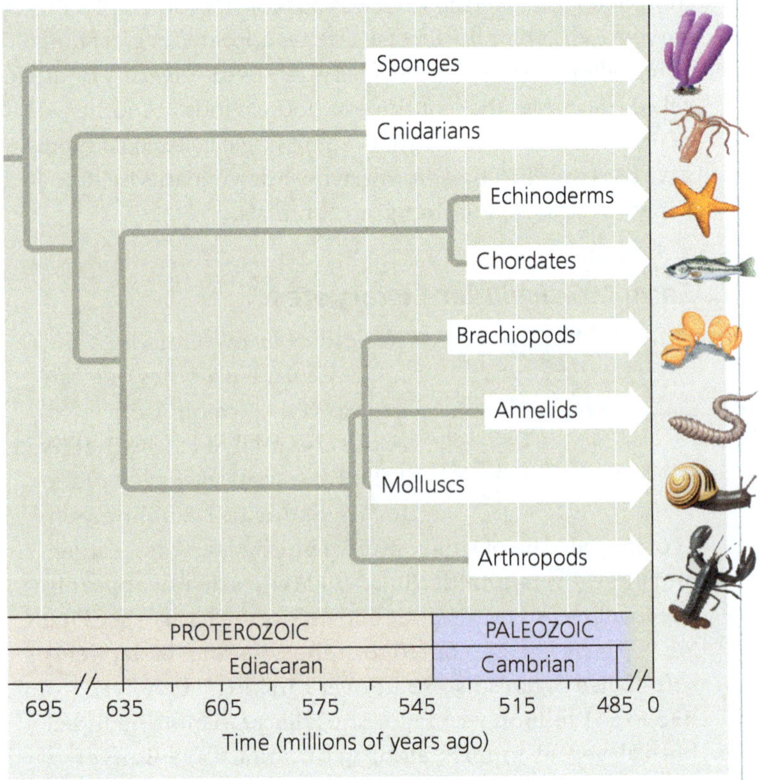

VISUAL SKILLS ➤ *Circle the branch point that represents the most recent common ancestor of chordates and annelids. What is a minimum estimate of that ancestor's age?*

The Colonization of Land

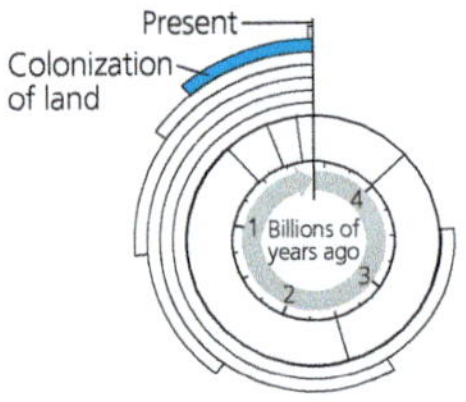

The colonization of land was another milestone in the history of life. There is fossil evidence that cyanobacteria and other photosynthetic prokaryotes coated damp terrestrial surfaces well over a billion years ago. However, larger forms of life, such as fungi, plants, and animals, did not begin to colonize land until about 500 million years ago. This gradual evolutionary venture out of aquatic environments was associated with adaptations that made it possible to reproduce on land and that helped prevent dehydration. For example, many land plants today have a vascular system for transporting materials internally and a waterproof coating of wax on their leaves that slows the loss of water to the air. Early signs of these adaptations were present 420 million years ago, at which time small plants (about 10 cm high) existed that had a vascular system but lacked true roots or leaves. By 40 million years later, plants had diversified greatly and included reeds and treelike plants with true roots and leaves.

Plants appear to have colonized land in the company of fungi. Even today, the roots of most plants are associated with fungi that aid in the absorption of water and minerals from the soil (see Concept 31.1). These root fungi (or *mycorrhizae*), in turn, obtain their organic nutrients from the plants. Such mutually beneficial associations of plants and fungi are evident in some of the oldest fossilized plants, dating this relationship back to the early spread of life onto land **(Figure 25.12)**.

Although many animal groups are now represented in terrestrial environments, the most widespread and diverse

Figure 25.12 An ancient symbiosis. This 405-million-year-old fossil stem (cross section) documents mycorrhizae in the early land plant *Aglaophyton major*. The inset shows an enlarged view of a cell containing a branched fungal structure called an arbuscule; the fossil arbuscule resembles those seen in plant cells today.

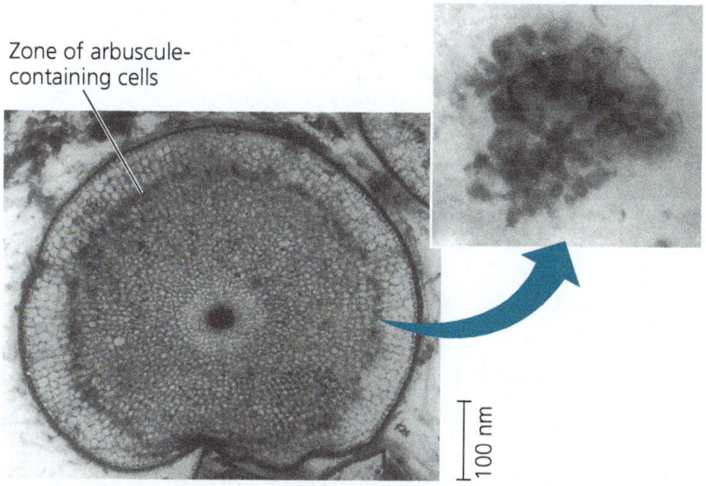

land animals are arthropods (particularly insects and spiders) and tetrapods. Arthropods were among the first animals to colonize land, roughly 450 million years ago. The earliest tetrapods found in the fossil record lived about 365 million years ago and appear to have evolved from a group of lobe-finned fishes (see Concept 34.3). Tetrapods include humans, although we are late arrivals on the scene. The human lineage diverged from other primates around 6–7 million years ago, and our species originated only about 195,000 years ago. If the clock of Earth's history were rescaled to represent an hour, humans appeared less than 0.2 second ago.

MB **Interview with Geerat Vermeij: What fossils reveal about the history of life**

CONCEPT CHECK 25.3

1. The first appearance of free oxygen in the atmosphere likely triggered a massive wave of extinctions among the prokaryotes of the time. Why?
2. What evidence supports the hypothesis that mitochondria preceded plastids in the evolution of eukaryotic cells?
3. **WHAT IF?** ➤ What would a fossil record of life today look like?

For suggested answers, see Appendix A.

CONCEPT 25.4

The rise and fall of groups of organisms reflect differences in speciation and extinction rates

From its beginnings, life on Earth has been marked by the rise and fall of groups of organisms. Anaerobic prokaryotes originated, flourished, and then declined as the oxygen content of the atmosphere rose. Billions of years later, the first tetrapods emerged from the sea, giving rise to several major new groups of organisms. One of these, the amphibians, went on to dominate life on land for 100 million years, until other tetrapods (including dinosaurs and, later, mammals) replaced them as the dominant terrestrial vertebrates.

Figure 25.13 How speciation and extinction affect diversity. The species diversity of an evolutionary lineage will increase when more new member species originate than are lost to extinction. In this hypothetical example, by 2 million years ago both lineage A and lineage B have given rise to four species, and no species have become extinct. Over the next 2 million years, however, lineage A experiences higher extinction rates than does lineage B (extinct species are denoted by a dagger symbol, †). As a result, after 4 million years (that is, by time 0), lineage A contains only one species, while lineage B contains eight species.

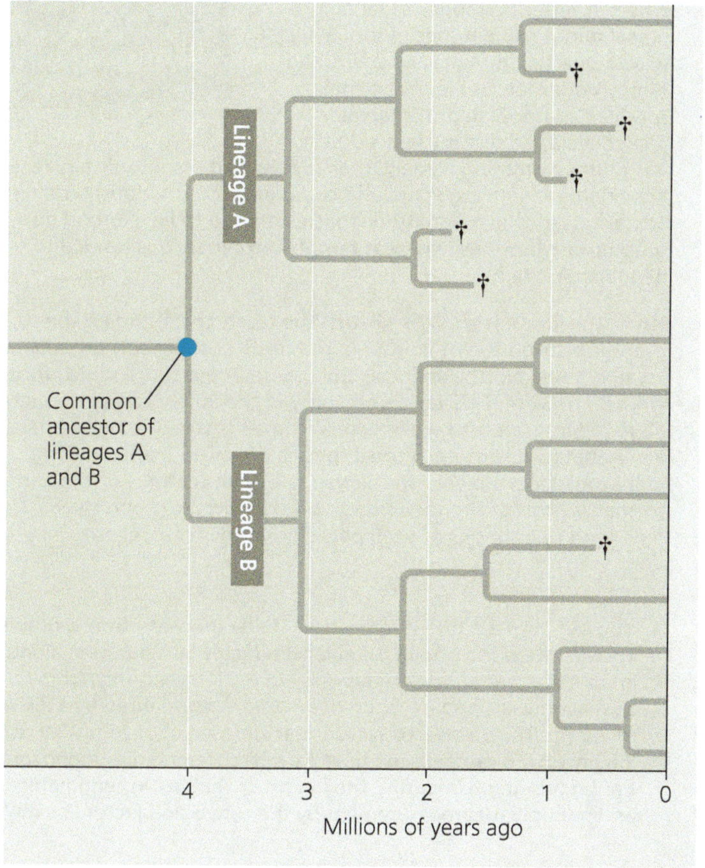

The rise and fall of these and other major groups of organisms have shaped the history of life. Narrowing our focus, we can also see that the rise or fall of any particular group is related to the speciation and extinction rates of its member species **(Figure 25.13)**. Just as a population increases in size when there are more births than deaths, the rise of a group of organisms occurs when more new species are produced than are lost to extinction. The reverse occurs when a group is in decline. In the **Scientific Skills Exercise**, you will interpret data from the fossil record about changes in a group of snail species in the early Paleogene period. Such changes in the fates of groups of organisms have been influenced by large-scale processes such as plate tectonics, mass extinctions, and adaptive radiations.

SCIENTIFIC SKILLS EXERCISE

Estimating Quantitative Data from a Graph and Developing Hypotheses

Do Ecological Factors Affect Evolutionary Rates? Researchers studied the fossil record to investigate whether differing modes of larval dispersal might explain species longevity within one taxon of marine snails, the family Volutidae. Some of the snail species had nonplanktonic larvae: They developed directly into adults without a swimming stage. Other species had planktonic larvae: They had a swimming stage and could disperse very long distances. The adults of these planktonic species tended to have broad geographic distributions, whereas nonplanktonic species tended to be more isolated.

How the Research Was Done The researchers studied the stratigraphic distribution of volutes in outcrops of sedimentary rocks located along North America's Gulf coast. These rocks, which formed from 66 to 37 million years ago, early in the Paleogene period, are an excellent source of well-preserved snail fossils. The researchers were able to classify each fossil species of volute snail as having planktonic or nonplanktonic larvae based on features of the earliest formed whorls of the snail's shell. Each bar in the graph shows how long one species of snail persisted in the fossil record.

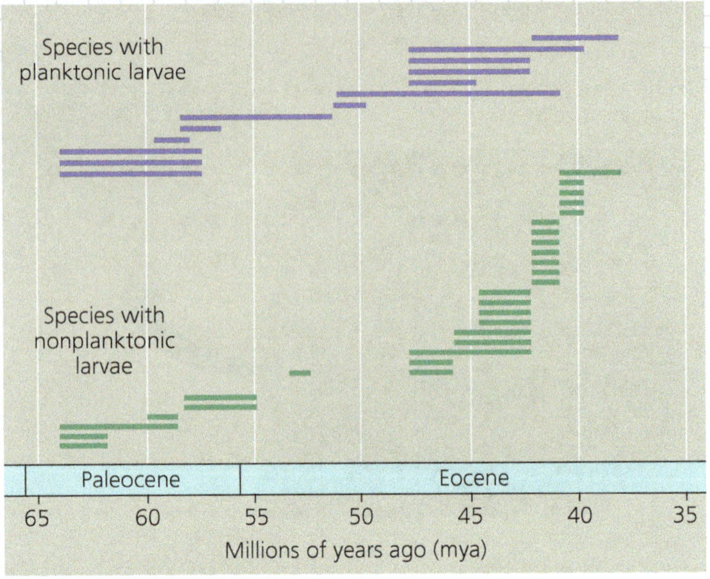

INTERPRET THE DATA

1. You can estimate quantitative data (fairly precisely) from a graph. The first step is to obtain a conversion factor by measuring along an axis that has a scale. In this case, 25 million years (my; from 60 to 35 million years ago [mya] on the *x*-axis) is represented by a distance of 7.0 cm. This yields a conversion factor (a ratio) of 25 my/7.0 cm = 3.6 my/cm. To estimate the time period represented by a horizontal bar on this graph, measure the length of that bar in centimeters and multiply that measurement by the conversion factor, 3.6 my/cm. For example, a bar that measures 1.1 cm on the graph represents a persistence time of 1.1 cm × 3.6 my/cm = 4 million years.
2. Calculate the mean (average) persistence times for species with planktonic larvae and species with nonplanktonic larvae.
3. Count the number of new species that form in each group beginning at 60 mya (the first three species in each group were present around 64 mya, the first time period sampled, so we don't know when those species first appear in the fossil record).
4. Propose a hypothesis to explain the differences in longevity of snail species with planktonic and nonplanktonic larvae.

Data from T. A. Hansen, Larval dispersal and species longevity in Lower Tertiary gastropods, *Science* 199:885–887 (1978). Reprinted with permission from AAAS.

Instructors: A version of this Scientific Skills Exercise can be assigned in MasteringBiology.

Plate Tectonics

If photographs of Earth were taken from space every 10,000 years and spliced together to make a movie, it would show something many of us find hard to imagine: The seemingly "rock solid" continents we live on move over time. Over the past billion years, there have been three occasions (1 billion, 600 million, and 250 million years ago) when most of the landmasses of Earth came together to form a supercontinent, then later broke apart. Each time, this breakup yielded a different configuration of continents. Based on the directions in which the continents are moving today, some geologists have estimated that a new supercontinent will form roughly 250 million years from now.

According to the theory of **plate tectonics**, the continents are part of great plates of Earth's crust that essentially float on the hot, underlying portion of the mantle **(Figure 25.14)**. Movements in the mantle cause the plates to move over time in a process called *continental drift*. Geologists can measure the rate at which the plates are moving now, usually only a few centimeters per year. They can also infer the past locations of the continents using the magnetic signal recorded in rocks at the time of their formation. This method works because as a continent shifts its position over time, the direction of magnetic north recorded in its newly formed rocks also changes.

▼ **Figure 25.14 Cutaway view of Earth.** The thickness of the crust is exaggerated here.

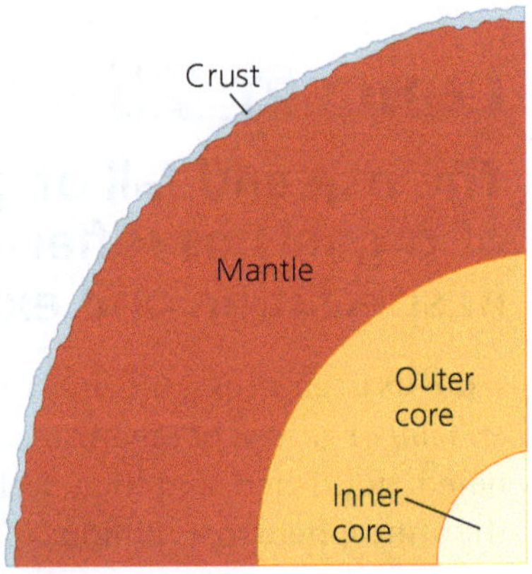

Earth's major tectonic plates are shown in **Figure 25.15**. Many important geologic processes, including the formation of mountains and islands, occur at plate boundaries. In some cases, two plates are moving away from each other, as are the North American and Eurasian

Figure 25.15 Earth's major tectonic plates. The arrows indicate direction of movement. The reddish orange dots represent zones of violent tectonic activity.

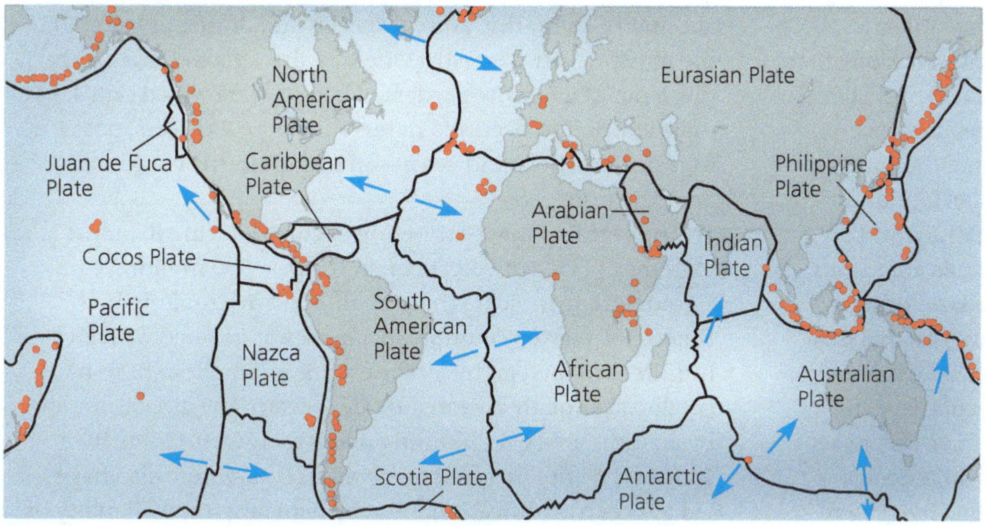

HHMI Animation: Plate Tectonics hhmi BioInteractive

plates, which are currently drifting apart at a rate of about 2 cm per year. In other cases, two plates are sliding past each other, forming regions where earthquakes are common. California's infamous San Andreas Fault is part of a border where two plates slide past each other. In still other cases, two plates collide, producing violent upheavals and forming new mountains along the plate boundaries. One spectacular example of this occurred 45 million years ago, when the Indian plate crashed into the Eurasian plate, starting the formation of the Himalayan mountains.

HHMI Video: Animated Life: Pangea, Wegener, and Continental Drift hhmi BioInteractive

Consequences of Continental Drift

Plate movements rearrange geography slowly, but their cumulative effects are dramatic. In addition to reshaping the physical features of our planet, continental drift also has a major impact on life on Earth.

One reason for this is that continental drift alters the habitats in which organisms live. Consider the changes shown in **Figure 25.16**. About 250 million years ago, plate movements brought previously separated landmasses together into a supercontinent named **Pangaea**. Ocean basins became deeper, which drained shallow coastal seas. At that time, as now, most marine species inhabited shallow waters, and the formation of Pangaea destroyed much of that habitat. Pangaea's interior was cold and dry, probably an even more severe environment than that of central Asia today. Overall, the formation of Pangaea greatly altered the physical environment and climate, which drove some species to extinction and provided new opportunities for groups of organisms that survived the crisis.

Organisms are also affected by the climate change that results when a continent shifts its location. The southern tip of Labrador, Canada, for example, once was located in the tropics but has moved 40° to the north over the last 200 million years. When faced with the changes in climate that such shifts in position entail, organisms adapt, move to a new location, or become extinct (this last outcome occurred for many organisms stranded on Antarctica, which separated from Australia 40 million years ago).

Continental drift also promotes allopatric speciation on a grand scale. When supercontinents break apart, regions that once were connected become isolated. As the continents drifted apart over the last 200 million years, each became a separate evolutionary arena, with lineages of plants and animals that diverged from those on other continents.

Figure 25.16 The history of continental drift during the Phanerozoic eon.

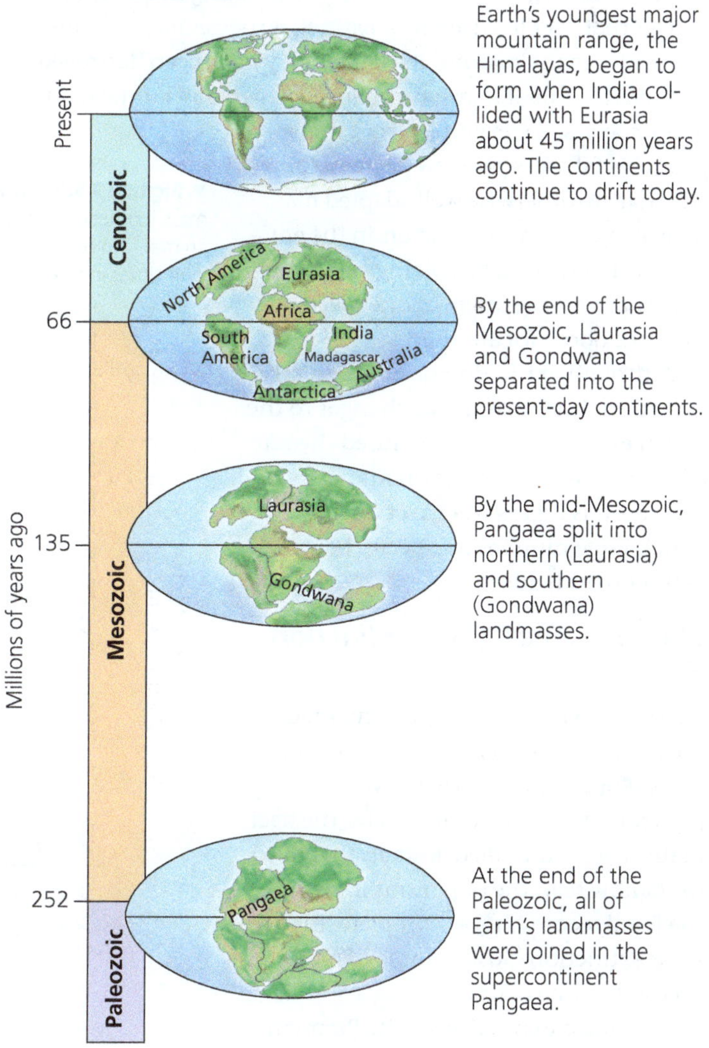

VISUAL SKILLS ➤ *Is the Australian plate's current direction of movement (see Figure 25.15) similar to the direction it traveled over the past 66 million years?*

Finally, continental drift can help explain puzzles about the geographic distribution of extinct organisms, such as why fossils of the same species of Permian freshwater reptiles have been discovered in both Brazil and the West African nation of Ghana. These two parts of the world, now separated by 3,000 km of ocean, were joined together when these reptiles were living. Continental drift also explains much about the current distributions of organisms, such as why Australian fauna and flora contrast so sharply with those of the rest of the world. Marsupial mammals fill ecological roles in Australia analogous to those filled by eutherians (placental mammals) on other continents (see Figure 22.18). Fossil evidence suggests that marsupials originated in what is now Asia and reached Australia via South America and Antarctica while the continents were still joined. The subsequent breakup of the southern continents set Australia "afloat," like a giant raft of marsupials. In Australia, marsupials diversified, and the few eutherians that lived there became extinct; on other continents, most marsupials became extinct, and the eutherians diversified.

Mass Extinctions

The fossil record shows that the overwhelming majority of species that ever lived are now extinct. A species may become extinct for many reasons. Its habitat may have been destroyed, or its environment may have changed in a manner unfavorable to the species. For example, if ocean temperatures fall by even a few degrees, species that are otherwise well adapted may perish. Even if physical factors in the environment remain stable, biological factors may change—the origin of one species can spell doom for another.

Although extinction occurs regularly, at certain times disruptive changes to the global environment have caused the rate of extinction to increase dramatically. The result is a **mass extinction**, in which large numbers of species become extinct worldwide.

The "Big Five" Mass Extinction Events

Five mass extinctions are documented in the fossil record over the past 500 million years **(Figure 25.17)**. These events are particularly well documented for the decimation of hard-bodied animals that lived in shallow seas, the organisms for which the fossil record is most complete. In each mass extinction, 50% or more of marine species became extinct.

Two mass extinctions—the Permian and the Cretaceous—have received the most attention. The Permian mass extinction, which defines the boundary between the Paleozoic and Mesozoic eras (252 million years ago), claimed about 96% of marine animal species and drastically altered life in the ocean. Terrestrial life was also affected. For example, 8 out of 27 known orders of insects were wiped out. This mass extinction occurred in less than 500,000 years, possibly in just a few thousand years—an instant in the context of geologic time.

The Permian mass extinction occurred during the most extreme episode of volcanism in the past 500 million years. Geologic data indicate that 1.6 million km^2 (roughly half the size of Western Europe) in Siberia was covered with lava hundreds of meters thick. The eruptions are thought to have produced enough carbon dioxide to warm the global climate by an estimated 6°C, harming many temperature-sensitive species. The rise in atmospheric CO_2 levels would also have led to ocean acidification, thereby reducing the availability of calcium carbonate, which is required by reef-building corals and many shell-building species (see Figure 3.12). The explosions would also have added nutrients such as phosphorus to marine ecosystems, stimulating the growth of microorganisms. Upon their deaths, these microorganisms would have provided food for bacterial decomposers. Bacteria use oxygen as they decompose the bodies of dead organisms, thus causing oxygen concentrations to drop. This would have harmed

Figure 25.17 Mass extinction and the diversity of life. The five generally recognized mass extinction events, indicated by red arrows, represent peaks in the extinction rate of marine animal families (red line and left vertical axis). These mass extinctions interrupted the overall increase, over time, in the number of extant families of marine animals (blue line and right vertical axis).

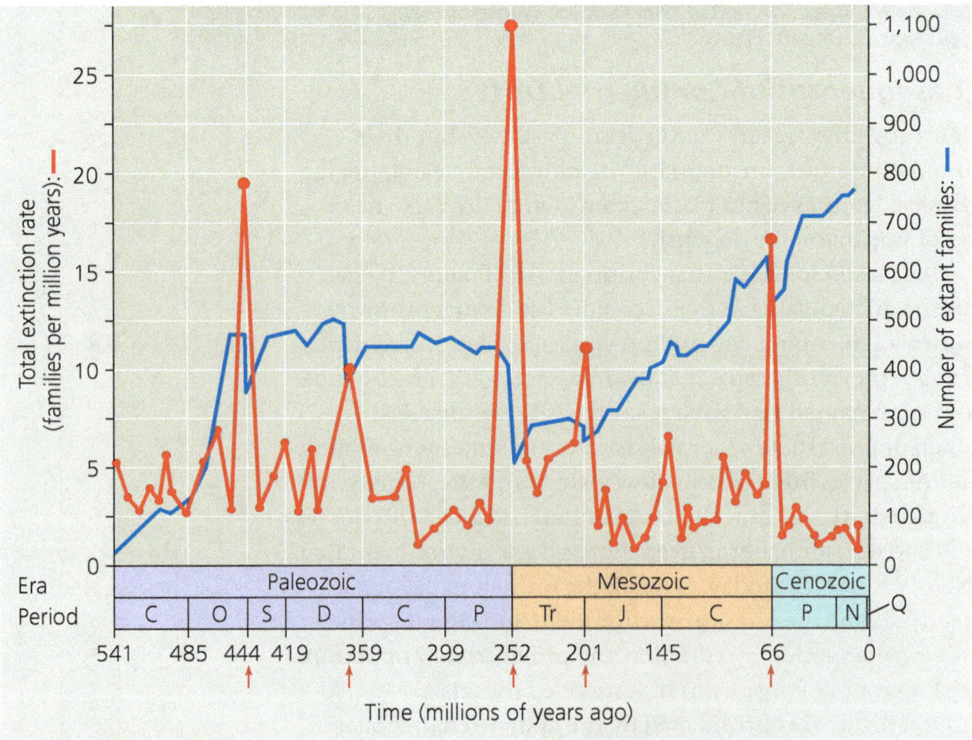

INTERPRET THE DATA ➤ *As mentioned in the text, 96% of marine animal species became extinct in the Permian mass extinction. Explain why the blue curve shows only a 50% drop at that time.*

oxygen-breathers and promoted the growth of anaerobic bacteria that emit a poisonous metabolic by-product, hydrogen sulfide (H_2S) gas. Overall, the volcanic eruptions may have triggered a series of catastrophic events that together resulted in the Permian mass extinction.

The Cretaceous mass extinction occurred 66 million years ago. This event extinguished more than half of all marine species and eliminated many families of terrestrial plants and animals, including all dinosaurs (except birds, which are members of the same group; see Figure 34.25). One clue to a possible cause of the Cretaceous mass extinction is a thin layer of clay enriched in iridium that dates to the time of the mass extinction. Iridium is an element that is very rare on Earth but common in many of the meteorites and other extraterrestrial objects that occasionally fall to Earth. As a result, researchers proposed that this clay is fallout from a huge cloud of debris that billowed into the atmosphere when an asteroid or large comet collided with Earth. This cloud would have blocked sunlight and severely disturbed the global climate for several months.

Is there evidence of such an asteroid or comet? Research has focused on the Chicxulub crater, a 66-million-year-old scar beneath sediments off the coast of Mexico **(Figure 25.18)**.

▼ **Figure 25.18 A trauma for Cretaceous life.** Beneath the Caribbean Sea, the 66-million-year-old Chicxulub crater measures 180 km across. The horseshoe shape of the crater and the pattern of debris in sedimentary rocks indicate that an asteroid or comet struck at a low angle from the southeast. This drawing represents the impact and its immediate effect: a cloud of hot vapor and debris that could have killed many of the plants and animals in North America within hours.

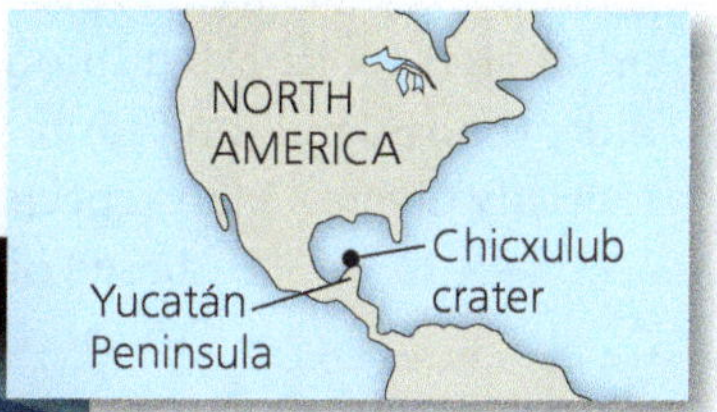

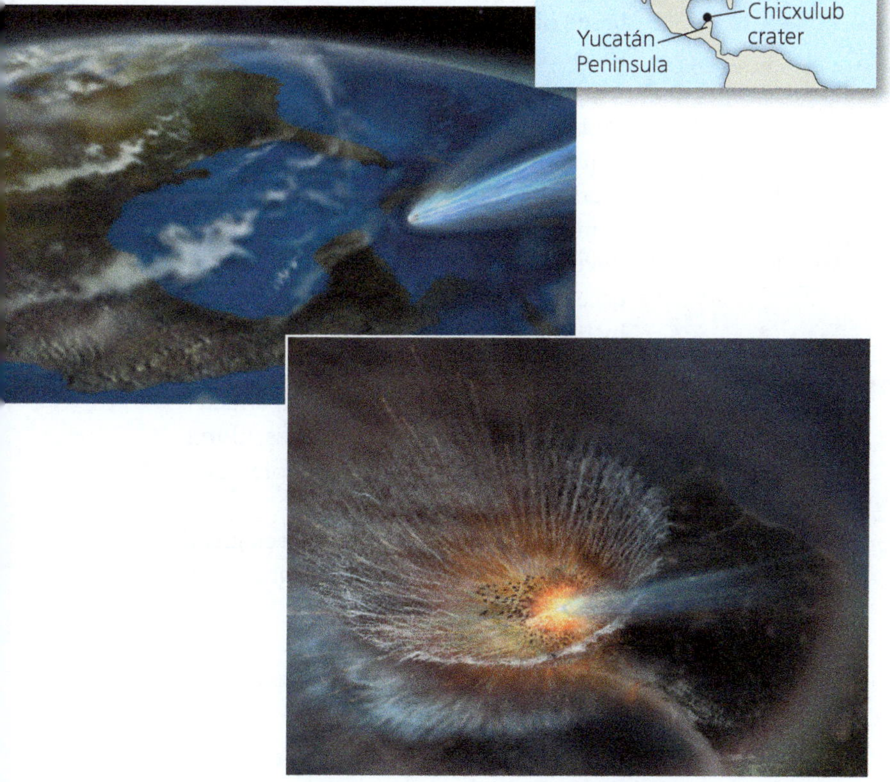

The crater is the right size to have been caused by an object with a diameter of 10 km. Critical evaluation of this and other hypotheses for mass extinctions continues.

MB HHMI Video: The Day the Mesozoic Died

Is a Sixth Mass Extinction Under Way?

As you will read further in Concept 56.1, human actions, such as habitat destruction, are modifying the global environment to such an extent that many species are threatened with extinction. More than 1,000 species have become extinct in the last 400 years. Scientists estimate that this rate is 100 to 1,000 times the typical background rate seen in the fossil record. Is a sixth mass extinction now in progress?

This question is difficult to answer, in part because it is hard to document the total number of extinctions occurring today. Tropical rain forests, for example, harbor many undiscovered species. As a result, destroying tropical forest may drive species to extinction before we even learn of their existence. Such uncertainties make it hard to assess the full extent of the current extinction crisis. Even so, it is clear that losses to date have not reached those of the "big five" mass extinctions, in which large percentages of Earth's species became extinct. This does not in any way discount the seriousness of today's situation. Monitoring programs show that many species are declining at an alarming rate due to habitat loss, introduced species, overharvesting, and other factors. Recent studies on a variety of organisms, including lizards, pine trees, and polar bears, suggest that climate change may hasten some of these declines. The fossil record also highlights the potential importance of climate change: Over the last 500 million years, extinction rates have tended to increase when global temperatures were high **(Figure 25.19)**.

▼ **Figure 25.19 Fossil extinctions and temperature.** Extinction rates increased when global temperatures were high. Temperatures were estimated using ratios of oxygen isotopes and converted to an index in which 0 is the overall average temperature.

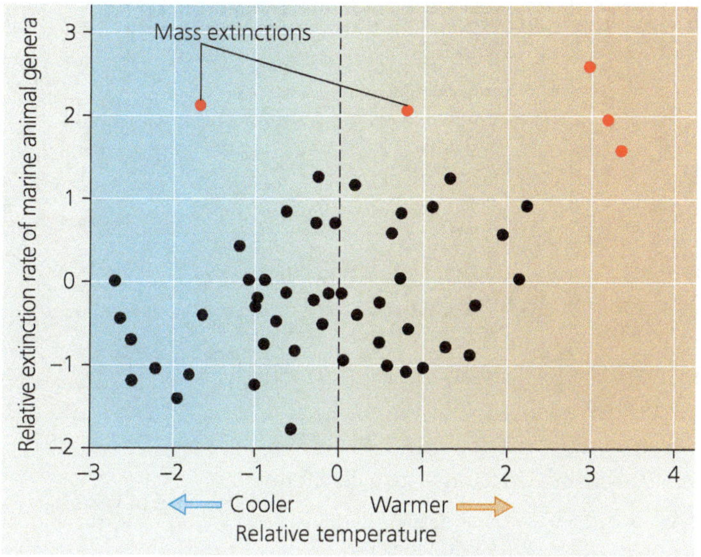

Overall, the evidence suggests that unless dramatic actions are taken, a sixth, human-caused mass extinction is likely to occur within the next few centuries or millennia.

Consequences of Mass Extinctions

Mass extinctions have significant and long-term effects. By eliminating large numbers of species, a mass extinction can reduce a thriving and complex ecological community to a pale shadow of its former self. And once an evolutionary lineage disappears, it cannot reappear. The course of evolution is changed forever. Consider what would have happened if the early primates living 66 million years ago had died out in the Cretaceous mass extinction. Humans would not exist, and life on Earth would differ greatly from what it is today.

The fossil record shows that it typically takes 5–10 million years for the diversity of life to recover to previous levels after a mass extinction. In some cases, it has taken much longer than that: It took about 100 million years for the number of marine families to recover after the Permian mass extinction (see Figure 25.17). These data have sobering implications. If current trends continue and a sixth mass extinction occurs, it will take millions of years for life on Earth to recover.

Mass extinctions can also alter ecological communities by changing the types of organisms residing there. For example, after the Permian and Cretaceous mass extinctions, the percentage of marine organisms that were predators increased substantially **(Figure 25.20)**. A rise in the number of predators can increase both the risks faced by prey and the competition among predators for food. In addition, mass extinctions can curtail lineages with novel and advantageous features. For example, in the late Triassic period, a group of gastropods (snails and their relatives) arose that could drill through the shells of bivalves (such as clams) and feed on the animals inside. Although shell drilling provided access to a new and abundant source of food, this newly formed group was wiped out during the mass extinction at the end of the Triassic (about 200 million years ago). Another 120 million years passed before another group of gastropods (the oyster drills) exhibited the ability to drill through shells. As their predecessors might have done if they had not originated at an unfortunate time, oyster drills have since diversified into many new species. Finally, by eliminating so many species, mass extinctions can pave the way for adaptive radiations, in which new groups of organisms proliferate.

▼ Figure 25.20 Mass extinctions and ecology. The Permian and Cretaceous mass extinctions (indicated by red arrows) altered the ecology of the oceans by increasing the percentage of marine genera that were predators.

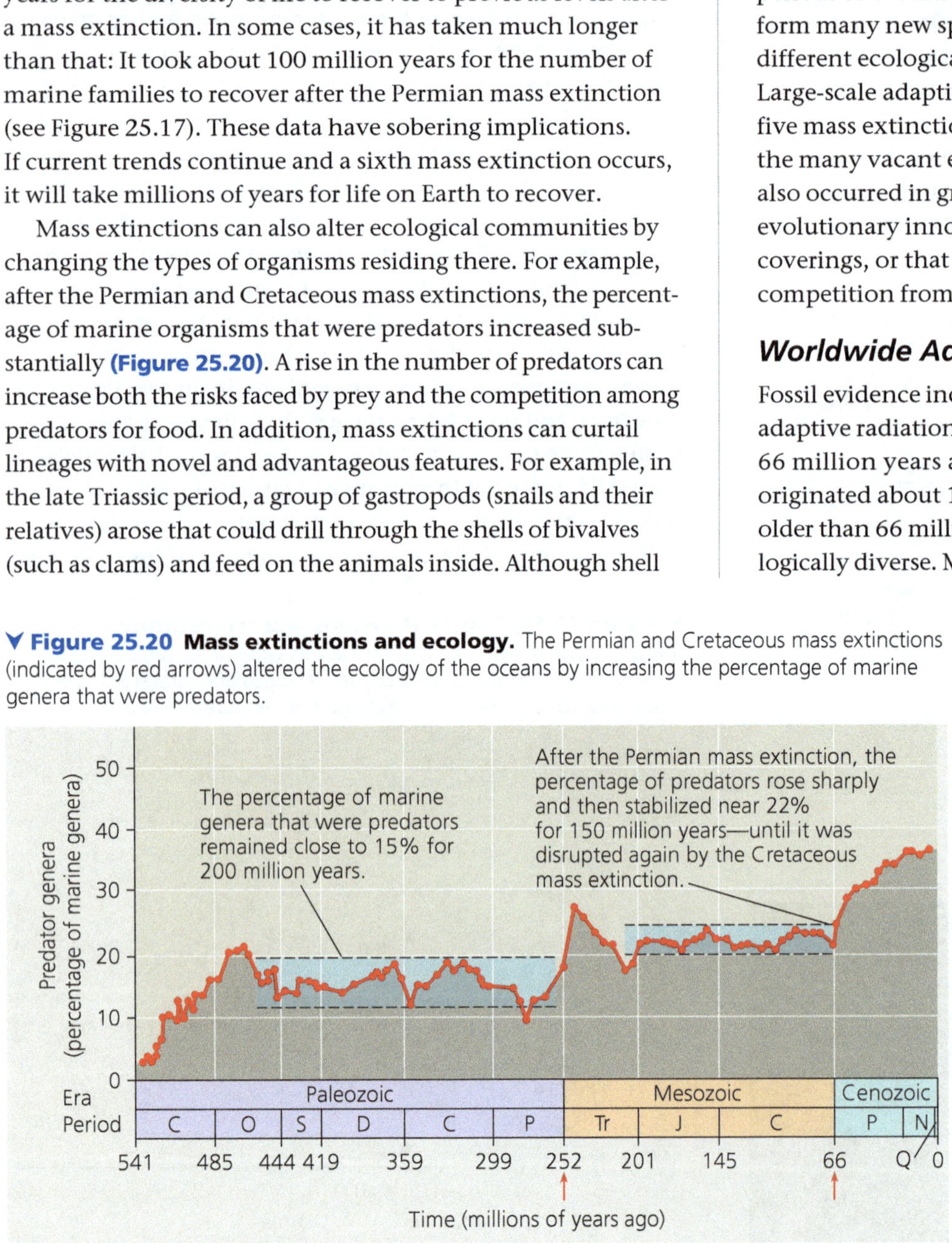

Adaptive Radiations

The fossil record shows that the diversity of life has increased over the past 250 million years (see blue line in Figure 25.17). This increase has been fueled by **adaptive radiations**, periods of evolutionary change in which groups of organisms form many new species whose adaptations allow them to fill different ecological roles, or niches, in their communities. Large-scale adaptive radiations occurred after each of the big five mass extinctions, when survivors became adapted to the many vacant ecological niches. Adaptive radiations have also occurred in groups of organisms that possessed major evolutionary innovations, such as seeds or armored body coverings, or that colonized regions in which they faced little competition from other species.

Worldwide Adaptive Radiations

Fossil evidence indicates that mammals underwent a dramatic adaptive radiation after the extinction of terrestrial dinosaurs 66 million years ago **(Figure 25.21)**. Although mammals originated about 180 million years ago, the mammal fossils older than 66 million years are mostly small and not morphologically diverse. Many species appear to have been nocturnal based on their large eye sockets, similar to those in living nocturnal mammals. A few early mammals were intermediate in size, such as *Repenomamus giganticus*, a 1-m-long predator that lived 130 million years ago—but none approached the size of many dinosaurs. Early mammals may have been restricted in size and diversity because they were eaten or outcompeted by the larger and more diverse dinosaurs. With the disappearance of the dinosaurs (except for birds), mammals expanded greatly in both diversity and size, filling the ecological roles once occupied by terrestrial dinosaurs.

The history of life has also been greatly altered by radiations in which groups of organisms increased in diversity as they came to play entirely new ecological roles in their communities. As we'll explore

Figure 25.21 Adaptive radiation of mammals.

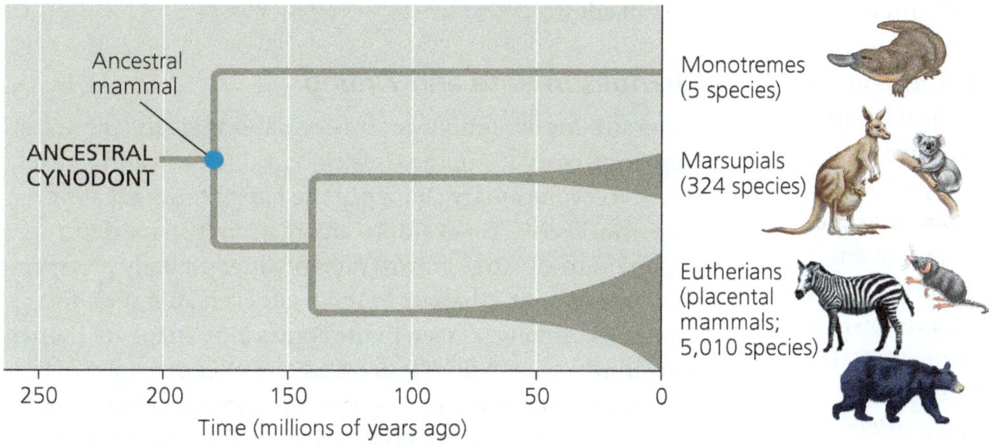

in later chapters, examples include the rise of photosynthetic prokaryotes, the evolution of large predators in the Cambrian explosion, and the radiations following the colonization of land by plants, insects, and tetrapods. Each of these last three radiations was associated with major evolutionary innovations that facilitated life on land. The radiation of land plants, for example, was associated with key adaptations, such as stems that support plants against gravity and a waxy coat that protects leaves from water loss. Finally, organisms that arise in an adaptive radiation can serve as a new source of food for still other organisms. In fact, the diversification of land plants stimulated a series of adaptive radiations in insects that ate or pollinated plants, one reason that insects are the most diverse group of animals on Earth today.

Regional Adaptive Radiations

Striking adaptive radiations have also occurred over more limited geographic areas. Such radiations can be initiated when a few organisms make their way to a new, often distant location in which they face relatively little competition from other organisms. The Hawaiian archipelago is one of the world's great showcases of this type of adaptive radiation **(Figure 25.22)**. Located about 3,500 km from the nearest continent, the volcanic islands are progressively older as one follows the chain toward the northwest; the youngest

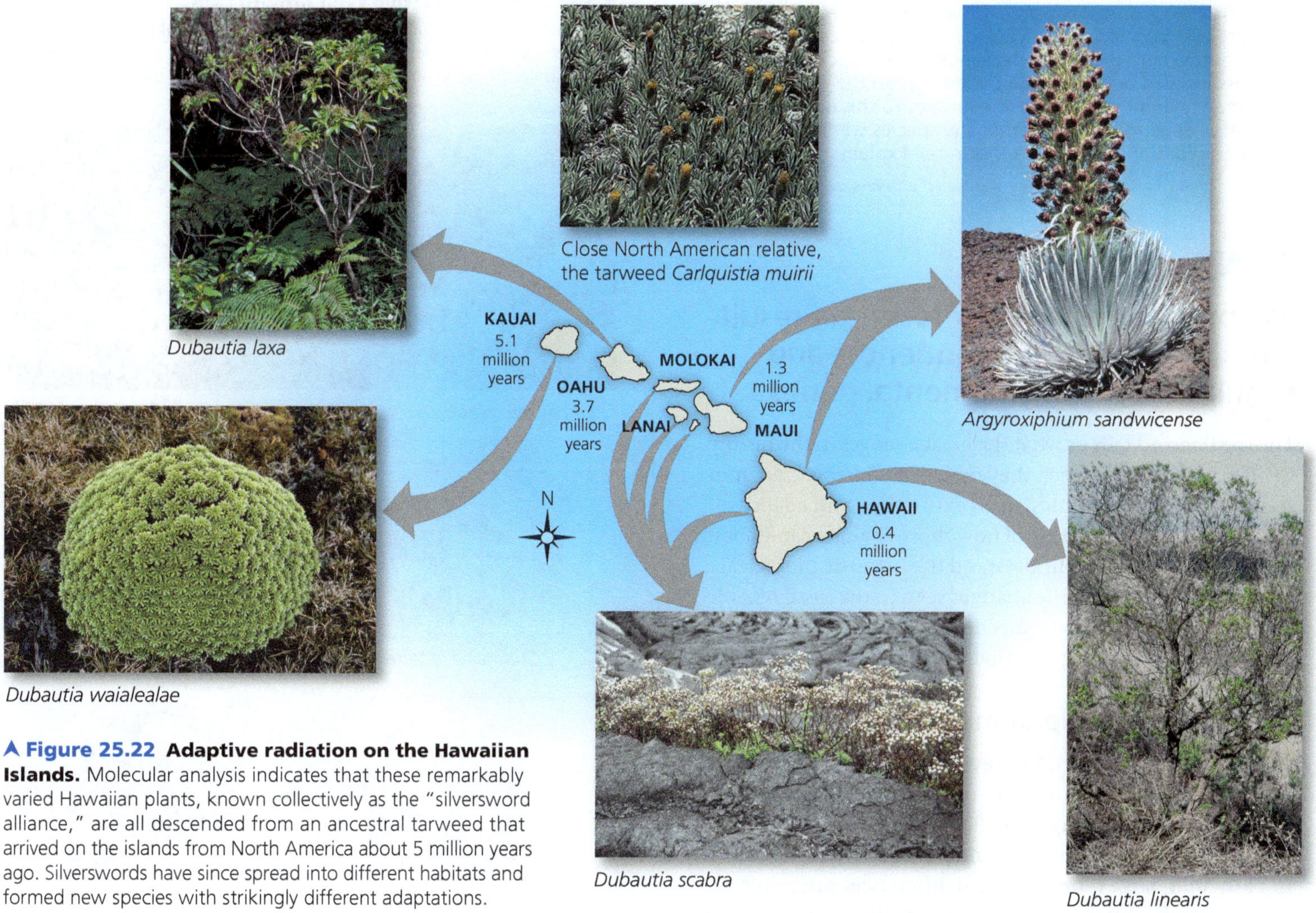

Figure 25.22 Adaptive radiation on the Hawaiian Islands. Molecular analysis indicates that these remarkably varied Hawaiian plants, known collectively as the "silversword alliance," are all descended from an ancestral tarweed that arrived on the islands from North America about 5 million years ago. Silverswords have since spread into different habitats and formed new species with strikingly different adaptations.

island, Hawaii, is less than a million years old and still has active volcanoes. Each island was born "naked" and was gradually populated by stray organisms that rode the ocean currents and winds either from far-distant land areas or from older islands of the archipelago itself. The physical diversity of each island, including immense variation in soil conditions, elevation, and rainfall, provides many opportunities for evolutionary divergence by natural selection. Multiple invasions followed by speciation events have ignited an explosion of adaptive radiation in Hawaii. As a result, thousands of species that inhabit the islands are found nowhere else on Earth. Among plants, for example, about 1,100 species are unique to the Hawaiian Islands. Unfortunately, many of these species are now facing an elevated risk of extinction due to human actions such as habitat destruction and the introduction of non-native plant species.

MB Animation: Overview of Macroevolution

CONCEPT CHECK 25.4

1. Explain the consequences of plate tectonics for life on Earth.
2. What factors promote adaptive radiations?
3. **WHAT IF?** ➤ Suppose that an invertebrate species was lost in a mass extinction caused by a sudden catastrophic event. Would the last appearance of this species in the fossil record necessarily be close to when the extinction actually occurred? Would the answer to this question differ depending on whether the species was common (abundant and widespread) or rare? Explain.

For suggested answers, see Appendix A.

CONCEPT 25.5

Major changes in body form can result from changes in the sequences and regulation of developmental genes

The fossil record tells us what the great changes in the history of life have been and when they occurred. Moreover, an understanding of plate tectonics, mass extinction, and adaptive radiation provides a picture of how those changes came about. But we can also seek to understand the intrinsic biological mechanisms that underlie changes seen in the fossil record. For this, we turn to genetic mechanisms of change, paying particular attention to genes that influence development.

Effects of Developmental Genes

As you read in Concept 21.6, "evo-devo"—research at the interface between evolutionary biology and developmental biology—is illuminating how slight genetic differences can produce major morphological differences between species. In particular, large morphological differences can result from genes that alter the rate, timing, and spatial pattern of change in an organism's form as it develops from a zygote into an adult.

Changes in Rate and Timing

Many striking evolutionary transformations are the result of **heterochrony** (from the Greek *hetero*, different, and *chronos*, time), an evolutionary change in the rate or timing of developmental events. For example, an organism's shape depends in part on the relative growth rates of different body parts during development. Changes to these rates can alter the adult form substantially, as seen in the contrasting shapes of human and chimpanzee skulls **(Figure 25.23)**. Other examples of the dramatic evolutionary effects of heterochrony include how increased growth rates of finger bones yielded the skeletal structure of wings in bats (see Figure 22.15) and how slowed growth of leg and pelvic bones led to the reduction and eventual loss of hind limbs in whales (see Figure 22.20).

Heterochrony can also alter the timing of reproductive development relative to the development of nonreproductive organs. If the development of reproductive organs accelerates compared to that of other organs, the sexually mature stage of

▼ **Figure 25.23 Relative skull growth rates.** In the human evolutionary lineage, mutations slowed the growth of the jaw relative to other parts of the skull. As a result, in humans the skull of an adult is more similar to the skull of an infant than is the case for chimpanzees.

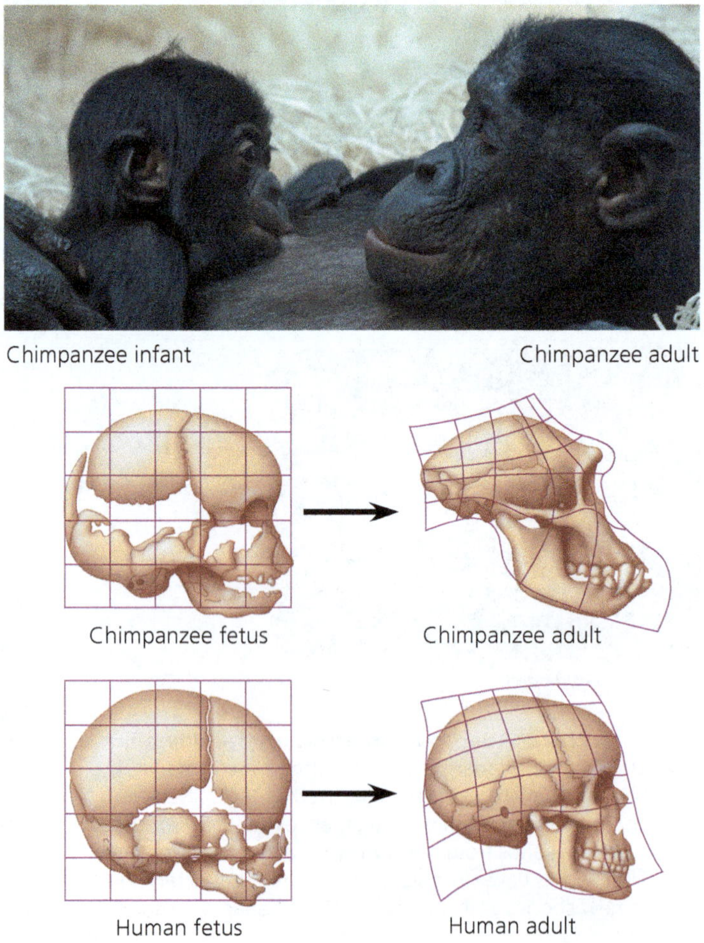

Figure 25.24 Paedomorphosis. The adults of some species retain features that were juvenile in ancestors. This salamander is an axolotl, an aquatic species becomes a sexually mature adult while retaining certain larval (tadpole) characteristics, including gills.

a species may retain body features that were juvenile structures in an ancestral species, a condition called **paedomorphosis** (from the Greek *paedos*, of a child, and *morphosis*, formation). For example, most salamander species have aquatic larvae that undergo metamorphosis in becoming adults. But some species grow to adult size and become sexually mature while retaining gills and other larval features **(Figure 25.24)**. Such an evolutionary alteration of developmental timing can produce animals that appear very different from their ancestors, even though the overall genetic change may be small. Indeed, recent evidence indicates that a change at a single locus was probably sufficient to bring about paedomorphosis in the axolotl salamander, although other genes may have contributed as well.

Changes in Spatial Pattern

Substantial evolutionary changes can also result from alterations in genes that control the spatial organization of body parts. For example, master regulatory genes called **homeotic genes** (see Concept 18.1) determine such basic features as where a pair of wings and legs will develop on a bird or how a plant's flower parts are arranged.

The products of one class of homeotic genes, the *Hox* genes, provide positional information in an animal embryo. This information prompts cells to develop into structures appropriate for a particular location. Changes in *Hox* genes or in how they are expressed can have a profound impact on morphology. For example, among crustaceans, a change in the location where two *Hox* genes (*Ubx* and *Scr*) are expressed correlates with the conversion of a swimming appendage to a feeding appendage. Similarly, when comparing plant species, changes to the expression of homeotic genes known as *MADS-box* genes can produce flowers that differ dramatically in form (see Concept 35.5).

The Evolution of Development

The 560-million-year-old fossils of Ediacaran animals in Figure 25.5 suggest that a set of genes sufficient to produce complex animals existed at least 25 million years *before* the Cambrian explosion. If such genes have existed for so long, how can we explain the astonishing increases in diversity seen during and since the Cambrian explosion?

Adaptive evolution by natural selection provides one answer to this question. As we've seen throughout this unit, by sorting among differences in the sequences of protein-encoding genes, selection can improve adaptations rapidly. In addition, new genes (created by gene duplication events) can take on new metabolic and structural functions, as can existing genes that are regulated in new ways.

Examples in the previous section suggest that developmental genes may have been particularly important. Thus, we'll turn next to how new morphological forms can arise from changes in the nucleotide sequences or regulation of developmental genes.

Changes in Gene Sequence

New developmental genes arising after gene duplication events probably facilitated the origin of novel morphological forms. But since other genetic changes also may have occurred at such times, it can be difficult to establish causal links between genetic and morphological changes that occurred in the past.

This difficulty was sidestepped in a study of developmental changes associated with the divergence of six-legged insects from crustacean ancestors that had more than six legs. (As discussed in Concept 33.4, insects arose from within a subgroup of the crustaceans, the traditional name for organisms such as shrimp, crabs, and lobsters.) Researchers noted differences between crustaceans and insects in the pattern of expression and the effects of the *Hox* gene *Ubx*: In particular, in insects, *Ubx* suppresses leg formation where it is expressed **(Figure 25.25)**.

Figure 25.25 Effects of the *Hox* gene *Ubx* on the insect body plan. In crustaceans, the *Hox* gene *Ubx* is expressed in the region shaded green, the body segments between the head and genital segments. In insects, *Ubx* is expressed in only a subset (shaded pink) of the homologous body segments, where it suppresses leg formation.

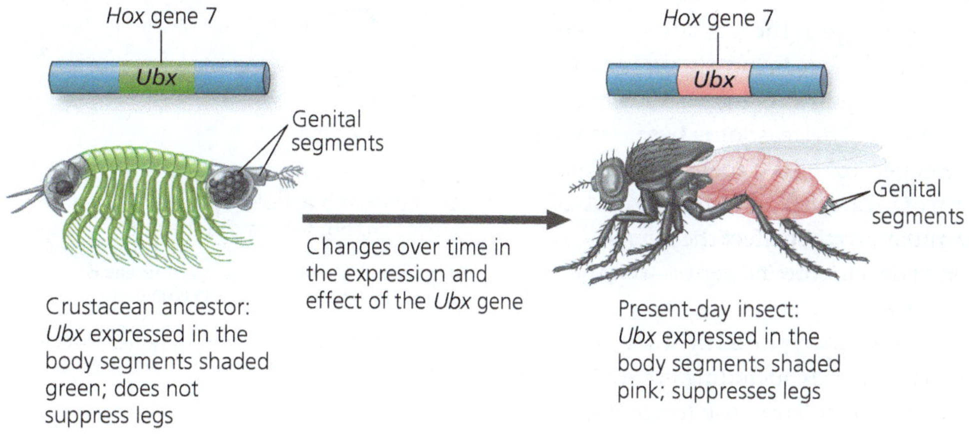

To examine the workings of this gene, researchers cloned the *Ubx* gene from an insect, the fruit fly *Drosophila*, and from a crustacean, the brine shrimp *Artemia*. Next, they genetically engineered fruit fly embryos to express either the *Drosophila Ubx* gene or the *Artemia Ubx* gene throughout their bodies. The *Drosophila* gene suppressed 100% of the limbs in the embryos, as expected, whereas the *Artemia* gene suppressed only 15%.

The researchers then sought to uncover key steps involved in the evolutionary transition from an ancestral *Ubx* gene to an insect *Ubx* gene. Their approach was to identify mutations that would cause the *Artemia Ubx* gene to suppress leg formation, thus making its gene act more like an insect *Ubx* gene. To do this, they constructed a series of "hybrid" *Ubx* genes, each of which contained known segments of the *Drosophila Ubx* gene and known segments of the *Artemia Ubx* gene. By inserting these hybrid genes into fruit fly embryos (one hybrid gene per embryo) and observing their effects on leg development, the researchers were able to pinpoint the exact amino acid changes responsible for the suppression of additional limbs in insects. In so doing, this study provided evidence that particular changes in the nucleotide sequence of a developmental gene contributed to a major evolutionary change: the origin of the six-legged insect body plan.

Changes in Gene Regulation

A change in the nucleotide sequence of a gene may affect its function wherever the gene is expressed, while changes in the regulation of gene expression can be limited to one cell type (see Concept 18.4). Thus, a change in the regulation of a developmental gene may have fewer harmful side effects than a change to the sequence of the gene. This reasoning has prompted researchers to suggest that changes in the form of organisms may often be caused by mutations that affect the regulation of developmental genes—not their sequences.

This idea is supported by studies of a variety of species, including threespine stickleback fish. These fish live in the open

▼ Figure 25.26

Inquiry What causes the loss of spines in lake stickleback fish?

Experiment Marine populations of the threespine stickleback fish (*Gasterosteus aculeatus*) have a set of protective spines on their lower (ventral) surface; however, these spines have been lost or reduced in some lake populations of this fish. Working at Stanford University, Michael Shapiro, David Kingsley, and colleagues performed genetic crosses and found that most of the reduction in spine size resulted from the effects of a single developmental gene, *Pitx1*. The researchers then tested two hypotheses about how *Pitx1* causes this morphological change.

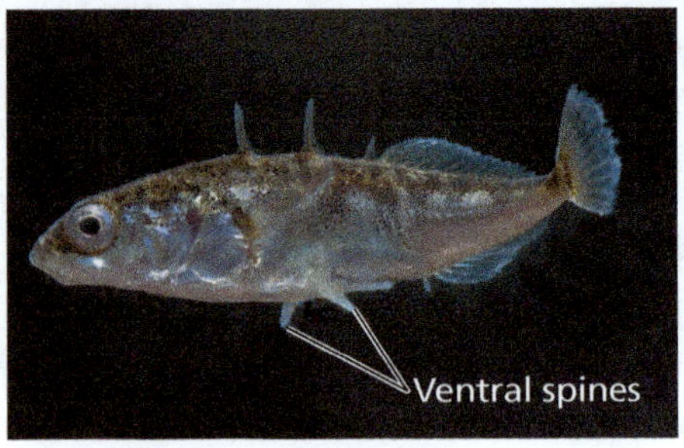

▲ Threespine stickleback (*Gasterosteus aculeatus*)

Hypothesis A: A change in the DNA sequence of *Pitx1* had caused spine reduction in lake populations. To test this idea, the team used DNA sequencing to compare the coding sequence of the *Pitx1* gene between marine and lake stickleback populations.

Hypothesis B: A change in the regulation of the expression of *Pitx1* had caused spine reduction. To test this idea, the researchers monitored where in the developing embryo the *Pitx1* gene was expressed. They conducted whole-body *in situ* hybridization experiments (see Concept 20.2) using *Pitx1* DNA as a probe to detect *Pitx1* mRNA in the fish.

Results

Test of Hypothesis A:	Are there differences in the coding sequence of the *Pitx1* gene in marine and lake stickleback fish?	**Result: No** →	The 283 amino acids of the *Pitx1* protein are identical in marine and lake stickleback populations.
Test of Hypothesis B:	Are there any differences in the regulation of expression of *Pitx1*?	**Result: Yes** →	Red arrows (→) indicate regions of *Pitx1* gene expression in the photographs below. *Pitx1* is expressed in the ventral spine and mouth regions of developing marine stickleback fish but only in the mouth region of developing lake stickleback fish.

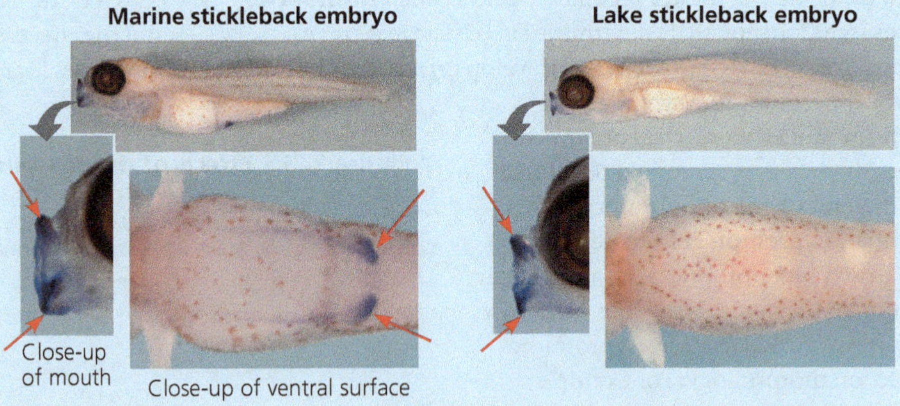

Conclusion The loss or reduction of ventral spines in lake populations of threespine stickleback fish appears to have resulted primarily from a change in the regulation of *Pitx1* gene expression, not from a change in the gene's sequence.

Data from M. D. Shapiro et al., Genetic and developmental basis of evolutionary pelvic reduction in three-spine sticklebacks, *Nature* 428:717–723 (2004).

WHAT IF? ➤ *Describe the set of results that would have led researchers to the conclusion that a change in the coding sequence of the* Pitx1 *gene was more important than a change in regulation of gene expression.*

ocean and in shallow, coastal waters. In western Canada, they also live in lakes formed when the coastline receded during the past 12,000 years. Marine stickleback fish have a pair of spines on their ventral (lower) surface, which deter some predators. These spines are often reduced or absent in stickleback fish living in lakes that lack predatory fishes and that are also low in calcium. Spines may have been lost in such lakes because they are not advantageous in the absence of predators, and the limited calcium is needed for purposes other than constructing spines.

At the genetic level, the developmental gene *Pitx1* was known to influence whether stickleback fish have ventral spines. Was the reduction of spines in some lake populations due to changes in the *Pitx1* gene or to changes in how the gene is expressed **(Figure 25.26)**? The researchers' results indicate that the regulation of gene expression has changed, not the DNA sequence. Moreover, lake stickleback fish do express the *Pitx1* gene in tissues not related to the production of spines (for example, the mouth), illustrating how morphological change can be caused by altering the expression of a developmental gene in some parts of the body but not others. In a 2010 follow-up study, researchers showed that changes to the *Pel* enhancer, a noncoding DNA region that affects expression of the *Pitx1* gene, resulted in the reduction of ventral spines in lake sticklebacks. Overall, results from studies on stickleback fish provide a clear and detailed example of how changes in gene regulation can alter the form of individual organisms and ultimately lead to evolutionary change in populations.

MB HHMI Video: The Making of the Fittest: Evolving Switches, Evolving Bodies (Stickleback) hhmi BioInteractive

CONCEPT CHECK 25.5

1. Explain how new body forms can originate by heterochrony.
2. Why is it likely that *Hox* genes have played a major role in the evolution of novel morphological forms?
3. **MAKE CONNECTIONS ➤** Given that changes in morphology are often caused by changes in the regulation of gene expression, predict whether noncoding DNA is likely to be affected by natural selection. See Concept 18.3 to review noncoding DNA and regulation of gene expression.

For suggested answers, see Appendix A.

CONCEPT 25.6

Evolution is not goal oriented

What does our study of macroevolution tell us about how evolution works? One lesson is that throughout the history of life, the origin of new species has been affected by both the small-scale factors described in Concept 23.3 (such as natural selection operating in populations) and the large-scale factors described in this chapter (such as continental drift promoting bursts of speciation throughout the globe). Moreover, to paraphrase the Nobel Prize–winning geneticist François Jacob, evolution is like tinkering—a process in which new forms arise by the modification of existing structures or existing developmental genes. Over time, such tinkering has led to the three key features of the natural world described on the opening page of Chapter 22: the striking ways in which organisms are suited for life in their environments, the many shared characteristics of life, and the rich diversity of life.

Evolutionary Novelties

François Jacob's view of evolution harkens back to Darwin's concept of descent with modification. As new species form, novel and complex structures can arise as gradual modifications of ancestral structures. In many cases, complex structures have evolved in increments from simpler versions that performed the same basic function. For example, consider the human eye, an intricate organ constructed from numerous parts that work together in forming an image and transmitting it to the brain. How could the human eye have evolved in gradual increments? Some argue that if the eye needs all of its components to function, a partial eye could not have been of use to our ancestors.

The flaw in this argument, as Darwin himself noted, lies in the assumption that only complicated eyes are useful. In fact, many animals depend on eyes that are far less complex than our own. The simplest eyes that we know of are patches of light-sensitive photoreceptor cells. These simple eyes appear to have had a single evolutionary origin and are now found in a variety of animals, including small molluscs called limpets. Such eyes have no equipment for focusing images, but they do enable the animal to distinguish light from dark. Limpets cling more tightly to their rock when a shadow falls on them, a behavioral adaptation that reduces the risk of being eaten **(Figure 25.27)**. Limpets have had a long evolutionary history, demonstrating that their "simple" eyes are quite adequate to support their survival and reproduction.

In the animal kingdom, complex eyes have evolved independently from such basic structures many times. Some molluscs, such as squids and octopuses, have eyes as complex as those

▼ Figure 25.27 Limpets (*Patella vulgata*), molluscs that can sense light and dark with a simple patch of photoreceptor cells.

Figure 25.28 A range of eye complexity among molluscs.

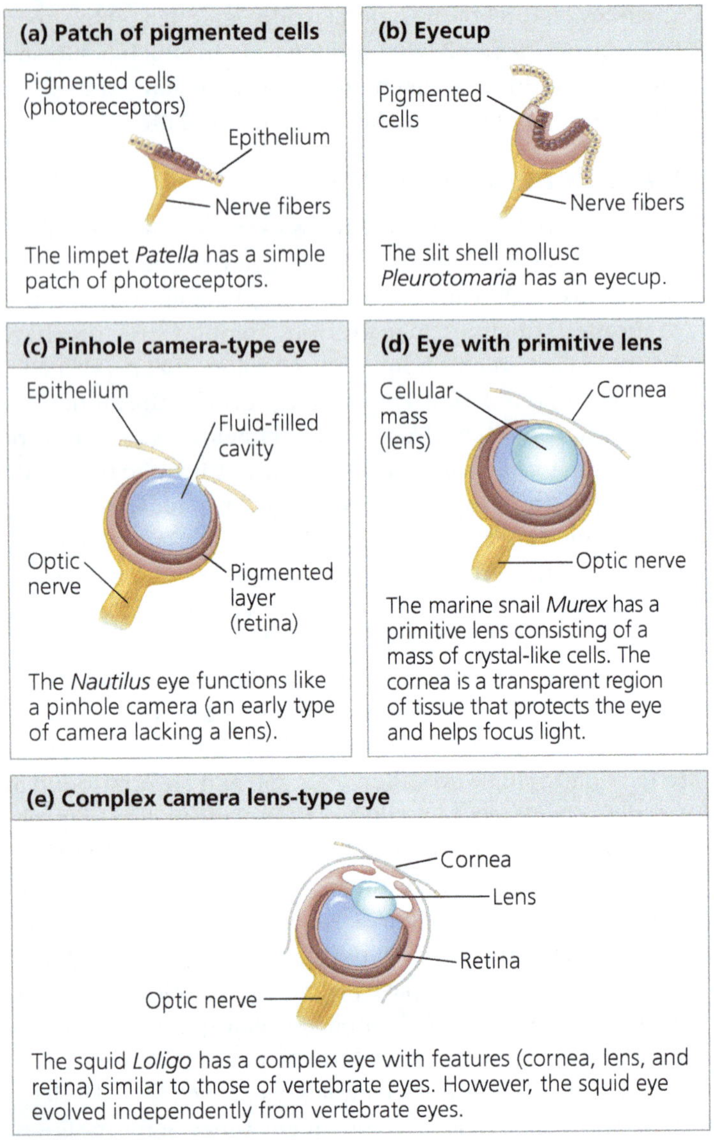

(a) The limpet *Patella* has a simple patch of photoreceptors.

(b) The slit shell mollusc *Pleurotomaria* has an eyecup.

(c) The *Nautilus* eye functions like a pinhole camera (an early type of camera lacking a lens).

(d) The marine snail *Murex* has a primitive lens consisting of a mass of crystal-like cells. The cornea is a transparent region of tissue that protects the eye and helps focus light.

(e) The squid *Loligo* has a complex eye with features (cornea, lens, and retina) similar to those of vertebrate eyes. However, the squid eye evolved independently from vertebrate eyes.

BBC Video: How Did Eyes Evolve?

of humans and other vertebrates **(Figure 25.28)**. Although complex mollusc eyes evolved independently of vertebrate eyes, both evolved from a simple cluster of photoreceptor cells present in a common ancestor. In each case, the complex eye evolved through a series of steps that benefited the eyes' owners at every stage. Evidence of their independent evolution can be found in their structure: Vertebrate eyes detect light at the back layer of the retina and conduct nerve impulses toward the front, while complex mollusc eyes do the reverse.

Throughout their evolutionary history, eyes retained their basic function of vision. But evolutionary novelties can also arise when structures that originally played one role gradually acquire a different one. For example, as cynodonts gave rise to early mammals, bones that formerly comprised the jaw hinge (the articular and quadrate; see Figure 25.7) were incorporated into the ear region of mammals, where they eventually took on a new function: the transmission of sound (see Concept 34.6). Structures that evolve in one context but become co-opted for another function are sometimes called *exaptations* to distinguish them from the adaptive origin of the original structure. Note that the concept of exaptation does not imply that a structure somehow evolves in anticipation of future use. Natural selection cannot predict the future; it can only improve a structure in the context of its *current* utility. Novel features, such as the new jaw hinge and ear bones of early mammals, can arise gradually via a series of intermediate stages, each of which has some function in the organism's current context.

Evolutionary Trends

What else can we learn from patterns of macroevolution? Consider evolutionary "trends" observed in the fossil record. For instance, some evolutionary lineages exhibit a trend toward larger or smaller body size. An example is the evolution of the present-day horse (genus *Equus*), a descendant of the 55-million-year-old *Hyracotherium* **(Figure 25.29)**. About the size of a large dog, *Hyracotherium* had four toes on its front feet, three toes on its hind feet, and teeth adapted for browsing on bushes and trees. In comparison, present-day horses are larger, have only one toe on each foot, and possess teeth modified for grazing on grasses.

Extracting a single evolutionary progression from the fossil record can be misleading, however; it is like describing a bush as growing toward a single point by tracing only the branches that lead to that twig. For example, by selecting certain species from the available fossils, it is possible to arrange a succession of animals intermediate between *Hyracotherium* and living horses that shows a trend toward large, single-toed species (follow the yellow highlighting in Figure 25.29). However, if we consider *all* fossil horses known today, this apparent trend vanishes. The genus *Equus* did not evolve in a straight line; it is the only surviving twig of an evolutionary tree that is so branched that it is more like a bush. *Equus* actually descended through a series of speciation episodes that included several adaptive radiations, not all of which led to large, one-toed, grazing horses. In fact, phylogenetic analyses suggest that all lineages that include grazers are closely related to *Parahippus*; the many other horse lineages, all of which are now extinct, remained multi-toed browsers for 35 million years.

Branching evolution *can* result in a real evolutionary trend even if some species counter the trend. One model of long-term trends views species as analogous to individuals: Speciation is their birth, extinction is their death, and new species that diverge from them are their offspring. In this model, just as populations of individual organisms undergo natural selection, species undergo *species selection*. The species that endure the longest and generate the most new offspring species determine the direction of major evolutionary trends. The species selection model suggests that "differential speciation success" plays a role in macroevolution similar to the role of differential reproductive success in microevolution. Evolutionary trends can

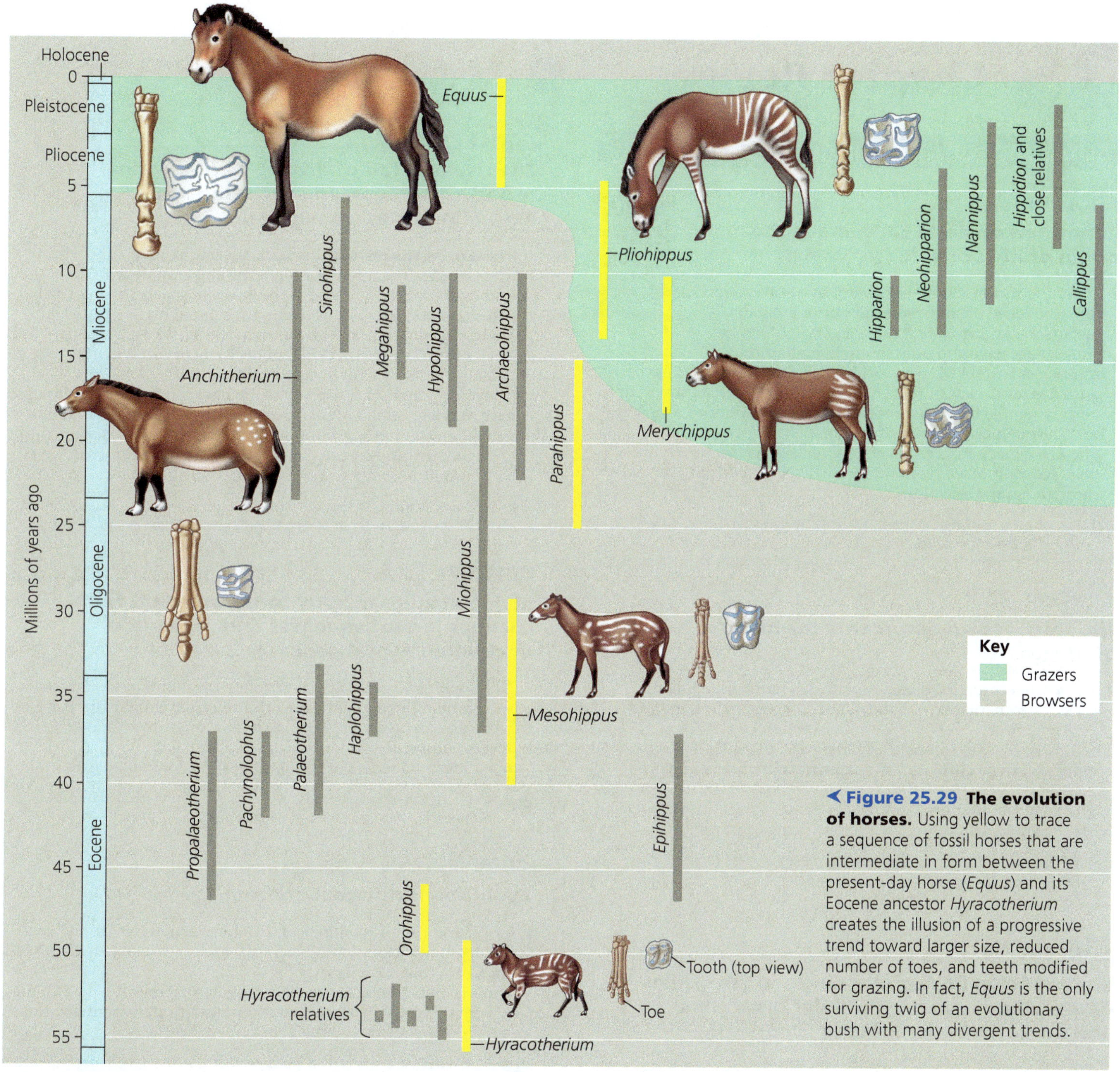

Figure 25.29 The evolution of horses. Using yellow to trace a sequence of fossil horses that are intermediate in form between the present-day horse (*Equus*) and its Eocene ancestor *Hyracotherium* creates the illusion of a progressive trend toward larger size, reduced number of toes, and teeth modified for grazing. In fact, *Equus* is the only surviving twig of an evolutionary bush with many divergent trends.

also result directly from natural selection. For example, when horse ancestors invaded the grasslands that spread during the mid-Cenozoic, there was strong selection for grazers that could escape predators by running faster. This trend would not have occurred without open grasslands.

Whatever its cause, an evolutionary trend does not imply that there is some intrinsic drive toward a particular phenotype. Evolution is the result of the interactions between organisms and their current environments; if environmental conditions change, an evolutionary trend may cease or even reverse itself. The cumulative effect of these ongoing interactions between organisms and their environments is enormous: It is through them that the staggering diversity of life—Darwin's "endless forms most beautiful"—has arisen.

CONCEPT CHECK 25.6

1. How can the Darwinian concept of descent with modification explain the evolution of such complex structures as the vertebrate eye?
2. **WHAT IF?** ➤ The myxoma virus kills up to 99.8% of infected European rabbits in populations with no previous exposure to the virus. The virus is transmitted between living rabbits by mosquitoes. Describe an evolutionary trend (in either the rabbit or virus) that might occur after a rabbit population first encounters the virus.

For suggested answers, see Appendix A.

25 Chapter Review

Go to **MasteringBiology**™ for Videos, Animations, Vocab Self-Quiz, Practice Tests, and more in the Study Area.

SUMMARY OF KEY CONCEPTS

CONCEPT 25.1

Conditions on early Earth made the origin of life possible *(pp. 524–526)*

VOCAB SELF-QUIZ goo.gl/6u55ks

- Experiments simulating possible early atmospheres have produced organic molecules from inorganic precursors. Amino acids, lipids, sugars, and nitrogenous bases have also been found in meteorites.
- Amino acids and RNA nucleotides polymerize when dripped onto hot sand, clay, or rock. Organic compounds can spontaneously assemble into **protocells**, membrane-bounded droplets that have some properties of cells.
- The first genetic material may have self-replicating, catalytic RNA. Early protocells containing such RNA would have increased through natural selection.

? *Describe the roles that montmorillonite clay and vesicles may have played in the origin of life.*

CONCEPT 25.2

The fossil record documents the history of life *(pp. 526–530)*

- The fossil record, based largely on fossils found in sedimentary rocks, documents the rise and fall of different groups of organisms over time.
- Sedimentary strata reveal the relative ages of fossils. The ages of fossils can be estimated by **radiometric dating** and other methods.
- The fossil record shows how new groups of organisms can arise via the gradual modification of preexisting organisms.

? *What are the challenges of estimating the ages of old fossils? Explain how these challenges may be overcome in some circumstances.*

CONCEPT 25.3

Key events in life's history include the origins of unicellular and multicellular organisms and the colonization of land *(pp. 530–535)*

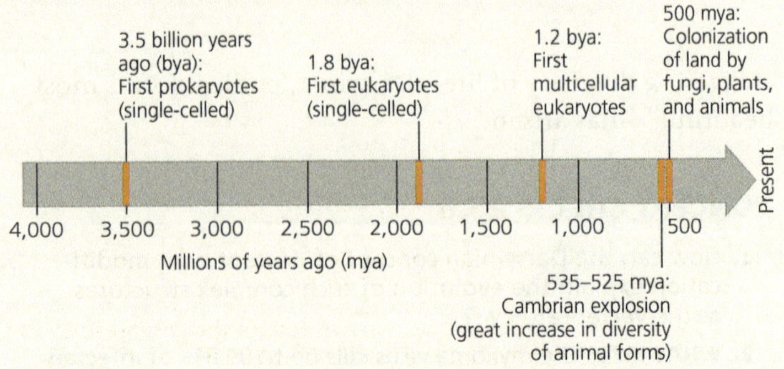

? *What is the "Cambrian explosion," and why is it significant?*

CONCEPT 25.4

The rise and fall of groups of organisms reflect differences in speciation and extinction rates *(pp. 535–542)*

- In **plate tectonics**, continental plates move gradually over time, altering the physical geography and climate of Earth, leading to extinctions in some groups and speciation in others.
- Evolutionary history has been punctuated by five **mass extinctions** that radically altered life's history. Possible causes for these extinctions include continental drift, volcanic activity, and impacts from comets.
- Large increases in the diversity of life have resulted from **adaptive radiations** that followed mass extinctions. Adaptive radiations have also occurred in groups of organisms that possessed major evolutionary innovations or that colonized new regions in which there was little competition from other organisms.

? *Explain how the broad evolutionary changes seen in the fossil record are the cumulative result of speciation and extinction events.*

CONCEPT 25.5

Major changes in body form can result from changes in the sequences and regulation of developmental genes *(pp. 542–545)*

- Developmental genes affect morphological differences between species by influencing the rate, timing, and spatial patterns of change in an organism's form as it develops into an adult.
- The evolution of new forms can be caused by changes in the nucleotide sequences or regulation of developmental genes.

? *How could changes in a single gene or DNA region ultimately lead to the origin of a new group of organisms?*

CONCEPT 25.6

Evolution is not goal oriented *(pp. 545–547)*

- Novel and complex biological structures can evolve through a series of incremental modifications, each of which benefits the organism that possesses it.
- Evolutionary trends can be caused by natural selection in a changing environment or species selection, resulting from interactions between organisms and their current environments.

? *Explain the reasoning behind the statement "Evolution is not goal oriented."*

TEST YOUR UNDERSTANDING

Level 1: Knowledge/Comprehension

PRACTICE TEST goo.gl/CUYGKD

1. Fossilized stromatolites
   (A) formed around deep-sea vents.
   (B) resemble structures formed by bacterial communities that are found today in some shallow marine bays.
   (C) provide evidence that plants moved onto land in the company of fungi around 500 million years ago.
   (D) contain the first undisputed fossils of eukaryotes.

2. The oxygen revolution changed Earth's environment dramatically. Which of the following took advantage of the presence of free oxygen in the oceans and atmosphere?
(A) the evolution of cellular respiration, which used oxygen to help harvest energy from organic molecules
(B) the persistence of some animal groups in anaerobic habitats
(C) the evolution of photosynthetic pigments that protected early algae from the corrosive effects of oxygen
(D) the evolution of chloroplasts after early protists incorporated photosynthetic cyanobacteria

3. Which factor most likely caused animals and plants in India to differ greatly from species in nearby southeast Asia?
(A) The species became separated by convergent evolution.
(B) The climates of the two regions are similar.
(C) India is in the process of separating from the rest of Asia.
(D) India was a separate continent until 45 million years ago.

4. Adaptive radiations can be a direct consequence of three of the following four factors. Select the exception.
(A) vacant ecological niches
(B) genetic drift
(C) colonization of an isolated region that contains suitable habitat and few competitor species
(D) evolutionary innovation

5. Which of the following steps has *not* yet been accomplished by scientists studying the origin of life?
(A) synthesis of small RNA polymers by ribozymes
(B) formation of molecular aggregates with selectively permeable membranes
(C) formation of protocells that use DNA to direct the polymerization of amino acids
(D) abiotic synthesis of organic molecules

Level 2: Application/Analysis

6. A genetic change that caused a certain *Hox* gene to be expressed along the tip of a vertebrate limb bud instead of farther back helped make possible the evolution of the tetrapod limb. This type of change is illustrative of
(A) the influence of environment on development.
(B) paedomorphosis.
(C) a change in a developmental gene or in its regulation that altered the spatial organization of body parts.
(D) heterochrony.

7. A swim bladder is a gas-filled sac that helps fish maintain buoyancy. The evolution of the swim bladder from the air-breathing organ (a simple lung) of an ancestral fish is an example of
(A) exaptation.
(B) changes in *Hox* gene expression.
(C) paedomorphosis.
(D) adaptive radiation.

Level 3: Synthesis/Evaluation

8. **EVOLUTION CONNECTION** Describe how gene flow, genetic drift, and natural selection all can influence macroevolution.

9. **SCIENTIFIC INQUIRY** Herbivory (plant eating) has evolved repeatedly in insects, typically from meat-eating or detritus-feeding ancestors (detritus is dead organic matter). Moths and butterflies, for example, eat plants, whereas their "sister group" (the insect group to which they are most closely related), the caddisflies, feed on animals, fungi, or detritus. As illustrated in the following phylogenetic tree, the combined moth/butterfly and caddisfly group shares a common ancestor with flies and fleas. Like caddisflies, flies and fleas are thought to have evolved from ancestors that did not eat plants.

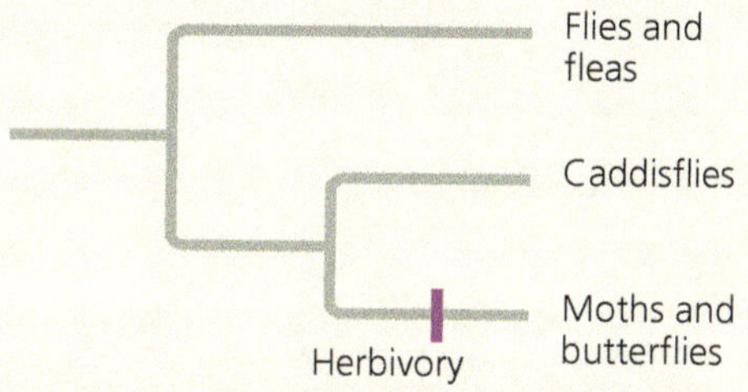

There are 140,000 species of moths and butterflies and 7,000 species of caddisflies. State a hypothesis about the impact of herbivory on adaptive radiations in insects. How could this hypothesis be tested?

10. **WRITE ABOUT A THEME: ORGANIZATION** You have seen many examples of how form fits function at all levels of the biological hierarchy. However, we can imagine forms that would function better than some forms actually found in nature. For example, if the wings of a bird were not formed from its forelimbs, such a hypothetical bird could fly yet also hold objects with its forelimbs. In a short essay (100–150 words), use the concept of "evolution as tinkering" to explain why there are limits to the functionality of forms in nature.

11. **SYNTHESIZE YOUR KNOWLEDGE**

In 2010, the Soufriere Hills volcano on the Caribbean island of Montserrat erupted violently, spewing huge clouds of ash and gases into the sky. Explain how the volcanic eruptions at the end of the Permian period and the formation of Pangaea, both of which occurred about 252 million years ago, set in motion events that altered evolutionary history.

For selected answers, see Appendix A.

For additional practice questions, check out the **Dynamic Study Modules** in MasteringBiology. You can use them to study on your smartphone, tablet, or computer anytime, anywhere!

APPENDIX A Answers

NOTE: Answers to Scientific Skills Exercises, Problem-Solving Exercises, Interpret the Data questions, and short-answer essay questions are available only for instructors in the Instructor Resources area of MasteringBiology. Scientific Skills Exercises, Problem-Solving Exercises, Interpret the Data questions, and additional questions for the Visualizing Figures can be assigned and automatically graded in MasteringBiology.

Chapter 1

Figure Questions

Figure 1.4 The scale bar is about 8.5 mm long, and it corresponds to 1 μm. The prokaryotic cell is about 2 cm = 20 mm long. Dividing by 8.5 mm/scale bar, the length of the prokaryotic cell is about 2.4 scale bars. Each scale bar represents 1 μm, so the prokaryotic cell is about 2.4 μm long. The eukaryotic cell is about 82 mm across (from lower left to upper right) divided by 8.5 mm/scale bar = 9.6 scale bars = 9.6 μm across. **Figure 1.10** The response to insulin is glucose uptake by cells and glucose storage in liver cells. The initial stimulus is high glucose levels, which are reduced when glucose is taken up by cells.
Figure 1.18

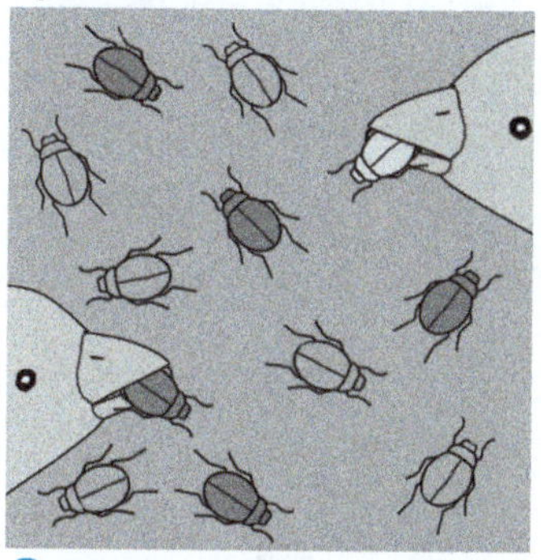

⑤ Environmental change resulting in survival of organisms with different traits

As the soil gradually becomes lighter brown, beetles that match the color of the soil will not be seen by birds and therefore will not be eaten. For example, when the soil is medium-colored, birds will be able to see and eat the darker beetles and any lighter beetles that arise. (Most or all of the lighter beetles will have been eaten earlier, but new light beetles will arise due to variation in the population.) Thus, over time, the population will become lighter as the soil becomes lighter.

Concept Check 1.1

1. Examples: A molecule consists of *atoms* bonded together. Each organelle has an orderly arrangement of *molecules*. Photosynthetic plant cells contain *organelles* called chloroplasts. A tissue consists of a group of similar *cells*. Organs such as the heart are constructed from several *tissues*. A complex multicellular organism, such as a plant, has several types of *organs*, such as leaves and roots. A population is a set of *organisms* of the same species. A community consists of *populations* of the various species inhabiting a specific area. An ecosystem consists of a biological *community* along with the nonliving factors important to life, such as air, soil, and water. The biosphere is made up of all of Earth's *ecosystems*. **2.** (a) New properties emerge at successive levels of biological organization: Structure and function are correlated. (b) Life's processes involve the expression and transmission of genetic information. (c) Life requires the transfer and transformation of energy and matter. **3.** Some possible answers: *Organization (Emergent properties)*: The ability of a human heart to pump blood requires an intact heart; it is not a capability of any of the heart's tissues or cells working alone. *Organization (Structure and function)*: The strong, sharp teeth of a wolf are well suited to grasping and dismembering its prey. *Information*: Human eye color is determined by the combination of genes inherited from the two parents. *Energy and Matter*: A plant, such as a grass, absorbs energy from the sun and transforms it into molecules that act as stored fuel. Animals can eat parts of the plant and use the food for energy to carry out their activities. *Interactions (Molecules)*: When your stomach is full, it signals your brain to decrease your appetite. *Interactions (Ecosystems)*: A mouse eats food, such as nuts or grasses, and deposits some of the food material as wastes (feces and urine). Construction of a nest rearranges the physical environment and may hasten degradation of some of its components. The mouse may also act as food for a predator.

Concept Check 1.2

1. The naturally occurring heritable variation in a population is "edited" by natural selection because individuals with heritable traits better suited to the environment survive and reproduce more successfully than others. Over time, better-suited individuals persist and their percentage in the population increases, while less well-suited individuals become less prevalent—a type of population editing. **2.** Here is one possible explanation: The ancestor species of the green warbler finch lived on an island where insects were a plentiful food source. Among individuals in the ancestor population, there was likely variation in beak shape and size. Individuals with slender, sharp beaks were likely more successful at picking up insects for food. Being well-nourished, they gave rise to more offspring than birds with thick, short beaks. Their many offspring inherited slender, sharp beaks (because of genetic information being passed from generation to generation, although Darwin didn't know this). In each generation, the offspring birds with the beaks of a shape best at picking up insects would eat more and have more offspring. Therefore, the green warbler finch of today has a slender beak that is very well matched (adapted) to its food source, insects.
3.

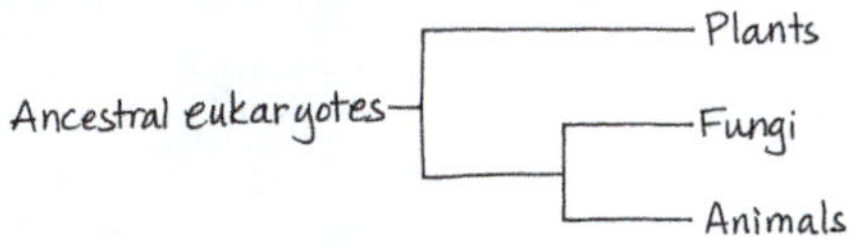

Concept Check 1.3

1. Mouse coat color matches the environment for both beach and inland populations. **2.** Inductive reasoning derives generalizations from specific cases; deductive reasoning predicts specific outcomes from general premises. **3.** Compared to a hypothesis, a scientific theory is usually more general and substantiated by a much greater amount of evidence. Natural selection is an explanatory idea that applies to all kinds of organisms and is supported by vast amounts of evidence of various kinds. **4.** Based on the mouse coloration in Figure 1.25, you might expect that the mice that live on the sandy soil would be lighter in color and those that live on the lava rock would be much darker. And in fact, that is what researchers have found. You would predict that each color of mouse would be less preyed upon in its native habitat than it would be in the other habitat. (Research results also support this prediction.) You could repeat the Hoekstra experiment with colored models, painted to resemble these two types of mouse. Or you could try transplanting some of each population to its non-native habitat and counting how many you can recapture over the next few days, then comparing the four samples as was done in Hoekstra's experiment. (The painted models are easier to recapture, of course!) In the live mouse transplantation experiment, you would have to do controls to eliminate the variable represented by the transplanted mice being in a new, unknown territory. You could control for the transplantation process by transplanting some dark mice from one area of lava rock to one far distant, and some light mice from one area of sandy soil to a distant area.

Concept Check 1.4

1. Science aims to understand natural phenomena and how they work, while technology involves application of scientific discoveries for a particular purpose or to solve a specific problem. **2.** Natural selection could be operating. Malaria is present in sub-Saharan Africa, so there might be an advantage to people with the sickle-cell disease form of the gene that makes them more able to survive and pass on their genes to offspring. Among those of African descent living in the United States, where malaria is absent, there would be no advantage, so they would be selected against more strongly, resulting in fewer individuals with the sickle-cell disease form of the gene.

Summary of Key Concepts Questions

1.1 Finger movements rely on the coordination of the many structural components of the hand (muscles, nerves, bones, etc.), each of which is composed of elements from lower levels of biological *organization* (cells, molecules). The development of the hand relies on the genetic *information* encoded in chromosomes found in cells throughout the body. To power the finger movements that result in a text message, muscle and nerve cells require chemical *energy* that they transform in powering muscle contraction or in propagating nerve impulses. Texting is in essence communication, an *interaction* that conveys information between organisms, in this case of the same species. **1.2** Ancestors of the beach mouse may have exhibited variations in their coat color. Because of the prevalence of visual predators, the better-camouflaged (lighter) mice in the beach habitat may have survived longer and been able to produce more offspring. Over time, a higher and higher proportion of individuals in the population would have had the adaptation of lighter fur that acted to camouflage the mouse in the beach habitat.
1.3 Gathering and interpreting data are core activities in the scientific process, and they are affected by, and affect in turn, three other arenas of the scientific process: exploration and discovery, community analysis and feedback, and societal benefits and outcomes. **1.4** Different approaches taken by scientists studying natural phenomena at different levels complement each other, so more is learned about each problem being studied. A diversity of backgrounds among scientists may lead to fruitful ideas in the same way that important innovations have often arisen where a mix of cultures coexist, due to multiple different viewpoints.

Test Your Understanding

1. B **2.** C **3.** C **4.** B **5.** C **6.** A **7.** D **8.** Your figure should show the following: (1) for the biosphere, the Earth with an arrow coming out of a tropical ocean; (2) for the ecosystem, a distant view of a coral reef; (3) for the community, a collection of reef animals and algae, with corals, fishes, some seaweed, and any other organisms you can think of; (4) for the population, a group of fish of the same species; (5) for the organism, one fish from your population; (6) for the organ, the fish's stomach; (7) for a tissue, a group of similar cells from the stomach; (8) for a cell, one cell from the tissue, showing its nucleus and a few other organelles; (9) for an organelle, the nucleus, where most of the cell's DNA is located; and (10) for a molecule, a DNA double helix. Your sketches can be very rough!

Chapter 2

Figure Questions

Figure 2.7 Atomic number = 12; 12 protons, 12 electrons; 3 electron shells; 2 valence electrons

Figure 2.14 One possible answer:

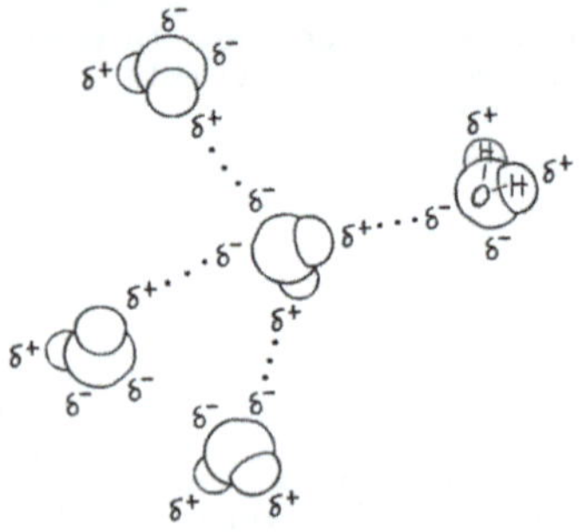

Figure 2.17

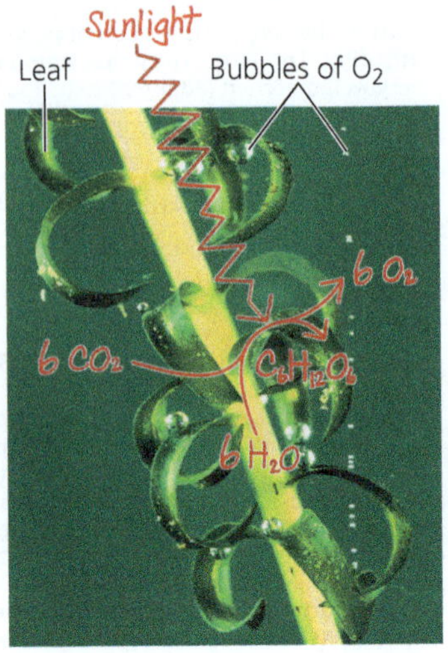

Concept Check 2.1

1. Table salt (sodium chloride) is made up of sodium and chlorine. We are able to eat the compound, showing that it has different properties from those of a metal (sodium) and a poisonous gas (chlorine). **2.** Yes, because an organism requires trace elements, even though only in small amounts **3.** A person with an iron deficiency will probably show fatigue and other effects of a low oxygen level in the blood. (The condition is called anemia and can also result from too few red blood cells or abnormal hemoglobin.) **4.** Variant ancestral plants that could tolerate elevated levels of the elements in serpentine soils could grow and reproduce there. (Plants that were well adapted to nonserpentine soils would not be expected to survive in serpentine areas.) The offspring of the variants would also vary, with those most capable of thriving under serpentine conditions growing best and reproducing most. Over many generations, this probably led to the serpentine-adapted species we see today.

Concept Check 2.2

1. 7 **2.** ${}^{15}_{7}N$ **3.** 9 electrons; two electron shells; 1*s*, 2*s*, 2*p* (three orbitals); 1 electron is needed to fill the valence shell. **4.** The elements in a row all have the same number of electron shells. In a column, all the elements have the same number of electrons in their valence shells.

Concept Check 2.3

1. In this structure, each carbon atom has only three covalent bonds instead of the required four. **2.** The attraction between oppositely charged ions, forming ionic bonds **3.** If you could synthesize molecules that mimic these shapes, you might be able to treat diseases or conditions caused by the inability of affected individuals to synthesize such molecules.

Concept Check 2.4

1.

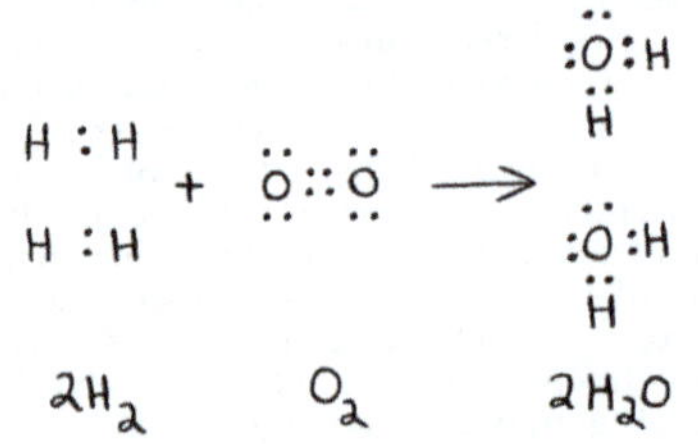

2. At equilibrium, the forward and reverse reactions occur at the same rate. **3.** $C_6H_{12}O_6 + 6\ O_2 \rightarrow 6\ CO_2 + 6\ H_2O$ + Energy. Glucose and oxygen react to form carbon dioxide and water, releasing energy. We breathe in oxygen because we need it for this reaction to occur, and we breathe out carbon dioxide because it is a by-product of this reaction. (This reaction is called cellular respiration, and you will learn more about it in Chapter 9.)

Summary of Key Concepts Questions

2.1 A compound is made up of two or more elements combined in a fixed ratio, while an element is a substance that cannot be broken down to other substances.

2.2

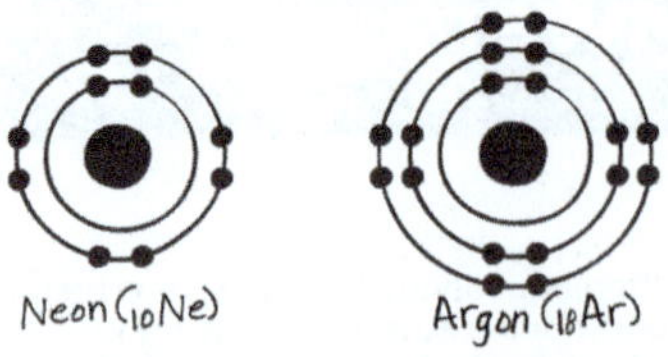

Both neon and argon have completed valence shells, containing 8 electrons. They do not have unpaired electrons that could participate in chemical bonds.

2.3 Electrons are shared equally between the two atoms in a nonpolar covalent bond. In a polar covalent bond, the electrons are drawn closer to the more electronegative atom. In the formation of ions, an electron is completely transferred from one atom to a much more electronegative atom. **2.4** The concentration of products would increase as the added reactants were converted to products. Eventually, an equilibrium would again be reached in which the forward and reverse reactions were proceeding at the same rate and the relative concentrations of reactants and products returned to where they were before the addition of more reactants.

Test Your Understanding

1. A **2.** D **3.** B **4.** A **5.** D **6.** B **7.** C **8.** D

9.

a. 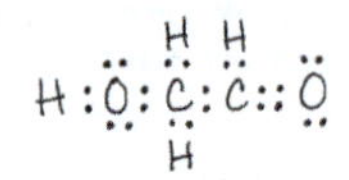

This structure makes sense because all valence shells are complete, and all bonds have the correct number of electrons.

b. 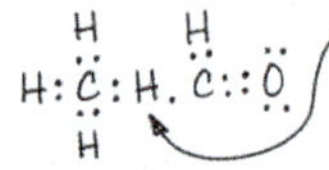

This structure doesn't make sense because H has only 1 electron to share, so it cannot form bonds with 2 atoms.

Chapter 3

Figure Questions

Figure 3.2 One possible answer:

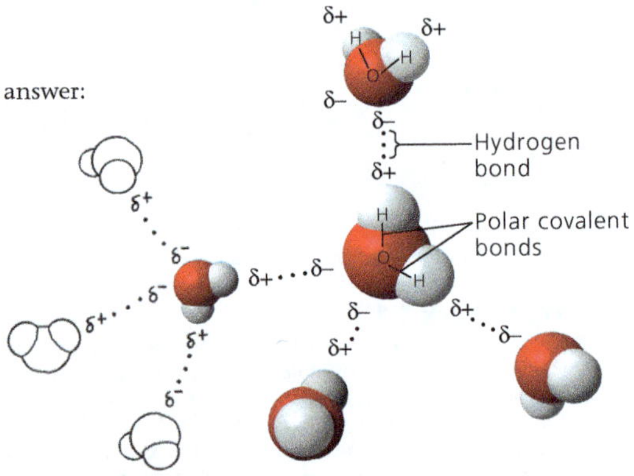

Figure 3.6 Without hydrogen bonds, water would behave like other small molecules, and the solid phase (ice) would be denser than liquid water. The ice would sink to the bottom and would no longer insulate the whole body of water, which would eventually freeze because of the freezing temperatures in the Southern Ocean near Antarctica. The krill would not survive. **Figure 3.8** Heating the solution would cause the water to evaporate faster than it is evaporating at room temperature. At a certain point, there wouldn't be enough water molecules to dissolve the salt ions. The salt would start coming out of solution and re-forming crystals. Eventually, all the water would evaporate, leaving behind a pile of salt like the original pile. **Figure 3.12** Adding excess CO_2 to the oceans ultimately reduces the rate at which calcification (by organisms) can occur.

Concept Check 3.1

1. Electronegativity is the attraction of an atom for the electrons of a covalent bond. Because oxygen is more electronegative than hydrogen, the oxygen atom in H_2O pulls electrons toward itself, resulting in a partial negative charge on the oxygen atom and partial positive charges on the hydrogen atoms. Atoms in neighboring water molecules with opposite partial charges are attracted to each other, forming a hydrogen bond. **2.** Due to its two polar covalent bonds, a water molecule has four regions of partial charge: two positive regions on the two hydrogens and two negative regions on the oxygen atom. Each of these can bind to a region of opposite partial charge on another water molecule. **3.** The hydrogen atoms of one molecule, with their partial positive charges, would repel the hydrogen atoms of the adjacent molecule. **4.** The covalent bonds of water molecules would not be polar, so no regions of the molecule would carry partial charges and water molecules would not form hydrogen bonds with each other.

Concept Check 3.2

1. Hydrogen bonds hold neighboring water molecules together. This cohesion helps chains of water molecules move upward against gravity in water-conducting cells as water evaporates from the leaves. Adhesion between water molecules and the walls of the water-conducting cells also helps counter gravity. **2.** High humidity hampers cooling by suppressing the evaporation of sweat. **3.** As water freezes, it expands because water molecules move farther apart in forming ice crystals. When there is water in a crevice of a boulder, expansion due to freezing may crack the boulder. **4.** The hydrophobic substance repels water, perhaps helping to keep the ends of the legs from becoming coated with water and breaking through the surface. If the legs were coated with a hydrophilic substance, water would be drawn up them, possibly making it more difficult for the water strider to walk on water.

Concept Check 3.3

1. 10^5, or 100,000 **2.** $[H^+] = 0.01\ M = 10^{-2}\ M$, so pH = 2. **3.** $CH_3COOH \rightarrow CH_3COO^- + H^+$. CH_3COOH is the acid (the H^+ donor), and CH_3COO^- is the base (the H^+ acceptor). **4.** The pH of the water should decrease from 7 to about 2 (as mentioned in the text); the pH of the acetic acid solution will decrease only a small amount, because as a weak acid, it acts (like carbonic acid) as a buffer. The reaction shown for question 3 will shift to the left, with CH_3COO^- accepting the influx of H^+ and becoming CH_3COOH molecules.

Summary of Key Concepts Questions

3.1

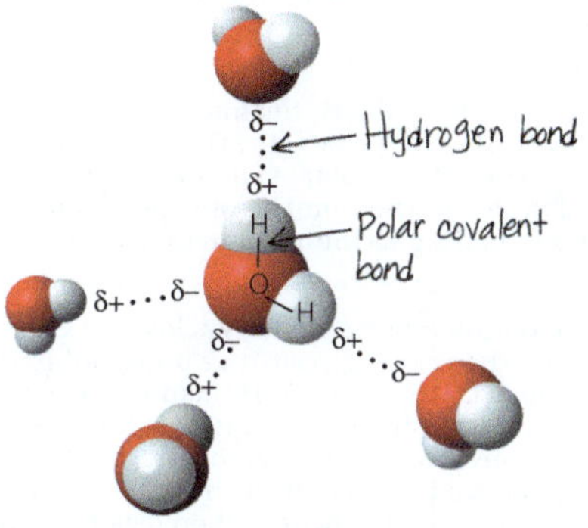

No. A covalent bond is a strong bond in which electrons are shared between two atoms. A hydrogen bond is a weak bond, which does not involve electron sharing, but is simply an attraction between two partial charges on neighboring atoms. **3.2** Ions dissolve in water when polar water molecules form a hydration shell around them, with partially charged regions of water molecules being attracted to ions of the opposite charge. Polar molecules dissolve as water molecules form hydrogen bonds with them and surround them. Solutions are homogeneous mixtures of solute and solvent. **3.3** The concentration of hydrogen ions (H^+) would be 10^{-11}, and the pH of the solution would be 11.

Test Your Understanding

1. C **2.** D **3.** C **4.** A **5.** D

6.

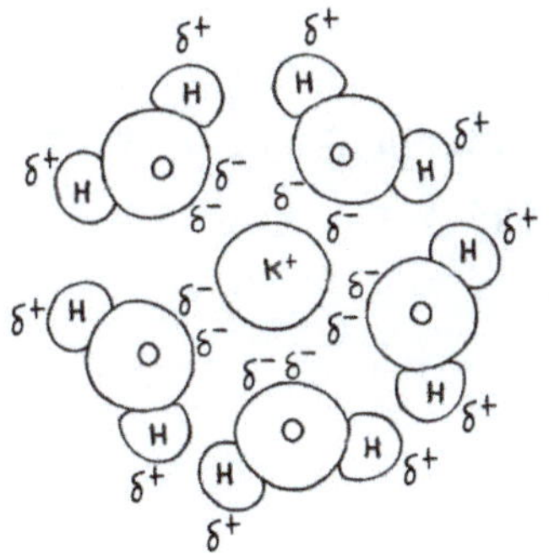

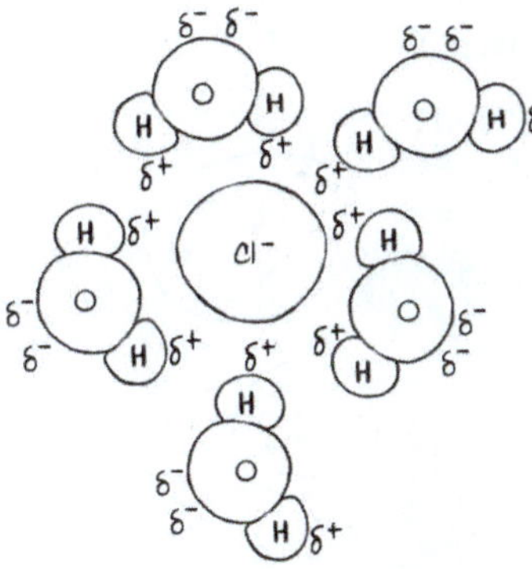

7. Due to intermolecular hydrogen bonds, water has a high specific heat (the amount of heat required to increase the temperature of water by 1°C). When water is heated, much of the heat is absorbed in breaking hydrogen bonds before the water molecules increase their motion and the temperature increases. Conversely, when water is cooled, many H bonds are formed, which releases a significant amount of heat. This release of heat can provide some protection against freezing of the plants' leaves, thus protecting the cells from damage. **8.** Both global warming and ocean acidification are caused by increasing levels of carbon dioxide in the atmosphere, the result of burning fossil fuels.

Chapter 4

Figure Questions

Figure 4.2 Because the concentration of the reactants influences the equilibrium (as discussed in Concept 2.4), there might have been more HCN relative to CH_2O, since there would have been a higher concentration of the reactant gas containing nitrogen.

Figure 4.4

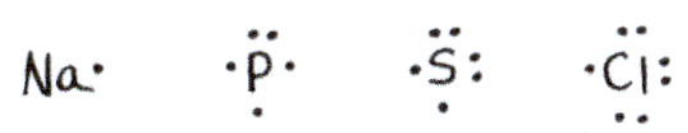

Figure 4.6 The tails of fats contain only carbon-hydrogen bonds, which are relatively nonpolar. Because the tails occupy the bulk of a fat molecule, they make the molecule as a whole nonpolar and therefore incapable of forming hydrogen bonds with water.

Figure 4.7

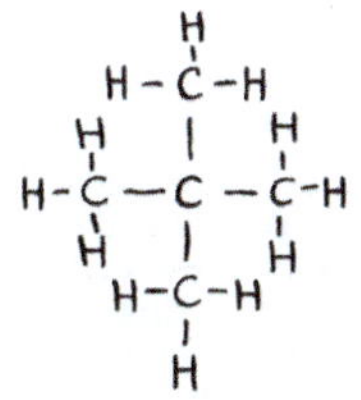

Concept Check 4.1

1. Prior to Wöhler's experiment, the prevailing view was that only living organisms could synthesize "organic" compounds. Wöhler made urea, an organic compound, without the involvement of living organisms. **2.** The sparks provided energy needed for the inorganic molecules in the atmosphere to react with each other. (You'll learn more about energy and chemical reactions in Chapter 8.)

Concept Check 4.2

1.

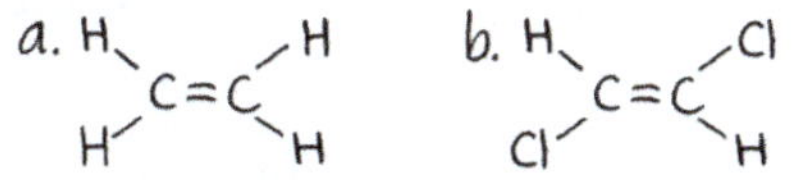

2. The forms of C_4H_{10} in (b) are structural isomers, as are the butenes (forms of C_4H_8) in (c). **3.** Both consist largely of hydrocarbon chains, which provide fuel—gasoline for engines and fats for plant embryos and animals. Reactions of both types of molecules release energy. **4.** No. There is not enough diversity in propane's atoms. It can't form structural isomers because there is only one way for three carbons to attach to each other (in a line). There are no double bonds, so *cis-trans* isomers are not possible. Each carbon has at least two hydrogens attached to it, so the molecule is symmetrical and cannot have enantiomers.

Concept Check 4.3

1. It has both an amino group ($—NH_2$), which makes it an amine, and a carboxyl group (—COOH), which makes it a carboxylic acid. **2.** The ATP molecule loses a phosphate, becoming ADP.

3.

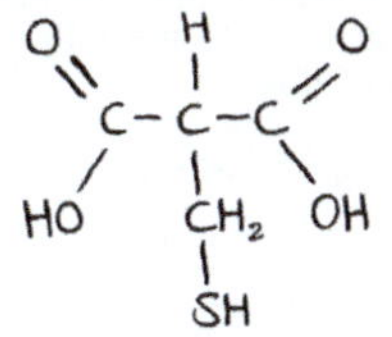

A chemical group that can act as a base has been replaced with a group that can act as an acid, increasing the acidic properties of the molecule. The shape of the molecule would also change, likely changing the molecules with which it can interact. The original cysteine molecule has an asymmetric carbon in the center. After replacement of the amino group with a carboxyl group, this carbon is no longer asymmetric.

Summary of Key Concepts Questions

4.1 Miller showed that organic molecules could form under the physical and chemical conditions estimated to have been present on early Earth. This abiotic synthesis of organic molecules would have been a first step in the origin of life. **4.2** Acetone and propanal are structural isomers. Acetic acid and glycine have no asymmetric carbons, whereas glycerol phosphate has one. Therefore, glycerol phosphate can exist as forms that are enantiomers, but acetic acid and glycine cannot. **4.3** The methyl group is nonpolar and not reactive. The other six groups are called functional groups because they can participate in chemical reactions. Also, all except the sulfhydryl group are hydrophilic, increasing the solubility of organic compounds in water.

Test Your Understanding

1. B **2.** B **3.** C **4.** C **5.** A **6.** B **7.** A **8.** The molecule on the right; the middle carbon is asymmetric.

9.

Silicon has 4 valence electrons, the same number as carbon. Therefore, silicon would be able to form long chains, including branches, that could act as skeletons for large molecules. It would clearly do this much better than neon (with no valence electrons) or aluminum (with 3 valence electrons).

Chapter 5

Figure Questions

Figure 5.3 Glucose and fructose are structural isomers.

Figure 5.4

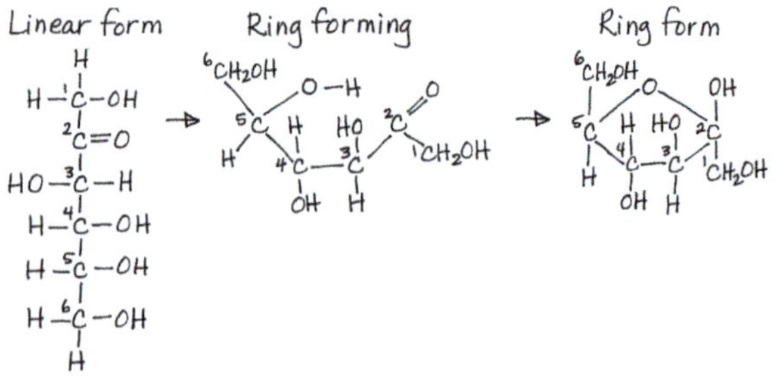

Note that the oxygen on carbon 5 lost its proton and that the oxygen on carbon 2, which used to be the carbonyl oxygen, gained a proton. Four carbons are in the fructose ring, and two are not. (The latter two carbons are attached to carbons 2 and 5, which are in the ring.) The fructose ring differs from the glucose ring, which has five carbons in the ring and one that is not. (Note that the orientation of this fructose molecule is flipped horizontally relative to that of the one in Figure 5.5b.)

Figure 5.5

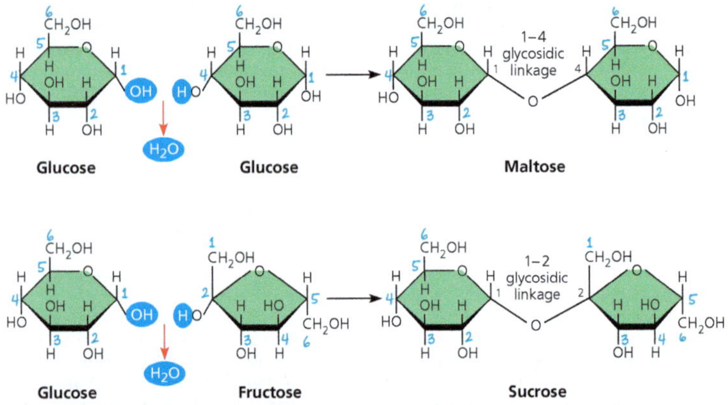

(a) In maltose, the linkage is called a 1–4 glycosidic linkage because the number 1 carbon in the left monosaccharide (glucose) is linked to the number 4 carbon in the right monosaccharide (also glucose). (b) In sucrose, the linkage is called a 1–2 glycosidic linkage because the number 1 carbon in the left monosaccharide (glucose) is linked to the number 2 carbon in the right monosaccharide (fructose). (Note that the fructose molecule is oriented differently from glucose in Figure 5.5b and from the fructose shown in the answer for Figure 5.4, above. In Figure 5.5b and in this answer, carbon 2 of fructose is close to carbon 1 of glucose.)

Figure 5.11

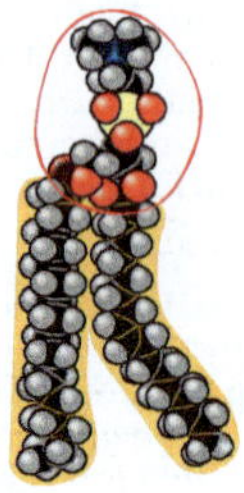

Figure 5.12

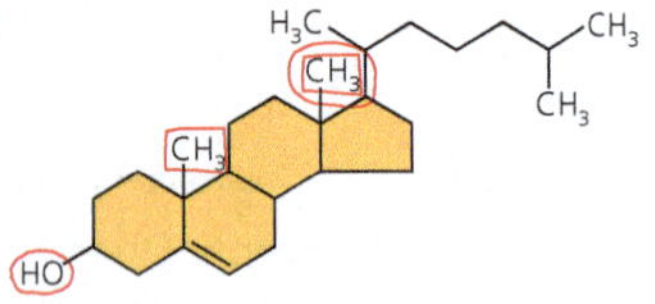

Figure 5.15

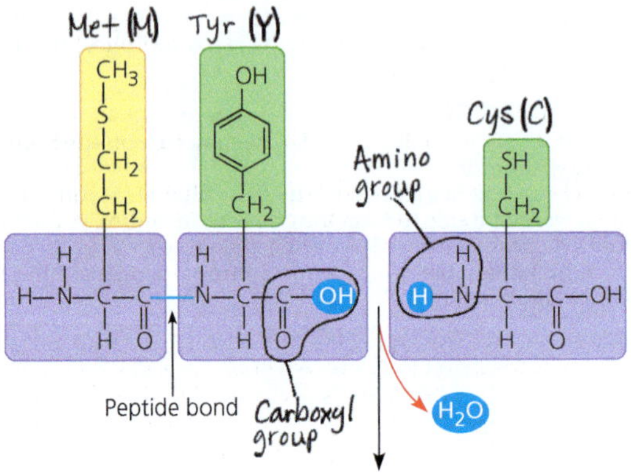

Figure 5.16 (1) The polypeptide backbone is most easily followed in the ribbon model.
(2)

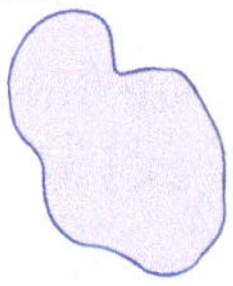

(3) The point of this diagram is to show that a pancreas cell secretes insulin proteins, so the shape is not important to the process being illustrated. **Figure 5.17** We can see that their complementary shapes allow the two proteins to fit together quite precisely. **Figure 5.19** The R group on glutamic acid is acidic and hydrophilic, whereas that on valine is nonpolar and hydrophobic. Therefore, it is unlikely that valine and glutamic acid participate in the same intramolecular interactions. A change in these interactions could (and does) cause a disruption of molecular structure. **Figure 5.26** Using a genomics approach allows us to use gene sequences to identify species and to learn about evolutionary relationships among any two species. This is because all species are related by their evolutionary history, and the evidence is in the DNA sequences. Proteomics—looking at proteins that are expressed—allows us to learn about how organisms or cells are functioning at a given time or in an association with another species.

Concept Check 5.1

1. The four main classes are proteins, carbohydrates, lipids, and nucleic acids. Lipids are not polymers. **2.** Nine, with one water molecule required to hydrolyze each connection between adjacent monomers **3.** The amino acids in the fish protein must be released in hydrolysis reactions and incorporated into other proteins in dehydration reactions.

Concept Check 5.2

1. $C_3H_6O_3$ **2.** $C_{12}H_{22}O_{11}$ **3.** The antibiotic treatment is likely to have killed the cellulose-digesting prokaryotes in the cow's gut. The absence of these prokaryotes would hamper the cow's ability to obtain energy from food and could lead to weight loss and possibly death. Thus, prokaryotic species are reintroduced, in appropriate combinations, in the gut culture given to treated cows.

Concept Check 5.3

1. Both have a glycerol molecule attached to fatty acids. The glycerol of a fat has three fatty acids attached, whereas the glycerol of a phospholipid is attached to two fatty acids and one phosphate group. **2.** Human sex hormones are steroids, a type of compound that is hydrophobic and thus classified as a lipid. **3.** The oil droplet membrane could consist of a single layer of phospholipids rather than a bilayer, because an arrangement in which the hydrophobic tails of the membrane phospholipids were in contact with the hydrocarbon regions of the oil molecules would be more stable.

Concept Check 5.4

1. Secondary structure involves hydrogen bonds between atoms of the polypeptide backbone. Tertiary structure involves interactions between atoms of the side chains of the amino acid subunits. **2.** The two ring forms of glucose are called α and β, depending on how the glycosidic bond dictates the position of a hydroxyl group. Proteins have α helices and β pleated sheets, two types of repeating structures found in polypeptides due to interactions between the repeating constituents of the chain (not the side chains). The hemoglobin molecule is made up of two types of polypeptides: It contains two molecules each of α-globin and β-globin. **3.** These are all nonpolar, hydrophobic amino acids, so you would expect this region to be located in the interior of the folded polypeptide, where it would not contact the aqueous environment inside the cell.

Concept Check 5.5

1.

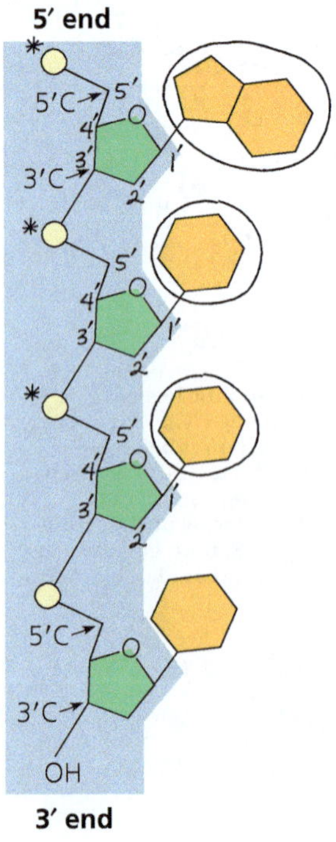

2.

5′–T A G G C C T–3′
3′–A T C C G G A–5′

Concept Check 5.6

1. The DNA of an organism encodes all of its proteins, and proteins are the molecules that carry out the work of cells, whether an organism is unicellular or multicellular. By knowing the DNA sequence of an organism, scientists would be able to catalog the protein sequences as well. **2.** Ultimately, the DNA sequence carries the information necessary to make the proteins that determine the traits of a particular species. Because the traits of the two species are similar, you would expect the proteins to be similar as well, and therefore the gene sequences should also have a high degree of similarity.

Summary of Key Concepts Questions

Concept 5.1 The polymers of large carbohydrates (polysaccharides), proteins, and nucleic acids are built from three different types of monomers (monosaccharides, amino acids, and nucleotides, respectively). **Concept 5.2** Both starch and cellulose are polymers of glucose, but the glucose monomers are in the α configuration in starch and the β configuration in cellulose. The glycosidic linkages thus have different geometries, giving the polymers different shapes and thus different properties. Starch is an energy-storage compound in plants; cellulose is a structural component of plant cell walls. Humans can hydrolyze starch to provide energy but cannot hydrolyze cellulose. Cellulose aids in the passage of food through the digestive tract. **Concept 5.3** Lipids are not polymers because they do not exist as a chain of linked monomers. They are not considered macromolecules because they do not reach the giant size of many polysaccharides, proteins, and nucleic acids. **Concept 5.4** A polypeptide, which may consist of hundreds of amino acids in a specific sequence (primary structure), has regions of coils and pleats (secondary structure), which are then folded into irregular contortions (tertiary structure) and may be noncovalently associated with other polypeptides (quaternary structure). The linear order of amino acids, with the varying properties of their side chains (R groups), determines what secondary and tertiary structures will form to produce a protein. The resulting unique three-dimensional shapes of proteins are key to their specific and diverse functions. **Concept 5.5** The complementary base pairing of the two strands of DNA makes possible the precise replication of DNA every time a cell divides, ensuring that genetic information is faithfully transmitted. In some types of RNA, complementary base pairing enables RNA molecules to assume specific three-dimensional shapes that facilitate diverse functions. **Concept 5.6** You would expect the human gene sequence to be most similar to that of the mouse (another mammal), then to that of the fish (another vertebrate), and least similar to that of the fruit fly (an invertebrate).

Test Your Understanding

1. D **2.** A **3.** B **4.** A **5.** B **6.** B **7.** C

8.

	Monomers or Components	Polymer or larger molecule	Type of linkage
Carbohydrates	Monosaccharides	Polysaccharides	Glycosidic linkages
Fats	Fatty acids	Triacylglycerols	Ester linkages
Proteins	Amino acids	Polypeptides	Peptide bonds
Nucleic acids	Nucleotides	Polynucleotides	Phosphodiester linkages

9.

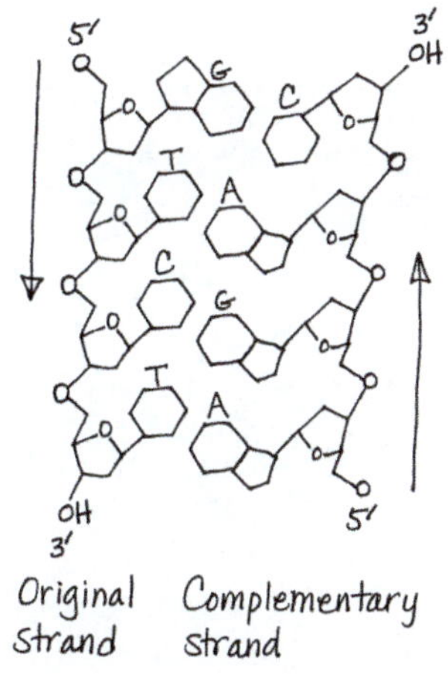

Chapter 6

Figure Questions

Figure 6.3 The cilia in the upper left were oriented lengthwise in the plane of the slice, while those on the right were oriented perpendicular to the plane of the slice. Therefore the former were cut in longitudinal section, and the latter in cross section. **Figure 6.4** You would use the pellet from the final fraction, which is rich in ribosomes. These are the sites of protein translation. **Figure 6.6** The dark bands in the TEM correspond to the hydrophilic heads of the phospholipids, while the light band corresponds to the hydrophobic fatty acid tails of the phospholipids. **Figure 6.9** The DNA in a chromosome dictates synthesis of a messenger RNA (mRNA) molecule, which then moves out to the cytoplasm. There, the information is used for the production, on ribosomes, of proteins that carry out cellular functions. **Figure 6.10** Any of the bound ribosomes (attached to the endoplasmic reticulum) could be circled, because any could be making a protein that will be secreted. **Figure 6.22** Each centriole has 9 sets of 3 microtubules, so the entire centrosome (two centrioles) has 54 microtubules. Each microtubule consists of a helical array of tubulin dimers (as shown in Table 6.1).

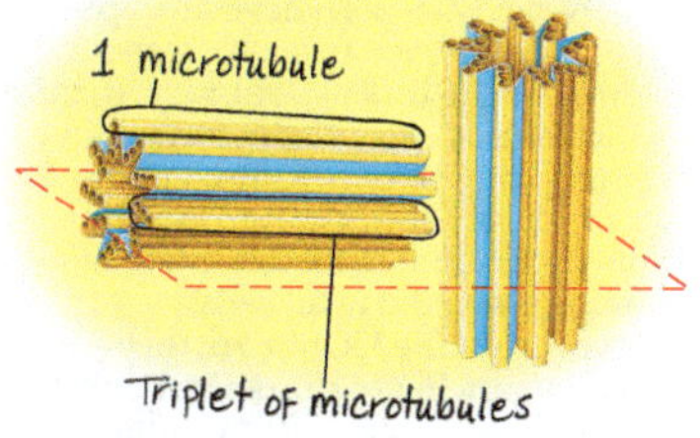

Figure 6.24 The two central microtubules terminate above the basal body, so they aren't present at the level of the cross section through the basal body, indicated by the lower red rectangle shown in the EM on the left.

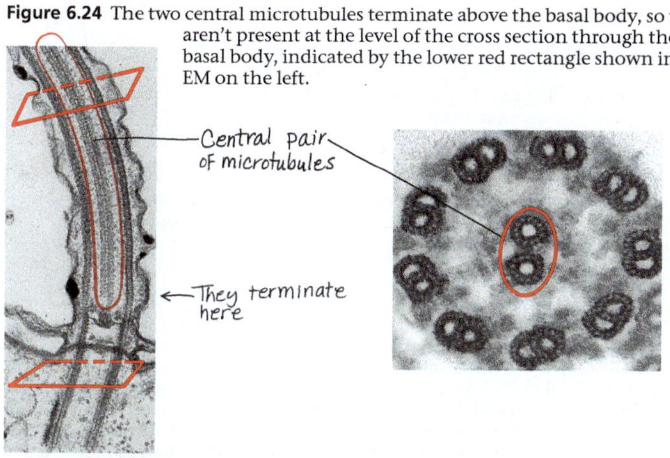

Figure 6.32 (1) nuclear pore, ribosome, proton pump, Cyt *c*. (2) As shown in the figure, the enzyme RNA polymerase moves along the DNA, transcribing the genetic information into an mRNA molecule. Given that RNA polymerase is somewhat larger than a nucleosome, the enzyme would not be able to fit between the histone proteins of the nucleosome and the DNA itself. Thus, the group of histone proteins must be separated from or moved along the DNA somehow in order for the RNA polymerase enzyme to access the DNA. (3) A mitochondrion.

Concept Check 6.1

1. Stains used for light microscopy are colored molecules that bind to cell components, affecting the light passing through, while stains used for electron microscopy involve heavy metals that affect the beams of electrons. **2.** (a) Light microscope, (b) scanning electron microscope

Concept Check 6.2

1. See Figure 6.8.
2.

This cell would have the same volume as the cells in columns 2 and 3 in Figure 6.7 but proportionally more surface area than that in column 2 and less than that in column 3. Thus, the surface-to-volume ratio should be greater than 1.2 but less than 6. To obtain the surface area, you would add the area of the six sides (the top, bottom, sides, and ends): $125 + 125 + 125 + 125 + 1 + 1 = 502$. The surface-to-volume ratio equals 502 divided by a volume of 125, or 4.0.

Concept Check 6.3

1. Ribosomes in the cytoplasm translate the genetic message, carried from the DNA in the nucleus by mRNA, into a polypeptide chain. **2.** Nucleoli consist of DNA and the ribosomal RNAs (rRNAs) made according to its instructions, as well as proteins imported from the cytoplasm. Together, the rRNAs and proteins are assembled into large and small ribosomal subunits. (These are exported through nuclear pores to the cytoplasm, where they will participate in polypeptide synthesis.) **3.** Each chromosome consists of one long DNA molecule attached to numerous protein molecules, a combination called chromatin. As a cell begins division, each chromosome becomes "condensed" as its diffuse mass of chromatin coils up.

Concept Check 6.4

1. The primary distinction between rough and smooth ER is the presence of bound ribosomes on the rough ER. Both types of ER make phospholipids, but membrane proteins and secretory proteins are all produced by the ribosomes on the rough ER. The smooth ER also functions in detoxification, carbohydrate metabolism, and storage of calcium ions. **2.** Transport vesicles move membranes and the substances they enclose between other components of the endomembrane system. **3.** The mRNA is synthesized in the nucleus and then passes out through a nuclear pore to the cytoplasm, where it is translated on a bound ribosome, attached to the rough ER. The protein is synthesized into the lumen of the ER and perhaps modified there. A transport vesicle carries the protein to the Golgi apparatus. After further modification in the Golgi, another transport vesicle carries it back to the ER, where it will perform its cellular function.

Concept Check 6.5

1. Both organelles are involved in energy transformation, mitochondria in cellular respiration and chloroplasts in photosynthesis. They both have multiple membranes that separate their interiors into compartments. In both organelles, the innermost membranes—cristae, or infoldings of the inner membrane, in mitochondria and the thylakoid membranes in chloroplasts—have large surface areas with embedded enzymes that carry out their main functions. **2.** Yes. Plant cells are able to make their own sugar by photosynthesis, but mitochondria in these eukaryotic cells are the organelles that are able to generate ATP molecules to be used for energy generation from sugars, a function required in all cells.
3. Mitochondria and chloroplasts are not derived from the ER, nor are they connected physically or via transport vesicles to organelles of the endomembrane system. Mitochondria and chloroplasts are structurally quite different from vesicles derived from the ER, which are bounded by a single membrane.

Concept Check 6.6

1. Dynein arms, powered by ATP, move neighboring doublets of microtubules relative to each other. Because they are anchored within the flagellum or cilium and with respect to one another, the doublets bend instead of sliding past each other. Synchronized bending of the nine microtubule doublets brings about bending of both cilia and flagella. **2.** Such individuals have defects in the microtubule-based movement of cilia and flagella. Thus, the sperm can't move because of malfunctioning or nonexistent flagella, and the airways are compromised because cilia that line the trachea malfunction or don't exist, and so mucus cannot be cleared from the lungs.

Concept Check 6.7

1. The most obvious difference is the presence of direct cytoplasmic connections between cells of plants (plasmodesmata) and animals (gap junctions). These connections result in the cytoplasm being continuous between adjacent cells. **2.** The cell would not be able to function properly and would probably soon die, as the cell wall or ECM must be permeable to allow the exchange of matter between the cell and its external environment. Molecules involved in energy production and use must be allowed entry, as well as those that provide information about the cell's environment. Other molecules, such as products synthesized by the cell for export and the by-products of cellular respiration, must be allowed to exit. **3.** The parts of the protein that face aqueous regions would be expected to have polar or charged (hydrophilic) amino acids, while the parts that go through the membrane would be expected to have nonpolar (hydrophobic) amino acids. You would predict polar or charged amino acids at each end (tail), in the region of the cytoplasmic loop, and in the regions of the two extracellular loops. You would predict nonpolar amino acids in the four regions that go through the membrane between the tails and loops.

Concept Check 6.8

1. *Colpidium colpoda* moves around in freshwater using cilia, projections from the plasma membrane that enclose microtubules in a "9 + 2" arrangement. The interactions between motor proteins and microtubules cause the cilia to bend synchronously, propelling the cell through the water. This is powered by ATP, obtained via breaking down sugars from food in a process that occurs in mitochondria. *C. colpoda* obtains bacteria as their food source, maybe via the same process (involving filopodia) the macrophage uses in Figure 6.31. This process uses actin filaments and other elements of the cytoskeleton to ingest the bacteria. Once ingested, the bacteria are broken down by enzymes in lysosomes. The proteins involved in all of these processes are encoded by genes on DNA in the nucleus of the *C. colpoda*.

Summary of Key Concepts Questions

6.1 Both light and electron microscopy allow cells to be studied visually, thus helping us understand internal cellular structure and the arrangement of cell components. Cell fractionation techniques separate out different groups of cell components, which can then be analyzed biochemically to determine their function. Performing microscopy on the same cell fraction helps to correlate the biochemical function of the cell with the cell component responsible.
6.2 The separation of different functions in different organelles has several advantages. Reactants and enzymes can be concentrated in one area instead of spread throughout the cell. Reactions that require specific conditions, such as a lower pH, can be compartmentalized. And enzymes for specific reactions are often embedded in the membranes that enclose or partition an organelle.
6.3 The nucleus contains the genetic material of the cell in the form of DNA, which codes for messenger RNA, which in turn provides instructions for the synthesis of proteins (including the proteins that make up part of the ribosomes). DNA also codes for ribosomal RNAs, which are combined with proteins in the nucleolus into the subunits of ribosomes. Within the cytoplasm, ribosomes join with mRNA to build polypeptides, using the genetic information in the mRNA. **6.4** Transport vesicles move proteins and membranes synthesized by the rough ER to the Golgi for further processing and then to the plasma membrane, lysosomes, or other locations in the cell, including back to the ER.
6.5 According to the endosymbiont theory, mitochondria originated from an oxygen-using prokaryotic cell that was engulfed by an ancestral eukaryotic cell. Over time, the host and endosymbiont evolved into a single unicellular organism. Chloroplasts originated when at least one of these eukaryotic cells containing mitochondria engulfed and then retained a photosynthetic prokaryote. **6.6** Inside the cell, motor proteins interact with components of the cytoskeleton to move cellular parts. Motor proteins "walk" vesicles along microtubules. The movement of cytoplasm within a cell involves interactions of the motor protein myosin and microfilaments (actin filaments). Whole cells can be moved by the rapid bending of flagella or cilia, which is caused by the motor-protein-powered sliding of microtubules within these structures. Cell movement can also occur when pseudopodia form at one end of a cell (caused by actin polymerization into a filamentous network), followed by contraction of the cell toward that end; this amoeboid movement is powered by interactions of microfilaments with myosin. Interactions of motor proteins and microfilaments in muscle cells can cause muscle contraction that can propel whole organisms (for example, by walking or swimming). **6.7** A plant cell wall is primarily composed of microfibrils of cellulose embedded in other polysaccharides and proteins. The ECM of animal cells is primarily composed of collagen and other protein fibers, such as fibronectin and other glycoproteins. These fibers are embedded in a network of carbohydrate-rich proteoglycans. A plant cell wall provides structural support for the cell and, collectively, for the plant body. In addition to giving support, the ECM of an animal cell allows for communication of environmental changes into the cell. **6.8** The nucleus houses the chromosomes; each is made up of proteins and a single DNA molecule. The genes that exist along the DNA carry the genetic information necessary to make the proteins involved in ingesting a bacterial cell, such as the actin of microfilaments that form pseudopodia (filopodia), the proteins in the mitochondria responsible for providing the necessary ATP, and the enzymes present in the lysosomes that will digest the bacterial cell.

Test Your Understanding

1. B **2.** C **3.** B **4.** A **5.** D **6.** See Figure 6.8.

Chapter 7

Figure Questions

Figure 7.2

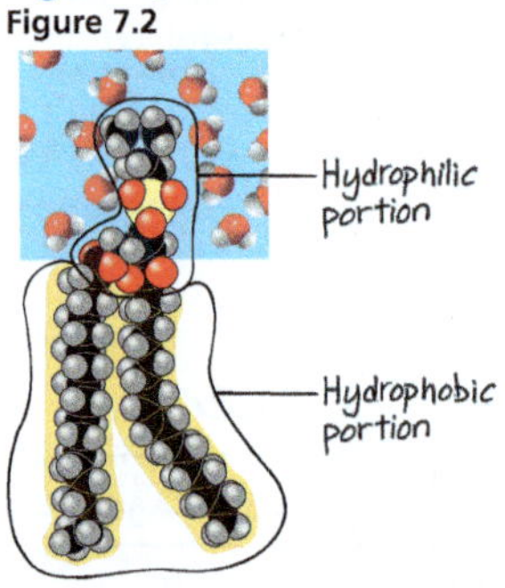

The hydrophilic portion is in contact with an aqueous environment (cytosol or extracellular fluid), and the hydrophobic portion is in contact with the hydrophobic portions of other phospholipids in the interior of the bilayer. **Figure 7.4** You couldn't rule out movement of proteins within membranes of the same species. You might propose that the membrane lipids and proteins from one species weren't able to mingle with those from the other species because of some incompatibility. **Figure 7.7** A transmembrane protein like the dimer in (f) might change its shape upon binding to a particular extracellular matrix (ECM) molecule. The new shape might enable the interior portion of the protein to bind to a second, cytoplasmic protein that would relay the message to the inside of the cell, as shown in (c). **Figure 7.8** The shape of a protein on the HIV surface is likely to be complementary to the shape of the receptor (CD4) and also to that of the co-receptor (CCR5). A molecule with a shape similar to that of the HIV surface protein could bind CCR5, blocking HIV binding. (Another answer would be a molecule that bound to CCR5 and changed the shape of CCR5 so it could no longer bind HIV; in fact, this is how maraviroc works.)
Figure 7.9

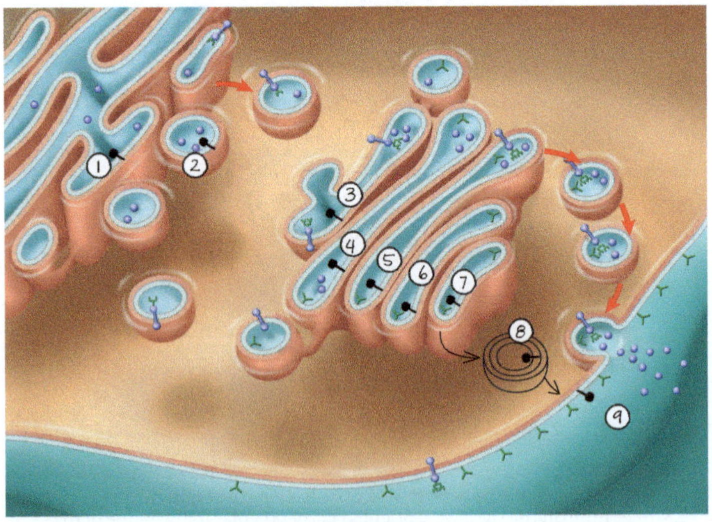

The protein would contact the extracellular fluid. (Because one end of the protein is in the ER membrane, no part of the protein extends into the cytoplasm.) The part of the protein not in the membrane extends into the ER lumen. Once the vesicle fuses with the plasma membrane, the "inside" of the ER membrane, facing the lumen, will become the "outside" of the plasma membrane, facing the extracellular fluid. **Figure 7.11** The orange dye would be evenly distributed throughout the solution on both sides of the membrane. The solution levels would not be affected because the orange dye can diffuse through the membrane and equalize its concentration. Thus, no additional osmosis would take place in either direction. **Figure 7.16** The diamond solutes are moving into the cell (down), and the round solutes are moving out of the cell (up); each is moving against its concentration gradient. **Figure 7.19** (a) In the micrograph of the algal cell, the diameter of the algal cell is about 2.3 times longer than the scale bar, which represents 5 µm, so the diameter of the algal cell is about 11.5 µm. (b) In the micrograph of the coated vesicle, the diameter of the coated vesicle is about 1.2 times longer than the scale bar, which represents 0.25 µm, so the diameter of the coated vesicle is about 0.3 µm. (c) Therefore, the food vacuole around the algal cell will be about 40× larger than the coated vesicle.

Concept Check 7.1

1. They are on the inside of the transport vesicle membrane. **2.** The grasses living in the cooler region would be expected to have more unsaturated fatty acids in their membranes because those fatty acids remain fluid at lower temperatures. The grasses living immediately adjacent to the hot springs would be expected to have more saturated fatty acids, which would allow the fatty acids to "stack" more closely, making the membranes less fluid and therefore helping them to stay intact at higher temperatures. (In plants, cholesterol is generally not used to moderate the effects of temperature on membrane fluidity because it is found at vastly lower levels in membranes of plant cells than in those of animal cells.)

Concept Check 7.2

1. O_2 and CO_2 are both nonpolar molecules that can easily pass through the hydrophobic interior of a membrane. **2.** Water is a polar molecule, so it cannot pass very rapidly through the hydrophobic region in the middle of a phospholipid bilayer. **3.** The hydronium ion is charged, while glycerol is not. Charge is probably more significant than size as a basis for exclusion by the aquaporin channel.

Concept Check 7.3

1. CO_2 is a nonpolar molecule that can diffuse through the plasma membrane. As long as it diffuses away so that the concentration remains low outside the cell, it will continue to exit the cell in this way. (This is the opposite of the case for O_2, described in this section of the text.) **2.** The activity of *Paramecium*'s contractile vacuole will decrease. The vacuole pumps out excess water that accumulates in the cell; this accumulation occurs only in a hypotonic environment.

Concept Check 7.4

1. These pumps use ATP. To establish a voltage, ions have to be pumped against their gradients, which requires energy. **2.** Each ion is being transported against its electrochemical gradient. If either ion were transported down its electrochemical gradient, this *would* be considered cotransport. **3.** The internal environment of a lysosome is acidic, so it has a higher concentration of H^+ than does the cytoplasm. Therefore, you might expect the membrane of the lysosome to have a proton pump such as that shown in Figure 7.17 to pump H^+ into the lysosome.

Concept Check 7.5

1. Exocytosis. When a transport vesicle fuses with the plasma membrane, the vesicle membrane becomes part of the plasma membrane.

2.

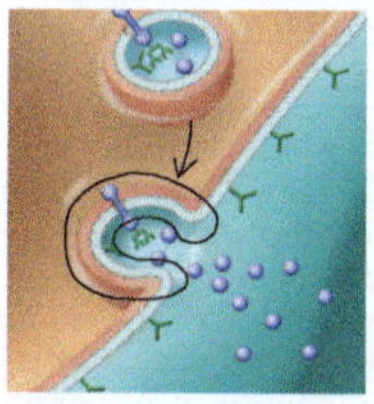

3. The glycoprotein would be synthesized in the ER lumen, move through the Golgi apparatus, and then travel in a vesicle to the plasma membrane, where it would undergo exocytosis and become part of the ECM.

Summary of Key Concepts Questions

7.1 Plasma membranes define the cell by separating the cellular components from the external environment. This allows conditions inside cells to be controlled by membrane proteins, which regulate entry and exit of molecules and even cell function (see Figure 7.7). The processes of life can be carried out inside the controlled environment of the cell, so membranes are crucial. In eukaryotes, membranes also function to subdivide the cytoplasm into different compartments where distinct processes can occur, even under differing conditions such as low or high pH. **7.2** Aquaporins are channel proteins that greatly increase the permeability of a membrane to water molecules, which are polar and therefore do not readily diffuse through the hydrophobic interior of the membrane. **7.3** There will be a net diffusion of water out of a cell into a hypertonic solution. The free water concentration is higher inside the cell than in the solution (where not as many water molecules are free, because many are clustered around the higher concentration of solute particles). **7.4** One of the solutes moved by the cotransporter is actively transported against its concentration gradient. The energy for this transport comes from the concentration gradient of the other solute, which was established by an electrogenic pump that used energy to transport the other solute across the membrane. Because energy is required overall to drive this process (because ATP is used to establish the concentration gradient), it is considered active transport. **7.5** In receptor-mediated endocytosis, specific molecules bind to receptors on the plasma membrane in a region where a coated pit develops. The cell can acquire bulk quantities of those specific molecules when the coated pit forms a vesicle and carries the bound molecules into the cell.

Test Your Understanding

1. B **2.** C **3.** A **4.** C **5.** B
6. (a)

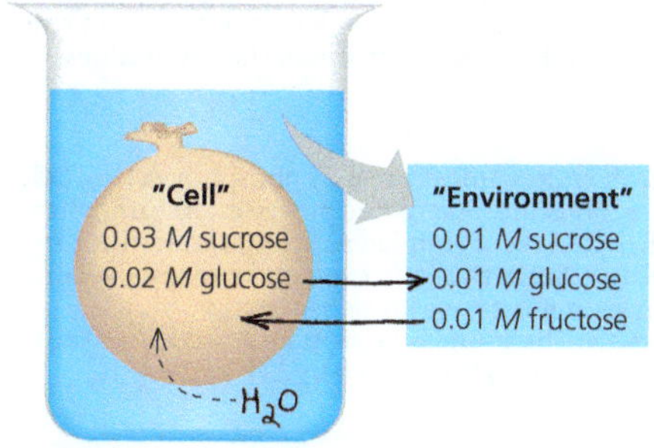

(b) The solution outside is hypotonic. It has less sucrose, which is a nonpenetrating solute. (c) See answer for (a). (d) The artificial cell will become more turgid. (e) Eventually, the two solutions will have the same solute concentrations. Even though sucrose can't move through the membrane, water flow (osmosis) will lead to isotonic conditions.

Chapter 8

Figure Questions

Figure 8.5 With a proton pump (Figure 7.17), the energy stored in ATP is used to pump protons across the membrane and build up a higher (nonrandom) concentration outside of the cell, so this process results in higher free energy. When solute molecules (analogous to hydrogen ions) are uniformly distributed, similar to the random distribution in the bottom of (b), the system has less free energy than it does in the top of (b). The system in the bottom can do no work. Because the concentration gradient created by a proton pump (Figure 7.17) represents higher free energy, this system has the potential to do work once there is a higher concentration of protons on one side of the membrane (as you will see in Figure 9.15). **Figure 8.10** Glutamic acid has a carboxyl group at the end of its R group. Glutamine has exactly the same structure as glutamic acid, except that there is an amino group in place of the —O⁻ on the R group. (The O atom on the R group leaves during the synthesis reaction.) Thus, in this figure, Gln is drawn as a Glu with an attached NH_2.

Figure 8.13

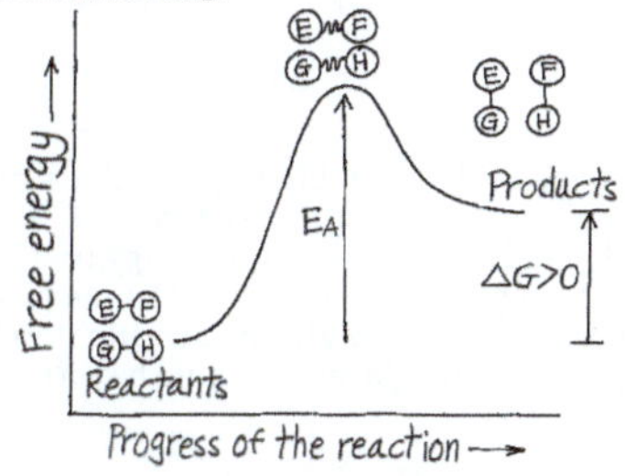

Figure 8.16

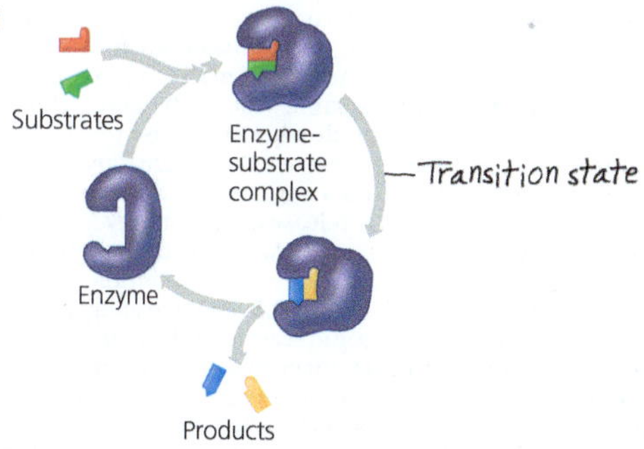

Concept Check 8.1

1. The second law is the trend toward randomization, or increasing entropy. When the concentrations of a substance on both sides of a membrane are equal, the distribution is more random than when they are unequal. Diffusion of a substance to a region where it is initially less concentrated increases entropy, making it an energetically favorable (spontaneous) process as described by the second law. This explains the process seen in Figure 7.10. **2.** The apple has potential energy in its position hanging on the tree, and the sugars and other nutrients it contains have chemical energy. The apple has kinetic energy as it falls from the tree to the ground. Finally, when the apple is digested and its molecules broken down, some of the chemical energy is used to do work, and the rest is lost as thermal energy. **3.** The sugar crystals become less ordered (entropy increases) as they dissolve and become randomly spread out in the water. Over time, the water evaporates, and the crystals form again because the water volume is insufficient to keep them in solution. While the reappearance of sugar crystals may represent a "spontaneous" increase in order (decrease in entropy), it is balanced by the decrease in order (increase in entropy) of the water molecules, which changed from a relatively compact arrangement as liquid water to a much more dispersed and disordered form as water vapor.

Concept Check 8.2

1. Cellular respiration is a spontaneous and exergonic process. The energy released from glucose is used to do work in the cell or is lost as heat.

2. Catabolism breaks down organic molecules, releasing their chemical energy and resulting in smaller products with more entropy, as when moving from the top to the bottom of Figure 8.5c. Anabolism consumes energy to synthesize larger molecules from simpler ones, as when moving from the bottom to the top of part (c). **3.** The reaction is exergonic because it releases energy—in this case, in the form of light. (This is a nonbiological version of the bioluminescence seen in Figure 8.1.)

Concept Check 8.3

1. ATP usually transfers energy to an endergonic process by phosphorylating (adding a phosphate group to) another molecule. (Exergonic processes, in turn, phosphorylate ADP to regenerate ATP.) **2.** A set of coupled reactions can transform the first combination into the second. Since this is an exergonic process overall, ΔG is negative and the first combination must have more free energy (see Figure 8.10). **3.** Active transport: The solute is being transported against its concentration gradient, which requires energy, provided by ATP hydrolysis.

Concept Check 8.4

1. A spontaneous reaction is a reaction that is exergonic. However, if it has a high activation energy that is rarely attained, the rate of the reaction may be low. **2.** Only the specific substrate(s) will fit properly into the active site of an enzyme, the part of the enzyme that carries out catalysis. **3.** In the presence of malonate, increase the concentration of the normal substrate (succinate) and see whether the rate of reaction increases. If it does, malonate is a competitive inhibitor.
4.

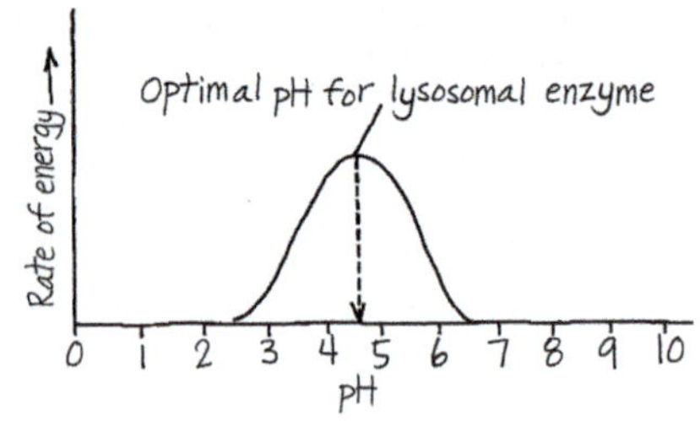

Concept Check 8.5

1. The activator binds in such a way that it stabilizes the active form of an enzyme, whereas the inhibitor stabilizes the inactive form. **2.** A catabolic pathway breaks down organic molecules, generating energy that is stored in ATP molecules. In feedback inhibition of such a pathway, ATP (one product) would act as an allosteric inhibitor of an enzyme catalyzing an early step in the catabolic process. When ATP is plentiful, the pathway would be turned off and no more would be made.

Summary of Key Concepts Questions

8.1 The process of "ordering" a cell's structure is accompanied by an increase in the entropy or disorder of the universe. For example, an animal cell takes in highly ordered organic molecules as the source of matter and energy used to build and maintain its structures. In the same process, however, the cell releases heat and the simple molecules of carbon dioxide and water to the surroundings. The increase in entropy of the latter process offsets the entropy decrease in the former. **8.2** A spontaneous reaction has a negative ΔG and is exergonic. For a chemical reaction to proceed with a net release of free energy ($-\Delta G$), the enthalpy or total energy of the system must decrease ($-\Delta H$), and/or the entropy or disorder must increase (yielding a more negative term, $-T\Delta S$). Spontaneous reactions supply the energy to perform cellular work. **8.3** The free energy released from the hydrolysis of ATP may drive endergonic reactions through the transfer of a phosphate group to a reactant molecule, forming a more reactive phosphorylated intermediate. ATP hydrolysis also powers the mechanical and transport work of a cell, often by powering shape changes in the relevant motor proteins. Cellular respiration, the catabolic breakdown of glucose, provides the energy for the endergonic regeneration of ATP from ADP and Ⓟ$_i$. **8.4** Activation energy barriers prevent the complex molecules of the cell, which are rich in free energy, from spontaneously breaking down to less ordered, more stable molecules. Enzymes permit a regulated metabolism by binding to specific substrates and forming enzyme-substrate complexes that selectively lower the E_A for the chemical reactions in a cell. **8.5** A cell tightly regulates its metabolic pathways in response to fluctuating needs for energy and materials. The binding of activators or inhibitors to regulatory sites on allosteric enzymes stabilizes either the active or inactive form of the subunits. For example, the binding of ATP to a catabolic enzyme in a cell with excess ATP would inhibit that pathway. Such types of feedback inhibition preserve chemical resources within a cell. If ATP supplies are depleted, binding of ADP to the regulatory site of catabolic enzymes would activate that pathway, generating more ATP.

Test Your Understanding

1. B **2.** C **3.** B **4.** A **5.** C **6.** D **7.** C

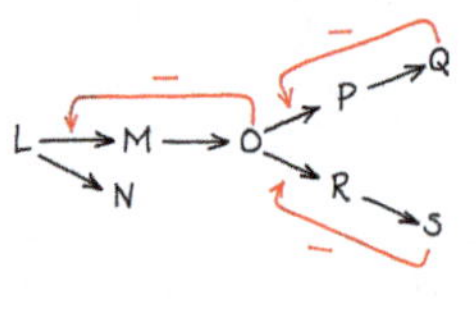

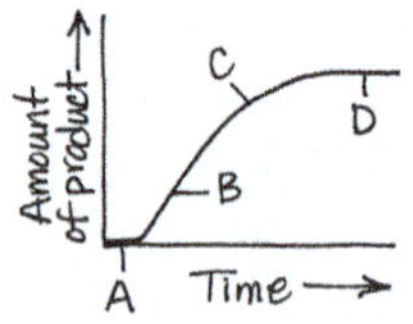

A. The substrate molecules are entering the pancreatic cells, so no product is made yet.
B. There is sufficient substrate, so the reaction is proceeding at a maximum rate.
C. As the substrate is used up, the rate decreases (the slope is less steep).
D. The line is flat because no new substrate remains and thus no new product appears.

Chapter 9

Figure Questions

Figure 9.4 The reduced form has an extra hydrogen, along with 2 electrons, bound to the carbon shown at the top of the nicotinamide (opposite the N). There are different numbers and positions of double bonds in the two forms: The oxidized form has three double bonds in the ring, while the reduced form has only two. (In organic chemistry you may have learned, or will learn, that three double bonds in a ring are able to "resonate," or act as a ring of electrons. Having three resonant double bonds is more "oxidized" than having only two double bonds in the ring.) In the oxidized form there is a + charge on the N (because it is sharing 4 electron pairs), whereas in the reduced form it is only sharing 3 electron pairs (having a pair of electrons to itself). **Figure 9.7** Because there is no external source of energy for the reaction, it must be exergonic, and the reactants must be at a higher energy level than the products. **Figure 9.9** The removal would probably stop glycolysis, or at least slow it down, since it would push the equilibrium for step 5 toward the bottom (toward DHAP). If less (or no) glyceraldehyde 3-phosphate were available, step 6 would slow down (or be unable to occur). **Figure 9.15** At first, some ATP could be made, since electron transport could proceed as far as complex III, and a small H^+ gradient could be built up. Soon, however, no more electrons could be passed to complex III because it could not be reoxidized by passing its electrons to complex IV. **Figure 9.16** First, there are 2 NADH from the oxidation of pyruvate plus 6 NADH from the citric acid cycle (CAC); 8 NADH $\times$ 2.5 ATP/NADH = 20 ATP. Second, there are 2 $FADH_2$ from the CAC; 2 $FADH_2$ $\times$ 1.5 ATP/$FADH_2$ = 3 ATP. Third, the 2 NADH from glycolysis enter the mitochondrion through one of two types of shuttle. They pass their electrons either to 2 FAD, which become $FADH_2$ and result in 3 ATP, or to 2 NAD^+, which become NADH and result in 5 ATP. Thus, 20 + 3 + 3 = 26 ATP, or 20 + 3 + 5 = 28 ATP from all NADH and $FADH_2$.

Concept Check 9.1

1. Both processes include glycolysis, the citric acid cycle, and oxidative phosphorylation. In aerobic respiration, the final electron acceptor is molecular oxygen (O_2); in anaerobic respiration, the final electron acceptor is a different substance.
2. $C_4H_6O_5$ would be oxidized and NAD^+ would be reduced.

Concept Check 9.2

1. NAD^+ acts as the oxidizing agent in step 6, accepting electrons from glyceraldehyde 3-phosphate (G3P), which thus acts as the reducing agent.

Concept Check 9.3

1. NADH and $FADH_2$; they will donate electrons to the electron transport chain.
2. CO_2 is released from the pyruvate that is the end product of glycolysis, and CO_2 is also released during the citric acid cycle. **3.** In both cases, the precursor molecule loses a CO_2 molecule and then donates electrons to an electron carrier in an oxidation step. Also, the product has been activated due to the attachment of a CoA group.

Concept Check 9.4

1. Oxidative phosphorylation would eventually stop entirely, resulting in no ATP production by this process. Without oxygen to "pull" electrons down the electron transport chain, H^+ would not be pumped into the mitochondrion's intermembrane space and chemiosmosis would not occur. **2.** Decreasing the pH means addition of H^+. This would establish a proton gradient even without the function of the electron transport chain, and we would expect ATP synthase to function and synthesize ATP. (In fact, it was experiments like this that provided support for chemiosmosis as an energy-coupling mechanism.) **3.** One of the components of the electron transport chain, ubiquinone (Q), must be able to diffuse within the membrane. It could not do so if the membrane components were locked rigidly into place.

Concept Check 9.5

1. A derivative of pyruvate, such as acetaldehyde during alcohol fermentation, or pyruvate itself during lactic acid fermentation; oxygen; another electron acceptor at the end of an electron transport chain, such as sulfate (SO_4^{2-}) **2.** The cell would need to consume glucose at a rate about 16 times the consumption rate in the aerobic environment (2 ATP are generated by fermentation versus up to 32 ATP by cellular respiration).

Concept Check 9.6

1. The fat is much more reduced; it has many $-CH_2-$ units, and in all these bonds the electrons are equally shared. The electrons present in a carbohydrate molecule are already somewhat oxidized (shared unequally in bonds; there are more C—O and O—H bonds), as quite a few of them are bound to oxygen. Electrons that are equally shared, as in fat, have a higher energy level than electrons that are unequally shared, as in carbohydrates. Thus, fat is a much better fuel than carbohydrate.
2. When we consume more food than necessary for metabolic processes, our body synthesizes fat as a way of storing energy for later use. **3.** AMP will accumulate, stimulating phosphofructokinase, and thus increasing the rate of glycolysis. Since oxygen is not present, the cell will convert pyruvate to lactate in lactic acid fermentation, providing a supply of ATP. **4.** When oxygen is present, the fatty acid chains

containing most of the energy of a fat are oxidized and fed into the citric acid cycle and the electron transport chain. During intense exercise, however, oxygen is scarce in muscle cells, so ATP must be generated by glycolysis alone. A very small part of the fat molecule, the glycerol backbone, can be oxidized via glycolysis, but the amount of energy released by this portion is insignificant compared to that released by the fatty acid chains. (This is why moderate exercise, staying below 70% maximum heart rate, is better for burning fat—because enough oxygen remains available to the muscles.)

Summary of Key Concepts Questions

9.1 Most of the ATP produced in cellular respiration comes from oxidative phosphorylation, in which the energy released from redox reactions in an electron transport chain is used to produce ATP. In substrate-level phosphorylation, an enzyme directly transfers a phosphate group to ADP from an intermediate substrate. All ATP production in glycolysis occurs by substrate-level phosphorylation; this form of ATP production also occurs at one step in the citric acid cycle. **9.2** The oxidation of the three-carbon sugar, glyceraldehyde 3-phosphate, yields energy. In this oxidation, electrons and H^+ are transferred to NAD^+, forming NADH, and a phosphate group is attached to the oxidized substrate. ATP is then formed by substrate-level phosphorylation when this phosphate group is transferred to ADP. **9.3** The release of six molecules of CO_2 represents the complete oxidation of glucose. During the processing of two pyruvates to acetyl CoA, the fully oxidized carboxyl groups ($—COO^-$) are given off as 2 CO_2. The remaining four carbons are released as CO_2 in the citric acid cycle as citrate is oxidized back to oxaloacetate. **9.4** The flow of H^+ through the ATP synthase complex causes the rotor and attached rod to rotate, exposing catalytic sites in the knob portion that produce ATP from ADP and Ⓟ$_i$. ATP synthases are found in the inner mitochondrial membrane, the plasma membrane of prokaryotes, and membranes within chloroplasts. **9.5** Anaerobic respiration yields more ATP. The 2 ATP produced by substrate-level phosphorylation in glycolysis represent the total energy yield of fermentation. NADH passes its "high-energy" electrons to pyruvate or a derivative of pyruvate, recycling NAD^+ and allowing glycolysis to continue. In anaerobic respiration, the NADH produced during glycolysis, as well as additional molecules of NADH produced as pyruvate is oxidized, are used to generate ATP molecules. An electron transport chain captures the energy of the electrons in NADH via a series of redox reactions; ultimately, the electrons are transferred to an electronegative molecule other than oxygen. **9.6** The ATP produced by catabolic pathways is used to drive anabolic pathways. Also, many of the intermediates of glycolysis and the citric acid cycle are used in the biosynthesis of a cell's molecules.

Test Your Understanding

1. C **2.** C **3.** A **4.** B **5.** D **6.** A **7.** B

8. Since the overall process of glycolysis results in net production of ATP, it would make sense for the process to slow down when ATP levels have increased substantially. Thus, we would expect ATP to allosterically inhibit phosphofructokinase.

9. The proton pump in Figures 7.17 and 7.18 is carrying out active transport, using ATP hydrolysis to pump protons against their concentration gradient. Because ATP is required, this is active transport of protons. The ATP synthase in Figure 9.14 is using the flow of protons down their concentration gradient to power ATP synthesis. Because the protons are moving down their concentration gradient, no energy is required, and this is passive transport.

10.

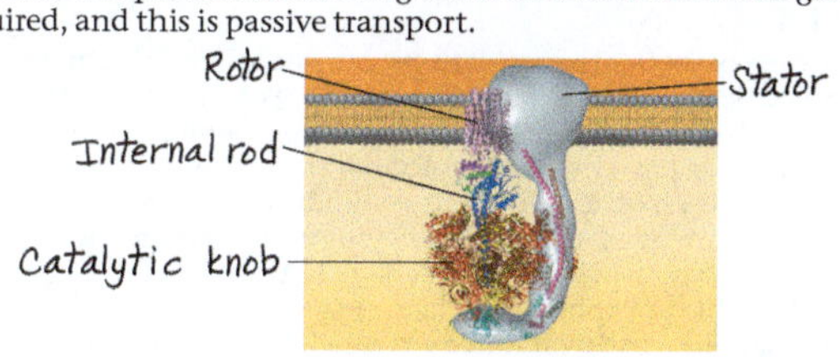

12.

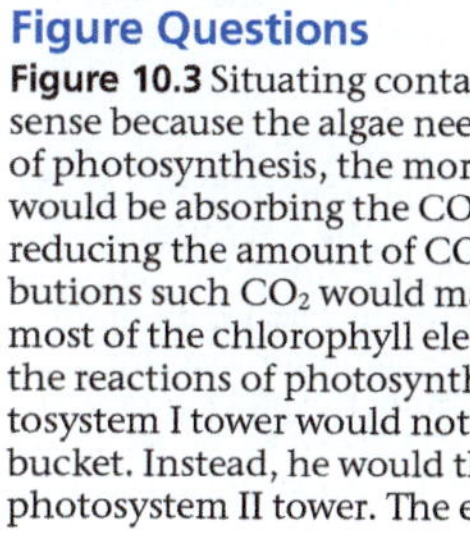

H^+ would continue to be pumped across the membrane into the intermembrane space, increasing the difference between the matrix pH and the intermembrane space pH. H^+ would not be able to flow back through ATP synthase, since the enzyme is inhibited by the poison, so rather than maintaining a constant difference across the membrane, the difference would continue to increase. (Ultimately, the H^+ concentration in the intermembrane space would be so high that no more H^+ would be able to be pumped against the gradient, but this isn't shown in the graph.)

Chapter 10

Figure Questions

Figure 10.3 Situating containers of algae near sources of CO_2 emissions makes sense because the algae need CO_2 to carry out photosynthesis. The higher their rate of photosynthesis, the more plant oil they will produce. At the same time, algae would be absorbing the CO_2 emitted from industrial plants or from car engines, reducing the amount of CO_2 entering the atmosphere—thus, lowering the contributions such CO_2 would make to global climate change. **Figure 10.12** In the leaf, most of the chlorophyll electrons excited by photon absorption are used to power the reactions of photosynthesis. **Figure 10.16** The person at the top of the photosystem I tower would not turn to his left and throw his electron into the NADPH bucket. Instead, he would throw it onto the top of the ramp at his right, next to the photosystem II tower. The electron would then roll down the ramp, get energized by a photon, and return to him. This cycle would continue as long as light was available. (This is why it's called cyclic electron flow.) **Figure 10.17** You would (a) decrease the pH outside the mitochondrion (thus increasing the H^+ concentration) and (b) increase the pH in the chloroplast stroma (thus decreasing the H^+ concentration). In both cases, this would generate an H^+ gradient across the membrane that would cause ATP synthase to synthesize ATP. **Figure 10.23** The gene encoding hexokinase is part of the DNA of a chromosome in the nucleus. There, the gene is transcribed into mRNA, which is transported to the cytoplasm where it is translated on a free ribosome into a polypeptide. The polypeptide folds into a functional protein with secondary and tertiary structure. Once functional, it carries out the first reaction of glycolysis in the cytoplasm.

Concept Check 10.1

1. CO_2 enters the leaves via stomata, and being a nonpolar molecule, can cross the leaf cell membrane and the chloroplast membranes to reach the stroma of the chloroplast. **2.** Using ^{18}O, a heavy isotope of oxygen, as a label, researchers were able to confirm van Niel's hypothesis that the oxygen produced during photosynthesis comes from water, not from carbon dioxide. **3.** The light reactions could *not* keep producing NADPH and ATP without the $NADP^+$, ADP, and Ⓟ$_i$ that the Calvin cycle generates. The two cycles are interdependent.

Concept Check 10.2

1. Green, because green light is mostly transmitted and reflected—not absorbed—by photosynthetic pigments **2.** Water (H_2O) is the initial electron donor; $NADP^+$ accepts electrons at the end of the electron transport chain, becoming reduced to NADPH. **3.** In this experiment, the rate of ATP synthesis would slow and eventually stop. Because the added compound would not allow a proton gradient to build up across the membrane, ATP synthase could not catalyze ATP production.

Concept Check 10.3

1. 6, 18, 12 **2.** The more potential energy and reducing power a molecule stores, the more energy and reducing power are required for the formation of that molecule. Glucose is a valuable energy source because it is highly reduced (lots of C—H bonds), storing lots of potential energy in its electrons. To reduce CO_2 to glucose, much energy and reducing power are required in the form of large numbers of ATP and NADPH molecules, respectively. **3.** The light reactions require ADP and $NADP^+$, which would not be formed in sufficient quantities from ATP and NADPH if the Calvin cycle stopped.

4.

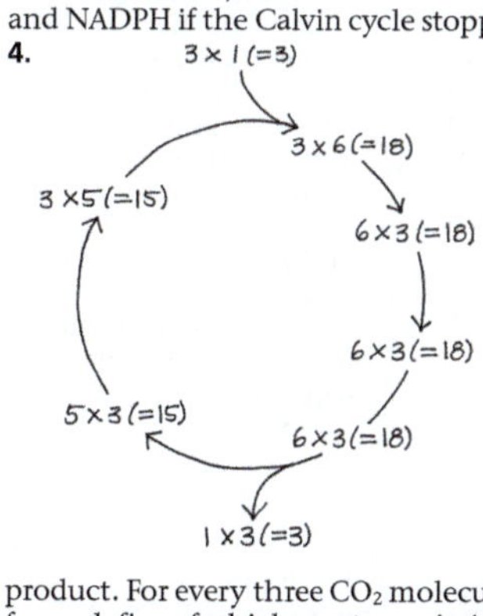

Three carbon atoms enter the cycle, one by one, as individual CO_2 molecules, and leave the cycle in one three-carbon molecule (G3P) per three turns of the cycle. **5.** In glycolysis, G3P acts as an intermediate. The 6-carbon sugar fructose 1,6-bisphosphate is cleaved into two 3-carbon sugars, one of which is G3P. The other is an isomer called dihydroxyacetone phosphate (DHAP), which can be converted to G3P by an isomerase. Because G3P is the substrate for the next enzyme, it is constantly removed, and the reaction equilibrium is pulled in the direction of conversion of DHAP to more G3P. In the Calvin cycle, G3P acts as both an intermediate and a product. For every three CO_2 molecules that enter the cycle, six G3P molecules are formed, five of which must remain in the cycle and become rearranged to regenerate three 5-carbon RuBP molecules. The one remaining G3P is a product, which can be thought of as the result of "reducing" the three CO_2 molecules that entered the cycle into a 3-carbon sugar that can later be used to generate energy.

Concept Check 10.4

1. Photorespiration decreases photosynthetic output by adding oxygen, instead of carbon dioxide, to the Calvin cycle. As a result, no sugar is generated (no carbon is fixed), and O_2 is used rather than generated. **2.** Without PS II, no O_2 is generated in bundle-sheath cells. This avoids the problem of O_2 competing with CO_2 for binding to rubisco in these cells. **3.** Both problems are caused by a drastic change in Earth's atmosphere due to burning of fossil fuels. The increase in CO_2 concentration affects ocean chemistry by decreasing pH, thus affecting calcification by marine organisms. On land, CO_2 concentration and air temperature are conditions that plants have become adapted to, and changes in these characteristics have a strong effect on photosynthesis by plants. Thus, alteration of these two fundamental factors could have critical effects on organisms all around the planet, in all different habitats. **4.** C_4 and CAM species would replace many of the C_3 species.

Concept Check 10.5

1. Yes, plants can break down the sugar (in the form of glucose) by cellular respiration, producing ATPs for various cellular processes such as endergonic chemical reactions, transport of substances across membranes, and movement of molecules in the cell. ATPs are also used for the movement of chloroplasts during cellular streaming in some plant cells (see Figure 6.26).

Summary of Key Concepts Questions

10.1 CO_2 and H_2O are the products of cellular respiration; they are the reactants in photosynthesis. In respiration, glucose is oxidized to CO_2 and electrons are passed through an electron transfer chain from glucose to O_2, producing H_2O. In photosynthesis, H_2O is the source of electrons, which are energized by light, temporarily stored in NADPH, and used to reduce CO_2 to carbohydrate. **10.2** The action spectrum of photosynthesis shows that some wavelengths of light that are not absorbed by chlorophyll *a* are still effective at promoting photosynthesis. The light-harvesting complexes of photosystems contain accessory pigments such as chlorophyll *b* and carotenoids, which absorb different wavelengths and pass the energy to chlorophyll *a*, broadening the spectrum of light usable for photosynthesis.

10.3

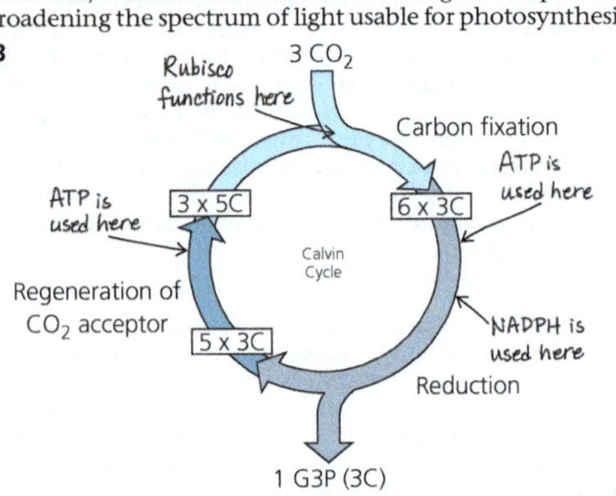

In the reduction phase of the Calvin cycle, ATP phosphorylates a three-carbon compound, and NADPH then reduces this compound to G3P. ATP is also used in the regeneration phase, when five molecules of G3P are converted to three molecules of the five-carbon compound RuBP. Rubisco catalyzes the first step of carbon fixation—the addition of CO_2 to RuBP.

10.4 Both C_4 photosynthesis and CAM photosynthesis involve initial fixation of CO_2 to produce a four-carbon compound (in mesophyll cells in C_4 plants and at night in CAM plants). These compounds are then broken down to release CO_2 (in the bundle-sheath cells in C_4 plants and during the day in CAM plants). ATP is required for recycling the molecule that is used initially to combine with CO_2. These pathways avoid the photorespiration that consumes ATP and reduces the photosynthetic output of C_3 plants when they close stomata on hot, dry, bright days. Thus, hot, arid climates would favor C_4 and CAM plants. **10.5** Photosynthetic organisms provide food (in the form of carbohydrates) to all other living organisms, either directly or indirectly. They do this by harnessing the energy of the sun to build carbohydrates, something that non-photosynthesizers cannot do. Photosynthetic organisms also produce oxygen (O_2), required by all aerobically respiring organisms.

Test Your Understanding

1. D **2.** B **3.** C **4.** A **5.** C **6.** B **7.** C

10.

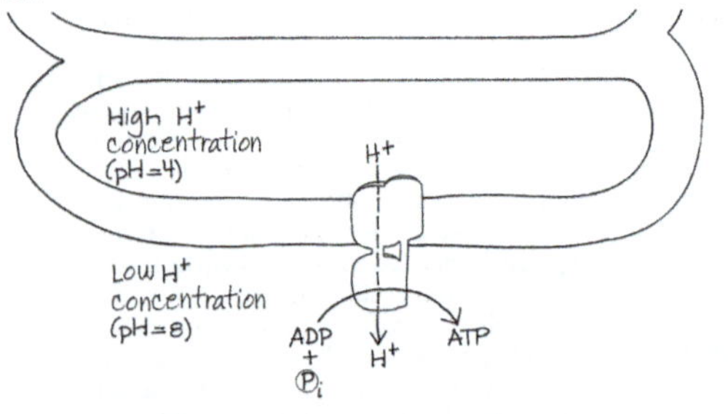

The ATP would end up outside the thylakoid. The thylakoids were able to make ATP in the dark because the researchers set up an artificial proton concentration gradient across the thylakoid membrane; thus, the light reactions were not necessary to establish the H^+ gradient required for ATP synthesis by ATP synthase.

Chapter 11

Figure Questions

Figure 11.6 Epinephrine is a signaling molecule; presumably, it binds to a cell-surface receptor protein. **Figure 11.8** This is an example of passive transport. The ion is moving down its concentration gradient, and no energy is required. **Figure 11.9** The aldosterone molecule, a steroid, is hydrophobic and can therefore pass directly through the hydrophobic lipid bilayer of the plasma membrane into the cell. (Hydrophilic molecules cannot do this.) **Figure 11.10** The entire phosphorylation cascade wouldn't operate. Regardless of whether or not the signaling molecule was bound, protein kinase 2 would always be inactive and would not be able to activate the purple-colored protein leading to the cellular response. **Figure 11.11** The signaling molecule (cAMP) would remain in its active form and would continue to signal.

Figure 11.12

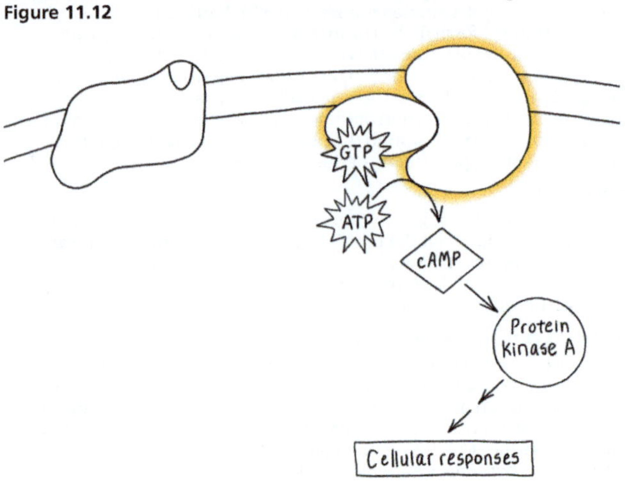

Figure 11.16 100,000,000 (one hundred million, or 10^8) glucose molecules are released. The first step results in 100× amplification (one epinephrine activates 100 G proteins); the next step does not amplify the response; the next step is a 100× amplification (10^2 active adenylyl cyclase molecules to 10^4 cyclic AMPs); the next step does not amplify; the next two steps are each 10× amplifications, and the final step is a 100× amplification. **Figure 11.17** The signaling pathway shown in **Figure 11.14** leads to the splitting of PIP_2 into the second messengers DAG and IP_3, which produce different responses. (The response elicited by DAG is mentioned but not shown.) The pathway shown for cell B is similar in that it branches and leads to two responses.

Concept Check 11.1

1. The two cells of opposite mating type (**a** and **α**) each secrete a certain signaling molecule, which can only be bound by receptors carried on cells of the opposite mating type. Thus, the **a** mating factor cannot bind to another **a** cell and cause it to grow toward the first **a** cell. Only an **α** cell can "receive" the signaling molecule and respond by directed growth. **2.** Glycogen phosphorylase acts in the third stage, the response to epinephrine signaling. **3.** Glucose 1-phosphate would not be generated because the activation of the enzyme requires an intact cell, with an intact receptor in the membrane and an intact signal transduction pathway. The enzyme cannot be activated directly by interaction with the signaling molecule in the cell-free mixture.

Concept Check 11.2

1. NGF is water-soluble (hydrophilic), so it cannot pass through the lipid membrane to reach intracellular receptors, as steroid hormones can. Therefore, you'd expect the NGF receptor to be in the plasma membrane—which is, in fact, the case. **2.** The cell with the faulty receptor would not be able to respond appropriately to the signaling molecule when it was present. This would most likely have dire consequences for the cell, since regulation of the cell's activities by this receptor would not occur appropriately. **3.** Binding of a ligand to a receptor changes the shape of the receptor, altering the ability of the receptor to transmit a signal. Binding of an allosteric regulator to an enzyme changes the shape of the enzyme, either promoting or inhibiting enzyme activity.

Concept Check 11.3

1. A protein kinase is an enzyme that transfers a phosphate group from ATP to a protein, usually activating that protein (often a second type of protein kinase). Many signal transduction pathways include a series of such interactions, in which each phosphorylated protein kinase in turn phosphorylates the next protein kinase in the series. Such phosphorylation cascades carry a signal from outside the cell to the cellular protein(s) that will carry out the response. **2.** Protein phosphatases reverse the effects of the kinases, and unless the signaling molecule is at a high enough concentration that it is continuously rebinding the receptor, the kinase molecules will all be returned to their inactive states by phosphatases. **3.** The signal that is being transduced is the *information* that a signaling molecule is bound to the cell-surface receptor. Information is transduced by way of sequential protein-protein interactions that change protein shapes, causing them to function in a way that passes the signal (the information) along. **4.** The IP_3-gated channel would open, allowing calcium ions to flow out of the ER and into the cytoplasm, which would raise the cytosolic Ca^{2+} concentration.

Concept Check 11.4

1. At each step in a cascade of sequential activations, one molecule or ion may activate numerous molecules functioning in the next step. This causes the response to be amplified at each such step and overall results in a large amplification of the original signal. **2.** Scaffolding proteins hold molecular components of signaling pathways in a complex with each other. Different scaffolding proteins would assemble different collections of proteins, facilitating different molecular interactions and leading to different cellular responses in the two cells. **3.** A malfunctioning protein phosphatase would not be able to dephosphorylate a particular receptor or relay protein. As a result, the signaling pathway, once activated, would not be able to be terminated. (In fact, one study found altered protein phosphatases in cells from 25% of colorectal tumors.)

Concept Check 11.5

1. In formation of the hand or paw in mammals, cells in the regions between the digits are programmed to undergo apoptosis. This serves to shape the digits of the hand or paw so that they are not webbed. (A lack of apoptosis in these regions in water birds results in webbed feet.) **2.** If a receptor protein for a death-signaling molecule were defective such that it was activated even in the absence of the death signal, this would lead to apoptosis when it wouldn't normally occur. Similar defects in any of the proteins in the signaling pathway would have the same effect if the defective proteins activated relay or response proteins in the absence of interaction with the previous protein or second messenger in the pathway. Conversely, if any protein in the pathway were defective in its ability to respond to an interaction with an early protein or other molecule or ion, apoptosis would not occur when it normally should. For example, a receptor protein for a death-signaling ligand might not be able to be activated, even when ligand was bound. This would stop the signal from being transduced into the cell.

Summary of Key Concepts Questions

11.1 A cell is able to respond to a hormone only if it has a receptor protein on the cell surface or inside the cell that can bind to the hormone. The response to a hormone depends on the specific signal transduction pathway within the cell, which will lead to the specific cellular response. The response can vary for different types of cells. **11.2** Both GPCRs and RTKs have an extracellular binding site for a signaling molecule (ligand) and one or more α-helical regions of the polypeptide that spans the membrane. A GPCR functions singly, while RTKs tend to dimerize or form larger groups of RTKs. GPCRs usually trigger a single transduction pathway, whereas the multiple activated tyrosines on an RTK dimer may trigger several different transduction pathways at the same time. **11.3** A protein kinase is an enzyme that adds a phosphate group to another protein. Protein kinases are often part of a phosphorylation cascade that transduces a signal. A second messenger is a small, nonprotein molecule or ion that rapidly diffuses and relays a signal throughout a cell. Both protein kinases and second messengers can operate in the same pathway. For example, the second messenger cAMP often activates protein kinase A, which then phosphorylates other proteins. **11.4** In G protein-coupled pathways, the GTPase portion of a G protein converts GTP to GDP and inactivates the G protein. Protein phosphatases remove phosphate groups from activated proteins, thus stopping a phosphorylation cascade of protein kinases. Phosphodiesterase converts cAMP to AMP, thus reducing the effect of cAMP in a signal transduction pathway. **11.5** The basic mechanism of controlled cell suicide evolved early in eukaryotic evolution, and the genetic basis for these pathways has been conserved during animal evolution. Such a mechanism is essential to the development and maintenance of all animals.

Test Your Understanding

1. D **2.** A **3.** B **4.** A **5.** C **6.** C **7.** C

8. This is one possible drawing of the pathway. (Similar drawings would also be correct.)

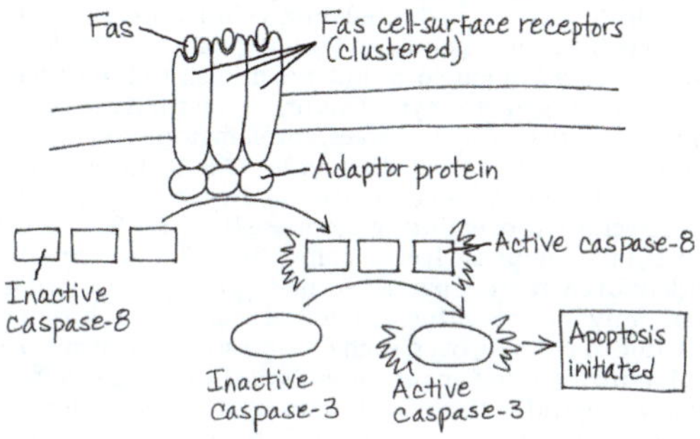

Chapter 12

Figure Questions

Figure 12.4

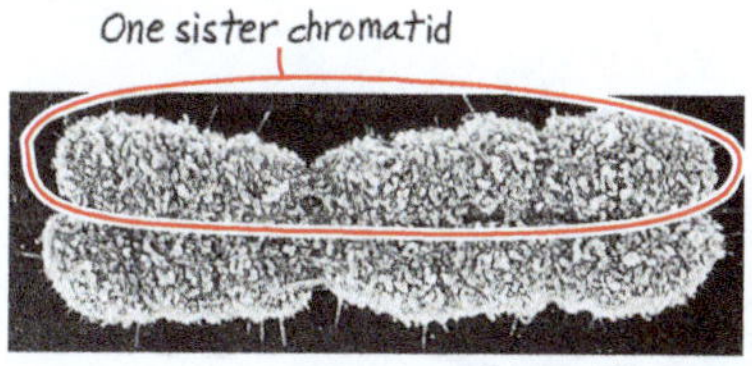

Circling the other chromatid instead would also be correct. **Figure 12.5** The chromosome has four arms. The single (duplicated) chromosome in step 2 becomes two (unduplicated) chromosomes in step 3. The duplicated chromosome in step 2 is considered one single chromosome. **Figure 12.7** 12; 2; 2; 1

Figure 12.8

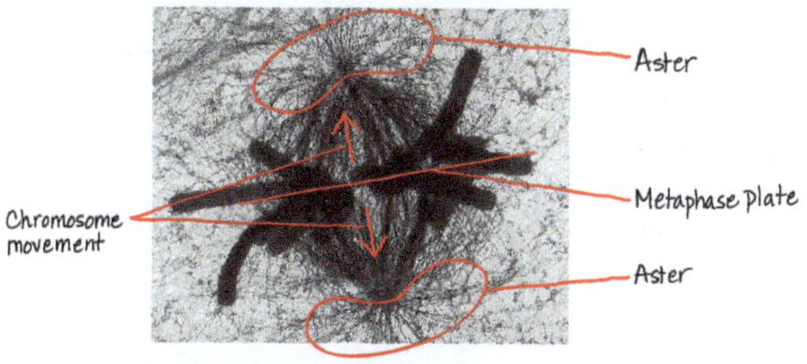

Figure 12.9 The mark would have moved toward the nearer pole. The lengths of fluorescent microtubules between that pole and the mark would have decreased, while the lengths between the chromosomes and the mark would have remained the same. **Figure 12.14** In both cases, the G_1 nucleus would have remained in G_1 until the time it normally would have entered the S phase. Chromosome condensation and spindle formation would not have occurred until the S and G_2 phases had been completed. **Figure 12.16** Passing the G_2 checkpoint in the diagram corresponds to the beginning of the "Time" axis of the graph, and entry into the mitotic phase (yellow background on the diagram) corresponds to the peaks of MPF activity and cyclin concentration on the graph (see the yellow M banner over the peaks). During G_1 and S phase in the diagram, Cdk is present without cyclin, so on the graph both cyclin concentration and MPF activity are low. The curved purple arrow in the diagram shows increasing cyclin concentration, seen on the graph during the end of S phase and throughout G_2 phase. Then the cell cycle begins again. **Figure 12.17** The cell would divide under conditions where it was inappropriate to do so. If the daughter cells and their descendants also ignored either of the checkpoints and divided, there would soon be an abnormal mass of cells. (This type of inappropriate cell division can contribute to the development of cancer.) **Figure 12.18** The cells in the vessel with PDGF would not be able to respond to the growth factor signal and thus would not divide. The culture would resemble that without the added PDGF.

Concept Check 12.1

1. 1; 1; 2 **2.** 39; 39; 78

Concept Check 12.2

1. 6 chromosomes; they are duplicated; 12 chromatids **2.** Following mitosis, cytokinesis results in two genetically identical daughter cells in both plant cells and animal cells. However, the mechanism of dividing the cytoplasm is different in animals and plants. In an animal cell, cytokinesis occurs by cleavage, which divides the parent cell in two with a contractile ring of actin filaments. In a plant cell, a cell plate forms in the middle of the cell and grows until its membrane fuses with the plasma membrane of the parent cell. A new cell wall grows inside the cell plate, thus eventually between the two new cells. **3.** From the end of S phase in interphase through the end of metaphase in mitosis **4.** During eukaryotic cell division, tubulin is involved in spindle formation and chromosome movement, while actin functions during cytokinesis. In bacterial binary fission, it's the opposite: Actin-like molecules are thought to move the daughter bacterial chromosomes to opposite ends of the cell, and tubulin-like molecules are thought to act in daughter cell separation. **5.** A kinetochore connects the spindle (a motor; note that it has motor proteins) to a chromosome (the cargo it will move). **6.** Microtubules made up of tubulin in the cell provide "rails" along which vesicles and other organelles can travel, based on interactions of motor proteins with tubulin in the microtubules. In muscle cells, actin in microfilaments interacts with myosin filaments to cause muscle contraction.

Concept Check 12.3

1. The nucleus on the right was originally in the G_1 phase; therefore, it had not yet duplicated its chromosomes. The nucleus on the left was in the M phase, so it had already duplicated its chromosomes. **2.** A sufficient amount of MPF has to exist for a cell to pass the G_2 checkpoint; this occurs through the accumulation of cyclin proteins, which combine with Cdk to form (active) MPF. MPF then phosphorylates other proteins, initiating mitosis. **3.** The intracellular receptor, once activated, would be able to act as a transcription factor in the nucleus, turning on genes that may cause the cell to pass a checkpoint and divide. The RTK receptor, when activated by a ligand, would form a dimer, and each subunit of the dimer would phosphorylate the other. This would lead to a series of signal transduction steps, ultimately turning on genes in the nucleus. As in the case of the estrogen receptor, the genes would code for proteins necessary to commit the cell to divide.

Summary of Key Concepts Questions

12.1 The DNA of a eukaryotic cell is packaged into structures called *chromosomes*. Each chromosome is a long molecule of DNA, which carries hundreds to thousands of genes, with associated proteins that maintain chromosome structure and help control gene activity. This DNA-protein complex is called *chromatin*. The chromatin of each chromosome is long and thin when the cell is not dividing. Prior to cell division, each chromosome is duplicated, and the resulting sister *chromatids* are attached to each other by proteins at the centromeres and, for many species, all along their lengths (a phenomenon called sister chromatid cohesion). **12.2** Chromosomes exist as single DNA molecules in G_1 of interphase and in anaphase and telophase of mitosis. During S phase, DNA replication produces sister chromatids, which persist during G_2 of interphase and through prophase, prometaphase, and metaphase of mitosis. **12.3** Checkpoints allow cellular surveillance mechanisms to determine whether the cell is prepared to go to the next stage. Internal and external signals move a cell past these checkpoints. The G_1 checkpoint determines whether a cell will proceed forward in the cell cycle or switch into the G_0 phase. The signals to pass this checkpoint often are external, such as growth factors. Passing the G_2 checkpoint requires sufficient numbers of active MPF complexes, which in turn orchestrate several mitotic events. MPF also initiates degradation of its cyclin component, terminating the M phase. The M phase will not begin again until sufficient cyclin is produced during the next S and G_2 phases. The signal to pass the M phase checkpoint is not activated until all chromosomes are attached to kinetochore fibers and are aligned at the metaphase plate. Only then will sister chromatid separation occur.

Test Your Understanding

1. B **2.** A **3.** C **4.** C **5.** A **6.** B **7.** A **8.** D **9.** See Figure 12.7 for a description of major events. Only one cell is indicated for each stage, but other correct answers are also present in this micrograph.

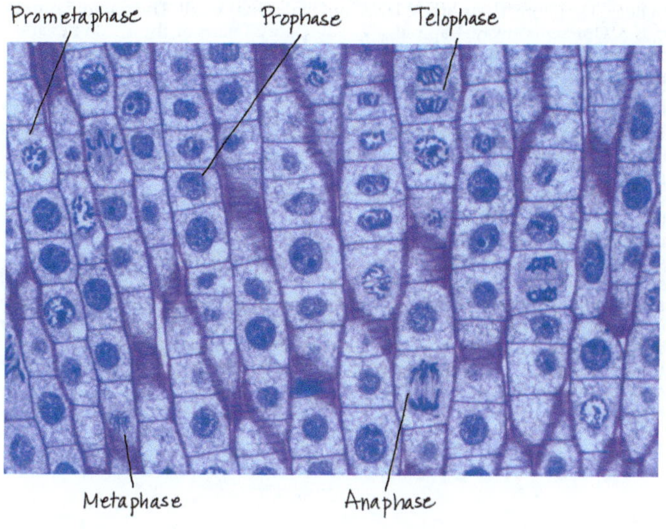

10.

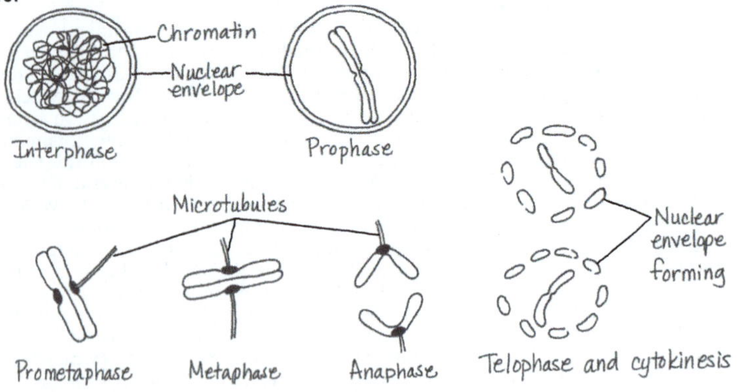

Chapter 13

Figure Questions

Figure 13.4 Two sets of chromosomes are present. Three pairs of homologous chromosomes are present. **Figure 13.6** In (a), haploid cells do not undergo mitosis. In (b), haploid spores undergo mitosis to form the gametophyte, and haploid cells of the gametophyte undergo mitosis to form gametes. In (c), haploid cells undergo mitosis to form either a multicellular haploid organism or a new unicellular haploid organism, and these haploid cells undergo mitosis to form gametes.
Figure 13.7

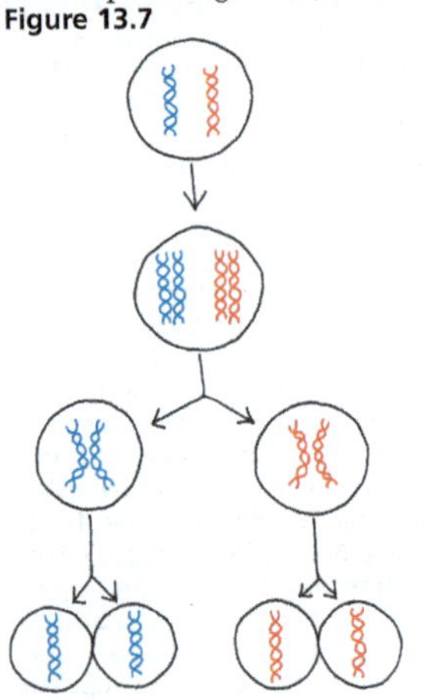

(A short strand of DNA is shown here for simplicity, but each chromosome or chromatid contains a very long coiled and folded DNA molecule.) **Figure 13.8** If a cell with six chromosomes undergoes two rounds of mitosis, each of the four resulting cells will have six chromosomes, while the four cells resulting from meiosis in Figure 13.8 each have three chromosomes. In mitosis, DNA replication (and thus chromosome duplication) precedes each prophase, ensuring that daughter cells have the same number of chromosomes as the parent cell. In meiosis, in contrast, DNA replication occurs only before prophase I (not prophase II). Thus, in two rounds of mitosis, the chromosomes duplicate twice and divide twice, while in meiosis, the chromosomes duplicate once and divide twice. **Figure 13.10** Yes. Each of the six chromosomes (three per cell) shown in telophase I has one nonrecombinant chromatid and one recombinant chromatid. Therefore, eight possible sets of chromosomes can be generated for the cell on the left and eight for the cell on the right.

Concept Check 13.1

1. Parents pass genes to their offspring; by dictating the production of messenger RNAs (mRNAs), the genes program cells to make specific enzymes and other proteins, whose cumulative action produces an individual's inherited traits.
2. Such organisms reproduce by mitosis, which generates offspring whose genomes are exact copies of the parent's genome (in the absence of mutation).
3. She should clone it. Crossbreeding it with another plant would generate offspring that have additional variation, which she no longer desires now that she has obtained her ideal orchid.

Concept Check 13.2

1. Each of the six chromosomes is duplicated, so each contains two DNA double helices. Therefore, there are 12 DNA molecules in the cell. The haploid number, n, is 3. One set is always haploid. **2.** There are 23 pairs of chromosomes and two sets. **3.** This organism has the life cycle shown in Figure 13.6c. Therefore, it must be a fungus or a protist, perhaps an alga.

Concept Check 13.3

1. The chromosomes are similar in that each is composed of two sister chromatids, and the individual chromosomes are positioned similarly at the metaphase plate. The chromosomes differ in that in a mitotically dividing cell, sister chromatids of each chromosome are genetically identical, but in a meiotically dividing cell, sister chromatids are genetically distinct because of crossing over in meiosis I. Moreover, the chromosomes in metaphase of mitosis can be a diploid set or a haploid set, but the chromosomes in metaphase of meiosis II always consist of a haploid set. **2.** If crossing over did not occur, the two homologs would not be associated in any way; each sister chromatid would be either all maternal or all paternal and would only be attached to its sister chromatid, not to a nonsister chromatid. This might result in incorrect arrangement of homologs during metaphase I and, ultimately, in formation of gametes with an abnormal number of chromosomes.

Concept Check 13.4

1. Mutations in a gene lead to the different versions (alleles) of that gene.
2. Without crossing over, independent assortment of chromosomes during meiosis I theoretically can generate 2^n possible haploid gametes, and random fertilization can produce $2^n \times 2^n$ possible diploid zygotes. Because the haploid number (n) of grasshoppers is 23 and that of fruit flies is 4, two grasshoppers would be expected to produce a greater variety of zygotes than would two fruit flies. **3.** If the segments of the maternal and paternal chromatids that undergo crossing over are genetically identical and thus have the same two alleles for every gene, then the recombinant chromosomes will be genetically equivalent to the parental chromosomes. Crossing over contributes to genetic variation only when it involves the rearrangement of different alleles.

Summary of Key Concepts Questions

13.1 Genes program specific traits, and offspring inherit their genes from each parent, accounting for similarities in their appearance to one or the other parent. Humans reproduce sexually, which ensures new combinations of genes (and thus traits) in the offspring. Consequently, the offspring are not clones of their parents (which would be the case if humans reproduced asexually). **13.2** Animals and plants both reproduce sexually, alternating meiosis with fertilization. Both have haploid gametes that unite to form a diploid zygote, which then goes on to divide mitotically, forming a diploid multicellular organism. In animals, haploid cells become gametes and don't undergo mitosis, while in plants, the haploid cells resulting from meiosis undergo mitosis to form a haploid multicellular organism, the gametophyte. This organism then goes on to generate haploid gametes. (In plants such as trees, the gametophyte is quite reduced in size and not obvious to the casual observer.) **13.3** At the end of meiosis I, the two members of a homologous pair end up in different cells, so they cannot pair up and undergo crossing over during prophase II. **13.4** First, during independent assortment in metaphase I, each pair of homologous chromosomes lines up independent of each other pair at the metaphase plate, so a daughter cell of meiosis I randomly inherits either a maternal or paternal chromosome of each pair. Second, due to crossing over, each chromosome is not exclusively maternal or paternal, but includes regions at the ends of the chromatid from a nonsister chromatid (a chromatid of the other homolog). (The nonsister segment can also be in an internal region of the chromatid if a second crossover occurs beyond the first one before the end of the chromatid.) This provides much additional diversity in the form of new combinations of alleles. Third, random fertilization ensures even more variation, since any sperm of a large number containing many possible genetic combinations can fertilize any egg of a similarly large number of possible combinations.

Test Your Understanding

1. A **2.** B **3.** A **4.** D **5.** C
6. (a)

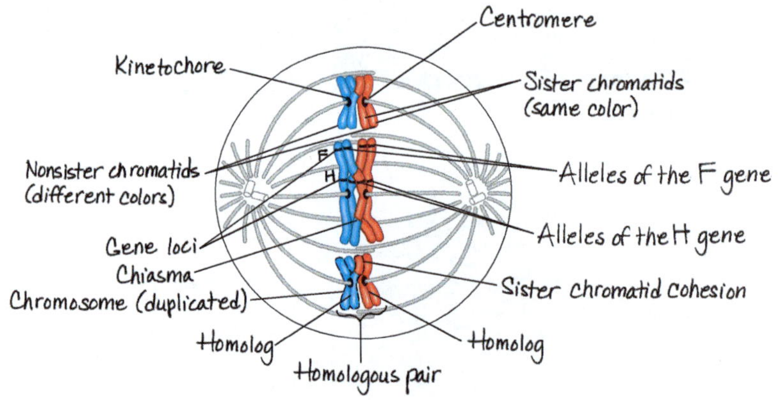

(b) A haploid set is made up of one long, one medium, and one short chromosome, no matter what combination of colors. For example, one red long, one blue medium, and one red short chromosome make up a haploid set. (In cases where crossovers have occurred, a haploid set of one color may include segments of chromatids of the other color.) All red and blue chromosomes together make up a diploid set. (c) Metaphase I **7.** This cell must be undergoing meiosis because the two homologs of a homologous pair are associated with each other at the metaphase plate; this does not occur in mitosis. Also, chiasmata are clearly present, meaning that crossing over has occurred, another process unique to meiosis.

Chapter 14

Figure Questions

Figure 14.3 All offspring would have purple flowers. (The ratio would be 1 purple: 0 white.) The P generation plants are true-breeding, so mating two purple-flowered plants produces the same result as self-pollination: All the offspring have the same trait. If Mendel had stopped after the F_1 generation, he could have concluded that the white factor had disappeared entirely and would not ever reappear.

Figure 14.8

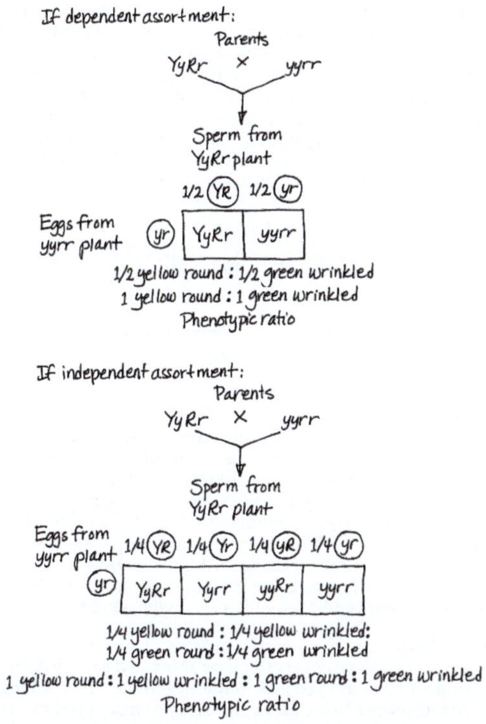

Yes, this cross would also have allowed Mendel to make different predictions for the two hypotheses, thereby allowing him to distinguish the correct one. **Figure 14.10** Your classmate would probably point out that the F_1 generation hybrids show an intermediate phenotype between those of the homozygous parents, which supports the blending hypothesis. You could respond that crossing the F_1 hybrids results in the reappearance of the white phenotype, rather than identical pink offspring, which fails to support the idea of traits blending during inheritance. **Figure 14.11** Both the I^A and I^B alleles are dominant to the *i* allele because the *i* allele results in no attached carbohydrate. The I^A and I^B alleles are codominant; both are expressed in the phenotype of I^AI^B heterozygotes, who have type AB blood. **Figure 14.12** In this cross, the final "3" and "1" of a standard cross are lumped together as a single phenotype. This occurs because in dogs that are *ee*, no pigment is deposited, thus the three dogs that have a *B* in their genotype (normally black) can no longer be distinguished from the dog that is *bb* (normally brown). **Figure 14.16** In the Punnett square, two of the three individuals with normal coloration are carriers, so the probability is ⅔. (Note that you must take into account everything you know when you calculate probability: You know she is not *aa*, so there are only three possible genotypes to consider.)

Concept Check 14.1

1. According to the law of independent assortment, 25 plants (1/16 of the offspring) are predicted to be *aatt*, or recessive for both characters. The actual result is likely to differ slightly from this value.

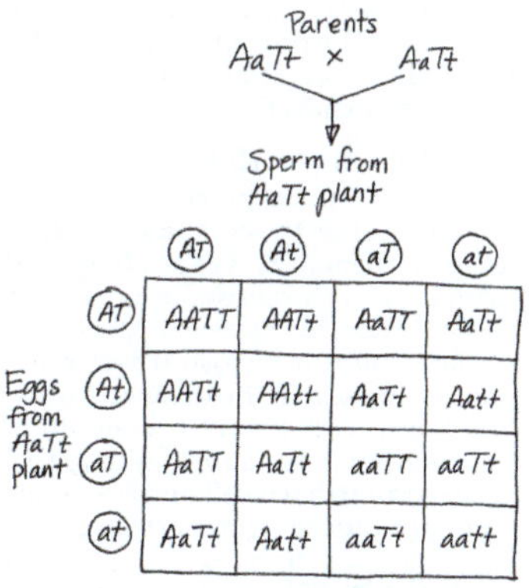

2. The plant could make eight different gametes (*YRI, YRi, YrI, Yri, yRI, yRi, yrI,* and *yri*). To fit all the possible gametes in a self-pollination, a Punnett square would need 8 rows and 8 columns. It would have spaces for the 64 possible unions of gametes in the offspring. **3.** Self-pollination is sexual reproduction because meiosis is involved in forming gametes, which unite during fertilization. As a result, the offspring in self-pollination are genetically different from the parent. (As mentioned in the footnote near the beginning of Concept 14.1, we have simplified the explanation in referring to the single pea plant as a parent. Technically, the gametophytes in the flower are the two "parents.")

Concept Check 14.2

1. ½ homozygous dominant (*AA*), 0 homozygous recessive (*aa*), and ½ heterozygous (*Aa*) **2.** ¼ *BBDD*; ¼ *BbDD*; ¼ *BBDd*; ¼ *BbDd* **3.** The genotypes that fulfill this condition are *ppyyIi, ppYyii, Ppyyii, ppYYii,* and *ppyyii*. Use the multiplication rule to find the probability of getting each genotype, and then use the addition rule to find the overall probability of meeting the conditions of this problem:

ppyyIi	½ (probability of pp) × ¼ (yy) × ½ (Ii)	= 1/16
ppYyii	½ (pp) × ½ (Yy) × ½ (ii)	= 2/16
Ppyyii	½ (Pp) × ¼ (yy) × ½ (ii)	= 1/16
ppYYii	½ (pp) × ¼ (YY) × ½ (ii)	= 1/16
ppyyii	½ (pp) × ¼ (yy) × ½ (ii)	= 1/16
Fraction predicted to have at least two recessive traits		= 6/16 or 3/8

Concept Check 14.3

1. Incomplete dominance describes the relationship between two alleles of a single gene, whereas epistasis relates to the genetic relationship between two genes (and the respective alleles of each). **2.** Half of the children would be expected to have type A blood and half type B blood. **3.** The black and white alleles are incompletely dominant, with heterozygotes being gray in color. A cross between a gray rooster and a black hen should yield approximately equal numbers of gray and black offspring.

Concept Check 14.4

1. ⅑ (Since cystic fibrosis is caused by a recessive allele, Beth and Tom's siblings who have CF must be homozygous recessive. Therefore, each parent must be a carrier of the recessive allele. Since neither Beth nor Tom has CF, this means they each have a ⅔ chance of being a carrier. If they are both carriers, there is a ¼ chance that they will have a child with CF. ⅔ × ⅔ × ¼ = ⅑); virtually 0 (Both Beth and Tom would have to be carriers to produce a child with the disease, unless a very rare mutation (change) occurred in the DNA of cells making eggs or sperm in a non-carrier that resulted in the CF allele.) **2.** In normal hemoglobin, the sixth amino acid is glutamic acid (Glu), which is acidic (has a negative charge on its side chain). In sickle-cell hemoglobin, Glu is replaced by valine (Val), which is a nonpolar amino acid, very different from Glu. The primary structure of a protein (its amino acid sequence) ultimately determines the shape of the protein and thus its function. The substitution of Val for Glu enables the hemoglobin molecules to interact with each other and form long fibers, leading to the protein's deficient function and the deformation of the red blood cell. **3.** Joan's genotype is *Dd*. Because the allele for polydactyly (*D*) is dominant to the allele for five digits per appendage (*d*), the trait is expressed in people with either the *DD* or *Dd* genotype. But because Joan's father does not have polydactyly, his genotype must be *dd*, which means that Joan inherited a *d* allele from him. Therefore, Joan, who does have the trait, must be heterozygous. **4.** In the monohybrid cross involving flower color, the ratio is 3.15 purple: 1 white, while in the human family in the pedigree, the ratio in the third generation is 1 can taste PTC: 1 cannot taste PTC. The difference is due to the small sample size (two offspring) in the human family. If the second-generation couple in this pedigree were able to have 929 offspring as in the pea plant cross, the ratio would likely be closer to 3:1. (Note that none of the pea plant crosses in Table 14.1 yielded *exactly* a 3:1 ratio.)

Summary of Key Concepts Questions

14.1 Alternative versions of genes, called alleles, are passed from parent to offspring during sexual reproduction. In a cross between purple- and white-flowered homozygous parents, the F_1 offspring are all heterozygous, each inheriting a purple allele from one parent and a white allele from the other. Because the purple allele is dominant, it determines the phenotype of the F_1 offspring to be purple, and the expression of the white allele is masked. Only in the F_2 generation is it possible for a white allele to exist in a homozygous state, which causes the white trait to be expressed.

14.2

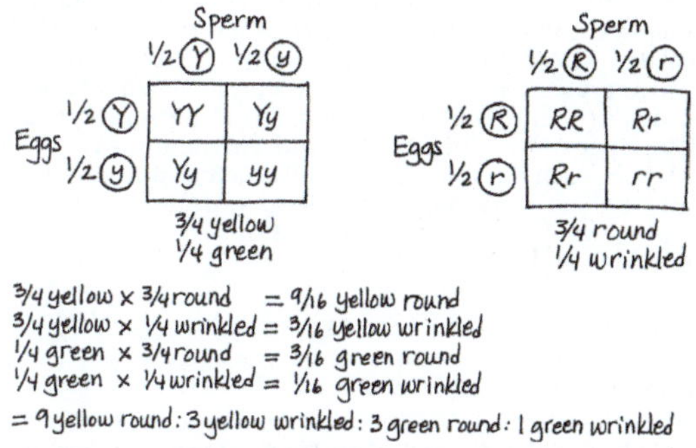

14.3 The ABO blood group is an example of multiple alleles because this single gene has more than two alleles (I^A, I^B, and *i*). Two of the alleles, I^A and I^B, exhibit codominance, since both carbohydrates (A and B) are present when these two alleles exist together in a genotype. I^A and I^B each exhibit complete dominance over the *i* allele. This situation is not an example of incomplete dominance because each allele affects the phenotype in a distinguishable way, so the result is not intermediate between the two phenotypes. Because this situation involves a single gene, it is not an example of epistasis or polygenic inheritance. **14.4** The chance of the fourth child having cystic fibrosis is ¼, as it was for each of the other children, because each birth is an independent event. We already know both parents are carriers, so whether their first three children are carriers or not has no bearing on the probability that their next child will have the disease. The parents' genotypes provide the only relevant information.

Test Your Understanding

1.

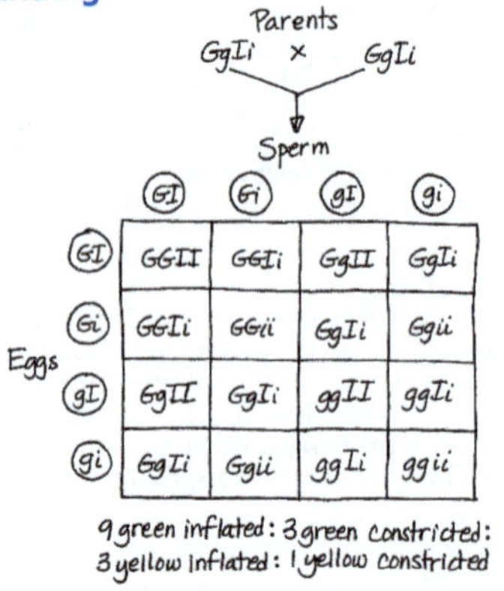

2. Man I^Ai; woman I^Bi; child *ii*. Genotypes for future children are predicted to be $\frac{1}{4}$ I^AI^B, $\frac{1}{4}$ I^Ai, $\frac{1}{4}$ I^Bi, $\frac{1}{4}$ *ii*. **3.** $\frac{1}{2}$ **4.** A cross of *Ii* × *ii* would yield offspring with a genotypic ratio of 1 *Ii*: 1 *ii* (2:2 is an equivalent answer) and a phenotypic ratio of 1 inflated: 1 constricted (2:2 is equivalent).

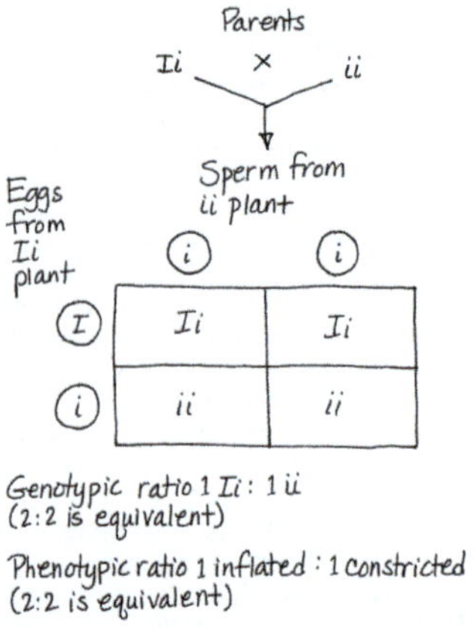

5. (a) $\frac{1}{64}$; (b) $\frac{1}{64}$; (c) $\frac{1}{8}$; (d) $\frac{1}{32}$ **6.** (a) $\frac{3}{4} \times \frac{3}{4} \times \frac{3}{4} = \frac{27}{64}$; (b) $1 - \frac{27}{64} = \frac{37}{64}$; (c) $\frac{1}{4} \times \frac{1}{4} \times \frac{1}{4} = \frac{1}{64}$; (d) $1 - \frac{1}{64} = \frac{63}{64}$ **7.** (a) $\frac{1}{256}$; (b) $\frac{1}{16}$; (c) $\frac{1}{256}$; (d) $\frac{1}{64}$; (e) $\frac{1}{128}$ **8.** (a) 1; (b) $\frac{1}{32}$; (c) $\frac{1}{8}$; (d) $\frac{1}{2}$ **9.** $\frac{1}{9}$ **10.** Matings of the original mutant cat with true-breeding noncurl cats will produce both curl and noncurl F_1 offspring if the curl allele is dominant, but only noncurl offspring if the curl allele is recessive. You would obtain some true-breeding offspring homozygous for the curl allele from matings between the F_1 cats resulting from the original curl × noncurl crosses whether the curl trait is dominant or recessive. You know that cats are true-breeding when curl × curl matings produce only curl offspring. As it turns out, the allele that causes curled ears is dominant. **11.** 25%, or $\frac{1}{4}$, will be cross-eyed; all (100%) of the cross-eyed offspring will also be white. **12.** The dominant allele *I* is epistatic to the *P/p* locus, and thus the genotypic ratio for the F_1 generation will be 9 *I-P-* (colorless) : 3 *I-pp* (colorless) : 3 *iiP-* (purple) : 1 *iipp* (red). Overall, the phenotypic ratio is 12 colorless : 3 purple : 1 red. **13.** Recessive. All affected individuals (Arlene, Tom, Wilma, and Carla) are homozygous recessive *aa*. George is *Aa*, since some of his children with Arlene are affected. Sam, Ann, Daniel, and Alan are each *Aa*, since they are all unaffected children with one affected parent. Michael also is *Aa*, since he has an affected child (Carla) with his heterozygous wife Ann. Sandra, Tina, and Christopher can each have either the *AA* or *Aa* genotype. **14.** $\frac{1}{6}$

Chapter 15

Figure Questions

Figure 15.2 The ratio would be 1 yellow round : 1 green round : 1 yellow wrinkled : 1 green wrinkled. **Figure 15.4** About $\frac{3}{4}$ of the F_2 offspring would have red eyes and about $\frac{1}{4}$ would have white eyes. About half of the white-eyed flies would be female and half would be male; similarly, about half of the red-eyed flies would be female and half would be male. (Note that the homologs with the eye color alleles would be the same shape in the Punnett square, and each offspring would inherit two alleles. The sex of the flies would be determined separately by inheritance of the sex chromosomes. Thus your Punnett square would have four possible combinations in sperm and four in eggs; it would have 16 squares altogether.) **Figure 15.7** All the males would be color-blind, and all the females would be carriers. (Another way to say this is that $\frac{1}{2}$ the offspring would be color-blind males, and $\frac{1}{2}$ the offspring would be carrier females.) **Figure 15.9** The two largest classes would still be the offspring with the phenotypes of the true-breeding P generation flies, but now would be gray vestigial and black normal, which is now the "parental type" because those were the specific allele combinations in the P generation. **Figure 15.10** The two chromosomes below, left, are like the two chromosomes inherited by the F_1 female, one from each P generation fly. They are passed by the F_1 female intact to the offspring and thus could be called "parental" chromosomes. The other two chromosomes result from crossing over during meiosis in the F_1 female. Because they have combinations of alleles not seen in either of the F_1 female's chromosomes, they can be called "recombinant" chromosomes. (Note that in this example, the alleles on the recombinant chromosomes, b^+ vg^+ and *b vg*, are the allele combinations that were on the parental chromosomes in the cross shown in Figures 15.9 and 15.10. The basis for calling them parental chromosomes is that they have the combination of alleles that was present on the P generation chromosomes.)

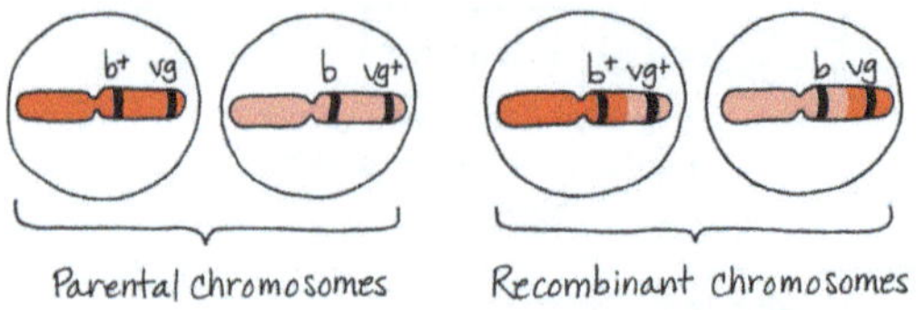

Concept Check 15.1

1. The law of segregation relates to the inheritance of alleles for a single character. The law of independent assortment of alleles relates to the inheritance of alleles for two characters. **2.** The physical basis for the law of segregation is the separation of homologs in anaphase I. The physical basis for the law of independent assortment is the alternative arrangements of all the different homologous chromosome pairs in metaphase I. **3.** To show the mutant phenotype, a male needs to possess only one mutant allele. If this gene had been on a pair of autosomes, the two alleles would both have had to be mutant in order for an individual to show the recessive mutant phenotype, a much less probable situation.

Concept Check 15.2

1. Because the gene for this eye color character is located on the X chromosome, all female offspring will be red-eyed and heterozygous ($X^{w^+}X^w$); all male offspring will inherit a Y chromosome from the father and be white-eyed (X^wY). (Another way to say this is that $\frac{1}{2}$ the offspring will be red-eyed heterozygous [carrier] females, and $\frac{1}{2}$ will be white-eyed males.) **2.** $\frac{1}{4}$ ($\frac{1}{2}$ chance that the child will inherit a Y chromosome from the father and be male × $\frac{1}{2}$ chance that he will inherit the X carrying the disease allele from his mother). If the child is a boy, there is a $\frac{1}{2}$ chance he will have the disease; a female would have zero chance (but $\frac{1}{2}$ chance of being a carrier). **3.** In a disorder caused by a dominant allele, there is no such thing as a "carrier," since those with the allele have the disorder. Because the allele is dominant, the females lose any "advantage" in having two X chromosomes, since one disorder-associated allele is sufficient to result in the disorder. All fathers who have the dominant allele will pass it along to *all* their daughters, who will also have the disorder. A mother who has the allele (and thus the disorder) will pass it to half of her sons and half of her daughters.

Concept Check 15.3

1. Crossing over during meiosis I in the heterozygous parent produces some gametes with recombinant genotypes for the two genes. Offspring with a recombinant phenotype arise from fertilization of the recombinant gametes by homozygous recessive gametes from the double-mutant parent. **2.** In each case, the alleles contributed by the female parent (in the egg) determine the phenotype of the offspring because the male in this cross contributes only recessive alleles. Thus, identifying the phenotype of the offspring tells you what alleles were in the egg. **3.** No. The order could be *A-C-B* or *C-A-B*. To determine which possibility is correct, you need to know the recombination frequency between *B* and *C*.

Concept Check 15.4

1. In meiosis, a combined 14-21 chromosome will behave as one chromosome. If a gamete receives the combined 14-21 chromosome and a normal copy of chromosome 21, trisomy 21 will result when this gamete combines with a normal gamete (with its own chromosome 21) during fertilization. **2.** No. The child can be either I^AI^Ai or I^Aii. A sperm of genotype I^AI^A could result from nondisjunction in the father during meiosis II, while an egg with the genotype *ii* could result from nondisjunction in the mother during either meiosis I or meiosis II. **3.** Activation of this gene could lead to the production of too much of this kinase. If the kinase is involved in a signaling pathway that triggers cell division, too much of it could trigger unrestricted cell division, which in turn could contribute to the development of a cancer (in this case, a cancer of one type of white blood cell).

Concept Check 15.5

1. Inactivation of an X chromosome in females and genomic imprinting. Because of X inactivation, the effective dose of genes on the X chromosome is the same in males and females. As a result of genomic imprinting, only one allele of certain genes is phenotypically expressed. **2.** The genes for leaf coloration are located in plastids within the cytoplasm. Normally, only the maternal parent transmits plastid genes to offspring. Since variegated offspring are produced only when the female parent is of the B variety, we can conclude that variety B contains both the wild-type and mutant alleles of pigment genes, producing variegated leaves. (Variety A must contain only the wild-type allele of pigment genes.) **3.** Each cell contains numerous mitochondria, and in affected individuals, most cells contain a variable mixture of normal and mutant mitochondria. The normal mitochondria carry out enough cellular respiration for survival. (The situation is similar for chloroplasts.)

Summary of Key Concepts Questions

15.1 Because the sex chromosomes are different from each other and because they determine the sex of the offspring, Morgan could use the sex of the offspring as a phenotypic character to follow the parental chromosomes. (He could also have followed them under a microscope, as the X and Y chromosomes look different.) At the same time, he could record eye color to follow the eye color alleles. **15.2** Males have only one X chromosome, along with a Y chromosome, while females have two X chromosomes. The Y chromosome has very few genes on it, while the X has about 1,000. When a recessive X-linked allele that causes a disorder is inherited by a male on the X from his mother, there isn't a second allele present on the Y (males are hemizygous), so the male has the disorder. Because females have two X chromosomes, they must inherit two recessive alleles in order to have the disorder, a rarer occurrence. **15.3** Crossing over results in new combinations of alleles. Crossing over is a random occurrence, and the more distance there is between two genes, the more chances there are for crossing over to occur, leading to new allele combinations. **15.4** In inversions and reciprocal translocations, the same genetic material is present in the same relative amount but just organized differently. In aneuploidy, duplications, deletions, and nonreciprocal translocations, the balance of genetic material is upset, as large segments are either missing or present in more than one copy. Apparently, this type of imbalance is very damaging to the organism. (Although it isn't lethal in the developing embryo, the reciprocal translocation that produces the Philadelphia chromosome can lead to a serious condition, cancer, by altering the expression of important genes.) **15.5** In these cases, the sex of the parent contributing an allele affects the inheritance pattern. For imprinted genes, either the paternal or the maternal allele is expressed, depending on the imprint. For mitochondrial and chloroplast genes, only the maternal contribution will affect offspring phenotype because the offspring inherit these organelles from the mother, via the egg cytoplasm.

Test Your Understanding

1. 0; ½; 1⁄16 **2.** Recessive; if the disorder were dominant, it would affect at least one parent of a child born with the disorder. The disorder's inheritance is sex-linked because it is seen only in boys. For a girl to have the disorder, she would have to inherit recessive alleles from *both* parents. This would be very rare, since males with the recessive allele on their X chromosome die in their early teens. **3.** 17%; yes, it is consistent. In Figure 15.9, the recombination frequency was also 17%. (You'd expect this to be the case since these are the very same two genes, and their distance from each other wouldn't change from one experiment to another.) **4.** Between *T* and *A*, 12%; between *A* and *S*, 5% **5.** Between *T* and *S*, 18%; sequence of genes is *T–A–S* **6.** 6%; wild-type heterozygous for normal wings and red eyes × recessive homozygous for vestigial wings and purple eyes **7.** Fifty percent of the offspring will show phenotypes resulting from crossovers. These results would be the same as those from a cross where *A* and *B* were *not* on the same chromosome, and you would interpret the results to mean that the genes are unlinked. (Further crosses involving other genes on the same chromosome would reveal the genetic linkage and map distances.) **8.** 450 each of blue oval and white round (parentals) and 50 each of blue round and white oval (recombinants) **9.** About one-third of the distance from the vestigial wing locus to the brown eye locus **10.** Because bananas are triploid, homologous pairs cannot line up during meiosis. Therefore, it is not possible to generate gametes that can fuse to produce a zygote with the triploid number of chromosomes. **12.** (a) For each pair of genes, you had to generate an F_1 dihybrid fly; let's use the *A* and *B* genes as an example. You obtained homozygous parental flies, either the first with dominant alleles of the two genes (*AABB*) and the second with recessive alleles (*aabb*), or the first with dominant alleles of gene *A* and recessive alleles of gene *B* (*AAbb*) and the second with recessive alleles of gene *A* and dominant alleles of gene *B* (*aaBB*). Breeding either of these pairs of P generation flies gave you an F_1 dihybrid, which you then testcrossed with a doubly homozygous recessive fly (*aabb*). You classed the offspring as parental or recombinant, based on the genotypes of the P generation parents (either of the two pairs described above). You added up the number of recombinant types and then divided by the total number of offspring. This gave you the recombination percentage (in this case, 8%), which you can translate into map units (8 map units) to construct your map.
(b)

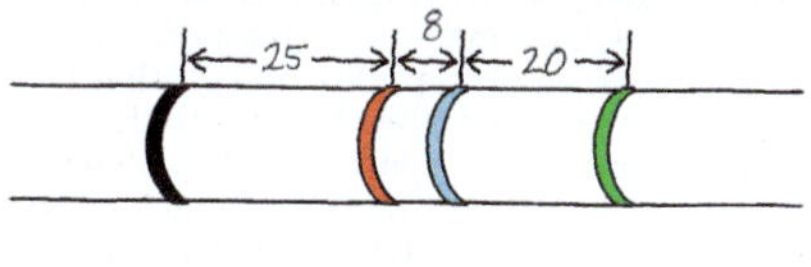

Chapter 16

Figure Questions

Figure 16.2 The living S cells found in the blood sample were able to reproduce to yield more S cells, indicating that the S trait is a permanent, heritable change, rather than just a one-time use of the dead S cells' capsules. **Figure 16.4** The radioactivity would have been found in the pellet when proteins were labeled (batch 1) because proteins would have had to enter the bacterial cells to program them with genetic instructions. It's hard for us to imagine now, but the DNA might have played a structural role that allowed some of the proteins to be injected while it remained outside the bacterial cell (thus no radioactivity in the pellet in batch 2). **Figure 16.7** (1) The nucleotides in a single DNA strand are held together by covalent bonds between an oxygen on the 3′ carbon of one nucleotide and the phosphate group on the 5′ carbon of the next nucleotide in the chain. Instead of covalent bonds, the bonds that hold the two strands together are hydrogen bonds between a nitrogenous base on one strand and the complementary nitrogenous base on the other strand. (Hydrogen bonds are weaker than covalent bonds, but there are so many hydrogen bonds in a DNA double helix that, together, they are enough to hold the two strands together.) (2) The left diagram shows the most detail. It shows that each sugar-phosphate backbone is made up of sugars (blue pentagons) and phosphates (yellow circles) joined by covalent bonds (black lines). The middle diagram doesn't show any detail in the backbone. Both the left and middle diagrams label the bases and represent their complementarity by the complementary shapes at the ends of the bases (curves/indents for G/C or V's/notches for T/A). The diagram on the right is the least detailed, implying that the base pairs pair up, but showing all bases as the same shape so not including the information about specificity and complementarity visible in the other two diagrams. The left and right diagrams show that the strand on the left was synthesized most recently, as indicated by the light blue color. All three diagrams show the 5′ and 3′ ends of the strands. **Figure 16.11** The tube from the first replication would look the same, with a middle band of hybrid ^{15}N–^{14}N DNA, but the second tube would not have the upper band of two light blue strands. Instead, it would have a bottom band of two dark blue strands, like the bottom band in the result predicted after one replication in the conservative model. **Figure 16.12** In the bubble at the top of the micrograph in (b), arrows should be drawn pointing left and right to indicate the two replication forks. **Figure 16.14** Looking at any of the DNA strands, we see that one end is called the 5′ end and the other the 3′ end. If we proceed from the 5′ end to the 3′ end on the left-most strand, for example, we list the components in this order: phosphate group → 5′ C of the sugar → 3′ C → phosphate → 5′ C → 3′ C. Going in the opposite direction on the same strand, the components proceed in the reverse order: 3′ C → 5′ C → phosphate. Thus, the two directions are distinguishable, which is what we mean when we say that the strands have directionality. (Review Figure 16.5 if necessary.)
Figure 16.17

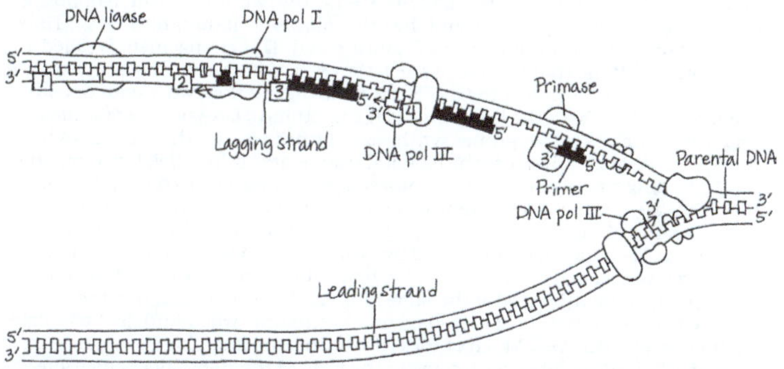

Figure 16.18

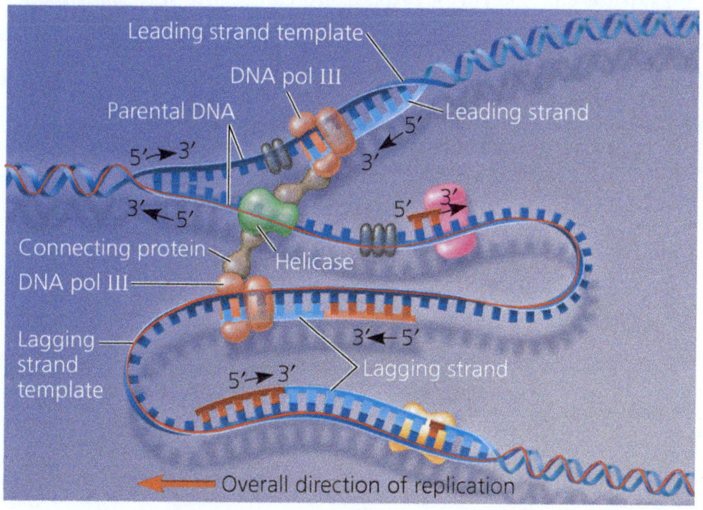

Figure 16.23 The two members of a homologous pair (which would be the same color) would be associated tightly together at the metaphase plate during metaphase I of meiosis I. In metaphase of mitosis, however, each chromosome would be lined up individually, so the two chromosomes of the same color would be in different places at the metaphase plate.

Concept Check 16.1

1. You can't tell which end is the 5′ end. You need to know which end has a phosphate group on the 5′ carbon (the 5′ end) or which end has an —OH group on the 3′ carbon (the 3′ end). **2.** He expected that the mouse injected with the mixture of heat-killed S cells and living R cells would survive, since neither type of cell alone would kill the mouse.

Concept Check 16.2

1. Complementary base pairing ensures that the two daughter molecules are exact copies of the parental molecule. When the two strands of the parental molecule separate, each serves as a template on which nucleotides are arranged, by the base-pairing rules, into new complementary strands. **2.** DNA pol III covalently adds nucleotides to new DNA strands and proofreads each added nucleotide for correct base pairing. **3.** In the cell cycle, DNA synthesis occurs during the S phase, between the G_1 and G_2 phases of interphase. DNA replication is therefore complete before the mitotic phase begins. **4.** Synthesis of the leading strand is initiated by an RNA primer, which must be removed and replaced with DNA, a task that could not be performed if the cell's DNA pol I were nonfunctional. In the overview box in Figure 16.17, just to the left of the top origin of replication, a functional DNA pol I would replace the RNA primer of the leading strand (shown in red) with DNA nucleotides (blue). The nucleotides would be added onto the 3′ end of the first Okazaki fragment of the upper lagging strand (the right half of the replication bubble).

Concept Check 16.3

1. A nucleosome is made up of eight histone proteins, two each of four different types, around which DNA is wound. Linker DNA runs from one nucleosome to the next. **2.** Euchromatin is chromatin that becomes less compacted during interphase and is accessible to the cellular machinery responsible for gene activity. Heterochromatin, on the other hand, remains quite condensed during interphase and contains genes that are largely inaccessible to this machinery. **3.** The nuclear lamina is a netlike array of protein filaments that provides mechanical support just inside the nuclear envelope and thus maintains the shape of the nucleus. Considerable evidence also supports the existence of a nuclear matrix, a framework of protein fibers extending throughout the nuclear interior.

Summary of Key Concepts Questions

16.1 Each strand in the double helix has polarity; the end with a phosphate group on the 5′ carbon of the sugar is called the 5′ end, and the end with an —OH group on the 3′ carbon of the sugar is called the 3′ end. The two strands run in opposite directions, one running 5′ → 3′ and the other alongside it running 3′ → 5′. Thus, each end of the molecule has both a 5′ and a 3′ end. This arrangement is called "antiparallel." If the strands were parallel, they would both run 5′ → 3′ in the same direction, so an end of the molecule would have either two 5′ ends or two 3′ ends. **16.2** On both the leading and lagging strands, DNA polymerase adds onto the 3′ end of an RNA primer synthesized by primase, synthesizing DNA in the 5′ → 3′ direction. Because the parental strands are antiparallel, however, only on the leading strand does synthesis proceed continuously into the replication fork. The lagging strand is synthesized bit by bit in the direction away from the fork as a series of shorter Okazaki fragments, which are later joined together by DNA ligase. Each fragment is initiated by synthesis of an RNA primer by primase as soon as a given stretch of single-stranded template strand is opened up. Although both strands are synthesized at the same rate, synthesis of the lagging strand is delayed because initiation of each fragment begins only when sufficient template strand is available. **16.3** Much of the chromatin in an interphase nucleus is present as the 30-nm fiber, with some in the form of the 10-nm fiber and some as looped domains of the 30-nm fiber. (These different levels of chromatin packing may reflect differences in gene expression occurring in these regions.) Also, a small percentage of the chromatin, such as that at the centromeres and telomeres, is highly condensed heterochromatin.

Test Your Understanding

1. C **2.** C **3.** B **4.** D **5.** A **6.** D **7.** B **8.** A

9. Like histones, the *E. coli* proteins would be expected to contain many basic (positively charged) amino acids, such as lysine and arginine, which can form weak bonds with the negatively charged phosphate groups on the sugar-phosphate backbone of the DNA molecule.

11.

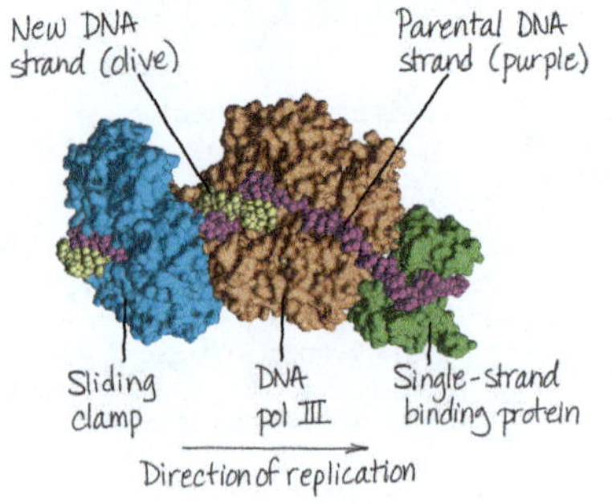

Chapter 17

Figure Questions

Figure 17.3 The previously presumed pathway would have been wrong. The new results would support this pathway: precursor → citrulline → ornithine → arginine. They would also indicate that class I mutants have a defect in the second step and class II mutants have a defect in the first step. **Figure 17.5** The mRNA sequence (5′-UGGUUUGGCUCA-3′) is the same as the nontemplate DNA strand sequence (5′-TGGTTTGGCTCA-3′), except there is a U in the mRNA wherever there is a T in the DNA. The nontemplate strand is probably used to represent a DNA sequence because it represents the mRNA sequence, containing codons. (This is why it's called the coding strand.) **Figure 17.6** Arg (or R)–Glu (or E)–Pro (or P)–Arg (or R) **Figure 17.8** The processes are similar in that polymerases form polynucleotides complementary to an antiparallel DNA template strand. In replication, however, both strands act as templates, whereas in transcription, only one DNA strand acts as a template. **Figure 17.9** The RNA polymerase would bind directly to the promoter, rather than being dependent on the previous binding of transcription factors.

Figure 17.12

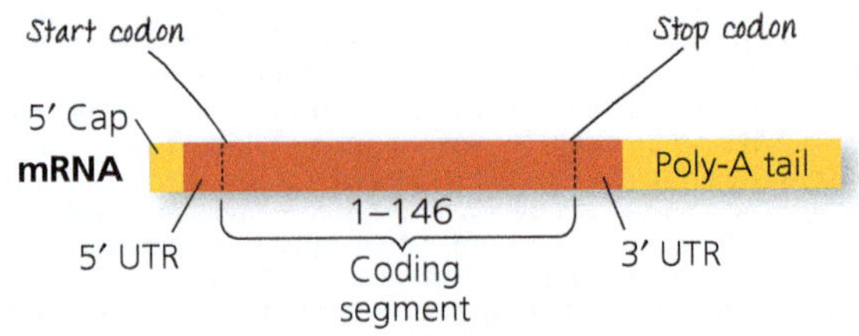

Figure 17.16 The anticodon on the tRNA is 3′-AAG-5′, so it would bind to the mRNA codon 5′-UUC-3′. This codon codes for phenylalanine, which is the amino acid this tRNA would carry. **Figure 17.22** It would be packaged in a vesicle, transported to the Golgi apparatus for further processing, and then transported via a vesicle to the plasma membrane. The vesicle would fuse with the membrane, releasing the protein outside the cell. **Figure 17.24** The mRNA farthest to the right (the longest one) started transcription first. The ribosome at the top, closest to the DNA, started translating first and thus has the longest polypeptide.

Concept Check 17.1

1. Recessive **2.** A polypeptide made up of 10 Gly (glycine) amino acids

3.

"Template sequence" (from nontemplate sequence in problem, written 3′ → 5′): 3′-ACGACTGAA-5′

mRNA sequence: 5′-UGCUGACUU-3′

Translated: Cys-STOP

If the nontemplate sequence could have been used as a template for transcribing the mRNA, the protein translated from the mRNA would have a completely different amino acid sequence and would most likely be nonfunctional. (It would also be shorter because of the UGA stop signal shown in the mRNA sequence above—and possibly others earlier in the mRNA sequence.)

Concept Check 17.2

1. A promoter is the region of DNA to which RNA polymerase binds to begin transcription. It is at the upstream end of the gene (transcription unit). **2.** In a bacterial cell, part of the RNA polymerase recognizes the gene's promoter and binds to it. In a eukaryotic cell, transcription factors must bind to the promoter first, then the RNA polymerase binds to them. In both cases, sequences in the promoter determine the precise binding of RNA polymerase so the enzyme is in the right location and orientation. **3.** The transcription factor that recognizes the TATA sequence would be unable to bind, so RNA polymerase could not bind and transcription of that gene probably would not occur.

Concept Check 17.3

1. Due to alternative splicing of exons, each gene can result in multiple different mRNAs and can thus direct synthesis of multiple different proteins. **2.** In watching a show recorded with a DVR, you watch segments of the show itself (exons) and fast-forward through the commercials, which are thus like introns. However, unlike introns, commercials remain in the recording, while the introns are cut out of the RNA transcript during RNA processing. **3.** Once the mRNA has exited the nucleus, the cap prevents it from being degraded by hydrolytic enzymes and facilitates its attachment to ribosomes. If the cap were removed from all mRNAs, the cell would no longer be able to synthesize any proteins and would probably die.

Concept Check 17.4

1. First, each aminoacyl-tRNA synthetase specifically recognizes a single amino acid and attaches it only to an appropriate tRNA. Second, a tRNA charged with its specific amino acid binds only to an mRNA codon for that amino acid. **2.** A signal peptide on the leading end of the polypeptide being synthesized is recognized by a signal-recognition particle that brings the ribosome to the ER membrane. There the ribosome attaches and continues to synthesize the polypeptide, depositing it in the ER lumen. **3.** Because of wobble, the tRNA could bind to either 5′-GCA-3′ or 5′-GCG-3′, both of which code for alanine (Ala). Alanine would be attached to the tRNA (see diagram, upper right). **4.** When one ribosome terminates translation and dissociates, the two subunits would be very close to the cap. This could facilitate their rebinding and initiating synthesis of a new polypeptide, thus increasing the efficiency of translation.

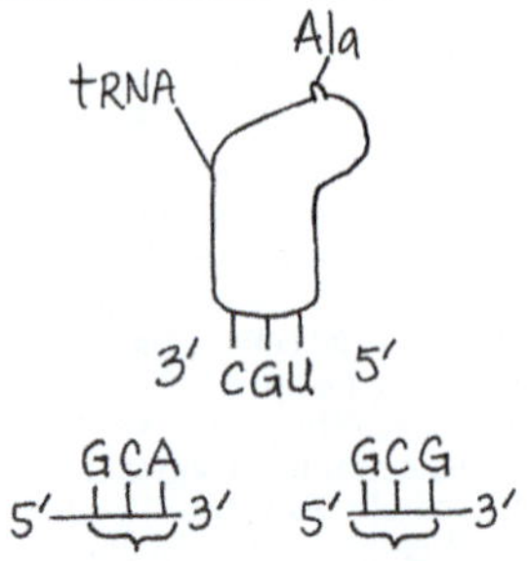

Concept Check 17.5

1. In the mRNA, the reading frame downstream from the deletion is shifted, leading to a long string of incorrect amino acids in the polypeptide, and in most cases, a stop codon will occur, leading to premature termination. The polypeptide will most likely be nonfunctional. **2.** Heterozygous individuals, said to have sickle-cell trait, have a copy each of the wild-type allele and the sickle-cell allele. Both alleles will be expressed, so these individuals will have both normal and sickle-cell hemoglobin molecules. Apparently, having a mix of the two forms of β-globin has no effect under most conditions, but during prolonged periods of low blood oxygen (such as at higher altitudes), these individuals can show some signs of sickle-cell disease.

3.

Normal DNA sequence (template strand is on top):	3′-TACTTGTCCGATATC-5′ 5′-ATGAACAGGCTATAG-3′
mRNA sequence:	5′-AUGAACAGGCUAUAG-3′
Amino acid sequence:	Met-Asn-Arg-Leu-STOP
Mutated DNA sequence (template strand is on top):	3′-TACTTGTCCAATATC-5′ 5′-ATGAACAGGTTATAG-3′
mRNA sequence:	5′-AUGAACAGGUUAUAG-3′
Amino acid sequence:	Met-Asn-Arg-Leu-STOP

No effect: The amino acid sequence is Met-Asn-Arg-Leu both before and after the mutation because the mRNA codons 5′-CUA-3′ and 5′-UUA-3′ both code for Leu. (The fifth codon is a stop codon.)

Summary of Key Concepts Questions

17.1 A gene contains genetic information in the form of a nucleotide sequence. The gene is first transcribed into an RNA molecule, and a messenger RNA molecule is ultimately translated into a polypeptide. The polypeptide makes up part or all of a protein, which performs a function in the cell and contributes to the phenotype of the organism. **17.2** Both bacterial and eukaryotic genes have promoters, regions where RNA polymerase ultimately binds and begins transcription. In bacteria, RNA polymerase binds directly to the promoter; in eukaryotes, transcription factors bind first to the promoter, and then RNA polymerase binds to the transcription factors and promoter together. **17.3** Both the 5′ cap and the 3′ poly-A tail help the mRNA exit from the nucleus and then, in the cytoplasm, help ensure mRNA stability and allow it to bind to ribosomes. **17.4** In the context of the ribosome, tRNAs function as translators between the nucleotide-based language of mRNA and the amino-acid-based language of polypeptides. A tRNA carries a specific amino acid, and the anticodon on the tRNA is complementary to the codon on the mRNA that codes for that amino acid. In the ribosome, the tRNA binds to the A site. Then the polypeptide being synthesized (currently on the tRNA in the P site) is joined to the new amino acid, which becomes the new (C-terminal) end of the polypeptide. Next, the tRNA in the A site moves to the P site. After the polypeptide is transferred to the new tRNA, thus adding the new amino acid, the now empty tRNA moves from the P site to the E site, where it exits the ribosome. **17.5** When a nucleotide base is altered chemically, its base-pairing characteristics may be changed. When that happens, an incorrect nucleotide is likely to be incorporated into the complementary strand during the next replication of the DNA, and successive rounds of replication will perpetuate the mutation. Once the gene is transcribed, the mutated codon may code for a different amino acid that inhibits or changes the function of a protein. If the chemical change in the base is detected and repaired by the DNA repair system before the next replication, no mutation will result.

Test Your Understanding

1. B **2.** C **3.** A **4.** A **5.** B **6.** C **7.** D **8.** No. Transcription and translation are separated in space and time in a eukaryotic cell, as a result of the eukaryotic cell's nuclear membrane.

9.

Type of RNA	Functions
Messenger RNA (mRNA)	Carries information specifying amino acid sequences of polypeptides from DNA to ribosomes
Transfer RNA (tRNA)	Serves as translator molecule in protein synthesis; translates mRNA codons into amino acids
Ribosomal RNA (rRNA)	In a ribosome, plays a structural role; as a ribozyme, plays a catalytic role (catalyzes peptide bond formation)
Primary transcript	Is a precursor to mRNA, rRNA, or tRNA, before being processed; some intron RNA acts as a ribozyme, catalyzing its own splicing
Small RNAs in spliceosome	Play structural and catalytic roles in spliceosomes, the complexes of protein and RNA that splice pre-mRNA

Chapter 18

Figure Questions

Figure 18.3 As the concentration of tryptophan in the cell falls, eventually there will be none bound to *trp* repressor molecules. These will then change into their inactive shapes and dissociate from the operator, allowing transcription of the operon to resume. The enzymes for tryptophan synthesis will be made, and they will again synthesize tryptophan in the cell. **Figure 18.9** Each of the two polypeptides has two regions—one that makes up part of MyoD's DNA-binding domain and one that makes up part of MyoD's activation domain. Each functional domain in the complete MyoD protein is made up of parts of both polypeptides. **Figure 18.11** In both types of cell, the albumin gene enhancer has the three control elements colored yellow, gray, and red. The sequences in the liver and lens cells would be identical, since the cells are in the same organism. **Figure 18.18** Even if the mutant MyoD protein couldn't activate the *myoD* gene, it could still turn on genes for the other proteins in the pathway (other transcription factors, which would turn on the genes for muscle-specific proteins, for example). Therefore, some differentiation would occur. But unless there were other activators that could compensate for the loss of the MyoD protein's activation of the *myoD* gene, the cell would not be able to maintain its differentiated state. **Figure 18.22** Normal Bicoid protein would be made in the anterior end and compensate for the presence of mutant *bicoid* mRNA put into the egg by the mother. Development should be normal, with a head present. (This is what was observed.) **Figure 18.25** The mutation is likely to be recessive because it is more likely to have an effect if both copies of the gene are mutated and code for nonfunctional proteins. If one normal copy of the gene is present, its product could inhibit the cell cycle. (However, there are also known cases of dominant *p53* mutations.) **Figure 18.27** Cancer is a disease in which cell division occurs without its usual regulation. Cell division can be stimulated by growth factors (see Figure 12.18), which bind to cell-surface receptors (see Figure 11.8). Cancer cells evade these normal controls and can often divide in the absence of growth factors (see Figure 12.19). This suggests that the receptor proteins or some other components in a signaling pathway are abnormal in some way (see, for example, the mutant Ras protein in Figure 18.24) or are expressed at abnormal levels, as seen for the receptors in this figure. Under some circumstances in the mammalian body, steroid hormones such as estrogen and progesterone can also promote cell division. These molecules also use cell-signaling pathways, as described in Concept 11.2 (see Figure 11.9). Because signaling receptors are involved in triggering cells to undergo cell division, it is not surprising that altered genes encoding these proteins might play a significant role in the development of cancer. Genes might be altered through either a mutation that changes the function of the protein product or a mutation that causes the gene to be expressed at abnormal levels that disrupt the overall regulation of the signaling pathway.

Concept Check 18.1

1. Binding by the *trp* corepressor (tryptophan) activates the *trp* repressor, which binds to the *trp* operator, shutting off transcription of the *trp* operon. Binding by the *lac* inducer (allolactose) inactivates the *lac* repressor, so that it can no longer bind to the *lac* operator, leading to transcription of the *lac* operon. **2.** When glucose is scarce, cAMP is bound to CRP and CRP is bound to the *lac* promoter, favoring the binding of RNA polymerase. However, in the absence of lactose, the *lac* repressor is bound to the *lac* operator, blocking RNA polymerase binding to the *lac* promoter. Therefore, the *lac* operon genes are not transcribed. **3.** The cell would continuously produce β-galactosidase and the two other enzymes for using lactose, even in the absence of lactose, thus wasting cell resources.

Concept Check 18.2

1. Histone acetylation is generally associated with gene expression, while DNA methylation is generally associated with lack of expression. **2.** The same enzyme could not methylate both a histone and a DNA base. Enzymes are very specific in structure, and an enzyme that could methylate an amino acid of a protein would not be able to fit the base of a DNA nucleotide into the same active site. **3.** General transcription factors function in assembling the transcription initiation complex at the promoters for all genes. Specific transcription factors bind to control elements associated with a particular gene and, once bound, either increase (activators) or decrease (repressors) transcription of that gene. **4.** Regulation of translation initiation, degradation of the mRNA, activation of the protein (by chemical modification, for example), and protein degradation **5.** The three genes should have some similar or identical sequences in the control elements of their enhancers. Because of this similarity, the same specific transcription factors in muscle cells could bind to the enhancers of all three genes and stimulate their expression coordinately.

Concept Check 18.3

1. Both miRNAs and siRNAs are small, single-stranded RNAs that associate with a complex of proteins and then can base-pair with mRNAs that have a complementary sequence. This base pairing leads to either degradation of the mRNA or blockage of its translation. In some yeasts, siRNAs associated with proteins in a different complex can bind back to centromeric chromatin, recruiting enzymes that cause condensation of that chromatin into heterochromatin. Both miRNAs and siRNAs are processed from double-stranded RNA precursors but have subtle variations in the structure of those precursors. **2.** The mRNA would persist and be translated into the cell division–promoting protein, and the cell would probably divide. If the intact miRNA is necessary for inhibition of cell division, then division of this cell might be inappropriate. Uncontrolled cell division could lead to formation of a mass of cells (tumor) that prevents proper functioning of the organism and could contribute to the development of cancer. **3.** The *XIST* RNA is transcribed from the *XIST* gene on the X chromosome that will be inactivated. It then binds to that chromosome and induces heterochromatin formation. A likely model is that *XIST* RNA somehow recruits chromatin modification enzymes that lead to formation of heterochromatin.

Concept Check 18.4

1. Cells undergo differentiation during embryonic development, becoming different from each other. Therefore, the adult organism is made up of many highly specialized cell types. **2.** By binding to a receptor on the receiving cell's surface and triggering a signal transduction pathway, involving intracellular molecules such as second messengers and transcription factors that affect gene expression **3.** The products of maternal effect genes, made and deposited into the egg by the mother, determine the head and tail ends, as well as the back and belly, of the egg and embryo (and eventually the adult fly). **4.** The lower cell is synthesizing signaling molecules because the gene encoding them is activated, meaning that the appropriate specific transcription factors are binding to the gene's enhancer. The genes encoding these specific transcription factors are also being expressed in this cell because the transcriptional activators that can turn them on were expressed in the precursor to this cell. A similar explanation also applies to the cells expressing the receptor proteins. This scenario began with specific cytoplasmic determinants localized in specific regions of the egg. These cytoplasmic determinants were distributed unevenly to daughter cells, resulting in cells going down different developmental pathways.

Concept Check 18.5

1. A cancer-causing mutation in a proto-oncogene usually makes the gene product overactive, whereas a cancer-causing mutation in a tumor-suppressor gene usually makes the gene product nonfunctional. **2.** When an individual has inherited an oncogene or a mutant allele of a tumor-suppressor gene **3.** Apoptosis is signaled by p53 protein when a cell has extensive DNA damage, so apoptosis plays a protective role in eliminating a cell that might contribute to cancer. If mutations in the genes in the apoptotic pathway blocked apoptosis, a cell with such damage could continue to divide and might lead to tumor formation.

Summary of Key Concepts Questions

18.1 A corepressor and an inducer are both small molecules that bind to the repressor protein in an operon, causing the repressor to change shape. In the case of a corepressor (like tryptophan), this shape change allows the repressor to bind to the operator, blocking transcription. In contrast, an inducer causes the repressor to dissociate from the operator, allowing transcription to begin. **18.2** The chromatin must not be tightly condensed because it must be accessible to transcription factors. The appropriate specific transcription factors (activators) must bind to the control elements in the enhancer of the gene, while repressors must not be bound. The DNA must be bent by a bending protein so the activators can contact the mediator proteins and form a complex with general transcription factors at the promoter. Then RNA polymerase must bind and begin transcription. **18.3** miRNAs do not "code" for the amino acids of a protein—they are never translated. Each miRNA associates with a group of proteins to form a complex. Binding of the complex to an mRNA with a complementary sequence causes that mRNA to be degraded or blocks its translation. This is considered gene regulation because it controls the amount of a particular mRNA that can be translated into a functional protein. **18.4** The first process involves cytoplasmic determinants, including mRNAs and proteins, placed into specific locations in the egg by maternal cells. The embryonic cells that are formed from different regions in the egg during early cell divisions will have different proteins in them, which will direct different programs of gene expression. The second process involves the cell in question responding to signaling molecules secreted by neighboring cells (induction). The signaling pathway in the responding cell also leads to a different pattern of gene expression. The coordination of these two processes results in each cell following a unique pathway in the developing embryo. **18.5** The protein product of a proto-oncogene is usually involved in a pathway that stimulates cell division. The protein product of a tumor-suppressor gene is usually involved in a pathway that inhibits cell division.

Test Your Understanding

1. C **2.** A **3.** B **4.** C **5.** C **6.** D **7.** A **8.** C **9.** B **10.** D
11. (a)

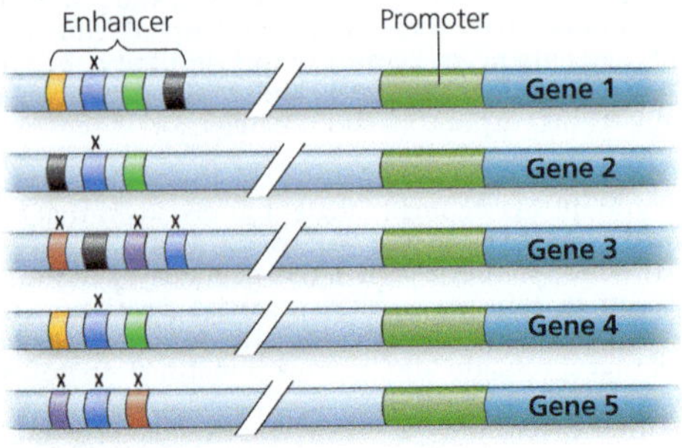

The purple, blue, and red activator proteins would be present.

(b)

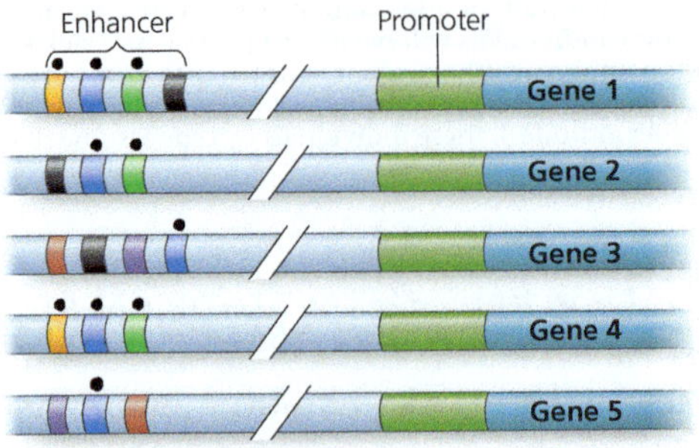

Only gene 4 would be transcribed.
(c) In nerve cells, the yellow, blue, green, and black activators would have to be present, thus activating transcription of genes 1, 2, and 4. In skin cells, the red, black, purple, and blue activators would have to be present, thus activating genes 3 and 5.

Chapter 19

Figure Questions

Figure 19.2 Beijerinck might have concluded that the agent was a toxin produced by the plant that was able to pass through a filter but that became more and more dilute. In this case, he would have concluded that the infectious agent could not replicate.
Figure 19.4

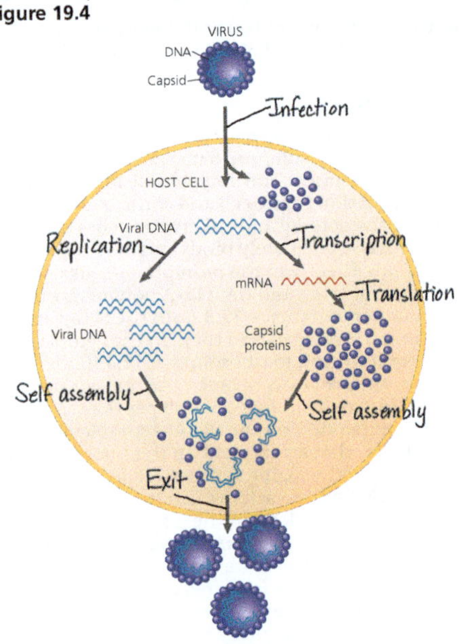

Figure 19.9 The main protein on the cell surface that HIV binds to is called CD4. However, HIV also requires a "co-receptor," which in many cases is a protein called CCR5. HIV binds to both of these proteins together and then is taken into the cell. Researchers discovered this requirement by studying individuals who seemed to be resistant to HIV infection despite multiple exposures. These individuals turned out to have mutations in the gene that encodes CCR5 such that the protein apparently cannot act as a co-receptor, and so HIV can't enter and infect cells.

Concept Check 19.1

1. TMV consists of one molecule of RNA surrounded by a helical array of proteins. The influenza virus has eight molecules of RNA, each associated with proteins and wound into a double helix. Another difference between the viruses is that the influenza virus has an outer envelope and TMV does not. **2.** The T2 phages were an excellent choice for use in the Hershey-Chase experiment because they consist of only DNA surrounded by a protein coat, and DNA and protein were the two candidates for macromolecules that carried genetic information. Hershey and Chase were able to radioactively label each type of molecule alone and follow it during separate infections of *E. coli* cells with T2. Only the DNA entered the bacterial cell during infection, and only labeled DNA showed up in some of the progeny phage. Hershey and Chase concluded that the DNA must carry the genetic information necessary for the phage to reprogram the cell and produce progeny phages.

Concept Check 19.2

1. Lytic phages can only carry out lysis of the host cell, whereas lysogenic phages may either lyse the host cell or integrate into the host chromosome. In the latter case, the viral DNA (prophage) is simply replicated along with the host chromosome. Under certain conditions, a prophage may exit the host chromosome and initiate a lytic cycle. **2.** Both the CRISPR-Cas system and miRNAs involve RNA molecules bound in a protein complex and acting as "homing devices" that enable the complex to bind a complementary sequence, but miRNAs are involved in regulating gene expression (by affecting mRNAs) and the CRISPR-Cas system protects bacterial cells from foreign invaders (infecting phages). Thus the CRISPR-Cas system is more like an immune system than are miRNAs. **3.** Both the viral RNA polymerase and the RNA polymerase in Figure 17.10 synthesize an RNA molecule complementary to a template strand. However, the RNA polymerase in Figure 17.10 uses one of the strands of the DNA double helix as a template, whereas the viral RNA polymerase uses the RNA of the viral genome as a template. **4.** HIV is called a retrovirus because it synthesizes DNA using its RNA genome as a template. This is the reverse ("retro") of the usual DNA → RNA information flow. **5.** There are many steps that could be interfered with: binding of the virus to the cell, reverse transcriptase function, integration into the host cell chromosome, genome synthesis (in this case, transcription of RNA from the integrated provirus), assembly of the virus inside the cell, and budding of the virus. (Many of these, if not all, are targets of actual medical strategies to block progress of the infection in HIV-infected people.)

Concept Check 19.3

1. Mutations can lead to a new strain of a virus that can no longer be effectively fought by the immune system, even if an animal had been exposed to the original strain; a virus can jump from one species to a new host; and a rare virus can spread if a host population becomes less isolated. **2.** In horizontal transmission, a plant is infected from an external source of virus, which enters through a break in the plant's epidermis due to damage by herbivores or other agents. In vertical transmission, a plant inherits viruses from its parent either via infected seeds (sexual reproduction) or via an infected cutting (asexual reproduction). **3.** Humans are not within the host range of TMV, so they can't be infected by the virus. (TMV can't bind to receptors on human cells and infect them.)

Summary of Key Concepts Questions

19.1 Viruses are generally considered nonliving, because they are not capable of replicating outside of a host cell and are unable to carry out the energy-transforming reactions of metabolism. To replicate and carry out metabolism, they depend completely on host enzymes and resources. **19.2** Single-stranded RNA viruses require an RNA polymerase that can make RNA using an RNA template. (Cellular RNA polymerases make RNA using a DNA template.) Retroviruses require reverse transcriptases to make DNA using an RNA template. (Once the first DNA strand has been made, the same enzyme can promote synthesis of the second DNA strand.) **19.3** The mutation rate of RNA viruses is higher than that of DNA viruses because RNA polymerase has no proofreading function, so errors in replication are not corrected. Their higher mutation rate means that RNA viruses change faster than DNA viruses, leading to their being able to have an altered host range and to evade immune defenses in possible hosts.

Test Your Understanding

1. C **2.** D **3.** C **4.** D **5.** B

6. As shown below, the viral genome would be translated into capsid proteins and envelope glycoproteins directly, rather than after a complementary RNA copy was made. A complementary RNA strand would still be made, however, that could be used as a template for many new copies of the viral genome.

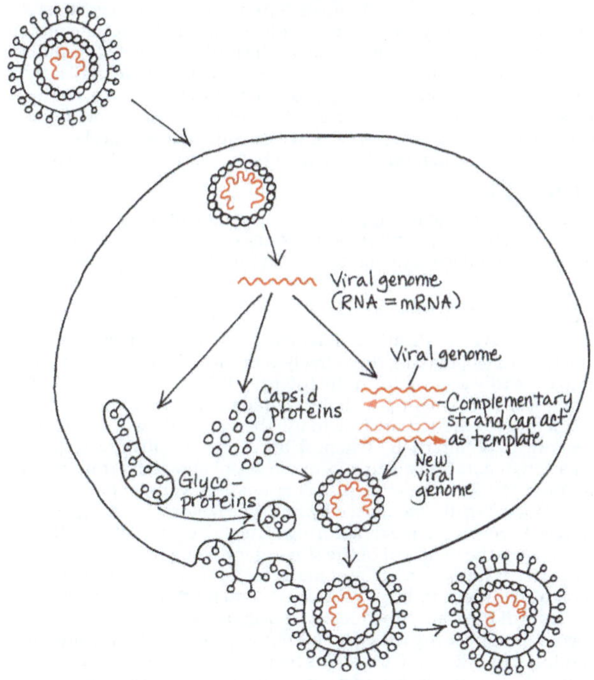

Chapter 20

Figure Questions

Figure 20.5

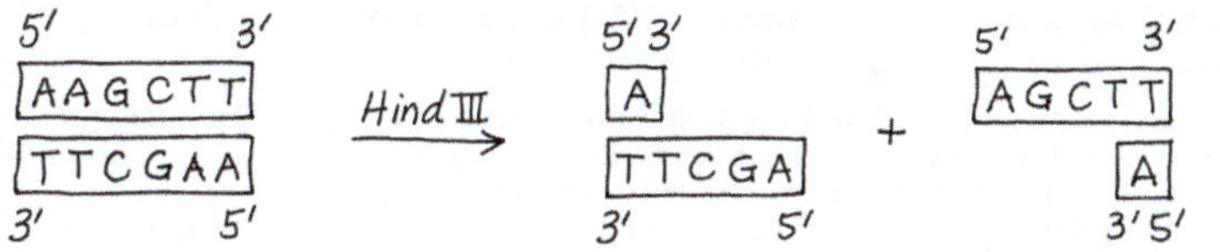

Figure 20.16 None of the eggs with the transplanted nuclei from the four-cell embryo at the upper left would have developed into a tadpole. Also, the result might include only some of the tissues of a tadpole, which might differ, depending on which nucleus was transplanted. (This assumes that there was some way to tell the four cells apart, as one can in some frog species.) **Figure 20.21** Using converted iPS cells would not carry the same risk, which is its major advantage. Because the donor cells would come from the patient, they would be perfectly matched. The patient's immune system would recognize them as "self" cells and would not mount an attack (which is what leads to rejection). On the other hand, cells that are rapidly dividing might carry a risk of inducing some type of tumor or contributing to development of cancer.

Concept Check 20.1

1. The covalent sugar-phosphate bonds of the DNA strands **2.** Yes, *Pvu*I will cut the molecule (at the position indicated by the dashed red line).

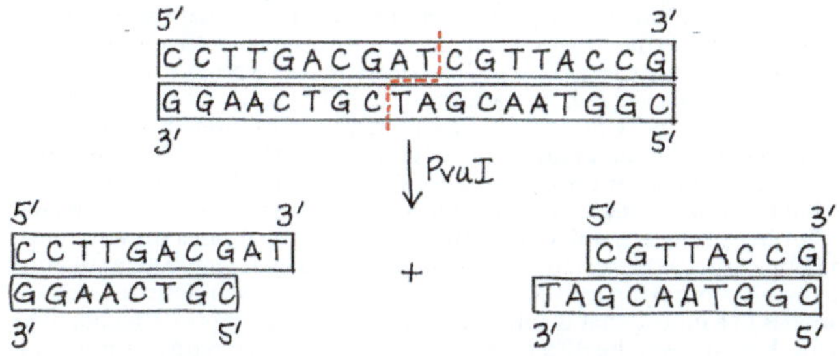

3. Some eukaryotic genes are too large to be incorporated into bacterial plasmids. Bacterial cells lack the means to process RNA transcripts into mRNA, and even if the need for RNA processing is avoided by using cDNA, bacteria lack enzymes to catalyze the post-translational processing that many eukaryotic proteins require to function properly. (This is often the case for human proteins, which are a focus of biotechnology.) **4.** During the replication of the ends of linear DNA molecules (see Figure 16.20), an RNA primer is used at the 5′ end of each new strand. The RNA must be replaced by DNA nucleotides, but DNA polymerase is incapable of starting from scratch at the 5′ end of a new DNA strand. During PCR, the primers are made of DNA nucleotides already, so they don't need to be replaced—they just remain as part of each new strand. Therefore, there is no problem with end replication during PCR, and the fragments don't shorten with each replication.

Concept Check 20.2

1. Complementary base pairing is involved in cDNA synthesis, which is required for the first three techniques: RT-PCR, DNA microarray analysis, and RNA sequencing. Reverse transcriptase uses mRNA as a template to synthesize the first strand of cDNA, adding nucleotides complementary to those on the mRNA. Complementary base pairing is also involved when DNA polymerase synthesizes the second strand of the cDNA. Furthermore, in RT-PCR, the primers must base-pair with their target sequences in the DNA mixture, locating one specific region among many. In DNA microarray analysis, the labeled cDNA probe binds only to the specific target sequence due to complementary nucleic acid hybridization (DNA-DNA hybridization). In RNA-seq, when sequencing the cDNAs, base complementarity plays a role in the sequencing process. During CRISPR-Cas9 editing, a guide RNA in the CRISPR-Cas9 complex must base-pair with its complementary sequence in the genome (in the target gene) before editing can occur. The repair system also uses complementarity of bases when using a template strand to repair breaks. **2.** As a researcher interested in cancer development, you would want to study genes represented by spots that are green or red because these are genes for which the expression level differs between the two types of tissues. Some of these genes may be expressed differently as a result of cancer, while others might play a role in causing cancer, so both would be of interest.

Concept Check 20.3

1. The state of chromatin modification in the nucleus from the intestinal cell was undoubtedly less similar to that of a nucleus from a fertilized egg, explaining why many fewer of these nuclei were able to be reprogrammed. In contrast, the chromatin in a nucleus from a cell at the four-cell stage would have been much more like that of a nucleus in a fertilized egg and therefore much more easily programmed to direct development. **2.** No, primarily because of subtle (and perhaps not so subtle) differences in the environment in which the clone develops and lives from that in which the original pet lived (see the differences noted in Figure 20.18). This does provoke ethical questions. To produce Dolly, also a mammal, several hundred embryos were cloned, but only one survived to adulthood. If any of the "reject" dog embryos survived to birth as defective dogs, would they be killed? Is it ethical to produce living animals that may be defective? You can probably think of other ethical issues as well. **3.** Given that muscle cell differentiation involves a master regulatory gene (*MyoD*), you might start by introducing either the MyoD protein or an expression vector carrying the *MyoD* gene into stem cells. (This is not likely to work, because the embryonic precursor cell in Figure 18.18 is more differentiated than the stem cells you are working with, and some other changes would have to be introduced as well. But it's a good way to start! And you may be able to think of others.)

Concept Check 20.4

1. Stem cells continue to reproduce themselves, ensuring that the corrective gene product will continue to be made. **2.** Herbicide resistance, pest resistance, disease resistance, salinity resistance, drought resistance, and delayed ripening **3.** Because hepatitis A is an RNA virus, you could isolate RNA from the blood and try to detect copies of hepatitis A RNA by RT-PCR. You would first reverse-transcribe the blood mRNA into cDNA and then use PCR to amplify the cDNA, using primers specific to hepatitis A sequences. If you then ran the products on an electrophoretic gel, the presence of a band of the appropriate size would support your hypothesis. Alternatively, you could use RNA-seq to sequence all the RNAs in your patient's blood and see whether any of the sequences match up with that of hepatitis A. (Since you are only seeking one sequence, though, RT-PCR is probably a better choice.)

Summary of Key Concepts Questions

20.1 A plasmid vector and a source of foreign DNA to be cloned are both cut with the same restriction enzyme, generating restriction fragments with sticky ends. These fragments are mixed together, ligated, and reintroduced into bacterial cells. The plasmid has a gene for resistance to an antibiotic. That antibiotic is added to the host cells, and only cells that have taken up a plasmid will grow. (Another technique allows researchers to select only the cells that have a recombinant plasmid, rather than the original plasmid without an inserted gene.) **20.2** The genes that are expressed in a given tissue or cell type determine the proteins (and noncoding RNAs) that are the basis of the structure and functions of that tissue or cell type. Understanding which groups of interacting genes establish particular structures and carry out certain functions will help us learn how the parts of an organism work together. We will also be better able to treat diseases that occur when faulty gene expression leads to malfunctioning tissues. **20.3** (1) Cloning a mouse involves transplanting a nucleus from a differentiated mouse cell into a mouse egg cell that has had its own nucleus removed. Activating the egg cell and promoting its development into an embryo in a surrogate mother results in a mouse that is genetically identical to the mouse that donated the nucleus. In this case, the differentiated nucleus has been reprogrammed by factors in the egg cytoplasm. (2) Mouse ES cells are generated from inner cells in mouse blastocysts, so in this case the cells are "naturally" reprogrammed by the process of reproduction and development. (Cloned mouse embryos can also be used as a source of ES cells.) (3) iPS cells can be generated without the use of embryos from a differentiated adult mouse cell by adding certain transcription factors into the cell. In this case, the transcription factors are reprogramming the cells to become pluripotent. **20.4** First, the disease must be caused by a single gene, and the molecular basis of the problem must be understood. Second, the cells that are going to be introduced into the patient must be cells that will integrate into body tissues and continue to multiply (and provide the needed gene product). Third, the gene must be able to be introduced into the cells in question in a safe way, as there have been instances of cancer resulting from some gene therapy trials. (Note that this will require testing the procedure in mice; moreover, the factors that determine a safe vector are not yet well understood. Maybe one of you will go on to solve this problem!)

Test Your Understanding

1. D **2.** B **3.** C **4.** B **5.** C **6.** B **7.** A **8.** B **9.** You would use PCR to amplify the gene. This could be done from genomic DNA. Alternatively, mRNA could be isolated from lens cells and reverse-transcribed by reverse transcriptase to make cDNA. This cDNA could then be used for PCR. In either case, the gene would then be inserted into an expression vector so you could produce the protein and study it. **10.** Crossing over, which causes recombination, is a random event. The chance of crossing over occurring between two loci increases as the distance between them increases. If a SNP is located very close to a disease-associated allele, it is said to be genetically linked. Crossing over will rarely occur between the SNP and the allele, so the SNP can be used as a genetic marker indicating the presence of the particular allele.

11.

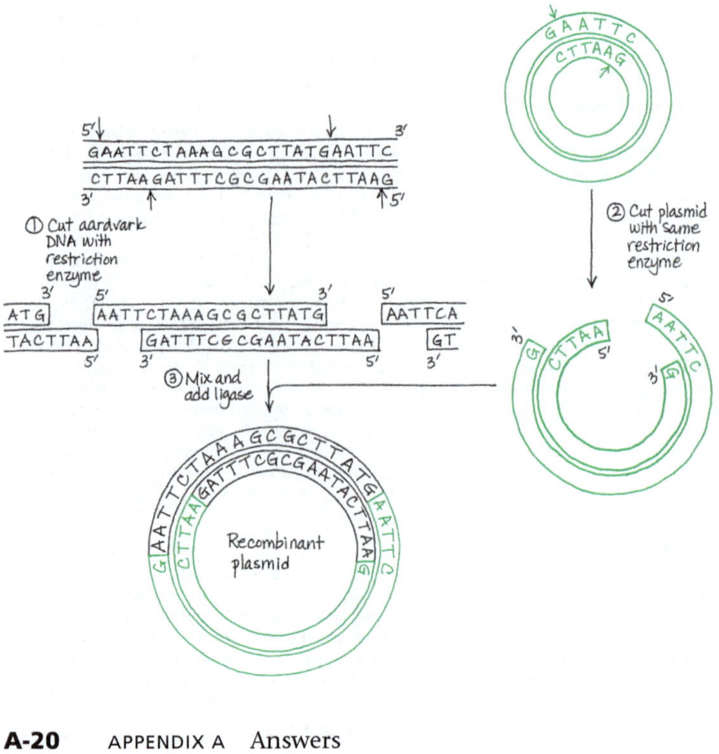

Chapter 21

Figure Questions

Figure 21.2 In step 2 of this figure, the order of the fragments relative to each other is not known and will be determined later by computer. The unordered nature of the fragments is reflected by their scattered arrangement in the diagram.
Figure 21.8 The transposon would be cut out of the DNA at the original site rather than copied, so the figure would show the original stretch of DNA without the transposon after the mobile transposon had been cut out. **Figure 21.10** The RNA transcripts extending from the DNA in each transcription unit are shorter on the left and longer on the right. This means that RNA polymerase must be starting on the left end of the unit and moving toward the right.

Figure 21.13

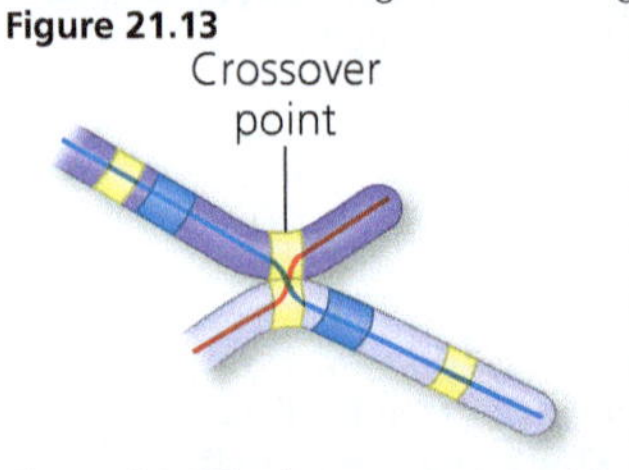

Figure 21.14 Pseudogenes are nonfunctional. They could have arisen by any mutations in the second copy that made the gene product unable to function. Examples are base changes that introduce stop codons in the sequence, alter amino acids, or change a region of the gene promoter so that the gene can no longer be expressed. **Figure 21.15** At position 5, there is an R (arginine) in lysozyme and a K (lysine) in α-lactalbumin; both of these are basic amino acids.
Figure 21.16 Let's say a transposable element (TE) existed in the intron to the left of the indicated EGF exon in the EGF gene, and the same TE was present in the intron to the right of the indicated F exon in the fibronectin gene. During meiotic recombination, these TEs could cause nonsister chromatids on homologous chromosomes to pair up incorrectly, as seen in Figure 21.13. One gene might end up with an F exon next to an EGF exon. Further mistakes in pairing over many generations might result in these two exons being separated from the rest of the gene and placed next to a single or duplicated K exon. In general, the presence of repeated sequences in introns and between genes facilitates these processes because it allows incorrect pairing of nonsister chromatids, leading to novel exon combinations. **Figure 21.18** Since you know that chimpanzees do not speak but humans do, you'd probably want to know how many amino acid differences there are between the human wild-type FOXP2 protein and that of the chimpanzee and whether these changes affect the function of the protein. (As we explain later in the text, there are two amino acid differences.) You know that humans with mutations in this gene have severe language impairment. You would want to learn more about the human mutations by checking whether they affect the same amino acids in the gene product that the chimpanzee sequence differences affect. If so, those amino acids might play an important role in the function of the protein in language. Going further, you could analyze the differences between the chimpanzee and mouse FOXP2 proteins. You might ask: Are they more similar than the chimpanzee and human proteins? (It turns out that the chimpanzee and mouse proteins have only one amino acid difference and thus are more similar than the chimpanzee and human proteins, which have two differences, and also are more similar than the human and mouse proteins, which have three differences.)

Concept Check 21.1

1. In the whole-genome shotgun approach, short fragments are generated by cutting the genome with multiple restriction enzymes. These fragments are cloned, sequenced, and then ordered by computer programs that identify overlapping regions.

Concept Check 21.2

1. The Internet allows centralization of databases such as GenBank and software resources such as BLAST, making them freely accessible. Having all the data in a central database, easily accessible on the Internet, minimizes the possibility of errors and of researchers working with different data. It streamlines the process of science, since all researchers are able to use the same software programs, rather than each having to obtain their own, possibly different, software. It speeds up dissemination of data and ensures as much as possible that errors are corrected in a timely fashion. These are just a few answers; you can probably think of more. **2.** Cancer is a disease caused by multiple factors. Focusing on a single gene or a single defect would mean ignoring other factors that may influence the cancer and even the behavior of the single gene being studied. The systems approach, because it takes into account many factors at the same time, is more likely to lead to an understanding of the causes and most useful treatments for cancer. **3.** Some of the transcribed region is accounted for by introns. The rest is transcribed into noncoding RNAs, including small RNAs, such as microRNAs (miRNAs), siRNAs, or piRNAs. These RNAs help regulate gene expression by blocking translation, causing degradation of mRNA, binding to the promoter and repressing transcription, or causing remodeling of chromatin structure. The longer noncoding RNAs (lncRNAs) may also contribute to gene regulation or to chromatin remodeling. **4.** Genome-wide association studies use the systems biology approach in that they consider the correlation of many single nucleotide polymorphisms (SNPs) with particular diseases, such as heart disease and diabetes, in an attempt to find patterns of SNPs that correlate with each disease.

Concept Check 21.3

1. Alternative splicing of RNA transcripts from a gene and post-translational processing of polypeptides **2.** At the top of the web page, you can see the number of genomes completed and those considered permanent drafts in a bar graph by year. Scrolling down, you can see the number of complete and incomplete sequencing projects by year, the number of projects by domain by year (the genomes of viruses and metagenomes are counted too, even though these are not "domains"), the phylogenetic distribution of bacterial genome projects, and projects by sequencing center. Finally, near the bottom, you can see a pie chart of the "Project Relevance of Bacterial

Genome Projects," which shows that about 47% have medical relevance. The web page ends with another pie chart showing the sequencing centers for archaeal and bacterial projects. **3.** Prokaryotes are generally smaller cells than eukaryotic cells, and they reproduce by binary fission. The evolutionary process involved is natural selection for more quickly reproducing cells: The faster they can replicate their DNA and divide, the more likely they will be able to dominate a population of prokaryotes. The less DNA they have to replicate, then, the faster they will reproduce.

Concept Check 21.4

1. The number of genes is higher in mammals, and the amount of noncoding DNA is greater. Also, the presence of introns in mammalian genes makes them larger, on average, than prokaryotic genes. **2.** The copy-and-paste transposon mechanism and retrotransposition **3.** In the rRNA gene family, identical transcription units for all three different RNA products are present in long arrays, repeated one after the other. The large number of copies of the rRNA genes enable organisms to produce the rRNA for enough ribosomes to carry out active protein synthesis, and the single transcription unit for the three rRNAs ensures that the relative amounts of the different rRNA molecules produced are correct—every time one rRNA is made, a copy of each of the other two is made as well. Rather than numerous identical units, each globin gene family consists of a relatively small number of nonidentical genes. The differences in the globin proteins encoded by these genes result in production of hemoglobin molecules adapted to particular developmental stages of the organism. **4.** The exons would be classified as exons (1.5%); the enhancer region containing the distal control elements, the region closer to the promoter containing the proximal control elements, and the promoter itself would be classified as regulatory sequences (5%); and the introns would be classified as introns (20%).

Concept Check 21.5

1. If meiosis is faulty, two copies of the entire genome can end up in a single cell. Errors in crossing over during meiosis can lead to one segment being duplicated while another is deleted. During DNA replication, slippage backward along the template strand can result in segment duplication. **2.** For either gene, a mistake in crossing over during meiosis could have occurred between the two copies of that gene, such that one ended up with a duplicated exon. (The other copy would have ended up with a deleted exon.) This could have happened several times, resulting in the multiple copies of a particular exon in each gene. **3.** Homologous transposable elements scattered throughout the genome provide sites where recombination can occur between different chromosomes. Movement of these elements into coding or regulatory sequences may change expression of genes, which can affect the phenotype in a way that is subject to natural selection. Transposable elements also can carry genes with them, leading to dispersion of genes and in some cases different patterns of expression. Transport of an exon during transposition and its insertion into a gene may add a new functional domain to the originally encoded protein, a type of exon shuffling. (For any of these changes to be heritable, they must happen in germ cells, cells that will give rise to gametes.) **4.** Because more offspring are born to women who have this inversion, it must provide some advantage during the process of reproduction and development. Because proportionally more offspring have this inversion, we would expect it to persist and spread in the population. (In fact, evidence in the study allowed the researchers to conclude that it has been increasing in proportion in the population. You'll learn more about population genetics in the next unit.)

Concept Check 21.6

1. Because both humans and macaques are primates, their genomes are expected to be more similar than the macaque and mouse genomes are. The mouse lineage diverged from the primate lineage before the human and macaque lineages diverged. **2.** Homeotic genes differ in their *non*homeobox sequences, which determine the interactions of homeotic gene products with other transcription factors and hence which genes are regulated by the homeotic genes. These nonhomeobox sequences differ in the two organisms, as do the expression patterns of the homeobox genes. **3.** *Alu* elements must have undergone transposition more actively in the human genome for some reason. Their increased numbers may have then allowed more recombination errors in the human genome, resulting in more or different duplications. The divergence of the organization and content of the two genomes presumably made the chromosomes of each genome less similar to those of the other, thus accelerating divergence of the two species by making matings less and less likely to result in fertile offspring due to the mismatch of genetic information.

Summary of Key Concepts Questions

21.1 One focus of the Human Genome Project was to improve sequencing technology in order to speed up the process. During the project, many advances in sequencing technology allowed faster reactions and detection of products, which were therefore less expensive. **21.2** The most significant finding is that more than 75% of the human genome appears to be transcribed at some point in at least one of the cell types studied. Also, at least 80% of the genome contains an element that is functional, participating in gene regulation or maintaining chromatin structure in some way. The project was expanded to include other species to further investigate the functions of these transcribed DNA elements. It is necessary to carry out this type of analysis on the genomes of species that can be used in laboratory experiments. **21.3** (a) In general, bacteria and archaea have smaller genomes, lower numbers of genes, and higher gene density than eukaryotes. (b) Among eukaryotes, there is no apparent systematic relationship between genome size and phenotype. The number of genes is often lower than would be expected from the size of the genome—in other words, the gene density is often lower in larger genomes. (Humans are an example.) **21.4** Transposable element–related sequences can move from place to place in the genome, and a subset of these sequences make a new copy of themselves when they do so. Thus, it is not surprising that they make up a significant percentage of the genome, and this percentage might be expected to increase over evolutionary time. **21.5** Chromosomal rearrangements within a species lead to some individuals having different chromosomal arrangements. Each of these individuals could still undergo meiosis and produce gametes, and fertilization involving gametes with different chromosomal arrangements could result in viable offspring. However, during meiosis in the offspring, the maternal and paternal chromosomes might not be able to pair up, causing gametes with incomplete sets of chromosomes to form. Most often, when zygotes are produced from such gametes, they do not survive. Ultimately, a new species could form if two different chromosomal arrangements became prevalent within a population and individuals could mate successfully only with other individuals having the same arrangement. **21.6** Comparing the genomes of two closely related species can reveal information about more recent evolutionary events, perhaps events that resulted in the distinguishing characteristics of the two species. Comparing the genomes of very distantly related species can tell us about evolutionary events that occurred a very long time ago. For example, genes that are shared between two distantly related species must have arisen before the two species diverged.

Test Your Understanding

1. B **2.** A **3.** C

4.
1. ATETI...PKSSD...TSSTT...NARRD
2. ATETI...PKSSE...TSSTT...NARRD
3. ATETI...PKSSD...TSSTT...NARRD
4. ATETI...PKSSD...TSSNT...SARRD
5. ATETI...PKSSD...TSSTT...NARRD
6. VTETI...PKSSD...TSSTT...NARRD

(a) Lines 1, 3, and 5 are the C, G, R species. (b) Line 4 is the human sequence. See the above figure for the differences between the human and C, G, R sequences—the underlined amino acids at which the human sequence has an N where the C, G, R sequences have a T, and an S where C, G, R have an N. (c) Line 6 is the O sequence. (d) See the above figure. There is one amino acid difference between the mouse (the circled E on line 2) and the C, G, R species (which have a D in that position). There are three amino acid differences between the mouse and the human. (The boxed E, T, and N in the mouse sequence are instead D, N, and S, respectively, in the human sequence.) (e) Because only one amino acid difference arose during the 60–100 million years since the mouse and C, G, R species diverged, it is somewhat surprising that two additional amino acid differences resulted during the 6 million years since chimpanzees and humans diverged. This indicates that the *FOXP2* gene has been evolving faster in the human lineage than in the lineages of other primates.

Chapter 22

Figure Questions

Figure 22.6 You should have circled the branch located at the far left of Figure 1.20. Although three of the descendants (*Certhidea olivacea, Camarhynchus pallidus*, and *Camarhynchus parvulus*) of this common ancestor did not eat insects, the other three species that descended from this ancestor did not eat insects. **Figure 22.8** The common ancestor lived about 5.5 million years ago. **Figure 22.12** The colors and body forms of these mantises allow them to blend into their surroundings, providing an example of how organisms are well suited for life in their environments. The mantises also share features with one another (and with other mantises), such as six legs, grasping forelimbs, and large eyes. These shared features illustrate another key observation about life: the unity that results from descent from a common ancestor. Over time, as these mantises diverged from a common ancestor, they accumulated different adaptations that made them well suited for life in their different environments. Eventually, as enough differences accumulated between mantis populations, new species were formed, thus contributing to the great diversity of life. **Figure 22.13** These results show that being reared from the egg stage on one plant species or the other did not result in the adult having a beak length appropriate for that host; instead, adult beak lengths were determined primarily by the population from which the eggs were obtained. Because an egg from a balloon vine population likely had long-beaked parents, while an egg from a goldenrain tree population likely had short-beaked parents, these results indicate that beak length is an inherited trait. **Figure 22.14** Both strategies should increase the time that it takes *S. aureus* to become resistant to a new drug. If a drug that harms *S. aureus* does not harm other bacteria, natural selection will not favor resistance to that drug in the other species. This would decrease the chance that *S. aureus* would acquire resistance genes from other bacteria—thus slowing the evolution of resistance. Similarly, selection for resistance to a drug that slows the growth but does not kill *S. aureus* is much weaker than selection for resistance to a drug that kills *S. aureus*—again slowing the evolution of resistance. **Figure 22.17** Based on this evolutionary tree, crocodiles are more closely related to birds than to lizards because they share a more recent common ancestor with birds (ancestor 5) than with lizards (ancestor 4). **Figure 22.20** Hind limb structure changed first. *Rodhocetus* lacked flukes, but its pelvic bones and hind limbs had changed substantially from how those bones were shaped and arranged in *Pakicetus*. For example, in *Rodhocetus*, the pelvis and hind limbs appear to be oriented for paddling, whereas they were oriented for walking in *Pakicetus*.

Concept Check 22.1

1. Hutton and Lyell proposed that geologic events in the past were caused by the same processes operating today, at the same gradual rate. This principle suggested that Earth must be much older than a few thousand years, the age that was widely accepted in the early 19th century. Hutton's and Lyell's ideas also stimulated Darwin to reason that the slow accumulation of small changes could ultimately produce the profound changes documented in the fossil record. In this context, the age of Earth was important to Darwin, because unless Earth was very old, he could not envision how there would have been enough time for evolution to occur.

2. By this criterion, Cuvier's explanation of the fossil record and Lamarck's hypothesis of evolution are both scientific. Cuvier thought that species did not evolve over time. He also suggested that sudden, catastrophic events caused extinctions in particular areas and that such regions were later repopulated by a different set of species that immigrated from other areas. These assertions can be tested against the fossil record. Lamarck's principle of use and disuse can be used to make testable predictions for fossils of groups such as whale ancestors as they adapted to a new habitat. Lamarck's principle of use and disuse and his associated principle of the inheritance of acquired characteristics can also be tested directly in living organisms.

Concept Check 22.2

1. Organisms share characteristics (the unity of life) because they share common ancestors; the great diversity of life occurs because new species have repeatedly formed when descendant organisms gradually adapted to different environments, thereby becoming different from their ancestors. **2.** The fossil mammal species (or its ancestors) would most likely have colonized the Andes from within South America, whereas ancestors of mammals currently found in Asian mountains would most likely have colonized those mountains from other parts of Asia. As a result, the Andes fossil species would share a more recent common ancestor with South American mammals than with mammals in Asia. Thus, for many of its traits, the fossil mammal species would probably more closely resemble mammals that live in South American jungles than mammals that live on Asian mountains. It is also possible, however, that the Andean fossil mammal could resemble a mammal from the mountains of Asia because similar environments had selected for similar adaptations (even though the fossil and Asian species were only distantly related to one another). **3.** As long as the white phenotype (encoded by the genotype *pp*) continues to be favored by natural selection, the frequency of the *p* allele will likely increase over time in the population. If the proportion of white individuals increases relative to purple individuals, the frequency of the recessive *p* allele will also increase relative to that of the *p* allele, which only appears in purple individuals (some of which also carry a *p* allele).

Concept Check 22.3

1. An environmental factor such as a drug does not create new traits, such as drug resistance, but rather selects for traits among those that are already present in the population. **2.** (a) Despite their different functions, the forelimbs of different mammals are structurally similar because they all represent modifications of a structure found in the common ancestor; thus, they are homologous structures. (b) In this case, the similar features of these mammals represent analogous features that arose by convergent evolution. The similarities between the sugar glider and flying squirrel indicate that similar environments selected for similar adaptations despite different ancestry. **3.** At the time that dinosaurs originated, Earth's landmasses formed a single large continent, Pangaea. Because many dinosaurs were large and mobile, it is likely that early members of these groups lived on many different parts of Pangaea. When Pangaea broke apart, fossils of these organisms would have moved with the rocks in which they were deposited. As a result, we would predict that fossils of early dinosaurs would have a broad geographic distribution (this prediction has been upheld).

Summary of Key Concepts Questions

Concept 22.1 Darwin thought that descent with modification occurred as a gradual, steplike process. The age of Earth was important to him because if Earth were only a few thousand years old (as conventional wisdom suggested), there wouldn't have been sufficient time for major evolutionary change. **Concept 22.2** All species have the potential to overreproduce—that is, to produce more offspring than can be supported by the environment. This ensures that there will be what Darwin called a "struggle for existence" in which many of the offspring are eaten, starved, diseased, or unable to reproduce for a variety of other reasons. Members of a population exhibit a range of heritable variations, some of which make it likely that their bearers will leave more offspring than other individuals (for example, the bearer may escape predators more effectively or be more tolerant of the physical conditions of the environment). Over time, natural selection resulting from factors such as predators, lack of food, or the physical conditions of the environment can increase the proportion of individuals with favorable traits in a population (evolutionary adaptation). **Concept 22.3** The hypothesis that cetaceans originated from a terrestrial mammal and are closely related to even-toed ungulates is supported by several lines of evidence. For example, fossils document that early cetaceans had hind limbs, as expected for organisms that descended from a land mammal; these fossils also show that cetacean hind limbs became reduced over time. Other fossils show that early cetaceans had a type of ankle bone that is otherwise found only in even-toed ungulates, providing strong evidence that even-toed ungulates are the land mammals to which cetaceans are most closely related. DNA sequence data also indicate that even-toed ungulates are the land mammals to which cetaceans are, most closely related.

Test Your Understanding

1. B **2.** D **3.** C **4.** B **5.** A

7. (a)

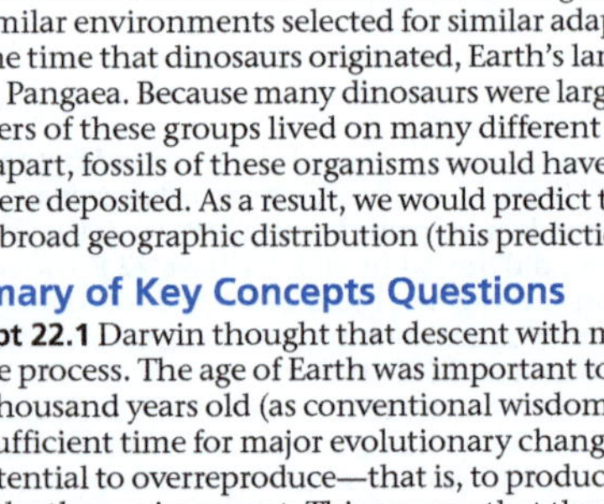

(b) The rapid rise in the percentage of mosquitoes resistant to DDT was most likely caused by natural selection in which mosquitoes resistant to DDT could survive and reproduce while other mosquitoes could not. (c) In India—where DDT resistance first appeared—natural selection would have caused the frequency of resistant mosquitoes to increase over time. If resistant mosquitoes then migrated from India (for example, transported by wind or in planes, trains, or ships) to other parts of the world, the frequency of DDT resistance would increase there as well. In addition, if resistance to DDT were to arise independently in mosquito populations outside of India, those populations would also experience an increase in the frequency of DDT resistance.

Chapter 23

Figure Questions

Figure 23.4 The genetic code is redundant, meaning that more than one codon can specify the same amino acid. As a result, a substitution at a particular site in a coding region of the *Adh* gene might change the codon but not the translated amino acid, and thus not the resulting protein encoded by the gene. One way an insertion in an exon would not affect the gene produced is if it occurs in an untranslated region of the exon. (This is the case for the insertion at location 1,703.) **Figure 23.7** There should be 24 red balls. **Figure 23.8** The predicted frequencies are 36% C^RC^R, 48% C^RC^W, and 16% C^WC^W. **Figure 23.9** Overall, by chance the frequency of the C^W allele first increases in generation 2 and then falls to zero in generation 3—causing the C^R allele to become fixed (reach a frequency of 100%). **Figure 23.12** The frequency of banded color patterns in island populations would probably increase. Since mainland populations did not decline in size, the number of individuals migrating from the mainland to the islands would probably not decline either. As a result, after island populations had decreased in size, alleles encoding banded coloration that were transferred from the mainland would comprise a larger proportion of the gene pool in island populations. This would cause the frequency of banded color patterns in island populations to increase. **Figure 23.13** Directional selection. Goldenrain tree has smaller fruit than does the native host, balloon vine. Thus, in soapberry bug populations feeding on goldenrain tree, bugs with shorter beaks had an advantage, resulting in directional selection for shorter beak length. **Figure 23.16** Crossing a single female's eggs with both an SC and an LC male's sperm allowed the researchers to directly compare the effects of the males' contribution to the next generation, since both batches of offspring had the same maternal contribution. This isolation of the male's impact enabled researchers to draw conclusions about differences in genetic "quality" between the SC and LC males. **Figure 23.18** Under prolonged low-oxygen conditions, some of the red blood cells of a heterozygote may sickle, leading to harmful effects. This does not occur in individuals with two wild-type hemoglobin alleles, suggesting that there may be selection against heterozygotes in malaria-free regions (where heterozygote advantage does not occur). However, since heterozygotes are healthy under most conditions, selection against them is unlikely to be strong.

Concept Check 23.1

1. Within a population, genetic differences among individuals provide the raw material on which natural selection and other mechanisms can act. Without such differences, allele frequencies could not change over time—and hence the population could not evolve. **2.** Many mutations occur in somatic cells, which do not produce gametes and so are lost when the organism dies. Of mutations that do occur in cell lines that produce gametes, many do not have a phenotypic effect on which natural selection can act. Others have a harmful effect and are thus unlikely to increase in frequency because they decrease the reproductive success of their bearers. **3.** Its genetic variation (whether measured at the level of the gene or at the level of nucleotide sequences) would probably drop over time. During meiosis, crossing over and the independent assortment of chromosomes produce many new combinations of alleles. In addition, a population contains a vast number of possible mating combinations, and fertilization brings together the gametes of individuals with different genetic backgrounds. Thus, via crossing over, independent assortment of chromosomes, and fertilization, sexual reproduction reshuffles alleles into fresh combinations each generation. Without sexual reproduction, the rate of forming new combinations of alleles would be vastly reduced, causing the overall amount of genetic variation to drop.

Concept Check 23.2

1. Each individual has two alleles, so the total number of alleles is 1,400. To calculate the frequency of allele *A*, note that each of the 85 individuals of genotype *AA* has two *A* alleles, each of the 320 individuals of genotype *Aa* has one *A* allele, and each of the 295 individuals of genotype *aa* has zero *A* alleles. Thus, the frequency (p) of allele *A* is

$$p = \frac{(2 \times 85) + (1 \times 320) + (0 \times 295)}{1{,}400} = 0.35$$

There are only two alleles (*A* and *a*) in our population, so the frequency of allele *a* must be $q = 1 - p = 0.65$. **2.** Because the frequency of allele *a* is 0.45, the frequency of allele *A* must be 0.55. Thus, the expected genotype frequencies are $p^2 = 0.3025$ for genotype *AA*, $2pq = 0.495$ for genotype *Aa*, and $q^2 = 0.2025$ for genotype *aa*. **3.** There are 120 individuals in the population, so there are 240 alleles. Of these, there are 124 *V* alleles—32 from the 16 *VV* individuals and 92 from the 92 *Vv* individuals. Thus, the frequency of the *V* allele is $p = 124/240 = 0.52$; hence, the frequency of the *v* allele is $q = 0.48$. Based on the Hardy-Weinberg equation, if the population were not evolving, the frequency of genotype *VV* should be $p^2 = 0.52 \times 0.52 = 0.27$; the frequency of genotype *Vv* should be $2pq = 2 \times 0.52 \times 0.48 = 0.5$; and the frequency of genotype *vv* should be $q^2 = 0.48 \times 0.48 = 0.23$. In a population of 120 individuals, these expected genotype frequencies lead us to predict that there would be 32 *VV* individuals (0.27×120), 60 *Vv* individuals (0.5×120), and 28 *vv* individuals (0.23×120). The actual numbers for the population (16 *VV*, 92 *Vv*, 12 *vv*) deviate from these expectations (fewer homozygotes and more heterozygotes than expected). This indicates that the population is not in Hardy-Weinberg equilibrium and hence may be evolving at this locus.

Concept Check 23.3

1. Natural selection is more "predictable" in that it alters allele frequencies in a nonrandom way: It tends to increase the frequency of alleles that increase the organism's reproductive success in its environment and decrease the frequency of alleles that decrease the organism's reproductive success. Alleles subject to genetic drift increase or decrease in frequency by chance alone, whether or not they are advantageous. **2.** Genetic drift results from chance events that cause allele frequencies to fluctuate at random from generation to generation; within a population, this process tends to decrease genetic variation over time. Gene flow is the transfer of alleles between populations, a process that can introduce new alleles to a population and hence may increase its genetic variation (albeit slightly, since rates of gene flow are often low). **3.** Selection is not important at this locus; furthermore, the populations are not small, and hence the effects of genetic drift should not be pronounced. Gene flow is occurring via the movement of pollen and seeds. Thus, allele and genotype frequencies in these populations should become more similar over time as a result of gene flow.

Concept Check 23.4

1. The relative fitness of a mule is zero, because fitness includes reproductive contribution to the next generation, and a sterile mule cannot produce offspring. **2.** Although both gene flow and genetic drift can increase the frequency of advantageous alleles in a population, they can also decrease the frequency of advantageous alleles or increase the frequency of harmful alleles. Only natural selection *consistently* results in an increase in the frequency of alleles that enhance survival or reproduction. Thus, natural selection is the only mechanism that consistently leads to adaptive evolution. **3.** The three modes of natural selection (directional, stabilizing, and disruptive) are defined in terms of the selective advantage of different *phenotypes*, not different genotypes. Thus, the type of selection represented by heterozygote advantage depends on the phenotype of the heterozygotes. In this question, because heterozygous individuals have a more extreme phenotype than either homozygote, heterozygote advantage represents directional selection.

Summary of Key Concepts Questions

23.1 Much of the nucleotide variability at a genetic locus occurs within introns. Nucleotide variation at these sites typically does not affect the phenotype because introns do not code for the protein product of the gene. (Note: In certain circumstances, it is possible that a change in an intron could affect RNA splicing and ultimately have some phenotypic effect on the organism, but such mechanisms are not covered in this introductory text.) There are also many variable nucleotide sites within exons. However, most of the variable sites within exons reflect changes to the DNA sequence that do not change the sequence of amino acids encoded by the gene (and hence may not affect the phenotype). **23.2** No, this is not an example of circular reasoning. Calculating p and q from observed genotype frequencies does not imply that those genotype frequencies must be in Hardy-Weinberg equilibrium. For example, consider a population that has 195 individuals of genotype *AA*, 10 of genotype *Aa*, and 195 of genotype *aa*. Calculating p and q from these values yields $p = q = 0.5$. Using the Hardy-Weinberg equation, the predicted equilibrium frequencies are $p^2 = 0.25$ for genotype *AA*, $2pq = 0.5$ for genotype *Aa*, and $q^2 = 0.25$ for genotype *aa*. Since there are 400 individuals in the population, these predicted genotype frequencies indicate that there should be 100 *AA* individuals, 200 *Aa* individuals, and 100 *aa* individuals—numbers that differ greatly from the values that we used to calculate p and q. **23.3** It is unlikely that two such populations would evolve in similar ways. Since their environments are very different, the alleles favored by natural selection would probably differ between the two populations. Although genetic drift may have important effects in each of these small populations, drift causes unpredictable changes in allele frequencies, so it is unlikely that drift would cause the populations to evolve in similar ways. Both populations are geographically isolated, suggesting that little gene flow would occur between them (again making it less likely that they would evolve in similar ways). **23.4** Compared to males, it is likely that the females of such species would be larger, more colorful, endowed with more elaborate ornamentation (for example, a large morphological feature such as the peacock's tail), and more apt to engage in behaviors intended to attract mates or prevent other members of their sex from obtaining mates.

Test Your Understanding

1. D **2.** C **3.** B **4.** A **5.** C

Chapter 24

Figure Questions

Figure 24.7 If this had not been done, the strong preference of "starch flies" and "maltose flies" to mate with like-adapted flies could have occurred simply because the flies could detect (for example, by sense of smell) what their potential mates had eaten as larvae—and preferred to mate with flies that had a similar smell to their own. **Figure 24.12** In murky waters where females distinguish colors poorly, females of each species might mate often with males of the other species. Hence, since hybrids between these species are viable and fertile, the gene pools of the two species might become more similar over time. **Figure 24.13** The graph indicates that there has been gene flow of some fire-bellied toad alleles into the range of the yellow-bellied toad. Otherwise, all individuals located to the left of the hybrid zone portion of the graph would have allele frequencies equal to 1. **Figure 24.14** Because the populations had only just begun to diverge from one another at this point in the process, it is likely that any existing barriers to reproduction would weaken over time. **Figure 24.18** Over time, the chromosomes of the experimental hybrids came to resemble those of *H. anomalus*. This occurred even though conditions in the laboratory differed greatly from conditions in the field, where *H. anomalus* is found, suggesting that selection for laboratory conditions was not strong. Thus, it is unlikely that the observed rise in the fertility of the experimental hybrids was due to selection for life under laboratory conditions. **Figure 24.19** The presence of *M. cardinalis* plants that carry the *M. lewisii yup* allele would make it more likely that bumblebees would transfer pollen between the two monkey flower species. As a result, we would expect the number of hybrid offspring to increase.

Concept Check 24.1

1. (a) All except the biological species concept can be applied to both asexual and sexual species because they define species on the basis of characteristics other than the ability to reproduce. In contrast, the biological species concept can be applied only to sexual species. (b) The easiest species concept to apply in the field would be the morphological species concept because it is based only on the appearance of the organism. Additional information about its ecological habits or reproduction is not required. **2.** Because these birds live in fairly similar environments and can breed successfully in captivity, the reproductive barrier in nature is probably prezygotic; given the species' differences in habitat preference, this barrier could result from habitat isolation.

Concept Check 24.2

1. In allopatric speciation, a new species forms while in geographic isolation from its parent species; in sympatric speciation, a new species forms in the absence of geographic isolation. Geographic isolation greatly reduces gene flow between populations, whereas ongoing gene flow is more likely in sympatric populations. As a result, allopatric speciation is more common than sympatric speciation. **2.** Gene flow between subsets of a population that live in the same area can be reduced in a variety of ways. In some species—especially plants—changes in chromosome number can block gene flow and establish reproductive isolation in a single generation. Gene flow can also be reduced in sympatric populations by habitat differentiation (as seen in the apple maggot fly, *Rhagoletis*) and sexual selection (as seen in Lake Victoria cichlids). **3.** Allopatric speciation would be less likely to occur on an island near a mainland than on a more isolated island of the same size. We expect this result because continued gene flow between mainland populations and those on a nearby island reduces the chance that enough genetic divergence will take place for allopatric speciation to occur. **4.** If all of the homologs failed to separate during anaphase I of meiosis, some gametes would end up with an extra set of chromosomes (and others would end up with no chromosomes). If a gamete with an extra set of chromosomes fused with a normal gamete, a triploid would result; if two gametes with an extra set of chromosomes fused with each other, a tetraploid would result.

Concept Check 24.3

1. Hybrid zones are regions in which members of different species meet and mate, producing some offspring of mixed ancestry. Such regions can be viewed as "natural laboratories" in which to study speciation because scientists can directly observe factors that cause (or fail to cause) reproductive isolation. **2.** (a) If hybrids consistently survived and reproduced poorly compared with the offspring of intraspecific matings, reinforcement could occur. If it did, natural selection would cause prezygotic barriers to reproduction between the parent species to strengthen over time, decreasing the production of unfit hybrids and leading to a completion of the speciation process. (b) If hybrid offspring survived and reproduced as well as the offspring of intraspecific matings, indiscriminate mating between the parent species would lead to the production of large numbers of hybrid offspring. As these hybrids mated with each other and with members of both parent species, the gene pools of the parent species could fuse over time, reversing the speciation process.

Concept Check 24.4

1. The time between speciation events includes (1) the length of time that it takes for populations of a newly formed species to begin diverging reproductively from one another and (2) the time it takes for speciation to be complete once this divergence begins. Although speciation can occur rapidly once populations have begun to diverge from one another, it may take millions of years for that divergence to begin. **2.** Investigators transferred alleles at the *yup* locus (which influences flower color) from each parent species to the other. *M. lewisii* plants with an *M. cardinalis yup* allele received many more visits from hummingbirds than usual; hummingbirds usually pollinate *M. cardinalis* but avoid *M. lewisii*. Similarly, *M. cardinalis* plants with an *M. lewisii yup* allele received many more visits from bumblebees than usual; bumblebees usually pollinate *M. lewisii* and avoid *M. cardinalis*. Thus, alleles at the *yup* locus can influence pollinator choice, which in these species provides the primary barrier to interspecific mating. Nevertheless, the experiment does not prove that the *yup* locus alone controls barriers to reproduction between *M. lewisii* and *M. cardinalis*; other genes might enhance the effect of the *yup* locus (by modifying flower color) or cause entirely different barriers to reproduction (for example, gametic isolation or a postzygotic barrier). **3.** Crossing over. If crossing over did not occur, each chromosome in an experimental hybrid would remain as in the F_1 generation: composed entirely of DNA from one parent species or the other.

Summary of Key Concepts Questions

24.1 According to the biological species concept, a species is a group of populations whose members interbreed and produce viable, fertile offspring; thus, gene flow occurs between populations of a species. In contrast, members of different species do not interbreed and hence no gene flow occurs between their populations. Overall, then, in the biological species concept, species can be viewed as designated by the *absence* of gene flow—making gene flow of central importance to the biological species concept. **24.2** Yes. Sympatric speciation can be promoted by factors such as polyploidy, sexual selection, and habitat shifts, all of which can reduce gene flow between the subpopulations of a larger population.

But such factors can also occur in allopatric populations and hence can also promote allopatric speciation. **24.3** If the hybrids are selected against, the hybrid zone could persist if individuals from the parent species regularly travel into the zone, where they mate to produce hybrid offspring. If hybrids are not selected against, there is no cost to the continued production of hybrids, and large numbers of hybrid offspring may be produced. However, natural selection for life in different environments may keep the gene pools of the two parent species distinct, thus preventing the loss (by fusion) of the parent species and once again causing the hybrid zone to be stable over time. **24.4** As the goatsbeard plant, Bahamas mosquitofish, and apple maggot fly illustrate, speciation continues to happen today. A new species can begin to form whenever gene flow is reduced between populations of the parent species. Such reductions in gene flow can occur in many ways: A new, geographically isolated population may be founded by a few colonists; some members of the parent species may begin to utilize a new habitat; or sexual selection may isolate formerly connected populations or subpopulations. These and many other such events are happening today.

Test Your Understanding

1. B **2.** C **3.** B **4.** A **5.** D **6.** C
7. Here is one possibility:

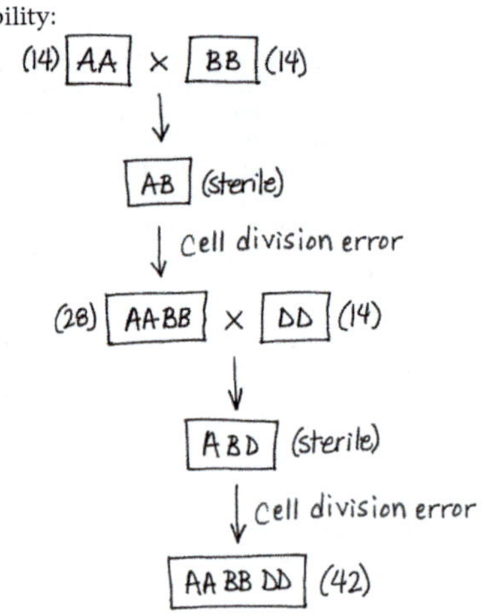

Chapter 25

Figure Questions

Figure 25.2 Proteins are almost always composed of the same 20 amino acids shown in Figure 5.14. However, many other amino acids could potentially form in this or any other experiment. For example, any molecule that had an R group that differed from those listed in Figure 5.14 would still be an amino acid as long as it also contained an α carbon, an amino group, and a carboxyl group—but that molecule would not be one of the 20 amino acids commonly found in nature. **Figure 25.4** The hydrophobic regions of such molecules are attracted to one another and excluded from water, whereas the hydrophilic regions have an affinity for water. As a result, the molecules can form a bilayer in which the hydrophilic regions are on the outside of the bilayer (facing water on each side of the bilayer) and the hydrophobic regions point toward each other (that is, toward the inside of the bilayer). **Figure 25.6** Because uranium-238 has a half-life of 4.5 billion years, the *x*-axis would be relabeled (in billions of years) as 4.5, 9, 13.5, and 18. **Figure 25.8** (1) The countdown timer and horizontal time scale indicate that prokaryotes originated 3.5 billion years ago and that the colonization of land took place 500 million years ago. On a 1-hour time scale, this indicates that prokaryotes appeared about 46 minutes ago, while the colonization of land took place less than 7 minutes ago. (2) From 3.5 billion years ago to 1.5 billion years ago, life on Earth consisted entirely of unicellular organisms. In fact, from 3.5 billion years ago to 1.8 billion years ago, all of Earth's organisms were prokaryotes; from 1.8 billion years ago until 1.5 billion years ago, these unicellular prokaryotes were joined by unicellular eukaryotes. The colonization of land did not occur until 500 million years ago. Hence, we can also infer that all or most of these unicellular organisms lived in the oceans or in freshwater environments for the first two billion years of life on Earth. **Figure 25.11** You should have circled the node, shown in the tree diagram at approximately 635 million years ago (mya), that leads to the echinoderm/chordate lineage and to the lineage that gave rise to brachiopods, annelids, molluscs, and arthropods. To determine a minimum estimate of the age of the ancestor represented by this node, note that the most recent common ancestor of chordates and annelids must be at least as old as any of its descendants. Since fossil molluscs date to about 560 mya, the common ancestor represented by the circled branch point must be at least 560 million years old. **Figure 25.16** The Australian plate's current direction of movement is roughly similar to the northeasterly direction the continent traveled over the past 66 million years. **Figure 25.26** The coding sequence of the *Pitx1* gene would differ between the marine and lake populations, but patterns of gene expression would not.

Concept Check 25.1

1. The hypothesis that conditions on early Earth could have permitted the synthesis of organic molecules from inorganic ingredients **2.** In contrast to random mingling of molecules in an open solution, segregation of molecular systems by membranes could concentrate organic molecules, assisting biochemical reactions. **3.** Today, genetic information usually flows from DNA to RNA, as when the DNA sequence of a gene is used as a template to synthesize the mRNA encoding a particular protein. However, the life cycle of retroviruses such as HIV shows that genetic information can flow in the reverse direction (from RNA to DNA). In these viruses, the enzyme reverse transcriptase uses RNA as a template for DNA synthesis, suggesting that a similar enzyme could have played a key role in the transition from an RNA world to a DNA world.

Concept Check 25.2

1. The fossil record shows that different groups of organisms dominated life on Earth at different points in time and that many organisms once alive are now extinct; specific examples of these points can be found in Figure 25.5. The fossil record also indicates that new groups of organisms can arise via the gradual modification of previously existing organisms, as illustrated by fossils that document the origin of mammals from their cynodont ancestors (see Figure 25.7). **2.** 22,920 years (four half-lives: 5,730 × 4)

Concept Check 25.3

1. Free oxygen attacks chemical bonds and can inhibit enzymes and damage cells. As a result, the appearance of oxygen in the atmosphere probably caused many prokaryotes that had thrived in anaerobic environments to survive and reproduce poorly, ultimately driving many of these species to extinction. **2.** All eukaryotes have mitochondria or remnants of these organelles, but not all eukaryotes have plastids. **3.** A fossil record of life today would include many organisms with hard body parts (such as vertebrates and many marine invertebrates), but might not include some species we are very familiar with, such as those that have small geographic ranges and/or small population sizes (for example, endangered species such as the giant panda, tiger, and several rhinoceros species).

Concept Check 25.4

1. The theory of plate tectonics describes the movement of Earth's continental plates, which alters the physical geography and climate of Earth, as well as the extent to which organisms are geographically isolated. Because these factors affect extinction and speciation rates, plate tectonics has a major impact on life on Earth. **2.** Mass extinctions; major evolutionary innovations; the diversification of another group of organisms (which can provide new sources of food); migration to new locations where few competitor species exist **3.** In theory, fossils of both common and rare species would be present right up to the time of the catastrophic event, then disappear. Reality is more complicated because the fossil record is not perfect. So the most recent fossil for a species might be a million years before the mass extinction—even though the species did not become extinct *until* the mass extinction. This complication is especially likely for rare species because few of their fossils will form and be discovered. Hence, for many rare species, the fossil record would not document that the species was alive immediately before the extinction (even if it was).

Concept Check 25.5

1. Heterochrony can cause a variety of morphological changes. For example, if the onset of sexual maturity changes, a retention of juvenile characteristics (paedomorphosis) may result. Paedomorphosis can be caused by small genetic changes that result in large changes in morphology, as seen in the axolotl salamander. **2.** In animal embryos, *Hox* genes influence the development of structures such as limbs and feeding appendages. As a result, changes in these genes—or in the regulation of these genes—are likely to have major effects on morphology. **3.** From genetics, we know that gene regulation is altered by how well transcription factors bind to noncoding DNA sequences called control elements. Thus, if changes in morphology are often caused by changes in gene regulation, portions of noncoding DNA that contain control elements are likely to be strongly affected by natural selection.

Concept Check 25.6

1. Complex structures do not evolve all at once, but in increments, with natural selection selecting for adaptive variants of the earlier versions. **2.** Although the myxoma virus is highly lethal, initially some of the rabbits are resistant (0.2% of infected rabbits are not killed). Thus, assuming resistance is an inherited trait, we would expect the rabbit population to show a trend for increased resistance to the virus. We would also expect the virus to show an evolutionary trend toward reduced lethality. We would expect this trend because a rabbit infected with a less lethal virus would be more likely to live long enough for a mosquito to bite it and hence potentially transmit the virus to another rabbit. (A virus that kills its rabbit host before a mosquito transmits the virus to another rabbit dies with its host.)

Summary of Key Concepts Questions

Concept 25.1 Particles of montmorillonite clay may have provided surfaces on which organic molecules became concentrated and hence were more likely to react with one another. Montmorillonite clay particles may also have facilitated the transport of key molecules, such as short strands of RNA, into vesicles. These vesicles can form spontaneously from simple precursor molecules, "reproduce" and "grow" on their own, and maintain internal concentrations of molecules that differ from those in the surrounding environment. These features of vesicles represent key steps in the emergence of protocells and (ultimately) the first living cells. **Concept 25.2** One challenge is that radioisotopes with very long half-lives are not used by organisms to build their bones or shells. As a result, fossils older than 75,000 years cannot be dated directly. Fossils are often found in sedimentary rock, but those rocks typically contain sediments of different ages, again posing a challenge when trying to date old fossils. To circumvent these challenges, geologists use radioisotopes with long half-lives to date layers of volcanic rock that surround old fossils. This approach provides minimum and maximum estimates for the ages of fossils sandwiched between two layers of volcanic rock. **Concept 25.3** The "Cambrian explosion" refers to a relatively short interval of time (535–525 million years ago) during which large forms of many present-day animal phyla first appear in the fossil record. The evolutionary changes that occurred during this time, such as the appearance of large predators and well-defended prey, were important because they set the stage for many of the key events in the history of life over the last 500 million years. **Concept 25.4** The broad evolutionary changes documented by the fossil record reflect the rise and fall of major groups of organisms. In turn, the rise or fall of any particular group results from a balance between speciation and extinction rates: A group increases in size when the rate at which its members produce new species is greater than the rate at which its member species are lost to extinction, while a group shrinks in size if extinction rates are greater than speciation rates. **Concept 25.5** A change in the sequence or regulation of a developmental gene can produce major morphological changes. In some cases, such morphological changes may enable organisms to perform new functions or live in new environments—thus potentially leading to an adaptive radiation and the formation of a new group of organisms. **Concept 25.6** Evolutionary change results from interactions between organisms and their current environments. No goal is involved in this process. As environments change over time, the features of organisms favored by natural selection may also change. When this happens, what once may have seemed like a "goal" of evolution (for example, improvements in the function of a feature previously favored by natural selection) may cease to be beneficial or may even be harmful.

Test Your Understanding

1. B **2.** A **3.** D **4.** B **5.** C **6.** C **7.** A

Chapter 26

Figure Questions

Figure 26.5 (1) Frogs are most closely related to a group consisting of lizards, chimps, and humans in this tree. (2) You should have circled the branch point splitting the frog lineage from the lineage leading to lizards, chimps, and humans. (3) Four: chimps–humans, lizards–chimps/humans; frogs–lizards/chimps/humans; and fishes–frogs/lizards/chimps/humans.
(4)

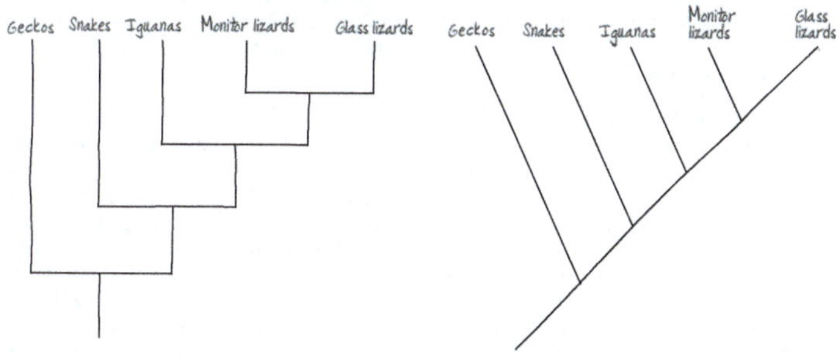

(5)

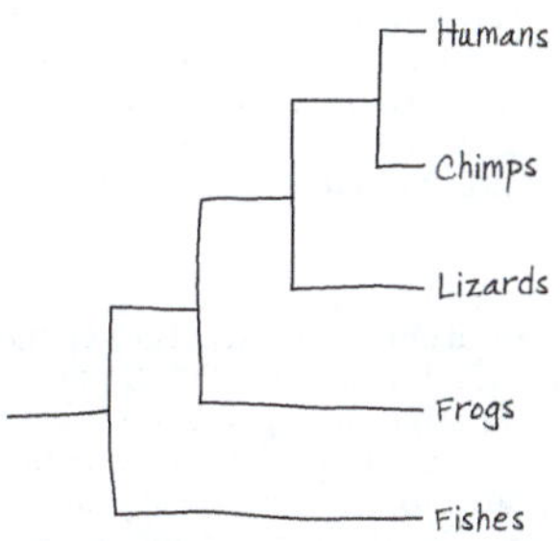

Each of the three trees identifies chimps and lizards as the two closest relatives of humans in these trees as they are the groups shown with whom we share the most recent common ancestors. **Figure 26.6** Unknown 1b (a portion of sample 1) and unknowns 9–13 all would have to be located on the branch of the tree that currently leads to Minke (Southern Hemisphere) and unknowns 1a and 2–8. **Figure 26.11** You should have circled the branch point that is drawn farthest to the left (the common ancestor of all taxa shown). Both cetaceans and seals descended from terrestrial lineages of mammals, indicating that the cetacean-seal common ancestor lacked a streamlined body form and hence would not be part of the cetacean-seal group. **Figure 26.12** Hinged jaws are a shared ancestral character for the group that includes frogs, turtles, and leopards. Thus, you should have circled the frog, turtle, and leopard lineages, along with their most recent common ancestor. **Figure 26.16** Crocodilians are the sister taxon to the dinosaur clade (which includes birds) because crocodilians and the dinosaur clade share an immediate common ancestor that is not shared by any other group. **Figure 26.21** This tree indicates that the sequences of rRNA and other genes in mitochondria are most closely related to those of proteobacteria, while the sequences of chloroplast genes are most closely related to those of cyanobacteria. These gene sequence relationships are what would be predicted from endosymbiont theory, which posits that both mitochondria and chloroplasts originated as engulfed prokaryotic cells.

Concept Check 26.1

1. We are classified the same from the domain level to the class level; both the leopard and human are mammals. Leopards belong to order Carnivora, whereas humans do not. **2.** The tree in (c) shows a different pattern of evolutionary relationships. In (c), C and B are sister taxa, whereas C and D are sister taxa in (a) and (b). **3.** The redrawn version of Figure 26.4 is shown below.

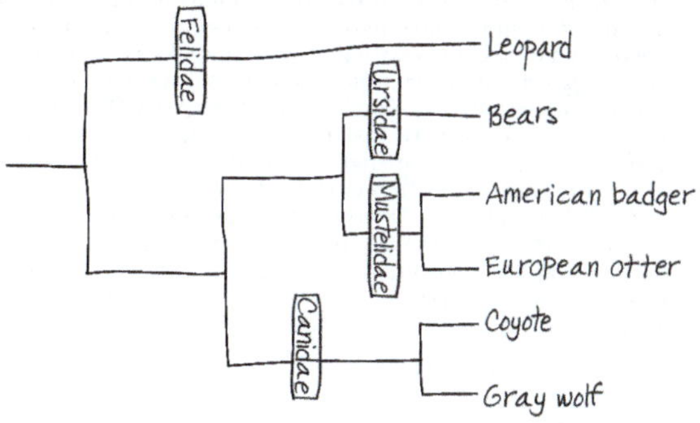

Concept Check 26.2

1. (a) Analogy, since porcupines and cacti are not closely related and since most other animals and plants do not have similar structures; (b) homology, since cats and humans are both mammals and have homologous forelimbs, of which the hand and paw are the lower part; (c) analogy, since owls and hornets are not closely related and since the structure of their wings is very different. **2.** Species B and C are more likely to be closely related. Small genetic changes (as between species B and C) can produce divergent physical appearances, but if many genes have diverged greatly (as in species A and B), then the lineages have probably been separate for a long time.

Concept Check 26.3

1. No; hair is a shared ancestral character common to all mammals and thus is not helpful in distinguishing different mammalian subgroups. **2.** The principle of maximum parsimony states that the hypothesis about nature we investigate first should be the simplest explanation found to be consistent with the facts. Actual evolutionary relationships may differ from those inferred by parsimony owing to complicating factors such as convergent evolution.
3.

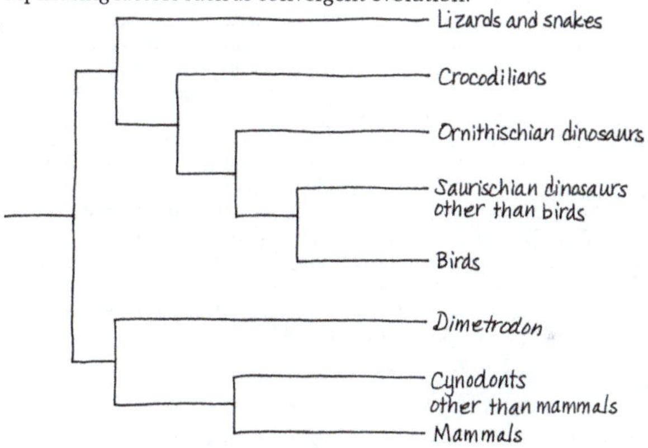

The traditional classification provides a poor match to evolutionary history, thus violating the basic principle of cladistics—that classification should be based on common descent. Both birds and mammals originated from groups traditionally designated as reptiles, making reptiles (as traditionally delineated) a paraphyletic group. These problems can be addressed by removing *Dimetrodon* and cynodonts from the reptiles and by regarding birds as a group of reptiles (specifically, as a group of dinosaurs).

Concept Check 26.4

1. Proteins are gene products. Their amino acid sequences are determined by the nucleotide sequences of the DNA that codes for them. Thus, differences between comparable proteins in two species reflect underlying genetic differences that have accumulated as the species diverged from one another. As a result, differences between the proteins can reflect the evolutionary history of the species. **2.** These observations suggest that the evolutionary lineages leading to species 1 and species 2 diverged from one another before a gene duplication event in species 1 produced gene B from gene A. **3.** In RNA processing, the exons or coding regions of a gene can be spliced together in different ways, yielding different mRNAs and hence different protein products. As a result, different proteins could potentially be produced from the same gene in different tissues, thereby enabling the gene to perform different functions in these different tissues.

Concept Check 26.5

1. A molecular clock is a method of estimating the actual time of evolutionary events based on numbers of base changes in orthologous genes. It is based on the assumption that the regions of genomes being compared evolve at constant rates. **2.** There are many portions of the genome that do not code for genes; mutations that alter the sequence of bases in such regions could accumulate through drift without affecting an organism's fitness. Even in coding regions of the genome, some mutations may not have a critical effect on genes or proteins. **3.** The gene (or genes) used for the molecular clock may have evolved more slowly in these two taxa than in the species used to calibrate the clock; as a result, the clock would underestimate the time at which the taxa diverged from each other.

Concept Check 26.6

1. The kingdom Monera included bacteria and archaea, but we now know that these organisms are in separate domains. Kingdoms are subsets of domains, so a single kingdom (like Monera) that includes taxa from different domains is not valid.
2. Because of horizontal gene transfer, some genes in eukaryotes are more closely related to bacteria, while others are more closely related to archaea; thus, depending on which genes are used, phylogenetic trees constructed from DNA data can yield conflicting results. **3.** Eukaryotes are hypothesized to have originated when a heterotrophic prokaryote (an archaeal host cell) engulfed a bacterium that would later become an organelle found in all eukaryotes—the mitochondrion. Over time, a fusion of organisms occurred as the archaeal host cell and its bacterial endosymbiont evolved to become a single organism. As a result, we would expect the cell of a eukaryote to include both archaeal DNA and bacterial DNA, making the origin of eukaryotes an example of horizontal gene transfer.

Summary of Key Concepts Questions

26.1 The fact that humans and chimpanzees are sister species indicates that we share a more recent common ancestor with chimpanzees than we do with any other living primate species. But that does not mean that humans evolved from chimpanzees, or vice versa; instead, it indicates that both humans and chimpanzees are descendants of that common ancestor. **26.2** Homologous characters result from shared ancestry. As organisms diverge over time, some of their homologous characters will also diverge. The homologous characters of organisms that diverged long ago typically differ more than do the homologous characters of organisms that diverged more recently. As a result, differences in homologous characters can be used to infer phylogeny. In contrast, analogous characters result from convergent evolution, not shared ancestry, and hence can give misleading estimates of phylogeny. **26.3** All features of organisms arose at some point in the history of life. In the group in which a new feature first arose, that feature is a shared derived character that is unique to that clade. The group in which each shared derived character first appeared can be determined, and the resulting nested pattern can be used to infer evolutionary history. **26.4** Orthologous genes should be used; for such genes, the homology results from speciation and hence reflects evolutionary history. **26.5** A key assumption of molecular clocks is that nucleotide substitutions occur at fixed rates, and hence the number of nucleotide differences between two DNA sequences is proportional to the time since the sequences diverged from each other. Some limitations of molecular clocks: No gene marks time with complete precision; natural selection can favor certain DNA changes over others; nucleotide substitution rates can change over long periods of time (causing molecular clock estimates of when events in the distant past occurred to be highly uncertain); and the same gene can evolve at different rates in different organisms. **26.6** Genetic data indicated that many prokaryotes differed as much from each other as they did from eukaryotes. This indicated that organisms should be grouped into three "super-kingdoms," or domains (Archaea, Bacteria, Eukarya). These data also indicated that the previous kingdom Monera (which had contained all the prokaryotes) did not make biological sense and should be abandoned. Later genetic and morphological data also indicated that the former kingdom Protista (which had primarily contained single-celled organisms) should be abandoned because some protists are more closely related to plants, fungi, or animals than they are to other protists.

Test Your Understanding

1. A **2.** C **3.** B **4.** C **5.** D **6.** A **7.** D
9.

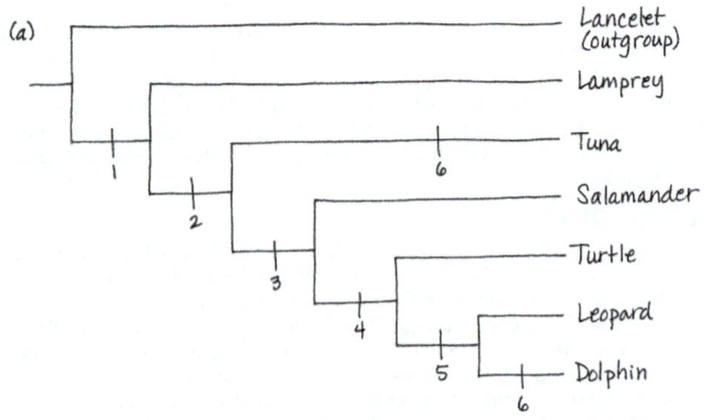

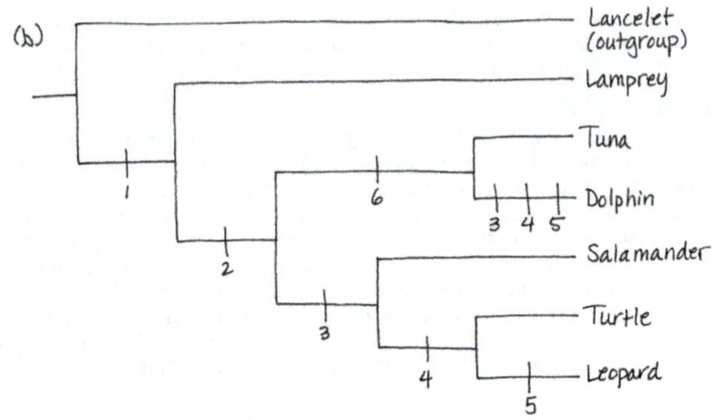

(c) The tree in (a) requires seven evolutionary changes, while the tree in (b) requires nine evolutionary changes. Thus, the tree in (a) is more parsimonious, since it requires fewer evolutionary changes.

Chapter 27

Figure Questions

Figure 27.7 The top ring, to which the hook is attached, is embedded within the interior, hydrophobic portion of the lipid bilayer of the outer membrane, suggesting that the top ring is hydrophobic. Likewise, the third ring down is embedded within the hydrophobic portion of the plasma membrane's lipid bilayer, suggesting that this ring also is hydrophobic. **Figure 27.10** It is likely that the expression or sequence of genes that affect glucose metabolism may have changed; genes for metabolic processes no longer needed by the cell also may have changed. **Figure 27.11** Transduction results in horizontal gene transfer when the host and recipient cells are members of different species. **Figure 27.15** Eukarya **Figure 27.17** Thermophiles live in very hot environments, so it is likely that their enzymes can continue to function normally at much higher temperatures than can the enzymes of other organisms. At low temperatures, however, the enzymes of thermophiles may not function as well as the enzymes of other organisms. **Figure 27.18** From the graph, plant uptake can be estimated as 0.72, 0.62, and 0.96 mg K^+ for strains 1, 2, and 3, respectively. These values average to 0.77 mg K^+. If bacteria had no effect, the average plant uptake of K^+ for strains 1, 2, and 3 should be close to 0.51 mg K^+, the value observed for plants grown in bacteria-free soil.

Concept Check 27.1

1. Adaptations include the capsule (shields prokaryotes from the host's immune system) and endospores (enable cells to survive harsh conditions and to revive when the environment becomes favorable). **2.** Prokaryotic cells lack the complex compartmentalization associated with the membrane-enclosed organelles of eukaryotic cells. Prokaryotic genomes have much less DNA than eukaryotic genomes, and most of this DNA is contained in a single ring-shaped chromosome located in the nucleoid rather than within a true membrane-enclosed nucleus. In addition, many prokaryotes also have plasmids, small ring-shaped DNA molecules containing a few genes. **3.** Plastids such as chloroplasts are thought to have evolved from an endosymbiotic photosynthetic prokaryote. More specifically, the phylogenetic tree shown in Figure 26.21 indicates that plastids are closely related to cyanobacteria. Hence, we can hypothesize that the thylakoid membranes of chloroplasts resemble those of cyanobacteria because chloroplasts evolved from an endosymbiotic cyanobacterium.

Concept Check 27.2

1. Prokaryotes can have extremely large population sizes, in part because they often have short generation times. The large number of individuals in prokaryotic populations makes it likely that in each generation there will be many individuals that have new mutations at any particular gene, thereby adding considerable genetic diversity to the population. **2.** In transformation, naked, foreign DNA from the environment is taken up by a bacterial cell. In transduction, phages carry bacterial genes from one bacterial cell to another. In conjugation, a bacterial cell directly transfers plasmid or chromosomal DNA to another cell via a mating bridge that temporarily connects the two cells. **3.** The population that includes individuals capable of conjugation would probably be more successful, since some of its members could form recombinant cells whose new gene combinations might be advantageous in a novel environment. **4.** Yes. Genes for antibiotic resistance could be transferred (by transformation, transduction, or conjugation) from the nonpathogenic bacterium to a pathogenic bacterium; this could make the pathogen an even greater threat to human health. In general, transformation, transduction, and conjugation tend to increase the spread of resistance genes.

Concept Check 27.3

1. A phototroph derives its energy from light, while a chemotroph gets its energy from chemical sources. An autotroph derives its carbon from CO_2, HCO_3^-, or related compounds, while a heterotroph gets its carbon from organic nutrients such as glucose. Thus, there are four nutritional modes: photoautotrophic, photoheterotrophic (unique to prokaryotes), chemoautotrophic (unique to prokaryotes), and chemoheterotrophic. **2.** Chemoheterotrophy; the bacterium must rely on chemical sources of energy, since it is not exposed to light, and it must be a heterotroph if it requires a source of carbon other than CO_2 (or a related compound, such as HCO_3^-). **3.** If humans could fix nitrogen, we could build proteins using atmospheric N_2 and hence would not need to eat high-protein foods such as meat, fish, or soy. Our diet would, however, need to include a source of carbon, along with minerals and water. Thus, a typical meal might consist of carbohydrates as a carbon source, along with fruits and vegetables to provide essential minerals (and additional carbon).

Concept Check 27.4

1. Molecular systematic studies indicate that some organisms once classified as bacteria are more closely related to eukaryotes and belong in a domain of their own: Archaea. Such studies have also shown that horizontal gene transfer is common and plays an important role in the evolution of prokaryotes. By not requiring that organisms be cultured in the laboratory, metagenomic studies have revealed an immense diversity of previously unknown prokaryotic species. Over time, the ongoing discovery of new species by metagenomic analyses may alter our understanding of prokaryotic phylogeny greatly. **2.** At present, all known methanogens are archaea in the clade Euryarchaeota; this suggests that this unique metabolic pathway probably arose in ancestral species within Euryarchaeota. Since Bacteria and Archaea have been separate evolutionary lineages for billions of years, the discovery of a methanogen from the domain Bacteria would suggest that adaptations that enabled the use of CO_2 to oxidize H_2 may have evolved twice—once in Archaea (within Euryarchaeota) and once in Bacteria. (It is also possible that a newly discovered bacterial methanogen could have acquired the genes for this metabolic pathway by horizontal gene transfer from a methanogen in domain Archaea. However, horizontal gene transfer is not a likely explanation because of the large number of genes involved and because gene transfers between species in different domains are relatively rare.)

Concept Check 27.5

1. Although prokaryotes are small, their large numbers and metabolic abilities enable them to play key roles in ecosystems by decomposing wastes, recycling chemicals, and affecting the concentrations of nutrients available to other organisms. **2.** Cyanobacteria produce oxygen when water is split in the light reactions of photosynthesis. The Calvin cycle incorporates CO_2 from the air into organic molecules, which are then converted to sugars.

Concept Check 27.6

1. Sample answers: eating fermented foods such as yogurt, sourdough bread, or cheese; receiving clean water from sewage treatment; taking medicines produced by bacteria **2.** No. If the poison is secreted as an exotoxin, live bacteria could be transmitted to another person. But the same is true if the poison is an endotoxin—only in this case, the live bacteria that are transmitted may be descendants of the (now-dead) bacteria that produced the poison. **3.** Some of the many different species of prokaryotes that live in the human gut compete with one another for resources (from the food that you eat). Because different prokaryotic species have different adaptations, a change in diet may alter which species can grow most rapidly, thus altering species abundance.

Summary of Key Concepts Questions

27.1 Specific structural features that enable prokaryotes to thrive in diverse environments include their cell walls (which provide shape and protection), flagella (which function in directed movement), and ability to form capsules or endospores (both of which can protect against harsh conditions). Prokaryotes also possess biochemical adaptations for growth in varied conditions, such as those that enable them to tolerate extremely hot or salty environments.
27.2 Many prokaryotic species can reproduce extremely rapidly, and their populations can number in the trillions. As a result, even though mutations are rare, every day many offspring are produced that have new mutations at particular gene loci. In addition, even though prokaryotes reproduce asexually and hence the vast majority of offspring are genetically identical to their parent, the genetic variation of their populations can be increased by transduction, transformation, and conjugation. Each of these (nonreproductive) processes can increase genetic variation by transferring DNA from one cell to another—even among cells that are of different species. **27.3** Prokaryotes have an exceptionally broad range of metabolic adaptations. As a group, prokaryotes perform all four modes of nutrition (photoautotrophy, chemoautotrophy, photoheterotrophy, and chemoheterotrophy), whereas eukaryotes perform only two of these (photoautotrophy and chemoheterotrophy). Prokaryotes are also able to metabolize nitrogen in a wide variety of forms (again unlike eukaryotes), and they frequently cooperate with other prokaryotic cells of the same or different species. **27.4** Phenotypic criteria such as shape, motility, and nutritional mode do not provide a clear picture of the evolutionary history of the prokaryotes. In contrast, molecular data have elucidated relationships among major groups of prokaryotes. Molecular data have also allowed researchers to sample genes directly from the environment; using such genes to construct phylogenies has led to the discovery of major new groups of prokaryotes. **27.5** Prokaryotes play key roles in the chemical cycles on which life depends. For example, prokaryotes are important decomposers, breaking down corpses and waste materials, thereby releasing nutrients to the environment where they can be used by other organisms. Prokaryotes also convert inorganic compounds to forms that other organisms can use. With respect to their ecological interactions, many prokaryotes form life-sustaining mutualisms with other species. In some cases, such as hydrothermal vent communities, the metabolic activities of prokaryotes provide an energy source on which hundreds of other species depend; in the absence of the prokaryotes, the community collapses. **27.6** Human well-being depends on our associations with mutualistic prokaryotes, such as the many species that live in our intestines and digest food that we cannot. Humans also can harness the remarkable metabolic capabilities of prokaryotes to produce a wide range of useful products and to perform key services such as bioremediation. Negative effects of prokaryotes result primarily from bacterial pathogens that cause disease.

Test Your Understanding

1. D **2.** A **3.** C **4.** C **5.** B **6.** A

Chapter 28

Figure Questions

Figure 28.2 The simplified version of the tree in Figure 28.2 and the modified tree showing unikonts as the sister group to all other eukaryotes would look as follows:

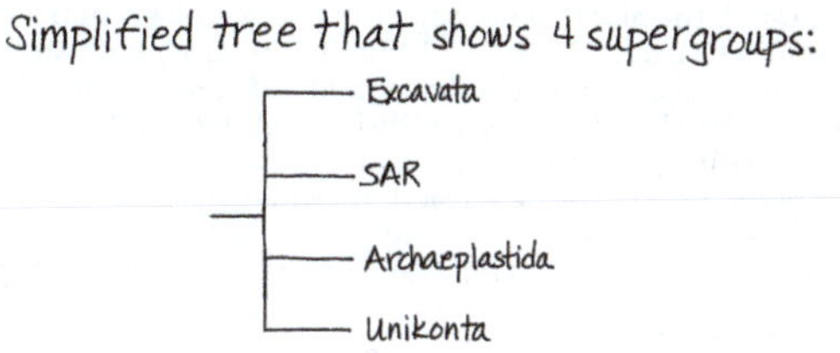

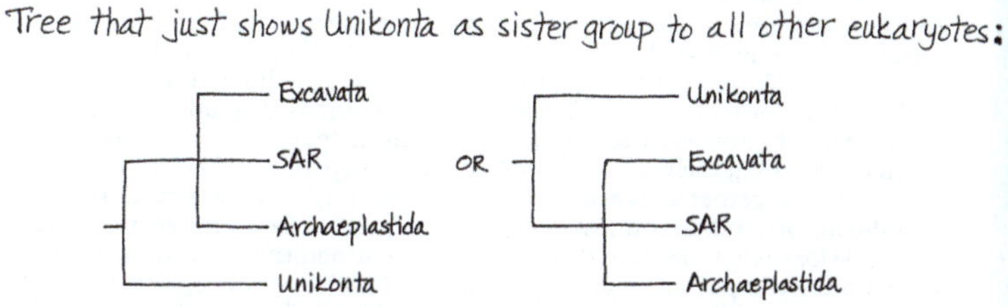

Figure 28.3 The diagram shows that a single secondary endosymbiosis event gave rise to the stramenopiles and alveolates—thus, these groups can trace their ancestry back to a single heterotrophic protist (shown in yellow) that ingested a red alga. In contrast, euglenids and chlorarachniophytes each descended from a different heterotrophic protist (one of which is shown in gray, the other in brown). Hence, it is likely that stramenopiles and alveolates are more closely related than are euglenids and chlorarachniophytes. **Figure 28.13** The sperm cells in the diagram are produced by the asexual (mitotic) division of cells in a single male gametophyte, which was itself produced by the asexual (mitotic) division of a single zoospore. Thus, the sperm cells are all derived from a single zoospore and so are genetically identical to one another. **Figure 28.16** Merozoites are produced by the asexual (mitotic) cell division of haploid sporozoites; similarly, gametocytes are produced by the asexual cell division of merozoites. Hence, it is likely that individuals in these three stages have the same complement of genes and that morphological differences between them result from changes in gene expression. **Figure 28.17** These events have a similar overall effect to fertilization. In both cases, haploid nuclei that were originally from two genetically different cells fuse to form a diploid nucleus. **Figure 28.23** The following stage should be circled: step 6, where a mature cell undergoes mitosis and forms four or more daughter cells. In step 7, the zoospores eventually grow into mature haploid cells, but they do not produce new daughter cells. Likewise, in step 2, a mature cell develops into a gamete, but it does not produce new daughter cells.
Figure 28.24 If the assumption is correct, then their results indicate that the fusion of the genes for DHFR and TS may be a derived trait shared by members of three supergroups of eukaryotes (Excavata, SAR, and Archaeplastida). However, if the assumption is not correct, the presence or absence of the gene fusion may tell little about phylogenetic history. For example, if the genes fused multiple times, groups could share the trait because of convergent evolution rather than common descent. If instead the genes were secondarily split, a group with such a split could be placed (incorrectly) in Unikonta rather than its correct placement in one of the other three supergroups. **Figure 28.26** They would be haploid because originally each of these cells was a haploid, solitary amoeba.

Concept Check 28.1

1. Sample response: Protists include unicellular, colonial, and multicellular organisms; photoautotrophs, heterotrophs, and mixotrophs; species that reproduce asexually, sexually, or both ways; and organisms with diverse physical forms and adaptations. **2.** Strong evidence shows that eukaryotes acquired mitochondria after a host cell (either an archaean or a close relative of the archaeans) first engulfed and then formed an endosymbiotic association with an alpha proteobacterium. Similarly, chloroplasts in red and green algae appear to have descended from a photosynthetic cyanobacterium that was engulfed by an ancient heterotrophic eukaryote. Secondary endosymbiosis also played an important role: Various protist lineages acquired plastids by engulfing unicellular red or green algae. **3.** Four. The first (and primary) genome is the DNA located in the chlorarachniophyte nucleus. A chlorarachniophyte also contain remnants of a green alga's nuclear DNA, located in the nucleomorph. Finally, mitochondria and chloroplasts contain DNA from the (different) bacteria from which they evolved. These two prokaryotic genomes comprise the third and fourth genomes contained within a chlorarachniophyte.

Concept Check 28.2

1. Their mitochondria do not have an electron transport chain and so cannot function in aerobic respiration. **2.** Since the unknown protist is more closely related to diplomonads than to euglenids, it must have originated after the lineage leading to the diplomonads and parabasalids diverged from the euglenozoans. In addition, since the unknown species has fully functional mitochondria—yet both diplomonads and parabasalids do not—it is likely that the unknown species originated *before* the last common ancestor of the diplomonads and parabasalids.

Concept Check 28.3

1. Because foram tests are hardened with calcium carbonate, they form long-lasting fossils in marine sediments and sedimentary rocks. **2.** The plastid DNA would likely be more similar to the chromosomal DNA of cyanobacteria based on the well-supported hypothesis that eukaryotic plastids (such as those found in the eukaryotic groups listed) originated by an endosymbiosis event in which a eukaryote engulfed a cyanobacterium. If the plastid is derived from the cyanobacterium, its DNA would be derived from the bacterial DNA. **3.** Figure 13.6b. Algae and plants with alternation of generations have a multicellular haploid stage *and* a multicellular diploid stage. In the other two life cycles, either the haploid stage or the diploid stage is unicellular. **4.** During photosynthesis, aerobic algae produce O_2 and use CO_2. O_2 is produced as a by-product of the light reactions, while CO_2 is used as an input to the Calvin cycle (the end products of which are sugars). Aerobic algae also perform cellular respiration, which uses O_2 as an input and produces CO_2 as a waste product.

Concept Check 28.4

1. Many red algae contain a photosynthetic pigment called phycoerythrin, which gives them a reddish color and allows them to carry out photosynthesis in relatively deep coastal water. Also unlike brown algae, red algae have no flagellated stages in their life cycle and must depend on water currents to bring gametes together for fertilization. **2.** *Ulva* contains many cells and its body is differentiated into leaflike blades and a rootlike holdfast. *Caulerpa*'s body is composed of multinucleate filaments without cross-walls, so it is essentially one large cell. **3.** Red algae have no flagellated stages in their life cycle and hence must depend on water currents to bring their gametes together. This feature of their biology might increase the difficulty of reproducing on land. In contrast, the gametes of green algae are flagellated, making it possible for them to swim in thin films of water. In addition, a variety of green algae contain compounds in their cytoplasm, cell wall, or zygote coat that protect against intense sunlight and other terrestrial conditions. Such compounds may have increased the chance that descendants of green algae could survive on land.

Concept Check 28.5

1. Amoebozoans have lobe- or tube-shaped pseudopodia, whereas forams have threadlike pseudopodia. **2.** Slime molds are fungus-like in that they produce fruiting bodies that aid in the dispersal of spores, and they are animal-like in that they are motile and ingest food. However, slime molds are more closely related to tubulinids and entamoebas than to fungi or animals.
3.

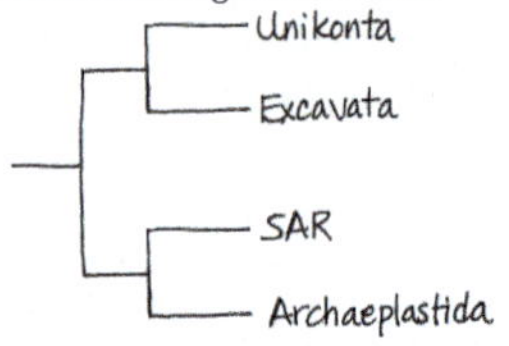

Concept Check 28.6

1. Because photosynthetic protists constitute the base of aquatic food webs, many aquatic organisms depend on them for food, either directly or indirectly. (In addition, a substantial percentage of the oxygen produced by photosynthesis is made by photosynthetic protists.) **2.** Protists form mutualistic and parasitic associations with other organisms. Examples include photosynthetic dinoflagellates that form a mutualistic symbiosis with coral polyps, parabasalids that form a mutualistic symbiosis with termites, and the stramenopile *Phytophthora ramorum*, a parasite of oak trees. **3.** Corals depend on their dinoflagellate symbionts for nourishment, so coral bleaching could cause the corals to die. As the corals died, less food would be available for fishes and other species that eat coral. As a result, populations of these species might decline, and that, in turn, might cause populations of their predators to decline. **4.** The two approaches differ in the evolutionary changes they may bring about. A strain of *Wolbachia* that confers resistance to infection by *Plasmodium* and does not harm mosquitoes would spread rapidly through the mosquito population. In this case, natural selection would favor any *Plasmodium* individuals that could overcome the resistance to infection conferred by *Wolbachia*. If insecticides are used, mosquitoes that are resistant to the insecticide would be favored by natural selection. Hence, use of *Wolbachia* could cause evolution in *Plasmodium* populations, while using insecticides could cause evolution in mosquito populations.

Summary of Key Concepts Questions

28.1 Sample response: Protists, plants, animals, and fungi are similar in that their cells have a nucleus and other membrane-enclosed organelles, unlike the cells of prokaryotes. These membrane-enclosed organelles make the cells of eukaryotes more complex than the cells of prokaryotes. Protists and other eukaryotes also differ from prokaryotes in having a well-developed cytoskeleton that enables them to have asymmetric forms and to change in shape as they feed, move, or grow. With respect to differences between protists and other eukaryotes, most protists are unicellular, unlike animals, plants, and most fungi. Protists also have greater nutritional diversity than other eukaryotes. **28.2** Unique cytoskeletal features are shared by many excavates. In addition, some members of Excavata have an "excavated" feeding groove for which the group was named. Moreover, recent genomic studies support the monophyly of the excavate supergroup. **28.3** Stramenopiles and alveolates are hypothesized to have originated by secondary endosymbiosis. Under this hypothesis, we can infer that the common ancestor of these two groups had a plastid, in this case of red algal origin. Thus, we would expect that apicomplexans (and alveolate or stramenopile protists) either would have plastids or would have lost their plastids over the course of evolution. **28.4** Red algae, green algae, and plants are placed in the same supergroup because considerable evidence indicates that these organisms all descended from the same ancestor, an ancient heterotrophic protist that acquired a cyanobacterial endosymbiont. **28.5** The unikonts are a diverse group of eukaryotes that includes many protists, along with animals and fungi. Most of the protists in Unikonta are amoebozoans, a clade of amoebas that have lobe- or tube-shaped pseudopodia (as opposed to the threadlike pseudopodia of rhizarians). Other protists in Unikonta include several groups that are closely related to fungi and several other groups that are closely related to animals. **28.6** Sample response: Ecologically important protists include photosynthetic dinoflagellates that provide essential sources of energy to their symbiotic partners, the corals that build coral reefs. Other important protistan symbionts include those that enable termites to digest wood and *Plasmodium*, the pathogen that causes malaria. Photosynthetic protists such as diatoms are among the most important producers in aquatic communities; as such, many other species in aquatic environments depend on them for food.

Test Your Understanding

1. D **2.** B **3.** B **4.** D **5.** D **6.** C
7.

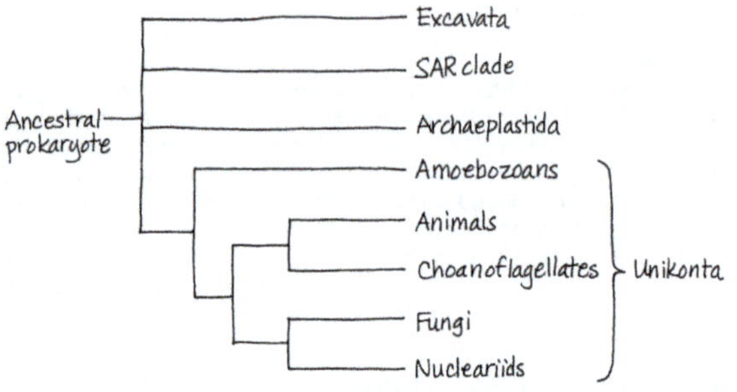

Pathogens that share a relatively recent common ancestor with humans will likely also share metabolic and structural characteristics with humans. Because drugs target the pathogen's metabolism or structure, developing drugs that harm the pathogen but not the patient should be most difficult for pathogens with whom we share the most recent evolutionary history. Working backward in time, we can use the phylogenetic tree to determine the order in which humans shared a common ancestor with pathogens in different taxa. This process leads to the prediction that it should be hardest to develop drugs to combat animal pathogens, followed by choanoflagellate pathogens, fungal and nucleariid pathogens, amoebozoans, other protists, and finally prokaryotes.

Chapter 29

Figure Questions

Figure 29.3 The life cycles of plants and some algae, shown in Figure 13.6b, have alternation of generations; other life cycles do not. Unlike in the animal life cycle (Figure 13.6a), in the plant/algal life cycle, meiosis produces spores, not gametes. These haploid spores then divide repeatedly by mitosis, ultimately forming a multicellular haploid individual that produces gametes. There is no multicellular haploid stage in the animal life cycle. An alternation of generations life cycle also has a multicellular diploid stage, whereas the life cycle of most fungi and some protists shown in Figure 13.6c does not. **Figure 29.6** Plants, vascular plants, and seed plants are monophyletic because each of these groups includes the common ancestor of the group and all of the descendants of that common ancestor. The other two categories of plants, the nonvascular plants and the seedless vascular plants, are paraphyletic: These groups do not include all of the descendants of the group's most recent common ancestor. **Figure 29.7** Yes. As shown in the diagram, the sperm cell and the egg cell that fuse each resulted from the mitotic division of spores produced by the same sporophyte. However, these spores would differ genetically from one another because they were produced by meiosis, a cell division process that generates genetic variation among the offspring cells. **Figure 29.9** Because the moss reduces nitrogen loss from the ecosystem, species that typically colonize the soils after the moss probably experience higher soil nitrogen levels than they otherwise would. The resulting increased availability of nitrogen may benefit these species because nitrogen is an essential nutrient that often is in short supply. **Figure 29.12** A fern that had wind-dispersed sperm would not require water for fertilization, thus removing a difficulty that ferns face when they live in arid environments. The fern would also be under strong selection to produce sperm above ground (as opposed to the current situation, where some fern gametophytes are located below ground).

Concept Check 29.1

1. Plants share some key traits only with charophytes: rings of cellulose-synthesizing complexes, similarity in sperm structure, and the formation of a phragmoplast in cell division. Comparisons of nuclear, chloroplast, and mitochondrial DNA sequences also indicate that certain groups of charophytes (such as *Zygnema* and *Coleochaete*) are the closest living relatives of plants. **2.** Spore walls toughened by sporopollenin (protects against harsh environmental conditions); multicellular, dependent embryos (provide nutrients and protection to the developing embryo); cuticle (reduces water loss); stomata (control gas exchange and reduce water loss) **3.** The multicellular diploid stage of the life cycle would not produce gametes. Instead, both males and females would produce haploid spores by meiosis. These spores would give rise to multicellular male and female haploid stages—a major change from the single-celled haploid stages (sperm and eggs) that we actually have. The multicellular haploid stages would produce gametes and reproduce sexually. An individual at the multicellular haploid stage of the human life cycle might look like us, or it might look completely different.

Concept Check 29.2

1. Most bryophytes do not have a vascular transport system, and their life cycle is dominated by gametophytes rather than sporophytes. **2.** Answers may include the following: Large surface area of protonema enhances absorption of water and minerals; the vase-shaped archegonia protect eggs during fertilization and transport nutrients to the embryos via placental transfer cells; the stalk-like seta conducts nutrients from the gametophyte to the capsule, where spores are produced; the peristome enables gradual spore discharge; stomata enable CO_2/O_2 exchange while minimizing water loss; lightweight spores are readily dispersed by wind.
3. Effects of global warming on peatlands could result in positive feedback, which occurs when an end product of a process increases its own production. In this case, global warming is expected to lower the water levels of some peatlands. This would expose peat to air and cause it to decompose, thereby releasing stored CO_2 to the atmosphere. The release of more stored CO_2 to the atmosphere could cause additional global warming, which in turn could cause further drops in water levels, the release of still more CO_2 to the atmosphere, additional warming, and so on: an example of positive feedback.

Concept Check 29.3

1. Lycophytes have microphylls, whereas seed plants and monilophytes (ferns and their relatives) have megaphylls. Monilophytes and seed plants also share other traits not found in lycophytes, such as the initiation of new root branches at various points along the length of an existing root. **2.** Both seedless vascular plants and bryophytes have flagellated sperm that require moisture for fertilization; this shared similarity poses challenges for these species in arid regions. With respect to key differences, seedless vascular plants have lignified, well-developed vascular tissue, a trait that enables the sporophyte to grow tall and that has transformed life on Earth (via the formation of forests). Seedless vascular plants also have true leaves and roots, which, when compared with bryophytes, provide increased surface area for photosynthesis and improve their ability to extract nutrients from soil. **3.** Three mechanisms contribute to the production of genetic variation in sexual reproduction: independent assortment of chromosomes, crossing over, and random fertilization. If fertilization were to occur between gametes from the same gametophyte, all of the offspring would be genetically identical. This would be the case because all of the cells produced by a gametophyte—including its sperm and egg cells—are the descendants of a single spore and hence are genetically identical. Although crossing over and the independent assortment of chromosomes would continue to generate genetic variation during the production of spores (which ultimately develop into gametophytes), overall the amount of genetic variation produced by sexual reproduction would drop.

Summary of Key Concepts Questions

29.1

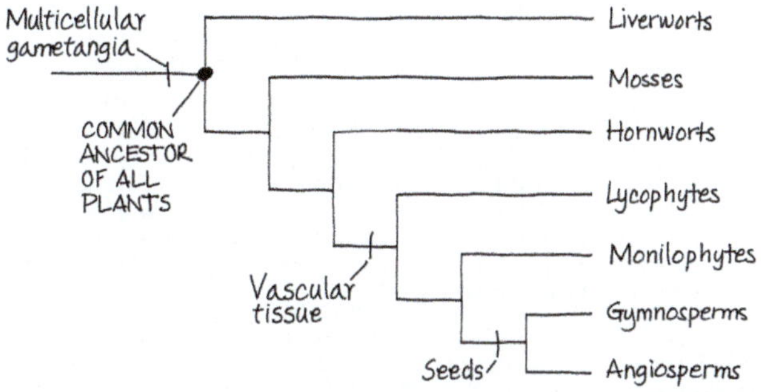

29.2 Some mosses colonize bare, sandy soils, leading to the increased retention of nitrogen in these otherwise low-nitrogen environments. Other mosses harbor nitrogen-fixing cyanobacteria that increase the availability of nitrogen in the ecosystem. The moss *Sphagnum* is often a major component of deposits of peat (partially decayed organic material). Boggy regions with thick layers of peat, known as peatlands, cover broad geographic regions and contain large reservoirs of carbon. By storing large amounts of carbon—in effect, removing CO_2 from the atmosphere—peatlands affect the global climate, making them of considerable ecological importance. **29.3** Lignified vascular tissue provided the strength needed to support a tall plant against gravity, as well as a means to transport water and nutrients to plant parts located high above ground. Roots were another key trait, anchoring the plant to the ground and providing additional structural support for plants that grew tall. Tall plants could shade shorter plants, thereby outcompeting them for light. Because the spores of a tall plant disperse farther than the spores of a short plant, it is also likely that tall plants could colonize new habitats more rapidly than short plants.

Test Your Understanding

1. B **2.** D **3.** C **4.** A **5.** B

6. (A) diploid; (B) haploid; (C) haploid; (D) diploid
7. Based on our current understanding of the evolution of major plant groups, the phylogeny has the four branch points shown here:

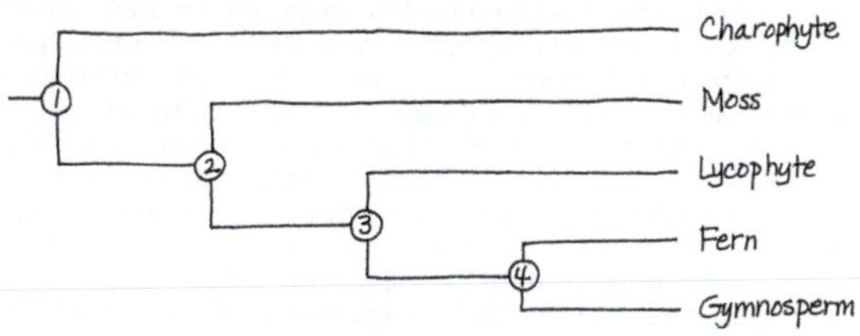

Derived characters unique to the charophyte and plant clade (indicated by branch point 1) include rings of cellulose-synthesizing complexes, flagellated sperm structure, and a phragmoplast. Derived characters unique to the plant clade (branch point 2) include alternation of generations; multicellular, dependent embryos; walled spores produced in sporangia; multicellular gametangia; and apical meristems. Derived characters unique to the vascular plant clade (branch point 3) include life cycles with dominant sporophytes, complex vascular systems (xylem and phloem), and well-developed roots and leaves. Derived characters unique to the monilophyte and seed plant clade (branch point 4) include megaphylls and roots that can branch at various points along the length of an existing root.
8. (a)

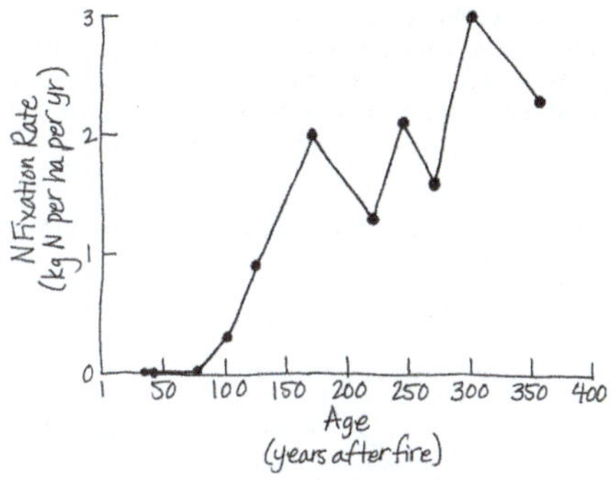

(b) In the first 40 years after a fire, nitrogen fixation rates were below 0.01 kg/(ha · yr), which was less than 1% of the amount of nitrogen deposited from the atmosphere. Thus, in the initial decades after a fire, the moss *Pleurozium* and the nitrogen-fixing bacteria it harbors had relatively little effect on the amount of nitrogen added to the forest. With time, however, *Pleurozium* and its symbiotic, nitrogen-fixing bacteria became increasingly important. By 170 years after a fire, the percentage of the ground surface covered by the moss had increased to about 70%, leading to a corresponding increase in populations of the symbiotic bacteria. As would be predicted from this result, in older forests considerably more nitrogen (130–300%) was added by nitrogen fixation than was deposited from the atmosphere.

Chapter 30

Figure Questions

Figure 30.2 Retaining the gametophyte within the sporophyte shields the egg-containing gametophyte from UV radiation. UV radiation is a mutagen. Hence, we would expect fewer mutations to occur in the egg cells produced by a gametophyte retained within the body of a sporophyte. Most mutations are harmful. Thus, the fitness of embryos should increase because fewer embryos would carry harmful mutations. **Figure 30.3** The seed contains cells from three generations: (1) the current sporophyte (cells of ploidy $2n$, found in the seed coat and in the megasporangium remnant that surrounds the spore wall), (2) the female gametophyte (cells of ploidy n, found in the food supply), and (3) the sporophyte of the next generation (cells of ploidy $2n$, found in the embryo). **Figure 30.4** Mitosis. A single haploid megaspore divides by mitosis to produce a multicellular, haploid female gametophyte. (Likewise, a single haploid microspore divides by mitosis to produce a multicellular male gametophyte.)

Figure 30.9

Figure 30.14 No. The branching order shown could still be correct if species on the lineages leading to basal angiosperms and magnolids had originated prior to 150 million years ago, but fossils of that age from those lineages had not yet been discovered. In such a situation, the 140-million-year-old date for the origin of the angiosperms shown on the phylogeny would be incorrect.

Concept Check 30.1

1. To reach the eggs, the flagellated sperm of seedless plants must swim through a film of water, usually over a distance of no more than a few centimeters. In contrast, the sperm of seed plants do not require water because they are produced within pollen grains that can be transported long distances by wind or by animal pollinators. Although flagellated in some species, the sperm of seed plants do not require mobility because pollen tubes convey them from the point at which the pollen grain is deposited (near the ovules) directly to the eggs. **2.** The reduced gametophytes of seed plants are nurtured by sporophytes and protected from stress, such as drought conditions and UV radiation. Pollen grains, with walls containing sporopollenin, provide protection during transport by wind or animals. Seeds have one or two layers of protective tissue, the seed coat, that improve survival by providing more protection from environmental stresses than do the walls of spores. Seeds also contain a stored supply of food, which provides nourishment for growth after dormancy is broken and the embryo emerges as a seedling. **3.** If a seed could not enter dormancy, the embryo would continue to grow after it was fertilized. As a result, the embryo might rapidly become too large to be dispersed, thus limiting its transport. The embryo's chance of survival might also be reduced because it could not delay growth until conditions become favorable.

Concept Check 30.2

1. Although gymnosperms are similar in not having their seeds enclosed in ovaries and fruits, their seed-bearing structures vary greatly. For instance, cycads have large cones, whereas some gymnosperms, such as *Ginkgo* and *Gnetum*, have small cones that look somewhat like berries, even though they are not fruits. Leaf shape also varies greatly, from the needles of many conifers to the palmlike leaves of cycads to *Gnetum* leaves that look like those of flowering plants. **2.** The pine life cycle illustrates heterospory, as ovulate cones produce megaspores and pollen cones produce microspores. The reduced gametophytes are evident in the form of the microscopic pollen grains that develop from microspores and the microscopic female gametophyte that develops from the megaspore. The egg is shown developing within an ovule, and a pollen tube is shown conveying the sperm. The figure also shows the protective and nutritive features of a seed.
3.

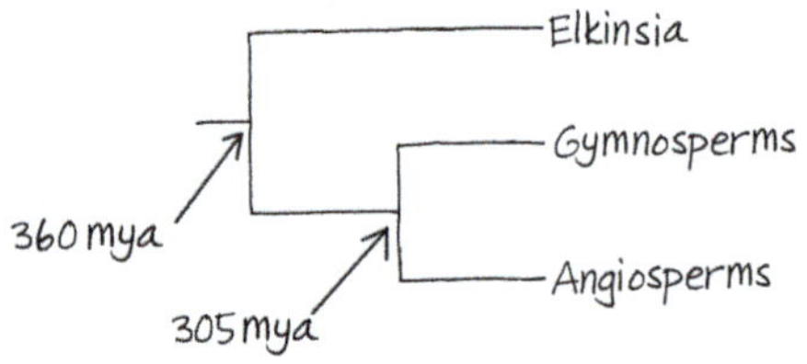

Concept Check 30.3

1. In the oak's life cycle, the tree (the sporophyte) produces flowers, which contain gametophytes in pollen grains and ovules; the eggs in ovules are fertilized; the mature ovaries develop into dry fruits called acorns. We can view the oak's life cycle as starting when the acorn seeds germinate, resulting in embryos giving rise to seedlings and finally to mature trees, which produce flowers—and then more acorns. **2.** Pine cones and flowers both have sporophylls, modified leaves that produce spores. Pine trees have separate pollen cones (with pollen grains) and ovulate cones (with ovules inside cone scales). In flowers, pollen grains are produced by the anthers of stamens, and ovules are within the ovaries of carpels. Unlike pine cones, many flowers produce both pollen and ovules. **3.** The fact that the clade with bilaterally symmetrical flowers had more species establishes a correlation between flower shape and the rate of plant speciation. Flower shape is not necessarily responsible for the result because the shape (that is, bilateral or radial symmetry) may have been correlated with another factor that was the actual cause of the observed result. Note, however, that flower shape was associated with increased speciation rates when averaged across 19 different pairs of plant lineages. Since these 19 lineage pairs were independent of one another, this association suggests—but does not establish—that differences in flower shape cause differences in speciation rates. In general, strong evidence for causation can come from controlled, manipulative experiments, but such experiments are usually not possible for studies of past evolutionary events.

Concept Check 30.4

1. Plant diversity can be considered a resource because plants provide many important benefits to humans; as a resource, plant diversity is nonrenewable because if a species is lost to extinction, that loss is permanent. **2.** A detailed phylogeny of the seed plants would identify many different monophyletic groups of seed plants. Using this phylogeny, researchers could look for clades that contained species in which medicinally useful compounds had already been discovered. Identification of such clades would allow researchers to concentrate their search for new medicinal compounds among clade members—as opposed to searching for new compounds in species that were selected at random from the more than 250,000 existing species of seed plants.

Summary of Key Concepts Questions

30.1 The integument of an ovule develops into the protective coat of a seed. The ovule's megaspore develops into a haploid female gametophyte, and two parts of the seed are related to that gametophyte: The food supply of the seed is derived from haploid gametophyte cells, and the embryo of the seed develops after the female gametophyte's egg cell is fertilized by a sperm cell. A remnant of the ovule's megasporangium surrounds the spore wall that encloses the seed's food supply and embryo. **30.2** Gymnosperms arose about 305 million years ago, making them a successful group in terms of their evolutionary longevity. Gymnosperms have the five derived traits common to all seed plants (reduced gametophytes, heterospory, ovules, pollen, and seeds), making them well adapted for life on land. Finally, because gymnosperms dominate immense geographic regions today, the group is also highly successful in geographic distribution. **30.3** Based on fossils known during his lifetime, Darwin was troubled by the relatively sudden and geographically widespread appearance of angiosperms in the fossil record. Recent fossil evidence shows that angiosperms arose and began to diversify over a period of 20–30 million years, a less rapid event than was suggested by the fossils known during Darwin's lifetime. Fossil discoveries have also uncovered extinct lineages of woody seed plants thought to have been more closely related to angiosperms than to gymnosperms; one such group, the Bennettitales, had flowerlike structures that may have been pollinated by insects. In addition, phylogenetic analyses have identified a woody species, *Amborella*, as the most basal lineage of extant angiosperms. The fact that both the extinct seed plant ancestors of angiosperms and the most basal taxon of extant angiosperms were woody suggests that the common ancestor of angiosperms also was woody. **30.4** The loss of tropical forests could contribute to global warming (which would have negative effects on many human societies). People also depend on Earth's biodiversity for many products and services and hence would be harmed by the loss of species that would occur if the world's remaining tropical forests were cut down. With respect to a possible mass extinction, tropical forests harbor at least 50% of the species on Earth. If the remaining tropical forests were destroyed, large numbers of these species could be driven to extinction, thus rivaling the losses that occurred in the five mass extinction events documented in the fossil record.

Test Your Understanding

1. C **2.** A **3.** B **4.** D **5.** C
6.

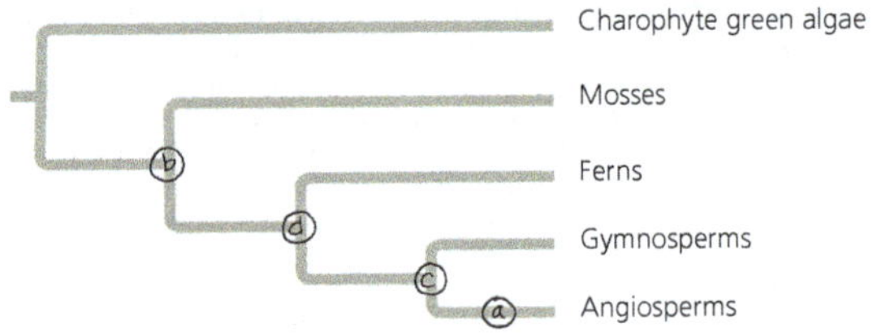

8. (a)

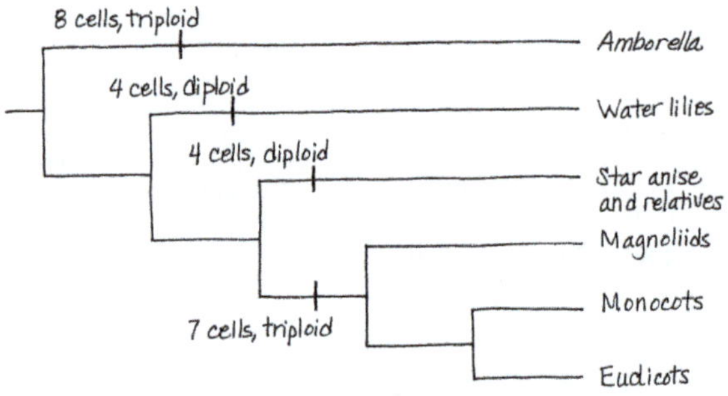

(b) The phylogeny indicates that basal angiosperms differed from other angiosperms in terms of the number of cells in female gametophytes and the ploidy of the endosperm. The ancestral state of the angiosperms cannot be determined from these data alone. It is possible that the common ancestor of angiosperms had seven-celled female gametophytes and triploid endosperm and hence that the eight-celled and four-celled conditions found in basal angiosperms represent derived traits for those lineages. Alternatively, either the eight-celled or four-celled condition may represent the ancestral state.

Chapter 31

Figure Questions

Figure 31.2 DNA from each of these mushrooms would be identical if each mushroom is part of a single hyphal network, as is likely. **Figure 31.5** The haploid spores produced in the sexual portion of the life cycle develop from haploid nuclei that were produced by meiosis; because genetic recombination occurs during meiosis, these spores will differ genetically from one another. In contrast, the haploid spores produced in the asexual portion of the life cycle develop from nuclei that were produced by mitosis; as a result, these spores are genetically identical to one another. **Figure 31.15** One or both of the following would apply to each species: DNA analyses would reveal that it is a member of the ascomycete clade, or aspects of its sexual life cycle would indicate that it is an ascomycete (for example, it would produce asci and ascospores). **Figure 31.16** The hypha is composed of cells that are haploid (n), as indicated by the teal-colored arrow behind it. **Figure 31.18** The mushroom is a basidiocarp, or fruiting body, of the dikaryotic mycelium, and so a cell from its stalk would be dikaryotic ($n + n$). **Figure 31.20** Two possible controls would be E–P– and E+P–. Results from an E–P– control could be compared with results from the E–P+ experiment, and results from an E+P– control could be compared with results from the E+P+ experiment. Together, these two comparisons would indicate whether the addition of the pathogen causes an increase in leaf mortality. Results from an E–P– experiment could also be compared with results from the second control (E+P–) to determine whether adding the fungal endophytes has a negative effect on the plant.

Concept Check 31.1

1. Both a fungus and a human are heterotrophs. Many fungi digest their food externally by secreting enzymes into the food and then absorbing the small molecules that result from digestion. Other fungi absorb such small molecules directly from their environment. In contrast, humans (and most other animals) ingest relatively large pieces of food and digest the food within their bodies. **2.** The ancestors of such a mutualist most likely secreted powerful enzymes to digest the body of their insect host. Since such enzymes would harm a living host, it is likely that the mutualist would not produce such enzymes or would restrict their secretion and use. **3.** Carbon that enters the plant through stomata is fixed into sugar through photosynthesis. Some of these sugars are absorbed by the fungus that partners with the plant to form mycorrhizae; others are transported within the plant body and used in the plant. Thus, the carbon may be deposited in either the body of the plant or the body of the fungus.

Concept Check 31.2

1. The majority of the fungal life cycle is spent in the haploid stage, whereas the majority of the human life cycle is spent in the diploid stage. **2.** The two mushrooms might be reproductive structures of the same mycelium (the same organism). Or they might be parts of two separate organisms that have arisen from a single parent organism through asexual reproduction (for example, from two genetically identical asexual spores) and thus carry the same genetic information.

Concept Check 31.3

1. DNA evidence indicates that fungi, animals, and their protistan relatives form a clade, the opisthokonts. Furthermore, some chytrids and other fungi thought to be members of basal lineages have posterior flagella, as do most other opisthokonts. This suggests that other fungal lineages lost their flagella after diverging from ancestors that had flagella. **2.** Mycorrhizae form extensive networks of hyphae through the soil, enabling nutrients to be absorbed more efficiently than a plant can do on its own; this is true today, and similar associations were probably very important for the earliest plants (which lacked roots). Evidence for the antiquity of mycorrhizal associations includes fossils showing arbuscular mycorrhizae in the early plant *Aglaophyton* and molecular results showing that genes required for the formation of mycorrhizae are present in liverworts and other basal plant lineages. **3.** Fungi are heterotrophs. Prior to the colonization of land by plants, terrestrial fungi would have lived where other organisms (or their remains) were present and provided a source of food. Thus, if fungi colonized land before plants, they could have fed on prokaryotes or protists that lived on land or by the water's edge—but not on the plants or animals on which many fungi feed today.

Concept Check 31.4

1. Flagellated spores; molecular evidence also suggests that chytrids include species that belong to lineages that diverged from other fungi early in the history of the group. **2.** Possible answers include the following: In zygomycetes, the sturdy, thick-walled zygosporangium can withstand harsh conditions and then undergo karyogamy and meiosis when the environment is favorable for reproduction. In glomeromycetes, the hyphae have a specialized morphology that enables the fungi to form arbuscular mycorrhizae with plant roots. In ascomycetes, the asexual spores (conidia) are often produced in chains or clusters at the tips of conidiophores, where they are easily dispersed by wind. The often cup-shaped ascocarps house the sexual spore-forming asci. In basidiomycetes, the basidiocarp supports and protects a large surface area of basidia, from which spores are dispersed. **3.** Such a change to the life cycle of an ascomycete would reduce the number and genetic diversity of ascospores that result from a mating event. Ascospore number would drop because a mating event would lead to the formation of only one ascus. Ascospore genetic diversity would also drop because in ascomycetes, one mating event leads to the formation of asci by many different dikaryotic cells. As a result, genetic recombination and meiosis occur independently many different times—which could not happen if only a single ascus was formed. It is also likely that if such an ascomycete formed an ascocarp, the shape of the ascocarp would differ considerably from that found in its close relatives.

Concept Check 31.5

1. A suitable environment for growth, retention of water and minerals, protection from intense sunlight, and protection from being eaten **2.** A hardy spore stage enables dispersal to host organisms through a variety of mechanisms; their ability to grow rapidly in a favorable new environment enables them to capitalize on the host's resources. **3.** Many different outcomes might have occurred. Organisms that currently form mutualisms with fungi might have gained the ability to perform the tasks currently done by their fungal partners, or they might have formed similar mutualisms with other organisms (such as bacteria). Alternatively, organisms that currently form mutualisms with fungi might be less effective at living in their present environments. For example, the colonization of land by plants might have been more difficult. And if plants did eventually colonize land without fungal mutualists, natural selection might have favored plants that formed more highly divided and extensive root systems (in part replacing mycorrhizae).

Summary of Key Concepts Questions

31.1 The body of a multicellular fungus typically consists of thin filaments called hyphae. These filaments form an interwoven mass (mycelium) that penetrates the substrate on which the fungus grows and feeds. Because the individual filaments are thin, the surface-to-volume ratio of the mycelium is maximized, making nutrient absorption highly efficient.

31.2

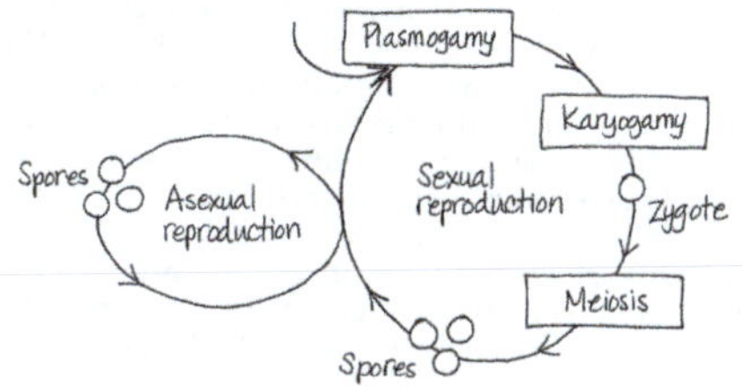

31.3 Phylogenetic analyses show that fungi and animals are more closely related to each other than either is to other multicellular eukaryotes (such as plants or multicellular algae). These analyses also show that fungi are more closely related to single-celled protists called nucleariids than they are to animals, whereas animals are more closely related to a different group of single-celled protists, the choanoflagellates, than they are to fungi. In combination, these results indicate that multicellularity evolved in fungi and animals independently, from different single-celled ancestors.

31.4

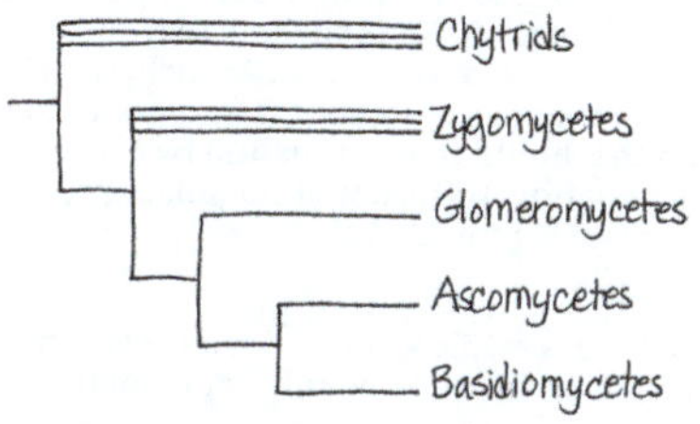

31.5 As decomposers, fungi break down the bodies of dead organisms, thereby recycling elements between the living and nonliving environments. Without the activities of fungi and bacterial decomposers, essential nutrients would remain tied up in organic matter, and life would cease. As an example of their key role as mutualists, fungi form mycorrhizal associations with plants. These associations improve the growth and survival of plants, thereby indirectly affecting the many other species (humans included) that depend on plants. As pathogens, fungi harm other species. In some cases, fungal pathogens have caused their host populations to decline across broad geographic regions, as seen for the American chestnut.

Test Your Understanding

1. B **2.** D **3.** A **4.** D

Chapter 32

Figure Questions

Figure 32.3 As described in 1 and 2, choanoflagellates and a broad range of animals have collar cells. Since collar cells have never been observed in plants, fungi, or non-choanoflagellate protists, this suggests that choanoflagellates may be more closely related to animals than to other eukaryotes. If choanoflagellates are more closely related to animals than to any other group of eukaryotes, choanoflagellates and animals should share other traits that are not found in other eukaryotes. The data described in 3 are consistent with this prediction. **Figure 32.10** The cells of an early embryo with deuterostome development typically are not committed to a particular developmental fate, whereas the cells of an early embryo with protostome development typically are committed to a particular developmental fate. As a result, an embryo with deuterostome development would be more likely to contain stem cells that could give rise to cells of any type. **Figure 32.11** Cnidaria is the sister phylum in this tree.

Concept Check 32.1

1. In most animals, the zygote undergoes cleavage, which leads to the formation of a blastula. Next, in gastrulation, one end of the embryo folds inward, producing layers of embryonic tissue. As the cells of these layers differentiate, a wide variety of animal forms are produced. Despite the diversity of animal forms, animal development is controlled by a similar set of *Hox* genes across a broad range of taxa. **2.** The imaginary plant would require tissues composed of cells that were analogous to the muscle and nerve cells found in animals: "Muscle" tissue would be necessary for the plant to chase prey, and "nerve" tissue would be required for the plant to coordinate its movements when chasing prey. To digest captured prey, the plant would need to either secrete enzymes into one or more digestive cavities (which could be modified leaves, as in a Venus flytrap) or secrete enzymes outside of its body and feed by absorption. To extract nutrients from the soil—yet be able to chase prey—the plant would need something other than fixed roots, perhaps retractable "roots" or a way to ingest soil. To conduct photosynthesis, the plant would require chloroplasts. Overall, such an imaginary plant would be very similar to an animal that had chloroplasts and retractable roots.

Concept Check 32.2

1. c, b, a, d **2.** The red-colored portion of the tree represents ancestors of animals that lived between 1 billion years ago and 770 million years ago. Although these ancestors are more closely related to animals than to fungi, they would not be classified as animals. One example of an ancestor represented by the red-colored portion of this tree is the most recent common ancestor shared by choanoflagellates and animals. **3.** In descent with modification, an organism shares characteristics with its ancestors (due to their shared ancestry), yet it also differs from its ancestors (because organisms accumulate differences over time as they adapt to their surroundings). As an example, consider the evolution of animal cadherin proteins, a key step in the origin of multicellular animals. These proteins illustrate both of these aspects of descent with modification: Animal cadherin proteins share many protein domains with a cadherin-like protein found in their choanoflagellate ancestors, yet they also have a unique "CCD" domain that is not found in choanoflagellates.

Concept Check 32.3

1. Grade-level characteristics are those that multiple lineages share regardless of evolutionary history. Some grade-level characteristics may have evolved multiple times independently. Features that unite clades are derived characteristics that originated in a common ancestor and were passed on to the various descendants. **2.** A snail has a spiral and determinate cleavage pattern; a human has radial, indeterminate cleavage. In a snail, the coelomic cavity is formed by splitting of mesoderm masses; in a human, the coelom forms from folds of archenteron. In a snail, the mouth forms from the blastopore; in a human, the anus develops from the blastopore. **3.** Most coelomate triploblasts have two openings to their digestive tract, a mouth and an anus. As such, their bodies have a structure that is analogous to that of a doughnut: The digestive tract (the hole of the doughnut) runs from the mouth to the anus and is surrounded by various tissues (the solid part of the doughnut). The doughnut analogy is most obvious at early stages of development (see Figure 32.10c).

Concept Check 32.4

1. Cnidarians possess tissues, while sponges do not. Also unlike sponges, cnidarians exhibit body symmetry, though it is radial and not bilateral as in most other animal phyla.
2.

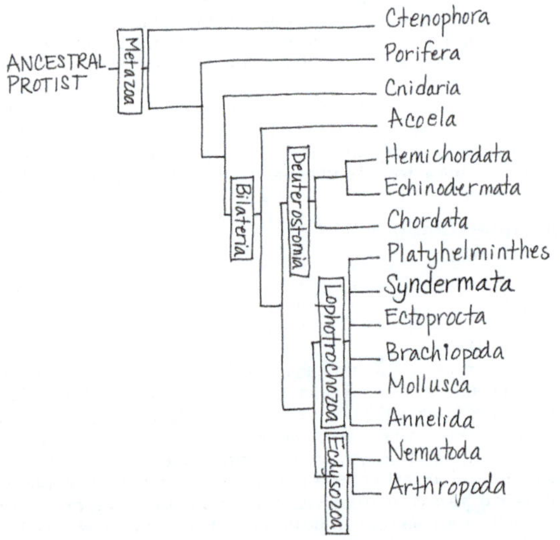

Under the hypothesis that ctenophores are basal metazoans, sponges (which lack tissues) would be nested within a clade whose other members all have tissues. As a result, a group composed of animals with tissues would not form a clade. **3.** The phylogeny in Figure 32.11 indicates that molluscs are members of Lophotrochozoa, one of the three main groups of bilaterians (the others being Deuterostomia and Ecdysozoa). As seen in Figure 25.11, the fossil record shows that molluscs were present tens of millions of years before the Cambrian explosion. Thus, long before the Cambrian explosion, the lophotrochoan clade had formed and was evolving independently of the evolutionary lineages leading to Deuterostomia and Ecdysozoa. Based on the phylogeny in Figure 32.11, we can also conclude that the lineages leading to Deuterostomia and Ecdysozoa were independent of one another before the Cambrian explosion. Since the lineages leading to the three main clades of bilaterians were evolving independently of one another prior to the Cambrian explosion, that explosion could be viewed as consisting of three "explosions," not one.

Summary of Key Concepts Questions

Concept 32.1 Unlike animals, which are heterotrophs that ingest their food, plants are autotrophs, and fungi are heterotrophs that grow on their food and feed by absorption. Animals lack cell walls, which are found in both plants and fungi. Animals also have muscle tissue and nerve tissue, which are not found in either plants or fungi. In addition, the sperm and egg cells of animals are produced by meiotic division, unlike what occurs in plants and fungi (where reproductive cells such as sperm and eggs are produced by mitotic division). Finally, animals regulate the development of body form with *Hox* genes, a unique group of genes that is not found in either plants or fungi. **Concept 32.2** Current hypotheses about the cause of the Cambrian explosion include new predator-prey relationships, an increase in atmospheric oxygen, and an increase in developmental flexibility provided by the origin of *Hox* genes and other genetic changes. **Concept 32.3** Body plans provide a helpful way to compare and contrast key features of organisms. However, phylogenetic analyses show that similar body plans have arisen independently in different groups of organisms. As such, similar body plans may have arisen by convergent evolution and hence may not be informative about evolutionary relationships. **Concept 32.4** Listed in order from the most to the least inclusive clade, humans belong to Metazoa, Eumetazoa, Bilateria, Deuterostomia, and Chordata.

Test Your Understanding

1. A **2.** D **3.** C **4.** B

Chapter 33

Figure Questions

Figure 33.8 The *Obelia* life cycle is most similar to the life cycle shown in Figure 13.6a. In *Obelia*, both the polyp and the medusa are diploid organisms. Typical of animals, only the single-celled gametes are haploid. By contrast, plants and some algae (Figure 13.6b) have a multicellular haploid generation and a multicellular diploid generation. *Obelia* also differs from fungi and some protists (Figure 13.6c) in that the diploid stage of those organisms is unicellular. **Figure 33.8** Both a feeding polyp and a medusa are diploid, as indicated by the pink arrow in the diagram. The medusa stage produces haploid gametes. **Figure 33.9** Possible examples might include the endoplasmic reticulum (flattening; increases area for biosynthesis), the cristae of mitochondria (folding; increases the surface area available for cellular respiration), root hairs (projections; increase area for absorption), or cardiovascular systems (branching; increase area for materials exchange in tissues). **Figure 33.11** Adding fertilizer to the water supply would probably increase the abundance of algae, and that, in turn, would likely increase the abundance of snails (which eat algae). If the water was also contaminated with infected human feces, an increase in the number of snails would likely lead to an increase in the abundance of blood flukes (which require snails as an intermediate host). As a result, the occurrence of schistosomiasis might increase. **Figure 33.22** The extinction of freshwater bivalves might lead to an increase in the abundance of photosynthetic protists and bacteria. Because these organisms are at the base of aquatic food webs, increases in their abundance could have major effects on aquatic communities (including both increases and decreases in the abundance of other species). **Figure 33.30** Such a result would be consistent with the origin of the *Ubx* and *abd-A Hox* genes having played a major role in the evolution of increased body segment diversity in arthropods. However, note that such a result would simply show that the presence of the *Ubx* and *abd-A Hox* genes was *correlated with* an increase in body segment diversity in arthropods; it would not provide direct experimental evidence that the origin of the *Ubx* and *abd-A* genes *caused* an increase in arthropod body segment diversity. **Figure 33.36** You should have circled the clade that includes the insects, remipedians, and other crustaceans, along with the branch point that represents their most recent common ancestor.

Concept Check 33.1

1. The flagella of choanocytes draw water through their collars, which trap food particles. The particles are engulfed by phagocytosis and digested, either by choanocytes or by amoebocytes. **2.** The collar cells of sponges bear a striking resemblance to a choanoflagellate cell. This suggests that the last common ancestor of animals and their protist sister group may have resembled a choanoflagellate. Nevertheless, mesomycetozoans could still be the sister group of animals. If this is the case, the lack of collar cells in mesomycetozoans would indicate that over time their structure evolved in ways that caused it to no longer resemble a choanoflagellate cell. It is also possible that choanoflagellates and sponges share similar-looking collar cells as a result of convergent evolution.

Concept Check 33.2

1. Both the polyp and the medusa are composed of an outer epidermis and an inner gastrodermis separated by a gelatinous layer, the mesoglea. The polyp is a cylindrical form that adheres to the substrate by its aboral end; the medusa is a flattened, mouth-down form that moves freely in the water. **2.** Cnidarian stinging cells (cnidocytes) function in defense and prey capture. They contain capsule-like organelles (cnidae), which in turn contain coiled threads. The threads either inject poison or stick to and entangle small prey. **3.** Evolution is not goal oriented; hence, it would not be correct to argue that cnidarians were not "highly evolved" simply because their form had changed relatively little over the past 560 million years. Instead, the fact that cnidarians have persisted for hundreds of millions of years indicates that the cnidarian body plan is a highly successful one.

Concept Check 33.3

1. Tapeworms can absorb food from their environment and release ammonia into their environment through their body surface because their body is very flat, due in part to the lack of a coelom. **2.** The inner tube is the alimentary canal, which runs the length of the body. The outer tube is the body wall. The two tubes are separated by the coelom. **3.** All molluscs have inherited a foot from their common ancestor. However, in different groups of molluscs, the structure of the foot has been modified over time by natural selection. In gastropods, the foot is used as a holdfast or to move slowly on the substrate. In cephalopods, the foot has been modified into part of the tentacles and into an excurrent siphon, through which water is propelled (resulting in movement in the opposite direction).

Concept Check 33.4

1. Nematodes lack body segments and a true coelom; annelids have both. **2.** The arthropod exoskeleton, which had already evolved in the ocean, allows terrestrial species to retain water and support their bodies on land. Wings allow insects to disperse quickly to new habitats and to find food and mates. The tracheal system allows for efficient gas exchange despite the presence of an exoskeleton. **3.** Yes. Under the traditional hypothesis, we would expect body segmentation to be controlled by similar *Hox* genes in annelids and arthropods. However, if annelids are in Lophotrochozoa and arthropods are in Ecdysozoa (as current evidence suggests), body segmentation may have evolved independently in these two groups. In such a case, we might expect that different *Hox* genes would control the development of body segmentation in the two clades.

Concept Check 33.5

1. Each tube foot consists of an ampulla and a podium. When the ampulla squeezes, it forces water into the podium, which causes the podium to expand and contact the substrate. Adhesive chemicals are then secreted from the base of the podium, thereby attaching the podium to the substrate. **2.** Both insects and nematodes are members of Ecdysozoa, one of the three major clades of bilaterians. Therefore, a characteristic shared by *Drosophila* and *Caenorhabditis* may be informative for other members of their clade—but not necessarily for members of Deuterostomia. Instead, Figure 33.2 suggests that a species within Echinodermata or Chordata might be a more appropriate invertebrate model organism from which to draw inferences about humans and other vertebrates. **3.** Echinoderms include species with a wide range of body forms. However, even echinoderms that look very different from one another, such as sea stars and sea cucumbers, share characteristics unique to their phylum, including a water vascular system and tube feet. The differences between echinoderm species illustrate the diversity of life, while the characteristics they share illustrate the unity of life. The match between organisms and their environments can be seen in such echinoderm features as the eversible stomachs of sea stars (enabling them to digest prey that are larger than their mouth) and the complex, jaw-like structure that sea urchins use to eat seaweed.

Summary of Key Concepts Questions

33.1 The sponge body consists of two layers of cells, both of which are in contact with water. As a result, gas exchange and waste removal occur as substances diffuse into and out of the cells of the body. Choanocytes and amoebocytes ingest food particles from the surrounding water. Choanocytes also release food particles to amoebocytes, which then digest the food particles and deliver nutrients to other cells. **33.2** The cnidarian body plan consists of a sac with a central digestive compartment, the gastrovascular cavity. The single opening to this compartment serves as both a mouth and an anus. The two main variations on this body plan are sessile polyps (which adhere to the substrate at the end of the body opposite to the mouth/anus) and motile medusae (which move freely through the water and resemble flattened, mouth-down versions of polyps). **33.3** No. Some lophotrochozoans have a crown of ciliated tentacles that function in feeding (called a lophophore), while others go through a distinctive developmental stage known as trochophore larvae. Many other lophotrochozoans do not have either of these features. As a result, the clade is defined primarily by DNA similarities, not morphological similarities. **33.4** Many nematode species live in soil and in sediments on the bottom of bodies of water. These free-living species play important roles in decomposition and nutrient cycling. Other nematodes are parasites, including many species that attack the roots of plants and some that attack animals (including humans). Arthropods have profound effects on all aspects of ecology. In aquatic environments, crustaceans play key roles as grazers (of algae), scavengers, and predators, and some species, such as krill, are important sources of food for whales and other vertebrates. On land, it is difficult to think of features of the natural world that are not affected in some way by insects and other arthropods, such as spiders and ticks. There are more than 1 million species of insects, many of which have enormous ecological effects as herbivores, predators, parasites, decomposers, and vectors of disease. Insects are also key sources of food for many organisms, including humans in some regions of the world. **33.5** Echinoderms and chordates are both members of Deuterostomia, one of the three main clades of bilaterian animals. As such, chordates (including humans) are more closely related to echinoderms than we are to animals in any of the other phyla covered in this chapter. Nevertheless, echinoderms and chordates have evolved independently for over 500 million years. This statement does not contradict the close relationship of echinoderms and chordates, but it does make clear that "close" is a relative term indicating that these two phyla are more closely related to each other than either is to animal phyla not in Deuterostomia.

Test Your Understanding

1. A **2.** C **3.** B **4.** D **5.** C **6.** D

Chapter 34

Figure Questions

Figure 34.2

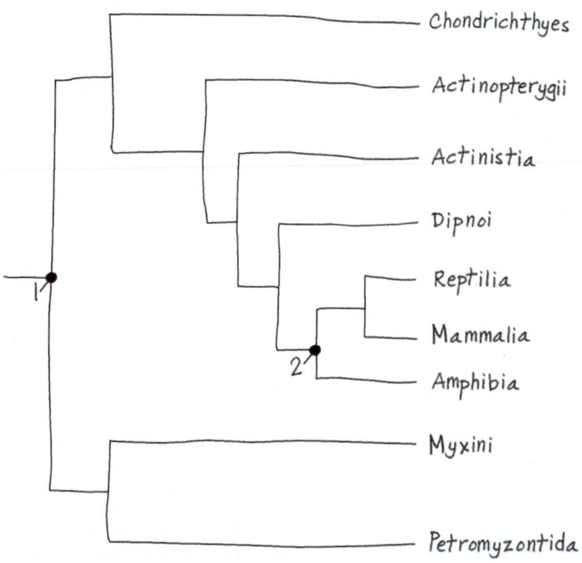

The redrawn tree shows mammals (including humans) as nested near the middle of the evolutionary tree of vertebrates. Showing the vertebrate tree in this way provides a visual illustration of the fact that the evolutionary history of vertebrates did not consist of a series of steps "leading to" humans. **Figure 34.6** The patterns in these figures suggest that specific *Hox* genes, as well as the order in which they are expressed, have been highly conserved over the course of evolution. **Figure 34.20** *Tiktaalik* was a lobe-fin fish that had both fish and tetrapod characters. Like a fish, *Tiktaalik* had fins, scales, and gills. As described by Darwin's concept of descent with modification, such shared characters can be attributed to descent from ancestral species—in this case, *Tiktaalik*'s descent from fish ancestors. *Tiktaalik* also had traits that were unlike a fish but like a tetrapod, including a flat skull, a neck, a full set of ribs, and the skeletal structure of its fin. These characters illustrate the second part of descent with modification, showing how ancestral features had become modified over time. **Figure 34.21** Sometime between 370 mya and 340 mya. We can infer this because amphibians must have originated after the most recent common ancestor of *Tulerpeton* and living tetrapods (and that ancestor is said to have originated 370 mya), but no later than the date of the earliest known fossils of amphibians (shown in the figure as 340 mya). **Figure 34.25** Pterosaurs did not descend from the common ancestor of all dinosaurs; hence, pterosaurs are not dinosaurs. However, birds are descendants of the common ancestor of the dinosaurs. As a result, a monophyletic clade of dinosaurs must include birds. In that sense, birds are dinosaurs. **Figure 34.37** In a catabolic pathway, like the aerobic processes of cellular respiration, water is released as a by-product when an organic compound such as glucose is mixed with oxygen. The kangaroo rat can retain and use that water, decreasing its need to drink water. **Figure 34.38** In general, the process of exaptation occurs as a structure that had one function acquires a different function via a series of intermediate stages. Each of these intermediate stages typically has some function in the organism in which it is found. The incorporation of articular and quadrate bones into the mammalian ear illustrates exaptation because these bones originally evolved as part of the jaw, where they functioned as the jaw hinge, but over time they became co-opted for another function, namely, the transmission of sound. **Figure 34.44** As shown in this phylogeny, chimpanzees and humans represent the tips of separate branches of evolution. As such, the human and chimpanzee lineages have evolved independently after they diverged from their common ancestor—an event that took place between 6 million and 7 million years ago. Hence, it is incorrect to say that humans evolved from chimpanzees (or vice versa). If humans had descended from chimpanzees, for example, the human lineage would be nested within the chimpanzee lineage, much as birds are nested within the reptile clade (see Figure 34.25). **Figure 34.51** Fossil evidence indicates that Neanderthals did not live in Africa; hence there would have been little opportunity for mating (gene flow) between Neanderthals and humans in Africa. However, as humans migrated from Africa, mating may have occurred between Neanderthals and humans in the first region where the two species encountered one another: the Middle East. Humans carrying Neanderthal genes may then have migrated to other locations, explaining why Neanderthals are equally related to humans from France, China, and Papua New Guinea.

Concept Check 34.1

1. The four characters are a notochord; a dorsal, hollow nerve chord; pharyngeal slits or clefts; and a muscular, post-anal tail. **2.** In humans, these characters are present only in the embryo. The notochord becomes disks between the vertebrae; the dorsal, hollow nerve cord develops into the brain and spinal cord; the pharyngeal clefts develop into various adult structures, and the tail is almost completely lost. **3.** You would expect the vertebrate groups Actinopterygii, Actinistia, Dipnoi, Amphibia, Reptilia, and Mammalia to have lungs or lung derivatives. All of these groups originate to the right of (evolved after) the hatch mark indicating the appearance of this derived character in their lineage.

Concept Check 34.2

1. Parasitic lampreys have a round, rasping mouth, which they use to attach to fish. Non-parasitic lampreys feed only as larvae; these larvae resemble lancelets and like them, are suspension feeders. Conodonts had two sets of mineralized dental elements, which may have been used to impale prey and cut it into smaller pieces.
2. Such a finding suggests that early organisms with a head were favored by natural selection in several different evolutionary lineages. However, while a logical argument can be made that having a head was advantageous, fossils alone do not constitute proof. **3.** In armored jawless vertebrates, bone served as external armor that may have provided protection from predators. Some species also had mineralized mouthparts, which could be used for either predation or scavenging.

Concept Check 34.3

1. Both are gnathostomes and have jaws, four clusters of *Hox* genes, enlarged forebrains, and lateral line systems. Shark skeletons consist mainly of cartilage, whereas tuna have bony skeletons. Sharks also have a spiral valve. Tuna have an operculum and a swim bladder, as well as flexible rays supporting their fins. **2.** Aquatic gnathostomes have jaws (an adaptation for feeding) and paired fins and a tail (adaptations for swimming). Aquatic gnathostomes also typically have streamlined bodies for efficient swimming and swim bladders or other mechanisms (such as oil storage in sharks) for buoyancy.
3.

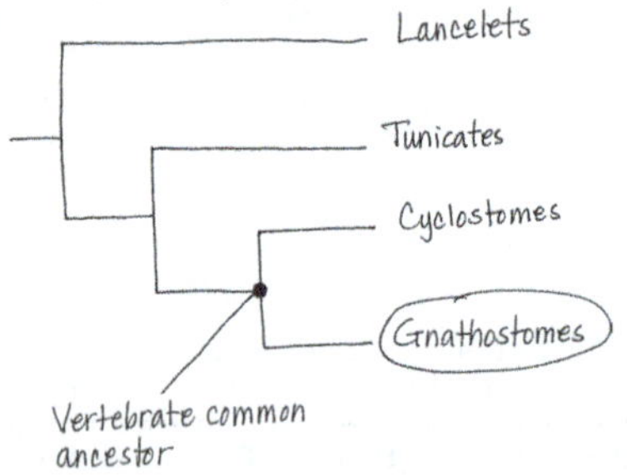

4. Yes, that could have happened. The paired appendages of aquatic gnathostomes other than the lobe-fins could have served as a starting point for the evolution of limbs. The colonization of land by aquatic gnathostomes other than the lobe-fins might have been facilitated in lineages that possessed lungs, as that would have enabled those organisms to breathe air.

Concept Check 34.4

1. Tetrapods are thought to have originated about 365 million years ago when the fins of some lobe-fins evolved into the limbs of tetrapods. In addition to their four limbs with digits—a key derived trait for which the group is named—other derived traits of tetrapods include a neck (consisting of vertebrae that separate the head from the rest of the body), and a pelvic girdle that is fused to the backbone. **2.** Some fully aquatic species are paedomorphic, retaining larval features for life in water as adults. Species that live in dry environments may avoid dehydration by burrowing or living under moist leaves, and they protect their eggs with foam nests, viviparity, and other adaptations. **3.** Many amphibians spend part of their life cycle in aquatic environments and part on land. Thus, they may be exposed to a wide range of environmental problems, including water and air pollution and the loss or degradation of aquatic and/or terrestrial habitats. In addition, amphibians have highly permeable skin, providing relatively little protection from external conditions, and their eggs do not have a protective shell.

Concept Check 34.5

1. The amniotic egg provides protection to the embryo and allows the embryo to develop on land, eliminating the necessity of a watery environment for reproduction. Another key adaptation is rib cage ventilation, which improves the efficiency of air intake and may have allowed early amniotes to dispense with breathing through their skin. Finally, not breathing through their skin allowed amniotes to develop relatively impermeable skin, thereby conserving water.
2. Yes. Although snakes lack limbs, they descended from lizards with legs. Some snakes retain vestigial pelvic and leg bones, providing evidence of their descent from an ancestor with legs. **3.** Birds have weight-saving modifications, including the absence of teeth, a urinary bladder, and a second ovary in females. The wings and feathers are adaptations that facilitate flight, as do efficient respiratory and circulatory systems that support a high metabolic rate.
4. (a) synapsids; (b) tuataras; (c) turtles

Concept Check 34.6

1. Monotremes lay eggs. Marsupials give birth to very small live young that attach to a nipple in the mother's pouch, where they complete development. Eutherians give birth to more developed live young. **2.** Hands and feet adapted for grasping, flat nails, large brain, forward-looking eyes on a flat face, parental care, and movable big toe and thumb **3.** Mammals are endothermic, enabling them to live in a wide range of habitats. Milk provides young with a balanced set of nutrients, and hair and a layer of fat under the skin help mammals retain heat. Mammals have differentiated teeth, enabling them to eat many different kinds of food. Mammals also have relatively large brains, and many species are capable learners. Following the mass extinction at the end of the Cretaceous period, the absence of large terrestrial dinosaurs may have opened many new ecological niches to mammals, promoting an adaptive radiation. Continental drift also isolated many groups of mammals from one another, promoting the formation of many new species.

Concept Check 34.7

1. Hominins are a clade within the ape clade that includes humans and all species more closely related to humans than to other apes. The derived characters of hominins include bipedal locomotion and relatively larger brains. **2.** In hominins, bipedal locomotion evolved long before large brain size. *Homo ergaster*, for example, was fully upright, bipedal, and as tall as modern humans, but its brain was significantly smaller than that of modern humans. **3.** Yes, both can be correct. *Homo sapiens* may have established populations outside of Africa as early as 115,000 years ago, as indicated by the fossil record. However, those populations may have left few or no descendants today. Instead, all living humans may have descended from Africans that spread from Africa roughly 50,000 years ago, as indicated by genetic data.

Summary of Key Concepts Questions

34.1 Lancelets are the most basal group of living chordates, and as adults they have key derived characters of chordates. This suggests that the chordate common ancestor may have resembled a lancelet in having an anterior end with a mouth along with the following four derived characters: a notochord; a dorsal, hollow nerve cord; pharyngeal slits or clefts; and a muscular, post-anal tail.
34.2 Conodonts, among the earliest vertebrates in the fossil record, were very abundant for 300 million years. While jawless, their well-developed teeth provide early signs of bone formation. Other species of jawless vertebrates developed armor on the outside of their bodies, which probably helped protect them from predators. Like lampreys, these species had paired fins for locomotion and an inner ear with semicircular canals that provided a sense of balance. There were many species of these armored jawless vertebrates, but they all became extinct by the close of the Devonian period, 359 million years ago. **34.3** The origin of jaws altered how fossil gnathostomes obtained food, which in turn had large effects on ecological interactions. Predators could use their jaws to grab prey or remove chunks of flesh, stimulating the evolution of increasingly sophisticated means of defense in prey species. Evidence for these changes can be found in the fossil record, which includes fossils of 10-m-long predators with remarkably powerful jaws, as well as lineages of well-defended prey species whose bodies were covered by armored plates. **34.4** Amphibians require water for reproduction; their bodies can lose water rapidly through their moist, highly permeable skin; and amphibian eggs do not have a shell and hence are vulnerable to desiccation. **34.5** Birds are descended from theropod dinosaurs, and dinosaurs are nested within the archosaur lineage, one of the two main reptile lineages. Thus, the other living archosaur reptiles, the crocodilians, are more closely related to birds than they are to non-archosaur reptiles such as lizards. As a result, birds are considered reptiles. (Note that if reptiles were defined as excluding birds, the reptiles would not form a clade; instead, the reptiles would be a paraphyletic group.) **34.6** Mammals are members of a group of amniotes called synapsids. Early (nonmammalian) synapsids laid eggs and had a sprawling gait. Fossil evidence shows that mammalian features arose gradually over a period of more than 100 million years. For example, the jaw was modified over time in nonmammalian synapsids, eventually coming to resemble that of a mammal. By 180 million years ago, the first mammals had appeared. There were many species of early mammals, but most of them were small, and they were not abundant or dominant members of their community. Mammals did not rise to ecological dominance until after the extinction of the dinosaurs. **34.7** The fossil record shows that from 4.5 to 2.5 million years ago, a wide range of hominin species walked upright but had relatively small brain sizes. About 2.5 million years ago, the first members of genus *Homo* emerged. These species used tools and had larger brains than those of earlier hominins. Fossil evidence indicates that multiple members of our genus were alive at any given point in time. Furthermore, until about 1.3 million years ago, these various *Homo* species also coexisted with members of earlier hominin lineages, such as *Paranthropus*. The different hominins alive at the same periods of time varied in body size, body shape, brain size, dental morphology, and the capacity for tool use. Ultimately, except for *Homo sapiens*, all of these species became extinct. Overall, human evolution can be viewed as an evolutionary tree with many branches—the only surviving lineage of which is our own.

Test Your Understanding

1. D **2.** C **3.** B **4.** C **5.** D **6.** A
8. (a) Because brain size tends to increase consistently in such lineages, we can conclude that natural selection favored the evolution of larger brains and hence that the benefits outweighed the costs. (b) As long as the benefits of brains that are large relative to body size are greater than the costs, large brains can evolve. Natural selection might favor the evolution of brains that are large relative to body size because such brains confer an advantage in obtaining mates and/or an advantage in survival.
(c)

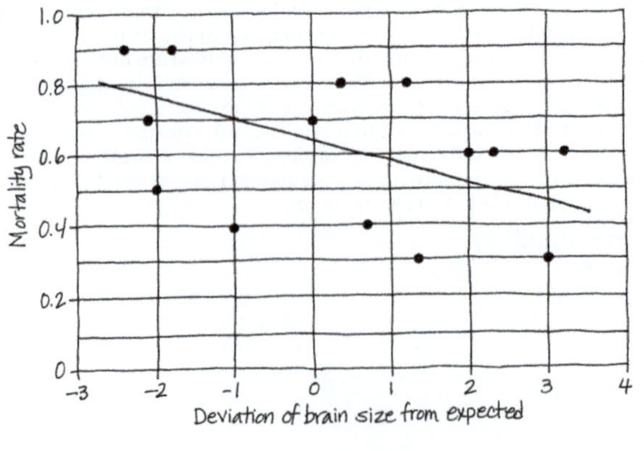

Adult mortality tends to be lower in birds with larger brains.

Chapter 35

Figure Questions

Figure 35.11
(1)

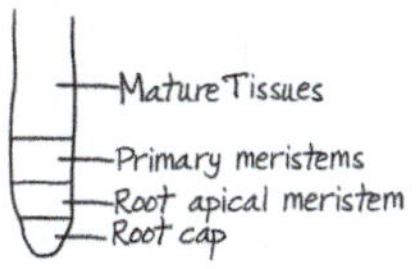

(2)

As a result of the addition of secondary xylem cells, the vascular cambium is pushed farther to the outside.

Figure 35.15

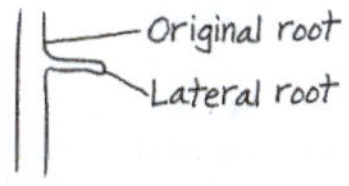

Figure 35.17 Pith and cortex are defined, respectively, as ground tissue that is internal and ground tissue that is external to vascular tissue. Since vascular bundles of monocot stems are scattered throughout the ground tissue, there is no clear distinction between internal and external relative to the vascular tissue. **Figure 35.19** The vascular cambium produces growth that increases the diameter of a stem or root. The tissues that are exterior to the vascular cambium cannot keep pace with the growth because their cells no longer divide. As a result, these tissues rupture. **Figure 35.23** Periderm (mainly cork and cork cambium), primary phloem, secondary phloem, vascular cambium, secondary xylem (sapwood and heartwood), primary xylem, and pith. At the base of ancient redwood that is many centuries old, the remnants of primary growth (primary phloem, primary xylem and pith would be quite insignificant. **Figure 35.33** Every root epidermal cell would develop a root hair. **Figure 35.35** Another example of homeotic gene mutation is the mutation in a *Hox* gene that causes legs to form in place of antennae in *Drosophila* (see Figure 18.20).
Figure 35.36
(a)

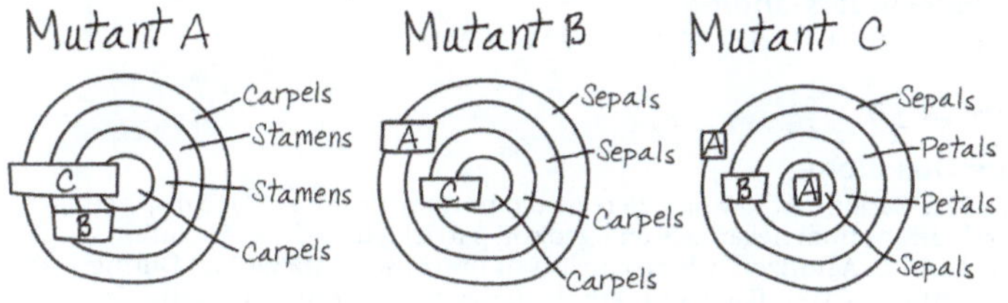

(b)

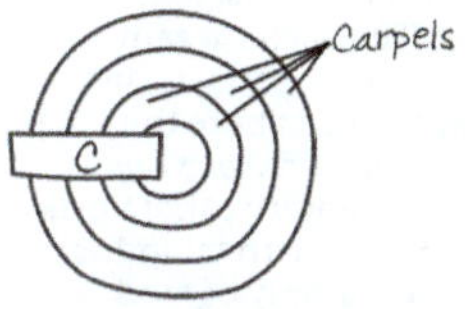

Concept Check 35.1

1. The vascular tissue system connects leaves and roots, allowing sugars to move from leaves to roots in the phloem and allowing water and minerals to move to the leaves in the xylem. **2.** To get sufficient energy from photosynthesis, we would need lots of surface area exposed to the sun. This large surface-to-volume ratio, however, would create a new problem—evaporative water loss. We would have to be permanently connected to a water source—the soil, also our source of minerals. In short, we would probably look and behave very much like plants.
3. As plant cells enlarge, they typically form a huge central vacuole that contains a dilute, watery sap. Central vacuoles enable plant cells to become large with only a minimal investment of new cytoplasm. The orientation of the cellulose microfibrils in plant cell walls affects the growth pattern of cells.

Concept Check 35.2

1. Yes. In a woody plant, secondary growth is occurring in the older parts of the stem and root, while primary growth is occurring at the root and shoot tips.
2. The largest, oldest leaves would be lowest on the shoot. Since they would probably be heavily shaded, they would not photosynthesize much regardless of their size. Determinate growth benefits the plant by keeping it from investing an ever-increasing amount of resources into organs that provide little photosynthetic product.
3. No. The carrot roots will probably be smaller at the end of the second year because the food stored in the roots will be used to produce flowers, fruits, and seeds.

Concept Check 35.3

1. In roots, primary growth occurs in three successive stages, moving away from the tip of the root: the zones of cell division, elongation, and differentiation. In shoots, it occurs at the tip of apical buds, with leaf primordia arising along the sides of an apical meristem. Most growth in length occurs in older internodes below the shoot tip. **2.** No. Because vertically oriented leaves, such as those of maize, can capture light equally well on both sides of the leaf, you would expect them to have mesophyll cells that are not differentiated into palisade and spongy layers. This is typically the case. Also, vertically oriented leaves usually have stomata on both leaf surfaces. **3.** Root hairs are cellular extensions that increase the surface area of the root epidermis, thereby enhancing the absorption of minerals and water. Microvilli are extensions that increase the absorption of nutrients by increasing the surface area of the gut.

Concept Check 35.4

1. The sign will still be 2 m above the ground because this part of the tree is no longer growing in length (primary growth); it is now growing only in thickness (secondary growth). **2.** Stomata must be able to close because evaporation is much more intensive from leaves than from the trunks of woody trees as a result of the higher surface-to-volume ratio in leaves. **3.** Since there is little seasonal temperature variation in the tropics, the growth rings of a tree from the tropics would be difficult to discern unless the tree came from an area that had pronounced wet and dry seasons. **4.** The tree would die slowly. Girdling removes an entire ring of secondary phloem (part of the bark), completely preventing transport of sugars and starches from the shoots to the roots. After several weeks, the roots would have used all of their stored carbohydrate reserves and would die.

Concept Check 35.5

1. Although all the living vegetative cells of a plant have the same genome, they develop different forms and functions because of differential gene expression.
2. Plants show indeterminate growth; juvenile and mature phases are found on the same individual plant; and cell differentiation in plants is more dependent on final position than on lineage. **3.** One hypothesis is that tepals arise if *B* gene activity is present in all three of the outer whorls of the flower.

Summary of Key Concepts Questions

35.1 Here are a few examples: The cuticle of leaves and stems protects these structures from desiccation. Collenchyma and sclerenchyma cells have thick walls that provide support for plants. Strong, branching root systems help anchor plants in the soil. **35.2** Primary growth arises from apical meristems and involves production and elongation of organs. Secondary growth arises from lateral meristems and adds to the diameter of roots and stems. **35.3** Lateral roots emerge from the pericycle and destroy plant cells as they emerge. In stems, branches arise from axillary buds and do not destroy any cells. **35.4** With the evolution of secondary growth, plants were able to grow taller and shade competitors. **35.5** The orientation of cellulose microfibrils in the innermost layers of the cell wall causes growth along one axis. Microtubules in the cell's outermost cytoplasm play a key role in regulating the axis of cell expansion because it is their orientation that determines the orientation of cellulose microfibrils.

Test Your Understanding

1. D **2.** C **3.** C **4.** A **5.** B **6.** D **7.** D
8.

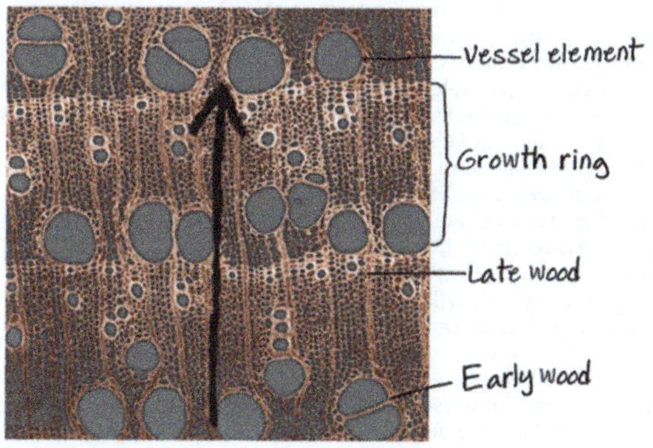

Chapter 36

Figure Questions

Figure 36.2 Cellular respiration is occurring in all parts of a growing plant at all times, with mitochondria continuously releasing CO_2 and consuming O_2. In photosynthetic cells, the CO_2 produced by mitochondria during the day is consumed by chloroplasts, which also consume CO_2 from the air. Meanwhile, the mitochondria obtain O_2 from the chloroplasts, which also release O_2 into the air. At night, when photosynthesis does not occur, the mitochondria must exchange gases with the air rather than with the chloroplasts. As a result, at night photosynthetic cells are *releasing* CO_2 into the air and *consuming* O_2 from the air, the opposite of what happens during the day. **Figure 36.3** The leaves are being produced in a counterclockwise spiral. The next leaf primordium will emerge approximately between and to the inside of leaves 8 and 13. **Figure 36.4** A higher leaf area index will not necessarily increase photosynthesis because of upper leaves shading lower leaves. **Figure 36.6** A proton pump inhibitor would depolarize (increase) the membrane potential because fewer hydrogen ions would be pumped out across the plasma membrane. The immediate effect of an inhibitor of the H^+/sucrose transporter would be to hyperpolarize (decrease) the membrane potential because fewer hydrogen ions would be leaking back into the cell through these cotransporters. An inhibitor of the H^+/NO_3^- cotransporter would have no effect on the membrane potential because the simultaneous cotransport of a positively charged ion and a negatively charged ion has no net effect on charge difference across the membrane. An inhibitor of the potassium ion channels would decrease the membrane potential because additional positively charged ions would not be accumulating outside the cell. **Figure 36.8** Few, if any, mesophyll cells are more than three cells from a vein. **Figure 36.9** The Casparian strip blocks water and minerals from moving between endodermal cells or moving around an endodermal cell via the cell's wall. Therefore, water and minerals must pass through an endodermal cell's plasma membrane. **Figure 36.18** Because the xylem is under negative pressure (tension), excising a stylet that had been inserted into a tracheid or vessel element would probably introduce air into the cell. No xylem sap would exude unless positive root pressure was predominant.

Concept Check 36.1

1. Vascular plants must transport minerals and water absorbed by the roots to all the other parts of the plant. They must also transport sugars from sites of production to sites of use. **2.** Increased stem elongation would raise the plant's upper leaves. Erect leaves and reduced lateral branching would make the plant less subject to shading by the encroaching neighbors. **3.** Pruning shoot tips removes apical dominance, resulting in lateral shoots (branches) growing from axillary buds (see Concept 35.3). This branching produces a bushier plant with a higher leaf area index.

Concept Check 36.2

1. The cell's Ψ_P is 0.7 MPa. In a solution with a Ψ of -0.4 MPa, the cell's Ψ_P at equilibrium would be 0.3 MPa. **2.** The cell would still adjust to changes in its osmotic environment, but its responses would be slower. Although aquaporins do not affect the water potential gradient across membranes, they allow for more rapid osmotic adjustments. **3.** If tracheids and vessel elements were alive at maturity, their cytoplasm would impede water movement, preventing rapid long-distance transport. **4.** The protoplasts would burst. Because the cytoplasm has many dissolved solutes, water would enter the protoplast continuously without reaching equilibrium. (When present, the cell wall prevents rupturing by limiting expansion of the protoplast.)

Concept Check 36.3

1. At dawn, a drop is exuded from the rooted stump because the xylem is under positive pressure due to root pressure. At noon, the xylem is under negative pressure (tension) when it is cut, and the xylem sap is pulled back into the rooted stump. Root pressure cannot keep pace with the increased rate of transpiration at noon. **2.** Perhaps greater root mass helps compensate for the lower water permeability of the plasma membranes. **3.** The Casparian strip and tight junctions both prevent movement of fluid between cells.

Concept Check 36.4

1. Stomatal opening at dawn is controlled mainly by light, CO_2 concentration, and a circadian rhythm. Environmental stresses such as drought, high temperature, and wind can stimulate stomata to close during the day. Water deficiency during the peak of the day can trigger release of the plant hormone abscisic acid, which signals guard cells to close stomata. **2.** The activation of the proton pumps of stomatal cells would cause the guard cells to take up K^+. The increased turgor of the guard cells would lock the stomata open and lead to extreme evaporation from the leaf. **3.** After the flowers are cut, transpiration from any leaves and from the petals (which are modified leaves) will continue to draw water up the xylem. If cut flowers are transferred directly to a vase, air pockets in xylem vessels prevent delivery of water from the vase to the flowers. Cutting stems again underwater, a few centimeters from the original cut, will sever the xylem above the air pocket. The water droplets prevent another air pocket from forming while the flowers are transferred to a vase. **4.** Water molecules are in constant motion, traveling at different speeds. If water molecules gain enough energy, the most energetic molecules near the liquid's surface will have sufficient speed, and therefore sufficient kinetic energy, to leave the liquid in the form of gaseous molecules (water vapor). As the molecules with the highest kinetic energy leave the liquid, the average kinetic energy of the remaining liquid decreases. Because a liquid's temperature is directly related to the average kinetic energy of its molecules, the temperature drops as evaporation proceeds.

Concept Check 36.5

1. In both cases, the long-distance transport is a bulk flow driven by a pressure difference at opposite ends of tubes. Pressure is generated at the source end of a sieve tube by the loading of sugar and resulting osmotic flow of water into the phloem, and this pressure *pushes* sap from the source end to the sink end of the tube. In contrast, transpiration generates a negative pressure potential (tension) that *pulls* the ascent of xylem sap. **2.** The main sources are fully grown leaves (producing sugar by photosynthesis) and fully developed storage organs (producing sugar by breakdown of starch). Roots, buds, stems, expanding leaves, and fruits are powerful sinks because they are actively growing. A storage organ may be a sink in the summer when accumulating carbohydrates but a source in the spring when breaking down starch into sugar for growing shoot tips. **3.** Positive pressure, whether it be in the xylem when root pressure predominates or in the sieve-tube elements of the phloem, requires active transport. Most long-distance transport in the xylem depends on bulk flow driven by the negative pressure potential generated ultimately by the evaporation of water from the leaf and does not require living cells. **4.** The spiral slash prevents optimal bulk flow of the phloem sap to the root sinks. Therefore, more phloem sap can move from the source leaves to the fruit sinks, making them sweeter.

Concept Check 36.6

1. Plasmodesmata, unlike gap junctions, have the ability to pass RNA, proteins, and viruses from cell to cell. **2.** Long-distance signaling is critical for the integrated functioning of all large organisms, but the speed of such signaling is much less critical to plants because their responses to the environment, unlike those of animals, do not typically involve rapid movements. **3.** Although this strategy would eliminate the systemic spread of viral infections, it would also severely impact the development of the plants.

Summary of Key Concepts Questions

36.1 Plants with tall shoots and elevated leaf canopies generally had an advantage over shorter competitors. A consequence of the selective pressure for tall shoots was the further separation of leaves from roots. This separation created problems for the transport of materials between root and shoot systems. Plants with xylem cells were more successful at supplying their shoot systems with soil resources (water and minerals). Similarly, those with phloem cells were more successful at supplying sugar sinks with carbohydrates. **36.2** Xylem sap is pulled up the plant by transpiration much more often than it is pushed up the plant by root pressure. **36.3** Hydrogen bonds are necessary for the cohesion of water molecules to each other and for the adhesion of water to other materials, such as cell walls. Both adhesion and cohesion of water molecules are involved in the ascent of xylem sap under conditions of negative pressure. **36.4** Although stomata account for most of the water lost from plants, they are necessary for exchange of gases—for example, for the uptake of carbon dioxide needed for photosynthesis. The loss of water through stomata also drives the long-distance transport of water that brings soil nutrients from roots to the rest of the plant. **36.5** Although the movement of phloem sap depends on bulk flow, the pressure gradient that drives phloem transport depends on the osmotic uptake of water in response to the loading of sugars into sieve-tube elements at sugar sources. Phloem loading depends on H^+ cotransport processes that ultimately depend on H^+ gradients established by active H^+ pumping. **36.6** Electrical signaling, cytoplasmic pH, cytoplasmic Ca^{2+} concentration, and viral movement proteins all affect symplastic communication, as do developmental changes in the number of plasmodesmata.

Test Your Understanding

1. A **2.** B **3.** B **4.** C **5.** B **6.** C **7.** A **8.** D

Chapter 37

Figure Questions

Figure 37.3 Cations. At low pH, there would be more protons (H^+) to displace mineral cations from negatively charged soil particles into the soil solution. **Figure 37.4** The A horizon, which consists of the topsoil. **Table 37.1** During photosynthesis, CO_2 is fixed into carbohydrates, which contribute to the dry mass. In cellular respiration, O_2 is reduced to H_2O and does not contribute to the dry mass. **Figure 37.10** Some other examples of mutualism are the following relationships. *Flashlight fish and bioluminescent bacteria:* The bacteria gain nutrients and protection from the fish, while the bioluminescence attracts prey and mates for the fish. *Flowering plants and pollinators:* Animals distribute the pollen and are rewarded by a meal of nectar or pollen. *Vertebrate herbivores and some bacteria in the digestive system:* Microorganisms in the alimentary canal break down cellulose to glucose and, in some cases, provide the animal with vitamins or amino acids. Meanwhile, the microorganisms have a steady supply of food and a warm environment. *Humans and some bacteria in the digestive system:* Some bacteria provide humans with vitamins, while the bacteria get nutrients from the digested food. **Figure 37.12** Both ammonium and nitrate. A decomposing animal would release amino acids into the soil that would be converted into ammonium by ammonifying bacteria. Some of this ammonium could be used directly by the plant. A large part of the ammonium, however, would be converted by nitrifying bacteria to form nitrate ions that could also be absorbed by the plant root system. **Figure 37.13** The legume plants benefit because the bacteria fix nitrogen that is absorbed by their roots. The bacteria benefit because they acquire photosynthetic products from the plants. **Figure 37.14** All three plant tissue systems are affected. Root hairs (dermal tissue) are modified to allow *Rhizobium* penetration. The cortex (ground tissue) and pericycle (vascular tissue) proliferate during nodule formation. The vascular tissue of the nodule connects to the vascular cylinder of the root to allow for efficient nutrient exchange.

Concept Check 37.1

1. Overwatering deprives roots of oxygen. Overfertilizing is wasteful and can lead to soil salinization and water pollution. **2.** As lawn clippings decompose, they restore mineral nutrients to the soil. If they are removed, the minerals lost from the soil must be replaced by fertilization. **3.** Because of their small size and negative charge, clay particles would increase the number of binding sites for cations and water molecules and would therefore increase cation exchange and water retention in the soil. **4.** Due to hydrogen bonding between water molecules, water expands when it freezes, and this causes mechanical fracturing of rocks. Water also coheres to many objects, and this cohesion combined with other forces, such as gravity, can help tug particles from rock. Finally, water, because it is polar, is an excellent solvent that allows many substances, including ions, to become dissolved in solution.

Concept Check 37.2

1. No. Even though macronutrients are required in greater amounts, all essential elements are necessary for the plant to complete its life cycle. **2.** No. The fact that the addition of an element results in an increase in the growth rate of a crop does not mean that the element is strictly required for the plant to complete its life cycle. **3.** Inadequate aeration of the roots of hydroponically grown plants would promote alcohol fermentation, which uses more energy and may lead to the accumulation of ethanol, a toxic by-product of fermentation.

Concept Check 37.3

1. The rhizosphere is the zone in the soil immediately adjacent to living roots. It harbors many rhizobacteria with which the root systems form beneficial mutualisms. Some rhizobacteria produce antibiotics that protect roots from disease. Others absorb toxic metals or make nutrients more available to roots. Still others convert gaseous nitrogen into forms usable by the plant or produce chemicals that stimulate plant growth. **2.** Soil bacteria and mycorrhizae enhance plant nutrition by making certain minerals more available to plants. For example, many types of soil bacteria are involved in the nitrogen cycle, and the hyphae of mycorrhizae provide a large surface area for the absorption of nutrients, particularly phosphate ions. **3.** Mixotrophy refers to the strategy of using photosynthesis and heterotrophy for nutrition. Euglenids are well-known mixotrophic protists. **4.** Saturating rainfall may deplete the soil of oxygen. A lack of soil oxygen would inhibit nitrogen fixation by the peanut root nodules and decrease the nitrogen available to the plants. Alternatively, heavy rain may leach nitrate from the soil. A symptom of nitrogen deficiency is yellowing of older leaves.

Summary of Key Concepts Questions

37.1 The term *ecosystem* refers to the communities of organisms within a given area and their interactions with the physical environment around them. Soil is teeming with many communities of organisms, including bacteria, fungi, animals, and the root systems of plants. The vigor of these individual communities depends on nonliving factors in the soil environment, such as minerals, oxygen, and water, as well as on interactions, both positive and negative, between different communities of organisms. **37.2** No. Plants can complete their life cycle when grown hydroponically, that is, in aerated salt solutions containing the proper ratios of all the minerals needed by plants. **37.3** No. Some parasitic plants obtain their energy by siphoning off carbon nutrients from other organisms.

Test Your Understanding

1. B **2.** B **3.** A **4.** D **5.** B **6.** B **7.** D **8.** C **9.** D
10.

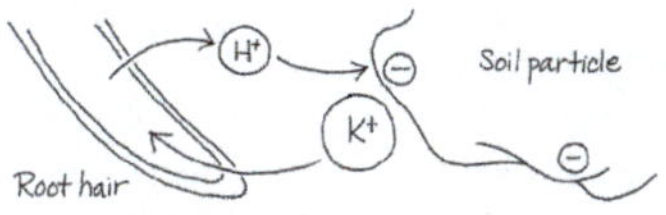

Chapter 38

Figure Questions

Figure 38.4 Having a specific pollinator is more efficient because less pollen gets delivered to flowers of the wrong species. However, it is also a risky strategy: If the pollinator population suffers to an unusual degree from predation, disease, or climate change, then the plant may not be able to produce seeds. **Figure 38.6** The part of the angiosperm life cycle characterized by the most mitotic divisions is the step between seed germination and the mature sporophyte. **Figure 38.8 Make Connections** In addition to having a single cotyledon, monocots have leaves with parallel leaf venation, scattered vascular bundles in their stems, a fibrous root system, floral parts in threes or multiples of threes, and pollen grains with only one opening. In contrast, dicots have two cotyledons, netlike leaf venation, vascular bundles in a ring, taproots, floral parts in fours or fives or multiples thereof, and pollen grains with three openings. **Figure 38.8 Visual Skills** The mature garden bean seed lacks an endosperm at maturity. Its endosperm was consumed during seed development, and its nutrients were stored anew in the cotyledons. **Figure 38.9** Beans use a hypocotyl hook to push through the soil. The delicate leaves and shoot apical meristem are also protected by being sandwiched between two large cotyledons. The coleoptile of maize seedlings helps protect the emerging leaves.

Concept Check 38.1

1. In angiosperms, pollination is the transfer of pollen from an anther to a stigma. Fertilization is the fusion of the egg and sperm to form the zygote; it cannot occur until after the growth of the pollen tube from the pollen grain. **2.** Long styles help to weed out pollen grains that are genetically inferior and not capable of successfully growing long pollen tubes. **3.** No. The haploid (gametophyte) generation of plants is multicellular and arises from spores. The haploid phase of the animal life cycles is a single-celled gamete (egg or sperm) that arises directly from meiosis: There are no spores.

Concept Check 38.2

1. Flowering plants can avoid self-fertilization by self-incompatibility, having male and female flowers on separate plants (dioecious species), or having stamens and styles of different heights on separate plants ("pin" and "thrum" flowers). **2.** Asexually propagated crops lack genetic diversity. Genetically diverse populations are less likely to become extinct in the face of an epidemic because there is a greater likelihood that a few individuals in the population are resistant. **3.** In the short term, selfing may be advantageous in a population that is so dispersed and sparse that pollen delivery is unreliable. In the long term, however, selfing is an evolutionary dead end because it leads to a loss of genetic diversity that may preclude adaptive evolution.

Concept Check 38.3

1. Traditional breeding and genetic engineering both involve artificial selection for desired traits. However, genetic engineering techniques facilitate faster gene transfer and are not limited to transferring genes between closely related varieties or species. **2.** *Bt* maize suffers less insect damage; therefore, *Bt* maize plants are less likely to be infected by fumonisin-producing fungi that infect plants through wounds. **3.** In such species, engineering the transgene into the chloroplast DNA would not prevent its escape in pollen; such a method requires that the chloroplast DNA be found only in the egg. An entirely different method of preventing transgene escape would therefore be needed, such as male sterility, apomixis, or self-pollinating closed flowers.

Summary of Key Concepts Questions

38.1 After pollination and fertilization, a flower changes into a fruit. The petals, sepals, and stamens typically fall off the flower. The stigma of the pistil withers, and the ovary begins to swell. The ovules (embryonic seeds) inside the ovary begin to mature. **38.2** Asexual reproduction can be advantageous in a stable environment because individual plants that are well suited to that environment pass on all their genes to offspring. Also, asexual reproduction generally results in offspring that are less fragile than the seedlings produced by sexual reproduction. However, sexual reproduction offers the advantage of dispersal of tough seeds. Moreover, sexual reproduction produces genetic variety, which may be advantageous in an unstable environment. The likelihood is better that at least one offspring of sexual reproduction will survive in a changed environment. **38.3** "Golden Rice," although not yet in commercial production, has been engineered to produce more vitamin A, thereby raising the nutritional value of rice. A protoxin gene from a soil bacterium has been engineered into *Bt* maize. This protoxin is lethal to invertebrates but harmless to vertebrates. *Bt* crops require less pesticide spraying and have lower levels of fungal infection and fungal toxins. The nutritional value of cassava is being increased in many ways by genetic engineering. Enriched levels of iron and beta-carotene (a vitamin A precursor) have been achieved, and cyanide-producing chemicals have been almost eliminated from the roots.

Test Your Understanding

1. A **2.** C **3.** C **4.** C **5.** D **6.** D **7.** D
8.

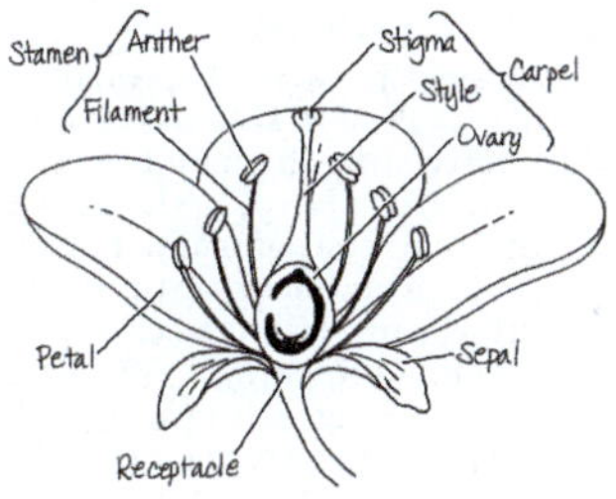

Chapter 39

Figure Questions

Figure 39.4 Panel B in Figure 11.17 shows a branching signal transduction pathway that resembles the branching phytochrome-dependent pathway involved in de-etiolation. **Figure 39.5** To determine which wavelengths of light are most effective in phototropism, you could use a glass prism to split white light into its component colors and see which colors cause the quickest bending (the answer is blue; see Figure 39.15). **Figure 39.6** No. Polar auxin transport depends on the distribution of auxin transport proteins at the basal ends of cells. **Figure 39.12** No. Since the *ein* mutation renders the seedling "blind" to ethylene, enhancing ethylene production by adding an *eto* mutation would have no effect on phenotype compared with the *ein* mutation alone. **Figure 39.16** Yes. The white light, which contains red light, would stimulate seed germination in all treatments. **Figure 39.20** Since far-red light, like darkness, causes an accumulation of the red-absorbing form (P_r) of phytochrome, single flashes of far-red light at night would have no effect on flowering beyond what the dark periods alone would have. **Figure 39.21** If this were true, florigen would be an inhibitor of flowering, not an inducer. **Figure 39.27** Photosynthetic adaptations can occur at the molecular level, as is apparent in the fact that C_3 plants use rubisco to fix carbon dioxide initially, whereas C_4 and CAM plants use PEP carboxylase. An adaptation at the tissue level is that plants have different stomatal densities based on their genotype and environmental conditions. At the organismal level, plants alter their shoot architectures to make photosynthesis more efficient. For example, self-pruning removes branches and leaves that respire more than they photosynthesize.

Concept Check 39.1

1. Dark-grown seedlings are etiolated: They have long stems, underdeveloped root systems, and unexpanded leaves, and their shoots lack chlorophyll. Etiolated growth is beneficial to seeds sprouting under the dark conditions they would encounter underground. By devoting more energy to stem elongation and less to leaf expansion and root growth, a plant increases the likelihood that the shoot will reach the sunlight before its stored foods run out. **2.** Cycloheximide should inhibit de-etiolation by preventing the synthesis of new proteins necessary for de-etiolation. **3.** No. Applying Viagra, like injecting cyclic GMP as described in the text, should cause only a partial de-etiolation response. Full de-etiolation would require activation of the calcium branch of the signal transduction pathway.

Concept Check 39.2

1. Fusicoccin's ability to cause an increase in plasma H^+ pump activity has an auxin-like effect and promotes stem cell elongation. **2.** The plant will exhibit a constitutive triple response. Because the kinase that normally prevents the triple response is dysfunctional, the plant will undergo the triple response regardless of whether ethylene is present or the ethylene receptor is functional. **3.** Since ethylene often stimulates its own synthesis, it is under positive-feedback regulation.

Concept Check 39.3

1. Not necessarily. Many environmental factors, such as temperature and light, change over a 24-hour period in the field. To determine whether the enzyme is under circadian control, a scientist would have to demonstrate that its activity oscillates even when environmental conditions are held constant. **2.** It is impossible to say. To establish that this species is a short-day plant, it would be necessary to establish the critical night length for flowering and that this species only flowers when the night is longer than the critical night length. **3.** According to the action spectrum of photosynthesis, red and blue light are the most effective in photosynthesis. Thus, it is not surprising that plants assess their light environment using blue- and red-light-absorbing photoreceptors.

Concept Check 39.4

1. A plant that overproduces ABA would undergo less evaporative cooling because its stomata would not open as widely. **2.** Plants close to the aisles may be more subject to mechanical stresses caused by passing workers and air currents. The plants nearer the center of the bench may also be taller as a result of shading and less evaporative stress. **3.** No. Because root caps are involved in sensing gravity, roots that have their root caps removed are almost completely insensitive to gravity.

Concept Check 39.5

1. Some insects increase plants' productivity by eating harmful insects or aiding in pollination. **2.** Mechanical damage breaches a plant's first line of defense against infection, its protective dermal tissue. **3.** No. Pathogens that kill their hosts would soon run out of victims and might themselves go extinct. **4.** Perhaps the breeze dilutes the local concentration of a volatile defense compound that the plants produce.

Summary of Key Concepts Questions

39.1 Signal transduction pathways often activate protein kinases, enzymes that phosphorylate other proteins. Protein kinases can directly activate certain preexisting enzymes by phosphorylating them, or they can regulate gene transcription (and enzyme production) by phosphorylating specific transcription factors. **39.2** Yes, there is truth to the old adage that one bad apple spoils the whole bunch. Ethylene, a gaseous hormone that stimulates ripening, is produced by damaged, infected, or overripe fruits. Ethylene can diffuse to healthy fruit in the "bunch" and stimulate their rapid ripening. **39.3** Plant physiologists proposed the existence of a floral-promoting factor (florigen) based on the fact that a plant induced to flower could induce flowering in a second plant to which it was grafted, even though the second plant was not in an environment that would normally induce flowering in that species. **39.4** Plants subjected to drought stress are often more resistant to freezing stress because the two types of stress are quite similar. Freezing of water in the extracellular spaces causes free water concentrations outside the cell to decrease. This, in turn, causes free water to leave the cell by osmosis, leading to the dehydration of cytoplasm, much like what is seen in drought stress. **39.5** Chewing insects make plants more susceptible to pathogen invasion by disrupting the waxy cuticle of shoots, thereby creating an opening for infection. Moreover, substances released from damaged cells can serve as nutrients for the invading pathogens.

Test Your Understanding

1. B **2.** C **3.** D **4.** C **5.** B **6.** B **7.** C
8.

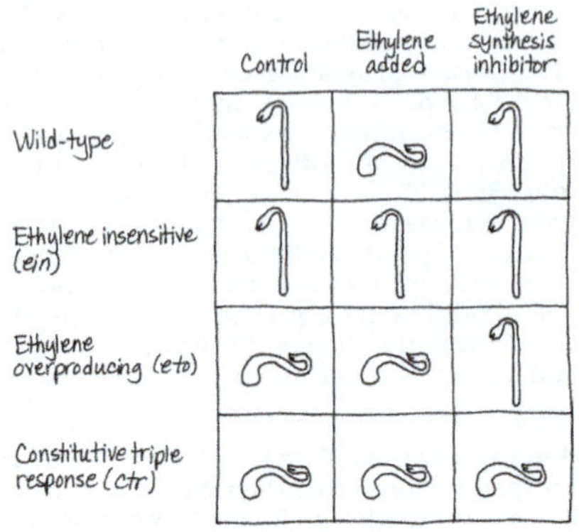

Chapter 40

Figure Questions

Figure 40.4 Such exchange surfaces are internal in the sense that they are inside the body. However, they are also continuous with openings on the external body surface that contact the environment. **Figure 40.6** Signals in the nervous system always travel on a direct route between the sending and receiving cell. In contrast, hormones that reach target cells can have an effect regardless of the path by which they arrive or how many times they travel through the circulatory system. **Figure 40.8** The stimuli (gray boxes) are the room temperature increasing in the top loop or decreasing in the bottom loop. The responses could include the heater turning off and the temperature decreasing in the top loop and the heater turning on and the temperature increasing in the bottom loop. The sensor/control center is the thermostat. The air conditioner would form a second control circuit, cooling the house when air temperature exceeded the set point. Such opposing, or antagonistic, pairs of control circuits increase the effectiveness of a homeostatic mechanism. **Figure 40.12** The conduction arrows would be in the opposite direction, transferring heat from the walrus to the ice because the walrus is warmer than the ice. **Figure 40.17** If a female Burmese python were not incubating eggs, her oxygen consumption would decrease with decreasing temperature, as for any other ectotherm. **Figure 40.18** The ice water would cool tissues in your head, including blood that would then circulate throughout your body. This effect would accelerate the return to a normal body temperature. If, however, the ice water reached the eardrum and cooled the blood vessel that supplies the hypothalamus, the hypothalamic thermostat would respond by inhibiting sweating and constricting blood vessels in the skin, slowing cooling elsewhere in the body. **Figure 40.19** The transport of nutrients across membranes and the synthesis of RNA and protein are coupled to ATP hydrolysis. These processes proceed spontaneously because there is an overall drop in free energy, with the excess energy given off as heat. Similarly, less than half of the free energy in glucose is captured in the coupled reactions of cellular respiration. The remainder of the energy is released as heat. **Figure 40.22** Nothing. Although genes that show a circadian variation in expression during euthermia exhibit constant RNA levels during hibernation, a gene that shows constant expression during hibernation might also show constant expression during euthermia. **Figure 40.23** In hot environments, both plants and animals experience evaporative cooling as a result of transpiration (in plants) or bathing, sweating, and panting (in animals); both plants and animals synthesize heat-shock proteins, which protect other proteins from heat stress; and animals also use various behavioral responses to minimize heat absorption. In cold environments, both plants and animals increase the proportion of unsaturated fatty acids in their membrane lipids and use antifreeze proteins that prevent or limit the formation of intracellular ice crystals; plants increase cytoplasmic levels of specific solutes that help reduce the loss of intracellular water during extracellular freezing; and animals increase metabolic heat production and use insulation, circulatory adaptations such as countercurrent exchange, and behavioral responses to minimize heat loss.

Concept Check 40.1

1. All types of epithelia consist of cells that line a surface, are tightly packed, are situated on top of a basal lamina, and form an active and protective interface with the external environment. **2.** An oxygen molecule must cross a plasma membrane when entering the body at an exchange surface in the respiratory system, in both entering and exiting the circulatory system, and in moving from the interstitial fluid to the cytoplasm of the body cell. **3.** You need the nervous system to perceive the danger and provoke a split-second muscular response to keep from falling. The nervous system, however, does not make a direct connection with blood vessels or glucose-storing cells in the liver. Instead, the nervous system triggers the release of a hormone (called epinephrine, or adrenaline) by the endocrine system, bringing about a change in these tissues in just a few seconds.

Concept Check 40.2

1. In thermoregulation, the product of the pathway (a change in temperature) decreases pathway activity by reducing the stimulus. In an enzyme-catalyzed biosynthetic process, the product of the pathway (in this case, isoleucine) inhibits the pathway that generated it. **2.** You would want to put the thermostat close to where you would be spending time, where it would be protected from environmental perturbations, such as direct sunshine, and not right in the path of the output of the heating system. Similarly, the sensors for homeostasis located in the human brain are separated from environmental influences and can monitor conditions in a vital and sensitive tissue. **3.** In convergent evolution, the same biological trait arises independently in two or more species. Gene analysis can provide evidence for an independent origin. In particular, if the genes responsible for the trait in one species lack significant sequence similarity to the corresponding genes in another species, scientists conclude that there is a separate genetic basis for the trait in the two species and thus an independent origin. In the case of circadian rhythms, the clock genes in cyanobacteria appear unrelated to those in humans.

Concept Check 40.3

1. "Wind chill" involves heat loss through convection, as the moving air contributes to heat loss from the skin surface. **2.** The hummingbird, being a very small endotherm, has a very high metabolic rate. If by absorbing sunlight certain flowers warm their nectar, a hummingbird feeding on these flowers is saved the metabolic expense of warming the nectar to its body temperature. **3.** To raise body temperature to the higher range of fever, the hypothalamus triggers heat generation by muscular contractions, or shivering. The person with a fever may in fact say that they feel cold, even though their body temperature is above normal.

Concept Check 40.4

1. The mouse would consume oxygen at a higher rate because it is an endotherm, so its basal metabolic rate is higher than the ectothermic lizard's standard metabolic rate. **2.** The house cat; smaller animals have a higher metabolic rate per unit body mass and a greater demand for food per unit body mass. **3.** The alligator's body temperature would decrease along with the air temperature. Its metabolic rate would therefore also decrease as chemical reactions slowed. In contrast, the lion's body temperature would not change. Its metabolic rate would increase as it shivered and produced heat to keep its body temperature constant.

Summary of Key Concepts Questions

40.1 Animals exchange materials with their environment across their body surface, and a spherical shape has the minimum surface area per unit volume. As body size increases, the ratio of surface area to body volume decreases. **40.2** No; an animal's internal environment fluctuates slightly around set points or within normal ranges. Homeostasis is a dynamic state. Furthermore, there are sometimes programmed changes in set points, such as those resulting in radical increases in hormone levels at particular times in development. **40.3** Heat exchange across the skin is a primary mechanism for the regulation of body core temperature, with the result that the skin is cooler than the body core. **40.4** Small animals have a higher BMR per unit mass and therefore consume more oxygen per unit mass than large animals. A higher breathing rate is required to support this increased oxygen consumption.

Test Your Understanding

1. B **2.** C **3.** A **4.** B **5.** C **6.** B **7.** D

8.

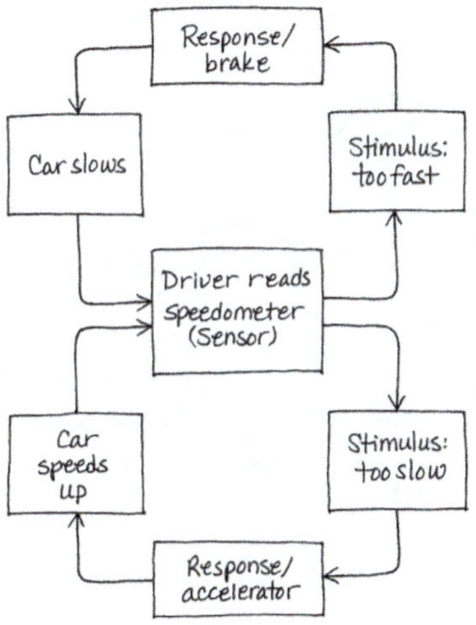

Chapter 41

Figure Questions

Figure 41.6 Your diagram should show food entering through the hydra's mouth and being digested into nutrients in the large portion of the gastrovascular cavity. The nutrients then diffuse into the extensions of that cavity that reach into the tentacles. There, nutrients would be absorbed by cells of the gastrodermis and transported to cells of the epidermis of a tentacle. **Figure 41.9** The airway must be open for exhaling to occur. If the epiglottis is up, milk entered the throat from the mouth encounters air forced out of the lungs and is carried along into the nasal cavity and out the nose. **Figure 41.11** Since enzymes are proteins, and proteins are hydrolyzed in the small intestine, the digestive enzymes in that compartment need to be resistant to enzymatic cleavage other than the cleavage required to activate them. **Figure 41.12** None. Since digestion is completed in the small intestine, tapeworms simply absorb predigested nutrients through their large body surface. **Figure 41.13** Yes. The exit of the chylomicrons involves exocytosis, an active process that consumes energy in the form of ATP. In contrast, the entry of monoglycerides and fatty acids into the cell by diffusion is a passive process that does not consume energy. **Figure 41.21** Both insulin and glucagon are involved in negative feedback circuits.

Concept Check 41.1

1. The only essential amino acids are those that an animal cannot synthesize from other molecules. **2.** Many vitamins serve as enzyme cofactors, which, like enzymes themselves, are unchanged by the chemical reactions in which they participate. Therefore, only very small amounts of vitamins are needed. **3.** To identify the essential nutrient missing from an animal's diet, a researcher could supplement the diet with individual nutrients one at a time and determine which nutrient eliminates the signs of malnutrition.

Concept Check 41.2

1. A gastrovascular cavity is a digestive pouch with a single opening that functions in both ingestion and elimination; an alimentary canal is a digestive tube with a separate mouth and anus at opposite ends. **2.** As long as nutrients are within the cavity of the alimentary canal, they are in a compartment that is continuous with the outside environment via the mouth and anus and have not yet crossed a membrane to enter the body. **3.** In both cases, high-energy fuels are consumed, complex molecules are broken down into simpler ones, and waste products are eliminated. In addition, gasoline, like food, is broken down in a specialized compartment, so that surrounding structures are protected from disassembly. Finally, just as food and wastes remain outside the body in a digestive tract, neither gasoline nor its waste products enter the passenger compartment of the automobile.

Concept Check 41.3

1. Because parietal cells in the stomach pump hydrogen ions into the stomach lumen where they combine with chloride ions to form HCl, a proton pump inhibitor reduces the acidity of chyme and thus the irritation that occurs when chyme enters the esophagus. **2.** By releasing sugars from starch or glycogen in the mouth, amylase might allow us to recognize foods that provide a ready source of energy. **3.** Proteins would be denatured and digested into peptides. Further digestion, to individual amino acids, would require enzymatic secretions found in the small intestine. No digestion of carbohydrates or lipids would occur.

Concept Check 41.4

1. The increased time for transit through the alimentary canal allows for more extensive processing, and the increased surface area of the canal provides greater opportunity for absorption. **2.** A mammal's digestive system provides mutualistic microorganisms with an environment that is protected against other microorganisms by saliva and gastric juice, that is held at a constant temperature conducive to enzyme action, and that provides a steady source of nutrients. **3.** For the yogurt treatment to be effective, the bacteria from yogurt would have to establish a mutualistic relationship with the small intestine, where disaccharides are broken down and sugars are absorbed. Conditions in the small intestine are likely to be very different from those in a yogurt culture. The bacteria might be killed before they reach the small intestine, or they might not be able to grow there in sufficient numbers to aid in digestion.

Concept Check 41.5

1. Over the long term, the body stores excess calories in fat, whether those calories come from fat, carbohydrate, or protein in food. **2.** In normal individuals, leptin levels decline during fasting. Individuals in the group with low levels of leptin are likely to be defective in leptin production, so leptin levels would remain low regardless of food intake. Individuals in the group with high leptin levels are likely to be defective in responding to leptin, but they still should shut off leptin production as fat stores are used up. **3.** The excess production of insulin will cause blood glucose levels to decrease below normal physiological levels. It will also trigger glycogen synthesis in the liver, further decreasing blood glucose levels. However, low blood glucose levels will stimulate the release of glucagon from alpha cells in the pancreas, which will trigger glycogen breakdown. Thus, there will be antagonistic effects in the liver.

Summary of Key Concepts Questions

41.1 Since the cofactor is necessary in all animals, those animals that do not require it in their diet must be able to synthesize it from other organic molecules. **41.2** A liquid diet containing glucose, amino acids, and other building blocks could be ingested and absorbed without the need for mechanical or chemical digestion. **41.3** The small intestine has a much larger surface area than the stomach. **41.4** The assortment of teeth in our mouth and the short length of our cecum suggest that our ancestors' digestive systems were not specialized for digesting plant material. **41.5** When mealtime arrives, nervous inputs from the brain signal the stomach to prepare to digest food through secretions and churning.

Test Your Understanding

1. B **2.** A **3.** B **4.** C **5.** B **6.** D **7.** B

8.

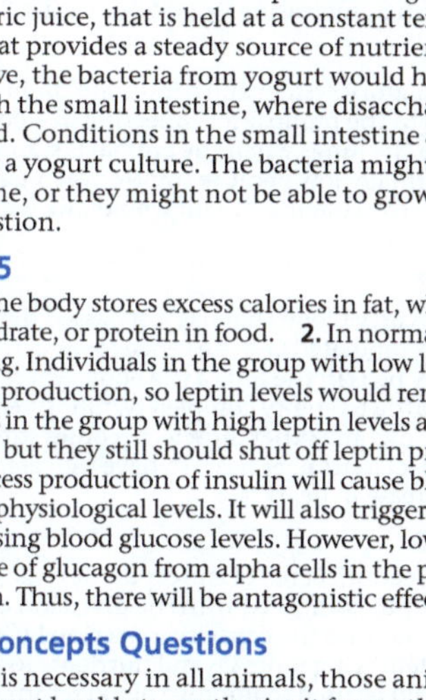

Chapter 42

Figure Questions

Figure 42.2 Although gas exchange might be improved by a steady, one-way flow of fluid, there would likely be inadequate time for food to be digested and nutrients absorbed if fluid flowed through the cavity in this manner. **Figure 42.5** Two capillary beds. The molecule of carbon dioxide would need to enter a capillary bed in the thumb before returning to the right atrium and ventricle, then travel to the lung and enter a capillary from which it could diffuse into an alveolus and be available to be exhaled. **Figure 42.8** Each feature of the ECG recording, such as the sharp upward spike, occurs once per cardiac cycle. Using the *x*-axis to measure the time in seconds between successive spikes and dividing that number by 60 would yield the heart rate as the number of cycles per minute. **Figure 42.25** The reduction in surface tension results from the presence of surfactant. Therefore, for all the infants who had died of RDS, you would expect the amount of surfactant to be near zero. For infants who had died of other causes, you would expect the amount of surfactant to be near zero for body masses less than 1,200 g but much greater than zero for body masses above 1,200 g. **Figure 42.27** Since exhalation is largely passive, the recoil of the elastic fibers in alveoli helps force air out of the lungs. When alveoli lose their elasticity, as occurs in the disease emphysema, less air is exhaled. Because more air is left in the lungs, less fresh air can be inhaled. With a smaller volume of air exchanged, there is a decrease in the partial pressure gradient that drives gas exchange. **Figure 42.28** Breathing at a rate greater than that needed to meet metabolic demand (hyperventilation) would lower blood CO_2 levels. Sensors in major blood vessels and the medulla would signal the breathing control center to decrease the rate of contraction of the diaphragm and rib muscles, decreasing the breathing rate and restoring normal CO_2 levels in the blood and other tissues. **Figure 42.29** The resulting increase in tidal volume would enhance ventilation within the lungs, increasing P_{O_2} and decreasing P_{CO_2} in the alveoli.

Concept Check 42.1

1. In both an open circulatory system and a fountain, fluid is pumped through a tube and then returns to the pump after collecting in a pool. **2.** The ability to shut off blood supply to the lungs when the animal is submerged **3.** The O_2 content would be abnormally low because some oxygen-depleted blood returned to the right atrium from the systemic circuit would mix with the oxygen-rich blood in the left atrium.

Concept Check 42.2

1. The pulmonary veins carry blood that has just passed through capillary beds in the lungs, where it accumulated O_2. The venae cavae carry blood that has just passed through capillary beds in the rest of the body, where it lost O_2 to the tissues. **2.** The delay allows the atria to empty completely, filling ventricles fully before they contract. **3.** The heart, like any other muscle, becomes stronger through regular exercise. You would expect a stronger heart to have a greater stroke volume, which would allow for the decrease in heart rate.

Concept Check 42.3

1. The large total cross-sectional area of the capillaries **2.** An increase in blood pressure and cardiac output combined with the diversion of more blood to the skeletal muscles would increase the capacity for action by increasing the rate of blood circulation and delivering more O_2 and nutrients to the skeletal muscles. **3.** Additional hearts could be used to improve blood return from the legs. However, it might be difficult to coordinate the activity of multiple hearts and to maintain adequate blood flow to hearts far from the gas exchange organs.

Concept Check 42.4

1. An increase in the number of white blood cells (leukocytes) may indicate that the person is combating an infection. **2.** Clotting factors do not initiate clotting but are essential steps in the clotting process. **3.** The chest pain results from inadequate blood flow in coronary arteries. Vasodilation promoted by nitric oxide from nitroglycerin increases blood flow, providing the heart muscle with additional oxygen and thus relieving the pain. **4.** Embryonic stem cells are pluripotent rather than multipotent, meaning that they can give rise to many rather than a few different cell types.

Concept Check 42.5

1. Their interior position helps gas exchange tissues stay moist. If the respiratory surfaces of lungs extended out into the terrestrial environment, they would quickly dry out, and diffusion of O_2 and CO_2 across these surfaces would stop. **2.** Earthworms need to keep their skin moist for gas exchange, but they need air outside this moist layer. If they stay in their waterlogged tunnels after a heavy rain, they will suffocate because they cannot get as much O_2 from water as from air. **3.** In fish, water passes over the gills in the direction opposite to that of blood flowing through the gill capillaries, maximizing the extraction of oxygen from the water along the length of the exchange surface. Similarly, in the extremities of some vertebrates, blood flows in opposite directions in neighboring veins and arteries; this countercurrent arrangement maximizes the recapture of heat from blood leaving the body core in arteries, which is important for thermoregulation in cold environments.

Concept Check 42.6

1. An increase in blood CO_2 concentration causes an increase in the rate of CO_2 diffusion into the cerebrospinal fluid, where the CO_2 combines with water to form carbonic acid. Dissociation of carbonic acid releases hydrogen ions, decreasing the pH of the cerebrospinal fluid. **2.** Increased heart rate increases the rate at which CO_2-rich blood is delivered to the lungs, where CO_2 is removed. **3.** A hole would allow air to enter the space between the inner and outer layers of the double membrane, resulting in a condition called a pneumothorax. The two layers would no longer stick together, and the lung on the side with the hole would collapse and cease functioning.

Concept Check 42.7

1. Differences in partial pressure between the capillaries and the surrounding tissues or medium; the net diffusion of a gas occurs from a region of higher partial pressure to a region of lower partial pressure. **2.** The Bohr shift causes hemoglobin to release more O_2 at a lower pH, such as found in the vicinity of tissues with high rates of cellular respiration and CO_2 release. **3.** The doctor is assuming that the rapid breathing is the body's response to low blood pH. Metabolic acidosis, the lowering of blood pH as a result of metabolism, can have many causes, including complications of certain types of diabetes, shock (extremely low blood pressure), and poisoning.

Summary of Key Concepts Questions

42.1 In a closed circulatory system, an ATP-driven muscular pump generally moves fluids in one direction on a scale of millimeters to meters. Exchange between cells and their environment relies on diffusion, which involves random movements of molecules. Concentration gradients of molecules across exchange surfaces can drive rapid net diffusion on a scale of 1 mm or less. **42.2** Replacement of a defective valve should increase stroke volume. A lower heart rate would therefore be sufficient to maintain the same cardiac output. **42.3** Blood pressure in the arm would fall by 25–30 mm Hg, the same difference as is normally seen between your heart and your brain. **42.4** One microliter of blood contains about 5 million erythrocytes and 5,000 leukocytes, so leukocytes make up only about 0.1% of the cells in the absence of infection. **42.5** Because CO_2 is such a small fraction of atmospheric gas (0.29 mm Hg/760 mm Hg, or less than 0.04%), the partial pressure gradient of CO_2 between the respiratory surface and the environment always strongly favors the release of CO_2 to the atmosphere. **42.6** Because the lungs do not empty completely with each breath, incoming and outgoing air mix. Lungs thus contain a mixture of fresh and stale air. **42.7** An enzyme speeds up a reaction without changing the equilibrium and without being consumed. Similarly, a respiratory pigment speeds up the exchange of gases between the body and the external environment without changing the equilibrium state and without being consumed.

Test Your Understanding

1. C **2.** A **3.** D **4.** C **5.** C **6.** A **7.** A
8.

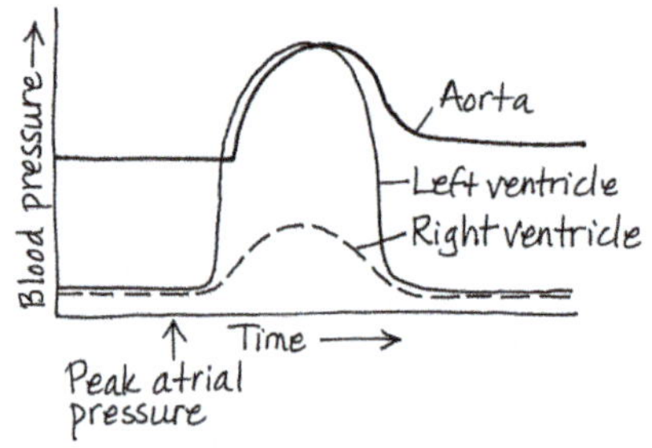

Chapter 43

Figure Questions

Figure 43.4 Dicer-2 binds double-stranded RNA without regard to size or sequence and then cuts that RNA into fragments, each 21 base pairs long. The Argo complex binds to double-stranded RNA fragments that are each 21 base pairs long, displaces one strand, and then uses the remaining strand to match to a particular target sequence in a single-stranded mRNA. **Figure 43.5** Cell-surface TLRs recognize molecules on the surface of pathogens, whereas TLRs in vesicles recognize internal molecules of pathogens after the pathogens are broken down. **Figure 43.7** Because the pain of a splinter stops almost immediately when you remove it from the skin, you can correctly deduce that the signals that mediate the inflammatory response are quite short-lived. **Figure 43.10** Part of the enzyme or antigen receptor provides a structural "backbone" that maintains overall shape, while interaction occurs at a surface with a close fit to the substrate or antigen. The combined effect of multiple noncovalent interactions at the active site or binding site is a high-affinity interaction of tremendous specificity. **Figure 43.13** After gene rearrangement, a lymphocyte and its daughter cells make a single version of the antigen receptor. In contrast, alternative splicing is not heritable and can give rise to diverse gene products in a single cell. **Figure 43.14** A single B cell has more than 100,000 identical antigen receptors on its surface, not four, and there are more than 1 million B cells differing in their antigen specificity, not three. **Figure 43.17** These receptors enable memory cells to present antigen on their cell surface to a helper T cell. This presentation of antigen is required to activate memory cells in a secondary immune response. **Figure 43.22** Primary response: arrows extending from Antigen (1st exposure), Antigen-presenting cell, Helper T cell, B cell, Plasma cells, Cytotoxic T cell, and Active cytotoxic T cells; secondary response: arrows extending from Antigen (2nd exposure), Memory helper T cells, Memory B cells, Memory cytotoxic T cells, Plasma cells, and Active cytotoxic T cells. **Figure 43.24** There would be no change in the results. Because the two antigen binding sites of an antibody have identical specificity, the two bacteriophages bound would have to display the same viral peptide.

Concept Check 43.1

1. Because pus contains white blood cells, fluid, and cell debris, it indicates an active and at least partially successful inflammatory response against invading pathogens. **2.** Whereas the ligand for the TLR receptor is a foreign molecule, the ligand for many signal transduction pathways is a molecule produced by the organism itself. **3.** Mounting an immune response would require recognition of some molecular feature of the wasp egg not found in the host. It might be that only some potential hosts have a receptor with the necessary specificity.

Concept Check 43.2

1. See Figure 43.9. The transmembrane regions lie within the C regions, which also form the disulfide bridges. In contrast, the antigen-binding sites are in the V regions. **2.** Generating memory cells ensures both that a receptor specific for a particular epitope will be present and that there will be more lymphocytes with this specificity than in a host that had never encountered the antigen. **3.** If each B cell produced two different light and heavy chains for its antigen receptor, different combinations would make four different receptors. If any one were self-reactive, the lymphocyte would be eliminated in the generation of self-tolerance. For this reason, many more B cells would be eliminated, and those that could respond to a foreign antigen would be less effective at doing so due to the variety of receptors (and antibodies) they express.

Concept Check 43.3

1. A child lacking a thymus would have no functional T cells. Without helper T cells to help activate B cells, the child would be unable to produce antibodies against extracellular bacteria. Furthermore, without cytotoxic T cells or helper T cells, the child's immune system would be unable to kill virus-infected cells. **2.** Since the antigen-binding site is intact, the antibody fragments could neutralize viruses and opsonize bacteria. **3.** If the handler developed immunity to proteins in the antivenin, another injection could provoke a severe immune response.

Concept Check 43.4

1. Myasthenia gravis is considered an autoimmune disease because the immune system produces antibodies against self molecules (certain receptors on muscle cells). **2.** A person with a cold is likely to produce oral and nasal secretions that facilitate viral transfer. In addition, since sickness can cause incapacitation or death, a virus that is programmed to exit the host when there is a physiological stress has the opportunity to find a new host at a time when the current host may cease to function. **3.** A person with a macrophage deficiency would have frequent infections. The causes would be poor innate responses, due to diminished phagocytosis and inflammation, and poor adaptive responses, due to the lack of macrophages to present antigens to helper T cells.

Summary of Key Concepts Questions

43.1 Lysozyme in saliva destroys bacterial cell walls; the viscosity of mucus helps trap bacteria; acidic pH in the stomach kills many bacteria; and the tight packing of cells lining the gut provides a physical barrier to infection. **43.2** Sufficient numbers of cells to mediate an innate immune response are always present, whereas an adaptive response requires selection and proliferation of an initially very small cell population specific for the infecting pathogen. **43.3** No. Immunological memory after a natural infection and that after vaccination are very similar. There may be minor differences in the particular antigens that can be recognized in a subsequent infection. **43.4** No. AIDS refers to a loss of immune function that can occur over time in an individual infected with HIV. However, certain multidrug combinations ("cocktails") or rare genetic variations usually prevent progression to AIDS in HIV-infected individuals.

Test Your Understanding

1. B **2.** C **3.** C **4.** B **5.** B **6.** B **7.** C **8.** One possible answer:

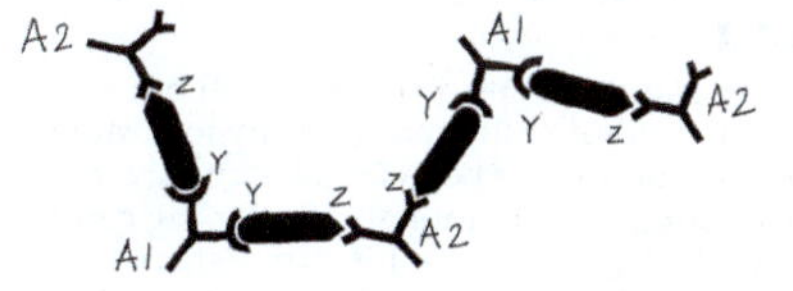

Chapter 44

Figure Questions

Figure 44.13 You would expect to find these cells lining tubules where they pass through the renal medulla. Because the extracellular fluid of the renal medulla has a very high osmolarity, production of organic solutes by tubule cells in this region keeps intracellular osmolarity high, with the result that these cells maintain normal volume. **Figure 44.14** Furosemide increases urine volume. The absence of ion transport in the ascending limb leaves the filtrate too concentrated for substantial volume reduction in the distal tubule and collecting duct. **Figure 44.17** When the concentration of an ion differs across a plasma membrane, the difference in the concentration of ions inside and outside represents chemical potential energy, while the resulting difference in charge inside and outside represents electrical potential energy. **Figure 44.20** The ADH levels would likely be elevated in both sets of patients with mutations because either defect prevents the recapture of water that restores blood osmolarity to normal levels. **Figure 44.21** Arrows that would be labeled "Secretion" are the arrows indicating secretion of aldosterone, angiotensinogen, and renin.

Concept Check 44.1

1. Because the salt is moved against its concentration gradient, from low concentration (fresh water) to high concentration (blood) **2.** A freshwater osmoconformer would have body fluids too dilute to carry out life's processes. **3.** Without a layer of insulating fur, the camel must use the cooling effect of evaporative water loss to maintain body temperature, thus linking thermoregulation and osmoregulation.

Concept Check 44.2

1. Because uric acid is largely insoluble in water, it can be excreted as a semisolid paste, thereby reducing an animal's water loss. **2.** Humans produce uric acid from purine breakdown, and reducing purines in the diet often lessens the severity of gout. Birds, however, produce uric acid as a waste product of general nitrogen metabolism. They would therefore need a diet low in all nitrogen-containing compounds, not just purines.

Concept Check 44.3

1. In flatworms, ciliated cells draw interstitial fluids containing waste products into protonephridia. In earthworms, waste products pass from interstitial fluids into the coelom. From there, cilia move the wastes into metanephridia via a funnel surrounding an internal opening to the metanephridia. In insects, the Malpighian tubules pump fluids from the hemolymph, which receives waste products during exchange with cells in the course of circulation. **2.** Filtrate is formed when the glomerulus filters blood from the renal artery within Bowman's capsule. Some of the filtrate contents are recovered, enter capillaries, and exit in the renal vein; the rest remain in the filtrate and pass out of the kidney in the ureter. **3.** The presence of Na^+ and other ions (electrolytes) in the dialysate would limit the extent to which they would be removed from the filtrate during dialysis. Adjusting the electrolytes in the starting dialysate can thus lead to the restoration of proper electrolyte concentrations in the plasma. Similarly, the absence of urea and other waste products in the starting dialysate facilitates their removal from the filtrate.

Concept Check 44.4

1. The numerous nephrons and well-developed glomeruli of freshwater fishes produce urine at a high rate, while the small numbers of nephrons and smaller glomeruli of marine fishes produce urine at a low rate. **2.** The kidney medulla would absorb less water; thus, the drug would increase the amount of water lost in the urine. **3.** A decline in blood pressure in the afferent arteriole would reduce the rate of filtration by moving less material through the vessels.

Concept Check 44.5

1. Alcohol inhibits the release of ADH, causing an increase in urinary water loss and increasing the chance of dehydration. **2.** The consumption of a very large amount of water in a short period of time, coupled with an absence of solute intake, can reduce sodium levels in the blood below tolerable levels. This condition, called hyponatremia, leads to disorientation and, sometimes, respiratory distress. It has occurred in some marathon runners who drink water rather than sports drinks. (It has also caused the death of a fraternity pledge as a consequence of a water hazing ritual and the death of a contestant in a water-drinking competition.) **3.** High blood pressure

Summary of Key Concepts Questions

44.1 Water moves into a cell by osmosis when the fluid outside the cells is hypoosmotic (has a lower solute concentration than the cytosol).
44.2

Waste Attribute	Ammonia	Urea	Uric Acid
Toxicity	High	Very low	Low
Energy cost to produce	Low	Moderate	High
Water loss to excretion	High	Moderate	Low

44.3 Filtration produces a fluid for exchange processes that is free of cells and large molecules, which are of benefit to the animal and could not readily be reabsorbed. **44.4** Both types of nephrons have proximal tubules that can reabsorb nutrients, but only juxtamedullary nephrons have loops of Henle that extend deep into the renal medulla. Thus, only kidneys containing juxtamedullary nephrons can produce urine that is more concentrated than the blood. **44.5** Patients who don't produce ADH have symptoms relieved by treatment with the hormone, but many patients with diabetes insipidus lack functional receptors for ADH.

Test Your Understanding

1. C **2.** A **3.** C **4.** D **5.** C **6.** B

Chapter 45

Figure Questions

Figure 45.4

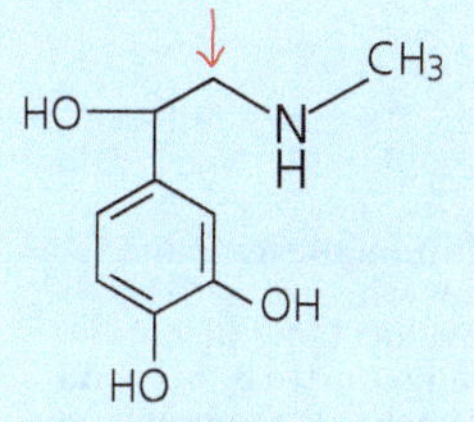

Epinephrine

Figure 45.5 The hormone is water-soluble and has a cell-surface receptor. Such receptors, unlike those for lipid-soluble hormones, can cause observable changes in cells without hormone-dependent gene transcription. **Figure 45.6** ATP is enzymatically converted to cAMP. The other steps represent binding reactions. **Figure 45.21** The embryonic gonad can become either a testis or an ovary. In contrast, the ducts either form a particular structure or degenerate, and the bladder forms in both males and females.

Concept Check 45.1

1. Water-soluble hormones, which cannot penetrate the plasma membrane, bind to cell-surface receptors. This interaction triggers an intracellular signal transduction pathway that ultimately alters the activity of a preexisting protein in the cytoplasm and/or changes transcription of specific genes in the nucleus. Steroid hormones are lipid-soluble and can cross the plasma membrane into the cell interior, where they bind to receptors located in the cytosol or nucleus. The hormone-receptor complex then functions directly as a transcription factor that changes transcription of specific genes. **2.** An exocrine gland, because pheromones are not secreted into interstitial fluid, but instead are typically released onto a body surface or into the environment **3.** Because receptors for water-soluble hormones are located on the cell surface, facing the extracellular space, injecting the hormone into the cytosol would not trigger a response.

Concept Check 45.2

1. Prolactin regulates milk production, and oxytocin regulates milk release. **2.** The posterior pituitary, an extension of the hypothalamus that contains the axons of neurosecretory cells, is the storage and release site for two neurohormones, oxytocin and antidiuretic hormone (ADH). The anterior pituitary contains endocrine cells that make at least six different hormones. Secretion of anterior pituitary hormones is controlled by hypothalamic hormones that travel via blood vessels to the anterior pituitary. **3.** The hypothalamus and pituitary glands function in many different endocrine pathways. Many defects in these glands, such as those affecting growth or organization, would therefore disrupt many hormone pathways. Only a very specific defect, such as a mutation affecting a particular hormone receptor, would alter just one endocrine pathway. The situation is quite different for the final gland in a pathway, such as the thyroid gland. In this case, a wide range of defects that disrupt gland function would disrupt only the one pathway or small set of pathways in which that gland functions. **4.** Both diagnoses could be correct. In one case, the thyroid gland may produce excess thyroid hormone despite normal hormonal input from the hypothalamus and anterior pituitary. In the other, abnormally elevated hormonal input (elevated TSH levels) may be the cause of the overactive thyroid gland.

Concept Check 45.3

1. If the function of the pathway is to provide a transient response, a short-lived stimulus would be less dependent on negative feedback. **2.** You would be exploiting the anti-inflammatory activity of glucocorticoids. Local injection avoids the effects on glucose metabolism that would occur if glucocorticoids were taken orally and transported throughout the body in the bloodstream. **3.** Both hormones produce opposite effects in different target tissues. In the fight-or-flight response, epinephrine increases blood flow to skeletal muscles and reduces blood flow to smooth muscles in the digestive system. In establishing apical dominance, auxin promotes the growth of apical buds and inhibits the growth of lateral buds.

Summary of Key Concepts Questions

45.1 As shown in Figure 43.16, helper T cell activation by cytokines acting as local regulators involves both autocrine and paracrine signaling. **45.2** The pancreas, parathyroid glands, and pineal gland **45.3** Both the pituitary and the adrenal glands are formed by fusion of neural and nonneural tissue. ADH is secreted by the neurosecretory portion of the pituitary gland, and epinephrine is secreted by the neurosecretory portion of the adrenal gland.

Test Your Understanding

1. C **2.** D **3.** D **4.** B **5.** B **6.** B **7.** A
8.

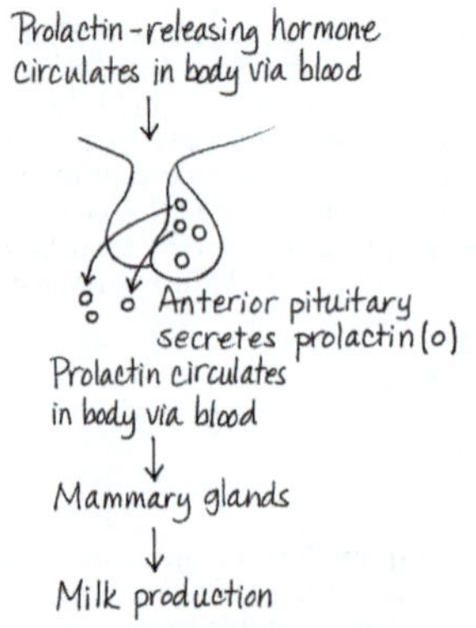

Chapter 46

Figure Questions

Figure 46.7 Newly formed sperm enter the seminal vesicle from the testis and exit via the ejaculatory duct during intercourse. Sperm enter the spermatheca after intercourse and, after storage, are released into the oviduct to fertilize an egg moving into the uterus. **Figure 46.8** When successfully courted by a second male, regardless of his genotype, about one-third of the females rid themselves of all sperm from the first mating. Thus, two-thirds retained some sperm from the first mating. We would therefore predict that two-thirds of those females would have some offspring exhibiting the small-eye phenotype of the dominant mutation carried by the males with which the females mated first. **Figure 46.11** The analysis would be informative because the polar bodies contain all of the maternal chromosomes that don't end up in the mature egg. For example, finding two copies of the disease gene in the polar bodies would indicate its absence in the egg. This method of genetic testing is sometimes carried out when oocytes collected from a female are fertilized with sperm in a laboratory dish. **Figure 46.15** The embryo normally implants about a week after conception, but it spends several days in the uterus before implanting, receiving nutrients from the endometrium. Therefore, the fertilized egg should be cultured for several days in liquid that is at normal body temperature and contains the same nutrients as those provided by the endometrium before implantation. **Figure 46.16** Testosterone can pass from fetal blood to maternal blood via the placental circulation, temporarily upsetting the hormonal balance in the mother. **Figure 46.18** Oxytocin would most likely induce labor, starting a positive-feedback loop that would direct labor to completion. Synthetic oxytocin is in fact frequently used to induce labor when prolonged pregnancy might endanger the mother or fetus.

Concept Check 46.1

1. The offspring of sexual reproduction are more genetically diverse. However, asexual reproduction can produce more offspring over multiple generations. **2.** Unlike other forms of asexual reproduction, parthenogenesis involves gamete production. By controlling whether or not haploid eggs are fertilized, species such as honeybees can readily switch between asexual and sexual reproduction. **3.** No. Owing to random assortment of chromosomes during meiosis, the offspring may receive the same copy or different copies of a particular parental chromosome from the sperm and the egg. Furthermore, genetic recombination during meiosis will result in reassortment of genes between pairs of parental chromosomes. **4.** Fragmentation occurs in both plants and animals. Also, budding in animals and the growth of adventitious from plant roots both involve emergence of new individuals from outgrowths of the parent.

Concept Check 46.2

1. Internal fertilization allows sperm to reach the egg without either gamete drying out. **2.** (a) Animals with external fertilization tend to release many gametes at once, resulting in the production of enormous numbers of zygotes. This increases the chances that some will survive to adulthood. (b) Animals with internal fertilization produce fewer offspring but generally exhibit greater care of the embryos and the young. **3.** Like the uterus of an insect, the ovary of a plant is the site of fertilization. Unlike the plant ovary, the uterus is not the site of egg production, which occurs in the insect ovary. In addition, the fertilized insect egg is expelled from the uterus, whereas the plant embryo develops within a seed in the ovary.

Concept Check 46.3

1. Spermatogenesis occurs normally only when the testicles are cooler than normal body temperature. Extensive use of a hot tub (or of very tight-fitting underwear) can cause a decrease in sperm quality and number. **2.** In humans, the secondary oocyte combines with a sperm before it finishes the second meiotic division. Thus, oogenesis is completed after, not before, fertilization. **3.** The only effect of sealing off each vas deferens is an absence of sperm in the ejaculate. Sexual response and ejaculate volume are unchanged. The cutting and sealing off of these ducts, a *vasectomy*, is a common surgical procedure for men who do not wish to produce any (more) offspring.

Concept Check 46.4

1. In the testis, FSH stimulates the Sertoli cells, which nourish developing sperm. LH stimulates the production of androgens (mainly testosterone), which in turn stimulate sperm production. In both females and males, FSH encourages the growth of cells that support and nourish developing gametes (follicle cells in females and Sertoli cells in males), and LH stimulates the production of sex hormones that promote gametogenesis (estrogens, primarily estradiol, in females and androgens, especially testosterone, in males). **2.** In estrous cycles, which occur in most female mammals, the endometrium is reabsorbed (rather than shed) if fertilization does not occur. Estrous cycles often occur just once or a few times a year, and the female is usually receptive to copulation only during the period around ovulation. Menstrual cycles are found only in humans and some other primates. They control the buildup and breakdown of the uterine lining, but not sexual receptivity. **3.** The combination of estradiol and progesterone would have a negative-feedback effect on the hypothalamus, blocking release of GnRH. This would interfere with LH secretion by the pituitary, thus preventing ovulation. This is in fact one basis of action of the most common hormonal contraceptives. **4.** In the viral replicative cycle, the production of new viral genomes is coordinated with capsid protein expression and with the production of phospholipids for viral coats. In the reproductive cycle of a human female, there is hormonally based coordination of egg maturation with the development of support tissues of the uterus.

Concept Check 46.5

1. The secretion of hCG by the early embryo stimulates the corpus luteum to make progesterone, which helps maintain the pregnancy. During the second trimester, however, hCG production drops, the corpus luteum disintegrates, and the placenta completely takes over progesterone production. **2.** Both tubal ligation and vasectomy block the movement of gametes from the gonads to a site where fertilization could take place. **3.** The introduction of a sperm nucleus directly into an oocyte bypasses the sperm's acquisition of motility in the epididymis, its swimming to meet the egg in the oviduct, and its fusion with the egg.

Summary of Key Concepts Questions

46.1 No. Because parthenogenesis involves meiosis, the mother would pass on to each offspring a random and therefore typically distinct combination of the chromosomes she inherited from her mother and father. **46.2** None **46.3** The small size and lack of cytoplasm characteristic of a sperm are adaptations well suited to its function as a delivery vehicle for DNA. The large size and rich cytoplasmic contents of eggs support the growth and development of the embryo. **46.4** Circulating anabolic steroids mimic the feedback regulation of testosterone, turning off pituitary signaling to the testes and thereby blocking the release of signals required for spermatogenesis. **46.5** Oxygen in maternal blood diffuses from pools in the endometrium into fetal capillaries in the chorionic villi of the placenta, and from there travels throughout the circulatory system of the fetus.

Test Your Understanding

1. D **2.** B **3.** A **4.** C **5.** A **6.** B **7.** C **8.** D
9.

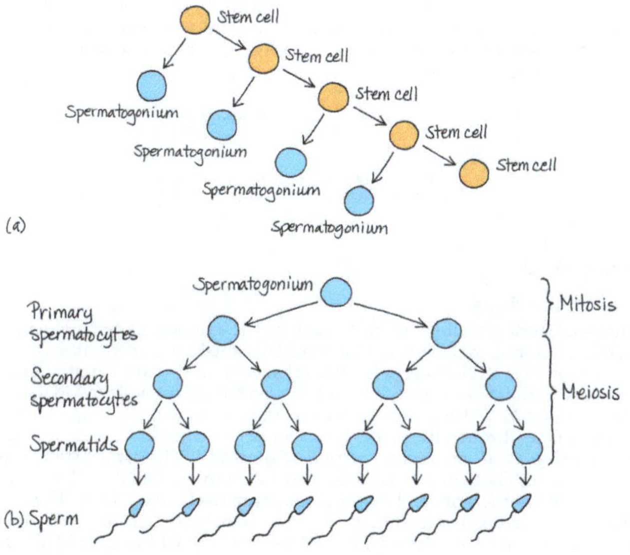

(c) The supply of stem cells would be used up, and spermatogenesis would not be able to continue.

Chapter 47

Figure Questions

Figure 47.4 You could inject the compound into an unfertilized egg, expose the egg to sperm, and see whether the fertilization envelope forms.
Figure 47.6 There would be fewer cells and they would be closer together.
Figure 47.8 (1) The blastocoel forms a single compartment that surrounds the gut, much like a doughnut surrounds a hole. (2) Ectoderm forms the outer covering of the animal, and endoderm lines the internal organs, such as the digestive tract. Mesoderm fills much of the space between these two layers.
Figure 47.19 Eight cell divisions are required to give rise to the intestinal cell closest to the mouth. **Figure 47.22** When the researchers allowed normal cortical rotation to occur, the "back-forming" determinants were activated. When they then forced the opposite rotation to occur, the back was established on the opposite side as well. Because the molecules on the normal side were already activated, forcing the opposite rotation apparently did not "cancel out" the establishment of the back side by the first rotation.
Figure 47.23 Draw It

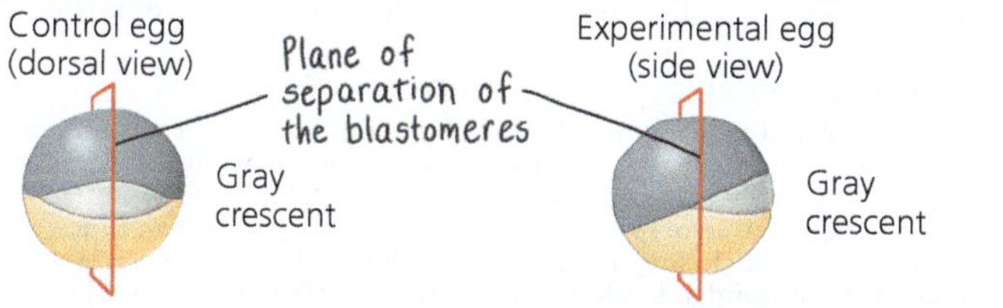

What If? In Spemann's control, the two blastomeres were physically separated, and each grew into a whole embryo. In Roux's experiment, remnants of the dead blastomere were still contacting the live blastomere, which developed into a half-embryo. Therefore, molecules present in the dead cell's remnants may have been signaling to the live cell, inhibiting it from making all the embryonic structures. **Figure 47.24** You could inject the isolated protein (or an mRNA encoding it) into ventral cells of an earlier gastrula. If dorsal structures form on the ventral side, that would support the idea that the protein is the signaling molecule secreted or presented by the dorsal lip. You should also do a control experiment to make sure the injection process alone did not cause dorsal structures to form. **Figure 47.26** Either Sonic hedgehog mRNA or protein can serve as a marker of the ZPA. The absence of either one after removal of the AER would support your hypothesis. You could also block FGF function and see whether the ZPA formed (by looking for Sonic hedgehog).

Concept Check 47.1

1. The fertilization envelope forms after cortical granules release their contents outside the egg, causing the vitelline membrane to rise and harden. The fertilization envelope serves as a barrier to fertilization by more than one sperm. **2.** The increased Ca^{2+} concentration in the egg would cause the cortical granules to fuse with the plasma membrane, releasing their contents and causing a fertilization envelope to form, even though no sperm had entered. This would prevent fertilization. **3.** You would expect it to fluctuate. The fluctuation of MPF drives the transition between DNA replication (S phase) and mitosis (M phase), which is still required in the abbreviated cleavage cell cycle.

Concept Check 47.2

1. The cells of the notochord migrate toward the midline of the embryo (converge), rearranging themselves so there are fewer cells across the notochord, which thus becomes longer overall (extends; see Figure 47.17). **2.** Because microfilaments would not be able to contract and decrease the size of one end of the cell, both the inward bending in the middle of the neural tube and the outward bending of the hinge regions at the edges would be blocked. Therefore, the neural tube probably would not form. **3.** Dietary intake of the vitamin folic acid dramatically reduces the frequency of neural tube defects.

Concept Check 47.3

1. Axis formation establishes the location and polarity of the three axes that provide the coordinates for development. Pattern formation positions particular tissues and organs in the three-dimensional space defined by those coordinates.
2. Morphogen gradients act by specifying cell fates across a field of cells through variation in the level of a determinant. Morphogen gradients thus act more globally than cytoplasmic determinants or inductive interactions between pairs of cells. **3.** Yes, a second embryo could develop because inhibiting BMP-4 activity would have the same effect as transplanting an organizer. **4.** The limb that developed probably would have a mirror-image duplication, with the most posterior digits in the middle and the most anterior digits at either end.

Summary of Key Concepts Questions

47.1 The binding of a sperm to a receptor on the egg surface is very specific and likely would not occur if the two gametes were from different species. Without sperm binding, the sperm and egg membranes would not fuse. **47.2** Apoptosis functions to eliminate structures required only in an immature form, nonfunctional cells from a pool larger than the number required, and tissues formed by a developmental program that is not adaptive for the organism as it has evolved. **47.3** Mutations that affected both limb and kidney development would be more likely to alter the function of monocilia because these organelles are important in several signaling pathways. Mutations that affected limb development but not kidney development would more likely alter a single pathway, such as Hedgehog signaling.

Test Your Understanding

1. A **2.** B **3.** D **4.** A **5.** D **6.** C **7.** B
8.

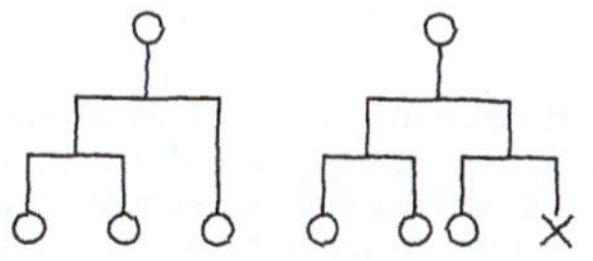

Chapter 48

Figure Questions

Figure 48.7 Adding chloride channels would make the membrane potential less positive. Adding potassium channels would have no effect because there are no potassium ions present. **Figure 48.9** In the absence of other forces, chemical concentration gradients govern net diffusion. In this case, ions are more concentrated outside of the cell and move in when the channel opens.
Figure 48.10

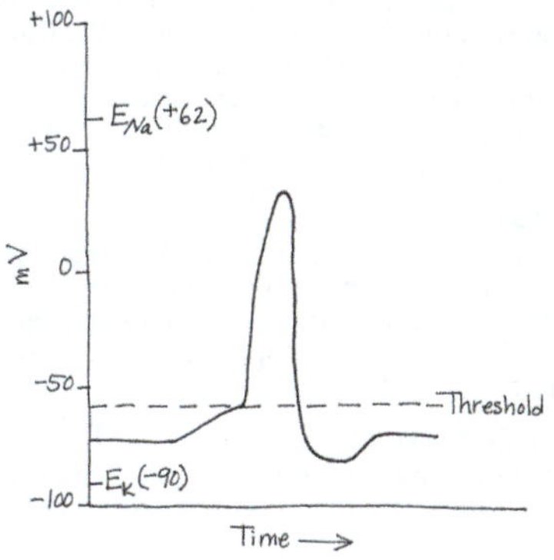

Appendix A Answers

Figure 48.11

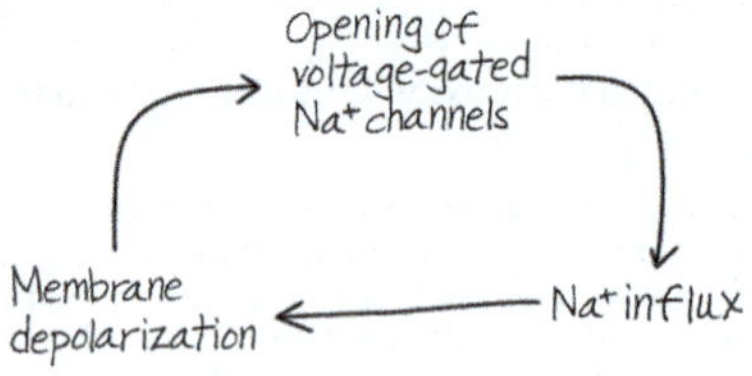

Figure 48.12

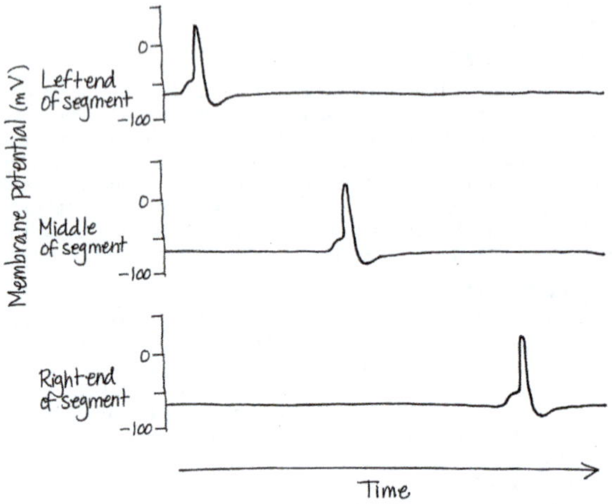

Figure 48.15 The production and transmission of action potentials would be unaffected. However, action potentials arriving at chemical synapses would be unable to trigger release of neurotransmitter. Signaling at such synapses would thus be blocked. **Figure 48.17** Summation only occurs if inputs occur simultaneously or nearly so. Thus, spatial summation, in which input is received from two different sources, is in effect also temporal summation.

Concept Check 48.1

1. Axons and dendrites extend from the cell body and function in information flow. Dendrites transfer information to the cell body, whereas axons transmit information from the cell body. A typical neuron has multiple dendrites and one axon. **2.** Sensors in your ear transmit information to your brain. There, the activity of interneurons in processing centers enables you to recognize your name. In response, signals transmitted via motor neurons cause contraction of muscles that turn your neck. **3.** Increased branching would allow control of a greater number of postsynaptic cells, enhancing coordination of responses to nervous system signals.

Concept Check 48.2

1. Ions can flow against a chemical concentration gradient if there is an opposing electrical gradient of greater magnitude. **2.** A decrease in permeability to K^+, an increase in permeability to Na^+, or both **3.** Charged dye molecules could equilibrate only if other charged molecules could also cross the membrane. If not, a membrane potential would develop that would counterbalance the chemical gradient.

Concept Check 48.3

1. A graded potential has a magnitude that varies with stimulus strength, whereas an action potential has an all-or-none magnitude that is independent of stimulus strength. **2.** Loss of the insulation provided by myelin sheaths leads to a disruption of action potential propagation along axons. Voltage-gated sodium channels are restricted to the nodes of Ranvier, and without the insulating effect of myelin, the inward current produced at one node during an action potential cannot depolarize the membrane to the threshold at the next node. **3.** Positive feedback is responsible for the rapid opening of many voltage-gated sodium channels, causing the rapid outflow of sodium ions responsible for the rising phase of the action potential. As the membrane potential becomes positive, voltage-gated potassium channels open in a form of negative feedback that helps bring about the falling phase of the action potential. **4.** The maximum frequency would decrease because the refractory period would be extended.

Concept Check 48.4

1. It can bind to different types of receptors, each triggering a specific response in postsynaptic cells. **2.** These toxins would prolong the EPSPs that acetylcholine produces because the neurotransmitter would remain longer in the synaptic cleft. **3.** Membrane depolarization, exocytosis, and membrane fusion each occur in fertilization and in neurotransmission.

Summary of Key Concepts Questions

48.1 It would prevent information from being transmitted away from the cell body along the axon. **48.2** There are very few open sodium channels in a resting neuron, so the resting potential either would not change or would become slightly more negative (hyperpolarization). **48.4** A given neurotransmitter can have many receptors that differ in their location and activity. Drugs that target receptor activity rather than neurotransmitter release or stability are therefore likely to exhibit greater specificity and potentially have fewer undesirable side effects.

Test Your Understanding

1. C **2.** C **3.** C **4.** B **5.** A **6.** D

7. The activity of the sodium-potassium pump is essential to maintain the resting potential. With the pump inactivated, the sodium and potassium concentration gradients would gradually disappear, resulting in a greatly reduced resting potential. **8.** Since GABA is an inhibitory neurotransmitter in the CNS, this drug would be expected to decrease brain activity. A decrease in brain activity might be expected to slow down or reduce behavioral activity. Many sedative drugs act in this fashion. **9.** As shown in this pair of drawings, a pair of action potentials would move outward in both directions from each electrode. (Action potentials are unidirectional only if they begin at one end of an axon.) However, because of the refractory period, the two action potentials between the electrodes both stop where they meet. Thus, only one action potential reaches the synaptic terminals.

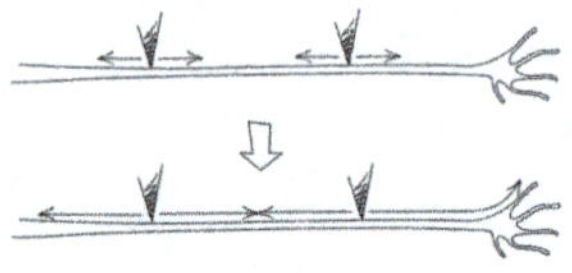

Chapter 49

Figure Questions

Figure 49.7 During swallowing, muscles along the esophagus alternately contract and relax, resulting in peristalsis. One model to explain this alternation is that each section of muscle receives nerve impulses that alternate between excitation and inhibition, just as the quadriceps and hamstring receive opposing signals in the knee-jerk reflex. **Figure 49.15** The gray areas have a different shape and pattern, indicating different planes through the brain. This fact indicates that the nucleus accumbens and the amygdala are in different planes. **Figure 49.17** The hand is shown larger than the forearm because the hand receives more innervation than the forearm for sensory input to the brain and motor output from the brain. **Figure 49.24** If the depolarization brings the membrane potential to or past threshold, it should initiate action potentials that cause dopamine release from the VTA neurons. This should mimic natural stimulation of the brain reward system, resulting in positive and perhaps pleasurable sensations.

Concept Check 49.1

1. The sympathetic division would likely be activated. It mediates the "fight-or-flight" response in stressful situations. **2.** Nerves contain bundles of axons, some that belong to motor neurons, which send signals outward from the CNS, and some that belong to sensory neurons, which bring signals into the CNS. Therefore, you would expect effects on both motor control and sensation.
3. Neurosecretory cells of the adrenal medulla secrete the hormones epinephrine and norepinephrine in response to preganglionic input from sympathetic neurons. These hormones travel in the circulation throughout the body, triggering responses in many tissues.

Concept Check 49.2

1. The cerebral cortex on the left side of the brain initiates voluntary movement of the right side of the body. **2.** Alcohol diminishes function of the cerebellum.
3. A coma reflects a disruption in the cycles of sleep and arousal regulated by communication between the midbrain and pons (reticular formation) and the cerebrum. You would expect this group to have damage to the midbrain, the pons, the cerebrum, or any part of the brain between these structures. Paralysis reflects an inability to carry out motor commands transmitted from the cerebrum to the spinal cord. You would expect this group to have damage to the portion of the CNS extending from the spinal cord up to but not including the midbrain and pons.

Concept Check 49.3

1. Brain damage that disrupts behavior, cognition, memory, or other functions provides evidence that the portion of the brain affected by the damage is important for the normal activity that is blocked or altered. **2.** Broca's area, which is active during the generation of speech, is located near the motor cortex, which controls skeletal muscles, including those in the face. Wernicke's area, which is active when speech is heard, is located in the posterior part of the temporal lobe, which is involved in hearing. **3.** Each cerebral hemisphere is specialized for different parts of this task—the right for face recognition and the left for language. Without an intact corpus callosum, neither hemisphere can take advantage of the other's processing abilities.

Concept Check 49.4

1. There can be an increase in the number of synapses between the neurons or an increase in the strength of existing synaptic connections. **2.** If consciousness is an emergent property resulting from the interaction of many different regions of the brain, then it is unlikely that localized brain damage will have a discrete effect on consciousness. **3.** The hippocampus is responsible for organizing newly acquired information. Without hippocampal function, the links necessary to retrieve information from the cerebral cortex will be lacking, and no functional memory, short- or long-term, will be formed.

Concept Check 49.5

1. Both are progressive brain diseases whose risk increases with advancing age. Both result from the death of brain neurons and are associated with the accumulation of peptide or protein aggregates. **2.** The symptoms of schizophrenia can be mimicked by a drug that stimulates dopamine-releasing neurons. The brain's reward system, which is involved in drug addiction, is composed of dopamine-releasing neurons that connect the ventral tegmental area to regions in the cerebrum. Parkinson's disease results from the death of dopamine-releasing neurons. **3.** Not necessarily. It might be that the plaques, tangles, and missing regions of the brain seen at death reflect secondary effects, the consequence of other unseen changes that are actually responsible for the alterations in brain function.

Summary of Key Concepts Questions

49.1 Because reflex circuits involve only a few neurons—the simplest consist of a sensory neuron and a motor neuron—the path for information transfer is short and simple, increasing the speed of the response. **49.2** The midbrain coordinates visual reflexes; the cerebellum controls coordination of movement that depends on visual input; the thalamus serves as a routing center for visual information; and the cerebrum is essential for converting visual input to a visual image. **49.3** You would expect the right side of the body to be paralyzed because it is controlled by the left cerebral hemisphere, where language generation and interpretation are localized. **49.4** Learning a new language likely requires the maintenance of synapses that are formed during early development but are otherwise lost prior to adulthood. **49.5** Whereas amphetamine stimulates dopamine release, PCP blocks glutamate receptors, suggesting that schizophrenia does not reflect a defect in the function of just one neurotransmitter.

Test Your Understanding

1. B **2.** B **3.** D **4.** C **5.** C **6.** A
7.

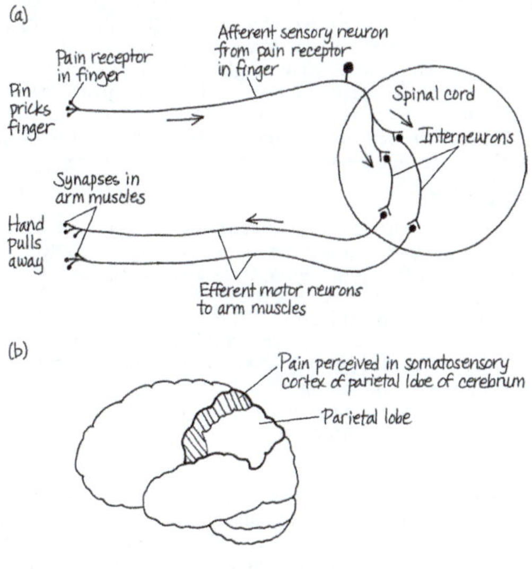

Chapter 50

Figure Questions

Figure 50.17

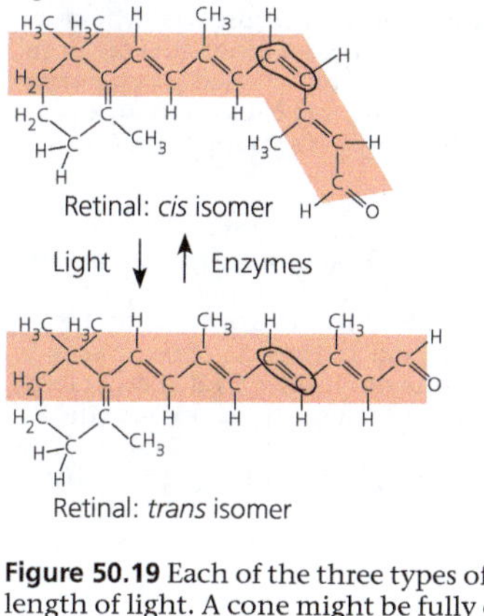

The hydrogen (H) atoms are on the same side of the carbon-carbon double bond in the *cis* isomer and the opposite side in the *trans* isomer.

Figure 50.19 Each of the three types of cones is most sensitive to a different wavelength of light. A cone might be fully depolarized when there is light present if the light is of a wavelength far from its optimum. **Figure 50.21** In humans, an X chromosome with a defect in the red or green opsin gene is much less common than a wild-type X chromosome. Color blindness therefore typically skips a generation as the defective allele passes from an affected male to a carrier daughter and back to an affected grandson. In squirrel monkeys, no X chromosome can confer full color vision. As a result, all males are color-blind and no unusual inheritance pattern is observed. **Figure 50.23** The results of the experiment would have been identical. What matters is the activation of particular sets of neurons, not the manner in which they are activated. Any signal from a bitter cell will be interpreted by the brain as a bitter taste, regardless of the nature of the compound and the receptor involved. **Figure 50.25** Only perception. Binding of an odorant to its receptor will cause action potentials to be sent to the brain. Although an excess of that odorant might cause a diminished response through adaptation, another odorant can mask the first only at the level of perception in the brain. **Figure 50.26** Both. A muscle fiber contains many myofibrils bundled together and divided lengthwise into many sarcomeres. A sarcomere is a contractile unit made up of portions of many myofibrils, and each myofibril is a part of many sarcomeres. **Figure 50.28** Hundreds of myosin heads participate in sliding each pair of thick and thin filaments past each other. Because cross-bridge formation and breakdown are not synchronized, many myosin heads are exerting force on the thin filaments at all times during muscle contraction. **Figure 50.33** By causing all of the motor neurons that control the muscle to generate action potentials at a rate high enough to produce tetanus in all of the muscle fibers

Concept Check 50.1

1. Electromagnetic receptors in general detect only external stimuli. Nonelectromagnetic receptors, such as chemoreceptors or mechanoreceptors, can act as either internal or external sensors. **2.** The capsaicin present in the peppers activates the thermoreceptor for high temperatures. In response to the perceived high temperature, the nervous system triggers sweating to achieve evaporative cooling. **3.** You would perceive the electrical stimulus as if the sensory receptors that regulate that neuron had been activated. For example, electrical stimulation of the sensory neuron controlled by the thermoreceptor activated by menthol would likely be perceived as a local cooling.

Concept Check 50.2

1. Otoliths detect the animal's orientation with respect to gravity, providing information that is essential in environments such as the tunnel habitat of the star-nosed mole, where light cues are absent. **2.** As a sound that changes gradually from a very low to a very high pitch **3.** The stapes and the other middle ear bones transmit vibrations from the tympanic membrane to the oval window. Fusion of these bones (as occurs in a disease called otosclerosis) would block this transmission and result in hearing loss. **4.** In animals, the statoliths are extracellular. In contrast, the statoliths of plants are found within an intracellular organelle. The methods for detecting their location also differ. In animals, detection is by means of mechanoreceptors on ciliated cells. In plants, the mechanism appears to involve calcium signaling.

Concept Check 50.3

1. Planarians have ocelli that cannot form images but can sense the intensity and direction of light, providing enough information to enable the animals to find protection in shaded places. Flies have compound eyes that form images and excel at detecting movement. **2.** The person can focus on distant objects but not close objects (without glasses) because close focusing requires the lens to become almost spherical. This problem is common after age 50. **3.** The signal produced by rod and cone cells is glutamate, and their release of glutamate decreases upon exposure to light. However, a decrease in glutamate production causes other retinal cells to increase the rate at which action potentials are sent to the brain, so that the brain receives more action potentials in light than in dark. **4.** Absorption of light by retinal converts retinal from its *cis* isomer to its *trans* isomer, initiating the process of light detection. In contrast, a photon absorbed by chlorophyll does not bring about isomerization, but instead boosts an electron to a higher energy orbital, initiating the electron flow that generates ATP and NADPH.

Concept Check 50.4

1. Both taste cells and olfactory cells have receptor proteins in their plasma membrane that bind certain substances, leading to membrane depolarization through a signal transduction pathway involving a G protein. However, olfactory cells are sensory neurons, whereas taste cells are not. **2.** Since animals rely on chemical signals for behaviors that include finding mates, marking territories, and avoiding dangerous substances, it is adaptive for the olfactory system to have a robust response to a very small number of molecules of a particular odorant. **3.** Because the sweet, bitter, and umami tastes involve GPCR proteins but the sour taste does not, you might predict that the mutation is in a molecule that acts in the signal transduction pathway common to the different GPCRs.

Concept Check 50.5

1. In a skeletal muscle fiber, Ca^{2+} binds to the troponin complex, which moves tropomyosin away from the myosin-binding sites on actin and allows cross-bridges to form. In a smooth muscle cell, Ca^{2+} binds to calmodulin, which activates an enzyme that phosphorylates the myosin head and thus enables cross-bridge formation. **2.** *Rigor mortis*, a Latin phrase meaning "stiffness of death," results from the complete depletion of ATP in skeletal muscle. Since ATP is required to release myosin from actin and to pump Ca^{2+} out of the cytosol, muscles become chronically contracted beginning about 3–4 hours after death. **3.** A competitive inhibitor binds to the same site as the substrate for the enzyme. In contrast, the troponin and tropomyosin complex masks, but does not bind to, the myosin-binding sites on actin.

Concept Check 50.6

1. The main problem in swimming is drag; a fusiform body minimizes drag. The main problem in flying is overcoming gravity; wings shaped like airfoils provide lift, and adaptations such as air-filled bones reduce body mass. **2.** In modeling peristalsis you would constrict the toothpaste tube at different points along its length, using your hand to encircle the tube and squeeze concentrically. To demonstrate movement of food through the digestive tract you would want the cap off the toothpaste tube, whereas you would want the cap on to show how peristalsis contributes to worm locomotion. **3.** When you grasp the sides of the chair, you are using a contraction of the triceps to keep your arms extended against the pull of gravity on your body. As you lower yourself slowly into the chair, you gradually decrease the number of motor units in the triceps that are contracted. Contracting your biceps would jerk you down, since you would no longer be opposing gravity.

Summary of Key Concepts Questions

50.1 Nociceptors overlap with other classes of receptors in the type of stimulus they detect. They differ from other receptors only in how a particular stimulus is perceived. **50.2** Volume is encoded by the frequency of action potentials transmitted to the brain; pitch is encoded by which axons are transmitting action potentials. **50.3** The major difference is that neurons in the retina integrate information from multiple sensory receptors (photoreceptors) before transmitting information to the central nervous system. **50.4** Our olfactory sense is responsible for most of what we describe as distinct tastes. A head cold or other source of congestion blocks odorant access to receptors lining portions of the nasal cavity. **50.5** Hydrolysis of ATP is required to convert myosin to a high-energy configuration for binding to actin and to power the Ca^{2+} pump that removes cytosolic Ca^{2+} during muscle relaxation. **50.6** Human body movements rely on the contraction of muscles anchored to a rigid endoskeleton. Tendons attach muscles to bones, which in turn are composed of fibers built up from a basic organizational unit, the sarcomere. The thin and thick filaments have separate points of attachment within the sarcomere. In response to nervous system motor output, the formation and breakdown of cross-bridges between myosin heads and actin ratchet the thin and thick filaments past each other. Because the filaments are anchored, this sliding movement shortens the muscle fibers. Furthermore, because the fibers themselves are part of the muscles attached at each end to bones, muscle contraction moves bones of the body relative to each other. In this way, the structural anchoring of muscles and filaments enables muscle function, such as the bending of an elbow by contraction of the biceps.

Test Your Understanding

1. D **2.** A **3.** B **4.** C **5.** B **6.** D
7.

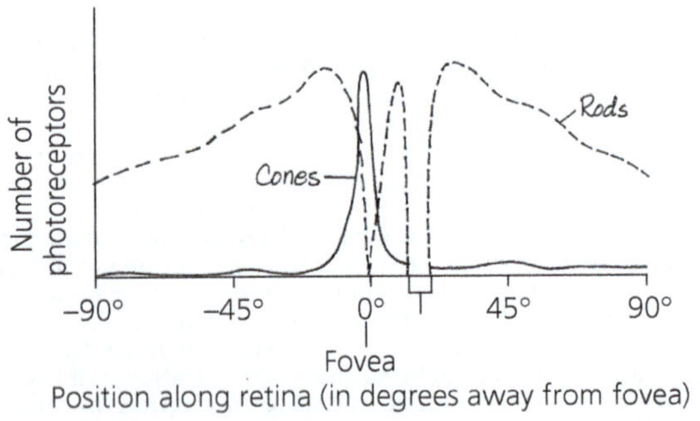

The answer shows the actual distribution of rods and cones in the human eye. Your graph may differ, but should have the following properties: only cones at the fovea; fewer cones and more rods at both ends of the *x*-axis; no photoreceptors in the optic disk.

Chapter 51

Figure Questions

Figure 51.2 The fixed action pattern based on the sign stimulus of a red belly ensures that the male will chase away any invading males of his species. By chasing away such males, the defender decreases the chance that another male will fertilize eggs laid in his nesting territory. **Figure 51.5** The straight-run portion conveys two pieces of information: direction, via the angle of that run relative to the wall of the hive, and distance, via the number of waggles performed during the straight run. At a minimum, the portions between the straight runs identify the activity as a waggle dance. Since they also provide contact with workers to one side and then the other, they may ensure transmission of information to a larger number of other bees. **Figure 51.7** There should be no effect. Imprinting is an innate behavior that is carried out anew in each generation. Assuming the nest was not disturbed, the offspring of the geese imprinted on a human would imprint on the mother goose. **Figure 51.8** Perhaps the wasp doesn't use visual cues. It might also be that wasps recognize objects native to their environment, but not foreign objects, such as the pinecones. Tinbergen addressed these ideas before carrying out the pinecone study. When he swept away the pebbles and sticks around the nest, the wasps could no longer find their nests. If he shifted the natural objects in their natural arrangement, the shift in the landmarks caused a shift in the site to which the wasps returned. Finally, if natural objects around the nest site were replaced with pinecones while the wasp was in the burrow, the wasp nevertheless found her way back to the nest site. **Figure 51.10** Switching the orientations of all three grids would control for an inherent preference for or against a particular orientation. If there were no inherent preference or bias, the experiment should work equally well after the switch. **Figure 51.24** It might be that the birds require stimuli during flight to exhibit their migratory preference. If this were true, the birds would show the same orientation in the funnel experiment despite their distinct genetic programming. **Figure 51.26** It holds true for some, but not all individuals. If a parent has more than one reproductive partner, the offspring of different partners will have a coefficient of relatedness less than 0.5.

Concept Check 51.1

1. The proximate explanation for this fixed action pattern might be that nudging and rolling are released by the sign stimulus of an object outside the nest, and the behavior is carried to completion once initiated. The ultimate explanation might be that ensuring that eggs remain in the nest increases the chance of producing healthy offspring. **2.** There might be selective pressure for other prey fish to detect an injured fish because the source of the injury might threaten them as well. Among predators, there might be selection for those that are attracted to the alarm substance because they would be more likely to encounter crippled prey. Fish with adequate defenses might show no change because they have a selective advantage if they do not waste energy responding to the alarm substance. **3.** In both cases, the detection of periodic variation in the environment results in a reproductive cycle timed to environmental conditions that optimize the opportunity for success.

Concept Check 51.2

1. Natural selection would tend to favor convergence in color pattern because a predator learning to associate a pattern with a sting or bad taste would avoid all other individuals with that same color pattern, regardless of species. **2.** You might move objects around to establish an abstract rule, such as "past landmark A, the same distance as A is from the starting point," while maintaining a minimum of fixed metric relationships, that is, avoiding having the food directly adjacent to or a set distance from a landmark. As you might surmise, designing an informative experiment of this kind is not easy. **3.** Learned behavior, just like innate behavior, can contribute to reproductive isolation and thus to speciation. For example, learned bird songs contribute to species recognition during courtship, thereby helping ensure that only members of the same species mate.

Concept Check 51.3

1. Certainty of paternity is higher with external fertilization. **2.** Balancing selection could maintain the two alleles at the *forager* locus if population density fluctuated from one generation to another. At times of low population density, the energy-conserving sitter larvae (carrying the for^s allele) would be favored, while at higher population density, the more mobile Rover larvae (for^R allele) would have a selective advantage. **3.** Because females would now be present in much larger numbers than males, all three types of males should have some reproductive success. Nevertheless, since the advantage that the blue-throats rely on—a limited number of females in their territory—will be absent, the yellow-throats are likely to increase in frequency in the short term.

Concept Check 51.4

1. Because this geographic variation corresponds to differences in prey availability between two garter snake habitats, it seems likely that snakes with characteristics enabling them to feed on the abundant prey in their locale would have had increased survival and reproductive success. In this way, natural selection would have resulted in the divergent foraging behaviors. **2.** The fact that the individual shares some genes with the offspring of its sibling (in the case of humans, with the individual's niece or nephew) means that the reproductive success of that niece or nephew increases the representation of those genes in the population (selects for them). **3.** The older individual cannot be the beneficiary because he or she cannot have extra offspring. However, the cost is low for an older individual performing the altruistic act because that individual has already reproduced (but perhaps is still caring for a child or grandchild). There can therefore be selection for an altruistic act by a postreproductive individual that benefits a young relative.

Summary of Key Concepts Questions

51.1 Circannual rhythms are typically based on the cycles of light and dark in the environment. As the global climate changes, animals that migrate in response to these rhythms may shift to a location before or after local environmental conditions are optimal for reproduction and survival. **51.2** For the goose, all that is acquired is an object at which the behavior is directed. In the case of the sparrow, learning takes place that will give shape to the behavior itself. **51.3** Because feeding the female is likely to improve her reproductive success, the genes from the sacrificed male are likely to appear in a greater number of progeny. **51.4** Studying the genetic basis of these behaviors reveals that changes in a single gene can have large-scale effects on even complex behaviors.

Test Your Understanding

1. C **2.** B **3.** B **4.** A **5.** C **6.** A

7.

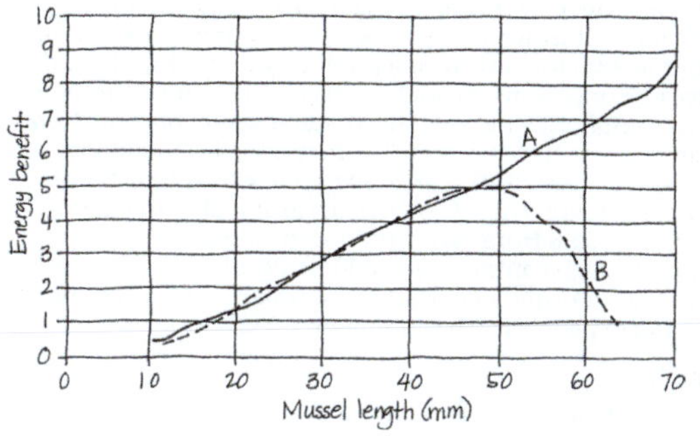

You could measure the size of mussels that oystercatchers successfully open and compare that with the size distribution in the habitat.

Chapter 52

Figure Questions

Figure 52.7 The species' distribution could be altered by dispersal limitations, the activities of people (such as a broad-scale conversion of forests to agriculture or selective harvesting), or many other factors, including those discussed later in the chapter (see Figure 52.17). **Figure 52.17** Some factors, such as fire, are relevant only for terrestrial systems. At first glance, water availability is primarily a terrestrial factor, too. However, species living along the intertidal zone of oceans or along the edge of lakes also suffer desiccation. Salinity stress is important for species in some aquatic and terrestrial systems. Oxygen availability is an important factor primarily for species in some aquatic systems and in soils and sediments.

Concept Check 52.1

1. In the tropics, high temperatures evaporate water and cause warm, moist air to rise. The rising air cools and releases much of its water as rain over the tropics. The remaining dry air descends at approximately 30° north and south, causing deserts to occur in those regions. **2.** The microclimate around the stream will be cooler, moister, and shadier than that around the unplanted agricultural field. **3.** Trees that require a long time to reach reproductive age are likely to evolve more slowly than annual plants in response to climate change, constraining the potential ability of such trees to respond to rapid climate change. **4.** Plants with C_4 photosynthesis are likely to expand their range globally as Earth's climate warms. C_4 photosynthesis minimizes photorespiration and enhances sugar production, an advantage that is especially useful in warmer regions where C_4 plants are found today.

Concept Check 52.2

1. The biggest difference between the two biomes is the higher amounts of precipitation that the forest receives. **2.** Answers will vary by location but should be based on the information and maps in Figure 52.12. How much your local area has been altered from its natural state will influence how much it reflects the expected characteristics of your biome, particularly the expected plants and animals. **3.** Northern coniferous forest is likely to replace tundra along the boundary between these biomes. To see why, note that northern coniferous forest is adjacent to tundra throughout North America, northern Europe, and Asia (see Figure 52.9) and that the temperature range for northern coniferous forest is just above that for tundra (see Figure 52.10).

Concept Check 52.3

1. In the oceanic pelagic zone, the ocean bottom lies below the photic zone, so there is too little light to support benthic algae or rooted plants. **2.** Aquatic organisms either gain or lose water by osmosis if the osmolarity of their environment differs from their internal osmolarity. Water gain can cause cells to swell, and water loss can cause them to shrink. To avoid excessive changes in cell volume, organisms that live in estuaries must be able to compensate for both water gain (under freshwater conditions) and water loss (under saltwater conditions). **3.** Oxygen serves as a reactant when decomposers break down the bodies of dead algae using aerobic respiration. Following an algal bloom, there are many dead algae; hence, decomposers may use a lot of oxygen to break down the bodies of dead algae, causing the lake's oxygen levels to drop.

Concept Check 52.4

1. (a) Humans might transplant a species to a new area that it could not previously reach because of a geographic barrier. (b) Humans might eliminate a predator or herbivore species, such as sea urchins, from an area. **2.** One test would be to build a fence around a plot of land in an area that has trees of that species, excluding all deer from the plot. You could then compare the abundance of tree seedlings inside and outside the fenced plot over time. **3.** Because the ancestor of the silverswords reached isolated Hawaii early in the islands' existence, it likely faced little competition and was able to occupy many unfilled niches. The cattle egret, in contrast, arrived in the Americas only recently and has to compete with a well-established group of species. Thus, its opportunities for adaptive radiation have probably been much more limited.

Concept Check 52.5

1. Changes in how organisms interact with one another and their environment can cause evolutionary change. In turn, an evolutionary change, such as an improvement in the ability of a predator to detect its prey, can alter ecological interactions. **2.** As cod adapt to the pressure of commercial fishing by reproducing at younger ages and smaller sizes, the number of offspring they produce each year will be lower. This may cause the population to decline as time goes on, thereby further reducing the population's ability to recover. If that happened, as the population becomes smaller over time, effects of genetic drift might become increasingly important. Drift could, for example, lead to the fixation of harmful alleles, which would further hinder the ability of the cod population to recover from overfishing.

Summary of Key Concepts Questions

52.1 Because dry air would descend at the equator instead of at 30° north and south latitude (where deserts exist today), deserts would be more likely to exist along the equator (see Figure 52.3). **52.2** The dominant plants in savanna ecosystems tend to be adapted to fire and tolerant of seasonal droughts. The savanna biome is maintained by periodic fires, both natural and set by humans, but humans are also clearing savannas for agriculture and other uses. **52.3** An aphotic zone is most likely to be found in the deep waters of a lake, the oceanic pelagic zone, or the marine benthic zone. **52.4** You might arrange a flowchart that begins with abiotic limitations—first determining the physical and chemical conditions under which a species could survive—and then moves through the other factors listed in the flowchart. **52.5** Because the introduced species had few predators or parasites, it might outcompete native species and thereby increase in number and expand its range in the new location. As the introduced species increased in abundance, natural selection might cause evolution in populations of competing species, favoring individuals with traits that made them more effective competitors with the introduced species. Selection could also cause evolution in populations of potential predator or parasite species, in this case favoring individuals with traits that enabled them to take advantage of this new potential source of food. Such evolutionary changes could modify the outcome of ecological interactions, potentially leading to further evolutionary changes, and so on.

Test Your Understanding

1. B **2.** B **3.** C **4.** D **5.** C **6.** A **7.** A **8.** B

Chapter 53

Figure Questions

Figure 53.4 The dispersion of the penguins would likely appear clumped as you flew over densely populated islands and sparsely populated ocean. **Figure 53.5** Ten percent (100/1,000) of the females survive to be three years old. **Figure 53.7** #109 **Figure 53.8** The population with $r = 1.0$ (blue curve) reaches 1,500 individuals in about 7.5 generations, whereas the population with $r = 0.5$ (red curve) reaches 1,500 individuals in about 14.5 generations.
Figure 53.16

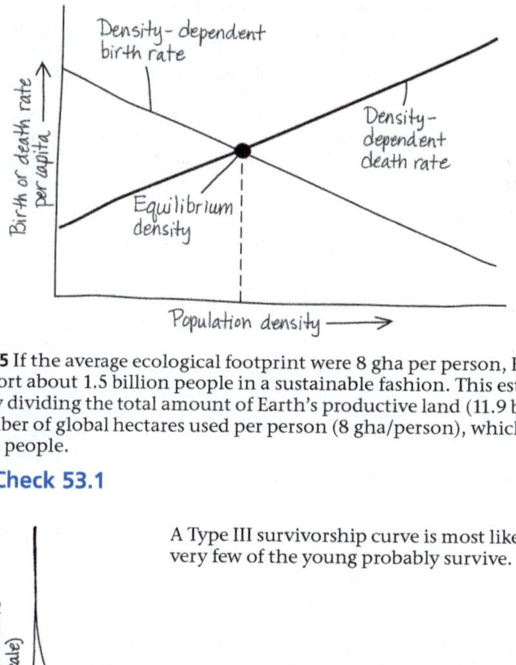

Figure 53.25 If the average ecological footprint were 8 gha per person, Earth could support about 1.5 billion people in a sustainable fashion. This estimate is obtained by dividing the total amount of Earth's productive land (11.9 billion gha) by the number of global hectares used per person (8 gha/person), which yields 1.49 billion people.

Concept Check 53.1

1.

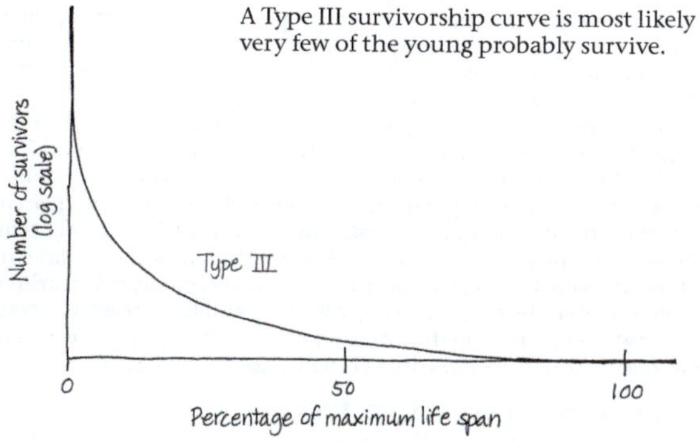

A Type III survivorship curve is most likely because very few of the young probably survive.

2. The proportion alive at the start of year 0–1 is 485/485 = 1.0. The proportion alive at the start of year 1–2 is 218/485 = 0.449. **3.** Male sticklebacks would likely have a uniform pattern of dispersion, with antagonistic interactions maintaining a relatively constant spacing between them.

Concept Check 53.2

1. Though r is constant, N, the population size, is increasing. As r is applied to an increasingly large N, population growth (rN) accelerates, producing the J-shaped curve. **2.** Exponential growth is more likely in the area where a forest was destroyed by fire. The first plants that found suitable habitat there would encounter an abundance of space, nutrients, and light. In the undisturbed forest, competition among plants for these resources would be intense. **3.** The equation for the number of people added to the population each year is $\Delta N/\Delta t = r_{\Delta t}N$. Therefore, the net population growth in 2014 was

$$\Delta N/\Delta t = 0.005 \times 320{,}000{,}000 = 1{,}600{,}000$$

or 1.6 million people. To determine whether the population is growing exponentially, you would need to determine whether $r > 0$ and if it is constant through time (across multiple years).

Concept Check 53.3

1. When N (population size) is small, there are relatively few individuals producing offspring. When N is large, near the carrying capacity, the per capita growth rate is relatively small because it is limited by available resources. The steepest part of the logistic growth curve corresponds to a population with a number of reproducing individuals that is substantial but not yet near carrying capacity. **2.** All else being equal, you would expect a plant species to have a larger carrying capacity at the equator than at high latitudes because there is more incident sunlight near the equator. **3.** The sudden change in environmental conditions might alter the phenotypic traits favored by natural selection. Assuming the newly favored traits were encoded at least in part by genes, natural selection might alter gene frequencies in this population. In addition, a substantial drop in the carrying capacity of the population could cause the size of the population to drop considerably. If this occurred, effects of genetic drift could become more pronounced—and that in turn could lead to the fixation of harmful alleles, hindering the ability of the population to rebound in size.

Concept Check 53.4

1. Three key life history traits are when reproduction begins, how often reproduction occurs, and how many offspring are produced per reproductive episode. Organisms differ widely for each of these traits. For example, the age of first reproduction is typically 3–4 years in coho salmon compared to 30 years in loggerhead turtles. Similarly, an agave reproduces only once during its lifetime, whereas an oak tree reproduces many times. Finally, the white rhinoceros produces a single calf when it reproduces, while most insects produce many offspring each time they reproduce. **2.** By preferentially investing in the eggs it lays in the nest, the peacock wrasse increases the chance those eggs will survive. The eggs it disperses widely and does not provide care for are less likely to survive, at least some of the time, but require a lower investment by the adults. (In this sense, the adults avoid the risk of placing all their eggs in one basket.) **3.** If a parent's survival is compromised greatly by bearing young during times of stress, the animal's fitness may increase if it abandons its current young and survives to produce healthier young at a later time.

Concept Check 53.5

1. Three attributes are the size, quality, and isolation of patches. A patch that is larger or of higher quality is more likely to attract individuals and to be a source of individuals for other patches. A patch that is relatively isolated will undergo fewer exchanges of individuals with other patches. **2.** You would need to study the population for more than one cycle (longer than 10 years and probably at least 20) before having sufficient data to examine changes through time. Otherwise, it would be impossible to know whether an observed decrease in the population size reflected a long-term trend or was part of the normal cycle. **3.** In negative feedback, the output, or product, of a process slows that process. In populations that have a density-dependent birth rate, such as dune fescue grass, an accumulation of product (more individuals, resulting in a higher population density) slows the process (population growth) by decreasing the birth rate.

Concept Check 53.6

1. A bottom-heavy age structure, with a disproportionate number of young people, portends continuing growth of the population as these young people begin reproducing. In contrast, a more evenly distributed age structure predicts a more stable population size, and a top-heavy age structure predicts a decrease in population size because relatively fewer young people are reproducing. **2.** The growth rate of Earth's human population has dropped by half since the 1960s, from 2.2% in 1962 to 1.1% today. Nonetheless, the yearly increase in population size has not slowed as much because the smaller growth rate is counterbalanced by increased population size; hence, the number of additional people on Earth each year remains enormous—approximately 78 million. **3.** Each student will calculate his or her own ecological footprint. Each of us influences our ecological footprint by how we live—what we eat, how much energy we use, and the amount of waste we generate—as well as by how many children we have. Making choices that reduce our demand for resources makes our ecological footprint smaller.

Summary of Key Concepts Questions

53.1 Ecologists can potentially estimate birth rates by counting the number of young born each year, and they can estimate death rates by seeing how the number of adults changes each year. **53.2** Under the exponential model, both populations will continue to grow to infinite size, regardless of the specific value of r (see Figure 53.8). **53.3** There are many things you can do to increase the carrying capacity of the species, including increasing its food supply, protecting it from predators, and providing more sites for nesting or reproduction. **53.4** Ecological trade-offs are common because organisms do not have access to unlimited amounts of energy and resources. As a result, the use of energy or resources for one function (such as reproduction) can decrease the energy or resources available to support other functions (such as growth or survival). **53.5** An example of a biotic factor is disease caused by a pathogen; natural disasters, such as earthquakes and floods, are examples of abiotic factors. **53.6** Humans are unique in our potential ability to reduce global population through contraception and family planning. Humans also are capable of consciously choosing their diet and personal lifestyle, and these choices influence the number of people Earth can support.

Test Your Understanding

1. B **2.** A **3.** A **4.** D **5.** C **6.** B **7.** C **8.** A **9.** A

Chapter 54

Figure Questions

Figure 54.3 Its realized and fundamental niches would be similar, unlike those of *Chthamalus*. **Figure 54.5** Individuals of a harmless species that resembled a distantly related harmful species might be attacked by predators less often than were other individuals that did not resemble the harmful species. As a result, individuals of the harmless species that resembled a harmful species would tend to contribute more offspring to the next generation than would other individuals of the harmless species. Over time, as natural selection by predators continued to favor those individuals of the harmless species that most closely resembled the harmful species, the resemblance of the harmless species to the harmful species would increase. However, selection is not the only process that could cause a harmless species to resemble a closely related harmful species. In this case, the two species could also resemble each other because they descended from a recent common ancestor and hence share many traits (including a resemblance to one another). **Figure 54.14** An increase in the abundance of carnivores that ate zooplankton might cause zooplankton abundance to drop, thereby causing phytoplankton abundance to increase. **Figure 54.15** The number of types of organisms eaten is zero for phytoplankton; one for copepods, crab-eater seals, and baleen whales; two for krill, carnivorous plankton, elephant seals, and sperm whales; three for squids, fishes, and leopard seals; and five for birds and smaller toothed whales. The two groups that both consume and are consumed by each other are fishes and squids. **Figure 54.18** The death of individuals of *Mytilus*, a dominant species, should open up space for other species and increase species richness even in the absence of *Pisaster*. **Figure 54.24** At the earliest stages of primary succession, free-living prokaryotes in the soil would reduce atmospheric N_2 to NH_3. Symbiotic nitrogen fixation could not occur until plants were present at the site. **Figure 54.28** We would expect that (a) population sizes would decrease because there would be fewer resources and less suitable habitat; (b) the extinction curve would rise more rapidly as the number of species on the island increased because small islands generally have fewer resources, less diverse habitats, and smaller population sizes; and (c) the predicted equilibrium species number would be smaller than shown in Figure 54.28. **Figure 54.31** Shrew populations in different locations and habitats might show substantial genetic variation in their susceptibility to the Lyme pathogen. As a result, there might be fewer infected ticks where shrew populations are less susceptible to the Lyme pathogen and more infected ticks where shrews are more susceptible.

Concept Check 54.1

1. Interspecific competition has negative effects on both species (−/−). In predation, the predator population benefits at the expense of the prey population; this is an example of exploitation (+/−). Mutualism is an interaction in which both species benefit (+/+). **2.** One of the competing species will become locally extinct because of the greater reproductive success of the more efficient competitor. **3.** By specializing in eating seeds of different plant species, individuals of the two finch species may be less likely to come into contact in the separate habitats, reinforcing a reproductive barrier to hybridization.

Concept Check 54.2

1. Species richness, the number of species in the community, and relative abundance, the proportions of the community represented by the various species, both contribute to species diversity. Compared to a community with a very high proportion of one species, one with a more even proportion of species is considered more diverse. **2.** A food chain presents a set of one-way transfers of food energy up to successively higher trophic levels. A food web documents how food chains are linked together, with many species weaving into the web at more than one trophic level. **3.** According to the bottom-up model, adding extra predators would have little effect on lower trophic levels, particularly vegetation. If the top-down model applied, increased bobcat numbers would decrease raccoon numbers, increase snake numbers, decrease mouse numbers, and increase grass biomass. **4.** A decrease in krill abundance might increase the abundance of organisms that krill eat (phytoplankton and copepods), while decreasing the abundance of organisms that eat krill (baleen whales, crabeater seals, birds, fishes, and carnivorous plankton); baleen whales and crabeater seals might be particularly at risk because they only eat krill. However, many of these possible changes could lead to other changes as well, making the overall outcome hard to predict. For example, a decrease in krill abundance could cause an increase in copepod abundance—but an increase in copepod abundance could counteract some of the other effects of decreased krill abundance (since like krill, copepods eat phytoplankton and are eaten by carnivorous plankton and fishes).

Concept Check 54.3

1. High levels of disturbance are generally so disruptive that they eliminate many species from communities, leaving the community dominated by a few tolerant species. Low levels of disturbance permit competitively dominant species to exclude other species from the community. On the other hand, moderate levels of disturbance can facilitate coexistence of a greater number of species in a community by preventing competitively dominant species from becoming abundant enough to eliminate other species from the community. **2.** Early successional species can facilitate the arrival of other species in many ways, including increasing the fertility or water-holding capacity of soils or providing shelter to seedlings from wind and intense sunlight. **3.** The absence of fire for 100 years would represent a change to a low level of disturbance. According to the intermediate disturbance hypothesis, this change should cause diversity to decline as competitively dominant species gain sufficient time to exclude less competitive species.

Concept Check 54.4

1. Ecologists propose that the greater species richness of tropical regions is the result of their longer evolutionary history and the greater solar energy input and water availability in tropical regions. **2.** Immigration of species to islands declines with distance from the mainland and increases with island area. Extinction of species is lower on larger islands and on less isolated islands. Since the number of species on islands is largely determined by the difference between rates of immigration and extinction, the number of species will be highest on large islands near the mainland and lowest on small islands far from the mainland. **3.** Because of their greater mobility, birds disperse to islands more often than snakes and lizards, so birds should have greater richness.

Concept Check 54.5

1. Pathogens are microorganisms, viruses, viroids, or prions that cause disease. **2.** To keep the rabies virus out, you could ban imports of all mammals, including pets. Potentially, you could also attempt to vaccinate all dogs in the British Isles against the virus. A more practical approach might be to quarantine all pets brought into the country that are potential carriers of the disease, the approach the British government actually takes.

Summary of Key Concepts Questions

54.1 Note: Sample answers follow; other answers could also be correct. Competition: a fox and a bobcat competing for prey. Predation: an orca eating a sea otter. Herbivory: a bison grazing in a prairie. Parasitism: a parasitoid wasp that lays its eggs on a caterpillar. Mutualism: a fungus and an alga that make up a lichen. Commensalism: a wildflower that grows in a maple forest and a maple tree. **54.2** Not necessarily if the more species-rich community is dominated by only one or a few species **54.3** Similar to clear-cutting a forest or plowing a field, some species would be present initially. As a result, the disturbance would initiate secondary succession in spite of its severe appearance. **54.4** Glaciations are major disturbances that can completely destroy communities found in temperate and polar regions. As a result, tropical communities are older than temperate or polar communities. This can cause species diversity to be high in the tropics simply because there has been more time for speciation to occur. **54.5** A keystone species is one with a pivotal ecological role. Hence, a pathogen that reduces the abundance or otherwise harms a keystone species could greatly alter the structure of the community. For example, if a novel pathogen drove a keystone species to local extinction, drastic changes in species diversity could occur.

Test Your Understanding

1. D **2.** C **3.** C **4.** C **5.** B **6.** C **7.** D **8.** B
9. Community 1: $H = -(0.05 \ln 0.05 + 0.05 \ln 0.05 + 0.85 \ln 0.85 + 0.05 \ln 0.05) = 0.59$. Community 2: $H = -(0.30 \ln 0.30 + 0.40 \ln 0.40 + 0.30 \ln 0.30) = 1.1$. Community 2 is more diverse. **10.** Crab numbers should increase, reducing the abundance of eelgrass.

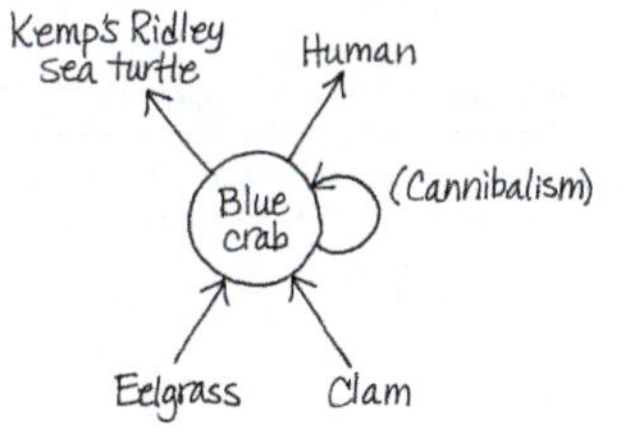

Chapter 55

Figure Questions

Figure 55.4 The blue arrow leading to *Primary consumers* could represent a grasshopper feeding on a plant. The blue arrow leading from *Primary consumers* to *Detritus* could represent the remains of a dead primary consumer (such as a grasshopper) becoming part of the detritus found in the ecosystem. The blue arrow leading from *Primary consumers* to *Secondary and tertiary consumers* could represent a bird (the secondary consumer) eating a grasshopper (the primary consumer). Finally, the blue arrow leading from *Primary consumers* to *Primary producers* could represent CO_2 released by a grasshopper in cellular respiration. **Figure 55.5** The map does not accurately reflect the productivity of wetlands, coral reefs, and coastal zones because these habitats cover areas that are too small to show up clearly on global maps. **Figure 55.6** New duck farms would add extra nitrogen and phosphorus to the water samples used in the experiment. We would expect that the extra phosphorus from these new duck farms would not alter the results (because in the original experiment, phosphorus levels were *already* so high that adding phosphorus did not increase phytoplankton growth). However, the new duck farms might increase nitrogen levels to the point where adding extra nitrogen in an experiment would not increase phytoplankton density. **Figure 55.12** The availability of water and exposure to light are other factors that may have varied across the sites. Factors such as these that are not included in the experimental design could make the results more difficult to interpret. Multiple factors can also be correlated to each other in nature, so ecologists must be careful that the factor they are studying is actually causing the observed response and is not just correlated with it.
Figure 55.13 (1) If the rate of decomposition slowed, more organic materials would be transferred from reservoir A to reservoir B; eventually, this might lead to more organic material becoming fossilized into fossil fuels. In addition, a decrease in decomposition rate would cause fewer inorganic materials to become available as nutrients in reservoir C, which would ultimately slow the rates of nutrient uptake and photosynthesis by living organisms. (2) Materials move into and out of reservoir A on a much shorter time scale than they move into reservoir B. Materials may remain in reservoir B for a very long time, or humans may remove them at a rapid pace by excavating and burning fossil fuels. **Figure 55.19** Populations evolve as organisms interact with each other and with the physical and chemical conditions of their environment. As a result, any human action that alters the environment has the potential to cause evolutionary change. In particular, since climate change has greatly affected arctic ecosystems, we would expect that climate change will cause evolution in arctic tundra populations.

Concept Check 55.1

1. Energy passes through an ecosystem, entering as sunlight and leaving as heat. It is not recycled within the ecosystem. **2.** You would need to know how much biomass the wildebeests ate from your plot and how much nitrogen was contained in that biomass. You would also need to know how much nitrogen they deposited in urine or feces. **3.** The second law states that in any energy transfer or transformation, some of the energy is dissipated to the surroundings as heat. For the ecosystem to remain intact, this "escape" of energy from the ecosystem must be offset by the continuous influx of solar radiation.

Concept Check 55.2

1. Only a fraction of solar radiation strikes plants or algae, only a portion of that fraction is of wavelengths suitable for photosynthesis, and much energy is lost as a result of reflection or heating of plant tissue. **2.** By manipulating the level of the factors of interest, such as phosphorus availability or soil moisture, and measuring responses by primary producers **3.** It is likely that NEP would decline after the fire. To see why, recall that $NEP = GPP - R_T$, where GPP is gross primary production and R_T is the total amount of cellular respiration in the ecosystem. By killing trees and other plants, the fire would cause GPP to decline from its pre-fire levels. In addition, as decomposers broke down the remains of trees killed by fire, the overall amount of cellular respiration (R_T) in the ecosystem could increase (because of increased cellular respiration by decomposers).
4. The enzyme rubisco, which catalyzes the first step in the Calvin cycle, is the most abundant protein on Earth. Like all proteins, rubisco contains nitrogen, and because photosynthetic organisms require so much rubisco, they also require considerable nitrogen to make it. Phosphorus is also needed as a component of several metabolites in the Calvin cycle and as a component of both ATP and NADPH (see Figure 10.19).

Concept Check 55.3

1. 20 J; 40% **2.** Nicotine protects the plant from herbivores. **3.** Total net primary production is $10{,}000 + 1{,}000 + 100 + 10\text{ J} = 11{,}110\text{ J}$. This is the amount of energy theoretically available to detritivores.

Concept Check 55.4

1. For example, for the carbon cycle:

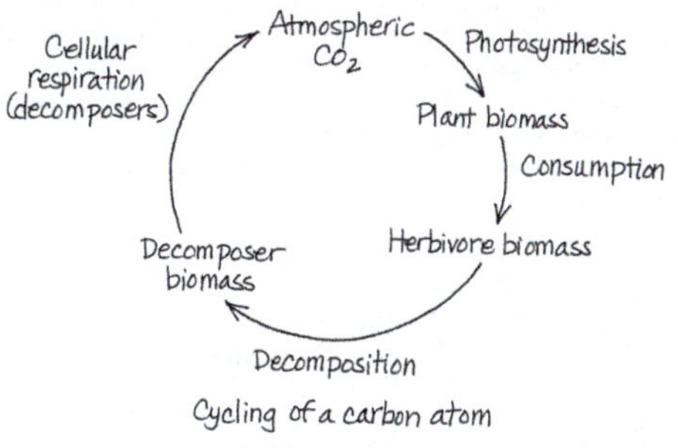

2. Removal of the trees stops nitrogen uptake from the soil, allowing nitrate to accumulate there. The nitrate is washed away by precipitation and enters the streams. **3.** Most of the nutrients in a tropical rain forest are contained in the trees, so removing the trees by logging rapidly depletes nutrients from the ecosystem. The nutrients that remain in the soil are quickly carried away into streams and groundwater by the abundant precipitation.

Concept Check 55.5

1. The main goal is to restore degraded ecosystems to a more natural state. **2.** The Kissimmee River project returns the flow of water to the original channel and restores natural flow, a self-sustaining outcome. Ecologists at the Maungatautari reserve will need to maintain the integrity of the fence indefinitely, an outcome that is not self-sustaining in the long term.

Summary of Key Concepts Questions

55.1 Because energy conversions are inefficient, with some energy inevitably lost as heat, you would expect that a given mass of primary producers would support a smaller biomass of secondary producers. **55.2** For estimates of NEP, you need to measure the respiration of all organisms in an ecosystem, not just the respiration of primary producers. In a sample of ocean water, primary producers and other organisms are usually mixed together, making their respective respirations hard to separate. **55.3** Runners use much more energy in respiration when they are running than when they are sedentary, reducing their production efficiency. **55.4** Factors other than temperature, including a shortage of water and nutrients, slow decomposition in hot deserts. **55.5** If the topsoil and deeper soil are kept separate, the engineers could return the deeper soil to the site first and then apply the more fertile topsoil to improve the success of revegetation and other restoration efforts.

Test Your Understanding

1. C **2.** B **3.** A **4.** C **5.** A **6.** B **7.** D **8.** D
9. (a)

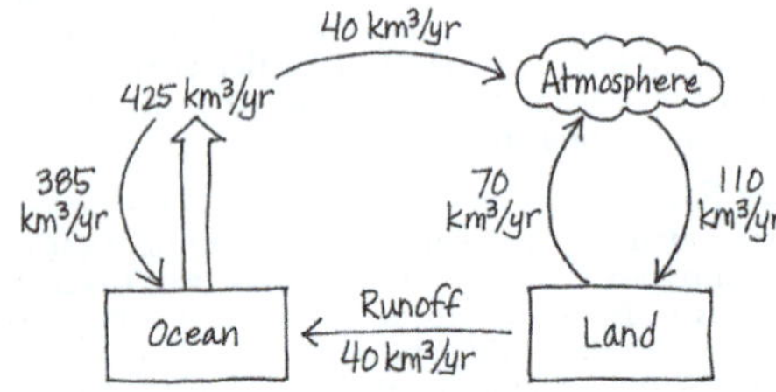

(b) On average, the ratio is 1, with equal amounts of water moving from the ocean to land as precipitation and moving from land to ocean in runoff.
(c) During an ice age, the amount of ocean evaporation falling on land as precipitation would be greater than the amount returning to the oceans in runoff; thus, the ratio would be >1. The difference would build up on land as ice.

Chapter 56

Figure Questions

Figure 56.4 You would need to know the complete range of the species and that it is missing across all of that range. You would also need to be certain that the species isn't hidden, as might be the case for an animal that is hibernating underground or a plant that is present in the form of seeds or spores. **Figure 56.9** The two examples are similar in that segments of DNA from the harvested samples were analyzed and compared with segments from specimens of known origin. One difference is that the whale researchers investigated relatedness at species and population levels to determine whether illegal activity had occurred, whereas the elephant investigators determined relatedness at the population level to determine the precise location of the poaching. Another difference is that mtDNA was used for the whale study, whereas nuclear DNA was used for the elephant study. The primary limitations of such approaches are the need to have (or generate) a reference database and the requirement that the organisms have sufficient variation in their DNA to reveal the relatedness of samples. **Figure 56.11** The higher the pH, the lower the acidity. Thus, the precipitation in this forest is becoming less acidic. **Figure 56.13** Answers may vary, but there are two reasons not to support transplanting additional birds. First, the Illinois population has a different genetic makeup than birds in other regions, and you would want to maintain to the greatest extent possible the frequency of beneficial genes or alleles found only in the Illinois population. Second, the translocation of birds from other states already caused the percentage of hatched eggs to increase dramatically, indicating that the transplantation of additional birds is not necessary. **Figure 56.15** The natural disturbance regime in this habitat includes frequent fires that clear undergrowth but do not kill mature pine trees. Without these fires, the undergrowth quickly fills in and the habitat becomes unsuitable for red-cockaded woodpeckers. **Figure 56.16** The photo shows edges between forest and grassland ecosystems, grassland and river ecosystems, and grassland and lake ecosystems. **Figure 56.25** The PCB concentration increased by a factor of 4.9 from phytoplankton to zooplankton, 41.6 from phytoplankton to smelt, 8.5 from zooplankton to smelt, 4.6 from smelt to lake trout, 119.2 from smelt to herring gull eggs, and 25.7 from lake trout to herring gull eggs. **Figure 56.30** Ocean acidification reduces the availability of carbonate ions (CO_3^{2-}). Corals and many other marine organisms require carbonate ions to build their shells. Since shell-building organisms depend upon their shells for survival, scientists have predicted that ocean acidification will cause many shell-building organisms to die. In turn, increased mortality rates of organisms that build shells would cause many other changes to ecological communities. For example, increased mortality rates of corals would harm the many other species that seek protection in coral reefs or that feed upon the species living there.

Concept Check 56.1

1. In addition to species loss, the biodiversity crisis includes the loss of genetic diversity within populations and species and the degradation of entire ecosystems. **2.** Habitat destruction, such as deforestation, channelizing of rivers, or conversion of natural ecosystems to agriculture or cities, deprives species of places to live. Introduced species, which are transported by humans to regions outside their native range, where they are not controlled by their natural pathogens or predators, often reduce the population sizes of native species through competition or predation. Overharvesting has reduced populations of plants and animals or driven them to extinction. Finally, global change is altering the environment to the extent that it reduces the capacity of Earth to sustain life. **3.** If both populations breed separately, then gene flow between the populations would not occur and genetic differences between them would be greater. As a result, the loss of genetic diversity would be greater than if the populations interbreed.

Concept Check 56.2

1. Reduced genetic variation decreases the capacity of a population to evolve in the face of change. **2.** The effective population size, N_e, would be $4(30 \times 10)/(30 + 10) = 30$ birds. **3.** Because millions of people use the greater Yellowstone ecosystem each year, it would be impossible to eliminate all contact between people and bears. Instead, you might try to reduce the kinds of encounters where bears are killed. You might recommend lower speed limits on roads in the park, adjust the timing or location of hunting seasons (where hunting is allowed outside the park) to minimize contact with mother bears and cubs, and provide financial incentives for livestock owners to try alternative means of protecting livestock, such as using guard dogs.

Concept Check 56.3

1. A small area supporting numerous endemic species as well as a large number of endangered and threatened species **2.** Zoned reserves may provide sustained supplies of forest products, water, hydroelectric power, educational opportunities, and income from tourism. **3.** Habitat corridors can increase the rate of movement or dispersal of organisms between habitat patches and thus the rate of gene flow between subpopulations. They thus help prevent a decrease in fitness attributable to inbreeding. They can also minimize interactions between organisms and humans as the organisms disperse; in cases involving potential predators, such as bears or large cats, minimizing such interactions is desirable.

Concept Check 56.4

1. Adding nutrients causes population explosions of algae and the organisms that feed on them. Increased respiration by algae and consumers, including detritivores, depletes the lake's oxygen, which the fish require. **2.** Decomposers are consumers that use nonliving organic matter as fuel for cellular respiration, which releases CO_2 as a by-product. Because higher temperatures lead to faster decomposition, organic matter in these soils could be decomposed to CO_2 more rapidly, thereby speeding up global warming. **3.** Reduced concentrations of ozone in the atmosphere increase the amount of UV radiation that reaches Earth's surface and the organisms living there. UV radiation can cause mutations by producing disruptive thymine dimers in DNA.

Concept Check 56.5

1. Sustainable development is an approach to development that works toward the long-term prosperity of human societies and the ecosystems that support them, which requires linking the biological sciences with the social sciences, economics, and humanities. **2.** Biophilia, our sense of connection to nature and all forms of life, may act as a significant motivation for the development of an environmental ethic that resolves not to allow species to become extinct or ecosystems to be destroyed. Such an ethic is necessary if we are to become more attentive and effective custodians of the environment. **3.** At a minimum, you would want to know the size of the population and the average reproductive rate of the individuals in it. To develop the fishery sustainably, you would seek a harvest rate that maintains the population near its original size and maximizes its harvest in the long term rather than the short term.

Summary of Key Concepts Questions

56.1 Nature provides us with many beneficial services, including a supply of reliable, clean water, the production of food and fiber, and the dilution and detoxification of our pollutants. **56.2** A more genetically diverse population is better able to withstand pressures from disease or environmental change, making it less likely to become extinct over a given period of time. **56.3** Habitat fragmentation can isolate populations, leading to inbreeding and genetic drift, and it can make populations more susceptible to local extinctions resulting from edge effects, including a change in physical conditions and an increase in competition or predation with edge-adapted species. **56.4** It's healthier to feed at a lower trophic level because biological magnification increases the concentration of toxins at higher levels. **56.5** One goal of conservation biology is to preserve as many species as possible. Sustainable approaches that maintain the quality of habitats are required for the long-term survival of organisms.

Test Your Understanding

1. C **2.** D **3.** B **4.** A **5.** B **6.** A
7.

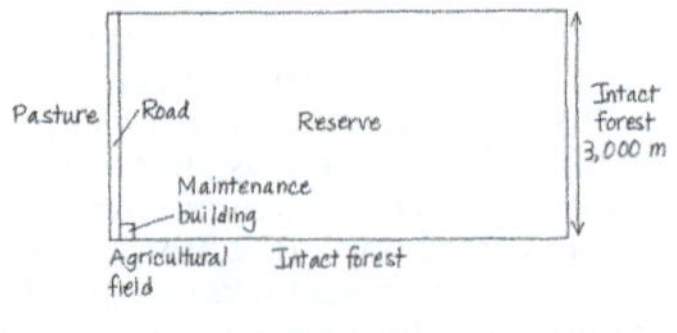

To minimize the area of forest into which the cowbirds penetrate, you should locate the road along the west edge of the reserve (since that edge abuts deforested pasture and an agricultural field). Any other location would increase the area of affected habitat. Similarly, the maintenance building should be in the southwest corner of the reserve to minimize the area susceptible to cowbirds.

APPENDIX B Periodic Table of the Elements

Atomic number (number of protons) → 6
Element symbol → C
Atomic mass (number of protons plus number of neutrons averaged over all isotopes) → 12.01

Metals (blue) | Metalloids (green) | Nonmetals (yellow)

Representative elements: Groups 1–2 (Alkali metals, Alkaline earth metals) and Groups 13–18 (Halogens, Noble gases)

Groups: Elements in a vertical column have the same number of electrons in their valence (outer) shell and thus have similar chemical properties.

Periods: Each horizontal row contains elements with the same total number of electron shells. Across each period, elements are ordered by increasing atomic number.

Transition elements: Groups 3–12

Period number	1 Group 1A (Alkali metals)	2 Group 2A (Alkaline earth metals)	3 3B	4 4B	5 5B	6 6B	7 7B	8 8B	9 8B	10 8B	11 1B	12 2B	13 Group 3A	14 Group 4A	15 Group 5A	16 Group 6A	17 Group 7A (Halogens)	18 Group 8A (Noble gases)
1	1 H 1.008																	2 He 4.003
2	3 Li 6.941	4 Be 9.012											5 B 10.81	6 C 12.01	7 N 14.01	8 O 16.00	9 F 19.00	10 Ne 20.18
3	11 Na 22.99	12 Mg 24.31											13 Al 26.98	14 Si 28.09	15 P 30.97	16 S 32.06	17 Cl 35.45	18 Ar 39.95
4	19 K 39.10	20 Ca 40.08	21 Sc 44.96	22 Ti 47.87	23 V 50.94	24 Cr 52.00	25 Mn 54.94	26 Fe 55.85	27 Co 58.93	28 Ni 58.69	29 Cu 63.55	30 Zn 65.38	31 Ga 69.72	32 Ge 72.64	33 As 74.92	34 Se 78.96	35 Br 79.90	36 Kr 83.80
5	37 Rb 85.47	38 Sr 87.62	39 Y 88.91	40 Zr 91.22	41 Nb 92.91	42 Mo 95.95	43 Tc (98)	44 Ru 101.1	45 Rh 102.9	46 Pd 106.4	47 Ag 107.9	48 Cd 112.4	49 In 114.8	50 Sn 118.7	51 Sb 121.8	52 Te 127.6	53 I 126.9	54 Xe 131.3
6	55 Cs 132.9	56 Ba 137.3	57* La 138.9	72 Hf 178.5	73 Ta 180.9	74 W 183.8	75 Re 186.2	76 Os 190.2	77 Ir 192.2	78 Pt 195.1	79 Au 197.0	80 Hg 200.6	81 Tl 204.4	82 Pb 207.2	83 Bi 209.0	84 Po (209)	85 At (210)	86 Rn (222)
7	87 Fr (223)	88 Ra (226)	89† Ac (227)	104 Rf (267)	105 Db (268)	106 Sg (271)	107 Bh (272)	108 Hs (277)	109 Mt (276)	110 Ds (281)	111 Rg (280)	112 Cn (285)	113 Nh (284)	114 Fl (289)	115 Mc (288)	116 Lv (293)	117 Ts (293)	118 Og (294)

*Lanthanides	58 Ce 140.1	59 Pr 140.9	60 Nd 144.2	61 Pm (145)	62 Sm 150.4	63 Eu 152.0	64 Gd 157.3	65 Tb 158.9	66 Dy 162.5	67 Ho 164.9	68 Er 167.3	69 Tm 168.9	70 Yb 173.0	71 Lu 175.0
†Actinides	90 Th 232.0	91 Pa 231.0	92 U 238.0	93 Np (237)	94 Pu (244)	95 Am (243)	96 Cm (247)	97 Bk (247)	98 Cf (251)	99 Es (252)	100 Fm (257)	101 Md (258)	102 No (259)	103 Lr (262)

Name (Symbol)	Atomic Number	Name (Symbol)	Atomic Number	Name (Symbol)	Atomic Number	Name (Symbol)	Atomic Number	Name (Symbol)	Atomic Number
Actinium (Ac)	89	Copper (Cu)	29	Iron (Fe)	26	Osmium (Os)	76	Silicon (Si)	14
Aluminum (Al)	13	Curium (Cm)	96	Krypton (Kr)	36	Oxygen (O)	8	Silver (Ag)	47
Americium (Am)	95	Darmstadtium (Ds)	110	Lanthanum (La)	57	Palladium (Pd)	46	Sodium (Na)	11
Antimony (Sb)	51	Dubnium (Db)	105	Lawrencium (Lr)	103	Phosphorus (P)	15	Strontium (Sr)	38
Argon (Ar)	18	Dysprosium (Dy)	66	Lead (Pb)	82	Platinum (Pt)	78	Sulfur (S)	16
Arsenic (As)	33	Einsteinium (Es)	99	Lithium (Li)	3	Plutonium (Pu)	94	Tantalum (Ta)	73
Astatine (At)	85	Erbium (Er)	68	Livermorium (Lv)	116	Polonium (Po)	84	Technetium (Tc)	43
Barium (Ba)	56	Europium (Eu)	63	Lutetium (Lu)	71	Potassium (K)	19	Tellurium (Te)	52
Berkelium (Bk)	97	Fermium (Fm)	100	Magnesium (Mg)	12	Praseodymium (Pr)	59	Tennessine (Ts)	117
Beryllium (Be)	4	Flerovium (Fl)	114	Manganese (Mn)	25	Promethium (Pm)	61	Terbium (Tb)	65
Bismuth (Bi)	83	Fluorine (F)	9	Meitnerium (Mt)	109	Protactinium (Pa)	91	Thallium (Tl)	81
Bohrium (Bh)	107	Francium (Fr)	87	Mendelevium (Md)	101	Radium (Ra)	88	Thorium (Th)	90
Boron (B)	5	Gadolinium (Gd)	64	Mercury (Hg)	80	Radon (Rn)	86	Thulium (Tm)	69
Bromine (Br)	35	Gallium (Ga)	31	Molybdenum (Mo)	42	Rhenium (Re)	75	Tin (Sn)	50
Cadmium (Cd)	48	Germanium (Ge)	32	Moscovium (Mc)	115	Rhodium (Rh)	45	Titanium (Ti)	22
Calcium (Ca)	20	Gold (Au)	79	Neodymium (Nd)	60	Roentgenium (Rg)	111	Tungsten (W)	74
Californium (Cf)	98	Hafnium (Hf)	72	Neon (Ne)	10	Rubidium (Rb)	37	Uranium (U)	92
Carbon (C)	6	Hassium (Hs)	108	Neptunium (Np)	93	Ruthenium (Ru)	44	Vanadium (V)	23
Cerium (Ce)	58	Helium (He)	2	Nickel (Ni)	28	Rutherfordium (Rf)	104	Xenon (Xe)	54
Cesium (Cs)	55	Holmium (Ho)	67	Nihonium (Nh)	113	Samarium (Sm)	62	Ytterbium (Yb)	70
Chlorine (Cl)	17	Hydrogen (H)	1	Niobium (Nb)	41	Scandium (Sc)	21	Yttrium (Y)	39
Chromium (Cr)	24	Indium (In)	49	Nitrogen (N)	7	Seaborgium (Sg)	106	Zinc (Zn)	30
Cobalt (Co)	27	Iodine (I)	53	Nobelium (No)	102	Selenium (Se)	34	Zirconium (Zr)	40
Copernicium (Cn)	112	Iridium (Ir)	77	Oganesson (Og)	118				

APPENDIX C The Metric System

Metric Prefixes:			
	10^9 = giga (G)	10^{-2} = centi (c)	10^{-9} = nano (n)
	10^6 = mega (M)	10^{-3} = milli (m)	10^{-12} = pico (p)
	10^3 = kilo (k)	10^{-6} = micro (μ)	10^{-15} = femto (f)

Measurement	Unit and Abbreviation	Metric Equivalent	Metric-to-English Conversion Factor	English-to-Metric Conversion Factor
Length	1 kilometer (km)	= 1,000 (10^3) meters	1 km = 0.62 mile	1 mile = 1.61 km
	1 meter (m)	= 100 (10^2) centimeters = 1,000 millimeters	1 m = 1.09 yards 1 m = 3.28 feet 1 m = 39.37 inches	1 yard = 0.914 m 1 foot = 0.305 m
	1 centimeter (cm)	= 0.01 (10^{-2}) meter	1 cm = 0.394 inch	1 foot = 30.5 cm 1 inch = 2.54 cm
	1 millimeter (mm)	= 0.001 (10^{-3}) meter	1 mm = 0.039 inch	
	1 micrometer (μm) (formerly micron, μ)	= 10^{-6} meter (10^{-3} mm)		
	1 nanometer (nm) (formerly millimicron, mμ)	= 10^{-9} meter (10^{-3} μm)		
	1 angstrom (Å)	= 10^{-10} meter (10^{-4} μm)		
Area	1 hectare (ha)	= 10,000 square meters	1 ha = 2.47 acres	1 acre = 0.405 ha
	1 square meter (m^2)	= 10,000 square centimeters	1 m^2 = 1.196 square yards 1 m^2 = 10.764 square feet	1 square yard = 0.8361 m^2 1 square foot = 0.0929 m^2
	1 square centimeter (cm^2)	= 100 square millimeters	1 cm^2 = 0.155 square inch	1 square inch = 6.4516 cm^2
Mass	1 metric ton (t)	= 1,000 kilograms	1 t = 1.103 tons	1 ton = 0.907 t
	1 kilogram (kg)	= 1,000 grams	1 kg = 2.205 pounds	1 pound = 0.4536 kg
	1 gram (g)	= 1,000 milligrams	1 g = 0.0353 ounce 1 g = 15.432 grains	1 ounce = 28.35 g
	1 milligram (mg)	= 10^{-3} gram	1 mg = approx. 0.015 grain	
	1 microgram (μg)	= 10^{-6} gram		
Volume (solids)	1 cubic meter (m^3)	= 1,000,000 cubic centimeters	1 m^3 = 1.308 cubic yards 1 m^3 = 35.315 cubic feet	1 cubic yard = 0.7646 m^3 1 cubic foot = 0.0283 m^3
	1 cubic centimeter (cm^3 or cc)	= 10^{-6} cubic meter	1 cm^3 = 0.061 cubic inch	1 cubic inch = 16.387 cm^3
	1 cubic millimeter (mm^3)	= 10^{-9} cubic meter = 10^{-3} cubic centimeter		
Volume (liquids and gases)	1 kiloliter (kL or kl)	= 1,000 liters	1 kL = 264.17 gallons	
	1 liter (L or l)	= 1,000 milliliters	1 L = 0.264 gallon 1 L = 1.057 quarts	1 gallon = 3.785 L 1 quart = 0.946 L
	1 milliliter (mL or ml)	= 10^{-3} liter = 1 cubic centimeter	1 mL = 0.034 fluid ounce 1 mL = approx. ¼ teaspoon 1 mL = approx. 15–16 drops (gtt.)	1 quart = 946 mL 1 pint = 473 mL 1 fluid ounce = 29.57 mL 1 teaspoon = approx. 5 mL
	1 microliter (μL or μl)	= 10^{-6} liter (10^{-3} milliliter)		
Pressure	1 megapascal (MPa)	= 1,000 kilopascals	1 MPa = 10 bars	1 bar = 0.1 MPa
	1 kilopascal (kPa)	= 1,000 pascals	1 kPa = 0.01 bar	1 bar = 100 kPa
	1 pascal (Pa)	= 1 newton/m^2 (N/m^2)	1 Pa = 1.0×10^{-5} bar	1 bar = 1.0×10^5 Pa
Time	1 second (s or sec)	= 1/60 minute		
	1 millisecond (ms or msec)	= 10^{-3} second		
Temperature	Degrees Celsius (°C) (0 K [Kelvin] = −273.15°C)		°F = 9/5°C + 32	°C = 5/9 (°F − 32)

APPENDIX E Classification of Life

This appendix presents a taxonomic classification for the major extant groups of organisms discussed in this text; not all phyla are included. The classification presented here is based on the three-domain system, which assigns the two major groups of prokaryotes, bacteria and archaea, to separate domains (with eukaryotes making up the third domain).

Various alternative classification schemes are discussed in Unit Five of the text. The taxonomic turmoil includes debates about the number and boundaries of kingdoms and about the alignment of the Linnaean classification hierarchy with the findings of modern cladistic analysis. In this review, asterisks (*) indicate currently recognized phyla thought by some systematists to be paraphyletic.

DOMAIN BACTERIA

- **Proteobacteria**
- **Chlamydia**
- **Spirochetes**
- **Cyanobacteria**
- **Gram-Positive Bacteria**

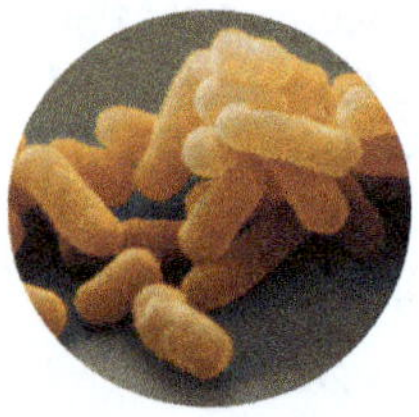

DOMAIN ARCHAEA

- **Euryarchaeota**
- **Thaumarchaeota**
- **Aigarchaeota**
- **Crenarchaeota**
- **Korarchaeota**

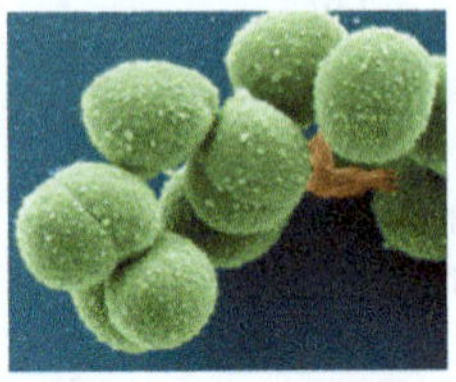

DOMAIN EUKARYA

In the phylogenetic hypothesis we present in Chapter 28, major clades of eukaryotes are grouped together in the four "supergroups" listed in bold type below and on the facing page. Formerly, all the eukaryotes generally called protists were assigned to a single kingdom, Protista. However, advances in systematics have made it clear that some protists are more closely related to plants, fungi, or animals than they are to other protists. As a result, the kingdom Protista has been abandoned.

Excavata

- Diplomonadida (diplomonads)
- Parabasala (parabasalids)
- Euglenozoa (euglenozoans)
  - Kinetoplastida (kinetoplastids)
  - Euglenophyta (euglenids)

SAR

- Stramenopila (stramenopiles)
  - Chrysophyta (golden algae)
  - Phaeophyta (brown algae)
  - Bacillariophyta (diatoms)

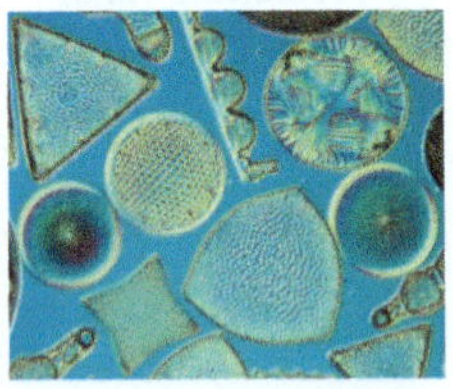

- Alveolata (alveolates)
  - Dinoflagellata (dinoflagellates)
  - Apicomplexa (apicomplexans)
  - Ciliophora (ciliates)
- Rhizaria (rhizarians)
  - Radiolaria (radiolarians)
  - Foraminifera (forams)
  - Cercozoa (cercozoans)

Archaeplastida

- Rhodophyta (red algae)
- Chlorophyta (green algae: chlorophytes)
- Charophyta (green algae: charophytes)
- Plantae
  - Nonvascular plants (bryophytes):
    - Phylum Hepatophyta (liverworts)
    - Phylum Bryophyta (mosses)
    - Phylum Anthocerophyta (hornworts)
  - Seedless vascular plants:
    - Phylum Lycophyta (lycophytes)
    - Phylum Monilophyta (ferns, horsetails, whisk ferns)
  - Seed plants:
    - Gymnosperms:
      - Phylum Ginkgophyta (ginkgo)
      - Phylum Cycadophyta (cycads)
      - Phylum Gnetophyta (gnetophytes)
      - Phylum Coniferophyta (conifers)
    - Angiosperms:
      - Phylum Anthophyta (flowering plants)

DOMAIN EUKARYA, continued

Unikonta

- Amoebozoa (amoebozoans)
  - Myxogastrida (plasmodial slime molds)
  - Dictyostelida (cellular slime molds)
  - Tubulinea (tubulinids)
  - Entamoeba (entamoebas)
- Nucleariida (nucleariids)
- Fungi
  - *Phylum Chytridiomycota (chytrids)
  - *Phylum Zygomycota (zygomycetes)
  - Phylum Glomeromycota (glomeromycetes)
  - Phylum Ascomycota (ascomycetes)
  - Phylum Basidiomycota (basidiomycetes)

- Choanoflagellata (choanoflagellates)
- Animalia
  - Phylum Porifera (sponges)
  - Phylum Ctenophora (comb jellies)
  - Phylum Cnidaria (cnidarians)
    - Medusozoa (hydrozoans, jellies, box jellies)
    - Anthozoa (sea anemones and most corals)
  - Phylum Acoela (acoel flatworms)
  - Phylum Placozoa (placozoans)
  - Lophotrochozoa (lophotrochozoans)
    - Phylum Platyhelminthes (flatworms)
      - Catenulida (chain worms)
      - Rhabditophora (planarians, flukes, tapeworms)
    - Phylum Nemertea (proboscis worms)
    - Phylum Ectoprocta (ectoprocts)
    - Phylum Brachiopoda (brachiopods)
    - Phylum Syndermata (rotifers and spiny-headed worms)
    - Phylum Cycliophora (cycliophorans)
    - Phylum Mollusca (molluscs)
      - Polyplacophora (chitons)
      - Gastropoda (gastropods)
      - Bivalvia (bivalves)
      - Cephalopoda (cephalopods)
    - Phylum Annelida (segmented worms)
      - Errantia (errantians)
      - Sedentaria (sedentarians)
  - Ecdysozoa (ecdysozoans)
    - Phylum Loricifera (loriciferans)
    - Phylum Priapula (priapulans)
    - Phylum Nematoda (roundworms)
    - Phylum Arthropoda (This survey groups arthropods into a single phylum, but some zoologists now split the arthropods into multiple phyla.)
      - Chelicerata (horseshoe crabs, arachnids)
      - Myriapoda (millipedes, centipedes)
      - Pancrustacea (crustaceans, insects)
    - Phylum Tardigrada (tardigrades)
    - Phylum Onychophora (velvet worms)
  - Deuterostomia (deuterostomes)
    - Phylum Hemichordata (hemichordates)
    - Phylum Echinodermata (echinoderms)
      - Asteroidea (sea stars, sea daisies)
      - Ophiuroidea (brittle stars)
      - Echinoides (sea urchins, sand dollars)
      - Crinoidea (sea lilies)
      - Holothuroidea (sea cucumbers)
    - Phylum Chordata (chordates)
      - Cephalochordata (cephalochordates: lancelets)
      - Urochordata (urochordates: tunicates)
      - Vertebrates:
        - Cyclostomata (cyclostomes)
          - Myxini (hagfishes)
          - Petromyzontida (lampreys)
        - Gnathostomata (gnathostomes)
          - Chondrichthyes (sharks, rays, chimaeras)
          - Actinopterygii (ray-finned fishes)
          - Actinistia (coelacanths)
          - Dipnoi (lungfishes)
          - Amphibia (amphibians: frogs, salamanders, caecilians)
          - Reptilia (reptiles: tuataras, lizards, snakes, turtles, crocodilians, birds)
          - Mammalia (mammals)

APPENDIX **D**

A Comparison of the Light Microscope and the Electron Microscope

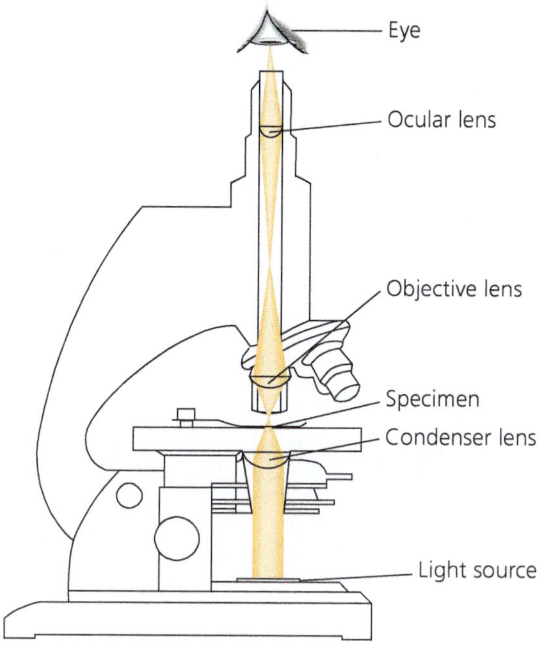

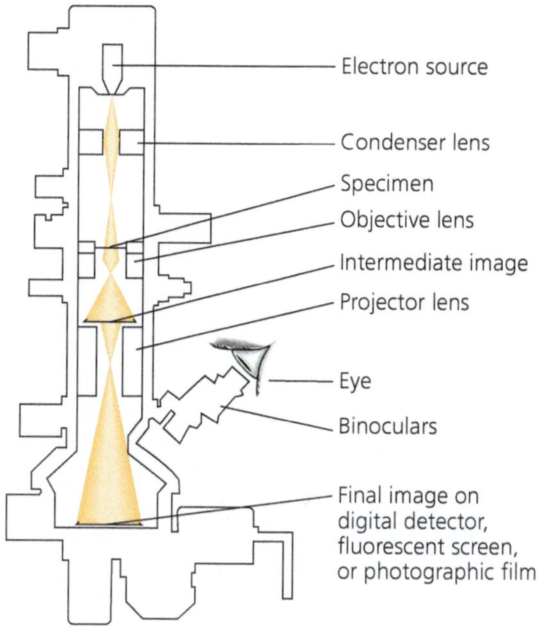

Light Microscope

In light microscopy, light is focused on a specimen by a glass condenser lens; the image is then magnified by an objective lens and an ocular lens for projection on the eye, digital camera, digital video camera, or photographic film.

Electron Microscope

In electron microscopy, a beam of electrons (top of the microscope) is used instead of light, and electromagnets are used instead of glass lenses. The electron beam is focused on the specimen by a condenser lens; the image is magnified by an objective lens and a projector lens for projection on a digital detector, fluorescent screen, or photographic film.

APPENDIX F Scientific Skills Review

Graphs

Graphs provide a visual representation of numerical data. They may reveal patterns or trends in the data that are not easy to recognize in a table. A graph is a diagram that shows how one variable in a data set is related (or perhaps not related) to another variable. The **independent variable** is the factor that is manipulated or changed by the researchers. The **dependent variable** is the factor that the researchers are measuring in relationship to the independent variable. The independent variable is typically plotted on the *x*-axis and the dependent variable on the *y*-axis. Types of graphs that are frequently used in biology include scatter plots, line graphs, bar graphs, and histograms.

➤ A **scatter plot** is used when the data for all variables are numerical and continuous. Each piece of data is represented by a point. In a **line graph**, each data point is connected to the next point in the data set with a straight line, as in the graph to the right. (To practice making and interpreting scatter plots and line graphs, see the Scientific Skills Exercises in Chapters 2, 3, 7, 8, 10, 13, 19, 24, 34, 43, 47, 49, 50, 52, 54, and 56.)

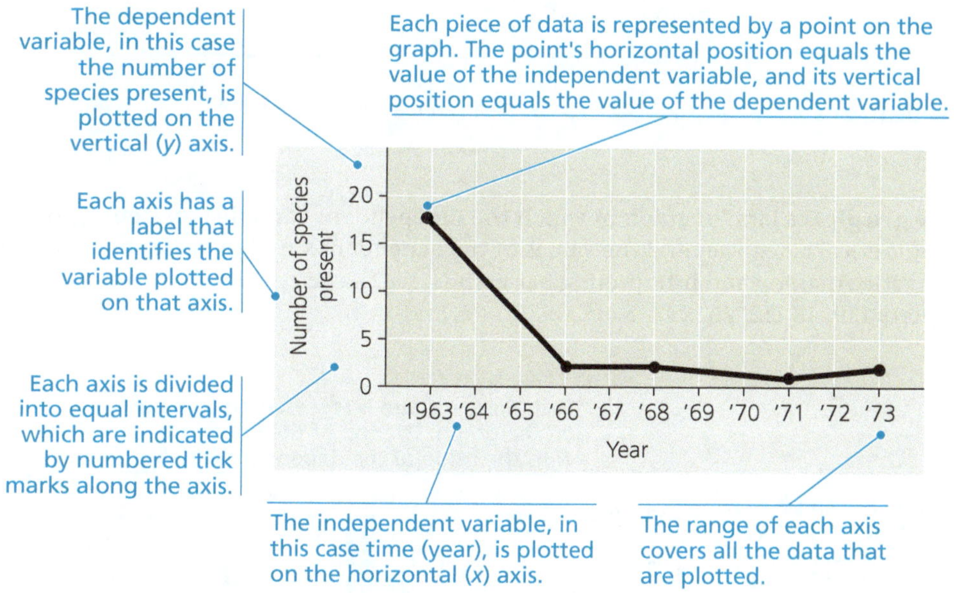

▼ Two or more data sets can be plotted on the same line graph to show how two dependent variables are related to the same independent variable. (To practice making and interpreting line graphs with two or more data sets, see the Scientific Skills Exercises in Chapters 7, 43, 47, 49, 50, 52, and 56.)

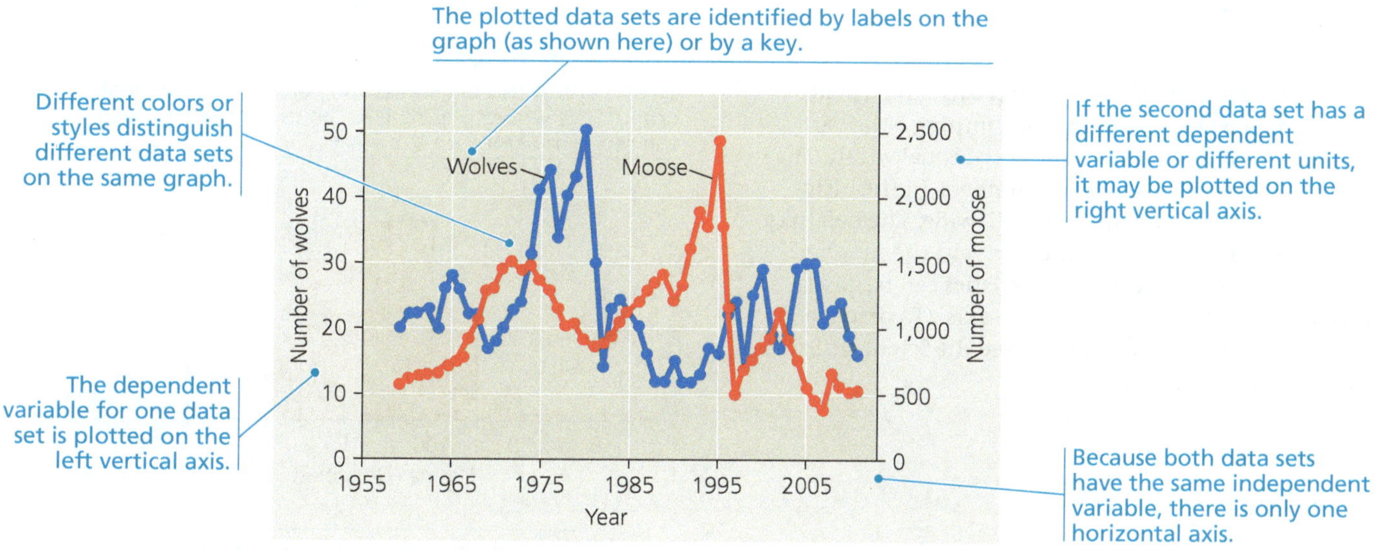

▼ In some scatter plot graphs, a straight or curved line is drawn through the entire data set to show the general trend in the data. A straight line that mathematically best fits the data is called a *regression line*. Alternatively, a mathematical function that best fits the data may describe a curved line, often termed a *best-fit curve*. (To practice making and interpreting regression lines, see the Scientific Skills Exercises in Chapters 3, 10, and 34.)

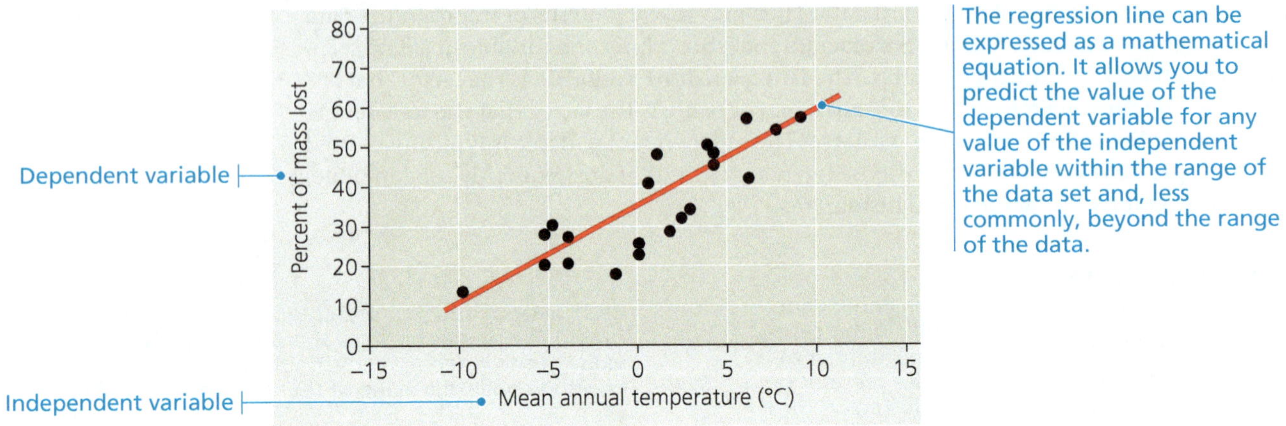

▼ A **bar graph** is a kind of graph in which the independent variable represents groups or nonnumerical categories and the values of the dependent variable(s) are shown by bars. (To practice making and interpreting bar graphs, see the Scientific Skills Exercises in Chapters 1, 9, 18, 22, 25, 29, 33, 35, 39, 51, 52, and 54.)

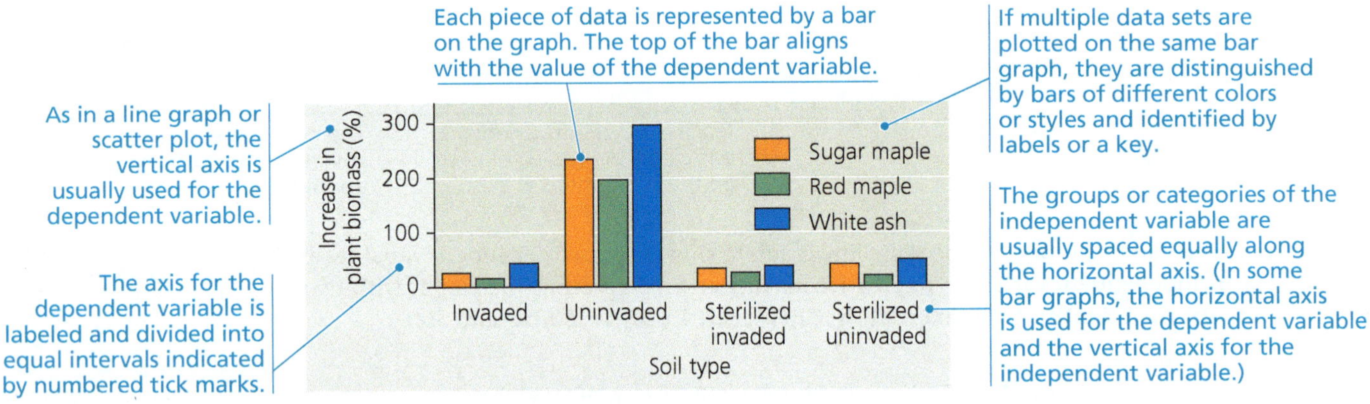

➤ A variant of a bar graph called a **histogram** can be made for numeric data by first grouping, or "binning," the variable plotted on the *x*-axis into intervals of equal width. The "bins" may be integers or ranges of numbers. In the histogram at right, the intervals are 25 mg/dL wide. The height of each bar shows the percent (or, alternatively, the number) of experimental subjects whose characteristics can be described by one of the intervals plotted on the *x*-axis. (To practice making and interpreting histograms, see the Scientific Skills Exercises in Chapters 12, 14, and 42.)

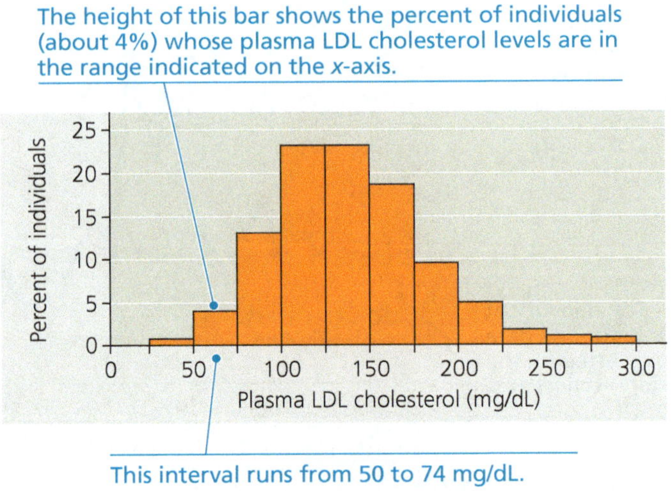

Glossary of Scientific Inquiry Terms

See Concept 1.3 for more discussion of the process of scientific inquiry.

control group In a controlled experiment, a set of subjects that lacks (or does not receive) the specific factor being tested. Ideally, the control group should be identical to the experimental group in other respects.

controlled experiment An experiment designed to compare an experimental group with a control group; ideally, the two groups differ only in the factor being tested.

data Recorded observations.

deductive reasoning A type of logic in which specific results are predicted from a general premise.

dependent variable A factor whose value is measured during an experiment to see whether it is influenced by changes in another factor (the independent variable).

experiment A scientific test. Often carried out under controlled conditions that involve manipulating one factor in a system in order to see the effects of changing that factor.

experimental group A set of subjects that has (or receives) the specific factor being tested in a controlled experiment. Ideally, the experimental group should be identical to the control group for all other factors.

hypothesis A testable explanation for a set of observations based on the available data and guided by inductive reasoning. A hypothesis is narrower in scope than a theory.

independent variable A factor whose value is manipulated or changed during an experiment to reveal possible effects on another factor (the dependent variable).

inductive reasoning A type of logic in which generalizations are based on a large number of specific observations.

inquiry The search for information and explanation, often focusing on specific questions.

model A physical or conceptual representation of a natural phenomenon.

prediction In deductive reasoning, a forecast that follows logically from a hypothesis. By testing predictions, experiments may allow certain hypotheses to be rejected.

theory An explanation that is broader in scope than a hypothesis, generates new hypotheses, and is supported by a large body of evidence.

variable A factor that varies during an experiment.

Chi-Square (χ^2) Distribution Table

To use the table, find the row that corresponds to the degrees of freedom in your data set. (The degrees of freedom is the number of categories of data minus 1.) Move along that row to the pair of values that your calculated χ^2 value lies between. Move up from those numbers to the probabilities at the top of the columns to find the probability range for your χ^2 value. A probability of 0.05 or less is generally considered significant. (To practice using the chi-square test, see the Scientific Skills Exercise in Chapter 15.)

Degrees of Freedom (df)	Probability										
	0.95	0.90	0.80	0.70	0.50	0.30	0.20	0.10	0.05	0.01	0.001
1	0.004	0.02	0.06	0.15	0.45	1.07	1.64	2.71	3.84	6.64	10.83
2	0.10	0.21	0.45	0.71	1.39	2.41	3.22	4.61	5.99	9.21	13.82
3	0.35	0.58	1.01	1.42	2.37	3.66	4.64	6.25	7.82	11.34	16.27
4	0.71	1.06	1.65	2.19	3.36	4.88	5.99	7.78	9.49	13.28	18.47
5	1.15	1.61	2.34	3.00	4.35	6.06	7.29	9.24	11.07	15.09	20.52
6	1.64	2.20	3.07	3.83	5.35	7.23	8.56	10.64	12.59	16.81	22.46
7	2.17	2.83	3.82	4.67	6.35	8.38	9.80	12.02	14.07	18.48	24.32
8	2.73	3.49	4.59	5.53	7.34	9.52	11.03	13.36	15.51	20.09	26.12
9	3.33	4.17	5.38	6.39	8.34	10.66	12.24	14.68	16.92	21.67	27.88
10	3.94	4.87	6.18	7.27	9.34	11.78	13.44	15.99	18.31	23.21	29.59

Mean and Standard Deviation

The **mean** is the sum of all data points in a data set divided by the number of data points. The mean (or average) represents a "typical" or central value around which the data points are clustered. The mean of a variable x (denoted by $\bar{x}$) is calculated from the following equation:

$$\bar{x} = \frac{1}{n}\sum_{i=1}^{n} x_i$$

In this formula, n is the number of observations, and x_i is the value of the ith observation of variable x; the "Σ" symbol indicates that the n values of x_i are to be summed. (To practice calculating the mean, see the Scientific Skills Exercises in Chapters 27, 32, and 34.)

The **standard deviation** provides a measure of the variation found in a set of data points. The standard deviation of a variable x (denoted s_x) is calculated from the following equation:

$$s = \sqrt{\frac{\sum_{i=1}^{n}(x_i - \bar{x})^2}{n-1}}$$

In this formula, n is the number of observations, x_i is the value of the ith observation of variable x, and $\bar{x}$ is the mean of x; the "Σ" symbol indicates that the n values of $(x_i - \bar{x})^2$ are to be summed. (To practice calculating standard deviation, see the Scientific Skills Exercises in Chapters 27, 32, and 34.)

Credits

Photo Credits

p. i and p. iv sunflower Radius Images/Getty Images; **p. vii 23.12 top** Kristin Stanford, Stone Laboratory, Ohio State University, Columbus; **23.12 bottom** Kent Bekker/United States Fish and Wildlife Service; **p. vii 34.53** From: *Homo naledi*, a new species of the genus *Homo* from the Dinaledi Chamber, South Africa. L. R. Berger et al. eLife 2015;4:e09560. Figs. 6 and 9.; **p. viii 55.8 fire** A. T. Willett/Alamy Stock Photo; **p. ix sunflower** Radius Images/Getty Images; **p. ix students by board** Rebecca Orr; **student at tables** Pearson Education; **p. x moth** William Mullins/Alamy Stock Photo; **larva** Reinaldo Aguilar, Vascular Plants of the Osa Peninsula, Costa Rica; **phone** Nik Merkulov/Shutterstock; **p. xi ant** John Cancalosi/Nature Picture Library; **plant 17.7a** Keith V Wood; **pig 17.7b** Simon Lin/AP Images; **p. xiii tablet** Designsstock/Shutterstock; **pondweed 2.17** Nigel Cattlin/Science Source; **p. xiv red blood cell 23.18** Eye of Science/Science Source; **sickle cell patient** Caroline Penn/Alamy Stock Photo; **p. xv mosquito** Kletr/Shutterstock; **p. xvi 7 Scientific Skills Exercise guinea pigs** Photo Fun/Shutterstock; **p. xvii price marker** ABC News Video; **p. xviii molecular model and DNA** Pearson Education; **laptop art** jazz/Shutterstock; **flower with flies 38.1** Ch'ien Lee; **eel 54.1** Jeremy Brown/123RF; **p. xxvi Lovell Jones** Lovell A. Jones; **Elba Serrano** Darren Phillips/New Mexico State University; **Shirley Tilghman** Denise Applewhite, Office of Communications, Princeton University; **Jack Szostak** Li Huang; **Nancy Moran** Courtesy of Nancy Moran, University of Texas, Austin; **Philip Benfey** Jie Huang; **Harald zur Hausen** DKFZ (German Cancer Research Center, Heidelberg)/T. Schwerdt; **Tracy Langkilde** Patrick Mansell/Penn State; **p. xxxii 1.17a hawk** Steve Byland/Fotolia; **guillemots** Erni/Fotolia; **p. xxxiii 5.21b** Photo Researchers, Inc./Science Source; ***Parmecium*** M. I. Walker/Science Source; **p. xxxiv top left 8.17** Jack Dykinga/Nature Picture Library; **bottom right** George Grall Agency/National Geographic Creative/Alamy Stock Photo; **p. xxxv left** Jane Stout and Claire Walczak, Indiana University, Winner of the GE Healthcare Life Sciences' 2012 Cell Imaging Competition; **right** Don W. Fawcett/Science Source; **p. xxxvi left 15.3** Martin Shields/Alamy Stock Photo; **right 17.1** ANGELO CUCCA/AFP/Getty Images; **p. xxxvii top left 19.10** Lei Sun, Richard J. Kuhn and Michael G. Rossmann, Purdue University, West Lafayette; **center** Image Quest Marine; **bottom** WaterFrame/Alamy Stock Photo; **p. xxxviii left 22.1** William Mullins/Alamy Stock Photo; **right** Karin Duthie/Alamy Stock Photo; **p. xxxix left 26.1** Trapp/blickwinkel/Alamy Stock Photo; **right 28.10** Steve Gschmeissner/Science Source; **p. xl left** Glam/Shutterstock; **top** Matthijs Wetterauw/Alamy Stock Photo; **right** Nigel Downer/NHPA/Photoshot; **p. xli left 33.1** Paul Anthony Stewart; **right 35.4** Dana Tezarr/Photodisc/GettyImages; **p. xlii left 37.16** Chris Mattison/Nature Picture Library; **right 38.12** California Department of Food and Agriculture's Plant Health and Pest Prevention Services; **p. xliii 42.1** John Cancalosi/Alamy Stock Photo; **p. xliv left** Juergen Berger/Science Source; **right 45.11** Cathy Keifer/123RF; **p. xlv top 46.1** Auscape/UIG/Getty Images; **bottom 47.13** Alejandro Díaz Díez/AGE Fotostock/Alamy Stock Photo; **p. xlvi left** Ivan Kuzmin/Alamy Stock Photo; **top 52.18** Sylvain Oliveira/Alamy Stock Photo; **right 53.9** Villiers Steyn/Shutterstock; **p. xlvii top 54.1** Jeremy Brown/123RF; **bottom** Eerika Schultz

Chapter 1 **1.1** J. B. Miller/Florida Park Service; **p. 2 bottom** Shawn P. Carey/Migration Productions; **1.2 top left** John Foxx/Image State Media Partners; **top center** R. Dirscherl/OceanPhoto/Frank Lane Picture Agency; **top right** Joe McDonald/Encyclopedia/Corbis; **1.2 bottom left** Toshiaki Ono/AmanaImages Inc./Alamy Stock Photo; **bottom left center** Frederic Didillon/Garden Picture Library/Getty Images; **bottom right center** Maximilian Weinzierl/Alamy Stock Photo; **bottom right** Malcolm Schuyl/Frank Lane Picture Agency; **1.3 biosphere** Leonello Calvetti/Stocktrek Images/Getty Images; **ecosystems** Terry Donnelly/Alamy Stock Photo; **communities; populations** Floris van Breugel/Nature Picture Library; **organisms** Greg Vaughn/Alamy Stock Photo; **organs** Pat Burner/Pearson Education; **tissue** Science Source; **cell** Andreas Holzenburg/Stanislav Vitha, Dept. of Biology and Microscopy, Imaging Center, Texas A&M University, College Station; **organelles** Jeremy Burgess/Science Source; **p. 6 bottom left** Survivalphotos/Alamy Stock Photo; **1.4 top** Steve Gschmeissner/Science Source; **bottom** A. Barry Dowsett/Science Source; **1.5 left, right** Conly L. Rieder; **1.6 baby** Gelpi/Fotolia; **1.7a** Photodisc/Getty Images; **1.8a top** Carol Yepes/Moment/Getty Images; **bottom** Ralf Dahm/Max Planck Institute of Neurobiology; **1.11 elephant** James Balog/Aurora/Getty Images; **1.12** Rod Williams/Nature Picture Library; **1.13a, 1.13b** Eye of Science/Science Source; **1.13c left** Kunst Scheidulin/AGE Fotostock; **left center** Daksel/Fotolia; **right center** Anup Shah/Nature Picture Library; **right** M. I. Walker/Science Source; **1.14 top left** Basel101658/Shutterstock; **bottom left** SPL/Science Source; **center** W. L. Dentler/Biological Photo Service; **right** Omikron/Science Source; **1.15** Dede Randrianarisata/Macalester College; **1.16 left** ARCHIV/Science Source; **right** Science Source; **1.17 hawk** Steve Byland/Fotolia; **robin** Sebastian Knight/Shutterstock; **flamingo** Zhaoyan/Shutterstock; **penguin** Volodymyr Goinyk/Shutterstock; **1.19** Frank Greenaway/Dorling Kindersley, Ltd; **1.21** Karl Ammann/Terra/Corbis; **inset** Tim Ridley/Dorling Kindersley, Ltd; **1.23 top** Martin Shields/Alamy Stock Photo; **center** xPACIFICA/The Image Bank/Getty Images; **bottom right** Maureen Spuhler/Pearson Education; **bottom left** All Canada Photos/Alamy Stock Photo; **1.24 beach** From: Darwin to DNA: The Genetic Basis of Color Adaptations. In Losos, J. *In the Light of Evolution: Essays from the Laboratory and Field*, Roberts and Co. Photo by Sacha Vignieri; **white mouse** Hopi Hoekstra, Harvard University; **inland** Shawn P. Carey/Migration Productions; **inland mouse** Vignieri Sacha; **1.25 left, left center, right center, right** From: The selective advantage of cryptic coloration in mice. Vignieri, S. N., J. Larson, and H. E. Hoekstra. 2010. *Evolution* 64:2153–2158. Fig. 1; **1 Scientific Skills Exercise** Imagebroker/Frank Lane Picture Agency; **1.26** Jay Janner/Austin American-Statesman/AP Images; **p. 25 top left** Photodisc/Getty Images; **p. 25 top right** James Balog/Aurora/Getty Images; **p. 26** Chris Mattison/Alamy Stock Photo

Unit One Interview Lovell A. Jones

Chapter 2 **2.1** Paul Quagliana/Bournemouth News & Picture Service; **p. 28 bottom** Paul Quagliana/Bournemouth News & Picture Service; **2.2 left** Chip Clark; **center, right** Stephen Frisch/Pearson Education; **2.3 left** C. Michael Hogan; **right** Rick York/California Native Plant Society; **bottom** Andrew Alden; **2.5** National Library of Medicine; **2 Scientific Skills Exercise** Pascal Goetgheluck/Science Source; **2.13 left** Stephen Frisch/Pearson Education; **p. 39** Martin Harvey/Photolibrary/Getty Images; **2.17** Nigel Cattlin/Science Source; **p. 43 top** Rolf Nussbaumer/Nature Picture Library; **p. 43 bottom** Thomas Eisner

Chapter 3 **3.1** Jeff Schmaltz/MODIS Rapid Response Team/NASA; **p. 44 bottom** Erni/Fotolia; **3.3 center** N.C. Brown Center for Ultrastructure Studies, SUNY, Syracuse; **3.4** Alasdair James/E+/Getty Images; **3.6 center** Jan van Franeker, Alfred Wegener Institute fur Polar und Meeresforschung, Germany; **3.10** NASA/JPL-Caltech/University of Arizona; **3.11 lemon** Paulista/Fotolia; **cola** Fotofermer/Fotolia; **blood cells** SCIEPRO/SPL/AGE Fotostock; **bleach** Beth Van Trees/Shutterstock; **3 Scientific Skills Exercise** Vlad61/Shutterstock; **p. 55** Eric Guilloret/Science Source

Chapter 4 **4.1** Florian Möllers/Nature Picture Library; **4 Scientific Skills Exercise** The Register of Stanley Miller Papers (Laboratory Notebook 2, page 114, Serial number 655, MSS642, Box 122), Mandeville Special Collections Library, UC San Diego.; **4 Scientific Skills Exercise inset** Jeffrey Bada, Scripps Institution of Oceanography, University of California, San Diego; **4.6 left** David M. Phillips/Science Source; **p. 65** George Sanker/Nature Picture Library

Chapter 5 **5.1** Mark J. Winter/Science Source; **p. 66 bottom** T. Naeser, Patrick Cramer Laboratory, Gene Center Munich, Ludwig-Maximilians-Universität München, Munich, Germany; **5.6 top left** Dougal Waters/Photodisc/Getty Images; **5.6a** Biological Photo Service; **5.6b** Paul B. Lazarow; **5.6c** Biophoto Associates/Science Source; **5.6 bottom left** John Durham/Science Source; **5.8 left** blickwinkel/Alamy Stock Photo; **5.10a** Dorling Kindersley, Ltd; **5.10b** David Murray/Dorling Kindersley, Ltd; **5.13 egg** Andrey Stratilatov/Shutterstock; **muscle; tissue** Nina Zanetti/Pearson Education; **5.16 right** Pearson Education; **5.17** Peter M. Colman; **p. 80** Dieter Hopf/Imagebroker/AGE Fotostock; **p. 81** SCIEPRO/SPL/AGE Fotostock; **5.19 top** Eye of Science/Science Source; **bottom** Eye of Science/Science Source; **5.21a top** CC-BY-3.0 Photo by Dsrjsr/Jane Shelby Richardson, Duke University; **5.21b** Laguna Design/Science Source; **5.25** P. Morris/Garvan Institute of Medical Research; **5.26 DNA** Alfred Pasieka/Science Source; **Neanderthal** Viktor Deak; **consultation** BSIP SA/Alamy Stock Photo; **plant** David Read, Department of Animal and Plant Sciences, University of Sheffield, UK; **whale** WaterFrame/Alamy Stock Photo; **elephants** ImageBroker/Frank Lane Picture Agency; **hippopotamus** Frontline Photography/Alamy Stock Photo; **p. 89 bottom** ABC News Video; **5 Scientific Skills Exercise left** lanych/Shutterstock; **center** David Bagnall/Alamy Stock Photo; **right** Eric Isselee/Shutterstock; **p. 90 butter** Dorling Kindersley, Ltd.; **oil** David Murray/Dorling Kindersley, Ltd.; **p. 91** Africa Studio/Shutterstock

Unit Two Interview **top** Darren Phillips, New Mexico State University; **bottom** Elba Serrano

Chapter 6 **6.1** Don W. Fawcett/Science Source; **p. 93 bottom** M. I. Walker/Science Source; **6.3 fluorescence** Michael W. Davidson/The Florida State University Research Foundation; **confocal** Karl Garsha; **deconvolution** Data courtesy of James G. Evans, Whitehead Institute, MIT, Boston and Hans van der Voort SVI.; **super-resolution** From: STED microscopy reveals that synaptotagmin remains clustered after synaptic vesicle exocytosis. Katrin I. Willig, Silvio O. Rizzoli, Volker Westphal, Reinhard Jahn & Stefan W. Hell. *Nature*, 440 (13) Apr 2006. Fig. 1d.; **SEM** J.L. Carson Custom Medical Stock Photo/Newscom; **brightfield, phase-contrast, DIC** Elisabeth Pierson, Pearson Education; **TEM top** William Dentler/Biological Photo Service; **TEM bottom** CNRI/Science Source; **6.6 top left** Daniel S. Friend; **6 Scientific Skills Exercise** Kelly Tatchell; **6.8 animal cell bottom left** S. Cinti/Science Source; **fungal cell** SPL/Science Source; **p. 100 cell bottom right** A. Barry Dowsett/Science Source; **6.8 plant cell bottom left** Biophoto Associates/Science Source; **unicellular eukaryotes** SPL/Science Source; **p.101 cell bottom right** Flagellar microtubule dynamics in *Chlamydomonas*: cytochalasin D induces periods of microtubule shortening and elongation; and colchicine induces disassembly of the distal, but not proximal, half of the flagellum, W. L. Dentler, C. Adams. *J Cell Biol.* 1992 Jun;117(6):1289-98. Fig. 10d.; **p. 102** Thomas Deerinck/Mark Ellisman/NCMIR; **6.9 top left** Reproduced with permission from *Freeze-Etch Histology*, by L. Orci and A. Perrelet, Springer-Verlag, Heidelberg, 1975.; Plate 25, page 53. © 1975 by Springer-Verlag GmbH & Co KG; **left center** Don W Fawcett/Science Source; **center** Ueli Aebi; **6.10 bottom left** Don W. Fawcett/Science Source; **bottom right** Harry Noller; **6.11 bottom** R. W. Bolender; Don W. Fawcett/Science Source; **6.12 right** Don W. Fawcett/Science Source; **6.13a; 6.13b** Daniel S. Friend; **6.14 bottom** Eldon H. Newcomb; **6.17a right** Daniel S. Friend; **6.17b** From: The shape of mitochondria and the number of mitochondrial nucleoids during the cell cycle of *Euglena gracilis*. Y. Hayashi and K. Ueda. *Journal of Cell Science*, 93:565–570, Fig. 3. © 1989 by Company of Biologists; **6.18a right** Jeremy Burgess/Mary Martin/Science Source; **6.18b** Franz Golig/Philipps University, Marburg, Germany; **6.19** Eldon H. Newcomb; **6.20** Albert Tousson; **6.21b** Bruce J. Schnapp; **Table 6.1 left** Mary Osborn; **center** Frank Solomon; **right** Mark Ladinsky; **6.22 bottom** Kent L. McDonald; **6.23a** Biophoto Associates/Science Source; **6.23b** Oliver Meckes/Nicole Ottawa/Science Source; **6.24a** OMIKRON/Science Source; **6.24b** Dartmouth College Electron Microscope Facility; **6.24c** From: Functional protofilament numbering of ciliary, flagellar, and centriolar microtubules. R. W. Linck, R. E. Stephens. *Cell Motil Cytoskeleton.* 2007 Jul; 64(7):489–95, Fig. 1B.; **6.25** From: Cross-linker system between neurofilaments, microtubules, and membranous organelles in frog axons revealed by the quick-freeze, deep-etching method. Hirokawa Nobutaka. *Journal of Cell Biology* 94(1): 129–142, 1982. Reproduced by permission of Rockefeller University Press.; **6.26a** Clara Franzini-Armstrong/University of Pennsylvania; **6.26b** M. I. Walker/Science Source;

6.26c Michael Clayton/University of Wisconsin; **6.27** G. F. Leedale/Biophoto Associates/Science Source; **6.29** Wm. P. Wergin, courtesy of Eldon H. Newcomb; **6.30 right** Reproduced with permission from *Freeze-Etch Histology*, by L. Orci and A. Perrelet, Springer-Verlag, Heidelberg, 1975. Plate 32. Page 68. © 1975 by Springer-Verlag GmbH & Co KG; **center** From: Fine structure of desmosomes., hemidesmosomes, and an adepidermal globular layer in developing newt epidermis. DE Kelly. *Journal of Cell Biology* 1966 Jan; 28(1):51–72. Fig. 7.; **bottom** From: Low resistance junctions in crayfish. Structural changes with functional uncoupling. C. Peracchia and A. F. Dulhunty, *The Journal of Cell Biology*. 1976 Aug; 70(2 pt 1):419–39. Fig. 6. Reproduced by permission of Rockefeller University Press.; **6.31** Eye of Science/Science Source; **p. 124 vacuole** Eldon H. Newcomb; **p. 125 peroxisome** Eldon H. Newcomb; **p. 125 bottom** Susumu Nishinaga/Science Source

Chapter 7 **7.1** Bert L. de Groot; **p. 126 bottom** Crystal structure of a mammalian voltage-dependent Shaker family K^+ channel. S. B. Long, et al. *Science*. 2005 Aug 5; 309 (5736):897–903. Epub 2005 Jul 7; **p. 129 bottom** camerawithlegs/Fotolia; **7.13** Michael Abbey/Science Source; **7 Scientific Skills Exercise** teddies Photo Fun/Shutterstock; **7.19 amoeba** Biophoto Associates/Science Source; **vesicles** Don W. Fawcett/Science Source; **coated pit; coated vesicle** From: M.M. Perry and A.B. Gilbert, *Journal of Cell Science* 39: 257–272, Fig. 11 (1979). © 1979 The Company of Biologists Ltd.; **p.142** Kristoffer Tripplaar/Alamy Stock Photo

Chapter 8 **8.1** Doug Perrine/Nature Picture Library; **p. 143 bottom** James Jordan Photography/Getty Images; **8.2** Stephen Simpson/Getty Images; **8.3a** Robert N. Johnson/RnJ Photography; **8.3b** Robert N. Johnson/RnJ Photography; **8.4 left** Image Quest Marine; **right** asharkyu/Shutterstock; **8.11b** Bruce J. Schnapp; **8 Scientific Skills Exercise** Fer Gregory/Shutterstock; **8.15a; 8.15b** Thomas Steitz; **8.17** Jack Dykinga/Nature Picture Library; **8.22** Nicolae Simionescu; **p. 163** PayPal/Getty Images

Chapter 9 **9.1** Sue Heaton/Alamy Stock Photo; **p. 164 bottom** Paul R. Sterry/Nature Photographers Ltd/Alamy Stock Photo; **9.4 center** Dionisvera/Fotolia; **9 Scientific Skills Exercise** Thomas Kitchin & Victoria Hurst/Design Pics Inc./Alamy Stock Photo; **p. 186 bottom** Stephen Rees/Shutterstock

Chapter 10 **10.1** Aflo/Nature Picture Library; **p. 187** George Grall/National Geographic Creative/Alamy Stock Photo; **10.2a** STILLFX/Shutterstock; **10.2b** Lawrence Naylor/Science Source; **10.2c** M. I. Walker/Science Source; **10.2d** Susan M. Barns; **10.2e** National Library of Medicine; **10.3** Qiang Hu; **10.4 top** Andreas Holzenburg and Stanislav Vitha, Dept. of Biology and Microscopy & Imaging Center, Texas A&M University, College Station; **bottom** Jeremy Burgess/Science Source; W.P. Wergin/Biological Photo Service; **10.12b** Christine Case; **10 Scientific Skills Exercise** Ohio State Weed Lab Archive, The Ohio State University, Bugwood.org; **10.21 top left** Doukdouk/Alamy Stock Photo; **top right** Keysurfing/Shutterstock; **10.22 tree** Andreas Holzenburg and Stanislav Vitha, Dept. of Biology and Microscopy & Imaging Center, Texas A&M University, College Station; **p. 211 bottom** gary yim/Shutterstock

Chapter 11 **11.1** Federico Veronesi/Gallo Images/Alamy Stock Photo; **p. 212 bottom** molekuul.be/Shutterstock; **11.3 top 3** Dale A. Kaiser; **11.3 bottom** Michiel Vos; **p. 214 left** Bruno Coignard/Jeff Hageman/CDC; **11.7** The Scripps Research Institute; **11.19** Gopal Murti/Science Source; **11.21 left; center; right** William Wood; **p. 233** Maureen Spuhler/Pearson Education

Chapter 12 **12.1** George von Dassow; **p. 234 bottom** Jane Stout and Claire Walczak, Indiana University, Winner of the GE Healthcare Life Sciences' 2012 Cell Imaging Competition.; **12.2a** Biophoto Associates/Science Source; **12.2b** Biology Pics/Science Source; **12.2c** Biophoto/Science Source; **12.3** John M. Murray, School for Medicine, University of Pennsylvania, Philadelphia.; **12.4** Biophoto/Science Source; **12.5 center** Biophoto/Science Source; **12.7** Conly L. Rieder; **12.8 right** J. Richard McIntosh, University of Colorado, Boulder; **left** Reproduced by permission from Matthew Schibler, from *Protoplasma* 137. © 1987: 29–44 by Springer-Verlag GmbH & Co KG; **12.10a** Don W. Fawcett/Science Source; **12.10b** Eldon H Newcomb; **12.11 left to right** Elisabeth/Pearson Education; **12.18 bottom** Guenter Albrecht-Buehler, Northwestern University, Chicago; **12.19a; 12.19b** Lan Bo Chen; **12.20 bottom right** Anne Weston/Wellcome Institute Library; **12 Scientific Skills Exercise** Mike Davidson; **p. 252 left** J.L. Carson/Newscom; **p. 252 right** Steve Gschmeissner/Science Source

Unit Three Interview **top** Denise Applewhite, Office of Communications, Princeton University; **bottom** Shirley Tilghman, Princeton University

Chapter 13 **13.1** Mango Productions/Getty Images; **p. 254 bottom** Don W. Fawcett/Science Source; **13.2a** Roland Birke/Science Source; **13.2b** George Ostertag/SuperStock; **13.3 top** Ermakoff/Science Source; **bottom** CNRI/Science Source; **13 Scientific Skills Exercise** SciMAT/Science Source; **13.12** Mark Petronczki and Maria Siomos; **13.13** John Walsh, Micrographia; **p. 268** Randy Ploetz

Chapter 14 **14.1** John Swithinbank/AGE Fotostock; **p. 269 bottom** Mendel Museum, Augustinian Abbey, Brno; **14.14a** Maximilian Weinzierl/Alamy Stock Photo; **14.14b** Paul Dymond/Alamy Stock Photo; **14 Scientific Skills Exercise** Apomares/E+/Getty Images; **14.15a; 14.15b** Pearson Education; **14.16** Rick Guidotti/Positive Exposure; **14.18** Michael Ciesielski Photography; **14.19** CNRI/Science Source; **p. 291 top** Pearson Education; **p. 293 left** Norma JubinVille/Patricia Speciale; **right** Rene Maletete/Gamma-Rapho/Getty Images

Chapter 15 **15.1** Peter Lichter and David Ward, *Science* 247 (1990). © 1990 American Association for the Advancement of Science; **p. 294** Zellsubstanz, Kern und Zelltheilung, by Walther Flemming, 1882, Courtesy of Yale University, Harvey Cushing/John Hay Whitney Medical Library; **15.3** Martin Shields/Alamy Stock Photo; **15.5** Andrew Syred/Science Source; **15.6b** Li Jingwang/E+/Getty Images; **15.6c** Kosam/Shutterstock; **15.6d** Creative Images/Fotolia; **15.8** Jagodka/Shutterstock; **15 Scientific Skills Exercise** Oliver911119/Shutterstock; **15.15 left** CNRI/Science Source; **right** Denys_Kuvaiev/Fotolia; **15.18** Phomphan/Shutterstock; **p. 313** James K Adams

Chapter 16 **16 Scientific Skills Exercise** Scott Ling **16.1** Andrey Prokhorov/E+/Getty Images; **p. 314 bottom** A. Barrington Brown/Science Source; **16.3** Oliver Meckes/Science Source; **16.6 left** Library of Congress Prints and Photographs Division; **right** Cold Spring Harbor Laboratory Archives; **16.12a** Micrograph by Jerome Vinograd. From: *Molecular Biology of the Cell*. 4th edition. DNA Replication Mechanisms. Figure 5–6.; **16.12b** From: Enrichment and visualization of small replication units from cultured mammalian cells. DJ Burks et al. *Journal of Cell Biology*. 1978 Jun;77(3):762-73. Fig. 6A.; **16.21** Peter Lansdorp; **16.22 left to right** Gopal Murti/Science Source, Victoria E. Foe, Barbara Hamkalo, U Laemmli/Science Source, Biophoto/Science Source; **16.23a** Thomas Reid, Genetics Branch/CCR/NCI/NIH; **16.23b** Michael R. Speicher/Medical University of Graz; **p. 334 left and right** Thomas A. Steitz/Yale University

Chapter 17 **17.1** ANGELO CUCCA/AFP/Getty Images; **p. 335 bottom** Richard Stockwell; **17.7a** Keith V Wood; **17.7b** Simon Lin/AP Images; **17.17 bottom right** Joachim Frank; **17.23b** Barbara Hamkalo; **17.24** Oscar L. Miller/SPL/Science Source; **17.26 cell** Eye of Science/Science Source; **p. 359** ABC News Video; **p. 362** Vasiliy Koval/Shutterstock

Chapter 18 **18.1** Andreas Werth; **p. 363 bottom** Gallimaufry/Shutterstock; **18 Scientific Skills Exercise** hidesy/E+/Getty Images; **18.12** Medical University of Graz; **18.16a; 18.16b** Mike Wu; **18.20 left; right** F. Rudolf Turner, Indiana University; **18.21 top; bottom** Wolfgang Driever, University of Freiburg, Freiburg, Germany; **18.22 top left** Ruth Lahmann, The Whitehead Institution; **18.27** Bloomberg/Getty Images; **p. 395 right** Peter Herring/Image Quest Marine

Chapter 19 **19.1** Richard Bizley/Science Source; **p. 396** Thomas Deerinck, NCMIR/Science Source; **19.2 left; right; bottom** Peter von Sengbusch, Botanik; **19.3a** Science Source; **19.3b** Linda M. Stannard, University of Cape Town/Science Source; **19.3c** Hazel Appleton, Health Protection Agency Centre for Infections/Science Source; **19.3d** Ami Images/Science Source; **19.7** molekuul.be/Fotolia; **19.9 top 2** Charles Dauguet/Science Source; **bottom 3** Petit Format/Science Source; **19.10a** National Institute of Allergy and Infectious Diseases (NIAID); **19.10b** Cynthia Goldsmith/Centers for Disease Control; **19.10c** Lei Sun, Richard J. Kuhn and Michael G. Rossmann, Purdue University, West Lafayette; **19 Scientific Skills Exercise** Dong yanjun/Imaginechina/AP Images; **19.11** Olivier Asselin/Alamy Stock Photo; **inset** James Gathany/Centers for Disease Control and Prevention; **19.12** Nigel Cattlin/Alamy Stock Photo; **p. 412 bottom** Nelson Hale/Shutterstock

Chapter 20 **20.1** Ian Derrington; **p. 413 bottom** John Elk III/Alamy Stock Photo; **20.2** P. Morris, Garvan Institute of Medical Research; **20.6b** Repligen Corporation; **20.9** Ethan Bier; **20.12** George S. Watts and Bernard W. Futscher, University of Arizona Cancer Center, Phoenix.; **20.17** Roslin Institute; **20.18** Pat Sullivan/AP Images; **20.19 fat** Steve Gschmeissner/Science Source; **bone** SPL/Science Source; **blood** Steve Gschmeissner/Science Source; **20.23 left; right** Brad DeCecco Photography; **20.24** Steve Helber/AP Images; **p. 439** Galyna Andrushko/Shutterstock

Chapter 21 **21.1** Karen Huntt/Corbis; **p. 440 bottom** Image Quest Marine; **21.4** University of Toronto Lab; **21.5** GeneChip Human Genome U133 Plus 2.0 Array, courtesy of Affymetrix; **21.7 left** AP Images; **21.7 right** Virginia Walbot; **21.10 top.**, Oscar L Miller Jr.; **21.18 top right** Nicholas Bergkessel, Jr./Science Source; **bottom left to right** From: Altered ultrasonic vocalization in mice with a disruption in the *Foxp2* gene. W. Shu et al. *Proc Natl Acad Sci U S A*. 2005 Jul 5; 102(27): 9643–9648. Fig. 3.; **p. 462 left** WaterFrame/Alamy Stock Photo; **p. 464 right** Patrick Landmann/Science Source

Unit Four Interview **top** Li Huang; **bottom** Janet Iwasa/Jack Szostak

Chapter 22 **22.1** William Mullins/Alamy Stock Photo; **p. 466 bottom** Reinaldo Aguilar, Vascular Plants of the Osa Peninsula, Costa Rica; **22.4** Karen Moskowitz/Stone/Getty Images; **22.5 inset left** ARCHIV/Science Source; **inset right** Science Source; **22.6a** Michael Gunther/Science Source; **22.6b** David Hosking/Frank Lane Picture Agency; **22.6c** David Hosking/Alamy Stock Photo; **22.7** Darwin's Tree of Life sketch, MS. DAR.121:p36. Reproduced with permission of the Cambridge University Library; **22.9 brussels sprouts** Arena Photo UK/Fotolia; **kale** Željko Radojko/Fotolia; **cabbage** Guy Shapira/Shutterstock; **wild mustard** Gerhard Schulz/Naturephoto; **broccoli** YinYang/E+/Getty Images; **kohlrabi** Motorolka/Shutterstock; **22.10** Laura Jesse; **22.11** Richard Packwood/Oxford Scientific/Getty Images; **22.12 top** Lighthouse/UIG/AGE Fotostock; **bottom** Gallo Images/Brand X Pictures/Getty Images; **22.13 top** Scott P. Carroll; **22.16 left** Keith Wheeler/Science Source; **right** Omikron/Science Source; **22.18 left** ant Photo Library/Science Source; **right** Steve Bloom Images/Alamy Stock Photo; **22.19a; 22.19b; 22.19c; 22.19d** Chris Linz, Thewissen lab, Northeastern Ohio Universities College of Medicine (NEOUCOM); **22.2 top left** Recherches sur les ossemens fossiles, Atlas G Cuvier, pl. 17 1836; **bottom left** Wayne Lynch/All Canada Photos/AGE Fotostock; **top center** Alfred Russel Wallace Memorial Fund; **top right** The Natural History Museum/Alamy Stock Photo; **bottom right** American Museum of Natural History; **p. 483** John Cancalosi/Nature Picture Library

Chapter 23 **23.1** Sylvain Cordier/Science Source; **p. 484 bottom** Rosemary B. Grant; **23.3** David Stoecklein/Lithium/AGE Fotostock; **23.5a; 23.5b** Erick Greene; **23.6 left** Gary Schultz/Photoshot; **right** Patrick Valkenburg/Alaska Department of Fish and Game; **23.Scientific Skills Exercise01** DLeonis/Fotolia; **23.11** William Ervin/Science Source; **23.12 inset top** Kristin Stanford, Stone Laboratory, Ohio State University, Columbus; **inset bottom** Kent Bekker, United States Fish and Wildlife Service; **23.14** John Visser/Bruce Coleman/Photoshot; **23.15** Dave Blackey/All Canada Photos/AGE Fotostock; **23.18 blood cells** Eye of Science/Science Source; **23.18 people** Caroline Penn/Alamy Stock Photo; **mosquito** Kletr/Shutterstock; **p. 503** Thomas & Pat Leeson/Science Source

Chapter 24 **24.1** Joel Sartore/National Geographic/Getty Images; **p. 504 bottom** Karin Duthie/Alamy Stock Photo; **24.2a left** Malcolm Schuyl/Alamy Stock Photo; **24.2a right** Wave RF/Getty Images; **24.2b top left** Robert Kneschke/Kalium/AGE Fotostock; **top center** Justin Horrocks/E+/Getty Images; **top right** Ryan McVay/Stockbyte/Getty Images; **24.2 bottom left** Dragon Images/Shutterstock; **bottom center** arek_malang/Shutterstock; **24.2** jaki good/Moment Open/Getty Images; **24.3a,b** Phil Huntley Franck; **24.3c** Hogle Zoo; **24.3d** Jerry A. Payne, USDA Agricultural Research Service, Bugwood.org; **24.3e** Imagebroker/Alamy Stock Photo; **24.3f** Reprinted by permission from: Evolution: single-gene speciation by left-right reversal. Ueshima R, Asami T. *Nature*. 2003. Oct 16; 425(6959):679; Fig.1 © 2003 Macmillan Magazines Limited.; **24.3g** William E. Ferguson; **24.3h** Charles W. Brown; **24.3i** Eyewire Collection/Getty Images; **24.3j** Corbis; **24.3k** Dawn YL/Fotolia; **24.3l** Kazutoshi Okuno; **24.4 top** CLFProductions/Shutterstock; **center** Boris Karpinski/Alamy Stock Photo; **bottom** Troy Maben/AP Images; **24.6** Brian Langerhans; **24.8 maps** Earth Observing System, NASA; **shrimp** Arthur Anker, Florida Museum of

Natural History; **24 Scientific Skills Exercise** John Shaw/Photoshot; **24.11** Pam Soltis; **24.12 top, bottom** Ole Seehausen; **24.13** Jeroen Speybroeck, Research Institute for Nature and Forest; **p. 516** Philimon Bulawayo/Reuters; **24.15 top left, top right, 24.16 bottom** Ole Seehausen; **24.17** Jason Rick and Loren Rieseberg; **24.19a–d** Reprinted by permission from: Allele substitution at a flower colour locus produces a pollinator shift in monkeyflowers. Bradshaw HD, Schemske DW. *Nature*. 2003 Nov 12; 426(6963):176–8. Fig. 1. © 2003. Macmillan Magazines Limited.; **p. 522** Thomas Marent/Rolf Nussbaumer Photography/Alamy Stock Photo

Chapter 25 **25.1** Juergen Ritterbach/Alamy Stock Photo; **p. 523 bottom** B. O'Kane/Alamy Stock Photo; **25.3 left** NASA; **right** Deborah S. Kelley; **25.4b** F. M. Menger and Kurt Gabrielson; **25.4c** Experimental models of primitive cellular compartments: encapsulation, growth, and division. MM Hanczyc et al. *Science*. 2003 Oct 24;302(5645):618–22. Fig.2i.; **25.5** ***Dimetrodon*** Maureen Spuhler/Pearson Education; **stromatolites** Roger Jones; **close-up stromatolites** S.M. Awramik/Biological Photo Service; **fossil** Sinclair Stammers/Science Source; **plesiosaur** Franz Xaver Schmidt; ***Hallucigenia*** Ted Daeschler/Academy of Natural Sciences; ***Dickensonia*** Chip Clark; ***Tappania*** Lisa-Ann Gershwin/Museum of Paleontology; **cross section** Andrew H. Knoll; **25.12a&b** From: Four hundred-million-year-old vesicular arbuscular mycorrhizae. Remy W, Taylor TN, Hass H, Kerp H. *Proc Natl Acad Sci U S A*. 1994 Dec 6;91(25):11841-3. Figure 1 and 4.; **25 Scientific Skills Exercise** Biophoto Associates/Science Source; **25.22** ***Dubautia laxa, sandwicense, linearis, scabra, waialealae*** Gerald D. Carr; **tarweed** Bruce G. Baldwin; **25.23** Jean Kern; **25.24** Juniors Bildarchiv GmbH/Alamy Stock Photo; **25.26 top** David Horsley; **bottom** From: Genetic and developmental basis of evolutionary pelvic reduction in threespine sticklebacks. MD Shapiro et al. *Nature*. Erratum. 2006 Feb 23; 439(7079):1014; Fig1.; **25.27** Sinclair Stammers/Science Source

Unit Five Interview Nancy Moran, University of Texas, Austin

Chapter 26 **26.1** Trapp/blickwinkel/Alamy Stock Photo; **26.17a** Mick Ellison; **26.17b** Ed Heck; **26.22 inset** Gerald Schoenknecht; **26.22** Gary Crabbe/Enlightened Images/Alamy Stock Photo; **26 Scientific Skills Exercise** Nigel Cattlin/Alamy Stock Photo; **p. 570** David Fleetham/Nature Picture Library

Chapter 27 **27.1** Zastolskiy Viktor/Shutterstock; **p. 571** NASA; **27.2a** Janice Haney Carr/Centers for Disease Control and Prevention; **27.2b** Kari Lounatmaa/Science Source; **27.2c** Stem Jems/Science Source; **27.3 center** L. Brent Selinger/Pearson; **27.4** Immo Rantala/SPL/Science Source; **27.5** H S Pankratz/T C Beaman/Biological Photo Service; **27.6** Kwangshin Kim/Science Source; **27.7 right** Julius Adler; **27.8a** From: Taxonomic Considerations of the Family Nitrobacteraceae Buchanan: Requests for Opinions. Stanley W. Watson, *IJSEM* (*International Journal of Systematic and Evolutionary Microbiology* formerly (in 1971) *Intl. Journal of Systematic Bacteriology*), July 1971 vol. 21 no. 3, 254-270. Fig. 14; **27.8b** Biological Photo Service; **27.9** Huntington Potter; **27.12** Charles C. Brinton, Jr.; **27.14** Susan M. Barns; **27.16** ***Rhizobium*** Biological Photo Service; ***Nitrosomonas*** Yuichi Suwa; ***Thiomargarita namibiensis*** National Library of Medicine; **proteobacteria** Patricia Grilione; ***Heliobacter pylori*** A Barry Dowsett/Science Source; ***Chlamydia*** Moredon Animal Health/SPL/Science Source; ***Leptospira* spirochetes** CNRI/SPL/Science Source; ***Oscillatoria*** CCALA/Institute of Botany CAS; ***Streptomyces*** Paul Alan Hoskisson; **mycoplasmas** David M. Phillips/Science Source; **27.17** Shaeri Mukherjee; **27.18** Pascale Frey-Klett; **27.19** Ken Lucas/Biological Photo Service; **27.20 left** Scott Camazine/Science Source; **center** David M. Phillips/Science Source; **right** James Gathany/Centers for Disease Control and Prevention; **27 Scientific Skills Exercise** Slava Epstein; **27.21ab** RNA-directed gene editing specifically eradicates latent and prevents new HIV-1 infection. W. Hu et al. *Proc Natl Acad Sci U S A*. 2014 Aug 5;111(31):11461-6. Fig. 3D.; **27.22** Metabolix Media; **27.23** Accent Alaska/Alamy Stock Photo; **p. 590** Biophoto Associates/Science Source

Chapter 28 **28.1** From: The molecular ecology of microbial eukaryotes unveils a hidden world, Moreira D, López-García P. *Trends Microbiol*. 2002 Jan;10(1):31–8. Fig. 4. Photo by Brian S. Leander and Mark Farmer.; **p. 591 bottom** Eric V. Grave/Science Source; **28 Scientific Skills Exercise** Shutterstock; **p. 594 right** Joel Mancuso, University of California, Berkeley; **p. 595 top left** M.I. Walker/Photoshot; **top right** Frank Fox/Science Source; **top right inset** David J. Patterson; **bottom left** Howard J. Spero; **bottom left inset** National Oceanic and Atmospheric Administration; **bottom right** Michael Abbey/Science Source; **28.4** Ken Ishida; **28.5** David M. Phillips/Science Source; **28.6** David J. Patterson; **28.7** Oliver Meckes/Science Source; **28.8** David J. Patterson; **28.9** Centers for Disease Control and Prevention; **28.10** Steve Gschmeissner/Science Source; **28.11** Stephen Durr; **28.12** Colin Bates; **28.13** J. R. Waaland/Biological Photo Service; **28.14** Guy Brugerolle; **28.15a** Virginia Institute of Marine Science; **28.15b** Science Source; **28.16** Masamichi Aikawa; **28.17a** M. I. Walker/Science Source; **28.18** Robert Brons/Biological Photo Service; **28.19** Nature Picture Library; **28.20** Eva Nowack; **28.21a** D. P. Wilson; Eric Hosking; David Hosking/Science Source; **28.21b** Michael Guiry; **28.21c** Biophoto Associates/Science Source; **28.21d** David Murray/Dorling Kindersley, Ltd.; **28.22a** M. I. Walker/Science Source; **28.22b** Laurie Campbell/Photoshot License Limited; **28.22c** David L. Ballantine; **28.23** William L. Dentler; **28.25** Ken Hickman; **28.26 left, bottom** Robert Kay; **28.27** Kevin Carpenter and Patrick Keeling; **28.28** David Rizzo; **p. 615** Greg Antipa/Biophoto Associates/Science Source

Chapter 29 **29.1** Exactostock/SuperStock; **p. 616 bottom** Belinda Images/SuperStock; **p. 617 center** R. Malcolm Brown, Jr.; **29.3 embryo** Linda E. Graham; **wall ingrowths** Karen S. Renzaglia; **wall spores left** Arterra Picture Library/Alamy Stock Photo; **wall spores right** Mike Peres RBP SPAS/CMSP Biology/Newscom; **liverwort** David John Jones; **root** Ed Reschke/Getty Images; **shoot** Ed Reschke/Getty Images; **29.4ab** Charles H Wellman; **29.5** From: A vascular conducting strand in the early land plant *Cooksonia*, D. Edwards, K. L. Davies & L. Axe. *Nature* 357, 683–685 (25 June 1992). Figure 1A.; **29.8 top left** Alvin E. Staffan/Science Source; **top left inset** Graham, Linda E.; **top right** The Hidden Forest; **bottom left** The Hidden Forest; **bottom right** Tony Wharton/Fundamental Photographs; **p. 625** From: *Mosses and Other Bryophytes, an Illustrated Glossary* (2006), Bill and Nancy Malcolm, Micro-optics Press; **29.10a** John Warburton-Lee Photography/Alamy Stock Photo; **29.10b** Thierry Lauzun/Iconotec/Alamy Stock Photo; **29.11** Hans Kerp; **29 Scientific Skills Exercise** Richard Becker/Fundamental Photographs; **29.12 top** Maureen Spuhler/Pearson Education; **bottom** FloralImages/Alamy Stock Photo; **29.13a** Maureen Spuhler/Pearson Education; **29.13b** FloralImages/Alamy Stock Photo; **29.14 top left** Jody Banks, Purdue University, West Lafayette; **top center** Murray Fagg/Australian National Botanic Gardens; **top right** Helga Rasbach; Kurt Rasbach; **bottom left** John Martin/Alamy Stock Photo; **bottom center** Stephen P. Parker/Science Source; **bottom right** Francisco Javier Yeste Garcia; **29.15** Open University, Department of Earth Sciences; **bottom right** Mike Peres RBP SPAS/CMSP Biology/Newscom; **p. 633** Wilhelm Barthlott

Chapter 30 **30.1** Lyn Topinka, USGS, U.S. Geological Survey Library; **p. 634 bottom** Marlin Harms; **p. 634 inset** Marlin Harms; **30 Scientific Skills Exercise** Guy Eisner; **30.5** Rudolph Serbet, Natural History and Biodiversity Institute, University of Kansas, Lawrence; **30.6** Copyright ESRF/PNAS/C. Soriano; **30.7 Cycadophyta** Warren Price Photography/Shutterstock; **Ginkgophyta** Travis Amos/Pearson; **Ginkgophyta inset** Kurt Stueber; ***Welwitschia*** Jeroen Peys/Getty Images; ***Welwitschia* cones** Thomas Schoepke; ***Gnetum*** Michael Clayton; ***Ephedra*** Bob Gibbons/Frank Lane Picture Agency Limited; **fir** vincentlouis/Fotolia; **juniper** Svetlana Tikhonova/Shutterstock; **larch** Adam Jones/Getty Images; **sequoia** Daniel Acevedo/AGE Fotostock/Alamy Stock Photo; ***Wollemia* pine fossil** Jaime Plaza, Royal Botanic Gardens Sydney; ***Wollemia* pine** Jaime Plaza/Wildlight Photo Agency/Alamy Stock Photo; **bristlecone pine** Russ Bishop/Alamy Stock Photo; **30.9 top** Zee/Fotolia; **bottom** Paul Atkinson/Shutterstock; **30.10 tomato** Dave King/Dorling Kindersley, Ltd.; **grapefruit** Andy Crawford/Dorling Kindersley, Ltd.; **nectarine** Dave King/Dorling Kindersley, Ltd.; **hazelnuts** Diana Taliun/Fotolia; **milkweed** Maria Dryfhout/123RF; **seeds** Mike Davis; **wings** PIXTAL/AGE Fotostock; **mouse** Eduard Kyslynskyy/Shutterstock; **cocklebur** Derek Hall/Dorling Kindersley, Ltd.; **dog** Scott Camazine/Science Source; **30.13** David L. Dilcher and Ge Sun; **30.15** D. Wilder; **30.17 lily** Howard Rice/Dorling Kindersley, Ltd.; **star anise** Floridata.com; ***Amborella*** Jack Scheper/Floridata.com; **magnolia** Andrew Butler/Dorling Kindersley, Ltd.; **orchid** Eric Crichton/Dorling Kindersley, Ltd.; **palm** John Dransfield; **barley** kenjii/Fotolia; **snow pea** Maria Dattola/Getty Images; **dog rose** Glam/Shutterstock; **oak** Matthew Ward/Dorling Kindersley, Ltd.; **30.18** NASA; **p. 651** Martin Turner/Getty Images

Chapter 31 **31.1** vvuls/123RF; **p. 652 bottom** Matthijs Wetterauw/Alamy Stock Photo; **31.2 top** Nata-Lia/Shutterstock; **bottom** Fred Rhoades; **inset** George L. Barron; **31.4a** G. L. Barron and N. Allin, University of Guelph/Biological Photo Service; **31 Scientific Skills Exercise** U.S. Department of Energy/DOE Photo; **31.6 left** Olga Popova/123RF; **right** Biophoto Associates/Science Source; **31.7** Stephen J. Kron; **31.9** Dirk Redecker; **31.10 chytrids** John Taylor; **zygomycete** Ray Watson; **glomeromycete** Mutualistic stability in the arbuscular mycorrhizal symbiosis: exploring hypotheses of evolutionary cooperation. E. T. Kiers, M. G. van der Heijden. *Ecology*. 2006 Jul; 87(7):1627-36. Fig. 1a. Image by Marcel van der Heijden, Swiss Federal Research Station for Agroecology and Agriculture.; **peels** blickwinkel/Alamy Stock Photo; **fungus** Science Source; **31.11** William E. Barstow; **31.12 mold** Antonio D'Albore/Getty Images; ***Rhizopus*** Alena Kubatova/Culture Collection of Fungi (CCF); **sporangia** George L. Barron; **zygosporangium** Ed Reschke/Getty Images; **31.13** George L. Barron/Biological Photo Service; **31.14** Biological Photo Service; **31.15 left** Bryan Eastham/Fotolia; **31.15 right** Science Source; **31.16** Fred Spiegel; **31.17 top** Frank Paul/Alamy Stock Photo; **center** kichigin19/Fotolia; **bottom** Fletcher and Baylis/Science Source; **31.18** Biophoto Associates/Science Source; **31.19** University of Tennessee Entomology and Plant Pathology; **31.21** Mark Bowler/Science Source; **31.22 top** Benvie/Wild Wonders of Europe/Nature Picture Library; **center** Geoff Simpson/Nature Picture Library; **bottom** Ralph Lee Hopkins/National Geographic/Getty Images; **31.23** Eye of Science/Science Source; **31.24a** Scott Camazine/Alamy Stock Photo; **31.24b** Peter Chadwick/Dorling Kindersley, Ltd.; **31.24c** Blickwinkel/Alamy Stock Photo; **31.25** Vance T. Vredenburg/San Francisco State University; **31.26** Gary Strobel; **p. 670** Erich G. Vallery/USDA Forest Service

Chapter 32 **32.1** Thomas Marent/Rolf Nussbaumer Photography/Alamy Stock Photo; **p. 671 bottom** Nigel Downer/Photoshot; **32.5a** Lisa-Ann Gershwin/Museum of Paleontology; **32.5b** Ediacaran fossils-*Kimberella* From: The Late Precambrian fossil *Kimberella* is a mollusc-like bilaterian organism. Mikhail A. Fedonkin and Benjamin M. Waggoner. *Nature* 388, 28 Aug 1997, 868-871 Fig. 1.; **32.6** From Predatorial borings in late precambrian mineralized exoskeletons. Bengtson S, Zhao Y. *Science*. 1992 Jul 17;257(5068):367-9. Reprinted with permission from AAAS; **32.7 left** Chip Clark; **right** The Natural History Museum Trading Company Ltd; **32.12** Blickwinkel/Alamy Stock Photo; **p. 683** WaterFrame/Alamy Stock Photo

Chapter 33 **33.1** Paul Anthony Stewart; **sponge** Andrew J. Martinez/Science Source; **jelly** Robert Brons/Biological Photo Service; **flatworms** Teresa Zuberbühler; **Placozoa** Stephen Dellaporta; **comb jelly** Gregory G. Dimijian/Science Source; **marine flatworm** Ed Robinson/Perspectives/Getty images; ***Plumatella repens*** Hecker/blickwinkel/Alamy Stock Photo; ***Brachionus*** M. I. Walker/Science Source; **lampshell** Image Quest Marine; **Gastrotricha** Sinclair Stammers/Nature Picture Library; **ribbon worm** Erling Svensen/UWPhoto ANS; **SEM** Peter Funch; **marine annelid** Fredrik Pleijel; **octopus** Photonimo/Shutterstock; **loriciferan** Reinhart Mobjerg Kristensen; **priapulan** Erling Svensen/UWPhoto ANS; **onychophoran** Thomas Stromberg; **roundworm** London Scientific Films/Oxford Scientific/Getty Images; **water bears** Andrew Syred/Science Source; **spider** Reinhard Hölzl/ImageBroker/AGE Fotostock; **acorn worm** Leslie Newman & Andrew Flowers/Science Source; **tunicate** Robert Brons/Biological Photo Service; **sea urchin** Louise Murray/Robert Harding World Imagery; **33.4** Andrew J. Martinez/Science Source; **33.7a left** Robert Brons/Biological Photo Service; **right** David Doubilet/National Geographic Creative/Getty Images; **33.7b left** Neil G. McDaniel/Science Source; **right** Mark Conlin/V&W/Image Quest Marine; **33.8** Robert Brons/Biological Photo Service; **33.9 flatworm** Amar and Isabelle Guillen, Guillen Photo LLC/Alamy Stock Photo; **mushroom mycelium** blickwinkel/Alamy Stock Photo; **chloroplasts** From: Cytochemical localization of catalase in leaf microbodies (peroxisomes). SE Frederick, EH Newcomb. *Journal of Cell Biology* 1969 Nov; 43(2):343-53. Fig. 6.; **villi** MedicalRF.com/AGE Fotostock; **33.11** Centers for Disease Control and Prevention; **33.12** Eye of Science/Science Source; **33.13** M. I. Walker/Science Source; **33.14** Holger Herlyn, University of Mainz, Germany; **33.15a** blickwinkel/Alamy Stock Photo; **33.15b** Image Quest Marine; **33.17** Image Quest Marine; **33.18a** Lubos Chlubny/Fotolia; **33.18b** Robert Marien/Corbis; **33 Scientific Skills Exercise** Christophe Courteau/Nature Picture Library; **33.19** Harold W. Pratt/Biological Photo Service; **33.21 top** Image Quest

Marine; **center** Photonimo/Shutterstock; **bottom** Jonathan Blair/Corbis; **33.22 left** Dave Clarke/Zoological Society of London; **right** The U.S. Bureau of Fisheries; **33.23** Fredrik Pleijel; **33.24** Wolcott Henry/National Geographic Creative/Getty Images; **33.25** Astrid Michler, Hanns-Frieder Michler/Science Source; **33.26** Photoshot; **33.27** London Scientific Films/Oxford Scientific/Getty Images; **33.28** Power and Syred/Science Source; **33.29** Dan Cooper; **33.30** Stephen Paddock; **33.32** Mark Newman/Frank Lane Picture Agency; **33.33 top** Tim Flach/The Image Bank/Getty Images; **center** Andrew Syred/Science Source; **bottom** Reinhard Hölzl/Imagebroker/AGE Fotostock; **33.34** Tim Flach/The Image Bank/Getty Images; **33.35a** PREMAPHOTOS/Nature Picture Library; **33.35b** Tom McHugh/Science Source; **33.37** Maximilian Weinzierl/Alamy Stock Photo; **33.38** Peter Herring/Image Quest Marine; **33.39** Peter Parks/Image Quest Marine; **33.41** Meul/ARCO/Nature Picture Library; **33.42a** Cathy Keifer/Shutterstock; **33.42b** Cathy Keifer/Shutterstock; **33.42c** Jim Zipp/Science Source; **33.42d** Cathy Keifer/Shutterstock; **33.42e** Cathy Keifer/Shutterstock; **33.43 bristletail** Kevin Murphy; **silverfish** Perry Babin; **weevil** PREMAPHOTOS/Nature Picture Library; **red tachinid** Bruce Marlin; **wasp** John Cancalosi/Nature Picture Library; **hummingbird** Hans Christoph Kappel/Nature Picture Library; **homopteran** Dante Fenolio/Science Source; **katydid** Chris Mattison/Alamy Stock Photo; **33.44** Andrey Nekrasov/Image Quest Marine; **33.45** Daniel Janies; **33.46** Jeff Rotman/Science Source; **33.47** Louise Murray/Robert Harding World Imagery; **33.48** Jurgen Freund/Nature Picture Library; **33.49** Hal Beral/Corbis; **p. 715** Lucy Arnold

Chapter 34 **34.1** Derek Siveter; **34.4** Natural Visions/Alamy Stock Photo; **34.5** Biological Photo Service; **34.8** Tom McHugh/Science Source; **34.9** Marevision/AGE Fotostock; **inset** A Hartl/AGE Fotostock; **34.10** Nanjing Institute of Geology and Palaeontology; **34.14** Field Museum Library/Premium Archive/Getty Images; **34.15a** Carlos Villoch/Image Quest Marine; **34.15b** Masa Ushioda/Image Quest Marine; **34.15c** Andy Murch/Image Quest Marine; **34.17 tuna** James D. Watt/Image Quest Marine; **lionfish** Jez Tryner/Image Quest Marine; **sea horse** George Grall/National Geographic/Getty Images; **eel** Fred McConnaughey/Science Source; **34.18** From: The oldest articulated osteichthyan reveals mosaic gnathostome characters. M. Zhu. *Nature*. 2009 Mar 26;458(7237):469-74. doi: 10.1038/nature07855. Fig. 2.; **34.19** Laurent Ballesta/www.blancpain-ocean-commitment.com/www.andromede-ocean.com and iSimangaliso Wetland Park Authority; **34.20 head; ribs; scales** Ted Daeschler, Academy of Natural Sciences; **skeleton** Kalliopi Monoyios Studio; **34.22a** Alberto Fernandez/AGE Fotostock; **34.22b** Paul A. Zahl/Science Source; **34.22c** Zeeshan Mirza/photocorp/Alamy Stock Photo; **34.23a** DP Wildlife Vertebrates/Alamy Stock Photo; **34.23b** FLPA/Alamy Stock Photo; **34.23c** John Cancalosi/Photolibrary/Getty Images; **34.24** Hinrich Kaiser/Victor Valley College; **p. 751 left** David L. Brill Photography; **34.27** Nobumichi Tamura; **34.28** Chris Mattison/Alamy Stock Photo; **34.29a** Heather Angel/Natural Visions/Alamy Stock Photo; **34.29b** Matt T. Lee; **34.29c** Matt T. Lee; **34.29d** Nick Garbutt/Nature Picture Library; **34.29e** Carl & Ann Purcell/Corbis; **34.30a** Visceralimage/Fotolia; **34.30b** The Natural History Museum/Alamy Stock Photo; **34.32** Boris Karpinski/Alamy Stock Photo; **34.33** DLILLC/Corbis; **34.34** Mariusz Blach/Fotolia; **34.35** The Africa Image Library/Alamy Stock Photo; **inset** Paolo Barbanera/AGE Fotostock; **34.39** clearviewstock/Shutterstock; **inset** Mervyn Griffiths, Commonwealth Scientific and Industrial Research Organization; **34.36** Gianpiero Ferrari/Frank Lane Picture Agency Limited; **34.40a** John Cancalosi/Alamy Stock Photo; **34.40b** Martin Harvey/Alamy Stock Photo; **34.43** Imagebroker/Alamy Stock Photo; **34.45a** Kevin Schafer/AGE Fotostock; **34.45b** J & C Sohns/Picture Press/Getty Images; **34.46a** Morales/AGE Fotostock; **34.46b** Juniors Bildarchiv GmbH/Alamy Stock Photo; **34.46c** T.J. RICH/Nature Picture Library; **34.46d** E.A. Janes/AGE Fotostock; **34.46e** Martin Harvey/Photolibrary/Getty Images; **34.48** David L. Brill Photography; **34.49a** John Reader/Science Source; **34.49b** John Gurche Studios; **34 Scientific Skills Exercise** Golfx/Shutterstock; **34.50** Alan Walker; **34.52** Erik Trinkaus; **34.53** *Homo naledi*, a new species of the genus *Homo* from the Dinaledi Chamber, South Africa. L. R. Berger et al. eLife 2015;4:e09560, Fig. 6 and 9.; **34.54** C. Henshilwood; **p. 754** Tony Heald/Nature Picture Library

Unit Six Interview **top** Jie Huang, Duke University; **bottom** Philip N. Benfey, Duke University & HHMI

Chapter 35 **35.1** O.Bellini/Shutterstock; **p. 756 bottom** John Walker; **35.3** Jeremy Burgess/Science Source; **35.4 prop roots** Natalie Bronstein; **beet root** Rob Walls/Alamy Stock Photo; **mangrove roots** Geoff Tompkinson/SPL/Science Source; **strangling roots** Dana Tezarr/Photodisc/Getty Images; **buttress roots** Karl Weidmann/Science Source; **35.5 rhizomes** Donald Gregory Clever; **stolons** Dorling Kindersley, Ltd.; **potato** Imagenavi/sozaijiten/AGE Fotostock; **35.7 top right** Neil Cooper/Alamy Stock Photo; **top left** Martin Ruegner/Photodisc/Getty Images; **bottom right** Gusto Production/Science Source; **bottom left** Jerome Wexler/Science Source; **35 Scientific Skills Exercise** Matthew Ward/Dorling Kindersley, Ltd.; **35.9** Steve Gschmeissner/SPL/AGE Fotostock; **35.10 parenchyma cells** M I Spike Walker/Alamy Stock Photo; **collenchyma cells** Clouds Hill Imaging/Last Refuge Limited; **sclereids; fiber cells** Graham Kent/Pearson Education; **vessels** N.C. Brown Center for Ultrastructure Studies; **TEM** Brian Gunning; **tube** Ray F. Evert; **plate** Graham Kent/Pearson Education; **mitotic** From: *Arabidopsis* TCP20 links regulation of growth and cell division control pathways. C. Li et al. *Proc Natl Acad Sci U S A*. 2005 Sep 6;102(36):12978-83. Epub 2005 Aug 25. Photo: Peter Doerner; **35.14ab** Ed Reschke; **35.14c** Natalie Bronstein; **35.15** Michael Clayton; **35.16** Michael Clayton; **35.17b** Ed Reschke; **35.18b** Ed Reschke; **35.18c** Ed Reschke; **35.20a** Michael Clayton; **35.20b** Alison W. Roberts; **35.23** California Historical Society Collection (CHS-1177), University of Southern California on behalf of the USC Specialized Libraries and Archival Collections; **35.24** WILDLIFE GmbH/Alamy Stock Photo; **35.26** From: Natural variation in *Arabidopsis*: from molecular genetics to ecological genomics. D. Weigel. *Plant Physiol*. 2012 Jan;158(1):2-22. doi: 10.1104/pp.111. Fig. 1A.; **35.27** From: Microtubule plus-ends reveal essential links between intracellular polarization and localized modulation of endocytosis during division-plane establishment in plant cells. P. Dhonukshe et al. *BMC Biol*. 2005 Apr 14;3:11. Fig. 4M.; **35.28** University of California, San Diego; **35.30** From: U. Mayer et al, *Development* 117 (1): 149-162. Fig. 1a © 1993 The Company of Biologists, Ltd; **35.31 left** B. Wells and K. Roberts; **right** From: Microtubule plus-ends reveal essential links between intracellular polarization and localized modulation of endocytosis during division-plane establishment in plant cells. P. Dhonukshe et al. *BMC Biol*. 2005 Apr 14;3:11. Fig. 4B.; **35.32** From: The dominant developmental mutants of tomato, Mouse-ear and Curl, are associated with distinct modes of abnormal transcriptional regulation of a Knotted gene. A. Parnis et al. *Plant Cell*. 1997 Dec;9(12):2143-58. Fig. 1.; **35.33** Reproduced by permission from Hung et al, *Plant Physiology* 117:73-84, Fig. 2g. © 1998 American Society of Plant Biologists. Image courtesy of John Schiefelbein/University of Michigan; **35.34** James B. Friday; **35.35** From: Genetic interactions among floral homeotic genes of *Arabidopsis*. JL Bowman, DR Smyth, EM Meyerowitz. *Development*. 1991 May;112(1):1-20; Fig. 1A.; **p. 781 left** From: Anatomy of the vessel network within and between tree rings of *Fraxinus lanuginosa* (Oleaceae). Peter B. Kitin, Tomoyuki Fujii, Hisashi Abe and Ryo Funada. *American Journal of Botany*. 2004;91:779-788.; **p. 781 right** Biophoto Associates/Science Source

Chapter 36 **36.1** Dennis Frates/Alamy Stock Photo; **p. 782 bottom** Bjorn Svensson/Science Source; **36.3a** Rolf Rutishauser; **36.3b** Rolf Rutishauser; **p. 788** Nigel Cattlin/Science Source; **36.8** Holger Herlyn, University of Mainz, Germany; **inset** Benjamin Blonder and David Elliott; **36.10** Scott Camazine/Science Source; **36.13a; 36.13b** Power and Syred/Science Source; **36.15 ocotillo** Gerald Holmes, California Polytechnic State University at San Luis Obispo, Bugwood.org; **inset** Steven Baskauf, Nature Conservancy Tennessee Chapter Headquarters; **leafless ocotillo** Kate Shane, Southwest School of Botanical Medicine; **oleander cross section** Natalie Bronstein; **oleander** Andrew de Lory/Dorling Kindersley, Ltd.; **cactus** Danita Delimont/Alamy Stock Photo; **36.18 left, center, right** M. H. Zimmerman, from P. B. Tomlinson, Harvard University; **36.19** From: A coiled-coil interaction mediates cauliflower mosaic virus cell-to-cell movement. L. Stavolone et al. *Proc Natl Acad Sci U S A*. 2005 Apr 26;102(17):6219-24. Fig. 5C.; **p. 802** Catalin Petolea/Alamy Stock Photo

Chapter 37 **37.1** Noah Elhardt; **inset** Wilhelm Barthlott; **p. 803 bottom** Bartosz Plachno; **37.2** ARS/USDA; **37.4** National Oceanic and Atmospheric Administration (NOAA); **37.5** USGS Menlo Park; **37.6** Kevin Horan/The Image Bank/Getty Images; **37.8 healthy** View Stock RF/AGE Fotostock; **nitrogen deficient** Guillermo Roberto Pugliese/International Plant Nutrition Institute (IPNI); **phosphate deficient** C. Witt/International Plant Nutrition Institute (IPNI); **potassium deficient** M.K. Sharma and P. Kumar/International Plant Nutrition Institute (IPNI); **p. 810 left** Nigel Cattlin/Science Source; **37.9** From: Changes in gene expression in *Arabidopsis* shoots during phosphate starvation and the potential for developing smart plants, J. P. Hammond et al. *Plant Physiol*. 2003 Jun;132(2):578-96. Fig. 4.; **37.10 lichen** David T. Webb/University of Montana; **photosynthetic bacteria** Ralf Wagner; **striped puffer** Andrey Nekrasov/Pixtal/AGE Fotostock; **floating fern** Daniel L Nickrent; **ants** Tim Flach/The Image Bank/Getty Images; **sorghum plant** USDA/Science Source; ***Azolla*** Alex Wild; **leaf cutter ants** Martin Dohrn/Nature Picture Library; **37.11** Sarah Lydia Lebeis; **37.13** Scimat/Science Source; **37.15 top left** Mark Brundrett; **top right** Hugues B. Massicotte/University of Northern British Columbia Ecosystem and Management Program, Prince George, BC, Canada; **bottom left** Mark Brundrett; **37.16 staghorn fern** David Wall/Alamy Stock Photo; **mistletoe** Peter Lane/Alamy Stock Photo; **dodder** Emilio Ereza/Alamy Stock Photo; **Indian pipe** Martin Shields/Alamy Stock Photo; **sundews** Fritz Polking/Frank Lane Picture Agency Limited; **pitcher plants** Philip Blenkinsop/Dorling Kindersley, Ltd.; **ant** Paul Zahl/Science Source; **Venus flytrap** Chris Mattison/Nature Picture Library; **p. 819** Mode Images/Alamy Stock Photo

Chapter 38 **38.1** Ch'ien Lee; **38.4 hazel** Wildlife GmbH/Alamy Stock Photo; **hazel carpellate** Friedhelm Adam/Imagebroker/Getty Images; **dandelion** Bjorn Rorslett; **under ultraviolet light** Bjorn Rorslett; **yucca flower** Doug Backlund/WildPhotos-Photography.com; ***Stapelia*** Kjell B. Sandved/Science Source; **long-nosed bat** Rolf Nussbaumer/Imagebroker/AGE Fotostock; **hummingbird** Rolf Nussbaumer/Nature Picture Library; **38.5** W. Barthlott and W. Rauh, Nees Institute for Biodiversity of Plants; **38.10** Blickwinkel/Alamy Stock Photo; **38.12 coconut** Kevin Schafer/Alamy Stock Photo; **Zanonia** Aquiya/Fotolia; **dandelion** Steve Bloom Images/Alamy Stock Photo; **maple seeds** Chrispo/Fotolia; **tumbleweed** Nurlan Kalchinov/Alamy Stock Photo; ***Tribulus terrestris*** California Department of Food and Agriculture's Plant Health and Pest Prevention Services; **seeds** Kim A Cabrera; **squirrel** Alan Williams/Alamy Stock Photo; **ant** Benoit Guénard; **38.13** Dennis Frates/Alamy Stock Photo; **p. 832** Toby Bradshaw; **38.14a left** Marcel Dorken; **right** Marcel Dorken; **38.14b** Nobumitsu Kawakubo; **38.15a-c** Meriel G. Jones/University of Liverpool School of Biological Sciences; **38.16** Andrew McRobb/Dorling Kindersley, Ltd.; **38.17** Gary P. Munkvold; **38.18** Ton Koene/Lineair/Still Pictures/Robert Harding World Imagery; **p. 839** Dartmouth College Electron Microscope Facility

Chapter 39 **39.1** Courtesy of the De Moraes and Mescher labs; **p. 840 bottom** Emilio Ereza/Alamy Stock Photo; **39.2a** Natalie Bronstein; **39.2b** Natalie Bronstein; **39.6ab** From: Regulation of polar auxin transport by AtPIN1 in *Arabidopsis* vascular tissue. L. Gälweiler et al. *Science*. 1998 Dec 18;282(5397):2226-30; Fig. 4ac.; **39.9a** Richard Amasino; **39.9b** Fred Jensen, Kearney Agricultural Center; **39.11 left** Mia Molvray; **right** Karen E. Koch; **39.13a** Kurt Stepnitz; **39.13b** Joseph J. Kieber; **39.14** Ed Reschke; **39.16 left** Nigel Cattlin/Alamy Stock Photo; **top and right** Nigel Cattlin/Alamy Stock Photo; **39.18 left** Martin Shields/Alamy Stock Photo; **right** Martin Shields/Alamy Stock Photo; **39.22a-d** Michael L. Evans, Ohio State University, Columbus; **39.23** Gregory Jensen and Elizabeth Haswell; **39.24a** Martin Shields/Science Source; **39.24b** Martin Shields/Science Source; **39.25a** J. L. Basq/M. C. Drew; **39.25b** J. L. Basq/M. C. Drew; **39.26** New York State Agricultural Experiment Station/Cornell University; **39.27 opium** Johan De Meester/Arterra Picture Library/Alamy Stock Photo; **crystals** David T. Webb; **cross section** Steve Gschmeissner/Science Source; **bristles** Susumu Nishinaga/Science Source; **snowflake leaf** Giuseppe Mazza; **leaf** Lawrence E. Gilbert/University of Texas, Austin; **hummingbird** Danny Kessler; **bamboo** Kim Jackson/Mode Images/Alamy Stock Photo; **caterpillar** Custom Life Science Images/Alamy Stock Photo; **inset** Custom Life Science Images/Alamy Stock Photo; **p. 869** Gary Crabbe/Alamy Stock Photo

Unit Seven Interview **top** T. Schwerdt, DKFZ (German Cancer Research Center in Heidelberg); **bottom** Stephen C. Harrison, The Laboratory of Structural Cell Biology, Harvard Medical School

Chapter 40 **40.1** Matthias Wittlinger; **p. 871 bottom** Premaphotos/Alamy Stock Photo; **40.2 top** Dave Fleetham/Robert Harding World Imagery; **center** Duncan Usher/Alamy Stock Photo; **bottom** Andre Seale/Image Quest Marine; **40.4 left** Eye of Science/

Science Source; **top right** Susumu Nishinaga/Science Source; **bottom right** Susumu Nishinaga/Science Source; **40.5 p. 875** Steve Downing/Pearson Education; **left top to bottom** Nina Zanetti/Pearson Education; Ed Reschke/Peter Arnold/Alamy Stock Photo; Nina Zanetti/Pearson Education; **40.5 p. 876 center** Nina Zanetti/Pearson Education; **40.5 p. 876 right top to bottom** Gopal Murti/Science Source; Chuck Brown/Science Source; **40.5 p. 877 top left to right** Nina Zanetti/Pearson Education; Ed Reschke/Photolibrary/Getty Images; Ed Reschke/Photolibrary/Getty Images; Ulrich Gartner; Thomas Deerinck; **40.7 top** Jeffrey Lepore/Science Source; **40.7 bottom** Neil McNicoll/Alamy Stock Photo; **40.10** Meiqianbao/Shutterstock; **40.11a** John Shaw/Science Source; **40.11b** Matt T. Lee; **40.14** Carol Walker/Nature Picture Library; **40.15** Robert Ganz; **40.16** From: Assessment of oxidative metabolism in brown fat using PET imaging. Otto Muzik, Thomas J. Mangner and James G. Granneman. *Front. Endocrinol.*, 08 Feb 2012 | http://dx.doi.org/10.3389/fendo.2012.00015 Fig. 2.; **40.20** Jeff Rotman/Alamy Stock Photo; **p. 891** Andrew Cooper/Nature Picture Library; **40.23 p. 892 leaves** Irin-K/Shutterstock; **bobcat** Thomas Kitchin/Victoria Hurst/All Canada Photos/AGE Fotostock; **sunflowers** Phil_Good/Fotolia; **fly eyes** WildPictures/Alamy Stock Photo; **plant** Bogdan Wankowicz/Shutterstock; **molting** Nature's Images/Science Source; **40.23 p. 893 xylem** Last Refuge/Robert Harding Picture Library Ltd/Alamy Stock Photo; **blood vessels** Susumu Nishinaga/Science Source; **root hairs** Rosanne Quinnell © The University of Sydney. eBot http://hdl.handle.net/102.100.100/1463; **intestinal lining** David M. Martin/Science Source; **peas** Scott Rothstein/Shutterstock; **piglets** Walter Hodges/Lithium/AGE Fotostock; **spongy mesophyll** Rosanne Quinnell; **alveoli** David M. Phillips/Science Source; **p. 895** Oliver Wien/epa European Pressphoto Agency creative account/Alamy Stock Photo

Chapter 41 **41.1** Jeff Foott/Discovery Channel Images/Getty Images; **41.3** Stefan Huwiler/Rolf Nussbaumer Photography/Alamy Stock Photo; **41.5 whale** Hervey Bay Whale Watch; **caterpillar** Thomas Eisner; **fly** Peter Parks/Image Quest Marine; **python** Gunter Ziesler/Photolibrary/Getty Images; **41.16 left** Fritz Polking/Alamy Stock Photo; **right** Tom Brakefield/Stockbyte/Getty Images; **41.18** Juergen Berger/Science Source; **p. 912 left** Peter Batson/Image Quest Marine; **41 Scientific Skills Exercise** The Jackson Laboratory; **p. 918** Jack Moskovita

Chapter 42 **42.1** John Cancalosi/Alamy Stock Photo; **42.2a** Reinhard Dirscherl/WaterFrame/Getty Images; **42.2b** Eric Grave/Science Source; **42.9 top** Indigo Instruments; **42.9 bottom** Ed Reschke/Photolibrary/Getty Images; **42.18** Eye of Science/Science Source; **42.19** Image Source/Exactostock.1598/SuperStock; **42 Scientific Skills Exercise** Fotolia; **42.21a** Peter Batson/Image Quest Marine; **42.21b** Olgysha/Shutterstock; **42.21c** Jez Tryner/Image Quest Marine; **42.23** Hong Y. Yan, University of Kentucky and Peng Chai, University of Texas; **42.24** Motta/Macchiarelli, Anatomy Dept./Univ. La Sapienza, Rome/SPL/Science Source; **42.26** Hans-Rainer Duncker, Institute of Anatomy and Cell Biology, Justus-Liebig-University Giessen; **p. 947** Doug Allan/Nature Picture Library; **p. 949** Stefan Hetz/WENN.com/Newscom

Chapter 43 **43.1** SPL/Science Source; **p. 950 bottom** Juergen Berger/Science Source; **p. 962** Steve Gschmeissner/Science Source; **43.26** CNRI/Science Source; **43 Scientific Skills Exercise** Eye of Science/Science Source; **43.28** Stephen C. Harrison/The Laboratory of Structural Cell Biology/Harvard Medical School; **p. 974** Tatan YUFLANA/AP Images

Chapter 44 **44.1** David Wall/Alamy Stock Photo; **p. 975 bottom** David Wall/Alamy Stock Photo; **44.3** Mark Conlin/Image Quest Marine; **44.5 left** Eye of Science/Science Source; **right** Eye of Science/Science Source; **44 Scientific Skills Exercise** Jiri Lochman/Lochman Transparencies; **44.7 left** GeorgePeters/E+/Getty Images; **center** Eric Isselée/Fotolia; **right** Maksym Gorpenyuk/Shutterstock; **44.12 right** Steve Gschmeissner/Science Source; **44.15** Michael Lynch/Shutterstock; **44.16** v_blinov/Fotolia; **44.17 cod** Roger Steene/Image Quest Marine; **frog** F. Rauschenbach/F1online digitale Bildagentur GmbH/Alamy Stock Photo; **stomata** Power and Syred/Science Source; **bacteria** Eye of Science/Science Source; **p. 996** Steven A. Wasserman

Chapter 45 **45.1** Phillip Colla/Oceanlight.com; **p. 997 bottom** Craig K. Lorenz/Science Source; **45.3** Volker Witte/Ludwig-Maximilians-Universitat Munchen; **45.11** Cathy Keifer/123RF; **p. 1008** angellodeco/Fotolia; **45.17** AP Images; **45.22 left** Blickwinkel/Alamy Stock Photo; **45.22 right** Jurgen and Christine Sohns/Frank Lane Picture Agency Limited; **p. 1016** Eric Roubos

Chapter 46 **46.1** Auscape/UIG/Getty Images; **46.2** Colin Marshall/Frank Lane Picture Agency; **46.3** P. de Vries; **46.5** Andy Sands/Nature Picture Library; **46.6** John Cancalosi/Alamy Stock Photo; **46 Scientific Skills Exercise** Tierbild Okapia/Science Source; **46.12** Design Pics Inc./Alamy Stock Photo; **46.17ab** Lennart Nilsson/Scanpix; **46.21** Phanie/SuperStock; **p. 1040** Dave Thompson/AP Images

Chapter 47 **47.1** Brad Smith/Stamps School of Art & Design, University of Michigan; **p. 1041 bottom** Oxford Scientific/Getty Images; **47.4abcd** From: Methods for quantitating sea urchin sperm-egg binding. V D Vacquier and J E Payne. *Exp Cell Res.* 1973 Nov; 82(1):227-35.; **47.4efgh** From: Wave of Free Calcium at Fertilization in the Sea Urchin Egg Visualized with Fura-2. M. Hafner et al. *Cell Motil. Cytoskel.*, 1988; 9:271-277; **47.6a-d** George von Dassow; **47.7 top** Jurgen Berger; **bottom** Andrew J. Ewald; **47.8a** Charles A. Ettensohn; **47.13b** Alejandro Díaz Díez/AGE Fotostock/Alamy Stock Photo; **47.14 left** Huw Williams; **right** Thomas Poole; **47.15b** Keith Wheeler/Science Source; **47.18b left** Hiroki Nishida; **right** Hiroki Nishida; **47.19** Medical Research Council; **47.20 & 21** MDC Biology Sinsheimer Labs; **47.24** From: Dorsal-ventral patterning and neural induction in *Xenopus* embryos. E. M. De Robertis and H. Kuroda. *Annu Rev Cell Dev Biol.* 2004;20:285-308. Fig. 1.; **47.25a** Kathryn W. Tosney; **47.26** Dennis Summerbell; **p. 1064** James Gerholdt/Getty Images

Chapter 48 **48.1** Franco Banfi/Science Source; **48.3** Thomas Deerinck; **48.13** Alan Peters; **48.16** Edwin R. Lewis, Y. Y. Zeevi and T. E, Everhart, University of California, Berkeley; **p. 1082** B.A.E. Inc./Alamy Stock Photo

Chapter 49 **49.1** Tamily Weissman; **49.4** Image by Sebastian Jessberger. Fred H. Gage, Laboratory of Genetics LOG-G, The Salk Institute for Biological Studies; **49.11** Larry Mulvehill/Corbis; **49.15ab** From: A functional MRI study of happy and sad affective states induced by classical music, M. T. Mitterschiffthaler et al. *Hum Brain Mapp.* 2007 Nov;28(11):1150-62. Fig. 1.; **49.18**, Marcus E Raichle; **49.19** From: Dr. Harlow's Case of Recovery from the passage of an Iron Bar through the Head, H. Bigelow. *Am. J of the Med. Sci.* July 1850;XXXIX. Images from the History of Medicine (NLM).; **49.25** Martin M. Rotker/Science Source; **p. 1104** Eric Delmar/Getty Images

Chapter 50 **50.1** Kenneth Catania; **50.6 top** CSIRO Publishing; **bottom** R. A. Steinbrecht; **50.7a** Michael Nolan/Robert Harding World Imagery; **50.7b** Grischa Georgiew/AGE Fotostock; **50.9** Richard Elzinga; **50.10** SPL//Science Source; **50.16a** USDA/APHIS Animal and Plant Health Inspection Service; **50.17** Steve Gschmeissner/Science Source; **50.21** Neitz Laboratory, University of Washington Medical School, Seattle; **50.26** Clara Franzini-Armstrong; **50.27** H. E. Huxley; **50.34** George Cathcart Photography; **50.39** Dave Watts/NHPA/Science Source; **50 Scientific Skills Exercise** Vance A. Tucker; **p. 1136** Dogs/Fotolia

Chapter 51 **51.1** Manamana/Shutterstock; **p. 1137 bottom** Ivan Kuzmin/Alamy Stock Photo; **51.2** Martin Harvey/Photolibrary/Getty Images; **51.3** Denis-Huot/Hemis/Alamy Stock Photo; **51.5** Kenneth Lorenzen; **p. 1142** Dustin Finkelstein/Getty Images; **51.7** Thomas D. McAvoy/The LIFE Picture Collection/Getty Images; **51.9** Lincoln Brower/Sweet Briar College; **51.11** Clive Bromhall/Oxford Scientific/Getty Images; **51.12** Richard Wrangham; **inset** Alissa Crandall/Encyclopedia/Corbis; **51 Scientific Skills Exercise** Matt Goff; **51.14a** Matt T. Lee; **51.14b** David Osborn/Alamy Stock Photo; **51.14c** David Tipling/Frank Lane Picture Agency Limited; **51.15** James D Watt/Image Quest Marine; **51.16** Gerald S. Wilkinson; **51.17** Cyril Laubscher/Dorling Kindersley, Ltd.; **51.20** Martin Harvey/Photolibrary/Getty Images; **51.21** Erik Svensson/Lund University, Sweden; **51.22** Lowell Getz; **51.23** Rory Doolin; **51.25** Jennifer Jarvis; **51.27** Marie Read/NHPA/Photoshot; **51.28** Jupiterimages/Creatas/Thinkstock/Getty Images; **p. 1160** William Leaman/Alamy Stock Photo

Unit Eight Interview **top** Patrick Mansell/Penn State; **bottom** Tracy Langkilde and Travis Robbins

Chapter 52 **52.1** Christopher Austin; **p. 1162 bottom** Christopher Austin; **52.2 top to bottom** Peter Blackwell/Nature Picture Library; Barrie Britton/Nature Picture Library; Oleg Znamenskiy/Fotolia; Juan Carlos Muñoz/AGE Fotostock; John Downer/Nature Picture Library; 1xpert/Fotolia; **p. 1168 American beech** Rick Koval; **52.8** Susan Carpenter; **52.10 desert** Anton Foltin/123RF; **grassland** David Halbakken/AGE Fotostock; **broadleaf** Gary718/Shutterstock; **tropical** Siepmann/Imagebroker/Alamy Stock Photo; **coniferous** Bent Nordeng/Shutterstock; **tundra** Juan Carlos Munoz/Nature Picture Library; **52.11 left** JTB Media Creation, Inc./Alamy Stock Photo; **right** Krystyna Szulecka/Alamy Stock Photo; **52.12 tropical** Siepmann/Imagebroker/Alamy Stock Photo; **desert** Anton Foltin/123RF; **savanna** Robert Harding Picture Library Ltd/Alamy Stock Photo; **chaparral** The California Chaparral Institute; **grassland** David Halbakken/AGE Fotostock; **coniferous** Bent Nordeng/Shutterstock; **broadleaf** Gary718/Shutterstock; **tundra** Juan Carlos Munoz/Nature Picture Library; **lake left** Susan Lee Powell; **lake right** AfriPics.com/Alamy Stock Photo; **wetlands** David Tipling/Nature Picture Library; **streams left** Ron Watts/Corbis; **streams right** Photononstop/SuperStock; **estuaries** Juan Carlos Munoz/AGE Fotostock; **intertidal** Stuart Westmorland/Danita Delimont/Alamy Stock Photo; **oceanic** Tatonka/Shutterstock; **reef** Digital Vision/Photodisc/Getty Images; **benthic** William Lange/Woods Hole Oceanographic Institute; **52.16** JLV Image Works/Fotolia; **52.18** Sylvain Oliveira/Alamy Stock Photo; **52.19** Scott Ling; **52 Scientific Skills Exercise left** John W. Bova/Science Source; **right** Dave Bevan/Alamy Stock Photo; **52.21** Daniel Mosquin; **p. 1187** Daryl Balfour/The Image Bank/Getty Images

Chapter 53 **53.1** Harpe/Robert Harding World Imagery; **p. 1188 bottom** Luiz Claudio Marigo/Nature Picture Library; **53.2** Todd Pusser/Nature Picture Library; **53.4a** Bernard Castelein/Nature Picture Library/Alamy Stock Photo; **53.4b** Michael S Nolan/AGE Fotostock; **53.4c** Alexander Chaikin/Shutterstock; **p. 1191 top** Jill M. Mateo; **p. 1191 bottom** Jennifer Dever; **53.9** Villiers Steyn/Shutterstock; **53.12** Photolibrary/Getty Images; **53 Scientific Skills Exercise** Laguna Design/Science Source; **53.13a** Stone Nature Photography/Alamy Stock Photo; **53.13b left** Kent Foster/Science Source; **inset** Robert D. and Jane L. Dorn; **53.14 inset** Dietmar Nill/Nature Picture Library; **53.15a** Steve Bloom Images/Alamy Stock Photo; **53.15b inset** Fernanda Preto/Alamy Stock Photo; **53.15b** Edward Parker/Alamy Stock Photo; **53.17 inset left** Chris Menjou; **inset right** Peter Brueggeman; **53.18 wheat** FotoVoyager/E+/Getty Images; **crowd** Jorge Dan/Reuters; **owl** Hellio & Van Ingen/NHPA/Photoshot; **cheetah** Gregory G. Dimijian/Science Source; **mice** Nicholas Bergkessel Jr./Science Source; **yeast** Andrew Syred/Science Source; **53.20** Alan & Sandy Carey/Science Source; **53.21 inset** Niclas Fritzen; **p. 1204** From: Tracking butterfly movements with harmonic radar reveals an effect of population age on movement distance. O. Ovaskainen et al. *Proc Natl Acad Sci U S A.* 2008 Dec 9;105(49):19090-5. doi: 10.1073/pnas.0802066105. Epub 2008 Dec 5. Fig. 1.; **53.26** NASA; **p. 1211** Reuters

Chapter 54 **54.1** Jeremy Brown/123RF; **p. 1212 bottom** Kristina Vackova/Shutterstock; **54.2 inset left** Joseph T. Collins/Science Source; **inset right** National Museum of Natural History/Smithsonian Institution; **p. 1214 bottom** Frank W Lane/Frank Lane Picture Agency Limited; **54 Scientific Skills Exercise** Johan Larson/Shutterstock; **54.5a** Tony Heald/Nature Picture Library; **54.5b** Tom Brakefield/Getty Images; **54.5c** Dante Fenolio/Science Source; **54.5d** Barry Mansell/Nature Picture Library; **54.5e left** Dante Fenolio/Science Source; **right** Robert Pickett/Papilio/Alamy Stock Photo; **54.5f left** Edward S. Ross; **right** James K. Lindsey; **54.6a-c** Roger Steene/Image Quest Marine; **54.7** Douglas Faulkner/Science Source; **54.8a** Bazzano Photography/Alamy Stock Photo; **54.8b** Nicholas Smythe/Science Source; **54.9** Daryl Balfour/Photoshot; **54.10a** Sally D. Hacker; **p. 1220** Gary W. Saunders; **54.12 top** Dung Vo Trung/Science Source; **54.13** Cedar Creek Ecosystem Science Reserve, University of Minnesota; **54.18** Genny Anderson; **54.19** Adam Welz; **p. 1225** DEA/T e G BALDIZZ/AGE Fotostock; **54.22a** National Park Service; **54.22b** National Park Service; **54.23 clockwise from top left** Charles D. Winters/Science Source; Keith Boggs; Terry Donnelly/Mary Liz Austin; Glacier Bay National Park and Preserve; **p. 1228** Custom Life Science Images/Alamy Stock Photo; **54.24 left to right** Charles D. Winters/Science Source; Keith Boggs; Terry Donnelly; Mary Liz Austin; Glacier Bay National Park and Preserve; **54.25 top** R. Grant Gilmore/Dynamac Corporation; **54.25 bottom** Lance Horn, National Undersea Research Center, University of North Carolina, Wilmington; **54.29** Tim Laman/National Geographic/Getty Images; **54.31** Nelish Pradhan, Bates College, Lewiston; **54.32** Josh Spice; **p. 1235** Jacques Rosès/Science Source

Chapter 55 **55.1** Steven Kazlowski/Nature Picture Library; **p. 1236 bottom** AGE Fotostock/Alamy Stock Photo; **55.2** Stone Nature Photography/Alamy Stock Photo; **55.3 left** Scimat/Science Source; **right** Justus de Cuveland/imagebroker/AGE Fotostock; **55.5** Image by Reto Stöckli, based on data provided by the MODIS Science Team/Earth Observatory/NASA; **55.8** A. T. Willett/Alamy Stock Photo; **p. 1243** Steven Katovich, USDA Forest Service, Bugwood.org; **inset** British Columbia Ministry of Forests, Lands and Natural Resource Operations; **55 Scientific Skills Exercise** David R. Frazier Photolibrary, Inc./Science Source; **55.15a-b** Hubbard Brook Research Foundation/USDA Forest Service; **55.16a-b** Mark Gallagher/Princeton Hydro, LLC/Ringoes, NJ; **55.17a** Kissimmee Division, South Florida Water Management District; **55.17b** Jean Hall/Holt Studio/Science Source; **55.17c** Tim Day, Xcluder Pest Proof Fencing Company; **55.17d** Kenji Morita/Environment Division, Tokyo Kyuei Co., Ltd; **55.18** U.S. Department of Energy; **Karoo** Jean Hall/Holt Studio/Science Source; **Maungatautari** Tim Day/Xcluder Pest Proof Fencing Company; **diver** Kenji Morita, Environment Division/Tokyo Kyuei Co., Ltd/ Morita, Kenji; **p. 1257** Eckart Pott/NHPA/Photoshot

Chapter 56 **56.1** Phung My Trung, vncreatures.net; **p. 1258 bottom** Eerika Schultz; **56.2** Trinh Le Nguyen/Shutterstock; **56.4 top** Neil Lucas/Nature Picture Library; **bottom** Mark Carwardine/Photolibrary/Getty Images; **56.5** Merlin D. Tuttle/Science Source; **56.6** Scott Camazine/Science Source; **56.7** Michael Edwards/The Image Bank/ Getty Images; **56.8 Chuck Pratt**/Alamy Stock Photo; **56.9** Benezeth Mutayoba University of Washington Center for Conservation Biology; **56.10** Travel Pictures/Alamy Stock Photo; **56.13** William Ervin/Science Source; **56.14** Craighead Institute; **56.15a** Chuck Bargeron, University of Georgia, Bugwood.org; **inset** William Leaman/Alamy Stock Photo; **56.15b** William D. Boyer/USDA; **56.16** Vladimir Melnikov/Fotolia; **56.17** R. O. Bierregaard, Jr., Biology Department, University of North Carolina, Charlotte; **56.18b** Frans Lemmens/Alamy Stock Photo; **56.20b** Edwin Giesbers/Nature Picture Library; **56.21** Mark Chiappone; **56.22** Lower Mainland Green Team; **56.23** Nigel Cattlin/Science Source; **56.24** NASA/Goddard Space Flight Center; **56.26** Erich Hartmann/ Magnum Photos; **56 Scientific Skills Exercise** Hank Morgan/Science Source; **56.30 resin canal** Biophoto Associates/Science Source; **tunnels** Ladd Livingston, Idaho Department of Lands, Bugwood.org; **dead trees** Dezene Huber, University of Northern British Columbia, Canada; **pika** Becka Barkley, courtesy of Chris Ray, University of Colorado, Boulder; **caribou *Rangifer tarandus*** E.A. Janes/Robert Harding World Imagery; ***Cerastium alpinum*** Gilles Delacroix/Garden World Images/AGE Fotostock; **urchin barrens** Scott Ling; **56.33 left** NASA; **right** NASA; **56.35a** Serge de Sazo/Science Source; **56.35b** Javier Trueba/MSF/Science Source; **56.35c** Gabriel Rojo/Nature Picture Library; **56.35d** Titus Lacoste/The Image Bank/Getty Images; **p. 1284** Edwin Giesbers/Nature Picture Library

Appendix A **p. A-5 6.24 left** OMIKRON/Science Source; **p. A-5 6.24 right** Dartmouth College Electron Microscope Facility; **p. A-11 12.4** Science Source; **p. A-11 12.8** J. Richard McIntosh; **p. A-11 bottom right** J.L. Carson "CMSP Biology"/Newscom; **p. A-16** Thomas A. Steitz, Yale University, New Haven; **p. A-29 30.9** Paul Atkinson/Shutterstock; **p. A-35 right 01** From: Anatomy of the vessel network within and between tree rings of *Fraxinus lanuginosa* (Oleaceae). Peter B. Kitin, Tomoyuki Fujii, Hisashi Abe and Ryo Funada. *American Journal of Botany*. 2004;91:779-788.

Appendix E **p. E-2 bacteria; archaea** Eye of Science/Science Source; **diatoms** M.I. Walker/Photoshot; **lily** Howard Rice/Dorling Kindersley, Ltd.; **p. E-3 fungus** Phil A. Dotson/Science Source; **macaque** Oliver Wien/European Pressphoto Agency creative account/Alamy Stock Photo

Illustration and Text Credits

Chapter 1 **1.23** Adapted from *The Real Process of Science* (2013), Understanding Science website. The University of California Museum of Paleontology, Berkeley, and the Regents of the University of California. Retrieved from http://undsci.berkeley.edu/article/howscienceworks_02; **1.25** Data from S. N. Vignieri, J. G. Larson, and H. E. Hoekstra, The Selective Advantage of Crypsis in Mice, *Evolution* 64:2153–2158 (2010); **1 Scientific Skills Exercise** Data from D. W. Kaufman, Adaptive Coloration in *Peromyscus polionotus*: Experimental Selection by Owls, *Journal of Mammalogy* 55:271–283 (1974).

Chapter 2 **2 Scientific Skills Exercise** Data from R. Pinhasi et al., Revised Age of late Neanderthal Occupation and the End of the Middle Paleolithic in the Northern Caucasus, *Proceedings of the National Academy of Sciences USA* 147:8611–8616 (2011). doi 10.1073/pnas.1018938108.

Chapter 3 **3.7** Republished with permission of American Association for the Advancement of Science, from Boom & Bust in the Great White North *Science* by Eli Kintisch. Vol. 349, Issue 6248, Pages 578–581; permission conveyed through Copyright Clearance Center, Inc.; adapted from figure on page 580; **3.7 map** Based on NOAA Fisheries. Bowhead Whale (*Balaena mysticetus*); **3.9** Based on Simulating Water and the Molecules of Life by Mark Gerstein and Michael Levitt, from *Scientific American*, November 1998; **3 Scientific Skills Exercise** Data from C. Langdon et al., Effect of Calcium Carbonate Saturation State on the Calcification Rate of an Experimental Coral Reef, *Global Biogeochemical Cycles* 14:639–654 (2000).

Chapter 4 **4.2** Data from S. L. Miller, A Production of Amino Acids Under Possible Primitive Earth Conditions, *Science* 117:528–529 (1953); **4 Scientific Skills Exercise** Data from E. T. Parker et al., Primordial Synthesis of Amines and Amino Acids in a 1958 Miller H_2S-rich Spark Discharge Experiment, *Proceedings of the National Academy of Sciences USA* 108:5526–5531 (2011). www.pnas.org/cgi/doi/10.1073/pnas.1019191108; **4.7** Adapted from Becker, Wayne M.; Reece, Jane B.; Poenie, Martin F., *The World of the Cell*, 3rd Ed., ©1996. Reprinted and electronically reproduced by permission of Pearson Education, Inc., Upper Saddle River, New Jersey.

Chapter 5 **5.11** Adapted from Wallace/Sanders/Ferl, *Biology: The Science of Life*, 3rd Ed., ©1991. Reprinted and electronically reproduced by permission of Pearson Education, Inc., Upper Saddle River, New Jersey; **5.13 Collagen** Data from Protein Data Bank ID 1CGD: "Hydration Structure of a Collagen Peptide" by Jordi Bella et al., from *Structure*, September 1995, Volume 3(9); **5.16 Space-filling model, Ribbon model** Data from PDB ID 2LYZ: R. Diamond. Real-space Refinement of the Structure of Hen Egg-white Lysozyme. *Journal of Molecular Biology* 82(3):371–91 (Jan. 25, 1974); **5.18 Transthyretin** Data from PDB ID 3GS0: S.K. Palaninathan, N.N. Mohamedmohaideen, E. Orlandini, G. Ortore, S. Nencetti, A. Lapucci, A. Rossello, J.S. Freundlich, J.C. Sacchettini. Novel Transthyretin Amyloid Fibril Formation Inhibitors: Synthesis, Biological Evaluation, and X-ray Structural Analysis. *Public Library of Science ONE* 4:e6290–e6290 (2009); **5.18 Collagen** Data from PDB ID 1CGD: J. Bella, B. Brodsky, and H.M. Berman. Hydration Structure of a Collagen Peptide, *Structure* 3:893–906 (1995); **5.18 Hemoglobin** Data from PDB ID 2HHB: G. Fermi, M.F. Perutz, B. Shaanan, R. Fourme. The Crystal Structure of Human Deoxyhaemoglobin at 1.74 Å resolution. *J. Mol. Biol.* 175:159–174 (1984).

Chapter 6 **6.6** Adapted from Becker, Wayne M.; Reece, Jane B.; Poenie, Martin F., *The World of the Cell*, 3rd Ed., ©1996. Reprinted and electronically reproduced by permission of Pearson Education, Inc. Upper Saddle River, New Jersey; **6.8 Animal cell** Adapted from Marieb, Elaine N.; Hoehn, Katja, *Human Anatomy and Physiology*, 8th Ed., © 2010. Printed and electronically reproduced by permission of Pearson Education, Inc., Upper Saddle River, New Jersey; **6.9 Small cell** Adapted from Marieb, Elaine N.; Hoehn, Katja, *Human Anatomy and Physiology*, 8th Ed., ©2010. Printed and electronically reproduced by permission of Pearson Education, Inc., Upper Saddle River, New Jersey; **6.10 Small cell** Adapted from Marieb, Elaine N.; Hoehn, Katja, *Human Anatomy and Physiology*, 8th Ed., ©2010. Printed and electronically reproduced by permission of Pearson Education, Inc., Upper Saddle River, New Jersey; **6.11 Small cell** Adapted from Marieb, Elaine N.; Hoehn, Katja, *Human Anatomy and Physiology*, 8th Ed., ©2010. Printed and electronically reproduced by permission of Pearson Education, Inc., Upper Saddle River, New Jersey; **6.12 Small cell** Adapted from Marieb, Elaine N.; Hoehn, Katja, *Human Anatomy and Physiology*, 8th Ed., ©2010. Printed and electronically reproduced by permission of Pearson Education, Inc., Upper Saddle River, New Jersey; **6.13 Small cell** Adapted from Marieb, Elaine N.; Hoehn, Katja, *Human Anatomy and Physiology*, 8th Ed., ©2010. Printed and electronically reproduced by permission of Pearson Education, Inc., Upper Saddle River, New Jersey; **6.15** Adapted from Marieb, Elaine N.; Hoehn, Katja, *Human Anatomy and Physiology*, 8th Ed., ©2010. Printed and electronically reproduced by permission of Pearson Education, Inc., Upper Saddle River, New Jersey; **6.17 Small cell** Adapted from Marieb, Elaine N.; Hoehn, Katja, *Human Anatomy and Physiology*, 8th Ed., ©2010. Printed and electronically reproduced by permission of Pearson Education, Inc., Upper Saddle River, New Jersey; **Table 6.1** Adapted from Hardin Jeff; Bertoni Gregory Paul, Kleinsmith, Lewis J., Becker's *World of the Cell*, 8th Edition, © 2012, p.423. Reprinted and electronically reproduced by permission of Pearson Education, Inc. Upper Saddle River, New Jersey; **6.22 Small cell** Adapted from Marieb, Elaine N.; Hoehn, Katja, *Human Anatomy and Physiology*, 8th Ed., ©2010. Printed and electronically reproduced by permission of Pearson Education, Inc., Upper Saddle River, New Jersey; **6.24 Small cell** Adapted from Marieb, Elaine N.; Hoehn, Katja, *Human Anatomy and Physiology*, 8th Ed., ©2010. Printed and electronically reproduced by permission of Pearson Education, Inc., Upper Saddle River, New Jersey; **6.32** Data from: Proton pump: PDB ID 3B8C: Crystal Structure of the Plasma Membrane Proton Pump, Pedersen, B.P., Buch-Pedersen, M.J., Morth, J.P., Palmgren, M.G., Nissen, P. (2007) *Nature* 450: 1111–1114; Calcium channel: PDB ID 5E1J: Structure of the Voltage-Gated Two-Pore Channel TPC1 from *Arabidopsis thaliana*, Guo, J., Zeng, W., Chen, Q., Lee, C., Chen, L., Yang, Y., Cang, C., Ren, D., Jiang, Y. (2016) *Nature* 531: 196–201; Aquaporin: PDB ID 5I32: Crystal Structure of an Ammonia-Permeable Aquaporin, Kirscht, A., Kaptan, S.S., Bienert, G.P., Chaumont, F., Nissen, P., de Groot, B.L., Kjellbom, P., Gourdon, P., Johanson, U. (2016) *Plos Biol.* 14: e1002411–e1002411; BRI1 and SERK1 co-receptors: PDB ID 4LSX: Molecular mechanism for plant steroid receptor activation by somatic embryogenesis co-receptor kinases, Santiago, J., Henzler, C., Hothorn, M. (2013) *Science* 341: 889–892; BRI1 kinase domain: PDB ID 4OAC: Crystal structures of the phosphorylated BRI1 kinase domain and implications for brassinosteroid signal initiation, Bojar, D., Martinez, J., Santiago, J., Rybin, V., Bayliss, R., Hothorn, M. (2014) *Plant J.* 78: 31–43; BAK1 kinase domain: PDB ID 3UIM: Structural basis for the impact of phosphorylation on the activation of plant receptor-like kinase BAK1, Yan, L., Ma, Y.Y., Liu, D., Wei, X., Sun, Y., Chen, X., Zhao, H., Zhou, J., Wang, Z., Shui, W., Lou, Z.Y. (2012) *Cell Res.* 22: 1304–1308; BSK8 pseudokinase: PDB ID: 4I92 Structural Characterization of the RLCK Family Member BSK8: A Pseudokinase with an Unprecedented Architecture, Grutter, C., Sreeramulu, S., Sessa, G., Rauh, D. (2013) *J. Mol. Biol.* 425: 4455–4467; ATP synthase PDB ID 1E79: The Structure of the Central Stalk in Bovine F(1)-ATPase at 2.4 Å Resolution, Gibbons, C., Montgomery, M.G., Leslie, A.G.W., Walker, J.E. (2000) *Nat. Struct. Biol.* 7: 1055; ATP synthase PDB ID 1C17: Structural changes linked to proton translocation by subunit c of the ATP synthase, Rastogi, V.K., Girvin, M.E. (1999) *Nature* 402: 263–268; ATP synthase PDB ID 1L2P: The "Second Stalk" of *Escherichia coli* ATP Synthase: Structure of the Isolated Dimerization Domain, Del Rizzo, P.A., Bi, Y., Dunn, S.D., Shilton, B.H. (2002) *Biochemistry* 41: 6875–6884; ATP synthase PDB ID 2A7U: Structural Characterization of the Interaction of the Delta and Alpha Subunits of the *Escherichia coli* F(1)F(0)-ATP Synthase by NMR Spectroscopy, Wilkens, S., Borchardt, D., Weber, J., Senior, A.E. (2005) *Biochemistry* 44: 11786–11794; Phosphofructokinase: PDB ID 1PFK: Crystal Structure of the Complex of Phosphofructokinase from *Escherichia coli* with Its Reaction Products, Shirakihara, Y., Evans, P.R. (1988) *J. Mol. Biol.* 204: 973–994; Hexokinase: PDB ID 4QS8: Biochemical and Structural Study of *Arabidopsis* Hexokinase 1, Feng, J., Zhao, S., Chen, X., Wang, W., Dong, W., Chen, J., Shen, J.-R., Liu, L., Kuang, T. (2015) *Acta Crystallogr., Sect. D* 71: 367–375; Isocitrate dehydrogenase: PDB ID 3BLW: Allosteric Motions in Structures of Yeast NAD^+-specific Isocitrate Dehydrogenase, Taylor, A.B., Hu, G., Hart, P.J., McAlister-Henn, L. (2008) *J. Biol. Chem.* 283:10872–10880; NADH-quinone oxidoreductase: PDB ID 3M9S: The architecture of respiratory complex I, Efremov, R.G., Baradaran, R., Sazanov, L.A. (2010) *Nature* 465: 441–445; NADH-quinone oxidore-ductase: PDB ID 3RKO: Structure of the membrane domain of respiratory complex I, Efremov,

R.G., Sazanov, L.A. (2011) *Nature* 476: 414–420; Succinate dehydrogenase: PDB ID 1NEK: Architecture of Succinate Dehydrogenase and Reactive Oxygen Species Generation, Yankovskaya, V., Horsefield, R., Tornroth, S., Luna-Chavez, C., Miyoshi, H., Leger, C., Byrne, B., Cecchini, G., Iwata, S. (2003) *Science* 299: 700–704; Ubiquinone: http://www.proteopedia.org/wiki/index.php/Image: Coenzyme_Q10.pdb; Cytochrome *bc1*: PDB ID 1BGY: Complete structure of the 11-subunit bovine mitochondrial cytochrome *bc1* complex, Iwata, S., Lee, J.W., Okada, K., Lee, J.K., Iwata, M., Rasmussen, B., Link, T.A., Ramaswamy, S., Jap, B.K. (1998) *Science* 281: 64–71; Cytochrome *c*: PDB ID 3CYT: Redox Conformation Changes in Refined Tuna Cytochrome *c*, Takano, T., Dickerson, R.E. (1980) *Proc. Natl. Acad. Sci. USA* 77: 6371–6375; Cytochrome *c* oxidase: PDB ID 1OCO: Redox-Coupled Crystal Structural Changes in Bovine Heart Cytochrome *c* Oxidase, Yoshikawa, S., Shinzawa-Itoh, K., Nakashima, R., Yaono, R., Yamashita, E., Inoue, N., Yao, M., Fei, M.J., Libeu, C.P., Mizushima, T., Yamaguchi, H., Tomizaki, T., Tsukihara, T. (1998) *Science* 280: 1723–1729; Rubisco: PDB ID 1RCX: The Structure of the Complex between Rubisco and its Natural Substrate Ribulose 1,5-Bisphosphate, Taylor, T.C., Andersson, I. (1997) *J. Mol. Biol.* 265: 432–444; Photosystem II: PDB ID 1S5L: Architecture of the Photosynthetic Oxygen-Evolving Center, Ferreira, K.N., Iverson, T.M., Maghlaoui, K., Barber, J., Iwata, S. (2004) *Science* 303: 1831–1838; Plastoquinone: http://www.rcsb.org/pdb/ligand/ligandsummary.do?hetId=PL9; Photosystem I: PDB ID 1JB0: Three-dimensional Structure of Cyanobacterial Photosystem I at 2.5 Å Resolution, Jordan, P., Fromme, P., Witt, H.T., Klukas, O., Saenger, W., Krauss, N. (2001) *Nature* 411: 909–917; Ferredoxin-NADP$^+$ reductase: PDB ID 3W5V: Concentration-Dependent Oligomerization of Cross-Linked Complexes between Ferredoxin and Ferredoxin-NADP$^{(+)}$ Reductase; DNA:PDB ID 1BNA: Structure of a B-DNA Dodecamer: Conformation and Dynamics, Drew, H.R., Wing, R.M., Takano, T., Broka, C., Tanaka, S., Itakura, K., Dickerson, R.E. (1981) *Proc. Natl. Acad. Sci. USA* 78: 2179–2183; RNA polymerase: PDB ID 2E2I: Structural basis of transcription: role of the trigger loop in substrate specificity and catalysis, Wang, D., Bushnell, D.A., Westover, K.D., Kaplan, C.D., Kornberg, R.D. (2006) *Cell* (Cambridge, Mass.) 127: 941–954; Nucleosome: PDB ID 1AOI: Crystal Structure of the Nucleosome Core Particle at 2.8 Å Resolution, Luger, K., Mader, A.W., Richmond, R.K., Sargent, D.F., Richmond, T.J. (1997) *Nature* 389: 251–260; tRNA: PDB ID 4TNA: Further refinement of the structure of yeast tRNAPhe, Hingerty, B., Brown, R.S., Jack, A. (1978) *J. Mol. Biol.* 124: 523–534; Ribosome: PDB ID 1FJF: Structure of the 30S Ribosomal Subunit, Wimberly, B.T., Brodersen, D.E., Clemons Jr., W.M., Morgan-Warren, R.J., Carter, A.P., Vonrhein, C., Hartsch, T., Ramakrishnan, V. (2000) *Nature* 407: 327–339; Ribosome: PDB ID 1JJ2: The Kink-Turn: A New RNA Secondary Structure Motif, Klein, D.J., Schmeing, T.M., Moore, P.B., Steitz, T.A. (2001) *EMBO J.* 20: 4214–4221; Microtubule: PDB ID 3J2U: Structural Model for Tubulin Recognition and Deformation by Kinesin-13 Microtubule Depolymerases, Asenjo, A.B., Chatterjee, C., Tan, D., Depaoli, V., Rice, W.J., Diaz-Avalos, R., Silvestry, M., Sosa, H. (2013) *Cell Rep.* 3: 759–768; Actin microfilament: PDB ID 1ATN: Atomic Structure of the Actin:DNase I Complex. Kabsch, W., Mannherz, H.G., Suck, D., Pai, E.F., Holmes, K.C. (1990) *Nature* 347: 37–44; Myosin: PDB ID 1M8Q: Molecular Modeling of Averaged Rigor Crossbridges from Tomograms of Insect Flight Muscle, Chen, L.F., Winkler, H., Reedy, M.K., Reedy, M.C., Taylor, K.A. (2002) *J. Struct. Biol.* 138: 92–104; Phosphoglucose Isomerase: PDB ID 1IAT: The Crystal Structure of Human Phosphoglucose Isomerase at 1.6 Å Resolution: Implications for Catalytic Mechanism, Cytokine Activity and Haemolytic Anaemia, Read, J., Pearce, J., Li, X., Muirhead, H., Chirgwin, J., Davies, C. (2001) *J. Mol. Biol.* 309: 447–463; Aldolase: PDB ID 1ALD: Activity and Specificity of Human Aldolases, Gamblin, S.J., Davies, G.J., Grimes, J.M., Jackson, R.M., Littlechild, J.A., Watson, H.C. (1991) *J. Mol. Biol.* 219: 573–576; Triosephosphate Isomerase: PDB ID 7TIM: Structure of the Triosephosphate Isomerase-Phosphoglycolohydroxamate Complex: An Analogue of the Intermediate on the Reaction Pathway, Davenport, R.C., Bash, P.A., Seaton, B.A., Karplus, M., Petsko, G.A., Ringe, D. (1991) *Biochemistry* 30: 5821–5826; Glyceraldehyde-3-Phosphate Dehydrogenase: PDB ID 3GPD: Twinning in Crystals of Human Skeletal Muscle D-Glyceraldehyde-3-Phosphate Dehydrogenase, Mercer, W.D., Winn, S.I., Watson, H.C. (1976) *J. Mol. Biol.* 104: 277–283; Phosphoglycerate Kinase: PDB ID 3PGK: Sequence and Structure of Yeast Phosphoglycerate Kinase, Watson, H.C., Walker, N.P., Shaw, P.J., Bryant, T.N., Wendell, P.L., Fothergill, L.A., Perkins, R.E., Conroy, S.C., Dobson, M.J., Tuite, M.F. (1982) *Embo J.* 1: 1635–1640; Phosphoglycerate Mutase: PDB ID 3PGM: Structure and Activity of Phosphoglycerate Mutase, Winn, S.I., Watson, H.C., Harkins, R.N., Fothergill, L.A. (1981) *Philos. Trans. R. Soc. London, Ser. B* 293: 121–130; Enolase: PDB ID 5ENL: Inhibition of Enolase: The Crystal Structures of Enolase-Ca$^{2(+)}$- 2-Phosphoglycerate and Enolase-Zn$^{2(+)}$-Phosphoglycolate Complexes at 2.2-Å Resolution, Lebioda, L., Stec, B., Brewer, J.M., Tykarska, E. (1991) *Biochemistry* 30: 2823–2827; Pyruvate Kinase: PDB ID 1A49: Structure of the Bis(Mg^{2+})-ATP-Oxalate Complex of the Rabbit Muscle Pyruvate Kinase at 2.1 Å Resolution: ATP Binding over a Barrel, Larsen, T.M., Benning, M.M., Rayment, I., Reed, G.H. (1998) *Biochemistry* 37: 6247–6255; Citrate Synthase: PDB ID 1CTS: Crystallographic Refinement and Atomic Models of Two Different Forms of Citrate Synthase at 2.7 and 1.7 Å Resolution, Remington, S., Wiegand, G., Huber, R. (1982) *J. Mol. Biol.* 158: 111–152; Succinyl-CoA Synthetase: PDB ID 2FP4: Interactions of GTP with the ATP-Grasp Domain of GTP-Specific Succinyl-CoA Synthetase, Fraser, M.E., Hayakawa, K., Hume, M.S., Ryan, D.G., Brownie, E.R. (2006) *J. Biol. Chem.* 281: 11058-11065; Malate Dehydrogenase: PDB ID 4WLE: Crystal Structure of Citrate Bound MDH2, Eo, Y.M., Han, B.G., Ahn, H.C. To Be Published; **6 Summary art nucleus** Adapted from Marieb, Elaine N.; Hoehn, Katja, *Human Anatomy and Physiology*, 8th Ed., ©2010. Printed and electronically reproduced by permission of Pearson Education, Inc., Upper Saddle River, New Jersey; **6 Summary art endoplasmic reticulum** Adapted from Marieb, Elaine N.; Hoehn, Katja, *Human Anatomy and Physiology*, 8th Ed., ©2010. Printed and electronically reproduced by permission of Pearson Education, Inc., Upper Saddle River, New Jersey; **6 Summary art Golgi apparatus** Adapted from Marieb, Elaine N.; Hoehn, Katja, *Human Anatomy and Physiology*, 8th Ed., ©2010. Printed and electronically reproduced by permission of Pearson Education, Inc., Upper Saddle River, New Jersey.

Chapter 7 **7.4** Data from L. D. Frye and M. Edidin, The Rapid Intermixing of Cell Surface Antigens after Formation of Mouse-human Heterokaryons, *Journal of Cell Science* 7:319 (1970); **7.6** Based on Similar Energetic Contributions of Packing in the Core of Membrane and Water-Soluble Proteins by Nathan H. Joh et al., from *Journal of the American Chemical Society*, Volume 131(31); **7 Scientific Skills Exercise** Data from Figure 1 in T. Kondo and E. Beutler, Developmental Changes in Glucose Transport of Guinea Pig Erythrocytes, *Journal of Clinical Investigation* 65:1–4 (1980).

Chapter 8 **8 Scientific Skills Exercise** Data from S. R. Commerford et al., Diets Enriched in Sucrose or Fat Increase Gluconeogenesis and G-6-pase but not Basal Glucose Production in Rats, *American Journal of Physiology—Endocrinology and Metabolism* 283:E545–E555 (2002); **8.19** Data from Protein Data Bank ID 3e1f: "Direct and Indirect Roles of His-418 in Metal Binding and in the Activity of Beta-Galactosidase (*E. coli*)" by Douglas H. Juers et al., from *Protein Science*, June 2009, Volume 18(6); **8.20** Data from Protein Data Bank ID 1MDYO: "Crystal Structure of MyoD bHLH Domain-DNA Complex: Perspectives on DNA Recognition and Implications for Transcriptional Activation" from *Cell*, May 1994, Volume 77(3); **8.22 Small cell** Adapted from Marieb, Elaine N.; Hoehn, Katja, *Human Anatomy and Physiology*, 8th Ed., ©2010. Printed and electronically reproduced by permission of Pearson Education, Inc., Upper Saddle River, New Jersey.

Chapter 9 **9.5** Adaptation of Figure 2.69 from *Molecular Biology of the Cell*, 4th Edition, by Bruce Alberts et al. Garland Science/Taylor & Francis LLC; **9.9** Figure adapted from *Biochemistry*, 4th Edition, by Christopher K. Mathews et al. Pearson Education, Inc.; **9 Scientific Skills Exercise** Data from M. E. Harper and M. D. Brand, The Quantitative Contributions of Mitochondrial Proton Leak and ATP Turnover Reactions to the Changed Respiration Rates of Hepatocytes from Rats of Different Thyroid Status, *Journal of Biological Chemistry* 268:14850–14860 (1993).

Chapter 10 **10.10** Data from T. W. Engelmann, Bacterium *Photometricum. Ein Beitrag zur Vergleichenden Physiologie des Licht-und Farbensinnes, Archiv. für Physiologie* 30:95–124 (1883); **10.13b** Data from Architecture of the Photosynthetic Oxygen-Evolving Center by Kristina N. Ferreira et al., from *Science*, March 2004, Volume 303(5665); **10.15** Adaptation of Figure 4.1 from *Energy, Plants, and Man*, by Richard Walker and David Alan Walker. Copyright © 1992 by Richard Walker and David Alan Walker. Reprinted with permission of Richard Walker; **10 Scientific Skills Exercise** Data from D. T. Patterson and E. P. Flint, Potential Effects of Global Atmospheric CO_2 Enrichment on the Growth and Competitiveness of C_3 and C_4 Weed and Crop Plants, *Weed Science* 28(1):71–75 (1980).

Chapter 11 **11 Problem-Solving Exercise** Data from N. Balaban et al., Treatment of *Staphylococcus aureus* Biofilm Infection by the Quorum-Sensing Inhibitor RIP, *Antimicrobial Agents and Chemotherapy*, 51:2226–2229 (2007); **11.8** Adapted from Becker, Wayne M.; Reece, Jane B.; Poenie, Martin F., *The World of the Cell*, 3rd Edition, © 1996. Reprinted and electronically reproduced by permission of Pearson Education, Inc., Upper Saddle River, New Jersey; **11.12** Adapted from Becker, Wayne M.; Reece, Jane B.; Poenie, Martin F., *The World of the Cell*, 3rd Edition, © 1996. Reprinted and electronically reproduced by permission of Pearson Education, Inc., Upper Saddle River, New Jersey.

Chapter 12 **12.9** Data from G. J. Gorbsky, P. J. Sammak, and G. G. Borisy, Chromosomes Move Poleward in Anaphase along Stationary Microtubules that Coordinately Disassemble from their Kinetochore Ends, *Journal of Cell Biology* 104:9–18 (1987); **12.13** Adaptation of Figure 18.41 from *Molecular Biology of the Cell*, 4th Edition, by Bruce Alberts et al. Garland Science/Taylor & Francis LLC; **12.14** Data from R. T. Johnson and P. N. Rao, Mammalian Cell Fusion: Induction of Premature Chromosome Condensation in Interphase Nuclei, *Nature* 226:717–722 (1970); **12 Scientific Skills Exercise** Data from K. K. Velpula et al., Regulation of Glioblastoma Progression by Cord Blood Stem Cells is Mediated by Downregulation of Cyclin D1, *PLoS ONE* 6(3): e18017 (2011).

Chapter 14 **14.3** Data from G. Mendel, Experiments in Plant Hybridization, *Proceedings of the Natural History Society of Brünn* 4:3–47 (1866); **14.8** Data from G. Mendel, Experiments in Plant Hybridization, *Proceedings of the Natural History Society of Brünn* 4:3–47 (1866).

Chapter 15 **15.4** Data from T. H. Morgan, Sex-limited inheritance in *Drosophila, Science* 32:120–122 (1910); **15.9** Based on the data from "The Linkage of Two Factors in Drosophila That Are Not Sex-Linked" by Thomas Hunt Morgan and Clara J. Lynch, from *Biological Bulletin*, August 1912, Volume 23(3).

Chapter 16 **16.2** Data from F. Griffith, The Significance of Pneumococcal Types, *Journal of Hygiene* 27:113–159 (1928); **16.4** Data from A. D. Hershey and M. Chase, Independent Functions of Viral Protein and Nucleic Acid in Growth of Bacteriophage, *Journal of General Physiology* 36:39–56 (1952); **16 Scientific Skills Exercise** Data from several papers by Chargaff: for example, E. Chargaff et al., Composition of the Desoxypentose Nucleic Acids of Four Genera of Sea-urchin, *Journal of Biological Chemistry* 195:155–160 (1952); **pp. 320–321 Quote** J. D. Watson and F. H. C. Crick, Genetical Implications of the Structure of Deoxyribonucleic Acid, *Nature* 171:964–967 (1953); **16.11** Data from M. Meselson and F. W. Stahl, The Replication of DNA in *Escherichia coli, Proceedings of the National Academy of Sciences USA* 44:671–682 (1958).

Chapter 17 **17.3** Data from A. M. Srb and N. H. Horowitz, The Ornithine Cycle in *Neurospora* and Its Genetic Control, *Journal of Biological Chemistry* 154:129–139 (1944); **17.12** Adapted from Becker, Wayne M.; Reece, Jane B.; Poenie, Martin F., *The World of the Cell*, 3rd Edition, © 1996. Reprinted and electronically reproduced by permission of Pearson Education, Inc., Upper Saddle River, New Jersey; **17.14** Adapted from Kleinsmith, Lewis J., Kish, Valerie M.; *Principles of Cell and Molecular Biology*. Reprinted and electronically reproduced by permissions of Pearson Education, Inc., Upper Saddle River, New Jersey; **17.18** Adapted from Mathews, Christopher K.; Van Holde, Kensal E., *Biochemistry*, 2nd ed., ©1996. Reprinted and electronically reproduced by permission of Pearson Education, Inc. Upper Saddle River, New Jersey; **17 Scientific Skills Exercise** Material provided courtesy of Dr. Thomas Schneider, National Cancer Institute,

National Institutes of Health, 2012; **17 Problem-Solving Exercise** Data from N. Nishi and K. Nanjo, Insulin Gene Mutations and Diabetes, *Journal of Diabetes Investigation* Vol. 2: 92–100 (2011).

Chapter 18 **18.9** Data from PDB ID 1MDY: P. C. Ma et al. Crystal structure of MyoD bHLH Domain-DNA Complex: Perspectives on DNA Recognition and Implications for Transcriptional Activation, *Cell* 77:451–459 (1994); **18 Scientific Skills Exercise** Data from J. N. Walters et al., Regulation of Human Microsomal Prostaglandin E Synthase-1 by IL-1b Requires a Distal Enhancer Element with a Unique Role for C/EBPb, *Biochemical Journal* 443:561–571 (2012); **18.26** Adapted from Becker, Wayne M.; Reece, Jane B.; Poenie, Martin F., *The World of the Cell,* 3rd Edition, © 1996. Reprinted and electronically reproduced by permission of Pearson Education, Inc., Upper Saddle River, New Jersey.

Chapter 19 **19.2** Data from M. J. Beijerinck, Concerning a Contagium Vivum Fluidum as Cause of the Spot Disease of Tobacco Leaves, *Verhandelingen der Koninkyke Akademie Wettenschappen te Amsterdam* 65:3–21 (1898). Translation published in English as *Phytopathological Classics Number 7* (1942), American Phytopathological Society Press, St. Paul, MN; **19 Scientific Skills Exercise** Data from J.-R. Yang et al., New Variants and Age Shift to High Fatality Groups Contribute to Severe Successive Waves in the 2009 Influenza Pandemic in Taiwan, *PLoS ONE* 6(11): e28288 (2011).

Chapter 20 **20.7** Adapted from Becker, Wayne M.; Reece, Jane B.; Poenie, Martin F., *The World of the Cell,* 3rd Edition, © 1996. Reprinted and electronically reproduced by permission of Pearson Education, Inc., Upper Saddle River, New Jersey; **20.16** Data from J. B. Gurdon et al., The Developmental Capacity of Nuclei Transplanted from Keratinized Cells of Adult Frogs, *Journal of Embryology and Experimental Morphology* 34:93–112 (1975); **20.21** Data from K. Takahashi et al., Induction of pluripotent stem cells from adult human fibroblasts by defined factors, *Cell* 131:861–872 (2007).

Chapter 21 **21.3** Simulated screen shots based on Mac OS X and from data found at NCBI, U.S. National Library of Medicine using Conserved Domain Database, Sequence Alignment Viewer, and Cn3D; **21.8** Adapted from Becker, Wayne M.; Reece, Jane B.; Poenie, Martin F., *The World of the Cell,* 3rd Edition, © 1996. Reprinted and electronically reproduced by permission of Pearson Education, Inc., Upper Saddle River, New Jersey; **21.9** Adapted from Becker, Wayne M.; Reece, Jane B.; Poenie, Martin F., *The World of the Cell,* 3rd Edition, © 1996. Reprinted and electronically reproduced by permission of Pearson Education, Inc., Upper Saddle River, New Jersey; **21.10 Hemoglobin** Data from PDB ID 2HHB: G. Fermi, M.F. Perutz, B. Shaanan, and R. Fourme. The Crystal Structure of Human Deoxyhaemoglobin at 1.74 Å resolution. *J. Mol. Biol.* 175:159–174 (1984); **21.15a** Drawn from data in Protein Data Bank ID 1LZ1: "Refinement of Human Lysozyme at 1.5 Å Resolution Analysis of Non-bonded and Hydrogen-bond Interactions" by P. J. Artymiuk and C. C. Blake, from *Journal of Molecular Biology*, 1981, 152:737–762; **21.15b** Drawn from data in Protein Data Bank ID 1A4V: "Structural Evidence for the Presence of a Secondary Calcium Binding Site in Human Alpha-Lactalbumin" by N. Chandra et al., from *Biochemistry*, 1998, 37:4767–4772; **21 Hemoglobin in Scientific Skills Exercise** PDB ID 2HHB: G. Fermi, M.F. Perutz, B. Shaanan, and R. Fourme. The Crystal Structure of Human Deoxyhaemoglobin at 1.74 Å Resolution. *J. Mol. Biol.* 175:159–174 (1984); **21 Scientific Skills Exercise** Compiled using data from NCBI; **21.18** Data from W. Shu et al., Altered ultrasonic vocalization in mice with a disruption in the *Foxp2* gene, *Proceedings of the National Academy of Sciences USA* 102:9643–9648 (2005); **21.19** Adapted from *The Homeobox: Something Very Precious That We Share with Flies, From Egg to Adult* by Peter Radetsky, © 1992. Reprinted by permission from William McGinnis; **21.20** Adaptation from "*Hox* Genes and the Evolution of Diverse Body Plans" by Michael Akam, from *Philosophical Transactions of the Royal Society B: Biological Sciences*, September 29, 1995, Volume 349(1329): 313–319. Reprinted by permission from The Royal Society.

Chapter 22 **22.8** Artwork by Utako Kikutani (as appeared in "What Can Make a Four-Ton Mammal a Most Sensitive Beast?" by Jeheskel Shoshani, from *Natural History*, November 1997, Volume 106(1), 36–45). Copyright © 1997 by Utako Kikutani. Reprinted with permission of the artist; **22.13** Data from "Host Race Radiation in the Soapberry Bug: Natural History with the History" by Scott P. Carroll and Christin Boyd, from *Evolution*, 1992, Volume 46(4); **22.14** Figure created by Dr. Binh Diep on request of Michael Cain. Copyright © 2011 by Binh Diep. Reprinted with permission; **22 Scientific Skills Exercise** Data from J. A. Endler, Natural Selection on Color Patterns in *Poecilia reticulata, Evolution* 34:76–91 (1980); **22 Test Your Understanding Question 7** Data from C. F. Curtis et al., Selection for and Against Insecticide Resistance and Possible Methods of Inhibiting the Evolution of Resistance in Mosquitoes, *Ecological Entomology* 3:273–287 (1978).

Chapter 23 **23.4** Based on the data from *Evolution*, by Douglas J. Futuyma. Sinauer Associates, 2006; and Nucleotide Polymorphism at the Alcohol Dehydrogenase Locus of *Drosophila melanogaster* by Martin Kreitman, from *Nature*, August 1983, Volume 304(5925); **23.11a Maps** Adapted Figure 20.6 from *Discover Biology*, 2nd Edition, edited by Michael L. Cain, Hans Damman, Robert A. Lue, and Carol Kaesuk Loon. W. W. Norton & Company, Inc.; **23.12** Data from Joseph H. Camin and Paul R. Ehrlich, Natural Selection in Water Snakes (*Natrix sipedon L.*) on Islands in Lake Erie, *Evolution* 12:504–511 (1958); **23.14** Based on many sources: *Evolution* by Douglas J. Futuyma. Sinauer Associates 2005; and *Vertebrate Paleontology and Evolution* by Robert L. Carroll. W.H. Freeman & Co., 1988; **23.16** Data from A. M. Welch et al., Call Duration as an Indicator of Genetic Quality in Male Gray Tree Frogs, *Science* 280:1928–1930 (1998); **23.17** Adapted from Frequency-Dependent Natural Selection in the Handedness of Scale-Eating Cichlid Fish by Michio Hori, from *Science*, April 1993, Volume 260(5105); **23 Test Your Understanding Question 7** Data from R. K. Koehn and T. J. Hilbish, The Adaptive Importance of Genetic Variation, *American Scientist* 75:134–141 (1987).

Chapter 24 **24.6** Original unpublished graph created by Brian Langerhans; **24.7** Data from D. M. B. Dodd, Reproductive Isolation as a Consequence of Adaptive Divergence in *Drosophila pseudoobscura, Evolution* 43:1308–1311 (1989); **24 Scientific Skills Exercise** Data from S. G. Tilley, A. Verrell, and S. J. Arnold, Correspondence between Sexual Isolation and Allozyme Differentiation: A Test in the Salamander *Desmognathus ochrophaeus, Proceedings of the National Academy of Sciences USA* 87:2715–2719 (1990); **24.12** Data from O. Seehausen and J. J. M. van Alphen, The Effect of Male Coloration on Female Mate Choice in Closely Related Lake Victoria Cichlids (*Haplochromis nyererei* complex), *Behavioral Ecology and Sociobiology* 42:1–8 (1998); **24.13d** Based on *Hybrid Zone and the Evolutionary Process*, edited by Richard G. Harrison. Oxford University Press; **24.19b** Data from Role of Gene Interactions in Hybrid Speciation: Evidence from Ancient and Experimental Hybrids by Loren H. Rieseberg et al., from *Science*, May 1996, Volume 272(5262).

Chapter 25 **25.2** Based on data from The Miller Volcanic Spark Discharge Experiment by Adam P. Johnson et al., from *Science*, October 2008, Volume 322(5900); **25.4** Based on "Experimental Models of Primitive Cellular Compartments: Encapsulation, Growth, and Division" by Martin M. Hanczyc, Shelly M. Fujikawa, and Jack W. Szostak, from *Science*, October 2003, Volume 302(5645); **25.6** Eicher, D. L, *Geologic Time*, 2nd Ed., ©1976, p. 119. Adapted and electronically reproduced by permission of Pearson Education, Inc., Upper Saddle River, New Jersey; **25.7 First four skulls** Adapted from many sources including D.J. Futuyma, *Evolution*, Fig. 4.10, Sunderland, MA: Sinauer Associates, Sunderland, MA (2005) and from R.L. Carroll, *Vertebrate Paleontology and Evolution*. W.H. Freeman & Co. (1988); **25.7 Last skull** Adapted from Z. Luo et al., A New Mammaliaform from the Early Jurassic and Evolution of Mammalian Characteristics, *Science* 292:1535 (2001); **25.8** Adapted from When Did Photosynthesis Emerge on Earth? by David J. Des Marais, from *Science*, September 2000, Volume 289(5485).**25.9** Adapted from The Rise of Atmospheric Oxygen by Lee R. Kump, from *Nature*, January 2008, Volume 451(7176); **25 Scientific Skills Exercise** Data from T. A. Hansen, Larval Dispersal and Species Longevity in Lower Tertiary Gastropods, *Science* 199:885–887 (1978); **25.15** Based on *Earthquake Information Bulletin*, December 1977, Volume 9(6), edited by Henry Spall; **25.17** Based on many sources: D.M. Raup and J.J. Sepkoski, Jr., Mass Extinctions in the Marine Fossil Record, *Science* 215:1501–1503 (1982); J.J. Sepkoski, Jr., A Kinetic Model of Phanerozoic Taxonomic Diversity. III. Post-Paleozoic Families and Mass Extinctions, *Paleobiology* 10:246–267 (1984); and D.J. Futuyma, *The Evolution of Biodiversity*, p. 143, Fig. 7.3a and p. 145, Fig. 7.6, Sinauer Associates, Sunderland, MA; **25.19** Based on data from A Long-Term Association between Global Temperature and Biodiversity, Origination and Extinction in the Fossil Record by P.J. Mayhew, G.B. Jenkins and T.G. Benton, *Proceedings of the Royal Society B: Biological Sciences* 275(1630):47–53. The Royal Society, 2008; **25.20** Adapted from Anatomical and Ecological Constraints on Phanerozoic Animal Diversity in the Marine Realm by Richard K. Bambach et al., from *Proceedings of the National Academy of Sciences USA*, May 2002, Volume 99(10); **25.25** Based on data from The Miller Volcanic Spark Discharge Experiment by Adam P. Johnson et al., from *Science*, October 2008, Volume 322(5900); **25.26** Data from Genetic and Developmental Basis of Evolutionary Pelvic Reduction in Threespine Sticklebacks by Michael D. Shapiro et al., from *Nature*, April 2004, Volume 428(6984); **25.28** Adaptations of Figure 3-1(a–d, f) from *Evolution*, 3rd Edition, by Monroe W. Strickberger. Jones & Bartlett Learning, Burlington, MA.

Chapter 26 **26.6** Data from C. S. Baker and S. R. Palumbi, Which Whales Are Hunted? A Molecular Genetic Approach to Monitoring Whaling, *Science* 265:1538–1539 (1994); **26.13** Based on The Evolution of the Hedgehog Gene Family in Chordates: Insights from Amphioxus Hedgehog by Sebastian M. Shimeld, from *Developmental Genes and Evolution*, January 1999, Volume 209(1); **26.19** Based on *Molecular Markers, Natural History, and Evolution*, 2nd ed., by J.C. Advise. Sinauer Associates, 2004; **26.20** Adapted from Timing the Ancestor of the HIV-1 Pandemic Strains by B. Korber et al., *Science* 288(5472):1789–1796 (6/9/00); **26.23** Adapted from Phylogenetic Classification and the Universal Tree by W.F. Doolittle, *Science* 284(5423):2124–2128 (6/25/99); **26 Scientific Skills Exercise** Data from Nancy A. Moran, Yale University. See N. A. Moran and T. Jarvik, Lateral transfer of genes from fungi underlies carotenoid production in aphids, *Science* 328:624–627 (2010).

Chapter 27 **27.10 Graph** Data from V. S. Cooper and R. E. Lenski, The Population Genetics of Ecological Specialization in Evolving *Escherichia coli* Populations, *Nature* 407:736–739 (2000); **27.18** Data from Root-Associated Bacteria Contribute to Mineral Weathering and to Mineral Nutrition in Trees: A Budgeting Analysis by Christophe Calvaruso et al., *Applied and Environmental Microbiology*, February 2006, Volume 72(2); **27 Scientific Skills Exercise** Data from L. Ling et al. A New Antibiotic Kills Pathogens without Detectable Resistance, *Nature* 517:455–459 (2015); **27 Test Your Understanding Question 8** Data from J. J. Burdon et al., Variation in the Effectiveness of Symbiotic Associations between Native Rhizobia and Temperate Australian *Acacia*: Within Species Interactions, *Journal of Applied Ecology* 36:398–408 (1999).

Chapter 28 **28 Scientific Skills Exercise** Data from D. Yang et al., Mitochondrial Origins, *Proceedings of the National Academy of Sciences USA* 82:4443–4447 (1985); **28.17** Adaptation of illustration by Kenneth X. Probst, from *Microbiology* by R.W. Bauman. Copyright © 2004 by Kenneth X. Probst; **28.24** Data from A. Stechmann and T. Cavalier-Smith, Rooting the Eukaryote Tree by Using a Derived Gene Fusion, *Science* 297:89–91 (2002); **28.30** Based on Global Phytoplankton Decline over the Past Century by Daniel G. Boyce et al., from *Nature*, July 29, 2010, Volume 466(7306); and authors' personal communications.

Chapter 29 **29.9** Data from "Inputs, Outputs, and Accumulation of Nitrogen in an Early Successional Moss (*Polytrichum*) Ecosystem" by Richard D. Bowden, from *Ecological Monographs*, June 1991, Volume 61(2); **29 Scientific Skills Exercise** Data from T.M. Lenton et al, First Plants Cooled the Ordovician. *Nature Geoscience* 5:86–89 (2012); **29 Test Your Understanding Question 8** Data from O. Zackrisson et al., Nitrogen Fixation Increases with Successional Age in Boreal Forests, *Ecology* 85:3327–3334 (2006).

Chapter 30 **30 Scientific Skills Exercise** Data from S. Sallon et al, Germination, Genetics, and Growth of an Ancient Date Seed. *Science* 320:1464 (2008); **30.14a** Adapted from "A Revision of *Williamsoniella*" by T. M. Harris, from *Proceedings of the Royal Society B: Biological Sciences*, October 1944, Volume 231(583): 313–328; **30.14b** Adaptation of Figure 2.3, *Phylogeny and Evolution of Angiosperm,* 2nd Edition, by Douglas E. Soltis et al. (2005). Sinauer Associates, Inc.

Chapter 31 **31 Scientific Skills Exercise** Data from F. Martin et al., The genome of *Laccaria bicolor* provides insights into mycorrhizal symbiosis, *Nature* 452:88–93 (2008);

31.20 Data from A. E. Arnold et al., Fungal Endophytes Limit Pathogen Damage in a Tropical Tree, *Proceedings of the National Academy of Sciences USA* 100:15649–15654 (2003); **31.25** Adaption of Figure 1 from "Reversing Introduced Species Effects: Experimental Removal of Introduced Fish Leads to Rapid Recovery of a Declining Frog" by Vance T. Vredenburg, from *Proceedings of the National Academy of Sciences USA*, May 2004, Volume 101(20). Copyright (2004) National Academy of Sciences, U.S.A.; **31 Test Your Understanding Question 5** Data from R. S. Redman et al., Thermotolerance Generated by Plant/Fungal Symbiosis, *Science* 298:1581 (2002).

Chapter 32 **32 Scientific Skills Exercise** Data from Bradley Deline, University of West Georgia, and Kevin Peterson, Dartmouth College, 2013; **32 Test Your Understanding Question 6** Data from A. Hejnol and M. Martindale, The Mouth, the Anus, and the Blastopore—Open Questions About Questionable Openings. In *Animal Evolution: Genomes, Fossils and Trees*, eds. D. T. J. Littlewood and M. J. Telford, Oxford University Press, pp. 33–40 (2009).

Chapter 33 **33 Scientific Skills Exercise** Data from R. Rochette et al., Interaction between an Invasive Decapod and a Native Gastropod: Predator Foraging Tactics and Prey Architectural Defenses, *Marine Ecology Progress Series* 330:179–188 (2007); **33.22** Adaptation of Figure 3 from "The Global Decline of Nonmarine Mollusks" by Charles Lydeard et al., from *Bioscience*, April 2004, Volume 54(4). American Institute of Biological Sciences. Oxford University Press; **33.30 Tree** Data from J. K. Grenier et al., Evolution of the Entire Arthropod *Hox* Gene Set Predated the Origin and Radiation of the Onychophoran/Arthropod Clade, *Current Biology* 7:547–553 (1997).

Chapter 34 **34.10** Adaptation of Figure 1a from "Fossil Sister Group of Craniates: Predicted and Found" by Jon Mallatt and Jun-yuan Chen, from *Journal of Morphology*, May 15, 2003, Volume 258(1). John Wiley & Sons, Inc.; **34.12** Adapted from *Vertebrates: Comparative Anatomy, Function, Evolution* (2002) by Kenneth Kardong. The McGraw-Hill Companies, Inc.; **34.18** Adaptation of Figure 3 from "The Oldest Articulated Osteichthyan Reveals Mosaic Gnathostome Characters" by Min Zhu et al., from *Nature*, March 26, 2009, Volume 458(7237); **34.21** Adaptation of Figure 4 from "The Pectoral Fin of *Tiktaalik roseae* and the Origin of the Tetrapod Limb" by Neil H. Shubin et al., from *Nature*, April 6, 2006, Volume 440(7085). Macmillan Publishers Ltd.; **34.21** Acanthostega Adaptation of Figure 27 from "The Devonian Tetrapod *Acanthostega gunnari Jarvik:* Postcranial Anatomy, Basal Tetrapod Relationships and Patterns of Skeletal Evolution" by Michael I. Coates, from *Transactions of the Royal Society of Edinburgh: Earth Sciences*, Volume 87: 398; **34.38a** Based on many sources including Figure 4.10 from *Evolution*, by Douglas J. Futuyma. Sinauer Associates, 2005; and *Vertebrate Paleontology and Evolution* by Robert L. Carroll. W.H. Freeman & Co., 1988; **34.47** Based on many photos of fossils. Some sources are *O. tugenensis* photo in "Early Hominid Sows Division" by Michael Balter, from *ScienceNow*, Feb. 22, 2001; *A. garhi* and *H. neanderthalensis* based on *The Human Evolution Coloring Book* by Adrienne L. Zihlman and Carla J. Simmons. HarperCollins, 2001; *K. platyops* based on photo in "New Hominin Genus from Eastern Africa Shows Diverse Middle Pliocene Lineages" by Meave Leakey et al., from *Nature*, March 2001, Volume 410(6827); *P. boisei* based on a photo by David Bill; *H. ergaster* based on a photo at www.museumsinhand.com; *S. tchadensis* based on Figure 1b from "A New Hominid from the Upper Miocene of Chad, Central Africa" by Michel Brunet et al., from *Nature*, July 2002, Volume 418(6894); **34 Scientific Skills Exercise** Data from Dean Falk, Florida State University, 2013; **34.51** Data from R. E. Green et al., A Draft Sequence of the Neanderthal Genome, *Science* 328:710–722 (2010); **34 Test Your Understanding Question 8** Data from D. Sol et al., Big-Brained Birds Survive Better in Nature, *Proceedings of the Royal Society B* 274:763–769 (2007).

Chapter 35 **35 Scientific Skills Exercise** Data from D. L. Royer et al., Phenotypic Plasticity of Leaf Shape Along a Temperature Gradient in *Acer rubrum, PLOS ONE* 4(10):e7653 (2009); **35.21** Data from "Mongolian Tree Rings and 20th-Century Warming" by Gordon C. Jacoby, et al., from *Science*, August 9, 1996, Volume 273(5276): 771–773.

Chapter 36 **36 Scientific Skills Exercise** Data from J. D. Murphy and D. L. Noland, Temperature Effects on Seed Imbibition and Leakage Mediated by Viscosity and Membranes, *Plant Physiology* 69:428–431 (1982); **36.18** Data from S. Rogers and A. J. Peel, Some Evidence for the Existence of Turgor Pressure in the Sieve Tubes of Willow (*Salix*), *Planta* 126:259–267 (1975).

Chapter 37 **37.11b** Data from D.S. Lundberg et al., Defining the Core *Arabidopsis thaliana* Root Microbiome, *Nature* 488:86–94 (2012).

Chapter 38 **38 Scientific Skills Exercise** Data from S. Sutherland and R. K. Vickery, Jr. Trade-offs between Sexual and Asexual Reproduction in the Genus *Mimulus*. *Oecologia* 76:330–335 (1998).

Chapter 39 **39.5** Data from C. R. Darwin, *The Power of Movement in Plants*, John Murray, London (1880). P. Boysen-Jensen, Concerning the Performance of Phototropic Stimuli on the Avenacoleoptile, *Berichte der Deutschen Botanischen Gesellschaft (Reports of the German Botanical Society)* 31:559–566 (1913); **39.6** Data from L. Gälweiler et al., Regulation of Polar Auxin Transport by AtPIN1 in *Arabidopsis* Vascular Tissue, *Science* 282:2226–2230 (1998); **39.15a** Based on *Plantwatching: How Plants Remember, Tell Time, Form Relationships and More* by Malcolm Wilkins. Facts on File, 1988; **39.16** Data from H. Borthwick et al., A Reversible Photo Reaction Controlling Seed Germination, *Proceedings of the National Academy of Sciences USA* 38:662–666 (1952); **39 Problem-Solving Exercise** Map data from Camilo Mora et al. Days for Plant Growth Disappear under Projected Climate Change: Potential Human and Biotic Vulnerability. *PLoS Biol.* 13(6): e1002167 (2015); **39 Scientific Skills Exercise** Data from O. Falik et al., Rumor Has It ...: Relay Communication of Stress Cues in Plants, *PLoS ONE* 6(11):e23625 (2011).

Chapter 40 **40.17** Data from V. H. Hutchison, H. G. Dowling, and A. Vinegar, Thermoregulation in a Brooding Female Indian Python, *Python molurus bivittatus, Science* 151:694–696 (1966); **40 Scientific Skills Exercise** Based on the data from M. A. Chappell et al., Energetics of Foraging in Breeding Adélie Penguins, *Ecology* 74:2450–2461 (1993); M. A. Chappell et al., Voluntary Running in Deer Mice: Speed, Distance, Energy Costs, and Temperature Effects, *Journal of Experimental Biology* 207:3839–3854 (2004); T. M. Ellis and M. A. Chappell, Metabolism, Temperature Relations, Maternal Behavior, and Reproductive Energetics in the Ball Python (*Python regius), Journal of Comparative Physiology B* 157:393–402 (1987); **40.22** Data from F. G. Revel et al., The Circadian Clock Stops Ticking During Deep Hibernation in the European Hamster, *Proceedings of the National Academy of Sciences USA* 104:13816–13820 (2007).

Chapter 41 **41.4** Data from R. W. Smithells et al., Possible Prevention of Neural-Tube Defects by Periconceptional Vitamin Supplementation, *Lancet* 315:339–340 (1980);**41.8** Adapted from Marieb, Elaine; Hoehn, Katja, *Human Anatomy and Physiology*, 8th Edition, 2010, p. 852, Reprinted and electronically reproduced by permission of Pearson Education, Upper Saddle River, New Jersey; **41.17** Adapted from Ottman N., Smidt H., de Vos W.M. and Belzer C. (2012) The function of our microbiota: who is out there and what do they do? *Front. Cell. Inf. Microbiol.* 2:104. doi: 10.3389/fcimb.2012.00104; **41.22** Republished with permission of American Association for the Advancement of Science, from Cellular Warriors at the Battle of the Bulge by Kathleen Sutliff and Jean Marx, from *Science*, February 2003, Volume 299(5608); permission conveyed through Copyright Clearance Center, Inc.; **41 Scientific Skills Exercise** Based on the data from D. L. Coleman, Effects of Parabiosis of Obese Mice with Diabetes and Normal Mice, *Diabetologia* 9:294–298 (1973).

Chapter 42 **42 Scientific Skills Exercise** Data from J. C. Cohen et al., Sequence Variations in PCSK9, Low LDL, and Protection Against Coronary Heart Disease, *New England Journal of Medicine* 354:1264–1272 (2006); **42.25** Data from M. E. Avery and J. Mead, Surface Properties in Relation to Atelectasis and Hyaline Membrane Disease, *American Journal of Diseases of Children* 97:517–523 (1959).

Chapter 43 **43.6** Adapted from Marieb, Elaine N.; Hoehn, Katja, *Human Anatomy and Physiology*, 8th Ed., © 2010. Reprinted and electronically reproduced by permission of Pearson Education, Inc., Upper Saddle River, New Jersey; **43.7**Adapted from *Microbiology: An Introduction*, 11th Edition, by Gerard J. Tortora, Berdell R. Funke, and Christine L. Case. Pearson Education, Inc.; **43.23** Based on multiple sources: *WHO/UNICEF Coverage Estimates 2014 Revision*. July 2015. Map Production: Immunization Vaccines and Biologicals (IVB). World Health Organization, 16 July 2015; *Our Progress Against Polio*, May 1, 2014. Centers for Disease Control and Prevention; **43 Scientific Skills Exercise** Data from sources: L. J. Morrison et al., Probabilistic Order in Antigenic Variation of *Trypanosoma brucei, International Journal for Parasitology* 35:961-972 (2005); and L. J. Morrison et al., Antigenic Variation in the African Trypanosome: Molecular Mechanisms and Phenotypic Complexity, *Cellular Microbiology* 1: 1724–1734 (2009).

Chapter 44 **44 Scientific Skills Exercise** Data from R. E. MacMillen et al., Water Economy and Energy Metabolism of the Sandy Inland Mouse, *Leggadina hermannsburgensis, Journal of Mammalogy* 53:529–539 (1972); **44.7** Adapted from Mitchell, Lawrence G., *Zoology*, © 1998. Reprinted and electronically reproduced by permission of Pearson Education, Inc., Upper Saddle River, New Jersey; **44.12 Kidney structure** Adapted from Marieb, Elaine N.; Hoehn, Katja, *Human Anatomy and Physiology*, 8th Ed., 2010. Reprinted and electronically reproduced by permission of Pearson Education, Inc., Upper Saddle River, New Jersey; **44.13a Kidney structure** Adapted from Marieb, Elaine N.; Hoehn, Katja, *Human Anatomy and Physiology*, 8th Ed., 2010. Reprinted and electronically reproduced by permission of Pearson Education, Inc., Upper Saddle River, New Jersey; **44.20** Data in tables from P. M. Deen et al., Requirement of Human Renal Water Channel Aquaporin-2 for Vasopressin-Dependent Concentration of Urine, *Science* 264:92–95 (1994); **44 Summary Figure** Adapted from Beck, *Life: An Introduction to Biology*, 3rd Ed., ©1991, p. 643. Reprinted and electronically reproduced by permission of Pearson Education, Inc., Upper Saddle River, New Jersey; **44 Test Your Understanding Question 7** Data for kangaroo rat from *Animal Physiology: Adaptation and Environment* by Knut Schmidt-Nielsen. Cambridge University Press, 1991.

Chapter 45 **45 Scientific Skills Exercise** Data from J. Born et al., Timing the End of Nocturnal Sleep, *Nature* 397:29–30 (1999).

Chapter 46 **46.8** Data from R. R. Snook and D. J. Hosken, Sperm Death and Dumping in *Drosophila, Nature* 428:939–941 (2004); **46 Scientific Skills Exercise** Data from A. Jost, *Recherches Sur la Differenciation Sexuelle de l'embryon de Lapin* (Studies on the Sexual Differentiation of the Rabbit Embryo), *Archives d'Anatomie Microscopique et de Morphologie Experimentale* 36:271–316 (1947); **46.16** Adapted from Marieb, Elaine N., Hoehn, Katja, *Human Anatomy and Physiology*, 8th Ed., 2010. Reprinted and electronically reproduced by permission of Pearson Education, Inc., Upper Saddle River, New Jersey.

Chapter 47 **47.4** Data from "Intracellular Calcium Release at Fertilization in the Sea Urchin Egg" by R. Steinhardt et al., from *Developmental Biology*, July 1977, Volume 58(1); **47 Scientific Skills Exercise** Data from J. Newport and M. Kirschner, A Major Developmental Transition in Early *Xenopus* Embryos: I. Characterization and Timing of Cellular Changes at the Midblastula Stage, *Cell* 30:675–686 (1982); **47.10** Adapted from Keller, R. E. 1986. The Cellular Basis of Amphibian Gastrulation. In L. Browder (ed.), *Developmental Biology: A Comprehensive Synthesis*, Vol. 2. Plenum, New York, pp. 241–327; **47.14** Based on "Cell Commitment and Gene Expression in the Axolotl Embryo" by T. J. Mohun et al., from *Cell*, November 1980, Volume 22(1); **47.17** *Principles of Development*, 2nd Edition by Wolpert (2002), Fig. 8.26, p. 275. By permission of Oxford University Press; **47.19** Republished with permission of Garland Science, Taylor & Francis Group, from *Molecular Biology of the Cell*, Bruce Alberts et al., 4th Edition, ©2002; permission conveyed through Copyright Clearance Center, Inc.; **47.23** Data from H. Spemann, *Embryonic Development and Induction*, Yale University Press, New Haven, CT (1938); **47.24** Data from H. Spemann and H. Mangold, Induction of Embryonic Primordia by Implantation of Organizers from a Different Species, Trans. V. Hamburger (1924). Reprinted in *International Journal of Developmental Biology* 45:13–38 (2001); **47.26** Data from L. S. Honig and D. Summerbell, Maps of strength of positional signaling activity in the developing chick wing bud, *Journal of Embryology and Experimental Morphology* 87:163–174 (1985); **47.27** Adapted from Marieb, Elaine N.; Hoehn, Katja, *Human Anatomy and Physiology*, 8th Edition, 2010. Reprinted and electronically reproduced by permission of Pearson Education, Inc., Upper Saddle River, New Jersey.

Chapter 48 **48.11 Graph** Based on Figure 6-2d from *Cellular Physiology of Nerve and Muscle*, 4th Edition, by Gary G. Matthews. Wiley-Blackwell, 2003; **48 Scientific Skills Exercise** Data from C. B. Pert and S. H. Snyder, Opiate Receptor: Demonstration in Nervous Tissue, *Science* 179:1011–1014 (1973).

Chapter 49 **49.9** Adapted from Marieb, Elaine N.; Hoehn, Katja, *Human Anatomy and Physiology*, 8th Ed., © 2010. Reprinted and electronically reproduced by permission of Pearson Education, Inc., Upper Saddle River, New Jersey; **49.13** Based on "Sleep in Marine Mammals" by L. M. Mukhametov, from *Sleep Mechanisms*, edited by Alexander A. Borberly and J. L. Valatx. Springer; **49 Scientific Skills Exercise** Data from M. R. Ralph et al., Transplanted Suprachiasmatic Nucleus Determines Circadian Period, *Science* 247:975–978 (1990); **49.20** Adaptation of Figure 1c from "Avian Brains and a New Understanding of Vertebrate Brain Evolution" by Erich D. Jarvis et al., from *Nature Reviews Neuroscience*, February 2005, Volume 6(2); **49.23** Adaptation of Figure 10 from *Schizophrenia Genesis: The Origins of Madness* by Irving I. Gottesman. Worth Publishers.

Chapter 50 **50.12a** Adapted from Marieb, Elaine N; Hoehn, Katja, *Human Anatomy and Physiology*, 8th Ed., © 2010 Reprinted and electronically reproduced by Permission of Pearson Education, Inc., Upper Saddle River, New Jersey; **50.13** Adapted from Marieb, Elaine N; Hoehn, Katja, *Human Anatomy and Physiology*, 8th Ed., © 2010 Reprinted and electronically reproduced by Permission of Pearson Education, Inc., Upper Saddle River, New Jersey; **50.17 Eye structure** Adapted from Marieb, Elaine N; Hoehn, Katja, *Human Anatomy and Physiology*, 8th Ed., © 2010 Reprinted and electronically reproduced by Permission of Pearson Education, Inc., Upper Saddle River, New Jersey; **50.23** Data from K. L. Mueller et al., The receptors and coding logic for bitter taste, *Nature* 434:225–229 (2005); **50.24a** Adapted from Marieb, Elaine N; Hoehn, Katja, *Human Anatomy and Physiology*, 8th Ed., © 2010 Reprinted and electronically reproduced by Permission of Pearson Education, Inc., Upper Saddle River, New Jersey; **50.26** Adapted from Marieb, Elaine N; Hoehn, Katja, *Human Anatomy and Physiology*, 8th Ed., © 2010 Reprinted and electronically reproduced by Permission of Pearson Education, Inc., Upper Saddle River, New Jersey; **50.31** Adapted from Marieb, Elaine N; Hoehn, Katja, *Human Anatomy and Physiology*, 8th Ed., © 2010 Reprinted and electronically reproduced by Permission of Pearson Education, Inc., Upper Saddle River, New Jersey; **50.35 Grasshopper** Based on Hickman et al., *Integrated Principles of Zoology*, 9th ed., p. 518, Fig. 22.6, McGraw-Hill Higher Education, NY (1993); **50 Scientific Skills Exercise** Data from K. Schmidt-Nielsen, Locomotion: Energy Cost of Swimming, Flying, and Running, *Science* 177:222–228 (1972).

Chapter 51 **51.4** Based on "*Drosophila*: Genetics Meets Behavior" by Marla B. Sokolowski, from *Nature Reviews: Genetics*, November 2001, Volume 2(11); **51.8** Data from *The Study of Instinct*, N. Tinbergen, Clarendon Press, Oxford (1951); **51.10** Adapted from "Prospective and Retrospective Learning in Honeybees" by Martin Giurfa and Julie Bernard, from *International Journal of Comparative Psychology*, 2006, Volume 19(3); **51.13** Adapted from Evolution of Foraging Behavior in *Drosophila* by Density Dependent Selection by Maria B. Sokolowski et al., from *Proceedings of the National Academy of Sciences USA*, July 8, 1997, Volume 94(14); **51 Scientific Skills Exercise** Data from Shell Dropping: Decision-Making and Optimal Foraging in North-western Crows by Reto Zach, from *Behaviour*, 1979, Volume 68(1–2); 51; **51.18** Reprinted by permission from Klaudia Witte; **51.24 Illustration** Adaptations of photograph by Jonathan Blair, Figure/PhotoID: 3.14, as appeared in *Animal Behavior: An Evolutionary Approach*, 8th Edition, Editor: John Alcock, p. 88. Reprinted by permission; **51.24 Map** Data from "Rapid Microevolution of Migratory Behaviour in a Wild Bird Species" by P. Berthold et al., from *Nature*, December 1992, Volume 360(6405); **51 Art for Concept 51.2 Summary:** Data from *The Study of Instinct*, N. Tinbergen, Clarendon Press, Oxford (1951).

Chapter 52 **52.18** Based on the data from *Ecology and Field Biology* by Robert L. Smith. Pearson Education, 1974; and *Sibley Guide to Birds* by David Allen Sibley. Random House, 2000; **52.19** Based on the data from W.J. Fletcher, Interactions among Subtidal Australian Sea Urchins, Gastropods and Algae: Effects of Experimental Removals, *Ecological Monographs* 57:89–109 (1987); **52.20** Based on S. D. Ling et al. Climate-Driven Range Extension of a Sea Urchin: Inferring Future Trends by Analysis of Recent Population Dynamics, *Global Change Biology* (2009) 15, 719–731, doi: 10.1111/j.1365-2486.2008.01734.x; **52 Scientific Skills Exercise** Based on the data from C. M. Crain et al., Physical and Biotic Drivers of Plant Distribution Across Estuarine Salinity Gradients, *Ecology* 85:2539–2549 (2004); **52.23** Based on Rana W. El-Sabaawi et al, Assessing the Effects of Guppy Life History Evolution on Nutrient Recycling: From Experiments to the Field, *Freshwater Biology* (2015) 60, 590–601, doi:10.1111/fwb.12507; **52 Graph for Test Your Understanding Question 11** Based on the data from J. Clausen et al., Experimental Studies on the Nature of Species. III. *Environmental Responses of Climatic Races of Achillea*, Carnegie Institution of Washington Publication No. 581 (1948).

Chapter 53 **53.2** Data from A. M. Gormley et al., Capture-Recapture Estimates of Hector's Dolphin Abundance at Banks Peninsula, New Zealand, *Marine Mammal Science* 21:204–216 (2005); Table 53.1 Data from P. W. Sherman and M. L. Morton, Demography of Belding's Ground Squirrel, *Ecology* 65:1617–1628 (1984); **53.5** Based on Demography of Belding's Ground Squirrels by Paul W. Sherman and Martin L. Morton, from *Ecology*, October 1984, Volume 65(5); **53.14** Data from Brood Size Manipulations in the Kestrel (*Falco tinnunculus*): Effects on Offspring and Parent Survival by C. Dijkstra et al., from *Journal of Animal Ecology*, 1990, Volume 59(1); **53.16** Based on Climate and Population Regulation: The Biogeographer's Dilemma by J. T. Enright, from *Oecologia*, 1976, Volume 24(4); **53.17** Based on the data from Predator Responses, Prey Refuges, and Density-Dependent Mortality of a Marine Fish by T.W. Anderson, *Ecology* 82(1):245–257 (2001); **53.19** Based on the Data Provided by Dr. Rolf O. Peterson; **53.22** Based on U.S. Census Bureau International Data Base; **53.23** Based on U.S. Census Bureau International Data Base; **53.24** Based on U.S. Census Bureau International Data Base; **53.25** Based on Ewing B., D. Moore, S. Goldfinger, A. Oursler, A. Reed, and M. Wackernagel. 2010. *The Ecological Footprint Atlas 2010*. Oakland: Global Footprint Network, p. 33 (www.footprintnetwork.org).

Chapter 54 **54.2** Based on A. Stanley Rand and Ernest E. Williams. The Anoles of La Palma: Aspects of Their Ecological Relationships, *Breviora*, Volume 327: 1–19. Museum of Comparative Zoology, Harvard University; **54.3** Data from J. H. Connell, The Influence of Interspecific Competition and Other Factors on the Distribution of the Barnacle *Chthamalus stellatus*, *Ecology* 42:710–723 (1961); **54 Scientific Skills Exercise** Based on the data from B. L. Phillips and R. Shine, An Invasive Species Induces Rapid Adaptive Change in a Native Predator: Cane Toads and Black Snakes in Australia, *Proceedings of the Royal Society B* 273:1545–1550 (2006); **54.10** Based on the data from Sally D. Hacker and Mark D. Bertness, Experimental Evidence for Factors Maintaining Plant Species Diversity in a New England Salt Marsh. *Ecology*, September 1999, Volume 80(6); **54.12 Graph** Data from N. Fierer and R. B. Jackson, The Diversity and Biogeography of Soil Bacterial Communities, *Proceedings of the National Academy of Sciences USA* 103:626–631 (2006); **54.15** Based on George A. Knox. Antarctic Marine Ecosystems, from *Antarctic Ecology*, Volume 1, edited by Martin W. Holdgate. Academic Press, 1970; **54.16** Adapted from Denise L. Breitburg et al., Varying Effects of Low Dissolved Oxygen on Trophic Interactions in an Estuarine Food Web. *Ecological Monographs*, November 1997, Volume 67(4). Used by permission of the Ecological Society of America; **54.17** Based on B. Jenkins et al., Productivity, Disturbance and Food Web Structure at a Local Spatial Scale in Experimental Container Habitats. *OIKOS*, November 1992, Volume 65(2); **54.18 Graph** Data from R. T. Paine, Food web complexity and species diversity, *American Naturalist* 100:65–75 (1966); **54.21** Based on the data from C.R. Townsend, M.R. Scarsbrook, and S. Doledec, The Intermediate Disturbance Hypothesis, Refugia, and Biodiversity in Streams, *Limnology and Oceanography* 42:938–949 (1997); **54.23** Based on Robert L. Crocker and Jack Major. Soil Development in Relation to Vegetation and Surface Age at Glacier Bay, Alaska. *Journal of Ecology*, July 1955, Volume 43(2); **54.24** Adapted from F. Stuart Chapin et al., Mechanisms of Primary Succession Following Deglaciation at Glacier Bay. *Ecological Monographs*, May 1994, Volume 64(2). Ecological Society of America; **54.26** Adapted from D. J. Currie. Energy and Large-Scale Patterns of Animal- and Plant-Species Richness. *American Naturalist*, January 1991, Volume 137(1): 27–49; **54.27** Adapted from Robert H. MacArthur and Edward O. Wilson, An Equilibrium Theory of Insular Zoogeography. *Evolution*, December 1963, Volume 17(4). Society for the Study of Evolution; **54.30** Based on Daniel S. Simberloff and Edward O. Wilson. 1969. Experimental Zoogeography of Islands: The Colonization of Empty Islands. *Ecology*, Vol. 50, No. 2 (Mar., 1969), pp. 278–296.

Chapter 55 **55.4** Based on Figure 1.2 from Donald L. DeAngelis (1992), *Dynamics of Nutrient Cycling and Food Webs*. Taylor & Francis; **55.6** Data from J. H. Ryther and W. M. Dunstan, Nitrogen, Phosphorus, and Eutrophication in the Coastal Marine Environment, *Science* 171:1008–1013 (1971); Table 55.1 Data from D. W. Menzel and J. H. Ryther, Nutrients Limiting the Production of Phytoplankton in the Sargasso Sea, with Special Reference to Iron, *Deep Sea Research* 7:276–281 (1961); **55.7** Based on the data from Fig. 4.1, p. 82, in R.H. Whittaker (1970), *Communities and Ecosystems*. Macmillan, New York; **55.8** Based on Fig. 3c and 3d from Temperate Forest Health in an Era of Emerging Megadisturbance, Constance I. Millar and Nathan L. Stephenson, *Science* 349, 823 (2015); doi: 10.1126/science.aaa9933; **55 Scientific Skills Exercise** Data from J. M. Teal, Energy Flow in the Salt Marsh Ecosystem of Georgia, *Ecology* 43:614–624 (1962); **55.12** Data from J. A. Trofymow and the CIDET Working Group, *The Canadian Intersite Decomposition Experiment: Project and Site Establishment Report* (Information Report BC-X-378), Natural Resources Canada, Canadian Forest Service, Pacific Forestry Centre (1998) and T. R. Moore et al., Litter decomposition rates in Canadian forests, *Global Change Biology* 5:75–82 (1999); **55.14** Adapted Figure 7.4 from Robert E. Ricklefs (2001), *The Economy of Nature*, 5th edition. W.H. Freeman and Company; **55.18b** Based on the data from Wei-Min Wu et al. (2006), Pilot-Scale in Situ Bioremediation of Uranium in a Highly Contaminated Aquifer. 2. Reduction of U(VI) and Geochemical Control of U(VI) Bioavailability. *Environmental Science Technology* 40 (12):3986–3995 (5/13/06); **55 Art for Concept 55.1 Summary** Based on Figure 1.2 from Donald L. DeAngelis (1992). *Dynamics of Nutrient Cycling and Food Webs*. Taylor & Francis.

Chapter 56 **56.11** Based on data from Gene Likens; **56.12** Krebs, Charles J., *Ecology: The Experimental Analysis of Distribution and Abundance*, 5th Ed., © 2001. Reprinted and electronically reproduced by permission of Pearson Education, Inc., Upper Saddle River, New Jersey; **56.13** Data from "*Tracking the Long-Term Decline and Recovery of an Isolated Population*" by R.L. Westemeier et al., *Science* Volume 282(5394):1695–1698 (11/27/98), AAAS; **56.19** Adapted from Norman Myers et al. (2000). Biodiversity Hotspots for Conservation Priorities, *Nature*, February 24, 2000, Volume 403(6772); **56.28** Based on CO_2 data from www.esrl.noaa.gov/gmd/ccgg/trends. Temperature data from www.giss.nasa.gov/gistemps/graphs/Fig.A.lrg.gif; **56 Scientific Skills Exercise** Based on data from National Oceanic & Atmospheric Administration, Earth System Research Laboratory, Global Monitoring Division; **56.31** Based on the data from "History of the Ozone Hole," from NASA website, February 26, 2013; and "Antarctic Ozone," from British Antarctic Society website, June 7, 2013; **56.34** Based on the data from Instituto Nacional de Estadistica y Censos de Costa Rica and Centro Centroamericano de Poblacion, Universidad de Costa Rica.

Appendix A Figure 5.11 Wallace/Sanders/Ferl, *Biology: The Science of Life*, 3rd Ed., ©1991. Reprinted and electronically reproduced by permission of Pearson Education, Inc., Upper Saddle River, New Jersey.

Glossary

Pronunciation Key

ā	ace
a/ah	ash
ch	chose
ē	meet
e/eh	bet
g	game
ī	ice
i	hit
ks	box
kw	quick
ng	song
ō	robe
o	ox
oy	boy
s	say
sh	shell
th	thin
ū	boot
u/uh	up
z	zoo

′ = primary accent
′ = secondary accent

5′ cap A modified form of guanine nucleotide added onto the 5′ end of a pre-mRNA molecule.

ABC hypothesis A model of flower formation identifying three classes of organ identity genes that direct formation of the four types of floral organs.

abiotic (ā′-bī-ot′-ik) Nonliving; referring to the physical and chemical properties of an environment.

abortion The termination of a pregnancy in progress.

abscisic acid (ABA) (ab-sis′-ik) A plant hormone that slows growth, often antagonizing the actions of growth hormones. Two of its many effects are to promote seed dormancy and facilitate drought tolerance.

absorption The third stage of food processing in animals: the uptake of small nutrient molecules by an organism's body.

absorption spectrum The range of a pigment's ability to absorb various wavelengths of light; also a graph of such a range.

abyssal zone (uh-bis′-ul) The part of the ocean's benthic zone between 2,000 and 6,000 m deep.

acanthodian (ak′-an-thō′-dē-un) Any of a group of ancient jawed aquatic vertebrates from the Silurian and Devonian periods.

accessory fruit A fruit, or assemblage of fruits, in which the fleshy parts are derived largely or entirely from tissues other than the ovary.

acclimatization (uh-klī′-muh-tī -zā′-shun) Physiological adjustment to a change in an environmental factor.

acetyl CoA Acetyl coenzyme A; the entry compound for the citric acid cycle in cellular respiration, formed from a two-carbon fragment of pyruvate attached to a coenzyme.

acetylcholine (as′-uh-til-kō′-lēn) One of the most common neurotransmitters; functions by binding to receptors and altering the permeability of the postsynaptic membrane to specific ions, either depolarizing or hyperpolarizing the membrane.

acid A substance that increases the hydrogen ion concentration of a solution.

acoelomate (uh-sē′-lō-māt) A solid-bodied animal lacking a cavity between the gut and outer body wall.

acquired immunodeficiency syndrome (AIDS) The symptoms and signs present during the late stages of HIV infection, defined by a specified reduction in the number of T cells and the appearance of characteristic secondary infections.

acrosomal reaction (ak′-ruh-sōm′-ul) The discharge of hydrolytic enzymes from the acrosome, a vesicle in the tip of a sperm, when the sperm approaches or contacts an egg.

acrosome (ak′-ruh-sōm) A vesicle in the tip of a sperm containing hydrolytic enzymes and other proteins that help the sperm reach the egg.

actin (ak′-tin) A globular protein that links into chains, two of which twist helically about each other, forming microfilaments (actin filaments) in muscle and other kinds of cells.

action potential An electrical signal that propagates (travels) along the membrane of a neuron or other excitable cell as a nongraded (all-or-none) depolarization.

action spectrum A graph that profiles the relative effectiveness of different wavelengths of radiation in driving a particular process.

activation energy The amount of energy that reactants must absorb before a chemical reaction will start; also called free energy of activation.

activator A protein that binds to DNA and stimulates gene transcription. In prokaryotes, activators bind in or near the promoter; in eukaryotes, activators generally bind to control elements in enhancers.

active immunity Long-lasting immunity conferred by the action of B cells and T cells and the resulting B and T memory cells specific for a pathogen. Active immunity can develop as a result of natural infection or immunization.

active site The specific region of an enzyme that binds the substrate and that forms the pocket in which catalysis occurs.

active transport The movement of a substance across a cell membrane against its concentration or electrochemical gradient, mediated by specific transport proteins and requiring an expenditure of energy.

adaptation Inherited characteristic of an organism that enhances its survival and reproduction in a specific environment.

adaptive evolution Evolution that results in a better match between organisms and their environment.

adaptive immunity A vertebrate-specific defense that is mediated by B lymphocytes (B cells) and T lymphocytes (T cells) and that exhibits specificity, memory, and self-nonself recognition; also called acquired immunity.

adaptive radiation Period of evolutionary change in which groups of organisms form many new species whose adaptations allow them to fill different ecological roles in their communities.

addition rule A rule of probability stating that the probability of any one of two or more mutually exclusive events occurring can be determined by adding their individual probabilities.

adenosine triphosphate *See* ATP (adenosine triphosphate).

adenylyl cyclase (uh-den′-uh-lil) An enzyme that converts ATP to cyclic AMP in response to an extracellular signal.

adhesion The clinging of one substance to another, such as water to plant cell walls by means of hydrogen bonds.

adipose tissue A connective tissue that insulates the body and serves as a fuel reserve; contains fat-storing cells called adipose cells.

adrenal gland (uh-drē′-nul) One of two endocrine glands located adjacent to the kidneys in mammals. Endocrine cells in the outer portion (cortex) respond to adrenocorticotropic hormone (ACTH) by secreting steroid hormones that help maintain homeostasis during long-term stress. Neurosecretory cells in the central portion (medulla) secrete epinephrine and norepinephrine in response to nerve signals triggered by short-term stress.

aerobic respiration A catabolic pathway for organic molecules, using oxygen (O_2) as the final electron acceptor in an electron transport chain and ultimately producing ATP. This is the most efficient catabolic pathway and is carried out in most eukaryotic cells and many prokaryotic organisms.

age structure The relative number of individuals of each age in a population.

aggregate fruit A fruit derived from a single flower that has more than one carpel.

AIDS (acquired immunodeficiency syndrome) The symptoms and signs present during the late stages of HIV infection, defined by a specified reduction in the number of T cells and the appearance of characteristic secondary infections.

alcohol fermentation Glycolysis followed by the reduction of pyruvate to ethyl alcohol, regenerating NAD^+ and releasing carbon dioxide.

alga (plural, **algae**) A general term for any species of photosynthetic protist, including both unicellular and multicellular forms. Algal species are included in three eukaryote supergroups (Excavata, SAR, and Archaeplastida).

alimentary canal (al′-uh-men′-tuh-rē) A complete digestive tract, consisting of a tube running between a mouth and an anus.

alkaline vent A deep-sea hydrothermal vent that releases water that is warm (40–90°C) rather than hot and that has a high pH (is basic). These vents consist of tiny pores lined with iron and other catalytic minerals that some scientists hypothesize might have been the location of the earliest abiotic synthesis of organic compounds.

allele (uh-lē′-ul) Any of the alternative versions of a gene that may produce distinguishable phenotypic effects.

allopatric speciation (al′-uh-pat′-rik) The formation of new species in populations that are geographically isolated from one another.

allopolyploid (al′-ō-pol′-ē-ployd) A fertile individual that has more than two chromosome sets as a result of two different species interbreeding and combining their chromosomes.

allosteric regulation The binding of a regulatory molecule to a protein at one site that affects the function of the protein at a different site.

alpha (α) helix (al′-fuh hē′-liks) A coiled region constituting one form of the secondary structure of proteins, arising from a specific pattern of hydrogen bonding between atoms of the polypeptide backbone (not the side chains).

alternation of generations A life cycle in which there is both a multicellular diploid form, the sporophyte, and a multicellular haploid form, the gametophyte; characteristic of plants and some algae.

alternative RNA splicing A type of eukaryotic gene regulation at the RNA-processing level in which different mRNA molecules are produced from the same primary transcript, depending on which RNA segments are treated as exons and which as introns.

altruism (al′-trū-iz-um) Selflessness; behavior that reduces an individual's fitness while increasing the fitness of another individual.

alveolates (al-vē′-uh-lets) One of the three major subgroups for which the SAR eukaryotic supergroup is named. This clade arose by secondary endosymbiosis, and its members have membrane-enclosed sacs (alveoli) located just under the plasma membrane.

alveolus (al-vē′-uh-lus) (plural, **alveoli**) One of the dead-end air sacs where gas exchange occurs in a mammalian lung.

Alzheimer's disease (alts′-hī-merz) An age-related dementia (mental deterioration) characterized by confusion and memory loss.

amino acid (uh-mēn′-ō) An organic molecule possessing both a carboxyl and an amino group. Amino acids serve as the monomers of polypeptides.

amino group (uh-mēn′-ō) A chemical group consisting of a nitrogen atom bonded to two hydrogen atoms; can act as a base in solution, accepting a hydrogen ion and acquiring a charge of 1 + .

aminoacyl-tRNA synthetase An enzyme that joins each amino acid to the appropriate tRNA.

ammonia A small, toxic molecule (NH_3) produced by nitrogen fixation or as a metabolic waste product of protein and nucleic acid metabolism.

ammonite A member of a group of shelled cephalopods that were important marine predators for hundreds of millions of years until their extinction at the end of the Cretaceous period (65.5 million years ago).

amniocentesis (am′-nē-ō-sen-tē′-sis) A technique associated with prenatal diagnosis in which amniotic fluid is obtained by aspiration from a needle inserted into the uterus. The fluid and the fetal cells it contains are analyzed to detect certain genetic and congenital defects in the fetus.

amniote (am′-nē-ōt) A member of a clade of tetrapods named for a key derived character, the amniotic egg, which contains specialized membranes, including the fluid-filled amnion, that protect the embryo. Amniotes include mammals as well as birds and other reptiles.

amniotic egg An egg that contains specialized membranes that function in protection, nourishment, and gas exchange. The amniotic egg was a major evolutionary innovation, allowing embryos to develop on land in a fluid-filled sac, thus reducing the dependence of tetrapods on water for reproduction.

amoeba (uh-mē′-buh) A protist characterized by the presence of pseudopodia.

amoebocyte (uh-mē′-buh-sīt′) An amoeba-like cell that moves by pseudopodia and is found in most animals. Depending on the species, it may digest and distribute food, dispose of wastes, form skeletal fibers, fight infections, or change into other cell types.

amoebozoan (uh-mē′-buh-zō′-an) A protist in a clade that includes many species with lobe- or tube-shaped pseudopodia.

amphibian A member of the clade of tetrapods that includes salamanders, frogs, and caecilians.

amphipathic (am′-fē-path′-ik) Having both a hydrophilic region and a hydrophobic region.

amplification The strengthening of stimulus energy during transduction.

amygdala (uh-mig′-duh-luh) A structure in the temporal lobe of the vertebrate brain that has a major role in the processing of emotions.

amylase (am′-uh-lās′) An enzyme that hydrolyzes starch (a glucose polymer from plants) and glycogen (a glucose polymer from animals) into smaller polysaccharides and the disaccharide maltose.

anabolic pathway (an′-uh-bol′-ik) A metabolic pathway that consumes energy to synthesize a complex molecule from simpler molecules.

anaerobic respiration (an-er-ō′-bik) A catabolic pathway in which inorganic molecules other than oxygen accept electrons at the "downhill" end of electron transport chains.

analogous Having characteristics that are similar because of convergent evolution, not homology.

analogy (an-al′-uh-jē) Similarity between two species that is due to convergent evolution rather than to descent from a common ancestor with the same trait.

anaphase The fourth stage of mitosis, in which the chromatids of each chromosome have separated and the daughter chromosomes are moving to the poles of the cell.

anatomy The structure of an organism.

anchorage dependence The requirement that a cell must be attached to a substratum in order to initiate cell division.

androgen (an′-drō-jen) Any steroid hormone, such as testosterone, that stimulates the development and maintenance of the male reproductive system and secondary sex characteristics.

aneuploidy (an′-yū-ploy′-dē) A chromosomal aberration in which one or more chromosomes are present in extra copies or are deficient in number.

angiosperm (an′-jē-ō-sperm) A flowering plant, which forms seeds inside a protective chamber called an ovary.

anhydrobiosis (an-hī′-drō-bī-ō′-sis) A dormant state involving loss of almost all body water.

animal pole The point at the end of an egg in the hemisphere where the least yolk is concentrated; opposite of vegetal pole.

anion (an′-ī-on) A negatively charged ion.

anterior Pertaining to the front, or head, of a bilaterally symmetrical animal.

anterior pituitary A portion of the pituitary gland that develops from nonneural tissue; consists of endocrine cells that synthesize and secrete several tropic and nontropic hormones.

anther In an angiosperm, the terminal pollen sac of a stamen, where pollen grains containing sperm-producing male gametophytes form.

antheridium (an-thuh-rid′-ē-um) (plural, **antheridia**) In plants, the male gametangium, a moist chamber in which gametes develop.

anthropoid (an′-thruh-poyd) A member of a primate group made up of the monkeys and the apes (gibbons, orangutans, gorillas, chimpanzees, bonobos, and humans).

antibody A protein secreted by plasma cells (differentiated B cells) that binds to a particular antigen; also called immunoglobulin. All antibodies have the same Y-shaped structure and in their monomer form consist of two identical heavy chains and two identical light chains.

anticodon (an′-tī-kō′-don) A nucleotide triplet at one end of a tRNA molecule that base-pairs with a particular complementary codon on an mRNA molecule.

antidiuretic hormone (ADH) (an′-tī-dī-yū-ret′-ik) A peptide hormone, also called vasopressin, that promotes water retention by the kidneys. Produced in the hypothalamus and released from the posterior pituitary, ADH also functions in the brain.

antigen (an′-ti-jen) A substance that elicits an immune response by binding to receptors of B or T cells.

antigen presentation (an′-ti-jen) The process by which an MHC molecule binds to a fragment of an intracellular protein antigen and carries it to the cell surface, where it is displayed and can be recognized by a T cell.

antigen-presenting cell (an′-ti-jen) A cell that upon ingesting pathogens or internalizing pathogen proteins generates peptide fragments that are bound by class II MHC molecules and subsequently displayed on the cell surface to T cells. Macrophages, dendritic cells, and B cells are the primary antigen-presenting cells.

antigen receptor (an′-ti-jen) The general term for a surface protein, located on B cells and T cells, that binds to antigens, initiating adaptive immune responses. The antigen receptors on B cells are called B cell receptors, and the antigen receptors on T cells are called T cell receptors.

antiparallel Referring to the arrangement of the sugar-phosphate backbones in a DNA double helix (they run in opposite 5′ S 3′ directions).

aphotic zone (ā′-fō′-tik) The part of an ocean or lake beneath the photic zone, where light does not penetrate sufficiently for photosynthesis to occur.

apical bud (ā′-pik-ul) A bud at the tip of a plant stem; also called a terminal bud.

apical dominance (ā′-pik-ul) Tendency for growth to be concentrated at the tip of a plant shoot because the apical bud partially inhibits axillary bud growth.

apical ectodermal ridge (AER) (ā′-pik-ul) A thickened area of ectoderm at the tip of a limb bud that promotes outgrowth of the limb bud.

apical meristem (ā′-pik-ul mār′-uh-stem) A localized region at a growing tip of a plant body where one or more cells divide repeatedly. The dividing cells of an apical meristem enable the plant to grow in length.

apicomplexan (ap′-ē-kom-pleks′-un) A group of alveolate protists, this clade includes many species that parasitize animals. Some apicomplexans cause human disease.

apomixis (ap′-uh-mik′-sis) The ability of some plant species to reproduce asexually through seeds without fertilization by a male gamete.

apoplast (ap′-ō-plast) Everything external to the plasma membrane of a plant cell, including cell walls, intercellular spaces, and the space within dead structures such as xylem vessels and tracheids.

apoptosis (ā-puh-tō′-sus) A type of programmed cell death, which is brought about by activation of enzymes that break down many chemical components in the cell.

aposematic coloration (ap′-ō-si-mat′-ik) The bright warning coloration of many animals with effective physical or chemical defenses.

appendix A small, finger-like extension of the vertebrate cecum; contains a mass of white blood cells that contribute to immunity.

aquaporin A channel protein in a cellular membrane that specifically facilitates osmosis, the diffusion of free water across the membrane.

aqueous solution (ā′-kwē-us) A solution in which water is the solvent.

arachnid A member of a subgroup of the major arthropod clade Chelicerata. Arachnids have six pairs of appendages, including four pairs of walking legs, and include spiders, scorpions, ticks, and mites.

arbuscular mycorrhiza (ar-bus′-kyū-lur mī′-kō-rī′-zuh) Association of a fungus with a plant root system in which the fungus causes the invagination of the host (plant) cells' plasma membranes.

arbuscular mycorrhizal fungus (ar-bus′-kyū-lur) A symbiotic fungus whose hyphae grow through the cell wall of plant roots and extend into the root cell (enclosed in tubes formed by invagination of the root cell plasma membrane).

arbuscules Specialized branching hyphae that are found in some mutualistic fungi and exchange nutrients with living plant cells.

Archaea (ar′-kē′-uh) One of two prokaryotic domains, the other being Bacteria.

Archaeplastida (ar′-kē-plas′-tid-uh) One of four supergroups of eukaryotes proposed in a current hypothesis of the evolutionary history of eukaryotes. This monophyletic group, which includes red algae, green algae, and plants, descended from an ancient protistan ancestor that engulfed a cyanobacterium. *See also* Excavata, SAR, and Unikonta.

archegonium (ar-ki-gō′-nē-um) (plural, **archegonia**) In plants, the female gametangium, a moist chamber in which gametes develop.

archenteron (ar-ken′-tuh-ron) The endoderm-lined cavity, formed during gastrulation, that develops into the digestive tract of an animal.

archosaur (ar′-kō-sōr) A member of the reptilian group that includes crocodiles, alligators and dinosaurs, including birds.

arteriole (ar-ter′-ē-ōl) A vessel that conveys blood between an artery and a capillary bed.

artery A vessel that carries blood away from the heart to organs throughout the body.

arthropod A segmented ecdysozoan with a hard exoskeleton and jointed appendages. Familiar examples include insects, spiders, millipedes, and crabs.

artificial selection The selective breeding of domesticated plants and animals to encourage the occurrence of desirable traits.

ascocarp The fruiting body of a sac fungus (ascomycete).

ascomycete (as′-kuh-mī′-sēt) A member of the fungal phylum Ascomycota, commonly called sac fungus. The name comes from the saclike structure in which the spores develop.

ascus (plural, **asci**) A saclike spore capsule located at the tip of a dikaryotic hypha of a sac fungus.

asexual reproduction The generation of offspring from a single parent that occurs without the fusion of gametes. In most cases, the offspring are genetically identical to the parent.

A site One of a ribosome's three binding sites for tRNA during translation. The A site holds the tRNA carrying the next amino acid to be added to the polypeptide chain. (A stands for aminoacyl tRNA.)

assisted migration The translocation of a species to a favorable habitat beyond its native range for the purpose of protecting the species from human-caused threats.

associative learning The acquired ability to associate one environmental feature (such as a color) with another (such as danger).

atherosclerosis A cardiovascular disease in which fatty deposits called plaques develop in the inner walls of the arteries, obstructing the arteries and causing them to harden.

atom The smallest unit of matter that retains the properties of an element.

atomic mass The total mass of an atom, numerically equivalent to the mass in grams of 1 mole of the atom. (For an element with more than one isotope, the atomic mass is the average mass of the naturally occurring isotopes, weighted by their abundance.)

atomic nucleus An atom's dense central core, containing protons and neutrons.

atomic number The number of protons in the nucleus of an atom, unique for each element and designated by a subscript.

ATP (adenosine triphosphate) (a-den′-ō-sēn trī-fos′-fāt) An adenine-containing nucleoside triphosphate that releases free energy when its phosphate bonds are hydrolyzed. This energy is used to drive endergonic reactions in cells.

ATP synthase A complex of several membrane proteins that functions in chemiosmosis with adjacent electron transport chains, using the energy of a hydrogen ion (proton) concentration gradient to make ATP. ATP synthases are found in the inner mitochondrial membranes of eukaryotic cells and in the plasma membranes of prokaryotes.

atrial natriuretic peptide (ANP) (ā′-trē-ul na′-trē-yū-ret′-ik) A peptide hormone secreted by cells of the atria of the heart in response to high blood pressure. ANP's effects on the kidney alter ion and water movement and reduce blood pressure.

atrioventricular (AV) node A region of specialized heart muscle tissue between the left and right atria where electrical impulses are delayed for about 0.1 second before spreading to both ventricles and causing them to contract.

atrioventricular (AV) valve A heart valve located between each atrium and ventricle that prevents a backflow of blood when the ventricle contracts.

atrium (ā′-trē-um) (plural, **atria**) A chamber of the vertebrate heart that receives blood from the veins and transfers blood to a ventricle.

autocrine Referring to a secreted molecule that acts on the cell that secreted it.

autoimmune disease An immunological disorder in which the immune system turns against self.

autonomic nervous system (ot′-ō-nom′-ik) An efferent branch of the vertebrate peripheral nervous system that regulates the internal environment; consists of the sympathetic, parasympathetic, and enteric divisions.

autopolyploid (ot′-ō-pol′-ē-ployd) An individual that has more than two chromosome sets that are all derived from a single species.

autosome (ot′-ō-sōm) A chromosome that is not directly involved in determining sex; not a sex chromosome.

autotroph (ot′-ō-trōf) An organism that obtains organic food molecules without eating other organisms or substances derived from other organisms. Autotrophs use energy from the sun or from oxidation of inorganic substances to make organic molecules from inorganic ones.

auxin (ôk′-sin) A term that primarily refers to indoleacetic acid (IAA), a natural plant hormone that has a variety of effects, including cell elongation, root formation, secondary growth, and fruit growth.

axillary bud (ak′-sil-ār-ē) A structure that has the potential to form a lateral shoot, or branch. The bud appears in the angle formed between a leaf and a stem.

axon (ak′-son) A typically long extension, or process, of a neuron that carries nerve impulses away from the cell body toward target cells.

B cells The lymphocytes that complete their development in the bone marrow and become effector cells for the humoral immune response.

Bacteria One of two prokaryotic domains, the other being Archaea.

bacteriophage (bak-tēr′-ē-ō-fāj) A virus that infects bacteria; also called a phage.

bacteroid A form of the bacterium *Rhizobium* contained within the vesicles formed by the root cells of a root nodule.

balancing selection Natural selection that maintains two or more phenotypic forms in a population.

bar graph A graph in which the independent variable represents groups or nonnumerical categories and the values of the dependent variable(s) are shown by bars.

bark All tissues external to the vascular cambium, consisting mainly of the secondary phloem and layers of periderm.

Barr body A dense object lying along the inside of the nuclear envelope in cells of female mammals, representing a highly condensed, inactivated X chromosome.

basal angiosperm A member of one of three clades of early-diverging lineages of extant flowering plants. Examples are *Amborella*, water lilies, and star anise and its relatives.

basal body (bā′-sul) A eukaryotic cell structure consisting of a "9 + 0" arrangement of microtubule triplets. The basal body may organize the microtubule assembly of a cilium or flagellum and is structurally very similar to a centriole.

basal metabolic rate (BMR) The metabolic rate of a resting, fasting, and nonstressed endotherm at a comfortable temperature.

basal taxon In a specified group of organisms, a taxon whose evolutionary lineage diverged early in the history of the group.

base A substance that reduces the hydrogen ion concentration of a solution.

basidiocarp Elaborate fruiting body of a dikaryotic mycelium of a club fungus.

basidiomycete (buh-sid′-ē-ō-mī′-sēt) A member of the fungal phylum Basidiomycota, commonly called club fungus. The name comes from the club-like shape of the basidium.

basidium (plural, **basidia**) (buh-sid′-ē-um, buh-sid′-ē-ah) A reproductive appendage that produces sexual spores on the gills of mushrooms (club fungi).

Batesian mimicry (bāt′-zē-un mim′-uh-krē) A type of mimicry in which a harmless species resembles an unpalatable or harmful species to which it is not closely related.

behavior Individually, an action carried out by muscles or glands under control of the nervous system in response to a stimulus; collectively, the sum of an animal's responses to external and internal stimuli.

behavioral ecology The study of the evolution of and ecological basis for animal behavior.

benign tumor A mass of abnormal cells with specific genetic and cellular changes such that the cells are not capable of surviving at a new site and generally remain at the site of the tumor's origin.

benthic zone The bottom surface of an aquatic environment.

benthos (ben′-thōz) The communities of organisms living in the benthic zone of an aquatic biome.

beta (β) pleated sheet One form of the secondary structure of proteins in which the polypeptide chain folds back and forth. Two regions of the chain lie parallel to each other and are held together by hydrogen bonds between atoms of the polypeptide backbone (not the side chains).

beta oxidation A metabolic sequence that breaks fatty acids down to two-carbon fragments that enter the citric acid cycle as acetyl CoA.

bicoid A maternal effect gene that codes for a protein responsible for specifying the anterior end in *Drosophila melanogaster*.

bilateral symmetry Body symmetry in which a central longitudinal plane divides the body into two equal but opposite halves.

bilaterian (bī′-luh-ter′-ē-uhn) A member of a clade of animals with bilateral symmetry and three germ layers.

bile A mixture of substances that is produced in the liver and stored in the gallbladder; enables formation of fat droplets in water as an aid in the digestion and absorption of fats.

binary fission A method of asexual reproduction in single-celled organisms in which the cell grows to roughly double its size and then divides into two cells. In prokaryotes, binary fission does not involve mitosis, but in single-celled eukaryotes that undergo binary fission, mitosis is part of the process.

binomial A common term for the two-part, latinized format for naming a species, consisting of the genus and specific epithet; also called a binomen.

biodiversity hot spot A relatively small area with numerous endemic species and a large number of endangered and threatened species.

bioenergetics (1) The overall flow and transformation of energy in an organism. (2) The study of how energy flows through organisms.

biofilm A surface-coating colony of one or more species of prokaryotes that engage in metabolic cooperation.

biofuel A fuel produced from biomass.

biogeochemical cycle Any of the various chemical cycles, which involve both biotic and abiotic components of ecosystems.

biogeography The scientific study of the past and present geographic distributions of species.

bioinformatics The use of computers, software, and mathematical models to process and integrate biological information from large data sets.

biological augmentation An approach to restoration ecology that uses organisms to add essential materials to a degraded ecosystem.

biological clock An internal timekeeper that controls an organism's biological rhythms. The biological clock marks time with or without environmental cues but often requires signals from the environment to remain tuned to an appropriate period. *See also* circadian rhythm.

biological magnification A process in which retained substances become more concentrated at each higher trophic level in a food chain.

biological species concept Definition of a species as a group of populations whose members have the potential to interbreed in nature and produce viable, fertile offspring but do not produce viable, fertile offspring with members of other such groups.

biology The scientific study of life.

biomanipulation An approach that applies the top-down model of community organization to alter ecosystem characteristics. For example, ecologists can prevent algal blooms and eutrophication by altering the density of higher-level consumers in lakes instead of by using chemical treatments.

biomass The total mass of organic matter comprising a group of organisms in a particular habitat.

biome (bī′-ōm) Any of the world's major ecosystem types, often classified according to the predominant vegetation for terrestrial biomes and the physical environment for aquatic biomes and characterized by adaptations of organisms to that particular environment.

bioremediation The use of organisms to detoxify and restore polluted and degraded ecosystems.

biosphere The entire portion of Earth inhabited by life; the sum of all the planet's ecosystems.

biotechnology The manipulation of organisms or their components to produce useful products.

biotic (bī-ot′-ik) Pertaining to the living factors—the organisms—in an environment.

bipolar disorder A depressive mental illness characterized by swings of mood from high to low; also called manic-depressive disorder.

birth control pill A hormonal contraceptive that inhibits ovulation, retards follicular development, or alters a woman's cervical mucus to prevent sperm from entering the uterus.

blade (1) A leaflike structure of a seaweed that provides most of the surface area for photosynthesis. (2) The flattened portion of a typical leaf.

blastocoel (blas′-tuh-sēl) The fluid-filled cavity that forms in the center of a blastula.

blastocyst (blas′-tuh-sist) The blastula stage of mammalian embryonic development, consisting of an inner cell mass, a cavity, and an outer layer, the trophoblast. In humans, the blastocyst forms 1 week after fertilization.

blastomere An early embryonic cell arising during the cleavage stage of an early embryo.

blastopore (blas′-tō-pōr) In a gastrula, the opening of the archenteron that typically develops into the anus in deuterostomes and the mouth in protostomes.

blastula (blas′-tyū-luh) A hollow ball of cells that marks the end of the cleavage stage during early embryonic development in animals.

blood A connective tissue with a fluid matrix called plasma in which red blood cells, white blood cells, and cell fragments called platelets are suspended.

blue-light photoreceptor A type of light receptor in plants that initiates a variety of responses, such as phototropism and slowing of hypocotyl elongation.

body cavity A fluid- or air-filled space between the digestive tract and the body wall.

body plan In multicellular eukaryotes, a set of morphological and developmental traits that are integrated into a functional whole—the living organism.

Bohr shift A lowering of the affinity of hemoglobin for oxygen, caused by a drop in pH. It facilitates the release of oxygen from hemoglobin in the vicinity of active tissues.

bolus A lubricated ball of chewed food.

bone A connective tissue consisting of living cells held in a rigid matrix of collagen fibers embedded in calcium salts.

book lung An organ of gas exchange in spiders, consisting of stacked plates contained in an internal chamber.

bottleneck effect Genetic drift that occurs when the size of a population is reduced, as by a natural disaster or human actions. Typically, the surviving population is no longer genetically representative of the original population.

bottom-up model A model of community organization in which mineral nutrients influence community organization by controlling plant or phytoplankton numbers, which in turn control herbivore numbers, which in turn control predator numbers.

Bowman's capsule (bō′-munz) A cup-shaped receptacle in the vertebrate kidney that is the initial, expanded segment of the nephron, where filtrate enters from the blood.

brachiopod (bra′-kē-uh-pod′) A marine lophophorate with a shell divided into dorsal and ventral halves; also called lamp shells.

brain Organ of the central nervous system where information is processed and integrated.

brainstem A collection of structures in the vertebrate brain, including the midbrain, the pons, and the medulla oblongata; functions in homeostasis, coordination of movement, and conduction of information to higher brain centers.

branch point The representation on a phylogenetic tree of the divergence of two or more taxa from a common ancestor. A branch point is usually shown as a dichotomy in which a branch representing the ancestral lineage splits (at the branch point) into two branches, one for each of the two descendant lineages.

brassinosteroid A steroid hormone in plants that has a variety of effects, including inducing cell elongation, retarding leaf abscission, and promoting xylem differentiation.

breathing Ventilation of the lungs through alternating inhalation and exhalation.

bronchiole (brong′-kē-ōl′) A fine branch of the bronchi that transports air to alveoli.

bronchus (brong′-kus) (plural, **bronchi**) One of a pair of breathing tubes that branch from the trachea into the lungs.

brown alga A multicellular, photosynthetic protist with a characteristic brown or olive color that results from carotenoids in its plastids. Most brown algae are marine, and some have a plantlike body.

bryophyte (brī′-uh-fīt) An informal name for a moss, liverwort, or hornwort; a nonvascular plant that lives on land but lacks some of the terrestrial adaptations of vascular plants.

buffer A solution that contains a weak acid and its corresponding base. A buffer minimizes changes in pH when acids or bases are added to the solution.

bulk feeder An animal that eats relatively large pieces of food.

bulk flow The movement of a fluid due to a difference in pressure between two locations.

bundle-sheath cell In C_4 plants, a type of photosynthetic cell arranged into tightly packed sheaths around the veins of a leaf.

C_3 plant A plant that uses the Calvin cycle for the initial steps that incorporate CO_2 into organic material, forming a three-carbon compound as the first stable intermediate.

C_4 plant A plant in which the Calvin cycle is preceded by reactions that incorporate CO_2 into a four-carbon compound, the end product of which supplies CO_2 for the Calvin cycle.

calcitonin (kal′-si-tō′-nin) A hormone secreted by the thyroid gland that lowers blood calcium levels by promoting calcium deposition in bone and calcium excretion from the kidneys; nonessential in adult humans.

callus A mass of dividing, undifferentiated cells growing at the site of a wound or in culture.

calorie (cal) The amount of heat energy required to raise the temperature of 1 g of water by 1°C; also the amount of heat energy that 1 g of water releases when it cools by 1°C. The Calorie (with a capital C), usually used to indicate the energy content of food, is a kilocalorie.

Calvin cycle The second of two major stages in photosynthesis (following the light reactions), involving fixation of atmospheric CO_2 and reduction of the fixed carbon into carbohydrate.

Cambrian explosion A relatively brief time in geologic history when many present-day phyla of animals first appeared in the fossil record. This burst of evolutionary change occurred about 535–525 million years ago and saw the emergence of the first large, hard-bodied animals.

CAM plant A plant that uses crassulacean acid metabolism, an adaptation for photosynthesis in arid conditions. In this process, CO_2 entering open stomata during the night is converted to organic acids, which release CO_2 for the Calvin cycle during the day, when stomata are closed.

canopy The uppermost layer of vegetation in a terrestrial biome.

capillary (kap′-il-ār′-ē) A microscopic blood vessel that penetrates the tissues and consists of a single layer of endothelial cells that allows exchange between the blood and interstitial fluid.

capillary bed (kap′-il-ār′-ē) A network of capillaries in a tissue or organ.

capsid The protein shell that encloses a viral genome. It may be rod-shaped, polyhedral, or more complex in shape.

capsule (1) In many prokaryotes, a dense and well-defined layer of polysaccharide or protein that surrounds the cell wall and is sticky, protecting the cell and enabling it to adhere to substrates or other cells. (2) The sporangium of a bryophyte (moss, liverwort, or hornwort).

carbohydrate (kar′-bō-hī′-drāt) A sugar (monosaccharide) or one of its dimers (disaccharides) or polymers (polysaccharides).

carbon fixation The initial incorporation of carbon from CO_2 into an organic compound by an autotrophic organism (a plant, another photosynthetic organism, or a chemoautotrophic prokaryote).

carbonyl group (kar′-buh-nil) A chemical group present in aldehydes and ketones and consisting of a carbon atom double-bonded to an oxygen atom.

carboxyl group (kar-bok′-sil) A chemical group present in organic acids and consisting of a single carbon atom double-bonded to an oxygen atom and also bonded to a hydroxyl group.

cardiac cycle (kar′-dē-ak) The alternating contractions and relaxations of the heart.

cardiac muscle (kar′-dē-ak) A type of striated muscle that forms the contractile wall of the heart. Its cells are joined by intercalated disks that relay the electrical signals underlying each heartbeat.

cardiac output (kar′-dē-ak) The volume of blood pumped per minute by each ventricle of the heart.

cardiovascular system A closed circulatory system with a heart and branching network of arteries, capillaries, and veins. The system is characteristic of vertebrates.

carnivore An animal that mainly eats other animals.

carotenoid (kuh-rot′-uh-noyd′) An accessory pigment, either yellow or orange, in the chloroplasts of plants and in some prokaryotes. By absorbing wavelengths of light that chlorophyll cannot, carotenoids broaden the spectrum of colors that can drive photosynthesis.

carpel (kar′-pul) The ovule-producing reproductive organ of a flower, consisting of the stigma, style, and ovary.

carrier In genetics, an individual who is heterozygous at a given genetic locus for a recessively inherited disorder. The heterozygote is generally phenotypically normal for the disorder but can pass on the recessive allele to offspring.

carrying capacity The maximum population size that can be supported by the available resources, symbolized as *K*.

cartilage (kar′-til-ij) A flexible connective tissue with an abundance of collagenous fibers embedded in chondroitin sulfate.

Casparian strip (ka-spār′-ē-un) A water-impermeable ring of wax in the endodermal cells of plants that blocks the passive flow of water and solutes into the stele by way of cell walls.

catabolic pathway (kat′-uh-bol′-ik) A metabolic pathway that releases energy by breaking down complex molecules to simpler molecules.

catalysis (kuh-ta′-luh-sis) A process by which a chemical agent called a catalyst selectively increases the rate of a reaction without being consumed by the reaction.

catalyst (kat′-uh-list) A chemical agent that selectively increases the rate of a reaction without being consumed by the reaction.

cation (cat′-ī′-on) A positively charged ion.

cation exchange (cat′-ī′-on) A process in which positively charged minerals are made available to a plant when hydrogen ions in the soil displace mineral ions from the clay particles.

cecum (sē′-kum) (plural, **ceca**) The blind pouch forming one branch of the large intestine.

cell Life's fundamental unit of structure and function; the smallest unit of organization that can perform all activities required for life.

cell body The part of a neuron that houses the nucleus and most other organelles.

cell cycle An ordered sequence of events in the life of a cell, from its origin in the division of a parent cell until its own division into two. The eukaryotic cell cycle is composed of interphase (including G_1, S, and G_2 phases) and M phase (including mitosis and cytokinesis).

cell cycle control system A cyclically operating set of molecules in the eukaryotic cell that both triggers and coordinates key events in the cell cycle.

cell division The reproduction of cells.

cell fractionation The disruption of a cell and separation of its parts by centrifugation at successively higher speeds.

cell-mediated immune response The branch of adaptive immunity that involves the activation of cytotoxic T cells, which defend against infected cells.

cell plate A membrane-bounded, flattened sac located at the midline of a dividing plant cell, inside which the new cell wall forms during cytokinesis.

cellular respiration The catabolic pathways of aerobic and anaerobic respiration, which break down organic molecules and use an electron transport chain for the production of ATP.

cellulose (sel′-yū-lōs) A structural polysaccharide of plant cell walls, consisting of glucose monomers joined by β glycosidic linkages.

cell wall A protective layer external to the plasma membrane in the cells of plants, prokaryotes, fungi, and some protists. Polysaccharides such as cellulose (in plants and some protists), chitin (in fungi), and peptidoglycan (in bacteria) are important structural components of cell walls.

central nervous system (CNS) The portion of the nervous system where signal integration occurs; in vertebrate animals, the brain and spinal cord.

central vacuole In a mature plant cell, a large membranous sac with diverse roles in growth, storage, and sequestration of toxic substances.

centriole (sen′-trē-ōl) A structure in the centrosome of an animal cell composed of a cylinder of microtubule triplets arranged in a "9 + 0" pattern. A centrosome has a pair of centrioles.

centromere (sen′-trō-mēr) In a duplicated chromosome, the region on each sister chromatid where it is most closely attached to its sister chromatid by proteins that bind to the centromeric DNA. Other proteins condense the chromatin in that region, so it appears as a narrow "waist" on the duplicated chromosome. (An unduplicated chromosome has a single centromere, identified by the proteins bound there.)

centrosome (sen′-trō-sōm) A structure present in the cytoplasm of animal cells that functions as a microtubule-organizing center and is important during cell division. A centrosome has two centrioles.

cercozoan An amoeboid or flagellated protist that feeds with threadlike pseudopodia.

cerebellum (sār′-ruh-bel′-um) Part of the vertebrate hindbrain located dorsally; functions in unconscious coordination of movement and balance.

cerebral cortex (suh-rē′-brul) The surface of the cerebrum; the largest and most complex part of the mammalian brain, containing nerve cell bodies of the cerebrum; the part of the vertebrate brain most changed through evolution.

cerebrum (suh-rē′-brum) The dorsal portion of the vertebrate forebrain, composed of right and left hemispheres; the integrating center for memory, learning, emotions, and other highly complex functions of the central nervous system.

cervix (ser′-viks) The neck of the uterus, which opens into the vagina.

chaparral A scrubland biome of dense, spiny evergreen shrubs found at midlatitudes along coasts where cold ocean currents circulate offshore; characterized by mild, rainy winters and long, hot, dry summers.

chaperonin (shap′-er-ō′-nin) A protein complex that assists in the proper folding of other proteins.

character An observable heritable feature that may vary among individuals.

character displacement The tendency for characteristics to be more divergent in sympatric populations of two species than in allopatric populations of the same two species.

checkpoint A control point in the cell cycle where stop and go-ahead signals can regulate the cycle.

chelicera (kē-lih′-suh-ruh) (plural, **chelicerae**) One of a pair of clawlike feeding appendages characteristic of chelicerates.

chelicerate (kē-lih-suh′-rāte) An arthropod that has chelicerae and a body divided into a cephalothorax and an abdomen. Living chelicerates include sea spiders, horseshoe crabs, scorpions, ticks, and spiders.

chemical bond An attraction between two atoms, resulting from a sharing of outer-shell electrons or the presence of opposite charges on the atoms. The bonded atoms gain complete outer electron shells.

chemical energy Energy available in molecules for release in a chemical reaction; a form of potential energy.

chemical equilibrium In a chemical reaction, the state in which the rate of the forward reaction equals the rate of the reverse reaction, so that the relative concentrations of the reactants and products do not change with time.

chemical reaction The making and breaking of chemical bonds, leading to changes in the composition of matter.

chemiosmosis (kem′-ē-oz-mō′-sis) An energy-coupling mechanism that uses energy stored in the form of a hydrogen ion gradient across a membrane to drive cellular work, such as the synthesis of ATP. Under aerobic conditions, most ATP synthesis in cells occurs by chemiosmosis.

chemoautotroph (kē′-mō-ot′-ō-trōf) An organism that obtains energy by oxidizing inorganic substances and needs only carbon dioxide as a carbon source.

chemoheterotroph (kē′-mō-het′-er-ō-trōf) An organism that requires organic molecules for both energy and carbon.

chemoreceptor A sensory receptor that responds to a chemical stimulus, such as a solute or an odorant.

chiasma (plural, **chiasmata**) (kī-az′-muh, kī-az′-muh-tuh) The X-shaped, microscopically visible region where crossing over has occurred earlier in prophase I between homologous nonsister chromatids. Chiasmata become visible after synapsis ends, with the two homologs remaining associated due to sister chromatid cohesion.

chitin (kī′-tin) A structural polysaccharide, consisting of amino sugar monomers, found in many fungal cell walls and in the exoskeletons of all arthropods.

chlorophyll (klōr′-ō-fil) A green pigment located in membranes within the chloroplasts of plants and algae and in the membranes of certain prokaryotes. Chlorophyll *a* participates directly in the light reactions, which convert solar energy to chemical energy.

chlorophyll *a* (klōr′-ō-fil) A photosynthetic pigment that participates directly in the light reactions, which convert solar energy to chemical energy.

chlorophyll *b* (klōr′-ō-fil) An accessory photosynthetic pigment that transfers energy to chlorophyll *a*.

chloroplast (klōr′-ō-plast) An organelle found in plants and photosynthetic protists that absorbs sunlight and uses it to drive the synthesis of organic compounds from carbon dioxide and water.

choanocyte (kō-an′-uh-sīt) A flagellated feeding cell found in sponges. Also called a collar cell, it has a collar-like ring that traps food particles around the base of its flagellum.

cholesterol (kō-les′-tuh-rol) A steroid that forms an essential component of animal cell membranes and acts as a precursor molecule for the synthesis of other biologically important steroids, such as many hormones.

chondrichthyan (kon-drik′-thē-an) A member of the clade Chondrichthyes, vertebrates with skeletons made mostly of cartilage, such as sharks and rays.

chordate A member of the phylum Chordata, animals that at some point during their development have a notochord; a dorsal, hollow nerve cord; pharyngeal slits or clefts; and a muscular, post-anal tail.

chorionic villus sampling (CVS) (kōr′-ē-on′-ik vil′-us) A technique associated with prenatal diagnosis in which a small sample of the fetal portion of the placenta is removed for analysis to detect certain genetic and congenital defects in the fetus.

chromatin (krō′-muh-tin) The complex of DNA and proteins that makes up eukaryotic chromosomes. When the cell is not dividing, chromatin exists in its dispersed form, as a mass of very long, thin fibers that are not visible with a light microscope.

chromosome (krō′-muh-sōm) A cellular structure consisting of one DNA molecule and associated protein molecules. (In some contexts, such as genome sequencing, the term may refer to the DNA alone.) A eukaryotic cell typically has multiple, linear chromosomes, which are located in the nucleus. A prokaryotic cell often has a single, circular chromosome, which is found in the nucleoid, a region that is not enclosed by a membrane. *See also* chromatin.

chromosome theory of inheritance (krō′-muh-sōm) A basic principle in biology stating that genes are located at specific positions (loci) on chromosomes and that the behavior of chromosomes during meiosis accounts for inheritance patterns.

chylomicron (kī′-lō-mī′-kron) A lipid transport globule composed of fats mixed with cholesterol and coated with proteins.

chyme (kīm) The mixture of partially digested food and digestive juices formed in the stomach.

chytrid (kī′-trid) A member of the fungal phylum Chytridiomycota, mostly aquatic fungi with flagellated zoospores that represent an early-diverging fungal lineage.

ciliate (sil′-ē-it) A type of protist that moves by means of cilia.

cilium (sil′-ē-um) (plural, **cilia**) A short appendage containing microtubules in eukaryotic cells. A motile cilium is specialized for locomotion or moving fluid past the cell; it is formed from a core of nine outer doublet microtubules and two inner single microtubules (the "9 + 2" arrangement) ensheathed in an extension of the plasma membrane. A primary cilium is usually nonmotile and plays a sensory and signaling role; it lacks the two inner microtubules (the "9 + 0" arrangement).

circadian rhythm (ser-kā′-dē-un) A physiological cycle of about 24 hours that persists even in the absence of external cues.

***cis-trans* isomer** One of several compounds that have the same molecular formula and covalent bonds between atoms but differ in the spatial arrangements of their atoms owing to the inflexibility of double bonds; formerly called a geometric isomer.

citric acid cycle A chemical cycle involving eight steps that completes the metabolic breakdown of glucose molecules begun in glycolysis by oxidizing acetyl CoA (derived from pyruvate) to carbon dioxide; occurs within the mitochondrion in eukaryotic cells and in the cytosol of prokaryotes; together with pyruvate oxidation, the second major stage in cellular respiration.

clade (klād) A group of species that includes an ancestral species and all of its descendants. A clade is equivalent to a monophyletic group.

cladistics (kluh-dis′-tiks) An approach to systematics in which organisms are placed into groups called clades based primarily on common descent.

class In Linnaean classification, the taxonomic category above the level of order.

cleavage (1) The process of cytokinesis in animal cells, characterized by pinching of the plasma membrane. (2) The succession of rapid cell divisions without significant growth during early embryonic development that converts the zygote to a ball of cells.

cleavage furrow The first sign of cleavage in an animal cell; a shallow groove around the cell in the cell surface near the old metaphase plate.

climate The long-term prevailing weather conditions at a given place.

climate change A directional change in temperature, precipitation, or other aspect of the global climate that lasts for three decades or more.

climograph A plot of the temperature and precipitation in a particular region.

clitoris (klit′-uh-ris) An organ at the upper intersection of the labia minora that engorges with blood and becomes erect during sexual arousal.

cloaca (klō-ā′-kuh) A common opening for the digestive, urinary, and reproductive tracts found in many nonmammalian vertebrates but in few mammals.

clonal selection The process by which an antigen selectively binds to and activates only those lymphocytes bearing receptors specific for the antigen. The selected lymphocytes proliferate and differentiate into a clone of effector cells and a clone of memory cells specific for the stimulating antigen.

clone (1) A lineage of genetically identical individuals or cells. (2) In popular usage, an individual that is genetically identical to another individual. (3) As a verb, to make one or more genetic replicas of an individual or cell. *See also* gene cloning.

cloning vector In genetic engineering, a DNA molecule that can carry foreign DNA into a host cell and replicate there. Cloning vectors include plasmids and bacterial artificial chromosomes (BACs), which move recombinant DNA from a test tube back into a cell, and viruses that transfer recombinant DNA by infection.

closed circulatory system A circulatory system in which blood is confined to vessels and is kept separate from the interstitial fluid.

cnidocyte (nī′-duh-sīt) A specialized cell unique to the phylum Cnidaria; contains a capsule-like organelle housing a coiled thread that, when discharged, explodes outward and functions in prey capture or defense.

cochlea (kok′-lē-uh) The complex, coiled organ of hearing that contains the organ of Corti.

coding strand Nontemplate strand of DNA, which has the same sequence as the mRNA except it has thymine (T) instead of uracil (U).

codominance The situation in which the phenotypes of both alleles are exhibited in the heterozygote because both alleles affect the phenotype in separate, distinguishable ways.

codon (kō′-don) A three-nucleotide sequence of DNA or mRNA that specifies a particular amino acid or termination signal; the basic unit of the genetic code.

coefficient of relatedness The fraction of genes that, on average, are shared by two individuals.

coelom (sē′-lōm) A body cavity lined by tissue derived only from mesoderm.

coelomate (sē′-lō-māt) An animal that possesses a true coelom (a body cavity lined by tissue completely derived from mesoderm).

coenocytic fungus (sē′-no-si′-tic) A fungus that lacks septa and hence whose body is made up of a continuous cytoplasmic mass that may contain hundreds or thousands of nuclei.

coenzyme (kō-en′-zīm) An organic molecule serving as a cofactor. Most vitamins function as coenzymes in metabolic reactions.

coevolution The joint evolution of two interacting species, each in response to selection imposed by the other.

cofactor Any nonprotein molecule or ion that is required for the proper functioning of an enzyme. Cofactors can be permanently bound to the active site or may bind loosely and reversibly, along with the substrate, during catalysis.

cognition The process of knowing that may include awareness, reasoning, recollection, and judgment.

cognitive map A neural representation of the abstract spatial relationships between objects in an animal's surroundings.

cohesion The linking together of like molecules, often by hydrogen bonds.

cohesion-tension hypothesis The leading explanation of the ascent of xylem sap. It states that transpiration exerts pull on xylem sap, putting the sap under negative pressure, or tension, and that the cohesion of water molecules transmits this pull along the entire length of the xylem from shoots to roots.

cohort A group of individuals of the same age in a population.

coleoptile (kō′-lē-op′-tul) The covering of the young shoot of the embryo of a grass seed.

coleorhiza (kō′-lē-uh-rī′-zuh) The covering of the young root of the embryo of a grass seed.

collagen A glycoprotein in the extracellular matrix of animal cells that forms strong fibers, found extensively in connective tissue and bone; the most abundant protein in the animal kingdom.

collecting duct The location in the kidney where processed filtrate, called urine, is collected from the renal tubules.

collenchyma cell (kō-len′-kim-uh) A flexible plant cell type that occurs in strands or cylinders that support young parts of the plant without restraining growth.

colon (kō′-len) The largest section of the vertebrate large intestine; functions in water absorption and formation of feces.

commensalism (kuh-men′-suh-lizm) A +/0 ecological interaction in which one organism benefits but the other is neither helped nor harmed.

communication In animal behavior, a process involving transmission of, reception of, and response to signals. The term is also used in connection with other organisms, as well as individual cells of multicellular organisms.

community All the organisms that inhabit a particular area; an assemblage of populations of different species living close enough together for potential interaction.

community ecology The study of how interactions between species affect community structure and organization.

companion cell A type of plant cell that is connected to a sieve-tube element by many plasmodesmata and whose nucleus and ribosomes may serve one or more adjacent sieve-tube elements.

competition A −/− interaction that occurs when individuals of different species compete for a resource that limits the survival and reproduction of each species.

competitive exclusion The concept that when populations of two similar species compete for the same limited resources, one population will use the resources more efficiently and have a reproductive advantage that will eventually lead to the elimination of the other population.

competitive inhibitor A substance that reduces the activity of an enzyme by entering the active site in place of the substrate, whose structure it mimics.

complement system A group of about 30 blood proteins that may amplify the inflammatory response, enhance phagocytosis, or directly lyse extracellular pathogens.

complementary DNA (cDNA) A double-stranded DNA molecule made *in vitro* using mRNA as a template and the enzymes reverse transcriptase and DNA polymerase. A cDNA molecule corresponds to the exons of a gene.

complete dominance The situation in which the phenotypes of the heterozygote and dominant homozygote are indistinguishable.

complete flower A flower that has all four basic floral organs: sepals, petals, stamens, and carpels.

complete metamorphosis The transformation of a larva into an adult that looks very different, and often functions very differently in its environment, than the larva.

compound A substance consisting of two or more different elements combined in a fixed ratio.

compound eye A type of multifaceted eye in insects and crustaceans consisting of up to several thousand light-detecting, focusing ommatidia.

concentration gradient A region along which the density of a chemical substance increases or decreases.

conception The fertilization of an egg by a sperm in humans.

cone A cone-shaped cell in the retina of the vertebrate eye, sensitive to color.

conformer An animal for which an internal condition conforms to (changes in accordance with) changes in an environmental variable.

conidium (plural, **conidia**) A haploid spore produced at the tip of a specialized hypha in ascomycetes during asexual reproduction.

conifer A member of the largest gymnosperm phylum. Most conifers are cone-bearing trees, such as pines and firs.

conjugation (kon′-jū-gā′-shun) (1) In prokaryotes, the direct transfer of DNA between two cells that are temporarily joined. When the two cells are members of different species, conjugation results in horizontal gene transfer. (2) In ciliates, a sexual process in which two cells exchange haploid micronuclei but do not reproduce.

connective tissue Animal tissue that functions mainly to bind and support other tissues, having a sparse population of cells scattered through an extracellular matrix.

conodont An early, soft-bodied vertebrate with prominent eyes and dental elements.

conservation biology The integrated study of ecology, evolutionary biology, physiology, molecular biology, and genetics to sustain biological diversity at all levels.

consumer An organism that feeds on producers, other consumers, or nonliving organic material.

contraception The deliberate prevention of pregnancy.

contractile vacuole A membranous sac that helps move excess water out of certain freshwater protists.

control element A segment of noncoding DNA that helps regulate transcription of a gene by serving as a binding site for a transcription factor. Multiple control elements are present in a eukaryotic gene's enhancer.

control group In a controlled experiment, a set of subjects that lacks (or does not receive) the specific factor being tested. Ideally, the control group should be identical to the experimental group in other respects.

controlled experiment An experiment designed to compare an experimental group with a control group; ideally, the two groups differ only in the factor being tested.

convergent evolution The evolution of similar features in independent evolutionary lineages.

convergent extension A process in which the cells of a tissue layer rearrange themselves in such a way that the sheet of cells becomes narrower (converges) and longer (extends).

cooperativity A kind of allosteric regulation whereby a shape change in one subunit of a protein caused by substrate binding is transmitted to all the other subunits, facilitating binding of additional substrate molecules to those subunits.

coral reef Typically a warm-water, tropical ecosystem dominated by the hard skeletal structures secreted primarily by corals. Some coral reefs also exist in cold, deep waters.

corepressor A small molecule that binds to a bacterial repressor protein and changes the protein's shape, allowing it to bind to the operator and switch an operon off.

cork cambium (kam′-bē-um) A cylinder of meristematic tissue in woody plants that replaces the epidermis with thicker, tougher cork cells.

corpus callosum (kor′-pus kuh-lō′-sum) The thick band of nerve fibers that connects the right and left cerebral hemispheres in mammals, enabling the hemispheres to process information together.

corpus luteum (kor′-pus lū′-tē-um) A secreting tissue in the ovary that forms from the collapsed follicle after ovulation and produces progesterone.

cortex (1) The outer region of cytoplasm in a eukaryotic cell, lying just under the plasma membrane, that has a more gel-like consistency than the inner regions due to the presence of multiple microfilaments. (2) In plants, ground tissue that is between the vascular tissue and dermal tissue in a root or eudicot stem.

cortical nephron In mammals and birds, a nephron with a loop of Henle located almost entirely in the renal cortex.

cotransport The coupling of the "downhill" diffusion of one substance to the "uphill" transport of another against its own concentration gradient.

cotyledon (kot′-uh-lē′-dun) A seed leaf of an angiosperm embryo. Some species have one cotyledon, others two.

countercurrent exchange The exchange of a substance or heat between two fluids flowing in opposite directions. For example, blood in a fish gill flows in the opposite direction of water passing over the gill, maximizing diffusion of oxygen into and carbon dioxide out of the blood.

countercurrent multiplier system A countercurrent system in which energy is expended in active transport to facilitate exchange of materials and generate concentration gradients.

covalent bond (kō-vā′-lent) A type of strong chemical bond in which two atoms share one or more pairs of valence electrons.

crassulacean acid metabolism (CAM) (crass-yū-lā′-shen) An adaptation for photosynthesis in arid conditions, first discovered in the family Crassulaceae. In this process, a plant takes up CO_2 and incorporates it into a variety of organic acids at night; during the day, CO_2 is released from organic acids for use in the Calvin cycle.

CRISPR-Cas9 system A technique for editing genes in living cells, involving a bacterial protein called Cas9 associated with a guide RNA complementary to a gene sequence of interest.

crista (plural, **cristae**) (kris′-tuh, kris′-tē) An infolding of the inner membrane of a mitochondrion. The inner membrane houses electron transport chains and molecules of the enzyme catalyzing the synthesis of ATP (ATP synthase).

critical load The amount of added nutrient, usually nitrogen or phosphorus, that can be absorbed by plants without damaging ecosystem integrity.

crop rotation The practice of growing different crops in succession on the same land chiefly to preserve the productive capacity of the soil.

cross-fostering study A behavioral study in which the young of one species are placed in the care of adults from another species.

crossing over The reciprocal exchange of genetic material between nonsister chromatids during prophase I of meiosis.

cross-pollination In angiosperms, the transfer of pollen from an anther of a flower on one plant to the stigma of a flower on another plant of the same species.

cryptic coloration Camouflage that makes a potential prey difficult to spot against its background.

culture A system of information transfer through social learning or teaching that influences the behavior of individuals in a population.

cuticle (kyū′-tuh-kul) (1) A waxy covering on the surface of stems and leaves that prevents desiccation in terrestrial plants. (2) A tough coat that covers the body of a nematode.

cyclic AMP (cAMP) Cyclic adenosine monophosphate, a ring-shaped molecule made from ATP that is a common intracellular signaling molecule (second messenger) in eukaryotic cells. It is also a regulator of some bacterial operons.

cyclic electron flow A route of electron flow during the light reactions of photosynthesis that involves only one photosystem and that produces ATP but not NADPH or O_2.

cyclin (sī′-klin) A cellular protein that occurs in a cyclically fluctuating concentration and that plays an important role in regulating the cell cycle.

cyclin-dependent kinase (Cdk) (sī′-klin) A protein kinase that is active only when attached to a particular cyclin.

cyclostome (sī′-cluh-stōm) Member of one of the two main clades of vertebrates; cyclostomes lack jaws and include lampreys and hagfishes. *See also* gnathostome.

cystic fibrosis (sis′-tik fī-brō′-sis) A human genetic disorder caused by a recessive allele for a chloride channel protein; characterized by an excessive secretion of mucus and consequent vulnerability to infection; fatal if untreated.

cytochrome (sī′-tō-krōm) An iron-containing protein that is a component of electron transport chains in the mitochondria and chloroplasts of eukaryotic cells and the plasma membranes of prokaryotic cells.

cytokinesis (sī′-tō-kuh-nē′-sis) The division of the cytoplasm to form two separate daughter cells immediately after mitosis, meiosis I, or meiosis II.

cytokinin (sī′-tō-kī′-nin) Any of a class of related plant hormones that retard aging and act in concert with auxin to stimulate cell division, influence the pathway of differentiation, and control apical dominance.

cytoplasm (sī′-tō-plaz-um) The contents of the cell bounded by the plasma membrane; in eukaryotes, the portion exclusive of the nucleus.

cytoplasmic determinant A maternal substance, such as a protein or RNA, that when placed into an egg influences the course of early development by regulating the expression of genes that affect the developmental fate of cells.

cytoplasmic streaming A circular flow of cytoplasm, involving interactions of myosin and actin filaments, that speeds the distribution of materials within cells.

cytoskeleton A network of microtubules, microfilaments, and intermediate filaments that extend throughout the cytoplasm and serve a variety of mechanical, transport, and signaling functions.

cytosol (sī′-tō-sol) The semifluid portion of the cytoplasm.

cytotoxic T cell A type of lymphocyte that, when activated, kills infected cells as well as certain cancer cells and transplanted cells.

dalton A measure of mass for atoms and subatomic particles; the same as the atomic mass unit, or amu.

data Recorded observations.

day-neutral plant A plant in which flower formation is not controlled by photoperiod or day length.

decomposer An organism that absorbs nutrients from nonliving organic material such as corpses, fallen plant material, and the wastes of living organisms and converts them to inorganic forms; a detritivore.

deductive reasoning A type of logic in which specific results are predicted from a general premise.

de-etiolation The changes a plant shoot undergoes in response to sunlight; also known informally as greening.

dehydration reaction A chemical reaction in which two molecules become covalently bonded to each other with the removal of a water molecule.

deletion (1) A deficiency in a chromosome resulting from the loss of a fragment through breakage. (2) A mutational loss of one or more nucleotide pairs from a gene.

demographic transition In a stable population, a shift from high birth and death rates to low birth and death rates.

demography The study of changes over time in the vital statistics of populations, especially birth rates and death rates.

denaturation (dē-nā′-chur-ā′-shun) In proteins, a process in which a protein loses its native shape due to the disruption of weak chemical bonds and interactions, thereby becoming biologically inactive; in DNA, the separation of the two strands of the double helix. Denaturation occurs under extreme (noncellular) conditions of pH, salt concentration, or temperature.

dendrite (den′-drīt) One of usually numerous, short, highly branched extensions of a neuron that receive signals from other neurons.

dendritic cell An antigen-presenting cell, located mainly in lymphatic tissues and skin, that is particularly efficient in presenting antigens to helper T cells, thereby initiating a primary immune response.

density The number of individuals per unit area or volume.

density dependent Referring to any characteristic that varies with population density.

density-dependent inhibition The phenomenon observed in normal animal cells that causes them to stop dividing when they come into contact with one another.

density independent Referring to any characteristic that is not affected by population density.

deoxyribonucleic acid (DNA) (dē-ok′-sē-rī′-bō-nū-klā′-ik) A nucleic acid molecule, usually a double-stranded helix, in which each polynucleotide strand consists of nucleotide monomers with a deoxyribose sugar and the nitrogenous bases adenine (A), cytosine (C), guanine (G), and thymine (T); capable of being replicated and determining the inherited structure of a cell's proteins.

deoxyribose (dē-ok′-si-rī′-bōs) The sugar component of DNA nucleotides, having one fewer hydroxyl group than ribose, the sugar component of RNA nucleotides.

dependent variable A factor whose value is measured during an experiment or other test to see whether it is influenced by changes in another factor (the independent variable).

depolarization A change in a cell's membrane potential such that the inside of the membrane is made less negative relative to the outside. For example, a neuron membrane is depolarized if a stimulus decreases its voltage from the resting potential of -70 mV in the direction of zero voltage.

dermal tissue system The outer protective covering of plants.

desert A terrestrial biome characterized by very low precipitation.

desmosome A type of intercellular junction in animal cells that functions as a rivet, fastening cells together.

determinate cleavage A type of embryonic development in protostomes that rigidly casts the developmental fate of each embryonic cell very early.

determinate growth A type of growth characteristic of most animals and some plant organs, in which growth stops after a certain size is reached.

determination The progressive restriction of developmental potential in which the possible fate of each cell becomes more limited as an embryo develops. At the end of determination, a cell is committed to its fate.

detritivore (deh-trī′-tuh-vōr) A consumer that derives its energy and nutrients from nonliving organic material such as corpses, fallen plant material, and the wastes of living organisms; a decomposer.

detritus (di-trī′-tus) Dead organic matter.

deuteromycete (dū′-tuh-rō-mī′-sēt) Traditional classification for a fungus with no known sexual stage.

deuterostome development (dū′-tuh-rō-stōm′) In animals, a developmental mode distinguished by the development of the anus from the blastopore; often also characterized by radial cleavage and by the body cavity forming as outpockets of mesodermal tissue.

Deuterostomia (dū′-tuh-rō-stōm′-ē-uh) One of the three main lineages of bilaterian animals. *See also* Ecdysozoa and Lophotrochozoa.

development The events involved in an organism's changing gradually from a simple to a more complex or specialized form.

diabetes mellitus (dī′-uh-bē′-tis mel′-uh-tus) An endocrine disorder marked by an inability to maintain glucose homeostasis. The type 1 form results from autoimmune destruction of insulin-secreting cells; treatment usually requires daily insulin injections. The type 2 form most commonly results from reduced responsiveness of target cells to insulin; obesity and lack of exercise are risk factors.

diacylglycerol (DAG) (dī-a′-sil-glis′-er-ol) A second messenger produced by the cleavage of the phospholipid PIP_2 in the plasma membrane.

diaphragm (dī′-uh-fram′) (1) A sheet of muscle that forms the bottom wall of the thoracic cavity in mammals. Contraction of the diaphragm pulls air into the lungs. (2) A dome-shaped rubber cup fitted into the upper portion of the vagina before sexual intercourse. It serves as a physical barrier to the passage of sperm into the uterus.

diapsid (dī-ap′-sid) A member of an amniote clade distinguished by a pair of holes on each side of the skull. Diapsids include the lepidosaurs and archosaurs.

diastole (dī-as′-tō-lē) The stage of the cardiac cycle in which a heart chamber is relaxed and fills with blood.

diastolic pressure Blood pressure in the arteries when the ventricles are relaxed.

diatom Photosynthetic protist in the stramenopile clade; diatoms have a unique glass-like wall made of silicon dioxide embedded in an organic matrix.

dicot A term traditionally used to refer to flowering plants that have two embryonic seed leaves, or cotyledons. Recent molecular evidence indicates that dicots do not form a clade; species once classified as dicots are now grouped into eudicots, magnoliids, and several lineages of basal angiosperms.

differential gene expression The expression of different sets of genes by cells with the same genome.

differentiation The process by which a cell or group of cells becomes specialized in structure and function.

diffusion The random thermal motion of particles of liquids, gases, or solids. In the presence of a concentration or electrochemical gradient, diffusion results in the net movement of a substance from a region where it is more concentrated to a region where it is less concentrated.

digestion The second stage of food processing in animals: the breaking down of food into molecules small enough for the body to absorb.

dihybrid (dī′-hī′-brid) An organism that is heterozygous with respect to two genes of interest. All the offspring from a cross between parents doubly homozygous for different alleles are dihybrids. For example, parents of genotypes *AABB* and *aabb* produce a dihybrid of genotype *AaBb*.

dihybrid cross (dī′-hī′-brid) A cross between two organisms that are each heterozygous for both of the characters being followed (or the self-pollination of a plant that is heterozygous for both characters).

dikaryotic (dī′-kār-ē-ot′-ik) Referring to a fungal mycelium with two haploid nuclei per cell, one from each parent.

dinoflagellate (dī′-nō-flaj′-uh-let) A member of a group of mostly unicellular photosynthetic algae with two flagella situated in perpendicular grooves in cellulose plates covering the cell.

dinosaur A member of an extremely diverse clade of reptiles varying in body shape, size, and habitat. Birds are the only extant dinosaurs.

dioecious (dī-ē′-shus) In plant biology, having the male and female reproductive parts on different individuals of the same species.

diploblastic Having two germ layers.

diploid cell (dip′-loyd) A cell containing two sets of chromosomes ($2n$), one set inherited from each parent.

diplomonad A protist that has modified mitochondria, two equal-sized nuclei, and multiple flagella.

directional selection Natural selection in which individuals at one end of the phenotypic range survive or reproduce more successfully than do other individuals.

disaccharide (dī-sak′-uh-rīd) A double sugar, consisting of two monosaccharides joined by a glycosidic linkage formed by a dehydration reaction.

dispersal The movement of individuals or gametes away from their parent location. This movement sometimes expands the geographic range of a population or species.

dispersion The pattern of spacing among individuals within the boundaries of a population.

disruptive selection Natural selection in which individuals on both extremes of a phenotypic range survive or reproduce more successfully than do individuals with intermediate phenotypes.

distal tubule In the vertebrate kidney, the portion of a nephron that helps refine filtrate and empties it into a collecting duct.

disturbance A natural or human-caused event that changes a biological community and usually removes organisms from it. Disturbances, such as fires and storms, play a pivotal role in structuring many communities.

disulfide bridge A strong covalent bond formed when the sulfur of one cysteine monomer bonds to the sulfur of another cysteine monomer.

DNA (deoxyribonucleic acid) (dē-ok′-sē-rī′-bō-nū-klā′-ik) A nucleic acid molecule, usually a double-stranded helix, in which each polynucleotide strand consists of nucleotide monomers with a deoxyribose sugar and the nitrogenous bases adenine (A), cytosine (C), guanine (G), and thymine (T); capable of being replicated and determining the inherited structure of a cell's proteins.

DNA cloning The production of multiple copies of a specific DNA segment.

DNA ligase (lī′-gās) A linking enzyme essential for DNA replication; catalyzes the covalent bonding of the 3′ end of one DNA fragment (such as an Okazaki fragment) to the 5′ end of another DNA fragment (such as a growing DNA chain).

DNA methylation The presence of methyl groups on the DNA bases (usually cytosine) of plants, animals, and fungi. (The term also refers to the process of adding methyl groups to DNA bases.)

DNA microarray assay A method to detect and measure the expression of thousands of genes at one time. Tiny amounts of a large number of single-stranded DNA fragments representing different genes are fixed to a glass slide and tested for hybridization with samples of labeled cDNA.

DNA polymerase (puh-lim′-er-ās) An enzyme that catalyzes the elongation of new DNA (for example, at a replication fork) by the addition of nucleotides to the 3′ end of an existing chain. There are several different DNA polymerases; DNA polymerase III and DNA polymerase I play major roles in DNA replication in *E. coli*.

DNA replication The process by which a DNA molecule is copied; also called DNA synthesis.

DNA sequencing Determining the complete nucleotide sequence of a gene or DNA segment.

DNA technology Techniques for sequencing and manipulating DNA.

domain (1) A taxonomic category above the kingdom level. The three domains are Archaea, Bacteria, and Eukarya. (2) A discrete structural and functional region of a protein.

dominant allele An allele that is fully expressed in the phenotype of a heterozygote.

dominant species A species with substantially higher abundance or biomass than other species in a community. Dominant species exert a powerful control over the occurrence and distribution of other species.

dormancy A condition typified by extremely low metabolic rate and a suspension of growth and development.

dorsal Pertaining to the top of an animal with radial or bilateral symmetry.

dorsal lip The region above the blastopore on the dorsal side of the amphibian embryo.

double bond A double covalent bond; the sharing of two pairs of valence electrons by two atoms.

double circulation A circulatory system consisting of separate pulmonary and systemic circuits, in which blood passes through the heart after completing each circuit.

double fertilization A mechanism of fertilization in angiosperms in which two sperm cells unite with two cells in the female gametophyte (embryo sac) to form the zygote and endosperm.

double helix The form of native DNA, referring to its two adjacent antiparallel polynucleotide strands wound around an imaginary axis into a spiral shape.

Down syndrome A human genetic disease usually caused by the presence of an extra chromosome 21; characterized by developmental delays and heart and other defects that are generally treatable or non-life-threatening.

Duchenne muscular dystrophy (duh-shen′) A human genetic disease caused by a sex-linked recessive allele; characterized by progressive weakening and a loss of muscle tissue.

duodenum (dū′-uh-dēn′-um) The first section of the small intestine, where chyme from the stomach mixes with digestive juices from the pancreas, liver, and gallbladder as well as from gland cells of the intestinal wall.

duplication An aberration in chromosome structure due to fusion with a fragment from a homologous chromosome, such that a portion of a chromosome is duplicated.

dynein (dī′-nē-un) In cilia and flagella, a large motor protein extending from one microtubule doublet to the adjacent doublet. ATP hydrolysis drives changes in dynein shape that lead to bending of cilia and flagella.

E site One of a ribosome's three binding sites for tRNA during translation. The E site is the place where discharged tRNAs leave the ribosome. (E stands for exit.)

Ecdysozoa (ek′-dē-sō-zō′-uh) One of the three main lineages of bilaterian animals; many ecdysozoans are molting animals. *See also* Deuterostomia and Lophotrochozoa.

echinoderm (i-kī′-nō-derm) A slow-moving or sessile marine deuterostome with a water vascular system and, in larvae, bilateral symmetry. Echinoderms include sea stars, brittle stars, sea urchins, feather stars, and sea cucumbers.

ecological footprint The aggregate land and water area required by a person, city, or nation to produce all of the resources it consumes and to absorb all of the waste it generates.

ecological niche (nich) The sum of a species' use of the biotic and abiotic resources in its environment.

ecological species concept Definition of a species in terms of ecological niche, the sum of how members of the species interact with the nonliving and living parts of their environment.

ecological succession Transition in the species composition of a community following a disturbance; establishment of a community in an area virtually barren of life.

ecology The study of how organisms interact with each other and their environment.

ecosystem All the organisms in a given area as well as the abiotic factors with which they interact; one or more communities and the physical environment around them.

ecosystem ecology The study of energy flow and the cycling of chemicals among the various biotic and abiotic components in an ecosystem.

ecosystem engineer An organism that influences community structure by causing physical changes in the environment.

ecosystem service A function performed by an ecosystem that directly or indirectly benefits humans.

ecotone The transition from one type of habitat or ecosystem to another, such as the transition from a forest to a grassland.

ectoderm (ek′-tō-durm) The outermost of the three primary germ layers in animal embryos; gives rise to the outer covering and, in some phyla, the nervous system, inner ear, and lens of the eye.

ectomycorrhiza (plural, **ectomycorrhizae**) (ek′-tō-mī′-kō-rī′-zuh, ek′-tō-mī′-kō-rī′-zē) Association of a fungus with a plant root system in which the fungus surrounds the roots but does not cause invagination of the host (plant) cell's plasma membrane.

ectomycorrhizal fungus A symbiotic fungus that forms sheaths of hyphae over the surface of plant roots and also grows into extracellular spaces of the root cortex.

ectoparasite A parasite that feeds on the external surface of a host.

ectopic Occurring in an abnormal location.

ectoproct A sessile, colonial lophophorate; also called a bryozoan.

ectothermic Referring to organisms for which external sources provide most of the heat for temperature regulation.

Ediacaran biota (ē′-dē-uh-keh′-run bī-ō′-tuh) An early group of macroscopic, soft-bodied, multicellular eukaryotes known from fossils that range in age from 635 million to 535 million years old.

effective population size An estimate of the size of a population based on the numbers of females and males that successfully breed; generally smaller than the total population.

effector A pathogen-encoded protein that cripples the host's innate immune system.

effector cell (1) A muscle cell or gland cell that carries out the body's response to stimuli as directed by signals from the brain or other processing center of the nervous system. (2) A lymphocyte that has undergone clonal selection and is capable of mediating an adaptive immune response.

egg The female gamete.

ejaculation The propulsion of sperm from the epididymis through the muscular vas deferens, ejaculatory duct, and urethra.

electrocardiogram (ECG or EKG) A record of the electrical impulses that travel through heart muscle during the cardiac cycle.

electrochemical gradient The diffusion gradient of an ion, which is affected by both the concentration difference of an ion across a membrane (a chemical force) and the ion's tendency to move relative to the membrane potential (an electrical force).

electrogenic pump An active transport protein that generates voltage across a membrane while pumping ions.

electromagnetic receptor A receptor of electromagnetic energy, such as visible light, electricity, or magnetism.

electromagnetic spectrum The entire spectrum of electromagnetic radiation, ranging in wavelength from less than a nanometer to more than a kilometer.

electron A subatomic particle with a single negative electrical charge and a mass about 1/2,000 that of a neutron or proton. One or more electrons move around the nucleus of an atom.

electron microscope (EM) A microscope that uses magnets to focus an electron beam on or through a specimen, resulting in a practical resolution that is 100-fold greater than that of a light microscope using standard techniques. A transmission electron microscope (TEM) is used to study the internal structure of thin sections of cells. A scanning electron microscope (SEM) is used to study the fine details of cell surfaces.

electron shell An energy level of electrons at a characteristic average distance from the nucleus of an atom.

electron transport chain A sequence of electron carrier molecules (membrane proteins) that shuttle electrons down a series of redox reactions that release energy used to make ATP.

electronegativity The attraction of a given atom for the electrons of a covalent bond.

electroporation A technique to introduce recombinant DNA into cells by applying a brief electrical pulse to a solution containing the cells. The pulse creates temporary holes in the cells' plasma membranes, through which DNA can enter.

element Any substance that cannot be broken down to any other substance by chemical reactions.

elimination The fourth and final stage of food processing in animals: the passing of undigested material out of the body.

embryo sac (em′-brē-ō) The female gametophyte of angiosperms, formed from the growth and division of the megaspore into a multicellular structure that typically has eight haploid nuclei.

embryonic lethal A mutation with a phenotype leading to death of an embryo or larva.

embryophyte Alternate name for land plants that refers to their shared derived trait of multicellular, dependent embryos.

emergent properties New properties that arise with each step upward in the hierarchy of life, owing to the arrangement and interactions of parts as complexity increases.

emigration The movement of individuals out of a population.

enantiomer (en-an′-tē-ō-mer) One of two compounds that are mirror images of each other and that differ in shape due to the presence of an asymmetric carbon.

endangered species A species that is in danger of extinction throughout all or a significant portion of its range.

endemic (en-dem′-ik) Referring to a species that is confined to a specific geographic area.

endergonic reaction (en′-der-gon′-ik) A nonspontaneous chemical reaction in which free energy is absorbed from the surroundings.

endocrine gland (en′-dō-krin) A ductless gland that secretes hormones directly into the interstitial fluid, from which they diffuse into the bloodstream.

endocrine system (en′-dō-krin) In animals, the internal system of communication involving hormones, the ductless glands that secrete hormones, and the molecular receptors on or in target cells that respond to hormones; functions in concert with the nervous system to effect internal regulation and maintain homeostasis.

endocytosis (en′-dō-sī-tō′-sis) Cellular uptake of biological molecules and particulate matter via formation of vesicles from the plasma membrane.

endoderm (en′-dō-durm) The innermost of the three primary germ layers in animal embryos; lines the archenteron and gives rise to the liver, pancreas, lungs, and the lining of the digestive tract in species that have these structures.

endodermis In plant roots, the innermost layer of the cortex that surrounds the vascular cylinder.

endomembrane system The collection of membranes inside and surrounding a eukaryotic cell, related either through direct physical contact or by the transfer of membranous vesicles; includes the plasma membrane, the nuclear envelope, the smooth and rough endoplasmic reticulum, the Golgi apparatus, lysosomes, vesicles, and vacuoles.

endometriosis (en′-dō-mē-trē-ō′-sis) The condition resulting from the presence of endometrial tissue outside of the uterus.

endometrium (en′-dō-mē′-trē-um) The inner lining of the uterus, which is richly supplied with blood vessels.

endoparasite A parasite that lives within a host.

endophyte A harmless fungus, or occasionally another organism, that lives between cells of a plant part or multicellular alga.

endoplasmic reticulum (ER) (en′-dō-plaz′-mik ruh-tik′-yū-lum) An extensive membranous network in eukaryotic cells, continuous with the outer nuclear membrane and composed of ribosome-studded (rough) and ribosome-free (smooth) regions.

endorphin (en-dōr′-fin) Any of several hormones produced in the brain and anterior pituitary that inhibit pain perception.

endoskeleton A hard skeleton buried within the soft tissues of an animal.

endosperm In angiosperms, a nutrient-rich tissue formed by the union of a sperm with two polar nuclei during double fertilization. The endosperm provides nourishment to the developing embryo in angiosperm seeds.

endospore A thick-coated, resistant cell produced by some bacterial cells when they are exposed to harsh conditions.

endosymbiont theory The theory that mitochondria and plastids, including chloroplasts, originated as prokaryotic cells engulfed by a host cell. The engulfed cell and its host cell then evolved into a single organism. *See also* endosymbiosis.

endosymbiosis A relationship between two species in which one organism lives inside the cell or cells of another organism. *See also* endosymbiont theory.

endothelium (en′-dō-thē′-lē-um) The simple squamous layer of cells lining the lumen of blood vessels.

endothermic Referring to organisms that are warmed by heat generated by their own metabolism. This heat usually maintains a relatively stable body temperature higher than that of the external environment.

endotoxin A toxic component of the outer membrane of certain gram-negative bacteria that is released only when the bacteria die.

energetic hypothesis The concept that the length of a food chain is limited by the inefficiency of energy transfer along the chain.

energy The capacity to cause change, especially to do work (to move matter against an opposing force).

energy coupling In cellular metabolism, the use of energy released from an exergonic reaction to drive an endergonic reaction.

enhancer A segment of eukaryotic DNA containing multiple control elements, usually located far from the gene whose transcription it regulates.

enteric nervous system A distinct network of neurons that exerts direct and partially independent control over the digestive tract, pancreas, and gallbladder.

entropy A measure of molecular disorder, or randomness.

enzyme (en′-zīm) A macromolecule serving as a catalyst, a chemical agent that increases the rate of a reaction without being consumed by the reaction. Most enzymes are proteins.

enzyme-substrate complex (en′-zīm) A temporary complex formed when an enzyme binds to its substrate molecule(s).

epicotyl (ep′-uh-kot′-ul) In an angiosperm embryo, the embryonic axis above the point of attachment of the cotyledon(s) and below the first pair of miniature leaves.

epidemic A widespread outbreak of a disease.

epidermis (1) The dermal tissue system of nonwoody plants, usually consisting of a single layer of tightly packed cells. (2) The outermost layer of cells in an animal.

epididymis (ep′-uh-did′-uh-mus) A coiled tubule located adjacent to the mammalian testis where sperm are stored.

epigenetic inheritance Inheritance of traits transmitted by mechanisms that do not involve the nucleotide sequence.

epinephrine (ep′-i-nef′-rin) A catecholamine that, when secreted as a hormone by the adrenal medulla, mediates "fight-or-flight" responses to short-term stresses; also released by some neurons as a neurotransmitter; also called adrenaline.

epiphyte (ep′-uh-fīt) A plant that nourishes itself but grows on the surface of another plant for support, usually on the branches or trunks of trees.

epistasis (ep′-i-stā′-sis) A type of gene interaction in which the phenotypic expression of one gene alters that of another independently inherited gene.

epithelial tissue (ep′-uh-thē′-lē-ul) Sheets of tightly packed cells that line organs and body cavities as well as external surfaces.

epithelium An epithelial tissue.

epitope A small, accessible region of an antigen to which an antigen receptor or antibody binds.

equilibrium potential (E_{ion}) The magnitude of a cell's membrane voltage at equilibrium; calculated using the Nernst equation.

erythrocyte (eh-rith′-ruh-sīt) A blood cell that contains hemoglobin, which transports oxygen; also called a red blood cell.

erythropoietin (EPO) (eh-rith′-rō-poy′-uh-tin) A hormone that stimulates the production of erythrocytes. It is secreted by the kidney when body tissues do not receive enough oxygen.

esophagus (eh-sof′-uh-gus) A muscular tube that conducts food, by peristalsis, from the pharynx to the stomach.

essential amino acid An amino acid that an animal cannot synthesize itself and must be obtained from food in prefabricated form.

essential element A chemical element required for an organism to survive, grow, and reproduce.

essential fatty acid An unsaturated fatty acid that an animal needs but cannot make.

essential nutrient A substance that an organism cannot synthesize from any other material and therefore must absorb in preassembled form.

estradiol (es′-truh-dī′-ol) A steroid hormone that stimulates the development and maintenance of the female reproductive system and secondary sex characteristics; the major estrogen in mammals.

estrogen (es′-trō-jen) Any steroid hormone, such as estradiol, that stimulates the development and maintenance of the female reproductive system and secondary sex characteristics.

estrous cycle (es′-trus) A reproductive cycle characteristic of female mammals except humans and certain other primates, in which the endometrium is reabsorbed in the absence of pregnancy and sexual response occurs only during a mid-cycle point known as estrus.

estuary The area where a freshwater stream or river merges with the ocean.

ethylene (eth′-uh-lēn) A gaseous plant hormone involved in responses to mechanical stress, programmed cell death, leaf abscission, and fruit ripening.

etiolation Plant morphological adaptations for growing in darkness.

euchromatin (yū-krō′-muh-tin) The less condensed form of eukaryotic chromatin that is available for transcription.

eudicot (yū-dī′-kot) A member of a clade that contains the vast majority of flowering plants that have two embryonic seed leaves, or cotyledons.

euglenid (yū′-glen-id) A protist, such as *Euglena* or its relatives, characterized by an anterior pocket from which one or two flagella emerge.

euglenozoan A member of a diverse clade of flagellated protists that includes predatory heterotrophs, photosynthetic autotrophs, and pathogenic parasites.

Eukarya (yū-kar′-ē-uh) The domain that includes all eukaryotic organisms.

eukaryotic cell (yū′-ker-ē-ot′-ik) A type of cell with a membrane-enclosed nucleus and membrane-enclosed organelles. Organisms with eukaryotic cells (protists, plants, fungi, and animals) are called eukaryotes.

eumetazoan (yū′-met-uh-zō′-un) A member of a clade of animals with true tissues. All animals except sponges and a few other groups are eumetazoans.

eurypterid (yur-ip′-tuh-rid) An extinct carnivorous chelicerate; also called a water scorpion.

Eustachian tube (yū-stā′-shun) The tube that connects the middle ear to the pharynx.

eutherian (yū-thēr′-ē-un) Placental mammal; mammal whose young complete their embryonic development within the uterus, joined to the mother by the placenta.

eutrophic lake (yū-trōf′-ik) A lake that has a high rate of biological productivity supported by a high rate of nutrient cycling.

eutrophication A process by which nutrients, particularly phosphorus and nitrogen, become highly concentrated in a body of water, leading to increased growth of organisms such as algae or cyanobacteria.

evaporative cooling The process in which the surface of an object becomes cooler during evaporation, a result of the molecules with the greatest kinetic energy changing from the liquid to the gaseous state.

evapotranspiration The total evaporation of water from an ecosystem, including water transpired by plants and evaporated from a landscape, usually measured in millimeters and estimated for a year.

evo-devo Evolutionary developmental biology; a field of biology that compares developmental processes of different multicellular organisms to understand how these processes have evolved and how changes can modify existing organismal features or lead to new ones.

evolution Descent with modification; the idea that living species are descendants of ancestral species that were different from the present-day ones; also defined more narrowly as the change in the genetic composition of a population from generation to generation.

evolutionary tree A branching diagram that reflects a hypothesis about evolutionary relationships among groups of organisms.

Excavata (ex′-kuh-vah′-tuh) One of four supergroups of eukaryotes proposed in a current hypothesis of the evolutionary history of eukaryotes. Excavates have unique cytoskeletal features, and some species have an "excavated" feeding groove on one side of the cell body. *See also* SAR, Archaeplastida, and Unikonta.

excitatory postsynaptic potential (EPSP) An electrical change (depolarization) in the membrane of a postsynaptic cell caused by the binding of an excitatory neurotransmitter from a presynaptic cell to a postsynaptic receptor; makes it more likely for a postsynaptic cell to generate an action potential.

excretion The disposal of nitrogen-containing metabolites and other waste products.

exergonic reaction (ek′-ser-gon′-ik) A spontaneous chemical reaction in which there is a net release of free energy.

exocytosis (ek′-sō-sī-tō′-sis) The cellular secretion of biological molecules by the fusion of vesicles containing them with the plasma membrane.

exon A sequence within a primary transcript that remains in the RNA after RNA processing; also refers to the region of DNA from which this sequence was transcribed.

exoskeleton A hard encasement on the surface of an animal, such as the shell of a mollusc or the cuticle of an arthropod, that provides protection and points of attachment for muscles.

exotoxin (ek′-sō-tok′-sin) A toxic protein that is secreted by a prokaryote or other pathogen and that produces specific symptoms, even if the pathogen is no longer present.

expansin Plant enzyme that breaks the cross-links (hydrogen bonds) between cellulose microfibrils and other cell wall constituents, loosening the wall's fabric.

experiment A scientific test. Often carried out under controlled conditions that involve manipulating one factor in a system in order to see the effects of changing that factor.

experimental group A set of subjects that has (or receives) the specific factor being tested in a controlled experiment. Ideally, the experimental group should be identical to the control group for all other factors.

exploitation A +/− ecological interaction in which one species benefits by feeding on the other species, which is harmed. Exploitative interactions include predation, herbivory, and parasitism.

exponential population growth Growth of a population in an ideal, unlimited environment, represented by a J-shaped curve when population size is plotted over time.

expression vector A cloning vector that contains a highly active bacterial promoter just upstream of a restriction site where a eukaryotic gene can be inserted, allowing the gene to be expressed in a bacterial cell. Expression vectors are also available that have been genetically engineered for use in specific types of eukaryotic cells.

extinction vortex A downward population spiral in which inbreeding and genetic drift combine to cause a small population to shrink and, unless the spiral is reversed, become extinct.

extracellular matrix (ECM) The meshwork surrounding animal cells, consisting of glycoproteins, polysaccharides, and proteoglycans synthesized and secreted by cells.

extraembryonic membrane One of four membranes (yolk sac, amnion, chorion, and allantois) located outside the embryo that support the developing embryo in reptiles and mammals.

extreme halophile An organism that lives in a highly saline environment, such as the Great Salt Lake or the Dead Sea.

extreme thermophile An organism that thrives in hot environments (often 60–80°C or hotter).

extremophile An organism that lives in environmental conditions so extreme that few other species can survive there. Extremophiles include extreme halophiles ("salt lovers") and extreme thermophiles ("heat lovers").

F_1 generation The first filial, hybrid (heterozygous) offspring arising from a parental (P generation) cross.

F_2 generation The offspring resulting from interbreeding (or self-pollination) of the hybrid F_1 generation.

facilitated diffusion The passage of molecules or ions down their electrochemical gradient across a biological membrane with the assistance of specific transmembrane transport proteins, requiring no energy expenditure.

facultative anaerobe (fak′-ul-tā′-tiv an′-uh-rōb) An organism that makes ATP by aerobic respiration if oxygen is present but that switches to anaerobic respiration or fermentation if oxygen is not present.

family In Linnaean classification, the taxonomic category above genus.

fast-twitch fiber A muscle fiber used for rapid, powerful contractions.

fat A lipid consisting of three fatty acids linked to one glycerol molecule; also called a triacylglycerol or triglyceride.

fate map A territorial diagram of embryonic development that displays the future derivatives of individual cells and tissues.

fatty acid A carboxylic acid with a long carbon chain. Fatty acids vary in length and in the number and location of double bonds; three fatty acids linked to a glycerol molecule form a fat molecule, also called triacylglycerol or triglyceride.

feces (fē′-sēz) The wastes of the digestive tract.

feedback inhibition A method of metabolic control in which the end product of a metabolic pathway acts as an inhibitor of an enzyme within that pathway.

feedback regulation The regulation of a process by its output or end product.

fermentation A catabolic process that makes a limited amount of ATP from glucose (or other organic molecules) without an electron transport chain and that produces a characteristic end product, such as ethyl alcohol or lactic acid.

fertilization (1) The union of haploid gametes to produce a diploid zygote. (2) The addition of mineral nutrients to the soil.

fetus (fē′-tus) A developing mammal that has all the major structures of an adult. In humans, the fetal stage lasts from the 9th week of gestation until birth.

F factor In bacteria, the DNA segment that confers the ability to form pili for conjugation and associated functions required for the transfer of DNA from donor to recipient. The F factor may exist as a plasmid or be integrated into the bacterial chromosome.

fiber A lignified cell type that reinforces the xylem of angiosperms and functions in mechanical support; a slender, tapered sclerenchyma cell that usually occurs in bundles.

fibroblast (fī′-brō-blast) A type of cell in loose connective tissue that secretes the protein ingredients of the extracellular fibers.

fibronectin An extracellular glycoprotein secreted by animal cells that helps them attach to the extracellular matrix.

filament In an angiosperm, the stalk portion of the stamen, the pollen-producing reproductive organ of a flower.

filter feeder An animal that feeds by using a filtration mechanism to strain small organisms or food particles from its surroundings.

filtrate Cell-free fluid extracted from the body fluid by the excretory system.

filtration In excretory systems, the extraction of water and small solutes, including metabolic wastes, from the body fluid.

fimbria (plural, **fimbriae**) A short, hairlike appendage of a prokaryotic cell that helps it adhere to the substrate or to other cells.

first law of thermodynamics The principle of conservation of energy: Energy can be transferred and transformed, but it cannot be created or destroyed.

fission The separation of an organism into two or more individuals of approximately equal size.

fixed action pattern In animal behavior, a sequence of unlearned acts that is essentially unchangeable and, once initiated, usually carried to completion.

flaccid (flas′-id) Limp. Lacking turgor (stiffness or firmness), as in a plant cell in surroundings where there is a tendency for water to leave the cell. (A walled cell becomes flaccid if it has a higher water potential than its surroundings, resulting in the loss of water.)

flagellum (fluh-jel′-um) (plural, **flagella**) A long cellular appendage specialized for locomotion. Like motile cilia, eukaryotic flagella have a core with nine outer doublet microtubules and two inner single microtubules (the "9 + 2" arrangement) ensheathed in an extension of the plasma membrane. Prokaryotic flagella have a different structure.

florigen A flowering signal, probably a protein, that is made in leaves under certain conditions and that travels to the shoot apical meristems, inducing them to switch from vegetative to reproductive growth.

flower In an angiosperm, a specialized shoot with up to four sets of modified leaves, bearing structures that function in sexual reproduction.

fluid feeder An animal that lives by sucking nutrient-rich fluids from another living organism.

fluid mosaic model The currently accepted model of cell membrane structure, which envisions the membrane as a mosaic of protein molecules drifting laterally in a fluid bilayer of phospholipids.

follicle (fol′-uh-kul) A microscopic structure in the ovary that contains the developing oocyte and secretes estrogens.

follicle-stimulating hormone (FSH) (fol′-uh-kul) A tropic hormone that is produced and secreted by the anterior pituitary and that stimulates the production of eggs by the ovaries and sperm by the testes.

food chain The pathway along which food energy is transferred from trophic level to trophic level, beginning with producers.

food vacuole A membranous sac formed by phagocytosis of microorganisms or particles to be used as food by the cell.

food web The interconnected feeding relationships in an ecosystem.

foot (1) The portion of a bryophyte sporophyte that gathers sugars, amino acids, water, and minerals from the parent gametophyte via transfer cells. (2) One of the three main parts of a mollusc; a muscular structure usually used for movement. *See also* mantle and visceral mass.

foraging The seeking and obtaining of food.

foram (foraminiferan) An aquatic protist that secretes a hardened shell containing calcium carbonate and extends pseudopodia through pores in the shell.

forebrain One of three ancestral and embryonic regions of the vertebrate brain; develops into the thalamus, hypothalamus, and cerebrum.

fossil A preserved remnant or impression of an organism that lived in the past.

founder effect Genetic drift that occurs when a few individuals become isolated from a larger population and form a new population whose gene pool composition is not reflective of that of the original population.

fovea (fō′-vē-uh) The place on the retina at the eye's center of focus, where cones are highly concentrated.

F plasmid The plasmid form of the F factor.

fragmentation A means of asexual reproduction whereby a single parent breaks into parts that regenerate into whole new individuals.

frameshift mutation A mutation occurring when nucleotides are inserted in or deleted from a gene and the number inserted or deleted is not a multiple of three, resulting in the improper grouping of the subsequent nucleotides into codons.

free energy The portion of a biological system's energy that can perform work when temperature and pressure are uniform throughout the system. The change in free energy of a system (ΔG) is calculated by the equation $\Delta G = \Delta H - T\Delta S$, where ΔH is the change in enthalpy (in biological systems, equivalent to total energy), ΔT is the absolute temperature, and ΔS is the change in entropy.

frequency-dependent selection Selection in which the fitness of a phenotype depends on how common the phenotype is in a population.

fruit A mature ovary of a flower. The fruit protects dormant seeds and often functions in their dispersal.

functional group A specific configuration of atoms commonly attached to the carbon skeletons of organic molecules and involved in chemical reactions.

fusion In evolutionary biology, a process in which gene flow between two species that can form hybrid offspring weakens barriers to reproduction between the species. This process causes their gene pools to become increasingly alike and can cause the two species to fuse into a single species.

G_0 phase A nondividing state occupied by cells that have left the cell cycle, sometimes reversibly.

G_1 phase The first gap, or growth phase, of the cell cycle, consisting of the portion of interphase before DNA synthesis begins.

G_2 phase The second gap, or growth phase, of the cell cycle, consisting of the portion of interphase after DNA synthesis occurs.

gallbladder An organ that stores bile and releases it as needed into the small intestine.

game theory An approach to evaluating alternative strategies in situations where the outcome of a particular strategy depends on the strategies used by other individuals.

gametangium (gam′-uh-tan′-jē-um) (plural, **gametangia**) Multicellular plant structure in which gametes are formed. Female gametangia are called archegonia, and male gametangia are called antheridia.

gamete (gam′-ēt) A haploid reproductive cell, such as an egg or sperm, that is formed by meiosis or is the descendant of cells formed by meiosis. Gametes unite during sexual reproduction to produce a diploid zygote.

gametogenesis (guh-mē′-tō-gen′-uh-sis) The process by which gametes are produced.

gametophore (guh-mē′-tō-fōr) The mature gamete-producing structure of a moss gametophyte.

gametophyte (guh-mē′-tō-fīt) In organisms (plants and some algae) that have alternation of generations, the multicellular haploid form that produces haploid gametes by mitosis. The haploid gametes unite and develop into sporophytes.

ganglion (gan′-glē-uhn) (plural, **ganglia**) A cluster (functional group) of nerve cell bodies.

gap junction A type of intercellular junction in animal cells, consisting of proteins surrounding a pore that allows the passage of materials between cells.

gas exchange The uptake of molecular oxygen from the environment and the discharge of carbon dioxide to the environment.

gastric juice A digestive fluid secreted by the stomach.

gastrovascular cavity A central cavity with a single opening in the body of certain animals, including cnidarians and flatworms, that functions in both the digestion and distribution of nutrients.

gastrula (gas′-trū-luh) An embryonic stage in animal development encompassing the formation of three layers: ectoderm, mesoderm, and endoderm.

gastrulation (gas′-trū-lā′-shun) In animal development, a series of cell and tissue movements in which the blastula-stage embryo folds inward, producing a three-layered embryo, the gastrula.

gated channel A transmembrane protein channel that opens or closes in response to a particular stimulus.

gated ion channel A gated channel for a specific ion. The opening or closing of such channels may alter a cell's membrane potential.

gel electrophoresis (ē-lek′-trō-fōr-ē′-sis) A technique for separating nucleic acids or proteins on the basis of their size and electrical charge, both of which affect their rate of movement through an electric field in a gel made of agarose or another polymer.

gene A discrete unit of hereditary information consisting of a specific nucleotide sequence in DNA (or RNA, in some viruses).

gene annotation Analysis of genomic sequences to identify protein-coding genes and determine the function of their products.

gene cloning The production of multiple copies of a gene.

gene drive A process that biases inheritance such that a particular allele is more likely to be inherited than are other alleles, causing the favored allele to spread (be "driven") through the population.

gene expression The process by which information encoded in DNA directs the synthesis of proteins or, in some cases, RNAs that are not translated into proteins and instead function as RNAs.

gene flow The transfer of alleles from one population to another, resulting from the movement of fertile individuals or their gametes.

gene pool The aggregate of all copies of every type of allele at all loci in every individual in a population. The term is also used in a more restricted sense as the aggregate of alleles for just one or a few loci in a population.

gene therapy The introduction of genes into an afflicted individual for therapeutic purposes.

genetic drift A process in which chance events cause unpredictable fluctuations in allele frequencies from one generation to the next. Effects of genetic drift are most pronounced in small populations.

genetic engineering The direct manipulation of genes for practical purposes.

genetic map An ordered list of genetic loci (genes or other genetic markers) along a chromosome.

genetic profile An individual's unique set of genetic markers, detected most often today by PCR or, previously, by electrophoresis and nucleic acid probes.

genetic recombination General term for the production of offspring with combinations of traits that differ from those found in either parent.

genetic variation Differences among individuals in the composition of their genes or other DNA segments.

genetically modified organism (GMO) An organism that has acquired one or more genes by artificial means; also called a transgenic organism.

genetics The scientific study of heredity and hereditary variation.

genome (jē′-nōm) The genetic material of an organism or virus; the complete complement of an organism's or virus's genes along with its noncoding nucleic acid sequences.

genome-wide association study (jē′-nōm) A large-scale analysis of the genomes of many people having a certain phenotype or disease, with the aim of finding genetic markers that correlate with that phenotype or disease.

genomic imprinting (juh-nō′-mik) A phenomenon in which expression of an allele in offspring depends on whether the allele is inherited from the male or female parent.

genomics (juh-nō′-miks) The systematic study of whole sets of genes (or other DNA) and their interactions within a species, as well as genome comparisons between species.

genotype (jē′-nō-tīp) The genetic makeup, or set of alleles, of an organism.

genus (jē′-nus) (plural, **genera**) A taxonomic category above the species level, designated by the first word of a species' two-part scientific name.

geologic record A standard time scale dividing Earth's history into time periods, grouped into four eons—Hadean, Archaean, Proterozoic, and Phanerozoic—and further subdivided into eras, periods, and epochs.

germ layer One of the three main layers in a gastrula that will form the various tissues and organs of an animal body.

gestation (jes-tā'-shun) *See* pregnancy.

gibberellin (jib'-uh-rel'-in) Any of a class of related plant hormones that stimulate growth in the stem and leaves, trigger the germination of seeds and breaking of bud dormancy, and (with auxin) stimulate fruit development.

glans The rounded structure at the tip of the clitoris or penis that is involved in sexual arousal.

glia (glial cells) Cells of the nervous system that support, regulate, and augment the functions of neurons.

global ecology The study of the functioning and distribution of organisms across the biosphere and how the regional exchange of energy and materials affects them.

glomeromycete (glō'-mer-ō-mī'-sēt) A member of the fungal phylum Glomeromycota, characterized by a distinct branching form of mycorrhizae called arbuscular mycorrhizae.

glomerulus (glō-mār'-yū-lus) A ball of capillaries surrounded by Bowman's capsule in the nephron and serving as the site of filtration in the vertebrate kidney.

glucocorticoid A steroid hormone that is secreted by the adrenal cortex and that influences glucose metabolism and immune function.

glucagon (glū'-kuh-gon) A hormone secreted by the pancreas that raises blood glucose levels. It promotes glycogen breakdown and release of glucose by the liver.

glyceraldehyde 3-phosphate (G3P) (glis'-er-al'-de-hīd) A three-carbon carbohydrate that is the direct product of the Calvin cycle; it is also an intermediate in glycolysis.

glycogen (glī'-kō-jen) An extensively branched glucose storage polysaccharide found in the liver and muscle of animals; the animal equivalent of starch.

glycolipid A lipid with one or more covalently attached carbohydrates.

glycolysis (glī-kol'-uh-sis) A series of reactions that ultimately splits glucose into pyruvate. Glycolysis occurs in almost all living cells, serving as the starting point for fermentation or cellular respiration.

glycoprotein A protein with one or more covalently attached carbohydrates.

glycosidic linkage A covalent bond formed between two monosaccharides by a dehydration reaction.

gnathostome (na'-thu-stōm) Member of one of the two main clades of vertebrates; gnathostomes have jaws and include sharks and rays, ray-finned fishes, coelacanths, lungfishes, amphibians, reptiles, and mammals. *See also* cyclostome.

golden alga A biflagellated, photosynthetic protist named for its color, which results from its yellow and brown carotenoids.

Golgi apparatus (gol'-jē) An organelle in eukaryotic cells consisting of stacks of flat membranous sacs that modify, store, and route products of the endoplasmic reticulum and synthesize some products, notably noncellulose carbohydrates.

gonad (gō'-nad) A male or female gamete-producing organ.

G protein A GTP-binding protein that relays signals from a plasma membrane signal receptor, known as a G protein-coupled receptor, to other signal transduction proteins inside the cell.

G protein-coupled receptor (GPCR) A signal receptor protein in the plasma membrane that responds to the binding of a signaling molecule by activating a G protein. Also called a G protein-linked receptor.

graded potential In a neuron, a shift in the membrane potential that has an amplitude proportional to signal strength and that decays as it spreads.

Gram stain A staining method that distinguishes between two different kinds of bacterial cell walls; may be used to help determine medical response to an infection.

gram-negative Describing the group of bacteria that have a cell wall that is structurally more complex and contains less peptidoglycan than the cell wall of gram-positive bacteria. Gram-negative bacteria are often more toxic than gram-positive bacteria.

gram-positive Describing the group of bacteria that have a cell wall that is structurally less complex and contains more peptidoglycan than the cell wall of gram-negative bacteria. Gram-positive bacteria are usually less toxic than gram-negative bacteria.

granum (gran'-um) (plural, **grana**) A stack of membrane-bounded thylakoids in the chloroplast. Grana function in the light reactions of photosynthesis.

gravitropism (grav'-uh-trō'-pizm) A response of a plant or animal to gravity.

gray matter Regions of clustered neuron cell bodies within the CNS.

green alga A photosynthetic protist, named for green chloroplasts that are similar in structure and pigment composition to the chloroplasts of plants. Green algae are a paraphyletic group; some members are more closely related to plants than they are to other green algae.

greenhouse effect The warming of Earth due to the atmospheric accumulation of carbon dioxide and certain other gases, which absorb reflected infrared radiation and reradiate some of it back toward Earth.

gross primary production (GPP) The total primary production of an ecosystem.

ground tissue system Plant tissues that are neither vascular nor dermal, fulfilling a variety of functions, such as storage, photosynthesis, and support.

growth factor (1) A protein that must be present in the extracellular environment (culture medium or animal body) for the growth and normal development of certain types of cells. (2) A local regulator that acts on nearby cells to stimulate cell proliferation and differentiation.

growth hormone (GH) A hormone that is produced and secreted by the anterior pituitary and that has both direct (nontropic) and tropic effects on a wide variety of tissues.

guard cells The two cells that flank the stomatal pore and regulate the opening and closing of the pore.

gustation The sense of taste.

guttation The exudation of water droplets from leaves, caused by root pressure in certain plants.

gymnosperm (jim'-nō-sperm) A vascular plant that bears naked seeds—seeds not enclosed in protective chambers.

hagfish Marine jawless vertebrates that have highly reduced vertebrae and a skull made of cartilage; most hagfishes are bottom-dwelling scavengers.

hair cell A mechanosensory cell that alters output to the nervous system when hairlike projections on the cell surface are displaced.

half-life The amount of time it takes for 50% of a sample of a radioactive isotope to decay.

halophile *See* extreme halophile.

Hamilton's rule The principle that for natural selection to favor an altruistic act, the benefit to the recipient, devalued by the coefficient of relatedness, must exceed the cost to the altruist.

haploid cell (hap'-loyd) A cell containing only one set of chromosomes (n).

Hardy-Weinberg equilibrium The state of a population in which frequencies of alleles and genotypes remain constant from generation to generation, provided that only Mendelian segregation and recombination of alleles are at work.

heart A muscular pump that uses metabolic energy to elevate the hydrostatic pressure of the circulatory fluid (blood or hemolymph). The fluid then flows down a pressure gradient through the body and eventually returns to the heart.

heart attack The damage or death of cardiac muscle tissue resulting from prolonged blockage of one or more coronary arteries.

heart murmur A hissing sound that most often results from blood squirting backward through a leaky valve in the heart.

heart rate The frequency of heart contraction (in beats per minute).

heat Thermal energy in transfer from one body of matter to another.

heat of vaporization The quantity of heat a liquid must absorb for 1 g of it to be converted from the liquid to the gaseous state.

heat-shock protein A protein that helps protect other proteins during heat stress. Heat-shock proteins are found in plants, animals, and microorganisms.

heavy chain One of the two types of polypeptide chains that make up an antibody molecule and B cell receptor; consists of a variable region, which contributes to the antigen-binding site, and a constant region.

helicase An enzyme that untwists the double helix of DNA at replication forks, separating the two strands and making them available as template strands.

helper T cell A type of T cell that, when activated, secretes cytokines that promote the response of B cells (humoral response) and cytotoxic T cells (cell-mediated response) to antigens.

hemoglobin (hē'-mō-glō'-bin) An iron-containing protein in red blood cells that reversibly binds oxygen.

hemolymph (hē′-mō-limf′) In invertebrates with an open circulatory system, the body fluid that bathes tissues.

hemophilia (hē′-muh-fil′-ē-uh) A human genetic disease caused by a sex-linked recessive allele resulting in the absence of one or more blood-clotting proteins; characterized by excessive bleeding following injury.

hepatic portal vein A large vessel that conveys nutrient-laden blood from the small intestine to the liver, which regulates the blood's nutrient content.

herbivore (hur′-bi-vōr′) An animal that mainly eats plants or algae.

herbivory An interaction in which an organism eats part of a plant or alga.

heredity The transmission of traits from one generation to the next.

hermaphrodite (hur-maf′-ruh-dīt′) An individual that functions as both male and female in sexual reproduction by producing both sperm and eggs.

hermaphroditism (hur-maf′-rō-dī-tizm) A condition in which an individual has both female and male gonads and functions as both a male and a female in sexual reproduction by producing both sperm and eggs.

heterochromatin (het′-er-ō-krō′-muh-tin) Eukaryotic chromatin that remains highly compacted during interphase and is generally not transcribed.

heterochrony (het′-uh-rok′-ruh-nē) Evolutionary change in the timing or rate of an organism's development.

heterocyst (het′-er-ō-sist) A specialized cell that engages in nitrogen fixation in some filamentous cyanobacteria; also called a heterocyte.

heterokaryon (het′-er-ō-kār′-ē-un) A fungal mycelium that contains two or more haploid nuclei per cell.

heteromorphic (het′-er-ō-mōr′-fik) Referring to a condition in the life cycle of plants and certain algae in which the sporophyte and gametophyte generations differ in morphology.

heterosporous (het-er-os′-pōr-us) Referring to a plant species that has two kinds of spores: microspores, which develop into male gametophytes, and megaspores, which develop into female gametophytes.

heterotroph (het′-er-ō-trōf) An organism that obtains organic food molecules by eating other organisms or substances derived from them.

heterozygote An organism that has two different alleles for a gene (encoding a character).

heterozygote advantage Greater reproductive success of heterozygous individuals compared with homozygotes; tends to preserve variation in a gene pool.

heterozygous (het′-er-ō-zī′-gus) Having two different alleles for a given gene.

hibernation A long-term physiological state in which metabolism decreases, the heart and respiratory system slow down, and body temperature is maintained at a lower level than normal.

high-density lipoprotein (HDL) A particle in the blood made up of thousands of cholesterol molecules and other lipids bound to a protein. HDL scavenges excess cholesterol.

hindbrain One of three ancestral and embryonic regions of the vertebrate brain; develops into the medulla oblongata, pons, and cerebellum.

histamine (his′-tuh-mēn) A substance released by mast cells that causes blood vessels to dilate and become more permeable in inflammatory and allergic responses.

histogram A variant of a bar graph that is made for numeric data by first grouping, or "binning," the variable plotted on the *x*-axis into intervals of equal width. The "bins" may be integers or ranges of numbers. The height of each bar shows the percent or number of experimental subjects whose characteristics can be described by one of the intervals plotted on the *x*-axis.

histone (his′-tōn) A small protein with a high proportion of positively charged amino acids that binds to the negatively charged DNA and plays a key role in chromatin structure.

histone acetylation (his′-tōn) The attachment of acetyl groups to certain amino acids of histone proteins.

HIV (human immunodeficiency virus) The infectious agent that causes AIDS. HIV is a retrovirus.

holdfast A rootlike structure that anchors a seaweed.

homeobox (hō′-mē-ō-boks′) A 180-nucleotide sequence within homeotic genes and some other developmental genes that is widely conserved in animals. Related sequences occur in plants and yeasts.

homeostasis (hō′-mē-ō-stā′-sis) The steady-state physiological condition of the body.

homeotic gene (hō-mē-o′-tik) Any of the master regulatory genes that control placement and spatial organization of body parts in animals, plants, and fungi by controlling the developmental fate of groups of cells.

hominin (hō′-mi-nin) A group consisting of humans and the extinct species that are more closely related to us than to chimpanzees.

homologous chromosomes (or homologs) (hō-mol′-uh-gus) A pair of chromosomes of the same length, centromere position, and staining pattern that possess genes for the same characters at corresponding loci. One homologous chromosome is inherited from the organism's father, the other from the mother. Also called a homologous pair.

homologous pair *See* homologous chromosomes.

homologous structures (hō-mol′-uh-gus) Structures in different species that are similar because of common ancestry.

homologs *See* homologous chromosomes.

homology (hō-mol′-ō-jē) Similarity in characteristics resulting from a shared ancestry.

homoplasy (hō′-muh-play′-zē) A similar (analogous) structure or molecular sequence that has evolved independently in two species.

homosporous (hō-mos′-puh-rus) Referring to a plant species that has a single kind of spore, which typically develops into a bisexual gametophyte.

homozygote An organism that has a pair of identical alleles for a gene (encoding a character).

homozygous (hō′-mō-zī′-gus) Having two identical alleles for a given gene.

horizontal gene transfer The transfer of genes from one genome to another through mechanisms such as transposable elements, plasmid exchange, viral activity, and perhaps fusions of different organisms.

hormone In multicellular organisms, one of many types of secreted chemicals that are formed in specialized cells, travel in body fluids, and act on specific target cells in other parts of the organism, changing the target cells' functioning.

hornwort A small, herbaceous, nonvascular plant that is a member of the phylum Anthocerophyta.

host The larger participant in a symbiotic relationship, often providing a home and food source for the smaller symbiont.

host range The limited number of species whose cells can be infected by a particular virus.

Human Genome Project An international collaborative effort to map and sequence the DNA of the entire human genome.

human immunodeficiency virus (HIV) The infectious agent that causes AIDS (acquired immunodeficiency syndrome). HIV is a retrovirus.

humoral immune response (hyū′-mer-ul) The branch of adaptive immunity that involves the activation of B cells and that leads to the production of antibodies, which defend against bacteria and viruses in body fluids.

humus (hyū′-mus) Decomposing organic material that is a component of topsoil.

Huntington's disease A human genetic disease caused by a dominant allele; characterized by uncontrollable body movements and degeneration of the nervous system; usually fatal 10 to 20 years after the onset of symptoms.

hybrid Offspring that results from the mating of individuals from two different species or from two true-breeding varieties of the same species.

hybrid zone A geographic region in which members of different species meet and mate, producing at least some offspring of mixed ancestry.

hybridization In genetics, the mating, or crossing, of two true-breeding varieties.

hydration shell The sphere of water molecules around a dissolved ion.

hydrocarbon An organic molecule consisting only of carbon and hydrogen.

hydrogen bond A type of weak chemical bond that is formed when the slightly positive hydrogen atom of a polar covalent bond in one molecule is attracted to the slightly negative atom of a polar covalent bond in another molecule or in another region of the same molecule.

hydrogen ion A single proton with a charge of 1 + . The dissociation of a water molecule (H_2O) leads to the generation of a hydroxide ion (OH^-) and a hydrogen ion (H^+); in water, H^+ is not found alone but associates with a water molecule to form a hydronium ion.

hydrolysis (hī-drol′-uh-sis) A chemical reaction that breaks bonds between two molecules by the addition of water; functions in disassembly of polymers to monomers.

hydronium ion A water molecule that has an extra proton bound to it; H_3O^+, commonly represented as H^+.

hydrophilic (hī′-drō-fil′-ik) Having an affinity for water.

hydrophobic (hī′-drō-fō′-bik) Having no affinity for water; tending to coalesce and form droplets in water.

hydrophobic interaction (hī′-drō-fō′-bik) A type of weak chemical interaction caused when molecules that do not mix with water coalesce to exclude water.

hydroponic culture A method in which plants are grown in mineral solutions rather than in soil.

hydrostatic skeleton A skeletal system composed of fluid held under pressure in a closed body compartment; the main skeleton of most cnidarians, flatworms, nematodes, and annelids.

hydrothermal vent An area on the seafloor where heated water and minerals from Earth's interior gush into the seawater, producing a dark, hot, oxygen-deficient environment. The producers in a hydrothermal vent community are chemoautotrophic prokaryotes.

hydroxide ion A water molecule that has lost a proton; OH^-.

hydroxyl group (hī-drok′-sil) A chemical group consisting of an oxygen atom joined to a hydrogen atom. Molecules possessing this group are soluble in water and are called alcohols.

hyperpolarization A change in a cell's membrane potential such that the inside of the membrane becomes more negative relative to the outside. Hyperpolarization reduces the chance that a neuron will transmit a nerve impulse.

hypersensitive response A plant's localized defense response to a pathogen, involving the death of cells around the site of infection.

hypertension A disorder in which blood pressure remains abnormally high.

hypertonic Referring to a solution that, when surrounding a cell, will cause the cell to lose water.

hypha (plural, **hyphae**) (hī′-fuh, hī′-fē) One of many connected filaments that collectively make up the mycelium of a fungus.

hypocotyl (hī′-puh-cot′-ul) In an angiosperm embryo, the embryonic axis below the point of attachment of the cotyledon(s) and above the radicle.

hypothalamus (hī′-pō-thal′-uh-mus) The ventral part of the vertebrate forebrain; functions in maintaining homeostasis, especially in coordinating the endocrine and nervous systems; secretes hormones of the posterior pituitary and releasing factors that regulate the anterior pituitary.

hypothesis (hī-poth′-uh-sis) A testable explanation for a set of observations based on the available data and guided by inductive reasoning. A hypothesis is narrower in scope than a theory.

hypotonic Referring to a solution that, when surrounding a cell, will cause the cell to take up water.

imbibition The uptake of water by a seed or other structure, resulting in swelling.

immigration The influx of new individuals into a population from other areas.

immune system An organism's system of defenses against agents that cause disease.

immunization The process of generating a state of immunity by artificial means. In vaccination, an inactive or weakened form of a pathogen is administered, inducing B and T cell responses and immunological memory. In passive immunization, antibodies specific for a particular pathogen are administered, conferring immediate but temporary protection.

immunoglobulin (Ig) (im′-yū-nō-glob′-yū-lin) *See* antibody.

imprinting In animal behavior, the formation at a specific stage in life of a long-lasting behavioral response to a specific individual or object. *See also* genomic imprinting.

inclusive fitness The total effect an individual has on proliferating its genes by producing its own offspring and by providing aid that enables other close relatives to increase production of their offspring.

incomplete dominance The situation in which the phenotype of heterozygotes is intermediate between the phenotypes of individuals homozygous for either allele.

incomplete flower A flower in which one or more of the four basic floral organs (sepals, petals, stamens, or carpels) are either absent or nonfunctional.

incomplete metamorphosis A type of development in certain insects, such as grasshoppers, in which the young (called nymphs) resemble adults but are smaller and have different body proportions. The nymph goes through a series of molts, each time looking more like an adult, until it reaches full size.

independent variable A factor whose value is manipulated or changed during an experiment to reveal possible effects on another factor (the dependent variable).

indeterminate cleavage A type of embryonic development in deuterostomes in which each cell produced by early cleavage divisions retains the capacity to develop into a complete embryo.

indeterminate growth A type of growth characteristic of plants, in which the organism continues to grow as long as it lives.

induced fit Caused by entry of the substrate, the change in shape of the active site of an enzyme so that it binds more snugly to the substrate.

inducer A specific small molecule that binds to a bacterial repressor protein and changes the repressor's shape so that it cannot bind to an operator, thus switching an operon on.

induction A process in which a group of cells or tissues influences the development of another group through close-range interactions.

inductive reasoning A type of logic in which generalizations are based on a large number of specific observations.

inflammatory response An innate immune defense triggered by physical injury or infection of tissue involving the release of substances that promote swelling, enhance the infiltration of white blood cells, and aid in tissue repair and destruction of invading pathogens.

inflorescence A group of flowers tightly clustered together.

ingestion The first stage of food processing in animals: the act of eating.

ingroup A species or group of species whose evolutionary relationships are being examined in a given analysis.

inhibitory postsynaptic potential (IPSP) An electrical change (usually hyperpolarization) in the membrane of a postsynaptic neuron caused by the binding of an inhibitory neurotransmitter from a presynaptic cell to a postsynaptic receptor; makes it more difficult for a postsynaptic neuron to generate an action potential.

innate behavior Animal behavior that is developmentally fixed and under strong genetic control. Innate behavior is exhibited in virtually the same form by all individuals in a population despite internal and external environmental differences during development and throughout their lifetimes.

innate immunity A form of defense common to all animals that is active immediately upon exposure to a pathogen and that is the same whether or not the pathogen has been encountered previously.

inner cell mass An inner cluster of cells at one end of a mammalian blastocyst that subsequently develops into the embryo proper and some of the extraembryonic membranes.

inner ear One of the three main regions of the vertebrate ear; includes the cochlea (which in turn contains the organ of Corti) and the semicircular canals.

inositol trisphosphate (IP_3) (in-ō′-suh-tol) A second messenger that functions as an intermediate between certain signaling molecules and a subsequent second messenger, Ca^{2+}, by causing a rise in cytoplasmic Ca^{2+} concentration.

inquiry The search for information and explanation, often focusing on specific questions.

insertion A mutation involving the addition of one or more nucleotide pairs to a gene.

***in situ* hybridization** A technique using nucleic acid hybridization with a labeled probe to detect the location of a specific mRNA in an intact organism.

insulin (in′-suh-lin) A hormone secreted by pancreatic beta cells that lowers blood glucose levels. It promotes the uptake of glucose by most body cells and the synthesis and storage of glycogen in the liver and also stimulates protein and fat synthesis.

integral protein A transmembrane protein with hydrophobic regions that extend into and often completely span the hydrophobic interior of the membrane and with hydrophilic regions in contact with the aqueous solution on one or both sides of the membrane (or lining the channel in the case of a channel protein).

integrin (in′-tuh-grin) In animal cells, a transmembrane receptor protein with two subunits that interconnects the extracellular matrix and the cytoskeleton.

integument (in-teg′-yū-ment) Layer of sporophyte tissue that contributes to the structure of an ovule of a seed plant.

integumentary system The outer covering of a mammal's body, including skin, hair, and nails, claws, or hooves.

interferon (in′-ter-fēr′-on) A protein that has antiviral or immune regulatory functions. For example, interferons secreted by virus-infected cells help nearby cells resist viral infection.

intermediate disturbance hypothesis The concept that moderate levels of disturbance can foster greater species diversity than low or high levels of disturbance.

intermediate filament A component of the cytoskeleton that includes filaments intermediate in size between microtubules and microfilaments.

interneuron An association neuron; a nerve cell within the central nervous system that forms synapses with sensory and/or motor neurons and integrates sensory input and motor output.

internode A segment of a plant stem between the points where leaves are attached.

interphase The period in the cell cycle when the cell is not dividing. During interphase, cellular metabolic activity is high, chromosomes and organelles are duplicated, and cell size may increase. Interphase often accounts for about 90% of the cell cycle.

intersexual selection A form of natural selection in which individuals of one sex (usually the females) are choosy in selecting their mates from the other sex; also called mate choice.

interspecific interaction A relationship between individuals of two or more species in a community.

interstitial fluid The fluid filling the spaces between cells in most animals.

intertidal zone The shallow zone of the ocean adjacent to land and between the high- and low-tide lines.

intrasexual selection A form of natural selection in which there is direct competition among individuals of one sex for mates of the opposite sex.

intrinsic rate of increase (*r*) In population models, the per capita rate at which an exponentially growing population increases in size at each instant in time.

introduced species A species moved by humans, either intentionally or accidentally, from its native location to a new geographic region; also called non-native or exotic species.

intron (in′-tron) A noncoding, intervening sequence within a primary transcript that is removed from the transcript during RNA processing; also refers to the region of DNA from which this sequence was transcribed.

invasive species A species, often introduced by humans, that takes hold outside its native range.

inversion An aberration in chromosome structure resulting from reattachment of a chromosomal fragment in a reverse orientation to the chromosome from which it originated.

invertebrate An animal without a backbone. Invertebrates make up 95% of animal species.

***in vitro* fertilization (IVF)** (vē′-trō) Fertilization of oocytes in laboratory containers followed by artificial implantation of the early embryo in the mother's uterus.

***in vitro* mutagenesis** A technique used to discover the function of a gene by cloning it, introducing specific changes into the cloned gene's sequence, reinserting the mutated gene into a cell, and studying the phenotype of the mutant.

ion (ī′-on) An atom or group of atoms that has gained or lost one or more electrons, thus acquiring a charge.

ion channel (ī′-on) A transmembrane protein channel that allows a specific ion to diffuse across the membrane down its concentration or electrochemical gradient.

ionic bond (ī-on′-ik) A chemical bond resulting from the attraction between oppositely charged ions.

ionic compound (ī-on′-ik) A compound resulting from the formation of an ionic bond; also called a salt.

iris The colored part of the vertebrate eye, formed by the anterior portion of the choroid.

isomer (ī′-sō-mer) One of two or more compounds that have the same numbers of atoms of the same elements but different structures and hence different properties.

isomorphic Referring to alternating generations in plants and certain algae in which the sporophytes and gametophytes look alike, although they differ in chromosome number.

isotonic (ī′-sō-ton′-ik) Referring to a solution that, when surrounding a cell, causes no net movement of water into or out of the cell.

isotope (ī′-sō-tōp′) One of several atomic forms of an element, each with the same number of protons but a different number of neutrons, thus differing in atomic mass.

iteroparity Reproduction in which adults produce offspring over many years; also called repeated reproduction.

jasmonate Any of a class of plant hormones that regulate a wide range of developmental processes in plants and play a key role in plant defense against herbivores.

joule (J) A unit of energy: 1 J = 0.239 cal; 1 cal = 4.184 J.

juxtaglomerular apparatus (JGA) (juks′-tuh-gluh-mār′-yū-ler) A specialized tissue in nephrons that releases the enzyme renin in response to a drop in blood pressure or volume.

juxtamedullary nephron In mammals and birds, a nephron with a loop of Henle that extends far into the renal medulla.

karyogamy (kār′-ē-og′-uh-mē) In fungi, the fusion of haploid nuclei contributed by the two parents; occurs as one stage of sexual reproduction, preceded by plasmogamy.

karyotype (kār′-ē-ō-tīp) A display of the chromosome pairs of a cell arranged by size and shape.

keystone species A species that is not necessarily abundant in a community yet exerts strong control on community structure by the nature of its ecological role or niche.

kidney In vertebrates, one of a pair of excretory organs where blood filtrate is formed and processed into urine.

kilocalorie (kcal) A thousand calories; the amount of heat energy required to raise the temperature of 1 kg of water by 1°C.

kinetic energy (kuh-net′-ik) The energy associated with the relative motion of objects. Moving matter can perform work by imparting motion to other matter.

kinetochore (kuh-net′-uh-kōr) A structure of proteins attached to the centromere that links each sister chromatid to the mitotic spindle.

kinetoplastid A protist, such as a trypanosome, that has a single large mitochondrion that houses an organized mass of DNA.

kingdom A taxonomic category, the second broadest after domain.

kin selection Natural selection that favors altruistic behavior by enhancing the reproductive success of relatives.

***K*-selection** Selection for life history traits that are sensitive to population density; also called density-dependent selection.

labia majora A pair of thick, fatty ridges that encloses and protects the rest of the vulva.

labia minora A pair of slender skin folds that surrounds the openings of the vagina and urethra.

lacteal (lak′-tē-ul) A tiny lymph vessel extending into the core of an intestinal villus and serving as the destination for absorbed chylomicrons.

lactic acid fermentation Glycolysis followed by the reduction of pyruvate to lactate, regenerating NAD^+ with no release of carbon dioxide.

lagging strand A discontinuously synthesized DNA strand that elongates by means of Okazaki fragments, each synthesized in a $5' \rightarrow 3'$ direction away from the replication fork.

lamprey Any of the jawless vertebrates with highly reduced vertebrae that live in freshwater and marine environments. Almost half of extant lamprey species are parasites that feed by clamping their round, jawless mouth onto the flank of a live fish; nonparasitic lampreys are suspension feeders that feed only as larvae.

lancelet A member of the clade Cephalochordata, small blade-shaped marine chordates that lack a backbone.

landscape An area containing several different ecosystems linked by exchanges of energy, materials, and organisms.

landscape ecology The study of how the spatial arrangement of habitat types affects the distribution and abundance of organisms and ecosystem processes.

large intestine The portion of the vertebrate alimentary canal between the small intestine and the anus; functions mainly in water absorption and the formation of feces.

larva (lar′-vuh) (plural, **larvae**) A free-living, sexually immature form in some animal life cycles that may differ from the adult animal in morphology, nutrition, and habitat.

larynx (lār′-inks) The portion of the respiratory tract containing the vocal cords; also called the voice box.

lateralization Segregation of functions in the cortex of the left and right cerebral hemispheres.

lateral line system A mechanoreceptor system consisting of a series of pores and receptor units along the sides of the body in fishes and aquatic amphibians; detects water movements made by the animal itself and by other moving objects.

lateral meristem (mār′-uh-stem) A meristem that thickens the roots and shoots of woody plants. The vascular cambium and cork cambium are lateral meristems.

lateral root A root that arises from the pericycle of an established root.

law of conservation of mass A physical law stating that matter can change form but cannot be created or destroyed. In a closed system, the mass of the system is constant.

law of independent assortment Mendel's second law, stating that each pair of alleles segregates, or assorts, independently of each other pair during gamete formation; applies when genes for two characters are located on different pairs of homologous chromosomes or when they are far enough apart on the same chromosome to behave as though they are on different chromosomes.

law of segregation Mendel's first law, stating that the two alleles in a pair segregate (separate from each other) into different gametes during gamete formation.

leading strand The new complementary DNA strand synthesized continuously along the template strand toward the replication fork in the mandatory $5' \rightarrow 3'$ direction.

leaf The main photosynthetic organ of vascular plants.

leaf primordium (plural, **primordia**) A finger-like projection along the flank of a shoot apical meristem, from which a leaf arises.

learning The modification of behavior as a result of specific experiences.

lens The structure in an eye that focuses light rays onto the photoreceptors.

lenticel (len′-ti-sel) A small raised area in the bark of stems and roots that enables gas exchange between living cells and the outside air.

lepidosaur (leh-pid′-uh-sōr) A member of the reptilian group that includes lizards, snakes, and two species of New Zealand animals called tuataras.

leukocyte (lū′-kō-sīt′) A blood cell that functions in fighting infections; also called a white blood cell.

lichen The mutualistic association between a fungus and a photosynthetic alga or cyanobacterium.

life cycle The generation-to-generation sequence of stages in the reproductive history of an organism.

life history The traits that affect an organism's schedule of reproduction and survival.

life table A summary of the age-specific survival and reproductive rates of individuals in a population.

ligament A fibrous connective tissue that joins bones together at joints.

ligand (lig′-und) A molecule that binds specifically to another molecule, usually a larger one.

ligand-gated ion channel (lig′-und) A transmembrane protein containing a pore that opens or closes as it changes shape in response to a signaling molecule (ligand), allowing or blocking the flow of specific ions; also called an ionotropic receptor.

light chain One of the two types of polypeptide chains that make up an antibody molecule and B cell receptor; consists of a variable region, which contributes to the antigen-binding site, and a constant region.

light-harvesting complex A complex of proteins associated with pigment molecules (including chlorophyll *a*, chlorophyll *b*, and carotenoids) that captures light energy and transfers it to reaction-center pigments in a photosystem.

light microscope (LM) An optical instrument with lenses that refract (bend) visible light to magnify images of specimens.

light reactions The first of two major stages in photosynthesis (preceding the Calvin cycle). These reactions, which occur on the thylakoid membranes of the chloroplast or on membranes of certain prokaryotes, convert solar energy to the chemical energy of ATP and NADPH, releasing oxygen in the process.

lignin (lig′-nin) A strong polymer embedded in the cellulose matrix of the secondary cell walls of vascular plants that provides structural support in terrestrial species.

limiting nutrient An element that must be added for production to increase in a particular area.

limnetic zone In a lake, the well-lit, open surface waters far from shore.

linear electron flow A route of electron flow during the light reactions of photosynthesis that involves both photosystems (I and II) and produces ATP, NADPH, and O_2. The net electron flow is from H_2O to $NADP^+$.

line graph A graph in which each data point is connected to the next point in the data set with a straight line.

linkage map A genetic map based on the frequencies of recombination between markers during crossing over of homologous chromosomes.

linked genes Genes located close enough together on a chromosome that they tend to be inherited together.

lipid (lip′-id) Any of a group of large biological molecules, including fats, phospholipids, and steroids, that mix poorly, if at all, with water.

littoral zone In a lake, the shallow, well-lit waters close to shore.

liver A large internal organ in vertebrates that performs diverse functions, such as producing bile, maintaining blood glucose level, and detoxifying poisonous chemicals in the blood.

liverwort A small, herbaceous, nonvascular plant that is a member of the phylum Hepatophyta.

loam The most fertile soil type, made up of roughly equal amounts of sand, silt, and clay.

lobe-fin Member of a clade of osteichthyans having rod-shaped muscular fins. The group includes coelacanths, lungfishes, and tetrapods.

local regulator A secreted molecule that influences cells near where it is secreted.

locomotion Active motion from place to place.

locus (lō′-kus) (plural, **loci**) (lō′-sī) A specific place along the length of a chromosome where a given gene is located.

logistic population growth Population growth that levels off as population size approaches carrying capacity.

long-day plant A plant that flowers (usually in late spring or early summer) only when the light period is longer than a critical length.

long noncoding RNA (lncRNA) An RNA between 200 and hundreds of thousands of nucleotides in length that does not code for protein but is expressed at significant levels.

long-term memory The ability to hold, associate, and recall information over one's lifetime.

long-term potentiation (LTP) An enhanced responsiveness to an action potential (nerve signal) by a receiving neuron.

loop of Henle (hen′-lē) The hairpin turn, with a descending and ascending limb, between the proximal and distal tubules of the vertebrate kidney; functions in water and salt reabsorption.

lophophore (lof′-uh-fōr) In some lophotrochozoan animals, including brachiopods, a crown of ciliated tentacles that surround the mouth and function in feeding.

Lophotrochozoa (lo-phah′-truh-kō-zō′-uh) One of the three main lineages of bilaterian animals; lophotrochozoans include organisms that have lophophores or trochophore larvae. *See also* Deuterostomia and Ecdysozoa.

low-density lipoprotein (LDL) A particle in the blood made up of thousands of cholesterol molecules and other lipids bound to a protein. LDL transports cholesterol from the liver for incorporation into cell membranes.

lung An infolded respiratory surface of a terrestrial vertebrate, land snail, or spider that connects to the atmosphere by narrow tubes.

luteinizing hormone (LH) (lū′-tē-uh-nī′-zing) A tropic hormone that is produced and secreted by the anterior pituitary and that stimulates ovulation in females and androgen production in males.

lycophyte (lī′-kuh-fīt) An informal name for a member of the phylum Lycophyta, which includes club mosses, spike mosses, and quillworts.

lymph The colorless fluid, derived from interstitial fluid, in the lymphatic system of vertebrates.

lymph node An organ located along a lymph vessel. Lymph nodes filter lymph and contain cells that attack viruses and bacteria.

lymphatic system A system of vessels and nodes, separate from the circulatory system, that returns fluid, proteins, and cells to the blood.

lymphocyte A type of white blood cell that mediates immune responses. The two main classes are B cells and T cells.

lysogenic cycle (lī′-sō-jen′-ik) A type of phage replicative cycle in which the viral genome becomes incorporated into the bacterial host chromosome as a prophage, is replicated along with the chromosome, and does not kill the host.

lysosome (lī′-suh-sōm) A membrane-enclosed sac of hydrolytic enzymes found in the cytoplasm of animal cells and some protists.

lysozyme (lī′-sō-zīm) An enzyme that destroys bacterial cell walls; in mammals, it is found in sweat, tears, and saliva.

lytic cycle (lit′-ik) A type of phage replicative cycle resulting in the release of new phages by lysis (and death) of the host cell.

macroevolution Evolutionary change above the species level. Examples of macroevolutionary change include the origin of a new group of organisms through a series of speciation events and the impact of mass extinctions on the diversity of life and its subsequent recovery.

macromolecule A giant molecule formed by the joining of smaller molecules, usually by a dehydration reaction. Polysaccharides, proteins, and nucleic acids are macromolecules.

macronutrient An essential element that an organism must obtain in relatively large amounts. *See also* micronutrient.

macrophage (mak′-rō-fāj) A phagocytic cell present in many tissues that functions in innate immunity by destroying microbes and in acquired immunity as an antigen-presenting cell.

magnoliid A member of the angiosperm clade that is most closely related to the combined eudicot and monocot clades. Extant examples are magnolias, laurels, and black pepper plants.

major depressive disorder A mood disorder characterized by feelings of sadness, lack of self-worth, emptiness, or loss of interest in nearly all things.

major histocompatibility complex (MHC) molecule A host protein that functions in antigen presentation. Foreign MHC molecules on transplanted tissue can trigger T cell responses that may lead to rejection of the transplant.

malignant tumor A cancerous tumor containing cells that have significant genetic and cellular changes and are capable of invading and surviving in new sites. Malignant tumors can impair the functions of one or more organs.

Malpighian tubule (mal-pig′-ē-un) A unique excretory organ of insects that empties into the digestive tract, removes nitrogenous wastes from the hemolymph, and functions in osmoregulation.

mammal A member of the clade Mammalia, amniotes that have hair and mammary glands (glands that produce milk).

mammary gland An exocrine gland that secretes milk for nourishing the young. Mammary glands are characteristic of mammals.

mantle One of the three main parts of a mollusc; a fold of tissue that drapes over the mollusc's visceral mass and may secrete a shell. *See also* foot and visceral mass.

mantle cavity A water-filled chamber that houses the gills, anus, and excretory pores of a mollusc.

map unit A unit of measurement of the distance between genes. One map unit is equivalent to a 1% recombination frequency.

marine benthic zone The ocean floor.

mark-recapture method A sampling technique used to estimate the size of animal populations.

marsupial (mar-sū′-pē-ul) A mammal, such as a koala, kangaroo, or opossum, whose young complete their embryonic development inside a maternal pouch called the marsupium.

mass extinction The elimination of a large number of species throughout Earth, the result of global environmental changes.

mass number The total number of protons and neutrons in an atom's nucleus.

mate-choice copying Behavior in which individuals in a population copy the mate choice of others, apparently as a result of social learning.

maternal effect gene A gene that, when mutant in the mother, results in a mutant phenotype in the offspring, regardless of the offspring's genotype. Maternal effect genes, also called egg-polarity genes, were first identified in *Drosophila melanogaster*.

matter Anything that takes up space and has mass.

maximum likelihood As applied to DNA sequence data, a principle that states that when considering multiple phylogenetic hypotheses, one should take into account the hypothesis that reflects the most likely sequence of evolutionary events, given certain rules about how DNA changes over time.

maximum parsimony A principle that states that when considering multiple explanations for an observation, one should first investigate the simplest explanation that is consistent with the facts.

mean The sum of all data points in a data set divided by the number of data points.

mechanoreceptor A sensory receptor that detects physical deformation in the body's environment associated with pressure, touch, stretch, motion, or sound.

medulla oblongata (meh-dul′-uh ob′-long-go′-tuh) The lowest part of the vertebrate brain, commonly called the medulla; a swelling of the hindbrain anterior to the spinal cord that controls autonomic, homeostatic functions, including breathing, heart and blood vessel activity, swallowing, digestion, and vomiting.

medusa (plural, **medusae**) (muh-dū′-suh) The floating, flattened, mouth-down version of the cnidarian body plan. The alternate form is the polyp.

megapascal (MPa) (meg′-uh-pas-kal′) A unit of pressure equivalent to about 10 atmospheres of pressure.

megaphyll (meh′-guh-fil) A leaf with a highly branched vascular system, found in almost all vascular plants other than lycophytes. *See also* microphyll.

megaspore A spore from a heterosporous plant species that develops into a female gametophyte.

meiosis (mī-ō′-sis) A modified type of cell division in sexually reproducing organisms consisting of two rounds of cell division but only one round of DNA replication. It results in cells with half the number of chromosome sets as the original cell.

meiosis I (mī-ō′-sis) The first division of a two-stage process of cell division in sexually reproducing organisms that results in cells with half the number of chromosome sets as the original cell.

meiosis II (mī-ō′-sis) The second division of a two-stage process of cell division in sexually reproducing organisms that results in cells with half the number of chromosome sets as the original cell.

melanocyte-stimulating hormone (MSH) A hormone produced and secreted by the anterior pituitary with multiple activities, including regulating the behavior of pigment-containing cells in the skin of some vertebrates.

melatonin A hormone that is secreted by the pineal gland and that is involved in the regulation of biological rhythms and sleep.

membrane potential The difference in electrical charge (voltage) across a cell's plasma membrane due to the differential distribution of ions. Membrane potential affects the activity of excitable cells and the transmembrane movement of all charged substances.

memory cell One of a clone of long-lived lymphocytes, formed during the primary immune response, that remains in a lymphoid organ until activated by exposure to the same antigen that triggered its formation. Activated memory cells mount the secondary immune response.

menopause The cessation of ovulation and menstruation marking the end of a human female's reproductive years.

menstrual cycle (men′-strū-ul) In humans and certain other primates, the periodic growth and shedding of the uterine lining that occurs in the absence of pregnancy.

menstruation The shedding of portions of the endometrium during a uterine (menstrual) cycle.

meristem (mār′-uh-stem) Plant tissue that remains embryonic as long as the plant lives, allowing for indeterminate growth.

meristem identity gene (mār′-uh-stem) A plant gene that promotes the switch from vegetative growth to flowering.

mesoderm (mez′-ō-derm) The middle primary germ layer in a triploblastic animal embryo; develops into the notochord, the lining of the coelom, muscles, skeleton, gonads, kidneys, and most of the circulatory system in species that have these structures.

mesohyl (mez′-ō-hīl) A gelatinous region between the two layers of cells of a sponge.

mesophyll (mez′-ō-fil) Leaf cells specialized for photosynthesis. In C_3 and CAM plants, mesophyll cells are located between the upper and lower epidermis; in C_4 plants, they are located between the bundle-sheath cells and the epidermis.

messenger RNA (mRNA) A type of RNA, synthesized using a DNA template, that attaches to ribosomes in the cytoplasm and specifies the primary structure of a protein. (In eukaryotes, the primary RNA transcript must undergo RNA processing to become mRNA.)

metabolic pathway A series of chemical reactions that either builds a complex molecule (anabolic pathway) or breaks down a complex molecule to simpler molecules (catabolic pathway).

metabolic rate The total amount of energy an animal uses in a unit of time.

metabolism (muh-tab′-uh-lizm) The totality of an organism's chemical reactions, consisting of catabolic and anabolic pathways, which manage the material and energy resources of the organism.

metagenomics The collection and sequencing of DNA from a group of species, usually an environmental sample of microorganisms. Computer software sorts partial sequences and assembles them into genome sequences of individual species making up the sample.

metamorphosis (met′-uh-mōr′-fuh-sis) A developmental transformation that turns an animal larva into either an adult or an adult-like stage that is not yet sexually mature.

metanephridium (met′-uh-nuh-frid′-ē-um) (plural, **metanephridia**) An excretory organ found in many invertebrates that typically consists of tubules connecting ciliated internal openings to external openings.

metaphase The third stage of mitosis, in which the spindle is complete and the chromosomes, attached to microtubules at their kinetochores, are all aligned at the metaphase plate.

metaphase plate An imaginary structure located at a plane midway between the two poles of a cell in metaphase on which the centromeres of all the duplicated chromosomes are located.

metapopulation A group of spatially separated populations of one species that interact through immigration and emigration.

Glossary

metastasis (muh-tas′-tuh-sis) The spread of cancer cells to locations distant from their original site.

methanogen (meth-an′-ō-jen) An organism that produces methane as a waste product of the way it obtains energy. All known methanogens are in domain Archaea.

methyl group A chemical group consisting of a carbon bonded to three hydrogen atoms. The methyl group may be attached to a carbon or to a different atom.

microbiome The collection of microorganisms living in or on an organism's body, along with their genetic material.

microclimate Climate patterns on a very fine scale, such as the specific climatic conditions underneath a log.

microevolution Evolutionary change below the species level; change in the allele frequencies in a population over generations.

microfilament A cable composed of actin proteins in the cytoplasm of almost every eukaryotic cell, making up part of the cytoskeleton and acting alone or with myosin to cause cell contraction; also called an actin filament.

micronutrient An essential element that an organism needs in very small amounts. *See also* macronutrient.

microphyll (mī′-krō-fil) A small, usually spine-shaped leaf supported by a single strand of vascular tissue, found only in lycophytes.

micropyle A pore in the integuments of an ovule.

microRNA (miRNA) A small, single-stranded RNA molecule, generated from a double-stranded RNA precursor. The miRNA associates with one or more proteins in a complex that can degrade or prevent translation of an mRNA with a complementary sequence.

microspore A spore from a heterosporous plant species that develops into a male gametophyte.

microtubule A hollow rod composed of tubulin proteins that makes up part of the cytoskeleton in all eukaryotic cells and is found in cilia and flagella.

microvillus (plural, **microvilli**) One of many fine, finger-like projections of the epithelial cells in the lumen of the small intestine that increase its surface area.

midbrain One of three ancestral and embryonic regions of the vertebrate brain; develops into sensory integrating and relay centers that send sensory information to the cerebrum.

middle ear One of three main regions of the vertebrate ear; in mammals, a chamber containing three small bones (the malleus, incus, and stapes) that convey vibrations from the eardrum to the oval window.

middle lamella (luh-mel′-uh) In plants, a thin layer of adhesive extracellular material, primarily pectins, found between the primary walls of adjacent young cells.

migration A regular, long-distance change in location.

mineral In nutrition, a simple nutrient that is inorganic and therefore cannot be synthesized in the body.

mineralocorticoid A steroid hormone secreted by the adrenal cortex that regulates salt and water homeostasis.

minimum viable population (MVP) The smallest population size at which a species is able to sustain its numbers and survive.

mismatch repair The cellular process that uses specific enzymes to remove and replace incorrectly paired nucleotides.

missense mutation A nucleotide-pair substitution that results in a codon that codes for a different amino acid.

mitochondrial matrix The compartment of the mitochondrion enclosed by the inner membrane and containing enzymes and substrates for the citric acid cycle, as well as ribosomes and DNA.

mitochondrion (mī′-tō-kon′-drē-un) (plural, **mitochondria**) An organelle in eukaryotic cells that serves as the site of cellular respiration; uses oxygen to break down organic molecules and synthesize ATP.

mitosis (mī-tō′-sis) A process of nuclear division in eukaryotic cells conventionally divided into five stages: prophase, prometaphase, metaphase, anaphase, and telophase. Mitosis conserves chromosome number by allocating replicated chromosomes equally to each of the daughter nuclei.

mitotic (M) phase The phase of the cell cycle that includes mitosis and cytokinesis.

mitotic spindle An assemblage of microtubules and associated proteins that is involved in the movement of chromosomes during mitosis.

mixotroph An organism that is capable of both photosynthesis and heterotrophy.

model A physical or conceptual representation of a natural phenomenon.

model organism A particular species chosen for research into broad biological principles because it is representative of a larger group and usually easy to grow in a lab.

molarity A common measure of solute concentration, referring to the number of moles of solute per liter of solution.

mold Informal term for a fungus that grows as a filamentous fungus, producing haploid spores by mitosis and forming a visible mycelium.

mole (mol) The number of grams of a substance that equals its molecular or atomic mass in daltons; a mole contains Avogadro's number of the molecules or atoms in question.

molecular clock A method for estimating the time required for a given amount of evolutionary change, based on the observation that some regions of genomes evolve at constant rates.

molecular mass The sum of the masses of all the atoms in a molecule; sometimes called molecular weight.

molecule Two or more atoms held together by covalent bonds.

molting A process in ecdysozoans in which the exoskeleton is shed at intervals, allowing growth by the production of a larger exoskeleton.

monilophyte An informal name for a member of the phylum Monilophyta, which includes ferns, horsetails, and whisk ferns and their relatives.

monoclonal antibody (mon′-ō-klōn′-ul) Any of a preparation of antibodies that have been produced by a single clone of cultured cells and thus are all specific for the same epitope.

monocot A member of a clade consisting of flowering plants that have one embryonic seed leaf, or cotyledon.

monogamous (muh-nog′-uh-mus) Referring to a type of relationship in which one male mates with just one female.

monohybrid An organism that is heterozygous with respect to a single gene of interest. All the offspring from a cross between parents homozygous for different alleles are monohybrids. For example, parents of genotypes *AA* and *aa* produce a monohybrid of genotype *Aa*.

monohybrid cross A cross between two organisms that are heterozygous for the character being followed (or the self-pollination of a heterozygous plant).

monomer (mon′-uh-mer) The subunit that serves as the building block of a polymer.

monophyletic (mon′-ō-fī-let′-ik) Pertaining to a group of taxa that consists of a common ancestor and all of its descendants. A monophyletic taxon is equivalent to a clade.

monosaccharide (mon′-ō-sak′-uh-rīd) The simplest carbohydrate, active alone or serving as a monomer for disaccharides and polysaccharides. Also called simple sugars, monosaccharides have molecular formulas that are generally some multiple of CH_2O.

monosomic Referring to a diploid cell that has only one copy of a particular chromosome instead of the normal two.

monotreme An egg-laying mammal, such as a platypus or echidna. Like all mammals, monotremes have hair and produce milk, but they lack nipples.

morphogen A substance, such as Bicoid protein in *Drosophila*, that provides positional information in the form of a concentration gradient along an embryonic axis.

morphogenesis (mōr′-fō-jen′-uh-sis) The development of the form of an organism and its structures.

morphological species concept Definition of a species in terms of measurable anatomical criteria.

moss A small, herbaceous, nonvascular plant that is a member of the phylum Bryophyta.

motor neuron A nerve cell that transmits signals from the brain or spinal cord to muscles or glands.

motor protein A protein that interacts with cytoskeletal elements and other cell components, producing movement of the whole cell or parts of the cell.

motor system An efferent branch of the vertebrate peripheral nervous system composed of motor neurons that carry signals to skeletal muscles in response to external stimuli.

motor unit A single motor neuron and all the muscle fibers it controls.

movement corridor A series of small clumps or a narrow strip of quality habitat (usable by organisms) that connects otherwise isolated patches of quality habitat.

MPF Maturation-promoting factor (or M-phase-promoting factor); a protein complex required for a cell to progress from late interphase to mitosis. The active form consists of cyclin and a protein kinase.

mucus A viscous and slippery mixture of glycoproteins, cells, salts, and water that moistens and protects the membranes lining body cavities that open to the exterior.

Müllerian mimicry (myū-lār′-ē-un mim′-uh-krē) Reciprocal mimicry by two unpalatable species.

multifactorial Referring to a phenotypic character that is influenced by multiple genes and environmental factors.

multigene family A collection of genes with similar or identical sequences, presumably of common origin.

multiple fruit A fruit derived from an entire inflorescence.

multiplication rule A rule of probability stating that the probability of two or more independent events occurring together can be determined by multiplying their individual probabilities.

muscle tissue Tissue consisting of long muscle cells that can contract, either on its own or when stimulated by nerve impulses.

mutagen (myū′-tuh-jen) A chemical or physical agent that interacts with DNA and can cause a mutation.

mutation (myū-tā′-shun) A change in the nucleotide sequence of an organism's DNA or in the DNA or RNA of a virus.

mutualism (myū′-chū-ul-izm) A +/+ ecological interaction that benefits each of the interacting species.

mycelium (mī-sē′-lē-um) The densely branched network of hyphae in a fungus.

mycorrhiza (plural, **mycorrhizae**) (mī′-kō-rī′-zuh, mī′-kō-rī′-zē) A mutualistic association of plant roots and fungus.

mycosis (mī-kō′-sis) General term for a fungal infection.

myelin sheath (mī′-uh-lin) Wrapped around the axon of a neuron, an insulating coat of cell membranes from Schwann cells or oligodendrocytes. It is interrupted by nodes of Ranvier, where action potentials are generated.

myofibril (mī′-ō-fī′-bril) A longitudinal bundle in a muscle cell (fiber) that contains thin filaments of actin and regulatory proteins and thick filaments of myosin.

myoglobin (mī′-uh-glō′-bin) An oxygen-storing, pigmented protein in muscle cells.

myosin (mī′-uh-sin) A type of motor protein that associates into filaments that interact with actin filaments to cause cell contraction.

myriapod (mir′-ē-uh-pod′) A terrestrial arthropod with many body segments and one or two pairs of legs per segment. Millipedes and centipedes are the two major groups of living myriapods.

NAD^+ The oxidized form of nicotinamide adenine dinucleotide, a coenzyme that can accept electrons, becoming NADH. NADH temporarily stores electrons during cellular respiration.

NADH The reduced form of nicotinamide adenine dinucleotide that temporarily stores electrons during cellular respiration. NADH acts as an electron donor to the electron transport chain.

$NADP^+$ The oxidized form of nicotinamide adenine dinucleotide phosphate, an electron carrier that can accept electrons, becoming NADPH. NADPH temporarily stores energized electrons produced during the light reactions.

NADPH The reduced form of nicotinamide adenine dinucleotide phosphate; temporarily stores energized electrons produced during the light reactions. NADPH acts as "reducing power" that can be passed along to an electron acceptor, reducing it.

natural killer cell A type of white blood cell that can kill tumor cells and virus-infected cells as part of innate immunity.

natural selection A process in which individuals that have certain inherited traits tend to survive and reproduce at higher rates than other individuals because of those traits.

negative feedback A form of regulation in which accumulation of an end product of a process slows the process; in physiology, a primary mechanism of homeostasis, whereby a change in a variable triggers a response that counteracts the initial change.

negative pressure breathing A breathing system in which air is pulled into the lungs.

nematocyst (nem′-uh-tuh-sist′) In a cnidocyte of a cnidarian, a capsule-like organelle containing a coiled thread that when discharged can penetrate the body wall of the prey.

nephron (nef′-ron) The tubular excretory unit of the vertebrate kidney.

neritic zone The shallow region of the ocean overlying the continental shelf.

nerve A fiber composed primarily of the bundled axons of neurons.

nervous system In animals, the fast-acting internal system of communication involving sensory receptors, networks of nerve cells, and connections to muscles and glands that respond to nerve signals; functions in concert with the endocrine system to effect internal regulation and maintain homeostasis.

nervous tissue Tissue made up of neurons and supportive cells.

net ecosystem production (NEP) The gross primary production of an ecosystem minus the energy used by all autotrophs and heterotrophs for respiration.

net primary production (NPP) The gross primary production of an ecosystem minus the energy used by the producers for respiration.

neural crest In vertebrates, a region located along the sides of the neural tube where it pinches off from the ectoderm. Neural crest cells migrate to various parts of the embryo and form pigment cells in the skin and parts of the skull, teeth, adrenal glands, and peripheral nervous system.

neural tube A tube of infolded ectodermal cells that runs along the anterior-posterior axis of a vertebrate, just dorsal to the notochord. It will give rise to the central nervous system.

neurohormone A molecule that is secreted by a neuron, travels in body fluids, and acts on specific target cells, changing their functioning.

neuron (nyūr′-on) A nerve cell; the fundamental unit of the nervous system, having structure and properties that allow it to conduct signals by taking advantage of the electrical charge across its plasma membrane.

neuronal plasticity The capacity of a nervous system to change with experience.

neuropeptide A relatively short chain of amino acids that serves as a neurotransmitter.

neurotransmitter A molecule that is released from the synaptic terminal of a neuron at a chemical synapse, diffuses across the synaptic cleft, and binds to the postsynaptic cell, triggering a response.

neutral variation Genetic variation that does not provide a selective advantage or disadvantage.

neutron A subatomic particle having no electrical charge (electrically neutral), with a mass of about 1.7×10^{-24}g, found in the nucleus of an atom.

neutrophil The most abundant type of white blood cell. Neutrophils are phagocytic and tend to self-destruct as they destroy foreign invaders, limiting their life span to a few days.

nitric oxide (NO) A gas produced by many types of cells that functions as a local regulator and as a neurotransmitter.

nitrogen cycle The natural process by which nitrogen, either from the atmosphere or from decomposed organic material, is converted by soil bacteria to compounds assimilated by plants. This incorporated nitrogen is then taken in by other organisms and subsequently released, acted on by bacteria, and made available again to the nonliving environment.

nitrogen fixation The conversion of atmospheric nitrogen (N_2) to ammonia (NH_3). Biological nitrogen fixation is carried out by certain prokaryotes, some of which have mutualistic relationships with plants.

nociceptor (nō′-si-sep′-tur) A sensory receptor that responds to noxious or painful stimuli; also called a pain receptor.

node A point along the stem of a plant at which leaves are attached.

node of Ranvier (ron′-vē-ā′) Gap in the myelin sheath of certain axons where an action potential may be generated. In saltatory conduction, an action potential is regenerated at each node, appearing to "jump" along the axon from node to node.

nodule A swelling on the root of a legume. Nodules are composed of plant cells that contain nitrogen-fixing bacteria of the genus *Rhizobium*.

noncompetitive inhibitor A substance that reduces the activity of an enzyme by binding to a location remote from the active site, changing the enzyme's shape so that the active site no longer effectively catalyzes the conversion of substrate to product.

nondisjunction An error in meiosis or mitosis in which members of a pair of homologous chromosomes or a pair of sister chromatids fail to separate properly from each other.

nonequilibrium model A model that maintains that communities change constantly after being buffeted by disturbances.

nonpolar covalent bond A type of covalent bond in which electrons are shared equally between two atoms of similar electronegativity.

nonsense mutation A mutation that changes an amino acid codon to one of the three stop codons, resulting in a shorter and usually nonfunctional protein.

norepinephrine A catecholamine that is chemically and functionally similar to epinephrine and acts as a hormone or neurotransmitter; also called noradrenaline.

northern coniferous forest A terrestrial biome characterized by long, cold winters and dominated by cone-bearing trees.

no-till agriculture A plowing technique that minimally disturbs the soil, thereby reducing soil loss.

notochord (nō′-tuh-kord′) A longitudinal, flexible rod made of tightly packed mesodermal cells that runs along the anterior-posterior axis of a chordate in the dorsal part of the body.

nuclear envelope In a eukaryotic cell, the double membrane that surrounds the nucleus, perforated with pores that regulate traffic with the cytoplasm. The outer membrane is continuous with the endoplasmic reticulum.

nuclear lamina A netlike array of protein filaments that lines the inner surface of the nuclear envelope and helps maintain the shape of the nucleus.

nucleariid A member of a group of unicellular, amoeboid protists that are more closely related to fungi than they are to other protists.

nuclease An enzyme that cuts DNA or RNA, either removing one or a few bases or hydrolyzing the DNA or RNA completely into its component nucleotides.

nucleic acid (nū-klā′-ik) A polymer (polynucleotide) consisting of many nucleotide monomers; serves as a blueprint for proteins and, through the actions of proteins, for all cellular activities. The two types are DNA and RNA.

nucleic acid hybridization (nū-klā′-ik) The base pairing of one strand of a nucleic acid to the complementary sequence on a strand from *another* nucleic acid molecule.

nucleic acid probe (nū-klā′-ik) In DNA technology, a labeled single-stranded nucleic acid molecule used to locate a specific nucleotide sequence in a nucleic acid sample. Molecules of the probe hydrogen-bond to the complementary sequence wherever it occurs; radioactive, fluorescent, or other labeling of the probe allows its location to be detected.

nucleoid (nū′-klē-oyd) A non-membrane-enclosed region in a prokaryotic cell where its chromosome is located.

nucleolus (nū-klē′-ō-lus) (plural, **nucleoli**) A specialized structure in the nucleus, consisting of chromosomal regions containing ribosomal RNA (rRNA) genes along with ribosomal proteins imported from the cytoplasm; site of rRNA synthesis and ribosomal subunit assembly. *See also* ribosome.

nucleosome (nū′-klē-ō-sōm′) The basic, beadlike unit of DNA packing in eukaryotes, consisting of a segment of DNA wound around a protein core composed of two copies of each of four types of histone.

nucleotide (nū-klē-ō-tīd′) The building block of a nucleic acid, consisting of a five-carbon sugar covalently bonded to a nitrogenous base and one to three phosphate groups.

nucleotide excision repair (nū′-klē-ō-tīd′) A repair system that removes and then correctly replaces a damaged segment of DNA using the undamaged strand as a guide.

nucleotide-pair substitution (nū′-klē-ō-tīd′) A type of point mutation in which one nucleotide in a DNA strand and its partner in the complementary strand are replaced by another pair of nucleotides.

nucleus (1) An atom's central core, containing protons and neutrons. (2) The organelle of a eukaryotic cell that contains the genetic material in the form of chromosomes, made up of chromatin. (3) A cluster of neurons.

nutrition The process by which an organism takes in and makes use of food substances.

obligate aerobe (ob′-lig-et ār′-ōb) An organism that requires oxygen for cellular respiration and cannot live without it.

obligate anaerobe (ob′-lig-et an′-uh-rōb) An organism that carries out only fermentation or anaerobic respiration. Such organisms cannot use oxygen and in fact may be poisoned by it.

ocean acidification The process by which the pH of the ocean is lowered (made more acidic) when excess CO_2 dissolves in seawater and forms carbonic acid (H_2CO_3).

oceanic pelagic zone Most of the ocean's waters far from shore, constantly mixed by ocean currents.

odorant A molecule that can be detected by sensory receptors of the olfactory system.

Okazaki fragment (ō′-kah-zah′-kē) A short segment of DNA synthesized away from the replication fork on a template strand during DNA replication. Many such segments are joined together to make up the lagging strand of newly synthesized DNA.

olfaction The sense of smell.

oligodendrocyte A type of glial cell that forms insulating myelin sheaths around the axons of neurons in the central nervous system.

oligotrophic lake A nutrient-poor, clear lake with few phytoplankton.

ommatidium (ōm′-uh-tid′-ē-um) (plural, **ommatidia**) One of the facets of the compound eye of arthropods and some polychaete worms.

omnivore An animal that regularly eats animals as well as plants or algae.

oncogene (on′-kō-jēn) A gene found in viral or cellular genomes that is involved in triggering molecular events that can lead to cancer.

oocyte (ō′-uh-sīt) A cell in the female reproductive system that differentiates to form an egg.

oogenesis (ō′-uh-jen′-uh-sis) The process in the ovary that results in the production of female gametes.

oogonium (ō′-uh-gō′-nē-em) (plural, **oogonia**) A cell that divides mitotically to form oocytes.

open circulatory system A circulatory system in which fluid called hemolymph bathes the tissues and organs directly and there is no distinction between the circulating fluid and the interstitial fluid.

operator In bacterial and phage DNA, a sequence of nucleotides near the start of an operon to which an active repressor can attach. The binding of the repressor prevents RNA polymerase from attaching to the promoter and transcribing the genes of the operon.

operculum (ō-per′-kyuh-lum) In aquatic osteichthyans, a protective bony flap that covers and protects the gills.

operon (op′-er-on) A unit of genetic function found in bacteria and phages, consisting of a promoter, an operator, and a coordinately regulated cluster of genes whose products function in a common pathway.

opisthokont (uh-pis′-thuh-kont′) A member of an extremely diverse clade of eukaryotes that includes fungi, animals, and several closely related groups of protists.

opposable thumb A thumb that can touch the ventral surface (fingerprint side) of the fingertip of all four fingers of the same hand with its own ventral surface.

opsin A membrane protein bound to a light-absorbing pigment molecule.

optimal foraging model The basis for analyzing behavior as a compromise between feeding costs and feeding benefits.

oral cavity The mouth of an animal.

orbital The three-dimensional space where an electron is found 90% of the time.

order In Linnaean classification, the taxonomic category above the level of family.

organ A specialized center of body function composed of several different types of tissues.

organelle (ōr-guh-nel′) Any of several membrane-enclosed structures with specialized functions, suspended in the cytosol of eukaryotic cells.

organic chemistry The study of carbon compounds (organic compounds).

organ identity gene A plant homeotic gene that uses positional information to determine which emerging leaves develop into which types of floral organs.

organism An individual living thing, consisting of one or more cells.

organismal ecology The branch of ecology concerned with the morphological, physiological, and behavioral ways in which individual organisms meet the challenges posed by their biotic and abiotic environments.

organ of Corti (kor′-tē) The actual hearing organ of the vertebrate ear, located in the floor of the cochlear duct in the inner ear; contains the receptor cells (hair cells) of the ear.

organogenesis (ōr-gan′-ō-jen′-uh-sis) The process in which organ rudiments develop from the three germ layers after gastrulation.

organ system A group of organs that work together in performing vital body functions.

origin of replication Site where the replication of a DNA molecule begins, consisting of a specific sequence of nucleotides.

orthologous genes Homologous genes that are found in different species because of speciation.

osculum (os′-kyuh-lum) A large opening in a sponge that connects the spongocoel to the environment.

osmoconformer An animal that is isoosmotic with its environment.

osmolarity (oz′-mō-lār′-uh-tē) Solute concentration expressed as molarity.

osmoregulation Regulation of solute concentrations and water balance by a cell or organism.

osmoregulator An animal that controls its internal osmolarity independent of the external environment.

osmosis (oz-mō′-sis) The diffusion of free water across a selectively permeable membrane.

osteichthyan (os′-tē-ik′-thē-an) A member of a vertebrate clade with jaws and mostly bony skeletons.

outer ear One of the three main regions of the ear in reptiles (including birds) and mammals; made up of the auditory canal and, in many birds and mammals, the pinna.

outgroup A species or group of species from an evolutionary lineage that is known to have diverged before the lineage that contains the group of species being studied. An outgroup is selected so that its members are closely related to the group of species being studied, but not as closely related as any study-group members are to each other.

oval window In the vertebrate ear, a membrane-covered gap in the skull bone, through which sound waves pass from the middle ear to the inner ear.

ovarian cycle (ō-vār′-ē-un) The cyclic recurrence of the follicular phase, ovulation, and the luteal phase in the mammalian ovary, regulated by hormones.

ovary (ō′-vuh-rē) (1) In flowers, the portion of a carpel in which the egg-containing ovules develop. (2) In animals, the structure that produces female gametes and reproductive hormones.

oviduct (ō′-vuh-duct) A tube passing from the ovary to the vagina in invertebrates or to the uterus in vertebrates, where it is also called a fallopian tube.

oviparous (ō-vip′-uh-rus) Referring to a type of development in which young hatch from eggs laid outside the mother's body.

ovoviviparous (ō′-vō-vī-vip′-uh-rus) Referring to a type of development in which young hatch from eggs that are retained in the mother's uterus.

ovulation The release of an egg from an ovary. In humans, an ovarian follicle releases an egg during each uterine (menstrual) cycle.

ovule (o′-vyūl) A structure that develops within the ovary of a seed plant and contains the female gametophyte.

oxidation The complete or partial loss of electrons from a substance involved in a redox reaction.

oxidative phosphorylation (fos′-fōr-uh-lā′-shun) The production of ATP using energy derived from the redox reactions of an electron transport chain; the third major stage of cellular respiration.

oxidizing agent The electron acceptor in a redox reaction.

oxytocin (ok′-si-tō′-sen) A hormone produced by the hypothalamus and released from the posterior pituitary. It induces contractions of the uterine muscles during labor and causes the mammary glands to eject milk during nursing.

***p53* gene** A tumor-suppressor gene that codes for a specific transcription factor that promotes the synthesis of proteins that inhibit the cell cycle.

paedomorphosis (pē′-duh-mōr′-fuh-sis) The retention in an adult organism of the juvenile features of its evolutionary ancestors.

pain receptor A sensory receptor that responds to noxious or painful stimuli; also called a nociceptor.

paleoanthropology The study of human origins and evolution.

paleontology (pā′-lē-un-tol′-ō-jē) The scientific study of fossils.

pancreas (pan′-krē-us) A gland with exocrine and endocrine tissues. The exocrine portion functions in digestion, secreting enzymes and an alkaline solution into the small intestine via a duct; the ductless endocrine portion functions in homeostasis, secreting the hormones insulin and glucagon into the blood.

pancrustacean A member of a diverse arthropod clade that includes lobsters, crabs, barnacles and other crustaceans, as well as insects and their six-legged terrestrial relatives.

pandemic A global epidemic.

Pangaea (pan-jē′-uh) The supercontinent that formed near the end of the Paleozoic era, when plate movements brought all the landmasses of Earth together.

parabasalid A protist, such as a trichomonad, with modified mitochondria.

paracrine Referring to a secreted molecule that acts on a neighboring cell.

paralogous genes Homologous genes that are found in the same genome as a result of gene duplication.

paraphyletic (pār′-uh-fī-let′-ik) Pertaining to a group of taxa that consists of a common ancestor and some, but not all, of its descendants.

parareptile A basal group of reptiles, consisting mostly of large, stocky quadrupedal herbivores. Parareptiles died out in the late Triassic period.

parasite (pār′-uh-sīt) An organism that feeds on the cell contents, tissues, or body fluids of another species (the host) while in or on the host organism. Parasites harm but usually do not kill their host.

parasitism (pār′-uh-sit-izm) A +/− ecological interaction in which one organism, the parasite, benefits by feeding upon another organism, the host, which is harmed; some parasites live within the host (feeding on its tissues), while others feed on the host's external surface.

parasympathetic division One of three divisions of the autonomic nervous system; generally enhances body activities that gain and conserve energy, such as digestion and reduced heart rate.

parathyroid gland One of four small endocrine glands, embedded in the surface of the thyroid gland, that secrete parathyroid hormone.

parathyroid hormone (PTH) A hormone secreted by the parathyroid glands that raises blood calcium level by promoting calcium release from bone and calcium retention by the kidneys.

parenchyma cell (puh-ren′-ki-muh) A relatively unspecialized plant cell type that carries out most of the metabolism, synthesizes and stores organic products, and develops into a more differentiated cell type.

parental type An offspring with a phenotype that matches one of the true-breeding parental (P generation) phenotypes; also refers to the phenotype itself.

Parkinson's disease A progressive brain disease characterized by difficulty in initiating movements, slowness of movement, and rigidity.

parthenogenesis (par′-thuh-nō′-jen′-uh-sis) A form of asexual reproduction in which females produce offspring from unfertilized eggs.

partial pressure The pressure exerted by a particular gas in a mixture of gases (for instance, the pressure exerted by oxygen in air).

passive immunity Short-term immunity conferred by the transfer of antibodies, as occurs in the transfer of maternal antibodies to a fetus or nursing infant.

passive transport The diffusion of a substance across a biological membrane with no expenditure of energy.

pathogen An organism or virus that causes disease.

pathogen-associated molecular pattern (PAMP) A molecular sequence that is specific to a certain pathogen.

pattern formation The development of a multicellular organism's spatial organization, the arrangement of organs and tissues in their characteristic places in three-dimensional space.

peat Extensive deposits of partially decayed organic material often formed primarily from the wetland moss *Sphagnum*.

pedigree A diagram of a family tree with conventional symbols, showing the occurrence of heritable characters in parents and offspring over multiple generations.

pelagic zone The open-water component of aquatic biomes.

penis The copulatory structure of male mammals.

PEP carboxylase An enzyme that adds CO_2 to phosphoenolpyruvate (PEP) to form oxaloacetate in mesophyll cells of C_4 plants. It acts prior to photosynthesis.

pepsin An enzyme present in gastric juice that begins the hydrolysis of proteins.

pepsinogen The inactive form of pepsin secreted by chief cells located in gastric pits of the stomach.

peptide bond The covalent bond between the carboxyl group on one amino acid and the amino group on another, formed by a dehydration reaction.

peptidoglycan (pep′-tid-ō-glī′-kan) A type of polymer in bacterial cell walls consisting of modified sugars cross-linked by short polypeptides.

perception The interpretation of sensory system input by the brain.

pericycle The outermost layer in the vascular cylinder, from which lateral roots arise.

periderm (pār′-uh-derm′) The protective coat that replaces the epidermis in woody plants during secondary growth, formed of the cork and cork cambium.

peripheral nervous system (PNS) The sensory and motor neurons that connect to the central nervous system.

peripheral protein A protein loosely bound to the surface of a membrane or to part of an integral protein and not embedded in the lipid bilayer.

peristalsis (pār′-uh-stal′-sis) (1) Alternating waves of contraction and relaxation in the smooth muscles lining the alimentary canal that push food along the canal. (2) A type of movement on land produced by rhythmic waves of muscle contractions passing from front to back, as in many annelids.

peristome (pār′-uh-stōme′) A ring of interlocking, tooth-like structures on the upper part of a moss capsule (sporangium), often specialized for gradual spore discharge.

peritubular capillary One of the tiny blood vessels that form a network surrounding the proximal and distal tubules in the kidney.

peroxisome (puh-rok′-suh-sōm′) An organelle containing enzymes that transfer hydrogen atoms from various substrates to oxygen (O_2), producing and then degrading hydrogen peroxide (H_2O_2).

petal A modified leaf of a flowering plant. Petals are the often colorful parts of a flower that advertise it to insects and other pollinators.

petiole (pet′-ē-ōl) The stalk of a leaf, which joins the leaf to a node of the stem.

P generation The true-breeding (homozygous) parent individuals from which F_1 hybrid offspring are derived in studies of inheritance; P stands for "parental."

pH A measure of hydrogen ion concentration equal to $-\log[H^+]$ and ranging in value from 0 to 14.

phage (fāj) A virus that infects bacteria; also called a bacteriophage.

phagocytosis (fag′-ō-sī-tō′-sis) A type of endocytosis in which large particulate substances or small organisms are taken up by a cell. It is carried out by some protists and by certain immune cells of animals (in mammals, mainly macrophages, neutrophils, and dendritic cells).

pharyngeal cleft (fuh-rin′-jē-ul) In chordate embryos, one of the grooves that separate a series of arches along the outer surface of the pharynx and may develop into a pharyngeal slit.

pharyngeal slit (fuh-rin′-jē-ul) In chordate embryos, one of the slits that form from the pharyngeal clefts and open into the pharynx, later developing into gill slits in many vertebrates.

pharynx (far′-inks) (1) An area in the vertebrate throat where air and food passages cross. (2) In flatworms, the muscular tube that protrudes from the ventral side of the worm and ends in the mouth.

phase change (1) A shift from one developmental phase to another. (2) In plants, a morphological change that arises from a transition in shoot apical meristem activity.

phenotype (fē′-nō-tīp) The observable physical and physiological traits of an organism, which are determined by its genetic makeup.

pheromone (fār′-uh-mōn) In animals and fungi, a small molecule released into the environment that functions in communication between members of the same species. In animals, it acts much like a hormone in influencing physiology and behavior.

phloem (flō′-em) Vascular plant tissue consisting of living cells arranged into elongated tubes that transport sugar and other organic nutrients throughout the plant.

phloem sap (flō′-em) The sugar-rich solution carried through a plant's sieve tubes.

phosphate group A chemical group consisting of a phosphorus atom bonded to four oxygen atoms; important in energy transfer.

phospholipid (fos′-fō-lip′-id) A lipid made up of glycerol joined to two fatty acids and a phosphate group. The hydrocarbon chains of the fatty acids act as nonpolar, hydrophobic tails, while the rest of the molecule acts as a polar, hydrophilic head. Phospholipids form bilayers that function as biological membranes.

phosphorylated intermediate (fos′-fōr-uh-lā′-ted) A molecule (often a reactant) with a phosphate group covalently bound to it, making it more reactive (less stable) than the unphosphorylated molecule.

phosphorylation cascade (fos′-fōr-uh-lā′-shun) A series of chemical reactions during cell signaling mediated by enzymes (kinases), in which each kinase in turn phosphorylates and activates another, ultimately leading to phosphorylation of many proteins.

photic zone (fō′-tic) The narrow top layer of an ocean or lake, where light penetrates sufficiently for photosynthesis to occur.

photoautotroph (fō′-tō-ot′-ō-trōf) An organism that harnesses light energy to drive the synthesis of organic compounds from carbon dioxide.

photoheterotroph (fō′-tō-het′-er-ō-trōf) An organism that uses light to generate ATP but must obtain carbon in organic form.

photomorphogenesis Effects of light on plant morphology.

photon (fō′-ton) A quantum, or discrete quantity, of light energy that behaves as if it were a particle.

photoperiodism (fō′-tō-pēr′-ē-ō-dizm) A physiological response to photoperiod, the interval in a 24-hour period during which an organism is exposed to light. An example of photoperiodism is flowering.

photophosphorylation (fō′-tō-fos′-fōr-uh-lā′-shun) The process of generating ATP from ADP and phosphate by means of chemiosmosis, using a proton-motive force generated across the thylakoid membrane of the chloroplast or the membrane of certain prokaryotes during the light reactions of photosynthesis.

photoreceptor An electromagnetic receptor that detects the radiation known as visible light.

photorespiration A metabolic pathway that consumes oxygen and ATP, releases carbon dioxide, and decreases photosynthetic output. Photorespiration generally occurs on hot, dry, bright days, when stomata close and the O_2/CO_2 ratio in the leaf increases, favoring the binding of O_2 rather than CO_2 by rubisco.

photosynthesis (fō′-tō-sin′-thi-sis) The conversion of light energy to chemical energy that is stored in sugars or other organic compounds; occurs in plants, algae, and certain prokaryotes.

photosystem A light-capturing unit located in the thylakoid membrane of the chloroplast or in the membrane of some prokaryotes, consisting of a reaction-center complex surrounded by numerous light-harvesting complexes. There are two types of photosystems, I and II; they absorb light best at different wavelengths.

photosystem I (PS I) A light-capturing unit in a chloroplast's thylakoid membrane or in the membrane of some prokaryotes; it has two molecules of P700 chlorophyll *a* at its reaction center.

photosystem II (PS II) One of two light-capturing units in a chloroplast's thylakoid membrane or in the membrane of some prokaryotes; it has two molecules of P680 chlorophyll *a* at its reaction center.

phototropism (fō′-tō-trō′-pizm) Growth of a plant shoot toward or away from light.

phyllotaxy (fil′-uh-tak′-sē) The pattern of leaf attachment to the stem of a plant.

phylogenetic tree A branching diagram that represents a hypothesis about the evolutionary history of a group of organisms.

phylogeny (fī-loj′-uh-nē) The evolutionary history of a species or group of related species.

phylum (fī′-lum) (plural, **phyla**) In Linnaean classification, the taxonomic category above class.

physiology The processes and functions of an organism.

phytochrome (fī′-tuh-krōm) A type of light receptor in plants that mostly absorbs red light and regulates many plant responses, such as seed germination and shade avoidance.

phytoremediation An emerging technology that seeks to reclaim contaminated areas by taking advantage of some plant species' ability to extract heavy metals and other pollutants from the soil and to concentrate them in easily harvested portions of the plant.

pilus (plural, **pili**) (pī′-lus, pī′-lī) In bacteria, a structure that links one cell to another at the start of conjugation; also called a sex pilus or conjugation pilus.

pineal gland (pī′-nē-ul) A small gland on the dorsal surface of the vertebrate forebrain that secretes the hormone melatonin.

pinocytosis (pī′-nō-sī-tō′-sis) A type of endocytosis in which the cell ingests extracellular fluid and its dissolved solutes.

pistil A single carpel (a simple pistil) or a group of fused carpels (a compound pistil).

pith Ground tissue that is internal to the vascular tissue in a stem; in many monocot roots, parenchyma cells that form the central core of the vascular cylinder.

pituitary gland (puh-tū′-uh-tār′-ē) An endocrine gland at the base of the hypothalamus; consists of a posterior lobe, which stores and releases two hormones produced by the hypothalamus, and an anterior lobe, which produces and secretes many hormones that regulate diverse body functions.

placenta (pluh-sen′-tuh) A structure in the uterus of a pregnant eutherian mammal that nourishes the fetus with the mother's blood supply; formed from the uterine lining and embryonic membranes.

placoderm A member of an extinct group of fishlike vertebrates that had jaws and were enclosed in a tough outer armor.

planarian A free-living flatworm found in ponds and streams.

plasma (plaz′-muh) The liquid matrix of blood in which the blood cells are suspended.

plasma membrane (plaz′-muh) The membrane at the boundary of every cell that acts as a selective barrier, regulating the cell's chemical composition.

plasmid (plaz′-mid) A small, circular, double-stranded DNA molecule that carries accessory genes separate from those of a bacterial chromosome; in DNA cloning, plasmids are used as vectors carrying up to about 10,000 base pairs (10 kb) of DNA. Plasmids are also found in some eukaryotes, such as yeasts.

plasmodesma (plaz′-mō-dez′-muh) (plural, **plasmodesmata**) An open channel through the cell wall that connects the cytoplasm of adjacent plant cells, allowing water, small solutes, and some larger molecules to pass between the cells.

plasmogamy (plaz-moh′-guh-mē) In fungi, the fusion of the cytoplasm of cells from two individuals; occurs as one stage of sexual reproduction, followed later by karyogamy.

plasmolysis (plaz-mol′-uh-sis) A phenomenon in walled cells in which the cytoplasm shrivels and the plasma membrane pulls away from the cell wall; occurs when the cell loses water to a hypertonic environment.

plastid One of a family of closely related organelles that includes chloroplasts, chromoplasts, and amyloplasts. Plastids are found in cells of photosynthetic eukaryotes.

plate tectonics The theory that the continents are part of great plates of Earth's crust that float on the hot, underlying portion of the mantle. Movements in the mantle cause the continents to move slowly over time.

platelet A pinched-off cytoplasmic fragment of a specialized bone marrow cell. Platelets circulate in the blood and are important in blood clotting.

pleiotropy (plī′-o-truh-pē) The ability of a single gene to have multiple effects.

pluripotent Describing a cell that can give rise to many, but not all, parts of an organism.

point mutation A change in a single nucleotide pair of a gene.

polar covalent bond A covalent bond between atoms that differ in electronegativity. The shared electrons are pulled closer to the more electronegative atom, making it slightly negative and the other atom slightly positive.

polar molecule A molecule (such as water) with an uneven distribution of charges in different regions of the molecule.

polarity A lack of symmetry; structural differences in opposite ends of an organism or structure, such as the root end and shoot end of a plant.

pollen grain In seed plants, a structure consisting of the male gametophyte enclosed within a pollen wall.

pollen tube A tube that forms after germination of the pollen grain and that functions in the delivery of sperm to the ovule.

pollination (pol′-uh-nā′-shun) The transfer of pollen to the part of a seed plant containing the ovules, a process required for fertilization.

poly-A tail A sequence of 50–250 adenine nucleotides added onto the 3′ end of a pre-mRNA molecule.

polygamous Referring to a type of relationship in which an individual of one sex mates with several of the other.

polygenic inheritance (pol′-ē-jen′-ik) An additive effect of two or more genes on a single phenotypic character.

polymer (pol′-uh-mer) A long molecule consisting of many similar or identical monomers linked together by covalent bonds.

polymerase chain reaction (PCR) (puh-lim′-uh-rās) A technique for amplifying DNA *in vitro* by incubating it with specific primers, a heat-resistant DNA polymerase, and nucleotides.

polynucleotide (pol′-ē-nū′-klē-ō-tīd) A polymer consisting of many nucleotide monomers in a chain. The nucleotides can be those of DNA or RNA.

polyp The sessile variant of the cnidarian body plan. The alternate form is the medusa.

polypeptide (pol′-ē-pep′-tīd) A polymer of many amino acids linked together by peptide bonds.

polyphyletic (pol′-ē-fī-let′-ik) Pertaining to a group of taxa that includes distantly related organisms but does not include their most recent common ancestor.

polyploidy (pol′-ē-ploy′-dē) A chromosomal alteration in which the organism possesses more than two complete chromosome sets. It is the result of an accident of cell division.

polyribosome (polysome) (pol′-ē-rī′-buh-sōm′) A group of several ribosomes attached to, and translating, the same messenger RNA molecule.

polysaccharide (pol′-ē-sak′-uh-rīd) A polymer of many monosaccharides, formed by dehydration reactions.

polyspermy The fertilization of an egg by more than one sperm.

polytomy (puh-lit′-uh-mē) In a phylogenetic tree, a branch point from which more than two descendant taxa emerge. A polytomy indicates that the evolutionary relationships between the descendant taxa are not yet clear.

pons A portion of the brain that participates in certain automatic, homeostatic functions, such as regulating the breathing centers in the medulla.

population A group of individuals of the same species that live in the same area and interbreed, producing fertile offspring.

population dynamics The study of how complex interactions between biotic and abiotic factors influence variations in population size.

population ecology The study of populations in relation to their environment, including environmental influences on population density and distribution, age structure, and variations in population size.

positional information Molecular cues that control pattern formation in an animal or plant embryonic structure by indicating a cell's location relative to the organism's body axes. These cues elicit a response by genes that regulate development.

positive feedback A form of regulation in which an end product of a process speeds up that process; in physiology, a control mechanism in which a change in a variable triggers a response that reinforces or amplifies the change.

positive interaction A +/+ or +/0 ecological interaction in which at least one of the interacting species benefits and neither is harmed; positive interactions include mutualism and commensalism.

positive pressure breathing A breathing system in which air is forced into the lungs.

posterior Pertaining to the rear, or tail end, of a bilaterally symmetrical animal.

posterior pituitary An extension of the hypothalamus composed of nervous tissue that secretes oxytocin and antidiuretic hormone made in the hypothalamus; a temporary storage site for these hormones.

postzygotic barrier (pōst′-zī-got′-ik) A reproductive barrier that prevents hybrid zygotes produced by two different species from developing into viable, fertile adults.

potential energy The energy that matter possesses as a result of its location or spatial arrangement (structure).

predation An interaction between species in which one species, the predator, eats the other, the prey.

prediction In deductive reasoning, a forecast that follows logically from a hypothesis. By testing predictions, experiments may allow certain hypotheses to be rejected.

pregnancy The condition of carrying one or more embryos in the uterus; also called gestation.

prepuce (prē′-pyūs) A fold of skin covering the head of the clitoris or penis.

pressure potential (ψ_P) A component of water potential that consists of the physical pressure on a solution, which can be positive, zero, or negative.

prezygotic barrier (prē′-zī-got′-ik) A reproductive barrier that impedes mating between species or hinders fertilization if interspecific mating is attempted.

primary cell wall In plants, a relatively thin and flexible layer that surrounds the plasma membrane of a young cell.

primary consumer An herbivore; an organism that eats plants or other autotrophs.

primary electron acceptor In the thylakoid membrane of a chloroplast or in the membrane of some prokaryotes, a specialized molecule that shares the reaction-center complex with a pair of chlorophyll *a* molecules and that accepts an electron from them.

primary growth Growth produced by apical meristems, lengthening stems and roots.

primary immune response The initial adaptive immune response to an antigen, which appears after a lag of about 10–17 days.

primary meristems The three meristematic derivatives (protoderm, procambium, and ground meristem) of an apical meristem.

primary oocyte (ō′-uh-sīt) An oocyte prior to completion of meiosis I.

primary producer An autotroph, usually a photosynthetic organism. Collectively, autotrophs make up the trophic level of an ecosystem that ultimately supports all other levels.

primary production The amount of light energy converted to chemical energy (organic compounds) by the autotrophs in an ecosystem during a given time period.

primary structure The level of protein structure referring to the specific linear sequence of amino acids.

primary succession A type of ecological succession that occurs in an area where there were originally no organisms present and where soil has not yet formed.

primary transcript An initial RNA transcript from any gene; also called pre-mRNA when transcribed from a protein-coding gene.

primase An enzyme that joins RNA nucleotides to make a primer during DNA replication, using the parental DNA strand as a template.

primer A short polynucleotide with a free 3′ end, bound by complementary base pairing to the template strand and elongated with DNA nucleotides during DNA replication.

prion An infectious agent that is a misfolded version of a normal cellular protein. Prions appear to increase in number by converting correctly folded versions of the protein to more prions.

problem solving The cognitive activity of devising a method to proceed from one state to another in the face of real or apparent obstacles.

producer An organism that produces organic compounds from CO_2 by harnessing light energy (in photosynthesis) or by oxidizing inorganic chemicals (in chemosynthetic reactions carried out by some prokaryotes).

product A material resulting from a chemical reaction.

production efficiency The percentage of energy stored in assimilated food that is not used for respiration or eliminated as waste.

progesterone A steroid hormone that contributes to the menstrual cycle and prepares the uterus for pregnancy; the major progestin in mammals.

progestin Any steroid hormone with progesterone-like activity.

prokaryotic cell (prō′-kār′-ē-ot′-ik) A type of cell lacking a membrane-enclosed nucleus and membrane-enclosed organelles. Organisms with prokaryotic cells (bacteria and archaea) are called prokaryotes.

prolactin A hormone produced and secreted by the anterior pituitary with a great diversity of effects in different vertebrate species. In mammals, it stimulates growth of and milk production by the mammary glands.

prometaphase The second stage of mitosis, in which the nuclear envelope fragments and the spindle microtubules attach to the kinetochores of the chromosomes.

promoter A specific nucleotide sequence in the DNA of a gene that binds RNA polymerase, positioning it to start transcribing RNA at the appropriate place.

prophage (prō′-fāj) A phage genome that has been inserted into a specific site on a bacterial chromosome.

prophase The first stage of mitosis, in which the chromatin condenses into discrete chromosomes visible with a light microscope, the mitotic spindle begins to form, and the nucleolus disappears but the nucleus remains intact.

prostaglandin (pros′-tuh-glan′-din) One of a group of modified fatty acids that are secreted by virtually all tissues and that perform a wide variety of functions as local regulators.

prostate gland (pros′-tāt) A gland in human males that secretes an acid-neutralizing component of semen.

protease (prō′-tē-āz) An enzyme that digests proteins by hydrolysis.

protein (prō′-tēn) A biologically functional molecule consisting of one or more polypeptides folded and coiled into a specific three-dimensional structure.

protein kinase (prō′-tēn kī′-nās) An enzyme that transfers phosphate groups from ATP to a protein, thus phosphorylating the protein.

protein phosphatase (prō′-tēn fos′-fuh-tās) An enzyme that removes phosphate groups from (dephosphorylates) proteins, often functioning to reverse the effect of a protein kinase.

proteoglycan (prō′-tē-ō-glī′-kan) A large molecule consisting of a small core protein with many carbohydrate chains attached, found in the extracellular matrix of animal cells. A proteoglycan may consist of up to 95% carbohydrate.

proteome The entire set of proteins expressed by a given cell, tissue, or organism.

proteomics (prō′-tē-ō′-miks) The systematic study of sets of proteins and their properties, including their abundance, chemical modifications, and interactions.

protist An informal term applied to any eukaryote that is not a plant, animal, or fungus. Most protists are unicellular, though some are colonial or multicellular.

protocell An abiotic precursor of a living cell that had a membrane-like structure and that maintained an internal chemistry different from that of its surroundings.

proton (prō′-ton) A subatomic particle with a single positive electrical charge, with a mass of about 1.7×10^{-24}g, found in the nucleus of an atom.

protonema (prō′-tuh-nē′-muh) (plural, **protonemata**) A mass of green, branched, one-cell-thick filaments produced by germinating moss spores.

protonephridium (prō′-tō-nuh-frid′-ē-um) (plural, **protonephridia**) An excretory system, such as the flame bulb system of flatworms, consisting of a network of tubules lacking internal openings.

proton-motive force (prō′-ton) The potential energy stored in the form of a proton electrochemical gradient, generated by the pumping of hydrogen ions (H^+) across a biological membrane during chemiosmosis.

proton pump (prō′-ton) An active transport protein in a cell membrane that uses ATP to transport hydrogen ions out of a cell against their concentration gradient, generating a membrane potential in the process.

proto-oncogene (prō′-tō-on′-kō-jēn) A normal cellular gene that has the potential to become an oncogene.

protoplast The living part of a plant cell, which also includes the plasma membrane.

protostome development In animals, a developmental mode distinguished by the development of the mouth from the blastopore; often also characterized by spiral cleavage and by the body cavity forming when solid masses of mesoderm split.

provirus A viral genome that is permanently inserted into a host genome.

proximal tubule In the vertebrate kidney, the portion of a nephron immediately downstream from Bowman's capsule that conveys and helps refine filtrate.

pseudocoelomate (sū′-dō-sē′-lō-māt) An animal whose body cavity is lined by tissue derived from mesoderm and endoderm.

pseudogene (sū′-dō-jēn) A DNA segment that is very similar to a real gene but does not yield a functional product; a DNA segment that formerly functioned as a gene but has become inactivated in a particular species because of mutation.

pseudopodium (sū′-dō-pō′-dē-um) (plural, **pseudopodia**) A cellular extension of amoeboid cells used in moving and feeding.

P site One of a ribosome's three binding sites for tRNA during translation. The P site holds the tRNA carrying the growing polypeptide chain. (P stands for peptidyl tRNA.)

pterosaur Winged reptile that lived during the Mesozoic era.

pulse The rhythmic bulging of the artery walls with each heartbeat.

punctuated equilibria In the fossil record, long periods of apparent stasis, in which a species undergoes little or no morphological change, interrupted by relatively brief periods of sudden change.

Punnett square A diagram used in the study of inheritance to show the predicted genotypic results of random fertilization in genetic crosses between individuals of known genotype.

pupil The opening in the iris, which admits light into the interior of the vertebrate eye. Muscles in the iris regulate its size.

purine (pyū′-rēn) One of two types of nitrogenous bases found in nucleotides, characterized by a six-membered ring fused to a five-membered ring. Adenine (A) and guanine (G) are purines.

pyrimidine (puh-rim′-uh-dēn) One of two types of nitrogenous bases found in nucleotides, characterized by a six-membered ring. Cytosine (C), thymine (T), and uracil (U) are pyrimidines.

quantitative character A heritable feature that varies continuously over a range rather than in an either-or fashion.

quaternary structure (kwot′-er-nār′-ē) The particular shape of a complex, aggregate protein, defined by the characteristic three-dimensional arrangement of its constituent subunits, each a polypeptide.

radial cleavage A type of embryonic development in deuterostomes in which the planes of cell division that transform the zygote into a ball of cells are either parallel or perpendicular to the vertical axis of the embryo, thereby aligning tiers of cells one above the other.

radial symmetry Symmetry in which the body is shaped like a pie or barrel (lacking a left side and a right side) and can be divided into mirror-imaged halves by any plane through its central axis.

radicle An embryonic root of a plant.

radioactive isotope An isotope (an atomic form of a chemical element) that is unstable; the nucleus decays spontaneously, giving off detectable particles and energy.

radiolarian A protist, usually marine, with a shell generally made of silica and pseudopodia that radiate from the central body.

radiometric dating A method for determining the absolute age of rocks and fossils, based on the half-life of radioactive isotopes.

radula A straplike scraping organ used by many molluscs during feeding.

***ras* gene** A gene that codes for Ras, a G protein that relays a growth signal from a growth factor receptor on the plasma membrane to a cascade of protein kinases, ultimately resulting in stimulation of the cell cycle.

ratite (rat′-īt) A member of the group of flightless birds.

ray-finned fish A member of the clade Actinopterygii, aquatic osteichthyans with fins supported by long, flexible rays, including tuna, bass, and herring.

reabsorption In excretory systems, the recovery of solutes and water from filtrate.

reactant A starting material in a chemical reaction.

reaction-center complex A complex of proteins associated with a special pair of chlorophyll *a* molecules and a primary electron acceptor. Located centrally in a photosystem, this complex triggers the light reactions of photosynthesis. Excited by light energy, the pair of chlorophylls donates an electron to the primary electron acceptor, which passes an electron to an electron transport chain.

reading frame On an mRNA, the triplet grouping of ribonucleotides used by the translation machinery during polypeptide synthesis.

receptacle The base of a flower; the part of the stem that is the site of attachment of the floral organs.

reception In cellular communication, the first step of a signaling pathway in which a signaling molecule is detected by a receptor molecule on or in the cell.

receptor-mediated endocytosis (en′-dō-sī-tō′-sis) The movement of specific molecules into a cell by the infolding of vesicles containing proteins with receptor sites specific to the molecules being taken in; enables a cell to acquire bulk quantities of specific substances.

receptor potential An initial response of a receptor cell to a stimulus, consisting of a change in voltage across the receptor membrane proportional to the stimulus strength.

receptor tyrosine kinase (RTK) A receptor protein spanning the plasma membrane, the cytoplasmic (intracellular) part of which can catalyze the transfer of a phosphate group from ATP to a tyrosine on another protein. Receptor tyrosine kinases often respond to the binding of a signaling molecule by dimerizing and then phosphorylating a tyrosine on the cytoplasmic portion of the other receptor in the dimer.

recessive allele An allele whose phenotypic effect is not observed in a heterozygote.

reciprocal altruism Altruistic behavior between unrelated individuals, whereby the altruistic individual benefits in the future when the beneficiary reciprocates.

recombinant chromosome A chromosome created when crossing over combines DNA from two parents into a single chromosome.

recombinant DNA molecule A DNA molecule made *in vitro* with segments from different sources.

recombinant type (recombinant) An offspring whose phenotype differs from that of the true-breeding P generation parents; also refers to the phenotype itself.

rectum The terminal portion of the large intestine, where the feces are stored prior to elimination.

red alga A photosynthetic protist, named for its color, which results from a red pigment that masks the green of chlorophyll. Most red algae are multicellular and marine.

redox reaction (rē′-doks) A chemical reaction involving the complete or partial transfer of one or more electrons from one reactant to another; short for **red**uction-**ox**idation reaction.

reducing agent The electron donor in a redox reaction.

reduction The complete or partial addition of electrons to a substance involved in a redox reaction.

reflex An automatic reaction to a stimulus, mediated by the spinal cord or lower brain.

refractory period (rē-frakt′-ōr-ē) The short time immediately after an action potential in which the neuron cannot respond to another stimulus, owing to the inactivation of voltage-gated sodium channels.

regulator An animal for which mechanisms of homeostasis moderate internal changes in a particular variable in the face of external fluctuation of that variable.

regulatory gene A gene that codes for a protein, such as a repressor, that controls the transcription of another gene or group of genes.

reinforcement In evolutionary biology, a process in which natural selection strengthens prezygotic barriers to reproduction, thus reducing the chances of hybrid formation. Such a process is likely to occur only if hybrid offspring are less fit than members of the parent species.

relative abundance The proportional abundance of different species in a community.

relative fitness The contribution an individual makes to the gene pool of the next generation, relative to the contributions of other individuals in the population.

renal cortex The outer portion of the vertebrate kidney.

renal medulla The inner portion of the vertebrate kidney, beneath the renal cortex.

renal pelvis The funnel-shaped chamber that receives processed filtrate from the vertebrate kidney's collecting ducts and is drained by the ureter.

renin-angiotensin-aldosterone system (RAAS) A hormone cascade pathway that helps regulate blood pressure and blood volume.

repetitive DNA Nucleotide sequences, usually noncoding, that are present in many copies in a eukaryotic genome. The repeated units may be short and arranged tandemly (in series) or long and dispersed in the genome.

replication fork A Y-shaped region on a replicating DNA molecule where the parental strands are being unwound and new strands are being synthesized.

repressor A protein that inhibits gene transcription. In prokaryotes, repressors bind to the DNA in or near the promoter. In eukaryotes, repressors may bind to control elements within enhancers, to activators, or to other proteins in a way that blocks activators from binding to DNA.

reproductive isolation The existence of biological factors (barriers) that impede members of two species from producing viable, fertile offspring.

reptile A member of the clade of amniotes that includes tuataras, lizards, snakes, turtles, crocodilians, and birds.

reservoir In biogeochemical cycles, location of a chemical element, consisting of either organic or inorganic materials that are either available for direct use by organisms or unavailable as nutrients.

residual volume The amount of air that remains in the lungs after forceful exhalation.

resource partitioning The division of environmental resources by coexisting species such that the niche of each species differs by one or more significant factors from the niches of all coexisting species.

respiratory pigment A protein that transports oxygen in blood or hemolymph.

response (1) In cellular communication, the change in a specific cellular activity brought about by a transduced signal from outside the cell. (2) In feedback regulation, a physiological activity triggered by a change in a variable.

resting potential The membrane potential characteristic of a nonconducting excitable cell, with the inside of the cell more negative than the outside.

restriction enzyme An endonuclease (type of enzyme) that recognizes and cuts DNA molecules foreign to a bacterium (such as phage genomes). The enzyme cuts at specific nucleotide sequences (restriction sites).

restriction fragment A DNA segment that results from the cutting of DNA by a restriction enzyme.

restriction site A specific sequence on a DNA strand that is recognized and cut by a restriction enzyme.

retina (ret′-i-nuh) The innermost layer of the vertebrate eye, containing photoreceptor cells (rods and cones) and neurons; transmits images formed by the lens to the brain via the optic nerve.

retinal The light-absorbing pigment in rods and cones of the vertebrate eye.

retrotransposon (re′-trō-trans-pō′-zon) A transposable element that moves within a genome by means of an RNA intermediate, a transcript of the retrotransposon DNA.

retrovirus (re′-trō-vī′-rus) An RNA virus that replicates by transcribing its RNA into DNA and then inserting the DNA into a cellular chromosome; an important class of cancer-causing viruses.

reverse transcriptase (tran-skrip′-tās) An enzyme encoded by certain viruses (retroviruses) that uses RNA as a template for DNA synthesis.

reverse transcriptase–polymerase chain reaction (RT-PCR) A technique for determining expression of a particular gene. It uses reverse transcriptase and DNA polymerase to synthesize cDNA from all the mRNA in a sample and then subjects the cDNA to PCR amplification using primers specific for the gene of interest.

rhizarians (rī-za′-rē-uhns) One of the three major subgroups for which the SAR eukaryotic supergroup is named. Many species in this clade are amoebas characterized by threadlike pseudopodia.

rhizobacterium A soil bacterium whose population size is much enhanced in the rhizosphere, the soil region close to a plant's roots.

rhizoid (rī′-zoyd) A long, tubular single cell or filament of cells that anchors bryophytes to the ground. Unlike roots, rhizoids are not composed of tissues, lack specialized conducting cells, and do not play a primary role in water and mineral absorption.

rhizosphere The soil region close to plant roots and characterized by a high level of microbiological activity.

rhodopsin (rō-dop′-sin) A visual pigment consisting of retinal and opsin. Upon absorbing light, the retinal changes shape and dissociates from the opsin.

ribonucleic acid (RNA) (rī′-bō-nū-klā′-ik) A type of nucleic acid consisting of a polynucleotide made up of nucleotide monomers with a ribose sugar and the nitrogenous bases adenine (A), cytosine (C), guanine (G), and uracil (U); usually single-stranded; functions in protein synthesis, in gene regulation, and as the genome of some viruses.

ribose The sugar component of RNA nucleotides.

ribosomal RNA (rRNA) (rī′-buh-sō′-mul) RNA molecules that, together with proteins, make up ribosomes; the most abundant type of RNA.

ribosome (rī′-buh-sōm) A complex of rRNA and protein molecules that functions as a site of protein synthesis in the cytoplasm; consists of a large and a small subunit. In eukaryotic cells, each subunit is assembled in the nucleolus. *See also* nucleolus.

ribozyme (rī′-buh-zīm) An RNA molecule that functions as an enzyme, such as an intron that catalyzes its own removal during RNA splicing.

RNA interference (RNAi) A mechanism for silencing the expression of specific genes. In RNAi, double-stranded RNA molecules that match the sequence of a particular gene are processed into siRNAs that either block translation or trigger the degradation of the gene's messenger RNA. This happens naturally in some cells, and can be carried out in laboratory experiments as well.

RNA polymerase An enzyme that links ribonucleotides into a growing RNA chain during transcription, based on complementary binding to nucleotides on a DNA template strand.

RNA processing Modification of RNA primary transcripts, including splicing out of introns, joining together of exons, and alteration of the 5′ and 3′ ends.

RNA sequencing (RNA-seq) (RNA-sēk) A method of analyzing large sets of RNAs that involves making cDNAs and sequencing them.

RNA splicing After synthesis of a eukaryotic primary RNA transcript, the removal of portions of the transcript (introns) that will not be included in the mRNA and the joining together of the remaining portions (exons).

rod A rodlike cell in the retina of the vertebrate eye, sensitive to low light intensity.

root An organ in vascular plants that anchors the plant and enables it to absorb water and minerals from the soil.

root cap A cone of cells at the tip of a plant root that protects the apical meristem.

root hair A tiny extension of a root epidermal cell, growing just behind the root tip and increasing surface area for absorption of water and minerals.

root pressure Pressure exerted in the roots of plants as the result of osmosis, causing exudation from cut stems and guttation of water from leaves.

root system All of a plant's roots, which anchor it in the soil, absorb and transport minerals and water, and store food.

rooted Describing a phylogenetic tree that contains a branch point (often, the one farthest to the left) representing the most recent common ancestor of all taxa in the tree.

rough ER That portion of the endoplasmic reticulum with ribosomes attached.

round window In the mammalian ear, the point of contact where vibrations of the stapes create a traveling series of pressure waves in the fluid of the cochlea.

R plasmid A bacterial plasmid carrying genes that confer resistance to certain antibiotics.

***r*-selection** Selection for life history traits that maximize reproductive success in uncrowded environments; also called density-independent selection.

rubisco (rū-bis′-kō) Ribulose bisphosphate (RuBP) carboxylase-oxygenase, the enzyme that normally catalyzes the first step of the Calvin cycle (the addition of CO_2 to RuBP). When excess O_2 is present or CO_2 levels are low, rubisco can bind oxygen, resulting in photorespiration.

ruminant (rūh′-muh-nent) A cud-chewing animal, such as a cow or sheep, with multiple stomach compartments specialized for an herbivorous diet.

salicylic acid (sal′-i-sil′-ik) A signaling molecule in plants that may be partially responsible for activating systemic acquired resistance to pathogens.

salivary gland A gland associated with the oral cavity that secretes substances that lubricate food and begin the process of chemical digestion.

salt A compound resulting from the formation of an ionic bond; also called an ionic compound.

saltatory conduction (sol′-tuh-tōr′-ē) Rapid transmission of a nerve impulse along an axon, resulting from the action potential jumping from one node of Ranvier to another, skipping the myelin-sheathed regions of membrane.

SAR One of four supergroups of eukaryotes proposed in a current hypothesis of the evolutionary history of eukaryotes. This supergroup contains a large, extremely diverse collection of protists from three major subgroups: stramenopiles, alveolates, and rhizarians. *See also* Excavata, Archaeplastida, and Unikonta.

sarcomere (sar′-kō-mēr) The fundamental, repeating unit of striated muscle, delimited by the Z lines.

sarcoplasmic reticulum (SR) (sar′-kō-plaz′-mik ruh-tik′-yū-lum) A specialized endoplasmic reticulum that regulates the calcium concentration in the cytosol of muscle cells.

saturated fatty acid A fatty acid in which all carbons in the hydrocarbon tail are connected by single bonds, thus maximizing the number of hydrogen atoms that are attached to the carbon skeleton.

savanna A tropical grassland biome with scattered individual trees and large herbivores and maintained by occasional fires and drought.

scaffolding protein A type of large relay protein to which several other relay proteins are simultaneously attached, increasing the efficiency of signal transduction.

scanning electron microscope (SEM) A microscope that uses an electron beam to scan the surface of a sample, coated with metal atoms, to study details of its topography.

scatter plot A graph in which each piece of data is represented by a point. A scatter plot is used when the data for all variables are numerical and continuous.

schizophrenia (skit′-suh-frē′-nē-uh) A severe mental disturbance characterized by psychotic episodes in which patients have a distorted perception of reality.

Schwann cell A type of glial cell that forms insulating myelin sheaths around the axons of neurons in the peripheral nervous system.

science An approach to understanding the natural world.

scion (sī′-un) The twig grafted onto the stock when making a graft.

sclereid (sklār′-ē-id) A short, irregular sclerenchyma cell in nutshells and seed coats. Sclereids are scattered throughout the parenchyma of some plants.

sclerenchyma cell (skluh-ren′-kim-uh) A rigid, supportive plant cell type usually lacking a protoplast and possessing thick secondary walls strengthened by lignin at maturity.

scrotum A pouch of skin outside the abdomen that houses the testes; functions in maintaining the testes at the lower temperature required for spermatogenesis.

second law of thermodynamics The principle stating that every energy transfer or transformation increases the entropy of the universe. Usable forms of energy are at least partly converted to heat.

second messenger A small, nonprotein, water-soluble molecule or ion, such as a calcium ion (Ca^{2+}) or cyclic AMP, that relays a signal to a cell's interior in response to a signaling molecule bound by a signal receptor protein.

secondary cell wall In plant cells, a strong and durable matrix that is often deposited in several laminated layers around the plasma membrane and provides protection and support.

secondary consumer A carnivore that eats herbivores.

secondary endosymbiosis A process in eukaryotic evolution in which a heterotrophic eukaryotic cell engulfed a photosynthetic eukaryotic cell, which survived in a symbiotic relationship inside the heterotrophic cell.

secondary growth Growth produced by lateral meristems, thickening the roots and shoots of woody plants.

secondary immune response The adaptive immune response elicited on second or subsequent exposures to a particular antigen. The secondary immune response is more rapid, of greater magnitude, and of longer duration than the primary immune response.

Glossary

secondary oocyte (ō′-uh-sīt) An oocyte that has completed meiosis I.

secondary production The amount of chemical energy in consumers' food that is converted to their own new biomass during a given time period.

secondary structure Regions of repetitive coiling or folding of the polypeptide backbone of a protein due to hydrogen bonding between constituents of the backbone (not the side chains).

secondary succession A type of succession that occurs where an existing community has been cleared by some disturbance that leaves the soil or substrate intact.

secretion (1) The discharge of molecules synthesized by a cell. (2) The active transport of wastes and certain other solutes from the body fluid into the filtrate in an excretory system.

seed An adaptation of some terrestrial plants consisting of an embryo packaged along with a store of food within a protective coat.

seed coat A tough outer covering of a seed, formed from the outer coat of an ovule. In a flowering plant, the seed coat encloses and protects the embryo and endosperm.

seedless vascular plant An informal name for a plant that has vascular tissue but lacks seeds. Seedless vascular plants form a paraphyletic group that includes the phyla Lycophyta (club mosses and their relatives) and Monilophyta (ferns and their relatives).

selective permeability A property of biological membranes that allows them to regulate the passage of substances across them.

self-incompatibility The ability of a seed plant to reject its own pollen and sometimes the pollen of closely related individuals.

semelparity (seh′-mel-pār′-i-tē) Reproduction in which an organism produces all of its offspring in a single event; also called big-bang reproduction.

semen (sē′-mun) The fluid that is ejaculated by the male during orgasm; contains sperm and secretions from several glands of the male reproductive tract.

semicircular canals A three-part chamber of the inner ear that functions in maintaining equilibrium.

semiconservative model Type of DNA replication in which the replicated double helix consists of one old strand, derived from the parental molecule, and one newly made strand.

semilunar valve A valve located at each exit of the heart, where the aorta leaves the left ventricle and the pulmonary artery leaves the right ventricle.

seminal vesicle (sem′-i-nul ves′-i-kul) A gland in males that secretes a fluid component of semen that lubricates and nourishes sperm.

seminiferous tubule (sem′-i-nif′-er-us) A highly coiled tube in the testis in which sperm are produced.

senescence (se-nes′-ens) The growth phase in a plant or plant part (as a leaf) from full maturity to death.

sensitive period A limited phase in an animal's development when learning of particular behaviors can take place; also called a critical period.

sensor In homeostasis, a receptor that detects a stimulus.

sensory adaptation The tendency of sensory neurons to become less sensitive when they are stimulated repeatedly.

sensory neuron A nerve cell that receives information from the internal or external environment and transmits signals to the central nervous system.

sensory reception The detection of a stimulus by sensory cells.

sensory receptor A specialized structure or cell that responds to a stimulus from an animal's internal or external environment.

sensory transduction The conversion of stimulus energy to a change in the membrane potential of a sensory receptor cell.

sepal (sē′-pul) A modified leaf in angiosperms that helps enclose and protect a flower bud before it opens.

septum (plural, **septa**) One of the cross-walls that divide a fungal hypha into cells. Septa generally have pores large enough to allow ribosomes, mitochondria, and even nuclei to flow from cell to cell.

serial endosymbiosis A hypothesis for the origin of eukaryotes consisting of a sequence of endosymbiotic events in which mitochondria, chloroplasts, and perhaps other cellular structures were derived from small prokaryotes that had been engulfed by larger cells.

set point In homeostasis in animals, a value maintained for a particular variable, such as body temperature or solute concentration.

seta (sē′-tuh) (plural, **setae**) The elongated stalk of a bryophyte sporophyte.

sex chromosome A chromosome responsible for determining the sex of an individual.

sex-linked gene A gene located on either sex chromosome. Most sex-linked genes are on the X chromosome and show distinctive patterns of inheritance; there are very few genes on the Y chromosome.

sexual dimorphism (dī-mōr′-fizm) Differences between the secondary sex characteristics of males and females of the same species.

sexual reproduction Reproduction arising from fusion of two gametes.

sexual selection A process in which individuals with certain inherited characteristics are more likely than other individuals of the same sex to obtain mates.

Shannon diversity An index of community diversity symbolized by H and represented by the equation $H = -(p_A \ln p_A + p_B \ln p_B + p_C \ln p_C + \cdots)$, where A, B, C ... are species, p is the relative abundance of each species, and ln is the natural logarithm.

shared ancestral character A character, shared by members of a particular clade, that originated in an ancestor that is not a member of that clade.

shared derived character An evolutionary novelty that is unique to a particular clade.

shoot system The aerial portion of a plant body, consisting of stems, leaves, and (in angiosperms) flowers.

short tandem repeat (STR) Simple sequence DNA containing multiple tandemly repeated units of two to five nucleotides. Variations in STRs act as genetic markers in STR analysis, used to prepare genetic profiles.

short-day plant A plant that flowers (usually in late summer, fall, or winter) only when the light period is shorter than a critical length.

short-term memory The ability to hold information, anticipations, or goals for a time and then release them if they become irrelevant.

sickle-cell disease A recessively inherited human blood disorder in which a single nucleotide change in the α-globin gene causes hemoglobin to aggregate, changing red blood cell shape and causing multiple symptoms in afflicted individuals.

sieve plate An end wall in a sieve-tube element, which facilitates the flow of phloem sap in angiosperm sieve tubes.

sieve-tube element A living cell that conducts sugars and other organic nutrients in the phloem of angiosperms; also called a sieve-tube member. Connected end to end, they form sieve tubes.

sign stimulus An external sensory cue that triggers a fixed action pattern by an animal.

signal In animal behavior, transmission of a stimulus from one animal to another. The term is also used in the context of communication in other kinds of organisms and in cell-to-cell communication in all multicellular organisms.

signal peptide A sequence of about 20 amino acids at or near the leading (amino) end of a polypeptide that targets it to the endoplasmic reticulum or other organelles in a eukaryotic cell.

signal-recognition particle (SRP) A protein-RNA complex that recognizes a signal peptide as it emerges from a ribosome and helps direct the ribosome to the endoplasmic reticulum (ER) by binding to a receptor protein on the ER.

signal transduction The linkage of a mechanical, chemical, or electromagnetic stimulus to a specific cellular response.

signal transduction pathway A series of steps linking a mechanical, chemical, or electrical stimulus to a specific cellular response.

silent mutation A nucleotide-pair substitution that has no observable effect on the phenotype; for example, within a gene, a mutation that results in a codon that codes for the same amino acid.

simple fruit A fruit derived from a single carpel or several fused carpels.

simple sequence DNA A DNA sequence that contains many copies of tandemly repeated short sequences.

single bond A single covalent bond; the sharing of a pair of valence electrons by two atoms.

single circulation A circulatory system consisting of a single pump and circuit, in which blood passes from the sites of gas exchange to the rest of the body before returning to the heart.

single-lens eye The camera-like eye found in some jellies, polychaete worms, spiders, and many molluscs.

single nucleotide polymorphism (SNP) (snip) A single base-pair site in a genome where nucleotide variation is found in at least 1% of the population.

single-strand binding protein A protein that binds to the unpaired DNA strands during DNA replication, stabilizing them and holding them apart while they serve as templates for the synthesis of complementary strands of DNA.

sinoatrial (SA) node (sī′-nō-ā′-trē-uhl) A region in the right atrium of the heart that sets the rate and timing at which all cardiac muscle cells contract; the pacemaker.

sister chromatids Two copies of a duplicated chromosome attached to each other by proteins at the centromere and, sometimes, along the arms. While joined, two sister chromatids make up one chromosome. Chromatids are eventually separated during mitosis or meiosis II.

sister taxa Groups of organisms that share an immediate common ancestor and hence are each other's closest relatives.

skeletal muscle A type of striated muscle that is generally responsible for the voluntary movements of the body.

sliding-filament model The idea that muscle contraction is based on the movement of thin (actin) filaments along thick (myosin) filaments, shortening the sarcomere, the basic unit of muscle organization.

slow-twitch fiber A muscle fiber that can sustain long contractions.

small interfering RNA (siRNA) One of multiple small, single-stranded RNA molecules generated by cellular machinery from a long, linear, double-stranded RNA molecule. The siRNA associates with one or more proteins in a complex that can degrade or prevent translation of an mRNA with a complementary sequence.

small intestine The longest section of the alimentary canal, so named because of its small diameter compared with that of the large intestine; the principal site of the enzymatic hydrolysis of food macromolecules and the absorption of nutrients.

smooth ER That portion of the endoplasmic reticulum that is free of ribosomes.

smooth muscle A type of muscle lacking the striations of skeletal and cardiac muscle because of the uniform distribution of myosin filaments in the cells; responsible for involuntary body activities.

social learning Modification of behavior through the observation of other individuals.

sociobiology The study of social behavior based on evolutionary theory.

sodium-potassium pump A transport protein in the plasma membrane of animal cells that actively transports sodium out of the cell and potassium into the cell.

soil horizon A soil layer with physical characteristics that differ from those of the layers above or beneath.

solute (sol′-yūt) A substance that is dissolved in a solution.

solute potential (ψ_S) A component of water potential that is proportional to the molarity of a solution and that measures the effect of solutes on the direction of water movement; also called osmotic potential, it can be either zero or negative.

solution A liquid that is a homogeneous mixture of two or more substances.

solvent The dissolving agent of a solution. Water is the most versatile solvent known.

somatic cell (sō-mat′-ik) Any cell in a multicellular organism except a sperm or egg or their precursors.

somite One of a series of blocks of mesoderm that exist in pairs just lateral to the notochord in a vertebrate embryo.

soredium (suh-rē′-dē-um) (plural, **soredia**) In lichens, a small cluster of fungal hyphae with embedded algae.

sorus (plural, **sori**) A cluster of sporangia on a fern sporophyll. Sori may be arranged in various patterns, such as parallel lines or dots, which are useful in fern identification.

spatial learning The establishment of a memory that reflects the environment's spatial structure.

speciation (spē′-sē-ā′-shun) An evolutionary process in which one species splits into two or more species.

species (spē′-sēz) A population or group of populations whose members have the potential to interbreed in nature and produce viable, fertile offspring but do not produce viable, fertile offspring with members of other such groups.

species-area curve (spē′-sēz) The biodiversity pattern that shows that the larger the geographic area of a community is, the more species it has.

species diversity (spē′-sēz) The number and relative abundance of species in a biological community.

species richness (spē′-sēz) The number of species in a biological community.

specific heat The amount of heat that must be absorbed or lost for 1 g of a substance to change its temperature by 1°C.

spectrophotometer An instrument that measures the proportions of light of different wavelengths absorbed and transmitted by a pigment solution.

sperm The male gamete.

spermatheca (sper′-muh-thē′-kuh) (plural, **spermathecae**) In many insects, a sac in the female reproductive system where sperm are stored.

spermatogenesis (sper-ma′-tō-gen′-uh-sis) The continuous and prolific production of mature sperm in the testis.

spermatogonium (sper-ma′-tō-gō′-nē-um) (plural, **spermatogonia**) A cell that divides mitotically to form spermatocytes.

S phase The synthesis phase of the cell cycle; the portion of interphase during which DNA is replicated.

sphincter (sfink′-ter) A ringlike band of muscle fibers that controls the size of an opening in the body, such as the passage between the esophagus and the stomach.

spiral cleavage A type of embryonic development in protostomes in which the planes of cell division that transform the zygote into a ball of cells are diagonal to the vertical axis of the embryo. As a result, the cells of each tier sit in the grooves between cells of adjacent tiers.

spliceosome (splī′-sō-sōm) A large complex made up of proteins and RNA molecules that splices RNA by interacting with the ends of an RNA intron, releasing the intron and joining the two adjacent exons.

spongocoel (spon′-jō-sēl) The central cavity of a sponge.

spontaneous process A process that occurs without an overall input of energy; a process that is energetically favorable.

sporangium (spōr-an′-jē-um) (plural, **sporangia**) A multicellular organ in fungi and plants in which meiosis occurs and haploid cells develop.

spore (1) In the life cycle of a plant or alga undergoing alternation of generations, a haploid cell produced in the sporophyte by meiosis. A spore can divide by mitosis to develop into a multicellular haploid individual, the gametophyte, without fusing with another cell. (2) In fungi, a haploid cell, produced either sexually or asexually, that produces a mycelium after germination.

sporocyte (spō′-ruh-sīt) A diploid cell within a sporangium that undergoes meiosis and generates haploid spores; also called a spore mother cell.

sporophyll (spō′-ruh-fil) A modified leaf that bears sporangia and hence is specialized for reproduction.

sporophyte (spō-ruh-fīt·) In organisms (plants and some algae) that have alternation of generations, the multicellular diploid form that results from the union of gametes. Meiosis in the sporophyte produces haploid spores that develop into gametophytes.

sporopollenin (spōr-uh-pol′-eh-nin) A durable polymer that covers exposed zygotes of charophyte algae and forms the walls of plant spores, preventing them from drying out.

stability In evolutionary biology, a term referring to a hybrid zone in which hybrids continue to be produced; this causes the hybrid zone to be "stable" in the sense of persisting over time.

stabilizing selection Natural selection in which intermediate phenotypes survive or reproduce more successfully than do extreme phenotypes.

stamen (stā′-men) The pollen-producing reproductive organ of a flower, consisting of an anther and a filament.

standard deviation A measure of the variation found in a set of data points.

standard metabolic rate (SMR) Metabolic rate of a resting, fasting, and nonstressed ectotherm at a particular temperature.

starch A storage polysaccharide in plants, consisting entirely of glucose monomers joined by glycosidic linkages.

start point In transcription, the nucleotide position on the promoter where RNA polymerase begins synthesis of RNA.

statocyst (stat′-uh-sist′) A type of mechanoreceptor that functions in equilibrium in invertebrates by use of statoliths, which stimulate hair cells in relation to gravity.

statolith (stat′-uh-lith′) (1) In plants, a specialized plastid that contains dense starch grains and may play a role in detecting gravity. (2) In invertebrates, a dense particle that settles in response to gravity and is found in sensory organs that function in equilibrium.

stele (stēl) The vascular tissue of a stem or root.

stem A vascular plant organ consisting of an alternating system of nodes and internodes that support the leaves and reproductive structures.

stem cell Any relatively unspecialized cell that can produce, during a single division, two identical daughter cells or two more specialized daughter cells that can undergo further differentiation, or one cell of each type.

steroid A type of lipid characterized by a carbon skeleton consisting of four fused rings with various chemical groups attached.

sticky end A single-stranded end of a double-stranded restriction fragment.

stigma (plural, **stigmata**) The sticky part of a flower's carpel, which receives pollen grains.

stimulus In feedback regulation, a fluctuation in a variable that triggers a response.

stipe A stemlike structure of a seaweed.

stock The plant that provides the root system when making a graft.

stoma (stō′-muh) (plural, **stomata**) A microscopic pore surrounded by guard cells in the epidermis of leaves and stems that allows gas exchange between the environment and the interior of the plant.

stomach An organ of the digestive system that stores food and performs preliminary steps of digestion.

Stramenopiles (strah′-men-ō′-pē-lēs) One of the three major subgroups for which the SAR eukaryotic supergroup is named. This clade arose by secondary endosymbiosis and includes diatoms and brown algae.

stratum (strah′-tum) (plural, **strata**) A rock layer formed when new layers of sediment cover older ones and compress them.

strigolactone Any of a class of plant hormones that inhibit shoot branching, trigger the germination of parasitic plant seeds, and stimulate the association of plant roots with mycorrhizal fungi.

strobilus (strō-bī′-lus) (plural, **strobili**) The technical term for a cluster of sporophylls known commonly as a cone, found in most gymnosperms and some seedless vascular plants.

stroke The death of nervous tissue in the brain, usually resulting from rupture or blockage of arteries in the head.

stroke volume The volume of blood pumped by a heart ventricle in a single contraction.

stroma (strō′-muh) The dense fluid within the chloroplast surrounding the thylakoid membrane and containing ribosomes and DNA; involved in the synthesis of organic molecules from carbon dioxide and water.

stromatolite Layered rock that results from the activities of prokaryotes that bind thin films of sediment together.

structural isomer One of two or more compounds that have the same molecular formula but differ in the covalent arrangements of their atoms.

style The stalk of a flower's carpel, with the ovary at the base and the stigma at the top.

substrate The reactant on which an enzyme works.

substrate feeder An animal that lives in or on its food source, eating its way through the food.

substrate-level phosphorylation The enzyme-catalyzed formation of ATP by direct transfer of a phosphate group to ADP from an intermediate substrate in catabolism.

sugar sink A plant organ that is a net consumer or storer of sugar. Growing roots, shoot tips, stems, and fruits are examples of sugar sinks supplied by phloem.

sugar source A plant organ in which sugar is being produced by either photosynthesis or the breakdown of starch. Mature leaves are the primary sugar sources of plants.

sulfhydryl group A chemical group consisting of a sulfur atom bonded to a hydrogen atom.

summation A phenomenon of neural integration in which the membrane potential of the postsynaptic cell is determined by the combined effect of EPSPs or IPSPs produced in rapid succession at one synapse or simultaneously at different synapses.

suprachiasmatic nucleus (SCN) (sūp′-ruh-kē′-as-ma-tik) A group of neurons in the hypothalamus of mammals that functions as a biological clock.

surface tension A measure of how difficult it is to stretch or break the surface of a liquid. Water has a high surface tension because of the hydrogen bonding of surface molecules.

surfactant A substance secreted by alveoli that decreases surface tension in the fluid that coats the alveoli.

survivorship curve A plot of the number of members of a cohort that are still alive at each age; one way to represent age-specific mortality.

suspension feeder An animal that feeds by removing suspended food particles from the surrounding medium by a capture, trapping, or filtration mechanism.

sustainable agriculture Long-term productive farming methods that are environmentally safe.

sustainable development Development that meets the needs of people today without limiting the ability of future generations to meet their needs.

swim bladder In aquatic osteichthyans, an air sac that enables the animal to control its buoyancy in the water.

symbiont (sim′-bē-ont) The smaller participant in a symbiotic relationship, living in or on the host.

symbiosis An ecological relationship between organisms of two different species that live together in direct and intimate contact.

sympathetic division One of three divisions of the autonomic nervous system; generally increases energy expenditure and prepares the body for action.

sympatric speciation (sim-pat′-rik) The formation of new species in populations that live in the same geographic area.

symplast In plants, the continuum of cytoplasm connected by plasmodesmata between cells.

synapse (sin′-aps) The junction where a neuron communicates with another cell across a narrow gap via a neurotransmitter or an electrical coupling.

synapsid (si-nap′-sid) A member of an amniote clade distinguished by a single hole on each side of the skull. Synapsids include the mammals.

synapsis (si-nap′-sis) The pairing and physical connection of one duplicated chromosome to its homolog during prophase I of meiosis.

synaptonemal (si-nap′-tuh-nē′-muhl) **complex** A zipper-like structure composed of proteins, which connects a chromosome to its homolog tightly along their lengths during part of prophase I of meiosis.

systematics A scientific discipline focused on classifying organisms and determining their evolutionary relationships.

systemic acquired resistance A defensive response in infected plants that helps protect healthy tissue from pathogenic invasion.

systemic circuit The branch of the circulatory system that supplies oxygenated blood to and carries deoxygenated blood away from organs and tissues throughout the body.

systems biology An approach to studying biology that aims to model the dynamic behavior of whole biological systems based on a study of the interactions among the system's parts.

systole (sis′-tō-lē) The stage of the cardiac cycle in which a heart chamber contracts and pumps blood.

systolic pressure Blood pressure in the arteries during contraction of the ventricles.

taproot A main vertical root that develops from an embryonic root and gives rise to lateral (branch) roots.

tastant Any chemical that stimulates the sensory receptors in a taste bud.

taste bud A collection of modified epithelial cells on the tongue or in the mouth that are receptors for taste in mammals.

TATA box A DNA sequence in eukaryotic promoters crucial in forming the transcription initiation complex.

taxis (tak′-sis) An oriented movement toward or away from a stimulus.

taxon (plural, **taxa**) A named taxonomic unit at any given level of classification.

taxonomy (tak-son′-uh-mē) A scientific discipline concerned with naming and classifying the diverse forms of life.

Tay-Sachs disease A human genetic disease caused by a recessive allele for a dysfunctional enzyme, leading to accumulation of certain lipids in the brain. Seizures, blindness, and degeneration of motor and mental performance usually become manifest a few months after birth, followed by death within a few years.

T cells The class of lymphocytes that mature in the thymus; they include both effector cells for the cell-mediated immune response and helper cells required for both branches of adaptive immunity.

technology The application of scientific knowledge for a specific purpose, often involving industry or commerce but also including uses in basic research.

telomere (tel′-uh-mēr) The tandemly repetitive DNA at the end of a eukaryotic chromosome's DNA molecule. Telomeres protect the organism's genes from being eroded during successive rounds of replication. *See also* repetitive DNA.

telophase The fifth and final stage of mitosis, in which daughter nuclei are forming and cytokinesis has typically begun.

temperate broadleaf forest A biome located throughout midlatitude regions where there is sufficient moisture to support the growth of large, broadleaf deciduous trees.

temperate grassland A terrestrial biome that exists at midlatitude regions and is dominated by grasses and forbs.

temperate phage A phage that is capable of replicating by either a lytic or lysogenic cycle.

temperature A measure in degrees of the average kinetic energy (thermal energy) of the atoms and molecules in a body of matter.

template strand The DNA strand that provides the pattern, or template, for ordering, by complementary base pairing, the sequence of nucleotides in an RNA transcript.

tendon A fibrous connective tissue that attaches muscle to bone.

terminator In bacteria, a sequence of nucleotides in DNA that marks the end of a gene and signals RNA polymerase to release the newly made RNA molecule and detach from the DNA.

territoriality A behavior in which an animal defends a bounded physical space against encroachment by other individuals, usually of its own species.

tertiary consumer (ter′-shē-ār′-ē) A carnivore that eats other carnivores.

tertiary structure (ter′-shē-ār′-ē) The overall shape of a protein molecule due to interactions of amino acid side chains, including hydrophobic interactions, ionic bonds, hydrogen bonds, and disulfide bridges.

test In foram protists, a porous shell that consists of a single piece of organic material hardened with calcium carbonate.

testcross Breeding an organism of unknown genotype with a homozygous recessive individual to determine the unknown genotype. The ratio of phenotypes in the offspring reveals the unknown genotype.

testis (plural, **testes**) The male reproductive organ, or gonad, in which sperm and reproductive hormones are produced.

testosterone A steroid hormone required for development of the male reproductive system, spermatogenesis, and male secondary sex characteristics; the major androgen in mammals.

tetanus (tet′-uh-nus) The maximal, sustained contraction of a skeletal muscle, caused by a very high frequency of action potentials elicited by continual stimulation.

tetrapod A vertebrate clade whose members have limbs with digits. Tetrapods include mammals, amphibians, and birds and other reptiles.

thalamus (thal′-uh-mus) An integrating center of the vertebrate forebrain. Neurons with cell bodies in the thalamus relay neural input to specific areas in the cerebral cortex and regulate what information goes to the cerebral cortex.

theory An explanation that is broader in scope than a hypothesis, generates new hypotheses, and is supported by a large body of evidence.

thermal energy Kinetic energy due to the random motion of atoms and molecules; energy in its most random form. *See also* heat.

thermocline A narrow stratum of abrupt temperature change in the ocean and in many temperate-zone lakes.

thermodynamics (ther′-mō-dī-nam′-iks) The study of energy transformations that occur in a collection of matter. *See also* first law of thermodynamics and second law of thermodynamics.

thermophile *See* extreme thermophile.

thermoreceptor A receptor stimulated by either heat or cold.

thermoregulation The maintenance of internal body temperature within a tolerable range.

theropod A member of a group of dinosaurs that were bipedal carnivores.

thick filament A filament composed of staggered arrays of myosin molecules; a component of myofibrils in muscle fibers.

thigmomorphogenesis (thig′-mō-mor′-phō-gen′-uh-sis) A response in plants to chronic mechanical stimulation, resulting from increased ethylene production. An example is thickening stems in response to strong winds.

thigmotropism (thig-mō′-truh-pizm) A directional growth of a plant in response to touch.

thin filament A filament consisting of two strands of actin and two strands of regulatory protein coiled around one another; a component of myofibrils in muscle fibers.

threatened species A species that is considered likely to become endangered in the foreseeable future.

threshold The potential that an excitable cell membrane must reach for an action potential to be initiated.

thrombus A fibrin-containing clot that forms in a blood vessel and blocks the flow of blood.

thylakoid (thī′-luh-koyd) A flattened, membranous sac inside a chloroplast. Thylakoids often exist in stacks called grana that are interconnected; their membranes contain molecular "machinery" used to convert light energy to chemical energy.

thymus (thī′-mus) A small organ in the thoracic cavity of vertebrates where maturation of T cells is completed.

thyroid gland An endocrine gland, located on the ventral surface of the trachea, that secretes two iodine-containing hormones, triiodothyronine (T_3) and thyroxine (T_4), as well as calcitonin.

thyroid hormone Either of two iodine-containing hormones (triiodothyronine and thyroxine) that are secreted by the thyroid gland and that help regulate metabolism, development, and maturation in vertebrates.

thyroxine (T_4) One of two iodine-containing hormones that are secreted by the thyroid gland and that help regulate metabolism, development, and maturation in vertebrates.

tidal volume The volume of air a mammal inhales and exhales with each breath.

tight junction A type of intercellular junction between animal cells that prevents the leakage of material through the space between cells.

tissue An integrated group of cells with a common structure, function, or both.

tissue system One or more tissues organized into a functional unit connecting the organs of a plant.

Toll-like receptor (TLR) A membrane receptor on a phagocytic white blood cell that recognizes fragments of molecules common to a set of pathogens.

tonicity The ability of a solution surrounding a cell to cause that cell to gain or lose water.

top-down model A model of community organization in which predation influences community organization by controlling herbivore numbers, which in turn control plant or phytoplankton numbers, which in turn control nutrient levels; also called the trophic cascade model.

topoisomerase A protein that breaks, swivels, and rejoins DNA strands. During DNA replication, topoisomerase helps to relieve strain in the double helix ahead of the replication fork.

topsoil A mixture of particles derived from rock, living organisms, and decaying organic material (humus).

torpor A physiological state in which activity is low and metabolism decreases.

totipotent (tō′-tuh-pōt′-ent) Describing a cell that can give rise to all parts of the embryo and adult, as well as extraembryonic membranes in species that have them.

trace element An element indispensable for life but required in extremely minute amounts.

trachea (trā′-kē-uh) The portion of the respiratory tract that passes from the larynx to the bronchi; also called the windpipe.

tracheal system In insects, a system of branched, air-filled tubes that extends throughout the body and carries oxygen directly to cells.

tracheid (trā′-kē-id) A long, tapered water-conducting cell found in the xylem of nearly all vascular plants. Functioning tracheids are no longer living.

trait One of two or more detectable variants in a genetic character.

***trans* fat** An unsaturated fat, formed artificially during hydrogenation of oils, containing one or more *trans* double bonds.

transcription The synthesis of RNA using a DNA template.

transcription factor A regulatory protein that binds to DNA and affects transcription of specific genes.

transcription initiation complex The completed assembly of transcription factors and RNA polymerase bound to a promoter.

transcription unit A region of DNA that is transcribed into an RNA molecule.

transduction A process in which phages (viruses) carry bacterial DNA from one bacterial cell to another. When these two cells are members of different species, transduction results in horizontal gene transfer. *See also* signal transduction.

transfer RNA (tRNA) An RNA molecule that functions as a translator between nucleic acid and protein languages by picking up a specific amino acid and carrying it to the ribosome, where the tRNA recognizes the appropriate codon in the mRNA.

transformation (1) The process by which a cell in culture acquires the ability to divide indefinitely, similar to the division of cancer cells. (2) A change in genotype and phenotype due to the assimilation of external DNA by a cell. When the external DNA is from a member of a different species, transformation results in horizontal gene transfer.

transgenic Pertaining to an organism whose genome contains DNA introduced from another organism of the same or a different species.

translation The synthesis of a polypeptide using the genetic information encoded in an mRNA molecule. There is a change of "language" from nucleotides to amino acids.

translocation (1) An aberration in chromosome structure resulting from attachment of a chromosomal fragment to a nonhomologous chromosome. (2) During protein synthesis, the third stage in the elongation cycle, when the RNA carrying the growing polypeptide moves from the A site to the P site on the ribosome. (3) The transport of organic nutrients in the phloem of vascular plants.

transmission electron microscope (TEM) A microscope that passes an electron beam through very thin sections stained with metal atoms and is primarily used to study the internal structure of cells.

transpiration The evaporative loss of water from a plant.

transport epithelium One or more layers of specialized epithelial cells that carry out and regulate solute movement.

transport protein A transmembrane protein that helps a certain substance or class of closely related substances to cross the membrane.

transport vesicle A small membranous sac in a eukaryotic cell's cytoplasm carrying molecules produced by the cell.

transposable element A segment of DNA that can move within the genome of a cell by means of a DNA or RNA intermediate; also called a transposable genetic element.

transposon A transposable element that moves within a genome by means of a DNA intermediate.

transverse (T) tubule An infolding of the plasma membrane of skeletal muscle cells.

triacylglycerol (trī-as′-ul-glis′-uh-rol) A lipid consisting of three fatty acids linked to one glycerol molecule; also called a fat or triglyceride.

trichome An epidermal cell that is a highly specialized, often hairlike outgrowth on a plant shoot.

triple response A plant growth maneuver in response to mechanical stress, involving slowing of stem elongation, thickening of the stem, and a curvature that causes the stem to start growing horizontally.

triplet code A genetic information system in which a series of three-nucleotide-long words specifies a sequence of amino acids for a polypeptide chain.

triploblastic Possessing three germ layers: the endoderm, mesoderm, and ectoderm. All bilaterian animals are triploblastic.

trisomic Referring to a diploid cell that has three copies of a particular chromosome instead of the normal two.

trochophore larva (trō′-kuh-fōr) Distinctive larval stage observed in some lophotrochozoan animals, including some annelids and molluscs.

trophic efficiency The percentage of production transferred from one trophic level to the next higher trophic level.

trophic structure The different feeding relationships in an ecosystem, which determine the route of energy flow and the pattern of chemical cycling.

trophoblast The outer epithelium of a mammalian blastocyst. It forms the fetal part of the placenta, supporting embryonic development but not forming part of the embryo proper.

tropical dry forest A terrestrial biome characterized by relatively high temperatures and precipitation overall but with a pronounced dry season.

tropical rain forest A terrestrial biome characterized by relatively high precipitation and temperatures year-round.

tropics Latitudes between 23.5° north and south.

tropism A growth response that results in the curvature of whole plant organs toward or away from stimuli due to differential rates of cell elongation.

tropomyosin The regulatory protein that blocks the myosin-binding sites on actin molecules.

troponin complex The regulatory proteins that control the position of tropomyosin on the thin filament.

true-breeding Referring to organisms that produce offspring of the same variety over many generations of self-pollination.

tubal ligation A means of sterilization in which a woman's two oviducts (fallopian tubes) are tied closed and a segment of each is removed to prevent eggs from reaching the uterus.

tube foot One of numerous extensions of an echinoderm's water vascular system. Tube feet function in locomotion and feeding.

tumor-suppressor gene A gene whose protein product inhibits cell division, thereby preventing the uncontrolled cell growth that contributes to cancer.

tundra A terrestrial biome at the extreme limits of plant growth. At the northernmost limits, it is called arctic tundra, and at high altitudes, where plant forms are limited to low shrubby or matlike vegetation, it is called alpine tundra.

tunicate A member of the clade Urochordata, sessile marine chordates that lack a backbone.

turgid (ter′-jid) Swollen or distended, as in plant cells. (A walled cell becomes turgid if it has a lower water potential than its surroundings, resulting in entry of water.)

turgor pressure The force directed against a plant cell wall after the influx of water and swelling of the cell due to osmosis.

turnover The mixing of waters as a result of changing water-temperature profiles in a lake.

twin study A behavioral study in which researchers compare the behavior of identical twins raised apart with that of identical twins raised in the same household.

tympanic membrane Another name for the eardrum, the membrane between the outer and middle ear.

Unikonta (yū′-ni-kon′-tuh) One of four supergroups of eukaryotes proposed in a current hypothesis of the evolutionary history of eukaryotes. This clade, which is supported by studies of myosin proteins and DNA, consists of amoebozoans and opisthokonts. *See also* Excavata, SAR, and Archaeplastida.

unsaturated fatty acid A fatty acid that has one or more double bonds between carbons in the hydrocarbon tail. Such bonding reduces the number of hydrogen atoms attached to the carbon skeleton.

urban ecology The study of organisms and their environment in urban and suburban settings.

urea A soluble nitrogenous waste produced in the liver by a metabolic cycle that combines ammonia with carbon dioxide.

ureter (yū-rē′-ter) A duct leading from the kidney to the urinary bladder.

urethra (yū-rē′-thruh) A tube that releases urine from the mammalian body near the vagina in females and through the penis in males; also serves in males as the exit tube for the reproductive system.

uric acid A product of protein and purine metabolism and the major nitrogenous waste product of insects, land snails, and many reptiles. Uric acid is relatively nontoxic and largely insoluble in water.

urinary bladder The pouch where urine is stored prior to elimination.

uterine cycle The cyclic changes in the endometrium (uterine lining) of mammals that occur in the absence of pregnancy. In certain primates, including humans, the uterine cycle is a menstrual cycle.

uterus A female organ where eggs are fertilized and/or development of the young occurs.

vaccine A harmless variant or derivative of a pathogen that stimulates a host's immune system to mount defenses against the pathogen.

vacuole (vak′-yū-ōl′) A membrane-bounded vesicle whose specialized function varies in different kinds of cells.

vagina Part of the female reproductive system between the uterus and the outside opening; the birth canal in mammals. During copulation, the vagina accommodates the male's penis and receives sperm.

valence The bonding capacity of a given atom; the number of covalent bonds that an atom can form, which usually equals the number of unpaired electrons in its outermost (valence) shell.

valence electron An electron in the outermost electron shell.

valence shell The outermost energy shell of an atom, containing the valence electrons involved in the chemical reactions of that atom.

van der Waals interactions Weak attractions between molecules or parts of molecules that result from transient local partial charges.

variable A factor that varies in an experiment.

variation Differences between members of the same species.

vas deferens In mammals, the tube in the male reproductive system in which sperm travel from the epididymis to the urethra.

vasa recta The capillary system in the kidney that serves the loop of Henle.

vascular cambium A cylinder of meristematic tissue in woody plants that adds layers of secondary vascular tissue called secondary xylem (wood) and secondary phloem.

vascular plant A plant with vascular tissue. Vascular plants include all living plant species except liverworts, mosses, and hornworts.

vascular tissue Plant tissue consisting of cells joined into tubes that transport water and nutrients throughout the plant body.

vascular tissue system A transport system formed by xylem and phloem throughout a vascular plant. Xylem transports water and minerals; phloem transports sugars, the products of photosynthesis.

vasectomy The cutting and sealing of each vas deferens to prevent sperm from entering the urethra.

vasoconstriction A decrease in the diameter of blood vessels caused by contraction of smooth muscles in the vessel walls.

vasodilation An increase in the diameter of blood vessels caused by relaxation of smooth muscles in the vessel walls.

vasopressin *See* antidiuretic hormone (ADH).

vector An organism that transmits pathogens from one host to another.

vegetal pole The point at the end of an egg in the hemisphere where most yolk is concentrated; opposite of animal pole.

vegetative propagation Asexual reproduction in plants that is facilitated or induced by humans.

vegetative reproduction Asexual reproduction in plants.

vein (1) In animals, a vessel that carries blood toward the heart. (2) In plants, a vascular bundle in a leaf.

ventilation The flow of air or water over a respiratory surface.

ventral Pertaining to the underside, or bottom, of an animal with radial or bilateral symmetry.

ventricle (ven′-tri-kul) (1) A heart chamber that pumps blood out of the heart. (2) A space in the vertebrate brain, filled with cerebrospinal fluid.

venule (ven′-yūl) A vessel that conveys blood between a capillary bed and a vein.

vernalization The use of cold treatment to induce a plant to flower.

vertebrate A chordate animal with vertebrae, the series of bones that make up the backbone.

vesicle (ves′-i-kul) A membranous sac in the cytoplasm of a eukaryotic cell.

vessel A continuous water-conducting micropipe found in most angiosperms and a few nonflowering vascular plants.

vessel element A short, wide water-conducting cell found in the xylem of most angiosperms and a few nonflowering vascular plants. Dead at maturity, vessel elements are aligned end to end to form micropipes called vessels.

vestigial structure A feature of an organism that is a historical remnant of a structure that served a function in the organism's ancestors.

villus (plural, **villi**) (1) A finger-like projection of the inner surface of the small intestine. (2) A finger-like projection of the chorion of the mammalian placenta. Large numbers of villi increase the surface areas of these organs.

viral envelope A membrane, derived from membranes of the host cell, that cloaks the capsid, which in turn encloses a viral genome.

virulent phage A phage that replicates only by a lytic cycle.

virus An infectious particle incapable of replicating outside of a cell, consisting of an RNA or DNA genome surrounded by a protein coat (capsid) and, for some viruses, a membranous envelope.

visceral mass One of the three main parts of a mollusc; the part containing most of the internal organs. *See also* foot and mantle.

visible light That portion of the electromagnetic spectrum that can be detected as various colors by the human eye, ranging in wavelength from about 380 nm to about 750 nm.

vital capacity The maximum volume of air that a mammal can inhale and exhale with each breath.

vitamin An organic molecule required in the diet in very small amounts. Many vitamins serve as coenzymes or parts of coenzymes.

viviparous (vī-vip′-uh-rus) Referring to a type of development in which the young are born alive after having been nourished in the uterus by blood from the placenta.

voltage-gated ion channel A specialized ion channel that opens or closes in response to changes in membrane potential.

vulva Collective term for the female external genitalia.

water potential (ψ) The physical property predicting the direction in which water will flow, governed by solute concentration and applied pressure.

water vascular system A network of hydraulic canals unique to echinoderms that branches into extensions called tube feet, which function in locomotion and feeding.

wavelength The distance between crests of waves, such as those of the electromagnetic spectrum.

wetland A habitat that is inundated by water at least some of the time and that supports plants adapted to water-saturated soil.

white matter Tracts of axons within the CNS.

whole-genome shotgun approach Procedure for genome sequencing in which the genome is randomly cut into many overlapping short segments that are sequenced; computer software then assembles the complete sequence.

wild type The phenotype most commonly observed in natural populations; also refers to the individual with that phenotype.

wilting The drooping of leaves and stems as a result of plant cells becoming flaccid.

wobble Flexibility in the base-pairing rules in which the nucleotide at the 5′ end of a tRNA anticodon can form hydrogen bonds with more than one kind of base in the third position (3′ end) of a codon.

xerophyte (zir′-ō-fīt′) A plant adapted to an arid climate.

X-linked gene A gene located on the X chromosome; such genes show a distinctive pattern of inheritance.

X-ray crystallography A technique used to study the three-dimensional structure of molecules. It depends on the diffraction of an X-ray beam by the individual atoms of a crystallized molecule.

xylem (zī′-lum) Vascular plant tissue consisting mainly of tubular dead cells that conduct most of the water and minerals upward from the roots to the rest of the plant.

xylem sap (zī′-lum) The dilute solution of water and minerals carried through vessels and tracheids.

yeast Single-celled fungus. Yeasts reproduce asexually by binary fission or by the pinching of small buds off a parent cell. Many fungal species can grow both as yeasts and as a network of filaments; relatively few species grow only as yeasts.

yolk Nutrients stored in an egg.

zero population growth (ZPG) A period of stability in population size, when additions to the population through births and immigration are balanced by subtractions through deaths and emigration.

zona pellucida The extracellular matrix surrounding a mammalian egg.

zoned reserve An extensive region that includes areas relatively undisturbed by humans surrounded by areas that have been changed by human activity and are used for economic gain.

zone of polarizing activity (ZPA) A block of mesoderm located just under the ectoderm where the posterior side of a limb bud is attached to the body; required for proper pattern formation along the anterior-posterior axis of the limb.

zoonotic pathogen A disease-causing agent that is transmitted to humans from other animals.

zoospore Flagellated spore found in chytrid fungi and some protists.

zygomycete (zī′-guh-mī′-sēt) A member of the fungal phylum Zygomycota, characterized by the formation of a sturdy structure called a zygosporangium during sexual reproduction.

zygosporangium (zī′-guh-spōr-an′-jē-um) (plural, **zygosporangia**) In zygomycete fungi, a sturdy multinucleate structure in which karyogamy and meiosis occur.

zygote (zī′-gōt) The diploid cell produced by the union of haploid gametes during fertilization; a fertilized egg.

Index

NOTE: A page number in regular type indicates where a topic is discussed in the text; a **bold** page number indicates where a term is bold and defined; an *f* following a page number indicates a figure (the topic may also be discussed in the text on that page); a *t* following a page number indicates a table (the topic may also be discussed in the text on that page).

H

P

Q

R

S

V

W